www.mhhe.com/shier12

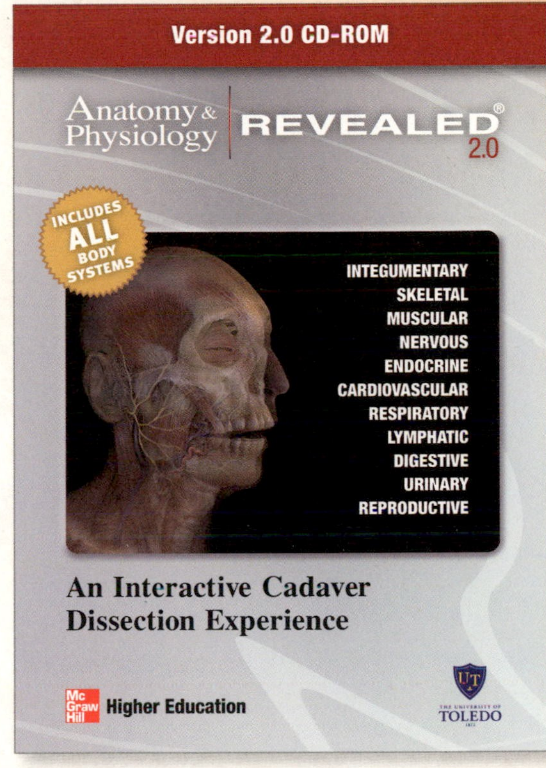

Anatomy and Physiology | Revealed

Available on CD-ROM or online at **www.aprevealed.com,** this amazing multimedia tool is designed to help students learn and review human anatomy using cadaver specimens. Detailed cadaver photographs blended together with a state-of-the-art layering technique provide a uniquely interactive dissection experience. This version features the following sections:

- **Dissection**—Peel away layers of the human body to reveal structures beneath the surface. Structures can be pinned and labeled, just like in a real dissection lab. Each labeled structure is accompanied by detailed information and an audio pronunciation. Dissection images can be captured and saved.
- **Animation**—Compelling animations demonstrate muscle actions, clarify anatomical relationships, or explain difficult concepts.
- **Histology**—Labeled light micrographs presented with each body system allow students to view tissues at their own pace.
- **Imaging**—Labeled X-ray, MRI, and CT images familiarize students with the appearance of key anatomical structures as seen through different medical imaging techniques.
- **Self-Test**—Challenging exercises let students test their ability to identify anatomical structures in a timed practical exam format or traditional multiple choice. A results page provides analysis of test scores plus links back to all incorrectly identified structures for review.
- **Anatomy Terms**—This visual glossary of general terms includes directional and regional terms, as well as planes and terms of movement.

Electronic Books

Go Green. Save Green.

McGraw-Hill eBooks.

GO GREEN!

Green. . . it's on everybody's minds these days. It's not only about saving trees; it's also about saving money. At 50% of the bookstore price, McGraw-Hill eBooks are an eco-friendly and cost-saving alternative to the traditional print textbook. So, you do some good for the environment. . . and you do some good for your wallet.

Visit **www.mhhe.com/ebooks** for details.

HOLE'S

HUMAN ANATOMY & PHYSIOLOGY

TWELFTH EDITION

DAVID SHIER
Washtenaw Community College

JACKIE BUTLER
Grayson County College

RICKI LEWIS
Alden March Bioethics Institute

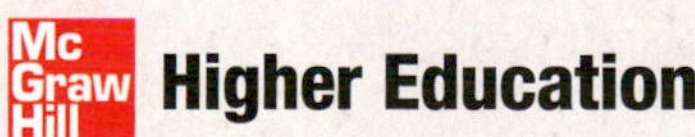

Boston Burr Ridge, IL Dubuque, IA New York San Francisco St. Louis
Bangkok Bogotá Caracas Kuala Lumpur Lisbon London Madrid Mexico City
Milan Montreal New Delhi Santiago Seoul Singapore Sydney Taipei Toronto

HOLE'S HUMAN ANATOMY & PHYSIOLOGY, TWELFTH EDITION

Published by McGraw-Hill, a business unit of The McGraw-Hill Companies, Inc., 1221 Avenue of the Americas, New York, NY 10020.

This book is printed on acid-free paper.

2 3 4 5 6 7 8 9 0 QPD/QPD 0 9

ISBN 978-0-07-352570-9
MHID 0-07-352570-7

Publisher: *Michelle Watnick*
Senior Sponsoring Editor: *James F. Connely*
Director of Development: *Kristine Tibbetts*
Developmental Editor: *Fran Schreiber*
Marketing Manager: *Lynn M. Breithaupt*
Senior Project Manager: *Jayne L. Klein*
Lead Production Supervisor: *Sandy Ludovissy*
Lead Media Project Manager: *Stacy A. Patch*
Designer: *Tara McDermott*
Cover/Interior Designer: *Elise Lansdon*
(USE) Cover Image: *© The McGraw-Hill Companies, Inc., Gerald Wofford, photographer (left photo); © Design Pics/PunchStock (middle photo); Scott Halleran/Getty Images (right photo)*
Senior Photo Research Coordinator: *John C. Leland*
Photo Research: *Danny Meldung/Photo Affairs, Inc.*
Supplement Producer: *Mary Jane Lampe*
Compositor: *Precision Graphics*
Typeface: *10/12 ITC Slimbach Std*
Printer: *Quebecor World Dubuque, IA*

The credits section for this book begins on page 977 and is considered an extension of the copyright page.

Library of Congress Cataloging-in-Publication Data

Shier, David.
Hole's human anatomy & physiology / David Shier, Jackie Butler, Ricki Lewis.—12th ed.
p. cm.
Includes index.
ISBN 978-0-07-352570-9 — ISBN 0-07-352570-7 (hard copy : alk. paper) 1. Human physiology. 2. Human anatomy. I. Butler, Jackie. II. Lewis, Ricki. III. Title. IV. Title: Hole's human anatomy and physiology. V. Title: Human anatomy & physiology.
QP34.5.S49 2010
612--dc22

2008033022

www.mhhe.com

BRIEF CONTENTS

About the Authors iv | Updates and Additions v | Learn, Practice, Assess vii |
Contents xv | Clinical Connections xx | Acknowledgments xxi |

FOUNDATIONS FOR SUCCESS XXII

UNIT I LEVELS OF ORGANIZATION 1

1 Introduction to Human Anatomy and Physiology 1
2 Chemical Basis of Life 50
3 Cells 75
4 Cellular Metabolism 114
5 Tissues 143

UNIT II SUPPORT AND MOVEMENT 170

6 Integumentary System 170
7 Skeletal System 192
8 Joints of the Skeletal System 260
9 Muscular System 284

UNIT III INTEGRATION AND COORDINATION 353

10 Nervous System I 353
11 Nervous System II 382
12 Nervous System III 437
13 Endocrine System 482

UNIT IV TRANSPORT 522

14 Blood 522
15 Cardiovascular System 552
16 Lymphatic System and Immunity 616

UNIT V ABSORPTION AND EXCRETION 651

17 Digestive System 651
18 Nutrition and Metabolism 698
19 Respiratory System 735
20 Urinary System 774
21 Water, Electrolyte, and Acid-Base Balance 810

UNIT VI THE HUMAN LIFE CYCLE 830

22 Reproductive Systems 830
23 Pregnancy, Growth, and Development 875
24 Genetics and Genomics 916

Appendixes 939
Glossary 951
Credits 977
Index 981

ABOUT THE AUTHORS

David Shier
Washtenaw Community College

My interest in physiology research and teaching began with a job as a research assistant at Harvard Medical School from 1976-1979. I completed my Ph.D. at the University of Michigan in 1984 and served on the faculty of the Medical College of Ohio from 1985-1989. I have been teaching Anatomy and Physiology and Pathophysiology full-time at Washtenaw Community College since 1990 and contributing as a member of the author team for the Hole texts since 1993.

Since the mid 1990s, when assessment of student academic achievement began to surface as a mandate for accreditation, I have become increasingly interested in the interrelationship between pedagogy and assessment. I think that we have all used some pedagogical tools (figures from the text, for example) on exams as part of assessment. Recently, in my own classroom, I have been using tools traditionally associated with assessment (e.g., lab quizzes) more and more as pedagogical tools, often in concert with group activities.

I also have interests outside of the classroom and away from the office! These include mountain biking, recorded music (vinyl!) and photography. My wife, Janet, is also an educator. We love to travel, but spend most of our time in Ann Arbor, Michigan, where we reside.

Jackie Butler
Grayson County College

My science career began in research at M.D. Anderson Hospital, where teaching was not one of my responsibilities. My masters committee at Texas A & M University quickly realized where my heart was. After I taught labs at Texas A & M for three years, they strongly recommended that I seek a teaching position when I relocated after graduation. As a result of their encouragement, I began teaching at Grayson County College in 1981. Many years later, I still feel excited and enthusiastic about being in the classroom.

John Hole's *Human Anatomy and Physiology,* Second Edition, was the book used at Grayson County College in 1981. We have continued teaching using this text through many editions. John Hole wrote a very well-organized, succinct text, appropriate for our student population. It has been a wonderful experience for me to be a part of this team that has worked to keep the text up-to-date and appropriate to the current student population. We have been selective in adding to the depth and detail of coverage in the text, so as to maintain Hole's original intent of readability and the desire not to overwhelm the student.

Outside the classroom, I enjoy traveling with my husband, Dale. Additional interests include: 6:00 AM walking with my friends (12–15 miles a week), quilting, and reading.

Ricki Lewis
Alden March Bioethics Institute

My career as a science communicator began with earning a PhD in genetics from Indiana University in 1980, and quickly blossomed into writing for newspapers and magazines, writing the introductory textbook *Life,* and teaching at several universities. Since then I have published many articles, the textbook *Human Genetics: Concepts and Applications,* an essay collection, and most recently my first novel. I love the challenge of being part of the Hole team.

Since 1984 I have been a genetic counselor for a large private ob/gyn practice. I also work with the Cure Huntington's Disease Initiative and write biotechnology market reports. As a hospice volunteer since 2005, I have learned about many disorders in a very personal manner. I also blog regularly at blog.bioethics.net.

When I'm not writing, I enjoy exercising, reading, and public speaking. I am also involved in launching a science center and teaching in an adult education program. My husband is a research chemist and we both are devoted to making science understandable to everyone. We have three daughters, many felines, a tortoise, and a hare, and reside in upstate New York and Martha's Vineyard.

UPDATES AND ADDITIONS

FROM THE AUTHORS

In biological evolution, a successful species becomes the best suited that it can be for a particular environment. In a similar manner, *Hole's Human Anatomy & Physiology* continues to evolve as a modern exploration of the human, from the cellular and molecular underpinnings of the functions of life to its interacting organ systems.

We are authors, but first and foremost we are teachers. What we and our reviewers do in class is reflected in each new edition. We are especially excited about the Learn, Practice, Assess approach to this new edition. Each chapter opens with Learning Outcomes, contains numerous opportunities to Practice throughout, and closes with Assessments that are closely tied to the learning outcomes.

Students have always come first in our approach to teaching and textbook authoring, but we now feel more excited than ever about the student-oriented, teacher-friendly quality of this text. We have never included detail for its own sake, but we have felt free to include extra detail if the end result is to clarify.

The level of this text is geared toward students in two-semester courses in anatomy and physiology who are pursuing careers in nursing and allied health fields and who have minimal background in physical and biological sciences. The first four chapters review chemistry and physiological processes. Students who have studied this material previously will view it as a welcomed review, but newcomers will not find it intimidating.

Remember that although you are working hard to successfully complete this course, you are not doing so for us, or even for your teacher. You are working for yourselves and for your future patients, as health care professionals. Your course is not so much a hurdle as a stepping stone, even more so a foundation. We have written this book to help prepare you for success along that path.

David Shier, Jackie Butler, Ricki Lewis

GLOBAL CHANGES

- Numbered A heads easily link to Learning Outcomes/Assessments
- Introductory sections for each chapter
- Practice questions dividing sections numbered sequentially for clearer/easier reference
- Added pronunciations
- New and updated boxes throughout illuminate new technologies, including biomarkers, reprogrammed (induced pluripotent) stem cells, DNA microarrays, nanotechnology, the metabolome and microbiome, microRNAs, brain banks, RNA interference, tissue engineering, vaccines, stem cell therapies, and direct-to-consumer genetic testing

Chapter 1

- New figure on directional terms
- Updated terminology to be consistent with Terminologia Anatomica
- Improved shading on figures depicting body cavities
- Gene expression profiling to flesh out anatomy and physiology ties in to final chapter

Chapter 3

- Vignette introduces new HIV drug
- Updated coverage of the mechanism of osmosis
- From Science to Technology box on tailoring stem cells to treat disease

Chapter 4

- Figure 4.20 on DNA replication includes the cell cycle, with reference to the changes in chromosome structure that occur during S phase
- From Science to Technology 4.1 discusses an innocence project case

Chapter 5

- New micrograph for chapter opener (better view of whole tissue)
- Moved intercellular junctions from chapter 3 to chapter 5 introduction
- Many new micrographs and accompanying line art
- New icon for figure 5.12*a* salivary glands instead of pancreas

Chapter 6

- New vignette highlights cryo-electron tomography view of proteins responsible for the skin's integrity
- Added to melanin production (from tyrosine in melanosomes)
- Moved skin color to melanin production earlier in chapter
- New micrographs and corresponding line art, and other new photos
- Clarified wound healing in the text and figure

Chapter 7

- Added scientific names to layers in epiphyseal plate
- Some labels added to figures to correlate to muscle attachments referenced in chapter 9
- Table of male/female skeletal differences reworked/expanded in side by side comparison

Chapter 8

- New vignette on glucosamine and chondroitin to treat arthritis
- New illustrations for joint movements using real people

Chapter 9

- New vignette on the muscular movements behind "texting"
- Piriformis and quadratus lumborum added to muscle coverage

Chapter 11

- New figure on brain and brain regions
- Updated discussion of Broca's area and Wernicke's area
- New figures add detail to brachial and lumbosacral nerve plexuses
- New Clinical Application on traumatic brain injury

Chapter 13

- Clinical Application updates performance enhancement
- Two boxes update progress in treating diabetes

Chapter 14

- New chapter opener photo
- Clinical Application case of a young editor with leukemia and the "miracle drug" Gleevec
- New Clinical Application on deep vein thrombosis
- New micrographs include the 5 types of white blood cells
- Moved up figure summarizing blood composition
- Improved figures 14.21 and 14.22
- Update of terminology (hematopoietic stem cell)
- New tables on ABO blood type frequencies and inherited blood disorders

Chapter 15

- Added coronal section of cadaver heart
- Figure 15.21 altered to emphasize depolarization/repolarization rather than valves
- Figure 15.24 added schematic of general reflex arc to correlate with the baroreceptor reflex control of heart rate
- Figures 15.53, 15.57, and 15.58 redrawn to depict paired veins in the upper and lower limbs

Chapter 16

- New micrographs
- Updated anti-rejection treatment protocols
- Added concept of herd immunity
- Figure 16.17*a* expanded to include cytotoxic and memory T cells
- Moved lymphocyte functions to T and B cell discussions for better flow

Chapter 17

- New vignette on gut microbiome
- Figure 17.4 rearranged into one column for better flow
- New micrographs and new corresponding line art
- Figure 17.17*b* revised labels
- Figure 17.19*a* new line art shows three layers of muscle
- Figure 17.19*b* new micrograph
- Figure 17.44 new radiograph of colon

Chapter 18

- Figure 18.1 expanded to include the effects of ghrelin on appetite, with the text reflecting the complexity of appetite control
- New photos for obesity/athlete/scurvy/anorexia
- Replaced Atkins diet food pyramid with Mediterranean diet pyramid
- Clinical Application 18.1 on obesity includes lap-band surgery and updated information on gastric bypass surgery
- Includes discussion of BMI

Chapter 19

- New vignette on secondhand smoke
- Updated coverage of respiratory control

Chapter 20

- New vignette on a medical mystery (Balkan nephropathy)
- New figures on nephron anatomy, including representation of the macula densa as part of the ascending limb of the nephron loop
- New table of developmental abnormalities of the urinary system

Chapter 22

- New chapter opener photo
- Moved meiosis to introduction, before details of spermatogenesis/oogenesis
- New micrographs
- Figure 22.8 more clearly explains number of chromatids per chromosome
- All new photos of birth control, including female condom and spermicides
- Updated STDs/sexually transmitted infections (STIs)

Chapter 23

- The conjoined twins in Clinical Application 23.2 are now teens!
- Clinical Application 23.4 on living to age 100
- Human embryonic stem cells de-emphasized to reflect other types of stem cells in use

Chapter 24

- Complete update and overhaul to reflect change in focus in field
- New vignette on direct-to-consumer (web-based) genetic testing
- New Clinical Application 24.1 introduces modes of inheritance through genetic counseling cases
- Final section on gene expression explaining anatomy and physiology brings the book full circle back to chapter 1

Learn, Practice, Assess!

A **major change** that you will notice in the 12th edition is a **new format.** The book is now organized with **Learning Outcomes and Assessments.**

LEARN

Learning Outcomes open the chapters, and are closely linked to Chapter Assessments and Integrative Assessments/Critical Thinking questions found at the end of each chapter.

LEARNING OUTCOMES

After you have studied this chapter, you should be able to:

5.1 Introduction
1 Describe a tissue, and explain the intercellular junctions in tissues.
2 List the four major tissue types in the body.

5.2 Epithelial Tissues
3 Describe the general characteristics and functions of epithelial tissue.
4 Name the types of epithelium and identify an organ in which each is found.
5 Explain how glands are classified.

5.3 Connective Tissues
6 Describe the general characteristics of connective tissue.
7 Compare and contrast the cellular components, structures, fibers, and extracellular matrix (where applicable) in each type of connective tissue. (p. 156)
8 Describe the major functions of each type of connective tissue.

5.4 Types of Membranes
9 Describe and locate each of the four types of membranes.

Learning tools to help you succeed...

Check out the **Chapter Preview, Foundations for Success,** on page xxiii. The Chapter Preview was specifically designed to help students **LEARN** how to study at the collegiate level and efficiently use the tools available to them. It provides helpful study tips.

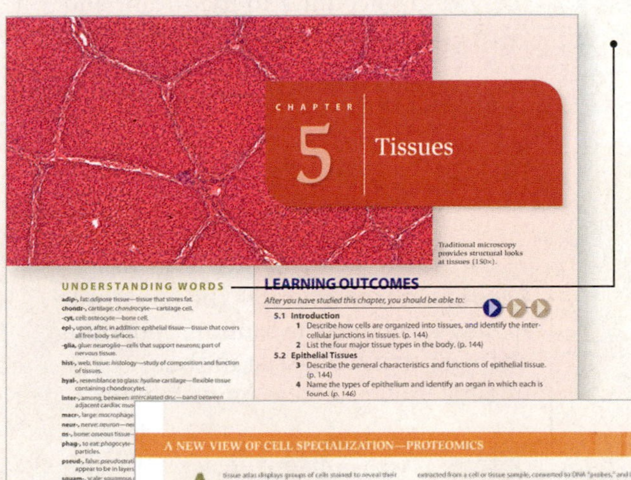

Understanding Words help you remember scientific word meanings. Examine root words, stems, prefixes, suffices, pronunciations, and build a solid anatomy and physiology vocabulary.

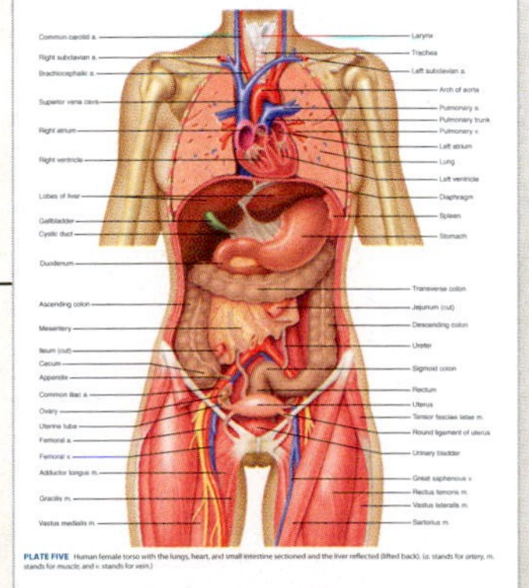

Reference Plates offer vibrant detail of body structures.

Chapter Opening Vignettes introduce each topic. Taken from headlines and scientific journal reports, they extend the student's view of the chapter content.

Learn, Practice, Assess!

PRACTICE

After each major section, a question or series of questions tests the student's understanding of the material. If he or she cannot answer these practice question(s), the student will want to reread that section.

PRACTICE 

1. What is a tissue?
2. What are the different types of intercellular junctions?
3. List the four major types of tissue.

Interesting applications help students practice and apply their knowledge...

Up to 90% of human cancers are *carcinomas,* growths that originate in epithelium. Most carcinomas begin on surfaces that contact the external environment, such as skin, linings of the airways in the respiratory tract, or linings of the stomach or intestines in the digestive tract. This observation suggests that the more common cancer-causing agents may not deeply penetrate tissues.

Boxed Information connects chapter ideas to clinical situations, discusses changes in organ structure and function, and introduces new medical technology or experiments.

5.1 FROM SCIENCE TO TECHNOLOGY

Nanotechnology meets the Blood-Brain Barrier

From Science to Technology previews the technological applications of knowledge in anatomy and physiology that students are likely to encounter in the future and explains how and why the technology was developed.

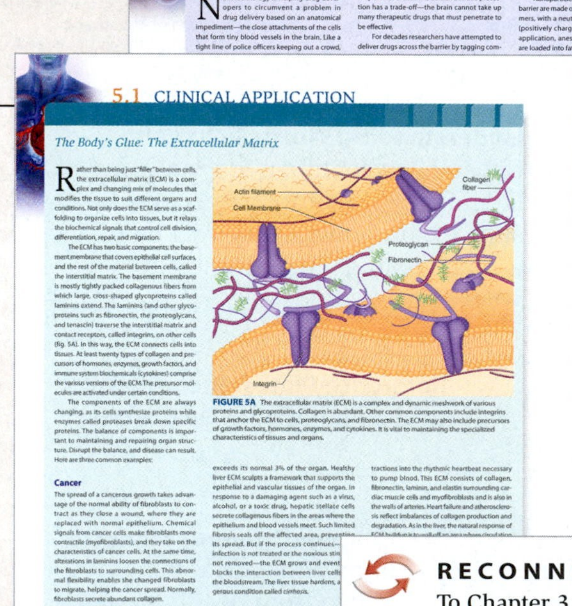
5.1 CLINICAL APPLICATION

The Body's Glue: The Extracellular Matrix

Clinical Applications encourage students to explore information on related pathology, historical insights, and clinical examples that they are likely to encounter in their careers.

Reconnect Icon prompts the student to review key concepts found in previous chapters that will assist in their understanding of new information.

RECONNECT
To Chapter 3, Movements Into and Out of the Cell, page 90.

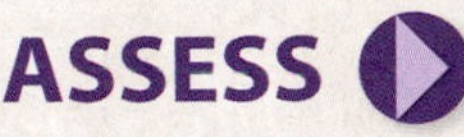

ASSESS

Tools to help students make the connection and master anatomy & physiology!

CHAPTER ASSESSMENTS

5.1 Introduction

1 Define *tissue*. (p. 144)

2 Describe three types of intercellular junctions. (p. 144)

3 Which of the following is a major tissue type in the body? (p. 144)

a. epithelial
b. nervous
c. muscle
d. connective
e. all of the above.

5.2 Epithelial Tissues

4 A general characteristic of epithelial tissues is that ________. (p. 145)

a. numerous blood vessels are present
b. cells are spaced apart
c. cells divid
d. there is m

8 Distinguish between an exocrine gland and an endocrine gland. (p. 150)

9 Describe how glands are classified according to the structure of their ducts and the organization of their cells. (p. 150)

10 A gland that secretes substances by exocytosis is a(n) ________ gland. (p. 150)

a. merocrine
b. apocrine
c. holocrine

5.3 Connective Tissues

11 Discuss the general characteristics of connective tissue. (p. 153)

12 Define *extracellular matrix* and *ground substance*. (p. 153)

13 Describe three major types of connective tissue cells. (p. 153)

Chapter Assessments found at the end of each chapter check student's understanding of the chapter's Learning Outcomes. The Chapter Assessment numbers correspond directly to the Learning Outcomes.

INTEGRATIVE ASSESSMENTS/CRITICAL THINKING

OUTCOMES 3.2, 3.6, 5.1, 5.2, 5.3, 5.5, 5.6

1. Tissue engineering combines living cells with synthetic materials to create functional substitutes for human tissues. What components would you use to engineer replacement (a) skin, (b) bone, (c) muscle, and (d) blood?

OUTCOMES 3.2, 5.2

2. In the lungs of smokers, a process called metaplasia occurs where normal lining cells of the lung are replaced by squamous metaplastic cells (many layers of squamous epithelial cells). Functionally, why is this an undesirable body reaction to tobacco smoke?

OUTCOMES 3.4, 3.5, 5.2, 5.3, 5.5, 5.6

3. Cancer-causing agents (carcinogens) usually act on dividing cells. Which of the four tissues would carcinogens most influence? Least influence?

OUTCOMES 5.2, 5.4

4. Sometimes, in response to irritants, mucous cells secrete excess mucus. What symptoms might this produce if it occurred in the (a) digestive tract or (b) respiratory passageway?

OUTCOME 5.3

5. Disorders of collagen are characterized by deterioration of connective tissues. Why would you expect such diseases to produce widely varying symptoms?

OUTCOME 5.3

6. Collagen and elastin are added to many beauty products. What type of tissues are they normally part of?

OUTCOME 5.3

7. Joints such as the shoulder, elbow, and knee contain considerable amounts of cartilage and dense regular connective tissue. How does this explain that joint injuries are often slow to heal?

Integrative Assessments/ Critical Thinking questions relate information from various Learning Outcomes within a chapter (and frequently from previous chapters) and apply that information.

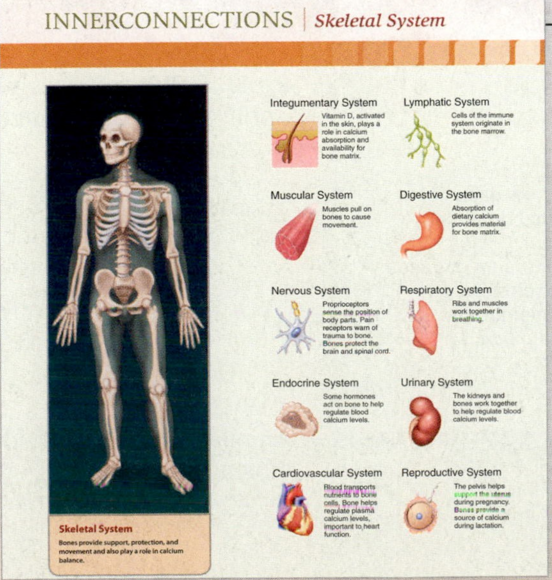

InnerConnections conceptually link the highlighted body system to every other system. These graphic representations review chapter concepts, make connections, and stress the "big picture" in learning and applying the concepts and facts of anatomy and physiology.

The Hole's Instructor Support Package
Can Help You...

- **Correlate ancillaries** that accompany your McGraw-Hill text
- **Incorporate engaging presentation materials** for lecture and lab
- **Measure your student's progress** with assessment tools and assignments
- **Improve performance** by providing self-study tools for students
- **Provide low-cost textbook alternatives** for your class

NEW for the twelfth edition **Ancillary Correlation Guide—** Instructors will find this guide invaluable. McGraw-Hill offers a variety of ancillary products to accompany our texts. The authors have gone through the ancillaries and correlated them to each Learning Outcome found at the beginning of the chapter! Here are the ancillaries that are correlated to the specific Learning Outcomes of *Hole's Human Anatomy & Physiology*, Twelfth Edition:

- Textbook
- Website—**www.mhhe.com/shier12**
- EZ Test Online
- Ph.I.L.S. 3.0
- MediaPhys 3.0
- Anatomy & Physiology Revealed (APR)
- Virtual Anatomy Dissection Review
- Student Study Guide

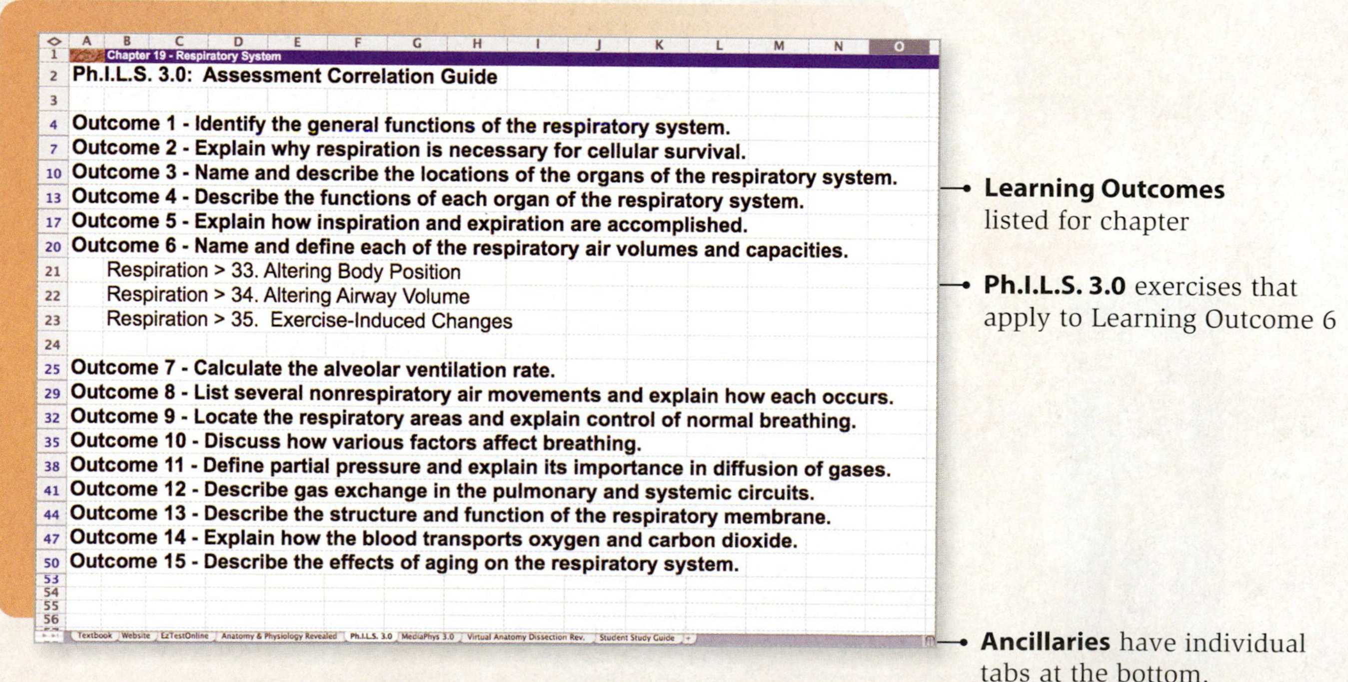

Chapter 19 - Respiratory System

Ph.I.L.S. 3.0: Assessment Correlation Guide

Row	
4	**Outcome 1 - Identify the general functions of the respiratory system.**
7	**Outcome 2 - Explain why respiration is necessary for cellular survival.**
10	**Outcome 3 - Name and describe the locations of the organs of the respiratory system.**
13	**Outcome 4 - Describe the functions of each organ of the respiratory system.**
17	**Outcome 5 - Explain how inspiration and expiration are accomplished.**
20	**Outcome 6 - Name and define each of the respiratory air volumes and capacities.**
21	Respiration > 33. Altering Body Position
22	Respiration > 34. Altering Airway Volume
23	Respiration > 35. Exercise-Induced Changes
24	
25	**Outcome 7 - Calculate the alveolar ventilation rate.**
29	**Outcome 8 - List several nonrespiratory air movements and explain how each occurs.**
32	**Outcome 9 - Locate the respiratory areas and explain control of normal breathing.**
35	**Outcome 10 - Discuss how various factors affect breathing.**
38	**Outcome 11 - Define partial pressure and explain its importance in diffusion of gases.**
41	**Outcome 12 - Describe gas exchange in the pulmonary and systemic circuits.**
44	**Outcome 13 - Describe the structure and function of the respiratory membrane.**
47	**Outcome 14 - Explain how the blood transports oxygen and carbon dioxide.**
50	**Outcome 15 - Describe the effects of aging on the respiratory system.**

Textbook | Website | EzTestOnline | Anatomy & Physiology Revealed | Ph.I.L.S. 3.0 | MediaPhys 3.0 | Virtual Anatomy Dissection Rev. | Student Study Guide

Learning Outcomes listed for chapter
Ph.I.L.S. 3.0 exercises that apply to Learning Outcome 6
Ancillaries have individual tabs at the bottom.

Incorporate Engaging Presentation Materials for Lecture and Lab

Incorporate customized lectures, visually enhanced tests and quizzes, compelling course websites, or attractive printed support materials using **McGraw-Hill's Presentation Tools.**

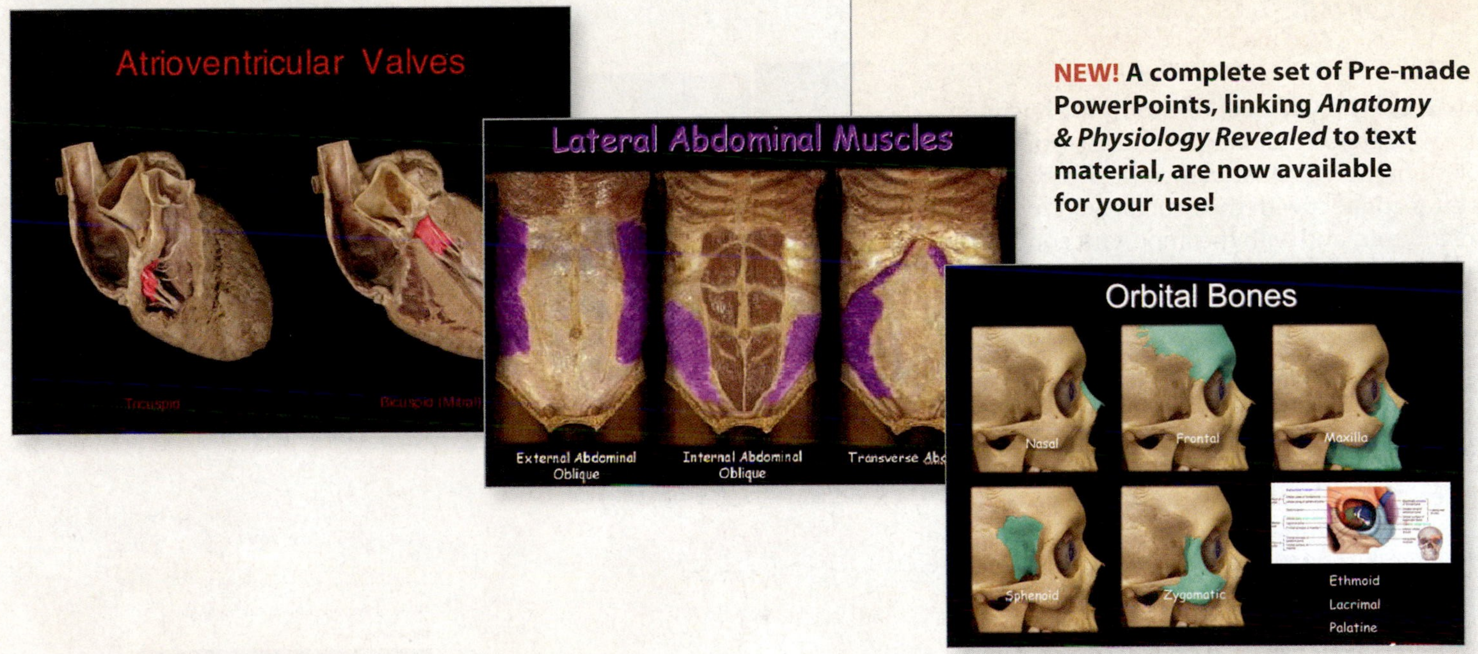

NEW! A complete set of Pre-made PowerPoints, linking *Anatomy & Physiology Revealed* to text material, are now available for your use!

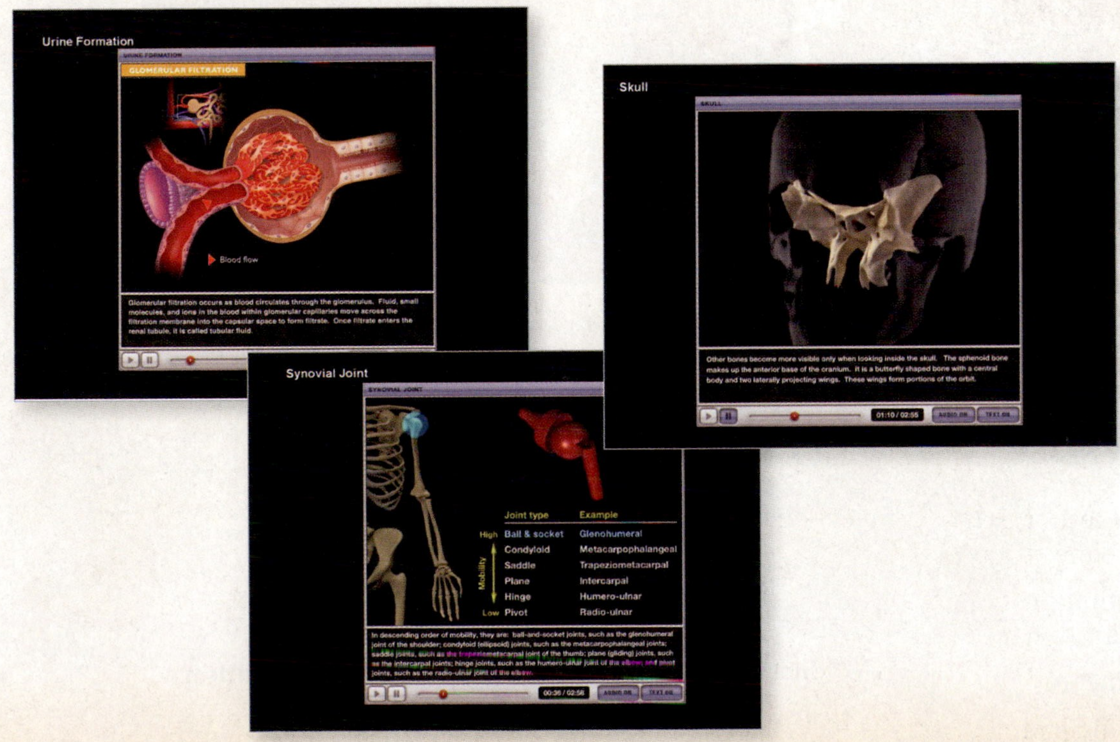

NEW! A complete set of animation embedded PowerPoint slides are now available.

Measure Your Student's Progress with Assessment Tools and Assignments

McGraw-Hill Connect Anatomy & Physiology is a web-based assignment and assessment platform that gives students the means to better connect with their coursework, with their instructors, and with the important concepts that they will need to know for success now and in the future. With Connect Anatomy & Physiology, instructors can deliver assignments, quizzes, and tests easily online. Students can practice important skills at their own pace and on their own schedule. With Connect Anatomy & Physiology Plus, students also get 24/7 online access to an eBook — an online edition of the text — to aid them in successfully completing their work, wherever and whenever they choose. www.mhhe.com/shier12

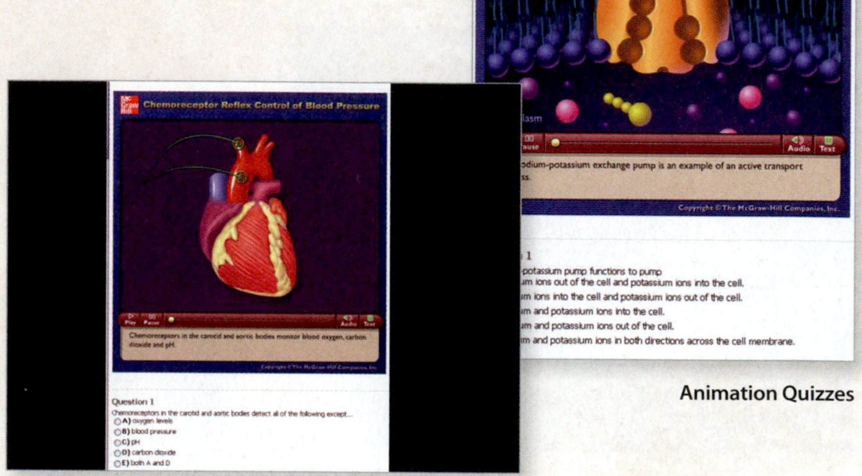

Animation Quizzes

Computerized Test Bank Edited by Author Team!

- Powered by McGraw-Hill's flexible electronic testing program EZ Test Online EZ TEST.
- Create paper and online tests or quizzes in one program!
- Create tests that can be easily shared with colleagues, adjuncts, WebCT, Blackboard, PageOut, and Apple's iQuiz.
- Sort questions by difficulty level or learning outcome.
- Create and access your test or quiz anywhere, at any time.
- Select questions from multiple McGraw-Hill test banks.
- Manage your tests online.
- Online automated scoring and reporting are also available.

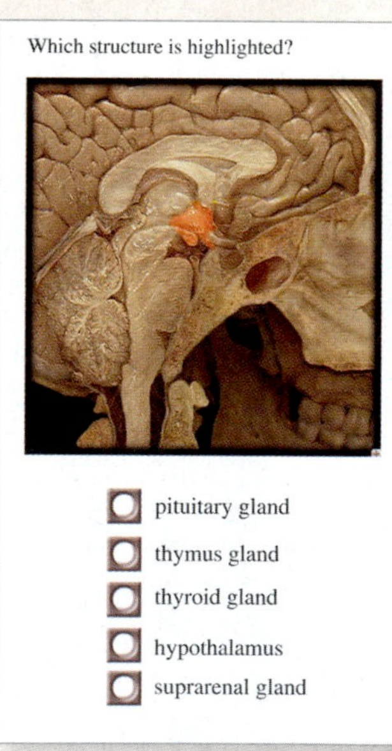

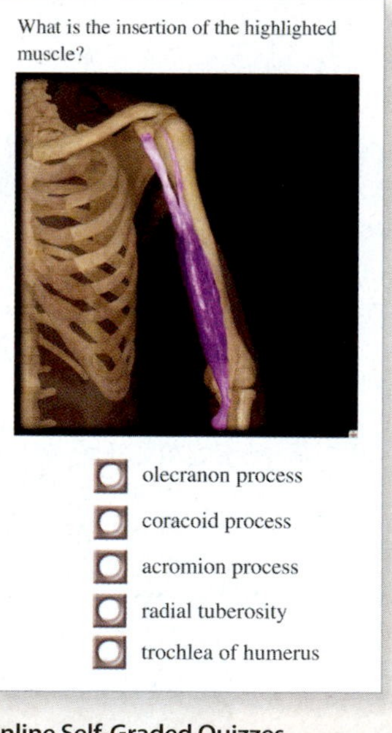

Online Self-Graded Quizzes

Instructor's Manual—*New* for the twelfth edition

Instructor's manual with all lecture outline suggestions and topical outlines tied to specific Learning Outcomes.

Improve Performance by Providing Self-Study Tools for Students

In a recent student survey:

96% **of students felt APR was fun to use!**

80% **of students reported they studied more often because of APR!**

94% **of students felt using APR helped improve their grade!**

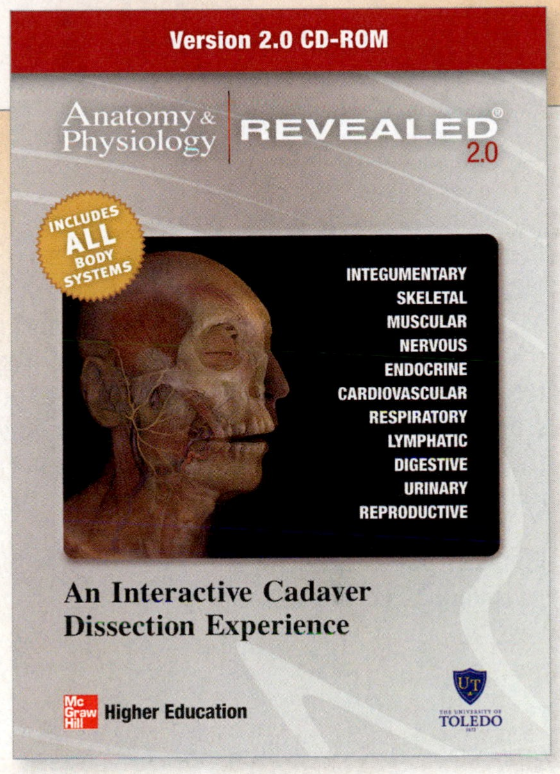

Anatomy & Physiology | Revealed 2.0

This amazing multimedia tool is designed to help students learn and review human anatomy using cadaver specimens. Detailed cadaver photographs blended together with a state-of-the-art layering technique provide a uniquely interactive dissection experience.

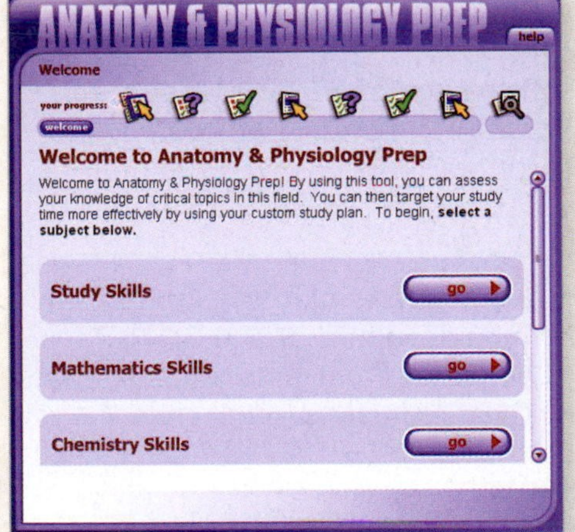

A&P Prep

A&P Prep, also available on the text website, helps students to prepare for their upcoming coursework in anatomy and physiology. This website enables students to perform self assessments, conduct self study sessions with tutorials, and perform a post assessment of their knowledge in the following areas:

- Biology Skills
- Mathematics Skills
- Chemistry Skills
- Physics Skills
- Study Skills
- Medical Terminology

Provide Low-Cost Textbook Alternatives for Your Class

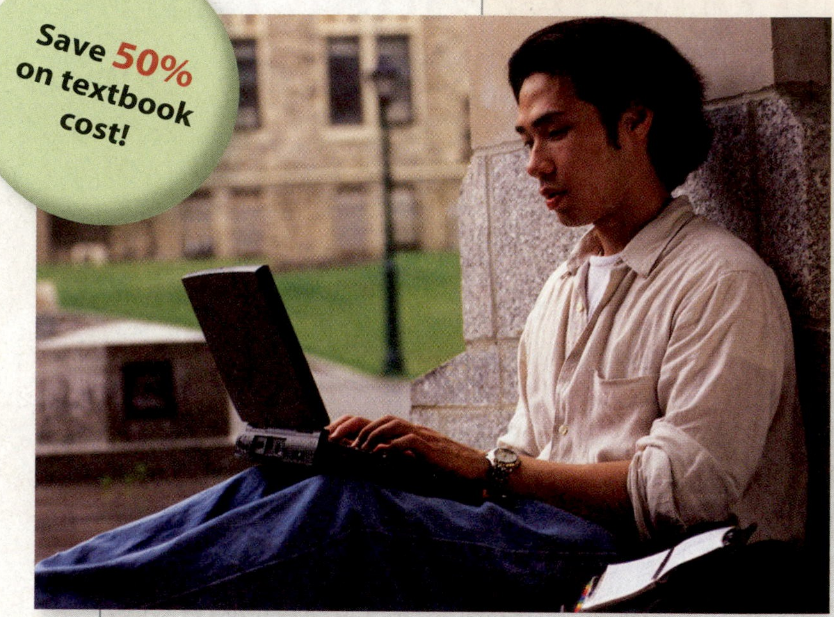

Electronic Books

If you or your students are ready for an alternative version of the traditional textbook, McGraw-Hill eBooks offer a cheaper and eco-friendly alternative to traditional textbooks. By purchasing eBooks from McGraw-Hill, students can save as much as 50% on selected titles delivered on the most advanced E-book platform available.

Contact your McGraw-Hill sales representative to discuss E-book packaging options.

Other Resources Available

Student Study Guide

by Nancy A. Sickels Corbett offers chapter overviews, chapter outcomes, focus questions, mastery tests, study activities, and mastery test answers.

Physiology Tutorials

MediaPhys offers detailed explanations, high quality illustrations, and animations to provide students with a thorough introduction to the world of physiology—giving them a virtual tour of physiological processes.

Physiology Interactive Lab Simulations

Ph.I.L.S. 3.0 offers 37 lab simulations that may be used to supplement or substitute for wet labs.

Clinical Applications Manual

This manual expands on *Anatomy & Physiology's* clinical themes, introduces new clinical topics, and provides test questions and case studies to develop students' abilities to apply knowledge to realistic situations. A print version is available for students.

Lab Manual Options to fit your course

Laboratory Manual for Hole's Human Anatomy and Physiology by Terry R. Martin, Kishwaukee College, is designed to accompany the twelfth edition of *Hole's Human Anatomy & Physiology*. **New** for the twelfth edition: This laboratory manual now comes in a cat or fetal pig version!

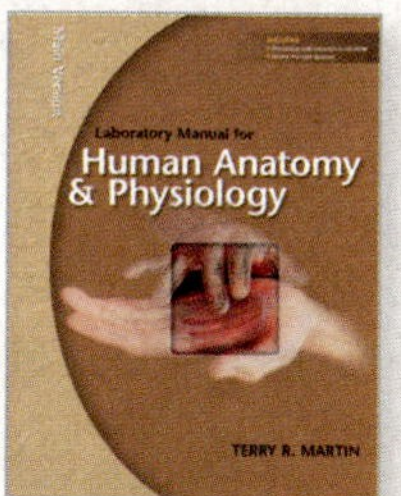

NEW! *The Laboratory Manual for Human Anatomy & Physiology* by Terry Martin of Kishwaukee College is written to coincide with any A&P textbook.

- 3 versions—main, cat, and fetal pig
- Includes Ph.I.L.S. 3.0 CD-ROM
- Outcomes and Assessments format
- Clear, concise writing style

Student Supplements

McGraw-Hill offers various tools and technology products to support the textbook. Students can order supplemental study materials by contacting their campus bookstore or online at **www.shopmcgraw-hill.com**.

Instructor Supplements

Instructors can obtain teaching aids by calling the McGraw-Hill Customer Service Department at 1-800-338-3987, visiting our online catalog at **www.mhhe.com**, or by contacting their local McGraw-Hill sales representative.

CONTENTS

FOUNDATIONS FOR SUCCESS XXIII

UNIT I **LEVELS OF ORGANIZATION 1**

CHAPTER 1

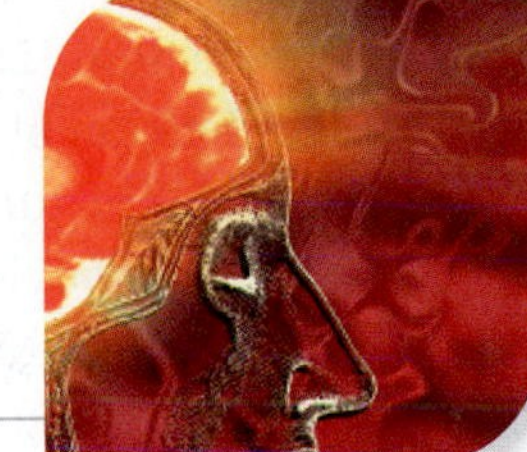

Introduction to Human Anatomy and Physiology 1

1.1 Introduction 3
1.2 Anatomy and Physiology 4
1.3 Levels of Organization 4
1.4 Characteristics of Life 6
1.5 Maintenance of Life 7
1.6 Organization of the Human Body 12
1.7 Life-Span Changes 20
1.8 Anatomical Terminology 20
Some Medical and Applied Sciences 24

CHAPTER SUMMARY 26
CHAPTER ASSESSMENTS 28
INTEGRATIVE ASSESSMENTS/CRITICAL THINKING 29

REFERENCE PLATES 1–25
The Human Organism 30

CHAPTER 2

Chemical Basis of Life 50

2.1 Introduction 51
2.2 Structure of Matter 51
2.3 Chemical Constituents of Cells 60

CHAPTER SUMMARY 72
CHAPTER ASSESSMENTS 73
INTEGRATIVE ASSESSMENTS/CRITICAL THINKING 74

CHAPTER 3

Cells 75

3.1 Introduction 76
3.2 A Composite Cell 76
3.3 Movements Into and Out of the Cell 90
3.4 The Cell Cycle 100
3.5 Control of Cell Division 103
3.6 Stem and Progenitor Cells 105
3.7 Cell Death 106

CHAPTER SUMMARY 109
CHAPTER ASSESSMENTS 111
INTEGRATIVE ASSESSMENTS/CRITICAL THINKING 113

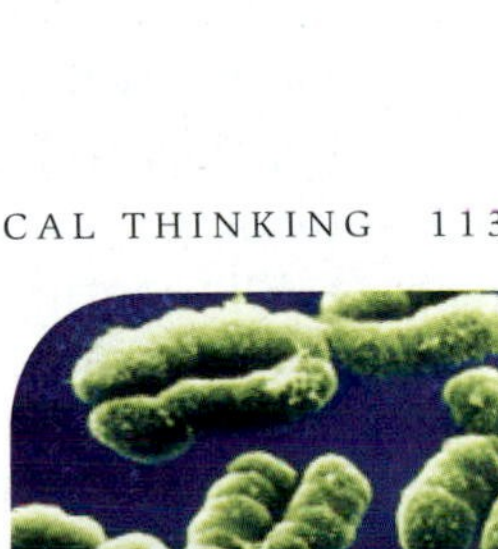

CHAPTER 4

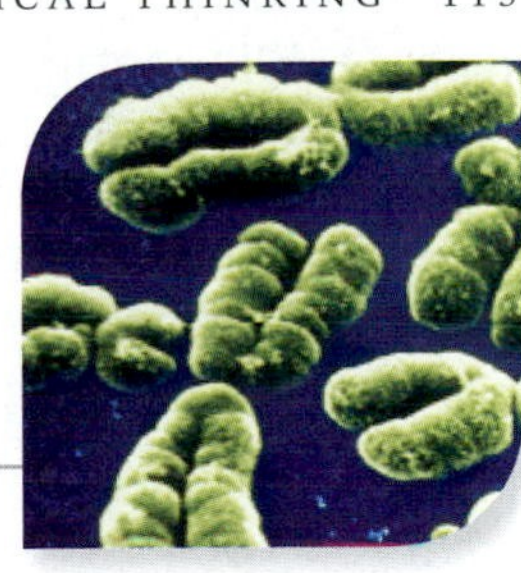

Cellular Metabolism 114

4.1 Introduction 115
4.2 Metabolic Processes 115
4.3 Control of Metabolic Reactions 117
4.4 Energy for Metabolic Reactions 119
4.5 Cellular Respiration 120
4.6 Nucleic Acids and Protein Synthesis 124
4.7 Changes in Genetic Information 135

CHAPTER SUMMARY 138
CHAPTER ASSESSMENTS 141
INTEGRATIVE ASSESSMENTS/CRITICAL THINKING 142

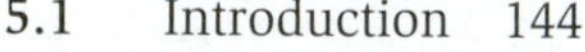

CHAPTER 5

Tissues 143

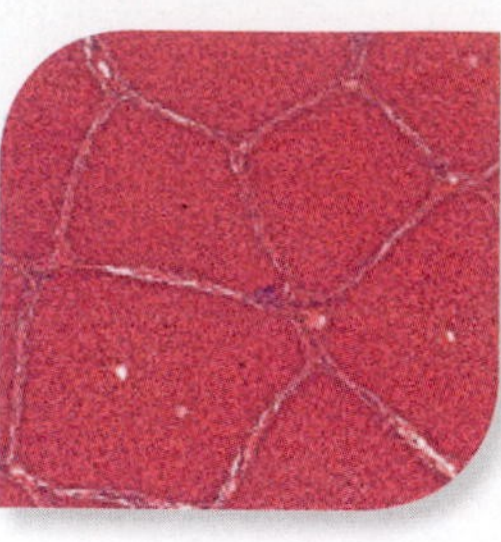

5.1 Introduction 144
5.2 Epithelial Tissues 144
5.3 Connective Tissues 152
5.4 Types of Membranes 163
5.5 Muscle Tissues 163
5.6 Nervous Tissues 164
CHAPTER SUMMARY 165
CHAPTER ASSESSMENTS 168
INTEGRATIVE ASSESSMENTS/CRITICAL THINKING 169

UNIT II SUPPORT AND MOVEMENT 170

CHAPTER 6

Integumentary System 170

6.1 Introduction 171
6.2 Skin and Its Tissues 171
6.3 Accessory Structures of the Skin 177
6.4 Regulation of Body Temperature 181
6.5 Healing of Wounds and Burns 183
6.6 Life-Span Changes 186
INNERCONNECTIONS: INTEGUMENTARY SYSTEM 188
CHAPTER SUMMARY 189
CHAPTER ASSESSMENTS 190
INTEGRATIVE ASSESSMENTS/CRITICAL THINKING 191

CHAPTER 7

Skeletal System 192

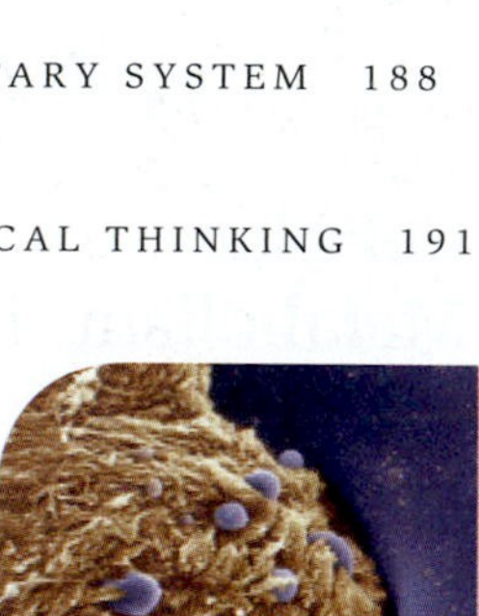

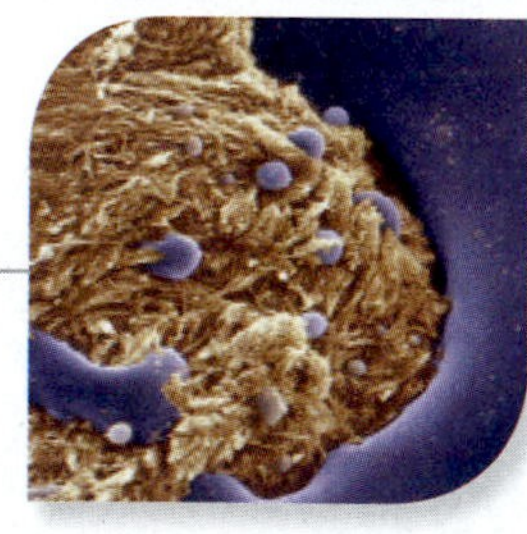

7.1 Introduction 193
7.2 Bone Structure 193
7.3 Bone Development and Growth 197
7.4 Bone Function 202
7.5 Skeletal Organization 205
7.6 Skull 206
7.7 Vertebral Column 218
7.8 Thoracic Cage 222
7.9 Pectoral Girdle 225
7.10 Upper Limb 226
7.11 Pelvic Girdle 231
7.12 Lower Limb 234
7.13 Life-Span Changes 238
INNERCONNECTIONS: SKELETAL SYSTEM 239
CHAPTER SUMMARY 240
CHAPTER ASSESSMENTS 243
INTEGRATIVE ASSESSMENTS/CRITICAL THINKING 244

REFERENCE PLATES 26–54

Human Skull 245

CHAPTER 8

Joints of the Skeletal System 260

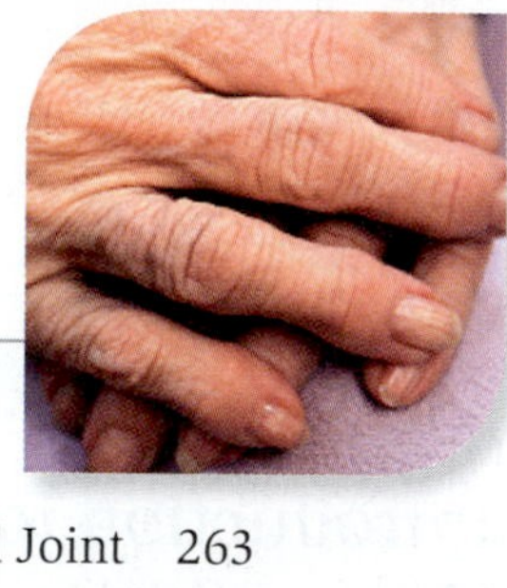

8.1 Introduction 261
8.2 Classification of Joints 261
8.3 General Structure of a Synovial Joint 263
8.4 Types of Synovial Joints 265
8.5 Types of Joint Movements 267
8.6 Examples of Synovial Joints 271
8.7 Life-Span Changes 278
CHAPTER SUMMARY 281
CHAPTER ASSESSMENTS 282
INTEGRATIVE ASSESSMENTS/CRITICAL THINKING 283

CHAPTER 9

Muscular System 284

9.1 Introduction 285
9.2 Structure of a Skeletal Muscle 285
9.3 Skeletal Muscle Contraction 289
9.4 Muscular Responses 296
9.5 Smooth Muscles 300
9.6 Cardiac Muscle 301
9.7 Skeletal Muscle Actions 301
9.8 Major Skeletal Muscles 305
9.9 Life-Span Changes 334
INNERCONNECTIONS: MUSCULAR SYSTEM 333
CHAPTER SUMMARY 334
CHAPTER ASSESSEMENTS 337
INTEGRATIVE ASSESSMENTS/CRITICAL THINKING 339

REFERENCE PLATES 55–75

Surface Anatomy and Cadaver Dissection 341

UNIT III INTEGRATION AND COORDINATION 353

CHAPTER 10

Nervous System I: Basic Structure and Function 353

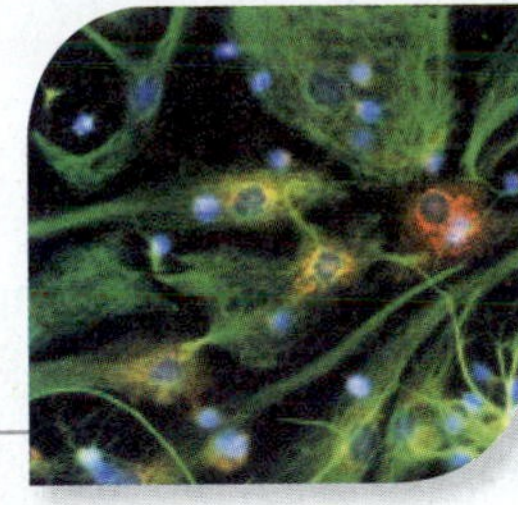

10.1 Introduction 354
10.2 General Functions of the Nervous System 355
10.3 Description of Cells of the Nervous System 356
10.4 Classification of Cells of the Nervous System 359
10.5 The Synapse 365
10.6 Cell Membrane Potential 365
10.7 Synaptic Transmission 371
10.8 Impulse Processing 374
CHAPTER SUMMARY 376
INNERCONNECTIONS: NERVOUS SYSTEM 378
CHAPTER ASSESSMENTS 380
INTEGRATIVE ASSESSMENTS/CRITICAL THINKING 381

CHAPTER 11

Nervous System II: Divisions of the Nervous System 382

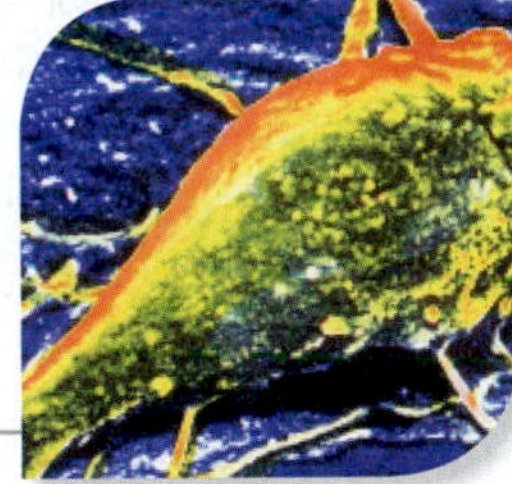

11.1 Introduction 384
11.2 Meninges 384
11.3 Ventricles and Cerebrospinal Fluid 385
11.4 Spinal Cord 387
11.5 Brain 397
11.6 Peripheral Nervous System 411
11.7 Autonomic Nervous System 424
11.8 Life-Span Changes 431
CHAPTER SUMMARY 432
CHAPTER ASSESSMENTS 434
INTEGRATIVE ASSESSMENTS/CRITICAL THINKING 436

CHAPTER 12

Nervous System III: Senses 437

12.1 Introduction 438
12.2 Receptors, Sensation, and Perception 438
12.3 General Senses 440
12.4 Special Senses 446
12.5 Life-Span Changes 476
CHAPTER SUMMARY 477
CHAPTER ASSESSMENTS 480
INTEGRATIVE ASSESSMENTS/CRITICAL THINKING 481

CHAPTER 13

Endocrine System 482

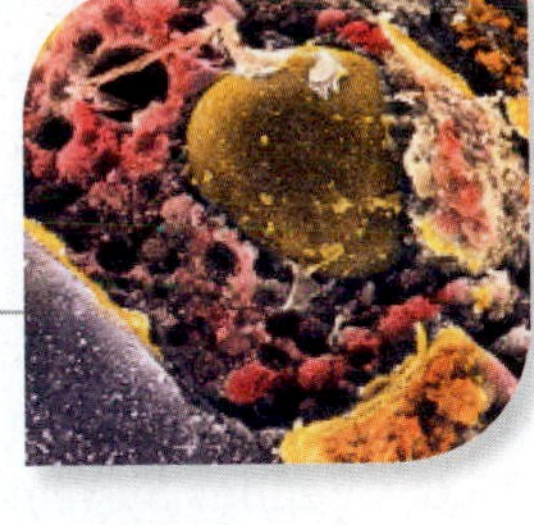

13.1 Introduction 483
13.2 General Characteristics of the Endocrine System 483
13.3 Hormone Action 484
13.4 Control of Hormonal Secretions 491
13.5 Pituitary Gland 492
13.6 Thyroid Gland 499
13.7 Parathyroid Glands 502
13.8 Adrenal Glands 504
13.9 Pancreas 509
13.10 Other Endocrine Glands 511
13.11 Stress and Its Effects 513
13.12 Life-Span Changes 515
INNERCONNECTIONS: ENDOCRINE SYSTEM 516
CHAPTER SUMMARY 517
CHAPTER ASSESSMENTS 519
INTEGRATIVE ASSESSMENTS/CRITICAL THINKING 520

UNIT IV TRANSPORT 522

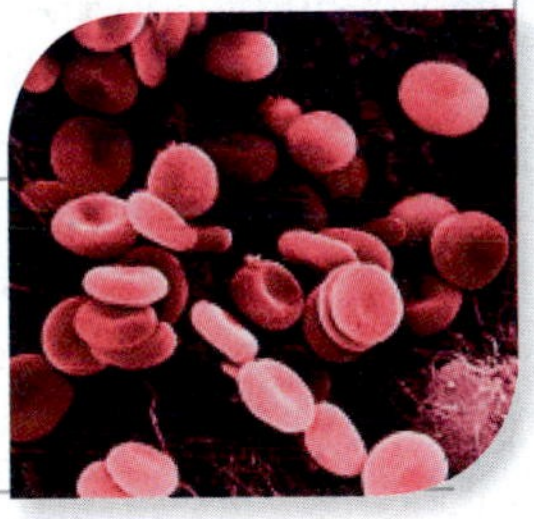

CHAPTER 14

Blood 522

14.1 Introduction 523
14.2 Blood Cells 524
14.3 Blood Plasma 535
14.4 Hemostasis 538
14.5 Blood Groups and Transfusions 544
CHAPTER SUMMARY 548
CHAPTER ASSESSMENTS 550
INTEGRATIVE ASSESSMENTS/CRITICAL THINKING 551

CHAPTER **15**

Cardiovascular System 552

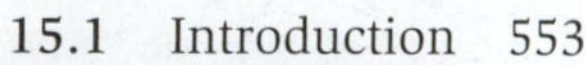

15.1 Introduction 553
15.2 Structure of the Heart 553
15.3 Heart Actions 564
15.4 Blood Vessels 573
15.5 Blood Pressure 580
15.6 Paths of Circulation 590
15.7 Arterial System 592
15.8 Venous System 600
15.9 Life-Span Changes 606
INNERCONNECTIONS: CARDIOVASCULAR SYSTEM 610
CHAPTER SUMMARY 611
CHAPTER ASSESSMENTS 614
INTEGRATIVE ASSESSMENTS/CRITICAL THINKING 615

CHAPTER **16**

Lymphatic System and Immunity 616

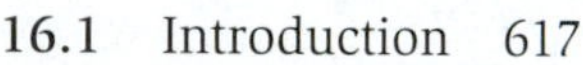

16.1 Introduction 617
16.2 Lymphatic Pathways 617
16.3 Tissue Fluid and Lymph 619
16.4 Lymph Movement 621
16.5 Lymph Nodes 621
16.6 Thymus and Spleen 623
16.7 Body Defenses Against Infection 625
16.8 Innate (Nonspecific) Defenses 626
16.9 Adaptive (Specific) Defenses or Immunity 628
16.10 Life-Span Changes 644
CHAPTER SUMMARY 644
INNERCONNECTIONS: LYMPHATIC SYSTEM 645
CHAPTER ASSESSMENTS 648
INTEGRATIVE ASSESSMENTS/CRITICAL THINKING 649

UNIT V ABSORPTION AND EXCRETION 651

CHAPTER **17**

Digestive System 651

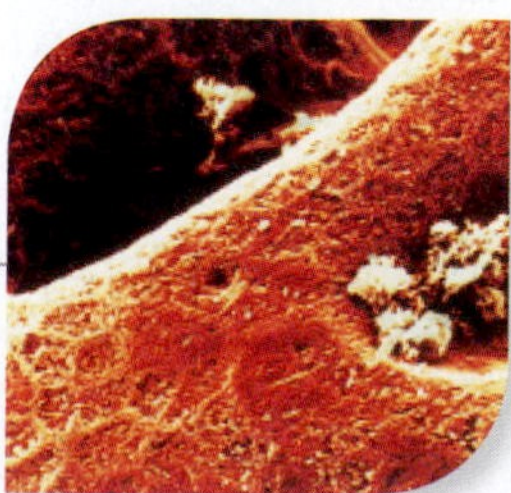

17.1 Introduction 652
17.2 General Characteristics of the Alimentary Canal 652
17.3 Mouth 656
17.4 Salivary Glands 660
17.5 Pharynx and Esophagus 661
17.6 Stomach 665
17.7 Pancreas 671
17.8 Liver 673
17.9 Small Intestine 678
17.10 Large Intestine 686
17.11 Life-Span Changes 690
INNERCONNECTIONS: DIGESTIVE SYSTEM 692
CHAPTER SUMMARY 693
CHAPTER ASSESSMENTS 696
INTEGRATIVE ASSESSMENTS/CRITICAL THINKING 697

CHAPTER **18**

Nutrition and Metabolism 698

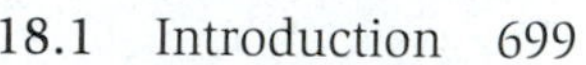

18.1 Introduction 699
18.2 Carbohydrates 700
18.3 Lipids 702
18.4 Proteins 704
18.5 Energy Expenditures 706
18.6 Vitamins 709
18.7 Minerals 717
18.8 Healthy Eating 722
18.9 Life-Span Changes 728
CHAPTER SUMMARY 729
CHAPTER ASSESSMENTS 732
INTEGRATIVE ASSESSMENTS/CRITICAL THINKING 733

CHAPTER **19**

Respiratory System 735

19.1 Introduction 736
19.2 Why We Breathe 736
19.3 Organs of the Respiratory System 737
19.4 Breathing Mechanism 747
19.5 Control of Breathing 755
19.6 Alveolar Gas Exchanges 759
19.7 Gas Transport 762
19.8 Life-Span Changes 767
INNERCONNECTIONS: RESPIRATORY SYSTEM 768
CHAPTER SUMMARY 769
CHAPTER ASSESSMENTS 771
INTEGRATIVE ASSESSMENTS/CRITICAL THINKING 773

CHAPTER 20
Urinary System 774

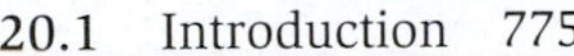
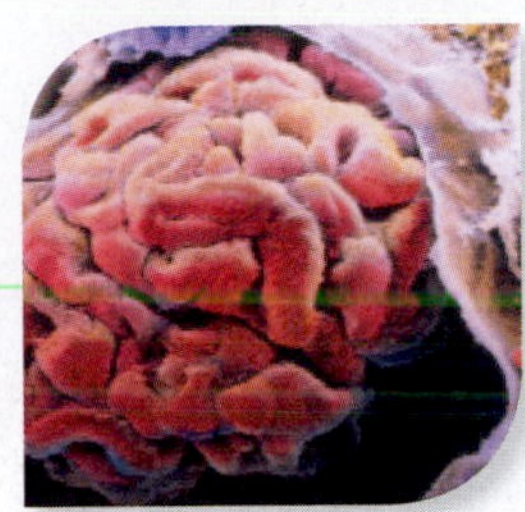

20.1 Introduction 775
20.2 Kidneys 776
20.3 Urine Formation 785
20.4 Elimination of Urine 798
20.5 Life-Span Changes 803
INNERCONNECTIONS: URINARY SYSTEM 804
CHAPTER SUMMARY 805
CHAPTER ASSESSMENTS 807
INTEGRATIVE ASSESSMENTS/CRITICAL THINKING 808

CHAPTER 21
Water, Electrolyte, and Acid-Base Balance 810

21.1 Introduction 811
21.2 Distribution of Body Fluids 811
21.3 Water Balance 813
21.4 Electrolyte Balance 815
21.5 Acid-Base Balance 819
21.6 Acid-Base Imbalances 824
CHAPTER SUMMARY 826
CHAPTER ASSESSMENTS 828
INTEGRATIVE ASSESSMENTS/CRITICAL THINKING 829

UNIT VI THE HUMAN LIFE CYCLE 830

CHAPTER 22
Reproductive Systems 830

22.1 Introduction 831
22.2 Organs of the Male Reproductive System 833
22.3 Hormonal Control of Male Reproductive Functions 845
22.4 Organs of the Female Reproductive System 846
22.5 Hormonal Control of Female Reproductive Functions 857
22.6 Mammary Glands 861
22.7 Birth Control 862
22.8 Sexually Transmitted Infections 867
INNERCONNECTIONS: REPRODUCTIVE SYSTEM 868
CHAPTER SUMMARY 869
CHAPTER ASSESSMENTS 873
INTEGRATIVE ASSESSMENTS/CRITICAL THINKING 874

CHAPTER 23
Pregnancy, Growth, and Development 875

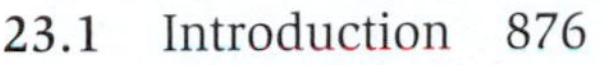

23.1 Introduction 876
23.2 Pregnancy 876
23.3 Prenatal Period 879
23.4 Postnatal Period 904
23.5 Aging 909
CHAPTER SUMMARY 911
CHAPTER ASSESSMENTS 914
INTEGRATIVE ASSESSMENTS/CRITICAL THINKING 915

CHAPTER 24
Genetics and Genomics 916

24.1 Introduction 917
24.2 Modes of Inheritance 918
24.3 Factors That Affect Expression of Single Genes 924
24.4 Multifactorial Traits 924
24.5 Matters of Sex 927
24.6 Chromosome Disorders 929
24.7 Gene Expression Explains Aspects of Anatomy and Physiology 933
CHAPTER SUMMARY 935
CHAPTER ASSESSMENTS 936
INTEGRATIVE ASSESSMENTS/CRITICAL THINKING 937

APPENDIX A Periodic Table of Elements 939
APPENDIX B Laboratory Tests of Clinical Importance 940
APPENDIX C Cellular Respiration 944
APPENDIX D A Closer Look at DNA and RNA Structures 948

Glossary *951*
Credits *977*
Index *981*

CLINICAL CONNECTIONS

Clinical Applications and From Science to Technology

1.1: Ultrasonography And Magnetic Resonance Imaging: A Tale Of Two Patients 6

2.1: Radioactive Isotopes Reveal Physiology 54
2.2: Ionizing Radiation: From the Cold War to Yucca Mountain 56
2.3: CT Scanning and PET Imaging 70

3.1: Faulty Ion Channels Cause Disease 82
3.2: Disease at the Organelle Level 87
3.1: Tailoring Stem Cells to Treat Disease 108

4.1: DNA Profiling Frees A Prisoner 130
4.2: Nucleic Acid Amplification 132
4.3: MicroRNAs and RNA Interference 136
4.4: The Human Metabolome 139

5.1: The Body's Glue: The Extracellular Matrix 154
5.2: Abnormalities of Collagen 157
5.1: Nanotechnology Meets the Blood-Brain Barrier 145
5.2: Tissue Engineering: Replacement Bladders and Hearts 166

6.1: Tanning and Skin Cancer 175
6.2: Hair Loss 179
6.3: Acne 181
6.4: Elevated Body Temperature 184

7.1: Fractures 202
7.2: Osteopenia and Osteoporosis: Preventing "Fragility Fractures" 204
7.3: Disorders of the Vertebral Column 225

8.1: Replacing Joints 276
8.2: Joint Disorders 278

9.1: Myasthenia Gravis 291
9.2: Use and Disuse of Skeletal Muscles 300
9.3: TMJ Syndrome 308

10.1: Migraine 356
10.2: Multiple Sclerosis 360
10.3: Factors Affecting Impulse Conduction 372
10.4: Opiates in the Human Body 375
10.5: Drug Addiction 377

11.1: Cerebrospinal Fluid Pressure 388
11.2: Uses of Reflexes 394
11.3: Spinal Cord Injuries 397
11.4: Traumatic Brain Injury 405
11.5: Parkinson Disease 406
11.6: Brain Waves 411
11.7: Spinal Nerve Injuries 424

12.1: Treating Pain 445
12.2: Mixed-Up Senses—Synesthesia 447
12.3: Smell and Taste Disorders 451
12.4: Getting a Cochlear Implant 455
12.5: Hearing Loss 459
12.6: Refraction Disorders 472

13.1: Using Hormones to Improve Athletic Performance 489
13.2: Growth Hormone Ups and Downs 497
13.3: Disorders of the Adrenal Cortex 509
13.4: Diabetes Mellitus 512
13.1: Treating Diabetes 513

14.1: King George III and Porphyria Variegata 530
14.2: Leukemia 536
14.3: Deep Vein Thrombosis 543
14.1: Blood Typing and Matching: From Serology to DNA Chips 546

15.1: Arrhythmias 572
15.2: Blood Vessel Disorders 581
15.3: Measurement of Arterial Blood Pressure 582
15.4: Space Medicine 585
15.5: Hypertension 588
15.6: Exercise and the Cardiovascular System 590
15.7: Molecular Causes of Cardiovascular Disease 608
15.8: Coronary Artery Disease 609
15.1: Replacing the Heart—From Transplants to Stem Cell Implants 566
15.2: Altering Angiogenesis 574

16.1: Immunity Breakdown: AIDS 642
16.1: Immunotherapy 636

17.1: Dental Caries 661
17.2: Oh, My Aching Stomach! 670
17.3: Hepatitis 677
17.4: Gallbladder Disease 679
17.5: Disorders of the Large Intestine 691
17.1: Replacing the Liver 676

18.1: Obesity 710
18.2: Dietary Supplements—Proceed with Caution 725
18.3: Nutrition and the Athlete 726

19.1: The Effects of Cigarette Smoking on the Respiratory System 739
19.2: Lung Irritants 749
19.3: Respiratory Disorders That Decrease Ventilation: Bronchial Asthma and Emphysema 756
19.4: Exercise and Breathing 760
19.5: Effects of High Altitude 763
19.6: Disorders That Impair Gas Exchange: Pneumonia, Tuberculosis, and Adult Respiratory Distress Syndrome 765

20.1: Chronic Kidney Failure 779
20.2: Glomerulonephritis 782
20.3: The Nephrotic Syndrome 793
20.4: Urinalysis: Clues to Health 803

21.1: Water Balance Disorders 816
21.2: Sodium and Potassium Imbalances 820

22.1: Prostate Enlargement 841
22.2: Male Infertility 842
22.3: Female Infertility 861
22.4: Treating Breast Cancer 863

23.1: Some Causes of Birth Defects 896
23.2: Joined For Life 900
23.3: Human Milk—The Perfect Food for Human Babies 904
23.4: Living to 100—And Beyond 910
23.1: Assisted Reproductive Technologies 878
23.2: Preimplantation Genetic Diagnosis 883

24.1: Genetic Counselors Communicate Modes of Inheritance 923
24.2: Down Syndrome 931

ACKNOWLEDGMENTS

Any textbook is the result of hard work by a large team. Although we directed the revision, many "behind-the scenes" people at McGraw-Hill were indispensable to the project. We would like to thank our editorial team of Marty Lange, Michelle Watnick, Jim Connely, and Fran Schreiber; Lynn Breithaupt, Marketing Manager, and our production team, which included Jayne Klein, Sandy Ludovissy, Tara McDermott, and John Leland; Kim Brucker, art director, Precision Graphics; and most of all, John Hole, for giving us the opportunity and freedom to continue his classic work. We also thank our wonderfully patient families for their support.

Thank you also to our McGraw-Hill "Champions": Clark Bierle, Grace Dueck, Kevin Fearns, Julie Halbritter, Laurie Helling, Paul Moorman, Barry Nitzberg, Kelly Post, Tracy Sawchuk, and Susan Vorwald.

David Shier, Jackie Butler, Ricki Lewis

REVIEWERS

We would like to acknowledge the valuable contributions of all professors and their students who have provided detailed recommendations for improving chapter content and illustrations throughout the revision process for each edition. Hundreds of professors from the U.S., Canada, and Europe have played a vital role in building a solid foundation for *Hole's Human Anatomy & Physiology.*

Emily Allen, *Gloucester County College*
Sharon D. Allen, *Rowan Cabarrus Community College*
Lynne Anderson, *Meridian Community College*
R. Michael Anson, *The Community College of Baltimore County*
Paul Allan Buttenhoff, *College of St. Catherine*
Jackie Carnegie, *University of Ottawa*
James T. Daniels, *Southern Arkansas University*
Cara L. Davies, *Ohio Northern University*
Sondra Dubowsky, *Allen County Community College*
Mary Catherine Flath, *Ashland Community and Technical College*
Juan C. Gutierrez, *North Harris College*
Michael J. Harman, *North Harris College*
Clare Hays, *Metropolitan State College of Denver*
Julie Huggins, *Arkansas State University*
J. Timothy Inglis, *University of British Columbia*
John Koons, *Jackson State Community College, Jackson Tennessee*
Tyjuanna R. LaBennett, *North Carolina Central University*
Joan Esterline Lafuze, *Indiana University East Richmond*
Barbara Mania-Farnell, *Purdue University Calumet*
Cynthia Conaway Mavroidis, *Northwest State Community College*
Jennifer McLeese, *University of Manitoba*
John P. McNamara, *Jefferson College of Health Sciences*
Elaine Orr, *Our Lady of Holy Cross College*
Ellen Ott-Reeves, *Blinn College*
Davonya J. Person, *Auburn University*
Danny M. Pincivero, *The University of Toledo*
Susan Rohde, *Triton College*
Marilyn Shopper, *Johnson County Community College*
Phillip D. Snider Jr., *Gadsden State Community College*
William Stewart, *Middle Tennessee State University*
John Stribley, *University of Michigan*
Yong Tang, *Front Range Community College*
Terry Thompson, *Wor-Wic Community College*
Bennett D. Tucker, Jr., *Gadsden State Community College*
Leticia Vosotros, *Ozarks Technical Community College*
James Earl Whaley, *Baker College*

We would especially like to thank the participants in a three-day symposium held during June 2007, during which the Learn, Practice, Assess model was fine tuned, the concept of an Ancillary Correlation Guide was developed, and the need for a Test Bank written by the authors was underscored. The attendees ranged from new instructors to veteran instructors and former instructors now in administration. Their input was invaluable, and we feel privileged to have had the opportunity to interact which such a talented and dedicated group of educators (not to mention a fun group to spend time with).

Abel Bult-Ito, *University of Alaska Fairbanks*
Ray D. Burkett, *Southwest Tennessee Community College*
Sandra I. Caudle, *John C. Calhoun Community College*
Sondra M. Evans, *Florida Community College – Jacksonville*
Jared R. Gilmore, *San Jacinto College - Central*
Carl F. Hirtzel, *Oklahoma City Community College*
Randy Lankford, *Galveston College*
Jason LaPres, *North Harris College*
John McNamara, *Jefferson College of Health Sciences*
Kamal Osman, *Baker College of Flint*
Ellen Ott-Reeves, *Blinn College*
Robin Robison, *Northwest Mississippi Community College*
Felicia Scott, *Macomb Community College – Clinton Twp*
Janet Steele, *University of Nebraska at Kearney*
Sanjay Tiwary, *Hinds Community College*
Marlena K. West, *Madisonville Community College*

A special thank you also goes to other focus group attendees. Their input was greatly appreciated.

Pegge Alciatore, *University of Louisiana*
Sara Brenizer, *Shelton State Community College*
Juville Dario-Becker, *Central Virginia Community College*
Deanna Ferguson, *Gloucester County College*
Pamela Fouche, *Walters State Community College*
Richard Griner, *August State University*
Carol Makravitz, SCC, *Luzerne County Community College*
Ronald A. Markle, *Northern Arizona University*
Joe Schiller, *Austin Peay State University*
Mark L. Wygoda, *McNeese State University*

Isaac Barjis, *New York City College of Technology*
Jerry Barton, *Tarrant County College – South Campus*
J. Gordon Betts, *Tyler Junior College*
Lois Brewer Borek, *Georgia State University*
Scott Dunham, *Illinois Central College*
Amy Harwell, *Oregon State University*
Alfredo Munoz, *University of Texas at Brownsville*
Margaret (Betsy) Ott, *Tyler Junior College*
Robert L. Pope, *Miami Dade College*
Gregory K. Reeder, *Broward Community College*
Hugo Rodriguez Uribe, *University of Texas at Brownsville*
Walied Samarrai, *New York City College of Technology*
Brad Sarchet, *Manatee Community College*
Mitzie Sowell, *Pensacola Junior College*
Eric Sun, *Macon State College*
Anupama Trzaska, *St. Louis Community College*
Anthony Weinhaus, *University of Minnesota*
Linda Wooter, *Bishop State Community College*

CHAPTER

PREVIEW

Foundations for Success

The Chapter Preview not only provides great study tips to offer a foundation for success, but it also offers tips on how to utilize this particular text. Those tips will be found in boxes just like this.

A photo on the opening page for each chapter generates interest.

UNDERSTANDING WORDS

This section introduces building blocks of words that your instructor may assign. They are good investments of your time, since they can be used over and over and apply to many of the terms you will use in your career. Inside the back cover and on the facing page is a comprehensive list of these prefixes, suffixes, and root words.

ana-, up: *ana*tomy—the study of breaking up the body into its parts.

multi-, many: *multi*tasking—performing several tasks simultaneously.

physio-, relationship to nature: *physio*logy—the study of how body parts function.

LEARNING OUTCOMES

Each chapter begins with a list of learning outcomes indicating the knowledge you should gain as you work through the chapter. (Note the blue learn arrow.) These are intended to help you master the similar outcomes set by your instructor. The outcomes will be tied directly to assessments of knowledge gained.

After you have studied this chapter, you should be able to:

P.1 Introduction

1 Explain the importance of an individualized approach to learning. (p. xxiv)

P.2 Strategies for Success

2 Summarize what you should do before attending class. (p. xxiv)

3 Identify student activities that enhance classroom experience. (p. xxviii)

4 Describe several study techniques that can facilitate learning new material. (p. xxviii)

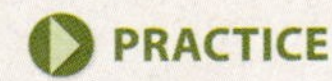

ASSESS

PAY ATTENTION

It is a beautiful day. You can't help but stare wistfully out the window, the scent of spring blooms and sounds of birds making it impossible to concentrate on what the instructor is saying. Gradually the lecture fades as you become aware of your own breathing, the beating of your heart, and the sheen of sweat that breaks out on your forehead in response to the radiant heat from the glorious day. Suddenly your reverie is cut short—the instructor has dropped a human anatomy and physiology textbook on your desk. You jump. Yelp. Your heart hammers and a flash of fear grips your chest, but you soon realize what has happened and recover.

The message is clear: pay attention. So you do, tuning out the great outdoors and focusing on the lecture. In this course, you will learn all about the events that you have just experienced, including your response to the sudden stimulation of the instructor's wake-up call. This is a good reason to learn how to stay focused in the course.

Opening Vignettes Beginning each chapter is a vignette that discusses current events or research news relating to the subject matter in the chapter. These demonstrate applications of the concepts learned in the study of anatomy and physiology.

P.1 INTRODUCTION

An overview tells you what to expect and why it is important.

Studying the human body can be overwhelming at times. The new terminology, used to describe body parts and how they work, can make it seem as if you are studying a foreign language. Learning all the parts of the body, along with the composition of each part, and how each part fits with the other parts to make the whole requires memorization. Understanding the way each body part works individually, as well as body parts working together, requires a higher level of knowledge, comprehension, and application. Identifying underlying structural similarities, from the macroscopic to the microscopic levels of body organization, taps more subtle critical thinking skills. This chapter will catalyze success in this active process of learning. (Remember that while the skills and tips discussed in this chapter relate to learning anatomy and physiology, they can be applied to other subjects.)

Students learn in different ways. Some students need to see the written word to remember it and the concept it describes or to actually write the words; others must hear the information or explain it to someone else. For some learners, true understanding remains elusive until a principle is revealed in a laboratory or clinical setting that provides a memorable context and engages all the senses.

After each major section, a question or series of questions tests your understanding of the material and enables you to practice using the new information. (Note the green practice arrow.) If you cannot answer the question(s) you should reread that section, being particularly on the lookout for the answer(s).

PRACTICE 

1 List some difficulties a student may experience when studying the human body.

P.2 STRATEGIES FOR SUCCESS

Major divisions within a chapter are called "A-heads." They are numbered sequentially in very large blue type and identify major content areas.

Many of the strategies for academic success are common sense, but it might help to review them. You may encounter new and helpful methods of learning.

Before Class

The major divisions are divided into no-less-important subdivisions called "B-heads," identified by large, gold type. These will help you organize the concepts upon which the major divisions are built.

Before attending class, prepare by reading and outlining or taking notes on the assigned pages of the text. If outlining, leave adequate space between entries to allow room for note-taking during lectures. Or, fold each page of notes taken before class in half so that class notes can be written on the blank side of the paper across from the reading notes on the same topic. This introduces the topics of the next class lecture, as well as new terms. Some students team a vocabulary list with each chapter's notes. The outline or notes from the reading can be taken to class and expanded during the lecture.

Sometimes in your reading you will be directed back to a related concept, discussed in an earlier chapter, to help you better understand the new concept that is being explained.

RECONNECT

To Chapter 11, Sympathetic Division, pages 424–426.

In a *hiatal hernia,* part of the stomach protrudes through a weakened area of the diaphragm, through the esophageal hiatus and into the thorax. Regurgitation (reflux) of gastric juice into the esophagus may inflame the esophageal mucosa, causing heartburn, difficulty in swallowing, or ulceration and blood loss. In response to the destructive action of gastric juice, columnar epithelium may replace the squamous epithelium that normally lines the esophagus (see chapter 5, pages 147–148). This condition, called *Barrett's esophagus,* increases the risk of developing esophageal cancer.

As you read, you may feel the need for a "study break." Sometimes you may just need to "chill out." Other times, you may just need to shift gears. Try the following! Throughout the book are shaded boxes that present sidelights to the main focus of the text. Indeed, some of these may cover topics that your instructor chooses to highlight. Read them! They are interesting, informative, and a change of pace.

2.2 FROM SCIENCE TO TECHNOLOGY

Ionizing Radiation: From the Cold War to Yucca Mountain

Alpha, beta, and gamma radiation are called ionizing radiation because their energy removes electrons from atoms (fig. 2C). Electrons dislodged by ionizing radiation can affect nearby atoms, disrupting physiology at the chemical level in a variety of ways—causing cancer, clouding the lens of the eye, and interfering with normal growth and development.

In the United States, some people are exposed to very low levels of ionizing radiation, mostly from background radiation, which originates from natural environmental sources (table 2A). For people who live near sites of atomic weapons manufacture, exposure is greater. Epidemiologists are investigating medical records that document illnesses linked to long-term exposure to ionizing radiation in a 1,200-square kilometer area in Germany.

The lake near Oberrothenback, Germany, which appears inviting, harbors enough toxins to kill thousands of people. It is polluted with heavy metals, low-level radioactive chemical waste, and 22,500 tons of arsenic. Radon, a radioactive by-product of uranium, permeates the soil. Many farm animals and pets that have drunk from the lake have died. Cancer rates and respiratory disorders among the human residents nearby are well above normal.

The lake in Oberrothenback was once a dump for a factory that produced "yellow cake," a term for processed uranium ore, used to build atomic bombs for the former Soviet Union. In the early 1950s, nearly half a million workers labored here and in surrounding areas in factories and mines. Records released in 1989, after the reunification of Germany, reveal that workers were given perks, such as alcoholic beverages and better wages, to work in the more dangerous areas. The workers paid a heavy price: many died of lung ailments.

Today, concern over the health effects of exposure to ionizing radiation centers on the u.s. government's plan to transport tens of thousands of metric tons of high-level nuclear waste from 109 reactors around the country for burial beneath yucca mountain, nevada, by 2021. The waste, currently stored near the reactors, will be buried in impenetrable containers under the mountain by robots. In the reactors, nuclear fuel rods contain uranium oxide, which produces electricity as it decays to plutonium, which gives off gamma rays. Periodically the fuel rods must be replaced, and the spent ones buried. Environmental groups are concerned that the waste could be exposed during transport and that the facility in the mountain may not adequately contain it. ■

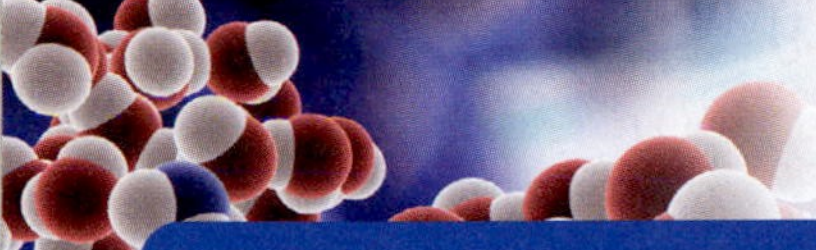

FIGURE 2C Ionizing radiation removes elecrons from atoms. *(a)* Ionizing radiation may dislodge an electron from an electrically neutral hydrogen atom. *(b)* Without its electron, the hydrogen atom becomes a positively-charged hydrogen ion (H^+).

TABLE 2A | Sources of Ionizing Radiation

Background (Natural environmental)	Cosmic rays from space
	Radioactive elements in earth's crust
	Rocks and clay in building materials
	Radioactive elements naturally in the body (potassium-40, carbon-14)
Medical and dental	X rays
	Radioactive substances
Other	Atomic and nuclear weapons
	Mining and processing radioactive minerals
	Radioactive fuels in nuclear power plants
	Radioactive elements in consumer products (luminescent dials, smoke detectors, color TV components)

Remember when you were very young and presented with a substantial book for the first time? You were likely intimidated by its length, but were reassured that there were "a lot of pictures." There are a lot of illustrations in this book as well, all designed to help you master the material.

Photographs and Line Art

Sometimes subdivisions have so many parts that the book goes to a third level, the "C-head." This information is presented in a slightly smaller, bold, black font that identifies a specific section with an example.

Photographs provide a realistic view of anatomy.

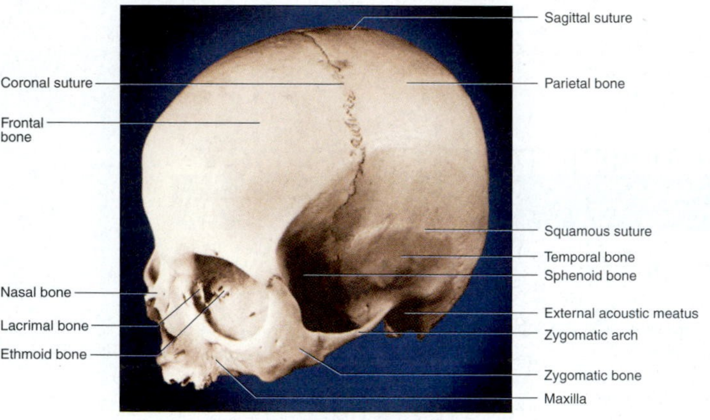

Since line art can be from different positions and layers, it can provide a unique view.

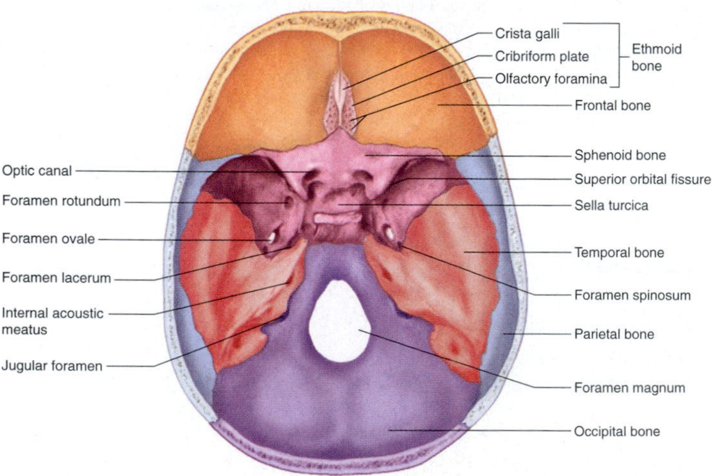

Macroscopic to Microscopic

Many figures show anatomical structures in a manner macroscopic to microscopic (or vice versa), both as electronic art and as photomicrographs.

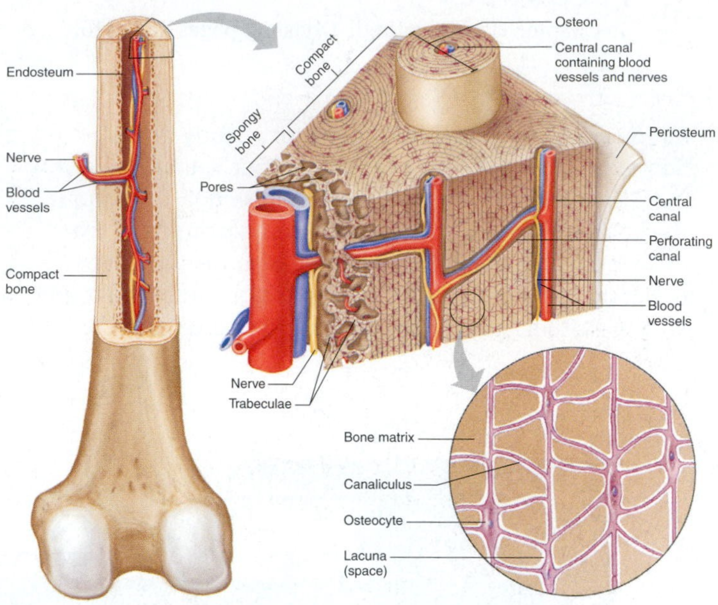

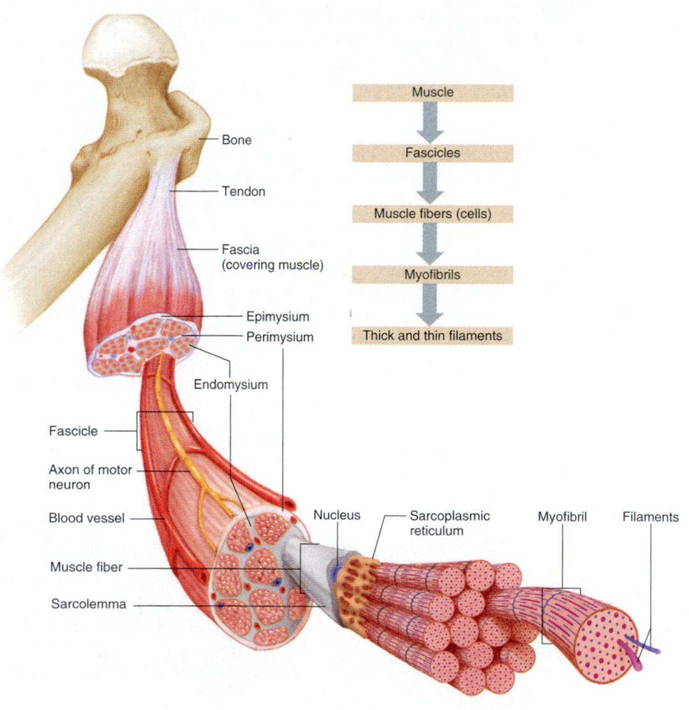

Flow Charts

Flow charts depict sequences of related events, steps of pathways, and complex concepts, easing comprehension. Other figures may show physiological processes.

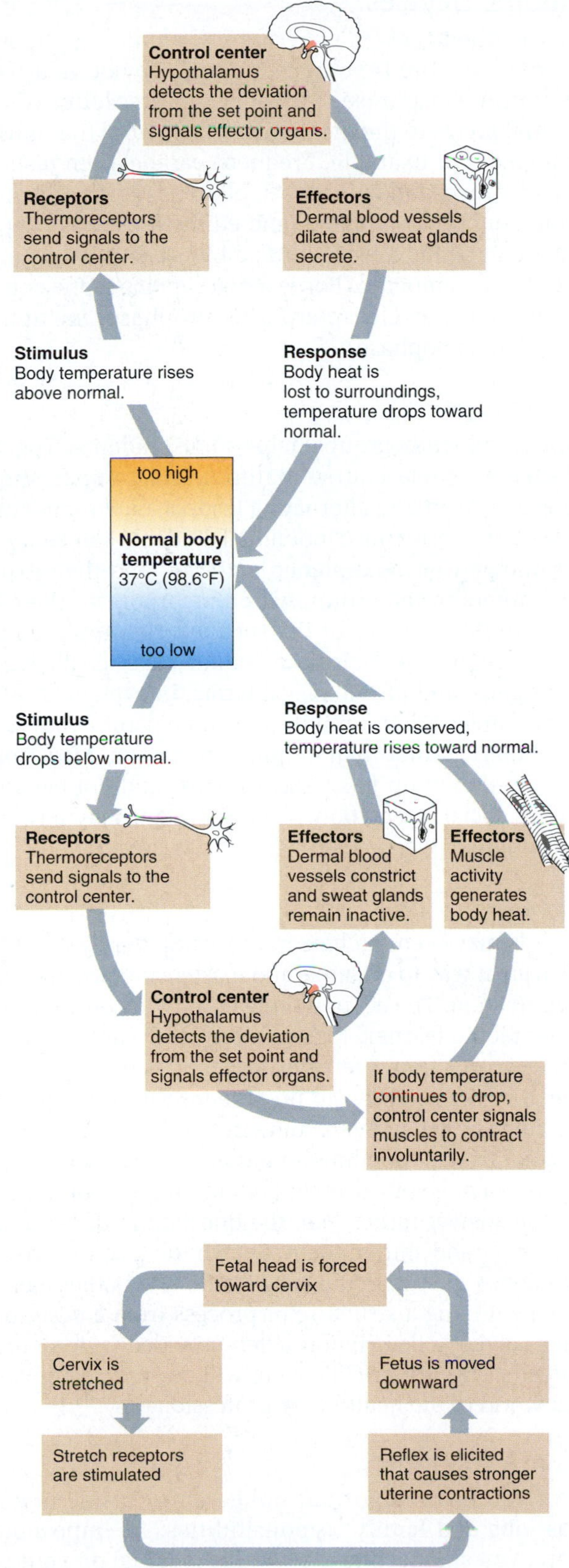

Control center
Hypothalamus detects the deviation from the set point and signals effector organs.
Receptors
Thermoreceptors send signals to the control center.
Effectors
Dermal blood vessels dilate and sweat glands secrete.
Stimulus
Body temperature rises above normal.
Response
Body heat is lost to surroundings, temperature drops toward normal.
too high
Normal body temperature
37°C (98.6°F)
too low
Stimulus
Body temperature drops below normal.
Response
Body heat is conserved, temperature rises toward normal.
Receptors
Thermoreceptors send signals to the control center.
Effectors
Dermal blood vessels constrict and sweat glands remain inactive.
Effectors
Muscle activity generates body heat.
Control center
Hypothalamus detects the deviation from the set point and signals effector organs.
If body temperature continues to drop, control center signals muscles to contract involuntarily.
Fetal head is forced toward cervix
Cervix is stretched
Fetus is moved downward
Stretch receptors are stimulated
Reflex is elicited that causes stronger uterine contractions

Anatomical Structures

Some figures illustrate the locations of anatomical structures.

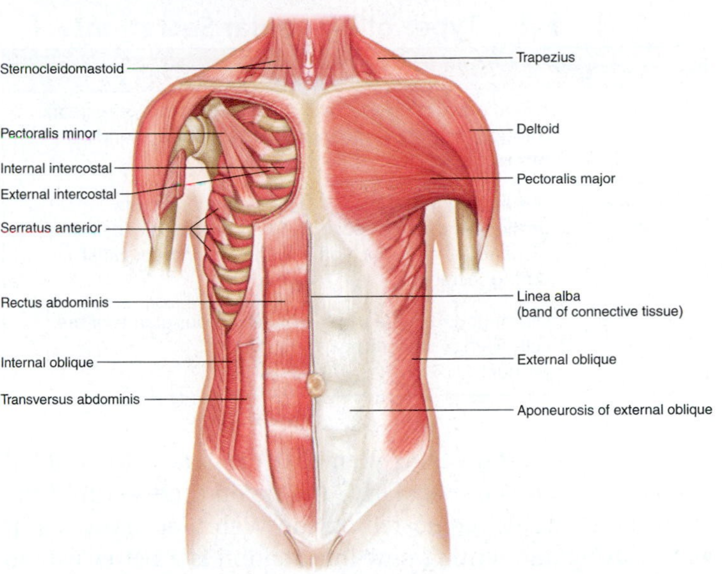

Sternocleidomastoid
Trapezius
Pectoralis minor
Deltoid
Internal intercostal
External intercostal
Pectoralis major
Serratus anterior
Rectus abdominis
Linea alba
(band of connective tissue)
Internal oblique
External oblique
Transversus abdominis
Aponeurosis of external oblique

Other figures illustrate the functional relationships of anatomical structures.

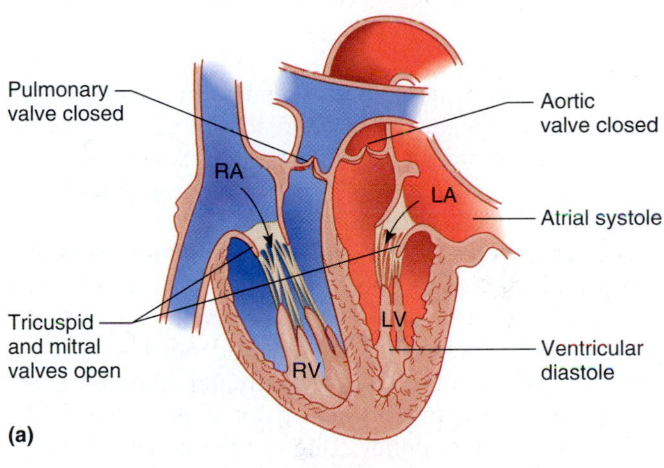

Pulmonary valve closed
Aortic valve closed
RA
LA
Atrial systole
Tricuspid and mitral valves open
LV
RV
Ventricular diastole
(a)

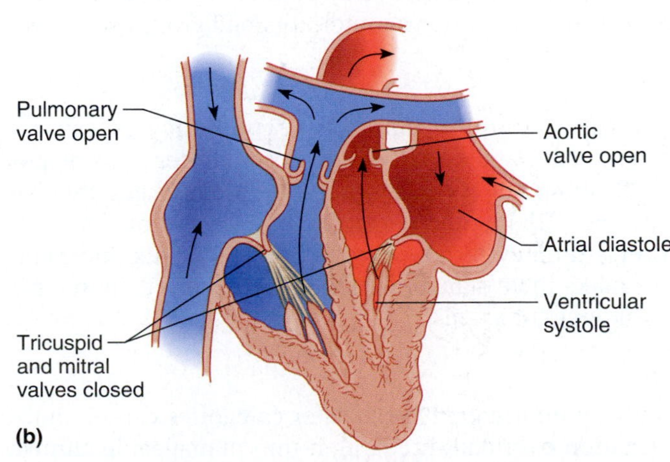

Pulmonary valve open
Aortic valve open
Atrial diastole
Ventricular systole
Tricuspid and mitral valves closed
(b)

Organizational Tables

Organizational tables can help "put it all together," but are not a substitute for reading the text or having good lecture notes.

TABLE 5.4 | **Types of Glandular Secretions**

Type	Description of Secretion	Example
Merocrine glands	A fluid product released through the cell membrane by exocytosis	Salivary glands, pancreatic glands, sweat glands of the skin
Apocrine glands	Cellular product and portions of the free ends of glandular cells pinch off during secretion	Mammary glands, ceruminous glands lining the external ear canal
Holocrine glands	Disintegrated entire cells filled with secretory products	Sebaceous glands of the skin

As many resources as your text provides, it is critical that you attend class regularly, and be on time—even if the instructor's notes are posted on the Web. For many learners, hearing and writing new information is a better way to retain facts than just scanning notes on a computer screen. Attending lectures and discussion sections also provides more detailed and applied analysis of the subject matter, as well as a chance to ask questions.

During Class

Be alert and attentive in class. Take notes by adding either to the outline or notes taken while reading. Auditory learners benefit from recording the lectures and listening to them while driving or doing chores. This is called **multitasking**—doing more than one activity at a time.

Participate in class discussions, asking questions of the instructor and answering questions he or she poses. All of the students are in the class to learn, and many will be glad someone asked a question others would not be comfortable asking. Such student response can alert the instructor to topics that are misunderstood or not understood at all. However, respect class policy. Due to time constraints and class size, asking questions may be more appropriate after a large lecture class or during tutorial (small group) sessions.

After Class

In learning complex material, expediency is critical. Organize, edit, and review notes as soon after class as possible, fleshing out sections where the lecturer got ahead of the listener. Highlighting or underlining (in color, for visual learners) the key terms, lists, important points and major topics make them stand out, which eases both daily reviews and studying for exams.

Lists

Organizing information into lists or categories can minimize information overload, breaking it into manageable chunks. For example, when studying the muscles of the thigh it is easier to learn the insertion, origin, action, and nerve supply of the four muscles making up the *quadriceps femoris* as a group, because they all have the same insertion, action, and nerve supply . . . they differ only in their origins.

Mnemonic Devices

Another method for remembering information is the **mnemonic device.** One type of mnemonic device is a list of words, forming a phrase, in which the first letter of each word corresponds to the first letter of each word that must be remembered. For example, *F*requent *p*arade *o*ften *t*ests *s*oldiers' *e*ndurance stands for the skull bones *f*rontal, *p*arietal, *o*ccipital, *t*emporal, *s*phenoid, and *e*thmoid. Another type of mnemonic device is a word formed by the first letters of the items to be remembered. For example, *ipmat* represents the stages in the cell cycle: *i*nterphase, *p*rophase, *m*etaphase, *a*naphase, and *t*elophase.

Study Groups

Forming small study groups helps some students. Together the students review course material and compare notes. Working as a team and alternating leaders allows students to verbalize the information. Individual students can study and master one part of the assigned material, and then explain it to the others in the group, which incorporates the information into the memory of the speaker. Hearing the material spoken aloud also helps the auditory learner. Be sure to use anatomical and physiological terms, in explanations and everyday conversation, until they become part of your working vocabulary, rather than intimidating jargon. Most important of all—the group must stay on task, and not become a vehicle for social interaction. Your instructor may have suggestions or guidelines for setting up study groups.

Flash Cards

Flash cards may seem archaic in this computer age, but they are still a great way to organize and master complex and abundant information. The act of writing or drawing on a note card helps the tactile learner. Master a few new cards each day, and review cards from previous days, and use them all again at the end of the semester to prepare for the comprehensive final exam. They may even come in handy later, such as in studying for exams for admission to medical school or graduate school. Divide your deck in half and flip half of the cards so that the answer rather than the question is showing. Mix them together and shuffle them. Switch them so that you see the questions rather than the answers from the other half. Get used to identifying a structure or process from a description as well as giving a description when provided with a process or structure. This is more like what will be expected of you in the real world of the health-care professional.

Manage Your Time

Many of you have important obligations outside of class, such as jobs and family responsibilities. As important as these are, you still need to master this material on your path to becoming a health-care professional. Good time management skills are therefore essential in your study of human anatomy and physiology. In addition to class, lab, and study time, multitask. Spend time waiting for a ride, or waiting in a doctor's office, reviewing notes or reading the text.

Daily repetition is helpful, so scheduling several short study periods each day can replace a last-minute crunch to cram for an exam. This does not take the place of time to prepare for the next class. Thinking about these suggestions for learning now can maximize study time throughout the semester, and, hopefully, lead to academic success. A working knowledge of the structure and function of the human body provides the foundation for all careers in the health sciences.

PRACTICE 

2 Why is it important to prepare before attending class?

3 Name two ways to participate in class discussions.

4 List several aids for remembering information.

CHAPTER SUMMARY

A summary of the chapter provides an outline to review major ideas and is a tool in organizing thoughts.

P.1 INTRODUCTION (PAGE XXIV)

Try a variety of methods to study the human body.

P.2 STRATEGIES FOR SUCCESS (PAGE XXIV)

While strategies for academic success seem to be common sense, you might benefit from reminders of study methods.

1. Before class
   Read the assigned text material prior to the corresponding class meeting.
   a. Reconnects refer back to helpful, previously discussed concepts.
   b. Shaded boxes present sidelights to the main focus of the text.
   c. Photographs, line art, flow charts, and organizational tables help in mastery of the materials.
2. During class
   Take notes and participate in class discussions.
3. After class
   a. Organize, edit, and review class notes.
   b. Mnemonic devices aid learning.
      (1) The first letters of the words to remember begin words of an easily recalled phrase.
      (2) The first letters of the items to be remembered form a word.
   c. Small study groups reviewing and vocalizing material can divide and conquer the learning task.
   d. Flash cards help the tactile learner.
   e. Time management skills encourage scheduled studying, including daily repetition instead of cramming for exams.

CHAPTER ASSESSMENTS

Chapter assessments that are tied directly to the learning outcomes allow you to self assess your mastery of the material. (Note the purple assess arrow.)

P.1 Introduction

1. Explain how students learn in different ways. (p. xxiv)

P.2 Strategies for Success

2. Methods to prepare for class include ____________. (p. xxiv)
   a. reading the chapter
   b. outlining the chapter
   c. taking notes on the assigned reading
   d. making a vocabulary list
   e. all of the above

3. Describe how you can participate in class discussions. (p. xxviii)

4. Forming the phrase "*I* passed *my* anatomy test." To remember the cell cycle (interphase, prophase, metaphase, anaphase, telophase) is an example of a ____________. (p. xxviii)

5. Explain the value of repetition in learning and preparation for exams. (p. xxviii)

6. Name a benefit and a drawback of small study groups. (p. xxviii)

INTEGRATIVE ASSESSMENTS/CRITICAL THINKING

Integrative assessments apply main concepts within the current chapter and from previous chapters to clinical or research situations and take the student beyond memorization to utilization of knowledge.

OUTCOMES P.1, P.2

1. Which study methods are most successful for you?

OUTCOME P.2

2. Design a personalized study schedule.

WEB CONNECTIONS

Be sure to visit the text website at **www.mhhe.com/shier12** for answers to chapter assessments, additional quizzes, and interactive learning exercises.

ANATOMY & PHYSIOLOGY REVEALED

Anatomy & Physiology Revealed® (APR) includes cadaver photos that allow you to peel away layers of the human body to reveal structures beneath the surface. This program also includes animations, radiologic imaging, audio pronunciations, and practice quizzing. Check out **www.aprevealed.com**. APR has been proven to help improve student grades!

UNIT ONE

CHAPTER 1

Introduction to Human Anatomy and Physiology

Our brain enables us to learn, to practice, and to assess our understanding—whether of a textbook, or how to handle a medical emergency.

UNDERSTANDING WORDS

append-, to hang something: *append*icular—pertaining to the upper limbs and lower limbs.

cardi-, heart: peri*cardi*um—membrane that surrounds the heart.

cerebr-, brain: *cerebr*um—largest part of the brain.

cran-, helmet: *cran*ial—pertaining to the part of the skull that surrounds the brain.

dors-, back: *dors*al—position toward the back of the body.

homeo-, same: *homeo*stasis—maintenance of a stable internal environment.

-logy, the study of: physio*logy*—study of body functions.

meta-, change: *meta*bolism—chemical changes that occur within the body.

nas-, nose: *nas*al—pertaining to the nose.

orb-, circle: *orb*ital—pertaining to the portion of skull that encircles an eye.

pariet-, wall: *pariet*al membrane—membrane that lines the wall of a cavity.

pelv-, basin: *pelv*ic cavity—basin-shaped cavity enclosed by the pelvic bones.

peri-, around: *peri*cardial membrane—membrane that surrounds the heart.

pleur-, rib: *pleur*al membrane—membrane that encloses the lungs within the rib cage.

-stasis, standing still: homeo*stasis*—maintenance of a stable internal environment.

super-, above: *super*ior—referring to a body part located above another.

-tomy, cutting: ana*tomy*—study of structure, which often involves cutting or removing body parts.

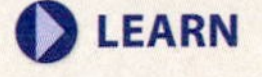 LEARN  PRACTICE  ASSESS

LEARNING OUTCOMES

After you have studied this chapter, you should be able to:

1.1 Introduction

1 Identify some of the early discoveries that lead to our current understanding of the human body. (p. 3)

1.2 Anatomy and Physiology

2 Explain how anatomy and physiology are related. (p. 4)

1.3 Levels of Organization

3 List the levels of organization in the human body and the characteristics of each. (p. 4)

1.4 Characteristics of Life

4 List and describe the major characteristics of life. (p. 6)

5 Give examples of *metabolism*. (p. 6)

1.5 Maintenance of Life

6 List and describe the major requirements of organisms. (p. 7)

7 Explain the importance of homeostasis to survival. (p. 9)

8 Describe the parts of a homeostatic mechanism and explain how they function together. (p. 9)

1.6 Organization of the Human Body

9 Identify the locations of the major body cavities. (p. 12)

10 List the organs located in each major body cavity. (p. 12)

11 Name and identify the locations of the membranes associated with the thoracic and abdominopelvic cavities. (p. 12)

12 Name the major organ systems, and list the organs associated with each. (p. 14)

13 Describe the general function of each organ system. (p. 14)

1.7 Life-Span Changes

14 For each decade of life, identify the levels of organization in the body at which aging occurs. (p. 20)

1.8 Anatomical Terminology

15 Properly use the terms that describe relative positions, body sections, and body regions. (p. 21)

EMERGENCY

Judith R. had not been wearing a seat belt when the accident occurred because she had to drive only a short distance. She hadn't anticipated the intoxicated driver in the oncoming lane who swerved right in front of her. Thrown several feet, she now lay near her wrecked car as emergency medical technicians immobilized her neck and spine. Terrified, Judith tried to assess her condition. She didn't think she was bleeding, and nothing hurt terribly, but she felt a dull ache in the upper right part of her abdomen.

Minutes later, in the emergency department, a nurse checked Judith's blood pressure, pulse and breathing rate, and other vital signs that reflect underlying metabolic activities necessary for life. Assessing vital signs is important in any medical decision. Judith's vital signs were stable, and she was alert, knew who and where she was, and didn't have obvious life-threatening injuries, so transfer to a trauma center was not necessary. However, Judith continued to report abdominal pain. The attending physician ordered abdominal X rays, knowing that about a third of patients with abdominal injuries show no outward sign of a problem. As part of standard procedure, Judith received oxygen and intravenous fluids, and a technician took several tubes of blood for testing.

A young physician approached and smiled at Judith as assistants snipped off her clothing. The doctor carefully looked and listened and gently poked and probed. She was looking for cuts; red areas called hematomas where blood vessels had broken; and treadmarks on the skin. Had Judith been wearing her seat belt, the doctor would have checked for characteristic "seat belt contusions," crushed bones or burst hollow organs caused by the twisting constrictions that can occur at the moment of impact when a person wears a seat belt. Had Judith been driving fast enough for the air bag to have deployed, she might have suffered abrasions from not having the seat belt on to hold her in a safe position. Finally, the doctor measured the girth of Judith's abdomen. If her abdomen swelled later on, this could indicate a complication, such as infection or internal bleeding.

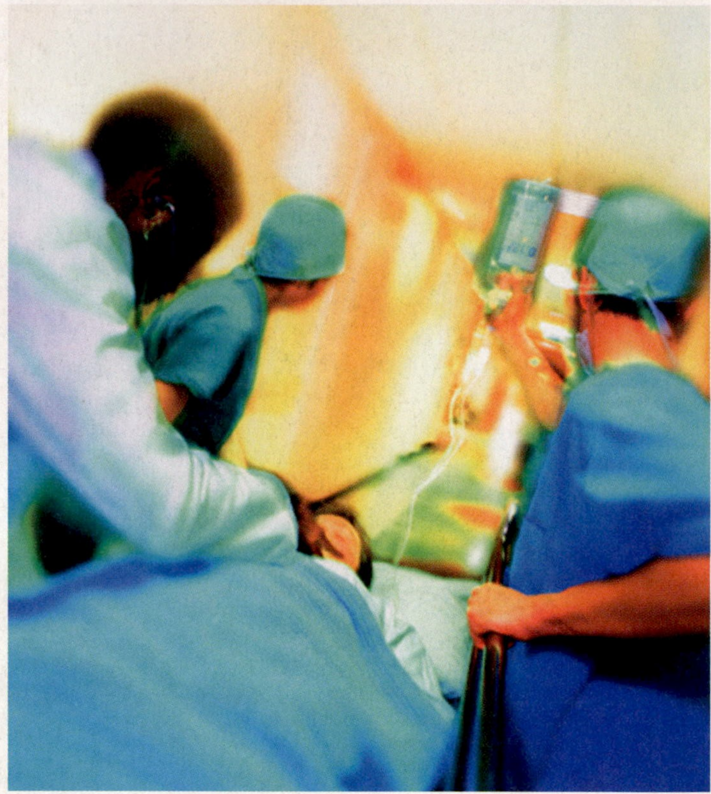

The difference between life and death may depend on a health-care professional's understanding of the human body.

On the basis of a hematoma in Judith's upper right abdomen and the continued pain coming from this area, the physician ordered a computed tomography (CT) scan. It revealed a lacerated liver. Judith underwent emergency surgery to remove the small torn portion of this vital organ.

When Judith awoke from surgery, a different physician was scanning her chart, looking up frequently. The doctor was studying her medical history for any notation of a disorder that might impede healing. Judith's history of slow blood clotting, he noted, might slow her recovery from surgery. Next, the physician looked and listened. A bluish discoloration of Judith's side might indicate bleeding from her pancreas, kidney, small intestine, or aorta (the artery leading from the heart). A bluish hue near the navel would indicate bleeding from the liver or spleen. Her umbilical area was somewhat discolored.

The doctor gently tapped Judith's abdomen and carefully listened to sounds from her digestive tract. A drumlike resonance could mean that a hollow organ had burst, whereas a dull sound might indicate internal bleeding. Judith's abdomen produced dull sounds throughout. In addition, her abdomen had become swollen and the pain intensified when the doctor gently pushed on the area. With Judith's heart rate increasing and blood pressure falling, bleeding from the damaged liver was a definite possibility.

Blood tests confirmed the doctor's suspicions. Blood is a complex mixture of cells and biochemicals, so it serves as a barometer of health. Injury or illness disrupts the body's maintenance of specific levels of various biochemicals, called homeostasis. Judith's blood tests revealed that her body had not yet recovered from the accident. Levels of clotting factors her liver produced were falling and blood oozed from her incision, a sign of impaired clotting. Judith's blood glucose level remained elevated, as it had been on arrival. Her body was still reacting to the injury.

Based on Judith's blood tests, heart rate, blood pressure, reports of pain, and the physical exam, the doctor sent her back to the operating room. Sure enough, the part of her liver where the injured portion had been removed was still bleeding. When the doctors placed packing material at the wound site, the oozing gradually stopped. Judith returned to the recovery room. When her condition stabilized, she continued recovering in a private room. This time, all went well, and a few days later, she was able to go home. The next time she drove, Judith wore her seat belt!

Imagine yourself as one of the health-care professionals who helped identify Judith R.'s injury and got her on the road back to health. How would you know what to look, listen, and feel for? How would you place the signs and symptoms into a bigger picture that would suggest the appropriate diagnosis? Nurses, doctors, technicians, and other integral members of health-care teams must have a working knowledge of the many intricacies of the human body. How can they begin to understand its astounding complexity? The study of human anatomy and physiology is a daunting, but fascinating and ultimately life-saving, challenge. ■

1.1 INTRODUCTION

Our understanding of the human body has a long and interesting history (fig. 1.1). Our earliest ancestors must have been curious about how their bodies worked. At first their interests most likely concerned injuries and illnesses, because healthy bodies demand little attention from their owners. Primitive people suffered aches and pains, injured themselves, bled, broke bones, developed diseases, and contracted infections.

The change from a hunter-gatherer to an agricultural lifestyle, which occurred from 6,000 to 10,000 years ago in various parts of the world, altered the spectrum of human illnesses. Before agriculture, isolated bands of peoples had little contact with each other, and so infectious diseases did not spread easily, as they do today with our global connections. These ancient peoples ate wild plants that provided chemicals that combated some parasitic infections. A man who died in the Austrian/Italian Alps 5,300 years ago and whose body was found frozen was carrying mushrooms that had antibiotic activity.

With agriculture came exposure to pinworms, tapeworms, and hookworms in excrement used as fertilizer, and less reliance on the natural drugs in wild plants.

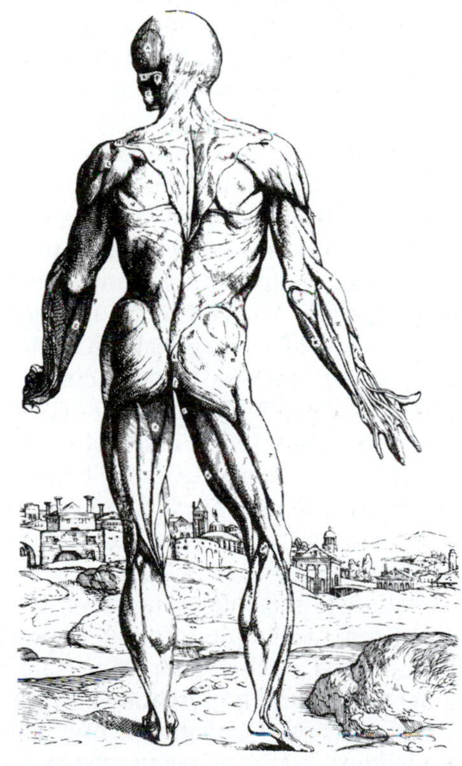

FIGURE 1.1 The study of the human body has a long history, as this illustration from the second book of *De Humani Corporis Fabrica* by Andreas Vesalius, issued in 1543, indicates. Note the similarity to the anatomical position (described on page 20).

Urbanization brought more infectious disease as well as malnutrition, as people became sedentary and altered their diets. Evidence from preserved bones and teeth chronicle these changes. Tooth decay, for example, affected 3% of samples from hunter-gatherers, but 8.7% from farmers, and 17% of samples from city residents. Preserved bones from children reflect increasing malnutrition as people moved from the grasslands to farms to cities. When a child starves or suffers from severe infection, the ends of the long bones stop growing. When health returns, growth resumes, but leaves behind telltale areas of dense bone.

The rise of medical science paralleled human prehistory and history. At first, healers relied heavily on superstitions and notions about magic. However, as they tried to help the sick, these early medical workers began to discover useful ways of examining and treating the human body. They observed the effects of injuries, noticed how wounds healed, and examined dead bodies to determine the causes of death. They also found that certain herbs and potions could treat coughs, headaches, and other common problems. These long-ago physicians began to wonder how these substances, the forerunners of modern drugs, affected body functions.

People began asking more questions and seeking answers, setting the stage for the development of modern medical science. Techniques for making accurate observations and performing careful experiments evolved, and knowledge of the human body expanded rapidly.

This new knowledge of the structure and function of the human body required a new, specialized language. Early medical providers devised many terms to name body parts, describe their locations, and explain their functions. These terms, most of which originated from Greek and Latin, formed the basis for the language of anatomy and physiology. (A list of some of the modern medical and applied sciences appears on pages 24–25.)

Although study of corpses was forbidden in Europe during the Middle Ages, dissection of dead bodies became a key part of medical education in the twentieth century. Today, cadaver dissection remains an important method to learn how the body functions and malfunctions, and autopsies are vividly depicted on television crime dramas. However, the traditional gross anatomy course in medical schools is sometimes supplemented with learning from body parts already dissected by instructors (in contrast to students doing this) as well as with computerized scans of cadavers, such as the Visible Human Project from the National Library of Medicine and Anatomy and Physiology Revealed available with this textbook.

PRACTICE

1. What factors probably stimulated an early interest in the human body?
2. How did human health change as lifestyle changed?
3. What types of activities helped promote the development of modern medical science?

1.2 ANATOMY AND PHYSIOLOGY

Two major areas of medical science, **anatomy** (ah-nat′o-me) and **physiology** (fiz″e-ol′o-je), address how the body maintains life. Anatomy, from the Greek for "a cutting up," examines the **structures,** or morphology, of body parts—their forms and organization. Physiology, from the Greek for "relationship to nature," considers the **functions** of body parts—what they do and how they do it. Although anatomists rely more on examination of the body and physiologists more on experimentation, together their efforts have provided a solid foundation for understanding how our bodies work.

It is difficult to separate the topics of anatomy and physiology because anatomical structures make possible their functions. Parts form a well-organized unit—the **human organism.** Each part contributes to the operation of the unit as a whole. This functional role arises from the way the part is constructed. For example, the arrangement of bones and muscles in the human hand, with its long, jointed fingers, makes grasping possible. The heart's powerful muscular walls contract and propel blood out of the chambers and into blood vessels, and heart valves keep blood moving in the proper direction. The shape of the mouth enables it to receive food; tooth shapes enable teeth to break solid foods into pieces; and the muscular tongue and cheeks are constructed in a way that helps mix food particles with saliva and prepare them for swallowing (fig. 1.2).

As ancient as the fields of anatomy and physiology are, we are always learning more. For example, researchers recently used imaging technology to identify a previously unrecognized part of the brain, the planum temporale, which enables people to locate sounds in space. Many discoveries today begin with investigations at the molecular or cellular level. In this way, researchers have discovered that certain cells in the small intestine bear the same taste receptor proteins found on the tongue—at both locations, the receptors detect the molecules that impart sweetness. The discovery of the planum temporale is anatomical; the discovery of sweet receptors in the intestine is physiological.

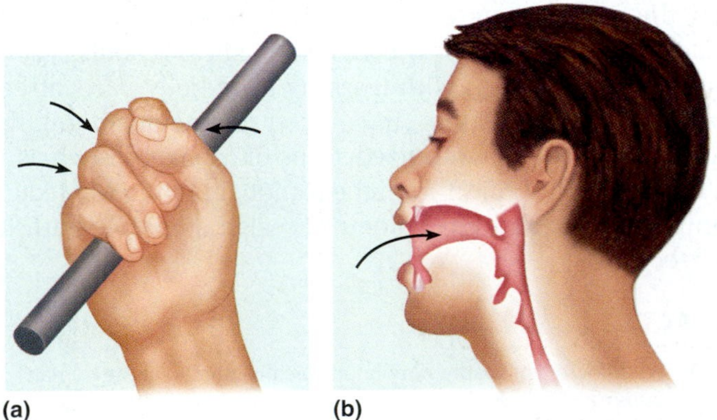

FIGURE 1.2 The structures of body parts make possible their functions: *(a)* The hand is adapted for grasping and *(b)* the mouth for receiving food. (Arrows indicate movements associated with these functions.)

Many nuances of physiology are being revealed with examination of the genes that function in particular cell types under particular conditions, leading to sometimes surprising findings. Using such "gene expression profiling," for example, researchers discovered that after a spinal cord injury, the damaged tissue releases a flood of proteins previously associated only with skin wounds. Finding these proteins in the aftermath of spinal cord injury suggests new drug targets.

PRACTICE 

4 What are the differences between anatomy and physiology?

5 Why is it difficult to separate the topics of anatomy and physiology?

6 List several examples that illustrate how the structure of a body part makes possible its function.

7 How are anatomy and physiology both old and new fields?

1.3 LEVELS OF ORGANIZATION

Early investigators, limited in their ability to observe small structures, such as cells and gene expression profiles focused their attention on larger body parts. Studies of small structures had to await invention of magnifying lenses and microscopes, about 400 years ago. These tools revealed that larger body structures were made up of smaller parts, which, in turn, were composed of even smaller ones.

Today, scientists recognize that all materials, including those that comprise the human body, are composed of chemicals. Chemicals consist of tiny particles called **atoms,** composed of even smaller **subatomic particles;** atoms can join to form larger **molecules;** small molecules may combine to form larger molecules called **macromolecules.**

In all organisms, including the human, the basic unit of structure and function is a **cell.** Although individual cells vary in size and shape, all share certain characteristics. Cells of complex organisms such as humans contain structures called **organelles** (or″gan-elz′) that carry on specific activities. Organelles are composed of assemblies of large molecules, including proteins, carbohydrates, lipids, and nucleic acids. Most human cells contain a complete set of genetic instructions, yet use only a subset of them, allowing cells to specialize. All cells share the same characteristics of life and must meet certain requirements to stay alive.

Specialized cells assemble into layers or masses that have specific functions. Such a group of cells forms a **tissue.** Groups of different tissues form **organs**—complex structures with specialized functions—and groups of organs that function closely together comprise **organ systems.** Interacting organ systems make up an **organism.**

A body part can be described at different levels. The heart, for example, consists of muscle, fat, and nervous tissue. These tissues, in turn, are constructed of cells, which contain organelles. All of the structures of life are, ultimately,

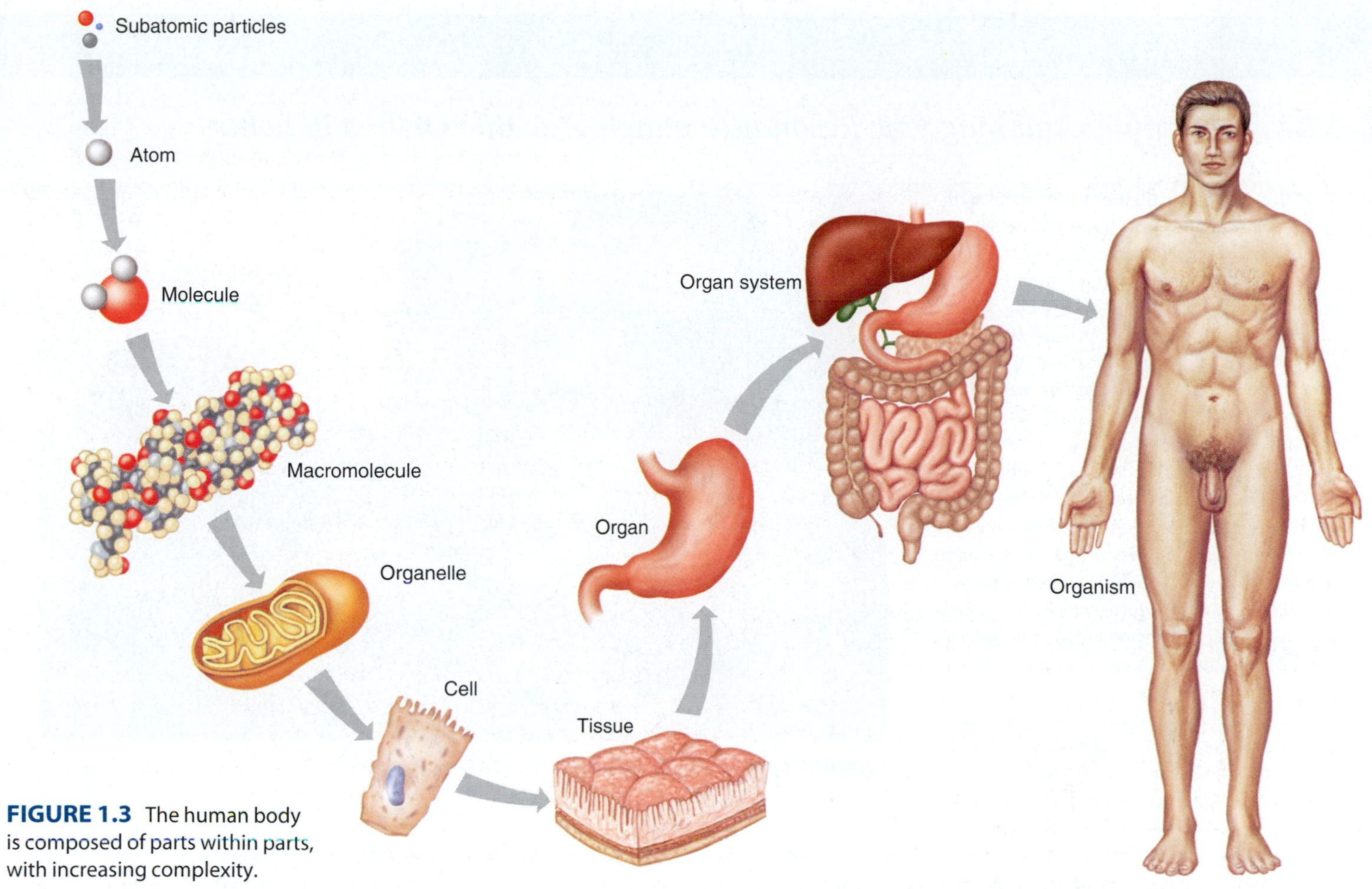

FIGURE 1.3 The human body is composed of parts within parts, with increasing complexity.

composed of chemicals (fig. 1.3). Clinical Application 1.1 describes two technologies used to visualize body parts based on body chemistry.

Chapters 2–6 discuss these levels of organization in more detail. Chapter 2 describes the atomic and molecular levels; chapter 3 presents organelles and cellular structures and functions; chapter 4 explores cellular metabolism; chapter 5 describes tissues; and chapter 6 presents the skin and its accessory organs as an example of an organ system. In the remaining chapters, the structures and functions of each of the other organ systems are described in detail. Table 1.1 lists the levels of organization and some corresponding illustrations in this textbook. Table 1.2 summarizes the organ systems, the major organs that comprise them, and their major functions in the order presented in this book. They are discussed in more detail later in this chapter (pages 14–18).

PRACTICE

8 How does the human body illustrate levels of organization?

9 What is an organism?

10 How do body parts at different levels of organization vary in complexity?

TABLE 1.1 | Levels of Organization

Level	Example	Illustration
Subatomic particles	Electrons, protons, neutrons	Figure 2.1
Atom	Hydrogen atom, lithium atom	Figure 2.3
Molecule	Water molecule, glucose molecule	Figure 2.7
Macromolecule	Protein molecule, DNA molecule	Figure 2.19
Organelle	Mitochondrion, Golgi apparatus, nucleus	Figure 3.3
Cell	Muscle cell, nerve cell	Figure 5.28
Tissue	Simple squamous epithelium, loose connective tissue	Figure 5.1
Organ	Skin, femur, heart, kidney	Figure 6.2
Organ system	Integumentary system, skeletal system, digestive system	Figure 1.13
Organism	Human	Figure 1.19

1.1 CLINICAL APPLICATION

Ultrasonography And Magnetic Resonance Imaging: A Tale Of Two Patients

The two patients enter the hospital medical scanning unit hoping for opposite outcomes. Vanessa Q., who has suffered several pregnancy losses, hopes that an ultrasound exam will reveal that her current pregnancy is progressing normally. Michael P., a sixteen-year-old who has excruciating headaches, is to undergo a magnetic resonance (MR) scan to assure his physician (and himself!) that the cause of the headache is not a brain tumor.

Ultrasound and magnetic resonance scans are noninvasive procedures that provide images of soft internal structures. Ultrasonography uses high-frequency sound waves beyond the range of human hearing. A technician gently presses a device called a transducer, which emits sound waves, against the skin and moves it slowly over the surface of the area being examined, which in this case is Vanessa's abdomen (fig. 1A).

Prior to the exam, Vanessa drank several glasses of water. Her filled bladder will intensify the contrast between her uterus (and its contents) and nearby organs because as the sound waves from the transducer travel into the body, some of the waves reflect back to the transducer when they reach a border between structures of slightly different densities. Other sound waves continue into deeper tissues, and some of them are reflected back by still other interfaces. As the reflected sound waves reach the transducer, they are converted into electrical impulses amplified and used to create a sectional image of the body's internal structure on a viewing screen. This image is a sonogram (fig. 1B).

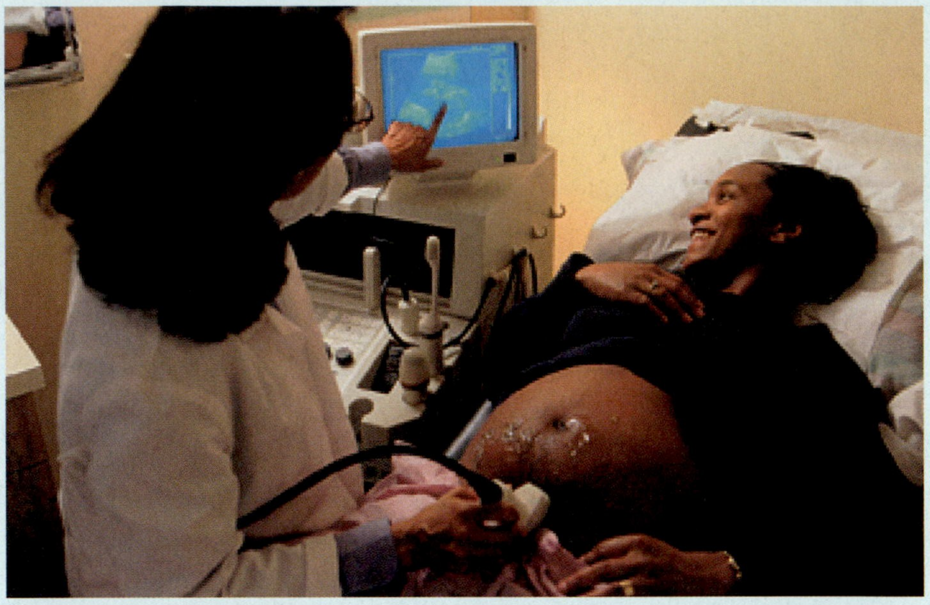

FIGURE 1A Ultrasonography uses reflected sound waves to visualize internal body structures.

Glancing at the screen, Vanessa smiles. The image reveals the fetus in her uterus, heart beating and already showing budlike structures that will develop into arms and legs. She happily heads home with a video of the fetus.

Vanessa's ultrasound exam takes only a few minutes, whereas Michael's MR scan takes an hour. First, Michael receives an injection of a dye that provides contrast so that a radiologist examining the scan can distinguish certain brain structures. Then, a nurse wheels the narrow bed on which Michael lies into a chamber surrounded by a powerful magnet and a special radio antenna. The chamber, which looks like a metal doughnut, is the MR imaging instrument. As Michael settles back, closes his eyes, and listens to the music through earphones, a technician activates the device.

1.4 CHARACTERISTICS OF LIFE

A scene such as Judith R.'s accident and injury underscores the delicate balance that must be maintained to sustain life. In those seconds at the limits of life—the birth of a baby, a trauma scene, or the precise instant of death following a long illness—we often think about just what combination of qualities constitutes this state that we call life. Indeed, although this text addresses the human body, the most fundamental characteristics of life are shared by all organisms.

As living organisms, we can respond to our surroundings. Our bodies grow, eventually becoming able to reproduce. We gain energy by ingesting (taking in), digesting (breaking down), absorbing, and assimilating the nutrients in food. The absorbed substances circulate throughout the internal environment of our bodies. We can then, by the process of respiration, use the energy in these nutrients for such vital functions as growth and repair of tissues. Finally, we excrete wastes. Taken together, these physical and chemical events that obtain, release, and use energy are a major part of **metabolism** (mĕ-tab′o-liz-m), all of the chemical reactions in cells. Table 1.3 summarizes the characteristics of life.

PRACTICE  

11 What are the characteristics of life?

12 Which physical and chemical events constitute metabolism?

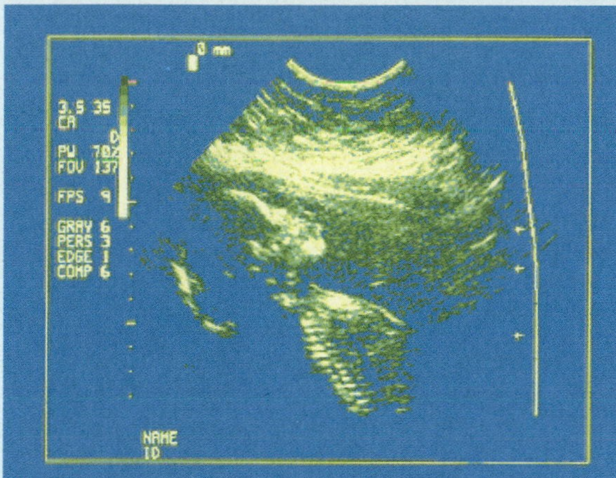

FIGURE 1B This image resulting from an ultrasonographic procedure reveals a fetus in the uterus.

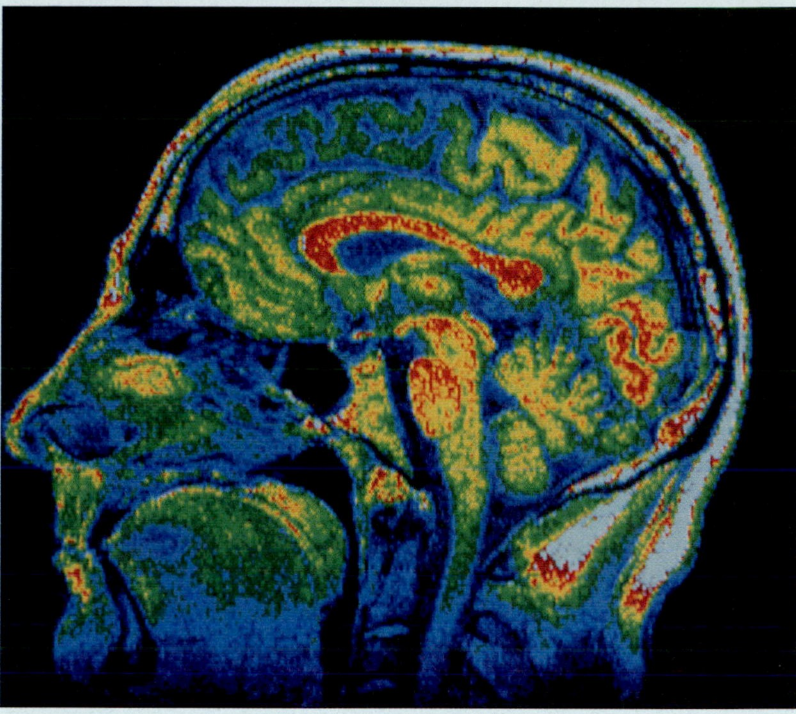

FIGURE 1C Falsely colored MR image of a human head and brain (sagittal section, see fig. 1.21).

The magnet generates a magnetic field that alters the alignment and spin of certain types of atoms within Michael's brain. At the same time, a second rotating magnetic field causes particular types of atoms (such as the hydrogen atoms in body fluids and organic compounds) to release weak radio waves with characteristic frequencies. The nearby antenna receives and amplifies the radio waves, which are then processed by a computer. Within a few minutes, the computer generates a sectional image based on the locations and concentrations of the atoms being studied (fig. 1C). The device continues to produce data, painting portraits of Michael's brain from different angles.

Michael and his parents nervously wait two days for the expert eyes of a radiologist to interpret the MR scan. Happily, the scan shows normal brain structure. Whatever is causing Michael's headaches, it is not a brain tumor—at least not one large enough to be imaged. ■

1.5 MAINTENANCE OF LIFE

With the exception of an organism's reproductive system, which perpetuates the species, all body structures and functions work in ways that maintain life.

Requirements of Organisms

Human life depends upon the following environmental factors:

1. **Water** is the most abundant substance in the body. It is required for a variety of metabolic processes, and it provides the environment in which most of them take place. Water also transports substances in organisms and is important in regulating body temperature.
2. **Food** refers to substances that provide organisms with necessary chemicals (nutrients) in addition to water. Nutrients supply energy and raw materials for building new living matter.
3. **Oxygen** is a gas that makes up about one-fifth of the air. It is used to release energy from nutrients. The energy, in turn, is used to drive metabolic processes.
4. **Heat** is a form of energy present in our environment. It is also a product of metabolic reactions, and it partly controls the rate at which these reactions occur. Generally, the more heat, the more rapidly chemical reactions take place. *Temperature* is a measure of the amount of heat present.

TABLE 1.2 | Organ Systems

Organ System	Major Organs	Major Functions
Integumentary	Skin, hair, nails, sweat glands, sebaceous glands	Protect tissues, regulate body temperature, support sensory receptors
Skeletal	Bones, ligaments, cartilages	Provide framework, protect soft tissues, provide attachments for muscles, produce blood cells, store inorganic salts
Muscular	Muscles	Cause movements, maintain posture, produce body heat
Nervous	Brain, spinal cord, nerves, sense organs	Detect changes, receive and interpret sensory information, stimulate muscles and glands
Endocrine	Glands that secrete hormones (pituitary gland, thyroid gland, parathyroid glands, adrenal glands, pancreas, ovaries, testes, pineal gland, and thymus)	Control metabolic activities of body structures
Cardiovascular	Heart, arteries, capillaries, veins	Move blood through blood vessels and transport substances throughout body
Lymphatic	Lymphatic vessels, lymph nodes, thymus, spleen	Return tissue fluid to the blood, carry certain absorbed food molecules, defend the body against infection
Digestive	Mouth, tongue, teeth, salivary glands, pharynx, esophagus, stomach, liver, gallbladder, pancreas, small and large intestines	Receive, break down, and absorb food; eliminate unabsorbed material
Respiratory	Nasal cavity, pharynx, larynx, trachea, bronchi, lungs	Intake and output of air, exchange of gases between air and blood
Urinary	Kidneys, ureters, urinary bladder, urethra	Remove wastes from blood, maintain water and electrolyte balance, store and transport urine
Reproductive	Male: scrotum, testes, epididymides, ductus deferentia, seminal vesicles, prostate gland, bulbourethral glands, urethra, penis	Produce and maintain sperm cells, transfer sperm cells into female reproductive tract
	Female: ovaries, uterine tubes, uterus, vagina, clitoris, vulva	Produce and maintain egg cells, receive sperm cells, support development of an embryo and function in birth process

TABLE 1.3 | Characteristics of Life

Process	Examples	Process	Examples
Movement	Change in position of the body or of a body part; motion of an internal organ	Digestion	Breakdown of food substances into simpler forms that can be absorbed and used
Responsiveness	Reaction to a change inside or outside the body	Absorption	Passage of substances through membranes and into body fluids
Growth	Increase in body size without change in shape	Circulation	Movement of substances in body fluids
Reproduction	Production of new organisms and new cells	Assimilation	Changing of absorbed substances into different chemical forms
Respiration	Obtaining oxygen, removing carbon dioxide, and releasing energy from foods (some forms of life do not use oxygen in respiration)	Excretion	Removal of wastes produced by metabolic reactions

5. **Pressure** is an application of force on an object or substance. For example, the force acting on the outside of a land organism due to the weight of air above it is called *atmospheric pressure*. In humans, this pressure plays an important role in breathing. Similarly, organisms living under water are subjected to *hydrostatic pressure*—a pressure a liquid exerts—due to the weight of water above them. In complex animals, such as humans, heart action produces blood pressure (another form of hydrostatic pressure), which keeps blood flowing through blood vessels.

Although the human organism requires water, food, oxygen, heat, and pressure, these factors alone are not enough to ensure survival. Both the quantities and the qualities of such factors are also important. Table 1.4 summarizes the major requirements of organisms.

Homeostasis

Most of the earth's residents are unicellular, or single-celled. The most ancient and abundant unicellular organisms are the bacteria. Their cells do not have membrane-bound organelles. Some unicellular organisms have organellas that are as complex as our own. This is the case for the amoeba

TABLE 1.4 | Requirements of Organisms

Factor	Characteristic	Use	Factor	Characteristic	Use
Water	A chemical substance	For metabolic processes, as a medium for metabolic reactions, to transport substances, and to regulate body temperature	Heat	A form of energy	To help regulate the rates of metabolic reactions
Food	Various chemical substances	To supply energy and raw materials for the production of necessary substances and for the regulation of vital reactions	Pressure	A force	Atmospheric pressure for breathing; hydrostatic pressure to help circulate blood
Oxygen	A chemical substance	To help release energy from food substances			

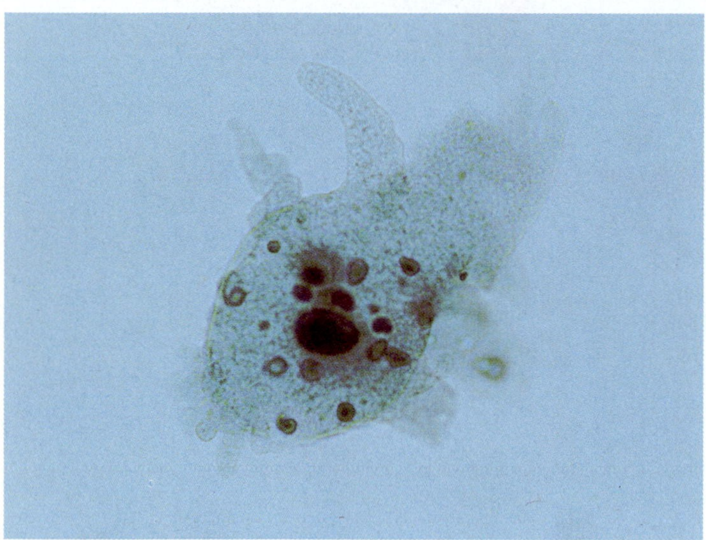

FIGURE 1.4 The amoeba is an organism consisting of a single, but complex, cell (100×).

(fig. 1.4). It survives and reproduces as long as its lake or pond environment is of a tolerable temperature and composition, and the amoeba can obtain food. With a limited ability to move, the amoeba depends upon the conditions in its lake or pond environment.

In contrast to the amoeba, humans are composed of 50 to 100 trillion cells in their own environment—our bodies. Our cells, as parts of organs and organ systems, interact in ways that keep this **internal environment** relatively constant, despite an ever-changing outside environment. Anatomically the internal environment is inside the body, but consists of fluid that surrounds cells, called the *extracellular fluid* (see chapter 21, p. 811). The internal environment protects our cells (and us!) from external changes that would kill isolated cells such as the amoeba (fig. 1.5). The body's maintenance of a stable internal environment is called **homeostasis** (ho″me-ō-sta′sis), and it is so important that it requires most of our metabolic energy. Many of the tests performed on Judith R. during her hospitalization (as described in this chapter's opening vignette on page 2) assessed her body's return to homeostasis.

The body maintains homeostasis through a number of self-regulating control systems, or **homeostatic mechanisms.** These mechanisms share the following three components (fig. 1.6):

1. **Receptors,** which provide information about specific conditions (stimuli) in the internal environment. A receptor may be a molecule or a cell.
2. A **control center,** which includes a **set point,** tells what a particular value should be (such as body temperature at 98.6°F).
3. **Effectors,** such as muscles or glands, which elicit responses that alter conditions in the internal environment.

A homeostatic mechanism works as follows. If the receptors measure deviations from the set point, effectors are activated that can return conditions toward normal. As conditions return toward normal, the deviation from the set point progressively lessens, and the effectors gradually shut down. Such a response is called a **negative feedback** (neg′ah-tiv fēd′bak) mechanism, both because the deviation from the set point is corrected (moves in the opposite or negative direction) and because the correction reduces the action of the effectors. This latter aspect is important because it prevents a correction from going too far.

To better understand this idea of maintaining a stable internal environment, imagine a room equipped with a furnace and an air conditioner. Suppose the room temperature is to remain near 20°C (68°F), so the thermostat is adjusted to a set point of 20°C. A thermostat is sensitive to temperature changes, so it will signal the furnace to start and the air conditioner to stop whenever the room temperature drops below the set point. If the temperature rises above the set point, the thermostat will cause the furnace to stop and the air conditioner to start. These actions maintain a relatively constant temperature in the room (fig. 1.7).

A similar homeostatic mechanism regulates body temperature in humans (fig. 1.8). The "thermostat" is a temperature-sensitive region in a control center of the brain called the hypothalamus. In healthy persons, the set point of this body thermostat is at or near 37°C (98.6°F).

If a person is exposed to a cold environment and the body temperature begins to drop, the hypothalamus senses this change and triggers heat-conserving and heat-generating activities. Blood vessels in the skin constrict, reducing blood

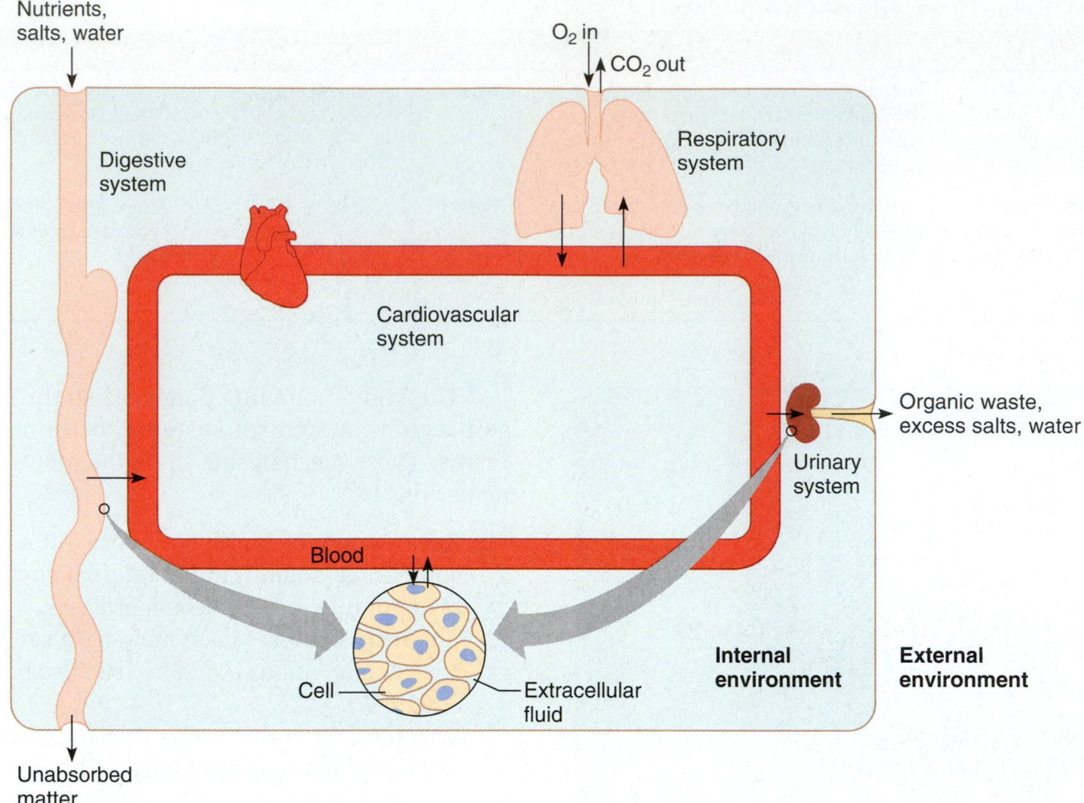

FIGURE 1.5 Our cells lie within an internal fluid environment (extracellular fluid). Concentrations of water, nutrients, and oxygen in the internal environment must be maintained within certain ranges to sustain life.

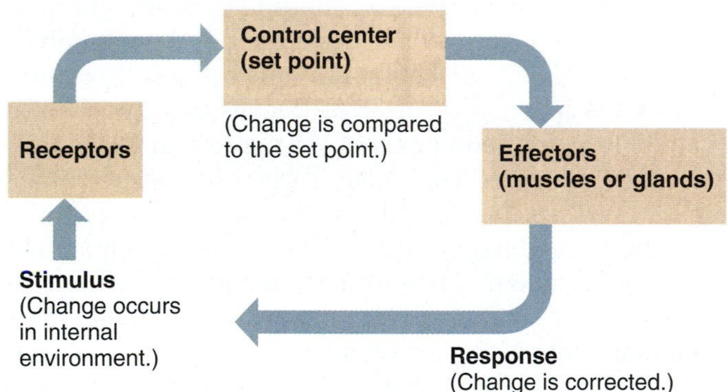

FIGURE 1.6 A homeostatic mechanism monitors a particular aspect of the internal environment and corrects any changes back to the value indicated by the set point.

flow and enabling deeper tissues to retain heat. At the same time, small groups of muscle cells may be stimulated to contract involuntarily, an action called shivering that produces heat, which helps warm the body.

If a person becomes overheated, the hypothalamus triggers a series of changes that dissipate body heat. Sweat glands in the skin secrete watery perspiration. Water evaporation from the surface carries away heat, cooling the skin. At the same time, blood vessels in the skin dilate. This allows the blood that carries heat from deeper tissues to reach the surface where more heat is lost to the outside. Chapter 6 discusses body temperature regulation in more detail (pp. 181 and 182).

Another homeostatic mechanism regulates the blood pressure in the blood vessels (arteries) leading away from the heart. In this instance, pressure-sensitive areas (sensory receptors) within the walls of these vessels detect changes in blood pressure and signal a pressure control center in the brain. If the blood pressure is above the pressure set point, the brain signals the heart, causing its chambers to contract more slowly and less forcefully. Because of decreased heart action, less blood enters the blood vessels, and the pressure inside the vessels decreases. If the blood pressure drops below the set point, the brain center signals the heart to contract more rapidly and with greater force, increasing the pressure in the vessels. Chapter 15 (pp. 585–587) discusses blood pressure regulation in more detail.

A homeostatic mechanism regulates the concentration of the sugar glucose in blood. In this case, cells of an organ called the pancreas determine the set point. If the concentration of blood glucose increases following a meal, the pancreas detects this change and releases a chemical (insulin) into the blood. Insulin allows glucose to move from the blood into various body cells and to be stored in the liver and muscles. As this occurs, the concentration of blood glucose decreases, and as it reaches the normal set point, the pancreas decreases its release of insulin. If, on the other hand, blood glucose concentration

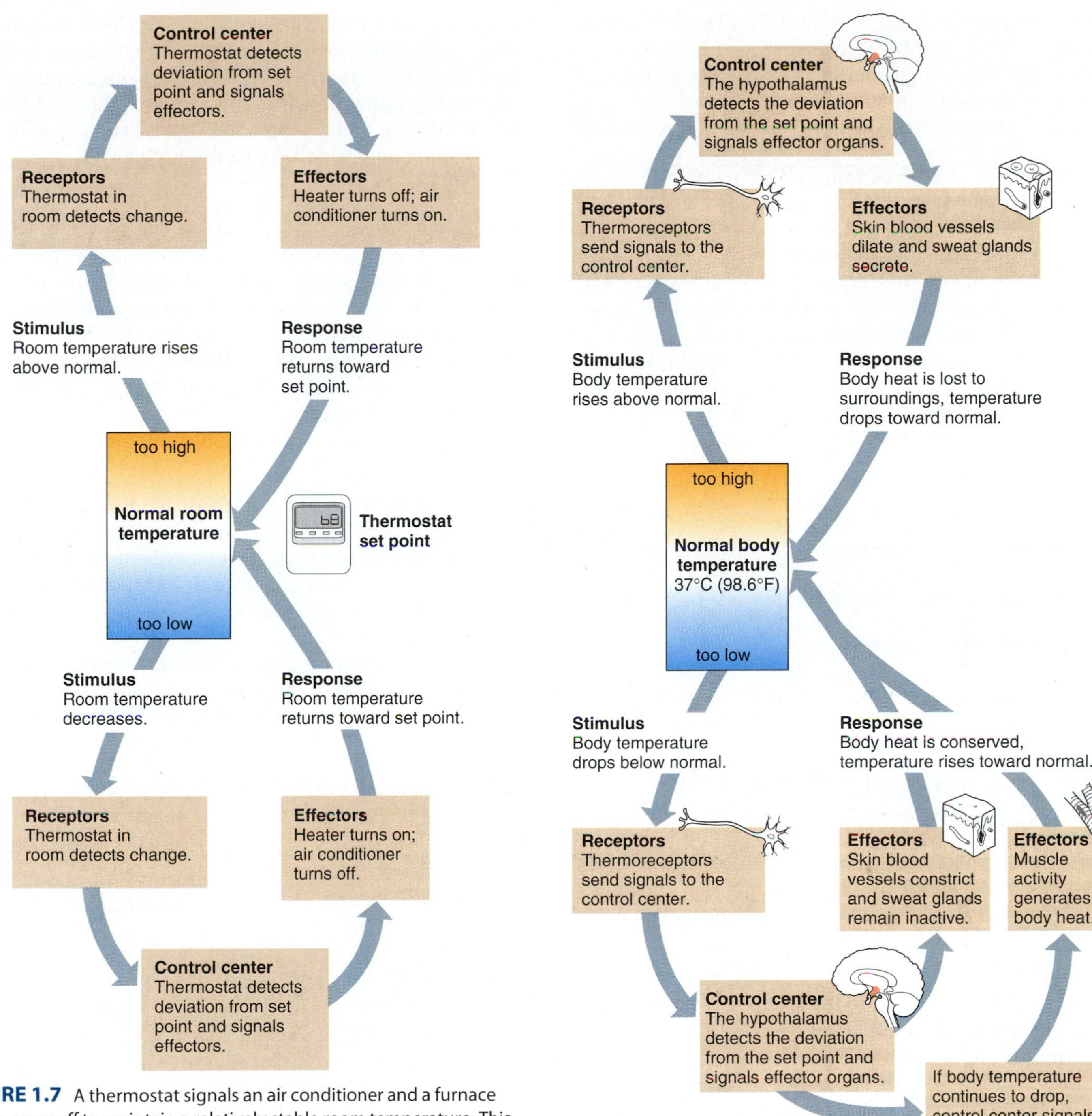
Control center
Thermostat detects deviation from set point and signals effectors.
Receptors
Thermostat in room detects change.
Effectors
Heater turns off; air conditioner turns on.
Stimulus
Room temperature rises above normal.
Response
Room temperature returns toward set point.
too high
Normal room temperature
too low
Thermostat set point
Stimulus
Room temperature decreases.
Response
Room temperature returns toward set point.
Receptors
Thermostat in room detects change.
Effectors
Heater turns on; air conditioner turns off.
Control center
Thermostat detects deviation from set point and signals effectors.
Control center
The hypothalamus detects the deviation from the set point and signals effector organs.
Receptors
Thermoreceptors send signals to the control center.
Effectors
Skin blood vessels dilate and sweat glands secrete.
Stimulus
Body temperature rises above normal.
Response
Body heat is lost to surroundings, temperature drops toward normal.
too high
Normal body temperature
37°C (98.6°F)
too low
Stimulus
Body temperature drops below normal.
Response
Body heat is conserved, temperature rises toward normal.
Receptors
Thermoreceptors send signals to the control center.
Effectors
Skin blood vessels constrict and sweat glands remain inactive.
Effectors
Muscle activity generates body heat.
Control center
The hypothalamus detects the deviation from the set point and signals effector organs.
If body temperature continues to drop, control center signals muscles to contract involuntarily.

FIGURE 1.7 A thermostat signals an air conditioner and a furnace to turn on or off to maintain a relatively stable room temperature. This system is an example of a homeostatic mechanism.

FIGURE 1.8 The homeostatic mechanism that regulates body temperature.

falls too low, the pancreas detects this change and secretes a different chemical (glucagon) that releases stored glucose into the blood. Chapter 13 (pp. 509–511) discusses regulation of blood glucose concentration in more detail (see fig. 13.36).

Human physiology offers many other examples of homeostatic mechanisms, which all work by the basic mechanism just described. Just as anatomical terms are used repeatedly throughout this book, so the basic principles of physiology apply in all organ systems.

Although most feedback mechanisms in the body are negative, some changes stimulate further change. A process that moves conditions away from the normal state is called a *positive feedback mechanism.*

Positive feedback mechanisms may be important to homeostasis and survival. In blood clotting, for example, certain chemicals stimulate more clotting, which minimizes bleeding (see chapter 14, pp. 539–541). Preventing blood loss following an injury is critical to sustaining life. Similarly, a positive feedback mechanism increases the strength of uterine contractions during childbirth.

Positive feedback mechanisms usually produce unstable conditions, which might not seem compatible with homeostasis. However, the few examples of positive feedback associated with health have very specific functions and are short-lived.

Homeostatic mechanisms maintain a relatively constant internal environment, yet physiological values may vary slightly in a person from time to time or from one person to the next. Therefore, both normal values for an individual and the idea of a **normal range** for the general population are clinically important. Numerous examples of homeostasis are presented throughout this book, and normal ranges for a number of physiological variables are listed in Appendix B, Laboratory Tests of Clinical Importance, pages 940–943.

PRACTICE

13 Which requirements of organisms does the external environment provide?

14 What is the relationship between oxygen use and heat production?

15 Why is homeostasis so important to survival?

16 Describe three homeostatic mechanisms.

1.6 ORGANIZATION OF THE HUMAN BODY

The human organism is a complex structure composed of many parts. The major features of the human body include cavities, various types of membranes, and organ systems.

Body Cavities

The human organism can be divided into an **axial** (ak′se-al) **portion,** which includes the head, neck, and trunk, and an **appendicular** (ap″en-dik′u-lar) **portion,** which includes the upper and lower limbs. Within the axial portion are the **cranial cavity,** which houses the brain; the **vertebral canal** (spinal cavity), which contains the spinal cord and is surrounded by sections of the backbone (vertebrae); the **thoracic** (tho-ras′ik) **cavity;** and the **abdominopelvic** (ab-dom′ĭ-no-pel′vik) **cavity.** The organs within these last two cavities are called **viscera** (vis′er-ah). Figure 1.9 shows these major body cavities.

The thoracic cavity is separated from the lower abdominopelvic cavity by a broad, thin muscle called the **diaphragm** (di′ah-fram). When it is at rest, this muscle curves upward into the thorax like a dome. When it contracts during inhalation, it presses down upon the abdominal viscera. The wall of the thoracic cavity is composed of skin, skeletal muscles, and bones. Within the thoracic cavity are the lungs and a region between the lungs, called the **mediastinum** (me″de-as-ti′num). The mediastinum separates the thorax into two compartments that contain the right and left lungs. The remaining thoracic viscera—heart, esophagus, trachea, and thymus—are within the mediastinum.

The abdominopelvic cavity, which includes an upper abdominal portion and a lower pelvic portion, extends from the diaphragm to the floor of the pelvis. Its wall primarily consists of skin, skeletal muscles, and bones. The viscera within the **abdominal cavity** include the stomach, liver, spleen, gallbladder, and the small and large intestines.

The **pelvic cavity** is the portion of the abdominopelvic cavity enclosed by the pelvic bones. It contains the terminal end of the large intestine, the urinary bladder, and the internal reproductive organs.

Smaller cavities within the head include the following (fig. 1.10):

1. *Oral cavity,* containing the teeth and tongue.
2. *Nasal cavity,* located within the nose and divided into right and left portions by a nasal septum. Several air-filled sinuses are connected to the nasal cavity. These include the sphenoidal and frontal sinuses (see fig. 7.25).
3. *Orbital cavities,* containing the eyes and associated skeletal muscles and nerves.
4. *Middle ear cavities,* containing the middle ear bones.

Thoracic and Abdominopelvic Membranes

Thin **serous membranes** line the walls of the thoracic and abdominal cavities and fold back to cover the organs within these cavities. These membranes secrete a slippery serous fluid that separates the layer lining the wall of the cavity (parietal layer) from the layer covering the organ (visceral layer). For example, the right and left thoracic compartments, which contain the lungs, are lined with a serous membrane called the *parietal pleura.* This membrane folds back to cover the lungs, forming the *visceral pleura.* A thin film of serous fluid separates the parietal and visceral **pleural** (ploo′ral) **membranes.** Although there is normally no space between these two membranes, the potential space between them is called the *pleural cavity.*

The heart, located in the broadest portion of the mediastinum, is surrounded by **pericardial** (per″ĭ-kar′de-al) **membranes.** A thin *visceral pericardium* (epicardium) covers the heart's surface and is separated from the *parietal pericardium* by a small volume of serous fluid. The potential space between these membranes is called the *pericardial cavity.* The parietal pericardium is covered by a much thicker third layer, the *fibrous pericardium.* Figure 1.11 shows the membranes associated with the heart and lungs.

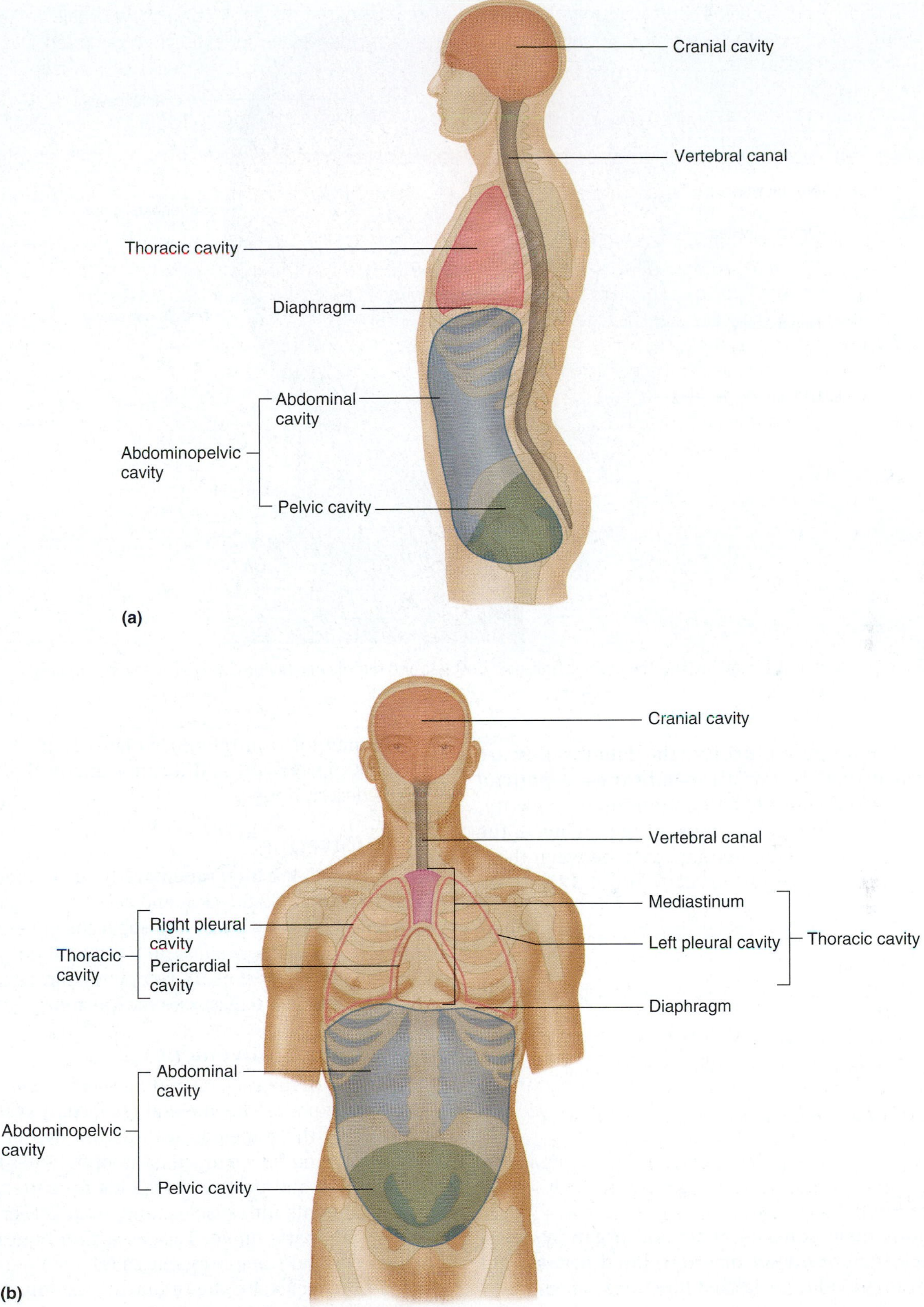

FIGURE 1.9 Major body cavities. (*a*) Lateral view. (*b*) Anterior view.

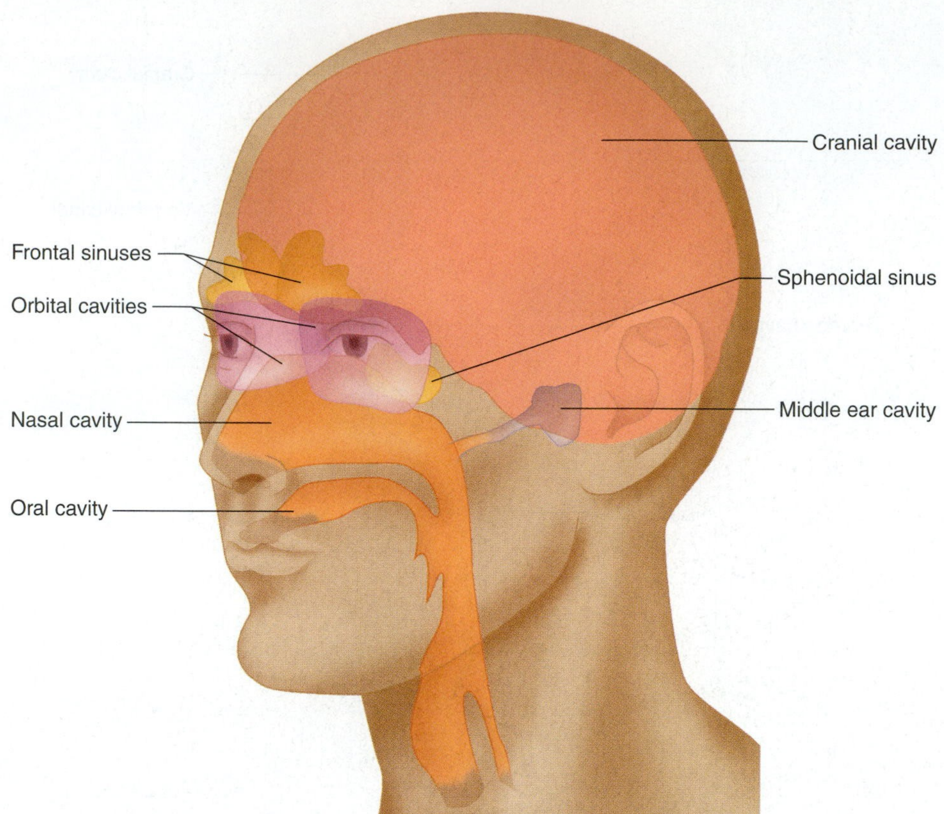

FIGURE 1.10 The cavities in the head include the cranial, oral, nasal, orbital, and middle ear cavities, as well as several sinuses.

In the abdominopelvic cavity, the membranes are called **peritoneal** (per″-ĭ-to-ne′al) **membranes.** A *parietal peritoneum* lines the wall of the abdominopelvic cavity, and a *visceral peritoneum* covers most of the organs in the abdominopelvic cavity. The potential space between these membranes is called the *peritoneal cavity* (fig. 1.12).

PRACTICE

17 What are the viscera?

18 Which organs occupy the thoracic cavity? The abdominal cavity? The pelvic cavity?

19 Name the cavities of the head.

20 Describe the membranes associated with the thoracic and abdominopelvic cavities.

21 Distinguish between the parietal and visceral peritoneum.

Organ Systems

The human organism consists of several organ systems, each of which includes a set of interrelated organs that work together to provide specialized functions. The maintenance of homeostasis depends on the coordination of organ systems. A figure called **"InnerConnections"** at the end of some chapters ties together the ways in which organ systems interact. As you read about each organ system, you may want to consult the illustrations and cadaver photos of the human torso in reference plates 1–25 at the end of this chapter (pp. 31–49) and locate some of the features listed in the descriptions.

Body Covering

The organs of the **integumentary** (in-teg-u-men′tar-e) **system** (fig. 1.13) include the skin and accessory organs such as the hair, nails, sweat glands, and sebaceous glands. These parts protect underlying tissues, help regulate body temperature, house a variety of sensory receptors, and synthesize certain products. Chapter 6 discusses the integumentary system.

Support and Movement

The organs of the skeletal and muscular systems support and move body parts. The **skeletal** (skel′ĕ-tal) **system** (fig. 1.14) consists of the bones as well as the ligaments and cartilages that bind bones together at joints. These parts provide frameworks and protective shields for softer tissues, serve as attachments for muscles, and act together with muscles when body parts move. Tissues within bones also produce blood cells and store inorganic salts.

The muscles are the organs of the **muscular** (mus′ku-lar) **system** (fig. 1.14). By contracting and pulling their ends closer together, muscles provide the forces that move body parts. Muscles also help maintain posture and are the primary source of body heat. Chapters 7, 8, and 9 discuss the skeletal and muscular systems.

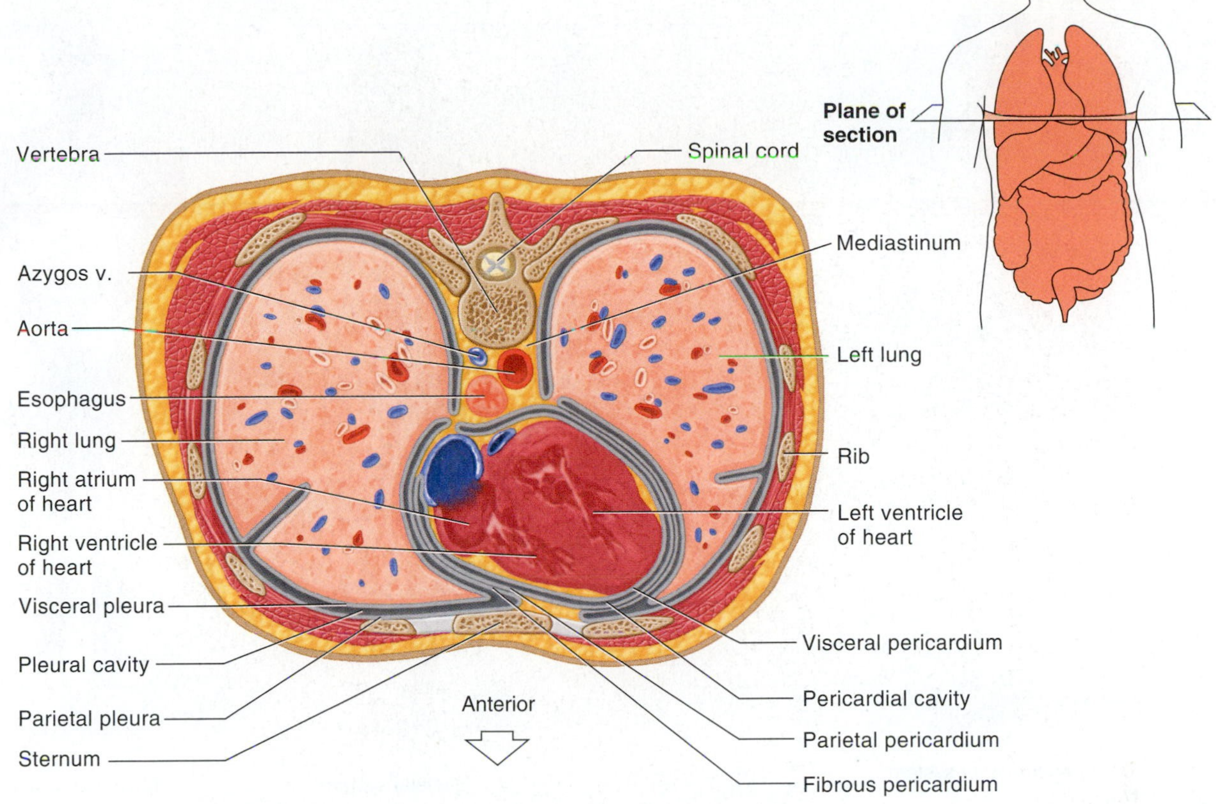

FIGURE 1.11 A transverse section through the thorax reveals the serous membranes associated with the heart and lungs (superior view).

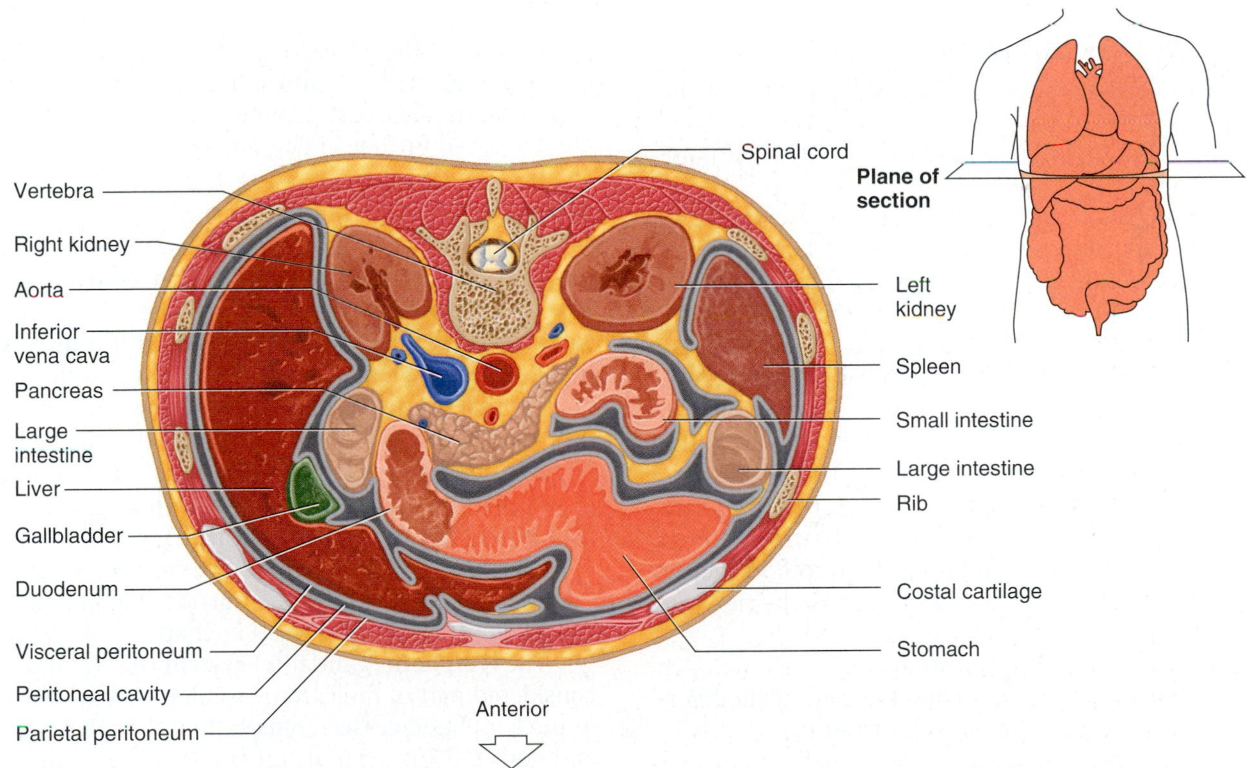

FIGURE 1.12 Transverse section through the abdomen (superior view).

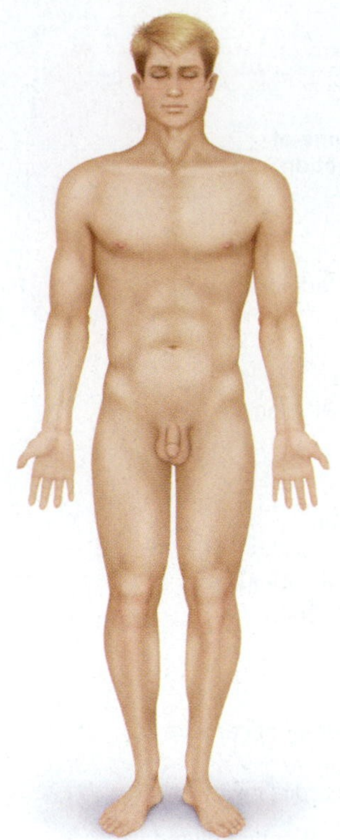

FIGURE 1.13 The integumentary system covers the body.

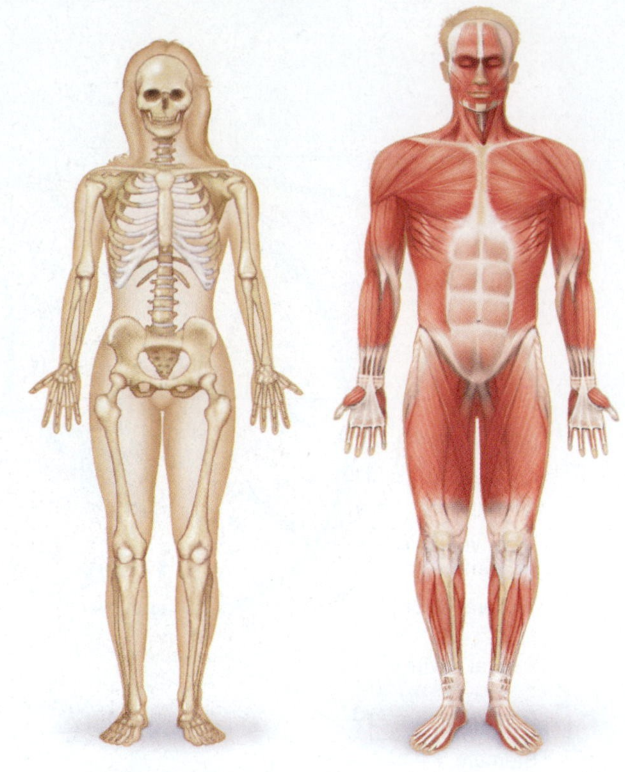

FIGURE 1.14 The skeletal and muscular systems provide support and movement.

Integration and Coordination

For the body to act as a unit, its parts must be integrated and coordinated. The nervous and endocrine systems control and adjust various organ functions from time to time, maintaining homeostasis.

The **nervous** (ner′vus) **system** (fig. 1.15) consists of the brain, spinal cord, nerves, and sense organs. Nerve cells within these organs use electrochemical signals called *nerve impulses* (action potentials) to communicate with one another and with muscles and glands. Each impulse produces a relatively short-term effect on its target. Some nerve cells act as specialized sensory receptors that can detect changes occurring inside and outside the body. Other nerve cells receive the impulses transmitted from these sensory units and interpret and act on the information. Still other nerve cells carry impulses from the brain or spinal cord to muscles or glands, stimulating them to contract or to secrete products. Chapters 10 and 11 discuss the nervous system, and chapter 12 discusses sense organs.

The **endocrine** (en′do-krin) **system** (fig. 1.15) includes all the glands that secrete chemical messengers, called *hormones.* Hormones, in turn, travel away from the glands in body fluids such as blood or tissue fluid. Usually a particular hormone affects only a particular group of cells, called its *target cells.* The effect of a hormone is to alter the metabolism of the target cells. Compared to nerve impulses, hormonal effects occur over a relatively long period.

Organs of the endocrine system include the pituitary, thyroid, parathyroid, and adrenal glands, as well as the pancreas, ovaries, testes, pineal gland, and thymus. These are discussed further in chapter 13.

Transport

Two organ systems transport substances throughout the internal environment. The **cardiovascular** (kahr″de-o-vas′ku-lur) **system** (fig. 1.16) includes the heart, arteries, capillaries, veins, and blood. The heart is a muscular pump that helps force blood through the blood vessels. Blood transports gases, nutrients, hormones, and wastes. It carries oxygen from the lungs and nutrients from the digestive organs to all body cells, where these substances are used in metabolic processes. Blood also transports hormones from endocrine glands to their target cells and carries wastes from body cells to the excretory organs, where the wastes are removed from the blood and released to the outside. Blood and the cardiovascular system are discussed in chapters 14 and 15.

The **lymphatic** (lim-fat′ik) **system** (fig. 1.16) is sometimes considered part of the cardiovascular system. It is composed of the lymphatic vessels, lymph fluid, lymph nodes, thymus, and spleen. This system transports some of the fluid from the spaces in tissues (tissue fluid) back to the bloodstream and carries certain fatty substances away from the digestive organs. Cells of the lymphatic system, called lymphocytes, defend the body against infections by removing pathogens

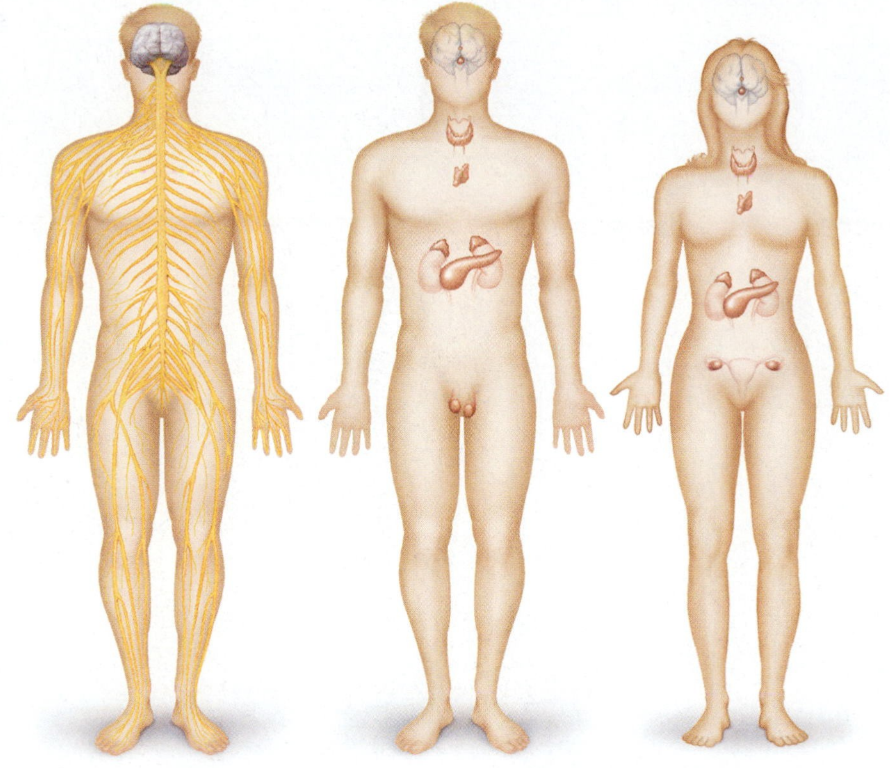

FIGURE 1.15 The nervous and endocrine systems integrate and coordinate body functions.

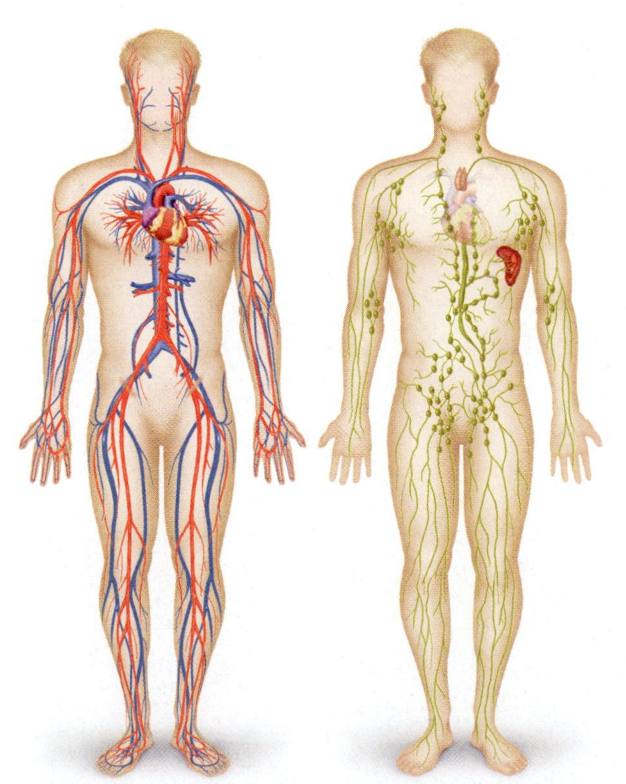

FIGURE 1.16 The cardiovascular and lymphatic systems transport fluids.

(disease-causing microorganisms and viruses) from tissue fluid. The lymphatic system is discussed in chapter 16.

Absorption and Excretion

Organs in several systems absorb nutrients and oxygen and excrete wastes. The organs of the **digestive** (di-jest′tiv) **system** (fig. 1.17), discussed in detail in chapter 17 receive foods and then break down food molecules into simpler forms that can be absorbed into the internal environment. Certain digestive organs (chapter 17, pp. 668, 671, 672) also produce hormones and thus function as parts of the endocrine system.

The digestive system includes the mouth, tongue, teeth, salivary glands, pharynx, esophagus, stomach, liver, gallbladder, pancreas, small intestine, and large intestine. Chapter 18 discusses nutrition and metabolism, considering the fate of foods in the body.

The organs of the **respiratory** (re-spi′rah-to″re) **system** (fig. 1.17) take air in and out and exchange gases between the blood and the air. More specifically, oxygen passes from air in the lungs into the blood, and carbon dioxide leaves the blood and enters the air. The nasal cavity, pharynx, larynx, trachea, bronchi, and lungs are parts of this system, discussed in chapter 19.

The **urinary** (u′rĭ-ner″e) **system** (fig. 1.17) consists of the kidneys, ureters, urinary bladder, and urethra. The kidneys remove wastes from blood and assist in maintaining the body's water and electrolyte balance. The product of these activities is urine. Other parts of the urinary system store urine and transport it outside the body. Chapter 20 discusses the urinary system. Sometimes the urinary system is called the

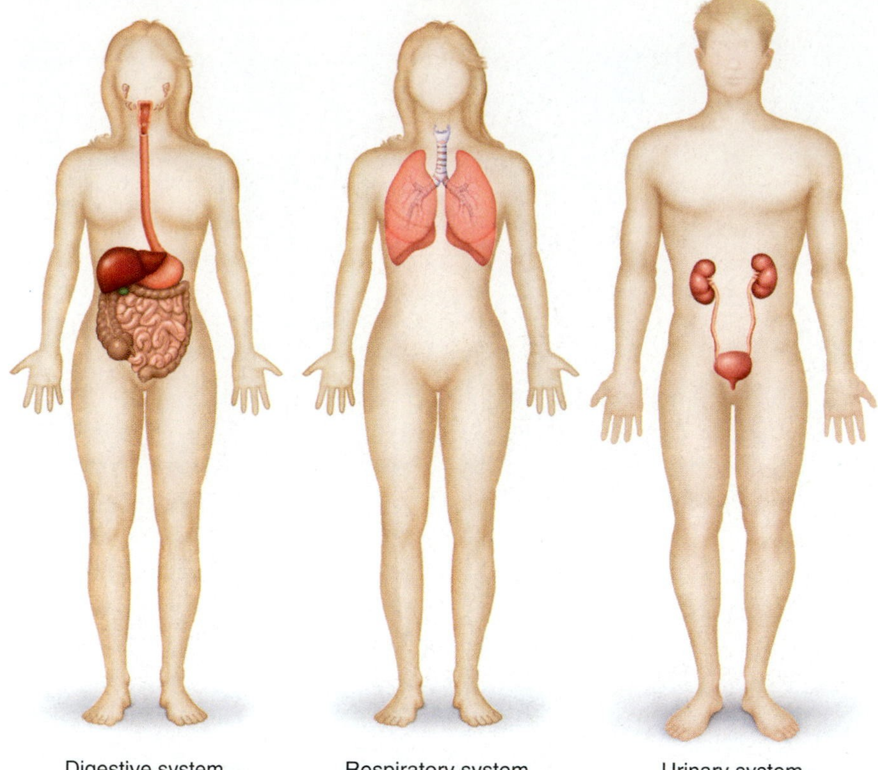

FIGURE 1.17 The digestive, respiratory, and urinary systems absorb nutrients, take in oxygen and release carbon dioxide, and excrete wastes.

excretory system. However, **excretion** (ek-skre′shun), or waste removal, is also a function of the respiratory system and, to a lesser extent, the digestive and integumentary systems.

Reproduction

Reproduction (re″pro-duk′shun) is the process of producing offspring (progeny). Cells reproduce when they divide and give rise to new cells. The **reproductive** (re″pro-duk′tiv) **system** (fig. 1.18) of an organism, however, produces whole new organisms like itself (see chapter 22).

The male reproductive system includes the scrotum, testes, epididymides, ductus deferentia, seminal vesicles, prostate gland, bulbourethral glands, urethra, and penis. These structures produce and maintain the male sex cells, or sperm cells (spermatozoa). The male reproductive system also transfers these cells into the female reproductive tract.

The female reproductive system consists of the ovaries, uterine tubes, uterus, vagina, clitoris, and vulva. These organs produce and maintain the female sex cell (egg cells or ova), transport the female's egg cell within the female reproductive system, and receive the male's sperm cells for the possibility of fertilizing an egg. The female reproductive system also supports development of embryos, carries a fetus to term, and functions in the birth process. Figure 1.19 illustrates the organ systems in humans.

PRACTICE 

22 Name the major organ systems and list the organs of each system.

23 Describe the general functions of each organ system.

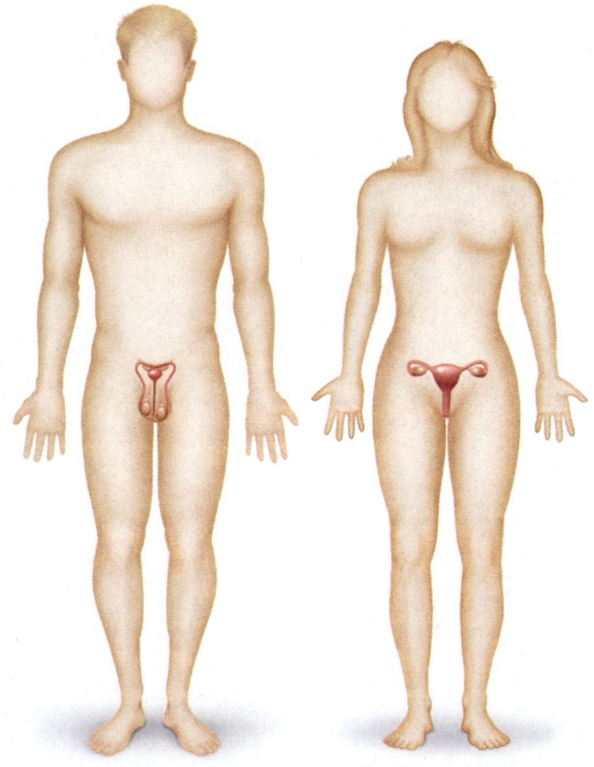

FIGURE 1.18 The reproductive systems manufacture and transport sex cells. The female reproductive system provides for prenatal development and childbirth.

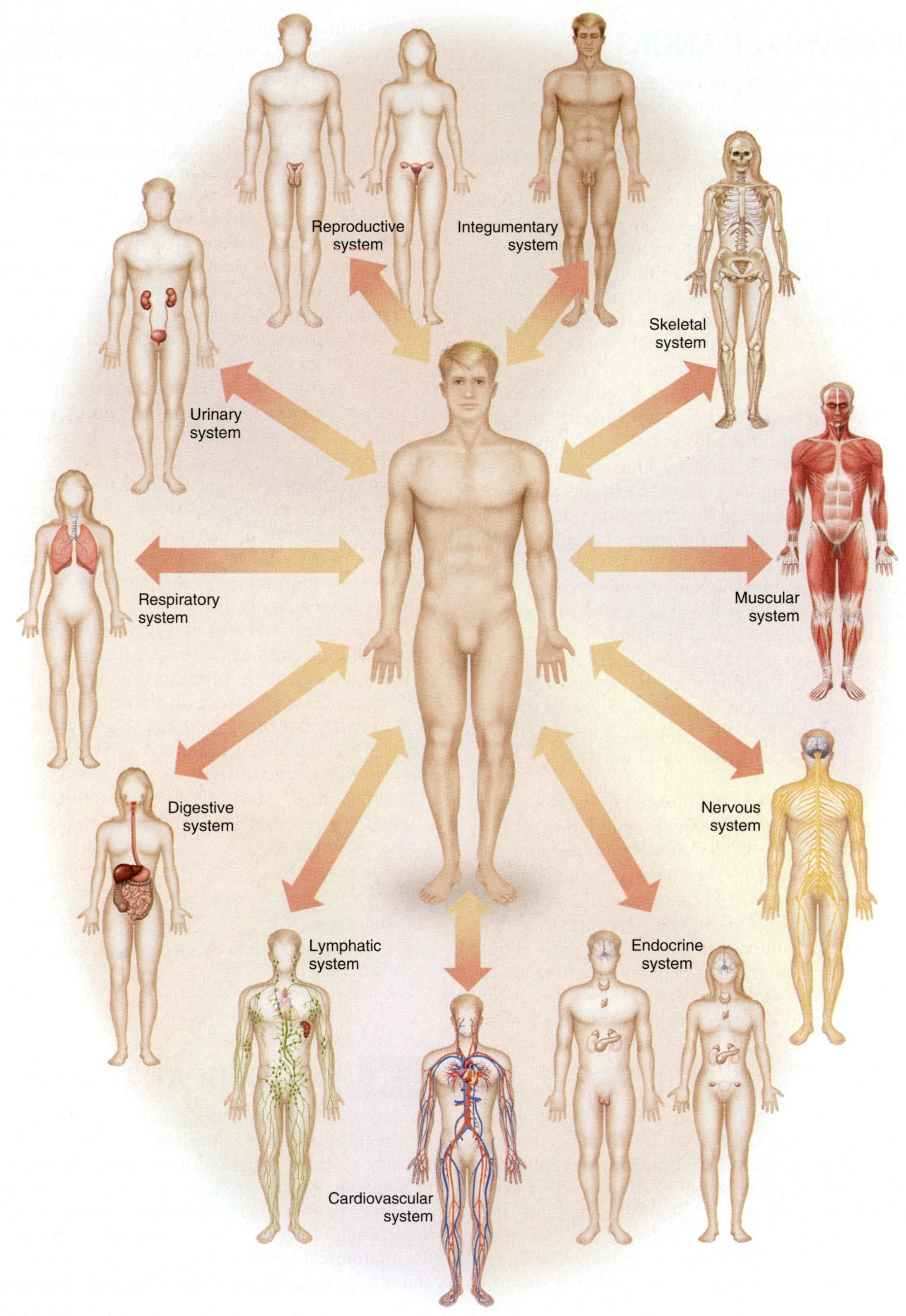

FIGURE 1.19 The organ systems in humans interact in ways that maintain homeostasis.

1.7 LIFE-SPAN CHANGES

Aging is a part of life. It is the process of becoming mature or old. **Aging** is the passage of time and the accompanying bodily changes. The passage of time is inevitable; so, too, is aging, despite common claims for the anti-aging properties of various diets, cosmetics, pills, and skin-care products.

Aging occurs from the microscopic to the whole-body level. Although programmed cell death begins in the fetus, we are usually not very aware of aging until the third decade of life, when a few gray hairs, faint lines etched into facial skin, and minor joint stiffness in the morning remind us that time marches on. A woman over the age of thirty-five attempting to conceive a child might be shocked to learn that she is of "advanced maternal age," because the chances of conceiving an offspring with an abnormal chromosome number increase with the age of the egg. In both sexes, by the fourth or fifth decade, as hair color fades and skin etches become wrinkles, the first signs of adult-onset disorders may appear, such as elevated blood pressure that one day may be considered hypertension, and slightly high blood glucose that could become type 2 diabetes mellitus. A person with a strong family history of heart disease, coupled with unhealthy diet and exercise habits, may be advised to change his or her lifestyle, and perhaps even begin taking a drug to lower serum cholesterol levels. The sixth decade sees grayer or whiter hair, more and deeper skin wrinkles, and a waning immunity that makes vaccinations against influenza and other infectious diseases important. Yet many, if not most, people in their sixties and older have sharp minds and are capable of all sorts of physical activities.

Changes at the tissue, cell, and molecular levels explain the familiar signs of aging. Decreased production of the connective tissue proteins collagen and elastin account for the stiffening of skin, and diminished levels of subcutaneous fat are responsible for wrinkling. Proportions of fat to water in the tissues change, with the percentage of fats increasing steadily in women, and increasing until about age sixty in men. These alterations explain why the elderly metabolize certain drugs at different rates than do younger people. As a person ages, tissues atrophy, and as a result, organs shrink.

Cells mark time too, many approaching the end of a limited number of predetermined cell divisions as their chromosome tips whittle down. Such cells reaching the end of their division days may enlarge or die. Some cells may be unable to build the apparatus that pulls apart replicated chromosomes in a cell on the verge of division. Impaired cell division slows wound healing, yet at the same time, the inappropriate cell division that underlies cancer becomes more likely. Certain subcellular functions lose efficiency, including repair of DNA damage and transport of substances into and out of cells. Aging cells are less efficient at extracting energy from nutrients and breaking down aged or damaged cell parts.

As changes at the tissue level cause organ-level signs of aging, certain biochemical changes fuel cellular aging. Lipofuscin and ceroid pigments accumulate as the cell can no longer prevent formation of damaging oxygen free radicals. A protein called beta amyloid may build up in the brain, contributing, in some individuals, to the development of Alzheimer disease. A generalized metabolic slowdown results from a dampening of thyroid gland function, impairing glucose use, the rate of protein synthesis, and production of digestive enzymes. At the whole-body level, we notice slowed metabolism as diminished tolerance to cold, weight gain, and fatigue.

Several investigations are identifying key characteristics, particularly gene variants, which people who live more than 100 years share. These fortunate individuals, called centenarians, fall into three broad groups: about 20 percent of them never get the diseases that kill most people; 40 percent get these diseases but at much older ages than average; and the other 40 percent live with and survive the more common disorders of aging. Environmental factors are important, too—another trait centenarians share is never having smoked.

Our organs and organ systems are interrelated, so aging-related changes in one influence the functioning of others. Several chapters in this book conclude with a "Life-Span Changes" section that discusses changes specific to particular organ systems. These changes reflect the natural breakdown of structure and function that accompanies the passage of time, as well as events in our genes ("nature") and symptoms or characteristics that might arise as a consequence of lifestyle choices and circumstances ("nurture").

PRACTICE 

24 Define aging.

25 List some aging-related changes at the microscopic and whole-body levels.

1.8 ANATOMICAL TERMINOLOGY

To communicate effectively with one another, investigators over the ages have developed a set of terms with precise meanings. Some of these terms concern the relative positions of body parts, others refer to imaginary planes along which cuts may be made, and still others describe body regions. When such terms are used, it is assumed that the body is in the **anatomical position**—standing erect; the face is forward; and the upper limbs are at the sides, with the palms forward.

Relative Position

Terms of relative position are used to describe the location of one body part with respect to another. They include the following (many of these terms are illustrated in figure 1.20):

1. **Superior** means a part is above another part, or closer to the head. (The thoracic cavity is superior to the abdominopelvic cavity.)
2. **Inferior** means a part is below another part, or toward the feet. (The neck is inferior to the head.)
3. **Anterior** (or ventral) means toward the front. (The eyes are anterior to the brain.)
4. **Posterior** (or dorsal) is the opposite of anterior; it means toward the back. (The pharynx is posterior to the oral cavity.)
5. **Medial** refers to an imaginary midline dividing the body into equal right and left halves. A part is medial if it is closer to this line than another part. (The nose is medial to the eyes.)
6. **Lateral** means toward the side with respect to the imaginary midline. (The ears are lateral to the eyes.)
7. **Bilateral** refers to paired structures, one on each side. (The lungs are bilateral.)
8. **Ipsilateral** refers to structures on the same side. (The right lung and the right kidney are ipsilateral.)
9. **Contralateral** refers to structures on the opposite side. (A patient with a fractured right leg would have to bear weight on the contralateral—in this case, left—lower limb.)
10. **Proximal** describes a part closer to the trunk of the body or closer to another specified point of reference than another part. (The elbow is proximal to the wrist.)
11. **Distal** is the opposite of proximal. It means a particular body part is farther from the trunk or farther from another specified point of reference than another part. (The fingers are distal to the wrist.)
12. **Superficial** means near the surface. (The epidermis is the superficial layer of the skin.) **Peripheral** also means outward or near the surface. It describes the location of certain blood vessels and nerves. (The nerves that branch from the brain and spinal cord are peripheral nerves.)
13. **Deep** describes more internal parts. (The dermis is the deep layer of the skin.)

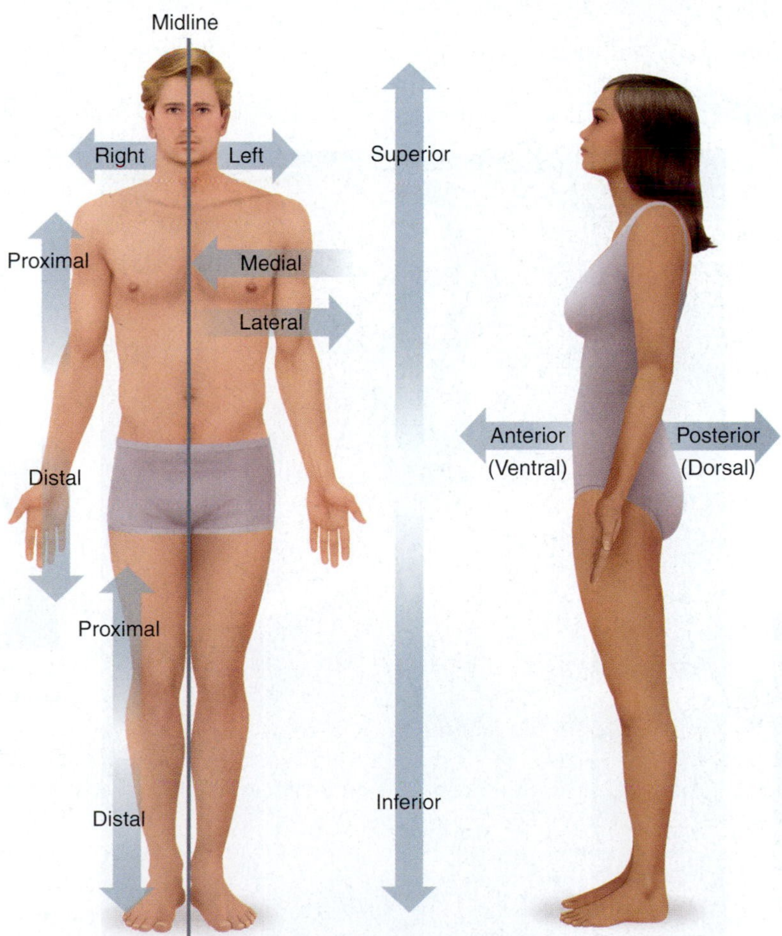

FIGURE 1.20 Relative positional terms describe a body part's location with respect to other body parts.

Body Sections

To observe the relative locations and arrangements of internal parts, it is necessary to cut, or section, the body along various planes (figs. 1.21 and 1.22). The following terms describe such planes and sections:

1. **Sagittal** refers to a lengthwise cut that divides the body into right and left portions. If a sagittal section passes along the midline and divides the body into equal parts, it is called median (midsagittal). A sagittal section lateral to midline is called parasagittal.
2. **Transverse** (or horizontal) refers to a cut that divides the body into superior and inferior portions.
3. **Frontal** (or coronal) refers to a section that divides the body into anterior and posterior portions.

Sometimes a cylindrical organ such as a blood vessel is sectioned. In this case, a cut across the structure is called a *cross section,* an angular cut is called an *oblique section,* and a lengthwise cut is called a *longitudinal section* (fig. 1.23).

Body Regions

A number of terms designate body regions. The abdominal area, for example, is subdivided into the following regions, as shown in figure 1.24*a:*

1. **Epigastric region** The upper middle portion.
2. **Left** and **right hypochondriac regions** On the left/right side of the epigastric region.
3. **Umbilical region** The central portion.
4. **Left** and **right lumbar regions** On the left/right side of the umbilical region.
5. **Hypogastric region** The lower middle portion.
6. **Left** and **right iliac (or inguinal) regions** On the left/right side of the hypogastric region.

The abdominal area may also be subdivided into the following four quadrants, as figure 1.24*b* illustrates:

1. **Right upper quadrant** (RUQ).
2. **Right lower quadrant** (RLQ).
3. **Left upper quadrant** (LUQ).
4. **Left lower quadrant** (LLQ).

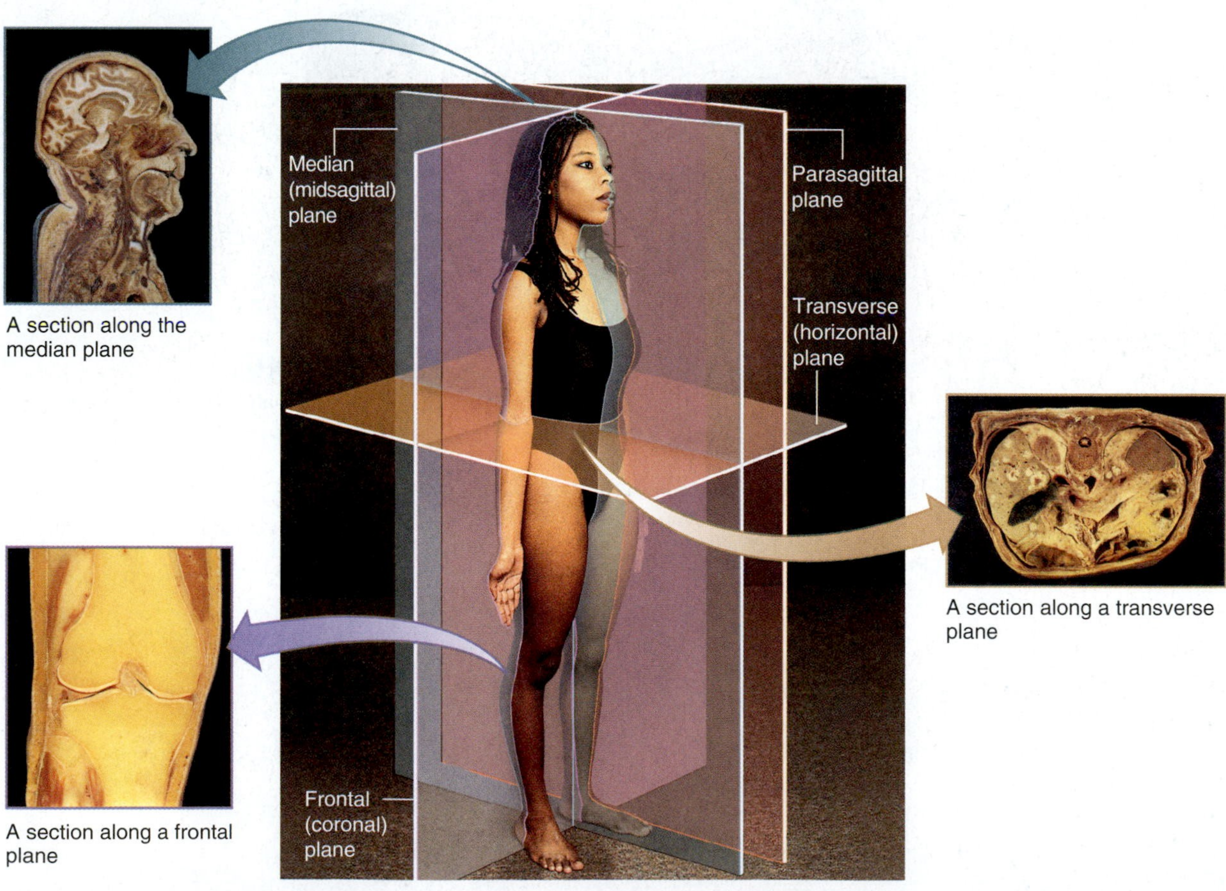

FIGURE 1.21 Observation of internal parts requires sectioning the body along various planes.

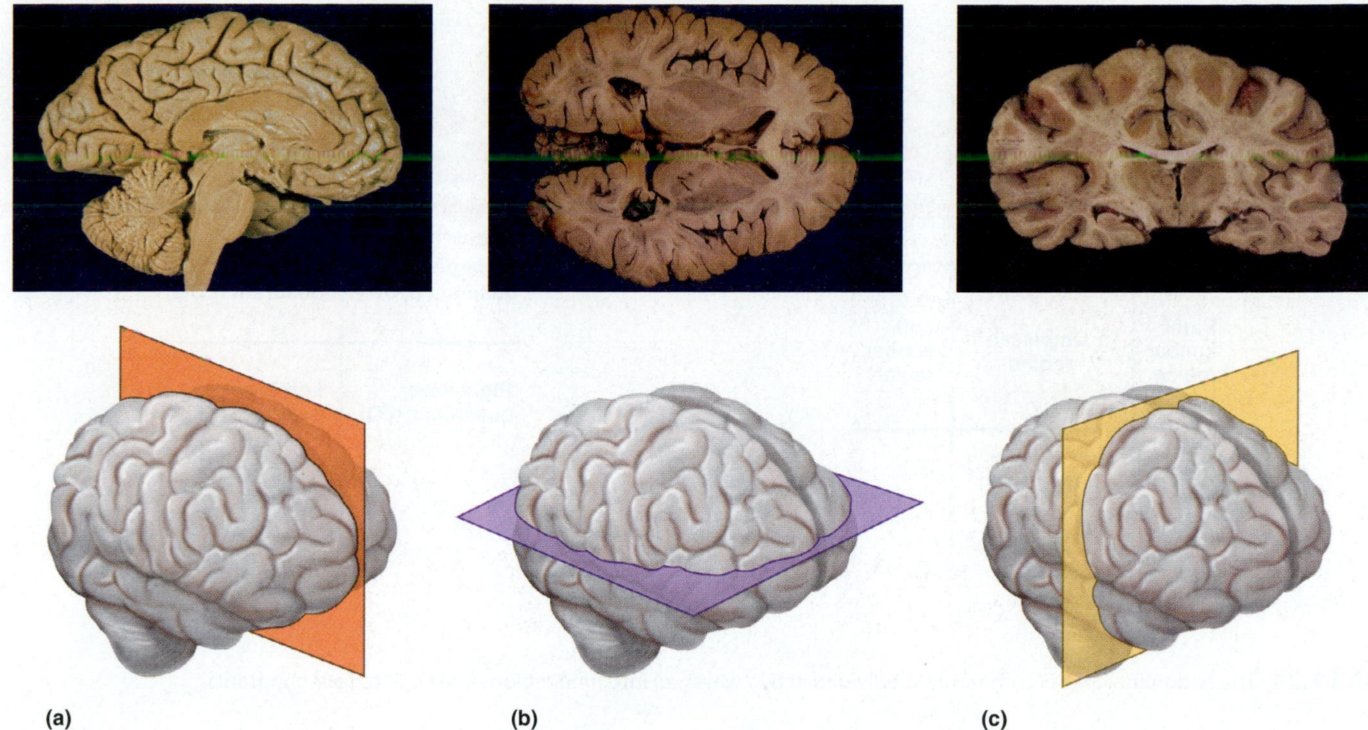

FIGURE 1.22 A human brain sectioned along (*a*) a sagittal plane, (*b*) a transverse plane, and (*c*) a frontal plane.

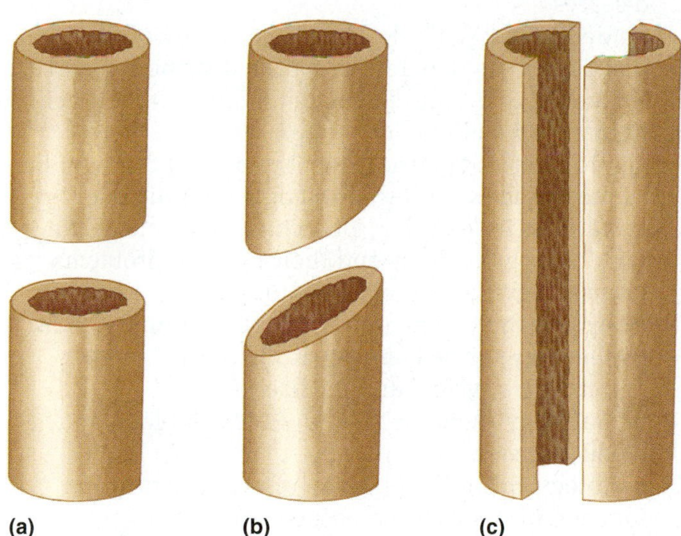

FIGURE 1.23 Cylindrical parts may be cut in (*a*) cross section, (*b*) oblique section, or (*c*) longitudinal section.

The following adjectives are commonly used when referring to various body regions. Figure 1.25 illustrates some of these regions.

abdominal (ab-dom′ĭ-nal) region between the thorax and pelvis
acromial (ah-kro′me-al) point of the shoulder
antebrachial (an″te-bra′ke-al) forearm
antecubital (an″te-ku′bĭ-tal) space in front of the elbow
axillary (ak′sĭ-ler″e) armpit
brachial (bra′ke-al) arm
buccal (buk′al) cheek
carpal (kar′pal) wrist
celiac (se′le-ak) abdomen
cephalic (sĕ-fal′ik) head
cervical (ser′vĭ-kal) neck
costal (kos′tal) ribs
coxal (kok′sal) hip
crural (kro͞or′al) leg
cubital (ku′bĭ-tal) elbow
digital (dij′ĭ-tal) finger or toe
dorsum (dor′sum) back
femoral (fem′or-al) thigh
frontal (frun′tal) forehead
genital (jen′i-tal) reproductive organs
gluteal (gloo′te-al) buttocks
inguinal (ing′gwĭ-nal) depressed area of the abdominal wall near the thigh (groin)
lumbar (lum′bar) region of the lower back between the ribs and the pelvis (loin)
mammary (mam′er-e) breast
mental (men′tal) chin
nasal (na′zal) nose
occipital (ok-sip′ĭ-tal) lower posterior region of the head
oral (o′ral) mouth
orbital (or′bi-tal) eye cavity
otic (o′tik) ear
palmar (pahl′mar) palm of the hand
patellar (pah-tel′ar) front of the knee

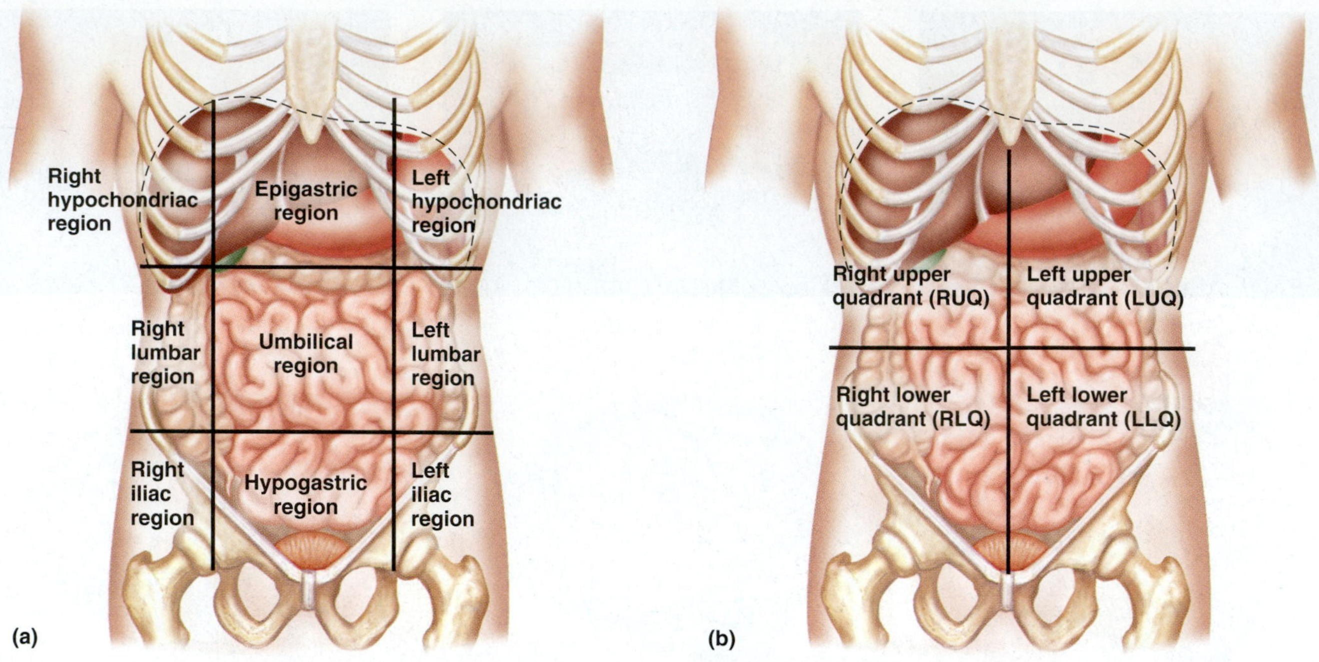

FIGURE 1.24 The abdominal area is commonly subdivided in two ways: (*a*) into nine regions and (*b*) into four quadrants

pectoral (pek′tor-al) chest
pedal (ped′al) foot
pelvic (pel′vik) pelvis
perineal (per″ĭ-ne′al) region between the anus and the external reproductive organs (perineum)
plantar (plan′tar) sole of the foot
popliteal (pop″lĭ-te′al) area behind the knee
sacral (sa′kral) posterior region between the hipbones
sternal (ster′nal) middle of the thorax, anteriorly
sural (su′ral) calf of the leg
tarsal (tahr′sal) instep of the foot (ankle)
umbilical (um-bil′ĭ-kal) navel
vertebral (ver′te-bral) spinal column

PRACTICE

26 Describe the anatomical position.
27 Using the appropriate terms, describe the relative positions of several body parts.
28 Describe three types of body sections.
29 Describe the nine regions of the abdomen.
30 Explain how the names of the abdominal regions describe their locations.

SOME MEDICAL AND APPLIED SCIENCES

cardiology (kar″de-ol′o-je) Branch of medical science dealing with the heart and heart diseases.
dermatology (der″mah-tol′o-je) Study of skin and its diseases.
endocrinology (en″do-krĭ-nol′o-je) Study of hormones, hormone-secreting glands, and associated diseases.
epidemiology (ep″ĭ-de″me-ol′o-je) Study of the factors that contribute to determining the distribution and frequency of health-related conditions within a defined human population.
gastroenterology (gas″tro-en″ter-ol′o-je) Study of the stomach and intestines, as well as their diseases.
geriatrics (jer″e-at′riks) Branch of medicine dealing with older individuals and their medical problems.
gerontology (jer″on-tol′o-je) Study of the process of aging and the various problems of older individuals.
gynecology (gi″nĕ-kol-o-je) Study of the female reproductive system and its diseases.
hematology (hem″ah-tol′o-je) Study of blood and blood diseases.
histology (his-tol′o-je) Study of the structure and function of tissues (microscopic anatomy).
immunology (im″u-nol′o-je) Study of the body's resistance to disease.
neonatology (ne″o-na-tol′o-je) Study of newborns and the treatment of their disorders.
nephrology (nĕ-frol′o-je) Study of the structure, function, and diseases of the kidneys.
neurology (nu-rol′o-je) Study of the nervous system in health and disease.
obstetrics (ob-stet′riks) Branch of medicine dealing with pregnancy and childbirth.
oncology (ong-kol′o-je) Study of cancers.
ophthalmology (of″thal-mol′o-je) Study of the eye and eye diseases.

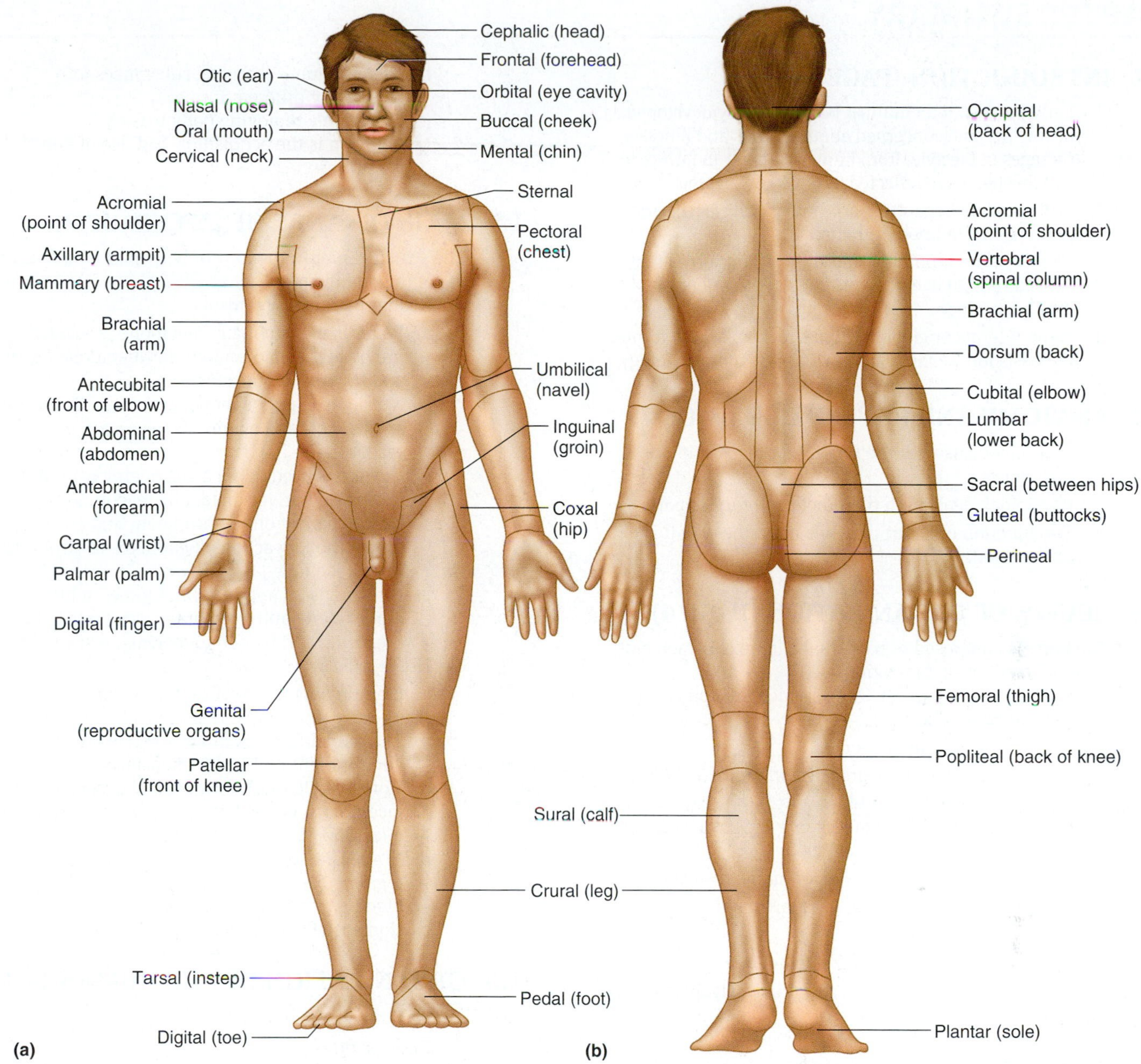

FIGURE 1.25 Some terms used to describe body regions. (*a*) Anterior regions. (*b*) Posterior regions.

orthopedics (or″tho-pe′diks) Branch of medicine dealing with the muscular and skeletal systems and their problems.
otolaryngology (o″to-lar″in-gol′o-je) Study of the ear, throat, larynx, and their diseases.
pathology (pah-thol′o-je) Study of structural and functional changes within the body associated with disease.
pediatrics (pe″de-at′riks) Branch of medicine dealing with children and their diseases.
pharmacology (fahr″mah-kol′o-je) Study of drugs and their uses in the treatment of diseases.
podiatry (po-di′ah-tre) Study of the care and treatment of the feet.
psychiatry (si-ki′ah-tre) Branch of medicine dealing with the mind and its disorders.
radiology (ra″de-ol′o-je) Study of X rays and radioactive substances, as well as their uses in diagnosing and treating diseases.
toxicology (tok″sĭ-kol′o-je) Study of poisonous substances and their effects on physiology.
urology (u-rol′o-je) Branch of medicine dealing with the urinary and male reproductive systems and their diseases.

CHAPTER SUMMARY

1.1 INTRODUCTION (PAGE 3)

1. Early interest in the human body probably developed as people became concerned about injuries and illnesses. Changes in lifestyle, from hunter-gatherer to farmer to city dweller, were reflected in types of illnesses.
2. Early doctors began to learn how certain herbs and potions affected body functions.
3. The idea that humans could understand forces that caused natural events led to the development of modern science.
4. A set of terms originating from Greek and Latin formed the basis for the language of anatomy and physiology.

1.2 ANATOMY AND PHYSIOLOGY (PAGE 4)

1. Anatomy deals with the form and organization of body parts.
2. Physiology deals with the functions of these parts.
3. The function of a part depends upon the way it is constructed.

1.3 LEVELS OF ORGANIZATION (PAGE 4)

The body is composed of parts that can be considered at different levels of organization.

1. Matter is composed of atoms, which are composed of subatomic particles.
2. Atoms join to form molecules.
3. Organelles consist of aggregates of interacting large molecules (macromolecules).
4. Cells, composed of organelles, are the basic units of structure and function of the body.
5. Cells are organized into layers or masses called tissues.
6. Tissues are organized into organs.
7. Organs form organ systems.
8. Organ systems constitute the organism.
9. These parts vary in complexity progressively from one level to the next.

1.4 CHARACTERISTICS OF LIFE (PAGE 6)

Characteristics of life are traits all organisms share.

1. These characteristics include
   a. Movement—changing body position or moving internal parts.
   b. Responsiveness—sensing and reacting to internal or external changes.
   c. Growth—increasing in size without changing in shape.
   d. Reproduction—producing offspring.
   e. Respiration—obtaining oxygen, using oxygen to release energy from foods, and removing gaseous wastes.
   f. Digestion—breaking down food substances into forms that can be absorbed.
   g. Absorption—moving substances through membranes and into body fluids.
   h. Circulation—moving substances through the body in body fluids.
   i. Assimilation—changing substances into chemically different forms.
   j. Excretion—removing body wastes.
2. Metabolism is the acquisition and use of energy by an organism.

1.5 MAINTENANCE OF LIFE (PAGE 7)

The structures and functions of body parts maintain the life of the organism.

1. Requirements of organisms
   a. Water is used in many metabolic processes, provides the environment for metabolic reactions, and transports substances.
   b. Nutrients supply energy, raw materials for building substances, and chemicals necessary in vital reactions.
   c. Oxygen is used in releasing energy from nutrients; this energy drives metabolic reactions.
   d. Heat is part of our environment and is a product of metabolic reactions; heat helps control rates of these reactions.
   e. Pressure is an application of force; in humans, atmospheric and hydrostatic pressures help breathing and blood movements, respectively.
2. Homeostasis
   a. If an organism is to survive, the conditions within its body fluids must remain relatively stable.
   b. The tendency to maintain a stable internal environment is called homeostasis.
   c. Homeostatic mechanisms involve sensory receptors, a control center with a set point, and effectors.
   d. Homeostatic mechanisms include those that regulate body temperature, blood pressure, and blood glucose concentration.
   e. Homeostatic mechanisms employ negative feedback.

1.6 ORGANIZATION OF THE HUMAN BODY (PAGE 12)

1. Body cavities
   a. The axial portion of the body contains the cranial cavity and vertebral canal, as well as the thoracic and abdominopelvic cavities, separated by the diaphragm.
   b. The organs within thoracic and abdominopelvic cavities are called viscera.
   c. Other body cavities include the oral, nasal, orbital, and middle ear cavities.
2. Thoracic and abdominopelvic membranes

   Parietal serous membranes line the walls of these cavities; visceral serous membranes cover organs within them. They secrete serous fluid.

   a. Thoracic membranes
      (1) Pleural membranes line the thoracic cavity and cover the lungs.
      (2) Pericardial membranes surround the heart and cover its surface.
      (3) The pleural and pericardial cavities are potential spaces between these membranes.

b. Abdominopelvic membranes
    (1) Peritoneal membranes line the abdominopelvic cavity and cover the organs inside.
    (2) The peritoneal cavity is a potential space between these membranes.

3. Organ systems

The human organism consists of several organ systems. Each system includes interrelated organs.

a. Integumentary system
    (1) The integumentary system covers the body.
    (2) It includes the skin, hair, nails, sweat glands, and sebaceous glands.
    (3) It protects underlying tissues, regulates body temperature, houses sensory receptors, and synthesizes substances.

b. Skeletal system
    (1) The skeletal system is composed of bones and the ligaments and cartilages that bind bones together.
    (2) It provides framework, protective shields, and attachments for muscles; it also produces blood cells and stores inorganic salts.

c. Muscular system
    (1) The muscular system includes the muscles of the body.
    (2) It moves body parts, maintains posture, and produces body heat.

d. Nervous system
    (1) The nervous system consists of the brain, spinal cord, nerves, and sense organs.
    (2) It receives impulses from sensory parts, interprets these impulses, and acts on them, stimulating muscles or glands to respond.

e. Endocrine system
    (1) The endocrine system consists of glands that secrete hormones.
    (2) Hormones help regulate metabolism by stimulating target tissues.
    (3) It includes the pituitary gland, thyroid gland, parathyroid glands, adrenal glands, pancreas, ovaries, testes, pineal gland, and thymus.

f. Digestive system
    (1) The digestive system receives foods, breaks down nutrients into forms that can pass through cell membranes, and eliminates unabsorbed materials.
    (2) Some digestive organs produce hormones.
    (3) The digestive system includes the mouth, tongue, teeth, salivary glands, pharynx, esophagus, stomach, liver, gallbladder, pancreas, small intestine, and large intestine.

g. Respiratory system
    (1) The respiratory system takes in and releases air and exchanges gases between the blood and the air.
    (2) It includes the nasal cavity, pharynx, larynx, trachea, bronchi, and lungs.

h. Cardiovascular system
    (1) The cardiovascular system includes the heart, which pumps blood, and the blood vessels, which carry blood to and from body parts.
    (2) Blood transports oxygen, nutrients, hormones, and wastes.

i. Lymphatic system
    (1) The lymphatic system is composed of lymphatic vessels, lymph nodes, thymus, and spleen.
    (2) It transports lymph from tissue spaces to the bloodstream and carries certain fatty substances away from the digestive organs. Lymphocytes defend the body against disease-causing agents.

j. Urinary system
    (1) The urinary system includes the kidneys, ureters, urinary bladder, and urethra.
    (2) It filters wastes from the blood and helps maintain fluid and electrolyte balance.

k. Reproductive systems
    (1) The reproductive system enables an organism to produce progeny.
    (2) The male reproductive system produces, maintains, and transports male sex cells. It includes the scrotum, testes, epididymides, ductus deferentia, seminal vesicles, prostate gland, bulbourethral glands, urethra, and penis.
    (3) The female reproductive system produces, maintains, and transports female sex cells. It includes the ovaries, uterine tubes, uterus, vagina, clitoris, and vulva.

1.7 LIFE-SPAN CHANGES (PAGE 20)

Aging occurs from conception on and has effects at the cell, tissue, organ, and organ system levels.

1. The first signs of aging are noticeable in one's thirties. Female fertility begins to decline during this time.
2. In the forties and fifties, adult-onset disorders may begin.
3. Skin changes reflect less elastin, collagen, and subcutaneous fat.
4. Older people may metabolize certain drugs at different rates than younger people.
5. Cells divide a limited number of times. As DNA repair falters, mutations may accumulate.
6. Oxygen free-radical damage produces certain pigments. Metabolism slows, and beta amyloid protein may build up in the brain.

1.8 ANATOMICAL TERMINOLOGY (PAGE 20)

Investigators use terms with precise meanings to effectively communicate with one another.

1. Relative position

    These terms describe the location of one part with respect to another part.

2. Body sections

    Body sections are planes along which the body may be cut to observe the relative locations and arrangements of internal parts.

3. Body regions

    Special terms designate various body regions.

CHAPTER ASSESSMENTS

1.1 Introduction

1 Describe how an early interest in the human body eventually led to the development of modern medical science. (p. 3)

1.2 Anatomy and Physiology

2 Distinguish between anatomy and physiology. (p. 4)

3 Explain the relationship between the form and function of body parts and give three examples. (p. 4)

1.3 Levels of Organization

4 Describe the relationship between each of the following pairs: molecules and cells, tissues and organs, organs and organ systems. (p. 4)

1.4 Characteristics of Life

5 Which characteristics of life can you identify in yourself? (p. 6)

6 Identify those characteristics of living organisms that depend on metabolism. (p. 6)

1.5 Maintenance of Life

7 Compare your own needs for survival with the requirements of organisms described in the chapter. (p. 7)

8 Explain the relationship between homeostasis and the internal environment. (p. 9)

9 Describe a general physiological control system, including the role of negative feedback. (p. 9)

10 Explain the control of body temperature. (p. 9)

11 Describe the homeostatic mechanisms that help regulate blood pressure and blood glucose—what do they have in common and how are they different? (p. 10)

1.6 Organization of the Human Body

12 Explain the difference between the axial and appendicular portions of the body. (p. 12)

13 Identify the cavities within the axial portion of the body. (p. 12)

14 Define *viscera*. (p. 12)

15 Describe the mediastinum and its contents. (p. 12)

16 List the cavities of the head and the contents of each cavity. (p. 12)

17 Name the body cavity that houses each of the following organs: (p. 12)

a. Stomach
b. Heart
c. Brain
d. Liver
e. Trachea
f. Rectum
g. Spinal cord
h. Esophagus
i. Spleen
j. Urinary bladder

18 Distinguish between a parietal and a visceral membrane. (p. 12)

19 Describe the general contribution of each of the organ systems to maintaining homeostasis. (p. 14)

20 List the major organs that compose each organ system and identify their functions. (p. 14)

1.7 Life-Span Changes

21 Describe physical changes associated with aging that occur during each decade past the age of 30. (p. 20)

22 List age-associated changes that occur at the molecular, cellular, tissue and/or organ levels. (p. 20)

1.8 Anatomical Terminology

23 Write complete sentences using each of the following terms to correctly describe the relative locations of specific body parts: (p. 21)

a. Superior
b. Inferior
c. Anterior
d. Posterior
e. Medial
f. Lateral
g. Bilateral
h. Ipsilateral
i. Contralateral
j. Proximal
k. Distal
l. Superficial
m. Peripheral
n. Deep

24 Sketch the outline of a human body, and use lines to indicate each of the following sections: (p. 22)

a. Sagittal
b. Transverse
c. Frontal

25 Sketch the abdominal area, and indicate the locations of the following regions: (p. 22)

a. Epigastric
b. Hypochondriac
c. Umbilical
d. Lumbar
e. Hypogastric
f. Iliac

26 Sketch the abdominal area, and indicate the location of the following regions: (p. 22)

a. Right upper quadrant
b. Right lower quadrant
c. Left upper quadrant
d. Left lower quadrant

27 Provide the common name for the region to which each of the following terms refers: (p. 23)

a. Acromial
b. Antebrachial
c. Axillary
d. Buccal
e. Celiac
f. Coxal
g. Crural
h. Femoral
i. Genital
j. Gluteal
k. Inguinal
l. Mental
m. Occipital
n. Orbital
o. Otic
p. Palmar
q. Pectoral
r. Pedal
s. Perineal
t. Plantar
u. Popliteal
v. Sacral
w. Sternal
x. Tarsal
y. Umbilical
z. Vertebral

INTEGRATIVE ASSESSMENTS/CRITICAL THINKING

OUTCOMES 1.2, 1.3, 1.4, 1.5

1. Which characteristics of life does a computer have? Why is a computer not alive?

OUTCOMES 1.2, 1.3, 1.4, 1.5, 1.6

2. Put the following in order, from smallest and simplest, to largest and most complex, and describe their individual roles in homeostasis: organ, molecule, organelle, atom, organ system, tissue, organism, cell, macromolecule.

OUTCOMES 1.4, 1.5

3. What environmental conditions would be necessary for a human to survive on another planet?

OUTCOMES 1.5, 1.6. 1.7

4. In health, body parts interact to maintain homeostasis. Illness can threaten the maintenance of homeostasis, requiring treatment. What treatments might be used to help control a patient's (*a*) body temperature, (*b*) blood oxygen level, and (*c*) blood glucose level?

OUTCOMES 1.5, 1.6, 1.7

5. How might health-care professionals provide the basic requirements of life to an unconscious patient? Describe the body parts involved in the treatment, using correct directional and regional terms.

OUTCOME 1.6

6. Suppose two individuals develop benign (noncancerous) tumors that produce symptoms because they occupy space and crowd adjacent organs. If one of these persons has the tumor in the thoracic cavity and the other has the tumor in the abdominopelvic cavity, which person would be likely to develop symptoms first? Why? Which might be more immediately serious? Why?

OUTCOME 1.6

7. If a patient complained of a "stomachache" and pointed to the umbilical region as the site of discomfort, which organs located in this region might be the source of the pain?

WEB CONNECTIONS

Be sure to visit the text website at **www.mhhe.com/shier12** for answers to chapter assessments, additional quizzes, and interactive learning exercises.

ANATOMY & PHYSIOLOGY REVEALED

Anatomy & Physiology Revealed® (APR) includes cadaver photos that allow you to peel away layers of the human body to reveal structures beneath the surface. This program also includes animations, radiologic imaging, audio pronunciations, and practice quizzing. Check out **www.aprevealed.com**. APR has been proven to help improve student grades!

THE HUMAN ORGANISM

The following series of plates includes illustrations of the major organs of the human torso and human cadaver photos. The first plate shows the anterior surface of the human torso and reveals the muscles on one side. Then, plates 2–7 expose deeper organs, including those in the thoracic, abdominal, and pelvic cavities. Plates 8–25 are photographs of sagittal sections and transverse sections of the torso of a human cadaver. These plates will help you visualize the proportional relationships between the major anatomical structures of actual specimens.

Variations exist in anatomical structures among humans. The illustrations in the textbook and the laboratory manual represent normal (normal means the most common variation) anatomy.

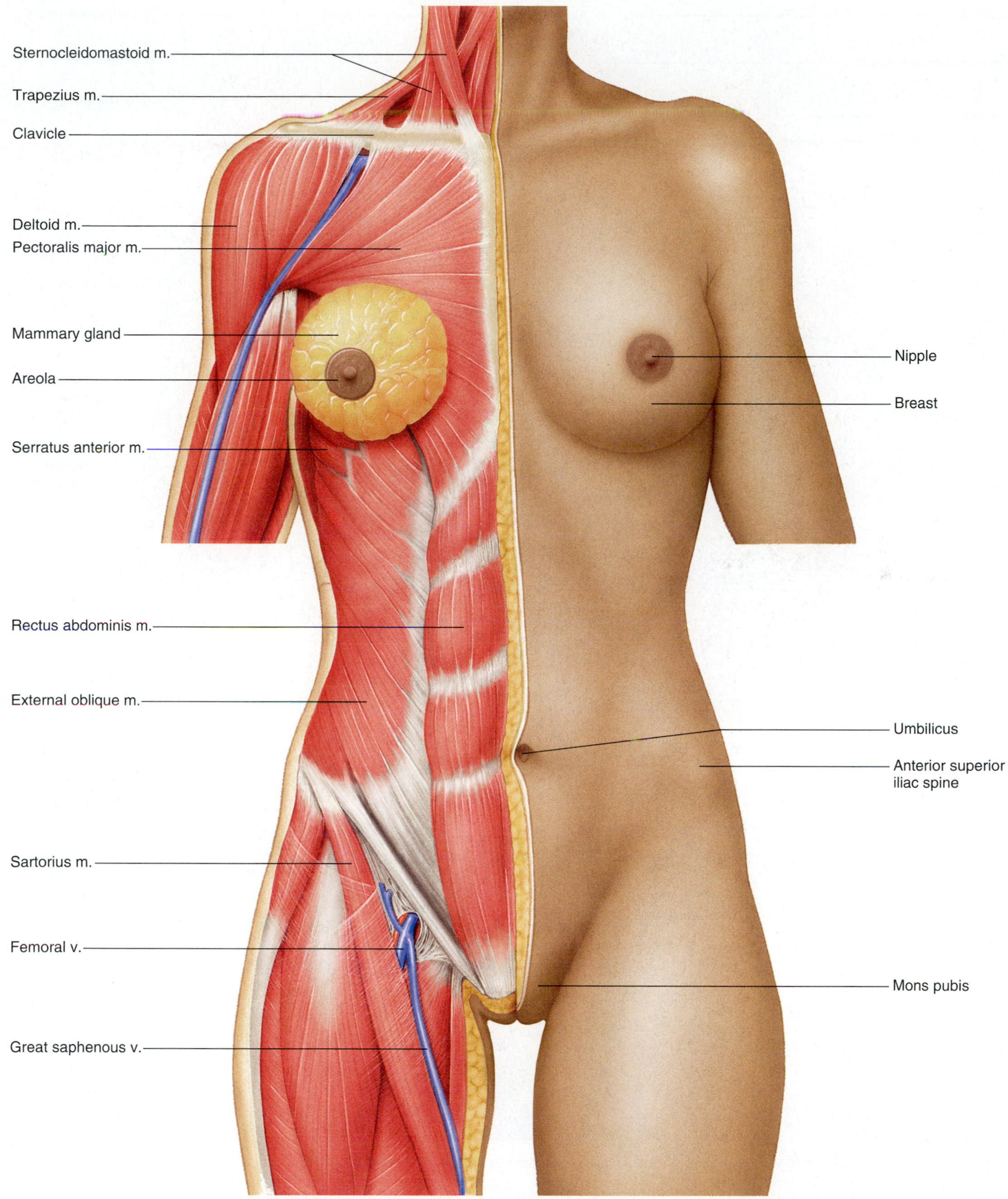

PLATE ONE Human female torso showing the anterior surface on one side and the superficial muscles exposed on the other side. (*m.* stands for *muscle*, and *v.* stands for *vein*.)

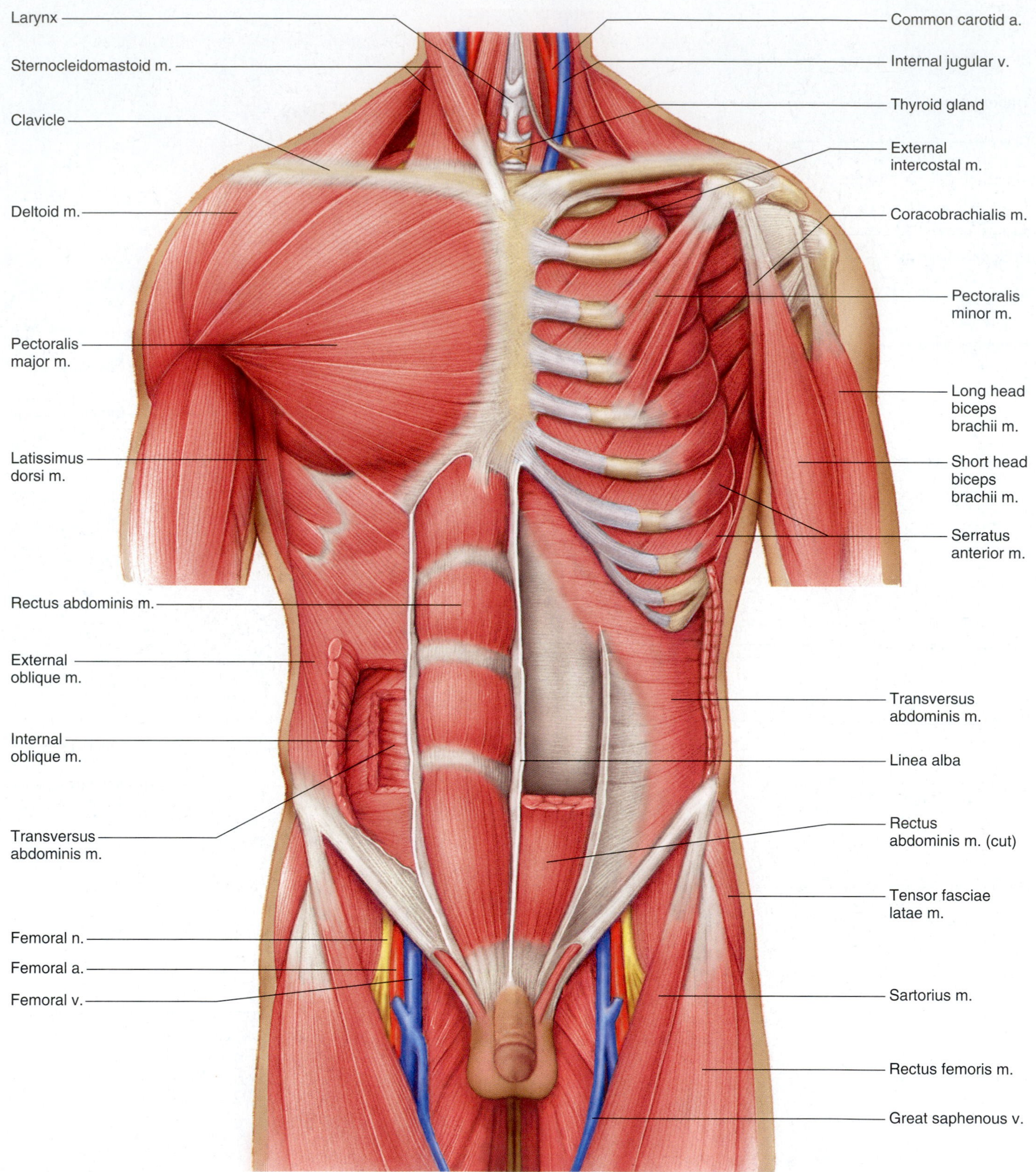

PLATE TWO Human male torso with the deeper muscle layers exposed. (*a.* stands for *artery*, *m.* stands for *muscle*, *n.* stands for *nerve*, and *v.* stands for *vein*.)

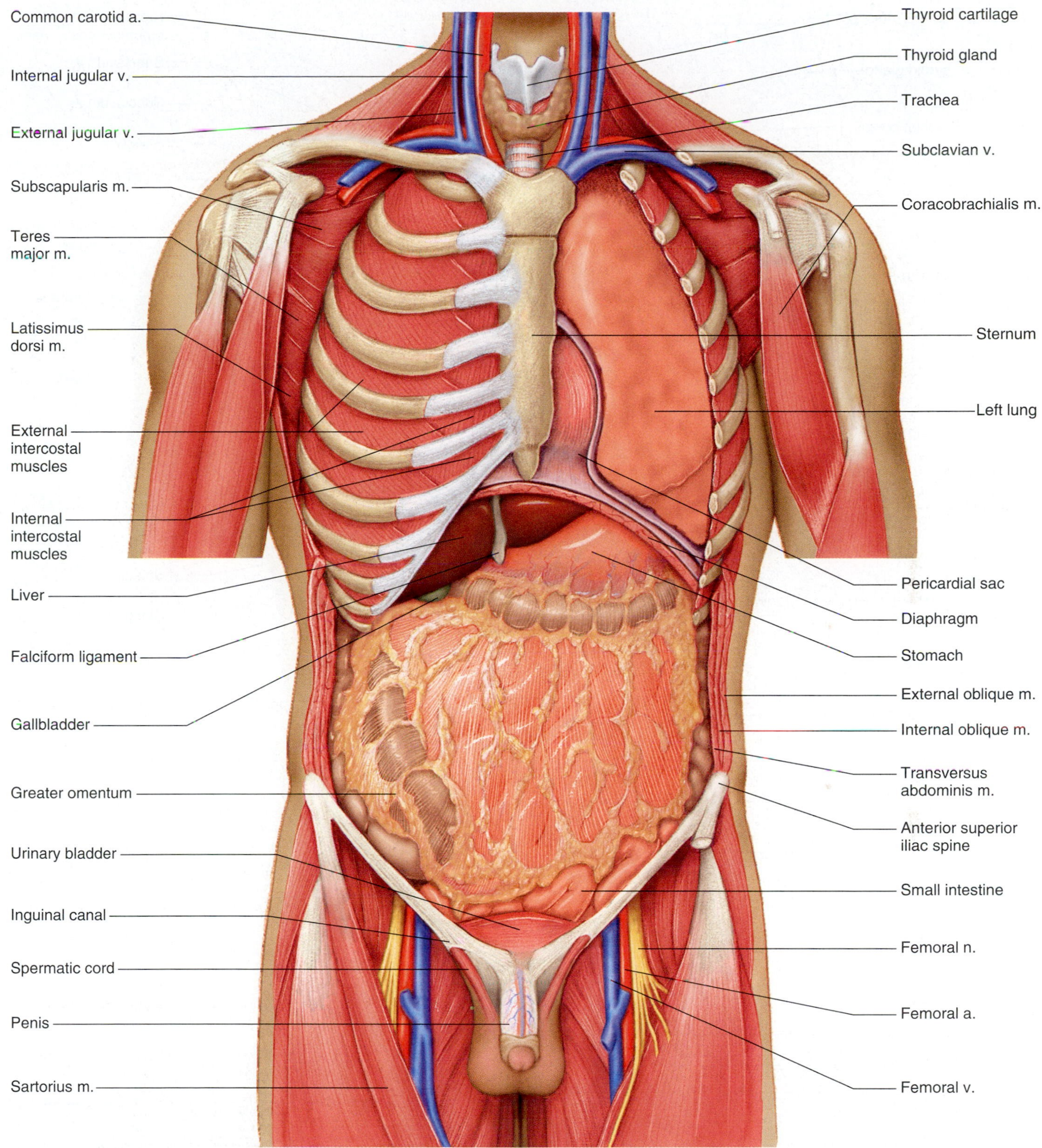

PLATE THREE Human male torso with the deep muscles removed and the abdominal viscera exposed. (*a.* stands for *artery, m.* stands for *muscle, n.* stands for *nerve,* and *v.* stands for *vein.*)

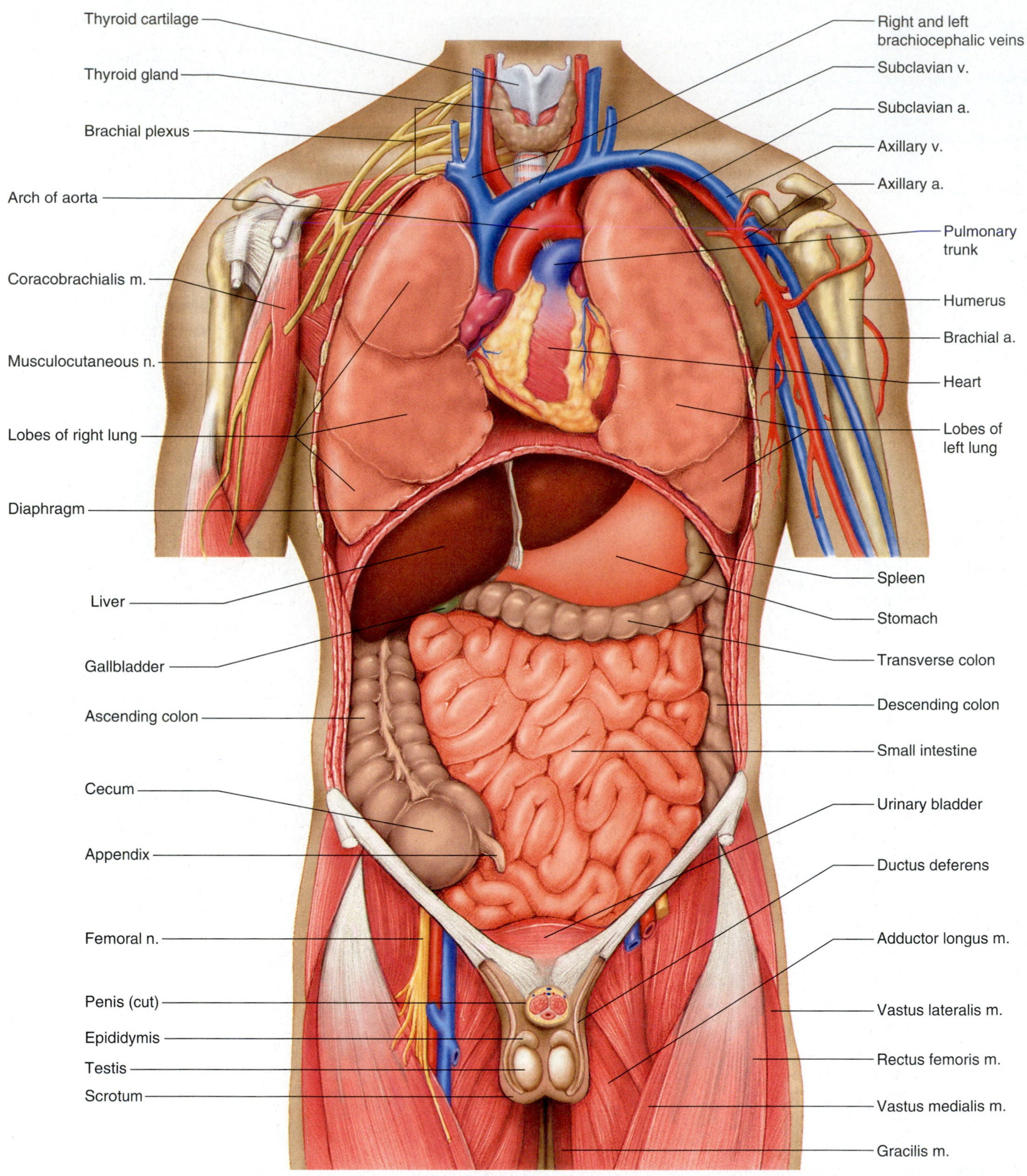

PLATE FOUR Human male torso with the thoracic and abdominal viscera exposed. (*a.* stands for *artery, m.* stands for *muscle, n.* stands for *nerve,* and *v.* stands for *vein.*)

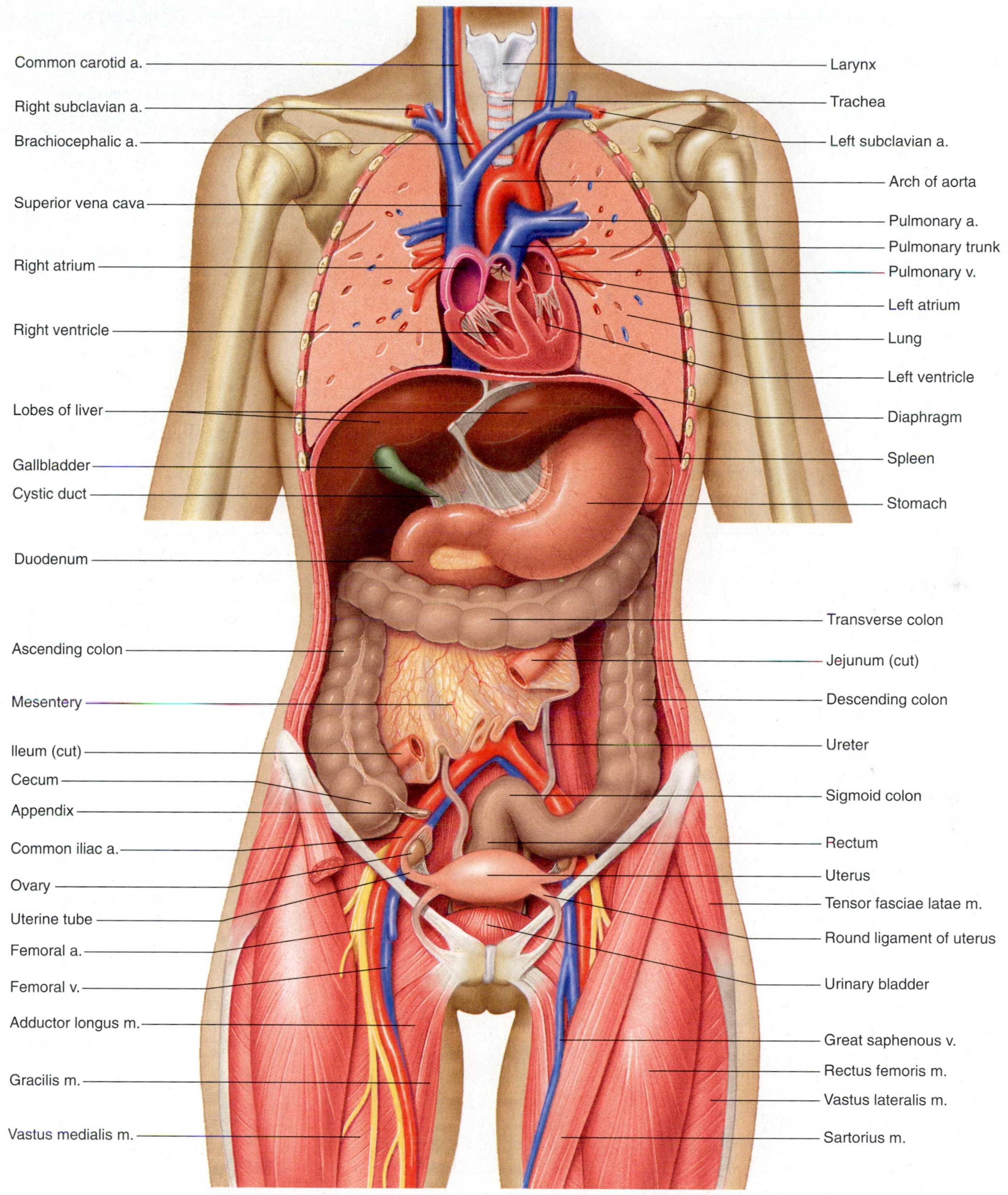

PLATE FIVE Human female torso with the lungs, heart, and small intestine sectioned and the liver reflected (lifted back). (*a.* stands for *artery*, *m.* stands for *muscle*, and *v.* stands for *vein*.)

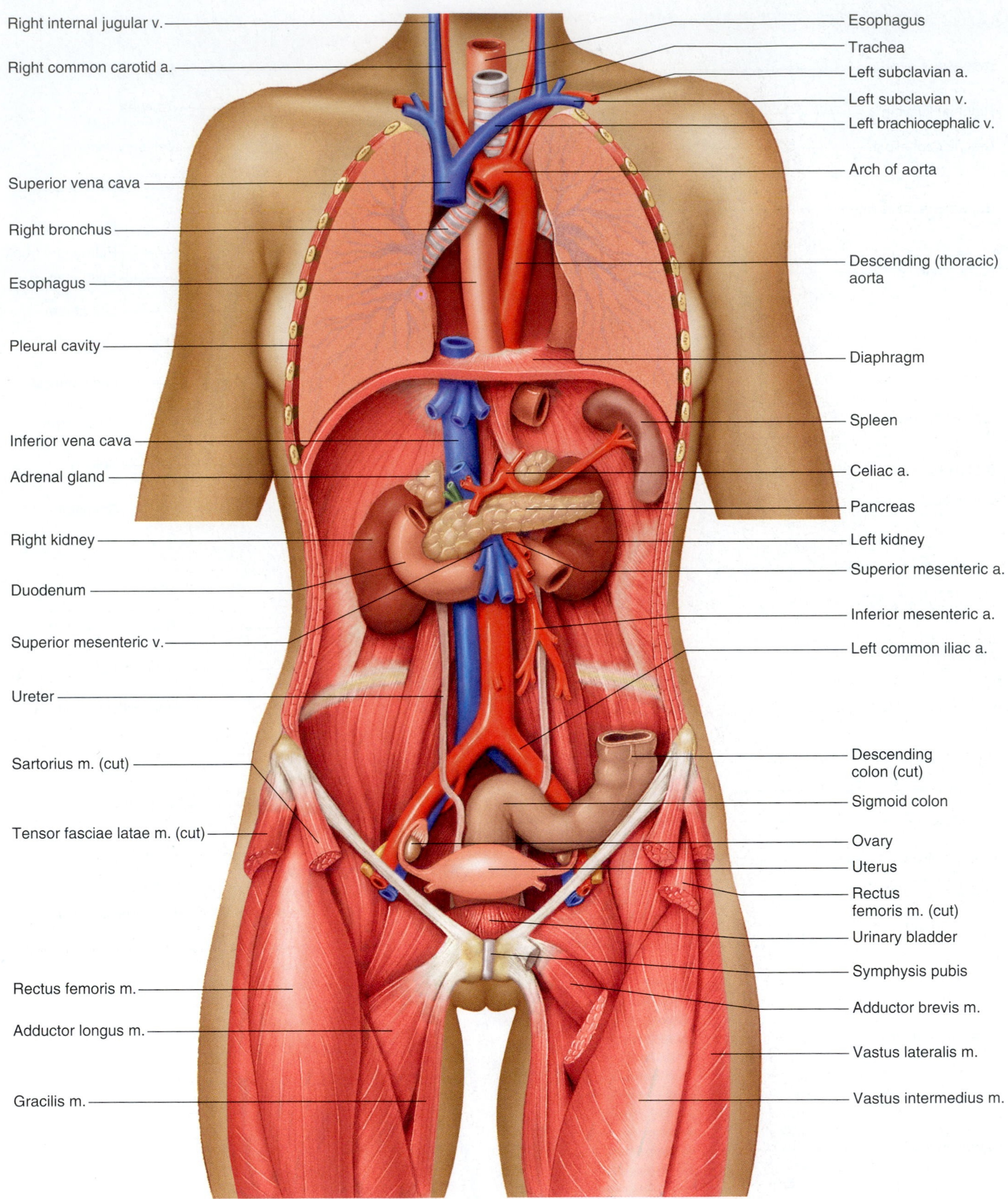

PLATE SIX Human female torso with the heart, stomach, liver, and parts of the intestine and lungs removed. (*a.* stands for *artery*, *m.* stands for *muscle*, and *v.* stands for *vein*.)

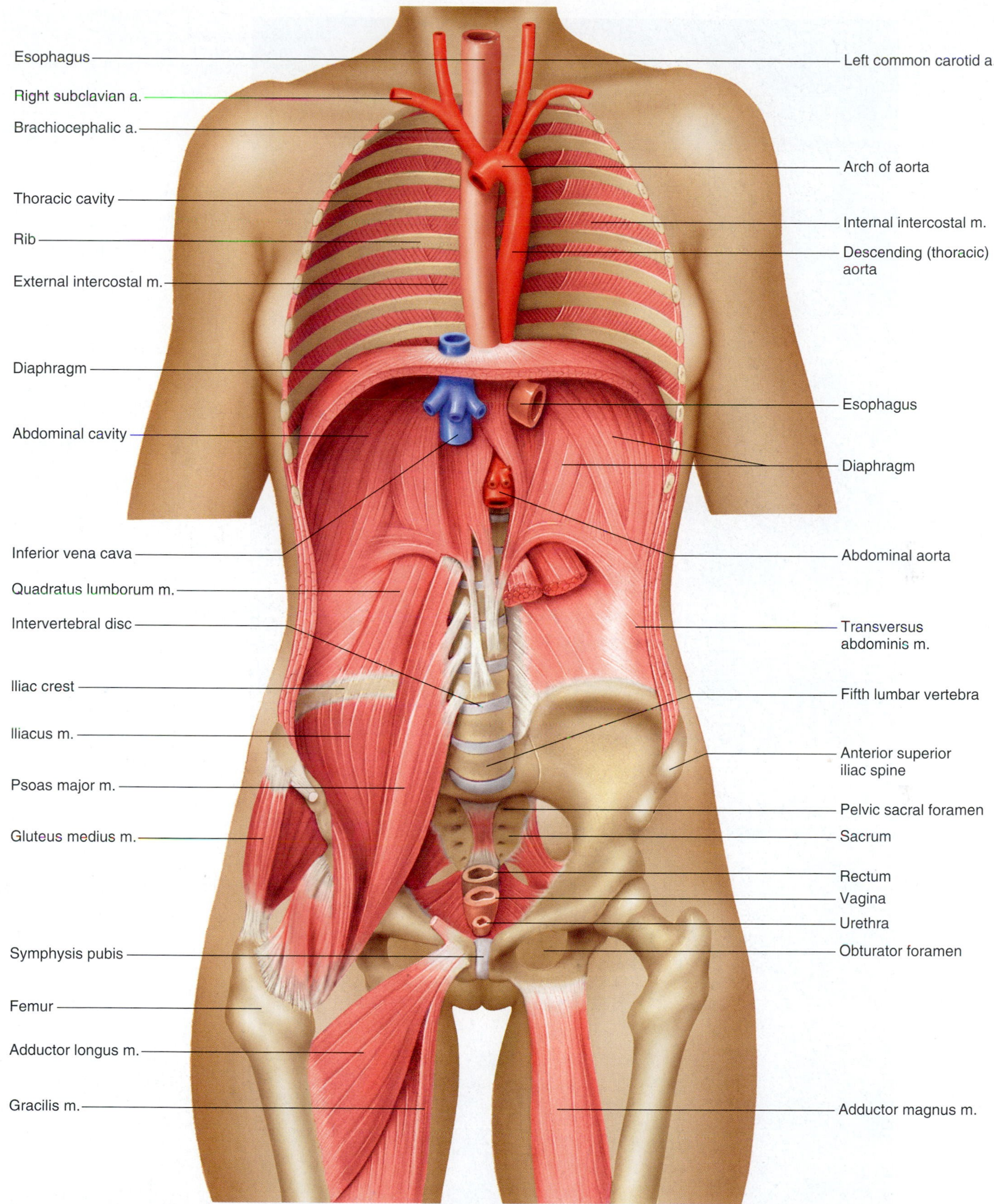

PLATE SEVEN Human female torso with the thoracic, abdominal, and pelvic viscera removed. (*a.* stands for *artery* and *m.* stands for *muscle.*)

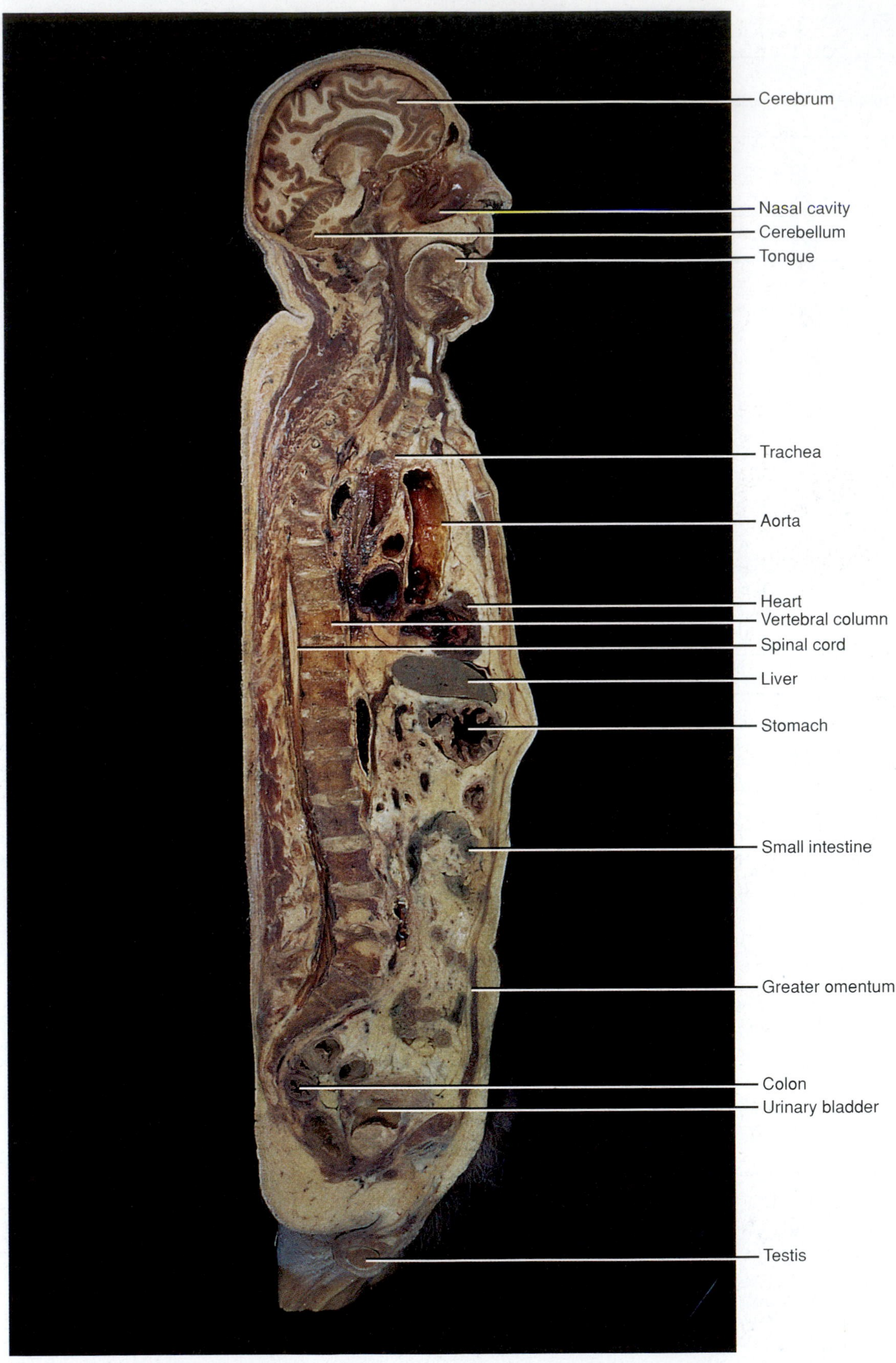

PLATE EIGHT Saggital section of the head and trunk.

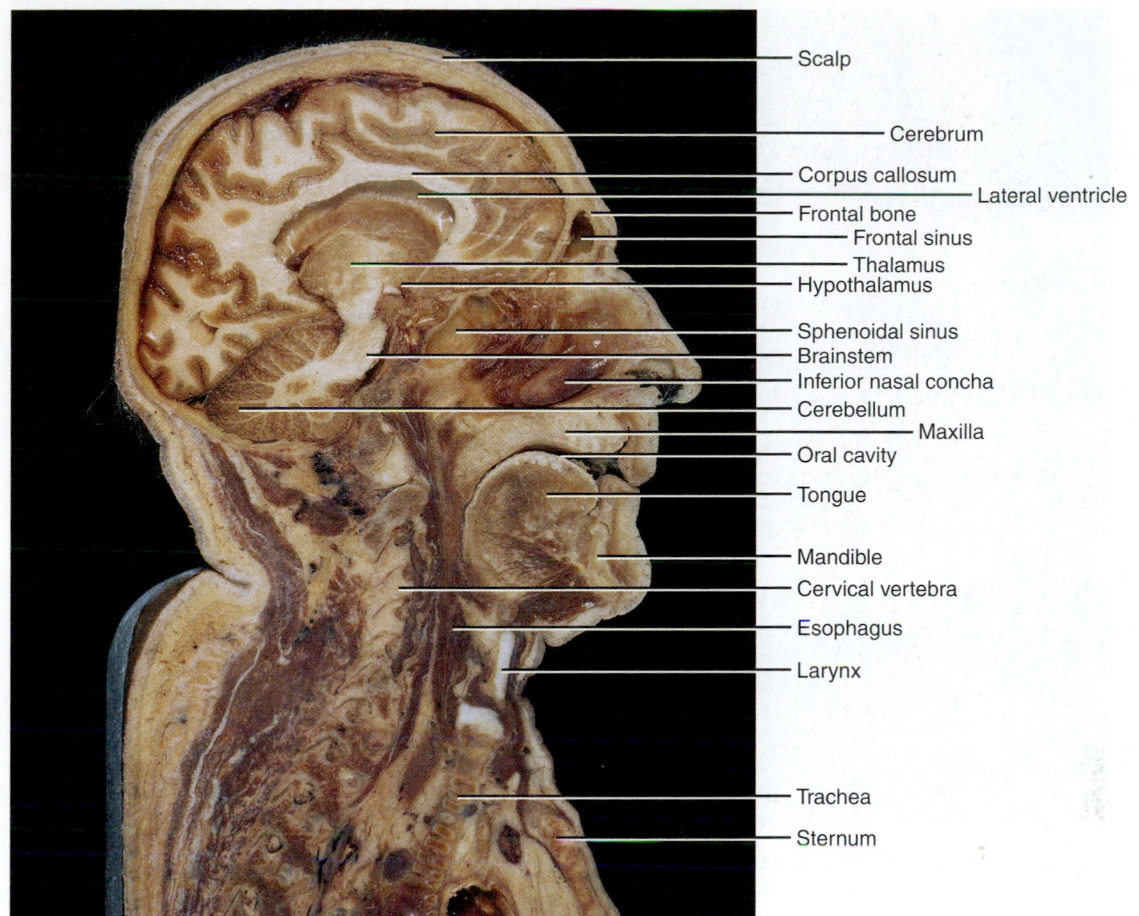

PLATE NINE Saggital section of the head and neck.

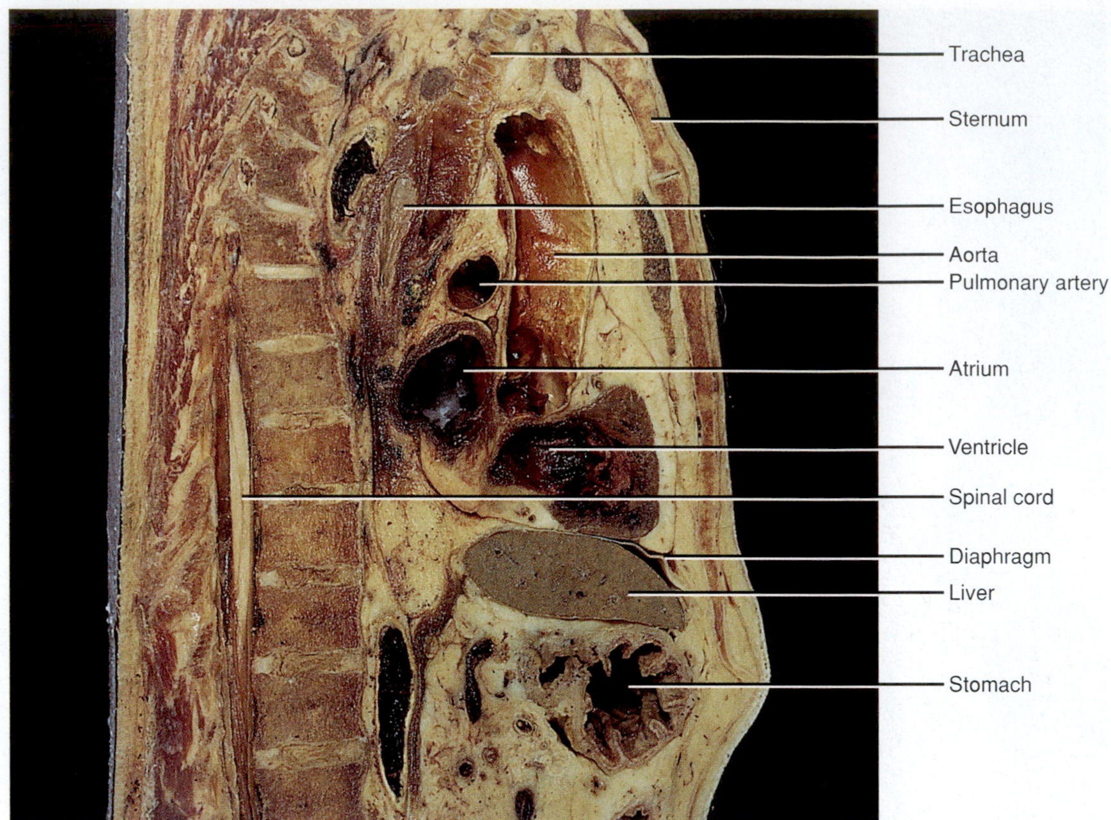

PLATE TEN Viscera of the thoracic cavity, sagittal section.

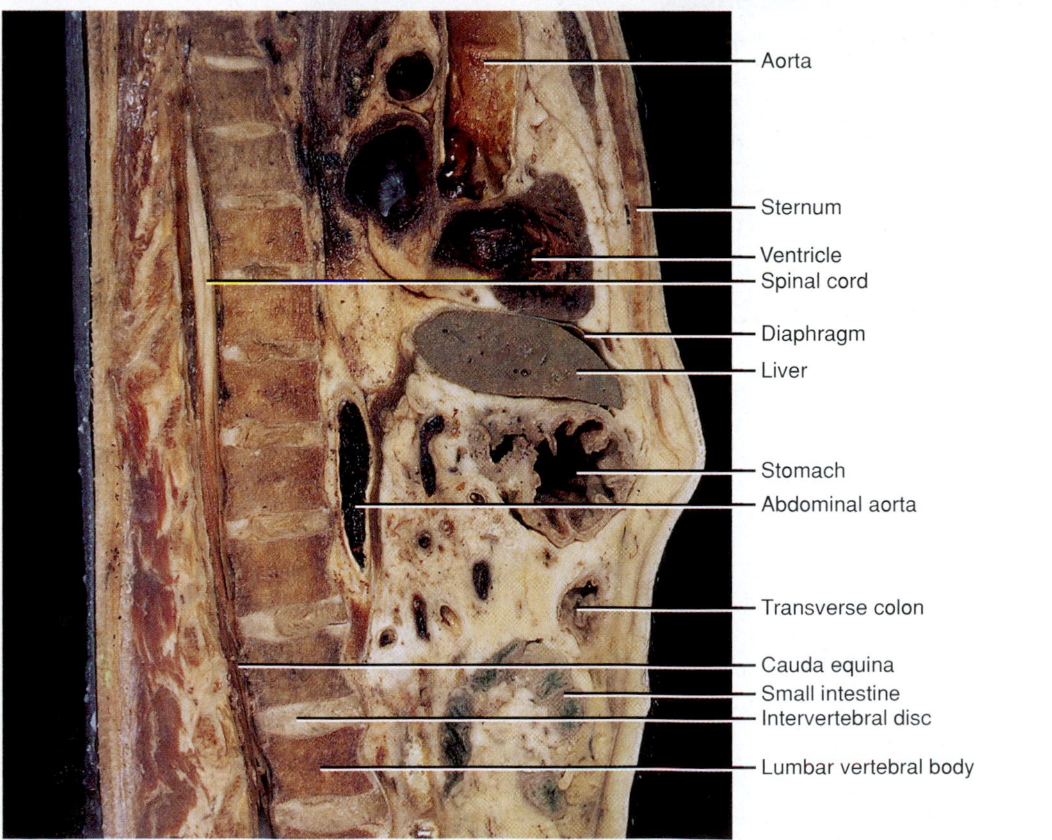

PLATE ELEVEN Viscera of the abdominal cavity, sagittal section.

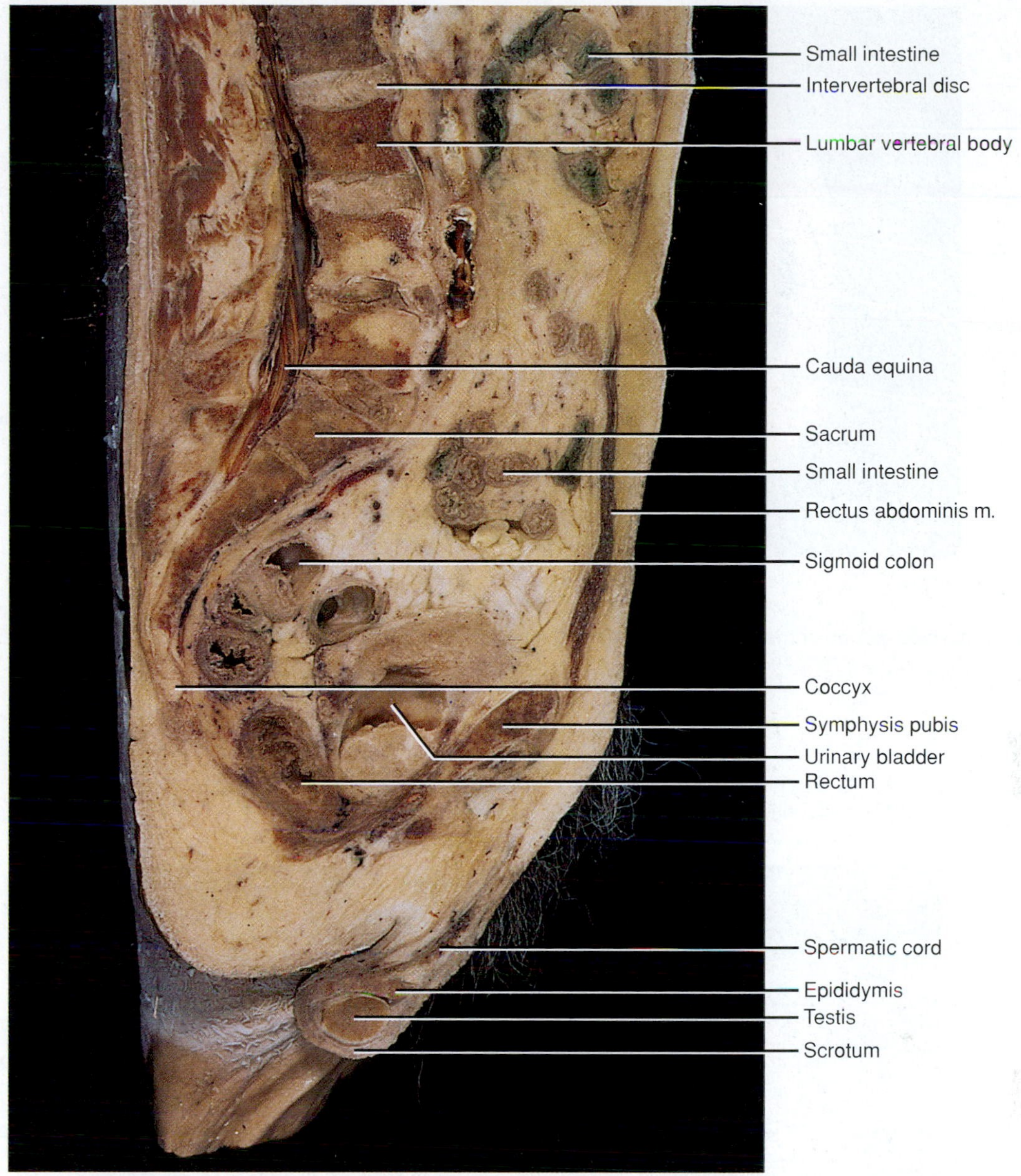

PLATE TWELVE Viscera of the pelvic cavity, sagittal section. (*m.* stands for muscle.)

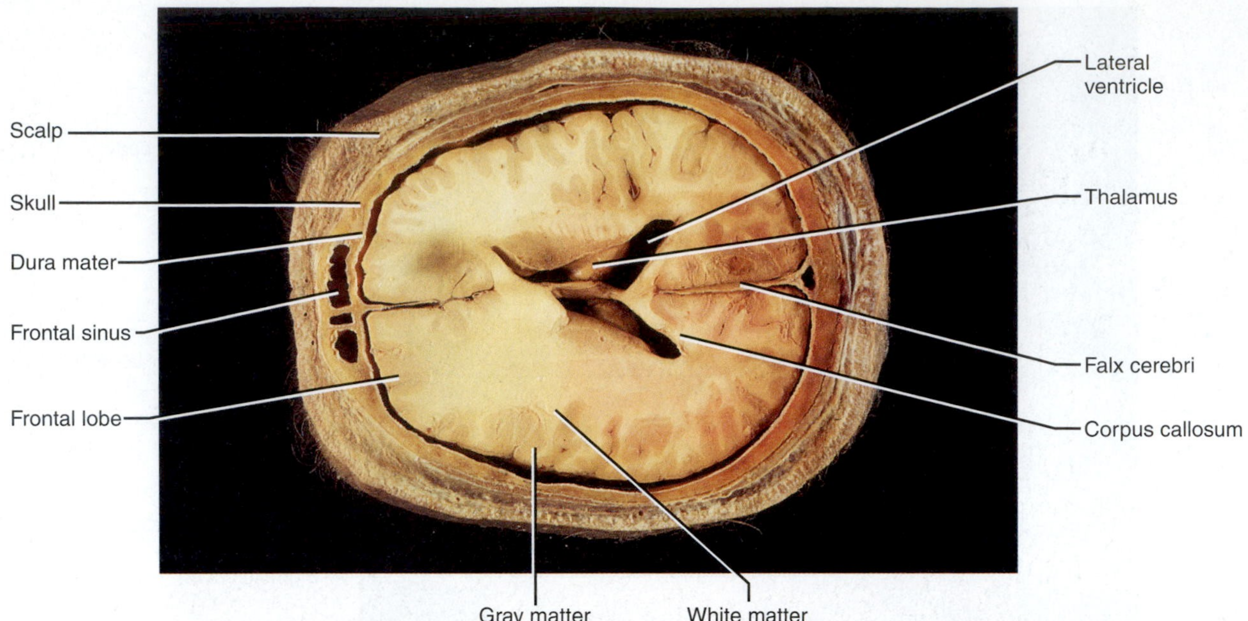

PLATE THIRTEEN Transverse section of the head above the eyes, superior view.

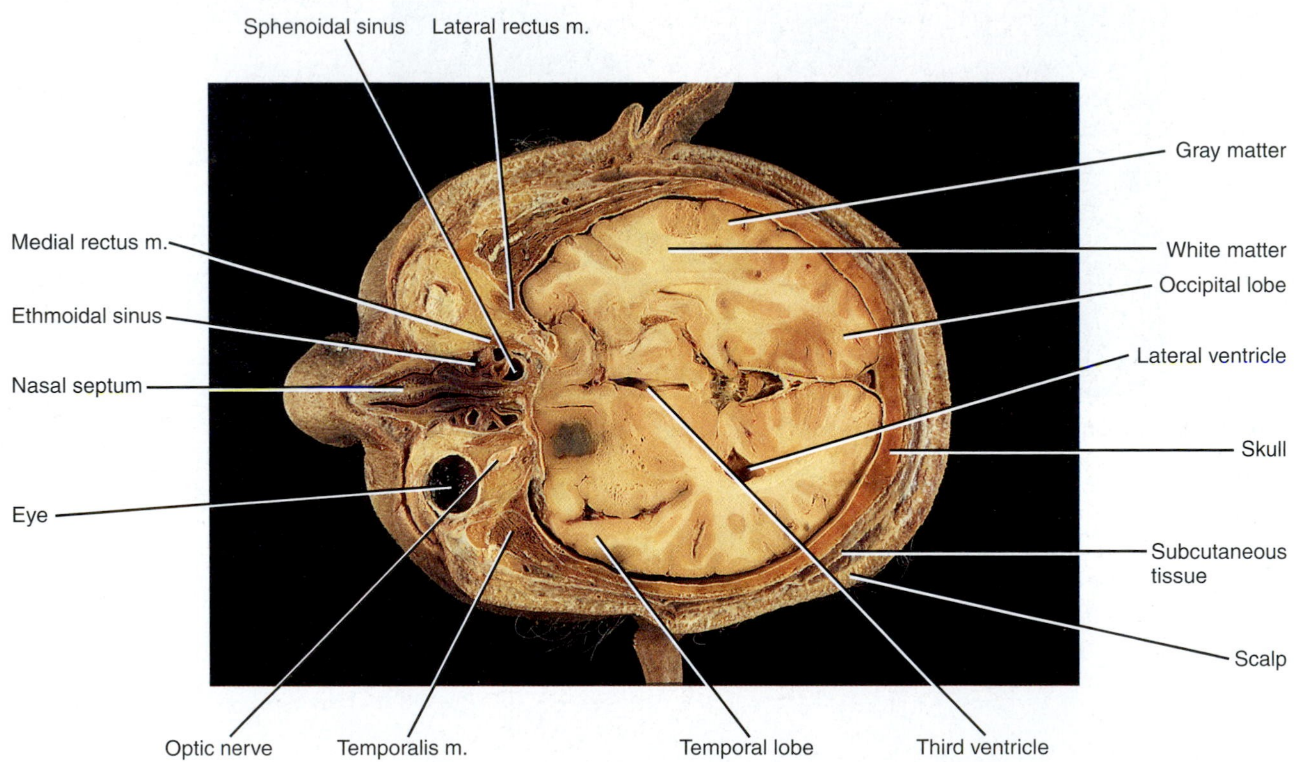

PLATE FOURTEEN Transverse section of the head at the level of the eyes, superior view. (*m.* stands for muscle.)

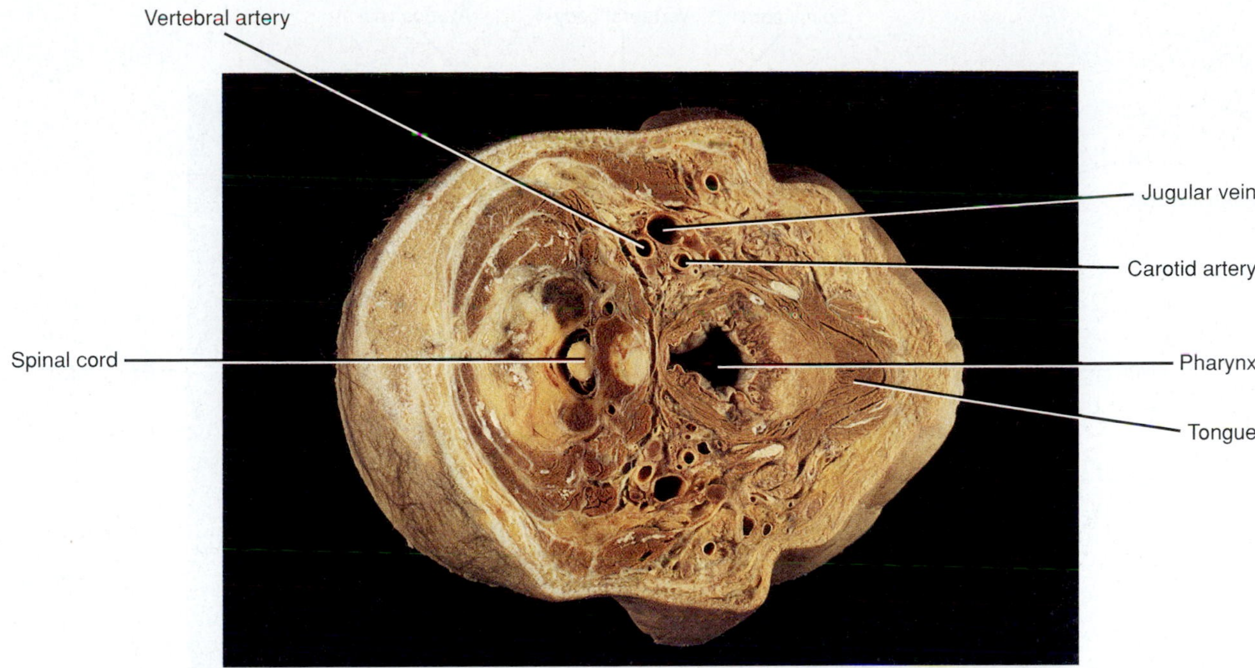

PLATE FIFTEEN Transverse section of the neck, inferior view.

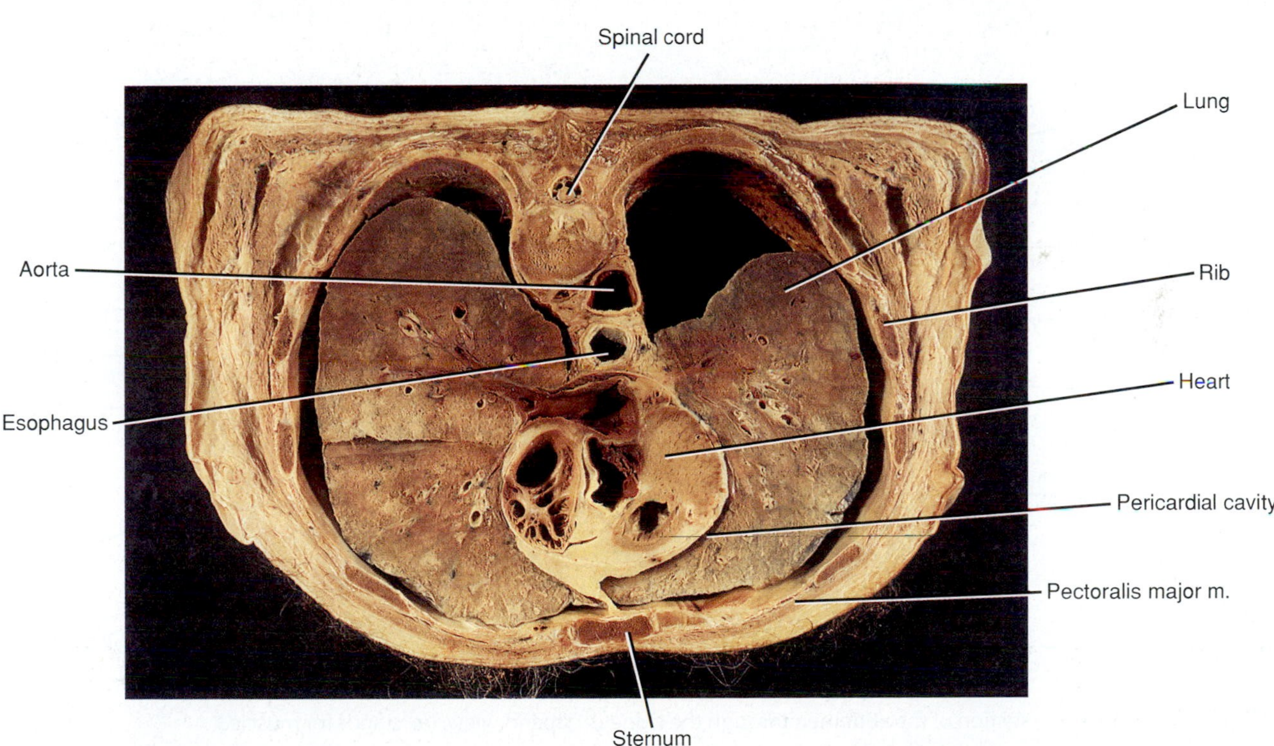

PLATE SIXTEEN Transverse section of the thorax through the base of the heart, superior view. (*m.* stands for muscle.)

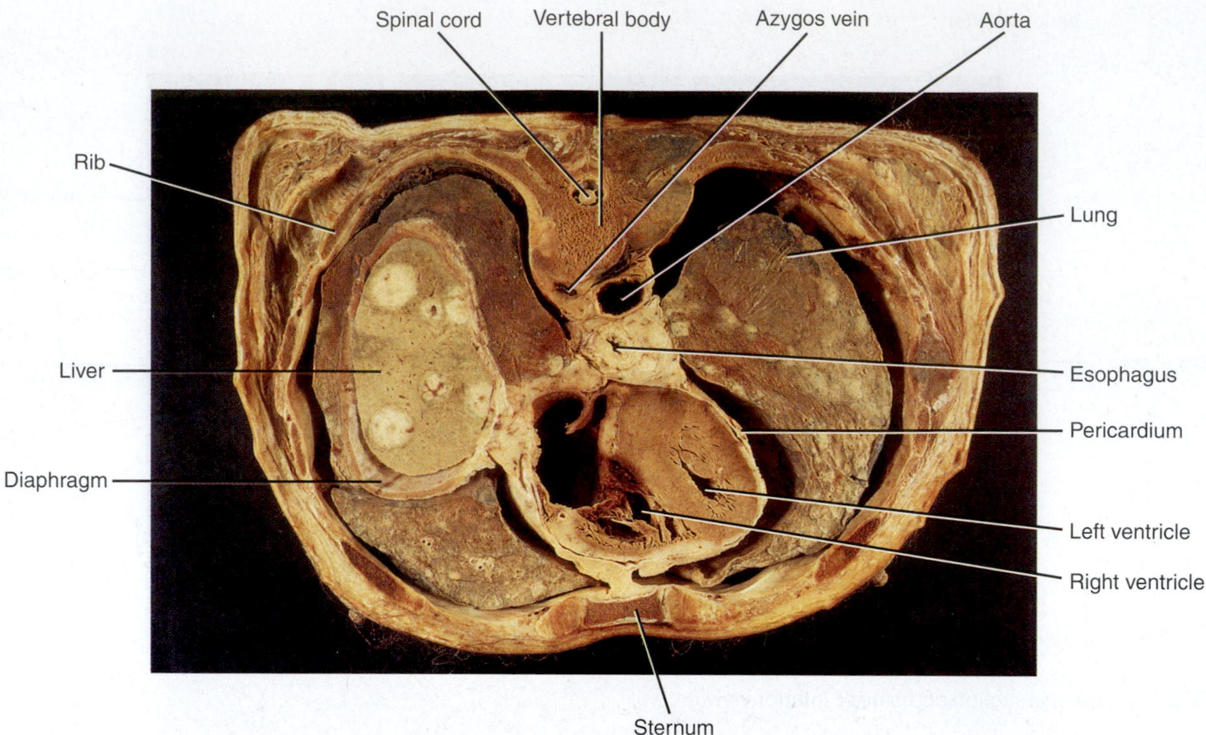

PLATE SEVENTEEN Transverse section of the thorax through the heart, superior view.

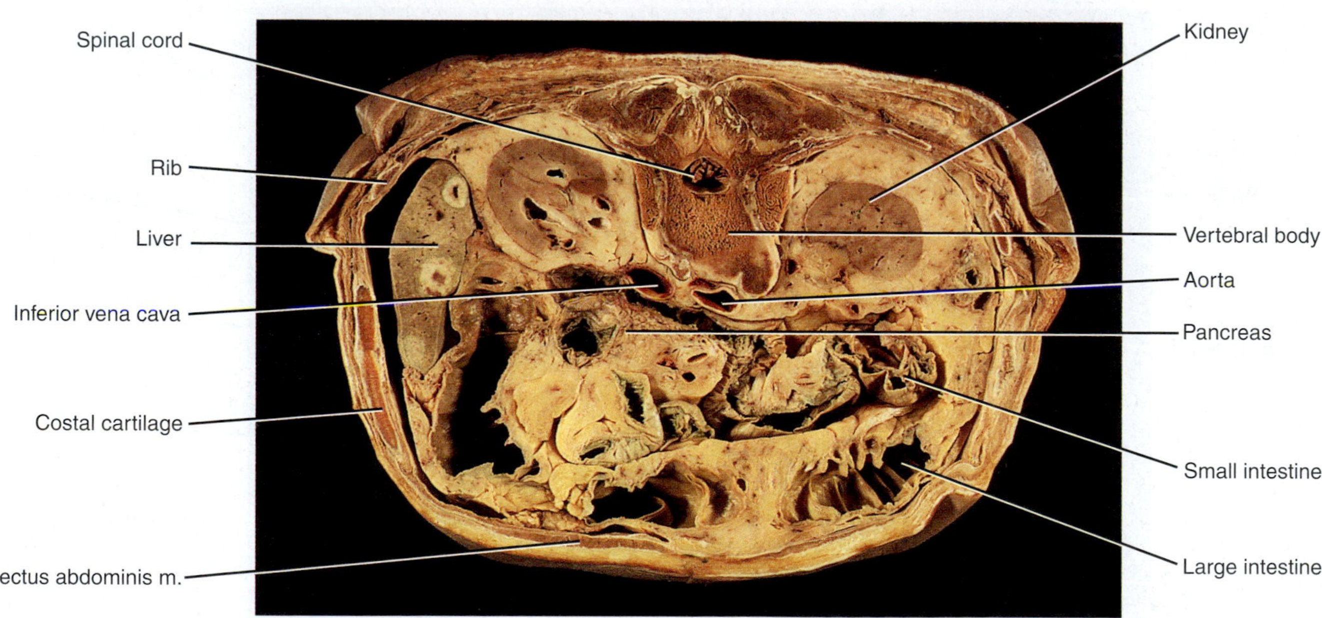

PLATE EIGHTEEN Transverse section of the abdomen through the kidneys, superior view. (*m.* stands for muscle.)

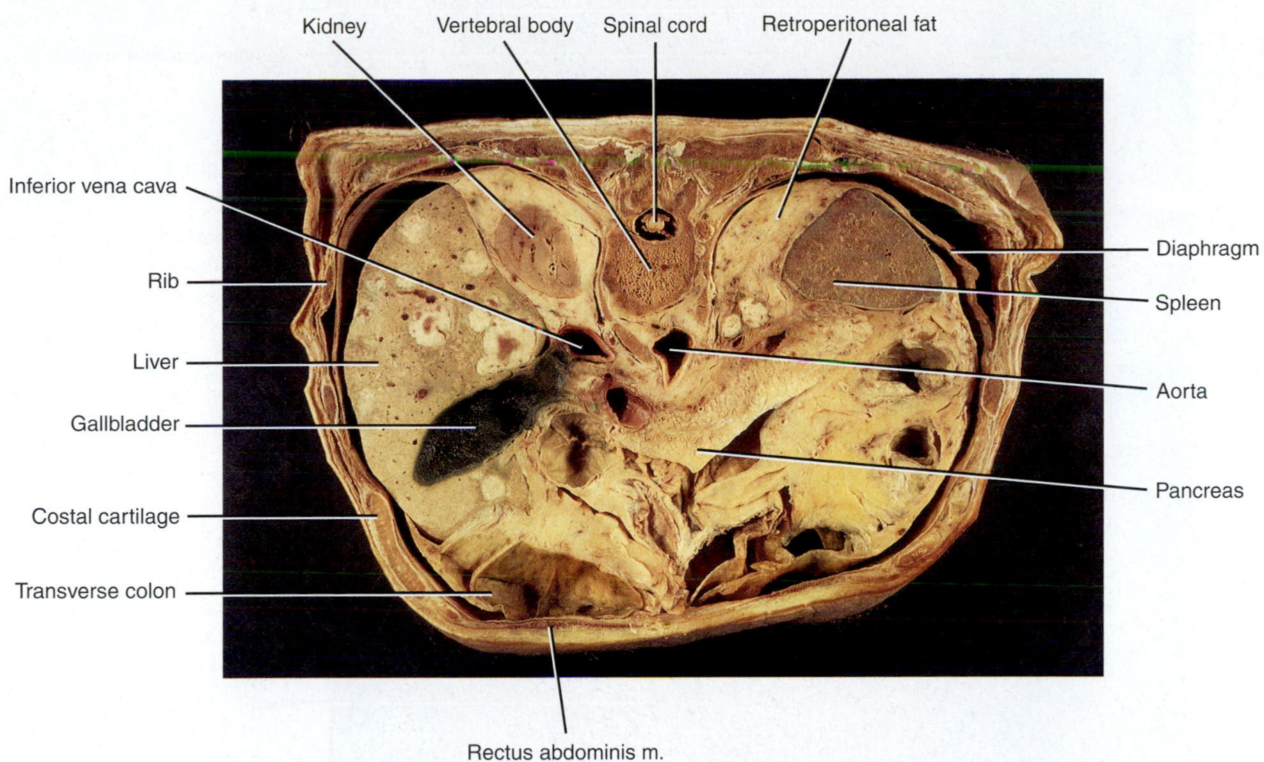

PLATE NINETEEN Transverse section of the abdomen through the pancreas, superior view. (*m.* stands for muscle.)

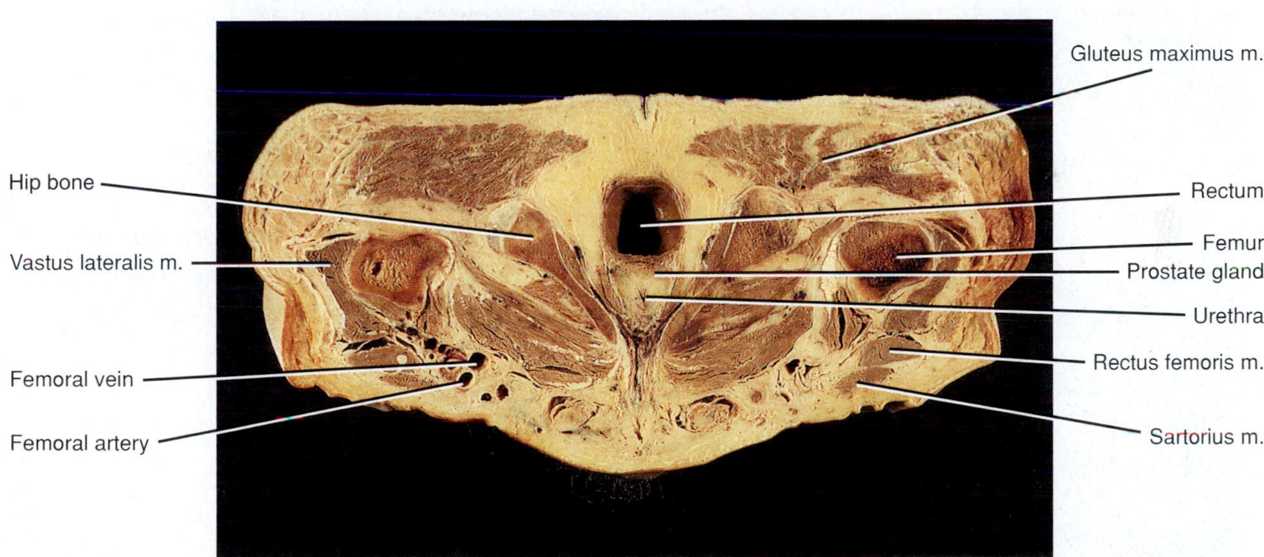

PLATE TWENTY Transverse section of the male pelvic cavity, superior view. (*m.* stands for muscle.)

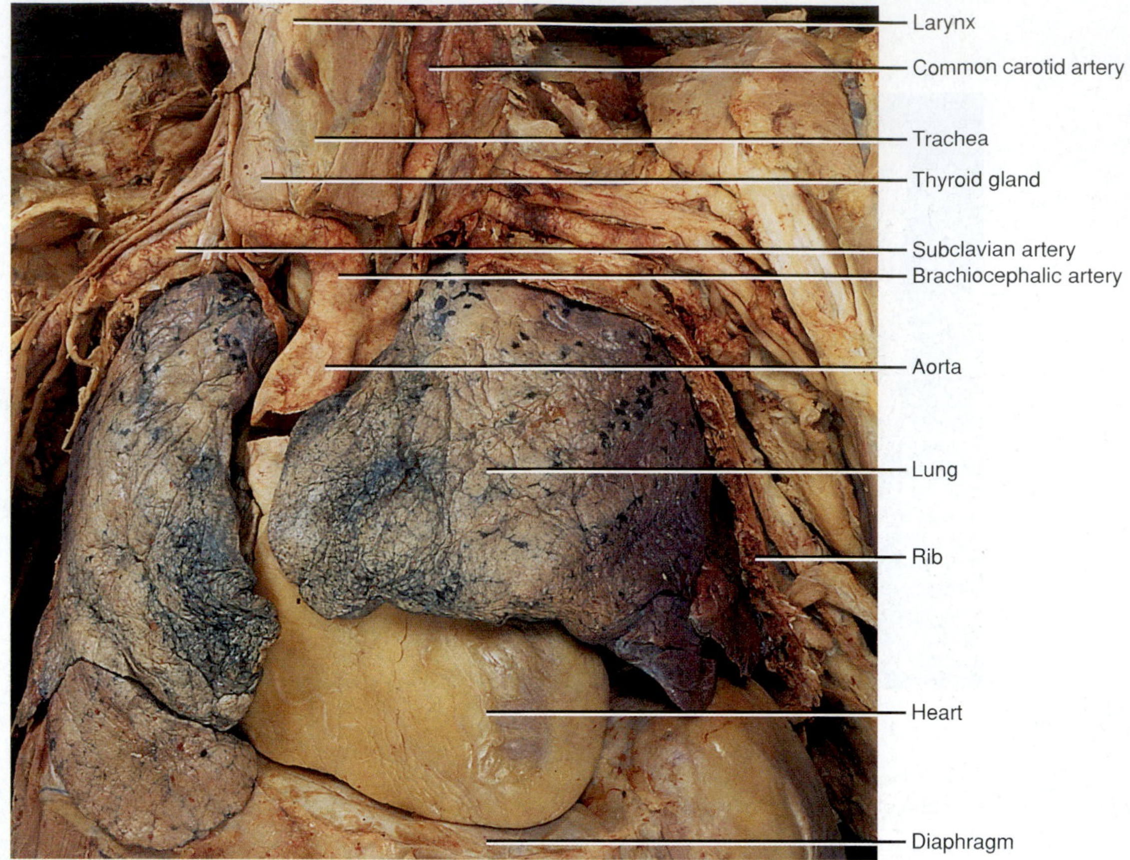

PLATE TWENTY-ONE Thoracic viscera, anterior view. (Brachiocephalic veins have been removed to better expose the brachiocephalic artery and the aorta.)

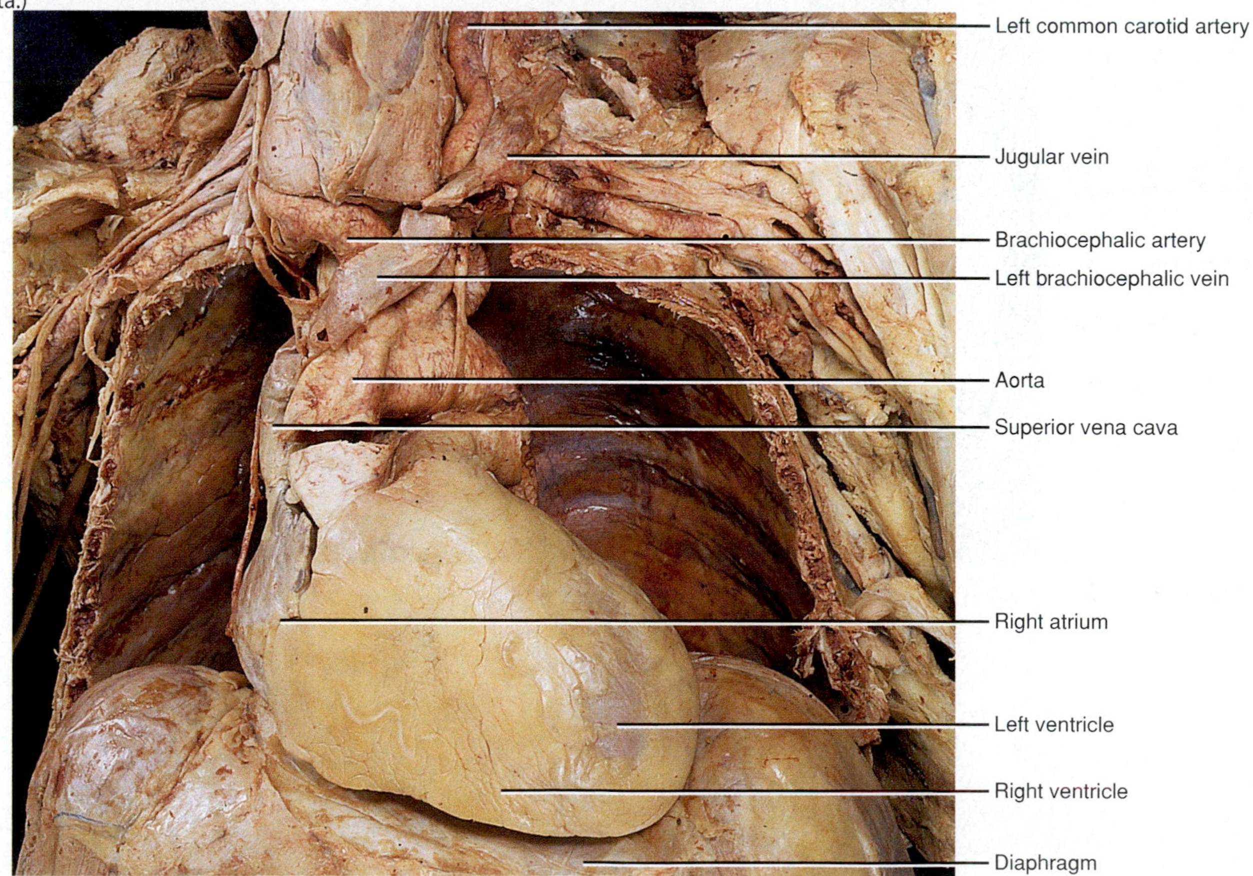

PLATE TWENTY-TWO Thorax with the lungs removed, anterior view.

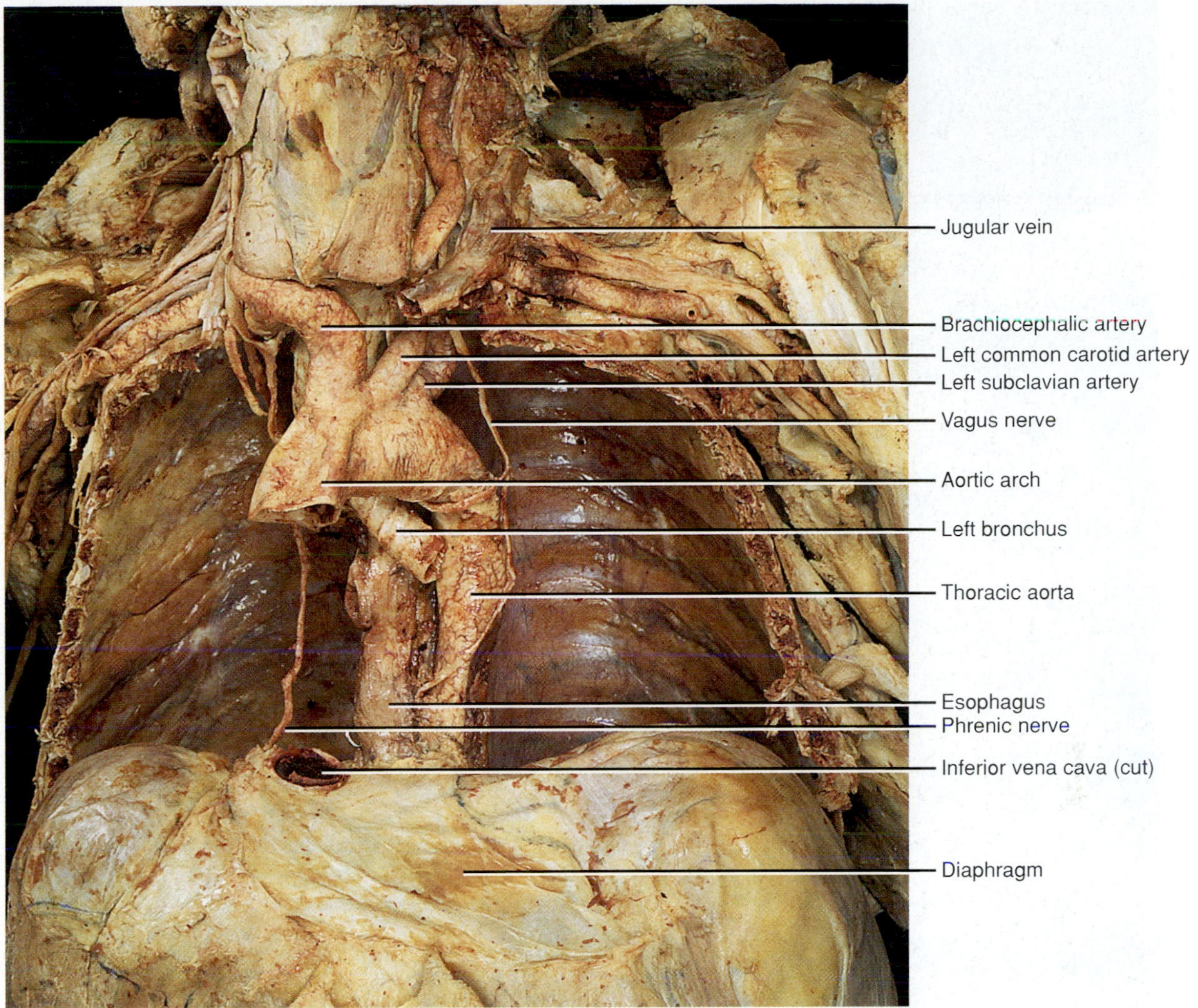

PLATE TWENTY-THREE Thorax with the heart and lungs removed, anterior view.

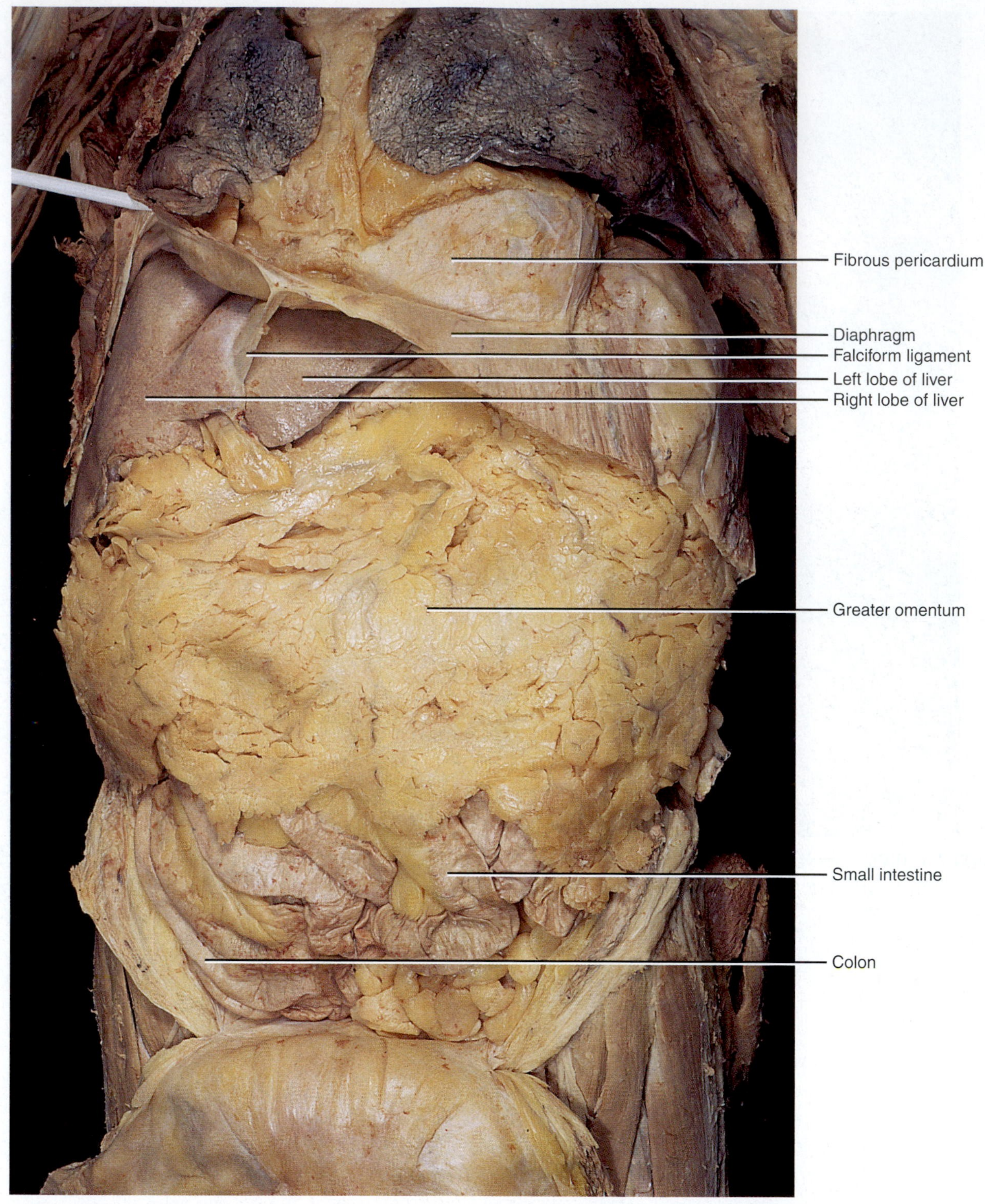

PLATE TWENTY-FOUR Abdominal viscera, anterior view.

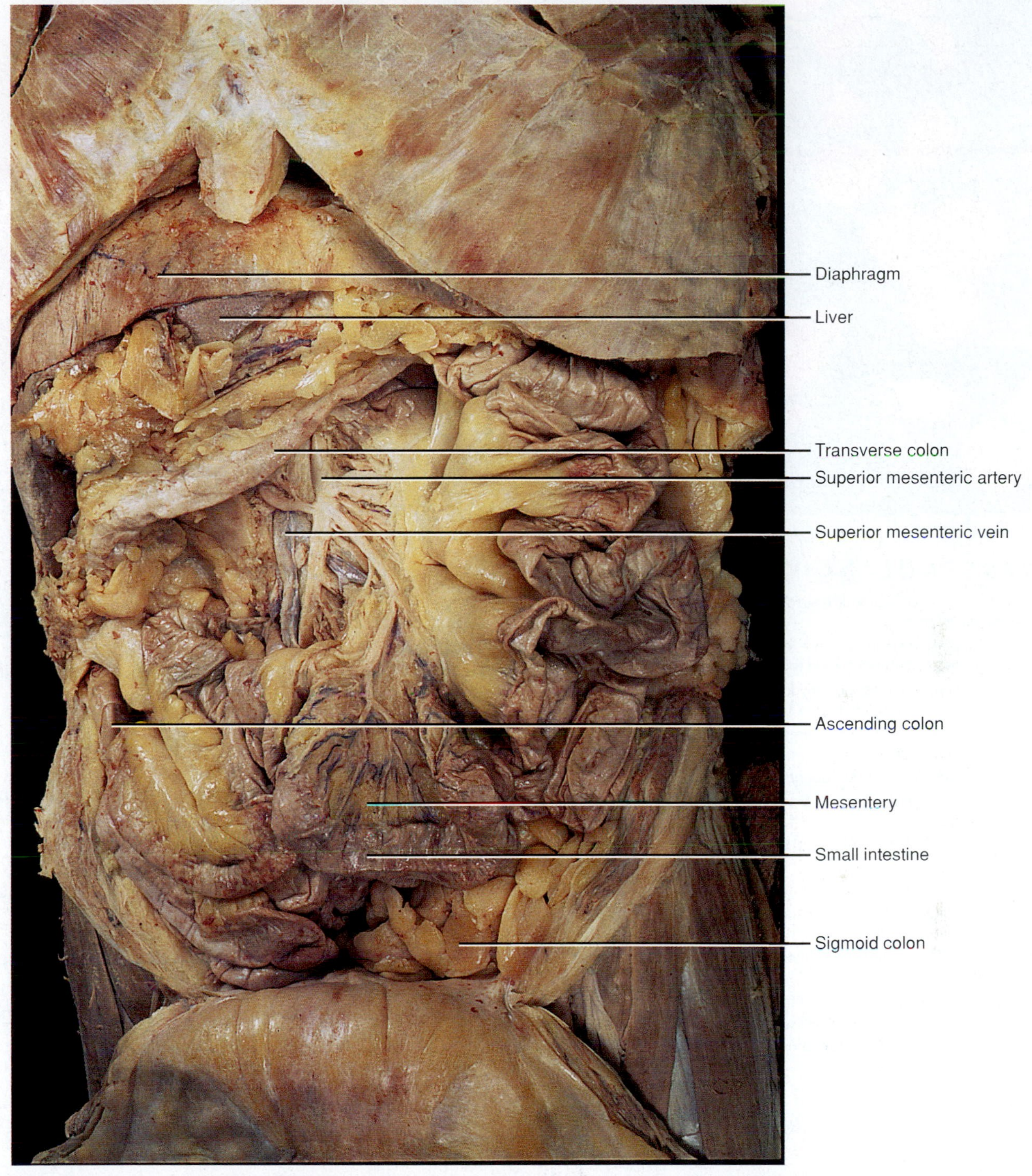

PLATE TWENTY-FIVE Abdominal viscera with the greater omentum removed, anterior view. (Small intestine has been displaced to the left.)

CHAPTER 2

Chemical Basis of Life

This partial model of DNA, the blueprint for a human, shows carbon atoms black, oxygen red, nitrogen blue, phosphorus yellow, and hydrogen white.

UNDERSTANDING WORDS

bio-, life: *bio*chemistry—branch of science dealing with the chemistry of life forms.

di-, two: *di*saccharide—compound whose molecules are composed of two sugar units bound together.

glyc-, sweet: *glyc*ogen—complex carbohydrate composed of glucose molecules bound together in a particular way.

iso-, equal: *iso*tope—atom that has the same atomic number as another atom but a different atomic weight.

lip-, fat: *lip*ids—group of organic compounds that includes fats.

-lyt, dissolvable: electro*lyt*e—substance that dissolves in water and releases ions.

mono-, one: *mono*saccharide—compound whose molecule consists of a single sugar unit.

nucle-, kernel: *nucle*us—central core of an atom.

poly-, many: *poly*unsaturated—molecule that has many double bonds between its carbon atoms.

sacchar-, sugar: mono*sacchar*ide—molecule consisting of a single sugar unit.

syn-, together: *syn*thesis—process by which substances are united to form a new type of substance.

-valent, having power: co*valent* bond—chemical bond produced when two atoms share electrons.

LEARNING OUTCOMES

After you have studied this chapter, you should be able to:

2.1 Introduction

1 Give examples of how the study of living materials requires an understanding of chemistry. (p. 51)

2.2 Structure of Matter

2 Describe the relationships among matter, atoms, and compounds. (p. 51)

3 Describe how atomic structure determines how atoms interact. (p. 52)

4 Explain how molecular and structural formulas symbolize the composition of compounds. (p. 53)

5 Describe three types of chemical reactions. (p. 58)

6 Describe the differences among acids, bases, and buffers. (p. 59)

7 Explain the pH scale. (p. 59)

2.3 Chemical Constituents of Cells

8 List the major groups of inorganic chemicals common in cells and explain the function(s) of each group. (p. 61)

9 Describe the general functions of the main classes of organic molecules in cells. (p. 61)

BIOMARKERS HERALD DISEASE

In our body fluids lie chemical clues to our health. A *biomarker* is a substance in the body that indicates a disease process or exposure to a toxin. Many medical tests that measure biomarkers are familiar, such as that for serum cholesterol. Such tests may indicate an increased risk of developing a particular disease or a presymptomatic stage, be used as a basis for diagnosis, or reflect response to a treatment.

Not just any chemical in the body can form the basis of a useful test. Biomarker tests should be simple and inexpensive and fulfill the following criteria:

- Sensitivity. The ability of a test to detect disease only when it is really present. The more sensitive a test, the fewer false positives.
- Specificity. The test's ability to exclude the disease in a patient who does not have it. The more specific a test, the fewer false negatives.
- Reproducibility. Results mean the same thing in different patients.
- Noninvasiveness: The test can be performed on an easily obtained body fluid.

The four major types of chemicals in a human body—carbohydrates, lipids, proteins, and nucleic acids—form the basis of many biomarker tests. Elevated levels of certain carbohydrates in the blood are signs of certain bacterial infections, and telltale carbohydrates in nipple fluid may indicate breast cancer. Cholesterol tests assess lipid levels in blood serum, which may be correlated to risk of cardiovascular disease. The prostate specific antigen (PSA) test measures the levels of a protein in the blood normally on cells of the prostate. An elevated level can indicate increased risk of prostate cancer. Genetic tests detect mutations in DNA that cause inherited disease or levels of RNA molecules characteristic of various disorders.

New biomarker tests evaluate levels of a number of chemicals in a body fluid sample, rather than one at a time. To assess exposure to tobacco smoke, for example, a biomarker panel measures carbon monoxide and biochemicals that the body produces (metabolites) as it breaks down several carcinogens in cigarette smoke. A biomarker test that measures levels of 120 different biochemicals in blood, including many immune system substances, hormones, growth factors, clotting factors, and proteins associated with cancer cells, provides risk estimates of cancer types.

Analysis of the human genome is allowing drug developers to assess the increasing and decreasing levels of many biochemicals in body fluids, and their research is providing information to develop new biomarker tests. However, the utility of such tests is important to consider. For example, a biomarker test would be useful if it led to earlier diagnosis or helped select a drug likely to help without causing adverse effects. It may be less useful for a patient who has a disease not currently treatable. ■

2.1 INTRODUCTION

Chemistry considers the composition of substances and how they change. It is possible to study anatomy without much reference to chemistry. However understanding the basics of chemistry is essential for understanding physiology, because body functions result from cellular functions that, in turn, result from chemical changes. The human body consists of chemicals, including salts, water, carbohydrates, lipids, proteins, and nucleic acids. The food that we eat, liquids that we drink, and medications that we take are chemicals.

As interest in the chemistry of living organisms grew and knowledge of the subject expanded, a field of life science called biological chemistry, or **biochemistry**, emerged. Biochemistry has been important not only in helping explain physiological processes but also in developing many new drugs and methods for treating diseases.

PRACTICE

1 Why is a knowledge of chemistry essential to understanding physiology?

2 What is biochemistry?

2.2 STRUCTURE OF MATTER

Matter is anything that has weight and takes up space. This includes all the solids, liquids, and gases in our surroundings as well as in our bodies. All matter consists of particles organized in specific ways. Table 2.1 lists some particles of matter and their characteristics.

Elements and Atoms

All matter is composed of fundamental substances called **elements** (el′ĕ-mentz). At present, 90 naturally occurring

TABLE 2.1 | Some Particles of Matter

Name	Characteristic	Name	Characteristic
Atom	Smallest particle of an element that has the properties of that element	Neutron (n^0)	Particle with about the same weight as a proton; uncharged and thus electrically neutral; found within an atomic nucleus
Electron (e^-)	Extremely small particle with almost no weight; carries a negative electrical charge and is in constant motion around an atomic nucleus	Ion	Particle that is electrically charged because it has gained or lost one or more electrons
Proton (p^+)	Relatively large atomic particle; carries a positive electrical charge and is found within an atomic nucleus	Molecule	Particle formed by the chemical union of two or more atoms

elements are known and at least 26 more have been created in the laboratory. Among these elements are such common materials as iron, copper, silver, gold, aluminum, carbon, hydrogen, and oxygen. Some elements exist in a pure form, but these and other elements are more commonly parts of chemical combinations called **compounds** (kom′powndz).

Elements the body requires in large amounts—such as carbon, hydrogen, oxygen, nitrogen, sulfur, and phosphorus—are termed **bulk elements.** These elements make up more than 95% (by weight) of the human body (table 2.2). Elements required in small amounts are called **trace elements.** Many trace elements are important parts of enzymes, proteins that regulate the rates of chemical reactions in living organisms. Some elements toxic in large amounts, such as arsenic, may be vital in very small amounts, and these are called **ultratrace elements.**

Elements are composed of particles called **atoms** (at′omz), the smallest complete units of the elements. The atoms that make up each element are chemically identical but they differ from the atoms that make up other elements. Atoms vary in size, weight, and the ways they interact with other atoms. Some atoms can combine with atoms like themselves or with other atoms by forming attractions called **chemical bonds,** while other atoms cannot form such bonds.

TABLE 2.2 | Major Elements in the Human Body (By Weight)

Major Elements	Symbol	Approximate Percentage of the Human Body	
Oxygen	O	65.0	99.9%
Carbon	C	18.5	
Hydrogen	H	9.5	
Nitrogen	N	3.2	
Calcium	Ca	1.5	
Phosphorus	P	1.0	
Potassium	K	0.4	
Sulfur	S	0.3	
Chlorine	Cl	0.2	
Sodium	Na	0.2	
Magnesium	Mg	0.1	
Trace Elements			
Cobalt	Co	less than 0.1%	
Copper	Cu		
Fluorine	F		
Iodine	I		
Iron	Fe		
Manganese	Mn		
Zinc	Zn		

Atomic Structure

An atom consists of a central portion called the **nucleus** (nu′kle-us) and one or more **electrons** (e-lek′tronz) that constantly move around the nucleus. The nucleus contains one or more relatively large particles, **protons** (pro′tonz) and usually **neutrons** (nu′tronz). Protons and neutrons are about equal in weight, but they are otherwise different (fig. 2.1).

Electrons, so small that they have almost no weight, carry a single, negative electrical charge (e^-). Each proton carries a single, positive electrical charge (p^+). Neutrons are uncharged and thus are electrically neutral (n^0).

The nucleus contains protons, so this part of an atom is always positively charged. However, the number of electrons outside the nucleus equals the number of protons, so a complete atom is said to have no net charge and is thus electrically neutral.

The atoms of different elements have different numbers of protons. The number of protons in the atoms of a particular element is called its **atomic number.** Hydrogen, for example, whose atoms have one proton, has atomic number 1; carbon, whose atoms have six protons, has atomic number 6.

The weight of an atom of an element is primarily due to the protons and neutrons in its nucleus, because the electrons are so light. For this reason, the number of protons plus the number of neutrons in each of an element's atoms essentially equals the **atomic weight** of that atom. The atomic weight of a hydrogen atom, which has only one proton and no neutrons, is approximately 1. The atomic weight of a carbon atom, with six protons and six neutrons, is approximately 12 (table 2.3).

Isotopes

All the atoms of a particular element have the same atomic number because they have the same number of protons and electrons. However, the atoms of an element vary in the number of neutrons in their nuclei; thus, they vary in atomic

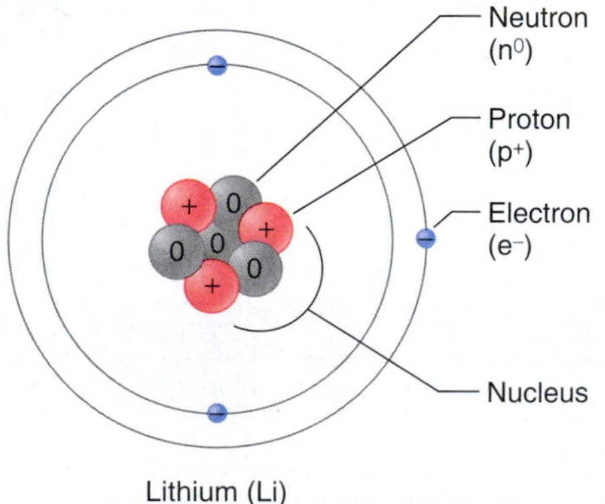

FIGURE 2.1 An atom consists of subatomic particles. In an atom of the element lithium, three electrons encircle a nucleus that consists of three protons and four neutrons.

TABLE 2.3 | Atomic Structure of Elements 1 Through 12

Element	Symbol	Number	Approximate Atomic Weight	Protons	Neutrons	Electrons in Shells: First	Second	Third
Hydrogen	H	1	1	1	0	1		
Helium	He	2	4	2	2	2 (inert)		
Lithium	Li	3	7	3	4	2	1	
Beryllium	Be	4	9	4	5	2	2	
Boron	B	5	11	5	6	2	3	
Carbon	C	6	12	6	6	2	4	
Nitrogen	N	7	14	7	7	2	5	
Oxygen	O	8	16	8	8	2	6	
Fluorine	F	9	19	9	10	2	7	
Neon	Ne	10	20	10	10	2	8 (inert)	
Sodium	Na	11	23	11	12	2	8	1
Magnesium	Mg	12	24	12	12	2	8	2

For more detail, see Appendix A, Periodic Table of the Elements.

weight. For example, all oxygen atoms have eight protons in their nuclei. Some, however, have eight neutrons (atomic weight 16), others have nine neutrons (atomic weight 17), and still others have ten neutrons (atomic weight 18). Atoms that have the same atomic numbers but different atomic weights are called **isotopes** (i′so-tōpz) of an element. A sample of an element is likely to include more than one isotope, so the atomic weight of the element is often considered to be the average weight of the isotopes present. (See Appendix A, Periodic Table of the Elements, p. 939.)

The ways atoms interact reflect their numbers of electrons. An atom has the same number of electrons and protons, so all the isotopes of a particular element have the same number of electrons and chemically react in the same manner. For example, any of the isotopes of oxygen can have the same function in metabolic reactions.

Isotopes of an element may be stable, or they may have unstable atomic nuclei that decompose, releasing energy or pieces of themselves until they reach a stable form. Such unstable isotopes are called *radioactive,* and the energy or atomic fragments they emit are called *atomic radiation*. Elements that have radioactive isotopes include oxygen, iodine, iron, phosphorus, and cobalt. Some radioactive isotopes are used to detect and treat disease (From Science to Technology 2.1).

Atomic radiation includes three common forms called alpha (α), beta (β), and gamma (γ). Each type of radioactive isotope produces one or more of these forms of radiation. Alpha radiation consists of particles from atomic nuclei, each of which includes two protons and two neutrons, that move slowly and cannot easily penetrate matter. Beta radiation consists of much smaller particles (electrons) that travel faster and more deeply penetrate matter. Gamma radiation is a form of energy similar to X-radiation and is the most penetrating form of atomic radiation. From Science to Technology 2.2 examines how radiation that moves electrons can affect human health.

PRACTICE

3 What is the relationship between matter and elements?

4 Which elements are most common in the human body?

5 Where are electrons, protons, and neutrons located within an atom?

6 What is an isotope?

7 What is atomic radiation?

Molecules and Compounds

Two or more atoms may combine to form a distinctive type of particle called a **molecule.** A **molecular formula** is shorthand used to depict the numbers and types of atoms in a molecule. It consists of the symbols of the elements in the molecule with numerical subscripts that indicate how many atoms of each element are present. For example, the molecular formula for water is H_2O, which indicates two atoms of hydrogen and one atom of oxygen in each molecule. The molecular formula for the sugar glucose, $C_6H_{12}O_6$, indicates six atoms of carbon, twelve atoms of hydrogen, and six atoms of oxygen.

If atoms of the same element combine, they produce molecules of that element. Gases of hydrogen (H_2), oxygen (O_2), and nitrogen (N_2) consist of such molecules. If atoms of different elements combine, molecules of compounds form. Two atoms of hydrogen, for example, can combine with one atom of oxygen to produce a molecule of the compound water

Radioactive Isotopes Reveal Physiology

Vicki L. arrived early at the nuclear medicine department of the health center. As she sat in an isolated cubicle, a doctor in full sterile dress approached with a small metal canister marked with warnings. The doctor carefully unscrewed the top, inserted a straw, and watched as the young woman sipped the fluid within. It tasted like stale water but was a solution containing a radioactive isotope, iodine-131.

Vicki's thyroid gland had been removed three months earlier, and this test was to determine whether any active thyroid tissue remained. The thyroid is the only part of the body to metabolize iodine, so if Vicki's body retained any of the radioactive drink, it would mean that some of her cancerous thyroid gland remained. By using a radioactive isotope, her physicians could detect iodide uptake using a scanning device called a scintillation counter (fig. 2A). Figure 2B illustrates iodine-131 uptake in a complete thyroid gland.

The next day, Vicki returned for the scan, which showed that a small amount of thyroid tissue was left and functioning. Another treatment would be necessary. Vicki would drink enough radioactive iodide to destroy the remaining tissue. This time, she drank the solution in an isolation room lined with paper to keep her from contaminating the floor, walls, and furniture. The same physician administered the radioactive iodide. Vicki's physician had this job because his thyroid had been removed many years earlier, and therefore, the radiation couldn't harm him.

After two days in isolation, Vicki went home with a list of odd instructions. She was to stay away from her children and pets, wash her clothing separately, use disposable utensils and plates, and flush the toilet three times each time she used it. These precautions would minimize her contaminating her family—mom was radioactive!

Iodine-131 is a medically useful radioactive isotope because it has a short half-life, a measurement of the time it takes for half of an amount of an isotope to decay to a nonradioactive form. The half-life of iodine-131 is 8.1 days. With the amount of radiation in Vicki's body dissipating by half every 8.1 days, after three months hardly any would be left. If all went well, any remaining cancer cells would leave her body along with the radioactive iodine.

Isotopes of other elements have different half-lives. The half-life of iron-59 is 45.1 days; that of phosphorus-32 is 14.3 days; that of cobalt-60 is 5.26 years; and that of radium-226 is 1,620 years.

A form of thallium-201 with a half-life of 73.5 hours is commonly used to detect disorders in the blood vessels supplying the heart muscle or to locate regions of damaged heart tissue after a heart attack.

Gallium-67, with a half-life of 78 hours, is used to detect and monitor the progress of certain cancers and inflammatory illnesses. These medical procedures inject the isotope into the blood and follow its path using detectors that record images on paper or film.

Radioactive isotopes are also used to assess kidney function, estimate the concentrations of hormones in body fluids, measure blood volume, and study changes in bone density. Cobalt-60 is a radioactive isotope used to treat some cancers. The cobalt emits radiation that damages cancer cells more readily than it does healthy cells. ■

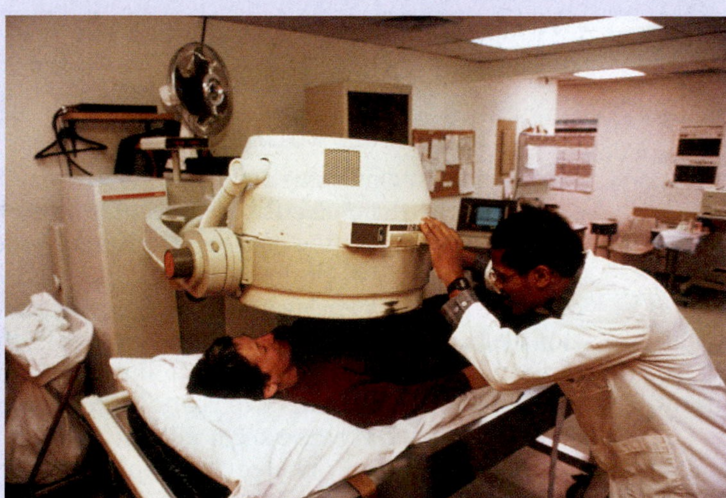

FIGURE 2A Scintillation counters detect radioactive isotopes.

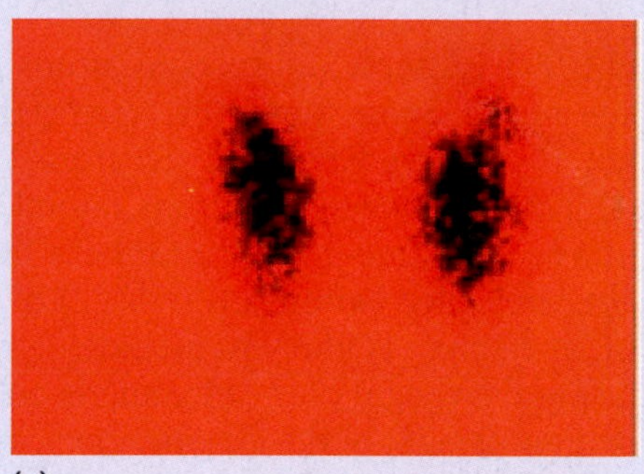

(a)

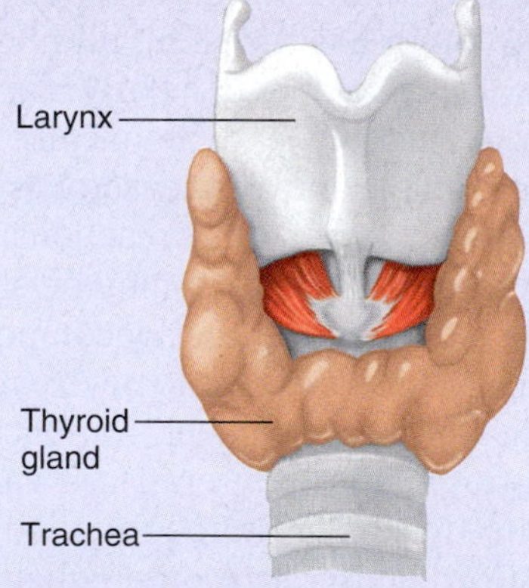

(b)

FIGURE 2B (*a*) A scan of the thyroid gland twenty-four hours after the patient receives radioactive iodine. Note how closely the scan in (*a*) resembles the shape of the thyroid gland shown in (*b*).

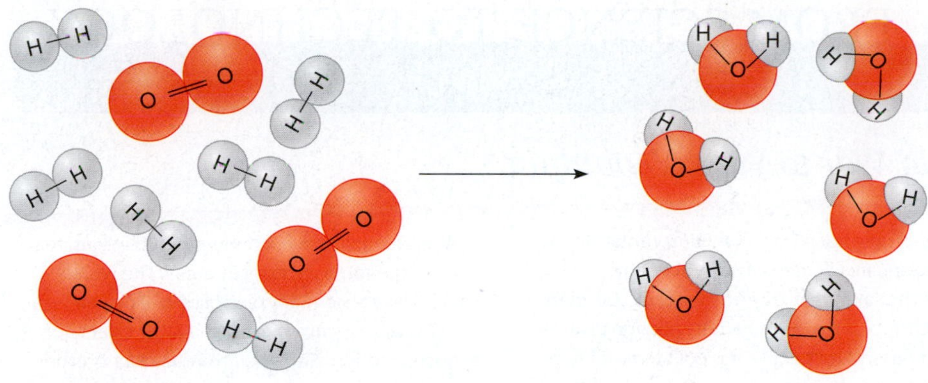

FIGURE 2.2 Under certain conditions, hydrogen molecules can combine with oxygen molecules, forming water molecules.

(H_2O), as figure 2.2 shows. Table sugar, baking soda, natural gas, beverage alcohol, and most drugs are compounds.

A molecule of a compound always consists of definite types and numbers of atoms. A molecule of water (H_2O), for instance, always has two hydrogen atoms and one oxygen atom. If two hydrogen atoms combine with two oxygen atoms, the compound formed is not water, but hydrogen peroxide (H_2O_2).

Bonding of Atoms

Atoms combine with other atoms by forming links called **bonds.** Chemical bonds result from interactions of electrons.

The electrons of an atom occupy one or more regions of space called **electron shells** that encircle the nucleus. Because electrons have a level of energy characteristic of the particular shell they are in, the shells are sometimes called *energy shells.* Each electron shell can hold a limited number of electrons. The maximum number of electrons that each of the first three shells can hold for elements of atomic number 18 and under is

First shell (closest to the nucleus)	*2 electrons*
Second shell	*8 electrons*
Third shell	*8 electrons*

More complex atoms may have as many as eighteen electrons in the third shell.

Simplified diagrams such as those in figure 2.3 are used to show electron configuration in atoms. The single electron of a hydrogen atom is in the first shell; the two electrons of a helium atom fill its first shell; and of the three electrons of a lithium atom, two are in the first shell and one is in the second shell. Lower energy shells, closer to the nucleus, must be filled first.

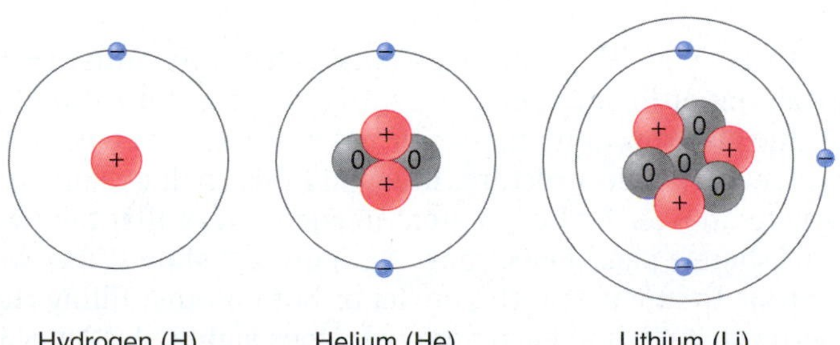

FIGURE 2.3 Electrons orbit the atomic nucleus. The single electron of a hydrogen atom moves within its first shell. The two electrons of a helium atom fill its first shell. Two of the three electrons of a lithium atom are in the first shell, and one is in the second shell.

The number of electrons in the outermost shell of an atom determines whether it will react with another atom. Atoms react in a way that leaves the outermost shell completely filled with electrons, achieving a more stable structure. This is sometimes called the **octet rule,** because, except for the first shell, it takes eight electrons to fill the shells in most of the atoms important in living organisms.

Atoms such as helium, whose outermost electron shells are filled, already have stable structures and are chemically inactive or **inert** (they cannot form chemical bonds). Atoms with incompletely filled outer shells, such as those of hydrogen or lithium, tend to gain, lose, or share electrons in ways that empty or fill their outer shells. In this way, they achieve stable structures.

Atoms that gain or lose electrons become electrically charged and are called **ions** (i′onz). An atom of sodium, for example, has eleven electrons: two in the first shell, eight in the second shell, and one in the third shell. This atom tends to lose the electron from its outer (third) shell, which leaves the second (now the outermost) shell filled and the new form stable (fig. 2.4*a*). In the process, sodium is left with eleven protons (11^+) in its nucleus and only ten electrons (10^-). As a result, the atom develops a net electrical charge of 1^+ and is called a sodium ion, symbolized Na^+.

A chlorine atom has seventeen electrons, with two in the first shell, eight in the second shell, and seven in the third shell. An atom of this type tends to accept a single electron, filling its outer (third) shell and becoming stable. In the process, the chlorine atom is left with seventeen protons (17^+) in its nucleus and eighteen electrons (18^-). As a result, the atom develops a net electrical charge of 1^- and is called a chloride ion, symbolized Cl^-.

Positively charged ions are called **cations** (kat′i-onz), and negatively charged ions are called **anions** (an′i-onz). Ions with opposite charges attract, forming **ionic bonds** (i-on′ik bondz). Sodium ions (Na^+) and chloride ions (Cl^-) uniting in this manner form the compound sodium chloride (NaCl), or table salt (fig. 2.4*b*). Similarly, hydrogen atoms

2.2 FROM SCIENCE TO TECHNOLOGY

Ionizing Radiation: From the Cold War to Yucca Mountain

Alpha, beta, and gamma radiation are called ionizing radiation because their energy removes electrons from atoms (fig. 2C). Electrons dislodged by ionizing radiation can affect nearby atoms, disrupting physiology at the chemical level in a variety of ways—causing cancer, clouding the lens of the eye, and interfering with normal growth and development.

In the United States, some people are exposed to very low levels of ionizing radiation, mostly from background radiation, which originates from natural environmental sources (table 2A). For people who live near sites of atomic weapons manufacture, exposure is greater. Epidemiologists are investigating medical records that document illnesses linked to long-term exposure to ionizing radiation in a 1,200-square kilometer area in Germany.

The lake near Oberrothenback, Germany, which appears inviting, harbors enough toxins to kill thousands of people. It is polluted with heavy metals, low-level radioactive chemical waste, and 22,500 tons of arsenic. Radon, a radioactive by-product of uranium, permeates the soil. Many farm animals and pets that have drunk from the lake have died. Cancer rates and respiratory disorders among the human residents nearby are well above normal.

The lake in Oberrothenback was once a dump for a factory that produced "yellow cake," a term for processed uranium ore, used to build atomic bombs for the former Soviet Union. In the early 1950s, nearly half a million workers labored here and in surrounding areas in factories and mines. Records released in 1989, after the reunification of Germany, reveal that workers were given perks, such as alcoholic beverages and better wages, to work in the more dangerous areas. The workers paid a heavy price: many died of lung ailments.

Today, concern over the health effects of exposure to ionizing radiation centers on the U.S. government's plan to transport tens of thousands of metric tons of high-level nuclear waste from 109 reactors around the country for burial beneath Yucca Mountain, Nevada, by 2021. The waste, currently stored near the reactors, will be buried in impenetrable containers under the mountain by robots. In the reactors, nuclear fuel rods contain uranium oxide, which produces electricity as it decays to plutonium, which gives off gamma rays. Periodically the fuel rods must be replaced, and the spent ones buried. Environmental groups are concerned that the waste could be exposed during transport and that the facility in the mountain may not adequately contain it. ■

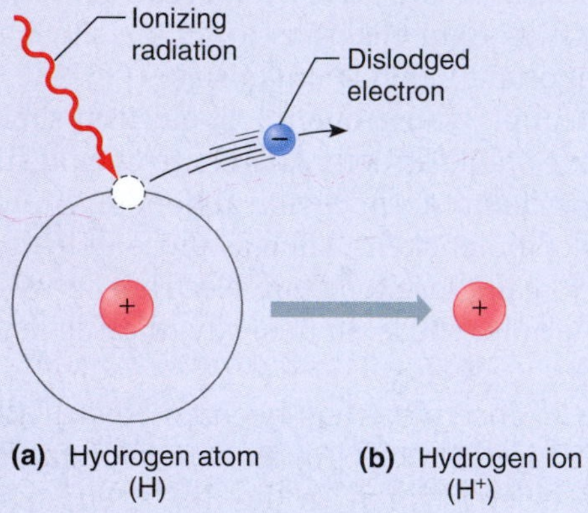

FIGURE 2C Ionizing radiation removes elecrons from atoms. (*a*) Ionizing radiation may dislodge an electron from an electrically neutral hydrogen atom. (*b*) Without its electron, the hydrogen atom becomes a positively-charged hydrogen ion (H^+).

TABLE 2A | Sources of Ionizing Radiation

Background (Natural environmental)	Cosmic rays from space
	Radioactive elements in earth's crust
	Rocks and clay in building materials
	Radioactive elements naturally in the body (potassium-40, carbon-14)
Medical and dental	X rays
	Radioactive substances
Other	Atomic and nuclear weapons
	Mining and processing radioactive minerals
	Radioactive fuels in nuclear power plants
	Radioactive elements in consumer products (luminescent dials, smoke detectors, color TV components)

may lose their single electrons and become hydrogen ions (H^+). Hydrogen ions can form ionic bonds with chloride ions (Cl^-) to form hydrogen chloride (HCl, which reacts in water to form hydrochloric acid).

Cations and anions attract each other in all directions, forming a three-dimensional structure. Ionically bound compounds do not form specific particles, so they do not exist as molecules. Rather, they form arrays, such as crystals of sodium chloride (fig. 2.4*c*). The molecular formulas for compounds such as sodium chloride (NaCl) give the relative amounts of each element.

Atoms may also bond by sharing electrons rather than by gaining or losing them. A hydrogen atom, for example, has one electron in its first shell but requires two electrons to achieve a stable structure. It may fill this shell by combining with another hydrogen atom in such a way that the two atoms share a pair of electrons. As figure 2.5 shows, the two electrons then encircle the nuclei of both atoms, filling the outermost shell, and each atom becomes stable. A chemical bond between atoms that share electrons is called a **covalent bond** (ko′va-lent bond).

Usually atoms of each element form a specific number of covalent bonds. Hydrogen atoms form single bonds, oxygen

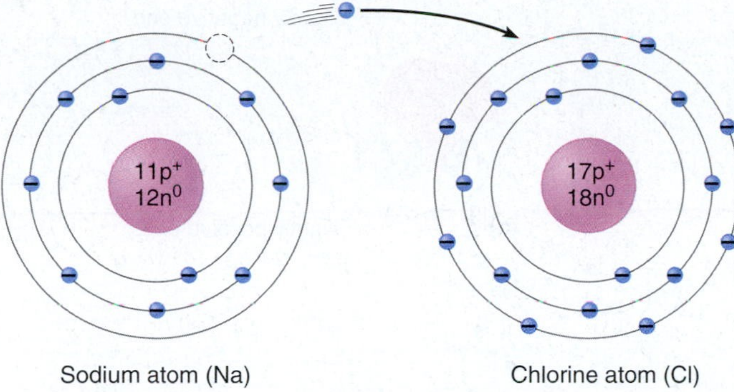

(a) Separate atoms
If a sodium atom loses an electron to a chlorine atom, the sodium atom becomes a sodium ion (Na^+), and the chlorine atom becomes a chloride ion (Cl^-).

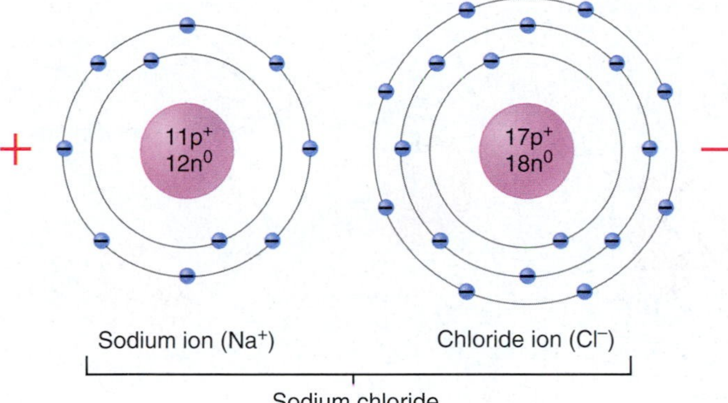

(b) Bonded ions
These oppositely charged particles attract electrically and join by an ionic bond.

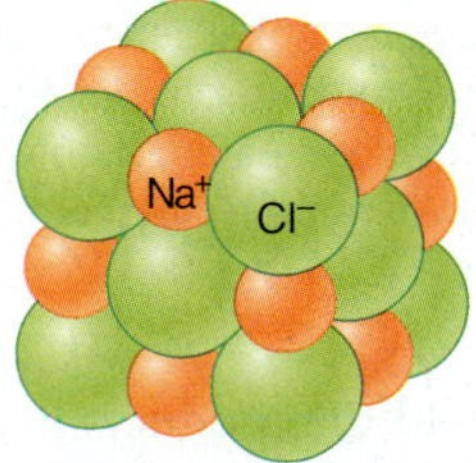

(c) Salt crystal
Ionically bonded substances form arrays such as a crystal of NaCl.

FIGURE 2.4 An ionic bond forms when (*a*) one atom gains and another atom loses electron(s) and then (*b*) oppositely charged ions attract. (*c*) Ionically bonded substances may form crystals.

atoms form two bonds, nitrogen atoms form three bonds, and carbon atoms form four bonds. Symbols and lines can be used to represent the bonding capacities of these atoms, as follows:

—H —O—

—N— (with one additional bond line below) —C— (with bond lines above and below)

Symbols and lines show how atoms bond and are arranged in various molecules. One pair of shared electrons, a single covalent bond, is depicted with a single line. Sometimes atoms may share two pairs of electrons (a double covalent bond), or even three pairs (a triple covalent bond), represented by two and three lines, respectively. Illustrations of this type, called **structural formulas** (fig. 2.6), are useful, but they cannot adequately capture the three-dimensional forms of molecules. In contrast, figure 2.7 shows a three-dimensional (space-filling) representation of a water molecule.

Different types of chemical bonds share electrons to different degrees. At one extreme is an ionic bond in which atoms gain or lose electrons. At the other extreme is a covalent bond that shares electrons equally. In between lies a covalent bond in which electrons are not shared equally, resulting in a molecule whose shape gives an uneven distribution of charges. Such a molecule is called **polar.** Unlike an ion, a polar molecule has an equal number of protons and electrons, but one end of the molecule has more than its share of electrons, becoming slightly negative, while the other end of the molecule has less than its share, becoming slightly positive. Typically, polar covalent bonds form where hydrogen atoms bond to oxygen or nitrogen atoms. Water is a polar molecule (fig. 2.8*a*).

The attraction of the positive hydrogen end of a polar molecule to the negative nitrogen or oxygen end of another polar molecule is called a **hydrogen bond.** These bonds are weak, particularly at body temperature. For example, below 0°C, the hydrogen bonds between water molecules shown in figure 2.8*b* are strong enough to form ice. As the temperature rises, increased molecular movement breaks the hydrogen bonds, and water becomes liquid. Even at body temperature,

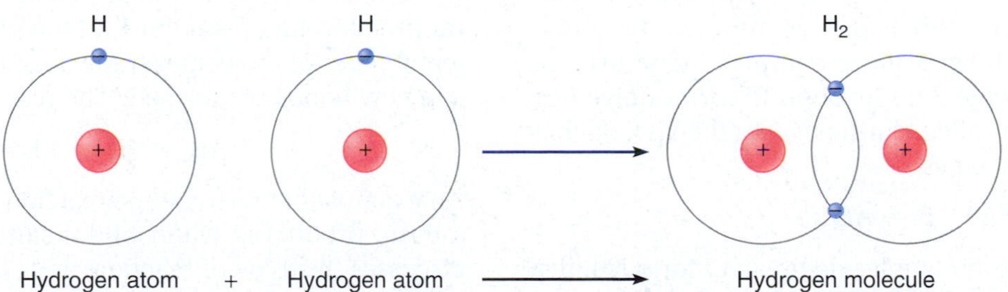

FIGURE 2.5 A hydrogen molecule forms when two hydrogen atoms share a pair of electrons. A covalent bond forms between the atoms.

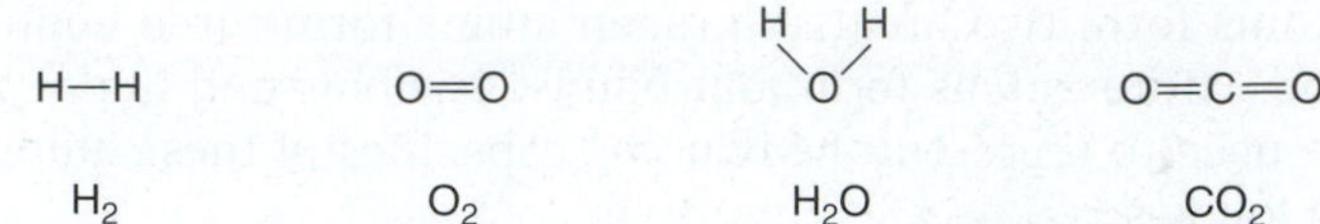

FIGURE 2.6 Structural and molecular formulas depict molecules of hydrogen, oxygen, water, and carbon dioxide. Note the double covalent bonds. (Triple covalent bonds are also possible between some atoms.)

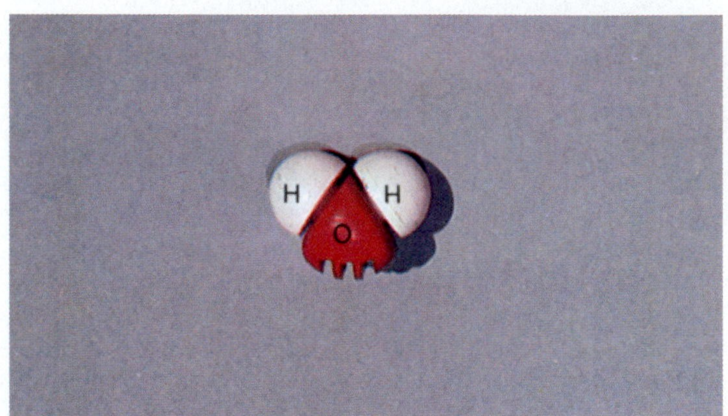

FIGURE 2.7 A three-dimensional model represents this water molecule (H_2O). The white parts represent the hydrogen atoms, and the red part represents oxygen.

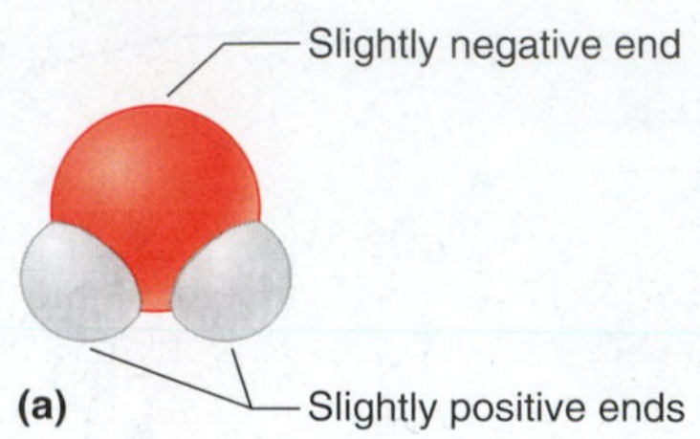

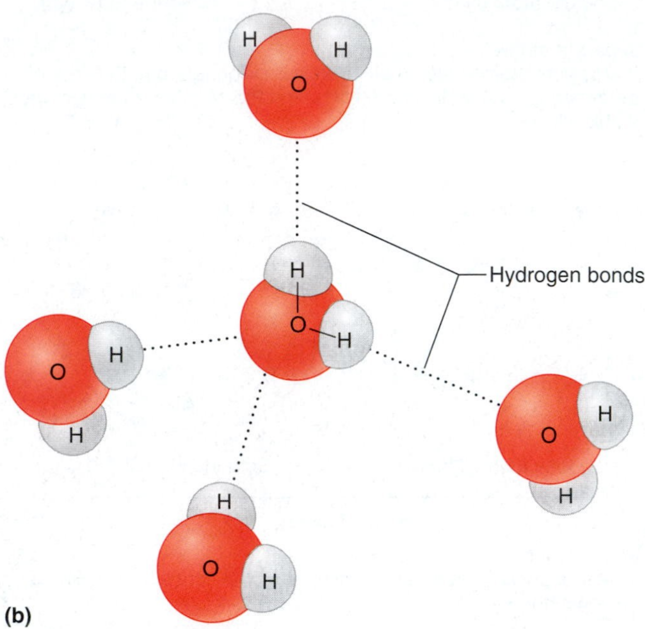

FIGURE 2.8 Water is a polar molecule. (*a*) Water molecules have equal numbers of electrons and protons but are polar because the electrons are shared unequally, creating slightly negative ends and slightly positive ends. (*b*) Water molecules form hydrogen bonds with each other.

hydrogen bonds are important in protein and nucleic acid structure. In these cases, hydrogen bonds form between polar regions of a single, very large molecule.

PRACTICE

8 Distinguish between a molecule and a compound.
9 What is an ion?
10 Describe two ways that atoms may combine with other atoms.
11 What is a molecular formula? A structural formula?
12 Distinguish between an ion and a polar molecule.

Chemical Reactions

Chemical reactions form or break bonds between atoms, ions, or molecules. The starting materials changed by the chemical reaction are called **reactants** (re-ak′tantz). The atoms, ions, or molecules formed at the reaction's conclusion are called **products.** When two or more atoms, ions, or molecules bond to form a more complex structure, as when hydrogen and oxygen atoms bond to form molecules of water, the reaction is called **synthesis** (sin′thĕ-sis). Such a reaction can be symbolized as

$$A + B \rightarrow AB$$

If the bonds of a reactant molecule break to form simpler molecules, atoms, or ions, the reaction is called **decomposition** (de-kom″po-zish′un). For example, molecules of water can decompose to yield the products hydrogen and oxygen. Decomposition is symbolized as

$$AB \rightarrow A + B$$

Synthetic reactions, which build larger molecules from smaller ones, are particularly important in growth of body parts and repair of worn or damaged tissues. Decomposition reactions digest nutrient molecules into molecules small enough to be absorbed into the bloodstream in the small intestine.

A third type of chemical reaction is an **exchange reaction** (replacement reaction). In this reaction, parts of two different types of molecules trade positions as bonds are broken and new bonds are formed. The reaction is symbolized as

$$AB + CD \rightarrow AD + CB$$

An example of an exchange reaction is an acid reacting with a base, producing water and a salt. The following section discusses this type of reaction.

Many chemical reactions are reversible. This means the product or products can change back to the reactant or reactants. A **reversible reaction** is symbolized using a double arrow:

$$A + B \rightleftarrows AB$$

Whether a reversible reaction proceeds in one direction or another depends on the relative proportions of reactant (or reactants) and product (or products) as well as the amount of energy available. **Catalysts** (kat′ah-listz) are molecules that influence the rates (not the direction) of chemical reactions but are not consumed in the process.

Acids, Bases, and Salts

When ionically bound substances are placed in water, the ions are attracted to the positive and negative ends of the water molecules and tend to leave each other, or dissociate. In this way, the polarity of water dissociates the salts in the internal environment. Sodium chloride (NaCl), for example, ionizes into sodium ions (Na^+) and chloride ions (Cl^-) in water (fig. 2.9). This reaction is represented as

$$NaCl \rightarrow Na^+ + Cl^-$$

The resulting solution has electrically charged particles (ions), so it conducts an electric current. Substances that release ions in water are, therefore, called **electrolytes** (e-lek′tro-lītz). Electrolytes that dissociate to release hydrogen ions (H^+) in water are called **acids** (as′idz). For example, in water, the compound hydrochloric acid (HCl) releases hydrogen ions (H^+) and chloride ions (Cl^-):

$$HCl \rightarrow H^+ + Cl^-$$

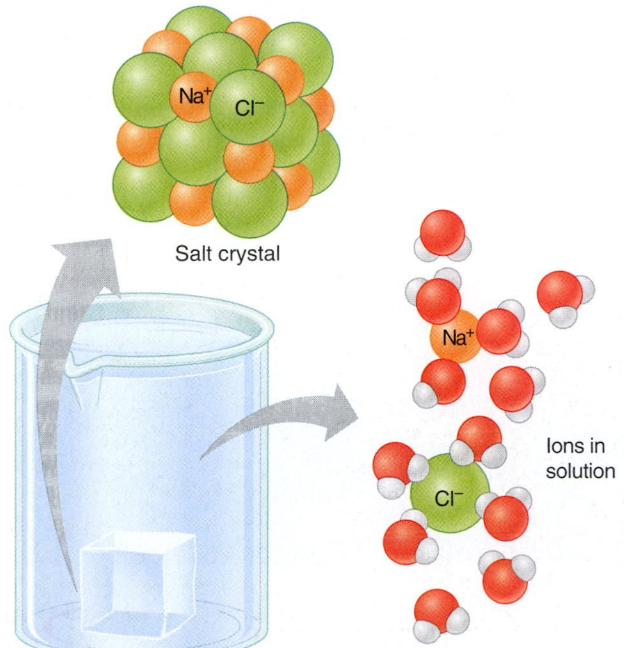

FIGURE 2.9 The polar nature of water molecules dissociates sodium chloride (NaCl) in water, releasing sodium ions (Na^+) and chloride ions (Cl^-).

Substances that combine with hydrogen ions are called **bases.** The compound sodium hydroxide (NaOH) in water releases hydroxide ions (OH^-), which can combine with hydrogen ions to form water. Thus, sodium hydroxide is a base:

$$NaOH \rightarrow Na^+ + OH^-$$

(*Note:* Some ions, such as OH^-, consist of two or more atoms. However, such a group usually behaves like a single atom and remains unchanged during a chemical reaction.)

Bases can react with acids to neutralize them, forming water and electrolytes called **salts.** For example, hydrochloric acid and sodium hydroxide react to form water and sodium chloride:

$$HCl + NaOH \rightarrow H_2O + NaCl$$

Table 2.4 summarizes the three types of electrolytes.

Acid and Base Concentrations

Concentrations of acids and bases affect the chemical reactions that constitute many life processes, such as those controlling breathing rate. Thus, the concentrations of these substances in body fluids are of special importance.

Hydrogen ion concentration can be measured in grams of ions per liter of solution. However, because hydrogen ion concentration can cover such a wide range (gastric juice has 0.01 grams H^+/liter; household ammonia has 0.00000000001 grams H^+/liter), a shorthand system called the **pH scale** is used. This system tracks the number of decimal places in a hydrogen ion concentration without writing them out. For example, a solution with a hydrogen ion concentration of 0.1 grams per liter has a pH of 1.0; a concentration of 0.01 g H^+/L has pH 2.0; 0.001 g H^+/L is pH 3.0; and so forth. Each whole number on the pH scale, which extends from 0 to 14, represents a tenfold difference in hydrogen ion concentration. As the hydrogen ion concentration increases, the pH number decreases. For example, a solution of pH 6 has ten times the hydrogen ion concentration as a solution with pH 7. Small changes in pH can reflect large changes in hydrogen ion concentration.

In pure water, which ionizes only slightly, the hydrogen ion concentration is 0.0000001 g/L, and the pH is 7.0. Water ionizes to release equal numbers of acidic hydrogen ions and basic hydroxide ions, so it is *neutral.*

$$H_2O \rightarrow H^+ + OH^-$$

TABLE 2.4 | Types of Electrolytes

	Characteristic	Examples
Acid	Substance that releases hydrogen ions (H^+)	Carbonic acid, hydrochloric acid, acetic acid, phosphoric acid
Base	Substance that releases ions that can combine with hydrogen ions	Sodium hydroxide, potassium hydroxide, magnesium hydroxide, sodium bicarbonate
Salt	Substance formed by the reaction between an acid and a base	Sodium chloride, aluminum chloride, magnesium sulfate

Many bases are present in body fluids, but because of the way bases react in water, the concentration of hydroxide ions is a good estimate of the total base concentration. The concentrations of hydrogen ions and hydroxide ions are always in balance such that if one increases, the other decreases, and vice versa. Solutions with more hydrogen ions than hydroxide ions are *acidic*. That is, acidic solutions have pH values less than 7.0 (fig. 2.10). Solutions with fewer hydrogen ions than hydroxide ions are *basic* (alkaline); they have pH values greater than 7.0.

Table 2.5 summarizes the relationship between hydrogen ion concentration and pH. Chapter 21 (p. 819) discusses the regulation of hydrogen ion concentrations in the internal environment.

Many fluids in the human body function within a narrow pH range. Illness results when pH changes. The normal pH of blood, for example, is 7.35 to 7.45. Blood pH of 7.5 to 7.8, called **alkalosis** (al″kah-lo′sis), makes one feel agitated and dizzy. This can be caused by breathing rapidly at high altitudes, taking too many antacids, high fever, anxiety, or mild to moderate vomiting that rids the body of stomach acid. **Acidosis** (as′ĭ-do′sis), in which blood pH falls to 7.0 to 7.3, makes one feel disoriented and fatigued, and breathing may become difficult. This condition can result from severe vomiting that empties the alkaline small intestinal contents, diabetes, brain damage, impaired breathing, and lung and kidney disease.

Buffers are chemicals that resist pH change. They combine with hydrogen ions when these ions are in excess, or they donate hydrogen ions when these ions are depleted. Buffers are discussed in chapter 21 (pp. 820–823).

PRACTICE

13 Describe three types of chemical reactions.

14 Compare the characteristics of an acid, a base, and a salt.

15 What does the pH scale measure?

16 What is a buffer?

TABLE 2.5 | Hydrogen Ion Concentrations and pH

Grams of H^+ per Liter	pH	
0.00000000000001	14	↑
0.0000000000001	13	
0.000000000001	12	
0.00000000001	11	Increasingly basic
0.0000000001	10	
0.000000001	9	
0.00000001	8	
0.0000001	7	Neutral—neither acidic nor basic
0.000001	6	
0.00001	5	
0.0001	4	
0.001	3	Increasingly acidic
0.01	2	
0.1	1	
1.0 0	0	↓

2.3 CHEMICAL CONSTITUENTS OF CELLS

Chemicals, including those that take part in metabolism (the cell's energy reactions), are of two general types. **Organic** (or-gan′ik) compounds have carbon and hydrogen. All other chemicals are **inorganic** (in′or-gan′ik). Many organic molecules have long chains or ring structures that can form because of a carbon atom's ability to form four covalent bonds.

Inorganic substances usually dissociate in water, forming ions; thus, they are *electrolytes*. Many organic compounds dissolve in water, but most dissolve in organic liquids such

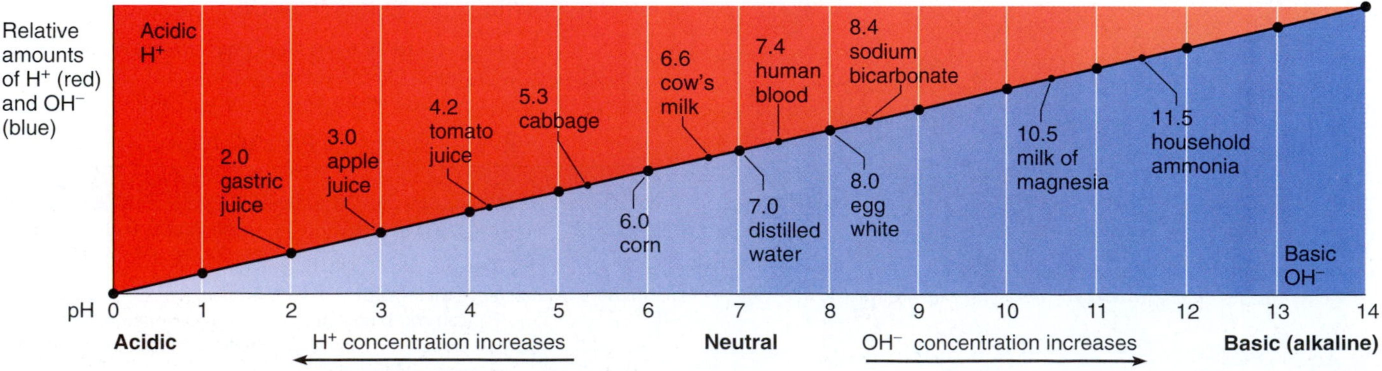

FIGURE 2.10 The pH scale reflects the hydrogen ion (H^+) concentration. As the concentration of H^+ increases, a solution becomes more acidic and the pH value decreases. As the concentration of ions that bond with H^+ (such as hydroxide ions) increases, a solution becomes more basic (alkaline) and the pH value increases. The pH values of some common substances are shown.

The word "organic" has several meanings. Technically, it refers to carbon-containing chemical compounds (except carbon dioxide), as opposed to inorganic compounds that do not contain carbon. "Organic" may also refer to a substance obtained from an organism or, even more generally, indicate a fundamental characteristic. In agriculture, "organic" refers to growing crops without the use of synthetic chemicals.

as ether or alcohol. Organic compounds that dissolve in water usually do not release ions and are therefore called *nonelectrolytes.*

Inorganic Substances

Common inorganic substances in cells include water, oxygen, carbon dioxide, and inorganic salts.

Water

Water (H_2O) is the most abundant compound in living material and accounts for about two-thirds of the weight of an adult human. It is the major component of blood and other body fluids, including fluids in cells.

When substances dissolve in water, the polar water molecules separate molecules of the substance, or even break them up into ions. These liberated particles are much more likely to react. Consequently, most metabolic reactions occur in water.

Water is also important in transporting chemicals in the body. Blood, mostly water, carries oxygen, sugars, salts, vitamins, and other vital substances from organs of the digestive and respiratory systems to cells. Blood also carries waste materials, such as carbon dioxide and urea, from cells to the lungs and kidneys, respectively, which remove them from the blood and release them outside the body.

Water absorbs and transports heat. Blood carries heat released from muscle cells during exercise from deeper parts of the body to the surface. At the same time, skin cells release water in the form of perspiration that can carry heat away by evaporation.

Oxygen

Molecules of oxygen gas (O_2) enter the internal environment through the respiratory organs and are transported throughout the body by the blood, especially by red blood cells. In cells, organelles use oxygen to release energy from nutrient molecules. The energy then drives the cell's metabolic activities. A continuing supply of oxygen is necessary for cell survival and, ultimately, for the survival of the person.

Carbon Dioxide

Carbon dioxide (CO_2) is a simple, carbon-containing inorganic compound. It is produced as a waste product when energy is released during certain metabolic reactions. As carbon dioxide moves from cells into surrounding body fluids and blood, most of it reacts with water to form a weak acid (carbonic acid, H_2CO_3). This acid ionizes, releasing hydrogen ions (H^+) and bicarbonate ions (HCO_3^-), which blood carries to the respiratory organs. There, the chemical reactions reverse, and carbon dioxide gas is produced and is eventually exhaled.

NO (nitric oxide) and CO (carbon monoxide) are two small chemicals that can harm health, yet are also important to normal physiology. NO is found in smog, cigarettes, and acid rain. CO is a colorless, odorless, gas that is deadly when it leaks from home heating systems or exhaust pipes in closed garages. However, NO and CO are important biological messenger molecules. NO is involved in digestion, memory, immunity, respiration, and circulation. CO functions in the spleen, which recycles old red blood cells, and in the parts of the brain that control memory, smell, and vital functions.

Inorganic Salts

Inorganic salts are abundant in body fluids. They are the sources of many necessary ions, including ions of sodium (Na^+), chloride (Cl^-), potassium (K^+), calcium (Ca^{+2}), magnesium (Mg^{+2}), phosphate (PO_4^{-2}), carbonate (CO_3^{-2}), bicarbonate (HCO_3^-), and sulfate (SO_4^{-2}). These ions play important roles in metabolism, helping to maintain proper water concentrations in body fluids, pH, blood clotting, bone development, energy transfer in cells, and muscle and nerve functions. The body regularly gains and loses these electrolytes, but they must be present in certain concentrations, both inside and outside cells, to maintain homeostasis. Such a condition is called **electrolyte balance.** Disrupted electrolyte balance occurs in certain diseases, and restoring it is a medical priority. Table 2.6 summarizes the functions of some of the inorganic components of cells.

PRACTICE

17 What are the general differences between an organic molecule and an inorganic molecule?

18 What is the difference between an electrolyte and a nonelectrolyte?

19 Define electrolyte balance.

Organic Substances

Important groups of organic chemicals in cells include carbohydrates, lipids, proteins, and nucleic acids.

Carbohydrates

Carbohydrates (kar′bo-hi′drātz) provide much of the energy that cells require. They also supply materials to build certain cell structures, and they often are stored as reserve energy supplies.

TABLE 2.6 | Inorganic Substances Common in Cells

Substance	Symbol or Formula	Functions
I. Inorganic Molecules		
Water	H_2O	Major component of body fluids (chapter 21, p. 811); medium in which most biochemical reactions occur; transports various chemical substances (chapter 14, p. 536); helps regulate body temperature (chapter 6, p. 182)
Oxygen	O_2	Used in release of energy from glucose molecules (chapter 4, p. 120)
Carbon dioxide	CO_2	Waste product that results from metabolism (chapter 4, p. 122); reacts with water to form carbonic acid (chapter 19, p. 766)
II. Inorganic Ions		
Bicarbonate ions	HCO_3^-	Help maintain acid-base balance (chapter 21, p. 821)
Calcium ions	Ca^{+2}	Necessary for bone development (chapter 7, p. 204); muscle contraction (chapter 9, p. 291), and blood clotting (chapter 14, fig. 14.19)
Carbonate ions	CO_3^{-2}	Component of bone tissue (chapter 7, p. 204)
Chloride ions	Cl^-	Help maintain water balance (chapter 21, p. 818)
Hydrogen ions	H^+	pH of the internal environment (chapters 19, pp. 757–758, and 766)
Magnesium ions	Mg^{+2}	Component of bone tissue (chapter 7, p. 204); required for certain metabolic processes (chapter 18, p. 720)
Phosphate ions	PO_4^{-3}	Required for synthesis of ATP, nucleic acids, and other vital substances (chapter 4, p. 125); component of bone tissue (chapter 7, p. 204); help maintain polarization of cell membranes (chapter 10, p. 366)
Potassium ions	K^+	Required for polarization of cell membranes (chapter 10, p. 366)
Sodium ions	Na^+	Required for polarization of cell membranes (chapter 10, p. 366); help maintain water balance (chapter 21, p. 818)
Sulfate ions	SO_4^{-2}	Help maintain polarization of cell membranes (chapter 10, p. 366) and acid-base balance (chapter 21, p. 818)

Carbohydrates are water-soluble molecules that include atoms of carbon, hydrogen, and oxygen. These molecules usually have twice as many hydrogen as oxygen atoms, the same ratio of hydrogen to oxygen as in water molecules (H_2O). This ratio is easy to see in the molecular formulas of the carbohydrates glucose ($C_6H_{12}O_6$) and sucrose ($C_{12}H_{22}O_{11}$).

Carbohydrates are classified by size. Simple carbohydrates, or **sugars,** include the **monosaccharides** (mon″o-sak′ah-rīdz) (single sugars) and **disaccharides** (di-sak′ah-rīdz) (double sugars). A monosaccharide may include from three to seven carbon atoms, in a straight chain or a ring (fig. 2.11). Monosaccharides include the five-carbon sugars ribose and deoxyribose, as well as the six-carbon sugars glucose, dextrose (a form of glucose), fructose, and galactose (fig. 2.12*a*). Disaccharides consist of two 6-carbon units (fig. 2.12*b*). Sucrose (table sugar) and lactose (milk sugar) are disaccharides.

Complex carbohydrates, also called **polysaccharides** (pol″e-sak′ah-rīdz), are built of simple carbohydrates (fig. 2.12*c*). Cellulose is a polysaccharide abundant in plants. It is made of many bonded glucose molecules. Humans cannot digest cellulose. It is considered to be dietary fiber, passing through the gastrointestinal tract without being broken down and absorbed into the bloodstream. Plant starch is another type of polysaccharide. Starch molecules consist of highly branched chains of glucose molecules connected differently than in cellulose. Humans easily digest starch.

Animals, including humans, synthesize a polysaccharide similar to starch called *glycogen,* stored in the liver and skeletal muscles. Its molecules also are branched chains of sugar units; each branch consists of up to a dozen glucose units.

Lipids

Lipids (lip′idz) are a group of organic chemicals that are insoluble in water but soluble in organic solvents, such as ether and chloroform. Lipids include a number of compounds, such as fats, phospholipids, and steroids, that have vital functions in cells and are important constituents of cell membranes (see chapter 3, p. 79). The most common lipids are the *fats,* primarily used to supply energy for cellular activities. Fat molecules can supply more energy gram for gram than can carbohydrate molecules.

Like carbohydrates, fat molecules are composed of carbon, hydrogen, and oxygen atoms. However, fats have a much smaller proportion of oxygen than do carbohydrates. The formula for the fat *tristearin,* $C_{57}H_{110}O_6$, illustrates these characteristic proportions.

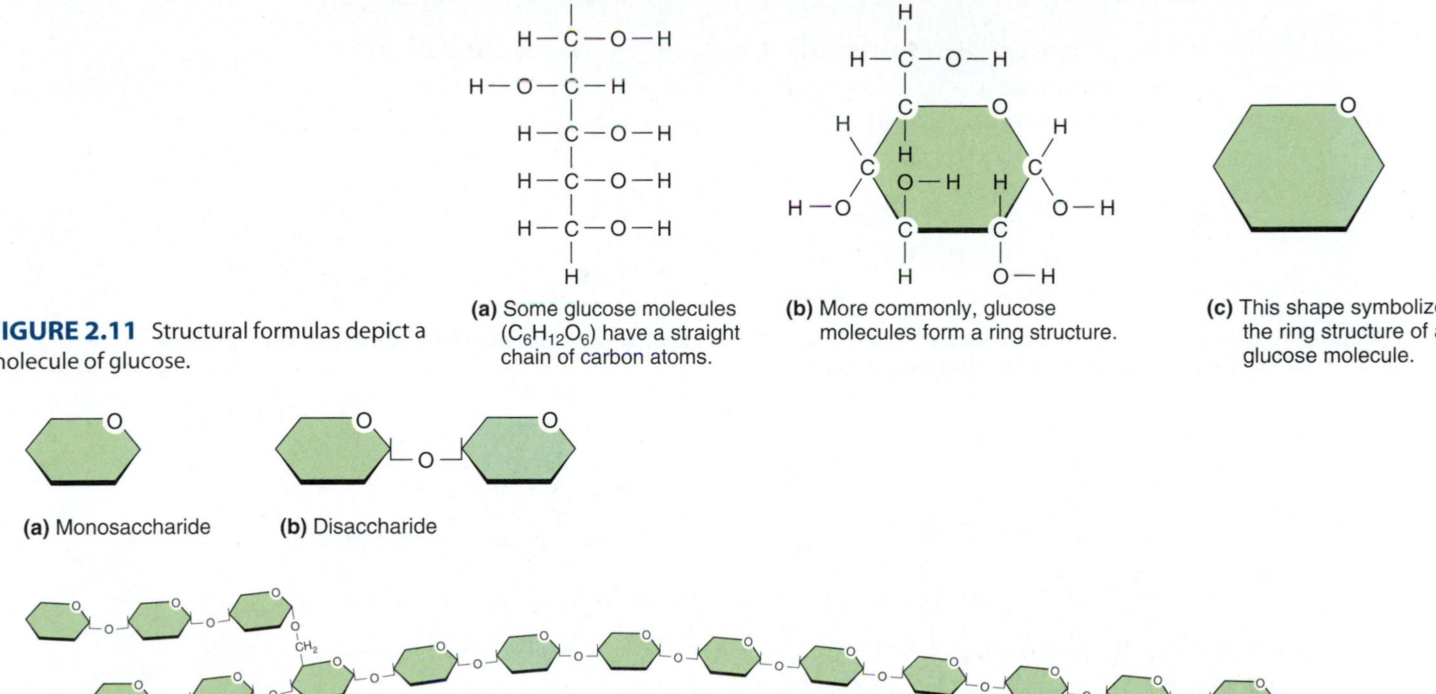

(a) Some glucose molecules ($C_6H_{12}O_6$) have a straight chain of carbon atoms.

(b) More commonly, glucose molecules form a ring structure.

(c) This shape symbolizes the ring structure of a glucose molecule.

FIGURE 2.11 Structural formulas depict a molecule of glucose.

(a) Monosaccharide

(b) Disaccharide

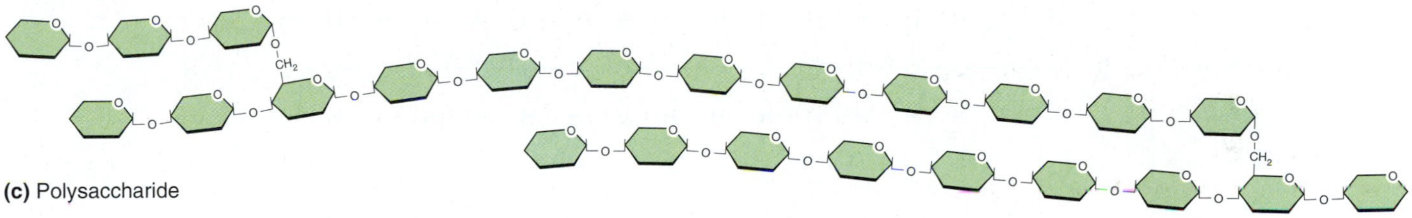

(c) Polysaccharide

FIGURE 2.12 Carbohydrate molecules vary in size. (*a*) A monosaccharide molecule consists of one 6-carbon atom building block. (*b*) A disaccharide molecule consists of two of these building blocks. (*c*) A polysaccharide molecule consists of many building blocks.

The building blocks of fat molecules are **fatty acids** (fat′e as′idz) and **glycerol** (glis′er-ol). Although the glycerol portion of every fat molecule is the same, there are many types of fatty acids and, therefore, many types of fats. All fatty acid molecules include a carboxyl group (—COOH) at the end of a chain of carbon atoms. Fatty acids differ in the lengths of their carbon atom chains, which usually have an even number of carbon atoms. The fatty acid chains also may vary in the ways the carbon atoms join. In some cases, single carbon-carbon bonds link all the carbon atoms. This type of fatty acid is called a **saturated fatty acid;** that is, each carbon atom binds as many hydrogen atoms as possible and is thus saturated with hydrogen atoms. Other fatty acid chains, **unsaturated fatty acids,** have one or more double bonds between carbon atoms. Fatty acids with one double bond are called *monounsaturated fatty acids,* and those with two or more double bonds are *polyunsaturated fatty acids* (fig. 2.13).

A glycerol molecule combines with three fatty acid molecules to form a single fat molecule, or *triglyceride* (fig. 2.14). The fatty acids of a triglyceride may have different lengths and degrees of saturation making the fats very diverse. Fat molecules that have only saturated fatty acids are called **saturated fats** (sat′u-rāt″ed fatz), and those that have unsaturated fatty acids are called **unsaturated fats** (unsat′u-rāted fatz). Each type of fat molecule has distinct properties.

A diet rich in saturated fat increases risk of atherosclerosis, which obstructs blood vessels. The risk is even greater if the diet is also high in refined carbohydrates, such as white flour and rice, because these raise triglyceride levels. Unsaturated, particularly monounsaturated, fats are healthier to eat than saturated fats. Monounsaturated fats include olive, canola, and macadamia nut oils.

Most saturated fats are solids at room temperature, such as butter, lard, and most other animal fats. Most unsaturated fats are liquids at room temperature, such as corn, sesame, peanut, sunflower, and soybean oils. Coconut and palm kernel oils are exceptions—they are relatively high in saturated fat.

A food-processing technique called hydrogenation adds hydrogens to an unsaturated fat, making it more solid and therefore useful in prepared foods. Margarine is an example. However, hydrogenation is an imperfect process. Some of the double bonds are changed to single bonds when hydrogens are forced onto the molecule, but some are not. Instead, the two hydrogens bonded to the two carbons that share a partially hydrogenated bond assume a "trans" configuration—that is, emanating in opposite directions from the carbons with respect to each other. (In the natural "cis" configuration, the two hydrogens lie on the same side of the carbon backbone.) Trans fats raise the risk of heart disease.

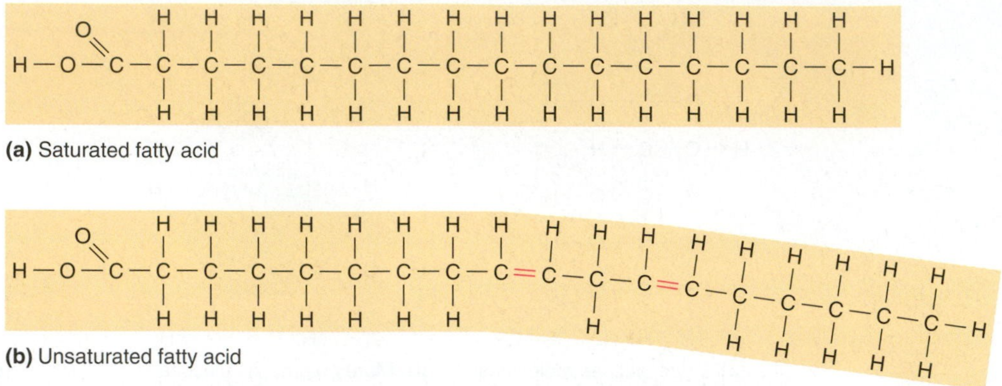

(a) Saturated fatty acid

(b) Unsaturated fatty acid

FIGURE 2.13 Fatty acids. (*a*) A molecule of saturated fatty acid and (*b*) a molecule of unsaturated fatty acid. Double bonds between carbon atoms are shown in red. They cause a "kink" in the shape of the molecule.

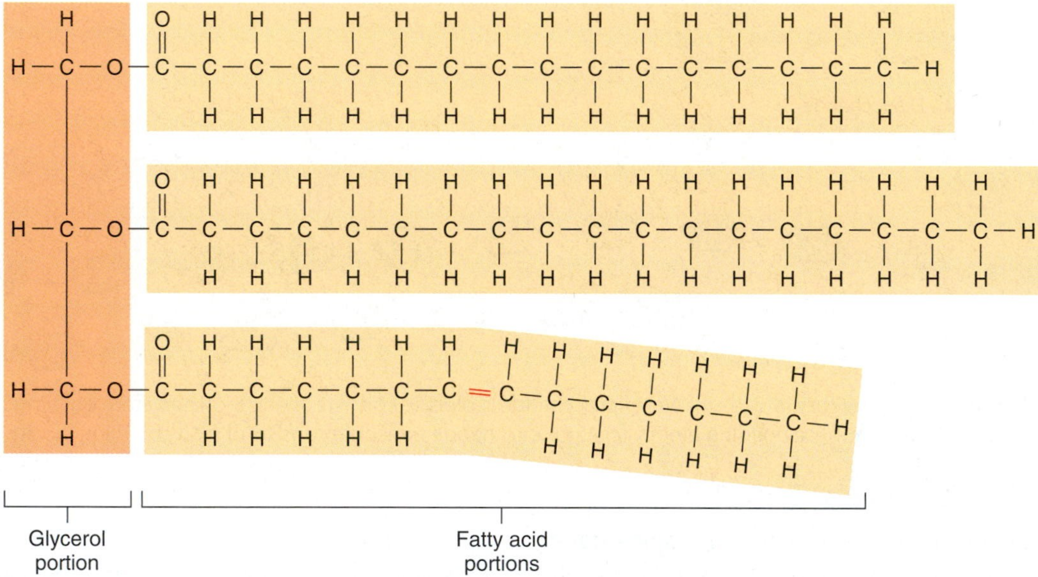

FIGURE 2.14 A triglyceride molecule (fat) consists of a glycerol portion and three fatty acid portions. This is an example of an unsaturated fat. The double bond between carbon atoms in the unsaturated fatty acid is shown in red.

A *phospholipid* molecule is similar to a fat molecule in that it includes a glycerol and fatty acid chains. The phospholipid, however, has only two fatty acid chains and, in place of the third, has a portion containing a phosphate group. This phosphate-containing part is soluble in water (hydrophilic) and forms the "head" of the molecule, whereas the fatty acid portion is insoluble in water (hydrophobic) and forms a "tail" (fig. 2.15).

Steroid molecules are complex structures that include connected rings of carbon atoms (fig. 2.16). Among the more important steroids are cholesterol, in all body cells and used to synthesize other steroids; sex hormones, such as estrogen, progesterone, and testosterone; and several hormones from the adrenal glands. Chapters 13, 18, 20, 21, and 22 discuss these steroids. Table 2.7 summarizes the molecular structures and characteristics of lipids.

Proteins

Proteins (pro′te-inz) have a great variety of functions. The human body has more than 200,000 types of proteins. They are structural materials, energy sources, and chemical messengers (hormones). Other proteins combine with carbohydrates (glycoproteins) and function as receptors on cell surfaces, allowing cells to respond to particular types of molecules that bind to them. Antibody proteins recognize and destroy substances foreign to the body, such as certain molecules on the surfaces of infecting bacteria. Proteins such as hemoglobin and myoglobin transport oxygen in the blood and muscles, respectively, and actin and myosin are contractile proteins that provide muscle action. Many proteins play vital roles in metabolism as **enzymes** (en′zīmz), catalysts in living systems. That is, they speed specific chemical reactions without being consumed. (Enzymes are discussed in chapter 4, p. 117.)

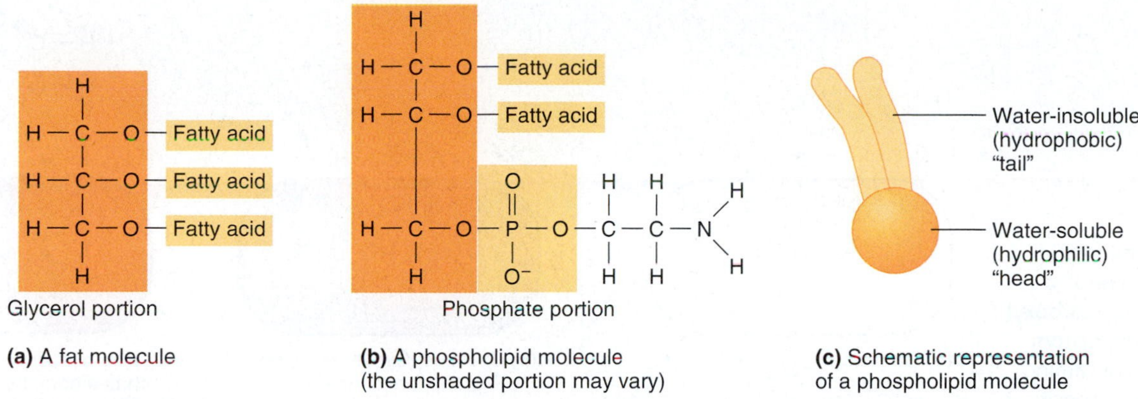

(a) A fat molecule

(b) A phospholipid molecule (the unshaded portion may vary)

(c) Schematic representation of a phospholipid molecule

FIGURE 2.15 Fats and phospholipids. (*a*) A fat molecule (triglyceride) consists of a glycerol and three fatty acids. (*b*) In a phospholipid molecule, a phosphate-containing group replaces one fatty acid. (*c*) Schematic representation of a phospholipid.

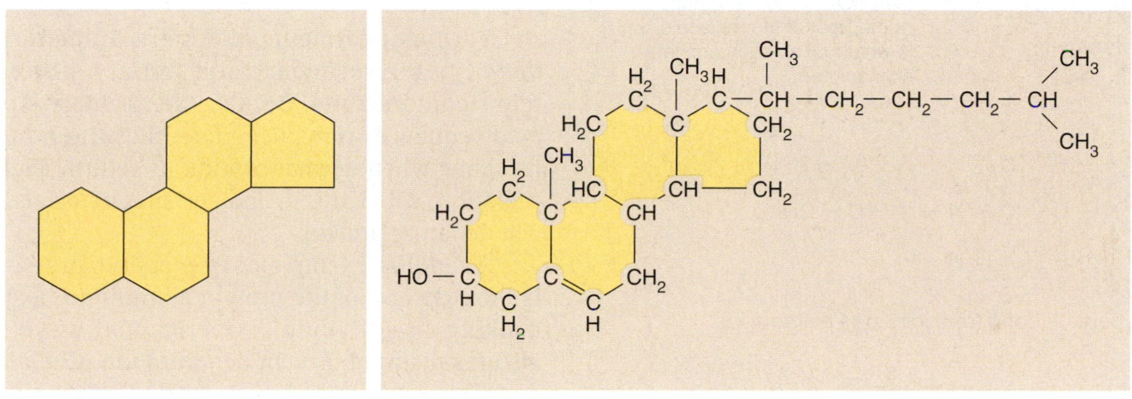

(a) General structure of a steroid

(b) Cholesterol

FIGURE 2.16 Steroid structure. (*a*) The general structure of a steroid. (*b*) The structural formula for cholesterol, a steroid widely distributed in the body and a component of cell membranes.

TABLE 2.7 | Important Groups of Lipids

Group	Basic Molecular Structure	Characteristics
Triglycerides	Three fatty acid molecules bound to a glycerol molecule	Most common lipid in the body; stored in fat tissue as an energy supply; fat tissue also provides insulation beneath the skin
Phospholipids	Two fatty acid molecules and a phosphate group bound to a glycerol molecule (may also include a nitrogen-containing molecule attached to the phosphate group)	Used as structural components in cell membranes; large amounts are in the liver and parts of the nervous system
Steroids	Four connected rings of carbon atoms	Widely distributed in the body with a variety of functions; includes cholesterol, sex hormones, and certain hormones of the adrenal glands

Like carbohydrates and lipids, proteins consist of atoms of carbon, hydrogen, and oxygen. In addition, proteins always include nitrogen atoms and sometimes sulfur atoms. The building blocks of proteins are **amino acids** (ah-me′no as′idz).

Twenty types of amino acids comprise proteins in organisms. Amino acid molecules have an amino group (—NH_2) at one end and a carboxyl group (—COOH) at the other end. Between these groups is a single carbon atom. This central carbon is bonded to a hydrogen atom and to another group of atoms called a *side chain* or *R group* ("R" may be thought of as the "Rest of the molecule"). The composition of the R group distinguishes one type of amino acid from another (fig. 2.17).

Proteins have complex three-dimensional shapes, called **conformations,** yet they are assembled from simple chains of amino acids connected by peptide bonds. These are covalent bonds that link the amino end of one amino acid with the carboxyl end of another. Figure 2.18 shows two amino acids connected by a peptide bond. The resulting molecule is a dipeptide. Adding a third amino acid creates a tripeptide. Many amino acids connected in this way constitute a polypeptide (fig. 2.19*a*).

Proteins have four levels of structure: primary, secondary, tertiary and quaternary. The *primary structure* is the amino acid sequence of the polypeptide chain. The primary

FIGURE 2.17 Amino acid structure. (*a*) An amino acid has an amino group, a carboxyl group, and a hydrogen atom that are common to all amino acid molecules, and a specific R group. (*b*) Some representative amino acids and their structural formulas. Each type of amino acid molecule has a particular shape due to its different R group.

(a) General structure of an amino acid. The portion common to all amino acids is within the oval. It includes the amino group (—NH_2) and the carboxyl group (—COOH). The "R" group, or the "rest of the molecule," is what makes each amino acid unique.

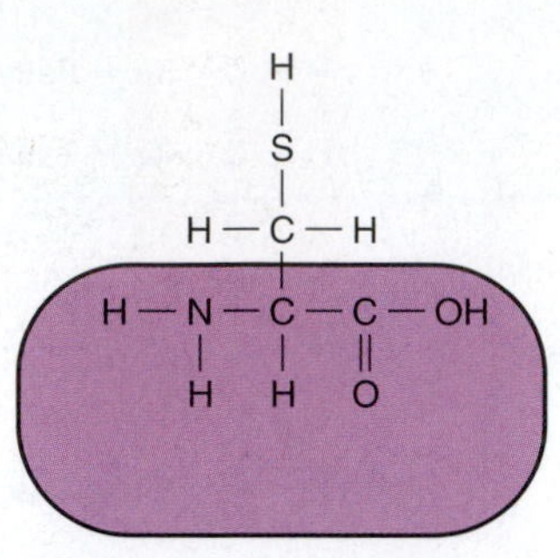

(b) Cysteine. Cysteine has an R group that contains sulfur.

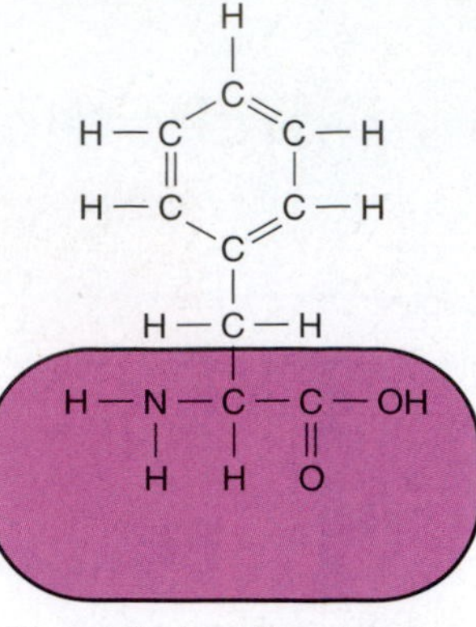

Phenylalanine. Phenylalanine has a complex R group. Improper metabolism of phenylalanine occurs in the disease *phenylketonuria*.

FIGURE 2.18 A peptide bond (red) joins two amino acids.

structure may range from fewer than 100 to more than 5,000 amino acids. The amino acid sequence is characteristic of a particular protein. Hemoglobin, actin, and an antibody protein have very different amino acid sequences.

In the *secondary structure* (fig. 2.19*b*), the polypeptide chain either forms a springlike coil (alpha helix) or folds back and forth on itself (beta-pleated sheet) or folds into other shapes. Secondary structure arises from hydrogen bonding. Recall that polar molecules result when electrons are not shared equally in certain covalent bonds. In amino acids, this results in slightly negative oxygen and nitrogen atoms and slightly positive hydrogen atoms. Hydrogen bonding between oxygen and hydrogen atoms in different parts of the molecule determines the secondary structure. A single polypeptide may include helices; sheets; and other localized shapes, called motifs.

Hydrogen bonding and even covalent bonding between atoms in different parts of a polypeptide can impart another, larger level of folding, the *tertiary structure.* Altogether, the primary, secondary, and tertiary structures contribute to a protein's distinct conformation (fig. 2.19*c*), which determines its function. Some proteins are long and fibrous, such as the keratins that form hair and the threads of fibrin that knit a blood clot. Myoglobin and hemoglobin are globular, as are many enzymes.

In many cases, slight, reversible changes in conformation may be part of the protein's normal function. For example, some of the proteins involved in muscle contraction exert a pulling force as a result of such a shape change, leading to movement. Such changes in shape are reversible, enabling the protein to function repeatedly.

Various treatments can more dramatically change or *denature* the secondary and tertiary structures of a protein's conformation. Because the primary structure (amino acid sequence) remains, sometimes the protein can regain its shape when normal conditions return. High temperature, radiation, pH changes, and certain chemicals (such as urea) can denature proteins.

A familiar example of irreversible protein denaturation is the response of the protein albumin to heat (for example, cooking an egg white). A permanent wave that curls hair also results from protein denaturation. Chemicals first break apart the tertiary structure formed when sulfur-containing amino acids attract each other within keratin molecules. This relaxes the hair. When the chemicals are washed out of the set hair, the sulfur bonds reform, but in different places. The appearance of the hair changes.

Not all proteins are single polypeptide chains. In some proteins, several polypeptide chains are connected in a fourth level, or *quaternary structure,* to form a very large structure (fig. 2.19*d*). Hemoglobin is a quaternary protein made up of four separate polypeptide chains.

A protein's conformation determines its function. The amino acid sequence and interactions among the amino acids in a protein determine the conformation. Thus, it is the amino acid sequence of a protein that determines its function in the body.

Protein misfolding can cause disease. In cystic fibrosis, for example, a protein cannot fold into its final form. It cannot anchor in the cell membrane, where it would normally control the flow of chloride ions. Certain body fluids dry up, which impairs respiration and digestion. A class of illnesses called transmissible spongiform encephalopathies, which includes "mad cow disease," results when a type of protein called a prion folds into an abnormal, infectious form—that is, it converts normal prion protein into the pathological form, which riddles the brain with holes. Alzheimer disease results from the cutting of a protein called beta amyloid into pieces of a certain size. The proteins misfold, attach, and accumulate, forming structures called plaques in parts of the brain controlling memory and cognition.

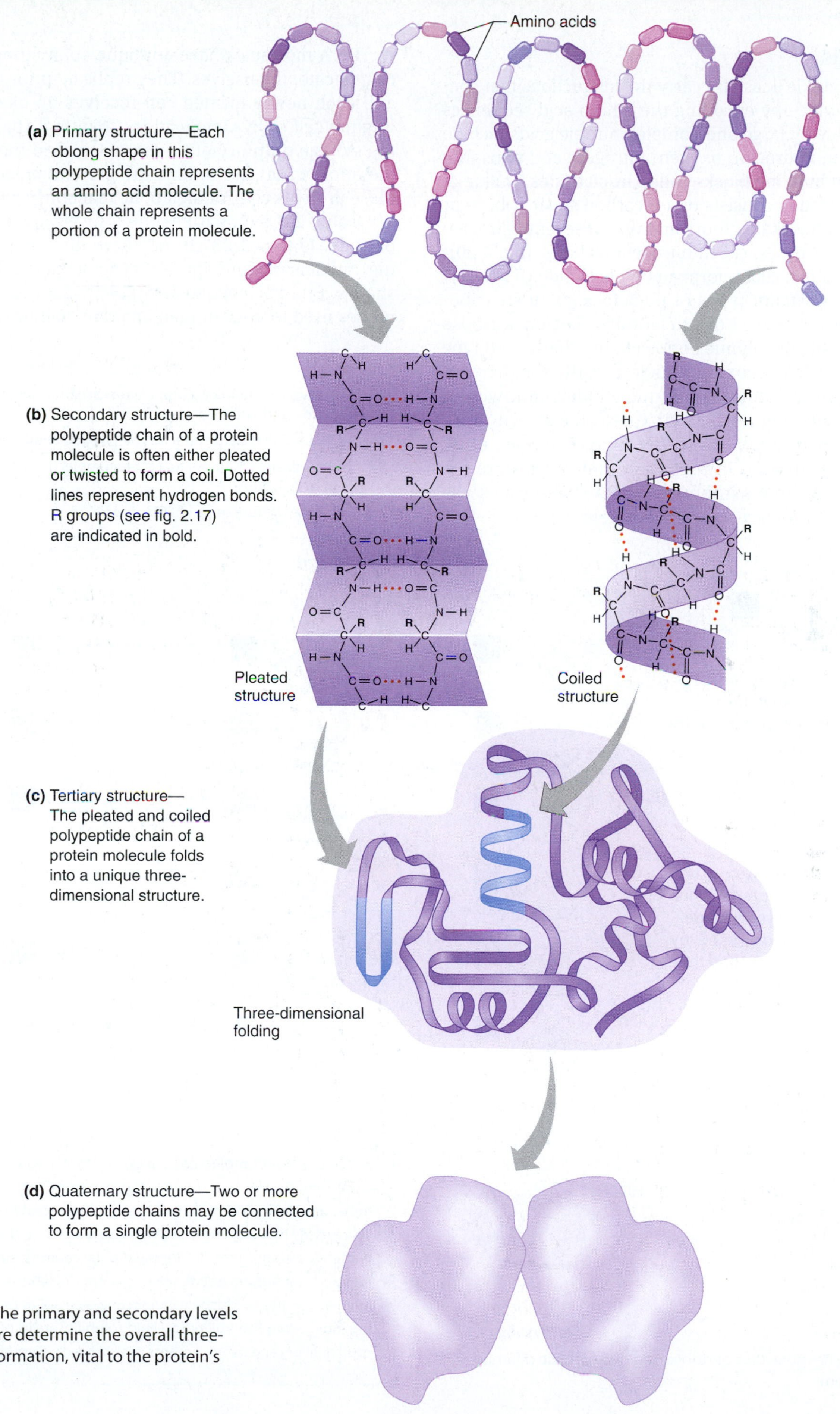

Amino acids
(a) Primary structure—Each oblong shape in this polypeptide chain represents an amino acid molecule. The whole chain represents a portion of a protein molecule.
(b) Secondary structure—The polypeptide chain of a protein molecule is often either pleated or twisted to form a coil. Dotted lines represent hydrogen bonds. R groups (see fig. 2.17) are indicated in bold.
Pleated structure
Coiled structure
(c) Tertiary structure—The pleated and coiled polypeptide chain of a protein molecule folds into a unique three-dimensional structure.
Three-dimensional folding
(d) Quaternary structure—Two or more polypeptide chains may be connected to form a single protein molecule.

FIGURE 2.19 The primary and secondary levels of protein structure determine the overall three-dimensional conformation, vital to the protein's function.

Nucleic Acids

Nucleic acids (nu-kle′ik as′idz) carry the instructions that control a cell's activities by encoding the amino acid sequences of proteins. The very large and complex nucleic acids include atoms of carbon, hydrogen, oxygen, nitrogen, and phosphorus, which form building blocks called **nucleotides** (nu′kle-o-tīdz). Each nucleotide consists of a 5-carbon sugar (ribose or deoxyribose), a phosphate group, and one of several nitrogen-containing organic bases, called nitrogenous bases (fig. 2.20). Such nucleotides, in a chain, form a polynucleotide (fig. 2.21).

There are two major types of nucleic acids. **RNA** (ribonucleic acid) is composed of nucleotides that have ribose sugar. RNA is a single polynucleotide chain. The second type of nucleic acid, **DNA** (deoxyribonucleic acid), has deoxyribose sugar. DNA is a double polynucleotide chain wound into a double helix. Figure 2.22 compares the structures of ribose and deoxyribose, which differ by one oxygen atom. DNA and RNA also differ in that DNA molecules store the information for protein synthesis and RNA molecules use this information to construct specific protein molecules.

DNA molecules have a unique ability to make copies of, or replicate, themselves. They replicate prior to cell division, and each newly formed cell receives an exact copy of the original cell's DNA molecules. Chapter 4 (p. 130) discusses the storage of information in nucleic acid molecules, use of the information to manufacture protein molecules, and how these proteins control metabolic reactions.

Table 2.8 summarizes the four groups of organic compounds. Figure 2.23 shows three-dimensional (space-filling) models of some important molecules, illustrating their shapes. From Science to Technology 2.3 describes two techniques used to view human anatomy and physiology.

PRACTICE 

20 Compare the chemical composition of carbohydrates, lipids, proteins, and nucleic acids.

21 How does an enzyme affect a chemical reaction?

22 What is likely to happen to a protein molecule exposed to intense heat or radiation?

23 What are the functions of DNA and RNA?

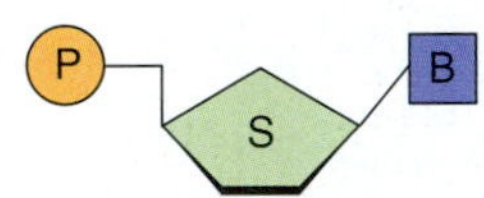

FIGURE 2.20 A nucleotide consists of a 5-carbon sugar (S = sugar), a phosphate group (P = phosphate), and a nitrogenous base (B = base).

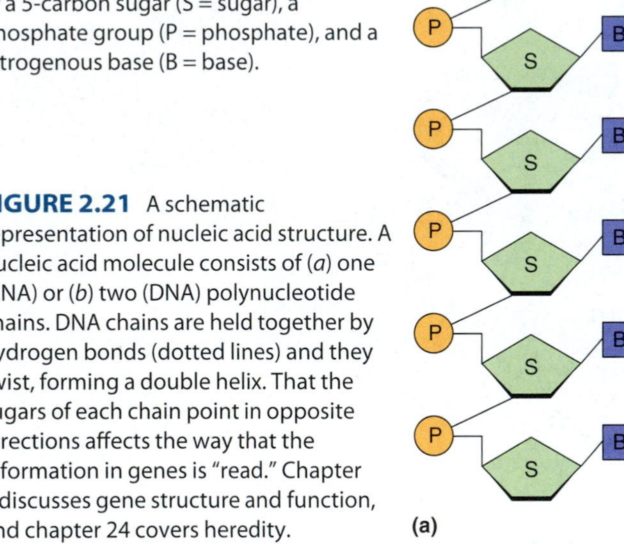

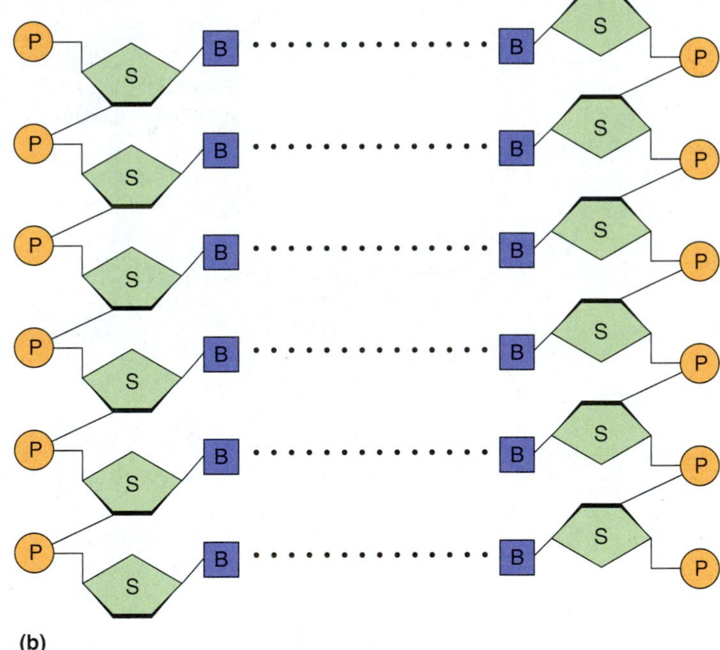

FIGURE 2.21 A schematic representation of nucleic acid structure. A nucleic acid molecule consists of (*a*) one (RNA) or (*b*) two (DNA) polynucleotide chains. DNA chains are held together by hydrogen bonds (dotted lines) and they twist, forming a double helix. That the sugars of each chain point in opposite directions affects the way that the information in genes is "read." Chapter 4 discusses gene structure and function, and chapter 24 covers heredity.

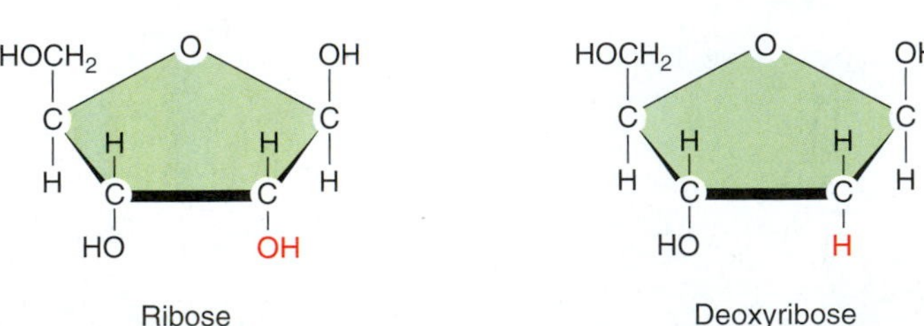

FIGURE 2.22 The molecules of ribose and deoxyribose differ by a single oxygen atom.

> Recall that water molecules are polar. Many larger molecules have polar regions where nitrogen or oxygen bond with hydrogen. Such molecules, including carbohydrates, proteins, and nucleic acids, dissolve easily in water. They are water soluble, or hydrophilic ("liking" water). Molecules that do not have polar regions, such as triglycerides and steroids, do not dissolve in water ("oil and water don't mix"). Such molecules do dissolve in lipid and are said to be lipophilic ("liking" lipid). Water solubility and lipid solubility are important factors in drug delivery and in movements of substances throughout the body.

TABLE 2.8 | Organic Compounds in Cells

Compound	Elements Present	Building Blocks	Functions	Examples
Carbohydrates	C,H,O	Simple sugar	Provide energy, cell structure	Glucose, starch
Lipids	C,H,O (often P)	Glycerol, fatty acids, phosphate groups	Provide energy, cell structure	Triglycerides, phospholipids, steroids
Proteins	C,H,O,N (often S)	Amino acids	Provide cell structure, enzymes, energy	Albumins, hemoglobin
Nucleic acids	C,H,O,N,P	Nucleotides	Store information for the synthesis of proteins, control cell activities	RNA, DNA

(a)

(b)

(c)

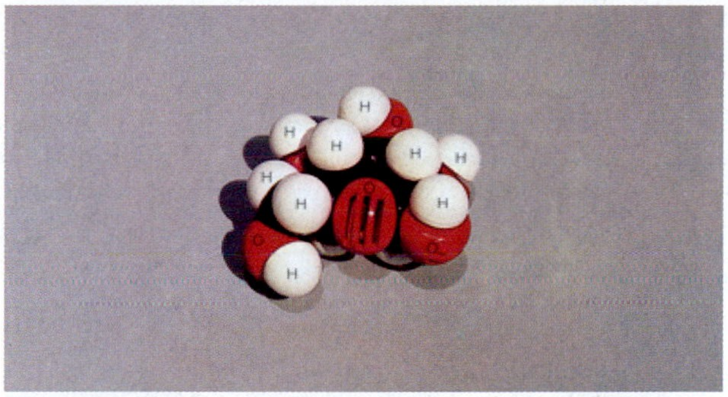

(d)

(e)

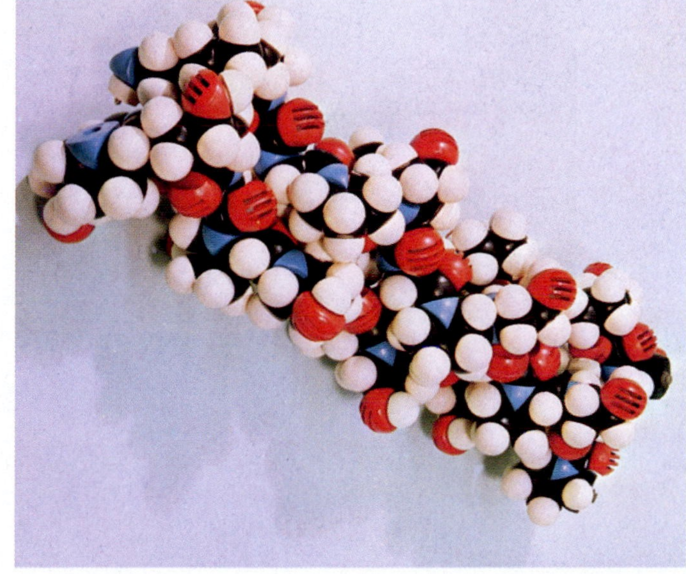
(f)

FIGURE 2.23 These three-dimensional (space-filling) models show the relative sizes of several important molecules: (*a*) water, (*b*) carbon dioxide, (*c*) glycine (an amino acid), (*d*) glucose (a monosaccharide), (*e*) a fatty acid, and (*f*) collagen (a protein). White = hydrogen, red = oxygen, blue = nitrogen, black = carbon.

CT Scanning and PET Imaging

Physicians use two techniques—computerized tomography (CT) scanning and positron emission tomography (PET) imaging—to paint portraits of anatomy and physiology, respectively.

In CT scanning, an X-ray-emitting device is positioned around the region of the body being examined. At the same time, an X-ray detector is moved in the opposite direction on the other side of the body. As these parts move, an X-ray beam passes through the body from hundreds of different angles. Tissues and organs of varying composition absorb X rays differently, so the intensity of X rays reaching the detector varies from position to position. A computer records the measurements made by the X-ray detector and combines them mathematically. This creates on a viewing screen a sectional image of the internal body parts (fig. 2D).

Ordinary X-ray techniques produce two-dimensional images known as radiographs, X rays, or films. A CT scan provides three-dimensional information. The CT scan can also clearly differentiate between soft tissues of slightly different densities, such as the liver and kidneys, which cannot be seen in a conventional X-ray image. In this way, a CT scan can detect abnormal tissue, such as a tumor. For example, a CT scan can tell whether a sinus headache that does not respond to antibiotic therapy is caused by a drug-resistant infection or by a tumor.

PET imaging uses radioactive isotopes that naturally emit positrons, atypical positively charged electrons, to detect biochemical activity in a specific body part. Useful isotopes in PET imaging include carbon-11, nitrogen-13, oxygen-15, and fluorine-18. When one of these isotopes releases a positron, it interacts with a nearby negatively charged electron. The two particles destroy each other in an event called annihilation. At the moment of destruction, two gamma rays form and move apart. Special equipment detects the gamma radiation.

To produce a PET image of biochemically active tissue, a person is injected with a metabolically active compound that includes a bound positron-emitting isotope. To study the brain, for example, a person is injected with glucose containing fluorine-18. After the brain takes up the isotope-tagged compound, the person rests the head within a circular array of radiation detectors. A device records each time two gamma rays are emitted simultaneously and travel in opposite directions (the result of annihilation). A computer collects and combines the data and generates a cross-sectional image. The image indicates the location and relative concentration of the radioactive isotope in different regions of the brain and can be used to study those parts metabolizing glucose.

PET images reveal the parts of the brain affected in such disorders as Huntington disease, Parkinson disease, epilepsy, and Alzheimer disease, and they are used to study blood flow in vessels supplying the brain and heart. The technology is invaluable for detecting the physiological bases of poorly understood behavioral disorders, such as obsessive-compulsive disorder. In this condition, a person repeatedly performs a certain behavior, such as washing hands, showering, locking doors, or checking to see that the stove is turned off. PET images of people with this disorder reveal intense activity in two parts

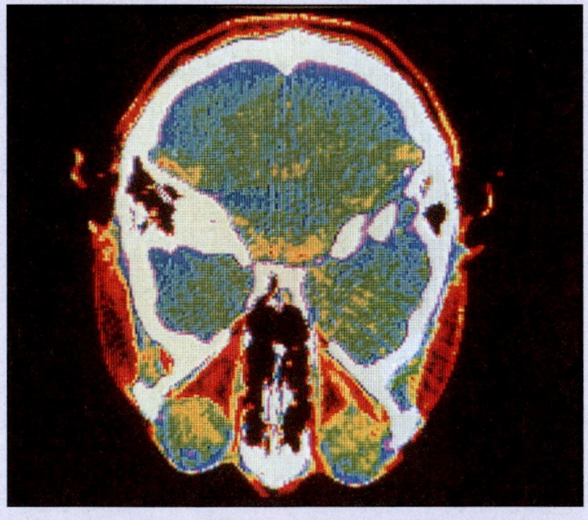

(a)

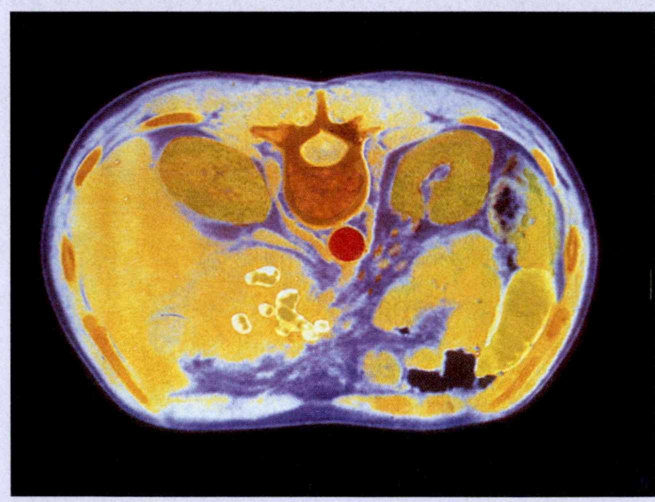

(b)

FIGURE 2D CT scans of (*a*) the head and (*b*) the abdomen.

of the brain that are quiet in the brains of unaffected individuals. Knowing the site of altered brain activity can help researchers develop more directed drug therapy.

In addition to highlighting biochemical activities behind illness, PET scans allow biologists to track normal brain physiology. Figure 2E shows that different patterns of brain activity are associated with learning and with reviewing something already learned.

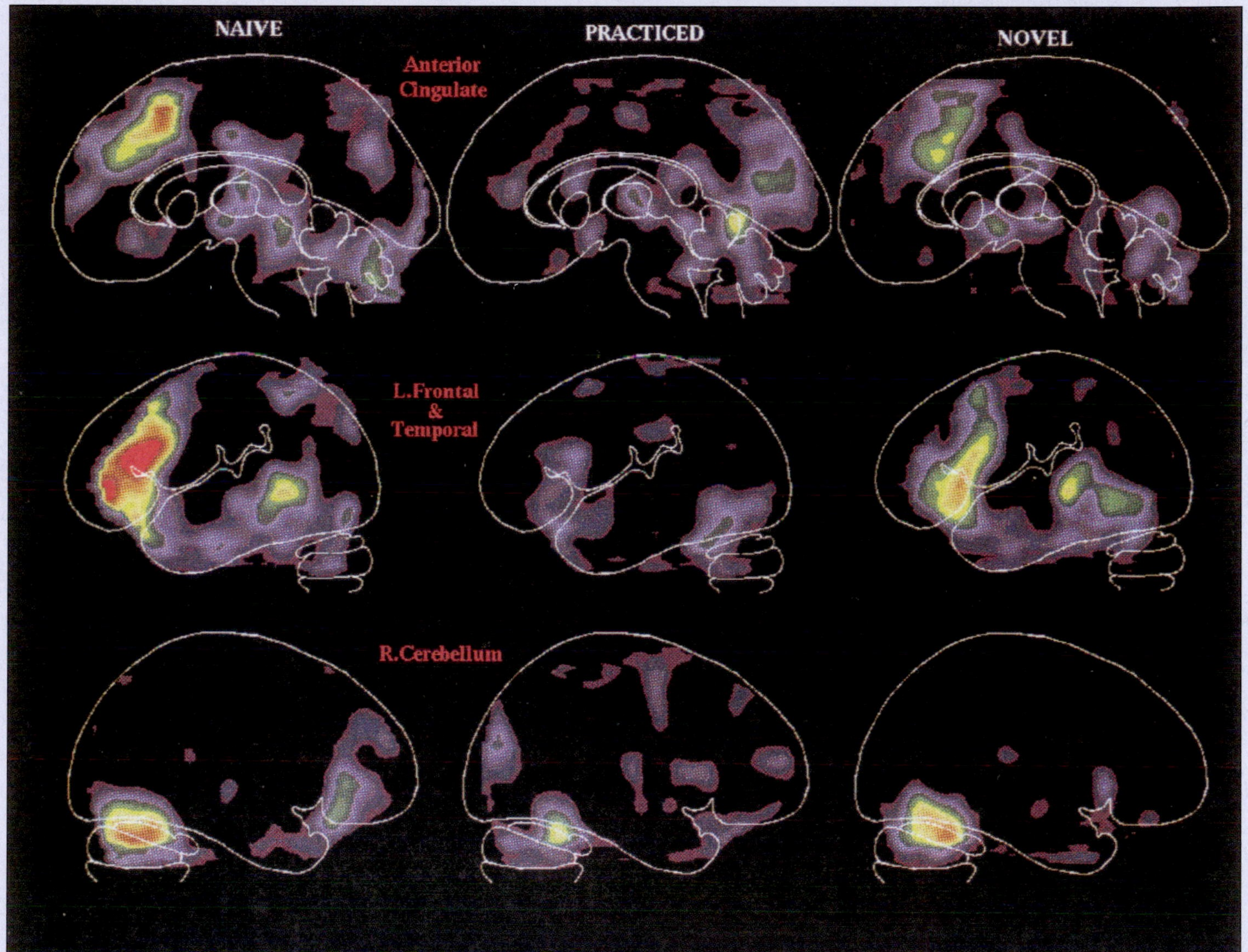

FIGURE 2E These PET images demonstrate brain changes that accompany learning. The top and bottom views show different parts of the same brain. The "naive" brain on the left has been given a list of nouns and asked to visualize each word. In the middle column, the person has practiced the task, so he can picture the nouns with less brain activity. In the third column, the person receives a new list of nouns. Learning centers in the brain show increased activity.

CHAPTER SUMMARY

2.1 INTRODUCTION (PAGE 51)

Chemistry deals with the composition of substances and changes in their composition. The human body is composed of chemicals. Biochemistry is the chemistry of living organisms.

2.2 STRUCTURE OF MATTER (PAGE 51)

Matter is anything that has weight and takes up space.

1. Elements and atoms
   a. Naturally occurring matter on earth is composed of ninety-two elements.
   b. Elements occur most frequently in chemical combinations called compounds.
   c. Elements are composed of atoms.
   d. Atoms of different elements vary in size, weight, and ways of interacting.
2. Atomic structure
   a. An atom consists of electrons surrounding a nucleus, which has protons and neutrons. The exception is hydrogen, which has only a proton in its nucleus.
   b. Electrons are negatively charged, protons positively charged, and neutrons uncharged.
   c. A complete atom is electrically neutral.
   d. The atomic number of an element is equal to the number of protons in each atom; the atomic weight is equal to the number of protons plus the number of neutrons in each atom.
3. Isotopes
   a. Isotopes are atoms with the same atomic number but different atomic weights (due to differing numbers of neutrons).
   b. All the isotopes of an element react chemically in the same manner.
   c. Some isotopes are radioactive and release atomic radiation.
4. Molecules and compounds
   a. Two or more atoms may combine to form a molecule.
   b. A molecular formula represents the numbers and types of atoms in a molecule.
   c. If atoms of the same element combine, they produce molecules of that element.
   d. If atoms of different elements combine, they form molecules of substances called compounds.
5. Bonding of atoms
   a. When atoms combine, they gain, lose, or share electrons.
   b. Electrons occupy space in areas called electron shells that encircle an atomic nucleus.
   c. Atoms with completely filled outer shells are inactive, whereas atoms with incompletely filled outer shells gain, lose, or share electrons and thus achieve stable structures.
   d. Atoms that lose electrons become positively charged; atoms that gain electrons become negatively charged.
   e. Ions with opposite charges attract and join by ionic bonds; atoms that share electrons join by covalent bonds.
   f. A structural formula represents the arrangement of atoms within a molecule.
   g. Polar molecules result from an unequal sharing of electrons.
   h. Hydrogen bonds occur between polar molecules.
6. Chemical reactions
   a. In a chemical reaction, bonds between atoms, ions, or molecules break or form.
   b. Three types of chemical reactions are synthesis, in which larger molecules form from smaller particles; decomposition, in which smaller particles form from breakdown of larger molecules; and exchange reactions, in which parts of two different molecules trade positions.
   c. Many reactions are reversible. The direction of a reaction depends upon the proportion of reactants and products, the energy available, and the presence or absence of catalysts.
7. Acids, bases, and salts
   a. Compounds that ionize when they dissolve in water are electrolytes.
   b. Electrolytes that release hydrogen ions are acids, and those that release hydroxide or other ions that react with hydrogen ions are bases.
   c. Acids and bases react to form water and electrolytes called salts.
8. Acid and base concentrations
   a. pH represents the concentration of hydrogen ions (H^+) and hydroxide ions (OH^-) in a solution.
   b. A solution with equal numbers of H^+ and OH^- is neutral and has a pH of 7.0; a solution with more H^+ than OH^- is acidic (pH less than 7.0); a solution with fewer H^+ than OH^- is basic (pH greater than 7.0).
   c. A tenfold difference in hydrogen ion concentration separates each whole number in the pH scale.
   d. Buffers are chemicals that resist pH change.

2.3 CHEMICAL CONSTITUENTS OF CELLS (PAGE 60)

Molecules containing carbon and hydrogen atoms are organic and are usually nonelectrolytes; other molecules are inorganic and are usually electrolytes.

1. Inorganic substances
   a. Water is the most abundant compound in cells. Many chemical reactions take place in water. Water transports chemicals and heat and helps release excess body heat.
   b. Oxygen releases energy needed for metabolic activities from glucose and other molecules.
   c. Carbon dioxide is produced when energy is released during metabolic processes.
   d. Inorganic salts provide ions needed in a variety of metabolic processes.
   e. Electrolytes must be present in certain concentrations inside and outside of cells.

2. Organic substances
   a. Carbohydrates provide much of the energy cells require; their building blocks are simple sugar molecules.
   b. Lipids, such as fats, phospholipids, and steroids, supply energy and are used to build cell parts; their building blocks are molecules of glycerol and fatty acids.
   c. Proteins serve as structural materials, energy sources, hormones, cell surface receptors, antibodies, and enzymes that initiate or speed chemical reactions without being consumed.
      (1) The building blocks of proteins are amino acids.
      (2) Proteins vary in the numbers and types of their constituent amino acids; the sequences of these amino acids; and their three-dimensional structures, or conformations.
      (3) The amino acid sequence determines the protein's conformation.
      (4) The protein's conformation determines its function.
      (5) Exposure to excessive heat, radiation, electricity, or certain chemicals can denature proteins.
   d. Nucleic acids constitute genes, the instructions that control cell activities, and direct protein synthesis.
      (1) The two types are RNA and DNA.
      (2) Nucleic acid molecules are composed of building blocks called nucleotides.
      (3) DNA molecules store information used by cell parts to construct specific types of protein molecules.
      (4) RNA molecules help synthesize proteins.
      (5) DNA molecules are replicated, and an exact copy of the original cell's DNA is passed to each of the newly formed cells resulting from cell division.

CHAPTER ASSESSMENTS

2.1 Introduction

1 Define *chemistry.* (p. 51)

2 Explain the difference between chemistry and biochemistry. (p. 51)

2.2 Structure of Matter

3 Define *matter.* (p. 51)

4 Define *compound.* (p. 52)

5 List the four most abundant elements in the human body. (p. 52)

6 Explain the relationship between elements and atoms. (p. 52)

7 Identify the major parts of an atom and where they are found within an atom. (p. 52)

8 Distinguish between protons and neutrons. (p. 52)

9 Explain why a complete atom is electrically neutral. (p. 52)

10 Distinguish between atomic number and atomic weight. (p. 52)

11 Define *isotope.* (p. 53)

12 Define *atomic radiation.* (p. 53)

13 Explain the relationship between molecules and compounds. (p. 53)

14 Explain how electrons are distributed within the electron shells of atoms. (p. 55)

15 Explain why some atoms are chemically inert. (p. 55)

16 An ionic bond forms when ______________. (p. 55)
   a. atoms share electrons
   b. positively-charged and negatively-charged parts of covalent molecules attract
   c. ions with opposite electrical charges attract
   d. two atoms exchange protons
   e. an element has two types of isotopes

17 A covalent bond forms when ______________. (p. 56)
   a. atoms share electrons
   b. positively-charged and negatively-charged parts of covalent molecules attract
   c. ions with opposite electrical charges attract
   d. two atoms exchange protons
   e. an element has two types of isotopes

18 Distinguish between a single covalent bond and a double covalent bond. (p. 57)

19 Show the difference between the molecular formula and the structural formula of a specific compound. (p. 57)

20 Explain how a hydrogen bond forms. (p. 57)

21 Identify two types of macromolecules in which hydrogen bonds are important parts of the structure. (p. 58)

22 Identify three major types of chemical reactions. (p. 58)

23 Define *reversible reaction.* (p. 59)

24 Define *catalyst.* (p. 59)

25 Define *electrolyte, acid, base,* and *salt.* (p. 59)

26 Explain pH and how to use the pH scale. (p. 59)

27 Define *buffer.* (p. 60)

2.3 Chemical Constituents of Cells

28 Distinguish between inorganic and organic substances. (p. 60)

29 Distinguish between electrolytes and nonelectrolytes. (p. 60)

30 Describe the functions of water and oxygen in the human body. (p. 61)

31 List several ions that cells require and identify their functions. (p. 61)

32 Define *electrolyte balance.* (p. 61)

33 Describe the general characteristics of carbohydrates. (p. 62)

34 Distinguish between simple and complex carbohydrates. (p. 62)

35 Describe the general characteristics of lipids. (p. 62)

36 List the three main types of lipids found in cells. (p. 62)

37 Explain the difference between saturated and unsaturated fats. (p. 63)

38 A hydrophilic molecule dissolves in ___________. (p. 64)
  a. lipid but not water
  b. water but not lipid
  c. neither lipid nor water
  d. both lipid and water

39 List at least three functions of proteins. (p. 64)

40 Describe the function of an enzyme. (p. 64)

41 Identify the four levels of protein structure. (p. 65)

42 Describe how the change in shape of a protein may be either abnormal or associated with normal function. (p. 66)

43 Describe the general characteristics of nucleic acids. (p. 68)

44 Explain the general functions of nucleic acids. (p. 68)

INTEGRATIVE ASSESSMENTS/CRITICAL THINKING

OUTCOME 2.2

1. The thyroid gland metabolizes iodine, the most common form of which has a molecular weight of 127 (iodine-127). A physician wants to use a radioactive isotope of iodine (iodine-131) to test whether a patient's thyroid gland is metabolizing normally. Based on what you know about how atoms react, do you think this physician's plan makes sense?

OUTCOME 2.2

2. How would you reassure a patient about to undergo CT scanning for evaluation of a tumor and fears becoming a radiation hazard to family members?

OUTCOMES 2.2, 2.3

3. What acidic and basic substances do you encounter in your everyday activities? What acidic foods do you eat regularly? What basic foods do you eat?

OUTCOME 2.3

4. A man on a very low-fat diet proclaims to his friend, "I'm going to get my cholesterol down to zero!" Is this desirable? Why or why not?

OUTCOME 2.3

5. How would you explain the importance of amino acids and proteins in a diet to a person following a diet composed primarily of carbohydrates?

OUTCOME 2.3

6. Explain why the symptoms of many inherited diseases result from abnormal protein function.

OUTCOME 2.3

7. A friend, while frying eggs, points to the change in the egg white (which contains a protein called albumin) and explains that if the conformation of a protein changes, it will no longer have the same properties and will lose its ability to function. Do you agree or disagree with this statement?

WEB CONNECTIONS

Be sure to visit the text website at **www.mhhe.com/shier12** for answers to chapter assessments, additional quizzes, and interactive learning exercises.

ANATOMY & PHYSIOLOGY REVEALED

Anatomy & Physiology Revealed® (APR) includes cadaver photos that allow you to peel away layers of the human body to reveal structures beneath the surface. This program also includes animations, radiologic imaging, audio pronunciations, and practice quizzing. Check out **www.aprevealed.com**. APR has been proven to help improve student grades!

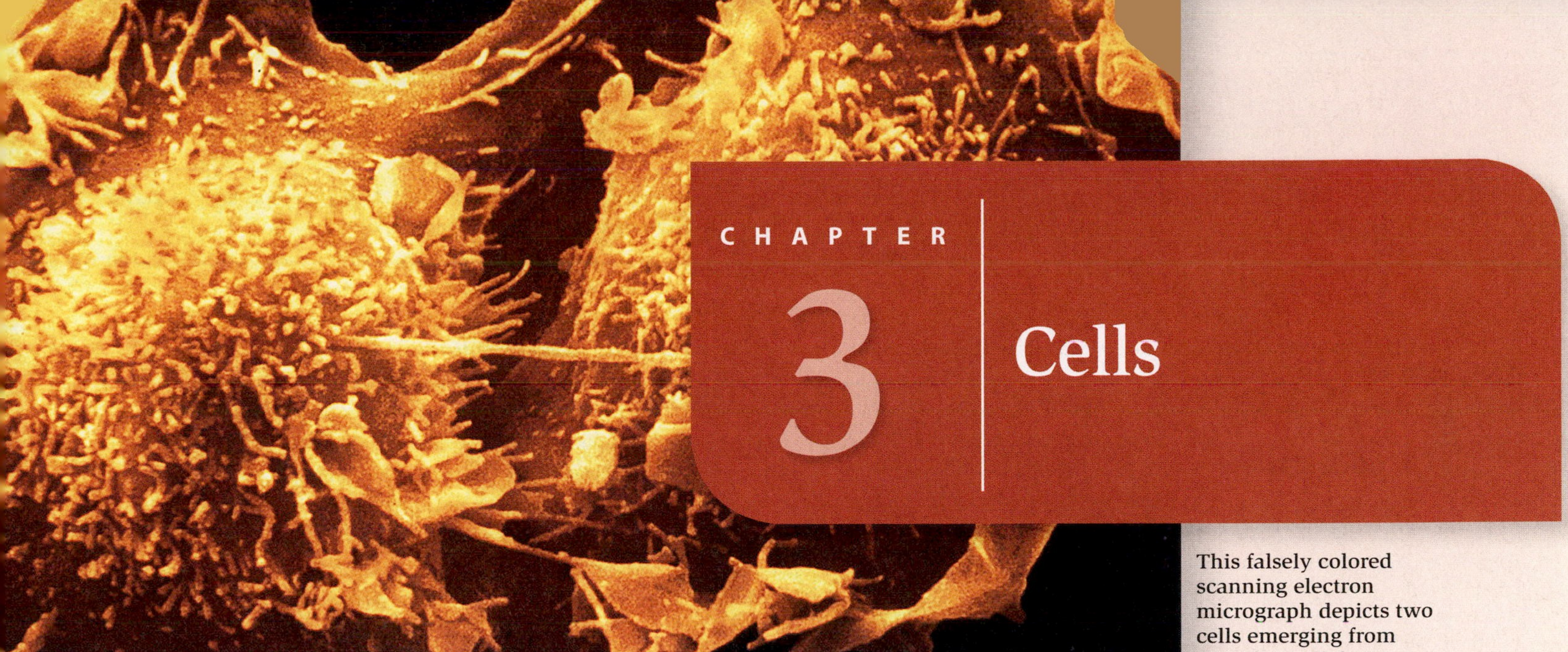

This falsely colored scanning electron micrograph depicts two cells emerging from one after cell division (5,100×).

UNDERSTANDING WORDS

apo-, away, off, apart: *apo*ptosis—a form of cell death in which cells are shed from a developing structure.

cyt-, cell: *cyto*plasm—fluid (cytosol) and organelles between the cell membrane and nuclear envelope.

endo-, within: *endo*plasmic reticulum—membranous complex in the cytoplasm.

hyper-, above: *hyper*tonic—solution that has a greater osmotic pressure than the cytosol.

hypo-, below: *hypo*tonic—solution that has a lesser osmotic pressure than the cytosol.

inter-, between: *inter*phase—stage between mitotic divisions of a cell.

iso-, equal: *iso*tonic—solution that has an osmotic pressure equal to that of the cytosol.

lys-, to break up: *lys*osome—organelle containing enzymes that break down proteins, carbohydrates, and nucleic acids.

mit-, thread: *mit*osis—stage of cell division when chromosomes condense.

phag-, to eat: *phago*cytosis—process by which a cell takes in solid particles.

pino-, to drink: *pino*cytosis—process by which a cell takes in tiny droplets of liquid.

pro-, before: *pro*phase—first stage of mitosis.

-som, body: ribo*some*—tiny, spherical organelle composed of protein and RNA that supports protein synthesis.

vesic-, bladder: *vesic*le—small, saclike organelle that contains substances to be transported within the cell or secreted.

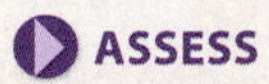
LEARN PRACTICE ASSESS

LEARNING OUTCOMES

After you have studied this chapter, you should be able to:

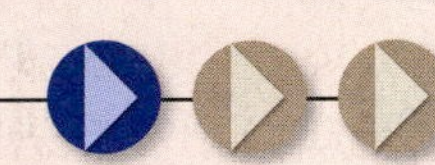

3.1 Introduction
1. Explain how cells differ from one another. (p. 76)

3.2 A Composite Cell
2. Describe the general characteristics of a composite cell. (p. 76)
3. Explain how the components of a cell's membrane provide its functions. (p. 79)
4. Describe each kind of cytoplasmic organelle and explain its function. (p. 82)
5. Describe the cell nucleus and its parts. (p. 89)

3.3 Movements Into and Out of the Cell
6. Explain how substances move into and out of cells. (p. 90)

3.4 The Cell Cycle
7. Describe the cell cycle. (p. 100)
8. Explain how a cell divides. (p. 100)

3.5 Control of Cell Division
9. Describe several controls of cell division. (p. 103)

3.6 Stem and Progenitor Cells
10. Explain how stem cells and progenitor cells make possible growth and repair of tissues. (p. 105)
11. Explain how two differentiated cell types can have the same genetic information, but different appearances and functions. (p. 106)

3.7 Cell Death
12. Discuss apoptosis. (p. 106)
13. Describe the relationship between apoptosis and mitosis. (p. 106)

FROM NATURAL PROTECTION AGAINST HIV TO A NEW DRUG

Understanding how HIV (Human Immunodeficiency Virus), the virus that causes AIDS (Acquired Immune Deficiency Syndrome), enters human cells has led to development of a new type of drug to treat the disease. In 1996, investigations of people in high-risk groups repeatedly exposed to the virus but never infected found that they were protected because HIV was unable to enter their cells. The reason: an inherited mutation blocks production of certain proteins that function as receptors, or doorways of sorts, on specific human cells. The virus was essentially kept out.

When infection begins, HIV typically enters CD4 helper T cells, which control many facets of the immune response. The viruses first bind to CD4 receptors on these cells, then also attach themselves to nearby receptors of another type, called CCR5. Only then can the virus enter the cell and begin reproducing itself at the cell's expense. Later during infection, the virus begins to use yet another type of receptor to enter cells.

The cells of about 1% of people of European ancestry lack CCR5 and are immune to HIV infection; the cells of another 13% make half the normal number of CCR5 receptors. If they become infected, it takes two years longer than the average for AIDS to develop. The mutation is rare in other population groups.

As soon as the mutation was discovered that could naturally protect against HIV infection, pharmaceutical companies began a race to mimic the effect in a drug. It took more than a decade, with the first drug approved in 2007 (Selzentry, or generic maraviroc). The drug is used for people, about 65,000 individuals in the U.S., who have developed resistance to the other three classes of anti-HIV drugs. It is a twice-daily pill, and patients must undergo testing to be certain that their strains of HIV use the CCR5 doorway. Studying the roles of cells in health and disease is a common route to developing drugs. ■

3.1 INTRODUCTION

An adult human body consists of about 50 to 100 trillion cells, the basic units of an organism. All cells have much in common, yet they come in at least 260 different varieties. Different cell types interact to build tissues, which interact to form organs.

Cells with specialized characteristics are termed **differentiated.** Such specialized cells form from less specialized cells that divide. A cell is like the Internet, harboring a vast store of information in its genome. However, like a person accessing only a small part of the Internet, a cell uses only some of the information in its genome as instructions for building its characteristic structures.

Cells vary considerably in size, which we measure in units called *micrometers* (mi′kro-me″terz). A micrometer equals one thousandth of a millimeter and is symbolized µm. A human egg cell is about 140 µm in diameter and is just barely visible to an unaided eye. This is large when compared to a red blood cell, about 7.5 µm in diameter, or the most common types of white blood cells, which are 10 to 12 µm in diameter. Smooth muscle cells are 20 to 500 µm long (fig. 3.1).

Differentiated cells have distinctive shapes that make possible their functions (fig. 3.2). For instance, nerve cells that have threadlike extensions many centimeters long transmit nerve impulses from one part of the body to another. Epithelial cells that line the inside of the mouth are thin, flattened, and tightly packed, somewhat like floor tiles. They form a barrier that shields underlying tissue. Muscle cells, slender and rod-like, contract and pull structures closer together.

3.2 A COMPOSITE CELL

It is not possible to describe a typical cell, because cells vary greatly in size, shape, content, and function. We can, however, consider a hypothetical composite cell that includes many known cell structures (fig. 3.3).

The three major parts of a cell—the **nucleus** (nu′kle-us), the **cytoplasm** (si′to-plazm), and the **cell membrane**—if appropriately stained are easily seen under the light microscope. In many cell types the nucleus is innermost and is enclosed by a thin membrane called the nuclear envelope. The nucleus contains the genetic material (DNA), which directs the cell's functions. The cytoplasm is composed of specialized structures called cytoplasmic organelles (organ-elz) suspended in a liquid called **cytosol** (si′to-sol). Organelles divide the labor in a cell by partitioning off certain areas or providing specific functions, such as dismantling debris or packaging secretions. The cytoplasm surrounds the nucleus and is contained by the cell membrane (also called a plasma membrane).

PRACTICE

1. What is a differentiated cell?
2. Name the major parts of a cell.
3. State the general functions of the cytoplasm and nucleus.

Cells with nuclei, such as those of the human body, are termed *eukaryotic,* meaning "true nucleus." In contrast are the *prokaryotic* ("before nucleus") cells of bacteria. Although bacterial cells lack nuclei and other membrane-bound organelles and are thus simpler than eukaryotic cells, the bacteria are widespread and have existed much longer than eukaryotic cells. Viruses are simpler than cells. They consist of genetic material in a protein coat and cannot reproduce outside of a host cell.

7.5 µm
(a)

12 µm
(b)

140 µm
(c)

200 µm
(d)

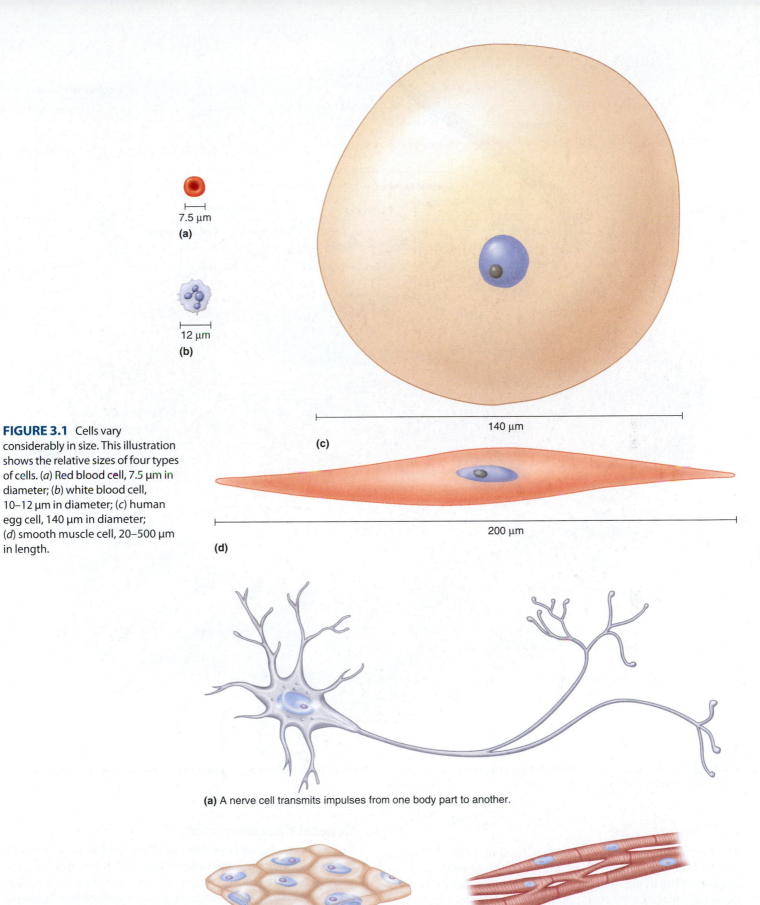

FIGURE 3.1 Cells vary considerably in size. This illustration shows the relative sizes of four types of cells. (*a*) Red blood cell, 7.5 µm in diameter; (*b*) white blood cell, 10–12 µm in diameter; (*c*) human egg cell, 140 µm in diameter; (*d*) smooth muscle cell, 20–500 µm in length.

(a) A nerve cell transmits impulses from one body part to another.

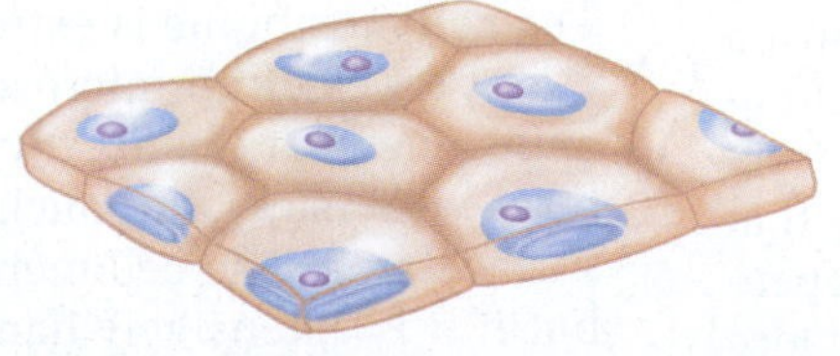

(b) Epithelial cells protect underlying cells.

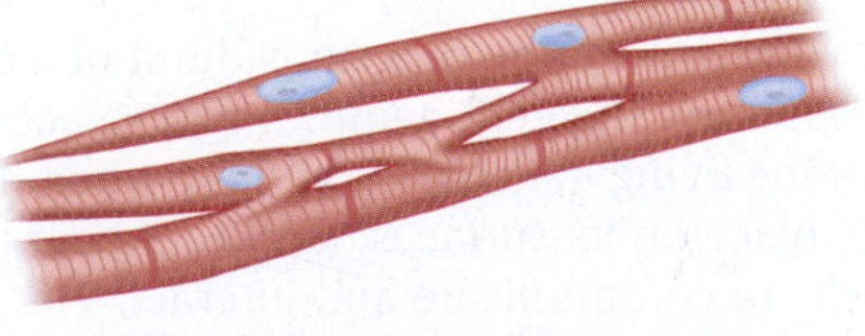

(c) Muscle cells contract, pulling structures closer together.

FIGURE 3.2 Cells vary in shape and function.

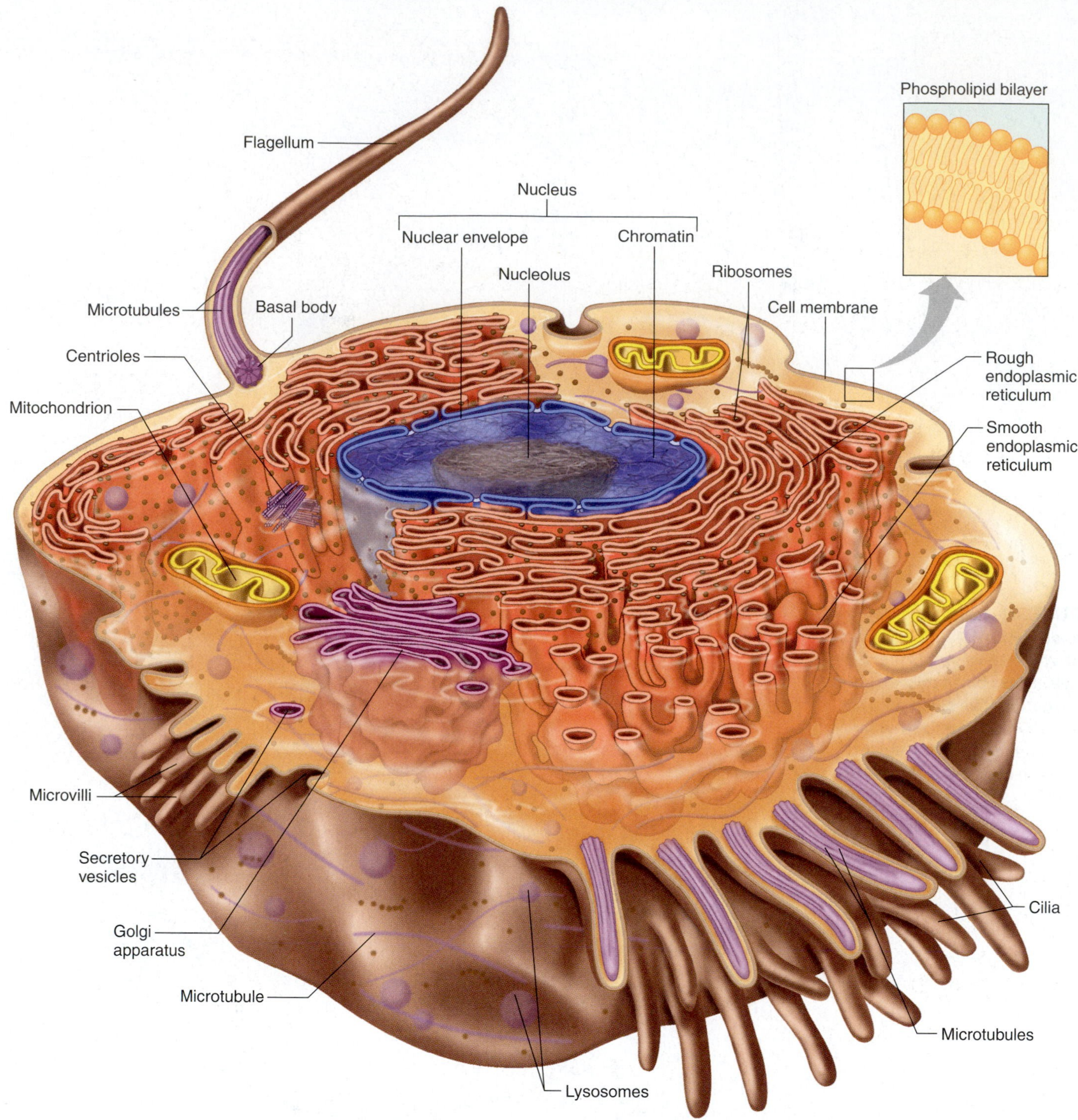

FIGURE 3.3 A composite cell illustrates the organelles and other structures found in cells. Specialized cells differ in the numbers and types of organelles, reflecting their functions. Organelles are not drawn to scale.

Cell Membrane

The cell membrane is the outermost limit of a cell. Not just a simple boundary, the cell membrane is an actively functioning part of the living material. Many important metabolic reactions take place on its surfaces, and it harbors molecules that enable cells to communicate and interact. The chapter opening vignette on page 76 offers one example of the importance of understanding the cell membrane in treating HIV infection.

General Characteristics

The cell membrane is extremely thin—visible only with the aid of an electron microscope (fig. 3.4)—but it is flexible and somewhat elastic. It typically has complex surface features with many outpouchings and infoldings that increase surface area. The cell membrane quickly seals tiny breaks, but if it is extensively damaged, cell contents exit and the cell dies.

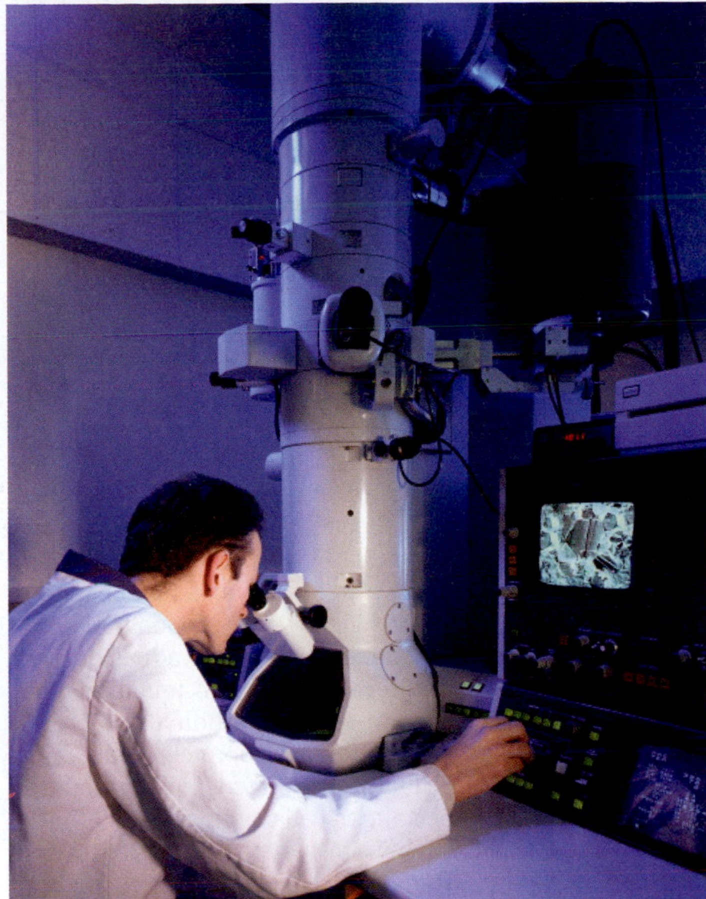

FIGURE 3.4 A transmission electron microscope.

In addition to maintaining the integrity of the cell, the cell membrane controls the entrance and exit of substances, allowing some in while excluding others. A membrane that functions in this manner is *selectively permeable* (per′me-ah-bl). The cell membrane is crucial because it is a conduit between the cell and the extracellular fluids in the body's internal environment. It allows the cell to receive and respond to incoming messages, in a process called **signal transduction.**

> The maximum effective magnification possible using a light microscope is about 5,000×. A confocal microscope is a type of light microscope that passes white or laser light through a pinhole and lens to impinge on the object, which greatly enhances resolution (ability to distinguish fine detail). A transmission electron microscope (TEM) provides an effective magnification of nearly 1,000,000×, whereas a scanning electron microscope (SEM) can provide about 50,000×. Photographs of microscopic objects (micrographs) produced using the light microscope and the transmission electron microscope are typically two-dimensional, but those obtained with the scanning electron microscope have a three-dimensional quality (fig. 3.5). Scanning probe microscopes work differently from light or electron microscopes. They move a probe over a surface and translate the distances into an image.

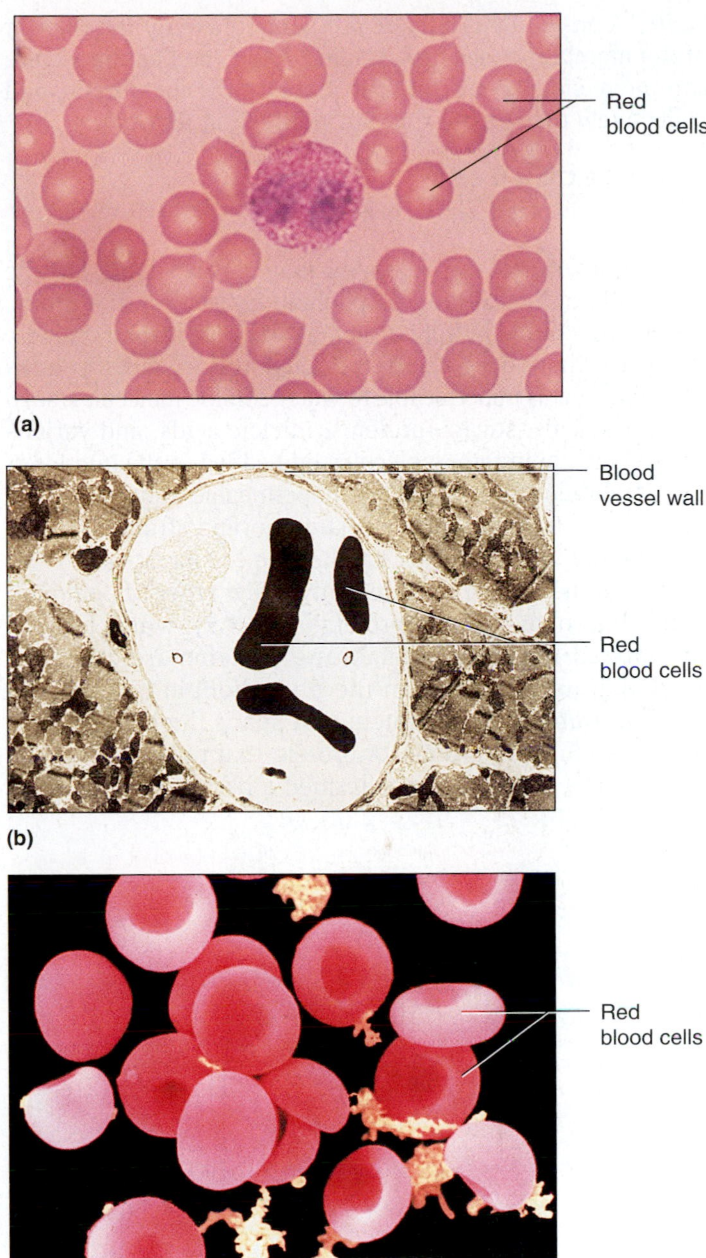

FIGURE 3.5 Human red blood cells as viewed using (*a*) a light microscope (1,200×), (*b*) a transmission electron microscope (2,500×), and (*c*) a scanning electron microscope (1,900×).

Membrane Structure

The cell membrane is mainly composed of lipids and proteins, with some carbohydrate. Its basic framework is a double layer (bilayer) of phospholipid molecules (see fig. 2.15) that self-assemble so that their water-soluble (hydrophilic) "heads," containing phosphate groups, form the surfaces of the membrane, and their water-insoluble (hydrophobic) "tails," consisting of fatty acid chains, make up the interior of the membrane (see figs. 3.3 and 3.6). The lipid molecules can move sideways within the plane of the membrane, and collectively they form a thin but stable fluid film.

RECONNECT

To chapter 2, Lipids, page 62.

The interior of the cell membrane consists largely of the fatty acid portions of the phospholipid molecules, so it is oily. Molecules soluble in lipids, such as oxygen, carbon dioxide, and steroid hormones, can pass through this layer easily; however, the layer is impermeable to water-soluble molecules, such as amino acids, sugars, proteins, nucleic acids, and various ions. Many cholesterol molecules embedded in the interior of the membrane also help make it impermeable to water-soluble substances. In addition, the relatively rigid structure of the cholesterol molecules helps stabilize the cell membrane.

A cell membrane includes only a few types of lipid molecules but many types of proteins (fig. 3.7), which provide specialized functions. Membrane proteins are classified by shape, locations within the phospholipid bilayer, and function (table 3.1). A protein that spans the membrane is termed an integral protein. A protein that projects from the membrane's outer surface is termed a peripheral protein. A protein that traverses the membrane and also extends from the outer surface is both an integral and a peripheral protein. A protein that extends outside the cell membrane at one end and dips into the cytoplasm on the interior is termed a transmembrane protein. Many transmembrane proteins are tightly coiled rods that function as *receptors*. They bind to specific incoming molecules, such as hormones, triggering responses from within the cell (see chapter 13, p. 483). CCR5, described in the vignette on page 76, is such a receptor.

Certain compact and globular proteins span the cell membrane and provide routes for small molecules and ions to cross the otherwise impermeable phospholipid bilayer. Some of these proteins form "pores" that admit water and others are highly selective and form channels that admit only particular ions. In nerve cells, for example, selective channels control the movements of sodium and potassium ions (see chapter 10, p. 365). Clinical Application 3.1 discusses how abnormal ion channels cause disease.

Peripheral proteins may also be enzymes (see chapter 4, p. 117), and many are part of signal transduction pathways. Other peripheral proteins function as **cellular adhesion molecules** (CAMs) that enable certain cells to touch or bind, discussed at the end of this section. Carbohydrate groups attached to peripheral proteins form glycoproteins that branch from a cell's surface, helping cells to recognize and bind to each other. This is important as cells aggregate to form tissues. Cell surface glycoproteins also mark the cells of an individual as "self," and distinguish particular differentiated cell

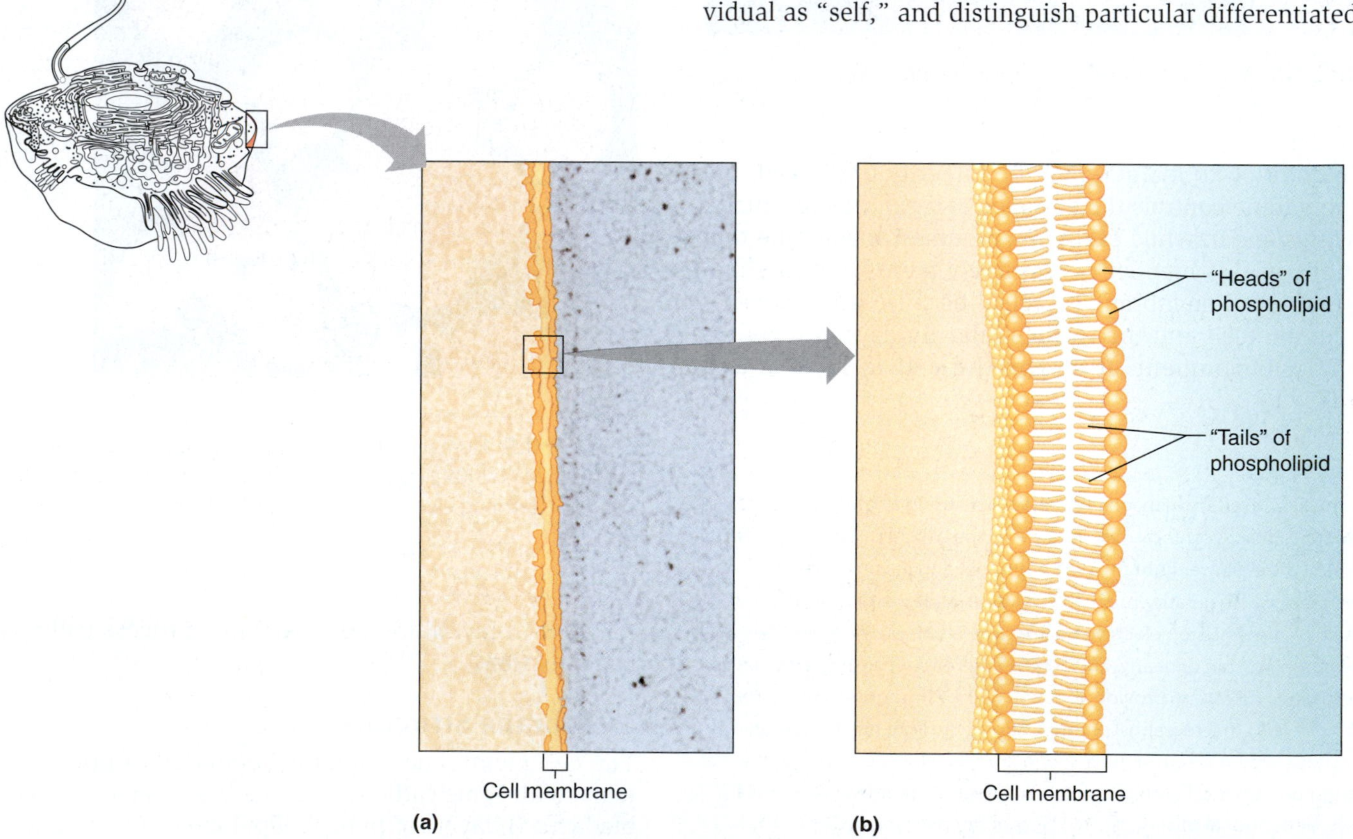

FIGURE 3.6 The cell membrane is a phospholipid bilayer. (*a*) A transmission electron micrograph of a cell membrane (600,000×); (*b*) the framework of the membrane consists of a double layer of phospholipid molecules. In actuality, many other molecules are embedded in and extend from the phospholipid bilayer.

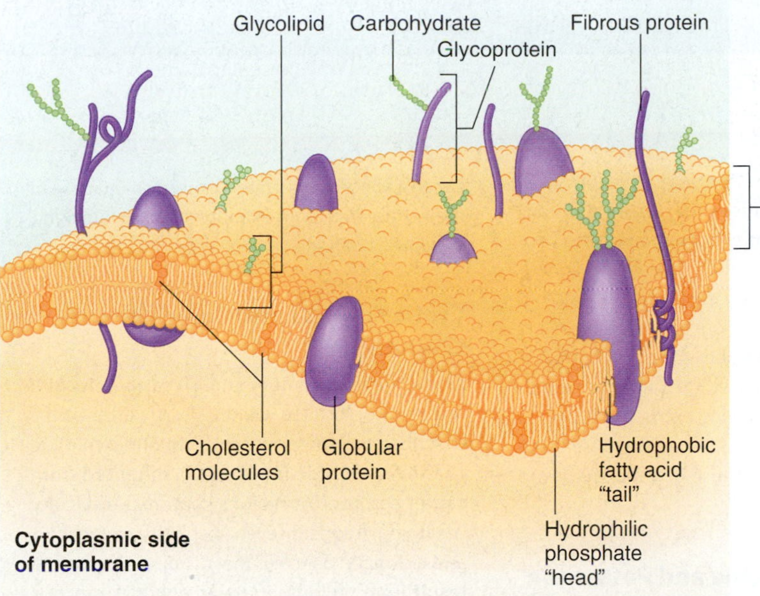

FIGURE 3.7 The cell membrane is composed primarily of phospholipids (and some cholesterol), with proteins scattered throughout the lipid bilayer and associated with its surfaces.

TABLE 3.1 | Types of Membrane Proteins

Protein Type	Function
Receptor proteins	Receive and transmit messages into a cell
Integral proteins	Form pores, channels, and carriers in cell membrane, transduce signals
Enzymes	Catalyze chemical reactions
Cellular adhesion molecules	Enable cells to stick to each other
Cell surface proteins	Establish self

types. The immune system can distinguish between "self" cell surfaces and "nonself" cell surfaces that may indicate a potential threat, such as the presence of infectious bacteria. Blood and tissue typing for transfusions or transplants consider the cell surface's protein and glycoprotein topography.

Cellular Adhesion Molecules

Often cells must interact dynamically and transiently, rather than form permanent attachments. Proteins called cellular adhesion molecules, or CAMs for short, guide cells on the move. Consider a white blood cell moving in the bloodstream to the site of an injury, where it is required to fight infection. Imagine that such a cell must reach a woody splinter embedded in a person's palm (fig. 3.8). Once near the splinter, the white blood cell must slow down in the turbulence of the bloodstream. A type of CAM called a *selectin* does this by coating the white blood cell and providing traction. The white blood cell slows to a roll and binds to carbohydrates on the inner capillary surface. Clotting blood, bacteria, and decaying tissue at the injury site release biochemicals (chemoattractants) that attract the white blood cell. Finally, a type of CAM called an *integrin* contacts an adhesion receptor protein protruding into the capillary space near the splinter and pushes up through the capillary cell membrane, grabbing the passing slowed white blood cell and directing it between the tilelike cells of the capillary wall. White blood cells collecting at an injury site produce inflammation and, with the dying bacteria, form pus. (The role of white blood cells in body defense is discussed further in chapter 14, pp. 531–532.)

Cellular adhesion is critical to many functions. CAMs guide cells surrounding an embryo to grow toward maternal cells and form the placenta, the supportive organ linking a pregnant woman to the fetus (see fig. 23.18). Sequences of CAMs help establish the connections between nerve cells that underlie learning and memory.

Abnormal cellular adhesion affects health. Lack of cellular adhesion, for example, eases the journey of cancer cells as they spread from one part of the body to another. Arthritis may occur when white blood cells are reined in by the wrong adhesion molecules and inflame a joint where there isn't an injury.

PRACTICE 

4 What is a selectively permeable membrane?
5 Describe the chemical structure of a cell membrane.
6 What are some functions of cell membrane proteins?
7 What are some of the events of cellular adhesion?

Cytoplasm

When viewed through a light microscope, cytoplasm usually appears clear with scattered specks. However, a transmission electron microscope (see fig. 3.4) reveals networks of membranes and organelles suspended in the cytosol. Cytoplasm also

3.1 CLINICAL APPLICATION

Faulty Ion Channels Cause Disease

What do abnormal pain intensity, irregular heartbeats, and cystic fibrosis have in common? All result from abnormal ion channels in cell membranes.

Ion channels are protein-lined tunnels in the phospholipid bilayer of a biological membrane. These passageways permit electrical signals to pass in and out of membranes as ions (charged particles). Many ion channels open or close like a gate in response to specific ions under specific conditions. These situations include a change in electrical forces across the membrane, binding of a molecule, or receiving biochemical messages from inside or outside the cell.

Ion channels are specific for calcium (Ca^{+2}), sodium (Na^{+}), potassium (K^{+}), or chloride (Cl^{-}). A cell membrane may have a few thousand ion channels specific for each of these ions. Ten million or more ions can pass through an ion channel in one second! Drugs may act by affecting ion channels (table 3A), and abnormal ion channels cause certain disorders, including the following:

Absent or Excess Pain

The ten-year-old boy amazed the people on the streets of the small northern Pakistani town. He was completely unable to feel pain and had become a performer, stabbing knives through his arms and walking on hot coals to entertain crowds. Several other people in this community, where relatives often married relatives, were also unable to feel pain. Researchers studied the connected families and discovered a mutation that alters sodium channels on certain nerve cells. The mutation blocks the channels so that the message to feel pain cannot be sent. The boy died at age thirteen from jumping off a roof. His genes could protect him from pain, but pain protects against injury by providing a warning.

A different mutation affecting the same sodium channels causes drastically different symptoms. In "burning man syndrome," the channels become hypersensitive, opening and flooding the body with pain easily, in response to exercise, an increase in room temperature, or just putting on socks. In another condition, "paroxysmal extreme pain disorder," the sodium channels stay open too long, causing excruciating pain in the rectum, jaw, and eyes. Researchers are using the information from these genetic studies to develop new painkillers.

Long-QT Syndrome and Potassium Channels

Four children in a Norwegian family were born deaf, and three of them died at ages four, five, and nine. All of the children had inherited from unaffected "carrier" parents "long-QT syndrome associated with deafness." They had abnormal potassium channels in the heart muscle and in the inner ear. In the heart, the malfunctioning channels disrupted electrical activity, causing a fatal disturbance to the heart rhythm. In the inner ear, the abnormal channels caused an increase in the extracellular concentration of potassium ions, impairing hearing.

Cystic Fibrosis and Chloride Channels

A seventeenth-century English saying, "A child that is salty to taste will die shortly after birth," described the consequence of abnormal chloride channels in cystic fibrosis (CF), inherited from carrier parents. The major symptoms—difficulty breathing, frequent severe respiratory infections, and a clogged pancreas that disrupts digestion—result from buildup of extremely thick mucous secretions.

Abnormal chloride channels in cells lining the lung passageways and ducts of the pancreas cause the symptoms of CF. The primary defect in the chloride channels also causes sodium channels to malfunction. The result: very salty sweat and abnormally thick mucus. Gene therapy is directed at supplying patients' lung-lining cells with the instructions to produce normal chloride channels. ■

TABLE 3A | Drugs That Affect Ion Channels

Target	Indication
Calcium channels	Antihypertensives Antiangina (chest pain)
Sodium channels	Antiarrhythmias, diuretics Local anesthetics
Chloride channels	Anticonvulsants Muscle relaxants
Potassium channels	Antihypertensives, antidiabetics (non-insulin-dependent)

contains abundant protein rods and tubules that form a supportive framework called the **cytoskeleton** (si′to-skel-ĕ-ton).

The activities of a cell occur largely in its cytoplasm, where nutrient molecules are received, processed, and used in metabolic reactions. The following cytoplasmic organelles have specific functions:

1. **Ribosomes.** Ribosomes (ri′bo-sōmz) are tiny, spherical structures composed of protein and RNA. They provide a structural support and enzymatic activity to link amino acids to form proteins (see chapter 4, p. 132). Unlike many of the other organelles, ribosomes are not composed of or contained in membranes. They are scattered in the cytoplasm and also bound to another organelle, the endoplasmic reticulum.
2. **Endoplasmic reticulum.** The endoplasmic reticulum (en-do-plaz′mik re-tik′u-lum) (ER) is a complex organelle composed of membrane-bound flattened sacs,

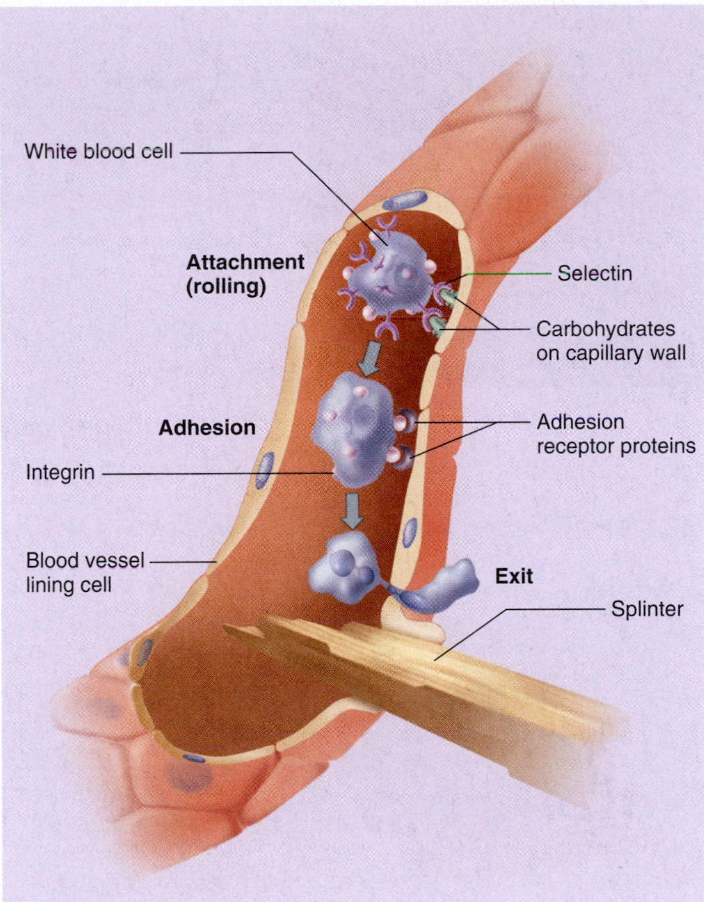

FIGURE 3.8 Cellular adhesion molecules (CAMs) direct white blood cells to injury sites, such as this splinter.

elongated canals, and fluid-filled vesicles (fig. 3.9). These parts are interconnected, and they interact with the cell membrane, the nuclear envelope, and certain other organelles. ER is widespread in the cytoplasm, providing a tubular transport system for molecules throughout the cell.

The ER participates in the synthesis of protein and lipid molecules. These molecules may leave the cell as secretions or be used within the cell for such functions as producing new ER or cell membrane as the cell grows.

The outer membranous surface of some ER is studded with many ribosomes that give the ER a textured appearance when viewed with an electron microscope. Such endoplasmic reticulum is termed *rough ER*. The ribosomes of rough ER are sites of protein synthesis. The proteins then move through the tubules of the endoplasmic reticulum to the Golgi apparatus for further processing.

ER that lacks ribosomes is called *smooth ER* (fig. 3.9). It contains enzymes important in synthesizing lipids, absorbing fats from the digestive tract, and breaking down drugs. Lipids are synthesized in the smooth ER and are added to proteins arriving from the rough ER. Smooth ER is especially abundant in liver cells that break down alcohol and drugs.

3. **Vesicles.** Vesicles (ves′ĭ-kelz) are membranous sacs that vary in size and contents. They may form when a portion of the cell membrane folds inward and pinches off. As a result, a tiny, bubblelike vesicle, containing some liquid or solid material formerly outside the cell, enters the cytoplasm. The Golgi apparatus and ER also form vesicles. Fleets of vesicles transport many substances into and out of cells in a process called vesicle trafficking.
4. **Golgi apparatus.** A Golgi apparatus (gol′je ap″ah-ra′tus) is a stack of half a dozen or so flattened, membranous sacs called *cisternae.* This organelle refines, packages, and delivers proteins synthesized on the rough ER (fig. 3.10).

   Proteins arrive at the Golgi apparatus enclosed in tiny vesicles composed of membrane from the ER. These sacs fuse to the membrane at the innermost end of the Golgi apparatus, specialized to receive proteins. Previously, in the ER sugar molecules were attached to these protein molecules, forming glycoproteins.

   As the glycoproteins pass from layer to layer through the Golgi stacks, they are modified chemically. For example, sugar molecules may be added or removed from them. When the altered glycoproteins reach the outermost layer, they are packaged in bits of Golgi apparatus membrane that bud off and form transport vesicles. Such a vesicle may then move to the cell membrane, where it fuses and releases its contents to the outside of the cell as a secretion. This is an example of a process called exocytosis (see page 98). Other vesicles may transport glycoproteins to organelles in the cell, as figure 3.11 shows for the process of milk secretion.

   Some cells, including certain liver cells and white blood cells (lymphocytes), secrete glycoprotein molecules as rapidly as they are synthesized. However, certain other cells, such as those that manufacture protein hormones, release vesicles containing newly synthesized molecules only when the cells are stimulated. Otherwise, the loaded vesicles remain in the cytoplasm. (Chapter 13, p. 493 discusses hormone secretion.)

   Secretory vesicles that originate in the ER not only release substances outside the cell, but also provide new cell membrane. This is especially important during cell growth.
5. **Mitochondria.** Mitochondria (mi″to-kon′dre-ah) are elongated, fluid-filled sacs 2–5 μm long. They often move slowly in the cytoplasm and can divide. A mitochondrion contains a small amount of DNA that encodes information for making a few types of proteins and specialized RNA. However, most proteins used in mitochondrial functions are encoded in the DNA of the nucleus. These proteins are synthesized elsewhere in the cell and then enter the mitochondria.

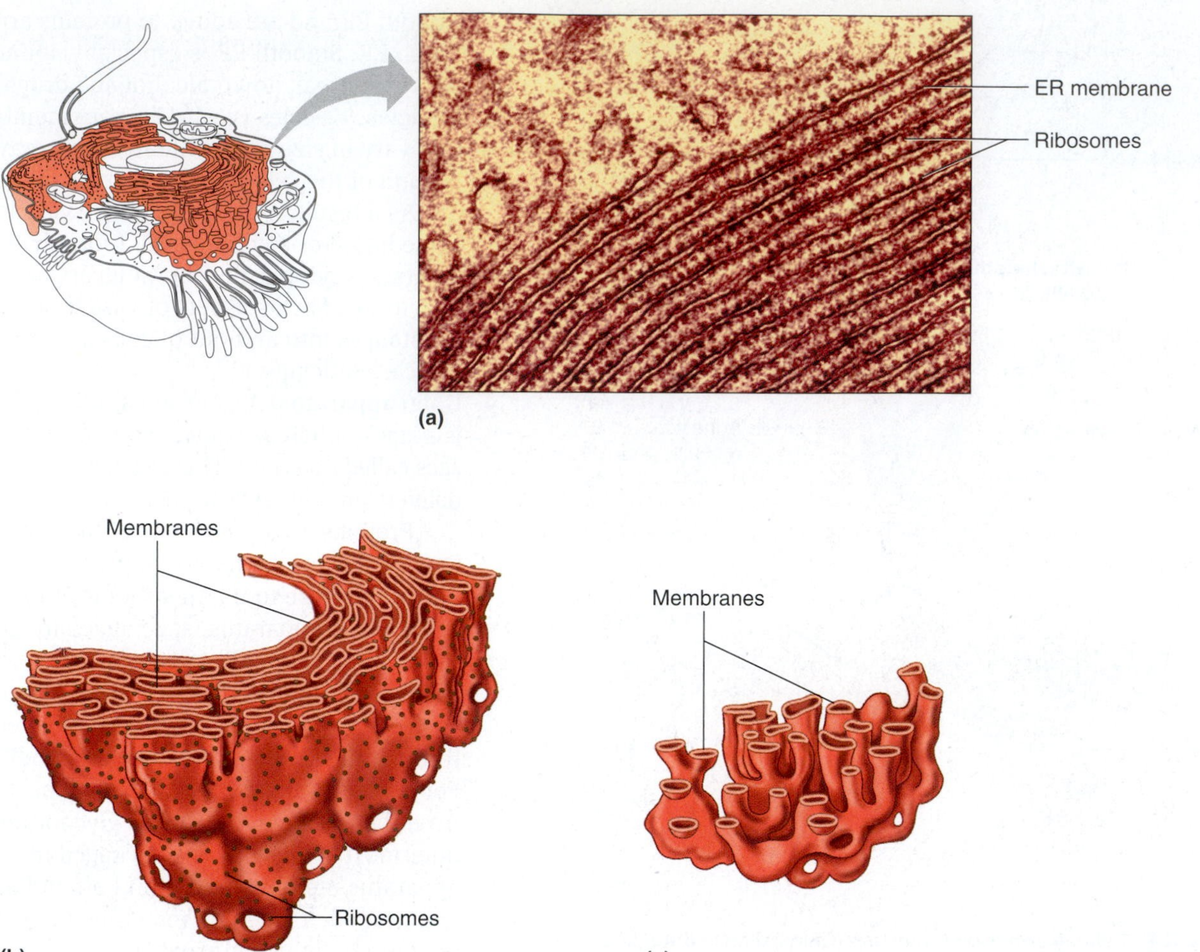

FIGURE 3.9 The endoplasmic reticulum is the site of protein and lipid synthesis, and serves as a transport system. (*a*) A transmission electron micrograph of rough endoplasmic reticulum (ER) (28,500×). (*b*) Rough ER is dotted with ribosomes, whereas (*c*) smooth ER lacks ribosomes.

A mitochondrion (mi″to-kon′dre-on) has two layers—an outer membrane and an inner membrane. The inner membrane is folded extensively in, forming shelflike partitions called *cristae* (fig. 3.12). This organization dramatically increases the surface area on which chemical reactions can occur. Small, stalked particles that contain enzymes are connected to the cristae. These enzymes and others dissolved in the fluid in the mitochondrion, called the matrix, control many of the chemical reactions that release energy from glucose and other nutrients. The mitochondrion captures and transfers this newly released energy into special chemical bonds of the molecule **adenosine triphosphate** (ATP), that cells can readily use (chapter 4, p. 119). For this reason, the mitochondrion is sometimes called the "powerhouse" of the cell.

A typical cell has about 1,700 mitochondria, but cells with very high energy requirements, such as skeletal muscle cells, have many thousands of mitochondria. This is why common symptoms of illnesses affecting mitochondria are exercise intolerance and weak, flaccid muscles. Some cells, such as red blood cells, lack mitochondria.

Mitochondria provide glimpses into the past. These organelles are passed to offspring from mothers only, because the mitochondria are excluded from the part of a sperm that enters an egg cell. Mitochondrial DNA sequences are consulted to trace human origins, back to a long-ago group of common ancestors of us all metaphorically called "mitochondrial Eve." Going even farther back, mitochondria are thought to be the remnants of once free-living bacterialike cells that entered primitive eukaryotic cells. These bacterial passengers remain in our cells today, where they participate in energy reactions. Mitochondria physically resemble bacteria.

6. **Lysosomes.** Lysosomes (li′so-sōmz) are the "garbage disposals" of the cell, where enzymes dismantle debris. Lysosomes can be difficult to identify because their shapes vary so greatly, but they often appear as tiny, membranous sacs (fig. 3.13). These sacs contain powerful enzymes that break down proteins,

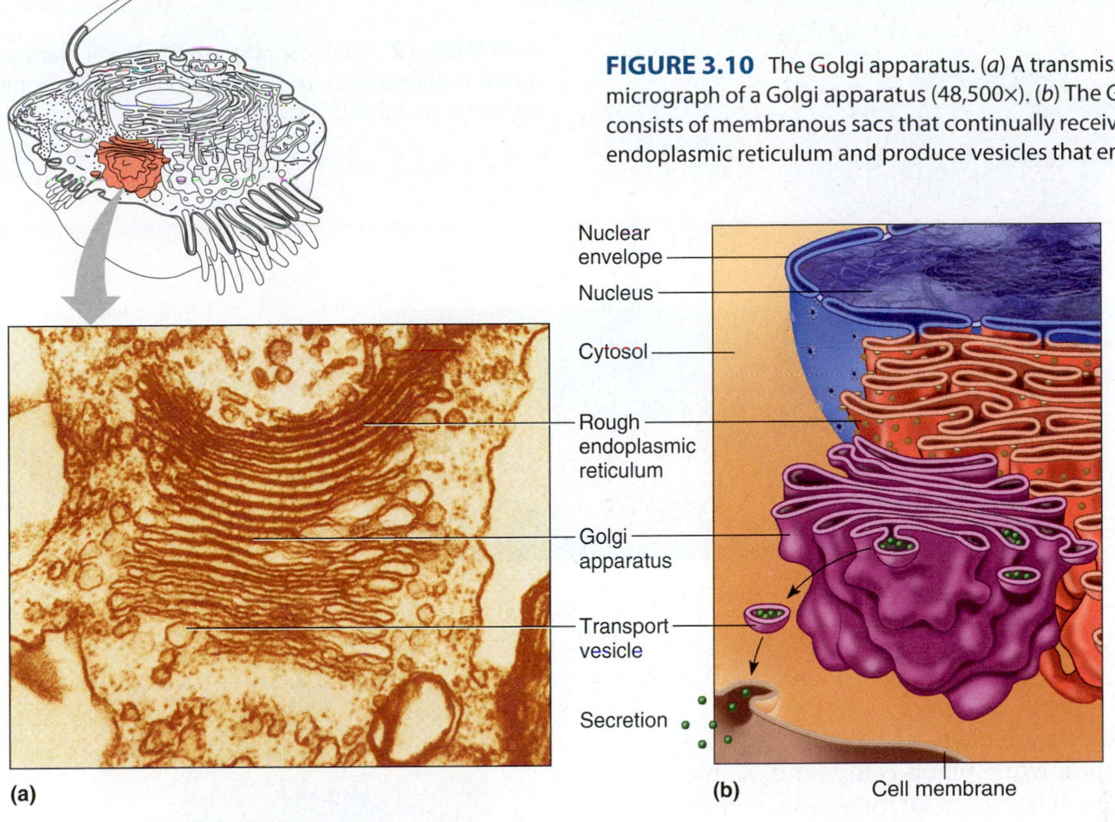

FIGURE 3.10 The Golgi apparatus. (*a*) A transmission electron micrograph of a Golgi apparatus (48,500×). (*b*) The Golgi apparatus consists of membranous sacs that continually receive vesicles from the endoplasmic reticulum and produce vesicles that enclose secretions.

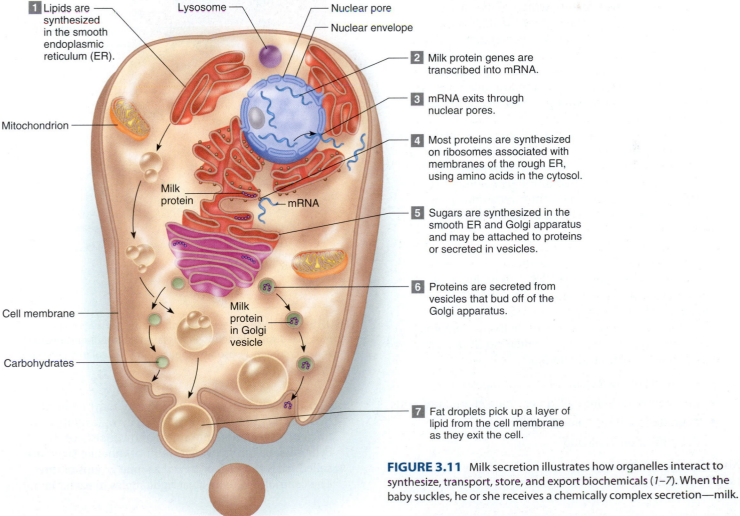

FIGURE 3.11 Milk secretion illustrates how organelles interact to synthesize, transport, store, and export biochemicals (*1–7*). When the baby suckles, he or she receives a chemically complex secretion—milk.

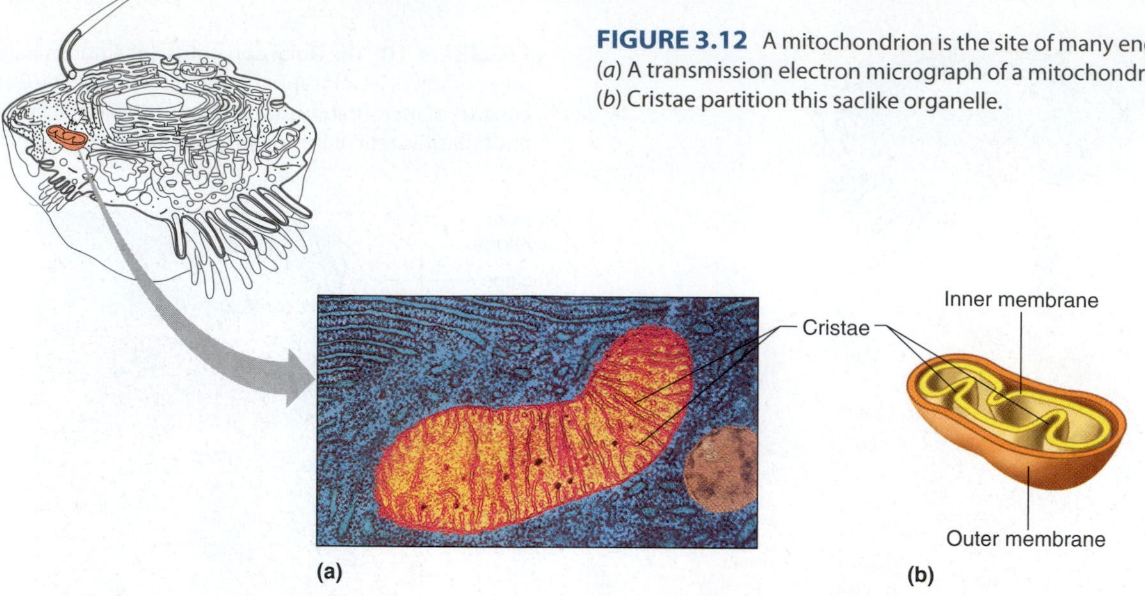

FIGURE 3.12 A mitochondrion is the site of many energy reactions. (*a*) A transmission electron micrograph of a mitochondrion (28,000×). (*b*) Cristae partition this saclike organelle.

carbohydrates, and nucleic acids, including foreign particles. Certain white blood cells, for example, engulf infecting bacteria, which lysosomal enzymes then digest.

Lysosomes also destroy worn cellular parts. Lysosomes in certain scavenger cells may engulf and digest entire body cells that have been damaged. How the lysosomal membrane is able to withstand being digested is not well understood, but this organelle sequesters enzymes that can function only under very acidic conditions, preventing them from destroying the cellular contents around them. Human lysosomes contain more than forty different types of enzymes. An abnormality in just one type of lysosomal enzyme can be devastating to health (Clinical Application 3.2).

7. **Peroxisomes** (pĕ-roks′ĭ-sōmz). Peroxisomes are membranous sacs that resemble lysosomes in size and shape. Although present in all human cells, peroxisomes are most abundant in cells of the liver and kidneys. Peroxisomes contain enzymes, called peroxidases, that catalyze metabolic reactions that release hydrogen peroxide (H_2O_2), which is toxic to cells. Peroxisomes also contain an enzyme called catalase, which decomposes hydrogen peroxide.

   The outer membrane of a peroxisome contains some forty types of enzymes, which catalyze a variety of biochemical reactions, including

   - synthesis of bile acids, used in fat digestion
   - breakdown of lipids called very long chain fatty acids
   - degradation of rare biochemicals
   - detoxification of alcohol

Abnormal peroxisomal enzymes can drastically affect health (see Clinical Application 3.2).

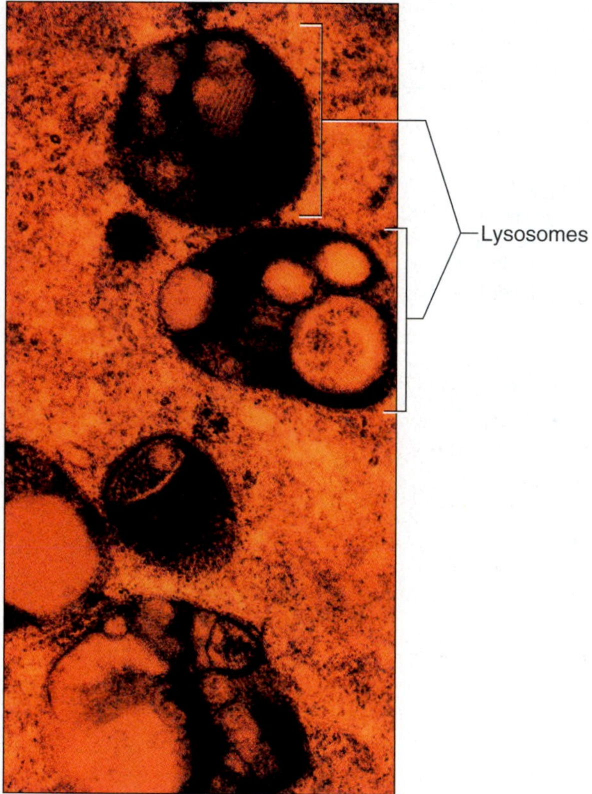

FIGURE 3.13 Lysosomes are membranous sacs that contain enzymes that dismantle debris. In this falsely colored transmission electron micrograph, lysosomes appear as dark circular structures (14,100×).

8. **Centrosome.** A centrosome (sen′tro-sōm) (central body) is a structure located in the cytoplasm near the nucleus. It is nonmembranous and consists of two cylinders, called *centrioles,* built of tubelike structures called microtubules organized as nine groups of three. The centrioles usually lie at right angles to each other.

3.2 CLINICAL APPLICATION

Disease at the Organelle Level

German physiologist Rudolph Virchow hypothesized cellular pathology—disease at the cellular level—in the 1850s. Today, treatments for many disorders are a direct result of understanding a disease process at the cellular level. Here, we examine how three abnormalities—in mitochondria, in lysosomes, and in peroxisomes—cause whole-body symptoms.

MELAS and Mitochondria

Sharon had always been small for her age, easily fatigued, slightly developmentally delayed, and had difficulty with schoolwork. She also had seizures. At age eleven, she suffered a stroke. An astute physician who observed Sharon's mother, Lillian, suspected that the girl's symptoms were all related, and the result of abnormal mitochondria, the organelles that house the biochemical reactions that extract energy from nutrients.

The doctor noticed that Lillian was uncoordinated and had numb hands. When she asked if Lillian ever had migraine headaches, she said that she suffered from them nearly daily, as did her two sisters and one brother. Lillian and her siblings also had diabetes mellitus and muscle weakness. The doctor ordered several blood tests for mother and daughter, which revealed that both had elevated levels of biochemicals (pyruvic acid and lactic acid) that indicated they were unable to extract maximal energy from nutrients. Their muscle cells had abnormal mitochondria. Accumulation of these mitochondria in smooth muscle cells in blood vessel walls in the brain caused Sharon's stroke, migraines, and seizures.

The affected family members had MELAS, which stands for the major symptoms—*m*itochondrial *e*ncephalomyopathy, *l*actic *a*cidosis, and *s*trokelike episodes. Their mitochondria cannot synthesize some of the proteins required to carry out the energy reactions. The mutant gene is part of the DNA in mitochondria, and Lillian's mother transmitted it to all of her children. But because mitochondria are inherited only from the mother, Lillian's brother will not pass MELAS to his children.

Tay-Sachs Disease and Lysosomes

Michael was a pleasant, happy infant who seemed to be developing normally until about six months of age. Able to roll over and sit for a few seconds, he suddenly lost those abilities. Soon, he no longer turned and smiled at his mother's voice, and he was no longer interested in his mobile. Concerned about Michael's reversals in development, his anxious parents took him to the pediatrician. It took exams by several other specialists to diagnose Tay-Sachs disease.

A neurologist saw telltale "cherry red spots" in Michael's eyes. His cells provided further clues—the lysosomes were swollen yet lacked one of the forty types of lysosomal enzymes, resulting in a "lysosomal storage disease" that built up fatty material on his nerve cells. Tests for the missing enzyme in the blood and tests for mutant genes confirmed the diagnosis. Michael's nervous system would continue to fail, and he would be paralyzed and unable to see or hear by the time he died, before the age of four years.

The cellular and molecular signs of Tay-Sachs disease—the swollen lysosomes and missing enzyme—had been present long before Michael began to lag developmentally. The next time his parents expected a child, they had the baby, a girl, tested before birth for the enzyme deficiency. They learned that she would be a carrier like themselves, but not ill.

Adrenoleukodystrophy (ALD) and Peroxisomes

For young Lorenzo Odone, the first sign of adrenoleukodystrophy was disruptive behavior in school. When he became lethargic, weak, and dizzy, his teachers and parents realized that his problem was not just temper tantrums. His skin darkened, blood sugar levels plummeted, heart rhythm altered, and the levels of electrolytes in his body fluids changed. He lost control over his limbs as his nervous system continued to deteriorate. Lorenzo's parents took him to many doctors. Finally, one of them tested the child's blood for an enzyme normally manufactured in peroxisomes.

Lorenzo's peroxisomes lacked the second most abundant protein in the outer membrane of this organelle, which normally transports an enzyme into the peroxisome. The enzyme controls breakdown of a type of very long chain fatty acid. Without the enzyme, the fatty acid builds up in cells in the brain and spinal cord, eventually stripping these cells of their lipid sheaths, made of a substance called myelin. Without the myelin sheaths, the nerve cells cannot transmit messages fast enough. Death comes in a few years. Boys inherit ALD from carrier mothers.

A 1992 film, *Lorenzo's Oil*, told the story of Lorenzo's parents' efforts to develop a mixture of oils to slow the buildup of the very long chain fatty acids. Although the oil is still a questionable treatment because of adverse effects, it did enable Lorenzo to live far longer than his doctors expected—he died at the age of 30 in 2008. At the time of his death, he could not talk or see and communicated with finger movements and eye blinks. Some boys with ALD have been cured with transplants of bone marrow stem cells, and an experimental gene therapy that delivers a functional version of the responsible mutant gene. ■

During cell division, the centrioles migrate to either side of the nucleus, where they form spindle fibers that pull on and distribute *chromosomes* (kro′mo-sōmz), which carry DNA information to the newly forming cells (fig. 3.14). Centrioles also form parts of hairlike cellular projections called cilia and flagella.

9. **Cilia** (sing., *cilium*)and **flagella** (sing., *flagellum*). Cilia and flagella are motile extensions of certain cells. They are structurally similar and differ mainly in their length and abundance. Both cilia and flagella consist of nine groups of three microtubules with two additional microtubules in the center, forming a distinct cylindrical pattern.

Cilia fringe the free surfaces of some epithelial cells. Each cilium is a hairlike structure about 10 μm long, which attaches just beneath the cell membrane to a modified centriole called a *basal body*. Cilia dot cells in precise patterns. They move in a coordinated "to-and-fro"manner so that rows of cilia beat one after the other, generating a wave that sweeps across the ciliated surface.

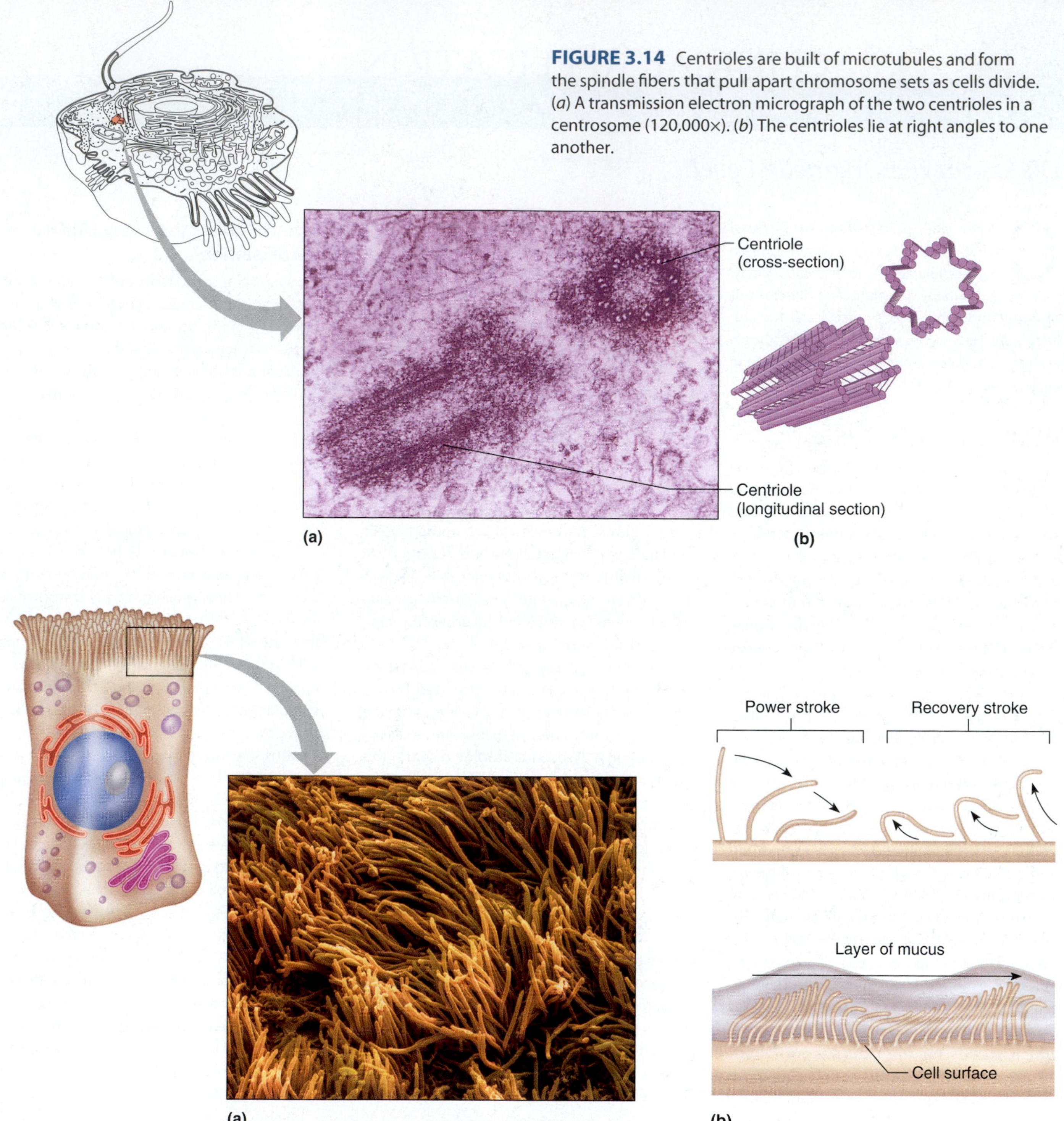

FIGURE 3.14 Centrioles are built of microtubules and form the spindle fibers that pull apart chromosome sets as cells divide. (*a*) A transmission electron micrograph of the two centrioles in a centrosome (120,000×). (*b*) The centrioles lie at right angles to one another.

FIGURE 3.15 Cilia are sweeping hairlike extensions. (*a*) They fringe certain cells, such as those forming the inner lining of the respiratory tract (5,400×). (*b*) Cilia have a power stroke and a recovery stroke that create a "to-and-fro" movement that sweeps fluids across the tissue surface.

For example, this action propels mucus over the lining of the respiratory tract (fig. 3.15). Chemicals in cigarette smoke destroy cilia, which impairs the respiratory tract's ability to expel bacteria. Infection may result.

A flagellum is much longer than a cilium, and a cell usually has only one. A flagellum begins its characteristic undulating, wavelike motion at its base. The tail of a sperm cell is a flagellum that generates swimming movements (fig. 3.16 and chapter 22, p. 837). It is the only known flagellum in humans.

10. **Microfilaments** and **microtubules.** Two types of threadlike structures in the cytoplasm are microfilaments and microtubules. They are distinguished by protein type, diameter, and how they assemble. Other proteins connect these components, creating the cytoskeleton that provides strength and the ability to resist force and maintain shape.

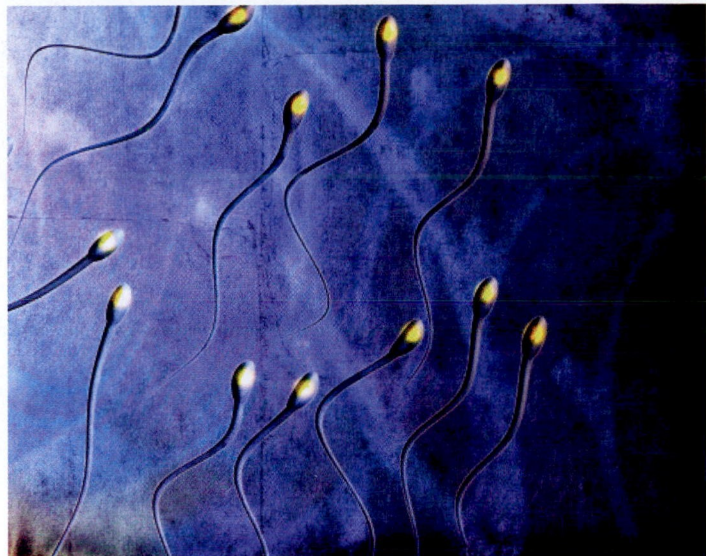

FIGURE 3.16 Flagella form the tails of these human sperm cells (1,400×).

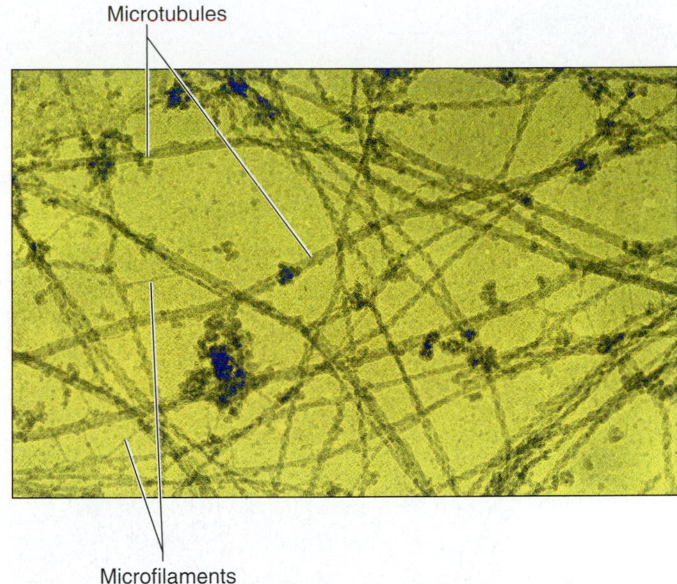

FIGURE 3.17 A falsely colored transmission electron micrograph of microfilaments and microtubules in the cytoplasm (35,000×).

Microfilaments are tiny rods of the protein actin that typically form meshworks or bundles and provide certain cellular movements. In muscle cells, for example, microfilaments constitute *myofibrils,* which shorten or contract these cells. In other cells, microfilaments associated with the inner surface of the cell membrane aid cell motility (fig. 3.17).

Microtubules are long, slender tubes with diameters two or three times greater than those of microfilaments. They are composed of the globular protein tubulin. Microtubules are usually somewhat rigid, which helps maintain the shape of the cell (fig. 3.18). In cilia and flagella, microtubule interactions provide movement (see figs. 3.15 and 3.16). Microtubules also move organelles and other cellular structures. For instance, microtubules form centrioles, and also provide conduits for organelles, like the tracks of a roller coaster.

All cells have microtubules and microfilaments, but some specialized cells have a third type of cytoskeletal component, intermediate filaments. These are composed of any of several types of proteins and take the general form of dimers (protein pairs) entwined into nested, coiled rods. Intermediate filaments are abundant in the actively dividing cells in the deepest part of the outer skin layer, the epidermis. Here they form a strong inner scaffolding that helps the cells attach and form a barrier.

11. **Other structures.** In addition to organelles, cytoplasm contains chemicals called *inclusions.* These usually are in a cell temporarily. Inclusions include stored nutrients, such as glycogen and lipids, and pigments, such as melanin in the skin.

In a group of inherited disorders called epidermolysis bullosa, the skin blisters easily as tissue layers separate due to abnormal intermediate filaments. A British documentary, called *The Boy Whose Skin Fell Off,* traces the life of a person with a severe form of the disease. Experimental stem cell therapy and gene therapy have had encouraging results in treating epidermolysis bullosa.

PRACTICE

8 What are the functions of the endoplasmic reticulum?

9 Describe how the Golgi apparatus functions.

10 Why are mitochondria called the "powerhouses" of cells?

11 How do lysosomes function?

12 Describe the functions of microfilaments and microtubules.

13 Distinguish between organelles and inclusions.

Cell Nucleus

A nucleus is a relatively large, usually spherical, structure that contains the genetic material (DNA) that directs the activities of the cell. The extremely long molecules of DNA are complexed with proteins to form dense, string-like chromatin fibers.

The nucleus is enclosed in a double-layered **nuclear envelope,** which consists of an inner and an outer lipid bilayer membrane. These two membranes have a narrow space between them, but are joined at places that surround openings called **nuclear pores.** These pores are not bare holes, but channels whose walls consist of more than 100 different types of proteins. Nuclear pores allow certain dissolved substances to move between the nucleus and the cytoplasm (fig. 3.19). Molecules of messenger RNA that carry genetic information exit the nucleus through nuclear pores.

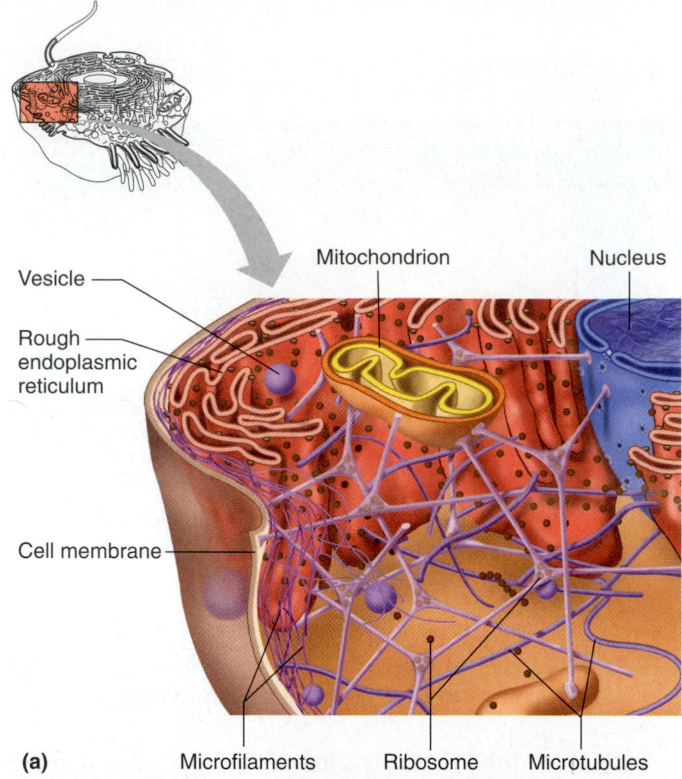

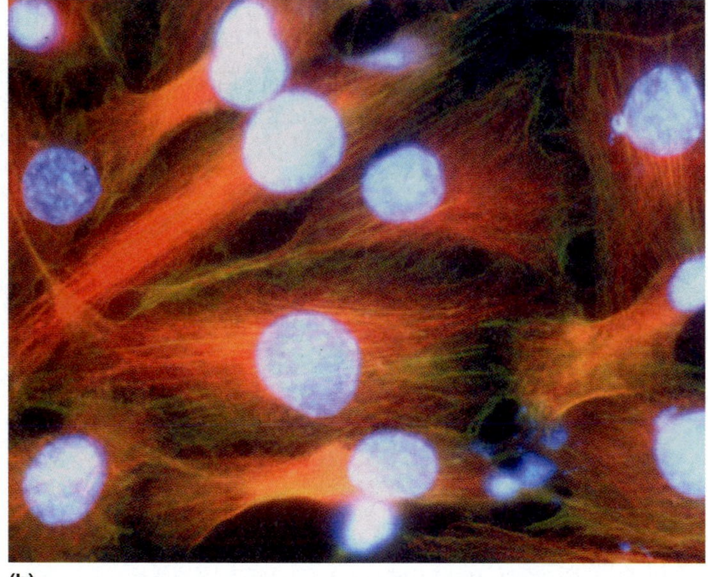

FIGURE 3.18 The cytoskeleton provides an inner scaffolding. (*a*) Microtubules and microfilaments help maintain the shape of a cell by forming an internal framework beneath the cell membrane and in the cytoplasm. (*b*) A falsely colored electron micrograph of cells showing the cytoskeletons (orange and yellow) (750×).

The nucleus contains a fluid (nucleoplasm) in which other structures are suspended. These structures include the following:

1. **Nucleolus.** A nucleolus (nu-kle′o-lus) ("little nucleus") is a small, dense body largely composed of RNA and protein. It has no surrounding membrane and is formed in specialized regions of certain chromosomes. The nucleolus is the site of ribosome production. Once ribosomes form, they migrate through the nuclear pores to the cytoplasm. A cell may have more than one nucleolus. The nuclei of cells that synthesize abundant protein, such as those of glands, may have especially large nucleoli.
2. **Chromatin.** Chromatin consists of loosely coiled fibers in the nuclear fluid. Chromatin fibers are composed of continuous DNA molecules wrapped around clusters of proteins called histones, giving the appearance of beads on a string (see fig. 4.19). Chromatin is the material that becomes organized and compacted to form **chromosomes.** When cell division begins, these fibers more tightly coil to form the rodlike chromosomes. The DNA molecules contain genes, the information for synthesis of proteins. The tightness in which chromatin is folded locally varies along the chromosomes, depending upon which genes are being accessed for their information at a particular time. "Chromatin" means colored substance, and "chromosome" means colored body.

Table 3.2 summarizes the structures and functions of cell parts.

PRACTICE 

14 How are the nuclear contents separated from the cytoplasm?

15 What is the function of the nucleolus?

16 What is chromatin?

3.3 MOVEMENTS INTO AND OUT OF THE CELL

The cell membrane is a barrier that controls which substances enter and leave the cell. Oxygen and nutrient molecules enter through this membrane, whereas carbon dioxide and other wastes leave through it. These movements involve *physical* (or passive) processes, such as diffusion, osmosis, facilitated diffusion, and filtration, and *physiological* (or active) processes, such as active transport, endocytosis, and exocytosis. Understanding the mechanisms that transport substances across the cell membrane is important for understanding many aspects of physiology.

Diffusion

Diffusion (dĭ-fu′zhun) (also called simple diffusion) is the tendency of atoms, molecules, and ions in a liquid or air solution to move from areas of higher concentration to areas of lower concentration, thus becoming more evenly distributed, or more *diffuse.* Diffusion occurs because atoms, molecules, and ions are in constant motion. Each particle travels in a separate path along a straight line until it collides with

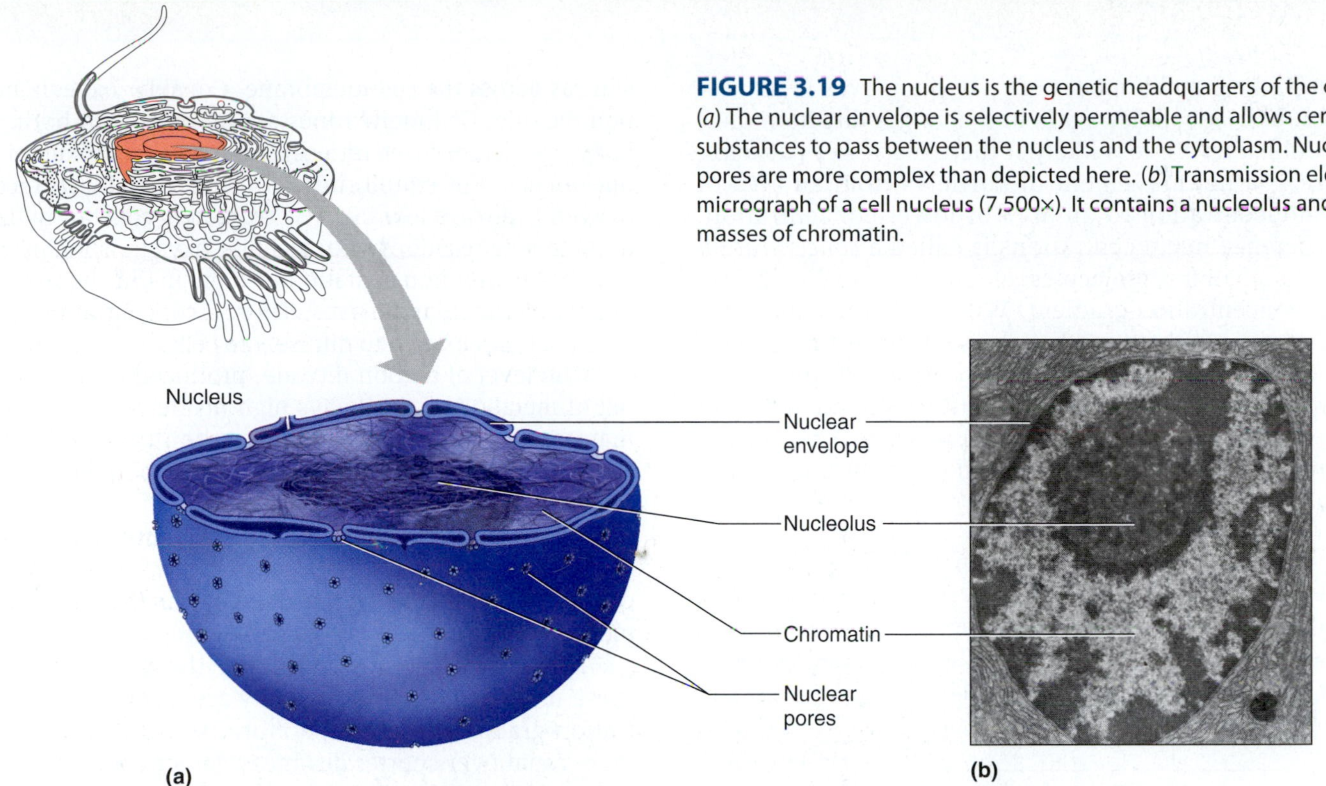

FIGURE 3.19 The nucleus is the genetic headquarters of the cell. (*a*) The nuclear envelope is selectively permeable and allows certain substances to pass between the nucleus and the cytoplasm. Nuclear pores are more complex than depicted here. (*b*) Transmission electron micrograph of a cell nucleus (7,500×). It contains a nucleolus and masses of chromatin.

TABLE 3.2 | Structures and Functions of Cell Parts

Cell Parts	Structure	Function
Cell membrane	Membrane mainly composed of protein and lipid molecules	Maintains integrity of the cell, controls the passage of materials into and out of the cell, and provides for signal transduction
Ribosomes	Particles composed of protein and RNA molecules	Synthesize proteins
Endoplasmic reticulum	Complex of connected, membrane-bound sacs, canals, and vesicles	Transports materials in the cell, provides attachment for ribosomes, and synthesizes lipids
Vesicles	Membranous sacs	Contain substances that recently entered the cell, store and transport newly synthesized molecules
Golgi apparatus	Group of flattened, membranous sacs	Packages and modifies protein molecules for transport and secretion
Mitochondria	Membranous sacs with inner partitions	Release energy from food molecules and convert the energy into a usable form
Lysosomes	Membranous sacs	Contain enzymes capable of digesting worn cellular parts or substances that enter cells
Peroxisomes	Membranous vesicles	Contain enzymes called peroxidases, important in the breakdown of many organic molecules
Centrosome	Nonmembranous structure composed of two rodlike centrioles	Helps distribute chromosomes to new cells during cell division, initiates formation of cilia
Cilia	Motile projections attached to basal bodies beneath the cell membrane	Propel fluids over cellular surface
Flagella	Motile projections attached to basal bodies beneath the cell membrane	Enable sperm cells to move
Microfilaments and microtubules	Thin rods and tubules	Support cytoplasm, help move substances and organelles within the cytoplasm
Nuclear envelope	Porous double membrane that separates the nuclear contents from the cytoplasm	Maintains the integrity of the nucleus, controls the passage of materials between the nucleus and cytoplasm
Nucleolus	Dense, nonmembranous body composed of protein and RNA molecules	Site of ribosome formation
Chromatin	Fibers composed of protein and DNA molecules	Carries information for synthesizing proteins

another particle and bounces off. Then it moves in its new direction until it collides again and changes direction once more. Collisions are less likely if there are fewer particles, so there is a net movement of particles from an area of higher concentration to an area of lower concentration. This difference in concentrations is called a *concentration gradient,* and atoms, molecules, and ions are said to diffuse down a concentration gradient. With time, the concentration of a given substance becomes uniform throughout a solution. This is the condition of *diffusional equilibrium* (dĭ-fu′zhun-ul e″kwi-lib′re-um). At diffusional equilibrium, although random movements continue, there is no further net movement, and the concentration of a substance is uniform throughout the solution.

> Random motion mixes molecules. At body temperature, small molecules such as water move more than a thousand miles per hour. However, the internal environment is crowded from a molecule's point of view. A single molecule may collide with other molecules a million times each second.

Sugar (a solute) put into a glass of water (a solvent), can be used to illustrate diffusion (fig. 3.20). The sugar at first remains in high concentration at the bottom of the glass. As the sugar molecules move, they may collide or miss each other. They are less likely to collide where there are fewer sugar molecules, so sugar molecules gradually diffuse from areas of higher concentration to areas of lower concentration (*down* the concentration gradient), and eventually become uniformly distributed in the water.

Diffusion of a substance across a membrane can occur only if (1) the cell membrane is permeable to that substance and (2) a concentration gradient exists such that the substance is at a higher concentration on one side of the membrane or the other (fig. 3.21). This principle applies to diffusion of substances across the cell membrane. Consider oxygen and carbon dioxide. Cell membranes are permeable to both. In the body, oxygen diffuses into cells and carbon dioxide diffuses out of cells, but equilibrium is never reached. Intracellular oxygen is always low because oxygen is constantly used up in metabolic reactions. Extracellular oxygen is maintained at a high level by homeostatic mechanisms in the respiratory and cardiovascular systems. Thus, a concentration gradient always allows oxygen to diffuse into cells.

The level of carbon dioxide, produced as a waste product of metabolism, is always high inside cells. Homeostasis maintains a lower extracellular carbon dioxide level, so a concentration gradient always favors carbon dioxide diffusing out of cells (fig. 3.22).

Diffusional equilibrium does not normally occur in organisms. The term *physiological steady state,* where concentrations of diffusing substances are unequal but stable, is more appropriate.

A number of factors influence the diffusion rate, but those most important in the body are distance, the concentration gradient, and temperature. In general, diffusion is more rapid over shorter distances, larger concentration gradients, and at higher temperatures. Homeostasis maintains all three of these factors at optimum levels.

Facilitated Diffusion

Some of the previous examples considered hypothetical membranes with specific permeabilities. For the cell membrane, permeability is more complex because of its selective nature. Lipid-soluble substances, such as oxygen, carbon dioxide, steroids, and general anesthetics, freely cross the cell membrane by simple diffusion. Small solutes that are not lipid-soluble, such as ions of sodium, potassium, and chloride, may diffuse through protein channels in the membrane, described earlier. (Water molecules also diffuse through similar channels, called pores.) This type of movement follows the concentration

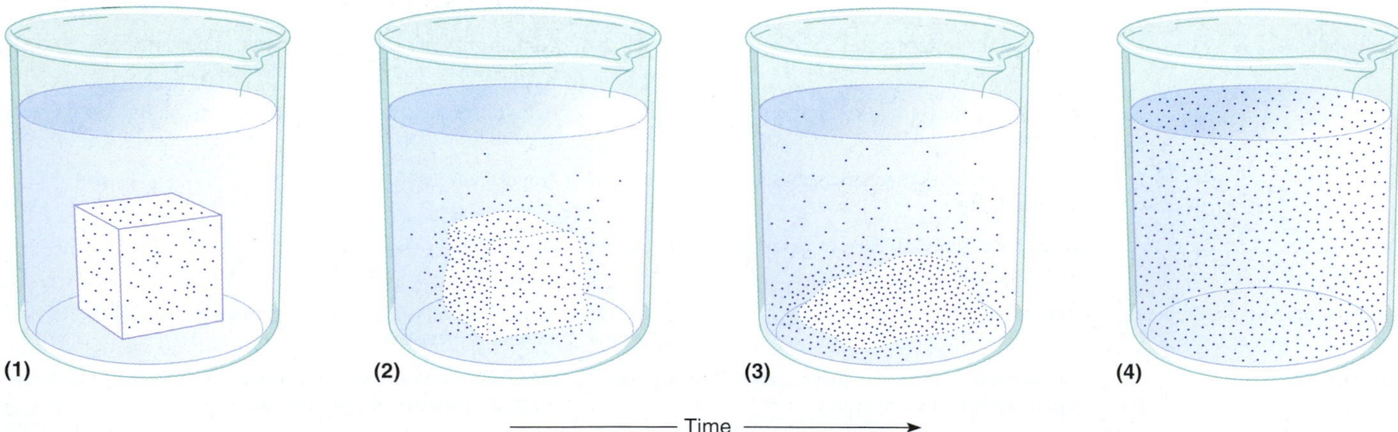

FIGURE 3.20 A dissolving sugar cube illustrates diffusion. *(1–3)* A sugar cube placed in water slowly disappears as the sugar molecules dissolve and then diffuse from regions where they are more concentrated toward regions where they are less concentrated. *(4)* Eventually, the sugar molecules are distributed evenly throughout the water.

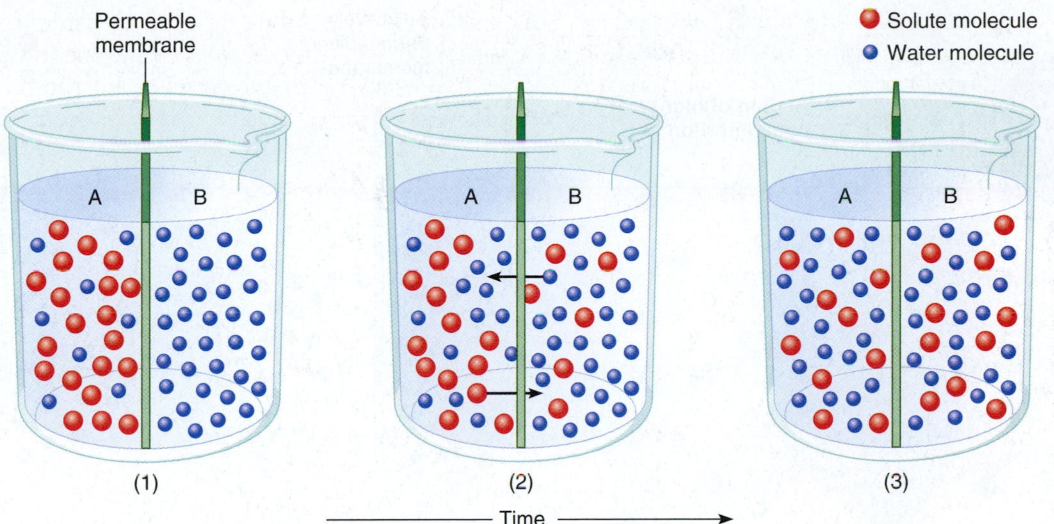

FIGURE 3.21 Diffusion is a passive movement of molecules. (*1*) A membrane permeable to water and solute molecules separates a container into two compartments. Compartment *A* contains both types of molecules, while compartment *B* contains only water molecules. (*2*) As a result of molecular motions, solute molecules tend to diffuse from compartment *A* into compartment *B*; water molecules tend to diffuse from compartment *B* into compartment *A*. (*3*) Eventually, equilibrium is reached.

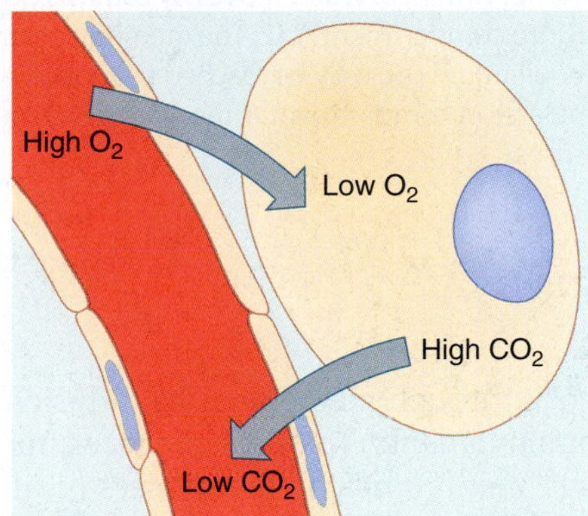

FIGURE 3.22 Oxygen enters cells and carbon dioxide leaves cells by diffusion.

gradient but uses membrane proteins as "carriers," so it is termed **facilitated diffusion** (fah-sil″ĭ-tāt′ed dĭ-fu′zhun). Facilitated diffusion is very important not only for ions, but for larger water-soluble molecules, such as glucose and amino acids.

Most sugars and amino acids are insoluble in lipids, and they are too large to pass through cell membrane pores. Facilitated diffusion includes not only protein channels, but also certain proteins that function as carriers to bring such molecules across the cell membrane. In the facilitated diffusion of glucose, for example, glucose combines with a protein carrier molecule at the surface of the cell membrane. This union of glucose and carrier molecule changes the shape of the carrier in a way that moves glucose to the inner surface of the membrane. The glucose portion is released, and the carrier molecule returns to its original shape to pick up another glucose molecule. The hormone *insulin*, discussed in chapter 13 (p. 510), promotes facilitated diffusion of glucose through the membranes of certain cells.

Facilitated diffusion is similar to simple diffusion in that it can move molecules only from regions of higher concentration toward regions of lower concentration. However, unlike simple diffusion, the number of carrier molecules in the cell membrane limits the rate of facilitated diffusion (fig. 3.23).

Osmosis

Osmosis (oz-mo′sis) is the movement of water across a selectively permeable membrane into a compartment containing solute that cannot cross the same membrane. What you may have heard is essentially true—"water follows salt"—although any impermeant solute will draw water by osmosis. The mechanism of osmosis is complex and beyond the scope of this discussion, but in part involves diffusion of water. Therefore, it might help to think of the entire process as diffusion of water down its concentration gradient. In the following example, assume that the membrane is permeable to water (the solvent) and impermeable to protein (the solute).

In solutions, a higher concentration of solute means a lower concentration of water; a lower concentration of solute means a higher concentration of water. This is because the solute molecules take up space that water molecules would otherwise occupy.

Like molecules of other substances, molecules of water will move from areas of higher concentration to areas of lower concentration. In figure 3.24, the greater concentration of protein in compartment *A* means that the water concentration there is less than the concentration of pure water in compartment *B*. Therefore, water moves from compartment *B* across the selectively permeable membrane and into compartment *A*. In other words, water moves from compartment *B* into compartment *A* by osmosis. Protein, on the other hand, cannot move out of compartment *A* because the selectively permeable membrane is impermeable to it. Note

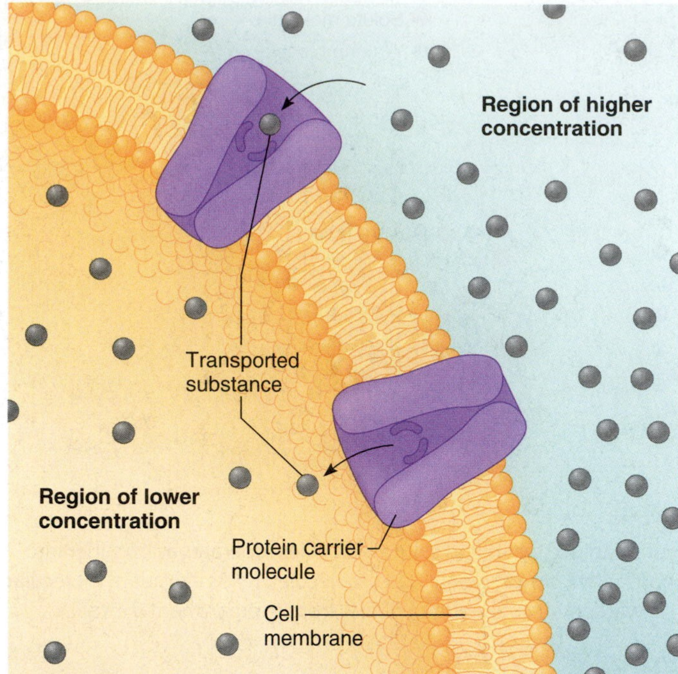

FIGURE 3.23 Facilitated diffusion uses carrier molecules to transport some substances into or out of cells, from a region of higher concentration to one of lower concentration.

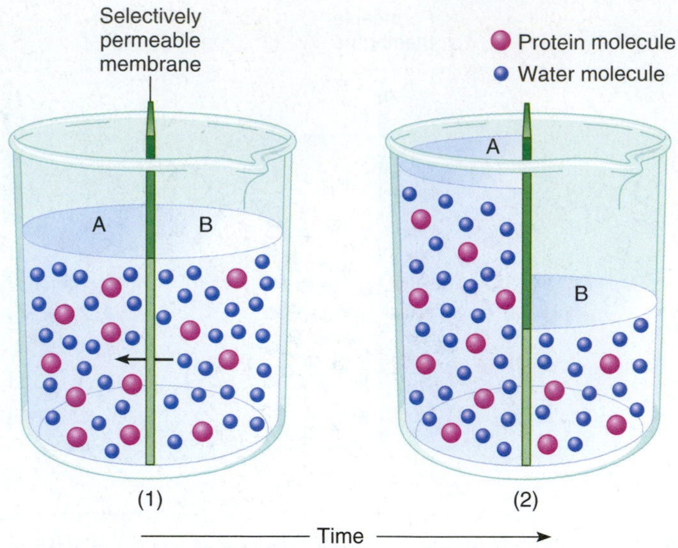

FIGURE 3.24 Osmosis. (*1*) A selectively permeable membrane separates the container into two compartments. At first, compartment *A* contains a higher concentration of protein (and a lower concentration of water) than compartment *B*. Water moves by osmosis from compartment *B* into compartment *A*. (*2*) The membrane is impermeable to proteins, so equilibrium can only be reached by movement of water. As water accumulates in compartment *A*, the water level on that side of the membrane rises.

in figure 3.24 that as osmosis occurs, the level of water on side *A* rises. This ability of osmosis to generate enough pressure to lift a volume of water is called *osmotic pressure.* Thus the osmotic movement of water alone achieves equilibrium.

The greater the concentration of impermeant solute particles (protein in this case) in a solution, the *lower* the water concentration of that solution and the *greater* the osmotic pressure. Water always tends to move toward solutions of greater osmotic pressure.

Cell membranes are generally permeable to water, so water equilibrates by osmosis throughout the body, and the concentration of water and solutes everywhere in the intracellular and extracellular fluids is essentially the same. Therefore, the osmotic pressure of the intracellular and extracellular fluids is the same. Any solution, such as a 0.9% NaCl solution (normal saline), that has the same osmotic pressure as body fluids is called **isotonic.** Cells will not change size in this solution.

Solutions that have a higher osmotic pressure than body fluids are called **hypertonic.** If cells are put into a hypertonic solution, there will be a net movement of water by osmosis out of the cells into the surrounding solution, and the cells shrink. Conversely, cells put into a **hypotonic** solution, which has a lower osmotic pressure than body fluids, gain water by osmosis and swell or possibly even burst (hemolyze). Although cell membranes are somewhat elastic, the cells may swell so much that they burst. Figure 3.25 illustrates the effects of the three types of solutions on red blood cells.

Filtration

Molecules move through membranes by diffusion because of their random movements. In other instances, molecules are forced through membranes by the process of **filtration** (fil-tra′shun).

Filtration is commonly used to separate solids from water. One method is to pour a mixture of solids and water onto filter paper in a funnel (fig. 3.26). The paper serves as a porous membrane through which the small water molecules can pass, leaving the larger solid particles behind. Hydrostatic pressure, created by the weight of water due to gravity, forces the water molecules through to the other side. An example of this is making coffee by the drip method.

In the body, tissue fluid forms when water and dissolved substances are forced out through the thin, porous walls of blood capillaries, but larger particles such as blood protein molecules are left inside (fig. 3.27). The force for this movement comes from blood pressure, generated largely by heart action, which is greater within the vessel than outside it. However, the impermeant proteins tend to hold water in blood vessels by osmosis, thus preventing the formation of excess tissue fluid, a condition called edema. (Although heart action is an active body process, filtration is considered passive because it can occur due to the pressure caused by gravity alone.) Filtration is discussed further in chapters 15 (p. 578) and 20 (p. 788).

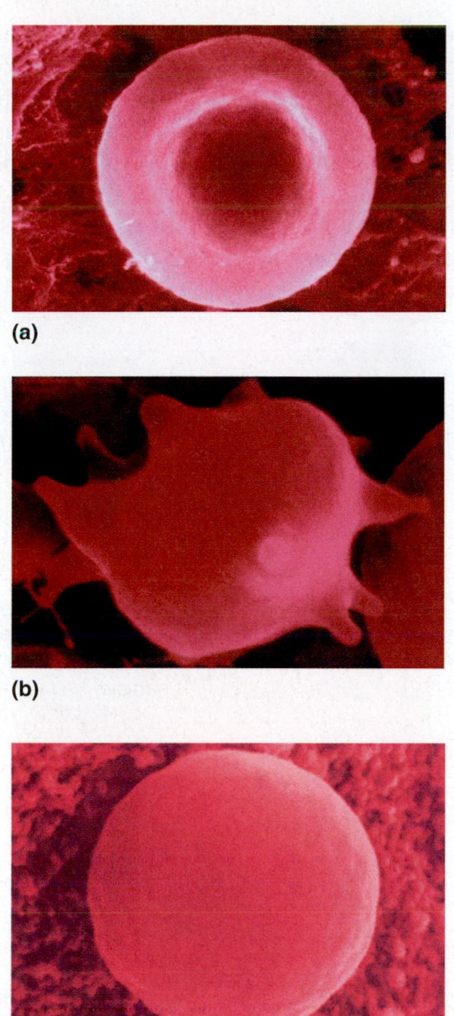

FIGURE 3.25 When red blood cells are placed (*a*) in an isotonic solution, equal volumes of water enter and leave the cells, and size and shape remain unchanged. (*b*) In a hypertonic solution, more water leaves than enters, so cells shrink. (*c*) In a hypotonic solution, more water enters than leaves, so cells swell and may burst (5,000×).

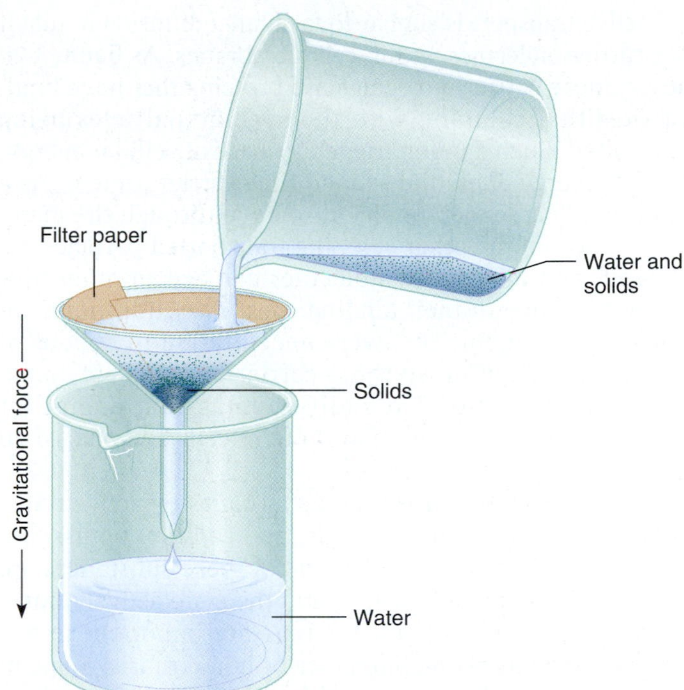

FIGURE 3.26 In filtration of water and solids, gravity forces water through filter paper, while tiny openings in the paper retain the solids. This process is similar to the drip method of preparing coffee.

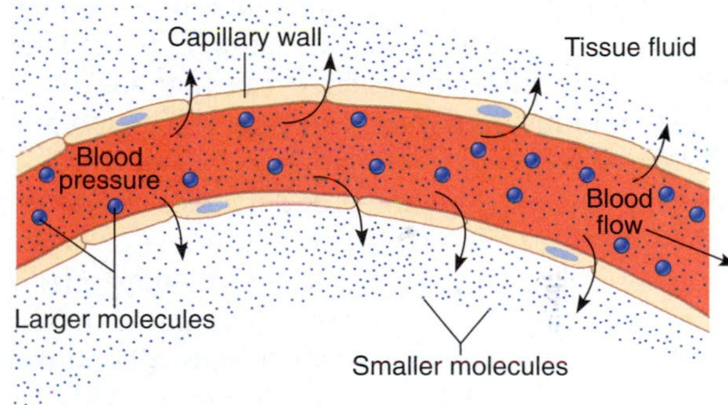

FIGURE 3.27 In filtration in the body, blood pressure forces smaller molecules through tiny openings in the capillary wall. The larger molecules remain inside.

PRACTICE

17 What kinds of substances most readily diffuse through a cell membrane?

18 Explain the differences among diffusion, facilitated diffusion, and osmosis.

19 Distinguish among isotonic, hypertonic, and hypotonic solutions.

20 Explain how filtration occurs in the body.

Active Transport

When molecules or ions pass through cell membranes by diffusion or facilitated diffusion their net movement is from regions of higher concentration to regions of lower concentration. Sometimes, however, the net movement of particles passing through membranes is in the opposite direction, from a region of lower concentration to one of higher concentration.

Sodium ions, for example, can diffuse slowly through cell membranes. Yet the concentration of these ions typically remains many times greater outside cells (in the extracellular fluid) than inside cells (in the intracellular fluid). This is because sodium ions are continually moved through the cell membrane from regions of lower concentration (inside) to regions of higher concentration (outside). Movement against a concentration gradient is called **active transport** (ak′tiv trans′port) and requires energy derived from cellular metabolism. Up to 40% of a cell's energy supply may be used for active transport of particles through its membranes.

Active transport is similar to facilitated diffusion in that it uses carrier molecules within cell membranes. As figure 3.28 shows, these carrier molecules are proteins that have binding sites that combine with the specific particles being transported. Such a union triggers release of cellular energy, and this energy alters the shape of the carrier protein. As a result, the "passenger" molecules move through the membrane. Once on the other side, the transported particles are released, and the carrier molecules can accept other passenger molecules at their binding sites. They transport substances from regions of lower concentration to regions of higher concentration, so these carrier proteins are sometimes called "pumps." A sodium/potassium pump, for example, transports sodium ions out of cells and potassium ions into cells.

Particles moved across cell membranes by active transport include sugars, amino acids, and sodium, potassium, calcium, and hydrogen ions. Some of these substances are actively transported into cells, and others are actively transported out. Movements of this type are important to cell survival, particularly maintenance of homeostasis. Some of these movements are described in subsequent chapters as they apply to specific organ systems.

Endocytosis

Cellular energy is used to move substances into or out of a cell without actually crossing the cell membrane. In **endocytosis** (en″do-si-to′sis), molecules or other particles that are too large to enter a cell by diffusion or active transport are conveyed in a vesicle that forms from a section of the cell membrane.

The three forms of endocytosis are pinocytosis, phagocytosis, and receptor-mediated endocytosis. In **pinocytosis** (pi″-no-si-to′sis), cells take in tiny droplets of liquid from their surroundings (fig. 3.29). When this happens, a small portion of cell membrane indents (invaginates). The open end of the tubelike part thus formed seals off and produces a small vesicle about 0.1 μm in diameter. This tiny sac detaches from the surface and moves into the cytoplasm. For a time, the vesicular membrane, part of the cell membrane, separates its contents from the rest of the cell; however, the membrane eventually breaks down, and the liquid inside becomes part of the cytoplasm. In this way, a cell is able to take in water and the particles dissolved in it, such as proteins, that otherwise might be too large to enter.

Phagocytosis (fag″o-si-to′sis) is similar to pinocytosis, but the cell takes in solids rather than liquid. Certain types of cells, including some white blood cells, are called **phagocytes** because they can take in solid particles such as bacteria and cellular debris. When a phagocyte first encounters such a particle, the particle attaches to the cell membrane. This stimulates a portion of the membrane to project outward, surround

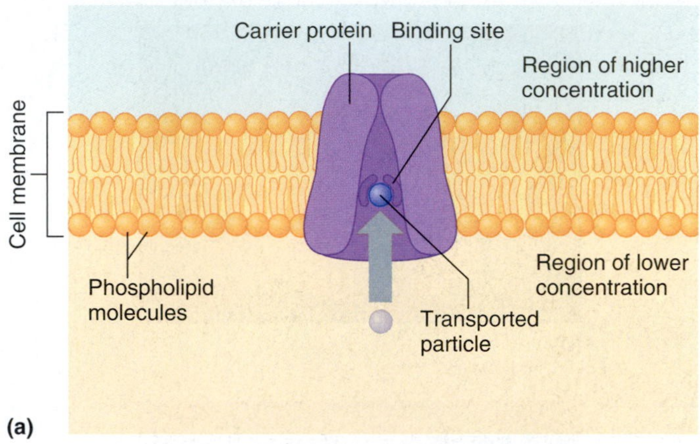

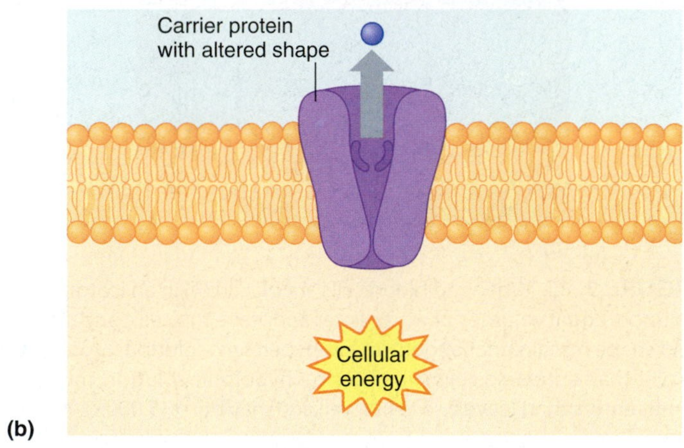

FIGURE 3.28 Active transport moves molecules against their concentration gradient. (*a*) During active transport, a molecule or an ion combines with a carrier protein, whose shape changes as a result. (*b*) This process, which requires cellular energy, transports the particle across the cell membrane.

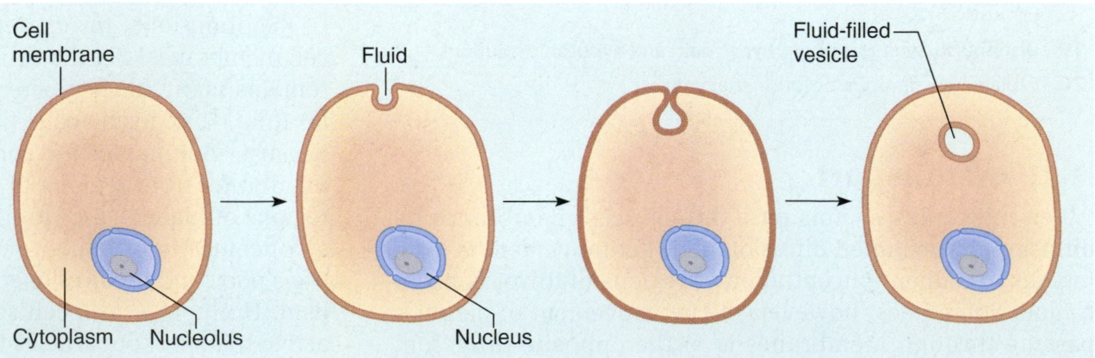

FIGURE 3.29 A cell may take in a tiny droplet of fluid from its surroundings by pinocytosis.

the particle, and slowly draw it inside the cell. The part of the membrane surrounding the solid detaches from the cell's surface, forming a vesicle containing the particle (fig. 3.30). Such a vesicle may be several micrometers in diameter.

Usually, a lysosome joins a newly formed vesicle, and lysosomal digestive enzymes decompose the contents (fig. 3.31). The products of this decomposition may then diffuse out of the lysosome and into the cytoplasm, where they may be used as raw materials in metabolic processes. Exocytosis may expel any remaining residue. In this way, phagocytic cells dispose of foreign objects, such as dust particles; remove damaged cells or cell parts that are no longer functional; or destroy disease-causing microorganisms. Phagocytosis is an important line of defense against infection.

Pinocytosis and phagocytosis engulf nonspecifically. In contrast is the more discriminating **receptor-mediated endocytosis,** which moves very specific types of particles into the cell. This process involves protein molecules that extend through the cell membrane and are exposed on its outer surface. These proteins are receptors to which specific molecules from the fluid surroundings of the cell can bind. Molecules that can bind to the receptor sites selectively enter the cell; other types of molecules are left outside (fig. 3.32). Molecules that bind specifically to receptors are called *ligands.*

Entry of cholesterol molecules into cells illustrates receptor-mediated endocytosis. Cholesterol molecules synthesized in liver cells are packaged into large spherical particles called *low-density lipoproteins* (LDL). An LDL particle has a coating that contains a binding protein called *apolipoprotein-B.* The membranes of various body cells (including liver cells)have receptors for apolipoprotein-B. When the liver releases LDL particles into the blood, cells with apolipoprotein-B receptors can recognize the LDL particles and bind them. Formation of such a receptor-ligand combination stimulates the cell membrane to indent and form a vesicle around the LDL particle. The vesicle carries the LDL particle to a lysosome, where enzymes digest it and release the cholesterol molecules for cellular use.

More than 25 million people in the United States take cholesterol-lowering drugs called statins. The drugs inhibit an enzyme, HMG-CoA reductase, which cells use to produce cholesterol—in addition to cholesterol we eat. Feedback is at play. When levels of the enzyme drop with taking the drug, liver cells are stimulated to make more LDL receptors. With statin use not only does the body make less cholesterol, but the more abundant LDL receptors remove cholesterol from the bloodstream more efficiently. Combined with a low-fat diet, taking a statin powerfully lowers blood serum cholesterol.

The idea to limit cholesterol synthesis inside the body came from studies of rare individuals with an inherited disease that prevents their cells from making LDL receptors. Excess cholesterol is deposited under the skin, appearing as yellowish lumps behind the knees. These individuals die of heart disease before age 20. Japanese researchers developed the first statin based on understanding the relationship between HMG-CoA reductase and the number of LDL receptors. The first statin was approved in the United States in 1987. Today the more than a dozen statin drugs differ by potency.

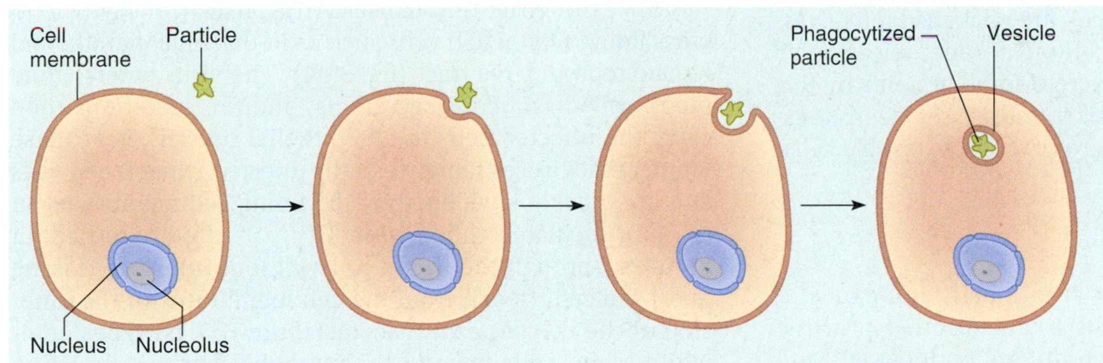

FIGURE 3.30 A cell may take in a solid particle from its surroundings by phagocytosis.

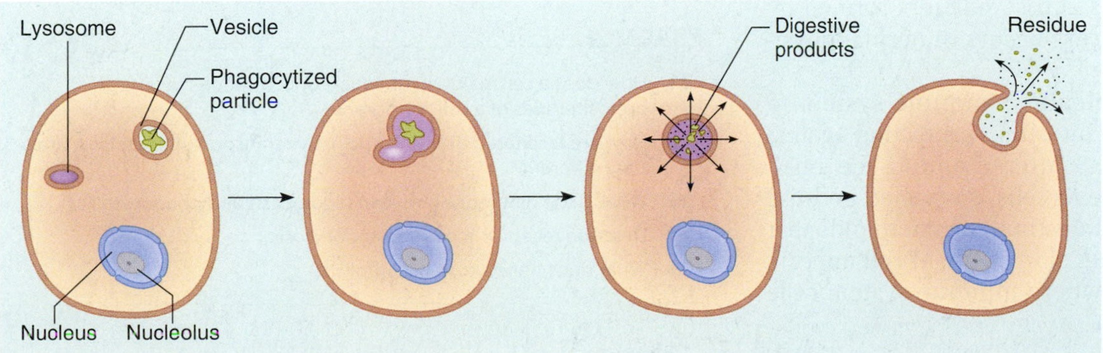

FIGURE 3.31 When a lysosome envelopes a vesicle that contains a phagocytized particle, its digestive enzymes may destroy the particle. The products of this intracellular digestion diffuse into the cytoplasm. Exocytosis may expel any residue.

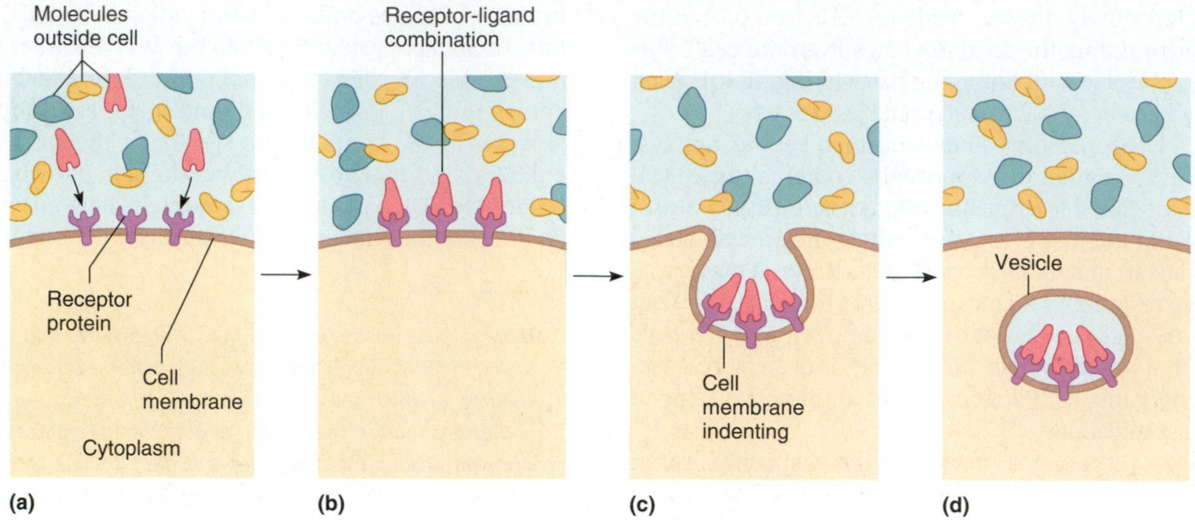

FIGURE 3.32 Receptor-mediated endocytosis. (*a, b*) A specific molecule binds to a receptor protein, forming a receptor-ligand combination. (*c*) The binding of the ligand to the receptor protein stimulates the cell membrane to indent. (*d*) Continued indentation forms a vesicle, which transports the molecule into the cytoplasm.

Receptor-mediated endocytosis is particularly important because it allows cells with the appropriate receptors to remove and process specific types of substances from their surroundings, even when these substances are present in very low concentrations. In short, receptor-mediated endocytosis provides specificity (fig. 3.32).

Exocytosis

Exocytosis (ex-o-si-to′sis) is essentially the reverse of endocytosis. Substances made in the cell are packaged into a vesicle, which then fuses with the cell membrane, releasing its contents outside the cell. Cells secrete some proteins by this process. Nerve cells use exocytosis to release the neurotransmitter chemicals that signal other nerve cells, muscle cells, or glands (fig. 3.33).

Transcytosis

Endocytosis brings a substance into a cell, and exocytosis transports a substance out of a cell. Another process, **transcytosis** (tranz-si-to′sis), combines endocytosis and exocytosis to selectively and rapidly transport a substance or particle from one end of a cell to the other (fig. 3.34). Transcytosis moves substances across barriers formed by tightly connected cells. The process occurs in normal physiology and in disease.

Transcytosis enables the healthy immune system to monitor pathogens in the small intestine, protecting against some forms of food poisoning. Scattered among the small intestinal epithelial cells are rare M cells, so-named because the cell side that faces into the intestine has microfolds that maximize surface area. The other side of the M cell appears punched in, forming a pocket where immune system cells gather. The M cell binds and takes in a bacterium from the intestinal side by endocytosis, then transports it through the cell to the side that faces the immune system cells, where it is released by exocytosis. The immune system cells bind parts of the bacterium, and, if they recognize surface features of a pathogen, they signal other cells to mature into antibody-producing cells. The antibodies are then secreted into the bloodstream and travel back to the small intestine, where they destroy the infecting bacteria.

HIV, the virus that causes AIDS, uses transcytosis to cross lining (epithelial) cells such as in the anus, mouth, and female reproductive tract (fig. 3.34). The virus enters white blood cells in mucous secretions, and the secretions then carry the infected cells to an epithelial barrier. Near these lining cells, viruses rapidly exit the infected white blood cells and are quickly enveloped by the lining cell membranes in receptor-mediated endocytosis. HIV particles are ferried, in vesicles, through the lining cell, without infecting (taking over) the cell, to exit from the cell membrane on the other side of the cell. After transcytosis, the HIV particles enter white blood cells beyond the epithelial barrier. Infection begins. Table 3.3 summarizes the types of movement into and out of the cell, including transcytosis.

PRACTICE 

21 How does a cell maintain unequal concentrations of ions on opposite sides of a cell membrane?

22 How are facilitated diffusion and active transport similar? How are they different?

23 What is the difference between pinocytosis and phagocytosis?

24 Describe receptor-mediated endocytosis.

25 What does transcytosis accomplish?

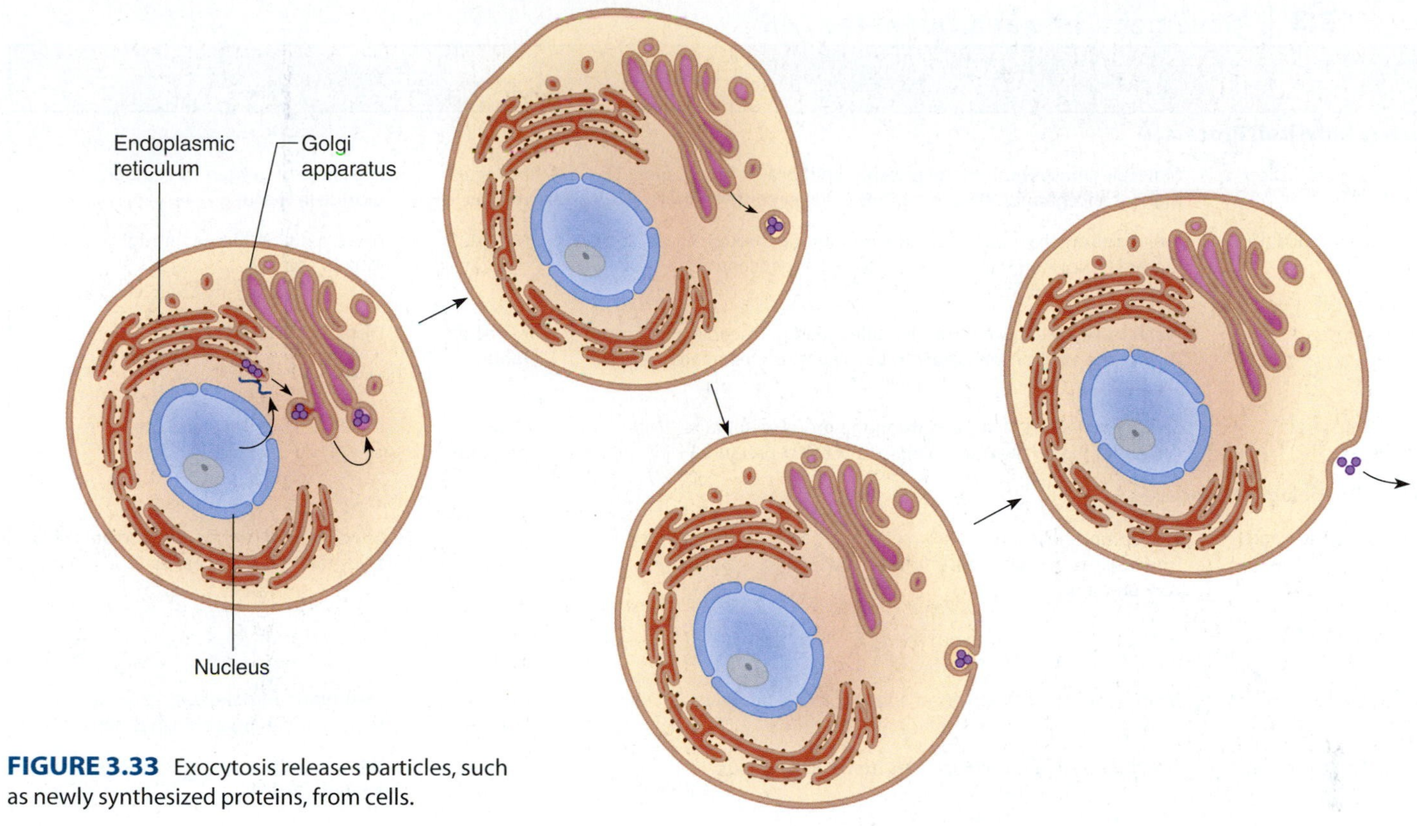

FIGURE 3.33 Exocytosis releases particles, such as newly synthesized proteins, from cells.

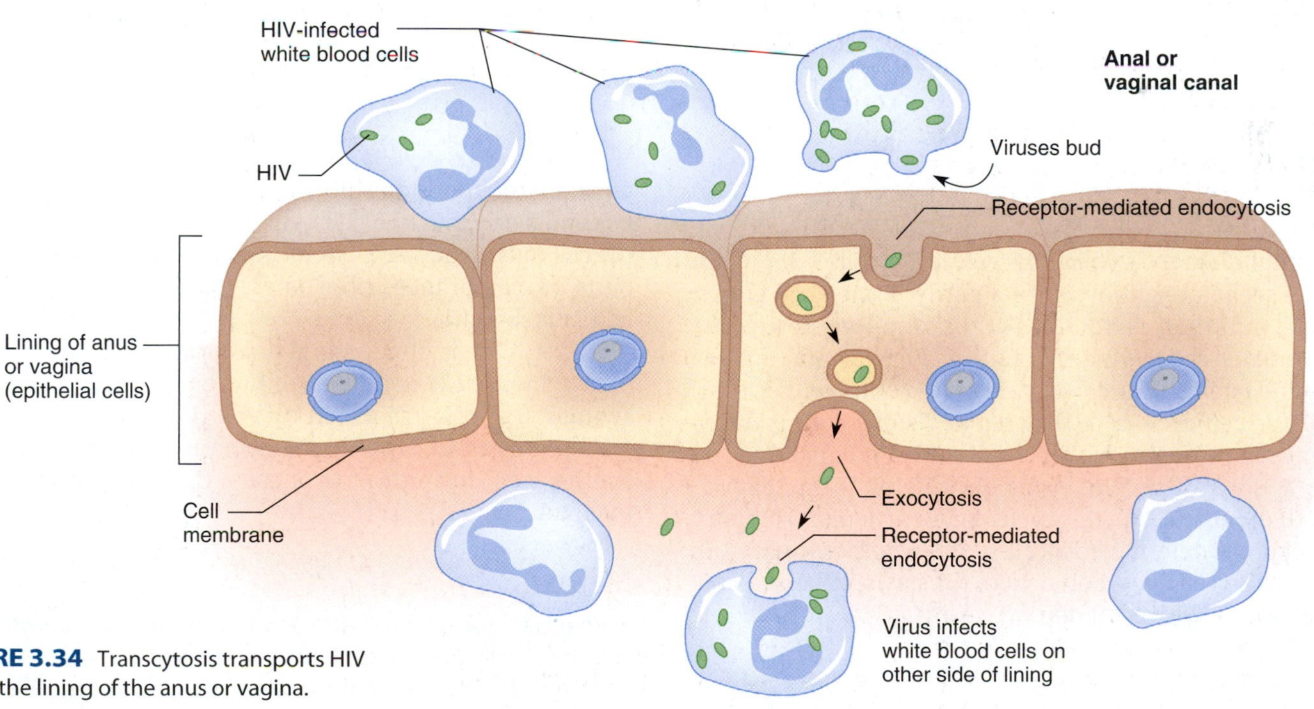

FIGURE 3.34 Transcytosis transports HIV across the lining of the anus or vagina.

TABLE 3.3 | Movements Into and Out of the Cell

Process	Characteristics	Source of Energy	Example
I. Passive (Physical) Processes			
A. Simple diffusion	Molecules move through the phospholipid bilayer from regions of higher concentration toward regions of lower concentration.	Molecular motion	Exchange of oxygen and carbon dioxide in the lungs
B. Facilitated diffusion	Molecules or ions move across the membrane through channels or by carrier molecules from a region of higher concentration to one of lower concentration.	Molecular motion	Movement of glucose through a cell membrane
C. Osmosis	Water molecules move through a selectively permeable membrane toward the solution with more impermeant solute (greater osmotic pressure).	Molecular motion	Distilled water entering a cell
D. Filtration	Smaller molecules are forced through porous membranes from regions of higher pressure to regions of lower pressure.	Hydrostatic pressure	Molecules leaving blood capillaries
II. Active (Physiological) Processes			
A. Active transport	Carrier molecules transport molecules or ions through membranes from regions of lower concentration toward regions of higher concentration.	Cellular energy	Movement of various ions and amino acids through membranes
B. Endocytosis			
1. Pinocytosis	Membrane engulfs droplets of liquid from surroundings.	Cellular energy	Membrane-forming vesicles containing large particles dissolved in water
2. Phagocytosis	Membrane engulfs solid particles from surroundings.	Cellular energy	White blood cell membrane engulfing bacterial cell
3. Receptor-mediated endocytosis	Membrane engulfs selected molecules combined with receptor proteins.	Cellular energy	Cell removing cholesterol-containing LDL particles from its surroundings
C. Exocytosis	Vesicles fuse with membrane and release contents outside of the cell.	Cellular energy	Protein secretion, neurotransmitter release
D. Transcytosis	Combines receptor-mediated endocytosis and exocytosis to ferry particles through a cell.	Cellular energy	HIV crossing a cell layer

3.4 THE CELL CYCLE

The series of changes that a cell undergoes, from the time it forms until it divides, is called the **cell cycle** (fig. 3.35). This cycle may seem straightforward—a newly formed cell grows for a time, and then divides in half to form two new cells, called daughter cells, which, in turn, may grow and divide. The specific events of the cycle are quite complex. For ease of study, the cell cycle is considered in distinct stages: interphase, mitosis, cytoplasmic division, and differentiation.

The actions of several types of proteins form "checkpoints" that control the cell cycle. One particularly important checkpoint determines a cell's fate, whether it will: (a) continue in the cell cycle and divide; (b) stay specialized and alive, yet not divide; or (c) die.

Interphase

Once thought to be a time of rest, **interphase** is actually a very active period. During interphase, the cell grows and maintains its routine functions as well as its contributions to the internal environment.

If the cell is developmentally programmed to divide, it must amass important biochemicals and duplicate much of its contents so that two cells can form from one. For example, the cell must replicate DNA and synthesize and assemble the parts of membranes, ribosomes, lysosomes, peroxisomes, and mitochondria.

Interphase is divided into phases based on the sequence of activities. DNA is replicated during S phase (S stands for synthesis) and is bracketed by two G phases, G_1 and G_2 (G stands for gap or growth). Structures other than DNA are synthesized during the G phases. Cellular growth occurs then, too (see fig. 3.35).

Mitosis

Mitosis is a form of cell division that occurs in somatic (nonsex) cells and produces two daughter cells from an original cell (fig. 3.36). These new cells are genetically identical, each with the full complement of 46 chromosomes. In contrast is *meiosis*, a second form of cell division that occurs only in the cells that give rise to sex cells (sperm and eggs). Meiosis halves the chromosome number. In this way, when

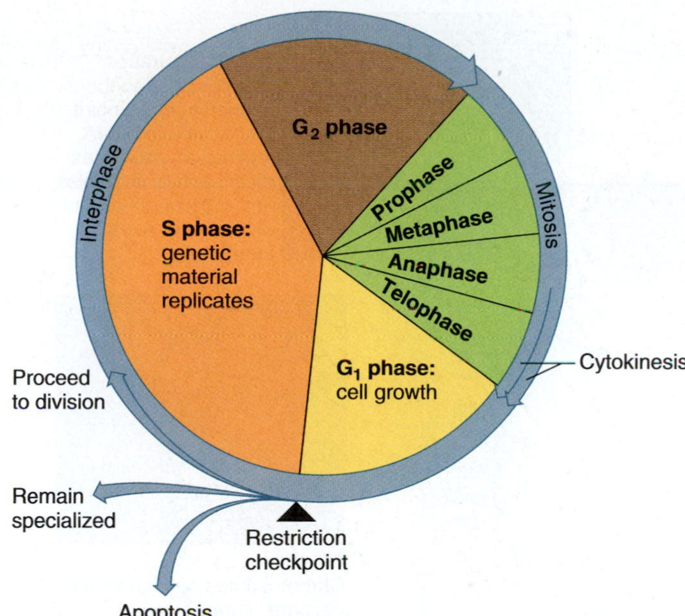

FIGURE 3.35 The cell cycle is divided into interphase, when cellular components duplicate, and cell division (mitosis and cytokinesis), when the cell splits in two, distributing its contents into two daughter cells. Interphase is divided into two gap phases (G_1 and G_2), when specific molecules and structures duplicate, and a synthesis phase (S), when DNA replicates. Mitosis can be considered in stages—prophase, metaphase, anaphase, and telophase.

a sperm fertilizes an egg, the total number of 46 chromosomes is restored. Chapter 22 (pp. 831–833) considers meiosis in detail.

During mitosis, the nuclear contents divide in an event called karyokinesis, which means "nucleus movement." Then the cytoplasm is apportioned into the two daughter cells in a process called cytokinesis, which means "cell movement." Mitosis must be very precise so that each new cell receives a complete copy of the genetic information. The chromosomes were duplicated in interphase, but it is in mitosis that the chromosome sets evenly distribute between the two forming cells.

Mitosis is a continuous process, but it is described in stages that indicate the sequence of major events, as follows:

1. **Prophase.** One of the first indications that a cell is going to divide is the condensation of chromatin fibers into tightly coiled rods. These are the chromosomes. During interphase, following DNA replication (discussed in chapter 4, page 127), each chromosome consists of two identical structures, called chromatids, temporarily attached by a region on each called a *centromere.*

   The centrioles of the centrosome replicate just before the onset of mitosis (fig. 3.36*a*), and during prophase, the two newly formed pairs of centrioles move to opposite sides of the cell. Soon the nuclear envelope and the nucleolus disperse and are no longer visible. Microtubules are assembled from tubulin proteins in the cytoplasm, and these structures associate with the centrioles and chromosomes. A spindle-shaped array of microtubules (spindle fibers) forms between the centrioles as they move apart (fig. 3.36*b*).
2. **Metaphase.** Spindle fibers attach to the centromeres so that a fiber accompanying one chromatid attaches to one centromere and a fiber accompanying the other chromatid attaches to its centromere (fig. 3.36*c*). The chromosomes move along the spindle fibers and are aligned about midway between the centrioles as a result of microtubule activity.
3. **Anaphase.** Soon the centromeres of the chromatids separate, and these identical chromatids are now considered individual chromosomes. The separated chromosomes move in opposite directions, and once again, the movement results from microtubule activity. The spindle fibers shorten and pull their attached chromosomes toward the centrioles at opposite sides of the cell (fig. 3.36*d*).
4. **Telophase.** The final stage of mitosis begins when the chromosomes complete their migration toward the centrioles. It is much like the reverse of prophase. As the identical sets of chromosomes approach their respective centrioles, they begin to elongate and unwind from rodlike structures to threadlike structures. A nuclear envelope forms around each chromosome set, and nucleoli become visible within the newly formed nuclei. Finally, the microtubules disassemble into free tubulin molecules (fig. 3.36*e*).

Table 3.4 summarizes the stages of mitosis.

> Mitosis is sometimes called cellular reproduction, because it results in two cells from one—the cell reproduces. This may be confusing, because meiosis is the prelude to human sexual reproduction. Both mitosis and meiosis are forms of cell *division,* with similar steps but different outcomes, and occurring in different types of cells.

Cytoplasmic Division

Cytoplasmic division (cytokinesis) begins during anaphase when the cell membrane starts to constrict around the middle, which it continues to do through telophase. The musclelike contraction of a ring of actin microfilaments pinches off two cells from one. The microfilaments assemble in the cytoplasm and attach to the inner surface of the cell membrane. The contractile ring forms at right angles to the microtubules that pulled the chromosomes to opposite ends of the cell during mitosis. As the ring pinches, it separates the two newly formed nuclei and apportions about half of the organelles into each of the daughter cells. The newly formed cells may differ slightly in size and number of organelles and inclusions, but they have identical chromosomes and thus contain

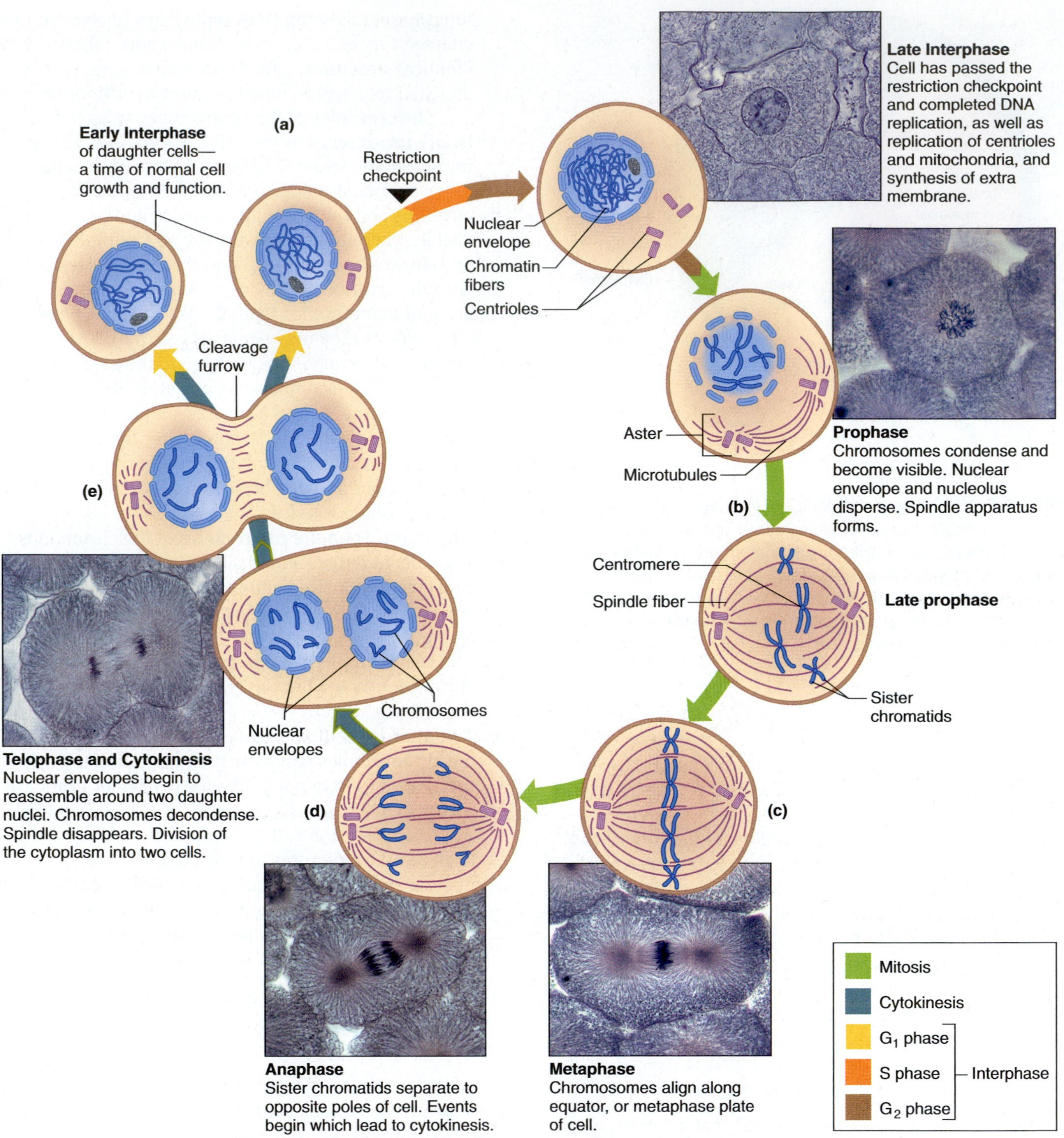
Early Interphase
of daughter cells—
a time of normal cell
growth and function.
(a)
Restriction
checkpoint
Nuclear
envelope
Chromatin
fibers
Centrioles
Late Interphase
Cell has passed the
restriction checkpoint
and completed DNA
replication, as well as
replication of centrioles
and mitochondria, and
synthesis of extra
membrane.
Cleavage
furrow
Aster
Microtubules
(b)
Prophase
Chromosomes condense and
become visible. Nuclear
envelope and nucleolus
disperse. Spindle apparatus
forms.
(e)
Centromere
Spindle fiber
Late prophase
Sister
chromatids
Chromosomes
Nuclear
envelopes
Telophase and Cytokinesis
Nuclear envelopes begin to
reassemble around two daughter
nuclei. Chromosomes decondense.
Spindle disappears. Division of
the cytoplasm into two cells.
(d)
(c)
Anaphase
Sister chromatids separate to
opposite poles of cell. Events
begin which lead to cytokinesis.
Metaphase
Chromosomes align along
equator, or metaphase plate
of cell.
Mitosis
Cytokinesis
G1 phase
S phase
G2 phase
Interphase

FIGURE 3.36 Mitosis and cytokinesis produce two cells from one. (*a*) During interphase, before mitosis, chromosomes are visible only as chromatin fibers. A single pair of centrioles is present, but not visible at this magnification. (*b*) In prophase, as mitosis begins, chromosomes have condensed and are easily visible when stained. The centrioles have replicated, and each pair moves to an opposite end of the cell. The nuclear envelope and nucleolus disappear, and spindle fibers associate with the centrioles and the chromosomes. (*c*) In metaphase, the chromosomes line up midway between the centrioles. (*d*) In anaphase, the centromeres are pulled apart by the spindle fibers, and the chromatids, now individual chromosomes, move in opposite directions. (*e*) In telophase, chromosomes complete their migration and become chromatin, the nuclear envelope reforms, and microtubules disassemble. Cytokinesis, which began during anaphase, continues during telophase. Not all chromosomes are shown in these drawings. (Micrographs approximately 360×)

TABLE 3.4 | Major Events in Mitosis

Stage	Major Events
Prophase	Chromatin condenses into chromosomes; centrioles move to opposite sides of cytoplasm; nuclear membrane and nucleolus disperse; microtubules assemble and associate with centrioles and chromatids of chromosomes.
Metaphase	Spindle fibers from the centrioles attach to the centromeres of each chromosome; chromosomes align midway between the centrioles.
Anaphase	Centromeres separate, and chromatids of the chromosomes separate; spindle fibers shorten and pull these new individual chromosomes toward centrioles.
Telophase	Chromosomes elongate and form chromatin threads; nuclear membranes form around each chromosome set; nucleoli form; microtubules break down.

identical DNA information (fig. 3.37). How that DNA is expressed (used to manufacture proteins) determines the specialization of the cell, a point we return to at the chapter's end (p. 106).

PRACTICE

26 Why is precise division of the genetic material during mitosis important?

27 Describe the events that occur during mitosis.

3.5 CONTROL OF CELL DIVISION

How often a cell divides is strictly controlled and varies with cell type. Skin cells, blood-forming cells, and cells that line the intestine, for example, divide often and continually. In contrast, the immature cells that give rise to neurons divide a specific number of times, and then cease—they become specialized and remain alive, but they no longer divide.

Most types of human cells divide from forty to sixty times when grown in the laboratory. Adherence to this limit can be startling. A connective tissue cell from a human fetus divides thirty-five to sixty-three times, the average being about fifty times. However, a similar cell from an adult divides only fourteen to twenty-nine times, as if the cell "knows" how many times it has already divided. In a body, however, signals from the immediate environment also influence mitotic potential.

A physical basis for this mitotic clock is the DNA at the tips of chromosomes, called *telomeres,* where the same six-nucleotide sequence repeats hundreds of times. Each mitosis removes up to 1,200 nucleotides. When the chromosome tips wear down to a certain point, this signals the cell to cease dividing. Studies show that severe psychological or emotional stress can hasten telomere shortening. This may be one way that stress can harm health.

Other external and internal factors influence the timing and frequency of mitosis. Within cells, waxing and waning levels of proteins called kinases and cyclins control the cell

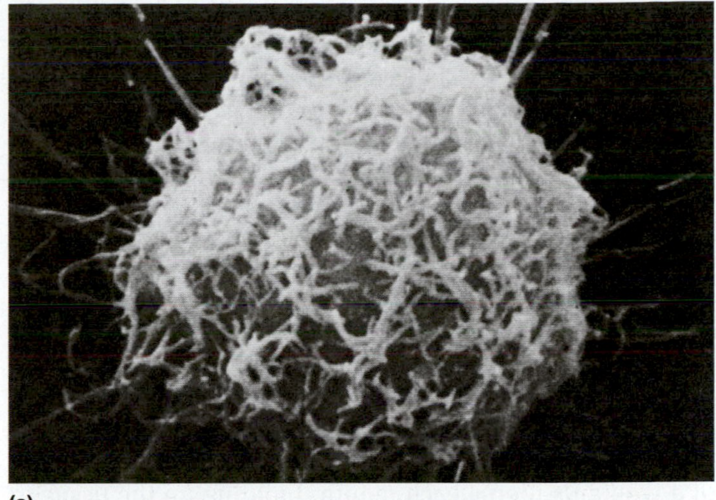

(a)

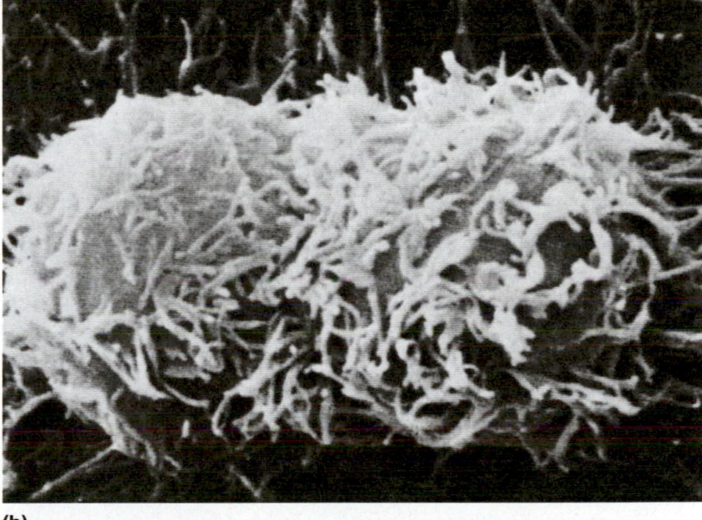

(b)

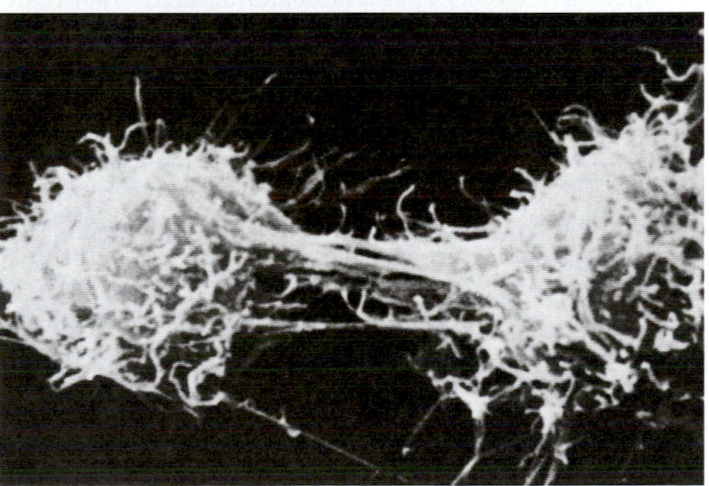

(c)

FIGURE 3.37 Cytoplasmic division is seen in these scanning electron micrographs (*a.* 3,750×; *b.* 3,750×; *c.* 3,190×). From *Scanning Electron Microscopy in Biology,* by R. G. Kessel and C. Y. Shih. © 1976 Springer-Verlag.

cycle. Another internal influence is cell size, specifically the ratio between the surface area the cell membrane provides and the cell volume. The larger the cell, the more nutrients it requires to maintain the activities of life. However, a cell's surface area limits the number of nutrient molecules that can enter. Volume increases faster than does surface area, so a cell can grow too large to efficiently obtain nutrients. Cell division solves this growth problem. The resulting daughter cells are smaller than the original cell and thus have a more favorable surface area-to-volume relationship. They require less energy and fewer nutrients, and diffusion is faster.

External controls of cell division include hormones and growth factors. Hormones are biochemicals manufactured in a gland and transported in the bloodstream to a site where they exert an effect. Hormones signal mitosis in the lining of a woman's uterus each month, building up the tissue to nurture a possible pregnancy. Similarly, a pregnant woman's hormones stimulate mitosis in her breasts when their function as milk-producing glands will soon be required.

Growth factors are like hormones in function but act closer to their sites of synthesis. Epidermal growth factor, for example, stimulates growth of new skin beneath the scab on a skinned knee. Salivary glands also produce this growth factor. This is why an animal's licking a wound may speed healing.

Many people with cancer benefit from drugs that affect growth factors. Granulocyte colony stimulating factor (G-CSF, sold under several brand names) is given as a drug to boost white blood cell counts, which plummet during chemotherapy. In contrast, anti-angiogenesis drugs work oppositely on vascular endothelial growth factor (VEGF), cutting off a tumor's blood supply.

Space availability is another external factor that influences the timing and rate of cell division. Healthy cells do not divide if they are surrounded by other cells, a phenomenon called contact (density dependent) inhibition.

Control of cell division is absolutely crucial to health. With too infrequent mitoses, an embryo could not develop, a child could not grow, and wounds would not heal. Too frequent mitoses or those that continue unabated produce an abnormal growth, or neoplasm, which may form a disorganized mass called a **tumor.**

Tumors are of two types. A *benign* tumor remains in place like a lump, eventually interfering with the function of healthy tissue. A *malignant,* or cancerous, tumor looks different—it is invasive, extending into surrounding tissue. A growing malignant tumor may roughly resemble a crab with outreaching claws. The word "cancer" comes from the Latin for "the crab." Cancer cells, if not stopped, eventually reach the circulation and spread, or metastasize, to other sites. Table 3.5 lists characteristics of cancer cells, and figure 3.38 illustrates how cancer cells infiltrate healthy tissue.

Cancer is a collection of disorders distinguished by their site of origin, the affected cell type, and differences in gene expression not always detectable by observing cancer cells under a microscope. Many cancers are treatable with surgery, radiation, chemicals (chemotherapy), or immune system substances used as drugs. A newer approach to treating cancer is to develop molecules that bind to receptors unique to, or unusually abundant on, cancer cells, blocking the cells from receiving signals to divide.

Two major types of genes cause cancer. **Oncogenes** are abnormal variants of genes that normally control the cell cycle, but are overexpressed, increasing cell division rate. **Tumor suppressor genes** normally hold mitosis in check. When tumor suppressor genes are removed or otherwise inactivated, this lifts control of the cell cycle, and uncontrolled cell division leading to cancer results (fig. 3.39). Cancer cells are said to be "immortal." Environmental factors, such as exposure to toxic

TABLE 3.5 | Characteristics of Cancer Cells

Characteristics of Cancer Cells
Loss of cell cycle control
Heritability (a cancer cell divides to form more cancer cells)
Transplantability (a cancer cell implanted into another individual will cause cancer to develop)
Dedifferentiation (loss of specialized characteristics)
Loss of contact inhibition
Ability to induce local blood vessel formation (angiogenesis)
Invasiveness
Ability to metastasize (spread)

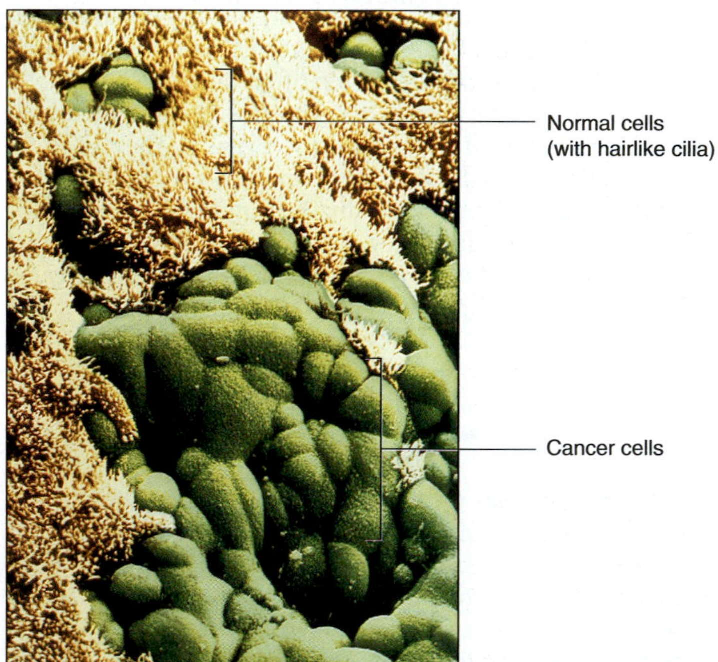

FIGURE 3.38 A cancer cell is rounder and less specialized than surrounding healthy cells. It secretes biochemicals that cut through nearby tissue (invasiveness) and other biochemicals that stimulate extension of blood vessels that nurture the tumor's growth (angiogenesis) (2,200×).

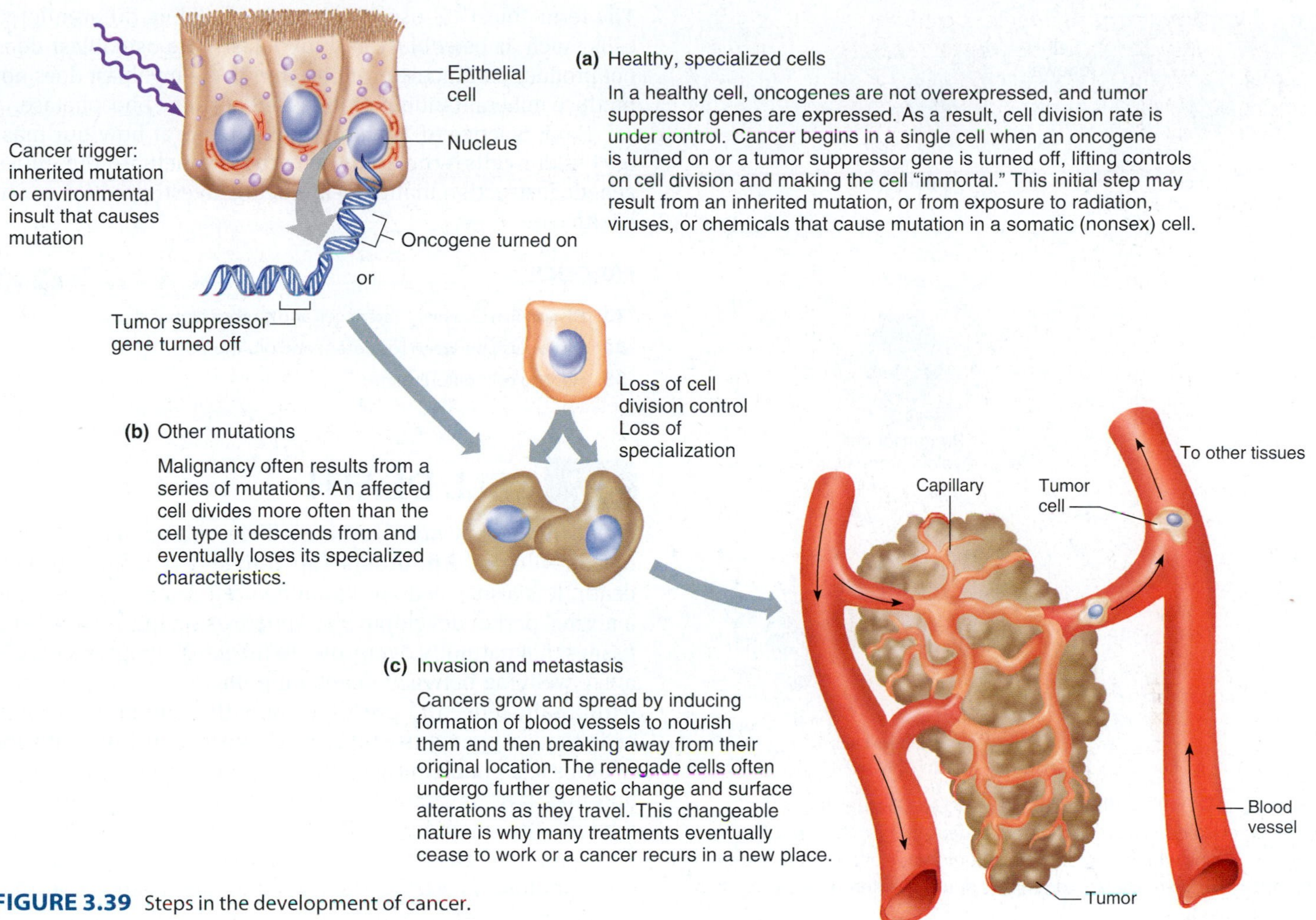

Epithelial cell
Nucleus
Cancer trigger: inherited mutation or environmental insult that causes mutation
Oncogene turned on
or
Tumor suppressor gene turned off
(a) Healthy, specialized cells
In a healthy cell, oncogenes are not overexpressed, and tumor suppressor genes are expressed. As a result, cell division rate is under control. Cancer begins in a single cell when an oncogene is turned on or a tumor suppressor gene is turned off, lifting controls on cell division and making the cell "immortal." This initial step may result from an inherited mutation, or from exposure to radiation, viruses, or chemicals that cause mutation in a somatic (nonsex) cell.
Loss of cell division control
Loss of specialization
(b) Other mutations
Malignancy often results from a series of mutations. An affected cell divides more often than the cell type it descends from and eventually loses its specialized characteristics.
Capillary
Tumor cell
To other tissues
(c) Invasion and metastasis
Cancers grow and spread by inducing formation of blood vessels to nourish them and then breaking away from their original location. The renegade cells often undergo further genetic change and surface alterations as they travel. This changeable nature is why many treatments eventually cease to work or a cancer recurs in a new place.
Blood vessel
Tumor

FIGURE 3.39 Steps in the development of cancer.

chemicals or radiation, may induce cancer by altering (mutating) oncogenes and tumor suppressor genes in body (somatic) cells. Cancer may also be the consequence of a failure of normal programmed cell death (apoptosis), resulting in overgrowth.

PRACTICE

28 How do cells vary in their rates of division?

29 Which factors control the number of times and the rate at which cells divide?

30 How can too infrequent or too frequent cell division affect health?

31 What is the difference between a benign and a malignant tumor?

32 What are two ways that genes cause cancer?

3.6 STEM AND PROGENITOR CELLS

Cells come from preexisting cells, by the processes of mitosis and cytokinesis. Cell division explains how a fertilized egg develops into an individual consisting of trillions of cells, of at least 260 specialized types. The process of specialization is called **differentiation.**

Cells that retain the ability to divide repeatedly enable the body to grow and injuries to heal (fig. 3.40). A **stem cell** divides mitotically to yield either two daughter cells like itself, or one daughter cell that is a stem cell and one that is partially specialized. One defining characteristic of a stem cell is its ability, called self-renewal, to divide to give rise to other stem cells. A stem cell can also differentiate as any of many cell types, given appropriate biochemical signals.

A partly specialized cell that is the daughter of a stem cell is intermediate between a stem cell and a fully differentiated cell and is termed a **progenitor cell.** A progenitor is said to be "committed" because its daughter cells can become any of a restricted number of cell types. For example, a neural stem cell divides to give rise to cells that become part of neural tissue (neurons and neuroglial), but not part of muscle or bone tissue. All of the differentiated cell types in a human body can be traced back through lineages of progenitor and stem cells.

Stem cells and progenitor cells are described in terms of their potential—according to the possible fates of their daughter cells. A fertilized egg and cells of the very early embryo, when it is a small ball of cells, are **totipotent,** which means that they can give rise to every cell type (fig. 3.41). In contrast,

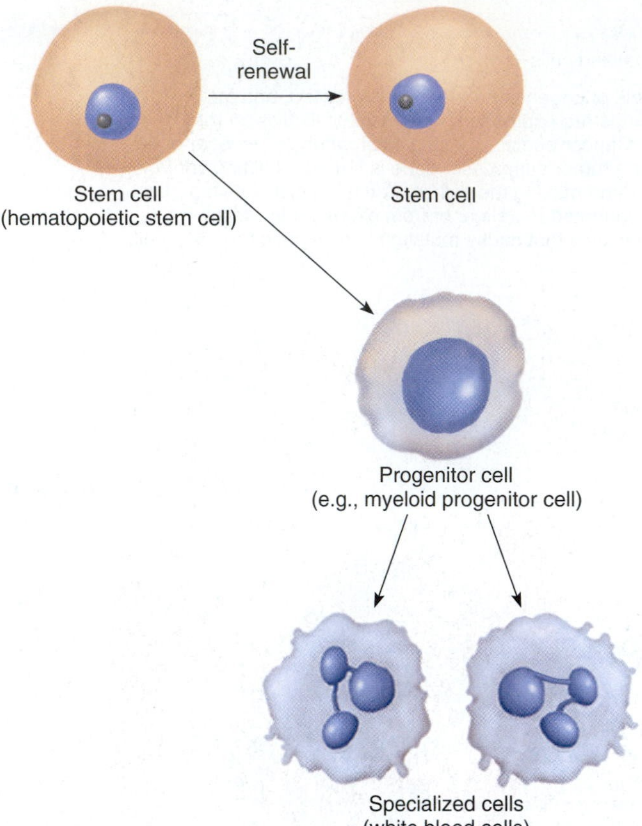

FIGURE 3.40 Stem cells and progenitor cells. A true stem cell divides mitotically to yield two stem cell daughters, or a stem cell and a progenitor cell, which may show the beginnings of differentiation. Progenitor cells give rise to progenitors or more differentiated cells of a restricted lineage.

stem cells present later in development as well as progenitor cells are **pluripotent,** which means that their daughter cells can follow any of several pathways, but not all of them.

Researchers are discovering that many, if not all, of the organs in an adult human body harbor very small populations of stem or progenitor cells activated when injury or illness occurs. For example, one in 10,000 to 15,000 bone marrow cells is a hematopoietic stem cell, which can give rise to blood and several other cell types. Stem cells in the adult body may have been set aside in the embryo or fetus, as repositories of future healing. Alternatively, or perhaps also, stem cells or progenitor cells may travel from bone marrow to replace damaged or dead cells in response to signals sent from injured or diseased tissues.

All cells in the human body (except red blood cells, which expel their nuclei), have the same set of genetic instructions, but as cells specialize, they use some genes and ignore others. For example, an immature bone cell (osteoblast) forms from a progenitor cell by manufacturing proteins necessary to bind bone mineral, as well as alkaline phosphatase, an enzyme required for bone formation. An immature muscle cell (myoblast), in contrast, forms from a muscle progenitor cell and accumulates the contractile proteins that define a muscle cell. The term "blast" is used to describe fledgling differentiated cells, such as osteoblast and myoblast. The osteoblast does not produce contractile proteins, just as the myoblast does not produce mineral-binding proteins and alkaline phosphatase.

From Science to Technology 3.1 looks at how our master builder cells—the stem cells that perpetuate the entire genetic instruction manual—are being investigated for use in health care.

PRACTICE 

33 Distinguish between a stem cell and a progenitor cell.

34 Distinguish between totipotent and pluripotent.

35 How do cells differentiate?

3.7 CELL DEATH

A cell that does not divide or differentiate has another option—death. **Apoptosis** (ap″o-to′sis) is a form of cell death. It is also called "programmed cell death" because it is a normal part of development. Apoptosis sculpts organs from tissues that naturally overgrow. In the fetus, apoptosis carves away webbing between developing fingers and toes, prunes extra brain cells, and preserves only those immune system cells that recognize the body's cell surface. If it weren't for apoptosis, a child's lung or liver couldn't grow to adult size and maintain its characteristic form. Apoptosis is also protective. After a sunburn, this form of cell death peels away damaged skin cells that might otherwise turn cancerous.

Apoptosis is a fast, orderly, contained destruction that packages cellular remnants into membrane-enclosed pieces that are then removed. It is a little like packaging up the content of a messy room into plastic bags. In contrast is necrosis, a disordered form of cell death associated with inflammation and injury.

Like mitosis, apoptosis is a continuous, stepwise process. It begins when a "death receptor" on the doomed cell's cell membrane receives a signal to die. Within seconds, enzymes called caspases are activated inside the cell, where they cut up various cell components. These enzymes:

- Destroy enzymes that replicate and repair DNA.
- Activate enzymes that cut DNA into similarly-sized pieces.
- Dismantle the cytoskeletal threads that support the nucleus, which collapses, condensing the DNA within.
- Fracture mitochondria, which release molecules that trigger further caspase activity, cut off the cell's energy supply, and destroy other organelles.
- Abolish the cell's ability to adhere to other cells.
- Transport certain phospholipids from the inner face of the cell membrane to the outside, where they attract phagocytes that break down debris.

A cell dying from apoptosis has a characteristic appearance (fig. 3.42). It rounds up as contacts with other cells are cut off, and the cell membrane undulates, forming bulges

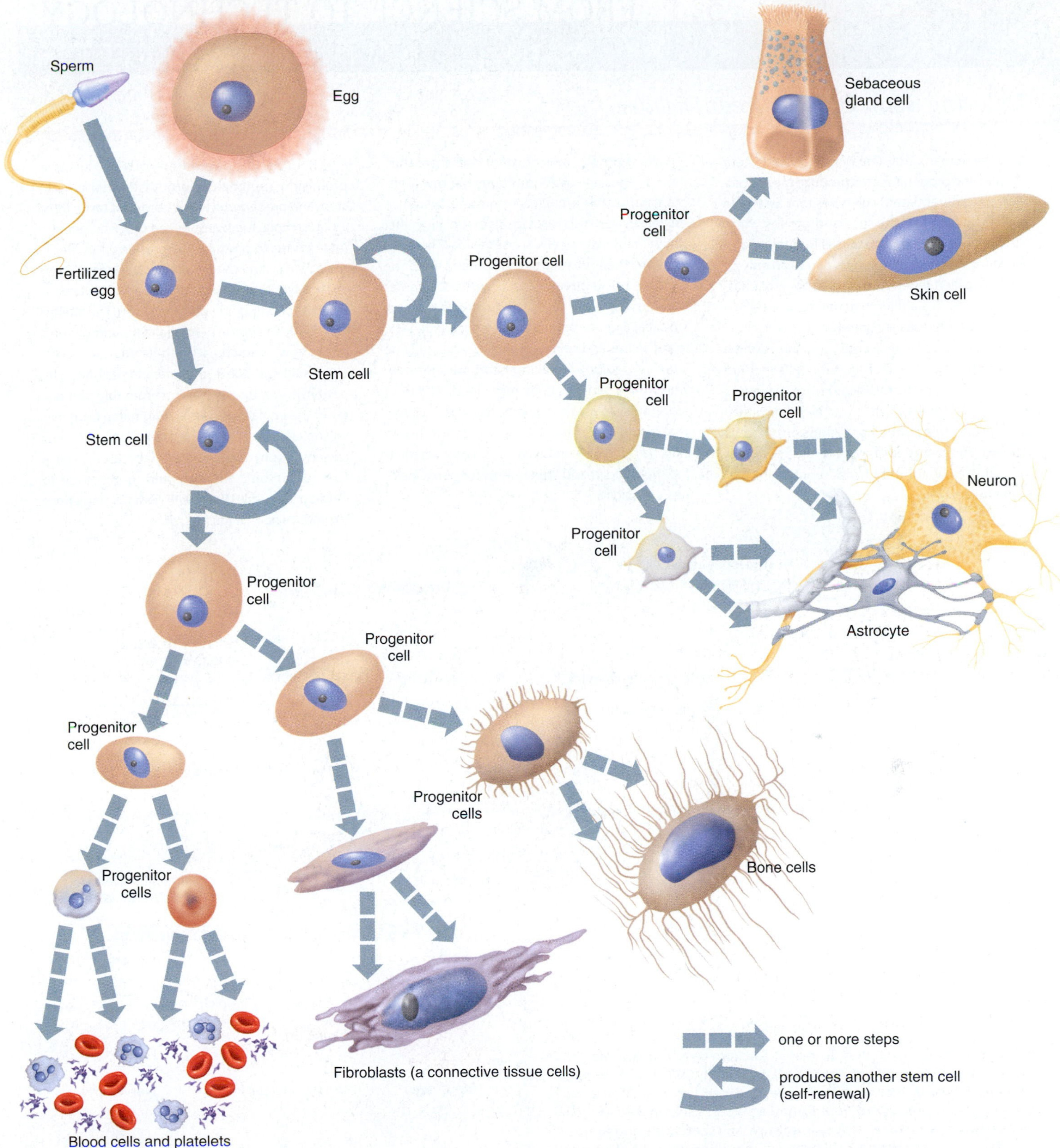

FIGURE 3.41 Cells specialize along cell lineage pathways. All cells in the human body ultimately descend from stem cells, through the processes of mitosis and differentiation. This simplified view depicts a few of the pathways that cells follow, grouping the cell types by the closeness of their lineages. A progenitor cell may yield daughter cells of the same type, such as bone cells, or daughter cells of different types, such as a neuron and an astrocyte. Imagine the complexity of the lineages of the more than 260 human cell types!

3.1 FROM SCIENCE TO TECHNOLOGY

Tailoring Stem Cells to Treat Disease

In the human body, lineages of dividing stem cells and progenitor cells produce the specialized (differentiated) cell types that assemble and interact to form tissues and organs. Stem and progenitor cells are essential for growth and healing. Stem cell technology is part of an emerging field, called regenerative medicine, that harnesses the body's ability to generate new cells to treat certain diseases and injuries.

Stem cells to treat disease come from donors or from the patient. Donor stem cells include umbilical cord stem cells saved from newborns and are used to treat a variety of blood disorders and certain metabolic conditions. Stem cells derived from a patient have two sources: their natural sites or cultured from "reprogrammed" differentiated cells.

An example of using stem cells from their natural site is an autologous bone marrow transplant, in which a person's immune system is essentially destroyed with drugs or radiation after the valuable stem cells are set aside. The stem cells are then infused to repopulate the bone marrow. This is already done. Future examples of using a patient's cells include directing neural stem cells in the brain to treat neurodegenerative diseases and spinal cord injury and applying stem cells to bolster failing heart muscle. Stem cells from a patient's body may one day be used to treat less serious conditions, too. The discovery that a single stem cell can divide to give rise to skin, hair, and oil glands suggests that manipulating them can provide treatments for burns, baldness, and acne.

Reprogramming differentiated cells is a promising approach to producing therapeutic stem and progenitor cells. A fibroblast taken from a skin sample, for example, can be given genetic instructions to produce key proteins that return the cell to a state that resembles a stem cell from an embryo. Then a cocktail of specific biochemicals is added to guide differentiation. The altered cell divides in culture, specializing and passing on its new characteristics to its daughter cells. The resulting tissue is implanted in the body. The patient's immune system presumably will not reject the implant because it originated from that person's skin cell. Fibroblasts from a boy with muscular dystrophy, for example, might be taken back to an embryonic-like state and then coaxed to develop as muscle, along with genetic instructions to produce normal muscle. ■

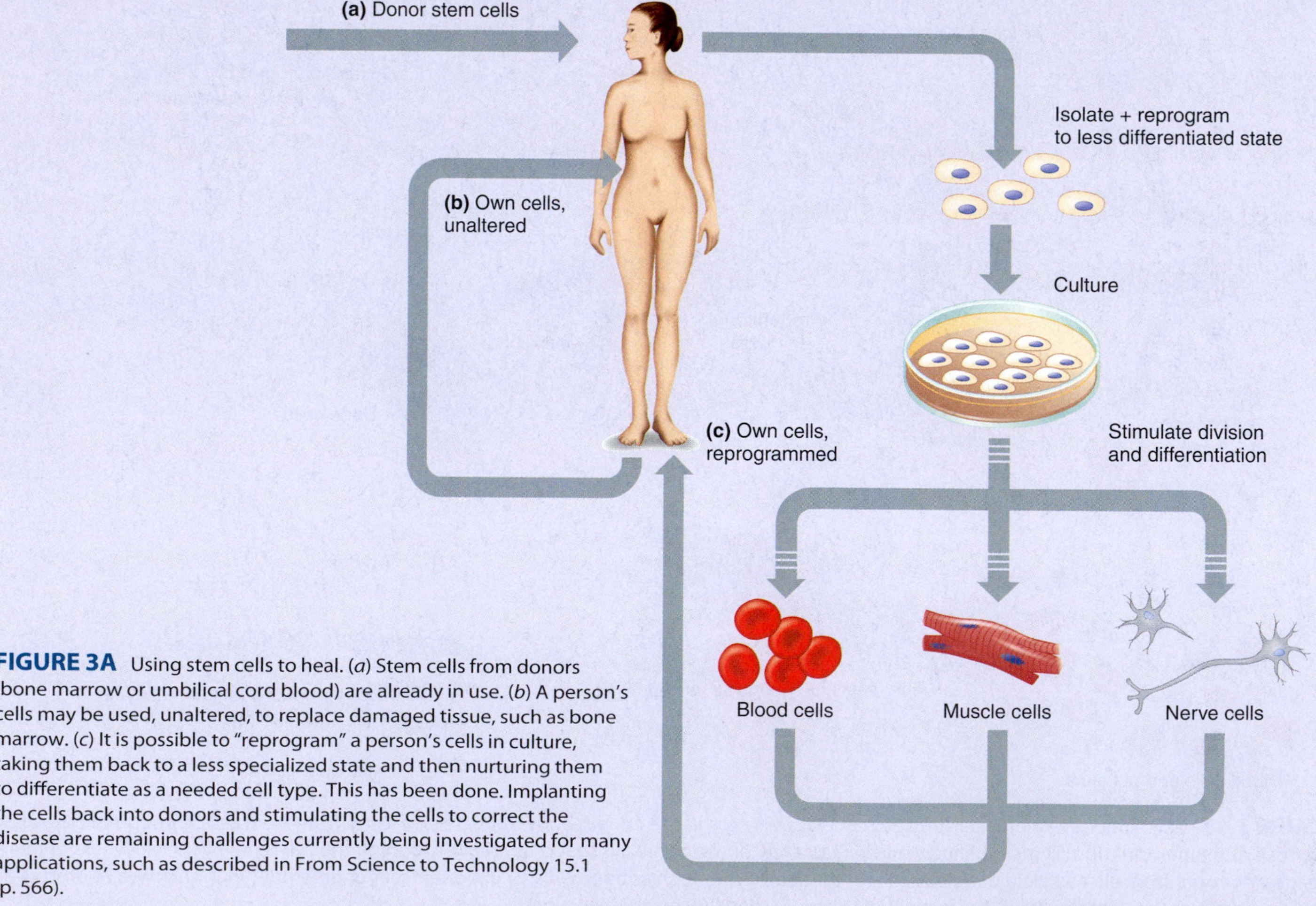

FIGURE 3A Using stem cells to heal. (*a*) Stem cells from donors (bone marrow or umbilical cord blood) are already in use. (*b*) A person's cells may be used, unaltered, to replace damaged tissue, such as bone marrow. (*c*) It is possible to "reprogram" a person's cells in culture, taking them back to a less specialized state and then nurturing them to differentiate as a needed cell type. This has been done. Implanting the cells back into donors and stimulating the cells to correct the disease are remaining challenges currently being investigated for many applications, such as described in From Science to Technology 15.1 (p. 566).

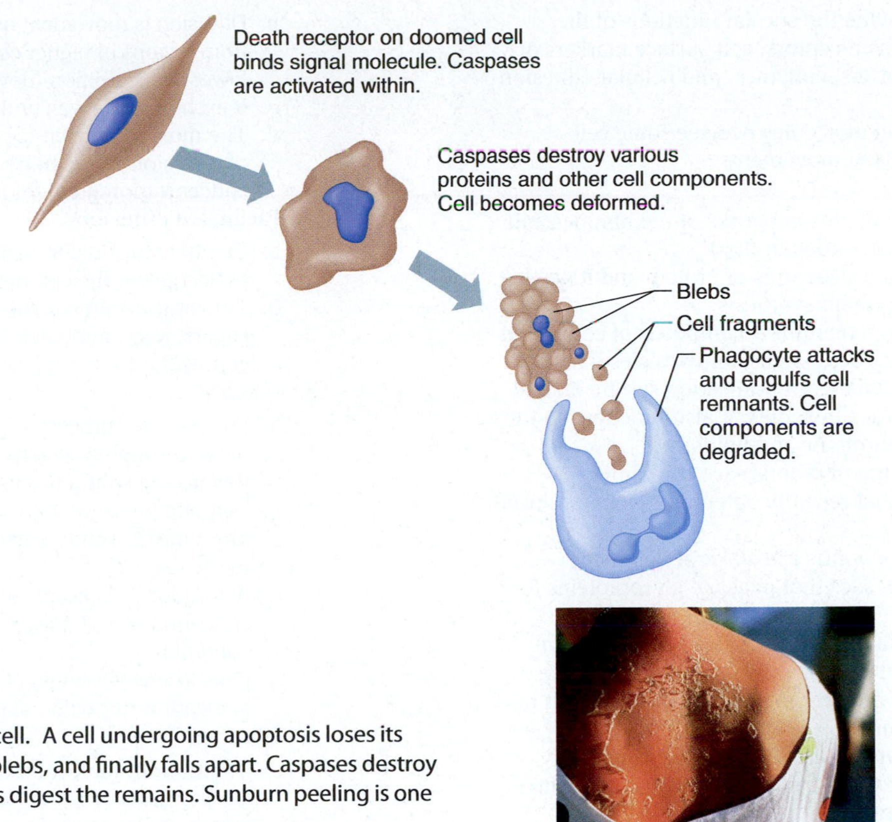
Death receptor on doomed cell binds signal molecule. Caspases are activated within.
Caspases destroy various proteins and other cell components. Cell becomes deformed.
Blebs
Cell fragments
Phagocyte attacks and engulfs cell remnants. Cell components are degraded.

FIGURE 3.42 Death of a cell. A cell undergoing apoptosis loses its characteristic shape, forms blebs, and finally falls apart. Caspases destroy the cell's insides. Phagocytes digest the remains. Sunburn peeling is one example of apoptosis.

called *blebs*. The nucleus bursts under the multiple strains, releasing same-sized DNA pieces. Mitochondria decompose. Finally, the cell shatters. Almost instantly, pieces of membrane encapsulate the fragments, which prevents the signaling that triggers inflammation. Within an hour of the first release of caspases, the cell that underwent apoptosis is gone.

Mitosis and apoptosis are synchronized throughout development, maturation, and aging, and as a result, tissues and organs neither overgrow nor shrink. Disruptions in either process can cause cancer.

PRACTICE 

36 What is apoptosis?

37 List two general functions of apoptosis.

38 List the steps of apoptosis.

CHAPTER SUMMARY

3.1 INTRODUCTION (PAGE 76)

1. Differentiated cells vary considerably in size, shape, and function.
2. The shapes of cells are important in determining their functions.
3. Specialized cells descend from less specialized cells.

3.2 A COMPOSITE CELL (PAGE 76)

1. A cell is a basic unit of an organism and includes a nucleus, cytoplasm, and a cell membrane.
2. Cytoplasmic organelles perform specific vital functions, but the nucleus controls the overall activities of the cell.
3. Cell membrane
    a. The cell membrane forms the outermost limit of the living material.
    b. It acts as a selectively permeable passageway that controls the movements of substances between the cell and its surroundings and thus is the site of signal transduction.
    c. It includes protein, lipid, and carbohydrate molecules.
    d. The cell membrane framework mainly consists of a double layer of phospholipid molecules.
    e. Molecules that are soluble in lipids pass through the membrane easily, but water-soluble molecules do not.
    f. Cholesterol molecules help stabilize the membrane.

g. Proteins provide the special functions of the membrane, as receptors, cell surface markers of self, transporters, enzymes, and cellular adhesion molecules.
h. Cell adhesion molecules oversee some cell interactions and movements.

4. Cytoplasm
   a. Cytoplasm contains networks of membranes and organelles suspended in fluid.
   b. Ribosomes are structures of protein and RNA that function in protein synthesis.
   c. Endoplasmic reticulum is composed of connected membranous sacs, canals, and vesicles that provide a tubular communication system and an attachment for ribosomes; it also functions in the synthesis of proteins and lipids.
   d. Vesicles are membranous sacs containing substances that recently entered or were produced in the cell.
   e. The Golgi apparatus is a stack of flattened, membranous sacs that package glycoproteins for secretion.
   f. Mitochondria are membranous sacs containing enzymes that catalyze the reactions that release energy from nutrient molecules and change it into a usable form.
   g. Lysosomes are membranous sacs containing digestive enzymes that destroy debris and worn-out organelles.
   h. Peroxisomes are membranous, enzyme-containing vesicles.
   i. The centrosome is a nonmembranous structure consisting of two centrioles that aid in the distribution of chromosomes during cell division.
   j. Cilia and flagella are motile extensions on some cell surfaces.
      (1) Cilia are tiny, hairlike structures that wave, moving fluids across cell surfaces.
      (2) Flagella are longer extensions.
   k. Microfilaments and microtubules are threadlike structures built of proteins that aid cellular movements and support and stabilize the cytoplasm.
   l. Cytoplasm may contain nonliving cellular products, such as nutrients and pigments, called inclusions.
5. Cell nucleus
   a. The nucleus is enclosed in a double-layered nuclear envelope that has nuclear pores that control movement of substances between the nucleus and cytoplasm.
   b. A nucleolus is a dense body of protein and RNA where ribosome synthesis occurs.
   c. Chromatin is composed of loosely coiled fibers of protein and DNA that condense into chromosomes during cell division.

3.3 MOVEMENTS INTO AND OUT OF THE CELL (PAGE 90)

Movement of substances into and out of the cell may use physical or physiological processes.

1. Diffusion
   a. Diffusion is due to the random movement of atoms, molecules, or ions in air or liquid solution.
   b. Diffusion is movement of atoms, molecules, or ions from regions of higher concentration toward regions of lower concentration (down a concentration gradient).
   c. It exchanges oxygen and carbon dioxide in the body.
   d. The most important factors determining the rate of diffusion in the body include distance, the concentration gradient, and temperature.
2. Facilitated diffusion
   a. Facilitated diffusion uses protein channels or carrier molecules in the cell membrane.
   b. This process moves substances such as ions, sugars, and amino acids from regions of higher concentration to regions of lower concentration.
3. Osmosis
   a. Osmosis is a process in which water molecules move through a selectively permeable membrane toward the solution with greater osmotic pressure.
   b. Osmotic pressure increases as the number of impermeant solute particles dissolved in a solution increases.
   c. A solution is isotonic when it contains the same concentration of dissolved particles as the cell contents.
   d. Cells lose water when placed in hypertonic solutions and gain water when placed in hypotonic solutions.
4. Filtration
   a. In filtration, molecules move through a membrane from regions of higher hydrostatic pressure toward regions of lower hydrostatic pressure.
   b. Blood pressure filters water and dissolved substances through porous capillary walls.
5. Active transport
   a. Active transport moves molecules or ions from regions of lower concentration to regions of higher concentration.
   b. It requires cellular energy and carrier molecules in the cell membrane.
6. Endocytosis
   a. In pinocytosis, a cell membrane engulfs tiny droplets of liquid.
   b. In phagocytosis, a cell membrane engulfs solid particles.
   c. In receptor-mediated endocytosis, receptor proteins combine with specific molecules in the cell surroundings. The membrane engulfs the combinations.
7. Exocytosis
   a. Exocytosis is the reverse of endocytosis.
   b. In exocytosis, vesicles containing secretions fuse with the cell membrane, releasing the substances to the outside.
8. Transcytosis
   a. Transcytosis combines endocytosis and exocytosis.
   b. In transcytosis, a substance or particle crosses a cell.
   c. Transcytosis is specific.

3.4 THE CELL CYCLE (PAGE 100)

1. The cell cycle includes interphase, mitosis, cytoplasmic division, and differentiation.
2. Interphase
   a. Interphase is the stage when a cell grows, DNA replicates, and new organelles form.
   b. It terminates when the cell begins mitosis.

3. Mitosis
   a. Mitosis is the division and distribution of DNA to daughter cells.
   b. The stages of mitosis include prophase, metaphase, anaphase, and telophase.
4. The cytoplasm divides into two portions with the completion of mitosis.

3.5 CONTROL OF CELL DIVISION (PAGE 103)

1. Cell division capacities vary greatly among cell types.
2. Chromosome tips that shorten with each mitosis provide a mitotic clock, usually limiting the number of divisions to fifty.
3. Cell division is limited and controlled by both internal and external factors.
4. As a cell grows, its surface area increases to a lesser degree than its volume, and eventually the area becomes inadequate for the requirements of the living material within the cell. When a cell divides, the daughter cells have more favorable surface area-volume relationships.
5. Growth factors and hormones also stimulate cell division.
6. Cancer is the consequence of a loss of cell cycle control.

3.6 STEM AND PROGENITOR CELLS (PAGE 105)

1. A stem cell divides to yield another stem cell and a partially differentiated progenitor cell.
2. Cells that give rise to any differentiated cell type are totipotent. Cells with more restricted fates are pluripotent.
3. Stem cells may be present in adult organs or migrate from the bone marrow to replace damaged cells—or both.
4. As cells specialize, they express different sets of genes that provide their distinct characteristics.

3.7 CELL DEATH (PAGE 106)

1. Apoptosis is a form of cell death that is part of normal development and growth.
2. It is a fast, orderly multistep process that begins when a cell surface receptor receives a signal to die. Caspases start a chain reaction that cuts up the cell into membrane-encapsulated pieces, and finally a phagocyte destroys the remains.
3. Apoptosis and mitosis are in balance.

CHAPTER ASSESSMENTS

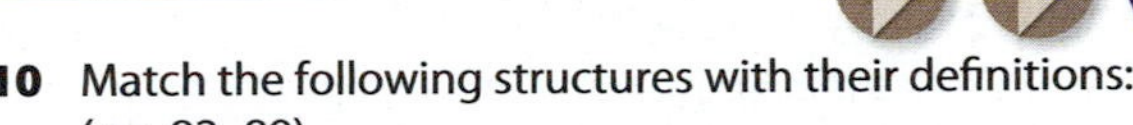

3.1 Introduction

1 An adult human body consists of about ______ cells. (p. 76)
   a. 2 billion
   b. 50 to 100 billion
   c. 50 to 100 trillion
   d. 8 quadrillion

2 Describe three types of differentiated cells. (p. 76)

3.2 A Composite Cell

3 The three major parts of a cell are ______________. (p. 76)
   a. the nucleus, the nucleolus, and the nuclear envelope
   b. the nucleus, the cytoplasm, and the cell membrane
   c. a nerve cell, an epithelial cell, and a muscle cell
   d. the endoplasmic reticulum, the Golgi apparatus, and ribosomes
   e. the cytoplasm, the organelles, and the chromatin

4 Distinguish between the cytoplasm and the cytosol of a cell. (p. 76)

5 Explain the general function of organelles. (p. 76)

6 Define *selectively permeable.* (p. 79)

7 Describe the structure of a cell membrane and explain how this structural organization provides the membrane's function. (p. 79)

8 List three functions of membrane proteins. (p. 80)

9 State a way that cellular adhesion is essential to health and a way that abnormal cellular adhesion harms health. (p. 81)

10 Match the following structures with their definitions: (pp. 82–89)

(1) Golgi apparatus
(2) mitochondria
(3) peroxisomes
(4) cilia
(5) endoplasmic reticulum
(6) cytoskeleton
(7) vesicles
(8) ribosomes

A. Sacs that contain enzymes that catalyze a variety of specific biochemical reactions
B. Structures on which protein synthesis occurs
C. Structures that house the reactions that release energy from nutrients
D. A network of microfilaments and microtubules that supports and shapes a cell
E. A structure that modifies, packages, and exports glycoproteins
F. Membrane-bounded sacs
G. A network of membranous channels and sacs where lipids and proteins are synthesized
H. Hairlike structures that extend from certain cell surfaces and wave about

11 Distinguish between organelles and inclusions. (p. 89)

12 List the parts of the nucleus and explain why each is important. (p. 89)

3.3 Movements Into and Out of the Cell

13 Distinguish between active and passive mechanisms of movement across cell membranes. (p. 90)

14 Match the transport mechanisms on the left with their descriptions on the right. (pp. 90–98)

(1) diffusion
(2) facilitated diffusion
(3) filtration
(4) active transport
(5) endocytosis
(6) exocytosis

A. The cell membrane engulfs a particle or substance, drawing it into the cell in a vesicle
B. Movement down the concentration gradient with a carrier protein, without energy input
C. Movement down the concentration gradient without a carrier protein or energy input
D. A particle or substance leaves a cell in a vesicle that merges with the cell membrane
E. Movement against the concentration gradient with energy input
F. Hydrostatic pressure forces substances through membranes

15 Define *osmosis.* (p. 93)

16 Distinguish between hypertonic, hypotonic, and isotonic solutions. (p. 94)

17 Explain how phagocytosis differs from receptor-mediated endocytosis. (p. 97)

18 Explain how transcytosis combines endocytosis and exocytosis. (p. 98)

3.4 The Cell Cycle

19 The period of the cell cycle when DNA replicates is____________. (p. 100)

a. G_1 phase
b. G_2 phase
c. S phase
d. prophase
e. telophase

20 Explain why interphase is not a period of rest for a cell. (p. 100)

21 Explain how meiosis differs from mitosis. (p. 100)

22 ____________________occur simultaneously. (p. 101)

a. G_1 phase and G_2 phase
b. Interphase and mitosis
c. Cytokinesis and telophase
d. Prophase and metaphase
e. Meiosis and mitotic metaphase

23 Describe the events of mitosis in sequence. (p. 101)

3.5 Control of Cell Division

24 List five factors that control when and if a cell divides. (p. 103)

25 Explain why it is important for the cell cycle to be highly regulated. (p. 104)

26 Discuss the consequences of too little cell division and too much cell division. (p. 104)

27 Distinguish between the ways that mutations in oncogenes and tumor suppressor genes cause cancer. (p. 104)

3.6 Stem and Progenitor Cells

28 Define *differentiation*. (p. 105)

29 A stem cell ______________. (p. 105)

a. self-renews
b. dies after fifty divisions
c. is differentiated
d. gives rise only to fully differentiated daughter cells
e. forms from a progenitor cell

30 Which of the following is true? (p. 105)

a. Progenitor cells are totiptent and stem cells are differentiated.
b. Stem cells are totipotent and progenitor cells are differentiated.
c. Differentiated cells are pluripotent until they specialize.
d. Stem cells in the early embryo are totipotent and progenitor cells are pluripotent.
e. Stem cells in the early embryo are pluripotent and progenitor cells are totipotent.

31 Describe a general function of stem cells in the body. (p. 106)

3.7 Cell Death

32 Explain how apoptosis (cell death) can be a normal part of development. (p. 106)

33 Provide an example of apoptosis. (p. 106)

34 List the steps of apoptosis. (p. 106)

35 Distinguish between necrosis and apoptosis. (p. 106)

36 Describe the relationship between apoptosis and mitosis. (p. 109)

INTEGRATIVE ASSESSMENTS/CRITICAL THINKING

OUTCOMES 2.3, 3.3

1. Liver cells are packed with glucose. What mechanism could transport more glucose into a liver cell? Why would only this mode of transport work?

OUTCOMES 2.3, 3.3

2. What characteristic of cell membranes may explain why fat-soluble substances such as chloroform and ether rapidly affect cells?

OUTCOMES 3.1, 3.2, 3.6

3. For experimental stem cell therapy, state the part of a cell reprogrammed to function like that of a stem cell and stimulated to differentiate in a particular way.

OUTCOME 3.2

4. Organelles compartmentalize a cell. What advantage does this offer a large cell? Cite two examples of organelles and the activities they compartmentalize.

OUTCOME 3.2

5. Exposure to tobacco smoke immobilizes and destroys cilia. How might this effect explain why smokers have an increased incidence of coughing and respiratory infections?

OUTCOME 3.3

6. Which process—diffusion, osmosis, or filtration—is used in the following situations?
   a. Injection of a drug hypertonic to the tissues stimulates pain.
   b. The urea concentration in the dialyzing fluid of an artificial kidney is decreased.
   c. A person with extremely low blood pressure stops producing urine.

OUTCOME 3.6

7. Reports in the media about stem cells usually state that they "turn into any kind of cell in the body." Explain why this statement is incorrect, including a description of what a stem cell really does.

WEB CONNECTIONS

Be sure to visit the text website at **www.mhhe.com/shier12** for answers to chapter assessments, additional quizzes, and interactive learning exercises.

ANATOMY & PHYSIOLOGY REVEALED

Anatomy & Physiology Revealed® (APR) includes cadaver photos that allow you to peel away layers of the human body to reveal structures beneath the surface. This program also includes animations, radiologic imaging, audio pronunciations, and practice quizzing. Check out **www.aprevealed.com**. APR has been proven to help improve student grades!

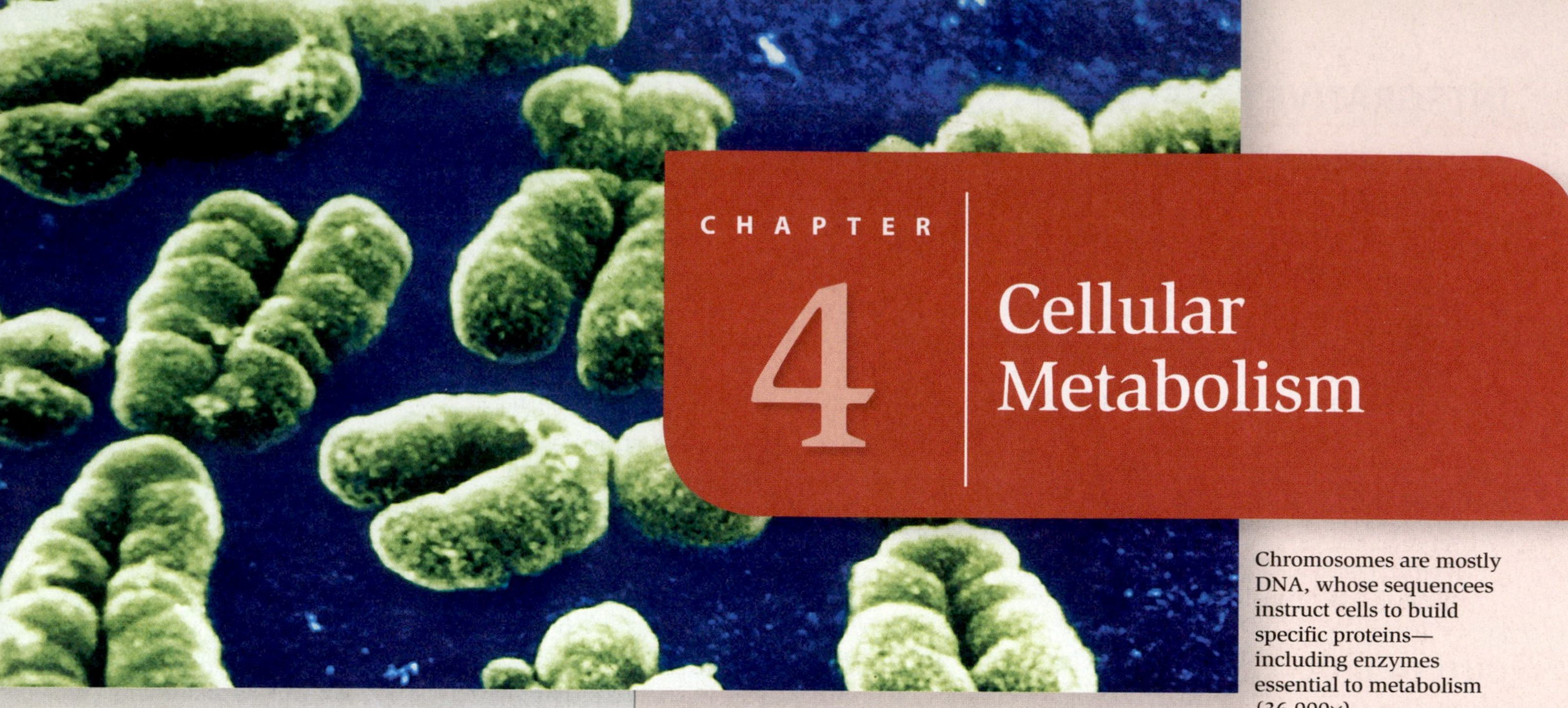

CHAPTER

4 Cellular Metabolism

Chromosomes are mostly DNA, whose sequencees instruct cells to build specific proteins—including enzymes essential to metabolism (36,000×).

UNDERSTANDING WORDS

aer-, air: *aer*obic respiration—respiratory process that requires oxygen.

an-, without: *an*aerobic respiration—respiratory process that does not require oxygen.

ana-, up: *ana*bolism—cellular processes in which smaller molecules are used to build up larger ones.

cata-, down: *cata*bolism—cellular processes in which larger molecules are broken down into smaller ones.

co-, with: *co*enzyme—substance that unites with a protein to complete the structure of an active enzyme molecule.

de-, undoing: *de*amination—process that removes nitrogen-containing portions of amino acid molecules.

mut-, change: *mut*ation—change in genetic information.

-strat, spread out: sub*strate*—substance upon which an enzyme acts.

sub-, under: *sub*strate—substance upon which an enzyme acts.

-zym, causing to ferment: en*zym*e—protein that speeds up a chemical reaction without itself being consumed.

LEARNING OUTCOMES

After you have studied this chapter, you should be able to:

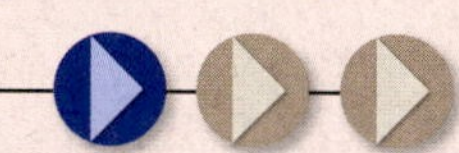

4.1 Introduction
1 Describe the linked pathways of metabolism. (p.115)

4.2 Metabolic Processes
2 Compare and contrast anabolism and catabolism. (p. 115)

4.3 Control of Metabolic Reactions
3 Describe how enzymes control metabolic reactions. (p.117)
4 Explain how metabolic pathways are regulated. (p.118)

4.4 Energy for Metabolic Reactions
5 Explain how ATP stores chemical energy and makes it available to a cell. (p. 119)

4.5 Cellular Respiration
6 Explain how the reactions of cellular respiration release chemical energy. (p.120)
7 Describe the general metabolic pathways of carbohydrate metabolism. (p.120)

4.6 Nucleic Acids and Protein Synthesis
8 Describe how DNA molecules store genetic information. (p. 124)
9 Describe how DNA molecules are replicated. (p. 127)
10 Explain how protein synthesis relies on genetic information. (p. 130)
11 Compare and contrast DNA and RNA. (p. 130)
12 Describe the steps of protein synthesis. (p. 131)

4.7 Changes in Genetic Information
13 Describe how genetic information can be altered. (p. 135)
14 Explain how a mutation may or may not affect an organism. (p. 136)

LEARN PRACTICE ASSESS

ARSENIC POISONING SHUTS DOWN METABOLISM

Disrupting the body's ability to extract energy from nutrients can drastically affect health. Arsenic is a chemical element that, if present in the body in excess, shuts down metabolism. It can do so suddenly or gradually.

Given in one large dose, arsenic poisoning causes chest pain, vomiting, diarrhea, shock, coma, and death. In contrast, many small doses cause dark skin lesions that feel as if they are burning, numb hands and feet, and eventually skin cancer. Such gradual poisoning, called arsenicosis, may occur from contact with pesticides or environmental pollutants. The world's largest outbreak of arsenicosis, however, is due to a natural exposure.

When the World Bank and UNICEF began tapping into aquifers in India and Bangladesh in the late 1960s, they were trying to supply clean water to areas ravaged by sewage and industrial waste released from rivers subject to cycles of floods and droughts. Millions of people had already perished from diarrheal diseases due to the poor sanitation. But digging wells to provide clean water backfired when workers unwittingly penetrated a layer of sediment naturally rich in arsenic. The chemical has since been leaching into the water in at least 2 million wells in Bangladesh, reaching levels fifty times the safety limit set by the World Health Organization. When effects on health began to appear years later, the people thought arsenicosis was contagious. Affected individuals not only suffered pain, but were shunned.

Arsenic damages the body by binding to bonds between sulfur atoms in proteins. It affects metabolism by impairing an enzyme that transports the breakdown products of glucose into mitochondria, where energy is extracted. The cell runs out of energy.

Today UNICEF is helping the people of India and Bangladesh to avoid arsenic poisoning. Workers are diagnosing and treating arsenicosis and providing tanks to collect and store rainwater. A vast education campaign has softened the stigma of arsenicosis. Although cases will continue to appear for a few more decades, the use of alternate water sources has finally slowed the progression of this public health problem. ■

4.1 INTRODUCTION

In every human cell, even in the most sedentary individual, thousands of chemical reactions essential to life take place every second. Special types of proteins called **enzymes** control the rate of each reaction. The sum total of chemical reactions in the cell constitutes **metabolism.**

Many metabolic reactions occur one after the other in a linked fashion, in which the products of one reaction are starting materials for the next. These reactions form pathways and cycles that may intersect where they share intermediate compounds, each step catalyzed by an enzyme. Metabolism in its entirety is complex. Individual pathways of metabolism reveal how cells function—in essence, how chemistry underlies biology.

> Metabolic reactions and pathways can be subgrouped. *Intermediary metabolism* refers to the processes that obtain, release, and use energy. Another way to classify metabolic reactions is by their necessity. *Primary metabolites* are products of metabolism essential to survival. *Secondary metabolites* are not essential to survival, but may provide an advantage or enhancement. Secondary metabolites are best studied in plants, where they usually help to defend against predators because they are toxins. Some of our most successful drugs are plant secondary metabolites. The vinca alkaloids, for example, protect the rosy Madagascar periwinkle that produces them by sickening animals that eat the vegetation, but we use these biochemicals to treat cancer. Their effect is to destabilize microtubule formation.

This chapter covers two complex and related subjects. The first is how metabolic pathways supply energy to a cell. Then, as an illustration of how cellular energy is used, and also of how proteins such as enzymes are produced, the second major topic considers how information in the building block sequences of DNA instructs the cell to assemble amino acids into proteins.

4.2 METABOLIC PROCESSES

Metabolic reactions and pathways are of two types. In **anabolism** (ăh-nab′o-liz″-ĕm), larger molecules are constructed from smaller ones, requiring input of energy. In **catabolism** (kă-tab′o-liz″-ĕm), larger molecules are broken down into smaller ones, releasing energy.

Anabolism

Anabolism provides all the materials required for cellular growth and repair. For example, a type of anabolic process called **dehydration synthesis** (de″hi-dra′shun sin′the-sis) joins many simple sugar molecules (monosaccharides) to form larger molecules of glycogen. When a runner consumes pasta the night before a race, digestion breaks down the complex carbohydrates in the prerace meal to monosaccharides. These are absorbed into the bloodstream, which carries the energy-rich molecules to body cells. Here, dehydration synthesis joins the monosaccharides to form glycogen, which stores energy that the runner may not need until later, as the finish line nears. When monosaccharide units join, an —OH (hydroxyl group) from one monosaccharide molecule and an —H (hydrogen atom) from an —OH group of another are removed. As the —H and —OH react to produce a water molecule, the monosaccharides are joined by a shared oxygen atom, as figure 4.1 shows (read from left to right). As the process repeats, the molecular chain extends, forming a polysaccharide.

Glycerol and fatty acid molecules also join by dehydration synthesis in fat (adipose tissue) cells to form fat molecules. In

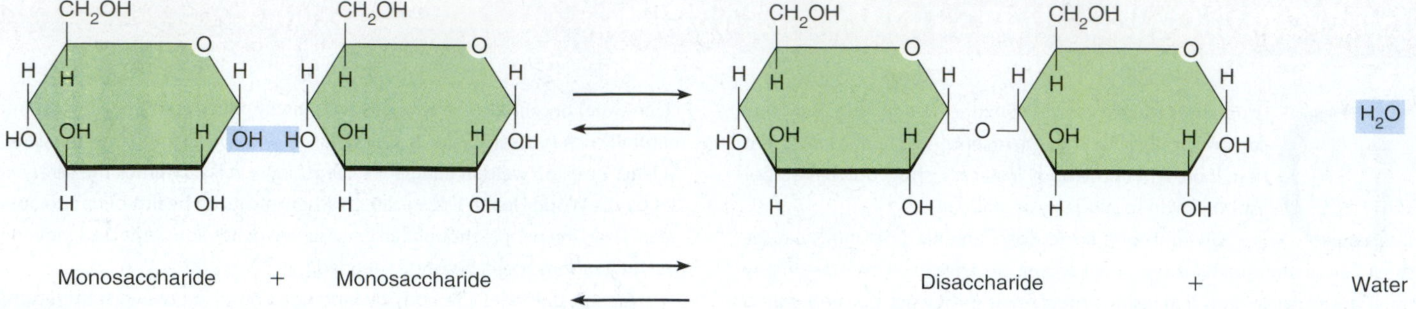

FIGURE 4.1 Building up and breaking down molecules. A disaccharide is formed from two monosaccharides in a dehydration synthesis reaction (arrows to the right). In the reverse reaction, hydrolysis, a disaccharide is broken down into two monosaccharides (arrows to the left).

this case, three hydrogen atoms are removed from a glycerol molecule, and an —OH group is removed from each of three fatty acid molecules, as figure 4.2 shows (read from left to right). The result is three water molecules and a single fat molecule whose glycerol and fatty acid portions are bound by shared oxygen atoms.

In cells, dehydration synthesis also builds protein molecules by joining amino acid molecules. When two amino acid molecules are united, an —OH from the —COOH group of one and an —H from the $—NH_2$ group of another are removed. A water molecule forms, and the amino acid molecules join by a bond between a carbon atom and a nitrogen atom (fig. 4.3; read from left to right). This type of bond, called a *peptide bond,* holds the amino acids together. Two such bound amino acids form a *dipeptide,* and many joined in a chain form a *polypeptide.* Generally, a polypeptide consisting of 100 or more amino acid molecules is called a *protein,* although the boundary between polypeptides and proteins is not precisely defined. Some proteins consist of more than one polypeptide chain.

Nucleic acids are also formed by dehydration synthesis. This process is described later in the chapter.

Catabolism

Metabolic processes that break down larger molecules into smaller ones constitute catabolism. An example of catabolism is **hydrolysis** (hi-drol′ĭ-sis), which can decompose carbohydrates, lipids, and proteins. A water molecule is used for each bond that is broken. Hydrolysis of a disaccharide, for instance, yields two monosaccharide molecules (see fig. 4.1; read from right to left). The bond between the simple sugars breaks, and the water molecule supplies a hydrogen atom to one sugar molecule and a hydroxyl group to the other. Hydrolysis is the reverse of dehydration synthesis.

Hydrolysis breaks down carbohydrates into monosaccharides; fats into glycerol and fatty acids (see fig. 4.2; read from right to left); proteins into amino acids (see fig. 4.3; read from right to left); and nucleic acids into nucleotides. It does not occur automatically, even though in the body water molecules are readily available to provide the necessary —H and —OH. For example, water-soluble substances such as the disaccharide sucrose (table sugar) *dissolve* in a glass of water but do not undergo hydrolysis. Like dehydration synthesis, hydrolysis requires specific enzymes, discussed in the next section, Control of Metabolic Reactions.

The reactions of metabolism are often reversible. However, the enzyme that speeds, or catalyzes, an anabolic reaction is often different from that which catalyzes the corresponding catabolic reaction.

Both catabolism and anabolism must be carefully controlled so that the breakdown or energy-releasing reactions occur at rates adjusted to the requirements of the building up or energy-utilizing reactions. Any disturbance in this balance is likely to damage or kill cells.

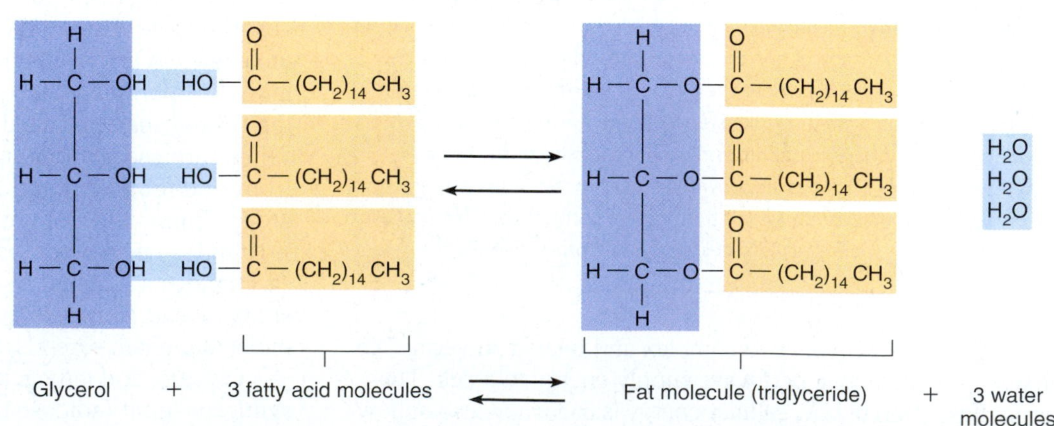

FIGURE 4.2 Forming a fat. A glycerol molecule and three fatty acid molecules react, yielding a fat molecule (triglyceride) in a dehydration synthesis reaction (arrows to the right). In the reverse reaction, hydrolysis, a triglyceride is broken down into three fatty acids and a glycerol (arrows to the left).

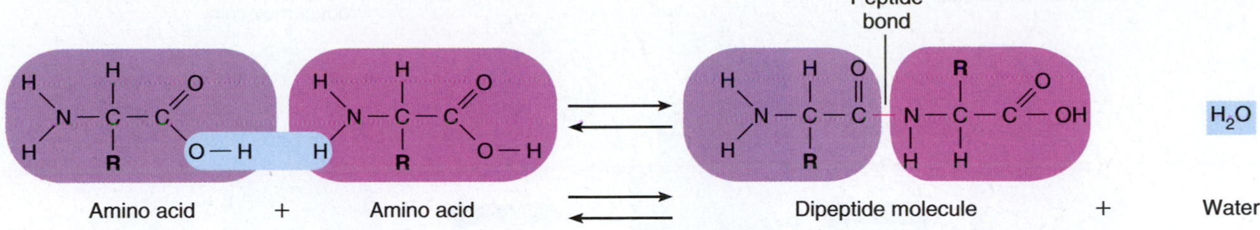

FIGURE 4.3 Peptide bonds link amino acids. When dehydration synthesis unites two amino acid molecules, a peptide bond forms between a carbon atom and a nitrogen atom, resulting in a dipeptide molecule (arrows to the right). In the reverse reaction, hydrolysis, a dipeptide molecule is broken down into two amino acids (arrows to the left).

PRACTICE 

1. What are the general functions of anabolism and catabolism?
2. What type of molecule is formed by the anabolism of monosaccharides? Of glycerol and fatty acids? Of amino acids?
3. Distinguish between dehydration synthesis and hydrolysis.

4.3 CONTROL OF METABOLIC REACTIONS

Different types of cells may conduct specialized metabolic processes, but all cells perform certain basic reactions, such as the buildup and breakdown of carbohydrates, lipids, proteins, and nucleic acids. These common reactions include hundreds of very specific chemical changes that must occur in particular sequences. Enzymes control the rates of these metabolic reactions.

Enzyme Action

Like other chemical reactions, metabolic reactions require energy *(activation energy)* before they proceed. This is why in laboratory experiments heat is used to increase the rates of chemical reactions. Heat energy increases the rate at which molecules move and the frequency of molecular collisions. These collisions increase the likelihood of interactions among the electrons of the molecules that can form new chemical bonds. The temperature conditions in cells are usually too mild to adequately promote the reactions of life. Enzymes make these reactions possible.

Most enzymes are globular proteins that catalyze specific chemical reactions in cells by lowering the activation energy required to start these reactions. Enzymes can speed metabolic reactions by a factor of a million or more.

Enzymes are required in small amounts, because as they work, they are not consumed and can, therefore, function repeatedly. Each enzyme is specific, acting only on a particular molecule, called its **substrate** (sub′strāt). For example, the substrate of an enzyme called catalase (found in the peroxisomes of liver and kidney cells) is hydrogen peroxide, a toxic by-product of certain metabolic reactions. This enzyme's only function is to decompose hydrogen peroxide into water and oxygen, an action that helps prevent accumulation of hydrogen peroxide, which damages cells.

> The action of the enzyme catalase is obvious when using hydrogen peroxide to cleanse a wound. Injured cells release catalase, and when hydrogen peroxide contacts them, bubbles of oxygen are set free. The resulting foam removes debris from inaccessible parts of the wound.

Each enzyme must be able to "recognize" its specific substrate. This ability to identify a substrate depends upon the shape of an enzyme molecule. That is, each enzyme's polypeptide chain twists and coils into a unique three-dimensional conformation that fits the particular shape of its substrate molecule.

RECONNECT
To Chapter 2, Proteins, page 65.

During an enzyme-catalyzed reaction, regions of the enzyme molecule called **active sites** temporarily combine with portions of the substrate, forming an enzyme-substrate complex. This interaction strains chemical bonds in the substrate in a way that makes a particular chemical reaction more likely to occur. When it does, the enzyme is released in its original form, able to bind another substrate molecule (fig. 4.4). Many enzyme-catalyzed reactions are reversible and in some cases the same enzyme catalyzes both directions.

Enzyme catalysis can be summarized as follows:

Substrate + Enzyme → Enzyme-substrate complex → Product + Enzyme (unchanged)

The speed of an enzyme-catalyzed reaction depends partly on the number of enzyme and substrate molecules in the cell. The reaction occurs more rapidly if the concentration of the enzyme or the concentration of the substrate increases. The efficiency of different types of enzymes varies greatly. Some enzymes can process only a few substrate molecules per second, whereas others can handle as many as hundreds of thousands.

Cellular metabolism includes hundreds of different chemical reactions, each controlled by a specific type of

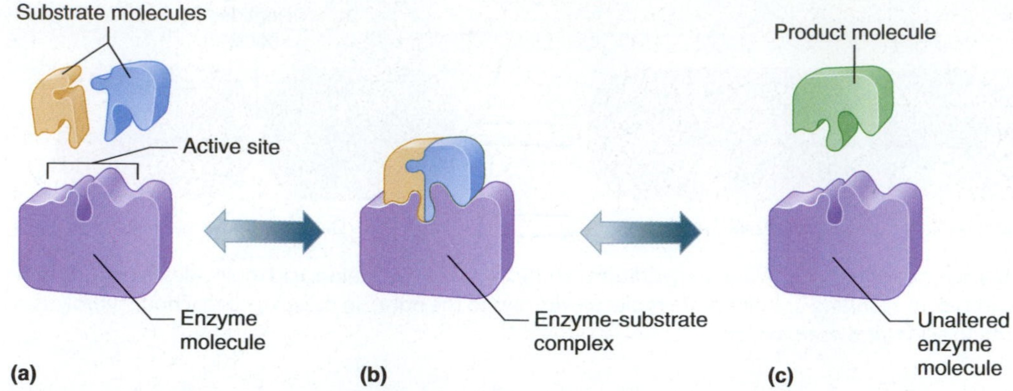

FIGURE 4.4 An enzyme-catalyzed reaction. (Many enzyme-catalyzed reactions, as depicted here, are reversible.) In the forward reaction (dark-shaded arrows), (*a*) the shapes of the substrate molecules fit the shape of the enzyme's active site. (*b*) When the substrate molecules temporarily combine with the enzyme, a chemical reaction occurs. (*c*) The result is a product molecule and an unaltered enzyme. The active site changes shape somewhat as the substrate binds, such that formation of the enzyme-substrate complex is more like a hand fitting into a glove, which has some flexibility, than a key fitting into a lock.

enzyme. Often sequences of enzyme-controlled reactions, called **metabolic pathways,** lead to synthesis or breakdown of particular biochemicals (fig. 4.5). Hundreds of different types of enzymes are present in every cell.

Enzyme names are often derived from the names of their substrates, with the suffix *-ase* added. For example, a lipid-splitting enzyme is called a *lipase,* a protein-splitting enzyme is a *protease,* and a starch (amylum)-splitting enzyme is an *amylase.* Similarly, *sucrase* is an enzyme that splits the sugar sucrose, *maltase* splits the sugar maltose, and *lactase* splits the sugar lactose.

Regulation of Metabolic Pathways

The rate at which a metabolic pathway functions is often determined by a regulatory enzyme that catalyzes one of its steps. The number of molecules of such a regulatory enzyme is limited. Consequently, these enzymes can become saturated when the substrate concentration exceeds a certain level. Once this happens, increasing the substrate concentration no longer affects the reaction rate. The enzyme becomes ineffectual at high substrate concentrations, so it is termed a **rate-limiting enzyme.**

Such an enzyme is often the first enzyme in a series (fig. 4.6). This position is important because an intermediate product of the pathway might accumulate if an enzyme occupying another position in the sequence were rate limiting.

Often the product of a metabolic pathway inhibits the rate-limiting regulatory enzyme. This type of control is an example of negative feedback. Accumulating product inhibits the pathway, and synthesis of the product falls. When the concentration of product decreases, the inhibition lifts, and more product is synthesized. In this way, a single enzyme can control a whole pathway, stabilizing the rate of production (fig. 4.6).

RECONNECT
To Chapter 1, Homeostasis, page 9.

Cofactors and Coenzymes

An enzyme may be inactive until it combines with a nonprotein component called a **cofactor** which helps the active site attain its appropriate shape or helps bind the enzyme to its

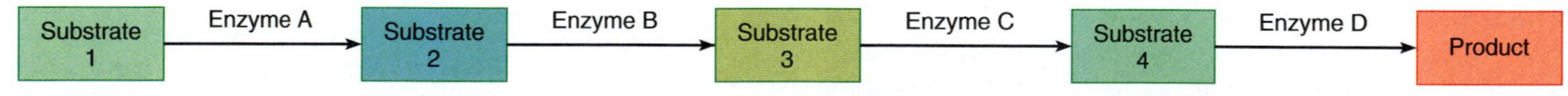

FIGURE 4.5 A metabolic pathway consists of a series of enzyme-controlled reactions leading to formation of a product.

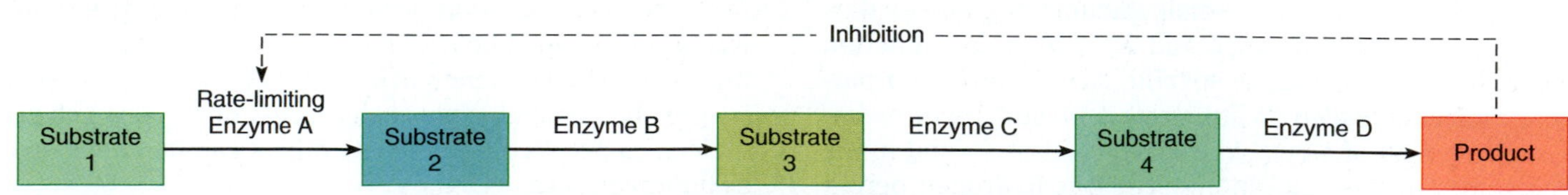

FIGURE 4.6 A negative feedback mechanism may control a rate-limiting enzyme in a metabolic pathway. The product of the pathway inhibits the enzyme.

substrate. A cofactor may be an ion of an element, such as copper, iron, or zinc, or a small organic molecule, called a **coenzyme** (ko-en′zīm). Many coenzymes are composed of vitamin molecules or incorporate altered forms of vitamin molecules into their structures.

Vitamins are essential organic molecules that human cells cannot synthesize (or may not synthesize in sufficient amounts) and therefore must come from the diet. Vitamins provide coenzymes that can, like enzymes, function repeatedly, so cells require small amounts of vitamins. An example is coenzyme A (derived from the vitamin pantothenic acid), which is necessary for one of the reactions of cellular respiration, discussed in the next section. Chapter 18 (pp. 710–716) discusses vitamins further.

Factors That Alter Enzymes

Almost all enzymes are proteins, and like other proteins, they can be denatured by exposure to excessive heat, radiation, electricity, certain chemicals, or fluids with extreme pH values. For example, many enzymes become inactive at 45°C, and nearly all of them are denatured at 55°C. Some poisons denature enzymes. Cyanide, for instance, can interfere with respiratory enzymes and damage cells by halting their energy-obtaining reactions.

> Certain microorganisms, colorfully called "extremophiles," live in conditions of extremely high or low heat, salinity, or pH. Their enzymes have evolved under these conditions and are useful in industrial processes too harsh to use other enzymes.

PRACTICE

4 How can an enzyme control the rate of a metabolic reaction?
5 How does an enzyme "recognize" its substrate?
6 How can a rate-limiting enzyme be an example of negative feedback control of a metabolic pathway?
7 What is the role of a cofactor?
8 What factors can denature enzymes?

4.4 ENERGY FOR METABOLIC REACTIONS

Energy is the capacity to change something; it is the ability to do work. Therefore, we recognize energy by what it can do. Common forms of energy are heat, light, sound, electrical energy, mechanical energy, and chemical energy.

Although energy cannot be created or destroyed, it can be changed from one form to another. An ordinary incandescent light bulb changes electrical energy to heat and light, and an automobile engine changes the chemical energy in gasoline to heat and mechanical energy.

Cellular respiration is the process that transfers energy from molecules such as glucose and makes it available for cellular use. The chemical reactions of cellular respiration must occur in a particular sequence, each one controlled by a different enzyme. Some of these enzymes are in the cell's cytosol, whereas others are in the mitochondria. Such precision of activity suggests that the enzymes are physically positioned in the exact sequence as that of the reactions they control. The enzymes responsible for some of the reactions of cellular respiration are located in tiny, stalked particles on the membranes (cristae) in the mitochondria (see chapter 3, p. 84).

Changes in the human body are a characteristic of life—whenever this happens, energy is being transferred. Thus, all metabolic reactions involve energy in some form.

ATP Molecules

Adenosine triphosphate (ATP) is a molecule that carries energy in a form that the cell can use. Each ATP molecule consists of three main parts—an adenine, a ribose, and three phosphates in a chain (fig. 4.7). The second and third phosphates of ATP are attached by high-energy bonds, and the chemical energy stored in one or both high-energy bonds may be quickly transferred to another molecule in a metabolic reaction. Energy from the breakdown of ATP powers cellular work such as skeletal muscle contraction, active transport across cell membranes, secretion, and many other functions.

An ATP molecule that loses its terminal phosphate becomes an **adenosine diphosphate (ADP)** molecule, which has only two phosphates. ATP can be resynthesized from an ADP by using energy released from cellular respiration to reattach a phosphate, in a process called **phosphorylation** (fos″fōr-ĭ-la′shun). Thus, as shown in figure 4.8, ATP and ADP molecules shuttle back and forth between the energy-transferring reactions of cellular respiration and the energy-transferring reactions of the cell.

ATP is the primary energy-carrying molecule in a cell. Even though there are other energy carriers, without enough ATP, cells quickly die.

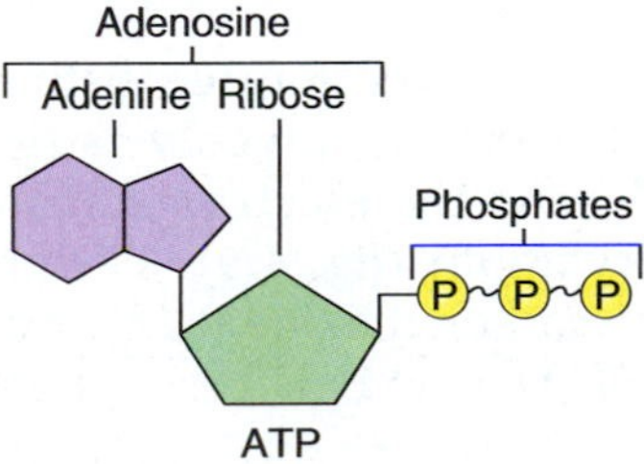

FIGURE 4.7 ATP provides cellular energy currency. An ATP (adenosine triphosphate) molecule consists of an adenine, a ribose, and three phosphates. The wavy lines connecting the last two phosphates represent high-energy chemical bonds.

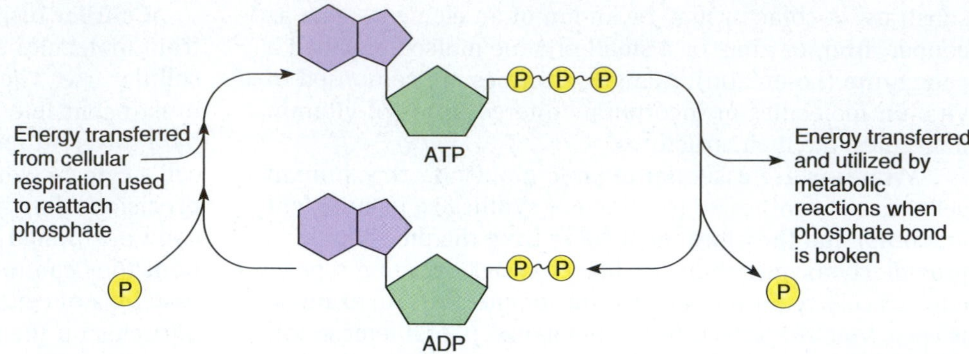
Energy transferred from cellular respiration used to reattach phosphate
P
ATP
P P P
ADP
P P
Energy transferred and utilized by metabolic reactions when phosphate bond is broken
P

FIGURE 4.8 ATP provides energy for metabolic reactions in cells. Cellular respiration generates ATP.

Release of Chemical Energy

Most metabolic processes require chemical energy stored in ATP. This form of energy is initially held in the chemical bonds that link atoms into molecules and is released when these bonds break. Burning a marshmallow over a campfire releases the chemical energy held in the bonds of the molecules that make up the marshmallow as heat and light. Similarly, when a marshmallow is eaten, digested, and absorbed, cells "burn" glucose molecules from that marshmallow in a process called **oxidation** (ok″sĭ-da′shun). The energy released by oxidation of glucose is harnessed to promote cellular metabolism.

Oxidation of substances inside cells and the burning of substances outside them have important differences. Burning in nonliving systems (such as starting a fire in a fireplace) usually requires a great deal of energy to begin, and most of the energy released escapes as heat or light. In cells, enzymes initiate oxidation by lowering the activation energy. Also, by transferring energy to ATP, cells are able to capture almost half of the energy released in the form of chemical energy. The rest escapes as heat, which helps maintain body temperature.

PRACTICE

9 What is energy?

10 Define cellular respiration.

11 How does cellular oxidation differ from burning?

4.5 CELLULAR RESPIRATION

Cellular respiration occurs in three distinct, yet interconnected, series of reactions: **glycolysis** (gli-kol′ĭ-sis), the **citric acid cycle,** and the **electron transport chain** (oxidative phosphorylation) (fig. 4.9). The products of these reactions include carbon dioxide (CO_2), water, and energy. Although most of the energy is lost as heat, almost half is captured as ATP.

Cellular respiration includes **aerobic** (a″er-ōb′ik), reactions which require oxygen, and **anaerobic** (an-a″er-ōb′ik) reactions, which do not require oxygen. For each glucose molecule decomposed completely by cellular respiration, up to thirty-eight molecules of ATP can be produced. All but two ATP molecules are formed by the aerobic reactions.

Glycolysis

Both aerobic and anaerobic pathways begin with glycolysis. Literally "the breaking of glucose," glycolysis is a series of ten enzyme-catalyzed reactions that break down the 6-carbon glucose molecule into two 3-carbon pyruvic acid molecules. Glycolysis occurs in the cytosol (see fig. 4.9), and because it does not require oxygen, it is sometimes referred to as the *anaerobic phase of cellular respiration.*

Three main events occur during glycolysis (fig. 4.10):

1. First, glucose is phosphorylated by the addition of two phosphate groups, one at each end of the molecule. Although this step requires ATP, it "primes" the molecule for some of the energy-releasing reactions that occur later.
2. Second, the 6-carbon glucose molecule is split into two 3-carbon molecules.
3. Third, the electron carrier NADH is produced, ATP is synthesized, and two 3-carbon pyruvic acid molecules result. Some of the reactions of glycolysis release hydrogen atoms. The electrons of these hydrogen atoms contain much of the energy associated with the chemical bonds of the original glucose molecule. To keep this energy in a form the cell can use, these hydrogen atoms are passed in pairs to molecules of the hydrogen carrier NAD^+ (nicotinamide adenine dinucleotide). In this reaction, two of the electrons and one hydrogen nucleus bind to NAD^+ to form NADH. The remaining hydrogen nucleus (a hydrogen ion) is released as follows:

$$\mathbf{NAD^+ + 2H \rightarrow NADH + H^+}$$

NADH delivers these high-energy electrons to the electron transport chain elsewhere in the mitochondria, where most of the ATP will be synthesized.

ATP is also synthesized directly in glycolysis. After subtracting the two ATP used in the priming step, this gives a net yield of two ATP per molecule of glucose.

Glycolysis

1 The 6-carbon sugar glucose is broken down in the cytosol into two 3-carbon pyruvic acid molecules with a net gain of 2 ATP and the release of high-energy electrons.

Citric Acid Cycle

2 The 3-carbon pyruvic acids generated by glycolysis enter the mitochondria. Each loses a carbon (generating CO_2) and is combined with a coenzyme to form a 2-carbon acetyl coenzyme A (acetyl CoA). More high-energy electrons are released.

3 Each acetyl CoA combines with a 4-carbon oxaloacetic acid to form the 6-carbon citric acid, for which the cycle is named. For each citric acid, a series of reactions removes 2 carbons (generating 2 CO_2's), synthesizes 1 ATP, and releases more high-energy electrons. The figure shows 2 ATP, resulting directly from 2 turns of the cycle per glucose molecule that enters glycolysis.

Electron Transport Chain

4 The high-energy electrons still contain most of the chemical energy of the original glucose molecule. Special carrier molecules bring the high-energy electrons to a series of enzymes that convert much of the remaining energy to more ATP molecules. The other products are heat and water. The function of oxygen as the final electron acceptor in this last step is why the overall process is called aerobic respiration.

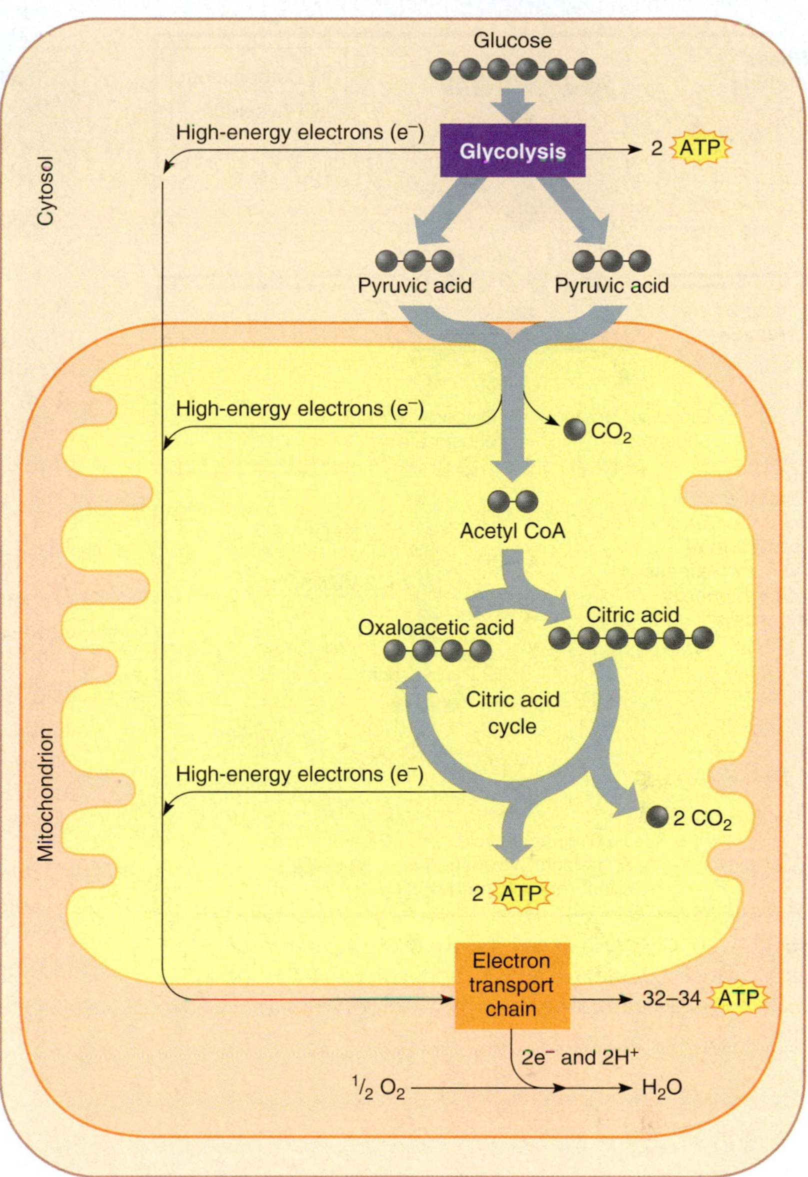

FIGURE 4.9 Glycolysis occurs in the cytosol and does not require oxygen. Aerobic respiration occurs in the mitochondria and only in the presence of oxygen. The products include ATP, heat, carbon dioxide, and water. Two ATP are generated by glycolysis, 2 result directly from the citric acid cycle, and 32–34 are generated by the electron transport chain. Thus, the total yield of ATP molecules per glucose molecule is 36–38, depending on the type of cell.

PRACTICE 

12 What are the final products of cellular respiration?

13 What are aerobic and anaerobic reactions?

14 What is the result of glycolysis?

Anaerobic Reactions

For glycolysis to continue, NADH + H^+ must be able to deliver electrons to the electron transport chain, replenishing the cellular supply of NAD^+. In the presence of oxygen, this is exactly what happens. Oxygen acts as the final electron acceptor at the end of the electron transport chain, enabling the chain to continue processing electrons and recycling NAD^+.

Under anaerobic conditions, however, the electron transport chain has nowhere to unload its electrons, and it can no longer accept new electrons from NADH. As an alternative, NADH + H^+ can give its electrons and hydrogens back to pyruvic acid in a reaction that forms **lactic acid.** Although this regenerates NAD^+, the buildup of lactic acid eventually inhibits glycolysis, and ATP production declines. The lactic acid diffuses into the blood, and when oxygen levels return to normal the liver converts the lactic acid back into pyruvic acid, which can finally enter the aerobic pathway.

PRACTICE 

15 What is the role of oxygen in cellular respiration?

16 Under what conditions does a cell produce lactic acid?

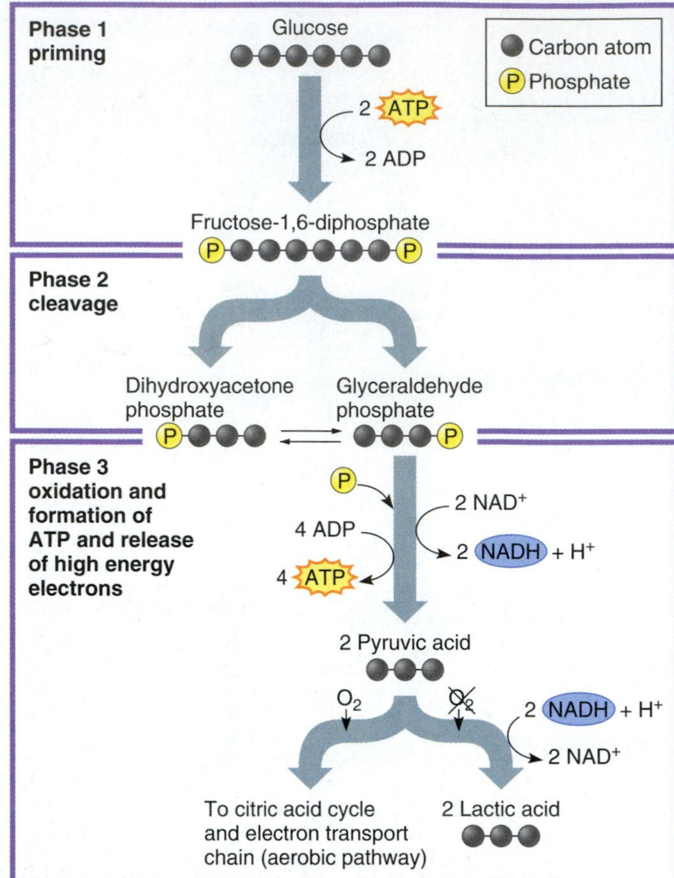

FIGURE 4.10 Glycolysis breaks down glucose in three stages: (1) phosphorylation, (2) splitting, and (3) production of NADH and ATP. Each glucose molecule broken down by glycolysis yields a net gain of 2 ATP.

Human muscle cells working so strenuously that their production of pyruvic acid exceeds the oxygen supply produce lactic acid. In this "oxygen debt," the muscle cells use solely the anaerobic pathway, which provides fewer ATPs per glucose molecule than do the aerobic reactions. The accumulation of lactic acid contributes to muscle fatigue and cramps. Walking after cramping can increase bloodflow that hastens depletion of lactic acid, easing the pain.

Aerobic Reactions

If enough oxygen is available, the pyruvic acid generated by glycolysis can continue through the aerobic pathways (see fig. 4.9). These reactions include the synthesis of **acetyl coenzyme A** (as′ĕ-til ko-en′zīm A) or acetyl CoA, the citric acid cycle, and the electron transport chain. In addition to carbon dioxide and water, the aerobic reactions yield up to thirty-six ATP molecules per glucose.

This book presents the theoretical yield of the aerobic reactions—up to 36 ATP per glucose molecule. In fact, more energy may be required to complete these reactions than once thought. Estimates taking this into account indicate a yield of ATP less than the theoretical maximum.

The aerobic reactions begin with pyruvic acid produced in glycolysis moving from the cytosol into the mitochondria (fig. 4.11). From each pyruvic acid molecule, enzymes inside the mitochondria remove two hydrogen atoms, a carbon atom, and two oxygen atoms, generating NADH and a CO_2 and leaving a 2-carbon acetic acid. The acetic acid then combines with a molecule of coenzyme A to form acetyl CoA. CoA "carries" the acetic acid into the citric acid cycle.

Citric Acid Cycle

The citric acid cycle begins when a 2-carbon acetyl CoA molecule combines with a 4-carbon oxaloacetic acid molecule to form the 6-carbon citric acid and CoA (fig. 4.11). The citric acid is changed through a series of reactions back into oxaloacetic acid. The CoA can be used again to combine with acetic acid to form acetyl CoA. The cycle repeats as long as the mitochondrion receives oxygen and pyruvic acid.

The citric acid cycle has three important consequences:

1. One ATP is produced directly for each citric acid molecule that goes through the cycle.
2. For each citric acid molecule, eight hydrogen atoms with high-energy electrons are transferred to the hydrogen carriers NAD^+ and the related FAD (flavine adenine dinucleotide):

$$\mathbf{NAD^+ + 2H \rightarrow NADH + H^+}$$

$$\mathbf{FAD + 2H \rightarrow FADH_2}$$

3. As the 6-carbon citric acid reacts to form the 4-carbon oxaloacetic acid, two carbon dioxide molecules are produced.

The carbon dioxide produced by the formation of acetyl CoA and in the citric acid cycle dissolves in the cytoplasm, diffuses from the cell, and enters the bloodstream. Eventually, the respiratory system excretes the carbon dioxide.

Electron Transport Chain

The hydrogen and high-energy electron carriers (NADH and $FADH_2$) generated by glycolysis and the citric acid cycle now hold most of the energy contained in the original glucose molecule. To couple this energy to ATP synthesis, the high-energy electrons are handed off to the electron transport chain, a series of enzyme complexes that carry and pass electrons along from one to another. These complexes dot the folds of the inner mitochondrial membranes (see chapter 3, p. 84), which, if stretched out, may be forty-five times as long as the cell membrane in some cells. The electron transport chain passes each electron along, gradually

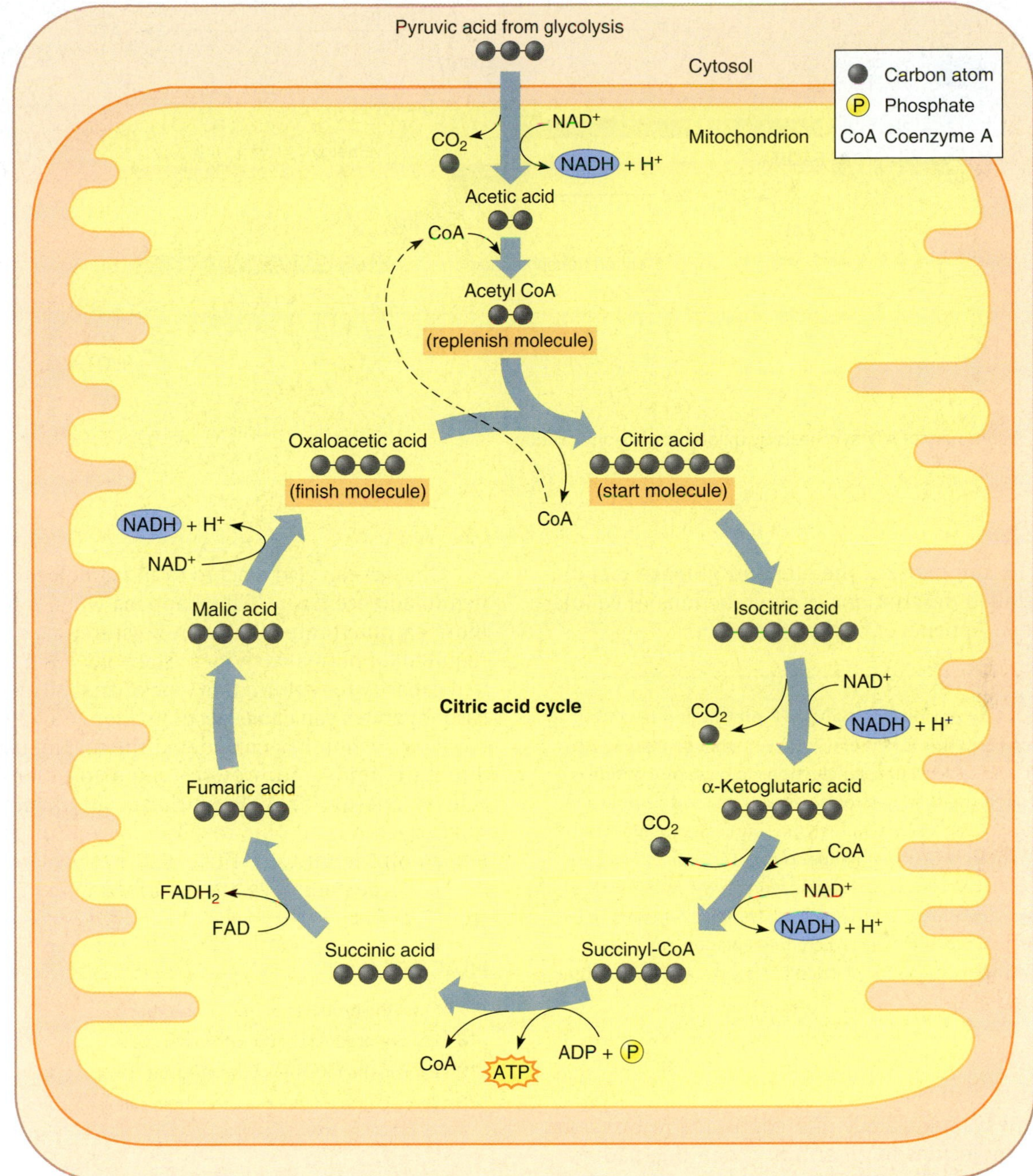

FIGURE 4.11 Each turn of the citric acid cycle (two "turns" or citric acids per glucose) produces one ATP directly, and two CO_2 molecules. Eight hydrogens with high-energy electrons are released.

lowering the electron's energy level and transferring that energy to ATP synthase, an enzyme complex that uses this energy to phosphorylate ADP to form ATP (fig. 4.12). These reactions, known as oxidation/reduction reactions, are described further in Appendix C, pages 944–947.

Neither glycolysis nor the citric acid cycle uses oxygen directly, although they are part of the aerobic metabolism of glucose. Instead, the final enzyme of the electron transport chain gives up a pair of electrons that combine with two hydrogen ions (provided by the hydrogen carriers) and an atom of oxygen to form a water molecule:

$$2e^- + 2H^+ + 1/2\ O_2 \rightarrow H_2O$$

Thus, oxygen is the final electron "carrier." In the absence of oxygen, electrons cannot continue to pass through the electron transport chain, and the aerobic reactions of cellular respiration grind to a halt.

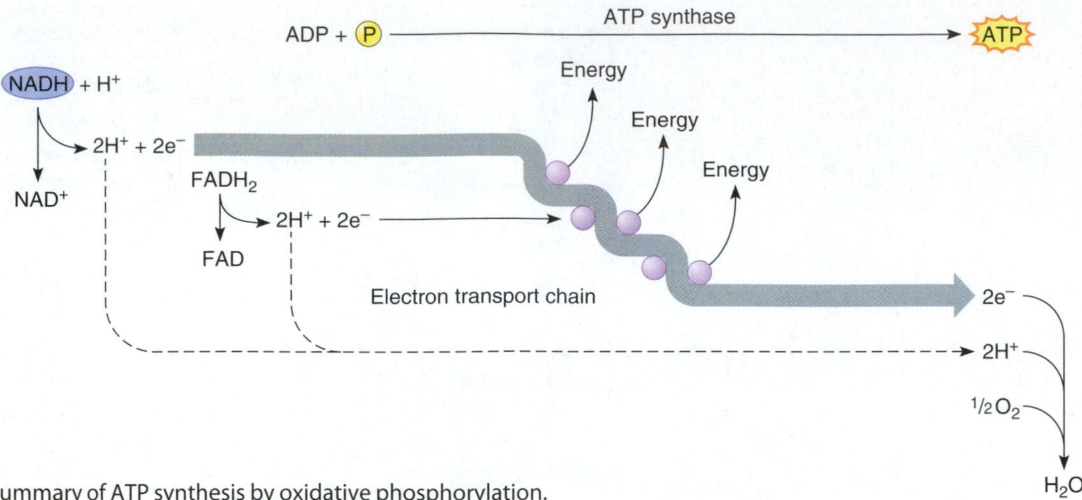

FIGURE 4.12 A summary of ATP synthesis by oxidative phosphorylation.

Figure 4.13 summarizes the steps in glucose metabolism. More detailed descriptions of the reactions of cellular respiration are in Appendix C, pages 944–947.

> Cyanide is a deadly poison that halts ATP production in cells. It binds to an iron atom that is part of the enzyme that enables NADH from the citric acid cycle to transfer electrons to oxygen. Cyanide is absorbed through the skin, gastrointestinal tract, and respiratory tract, and exposure can kill in minutes. One source of cyanide is bitter almonds (not the sweet type that people prefer), which produce a compound called amygdalin that an enzyme in the human small intestine breaks down, releasing the poison. Cyanide is encountered in certain industrial processes, including metal plating, gold extraction, and in the raw materials for plastics. Rat poison and fumigants also contain cyanide.

Carbohydrate Storage

Metabolic pathways are usually interconnected in ways that enable certain molecules to enter more than one pathway. For example, carbohydrate molecules from foods may enter catabolic pathways and be used to supply energy, or they may enter anabolic pathways and be stored or react to form some of the twenty different amino acids (fig. 4.14).

Excess glucose in cells may enter anabolic carbohydrate pathways and be linked into storage forms such as glycogen. Most cells can produce glycogen; liver and muscle cells store the greatest amounts. Following a meal, when blood glucose concentration is relatively high, liver cells obtain glucose from the blood and synthesize glycogen. Between meals, when blood glucose concentration is lower, the reaction reverses, and glucose is released into the blood. This mechanism ensures that cells throughout the body have a continual supply of glucose to support cellular respiration.

Glucose can also react to form fat molecules, later deposited in adipose tissue. This happens when a person takes in more carbohydrates than can be stored as glycogen or are required for normal activities. The body has an almost unlimited capacity to perform this type of anabolism, so overeating carbohydrates can cause accumulation of body fat.

This section has considered the metabolism of glucose, although lipids and proteins can also be broken down to release energy for ATP synthesis. In all three cases, the final process is aerobic respiration, and the most common entry point is into the citric acid cycle as acetyl CoA (fig. 4.15). These pathways are described in detail in chapter 18 (pp. 702–704).

PRACTICE 

17 State the products of the aerobic reactions.

18 List the products of the citric acid cycle.

19 Explain the function of the electron transport chain.

20 Discuss fates of glucose other than cellular respiration.

4.6 NUCLEIC ACIDS AND PROTEIN SYNTHESIS

Enzymes control the metabolic pathways that enable cells to survive, so cells must have information for producing these specialized proteins. Many other proteins are important in physiology as well, such as blood proteins, the proteins that form muscle and connective tissues, and the antibodies that protect against infection. The information that instructs a cell to synthesize a particular protein is held in the sequence of building blocks of **deoxyribonucleic acid (DNA),** the genetic material. As we will see later in this chapter, the correspondence between a unit of DNA information and a particular amino acid constitutes the **genetic code** (jĕ-net′ik kōd).

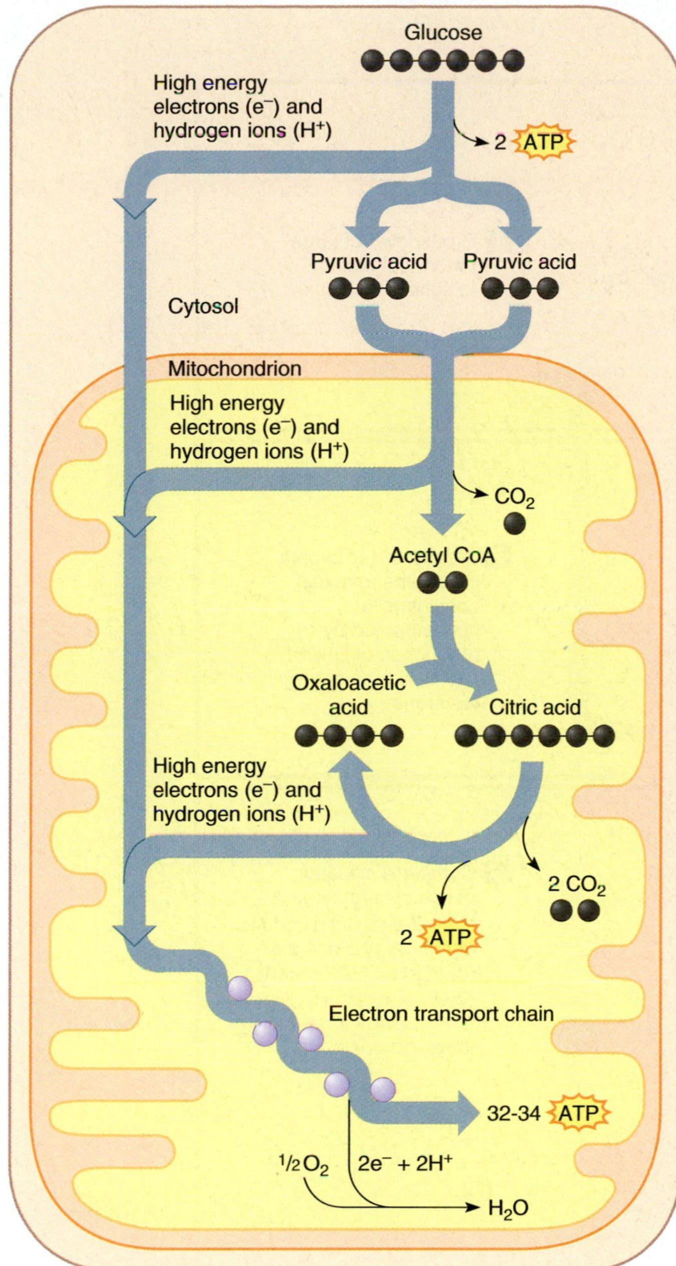

FIGURE 4.13 An overview of aerobic respiration, including the net yield of ATP at each step per molecule of glucose.

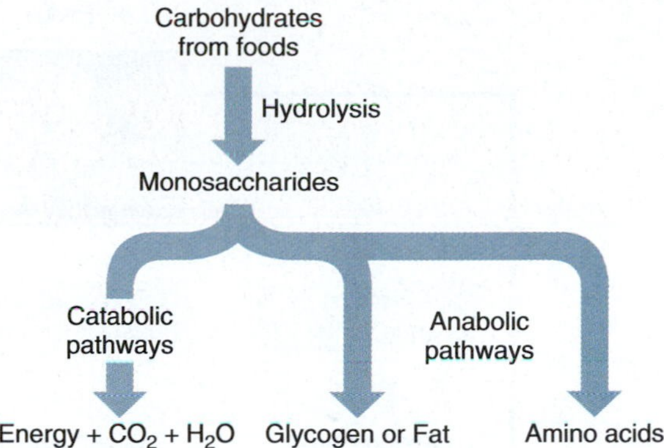

FIGURE 4.14 Hydrolysis breaks down carbohydrates from foods into monosaccharides. The resulting molecules may enter catabolic pathways and be used as energy sources, or they may enter anabolic pathways and be stored as glycogen or fat, or react to yield amino acids.

Genetic Information

Children resemble their parents because of inherited traits, but what passes from parents to a child is genetic information, in the form of DNA molecules from the parents' sex cells. Chromosomes are long molecules of DNA and associated proteins. As an offspring develops, mitosis passes the information in the DNA sequences of the chromosomes to new cells. Genetic information "tells" cells how to construct a great variety of protein molecules, each with a specific function. The portion of a DNA molecule that contains the genetic information for making a particular protein is called a **gene** (jēn). Enzymes control synthesis reactions, so all four groups of organic molecules—proteins, carbohydrates, lipids, and nucleic acids—depend on proteins, and thus require genetic instructions.

RECONNECT

To Chapter 3, Cell Nucleus, page 90.

The complete set of genetic instructions in a cell constitutes the **genome.** The "first draft" of the human genome sequence was announced in June 2000, following nearly fifteen years of discussion and work by thousands of researchers worldwide. Only a small part of the human genome encodes protein. The rest includes many controls over which proteins are produced in a particular cell under particular circumstances, called *gene expression*. Chapter 24 (p. 917) discusses the human genome.

Recall from chapter 2 (p. 68) that nucleotides are the building blocks of nucleic acids. A nucleotide consists of a 5-carbon sugar (ribose or deoxyribose), a phosphate group, and one of several nitrogenous bases (fig. 4.16). DNA and RNA nucleotides form long strands (polynucleotide chains) by alternately joining their sugar and phosphate portions by dehydration synthesis, which provides a "backbone" structure (fig. 4.17).

A DNA molecule consists of two polynucleotide chains, making it double-stranded. The nitrogenous bases project from the sugar-phosphate backbone of one strand and bind, or pair, by hydrogen bonds to the nitrogenous bases of the second strand (fig. 4.18). The resulting structure is somewhat like a ladder, in which the rails represent the sugar and phosphate backbones of the two strands and the rungs represent the paired nitrogenous bases. The sugars forming the two backbones point in opposite directions. For this reason, the two strands are called *antiparallel*.

Food

Proteins (egg white)

Carbohydrates (toast, hashbrowns)

Fats (butter)

Amino acids

Simple sugars (glucose)

Glycerol

Fatty acids

1 Breakdown of large macromolecules to simple molecules

Glycolysis

ATP

Pyruvic acid

Acetyl coenzyme A

2 Breakdown of simple molecules to acetyl coenzyme A accompanied by production of limited ATP and high energy electrons

Citric acid cycle

CO_2

ATP

High energy electrons carried by NADH and $FADH_2$

Electron transport chain

ATP

2e– and $2H^+$

$-NH_2$

$^1/_2 O_2$

H_2O

CO_2

Waste products

3 Complete oxidation of acetyl coenzyme A to H_2O and CO_2 produces high energy electrons (carried by NADH and $FADH_2$), which yield much ATP via the electron transport chain

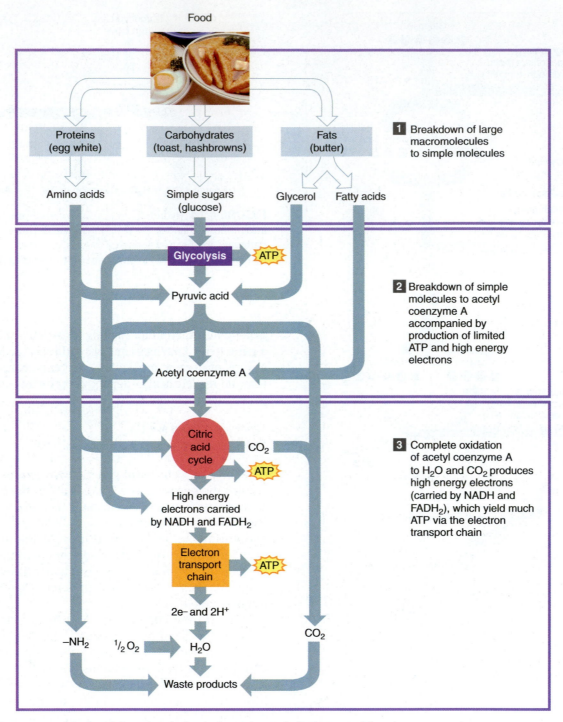

FIGURE 4.15 A summary of the breakdown (catabolism) of proteins, carbohydrates, and fats.

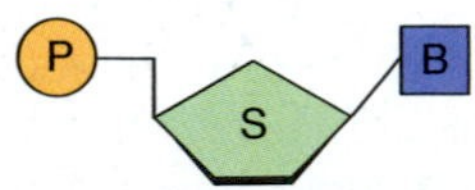

FIGURE 4.16 Each nucleotide of a nucleic acid consists of a 5-carbon sugar (S); a phosphate group (P); and an organic, nitrogenous base (B).

A DNA molecule is sleek and symmetrical because the bases pair in only two combinations, which maintains a constant width of the overall structure. In a DNA nucleotide, the base may be one of four types: adenine (A), thymine (T), cytosine (C), or guanine (G). A and G are **purines** (pu′rēnz), and they consist of two organic ring structures. T and C are **pyrimidines** (pe-rimi-denz), and they have a

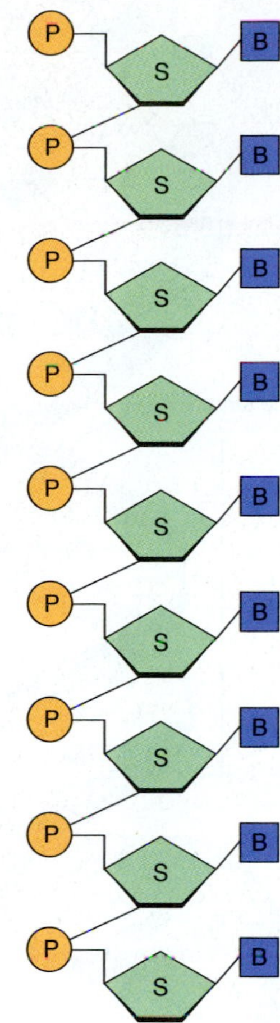

FIGURE 4.17 A polynucleotide chain consists of nucleotides connected by a sugar-phosphate backbone.

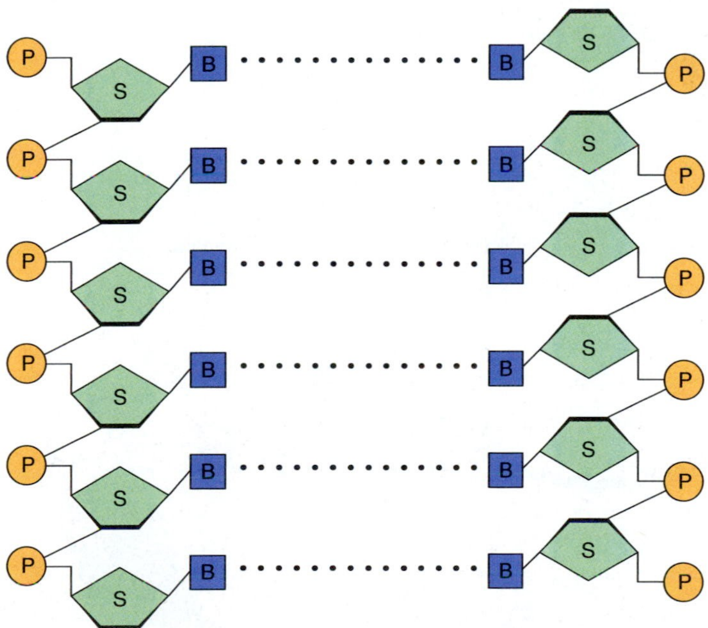

FIGURE 4.18 DNA is double-stranded, consisting of two polynucleotide chains. Hydrogen bonds (dotted lines) hold the nitrogenous bases of one strand to their partners on the other strand. The sugars point in opposite directions—that is, the strands are antiparallel.

single organic ring structure. A binds to T and G binds to C—that is, a purine always binds to a pyrimidine, and this is what establishes the constant width of the DNA molecule. These pairs—A with T, and G with C—are called **complementary base pairs** (fig. 4.19*a*). The sequence of one DNA strand can always be derived from the other by following the "base-pairing rules." If the sequence of one strand of the DNA molecule is G, A, C, T, then the complementary strand's sequence is C, T, G, A.

The double-stranded DNA molecule twists, forming a double helix, (fig. 4.19*b*). The human genome is 3.2 billion DNA bases long, dispersed over the 24 types of chromosomes. A single gene may be thousands or even millions of bases long. In the nucleus, DNA is wound around octets of proteins called *histones* to form chromatin (fig. 4.19*b*). Histones and other molecules come on and off different parts of the genome as some genes are accessed for their information to make proteins and others are silenced. During mitosis chromatin condenses to form chromosomes visible under the microscope (fig. 4.19*c*). Investigators can use DNA sequences to identify individuals (From Science to Technology 4.1). Appendix D, pages 948–949, has more detailed DNA structures.

DNA Replication

When a cell divides, each newly formed cell must have a copy of the original cell's genetic information (DNA) so it will be able to synthesize the proteins necessary to build cellular parts and metabolize. DNA **replication** (re″plĭ-ka′shun) is the process that creates an exact copy of a DNA molecule. It happens during interphase of the cell cycle.

RECONNECT

To Chapter 3, The Cell Cycle, page 100.

As DNA replication begins, hydrogen bonds break between the complementary base pairs of the double strands. Then the strands unwind and separate, exposing unpaired bases. New nucleotides pair with the exposed bases, forming hydrogen bonds. An enzyme, DNA polymerase, catalyzes this base pairing. Enzymes then knit together the new sugar-phosphate backbone. In this way, a new strand of complementary nucleotides extends along each of the old (original) strands. Two complete DNA molecules result, each with one new and one original strand (fig. 4.20). During mitosis, the two DNA molecules that form the two chromatids of each of the chromosomes separate so that one of these DNA molecules passes to each of the new cells.

From Science to Technology 4.2 discusses the polymerase chain reaction (PCR), a method for mass-producing, or amplifying, DNA. PCR has revolutionized biomedical science.

PRACTICE 

21 What is the function of DNA?

22 What is the structure of DNA?

23 How does DNA replicate?

FIGURE 4.19 DNA and chromosome structure. (*a*) The two polynucleotide chains of a DNA molecule point in opposite directions (antiparallel) and are held together by hydrogen bonds between complementary base pairs. (*b*) The molecular "ladder" of a DNA molecule twists into a double helix. (*c*) Histone proteins enable the long double helix to assume a compact form (chromosome) and move when sections of the DNA are accessed for gene expression.

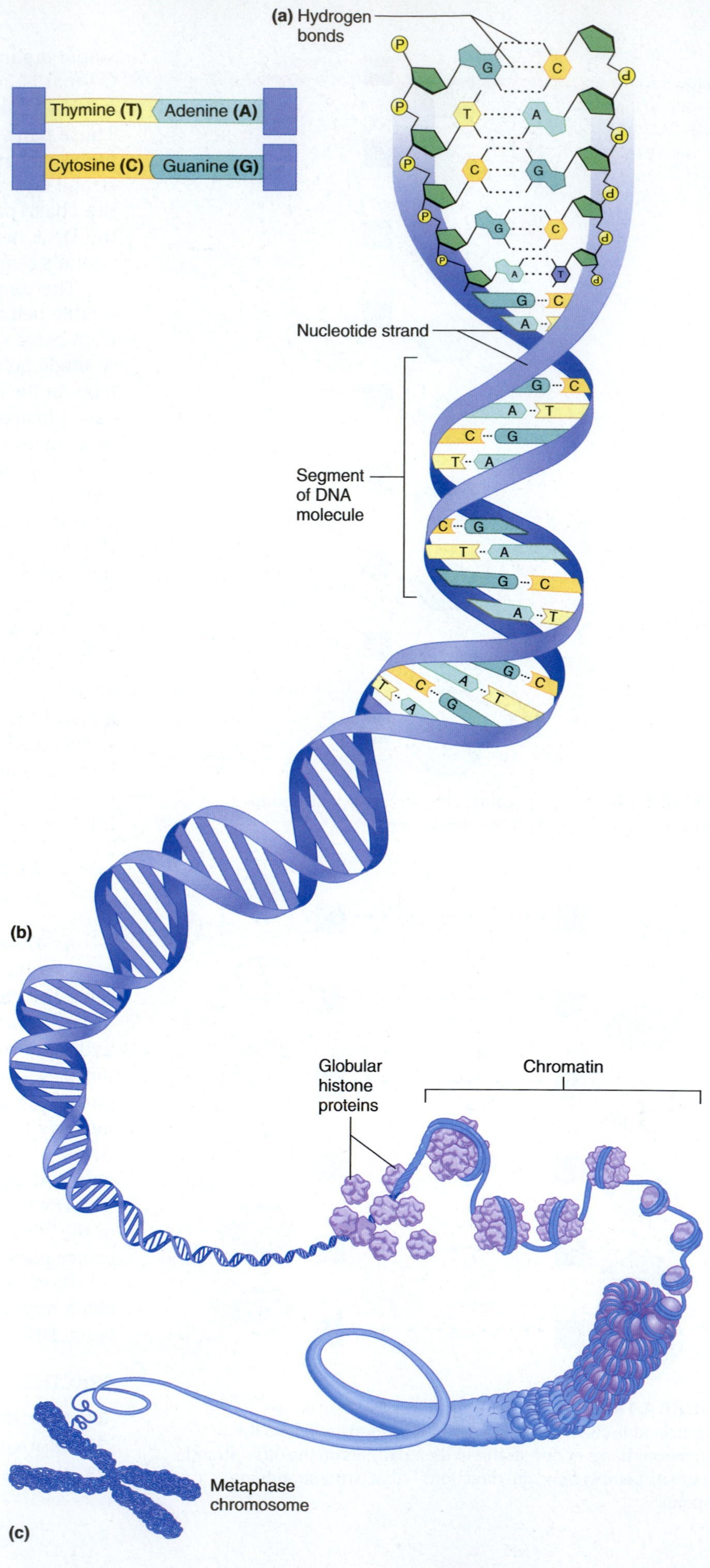

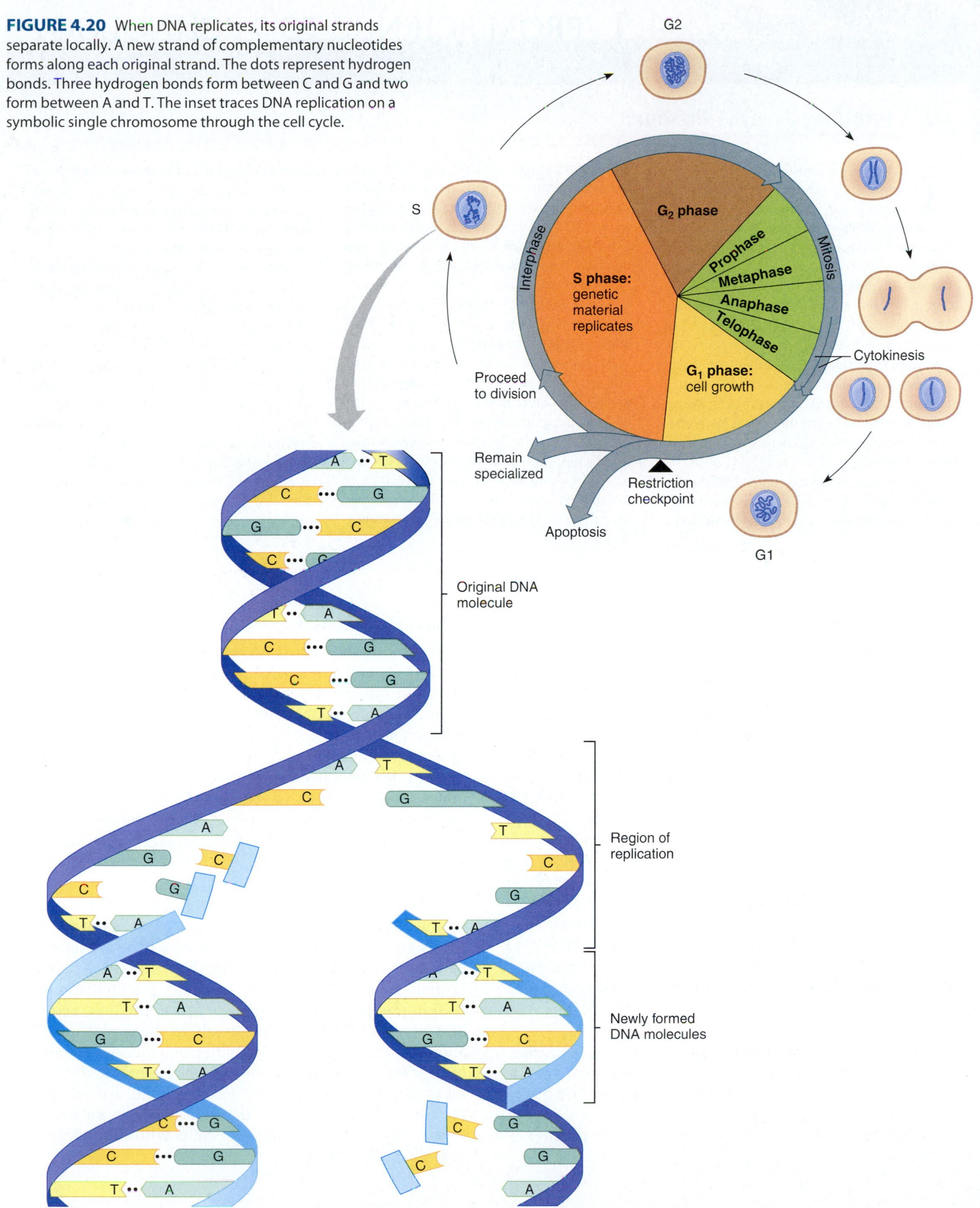

FIGURE 4.20 When DNA replicates, its original strands separate locally. A new strand of complementary nucleotides forms along each original strand. The dots represent hydrogen bonds. Three hydrogen bonds form between C and G and two form between A and T. The inset traces DNA replication on a symbolic single chromosome through the cell cycle.

4.1 FROM SCIENCE TO TECHNOLOGY

DNA Profiling Frees A Prisoner

The human genome sequence differs from person to person because it includes 3.2 billion bits of information. Techniques called DNA profiling (or fingerprinting) compare the most variable parts of the genome among individuals for several purposes—to identify remains at crime scenes or after natural disasters; to confirm or rule out "blood" relationships; and, increasingly, to establish innocence when other types of evidence are questionable. The Innocence Project is a national litigation and public policy organization that provides DNA testing to people who claim that they have been wrongfully convicted. So far the Innocence Project has exonerated more than 200 people. Among them is Josiah Sutton.

Sutton had served four and a half years of a twenty-five-year sentence for rape when DNA profiling established his innocence. He and a friend had become suspects after a woman in Houston identified them as the men who had raped and threatened her with a gun, leaving her in a field. The two young men supplied saliva and blood samples, from which DNA profiles were done and compared to DNA profiles from semen found in the victim and in her car. At the trial, an employee of the crime lab doing the DNA analysis testified that the probability that Sutton's DNA matched that of the evidence by chance was 1 in 694,000—a number so compelling that it led jurors to convict him, even though Sutton did not fit the victim's description of her assailant.

A DNA profile analyzes only 13 parts of the genome, known to vary in most populations. Usually this is sufficient information to rule out a suspect. Using these criteria, Sutton's DNA at first seemed to match the evidence. The problem, though, wasn't in the DNA, but in the population to which it was compared. Although Sutton's pattern may indeed have been very rare in the large population to which it was compared, among black men, it wasn't rare at all—1 in 16 black men have the exact same pattern!

Proclaiming his innocence all along, Sutton had asked right away for an independent DNA test, but was told he couldn't afford one. So while he was in prison, he read voraciously about DNA profiling and again, in a handwritten note, requested retesting. Then he got lucky. Two journalists began investigating the Houston crime laboratory. They sent information on a few cases to a professor of criminology, who immediately saw the errors made in Sutton's DNA analysis, claiming that the test wasn't even of the quality of a middle school science project. Retesting Sutton's DNA, and comparing it to a relevant population, proved his innocence. ■

Genetic Code

Genetic information specifies the correct sequence of amino acids in a polypeptide chain. Each of the twenty different types of amino acids is represented in a DNA molecule by a triplet code, consisting of sequences of three nucleotides. That is, the sequence C, G, T in a DNA strand represents one type of amino acid; the sequence G, C, A represents another type. Other sequences encode instructions for beginning or ending the synthesis of a protein molecule, and for determining which genes are accessed for their information.

> The genetic code is said to be universal because all species use the same DNA base triplets to specify the same amino acids. Researchers deciphered the code in the 1960s. When the media mentions an individual's genetic code, they really are referring to the sequence of DNA bases comprising a certain gene or genome—not the genetic code (the correspondence between DNA triplet and amino acid).

DNA molecules are in the nucleus and protein synthesis occurs in the cytoplasm. Because the cell must keep a permanent copy of the genetic instructions, genetic information must get from the nucleus into the cytoplasm for the cell to use it. RNA molecules accomplish this transfer of information.

RNA Molecules

RNA (ribonucleic acid) molecules differ from DNA molecules in several ways. RNA molecules are single-stranded, and their nucleotides have ribose rather than deoxyribose sugar. Like DNA, RNA nucleotides each have one of four nitrogenous bases, but whereas adenine, cytosine, and guanine nucleotides are part of both DNA and RNA, thymine nucleotides are only in DNA. In place of thymine nucleotides, RNA molecules have uracil (U) nucleotides (fig. 4.21 and Appendix D, p. 949). In RNA U pairs with A (fig. 4.22). Different types of RNA have different size ranges and functions. The process of copying DNA information into an RNA sequence is called **transcription** (trans-krip′-shun).

The first step in delivering information from the nucleus to the cytoplasm is the synthesis of **messenger RNA (mRNA).** RNA nucleotides form complementary base pairs with one of the two strands of DNA that encodes a particular protein. However, just as the words in a sentence must be read in the correct order to make sense, the base sequence of a strand of DNA must be "read" in the correct direction and from the correct starting point. Furthermore, only one of the two antiparallel strands of DNA contains the genetic message. An enzyme called RNA polymerase recognizes the correct DNA strand and the right direction for RNA synthesis. The "sentence" always begins with the mRNA base sequence AUG (fig. 4.23).

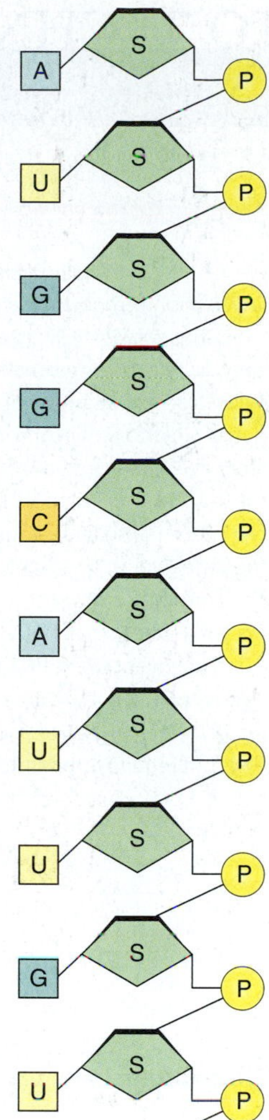

FIGURE 4.21 RNA differs from DNA in that it is single-stranded, contains ribose rather than deoxyribose, and has uracil (U) rather than thymine (T) as one of its four bases.

In mRNA synthesis, RNA polymerase binds to a promoter, a DNA base sequence that begins a gene. Then a section of the double-stranded DNA unwinds and pulls apart, exposing a portion of the gene. RNA polymerase moves along the strand, exposing other portions of the gene. At the same time, a molecule of mRNA forms as RNA nucleotides complementary to those along the DNA strand are strung together. For example, if the sequence of DNA bases is TACCCGAGG, the complementary bases in the developing mRNA molecule will be AUGGGCUCC, as figure 4.23 shows. For different genes, different strands of the DNA molecule may be used to manufacture RNA.

RNA polymerase continues to move along the DNA strand, exposing portions of the gene, until it reaches a special DNA base sequence (termination signal) that signals the end of the gene. At this point, the RNA polymerase releases the newly formed mRNA molecule and leaves the DNA. The DNA then rewinds and assumes its previous double helix structure.

Each amino acid in the protein to be synthesized was originally represented by a series of three bases in DNA.

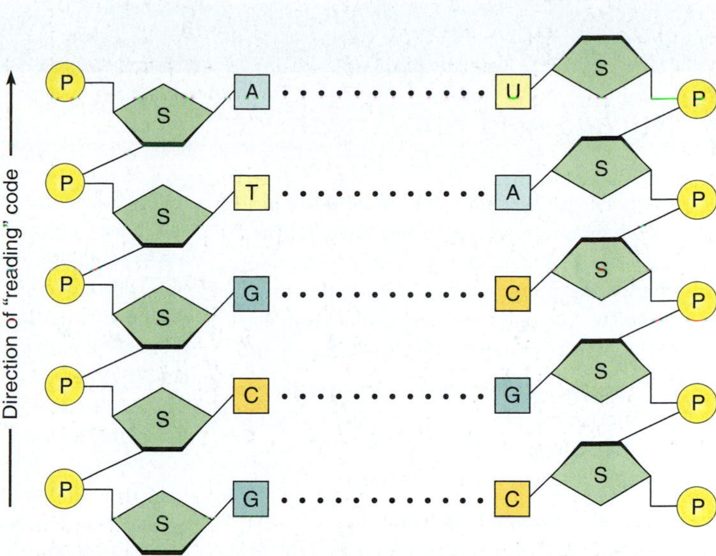

FIGURE 4.22 Transcription of RNA from DNA. When an RNA molecule is synthesized beside a strand of DNA, complementary nucleotides bond as in a double-stranded DNA molecule, with one exception: RNA contains uracil nucleotides (U) in place of thymine nucleotides (T).

Those amino acids, in the proper order, are now represented by a series of three base sequences, called **codons,** (ko′donz) in mRNA. To complete protein synthesis, mRNA must leave the nucleus and associate with a ribosome. There, the series of codons of the mRNA is translated from the "language" of nucleic acids to the "language" of amino acids. This process is fittingly called **translation** (see fig. 4.23). Table 4.1 compares DNA and RNA molecules.

Protein Synthesis

Synthesizing a protein molecule requires that the specified amino acid building blocks in the cytoplasm align in the proper sequence along an mRNA. A second type of RNA molecule, transcribed in the nucleus and called **transfer RNA (tRNA),** aligns amino acids in a way that enables them to bond to each other. A tRNA molecule consists of seventy to eighty nucleotides and has a complex three-dimensional shape, somewhat like a cloverleaf. The two ends of the tRNA molecule are important for the "connector" function (see fig. 4.23).

At one end, each tRNA molecule is a binding site for a particular amino acid. At least one type of tRNA specifies each of the twenty amino acids. An amino acid must be activated for a tRNA to pick it up. Special enzymes catalyze this step. ATP provides the energy for an amino acid and its tRNA to bond (fig. 4.24).

The other end of each transfer RNA molecule includes a specific three nucleotide sequence, called the **anticodon,** unique to that type of tRNA. An anticodon bonds only to the complementary mRNA codon. In this way, the appropriate tRNA carries its amino acid to the correct place in the mRNA sequence (fig. 4.24).

4.2 FROM SCIENCE TO TECHNOLOGY

Nucleic Acid Amplification

The polymerase chain reaction (PCR) is a procedure that borrows a cell's machinery for DNA replication, making many copies of a gene of interest. Developed in 1983, PCR was the first of several technologies called nucleic acid amplification. Starting materials for PCR include:

- two types of short DNA pieces known to bracket the gene of interest, called primers
- a large supply of DNA bases
- the enzymes that replicate DNA

Here's how it works. First, heat is used to separate the two strands of the target DNA—such as bacterial DNA in a body fluid sample from a person who has symptoms of an infection. Next, the temperature is lowered, and the two short DNA primers are added. The primers complementary base pair to the separated target strands. The third step adds DNA polymerase and bases. The DNA polymerase adds bases to the primers and builds a sequence complementary to the target sequence. The newly synthesized strands then act as templates in the next round of replication, which begins by raising the temperature. All of this is done in an automated device called a thermal cycler that controls the key temperature changes. The DNA polymerase can withstand the temperature shifts because it comes from a bacterium that lives in hot springs.

The pieces of DNA exponentially accumulate. The number of amplified pieces of DNA equals 2^n where n equals the number of temperature cycles. After just twenty cycles, 1 million copies of the original sequence are in the test tube. PCR has had many diverse applications, from detecting moose meat in hamburger to analysis of insect larvae in decomposing human corpses.

PCR's greatest strength is that it works on crude samples of rare and short DNA sequences, such as a bit of brain tissue on the bumper of a car, which in one criminal case led to identification of a missing person. PCR's greatest weakness, ironically, is its exquisite sensitivity. A blood sample submitted for diagnosis of an infection contaminated by leftover DNA from a previous run, or a stray eyelash dropped from the person running the reaction, can yield a false positive result. The technique is also limited in that a user must know the sequence to be amplified and that mutations can sometimes occur in the amplified DNA not present in the source DNA.

The invention of PCR inspired other nucleic acid amplification technologies. One, which copies DNA into RNA and then amplifies the RNA, does not require temperature shifts and produces 100 to 1,000 copies per cycle, compared to PCR's doubling. ■

The genetic code specifies more than enough information. Although only twenty types of amino acids need be encoded, the four types of bases can form sixty-four different mRNA codons. Therefore, some amino acids correspond to more than one codon (table 4.2). Three of the codons do not have a corresponding tRNA. They provide a "stop" signal, indicating the end of protein synthesis, much like the period at the end of this sentence. Sixty-one different tRNAs are specific for the remaining sixty-one codons, which means that more than one type of tRNA can correspond to the same amino acid type.

The binding of tRNA and mRNA occurs in close association with a ribosome. A ribosome is a tiny particle of two unequal-sized subunits composed of **ribosomal RNA (rRNA)** and protein molecules. The smaller subunit of a ribosome binds to a molecule of mRNA near the first codon. A tRNA with the complementary anticodon brings its attached amino acid into position, temporarily joining to the ribosome. A second tRNA, complementary to the second mRNA codon, then binds (with its activated amino acid) to an adjacent site on the ribosome. The first tRNA molecule releases its amino acid, providing the energy for a peptide bond to form between the two amino acids (see fig. 4.24). This process repeats as the ribosome moves along the mRNA, adding amino acids one at a time to the extending polypeptide chain. The enzymatic activity necessary for bonding of the amino acids comes from ribosomal proteins and some RNA molecules (ribozymes) in the larger subunit of the ribosome. This subunit also holds the growing chain of amino acids.

RECONNECT

To Chapter 3, A Composite Cell, page 82.

Protein synthesis is economical. A molecule of mRNA usually associates with several ribosomes at the same time. Thus, several copies of that protein, each in a different stage of formation, may be present at any given moment. As the polypeptide forms, proteins called *chaperones* fold it into its unique shape, and when the process is completed, the polypeptide is released as a separate functional molecule. The tRNA molecules, ribosomes, mRNA, and the enzymes can function repeatedly in protein synthesis.

ATP molecules provide the energy for protein synthesis. A protein may consist of many hundreds of amino acids and the energy from three ATP molecules is required to link each amino acid to the growing chain. This means that a large fraction of a cell's energy supply supports protein synthesis. Table 4.3 summarizes protein synthesis.

The number of molecules of a particular protein that a cell synthesizes is generally proportional to the number of corresponding mRNA molecules. The rate at which mRNA is transcribed from DNA in the nucleus and the rate at which enzymes (ribonucleases) destroy the mRNA in the cytoplasm therefore control protein synthesis.

Proteins called transcription factors activate certain genes, moving aside the surrounding histone proteins to expose the promoter DNA sequences that represent the start of a gene. These actions are called "chromatin remodeling,"

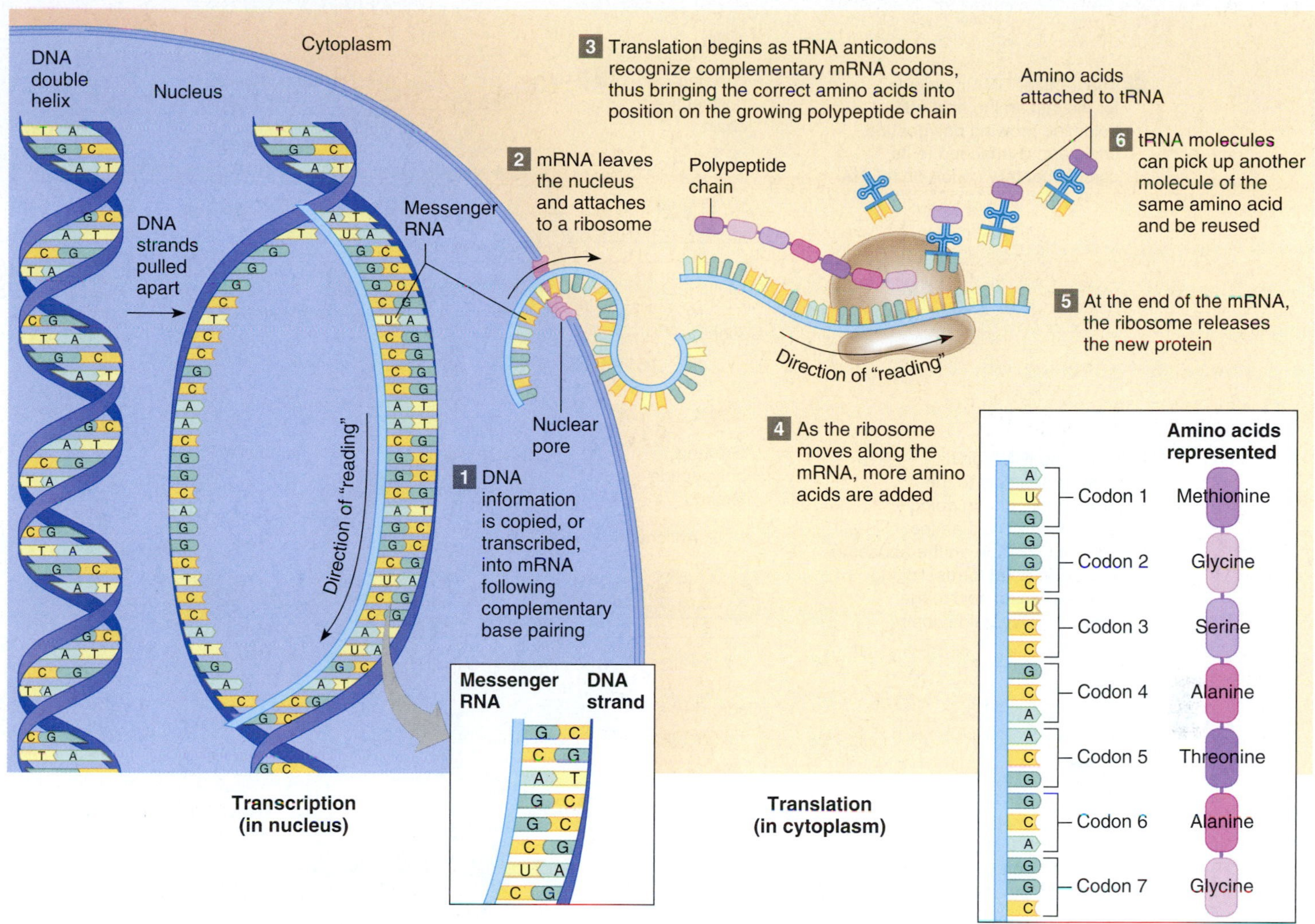

FIGURE 4.23 Protein synthesis. DNA information is transcribed into mRNA, which in turn is translated into a sequence of amino acids. The inset shows some examples of the correspondence between mRNA codons and the specific amino acids that they encode.

TABLE 4.1 | A Comparison of DNA and RNA Molecules

	DNA	RNA
Main location	Part of chromosomes, in nucleus	Cytoplasm
5-carbon sugar	Deoxyribose	Ribose
Basic molecular structure	Double-stranded	Single-stranded
Nitrogenous bases included	Cytosine, guanine, adenine, thymine	Cytosine, guanine, adenine, uracil
Major functions	Contains genetic code for protein synthesis, replicates prior to mitosis	Messenger RNA carries transcribed DNA information to cytoplasm and acts as template for synthesis of protein molecules; transfer RNA carries amino acids to messenger RNA; ribosomal RNA provides structure for ribosomes

and they control which proteins a cell produces and how many copies form under particular conditions. A connective tissue cell might have many mRNAs representing genes that encode the proteins collagen and elastin; a muscle cell would have abundant mRNAs encoding contractile proteins, such as actin and myosin. Extracellular signals such as hormones and growth factors activate transcription factors.

From Science to Technology 4.3 describes another type of transcriptional control—microRNAs.

PRACTICE

24 How is genetic information carried from the nucleus to the cytoplasm?

25 How are protein molecules synthesized?

26 How is gene expression controlled?

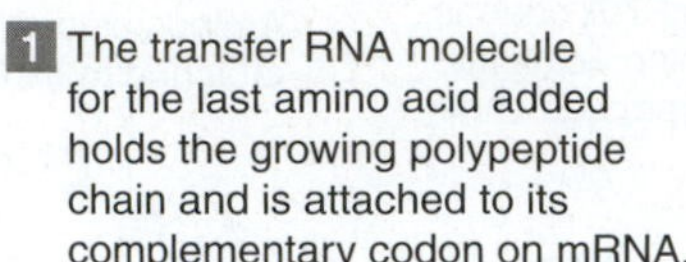

1 The transfer RNA molecule for the last amino acid added holds the growing polypeptide chain and is attached to its complementary codon on mRNA.

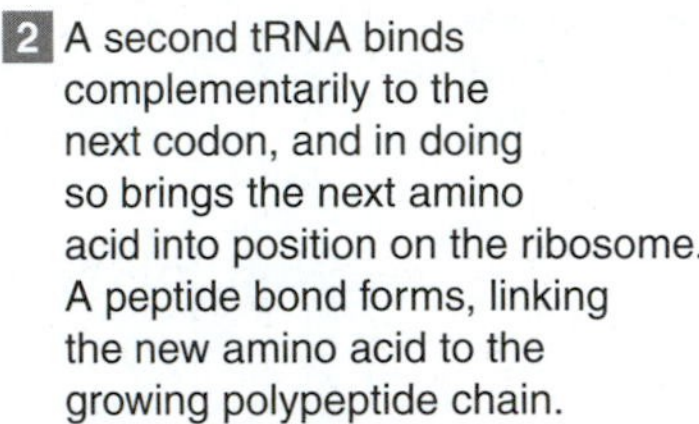

2 A second tRNA binds complementarily to the next codon, and in doing so brings the next amino acid into position on the ribosome. A peptide bond forms, linking the new amino acid to the growing polypeptide chain.

3 The tRNA molecule that brought the last amino acid to the ribosome is released to the cytoplasm, and will be used again. The ribosome moves to a new position at the next codon on mRNA.

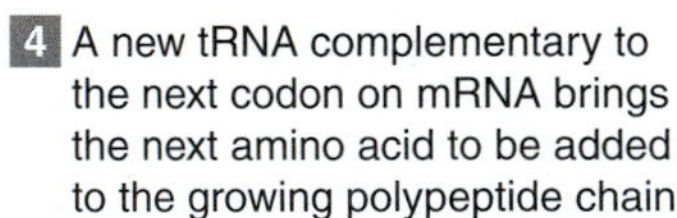

4 A new tRNA complementary to the next codon on mRNA brings the next amino acid to be added to the growing polypeptide chain.

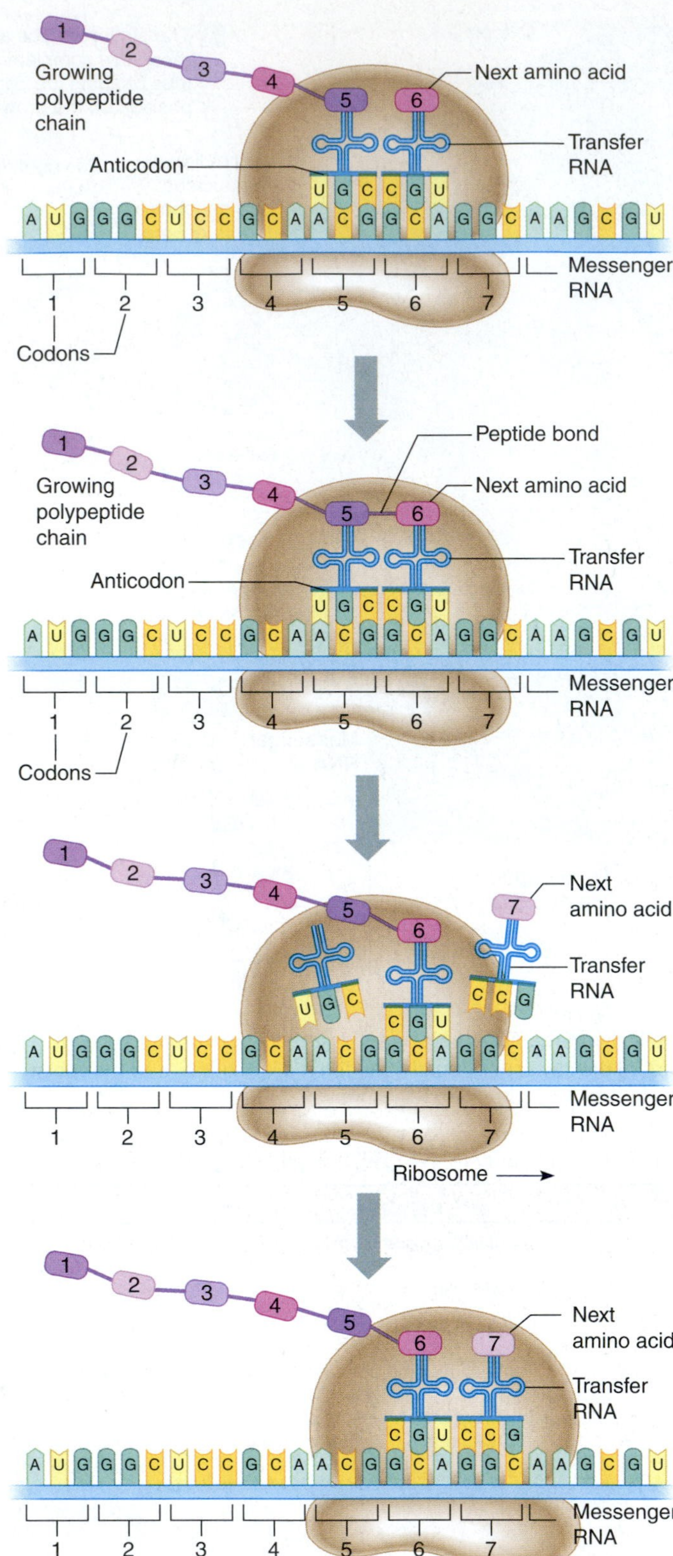

FIGURE 4.24 A closer look at protein synthesis. Molecules of transfer RNA (tRNA) attach to and carry specific amino acids, aligning them in the sequence determined by the codons of mRNA. These amino acids, connected by peptide bonds, form a polypeptide chain of a protein molecule. Protein synthesis occurs on ribosomes.

TABLE 4.2 | **Codons (mRNA Three Base Sequences)**

First Letter	SECOND LETTER: U		C		A		G		Third Letter
	U		C		A		G		
U	UUU	phenylalanine (phe)	UCU	serine (ser)	UAU	tyrosine (tyr)	UGU	cysteine (cys)	U
	UUC		UCC		UAC		UGC		C
	UUA	leucine (leu)	UCA		UAA	STOP	UGA	STOP	A
	UUG		UCG		UAG	STOP	UGG	tryptophan (trp)	G
C	CUU	leucine (leu)	CCU	proline (pro)	CAU	histidine (his)	CGU	arginine (arg)	U
	CUC		CCC		CAC		CGC		C
	CUA		CCA		CAA	glutamine (gln)	CGA		A
	CUG		CCG		CAG		CGG		G
A	AUU	isoleucine (ile)	ACU	threonine (thr)	AAU	asparagine (asn)	AGU	serine (ser)	U
	AUC		ACC		AAC		AGC		C
	AUA		ACA		AAA	lysine (lys)	AGA	arginine (arg)	A
	AUG	START methionine (met)	ACG		AAG		AGG		G
G	GUU	valine (val)	GCU	alanine (ala)	GAU	aspartic acid (asp)	GGU	glycine (gly)	U
	GUC		GCC		GAC		GGC		C
	GUA		GCA		GAA	glutamic acid (glu)	GGA		A
	GUG		GCG		GAG		GGG		G

TABLE 4.3 | **Protein Synthesis**

Transcription (In the Nucleus)
1. RNA polymerase binds to the DNA base sequence of a gene.
2. This enzyme unwinds a portion of the DNA molecule, exposing part of the gene.
3. RNA polymerase moves along one strand of the exposed gene and catalyzes synthesis of an mRNA, whose nucleotides are complementary to those of the strand of the gene.
4. When RNA polymerase reaches the end of the gene, the newly formed mRNA is released.
5. The DNA rewinds and closes the double helix.
6. The mRNA passes through a pore in the nuclear envelope and enters the cytoplasm.
Translation (In the Cytoplasm)
1. A ribosome binds to the mRNA near the codon at the beginning of the messenger strand.
2. A tRNA molecule that has the complementary anticodon brings its amino acid to the ribosome.
3. A second tRNA brings the next amino acid to the ribosome.
4. A peptide bond forms between the two amino acids, and the first tRNA is released.
5. This process is repeated for each codon in the mRNA sequence as the ribosome moves along its length, forming a chain of amino acids.
6. As the chain of amino acids grows, it folds, with the help of chaperone proteins, into the unique conformation of a functional protein molecule.
7. The completed protein molecule (polypeptide) is released. The mRNA molecule, ribosome, and tRNA molecules are recycled.

4.7 CHANGES IN GENETIC INFORMATION

Remarkably, we are more alike than different—human genome sequences are 99.9 percent the same among individuals. The tenth of a percent of the human genome that can vary from person-to-person includes rare DNA sequences that affect health or appearance, as well as common DNA base variations that do not exert any observable effects.

Nature of Mutations

The rare distinctions in DNA sequence that affect how we look or feel are called **mutations** (mu-ta′shunz) More common genetic variants with no detectable effects are called **single nucleotide polymorphisms,** abbreviated SNPs (pronounced "snips"). "Polymorphism" means "many forms."

To visualize the concept of genetic variability, imagine a simplified DNA sequence that is part of a particular genome region:

A A A A A A A A A A A A

A person with a mutation or polymorphism at the fourth base might have any of the following sequences for this portion of the genome, with the differences highlighted:

A A A C A A A A A A A A
A A A T A A A A A A A A
A A A G A A A A A A A A

4.3 FROM SCIENCE TO TECHNOLOGY

MicroRNAs and RNA Interference

The human genome provides blueprints for building a human body, and it also includes instructions for how to use the blueprints. Those instructions are so small—RNA molecules 21 or 22 bases long—that for many years researchers unwittingly threw them out. Today an entire industry is forming to adapt these natural controllers of gene expression, called microRNAs, into diagnostic tests and even new types of treatments for disease.

MicroRNAs belong to a class of RNA molecules called noncoding RNAs, so-named because they were not among the first three major classes of RNA described (mRNA, tRNA, and rRNA). The human genome probably has close to 1,000 microRNAs, about half of which have been discovered. The DNA sequences that encode microRNAs are found in parts of the genome accessed to produce proteins and also in the vast regions that do not encode protein and are less well understood.

Each microRNA binds to parts of the initial control regions (corresponding to DNA promoters) of a particular set of mRNAs, by complementary base pairing. When a microRNA binds a "target" mRNA, it turns off transcription. In this way, a single type of microRNA controls specific sets of genes. In turn, a single type of mRNA can bind several different microRNAs. To analyze these complex interactions, researchers use experiments as well as computational tools (bioinformatics).

Within the patterns of microRNA function may lie clues to developing new ways to fight disease, because these controls of gene expression have stood the test of evolutionary time. The first applications are in cancer, as certain microRNAs are either more or less abundant in cancer cells than in healthy cells of the same type from which the cancer cells formed. Restoring the levels of microRNAs that normally suppress the too-rapid cell cycling of cancer, or blocking production of microRNAs too abundant in cancer, could help to return cells to normal. The first microRNA-based diagnostic tests became available in 2008 and are used to distinguish types of lung cancer and for cancer that has spread and the original tumor cannot be identified by other means.

In a related technology called RNA interference (RNAi), small, synthetic RNA molecules are introduced into cells. They block gene expression in the same manner as the naturally occurring microRNAs. Many companies are developing RNAi-based drugs. The technological challenge is in directing where they affect the genome. ■

If the change affects the person in a noticeable or detectable way and occurs in less than one percent of the population, it is considered a mutation. If there is no detectable change from what is considered normal and the change is seen in more than one percent of the population, it is considered a SNP. These designations, however, are subjective. They depend upon what we can identify and what we consider normal. A more general and traditional use of the term "mutation" is as the mechanism of change in a DNA sequence.

The human genome has millions of SNPs. Association studies look at SNP combinations in populations and attempt to identify patterns found almost exclusively in people with a particular disorder. The correlations between SNP patterns and elevated disease risks can be used to guide clinical decision-making—for example, suggesting which patients might respond to one drug but not another. However, sometimes the associations are statistical flukes that vanish when more data are included. Still, several companies promote SNP-based tests direct-to-consumers on the Internet. These should be approached with caution, because the accuracy of using population-level data to diagnose disease in an individual has not been well-studied.

Another way that people differ in their DNA sequences is by the number of repeats of particular sequences, called copy number variants. Such a repeated sequence may range from only a few DNA bases to millions.

Mutations occur in two general ways—spontaneously or induced. They may happen spontaneously due to the chemical tendency of free nitrogenous bases to exist in two slightly different structures. For extremely short times, a base can be in an unstable form. If, by chance, such an unstable base is inserted into newly forming DNA, an error in sequence will be generated and perpetuated when the strand replicates. Another replication error that can cause mutation is when the existing (parental) DNA strand slips, adding nucleotides to or deleting nucleotides from the sequence.

In contrast to spontaneous mutations are induced mutations, a response to exposure to certain chemicals or radiation. Anything that causes mutation is termed a **mutagen** (mu′tah-jen). A familiar mutagen is ultraviolet radiation, part of sunlight. Prolonged exposure to ultraviolet radiation can form an extra bond between two adjacent thymine DNA bases that are part of the same DNA strand in a skin cell. This extra bond kinks the double helix, causing an incorrect base to be inserted during replication. The cell harboring such a mutation may not be affected, may be so damaged that it dies and peels off, or it may become cancerous. This is how too much sun exposure can cause skin cancer. Mutagens are also found in hair dye, food additives, smoked meats, and flame retardants.

Disease may result from a mutation, whether spontaneous or induced. If the mutation alters the amino acid sequence of the encoded protein so that it malfunctions or isn't produced at all, and health is impaired. For example, the muscle weakness of Duchenne muscular dystrophy results from a mutation in the gene encoding the protein dystrophin. This protein normally enables muscle cell membranes to withstand the force of contraction. The mutation may be a missing or changed nucleotide base or absence of part or all of the dystrophin gene. Lack of the normal protein causes muscle cells to collapse, and muscles throughout the body weaken and break down. Figure 4.25 shows how the change of one base causes another inherited illness, sickle cell disease.

Although mutations are commonly associated with diseases or otherwise considered abnormal, they also can confer an advantage. The opening vignette to chapter 3 (p. 76) describes one such helpful mutation that protects against HIV infection.

Once DNA changes, producing a mutation or a SNP or copy number variant, the change is transmitted every time the cell in which it originated divides. If that cell is an egg or sperm, then the change is passed to the next generation. We return to this point in the next section.

Protection Against Mutation

Cells detect many mutations and take action to correct the errors. Special "repair enzymes" recognize and remove mismatched nucleotides and fill the resulting gap with the accurate, complementary nucleotides. This mechanism, called the **DNA damage response,** restores the original structure of the double-stranded DNA molecule.

Disorders of the DNA damage response can make life difficult. Xeroderma pigmentosum, for example, causes extreme sun sensitivity. A child with the condition must completely cover up, swathing sunblock on any exposed skin to prevent freckles, sores, and cancer. Special camps and programs allow these children to play outdoors at night, away from the danger of the sun.

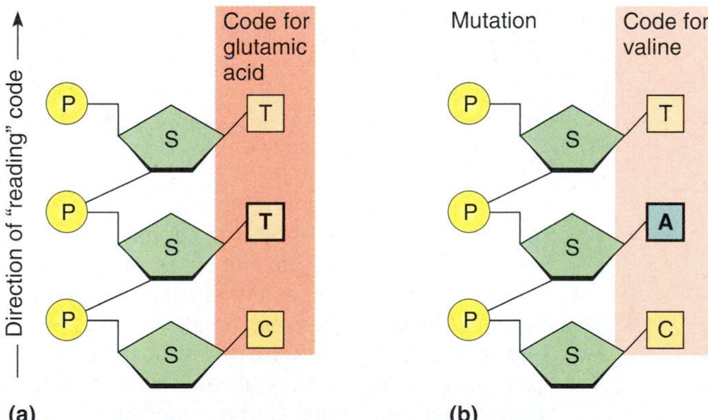

FIGURE 4.25 An example of mutation. (*a*) The DNA code for the amino acid glutamic acid is CTT. (*b*) If something happens to change the first T to A, the DNA code changes to CAT, which specifies the amino acid valine. The resulting mutation, when it occurs in the DNA that encodes the sixth amino acid in a subunit of the protein hemoglobin, causes sickle cell disease. The abnormal hemoglobin bends the red blood cells containing it into sickle shapes. The cells lodge in narrow blood vessels, blocking the circulation and causing great pain.

The nature of the genetic code protects against mutation, to a degree. Sixty-one codons specify the twenty types of amino acids, and therefore, some amino acids correspond to more than one codon type. Usually, two or three codons specifying the same amino acid differ only in the third base of the codon. A mutation that changes the third codon base can encode the same amino acid. For example, the DNA triplets GGA and GGG each specify the amino acid proline. If a mutation changes the third position of GGA to a G, the amino acid for that position in the encoded protein does not change—it is still proline.

If a mutation alters a base in the second position, the substituted amino acid is often similar in overall shape to the normal one, and the protein is not changed significantly enough to affect its function. This mutation, too, would go unnoticed. (An important exception is the mutation shown in fig. 4.25.) Yet another protection against mutation is that a person has two copies of each chromosome, and therefore of each gene. If one copy is mutated, the other may provide enough of the gene's normal function to maintain health. (This is more complicated for the sex chromosomes, X and Y, discussed in chapter 24, pp. 927–928.)

Timing of a mutation influences effects on health. A mutation in a sperm cell, egg cell, or fertilized ovum is repeated in every cell of the individual. A mutation in an embryo might be devastating because much of the body is still to develop, and many cells inherit the mutation. In contrast, a mutation in a body cell of an adult would most likely have no effect because it would be only one among trillions of cells that do not have the mutation. However, if such a somatic (body cell) mutation confers a faster cell cycle and therefore cells bearing the mutation have a division advantage, cancer can result.

Inborn Errors of Metabolism

The first part of the chapter discussed enzymes that catalyze the reactions of energy metabolism. Enzymes are also essential to many other reactions and pathways.

A type of disorder called an "inborn error of metabolism" results from inheriting a mutation that alters an enzyme. Such an enzyme block in a biochemical pathway has two general effects: the biochemical that the enzyme normally acts on builds up, and the biochemical resulting from the enzyme's normal action becomes scarce. It is similar to blocking a garden hose: water pressure builds up behind the block, but no water comes out after it.

The biochemical excesses and deficiencies that an inborn error of metabolism triggers can drastically affect health. The specific symptoms depend upon which pathways and biochemicals are affected. Figure 4.26 shows how blocks of different enzymes in one biochemical pathway lead to different sets of symptoms.

Understanding the pathways of metabolism and the many steps and controls of protein synthesis (gene expression) can be daunting. Advances in computational science, however, have vastly improved our ability to tease out the meanings from these complex processes that underlie our physiology. From Science to Technology 4.4 provides a glimpse of this new "systems biology" approach to dissecting the controls of how the human body functions.

PRACTICE 

27 Distinguish between a mutation and a SNP.

28 How do mutations arise?

29 How do mutations affect health or appearance?

30 Describe protections against mutation.

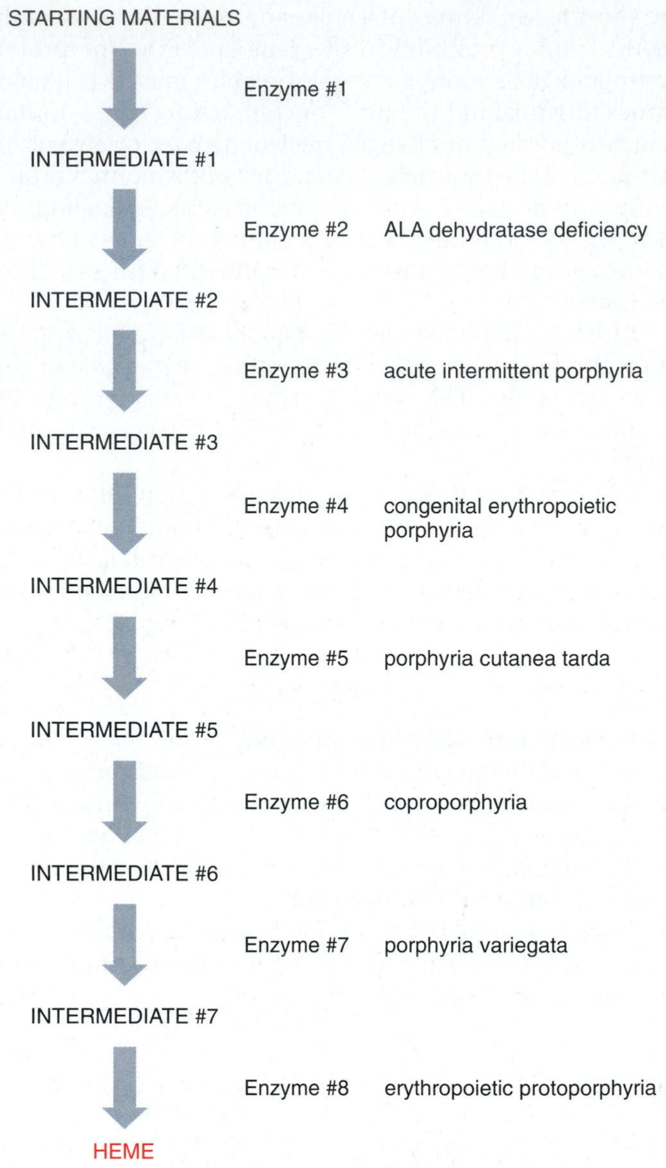

STARTING MATERIALS
↓ Enzyme #1
INTERMEDIATE #1
↓ Enzyme #2 — ALA dehydratase deficiency
INTERMEDIATE #2
↓ Enzyme #3 — acute intermittent porphyria
INTERMEDIATE #3
↓ Enzyme #4 — congenital erythropoietic porphyria
INTERMEDIATE #4
↓ Enzyme #5 — porphyria cutanea tarda
INTERMEDIATE #5
↓ Enzyme #6 — coproporphyria
INTERMEDIATE #6
↓ Enzyme #7 — porphyria variegata
INTERMEDIATE #7
↓ Enzyme #8 — erythropoietic protoporphyria
HEME

FIGURE 4.26 Seven related but distinct inborn errors of metabolism result from abnormal or missing enzymes that catalyze reactions of the pathway for the synthesis of heme, part of the hemoglobin molecule that is packed into red blood cells. In each disorder, the intermediate biochemical that a deficient enzyme would normally affect builds up. The excess is excreted in the urine or accumulates in blood, feces, or inside red blood cells. Some of the symptoms include reddish teeth, pink urine, excess hair, and photosensitivity.

CHAPTER SUMMARY

4.1 INTRODUCTION (PAGE 115)

A cell continuously carries on metabolic processes. Enzymes are critical to the reactions and pathways of metabolism.

4.2 METABOLIC PROCESSES (PAGE 115)

Metabolic processes include two types of reactions, anabolism and catabolism.

1. Anabolism
   a. Anabolism builds large molecules.
   b. In dehydration synthesis, hydrogen atoms and hydroxyl groups are removed, water forms, and smaller molecules bind by sharing atoms.
   c. Complex carbohydrates are synthesized from monosaccharides, fats are synthesized from glycerol and fatty acids, and proteins are synthesized from amino acids.
2. Catabolism
   a. Catabolism breaks down larger molecules.
   b. In hydrolysis, a water molecule supplies a hydrogen atom to one portion of a molecule and a hydroxyl group to a second portion; the bond between these two portions breaks.
   c. Complex carbohydrates are decomposed into monosaccharides, fats are decomposed into glycerol and fatty acids, and proteins are decomposed into amino acids.

4.3 CONTROL OF METABOLIC REACTIONS (PAGE 117)

Metabolic processes have many steps that occur in a specific sequence and are interconnected. A sequence of enzyme-controlled reactions is a metabolic pathway.

4.4 FROM SCIENCE TO TECHNOLOGY

The Human Metabolome

A generation ago, prehealth profession students had to memorize a frighteningly complex chart of biochemical pathways that represent all of the energy reactions in a cell. The cellular respiration pathways ran down the center, with branches radiating outward and in some places interconnecting into a giant web. Today, several technologies as well as the ability to store massive amounts of data have made possible the Human Metabolome Database (**www.hmdb.ca**).

"Metabolome" refers to all of the small molecules that are part of metabolism in a cell, tissue, organ, or an entire organism. The database is a vast, annotated catalog of those molecules. The government of Canada is supporting the effort to search all published papers and books that describe metabolites and link that information with experimental data. The techniques of electrophoresis and chromatography are used to separate metabolites, and mass spectrometry (MS) and nuclear magnetic resonance (NMR) spectroscopy describe their chemical characteristics.

Biochemists estimate that human cells have at least 2,500 different metabolites, but fewer than half have been identified. Far fewer have been analyzed for their concentrations in different cell types under different conditions. In the Human Metabolome Database, each entry has an electronic "MetaboCard" that includes 90 data fields, half with clinical data (such as associated diseases and drug interactions) and half with biochemical data (such as pathways and enzymes that interact with the metabolite). Each entry is also hyperlinked to other databases, interfacing with 1,500 drugs and 3,600 foods and food additives. The information in the Human Metabolome Database is being used in drug discovery, toxicology, transplant monitoring, clinical chemistry, disease diagnosis, and screening of newborns for metabolic disorders.

The "metabolome" is one of several "omes" now under intense study. The first was "genome," coined in 1920 to denote a complete set of genes. It was joined much more recently by "proteome" to denote the proteins in a cell or organism, and then "transcriptome" to list the RNA molecules in a cell type. The "omes" comprise the new field of systems biology, which examines the interactions and relationships among the parts of an organism. The genome, proteome, and transcriptome each describe a single type of molecule. The metabolome is the most complex set of biochemicals in a cell or organism. ■

1. Enzyme action
   a. Metabolic reactions require energy to start.
   b. Enzymes are proteins that increase the rate of specific metabolic reactions.
   c. An enzyme acts when its active site temporarily combines with the substrate, altering its chemical structure. This enables the substrate to react, forming a product. The enzyme is released in its original form.
   d. The rate of enzyme-controlled reactions depends upon the numbers of enzyme and substrate molecules and the efficiency of the enzyme.
   e. Enzymes are usually named according to their substrates, with *-ase* at the end.
2. Regulation of metabolic pathways
   a. A rate-limiting enzyme may regulate a metabolic pathway.
   b. A negative feedback mechanism in which the product of a pathway inhibits the regulatory enzyme may control the regulatory enzyme.
   c. The rate of product formation usually remains stable.
3. Cofactors and coenzymes
   a. Cofactors are additions to some enzymes that are necessary for their function.
   b. A cofactor may be an ion or a small organic molecule called a coenzyme.
   c. Vitamins, the sources of coenzymes, usually cannot be synthesized by human cells in adequate amounts.
4. Factors that alter enzymes
   a. Enzymes are proteins and can be denatured.
   b. Factors that may denature enzymes include heat, radiation, electricity, chemicals, and extreme pH values.

4.4 ENERGY FOR METABOLIC REACTIONS (PAGE 119)

Energy is a capacity to produce change or to do work. Common forms of energy include heat, light, sound, electrical energy, mechanical energy, and chemical energy. Metabolic energy is made available by the reactions of cellular respiration.

1. ATP molecules
   a. Energy is captured in the bond of the terminal phosphate of each ATP molecule.
   b. Captured energy is released when the terminal phosphate bond of an ATP molecule breaks.
   c. ATP that loses its terminal phosphate becomes ADP.
   d. ADP can be converted to ATP by capturing energy and a phosphate.
   e. ATP is the primary energy-carrying molecule in a cell.
2. Release of chemical energy
   a. Most metabolic processes use chemical energy released when molecular bonds break.
   b. The energy glucose releases during cellular respiration is used to promote metabolism.
   c. Enzymes in the cytoplasm and mitochondria control cellular respiration.

4.5 CELLULAR RESPIRATION (PAGE 120)

Cellular respiration transfers energy from molecules such as glucose and makes it available for cellular use. This process occurs in three distinct, interconnected series of reactions.

1. Glycolysis
   a. Glycolysis, the first step of glucose catabolism, occurs in the cytosol and does not require oxygen.
   b. Glycolysis can be divided into three stages, in which some of the energy released is transferred to ATP.
   c. Some of the energy released in glycolysis is in the form of high-energy electrons attached to hydrogen carriers.
2. Anaerobic reactions (absence of oxygen)
   a. Oxygen is the final electron acceptor in the aerobic reactions of cellular respiration.
   b. In the anaerobic reactions, NADH and H^+ instead donate electrons and hydrogens to pyruvic acid, generating lactic acid.
   c. Lactic acid builds up, eventually inhibiting glycolysis and ATP formation.
   d. When oxygen returns, in liver cells lactic acid reacts to form pyruvic acid.
3. Aerobic reactions (presence of oxygen)
   a. The second phase of glucose catabolism occurs in the mitochondria and requires oxygen.
   b. These reactions include the citric acid cycle and the electron transport chain.
   c. Considerably more energy is transferred to ATP during the aerobic reactions than during glycolysis.
   d. The products of the aerobic reactions of cellular respiration are heat, carbon dioxide, water, and energy.
   e. The citric acid cycle decomposes molecules, releases carbon dioxide, releases hydrogen atoms that have high-energy electrons, and forms ATP.
   f. High-energy electrons from hydrogen atoms enter an electron transport chain. Energy released from the chain is used to form ATP.
   g. Each metabolized glucose molecule yields up to thirty-eight ATP molecules.
   h. Excess carbohydrates may enter anabolic pathways and be polymerized into and stored as glycogen or react to produce fat.

4.6 NUCLEIC ACIDS AND PROTEIN SYNTHESIS (PAGE 124)

DNA molecules contain and maintain information that tells a cell how to synthesize proteins, including enzymes.

1. Genetic information
   a. DNA information specifies inherited traits.
   b. A gene is a portion of a DNA molecule that includes, in its nucleotide base sequence, the genetic information for making a protein.
   c. The DNA nucleotides from both strands pair in a complementary fashion, joining the two strands. A binds T, and G binds C.
2. DNA replication
   a. Each new cell requires a copy of the original cell's genetic information.
   b. DNA molecules are replicated during interphase of the cell cycle.
   c. Each new DNA molecule consists of one old strand and one new strand.
3. Genetic code
   a. Some of the sequence of nucleotides in a DNA molecule represents the sequence of amino acids in a protein molecule. This correspondence is the genetic code.
   b. RNA molecules transfer genetic information from the nucleus to the cytoplasm.
   c. RNA synthesis is transcription. Protein synthesis is translation.
4. RNA molecules
   a. RNA molecules are usually single-stranded, have ribose instead of deoxyribose, and uracil in place of thymine.
   b. Messenger RNA molecules, synthesized in the nucleus, have a nucleotide sequence complementary to that of an exposed strand of DNA.
   c. Messenger RNA molecules move into the cytoplasm, associate with ribosomes, and are templates for the synthesis of protein molecules.
5. Protein synthesis
   a. Molecules of tRNA position amino acids along a strand of mRNA.
   b. A ribosome binds to an mRNA and allows a tRNA to recognize its correct position on the mRNA.
   c. The ribosome has enzymes required for the synthesis of the protein and holds the protein until it is completed.
   d. As the protein forms, it folds into a unique shape.
   e. ATP provides the energy for protein synthesis.

4.7 CHANGES IN GENETIC INFORMATION (PAGE 135)

A DNA molecule contains a great amount of information. Mutation changes the genetic information. Not all changes to DNA are harmful.

1. Nature of mutations
   a. Mutations are rare and alter health or appearance.
   b. Single nucleotide polymorphisms are more common and have no observable effect.
   c. A protein synthesized from an altered DNA sequence may or may not function normally.
   d. Mutations may be spontaneous or induced.
   e. DNA changes are transmitted when the cell divides.
2. Protection against mutation
   a. Repair enzymes can correct some forms of DNA damage.
   b. The genetic code protects against some mutations.
   c. A mutation in a sex cell, fertilized egg, or embryo may have more effects than a later mutation because a greater proportion of cells bear the mutation.
3. Inborn errors of metabolism
   a. An enzyme deficiency may cause an inborn error of metabolism, in which a metabolic pathway is blocked.
   b. The substrate builds up and the product diminishes.

CHAPTER ASSESSMENTS

4.1 Introduction

1 Define *metabolism.* (p. 115)

2 Explain how metabolic pathways are linked and intersect. (p. 115)

4.2 Metabolic Processes

3 Distinguish between catabolism and anabolism. (p. 115)

4 Distinguish between dehydration synthesis and hydrolysis. (p. 115)

5 Give examples of a dehydration synthesis reaction and a hydrolysis reaction. (p. 115)

4.3 Control of Metabolic Reactions

6 Describe how an enzyme interacts with its substrate. (p. 117)

7 Define *active site.* (p. 117)

8 State two factors that control the rate of an enzyme-catalyzed reaction. (p. 117)

9 A cell has _________ types of enzymes and metabolic reactions. (p. 117)

a. 3
b. hundreds of
c. thousands of
d. millions of
e. 3 billion

10 Explain the importance of a rate-limiting enzyme. (p. 118)

11 Describe how negative feedback involving a rate-limiting enzyme controls a metabolic pathway. (p. 118)

12 Define *cofactor.* (p. 119)

13 Discuss the relationship between a coenzyme and a vitamin. (p. 119)

4.4 Energy for Metabolic Reactions

14 Define *energy.* (p. 119)

15 Explain the importance of ATP and its relationship to ADP. (p. 119)

16 Explain how the oxidation of molecules inside cells differs from the burning of substances outside cells. (p. 120)

4.5 Cellular Respiration

17 Define *cellular respiration.* (p. 120)

18 Distinguish between anaerobic and aerobic phases of cellular respiration. (p. 120)

19 Match the part of cellular respiration to the associated activities. (p. 120)

(1) electron transport chain
(2) glycolysis
(3) citric acid cycle

A. glucose molecules are broken down into pyruvic acid
B. carrier molecules and enzymes extract energy and store it as ATP, releasing water and heat
C. pyruvic acid molecules enter mitochondria, where CO_2 and high-energy electrons are released

20 Identify the final acceptor of the electrons released in the reactions of cellular respiration. (p. 123)

21 Excess glucose in cells may link and be stored as ____________. (p. 124)

4.6 Nucleic Acids and Protein Synthesis

22 The genetic code is ____________. (p. 124)

a. the bonding of purine to pyrimidine
b. the correspondence between DNA triplet and amino acid
c. the correspondence between DNA triplet and RNA triplet
d. the controls that determine where the instructions for a protein start and stop
e. unique in each individual

23 DNA information provides instructions for the cell to __________. (p. 125)

a. manufacture carbohydrate molecules
b. extract energy
c. manufacture RNA from amino acids
d. synthesize protein molecules

24 Distinguish between a gene and a genome. (p. 125)

25 Define *gene expression.* (p. 125)

26 If a DNA strand has the sequence ATGCGATCCGC then the sequence on the complementary DNA strand is ___________. (p. 127)

27 Explain why DNA replication is essential. (p. 127)

28 Describe the events of DNA replication. (p. 127)

29 Identify the part of a DNA molecule that encodes information. (p. 130)

30 List three ways that RNA differs from DNA. (p. 130)

31 If one strand of a DNA molecule has the sequence of ATTCTCGACTAT, the complementary mRNA has the sequence ____________. (p. 131)

a. ATTCTCGACTAT
b. AUUCUCGACUAU
c. TAAGAGCTGATA
d. UAAGAGCUGATA
e. Can't tell from given information.

32 Distinguish between transcription and translation. (p. 131)

33 Distinguish the functions of mRNA, rRNA, and tRNA. (p. 132)

34 List the steps of protein synthesis. (p. 132)

35 Describe the function of a ribosome in protein synthesis. (p. 132)

36 Calculate the number of amino acids that a DNA sequence of twenty-seven nucleotides encodes. (p. 132)

4.7 Changes in Genetic Information

37 Distinguish among mutations, SNPs, and copy number variants. (p. 135)

38 Discuss two major ways that mutation occurs. (p. 136)

39 Define *DNA damage response.* (p. 137)

40 Discuss three ways that the genetic code protects against the persistence of a mutation. (p. 137)

INTEGRATIVE ASSESSMENTS/CRITICAL THINKING

OUTCOMES 2.2, 4.2

1. How can the same molecule be both a reactant (starting material) and a product of a biochemical pathway?

OUTCOMES 2.3, 3.2, 4.3

2. What effect might changes in the pH of body fluids or body temperature that accompany illness have on cells?

OUTCOMES 2.3, 4.2, 4.5

3. Michael P. was very weak from birth, with poor muscle tone, difficulty breathing, and great fatigue. By his third month, he began having seizures. Michael's medical tests were normal except for one: his cerebrospinal fluid (the fluid that bathes the brain and spinal cord) was unusually high in glucose. Hypothesizing that the boy could not produce enough ATP, doctors tried an experimental treatment: they gave him a diet rich in certain fatty acids that caused the cellular respiration pathway to resume at the point of acetyl CoA formation. Michael rapidly improved. Explain what caused his symptoms.

OUTCOMES 4.4, 4.5

4. In cyanide poisoning, levels of ATP in the brain plummet, but levels of lactic acid increase markedly. Explain how both effects occur.

OUTCOMES 4.4, 4.5

5. A student is used to running 3 miles at a leisurely jogging pace. In a fitness class, she has to run a mile as fast as she can. Afterwards, she is winded and has sharp pains in her chest and leg muscles. What has she experienced, in terms of energy metabolism?

OUTCOME 4.6

6. Consider the following DNA sequence:

   TCGAGAATCTCGATT

   a. Write the sequence of the DNA strand that would be replicated from this one.
   b. Write the sequence of the RNA molecule that would be transcribed from the DNA strand.
   c. State how many codons the sequence specifies.
   d. State how many amino acids the sequence specifies.

OUTCOME 4.6

7. Some antibiotic drugs fight infection by interfering with DNA replication, transcription, or translation in bacteria. These processes are different enough in bacteria that the drugs do not harm us. Indicate whether each of the following effects is on replication, transcription, or translation.

   a. Rifampin binds to bacterial RNA polymerase.
   b. Streptomycin binds bacterial ribosomes, disabling them.
   c. Quinolone blocks an enzyme that prevents bacterial DNA from unwinding.

WEB CONNECTIONS

Be sure to visit the text website at **www.mhhe.com/shier12** for answers to chapter assessments, additional quizzes, and interactive learning exercises.

ANATOMY & PHYSIOLOGY REVEALED

Anatomy & Physiology Revealed® (APR) includes cadaver photos that allow you to peel away layers of the human body to reveal structures beneath the surface. This program also includes animations, radiologic imaging, audio pronunciations, and practice quizzing. Check out **www.aprevealed.com**. APR has been proven to help improve student grades!

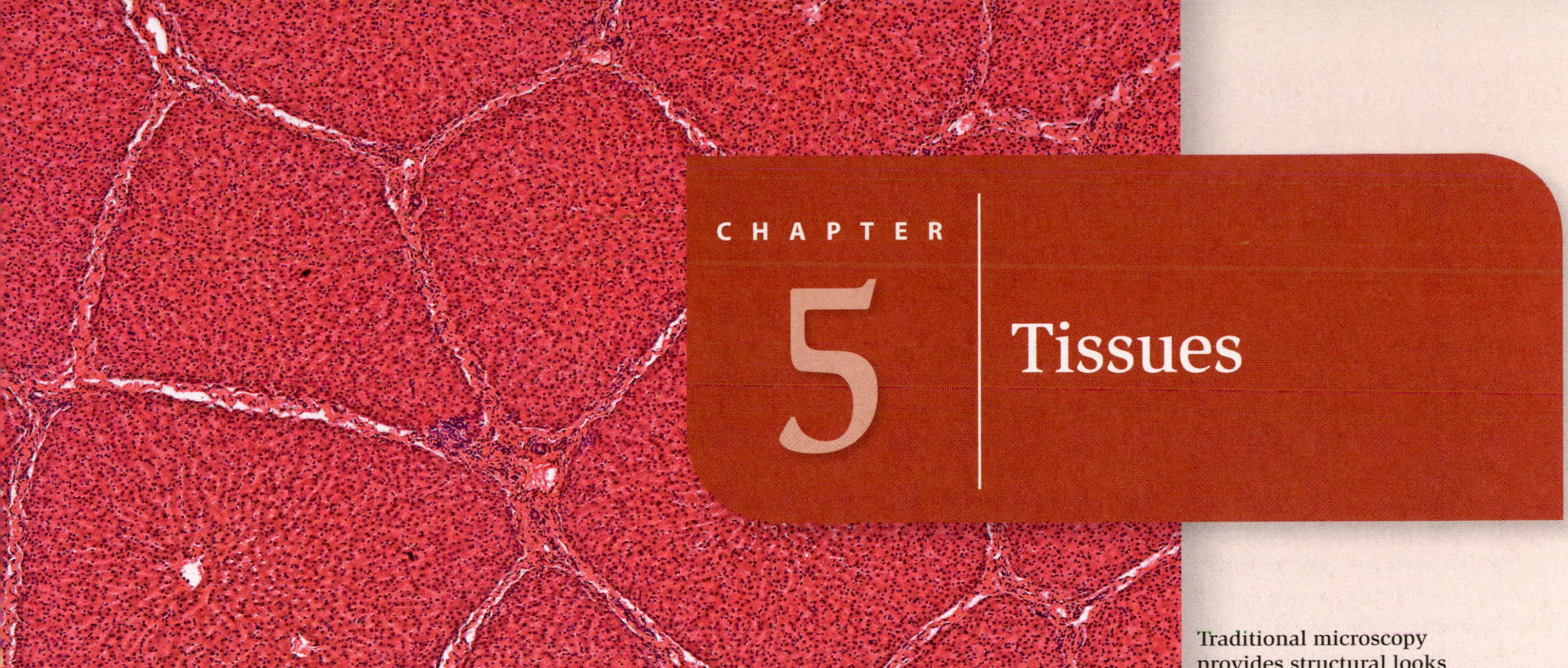

Traditional microscopy provides structural looks at tissues (150×).

UNDERSTANDING WORDS

adip-, fat: *adip*ose tissue—tissue that stores fat.

chondr-, cartilage: *chondr*ocyte—cartilage cell.

-cyt, cell: osteo*cyt*e—bone cell.

epi-, upon, after, in addition: *epi*thelial tissue—tissue that covers all free body surfaces.

-glia, glue: neuro*glia*—cells that support neurons; part of nervous tissue.

hist-, web, tissue: *hist*ology—study of composition and function of tissues.

hyal-, resemblance to glass: *hyal*ine cartilage—flexible tissue containing chondrocytes.

inter-, among, between: *inter*calated disc—band between adjacent cardiac muscle cells.

macr-, large: *macr*ophage—large phagocytic cell.

neur-, nerve: *neur*on—nerve cell.

os-, bone: *os*seous tissue—bone tissue.

phag-, to eat: *phag*ocyte—cell that engulfs and destroys foreign particles.

pseud-, false: *pseud*ostratified epithelium—tissue with cells that appear to be in layers, but are not.

squam-, scale: *squam*ous epithelium—tissue with flattened or scalelike cells.

strat-, layer: *strat*ified epithelium—tissue with cells in layers.

stria-, groove: *stria*ted muscle—tissue whose cells have alternating light and dark cross-markings.

 LEARN  PRACTICE  ASSESS

LEARNING OUTCOMES

After you have studied this chapter, you should be able to:

5.1 Introduction

1 Describe how cells are organized into tissues, and identify the intercellular junctions in tissues. (p. 144)

2 List the four major tissue types in the body. (p. 144)

5.2 Epithelial Tissues

3 Describe the general characteristics and functions of epithelial tissue. (p. 144)

4 Name the types of epithelium and identify an organ in which each is found. (p. 146)

5 Explain how glands are classified. (p. 150)

5.3 Connective Tissues

6 Describe the general characteristics of connective tissue. (p. 153)

7 Compare and contrast the cellular components, structures, fibers, and extracellular matrix (where applicable) in each type of connective tissue. (p. 156)

8 Describe the major functions of each type of connective tissue. (p. 156)

5.4 Types of Membranes

9 Describe and locate each of the four types of membranes. (p. 162)

5.5 Muscle Tissues

10 Distinguish among the three types of muscle tissue. (p. 163)

5.6 Nervous Tissues

11 Describe the general characteristics and functions of nervous tissue. (p. 164)

A NEW VIEW OF CELL SPECIALIZATION—PROTEOMICS

A tissue atlas displays groups of cells stained to reveal their specializations and viewed with the aid of a microscope. It's easy to tell skeletal muscle from adipose tissue from blood. A new way to look at tissues is to profile the proteins that their cells manufacture. These proteins are responsible for cell specializations and arise from the expression of subsets of the genome. Such an approach is called *proteomics.* A skeletal muscle cell, for example, transcribes messenger RNA molecules from genes that encode contractile proteins, whereas an adipose cell yields mRNAs whose protein products enable the cell to store massive amounts of fat. All cells also transcribe many mRNAs whose encoded proteins make life at the cellular level possible.

In the mid 1990s, technology was developed to display the genes expressed in particular cell types. The tool is a DNA microarray (also known as a gene chip). It is a square of glass or plastic smaller than a postage stamp to which thousands of small pieces of DNA of known sequence are bound, in a grid pattern, so that the position of each entrant is known. Then mRNAs are extracted from a cell or tissue sample, converted to DNA "probes," and labeled with a fluorescent dye. The grid positions where the probes bind fluoresce, which a laser scanner detects and converts to an image. The intensity of the fluorescence reveals the abundance of the mRNAs present. Probes representing two cell sources can be linked to different fluorescent tags so that their gene expression patterns can be directly compared—such as a healthy and cancerous version of the same cell type. A microarray can scan for activity in all genes or be customized to paint molecular portraits of specific functions.

Researchers are compiling DNA microarray patterns for the 260+ types of normal differentiated cells in a human body. A statistical analysis called hierarchy clustering groups cells by similarities in gene expression. The results generally agree with what is known of histology (the study of tissues) from microscopy, but sometimes reveal new proteins in specific cell types. Although DNA microarrays can fill in molecular details that cannot be seen under a microscope, a pair of discerning human eyes will always be necessary to see the bigger picture of how cells assemble into tissues. ■

5.1 INTRODUCTION

In all complex organisms, cells are organized into **tissues** (tish′uz), which are layers or groups of similar cells with a common function. Some cells, such as blood cells, are separated from each other in fluid-filled spaces or intercellular (in″ter-sell′u-lar) spaces. Many other cell types, however, are tightly packed, with structures called **intercellular junctions** that connect their cell membranes.

In one type of intercellular junction, called a *tight junction,* the membranes of adjacent cells converge and fuse. The area of fusion surrounds the cell like a belt, and the junction closes the space between the cells. Tight junctions typically join cells that form sheetlike layers, such as those that line the inside of the digestive tract. The linings of tiny blood vessels in the brain consist of cells held tightly together (From Science to Technology 5.1).

Another type of intercellular junction, called a *desmosome,* rivets or "spot welds" skin cells, enabling them to form a reinforced structural unit. The cell membranes of certain other cells, such as those in heart muscle and muscle of the digestive tract, are interconnected by tubular channels called *gap junctions.* These channels link the cytoplasm of adjacent cells and allow ions, nutrients (such as sugars, amino acids, and nucleotides), and other small molecules to move between them (fig. 5.1). Table 5.1 summarizes intercellular junctions.

Tissues can be distinguished from each other by variations in cell size, shape, organization, and function. The study of tissues, **histology,** will assist understanding in later discussions of the physiology of organs and organ systems. The tissues of the human body include four major types: *epithelial, connective, muscle,* and *nervous.* These tissues associate, assemble, and interact to form organs that have specialized functions. Table 5.2 compares the four major tissue types.

This chapter examines in detail epithelial and connective tissues and introduces muscle and nervous tissues. Throughout this chapter, simplified line drawings (for example, fig. 5.2*a*) are included with each micrograph (for example, fig. 5.2*b*) to emphasize the distinguishing characteristics of the specific tissue, as well as a locator icon (an example of where in the body that particular tissue may be found). Chapter 9 discusses muscle tissue in more detail, and chapters 10, 11, and 12 detail nervous tissue.

PRACTICE 

1 What is a tissue?
2 What are the different types of intercellular junctions?
3 List the four major types of tissue.

5.2 EPITHELIAL TISSUES

General Characteristics

Epithelial (ep″ĭ-the′le-al) **tissues** are found throughout the body. Epithelium covers the body surface and organs, forms the inner lining of body cavities, and lines hollow organs. It always has a *free (apical) surface* exposed to the outside or internally to an open space. A thin, nonliving layer called the **basement membrane** anchors epithelium to underlying connective tissue.

5.1 FROM SCIENCE TO TECHNOLOGY

Nanotechnology Meets the Blood-Brain Barrier

Nanotechnology is helping drug developers to circumvent a problem in drug delivery based on an anatomical impediment—the close attachments of the cells that form tiny blood vessels in the brain. Like a tight line of police officers keeping out a crowd, the blood-brain barrier is a vast network of capillaries in the brain whose cells are firmly attached by overlapping tight junctions. These cells also lack the scattered vesicles and windowlike clefts in other capillaries. In addition, star-shaped brain cells called astrocytes wrap around the barrier.

The 400-mile blood-brain barrier shields brain tissue from toxins and biochemical fluctuations that could be overwhelming. It also allows selective drug delivery. Certain antihistamines, for example, do not cause drowsiness because they cannot breach the barrier. But this protection has a trade-off—the brain cannot take up many therapeutic drugs that must penetrate to be effective.

For decades researchers have attempted to deliver drugs across the barrier by tagging compounds to substances that can cross, designing drugs to fit natural receptors in the cell membranes of the barrier, and injecting substances that temporarily relax the tight junctions. More recently, researchers have applied nanotechnology to the problem of circumventing the blood-brain barrier. Nanotechnology is the application of structures smaller than 100 billionths of a meter (100 nanometers) in at least one dimension.

Nanoparticles that can cross the blood-brain barrier are made of combinations of oils and polymers, with a neutral or slightly negative charge (positively charged particles are toxic). In one application, anesthetics or chemotherapeutics are loaded into fatty bubbles (liposomes) that are in turn placed in nanoparticles. This delivery system masks the part of the drug that cannot cross the barrier and slows release of the drug, which diminishes side effects.

In another application, insulin is delivered in inhaled nanoparticles 10 to 50 nanometers in diameter. Originally developed to provide insulin to people with diabetes instead of injecting it, clinical trials are showing that it is also helpful in maintaining memory in people who have mild cognitive impairment or Alzheimer disease. ■

Cancer cells secrete a substance that dissolves basement membranes, enabling the cells to invade tissue layers. Cancer cells also produce fewer adhesion proteins, or none at all, which allows them to spread into surrounding tissue.

As a rule, epithelial tissues lack blood vessels. However, nutrients diffuse to epithelium from underlying connective tissues, which have abundant blood vessels.

Epithelial cells readily divide, so injuries heal rapidly as new cells replace lost or damaged ones. Skin cells and the cells that line the stomach and intestines are continually being damaged and replaced.

Epithelial cells are tightly packed. In many places, desmosomes attach one to another, enabling these cells to form effective protective barriers in such structures as the outer layer of the skin and the inner lining of the mouth. Other epithelial functions include secretion, absorption, and excretion.

Epithelial tissues are classified according to the shape and number of layers of cells. Epithelial tissues composed of thin, flattened cells are *squamous;* those with cubelike cells are *cuboidal;* and those with elongated cells are *columnar.* Epithelium composed of a single layer of cells is *simple* and with two or more layers of cells, *stratified.* In the following descriptions, modifications of the free surfaces of epithelial cells reflect their specialized functions.

PRACTICE

4 List the general characteristics of epithelial tissue.

5 Explain how epithelial tissues are classified.

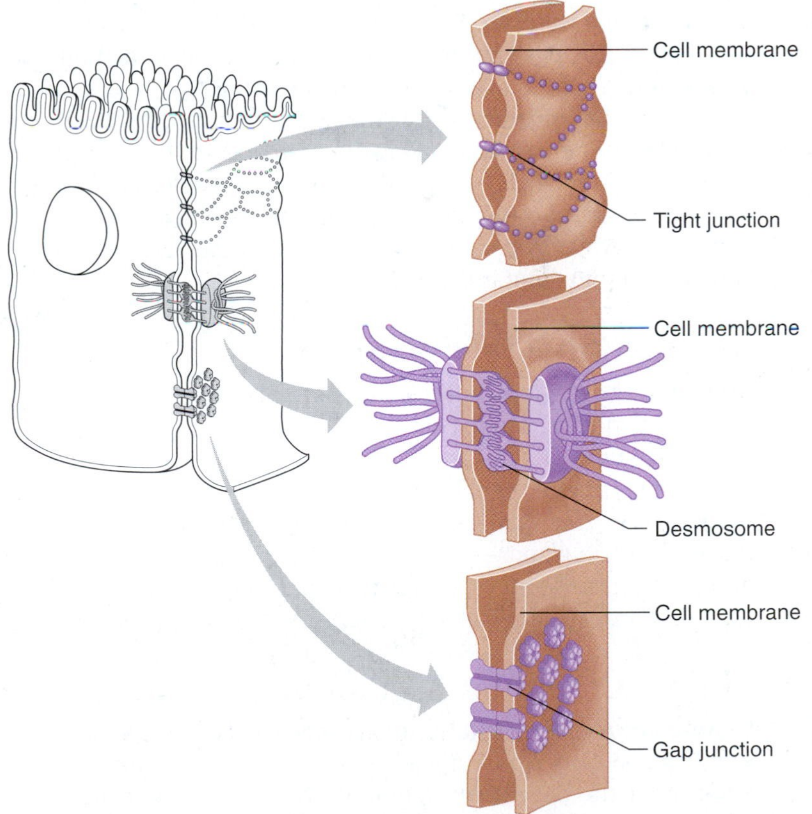

FIGURE 5.1 Some cells are joined by intercellular junctions, such as tight junctions that fuse neighboring cell membranes, desmosomes that serve as "spot welds," or gap junctions that allow small molecules to move between the cytoplasm of adjacent cells.

TABLE 5.1 | Types of Intercellular Junctions

Type	Characteristics	Example
Tight junctions	Close space between cells by fusing cell membranes	Cells that line the small intestine
Desmosomes	Bind cells by forming "spot welds" between cell membranes	Cells of the outer skin layer
Gap junctions	Form tubular channels between cells that allow exchange of substances	Muscle cells of the heart and digestive tract

TABLE 5.2 | Tissues

Type	Function	Location	Distinguishing Characteristics
Epithelial	Protection, secretion, absorption, excretion	Cover body surface, cover and line internal organs, compose glands	Lack blood vessels, cells readily divide, cells are tightly packed
Connective	Bind, support, protect, fill spaces, store fat, produce blood cells	Widely distributed throughout the body	Mostly have good blood supply, cells are farther apart than epithelial cells, with extracellular matrix in between
Muscle	Movement	Attached to bones, in the walls of hollow internal organs, heart	Able to contract in response to specific stimuli
Nervous	Transmit impulses for coordination, regulation, integration, and sensory reception	Brain, spinal cord, nerves	Cells communicate with each other and other body parts

Simple Squamous Epithelium

Simple squamous (skwa'mus) **epithelium** consists of a single layer of thin, flattened cells. These cells fit tightly together, somewhat like floor tiles, and their nuclei are usually broad and thin (fig. 5.2).

Substances pass rather easily through simple squamous epithelium. This tissue is common at sites of diffusion and filtration. Simple squamous epithelium lines the air sacs (alveoli) of the lungs where oxygen and carbon dioxide are exchanged. It also forms the walls of capillaries, lines the insides of blood and lymph vessels, and covers the membranes that line body cavities. However, because it is so thin and delicate, simple squamous epithelium is easily damaged.

Simple Cuboidal Epithelium

Simple cuboidal epithelium consists of a single layer of cube-shaped cells. These cells usually have centrally located, spherical nuclei (fig. 5.3).

Simple cuboidal epithelium lines the follicles of the thyroid gland, covers the ovaries, and lines the kidney tubules and ducts of certain glands—such as the salivary glands, pancreas, and liver. In the kidneys, it functions in tubular secretion and tubular reabsorption; in glands, it secretes glandular products.

Simple Columnar Epithelium

Simple columnar epithelium is composed of a single layer of elongated cells whose nuclei are usually at about the same level, near the basement membrane (fig. 5.4). The cells of this tissue can be ciliated or nonciliated. *Cilia,* 7 to 10 μm in length, extend from the free surfaces of the cells, and they move constantly. In the female, cilia aid in moving the egg cell through the uterine tube to the uterus.

Nonciliated simple columnar epithelium lines the uterus and portions of the digestive tract, including the stomach and small and large intestines. Its cells are elongated, so this tissue is thick, which enables it to protect underlying tissues. Simple columnar epithelium also secretes digestive fluids and absorbs nutrients from digested food.

Simple columnar cells, specialized for absorption, often have many tiny, cylindrical processes extending from their free surfaces. These processes, called *microvilli,* are from 0.5 to 1.0 μm long. They increase the surface area of the cell membrane where it is exposed to substances being absorbed (fig. 5.5).

Typically, specialized, flask-shaped glandular cells are scattered among the cells of simple columnar epithelium. These cells, called *goblet cells,* secrete a protective fluid called *mucus* onto the free surface of the tissue (see fig. 5.4).

Pseudostratified Columnar Epithelium

The cells of **pseudostratified** (soo″do-strat′ĭ-fīd) **columnar epithelium** appear stratified or layered, but they are not. A layered effect occurs because the nuclei are at two or more levels in the row of aligned cells. However, the cells, which vary in shape, all reach the basement membrane, even though some of them may not contact the free surface.

Pseudostratified columnar epithelial cells commonly have cilia, which extend from the free surfaces of the cells. Goblet cells scattered throughout this tissue secrete mucus, which the cilia sweep away (fig. 5.6).

Pseudostratified columnar epithelium lines the passages of the respiratory system. Here, the mucous-covered linings are sticky and trap dust and microorganisms that enter with the air. The cilia move the mucus and its captured particles upward and out of the airways.

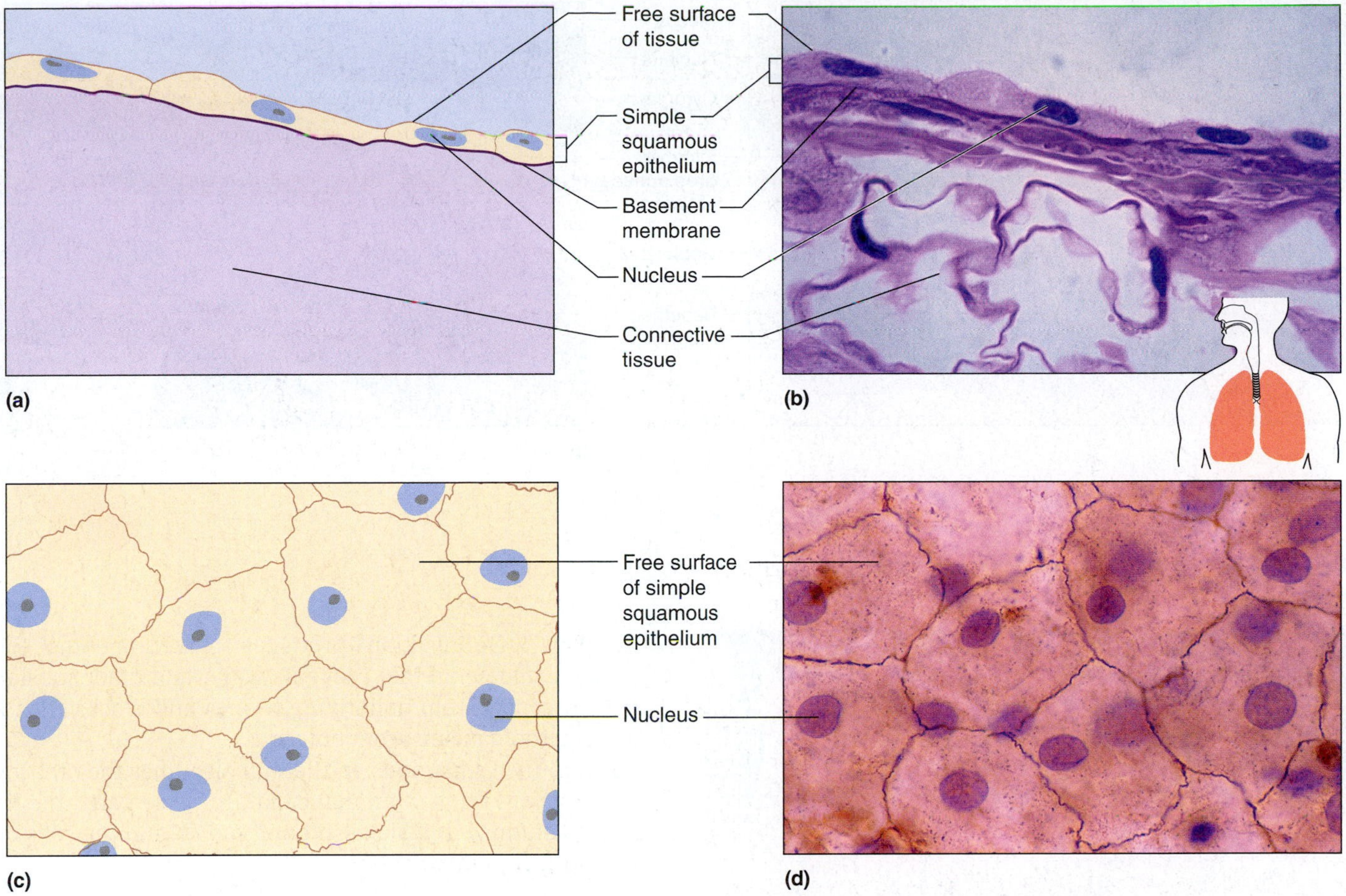

FIGURE 5.2 Simple squamous epithelium consists of a single layer of tightly packed, flattened cells (670×). (*a*) and (*b*) side view, (*c*) and (*d*) surface view.

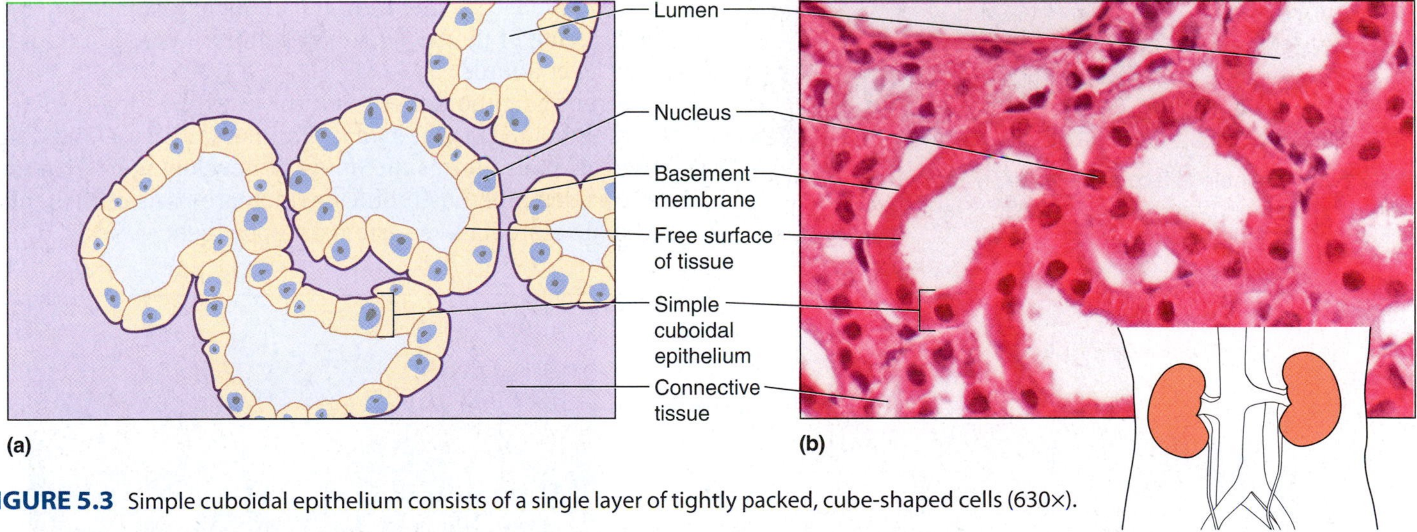

FIGURE 5.3 Simple cuboidal epithelium consists of a single layer of tightly packed, cube-shaped cells (630×).

Stratified Squamous Epithelium

Stratified epithelium is named for the shape of the cells forming the outermost layers. **Stratified squamous epithelium** consists of many layers of cells, making this tissue relatively thick. Cells nearest the free surface are flattened the most, whereas those in the deeper layers, where cell division occurs, are cuboidal or columnar. As the newer cells grow, older ones are pushed farther and farther outward, where they flatten (fig. 5.7).

The outermost layer of the skin (epidermis) is stratified squamous epithelium. As the older cells are pushed outward, they accumulate a protein called *keratin,* then harden and

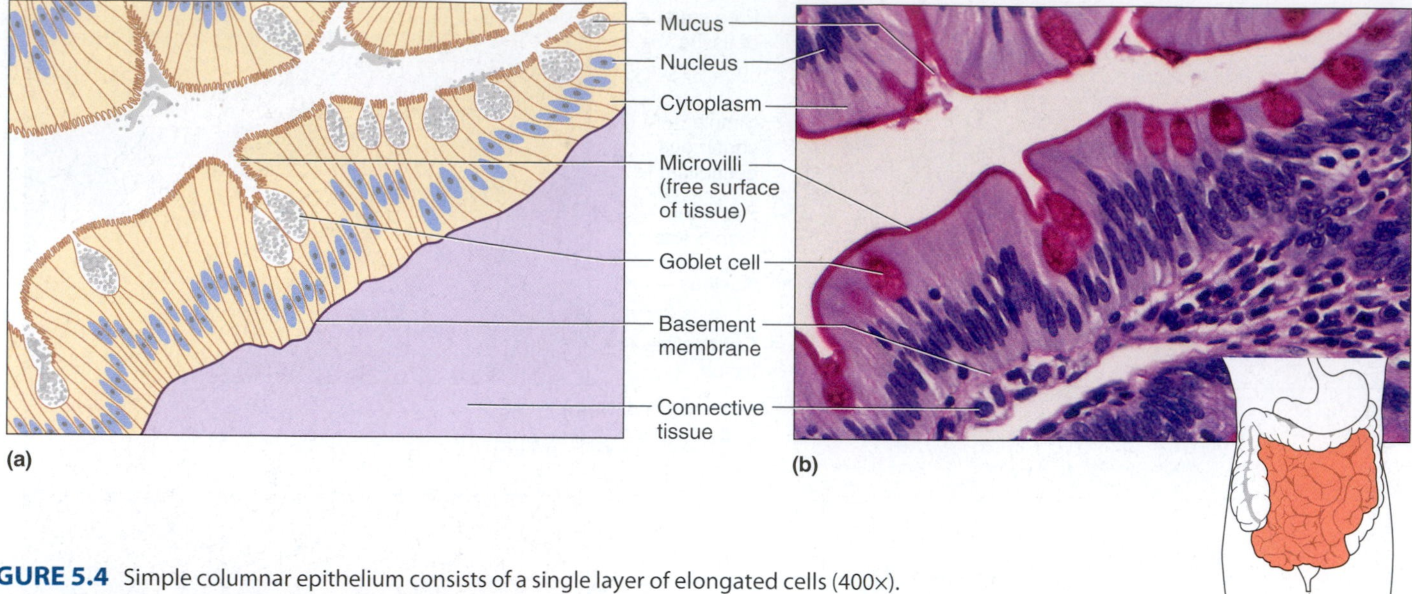

FIGURE 5.4 Simple columnar epithelium consists of a single layer of elongated cells (400×).

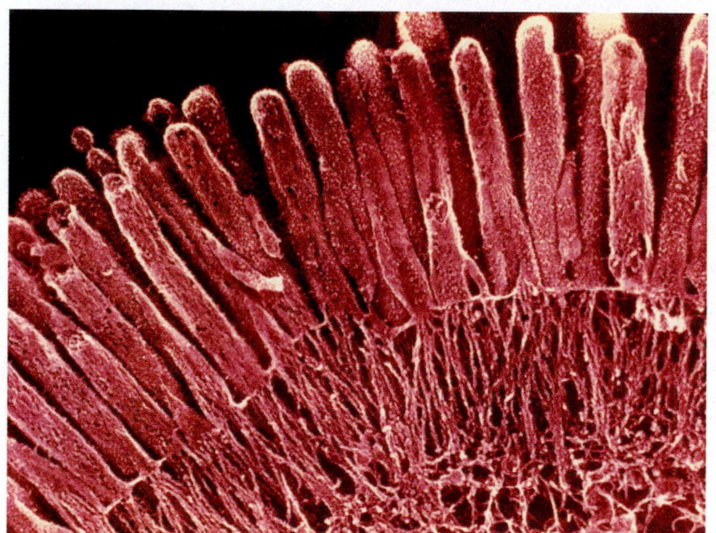

FIGURE 5.5 A scanning electron micrograph of microvilli, which fringe the free surfaces of some columnar epithelial cells (33,000×).

die. This "keratinization" produces a covering of dry, tough, protective material that prevents water and other substances from escaping from underlying tissues and blocks chemicals and microorganisms from entering.

Stratified squamous epithelium also lines the oral cavity, esophagus, vagina, and anal canal. In these parts, the tissue is not keratinized; it stays soft and moist, and the cells on its free surfaces remain alive.

Stratified Cuboidal Epithelium

Stratified cuboidal epithelium consists of two or three layers of cuboidal cells that form the lining of a lumen (fig. 5.8). The layering of the cells provides more protection than the single layer affords.

Stratified cuboidal epithelium lines the larger ducts of the mammary glands, sweat glands, salivary glands, and pancreas. It also forms the lining of developing ovarian follicles and seminiferous tubules, which are parts of the female and male reproductive systems, respectively.

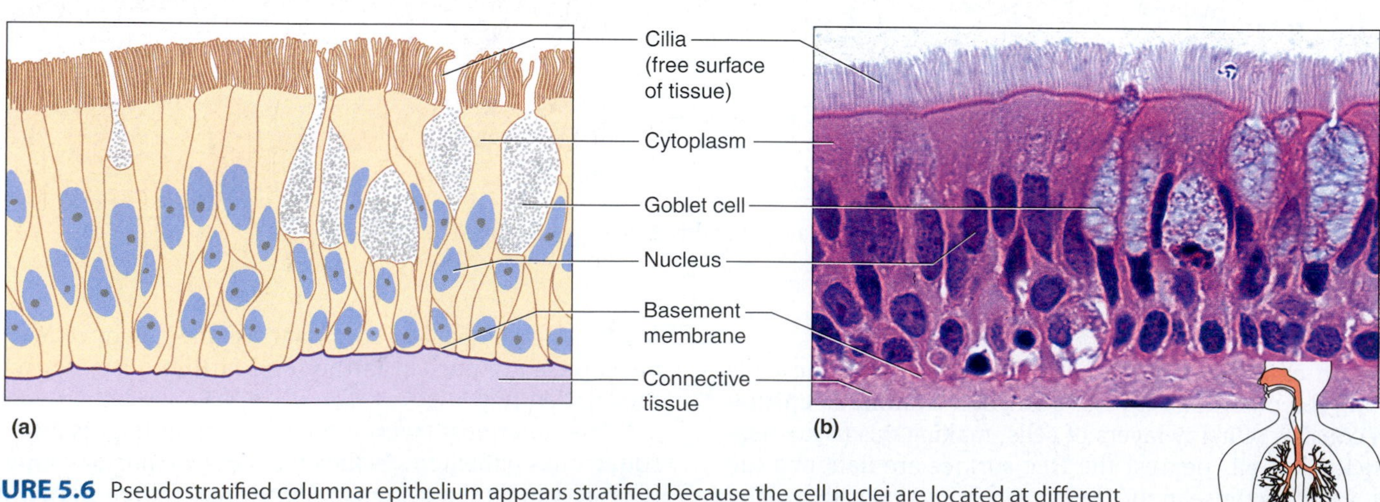

FIGURE 5.6 Pseudostratified columnar epithelium appears stratified because the cell nuclei are located at different levels (1,000×).

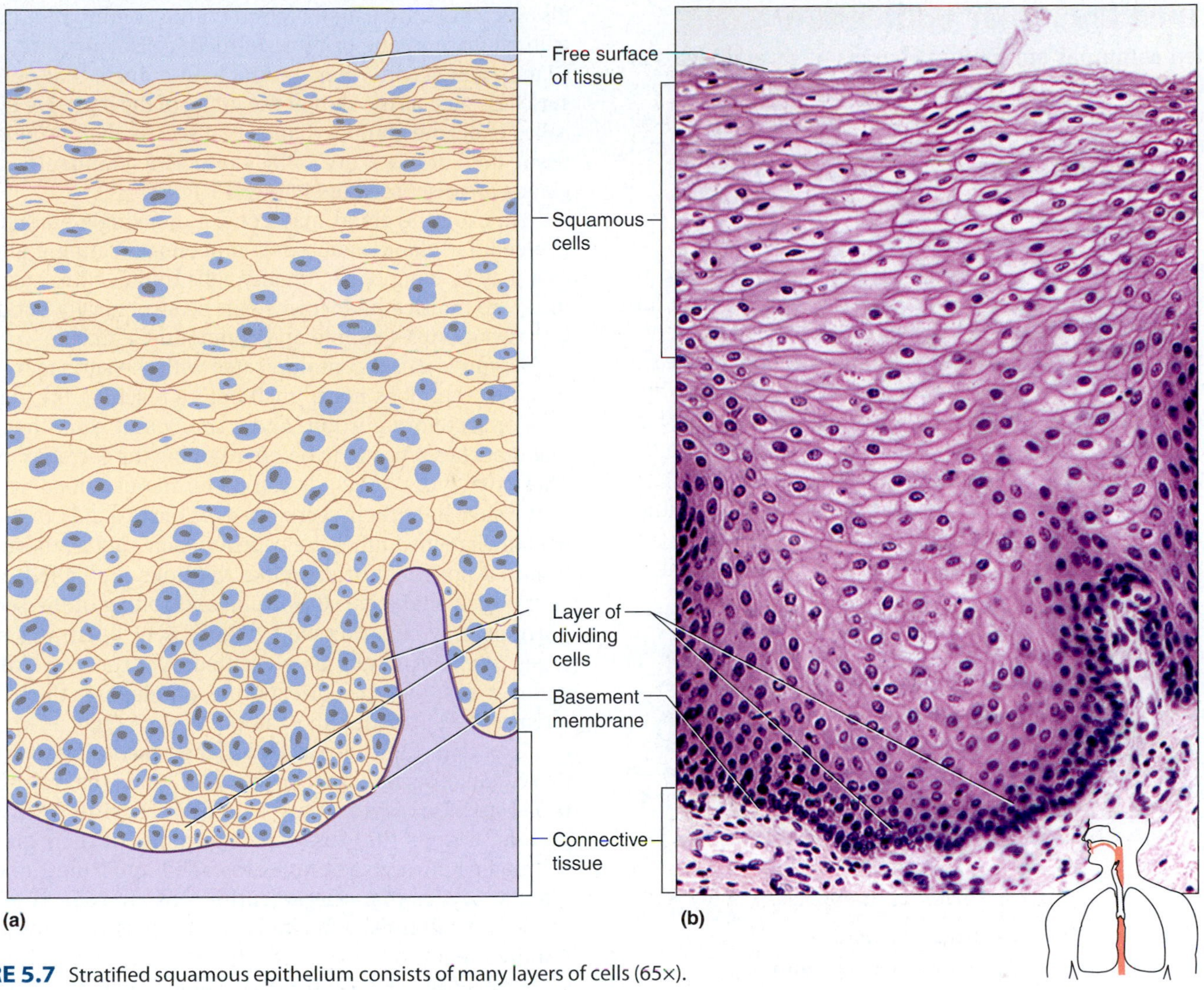

Free surface of tissue
Squamous cells
Layer of dividing cells
Basement membrane
Connective tissue
(a)
(b)

FIGURE 5.7 Stratified squamous epithelium consists of many layers of cells (65×).

Stratified cuboidal epithelium
Nucleus
Lumen
Free surface of tissue
Basement membrane
Connective tissue
(a)
(b)

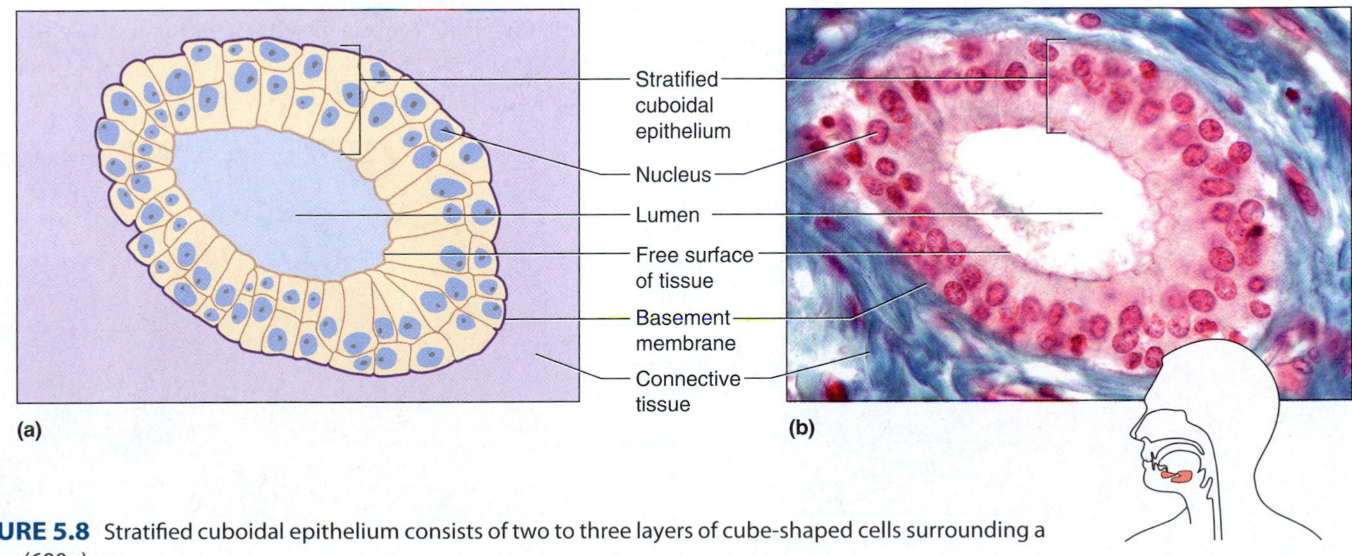

FIGURE 5.8 Stratified cuboidal epithelium consists of two to three layers of cube-shaped cells surrounding a lumen (600×).

Stratified Columnar Epithelium

Stratified columnar epithelium consists of several layers of cells (fig. 5.9). The superficial cells are elongated, whereas the basal layers consist of cube-shaped cells. Stratified columnar epithelium is found in part of the male urethra and ductus deferens and in parts of the pharynx.

Transitional Epithelium

Transitional epithelium (uroepithelium) is specialized to change in response to increased tension. It forms the inner lining of the urinary bladder and lines the ureters and the superior urethra. When the wall of one of these organs contracts, the tissue consists of several layers of cuboidal cells; however, when the organ is distended, the tissue stretches, and the physical relationships among the cells change. While distended, the tissue appears to contain only a few layers of cells (fig. 5.10). In addition to providing an expandable lining, transitional epithelium forms a barrier that helps prevent the contents of the urinary tract from diffusing back into the internal environment.

Up to 90% of human cancers are *carcinomas,* growths that originate in epithelium. Most carcinomas begin on surfaces that contact the external environment, such as skin, linings of the airways in the respiratory tract, or linings of the stomach or intestines in the digestive tract. This observation suggests that the more common cancer-causing agents may not deeply penetrate tissues.

PRACTICE

6 Describe the structure of each type of epithelium.

7 Describe the special functions of each type of epithelium.

Glandular Epithelium

Glandular epithelium is composed of cells specialized to produce and secrete substances into ducts or into body fluids. Such cells are usually found within columnar or cuboidal epithelium, and one or more of these cells constitute a *gland.* Glands that secrete their products into ducts that open onto surfaces, such as the skin or the lining of the digestive tract, are called **exocrine glands.** Glands that secrete their products into tissue fluid or blood are called **endocrine glands.** (Endocrine glands are discussed in chapter 13.)

An exocrine gland may consist of a single epithelial cell (unicellular gland), such as a mucous-secreting goblet cell, or it may be composed of many cells (multicellular gland). In turn, the multicellular forms can be structurally subdivided into two groups—simple and compound glands.

A *simple gland* communicates with the surface by means of a duct that does not branch before reaching the glandular cells or secretory portion, and a *compound gland* has a duct that branches repeatedly before reaching the secretory portion. These two types of glands can be further classified according to the shapes of their secretory portions. Glands that consist of epithelial-lined tubes are called *tubular glands;* those whose terminal portions form saclike dilations are called *alveolar glands* (acinar glands). Branching and coiling of the secretory portions may occur as well. Figure 5.11 illustrates several types of exocrine glands classified by structure. Table 5.3 summarizes the types of exocrine glands, lists their characteristics, and provides an example of each type.

Exocrine glands are also classified according to the ways these glands secrete their products. Glands that release fluid products by exocytosis are called **merocrine** (mer′o-krin) **glands.** Glands that lose small portions of their glandular cell bodies during secretion are called **apocrine** (ap′o-krin) **glands.** Glands that release entire cells are called **holocrine** (ho′lo-krin) **glands.** After release, the cells containing accumulated secretory products disintegrate, liberating their secretions (figs. 5.12 and 5.13). Table 5.4 summarizes these glands and their secretions.

RECONNECT

To Chapter 3, Movements Into and Out of the Cell, page 98.

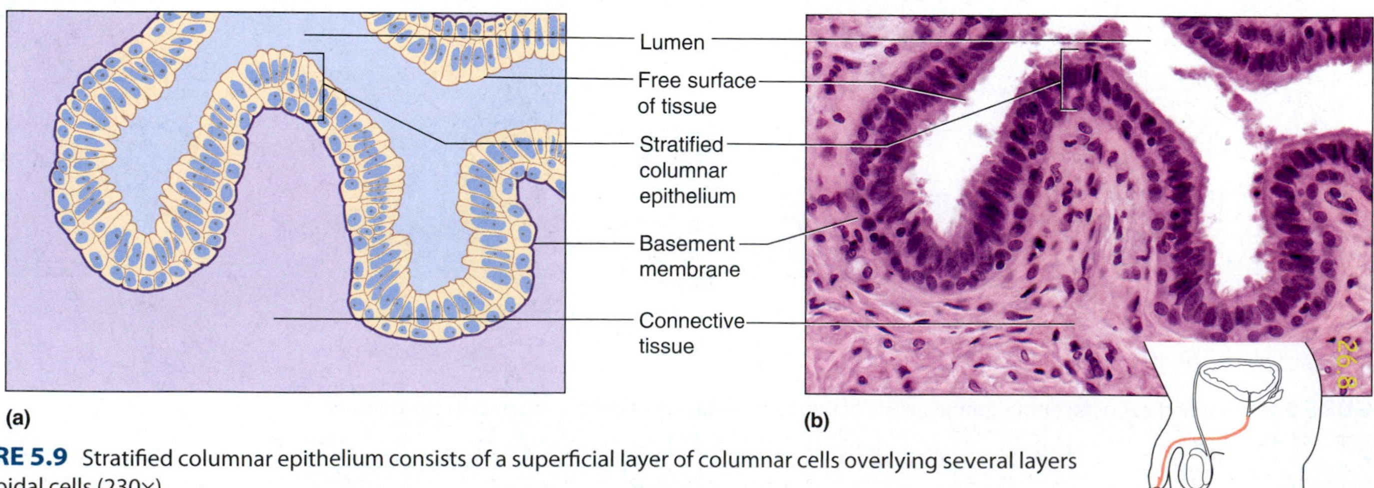

FIGURE 5.9 Stratified columnar epithelium consists of a superficial layer of columnar cells overlying several layers of cuboidal cells (230×).

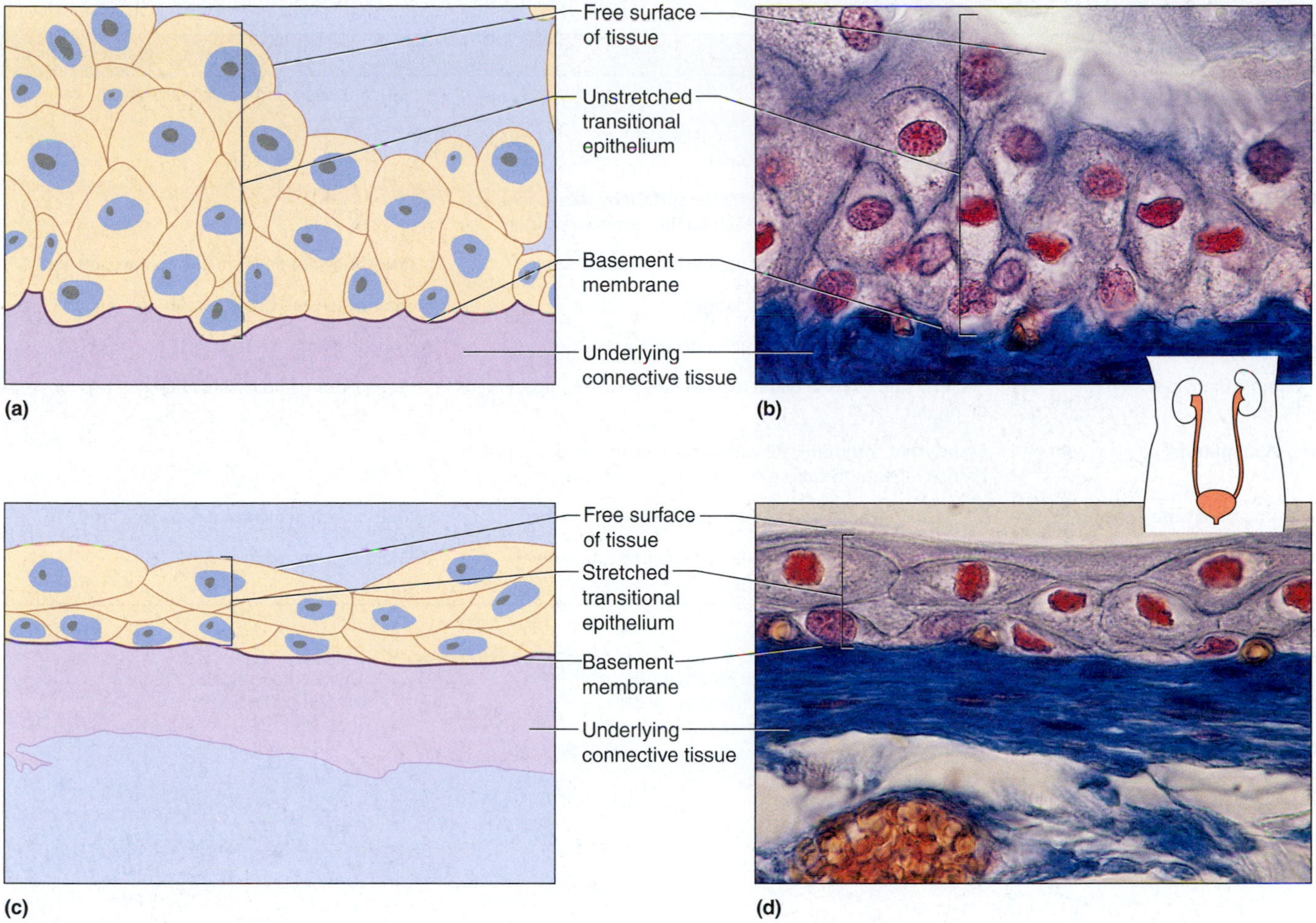

Free surface of tissue

Unstretched transitional epithelium

Basement membrane

Underlying connective tissue

(a)

(b)

Free surface of tissue

Stretched transitional epithelium

Basement membrane

Underlying connective tissue

(c)

(d)

FIGURE 5.10 Transitional epithelium. (*a* and *b*) When the organ wall contracts, transitional epithelium is unstretched and consists of many layers (675×). (*c* and *d*) When the organ is distended, the tissue stretches and appears thinner (675×).

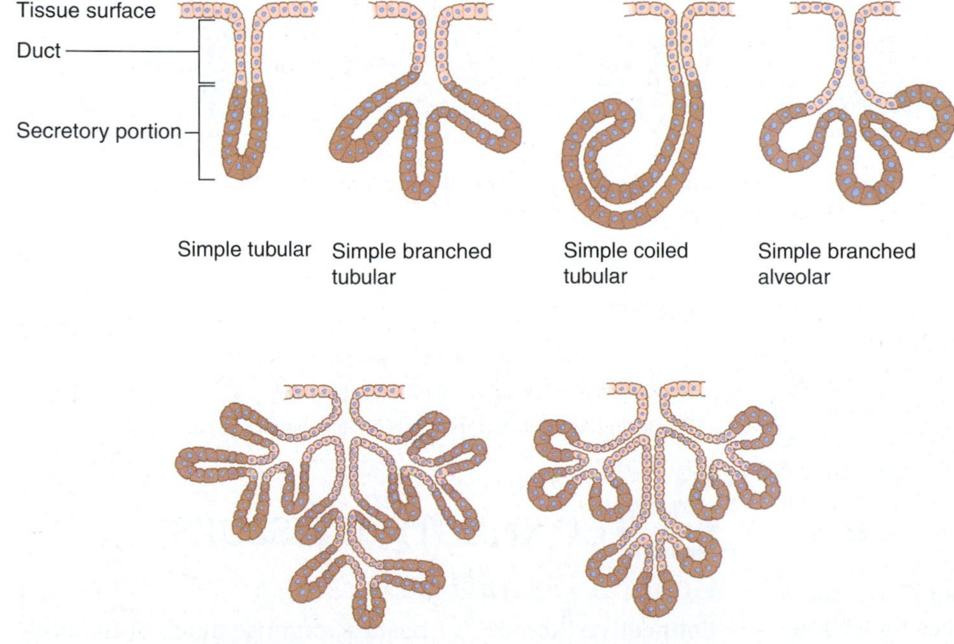

FIGURE 5.11 Structural types of exocrine glands.

TABLE 5.3 | Types of Exocrine Glands

Type	Characteristics	Example
Unicellular glands	A single secretory cell	Mucous-secreting goblet cell (see fig. 5.4)
Multicellular glands	Glands that consist of many cells	
Simple glands	Glands that communicate with the surface by means of ducts that do not branch before reaching the secretory portion	
1. Simple tubular gland	Straight tubelike gland that opens directly onto surface	Intestinal glands of small intestine (see fig. 17.3)
2. Simple coiled tubular gland	Long, coiled, tubelike gland; long duct	Eccrine (sweat) glands of skin (see fig. 6.11)
3. Simple branched tubular gland	Branched, tubelike gland; duct short or absent	Gastric glands (see fig. 17.19)
4. Simple branched alveolar gland	Secretory portions of gland expand into saclike compartments along duct	Sebaceous gland of skin (see fig. 5.13)
Compound glands	Glands that communicate with surface by means of ducts that branch repeatedly before reaching the secretory portion	
1. Compound tubular gland	Secretory portions are coiled tubules, usually branched	Bulbourethral glands of male (see fig. 22.4)
2. Compound alveolar gland	Secretory portions are irregularly branched tubules with numerous saclike outgrowths	Mammary glands (see fig. 23.30)

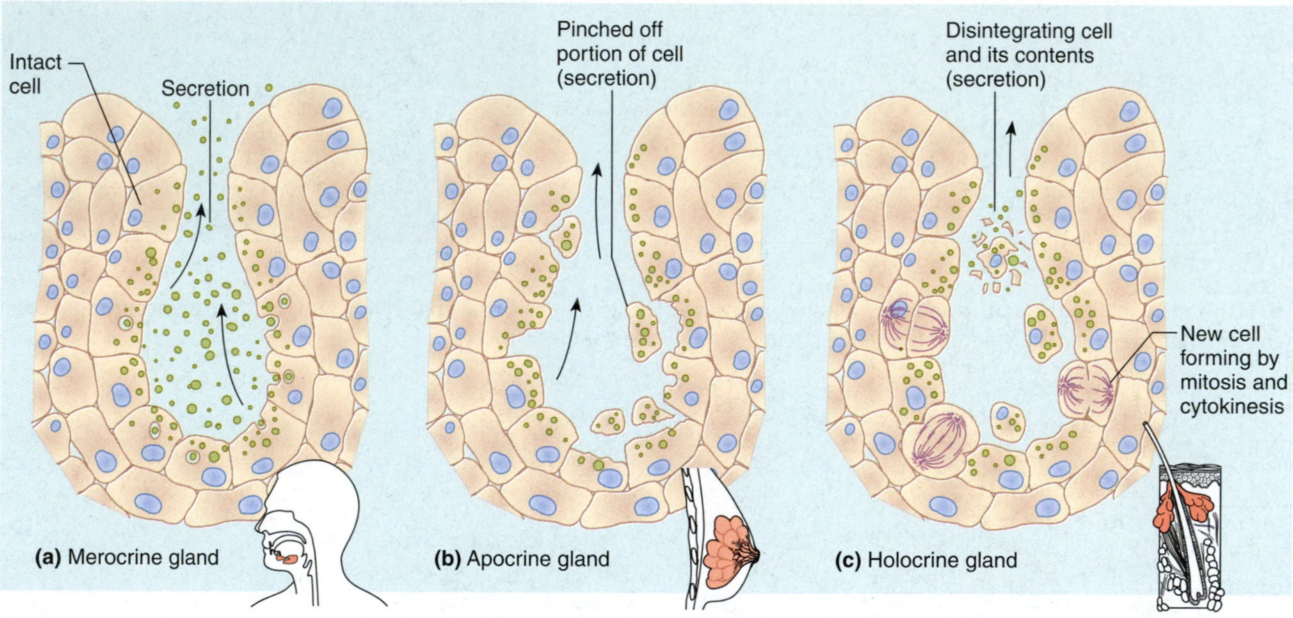

FIGURE 5.12 Glandular secretions. (*a*) Merocrine glands release secretions without losing cytoplasm. (*b*) Apocrine glands lose small portions of their cell bodies during secretion. (*c*) Holocrine glands release entire cells filled with secretory products.

Most exocrine secretory cells are merocrine, and they can be further subclassified based on their secretion of serous fluid or mucus. *Serous fluid* is typically watery and has a high concentration of enzymes. Serous cells secreting this fluid, which lubricates, are commonly associated with the visceral and parietal membranes of the thoracic and abdominopelvic cavities. The thicker fluid, *mucus,* is rich in the glycoprotein *mucin.* Cells in the inner linings of the digestive, respiratory, and reproductive systems secrete abundant mucus, which is protective. Mucous cells and goblet cells secrete mucus, but in different parts of the body. Table 5.5 summarizes the characteristics of the different types of epithelial tissues.

PRACTICE

8 Distinguish between exocrine and endocrine glands.

9 Explain how exocrine glands are classified.

10 Distinguish between a serous cell and a mucous cell.

5.3 CONNECTIVE TISSUES

General Characteristics

Connective (kŏ-nek′tiv) **tissues** comprise much of the body and are the most abundant type of tissue by weight. They

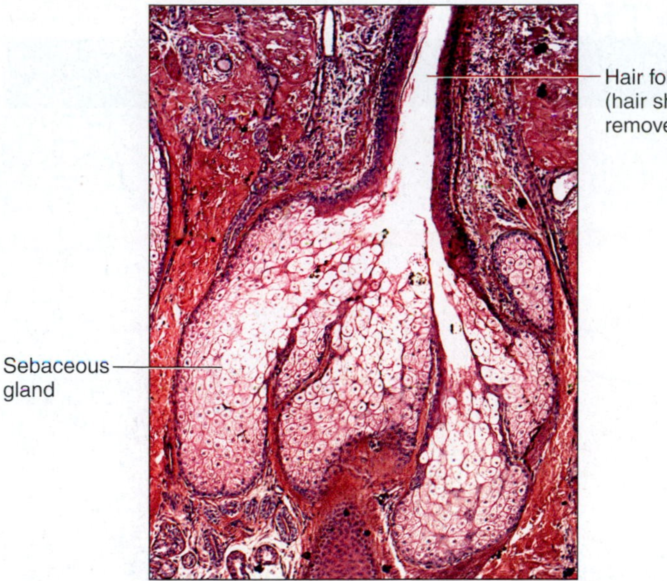

FIGURE 5.13 The sebaceous gland associated with a hair follicle is a simple-branched alveolar gland that secretes entire cells (40x).

bind structures, provide support and protection, serve as frameworks, fill spaces, store fat, produce blood cells, protect against infections, and help repair tissue damage.

Connective tissue cells are farther apart than epithelial cells, and they have an abundance of **extracellular matrix** (eks″trah-sel′u-lar ma′triks) between them. This extracellular matrix is composed of *protein fibers* and a *ground substance* consisting of nonfibrous protein and other molecules, and fluid. The consistency of the extracellular matrix varies from fluid to semisolid to solid. The ground substance binds, supports, and provides a medium through which substances may be transferred between the blood and cells of the tissue.

TABLE 5.4 | Types of Glandular Secretions

Type	Description of Secretion	Example
Merocrine glands	A fluid product released through the cell membrane by exocytosis	Salivary glands, pancreatic glands, sweat glands of the skin
Apocrine glands	Cellular product and portions of the free ends of glandular cells pinch off during secretion	Mammary glands, ceruminous glands lining the external ear canal
Holocrine glands	Disintegrated entire cells filled with secretory products	Sebaceous glands of the skin

Clinical Application 5.1 discusses the extracellular matrix and its relationship to disease.

Connective tissue cells can usually divide. These tissues have varying degrees of vascularity, but in most cases, they have good blood supplies and are well nourished. Some connective tissues, such as bone and cartilage, are rigid. Loose connective tissue and dense connective tissue are more flexible.

Major Cell Types

Connective tissues include a variety of cell types. Some of them are called *fixed cells* because they reside in the specific connective tissue type for an extended period. These include fibroblasts and mast cells. Other cells, such as macrophages, are *wandering cells*. They move through and appear in tissues temporarily, usually in response to an injury or infection.

Fibroblasts (fi′bro-blastz) are the most common type of fixed cell in connective tissues. These large, star-shaped cells produce fibers by secreting proteins into the extracellular matrix of connective tissues (fig. 5.14).

TABLE 5.5 | Epithelial Tissues

Type	Description	Function	Location
Simple squamous epithelium	Single layer, flattened cells	Filtration, diffusion, osmosis, covers surface	Air sacs of lungs, walls of capillaries, linings of blood and lymph vessels
Simple cuboidal epithelium	Single layer, cube-shaped cells	Secretion, absorption	Surface of ovaries, linings of kidney tubules, and linings of ducts of certain glands
Simple columnar epithelium	Single layer, elongated cells	Protection, secretion, absorption	Linings of uterus, stomach, and intestines
Pseudostratified columnar epithelium	Single layer, elongated cells	Protection, secretion, movement of mucus and substances	Linings of respiratory passages
Stratified squamous epithelium	Many layers, top cells flattened	Protection	Outer layer of skin, linings of oral cavity, vagina, and anal canal
Stratified cuboidal epithelium	2 to 3 layers, cube-shaped cells	Protection	Linings of larger ducts of mammary glands, sweat glands, salivary glands, and pancreas
Stratified columnar epithelium	Top layer of elongated cells, lower layers of cube-shaped cells	Protection, secretion	Part of the male urethra and parts of the pharynx
Transitional epithelium	Many layers of cube-shaped and elongated cells	Distensibility, protection	Inner lining of urinary bladder and linings of ureters and part of urethra
Glandular epithelium	Unicellular or multicellular	Secretion	Salivary glands, sweat glands, endocrine glands

5.1 CLINICAL APPLICATION

The Body's Glue: The Extracellular Matrix

Rather than being just "filler" between cells, the extracellular matrix (ECM) is a complex and changing mix of molecules that modifies the tissue to suit different organs and conditions. Not only does the ECM serve as a scaffolding to organize cells into tissues, but it relays the biochemical signals that control cell division, differentiation, repair, and migration.

The ECM has two basic components: the basement membrane that covers epithelial cell surfaces, and the rest of the material between cells, called the interstitial matrix. The basement membrane is mostly tightly packed collagenous fibers from which large, cross-shaped glycoproteins called laminins extend. The laminins (and other glycoproteins such as fibronectin, the proteoglycans, and tenascin) traverse the interstitial matrix and contact receptors, called integrins, on other cells (fig. 5A). In this way, the ECM connects cells into tissues. At least twenty types of collagen and precursors of hormones, enzymes, growth factors, and immune system biochemicals (cytokines) comprise the various versions of the ECM. The precursor molecules are activated under certain conditions.

The components of the ECM are always changing, as its cells synthesize proteins while enzymes called proteases break down specific proteins. The balance of components is important to maintaining and repairing organ structure. Disrupt the balance, and disease can result. Here are three common examples:

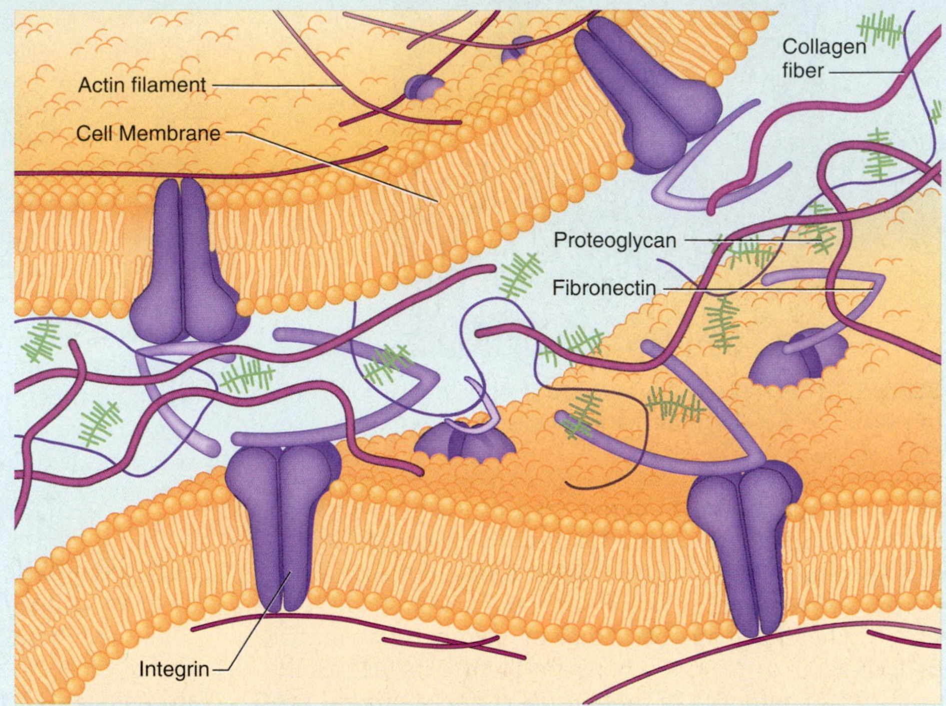

FIGURE 5A The extracellular matrix (ECM) is a complex and dynamic meshwork of various proteins and glycoproteins. Collagen is abundant. Other common components include integrins that anchor the ECM to cells, proteoglycans, and fibronectin. The ECM may also include precursors of growth factors, hormones, enzymes, and cytokines. It is vital to maintaining the specialized characteristics of tissues and organs.

Cancer

The spread of a cancerous growth takes advantage of the normal ability of fibroblasts to contract as they close a wound, where they are replaced with normal epithelium. Chemical signals from cancer cells make fibroblasts more contractile (myofibroblasts), and they take on the characteristics of cancer cells. At the same time, alterations in laminins loosen the connections of the fibroblasts to surrounding cells. This abnormal flexibility enables the changed fibroblasts to migrate, helping the cancer spread. Normally, fibroblasts secrete abundant collagen.

Liver Fibrosis

In fibrosis, a part of all chronic liver diseases, collagen deposition increases so that the ECM exceeds its normal 3% of the organ. Healthy liver ECM sculpts a framework that supports the epithelial and vascular tissues of the organ. In response to a damaging agent such as a virus, alcohol, or a toxic drug, hepatic stellate cells secrete collagenous fibers in the areas where the epithelium and blood vessels meet. Such limited fibrosis seals off the affected area, preventing its spread. But if the process continues—if an infection is not treated or the noxious stimulus not removed—the ECM grows and eventually blocks the interaction between liver cells and the bloodstream. The liver tissue hardens, a dangerous condition called *cirrhosis.*

Heart Failure and Atherosclerosis

The heart's ECM organizes cells into a three-dimensional network that coordinates their contractions into the rhythmic heartbeat necessary to pump blood. This ECM consists of collagen, fibronectin, laminin, and elastin surrounding cardiac muscle cells and myofibroblasts and is also in the walls of arteries. Heart failure and atherosclerosis reflect imbalances of collagen production and degradation. As in the liver, the natural response of ECM buildup is to wall off an area where circulation is blocked, but if it continues, the extra scaffolding stiffens the heart, which can lead to heart failure. In atherosclerosis, excess ECM accumulates on the interior linings of arteries, blocking blood flow. During a myocardial infarction (heart attack), collagen synthesis and deposition increase in affected and nonaffected heart parts, which is why damage can continue even after pain starts. From Science To Technology 5.2 (p. 166) and from Science to Technology 15.1 (p. 566) discuss engineering a semisynthetic replacement heart. ■

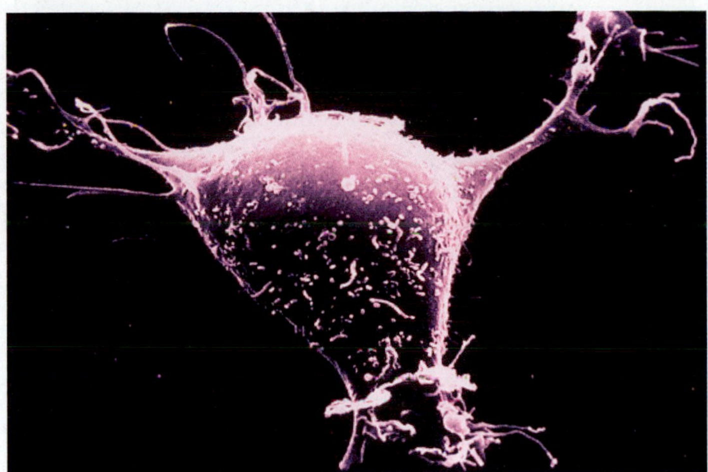

FIGURE 5.14 A scanning electron micrograph of a fibroblast (4,000×).

Macrophages (mak′ro-fājez), or histiocytes, originate as white blood cells (see chapter 14, p. 532) and are almost as numerous as fibroblasts in some connective tissues. They are usually attached to fibers but can detach and actively move about. Macrophages are specialized to carry on phagocytosis. They function as scavenger cells that can clear foreign particles from tissues, so macrophages are an important defense against infection (fig. 5.15). They also play a role in immunity (see chapter 16, p. 630).

Mast cells are large and widely distributed in connective tissues, where they are usually near blood vessels (fig. 5.16). They release *heparin,* a compound that prevents blood clotting. Mast cells also release *histamine,* a substance that promotes some of the reactions associated with inflammation and allergies, such as asthma and hay fever (see chapter 16, p. 639).

Release of histamine stimulates inflammation by dilating the small arterioles that feed capillaries, the tiniest blood vessels. The resulting swelling and redness is inhospitable to infectious bacteria and viruses and also dilutes toxins. Inappropriate histamine release as part of an allergic response can be most uncomfortable. Allergy medications called antihistamines counter this misplaced inflammation.

Connective Tissue Fibers

Fibroblasts produce three types of connective tissue fibers: collagenous fibers, elastic fibers, and reticular fibers. Of these, collagenous and elastic fibers are the most abundant.

Collagenous (kol-laj′ĕ-nus) **fibers** are thick threads of the protein **collagen** (kol′ah-jen), the major structural protein of the body. Collagenous fibers are grouped in long, parallel bundles, and they are flexible but only slightly elastic (fig. 5.17). More importantly, they have great tensile strength—that is, they can resist considerable pulling force. Thus, collagenous fibers are important components of body parts that hold structures together, such as **ligaments** (which connect bones to bones) and **tendons** (which connect muscles to bones).

Tissue containing abundant collagenous fibers is called *dense connective tissue.* Such tissue appears white, and for this reason collagenous fibers of dense connective tissue are sometimes called white fibers. *Loose connective tissue,* on the other hand, has sparse collagenous fibers. Clinical Application 5.2 describes disorders that result from abnormal collagen.

When skin is exposed to prolonged and intense sunlight, connective tissue fibers lose elasticity, and the skin stiffens and becomes leathery. In time, the skin may sag and wrinkle. Collagen injections may temporarily smooth out wrinkles. However, collagen applied as a cream to the skin does not combat wrinkles because collagen molecules are far too large to penetrate the skin.

Elastic fibers are composed of a springlike protein called **elastin.** These fibers branch, forming complex networks in various tissues. They are weaker than collagenous fibers but elastic. That is, they are easily stretched or deformed and will resume their original lengths and shapes when the force acting upon them is removed. Elastic fibers are common in body parts normally subjected to stretching, such as the vocal cords and air passages of the respiratory system. Elastic fibers are sometimes called yellow fibers, because tissues amply supplied with them appear yellowish (fig. 5.17).

Surgeons use elastin in foam, powder, or sheet form to prevent scar tissue adhesions from forming at the sites of tissue removal. Elastin is produced in bacteria that contain human genes that instruct them to manufacture the human protein. This is cheaper than synthesizing elastin chemically and safer than obtaining it from cadavers.

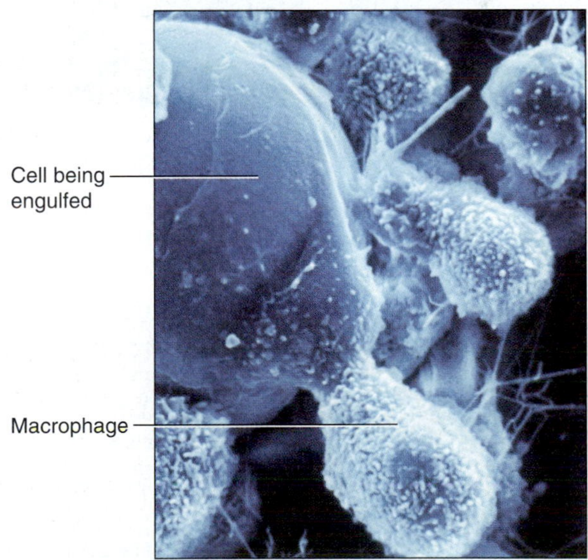

FIGURE 5.15 Macrophages are scavenger cells common in connective tissues. This scanning electron micrograph shows a number of macrophages engulfing parts of a larger cell (3,330×).

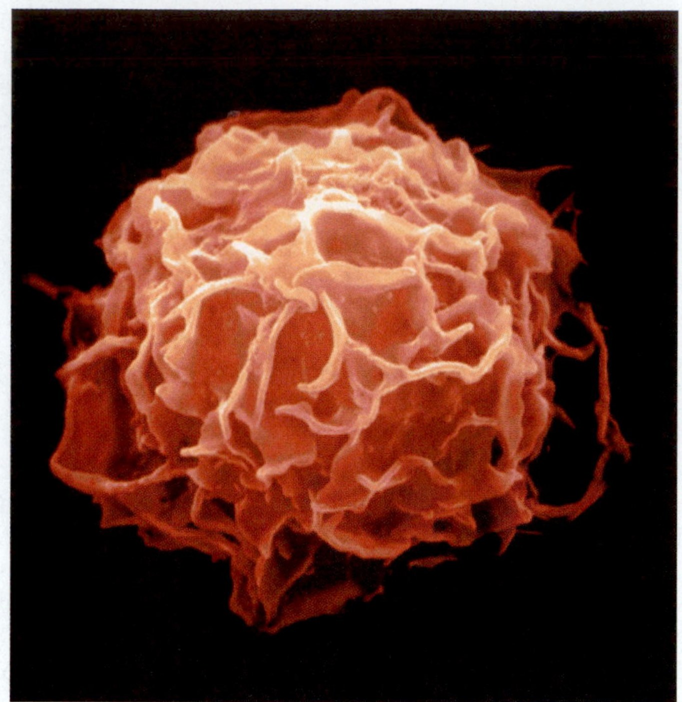

FIGURE 5.16 Scanning electron micrograph of a mast cell (6,600×).

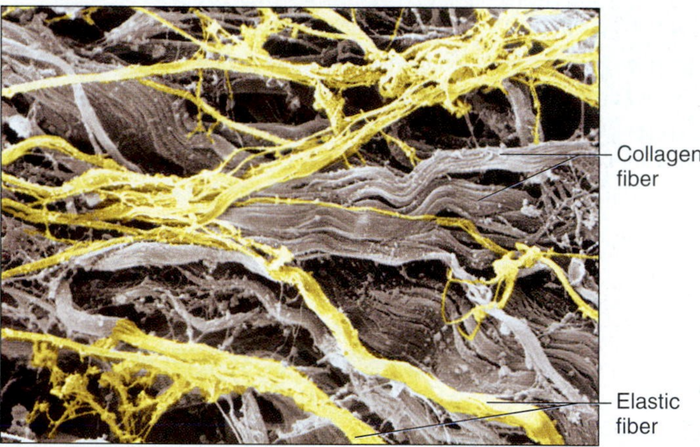

FIGURE 5.17 Scanning electron micrograph of collagenous fibers (shades of white to gray) and elastic fibers (yellow) (4,100×).

Reticular fibers are thin collagenous fibers. They are highly branched and form delicate supporting networks in a variety of tissues, including those of the spleen. Table 5.6 summarizes the components of connective tissue.

PRACTICE

11 What are the general characteristics of connective tissue?

12 What are the major types of cells in connective tissue?

13 What is the primary function of fibroblasts?

14 What are the characteristics of collagen and elastin?

Categories of Connective Tissues

Connective tissue is divided into two major categories. *Connective tissue proper* includes loose connective tissue (areolar, adipose, reticular) and dense connective tissue (dense regular, dense irregular, elastic). The *specialized connective tissues* include cartilage, bone, and blood. The following sections describe each type of connective tissue.

Areolar Tissue

Areolar (ah-re′o-lar) **tissue,** forms delicate, thin membranes throughout the body. The cells of this tissue, mainly fibroblasts, are located some distance apart and are separated by a gel-like ground substance that contains many collagenous and elastic fibers that fibroblasts secrete (fig. 5.18).

Areolar tissue binds the skin to the underlying organs and fills spaces between muscles. It underlies most layers of epithelium, where its many blood vessels nourish nearby epithelial cells.

Adipose Tissue

Adipose (ad′ĭ-pōs) **tissue,** or fat, develops when certain cells (adipocytes) store fat in droplets in their cytoplasm. At first, these cells resemble fibroblasts, but as they accumulate fat, they enlarge, and their nuclei are pushed to one side (fig. 5.19). When adipocytes become so abundant that they crowd out other cell types, they form adipose tissue. This tissue lies beneath the skin, in spaces between muscles, around the kidneys, behind the eyeballs, in certain abdominal membranes,

TABLE 5.6 | Components of Connective Tissue

Component	Characteristic	Function
Fibroblasts	Widely distributed, large, star-shaped cells	Secrete proteins that become fibers
Macrophages	Motile cells sometimes attached to fibers	Clear foreign particles from tissues by phagocytosis
Mast cells	Large cells, usually located near blood vessels	Release substances that may help prevent blood clotting and promote inflammation
Collagenous fibers (white fibers)	Thick, threadlike fibers of collagen with great tensile strength	Hold structures together
Elastic fibers (yellow fibers)	Bundles of microfibrils embedded in elastin	Provide elastic quality to parts that stretch
Reticular fibers	Thin fibers of collagen	Form supportive networks within tissues

5.2 CLINICAL APPLICATION

Abnormalities of Collagen

Much of the human body consists of the protein collagen. It accounts for more than 60% of the protein in bone and cartilage and provides 50% to 90% of the dry weight of skin, ligaments, tendons, and the dentin of teeth. Collagen is in the eyes, blood vessel linings, basement membranes, and connective tissue. It is not surprising that defects in collagen cause a variety of medical problems.

Collagen abnormalities are devastating because this protein has an extremely precise structure that is easily disrupted, even by slight alterations that might exert little noticeable effect in other proteins. Collagen is sculpted from a precursor molecule called procollagen. Three procollagen chains coil and entwine to form a regular triple helix.

Triple helices form as the procollagen is synthesized, but once secreted from the cell, the helices are trimmed. The collagen fibrils continue to associate outside the cell, building the networks that hold the body together. Collagen is rapidly synthesized and assembled into its rigid architecture. Many types of mutations can disrupt the protein's structure, including missing procollagen chains, kinks in the triple helix, failure to cut mature collagen, and defects in aggregation outside the cell.

Table 5A details some collagen disorders. Knowing which specific mutations cause disorders offers a way to identify the condition before symptoms arise. This can be helpful if early treatment can follow. A woman who has a high risk of developing hereditary osteoporosis, for example, might take calcium supplements before symptoms appear. Aortic aneurysm is a more serious connective tissue disorder that can be presymptomatically detected if the underlying mutation is discovered. In aortic aneurysm, a weakened aorta (the largest blood vessel in the body, which emerges from the heart) bursts. Knowing that the mutant gene has not been inherited can ease worries—and knowing that it has been inherited can warn affected individuals to have frequent ultrasound exams so that aortic weakening can be detected early enough to correct with surgery. ■

TABLE 5A | Collagen Disorders

Disorder	Molecular Defect	Signs and Symptoms
Chondrodysplasia	Collagen chains are too wide and asymmetric	Stunted growth; deformed joints
Dystrophic epidermolysis bullosa	Breakdown of collagen fibrils that attach skin layers to each other	Stretchy, easily scarred skin; lax joints
Hereditary osteoarthritis	Substituted amino acid in collagen chain alters shape	Painful joints
Marfan syndrome	Too little fibrillin, an elastic connective tissue protein	Long limbs, sunken chest, lens dislocation, spindly fingers, weakened aorta
Osteogenesis imperfecta type I	Too few collagen triple helices	Easily broken bones; deafness; blue sclera (whites of the eyes)
Stickler syndrome	Short collagen chains	Joint pain; degeneration of retina and fluid around it

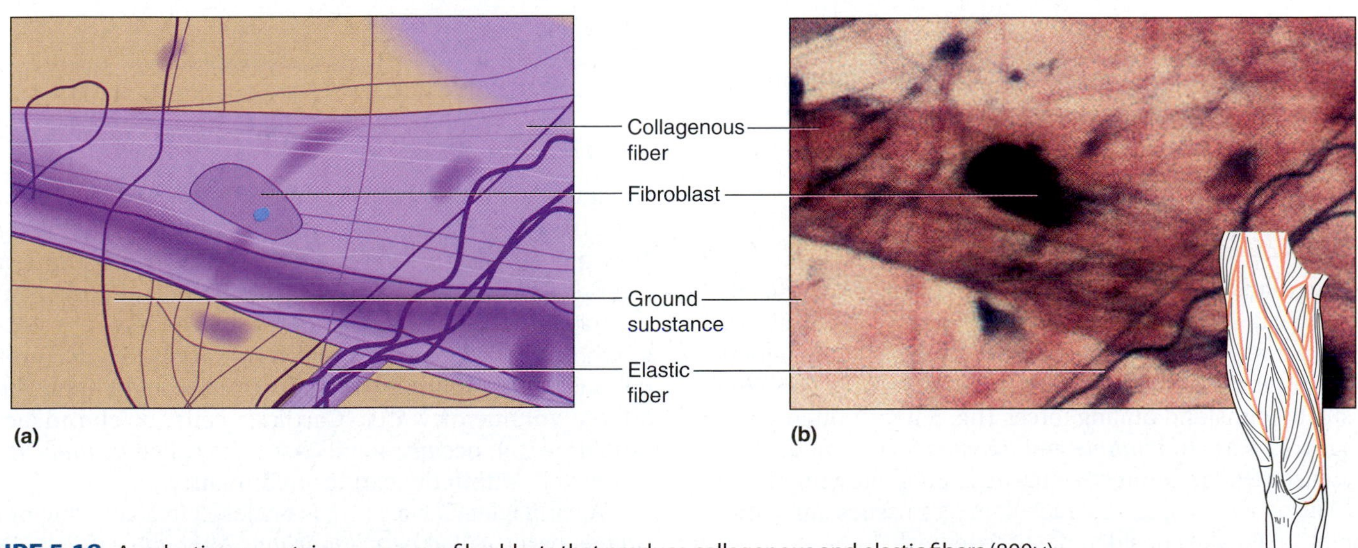

FIGURE 5.18 Areolar tissue contains numerous fibroblasts that produce collagenous and elastic fibers (800×).

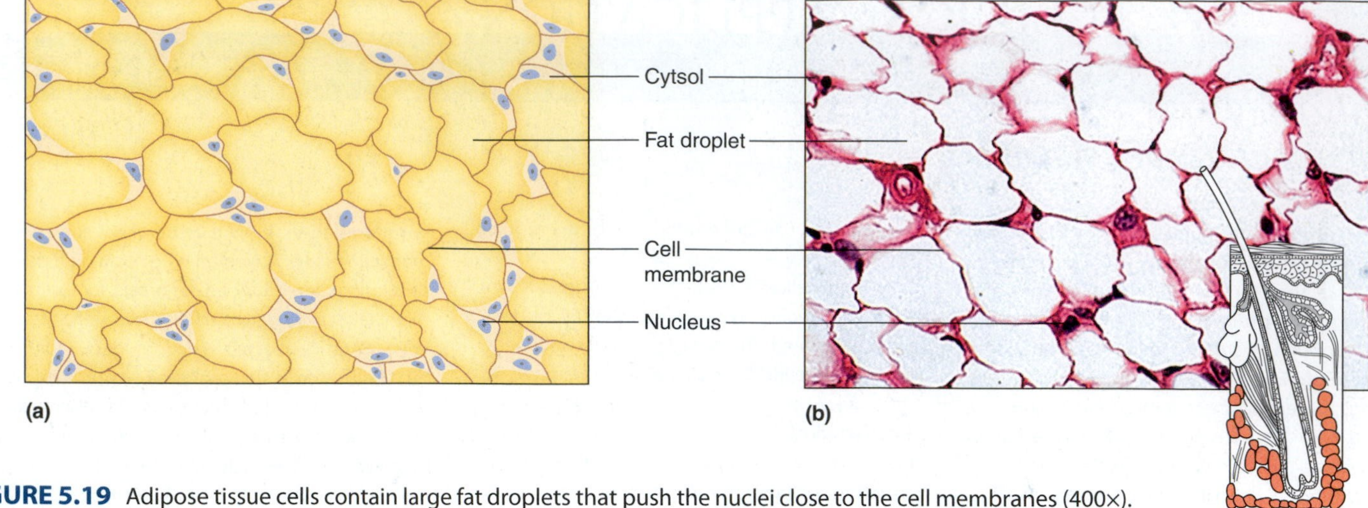

FIGURE 5.19 Adipose tissue cells contain large fat droplets that push the nuclei close to the cell membranes (400×).

on the surface of the heart, and around certain joints. Adipose tissue cushions joints and some organs, such as the kidneys. It also insulates beneath the skin, and it stores energy in fat molecules.

A person is born with a certain number of fat cells. Excess food calories are likely to be converted to fat and stored, so the amount of adipose tissue in the body reflects diet or an endocrine disorder. During a period of fasting, adipose cells may lose their fat droplets, shrink, and become more like fibroblasts again.

Infants and young children have a continuous layer of adipose tissue just beneath the skin, which gives their bodies a rounded appearance. In adults, this subcutaneous fat thins in some regions and remains thick in others. For example, in males, adipose tissue usually thickens in the upper back, arms, lower back, and buttocks; in females, it is more likely to develop in the breasts, buttocks, and thighs.

Reticular Connective Tissue

Reticular connective tissue is composed of thin, collagenous fibers in a three-dimensional network. It helps provide the framework of certain internal organs, such as the liver, spleen, and lymphatic organs (fig. 5.20).

Dense Regular Connective Tissue

Dense regular connective tissue consists of many closely packed, thick, collagenous fibers; a fine network of elastic fibers; and a few cells, mostly fibroblasts. Collagenous fibers of dense regular connective tissue are very strong, enabling the tissue to withstand pulling forces (fig. 5.21). It often binds body parts as parts of *tendons* and *ligaments*. The blood supply to dense regular connective tissue is poor, slowing tissue repair. This is why a sprain, which damages tissues surrounding a joint, may take considerable time to heal.

Dense Irregular Connective Tissue

Fibers of **dense irregular connective tissue** are thicker, interwoven, and more randomly organized. This allows the tissue to sustain tension exerted from many different directions. Dense irregular connective tissue is in the dermis, the inner skin layer.

Elastic Connective Tissue

Elastic connective tissue mainly consists of yellow, elastic fibers in parallel strands or in branching networks. Between these fibers are collagenous fibers and fibroblasts. This tissue is found in the attachments between bones of the spinal column (ligamenta flava). It is also in the layers within the walls of certain hollow internal organs, including the larger arteries; some portions of the heart; and the larger airways, where it imparts an elastic quality (fig. 5.22).

PRACTICE

15 Differentiate between loose connective tissue and dense connective tissue.

16 What are the functions of adipose tissue?

17 Distinguish between reticular and elastic connective tissues.

Cartilage

Cartilage (kar′ti-lij) is a rigid connective tissue. It provides support, frameworks, and attachments; protects underlying tissues; and forms structural models for many developing bones.

Cartilage extracellular matrix is abundant and is largely composed of collagenous fibers embedded in a gel-like ground substance. This ground substance is rich in a protein-polysaccharide complex (chondromucoprotein) and contains a large volume of water. Cartilage cells, or **chondrocytes** (kon′dro-sītz), occupy small chambers called *lacunae* and lie completely within the extracellular matrix.

A cartilaginous structure is enclosed in a covering of connective tissue called *perichondrium*. Although cartilage tissue

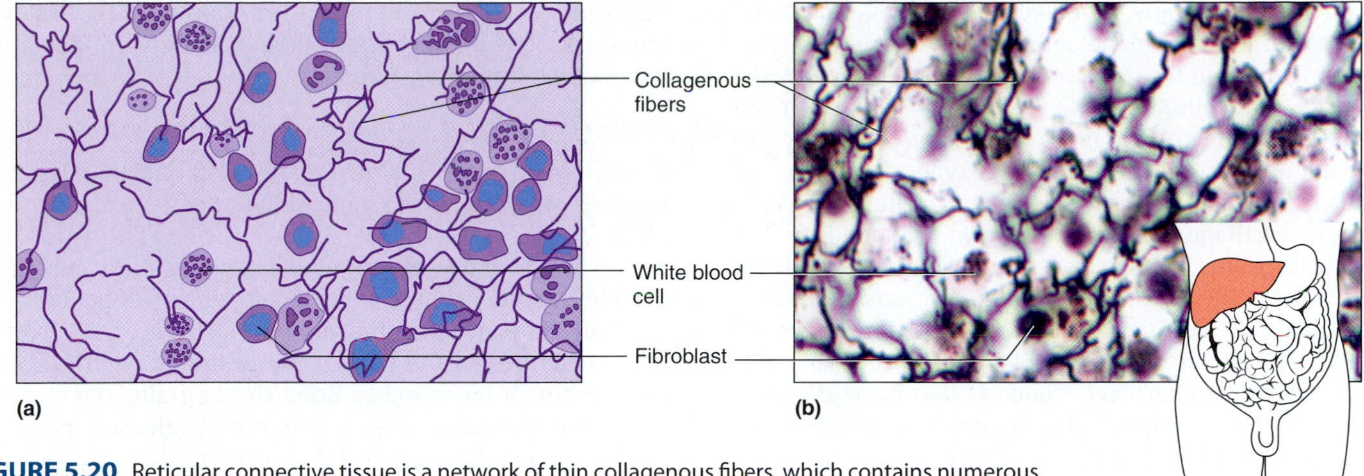

FIGURE 5.20 Reticular connective tissue is a network of thin collagenous fibers, which contains numerous fibroblasts and white blood cells (1,000×).

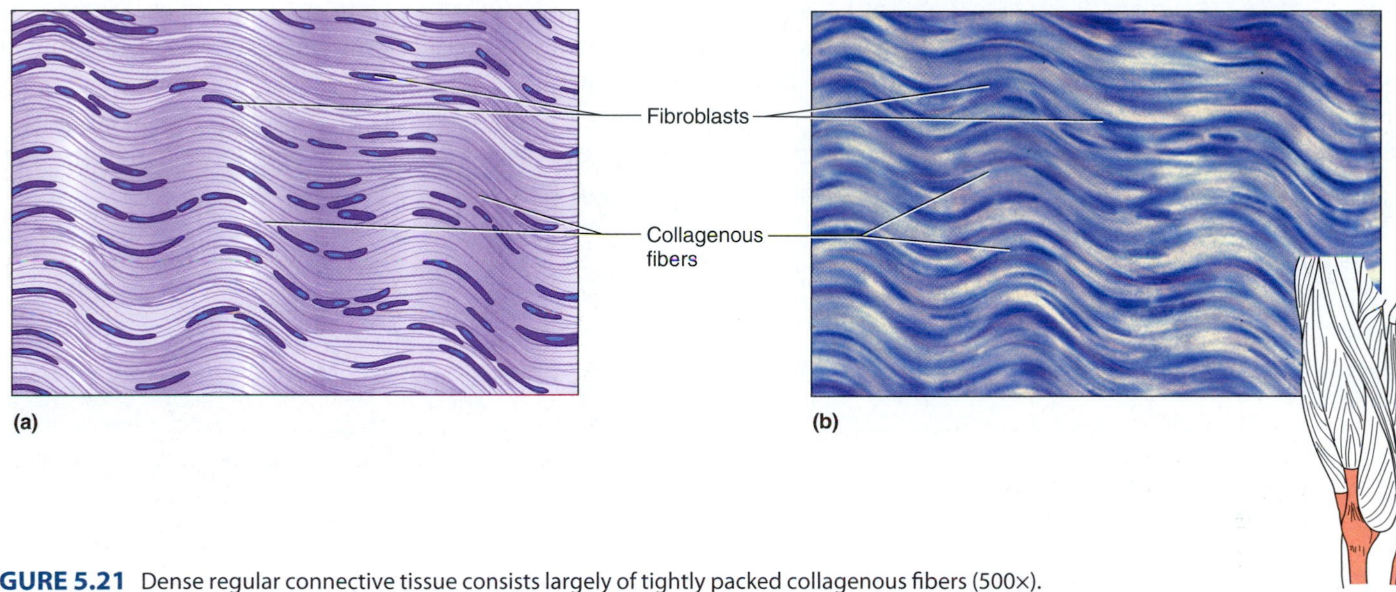

FIGURE 5.21 Dense regular connective tissue consists largely of tightly packed collagenous fibers (500×).

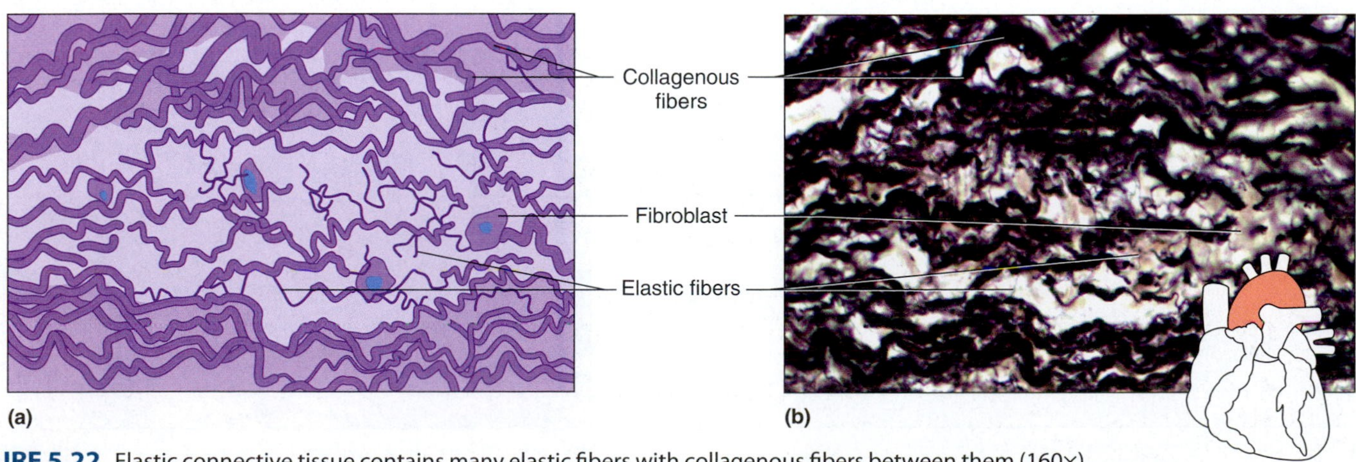

FIGURE 5.22 Elastic connective tissue contains many elastic fibers with collagenous fibers between them (160×).

lacks a direct blood supply, blood vessels are in the surrounding perichondrium. Cartilage cells near the perichondrium obtain nutrients from these vessels by diffusion, aided by the water in the extracellular matrix. This lack of a direct blood supply is why torn cartilage heals slowly and why chondrocytes do not divide frequently. The three types of cartilage are distinguished by their different types of extracellular matrix.

Hyaline cartilage (fig. 5.23), the most common type, has very fine collagenous fibers in its extracellular matrix and looks somewhat like white glass. It is found on the ends of bones in many joints, in the soft part of the nose, and in the supporting rings of the respiratory passages. Parts of an embryo's skeleton begin as hyaline cartilage "models" that bone gradually replaces. Hyaline cartilage is also important in the development and growth of most bones (see chapter 7, p. 197).

Elastic cartilage (fig. 5.24) is more flexible than hyaline cartilage because its extracellular matrix has a dense network of elastic fibers. It provides the framework for the external ears and parts of the larynx.

Fibrocartilage (fig. 5.25), a very tough tissue, has many collagenous fibers. It is a shock absorber for structures subjected to pressure. For example, fibrocartilage forms pads (intervertebral discs) between the individual bones (vertebrae) of the spinal column. It also cushions bones in the knees and in the pelvic girdle.

Bone

Bone (osseous tissue) is the most rigid connective tissue. Its hardness is largely due to mineral salts, such as calcium phosphate and calcium carbonate, between cells. This extracellular matrix also contains abundant collagenous fibers, which are flexible and reinforce the mineral components of bone.

Bone internally supports body structures. It protects vital structures in the cranial and thoracic cavities and is an attachment for muscles. Bone also contains red marrow, which forms blood cells. It stores and releases inorganic chemicals such as calcium and phosphorus.

Bone matrix is deposited by bone cells, called *osteoblasts,* in thin layers called *lamellae,* which form concentric patterns around capillaries located within tiny longitudinal tubes called *central,* or *Haversian, canals.* Once osteoblasts are in lacunae surrounded by matrix, they are called osteocytes and are rather evenly spaced within the lamellae. Consequently, osteocytes also form concentric circles (fig. 5.26).

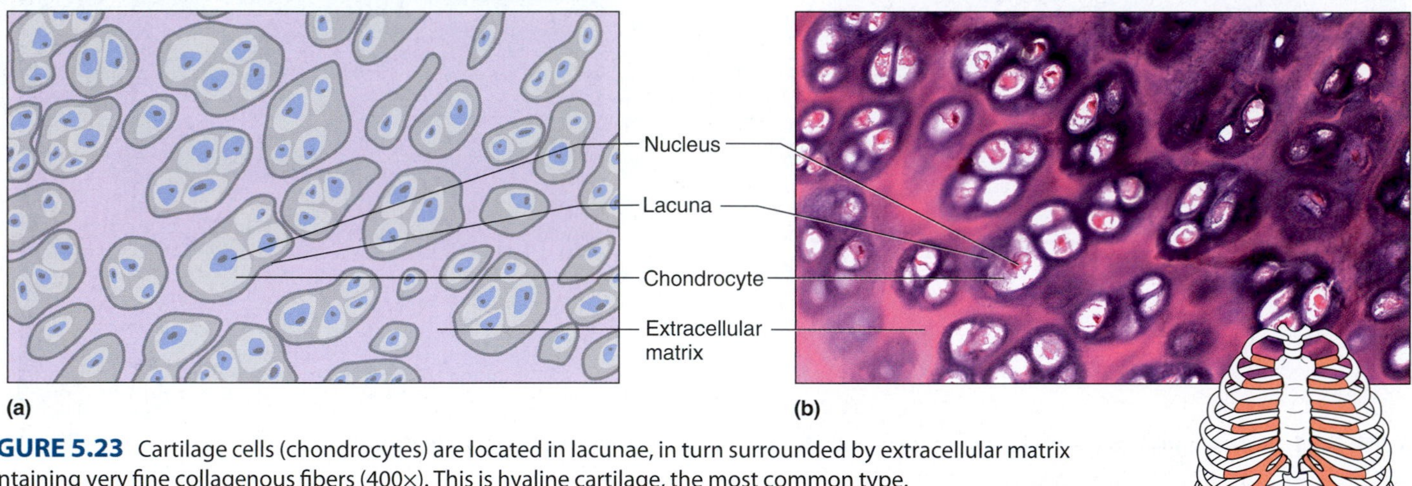

FIGURE 5.23 Cartilage cells (chondrocytes) are located in lacunae, in turn surrounded by extracellular matrix containing very fine collagenous fibers (400×). This is hyaline cartilage, the most common type.

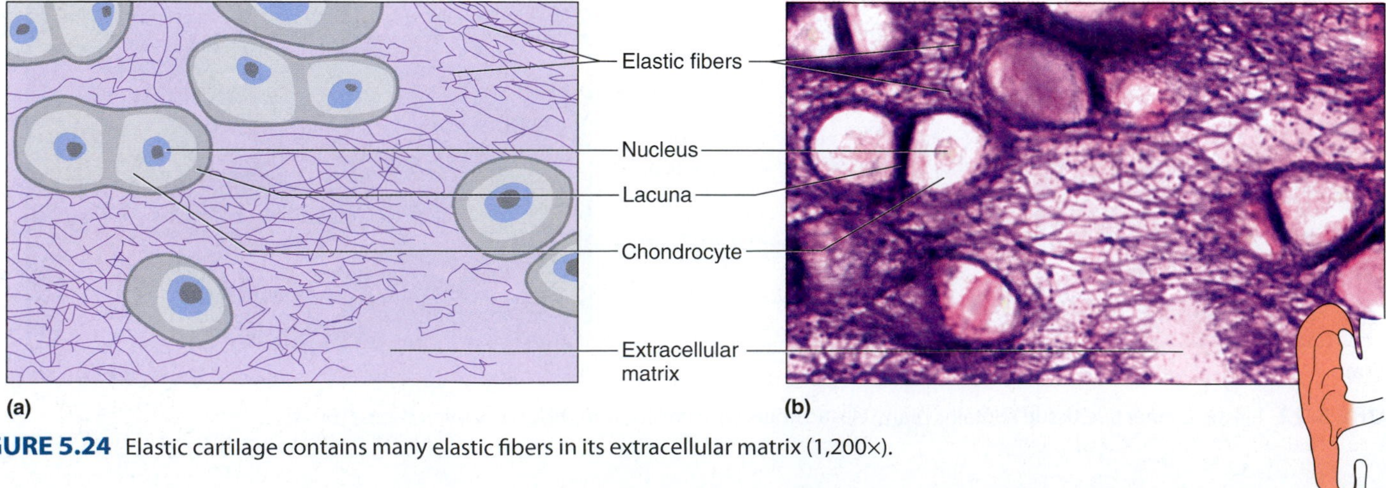

FIGURE 5.24 Elastic cartilage contains many elastic fibers in its extracellular matrix (1,200×).

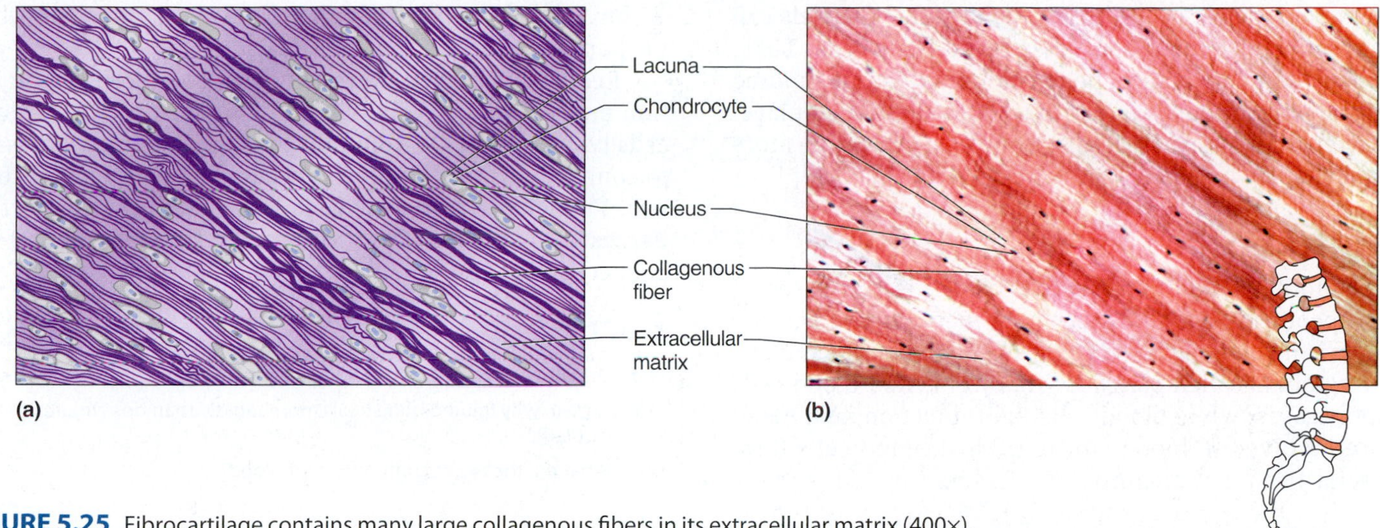

FIGURE 5.25 Fibrocartilage contains many large collagenous fibers in its extracellular matrix (400×).

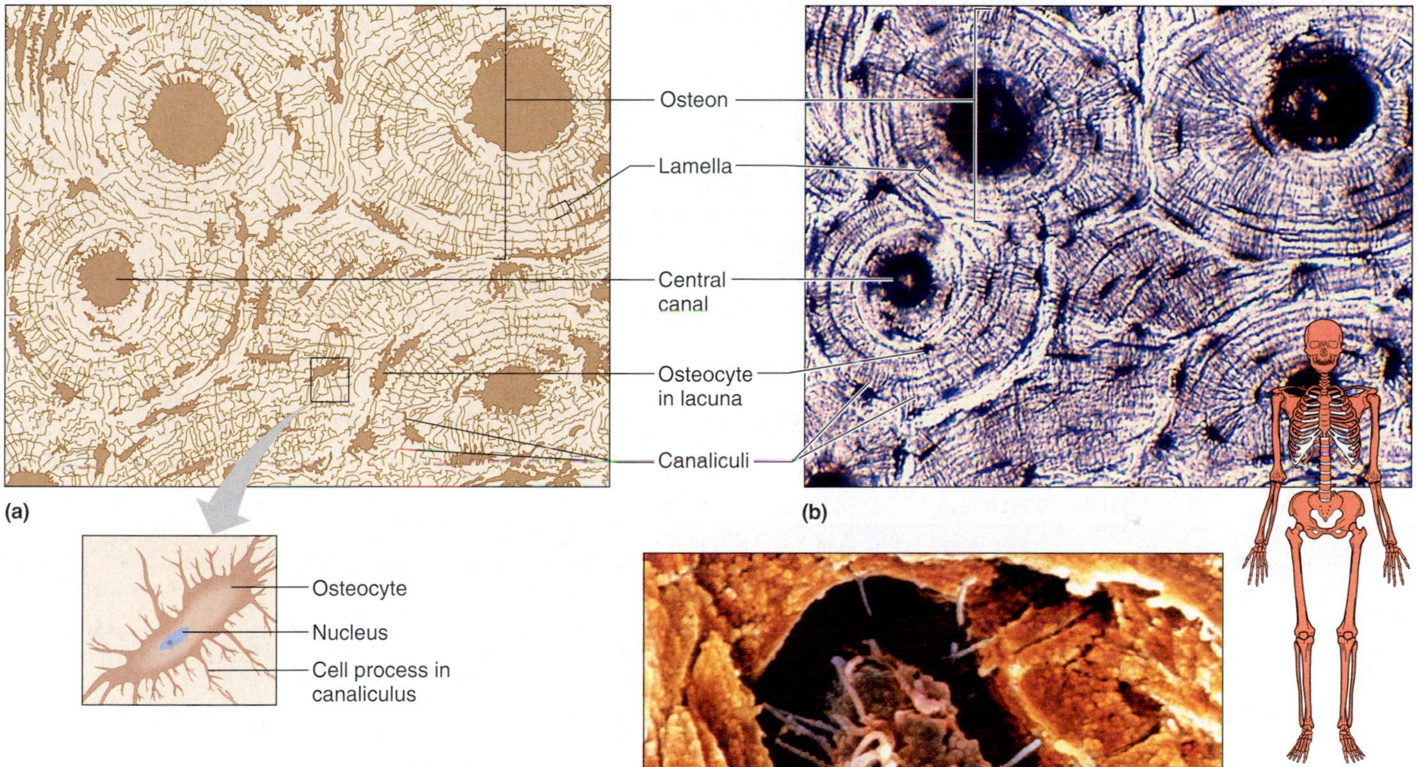

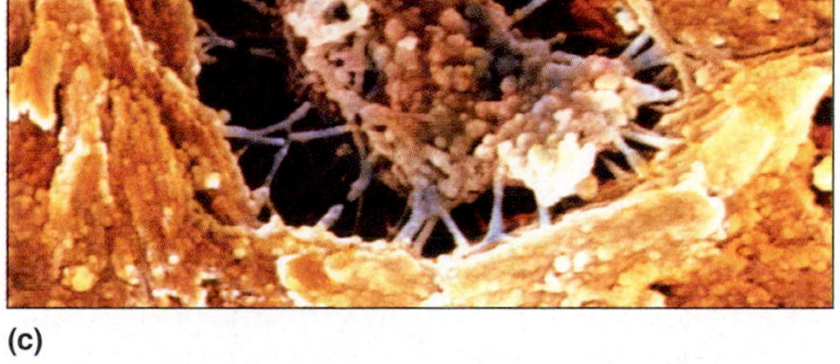

FIGURE 5.26 Bone tissue. (*a*) Bone matrix is deposited in concentric layers around central canals. (*b*) Micrograph of bone tissue (200×). (*c*) Artificially colored scanning electron micrograph of an osteocyte within a lacuna (6,000×).

In a bone, the osteocytes and layers of extracellular matrix, concentrically clustered around a central canal, form a cylinder-shaped unit called an *osteon,* or a Haversian system. Many of these units cemented together form the substance of bone (see chapter 7, p. 195).

Each central canal contains a blood vessel, so every bone cell is fairly close to a nutrient supply. In addition, the bone cells have many cytoplasmic processes that extend outward and pass through minute tubes in the extracellular matrix called *canaliculi.* Gap junctions attach these cellular processes to the membranes of nearby cells. As a result, materials can move rapidly between blood vessels and bone cells. Thus, despite its inert appearance, bone is an active tissue. Injured bone heals much more rapidly than does injured cartilage. (The microscopic structure of bone is described in more detail in chapter 7, p. 195.)

Blood

Blood, another type of connective tissue, is composed of cells suspended in a fluid extracellular matrix called *plasma*. These cells include *red blood cells, white blood cells,* and cellular fragments called *platelets* (fig. 5.27). Red blood cells transport gases; white blood cells fight infection; and platelets are involved in blood clotting. Most blood cells form in special tissues (hematopoietic tissues) in red marrow within the hollow parts of certain bones. Blood is described in chapter 14.

Red blood cells are the only type of blood cells that function entirely in the blood vessels. In contrast, white blood cells typically migrate from the blood through capillary walls to connective tissues, where they carry on their major activities. The white blood cells usually reside in the connective tissues until they die. Table 5.7 lists the characteristics of the connective tissues.

PRACTICE

18 Describe the general characteristics of cartilage.

19 Explain why injured bone heals more rapidly than does injured cartilage.

20 What are the major components of blood?

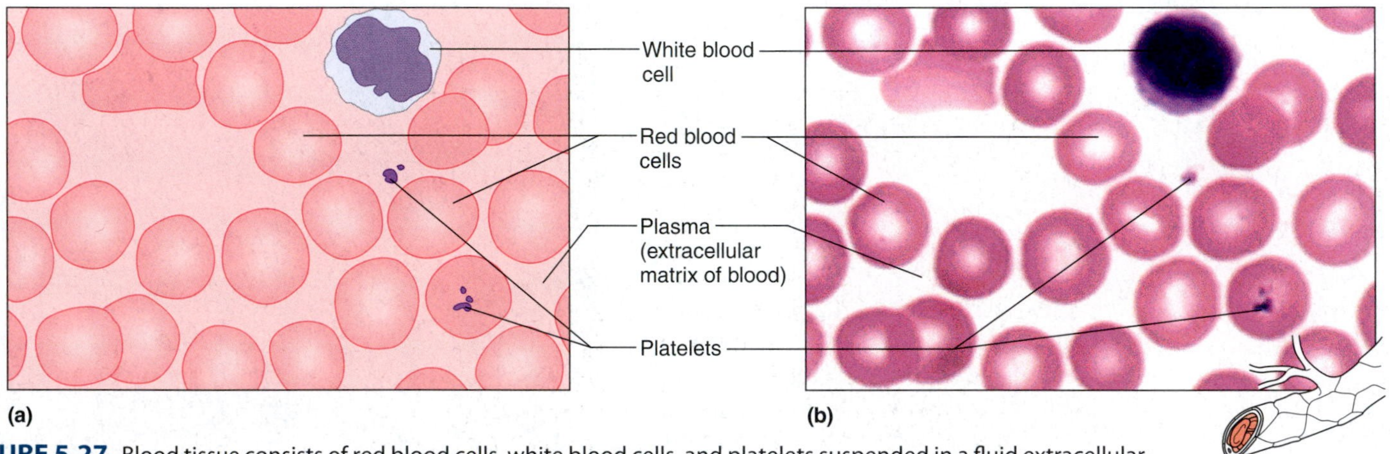

FIGURE 5.27 Blood tissue consists of red blood cells, white blood cells, and platelets suspended in a fluid extracellular matrix (1,000×).

TABLE 5.7 | Connective Tissues

Type	Description	Function	Location
Areolar connective tissue	Cells in fluid-gel matrix	Binds organs, holds tissue fluids	Beneath the skin, between muscles, beneath epithelial tissues
Adipose tissue	Cells in fluid-gel matrix	Protects, insulates, and stores fat	Beneath the skin, around the kidneys, behind the eyeballs, on the surface of the heart
Reticular connective tissue	Cells in fluid-gel matrix	Supports	Walls of liver, spleen, and lymphatic organs
Dense regular connective tissue	Cells in fluid-gel matrix	Binds body parts	Tendons, ligaments
Dense irregular connective tissue	Cells in fluid-gel matrix	Sustains tissue tension	Dermis
Elastic connective tissue	Cells in fluid-gel matrix	Provides elastic quality	Connecting parts of the spinal column, in walls of arteries and airways
Hyaline cartilage	Cells in solid-gel matrix	Supports, protects, provides framework	Ends of bones, nose, and rings in walls of respiratory passages
Elastic cartilage	Cells in solid-gel matrix	Supports, protects, provides flexible framework	Framework of external ear and part of larynx
Fibrocartilage	Cells in solid-gel matrix	Supports, protects, absorbs shock	Between bony parts of spinal column, parts of pelvic girdle, and knee
Bone	Cells in solid matrix	Supports, protects, provides framework	Bones of skeleton, middle ear
Blood	Cells and platelets in fluid matrix	Transports gases, defends against disease, clotting	Throughout the body in a closed system of blood vessels and heart chambers

5.4 TYPES OF MEMBRANES

After discussing epithelial and connective tissues, sheets of cells called membranes are better understood. **Epithelial membranes** are thin structures that are usually composed of epithelium and underlying connective tissue, covering body surfaces and lining body cavities. The three major types of epithelial membranes are *serous, mucous,* and *cutaneous.*

Serous (se′rus) **membranes** line the body cavities that do not open to the outside and reduce friction between the organs and cavity walls. They form the inner linings of the thorax and abdomen, and they cover the organs in these cavities (see figs. 1.11 and 1.12). A serous membrane consists of a layer of simple squamous epithelium (mesothelium) and a thin layer of loose connective tissue. Cells of a serous membrane secrete watery *serous fluid,* which helps lubricate membrane surfaces.

Mucous (mu′kus) **membranes** line the cavities and tubes that open to the outside of the body. These include the oral and nasal cavities and the tubes of the digestive, respiratory, urinary, and reproductive systems. A mucous membrane consists of epithelium overlying a layer of loose connective tissue. However, the type of epithelium varies with the location of the membrane. For example, stratified squamous epithelium lines the oral cavity, pseudostratified columnar epithelium lines part of the nasal cavity, and simple columnar epithelium lines the small intestine. Goblet cells within a mucous membrane secrete *mucus.*

Another epithelial membrane is the **cutaneous** (ku-ta′ne-us) **membrane,** more commonly called *skin.* It is part of the integumentary system described in detail in chapter 6.

A type of membrane composed entirely of connective tissues is a **synovial** (sĭ-no′ve-al) **membrane.** It lines joints and is discussed further in chapter 8 (p. 264).

PRACTICE

21 Name the four types of membranes, and explain how they differ.

5.5 MUSCLE TISSUES

General Characteristics

Muscle tissues are contractile; they can shorten and thicken. As they contract, muscle cells pull at their attached ends, which moves body parts. The cells that comprise muscle tissues are sometimes called *muscle fibers* because they are elongated. The three types of muscle tissue, skeletal, smooth, and cardiac, are introduced here and discussed further in chapter 9.

Skeletal Muscle Tissue

Skeletal muscle tissue (fig. 5.28) forms muscles that usually attach to bones and are controlled by conscious effort. For this reason, it is often called *voluntary* muscle tissue. Skeletal muscle cells are long—up to or more than 40 mm in length—and narrow—less than 0.1 mm in width. These threadlike cells have alternating light and dark cross-markings called *striations.* Each cell has many nuclei (multinucleate). A nerve cell can stimulate protein filaments in the muscle cell to slide past one another, which contracts the cell. The muscle cell relaxes when stimulation stops. Skeletal muscles move the head, trunk, and limbs and enable us to make facial expressions, write, talk, and sing, as well as chew, swallow, and breathe.

Smooth Muscle Tissue

Smooth muscle tissue (fig. 5.29) is called smooth because its cells lack striations. Smooth muscle cells are shorter than those of skeletal muscle and are spindle-shaped, each with a single, centrally located nucleus. This tissue comprises the walls of hollow internal organs, such as the stomach, intestines, urinary bladder, uterus, and blood vessels. Unlike skeletal muscle, smooth muscle usually cannot be stimulated to contract by conscious effort. Thus, its actions are *involuntary.* Smooth muscle tissue moves food through the digestive tract, constricts blood vessels, and empties the urinary bladder.

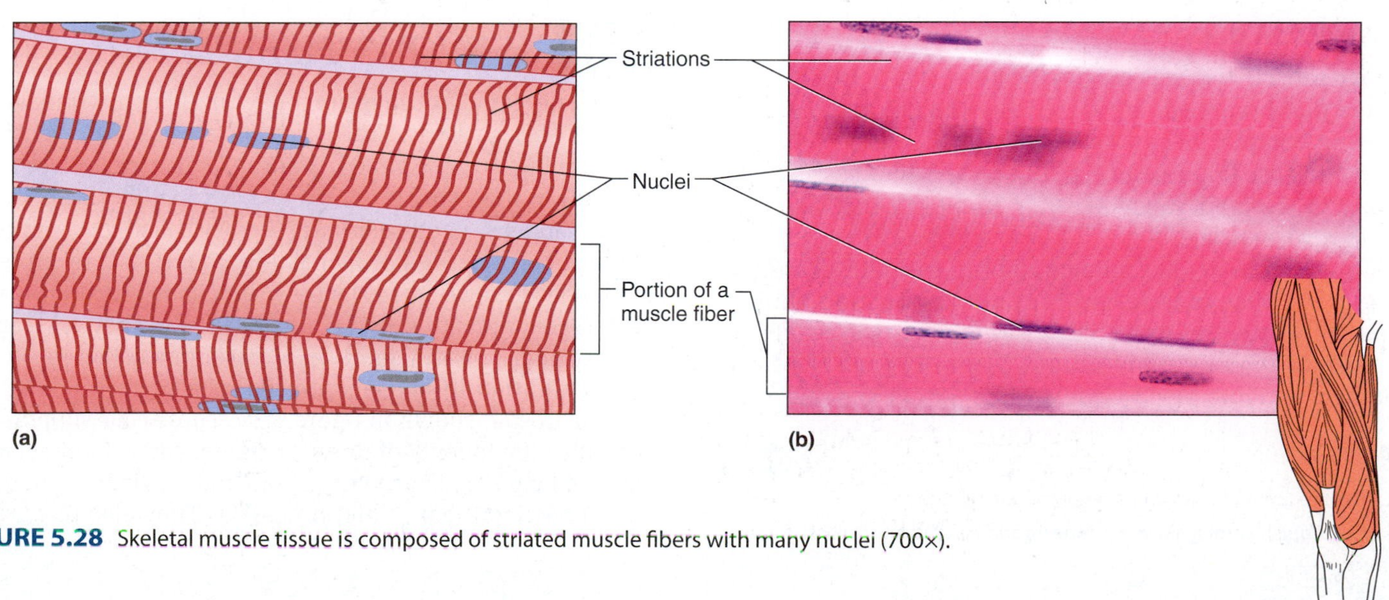

FIGURE 5.28 Skeletal muscle tissue is composed of striated muscle fibers with many nuclei (700×).

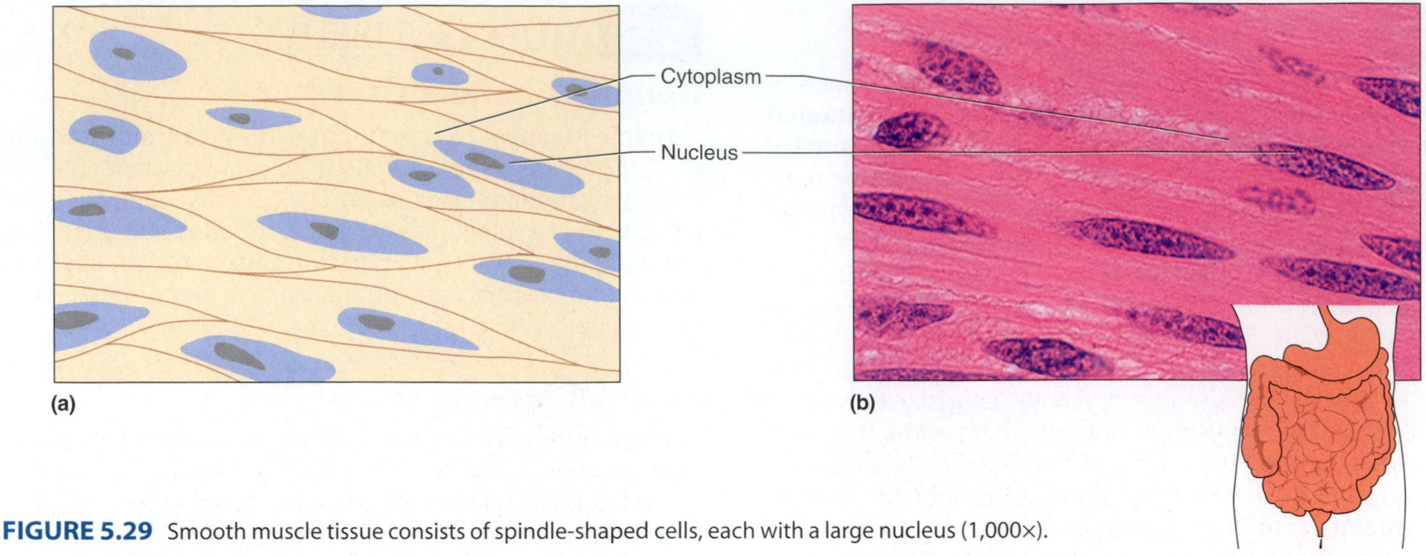

FIGURE 5.29 Smooth muscle tissue consists of spindle-shaped cells, each with a large nucleus (1,000×).

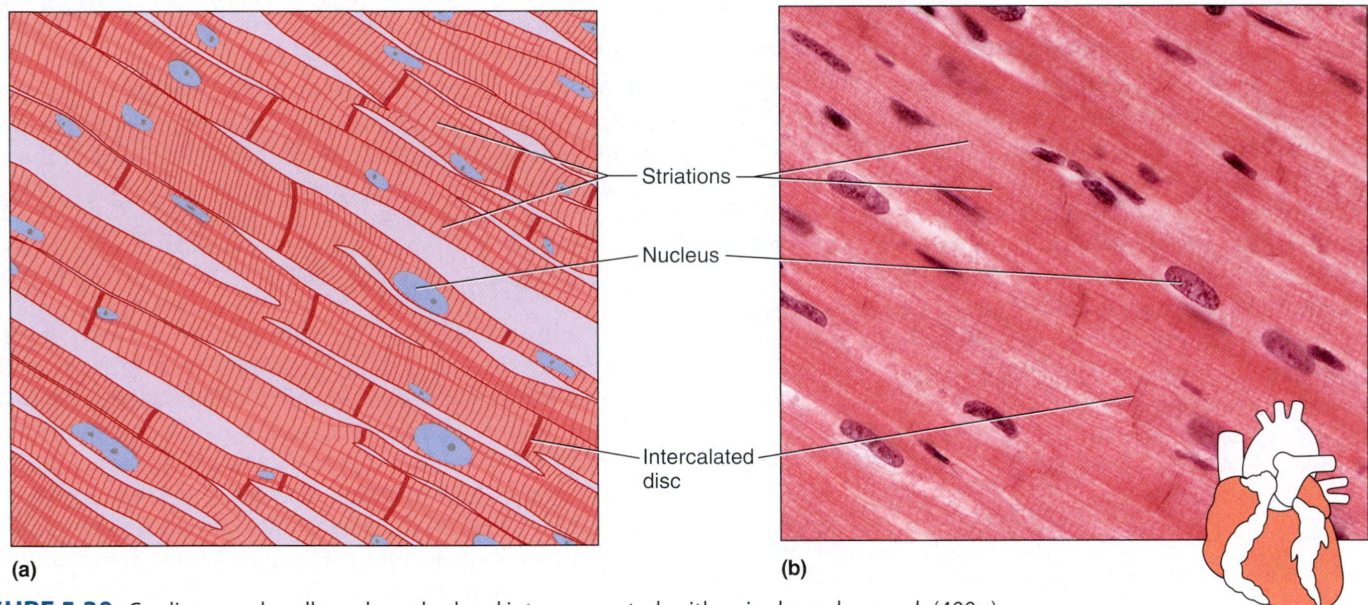

FIGURE 5.30 Cardiac muscle cells are branched and interconnected, with a single nucleus each (400×).

Cardiac Muscle Tissue

Cardiac muscle tissue is only in the heart (fig. 5.30). Its cells, striated and branched, are joined end-to-end, and interconnected in complex networks. Each cardiac muscle cell has a single nucleus. Where one cell touches another cell is a specialized intercellular junction called an *intercalated disc,* seen only in cardiac tissue.

Cardiac muscle, like smooth muscle, is controlled involuntarily. Cardiac muscle can continue to function without being stimulated by nerve impulses. This tissue makes up the bulk of the heart and pumps blood through the heart chambers and into blood vessels.

PRACTICE

22 List the general characteristics of muscle tissue.

23 Distinguish among skeletal, smooth, and cardiac muscle tissues.

5.6 NERVOUS TISSUES

Nervous (ner′vus) **tissues** are found in the brain, spinal cord, and peripheral nerves. The basic cells are called *neurons,* and they are highly specialized. Neurons sense certain types of changes in their surroundings and respond by transmitting nerve impulses along cellular processes called *axons* to other neurons or to muscles or glands (fig. 5.31). As a result of the extremely complex patterns by which neurons connect with each other and with muscle and gland cells, they can coordinate, regulate, and integrate many body functions.

In addition to neurons, nervous tissue includes abundant *neuroglia,* shown in figure 5.31. These cells support and bind the components of nervous tissue, carry on phagocytosis, and help supply growth factors and nutrients to neurons by connecting them to blood vessels. They also play a role in cell-to-cell communications. Chapter 10 discusses nervous tissue.

Table 5.8 summarizes the general characteristics of muscle and nervous tissues. From Science to Technology 5.2 discusses tissue engineering, part of a field called regenerative medicine.

PRACTICE 

24 Describe the general characteristics of nervous tissue.

25 Distinguish between neurons and neuroglia.

The cells of different tissues vary greatly in their abilities to divide. Cells that divide continuously include the epithelial cells of the skin, the inner lining of the digestive tract, and the connective tissue progenitor cells that form blood cells in red bone marrow. However, skeletal and cardiac muscle cells and nerve cells do not usually divide at all after differentiating.

Fibroblasts respond rapidly to injuries by increasing in number and fiber production. They are often the principal agents of repair in tissues that have limited abilities to regenerate. For instance, fibroblasts form scar tissue after a heart attack occurs.

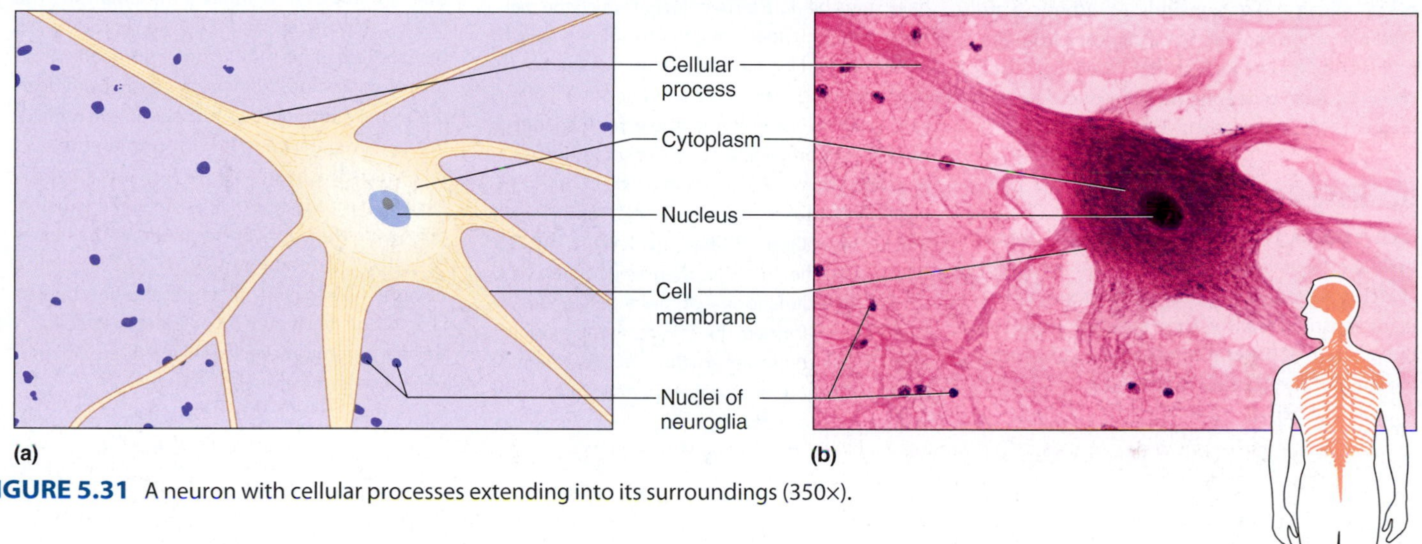

FIGURE 5.31 A neuron with cellular processes extending into its surroundings (350×).

TABLE 5.8 | Muscle and Nervous Tissues

Type	Description	Function	Location
Skeletal muscle tissue	Long, threadlike cells, striated, many nuclei	Voluntary movements of skeletal parts	Muscles usually attached to bones
Smooth muscle tissue	Shorter cells, single, central nucleus	Involuntary movements of internal organs	Walls of hollow internal organs
Cardiac muscle tissue	Branched cells, striated, single nucleus	Heart movements	Heart muscle
Nervous tissue	Cell with cytoplasmic extensions	Sensory reception and conduction of nerve impulses	Brain, spinal cord, and peripheral nerves

CHAPTER SUMMARY

5.1 INTRODUCTION (PAGE 144)

1. Cells are organized in layers or groups to form tissues.
2. Specialized intercellular junctions (tight junctions, desmosomes, and gap junctions) connect cells.
3. The study of tissues is called histology.
4. The four major types of human tissue are epithelial, connective, muscle, and nervous.

5.2 EPITHELIAL TISSUES (PAGE 144)

1. General characteristics
   a. Epithelial tissue covers all free body surfaces, forms the inner lining of body cavities, lines hollow organs, and is the major tissue of glands.
   b. A basement membrane anchors epithelium to connective tissue. Epithelial tissue lacks blood vessels, has cells that are tightly packed, and is continuously replaced.
   c. It functions in protection, secretion, absorption, and excretion.
2. Simple squamous epithelium
   a. This tissue consists of a single layer of thin, flattened cells through which substances pass easily.
   b. It functions in the exchange of gases in the lungs and lines blood vessels, lymph vessels, and membranes within the thorax and abdomen.
3. Simple cuboidal epithelium
   a. This tissue consists of a single layer of cube-shaped cells.

5.2 FROM SCIENCE TO TECHNOLOGY

Tissue Engineering: Replacement Bladders and Hearts

If an appliance part is damaged or fails, replacing it is simple. Not so for the human body. Donor organs and tissues for transplant are in short supply, so in the future spare parts may come from tissue engineering. In this technology, a patient's cells, extracellular matrix, and other biochemicals are grown with a synthetic scaffold to form an implant. The cells come from the patient, so the immune system does not reject them. Tissue engineering has provided skin, cartilage, and blood vessels.

Building a Better Bladder

Each year in the United States, about 10,000 people need their urinary bladders repaired or replaced. Typically a urologic surgeon replaces part of the bladder with part of the large intestine. However, the function of the intestine is to absorb, and the function of the bladder is to hold waste. Tissue engineering is providing a better replacement bladder. The natural organ is balloonlike, with smooth muscle on the outside and lining tissue (urothelium) and connective tissue on the inside.

Researchers created replacements for part of the bladder of seven children and teens who have spina bifida, a birth defect in which the malfunctioning bladder can harm the kidneys. Each patient donated a postage-stamp size sample of bladder tissue that consisted of about a million cells. From the samples, the researchers separated two types of progenitor cells—for smooth muscle and urothelium—and let them divide in culture in a specific "cocktail" of growth factors. Within seven weeks the million cells had divided to yield 1.5 billion cells. The cells were then seeded onto synthetic, three-dimensional domes. After confluent layers of cells formed, the domes were surgically attached to the lower portions of the patients' bladders, after removing the upper portions. The scaffolds degenerated over time, leaving new bladders built from the patients' own cells.

A Healed Heart

A heart is a considerably more complex organ than a bladder, essentially a muscular sac. The heart's architecture is difficult to reproduce, so researchers used a different approach called "decellularization." They took hearts from dead rats, removed the cells, and seeded the remaining extracellular matrix with progenitor cells taken from the hearts of newborn rats. Over the ensuing days, the cells divided and differentiated, occupying the nooks and crannies of the hearts' "skeletons" to rebuild the organ. On the eighth day, the hearts beat! To replace failing human hearts, one day it may be possible to decellularize hearts from cadavers and seed them with progenitor cells from patients. ■

   b. It carries on secretion and absorption in the kidneys and various glands.
4. Simple columnar epithelium
   a. This tissue is composed of elongated cells whose nuclei are near the basement membrane.
   b. It lines the uterus and digestive tract, where it functions in protection, secretion, and absorption.
   c. Absorbing cells often possess microvilli.
   d. This tissue usually contains goblet cells that secrete mucus.
5. Pseudostratified columnar epithelium
   a. This tissue appears stratified because the nuclei are at two or more levels.
   b. Its cells may have cilia that move mucus over the surface of the tissue.
   c. It lines tubes of the respiratory system.
6. Stratified squamous epithelium
   a. This tissue is composed of many layers of cells; the top layers are flattened.
   b. It protects underlying cells from harmful environmental effects.
   c. It is the outer layer of the skin and lines the oral cavity, esophagus, vagina, and anal canal.
7. Stratified cuboidal epithelium
   a. This tissue is composed of two or three layers of cube-shaped cells.
   b. It lines the larger ducts of the mammary glands, sweat glands, salivary glands, and pancreas.
   c. It functions in protection.
8. Stratified columnar epithelium
   a. The top layer of cells in this tissue contains elongated columns. Cube-shaped cells make up the bottom layers.
   b. It is in part of the male urethra and ductus deferens, and parts of the pharynx.
   c. This tissue functions in protection and secretion.
9. Transitional epithelium
   a. This tissue is specialized to become distended.
   b. It lines the urinary bladder, ureters, and superior urethra.
   c. It helps prevent the contents of the urinary passageways from diffusing out.
10. Glandular epithelium
    a. Glandular epithelium is composed of cells specialized to secrete substances.
    b. A gland consists of one or more cells.
       (1) Exocrine glands secrete into ducts.
       (2) Endocrine glands secrete into tissue fluid or blood.
    c. Exocrine glands are classified according to the organization of their cells.
       (1) Simple glands have ducts that do not branch before reaching the secretory portion.
       (2) Compound glands have ducts that branch repeatedly before the secretory portion.
       (3) Tubular glands consist of simple epithelium-lined tubes.
       (4) Alveolar glands consist of saclike dilations connected to the surface by narrowed ducts.

d. Exocrine glands are classified according to the composition of their secretions.
   (1) Merocrine glands secrete watery fluids without loss of cytoplasm. Most secretory cells are merocrine.
      (a) Serous cells secrete watery fluid with a high enzyme content.
      (b) Mucous cells secrete mucus.
   (2) Apocrine glands lose portions of their cells during secretion.
   (3) Holocrine glands release cells filled with secretions.

5.3 CONNECTIVE TISSUES (PAGE 152)

1. General characteristics
   a. Connective tissue connects, supports, protects, provides frameworks, fills spaces, stores fat, produces blood cells, protects against infection, and helps repair damaged tissues.
   b. Connective tissue cells usually have considerable extracellular matrix between them.
   c. This extracellular matrix consists of fibers, a ground substance, and fluid.
2. Major cell types
   a. Fibroblasts produce collagenous and elastic fibers.
   b. Macrophages are phagocytes.
   c. Mast cells release heparin and histamine.
3. Connective tissue fibers
   a. Collagenous fibers are composed of collagen and have great tensile strength.
   b. Elastic fibers are composed of elastin and are elastic.
   c. Reticular fibers are fine collagenous fibers.
4. Categories of connective tissues
   a. Connective tissue proper includes loose connective tissue (areolar, adipose, reticular) and dense connective tissue (dense regular, dense irregular, elastic).
   b. Specialized connective tissues include cartilage, bone, and blood.
5. Areolar tissue
   a. Areolar tissue forms thin membranes between organs and binds them.
   b. It is beneath the skin and between muscles.
6. Adipose tissue
   a. Adipose tissue is a specialized form of connective tissue that stores fat, cushions, and insulates.
   b. It is found beneath the skin; in certain abdominal membranes; and around the kidneys, heart, and various joints.
7. Reticular tissue
   a. Reticular connective tissue largely consists of thin, branched collagenous fibers.
   b. It supports the walls of the liver, spleen, and lymphatic organs.
8. Dense regular connective tissue

   Dense regular connective tissue is largely composed of strong, collagenous fibers that bind structures as parts of tendons and ligaments.
9. Dense irregular connective tissue

   Dense irregular connective tissue has thicker, randomly distributed collagenous fibers and is found in the dermis.
10. Elastic connective tissue

    Elastic connective tissue is mainly composed of elastic fibers and imparts an elastic quality to the walls of certain hollow internal organs such as the lungs and blood vessels.
11. Cartilage
    a. Cartilage provides a supportive framework for various structures.
    b. Its extracellular matrix is composed of fibers and a gel-like ground substance.
    c. It lacks a direct blood supply and is slow to heal.
    d. Most cartilaginous structures are enclosed in a perichondrium, which contains blood vessels.
    e. Major types are hyaline cartilage, elastic cartilage, and fibrocartilage.
    f. Cartilage is at the ends of various bones; in the ear; in the larynx; and in the pads between the bones of the spinal column, pelvic girdle, and knees.
12. Bone
    a. The extracellular matrix of bone contains mineral salts and collagen.
    b. Its cells usually form concentric circles around central canals. Canaliculi connect the cells.
    c. It is an active tissue that heals rapidly.
13. Blood
    a. Blood is composed of cells suspended in fluid.
    b. Blood cells are formed by special tissue in the hollow parts of certain bones.

5.4 TYPES OF MEMBRANES (PAGE 162)

1. Epithelial membranes
   a. Serous membranes
      (1) Serous membranes line body cavities that do not open to the outside.
      (2) They are composed of epithelium and loose connective tissue.
      (3) Cells of serous membranes secrete watery serous fluid that lubricates membrane surfaces.
   b. Mucous membranes
      (1) Mucous membranes line cavities and tubes opening to the outside of the body.
      (2) They are composed of epithelium and loose connective tissue.
      (3) Cells of mucous membranes secrete mucus.
   c. The cutaneous membrane is the external body covering commonly called skin.
2. Synovial membranes are composed of connective tissue only, and line joints.

5.5 MUSCLE TISSUES (PAGE 163)

1. General characteristics
   a. Muscle tissue contracts, moving structures attached to it.
   b. Three types are skeletal, smooth, and cardiac muscle tissues.
2. Skeletal muscle tissue
   a. Muscles containing this tissue usually attach to bones and are controlled by conscious effort.
   b. Muscle cells are long and threadlike, containing several nuclei, with alternating light and dark cross-markings (striations).

c. Muscle cells contract when stimulated by nerve impulses, then immediately relax when they are no longer stimulated.
3. Smooth muscle tissue
   a. This tissue of spindle-shaped cells, each with one nucleus, is in the walls of hollow internal organs.
   b. Usually it is involuntarily controlled.
4. Cardiac muscle tissue
   a. This tissue is found only in the heart.
   b. Striated cells, each with a single nucleus, are joined by intercalated discs and form branched networks.
   c. Cardiac muscle tissue is involuntarily controlled.

5.6 NERVOUS TISSUES (PAGE 164)

1. Nervous tissue is in the brain, spinal cord, and peripheral nerves.
2. Neurons
   a. Neurons sense changes and respond by transmitting nerve impulses to other neurons or to muscles or glands.
   b. They coordinate, regulate, and integrate body activities.
3. Neuroglia
   a. Some of these cells bind and support nervous tissue.
   b. Others carry on phagocytosis.
   c. Still others connect neurons to blood vessels.
   d. Some are involved in cell-to-cell communication.

CHAPTER ASSESSMENTS

5.1 Introduction

1 Define *tissue*. (p. 144)

2 Describe three types of intercellular junctions. (p. 144)

3 Which of the following is a major tissue type in the body? (p. 144)
- a. epithelial
- b. nervous
- c. muscle
- d. connective
- e. all of the above.

5.2 Epithelial Tissues

4 A general characteristic of epithelial tissues is that _______________. (p. 145)
- a. numerous blood vessels are present
- b. cells are spaced apart
- c. cells divide rapidly
- d. there is much extracellular matrix between cells

5 Distinguish between simple epithelium and stratified epithelium. (p. 145)

6 Explain how the structure of simple squamous epithelium provides its function. (p. 145)

7 Match the epithelial tissue on the left to an organ in which the tissue is found. (p. 146–152)

(1) simple squamous epithelium	A. lining of intestines
(2) simple cuboidal epithelium	B. lining of ducts of mammary glands
(3) simple columnar epithelium	C. lining of urinary bladder
(4) pseudostratified columnar epithelium	D. salivary glands
(5) stratified squamous epithelium	E. air sacs of lungs
(6) stratified cuboidal epithelium	F. respiratory passages
(7) stratified columnar epithelium	G. ductus deferens
(8) transitional epithelium	H. lining of kidney tubules
(9) glandular epithelium	I. outer layer of skin

8 Distinguish between an exocrine gland and an endocrine gland. (p. 150)

9 Describe how glands are classified according to the structure of their ducts and the organization of their cells. (p. 150)

10 A gland that secretes substances by exocytosis is a(n) _______________ gland. (p. 150)
- a. merocrine
- b. apocrine
- c. holocrine

5.3 Connective Tissues

11 Discuss the general characteristics of connective tissue. (p. 153)

12 Define *extracellular matrix* and *ground substance*. (p. 153)

13 Describe three major types of connective tissue cells. (p. 153)

14 _______________ are thick fibers that have great tensile strength and are flexible, but only slightly elastic fibers. (p. 155)
- a. Reticular
- b. Elastic
- c. Collagenous

15 Explain the difference between loose connective tissue and dense connective tissue. (p. 156)

16 Explain how the amount of adipose tissue in the body reflects diet. (p. 156)

17 Contrast dense regular and dense irregular connective tissues. (p. 158)

18 Explain why injured dense regular connective tissue and cartilage are usually slow to heal. (p. 158)

19 Distinguish between reticular and elastic connective tissues. (p. 158)

20 Name the major types of cartilage, and describe their differences and similarities. (p. 158)

21 Describe how bone cells are organized in bone tissue. (p. 160)

22 Explain how bone cells receive nutrients. (p. 160)

23 The fluid extracellular matrix of blood is called ________. (p. 161)

a. white blood cells
b. red blood cells
c. platelets
d. plasma
e. bone marrow

5.4 Types of Membranes

24 Describe the structure of epithelial membranes in contrast to synovial membranes. (p. 162)

25 Identify locations in the body of four types of membranes. (p. 163)

5.5 Muscle Tissues

26 Describe the general characteristics of muscle tissues. (p. 163)

27 Compare and contrast skeletal, smooth, and cardiac muscle tissues in terms of location, cell appearance, and control. (p. 163)

5.6 Nervous Tissues

28 Describe the general characteristics of nervous tissue. (p. 164)

29 Distinguish between the functions of neurons and neuroglia. (p. 164)

INTEGRATIVE ASSESSMENTS/CRITICAL THINKING

OUTCOMES 3.2, 3.6, 5.1, 5.2, 5.3, 5.5, 5.6

1. Tissue engineering combines living cells with synthetic materials to create functional substitutes for human tissues. What components would you use to engineer replacement (a) skin, (b) bone, (c) muscle, and (d) blood?

OUTCOMES 3.2, 5.2

2. In the lungs of smokers, a process called metaplasia occurs where normal lining cells of the lung are replaced by squamous metaplastic cells (many layers of squamous epithelial cells). Functionally, why is this an undesirable body reaction to tobacco smoke?

OUTCOMES 3.4, 3.5, 5.2, 5.3, 5.5, 5.6

3. Cancer-causing agents (carcinogens) usually act on dividing cells. Which of the four tissues would carcinogens most influence? Least influence?

OUTCOMES 5.2, 5.4

4. Sometimes, in response to irritants, mucous cells secrete excess mucus. What symptoms might this produce if it occurred in the (a) digestive tract or (b) respiratory passageway?

OUTCOME 5.3

5. Disorders of collagen are characterized by deterioration of connective tissues. Why would you expect such diseases to produce widely varying symptoms?

OUTCOME 5.3

6. Collagen and elastin are added to many beauty products. What type of tissues are they normally part of?

OUTCOME 5.3

7. Joints such as the shoulder, elbow, and knee contain considerable amounts of cartilage and dense regular connective tissue. How does this explain that joint injuries are often slow to heal?

WEB CONNECTIONS

Be sure to visit the text website at **www.mhhe.com/shier12** for answers to chapter assessments, additional quizzes, and interactive learning exercises.

ANATOMY & PHYSIOLOGY REVEALED

Anatomy & Physiology Revealed® (APR) includes cadaver photos that allow you to peel away layers of the human body to reveal structures beneath the surface. This program also includes animations, radiologic imaging, audio pronunciations, and practice quizzing. Check out **www.aprevealed.com**. APR has been proven to help improve student grades!

Falsely colored scanning electron micrograph of skin from the palm with sweat pores resembling mini craters (70×).

UNDERSTANDING WORDS

alb-, white: *alb*inism—condition characterized by a lack of pigment in skin, hair, and eyes.

cut-, skin: sub*cut*aneous—beneath the skin.

derm-, skin: *derm*is—inner layer of the skin.

epi-, upon, after, in addition: *epi*dermis—outer layer of the skin.

follic-, small bag: hair *follic*le—tubelike depression in which a hair develops.

hol-, entire, whole: *hol*ocrine gland—gland that discharges the entire cell containing the secretion.

kerat-, horn: *kerat*in—protein produced as epidermal cells die and harden.

melan-, black: *melan*in—dark pigment produced by certain cells.

por-, passage, channel: *por*e—opening by which a sweat gland communicates to the skin's surface.

seb-, grease: *seb*aceous gland—gland that secretes an oily substance.

LEARN PRACTICE ASSESS

LEARNING OUTCOMES

After you have studied this chapter, you should be able to:

6.1 Introduction

1 Define organ, and name the large organ of the integumentary system. (p. 171)

6.2 Skin and Its Tissues

2 List the general functions of the skin. (p. 171)
3 Describe the structure of the layers of the skin. (p. 173)
4 Summarize the factors that determine skin color. (p. 174)

6.3 Accessory Structures of the Skin

5 Describe the accessory structures associated with the skin. (p. 177)
6 Explain the functions of each accessory structure of the skin. (p. 177)

6.4 Regulation of Body Temperature

7 Explain how the skin helps regulate body temperature. (p. 182)

6.5 Healing of Wounds and Burns

8 Describe wound healing. (p. 183)
9 Distinguish among the types of burns, including a description of healing with each type. (p. 184)

6.6 Life-Span Changes

10 Summarize life-span changes in the integumentary system. (p. 186)

THE SECRET OF SKIN'S STRENGTH AND FLEXIBILITY

The skin is an amazing organ. It provides strength and flexibility, is waterproof, and covers our bodies in one smooth sheath. But investigating exactly what lies behind these properties has been hampered by limited technology—until now.

Preparing skin cells for visualization using a light microscope strips away important proteins. Preparing cells for imaging with an electron microscope requires harsh chemical treatment or coating with metal. As a result, these standard forms of microscopy do not provide three-dimensional close-ups of the junctions between cells as they are in the body. A technique called cryo-electron tomography images skin cells in their natural state. Skin samples taken from a healthy man's arm were flash-frozen and then probed from various angles with a special electron microscope. Zeroing in on the desmosomes (see figure 5.1), and especially the cadherin proteins that link the cells, revealed the secret of skin's strength and flexibility.

Cadherins emanate straight out from the cell membranes of squamous epithelium, yet maintain the ability to move about 20 degrees in any direction. This flexibility is essential for movements associated with growth and development that move and stretch the skin. Cadherin proteins also have a right-left orientation. They link to each other with alternating symmetry, a little like children holding hands, but alternating the direction in which each child faces. The strength of the skin comes from the fact that each cadherin protein binds not only to its neighboring cells in one plane, but to juxtaposed cells too—similar to sheets of stamps glued together. An individual cadherin is not very adhesive, but there is strength in numbers. When cadherins are aligned at the surface of a skin cell facing others on all sides, the combined integrity is formidable.

Although researchers have observed only static views of the skin using cryo-electron tomography, they hypothesize how this organ might develop.

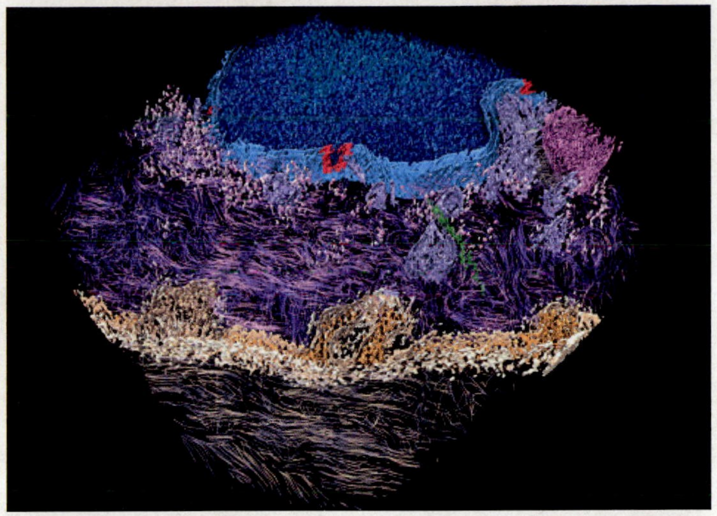

Cryo-electron tomography provides three-dimensional reconstructions of the junction between two skin cells (40,000×). The cells touch at the tan area. The nucleus is blue, set off by the light blue nuclear envelope. Nuclear pores are red. The purple structures are mitochondria, the green are microtubules, and the steel blue in the center is endoplasmic reticulum.

In a developing embryo, skin cells approach as they divide and join initially at sites where a few cadherin proteins bind. As time passes and incoming signals indicate that the skin is where it should be, more cadherins join at the sites of the original ones until the cells are strongly, but flexibly, attached. ■

6.1 INTRODUCTION

Two or more types of tissues grouped together and performing specialized functions constitute an **organ.** The skin, the largest organ in the body by weight, and its various accessory structures make up the **integumentary** (in-teg-u-men′tar-e) **system.** Skin is a strong yet flexible covering of our bodies.

6.2 SKIN AND ITS TISSUES

The skin is composed of several types of tissues (fig. 6.1). It is one of the more versatile organs of the body and is vital in maintaining homeostasis. A protective covering, the skin prevents many harmful substances, as well as microorganisms, from entering the body. Skin also retards water loss by diffusion from deeper tissues and helps regulate body temperature. It houses sensory receptors; synthesizes various chemicals, including vitamin D; contains immune system cells; and excretes small quantities of waste.

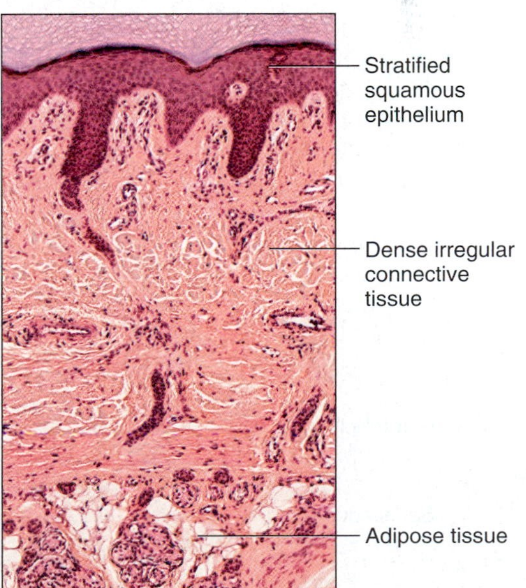

FIGURE 6.1 An organ, such as the skin, is composed of several types of tissues (30×).

Skin cells help produce vitamin D, necessary for normal bone and tooth development. This vitamin is ingested or it forms from a substance (dehydrocholesterol) synthesized by cells in the digestive system. When dehydrocholesterol (provitamin D) reaches the skin by means of the blood and is exposed to ultraviolet light from the sun, it is converted to vitamin D.

The skin, or cutaneous membrane, includes two distinct layers: epithelial tissue overlying connective tissue. The outer layer, called the **epidermis** (ep″i-der′mis), is composed of stratified squamous epithelium. The inner layer, or **dermis** (der′mis), is thicker than the epidermis and is made up of connective tissue containing collagen and elastic fibers, smooth muscle tissue, nervous tissue, and blood. A *basement membrane* anchored to the dermis by short fibrils separates the two skin layers.

Beneath the dermis, masses of areolar and adipose tissues bind the skin to underlying organs. These tissues are not part of the skin. They form the **subcutaneous layer** (sub″ku-ta′ne-us la′er), or hypodermis (fig. 6.2). The collagenous and elastic fibers of this layer are continuous with those of the dermis. Most of these fibers run parallel to the surface of the skin, extending in all directions. As a result, no sharp boundary separates the dermis and the subcutaneous layer.

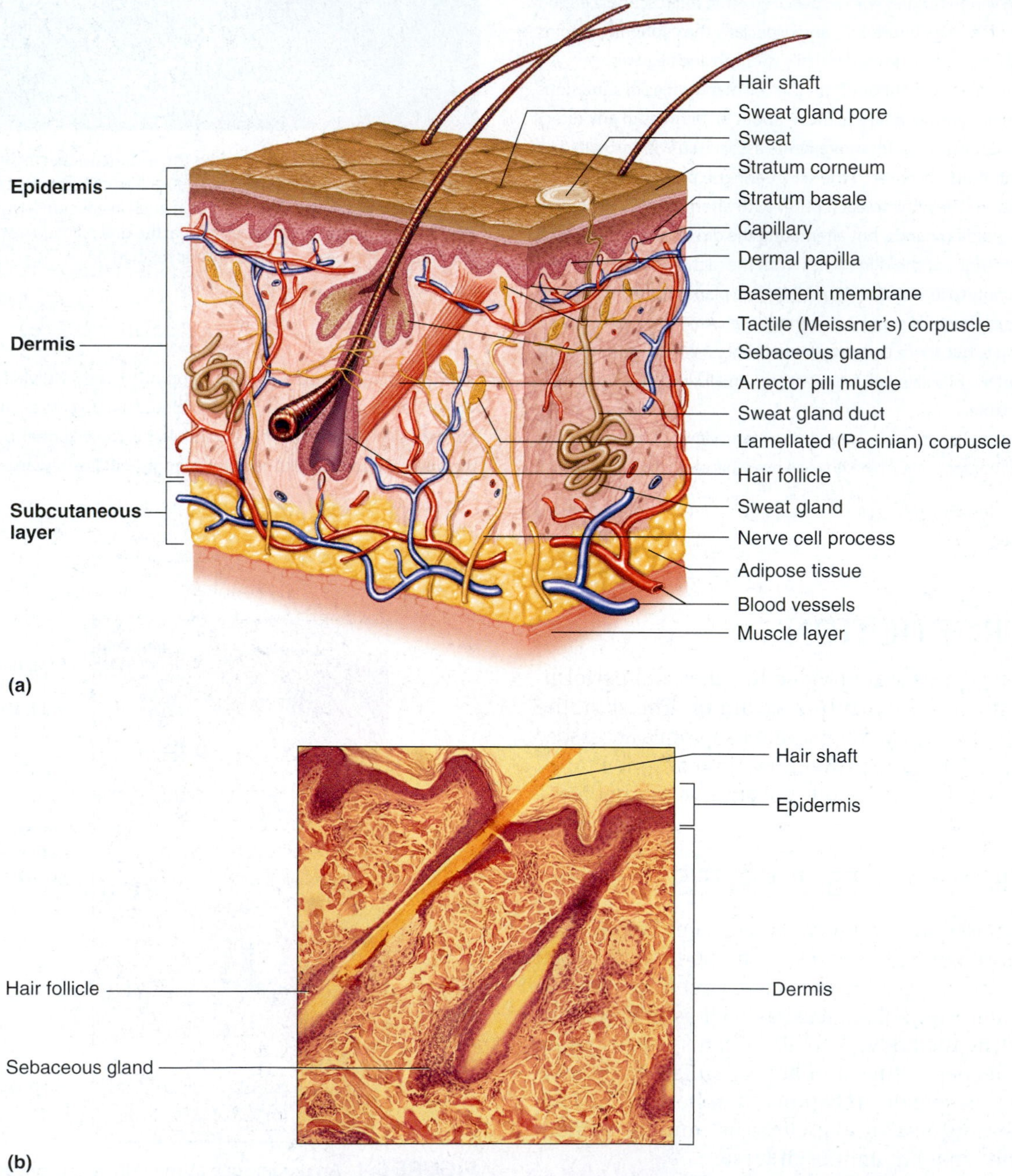

FIGURE 6.2 Skin. (*a*) A section of skin and the subcutaneous layer. (*b*) A light micrograph depicting the layered structure of the skin (75×).

The adipose tissue of the subcutaneous layer insulates, helping to conserve body heat and impeding the entrance of heat from the outside. The subcutaneous layer also contains the major blood vessels that supply the skin. Branches of these vessels form a network (rete cutaneum) between the dermis and the subcutaneous layer. They, in turn, give off smaller vessels that supply the dermis above and the underlying adipose tissue.

Drugs that cannot be taken orally are delivered through the skin in several ways: into the skin (intradermal injections), beneath the skin (subcutaneous injections), and into muscles (intramuscular injections). (Subcutaneous and intramuscular injections are also called hypodermic injections.) Another, less painful route is with an adhesive transdermal patch. The drug is in a small reservoir in the patch. It leaves through a permeable membrane at a known rate, diffuses through the epidermis, and enters blood vessels in the dermis. Drugs that alleviate chest pain, prevent motion sickness, lower blood pressure, and help people stop smoking are delivered with transdermal patches. A new type of transdermal patch may extend the approach to drugs that can be taken by mouth because they can use lower doses with fewer adverse effects. This technique uses "microneedles" to painlessly punch tiny holes in the stratum corneum.

PRACTICE 

1. List the general functions of the skin.
2. Name the tissue in the outer layer of the skin.
3. Name the tissues in the inner layer of the skin.
4. Name the tissues in the subcutaneous layer beneath the skin.
5. What are the functions of the subcutaneous layer?

Epidermis

The epidermis is composed entirely of stratified squamous epithelium, and therefore it lacks blood vessels. However, the deepest layer of epidermal cells, called the *stratum basale,* is close to the dermis and is nourished by dermal blood vessels, which enables the cells to divide and grow. As new cells enlarge, they push the older epidermal cells away from the dermis toward the surface of the skin. The farther the cells are moved, the poorer their nutrient supply becomes, and in time, they die.

The cell membranes of older skin cells (keratinocytes) thicken and develop many desmosomes that fasten them to each other (see chapter 5, p. 144 and the opening vignette to this chapter, p. 171). At the same time, the cells begin to harden, in a process called **keratinization** (ker″ah-tin″-za′shun). Strands of tough, fibrous, waterproof keratin proteins are synthesized and stored in the cell. As a result, many layers of tough, tightly packed dead cells accumulate in the epidermis, forming an outermost layer called the *stratum corneum.* These dead cells are eventually shed. Rubbing the skin briskly with a towel sheds dead cells.

The epidermis receives its nutrients from blood vessels in the dermis, so interference with blood flow may kill epidermal cells. For example, when a person lies in one position for a prolonged period, the weight of the body pressing against the bed blocks the skin's blood supply. If cells die, the tissues begin to break down (necrosis), and a pressure ulcer (also called a decubitus ulcer or bedsore) may appear.

Pressure ulcers usually form in the skin overlying bony projections, such as on the hip, heel, elbow, or shoulder. Frequently changing body position or massaging the skin to stimulate blood flow in regions associated with bony prominences can prevent pressure ulcers. For a paralyzed person who cannot feel pressure or respond to it by shifting position, caregivers must turn the body often to prevent pressure ulcers. Beds, wheelchairs, and other specialized equipment can periodically shift the patient, lowering the risk of developing pressure ulcers.

The structural organization of the epidermis varies from region to region. It is thickest on the palms of the hands and the soles of the feet, where it may be 0.8–1.4 mm thick. In most areas, only four layers are distinguishable. They are the *stratum basale* (stratum germinativum, or basal cell layer), the deepest layer; the *stratum spinosum;* the *stratum granulosum;* and the *stratum corneum,* a fully keratinized outermost layer. An additional layer, the *stratum lucidum* (between the stratum granulosum and the stratum corneum) is in the thickened skin of the palms and soles. The cells of these layers change shape as they are pushed toward the surface (fig. 6.3).

In body regions other than the palms and soles, the epidermis is usually thin, averaging 0.07–0.12 mm. The stratum lucidum may be missing where the epidermis is thin. Table 6.1 describes the characteristics of each layer of the epidermis.

In healthy skin, production of epidermal cells closely balances loss of dead cells from the stratum corneum. As a result, skin does not completely wear away. The rate of cell division increases where the skin is rubbed or pressed regularly, causing the growth of thickened areas called *calluses* on the palms and soles and keratinized conical masses on the toes called *corns.*

In psoriasis, a chronic skin disease, cells in the epidermis divide seven times more frequently than normal. Excess cells accumulate, forming bright red patches covered with silvery scales, which are keratinized cells. Medications used to treat cancer, such as methotrexate, are used to treat severe cases of psoriasis. Immune suppressing medications, such as topical corticosteroids, are used for treatment of chronic psoriasis. Five million people in the United States and 2% of people worldwide have psoriasis.

The epidermis has important protective functions. It shields the moist underlying tissues against excess water loss, mechanical injury, and the effects of harmful chemicals. When intact, the epidermis also keeps out disease-causing microorganisms (pathogens).

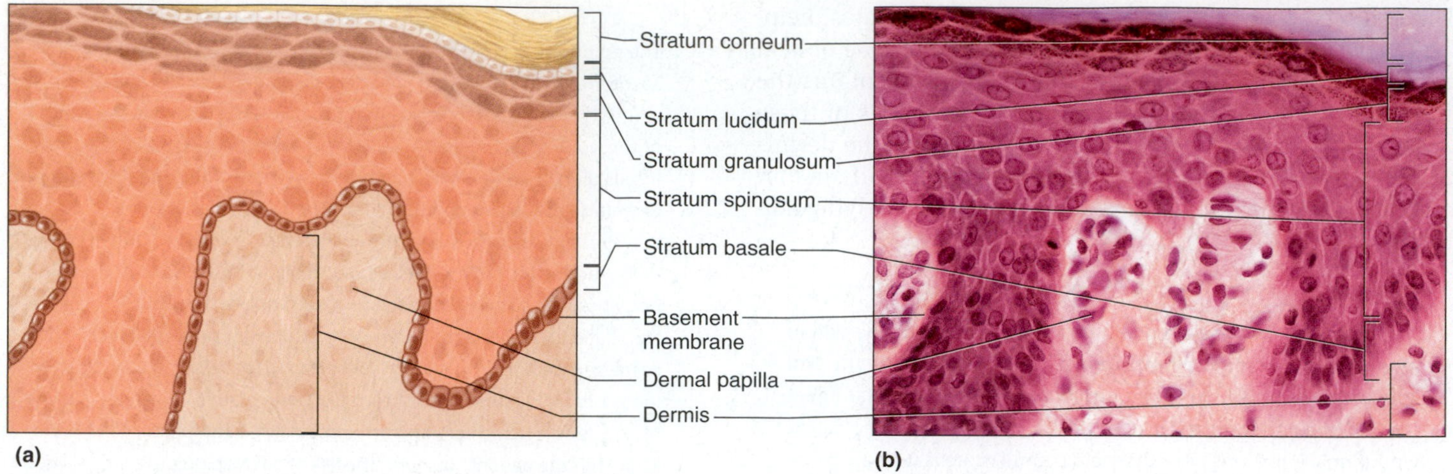

FIGURE 6.3 Epidermis of thick skin. (*a*) The layers of the epidermis are distinguished by changes in cells as they are pushed toward the surface of the skin. (*b*) Light micrograph of skin (120×).

TABLE 6.1 | Layers of the Epidermis

Layer	Location	Characteristics
Stratum corneum	Outermost layer	Many layers of keratinized, dead epithelial cells that are flattened and nonnucleated
Stratum lucidum	Between stratum corneum and stratum granulosum on soles and palms	Cells appear clear; nuclei, organelles, and cell membranes are no longer visible
Stratum granulosum	Beneath the stratum corneum	Three to five layers of flattened granular cells that contain shrunken fibers of keratin and shriveled nuclei
Stratum spinosum	Beneath the stratum granulosum	Many layers of cells with centrally located, large, oval nuclei and developing fibers of keratin; cells becoming flattened
Stratum basale (basal cell layer)	Deepest layer	A single row of cuboidal or columnar cells that divide and grow; this layer also includes melanocytes

PRACTICE 

6 Explain how the epidermis is formed.

7 Distinguish between the stratum basale and the stratum corneum.

8 List the protective functions of the epidermis.

Specialized cells in the epidermis called **melanocytes** produce the dark pigment **melanin** (mel′ah-nin) from the amino acid tyrosine in organelles called melanosomes. Melanin provides skin color (fig. 6.4*a*). Melanin also absorbs ultraviolet radiation in sunlight, which would otherwise cause mutations in the DNA of skin cells and other damaging effects. Clinical Application 6.1 discusses one consequence of excess sun exposure—skin cancer.

Melanocytes lie in the stratum basale of the epidermis. They are the only cells that can produce melanin, but the pigment gets into nearby epidermal cells. This happens because melanocytes have long, pigment-containing cellular extensions that pass upward between neighboring epidermal cells, and the extensions can transfer granules of melanin in melanosomes into keratinocytes, which may accumulate more melanin than melanocytes (fig. 6.4*b*). Certain keratinocytes, called pigment recipient cells, attract melanocytes and stimulate them to release melanin, causing pigment deposition. The pigment recipient cells, recently discovered, are thought to act like the outlines in a children's coloring book, delineating areas to be filled in with color.

Humans come in a wide variety of hues. Heredity and the environment determine skin color. Regardless of racial origin, all people have about the same number of melanocytes in their skin. Differences in skin color result from differences in the amount of melanin these cells produce. This is controlled by several genes. The more melanin, the darker the skin. The distribution and the size of pigment granules within melanocytes also influence skin color. The granules in very dark skin are single and large; those in lighter skin occur in clusters of two to four granules and are smaller. People who inherit mutant melanin genes have nonpigmented skin. This white skin is part of *albinism*. It affects people of all races and also many other species (fig. 6.5).

Environmental factors such as sunlight, ultraviolet light from sunlamps, and X rays affect skin color. These factors

6.1 Clinical Application

Tanning and Skin Cancer

Like cigarette smoking, a deep, dark tan was once desirable. In the 1960s, a teenager might have spent hours on a beach, skin glistening with oil, maybe even using a reflecting device to concentrate sun exposure on the face. Today, as they lather on sunblock, many of these people realize that the tans of yesterday may cause cancer tomorrow.

Tanning

Usually the DNA damage response, discussed in Chapter 4 (p. 137), protects against sun exposure. The solar radiation activates a gene that encodes a protein called p53 that normally mediates harmful effects of environmental insults in various tissues. In the skin, p53 stimulates a series of familiar responses to sunning: keratinocytes produce signaling molecules that promote the redness (erythema) and swelling of inflammation. Meanwhile, melanocytes further differentiate and increase their production of melanin, which melanosomes transfer to keratinocytes. The result is tanning.

Researchers hypothesize that the tanning response evolved about a million years ago, as our ancestors ventured from the forests onto the plains of Africa. Biology may also explain why we like to sunbathe—it stimulates keratinocytes to release beta endorphin, a molecule related to opiates that promotes a sense of well-being. However, like anything else, sun exposure should be done in moderation. Use of tanning booths is particularly dangerous because it bathes the skin in doses of ultraviolet radiation that can overwhelm the natural protection against cancer.

Skin Cancer

Cancer begins when the sun exposure overwhelms the ability of p53 to protect the skin. Usually, skin cancer arises in nonpigmented epithelial cells in the deep layer of the epidermis or from pigmented melanocytes. Skin cancers originating from epithelial cells are called *cutaneous carcinomas* (basal cell carcinoma or squamous cell carcinoma); those arising from melanocytes are *cutaneous melanomas* (melanocarcinomas or malignant melanomas) (fig. 6A).

Cutaneous carcinomas are the most common type of skin cancer, affecting mostly light-skinned people over forty years of age regularly exposed to sunlight. Such a cancer usually develops from a hard, dry, scaly growth with a reddish base. The lesion may be flat or raised and usually firmly adheres to the skin, appearing most often on the neck, face, or scalp. Fortunately, cutaneous carcinomas are typically slow growing and can usually be cured completely by surgical removal or radiation treatment.

A cutaneous melanoma is pigmented with melanin, often with a variety of colored areas—variegated brown, black, gray, or blue. A melanoma usually has irregular rather than smooth outlines and may feel bumpy. Melanoma accounts for only 4% of skin cancers but for 80% of skin cancer deaths.

People of any age may develop a cutaneous melanoma. These cancers seem to be caused by short, intermittent exposure to high-intensity sunlight. Thus, risk of melanoma increases in persons who stay indoors but occasionally sustain blistering sunburns.

Light-skinned people who burn rather than tan are at higher risk of developing a cutaneous melanoma. The cancer usually appears in the skin of the trunk, especially the back, or the limbs, arising from normal-appearing skin or from a mole (nevus). The lesion spreads horizontally through the skin, but eventually may thicken and grow downward into the skin, invading deeper tissues. Surgical removal during the horizontal growth phase can arrest the cancer. But once the lesion thickens and spreads into deeper tissues, it becomes more difficult to treat, and the survival rate is low. To reduce risk, avoid exposure to high-intensity sunlight, use sunscreens and sunblocks, and examine the skin regularly. Report any unusual lesions—particularly those that change in color, shape, or surface texture—to a physician. ■

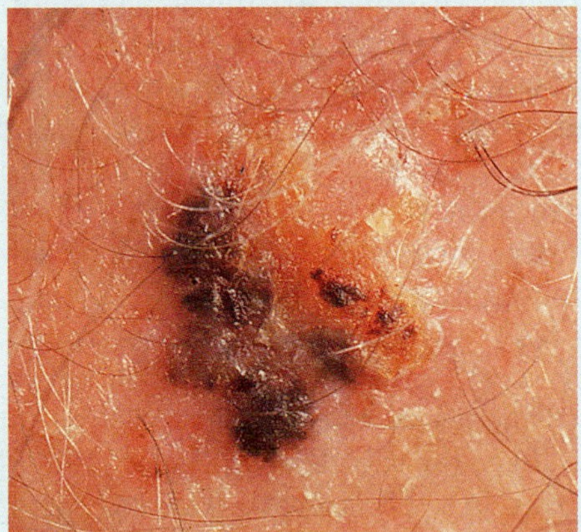

(a)

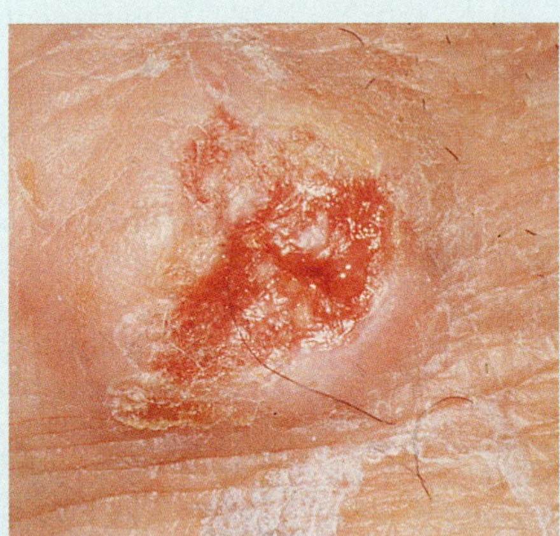

(b)

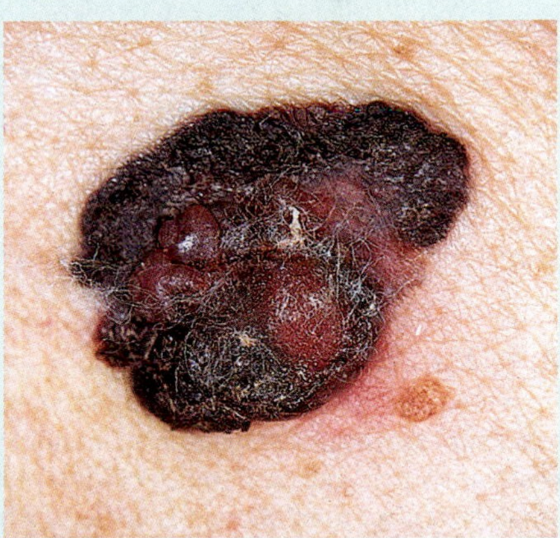

(c)

FIGURE 6A Skin cancer. (*a*) Squamous cell carcinoma. (*b*) Basal cell carcinoma. (*c*) Malignant melanoma.

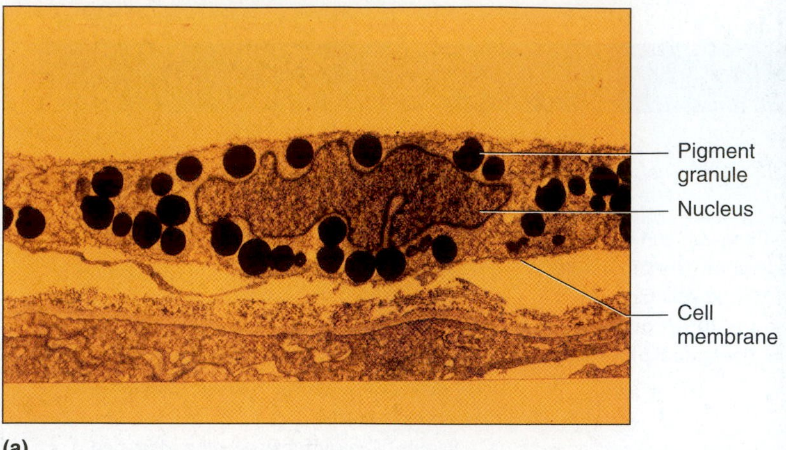

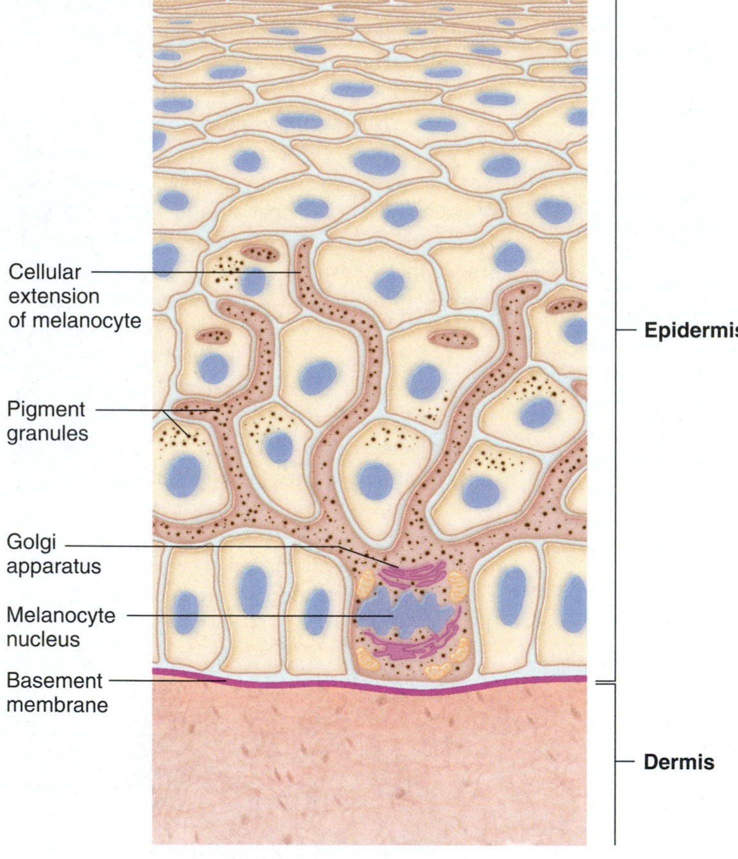

FIGURE 6.4 Melanocyte. (*a*) Transmission electron micrograph of a melanocyte with pigment-containing granules (10,600×). (*b*) A melanocyte may have pigment-containing extensions that pass between epidermal cells and transfer pigment into them. Much of the melanin is deposited above the nucleus, where the pigment can absorb UV radiation from outside before the DNA is damaged.

rapidly darken existing melanin, and they stimulate melanocytes to produce more pigment and transfer it to nearby epidermal cells within a few days. Unless exposure to sunlight continues, the tan fades as pigmented keratinocytes wear away.

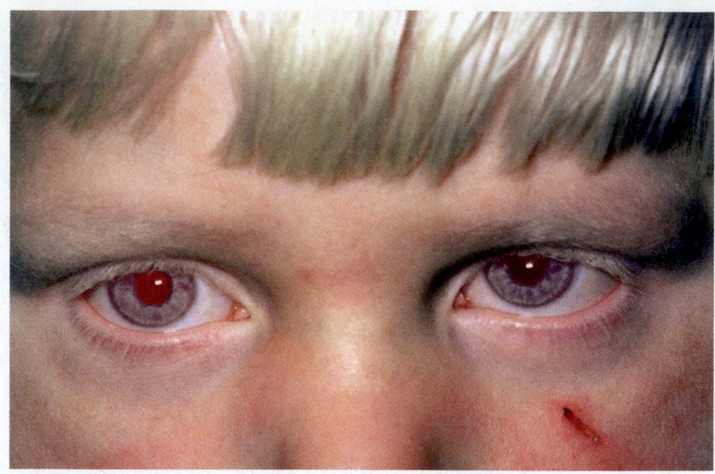

FIGURE 6.5 The pale or red eyes, skin, and hair of a person with albinism reflect lack of melanin. Albinism is inherited.

Worldwide, 1 in 110,000 people has albinism. Among the native Hopi people in Arizona, however, the incidence is 1 in 200. The reason for this is as much sociological as it is biological. Men with albinism help the women rather than risk severe sunburn in the fields with the other men. These men disproportionately contribute to the next generation because they have more sexual contact with women.

Blood in the dermal vessels adds color to the skin. When blood is well oxygenated, the blood pigment hemoglobin is bright red, making the skins of light-complexioned people appear pinkish. When the blood oxygen concentration is low, hemoglobin is dark red, and the skin appears bluish—a condition called *cyanosis.*

The state of the blood vessels also affects skin color. If the vessels are dilated, more blood enters the dermis, reddening the skin of a light-complexioned person. This may happen when a person is overheated, embarrassed, or under the influence of alcohol. Conversely, conditions that constrict blood vessels cause the skin to lose this reddish color. Thus, if body temperature drops abnormally or if a person is frightened, the skin may appear pale.

A yellow-orange plant pigment called *carotene,* found in yellow vegetables, can give skin a yellowish cast if a person consumes too much. This results from accumulation of carotene in the adipose tissue of the subcutaneous layer. Illnesses may also affect skin color. A yellowish skin tone can indicate *jaundice,* a consequence of liver malfunction.

PRACTICE

9 What is the function of melanin?
10 How do genetic factors influence skin color?
11 Which environmental factors influence skin color?
12 How do physiological factors influence skin color?

Some newborns develop the yellowish skin of jaundice shortly after birth. A blood incompatibility or an immature liver can cause jaundice. An observant British hospital nurse discovered a treatment for newborn jaundice in 1958. She liked to take her tiny charges out in the sun, and she noticed that a child whose skin had a yellow pallor developed normal pigmentation when he lay in sunlight. However, the part of the child's body covered by a diaper and therefore not exposed to the sun remained yellow. Further investigation showed that sunlight enables the body to break down bilirubin, the liver substance that accumulates in the skin. Today, newborns who develop persistently yellowish skin may have to lie under artificial "bili lights" for a few days, clad only in protective goggles.

Dermis

The boundary between the epidermis and dermis is usually uneven. This is because the epidermis has ridges projecting inward and the dermis has conical *dermal papillae* passing into the spaces between the ridges (see figs. 6.2 and 6.3). Dermal papillae increase the surface area where epidermal cells receive oxygen and nutrients from dermal capillaries.

Fingerprints form from these undulations of the skin at the distal end of the palmar surface of a finger. The undulations increase friction at the fingertips for grasping. Fingerprints may be used for purposes of identification because they are individually unique. The pattern of a fingerprint is genetically determined, and the prints form during fetal existence. However, during a certain time early in development, fetal movements can change the print pattern. No two fetuses move exactly alike, so even the fingerprints of identical twins are not exactly the same.

The dermis binds the epidermis to the underlying tissues. It is largely composed of dense irregular connective tissue that includes tough collagenous fibers and elastic fibers in a gel-like ground substance. Networks of these fibers give the skin toughness and elasticity. On the average, the dermis is 1.0–2.0 mm thick; however, it may be as thin as 0.5 mm or less on the eyelids or as thick as 3.0 mm on the soles of the feet.

The dermis also contains muscle fibers. Some regions, such as the skin that encloses the testes (scrotum), contain many smooth muscle cells that can wrinkle the skin when they contract. Other smooth muscles in the dermis are associated with accessory organs such as hair follicles and glands. Many skeletal muscle fibers are anchored to the dermis in the skin of the face. They help produce the voluntary movements associated with facial expressions.

Nerve cell processes are scattered throughout the dermis. Motor processes carry impulses to dermal muscles and glands, and sensory processes carry impulses away from specialized sensory receptors (see fig. 6.2).

One type of dermal sensory receptor, lamellated (Pacinian) corpuscles, is stimulated by heavy pressure, whereas another type, tactile (Meissner's) corpuscles, senses light touch. Still other receptors (free nerve endings) respond to temperature changes or to factors that can damage tissues and extend into the epidermis. Sensory receptors are discussed in chapter 12 (p. 440). The dermis also contains accessory structures including blood vessels, hair follicles, sebaceous glands, and sweat glands.

To create a tattoo, very fine needles inject inks into the dermis. The color is permanent, because dermis cells are not shed, as are cells of the epidermis. To remove a tattoo, a laser shatters the ink molecules, and the immune system removes the resulting debris. Before laser removal became available in the late 1980s, unwanted tattoos were scraped, frozen, or cut away—all painful procedures.

PRACTICE

13 What types of tissues make up the dermis?

14 What are the functions of these tissues?

6.3 ACCESSORY STRUCTURES OF THE SKIN

Accessory structures of the skin originate from the epidermis and include nails, hair follicles, and skin glands. As long as accessory structures remain intact, severely burned or injured dermis can regenerate.

Nails

Nails are protective coverings on the ends of the fingers and toes. Each nail consists of a *nail plate* that overlies a surface of skin called the *nail bed.* Specialized epithelial cells continuous with the epithelium of the skin produce the nail bed. The whitish, thickened, half-moon-shaped region (lunula) at the base of a nail plate is the most active growing region. The epithelial cells here divide, and the newly formed cells become keratinized. This gives rise to tiny, keratinized scales that become part of the nail plate, pushing it forward over the nail bed. In time, the plate extends beyond the end of the nail bed and with normal use gradually wears away (fig. 6.6).

Nail appearance mirrors health. Bluish nail beds may reflect a circulation problem. A white nail bed or oval depressions in a nail can indicate anemia. A pigmented spot under a nail that isn't caused by an injury may be a melanoma. Horizontal furrows may result from a period of serious illness or indicate malnutrition. Certain disorders of the lungs, heart, or liver may cause extreme curvature of the nails. Red streaks in noninjured nails may be traced to rheumatoid arthritis, ulcers, or hypertension.

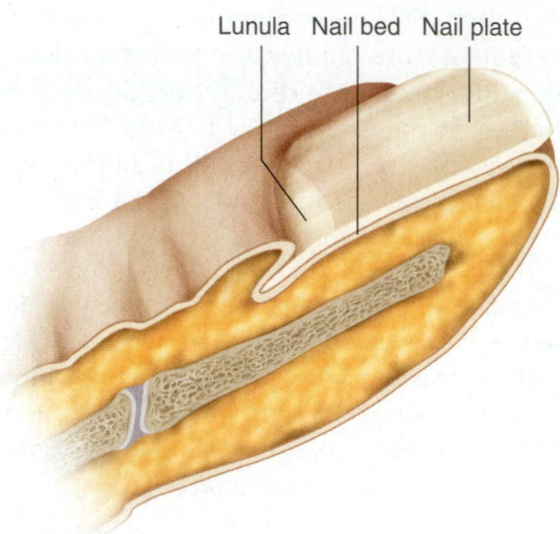

FIGURE 6.6 Nails grow from epithelial cells that divide and become as keratinized as the rest of the nail.

Hair Follicles

A healthy person loses from twenty to 100 hairs a day as part of the normal growth cycle of hair. A hair typically grows for two to six years, rests for two to three months, then falls out. A new hair grows in its place. At any time, 90% of hair is in the growth phase.

Hair is present on all skin surfaces except the palms, soles, lips, nipples, and parts of the external reproductive organs; however, it is not always well developed. For example, hair on the forehead is usually very fine.

Each hair develops from a group of epidermal cells at the base of a tubelike depression called a **hair follicle** (hār fol′ĭ-kl). This follicle extends from the surface into the dermis and contains the hair *root,* the portion of hair embedded in the skin. The epidermal cells at its base are nourished from dermal blood vessels in a projection of connective tissue (hair papilla) at the deep end of the follicle. As these epidermal cells divide and grow, older cells are pushed toward the surface. The cells that move upward and away from the nutrient supply become keratinized and die. Their remains constitute the structure of a developing *hair shaft* that extends away from the skin surface. In other words, a hair is composed of dead epidermal cells (figs. 6.7 and 6.8). Both hair and epidermal cells develop from the same types of stem cells.

Usually a hair grows for a time and then rests while it remains anchored in its follicle. Later, a new hair begins to grow from the base of the follicle, and the old hair is pushed outward and drops off. Sometimes, however, the hairs are not replaced. When this occurs in the scalp, the result is baldness, described in Clinical Application 6.2.

Genes determine hair color by directing the type and amount of pigment that epidermal melanocytes produce. Dark hair has more of the brownish-black **eumelanin,** while blonde hair and red hair have more of the reddish-yellow **pheomelanin.** The white hair of a person with *albinism* lacks melanin altogether. A mixture of pigmented hairs and unpigmented hairs usually appears gray.

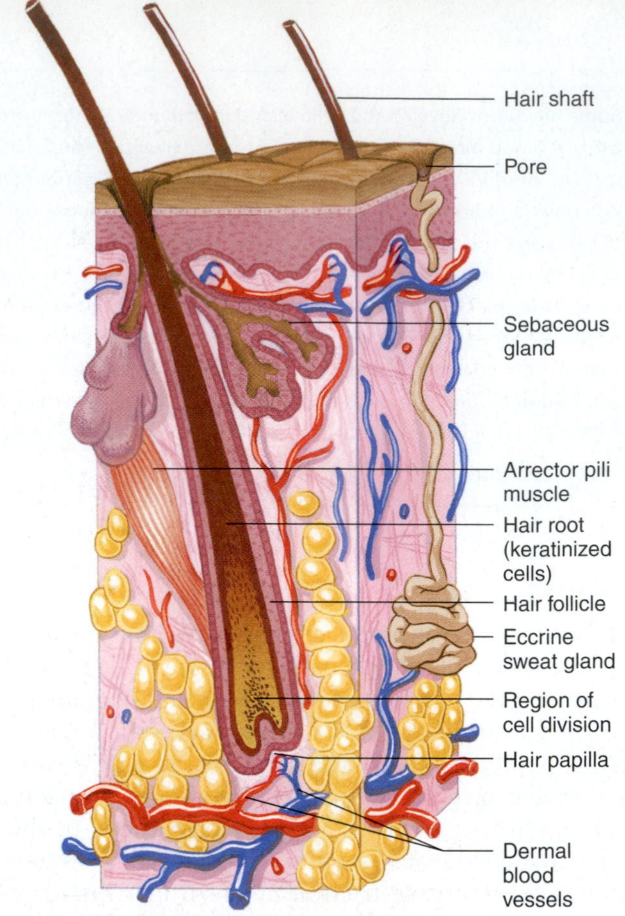

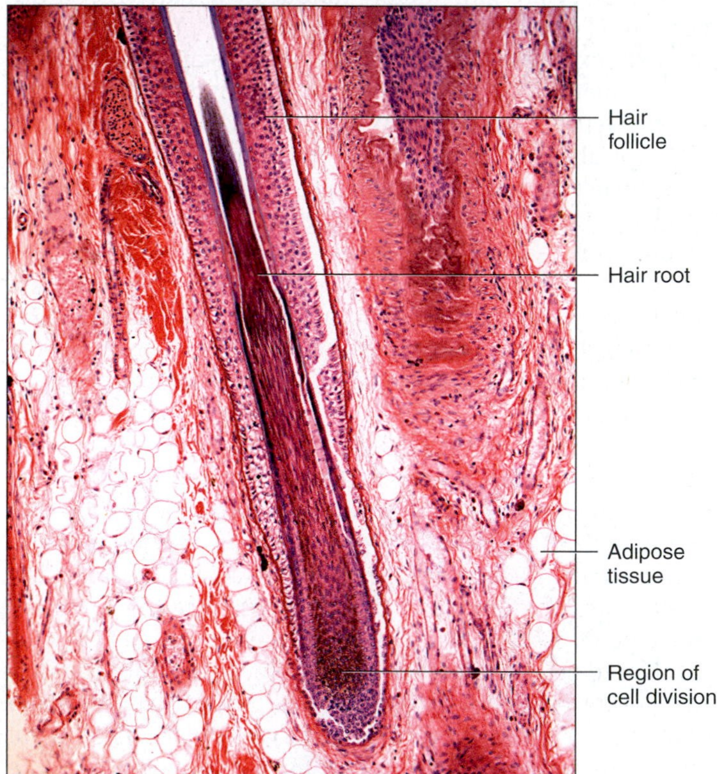

FIGURE 6.7 Hair follicle. (*a*) A hair grows from the base of a hair follicle when epidermal cells divide and older cells move outward and become keratinized. (*b*) Light micrograph of a hair follicle (175×).

6.2 CLINICAL APPLICATION

Hair Loss

About 1.4 billion people worldwide are bald. The most common type of baldness in adults is pattern baldness, in which the top of the head loses hair. Pattern baldness affects 35 million men and 20 million women in the United States and is also common elsewhere. The women tend to be past menopause, when lowered amounts of the hormone estrogen contribute to hair loss, which occurs more evenly on the scalp than it does in men. Pattern baldness is called *androgenic alopecia* because it is associated with testosterone, an androgenic (male) hormone. Variations in the androgen receptor gene, which determines the activity of androgens in hair follicles, may lie behind susceptibility to pattern baldness. Abnormal hormone levels that mimic menopause may cause hair loss in young women.

Another type of baldness is *alopecia areata,* in which the body manufactures antibodies that attack the hair follicles. This results in oval bald spots in mild cases but complete loss of scalp and body hair in severe cases. About 2.5 million people in the United States have alopecia areata.

Temporary hair loss has several causes. Lowered estrogen levels shortly before and after giving birth may cause a woman's hair to fall out in clumps. Taking birth control pills, cough medications, certain antibiotics, vitamin A derivatives, antidepressants, and many other medications can also cause temporary hair loss. A sustained high fever may prompt hair loss six weeks to three months later.

Many people losing their hair seek treatment (fig. 6B). One treatment is minoxidil (Rogaine), a drug originally used to lower high blood pressure. Rogaine causes new hair to grow in 10% to 14% of cases, and in 90% of people, it slows hair loss. However, when a person stops taking it, any new hair falls out. Hair transplants move hair follicles from a hairy body part to a bald part. They work. Several other approaches, however, can damage the scalp or lead to infection. These include suturing on hair pieces and implants of high-density artificial fibers. Products called "thinning hair supplements" are ordinary conditioners that make hair feel thicker. They are concoctions of herbs and the carbohydrate polysorbate.

A future approach to treating baldness may harness the ability of stem cells to divide and differentiate to give rise to new hair follicles. Stem cells that can produce hair as well as epidermal cells and sebaceous glands lie just above the "bulge" region at the base of a hair follicle. The first clue to the existence of these cells was that new skin in burn patients arises from hair follicles. Then, experiments in mice that mark stem cells and their descendants showed that the cells give rise to hair and skin. Manipulating stem cells could someday treat extreme hairiness (hirsutism) as well as baldness. ■

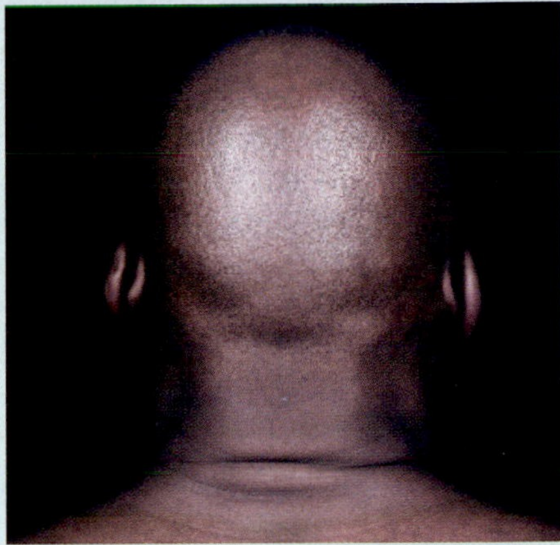

FIGURE 6B Being bald can be beautiful, but many people with hair loss seek ways to grow hair.

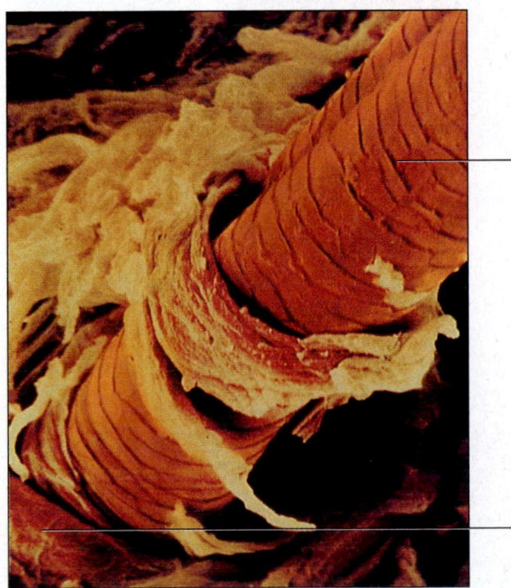

FIGURE 6.8 Scanning electron micrograph of a hair emerging from the epidermis (875×).

A single gene controls the proportions of eumelanin and pheomelanin in hair. Analysis of this gene in cells from arm bones of Neanderthals from about 45,000 years ago indicates that some of them had reddish hair and pale skin, in contrast to the common view of Neanderthals as having dark pigmentation.

A bundle of smooth muscle cells, forming the *arrector pili muscle* (see figs. 6.2*a* and 6.7*a*), attaches to each hair follicle. This muscle is positioned so that a short hair in the follicle stands on end when the muscle contracts. If a person is emotionally upset or very cold, nerve impulses may stimulate the arrector pili muscles to contract, raising gooseflesh, or goose bumps. Each hair follicle also has associated with it one or more sebaceous (oil-producing) glands.

Skin Glands

Sebaceous glands (se-ba′shus glandz) (see fig. 6.2) contain groups of specialized epithelial cells and are usually associated with hair follicles. They are holocrine glands (see chapter 5, p. 150), and their cells produce globules of a fatty material that accumulate, swelling and bursting the cells. The resulting mixture of fatty material and cellular debris is called *sebum.*

Sebum is secreted into hair follicles through short ducts and helps keep the hairs and the skin soft, pliable, and waterproof (fig. 6.9). Acne results from excess sebum secretion (Clinical Application 6.3).

Sebaceous glands are scattered throughout the skin but are not on the palms and soles. In some regions, such as the lips, the corners of the mouth, and parts of the external reproductive organs, sebaceous glands open directly to the surface of the skin rather than being connected to hair follicles.

Sweat (swet) **glands,** or sudoriferous glands, are widespread in the skin. Each gland consists of a tiny tube that originates as a ball-shaped coil in the deeper dermis or superficial subcutaneous layer. The coiled portion of the gland is closed at its deep end and is lined with sweat-secreting epithelial cells. The most numerous sweat glands, called **eccrine** (ek′rin) **glands,** respond throughout life to body temperature elevated by environmental heat or physical exercise (fig. 6.10). These glands are abundant on the forehead, neck, and back, where they produce profuse sweat on hot days or during intense physical activity. They also release the moisture that appears on the palms and soles when a person is emotionally stressed.

The fluid the eccrine sweat glands secrete is carried by a tube (duct) that opens at the surface as a *pore* (fig. 6.11). Sweat is mostly water, but it also contains small amounts of salts and wastes, such as urea and uric acid. Thus, sweating is also an excretory function.

The secretions of certain sweat glands, called **apocrine** (ap′o-krin) **glands,** develop a scent as skin bacteria metabolize them (see fig. 6.10). (Although these glands are currently called apocrine, they secrete by the same mechanism as eccrine glands—see merocrine glands described in chapter 5, p. 150.) Apocrine sweat glands become active at puberty and can wet certain areas of the skin when a person is emotionally upset, frightened, or in pain. Apocrine sweat glands are also active during sexual arousal. In adults,

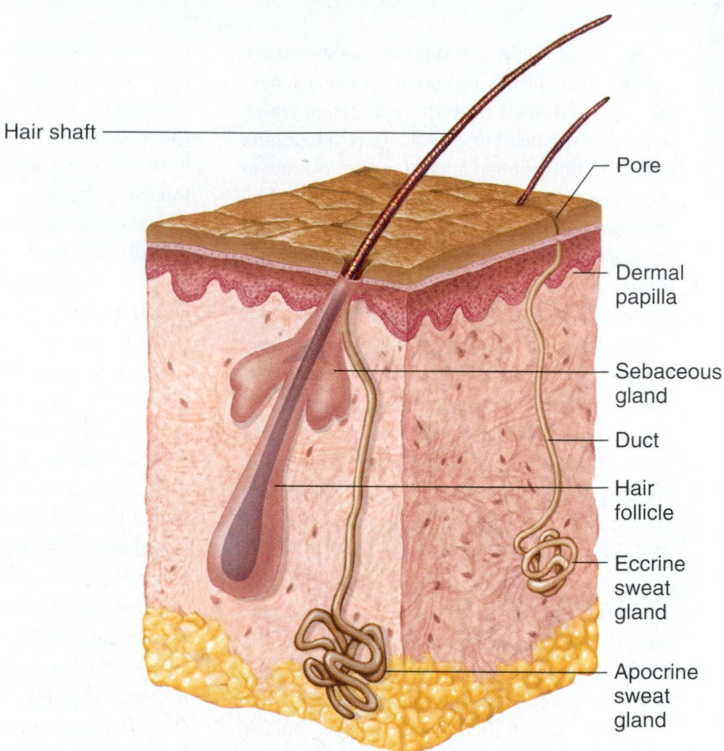

FIGURE 6.10 Note the difference in location of the ducts of the eccrine and apocrine sweat glands.

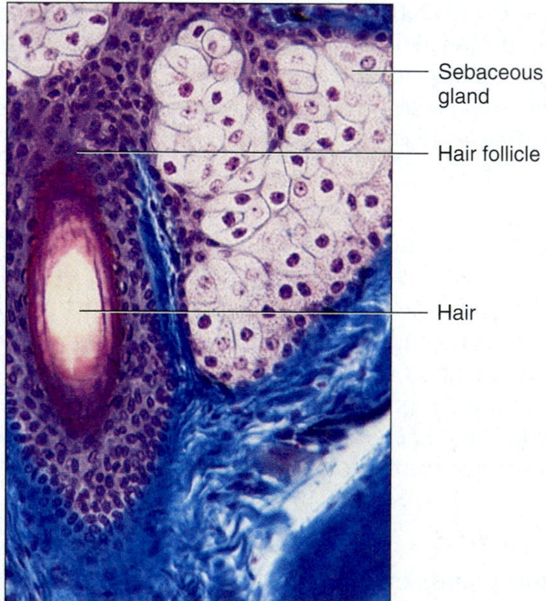

FIGURE 6.9 A sebaceous gland secretes sebum into a hair follicle, shown here in oblique section (300×).

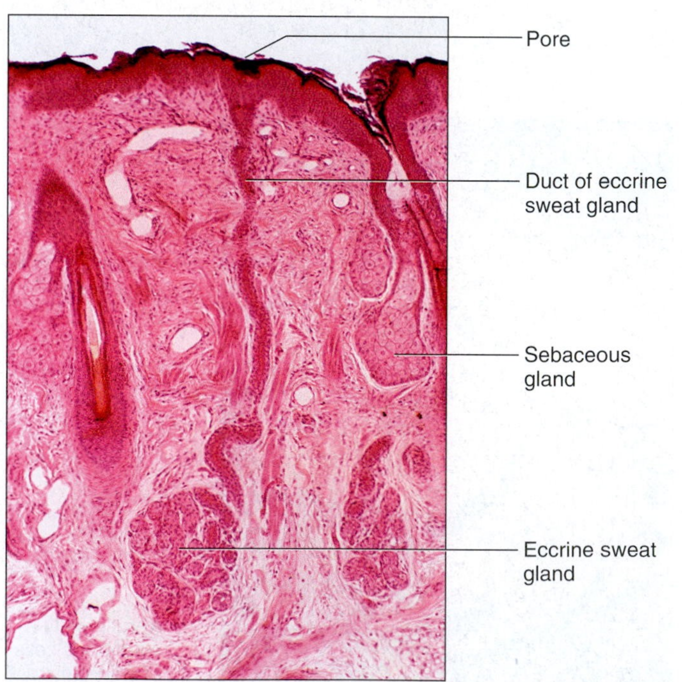

FIGURE 6.11 Light micrograph of the skin showing an eccrine sweat gland with its duct extending to a pore (30×).

6.3 CLINICAL APPLICATION

Acne

Many young people are all too familiar with *acne vulgaris*, a disorder of the sebaceous glands. Excess sebum and squamous epithelial cells clog the glands, producing blackheads and whiteheads (comedones). The blackness is not dirt but results from the accumulated cells blocking light. In addition, the clogged sebaceous gland provides an attractive environment for anaerobic bacteria. Their presence signals the immune system to trigger inflammation. The inflamed, raised area is a pimple (pustule).

A Hormonal Problem

Acne is the most common skin disease, affecting 80% of people at some time between the ages of eleven and thirty. It is usually hormonally induced. Just before puberty, the adrenal glands increase production of androgens, which stimulate increased secretion of sebum. At puberty, sebum production surges again. Acne usually develops because the sebaceous glands are extra responsive to androgens, but in some cases, androgens may be produced in excess.

Acne can cause skin blemishes far more serious than the perfect models in acne medication ads depict (fig. 6C). Scarring from acne can lead to emotional problems. Fortunately, several highly effective treatments are available.

What to Do—And Not Do

Acne is not caused by uncleanliness or eating too much chocolate or greasy food. Although cleansing products containing soaps, detergents, or astringents can remove surface sebum, they do not stop the flow of oil that contributes to acne. Abrasive products are harmful because they irritate the skin and increase inflammation.

Most acne treatments take weeks to months to work. Women with acne are sometimes prescribed certain types of birth control pills because the estrogens counter androgen excess. Isotretinoin is a very effective derivative of vitamin A but has side effects and causes birth defects. Systemic antibiotics can treat acne by clearing bacteria from sebaceous glands. Topical treatments include tretinoin (another vitamin A derivative), salicylic acid (an aspirin solution), and benzoyl peroxide.

Treatment for severe acne requires a doctor's care. Drug combinations are tailored to the severity of the condition (table 6A). ■

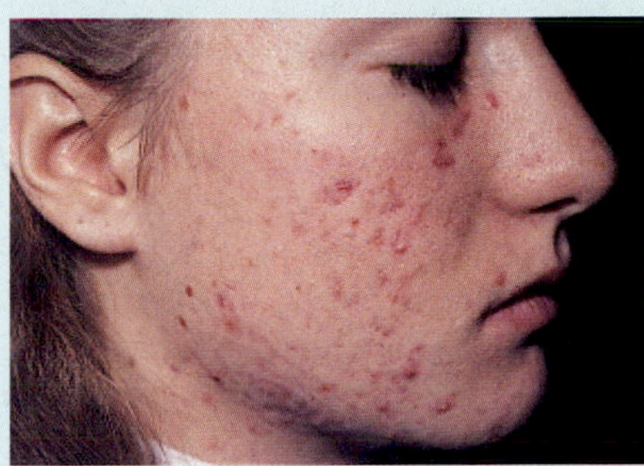

FIGURE 6C Acne is a common skin condition usually associated with a surge of androgen activity—not eating chocolate, as was once believed.

TABLE 6A | Acne Treatments (by Increasing Severity)

Condition	Treatment
Noninflammatory comedonal acne (blackheads and whiteheads)	Topical tretinoin or salicylic acid
Papular inflammatory acne	Topical antibiotic
Widespread blackheads and pustules	Topical tretinoin and systemic antibiotic
Severe cysts	Systemic isotretinoin
Explosive acne (ulcerated lesions, fever, joint pain)	Systemic corticosteroids

the apocrine glands are most numerous in axillary regions, the groin, and the area around the nipples. Ducts of these glands open into hair follicles.

Other sweat glands are structurally and functionally modified to secrete specific fluids, such as the ceruminous glands of the external ear canal that secrete ear wax (see chapter 12, p. 450) and the female mammary glands that secrete milk (see chapter 23, pp. 902–903). Table 6.2 summarizes skin glands.

PRACTICE 

15 How does the composition of a fingernail differ from that of a hair?

16 Explain how a hair forms.

17 What causes gooseflesh?

18 What is the function of the sebaceous glands?

19 Describe the locations of the sweat glands.

20 How do the functions of eccrine sweat glands and apocrine sweat glands differ?

6.4 REGULATION OF BODY TEMPERATURE

The regulation of body temperature is vitally important because even slight shifts can disrupt the rates of metabolic reactions. Normally, the temperature of deeper body parts remains close to a set point of 37°C (98.6°F). The maintenance of a stable temperature requires that the amount of heat the body loses be balanced by the amount it produces. The skin plays a key role in the homeostatic mechanism that regulates body temperature.

RECONNECT

To Chapter 1, Homeostasis, pages 9–10.

Heat Production and Loss

Heat is a product of cellular metabolism; thus, the more active cells of the body are the major heat producers. These

TABLE 6.2 | Skin Glands

Type	Description	Function	Location
Sebaceous glands	Groups of specialized epithelial cells	Keep hair soft, pliable, waterproof	Near or connected to hair follicles, everywhere but on palms and soles
Eccrine sweat glands	Abundant sweat glands with odorless secretion	Lower body temperature	Originate in deep dermis or subcutaneous layer and open to surface on forehead, neck, and back
Apocrine sweat glands	Less numerous sweat glands with secretions that develop odors	Wet skin during pain, fear, emotional upset, and sexual arousal	Near hair follicles in armpit, groin, around nipples
Ceruminous glands	Modified sweat glands	Secrete earwax	External ear canal
Mammary glands	Modified sweat glands	Secrete milk	Breasts

cells include skeletal and cardiac muscle cells and the cells of certain glands, such as the liver.

When body temperature rises above the set point, nerve impulses stimulate structures in the skin and other organs to release heat. For example, during physical exercise, active muscles release heat, which the blood carries away. The warmed blood reaches the part of the brain (the hypothalamus) that controls the body's temperature set point, which signals muscles in the walls of dermal blood vessels to relax. As these vessels dilate (vasodilation), more blood enters them, and some of the heat the blood carries escapes to the outside. At the same time, deeper blood vessels contract (vasoconstriction), diverting blood to the surface, and the skin reddens. The heart is stimulated to beat faster, moving more blood out of the deeper regions.

The primary means of body heat loss is **radiation** (ra-de-a′shun), by which infrared heat rays escape from warmer surfaces to cooler surroundings. These rays radiate in all directions, much like those from the bulb of a heat lamp.

Conduction and convection release less heat. In **conduction** (kon-duk′shun), heat moves from the body directly into the molecules of cooler objects in contact with its surface. For example, heat is lost by conduction into the seat of a chair when a person sits down. The heat loss continues as long as the chair is cooler than the body surface touching it. Heat is also lost by conduction to the air molecules that contact the body. As air becomes heated, it moves away from the body, carrying heat with it, and is replaced by cooler air moving toward the body. This type of continuous circulation of air over a warm surface is **convection** (kon-vek′shun).

Still another means of body heat loss is **evaporation** (e-vap″o-ra′shun). When the body temperature rises above normal, the nervous system stimulates eccrine sweat glands to release sweat onto the surface of the skin. As this fluid evaporates (changes from a liquid to a gas), it carries heat away from the surface, cooling the skin.

When body temperature drops below the set point, the brain triggers different responses in the skin structures. Muscles in the walls of dermal blood vessels are stimulated to contract; this decreases the flow of heat-carrying blood through the skin, which loses color, and helps reduce heat loss by radiation, conduction, and convection. At the same time, sweat glands remain inactive, decreasing heat loss by evaporation. If the body temperature continues to drop, the nervous system may stimulate muscle cells in the skeletal muscles throughout the body to contract slightly. This action requires an increase in the rate of cellular respiration and releases heat as a by-product. If this response does not raise the body temperature to normal, small groups of muscles may rhythmically contract with greater force, causing the person to shiver, generating more heat. Figure 6.12 summarizes the body's temperature-regulating mechanism, and Clinical Application 6.4 examines two causes of elevated body temperature.

Problems in Temperature Regulation

The body's temperature-regulating mechanism does not always operate satisfactorily, and the consequences may be dangerous. For example, air can hold only a limited volume of water vapor, so on a hot, humid day, the air may become nearly saturated with water. At such times, the sweat glands may be activated, but the sweat cannot quickly evaporate. The skin becomes wet, but the person remains hot and uncomfortable. Body temperature may rise, in a condition called hyperthermia. In addition, if the air temperature is high, heat loss by radiation is less effective. If the air temperature exceeds body temperature, the person may gain heat from the surroundings, elevating body temperature even higher.

Hypothermia, or lowered body temperature, can result from prolonged exposure to cold or as part of an illness. It can be extremely dangerous. Hypothermia begins with shivering and a feeling of coldness, but if not treated, progresses to mental confusion; lethargy; loss of reflexes and consciousness; and, eventually, a shutting down of major organs. If the temperature in the body's core drops just a few degrees, fatal respiratory failure or heart arrhythmia may result. However, the extremities can withstand drops of 20°F to 30°F below normal.

Certain people are at higher risk for developing hypothermia due to less adipose tissue in the subcutaneous layer beneath the skin (less insulation). These include the very old, very thin individuals, and the homeless. The very young with undeveloped nervous systems have difficulty regulating their body temperature. Dressing appropriately and staying active in the cold can prevent hypothermia. A person suffering from hypothermia must be warmed gradually so that respiratory and cardiovascular functioning remain stable.

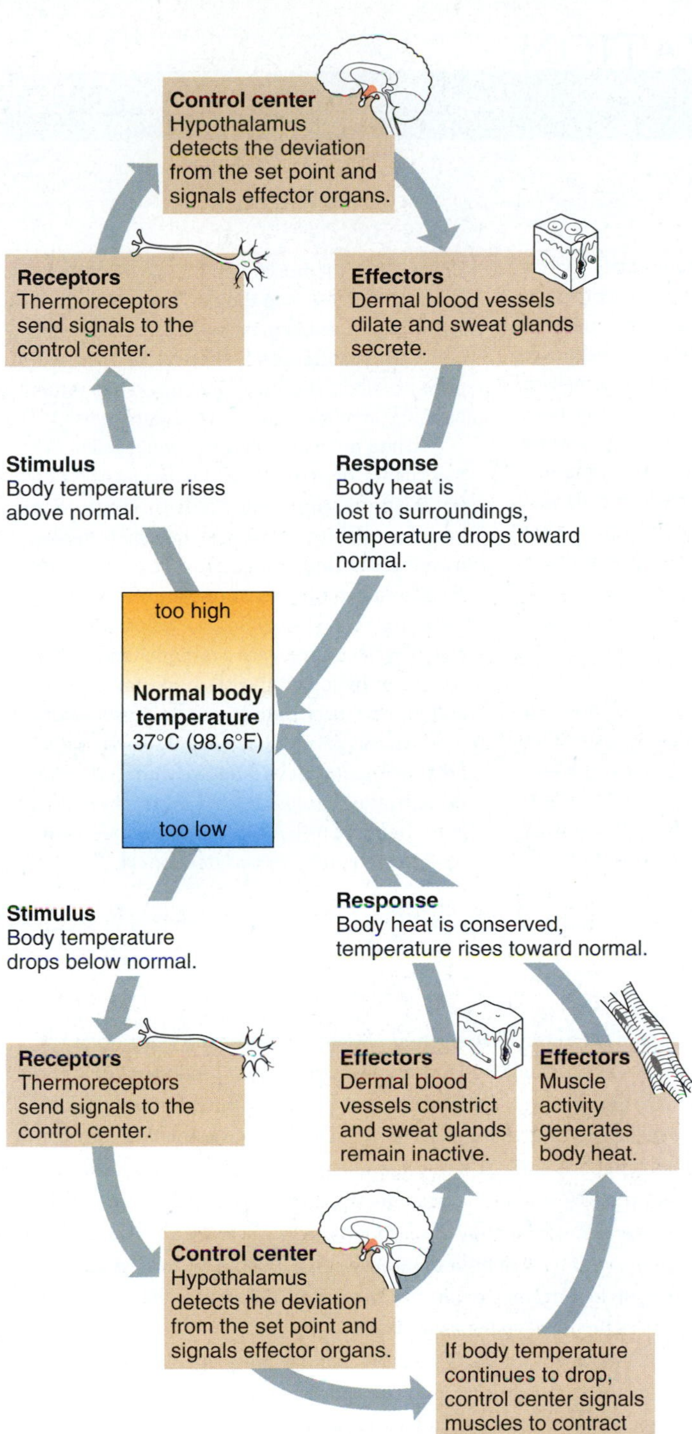
Control center
Hypothalamus detects the deviation from the set point and signals effector organs.
Receptors
Thermoreceptors send signals to the control center.
Effectors
Dermal blood vessels dilate and sweat glands secrete.
Stimulus
Body temperature rises above normal.
Response
Body heat is lost to surroundings, temperature drops toward normal.
too high
Normal body temperature
37°C (98.6°F)
too low
Stimulus
Body temperature drops below normal.
Response
Body heat is conserved, temperature rises toward normal.
Receptors
Thermoreceptors send signals to the control center.
Effectors
Dermal blood vessels constrict and sweat glands remain inactive.
Effectors
Muscle activity generates body heat.
Control center
Hypothalamus detects the deviation from the set point and signals effector organs.
If body temperature continues to drop, control center signals muscles to contract involuntarily.

FIGURE 6.12 Body temperature regulation is an example of homeostasis.

Hypothermia is intentionally induced during certain surgical procedures involving the heart, brain, or spinal cord. In heart surgery, body temperature may be lowered to between 78°F (26°C) and 89°F (32°C), which lowers the body's metabolic rate so that less oxygen is required. Hypothermia for surgery is accomplished by packing the patient in ice or by removing blood, cooling it, and returning it.

PRACTICE

21 Why is regulation of body temperature so important?

22 How is body heat produced?

23 How does the body lose excess heat?

24 How does the skin help regulate body temperature?

25 What are the dangers of hypothermia?

6.5 HEALING OF WOUNDS AND BURNS

Inflammation is a normal response to injury or stress. Blood vessels in affected tissues dilate and become more permeable, allowing fluids to leak into the damaged tissues. Inflamed skin may become reddened, swollen, warm, and painful to touch. However, the dilated blood vessels provide the tissues with more nutrients and oxygen, which aids healing. The specific events in the healing process depend on the nature and extent of the injury.

Cuts

If a break in the skin is shallow, epithelial cells along its margin are stimulated to divide more rapidly than usual. The newly formed cells fill the gap.

If an injury extends into the dermis or subcutaneous layer, blood vessels break, and the escaping blood forms a clot in the wound. A clot consists mainly of a fibrous protein (fibrin) that forms from another protein in the plasma, blood cells, and platelets trapped in the protein fibers. Tissue fluids seep into the area and dry. The blood clot and the dried fluids form a *scab* that covers and protects underlying tissues. Epithelial cells proliferate beneath the scab, bridging the wound. Before long, fibroblasts migrate into the injured region and begin secreting collagenous fibers that bind the edges of the wound. Suturing or otherwise closing a large break in the skin speeds this process. In addition, the connective tissue matrix releases *growth factors* that stimulate certain cells to divide and regenerate the damaged tissue.

As healing continues, blood vessels extend beneath the scab. Phagocytic cells remove dead cells and other debris. Eventually, the damaged tissues are replaced, and the scab sloughs off. If the wound is deep, extensive production of collagenous fibers may form an elevation above the normal epidermal surface, called a *scar.*

In large, open wounds, healing may be accompanied by formation of small, rounded masses called *granulations* that develop in the exposed tissues. A granulation consists of a new branch of a blood vessel and a cluster of collagen-secreting fibroblasts that the vessel nourishes. In time, some of the blood vessels are resorbed, and the fibroblasts move away, leaving a scar largely composed of collagenous fibers. Figure 6.13 shows the stages in the healing of a wound.

6.4 CLINICAL APPLICATION

Elevated Body Temperature

It was a warm June morning when the harried and hurried father strapped his five-month-old son Bryan into the backseat of his car and headed for work. Tragically, the father forgot to drop his son off at the babysitter's. When his wife called him at work late that afternoon to inquire why the child was not at the sitter's, the shocked father realized his mistake and hurried down to his parked car. But it was too late—Bryan had died. Left for ten hours in the car in the sun, all windows shut, the baby's temperature had quickly soared. Two hours after he was discovered, the child's temperature still exceeded 41°C (106°F).

Sarah L.'s case of elevated body temperature was more typical. She awoke with a fever of 40°C (104°F) and a terribly painful sore throat. At the doctor's office, a test revealed that Sarah had a *Streptococcus* infection. The fever was her body's attempt to fight the infection.

The true cases of Bryan and Sarah illustrate two reasons why body temperature may rise—inability of the temperature homeostatic mechanism to handle an extreme environment and an immune system response to infection. In Bryan's case, sustained exposure to very high heat overwhelmed the temperature-regulating mechanism, resulting in hyperthermia. Body heat built up faster than it could dissipate, and body temperature rose, even though the set point of the thermostat was normal. His blood vessels dilated so greatly in an attempt to dissipate the excess heat that after a few hours, his cardiovascular system collapsed.

Fever is a special case of hyperthermia in which temperature rises in response to an elevated set point. In fever, molecules on the surfaces of the infectious agents (usually bacteria or viruses) stimulate phagocytes to release a substance called interleukin-1 (also called endogenous pyrogen, meaning "fire maker from within"). The bloodstream carries interleukin-1 to the hypothalamus, where it raises the set point controlling temperature. In response, the brain signals skeletal muscles to increase heat production, blood flow to the skin to decrease, and sweat glands to decrease secretion. As a result, body temperature rises to the new set point, and fever develops. The increased body temperature helps the immune system kill the pathogens.

Rising body temperature requires different treatments, depending on the degree of elevation. Hyperthermia in response to exposure to intense, sustained heat should be rapidly treated by administering liquids to replace lost body fluids and electrolytes, sponging the skin with water to increase cooling by evaporation, and covering the person with a refrigerated blanket. Fever can be lowered with ibuprofen or acetaminophen, or aspirin in adults. Some health professionals believe that a slightly elevated temperature should not be reduced (with medication or cold baths) because it may be part of a normal immune response. A high or prolonged fever, however, requires medical attention. ■

Burns

Slightly burned skin, such as from a minor sunburn, may become warm and reddened (erythema) as dermal blood vessels dilate. This response may be accompanied by mild edema, and, in time, the surface layer of skin may be shed. A burn injuring only the epidermis is called a *superficial partial-thickness* (first-degree) *burn*. Healing usually occurs within a few days to two weeks, with no scarring.

A burn that destroys some epidermis as well as some underlying dermis is a *deep partial-thickness* (second-degree) *burn*. Fluid escapes from damaged dermal capillaries, and as it accumulates beneath the outer layer of epidermal cells, blisters appear. The injured region becomes moist and firm and may vary in color from dark red to waxy white. Such a burn most commonly occurs as a result of exposure to hot objects, hot liquids, flames, or burning clothing.

The healing of a deep partial-thickness burn depends upon stem cells that are associated with accessory structures of the skin. These structures include hair follicles, sweat glands, and sebaceous glands. They survive the injury because they are derived from the epidermis located deep in the dermis. During healing, the stem cells divide, and their daughter cells grow out onto the surface of the dermis, spread over it, and differentiate as new epidermis. In time, the skin usually completely recovers, and scar tissue does not develop unless an infection occurs.

Acute sunburn (solar erythema) is an inflammatory reaction of the skin to excessive exposure to ultraviolet radiation in sunlight. The skin becomes very red, swollen, and painful, with discomfort peaking between 6 and 48 hours after exposure. Within a few days the skin may peel, as surface cells die and are shed. Peeling, an example of apoptosis (programmed cell death), prevents cancer from developing by ridding the body of susceptible cells. Microscopic skin changes begin within a half hour of intense sun exposure, including damage to cells in the upper, epidermal layer of the skin, and swelling of blood vessels in the deeper, dermal layer.

Treatment for acute sunburn includes frequent cool baths, perhaps with oatmeal or baking soda added to soothe. Do not wash the area with a harsh soap, and avoid products with benzocaine, which can cause allergic reactions. Apply aloe for the first two days, but do not use petroleum jelly, ointments, or butters—these lock in the heat. Seek medical care if fever, blistering, dizziness, or visual disturbances develop, which are signs of sun poisoning.

To avoid sunburn, stay out of the sun between the hours of 10 A.M. and 3 P.M., and when exposed, apply sunblock with an SPF factor of at least 15—even on a cloudy day. Certain medications can hasten or intensify the skin's reaction to sun. Tanning lotions, reflectors, sunlamps, or tanning booths may pose a risk for sunburn.

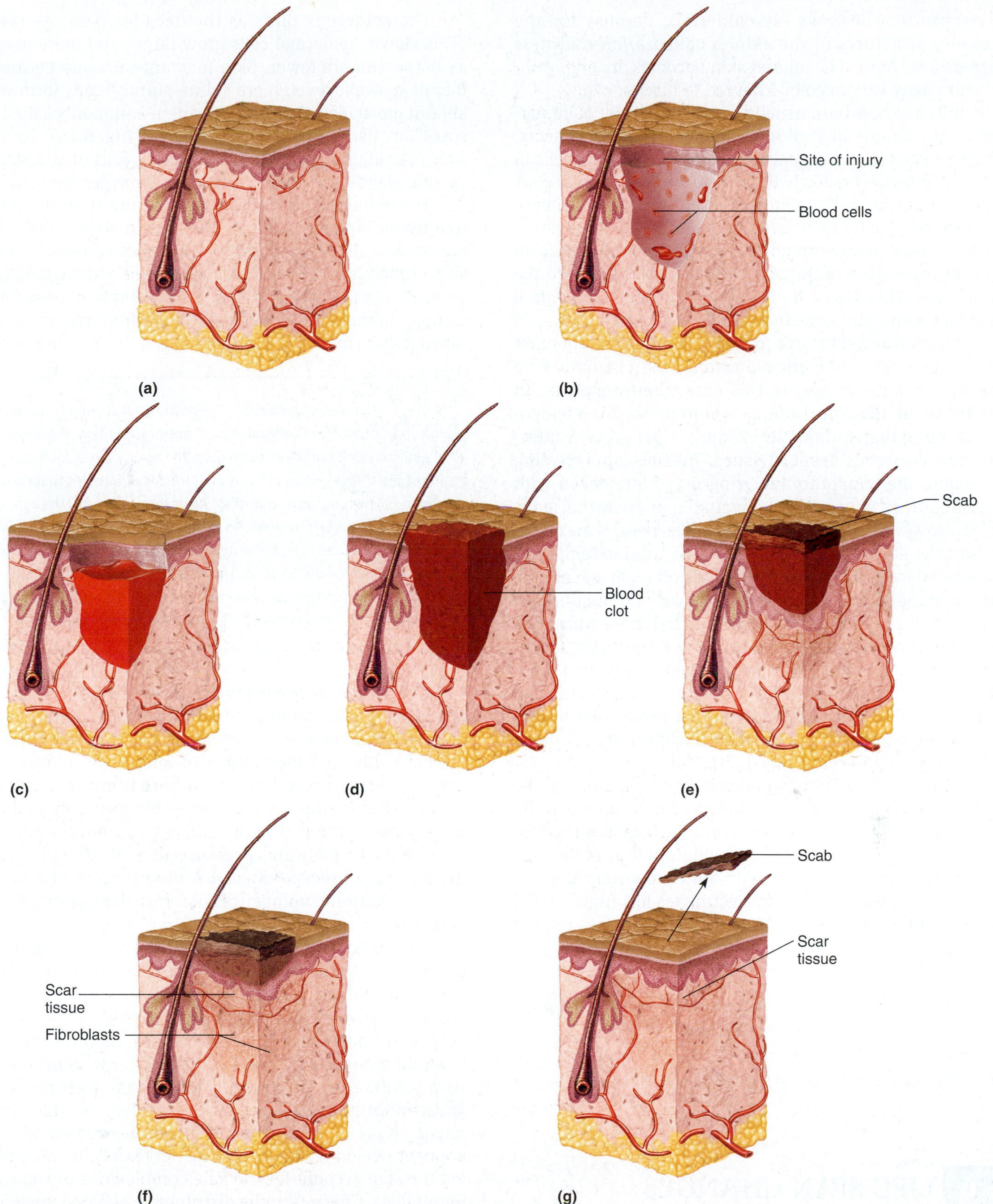

FIGURE 6.13 Healing of a wound. (*a*) If normal skin is (*b*) injured, (*c*) blood escapes from dermal blood vessels, and (*d*) a blood clot soon forms. (*e*) The blood clot and dried tissue fluid form a scab that protects the damaged region. (*f*) Later, blood vessels send out branches, and fibroblasts migrate into the area. The fibroblasts produce new connective tissue fibers, and (*g*)when the skin is mostly repaired, the scab sloughs off. Scar tissue continues to form, elevating the epidermal surface.

A burn that destroys the epidermis, dermis, and the accessory structures of the skin is called a *full-thickness* (third-degree) *burn.* The injured skin becomes dry and leathery, and it may vary in color from red to black to white.

A full-thickness burn usually occurs as a result of immersion in hot liquids or prolonged exposure to hot objects, flames, or corrosive chemicals. Most of the epithelial cells in the affected region are likely to be destroyed, so spontaneous healing can occur only by growth of epithelial cells inward from the margin of the burn. If the injury is extensive, treatment may involve removing a thin layer of skin from an unburned region of the body and transplanting it to the injured area. This procedure is an example of an *autograft,* a transplant within the same individual.

If a burn is too extensive to replace with skin from other parts of the body, cadaveric skin from a skin bank may be used to cover the injury. In this case, the transplant, an example of an *allograft* (from person to person) is a temporary covering that shrinks the wound, helps prevent infection, and preserves deeper tissues. In time, after healing has begun, the temporary covering may be replaced with an autograft, as skin becomes available in areas that have healed. However, skin grafts can leave extensive scars.

Various skin substitutes also may be used to temporarily cover extensive burns. These include amniotic membrane that surrounded a human fetus and artificial membranes composed of silicone, polyurethane, or nylon together with a network of collagenous fibers. Another type of skin substitute comes from cultured human epithelial cells. In a laboratory, a bit of human skin the size of a postage stamp can grow to the size of a bathmat in about three weeks. Skin substitutes are a major focus of tissue engineering, discussed in From Science to Technology 5.2 (p. 166).

The treatment of a burn patient requires estimating the extent of the body's affected surface. Physicians use the "rule of nines," subdividing the skin's surface into regions, each accounting for 9% (or some multiple of 9%) of the total surface area (fig. 6.14). This estimate is important in planning to replace body fluids and electrolytes lost from injured tissues and for covering the burned area with skin or skin substitutes.

PRACTICE

26 What is the tissue response to inflammation?

27 How does a scab slough off?

28 Which type of burn is most likely to leave a scar? Why?

6.6 LIFE-SPAN CHANGES

We are more aware of aging-related changes in skin than in other organ systems, because we can easily see them. Aging skin affects appearance, temperature regulation, and vitamin D activation.

The epidermis thins as the decades pass. As the cell cycle slows, epidermal cells grow larger and more irregular in shape, but are fewer. Skin may appear scaly because, at the microscopic level, more sulfur–sulfur bonds form within keratin molecules. Patches of pigment commonly called "age spots" or "liver spots" appear and grow (fig. 6.15). These are sites of oxidation of fats in the secretory cells of apocrine and eccrine glands and reflect formation of oxygen free radicals.

The dermis becomes reduced as synthesis of the connective tissue proteins collagen and elastin slows. The combination of a shrinking dermis and loss of some fat from the subcutaneous layer results in wrinkling and sagging of the skin. Fewer fibroblasts delay wound healing. Some of the changes in the skin's appearance result from specific deficits. The decrease in oil from sebaceous glands dries the skin.

> Various treatments temporarily smooth facial wrinkles. "Botox" is injection of a very dilute solution of botulinum toxin. Produced by the bacterium *Clostridium botulinum,* the toxin causes food poisoning. It also blocks nerve activation of the facial muscles that control smiling, frowning, and squinting. After three months, though, the facial nerves contact the muscles at different points, and the wrinkles return. (Botox used at higher doses to treat neuromuscular conditions can cause adverse effects.) Other anti-wrinkle treatments include chemical peels and dermabrasion to reveal new skin surface; collagen injections; and transplants of fat from the buttocks to the face.

The skin's accessory structures also show signs of aging. Slowed melanin production whitens hair as the follicle becomes increasingly transparent. Hair growth slows, the hairs thin, and the number of follicles decreases. Males may develop pattern baldness—hereditary, but not often expressed in females. A diminished blood supply to the nail beds impairs their growth, dulling and hardening them. Sensitivity to pain and pressure diminishes with age as the number of receptors falls. A ninety-year-old's skin has only one-third the number of such receptors as the skin of a young adult.

The ability to control temperature falters as the number of sweat glands in the skin falls, as the capillary beds that surround sweat glands and hair follicles shrink, and as the ability to shiver declines. In addition, the number of blood vessels in the deeper layers decreases, as does the ability to shunt blood toward the body's interior to conserve heat. As a result, an older person is less able to tolerate the cold and cannot regulate heat. In the winter, an older person might set the thermostat ten to fifteen degrees higher than a younger person would. Fewer blood vessels in and underlying the skin account for the pale complexions of some older individuals. Changes in the distribution of blood vessels also contribute to development of pressure sores in a bedridden person whose skin does not receive adequate circulation.

Aging of the skin is also related to skeletal health. The skin is the site of activation of vitamin D, which requires

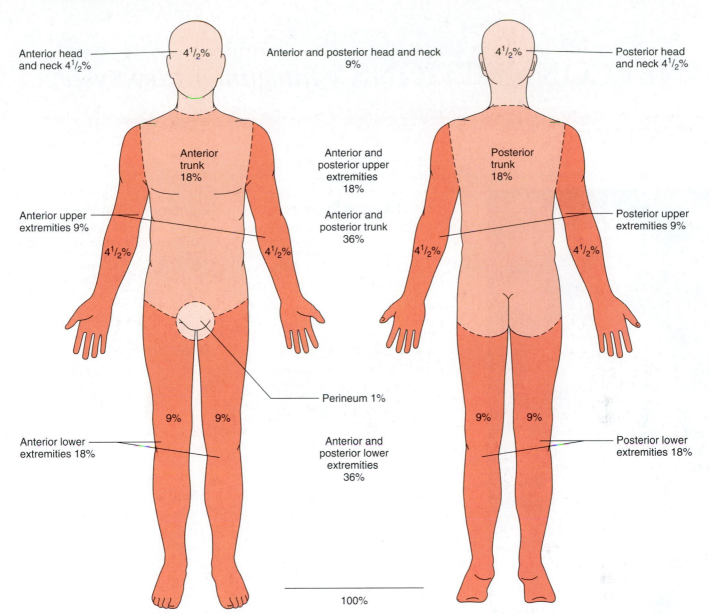

Anterior head and neck $4^{1}/_{2}\%$
$4^{1}/_{2}\%$
Anterior and posterior head and neck 9%
$4^{1}/_{2}\%$
Posterior head and neck $4^{1}/_{2}\%$
Anterior trunk 18%
Anterior and posterior upper extremities 18%
Posterior trunk 18%
Anterior upper extremities 9%
Anterior and posterior trunk 36%
Posterior upper extremities 9%
$4^{1}/_{2}\%$
$4^{1}/_{2}\%$
$4^{1}/_{2}\%$
$4^{1}/_{2}\%$
Perineum 1%
9%
9%
9%
9%
Anterior lower extremities 18%
Anterior and posterior lower extremities 36%
Posterior lower extremities 18%
100%

FIGURE 6.14 As an aid for estimating the extent of damage burns cause, the body is subdivided into regions, each representing 9% (or some multiple of 9%) of the total skin surface area.

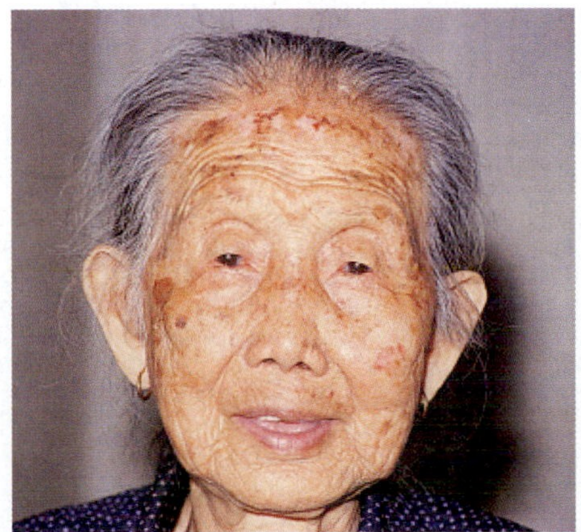

FIGURE 6.15 Aging-associated changes are obvious in the skin.

exposure to the sun. Vitamin D is necessary for absorption of calcium, needed for bone structure. Many older people do not get outdoors much, and the wavelengths of light that are important for vitamin D activation do not readily penetrate glass windows. In addition, older skin has a diminished ability to activate the vitamin. Therefore, homebound seniors can benefit from vitamin D supplements to help maintain bone structure.

PRACTICE

29 What changes occur with age in the epidermis and dermis?

30 How do the skin's accessory structures change over time?

31 Why do older people have more difficulty controlling body temperature than do younger people?

INNERCONNECTIONS | *Integumentary System*

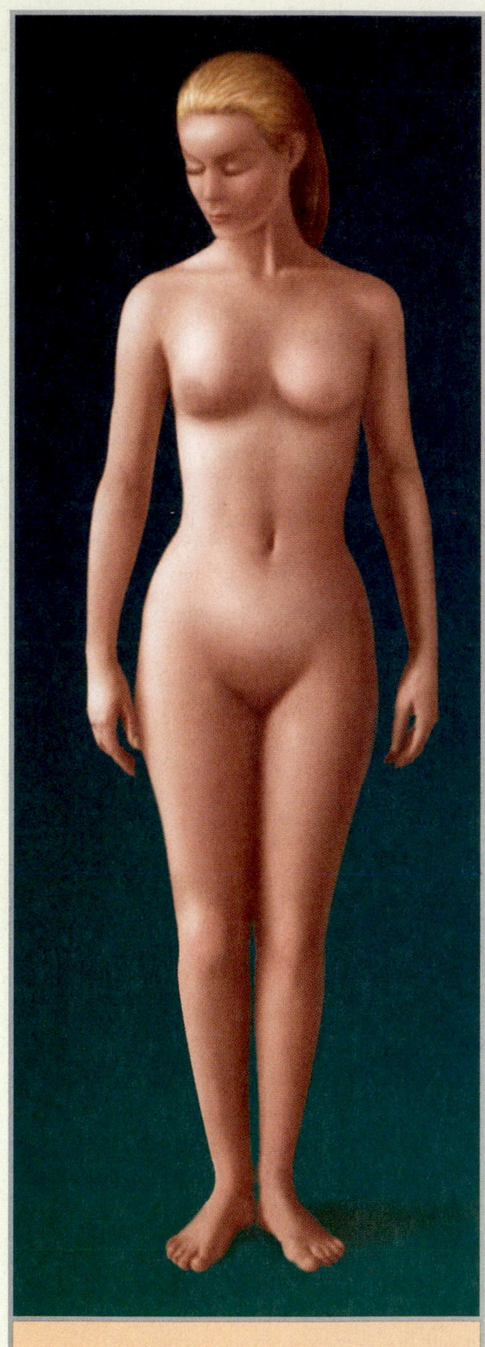

Integumentary System

The skin provides protection, contains sensory organs, and helps control body temperature.

Skeletal System

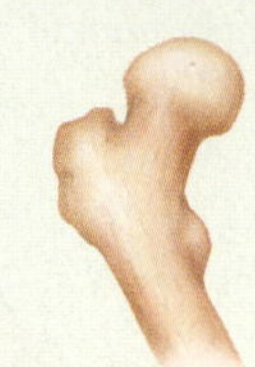

Vitamin D activated by the skin helps provide calcium for bone matrix.

Lymphatic System

The skin, acting as a barrier, provides an important first line of defense for the immune system.

Muscular System

Involuntary muscle contractions (shivering) work with the skin to control body temperature. Muscles act on facial skin to create expressions.

Digestive System

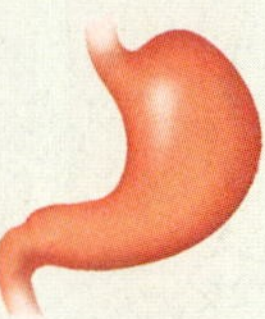

Excess calories may be stored as subcutaneous fat. Vitamin D activated by the skin stimulates dietary calcium absorption.

Nervous System

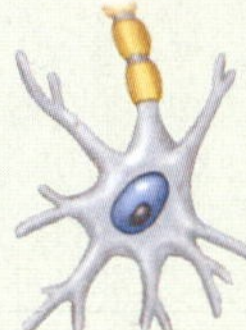

Sensory receptors provide information about the outside world to the nervous system. Nerves control the activity of sweat glands.

Respiratory System

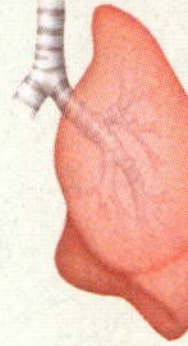

Stimulation of skin receptors may alter respiratory rate.

Endocrine System

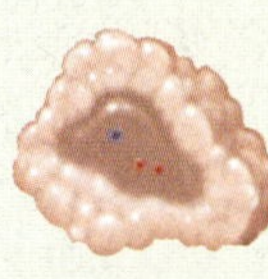

Hormones help to increase skin blood flow during exercise. Other hormones stimulate either the synthesis or the decomposition of subcutaneous fat.

Urinary System

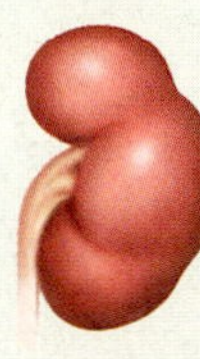

The kidneys help compensate for water and electrolytes lost in sweat.

Cardiovascular System

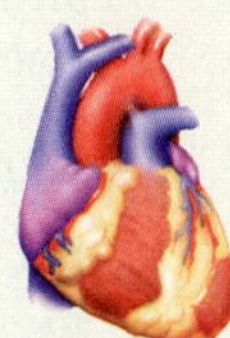

Skin blood vessels play a role in regulating body temperature.

Reproductive System

Sensory receptors play an important role in sexual activity and in the suckling reflex.

CHAPTER SUMMARY

6.1 INTRODUCTION (PAGE 171)

The skin, the largest organ in the body, and its accessory structures constitute the integumentary system.

6.2 SKIN AND ITS TISSUES (PAGE 171)

Skin is a protective covering, helps regulate body temperature, houses sensory receptors, synthesizes chemicals, and excretes wastes. It is composed of an epidermis and a dermis separated by a basement membrane. A subcutaneous layer, not part of the skin, lies beneath the dermis. The subcutaneous layer is composed of areolar tissue and adipose tissue that helps conserve body heat. This layer contains blood vessels that supply the skin.

1. Epidermis
   a. The epidermis is stratified squamous epithelium that lacks blood vessels.
   b. The deepest layer, called the stratum basale, contains cells that divide and grow.
   c. Epidermal cells undergo keratinization as they are pushed toward the surface.
   d. The outermost layer, called the stratum corneum, is composed of dead epidermal cells.
   e. Production of epidermal cells balances the rate at which they are lost at the surface.
   f. The epidermis protects underlying tissues against water loss, mechanical injury, and the effects of harmful chemicals.
   g. Melanin, a pigment produced from the amino acid tyrosine, provides skin color and protects underlying cells from the effects of ultraviolet light.
   h. Melanocytes transfer melanin to nearby epidermal cells.
   i. All humans have about the same concentration of melanocytes. Skin color is largely due to the amount of melanin in the epidermis.
      (1) Each person inherits genes for melanin production.
         (a) Dark skin is due to genes that cause large amounts of melanin to be produced; lighter skin is due to genes that cause lesser amounts of melanin to form.
         (b) Mutant genes may cause a lack of melanin in the skin.
      (2) Environmental factors that influence skin color include sunlight, ultraviolet light, and X rays. These factors darken existing melanin and stimulate additional melanin production.
      (3) Physiological factors influence skin color.
         (a) The oxygen content of the blood in dermal vessels may cause the skin of light-complexioned persons to appear pinkish or bluish.
         (b) Carotene in the subcutaneous layer may cause the skin to appear yellowish.
         (c) Disease may affect skin color.
2. Dermis
   a. The dermis is a layer composed of dense irregular connective tissue that binds the epidermis to underlying tissues.
   b. It also contains muscle cells, blood vessels, and nerve cell processes.
   c. Dermal blood vessels supply nutrients to all skin cells and help regulate body temperature.
   d. Nervous tissue is scattered throughout the dermis.
      (1) Some dermal nerve cell processes carry impulses to muscles and glands of the skin.
      (2) Other dermal nerve cell processes are associated with sensory receptors in the skin.

6.3 ACCESSORY STRUCTURES OF THE SKIN (PAGE 177)

1. Nails
   a. Nails are protective covers on the ends of fingers and toes.
   b. They consist of keratinized epidermal cells.
2. Hair follicles
   a. Hair covers nearly all regions of the skin.
   b. Each hair develops from epidermal cells at the base of a tubelike hair follicle.
   c. As newly formed cells develop and grow, older cells are pushed toward the surface and undergo keratinization.
   d. A hair usually grows for a while, rests, and then is replaced by a new hair.
   e. Hair color is determined by genes that direct the type and amount of pigment in hair cells.
   f. A bundle of smooth muscle cells and one or more sebaceous glands are attached to each hair follicle.
3. Skin glands
   a. Sebaceous glands secrete sebum, which softens and waterproofs both the skin and hair.
   b. Sebaceous glands are usually associated with hair follicles.
   c. Sweat glands are located in nearly all regions of the skin.
   d. Each sweat gland consists of a coiled tube.
   e. Eccrine sweat glands, located on the forehead, neck, back, palms, and soles, respond to elevated body temperature or emotional stress.
   f. Sweat is primarily water but also contains salts and waste products.
   g. Apocrine sweat glands, located in the axillary regions, groin, and around the nipples, moisten the skin when a person is emotionally upset, scared, in pain, or sexually aroused.

6.4 REGULATION OF BODY TEMPERATURE (PAGE 181)

Regulation of body temperature is vital because heat affects the rates of metabolic reactions. Normal temperature of deeper body parts is close to a set point of 37°C (98.6°F).

1. Heat production and loss
   a. Heat is a by-product of cellular respiration.
   b. When body temperature rises above normal, more blood enters dermal blood vessels and the skin reddens.

c. Heat is lost to the outside by radiation, conduction, convection, and evaporation.
d. Sweat gland activity increases heat loss by evaporation.
e. When body temperature drops below normal, dermal blood vessels constrict, causing the skin to lose color, and sweat glands become inactive.
f. If body temperature continues to drop, skeletal muscles involuntarily contract; this increases cellular respiration and produces additional heat.

2. Problems in temperature regulation
   a. Air can hold a limited volume of water vapor.
   b. When the air is saturated with water, sweat may fail to evaporate and body temperature may remain elevated.
   c. Hypothermia is lowered body temperature. It causes shivering, mental confusion, lethargy, loss of reflexes and consciousness, and eventually major organ failure.

6.5 HEALING OF WOUNDS AND BURNS (PAGE 183)

Skin injuries trigger inflammation. The affected area becomes red, warm, swollen, and tender.

1. A cut in the epidermis is filled in by dividing epithelial cells. Clots close deeper cuts, sometimes leaving a scar where connective tissue produces collagenous fibers, forming an elevation above the normal epidermal surface. Granulations form as part of the healing process in large, open wounds.
2. A superficial partial-thickness burn heals quickly with no scarring. The area is warm and red. A burn penetrating to the dermis is a deep partial-thickness burn. It blisters. Deeper skin structures help heal this more serious type of burn. A full-thickness burn is the most severe and may require a skin graft.

6.6 LIFE-SPAN CHANGES (PAGE 186)

1. Aging skin affects appearance as "age spots" or "liver spots" appear and grow, along with wrinkling and sagging.
2. Due to changes in the number of sweat glands and shrinking capillary beds in the skin, elderly people are less able to tolerate the cold and cannot regulate heat.
3. Older skin has a diminished ability to activate vitamin D necessary for skeletal health.

CHAPTER ASSESSMENTS

6.1 Introduction

1 Two or more tissues grouped together and performing specialized functions define a(n) _____________. (p. 171)
a. organelle
b. cell
c. organ
d. organ system

2 The largest organ(s) in the body is (are) the _____________. (p. 171)
a. liver
b. intestines
c. lungs
d. skin

6.2 Skin and Its Tissues

3 Functions of the skin include _____________. (p. 171)
a. retarding water loss
b. body temperature regulation
c. sensory reception
d. excretion
e. All of the above.

4 List the remaining functions of skin not mentioned in question 3. (p. 171)

5 The epidermis is composed of layers of _____________ tissue. (p. 172)

6 Distinguish between the epidermis and the dermis. (p. 172)

7 Explain the functions of the subcutaneous layer. (p. 172)

8 Explain what happens to epidermal cells as they undergo keratinization. (p. 173)

9 Place the layers of the epidermis in order (1–5) from the outermost layer to the layer attached to the dermis by the basement membrane. (p. 173)
________ stratum spinosum
________ stratum corneum
________ stratum basale
________ stratum spinosum
________ stratum granulosum

10 Describe the function of melanocytes. (p. 174)

11 Discuss the function of melanin, other than providing color to the skin. (p. 174)

12 Explain how environmental factors affect skin color. (p. 176)

13 Describe three physiological factors that affect skin color. (p. 176)

14 Name the tissue(s) of the dermis. (p. 177)

15 Review the functions of dermal nervous tissue. (p. 177)

6.3 Accessory Structures of the Skin

16 Describe how nails are formed. (p. 177)

17 Distinguish between a hair and a hair follicle. (p. 177)

18 Review how hair color is determined. (p. 178)

19 Explain the function of sebaceous glands. (p. 180)

20 The sweat glands that respond to elevated body temperature and are commonly found on the forehead, neck, and back are _____________ glands. (p. 180)
a. sebaceous
b. holocrine
c. eccrine
d. apocrine
e. ceruminous

6.4 Regulation of Body Temperature

21 Explain the importance of body temperature regulation. (p. 181)

22 Describe the role of the skin in promoting the loss of excess body heat. (p. 182)

23 Match each means of losing body heat with its description. (p. 182)

(1) radiation	A. fluid changes from liquid to a gas
(2) conduction	B. heat moves from body directly into molecules of cooler objects in contact with its surface
(3) convection	C. heat rays escape from warmer surfaces to cooler surroundings
(4) evaporation	D. continuous circulation of air over a warm surface

24 Describe the body's responses to decreasing body temperature. (p. 182)

25 Review how air saturated with water vapor may interfere with body temperature regulation. (p. 182)

6.5 Healing of Wounds and Burns

26 Distinguish between the healing of shallow and deeper breaks in the skin. (p. 183)

27 Distinguish among first-, second-, and third-degree burns. (p. 184)

28 Describe possible treatments for a third-degree burn. (p. 186)

6.6 Life-Span Changes

29 Discuss three affects of aging on skin. (p. 186)

INTEGRATIVE ASSESSMENTS/CRITICAL THINKING

OUTCOMES 1.5, 6.3, 6.4

1. What methods might be used to cool the skin of a child experiencing a high fever? For each method you list, identify the means by which it promotes heat loss—radiation, conduction, convection, or evaporation.

OUTCOMES 5.3, 6.2

2. Why would collagen and elastin added to skin creams be unlikely to penetrate the skin—as some advertisements imply they do?

OUTCOMES 5.3, 6.2, 6.4

3. A premature infant typically lacks subcutaneous adipose tissue. Also, the surface area of an infant's body is relatively large compared to its volume. How do these factors affect the ability of an infant to regulate its body temperature?

OUTCOME 6.2

4. Which of the following would result in the more rapid absorption of a drug: a subcutaneous injection or an intradermal injection? Why?

OUTCOME 6.2

5. Everyone's skin contains about the same number of melanocytes even though people come in many different colors. How is this possible?

OUTCOMES 6.2, 6.3, 6.4

6. What special problems would result from the loss of 50% of a person's functional skin surface? How might this person's environment be modified to compensate partially for such a loss?

OUTCOMES 6.2, 6.5

7. As a rule, a superficial partial-thickness burn is more painful than one involving deeper tissues. How would you explain this observation?

WEB CONNECTIONS

Be sure to visit the text website at **www.mhhe.com/shier12** for answers to chapter assessments, additional quizzes, and interactive learning exercises.

ANATOMY & PHYSIOLOGY REVEALED

Anatomy & Physiology Revealed® (APR) includes cadaver photos that allow you to peel away layers of the human body to reveal structures beneath the surface. This program also includes animations, radiologic imaging, audio pronunciations, and practice quizzing. Check out **www.aprevealed.com**. APR has been proven to help improve student grades!

CHAPTER

7 Skeletal System

Falsely colored scanning electron micrograph of an osteoclast (large blue cell), a type of bone cell involved in normal bone remodeling (1,240×).

UNDERSTANDING WORDS

acetabul-, vinegar cup: *acetabul*um—depression of the hip bone that articulates with the head of the femur.

ax-, axis: *ax*ial skeleton—upright portion of the skeleton that supports the head, neck, and trunk.

-blast, bud, a growing organism in early stages: osteo*blast*—cell that will form bone tissue.

canal-, channel: *canal*iculus—tubular passage.

carp-, wrist: *carp*als—wrist bones.

-clast, break: osteo*clast*—cell that breaks down bone tissue.

clav-, bar: *clav*icle—bone that articulates with the sternum and scapula.

condyl-, knob, knuckle: *condyl*e—rounded, bony process.

corac-, a crow's beak: *corac*oid process—beaklike process of the scapula.

cribr-, sieve: *cribr*iform plate—portion of the ethmoid bone with many small openings.

crist-, crest: *crist*a galli—bony ridge that projects upward into the cranial cavity.

fov-, pit: *fov*ea capitis—pit in the head of a femur.

glen-, joint socket: *glen*oid cavity—depression in the scapula that articulates with the head of a humerus.

inter-, among, between: *inter*vertebral disc—structure between vertebrae.

intra-, inside: *intra*membranous bone—bone that forms within sheetlike masses of connective tissue.

lamell-, thin plate: *lamell*a—thin, bony plate.

meat-, passage: auditory *meat*us—canal of the temporal bone that leads inward to parts of the ear.

odont-, tooth: *odont*oid process—toothlike process of the second cervical vertebra.

poie-, make, produce: hemato*poie*sis—process that forms blood cells.

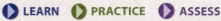

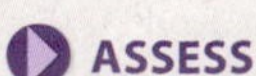
LEARN PRACTICE ASSESS

LEARNING OUTCOMES

After you have studied this chapter, you should be able to

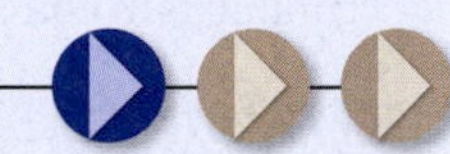

7.1 Introduction

1 Discuss the living tissues found in bone even though bone appears to be inert. (p. 193)

7.2 Bone Structure

2 Classify bones according to their shapes, and name an example from each group. (p. 193)

3 Describe the macroscopic and microscopic structure of a long bone, and list the functions of these parts. (p. 194)

7.3 Bone Development and Growth

4 Distinguish between intramembranous and endochondral bones, and explain how such bones develop and grow. (p. 197)

5 Describe the effects of sunlight, nutrition, hormonal secretions, and exercise on bone development and growth. (p. 200)

7.4 Bone Function

6 Discuss the major functions of bones. (p. 202)

7.5 Skeletal Organization

7 Distinguish between the axial and appendicular skeletons, and name the major parts of each. (p. 206)

7.6 Skull–7.12 Lower Limb

8 Locate and identify the bones and the major features of the bones that comprise the skull, vertebral column, thoracic cage, pectoral girdle, upper limb, pelvic girdle, and lower limb. (p. 206)

9 Describe the differences between male and female skeletons. (p. 234)

7.13 Life-Span Changes

10 Describe life-span changes in the skeletal system. (p. 238)

SKELETAL CLUES TO THE PAST

As a hard and enduring human tissue, bone provides important clues to the distant past. We have a glimpse of our ancestors from 156,000 years ago in skulls discovered near the town of Herto in Ethiopia. Driving by Herto after a season of punishing rains, paleoanthropologist Tim White of the University of California, Berkeley, spotted a skull jutting from the sand near the Awash River. The skull, from a hippo, bore cut marks indicating butchery. Returning with helpers, the researchers uncovered the fossilized remains of three human skulls, preserved because the rain had driven the modern-day residents and their cattle from Herto before they could trample the evidence.

The researchers named this earliest known member of the human family *Homo sapiens idaltu,* which means "elder" in the local Afar language. One adult skull was that of a young man and because it was only partially crushed, could be reconstructed; another was damaged beyond recognition. The third skull was that of a child about seven years old. It had been smashed into more than 200 pieces and scattered over a 400-square-meter area. Telltale clues suggested reverence for the dead: The skulls were smooth, as if they had been repeatedly handled, and they bore highly symmetrical cut marks. They were also found alone, with no other body parts, suggesting that they had been transported. Some societies treat skulls in this manner to honor the dead. Nearby were many stone blades, axes, and flaking tools. It looked like a band of early humans had lived near a shallow lake that formed when the river overflowed. Other preserved bones indicate that the lake was also home to hippos, catfish, and crocodiles, and that buffalo lived near the surrounding lush vegetation. This part of Ethiopia has unusual geological features that have preserved human ancestors from more than 6 million years ago to recent times.

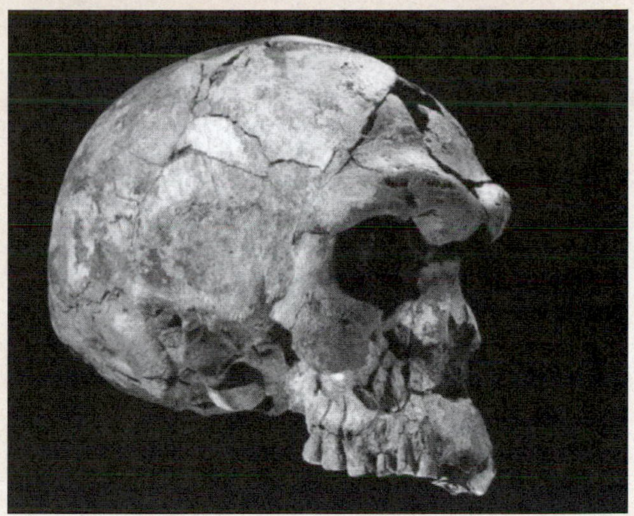

Fossilized skulls from 156,000 years ago provide glimpses of the oldest known anatomically modern humans, *Homo sapiens idaltu*.

It took an international team three years to assemble the human skulls from Herto and another three to analyze them sufficiently to publish preliminary results. The researchers compared the skull dimensions to 6,000 modern human skulls. *H. sapiens idaltu* had facial features much like ours and a slightly larger and longer head. It will be interesting to learn more about these ancestors. ■

7.1 INTRODUCTION

A bone may appear to be inert because of nonliving material in the extracellular matrix of bone tissue. However, bone also includes active, living tissues: bone tissue, cartilage, dense connective tissue, blood, and nervous tissue. Bones are not only alive, but also multifunctional. Bones, the organs of the **skeletal system,** support and protect softer tissues, provide points of attachment for muscles, house blood-producing cells, and store inorganic salts.

PRACTICE

1 List the living tissues in bone.

7.2 BONE STRUCTURE

The bones of the skeletal system vary greatly in size and shape. However, bones are similar in structure, development, and function.

Bone Classification

Bones are classified according to their shapes—long, short, flat, or irregular (fig. 7.1).

- **Long bones** have long longitudinal axes and expanded ends. Examples of long bones are the forearm and thigh bones.
- **Short bones** are cubelike, with roughly equal lengths and widths. The bones of the wrists and ankles are this type.
- **Flat bones** are platelike structures with broad surfaces, such as the ribs, scapulae, and some bones of the skull.
- **Irregular bones** have a variety of shapes and are usually connected to several other bones. Irregular bones include the vertebrae that comprise the backbone and many facial bones.

In addition to these four groups of bones, some authorities recognize a fifth group called **sesamoid bones,** or **round bones** (see fig. 7.45*c*). These bones are usually small and nodular and are embedded in tendons adjacent to joints, where the tendons are compressed. The kneecap (patella) is a sesamoid bone.

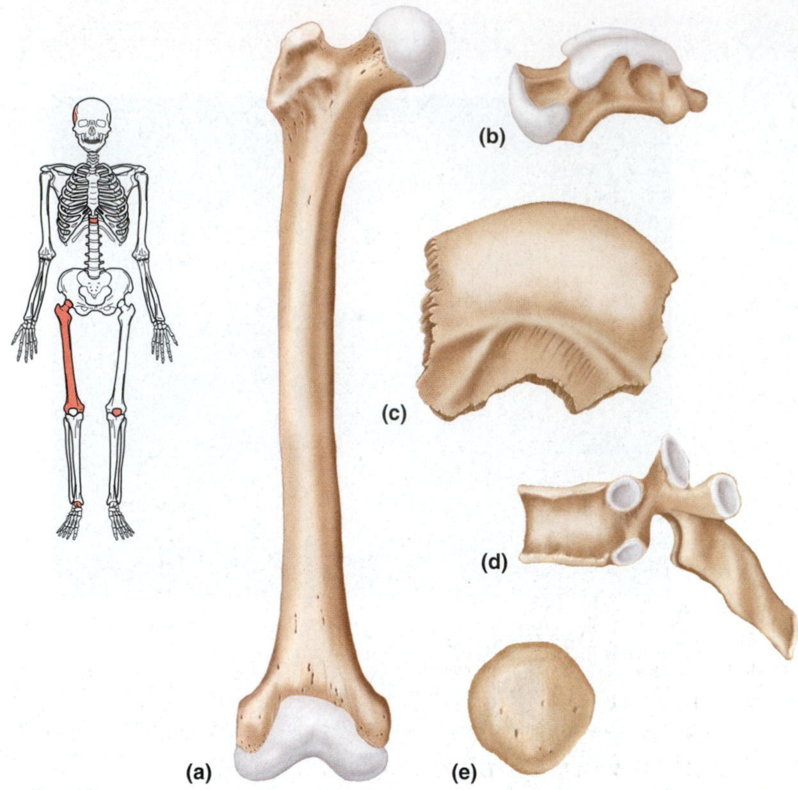

FIGURE 7.1 Bones are classified by shape. (*a*) The femur of the thigh is a long bone, (*b*) a tarsal bone of the ankle is a short bone, (*c*) a parietal bone of the skull is a flat bone, (*d*) a vertebra of the backbone is an irregular bone, and (*e*) the patella of the knee is a sesamoid bone. The whole-skeleton location icon highlights the bones used as examples for classification.

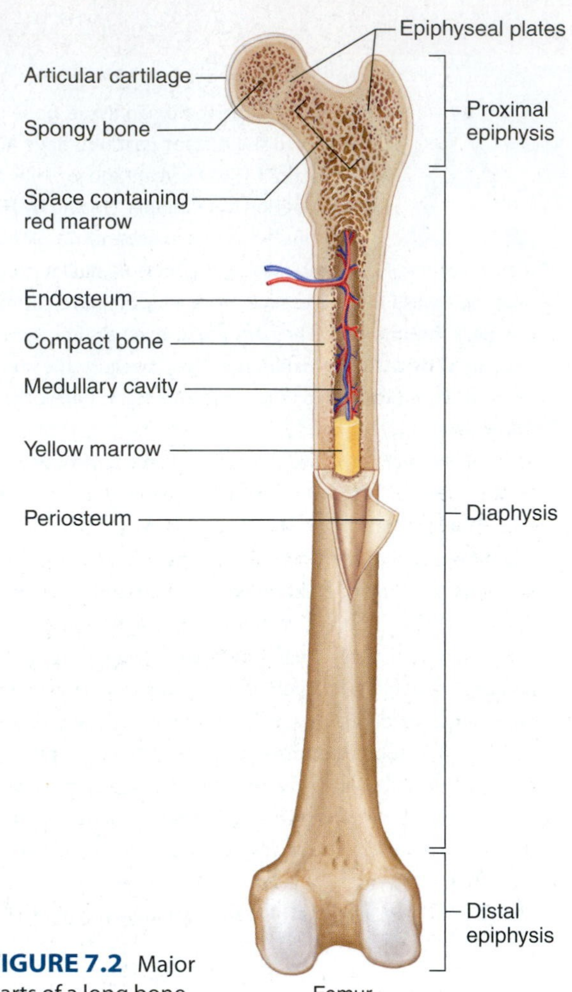

FIGURE 7.2 Major parts of a long bone.

Parts of a Long Bone

The femur, the long bone in the thigh, illustrates the structure of bone (fig. 7.2). At each end of such a bone is an expanded portion called an **epiphysis** (e-pif′ĭ-sis) (pl., *epiphyses*), which articulates (or forms a joint) with another bone. One epiphysis, called the proximal epiphysis, is nearest to the torso. The other, called the distal epiphysis, is farthest from the torso. On its outer surface, the articulating portion of the epiphysis is coated with a layer of hyaline cartilage called **articular cartilage** (ar-tik′u-lar kar′tĭ-lij). The shaft of the bone, between the epiphyses, is called the **diaphysis** (di-af′ĭ-sis).

A bone is enclosed by a tough, vascular covering of fibrous tissue called the **periosteum** (per″e-os′te-um), except for the articular cartilage on its ends. The periosteum is firmly attached to the bone, and the periosteal fibers are continuous with connected ligaments and tendons. The periosteum also helps form and repair bone tissue.

A bone's shape makes possible its functions. Bony projections called *processes,* for example, provide sites for attachment of ligaments and tendons; grooves and openings are passageways for blood vessels and nerves; and a depression of one bone might articulate with a process of another.

The wall of the diaphysis is mainly composed of tightly packed tissue called **compact bone** (kom′pakt bōn), or cortical bone. This type of bone has a continuous extracellular matrix with no gaps (fig. 7.3*a*). The epiphyses, on the other hand, are largely composed of **spongy bone** (spunj′e bōn), or cancellous bone, with thin layers of compact bone on their surfaces (fig. 7.3*b*). Spongy bone consists of many branching bony plates called **trabeculae** (trah-bek′u-le). Irregular connecting spaces between these plates help reduce the bone's weight. The bony plates are most highly developed in the regions of the epiphyses subjected to compressive forces. Both compact and spongy bone are strong and resist bending.

A bone usually has compact bone overlying spongy bone, with the relative amounts of each varying in the differently shaped bones. Short, flat, and irregular bones typically consist of a mass of spongy bone either covered by a layer of compact bone or sandwiched between plates of compact bone (fig. 7.3*c*).

Compact bone in the diaphysis of a long bone forms a semirigid tube with a hollow chamber called the **medullary cavity** (med′u-lār″e kav′ĭ-te) that is continuous with the spaces of the spongy bone. A thin membrane containing bone-forming cells, called **endosteum** (en-dos′tē-um), lines these

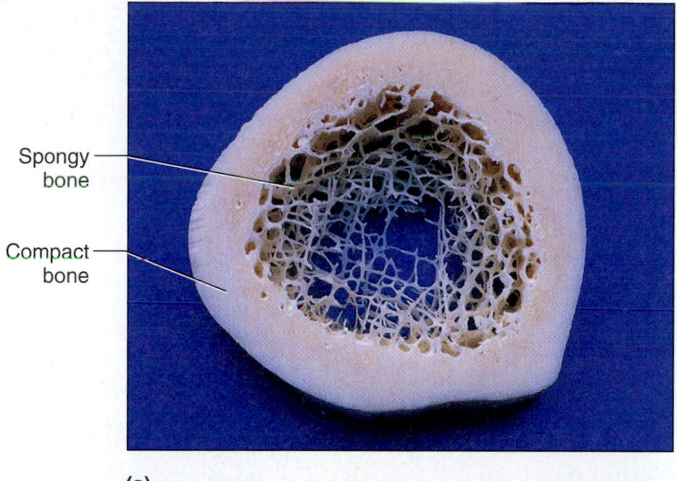

(a)

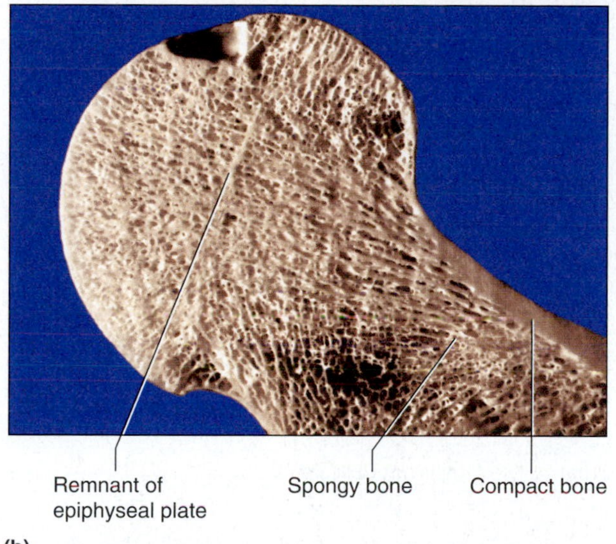

(b)

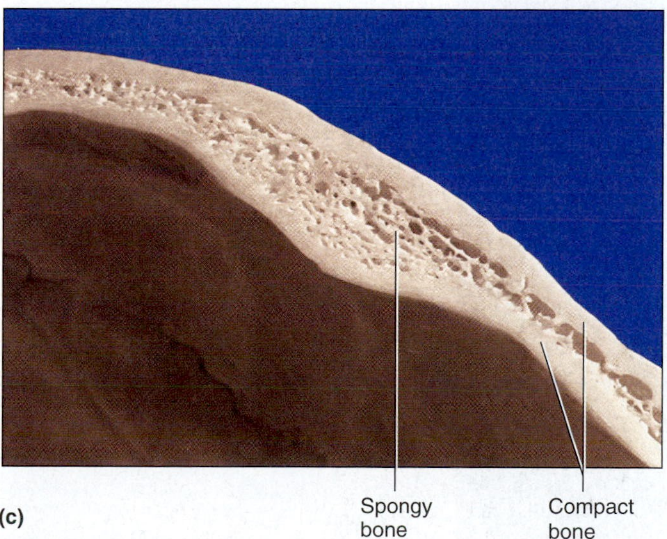

(c)

FIGURE 7.3 Compact bone and spongy bone. (*a*) In a femur, the wall of the diaphysis consists mostly of compact bone. (*b*) The epiphyses of the femur contain spongy bone enclosed by a thin layer of compact bone. (*c*) This skull bone contains a layer of spongy bone sandwiched between plates of compact bone.

areas, and a specialized type of soft connective tissue called **marrow** (mar′o) fills them. The two forms of marrow, red and yellow, are described later in this chapter (see also fig. 7.2).

Microscopic Structure

Recall from chapter 5 (p. 160) that bone cells called **osteocytes** (os′te-o-sītz) are in tiny, bony chambers called *lacunae*, which form concentric circles around *central canals* (Haversian canals). Osteocytes transport nutrients and wastes to and from nearby cells by means of cellular processes passing through *canaliculi*. The extracellular matrix of bone tissue is largely collagen and inorganic salts. Collagen gives bone its strength and resilience, and inorganic salts make it hard and resistant to crushing.

Compact Bone

In compact bone, the osteocytes and layers of extracellular matrix concentrically clustered around a central canal form a cylinder-shaped unit called an **osteon** (os′te-on), sometimes called an Haversian system (figs. 7.4 and 7.5). Many of these units cemented together form the substance of compact bone. The orientation of the osteons resists compressive forces.

Each central canal contains blood vessels and nerve fibers surrounded by loose connective tissue. Blood in these vessels nourishes bone cells associated with the central canal via gap junctions between osteocytes.

Central canals extend longitudinally through bone tissue, and transverse *perforating canals* (Volkmann's canals) connect them. Perforating canals contain larger blood vessels and nerves by which the smaller blood vessels and nerve fibers in central canals communicate with the surface of the bone and the medullary cavity (see fig. 7.4).

Spongy Bone

Spongy bone is also composed of osteocytes and extracellular matrix, but the bone cells do not aggregate around central canals. Instead, the cells lie within the trabeculae and get nutrients from substances diffusing into the canaliculi that lead to the surfaces of these thin, bony plates.

Severe bone pain is one symptom of sickle cell disease, which is inherited. Under low oxygen conditions, abnormal hemoglobin (an oxygen-carrying protein) bends the red blood cells that contain it into sickle shapes, obstructing circulation. Radiographs can reveal blocked arterial blood flow in bones of sickle cell disease patients.

PRACTICE 

2 Explain how bones are classified.
3 List five major parts of a long bone.
4 How do compact and spongy bone differ in structure?
5 Describe the microscopic structure of compact bone.

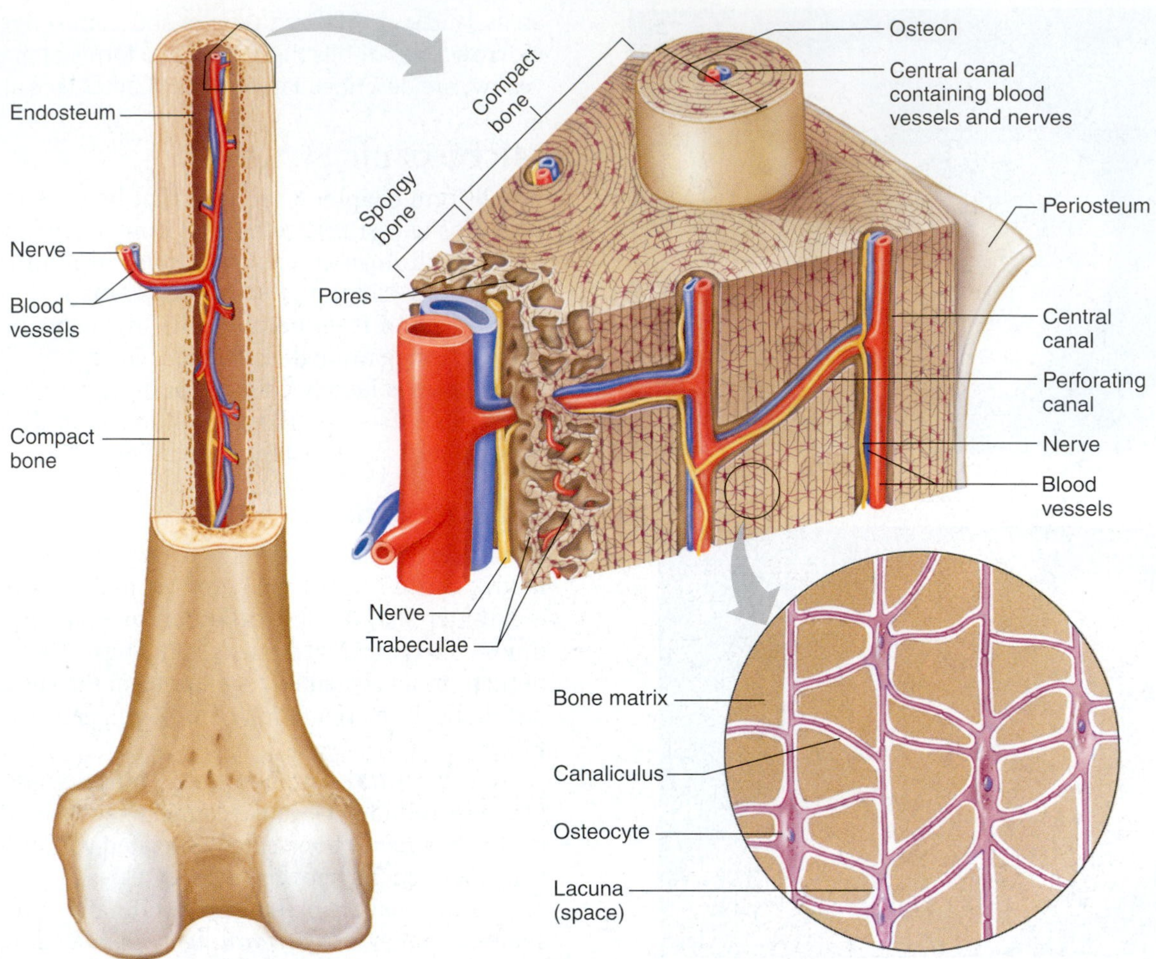

FIGURE 7.4 Compact bone is composed of osteons cemented together by bone matrix. Drawing is not to scale.

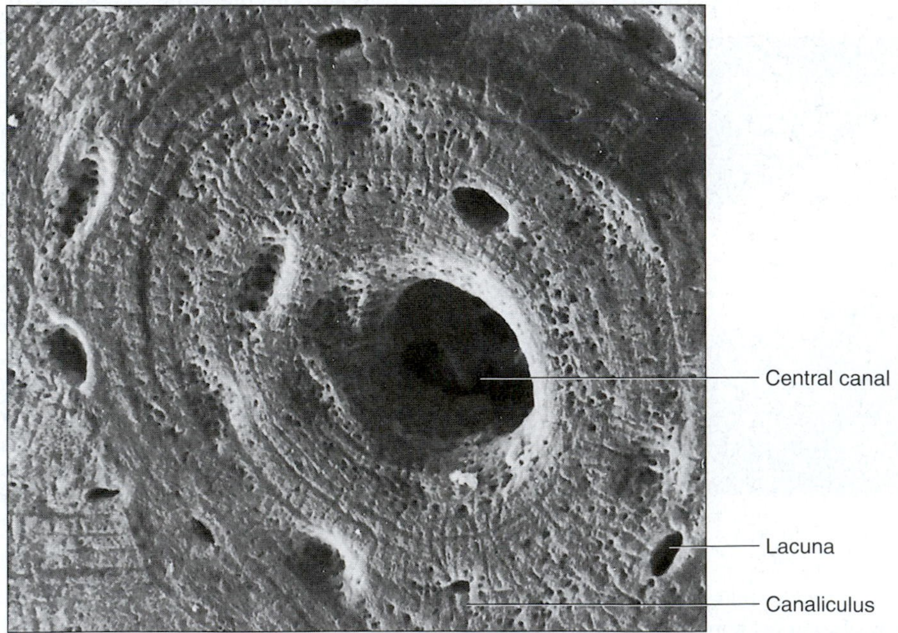

FIGURE 7.5 Scanning electron micrograph of a single osteon in compact bone (575×). *Tissues and Organs: A Text-Atlas of Scanning Electron Microscopy,* by R. G. Kessel and R. H. Kardon. © 1979 W. H. Freeman and Company.

7.3 BONE DEVELOPMENT AND GROWTH

Parts of the skeletal system begin to form during the first few weeks of prenatal development, and bony structures continue to grow and develop into adulthood. Bones form by replacing existing connective tissue in one of two ways. Some bones originate within sheetlike layers of connective tissues; they are called *intramembranous bones.* Others begin as masses of cartilage later replaced by bone tissue; they are called *endochondral bones* (fig. 7.6).

Intramembranous Bones

The broad, flat bones of the skull are **intramembranous bones** (in″trah-mem′brah-nus bōnz). During their development (osteogenesis), membranelike layers of unspecialized, or relatively undifferentiated, connective tissues appear at the sites of the future bones. Dense networks of blood vessels supply these connective tissue layers, which may form around the vessels. These partially differentiated progenitor cells enlarge and further differentiate into bone-forming cells called **osteoblasts** (os′te-o-blasts), which, in turn, deposit bony matrix around themselves. As a result, spongy bone forms in all directions along blood vessels within the layers of connective tissues. Later, some spongy bone may become compact bone as spaces fill with bone matrix.

As development continues, the osteoblasts may become completely surrounded by extracellular matrix, and in this manner, they become secluded within lacunae. At the same time, extracellular matrix enclosing the cellular processes of the osteoblasts gives rise to canaliculi. Once isolated in lacunae, these cells are called *osteocytes* (fig. 7.7).

Cells of the connective tissue that persist outside the developing bone give rise to the periosteum. Osteoblasts on the inside of the periosteum form a layer of compact bone over the surface of the newly formed spongy bone.

This process of replacing connective tissue to form an intramembranous bone is called *intramembranous ossification.* Table 7.1 lists the major steps of the process.

Endochondral Bones

Most of the bones of the skeleton are **endochondral bones** (en′do-kon′dral bōnz). They develop from masses of hyaline cartilage shaped like future bony structures. These cartilaginous models grow rapidly for a time and then begin to change extensively. Cartilage cells enlarge and their lacunae grow. The surrounding matrix breaks down, and soon the cartilage cells die and degenerate.

As the cartilage decomposes, a periosteum forms from connective tissue that encircles the developing structure. Blood vessels and partially differentiated connective tissue cells invade the disintegrating tissue. Some of the invading cells further differentiate into osteoblasts and begin to form spongy bone in the spaces previously housing the cartilage.

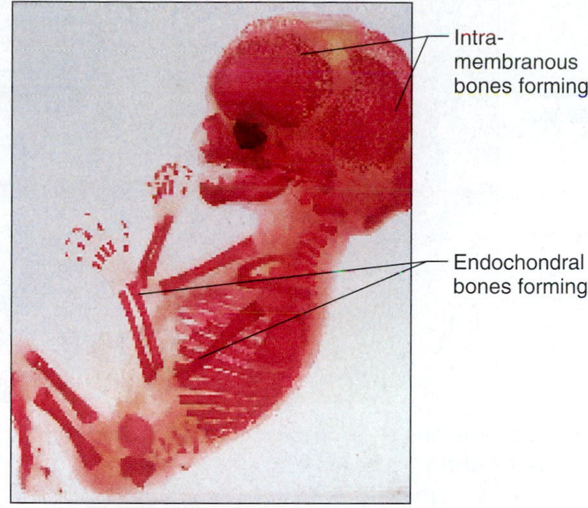

(a)

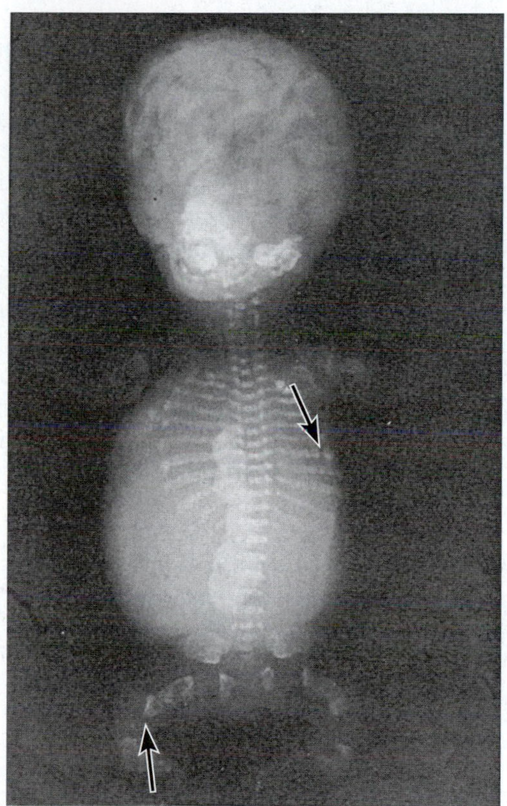

(b)

FIGURE 7.6 Fetal skeleton. (*a*) Note the stained bones of this fourteen-week fetus. (*b*) Bones can fracture even before birth. This fetus has numerous broken bones (arrows) because of an inherited defect in collagen called osteogenesis imperfecta, which causes brittle bones.

Once completely surrounded by the bony matrix, osteoblasts are called osteocytes. As ossification continues, osteoblasts beneath the periosteum deposit compact bone around the spongy bone.

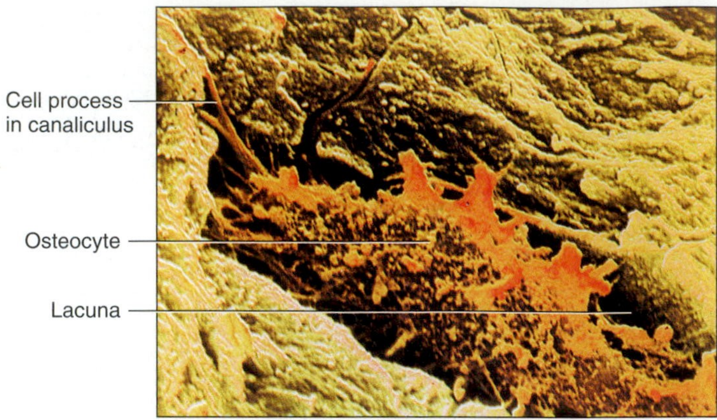

FIGURE 7.7 Scanning electron micrograph (falsely colored) of an osteocyte isolated in a lacuna (4,700×).

The process of forming an endochondral bone by the replacement of hyaline cartilage is called *endochondral ossification.* Its major steps are listed in table 7.1 and illustrated in figure 7.8.

In a long bone, bony tissue begins to replace hyaline cartilage in the center of the diaphysis. This region is called the *primary ossification center,* and bone develops from it toward the ends of the cartilaginous structure. Meanwhile, osteoblasts from the periosteum deposit a thin layer of compact bone around the primary ossification center. The epiphyses of the developing bone remain cartilaginous and continue to grow. Later, *secondary ossification centers* appear in the epiphyses, and spongy bone forms in all directions from them. As spongy bone is deposited in the diaphysis and in the epiphysis, a band of cartilage called the **epiphyseal plate** (ep″ĭ-fiz′e-al plāt) remains between the two ossification centers (see figs. 7.2, 7.3*b,* and 7.8).

Growth at the Epiphyseal Plate

In a long bone, the diaphysis is separated from the epiphysis by an epiphyseal plate. The cartilaginous cells of the epiphyseal plate form four layers, each of which may be several cells thick, as shown in figure 7.9. The first layer, or *zone of resting cartilage,* is closest to the end of the epiphysis. It is composed of resting cells that do not actively participate in growth. This layer anchors the epiphyseal plate to the bony tissue of the epiphysis.

The second layer of the epiphyseal plate, or *zone of proliferating cartilage,* includes rows of many young cells undergoing mitosis. As new cells appear and as extracellular matrix forms around them, the cartilaginous plate thickens.

The rows of older cells, left behind when new cells appear, form the third layer, or *zone of hypertrophic cartilage,* enlarging and thickening the epiphyseal plate still more. Consequently, the entire bone lengthens. At the same time, invading osteoblasts, which secrete calcium salts, accumulate in the extracellular matrix adjacent to the oldest cartilaginous cells, and as the extracellular matrix calcifies, the cells begin to die.

The fourth layer of the epiphyseal plate, or *zone of calcified cartilage,* is thin. It is composed of dead cells and calcified extracellular matrix.

In time, large, multinucleated cells called **osteoclasts** (os′te-o-klasts) break down the calcified matrix. These large cells originate from the fusion of single-nucleated white blood cells called monocytes (see chapter 14, p. 532). Osteoclasts secrete an acid that dissolves the inorganic component of the calcified matrix, and their lysosomal enzymes digest the organic components. Osteoclasts also phagocytize components of the bony matrix. After osteoclasts remove the extracellular matrix, bone-building osteoblasts invade the region and deposit bone tissue in place of the calcified cartilage.

In bone cancers, abnormally active osteoclasts destroy bone tissue. Interestingly, advanced cancer of the prostate gland can have the opposite effect. If such cancer cells reach the bone marrow, they stimulate osteoblast activity. This promotes formation of new bone on the surfaces of the bony trabeculae.

TABLE 7.1 | Major Steps in Bone Development

Intramembranous Ossification	Endochondral Ossification
1. Sheets of relatively undifferentiated connective tissue appear at sites of future bones.	1. Masses of hyaline cartilage form models of future bones.
2. Partially differentiated connective tissue cells collect around blood vessels in these layers.	2. Cartilage tissue breaks down. Periosteum develops.
3. Connective tissue cells further differentiate into osteoblasts, which deposit spongy bone.	3. Blood vessels and differentiating osteoblasts from the periosteum invade the disintegrating tissue.
4. Osteoblasts become osteocytes when bony matrix completely surrounds them.	4. Osteoblasts form spongy bone in the space occupied by cartilage.
5. Connective tissue on the surface of each developing structure forms a periosteum.	5. Osteoblasts beneath the periosteum deposit a thin layer of compact bone.
6. Osteoblasts on the inside of the periosteum deposit compact bone over the spongy bone.	6. Osteoblasts become osteocytes when bony matrix completely surrounds them.

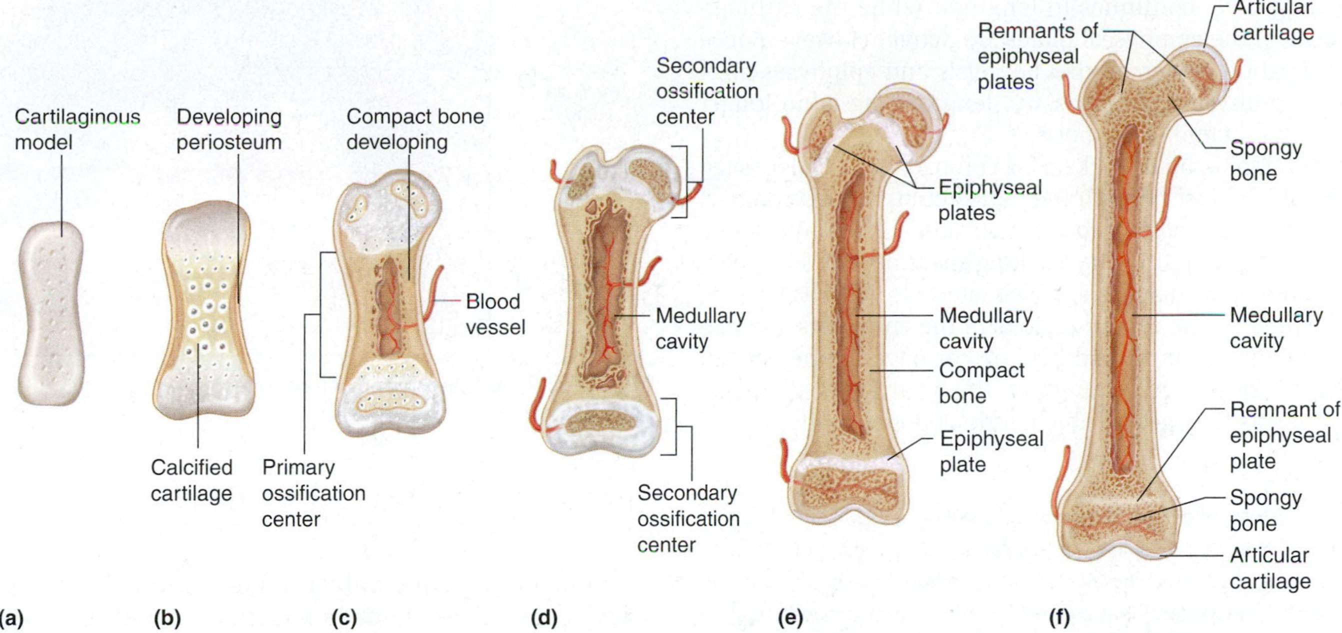

FIGURE 7.8 Major stages (*a–d* fetal, *e* child, *f* adult) in the development of an endochondral bone. (Relative bone sizes are not to scale.)

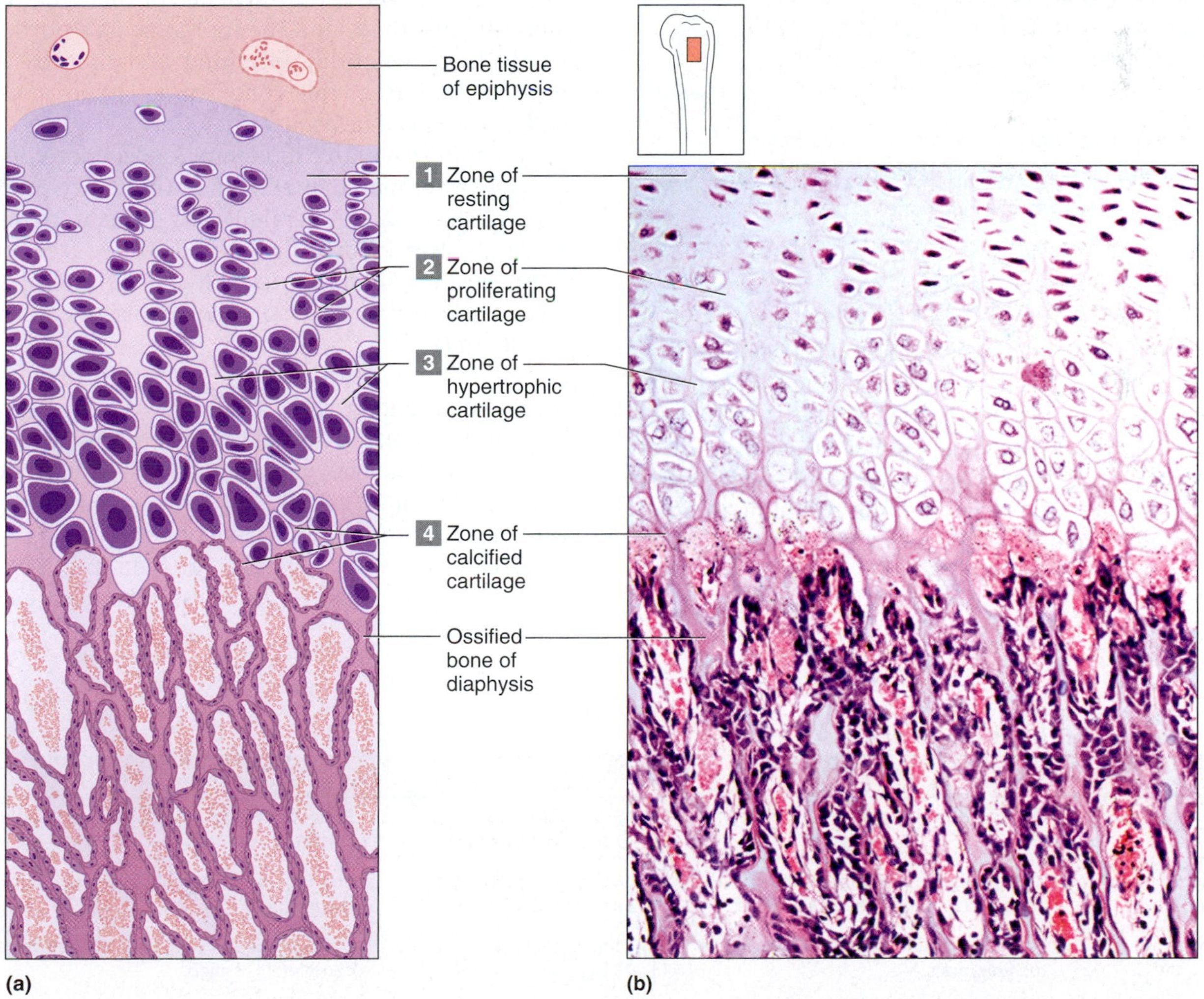

Bone tissue of epiphysis
1 Zone of resting cartilage
2 Zone of proliferating cartilage
3 Zone of hypertrophic cartilage
4 Zone of calcified cartilage
Ossified bone of diaphysis
(a)
(b)

FIGURE 7.9 Epiphyseal plate. (*a*) The cartilaginous cells of an epiphyseal plate lie in four layers, each of which may be several cells thick. (*b*) A micrograph of an epiphyseal plate (100×).

A long bone continues to lengthen while the cartilaginous cells of the epiphyseal plates are active. However, once the ossification centers of the diaphysis and epiphyses meet and the epiphyseal plates ossify, lengthening is no longer possible in that end of the bone.

A developing bone thickens as compact bone is deposited on the outside, just beneath the periosteum. As this compact bone forms on the surface, osteoclasts erode other bone tissue on the inside (fig. 7.10). The resulting space becomes the medullary cavity of the diaphysis, which later fills with marrow.

The bone in the central regions of the epiphyses and diaphysis remains spongy, and hyaline cartilage on the ends of the epiphyses persists throughout life as articular cartilage. Table 7.2 lists the ages at which various bones ossify.

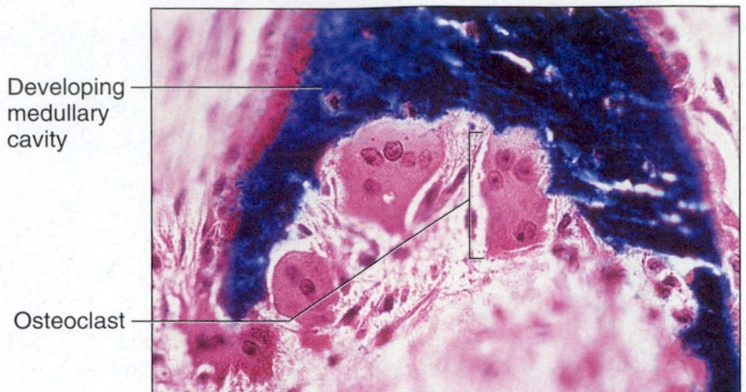

FIGURE 7.10 Micrograph of a bone-resorbing osteoclast (800×).

> If a child's long bones are still growing, a radiograph will reveal epiphyseal plates (fig. 7.11). If a plate is damaged as a result of a fracture before it ossifies, elongation of that long bone may prematurely cease, or if growth continues, it may be uneven. For this reason, injuries to the epiphyses of a young person's bones are of special concern. Surgery is used on an epiphysis to equalize growth of bones developing at very different rates.

PRACTICE

6 Describe the development of an intramembranous bone.

7 Explain how an endochondral bone develops.

8 List the steps in the growth of a long bone.

Homeostasis of Bone Tissue

After the intramembranous and endochondral bones form, the actions of osteoclasts and osteoblasts continually remodel them. **Bone remodeling** occurs throughout life as osteoclasts resorb bone tissue, and osteoblasts replace the bone. These opposing processes of *resorption* and *deposition* are highly regulated so that the total mass of bone tissue in an adult skeleton normally remains nearly constant, even though 3% to 5% of bone calcium is exchanged each year.

Factors Affecting Bone Development, Growth, and Repair

A number of factors influence bone development, growth, and repair. These include nutrition, exposure to sunlight, hormonal secretions, and physical exercise. For example, vitamin D is necessary for proper absorption of calcium in the small intestine. In the absence of this vitamin, calcium is poorly absorbed, and the inorganic salt portion of bone matrix lacks calcium, softening and thereby deforming bones. In children, this condition is called *rickets*, and in adults, it is called *osteomalacia*.

Vitamin D is scarce in natural foods, except for eggs, but it is readily available in milk and other dairy products fortified with vitamin D. Vitamin D also forms from dehydrocholesterol, produced by cells in the digestive tract or obtained in the diet. The blood carries dehydrocholesterol to the skin, where exposure to ultraviolet light from the sun converts it to vitamin D.

Vitamins A and C are also required for normal bone development and growth. Vitamin A is necessary for osteoblast and osteoclast activity during normal development. This is why deficiency of vitamin A may retard bone development. Vitamin C is required for collagen synthesis, so its lack may also inhibit bone development. In this case, osteoblasts cannot produce enough collagen in the extracellular matrix of the bone tissue. As a result, bones are abnormally slender and fragile.

TABLE 7.2 | Ossification Timetable

Age	Occurrence	Age	Occurrence
Third month of prenatal development	Ossification in long bones begins.	15 to 18 years (females) 17 to 20 years (males)	Bones of the upper limbs and scapulae completely ossified.
Fourth month of prenatal development	Most primary ossification centers have appeared in the diaphyses of bones.	16 to 21 years (females) 18 to 23 years (males)	Bones of the lower limbs and hip bones completely ossified.
Birth to 5 years	Secondary ossification centers appear in the epiphyses.	21 to 23 years (females) 23 to 25 years (males)	Bones of the sternum, clavicles, and vertebrae completely ossified.
5 to 12 years (females) 5 to 14 years (males)	Ossification rapidly spreads from the ossification centers.	By 23 years (females) By 25 years (males)	Nearly all bones completely ossified.

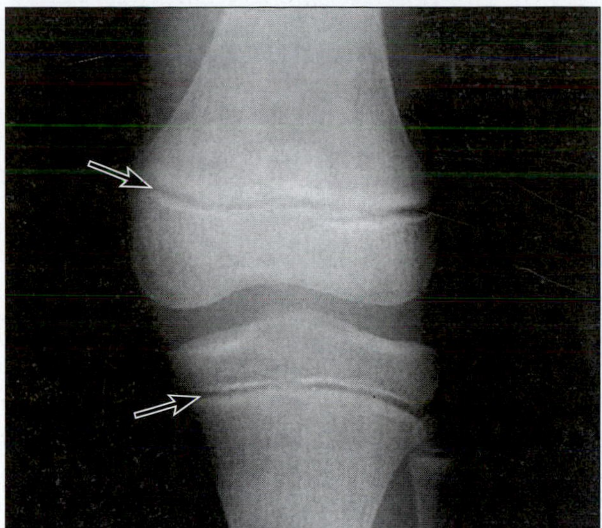

FIGURE 7.11 Radiograph showing epiphyseal plates (arrows) in a child's bones indicates that the bones are still lengthening.

About 90 % of the protein that is part of bone is collagen. Less abundant bone proteins are important too.

- Osteocalcin is activated by vitamin K to bind calcium, which in bone is part of the compound hydroxyapatite the main component of bone matrix.
- Osteonectin binds hydroxyapatite and collagen and stimulates mineral crystal deposition in bone.
- Osteopontin speeds bone remodeling.
- Bone morphogenetic proteins include growth factors that induce bone and cartilage formation. They are used in spinal fusion procedures.

Hormones secreted by the pituitary gland, thyroid gland, parathyroid glands, and ovaries or testes affect bone growth and development. The pituitary gland secretes **growth hormone,** which stimulates division of cartilage cells in the epiphyseal plates. In the absence of this hormone, the long bones of the limbs fail to develop normally, and the child has *pituitary dwarfism.* He or she is very short but has normal body proportions. If excess growth hormone is released before the epiphyseal plates ossify, height may exceed 8 feet—a condition called *pituitary gigantism.* In an adult, secretion of excess growth hormone causes *acromegaly,* in which the hands, feet, and jaw enlarge (see chapter 13, pp. 494–495).

The thyroid hormone thyroxine stimulates replacement of cartilage in the epiphyseal plates of long bones with bone tissue. This hormone increases cellular metabolism, including stimulating osteoblast activity. In contrast to the bone-forming activity of thyroid hormone, parathyroid hormone stimulates an increase in the number and activity of osteoclasts, which break down bone (see chapter 13, pp. 501–503).

Both male and female sex hormones (called testosterone and estrogens, respectively) from the testes and ovaries promote formation of bone tissue. Beginning at puberty, these hormones are abundant, causing the long bones to grow considerably (see chapter 22, pp. 846, 857). However, sex hormones also stimulate ossification of the epiphyseal plates, and consequently they stop bone lengthening at a relatively early age. The effect of estrogens on the epiphyseal plates is somewhat stronger than that of testosterone. For this reason, females typically reach their maximum heights earlier than males.

Physical stress also stimulates bone growth. For example, when skeletal muscles contract, they pull at their attachments on bones, and the resulting stress stimulates the bone tissue to thicken and strengthen (hypertrophy). Conversely, with lack of exercise, the same bone tissue wastes, becoming thinner and weaker (atrophy). This is why the bones of athletes are usually stronger and heavier than those of nonathletes (fig. 7.12). It is also why fractured bones immobilized in casts may shorten. Clinical Application 7.1 describes what happens when a bone breaks.

Astronauts experience a 1% loss of bone mass per month in space. Under microgravity conditions, osteoblast activity decreases and osteoclast activity increases, with greater loss in spongy compared to compact bone. Researchers predict that a 50% bone loss could occur on a several-year-long space flight, such as a mission to Mars.

PRACTICE

9 Explain how nutritional factors affect bone development.

10 What effects do hormones have on bone growth?

11 How does physical exercise affect bone structure?

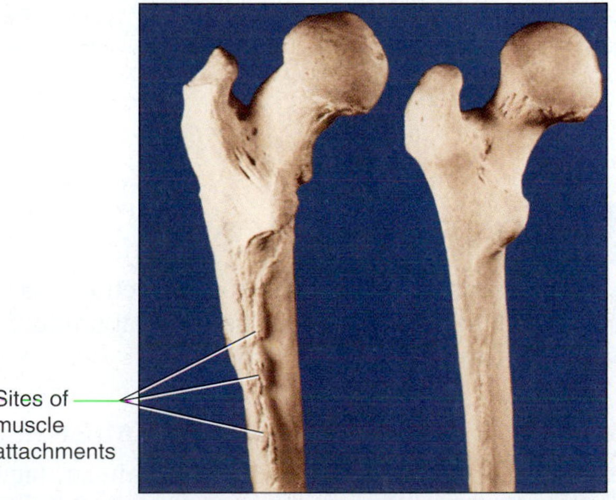

FIGURE 7.12 Note the increased amount of bone at the sites of muscle attachments in the femur on the left. The thickened bone is better able to withstand the forces resulting from muscle contraction.

7.1 CLINICAL APPLICATION

Fractures

When seven-year-old Jacob fell from the tree limb, he had been hanging about eight feet from the ground. He landed in a crumpled heap, crying, with his right leg at an abnormal angle. Emergency medical technicians immobilized the leg and took Jacob to the emergency department at the nearest hospital, where an X ray indicated a broken tibia. He spent the next six weeks in a cast, and the bone continued to heal over several months. By the next summer, Jacob was again climbing trees—but more carefully.

Many of us have experienced fractured, or broken, bones. A fracture is classified by its cause and the nature of the break. For example, a break due to injury is a traumatic fracture, whereas one resulting from disease is a *spontaneous,* or *pathologic, fracture.* A broken bone exposed to the outside by an opening in the skin is termed a *compound (open) fracture.* It has the added danger of infection, because microorganisms enter through the broken skin. A break protected by uninjured skin is a *closed fracture.* Figure 7A shows several types of traumatic fractures.

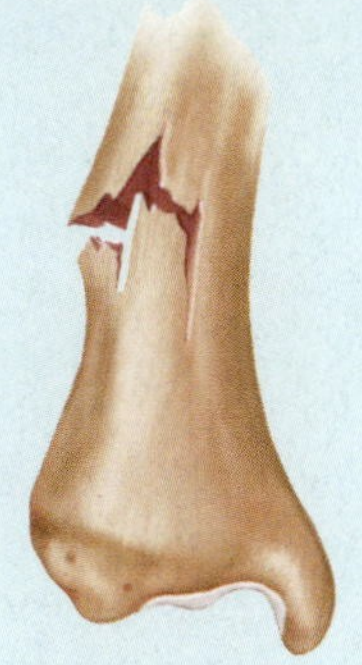

A *greenstick* fracture is incomplete, and the break occurs on the convex surface of the bend in the bone.

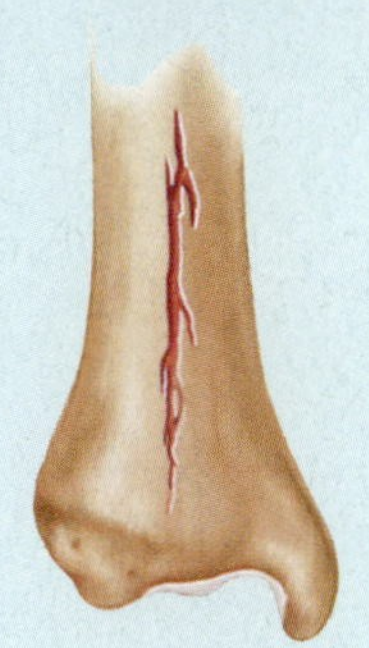

A *fissured* fracture involves an incomplete longitudinal break.

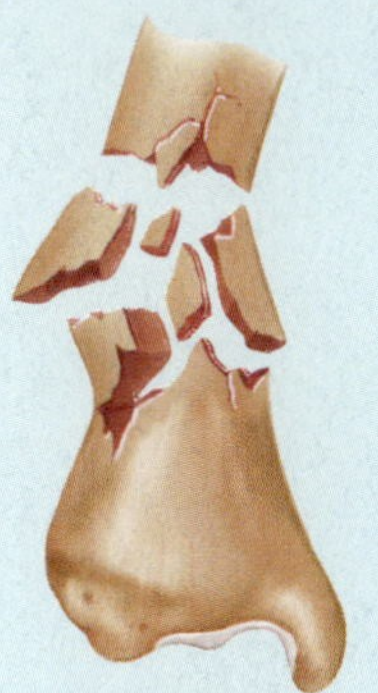

A *comminuted* fracture is complete and fragments the bone.

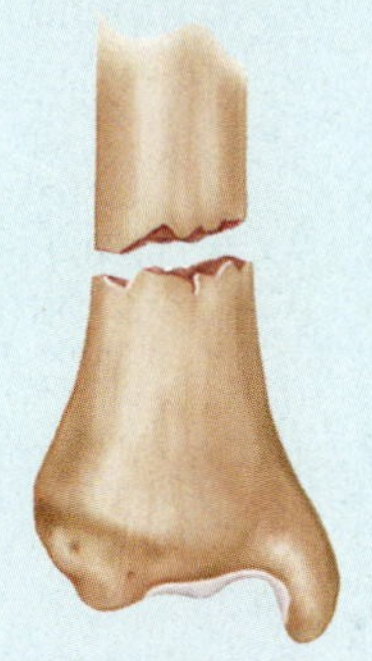

A *transverse* fracture is complete, and the break occurs at a right angle to the axis of the bone.

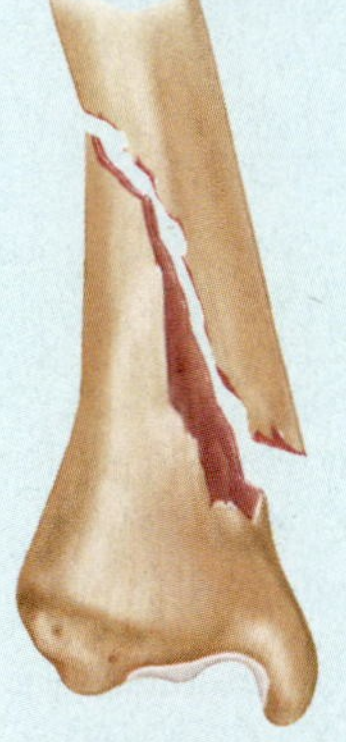

An *oblique* fracture occurs at an angle other than a right angle to the axis of the bone.

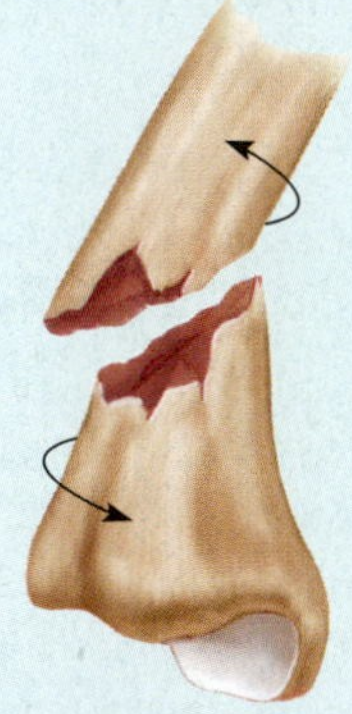

A *spiral* fracture is caused by twisting a bone excessively.

FIGURE 7A Various types of fractures.

Repair of a Fracture

When a bone breaks, blood vessels in it rupture, and the periosteum is likely to tear. Blood from the broken vessels spreads through the damaged area and soon forms a blood clot, or *hematoma.* Vessels in surrounding tissues dilate, swelling and inflaming tissues.

Within days or weeks, developing blood vessels and large numbers of osteoblasts originating from the periosteum invade the hematoma. The osteoblasts rapidly divide in the regions close to the new blood vessels, building spongy bone nearby. Granulation tissue develops, and in regions farther from a blood supply, fibroblasts produce masses of fibrocartilage. Meanwhile, phagocytic cells begin to remove the blood clot as well as any dead or damaged cells in the

7.4 BONE FUNCTION

Bones shape, support, and protect body structures, as well as aid body movements. They house tissues that produce blood cells and store various inorganic salts.

Support, Protection, and Movement

Bones give shape to structures such as the head, face, thorax, and limbs. They also support and protect. For example, the bones of the lower limbs, pelvis, and vertebral column support the body's weight. The bones of the skull protect the eyes, ears, and brain. Bones of the rib cage and shoulder girdle protect the heart and lungs, whereas bones of the pelvic girdle protect the lower abdominal and internal reproductive organs. Whenever limbs or other body parts move, bones and muscles interact.

Blood Cell Formation

The process of blood cell formation, called **hematopoiesis** (hem″ah-to-poi-e′sis), or hemopoiesis, begins in the yolk sac, which lies outside the embryo (see chapter 23, p. 891). Later in development, blood cells are manufactured in the liver and spleen, and still later, they form in bone marrow.

affected area. Osteoclasts also appear and resorb bone fragments, aiding in "cleaning up" debris.

In time, fibrocartilage fills the gap between the ends of the broken bone. This mass, termed a cartilaginous callus, is later replaced by bone tissue in much the same way that the hyaline cartilage of a developing endochondral bone is replaced. That is, the cartilaginous callus breaks down, blood vessels and osteoblasts invade the area, and a bony callus fills the space.

Typically, more bone is produced at the site of a healing fracture than is necessary to replace the damaged tissues. Osteoclasts remove the excess, and the result is a bone shaped much like the original. Figure 7B shows the steps in the healing of a fracture.

If the ends of a broken bone are close together, healing is faster than if they are far apart. Physicians can help the bone-healing process. The first casts to immobilize fractured bones were introduced in Philadelphia in 1876, and soon after, doctors began using screws and plates internally to align healing bone parts. Today, orthopedic surgeons also use rods, wires, and nails. These devices have become lighter and smaller; many are built of titanium. A new approach, called a hybrid fixator, treats a broken leg using metal pins internally to align bone pieces. The pins are anchored to a metal ring device worn outside the leg.

Some bones naturally heal more rapidly than others. The long bones of the upper limbs, for example, may heal in half the time required by the long bones of the lower limbs, as Jacob was unhappy to discover. However, his young age would favor quicker healing. ■

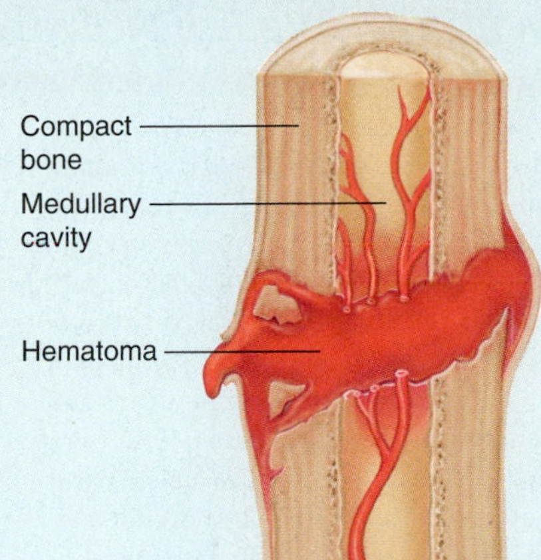

(a) Blood escapes from ruptured blood vessels and forms a hematoma.

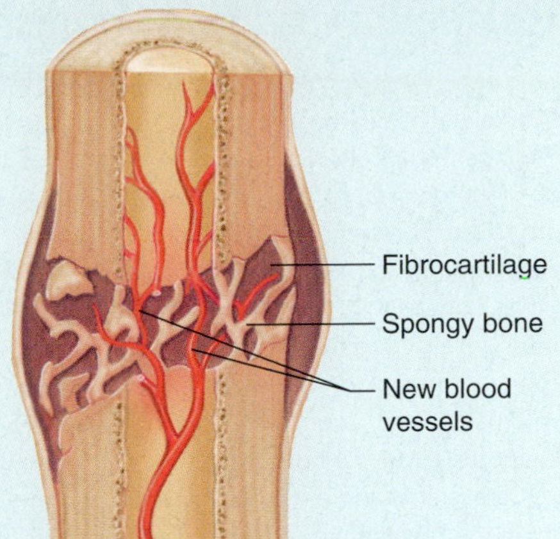

(b) Spongy bone forms in regions close to developing blood vessels, and fibrocartilage forms in more distant regions.

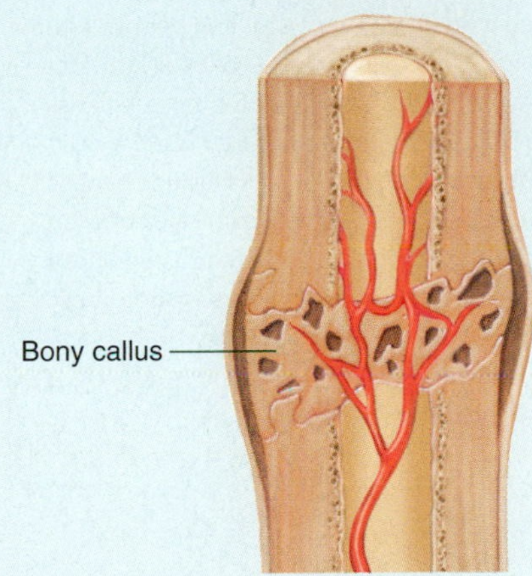

(c) A bony callus replaces fibrocartilage.

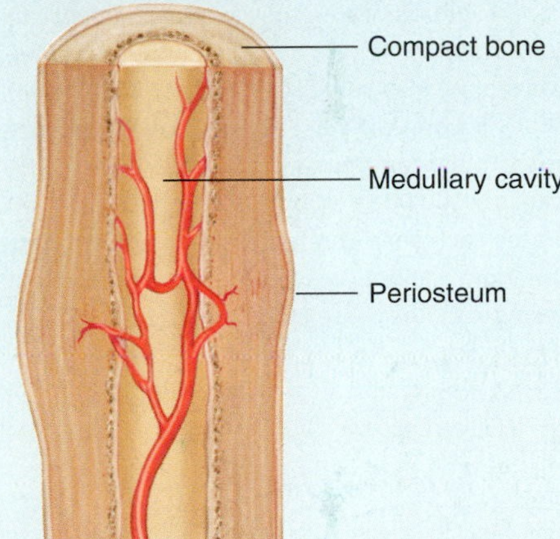

(d) Osteoclasts remove excess bony tissue, restoring new bone structure much like the original.

FIGURE 7B Major steps (*a–d*) in the repair of a fracture.

Marrow is a soft, netlike mass of connective tissue in the medullary cavities of long bones, in the irregular spaces of spongy bone, and in the larger central canals of compact bone tissue. The two types of marrow are red and yellow. *Red marrow* functions in the formation of red blood cells (erythrocytes), white blood cells (leukocytes), and blood platelets. It is red because of the red, oxygen-carrying pigment **hemoglobin** in red blood cells.

The distribution of marrow changes with age. In an infant, red marrow occupies the cavities of most bones. With increasing age, however, yellow marrow replaces much of it. *Yellow marrow* stores fat and is inactive in blood cell production. In an adult, red marrow is primarily found in the spongy bone of the skull, ribs, sternum, clavicles, vertebrae, and hip bones. If the blood cell supply is deficient, some yellow marrow may change back into red marrow and produce blood cells. Chapter 14 (p. 524) discusses blood cell formation.

Inorganic Salt Storage

Recall that the extracellular matrix of bone tissue includes collagen and inorganic mineral salts. The salts account for about 70% of the extracellular matrix by weight and are mostly small crystals of a type of calcium phosphate called *hydroxyapatite*. Clinical Application 7.2 discusses osteoporosis, a condition that results from loss of bone mineral.

7.2 CLINICAL APPLICATION

Osteopenia and Osteoporosis: Preventing "Fragility Fractures"

Skeletal health is a matter of balance. Before age thirty, cells that form new bone tissue (osteoblasts) counter cells that degrade it (osteoclasts), keeping living bone in a constant state of remodeling. Over time, the balance shifts so that bone is lost, especially in women past menopause due to falling estrogen levels. This imbalance may progress to osteopenia ("low bone mass") and, eventually, the more severe osteoporosis ("porous bones").

Osteopenia and osteoporosis are a continuum in the breakdown of the microarchitecture of bone tissue. By the time the declining bone mass is considered osteoporosis, trabeculae are lost and the bones develop spaces and canals, which enlarge and fill with fibrous and fatty tissues. Such bones easily fracture and may spontaneously break because they are no longer able to support body weight. Sections of the backbone (vertebrae) may collapse or the distal portion of a forearm bone (radius) snap as a result of minor stress. Most common is hip fracture, which happens to 200,000 senior citizens in the United States each year. The femur may begin to fracture from a minor movement before the fall that seems to be the cause of the break.

A "fragility fracture" is a telltale sign of dangerously low bone density. This is a fracture that happens after a fall from less than standing height, which a strong, healthy skeleton could resist. Fragility fractures occur in 1.5 million people in the United States each year, yet despite this warning sign, only one quarter to one third of them are followed up with bone scans and treatment to build new bone tissue. Since 1995, five new drugs have become available to treat osteoporosis. One class, the bisphosphonates, builds new bone. They are taken once a week or once a month.

Osteopenia and osteoporosis are common. The Surgeon General estimates that half of people over age fifty have either condition, which amounts to thirty-five million people with osteopenia and another ten million people with osteoporosis. Screening is advised for all individuals over age sixty-five, as well as for those with risk factors. The most telling predictor is a previous fragility fracture. Other risk factors include genetic predisposition, low dietary calcium, lack of exercise, smoking, drinking alcohol, recent weight loss, recent height loss (this could be an asymptomatic compression fracture), and older age.

Osteopenia and osteoporosis are assessed by measuring bone mineral density (BMD). This is most often done in the hip bone and lower spine with a technique called dual-energy X-ray absorptiometry. Osteopenia is defined as BMD at least 1 to 2.5 standard deviations below the mean. Osteoporosis is defined as BMD at least 2.5 standard deviations below the mean for young adults. These measurements produce T values. Another measurement, a Z value, compares BMD to other individuals of a person's age and is used to assess skeletal health in individuals under age sixty-five.

People approaching retirement age are not the only ones who should be concerned about osteopenia and osteoporosis, because these conditions can be delayed or prevented if dealt with early. Researchers think that what puts people at risk is failing to attain maximal possible bone density by age thirty. To keep bones as strong as possible for as long as possible, it is essential to get at least thirty minutes of exercise daily (some weight bearing), consume enough daily calcium (1,000–1,200 mg) and vitamin D (200 IU), and not smoke. There is much you can do to promote skeletal health—at any age. ■

The human body requires calcium for a number of vital metabolic processes, including muscle cell contraction, nerve impulse conduction, and blood clot formation. When the blood is low in calcium, parathyroid hormone stimulates osteoclasts to break down bone tissue, releasing calcium salts from the extracellular matrix into the blood. On the other hand, very high blood calcium inhibits osteoclast activity, and calcitonin from the thyroid gland stimulates osteoblasts to form bone tissue, storing excess calcium in the extracellular matrix (fig. 7.13). This response is particularly important in developing bone matrix in children. The details of this homeostatic mechanism are in chapter 13, pp. 501–503.

In addition to storing calcium and phosphorus (as calcium phosphate), bone tissue contains smaller amounts of magnesium, sodium, potassium, and carbonate ions. Bones also accumulate certain harmful metallic elements such as lead, radium, and strontium, which are not normally present in the body but sometimes accidentally ingested.

Biomineralization—the combining of minerals with organic molecules, as occurs in bones—is seen in many animal species. Ancient Mayan human skulls have teeth composed of nacre, also known as "mother-of-pearl" (found in clam shells) attached to human tooth roots. The Mayan dentists knew that the human body could somehow incorporate a biomineral from another species. Today, nacre is used to fill in bone lost in the upper jaw. The nacre not only does not evoke rejection by the immune system, but it also stimulates the person's osteoblasts to produce new bone tissue.

PRACTICE

12 Name the major functions of bones.

13 Distinguish between the functions of red marrow and yellow marrow.

14 Explain regulation of the concentration of blood calcium.

15 List the substances normally stored in bone tissue.

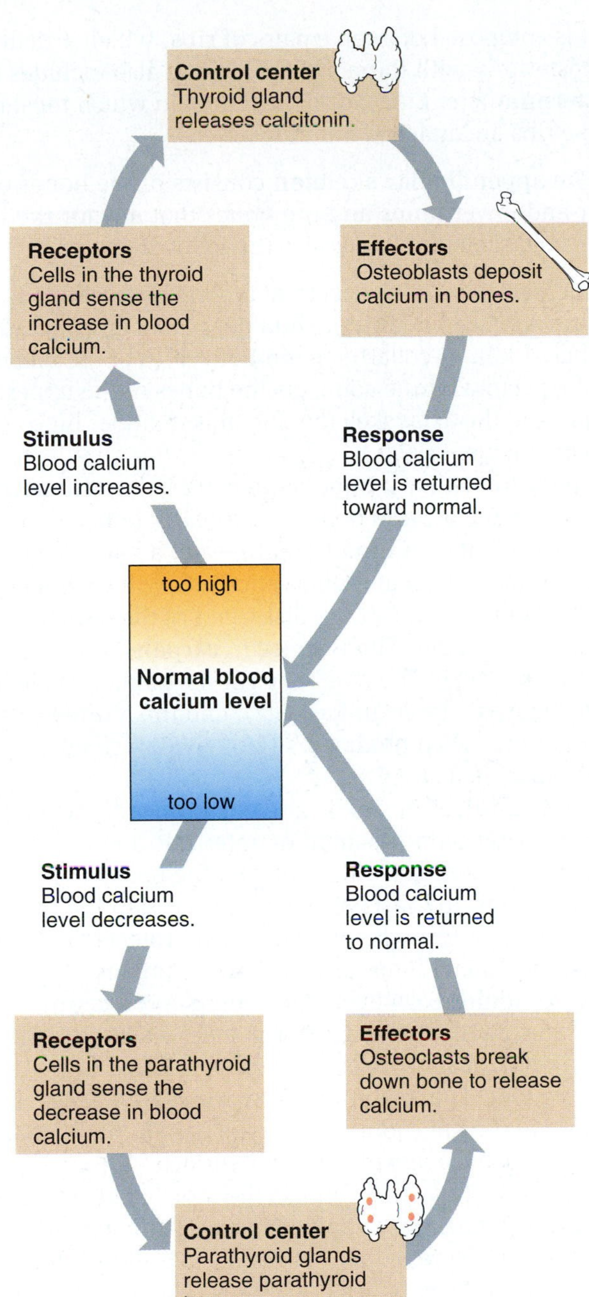

FIGURE 7.13 Hormonal regulation of bone calcium resorption and deposition.

7.5 SKELETAL ORGANIZATION

Number of Bones

The number of bones in a human skeleton is often reported to be 206 (table 7.3), but the number varies from person to person. People may lack certain bones or have extra ones. For example, the flat bones of the skull usually grow together and tightly join along irregular lines called *sutures*. Occasionally, extra bones called sutural bones (wormian bones) develop in

TABLE 7.3 | Bones of the Adult Skeleton

1. Axial Skeleton	
a. Skull	22 bones
8 cranial bones	
frontal 1	
parietal 2	
occipital 1	
temporal 2	
sphenoid 1	
ethmoid 1	
14 facial bones	
maxilla 2	
palatine 2	
zygomatic 2	
lacrimal 2	
nasal 2	
vomer 1	
inferior nasal concha 2	
mandible 1	
b. Middle ear bones	6 bones
malleus 2	
incus 2	
stapes 2	
c. Hyoid	1 bone
hyoid bone 1	
d. Vertebral column	26 bones
cervical vertebra 7	
thoracic vertebra 12	
lumbar vertebra 5	
sacrum 1	
coccyx 1	
e. Thoracic cage	25 bones
rib 24	
sternum 1	
2. Appendicular Skeleton	
a. Pectoral girdle	4 bones
scapula 2	
clavicle 2	
b. Upper limbs	60 bones
humerus 2	
radius 2	
ulna 2	
carpal 16	
metacarpal 10	
phalanx 28	
c. Pelvic girdle	2 bones
hip bone 2	
d. Lower limbs	60 bones
femur 2	
tibia 2	
fibula 2	
patella 2	
tarsal 14	
metatarsal 10	
phalanx 28	
Total	206 bones

these sutures (fig. 7.14). Extra small, round sesamoid bones may develop in tendons, where they reduce friction in places where tendons pass over bony prominences.

Divisions of the Skeleton

For purposes of study, it is convenient to divide the skeleton into two major portions—an axial skeleton and an appendicular skeleton (fig. 7.15). The **axial skeleton** consists of the bony and cartilaginous parts that support and protect the organs of the head, neck, and trunk. These parts include the following:

1. **Skull.** The skull is composed of the *cranium* (brain case) and the *facial bones.*
2. **Hyoid bone.** The hyoid (hi′oid) bone is located in the neck between the lower jaw and the larynx (fig. 7.16). It does not articulate with any other bones but is fixed in position by muscles and ligaments. The hyoid bone supports the tongue and is an attachment for certain muscles that help move the tongue during swallowing. It can be felt approximately a finger's width above the anterior prominence of the larynx.
3. **Vertebral column.** The vertebral column, or spinal column, consists of many vertebrae separated by cartilaginous *intervertebral discs.* This column forms the central axis of the skeleton. Near its distal end, five vertebrae fuse to form the **sacrum** (sa′krum), part of the pelvis. A small tailbone formed by the fusion of four vertebrae and called the **coccyx** (kok′siks) is attached to the end of the sacrum.
4. **Thoracic cage.** The thoracic cage protects the organs of the thoracic cavity and the upper abdominal cavity. It is composed of twelve pairs of **ribs,** which articulate posteriorly with thoracic vertebrae. It also includes the **sternum** (ster′num), or breastbone, to which most of the ribs are attached anteriorly.

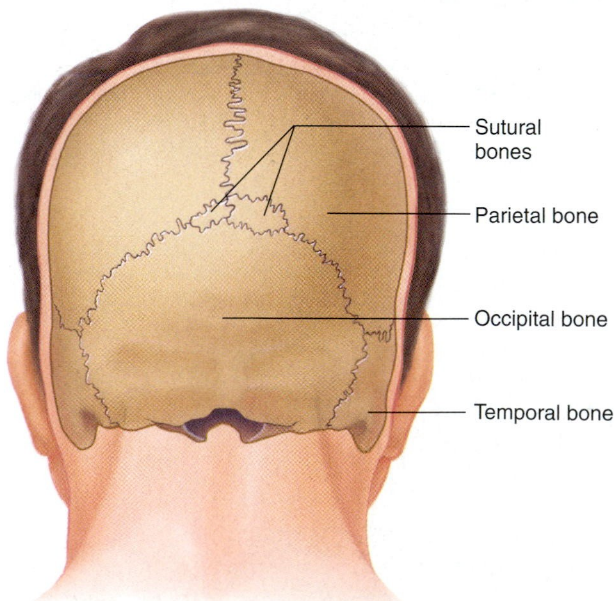

FIGURE 7.14 Sutural (wormian) bones are extra bones that sometimes develop in sutures between the flat bones of the skull.

The **appendicular skeleton** consists of the bones of the upper and lower limbs and the bones that anchor the limbs to the axial skeleton. It includes the following:

1. **Pectoral girdle.** The pectoral girdle is formed by a **scapula** (scap′u-lah), or shoulder blade, and a **clavicle** (klav′ĭ-k′l), or collarbone, on both sides of the body. The pectoral girdle connects the bones of the upper limbs to the axial skeleton and aids in upper limb movements.
2. **Upper limbs.** Each upper limb consists of a **humerus** (hu′mer-us), or arm bone; two forearm bones—a **radius** (ra′de-us) and an **ulna** (ul′nah)—and a hand. The humerus, radius, and ulna articulate with each other at the elbow joint. At the distal end of the radius and ulna is the hand. There are eight **carpals** (kar′palz), or wrist bones. The five bones of the palm are called **metacarpals** (met″ah-kar′palz), and the fourteen finger bones are called **phalanges** (fah-lan′jēz); singular, *phalanx,* (fa′lanks).
3. **Pelvic girdle.** The pelvic girdle is formed by two hip bones attached to each other anteriorly and to the sacrum posteriorly. They connect the bones of the lower limbs to the axial skeleton and, with the sacrum and coccyx, form the **pelvis,** which protects the lower abdominal and internal reproductive organs.
4. **Lower limbs.** Each lower limb consists of a **femur** (fe′mur), or thigh bone; two leg bones—a large **tibia** (tib′e-ah), or shin bone, and a slender **fibula** (fib′u-lah)—and a foot. The femur and tibia articulate with each other at the knee joint, where the **patella** (pah-tel′ah), or kneecap, covers the anterior surface. At the distal ends of the tibia and fibula is the foot. There are seven **tarsals** (tahr′salz), or ankle bones. The five bones of the instep are called **metatarsals** (met″ah-tar′salz), and the fourteen bones of the toes (like the fingers) are called **phalanges.** Table 7.4 defines some terms used to describe skeletal structures.

PRACTICE 

16 Distinguish between the axial and appendicular skeletons.

17 List the bones of the axial skeleton and of the appendicular skeleton.

7.6 SKULL

A human skull usually consists of twenty-two bones that, except for the lower jaw, are firmly interlocked along sutures. Eight of these interlocked bones make up the cranium and fourteen form the facial skeleton. The **mandible** (man′dĭ-b′l),

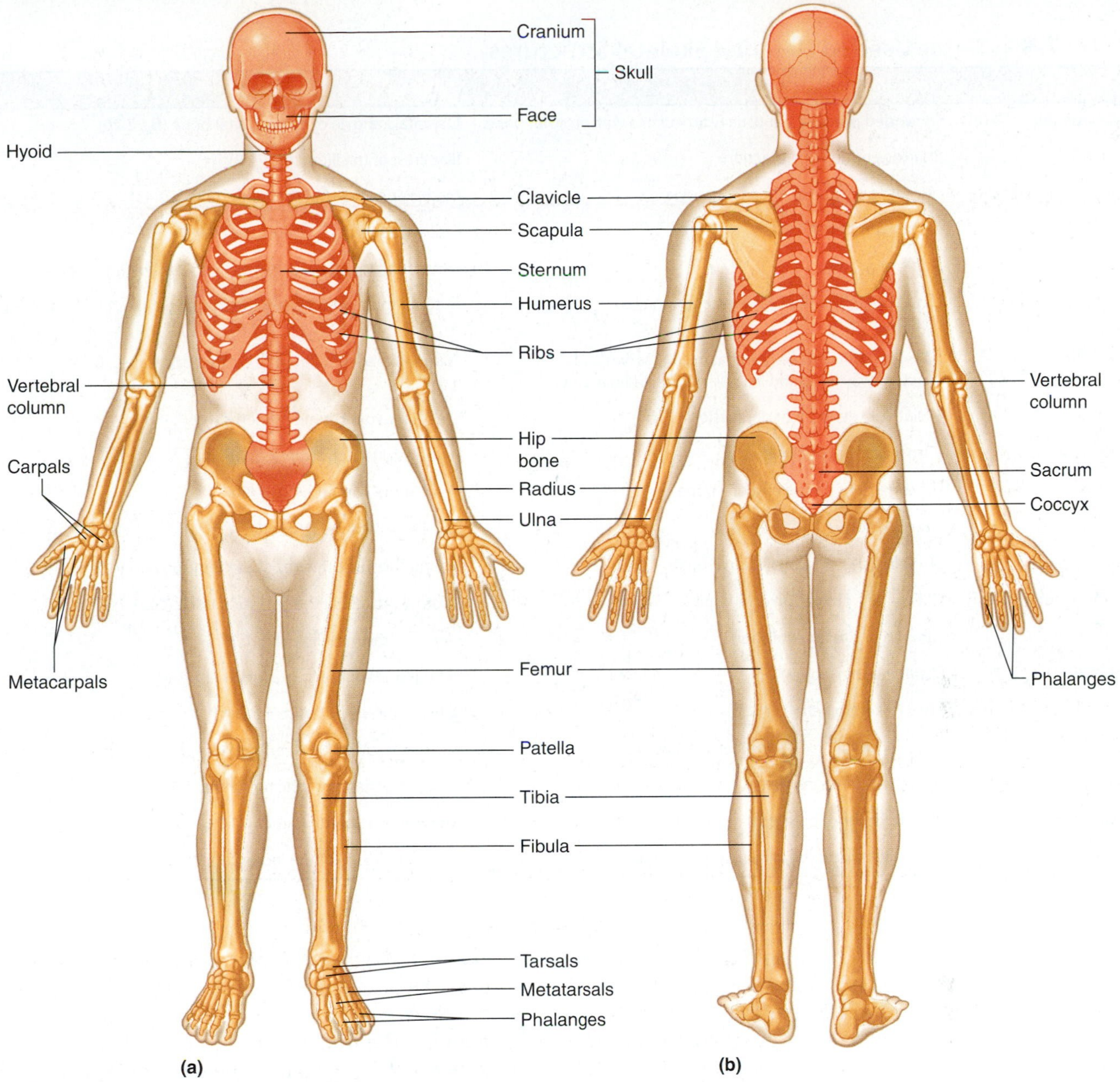

Cranium
Skull
Face
Hyoid
Clavicle
Scapula
Sternum
Humerus
Ribs
Vertebral column
Carpals
Metacarpals
Hip bone
Radius
Ulna
Femur
Patella
Tibia
Fibula
Tarsals
Metatarsals
Phalanges
(a)

Clavicle
Scapula
Humerus
Ribs
Vertebral column
Hip bone
Sacrum
Radius
Ulna
Coccyx
Phalanges
Femur
Tibia
Fibula
(b)

FIGURE 7.15 Major bones of the skeleton. (*a*) Anterior view. (*b*) Posterior view. The axial portion is shown in orange, and the appendicular portions are shown in yellow.

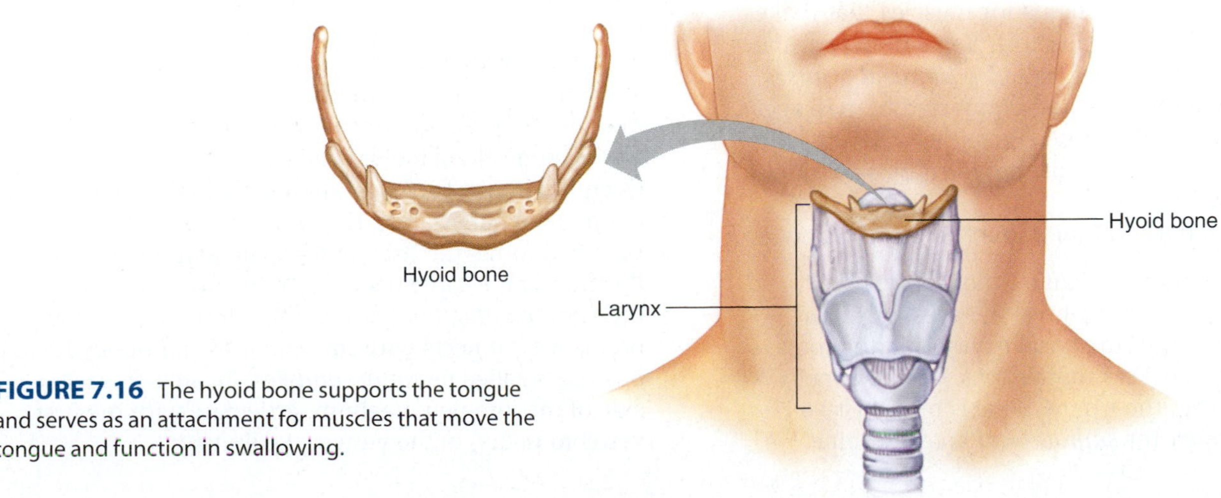

FIGURE 7.16 The hyoid bone supports the tongue and serves as an attachment for muscles that move the tongue and function in swallowing.

TABLE 7.4 | Terms Used to Describe Skeletal Structures

Term	Definition	Example
Condyle (kon′dīl)	Rounded process that usually articulates with another bone	Occipital condyle of the occipital bone (fig. 7.20)
Crest (krest)	Narrow, ridgelike projection	Iliac crest of the ilium (fig. 7.48)
Epicondyle (ep″ĭ-kon′dīl)	Projection situated above a condyle	Medial epicondyle of the humerus (fig. 7.43)
Facet (fas′et)	Small, nearly flat surface	Facet of a thoracic vertebra (fig. 7.36*b*)
Fissure (fish′ūr)	Cleft or groove	Inferior orbital fissure in the orbit of the eye (fig. 7.18)
Fontanel (fon″tah-nel′)	Soft spot in the skull where membranes cover the space between bones	Anterior fontanel between the frontal and parietal bones (fig. 7.31*a*)
Foramen (fo-ra′men)	Opening through a bone that usually serves as a passageway for blood vessels, nerves, or ligaments	Foramen magnum of the occipital bone (fig. 7.20)
Fossa (fos′ah)	Relatively deep pit or depression	Olecranon fossa of the humerus (fig. 7.43*b*)
Fovea (fo′ve-ah)	Tiny pit or depression	Fovea capitis of the femur (fig. 7.51*b*)
Head (hed)	Enlargement on the end of a bone	Head of the humerus (fig. 7.43)
Linea (lin′e-ah)	Narrow ridge	Linea aspera of the femur (fig. 7.51*b*)
Meatus (me-a′tus)	Tubelike passageway within a bone	External acoustic meatus of the temporal bone (fig. 7.19)
Process (pros′es)	Prominent projection on a bone	Mastoid process of the temporal bone (fig. 7.19)
Ramus (ra′mus)	Branch or similar extension	Ramus of the mandible (fig. 7.29*a*)
Sinus (si′nus)	Cavity within a bone	Frontal sinus of the frontal bone (fig. 7.25)
Spine (spīn)	Thornlike projection	Spine of the scapula (fig. 7.41*a, b*)
Suture (soo′cher)	Interlocking line of union between bones	Lambdoid suture between the occipital and parietal bones (fig. 7.19)
Trochanter (tro-kan′ter)	Relatively large process	Greater trochanter of the femur (fig. 7.51*a*)
Tubercle (tu′ber-kl)	Small, knoblike process	Tubercle of a rib (fig. 7.39)
Tuberosity (tu″bĕ-ros′ĭ-te)	Knoblike process usually larger than a tubercle	Radial tuberosity of the radius (fig. 7.44*a*)

or lower jawbone, is a movable bone held to the cranium by ligaments (figs. 7.17 and 7.19). Some facial and cranial bones together form the orbit of the eye (fig. 7.18). Plates 26–54 on pages 245–259 show a set of photographs of the human skull and its parts.

Cranium

The **cranium** (kra′ne-um) encloses and protects the brain, and its surface provides attachments for muscles that make chewing and head movements possible. Some of the cranial bones contain air-filled cavities called *paranasal sinuses,* lined with mucous membranes and connected by passageways to the nasal cavity. Sinuses reduce the weight of the skull and increase the intensity of the voice by serving as resonant sound chambers.

The eight bones of the cranium (table 7.5) are as follows:

1. **Frontal bone.** The frontal (frun′tal) bone forms the anterior portion of the skull above the eyes, including the forehead, the roof of the nasal cavity, and the roofs of the orbits (bony sockets) of the eyes. On the upper margin of each orbit, the frontal bone is marked by a *supraorbital foramen* (or *supraorbital notch* in some skulls) through which blood vessels and nerves pass to the tissues of the forehead. Within the frontal bone are two *frontal sinuses,* one above each eye near the midline. The frontal bone is a single bone in adults, but it develops in two parts (see fig. 7.31*b*). These halves grow together and usually completely fuse by the fifth or sixth year of life.
2. **Parietal bones.** One parietal (pah-ri′ĕ-tal) bone is located on each side of the skull just behind the frontal bone. Each is shaped like a curved plate and has four borders. Together, the parietal bones form the bulging sides and roof of the cranium. They are fused at the midline along the *sagittal suture,* and they meet the frontal bone along the *coronal suture.*
3. **Occipital bone.** The occipital (ok-sip′i-tal) bone joins the parietal bones along the *lambdoid* (lam′doid) *suture.* It forms the back of the skull and the base of the cranium. A large opening on its lower surface is the *foramen magnum,* where the inferior part of the brainstem connects with the spinal cord. Rounded processes called *occipital condyles,* located on each side of the foramen magnum, articulate with the first vertebra (atlas) of the vertebral column.

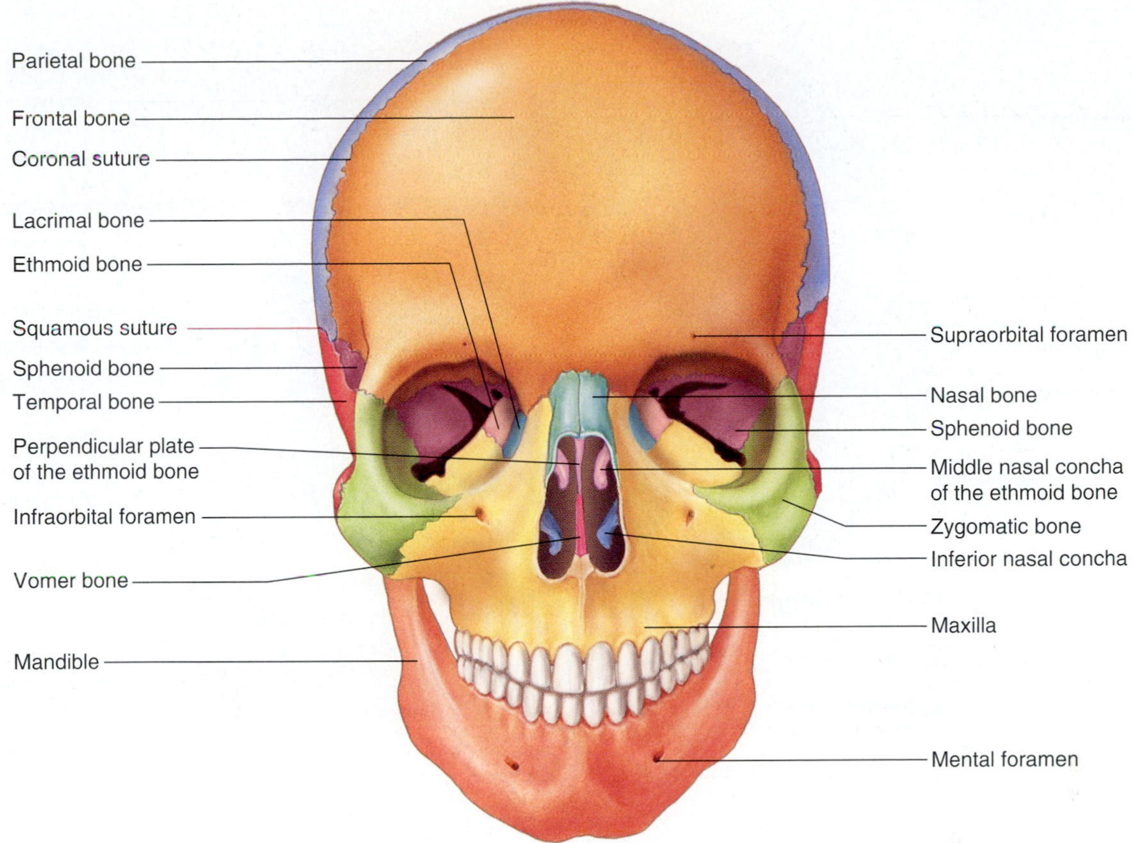

FIGURE 7.17 Anterior view of the skull.

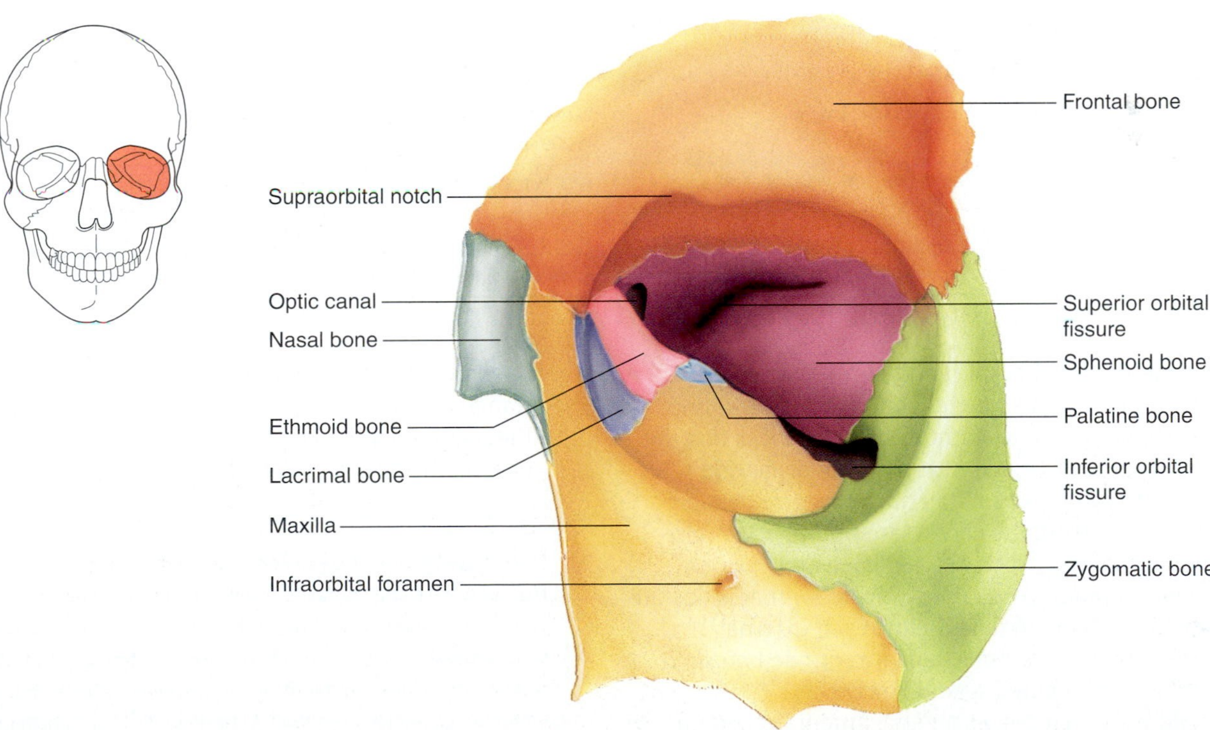

FIGURE 7.18 The orbit of the eye includes both cranial and facial bones.

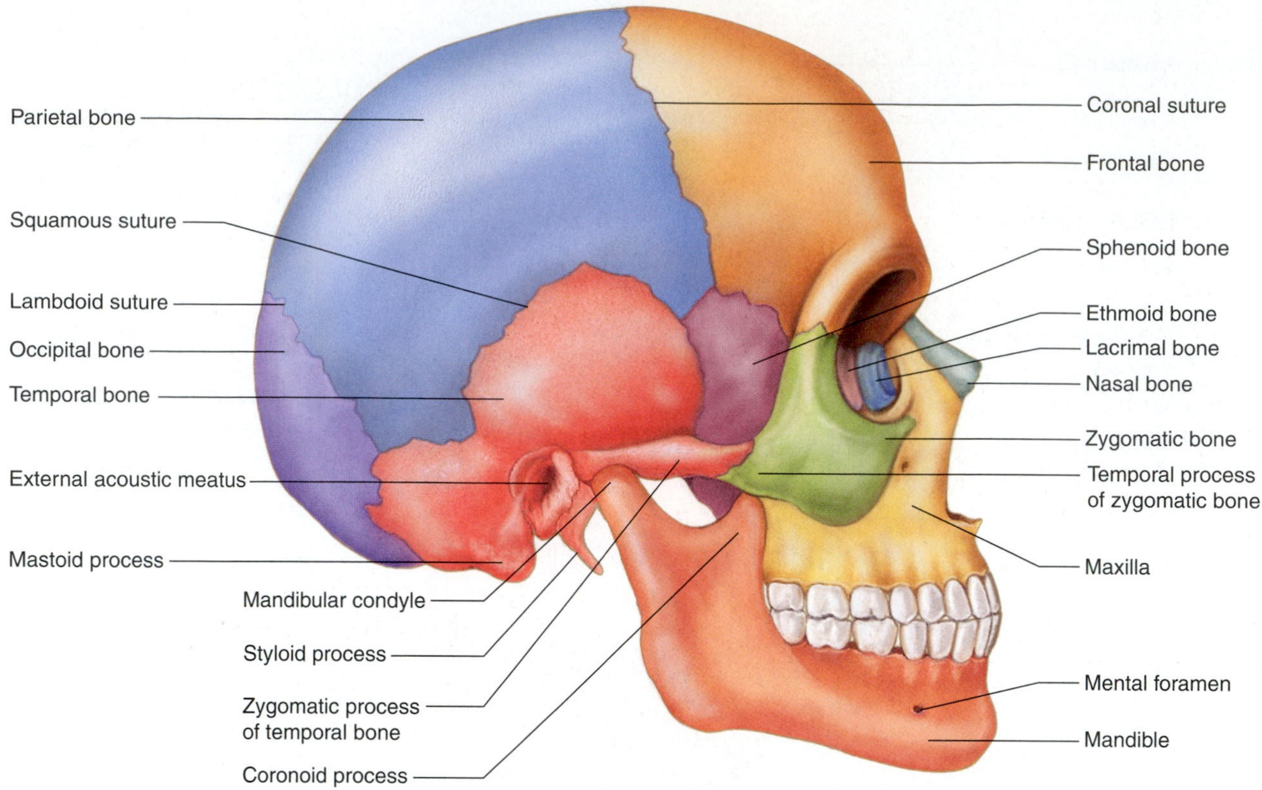

FIGURE 7.19 Right lateral view of the skull.

TABLE 7.5 | Cranial Bones

Name and Number	Description	Special Features
Frontal (1)	Forms forehead, roof of nasal cavity, and roofs of orbits	Supraorbital foramen, frontal sinuses
Parietal (2)	Form side walls and roof of cranium	Fused at midline along sagittal suture
Occipital (1)	Forms back of skull and base of cranium	Foramen magnum, occipital condyles
Temporal (2)	Form side walls and floor of cranium	External acoustic meatus, mandibular fossa, mastoid process, styloid process, zygomatic process
Sphenoid (1)	Forms parts of base of cranium, sides of skull, and floors and sides of orbits	Sella turcica, sphenoidal sinuses
Ethmoid (1)	Forms parts of roof and walls of nasal cavity, floor of cranium, and walls of orbits	Cribriform plates, perpendicular plate, superior and middle nasal conchae, ethmoidal sinuses, crista galli

4. **Temporal bones.** A temporal (tem′por-al) bone on each side of the skull joins the parietal bone along a *squamous suture.* The temporal bones form parts of the sides and the base of the cranium. Located near the inferior margin is an opening, the *external acoustic* (auditory) *meatus,* which leads inward to parts of the ear. The temporal bones also house the internal ear structures and have depressions called the *mandibular fossae* (glenoid fossae) that articulate with condyles of the mandible. Below each external acoustic meatus are two projections—a rounded *mastoid process* and a long, pointed *styloid process* (fig. 7.19). The mastoid process provides an attachment for certain muscles of the neck, whereas the styloid process anchors muscles associated with the tongue and pharynx. An opening near the mastoid process, the *carotid canal,* transmits the internal carotid artery. An opening between the temporal and occipital bones, the *jugular foramen,* accommodates the internal jugular vein (fig. 7.20).

The mastoid process may become infected. The tissues in this region of the temporal bone contain a number of interconnected air cells lined with mucous membranes that communicate with the middle ear. These spaces sometimes become inflamed when microorganisms spread into them from an infected middle ear *(otitis media).* The resulting mastoid infection, called *mastoiditis,* is of particular concern because nearby membranes that surround the brain may become infected.

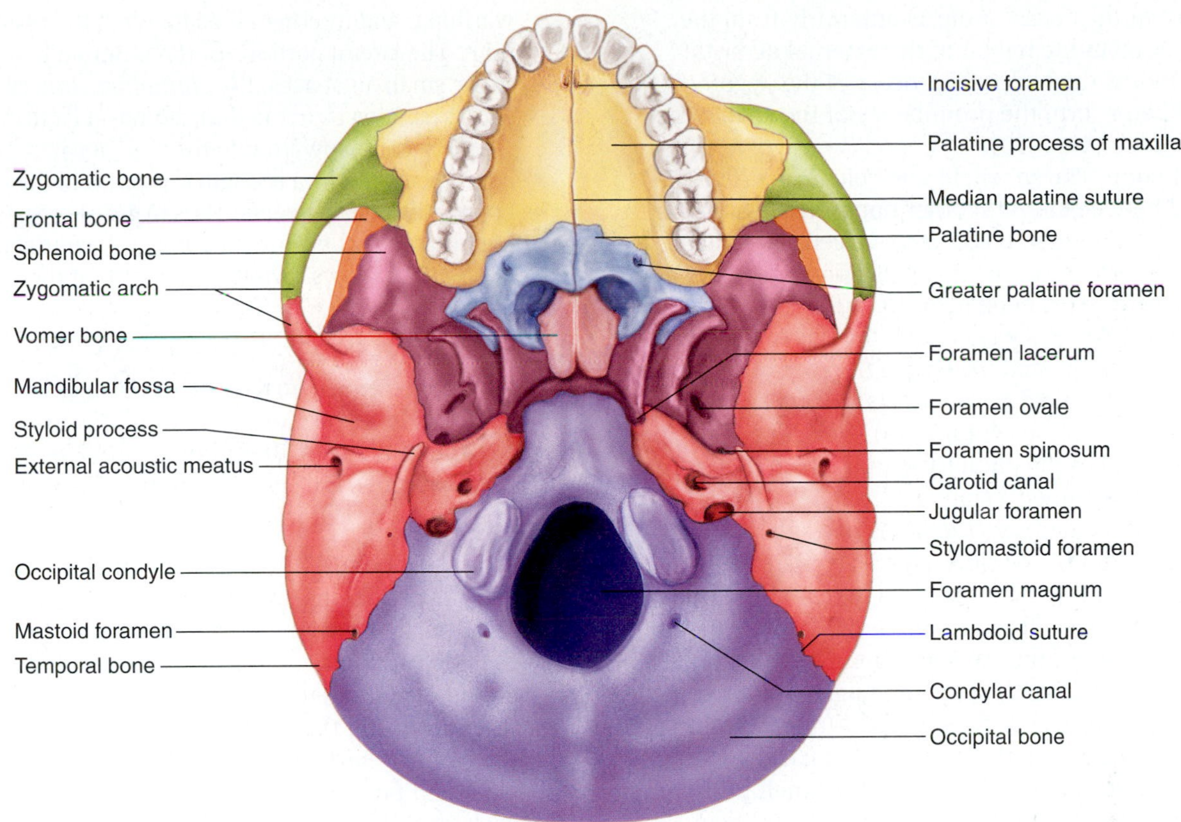

FIGURE 7.20 Inferior view of the skull.

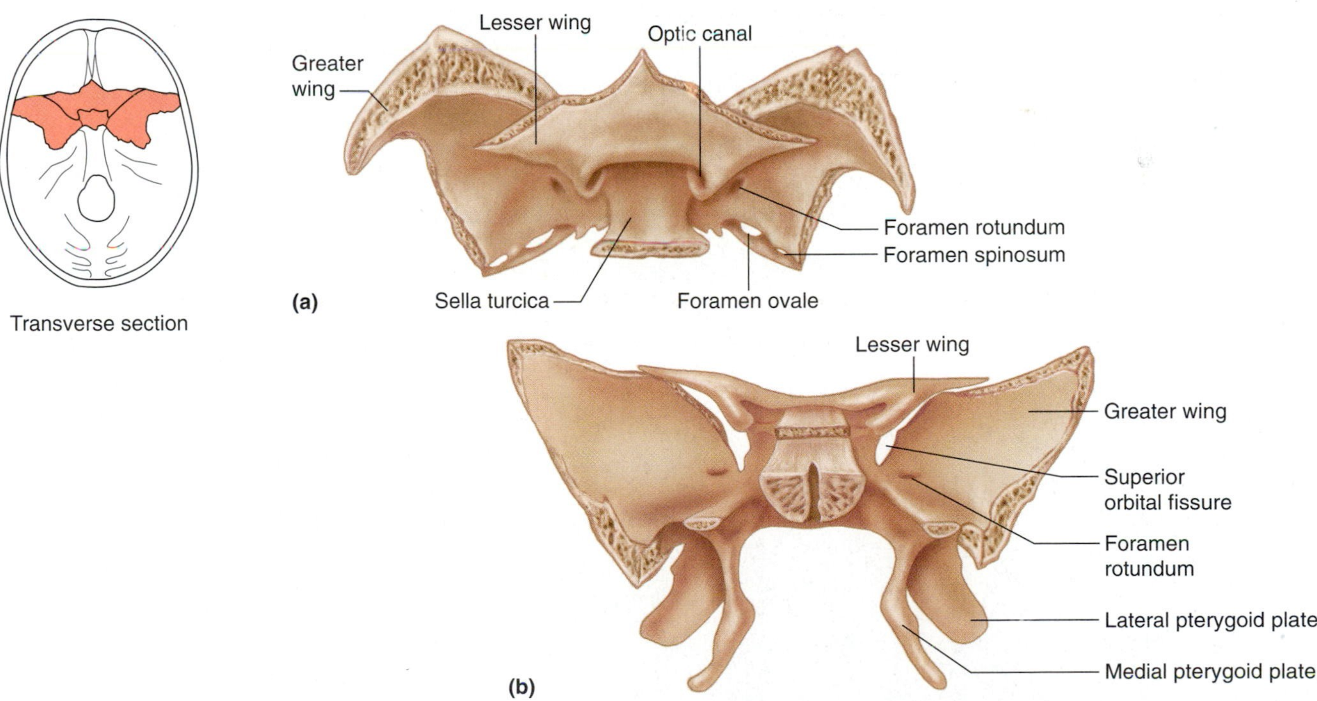

FIGURE 7.21 The sphenoid bone. (*a*) Superior view. (*b*) Posterior view. (The sphenoidal sinuses are within the bone and are not visible in this representation.)

A *zygomatic process* projects anteriorly from the temporal bone in the region of the external acoustic meatus. It joins the temporal process of the zygomatic bone and helps form the prominence of the cheek, the *zygomatic arch* (fig. 7.20).

5. **Sphenoid bone.** The sphenoid (sfe′noid) bone (fig. 7.21) is wedged between several other bones in the anterior portion of the cranium. It consists of a central part and two winglike structures that extend laterally toward each side of the skull. This bone helps form the base of the cranium, the sides of the skull, and the floors and sides of the orbits. Along the midline within the cranial cavity, a portion of the sphenoid bone indents to form the saddle-shaped *sella turcica* (sel′ah tur′si-ka) (Turk's saddle). In this depression lies the pituitary gland, which hangs from the base of the brain by a stalk.

   The sphenoid bone also contains two *sphenoidal sinuses.* These lie side by side and are separated by a bony septum that projects downward into the nasal cavity.

6. **Ethmoid bone.** The ethmoid (eth′moid) bone (fig. 7.22) is located in front of the sphenoid bone. It consists of two masses, one on each side of the nasal cavity, joined horizontally by thin *cribriform* (krib′rĭ-form) *plates.* These plates form part of the roof of the nasal cavity, and nerves associated with the sense of smell pass through tiny openings (*olfactory foramina*) in them. Portions of the ethmoid bone also form sections of the cranial floor, orbital walls, and nasal cavity walls. A *perpendicular plate* projects downward in the midline from the cribriform plates to form most of the nasal septum.

   Delicate, scroll-shaped plates called the *superior nasal concha* (kong′kah) and the *middle nasal concha* project inward from the lateral portions of the ethmoid bone toward the perpendicular plate. These bony plates support mucous membranes that line the nasal cavity. The mucous membranes, in turn, begin moistening, warming, and filtering air as it enters the respiratory tract. The lateral portions of the ethmoid bone contain many small air spaces, the *ethmoidal sinuses.* Figure 7.23 shows various structures in the nasal cavity.

   Projecting upward into the cranial cavity between the cribriform plates is a triangular process of the ethmoid bone called the *crista galli* (kris′tă gal′li) (cock's comb). Membranes that enclose the brain attach to this process. Figure 7.24 shows a view of the floor of the cranial cavity.

Facial Skeleton

The **facial skeleton** consists of thirteen immovable bones and a movable lower jawbone. In addition to forming the basic shape of the face, these bones provide attachments for muscles that move the jaw and control facial expressions.

The bones of the facial skeleton are as follows:

1. **Maxillary bones.** The maxillary (mak′sĭ-ler″e) bones (sing., maxilla, mak-sil′ah; pl., maxillae, mak-sĭl′e) form the upper jaw; together they form the keystone of the face, because the other immovable facial bones articulate with them.

   Portions of these bones comprise the anterior roof of the mouth (*hard palate*), the floors of the orbits, and the sides and floor of the nasal cavity. They also contain the sockets of the upper teeth. Inside the maxillae, lateral to the nasal cavity, are *maxillary sinuses.* These spaces are the largest of the sinuses, and they extend from the floor of the orbits to the roots of the upper teeth. Figure 7.25 shows the locations of the maxillary and other paranasal sinuses.

   During development, portions of the maxillary bones called *palatine processes* grow together and fuse along the midline, or median palatine suture. This forms the anterior section of the hard palate (see fig. 7.20).

   The inferior border of each maxillary bone projects downward, forming an *alveolar* (al-ve′o-lar) *process.*

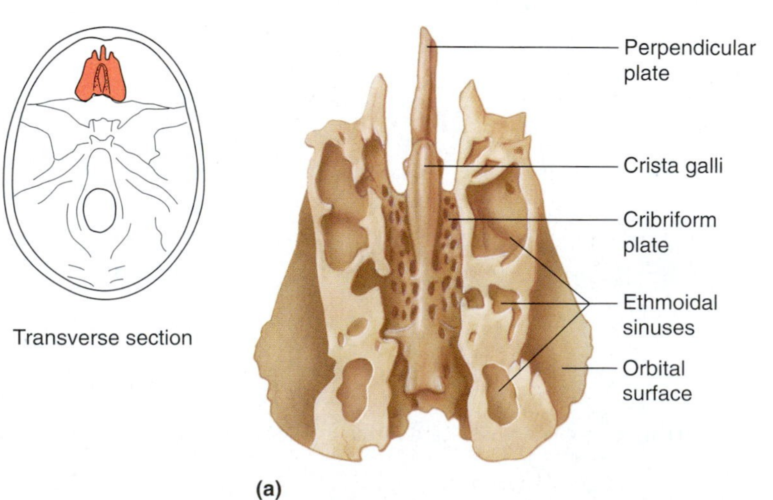

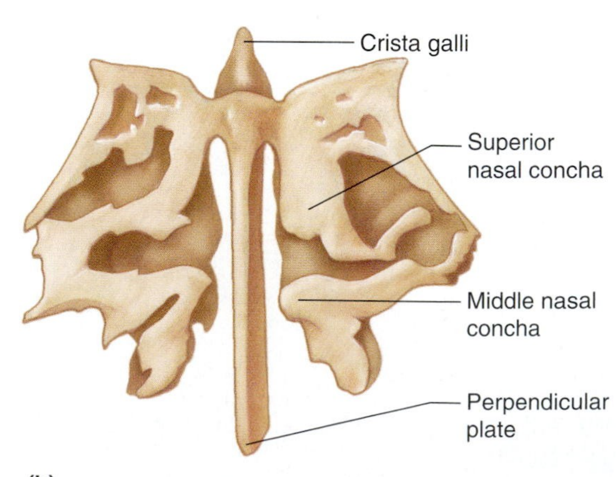

FIGURE 7.22 The ethmoid bone. (*a*) Superior view and (*b*) Posterior view.

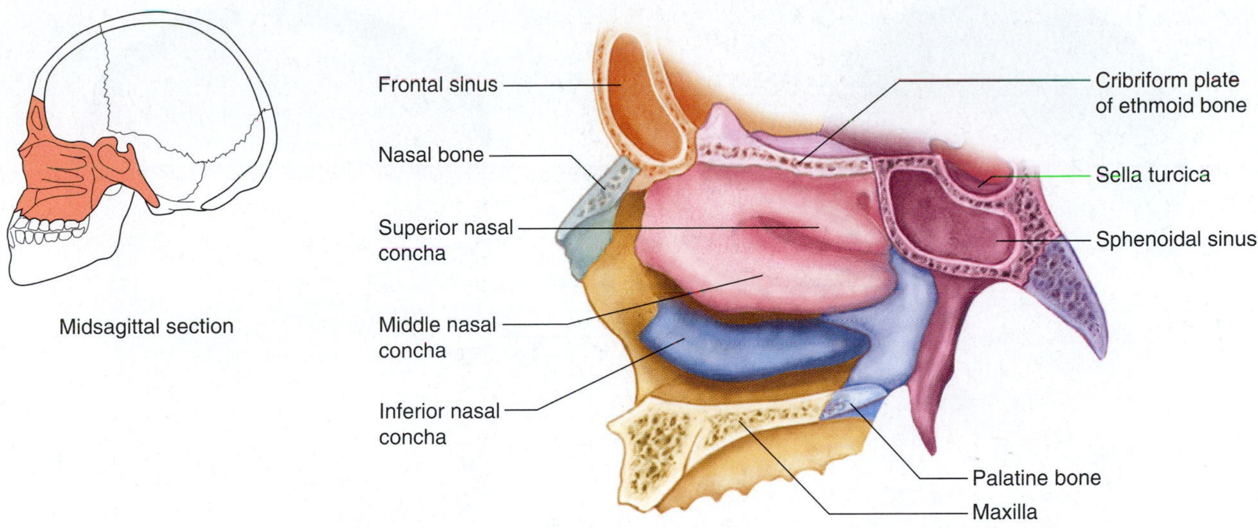

FIGURE 7.23 Lateral wall of the nasal cavity.

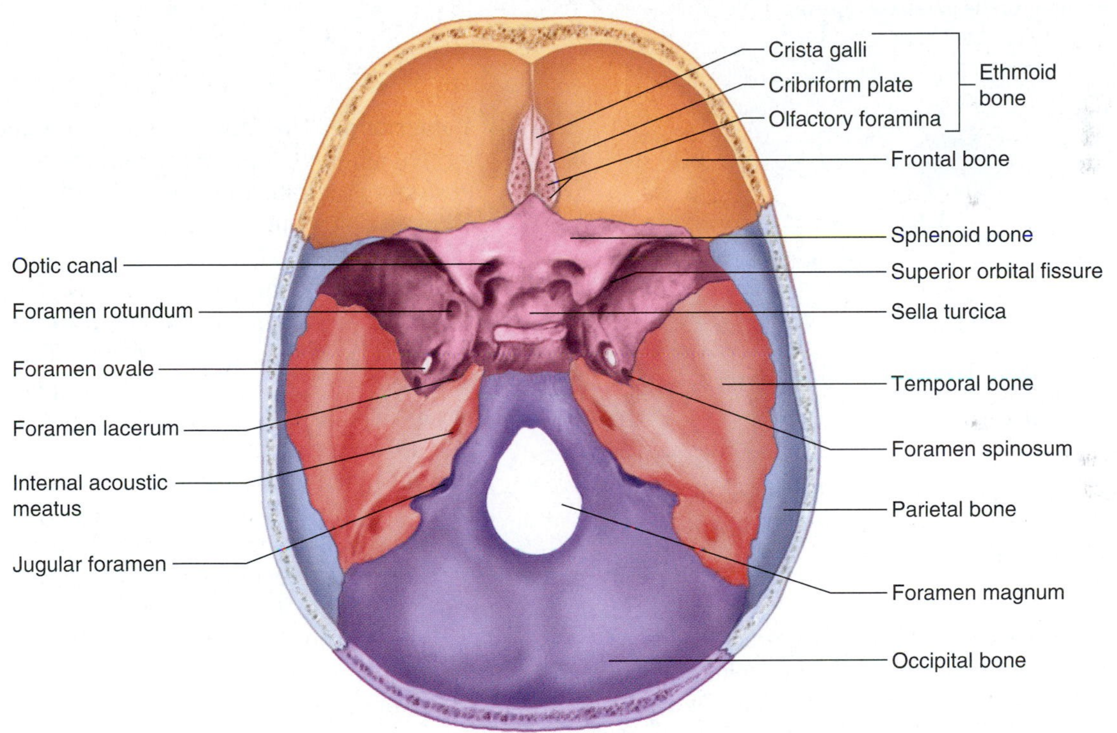

FIGURE 7.24 Floor of the cranial cavity, viewed from above.

Together these processes form a horseshoe-shaped *alveolar arch* (dental arch). Teeth occupy cavities in this arch (dental alveoli). Dense connective tissue binds teeth to the bony sockets (see chapter 17, p. 660).

2. **Palatine bones.** The L-shaped palatine (pal′ah-tīn) bones (fig. 7.26) are located behind the maxillae. The horizontal portions form the posterior section of the hard palate and the floor of the nasal cavity. The perpendicular portions help form the lateral walls of the nasal cavity.

Sometimes, fusion of the palatine processes of the maxillae is incomplete at birth; the result is a *cleft palate*. Infants with a cleft palate may have trouble suckling because of the opening between the oral and nasal cavities. A temporary prosthetic device (artificial palate) may be inserted into the mouth or a special type of nipple can be placed on bottles, so the child can eat and drink until surgery can be performed to correct the cleft.

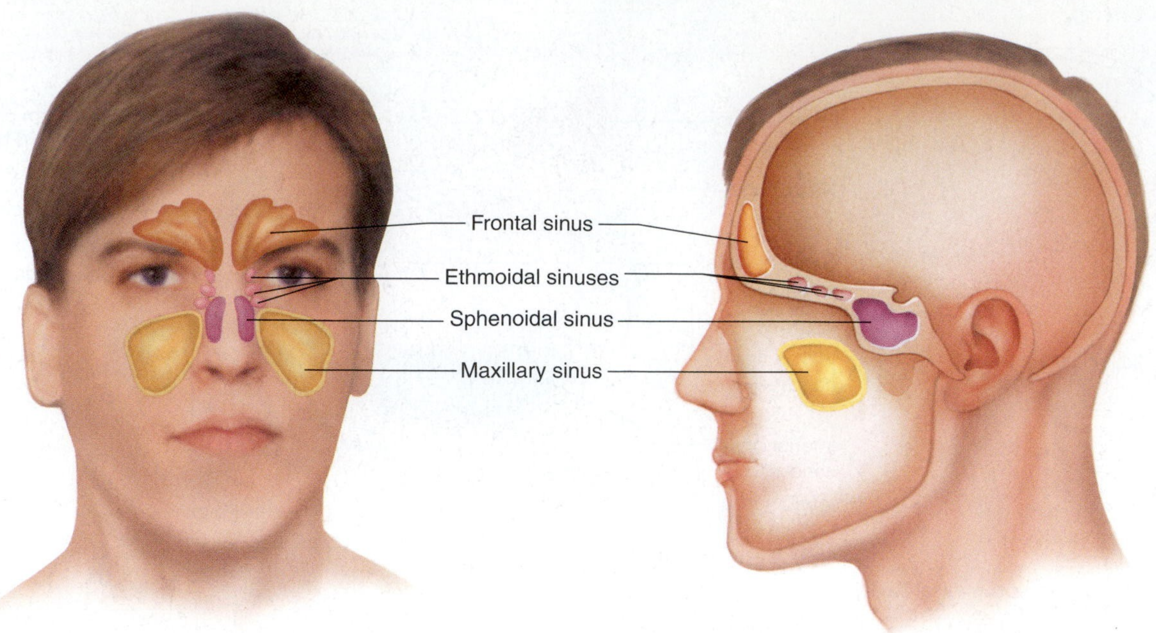

FIGURE 7.25 Locations of the paranasal sinuses.

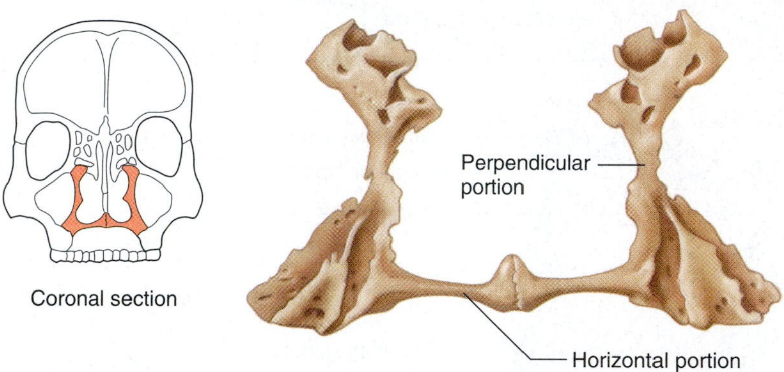

FIGURE 7.26 The horizontal portions of the palatine bones form the posterior section of the hard palate, and the perpendicular portions help form the lateral walls of the nasal cavity.

3. **Zygomatic bones.** The zygomatic (zi″go-mat′ik) bones are responsible for the prominences of the cheeks below and to the sides of the eyes. These bones also help form the lateral walls and the floors of the orbits. Each bone has a *temporal process,* which extends posteriorly to join the zygomatic process of a temporal bone (see fig. 7.19).
4. **Lacrimal bones.** A lacrimal (lak′rĭ-mal) bone is a thin, scalelike structure located in the medial wall of each orbit between the ethmoid bone and the maxilla (see fig. 7.19). A groove in its anterior portion leads from the orbit to the nasal cavity, providing a pathway for a channel that carries tears from the eye to the nasal cavity.
5. **Nasal bones.** The nasal (na′zal) bones are long, thin, and nearly rectangular (see fig. 7.17). They lie side by side and are fused at the midline, where they form the bridge of the nose. These bones are attachments for the cartilaginous tissues that form the shape of the nose.
6. **Vomer bone.** The thin, flat vomer (vo′mer) bone is located along the midline within the nasal cavity. Posteriorly, it joins the perpendicular plate of the ethmoid bone, and together they form the nasal septum (figs. 7.27 and 7.28).
7. **Inferior nasal conchae.** The inferior nasal conchae (kong′ke) are fragile, scroll-shaped bones attached to the lateral walls of the nasal cavity. They are the largest of the conchae and are below the superior and middle nasal conchae of the ethmoid bone (see figs. 7.17 and 7.23). Like the ethmoidal conchae, the inferior conchae support mucous membranes in the nasal cavity.
8. **Mandible.** The mandible (man′dĭ-b′l), or lower jawbone, is a horizontal, horseshoe-shaped body with a flat *ramus* projecting upward at each end. The rami are divided into a posterior *mandibular condyle* and an anterior *coronoid* (kor′o-noid) *process* (fig. 7.29). The mandibular condyles articulate with the mandibular fossae of the temporal bones, whereas the coronoid processes provide attachments for muscles used in chewing. Other large chewing muscles are inserted on the lateral surfaces of the rami. A curved bar of bone on the superior border of the mandible, the *alveolar border,* contains the hollow sockets (dental alveoli) that bear the lower teeth.

   On the medial side of the mandible, near the center of each ramus, is a *mandibular foramen.* This opening admits blood vessels and a nerve, which supply the roots of the lower teeth. Dentists inject anesthetic into the tissues near this foramen to temporarily block nerve impulse conduction and desensitize teeth on that side of the jaw. Branches of the blood vessels and the nerve emerge from the

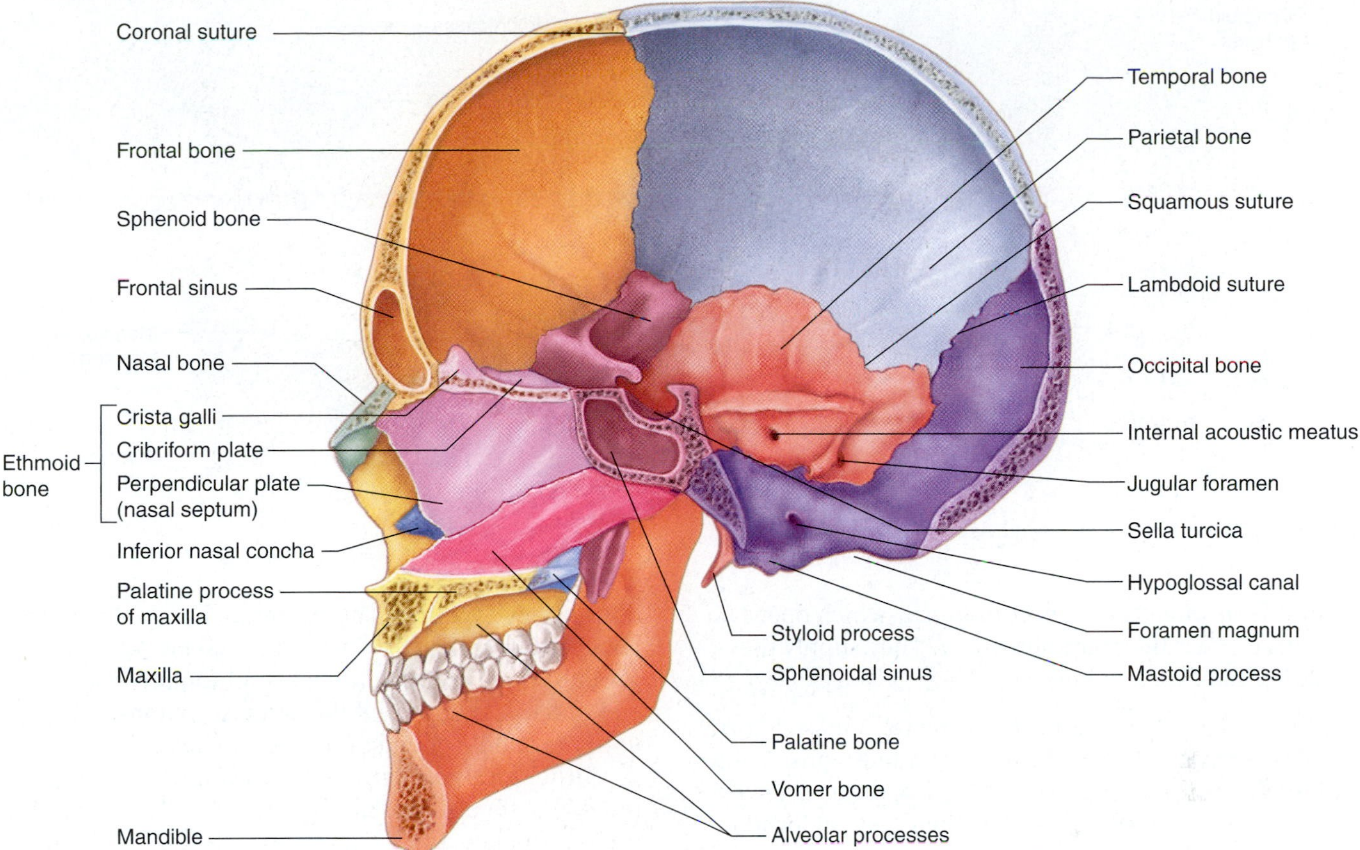

FIGURE 7.27 Sagittal section of the skull.

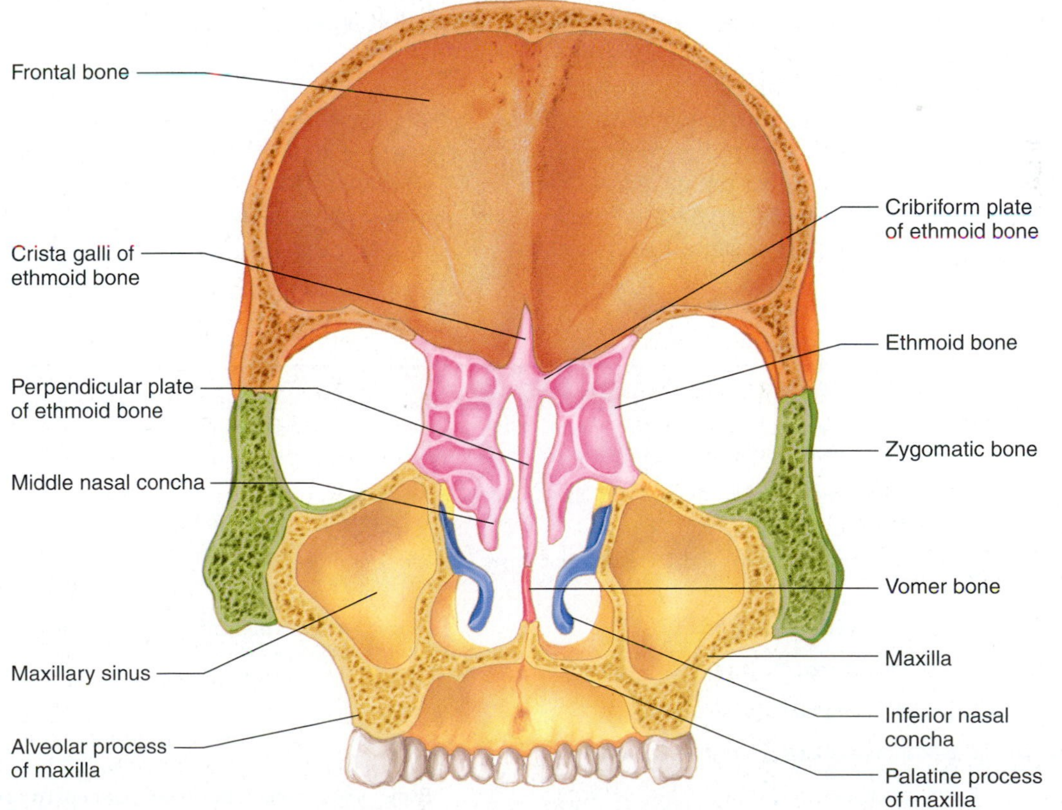

FIGURE 7.28 Coronal section of the skull (posterior view).

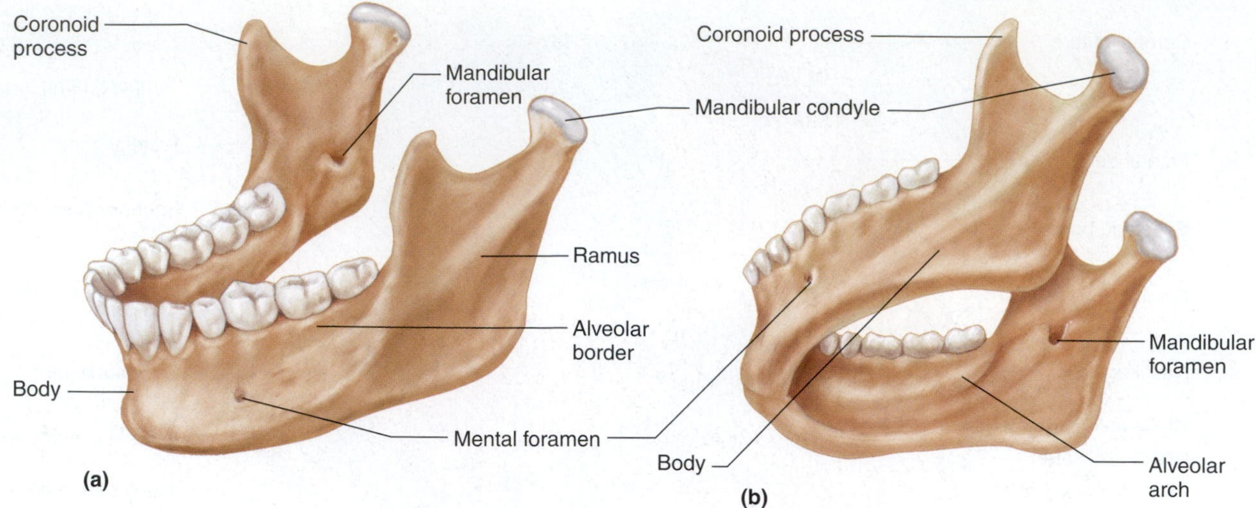

FIGURE 7.29 Mandible. (*a*) Left lateral view. (*b*) Inferior view.

mandible through the *mental foramen*, which opens on the outside near the point of the jaw. They supply the tissues of the chin and lower lip.

Table 7.6 describes the fourteen facial bones. Figure 7.30 shows features of these bones on radiographs. Table 7.7 lists the major openings *(foramina)* and passageways through bones of the skull, as well as their general locations and the structures that pass through them.

Infantile Skull

At birth, the skull is incompletely developed, with fibrous membranes connecting the cranial bones. These membranous areas are called **fontanels** (fon″tah-nel′z), or, more commonly, soft spots. They permit some movement between the bones so that the developing skull is partially compressible and can slightly change shape. This action, called *molding*, enables an infant's skull to more easily pass through the birth canal. Eventually, the fontanels close as the cranial bones grow together. The posterior fontanel usually closes about two months after birth; the sphenoidal fontanel closes at about three months; the mastoid fontanel closes near the end of the first year; and the anterior fontanel may not close until the middle or end of the second year.

Other characteristics of an infantile skull (fig. 7.31) include a small face with a prominent forehead and large orbits. The jaw and nasal cavity are small, the sinuses are incompletely formed, and the frontal bone is in two parts (reference plate 51). The skull bones are thin, but they are also somewhat flexible and thus are less easily fractured than adult bones.

In the infantile skull, a frontal suture (metopic suture) separates the two parts of the developing frontal bone in the midline. This suture usually closes before the sixth year; however, in a few adults, the frontal suture remains open.

TABLE 7.6 | Bones of the Facial Skeleton

Name and Number	Description	Special Features
Maxillary (2)	Form upper jaw, anterior roof of mouth, floors of orbits, and sides and floor of nasal cavity	Alveolar processes, maxillary sinuses, palatine process
Palatine (2)	Form posterior roof of mouth and floor and lateral walls of nasal cavity	
Zygomatic (2)	Form prominences of cheeks and lateral walls and floors of orbits	Temporal process
Lacrimal (2)	Form part of medial walls of orbits	Groove that leads from orbit to nasal cavity
Nasal (2)	Form bridge of nose	
Vomer (1)	Forms inferior portion of nasal septum	
Inferior nasal conchae (2)	Extend into nasal cavity from its lateral walls	
Mandible (1)	Forms lower jaw	Body, ramus, mandibular condyle, coronoid process, alveolar process, mandibular foramen, mental foramen

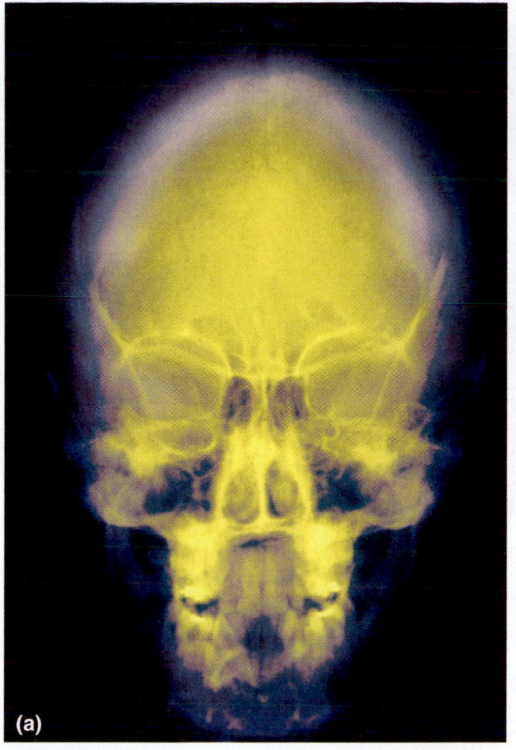

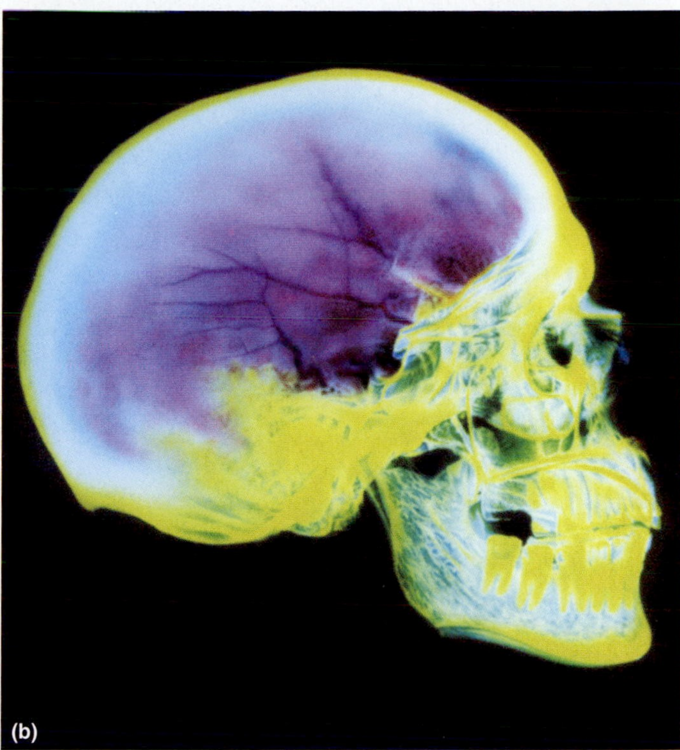

FIGURE 7.30 Falsely colored radiographs of the skull. (*a*) Anterior view. (*b*) Right lateral view.

TABLE 7.7 | Passageways Through Bones of the Skull

Passageway	Location	Major Stuctures Passing Through
Carotid canal (fig. 7.20)	Inferior surface of the temporal bone	Internal carotid artery, veins, and nerves
Foramen lacerum (fig. 7.20)	Floor of cranial cavity between temporal and sphenoid bones	Branch of pharyngeal artery (in life, opening is largely covered by fibrocartilage)
Foramen magnum (fig. 7.24)	Base of skull in occipital bone	Inferior part of brainstem connecting to spinal cord, also certain arteries
Foramen ovale (fig. 7.20)	Floor of cranial cavity in sphenoid bone	Mandibular division of trigeminal nerve and veins
Foramen rotundum (fig. 7.24)	Floor of cranial cavity in sphenoid bone	Maxillary division of trigeminal nerve
Foramen spinosum (fig. 7.24)	Floor of cranial cavity in sphenoid bone	Middle meningeal blood vessels and branch of mandibular nerve
Greater palatine foramen (fig. 7.20)	Posterior portion of hard palate in palatine bone	Palatine blood vessels and nerves
Hypoglossal canal (fig. 7.27)	Near margin of foramen magnum in occipital bone	Hypoglossal nerve
Incisive foramen (fig. 7.20)	Incisive fossa in anterior portion of hard palate	Nasopalatine nerves, openings of vomeronasal organ
Inferior orbital fissure (fig. 7.18)	Floor of the orbit	Maxillary nerve and blood vessels
Infraorbital foramen (fig. 7.18)	Below the orbit in maxillary bone	Infraorbital blood vessels and nerves
Internal acoustic meatus (fig. 7.24)	Floor of cranial cavity in temporal bone	Branches of facial and vestibulocochlear nerves and blood vessels
Jugular foramen (fig. 7.24)	Base of the skull between temporal and occipital bones	Glossopharyngeal, vagus and accessory nerves, and blood vessels
Mandibular foramen (fig. 7.29)	Inner surface of ramus of mandible	Inferior alveolar blood vessels and nerves
Mental foramen (fig. 7.29)	Near point of jaw in mandible	Mental nerve and blood vessels
Optic canal (fig. 7.18)	Posterior portion of orbit in sphenoid bone	Optic nerve and ophthalmic artery
Stylomastoid foramen (fig. 7.20)	Between styloid and mastoid processes	Facial nerve and blood vessels
Superior orbital fissure (fig. 7.18)	Lateral wall of orbit	Oculomotor, trochlear, and abducens nerves and ophthalmic division of trigeminal nerve
Supraorbital foramen (fig. 7.17)	Upper margin of orbit in frontal bone	Supraorbital blood vessels and nerves

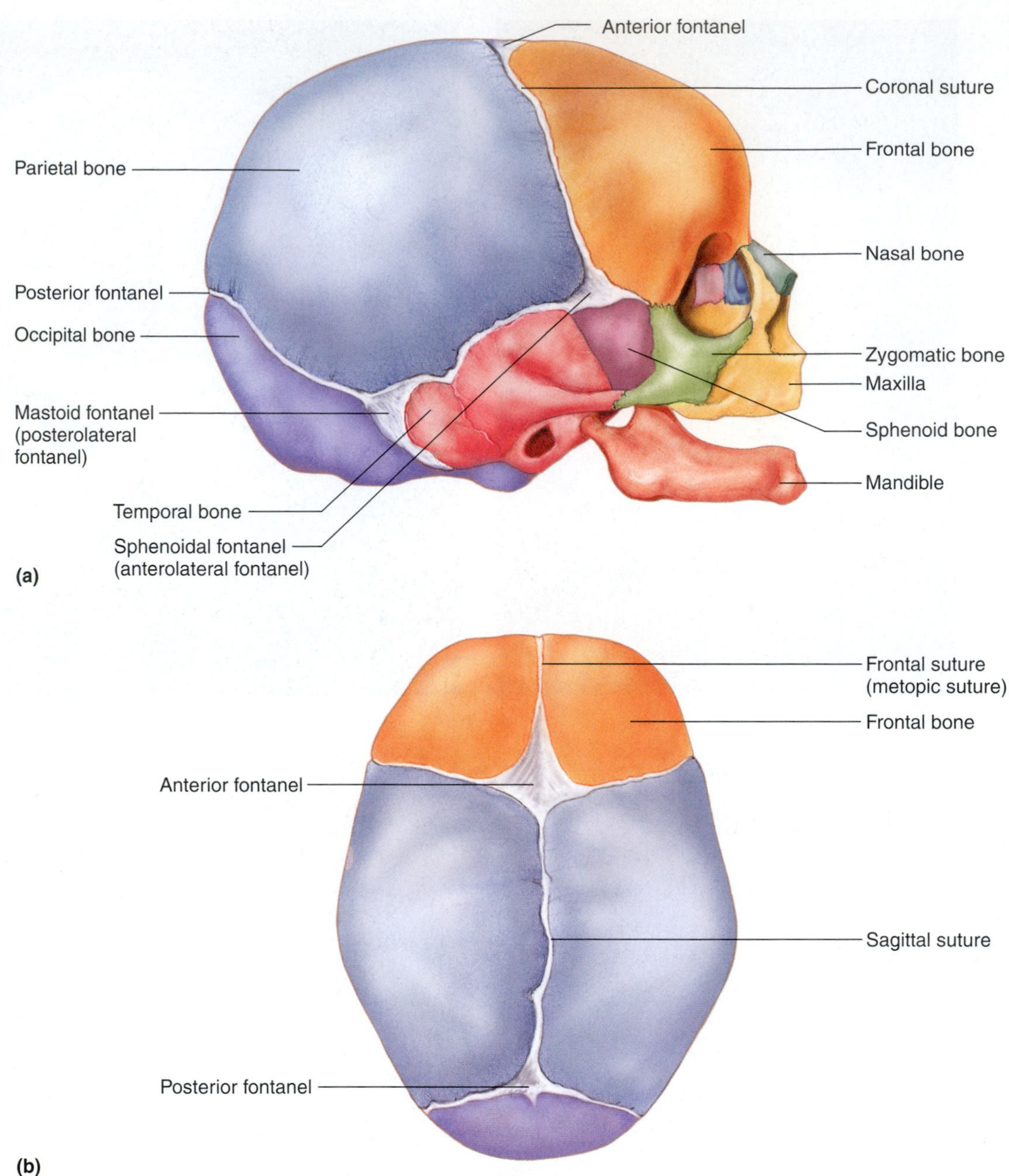

FIGURE 7.31 Fontanels. (*a*) Right lateral view and (*b*) superior view of the infantile skull.

PRACTICE

18 Locate and name each of the bones of the cranium.

19 Locate and name each of the facial bones.

20 Explain how an adult skull differs from that of an infant.

7.7 VERTEBRAL COLUMN

The **vertebral column** extends from the skull to the pelvis and forms the vertical axis of the skeleton (fig. 7.32). It is composed of many bony parts called **vertebrae** (ver′tĕ-bre) separated by masses of fibrocartilage called *intervertebral discs* and connected to one another by ligaments. The vertebral column supports the head and the trunk of the body, yet is flexible enough to permit movements, such as bending forward, backward, or to the side and turning or rotating on the central axis. It also protects the spinal cord, which passes through a *vertebral canal* formed by openings in the vertebrae.

An infant has thirty-three separate bones in the vertebral column. Five of these bones eventually fuse to form the sacrum, and four others join to become the coccyx. As a result, an adult vertebral column has twenty-six bones.

Normally, the vertebral column has four curvatures, which give it a degree of resiliency. The names of the curves

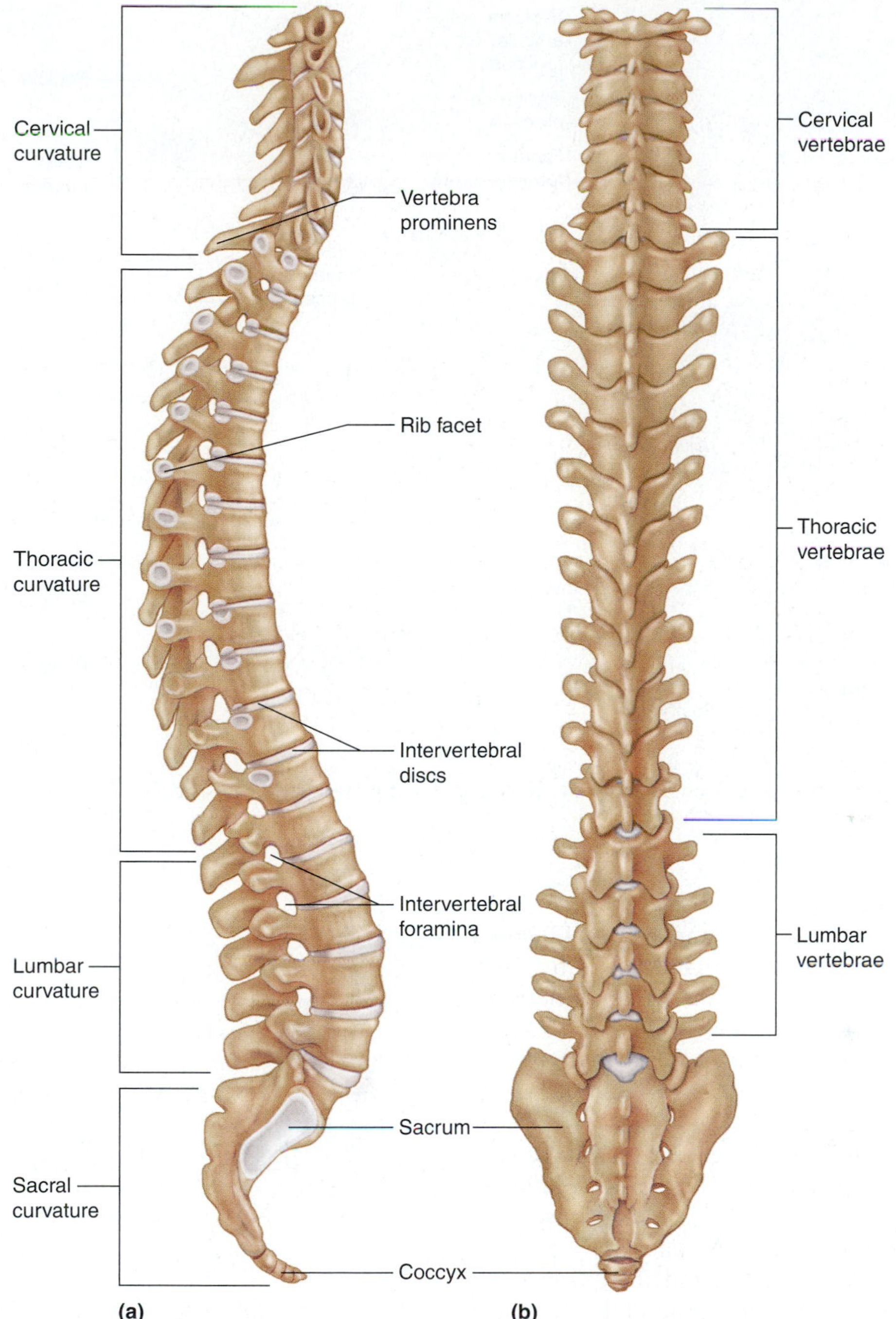

FIGURE 7.32 The curved vertebral column consists of many vertebrae separated by intervertebral discs. (*a*) Right lateral view. (*b*) Posterior view.

correspond to the regions in which they occur, as shown in figure 7.32. The *thoracic* and *sacral curvatures* are concave anteriorly and are called primary curves. The *cervical curvature* in the neck and the *lumbar curvature* in the lower back are convex anteriorly and are called secondary curves. The cervical curvature develops when a baby begins to hold up its head, and the lumbar curvature develops when the child begins to stand.

A Typical Vertebra

Although the vertebrae in different regions of the vertebral column have special characteristics, they also have features in common. A typical vertebra has a drum-shaped *body,* which forms the thick, anterior portion of the bone (fig. 7.33). A longitudinal row of these vertebral bodies supports the weight of the head and trunk. The intervertebral discs, which separate adjacent vertebrae, are fastened to the

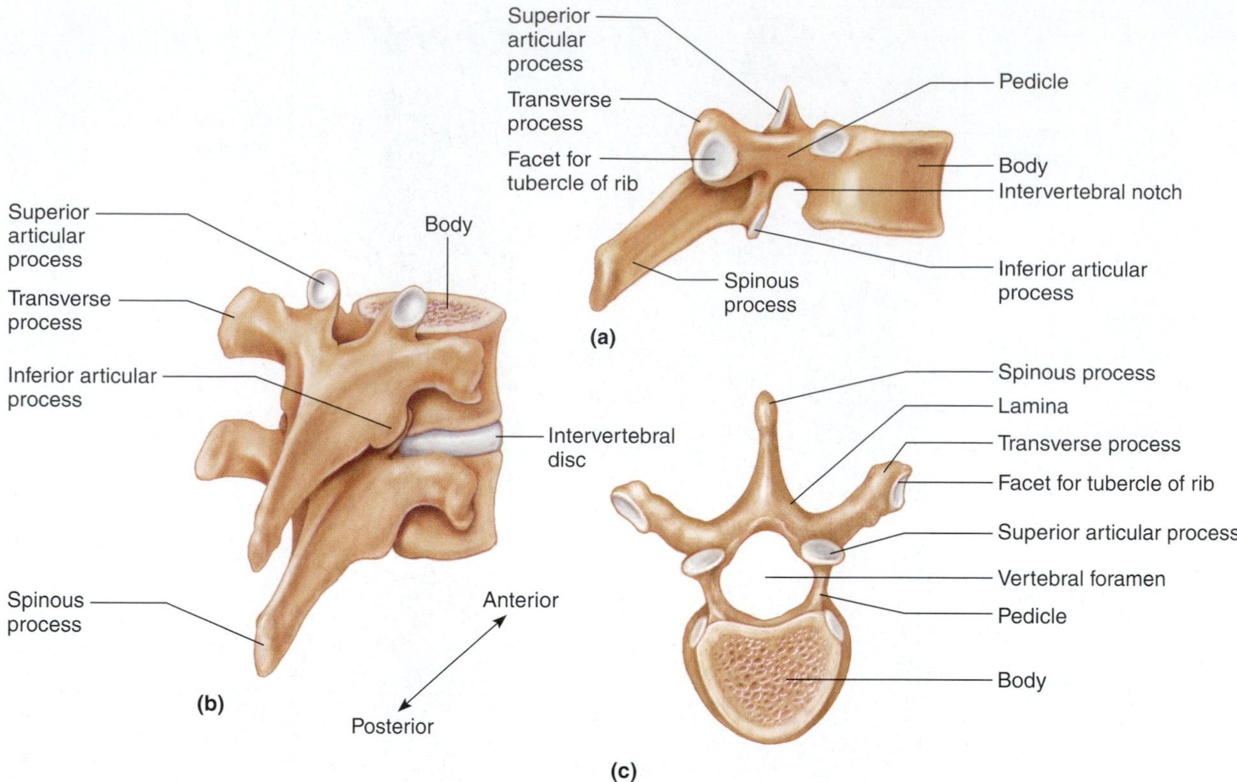

FIGURE 7.33 Typical thoracic vertebra. (*a*) Right lateral view. (*b*) Adjacent vertebrae join at their articular processes. (*c*) Superior view.

roughened upper and lower surfaces of the vertebral bodies. These discs cushion and soften the forces caused by such movements as walking and jumping, which might otherwise fracture vertebrae or jar the brain. The bodies of adjacent vertebrae are joined on their anterior surfaces by *anterior longitudinal ligaments* and on their posterior surfaces by *posterior longitudinal ligaments*.

Projecting posteriorly from each vertebral body are two short stalks called *pedicles* (ped′ĭ-k′lz). They form the sides of the *vertebral foramen*. Two plates called *laminae* (lam′ĭ-ne) arise from the pedicles and fuse in the back to become a *spinous process*. The pedicles, laminae, and spinous process together complete a bony *vertebral arch* around the vertebral foramen, through which the spinal cord passes.

Between the pedicles and laminae of a typical vertebra is a *transverse process*, which projects laterally and posteriorly. Various ligaments and muscles are attached to the dorsal spinous process and the transverse processes. Projecting upward and downward from each vertebral arch are *superior* and *inferior articulating processes*. These processes bear cartilage-covered facets by which each vertebra is joined to the one above and the one below.

On the lower surfaces of the vertebral pedicles are notches that align with adjacent vertebrae to help form openings called *intervertebral foramina* (in″ter-ver′tĕ-bral fo-ram′ĭ-nah). These openings provide passageways for spinal nerves.

Gymnasts, high jumpers, pole vaulters, and other athletes who hyperextend and rotate their vertebral columns and stress them with impact sometimes fracture the portion of the vertebra between the superior and inferior articulating processes (the pars interarticularis). Such damage to the vertebra is called spondylolysis, and it is most common at L5.

Cervical Vertebrae

Seven **cervical vertebrae** comprise the bony axis of the neck. These are the smallest of the vertebrae, but their bone tissues are denser than those in any other region of the vertebral column.

The transverse processes of the cervical vertebrae are distinctive because they have *transverse foramina*, passageways for arteries leading to the brain. Also, the spinous processes of the second through the sixth cervical vertebrae are uniquely forked (bifid). These processes provide attachments for muscles.

The spinous process of the seventh vertebra is longer and protrudes beyond the other cervical spines. It is called the *vertebra prominens*, and because it can be felt through the skin, it is a useful landmark for locating other vertebral parts (see fig. 7.32).

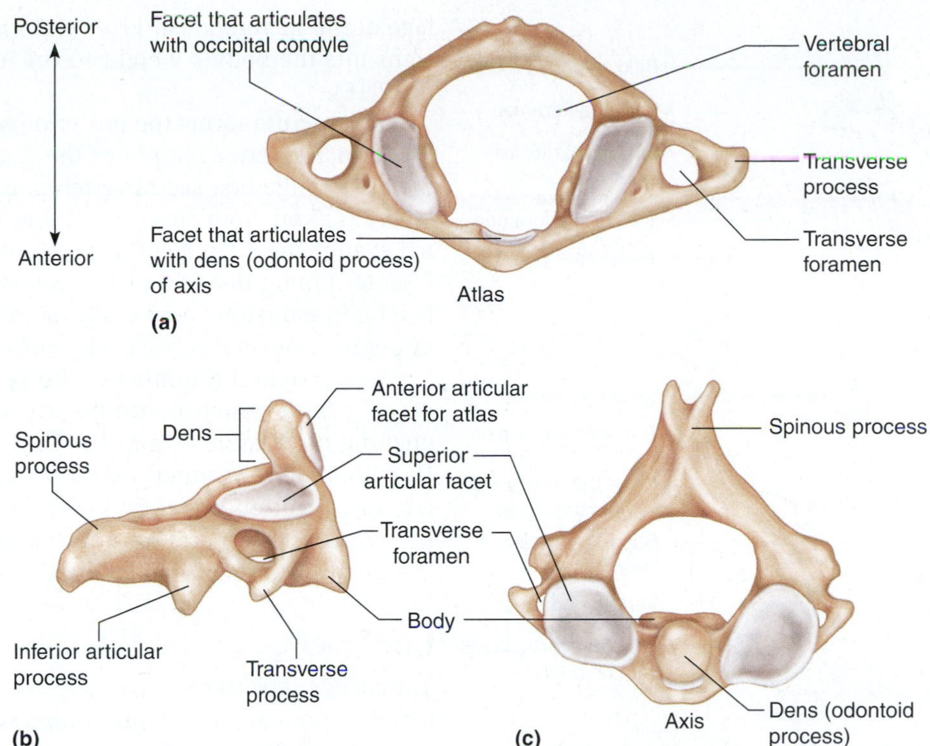

FIGURE 7.34 Atlas and axis. (*a*) Superior view of the atlas. (*b*) Right lateral view and (*c*) superior view of the axis.

Two of the cervical vertebrae, shown in figure 7.34, are of special interest. The first vertebra, or **atlas** (at′las), supports the head. It has practically no body or spine and appears as a bony ring with two transverse processes. On its superior surface, the atlas has two kidney-shaped *facets*, which articulate with the occipital condyles.

The second cervical vertebra, or **axis** (ak′sis), bears a toothlike *dens* (odontoid process) on its body. This process projects upward and lies in the ring of the atlas. As the head is turned from side to side, the atlas pivots around the dens (figs. 7.34 and 7.35).

Thoracic Vertebrae

The twelve **thoracic vertebrae** are larger than those in the cervical region. Their transverse processes project posteriorly at sharp angles. Each vertebra has a long, pointed spinous process, which slopes downward, and a facet on each side of its body, which articulates with a rib.

Beginning with the third thoracic vertebra and moving inferiorly, the bodies of these bones increase in size. Thus, they are adapted to bear increasing loads of body weight.

Lumbar Vertebrae

The five **lumbar vertebrae** in the small of the back (loin) support more weight than the superior vertebrae and have larger and stronger bodies. Compared to other types of vertebrae, the thinner transverse processes of these vertebrae project laterally, whereas their short, thick spinous processes project posteriorly nearly horizontal. Figure 7.36 compares the structures of the cervical, thoracic, and lumbar vertebrae.

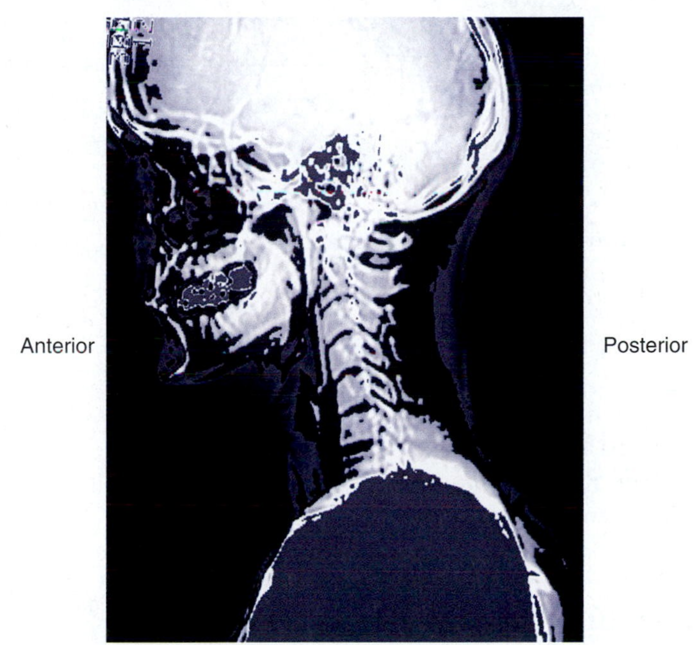

FIGURE 7.35 Radiograph of the cervical vertebrae.

The painful condition of spondylolisthesis occurs when a vertebra slips out of place over the vertebra below. Most commonly the fifth lumbar vertebra slides forward over the body of the sacrum. Persons with spondylolysis (see previous box) may be more likely to develop spondylolisthesis, as are gymnasts, football players, and others who flex or extend their vertebral columns excessively and forcefully.

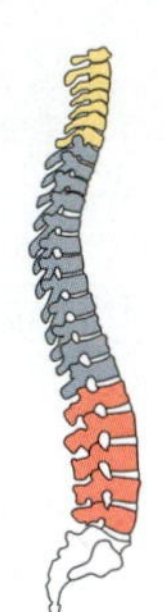

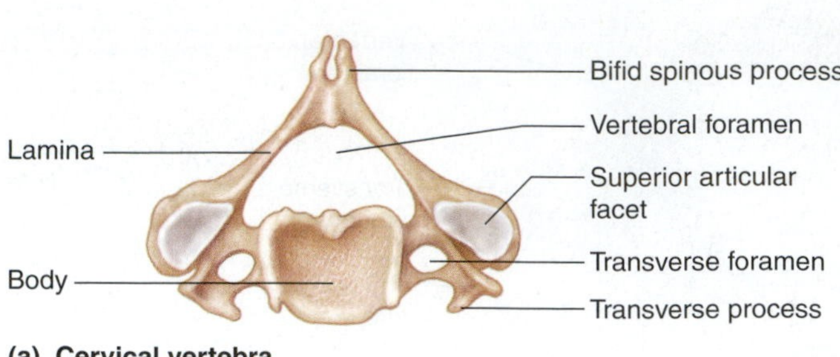

(a) Cervical vertebra

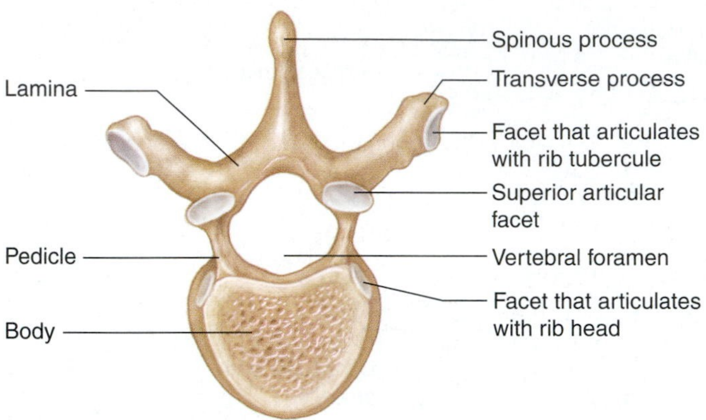

(b) Thoracic vertebra

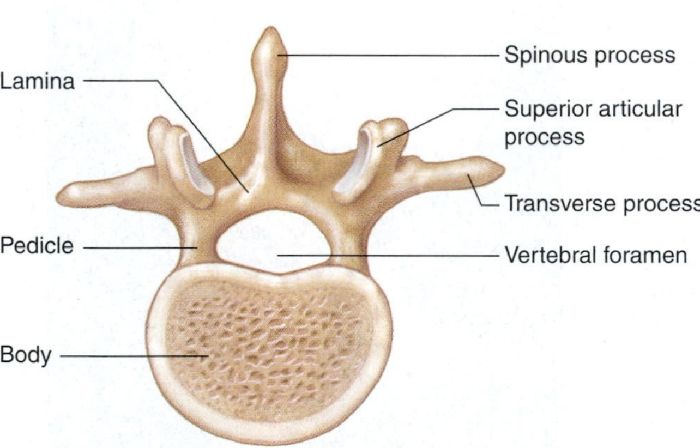

(c) Lumbar vertebra

FIGURE 7.36 Superior view of (*a*) a cervical vertebra, (*b*) a thoracic vertebra, and (*c*) a lumbar vertebra.

Sacrum

The **sacrum** (sa′krum) is a triangular structure at the base of the vertebral column. It is composed of five vertebrae that develop separately but gradually fuse between ages eighteen and thirty (fig. 7.37). Sometimes only four vertebrae fuse to form the sacrum and the fifth vertebra becomes a sixth lumbar vertebra. The spinous processes of these fused bones form a ridge of tubercles, the *median sacral crest.* Nerves and blood vessels pass through rows of openings, called the *posterior sacral foramina,* located to the sides of the tubercles.

The sacrum is wedged between the hip bones of the pelvis and joins them at its *auricular surfaces* by fibrocartilage of the *sacroiliac* (sa″kro-il′e-ak) *joints.* The pelvic girdle transmits the body's weight to the legs at these joints (see fig. 7.15).

The sacrum forms the posterior wall of the pelvic cavity. The upper anterior margin of the sacrum, which represents the body of the first sacral vertebra, is called the *sacral promontory* (sa′kral prom′on-to″re). A physician performing a vaginal examination can feel this projection and use it as a guide in determining the size of the pelvis. This measurement is helpful in estimating how easily an infant may be able to pass through a woman's pelvic cavity during childbirth.

The vertebral foramina of the sacral vertebrae form the *sacral canal,* which continues through the sacrum to an opening of variable size at the tip, called the *sacral hiatus* (hi-a′tus). This foramen exists because the laminae of the last sacral vertebra are not fused. On the ventral surface of the sacrum, four pairs of *anterior sacral foramina* provide passageways for nerves and blood vessels.

Coccyx

The **coccyx** (kok′siks), or tailbone, is the lowest part of the vertebral column and is usually composed of four vertebrae that fuse between the ages of twenty-five and thirty (fig. 7.37).

Variations in individuals include three to five coccygeal vertebrae with typically the last three fused. In the elderly, the coccyx may fuse to the sacrum. Ligaments attach the coccyx to the margins of the sacral hiatus. Sitting presses on the coccyx, and it moves forward, acting like a shock absorber. Sitting down with great force, as when slipping and falling on ice, can fracture or dislocate the coccyx. The coccyx also serves as an attachment for the muscles of the pelvic floor. Table 7.8 summarizes the bones of the vertebral column, and Clinical Application 7.3 discusses disorders of the vertebral column.

PRACTICE 

21 Describe the structure of the vertebral column.

22 Explain the difference between the vertebral column of an adult and that of an infant.

23 Describe a typical vertebra.

24 How do the structures of cervical, thoracic, and lumbar vertebrae differ?

7.8 THORACIC CAGE

The **thoracic cage** includes the ribs, the thoracic vertebrae, the sternum, and the costal cartilages that attach the ribs to the sternum. These bones support the pectoral girdle and upper limbs, protect the viscera in the thoracic and upper abdominal cavities, and play a role in breathing (fig. 7.38).

Ribs

The usual number of **ribs** is twenty-four—one pair attached to each of the twelve thoracic vertebrae. Some individuals have extra ribs associated with their cervical or lumbar vertebrae.

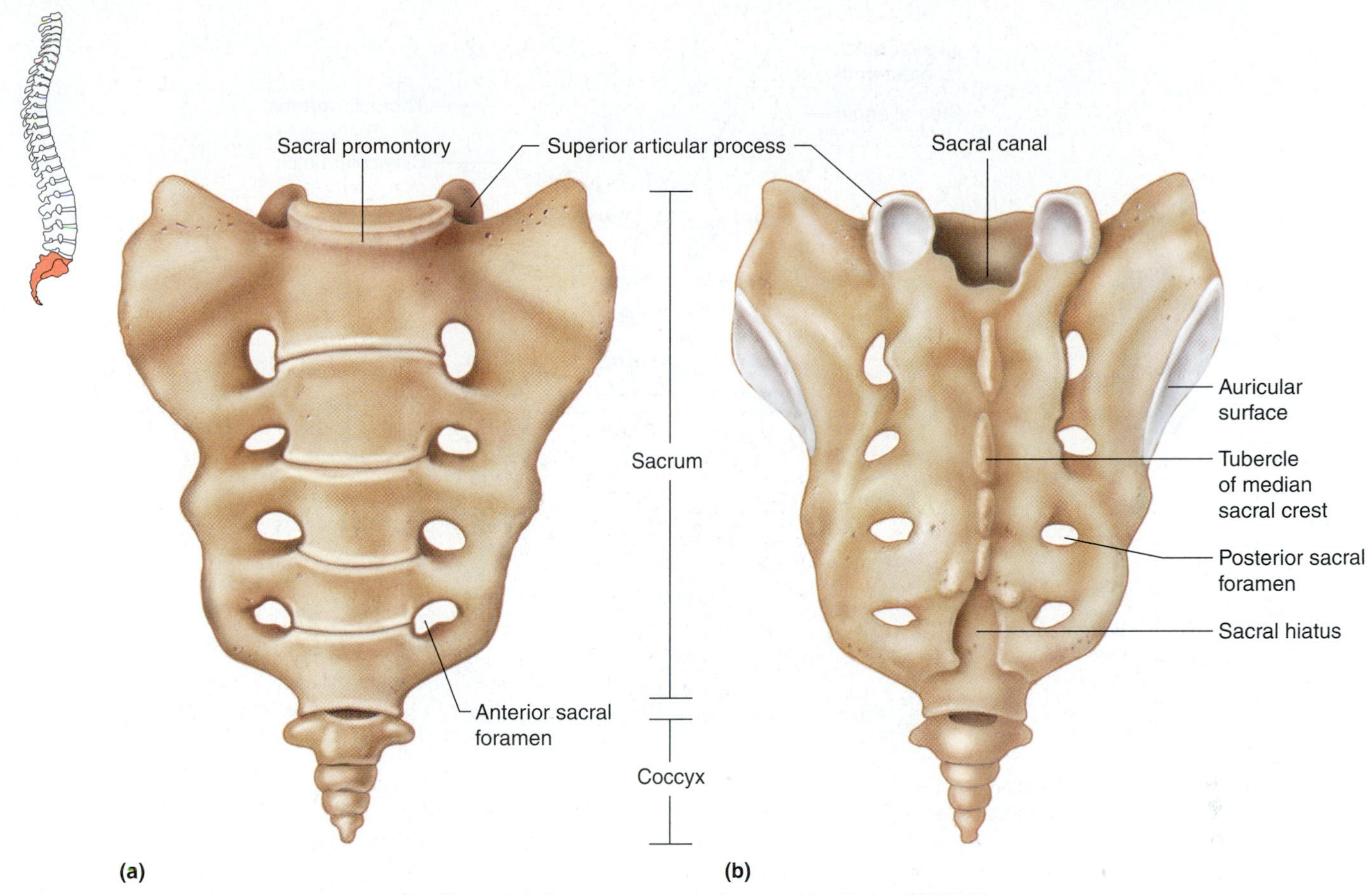

FIGURE 7.37 Sacrum and coccyx. (*a*) Anterior view. (*b*) Posterior view.

TABLE 7.8 | Bones of the Vertebral Column

Bones	Number	Special Features	Bones	Number	Special Features
Cervical vertebrae	7	Transverse foramina; facets of atlas articulate with occipital condyles of skull; dens of axis articulates with atlas; spinous processes of second through sixth vertebrae are bifid	Lumbar vertebrae	5	Large bodies; thinner transverse processes that project laterally; short, thick spinous processes that project posteriorly nearly horizontal
Thoracic vertebrae	12	Transverse processes project posteriorly at sharp angles; pointed spinous processes that slope downward; facets that articulate with ribs	Sacrum	4–5 vertebrae fused	Posterior sacral foramina, auricular surfaces, sacral promontory, sacral canal, sacral hiatus, anterior sacral foramina
			Coccyx	3–5 vertebrae fused	Attached by ligaments to the margins of the sacral hiatus

The first seven rib pairs, called the *true ribs* (vertebrosternal ribs), join the sternum directly by their costal cartilages. The remaining five pairs are called *false ribs* because their cartilages do not reach the sternum directly. Instead, the cartilages of the upper three false ribs (vertebrochondral ribs) join the cartilages of the seventh rib, whereas the last two rib pairs have no attachments to the sternum. These last two pairs (or sometimes the last three pairs) are called *floating ribs* (vertebral ribs).

A typical rib (fig. 7.39) has a long, slender shaft, which curves around the chest and slopes downward. On the posterior end is an enlarged *head* by which the rib articulates with a facet on the body of its own vertebra and with the body of the next higher vertebra. The neck of the rib is flattened, lateral to the head, where ligaments attach. A *tubercle,* close to the head of the rib, articulates with the transverse process of the vertebra.

The costal cartilages are composed of hyaline cartilage. They are attached to the anterior ends of the ribs and continue in line with them toward the sternum.

Sternum

The **sternum** (ster′num), or breastbone, is located along the midline in the anterior portion of the thoracic cage. It is a flat, elongated bone that develops in three parts—an

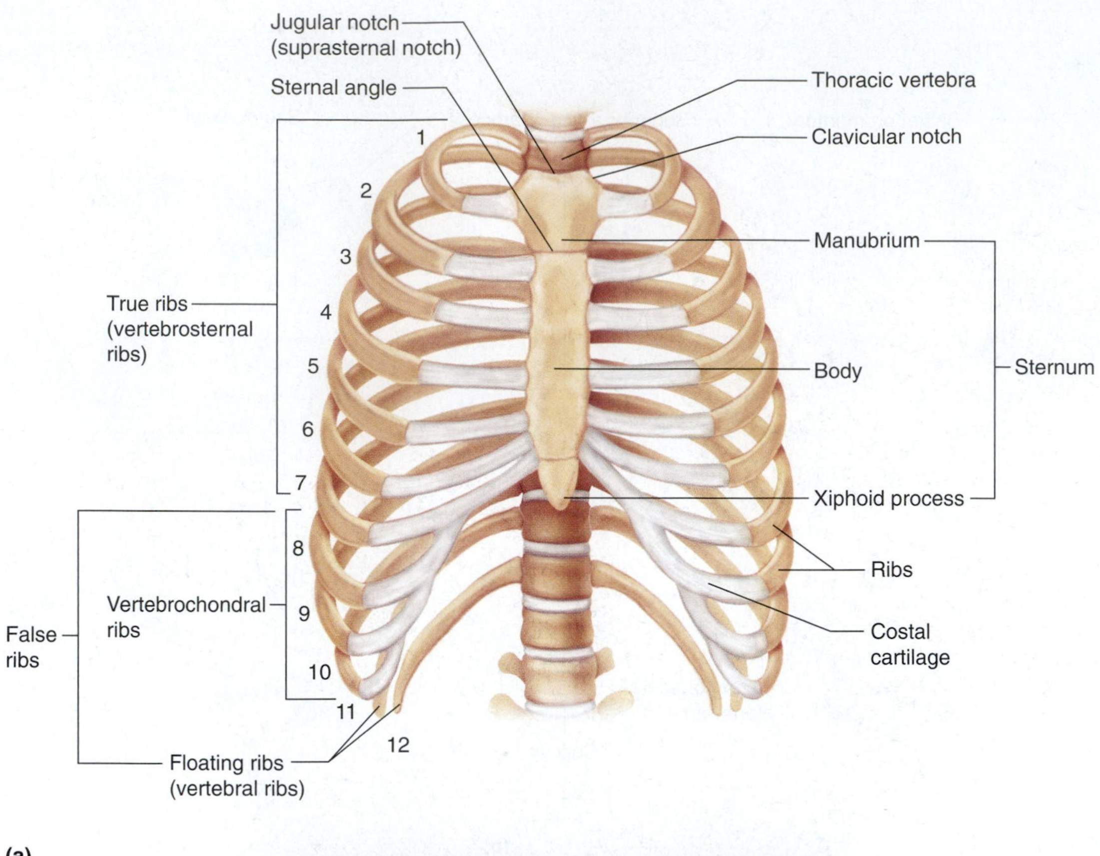

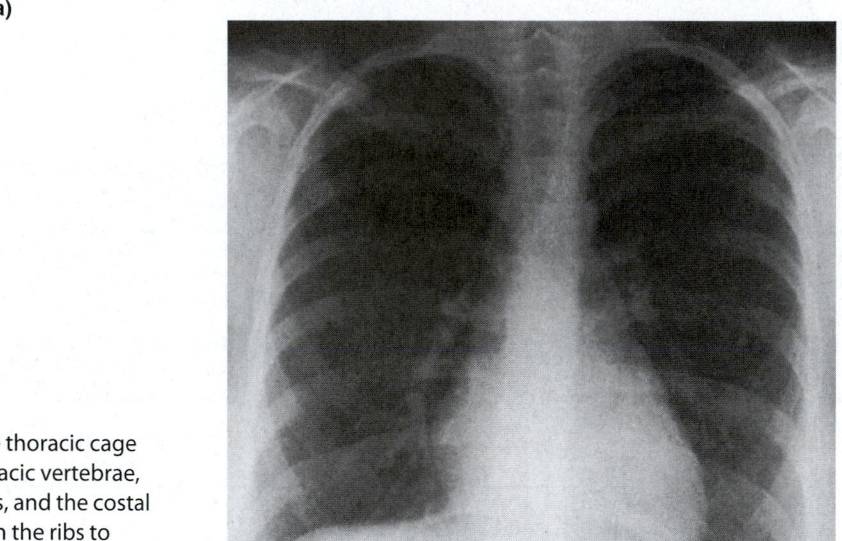

FIGURE 7.38 The thoracic cage includes (*a*) the thoracic vertebrae, the sternum, the ribs, and the costal cartilages that attach the ribs to the sternum. (*b*) Radiograph of the thoracic cage, anterior view. The light region behind the sternum and above the diaphragm is the heart.

upper *manubrium* (mah-nu′bre-um), a middle *body,* and a lower *xiphoid* (zif′oid) *process* that projects downward (see fig. 7.38).

The sides of the manubrium and the body are notched where they articulate with costal cartilages. The manubrium also articulates with the clavicles by facets on its superior border. It usually remains as a separate bone until middle age or later, when it fuses to the body of the sternum.

The manubrium and body of the sternum lie in different planes, so their line of union projects slightly forward. This projection, at the level of the second costal cartilage, is called the sternal angle (angle of Louis). It is commonly used as a clinical landmark to locate a particular rib (see fig. 7.38).

7.3 CLINICAL APPLICATION

Disorders of the Vertebral Column

Changes in the intervertebral discs may cause various problems. Each disc is composed of a tough, outer layer of fibrocartilage (annulus fibrosus) and an elastic central mass (nucleus pulposus). With age, these discs degenerate—the central masses lose firmness, and the outer layers thin and weaken, developing cracks. Extra pressure, as when a person falls or lifts a heavy object, can break the outer layers of the discs, squeezing out the central masses. Such a rupture may press on the spinal cord or on spinal nerves that branch from it. This condition, called a *ruptured,* or *herniated disc,* may cause back pain and numbness or loss of muscular function in the parts innervated by the affected spinal nerves.

A surgical procedure called a *laminectomy* may relieve the pain of a herniated disc by removing a portion of the posterior arch of a vertebra. This reduces the pressure on the affected nerve tissues. Alternatively, a protein-digesting enzyme (chymopapain) may be injected into the injured disc to shrink it.

Sometimes problems develop in the curvatures of the vertebral column because of poor posture, injury, or disease. An exaggerated thoracic curvature causes rounded shoulders and a hunchback. This condition, called kyphosis, occasionally develops in adolescents who undertake strenuous athletic activities. Unless corrected before bone growth completes, the condition can permanently deform the vertebral column.

Sometimes the vertebral column develops an abnormal lateral curvature, so that one hip or shoulder is lower than the other. This may displace or compress the thoracic and abdominal organs. With unknown cause, this condition, called scoliosis, is most common in adolescent females. It also may accompany such diseases as poliomyelitis, rickets, or tuberculosis. An accentuated lumbar curvature is called *lordosis,* or swayback.

As a person ages, the intervertebral discs shrink and become more rigid, and compression is more likely to fracture the vertebral bodies. Consequently, height may decrease, and the thoracic curvature of the vertebral column may be accentuated, bowing the back. ■

The xiphoid process begins as a piece of cartilage. It slowly ossifies, and by middle age it usually fuses to the body of the sternum.

Red marrow within the spongy bone of the sternum produces blood cells into adulthood. The sternum has a thin covering of compact bone and is easy to reach, so samples of its marrow may be removed to diagnose diseases. This procedure, a sternal puncture, suctions (aspirates) some marrow through a hollow needle. (Marrow may also be removed from the iliac crest of a hip bone.)

PRACTICE

25 Which bones comprise the thoracic cage?

26 Describe a typical rib.

27 What are the differences among true, false, and floating ribs?

28 Name the three parts of the sternum.

7.9 PECTORAL GIRDLE

The **pectoral** (pek′tor-al) **girdle,** or shoulder girdle, is composed of four parts—two clavicles (collarbones) and two scapulae (shoulder blades). Although the word *girdle* suggests a ring-shaped structure, the pectoral girdle is an incomplete ring. It is open in the back between the scapulae, and the sternum separates its bones in front. The pectoral girdle supports the upper limbs and is an attachment for several muscles that move them (fig. 7.40).

Clavicles

The **clavicles** (klav′ĭ-k′lz) are slender, rodlike bones with elongated S-shapes (fig. 7.40). Located at the base of the neck, they run horizontally between the manubrium and the scapulae. The sternal (or medial) ends of the clavicles articulate with the manubrium, and the acromial (or lateral) ends join processes of the scapulae.

The clavicles brace the freely movable scapulae, helping to hold the shoulders in place. They also provide attachments for muscles of the upper limbs, chest, and back. The clavicle is structurally weak because of its elongated double curve. If compressed lengthwise due to abnormal pressure on the shoulder, it is likely to fracture.

In the epic poem the *Iliad,* Homer describes a man whose "shoulders were bent and met over his chest." The man probably had a rare inherited condition, called cleidocranial dysplasia, in which certain bones do not grow. In the condition, the skull consists of small fragments joined by fibrous connective tissue, rather than the normal large, interlocking hard bony plates. The clavicles are stunted or missing.

Cleidocranial dysplasia was first reported in a child in the huge Arnold family, founded by a Chinese immigrant to South Africa. The child had been kicked by a horse, and X rays revealed that the fontanels atop the head had never closed. The condition became known as "Arnold head." It is caused not by a horse's kick, but by a malfunctioning gene that normally instructs differentiation of bone cells.

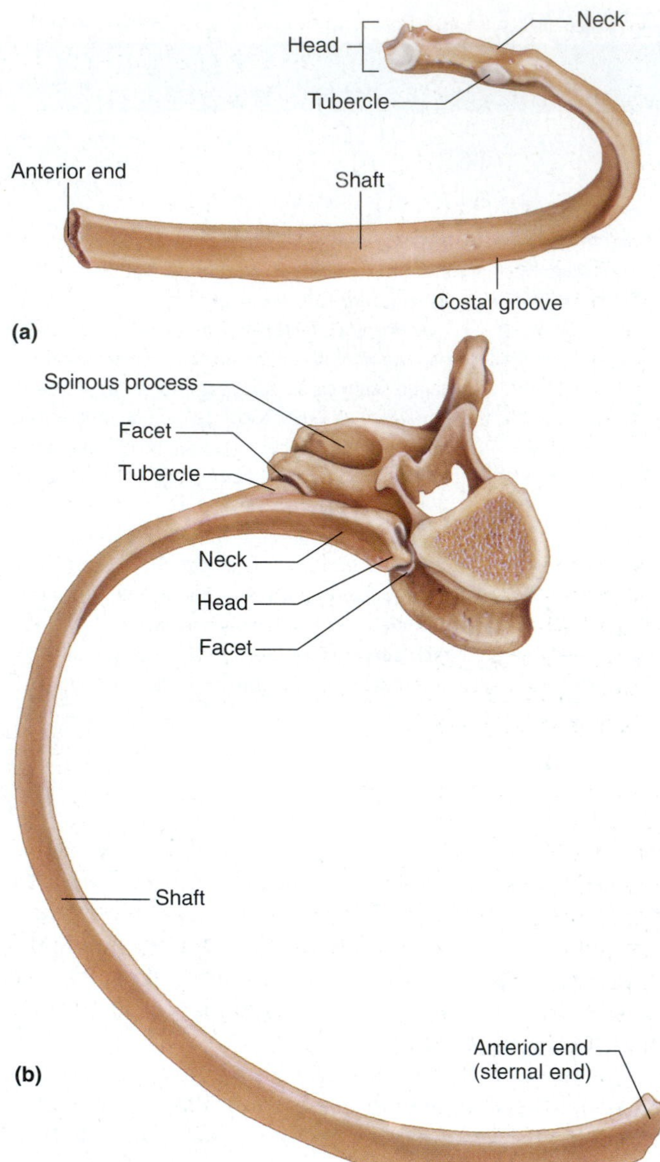

FIGURE 7.39 A typical rib. (*a*) Posterior view. (*b*) Articulations of a rib with a thoracic vertebra (superior view).

Scapulae

The **scapulae** (skap′u-le) are broad, somewhat triangular bones located on either side of the upper back. They have flat bodies with concave anterior surfaces (fig. 7.41). The posterior surface of each scapula is divided into unequal portions by a *spine.* Above the spine is the *supraspinous fossa,* and below the spine is the *infraspinous fossa.* This spine leads to an *acromion* (ah-kro′me-on) *process* that forms the tip of the shoulder. The acromion process articulates with the clavicle and provides attachments for muscles of the upper limb and chest. A *coracoid* (kor′ah-koid) *process* curves anteriorly and inferiorly to the acromion process. The coracoid process also provides attachments for upper limb and chest muscles. On the lateral surface of the scapula between the processes is a depression called the *glenoid cavity* (glenoid fossa of the scapula). It articulates with the head of the arm bone (humerus).

The scapula has three borders. The *superior border* is on the superior edge. The *axillary,* or *lateral border,* is directed toward the upper limb. The *vertebral,* or *medial border,* is closest to the vertebral column, about 5 cm away.

PRACTICE

29 Which bones form the pectoral girdle?

30 What is the function of the pectoral girdle?

7.10 UPPER LIMB

The bones of the upper limb form the framework of the arm, forearm, and hand. They also provide attachments for muscles and interact with muscles to move limb parts. These bones include a humerus, a radius, an ulna, carpals, metacarpals, and phalanges (fig. 7.42).

Humerus

The **humerus** is a long bone that extends from the scapula to the elbow. At its upper end is a smooth, rounded *head* that fits into the glenoid cavity of the scapula (fig. 7.43). Just below the head are two processes—a *greater tubercle* on the lateral side and a *lesser tubercle* on the anterior side. These tubercles provide attachments for muscles that move the upper limb at the shoulder. Between them is a narrow furrow, the *intertubercular groove,* through which a tendon passes from a muscle in the arm (biceps brachii) to the shoulder.

The narrow depression along the lower margin of the head that separates it from the tubercles is called the *anatomical neck.* Just below the head and the tubercles of the humerus is a tapering region called the *surgical neck,* so named because fractures commonly occur there. Near the middle of the bony shaft on the lateral side is a rough V-shaped area called the *deltoid tuberosity.* It provides an attachment for the muscle (deltoid) that raises the upper limb horizontally to the side.

At the lower end of the humerus are two smooth *condyles*—a knoblike *capitulum* (kah-pit′u-lum) on the lateral side and a pulley-shaped *trochlea* (trok′le-ah) on the medial side. The capitulum articulates with the radius at the elbow, whereas the trochlea joins the ulna.

Above the condyles on either side are *epicondyles,* which provide attachments for muscles and ligaments of the elbow. Between the epicondyles anteriorly is a depression, the *coronoid fossa,* that receives a process of the ulna (coronoid process) when the elbow bends. Another depression on the posterior surface, the *olecranon* (o″lek′ra-non) *fossa,* receives an olecranon process when the elbow straightens.

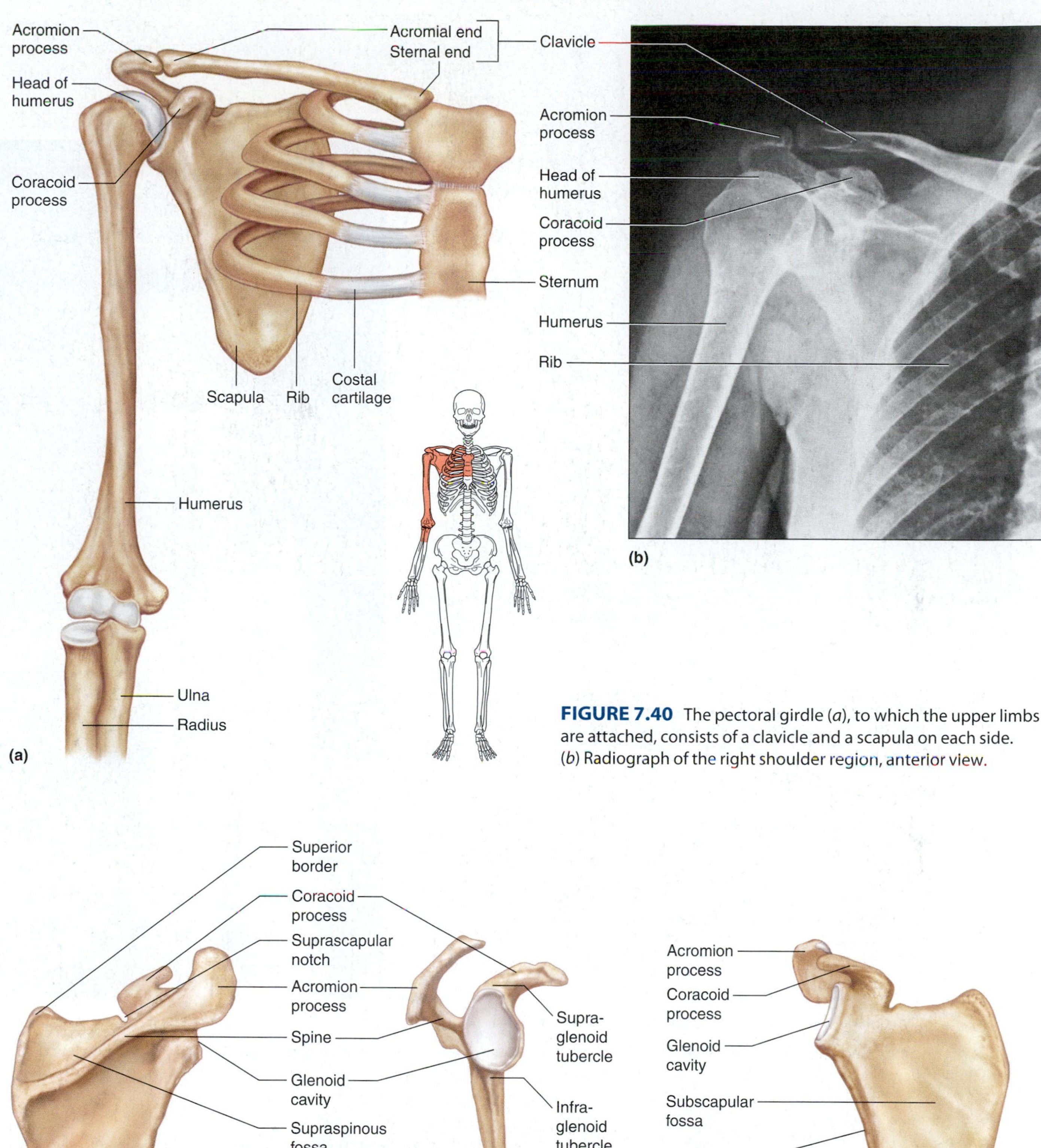

FIGURE 7.40 The pectoral girdle (*a*), to which the upper limbs are attached, consists of a clavicle and a scapula on each side. (*b*) Radiograph of the right shoulder region, anterior view.

FIGURE 7.41 Right scapula. (*a*) Posterior surface. (*b*) Lateral view showing the glenoid cavity that articulates with the head of the humerus. (*c*) Anterior surface.

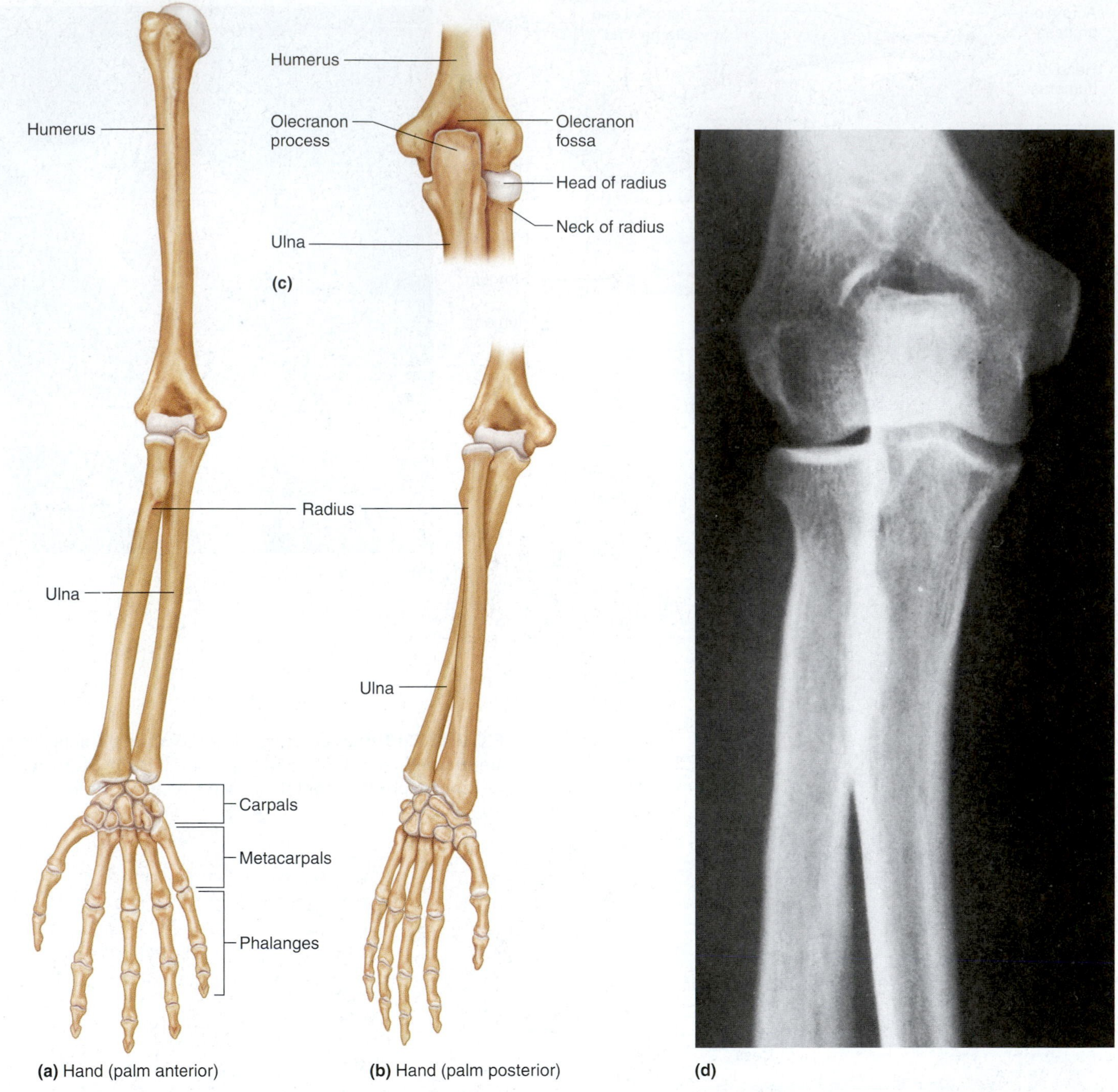

FIGURE 7.42 Right upper limb. (*a*) Anterior view with the hand, palm anterior and (*b*) with the hand, palm posterior. (*c*) Posterior view of the right elbow. (*d*) Radiograph of the right elbow and forearm, anterior view.

Radius

The **radius,** located on the thumb side of the forearm, is somewhat shorter than its companion, the ulna (fig. 7.44). The radius extends from the elbow to the wrist and crosses over the ulna when the hand is turned so that the palm faces backward.

A thick, disclike *head* at the upper end of the radius articulates with the capitulum of the humerus and a notch of the ulna (radial notch). This arrangement allows the radius to rotate.

On the radial shaft just below the head is a process called the *radial tuberosity.* It is an attachment for a muscle (biceps brachii) that bends the upper limb at the elbow. At the distal end of the radius, a lateral *styloid* (sti′loid) *process* provides attachments for ligaments of the wrist.

Ulna

The **ulna** is longer than the radius and overlaps the end of the humerus posteriorly. At its proximal end, the ulna has a

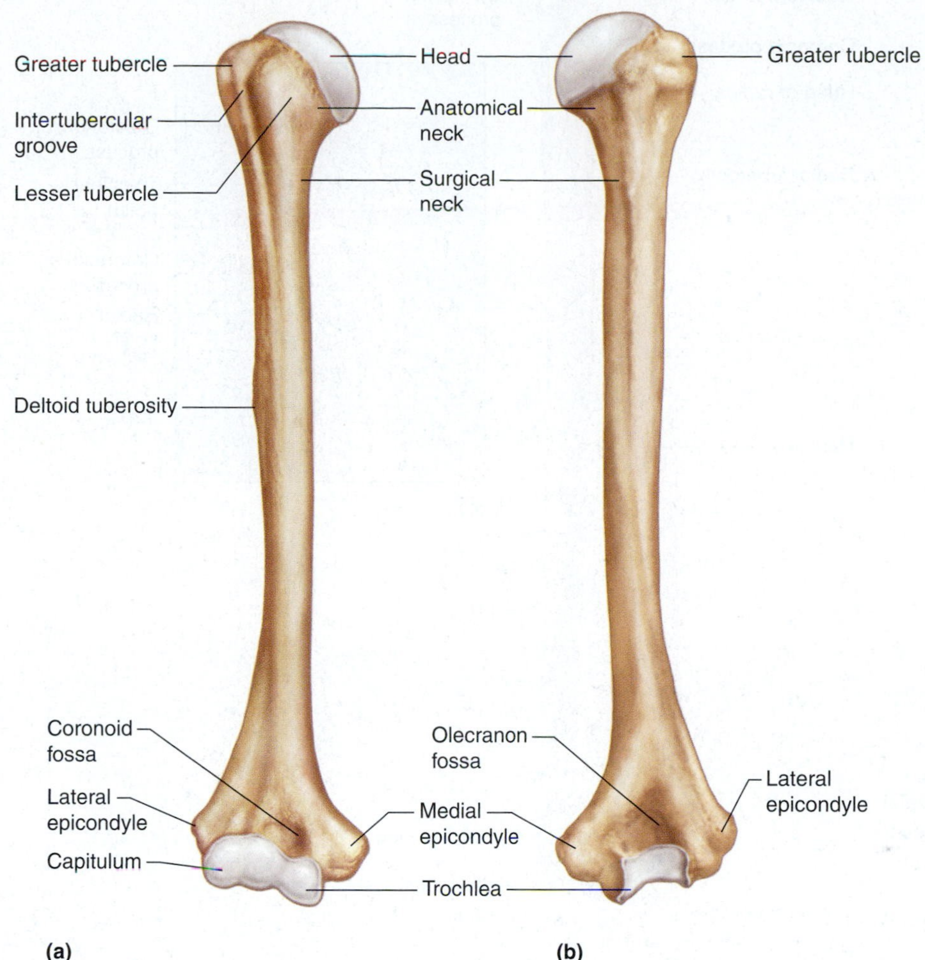

FIGURE 7.43 Right humerus. (*a*) Anterior surface. (*b*) Posterior surface.

wrenchlike opening, the *trochlear notch* (semilunar notch), that articulates with the trochlea of the humerus. A process lies on either side of this notch. The *olecranon process,* located above the trochlear notch, provides an attachment for the muscle (triceps brachii) that straightens the upper limb at the elbow. During this movement, the olecranon process of the ulna fits into the olecranon fossa of the humerus. Similarly, the *coronoid process,* just below the trochlear notch, fits into the coronoid fossa of the humerus when the elbow bends.

> Many a thirtyish parent of a young little leaguer or softball player becomes tempted to join in. But if he or she has not pitched in many years, sudden activity may break the forearm. Forearm pain while pitching is a signal that a fracture could happen. Medical specialists advise returning to the pitching mound gradually. Start with twenty pitches, five days a week, for two to three months before regular games begin. By the season's start, 120 pitches per daily practice session should be painless.

At the distal end of the ulna, its knoblike *head* articulates laterally with a notch of the radius (ulnar notch) and with a disc of fibrocartilage inferiorly (fig. 7.44). This disc, in turn, joins a wrist bone (triquetrum). A medial *styloid process* at the distal end of the ulna provides attachments for ligaments of the wrist.

Hand

The hand is made up of the wrist, palm, and fingers. The skeleton of the wrist consists of eight small **carpal bones** firmly bound in two rows of four bones each. The resulting compact mass is called a *carpus* (kar′pus). The carpus is rounded on its proximal surface, where it articulates with the radius and with the fibrocartilaginous disc on the ulnar side. The carpus is concave anteriorly, forming a canal through which tendons and nerves extend to the palm. Its distal surface articulates with the metacarpal bones. Figure 7.45 names the individual bones of the carpus.

Five **metacarpal bones,** one in line with each finger, form the framework of the palm or *metacarpus* (met″ah-kar′pus) of the hand. These bones are cylindrical, with rounded distal ends that form the knuckles of a clenched

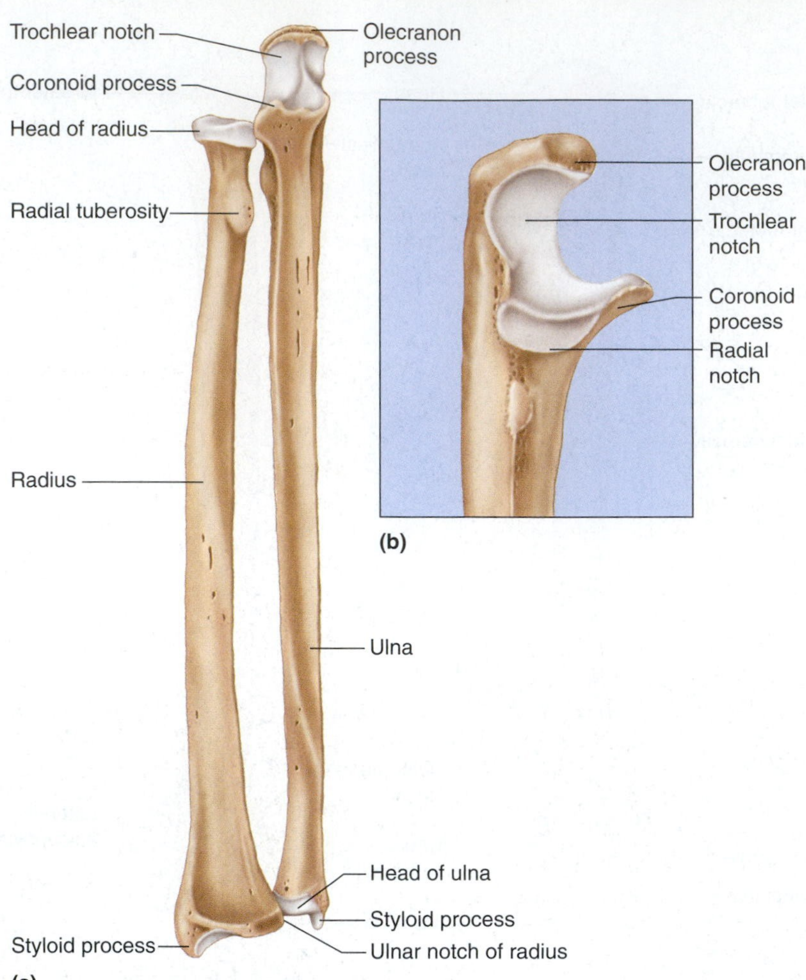

FIGURE 7.44 Right radius and ulna. (*a*) The head of the radius articulates with the radial notch of the ulna, and the head of the ulna articulates with the ulnar notch of the radius. (*b*) Lateral view of the proximal end of the ulna.

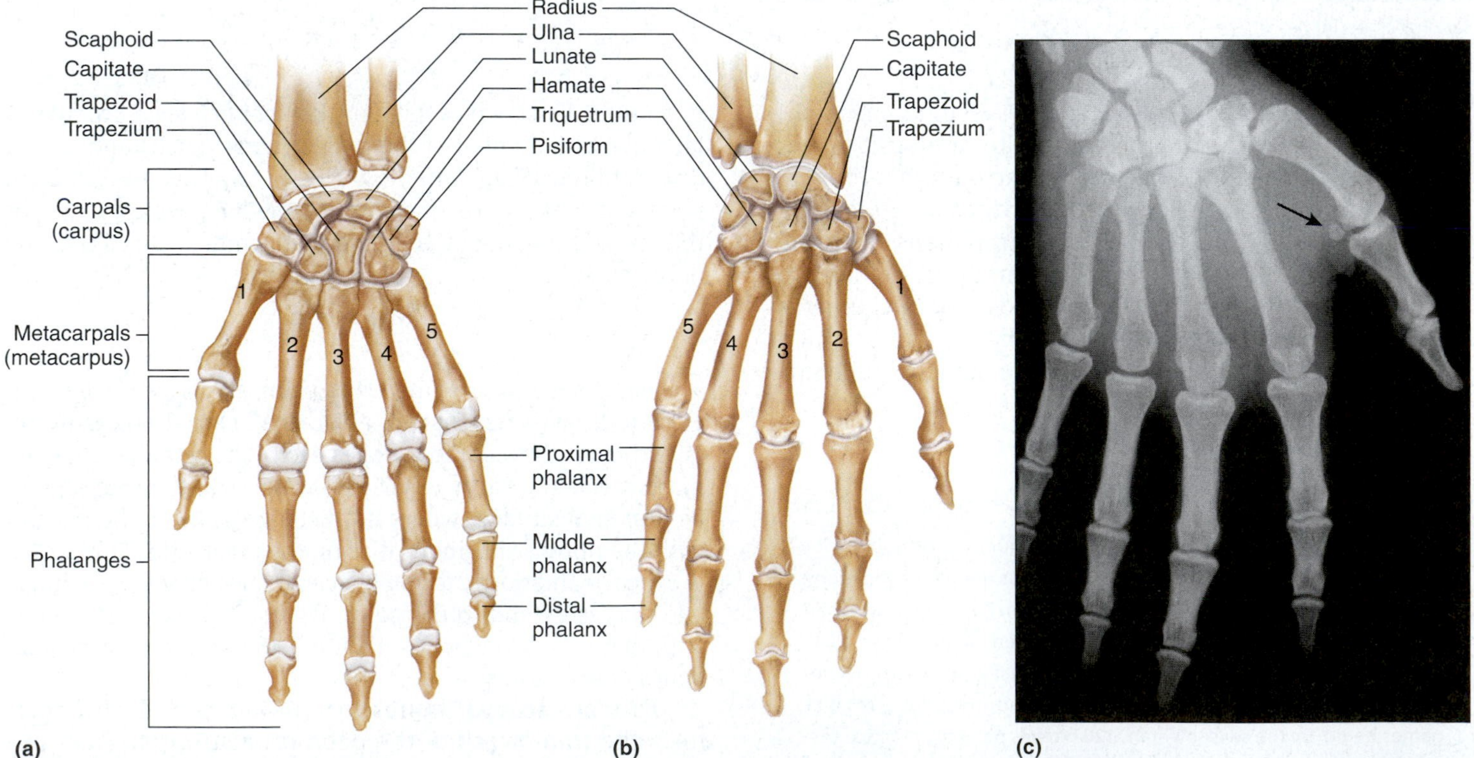

FIGURE 7.45 Right hand. (*a*) Anterior view. (*b*) Posterior view. (*c*) Radiograph, posterior view. Note the small sesamoid bone associated with the joint at the base of the thumb (arrow).

TABLE 7.9 | Bones of the Pectoral Girdle and Upper Limbs

Name and Number	Location	Special Features
Clavicle (2)	Base of neck between sternum and scapula	Sternal end, acromial end
Scapula (2)	Upper back, forming part of shoulder	Body, spine, acromion process, coracoid process, glenoid cavity
Humerus (2)	Arm, between scapula and elbow	Head, greater tubercle, lesser tubercle, intertubercular groove, anatomical neck, surgical neck, deltoid tuberosity, capitulum, trochlea, medial epicondyle, lateral epicondyle, coronoid fossa, olecranon fossa
Radius (2)	Lateral side of forearm, between elbow and wrist	Head, radial tuberosity, styloid process, ulnar notch
Ulna (2)	Medial side of forearm, between elbow and wrist	Trochlear notch, olecranon process, coronoid process, head, styloid process, radial notch
Carpal (16)	Wrist	Two rows of four bones each
Metacarpal (10)	Palm	One in line with each finger and thumb
Phalanx (28)	Finger	Three in each finger; two in each thumb

fist. The metacarpals articulate proximally with the carpals and distally with the phalanges. The metacarpal on the lateral side is the most freely movable; it permits the thumb to oppose the fingers when grasping something. These bones are numbered 1 to 5, beginning with the metacarpal of the thumb (fig. 7.45).

The **phalanges** are the finger bones. Three are in each finger—a proximal, a middle, and a distal phalanx—and two are in the thumb. (The thumb lacks a middle phalanx.) Thus, each hand has fourteen finger bones. Table 7.9 summarizes the bones of the pectoral girdle and upper limbs.

It is not uncommon for a baby to be born with an extra finger or toe, but because the extra digit is usually surgically removed early in life, hands like the ones in figure 7.46 are rare. Polydactyly ("many digits") is an inherited trait. It is common in cats. A lone but popular male cat brought the trait from England to colonial Boston. Polydactyly is also common among the Amish people.

PRACTICE

31 Locate and name each of the bones of the upper limb.

32 Explain how the bones of the upper limb articulate.

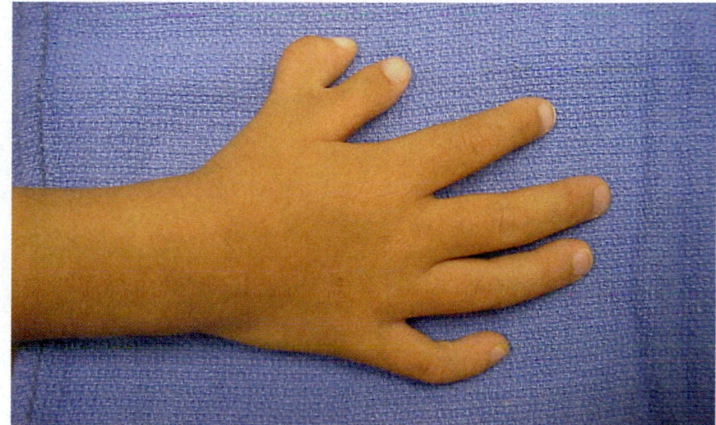

FIGURE 7.46 A person with polydactyly has extra digits.

7.11 PELVIC GIRDLE

The **pelvic girdle** consists of the two hip bones, also known as coxal bones, pelvic bones or innominate bones, which articulate with each other anteriorly and with the sacrum posteriorly (fig. 7.47). The sacrum, coccyx, and pelvic girdle form the bowl-shaped *pelvis.* The pelvic girdle supports the trunk of the body; provides attachments for the lower limbs; and protects the urinary bladder, the distal end of the large intestine, and the internal reproductive organs. The body's weight is transmitted through the pelvic girdle to the lower limbs and then onto the ground.

Hip Bones

Each hip bone develops from three parts—an ilium, an ischium, and a pubis (fig. 7.48). These parts fuse in the region of a cup-shaped cavity called the *acetabulum* (as″ĕ-tab′u-lum). This depression, on the lateral surface of the hip bone, receives the rounded head of the femur or thigh bone.

The **ilium** (il′e-um), the largest and most superior portion of the hip bone, flares outward, forming the prominence of the hip. The margin of this prominence is called the *iliac crest.* The smooth, concave surface on the anterior aspect of the ilium is the *iliac fossa.*

Posteriorly, the ilium joins the sacrum at the *sacroiliac joint.* Anteriorly, a projection of the ilium, the *anterior superior iliac spine,* can be felt lateral to the groin. This spine provides attachments for ligaments and muscles and is an important surgical landmark.

A common injury in contact sports such as football is bruising the soft tissues and bone associated with the anterior superior iliac spine. Wearing protective padding can prevent this painful injury, called a *hip pointer.*

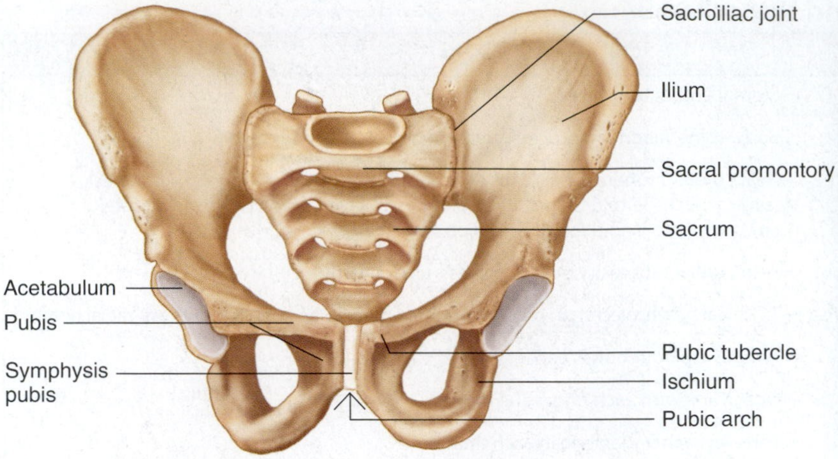

Sacroiliac joint
Ilium
Sacral promontory
Sacrum
Acetabulum
Pubis
Symphysis pubis
Pubic tubercle
Ischium
Pubic arch

(a)

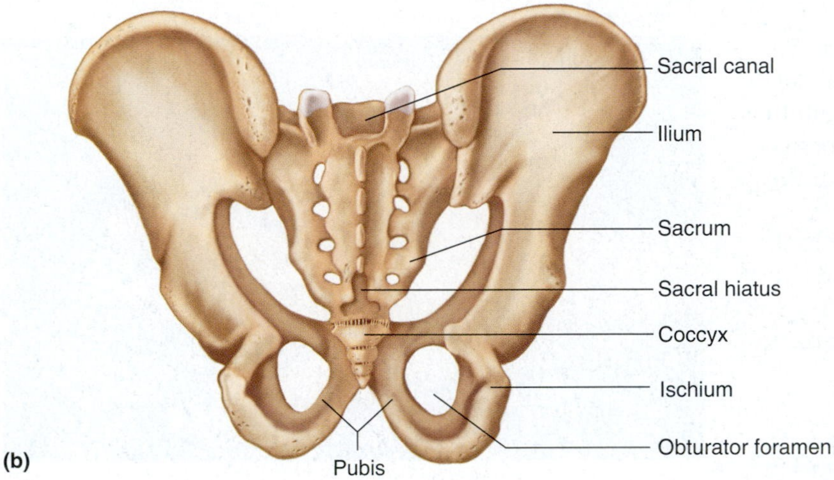

Sacral canal
Ilium
Sacrum
Sacral hiatus
Coccyx
Ischium
Obturator foramen
Pubis

(b)

(c)

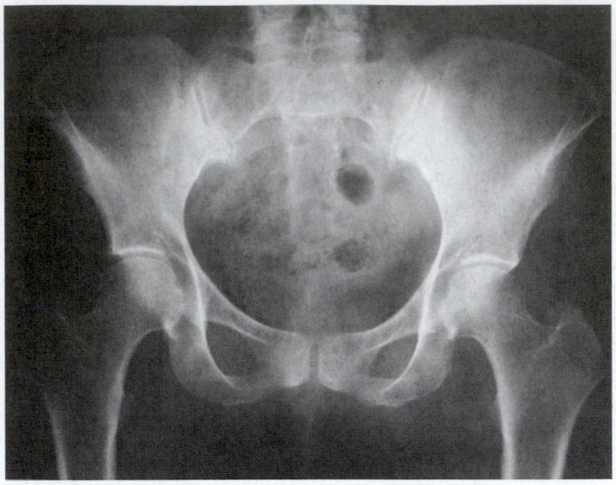

FIGURE 7.47 Pelvic girdle. (*a*) Anterior view. (*b*) Posterior view. This girdle provides an attachment for the lower limbs and together with the sacrum and coccyx forms the pelvis. (*c*) Radiograph of the pelvic girdle.

Iliac crest
Iliac fossa
Anterior superior iliac spine
Ilium
Anterior inferior iliac spine
Obturator foramen
Pubis
Ischium
Posterior superior iliac spine
Posterior inferior iliac spine
Greater sciatic notch
Ischial spine
Lesser sciatic notch
Ischial tuberosity
Iliac crest
Ilium
Acetabulum
Obturator foramen
Pubic crest
Pubic tubercle
Ischium
Pubis

(a) (b)

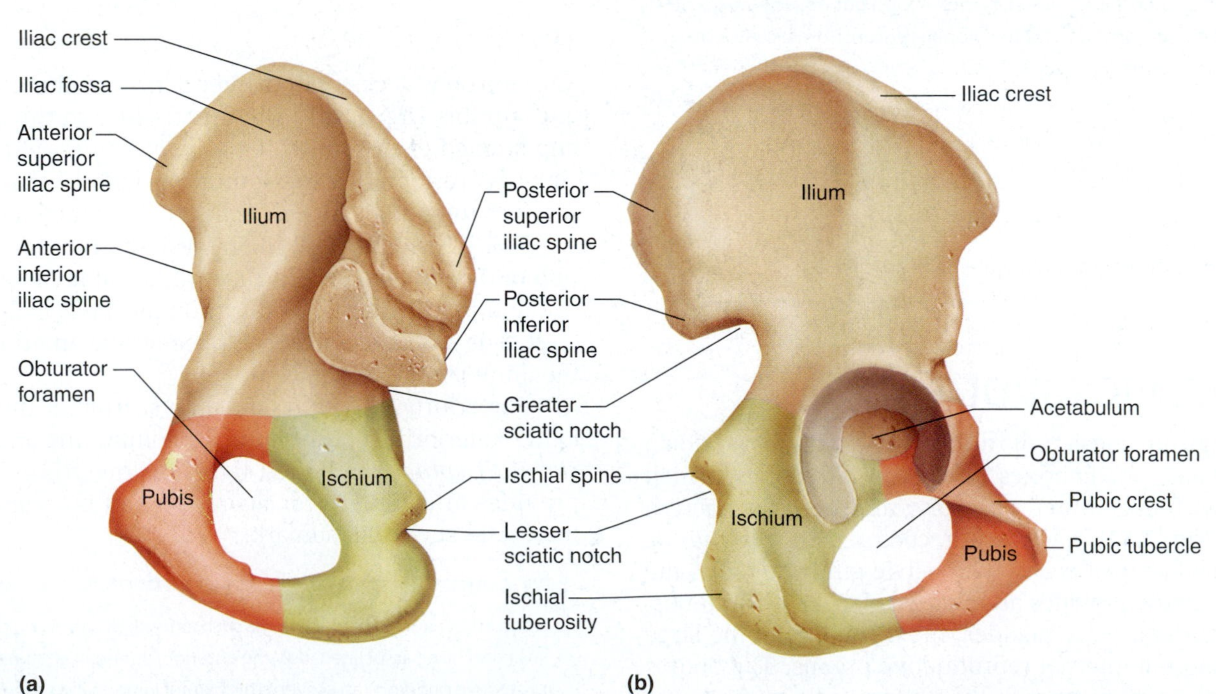

FIGURE 7.48 Right hip bone. (*a*) Medial surface. (*b*) Lateral view.

On the posterior border of the ilium is a *posterior superior iliac spine.* Below this spine is a deep indentation, the *greater sciatic notch,* through which a number of nerves and blood vessels pass.

The **ischium** (is′ke-um), which forms the lowest portion of the hip bone, is L-shaped, with its angle, the *ischial tuberosity,* pointing posteriorly and downward. This tuberosity has a rough surface that provides attachments for ligaments and lower limb muscles. It also supports the weight of the body during sitting. Above the ischial tuberosity, near the junction of the ilium and ischium, is a sharp projection called the *ischial spine.* Like the sacral promontory, this spine, which can be felt during a vaginal examination, is used as a guide for determining pelvis size. The distance between the ischial spines is the shortest diameter of the pelvic outlet.

The **pubis** (pu′bis) constitutes the anterior portion of the hip bone. The two pubic bones come together at the midline to form a joint called the *symphysis pubis* (sim′fĭ-sis pu′bis). The angle these bones form below the symphysis is the *pubic arch* (fig. 7.49).

A portion of each pubis passes posteriorly and downward to join an ischium. Between the bodies of these bones on either side is a large opening, the *obturator foramen,* which is the largest foramen in the skeleton. An obturator membrane covers and nearly closes this foramen (see figs. 7.47 and 7.48).

Greater and Lesser Pelves

If a line were drawn along each side of the pelvis from the sacral promontory downward and anteriorly to the upper margin of the symphysis pubis, it would mark the *pelvic brim* (linea terminalis). This margin separates the lower, or lesser (true), pelvis from the upper, or greater (false), pelvis (see fig. 7.49).

The *greater pelvis* is bounded posteriorly by the lumbar vertebrae, laterally by the flared parts of the iliac bones, and anteriorly by the abdominal wall. The greater pelvis helps support the abdominal organs.

The *lesser pelvis* is bounded posteriorly by the sacrum and coccyx and laterally and anteriorly by the lower ilium, ischium,

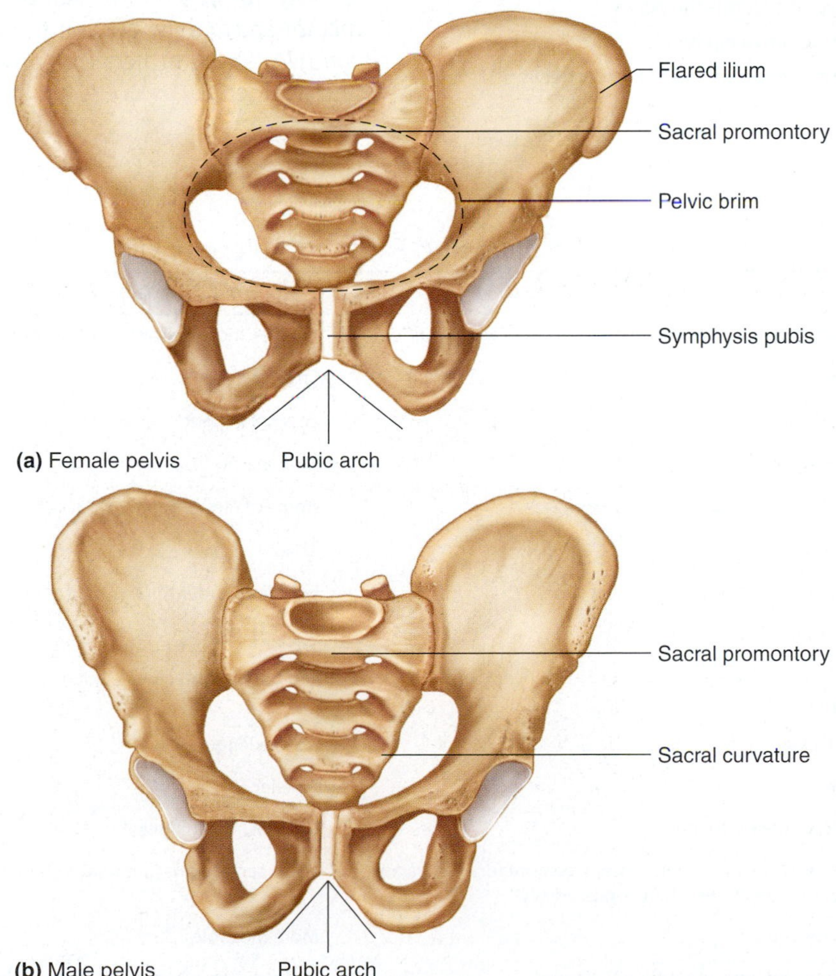

FIGURE 7.49 The female pelvis is usually wider in all diameters and roomier than that of the male. (*a*) Female pelvis. (*b*) Male pelvis.

and pubis bones. This portion of the pelvis surrounds a short, canal-like cavity that has an upper inlet and a lower outlet. An infant passes through this cavity during childbirth.

Differences Between Male and Female Pelves

Some basic structural differences distinguish the male and the female pelves, even though it may be difficult to find all of the "typical" characteristics in any one individual. These differences arise from the function of the female pelvis as a birth canal. Usually, the female iliac bones are more flared than those of the male, and consequently, the female hips are usually broader than the male's. The angle of the female pubic arch may be greater, there may be more distance between the ischial spines and the ischial tuberosities, and the sacral curvature may be shorter and flatter. Thus, the female pelvic cavity is usually wider in all diameters than that of the male. Also, the bones of the female pelvis are usually lighter, more delicate, and show less evidence of muscle attachments (fig. 7.49). Table 7.10 summarizes some of the differences between the male and female skeletons.

PRACTICE

33 Locate and name each bone that forms the pelvis.

34 Name the bones that fuse to form a hip bone.

35 Distinguish between the greater pelvis and the lesser pelvis.

36 How are male and female pelves different?

7.12 LOWER LIMB

The bones of the lower limb form the frameworks of the thigh, leg, and foot. They include a femur, a tibia, a fibula, tarsals, metatarsals, and phalanges (fig. 7.50).

Femur

The **femur,** or thigh bone, is the longest bone in the body and extends from the hip to the knee. A large, rounded *head* at its proximal end projects medially into the acetabulum of the hip bone (fig. 7.51). On the head, a pit called the *fovea capitis* marks the attachment of a ligament. Just below the head are a constriction, or *neck,* and two large processes—a superior, lateral *greater trochanter* and an inferior, medial *lesser trochanter.* These processes provide attachments for muscles of the lower limbs and buttocks. On the posterior surface in the middle third of the shaft is a longitudinal crest called the *linea aspera.* This rough strip is an attachment for several muscles.

At the distal end of the femur, two rounded processes, the *lateral* and *medial condyles,* articulate with the tibia of the leg. A patella also articulates with the femur on its distal anterior surface.

On the medial surface at its distal end is a prominent *medial epicondyle,* and on the lateral surface is a *lateral epicondyle.* These projections provide attachments for muscles and ligaments.

TABLE 7.10 | Differences Between the Male and Female Skeletons

Part	Male Differences	Female Differences
Skull	Larger, heavier, more conspicuous muscle attachment	Smaller, more delicate, less evidence of muscle attachment
mastoid process	Larger	Smaller
supraorbital ridge	More prominent	Less prominent
chin	More squared	More pointed
jaw angle	Angle of ramus about 90 degrees	Angle of ramus greater than 125 degrees
forehead	Shorter	Taller
orbit	Superior border thicker, blunt edge	Superior border thinner, sharp edge
palate	U-shaped	V-shaped
Pelvis	Hip bones heavier, thicker, more evidence of muscle attachment	Hip bones lighter, less evidence of muscle attachment
obturator foramen	More oval	More triangular
acetabulum	Larger	Smaller
pubic arch	Narrow, more V-shaped	Broader, more convex
sacrum	Narrow, sacral promontory projects more forward, sacral curvature bends less sharply posteriorly	Wide, sacral curvature bends sharply posteriorly
coccyx	Less movable	More movable
cavity	Narrow and long, more funnel-shaped	Wide, distance betweeen ischial spines and ischial tuberosities is greater

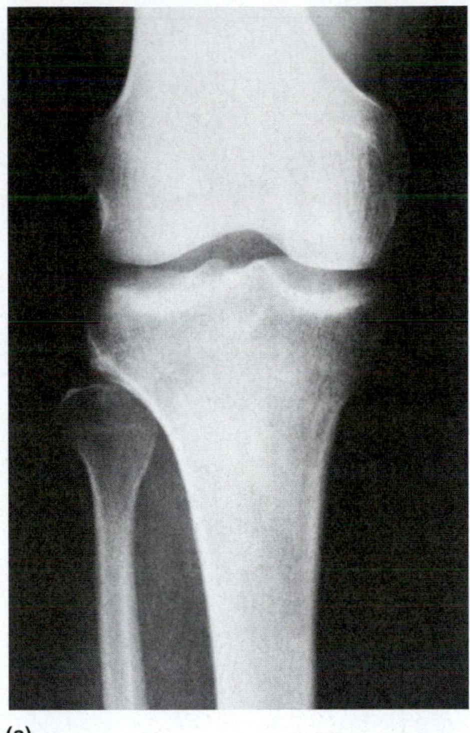

(a)

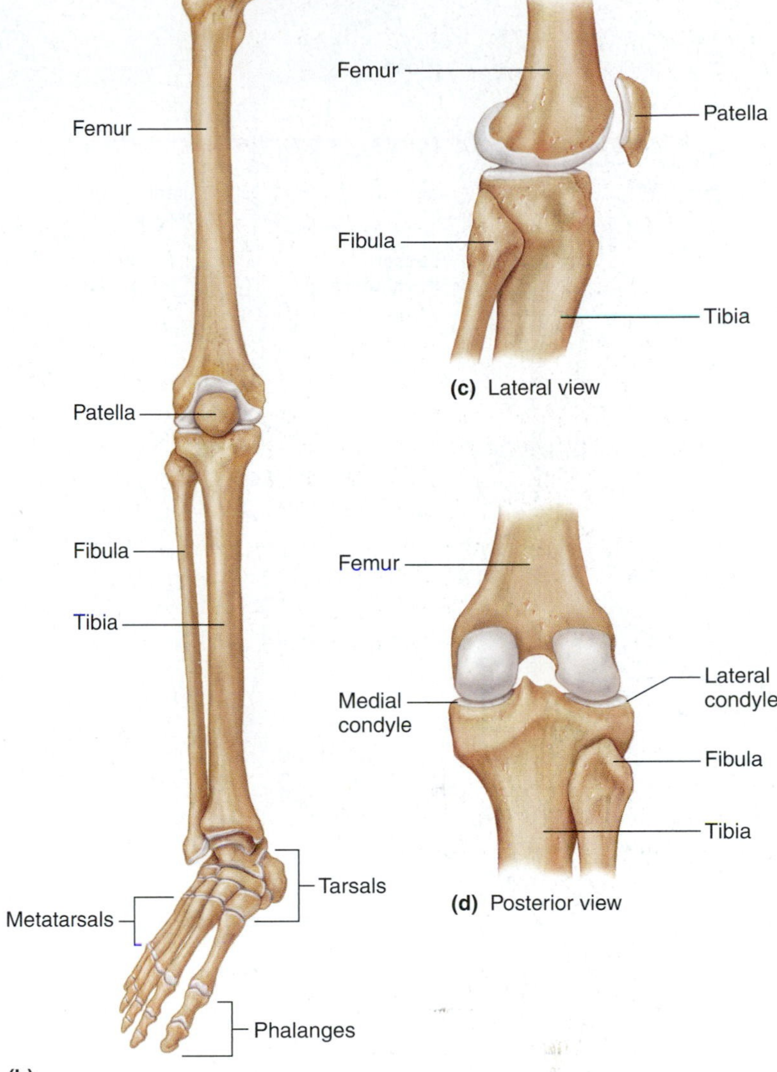

FIGURE 7.50 Parts of the right lower limb. (*a*) Radiograph of the knee (anterior view), showing the ends of the femur, tibia, and fibula. Thinner areas of bone, such as part of the head of the fibula and the patella, barely show in this radiograph. (*b*) Anterior view of the lower limb. (*c*) Lateral view of the knee. (*d*) Posterior view of the knee.

Patella

The **patella,** or kneecap, is a flat sesamoid bone located in a tendon that passes anteriorly over the knee (see fig. 7.50). The patella, because of its position, controls the angle at which this tendon continues toward the tibia, so it functions in lever actions associated with lower limb movements.

> As a result of a blow to the knee or a forceful unnatural movement of the leg, the patella sometimes slips to one side. This painful condition is called a *patellar dislocation.* Doing exercises that strengthen muscles associated with the knee and wearing protective padding can prevent knee displacement. Unfortunately, once the soft tissues that hold the patella in place are stretched, patellar dislocation tends to recur.

Tibia

The **tibia,** or shin bone, is the larger of the two leg bones and is located on the medial side. Its proximal end is expanded into *medial* and *lateral condyles,* which have concave surfaces and articulate with the condyles of the femur (fig. 7.52). Below the condyles, on the anterior surface, is a process called the *tibial tuberosity,* which provides an attachment for the *patellar ligament* (a continuation of the patella-bearing tendon). A prominent *anterior crest* extends downward from the tuberosity and attaches connective tissues in the leg.

At its distal end, the tibia expands to form a prominence on the inner ankle called the *medial malleolus* (mah-le′o-lus), an attachment for ligaments. On its lateral side is a depression that articulates with the fibula. The inferior surface of the tibia's distal end articulates with a large bone (the talus) in the ankle.

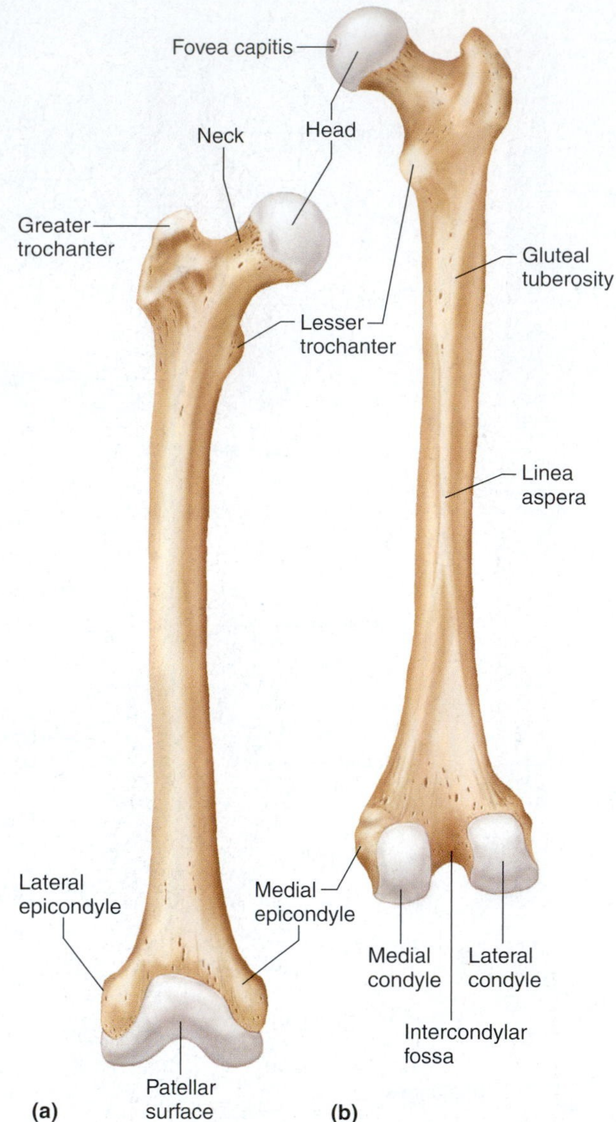

FIGURE 7.51 Right femur. (*a*) Anterior surface. (*b*) Posterior surface.

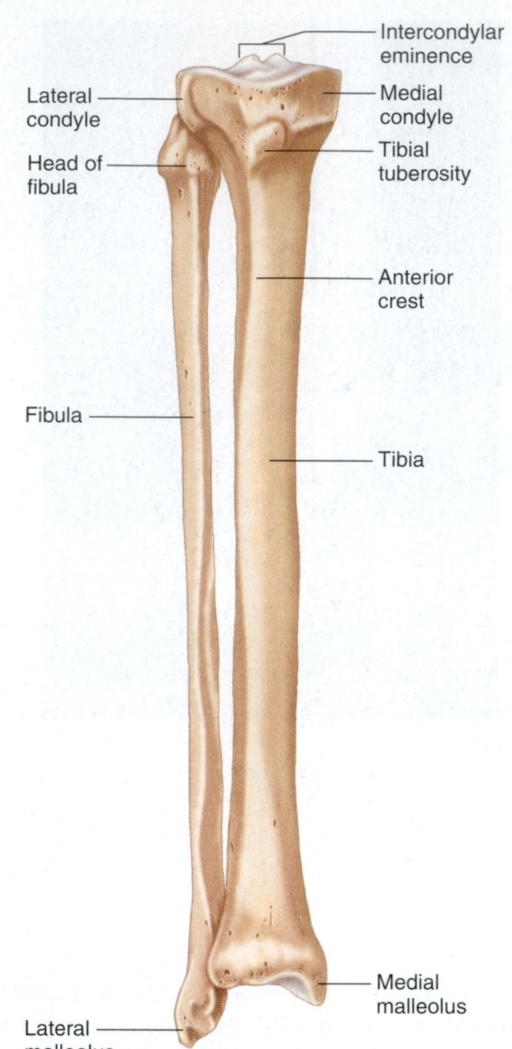

FIGURE 7.52 Bones of the right leg, anterior view.

The skeleton is particularly vulnerable to injury during the turbulent teen years, when bones grow rapidly. Athletic teens sometimes develop Osgood-Schlatter disease, which is a painful swelling of a bony projection of the tibia below the knee. Overusing the thigh muscles to straighten the lower limb irritates the area, causing the swelling. Usually a few months of rest and no athletic activity allows the bone to heal on its own. Rarely, a cast must be used to immobilize the knee.

Fibula

The **fibula** is a long, slender bone located on the lateral side of the tibia. Its ends are slightly enlarged into a proximal *head* and a distal *lateral malleolus* (fig. 7.52). The head articulates with the tibia just below the lateral condyle; however, it does not enter into the knee joint and does not bear any body weight. The lateral malleolus articulates with the ankle and protrudes on the lateral side.

Foot

The foot is made up of the ankle, the instep, and the toes. The ankle or *tarsus* (tahr′sus) is composed of seven **tarsal bones.** One of these bones, the **talus** (ta′lus), can move freely where it joins the tibia and fibula, forming the ankle. The other tarsal bones are firmly bound, supporting the talus. Figures 7.53 and 7.54 name the bones of the tarsus.

The largest of the tarsals, the **calcaneus** (kal-ka′ne-us), or heel bone, is below the talus where it projects backward to form the base of the heel. The calcaneus helps support body weight and provides an attachment, the *calcaneal tuberosity,* for muscles that move the foot.

The instep or *metatarsus* (met″ah-tahr′sus) consists of five elongated **metatarsal bones,** which articulate with the tarsus. They are numbered 1 to 5, beginning on the medial side

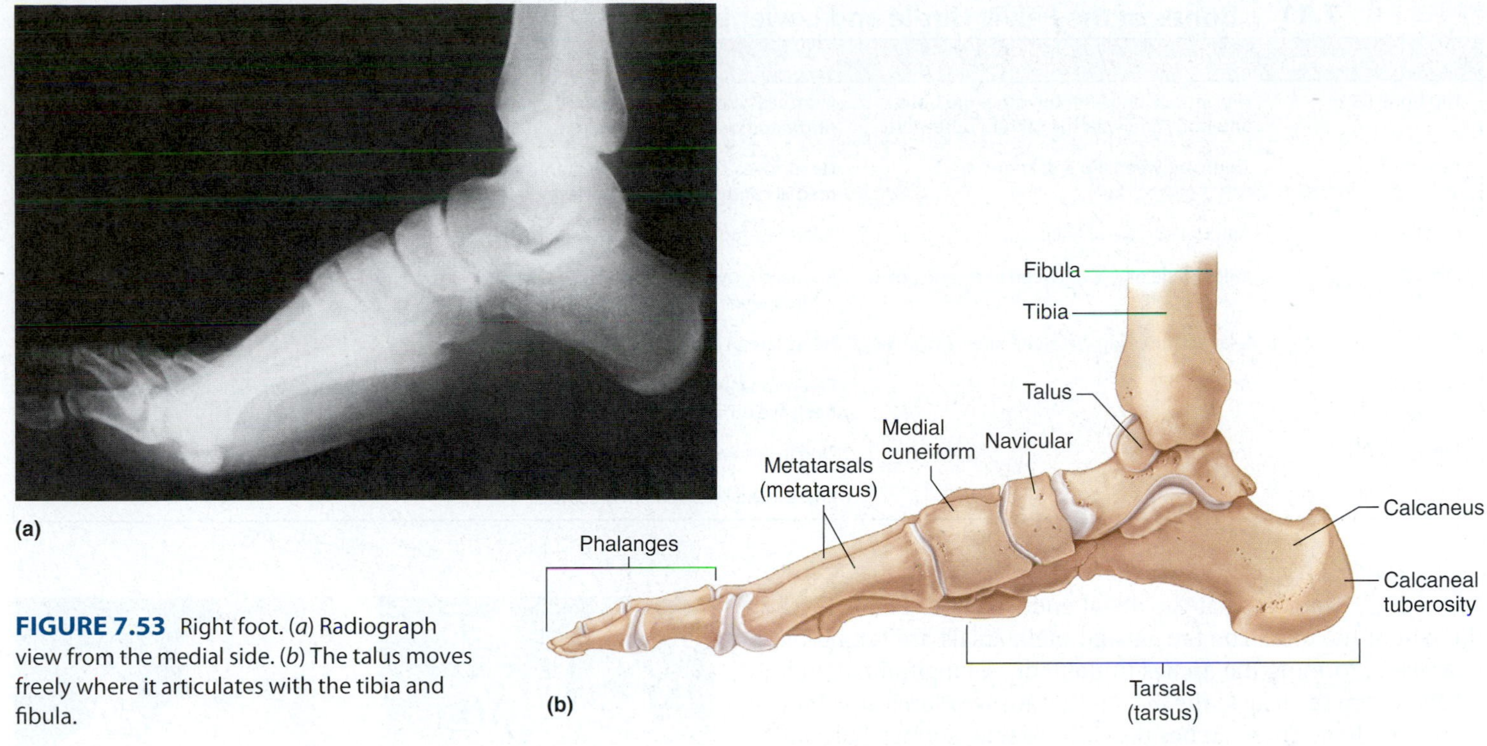

FIGURE 7.53 Right foot. (*a*) Radiograph view from the medial side. (*b*) The talus moves freely where it articulates with the tibia and fibula.

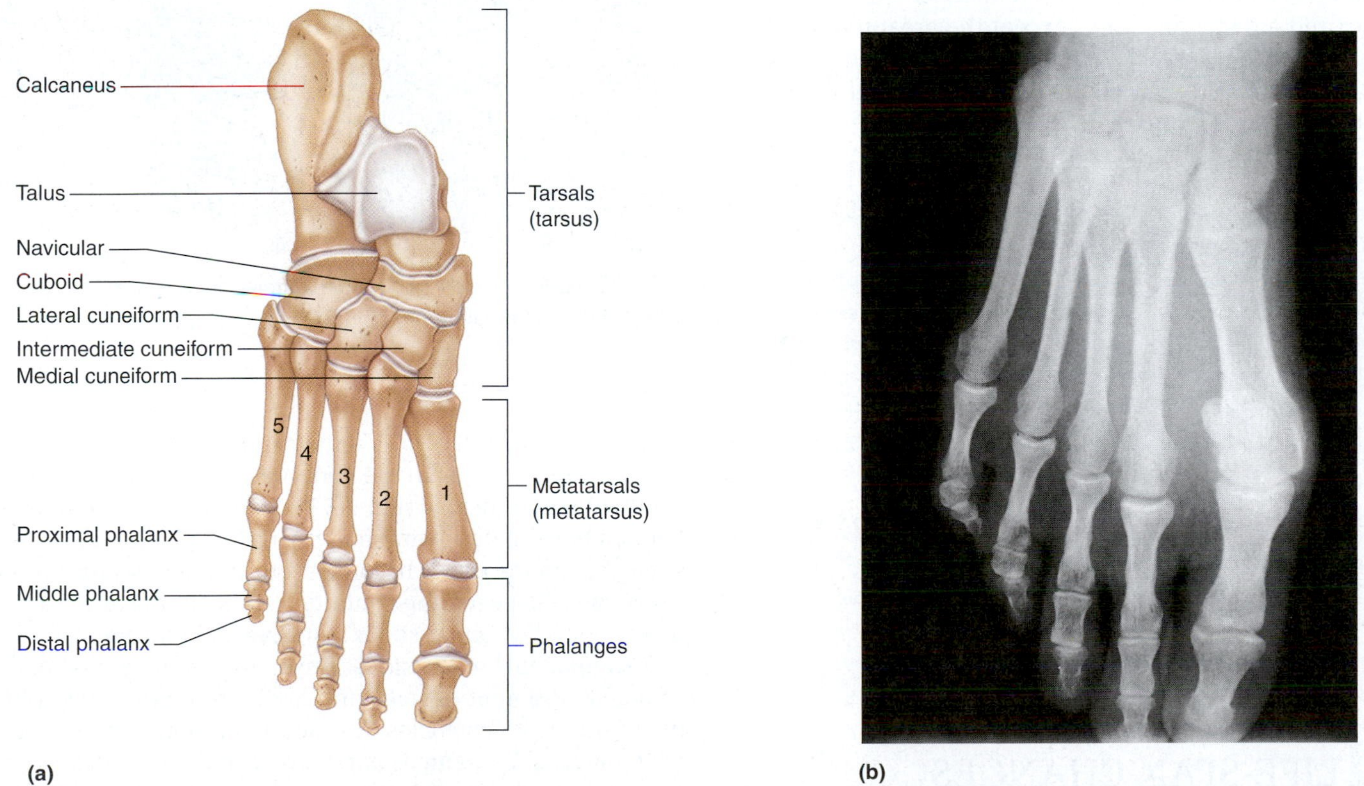

FIGURE 7.54 Right foot. (*a*) Viewed superiorly. (*b*) Radiograph of the foot viewed superiorly. Note: Sesamoid bone under first metatarsal in radiograph.

TABLE 7.11 | Bones of the Pelvic Girdle and Lower Limbs

Name and Number	Location	Special Features
Hip bone (2)	Hip, articulating with the other hip bone anteriorly and with the sacrum posteriorly	Ilium, iliac crest, anterior superior iliac spine, ischium, ischial tuberosity, ischial spine, obturator foramen, acetabulum, pubis
Femur (2)	Thigh, between hip and knee	Head, fovea capitis, neck, greater trochanter, lesser trochanter, linea aspera, lateral condyle, medial condyle, gluteal tuberosity, intercondylar fossa
Patella (2)	Anterior surface of knee	A flat sesamoid bone located within a tendon
Tibia (2)	Medial side of leg, between knee and ankle	Medial condyle, lateral condyle, tibial tuberosity, anterior crest, medial malleolus, intercondylar eminence
Fibula (2)	Lateral side of leg, between knee and ankle	Head, lateral malleolus
Tarsal (14)	Ankle	Freely movable talus that articulates with leg bones; calcaneus that forms the base of the heel; five other tarsal bones bound firmly together
Metatarsal (10)	Instep	One in line with each toe, bound by ligaments to form arches
Phalanx (28)	Toe	Three in each toe, two in great toe

(fig. 7.54). The heads at the distal ends of these bones form the ball of the foot. The tarsals and metatarsals are bound by ligaments, forming the arches of the foot. A longitudinal arch extends from the heel to the toe, and a transverse arch stretches across the foot. These arches provide a stable, springy base for the body. Sometimes, however, the tissues that bind the metatarsals weaken, producing fallen arches, or flat feet.

The **phalanges** of the toes are shorter but otherwise similar to those of the fingers and align and articulate with the metatarsals. Each toe has three phalanges—a proximal, a middle, and a distal phalanx—except the great toe, which has only two because it lacks the middle phalanx (fig. 7.54). Table 7.11 summarizes the bones of the pelvic girdle and lower limbs.

An infant with two casts on her feet is probably being treated for clubfoot, a common birth defect in which the foot twists out of its normal position, turning in, out, up, down, or some combination of these directions. Clubfoot probably results from arrested development during fetal existence, but the precise cause is not known. Clubfoot can almost always be corrected with special shoes, or surgery, followed by several months in casts to hold the feet in the correct position.

PRACTICE

37 Locate and name each of the bones of the lower limb.

38 Explain how the bones of the lower limb articulate with one another.

39 Describe how the foot is adapted to support the body.

7.13 LIFE-SPAN CHANGES

Aging-associated changes in the skeletal system are apparent at the cellular and whole-body levels. Most obvious is the incremental decrease in height that begins at about age thirty, with a loss of about 1/16 of an inch a year. In the later years, compression fractures in the vertebrae may contribute significantly to loss of height (fig. 7.55). Overall, as calcium levels fall and bone material gradually vanishes, the skeleton loses strength, and the bones become brittle and increasingly prone to fracture. However, the continued ability of fractures to heal reveals that the bone tissue is still alive and functional.

FIGURE 7.55 The bones change to different degrees and at different rates over a lifetime.

Components of the skeletal system and individual bones change to different degrees and at different rates over a lifetime. Gradually, osteoclasts come to outnumber osteoblasts, which means that bone is eaten away in the remodeling process at a faster rate than it is replaced—resulting in more spaces in bones. The bone thins, its strength waning. Bone matrix changes, with the ratio of mineral to protein increasing, making bones more brittle and prone to fracture. Beginning in

INNERCONNECTIONS | *Skeletal System*

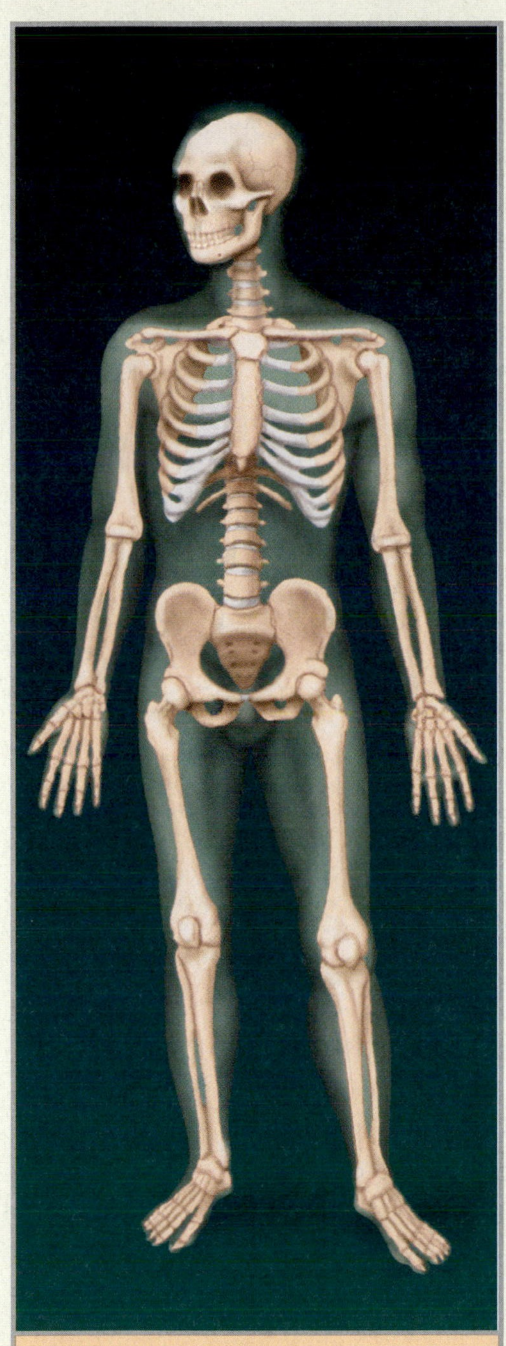

Skeletal System

Bones provide support, protection, and movement and also play a role in calcium balance.

Integumentary System

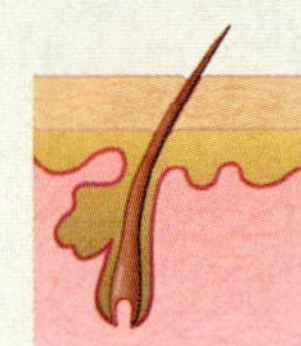

Vitamin D, activated in the skin, plays a role in calcium absorption and availability for bone matrix.

Lymphatic System

Cells of the immune system originate in the bone marrow.

Muscular System

Muscles pull on bones to cause movement.

Digestive System

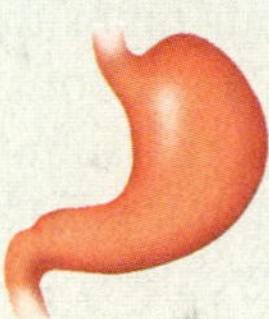

Absorption of dietary calcium provides material for bone matrix.

Nervous System

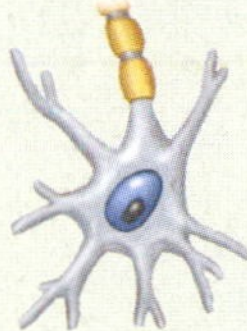

Proprioceptors sense the position of body parts. Pain receptors warn of trauma to bone. Bones protect the brain and spinal cord.

Respiratory System

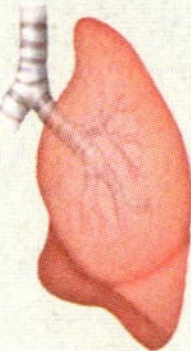

Ribs and muscles work together in breathing.

Endocrine System

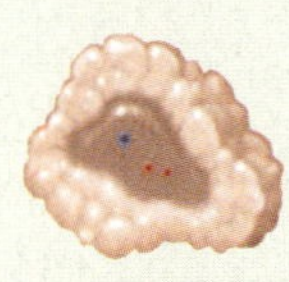

Some hormones act on bone to help regulate blood calcium levels.

Urinary System

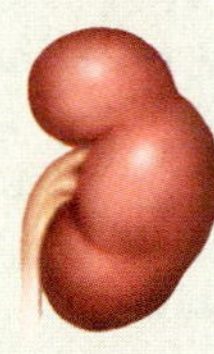

The kidneys and bones work together to help regulate blood calcium levels.

Cardiovascular System

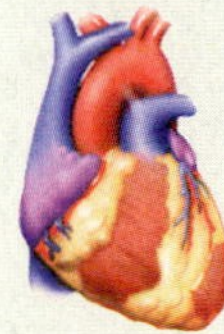

Blood transports nutrients to bone cells. Bone helps regulate plasma calcium levels, important to heart function.

Reproductive System

The pelvis helps support the uterus during pregnancy. Bones provide a source of calcium during lactation.

the third decade of life, bone matrix is removed faster than it is laid down. By age thirty-five, we start to lose bone mass.

Trabecular bone, due to its spongy, less compact nature, shows the changes of aging first, as it thins, increasing in porosity and weakening the overall structure. The vertebrae consist mostly of trabecular bone. It is also found in the upper part of the femur, whereas the shaft is more compact bone. That trabecular bone weakens sooner than compact bone destabilizes the femur, which is why it is a commonly broken bone among the elderly.

Compact bone loss begins at around age forty and continues at about half the rate of loss of trabecular bone. As remodeling continues throughout life, older osteons disappear as new ones are built next to them. With age, the osteons may coalesce, further weakening the overall structures as gaps form.

Bone loss is slow and steady in men, but in women, it is clearly linked to changing hormone levels. In the first decade following menopause, 15% to 20% of trabecular bone is lost, two to three times the rate of loss in men and premenopausal women. During the same time, compact bone loss is 10% to 15%, three to four times the rate of loss in men and premenopausal women. By about age seventy, both sexes are losing bone at about the same rate. By very old age, a woman may have only half the trabecular and compact bone mass as she did in her twenties, whereas a very elderly man may have one-third less bone mass.

Falls among the elderly are common and have many causes (see table 7.12). The most common fractures, after vertebral compression and hip fracture, are of the wrist, leg, and pelvis. Aging-related increased risk of fracture usually begins at about age fifty. Healing is slowed, so pain from a broken bone may persist for months. To preserve skeletal health, avoid falls, take calcium supplements, get enough vitamin D, avoid carbonated beverages (phosphates deplete bone), and get regular exercise.

PRACTICE

40 Why is bone lost faster with aging than it is replaced?

41 Which bones most commonly fracture in the elderly?

TABLE 7.12 | Possible Reasons for Falls Among the Elderly

Possible Reasons for Falls Among the Elderly
Overall frailty
Decreased muscle strength
Decreased coordination
Side effects of medication
Slowed reaction time due to stiffening joints
Poor vision and/or hearing
Disease (cancer, infection, arthritis)

CHAPTER SUMMARY

7.1 INTRODUCTION (PAGE 193)

Individual bones are the organs of the skeletal system. A bone contains active tissues. Bones support and protect soft tissues, provide attachment for muscles, house blood-producing cells, and store inorganic salts.

7.2 BONE STRUCTURE (PAGE 193)

Bone structure reflects its function.

1. Bone classification

   Bones are grouped according to their shapes—long, short, flat, irregular, or round (sesamoid).
2. Parts of a long bone
   a. Epiphyses at each end are covered with articular cartilage and articulate with other bones.
   b. The shaft of a bone is called the diaphysis.
   c. Except for the articular cartilage, a bone is covered by a periosteum.
   d. Compact bone has a continuous extracellular matrix with no gaps.
   e. Spongy bone has irregular interconnecting spaces between bony plates.
   f. Both compact and spongy bone are strong and resist bending.
   g. The diaphysis contains a medullary cavity filled with marrow.
3. Microscopic structure
   a. Compact bone contains osteons cemented together.
   b. Central canals contain blood vessels that nourish the cells of osteons.
   c. Perforating canals connect central canals transversely and communicate with the bone's surface and the medullary cavity.
   d. Diffusion from the surface of thin bony plates nourishes cells of spongy bones.

7.3 BONE DEVELOPMENT AND GROWTH (PAGE 197)

1. Intramembranous bones
   a. Certain flat bones of the skull are intramembranous bones.
   b. They develop from layers of connective tissues.
   c. Osteoblasts within the membranous layers form bone tissue.
   d. Osteoblasts surrounded by extracellular matrix are called osteocytes.
   e. Relatively unspecialized connective tissue gives rise to the periosteum.

2. Endochondral bones
   a. Most of the bones of the skeleton are endochondral.
   b. They develop as hyaline cartilage that bone tissue later replaces.
   c. The primary ossification center appears in the diaphysis, whereas secondary ossification centers appear in the epiphyses.
   d. An epiphyseal plate remains between the primary and secondary ossification centers.
3. Growth at the epiphyseal plate
   a. An epiphyseal plate consists of layers of cells: zone of resting cartilage, zone of proliferating cartilage, zone of hypertrophic cartilage, and zone of calcified cartilage.
   b. The epiphyseal plates are responsible for bone lengthening.
   c. Long bones continue to lengthen until the epiphyseal plates are ossified.
   d. Growth in bone thickness is due to ossification beneath the periosteum.
   e. The action of osteoclasts forms the medullary cavity.
4. Homeostasis of bone tissue
   a. Osteoclasts and osteoblasts continually remodel bone.
   b. The total mass of bone remains nearly constant.
5. Factors affecting bone development, growth, and repair
   a. Deficiencies of vitamin A, C, or D result in abnormal bone development.
   b. Insufficient secretion of pituitary growth hormone may result in dwarfism; excessive secretion may result in gigantism, or acromegaly.
   c. Deficiency of thyroid hormone delays bone growth.
   d. Male and female sex hormones promote bone formation and stimulate ossification of the epiphyseal plates.

7.4 BONE FUNCTION (PAGE 202)

1. Support, protection, and movement
   a. Bones shape and form body structures.
   b. Bones support and protect softer, underlying tissues.
   c. Bones and muscles interact, producing movement.
2. Blood cell formation
   a. At different ages, hematopoiesis occurs in the yolk sac, the liver, the spleen, and the red bone marrow.
   b. Red marrow houses developing red blood cells, white blood cells, and blood platelets.
3. Inorganic salt storage
   a. The extracellular matrix of bone tissue contains abundant calcium phosphate in the form of hydroxyapatite.
   b. When blood calcium ion concentration is low, osteoclasts resorb bone, releasing calcium salts.
   c. When blood calcium ion concentration is high, osteoblasts are stimulated to form bone tissue and store calcium salts.
   d. Bone stores small amounts of sodium, magnesium, potassium, and carbonate ions.
   e. Bone tissues may accumulate lead, radium, or strontium.

7.5 SKELETAL ORGANIZATION (PAGE 205)

1. Number of bones
   a. Usually a human skeleton has 206 bones, but the number may vary.
   b. Extra bones in sutures are called sutural bones.
2. Divisions of the skeleton
   a. The skeleton can be divided into axial and appendicular portions.
   b. The axial skeleton consists of the skull, hyoid bone, vertebral column, and thoracic cage.
   c. The appendicular skeleton consists of the pectoral girdle, upper limbs, pelvic girdle, and lower limbs.

7.6 SKULL (PAGE 206)

The skull consists of twenty-two bones, which include eight cranial bones and fourteen facial bones.

1. Cranium
   a. The cranium encloses and protects the brain and provides attachments for muscles.
   b. Some cranial bones contain air-filled paranasal sinuses that help reduce the weight of the skull.
   c. Cranial bones include the frontal bone, parietal bones, occipital bone, temporal bones, sphenoid bone, and ethmoid bone.
2. Facial skeleton
   a. Facial bones form the basic shape of the face and provide attachments for muscles.
   b. Facial bones include the maxillary bones, palatine bones, zygomatic bones, lacrimal bones, nasal bones, vomer bone, inferior nasal conchae, and mandible.
3. Infantile skull
   a. Incompletely developed bones, connected by fontanels, enable the infantile skull to change shape slightly during childbirth.
   b. Infantile skull bones are thin, somewhat flexible, and less easily fractured.

7.7 VERTEBRAL COLUMN (PAGE 218)

The vertebral column extends from the skull to the pelvis and protects the spinal cord. It is composed of vertebrae separated by intervertebral discs. An infant has thirty-three vertebral bones and an adult has twenty-six. The vertebral column has four curvatures—cervical, thoracic, lumbar, and sacral.

1. A typical vertebra
   a. A typical vertebra consists of a body, pedicles, laminae, spinous process, transverse processes, and superior and inferior articulating processes.
   b. Notches on the upper and lower surfaces of the pedicles on adjacent vertebrae form intervertebral foramina through which spinal nerves pass.
2. Cervical vertebrae
   a. Cervical vertebrae comprise the bones of the neck.
   b. Transverse processes have transverse foramina.
   c. The atlas (first vertebra) supports the head.
   d. The dens of the axis (second vertebra) provides a pivot for the atlas when the head turns from side to side.

3. Thoracic vertebrae
    a. Thoracic vertebrae are larger than cervical vertebrae.
    b. Their transverse processes project posteriorly at sharp angles.
    c. Their long spinous processes slope downward, and facets on the sides of bodies articulate with the ribs.
4. Lumbar vertebrae
    a. Vertebral bodies of lumbar vertebrae are large and strong.
    b. Their transverse processes project laterally, and their spinous processes project posteriorly nearly horizontal.
5. Sacrum
    a. The sacrum, formed of five fused vertebrae, is a triangular structure that has rows of dorsal sacral foramina.
    b. It is united with the hip bones at the sacroiliac joints.
    c. The sacral promontory provides a guide for determining the size of the pelvis.
6. Coccyx
    a. The coccyx, composed of four fused vertebrae, forms the lowest part of the vertebral column.
    b. It acts as a shock absorber when a person sits and is an attachment for muscles of the pelvic floor.

7.8 THORACIC CAGE (PAGE 222)

The thoracic cage includes the ribs, thoracic vertebrae, sternum, and costal cartilages. It supports the pectoral girdle and upper limbs, protects viscera, and functions in breathing.

1. Ribs
    a. Twelve pairs of ribs are attached to the twelve thoracic vertebrae.
    b. Costal cartilages of the true ribs join the sternum directly; those of the false ribs join indirectly or not at all.
    c. A typical rib has a shaft, head, and tubercles that articulate with the vertebrae.
2. Sternum
    a. The sternum consists of a manubrium, body, and xiphoid process.
    b. It articulates with costal cartilages and clavicles.

7.9 PECTORAL GIRDLE (PAGE 225)

The pectoral girdle is composed of two clavicles and two scapulae. It forms an incomplete ring that supports the upper limbs and provides attachments for muscles that move the upper limbs.

1. Clavicles
    a. Clavicles are rodlike bones that run horizontally between the sternum and shoulders.
    b. They hold the shoulders in place and provide attachments for muscles.
2. Scapulae
    a. The scapulae are broad, triangular bones with bodies, spines, acromion processes, coracoid processes, glenoid cavities, supraspinous and infraspinous fossae, superior borders, axillary borders, and vertebral borders.
    b. They articulate with the humerus of each upper limb and provide attachments for muscles of the upper limbs and chest.

7.10 UPPER LIMB (PAGE 226)

Limb bones form the framework and provide the attachments for muscles that move the limb.

1. Humerus
    a. The humerus extends from the scapula to the elbow.
    b. It has a head, greater tubercle, lesser tubercle, intertubercular groove, anatomical neck, surgical neck, deltoid tuberosity, capitulum, trochlea, epicondyles, coronoid fossa, and olecranon fossa.
2. Radius
    a. The radius is on the thumb side of the forearm between the elbow and wrist.
    b. It has a head, radial tuberosity, styloid process, and ulnar notch.
3. Ulna
    a. The ulna is longer than the radius and overlaps the humerus posteriorly.
    b. It has a trochlear notch, olecranon process, coronoid process, head, styloid process, and radial notch.
    c. It articulates with the radius laterally and with a disc of fibrocartilage inferiorly.
4. Hand
    a. The wrist has eight carpals.
    b. The palm has five metacarpals.
    c. The five fingers have fourteen phalanges.

7.11 PELVIC GIRDLE (PAGE 231)

The pelvic girdle consists of two hip bones that articulate with each other anteriorly and with the sacrum posteriorly. The sacrum, coccyx, and pelvic girdle form the pelvis. The girdle provides support for body weight and attachments for muscles and protects visceral organs.

1. Hip bones

    Each hip bone consists of an ilium, ischium, and pubis, fused in the region of the acetabulum.

    a. Ilium
        (1) The ilium, the largest portion of the hip bone, joins the sacrum at the sacroiliac joint.
        (2) It has an iliac crest with anterior and posterior superior iliac spines and iliac fossae.
    b. Ischium
        (1) The ischium is the lowest portion of the hip bone.
        (2) It has an ischial tuberosity and ischial spine.
    c. Pubis
        (1) The pubis is the anterior portion of the hip bone.
        (2) Pubis bones are fused anteriorly at the symphysis pubis.
2. Greater and lesser pelves
    a. The greater pelvis is above the pelvic brim; the lesser pelvis is below it.
    b. The greater pelvis helps support abdominal organs; the lesser pelvis functions as a birth canal.

3. Differences between male and female pelves
    a. Differences between male and female pelves reflect the function of the female pelvis as a birth canal.
    b. Usually the female pelvis is more flared; pubic arch is broader; distance between the ischial spines and the ischial tuberosities is greater; and sacral curvature is shorter.

7.12 LOWER LIMB (PAGE 234)

Bones of the lower limb provide the frameworks of the thigh, leg, ankle, and foot.

1. Femur
    a. The femur extends from the hip to the knee.
    b. It has a head, fovea capitis, neck, greater trochanter, lesser trochanter, linea aspera, lateral condyle, and medial condyle.
2. Patella
    a. The patella is a sesamoid bone in the tendon that passes anteriorly over the knee.
    b. It controls the angle of this tendon and functions in lever actions associated with lower limb movements.
3. Tibia
    a. The tibia is located on the medial side of the leg.
    b. It has medial and lateral condyles, tibial tuberosity, anterior crest, and medial malleolus.
    c. It articulates with the talus of the ankle.
4. Fibula
    a. The fibula is located on the lateral side of the tibia.
    b. It has a head and lateral malleolus that articulates with the ankle but does not bear body weight.
5. Foot
    a. The ankle includes the talus and six other tarsals.
    b. The instep has five metatarsals.
    c. The five toes have fourteen phalanges.

7.13 LIFE-SPAN CHANGES (PAGE 238)

Aging-associated changes in the skeleton are apparent at the cellular and whole-body levels.

1. Incremental decrease in height begins at about age thirty.
2. Gradually, bone loss exceeds bone replacement.
    a. In the first decade following menopause, bone loss occurs more rapidly in women than in men or premenopausal women. By age seventy, both sexes are losing bone at about the same rate.
    b. Aging increases risk of bone fractures.

CHAPTER ASSESSMENTS

7.1 Introduction

1 Active, living tissues found in bone include ______________________. (p. 193)

a. blood
b. nervous tissue
c. dense connective tissue
d. bone tissue
e. all of the above.

7.2 Bone Structure

2 List four groups of bones based on their shapes, and give an example from each group. (p. 193)

3 Sketch a typical long bone, and label its epiphyses, diaphysis, medullary cavity, periosteum, and articular cartilages. Designate the locations of compact and spongy bone. (p. 194)

4 Discuss the functions of the parts labeled in the sketch you made for question 3. (p. 194)

5 Distinguish between the microscopic structure of compact bone and spongy bone. (p. 195)

6 Explain how central canals and perforating canals are related. (p. 195)

7.3 Bone Development and Growth

7 Explain how the development of intramembranous bone differs from that of endochondral bone. (p. 197)

8 ____________________ are bone cells in lacunae, whereas ____________________ are bone-forming cells and ____________________ are bone-resorbing cells. (p. 197)

9 Explain the function of an epiphyseal plate. (p. 198)

10 Place the zones of cartilage in an epiphyseal plate in order (1–4), with the first zone attached to the epiphysis. (p. 198)

__________ zone of hypertrophic cartilage
__________ zone of calcified cartilage
__________ zone of resting cartilage
__________ zone of perforating cartilage

11 Explain how osteoblasts and osteoclasts regulate bone mass. (p. 200)

12 Describe the effects of vitamin deficiencies on bone development and growth. (p. 200)

13 Explain the causes of pituitary dwarfism and gigantism. (p. 201)

14 Describe the effects of thyroid and sex hormones on bone development and growth. (p. 201)

15 Physical exercise pulling on muscular attachments to bone, stimulates ______________________. (p. 201)

7.4 Bone Function

16 Provide several examples to illustrate how bones support and protect body parts. (p. 202)

17 Describe the functions of red and yellow bone marrow. (p. 203)

18 Explain the mechanism that regulates the concentration of blood calcium ions. (p. 204)

19 List three metallic elements that may be abnormally stored in bone. (p. 204)

7.5 Skeletal Organization

20 Bones of the head, neck, and trunk compose the ____________ skeleton; bones of the limbs and their attachments compose the ____________ skeleton. (p. 206)

7.6 Skull–7.12 Lower Limb

21 Name the bones of the cranium and the facial skeleton. (p. 208)

22 Explain the importance of fontanels. (p. 216)

23 Describe a typical vertebra, and distinguish among the cervical, thoracic, and lumbar vertebrae. (p. 219)

24 Describe the locations of the sacroiliac joint, the sacral promontory, and the sacral hiatus. (p. 222)

25 Name the bones that comprise the thoracic cage. (p. 222)

26 The clavicle and scapula form the ________________ girdle, whereas the hip bones and sacrum form the ________________ girdle. (p. 225)

27 Name the bones of the upper limb, and describe their locations. (p. 226)

28 Name the bones that comprise the hip bone. (p. 231)

29 Explain the major differences between the male and female skeletons. (p. 234)

30 Name the bones of the lower limb, and describe their locations. (p. 234)

31 Match the parts listed on the left with the bones listed on the right. (pp. 208–236)

(1) Coronoid process	A. Ethmoid bone
(2) Cribriform plate	B. Frontal bone
(3) Foramen magnum	C. Mandible
(4) Mastoid process	D. Maxillary bone
(5) Palatine process	E. Occipital bone
(6) Sella turcica	F. Temporal bone
(7) Supraorbital notch	G. Sphenoid bone
(8) Temporal process	H. Zygomatic bone
(9) Acromion process	I. Femur
(10) Deltoid tuberosity	J. Fibula
(11) Greater trochanter	K. Humerus
(12) Lateral malleolus	L. Radius
(13) Medial malleolus	M. Scapula
(14) Olecranon process	N. Sternum
(15) Radial tuberosity	O. Tibia
(16) Xiphoid process	P. Ulna

7.13 Life-Span Changes

32 Describe the changes, brought about by aging, in trabecular bone. (p. 240)

33 List factors that may preserve skeletal health. (p. 240)

INTEGRATIVE ASSESSMENTS/CRITICAL THINKING

OUTCOMES 1.6, 7.3, 7.4, 7.6

1. How might the condition of an infant's fontanels be used to evaluate skeletal development? How might the fontanels be used to estimate intracranial pressure (pressure in the cranial cavity)?

OUTCOMES 1.8, 7.2, 7.3

2. Why are incomplete, longitudinal fractures of bone shafts (greenstick fractures) more common in children than in adults?

OUTCOMES 5.3, 7.2, 7.6

3. How does the structure of a bone make it strong yet lightweight?

OUTCOMES 5.3, 7.3

4. If a young patient's forearm and elbow are immobilized by a cast for several weeks, what changes would you expect to occur in the bones of the upper limb?

OUTCOMES 7.3, 7.4, 7.10, 7.12

5. When a child's bone is fractured, growth may be stimulated at the epiphyseal plate. What problems might this extra growth cause in an upper or lower limb before the growth of the other limb compensates for the difference in length?

OUTCOMES 7.3, 7.11, 7.13

6. Archeologists discover skeletal remains of humanlike animals in Ethiopia. Examination of the bones suggests that the remains represent four types of individuals. Two of the skeletons have bone densities 30% less than those of the other two skeletons. The skeletons with the lower bone mass also have broader front pelvic bones. Within the two groups defined by bone mass, smaller skeletons have bones with evidence of epiphyseal plates, but larger bones have only a thin line where the epiphyseal plates should be. Give the age group and gender of the individuals in this find.

OUTCOMES 7.7, 7.13

7. Why do elderly persons often develop bowed backs and appear to lose height?

WEB CONNECTIONS

Be sure to visit the text website at **www.mhhe.com/shier12** for answers to chapter assessments, additional quizzes, and interactive learning exercises.

ANATOMY & PHYSIOLOGY REVEALED

Anatomy & Physiology Revealed® (APR) includes cadaver photos that allow you to peel away layers of the human body to reveal structures beneath the surface. This program also includes animations, radiologic imaging, audio pronunciations, and practice quizzing. Check out **www.aprevealed.com**. APR has been proven to help improve student grades!

HUMAN SKULL

The following set of reference plates will help you locate some of the more prominent features of the human skull. As you study these photographs, it is important to remember that individual human skulls vary in every characteristic. Also, the photographs in this set depict bones from several different skulls.

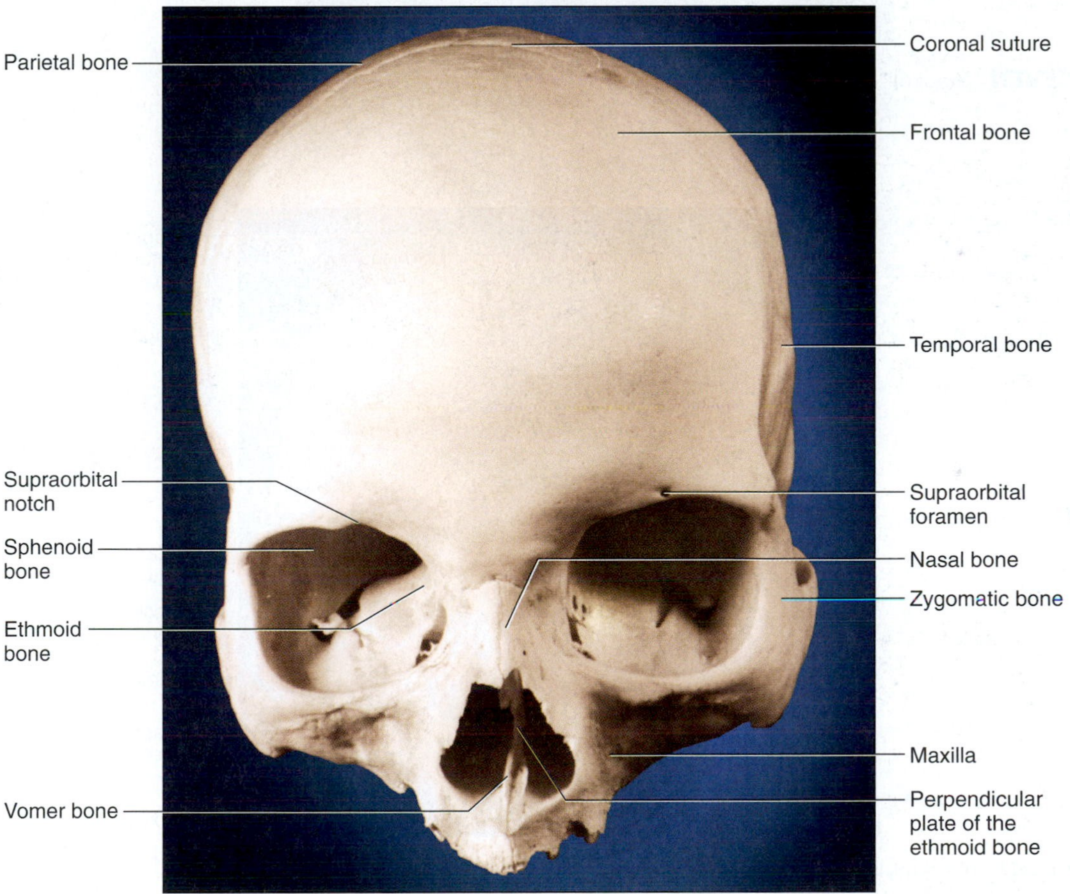

PLATE TWENTY-SIX The skull, frontal view.

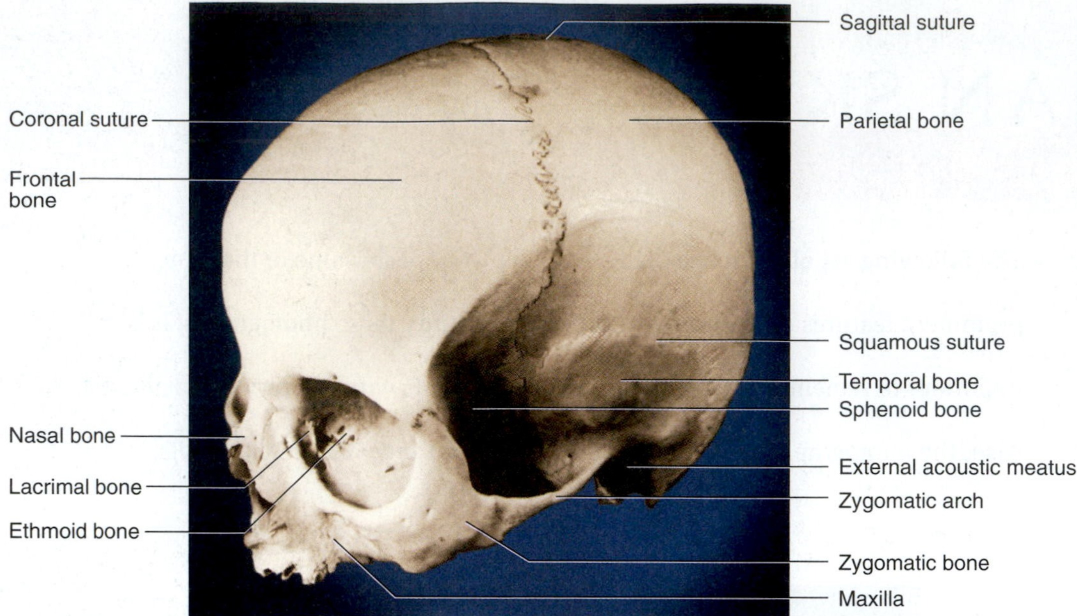

PLATE TWENTY-SEVEN The skull, left anterolateral view.

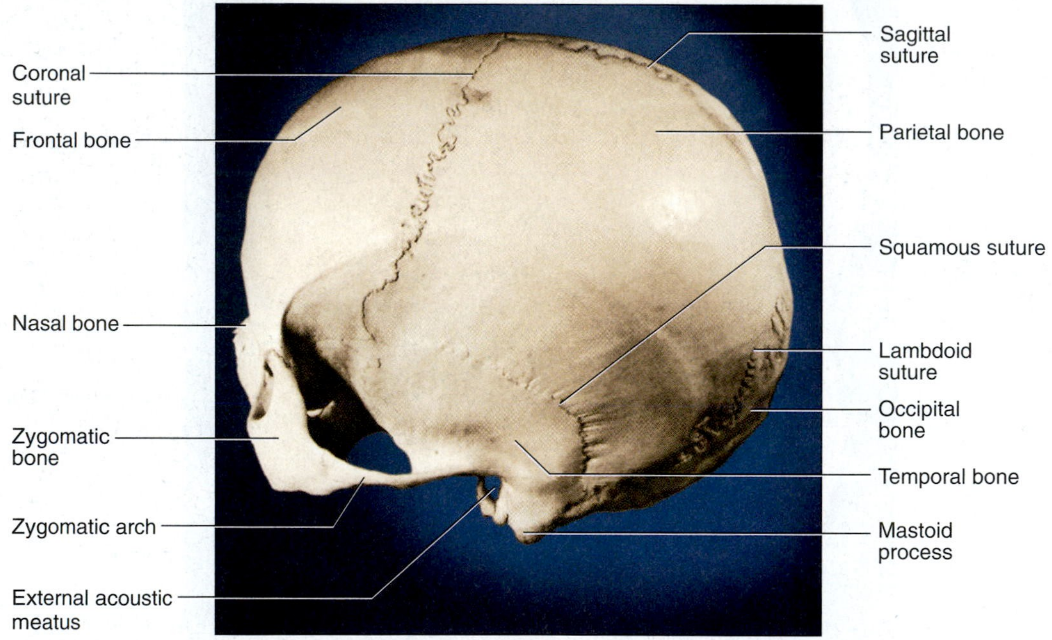

PLATE TWENTY-EIGHT The skull, left posterolateral view.

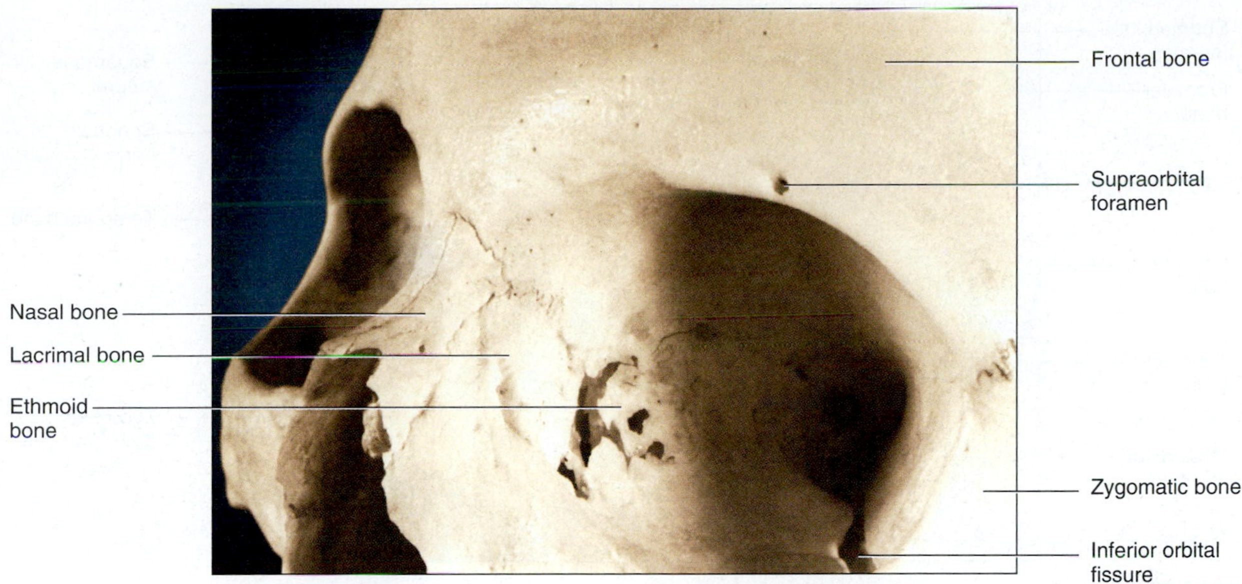

PLATE TWENTY-NINE Bones of the left orbital region.

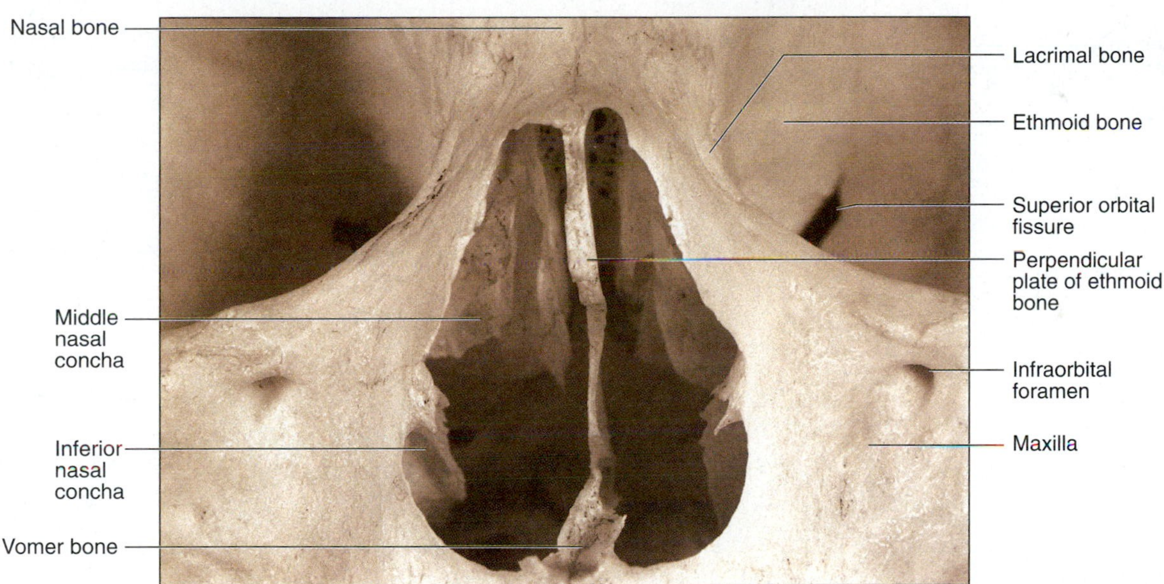

PLATE THIRTY Bones of the anterior nasal region.

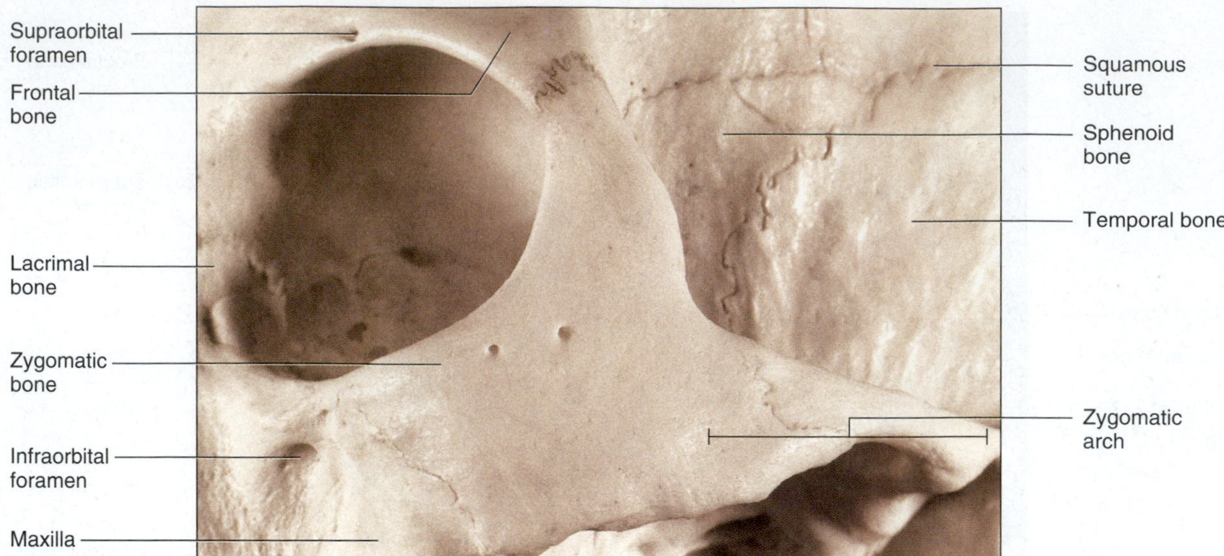

PLATE THIRTY-ONE Bones of the left zygomatic region.

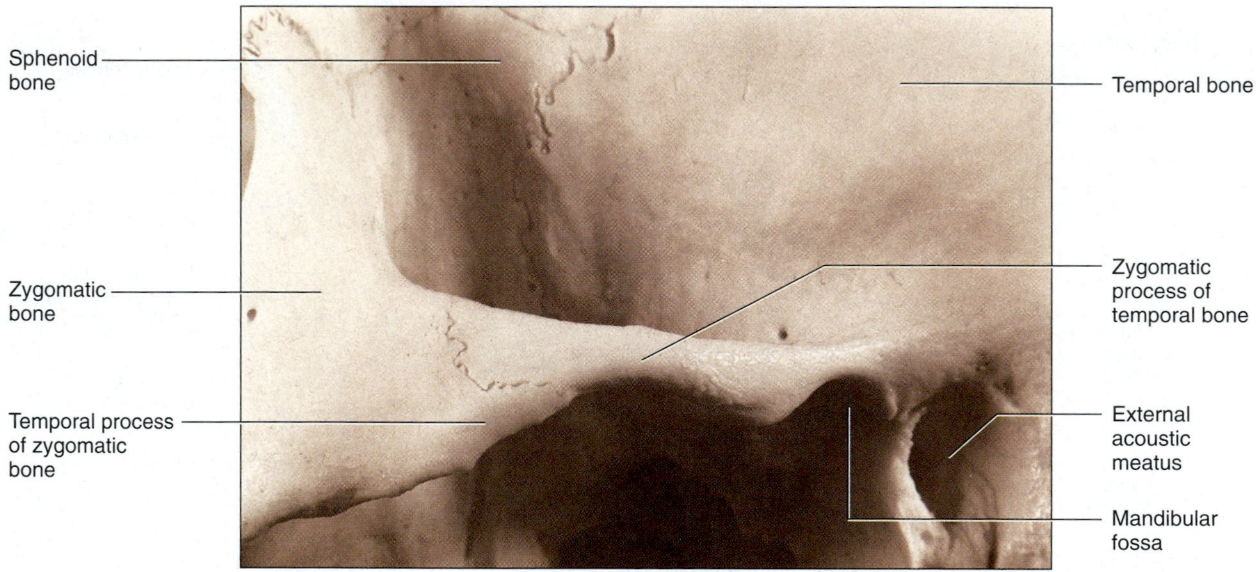

PLATE THIRTY-TWO Bones of the left temporal region.

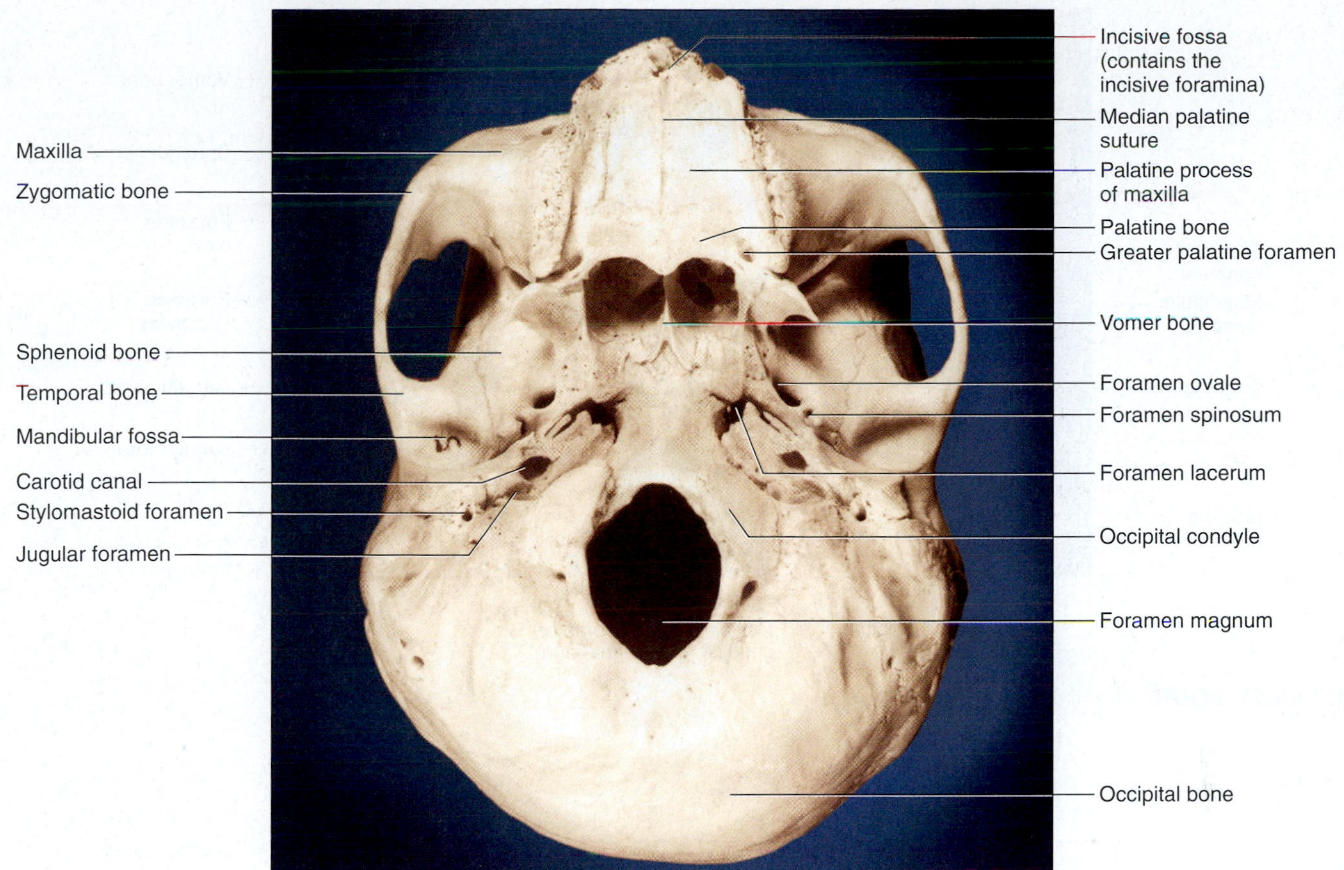

PLATE THIRTY-THREE The skull, inferior view.

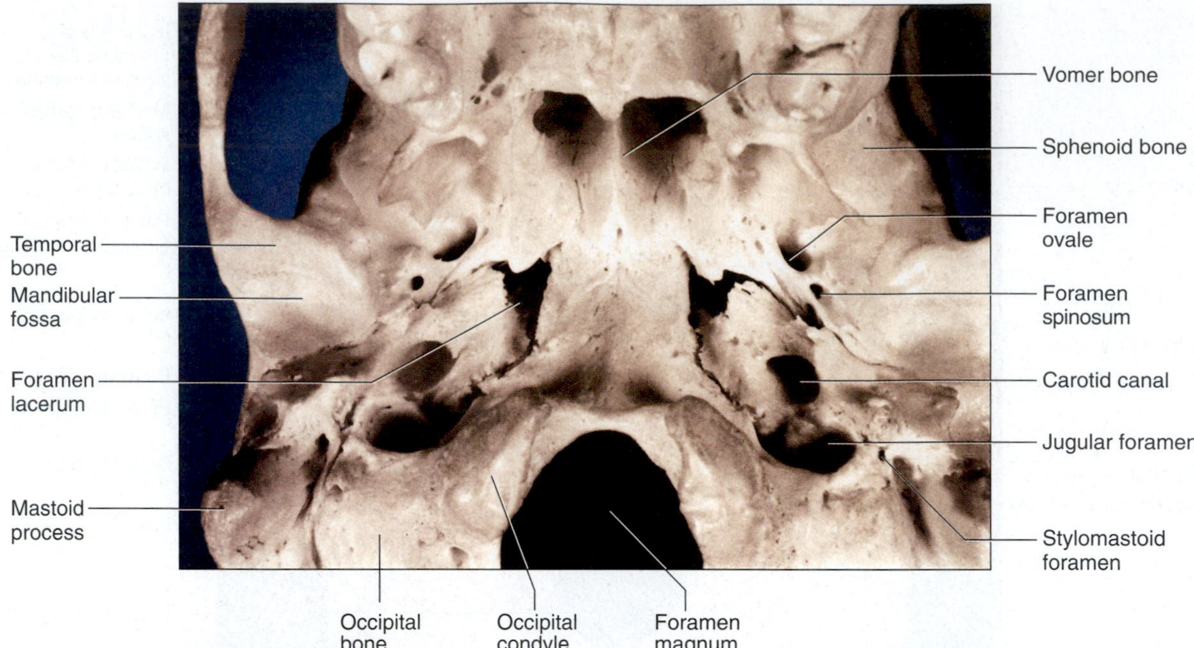

PLATE THIRTY-FOUR Base of the skull, sphenoid region.

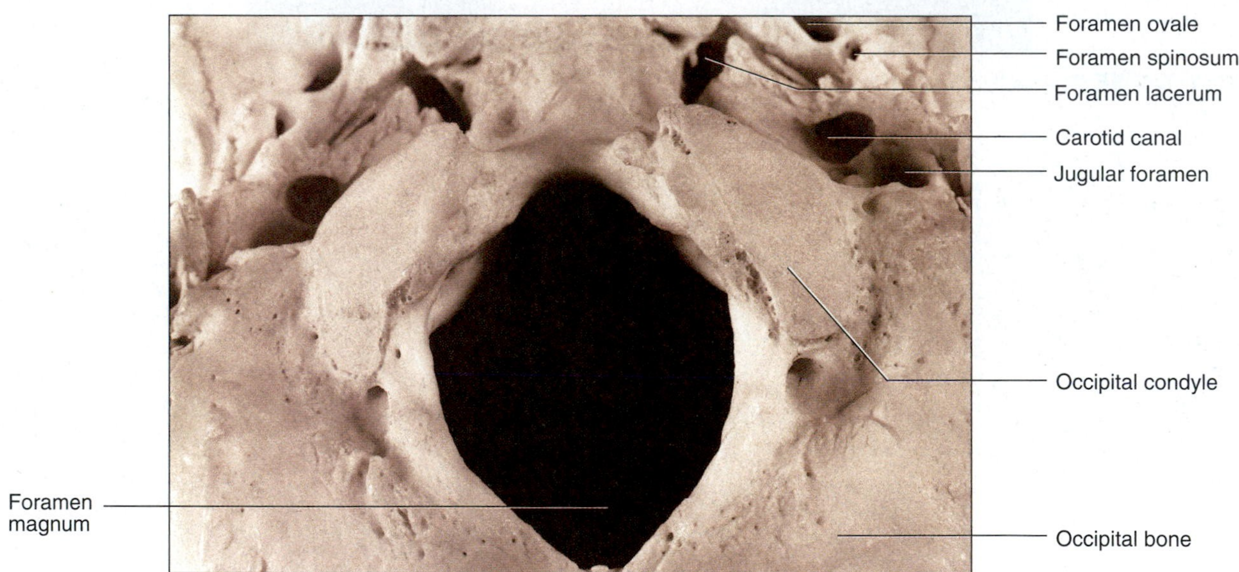

PLATE THIRTY-FIVE Base of the skull, occipital region.

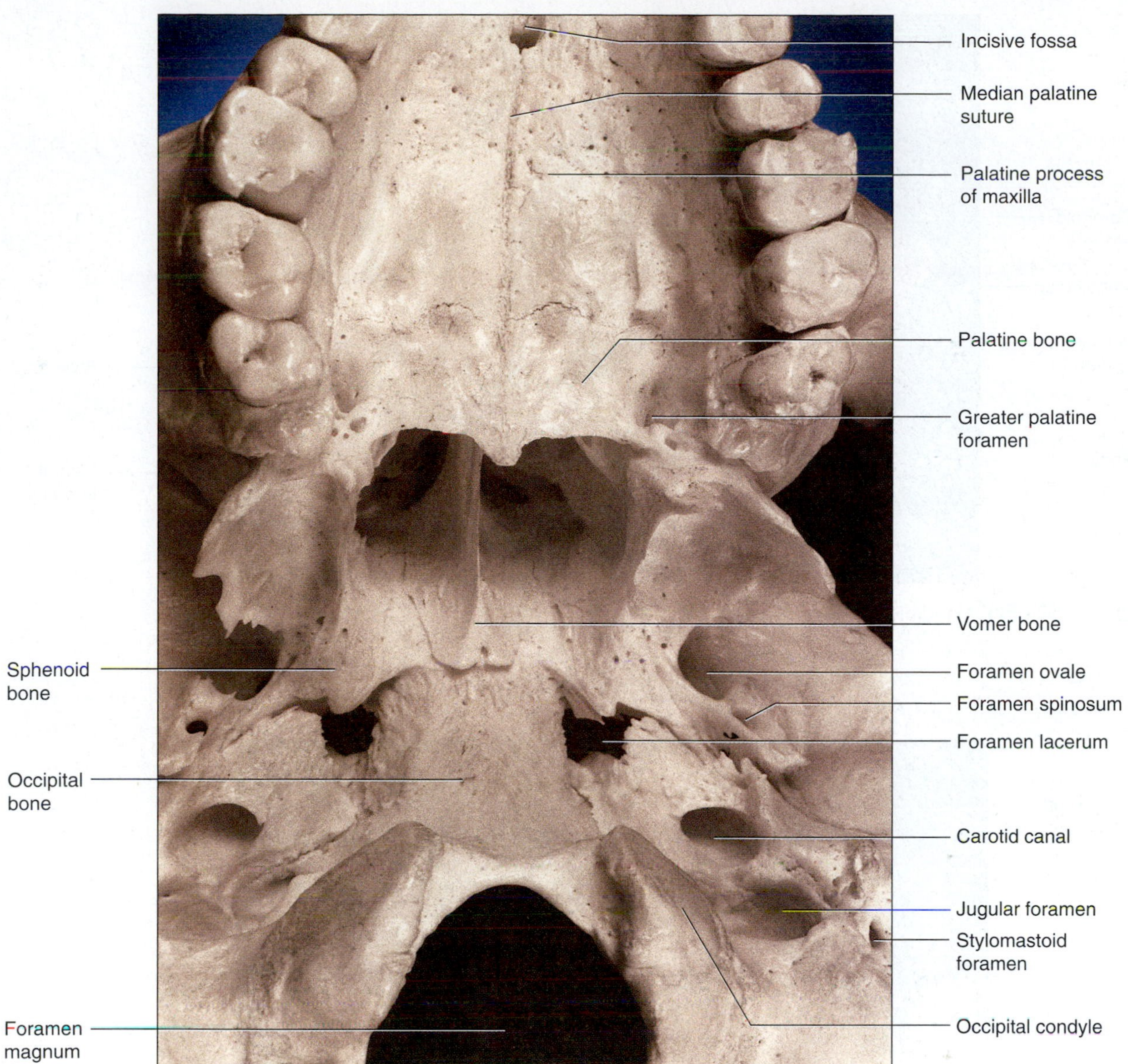

PLATE THIRTY-SIX Base of the skull, maxillary region.

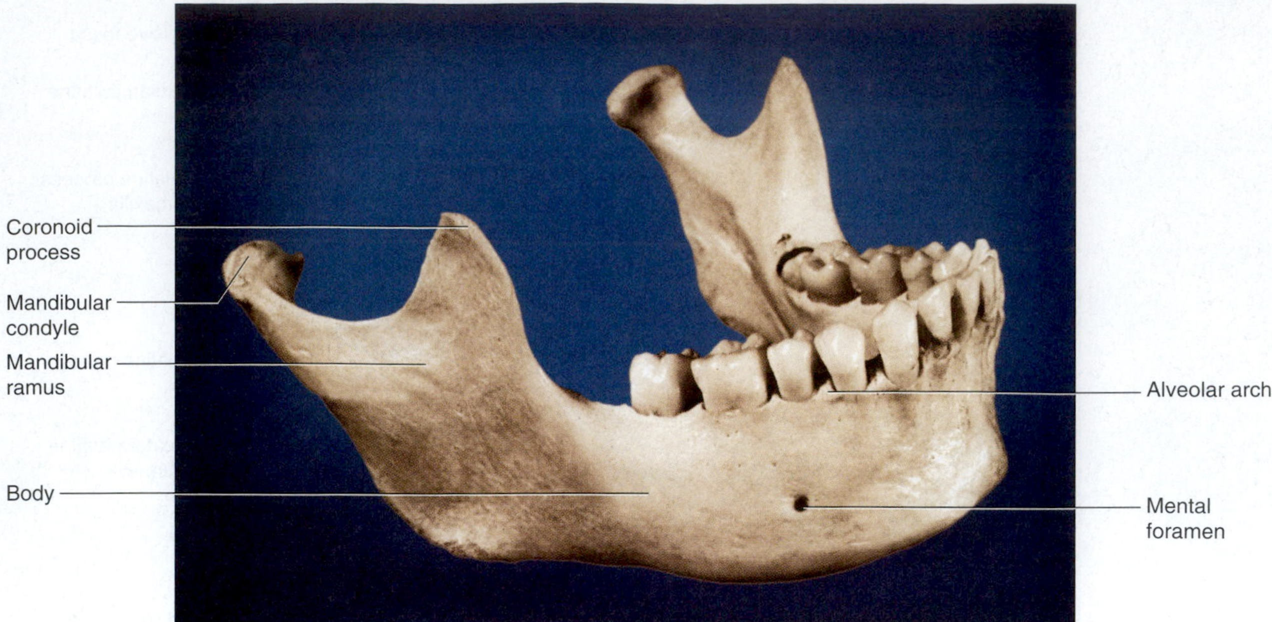

PLATE THIRTY-SEVEN Mandible, right lateral view.

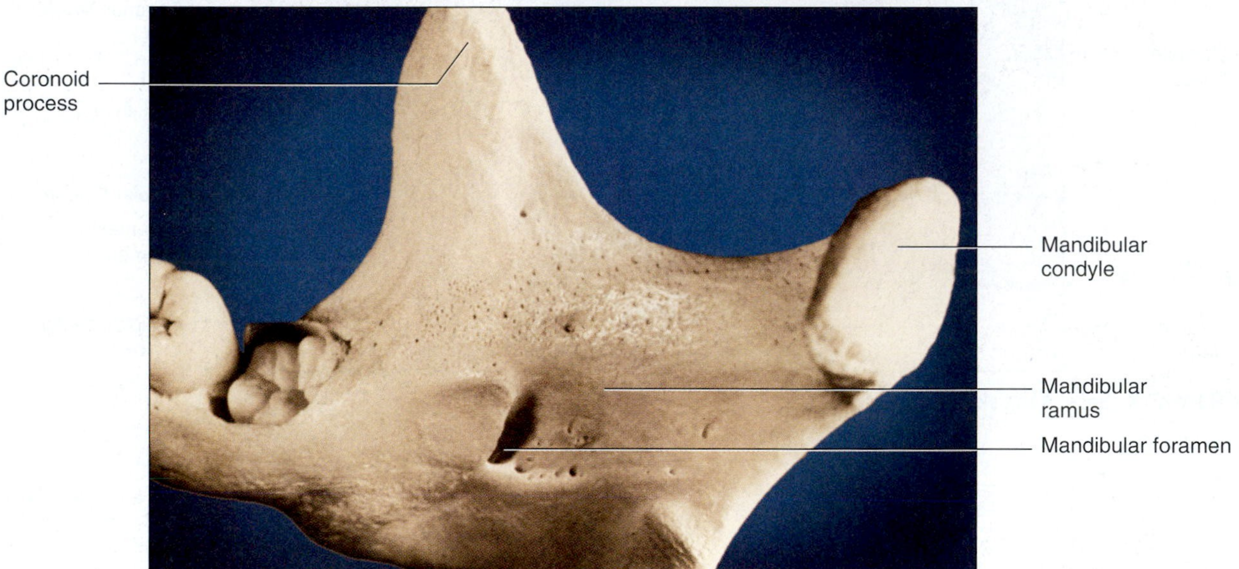

PLATE THIRTY-EIGHT Mandible, medial surface of right ramus.

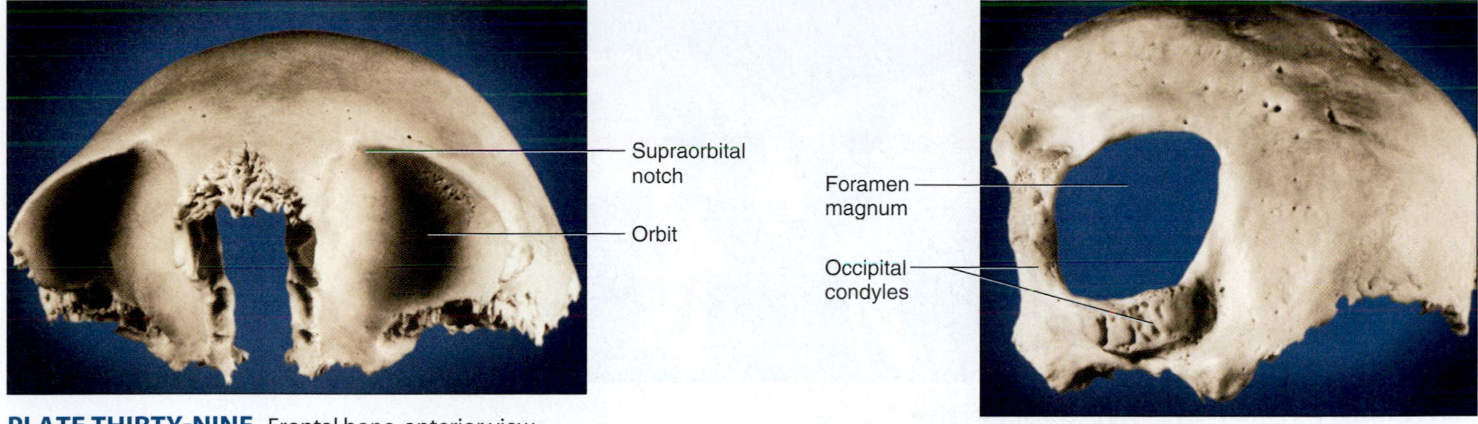

PLATE THIRTY-NINE Frontal bone, anterior view.

PLATE FORTY Occipital bone, inferior view.

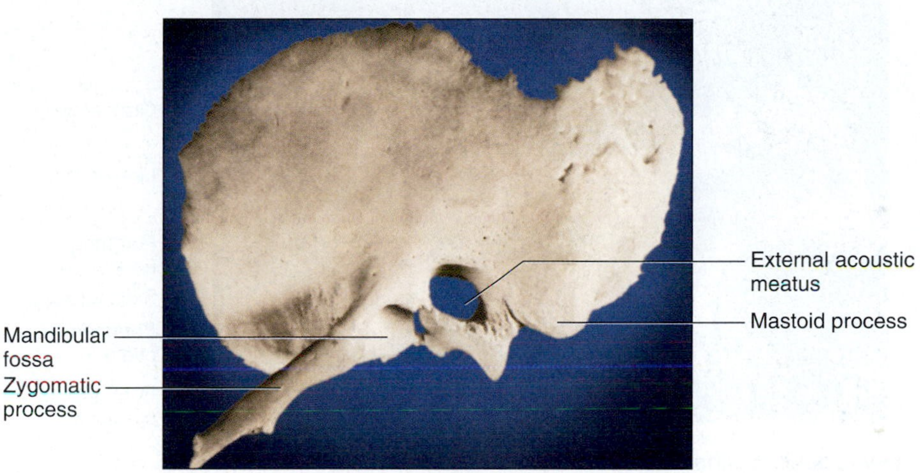

PLATE FORTY-ONE Temporal bone, left lateral view.

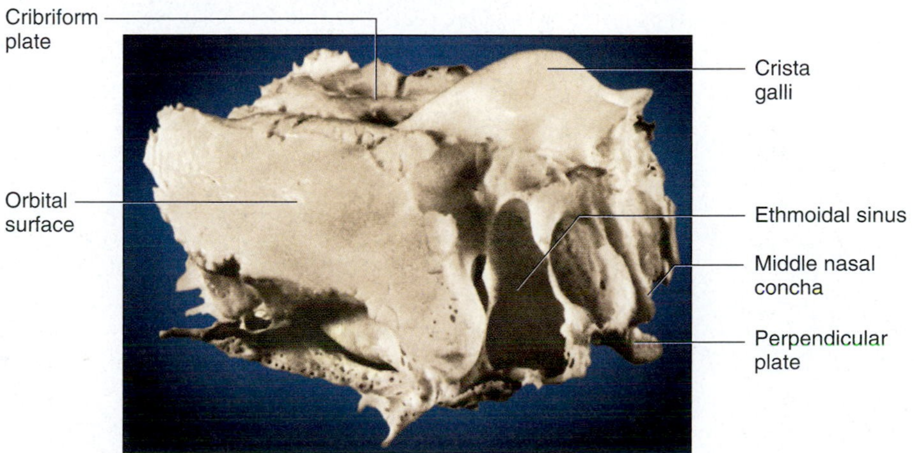

PLATE FORTY-TWO Ethmoid bone, right lateral view.

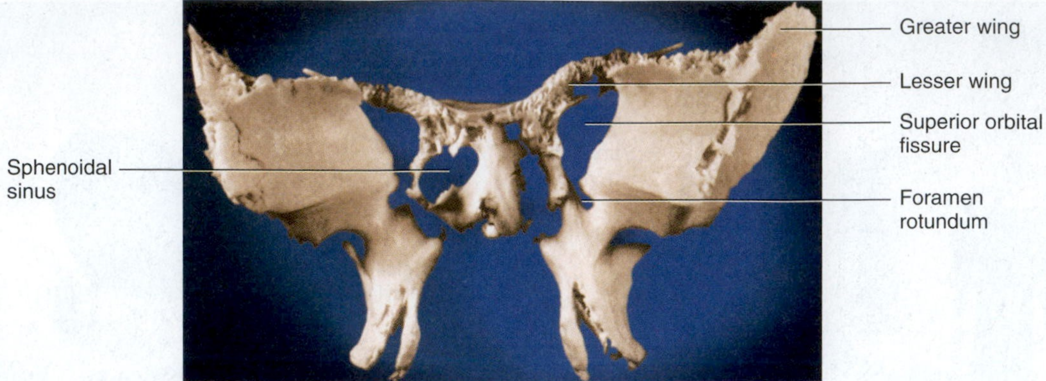

PLATE FORTY-THREE Sphenoid bone, anterior view.

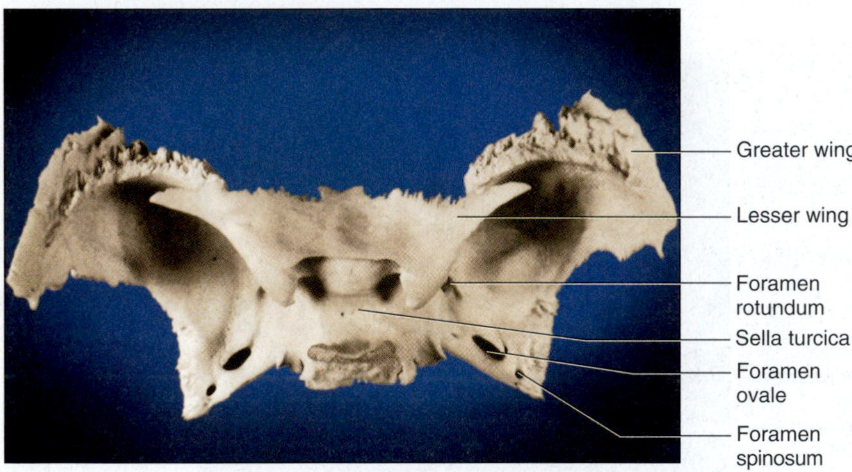

PLATE FORTY-FOUR Sphenoid bone, superior view.

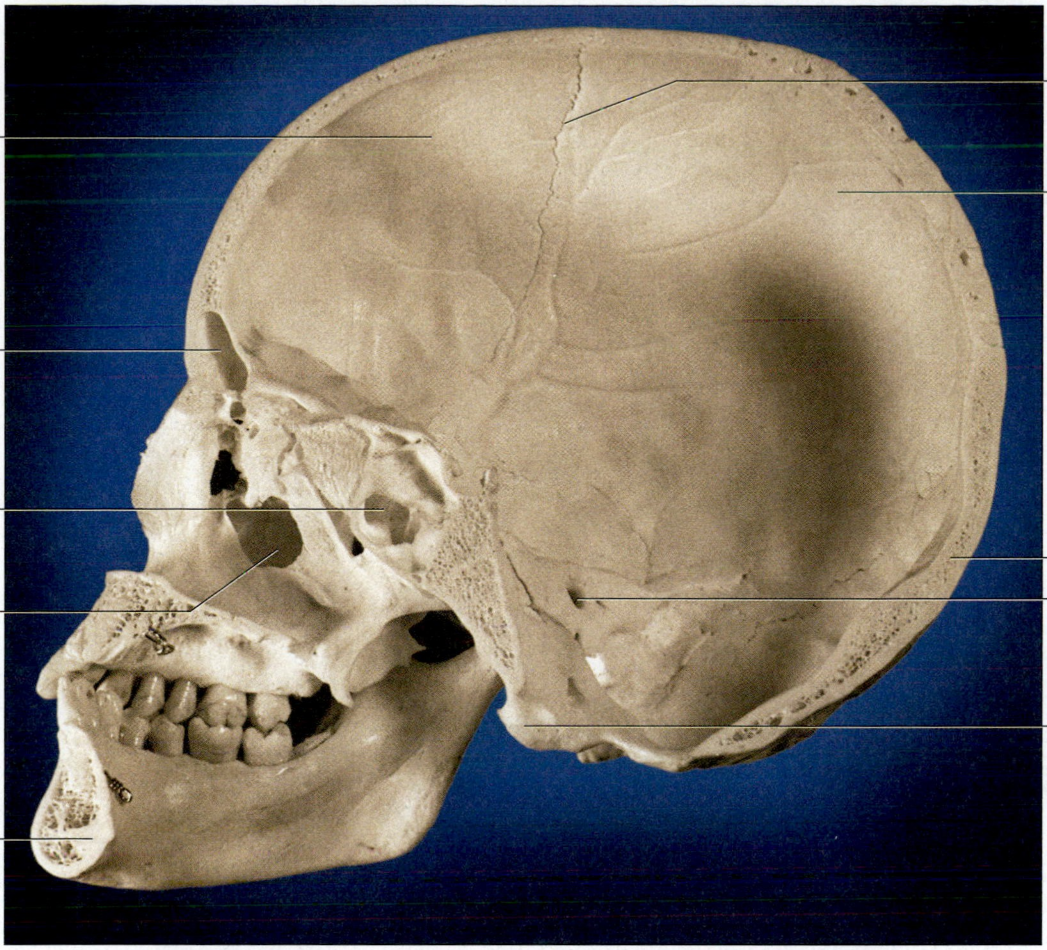

Frontal bone
Frontal sinus
Sphenoidal sinus
Maxillary sinus
Mandible
Coronal suture
Parietal bone
Occipital bone
Internal acoustic meatus
Occipital condyle

PLATE FORTY-FIVE The skull, sagittal section.

Frontal bone
Frontal sinus
Ethmoidal sinus
Maxillary sinus
Ethmoid bone
Sphenoid bone

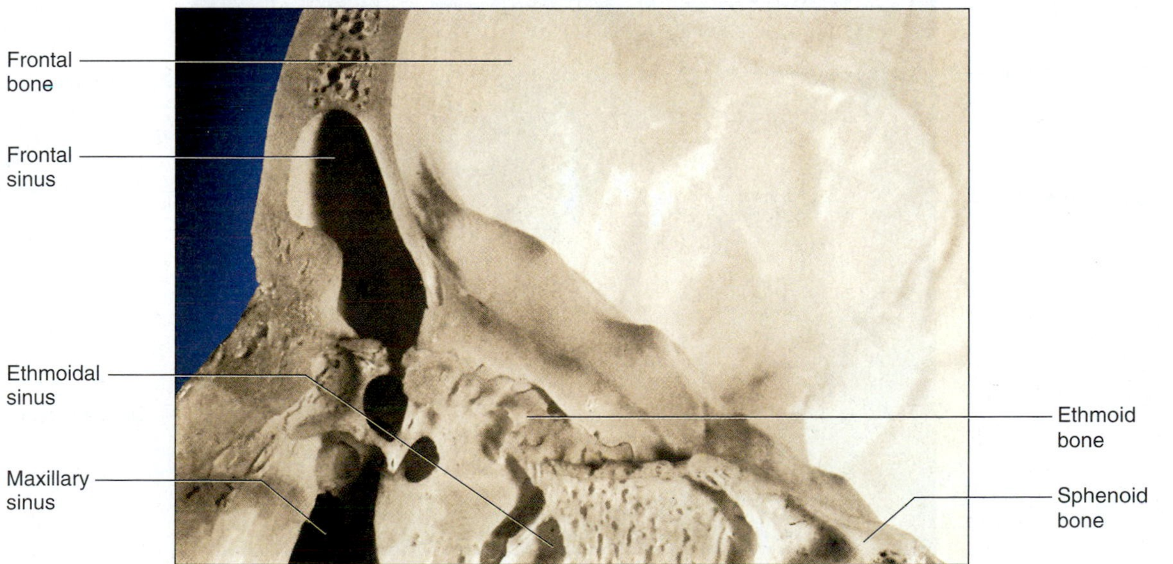

PLATE FORTY-SIX Ethmoidal region, sagittal section.

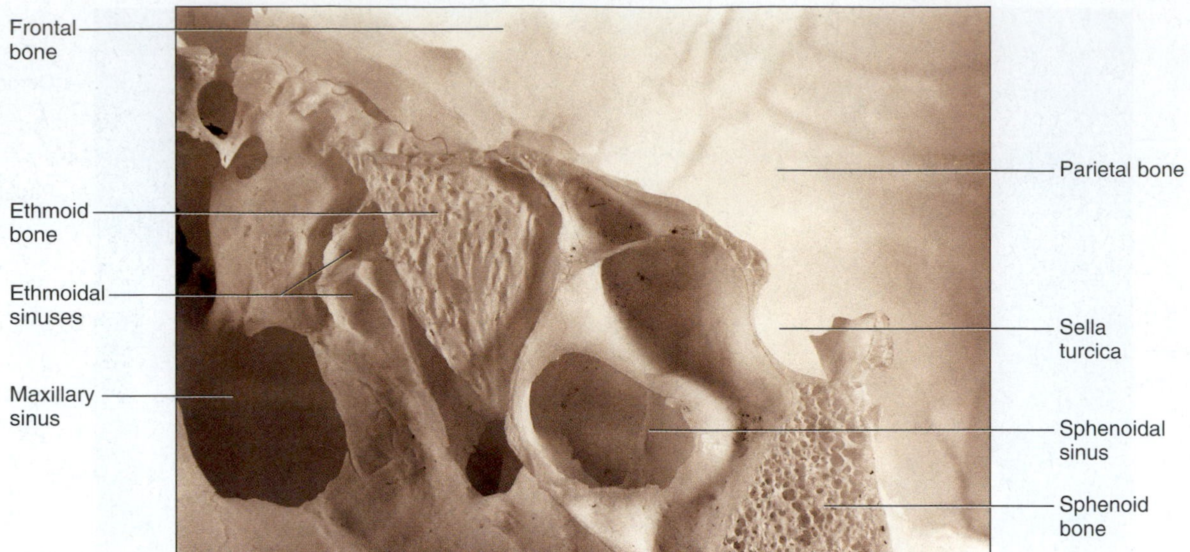

PLATE FORTY-SEVEN Sphenoidal region, sagittal section.

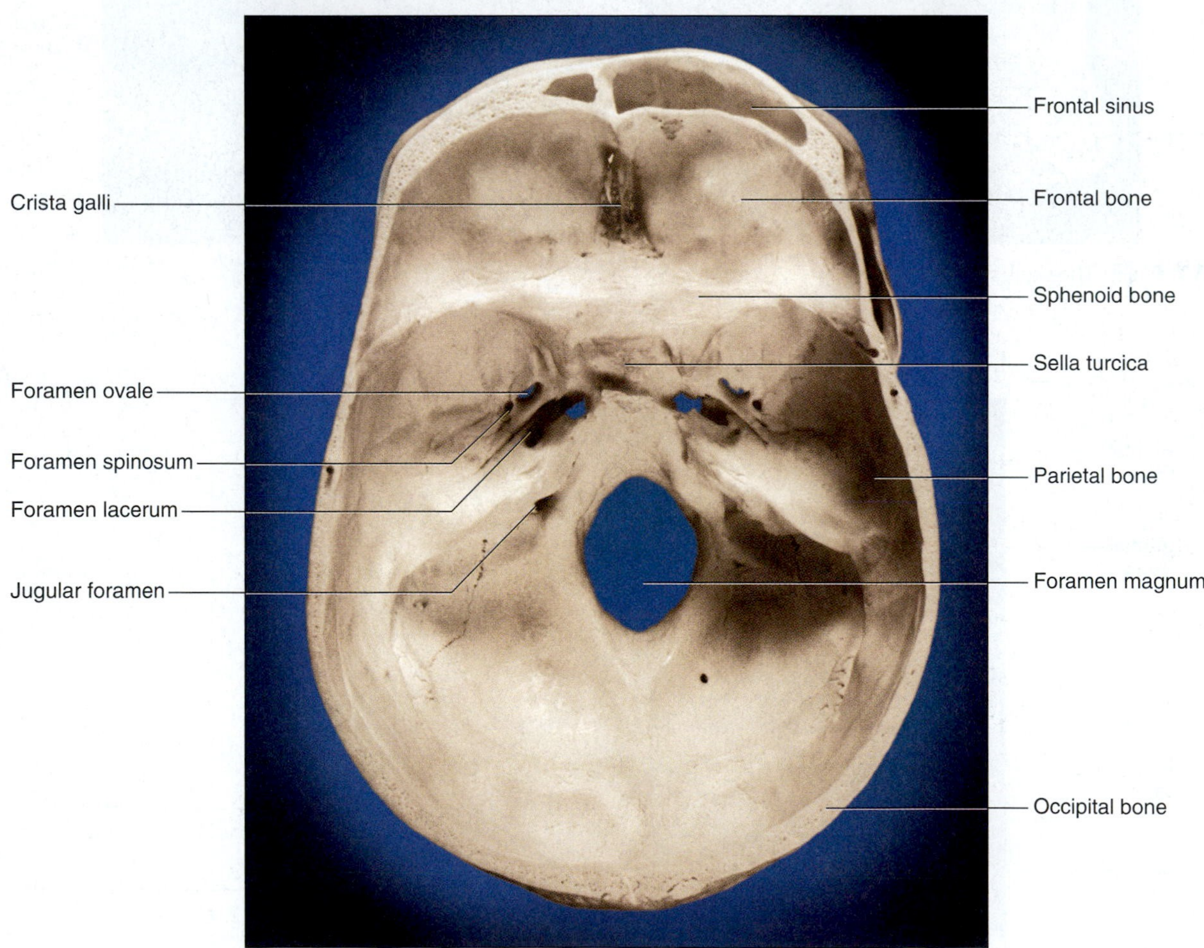

PLATE FORTY-EIGHT The skull, floor of the cranial cavity.

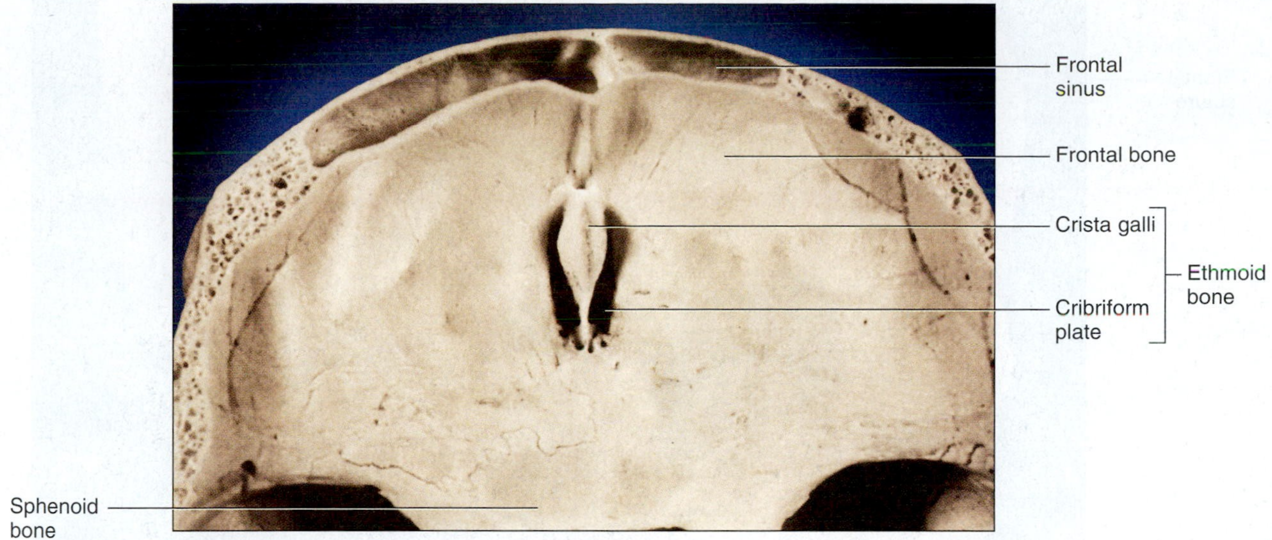

PLATE FORTY-NINE Frontal region, transverse section.

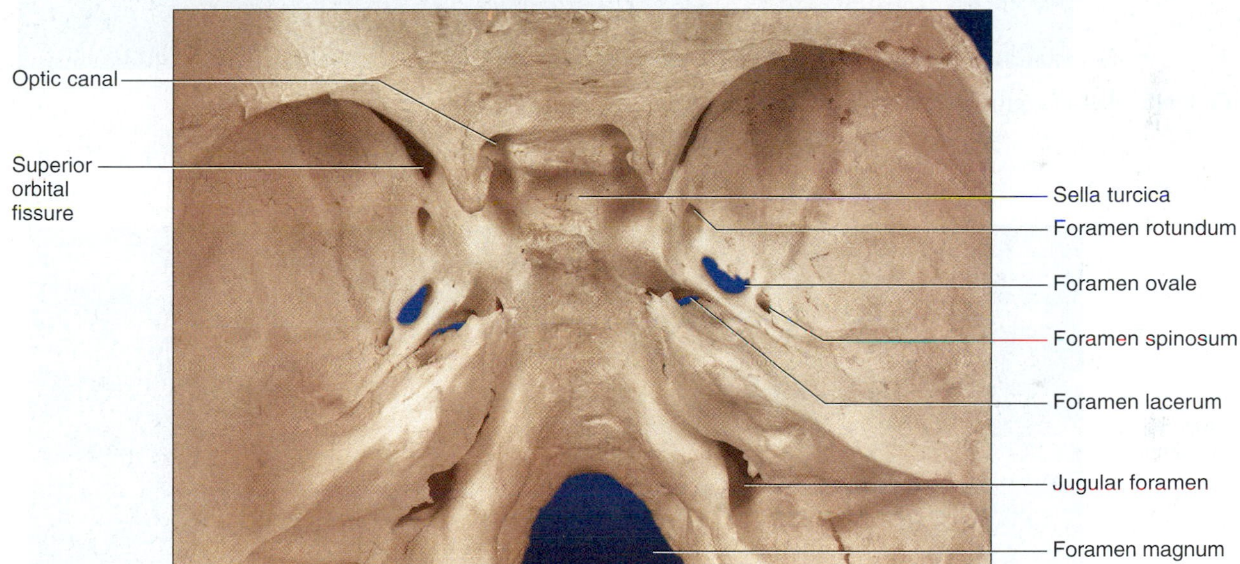

PLATE FIFTY Sphenoidal region, floor of the cranial cavity.

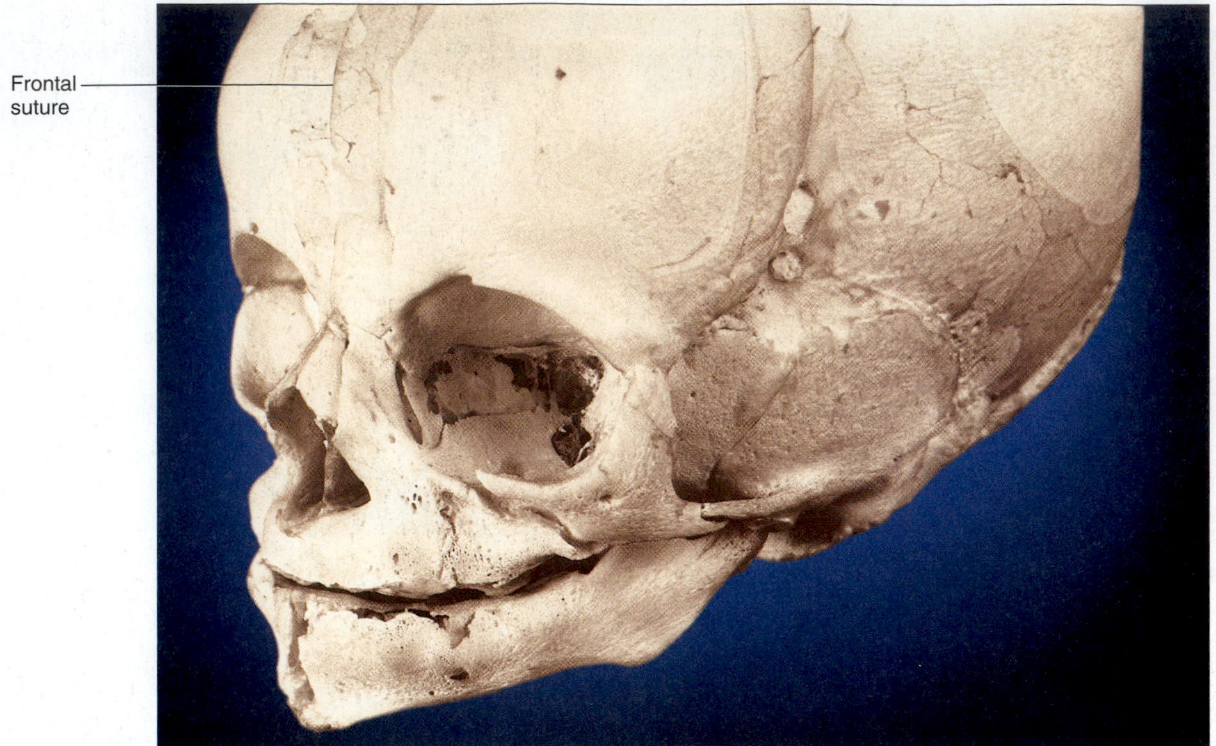

PLATE FIFTY-ONE Skull of a fetus, left anterolateral view.

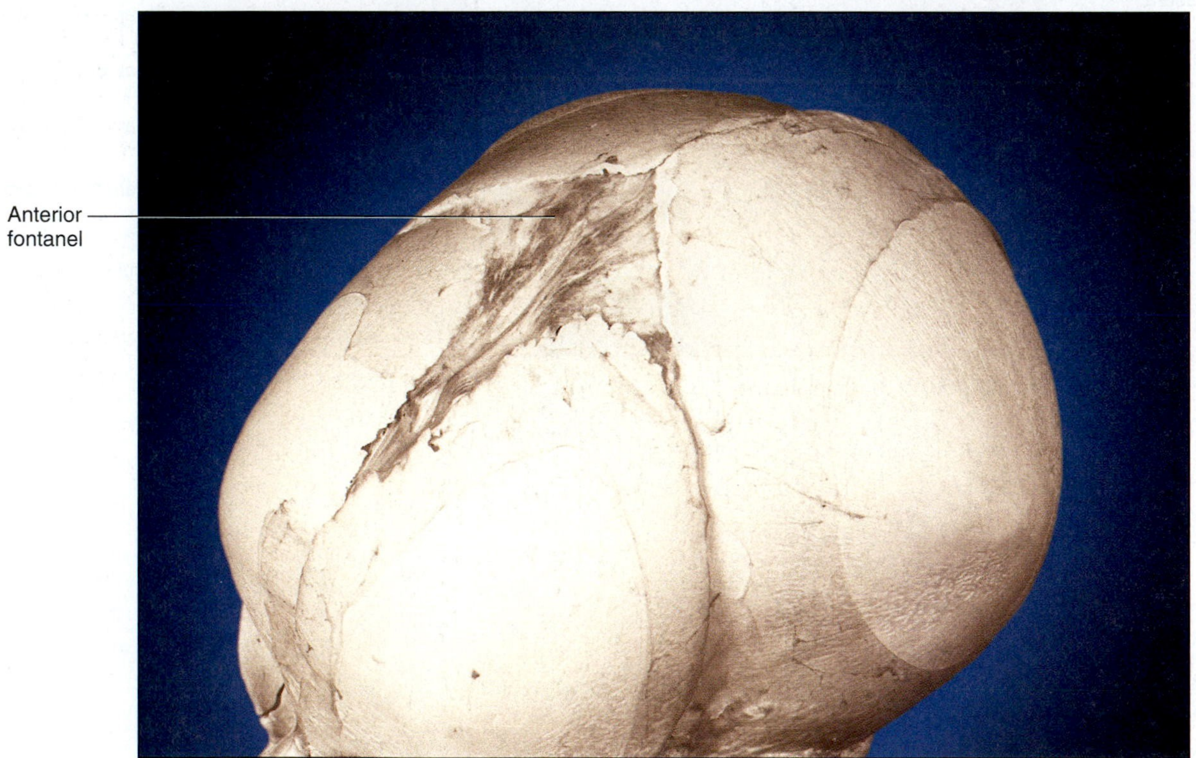

PLATE FIFTY-TWO Skull of a fetus, left superior view.

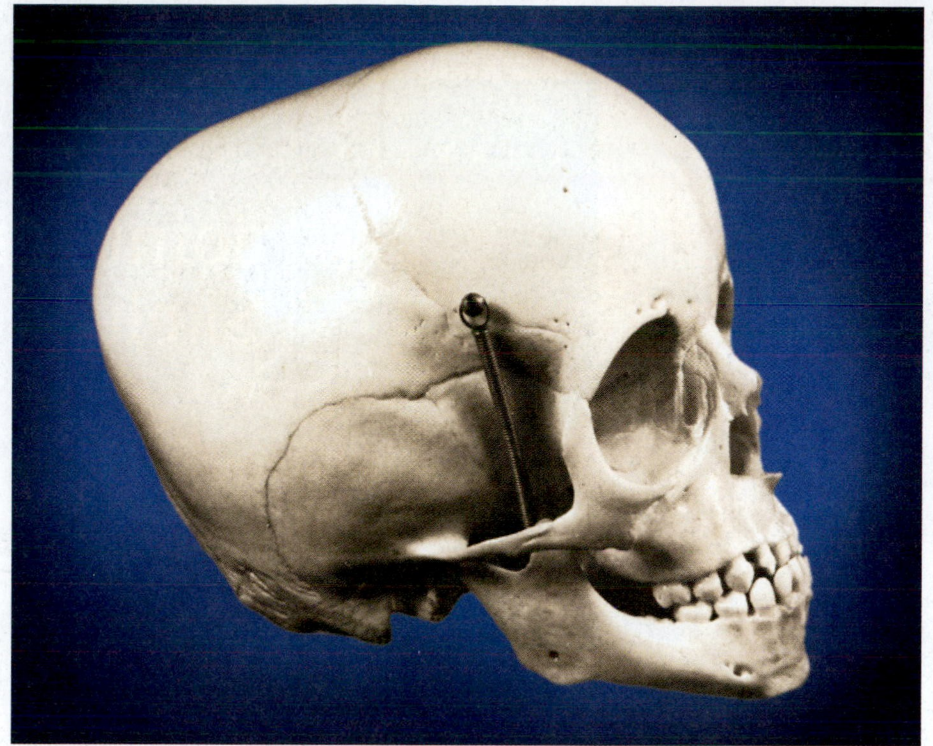

PLATE FIFTY-THREE Skull of a child, right lateral view.

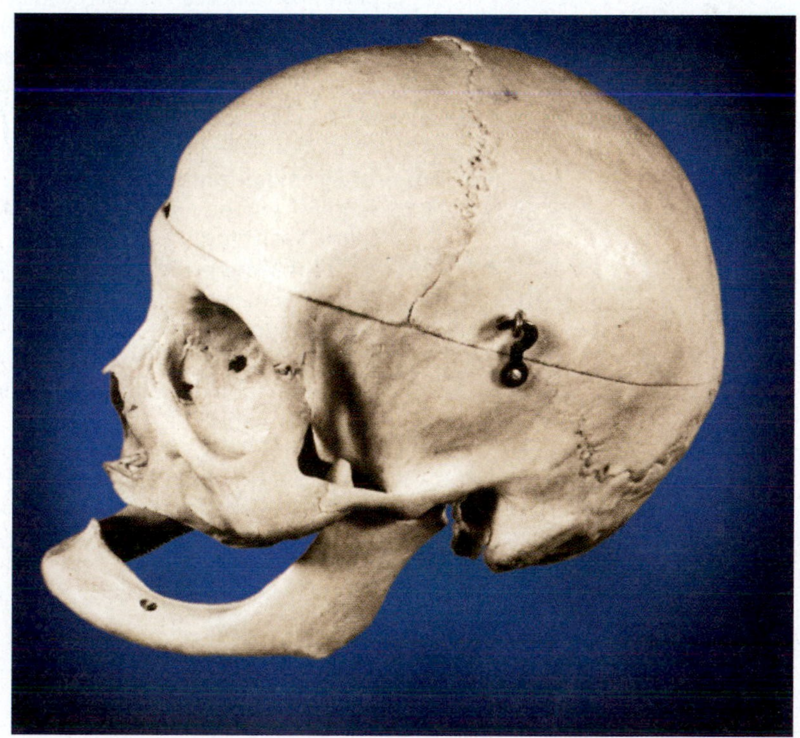

PLATE FIFTY-FOUR Skull of an aged person, left lateral view. (This skull has been cut postmortem to allow the removal of the cranium.)

CHAPTER 8 Joints of the Skeletal System

Arthritis has inflamed the joints in these fingers. Drugs and replacement joints are used to treat this painful condition.

UNDERSTANDING WORDS

anul-, ring: *anul*ar ligament—ring-shaped band of connective tissue below the elbow joint that encircles the head of the radius.

arth-, joint: *arth*rology—study of joints and ligaments.

burs-, bag, purse: prepatellar *burs*a—fluid-filled sac between the skin and the patella.

condyl-, knob: medial *condyl*e—rounded bony process at the distal end of the femur.

fov-, pit: *fov*ea capitis—pit in the head of the femur to which a ligament is attached.

glen-, joint socket: *glen*oid cavity—depression in the scapula that articulates with the head of the humerus.

labr-, lip: glenoidal *labr*um—rim of fibrocartilage attached to the margin of the glenoid cavity.

ov-, egglike: syn*ov*ial fluid—thick fluid in a joint cavity that resembles egg white.

sutur-, sewing: *sutur*e—type of joint in which flat bones are interlocked by a set of tiny bony processes.

syndesm-, binding together: *syndesm*osis—type of joint in which the bones are held together by long fibers of connective tissue.

 LEARN  PRACTICE ▶ ASSESS

LEARNING OUTCOMES

After you have studied this chapter, you should be able to:

8.1 Introduction
- **1** List the functions of joints. (p. 261)

8.2 Classification of Joints
- **2** Explain how joints can be classified according to the type of tissue that binds the bones together. (p. 261)
- **3** Describe how bones of fibrous joints are held together. (p. 261)
- **4** Describe how bones of cartilaginous joints are held together. (p. 262)

8.3 General Structure of a Synovial Joint
- **5** Describe the general structure of a synovial joint. (p. 263)

8.4 Types of Synovial Joints
- **6** Distinguish among the six types of synovial joints and name an example of each type. (p. 265)

8.5 Types of Joint Movements
- **7** Explain how skeletal muscles produce movements at joints, and identify several types of joint movements. (p. 267)

8.6 Examples of Synovial Joints
- **8** Describe the shoulder joint and explain how its articulating parts are held together. (p. 271)
- **9** Describe the elbow, hip, and knee joints and explain how their articulating parts are held together. (p. 272)

8.7 Life-Span Changes
- **10** Describe life-span changes in joints. (p. 278)

DO GLUCOSAMINE AND CHONDROITIN HELP ARTHRITIS PAIN?

Glucosamine and chondroitin are widely sold as "dietary supplements" to treat the joint pain and joint space narrowing of osteoarthritis. Anecdotal reports and many small studies indicate that these supplements are effective, but large, controlled clinical trials have yielded conflicting and confusing results.

Glucosamine and chondroitin are carbohydrates produced in the body, but in dietary supplements come from shells and cow cartilage, respectively. The recommended dose is 1,500 milligrams of glucosamine a day and 1,200 milligrams of chondroitin, usually taken in combined form two or three times a day. Dosages may be uneven, however, because these biochemicals are marketed as dietary supplements and not drugs, which are more precisely regulated.

The largest study completed so far is GAIT (Glucosamine/Chondroitin Arthritis Intervention Trial). Run by the Veterans Administration and the National Institutes of Health with results published in 2006, GAIT randomized 1,583 people with osteoarthritis of the knee into five treatment groups: glucosamine alone, chondroitin alone, glucosamine and chondroitin, celecoxib (a drug), and placebo. After tweny-four weeks, glucosamine and/or chondroitin did not improve arthritis symptoms any more than the placebo; the people taking the drug improved somewhat. However, when the researchers considered only those participants with moderate to severe arthritis, the dietary supplements (alone or in combination) did alleviate pain, although not as fast as the drug.

In a three-year study reported in 2001, people with knee arthritis taking the placebo had narrowing of the joint space, whereas participants taking glucosamine did not. Yet a two-year study published in 2008 on 222 people with hip arthritis showed that glucosamine worked only as well as the placebo in alleviating pain and stalling joint narrowing in that affected joint.

Two large ongoing clinical trials will add to the continuing evaluation of these popular dietary supplements. One investigation is assessing biochemical evidence of cartilage breakdown, so may provide more definitive results. Meanwhile, it is best to consult a physician before using glucosamine/chondroitin to alleviate the pain of arthritis. Many people report that the supplements help, and they have been studied enough to indicate that they seem safe, but they usually take two to three months to be effective—and we still do not know exactly what they do to our joints.

8.1 INTRODUCTION

Joints, or **articulations** (ar-tik″u-la′shunz), are functional junctions between bones. They bind parts of the skeletal system, make possible bone growth, permit parts of the skeleton to change shape during childbirth, and enable the body to move in response to skeletal muscle contractions.

8.2 CLASSIFICATION OF JOINTS

Joints vary considerably in structure and function. However, they can be classified by the type of tissue that binds the bones at each junction. Three general groups are fibrous joints, cartilaginous joints, and synovial joints.

Joints can also be grouped according to the degree of movement possible at the bony junctions. In this scheme, joints are classified as immovable (synarthrotic), slightly movable (amphiarthrotic), and freely movable (diarthrotic). At some diarthrotic joints, movement can occur over considerable distances, such as flexion and extension of the elbow. Whereas other, such as the joint between the saccrum and the ilium, move freely, but only for a short distance. The structural and functional classification schemes overlap somewhat. Currently, structural classification is the one most commonly used.

Fibrous Joints

Fibrous (fi′brus) **joints** are so named because the dense connective tissue holding them together includes many collagenous fibers. These joints are between bones in close contact. The three types of fibrous joints are

1. **Syndesmosis** (sin″des-mo′sis). In this type of joint, the bones are bound by a sheet (*interosseous membrane*) or bundle of dense connective tissue (*interosseous ligament*). This junction is flexible and may be twisted, so the joint may permit slight movement and thus is amphiarthrotic (am″fe-ar-thro′tik). A syndesmosis lies between the tibia and fibula (fig. 8.1).
2. **Suture** (soo′cher). Sutures are only between flat bones of the skull, where the broad margins of adjacent bones grow together and unite by a thin layer of dense

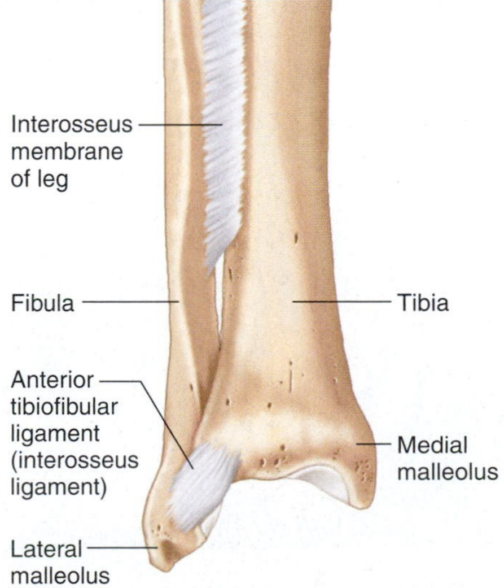

FIGURE 8.1 The articulation between the tibia and fibula is an example of a syndesmosis.

connective tissue called a *sutural ligament*. Recall from chapter 7 (p. 216) that the infantile skull is incompletely developed, with several of the bones connected by membranous areas called *fontanels* (see fig. 7.31). These areas allow the skull to change shape slightly during childbirth, but as the bones continue to grow, the fontanels close, and sutures replace them. With time, some of the bones at sutures interlock by tiny bony processes. Such a suture is in the adult human skull where the parietal and occipital bones meet to form the lambdoid suture. They are immovable, so sutures are synarthrotic (sin′ar-thro′tik) joints (figs. 8.2 and 8.3).

3. **Gomphosis** (gom-fo′sis). A gomphosis is a joint formed by the union of a cone-shaped bony process in a bony socket. The peglike root of a tooth fastened to a maxilla or the mandible by a *periodontal ligament* is such a joint. This ligament surrounds the root and firmly attaches it to the bone with bundles of thick collagenous fibers. A gomphosis is a synarthrotic joint (fig. 8.4).

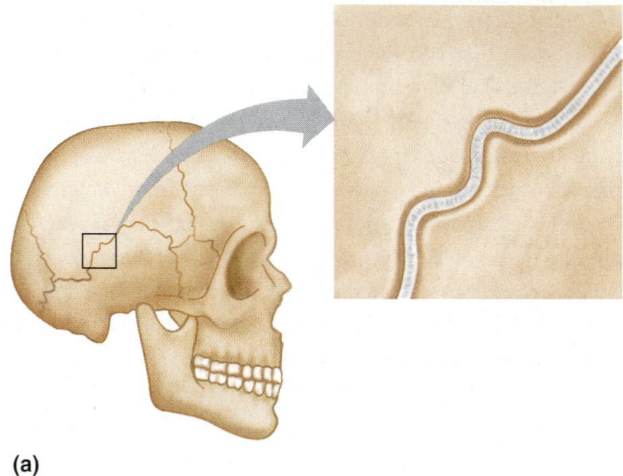

(a)

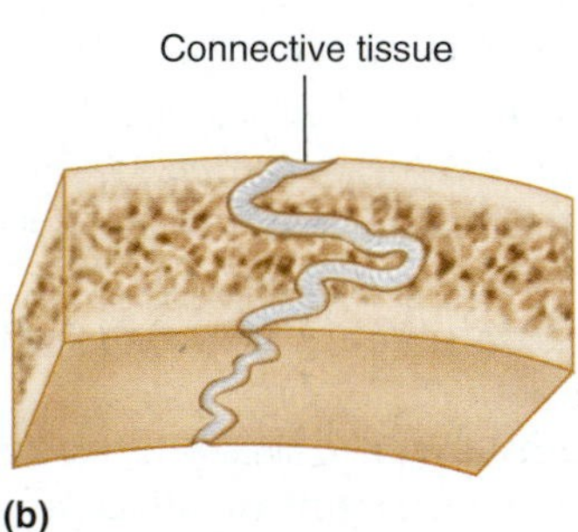

(b)

FIGURE 8.2 Fibrous joints. (*a*) The fibrous joints between the bones of the skull are immovable and are called sutures. (*b*) A thin layer of connective tissue connects the bones at the suture.

PRACTICE 

1. What is a joint?
2. How are joints classified?
3. Describe three types of fibrous joints.
4. What is the function of the fontanels?

Cartilaginous Joints

Hyaline cartilage or fibrocartilage connects the bones of **cartilaginous** (kar″tĭ-laj′ĭnus) **joints.** The two types are

1. **Synchondrosis** (sin″kon-dro′sis). In a synchondrosis, bands of hyaline cartilage unite the bones. Many of these joints are temporary structures that disappear during growth. An example is an immature long bone where a band of hyaline cartilage (the epiphyseal plate) connects an epiphysis to a diaphysis. This cartilage band participates in bone lengthening and, in time, is replaced with bone. When ossification completes, usually before the age of twenty-five years, the joint becomes a *synostosis,* a bony joint. The synostosis is synarthrotic (see fig. 7.11).

   Another synchondrosis lies between the manubrium and the first rib, directly united by costal cartilage (fig. 8.5). This joint is also synarthrotic, but permanent.

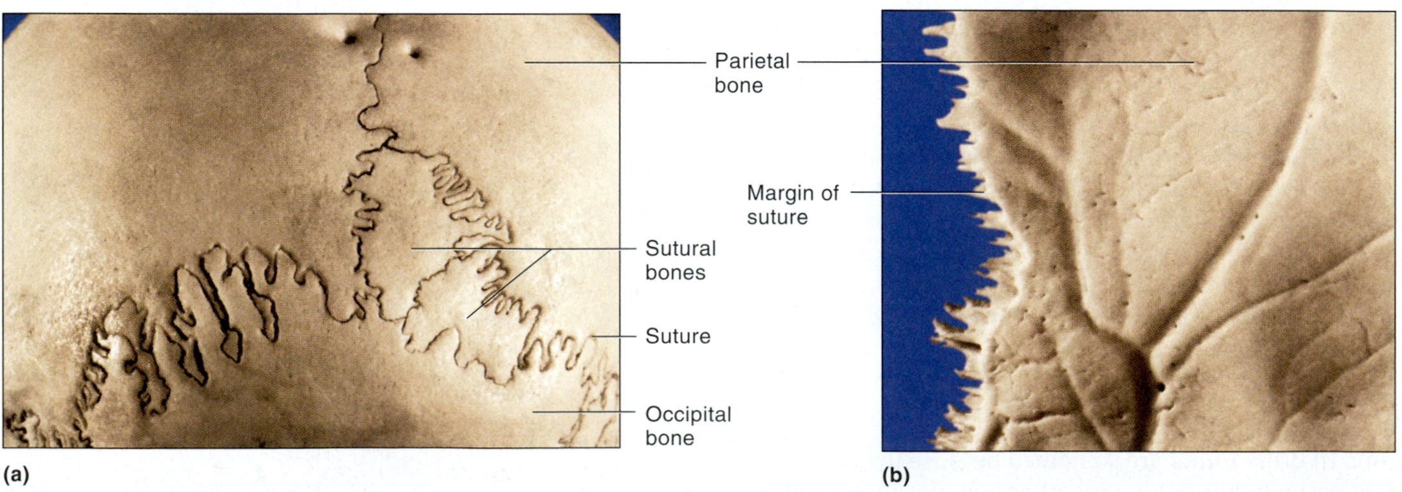

FIGURE 8.3 Cranial sutures. (*a*) Sutures between the parietal and occipital bones of the skull. (*b*) The inner margin of a parietal suture. The grooves on the inside of this parietal bone mark the paths of blood vessels near the brain's surface.

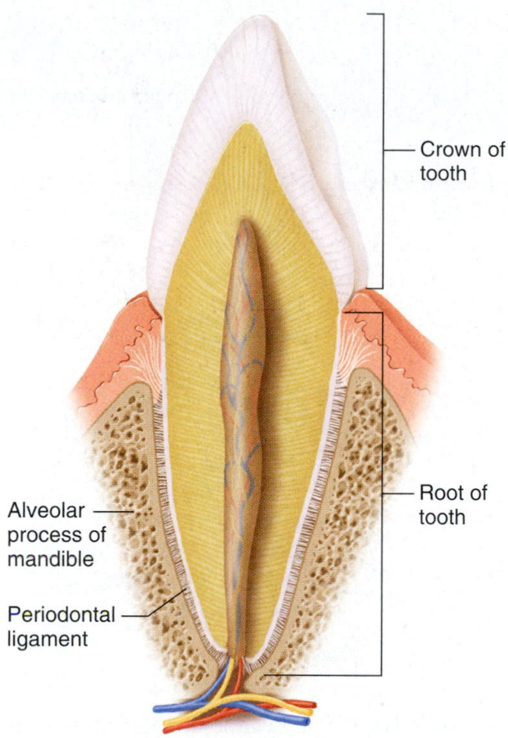

FIGURE 8.4 The articulation between the root of a tooth and the mandible is a gomphosis.

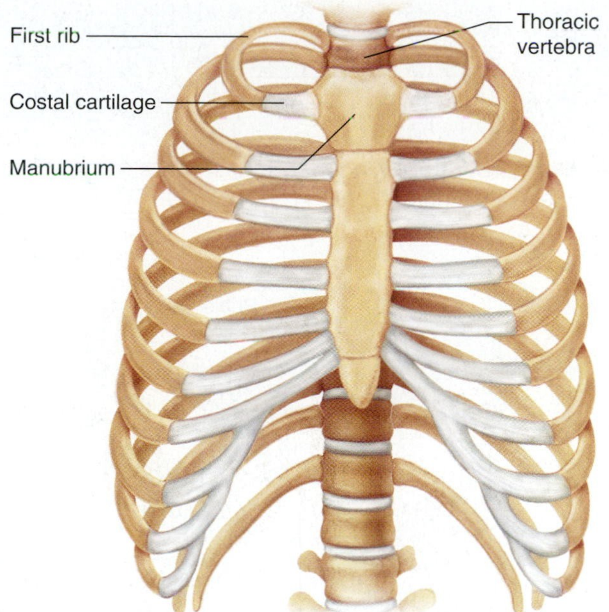

FIGURE 8.5 The articulation between the first rib and the manubrium is a synchondrosis.

The joints between the costal cartilages and the sternum of ribs 2 through 7 are usually synovial joints.

2. **Symphysis** (sim′fĭ-sis). The articular surfaces of the bones at a symphysis are covered by a thin layer of hyaline cartilage, and the cartilage, in turn, is attached to a pad of springy fibrocartilage. Limited movement occurs at such a joint whenever forces compress or deform the cartilaginous pad. An example of this type of joint is the symphysis pubis in the pelvis, which allows maternal pelvic bones to shift as an infant passes through the birth canal (fig. 8.6*a*).

    The joint formed by the bodies of two adjacent vertebrae separated by an intervertebral disc is also a symphysis (fig. 8.6*b* and reference plate 11, p. 40). Each intervertebral disc is composed of a band of fibrocartilage (annulus fibrosus) that surrounds a gelatinous core (nucleus pulposus). The disc absorbs shocks and helps equalize pressure between the vertebrae when the body moves. Each disc is slightly flexible, so the combined movement of many of the joints in the vertebral column allows the back to bend forward or to the side or to twist. They are amphiarthrotic joints because these joints allow slight movements.

Synovial Joints

Most joints of the skeletal system are **synovial** (sĭ-no′ve-al) **joints,** and because they allow free movement, they are diarthrotic (di″ar-thro′tik). These joints are more complex structurally than fibrous or cartilaginous joints. They consist of articular cartilage; a joint capsule; and a synovial membrane, which secretes synovial fluid.

Virtuoso violinist Niccolò Paganini (1782–1840) astounded concertgoers with his ability to reach three octaves across the bridge of his instrument. So lax were his joints that he could bend his thumb backward until the nail touched the back of his hand.

Paganini had "benign joint hypermobility syndrome," defined as a range of motion much greater than normal. Today the condition is studied in people whose professions make lax joints either a benefit or a liability. In athletes and dancers, for example, loose joints increase the risk of injury. Musicians are especially interesting. The nimble fingers, hands, and wrists of hypermobility syndrome help woodwind and string players, but lax joints also tend to cause back and knee problems. Rather than gaining strength from repetitive movements of playing instruments, these joints must bear weight from long hours of sitting in one position.

Perhaps rock guitarists make the best use of hypermobile joints. They stretch their fingers like Paganini while jumping about onstage to better distribute their weight on the other joints!

8.3 GENERAL STRUCTURE OF A SYNOVIAL JOINT

The articular ends of the bones in a synovial joint are covered with a thin layer of hyaline cartilage. This layer, the **articular cartilage,** resists wear and minimizes friction when it is compressed as the joint moves (fig. 8.7).

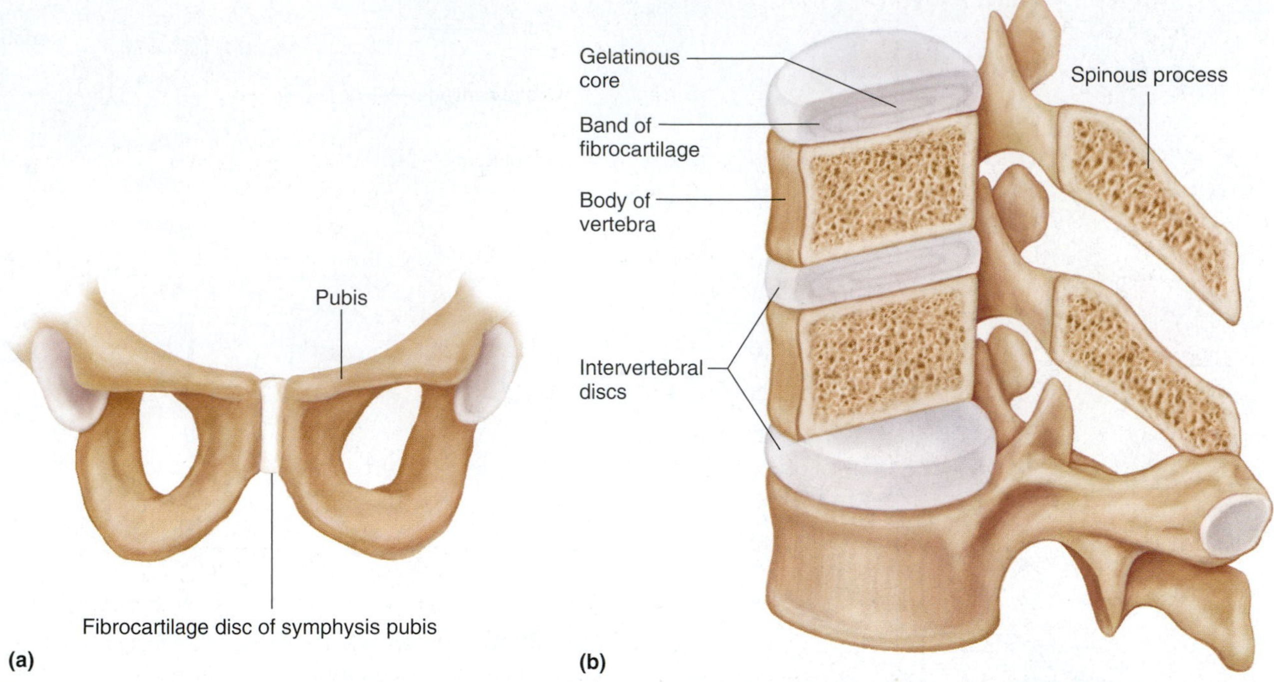

FIGURE 8.6 Fibrocartilage composes (*a*) the symphysis pubis that separates the pubic bones and (*b*) the intervertebral discs that separate vertebrae.

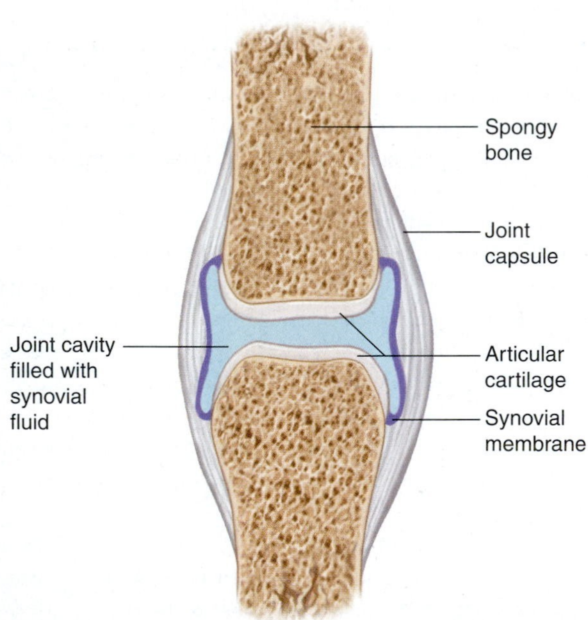

FIGURE 8.7 The generalized structure of a synovial joint.

A tubular **joint capsule** (articular capsule) that has two distinct layers holds together the bones of a synovial joint. The outer layer largely consists of dense connective tissue, whose fibers attach to the periosteum around the circumference of each bone of the joint near its articular end. Thus, the outer fibrous layer of the capsule completely encloses the other parts of the joint. It is, however, flexible enough to permit movement and strong enough to help prevent the articular surfaces from being pulled apart.

Bundles of strong, tough collagenous fibers called **ligaments** (lig′ah-mentz) reinforce the joint capsule and help bind the articular ends of the bones. Some ligaments appear as thickenings in the fibrous layer of the capsule, whereas others are *accessory structures* located outside the capsule. In either case, these structures help prevent excessive movement at the joint. That is, the ligament is relatively inelastic, and it tightens when the joint is stressed.

The inner layer of the joint capsule consists of a shiny, vascular lining of loose connective tissue called the **synovial membrane.** This membrane, only a few cells thick, covers all of the surfaces within the joint capsule, except the areas the articular cartilage covers. The synovial membrane surrounds a closed sac called the *synovial cavity,* into which the membrane secretes a clear, viscous fluid called **synovial fluid.** In some regions, the surface of the synovial membrane has villi as well as larger folds and projections that extend into the cavity. Besides filling spaces and irregularities of the joint cavity, these extensions increase the surface area of the synovial membrane. The membrane may also store adipose tissue and form movable fatty pads in the joint. This multifunctional membrane also reabsorbs fluid, which is important when a joint cavity is injured or infected. Synovial fluid contains stem cells, which may function in ligament regeneration following injury.

Synovial fluid has a consistency similar to uncooked egg white, and it moistens and lubricates the smooth cartilaginous surfaces of the joint. It also helps supply articular cartilage with nutrients obtained from blood vessels of the synovial membrane. The volume of synovial fluid in a joint cavity is usually just enough to cover the articulating surfaces with a thin film of fluid. The volume of synovial fluid in the cavity of the knee is 0.5 mL or less.

> A physician can determine the cause of joint inflammation or degeneration (arthritis) by aspirating a sample of synovial fluid from the affected joint using a procedure called arthrocentesis. Bloody fluid with lipid on top indicates a fracture extending into the joint. Clear fluid and an increase in stem cell number is found in osteoarthritis, a degeneration of collagen in the joint. Cloudy, yellowish fluid may indicate rheumatoid arthritis, and crystals in the synovial fluid signal gout. If the fluid is cloudy but red-tinged and containing pus, a bacterial infection may be present that requires prompt treatment. Normal synovial fluid has 180 or fewer leukocytes (white blood cells) per mL. If the fluid is infected, the leukocyte count exceeds 2,000.

Some synovial joints are partially or completely divided into two compartments by discs of fibrocartilage called **menisci** (me-nis′ke) (sing., *meniscus*) between the articular surfaces. Each meniscus attaches to the fibrous layer of the joint capsule peripherally, and its free surface projects into the joint cavity. In the knee joint, crescent-shaped menisci cushion the articulating surfaces and help distribute body weight onto these surfaces (fig. 8.8).

Fluid-filled sacs called **bursae** (ber′se) are associated with certain synovial joints. Each bursa has an inner lining of synovial membrane, which may be continuous with the synovial membrane of a nearby joint cavity. These sacs contain synovial fluid and are commonly located between the skin and underlying bony prominences, as in the case of the patella of the knee or the olecranon process of the elbow. Bursae cushion and aid the movement of tendons that glide over bony parts or over other tendons. The names of bursae indicate their locations. Figure 8.8 shows a *suprapatellar bursa,* a *prepatellar bursa,* and an *infrapatellar bursa.*

PRACTICE

5 Describe two types of cartilaginous joints.
6 What is the function of an intervertebral disc?
7 Describe the structure of a synovial joint.
8 What is the function of the synovial fluid?

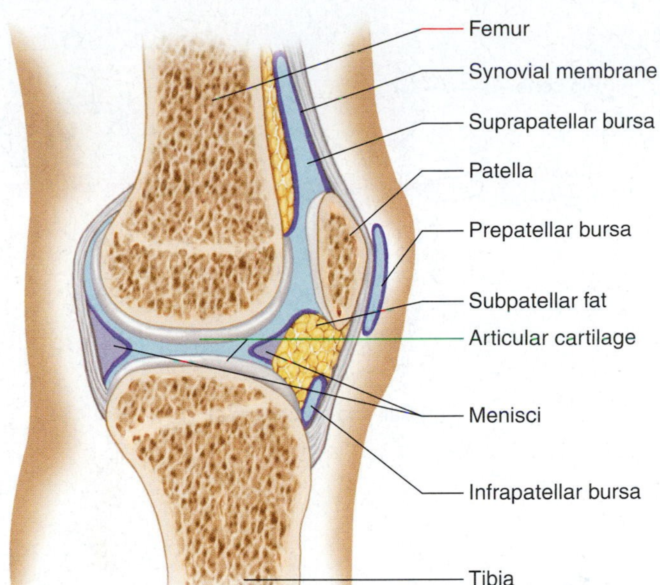

FIGURE 8.8 Menisci separate the articulating surfaces of the femur and tibia. Several bursae are associated with the knee joint.

> Articular cartilage, like other cartilaginous structures, lacks a direct blood supply (see chapter 5, p. 158). Surrounding synovial fluid supplies oxygen, nutrients, and other vital chemicals. Normal body movements force these substances into the joint cartilage. When a joint is immobilized or is not used for a long time, inactivity may cause degeneration of the articular cartilage. The degeneration may reverse when joint movements resume. However, it is important to avoid exercises that greatly compress the tissue during the period of regeneration. Otherwise, chondrocytes in the thinned cartilage may be injured, hindering repair.

8.4 TYPES OF SYNOVIAL JOINTS

The articulating bones of synovial joints have a variety of shapes that allow different types of movement. Based upon their shapes and the movements they permit, these joints can be classified into six major types—ball-and-socket joints, condylar joints, plane joints, hinge joints, pivot joints, and saddle joints.

1. A **ball-and-socket joint,** or **spheroidal joint,** consists of a bone with a globular or slightly egg-shaped head that articulates with the cup-shaped cavity of another bone. Such a joint allows a wider range of motion than does any other type, permitting movements in all planes, as well as rotational movement around a central axis. The hip and shoulder have joints of this type (fig. 8.9a).
2. In a **condylar joint,** or **ellipsoidal joint,** the ovoid condyle of one bone fits into the elliptical cavity of another bone, as in the joints between the metacarpals

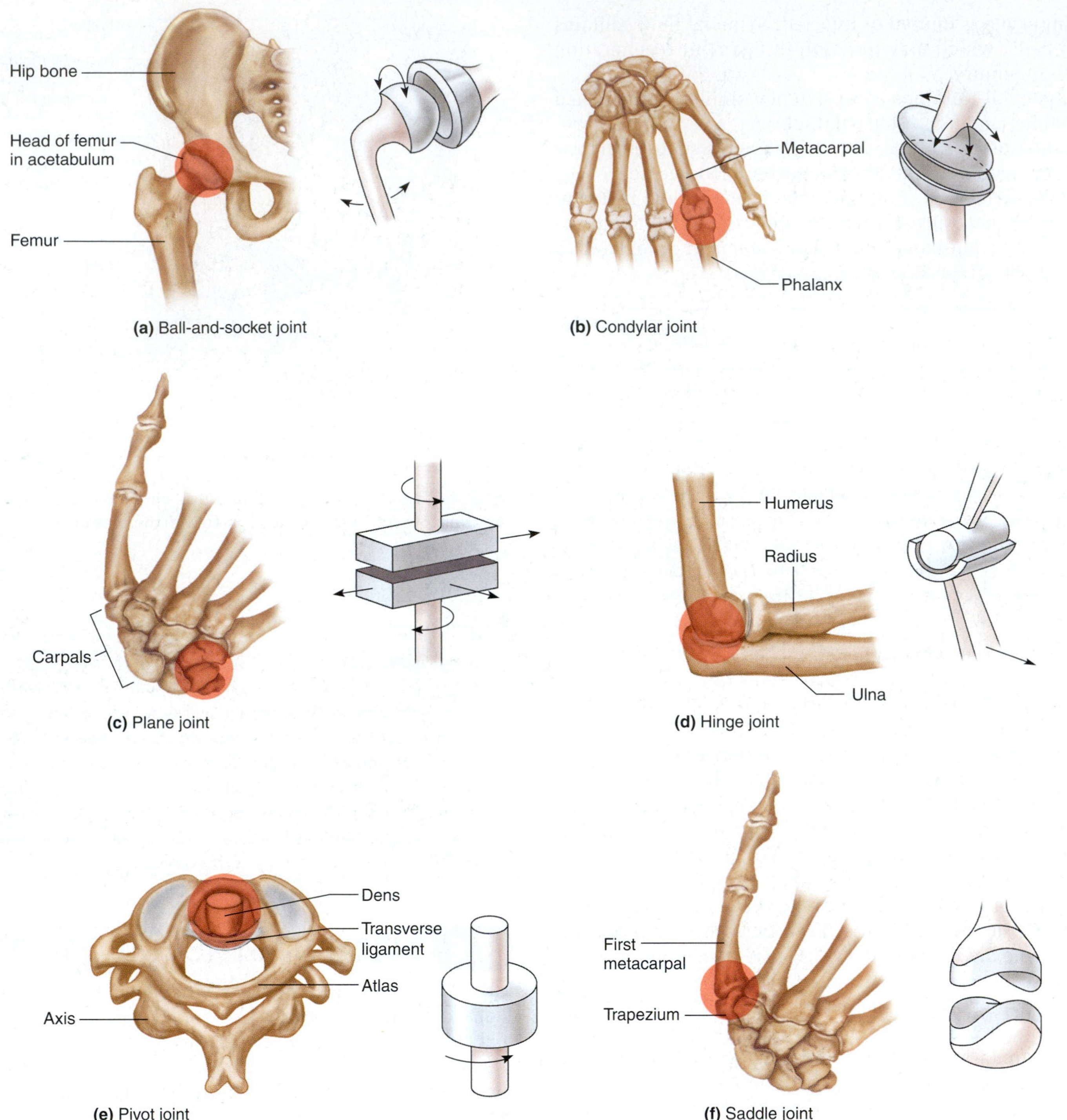

FIGURE 8.9 Types and examples of synovial (freely movable) joints.

and phalanges. This type of joint permits a variety of movements in different planes; rotational movement, however, is not possible (fig. 8.9*b*).

3. The articulating surfaces of **plane joints,** or **gliding joints,** are nearly flat or slightly curved. These joints allow sliding or back-and-forth motion and twisting movements. Most of the joints in the wrist and ankle, as well as those between the articular processes of vertebrae, belong to this group (fig. 8.9*c*). The sacroiliac joints and the joints formed by ribs 2 through 7 connecting with the sternum are also plane joints.
4. In a **hinge joint,** the convex surface of one bone fits into the concave surface of another, as in the elbow and the joints of the phalanges. Such a joint resembles the hinge of a door in that it permits movement in one plane only (fig. 8.9*d*).
5. In a **pivot joint,** or **trochoid joint,** the cylindrical surface of one bone rotates in a ring formed of bone

and a ligament. Movement at such a joint is limited to rotation around a central axis. The joint between the proximal ends of the radius and the ulna, where the head of the radius rotates in a ring formed by the radial notch of the ulna and a ligament (anular ligament), is of this type. Similarly, a pivot joint functions in the neck as the head turns from side to side. In this case, the ring formed by a ligament (transverse ligament) and the anterior arch of the atlas rotates around the dens of the axis (fig. 8.9*e*).

6. A **saddle joint,** or **sellar joint,** forms between bones whose articulating surfaces have both concave and convex regions. The surface of one bone fits the complementary surface of the other. This physical relationship permits a variety of movements, mainly in two planes, as in the case of the joint between the carpal (trapezium) and the metacarpal of the thumb (fig. 8.9*f*).

Table 8.1 summarizes the types of joints.

PRACTICE

9 Name six types of synovial joints.

10 Describe the structure of each type of synovial joint.

8.5 TYPES OF JOINT MOVEMENTS

Skeletal muscle action produces movements at synovial joints. Typically, one end of a muscle is attached to a relatively immovable or fixed part on one side of a joint, and the other end of the muscle is fastened to a movable part on the other side. When the muscle contracts, its fibers pull its movable end (*insertion*) toward its fixed end (*origin*), and a movement occurs at the joint.

The following terms describe movements at joints that occur in different directions and in different planes (figs. 8.10, 8.11, and 8.12):

flexion (flek′shun) Bending parts at a joint so that the angle between them decreases and the parts come closer together (bending the knee).

extension (ek-sten′shun) Straightening parts at a joint so that the angle between them increases and the parts move farther apart (straightening the knee).

hyperextension (hi″per-ek-sten′shun) Extension of the parts at a joint beyond the anatomical position (bending the head back beyond the upright position); often used

TABLE 8.1 | Types of Joints

Type of Joint	Description	Possible Movements	Example
Fibrous	Articulating bones fastened together by thin layer of dense connective tissue containing many collagenous fibers		
1. *Syndesmosis* (amphiarthrotic)	Bones bound by interosseous ligament	Joint flexible and may be twisted	Tibiofibular articulation
2. *Suture* (synarthrotic)	Flat bones united by sutural ligament	None	Parietal bones articulate at sagittal suture of skull
3. *Gomphosis* (synarthrotic)	Cone-shaped process fastened in bony socket by periodontal ligament	None	Root of tooth united with mandible
Cartilaginous	Articulating bones connected by hyaline cartilage or fibrocartilage		
1. *Synchondrosis* (synarthrotic)	Bones united by bands of hyaline cartilage	None	Joint between epiphysis and diaphysis of a long bone
2. *Symphysis* (amphiarthrotic)	Articular surfaces separated by thin layers of hyaline cartilage attached to a pad of fibrocartilage	Limited movement, as when the back is bent or twisted	Joints between bodies of vertebrae
Synovial (diarthrotic)	Articulating ends of bones surrounded by a joint capsule; articular bone ends covered by hyaline cartilage and separated by synovial fluid		
1. *Ball-and-socket*	Ball-shaped head of one bone articulates with cup-shaped socket of another	Movements in all planes, including rotation	Shoulder, hip
2. *Condylar*	Oval-shaped condyle of one bone articulates with elliptical cavity of another	Variety of movements in different planes, but no rotation	Joints between metacarpals and phalanges
3. *Plane*	Articulating surfaces are nearly flat or slightly curved	Sliding or twisting	Joints between various bones of wrist and ankle
4. *Hinge*	Convex surface of one bone articulates with concave surface of another	Flexion and extension	Elbow and joints of phalanges
5. *Pivot*	Cylindrical surface of one bone articulates with ring of bone and ligament	Rotation	Joint between proximal ends of radius and ulna
6. *Saddle*	Articulating surfaces have both concave and convex regions; surface of one bone fits the complementary surface of another	Variety of movements, mainly in two planes	Joint between carpal and metacarpal of thumb

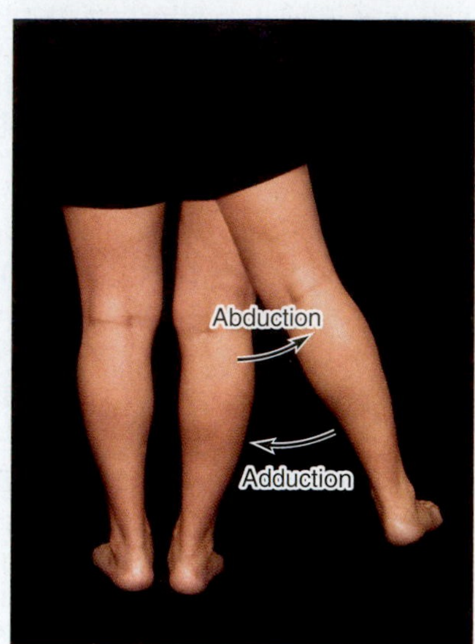

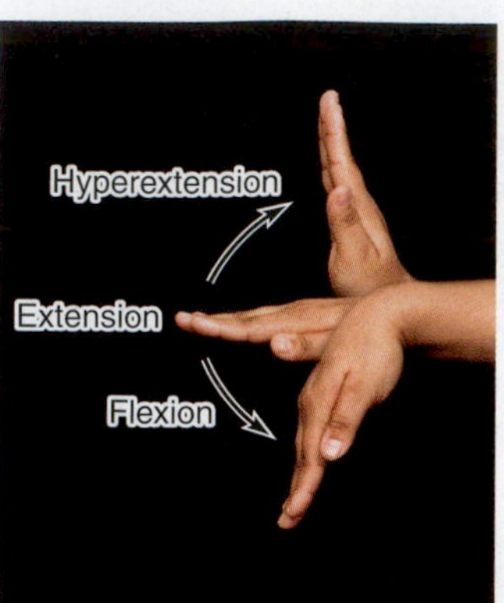

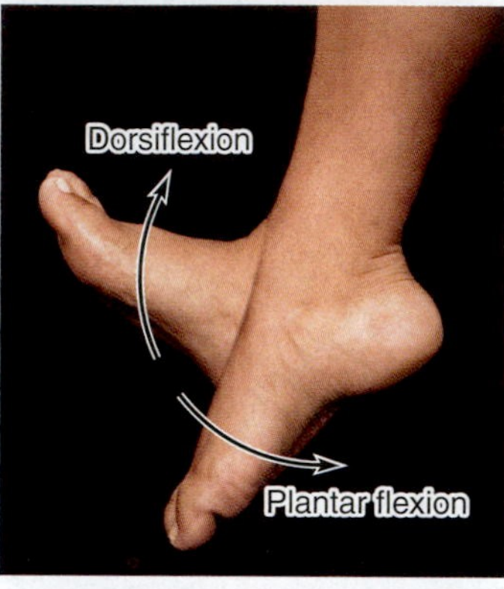

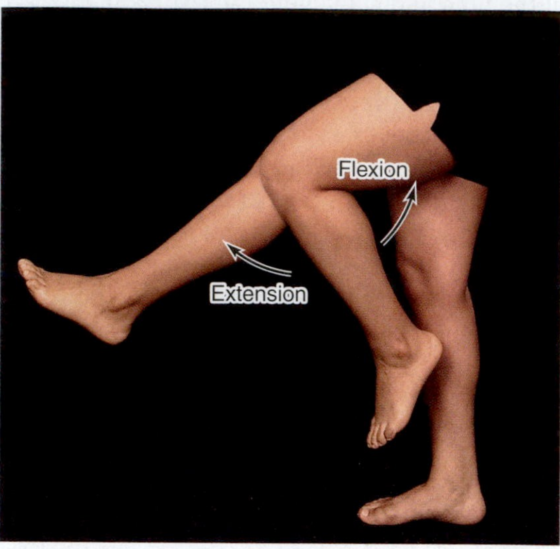

FIGURE 8.10 Joint movements illustrating adduction, abduction, dorsiflexion, plantar flexion, hyperextension, extension, and flexion.

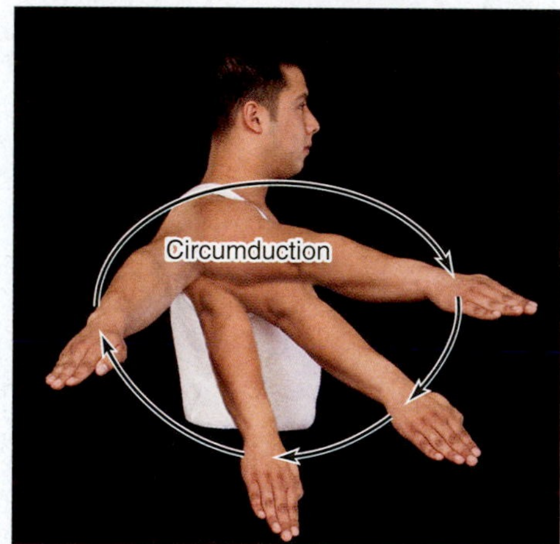

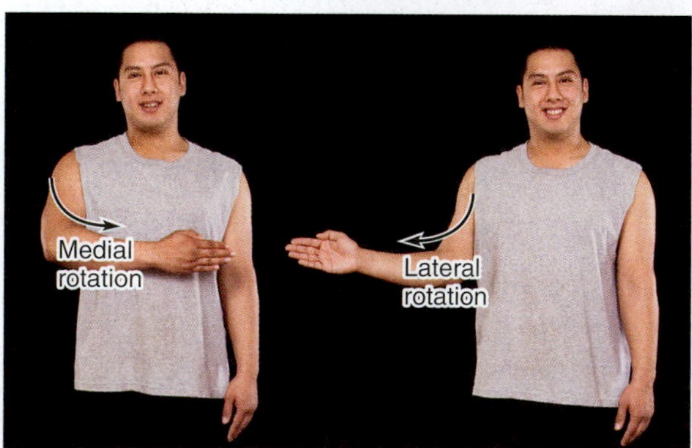

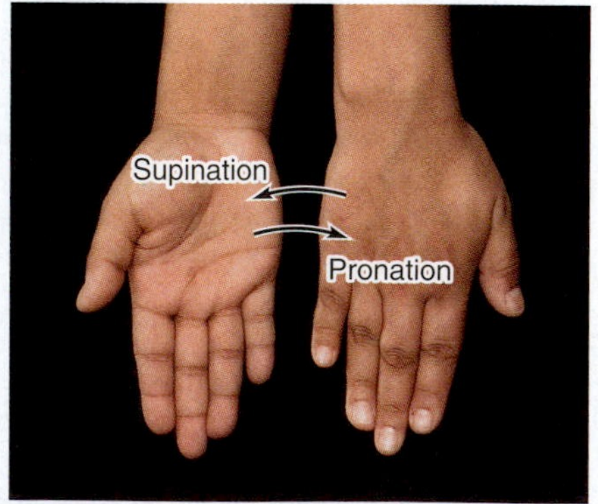

FIGURE 8.11 Joint movements illustrating rotation, circumduction, pronation, and supination.

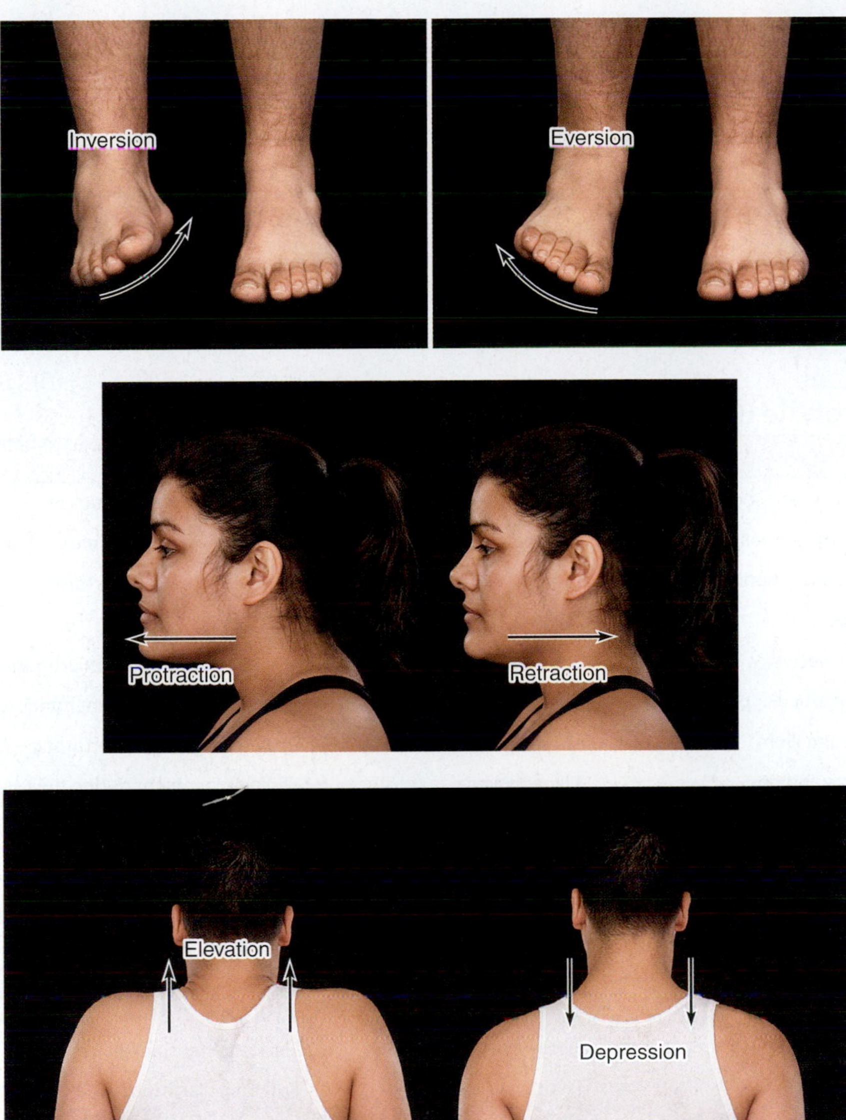

FIGURE 8.12 Joint movements illustrating eversion, inversion, retraction, protraction, elevation, and depression.

to describe an abnormal extension beyond the normal range of motion resulting in injury.

dorsiflexion (dor″si-flek′shun) Movement at the ankle that brings the foot closer to the shin (walking on heels).

plantar flexion (plan′tar flek′shun) Movement at the ankle that brings the foot farther from the shin (walking or standing on toes).

abduction (ab-duk′shun) Moving a part away from the midline (lifting the upper limb horizontally to form a right angle with the side of the body) or from the axial line of the limb (spreading the fingers or toes).

adduction (ah-duk′shun) Moving a part toward the midline (returning the upper limb from the horizontal position to the side of the body) or toward the axial line of the limb (closing in the fingers or toes).

rotation (ro-ta′shun) Moving a part around an axis (twisting the head from side to side). Medial (internal) rotation is movement toward the midline of the anterior surface, whereas lateral (external) rotation is movement in the opposite direction.

circumduction (ser″kum-duk′shun) Moving a part so that its end follows a circular path (moving the finger in a circular motion without moving the hand).

supination (soo″pĭ-na′shun) Turning the hand so the palm is upward or facing anteriorly.

pronation (pro-na′shun) Turning the hand so the palm is downward or facing posteriorly.

eversion (e-ver′zhun) Turning the foot so the plantar surface faces laterally.

inversion (in-ver′zhun) Turning the foot so the plantar surface faces medially.

protraction (pro-trak′shun) Moving a part forward (thrusting the head forward).

retraction (rĕ-trak′shun) Moving a part backward (pulling the head backward).

elevation (el-e-vā′shun) Raising a part (shrugging the shoulders).

depression (de-presh′un) Lowering a part (drooping the shoulders).

Description of movements of body parts is complex. At times, it will suffice to include the descriptive term of the movement followed by the part that is moving. For example, the deltoid muscle "abducts arm." Because the action of a muscle is at the insertion, the action of the biceps brachii muscle, for example, is sometimes described as "flexes forearm at the elbow." We have elected to use the more anatomically correct description of the change in geometry at the joint, "flexes elbow," to describe the action of the biceps brachii muscle. Table 8.2 lists information on several joints.

PRACTICE

11 Describe how movement occurs at a joint when a muscle contracts.

12 What terms describe movements at synovial joints?

TABLE 8.2 | Joints of the Body

Joint	Location	Type of Joint	Type of Movement
Skull	Cranial and facial bones	Suture, fibrous	Immovable, synarthrotic
Temporomandibular	Temporal bone, mandible	Modified hinge, synovial	Elevation, depression, protraction, retraction, diarthrotic
Atlanto-occipital	Atlas, occipital bone	Condylar, synovial	Flexion, extension, diarthrotic
Atlantoaxial	Atlas, axis	Pivot, synovial	Rotation
Intervertebral	Between vertebral bodies	Symphysis, cartilaginous	Slight movement, amphiarthrotic
Intervertebral	Between articular processes	Plane, synovial	Flexion, extension, slight rotation, diarthrotic
Sacroiliac	Sacrum and ilium	Plane, synovial	Sliding movement, diarthrotic
Vertebrocostal	Vertebrae and ribs	Plane, synovial	Sliding movement during breathing, diarthrotic
Sternoclavicular	Sternum and clavicle	Plane, synovial	Sliding movement when shrugging shoulders, diarthrotic
Sternocostal	Sternum and rib 1	Synchondrosis, cartilaginous	Immovable, synarthrotic
Sternocostal	Sternum and ribs 2–7	Plane, synovial	Sliding movement during breathing, diarthrotic
Acromioclavicular	Scapula and clavicle	Plane, synovial	Protraction, retraction, elevation, depression, rotation, diarthrotic
Shoulder (glenohumeral)	Humerus and scapula	Ball-and-socket, synovial	Flexion, extension, adduction, abduction, rotation, circumduction, diarthrotic
Elbow	Humerus and ulna	Hinge, synovial	Flexion, extension, diarthrotic
Proximal radioulnar	Radius and ulna	Pivot, synovial	Rotation, diarthrotic
Distal radioulnar	Radius and ulna	Pivot, synovial	Pronation, supination, diarthrotic
Wrist (radiocarpal)	Radius and carpals	Condylar, synovial	Flexion, extension, adduction, abduction, circumduction, diarthrotic
Intercarpal	Adjacent carpals	Plane, synovial	Sliding movement, adduction, abduction, flexion, extension, diarthrotic
Carpometacarpal	Carpal and metacarpal 1	Saddle, synovial	Flexion, extension, adduction, abduction, diarthrotic
Carpometacarpal	Carpals and metacarpals 2–5	Condylar, synovial	Flexion, extension, adduction, abduction, circumduction, diarthrotic
Metacarpophalangeal	Metacarpal and proximal phalanx	Condylar, synovial	Flexion, extension, adduction, abduction, circumduction, diarthrotic
Interphalangeal	Adjacent phalanges	Hinge, synovial	Flexion, extension, diarthrotic
Symphysis pubis	Pubic bones	Symphysis, cartilaginous	Slight movement, amphiarthrotic
Hip	Hip bone and femur	Ball-and-socket, synovial	Flexion, extension, adduction, abduction, rotation, circumduction, diarthrotic
Knee (tibiofemoral)	Femur and tibia	Modified hinge, synovial	Flexion, extension, slight rotation when flexed, diarthrotic
Knee (femoropatellar)	Femur and patella	Plane, synovial	Sliding movement, diarthrotic
Proximal tibiofibular	Tibia and fibula	Plane, synovial	Sliding movement, diarthrotic
Distal tibiofibular	Tibia and fibula	Syndesmosis, fibrous	Slight rotation during dorsiflexion, amphiarthrotic
Ankle (talocrural)	Talus, tibia, and fibula	Hinge, synovial	Dorsiflexion, plantar flexion, slight circumduction, diarthrotic
Intertarsal	Adjacent tarsals	Plane, synovial	Inversion, eversion, diarthrotic
Tarsometatarsal	Tarsals and metatarsals	Plane, synovial	Sliding movement, diarthrotic
Metatarsophalangeal	Metatarsal and proximal phalanx	Condylar, synovial	Flexion, extension, adduction, abduction, diarthrotic

8.6 EXAMPLES OF SYNOVIAL JOINTS

The shoulder, elbow, hip, and knee are large, freely movable joints. Although these joints have much in common, each has a unique structure that makes possible its specific function.

Shoulder Joint

The **shoulder joint** is a ball-and-socket joint that consists of the rounded head of the humerus and the shallow glenoid cavity of the scapula. The coracoid and acromion processes of the scapula protect these parts, and dense connective tissue and muscle hold them together.

The joint capsule of the shoulder is attached along the circumference of the glenoid cavity and the anatomical neck of the humerus. Although it completely envelops the joint, the capsule is very loose, and by itself is unable to keep the bones of the joint in close contact. However, muscles and tendons surround and reinforce the capsule, keeping together the articulating parts of the shoulder (fig. 8.13).

> The tendons of several muscles intimately blend with the fibrous layer of the shoulder joint capsule, forming the *rotator cuff,* which reinforces and supports the shoulder joint. Throwing a ball can create powerful decelerating forces that injure the rotator cuff.

The ligaments of the shoulder joint, some of which help prevent displacement of the articulating surfaces, include the following (fig. 8.14):

1. **Coracohumeral** (kor″ah-ko-hu′mer-al) **ligament.** This ligament is composed of a broad band of connective tissue that connects the coracoid process of the scapula to the greater tubercle of the humerus. It strengthens the superior portion of the joint capsule.
2. **Glenohumeral** (gle″no-hu′mer-al) **ligaments.** These include three bands of fibers that appear as thickenings in the ventral wall of the joint capsule. They extend from the edge of the glenoid cavity to the lesser tubercle and the anatomical neck of the humerus.
3. **Transverse humeral ligament.** This ligament consists of a narrow sheet of connective tissue fibers that runs between the lesser and the greater tubercles of the humerus. Together with the intertubercular groove of the humerus, the ligament forms a canal (retinaculum) through which the long head of the biceps brachii muscle passes.

The *glenoid labrum* (gle′noid la′brum) is composed of fibrocartilage. It is attached along the margin of the glenoid cavity and forms a rim with a thin, free edge that deepens the cavity.

Several bursae are associated with the shoulder joint. The major ones include the *subscapular bursa* between the joint capsule and the tendon of the subscapularis muscle, the *subdeltoid bursa* between the joint capsule and the deep surface

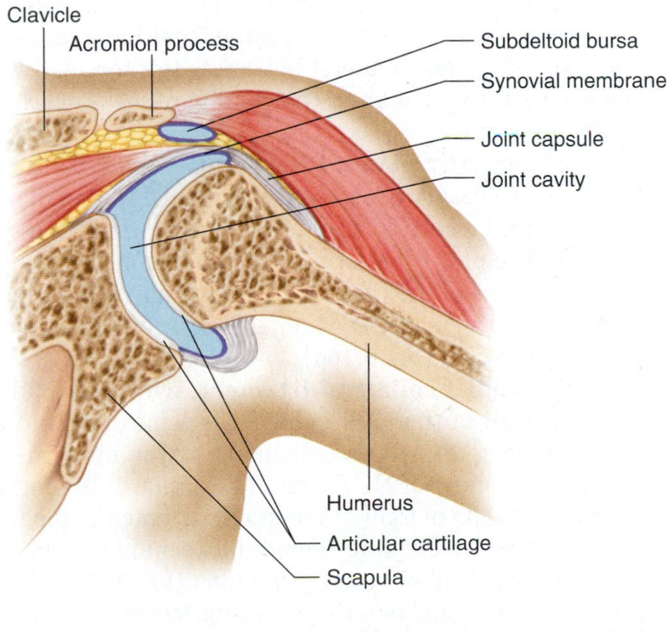

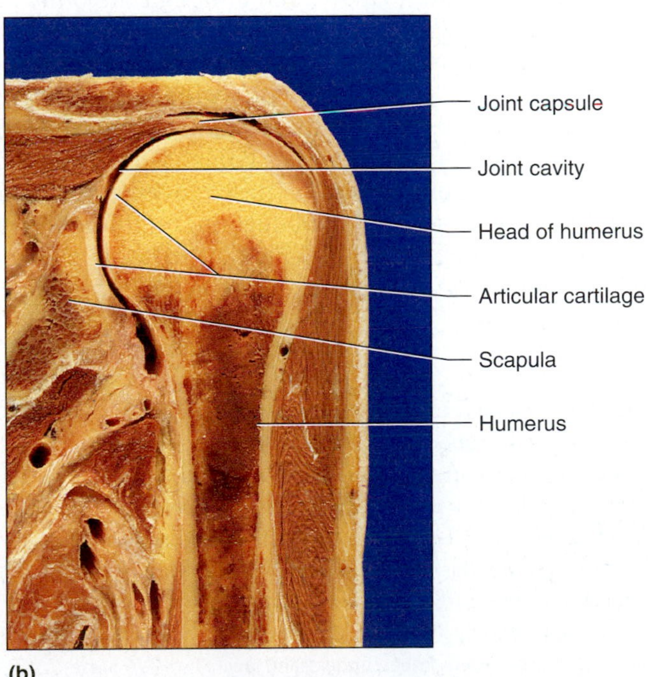

FIGURE 8.13 Shoulder joint. (*a*) The shoulder joint allows movements in all directions. A bursa is associated with this joint. (*b*) Photograph of the shoulder joint (frontal section).

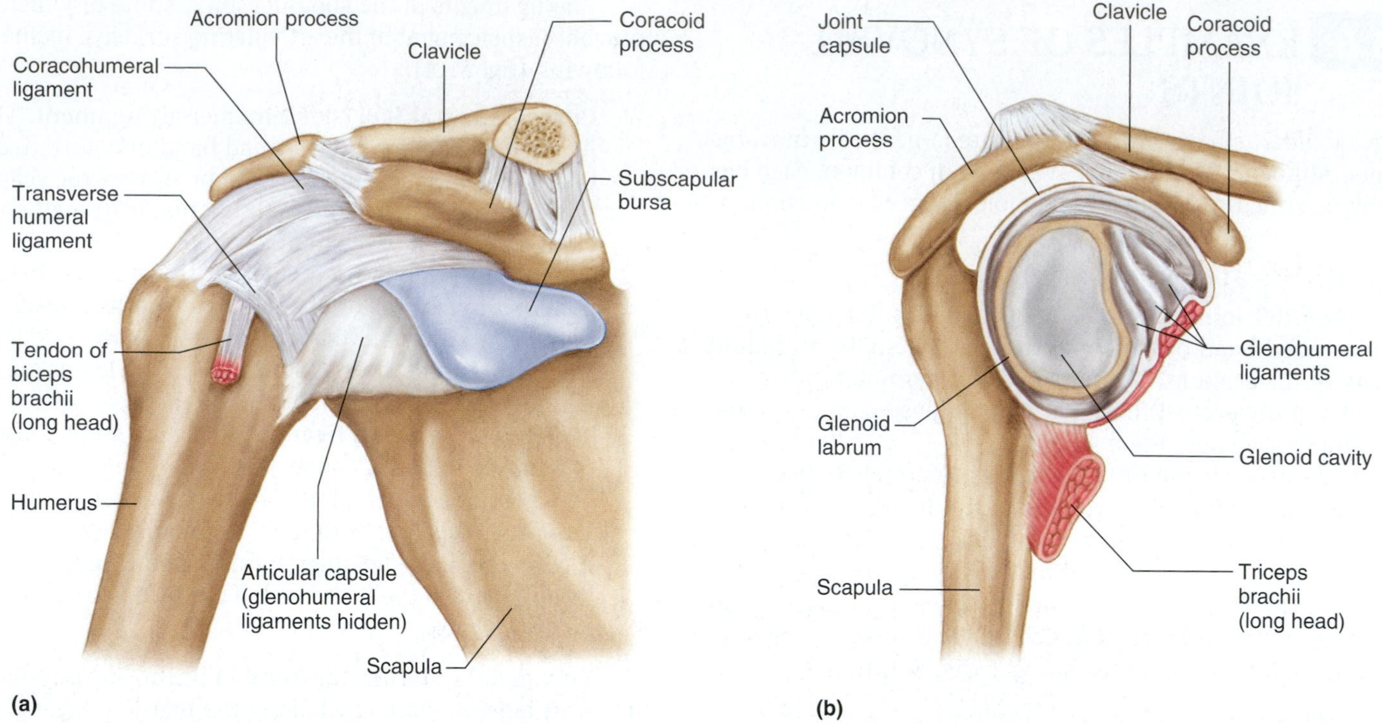

FIGURE 8.14 Ligaments associated with the shoulder joint. (*a*) Ligaments hold together the articulating surfaces of the shoulder. (*b*) The glenoid labrum is composed of fibrocartilage.

of the deltoid muscle, the *subacromial bursa* between the joint capsule and the undersurface of the acromion process of the scapula, and the *subcoracoid bursa* between the joint capsule and the coracoid process of the scapula. Of these, the subscapular bursa is usually continuous with the synovial cavity of the joint cavity, and although the others do not communicate with the joint cavity, they may be connected to each other (see figs. 8.13 and 8.14).

The shoulder joint is capable of a wide range of movement, due to the looseness of its attachments and the large articular surface of the humerus compared to the shallow depth of the glenoid cavity. These movements include flexion, extension, adduction, abduction, rotation, and circumduction. Motion occurring simultaneously in the joint formed between the scapula and the clavicle may also aid such movements.

The shoulder joint is somewhat weak because the bones are mainly held together by supporting muscles rather than by bony structures and strong ligaments. Consequently, the articulating surfaces may become displaced or dislocated easily. Such a *dislocation* can occur with a forceful impact during abduction, as when a person falls on an outstretched arm. This movement may press the head of the humerus against the lower part of the joint capsule where its wall is thin and poorly supported by ligaments. Dislocations most commonly affect joints of the shoulders, knees, fingers, and jaw.

Elbow Joint

The **elbow joint** is a complex structure that includes two articulations—a hinge joint between the trochlea of the humerus and the trochlear notch of the ulna and a plane joint between the capitulum of the humerus and a shallow depression (fovea) on the head of the radius. A joint capsule completely encloses and holds together these unions (fig. 8.15). Ulnar and radial collateral ligaments thicken the two joints, and fibers from a muscle (brachialis) in the arm reinforce its anterior surface.

The **ulnar collateral ligament,** a thick band of dense connective tissue, is located in the medial wall of the capsule. The anterior portion of this ligament connects the medial epicondyle of the humerus to the medial margin of the coronoid process of the ulna. Its posterior part is attached to the medial epicondyle of the humerus and to the olecranon process of the ulna (fig. 8.16*a*).

The **radial collateral ligament,** which strengthens the lateral wall of the joint capsule, is a fibrous band extending between the lateral epicondyle of the humerus and the *anular ligament of the radius.* The anular ligament, in turn, attaches to the margin of the trochlear notch of the ulna, and it encircles the head of the radius, keeping the head in contact with the radial notch of the ulna (fig. 8.16*b*). The elbow joint capsule encloses the resulting radioulnar joint so that its function is closely associated with the elbow.

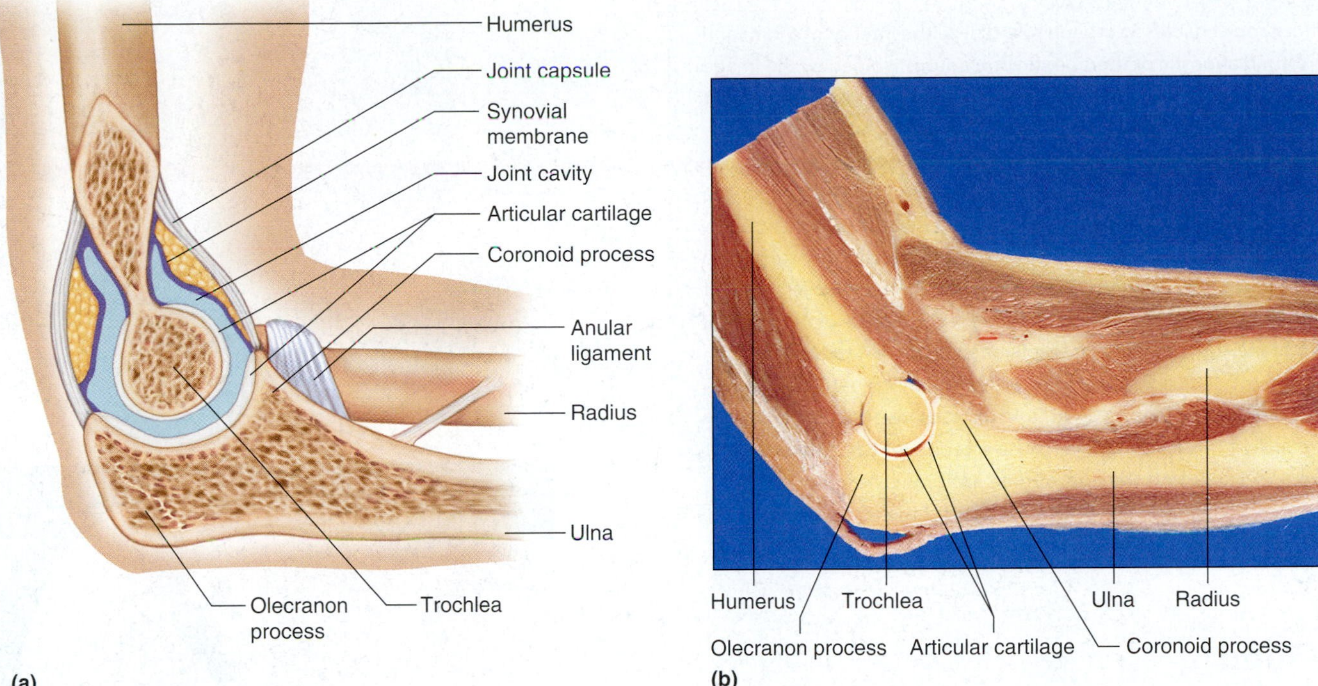

FIGURE 8.15 Elbow joint. (*a*) The elbow joint allows hinge movements, as well as pronation and supination of the hand. (*b*) Photograph of the elbow joint (sagittal section).

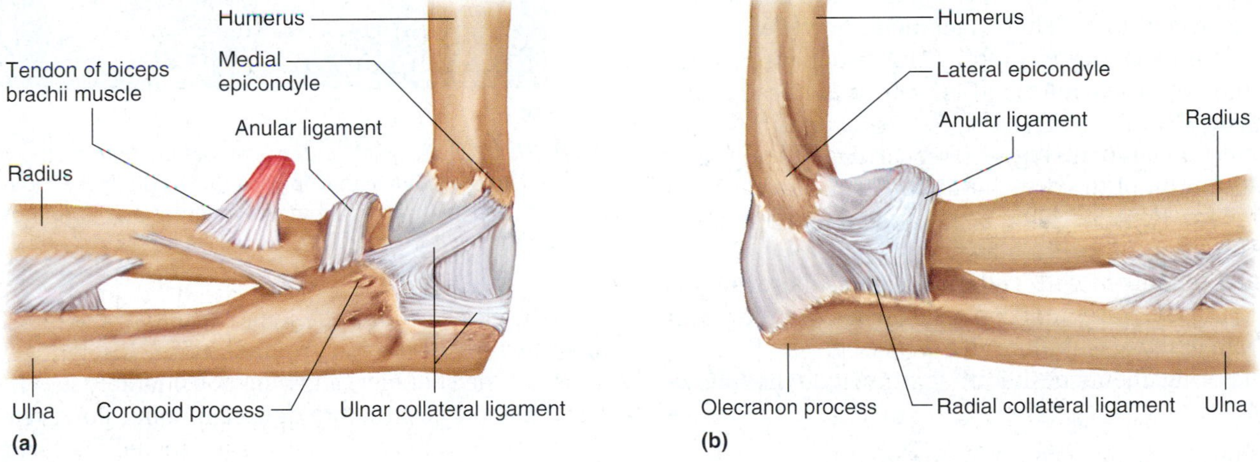

FIGURE 8.16 Ligaments associated with the elbow joint. (*a*) The ulnar collateral ligament, medial view, and (*b*) the radial collateral ligament strengthen the capsular wall of the elbow joint, lateral view.

The synovial membrane that forms the inner lining of the elbow capsule projects into the joint cavity between the radius and ulna and partially divides the joint into humerus–ulnar and humerus–radial portions. Also, varying amounts of adipose tissue form fatty pads between the synovial membrane and the fibrous layer of the joint capsule. These pads help protect nonarticular bony areas during joint movements.

The only movements that can occur at the elbow between the humerus and ulna are hinge-type movements—flexion and extension. The head of the radius, however, is free to rotate in the anular ligament. This movement allows pronation and supination of the hand.

PRACTICE  

13 Which parts help keep together the articulating surfaces of the shoulder joint?

14 What factors allow an especially wide range of motion in the shoulder?

15 Which structures form the hinge joint of the elbow?

16 Which parts of the elbow permit pronation and supination of the hand?

Arthroscopy enables a surgeon to visualize the interior of a joint and perform diagnostic or therapeutic procedures, guided by the image on a video screen. An arthroscope is a thin, tubular instrument about 25 cm long containing optical fibers that transmit an image. The surgeon inserts the device through a small incision in the joint capsule. It is far less invasive than conventional surgery. Many runners have undergone uncomplicated arthroscopy and raced just weeks later.

Arthroscopy is combined with a genetic technique called the polymerase chain reaction (PCR) to rapidly diagnose infection. Guided by an arthroscope, the surgeon samples a small piece of the synovial membrane. PCR detects and amplifies specific DNA sequences, such as those of bacteria. For example, the technique can rapidly diagnose Lyme disease by detecting DNA from the causative bacterium *Borrelia burgdorferi*. This is valuable because a variety of bacteria can infect joints, and choosing the appropriate antibiotic, based on knowing the type of bacterium, is crucial for fast and complete recovery.

Hip Joint

The **hip joint** is a ball-and-socket joint that consists of the head of the femur and the cup-shaped acetabulum of the hip bone. A ligament (ligamentum capitis) attaches to a pit (fovea capitis) on the head of the femur and to connective tissue in the acetabulum. This attachment, however, seems to have little importance in holding the articulating bones together, but rather carries blood vessels to the head of the femur (fig. 8.17).

A horseshoe-shaped ring of fibrocartilage (acetabular labrum) at the rim of the acetabulum deepens the cavity of the acetabulum. It encloses the head of the femur and helps hold it securely in place. In addition, a heavy, cylindrical joint capsule reinforced with still other ligaments surrounds the articulating structures and connects the neck of the femur to the margin of the acetabulum (fig. 8.18).

The major ligaments of the hip joint include the following (fig. 8.19):

1. **Iliofemoral** (il″e-o-fem′o-ral) **ligament.** This ligament consists of a Y-shaped band of strong fibers that connects the anterior inferior iliac spine of the hip bone to a bony line (intertrochanteric line) extending between the greater and lesser trochanters of the femur. The iliofemoral ligament is the strongest ligament in the body.
2. **Pubofemoral** (pu″bo-fem′o-ral) **ligament.** The pubofemoral ligament extends between the superior portion of the pubis and the iliofemoral ligament. Its fibers also blend with the fibers of the joint capsule.
3. **Ischiofemoral** (is″ke-o-fem′o-ral) **ligament.** This ligament consists of a band of strong fibers that originates on the ischium just posterior to the acetabulum and blends with the fibers of the joint capsule.

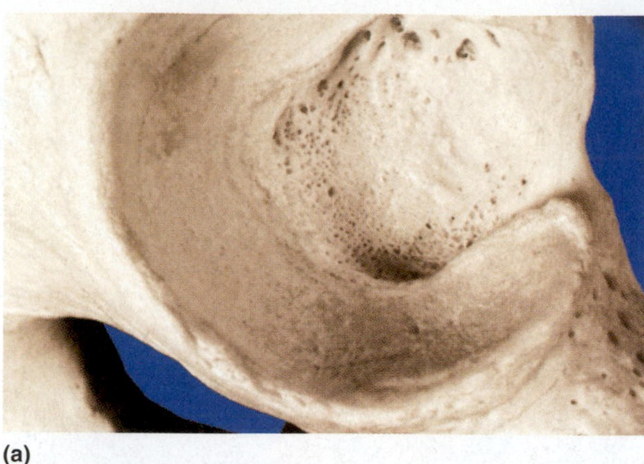

(a)

(b)

FIGURE 8.17 Hip joint. (*a*) The acetabulum provides the socket for the head of the femur in the hip joint. (*b*) The pit (fovea capitis) in the femur's head marks attachment of a ligament that surrounds blood vessels and nerves.

Muscles surround the joint capsule of the hip. The articulating parts of the hip are held more closely together than those of the shoulder, allowing considerably less freedom of movement. The structure of the hip joint, however, still permits a wide variety of movements, including flexion, extension, adduction, abduction, rotation, and circumduction. The hip is one of the joints most frequently replaced (Clinical Application 8.1).

Knee Joint

The **knee joint** is the largest and most complex of the synovial joints. It consists of the medial and lateral condyles at the distal end of the femur and the medial and lateral condyles at the proximal end of the tibia. In addition, the femur articulates anteriorly with the patella. Although the knee functions largely as a modified hinge joint (allowing flexion and extension), the articulations between the femur and tibia are condylar (allowing some rotation when the knee is flexed), and the joint between the femur and patella is a plane joint.

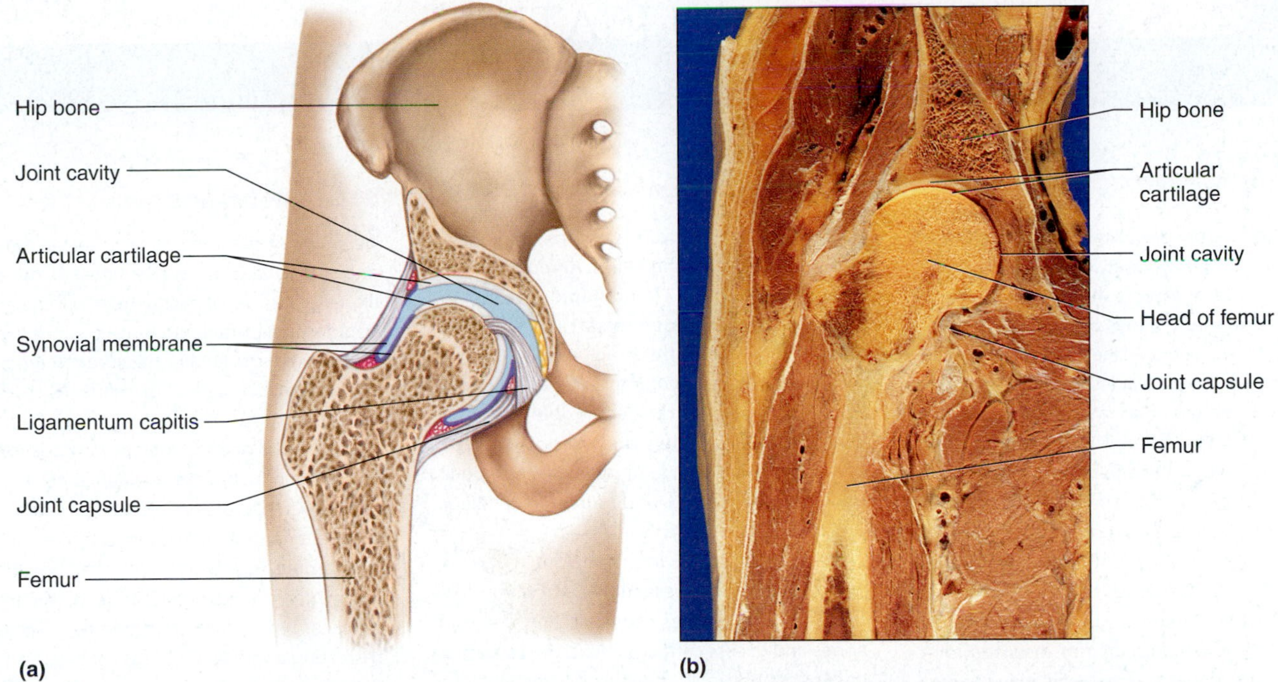

FIGURE 8.18 Hip joint. (*a*) A ring of cartilage in the acetabulum and a ligament-reinforced joint capsule hold together the hip joint. (*b*) Photograph of the hip joint (frontal section).

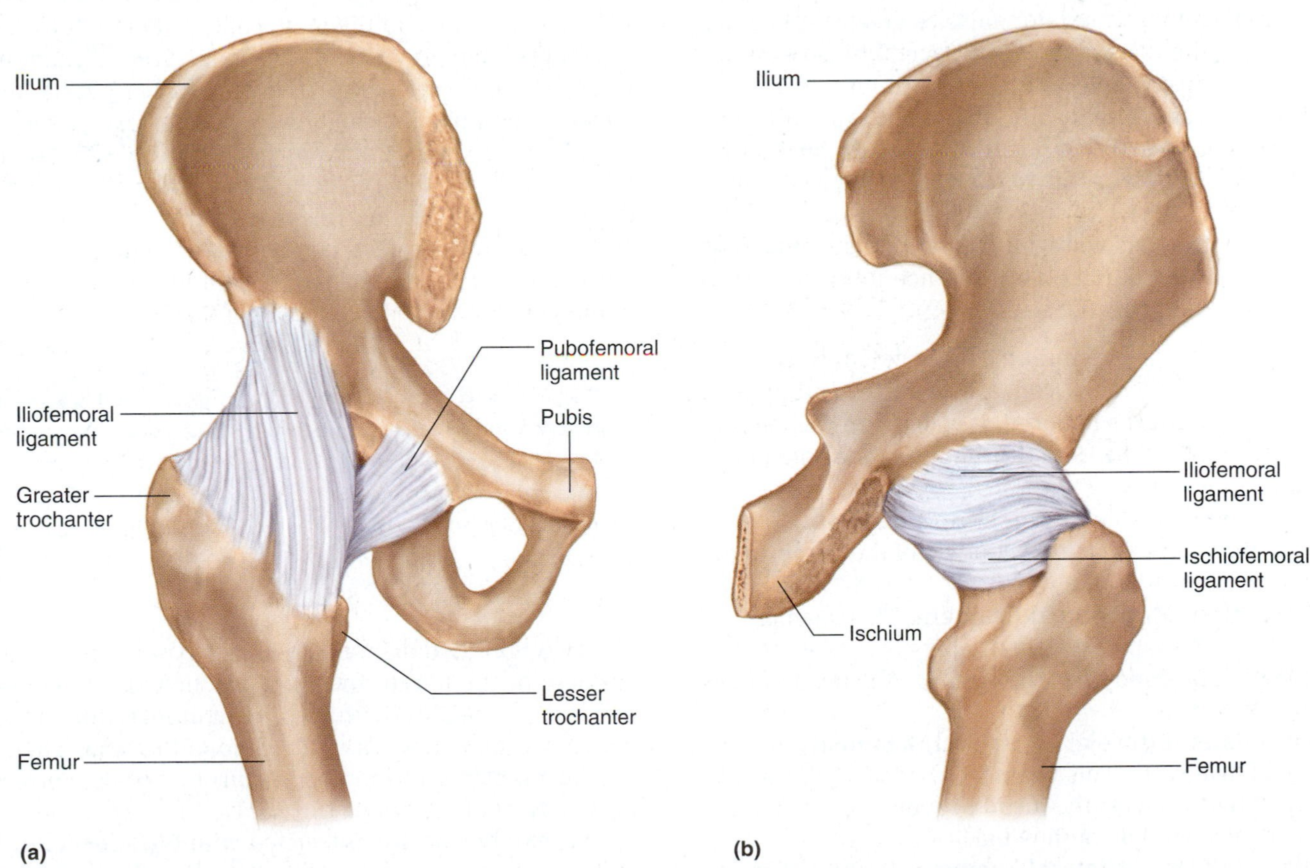

FIGURE 8.19 The major ligaments of the right hip joint. (*a*) Anterior view. (*b*) Posterior view.

8.1 CLINICAL APPLICATION

Replacing Joints

Surgeons use several synthetic materials to replace joints severely damaged by arthritis or injury. Metals such as cobalt, chromium, and titanium alloys are used to replace larger joints, whereas silicone polymers are more commonly used to replace smaller joints. Such artificial joints must be durable yet not provoke immune system rejection. They must also allow normal healing to occur and not move surrounding structures out of their normal positions. Ceramic materials are used in about 5% of hip replacements, but recipients sometimes complain of the joint squeaking.

Before the advent of joint replacements, surgeons removed damaged or diseased joint surfaces, hoping that scar tissue filling in the area would restore mobility. This type of surgery was rarely successful. In the 1950s, Alfred Swanson, an army surgeon in Grand Rapids, Michigan, invented the first joint implants using silicone polymers. By 1969, after much refinement, the first silicone-based joint implants became available. These devices provided flexible hinges for joints of the toes, fingers, and wrists. Since then, more than two dozen joint replacement models have been developed, and more than a million people have them, mostly in the hip.

A surgeon inserts a joint implant in a procedure called implant resection arthroplasty. The surgeon first removes the surface of the joint bones and excess cartilage. Next, the centers of the tips of abutting bones are hollowed out, and the stems of the implant are inserted here. The hinge part of the implant lies between the bones, aligning them yet allowing them to bend, as they would at a natural joint. Bone cement fixes the implant in place. Finally, the surgeon repairs the tendons, muscles, and ligaments. As the site of the implant heals, the patient must exercise the joint. A year of physical therapy may be necessary to fully benefit from replacement joints.

Newer joint replacements use materials that resemble natural body chemicals. Hip implants, for example, may bear a coat of hydroxyapatite, which interacts with natural bone. Instead of filling in spaces with bone cement, some investigators are testing a variety of porous coatings that allow bone tissue to grow into the implant area. ■

The *joint capsule* of the knee is relatively thin, but ligaments and the tendons of several muscles greatly strengthen it. For example, the fused tendons of several muscles in the thigh cover the capsule anteriorly. Fibers from these tendons descend to the patella, partially enclose it, and continue downward to the tibia. The capsule attaches to the margins of the femoral and tibial condyles as well as between these condyles (fig. 8.20).

The ligaments associated with the joint capsule that help keep the articulating surfaces of the knee joint in contact include the following (fig. 8.21):

1. **Patellar** (pah-tel′ar) **ligament.** This ligament is a continuation of a tendon from a large muscle group in the thigh (quadriceps femoris). It consists of a strong, flat band that extends from the margin of the patella to the tibial tuberosity.
2. **Oblique popliteal** (ŏ′blēk pop-lit′e-al) **ligament.** This ligament connects the lateral condyle of the femur to the margin of the head of the tibia.
3. **Arcuate** (ar′ku-āt) **popliteal ligament.** This ligament appears as a Y-shaped system of fibers that extends from the lateral condyle of the femur to the head of the fibula.
4. **Tibial collateral** (tib′e-al kŏ-lat′er-al) **ligament** (medial collateral ligament). This ligament is a broad, flat band of tissue that connects the medial condyle of the femur to the medial condyle of the tibia.
5. **Fibular** (fib′u-lar) **collateral ligament** (lateral collateral ligament). This ligament consists of a strong, round cord located between the lateral condyle of the femur and the head of the fibula.

In addition to the ligaments that strengthen the joint capsule, two ligaments in the joint, called **cruciate** (kroo′she-āt) **ligaments,** help prevent displacement of the articulating surfaces. These strong bands of fibrous tissue stretch upward between the tibia and the femur, crossing each other on the way. They are named according to their positions of attachment to the tibia. For example, the *anterior cruciate ligament* originates from the anterior intercondylar area of the tibia and extends to the lateral condyle of the femur. The *posterior cruciate ligament* connects the posterior intercondylar area of the tibia to the medial condyle of the femur.

The young soccer player, running at full speed, suddenly switches direction and is literally stopped in her tracks by a popping sound followed by a searing pain in her knee. Two hours after she veered toward the ball, her knee is swollen and painful, due to bleeding into the joint. She has torn the anterior cruciate ligament, a serious knee injury.

Two fibrocartilaginous *menisci* separate the articulating surfaces of the femur and tibia and help align them. Each meniscus is roughly C-shaped, with a thick rim and a thinner center, and attaches to the head of the tibia. The medial and lateral menisci form depressions that fit the corresponding condyles of the femur (fig. 8.21).

Several bursae are associated with the knee joint. These include a large extension of the knee joint cavity called the *suprapatellar bursa,* located between the anterior surface of the distal end of the femur and the muscle group (quadriceps femoris) above it; a large *prepatellar bursa* between

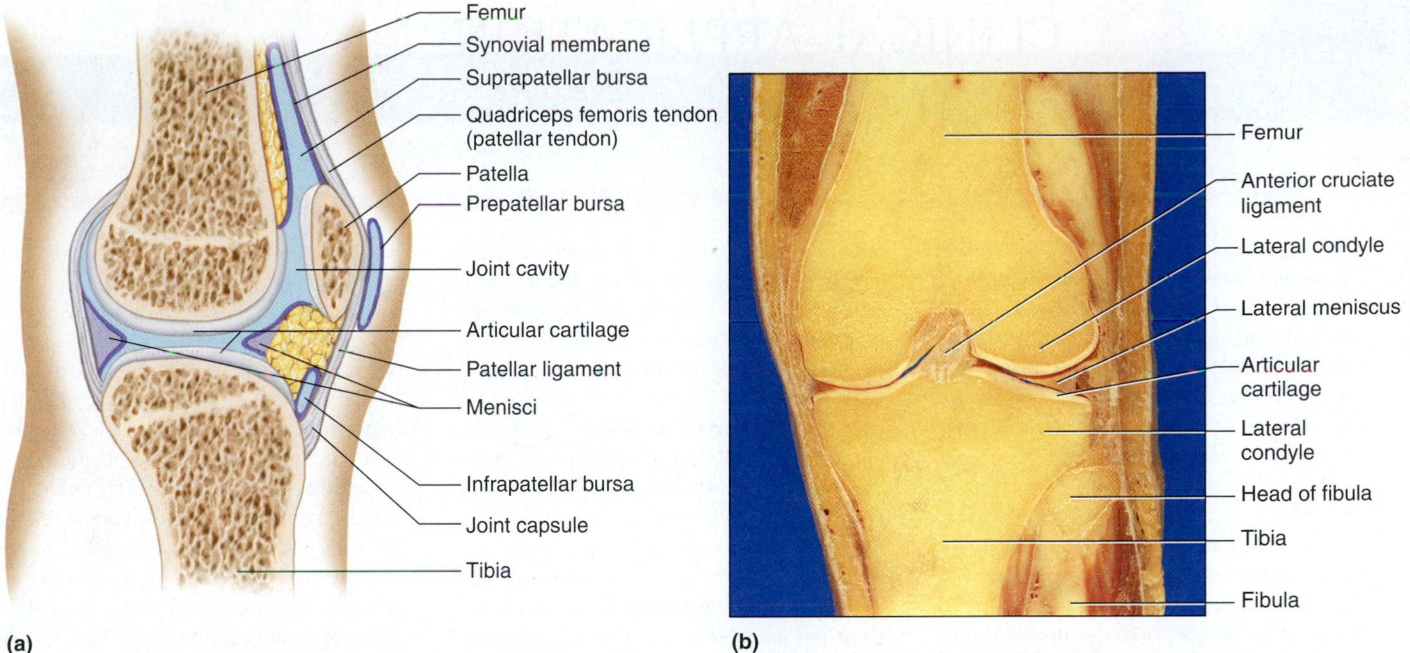

FIGURE 8.20 Knee joint. (*a*) The knee joint is the most complex of the synovial joints (sagittal section). (*b*) Photograph of the left knee joint (frontal section).

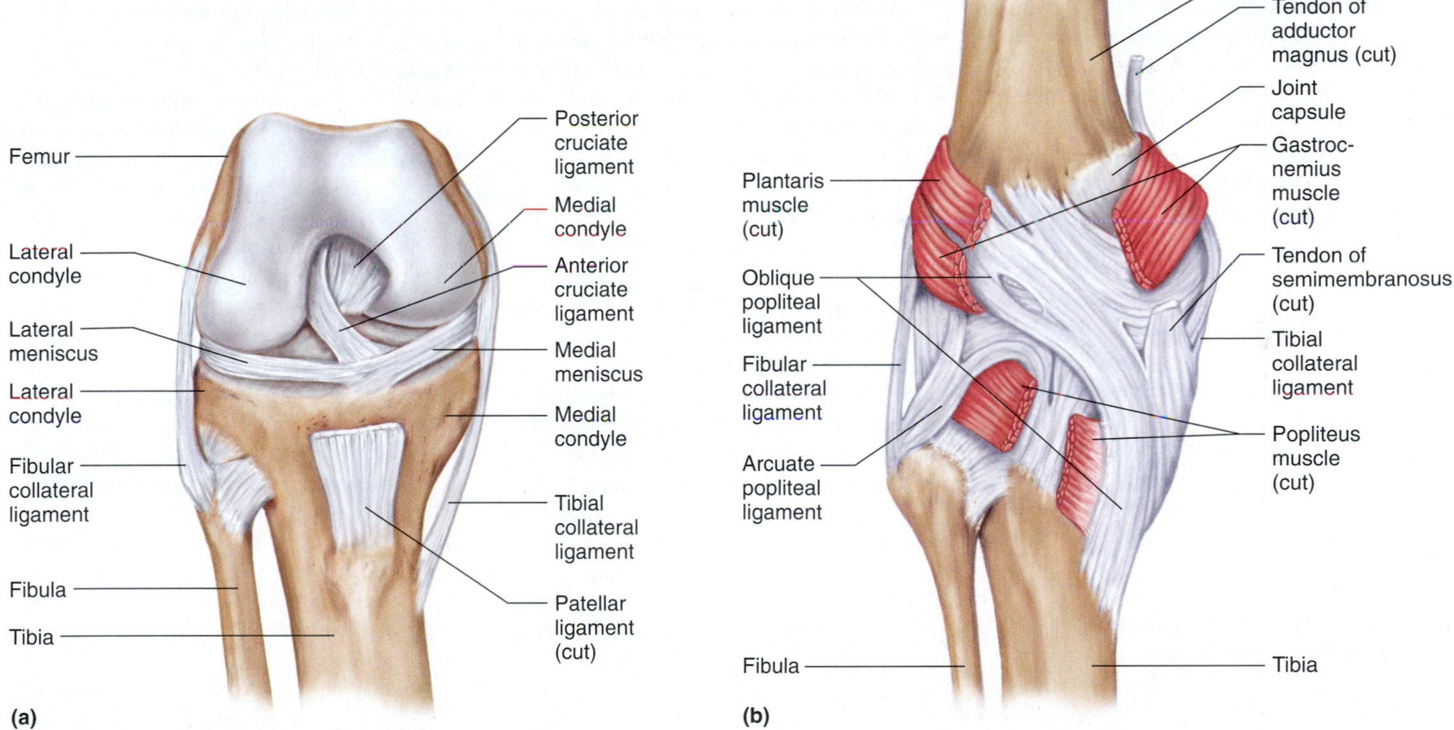

FIGURE 8.21 Ligaments within the knee joint help to strengthen it. (*a*) Anterior view of right bent knee (patella removed). (*b*) Posterior view of left knee.

the patella and the skin; and a smaller *infrapatellar bursa* between the proximal end of the tibia and the patellar ligament (see fig. 8.8).

As with a hinge joint, the basic structure of the knee joint permits flexion and extension. However, when the knee is flexed, rotation is also possible. Clinical Application 8.2 discusses some common joint disorders.

PRACTICE

17 Which structures help keep the articulating surfaces of the hip together?

18 What types of movement does the structure of the hip permit?

19 What types of joints are in the knee?

20 Which parts help hold together the articulating surfaces of the knee?

8.2 CLINICAL APPLICATION

Joint Disorders

Joints have a tough job. They must support weight, provide a great variety of body movements, and are used frequently. In addition to this normal wear and tear, these structures are sometimes subjected to injury from trauma, overuse, infection, a misplaced immune system attack, or degeneration. Here is a look at some common joint problems.

Sprains

Sprains result from overstretching or tearing the connective tissues, including cartilage, ligaments, and tendons, associated with a joint, but they do not dislocate the articular bones. Usually forceful wrenching or twisting sprains the wrist or ankles. For example, inverting an ankle too far can sprain it by stretching the ligaments on its lateral surface. Severe injuries may pull these tissues loose from their attachments.

A sprained joint is painful and swollen, restricting movement. Immediate treatment of a sprain is rest; more serious cases require medical attention. However, immobilization of a joint, even for a brief period, causes bone resorption and weakens ligaments. Consequently, exercise may help strengthen the joint.

Bursitis

Overuse of a joint or stress on a bursa may cause *bursitis*, an inflammation of a bursa. The bursa between the heel bone (calcaneus) and the Achilles tendon may become inflamed as a result of a sudden increase in physical activity using the feet. Similarly, a form of bursitis called tennis elbow affects the bursa between the olecranon process and the skin. Bursitis is treated with rest. Medical attention may be necessary.

Arthritis

Arthritis is a disease that causes inflamed, swollen, and painful joints. More than a hundred different types of arthritis affect 50 million people in the United States. Arthritis can also be part of other syndromes (table 8A). The most common types of arthritis are rheumatoid arthritis (RA), osteoarthritis, and Lyme arthritis.

Rheumatoid Arthritis (RA)

Rheumatoid arthritis, an autoimmune disorder (a condition in which the immune system attacks the body's healthy tissues), is painful and debilitating. The synovial membrane of a joint becomes inflamed and thickens, forming a mass called a pannus. Then, the articular cartilage is damaged, and fibrous tissue infiltrates, interfering with joint movements. In time, the joint may ossify, fusing the articulating bones (bony ankylosis). Joints severely damaged by RA may be surgically replaced.

RA may affect many joints or only a few. It is usually a systemic illness, accompanied by fatigue, muscular atrophy, anemia, and osteoporosis, as well as changes in the skin, eyes, lungs, blood vessels, and heart. RA usually affects adults, but there is a juvenile form.

Osteoarthritis

Osteoarthritis, a degenerative disorder, is the most common type of arthritis (fig. 8A). It usually occurs with aging, but an inherited form may appear as early as one's thirties. A person may first become aware of osteoarthritis when a blow to the affected joint produces pain much more intense than normal. Gradually, the area of the affected joint deforms. Arthritic fingers take on a gnarled appearance, or a knee may bulge.

In osteoarthritis, articular cartilage softens and disintegrates gradually, roughening the articular surfaces. Joints become painful, with restricted movement. For example, arthritic fingers may lock into place while a person is playing the guitar or tying a shoelace. Osteoarthritis most often affects joints used the most over a lifetime, such as those of the fingers, hips, knees, and the lower parts of the vertebral column.

Nonsteroidal anti-inflammatory drugs (NSAIDs) have been used for many years to control osteoarthritis symptoms. NSAIDs called COX-2 inhibitors relieve inflammation without the gastrointestinal side effects of older drugs, but they are prescribed only to people who do not have risk factors for cardiovascular disease, to which some of these drugs are linked. The COX-2 inhibitors are generally not more effective at relieving joint pain than other NSAIDs. Exercise can keep osteoarthritic joints more flexible.

Lyme Arthritis

Lyme disease, a bacterial infection passed in a tick bite, causes intermittent arthritis of several joints, usually weeks after the initial symptoms of rash, fatigue, and flulike aches and pains. Lyme arthritis was first observed in Lyme, Connecticut, where an astute woman kept a journal after noticing that many of her young neighbors had what appeared to be the very rare juvenile form of rheumatoid arthritis. Her observations led Allen Steere, a Yale University rheumatologist, to trace the illness to a tick-borne bacterial infection. Antibiotic treatment beginning as soon as the early symptoms of Lyme disease are recognized can prevent development of the associated arthritis.

Other types of bacteria that cause arthritis include common *Staphylococcus* and *Streptococcus* species, *Neisseria gonorrhoeae* (which causes the sexually transmitted disease gonorrhea), and *Mycobacterium* (which causes tuberculosis). Arthritis may also be associated with AIDS, because the immune system breakdown raises the risk of infection by bacteria that can cause arthritis. ■

Tearing or displacing a meniscus is a common knee injury, usually resulting from forcefully twisting the knee when the leg is flexed (fig. 8.22). The meniscus is composed of fibrocartilage, so this type of injury heals slowly. Also, a torn and displaced portion of cartilage jammed between the articulating surfaces impedes movement of the joint. Following such a knee injury, the synovial membrane may become inflamed (acute synovitis) and secrete excess fluid, distending the joint capsule so that the knee swells above and on the sides of the patella.

8.7 LIFE-SPAN CHANGES

Joint stiffness is an early sign of aging. By the fourth decade, a person may notice that the first steps each morning become difficult. Changes in collagen structure lie behind the increasing stiffness (fig. 8.23). Range of motion may diminish. However, joints age slowly, and exercise can lessen or forestall stiffness.

TABLE 8A | **Different Types of Arthritis**

Type	Incidence in the United States	Age of Onset	Symptoms
Some More-Common Forms of Arthritis			
Osteoarthritis	20.7 million		
Rheumatoid arthritis	2.1 million		
Spondyloarthropathies	2.5 million		
Some Less-Common Forms of Arthritis			
Gout	1.6 million (85% male)	>40	Sudden onset of extreme pain and swelling of a large joint
Juvenile rheumatoid arthritis	100,000	<18	Joint stiffness, often in knee
Scleroderma	300,000	30–50	Skin hardens and thickens
Systemic lupus erythematosus	500,000 (>90% female)	teens–50s	Fever, weakness, upper body rash, joint pain
Kawasaki disease	Hundreds of cases in local outbreaks	6 months–11 years	Fever, joint pain, red rash on palms and soles, heart complications
Strep A infection	100,000	any age	Confusion, body aches, shock, low blood pressure, dizziness, arthritis, pneumonia
Lyme disease	15,000	any age	Arthritis, malaise, neurologic and cardiac manifestations

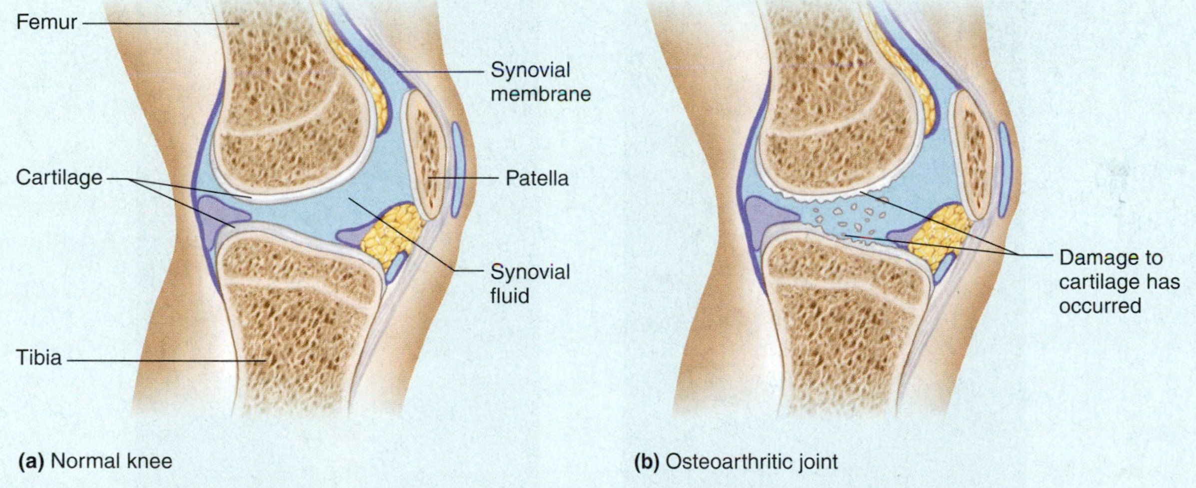

FIGURE 8A In osteoarthritis, an inherited defect in collagen, trauma, or prolonged wear and tear destroys joints.

The fibrous joints are the first to change, as the four types of fontanels close the bony plates of the skull at two, three, twelve, and eighteen to twenty-four months of age. Other fibrous joints may accumulate bone matrix over time, bringing bones closer together, even fusing them. Fibrous joints strengthen over a lifetime.

Synchondroses that connect epiphyses to diaphyses in long bones disappear as the skeleton grows and develops. Another synchondrosis is the joint that links the first rib to the manubrium (sternum). As water content decreases and deposition of calcium salts increases, this cartilage stiffens. Ligaments lose their elasticity as the collagen fibers become more tightly cross-linked. Breathing may become labored, and movement more restrained.

Aging also affects symphysis joints, which consist of a pad of fibrocartilage sandwiched between thin layers of hyaline

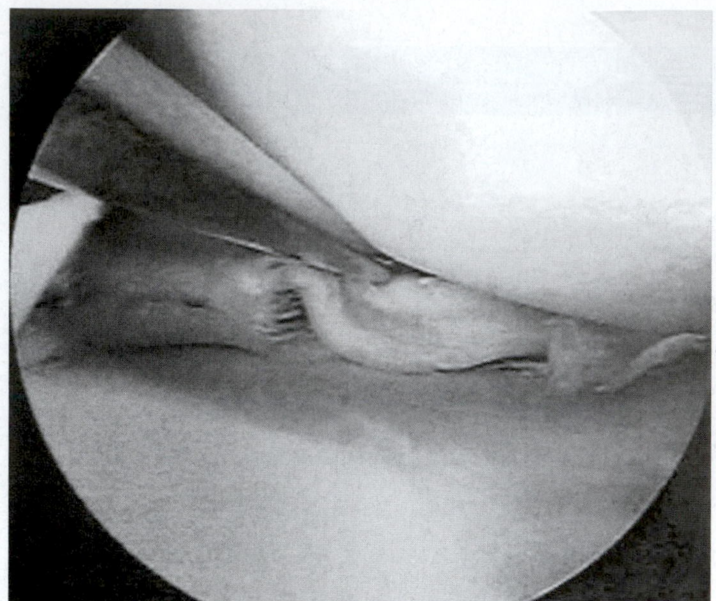

FIGURE 8.22 Arthroscopic view of a torn meniscus in the knee and arthroscopic scissors. Fibrocartilage does not heal well, so in many cases of torn meniscus the only treatment option is to cut out the damaged portion.

cartilage. In the intervertebral discs, less water diminishes the flexibility of the vertebral column and impairs the ability of the soft centers of the discs to absorb shocks. The discs may even collapse on themselves slightly, contributing to the loss of height in the elderly. The stiffening spine gradually restricts the range of motion.

Loss of function in synovial joints begins in the third decade of life, but progresses slowly. Fewer capillaries serving the synovial membrane slows the circulation of synovial fluid, and the membrane may become infiltrated with fibrous material and cartilage. As a result, the joint may lose elasticity, stiffening. More collagen cross-links shorten and stiffen ligaments, affecting the range of motion. This may, in turn, upset balance and retard the ability to respond in a protective way to falling, which may explain why older people are more likely to be injured in a fall than younger individuals.

Using joints, through activity and exercise, can keep them functional longer. Disuse hampers the blood supply to joints, which hastens stiffening. Paradoxically, this can keep people from exercising, when this is exactly what they should be doing.

PRACTICE

21 Which type of joint is the first to show signs of aging?

22 Describe the loss of function in synovial joints as a progressive process.

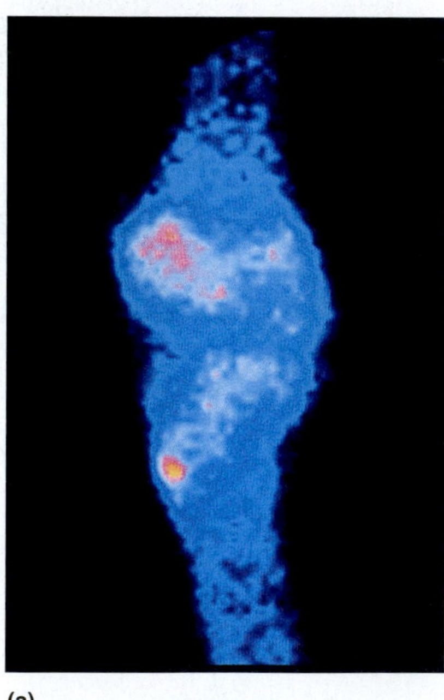

(a)

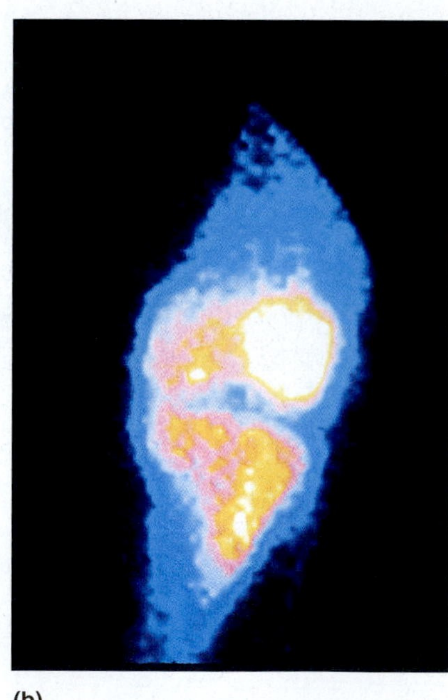

(b)

FIGURE 8.23 Nuclear scan of (*a*) a healthy knee and (*b*) an arthritic knee. The different colors in (*b*) indicate changes within the tissues associated with degeneration.

CHAPTER SUMMARY

8.1 INTRODUCTION (PAGE 261)

A joint forms wherever two or more bones meet. Joints bind parts of the skeleton, allow for bone growth, permit skeletal parts to change shape during childbirth, and enable movement in response to skeletal muscle contractions.

8.2 CLASSIFICATION OF JOINTS (PAGE 261)

Joints are classified according to the type of tissue that binds the bones together.

1. Fibrous joints
   a. Bones at fibrous joints are tightly fastened to each other by a layer of dense connective tissue with many collagenous fibers.
   b. There are three types of fibrous joints.
      (1) A syndesmosis has bones bound by long connective tissue fibers.
      (2) A suture is where flat bones are united by a thin layer of connective tissue and are interlocked by a set of bony processes.
      (3) A gomphosis is formed by the union of a cone-shaped bony process with a bony socket.
2. Cartilaginous joints
   a. A layer of cartilage holds together bones of cartilaginous joints.
   b. There are two types of cartilaginous joints.
      (1) A synchondrosis occurs where bones are united by hyaline cartilage that may disappear as a result of growth.
      (2) A symphysis occurs where articular surfaces of the bones are covered by hyaline cartilage and the cartilage is attached to a pad of fibrocartilage.
3. Synovial joints
   a. Synovial joints have a more complex structure than other types of joints.
   b. These joints include articular cartilage, a joint capsule, and a synovial membrane.

8.3 GENERAL STRUCTURE OF A SYNOVIAL JOINT (PAGE 263)

1. Articular cartilage covers articular ends of bones in a synovial joint.
2. A joint capsule strengthened by ligaments holds bones together.
3. A synovial membrane that secretes synovial fluid lines the inner layer of a joint capsule.
4. Synovial fluid moistens, provides nutrients, and lubricates the articular surfaces.
5. Menisci divide some synovial joints into compartments.
6. Some synovial joints have fluid-filled bursae.
   a. Bursae are usually located between the skin and underlying bony prominences.
   b. Bursae cushion and aid movement of tendons over bony parts.
   c. Bursae are named according to their locations.

8.4 TYPES OF SYNOVIAL JOINTS (PAGE 267)

1. Ball-and-socket joints
   a. In a ball-and-socket joint, the globular head of a bone fits into the cup-shaped cavity of another bone.
   b. These joints permit a wide variety of movements.
   c. The hip and shoulder are ball-and-socket joints.
2. Condylar joints
   a. A condylar joint consists of an ovoid condyle of one bone fitting into an elliptical cavity of another bone.
   b. This joint permits a variety of movements.
   c. The joints between the metacarpals and phalanges are condylar.
3. Plane joints
   a. Articular surfaces of plane joints are nearly flat.
   b. These joints permit the articular surfaces to slide back and forth.
   c. Most of the joints of the wrist and ankle are plane joints.
4. Hinge joints
   a. In a hinge joint, the convex surface of one bone fits into the concave surface of another bone.
   b. This joint permits movement in one plane only.
   c. The elbow and the joints of the phalanges are the hinge type.
5. Pivot joints
   a. In a pivot joint, a cylindrical surface of one bone rotates within a ring of bone and ligament.
   b. This joint permits rotational movement.
   c. The articulation between the proximal ends of the radius and the ulna is a pivot joint.
6. Saddle joints
   a. A saddle joint forms between bones that have complementary surfaces with both concave and convex regions.
   b. This joint permits a variety of movements.
   c. The articulation between the carpal and metacarpal of the thumb is a saddle joint.

8.5 TYPES OF JOINT MOVEMENTS (PAGE 267)

1. Muscles acting at synovial joints produce movements in different directions and in different planes.
2. Joint movements include flexion, extension, hyperextension, dorsiflexion, plantar flexion, abduction, adduction, rotation, circumduction, supination, pronation, eversion, inversion, protraction, retraction, elevation, and depression.

8.6 EXAMPLES OF SYNOVIAL JOINTS (PAGE 271)

1. Shoulder joint
   a. The shoulder joint is a ball-and-socket joint that consists of the head of the humerus and the glenoid cavity of the scapula.

b. A cylindrical joint capsule envelops the joint.
   (1) The capsule is loose and by itself cannot keep the articular surfaces together.
   (2) It is reinforced by surrounding muscles and tendons.
c. Several ligaments help prevent displacement of the bones.
d. Several bursae are associated with the shoulder joint.
e. Its parts are loosely attached, so the shoulder joint permits a wide range of movements.

2. Elbow joint
   a. The elbow has a hinge joint between the humerus and the ulna and a plane joint between the humerus and the radius.
   b. Collateral ligaments reinforce the joint capsule.
   c. A synovial membrane partially divides the joint cavity into two portions.
   d. The joint between the humerus and the ulna permits flexion and extension only.
3. Hip joint
   a. The hip joint is a ball-and-socket joint between the femur and the hip bone.
   b. A ring of fibrocartilage deepens the cavity of the acetabulum.
   c. The articular surfaces are held together by a heavy joint capsule reinforced by ligaments.
   d. The hip joint permits a wide variety of movements.
4. Knee joint
   a. The knee joint includes two condylar joints between the femur and the tibia and a plane joint between the femur and the patella.
   b. Ligaments and tendons strengthen the thin joint capsule.
   c. Several ligaments, some in the joint capsule, bind articular surfaces.
   d. Two menisci separate the articulating surfaces of the femur and the tibia.
   e. Several bursae are associated with the knee joint.
   f. The knee joint permits flexion and extension; when the lower limb is flexed at the knee, some rotation is possible.

8.7 LIFE-SPAN CHANGES (PAGE 278)

1. Joint stiffness is often the earliest sign of aging.
   a. Collagen changes cause the feeling of stiffness.
   b. Regular exercise can lessen the effects.
2. Fibrous joints are the first to begin to change and strengthen over a lifetime.
3. Synchondroses of the long bones disappear with growth and development.
4. Changes in symphysis joints of the vertebral column diminish flexibility and decrease height.
5. Over time, synovial joints lose elasticity.

CHAPTER ASSESSMENTS

8.1 Introduction

1 Functions of joints include ________________. (p. 261)
   a. binding skeletal parts
   b. allowing for bone growth
   c. permitting the skeleton to change shape during childbirth
   d. enabling movement in response to skeletal muscle contractions
   e. all of the above.

8.2 Classification of Joints

2 Describe how joints are classified. (p. 261)

3 Compare the structure of a fibrous joint with that of a cartilaginous joint. (p. 261)

4 A ____________ is a fibrous joint with bones bound by long connective tissue fibers, whereas a ____________ is a fibrous joint where flat bones are united by a thin layer of connective tissue. (p. 261)

5 Describe a gomphosis, and name an example. (p. 262)

6 Compare the structures of a synchondrosis and a symphysis. (p. 262)

7 Explain how the joints between vertebrae permit movement. (p. 262)

8.3 General Structure of a Synovial Joint

8 Draw the general structure of a synovial joint, labeling all the main parts. (p. 263)

9 Describe how a joint capsule may be reinforced. (p. 264)

10 Explain the function of a synovial membrane. (p. 264)

11 Explain the function of synovial fluid. (p. 264)

12 Define *meniscus*. (p. 265)

13 Define *bursa*. (p. 265)

8.4 Types of Synovial Joints

14 Describe the six types of synovial joints, and name an example of each type. (p. 265)

15 Describe the movements permitted by each type of synovial joint. (p. 265)

8.5 Types of Joint Movements

16 Joint movements occur when a muscle contracts and the muscle fibers pull the muscle's movable end of attachment, the ________________, toward its fixed end, the ________________. (p. 267)

17 Match the movements listed on the left with the descriptions listed on the right. (pp. 267–270)

(1) Rotation	A. turning palm upward
(2) Supination	B. decreasing the angle between parts
(3) Extension	C. moving part forward
(4) Eversion	D. moving part around an axis
(5) Protraction	E. moving part toward midline
(6) Flexion	F. turning the foot so the plantar surface faces laterally
(7) Pronation	G. increasing angle between parts
(8) Abduction	H. lowering a part
(9) Depression	I. turning palm downward
(10) Adduction	J. moving part away from midline

8.6 Examples of Synovial Joints

18 Name the parts that comprise the shoulder joint. (p. 271)

19 Name the major ligaments associated with the shoulder joint. (p. 271)

20 Explain why the shoulder joint permits a wide range of movements. (p. 271)

21 Name the parts that comprise the elbow joint. (p. 272)

22 Name the major ligaments associated with the elbow joint. (p. 272)

23 Describe the movements permitted by the elbow joint. (p. 273)

24 Name the parts that comprise the hip joint. (p. 274)

25 Describe how the articular surfaces of the hip joint are held together. (p. 274)

26 Explain why there is less freedom of movement in the hip joint than in the shoulder joint. (p. 274)

27 Name the parts that comprise the knee joint. (p. 274)

28 Describe the major ligaments associated with the knee joint. (p. 274)

29 Explain the function of the menisci of the knee. (p. 276)

30 Describe the locations of the bursae associated with the knee joint. (p. 276)

8.7 Life-Span Changes

31 Describe the process of aging as it contributes to the stiffening of fibrous, cartilaginous, and synovial joints. (p. 279)

INTEGRATIVE ASSESSMENTS/CRITICAL THINKING

OUTCOMES 5.3, 8.3, 8.6, 8.7

1. How would you explain to an athlete why damaged joint ligaments and cartilages are so slow to heal following an injury?

OUTCOMES 8.3, 8.4, 8.6

2. Based upon your knowledge of joint structures, which do you think could be more satisfactorily replaced by a prosthetic device, a hip joint or a knee joint? Why?

OUTCOMES 8.3, 8.5, 8.6, 8.7

3. How would you explain to a person with a dislocated shoulder that the shoulder is likely to become more easily dislocated in the future?

OUTCOMES 8.3, 8.5, 8.6, 8.7

4. Why is it important to encourage an inactive patient to keep all joints mobile, even if it is necessary to have another person or a device move the joints (passive movement)?

OUTCOMES 8.4, 8.6

5. Compared to the shoulder and hip joints, in what way is the knee joint poorly protected and thus especially vulnerable to injuries?

WEB CONNECTIONS

Be sure to visit the text website at **www.mhhe.com/shier12** for answers to chapter assessments, additional quizzes, and interactive learning exercises.

ANATOMY & PHYSIOLOGY REVEALED

Anatomy & Physiology Revealed® (APR) includes cadaver photos that allow you to peel away layers of the human body to reveal structures beneath the surface. This program also includes animations, radiologic imaging, audio pronunciations, and practice quizzing. Check out **www.aprevealed.com**. APR has been proven to help improve student grades!

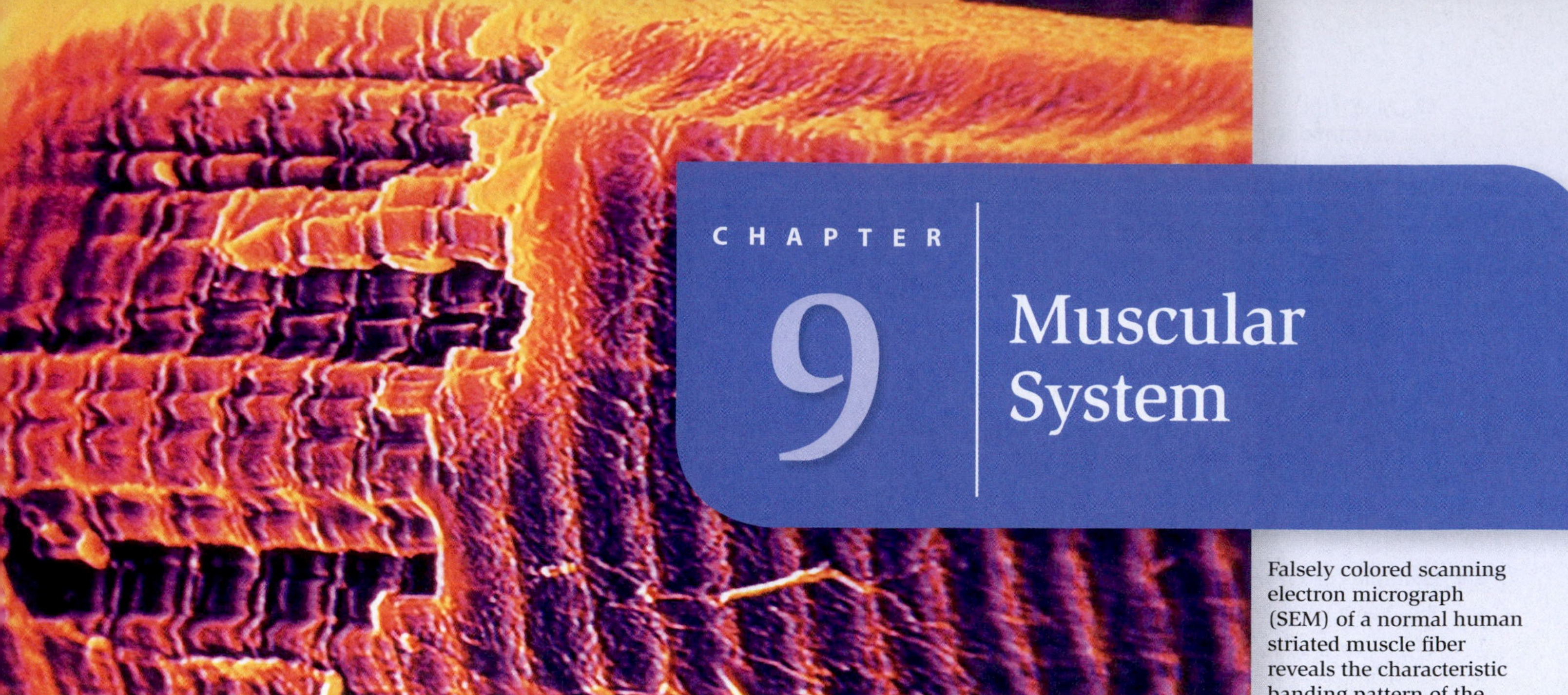

CHAPTER

9 Muscular System

Falsely colored scanning electron micrograph (SEM) of a normal human striated muscle fiber reveals the characteristic banding pattern of the myofibrils (3,000×).

UNDERSTANDING WORDS

calat-, something inserted: inter*calat*ed disc—membranous band that connects cardiac muscle cells.

erg-, work: syn*erg*ist—muscle that works with a prime mover, producing a movement.

fasc-, bundle: *fasc*iculus—bundle of muscle fibers.

-gram, something written: myo*gram*—recording of a muscular contraction.

hyper-, over, more: muscular *hyper*trophy—enlargement of muscle fibers.

inter-, between: *inter*calated disc—membranous band that connects cardiac muscle cells.

iso-, equal: *iso*tonic contraction—contraction during which the tension in a muscle remains unchanged.

laten-, hidden: *laten*t period—period between a stimulus and the beginning of a muscle contraction.

myo-, muscle: *myo*fibril—contractile fiber of a muscle cell.

reticul-, a net: sarcoplasmic *reticul*um—network of membranous channels within a muscle fiber.

sarco-, flesh: *sarco*plasm—substance (cytoplasm) within a muscle fiber.

syn-, together: *syn*ergist—muscle that works with a prime mover, producing a movement.

tetan-, stiff: *tetan*ic contraction—sustained muscular contraction.

-tonic, stretched: iso*tonic* contraction—contraction during which the tension of a muscle remains unchanged.

-troph, well fed: muscular hyper*trophy*—enlargement of muscle fibers.

voluntar-, of one's free will: *voluntar*y muscle—muscle that can be controlled by conscious effort.

 LEARN  PRACTICE  ASSESS

LEARNING OUTCOMES

After you have studied this chapter, you should be able to:

9.1 Introduction
1. List various outcomes of muscle actions (p. 285)

9.2 Structure of a Skeletal Muscle
2. Describe the structure of a skeletal muscle. (p. 285)
3. Name the major parts of a skeletal muscle fiber and describe the functions of each. (p. 287)

9.3 Skeletal Muscle Contraction
4. Describe the neural control of skeletal muscle contraction (p. 289)
5. Identify the major events of skeletal muscle fiber contraction. (p. 291)
6. List the energy sources for skeletal muscle fiber contraction. (p. 294)
7. Describe oxygen debt. (p. 295)
8. Describe how a muscle may become fatigued. (p. 296)

9.4 Muscular Responses
9. Distinguish between a twitch and a sustained contraction. (p. 296)
10. Explain how various types of muscular contractions produce body movements and help maintain posture. (p. 298)
11. Distinguish between fast and slow twitch muscle fibers. (p. 299)

9.5 Smooth Muscles
12. Distinguish between the structures and functions of multiunit smooth muscle and visceral smooth muscle. (p. 300)
13. Compare the contraction mechanisms of skeletal and smooth muscle fibers. (p. 301)

9.6 Cardiac Muscle
14. Compare the contraction mechanisms of skeletal and cardiac muscle fibers. (p. 301)

9.7 Skeletal Muscle Actions
15. Explain how the attachments, locations, and interactions of skeletal muscles make possible certain movements. (p. 301)

9.8 Major Skeletal Muscles
16. Identify and locate the skeletal muscles of each body region and describe the action(s) of each muscle. (p. 305)

9.9 Life-Span Changes
17. Describe aging-related changes in the muscular system. (p. 334)
18. Discuss how exercise can help maintain a healthy muscular system as the body ages. (p. 334)

THE MUSCULAR MOVEMENTS BEHIND "TEXTING"

Our musculoskeletal systems can rapidly adapt to new challenges. Consider sending text messages on a handheld device ("texting") or other movements that require the fingers to rapidly press precise sequences of very small buttons. Texting is similar to other challenges to dexterity, such as manipulating buttons on clothing or slicing or dicing foods. Loss of this dexterity may be an early sign of a disease that affects the muscles, such as amyotrophic lateral sclerosis (Lou Gehrig's disease).

Fingertip dexterity and hand movements are more complex than it may seem, altogether involving more than 30 muscles. To track the exact movements required for sending a text message, researchers recorded the electrical activity (using a measure called an electromyogram) and fingertip force in seven muscles of the index fingers of volunteers as they pushed their fingers against a surface. The researchers used an algorithm to assess the coordination and movements of the hand as the finger pressed the pad. They saw two clearly different patterns of muscle activation, indicating two different types of movement—light tapping from an angle versus direct downward pressure on one key. The act of texting entails a key-locating "tap" followed by a more direct push (static force). The switch from one type of movement to another is so fast and fluid that we usually are not aware of it.

Understanding the complexity of these dual tasks helps to explain why it takes years for children to master fine-hand coordination, as well as why these skills are often the first to be noticeably lost in neuromuscular disease. Our finger dexterity enabled our distant ancestors to live in the trees and then come down from them. Practical applications of the findings include guidance of prosthetic design, suggesting physical therapy techniques, and assisting the design of machines and electronic devices to be compatible with our natural finger and hand movements. ■

9.1 INTRODUCTION

Talking and walking, breathing and sneezing—all movements—require muscles. Muscles are organs composed of specialized cells that use the chemical energy stored in nutrients to exert a pulling force on structures to which they are attached. Muscular actions also provide muscle tone, propel body fluids and food, generate the heartbeat, and distribute heat.

Muscles are of three types—skeletal muscle, smooth muscle, and cardiac muscle, as described in chapter 5 (pp. 163–164). This chapter focuses mostly on skeletal muscle, which attaches to bones and to the skin of the face and is under conscious control. Smooth muscle and cardiac muscle are discussed briefly.

9.2 STRUCTURE OF A SKELETAL MUSCLE

A skeletal muscle is an organ of the muscular system. It is composed primarily of skeletal muscle tissue, nervous tissue, blood and other connective tissues.

Connective Tissue Coverings

An individual skeletal muscle is separated from adjacent muscles and held in position by layers of dense connective tissue called **fascia** (fash′e-ah). This connective tissue surrounds each muscle and may project beyond the ends of its muscle fibers, forming a cordlike **tendon.** Fibers in a tendon may intertwine with those in the periosteum of a bone, attaching the muscle to the bone. Or, the connective tissues associated with a muscle form broad, fibrous sheets called **aponeuroses** (ap″o-nu-ro′sēz), which may attach to bone or the coverings of adjacent muscles(figs. 9.1 and 9.2).

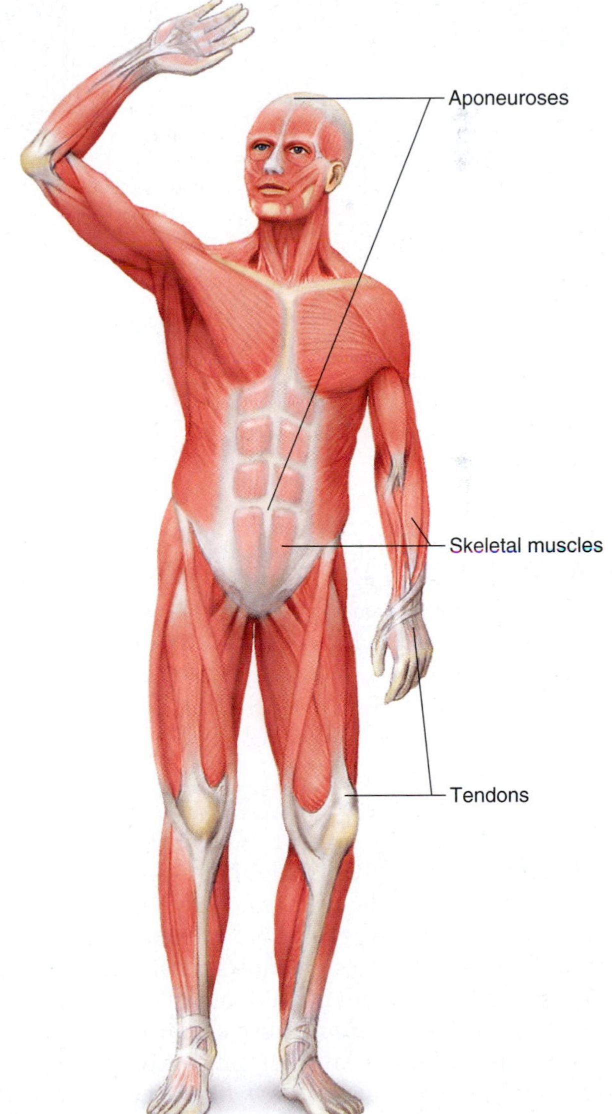

FIGURE 9.1 Tendons attach muscles to bones, whereas aponeuroses attach muscles to other muscles.

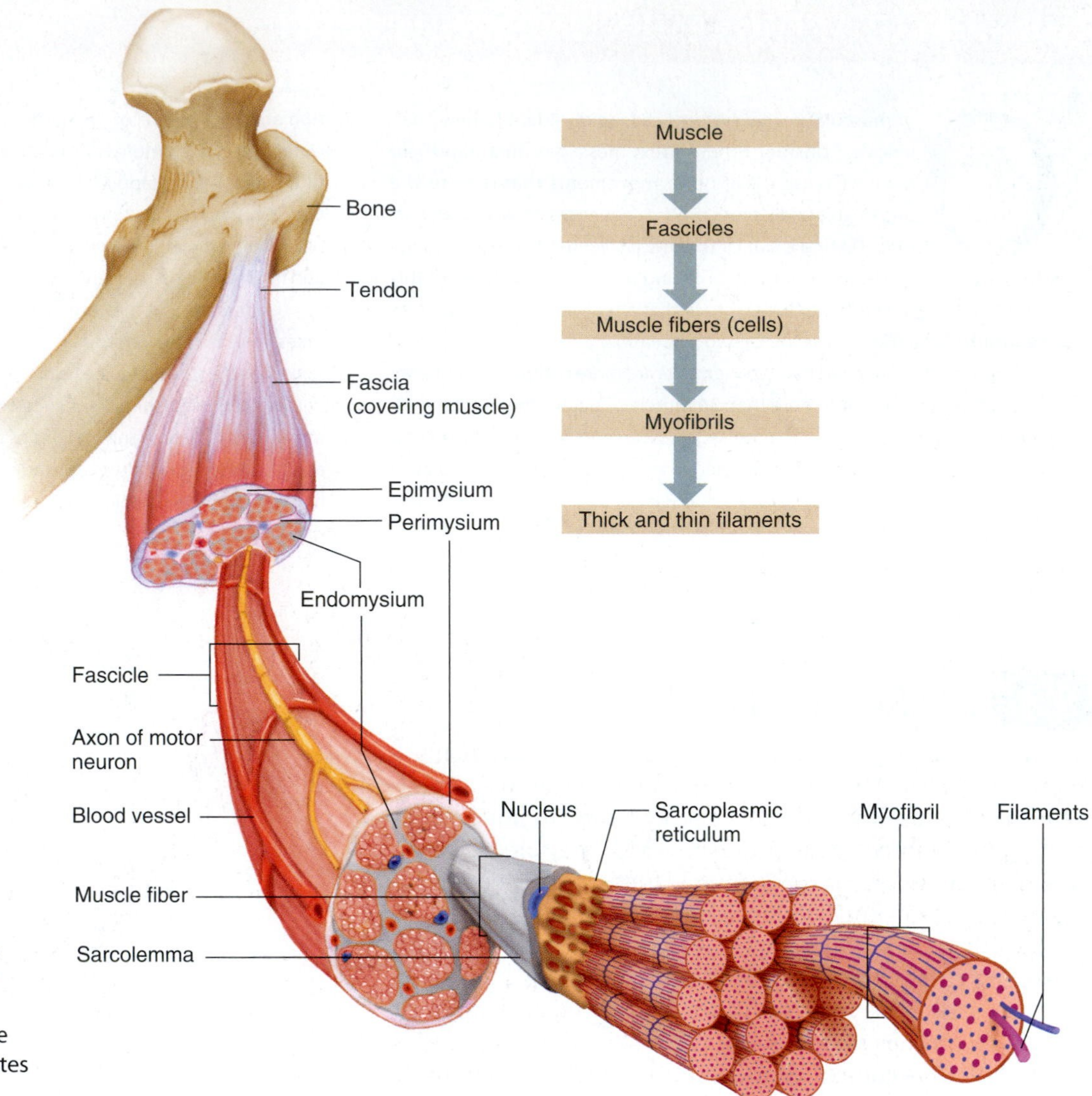

FIGURE 9.2 A skeletal muscle is composed of a variety of tissues, including layers of connective tissue. Fascia covers the surface of the muscle, epimysium lies beneath the fascia, and perimysium extends into the structure of the muscle where it separates muscle cells into fascicles. Endomysium separates individual muscle fibers.

A tendon or the connective tissue sheath of a tendon (tenosynovium) may become painfully inflamed and swollen following an injury or the repeated stress of athletic activity. These conditions are called *tendinitis* and *tenosynovitis*, respectively. The tendons most commonly affected are those associated with the joint capsules of the shoulder, elbow, hip, and knee and those involved with moving the wrist, hand, thigh, and foot.

The layer of connective tissue that closely surrounds a skeletal muscle is called the *epimysium*. Another layer of connective tissue, called the *perimysium*, extends inward from the epimysium and separates the muscle tissue into small sections. These sections contain bundles of skeletal muscle fibers called *fascicles* (fasciculi). Each muscle fiber within a fascicle (fasciculus) lies within a layer of connective tissue in the form of a thin covering called *endomysium* (figs. 9.2 and 9.3).

Layers of connective tissue, therefore, enclose and separate all parts of a skeletal muscle. This organization allows the parts to move somewhat independently. Also, many blood vessels and nerves pass through these layers.

A *compartment* is the space that contains a particular group of muscles, blood vessels, and nerves, all tightly enclosed by fascia. The limbs have many such compartments. If an injury causes fluid, such as blood from an internal hemorrhage, to accumulate in a compartment, the pressure inside will rise. The increased pressure, in turn, may interfere with blood flow into the region, reducing the supply of oxygen and nutrients to the affected tissues. This condition, called compartment syndrome, often produces severe, unrelenting pain. Persistently elevated compartmental pressure may irreversibly damage the enclosed muscles and nerves. Treatment for compartment syndrome may require an immediate surgical incision through the fascia (fasciotomy) to relieve the pressure and restore circulation.

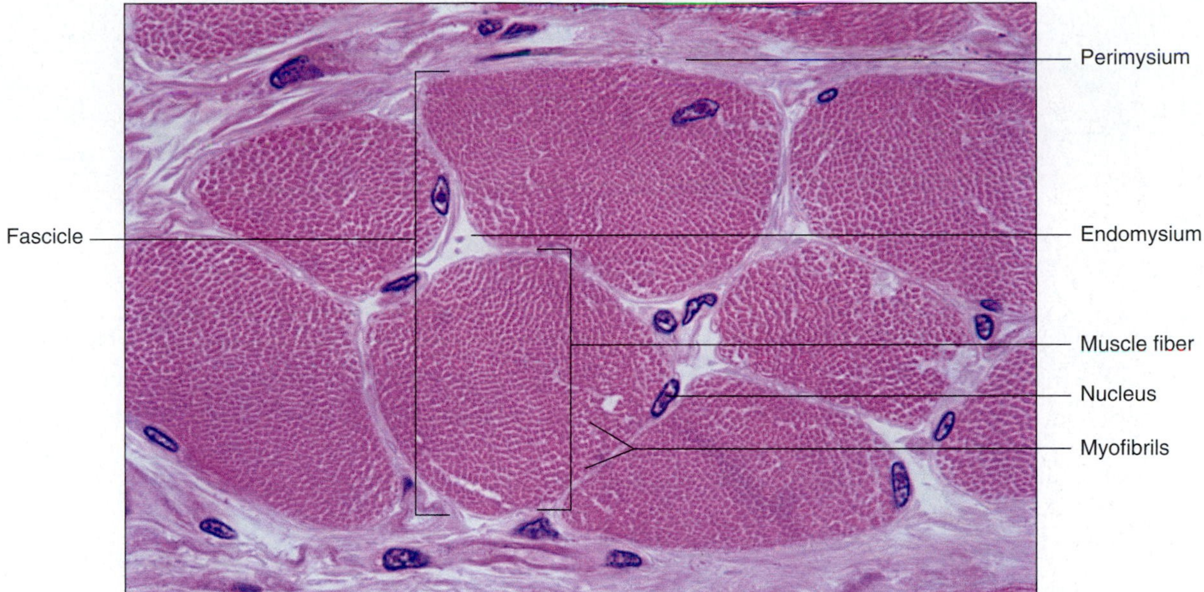

FIGURE 9.3 Scanning electron micrograph of a fascicle (fasciculus) surrounded by its connective tissue sheath, the perimysium. Muscle fibers within the fascicle are surrounded by endomysium (320×).

The fascia associated with each individual organ of the muscular system is part of a complex network of fasciae that extends throughout the body. The portion of the network that surrounds the muscles is called *deep fascia.* It is continuous with the *subcutaneous fascia* that lies just beneath the skin, forming the subcutaneous layer described in chapter 6 (p. 172). The network is also continuous with the *subserous fascia* that forms the connective tissue layer of the serous membranes covering organs in various body cavities and lining those cavities (see chapter 5, p. 163).

Skeletal Muscle Fibers

Recall from chapter 5 (p. 164) that a skeletal muscle fiber is a single muscle cell (see fig. 5.28). Each fiber forms from many undifferentiated cells that fuse during development. The resulting multinucleated muscle fiber is a thin, elongated cylinder with rounded ends that attach to the connective tissues associated with a muscle. Just beneath the muscle cell membrane (*sarcolemma*), the cytoplasm (*sarcoplasm*) of the fiber contains many small, oval nuclei and mitochondria. The sarcoplasm also has abundant, parallel, threadlike structures called **myofibrils** (mi″o-fi′-brilz) (fig. 9.4*a*).

The myofibrils play a fundamental role in the muscle contraction mechanism. They consist of two types of protein filaments: thick filaments composed of the protein **myosin** (mi′o-sin), and thin filaments composed primarily of the protein **actin** (ak′tin). (Two other thin filament proteins, troponin and tropomyosin, will be discussed later.) The organization of these filaments produces the alternating light and dark striations characteristic of skeletal muscle (and cardiac muscle) fibers. The striations form a repeating pattern of units called **sarcomeres** (sar′ko-mērz) along each muscle fiber. The myofibrils may be thought of as sarcomeres joined end to end. (fig. 9.4*a*). Muscle fibers, and in a way muscles themselves, are basically collections of sarcomeres, discussed later in this chapter as the functional units of muscle contraction.

The striation pattern of skeletal muscle has two main parts. The first, the *I bands* (the light bands), are composed of thin actin filaments held by direct attachments to structures called *Z lines,* which appear in the center of the I bands. The second part of the striation pattern consists of the *A bands* (the dark bands), composed of thick myosin filaments overlapping thin actin filaments (fig. 9.4*b*).

The A band consists not only of a region where thick and thin filaments overlap, but also a slightly lighter central region (*H zone*) consisting only of thick filaments. The A band includes a thickening known as the *M line,* which consists of proteins that help hold the thick filaments in place (fig. 9.4*b*). The myosin filaments are also held in place by the Z lines but are attached to them by a large protein called **titin** (connectin) (fig. 9.5). A sarcomere extends from one Z line to the next.

Thick filaments are composed of many molecules of myosin. Each myosin molecule consists of two twisted protein strands with globular parts called *cross-bridges* (heads) that project outward along their lengths. Thin filaments consist of double strands of actin twisted into a helix. Actin molecules are globular, and each has a binding site to which the cross-bridges of a myosin molecule can attach (fig. 9.6).

Two other types of protein, **troponin** and **tropomyosin,** associate with actin filaments. Troponin molecules have three protein subunits and are attached to actin. Tropomyosin molecules are rod-shaped and occupy the longitudinal grooves of the actin helix. Each tropomyosin is held in place by a troponin molecule, forming a troponin-tropomyosin complex (see fig. 9.6).

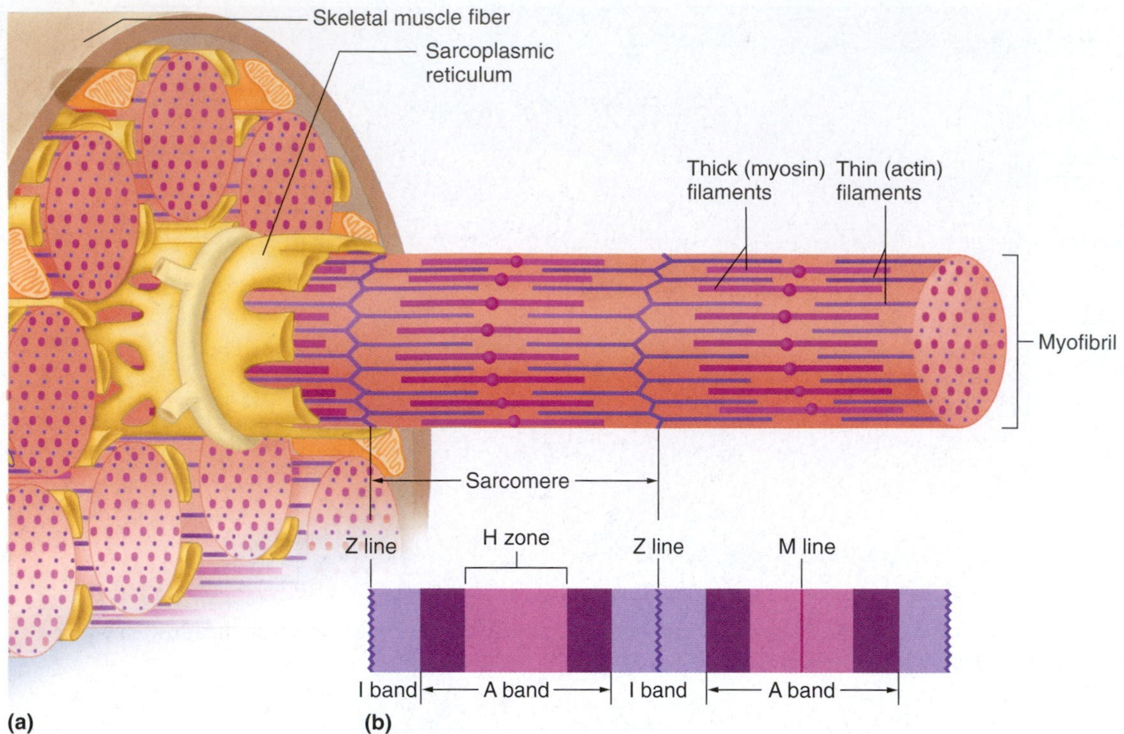

FIGURE 9.4 Skeletal muscle fiber. (*a*) A skeletal muscle fiber contains numerous myofibrils, each consisting of (*b*) repeating units called sarcomeres. The characteristic striations of a sarcomere reflect the organization of actin and myosin filaments.

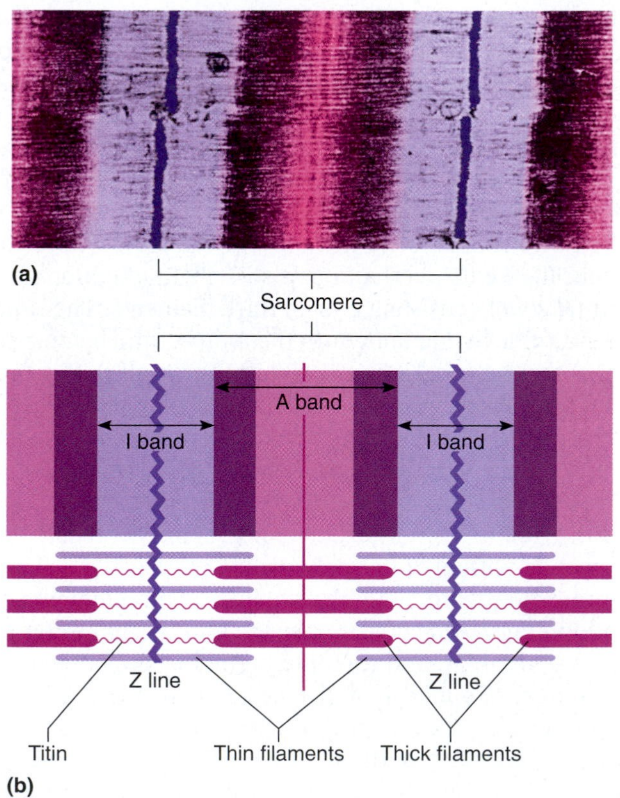

FIGURE 9.5 A sarcomere. (*a*) Micrograph (16,000×). (*b*) The relationship of thin and thick filaments in a sarcomere. The size of the H zone may change depending on the degree of filament overlap. Compare with the size of the H zone and filament overlap in figure 9.4*a* and *b*.

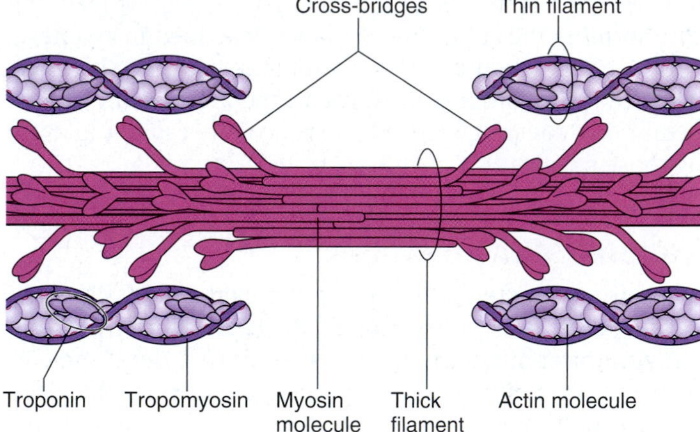

FIGURE 9.6 Thick filaments are composed of the protein myosin, and thin filaments are primarily composed of the protein actin. Myosin molecules have cross-bridges that extend toward nearby actin filaments.

Within the sarcoplasm of a muscle fiber is a network of membranous channels that surrounds each myofibril and runs parallel to it. These membranes form the **sarcoplasmic reticulum,** which corresponds to the endoplasmic reticulum of other cells (see figs. 9.2 and 9.4). A set of membranous channels, the **transverse tubules** (T-tubules), extends into the sarcoplasm as invaginations continuous with the sarcolemma and contains extracellular fluid. Each transverse tubule lies between two enlarged portions of the sarcoplasmic reticulum called **cisternae.** These three structures form a **triad** near the region where the actin and myosin filaments overlap (fig. 9.7).

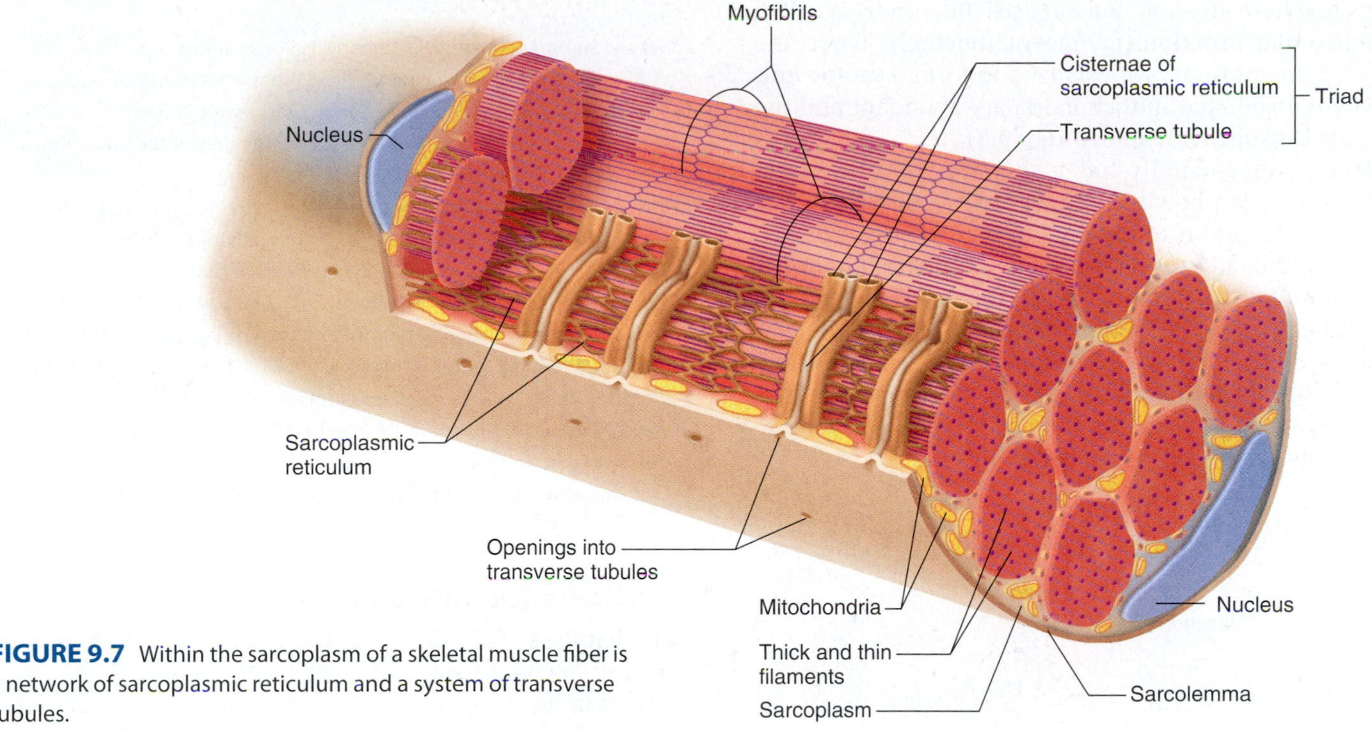

FIGURE 9.7 Within the sarcoplasm of a skeletal muscle fiber is a network of sarcoplasmic reticulum and a system of transverse tubules.

Although muscle fibers and the connective tissues associated with them are flexible, they can tear if overstretched. This type of injury, common in athletes, is called a *muscle strain*. The seriousness of the injury depends on the degree of damage the tissues sustain. In a mild strain, only a few muscle fibers are injured, the fascia remains intact, and little function is lost. In a severe strain, many muscle fibers as well as fascia tear, and muscle function may be lost completely. A severe strain is very painful and is accompanied by discoloration and swelling of tissues due to ruptured blood vessels. Surgery may be required to reconnect the separated tissues.

Actin, myosin, troponin, and tropomyosin are abundant in muscle cells. Scarcer proteins are also vital to muscle function. This is the case for a rod-shaped muscle protein called *dystrophin*. It accounts for only 0.002% of total muscle protein in skeletal muscle, but its absence causes the devastating inherited disorder Duchenne muscular dystrophy, a disease that only affects males. Dystrophin binds to the inside face of muscle cell membranes, supporting them against the powerful force of contraction. Without even these minute amounts of dystrophin, muscle cells burst and die. Other forms of muscular dystrophy result from abnormalities of proteins to which dystrophin attaches.

PRACTICE

1. Describe how connective tissue is associated with a skeletal muscle.
2. Describe the general structure of a skeletal muscle fiber.
3. Explain why skeletal muscle fibers appear striated.
4. Explain the physical relationship between the sarcoplasmic reticulum and the transverse tubules.

9.3 SKELETAL MUSCLE CONTRACTION

A muscle fiber contraction is a complex interaction of several cellular and chemical constituents. The result is a movement within the myofibrils in which the filaments of actin and myosin slide past one another, shortening the sarcomeres. When this happens, the muscle fiber shortens and pulls on its attachments.

Neuromuscular Junction

Recall from chapter 5 (p. 164) that neurons establish communication networks throughout the body. Each neuron has a process called an axon, which extends from the cell body and is capable of conducting a nerve impulse. Neurons that control effectors, including skeletal muscle, are called **motor neurons.**

Each skeletal muscle fiber is functionally (but not physically) connected to an axon of a motor neuron that passes outward from the brain or the spinal cord, in much the same way that you can talk into a cell phone although your mouth is not in direct physical contact with it. The site of this functional connection is called a **synapse.** It is a space through which information can pass. Neurons communicate with the cells that they control by releasing chemicals, called **neurotransmitters** (nu″ro-trans-mit-erz), at a synapse. Normally a skeletal muscle fiber contracts only upon stimulation by a motor neuron.

The site where an axon and a muscle fiber meet is called a **neuromuscular junction** (myoneural junction). There, the muscle fiber membrane is specialized to form a **motor end plate,** where nuclei and mitochondria are abundant and the sarcolemma is extensively folded (fig. 9.8).

A muscle fiber usually has a single motor end plate. Motor neuron axons, however, are densely branched, which enables one such axon to connect to many muscle fibers. Together, a motor neuron and the muscle fibers it controls constitute a **motor unit** (mo′tor u′nit) (fig. 9.9).

A small gap called the **synaptic cleft** separates the membrane of the neuron and the membrane of the muscle fiber. The cytoplasm at the distal ends of the nerve fiber is rich in mitochondria and contains many tiny vesicles (synaptic vesicles) that store neurotransmitters.

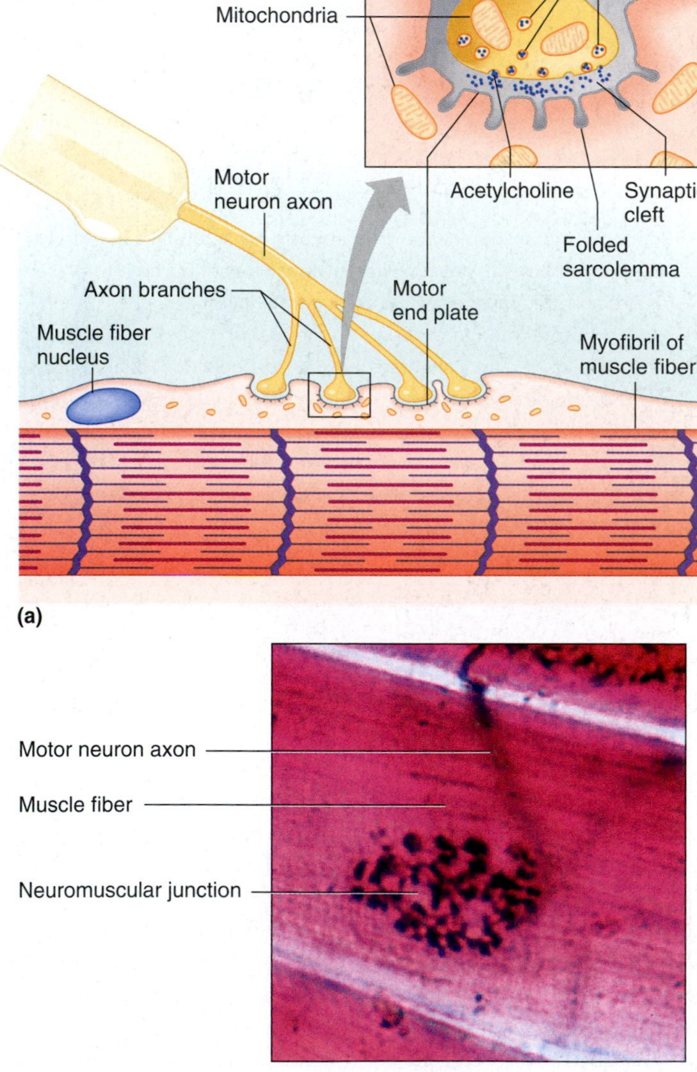

FIGURE 9.8 Neuromuscular junction. (*a*) A neuromuscular junction includes the end of a motor neuron and the motor end plate of a muscle fiber. (*b*) Micrograph of a neuromuscular junction (500×).

In the summer months of the early 1950s, millions of children contracted poliomyelitis, a viral infection that attacks motor neurons. Fever, headache, and nausea rapidly progressed to a stiffened back and neck, drowsiness, and then paralysis, usually of the lower limbs or muscles that control breathing or swallowing.

Vaccines introduced in the middle 1950s vanquished polio in many nations, but the disease resurged in Nigeria in 2003, where rumors that the vaccine causes female infertility led to a boycott of the World Health Organization's Global Polio Eradication Initiative. Polio has spread to neighboring nations and to as far away as Indonesia.

In the United States, a third of the 1.6 million polio survivors suffer the fatigue, muscle weakness and atrophy, and difficulty breathing of postpolio syndrome. Researchers think that in this condition, surviving motor neurons that grew extra axon branches to compensate for neurons lost during polio degenerate from years of overuse.

Stimulus for Contraction

Acetylcholine (ACh) is the neurotransmitter that motor neurons use to control skeletal muscle contraction. ACh is synthesized in the cytoplasm of the motor neuron and is stored in synaptic vesicles near the distal end of its axon. When a nerve impulse (a series of action potentials, described in chapter 10, pp. 369–370) reaches the end of the axon, some of these vesicles release acetylcholine into the synaptic cleft (see fig. 9.8).

Acetylcholine diffuses rapidly across the synaptic cleft and binds to specific protein molecules (receptors) in the muscle fiber membrane, increasing the membrane permeability to sodium ions. The entry of these charged particles into the muscle cell stimulates a **muscle impulse** (a series of action potentials), an electrical signal very much like a nerve impulse. A muscle impulse changes the muscle cell membrane in a way that transmits the impulse in all directions along and around the muscle cell, into the transverse tubules, into the sarcoplasm, and ultimately to the sarcoplasmic reticulum and the cisternae. Clinical Application 9.1 discusses myasthenia gravis, in which the immune system attacks certain neuromuscular junctions.

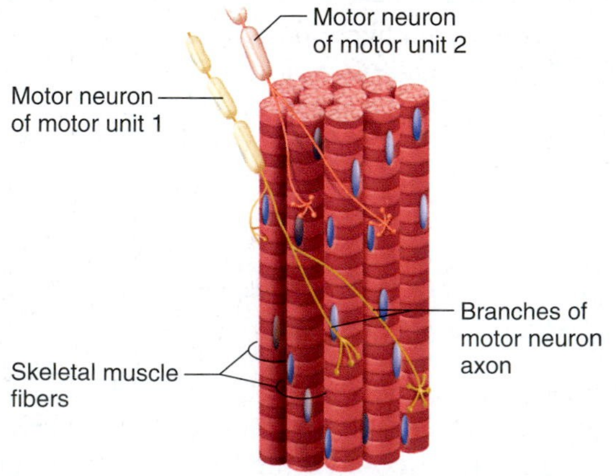

FIGURE 9.9 Two motor units. The muscle fibers of a motor unit are innervated by a single motor neuron and may be distributed throughout the muscle.

9.1 CLINICAL APPLICATION

Myasthenia Gravis

In an autoimmune disorder, the immune system attacks part of the body. In myasthenia gravis (MG), that part is the muscular system. The body produces antibodies that target receptors for the neurotransmitter acetylcholine on muscle cells at neuromuscular junctions. People with MG have one-third the normal number of acetylcholine receptors here. On a whole-body level, this causes weak and easily fatigued muscles.

MG affects hundreds of thousands of people worldwide, mostly women beginning in their twenties or thirties, and men in their sixties and seventies. The specific symptoms depend upon the site of attack. For 85% of patients, the disease causes generalized muscle weakness. Many people develop a characteristic flat smile and nasal voice and have difficulty chewing and swallowing due to affected facial and neck muscles. Many have limb weakness. About 15% of patients experience the illness only in the muscles surrounding their eyes. The disease reaches crisis level when respiratory muscles are affected, requiring a ventilator to support breathing. MG does not affect sensation or reflexes.

MG can usually be controlled, thanks to a combination of the following treatments:

- Drugs that inhibit acetylcholinesterase, the enzyme that normally breaks down acetylcholine, thus increasing levels of the neurotransmitter increase.
- Removing the thymus gland, which oversees much of the immune response.
- Immunosuppressant drugs that decrease production of antibodies.
- Intravenous antibodies that bind and inactivate the ones causing the damage.
- Plasma exchange, which rapidly removes the damaging antibodies from the circulation, helping people in crisis. ■

When the bacterium *Clostridium botulinum* grows in an anaerobic (oxygen-poor) environment, such as in a can of unrefrigerated food, it produces a toxin that prevents the release of acetylcholine from nerve terminals if ingested by a person. Symptoms include nausea, vomiting, and diarrhea; headache, dizziness, and blurred or double vision; and finally, weakness, hoarseness, and difficulty swallowing and, eventually, breathing. Physicians can administer an antitoxin substance that binds to and inactivates botulinum toxin in the bloodstream, stemming further symptoms, although not correcting damage already done. Small amounts of botulinum toxin are used to treat migraine headaches and to temporarily paralyze selected facial muscles, smoothing wrinkles.

Excitation Contraction Coupling

The sarcoplasmic reticulum has a high concentration of calcium ions compared to the cytosol. This is due to active transport of calcium ions (calcium pump) in the membrane of the sarcoplasmic reticulum. In response to a muscle impulse, the membranes of the cisternae become more permeable to these ions, and the calcium ions diffuse out of the cisternae into the cytosol of the muscle fiber (see fig. 9.7).

RECONNECT
To Chapter 3, Active Transport, page 95.

When a muscle fiber is at rest, the troponin-tropomyosin complexes block the binding sites on the actin molecules and thus prevent the formation of linkages with myosin cross-bridges (fig. 9.10 1). As the concentration of calcium ions in the cytosol rises, however, the calcium ions bind to the troponin, changing its shape (conformation) and altering the position of the tropomyosin. The movement of the tropomyosin molecules exposes the binding sites on the actin filaments, allowing linkages to form between myosin cross-bridges and actin (fig. 9.10 2).

RECONNECT
To Chapter 2, Proteins, pages 64–66.

The Sliding Filament Model of Muscle Contraction

The sarcomere is considered the functional unit of skeletal muscles because contraction of an entire skeletal muscle can be described in terms of the shortening of the sarcomeres of its muscle fibers. According to the **sliding filament model,** when sarcomeres shorten, the thick and thin filaments do not change length. Rather, they slide past one another, with the thin filaments moving toward the center of the sarcomere from both ends. As this occurs, the H zones and the I bands narrow; the regions of overlap widen; and the Z lines move closer together, shortening the sarcomere (fig. 9.11).

Cross-Bridge Cycling

The force that shortens the sarcomeres comes from cross-bridges pulling on the thin filaments. A myosin cross-bridge can attach to an actin binding site and bend slightly, pulling on the actin filament. Then the head can release, straighten, combine with another binding site further down the actin filament, and pull again (see fig. 9.10 2–6).

Myosin cross-bridges contain the enzyme **ATPase,** which catalyzes the breakdown of ATP to ADP and phosphate. This

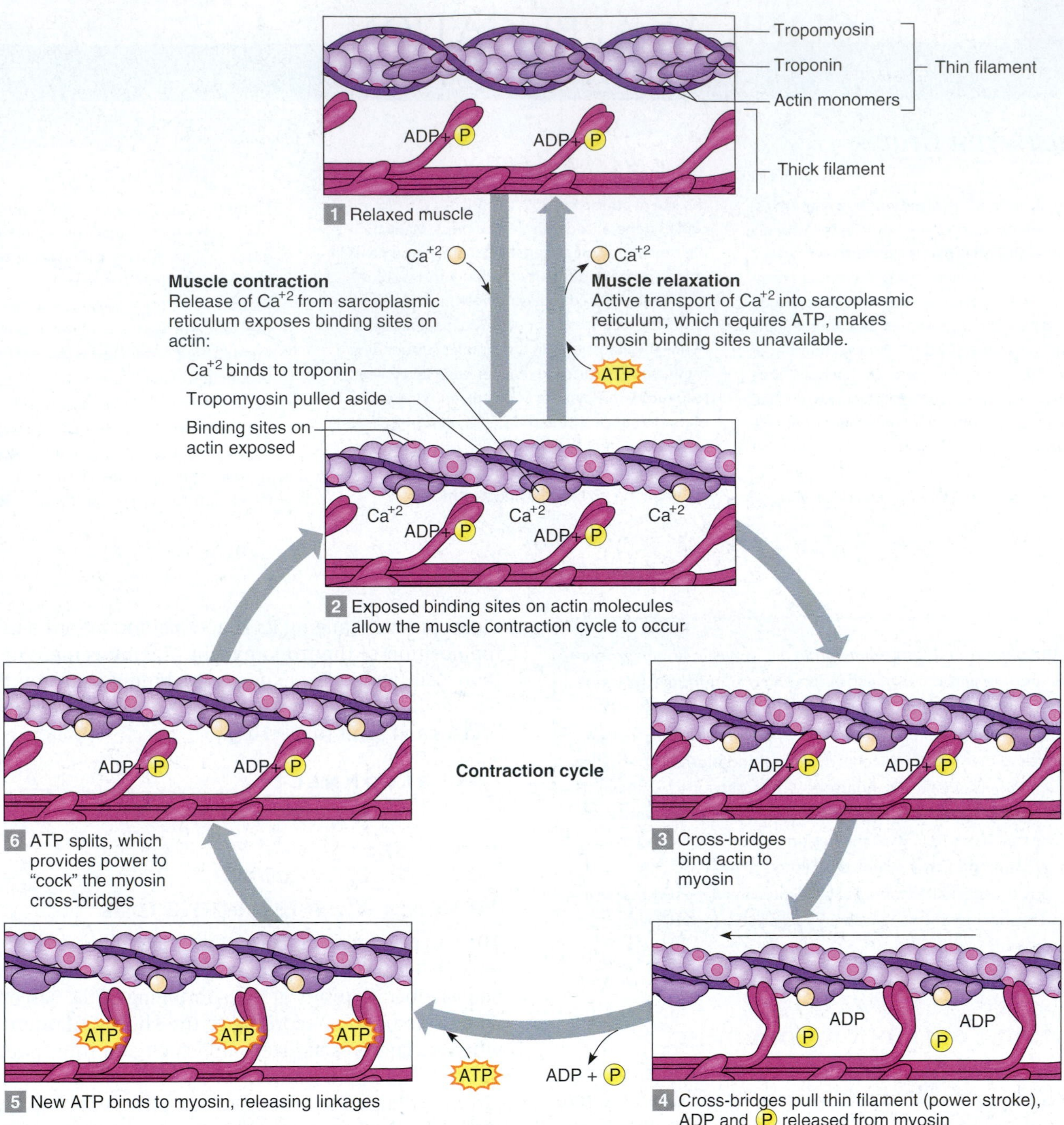

FIGURE 9.10 According to the sliding filament theory (*1–3*) when calcium ion concentration rises, binding sites on actin filaments open, and cross-bridges attach. (*4*) Upon binding to actin, cross-bridges spring from the cocked position and pull on actin filaments. (*5*) ATP binds to the cross-bridge (but is not yet broken down), releasing it from the actin filament. (*6*) ATP breakdown provides energy to "cock" the unattached myosin cross-bridge. As long as ATP and calcium ions are present, the cycle continues. When calcium ion concentration is low in the cytosol, the muscle cell remains relaxed. Not all cross-bridges form and release simultaneously.

reaction releases energy (see chapter 4, p. 119) that provides the force for muscle contraction. Breakdown of ATP puts the myosin cross-bridge in a "cocked" position (see fig. 9.10 6). When a muscle is stimulated to contract, a cocked cross-bridge attaches to actin (see fig. 9.10 3) and pulls the actin filament toward the center of the sarcomere (see fig. 9.10 4). This causes a greater overlap of the actin and myosin filaments, shortens the sarcomere and thus shortens the muscle (fig. 9.11). When another ATP binds, the cross-bridge is first released from the actin binding site (see fig. 9.10 5), then breaks down the ATP to return to the cocked position (see fig. 9.10 6). This cross-bridge cycle may repeat as long as ATP is present and nerve impulses release ACh at that neuromuscular junction.

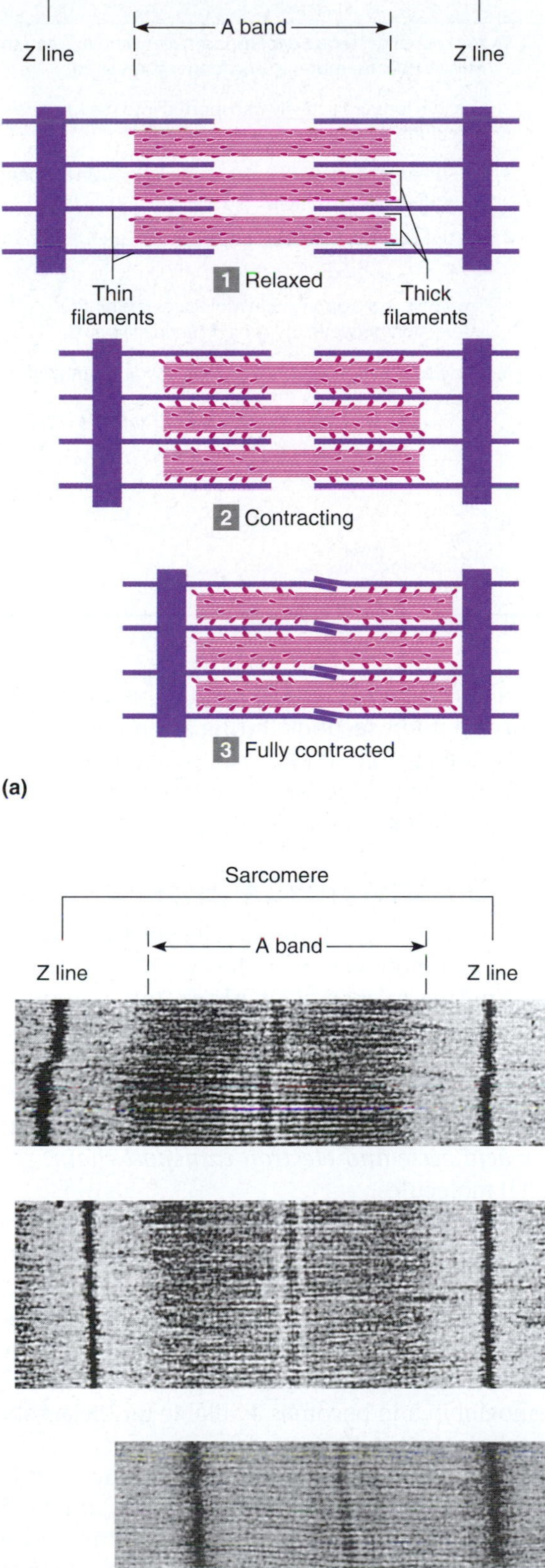

FIGURE 9.11 When a skeletal muscle contracts (*a*), individual sarcomeres shorten as thick and thin filaments slide past one another. (*b*) Transmission electron micrograph showing a sarcomere shortening during muscle contraction (23,000×).

Relaxation

When nerve impulses cease, two events relax the muscle fiber. First, an enzyme called **acetylcholinesterase** rapidly decomposes acetylcholine remaining in the synapse. This enzyme, present in the synapse and on the membranes of the motor end plate, prevents a single nerve impulse from continuously stimulating a muscle fiber. Second, when ACh breaks down, the stimulus to the sarcolemma and the membranes of the muscle fiber ceases. The calcium pump (which requires ATP) quickly moves calcium ions back into the sarcoplasmic reticulum, decreasing the calcium ion concentration of the cytosol. The cross-bridge linkages break (see fig. 9.10 6—this also requires ATP, although it is not broken down in this step), and tropomyosin rolls back into its groove, preventing cross-bridge attachment (see fig. 9.10 7). Consequently, the muscle fiber relaxes. Table 9.1 summarizes the major events leading to muscle contraction and relaxation.

If acetylcholine receptors at the motor end plate are too few, or blocked, muscles cannot receive the signal to contract. This may occur as the result of a disease, such as myasthenia gravis, or exposure to a poison, such as nerve gas. A drug called pyridostigmine bromide is used to treat myasthenia gravis. The drug inhibits the enzyme (acetylcholinesterase) that normally breaks down acetylcholine, keeping the neurotransmitter around longer. It was given to veterans of the first Gulf War who reported muscle aches in the months following their military service. Health officials reasoned that the drug's effect on myasthenia gravis might also help restore muscle function if the veterans' symptoms arose from exposure to nerve gas during the war. Acetylcholinesterase inhibitors are also used as insecticides. The buildup of acetylcholine causes an insect to twitch violently, then die.

It is important to remember that ATP is necessary for both muscle contraction and for muscle relaxation. The trigger for contraction is the increase in cytosolic calcium in response to stimulation by ACh from a motor neuron.

A few hours after death, the skeletal muscles partially contract, fixing the joints. This condition, called *rigor mortis,* may continue for seventy-two hours or more. It results from an increase in membrane permeability to calcium ions, which promotes cross-bridge attachment, and a decrease in availability of ATP in the muscle fibers, which prevents cross-bridge release from actin. Thus, the actin and myosin filaments of the muscle fibers remain linked until the muscles begin to decompose.

PRACTICE

5 Describe a neuromuscular junction.

6 Define motor unit.

7 List four proteins associated with myofibrils, and explain their structural and functional relationships.

8 Explain how the filaments of a myofibril interact during muscle contraction.

9 Explain how a motor nerve impulse can trigger a muscle contraction.

TABLE 9.1 | Major Events of Muscle Contraction and Relaxation

Muscle Fiber Contraction	Muscle Fiber Relaxation
1. A nerve impulse travels down a motor neuron axon.	1. Acetylcholinesterase decomposes acetylcholine, and the muscle fiber membrane is no longer stimulated.
2. The motor neuron terminal releases the neurotransmitter acetylcholine (ACh).	2. Calcium ions are actively transported into the sarcoplasmic reticulum.
3. ACh binds to ACh receptors on the muscle fiber.	3. ATP breaks linkages between actin and myosin filaments without breakdown of the ATP itself.
4. The sarcolemma is stimulated, and a muscle impulse travels over the surface of the muscle fiber and deep into the fiber through the transverse tubules.	4. Breakdown of ATP "cocks" the cross-bridges.
5. The muscle impulse reaches the sarcoplasmic reticulum, and calcium channels open.	5. Troponin and tropomyosin molecules inhibit the interaction between myosin and actin filaments.
6. Calcium ions diffuse from the sarcoplasmic reticulum into the sarcoplasm and bind to troponin molecules.	6. Muscle fiber remains relaxed, yet ready until stimulated again.
7. Tropomyosin molecules move and expose specific sites on actin.	
8. Actin and myosin form linkages.	
9. Thin (actin) filaments are pulled toward the center of the sarcomere by myosin cross-bridges increasing the overlap of the thin and thick filaments.	
10. The muscle fiber contracts.	

Energy Sources for Contraction

The energy used to power the interaction between actin and myosin filaments during muscle fiber contraction comes from ATP molecules. However, a muscle fiber has only enough ATP to contract briefly. Therefore, an active fiber requires regeneration of ATP.

The initial source of energy available to regenerate ATP from ADP and phosphate is **creatine phosphate.** Like ATP, creatine phosphate includes a high-energy phosphate bond, and this molecule is four to six times more abundant in muscle fibers than ATP. Creatine phosphate, however, cannot directly supply energy to a cell. Instead, it stores energy released from mitochondria. Whenever sufficient ATP is present, an enzyme in the mitochondria (creatine phosphokinase) promotes the synthesis of creatine phosphate, which stores excess energy in its phosphate bond (fig. 9.12).

As ATP is decomposed to ADP, the energy from creatine phosphate molecules is transferred to these ADP molecules, quickly phosphorylating them back into ATP. The amount of ATP and creatine phosphate in a skeletal muscle, however, is usually not sufficient to support maximal muscle activity for more than about ten seconds during an intense contraction. As a result, the muscle fibers in an active muscle soon use cellular respiration of glucose to synthesize ATP. Typically, a muscle stores glucose in the form of glycogen.

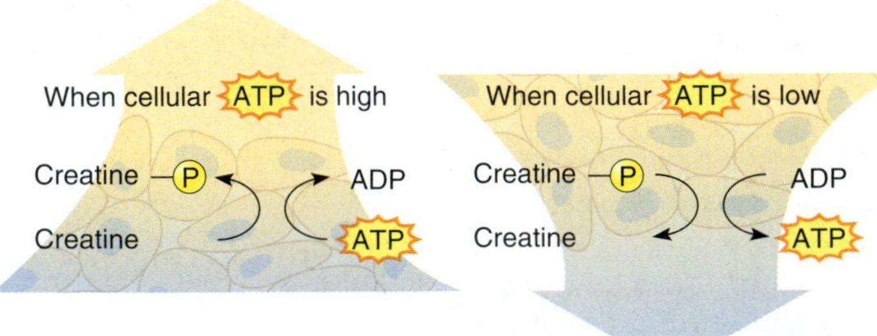

FIGURE 9.12 Creatine phosphate may be used to replenish ATP stores when ATP levels in a muscle cell are low.

Oxygen Supply and Cellular Respiration

Recall from chapter 4 (p. 120) that glycolysis, the early phase of cellular respiration, occurs in the cytoplasm and is *anaerobic,* not requiring oxygen. This phase only partially breaks down energy-supplying glucose and releases only a few ATP molecules. The complete breakdown of glucose occurs in the mitochondria and is *aerobic,* requiring oxygen. This process, which includes the complex series of reactions of the *citric acid cycle and electron transport chain,* produces many ATP molecules.

Blood carries the oxygen necessary to support the aerobic reactions of cellular respiration from the lungs to body cells. Oxygen is transported in red blood cells, where it is loosely bound to molecules of hemoglobin, the pigment responsible for the red color of blood. In regions of the body where the oxygen concentration is low, oxygen is released from hemoglobin and becomes available for the aerobic reactions of cellular respiration.

Another pigment, **myoglobin,** is synthesized in muscle cells and imparts the reddish brown color of skeletal muscle tissue. Like hemoglobin, myoglobin can loosely bind oxygen and, in fact, has a greater attraction for oxygen than does hemoglobin. Myoglobin can temporarily store oxygen in muscle tissue, which reduces a muscle's requirement for a continuous blood supply during contraction. This oxygen storage is important because blood flow

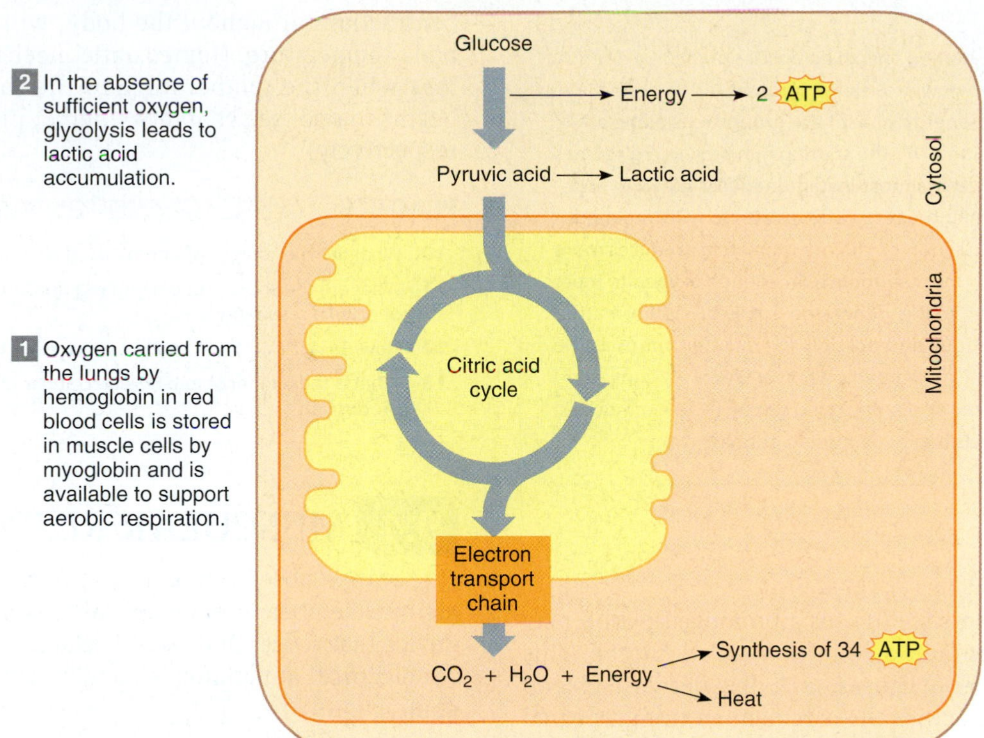

FIGURE 9.13 The oxygen required to support the aerobic reactions of cellular respiration is carried in the blood and stored in myoglobin. In the absence of sufficient oxygen, anaerobic reactions use pyruvic acid to produce lactic acid. The maximum number of ATPs generated per glucose molecule varies with cell type; in skeletal muscle, it is 36 (2 + 34).

may decrease during muscular contraction when contracting muscle fibers compress blood vessels (fig. 9.13).

Oxygen Debt

When a person is resting or moderately active, the respiratory and cardiovascular systems can usually supply sufficient oxygen to the skeletal muscles to support the aerobic reactions of cellular respiration. However, when skeletal muscles are used more strenuously, these systems may not be able to supply enough oxygen to sustain the aerobic reactions of cellular respiration.

Chapter 4 (pp. 120–122) discussed how the anaerobic reactions break down glucose into pyruvic acid, which then reacts to produce lactic acid. This shift in metabolism is referred to as the anaerobic threshold, or the **lactic acid threshold.** The lactic acid diffuses out of the muscle fibers and is carried in the bloodstream to the liver. Liver cells can react the lactic acid to form *glucose,* but this requires energy from ATP (fig. 9.14). During strenuous exercise, available oxygen is primarily used to synthesize ATP for muscle contraction rather than to make ATP for reacting lactic acid to yield glucose. Consequently, as lactic acid accumulates, a person develops an **oxygen debt** that must be repaid at a later time. The amount of oxygen debt roughly equals the amount of oxygen that liver cells require to use the accumulated lactic acid to produce glucose, plus the amount that the muscle cells require to resynthesize sufficient ATP and creatine phosphate to restore their original concentrations. The degree of oxygen debt also reflects the oxygen required to restore blood and tissue oxygen levels to preexercise levels.

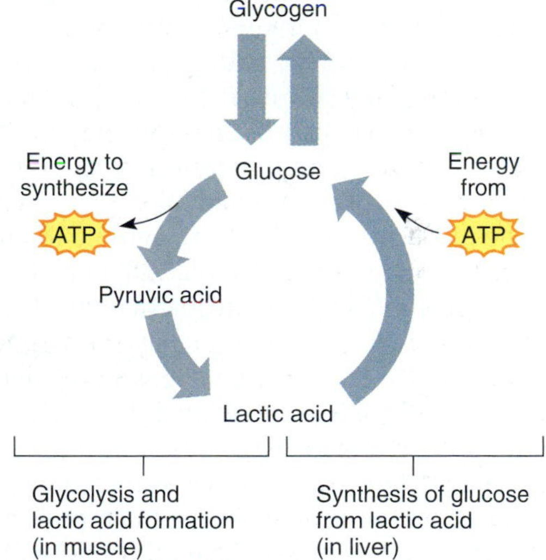

FIGURE 9.14 Liver cells can react lactic acid generated by muscles anaerobically to produce glucose.

The metabolic capacity of a muscle may change with athletic training. With high-intensity exercise, which depends more on glycolysis for ATP, a muscle will synthesize more glycolytic enzymes, and its capacity for glycolysis will increase. With aerobic exercise, more capillaries and mitochondria develop, and the muscles' capacity for the aerobic reactions of cellular respiration increases.

The runners are on the starting line, their muscles primed for a sprint. Glycogen will be broken down to release glucose, and creatine phosphate will supply high-energy phosphate groups to replenish ATP stores by phosphorylating ADP. The starting gun fires. Energy comes first from residual ATP, but almost instantaneously, creatine phosphate begins donating high-energy phosphates to ADP, regenerating ATP. Meanwhile, oxidation of glucose ultimately produces more ATP. But because the runner cannot take in enough oxygen to meet the high demand, most ATP is generated in glycolysis. Formation of lactic acid causes fatigue and possibly leg muscle cramps as the runner crosses the finish line. Already, her liver is actively converting lactic acid back to pyruvic acid and storing glycogen. In her muscles, creatine phosphate levels begin to return to normal.

Muscle Fatigue

A muscle exercised persistently for a prolonged period may lose its ability to contract, a condition called *fatigue*. This condition has a number of causes, including decreased blood flow, ion imbalances across the sarcolemma from repeated stimulation, and psychological loss of the desire to continue the exercise. However, muscle fatigue is most likely to arise from accumulation of lactic acid in the muscle from anaerobic ATP production. The lowered pH from the lactic acid prevents muscle fibers from responding to stimulation.

Occasionally a muscle fatigues and cramps at the same time. A cramp is a sustained, painful, involuntary muscle contraction. Cramps may result when changes, particularly a decreased electrolyte concentration, occurring in the extracellular fluid surrounding the muscle fibers and their motor neurons trigger uncontrolled stimulation of the muscle.

As muscle metabolism shifts from aerobic to anaerobic ATP production, lactic acid begins to accumulate in muscles and to appear in the bloodstream (lactic acid threshold). This leads to muscle fatigue. How quickly this happens varies among individuals, although people who regularly exercise aerobically produce less lactic acid than those who do not. The strenuous exercise of aerobic training stimulates new capillaries to extend into muscles, supplying more oxygen and nutrients to the muscle fibers. Such physical training also adds mitochondria, increasing the ability of muscle fibers to produce ATP aerobically. Some muscle fibers may be more likely to accumulate lactic acid than others, as described in a later section entitled "Fast- and Slow-Twitch Muscle Fibers."

Heat Production

All active cells generate heat, which is a by-product of cellular respiration. Muscle tissue constitutes such a large proportion of total body mass, that it is a major source of heat.

Less than half of the energy released in cellular respiration is available for use in metabolic processes; the rest becomes heat. Blood transports the heat from muscle contraction throughout the body, which helps to maintain body temperature. Homeostatic mechanisms promote heat loss when the temperature of the internal environment begins to rise (see chapters 1 and 6, pp. 9–10 and 181–183, respectively).

PRACTICE 

10 What are the sources of energy used to regenerate ATP?

11 What are the sources of oxygen required for the aerobic reactions of cellular respiration?

12 How do lactic acid and oxygen debt relate to muscle fatigue?

13 What is the relationship between cellular respiration and heat production?

9.4 MUSCULAR RESPONSES

One way to observe muscle contraction is to remove a single muscle fiber from a skeletal muscle and connect it to a device that senses and records changes in the fiber's length. An electrical stimulator is usually used to promote muscle contraction.

Threshold Stimulus

When an isolated muscle fiber is exposed to a series of stimuli of increasing strength, the fiber remains unresponsive until a certain strength of stimulation called the **threshold stimulus** (thresh'old stim'u-lus) is applied. Once threshold is reached, an action potential is generated, resulting in a muscle impulse that spreads throughout the muscle fiber, releasing enough calcium ions from the sarcoplasmic reticulum to activate cross-bridge binding and contract that fiber. A single nerve impulse in a motor neuron normally releases enough ACh to bring the muscle fibers in its motor unit to threshold, generating a muscle impulse in each muscle fiber.

Recording of a Muscle Contraction

The contractile response of a single muscle fiber to a muscle impulse is called a **twitch.** A twitch consists of a period of contraction, during which the fiber pulls at its attachments, followed by a period of relaxation, during which the pulling force declines. These events can be recorded in a pattern called a myogram (fig. 9.15). A twitch has a brief delay between the time of stimulation and the beginning of contraction. This is the **latent period,** which in human muscle may be less than 2 milliseconds.

The length to which a muscle fiber is stretched before stimulation affects the force it will develop. If a muscle fiber is stretched well beyond its normal resting length, the force will decrease. This is because sarcomeres of that fiber become so extended that myosin cross-bridges cannot reach binding sites on the thin filaments and cannot contribute to contraction. Conversely, at very short fiber lengths, the sarcomeres become compressed, and further shortening is not possible (fig. 9.16). During normal activities, muscle fibers contract at their optimal lengths. Some activities, such as walking up stairs two at a time or lifting something from an awkward position, put fibers at a disadvantageous length and compromise muscle performance.

> A muscle fiber brought to threshold under a given set of conditions contracts completely, and each twitch generates equal force. This is an *all-or-none* response. However, "all-or-none" is misleading, because in normal use of muscles, the force generated by muscle fibers and by whole muscles must vary.

Understanding the contraction of individual muscle fibers is important for understanding how muscles work, but such contractions by themselves are of little significance in day-to-day activities. Rather, the actions we need to perform usually require the contribution of multiple muscle fibers simultaneously. To record how a whole muscle responds to stimulation, a skeletal muscle can be removed from a frog or other small animal and mounted on a special device. The muscle is then electrically stimulated, and when it contracts, it pulls on a lever. The lever's movement is recorded as a myogram. The myogram results from the combined twitches of muscle fibers taking part in the contraction, so it looks essentially the same as the twitch contraction depicted in figure 9.15.

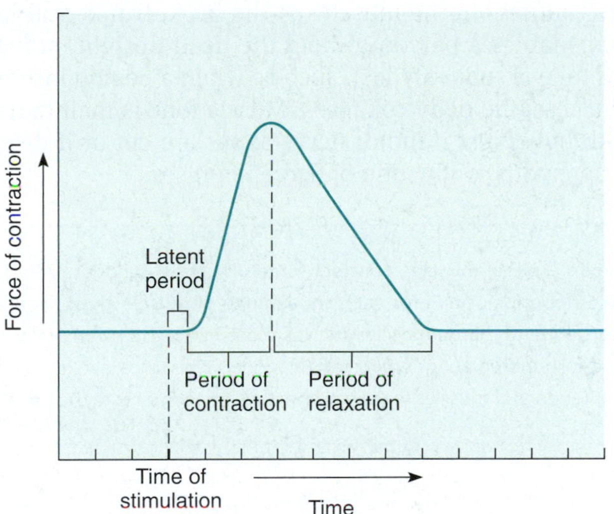

FIGURE 9.15 A myogram of a single muscle twitch.

Sustained contractions of whole muscles enable us to perform everyday activities, but the force generated by those contractions must be controlled. For example, holding a styrofoam cup of coffee firmly enough that it does not slip through our fingers, but not so forcefully as to crush it, requires precise control of contractile force. In the whole muscle, the force developed reflects (1) the frequency at which individual muscle fibers are stimulated and (2) how many fibers take part in the overall contraction of the muscle.

Summation

The force that a muscle fiber can generate is not limited to the maximum force of a single twitch (fig. 9.17*a*). A muscle fiber exposed to a series of stimuli of increasing frequency reaches a point when it is unable to completely relax before the next stimulus in the series arrives. When this happens, the individual twitches begin to combine, and the contraction becomes sustained. In such a *sustained contraction,* the force of individual twitches combines by the process of **summation** (fig. 9.17*b*). When the resulting forceful, sustained contraction lacks even partial relaxation, it is called a **tetanic** (tĕ-tan′ik) **contraction** (tetanus) (fig. 9.17*c*).

Recruitment of Motor Units

The number of muscle fibers in a motor unit varies considerably. The fewer muscle fibers in the motor units, however, the more precise the movements that can be produced in a particular muscle. For example, the motor units of the muscles that move the eyes may include fewer than ten muscle

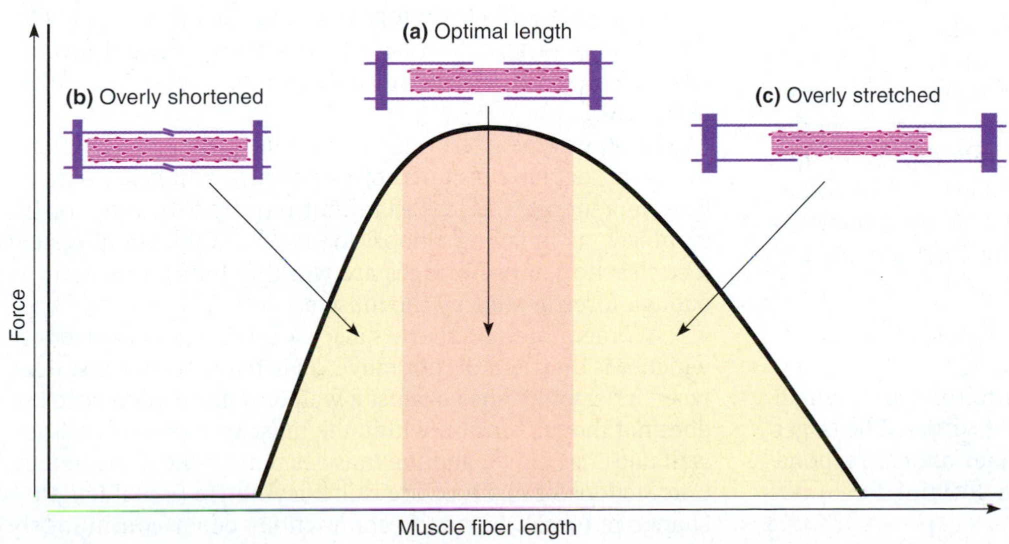

FIGURE 9.16 The force a muscle fiber can generate depends on the length to which it is stretched when stimulated.

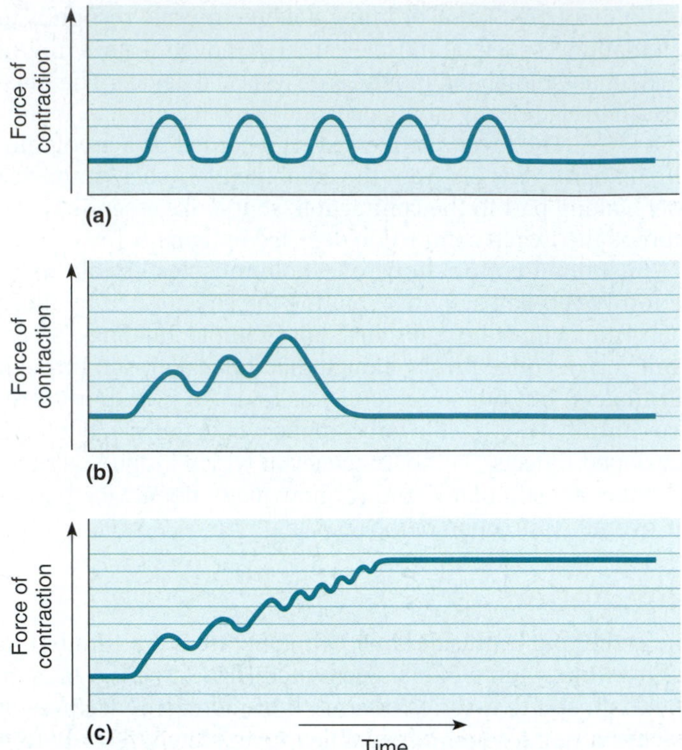

FIGURE 9.17 Myograms of (*a*) a series of twitches, (*b*) summation, and (*c*) a tetanic contraction. Stimulation frequency increases from one myogram to the next.

fibers per motor unit and can produce very slight movements. Conversely, the motor units of the large muscles in the back may include a hundred or more muscle fibers. When these motor units are stimulated, the movements that result are less gradual compared to those of the eye.

Anatomically, the muscle fibers of a muscle are organized into motor units, each of which is controlled by a single motor neuron. Each motor unit is also a functional unit, because a nerve impulse in its motor neuron will contract all of the fibers in that motor unit at the same time. A whole muscle is composed of many such motor units controlled by different motor neurons, which respond to different thresholds of stimulation. If only the more easily stimulated motor neurons are involved, few motor units contract. At higher intensities of stimulation, other motor neurons respond, and more motor units are activated. Such an increase in the number of activated motor units is called *multiple motor unit summation,* or **recruitment** (re-krōō-t′ment). As the intensity of stimulation increases, recruitment of motor units continues until finally all possible motor units are activated in that muscle.

Sustained Contractions

During sustained contractions, smaller motor units, which have smaller diameter axons, are recruited earlier. The larger motor units, which include larger diameter axons, respond later and more forcefully. The result is a sustained contraction of increasing strength.

Typically, many action potentials are triggered in a motor neuron, and so individual twitches do not normally occur. Tetanic contractions of muscle fibers are common. On the whole-muscle level, contractions are smooth rather than irregular or jerky because the spinal cord stimulates contractions in different sets of motor units at different moments.

Tetanic contractions occur frequently in skeletal muscles during everyday activities, often in only a portion of a muscle. For example, when a person lifts a weight or walks, sustained contractions are maintained in the upper limb or lower limb muscles for varying lengths of time. These contractions are responses to a rapid series of stimuli transmitted from the brain and spinal cord on motor neurons.

Even when a muscle appears to be at rest, its fibers undergo a certain degree of sustained contraction. This is called **muscle tone** (tonus), and it is a response to nerve impulses originating repeatedly in the spinal cord and traveling to a few muscle fibers. The result is a continuous state of partial contraction.

Muscle tone is particularly important in maintaining posture. Tautness in the muscles of the neck, trunk, and lower limbs enables a person to hold the head upright, stand, or sit. If tone is suddenly lost, such as when a person loses consciousness, the body collapses. Muscle tone is maintained in health but is lost if motor nerve axons are cut or if diseases interfere with conduction of nerve impulses.

> When skeletal muscles contract forcefully, they may generate up to 50 pounds of pull for each square inch of muscle cross section. Consequently, large muscles such as those in the thigh can pull with several hundred pounds of force. Occasionally, this force is so great that the tendons of muscles tear away from their attachments to the bones.

Types of Contractions

Sometimes muscles shorten when they contract. For example, if a person lifts an object, the muscles remain taut, their attached ends pull closer together, and the object is moved. This type of contraction is termed **isotonic** (equal force—change in length), and because shortening occurs, it is called **concentric.**

Another type of isotonic contraction, called a lengthening or an **eccentric contraction,** occurs when the force a muscle generates is less than that required to move or lift an object, as in laying a book down on a table. Even in such a contraction, cross-bridges are working but not generating enough force to shorten the muscle.

At other times, a skeletal muscle contracts, but the parts to which it is attached do not move. This happens, for instance, when a person pushes against a wall or holds a yoga pose but does not move. Tension within the muscles increases, but the wall does not move, and the muscles remain the same length. Contractions of this type are called **isometric** (equal length—change in force). Isometric contractions occur continuously in postural muscles that stabilize skeletal parts and hold the body upright. Figure 9.18 illustrates isotonic and isometric contractions.

Most body actions require both isotonic and isometric contractions. In walking, for instance, certain leg and thigh muscles contract isometrically and keep the limb stiff as it touches the ground, while other muscles contract isotonically, bending and lifting the limb. Similarly, walking down stairs requires eccentric contraction of certain thigh muscles.

Fast- and Slow-Twitch Muscle Fibers

Muscle fibers vary in contraction speed (slow-twitch or fast-twitch) and in whether they produce ATP oxidatively or glycolytically. At least three combinations of these characteristics are found in humans. Slow-twitch fibers (type I) are always oxidative and are therefore resistant to fatigue. Fast-twitch fibers (type II) may be primarily glycolytic (fatigable) or primarily oxidative (fatigue resistant).

Slow-twitch (type I) fibers, such as those in the long muscles of the back, are often called *red fibers* because they contain the red, oxygen-storing pigment myoglobin. These fibers are well supplied with oxygen-carrying blood. In addition, red fibers contain many mitochondria, an adaptation for the aerobic reactions of cellular respiration. These fibers have a high respiratory capacity and can generate ATP fast enough to keep up with the ATP breakdown that occurs when they contract. For this reason, these fibers can contract for long periods without fatiguing.

Fast-twitch glycolytic fibers (type IIb) are also called *white fibers* because they have less myoglobin and have a poorer blood supply than red fibers. They include fibers in certain hand muscles as well as in muscles that move the eye. These fibers have fewer mitochondria and thus have a reduced respiratory capacity. However, they have a more extensive sarcoplasmic reticulum to store and reabsorb calcium ions, and their ATPase is faster than that of red fibers. White muscle fibers can contract rapidly because of these factors, although they fatigue as lactic acid accumulates and as the ATP and the biochemicals to regenerate ATP are depleted.

A type of white fiber, the fast-twitch fatigue-resistant fibers (type IIa), are also called *intermediate fibers*. These fibers have the fast-twitch speed associated with white fibers with a substantial oxidative capacity more characteristic of red fibers.

While some muscles may have mostly one fiber type or another, all muscles include a combination of fiber types. The speed of contraction and aerobic capacities of the fibers reflect the specialized functions of the muscle. For example, muscles that move the eyes contract about ten times faster than those that maintain posture, and the muscles that move the limbs contract at intermediate rates. Slowing of eye movements is an early sign of certain neurological diseases. Clinical Application 9.2 discusses noticeable effects of muscle use and disuse.

Birds that migrate long distances have abundant dark, slow-twitch muscles—this is why their flesh is dark. In contrast, chickens that can only flap around the barnyard have abundant fast-twitch muscles and mostly white flesh.

World-class distance runners are the human equivalent of the migrating bird. Their muscles may have more than 90% slow-twitch fibers! In some European nations, athletic coaches measure slow-twitch to fast-twitch muscle fiber ratios to predict who will excel at long-distance events and who will fare better in sprints.

PRACTICE

14 Define *threshold stimulus.*

15 Distinguish between a twitch and a sustained contraction.

16 Define *muscle tone.*

17 Explain the differences between isometric and isotonic contractions.

18 Distinguish between fast-twitch and slow-twitch muscles fibers.

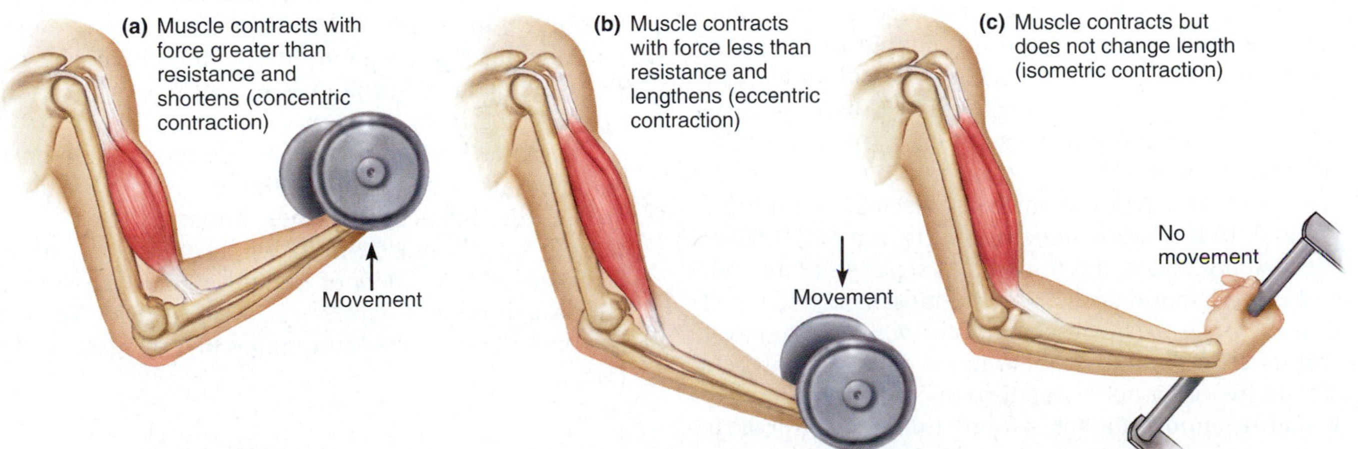
(a) Muscle contracts with force greater than resistance and shortens (concentric contraction)
Movement
(b) Muscle contracts with force less than resistance and lengthens (eccentric contraction)
Movement
(c) Muscle contracts but does not change length (isometric contraction)
No movement

FIGURE 9.18 Types of muscle contractions. (*a* and *b*) Isotonic contractions include concentric and eccentric contractions. (*c*) Isometric contractions occur when a muscle contracts but does not shorten or lengthen.

9.2 CLINICAL APPLICATION

Use and Disuse of Skeletal Muscles

Skeletal muscles respond to use and disuse. Forcefully exercised muscles enlarge, or *hypertrophy*. Unused muscles *atrophy*, decreasing in size and strength.

The way a muscle responds to use also depends on the type of exercise. Weak contractions, such as in swimming and running, activate slow, fatigue-resistant red fibers. In response, these fibers develop more mitochondria and more extensive capillary networks, which increase fatigue-resistance during prolonged exercise, although sizes and strengths of the muscle fibers may not change.

In forceful exercise, such as weightlifting, a muscle exerts more than 75% of its maximum tension, using predominantly the muscle's fast, fatigable white fibers. In response, existing muscle fibers synthesize new filaments of actin and myosin, and as their diameters increase, the entire muscle enlarges. However, no new muscle fibers are produced during hypertrophy.

The strength of a contraction is directly proportional to the diameter of the muscle fibers, so an enlarged muscle can contract more strongly than before. However, such a change does not increase the muscle's ability to resist fatigue during activities such as running or swimming.

Microscopic muscle damage can occur with too-frequent weight lifting (strength training). This is why trainers advise lifting weights every other day, rather than daily.

If regular exercise stops, capillary networks shrink, and muscle fibers lose some mitochondria. Actin and myosin filaments diminish, and the entire muscle atrophies. Injured limbs immobilized in casts, or accidents or diseases that interfere with motor nerve impulses, cause muscle atrophy. An unexercised muscle may shrink to less than one-half its usual size in a few months. Muscle fibers whose motor neurons are severed not only shrink but also may fragment, and in time fat or fibrous tissue replaces them. However, reinnervation of such a muscle within the first few months following an injury can restore function.

New technologies can compensate for some muscle loss. "Targeted muscle reinnervation," for example, can tap into the neuromuscular system to assist a person who has lost an upper limb. A surgeon reattaches muscles from a severed arm to the patient's chest wall, then uses electromyography to detect the electrical activity that still reaches those muscles. The information is sent to a microprocessor built into an attached prosthetic arm, where a "neural-machine interface" enables the patient to move the replacement arm at will, just as he or she would consciously direct the movement of the missing part. ■

9.5 SMOOTH MUSCLES

The contractile mechanisms of smooth and cardiac muscles are essentially the same as those of skeletal muscles. However, the cells of these tissues have important structural and functional distinctions.

Smooth Muscle Fibers

Recall from chapter 5 (p. 163) that smooth muscle cells are shorter than the fibers of skeletal muscle, and they have single, centrally located nuclei. Smooth muscle cells are elongated with tapering ends and contain filaments of actin and myosin in myofibrils that extend throughout their lengths. However, the filaments are thin and more randomly distributed than those in skeletal muscle fibers. Smooth muscle cells lack striations and transverse tubules, and their sarcoplasmic reticula are not well developed.

The two major types of smooth muscles are multiunit and visceral. In **multiunit smooth muscle,** the muscle fibers are less well organized and function as separate units, independent of neighboring cells. Smooth muscle of this type is found in the irises of the eyes and in the walls of blood vessels. Typically, multiunit smooth muscle contracts only after stimulation by motor nerve impulses or certain hormones.

Visceral smooth muscle (single-unit smooth muscle) is composed of sheets of spindle-shaped cells held in close contact by gap junctions. The thick portion of each cell lies next to the thin parts of adjacent cells. Fibers of visceral smooth muscle respond as a single unit. When one fiber is stimulated, the impulse moving over its surface may excite adjacent fibers that, in turn, stimulate others. Some visceral smooth muscle cells also display *rhythmicity*—a pattern of spontaneous repeated contractions.

These two features of visceral smooth muscle—transmission of impulses from cell to cell and rhythmicity—are largely responsible for the wavelike motion called **peristalsis** of certain tubular organs (see chapter 17, pp. 654 and 656). Peristalsis consists of alternate contractions and relaxations of the longitudinal and circular muscles. These movements help force the contents of a tube along its length. In the intestines, for example, peristaltic waves move masses of partially digested food and help to mix them with digestive fluids. Peristalsis in the ureters moves urine from the kidneys to the urinary bladder.

Visceral smooth muscle is the more common type of smooth muscle and is found in the walls of hollow organs, such as the stomach, intestines, urinary bladder, and uterus. Usually smooth muscle in the walls of these organs has two thicknesses. The fibers of the outer coats are longitudinal, whereas those of the inner coats are circular. The muscular layers change the sizes and shapes of the organs as they contract and relax.

Smooth Muscle Contraction

Smooth muscle contraction resembles skeletal muscle contraction in a number of ways. Both mechanisms reflect reactions of actin and myosin; both are triggered by membrane impulses and release of calcium ions; and both use energy from ATP molecules. However, smooth and skeletal muscle action also differs. For example, smooth muscle fibers lack troponin, the protein that binds to calcium ions in skeletal muscle. Instead, smooth muscle uses a protein called *calmodulin*, which binds to calcium ions released when its fibers are stimulated, activating contraction. In addition, much of the calcium necessary for smooth muscle contraction diffuses into the cell from the extracellular fluid.

Acetylcholine, the neurotransmitter in skeletal muscle, as well as *norepinephrine*, affect smooth muscle. Each of these neurotransmitters stimulates contractions in some smooth muscles and inhibits contractions in others. The discussion of the autonomic nervous system in chapter 11 (p. 424) describes these actions in greater detail.

Hormones affect smooth muscles by stimulating or inhibiting contraction in some cases and altering the degree of response to neurotransmitters in others. For example, during the later stages of childbirth, the hormone oxytocin stimulates smooth muscles in the wall of the uterus to contract (see chapter 23, pp. 899–901).

Stretching of smooth muscle fibers can also trigger contractions. This response is particularly important to the function of visceral smooth muscle in the walls of certain hollow organs, such as the urinary bladder and the intestines. For example, when partially digested food stretches the wall of the intestine, contractions move the contents further along the intestine.

Smooth muscle is slower to contract and relax than skeletal muscle, yet smooth muscle can forcefully contract longer with the same amount of ATP. Unlike skeletal muscle, smooth muscle fibers can change length without changing tautness; because of this, smooth muscles in the stomach and intestinal walls can stretch as these organs fill, holding the pressure inside the organs constant.

PRACTICE

19 Describe the two major types of smooth muscle.

20 What special characteristics of visceral smooth muscle make peristalsis possible?

21 How is smooth muscle contraction similar to skeletal muscle contraction?

22 How do the contraction mechanisms of smooth and skeletal muscles differ?

9.6 CARDIAC MUSCLE

Cardiac muscle appears only in the heart. It is composed of striated cells joined end to end, forming fibers interconnected in branching, three-dimensional networks. Each cell contains a single nucleus and many filaments of actin and myosin similar to those in skeletal muscle. A cardiac muscle cell also has a well-developed sarcoplasmic reticulum, a system of transverse tubules, and many mitochondria. However, the cisternae of the sarcoplasmic reticulum of a cardiac muscle fiber are less developed and store less calcium than those of a skeletal muscle fiber. On the other hand, the transverse tubules of cardiac muscle fibers are larger than those in skeletal muscle, and they release many calcium ions into the sarcoplasm in response to a single muscle impulse.

The calcium ions in transverse tubules come from the fluid outside the muscle fiber. In this way, extracellular calcium partially controls the strength of cardiac muscle contraction and enables cardiac muscle fibers to contract longer than skeletal muscle fibers can.

Drugs called calcium channel blockers are used to treat irregular heart rhythms. They do this by blocking ion channels that admit extracellular calcium into cardiac muscle cells.

The opposing ends of cardiac muscle cells are connected by cross-bands called *intercalated discs*. These bands are complex membrane junctions. Not only do they help join cells and transmit the force of contraction from cell to cell, but the intercellular junctions of the fused membranes of intercalated discs allow ions to diffuse between the cells. This allows muscle impulses to travel rapidly from cell to cell (see figs. 5.30 and 9.19).

When one portion of the cardiac muscle network is stimulated, the impulse passes to other fibers of the network, and the whole structure contracts as a unit (a *syncytium*); that is, the network responds to stimulation in an all-or-none manner. Cardiac muscle is also self-exciting and rhythmic. Consequently, a pattern of contraction and relaxation repeats, generating the rhythmic contraction of the heart. Also, the refractory period of cardiac muscle is longer than in skeletal muscle and lasts until the contraction ends. Thus, sustained or tetanic contractions do not occur in the heart muscle. Table 9.2 summarizes characteristics of the three types of muscles.

PRACTICE

23 How is cardiac muscle similar to skeletal muscle?

24 How does cardiac muscle differ from skeletal muscle?

25 What is the function of intercalated discs?

26 What characteristic of cardiac muscle causes the heart to contract as a unit?

9.7 SKELETAL MUSCLE ACTIONS

Skeletal muscles generate a great variety of body movements. The action of each muscle mostly depends upon the type of joint it is associated with and the way the muscle is attached on either side of that joint.

Body Movement

Whenever limbs or other body parts move, bones and muscles interact as simple mechanical devices called **levers** (lev'erz).

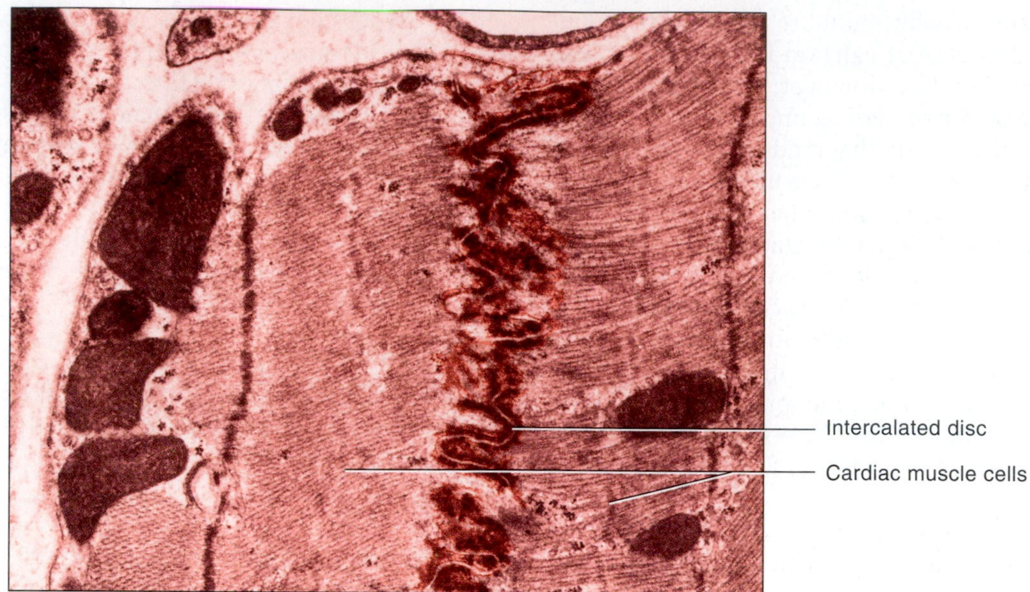

FIGURE 9.19 The intercalated discs of cardiac muscle, shown in this transmission electron micrograph, bind adjacent cells and allow ions to move between cells (12,500×).

TABLE 9.2 | Characteristics of Muscle Tissues

	Skeletal	Smooth	Cardiac
Dimensions			
Length	Up to 30 cm	30–200 μm	50–100 μm
Diameter	10–100 μm	3–6 μm	14 μm
Major location	Skeletal muscles	Walls of hollow organs	Wall of the heart
Major function	Movement of bones at joints; maintenance of posture	Movement of walls of hollow organs; peristalsis; vasoconstriction	Pumping action of the heart
Cellular characteristics			
Striations	Present	Absent	Present
Nucleus	Multiple nuclei	Single nucleus	Single nucleus
Special features	Transverse tubule system is well developed	Lacks transverse tubules	Transverse tubule system is well developed; intercalated discs separate cells
Mode of control	Voluntary	Involuntary	Involuntary
Contraction characteristics	Contracts and relaxes relatively rapidly	Contracts and relaxes relatively slowly; some types self-exciting; rhythmic	Network of fibers contracts as a unit; self-exciting; rhythmic; remains refractory until contraction ends

A lever has four basic components: (1) a rigid bar or rod, (2) a fulcrum or pivot on which the bar turns, (3) an object moved against resistance, and (4) a force that supplies energy for the movement of the bar.

A pair of scissors is a lever. The handle and blade form a rigid bar that rocks on a fulcrum near the center (the screw). The material to be cut by the blades represents the resistance, while the person on the handle end supplies the force needed for cutting the material.

Figure 9.20 shows the three types of levers, which differ in their arrangements. A first-class lever's parts are like those of a pair of scissors. Its fulcrum is located between the resistance and the force, making the sequence of components resistance–fulcrum–force. Other examples of first-class levers are seesaws and hemostats (devices used to clamp blood vessels).

The parts of a second-class lever are in the sequence fulcrum–resistance–force, as in a wheelbarrow. The parts of a third-class lever are in the sequence resistance–force–fulcrum. Eyebrow tweezers or forceps used to grasp an object illustrate this type of lever.

The actions of bending and straightening the upper limb at the elbow illustrate bones and muscles functioning as levers. When the upper limb bends, the forearm bones represent the rigid bar; the elbow joint is the fulcrum; the hand is moved against the resistance provided by the weight; and the force is supplied by muscles on the anterior side of the arm (fig. 9.21*a*). One of these muscles, the *biceps brachii*, is

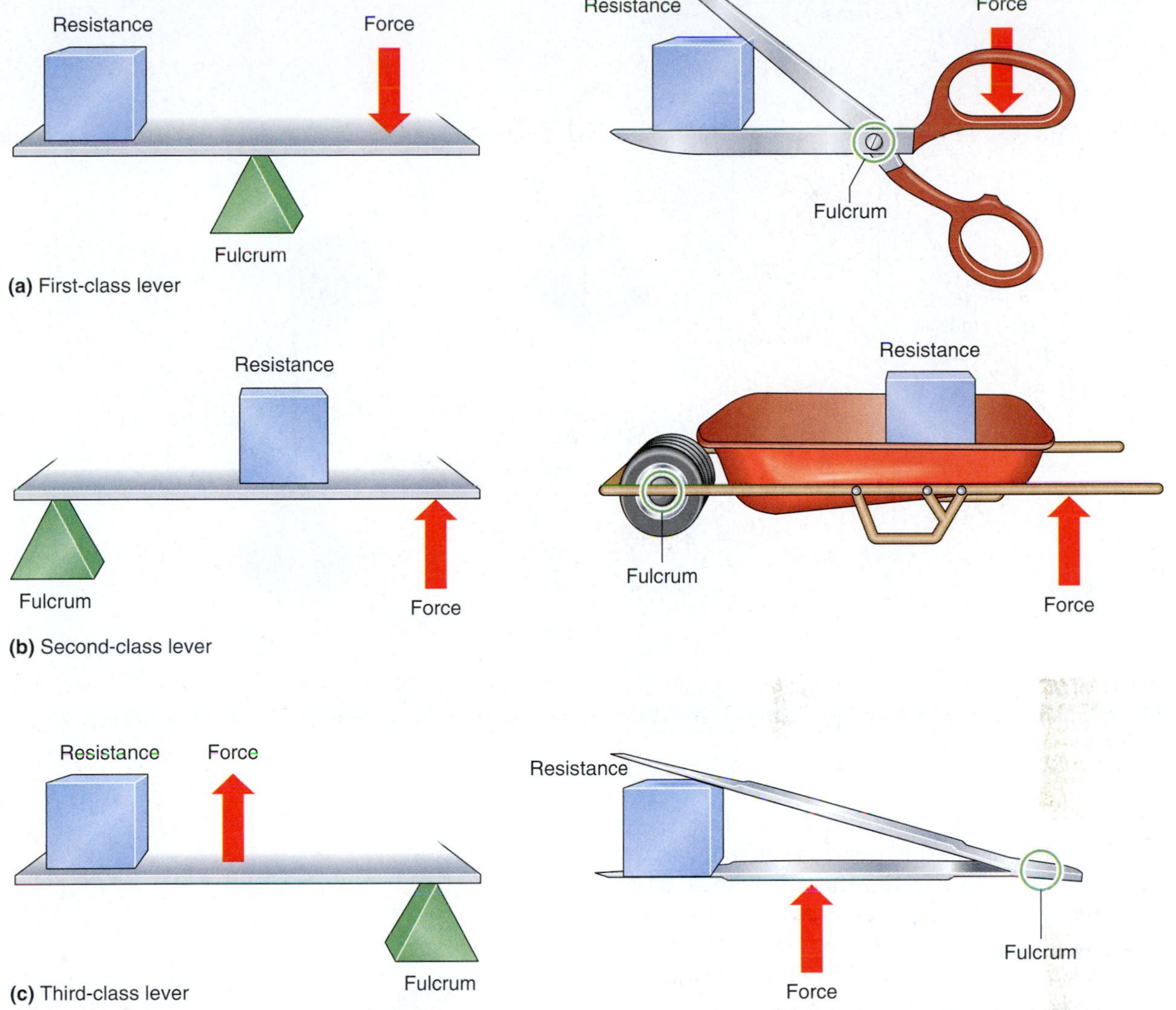

FIGURE 9.20 Three types of levers. (*a*) A first-class lever is used in a pair of scissors, (*b*) a second-class lever is used in a wheelbarrow, and (*c*) a third-class lever is used in a pair of forceps.

attached by a tendon to a projection (radial tuberosity) on the *radius* bone in the forearm, a short distance below the elbow. The parts of this lever are arranged in the sequence resistance–force–fulcrum, so it is a third-class lever.

When the upper limb straightens at the elbow, the forearm bones again serve as the rigid bar, the hand moves against the resistance by pulling on the rope to raise the weight (fig. 9.21*b*), and the elbow joint serves as the fulcrum. However, this time the *triceps brachii,* a muscle located on the posterior side of the arm, supplies the force. A tendon of this muscle attaches to a projection (olecranon process) of the ulna bone at the point of the elbow. The parts of the lever are arranged resistance–fulcrum–force, so it is a first-class lever.

A second-class lever (fulcrum–resistance–force) is also demonstrated in the human body. The fulcrum is the temporomandibular joint; muscles supply the resistance, attaching to a projection (coronoid process) and body of the mandible, that resist or oppose opening the mouth. The muscles attached to the chin area of the mandible provide the force that opens the mouth.

Levers provide a range of movements. Levers that move limbs, for example, produce rapid motions, whereas others, such as those that move the head, help maintain posture with minimal effort.

Origin and Insertion

Recall from chapter 8 (p. 267) that one end of a skeletal muscle is usually fastened to a relatively immovable or fixed part, and the other end is connected to a movable part on the other side of a joint. The immovable end is called the **origin** of the muscle, and the movable end is called its **insertion.** When a muscle contracts, its insertion is pulled toward its origin (fig. 9.22). The head of a muscle is the part nearest its origin.

Some muscles have more than one origin or insertion. The *biceps brachii* in the arm, for example, has two origins.

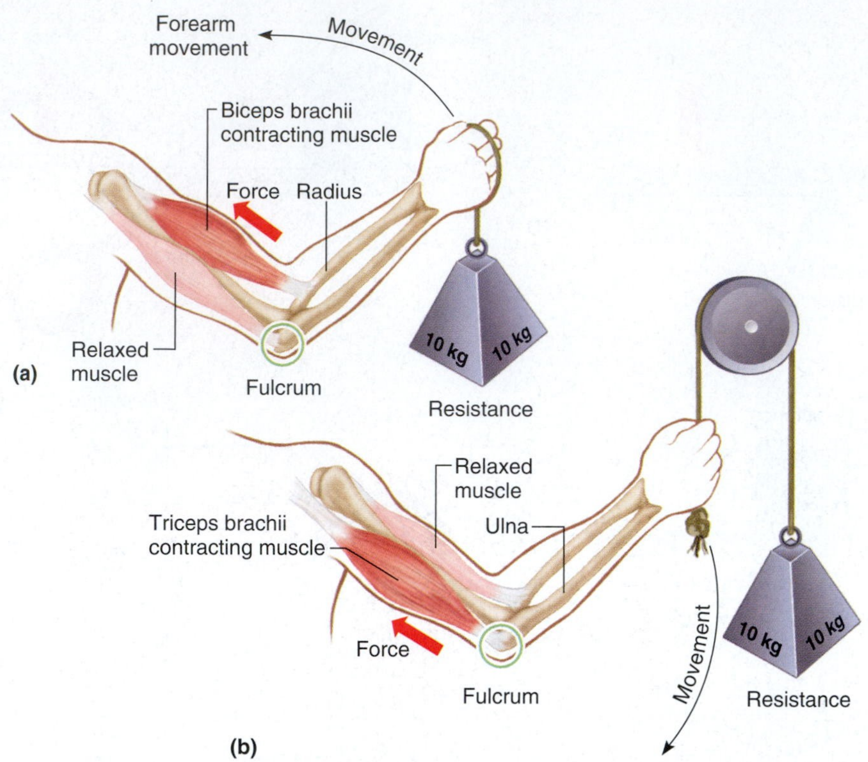

FIGURE 9.21 Levers and movement. (*a*) When the forearm bends at the elbow or (*b*) when the forearm straightens at the elbow, the bones and muscles function as a lever.

This is reflected in its name *biceps,* meaning "two heads." As figure 9.22 shows, one head of the muscle is attached to the coracoid process of the scapula, and the other head arises from a tubercle above the glenoid cavity of the scapula. The muscle extends along the anterior surface of the humerus and is inserted by a single tendon on the radial tuberosity of the radius. When the biceps brachii contracts, its insertion is pulled toward its origin, and the elbow bends.

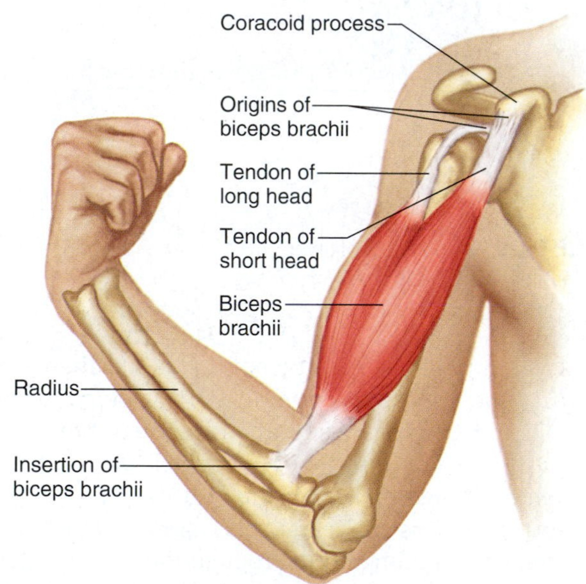

FIGURE 9.22 The biceps brachii has two heads that originate on the scapula. A tendon inserts this muscle on the radius.

Interaction of Skeletal Muscles

Skeletal muscles almost always function in groups. As a result, when a particular body part moves, a person must do more than contract a single muscle; instead, after learning to make a particular movement, the person wills the movement to occur, and the nervous system stimulates the appropriate group of muscles.

By carefully observing body movements, it is possible to determine the roles of particular muscles. For instance, abduction of the arm requires contracting the *deltoid* muscle, said to be the **prime mover** or **agonist.** A prime mover is the muscle primarily responsible for producing an action. However, while a prime mover is acting, certain nearby muscles also contract. When a deltoid muscle contracts, nearby muscles help hold the shoulder steady and in this way make the action of the prime mover more effective. Muscles that contract and assist a prime mover are called **synergists** (sin′er-jists).

Still other muscles act as **antagonists** (an-tag′o-nists) to prime movers. These muscles can resist a prime mover's action and cause movement in the opposite direction—the antagonist of the prime mover that raises the upper limb can lower the upper limb, or the antagonist of the prime mover that bends the upper limb can straighten it. If both a prime mover and its antagonist contract simultaneously, the structure they act upon remains rigid. Similarly, smooth body movements depend upon the antagonists' relaxing and giving way to the prime movers whenever the prime movers contract. Once again, the nervous system coordinates these complex actions, as described in chapter 11 (p. 410).

Sometimes the relationship between two muscles changes. For example, the pectoralis major and latissimus dorsi are antagonistic for flexion and extension of the shoulder. However, they are synergistic for medial rotation of the shoulder. Thus, the role of a muscle must be learned in the context of a particular movement.

The movements termed "flexion" and "extension" describe changes in the angle between bones that meet at a joint. For example, flexion of the elbow joint refers to a movement of the forearm that decreases the angle at the elbow joint. Alternatively, one could say that flexion at the elbow results from the action of the biceps brachii on the radius of the forearm.

Students find it helpful to think of movements in terms of the specific actions of the muscles involved, so we may also describe flexion and extension in these terms. Thus, the action of the biceps brachii may be described as "flexion of the forearm at the elbow" and the action of the quadriceps group as "extension of the leg at the knee." We believe that this occasional departure from strict anatomical terminology eases understanding and learning.

PRACTICE

27 Explain how parts of the upper limb form a first-class lever and a third-class lever.

28 Distinguish between the origin and the insertion of a muscle.

29 Define prime mover.

30 What is the function of a synergist? An antagonist?

9.8 MAJOR SKELETAL MUSCLES

This section discusses the locations, actions, origins, and insertions of some of the major skeletal muscles. The tables that summarize the information concerning groups of these muscles also include the names of nerves that supply the individual muscles in each group. Chapter 11 (pp. 414–423) presents the origins and pathways of these nerves.

Figures 9.23 and 9.24 show the locations of superficial skeletal muscles—that is, those near the surface. The names of muscles often describe them. A name may indicate a muscle's size, shape, location, action, number of attachments, or the direction of its fibers, as in the following examples:

pectoralis major A muscle of large size (major) in the pectoral region (chest).
deltoid Shaped like a delta or triangle.
extensor digitorum Extends the digits (fingers or toes).
biceps brachii A muscle with two heads (biceps), or points of origin, in the brachium (arm).

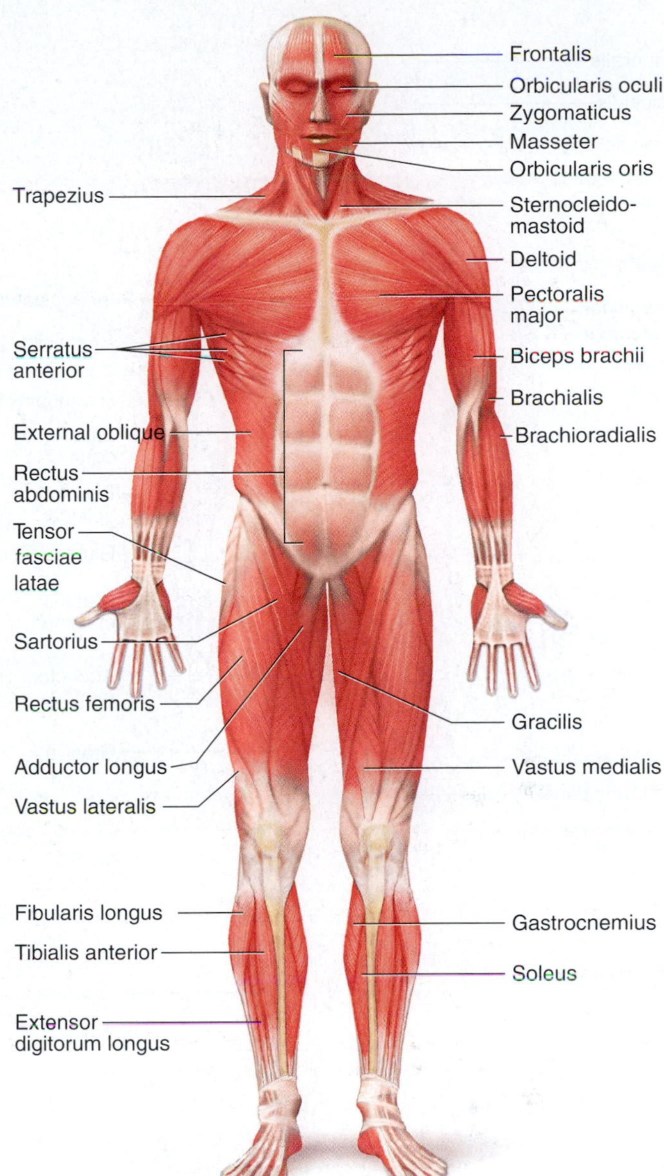

FIGURE 9.23 Anterior view of superficial skeletal muscles.

sternocleidomastoid Attached to the sternum, clavicle, and mastoid process.
external oblique Located near the outside, with fibers that run obliquely or in a slanting direction.

Muscles of Facial Expression

A number of small muscles beneath the skin of the face and scalp enable us to communicate feelings through facial expression. Many of these muscles are located around the eyes and mouth, and they make possible such expressions as surprise, sadness, anger, fear, disgust, and pain. As a group, the muscles of facial expression connect the bones of the skull to connective tissue in regions of the overlying skin.

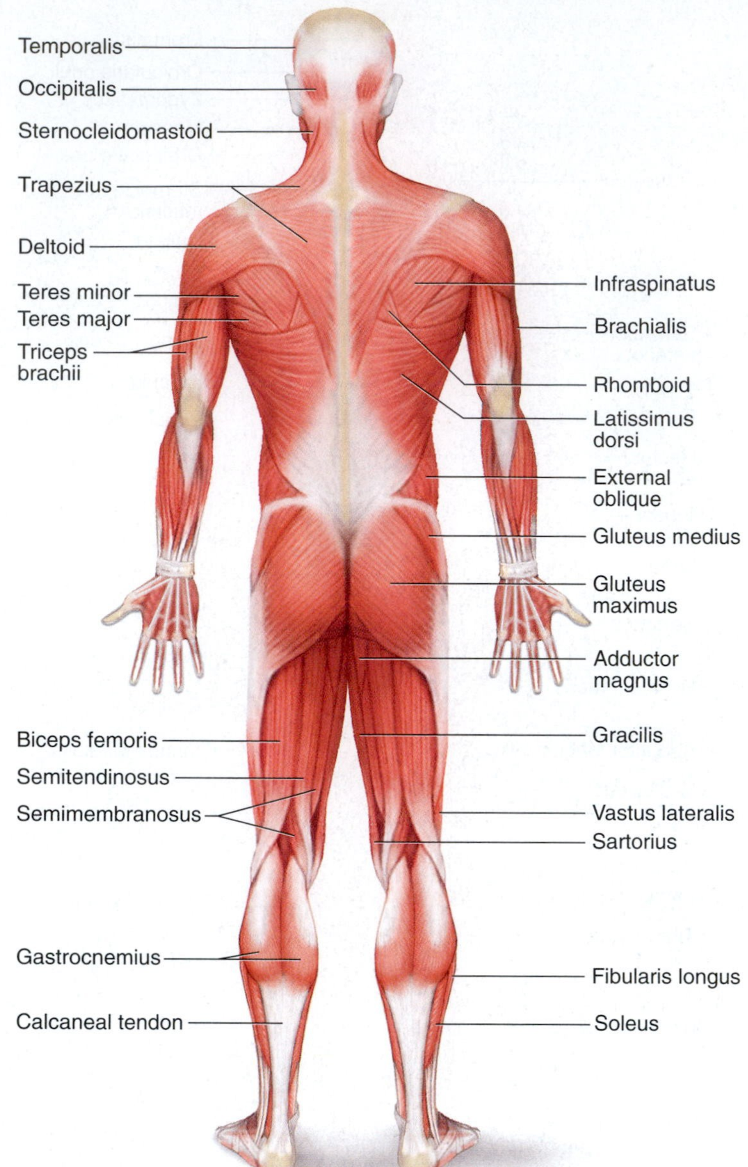

FIGURE 9.24 Posterior view of superficial skeletal muscles.

Figure 9.25 and reference plate 66 show these muscles, and table 9.3 lists them. The muscles of facial expression include the following:

Epicranius	Zygomaticus major
Orbicularis oculi	Zygomaticus minor
Orbicularis oris	Platysma
Buccinator	

The **epicranius** (ep″ĭ-kra′ne-us) covers the upper part of the cranium and consists of two muscular parts—the *frontalis* (frun-ta′lis), which lies over the frontal bone, and the *occipitalis* (ok-sip″ĭ-ta′lis), which lies over the occipital bone. These muscles are united by a broad, tendinous membrane called the *epicranial aponeurosis,* which covers the cranium like a cap. Contraction of the epicranius raises the eyebrows and horizontally wrinkles the skin of the forehead, as when a person expresses surprise. Headaches often result from sustained contraction of this muscle.

The **orbicularis oculi** (or-bik′u-la-rus ok′u-li) is a ringlike band of muscle, called a *sphincter muscle,* that surrounds the eye. It lies in the subcutaneous tissue of the eyelid and closes or blinks the eye. At the same time, it compresses the nearby tear gland, or *lacrimal gland,* aiding the flow of tears over the surface of the eye. Contraction of the orbicularis oculi also causes the folds, or crow's feet, that radiate laterally from the corner of the eye. The muscles that move the eye are described in chapter 12 (pp. 464–465).

The **orbicularis oris** (or-bik′u-la-rus o′ris) is a sphincter muscle that encircles the mouth. It lies between the skin and the mucous membranes of the lips, extending upward to the nose and downward to the region between the lower lip and chin. The orbicularis oris is sometimes called the kissing muscle because it closes and puckers the lips.

The **buccinator** (buk′sĭ-na″tor) is located in the wall of the cheek. Its fibers are directed forward from the bones of the jaws to the angle of the mouth, and when they contract, the cheek is compressed inward. This action helps hold food in contact with the teeth when a person is chewing. The buccinator also aids in blowing air out of the mouth, and for this reason, it is sometimes called the trumpeter muscle.

The **zygomaticus** (zi″go-mat′ik-us) **major** and **minor** extend from the zygomatic arch downward to the corner of the mouth. When they contract, the corner of the mouth is drawn upward, as in smiling or laughing.

TABLE 9.3 | Muscles of Facial Expression

Muscle	Origin	Insertion	Action	Nerve Supply
Epicranius	Occipital bone	Skin and muscles around eye	Raises eyebrow as when surprised	Facial n.
Orbicularis oculi	Maxillary and frontal bones	Skin around eye	Closes eye as in blinking	Facial n.
Orbicularis oris	Muscles near the mouth	Skin of central lip	Closes lips, protrudes lips as for kissing	Facial n.
Buccinator	Outer surfaces of maxilla and mandible	Orbicularis oris	Compresses cheeks inward as when blowing air	Facial n.
Zygomaticus major	Zygomatic bone	Corner of mouth	Raises corner of mouth as when smiling	Facial n.
Zygomaticus minor	Zygomatic bone	Corner of mouth	Raises corner of mouth as when smiling	Facial n.
Platysma	Fascia in upper chest	Lower border of mandible	Draws angle of mouth downward as when pouting	Facial n.

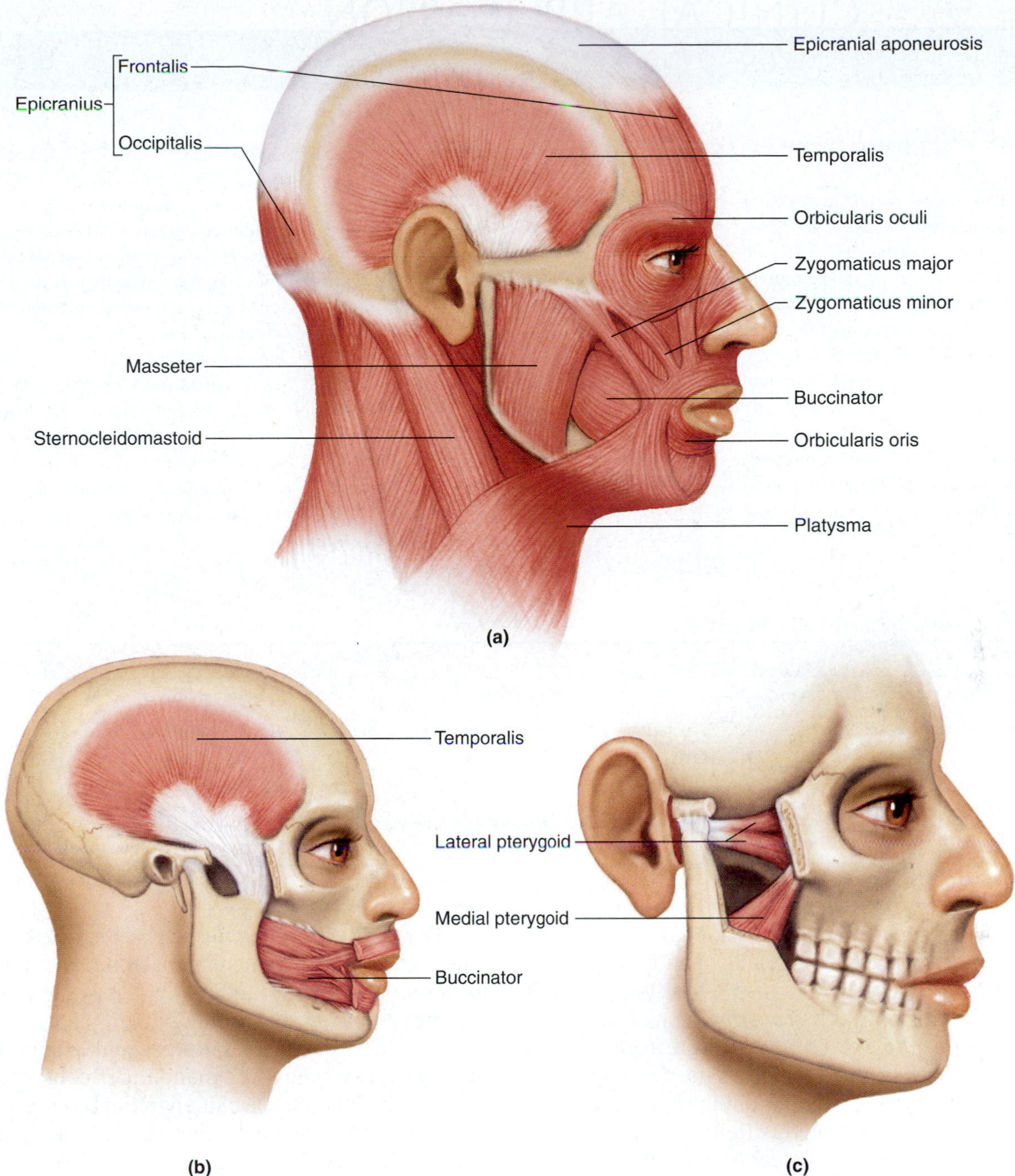

FIGURE 9.25 Muscles of the head and face. (*a*) Muscles of facial expression and mastication; isolated views of (*b*) the temporalis and buccinator muscles and (*c*) the lateral and medial pterygoid muscles.

The **platysma** (plah-tiz′mah) is a thin, sheetlike muscle whose fibers extend from the chest upward over the neck to the face. It pulls the angle of the mouth downward, as in pouting. The platysma also helps lower the mandible.

Muscles of Mastication

Four pairs of muscles attached to the mandible produce chewing movements. Three pairs of these muscles close the lower jaw, as in biting; the fourth pair can lower the jaw; cause side-to-side grinding motions of the mandible; and pull the mandible forward, causing it to protrude. The muscles of mastication are shown in figure 9.25 and reference plate 66 and are listed in table 9.4. They include the following:

Masseter	Medial pterygoid
Temporalis	Lateral pterygoid

The **masseter** (mas-se′ter) is a thick, flattened muscle that can be felt just in front of the ear when the teeth are clenched. Its fibers extend downward from the zygomatic

9.3 CLINICAL APPLICATION

TMJ Syndrome

Facial pain, headache, ringing in the ears, a clicking jaw, insomnia, teeth sensitive to heat or cold, backache, dizziness, and pain in front of the ears are aches and pains that may all result from temporomandibular joint (TMJ) syndrome. A misaligned jaw or grinding or clenching the teeth can cause TMJ by stressing the temporomandibular joint, the articulation between the mandibular condyle of the mandible and the mandibular fossa of the temporal bone. Loss of coordination of these structures affects the nerves that pass through the neck and jaw region, causing the symptoms. In TMJ syndrome, tensing a muscle in the forehead can cause a headache, or a spasm in the muscle that normally opens the auditory tubes during swallowing can impair ability to clear the ears.

Many cases of TMJ can be treated without medical intervention if a cause is identified. Getting enough sleep, drinking enough water, and learning relaxation techniques to combat stress can help. Massaging affected muscles can alleviate symptoms. Posture is sometimes the culprit—sitting for long hours in one position in front of a computer screen can cause or worsen TMJ.

Doctors diagnose TMJ syndrome using an electromyograph, in which electrodes record muscle activity in four pairs of head and neck muscle groups. One treatment is transcutaneous electrical nerve stimulation (TENS), which stimulates the facial muscles for up to an hour. Another treatment is an orthotic device fitted by a dentist. Worn for three to six months, the device fine-tunes the action of jaw muscles to form a more comfortable bite. A dentist can use bonding materials to alter shapes of certain teeth to provide a more permanent treatment for TMJ syndrome. ■

TABLE 9.4 | Muscles of Mastication

Muscle	Origin	Insertion	Action	Nerve Supply
Masseter	Lower border of zygomatic arch	Lateral surface of mandible	Elevates mandible	Trigeminal n.
Temporalis	Temporal bone	Coronoid process and anterior ramus of mandible	Elevates mandible	Trigeminal n.
Medial pterygoid	Sphenoid, palatine, and maxillary bones	Medial surface of mandible	Elevates mandible and moves it from side to side	Trigeminal n.
Lateral pterygoid	Sphenoid bone	Anterior surface of mandibular condyle	Depresses and protracts mandible and moves it from side to side	Trigeminal n.

arch to the mandible. The masseter raises the jaw, but it can also control the rate at which the jaw falls open in response to gravity (fig. 9.25*a*).

The **temporalis** (tem-po-ra′lis) is a fan-shaped muscle located on the side of the skull above and in front of the ear. Its fibers, which also raise the jaw, pass downward beneath the zygomatic arch to the mandible (fig. 9.25*a* and *b*). Tensing this muscle is associated with temporomandibular joint syndrome, discussed in Clinical Application 9.3.

When two dentists examined an eyeless cadaver's skull from an unusual perspective, they discovered a then-unknown muscle. Named the sphenomandibularis, the muscle extends about an inch and a half from behind the eyes to the inside of the jawbone and may assist chewing movements. In traditional dissection from the side, the new muscle's origin and insertion are not visible, so it may have appeared to be part of the larger and overlying temporalis muscle. Although the sphenomandibularis inserts on the inner side of the jawbone, as does the temporalis, it originates differently, on the sphenoid bone.

The **medial pterygoid** (ter′ĭ-goid) extends back and downward from the sphenoid, palatine, and maxillary bones to the ramus of the mandible. It closes the jaw (fig. 9.25*c*) and moves it from side to side.

The fibers of the **lateral pterygoid** extend forward from the region just below the mandibular condyle to the sphenoid bone. This muscle can open the mouth, pull the mandible forward to make it protrude, and move the mandible from side to side (fig. 9.25*c*).

Muscles That Move the Head and Vertebral Column

Paired muscles in the neck and back flex, extend, and rotate the head and hold the torso erect (figs. 9.26 and 9.28 and table 9.5). They include the following:

Sternocleidomastoid
Splenius capitis
Semispinalis capitis
Quadratus lumborum
Erector spinae

The **sternocleidomastoid** (ster″no-kli″do-mas′toid) is a long muscle in the side of the neck that extends upward from the thorax to the base of the skull behind the ear. When

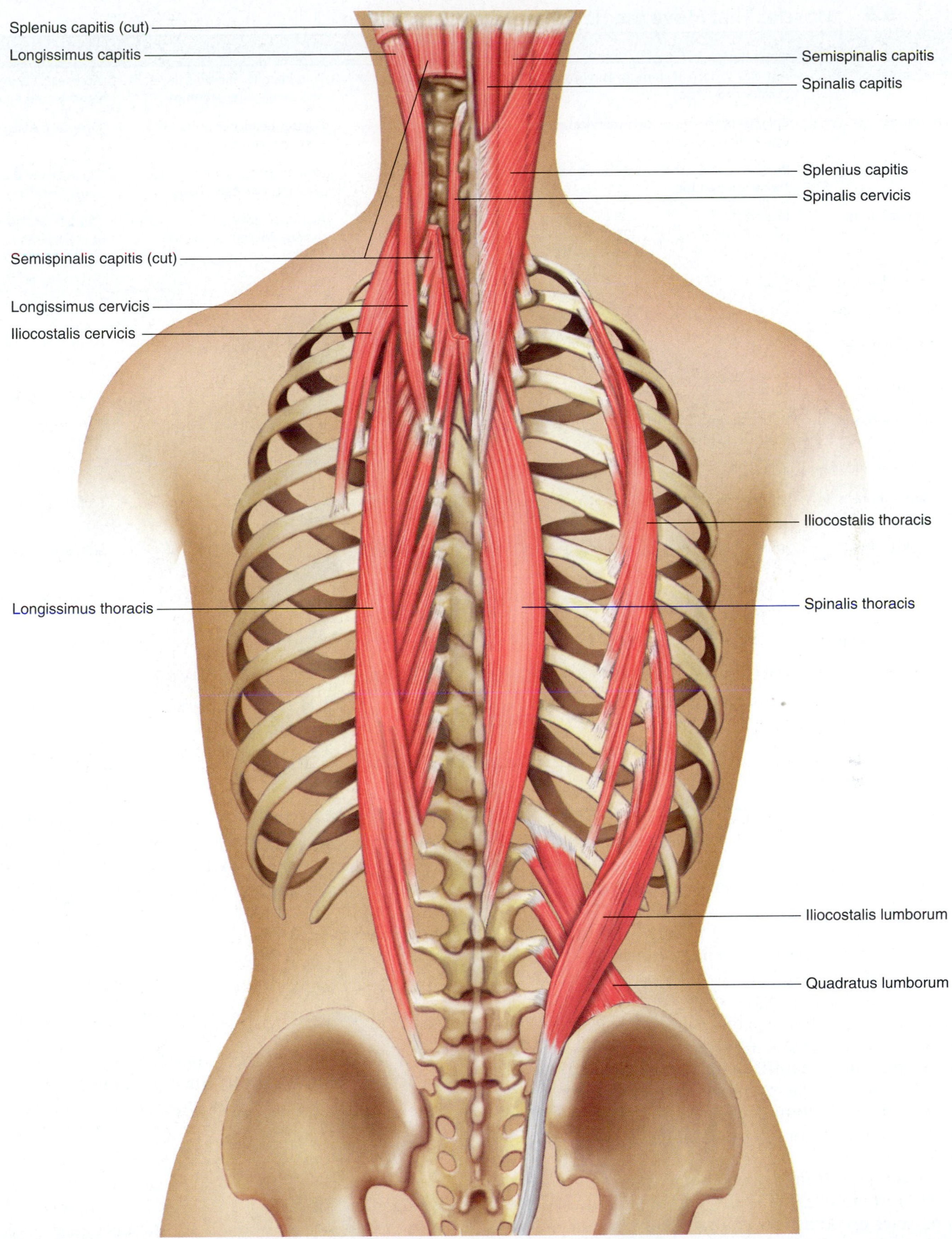

FIGURE 9.26 Deep muscles of the back and the neck help move the head (posterior view) and hold the torso erect. The splenius capitis and semispinalis capitis are removed on the left to show underlying muscles.

TABLE 9.5 | Muscles That Move the Head and Vertebral Column

Muscle	Origin	Insertion	Action	Nerve Supply
Sternocleidomastoid	Anterior surface of sternum and upper surface of clavicle	Mastoid process of temporal bone	Pulls head to one side, flexes neck or elevates sternum	Accessory, C2 and C3 cervical nerves
Splenius capitis	Spinous processes of lower cervical and upper thoracic vertebrae	Occipital bone	Rotates head, bends head to one side, or extends neck	Cervical nerves
Semispinalis capitis	Processes of lower cervical and upper thoracic vertebrae	Occipital bone	Extends head, bends head to one side, or rotates head	Cervical and thoracic spinal nerves
Quadratus lumborum	Iliac crest	Upper lumbar vertebrae and twelfth rib	Aids in breathing, extends lumbar region of vertebral column	Thoracic and lumbar spinal nerves
Erector spinae (divides into three columns)				
Iliocostalis (lateral) group				
Iliocostalis lumborum	Iliac crest	Lower six ribs	Extends lumbar region of vertebral column	Lumbar spinal nerves
Iliocostalis thoracis	Lower six ribs	Upper six ribs	Holds spine erect	Thoracic spinal nerves
Iliocostalis cervicis	Upper six ribs	Fourth through sixth cervical vertebrae	Extends cervical region of vertebral column	Cervical spinal nerves
Longissimus (intermediate) group				
Longissimus thoracis	Lumbar vertebrae	Thoracic and upper lumbar vertebrae and ribs 9 and 10	Extends thoracic region of vertebral column	Spinal nerves
Longissimus cervicis	Fourth and fifth thoracic vertebrae	Second through sixth cervical vertebrae	Extends cervical region of vertebral column	Spinal nerves
Longissimus capitis	Upper thoracic and lower cervical vertebrae	Mastoid process of temporal bone	Extends and rotates head	Cervical spinal nerves
Spinalis (medial) group				
Spinalis thoracis	Upper lumbar and lower thoracic vertebrae	Upper thoracic vertebrae	Extends vertebral column	Spinal nerves
Spinalis cervicis	Ligamentum nuchae and seventh cervical vertebra	Axis	Extends vertebral column	Spinal nerves
Spinalis capitis	Upper thoracic and lower cervical vertebrae	Occipital bone	Extends vertebral column	Spinal nerves

the sternocleidomastoid on one side contracts, the face turns to the opposite side. When both muscles contract, the head bends toward the chest. If other muscles fix the head in position, the sternocleidomastoids can raise the sternum, aiding forceful inhalation (see fig. 9.28 and table 9.5).

The **splenius capitis** (sple′ne-us kap′ĭ-tis) is a broad, straplike muscle in the back of the neck. It connects the base of the skull to the vertebrae in the neck and upper thorax. A splenius capitis acting singly rotates the head and bends it toward one side. Acting together, these muscles bring the head into an upright position (see fig. 9.26 and table 9.5).

The **semispinalis capitis** (sem″e-spi-na′lis kap′ĭ-tis) is a broad, sheetlike muscle extending upward from the vertebrae in the neck and thorax to the occipital bone. It extends the head, bends it to one side, or rotates it (see fig. 9.26 and table 9.5).

The **quadratus lumborum** (kwod-ra′tus lum-bo′rum) is located in the lumbar region. When the quadratus lumborum muscles on both sides contract the vertebral column is extended. When the muscle on only one side contracts, the vertebral column is flexed laterally.

Erector spinae muscles run longitudinally along the back, with origins and insertions at many places on the axial skeleton. These muscles extend and rotate the head and maintain the erect position of the vertebral column. Erector spinae can be subdivided into lateral, intermediate, and medial groups (table 9.5).

Muscles That Move the Pectoral Girdle

The muscles that move the pectoral girdle are closely associated with those that move the arm. A number of these chest and shoulder muscles connect the scapula to nearby bones and move the scapula upward, downward, forward, and backward (figs. 9.27, 9.28, 9.29; reference plates 68, 69; table 9.6). Muscles that move the pectoral girdle include the following:

- Trapezius
- Rhomboid major
- Rhomboid minor
- Levator scapulae
- Serratus anterior
- Pectoralis minor

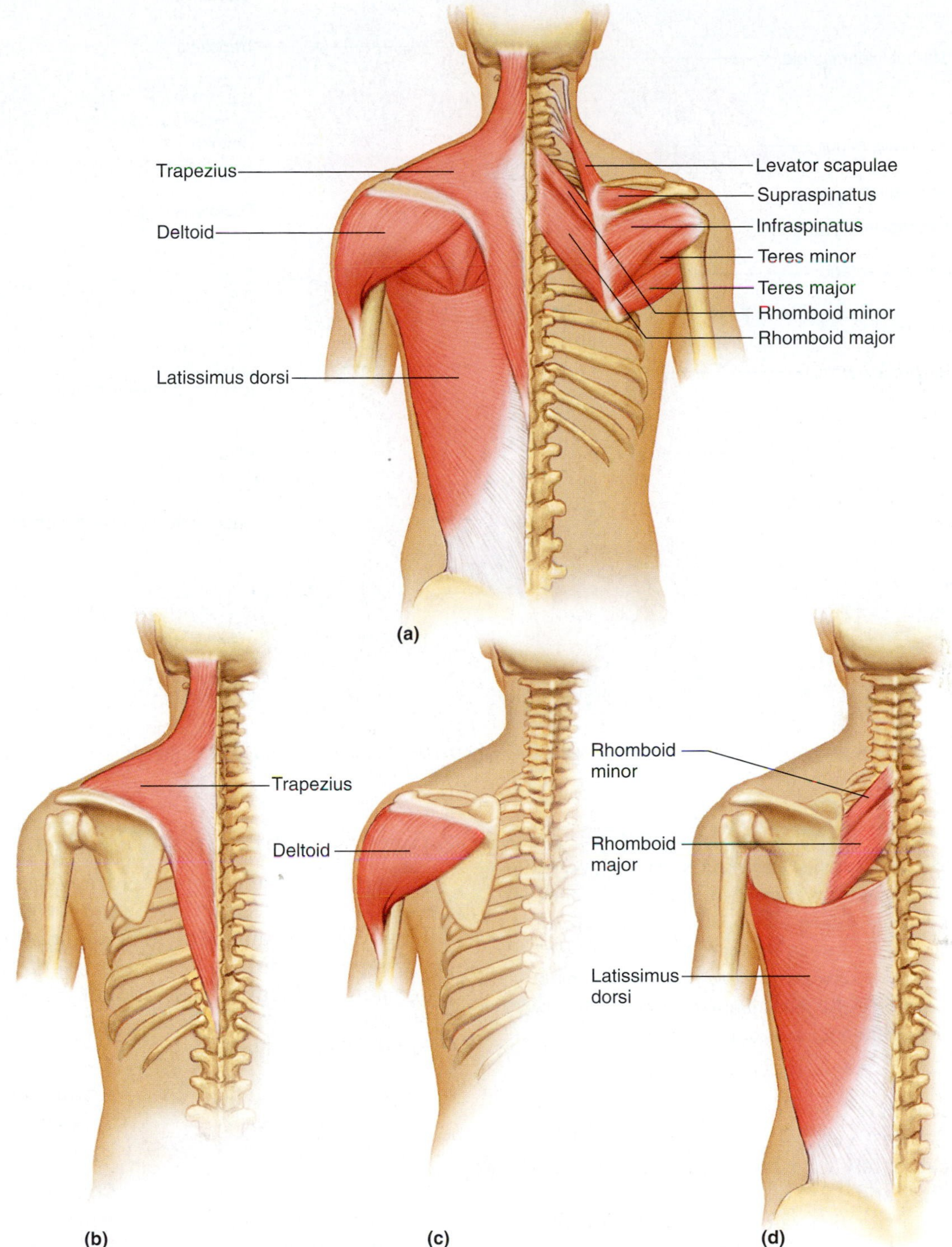

FIGURE 9.27 Muscles of the shoulder and back. (*a*) Muscles of the posterior shoulder. The right trapezius is removed to show underlying muscles. Isolated views of (*b*) trapezius, (*c*) deltoid, and (*d*) rhomboid and latissimus dorsi muscles.

The **trapezius** (trah-pe′ze-us) is a large, triangular muscle in the upper back that extends horizontally from the base of the skull and the cervical and thoracic vertebrae to the shoulder. Its fibers are organized into three groups—upper, middle, and lower. Together these fibers rotate the scapula. The upper fibers acting alone raise the scapula and shoulder, as when the shoulders are shrugged to express a feeling of indifference. The middle fibers pull the scapula toward the vertebral column, and the lower fibers draw the scapula and shoulder downward. When other muscles fix the shoulder in position, the trapezius can pull the head backward or to one side (see fig. 9.24).

Rhomboid (rom-boid′) **major** and **minor** connect the vertebral column to the scapula. Both retract and elevate the scapula. Rhomboid major can also rotate the scapula downward (see fig. 9.27).

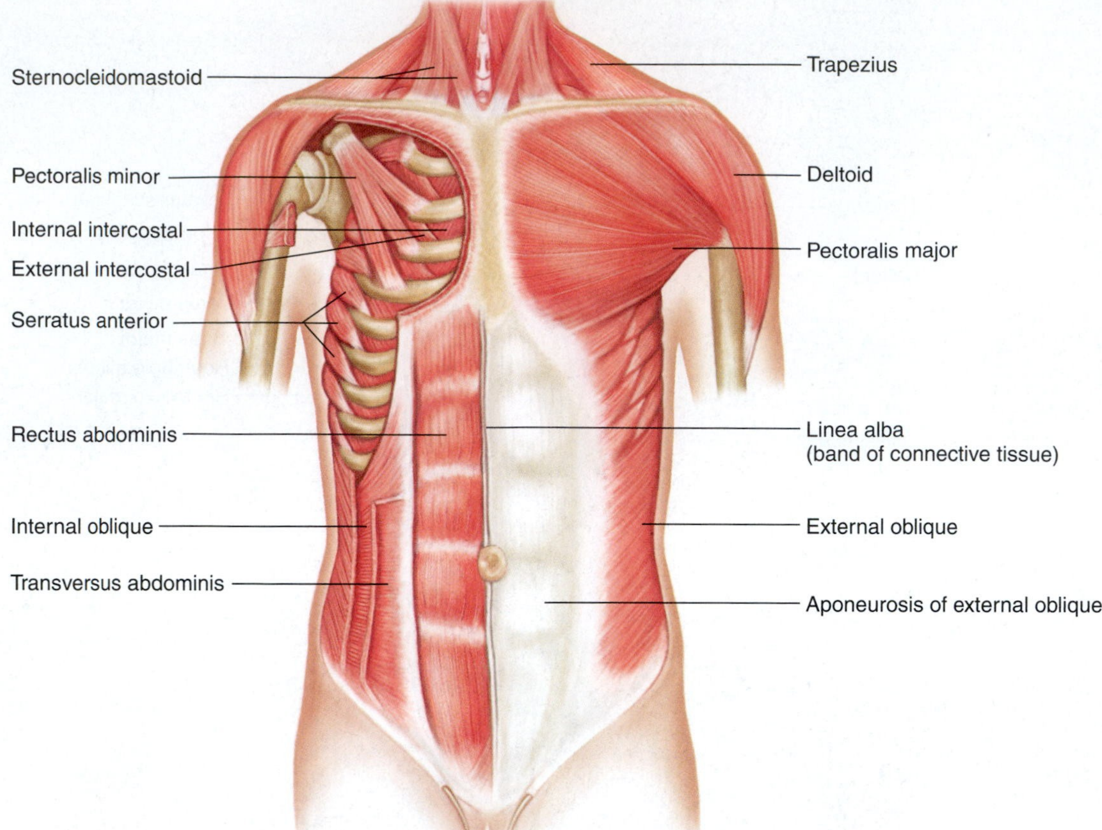

FIGURE 9.28 Muscles of the anterior chest and abdominal wall. The right pectoralis major is removed to show the pectoralis minor.

> A small, triangular region, called the *triangle of auscultation*, is located in the back where the trapezius overlaps the superior border of the latissimus dorsi and the underlying rhomboideus major. This area, near the medial border of the scapula, enlarges when a person bends forward with the arms folded across the chest. By placing the bell of a stethoscope in the triangle of auscultation, a physician can usually clearly hear the sounds of the respiratory organs.

The **levator scapulae** (le-va′tor scap′u-lē) is a straplike muscle that runs almost vertically through the neck, connecting the cervical vertebrae to the scapula. It elevates the scapula (see figs. 9.27 and 9.29).

The **serratus anterior** (ser-ra′tus an-te′re-or) is a broad, curved muscle located on the side of the chest. It arises as fleshy, narrow strips on the upper ribs and extends along the medial wall of the axilla to the ventral surface of the scapula. It pulls the scapula downward and anteriorly and is used to thrust the shoulder forward, as when pushing something (see fig. 9.28).

The **pectoralis** (pek′tor-a′lis) **minor** is a thin, flat muscle that lies beneath the larger pectoralis major. It extends laterally and upward from the ribs to the scapula and pulls the scapula forward and downward. When other muscles fix the scapula in position, the pectoralis minor can raise the ribs and thus aid forceful inhalation (see fig. 9.28).

Muscles That Move the Arm

The arm is one of the more freely movable parts of the body because muscles connect the humerus to regions of the pectoral girdle, ribs, and vertebral column. These muscles can be grouped according to their primary actions—flexion, extension, abduction, and rotation (figs. 9.29, 9.30, 9.31; reference plates 67, 68, 69; table 9.7). Muscles that move the arm include the following:

Flexors	***Abductors***
Coracobrachialis	Supraspinatus
Pectoralis major	Deltoid
Extensors	***Rotators***
Teres major	Subscapularis
Latissimus dorsi	Infraspinatus
	Teres minor

Flexors

The **coracobrachialis** (kor″ah-ko-bra′ke-al-is) extends from the scapula to the middle of the humerus along its medial surface. It flexes and adducts the arm (see figs. 9.30 and 9.31).

The **pectoralis major** is a thick, fan-shaped muscle in the upper chest. Its fibers extend from the center of the thorax through the armpit to the humerus. This muscle primarily pulls the arm forward and across the chest. It can also rotate the humerus medially and adduct the arm from a raised position (see fig. 9.28).

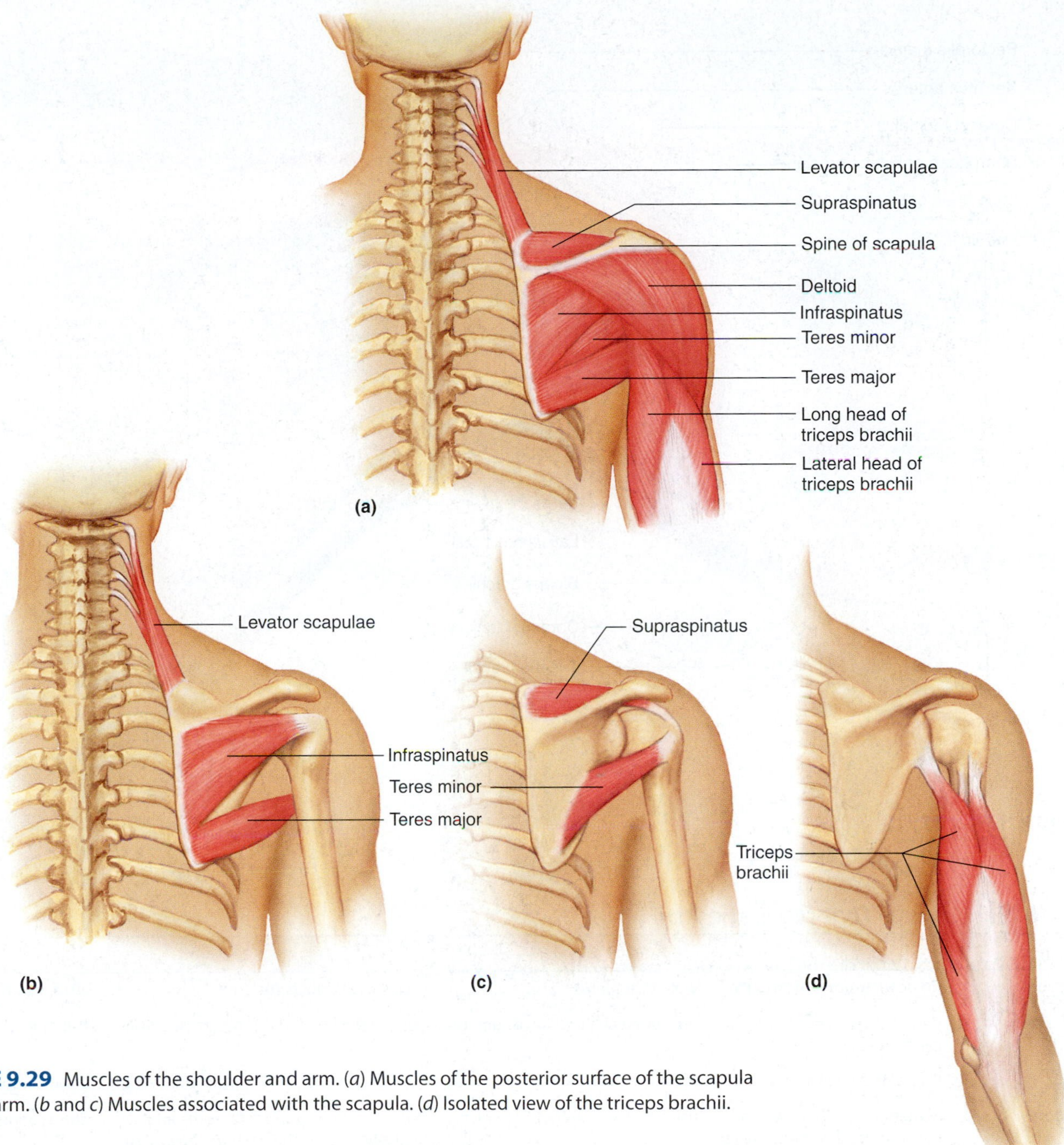

FIGURE 9.29 Muscles of the shoulder and arm. (*a*) Muscles of the posterior surface of the scapula and the arm. (*b* and *c*) Muscles associated with the scapula. (*d*) Isolated view of the triceps brachii.

TABLE 9.6 | Muscles That Move the Pectoral Girdle

Muscle	Origin	Insertion	Action	Nerve Supply
Trapezius	Occipital bone and spines of cervical and thoracic vertebrae	Clavicle, spine, and acromion process of scapula	Rotates scapula; various fibers raise scapula, pull scapula medially, or pull scapula and shoulder downward	Accessory n.
Rhomboid major	Spines of upper thoracic vertebrae	Medial border of scapula	Retracts, elevates, and rotates scapula	Dorsal scapular n.
Rhomboid minor	Spines of lower cervical vertebrae	Medial border of scapula	Retracts and elevates scapula	Dorsal scapular n.
Levator scapulae	Transverse processes of cervical vertebrae	Medial margin of scapula	Elevates scapula	Dorsal scapular and cervical nerves
Serratus anterior	Outer surfaces of upper ribs	Ventral surface of scapula	Pulls scapula anteriorly and downward	Long thoracic n.
Pectoralis minor	Sternal ends of upper ribs	Coracoid process of scapula	Pulls scapula forward and downward or raises ribs	Pectoral n.

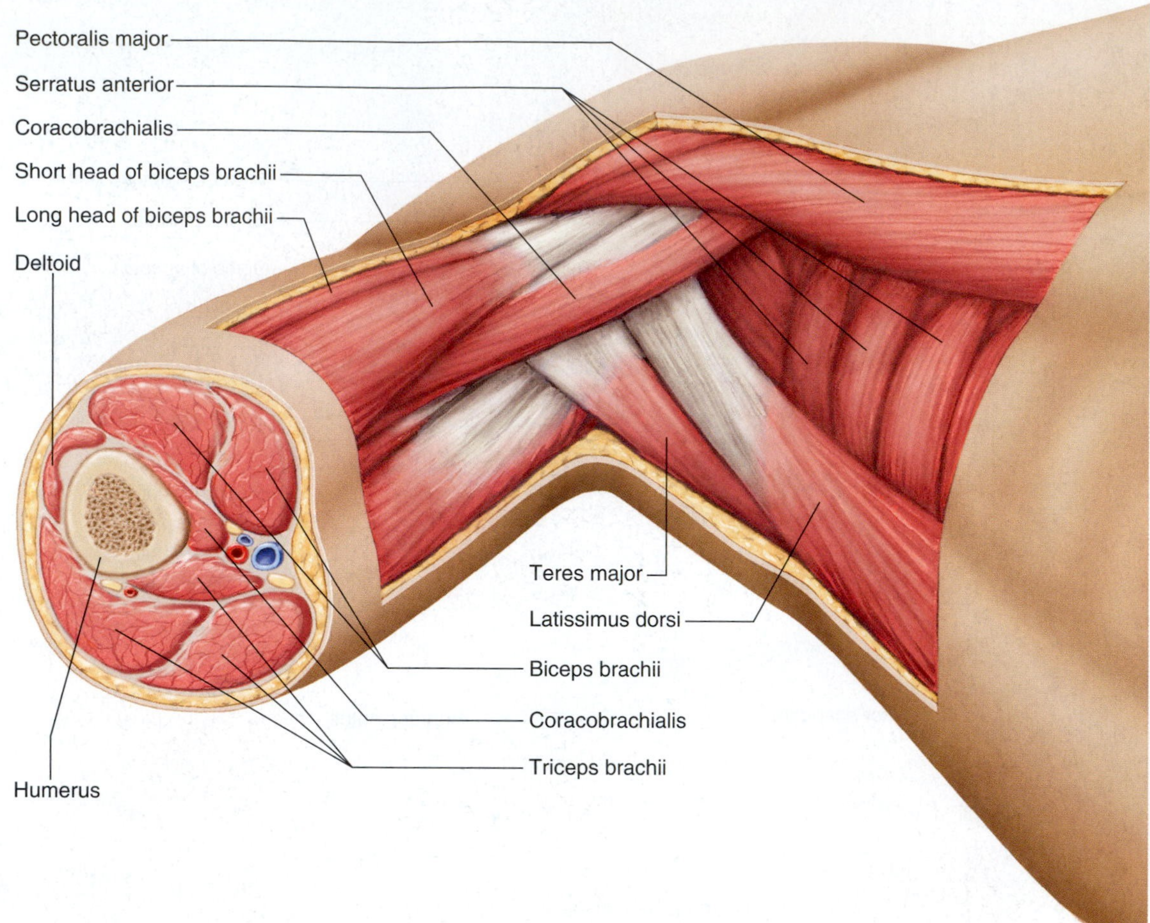

FIGURE 9.30 Cross section of the arm.

TABLE 9.7 | Muscles That Move the Arm

Muscle	Origin	Insertion	Action	Nerve Supply
Coracobrachialis	Coracoid process of scapula	Shaft of humerus	Flexes and adducts the arm	Musculocutaneus n.
Pectoralis major	Clavicle, sternum, and costal cartilages of upper ribs	Intertubercular groove of humerus	Flexes, adducts, and rotates arm medially	Pectoral n.
Teres major	Lateral border of scapula	Intertubercular groove of humerus	Extends, adducts, and rotates arm medially	Lower subscapular n.
Latissimus dorsi	Spines of sacral, lumbar, and lower thoracic vertebrae, iliac crest, and lower ribs	Intertubercular groove of humerus	Extends, adducts, and rotates the arm medially, or pulls the shoulder downward and back	Thoracodorsal n.
Supraspinatus	Posterior surface of scapula above spine	Greater tubercle of humerus	Abducts the arm	Suprascapular n.
Deltoid	Acromion process, spine of the scapula, and the clavicle	Deltoid tuberosity of humerus	Abducts, extends, and flexes arm	Axillary n.
Subscapularis	Anterior surface of scapula	Lesser tubercle of humerus	Rotates arm medially	Subscapular n.
Infraspinatus	Posterior surface of scapula below spine	Greater tubercle of humerus	Rotates arm laterally	Suprascapular n.
Teres minor	Lateral border of scapula	Greater tubercle of humerus	Rotates arm laterally	Axillary n.

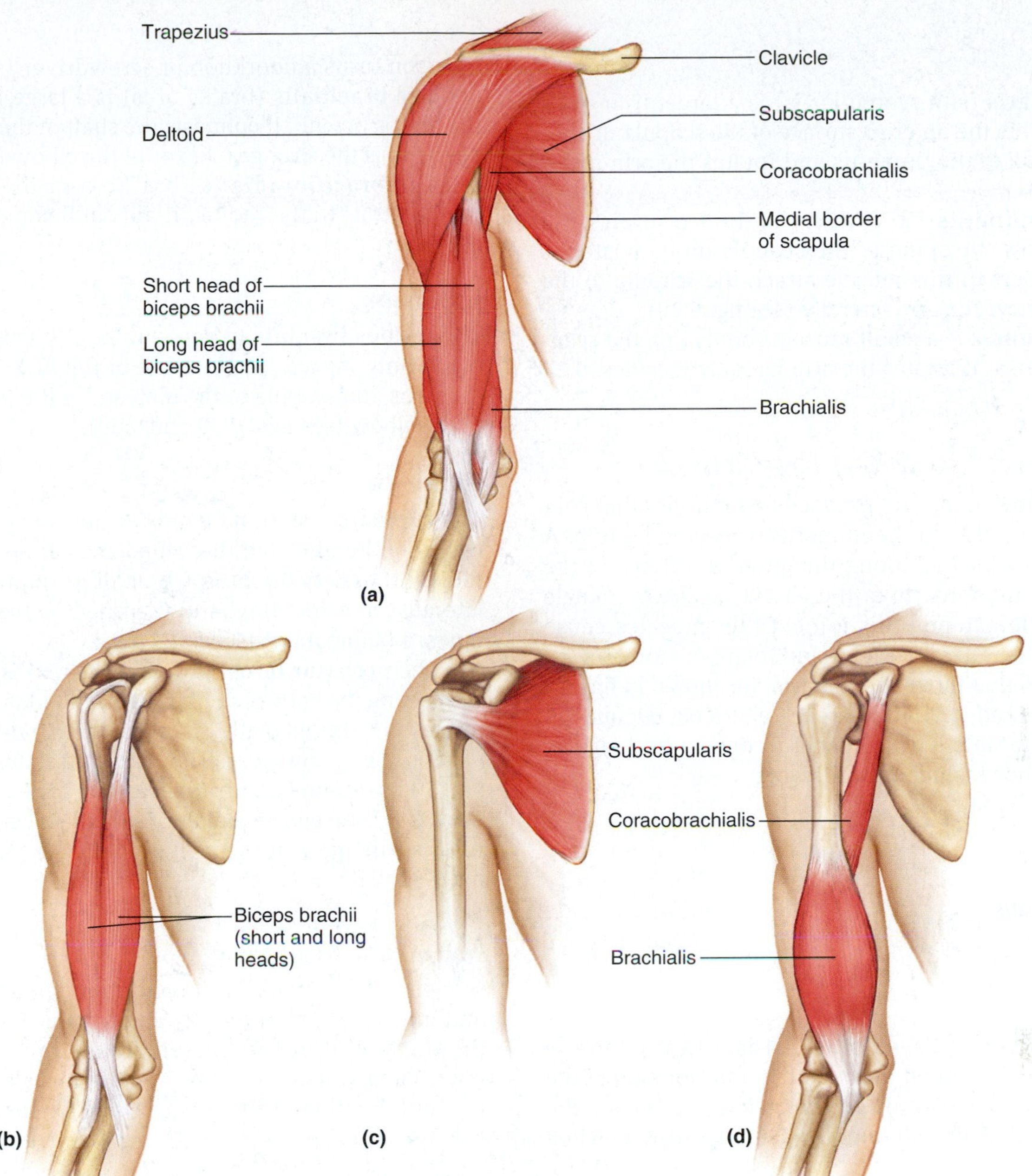

FIGURE 9.31 Muscles of the shoulder and arm. (*a*) Muscles of the anterior shoulder and the arm, with the rib cage removed. (*b, c,* and *d*) Isolated views of muscles associated with the arm.

Extensors

The **teres** (te′rēz) **major** connects the scapula to the humerus. It extends the humerus and can also adduct and rotate the arm medially (see figs. 9.27 and 9.29).

The **latissimus dorsi** (lah-tis′ĭ-mus dor′si) is a wide, triangular muscle that curves upward from the lower back, around the side, and to the armpit. It can extend and adduct the arm and rotate the humerus medially. It also pulls the shoulder downward and back. This muscle is used to pull the arm back in swimming, climbing, and rowing (see figs. 9.27 and 9.30).

Abductors

The **supraspinatus** (su″prah-spi′na-tus) is located in the depression above the spine of the scapula on its posterior surface. It connects the scapula to the greater tubercle of the humerus and abducts the arm (see figs. 9.27 and 9.29).

The **deltoid** (del′toid) is a thick, triangular muscle that covers the shoulder joint. It connects the clavicle and scapula to the lateral side of the humerus and abducts the arm. The deltoid's posterior fibers can extend the humerus, and its anterior fibers can flex the humerus (see fig. 9.27).

A humerus fractured at its surgical neck may damage the axillary nerve that supplies the deltoid muscle (see fig. 7.43). If this occurs, the muscle is likely to shrink and weaken. To test the deltoid for such weakness, a physician may ask a patient to abduct the arm against some resistance and maintain that posture for a time.

Rotators

The **subscapularis** (sub-scap′u-lar-is) is a large, triangular muscle that covers the anterior surface of the scapula. It connects the scapula to the humerus and rotates the arm medially (see fig. 9.31).

The **infraspinatus** (in″frah-spi′na-tus) occupies the depression below the spine of the scapula on its posterior surface. The fibers of this muscle attach the scapula to the humerus and rotate the arm laterally (see fig. 9.29).

The **teres minor** is a small muscle connecting the scapula to the humerus. It rotates the arm laterally (see figs. 9.27 and 9.29).

Muscles That Move the Forearm

Most forearm movements are produced by muscles that connect the radius or ulna to the humerus or pectoral girdle. A group of muscles located along the anterior surface of the humerus flexes the forearm at the elbow, whereas a single posterior muscle extends this joint. Other muscles cause movements at the radioulnar joint and rotate the forearm.

The muscles that move the forearm are shown in figures 9.31, 9.32, 9.33, and 9.34, in reference plates 68, 69, and 70, and are listed in table 9.8, grouped according to their primary actions. They include the following:

Flexors	***Extensor***	***Rotators***
Biceps brachii	Triceps brachii	Supinator
Brachialis		Pronator teres
Brachioradialis		Pronator quadratus

Flexors

The **biceps brachii** (bi′seps bra′ke-i) is a fleshy muscle that forms a long, rounded mass on the anterior side of the arm. It connects the scapula to the radius and flexes the elbow and rotates the hand laterally (supination), as when a person turns a doorknob or screwdriver (see fig. 9.31).

The **brachialis** (bra′ke-al-is) is a large muscle beneath the biceps brachii. It connects the shaft of the humerus to the ulna and is the strongest flexor of the elbow (see fig. 9.31).

The **brachioradialis** (bra″ke-o-ra″de-a′lis) connects the humerus to the radius. It aids in flexing the elbow (see fig. 9.32).

Extensor

The **triceps brachii** (tri′seps bra′ke-i) has three heads and is the only muscle on the back of the arm. It connects the humerus and scapula to the ulna and is the primary extensor of the elbow (see figs. 9.29 and 9.30).

Rotators

The **supinator** (su′pĭ-na-tor) is a short muscle whose fibers run from the ulna and the lateral end of the humerus to the radius. It assists the biceps brachii in rotating the forearm laterally, as when the hand is turned so the palm is facing upward (supination) (see fig. 9.32).

The **pronator teres** (pro-na′tor te′rēz) is a short muscle connecting the ends of the humerus and ulna to the radius. It rotates the arm medially, as when the hand is turned so the palm is facing downward (pronation) (see fig. 9.32).

The **pronator quadratus** (pro-na′tor kwod-ra′tus) runs from the distal end of the ulna to the distal end of the radius. It assists the pronator teres in rotating the arm medially (see fig. 9.32).

Muscles That Move the Hand

Movements of the hand include movements of the wrist and fingers. Many of the implicated muscles originate from the distal end of the humerus and from the radius and ulna. The two major groups of these muscles are flexors on the anterior side of the forearm and extensors on the posterior side. Figures 9.32, 9.33, 9.34, reference plate 70, and

TABLE 9.8 | Muscles That Move the Forearm

Muscle	Origin	Insertion	Action	Nerve Supply
Biceps brachii	Coracoid process and tubercle above glenoid cavity of scapula	Radial tuberosity of radius	Flexes elbow and rotates hand laterally	Musculocutaneous n.
Brachialis	Anterior shaft of humerus	Coronoid process of ulna	Flexes elbow	Musculocutaneous, median, and radial nerves
Brachioradialis	Distal lateral end of humerus	Lateral surface of radius above styloid process	Flexes elbow	Radial n.
Triceps brachii	Tubercle below glenoid cavity and lateral and medial surfaces of humerus	Olecranon process of ulna	Extends elbow	Radial n.
Supinator	Lateral epicondyle of humerus and crest of ulna	Lateral surface of radius	Rotates forearm laterally and supinates hand	Radial n.
Pronator teres	Medial epicondyle of humerus and coronoid process of ulna	Lateral surface of radius	Rotates forearm medially and pronates hand	Median n.
Pronator quadratus	Anterior distal end of ulna	Anterior distal end of radius	Rotates forearm medially and pronates hand	Median n.

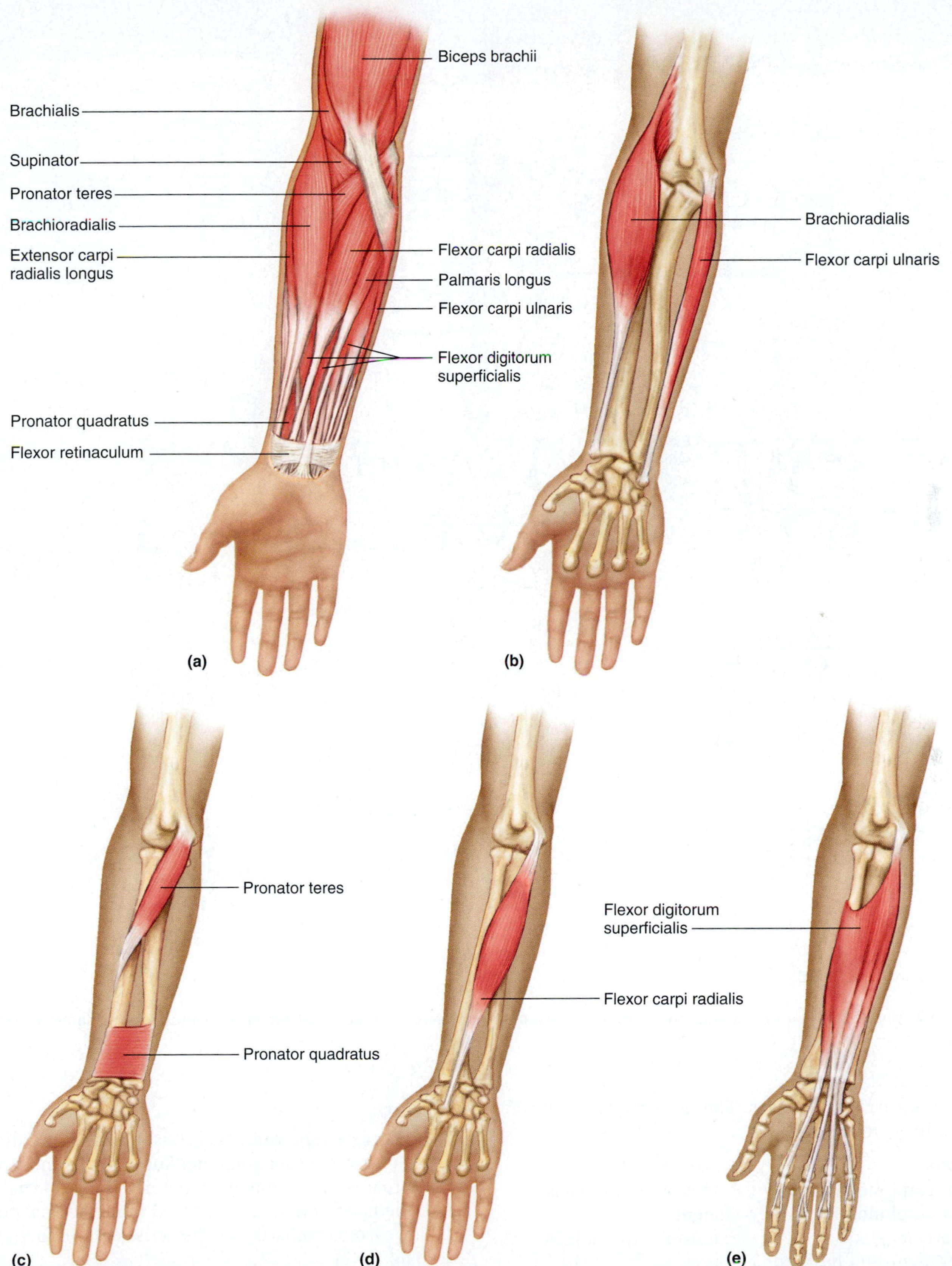

FIGURE 9.32 Muscles of the arm and forearm. (*a*) Muscles of the anterior forearm. (*b–e*) Isolated views of muscles associated with the anterior forearm.

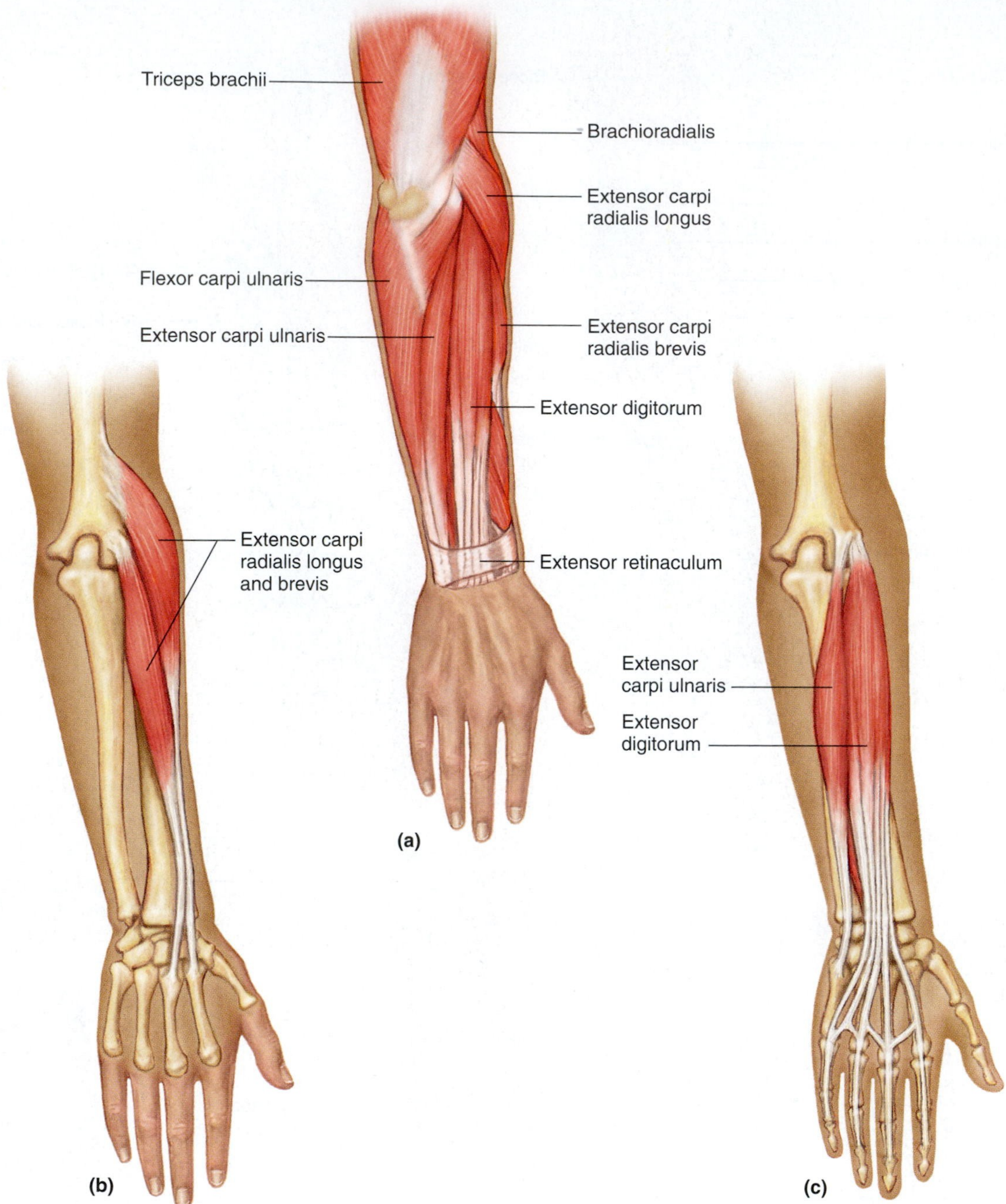

FIGURE 9.33 Muscles of the arm and forearm. (*a*) Muscles of the posterior forearm. (*b* and *c*) Isolated views of muscles associated with the posterior forearm.

table 9.9 concern these muscles. The muscles that move the hand include the following:

Flexors	***Extensors***
Flexor carpi radialis	Extensor carpi radialis longus
Flexor carpi ulnaris	Extensor carpi radialis brevis
Palmaris longus	Extensor carpi ulnaris
Flexor digitorum profundus	Extensor digitorum
Flexor digitorum superficialis	

Flexors

The **flexor carpi radialis** (flek′sor kar-pi′ra″de-a′lis) is a fleshy muscle that runs medially on the anterior side of the forearm. It extends from the distal end of the humerus into the hand, where it is attached to metacarpal bones. The flexor carpi radialis flexes the wrist and abducts the hand (see fig. 9.32).

The **flexor carpi ulnaris** (flek′sor kar-pi′ ul-na′ris) is located along the medial border of the forearm. It connects the distal end of the humerus and the proximal end of the

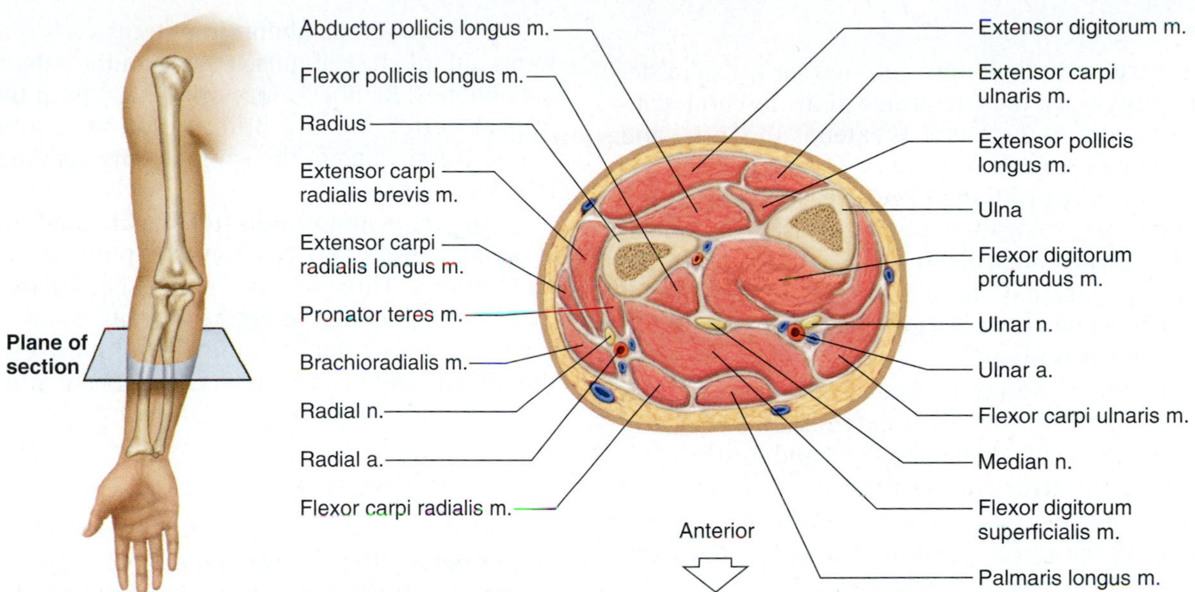

FIGURE 9.34 A cross section of the forearm (superior view). (*a.* stands for artery, *m.* stands for muscle, and *n.* stands for nerve.)

TABLE 9.9 | Muscles That Move the Hand

Muscle	Origin	Insertion	Action	Nerve Supply
Flexor carpi radialis	Medial epicondyle of humerus	Base of second and third metacarpals	Flexes wrist and abducts hand	Median n.
Flexor carpi ulnaris	Medial epicondyle of humerus and olecranon process	Carpal and metacarpal bones	Flexes wrist and adducts hand	Ulnar n.
Palmaris longus	Medial epicondyle of humerus	Fascia of palm	Flexes wrist	Median n.
Flexor digitorum profundus	Anterior surface of ulna	Bases of distal phalanges in fingers 2–5	Flexes distal joints of fingers	Median and ulnar nerves
Flexor digitorum superficialis	Medial epicondyle of humerus, coronoid process of ulna, and radius	Tendons of fingers	Flexes fingers and wrist	Median n.
Extensor carpi radialis longus	Distal end of humerus	Base of second metacarpal	Extends wrist and abducts hand	Radial n.
Extensor carpi radialis brevis	Lateral epicondyle of humerus	Base of second and third metacarpals	Extends wrist and abducts hand	Radial n.
Extensor carpi ulnaris	Lateral epicondyle of humerus	Base of fifth metacarpal	Extends wrist and adducts hand	Radial n.
Extensor digitorum	Lateral epicondyle of humerus	Posterior surface of phalanges in fingers 2–5	Extends fingers	Radial n.

ulna to carpal and metacarpal bones. It flexes the wrist and adducts the hand (see fig. 9.32).

The **palmaris longus** (pal-ma′ris long′gus) is a slender muscle located on the medial side of the forearm between the flexor carpi radialis and the flexor carpi ulnaris. It connects the distal end of the humerus to fascia of the palm and flexes the wrist (see fig. 9.32).

The **flexor digitorum profundus** (flek′sor dij″ĭ-to′rum pro-fun′dus) is a large muscle that connects the ulna to the distal phalanges. It flexes the distal joints of the fingers, as when a fist is made (see fig. 9.34).

The **flexor digitorum superficialis** (flek′sor dij″ĭ-to′rum su″per-fish″e-a′lis) is a large muscle located beneath the flexor carpi ulnaris. It arises by three heads—one from the medial epicondyle of the humerus, one from the medial side of the ulna, and one from the radius. It is inserted in the tendons of the fingers and flexes the fingers and, by a combined action, flexes the wrist (see fig. 9.32).

Some of the first signs of Parkinson disease appear in the hands. In this disorder, certain brain cells degenerate and damage nerve cells that control muscles. Once called "shaking palsy," the disease often begins with a hand tremor that resembles the motion of rolling a marble between the thumb and forefinger. Another sign is called "cogwheel rigidity." When a doctor rotates the patient's hand in an arc, the hand resists the movement and then jerks, like the cogs in a gear.

Extensors

The **extensor carpi radialis longus** (eks-ten′sor kar-pi′ra″de-a′lis long′gus) runs along the lateral side of the forearm, connecting the humerus to the hand. It extends the wrist and assists in abducting the hand (see figs. 9.33 and 9.34).

The **extensor carpi radialis brevis** (eks-ten′sor kar-pi′ ra″de-a′lis brev′ĭs) is a companion of the extensor carpi radialis longus and is located medially to it. This muscle runs from the humerus to metacarpal bones and extends the wrist. It also assists in abducting the hand (see figs. 9.33 and 9.34).

The **extensor carpi ulnaris** (eks-ten′sor kar-pi′ ul-na′ris) is located along the posterior surface of the ulna and connects the humerus to the hand. It extends the wrist and assists in adducting the hand (see figs. 9.33 and 9.34).

The **extensor digitorum** (eks-ten′sor dij″ĭ-to′rum) runs medially along the back of the forearm. It connects the humerus to the posterior surface of the phalanges and extends the fingers (see figs. 9.33 and 9.34).

A structure called the *extensor retinaculum* consists of a group of heavy connective tissue fibers in the fascia of the wrist (see fig. 9.33). It connects the lateral margin of the radius with the medial border of the styloid process of the ulna and certain bones of the wrist. The retinaculum gives off branches of connective tissue to the underlying wrist bones, creating a series of sheathlike compartments through which the tendons of the extensor muscles pass to the wrist and fingers.

Muscles of the Abdominal Wall

The walls of the chest and pelvic regions are supported directly by bone, but those of the abdomen are not. Instead, the anterior and lateral walls of the abdomen are composed of layers of broad, flattened muscles. These muscles connect the rib cage and vertebral column to the pelvic girdle. A band of tough connective tissue, called the **linea alba** (lin′e-ah al′bah), extends from the xiphoid process of the sternum to the symphysis pubis. It is an attachment for some of the abdominal wall muscles.

Contraction of these muscles decreases the volume of the abdominal cavity and increases the pressure inside. This action helps force air out of the lungs during forceful exhalation and also aids in defecation, urination, vomiting, and childbirth.

The abdominal wall muscles are shown in figure 9.35, reference plate 67, and are listed in table 9.10. They include the following:

External oblique	Transversus abdominis
Internal oblique	Rectus abdominis

The **external oblique** (eks-ter′nal ŏ-blēk) is a broad, thin sheet of muscle whose fibers slant downward from the lower ribs to the pelvic girdle and the linea alba. When this muscle contracts, it tenses the abdominal wall and compresses the contents of the abdominal cavity.

Similarly, the **internal oblique** (in-ter′nal ŏ-blēk) is a broad, thin sheet of muscle located beneath the external oblique. Its fibers run up and forward from the pelvic girdle to the lower ribs. Its function is similar to that of the external oblique.

The **transversus abdominis** (trans-ver′sus ab-dom′ĭ-nis) forms a third layer of muscle beneath the external and internal obliques. Its fibers run horizontally from the lower ribs, lumbar vertebrae, and ilium to the linea alba and pubic bones. It functions in the same manner as the external and internal obliques.

The **rectus abdominis** (rek′tus ab-dom′ĭ-nis) is a long, straplike muscle that connects the pubic bones to the ribs and sternum. Three or more fibrous bands cross the muscle transversely, giving it a segmented appearance. The muscle functions with other abdominal wall muscles to compress the contents of the abdominal cavity, and it also helps to flex the vertebral column.

Muscles of the Pelvic Outlet

Two muscular sheets span the outlet of the pelvis—a deeper **pelvic diaphragm** and a more superficial **urogenital diaphragm.** The pelvic diaphragm forms the floor of the pelvic cavity, and the urogenital diaphragm fills the space within the pubic arch. Figure 9.36 and table 9.11 show the muscles of the male and female pelvic outlets. They include the following:

Pelvic Diaphragm	*Urogenital Diaphragm*
Levator ani	Superficial transversus perinei
Coccygeus	Bulbospongiosus
	Ischiocavernosus
	Sphincter urethrae

Pelvic Diaphragm

The **levator ani** (le-va′tor ah-ni′) muscles form a thin sheet across the pelvic outlet. They are connected at the midline posteriorly by a ligament that extends from the tip of the coccyx to the anal canal. Anteriorly, they are separated in the male by the urethra and the anal canal, and in the female by the urethra, vagina, and anal canal. These muscles help support the pelvic viscera and provide sphincterlike action in the anal canal and vagina.

An *external anal sphincter* under voluntary control and an *internal anal sphincter* formed of involuntary muscle fibers of the intestine encircle the anal canal and keep it closed.

The **coccygeus** (kok-sij′e-us) is a fan-shaped muscle that extends from the ischial spine to the coccyx and sacrum. It aids the levator ani.

Urogenital Diaphragm

The **superficial transversus perinei** (soo′per-fish′al trans-ver′sus per″ĭ-ne′i) consists of a small bundle of muscle fibers that passes medially from the ischial tuberosity along the posterior border of the urogenital diaphragm. It assists other muscles in supporting the pelvic viscera.

In males, the **bulbospongiosus** (bul″bo-spon″je-o′sus) muscles are united surrounding the base of the penis. They assist in emptying the urethra. In females, these muscles are separated medially by the vagina and constrict the vaginal opening. They can also retard the flow of blood in veins,

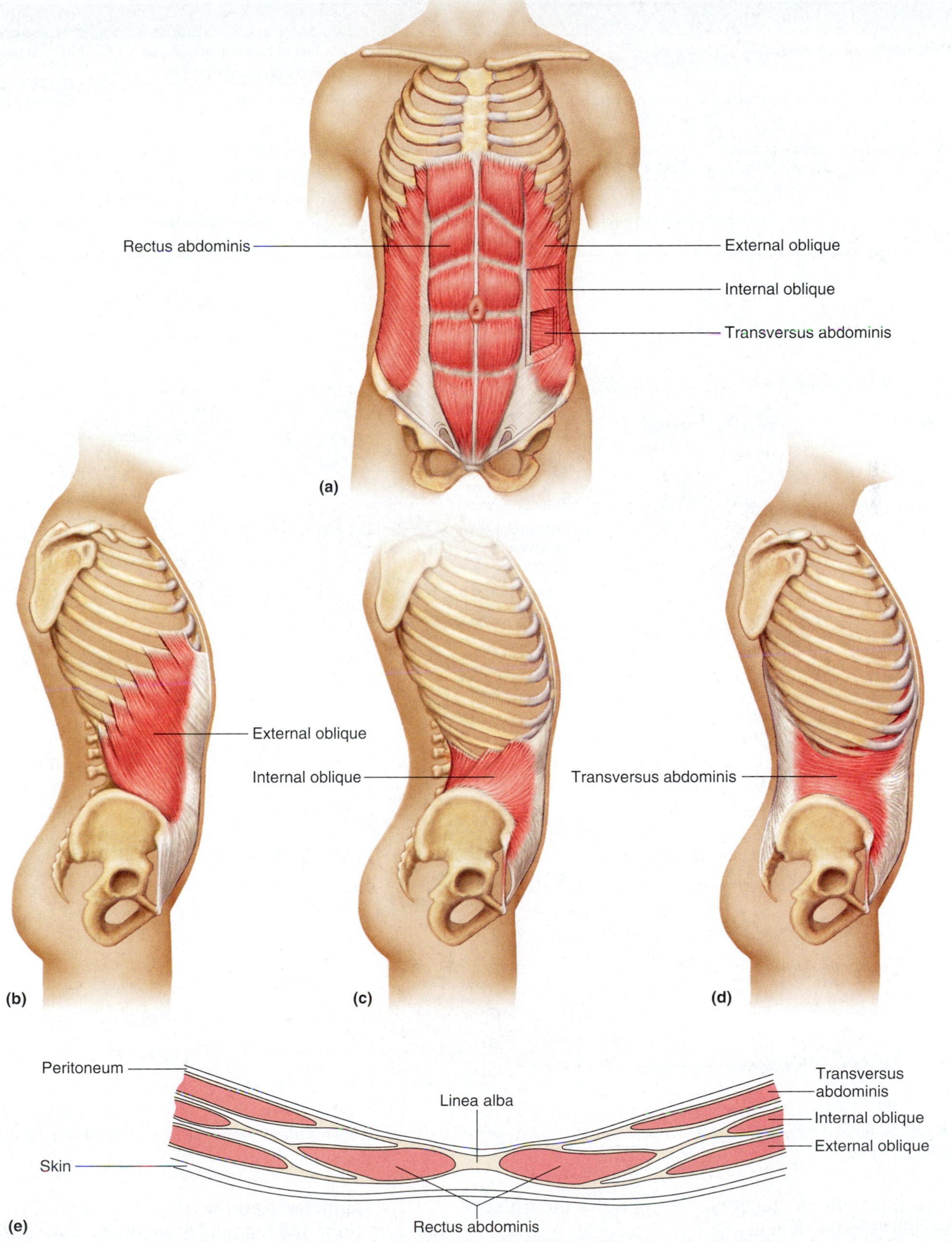

FIGURE 9.35 Muscles of the abdominal wall. (*a–d*) Isolated muscles of the abdominal wall. (*e*) Transverse section through the abdominal wall.

TABLE 9.10 | Muscles of the Abdominal Wall

Muscle	Origin	Insertion	Action	Nerve Supply
External oblique	Outer surfaces of lower ribs	Outer lip of iliac crest and linea alba	Tenses abdominal wall and compresses abdominal contents	Intercostal nerves 7–12
Internal oblique	Crest of ilium and inguinal ligament	Cartilages of lower ribs, linea alba, and crest of pubis	Same as above	Intercostal nerves 7–12
Transversus abdominis	Costal cartilages of lower ribs, processes of lumbar vertebrae, lip of iliac crest, and inguinal ligament	Linea alba and crest of pubis	Same as above	Intercostal nerves 7–12
Rectus abdominis	Crest of pubis and symphysis pubis	Xiphoid process of sternum and costal cartilages	Same as above; also flexes vertebral column	Intercostal nerves 7–12

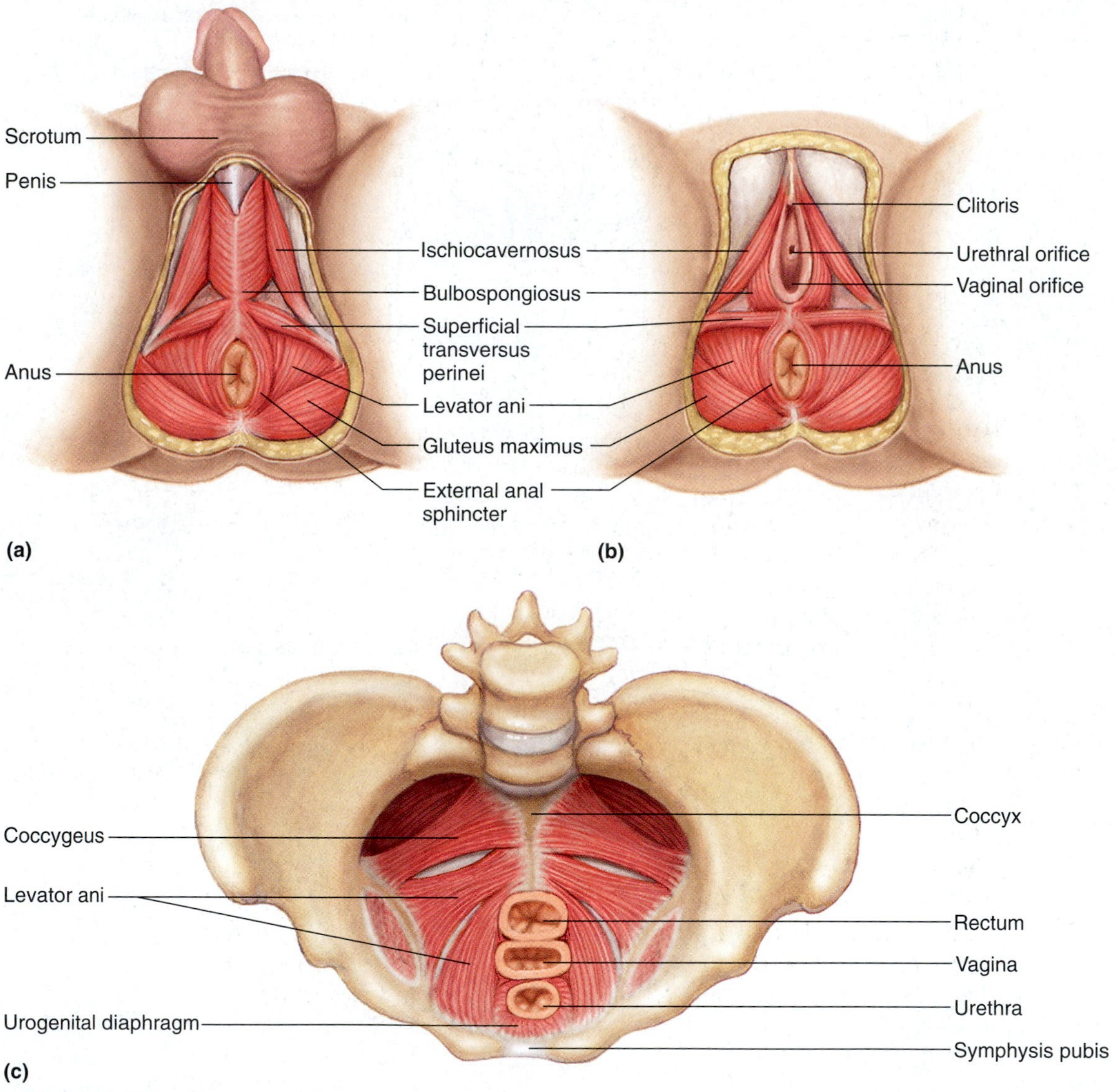

FIGURE 9.36 External view of muscles of (*a*) the male pelvic outlet and (*b*) the female pelvic outlet. (*c*) Internal view of female pelvic and urogenital diaphragms.

which helps maintain an erection of the penis in the male and of the clitoris in the female.

The **ischiocavernosus** (is″ke-o-kav″er-no′sus) muscle is a tendinous structure that extends from the ischial tuberosity to the margin of the pubic arch. It assists erection of the penis in males and the clitoris in females.

The **sphincter urethrae** (sfingk′ter u-re′thrē) are muscles that arise from the margins of the pubic and ischial bones. Each arches around the urethra and unites with the one on the other side. Together they act as a sphincter that closes the urethra by compression and opens it by relaxation, thus helping control the flow of urine.

TABLE 9.11 | Muscles of the Pelvic Outlet

Muscle	Origin	Insertion	Action	Nerve Supply
Levator ani	Pubic bone and ischial spine	Coccyx	Supports pelvic viscera and provides sphincterlike action in anal canal and vagina	Pudendal n.
Coccygeus	Ischial spine	Sacrum and coccyx	Same as above	S4 and S5 nerves
Superficial transversus perinei	Ischial tuberosity	Central tendon	Supports pelvic viscera	Pudendal n.
Bulbospongiosus	Central tendon	Males: Urogenital diaphragm and fascia of penis Females: Pubic arch and root of clitoris	Males: Assists emptying of urethra and assists in erection of penis Females: Constricts vagina and assists in erection of cltoris	Pudendal n.
Ischiocavernosus	Ischial tuberosity	Pubic arch	Males: Erects penis Females: Erects clitoris	Pudendal n.
Sphincter urethrae	Margins of pubis and ischium	Fibers of each unite with those from other side	Opens and closes urethra	Pudendal n.

Muscles That Move the Thigh

The muscles that move the thigh are attached to the femur and to some part of the pelvic girdle. (An important exception is the sartorius, described later.) They can be separated into anterior and posterior groups. The muscles of the anterior group primarily flex the thigh; those of the posterior group extend, abduct, or rotate it. The muscles in these groups are shown in figures 9.37, 9.38, 9.39, and 9.40, in reference plates 71 and 72, and are listed in table 9.12. Muscles that move the thigh include the following:

Anterior Group	*Posterior Group*
Psoas major	Gluteus maximus
Iliacus	Gluteus medius
	Gluteus minimus
	Piriformis
	Tensor fasciae latae

Another group of muscles, attached to the femur and pelvic girdle, adducts the thigh. This group includes the following:

Pectineus
Adductor brevis
Adductor longus
Adductor magnus
Gracilis

Anterior Group

The **psoas** (so′as) **major** is a long, thick muscle that connects the lumbar vertebrae to the femur. It flexes the thigh (see fig. 9.37).

The **iliacus** (il′e-ak-us), a large, fan-shaped muscle, lies along the lateral side of the psoas major. The iliacus and the psoas major are the primary flexors of the thigh, and they advance the lower limb in walking movements (see fig. 9.37).

Posterior Group

The **gluteus maximus** (gloo′te-us mak′si-mus) is the largest muscle in the body and covers a large part of each buttock. It connects the ilium, sacrum, and coccyx to the femur by fascia of the thigh and extends the thigh. The gluteus maximus helps to straighten the lower limb at the hip when a person walks, runs, or climbs. It is also used to raise the body from a sitting position (see fig. 9.38).

The **gluteus medius** (gloo′te-us me′de-us) is partly covered by the gluteus maximus. Its fibers extend from the ilium to the femur, and they abduct the thigh and rotate it medially (see fig. 9.38).

The **gluteus minimus** (gloo′te-us min′ĭ-mus) lies beneath the gluteus medius and is its companion in attachments and functions (see fig. 9.38).

The **piriformis** (pir-ĭ-for′mis) is shaped like a pyramid and located inferior to the gluteus minimus. It abducts and laterally rotates the thigh and is part of the posterior group of muscles that stabilizes the hip.

The **tensor fasciae latae** (ten′sor fash′e-e lah-tē) connects the ilium to the iliotibial tract (fascia of the thigh), which continues downward to the tibia. This muscle abducts and flexes the thigh and rotates it medially (see fig. 9.38).

The gluteus medius and gluteus minimus help support and maintain the normal position of the pelvis. If these muscles are paralyzed as a result of injury or disease, the pelvis tends to drop to one side whenever the foot on that side is raised. Consequently, the person walks with a waddling limp called the *gluteal gait*.

Thigh Adductors

The **pectineus** (pek-tin′e-us) muscle runs from the spine of the pubis to the femur. It flexes and adducts the thigh (see fig. 9.37).

The **adductor brevis** (ah-duk′tor brev′ĭs) is a short, triangular muscle that runs from the pubic bone to the femur. It adducts the thigh and assists in flexing and rotating it laterally (see fig. 9.37).

The **adductor longus** (ah-duk′tor long′gus) is a long, triangular muscle that runs from the pubic bone to the femur. It adducts the thigh and assists in flexing and rotating it laterally (see fig. 9.37).

The **adductor magnus** (ah-duk′tor mag′nus) is the largest adductor of the thigh. It is a triangular muscle that connects

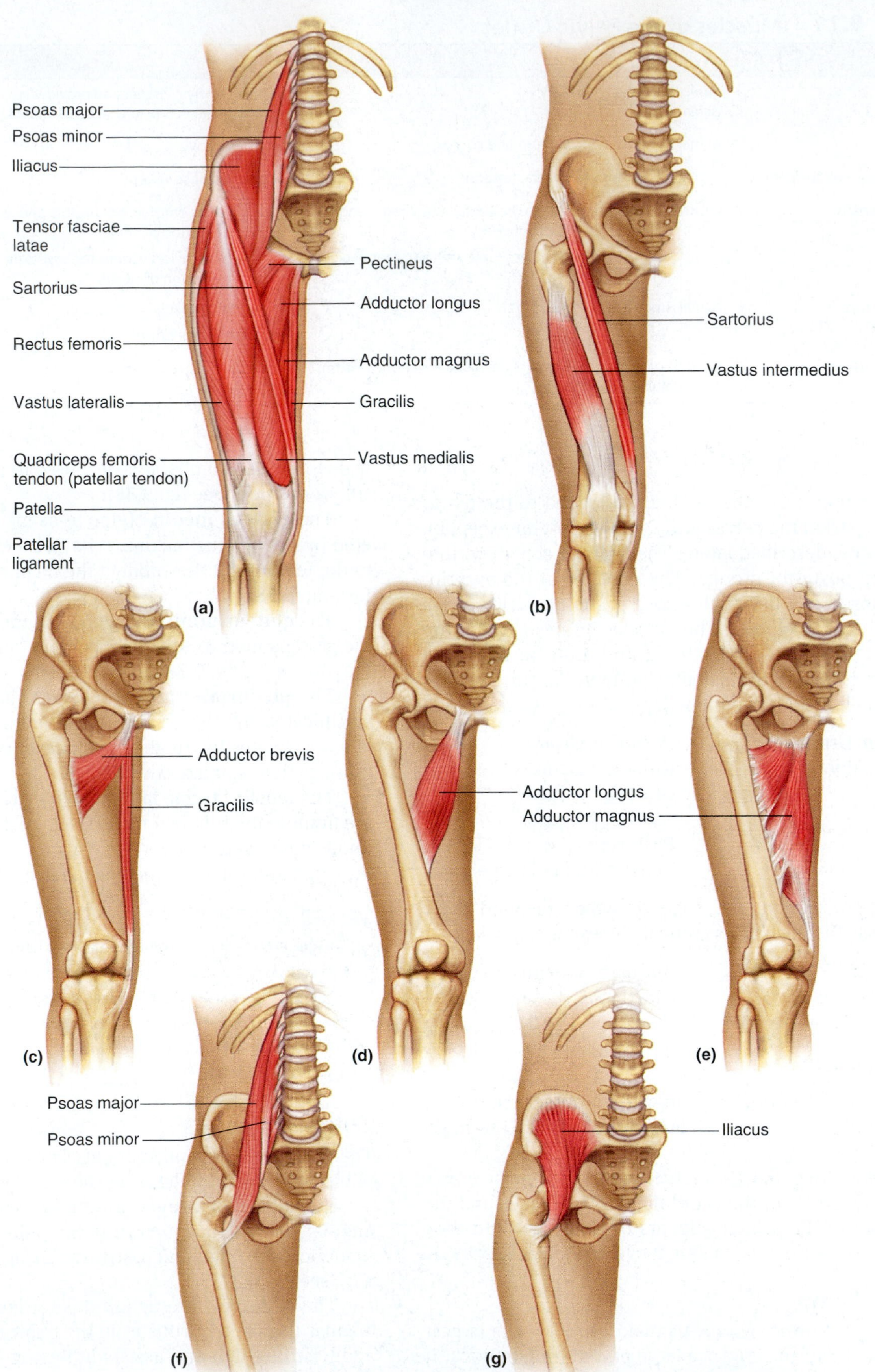

FIGURE 9.37 Muscles of the thigh and leg. (*a*) Muscles of the anterior right thigh. Isolated views of (*b*) the vastus intermedius, (*c–e*) adductors of the thigh, (*f–g*) flexors of the thigh.

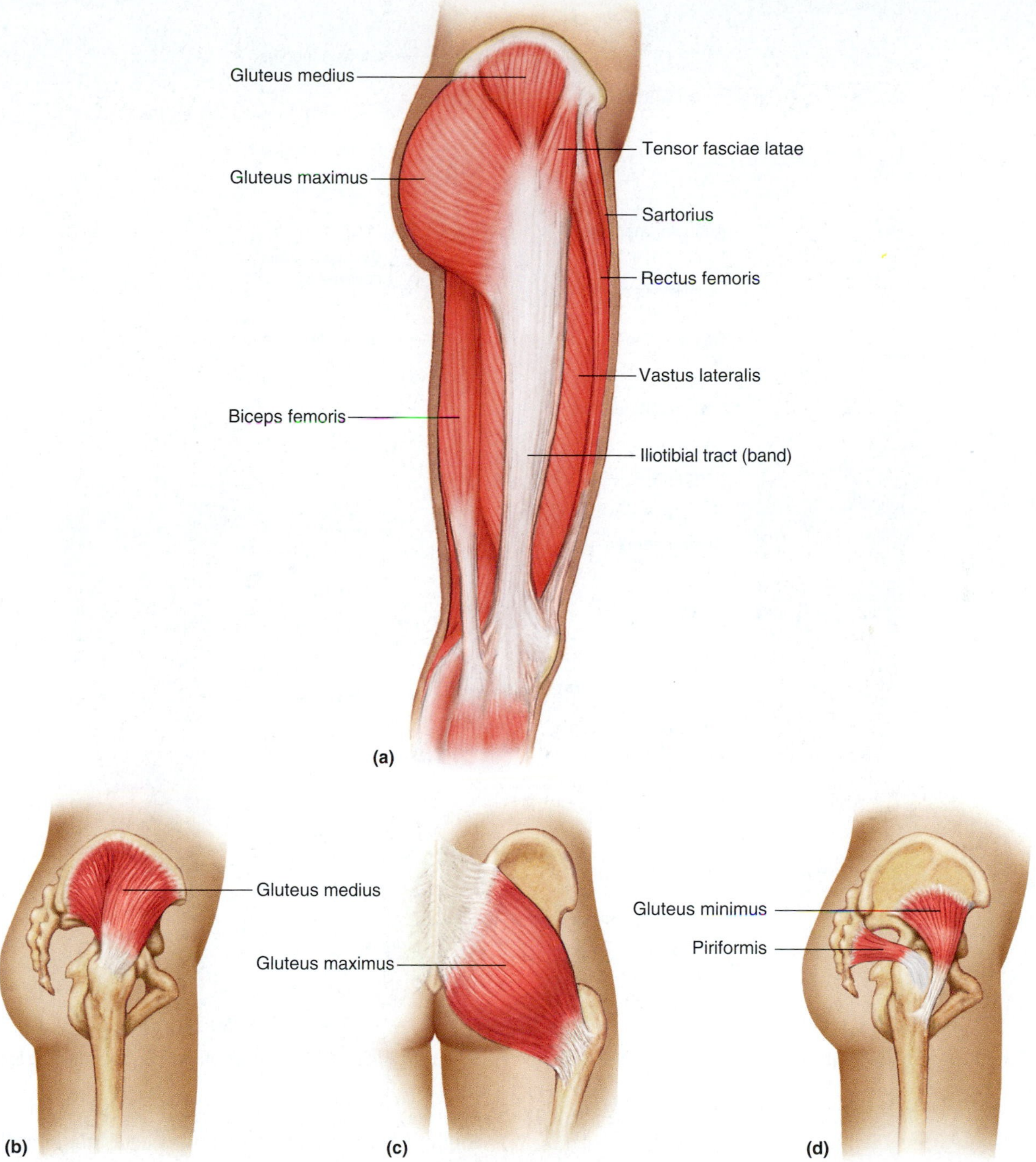

FIGURE 9.38 Muscles of the thigh and leg. (*a*) Muscles of the lateral right thigh. (*b–d*) Isolated views of the gluteal muscles.

the ischium to the femur. It adducts the thigh and portions assist in flexing and extending the thigh (see fig. 9.37).

The **gracilis** (gras′il-is) is a long, straplike muscle that passes from the pubic bone to the tibia. It adducts the thigh and flexes the leg at the knee (see fig. 9.37).

Muscles That Move the Leg

The muscles that move the leg connect the tibia or fibula to the femur or to the pelvic girdle. They fall into two major groups—those that flex the knee and those that extend it. The muscles of these groups are shown in figures 9.37, 9.38, 9.39, 9.40, in reference plates 71 and 72, and are listed in table 9.13. Muscles that move the leg include the following:

Flexors	*Extensor*
Biceps femoris	Quadriceps femoris group
Semitendinosus	
Semimembranosus	
Sartorius	

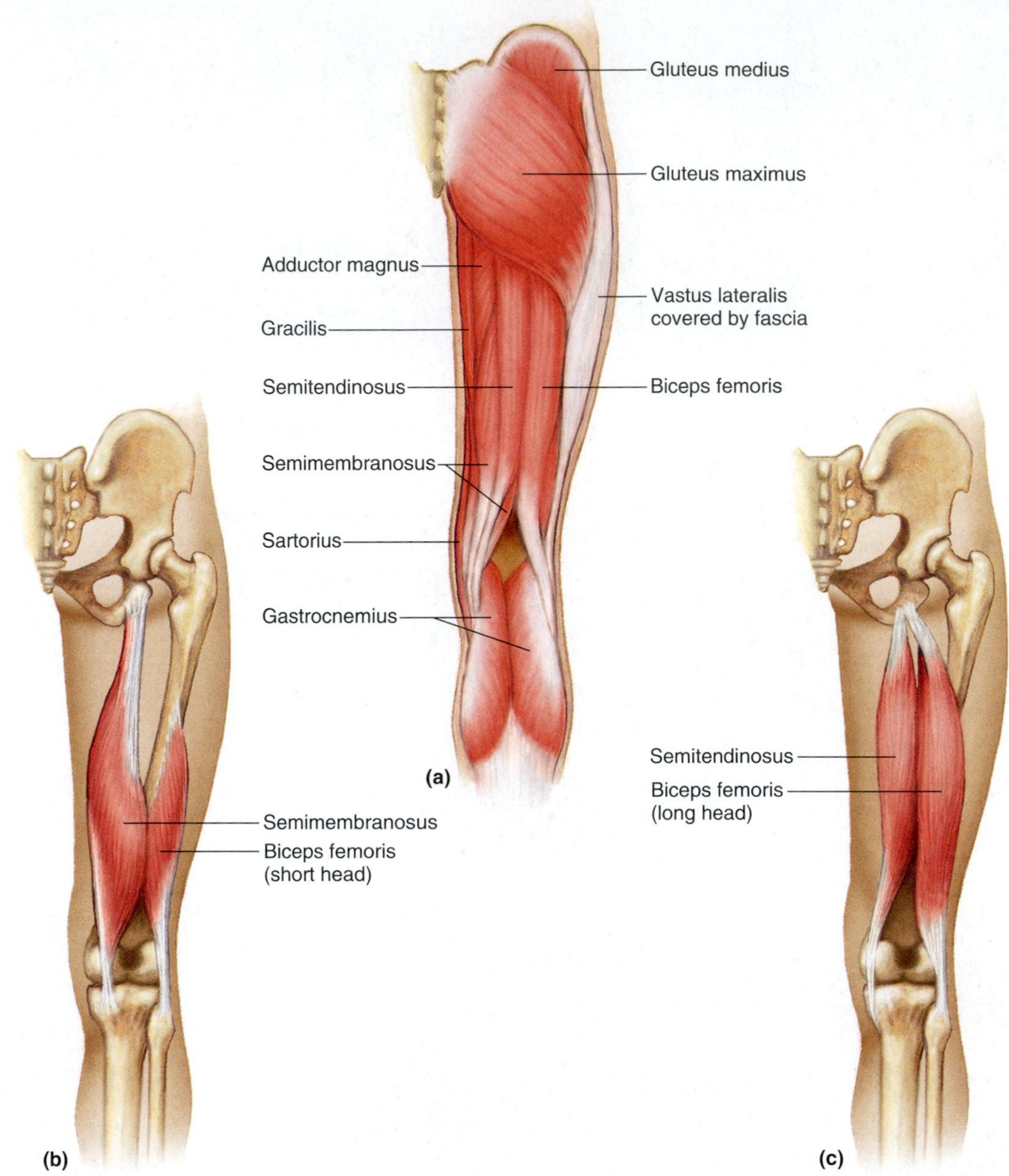

FIGURE 9.39 Muscles of the thigh and leg. (*a*) Muscles of the posterior right thigh. (*b* and *c*) Isolated views of muscles that flex the leg at the knee.

Flexors

As its name implies, the **biceps femoris** (bi′seps fem′or-is) has two heads, one attached to the ischium and the other attached to the femur. This muscle passes along the back of the thigh on the lateral side and connects to the proximal ends of the fibula and tibia. The biceps femoris is one of the hamstring muscles, and its tendon (hamstring) can be felt as a lateral ridge behind the knee. This muscle flexes and rotates the leg laterally and extends the thigh (see figs. 9.38 and 9.39).

The **semitendinosus** (sem″e-ten′dĭ-no-sus) is another hamstring muscle. It is a long, bandlike muscle on the back of the thigh toward the medial side, connecting the ischium to the proximal end of the tibia. The semitendinosus is so named because it becomes tendinous in the middle of the thigh, continuing to its insertion as a long, cordlike tendon. It flexes and rotates the leg medially and extends the thigh (see fig. 9.39).

The **semimembranosus** (sem″e-mem′brah-no-sus) is the third hamstring muscle and is the most medially located muscle in the back of the thigh. It connects the ischium to the tibia and flexes and rotates the leg medially and extends the thigh (see fig. 9.39).

Strenuous running or kicking motions can tear the tendinous attachments of the hamstring muscles to the ischial tuberosity. Internal bleeding from damaged blood vessels that supply the muscles usually occurs with this painful injury, commonly called "pulled hamstrings."

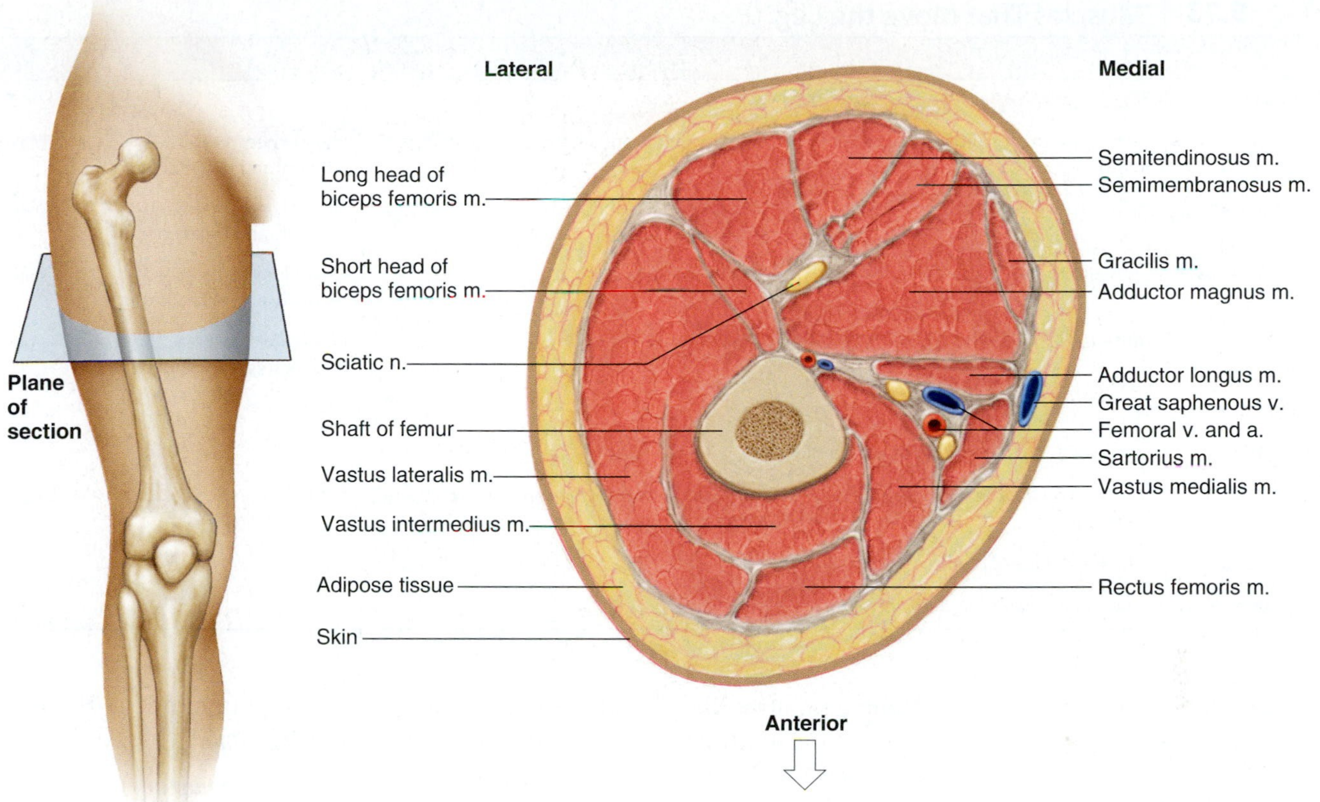

FIGURE 9.40 A cross section of the thigh (superior view). (*a.* stands for artery, *m.* stands for muscle, *n.* stands for nerve, and *v.* stands for vein.)

TABLE 9.12 | Muscles That Move The Thigh

Muscle	Origin	Insertion	Action	Nerve Supply
Psoas major	Lumbar intervertebral discs; bodies and transverse processes of lumbar vertebrae	Lesser trochanter of femur	Flexes thigh	Branches of L1-3 nerves
Iliacus	Iliac fossa of ilium	Lesser trochanter of femur	Flexes thigh	Femoral n.
Gluteus maximus	Sacrum, coccyx, and posterior surface of ilium	Posterior surface of femur and fascia of thigh	Extends hip	Inferior gluteal n.
Gluteus medius	Lateral surface of ilium	Greater trochanter of femur	Abducts and rotates thigh medially	Superior gluteal n.
Gluteus minimus	Lateral surface of ilium	Greater trochanter of femur	Same as gluteus medius	Superior gluteal n.
Piriformis	Anterior surface of sacrum	Greater trochanter of femur	Abducts and rotates thigh laterally	L5, S1, and S2 nerves
Tensor fasciae latae	Anterior iliac crest	Iliotibial tract (fascia of thigh)	Abducts, flexes, and rotates thigh medially	Superior gluteal n.
Pectineus	Spine of pubis	Femur distal to lesser trochanter	Flexes and adducts thigh	Obturator and femoral nerves
Adductor brevis	Pubic bone	Posterior surface of femur	Adducts, flexes, and rotates thigh laterally	Obturator n.
Adductor longus	Pubic bone near symphysis pubis	Posterior surface of femur	Adducts, flexes, and rotates thigh laterally	Obturator n.
Adductor magnus	Ischial tuberosity	Posterior surface of femur	Adducts thigh, posterior portion extends and anterior portion flexes thigh	Obturator and branch of sciatic nerves
Gracilis	Lower edge of symphysis pubis	Medial surface of tibia	Adducts thigh and flexes knee	Obturator n.

TABLE 9.13 | Muscles That Move the Leg

Muscle	Origin	Insertion	Action	Nerve Supply
Hamstring Group				
Biceps femoris	Ischial tuberosity and linea aspera of femur	Head of fibula and lateral condyle of tibia	Flexes knee, rotates leg laterally and extends thigh	Tibial n.
Semitendinosus	Ischial tuberosity	Medial surface of tibia	Flexes knee, rotates leg medially and extends thigh	Tibial n.
Semimembranosus	Ischial tuberosity	Medial condyle of tibia	Flexes knee, rotates leg medially and extends thigh	Tibial n.
Sartorius	Anterior superior iliac spine	Medial surface of tibia	Flexes knee and hip, abducts and rotates thigh laterally	Femoral n.
Quadriceps Femoris Group				
Rectus femoris	Spine of ilium and margin of acetabulum			
Vastus lateralis	Greater trochanter and posterior surface of femur	Patella by common tendon, which continues as patellar ligament to tibial tuberosity	Extends knee	Femoral n.
Vastus medialis	Medial surface of femur			
Vastus intermedius	Anterior and lateral surfaces of femur			

The **sartorius** (sar-to′re-us) is an elongated, straplike muscle that passes obliquely across the front of the thigh and then descends over the medial side of the knee. It connects the ilium to the tibia and flexes the leg and the thigh. It can also abduct the thigh and rotate it laterally (see figs. 9.37 and 9.38).

Extensor

The large, fleshy muscle group called the **quadriceps femoris** (kwod′rĭ-seps fem′or-is) occupies the front and sides of the thigh and is the primary extensor of the knee. It is composed of four parts—*rectus femoris, vastus lateralis, vastus medialis,* and *vastus intermedius* (see figs. 9.38 and 9.40). These parts connect the ilium and femur to a common *patellar tendon,* which passes over the front of the knee and attaches to the patella. This tendon then continues as the *patellar ligament* to the tibia.

> In a traumatic injury that compresses a muscle, such as the quadriceps femoris, against an underlying bone, new bone tissue may begin to develop in the damaged muscle. This condition is called *myositis ossificans*. Surgery can remove the newly formed bone when it matures several months after the injury.

Muscles That Move the Foot

Movements of the foot include movements of the ankle and toes. A number of muscles that move the foot are in the leg. They attach the femur, tibia, and fibula to bones of the foot and move the foot upward (dorsiflexion) or downward (plantar flexion) and turn the foot so the plantar surface faces medially (inversion) or laterally (eversion). These muscles are shown in figures 9.41, 9.42, 9.43, and 9.44, in reference plates 73, 74, 75, and are listed in table 9.14. Muscles that move the foot include the following:

Dorsal Flexors
Tibialis anterior
Fibularis tertius
Extensor digitorum longus
Extensor hallucis longus

Invertor
Tibialis posterior

Plantar Flexors
Gastrocnemius
Soleus
Plantaris
Flexor digitorum longus

Evertor
Fibularis longus

Dorsal Flexors

The **tibialis anterior** (tib″e-a′lis ante′re-or) is an elongated, spindle-shaped muscle located on the front of the leg. It arises from the surface of the tibia, passes medially over the distal end of the tibia, and attaches to bones of the foot. Contraction of the tibialis anterior causes dorsiflexion and inversion of the foot (see fig. 9.41).

The **fibularis** (peroneus) **tertius** (fib″u-la′ris ter′shus) is a muscle of variable size that connects the fibula to the lateral side of the foot. It functions in dorsiflexion and eversion of the foot (see fig. 9.41).

The **extensor digitorum longus** (eks-ten′sor dij″ĭ-to′rum long′gus) is situated along the lateral side of the leg just behind the tibialis anterior. It arises from the proximal end of the tibia and the shaft of the fibula. Its tendon divides into four parts as it passes over the front of the ankle. These parts continue over the surface of the foot and attach to the four lateral toes. The actions of the extensor digitorum longus include dorsiflexion of the foot, eversion of the foot, and extension of the toes (see figs. 9.41 and 9.42).

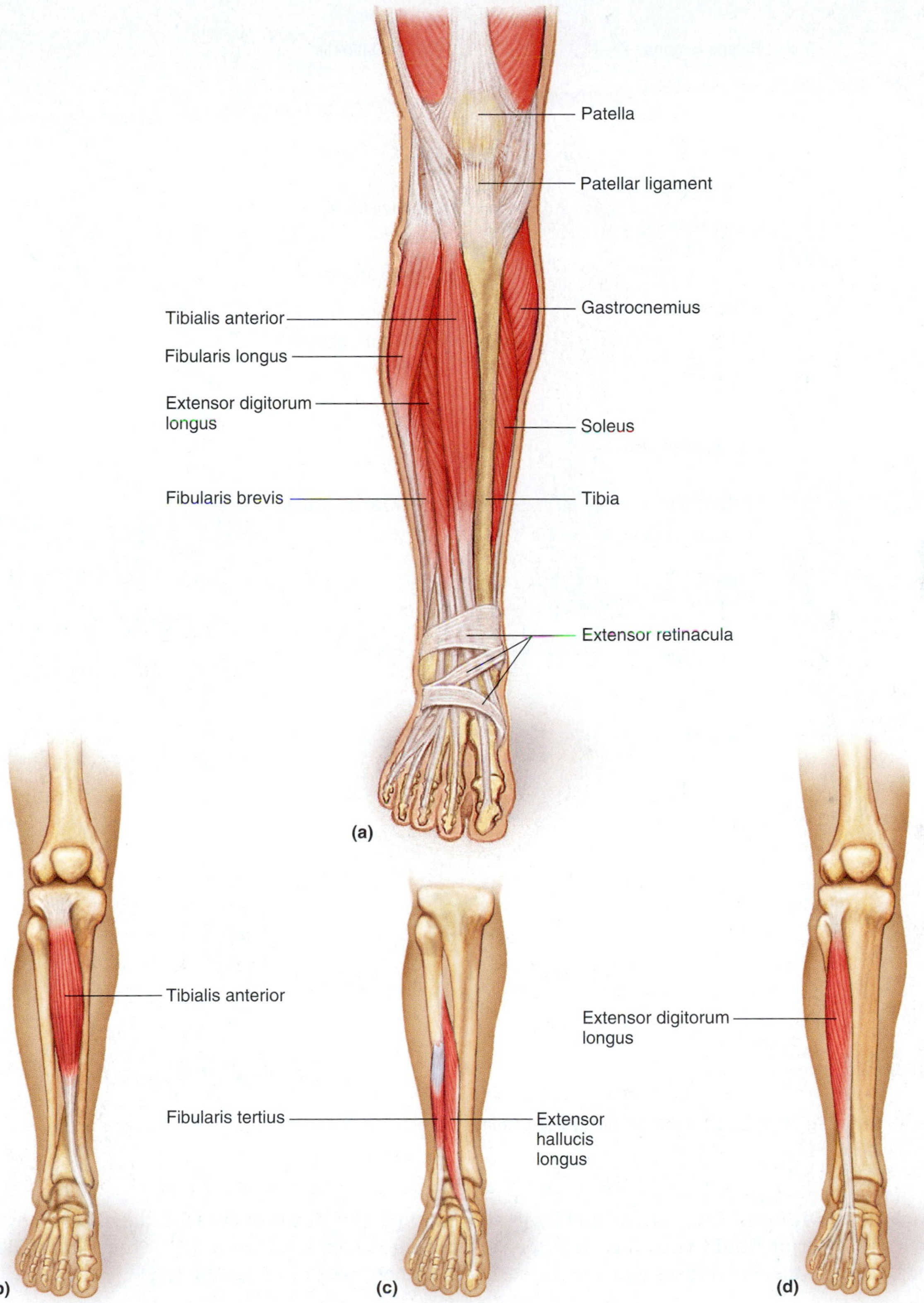

FIGURE 9.41 Muscles of the leg. (*a*) Muscles of the anterior right leg. (*b–d*) Isolated views of muscles associated with the anterior right leg.

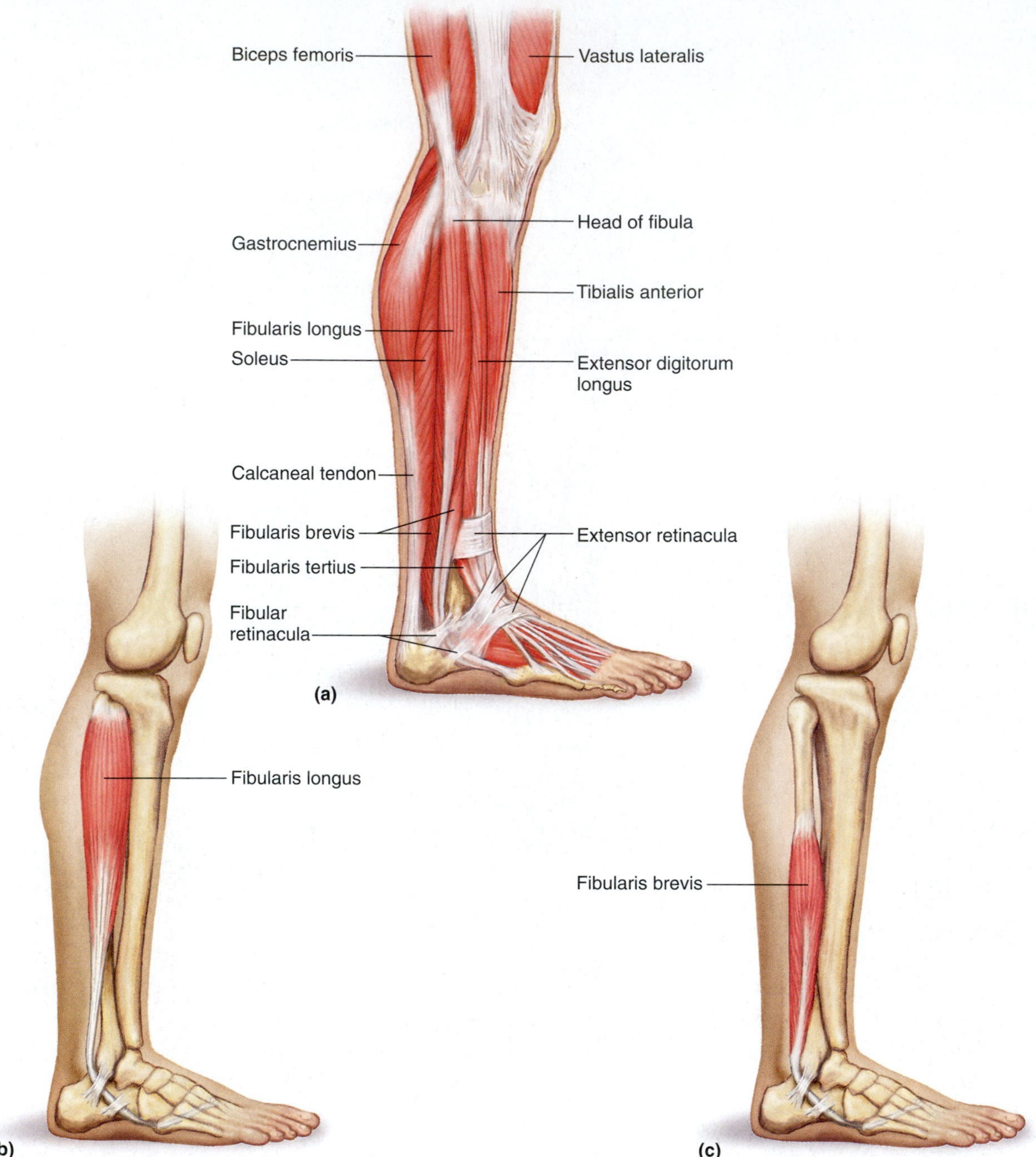

FIGURE 9.42 Muscles of the leg. (*a*) Muscles of the lateral right leg. Isolated views of (*b*) fibularis longus and (*c*) fibularis brevis.

The **extensor hallucis longus** (eks-ten′sor hal′lu-sis long′gus) connects the anterior fibula with the great toe. Contraction extends the great toe, dorsiflexes and inverts the foot (see fig. 9.41).

Plantar Flexors

The **gastrocnemius** (gas″trok-ne′me-us) on the back of the leg forms part of the calf. It arises by two heads from the femur. The distal end of this muscle joins the strong *calcaneal tendon* (Achilles tendon), which descends to the heel and attaches to the calcaneus. The gastrocnemius is a powerful plantar flexor of the foot that aids in pushing the body forward when a person walks or runs. It also flexes the leg at the knee (see figs. 9.42 and 9.43).

Strenuous athletic activity may partially or completely tear the calcaneal (Achilles) tendon. This injury occurs most frequently in middle-aged athletes who run or play sports that involve quick movements and directional changes. A torn calcaneal tendon usually requires surgical treatment.

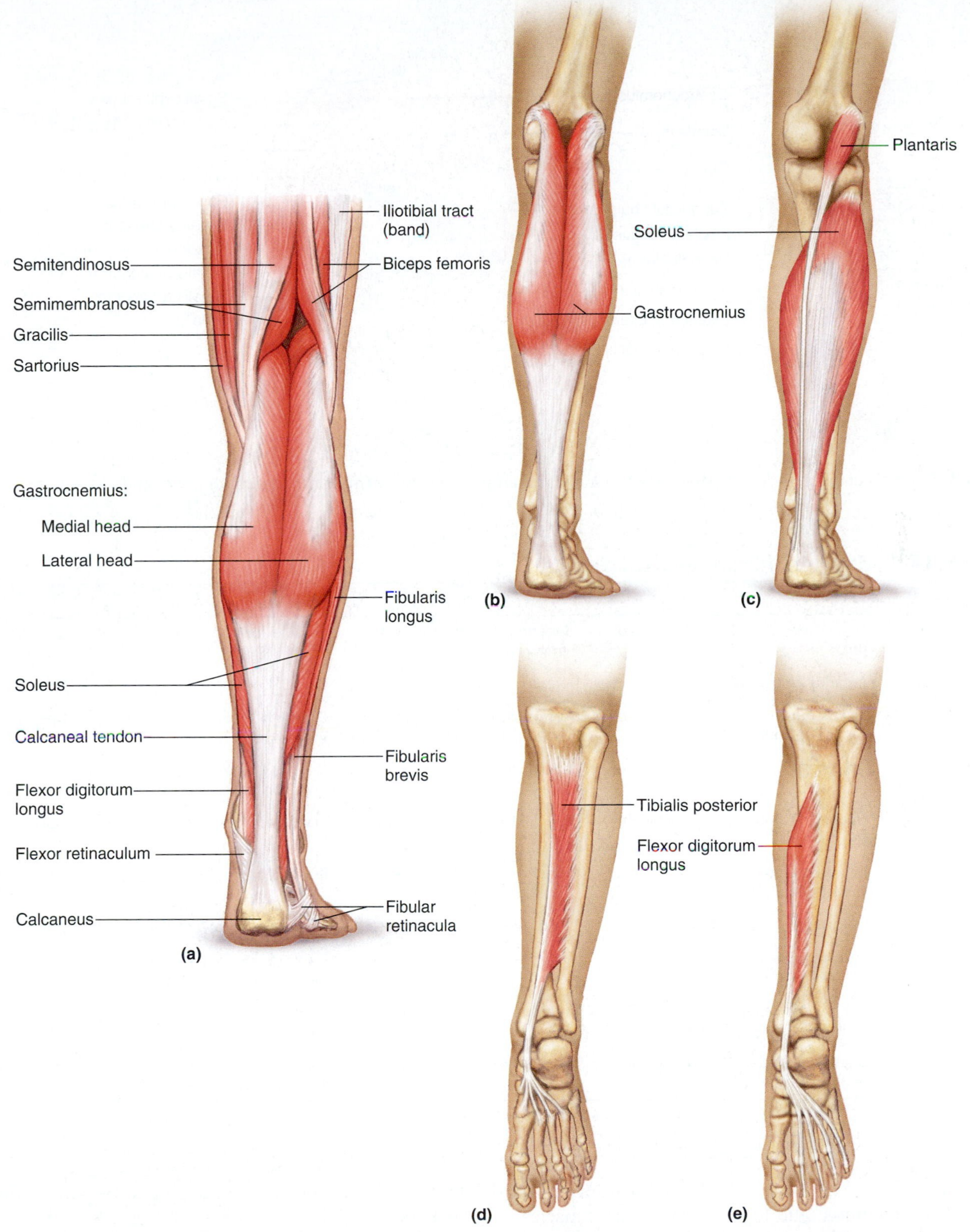

FIGURE 9.43 Muscles of the leg. (*a*) Muscles of the posterior right leg. (*b–e*) Isolated views of muscles associated with the posterior right leg.

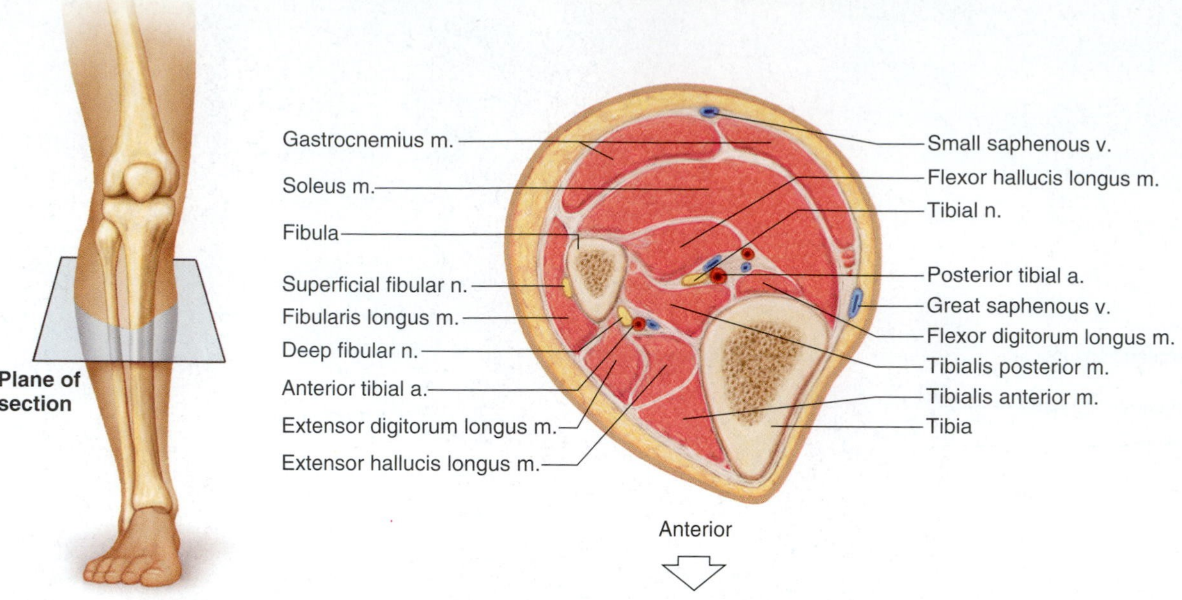

FIGURE 9.44 A cross section of the leg (superior view). (*a.* stands for artery, *m.* stands for muscle, *n.* stands for nerve, and *v.* stands for vein.)

TABLE 9.14 | Muscles That Move the Foot

Muscle	Origin	Insertion	Action	Nerve Supply
Tibialis anterior	Lateral condyle and lateral surface of tibia	Tarsal bone (cuneiform) and first metatarsal	Dorsiflexion and inversion of foot	Deep fibular n.
Fibularis tertius	Anterior surface of fibula	Dorsal surface of fifth metatarsal	Dorsiflexion and eversion of foot	Deep fibular n.
Extensor digitorum longus	Lateral condyle of tibia and anterior surface of fibula	Dorsal surfaces of second and third phalanges of four lateral toes	Dorsiflexion and eversion of foot, extends toes	Deep fibular n.
Extensor hallucis longus	Anterior surface of fibula	Distal phalanx of the great toe	Extends great toe, dorsiflexion and inversion of foot	Deep fibular n.
Gastrocnemius	Lateral and medial condyles of femur	Posterior surface of calcaneus	Plantar flexion of foot, flexes knee	Tibial n.
Soleus	Head and shaft of fibula and posterior surface of tibia	Posterior surface of calcaneus	Plantar flexion of foot	Tibial n.
Plantaris	Femur	Calcaneus	Plantar flexion of foot, flexes knee	Tibial n.
Flexor digitorum longus	Posterior surface of tibia	Distal phalanges of four lateral toes	Plantar flexion and inversion of foot, flexes four lateral toes	Tibial n.
Tibialis posterior	Lateral condyle and posterior surface of tibia and posterior surface of fibula	Tarsal and metatarsal bones	Plantar flexion and inversion of foot	Tibial n.
Fibularis longus	Lateral condyle of tibia and head and shaft of fibula	Tarsal and metatarsal bones	Plantar flexion and eversion of foot, also supports arch	Superficial fibular n.

The **soleus** (so′le-us) is a thick, flat muscle located beneath the gastrocnemius, and together these two muscles form the calf of the leg. The soleus arises from the tibia and fibula, and it extends to the heel by way of the calcaneal tendon. It acts with the gastrocnemius to cause plantar flexion of the foot (see figs. 9.42 and 9.43).

The **plantaris** (plan-ta′ris) connects the femur to the heel, where it inserts with the gastrocnemius and soleus via the calcaneal tendon. When the plantaris contracts it flexes the foot, and because it crosses the knee joint, it also flexes the knee.

The **flexor digitorum longus** (flek′sor dij″ĭ-to′rum long′gus) extends from the posterior surface of the tibia to the foot. Its tendon passes along the plantar surface of the foot. There the tendon divides into four parts that attach to the terminal bones of the four lateral toes. This muscle assists in plantar flexion of the foot, flexion of the four lateral toes, and inversion of the foot (see fig. 9.43).

INNERCONNECTIONS | *Muscular System*

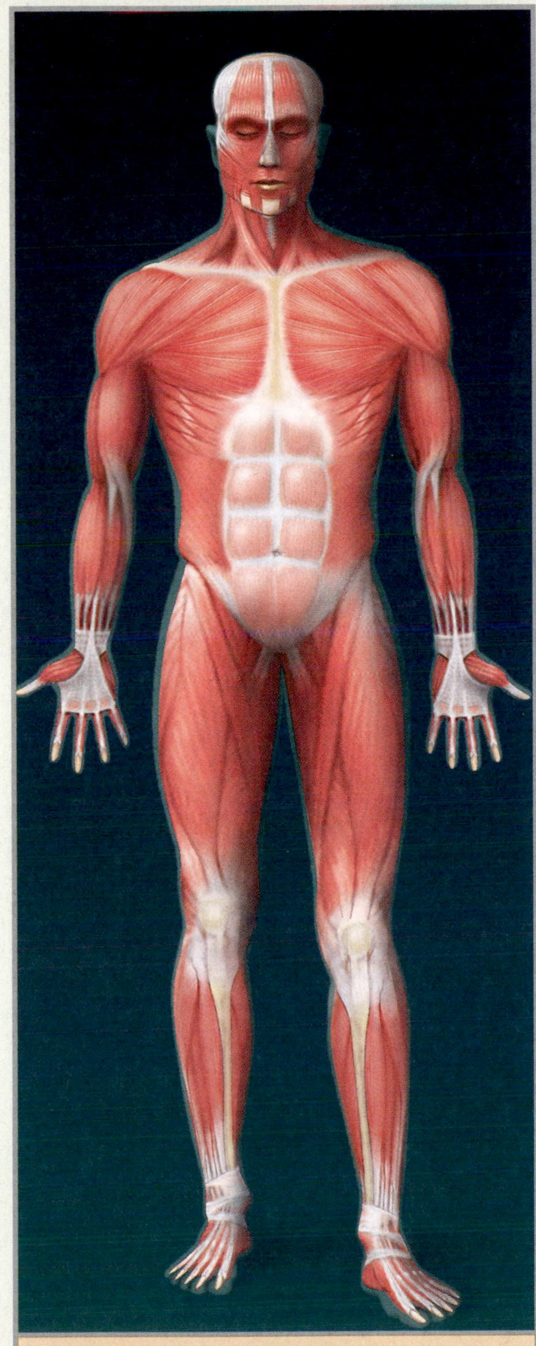

Muscular System

Muscles provide the force for moving body parts.

Integumentary System

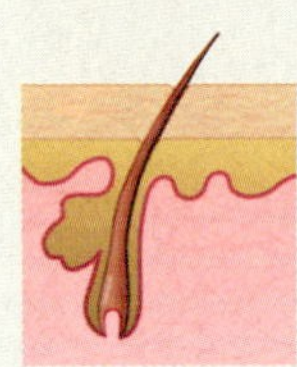

The skin increases heat loss during skeletal muscle activity. Sensory receptors function in the reflex control of skeletal muscles.

Lymphatic System

Muscle action pumps lymph through lymphatic vessels.

Skeletal System

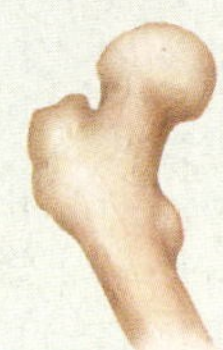

Bones provide attachments that allow skeletal muscles to cause movement.

Digestive System

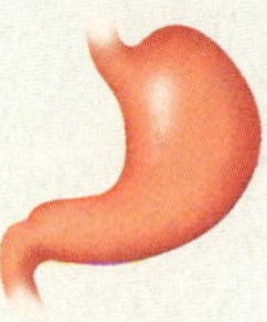

Skeletal muscles are important in swallowing. Smooth muscle moves food through the digestive tract. The digestive system absorbs needed nutrients.

Nervous System

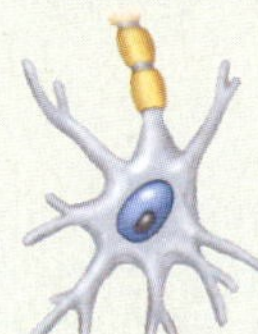

Neurons control muscle contractions.

Respiratory System

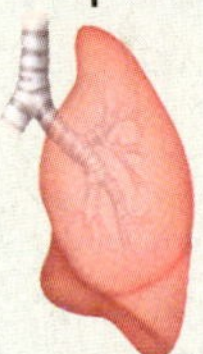

Breathing depends on skeletal muscles. The lungs provide oxygen for energy releasing reactions in muscle cells and excrete carbon dioxide waste from those reactions.

Endocrine System

Hormones help increase blood flow to exercising skeletal muscles.

Urinary System

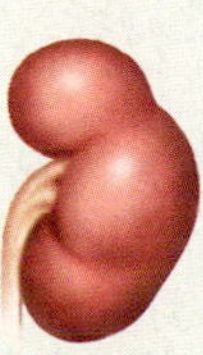

Skeletal muscles help control expulsion of urine from the urinary bladder.

Cardiovascular System

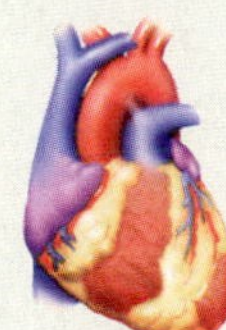

Blood flow delivers oxygen and nutrients and removes wastes. Cardiac muscle pumps blood, smooth muscle in vessel walls enables vasoconstriction, vasodilation.

Reproductive System

Skeletal muscles are important in sexual activity.

Invertor

The **tibialis posterior** (tib″e-a′lis pos-tēr′e-or) is the deepest of the muscles on the back of the leg. It connects the fibula and tibia to the ankle bones by means of a tendon that curves under the medial malleolus. This muscle assists in inversion and plantar flexion of the foot (see fig. 9.43). The dorsiflexor *extensor hallucis longus,* because it pulls up on the medial portion, also inverts the foot (see fig. 9.41).

Evertor

The **fibularis** (peroneus) **longus** (fib″u-la′ris long′gus) is a long, straplike muscle located on the lateral side of the leg. It connects the tibia and the fibula to the foot by means of a stout tendon that passes behind the lateral malleolus. It everts the foot, assists in plantar flexion, and helps support the arch of the foot (see figs. 9.42 and 9.44).

As in the wrist, fascia in various regions of the ankle thicken to form retinacula. Anteriorly, for example, *extensor retinacula* connect the tibia and fibula as well as the calcaneus and fascia of the sole. These retinacula form sheaths for tendons crossing the front of the ankle (see figs. 9.41 and 9.42).

Posteriorly, on the inside, a *flexor retinaculum* runs between the medial malleolus and the calcaneus and forms sheaths for tendons passing beneath the foot (see fig. 9.43). *Fibular retinacula* connect the lateral malleolus and the calcaneus, providing sheaths for tendons on the lateral side of the ankle (see fig. 9.42).

9.9 LIFE-SPAN CHANGES

Signs of aging in the muscular system begin to appear in one's forties, although a person can still be active. At a microscopic level, supplies of the molecules that enable muscles to function—myoglobin, ATP, and creatine phosphate—decline. The diameters of some muscle fibers may shrink, as the muscle layers in the walls of veins thicken, making the vessels more rigid and less elastic. Gradually, the muscles become smaller, drier, and capable of less forceful contraction. Connective tissue and adipose cells begin to replace some muscle tissue. By age eighty, nearly half the muscle mass has atrophied, due to a decline in motor neuron activity. Diminishing muscular strength slows reflexes.

Exercise can help maintain a healthy muscular system throughout life, countering the less effective oxygen delivery that results from the decreased muscle mass that accompanies aging. Exercise can even lead to formation of new muscle by stimulating skeletal muscle cells to release interleukin-6 (IL-6), a type of proinflammatory molecule called a cytokine. The IL-6 stimulates satellite cells, which function as muscle stem cells. They divide and migrate, becoming incorporated into the muscle fiber. Exercise also maintains the flexibility of blood vessels, which helps to keep blood pressure at healthy levels. A physician should be consulted before starting any exercise program.

According to the National Institute on Aging, exercise should include strength training and aerobics, with stretching before and after. Strength training consists of weight lifting or using a machine that works specific muscles against a resistance, performed so that the same muscle is not exercised on consecutive days. Strength training increases muscle mass, and the resulting stronger muscles can alleviate pressure on the joints, which may lessen arthritis pain. Aerobic exercise improves oxygen use by muscles and increases endurance. Stretching increases flexibility and decreases muscle strain, while improving blood flow to all muscles. A side benefit of regular exercise, especially among older individuals, is fewer bouts of depression.

PRACTICE

31 What changes are associated with an aging muscular system?

32 Describe two types of recommended exercise.

CHAPTER SUMMARY

9.1 INTRODUCTION (PAGE 285)

All movements require muscles. The three types of muscle tissue are skeletal, smooth, and cardiac.

9.2 STRUCTURE OF A SKELETAL MUSCLE (PAGE 285)

Skeletal muscles are composed of nervous, vascular and various other connective tissues, as well as skeletal muscle tissue.

1. Connective tissue coverings
   a. Fascia covers each skeletal muscle.
   b. Other connective tissues surround cells and groups of cells within the muscle's structure (epimysium, perimysium, endomysium).
   c. Fascia is part of a complex network of connective tissue that extends throughout the body.
2. Skeletal muscle fibers
   a. Each skeletal muscle fiber is a single muscle cell, the unit of contraction.
   b. Muscle fibers are cylindrical cells with many nuclei.
   c. The cytoplasm contains mitochondria, sarcoplasmic reticulum, and myofibrils of actin and myosin.
   d. The arrangement of the actin and myosin filaments causes striations. (I bands, Z lines, A bands, H zone and M line)
   e. Cross-bridges of myosin filaments form linkages with actin filaments. The reaction between actin and myosin filaments provides the basis for contraction.

f. When a fiber is at rest, troponin and tropomyosin molecules interfere with linkage formation. Calcium ions remove the inhibition.
g. Transverse tubules extend from the cell membrane into the cytoplasm and are associated with the cisternae of the sarcoplasmic reticulum.

9.3 SKELETAL MUSCLE CONTRACTION (PAGE 289)

Muscle fiber contraction results from a sliding movement of actin and myosin filaments overlapping that shortens the muscle fiber.

1. Neuromuscular junction
   a. Motor neurons stimulate muscle fibers to contract.
   b. The motor end plate of a muscle fiber lies on one side of a neuromuscular junction.
   c. One motor neuron and the muscle fibers associated with it constitute a motor unit.
   d. In response to a nerve impulse, the end of a motor nerve fiber secretes a neurotransmitter, which diffuses across the junction and stimulates the muscle fiber.
2. Stimulus for contraction
   a. Acetylcholine released from the end of a motor nerve fiber stimulates a muscle fiber.
   b. Acetylcholinesterase decomposes acetylcholine, preventing continuous stimulation.
   c. Stimulation causes a muscle fiber to conduct an impulse that travels over the surface of the sarcolemma and reaches the deep parts of the fiber by means of the transverse tubules.
3. Excitation contraction coupling
   a. A muscle impulse signals the sarcoplasmic reticulum to release calcium ions.
   b. Calcium ions combine with troponin, causing the tropomyosin to shift and expose active sites on the actin for myosin binding.
   c. Linkages form between myosin and actin, and the actin filaments move inward, shortening the sarcomere.
4. The sliding filament model of muscle contraction
   a. The sarcomere, defined by striations, is the functional unit of skeletal muscle.
   b. When the overlapping thick and thin myofilaments slide past one another, the sarcomeres shorten. The muscle contracts.
5. Cross-bridge cycling
   a. A myosin cross-bridge can attach to an actin binding site and pull on the actin filament. The myosin head can then release the actin and combine with another active binding site farther down the actin filament and pull again.
   b. The breakdown of ATP releases energy that provides the repetition of the cross-bridge cycle.
6. Relaxation
   a. Acetylcholinesterase rapidly decomposes acetylcholine remaining in the synapse preventing continuous stimulation of a muscle fiber.
   b. The muscle fiber relaxes when calcium ions are transported back into the sarcoplasmic reticulum.
   c. Cross-bridge linkages break and do not re-form—the muscle fiber relaxes.
7. Energy sources for contraction
   a. ATP supplies the energy for muscle fiber contraction.
   b. Creatine phosphate stores energy that can be used to synthesize ATP as it is decomposed.
   c. Active muscles require cellular respiration for energy.
8. Oxygen supply and cellular respiration
   a. Anaerobic reactions of cellular respiration yield few ATP molecules, whereas aerobic reactions of cellular respiration provide many ATP molecules.
   b. Hemoglobin in red blood cells carries oxygen from the lungs to body cells.
   c. Myoglobin in muscle cells temporarily stores some oxygen.
9. Oxygen debt
   a. During rest or moderate exercise, oxygen is sufficient to support the aerobic reactions of cellular respiration.
   b. During strenuous exercise, oxygen deficiency may develop, and lactic acid may accumulate as a result of the anaerobic reactions of cellular respiration.
   c. The amount of oxygen required to react accumulated lactic acid to form glucose and to restore supplies of ATP and creatine phosphate is called oxygen debt.
10. Muscle fatigue
   a. A fatigued muscle loses its ability to contract.
   b. Muscle fatigue is usually due to the effects of accumulation of lactic acid.
   c. Athletes usually produce less lactic acid than nonathletes because of their increased ability to supply oxygen and nutrients to muscles.
11. Heat production
   a. Muscular contraction generates body heat.
   b. Most of the energy released by cellular respiration is lost as heat.

9.4 MUSCULAR RESPONSES (PAGE 296)

1. Threshold stimulus is the minimal stimulus needed to elicit a muscular contraction.
2. Recording of a muscle contraction
   a. A twitch is a single, short contraction of a muscle fiber.
   b. A myogram is a recording of the contraction of an electrically stimulated isolated muscle or muscle fiber.
   c. The latent period is the time between stimulus and responding contraction.
   d. During the refractory period immediately following contraction, a muscle fiber cannot respond.
   e. The length to which a muscle is stretched before stimulation affects the force it will develop.
      (1) Normal activities occur at optimal length.
      (2) Too long or too short decreases force.
   f. Sustained contractions are more important than twitch contractions in everyday activities.
3. Summation
   a. A rapid series of stimuli may produce summation of twitches and sustained contraction.
   b. Forceful, sustained contraction without relaxation is a tetanic contraction.

4. Recruitment of motor units
    a. Muscles whose motor units have few muscle fibers produce finer movements.
    b. Motor units respond in an all-or-none manner.
    c. At low intensity of stimulation, relatively few motor units contract.
    d. At increasing intensities of stimulation, other motor units are recruited until the muscle contracts with maximal tension.
5. Sustained contractions
    a. Tetanic contractions are common in everyday activities.
    b. Even when a whole muscle appears at rest, some of its fibers undergo sustained contraction. This is called muscle tone.
6. Types of contractions
    a. One type of isotonic contraction occurs when a muscle contracts and its ends are pulled closer together. Because the muscle shortens, it is called a concentric contraction.
    b. In another type of isotonic contraction, the force a muscle generates is less than that required to move or lift an object. This lengthening contraction is an eccentric contraction.
    c. When a muscle contracts but its attachments do not move, the contraction is isometric.
    d. Most body movements involve both isometric and isotonic contractions.
7. Fast-and-slow twitch muscle fibers
    a. The speed of contraction is related to a muscle's specific function.
    b. Slow-contracting, or red, muscles can generate ATP fast enough to keep up with ATP breakdown and can contract for long periods.
    c. Fast-contracting, or white, muscles have reduced ability to carry on the aerobic reactions of cellular respiration and tend to fatigue rapidly.

9.5 SMOOTH MUSCLES (PAGE 300)

The contractile mechanisms of smooth and cardiac muscles are similar to those of skeletal muscle.

1. Smooth muscle fibers
    a. Smooth muscle cells contain filaments of myosin and actin.
    b. They lack transverse tubules, and the sarcoplasmic reticula are not well developed.
    c. Types include multiunit smooth muscle and visceral smooth muscle.
    d. Visceral smooth muscle displays rhythmicity.
    e. Peristalsis aids movement of material through hollow organs.
2. Smooth muscle contraction
    a. In smooth muscles, calmodulin binds to calcium ions and activates the contraction mechanism.
    b. Both acetylcholine and norepinephrine are neurotransmitters for smooth muscles.
    c. Hormones and stretching affect smooth muscle contractions.
    d. With a given amount of energy, smooth muscle can maintain a contraction longer than skeletal muscle.
    e. Smooth muscles can change length without changing tautness.

9.6 CARDIAC MUSCLE (PAGE 301)

1. Cardiac muscle contracts for a longer time than skeletal muscle because transverse tubules supply extra calcium ions.
2. Intercalated discs connect the ends of adjacent cardiac muscle cells and hold the cells together.
3. A network of fibers contracts as a unit and responds to stimulation in an all-or-none manner.
4. Cardiac muscle is self-exciting, rhythmic, and remains refractory until a contraction is completed.

9.7 SKELETAL MUSCLE ACTIONS (PAGE 301)

1. Body movement
    a. Bones and muscles function together as levers.
    b. A lever consists of a rod, a fulcrum (pivot), a resistance, and a force that supplies energy.
    c. Parts of a first-class lever are arranged resistance–fulcrum–force; of a second-class lever, fulcrum–resistance–force; and of a third-class lever, resistance–force–fulcrum.
2. Origin and insertion
    a. The movable end of attachment of a skeletal muscle to a bone is its insertion, and the immovable end is its origin.
    b. Some muscles have more than one origin or insertion.
3. Interaction of skeletal muscles
    a. Skeletal muscles function in groups.
    b. A prime mover is responsible for most of a movement; synergists aid prime movers; antagonists can resist the movement of a prime mover.
    c. Smooth movements depend upon antagonists giving way to the actions of prime movers.

9.8 MAJOR SKELETAL MUSCLES (PAGE 305)

Muscle names often describe sizes, shapes, locations, actions, number of attachments, or direction of fibers.

1. Muscles of facial expression
    a. These muscles lie beneath the skin of the face and scalp and are used to communicate feelings through facial expression.
    b. They include the epicranius, orbicularis oculi, orbicularis oris, buccinator, zygomaticus major, zygomaticus minor, and platysma.
2. Muscles of mastication
    a. These muscles are attached to the mandible and are used in chewing.
    b. They include the masseter, temporalis, medial pterygoid, and lateral pterygoid.
3. Muscles that move the head and vertebral column
    a. Muscles in the neck and back move the head.
    b. They include the sternocleidomastoid, splenius capitis, semispinalis capitis, quadratus lumborum, and erector spinae.
4. Muscles that move the pectoral girdle
    a. Most of these muscles connect the scapula to nearby bones and are closely associated with muscles that move the arm.
    b. They include the trapezius, rhomboid major, rhomboid minor, levator scapulae, serratus anterior, and pectoralis minor.

5. Muscles that move the arm
   a. These muscles connect the humerus to various regions of the pectoral girdle, ribs, and vertebral column.
   b. They include the coracobrachialis, pectoralis major, teres major, latissimus dorsi, supraspinatus, deltoid, subscapularis, infraspinatus, and teres minor.
6. Muscles that move the forearm
   a. These muscles connect the radius and ulna to the humerus and pectoral girdle.
   b. They include the biceps brachii, brachialis, brachioradialis, triceps brachii, supinator, pronator teres, and pronator quadratus.
7. Muscles that move the hand
   a. These muscles arise from the distal end of the humerus and from the radius and ulna.
   b. They include the flexor carpi radialis, flexor carpi ulnaris, palmaris longus, flexor digitorum profundus, flexor digitorum superficialis, extensor carpi radialis longus, extensor carpi radialis brevis, extensor carpi ulnaris, and extensor digitorum.
   c. An extensor retinaculum forms sheaths for tendons of the extensor muscles.
8. Muscles of the abdominal wall
   a. These muscles connect the rib cage and vertebral column to the pelvic girdle.
   b. They include the external oblique, internal oblique, transversus abdominis, and rectus abdominis.
9. Muscles of the pelvic outlet
   a. These muscles form the floor of the pelvic cavity and fill the space of the pubic arch.
   b. They include the levator ani, coccygeus, superficial transversus perinei, bulbospongiosus, ischiocavernosus, and sphincter urethrae.
10. Muscles that move the thigh
   a. These muscles are attached to the femur and to some part of the pelvic girdle.
   b. They include the psoas major, iliacus, gluteus maximus, gluteus medius, gluteus minimus, piriformis, tensor fasciae latae, pectineus, adductor brevis, adductor longus, adductor magnus, and gracilis.
11. Muscles that move the leg
   a. These muscles connect the tibia or fibula to the femur or pelvic girdle.
   b. They include the biceps femoris, semitendinosus, semimembranosus, sartorius, rectus femoris, vastus lateralis, vastus medialis, and vastus intermedius.
12. Muscles that move the foot
   a. These muscles attach the femur, tibia, and fibula to various bones of the foot.
   b. They include the tibialis anterior, fibularis tertius, extensor digitorum longus, extensor hallucis longus, gastrocnemius, soleus, plantaris, flexor digitorum longus, tibialis posterior, and fibularis longus.
   c. Retinacula form sheaths for tendons passing to the foot.

9.9 LIFE-SPAN CHANGES (PAGE 334)

1. Beginning in one's forties, supplies of ATP, myoglobin, and creatine phosphate begin to decline.
2. By age eighty, muscle mass may be halved. Reflexes slow. Adipose cells and connective tissue replace some muscle tissue.
3. Exercise is beneficial in maintaining muscle function.

CHAPTER ASSESSMENTS

PART A

9.1 Introduction

1 List three outcomes of muscle actions. (p. 285)

9.2 Structure of a Skeletal Muscle

2 Describe the difference between a tendon and an aponeurosis. (p. 285)

3 Describe how connective tissue is part of the structure of a skeletal muscle. (p. 286)

4 Distinguish among deep fascia, subcutaneous fascia, and subserous fascia. (p. 287)

5 Identify the major parts of a skeletal muscle fiber and describe the functions of each. (p. 287)

9.3 Skeletal Muscle Contraction

6 Describe the neuromuscular junction. (p. 289)

7 Define *motor unit* and explain how the number of fibers in a unit affects muscular contractions. (p. 290)

8 Describe the neural control of skeletal muscle contraction. (p. 290)

9 A neurotransmitter ________________. (p. 290)

a. binds actin filaments, causing them to slide
b. diffuses across a synaptic cleft from a neuron to a muscle cell
c. transports ATP across the synaptic cleft
d. breaks down acetylcholine at the synapse
e. is a contractile protein in the muscle fiber

10 Identify the major events that occur during skeletal muscle fiber contraction. (p. 291)

11 Explain how ATP and creatine phosphate function in skeletal muscle fiber contraction. (p. 292)

12 Describe how oxygen is supplied to skeletal muscle. (p. 294)

13 Describe how oxygen debt may develop. (p. 295)

14 Explain how a muscle may become fatigued, and how a person's physical condition may affect tolerance to fatigue. (p. 296)

9.4 Muscular Responses

15 Distinguish between a twitch and a sustained contraction. (p. 296)

16 Define *threshold stimulus*. (p. 296)

17 Which of the following describes addition of muscle fibers to take part in a contraction? (p. 297)

a. summation
b. recruitment
c. tetany
d. twitch
e. relaxation

18 Explain how a skeletal muscle can be stimulated to produce a sustained contraction. (p. 297)

19 Distinguish between a tetanic contraction and muscle tone. (p. 297)

20 Distinguish between concentric and eccentric contractions, and explain how each is used in body movements. (p. 298)

21 Distinguish between fast- and slow-twitch muscle fibers. (p. 299)

9.5 Smooth Muscle

22 Distinguish between multiunit smooth muscle and visceral smooth muscle. (p. 300)

23 Define *peristalsis* and explain its function. (p. 300)

24 Compare the characteristics of skeletal and smooth muscle fiber contractions. (p. 301)

9.6 Cardiac Muscle

25 Compare the characteristics of skeletal and cardiac muscle fiber contractions. (p. 301)

9.7 Skeletal Muscle Actions

26 Describe a lever, and explain how its parts may be arranged to form first-class, second-class, and third-class levers. (p. 302)

27 Explain how limb movements function as levers. (p. 302)

28 Distinguish between a muscle's origin and its insertion. (p. 303)

29 Define *prime mover, agonist, synergist,* and *antagonist*. (p. 304)

PART B

9.8 Major Skeletal Muscles

30 Match the muscle with its description or action. (pp. 306–330)

(1) Buccinator
(2) Epicranius
(3) Lateral pterygoid
(4) Platysma
(5) Rhomboideus major
(6) Splenius capitis
(7) Temporalis
(8) Zygomaticus major
(9) Biceps brachii
(10) Brachialis
(11) Deltoid
(12) Latissimus dorsi
(13) Pectoralis major
(14) Pronator teres
(15) Teres minor
(16) Triceps brachii
(17) Biceps femoris
(18) External oblique
(19) Gastrocnemius
(20) Gluteus maximus
(21) Gluteus medius
(22) Gracilis
(23) Rectus femoris
(24) Tibialis anterior

A. Inserted on the coronoid process of the mandible
B. Draws the corner of the mouth upward
C. Can raise and adduct the scapula
D. Can pull the head into an upright position
E. Consists of two parts—the frontalis and the occipitalis
F. Compresses the cheeks
G. Extends over the neck from the chest to the face
H. Pulls the jaw from side to side
I. Primary extensor of the elbow
J. Pulls the shoulder back and downward
K. Abducts the arm
L. Rotates the arm laterally
M. Pulls the arm forward and across the chest
N. Rotates the arm medially
O. Strongest flexor of the elbow
P. Strongest supinator of the forearm
Q. Inverts the foot
R. A member of the quadriceps femoris group
S. A plantar flexor of the foot
T. Compresses the contents of the abdominal cavity
U. Largest muscle in the body
V. A hamstring muscle
W. Adducts the thigh
X. Abducts the thigh

PART C

31 Label as many muscles as you can identify in these photos of a model whose muscles are enlarged by exercise. Describe the action of each muscle identified. (pp. 306–334)

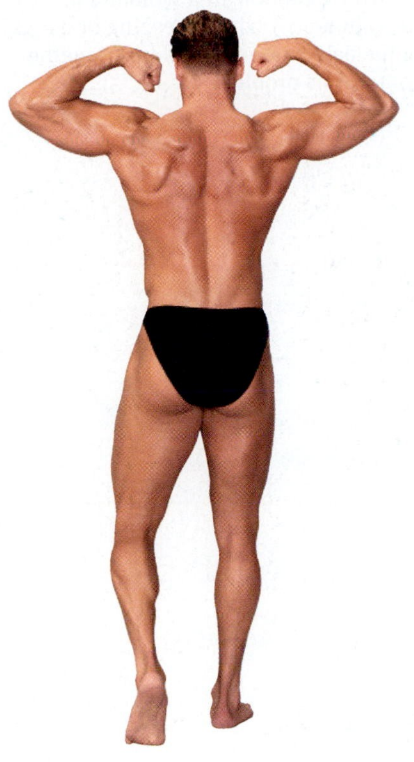

9.9 Life-Span Changes

32 Describe three aging-related changes in the muscular system. (p. 334)

33 Explain the benefits of exercise for maintaining muscular health while aging. (p. 334)

INTEGRATIVE ASSESSMENTS / CRITICAL THINKING

OUTCOMES 7.11, 9.8

1. Several important nerves and blood vessels course through the muscles of the gluteal region. To avoid the possibility of damaging such parts, intramuscular injections are usually made into the lateral, superior portion of the gluteus medius. What landmarks would help you locate this muscle in a patient?

OUTCOMES 9.2, 9.3

2. Millions of people take drugs called statins to lower serum cholesterol levels. In a small percentage of people taking these drugs, muscle pain, termed myopathy, is an adverse effect. In a small percentage of these individuals, the condition progresses to rhabdomyolysis, in which the sarcolemma breaks down.
   a. Describe the structure and state the function of the sarcolemma.
   b. Physicians can measure a patient's levels of creatine phosphokinase to track the safety of using a statin. Enzyme levels that exceed 10 times normal indicate possible rhabdomyolysis. Explain what this elevated enzyme level indicates about the physiology of the muscle cell.
   c. Explain why a dusky, dark color in the urine, resulting from the presence of myoglobin, also indicates muscle breakdown.

OUTCOMES 9.2, 9.3, 9.4

3. What steps might be taken to minimize atrophy of skeletal muscles in patients confined to bed for prolonged times?

OUTCOME 9.3

4. As lactic acid and other substances accumulate in an active muscle, they stimulate pain receptors, and the muscle may feel sore. How might the application of heat or substances that dilate blood vessels help relieve such soreness?

OUTCOMES 9.3, 9.4, 9.6

5. Why do you think athletes generally perform better if they warm up by exercising lightly before a competitive event?

OUTCOMES 9.3, 9.4, 9.7

6. Following an injury to a nerve, the muscles it supplies with motor nerve fibers may become paralyzed. How would you explain to a patient the importance of moving the disabled muscles passively or contracting them with electrical stimulation?

OUTCOMES 9.4, 9.8

7. Following childbirth, a woman may lose urinary control (incontinence) when sneezing or coughing. Which muscles of the pelvic floor should be strengthened by exercise to help control this problem?

WEB CONNECTIONS

Be sure to visit the text website at **www.mhhe.com/shier12** for answers to chapter assessments, additional quizzes, and interactive learning exercises.

ANATOMY & PHYSIOLOGY REVEALED

Anatomy & Physiology Revealed® (APR) includes cadaver photos that allow you to peel away layers of the human body to reveal structures beneath the surface. This program also includes animations, radiologic imaging, audio pronunciations, and practice quizzing. Check out **www.aprevealed.com**. APR has been proven to help improve student grades!

SURFACE ANATOMY AND CADAVER DISSECTION

The following set of reference plates, made up of surface anatomy photos and cadaver dissection photos, is presented to help you locate some of the more prominent surface features in various regions of the body. For the most part, the labeled structures on the surface anatomy photos are easily seen or palpated (felt) through the skin. As a review, you may want to locate as many of these features as possible on your own body. The cadaver dissection photos reveal the structures located beneath the skin.

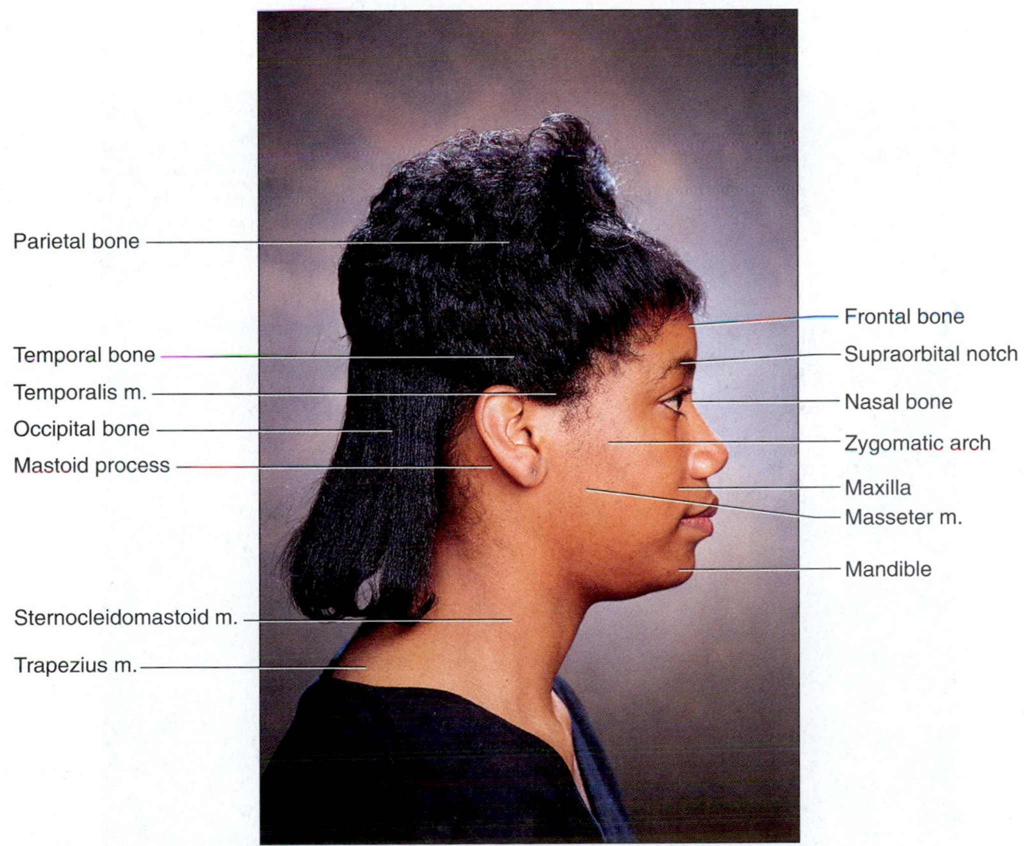

PLATE FIFTY-FIVE Surface anatomy of head and neck, lateral view. (*m.* stands for muscle.)

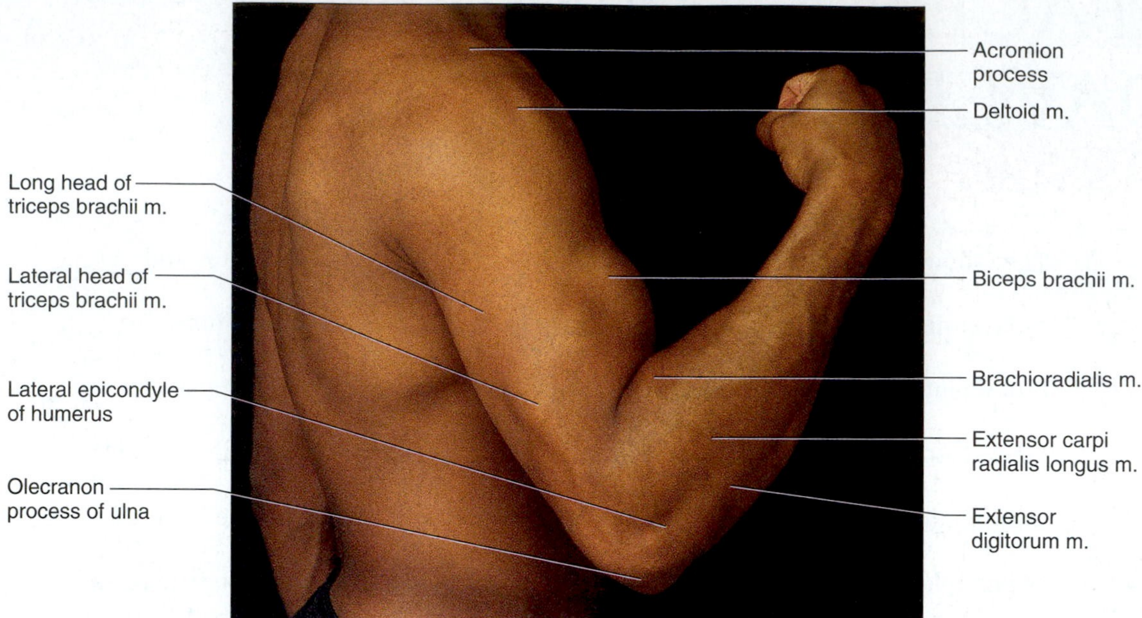

PLATE FIFTY-SIX Surface anatomy of upper limb and thorax, lateral view. (*m.* stands for muscle.)

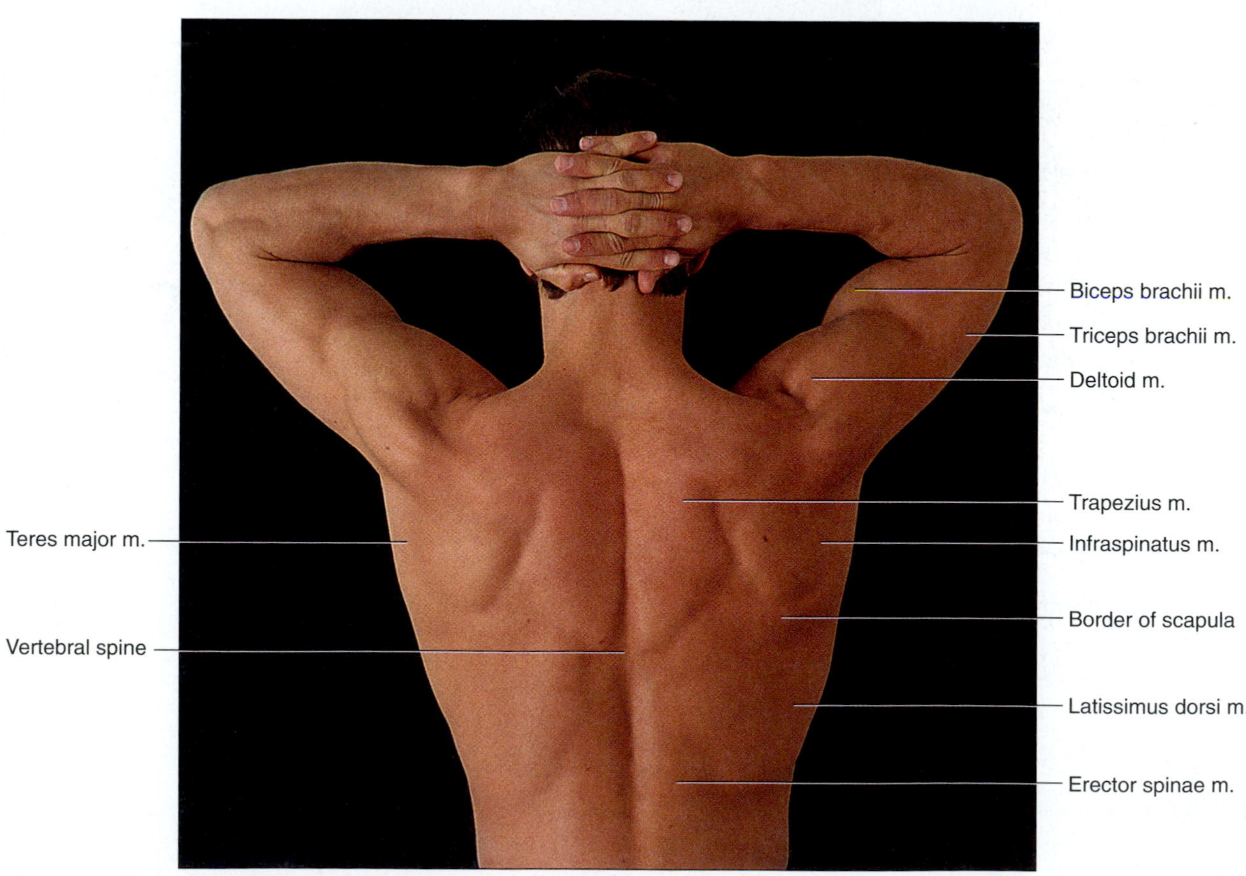

PLATE FIFTY-SEVEN Surface anatomy of back and upper limbs, posterior view. (*m.* stands for muscle.)

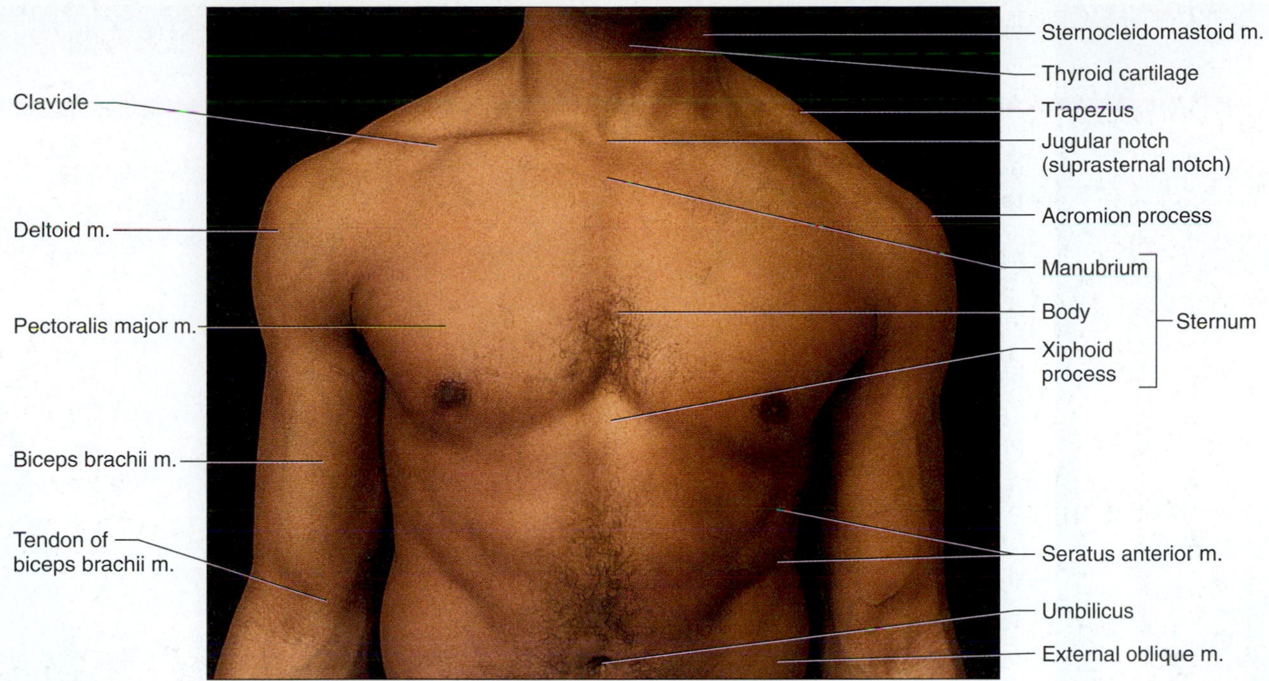

PLATE FIFTY-EIGHT Surface anatomy of torso and arms, anterior view. (*m.* stands for muscle.)

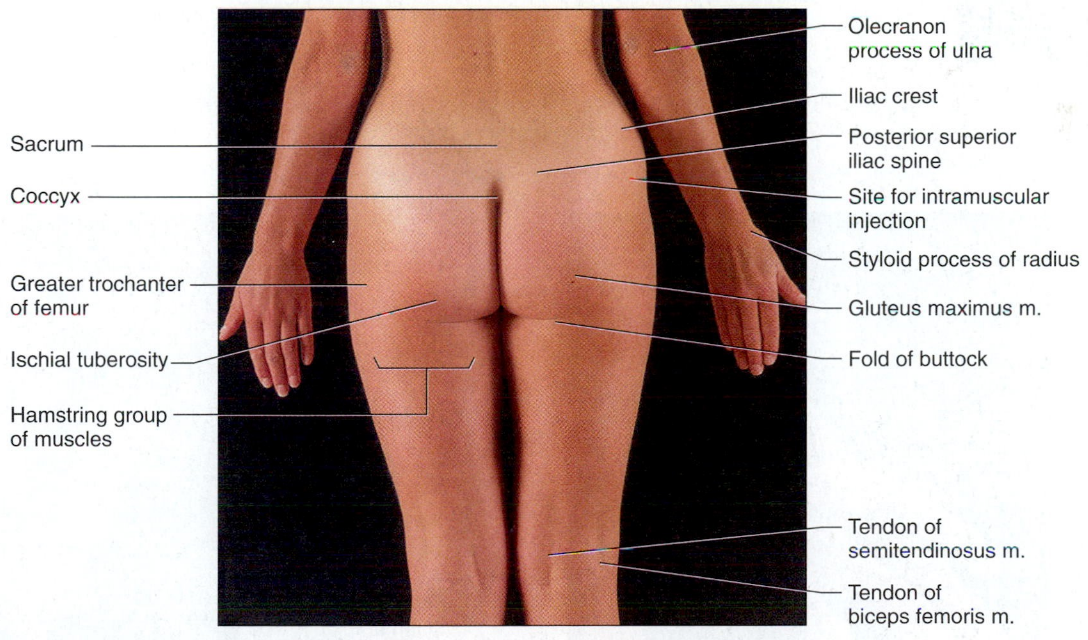

PLATE FIFTY-NINE Surface anatomy of torso and thighs, posterior view. (*m.* stands for muscle.)

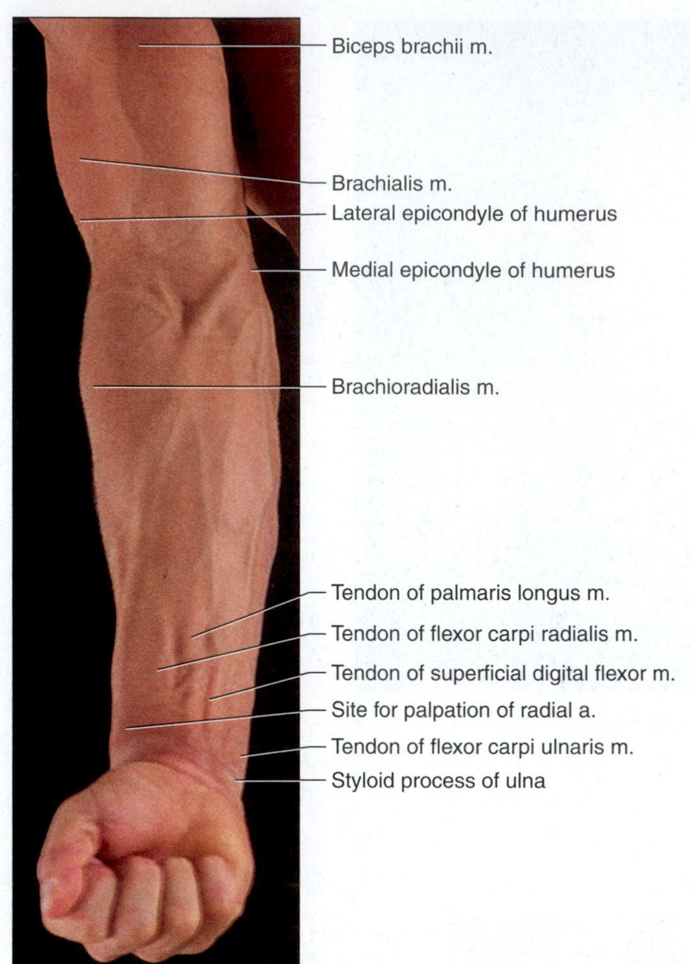

PLATE SIXTY Surface anatomy of right forearm, anterior view. (*m.* stands for muscle and *a.* stands for artery.)

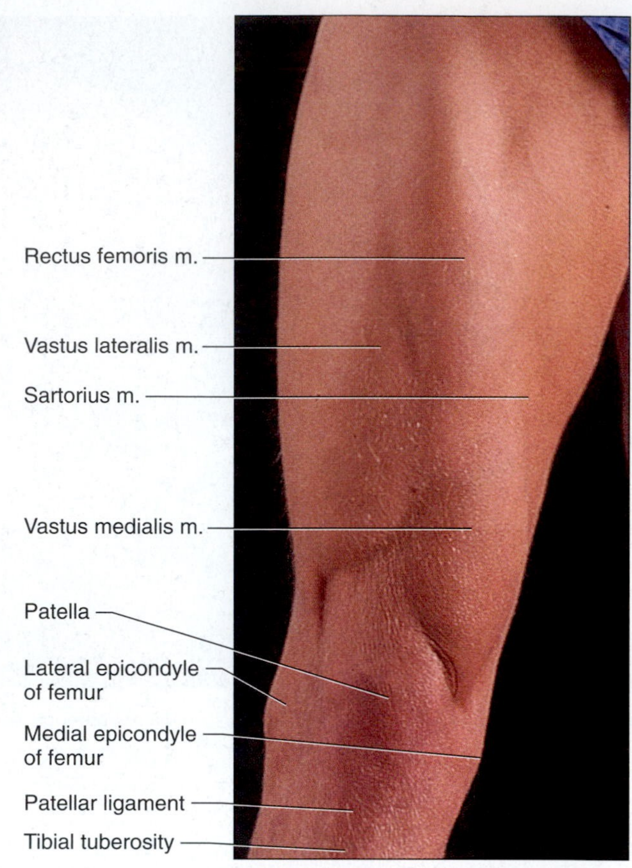

PLATE SIXTY-TWO Surface anatomy of right knee and surrounding area, anterior view. (*m.* stands for muscle.)

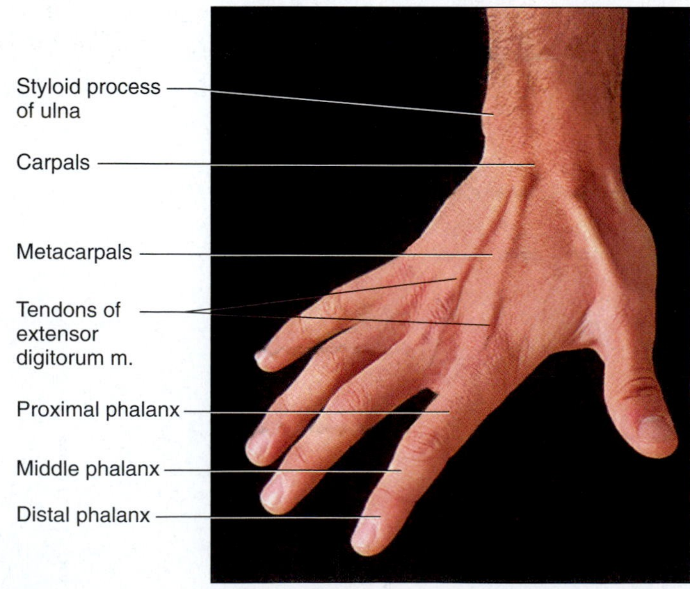

PLATE SIXTY-ONE Surface anatomy of the right hand. (*m.* stands for muscle.)

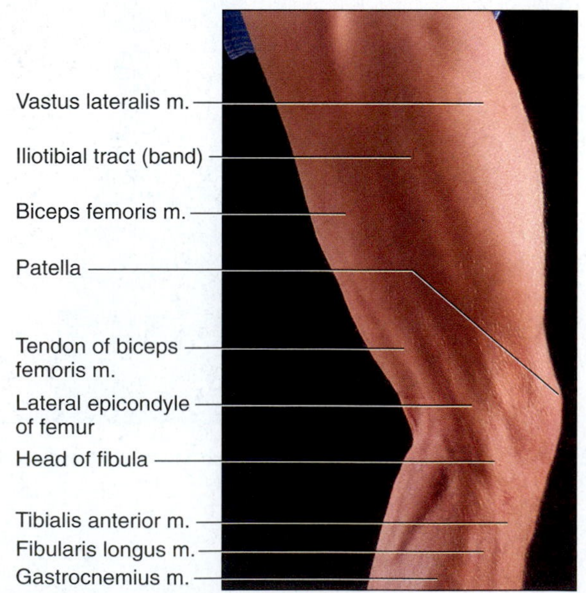

PLATE SIXTY-THREE Surface anatomy of right knee and surrounding area, lateral view. (*m.* stands for muscle.)

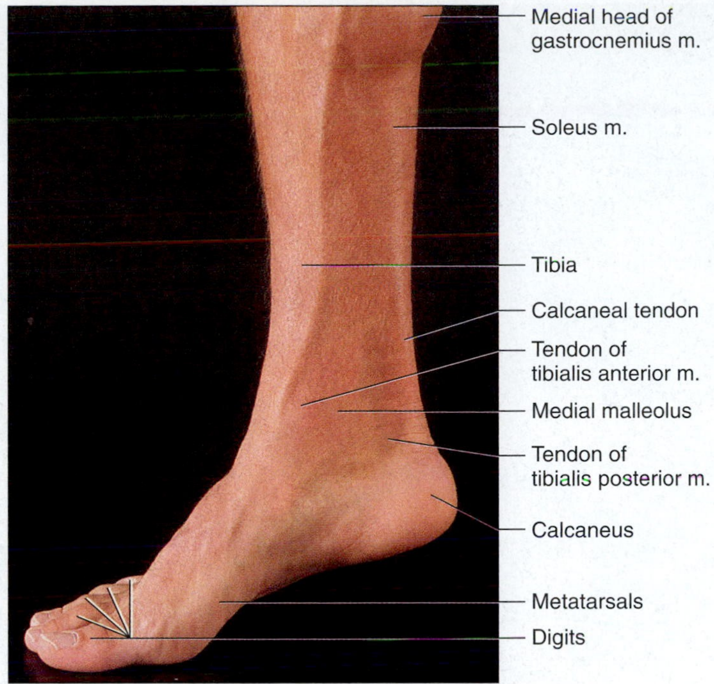

PLATE SIXTY-FOUR Surface anatomy of right foot and leg, medial view. (*m.* stands for muscle.)

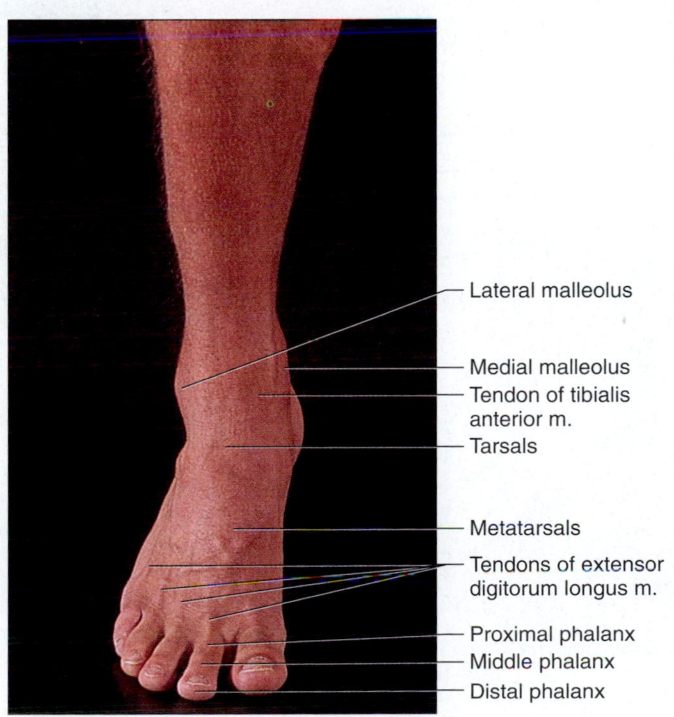

PLATE SIXTY-FIVE Surface anatomy of right foot. (*m.* stands for muscle.)

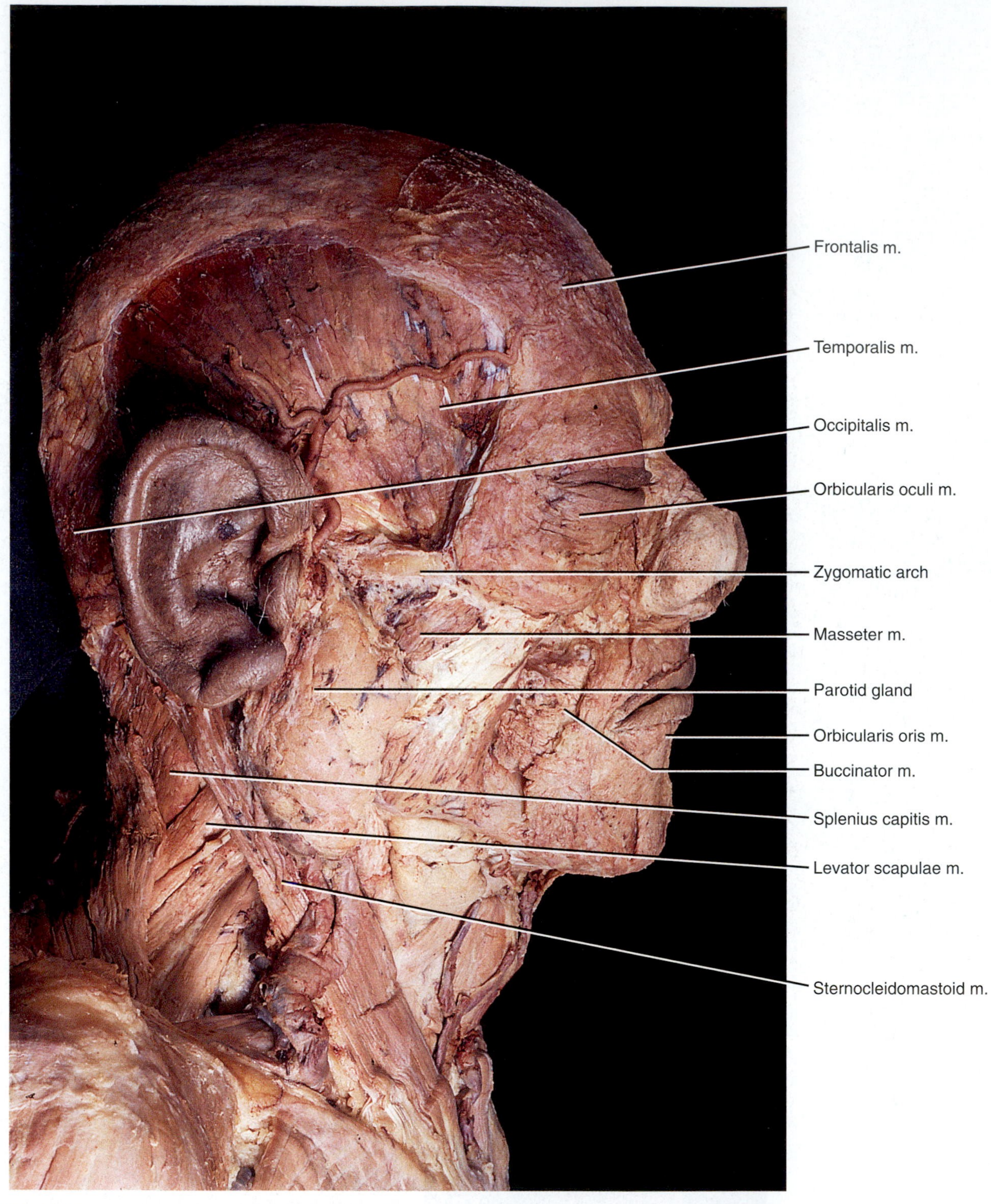

PLATE SIXTY-SIX Lateral view of the head. (*m.* stands for muscle.)

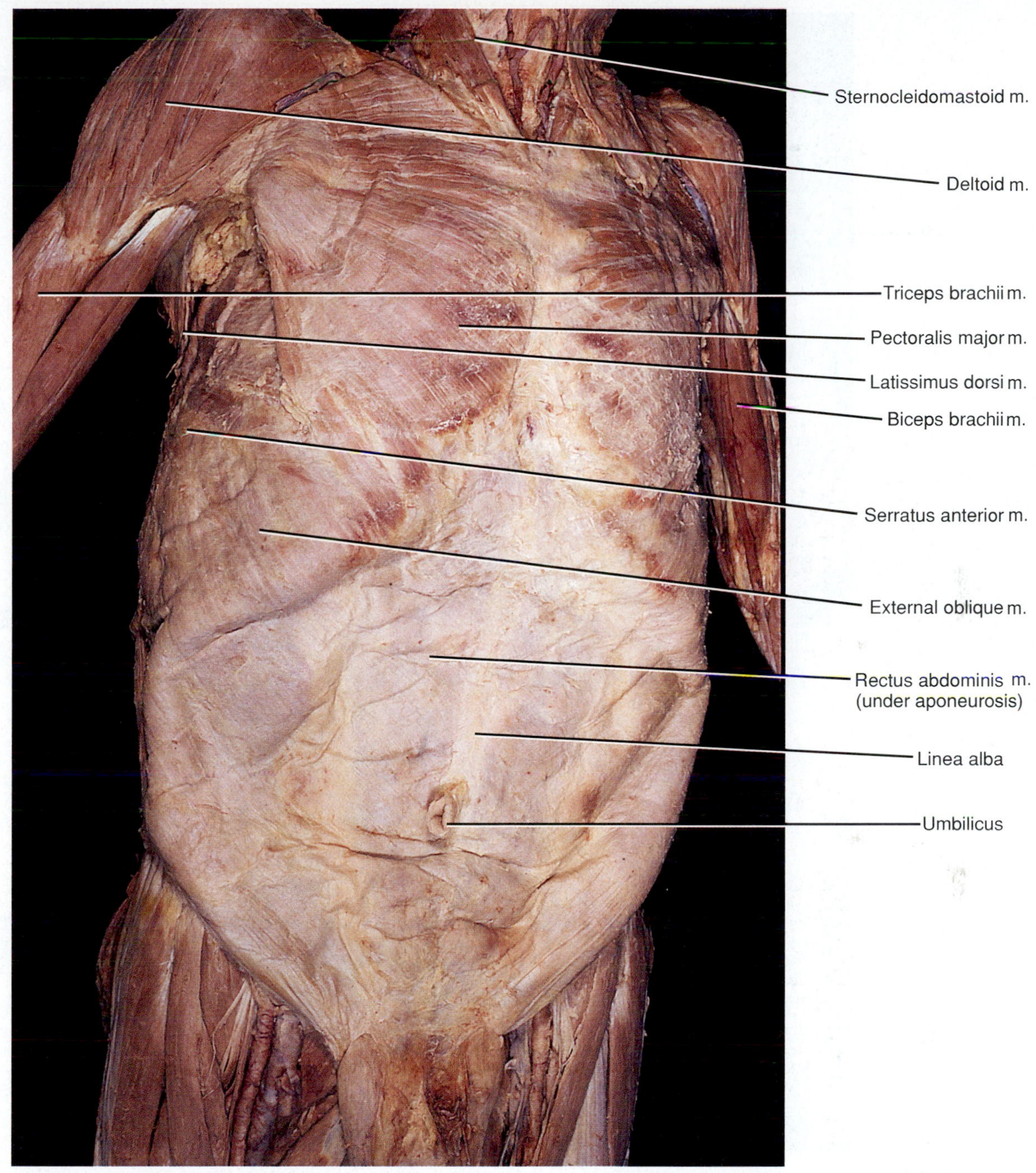

PLATE SIXTY-SEVEN Anterior view of the trunk. (*m.* stands for muscle.)

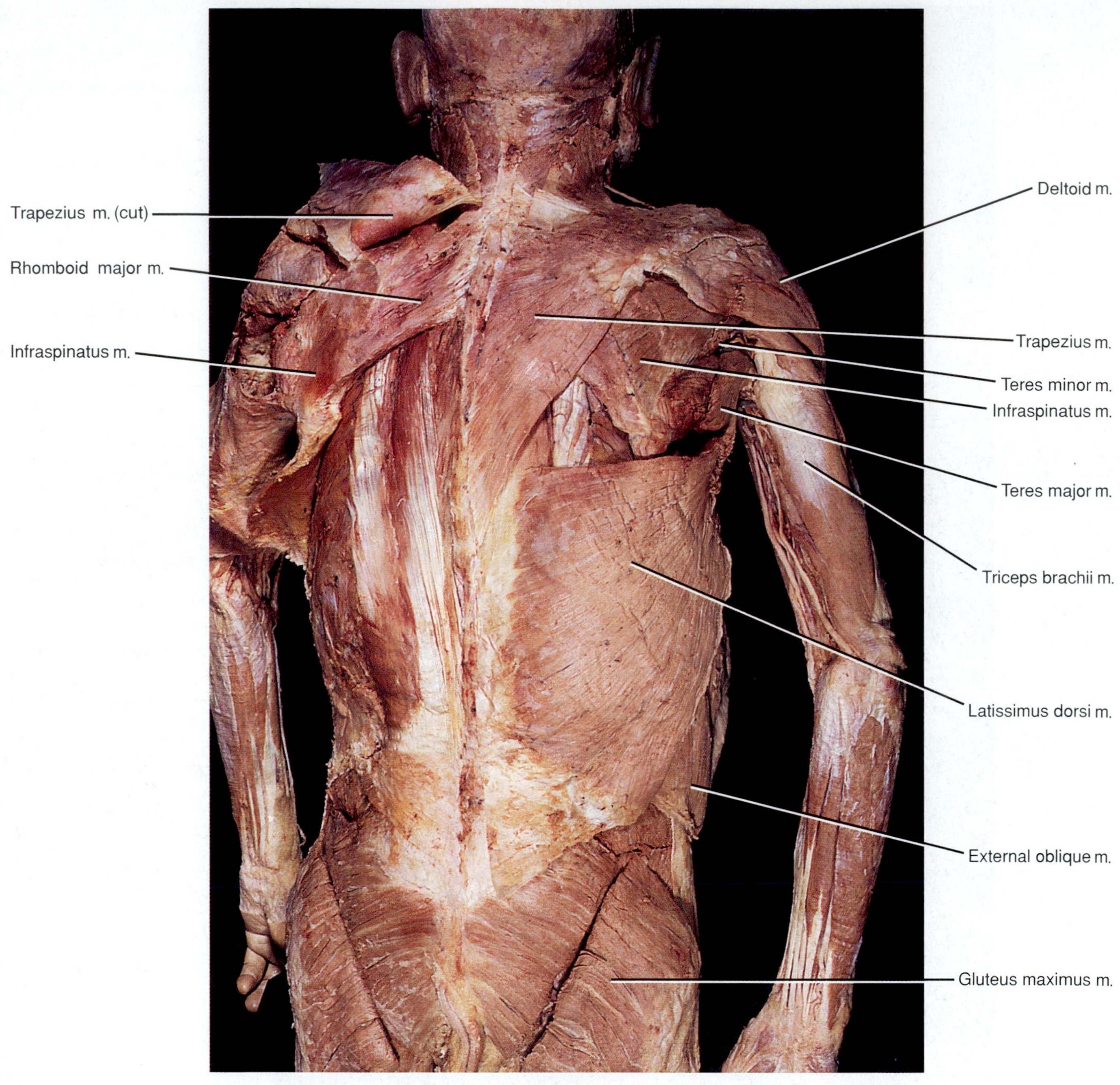

PLATE SIXTY-EIGHT Posterior view of the trunk, with deep thoracic muscles exposed on the left. (*m.* stands for muscle.)

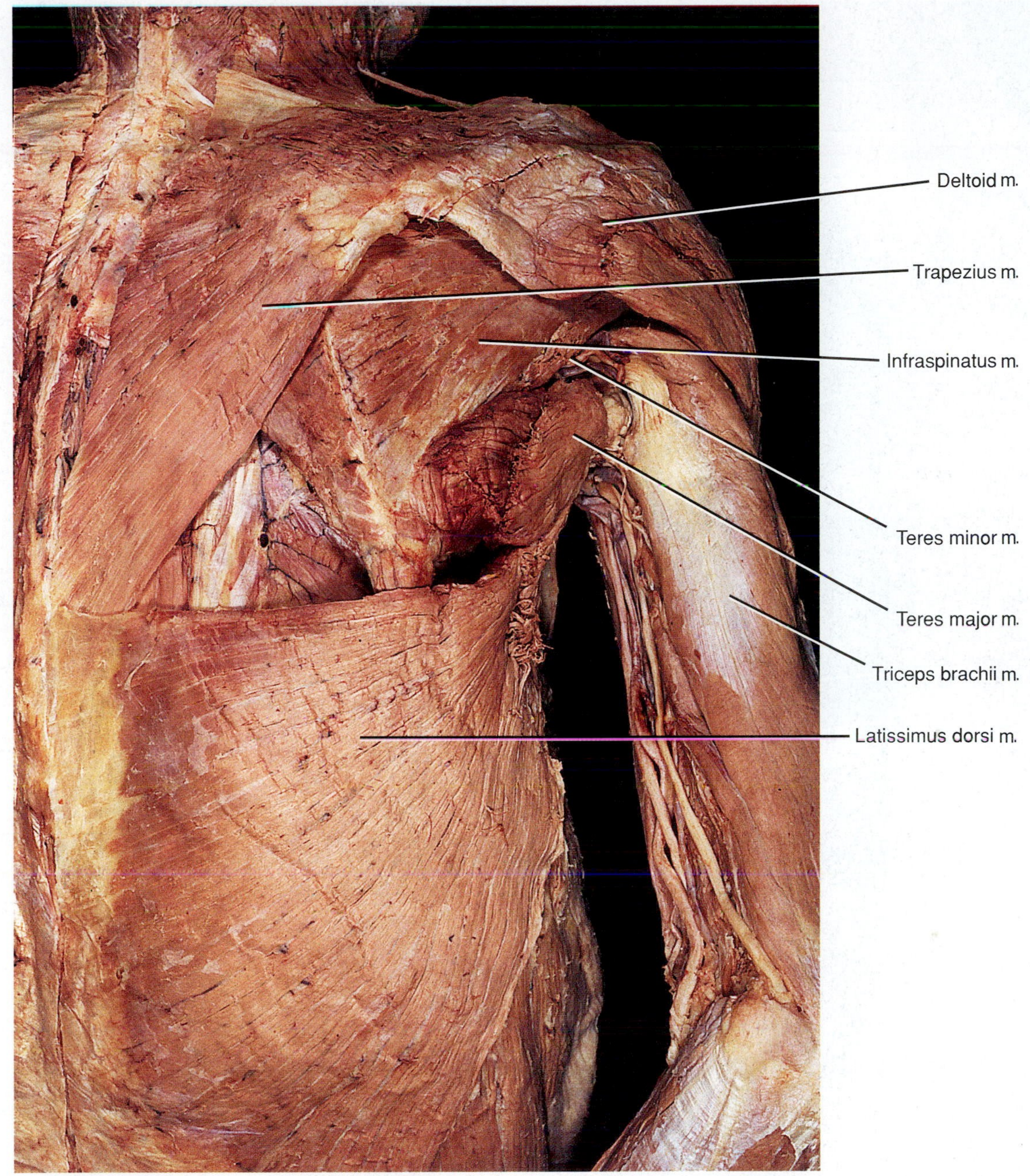

PLATE SIXTY-NINE Posterior view of the right thorax and arm. (*m.* stands for muscle.)

Triceps brachii m.
Brachioradialis m.
Brachialis m.
Biceps brachii m.
Extensor carpi radialis longus m.
Extensor digitorum m.
Extensor carpi radialis brevis m.
Extensor carpi ulnaris m.
Extensor retinaculum

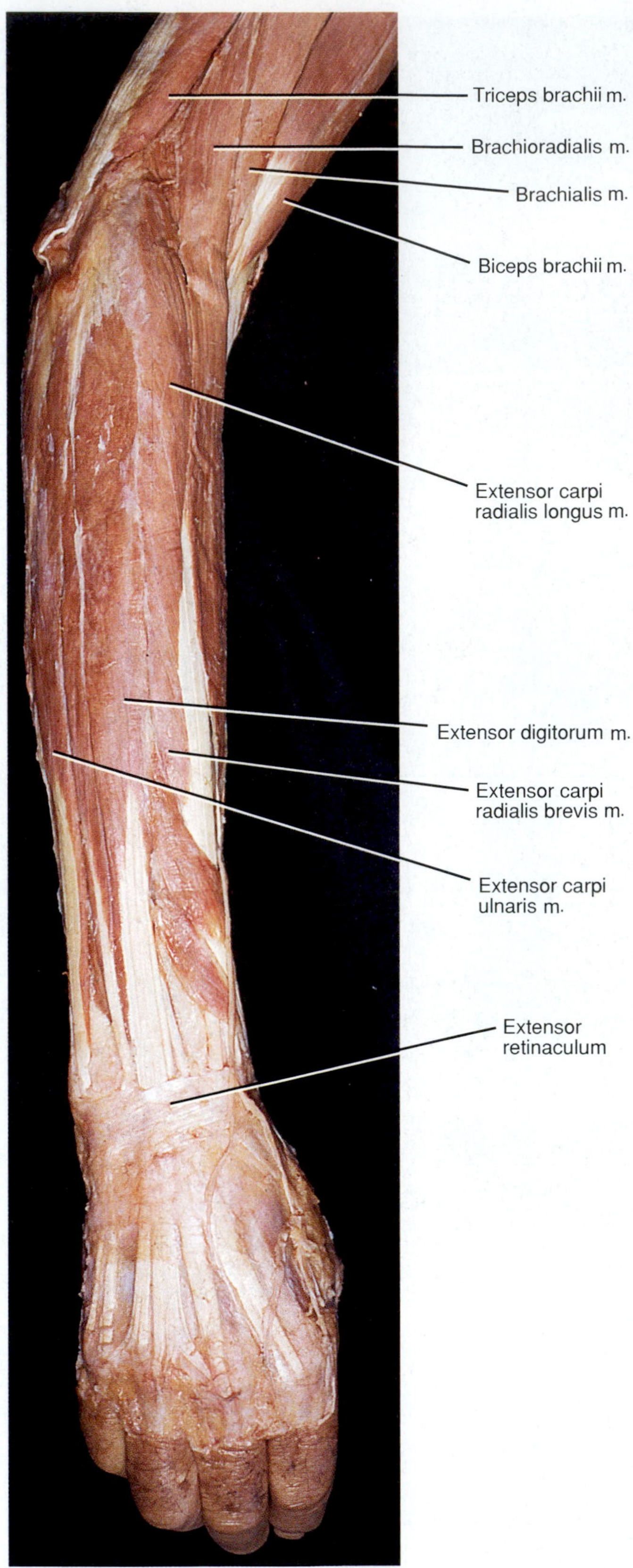

PLATE SEVENTY Posterior view of the right forearm and hand. (*m.* stands for muscle.)

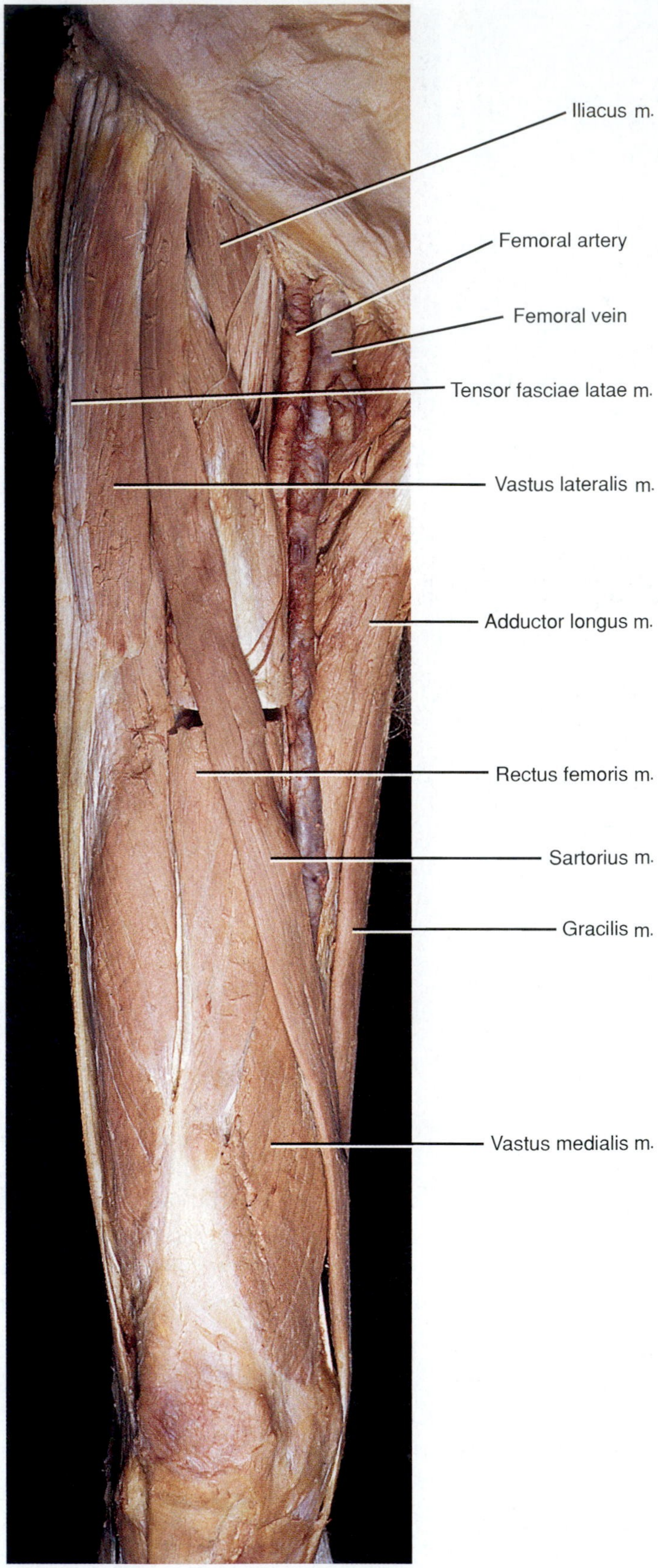

Iliacus m.
Femoral artery
Femoral vein
Tensor fasciae latae m.
Vastus lateralis m.
Adductor longus m.
Rectus femoris m.
Sartorius m.
Gracilis m.
Vastus medialis m.

PLATE SEVENTY-ONE Anterior view of the right thigh. (*m.* stands for muscle.)

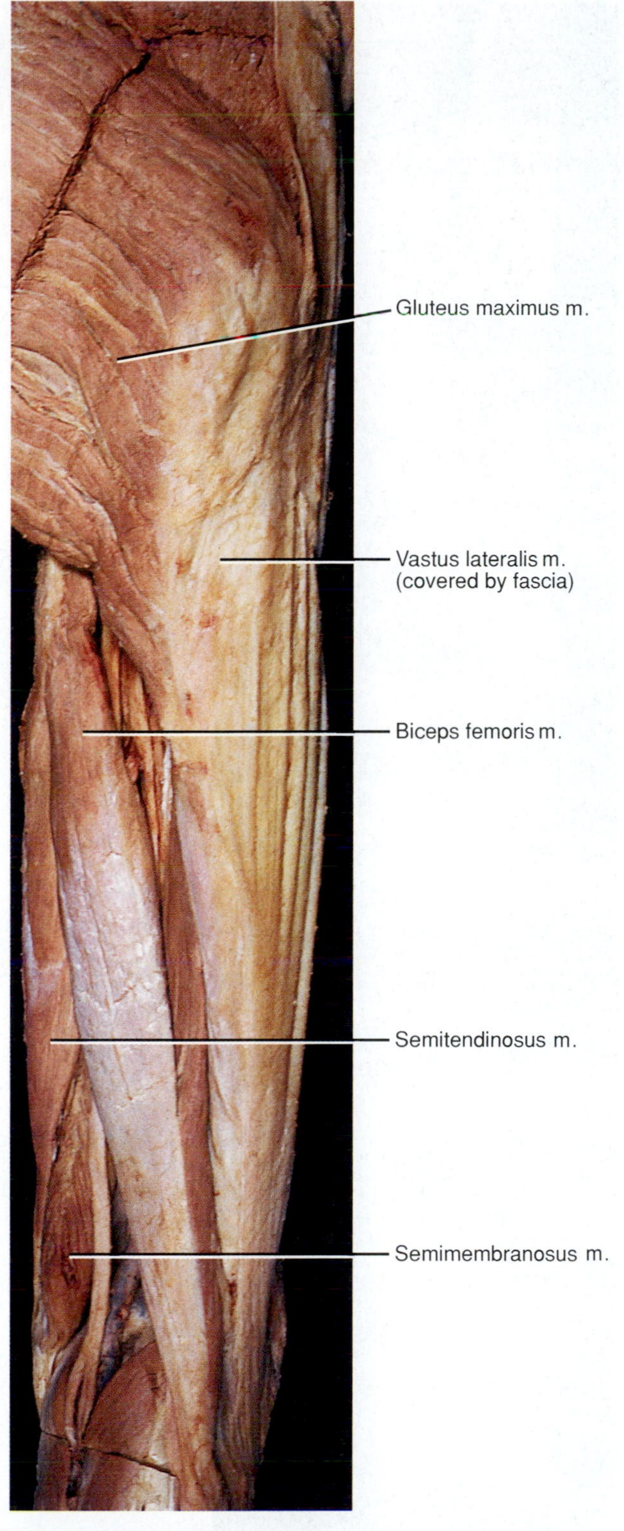

PLATE SEVENTY-TWO Posterior view of the right thigh. (*m.* stands for muscle.)

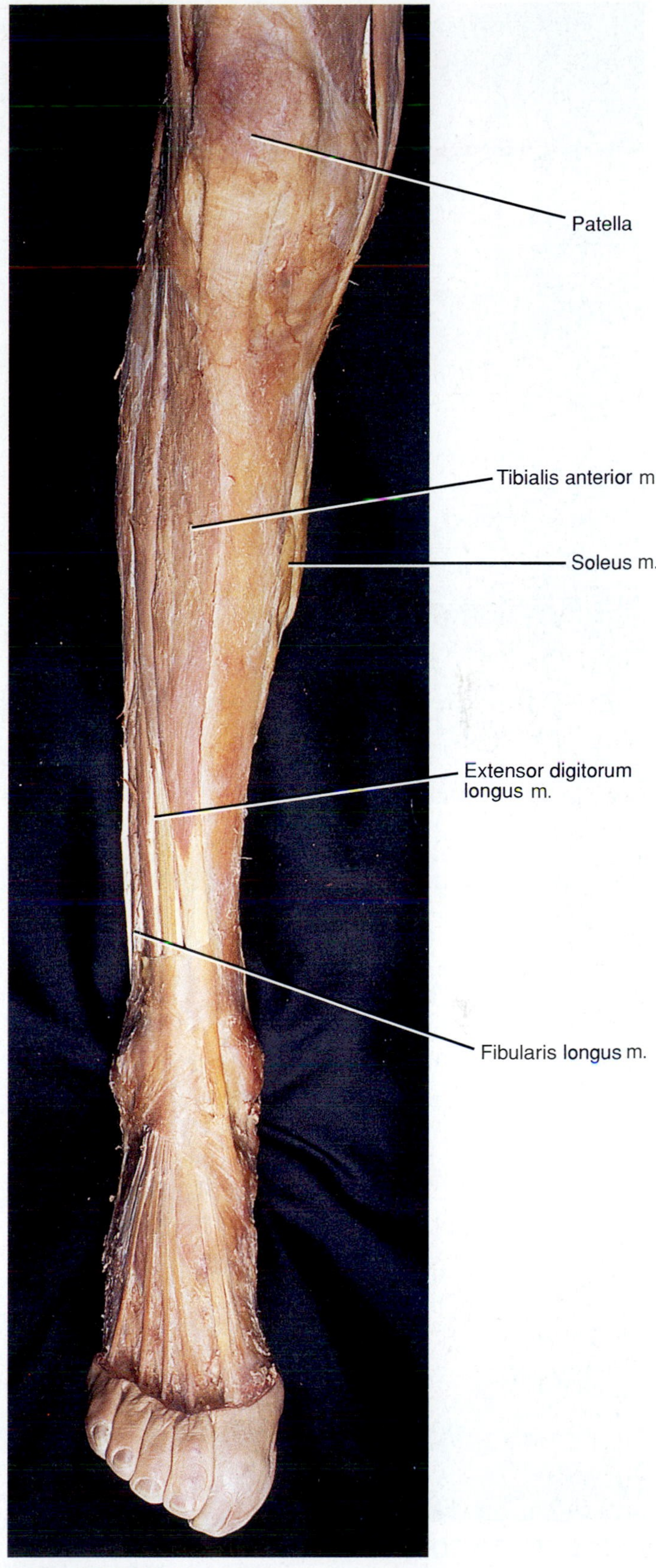

PLATE SEVENTY-THREE Anterior view of the right leg. (*m.* stands for muscle.)

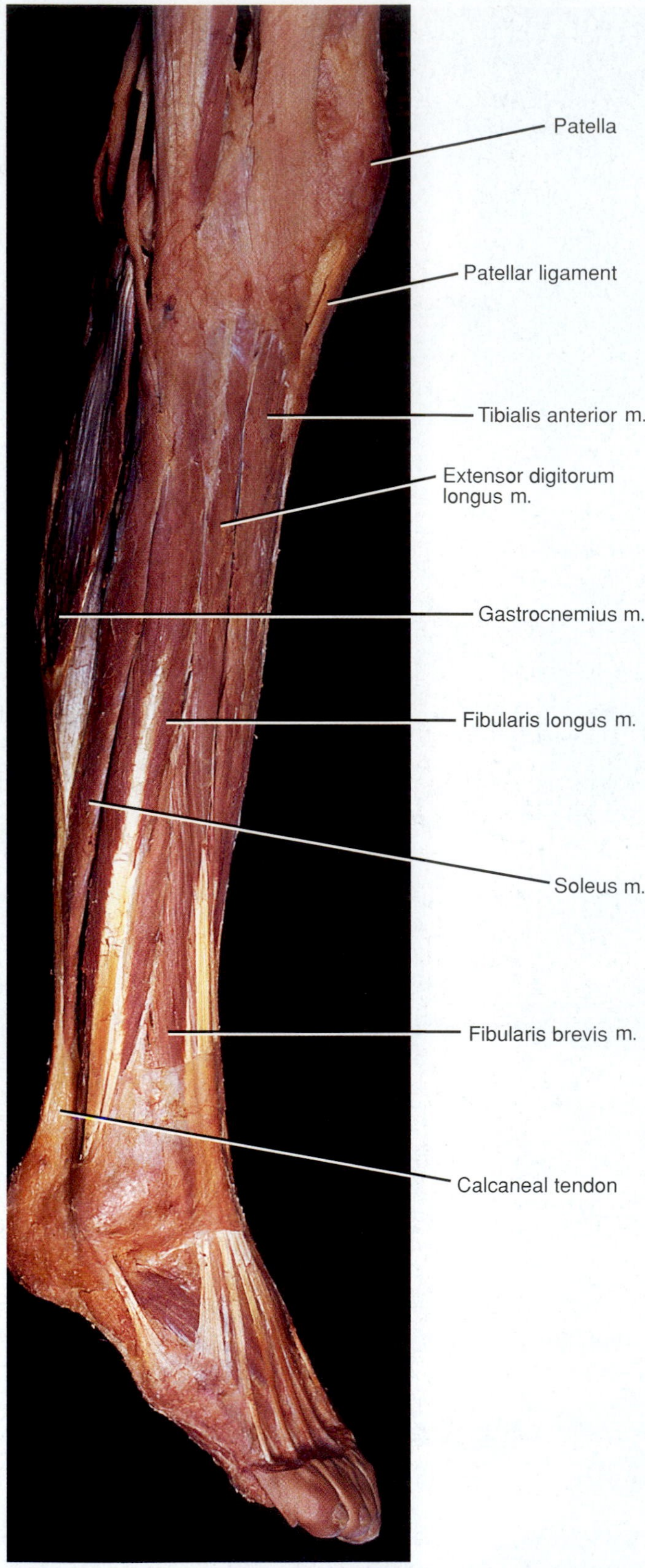

PLATE SEVENTY-FOUR Lateral view of the right leg. (*m.* stands for muscle.)

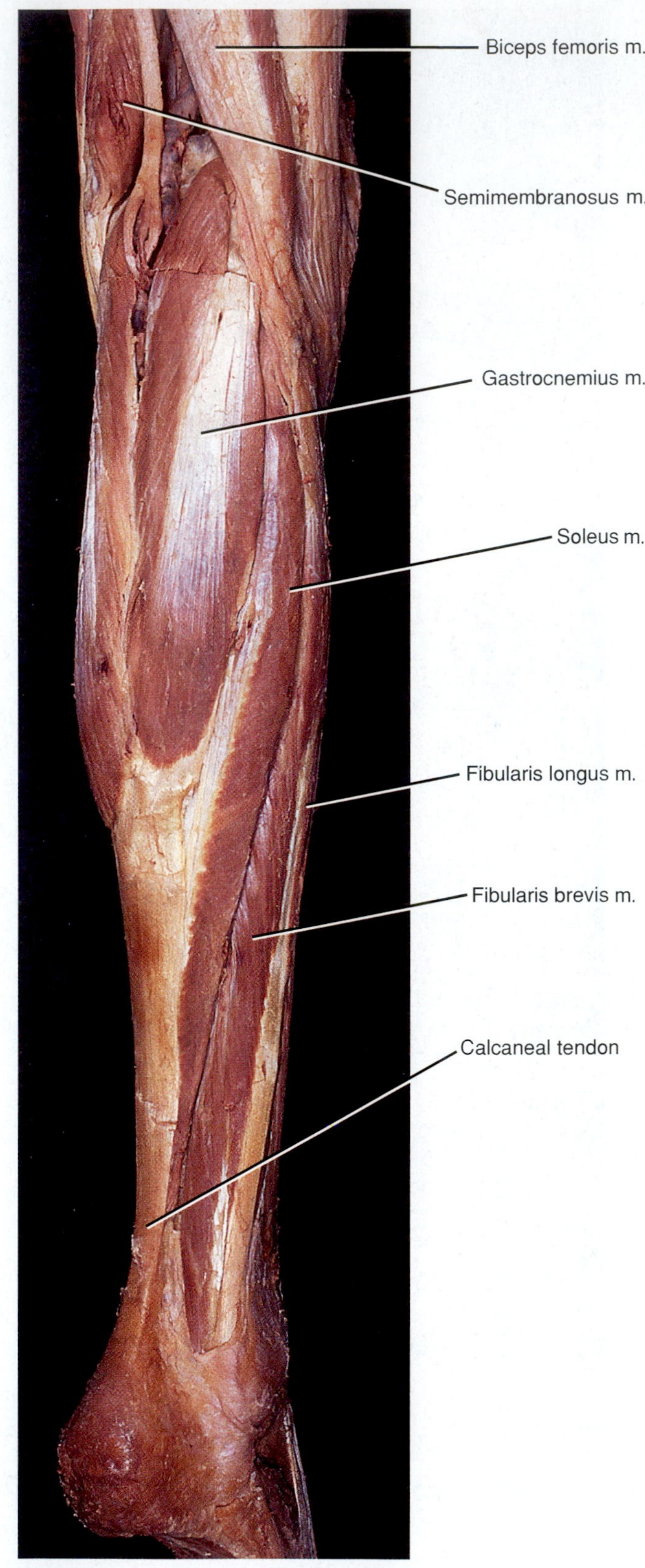

PLATE SEVENTY-FIVE Posterior view of the right leg. (*m.* stands for muscle.)

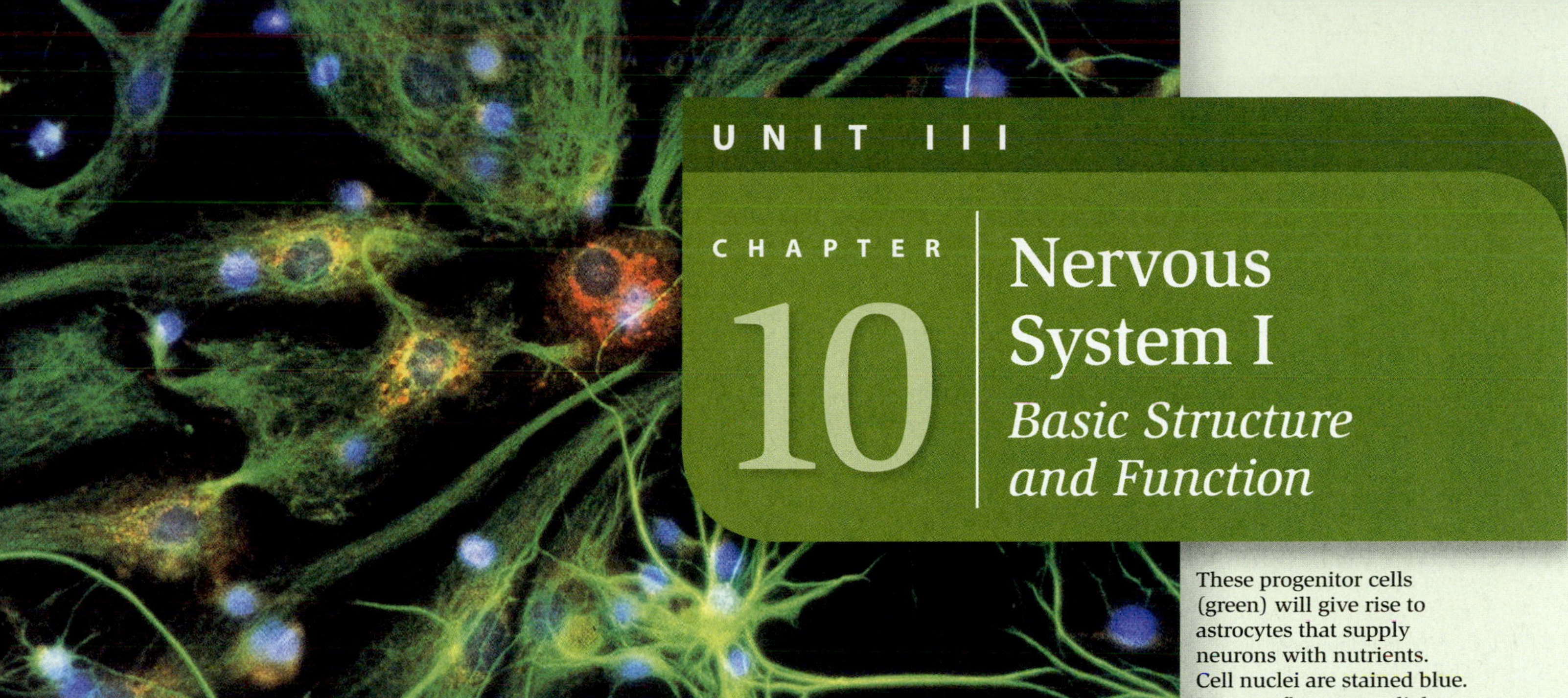

UNIT III

CHAPTER 10

Nervous System I

Basic Structure and Function

These progenitor cells (green) will give rise to astrocytes that supply neurons with nutrients. Cell nuclei are stained blue. Immunofluorescent light micrograph (1,150×).

UNDERSTANDING WORDS

astr-, starlike: *astro*cyte—star-shaped neuroglial cell.

ax-, axle: *ax*on—cylindrical nerve process that carries impulses away from a neuron cell body.

bi-, two: *bi*polar neuron—neuron with two processes extending from the cell body.

dendr-, tree: *dendr*ite—branched nerve process that serves as the receptor surface of a neuron.

ependym-, tunic: *ependym*a—neuroglial cells that line spaces in the brain and spinal cord.

-lemm, rind or peel: neuri*lemm*a—sheath that surrounds the myelin of a nerve cell process.

moto-, moving: *moto*r neuron—neuron that stimulates a muscle to contract or a gland to release a secretion.

multi-, many: *multi*polar neuron—neuron with many processes extending from the cell body.

oligo-, few: *oligo*dendrocyte—small neuroglial cell with few cellular processes.

peri-, all around: *peri*pheral nervous system—portion of the nervous system that consists of the nerves branching from the brain and spinal cord.

saltator-, a dancer: *saltator*y conduction—nerve impulse conduction in which the impulse seems to jump from node to node along the nerve fiber.

sens-, feeling: *sens*ory neuron—neuron that can be stimulated by a sensory receptor and conducts impulses into the brain or spinal cord.

syn-, together: *syn*apse—junction between two neurons.

uni-, one: *uni*polar—neuron with only one process extending from the cell body.

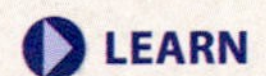 LEARN 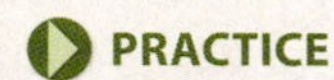 PRACTICE  ASSESS

LEARNING OUTCOMES

After you have studied this chapter, you should be able to

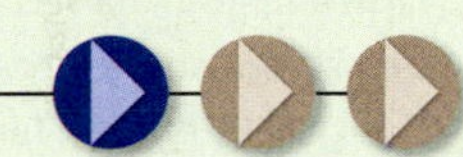

10.1 Introduction
1. Describe the general functions of the nervous system. (p. 354)
2. Identify the two types of cells that comprise nervous tissue. (p. 354)
3. Identify the two major groups of nervous system organs. (p. 354)

10.2 General Functions of the Nervous System
4. List the functions of sensory receptors. (p. 355)
5. Describe how the nervous system responds to stimuli. (p. 355)

10.3 Description of Cells of the Nervous System
6. Describe the parts of a neuron. (p. 356)
7. Describe the relationships among myelin, the neurilemma, and nodes of Ranvier. (p. 358)
8. Distinguish between the sources of white matter and gray matter. (p. 358)

10.4 Classification of Cells of the Nervous System
9. Identify structural and functional differences among neurons. (p. 359)
10. Identify the types of neuroglia in the central nervous system and their functions. (p. 361)
11. Describe the role of Schwann cells in the peripheral nervous system. (p. 363)

10.5 The Synapse
12. Explain how information passes from a presynaptic neuron to a postsynaptic cell. (p. 365)

10.6 Cell Membrane Potential
13. Explain how a cell membrane becomes polarized. (p. 365)
14. Describe the events leading to and the conduction of a nerve impulse. (p. 368)
15. Compare nerve impulse conduction in myelinated and unmyelinated neurons. (p. 371)

10.7 Synaptic Transmission
16. Identify the changes in membrane potential associated with excitatory and inhibitory neurotransmitters. (p. 371)
17. Explain what prevents a postsynaptic cell from being continuously stimulated. (p. 374)

10.8 Impulse Processing
18. Describe the basic ways in which the nervous system processes information. (p. 374)

BRAIN BANKS

In a large room at the Croatian Institute for Brain Research, rows of shelves hold a variety of fluid-filled jars, a human brain suspended in each. Their sizes differ, reflecting their origins from embryos up to the elderly. Researchers can use the more than 1,000 brains and more than 130,000 histological slides at the bank to investigate brain-based diseases and injuries that affect many millions of people worldwide and also to better understand the functioning of the normal human brain.

In the United States, several brain banks offer tissue sections from thousands of people who willed their brains to science. Unlike donated hearts, lungs, or corneas, which directly help other people, donated brains go to research labs.

Many brain banks are specialized. The bank at Harvard University is devoted to neurodegenerative diseases, such as Alzheimer and Parkinson diseases, while the resource at the University of Maryland in Baltimore focuses on developmental disorders, including Down syndrome and autism. The brain bank at the University of Miami has brains from people who had schizophrenia, depression, amyotrophic lateral sclerosis, and several other disorders, as well as undiseased brains for comparison.

Brains must be removed from the skull within twelve hours of death. Then they are halved and cut into one-centimeter thick sections and frozen in plastic bags. The specimens are provided free to researchers.

Study of brain function and malfunction is also possible at the cellular level. The National Human Neural Stem Cell Resource provides neural stem cells, which function after death longer than neurons because their energetic and oxygen requirements are not as high as those of the more specialized cells. Hospitals collect brain material upon autopsy and send it to the facility, where a special protocol is used to obtain and preserve the cells from several brain areas. These techniques were perfected on the brains of pigs, cats, and sheep. Investigators use the human neural stem cells to study neurodegenerative disorders, stroke, traumatic brain injury, rare inborn errors of metabolism, as well as the development of the incredibly complex human brain from initial stem and progenitor cells. The material in brain and stem cell banks is also being used in drug discovery and in developing new treatments based on cell implants. The chapter opening image shows neural progenitor cells and the photo accompanying this essay shows neural stem cells. ■

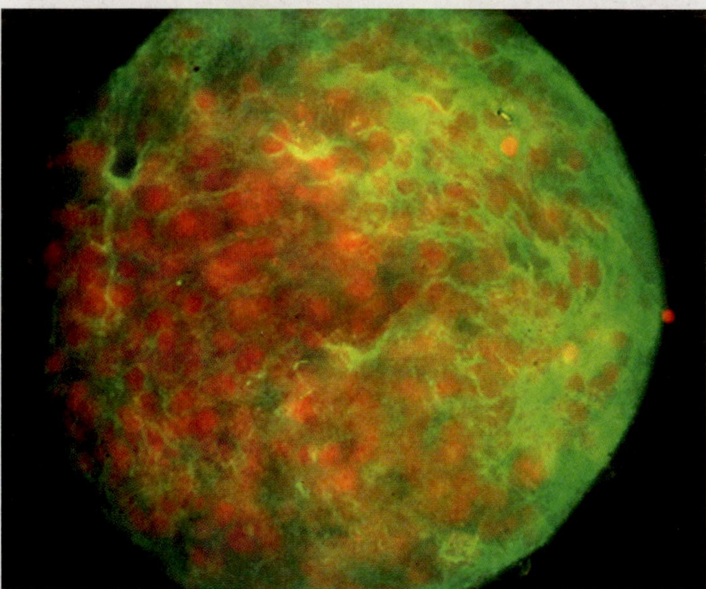

Neurospheres cultured in the laboratory consist of neural stem cells. These cells can divide and differentiate to give rise to neural progenitor cells, which in turn divide and differentiate, yielding neurons and neuroglia. In the brain, neural stem cells occupy certain areas but are exceedingly rare. Researchers are attempting to harness the natural ability of neural stem and progenitor cells to divide and replace damaged or diseased neural tissue.

10.1 INTRODUCTION

The nervous system oversees all that we do and largely determines who we are. Through a vast communicating network of cells and the biochemicals that they send and receive, the nervous system can detect changes in the body, make decisions on the basis of the information received, and stimulate muscles or glands to respond. Typically, these responses counteract the effects of the changes, and in this way, the nervous system helps maintain homeostasis. Clinical Application 10.1 discusses how environmental changes may trigger migraine headaches, a common medical problem attributed to the nervous system that may involve its blood supply as well as neurons.

The nervous system is composed predominantly of neural tissue, but also includes blood vessels and connective tissue. Neural tissue consists of two cell types: nerve cells, or **neurons** (nu′ronz), and **neuroglia** (nu-ro′gle-ah) (or neuroglial cells). Neurons are specialized to react to physical and chemical changes in their surroundings. Small cellular processes called **dendrites** (den′drītz) receive the input, and a longer process called an **axon** (ak′son), or nerve fiber, carries the information away from the cell in the form of bioelectric signals called **nerve impulses** (fig. 10.1). **Nerves** are bundles of axons. Neuroglia were once thought only to fill spaces and surround or support neurons. Today, we know that they have many other functions, including nourishing neurons and perhaps even sending and receiving messages.

An important part of the nervous system at the cellular level is not a cell at all, but the small space between a neuron and the cell(s) with which it communicates called a **synapse** (sin′aps). Much of the work of the nervous system is to send and receive electrochemical messages between neurons and other cells at synapses. Biological messenger molecules called **neurotransmitters** (nu″ro-trans-mit′erz) are the actual conveyors of this neural information.

The organs of the nervous system can be divided into two groups. One group, consisting of the brain and spinal cord, forms the **central nervous system (CNS)**, and the other, composed of the nerves (cranial and spinal nerves) that connect the central nervous system to other body parts, is called the **peripheral nervous system (PNS)** (fig. 10.2).

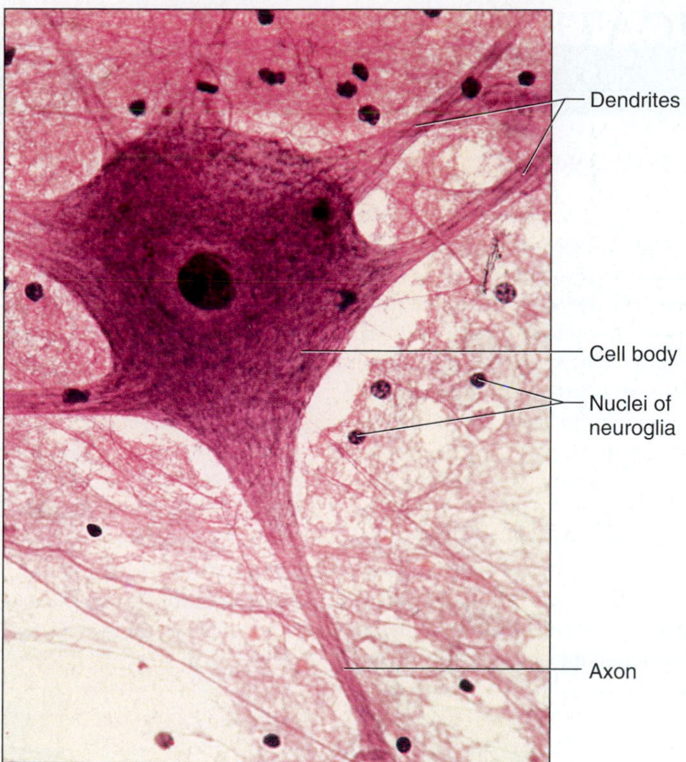

FIGURE 10.1 Neurons are the structural and functional units of the nervous system (600×). Neuroglia are cells that surround and support a neuron, appearing as dark dots. Note the locations of the neuron processes (dendrites and a single axon).

10.2 GENERAL FUNCTIONS OF THE NERVOUS SYSTEM

The three general functions of the nervous system—receiving, deciding, and reacting to stimuli—are termed sensory, integrative, and motor. Structures called **sensory receptors** at the ends of peripheral neurons provide the sensory function of the nervous system (see chapter 11, p. 389). These receptors gather information by detecting changes inside and outside the body. They monitor external environmental factors such as light and sound intensities as well as the temperature, oxygen concentration, and other conditions of the body's internal environment.

Sensory receptors convert (or transduce) their information into nerve impulses, which are then transmitted over peripheral nerves to the CNS. There the signals are integrated—that is, they are brought together, creating sensations, adding to memory, or helping produce thoughts. Following integration, conscious or subconscious decisions are made and then acted upon by means of motor functions.

The motor functions of the nervous system are carried out by neurons that carry impulses from the CNS to responsive structures called *effectors*. These effectors are outside the nervous system and include muscles that contract in response to nerve impulse stimulation and glands that secrete when stimulated. The motor portion of the PNS

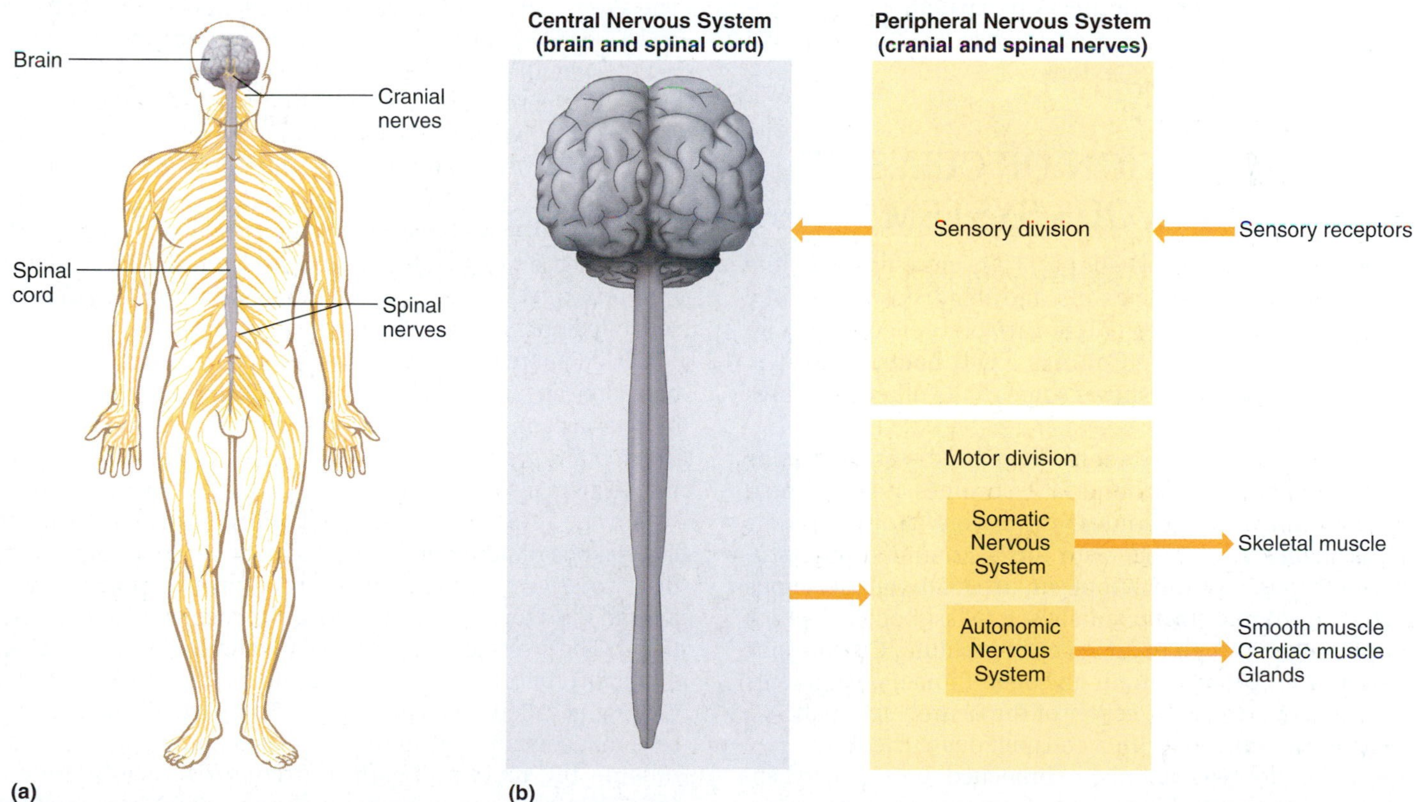

FIGURE 10.2 A diagrammatic representation of the nervous system. (*a*) The nervous system includes the central nervous system (brain and spinal cord) and the peripheral nervous system (cranial nerves and spinal nerves). (*b*) The nervous system receives information from sensory receptors and initiates responses through effector organs (muscles and glands).

10.1 CLINICAL APPLICATION

Migraine

The signs of a migraine are unmistakable—a pounding head, waves of nausea, sometimes shimmering images in the peripheral visual field, and extreme sensitivity to light or sound. Inherited susceptibilities and environmental factors probably cause migraines. Environmental triggers include sudden exposure to bright light, eating a particular food (chocolate, red wine, nuts, and processed meats top the list), lack of sleep, stress, high altitude, stormy weather, and excessive caffeine or alcohol intake. Hormonal influences may also be involved, because two-thirds of the 300 million people who suffer from migraines worldwide are women between the ages of 15 and 55.

A migraine attack may last only a few hours, or days. It is due to a phenomenon called "cortical spreading depression," in which an intense wave of excitation followed by a brief period of unresponsiveness in certain neurons stimulates the trigeminal nucleus at the base of the brain to produce pain sensations. The excitation and dampening of the activity level of these neurons also triggers changes in blood flow in the brain that were once thought to be the direct cause of migraine.

Drugs called triptans can very effectively halt a migraine attack, but must be taken as soon as symptoms begin. Triptans block the release of neurotransmitter from the trigeminal nerves. Because triptans constrict blood vessels throughout the body, making them dangerous for some people, newer migraine drugs have been developed that block the specific neurotransmitter that the trigeminal nerves release (calcitonin gene-related peptide), better targeting the therapeutic effect.

Several drugs developed to treat other conditions are used on a long-term, daily basis to lessen the frequency of migraines. These drugs include certain antidepressants, anticonvulsants, and drugs used to treat high blood pressure (calcium channel blockers and beta blockers). A physician must consider an individual's family and health history before prescribing these drugs to prevent migraine. ■

can be subdivided into the somatic and the autonomic nervous systems. Generally the **somatic nervous system** oversees conscious (voluntary) activities, such as skeletal muscle contraction. The **autonomic nervous system** controls viscera, such as the heart and various glands, and thus controls subconscious (involuntary) actions.

10.3 DESCRIPTION OF CELLS OF THE NERVOUS SYSTEM

Neurons vary in size and shape. They may differ in the lengths and sizes of their axons and dendrites and in the number of processes. Despite this variability, neurons share certain features. Every neuron has a **cell body,** dendrites, and an axon. Figure 10.3 shows some of the other structures common to neurons.

A neuron's cell body (soma or perikaryon) contains granular cytoplasm, mitochondria, lysosomes, a Golgi apparatus, and many microtubules. A network of fine threads called **neurofibrils** extends into the axon and supports it. Scattered throughout the cytoplasm are many membranous packets of **chromatophilic substance** (Nissl bodies), which consist mainly of rough endoplasmic reticulum. Cytoplasmic inclusions in neurons contain glycogen, lipids, or pigments such as melanin. Near the center of the neuron cell body is a large, spherical nucleus with a conspicuous nucleolus.

Dendrites are typically highly branched, providing receptive surfaces with which processes from other neurons communicate. (In some types of neurons, the cell body provides such a receptive surface.) Some dendrites have tiny, thornlike spines (dendritic spines) on their surfaces, which are contact points for other neurons.

A neuron may have many dendrites, but only one axon. The axon, which often arises from a slight elevation of the cell body (axonal hillock), is a slender, cylindrical process with a nearly smooth surface and uniform diameter. It is specialized to conduct nerve impulses away from the cell body. The cytoplasm of the axon includes many mitochondria, microtubules, and neurofibrils (ribosomes are found only in the cell body). The axon may give off branches, called *collaterals.* Near its end, an axon may have many fine extensions, each with a specialized ending called an *axon terminal.* This ends as a *synaptic knob* close to the receptive surface of another cell, separated only by a space called the **synaptic cleft.**

In addition to conducting nerve impulses, an axon conveys biochemicals produced in the neuron cell body, which can be quite a task in these long cells. In this activity, called *axonal transport,* vesicles, mitochondria, ions, nutrients, and neurotransmitters move from the cell body to the ends of the axon.

In the PNS, neuroglia called **Schwann cells** encase the large axons of peripheral neurons in lipid-rich sheaths. These tight coverings form as layers of cell membrane and wind around the axons somewhat like a bandage wrapped around a finger. The layers are composed of **myelin** (mi′ĕ-lin), which has a higher proportion of lipid than other cell membranes. This coating is called a *myelin sheath.* The parts of the Schwann cells that contain most of the cytoplasm and the nuclei remain outside the myelin sheath and comprise a **neurilemma** (nur″ĭlem′ah), or *neurilemmal sheath,* which surrounds the myelin sheath. Narrow gaps in the myelin sheath between Schwann cells are called **nodes of Ranvier** (fig. 10.4).

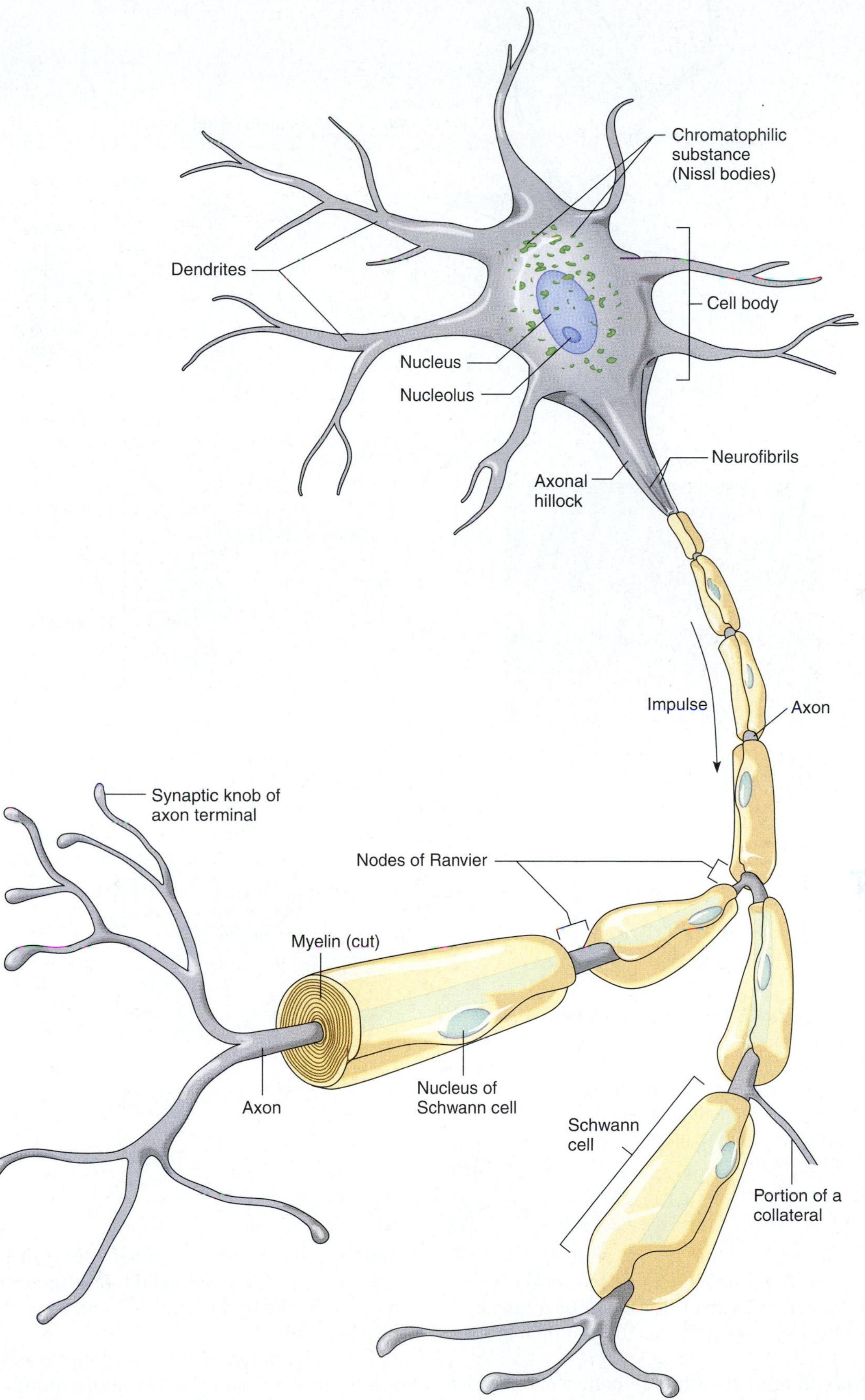

FIGURE 10.3 A common neuron.

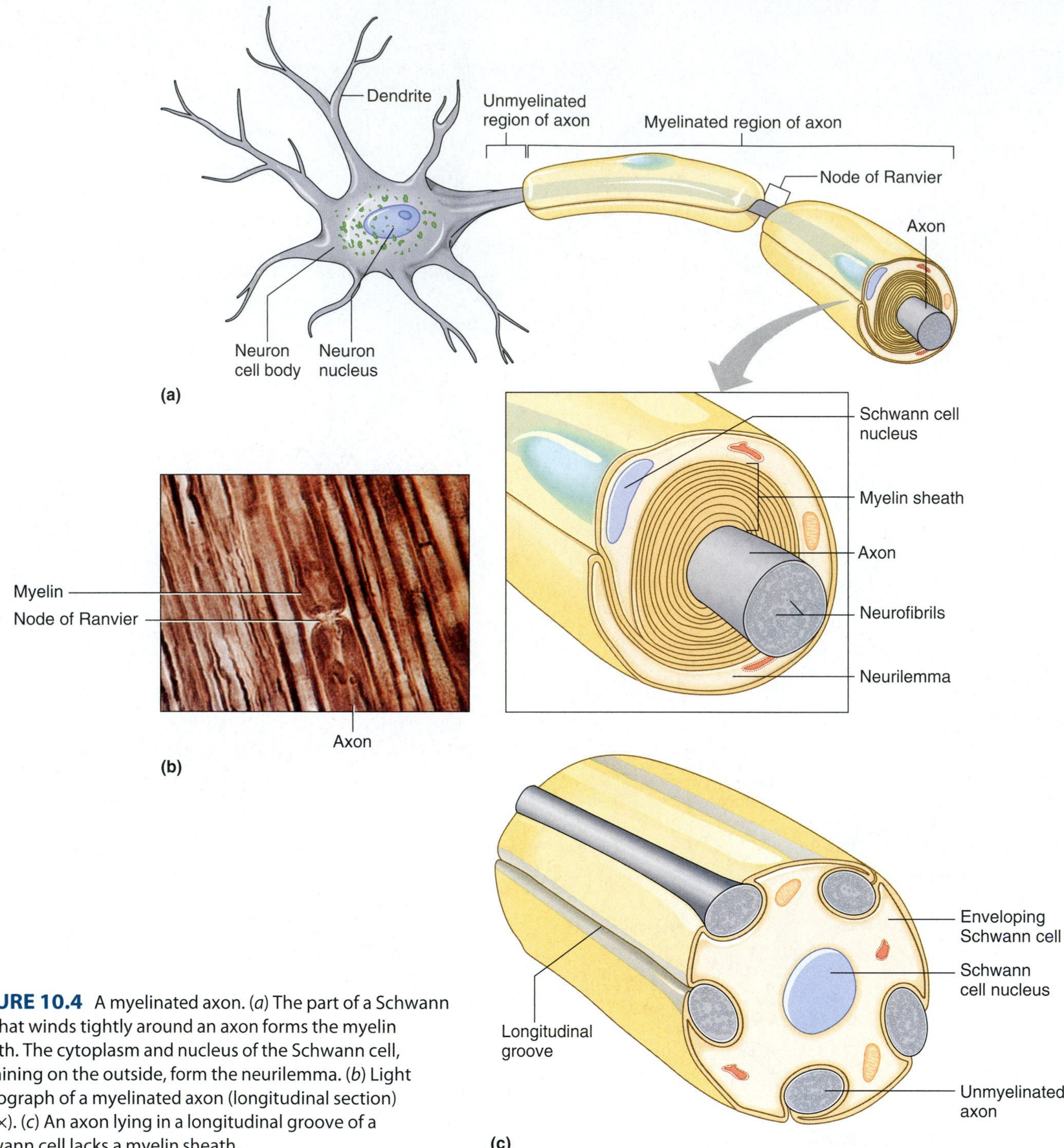

FIGURE 10.4 A myelinated axon. (*a*) The part of a Schwann cell that winds tightly around an axon forms the myelin sheath. The cytoplasm and nucleus of the Schwann cell, remaining on the outside, form the neurilemma. (*b*) Light micrograph of a myelinated axon (longitudinal section) (650×). (*c*) An axon lying in a longitudinal groove of a Schwann cell lacks a myelin sheath.

Schwann cells also enclose, but do not wind around, the smallest axons of peripheral neurons. Consequently, these axons lack myelin sheaths. Instead, the axon or a group of axons may lie partially or completely in a longitudinal groove of Schwann cells.

Axons that have myelin sheaths are called *myelinated* (medullated) axons, and those that lack these sheaths are *unmyelinated axons* (fig. 10.5). Groups of myelinated axons appear white. Masses of such axons impart color to the *white matter* in the brain and spinal cord, but in the CNS another type of neuroglial cell called an **oligodendrocyte** produces myelin. In the brain and spinal cord, myelinated axons lack neurilemmae.

Unmyelinated nerve tissue appears gray. Thus, the *gray matter* in the CNS contains many unmyelinated axons and neuron cell bodies. Clinical Application 10.2 discusses multiple sclerosis, in which neurons in the brain and spinal cord lose their myelin.

PRACTICE

1 List the general functions of the nervous system.
2 Describe a neuron.
3 Explain how an axon in the peripheral nervous system becomes myelinated.

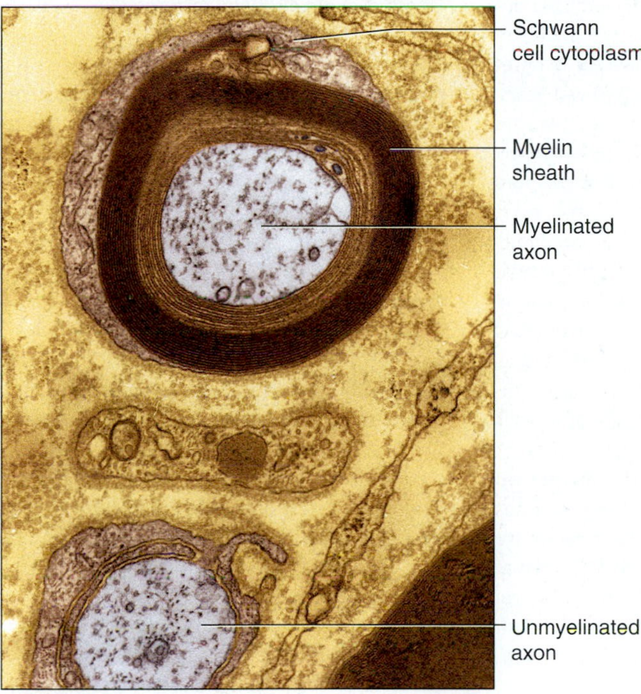

FIGURE 10.5 A falsely colored transmission electron micrograph of myelinated and unmyelinated axons in cross section (30,000×).

Myelin begins to form on axons during the fourteenth week of prenatal development. By the time of birth, many axons are not completely myelinated. All myelinated axons have begun to develop sheaths by the time a child starts to walk, and myelination continues into adolescence.

Excess myelin seriously impairs nervous system functioning. In Tay-Sachs disease, an inherited defect in a lysosomal enzyme causes myelin to accumulate, burying neurons in fat. The affected child begins to show symptoms by six months of age, gradually losing sight, hearing, and muscle function until death occurs by age four. Thanks to genetic screening among people of eastern European descent who are most likely to carry this gene, Tay-Sachs disease is extremely rare.

10.4 CLASSIFICATION OF CELLS OF THE NERVOUS SYSTEM

Neurons and neuroglia are intimately related. They descend from the same neural stem cells and remain associated throughout their existence.

Classification of Neurons

Neurons vary in size and shape and may differ in the lengths and sizes of their axons and dendrites and in the number of dendrites. Based on *structural differences,* neurons can be classified into three major groups, as figure 10.6 shows. Each

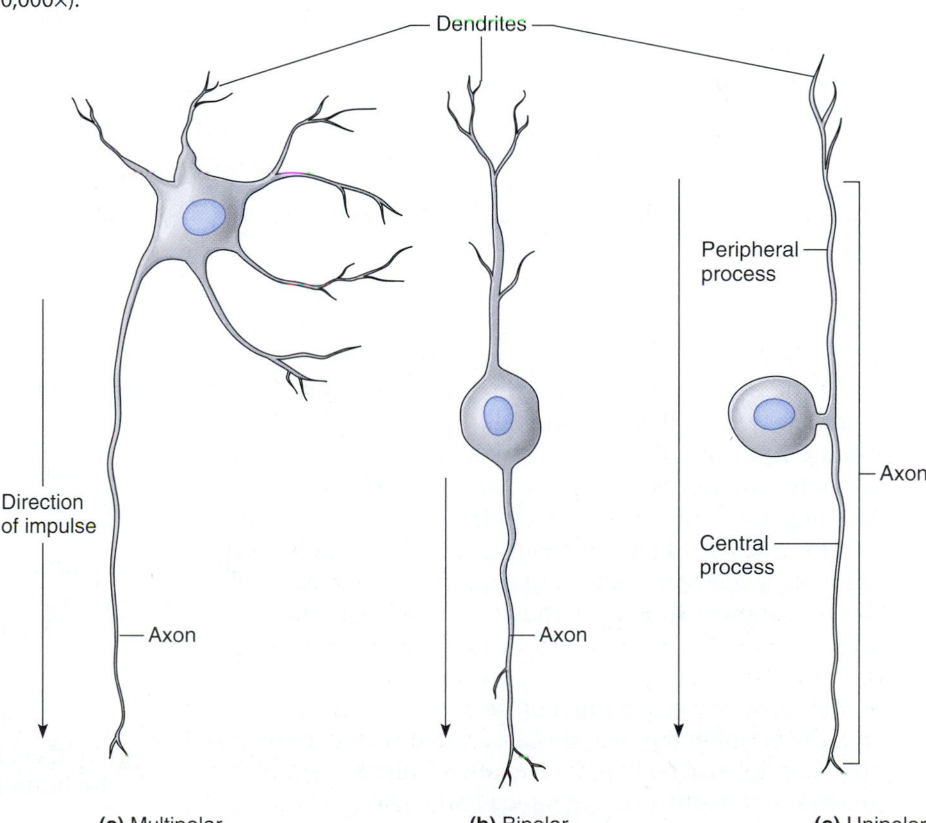

FIGURE 10.6 Structural types of neurons include (*a*) the multipolar neuron, (*b*) the bipolar neuron, and (*c*) the unipolar neuron.

10.2 CLINICAL APPLICATION

Multiple Sclerosis

Multiple sclerosis (MS) is a disorder of the CNS that affects 2.5 million people worldwide, and 400,000 in North America. In addition to overt nervous system symptoms, affected individuals experience disability, mood problems such as depression, and great fatigue. Four subtypes of MS are recognized, based on the pattern of symptomatic periods over time.

In MS, the myelin coating in various sites through the brain and spinal cord becomes inflamed due to an immune response and is eventually destroyed, leaving hard scars, called scleroses, that block the underlying neurons from transmitting messages. Muscles that no longer receive input from motor neurons stop contracting, and eventually, they atrophy. Symptoms reflect the specific neurons affected. Shortcircuiting in one part of the brain may affect fine coordination in one hand; if another brain part is affected, vision may be altered.

The first symptoms of MS are often blurred vision and numb legs or arms, but because in many cases these are intermittent, diagnosis may take awhile. Diagnosis is based on symptoms and repeated magnetic resonance (MR) scans, which can track development of lesions. A diagnostic work-up for MS might also include a lumbar puncture to rule out infection and an evoked potential test to measure electrical signals sent from the brain. About 70% of affected individuals first notice symptoms between the ages of twenty and forty; the earliest known age of onset is three years, and the latest, sixty-seven years. Some affected individuals eventually become permanently paralyzed. Women are twice as likely to develop MS as men, and Caucasians are more often affected than people of other races.

MS may develop when certain infections in certain individuals stimulate T cells (a type of white blood cell that takes part in immune responses) in the periphery, which then cross the blood-brain barrier. Here, the T cells attack myelin-producing cells through a flood of inflammatory molecules and by stimulating other cells to produce antibodies against myelin.

A virus may lie behind the misplaced immune attack that is MS. Evidence includes the observations that viral infection can cause repeated bouts of symptoms, as can MS, and that MS is much more common in some geographical regions (the temperate zones of Europe, South America, and North America) than others, suggesting a pattern of infection.

Various drugs are used to manage MS. Drugs to decrease bladder spasms can temper problems of urinary urgency and incontinence. Antidepressants are sometimes prescribed, and short-term steroid drugs are used to shorten the length of acute disabling relapses. Muscle relaxants ease stiffness and spasms.

Several drugs are used for long-term treatment of MS. Beta interferons are immune system biochemicals that are widely prescribed, even after only one attack if MS seems a likely diagnosis. This treatment diminishes the intensity of flare-ups, but effects on the course of illness over time are not yet known. Beta interferons may cause flulike adverse effects. They are self-injected once to several times a week.

Glatiramer is an alternative to beta interferons. It is prescribed if the course of the disease is "relapsing remitting," with periodic flare-ups. Glatiramer is self-injected daily and dampens the immune system's attack on myelin. It consists of part of myelin basic protein, the most abundant protein of myelin. In response, T cells decrease inflammation. Glatiramer also stimulates increased production of brain-derived neurotrophic factor, which protects axons.

Mitoxantrone is another drug that halts the immune response against CNS myelin, but because it can damage the heart, it is typically used in severe cases of MS and only up to two years. Another drug, natalizumab, prevents T cells from binding blood vessels in the brain, also quelling the abnormal immune response against myelin. It too may have rare but serious adverse effects. ■

type of neuron is specialized to send a nerve impulse in one direction.

1. **Multipolar neurons.** Multipolar neurons have many processes arising from their cell bodies. Only one is an axon; the rest are dendrites. Most neurons whose cell bodies lie within the brain or spinal cord are of this type. The neuron illustrated in figure 10.3 is multipolar.
2. **Bipolar neurons.** The cell body of a bipolar neuron has only two processes, one arising from either end. Although these processes are similar in structure, one is an axon and the other is a dendrite. Bipolar neurons are found in specialized parts of the eyes, nose, and ears.
3. **Unipolar neurons.** Each unipolar neuron has a single process extending from its cell body. A short distance from the cell body, this process divides into two branches, which really function as a single axon: One branch (peripheral process) is associated with dendrites near a peripheral body part. The other branch (central process) enters the brain or spinal cord. The cell bodies of some unipolar neurons aggregate in specialized masses of nerve tissue called **ganglia,** located outside the brain and spinal cord.

Neurons can also be classified by *functional differences* into the following groups, depending on whether they carry information into the CNS, completely within the CNS, or out of the CNS (fig. 10.7).

1. **Sensory neurons** (afferent neurons) carry nerve impulses from peripheral body parts into the brain or spinal cord. At their distal ends, the dendrites of these neurons or specialized structures associated with them act as sensory receptors, detecting changes in the outside world (for example, eyes, ears, or touch receptors in the skin) or in the body (for example, temperature or blood pressure receptors). When sufficiently stimulated, sensory receptors trigger impulses that travel on sensory neuron axons into the brain or spinal cord. Most sensory neurons are

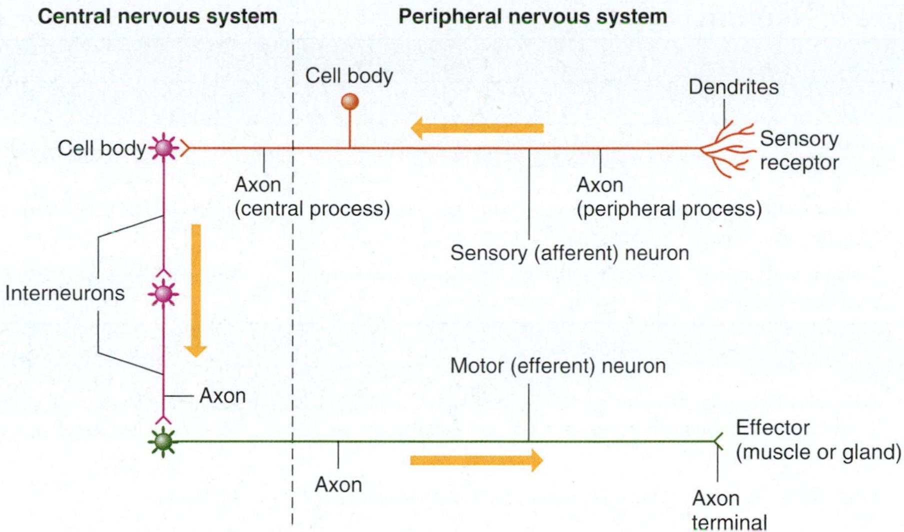

FIGURE 10.7 Neurons are classified by function as well as structure. Sensory (afferent) neurons carry information into the central nervous system (CNS), interneurons are completely within the CNS, and motor (efferent) neurons carry instructions to effectors.

unipolar, as shown in figure 10.7, although some are bipolar.

2. **Interneurons** (also called association or internuncial neurons) lie within the brain or spinal cord. They are multipolar and form links between other neurons. Interneurons transmit impulses from one part of the brain or spinal cord to another. That is, they may direct incoming sensory impulses to appropriate regions for processing and interpreting. Other incoming impulses are transferred to motor neurons.
3. **Motor neurons** (efferent neurons) are multipolar and carry nerve impulses out of the brain or spinal cord to effectors—structures that respond, such as muscles or glands. For example, when motor impulses reach muscles, they contract; when motor impulses reach glands, they release secretions.

   Motor neurons of the somatic nervous system (see fig. 10.2) that control skeletal muscle contraction are under voluntary (conscious) control. Those that control cardiac and smooth muscle contraction and the secretions of glands are part of the autonomic nervous system and are largely under involuntary control.

Table 10.1 summarizes the classification of neurons.

Classification of Neuroglia

Neuroglia were once thought to be mere bystanders to neural function, providing scaffolding and controlling the sites at which neurons contact one another (figs. 10.8 and 10.9). These important cells have additional functions. In the embryo, neuroglia guide neurons to their positions and may stimulate them to specialize. Neuroglia also produce the growth factors that nourish neurons and remove ions and neurotransmitters that accumulate between neurons, enabling them to continue transmitting information. In cell culture experiments, certain types of neuroglia (astrocytes) signal neurons to form and maintain synapses.

Neuroglia of the CNS

The four types of CNS neuroglia are astrocytes, oligodendrocytes, microglia, and ependyma:

1. **Astrocytes.** As their name implies, astrocytes are star-shaped cells. They are commonly found between neurons and blood vessels, where they provide support and hold structures together with abundant cellular processes. Astrocytes aid metabolism of certain substances, such as glucose, and they may help regulate the concentrations of important ions, such as potassium ions, in the interstitial space of nervous tissue. Astrocytes also respond to injury of brain tissue and form a special type of scar tissue, which fills spaces and closes gaps in the CNS. These multifunctional cells also have a nutritive function, regulating movement of substances from blood vessels to neurons and bathing nearby neurons in growth factors. Astrocytes play an important role in the blood-brain barrier, which restricts movement of substances between the blood and the CNS (see Clinical Application 5.1, p. 145). Gap junctions link astrocytes to one another, forming protein-lined channels through which calcium ions travel, possibly stimulating neurons.
2. **Oligodendrocytes.** Oligodendrocytes resemble astrocytes but are smaller and have fewer processes. They form in rows along myelinated axons, and produce myelin in the brain and spinal cord.

   Unlike the Schwann cells of the PNS, oligodendrocytes can send out a number of processes, each of which forms a myelin sheath around a nearby axon. In this way,

TABLE 10.1 | Types of Neurons

A. Classified by Structure		
Type	**Structural Characteristics**	**Location**
1. Multipolar neuron	Cell body with many processes, one of which is an axon, the rest dendrites	Most common type of neuron in the brain and spinal cord
2. Bipolar neuron	Cell body with a process, arising from each end, one axon and one dendrite	In specialized parts of the eyes, nose, and ears
3. Unipolar neuron	Cell body with a single process that divides into two branches and functions as an axon	Found in ganglia outside the brain or spinal cord
B. Classified by Function		
Type	**Functional Characteristics**	**Structural Characteristics**
1. Sensory neuron	Conducts nerve impulses from receptors in peripheral body parts into the brain or spinal cord	Most unipolar; some bipolar
2. Interneuron	Transmits nerve impulses between neurons in the brain and spinal cord	Multipolar
3. Motor neuron	Conducts nerve impulses from the brain or spinal cord out to effectors—muscles or glands	Multipolar

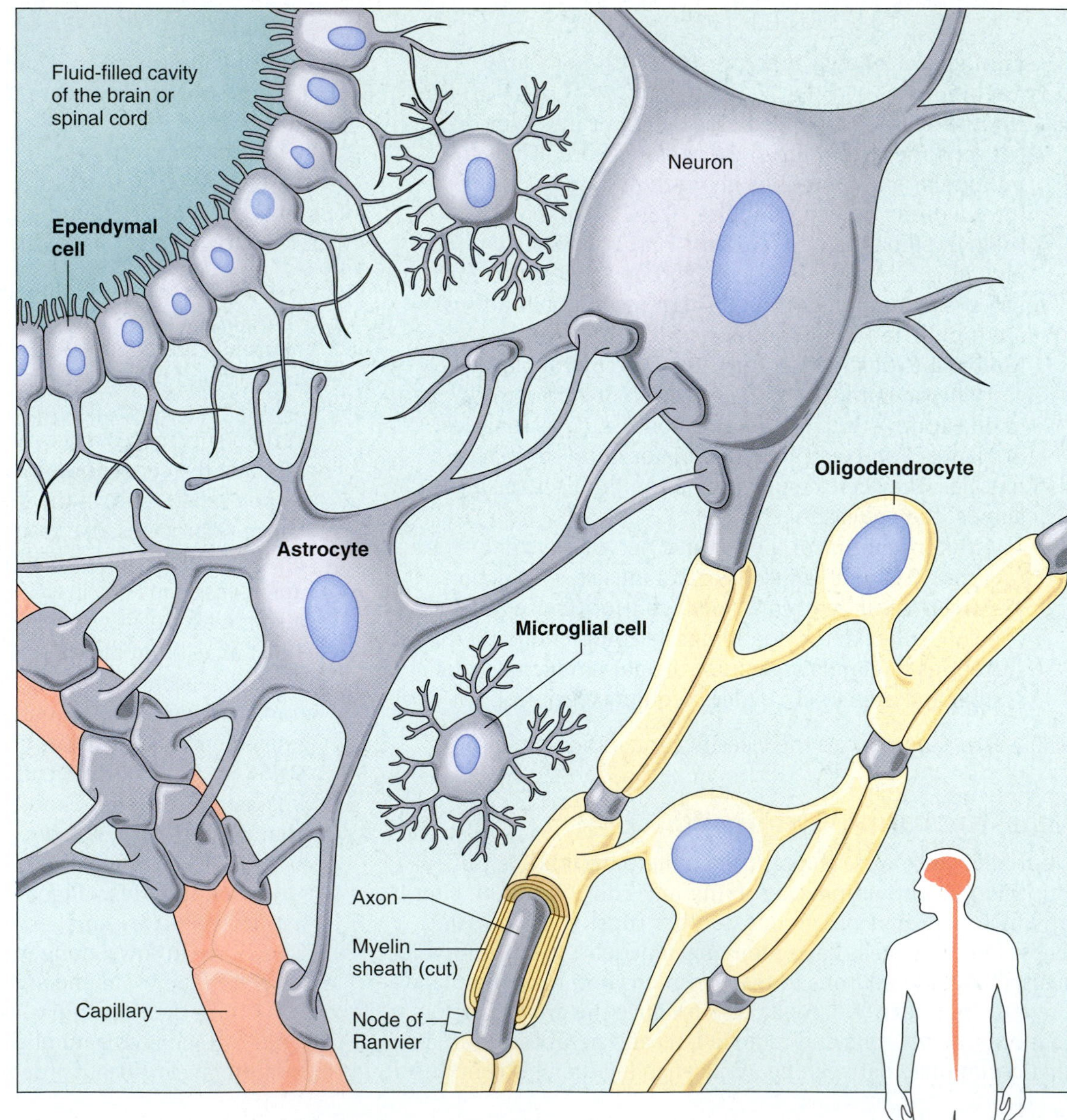

FIGURE 10.8 Types of neuroglia in the central nervous system include the astrocyte, oligodendrocyte, microglial cell, and ependymal cell. Cilia are on most ependymal cells during development and early childhood, but in the adult are mostly on ependymal cells in the ventricles of the brain.

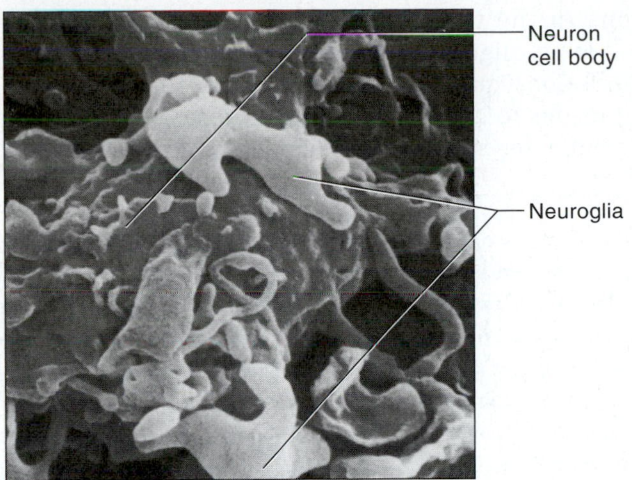

FIGURE 10.9 A scanning electron micrograph of a neuron cell body and some of the neuroglia associated with it (10,000×). (Tissues and Organs: *A Text-Atlas of Scanning Electron Microscopy,* by R. G. Kessel and R. H. Kardon, (c) 1979 W. H. Freeman and Company.)

a single oligodendrocyte may provide myelin for many axons. However, these cells do not form neurilemmae.

3. **Microglia.** Microglial cells are small and have fewer processes than other types of neuroglia. These cells are scattered throughout the CNS, where they help support neurons and phagocytize bacterial cells and cellular debris. They usually proliferate whenever the brain or spinal cord is inflamed because of injury or disease.
4. **Ependyma.** Ependymal cells are cuboidal or columnar in shape and may have cilia. They form the inner lining of the *central canal* that extends downward through the spinal cord. Ependymal cells also form a one-cell-thick epithelial-like membrane that covers the inside of spaces in the brain called *ventricles* (see chapter 11, pp. 385–386). Here, gap junctions join ependymal cells, forming a porous layer through which substances diffuse freely between the interstitial fluid of the brain tissues and the fluid (cerebrospinal fluid) in the ventricles.

    Ependymal cells also cover the specialized capillaries called *choroid plexuses* associated with the ventricles of the brain. Here they help regulate the composition of the cerebrospinal fluid.

Neuroglia, which comprise more than half of the volume of the brain and outnumber neurons 10 to 1, are critical to neuron function.

Abnormal neuroglia are associated with certain disorders. Most brain tumors, for example, consist of neuroglia that divide too often. Neuroglia that produce toxins may lie behind some neurodegenerative disorders. In one familial form of amyotrophic lateral sclerosis (ALS, or Lou Gehrig's disease), astrocytes release a toxin that destroys motor neurons, causing progressive weakness. In Huntington disease (HD), which causes uncontrollable movements and cognitive impairment, microglia in the brain release a toxin that damages neurons. In both ALS and HD, only specific sets of neurons are affected. Identifying the unexpected roles of neuroglia in nervous system disorders suggests new targets for treatments.

Neuroglia of the PNS

The two types of neuroglia in the peripheral nervous system are Schwann cells and **satellite cells:**

1. Schwann cells produce the myelin on peripheral myelinated neurons, as described earlier.
2. Satellite cells support clusters of neuron cell bodies called *ganglia,* in the PNS.

Table 10.2 summarizes the characteristics and functions of neuroglia.

TABLE 10.2 | Types of Neuroglia

Type	Characteristics	Functions
CNS		
Astrocytes	Star-shaped cells between neurons and blood vessels	Structural support, formation of scar tissue, transport of substances between blood vessels and neurons, communicate with one another and with neurons, mop up excess ions and neurotransmitters, induce synapse formation
Oligodendrocytes	Shaped like astrocytes, but with fewer cellular processes, occur in rows along axons	Form myelin sheaths in the brain and spinal cord, produce nerve growth factors
Microglia	Small cells with few cellular processes and found throughout the CNS	Structural support and phagocytosis (immune protection)
Ependyma	Cuboidal and columnar cells in the inner lining of the ventricles of the brain and the central canal of the spinal cord	Form a porous layer through which substances diffuse between the interstitial fluid of the brain and spinal cord and the cerebrospinal fluid
PNS		
Schwann cells	Cells with abundant, lipid-rich membranes that wrap tightly around the axons of peripheral neurons	Speed neurotransmission
Satellite cells	Small, cuboidal cells that surround cell bodies of neurons in ganglia	Support ganglia in the PNS

Neuroglia and Axonal Regeneration

Injury to the cell body usually kills the neuron, and because mature neurons do not divide, the destroyed cell is not replaced unless neural stem cells become stimulated to proliferate. However, a damaged peripheral axon may regenerate. For example, if injury or disease separates an axon in a peripheral nerve from its cell body, the distal portion of the axon and its myelin sheath deteriorate within a few weeks. Macrophages remove the fragments of myelin and other cellular debris. The proximal end of the injured axon develops sprouts shortly after the injury. Influenced by nerve growth factors that nearby neuroglia secrete, one of these sprouts may grow into a tube formed by remaining basement membrane and connective tissue. At the same time, any remaining Schwann cells proliferate along the length of the degenerating portion and form new myelin around the growing axon.

Growth of a regenerating axon is slow (3 to 4 millimeters per day), but eventually the new axon may reestablish the former connection (fig. 10.10). Nerve growth factors, secreted by neuroglia, may help direct the growing axon. However, the regenerating axon may still end up in the wrong place, so full function often does not return.

If an axon of a neuron within the CNS is separated from its cell body, the distal portion of the axon will degenerate, but more slowly than a separated axon in the PNS. However, axons in the CNS lack neurilemmae, and the myelin-producing oligodendrocytes do not proliferate following injury. Consequently, the proximal end of a damaged axon that begins to grow has no tube of sheath cells to guide it. Therefore, regeneration is unlikely.

If a peripheral nerve is severed, it is important that the two cut ends be connected as soon as possible so that the regenerating sprouts of the axons can more easily reach the tubes formed by the basement membranes and connective tissues on the distal side of the gap. When the gap exceeds 3 millimeters, the regenerating axons may form a tangled mass called a neuroma. It is composed of sensory axons and is painfully sensitive to pressure. Neuromas sometimes complicate a patient's recovery following limb amputation.

PRACTICE 

4 What are neuroglia?

5 Name and describe four types of neuroglia.

6 What are some functions of neuroglia?

7 Explain how an injured peripheral axon might regenerate.

8 Explain why functionally significant regeneration is unlikely in the CNS.

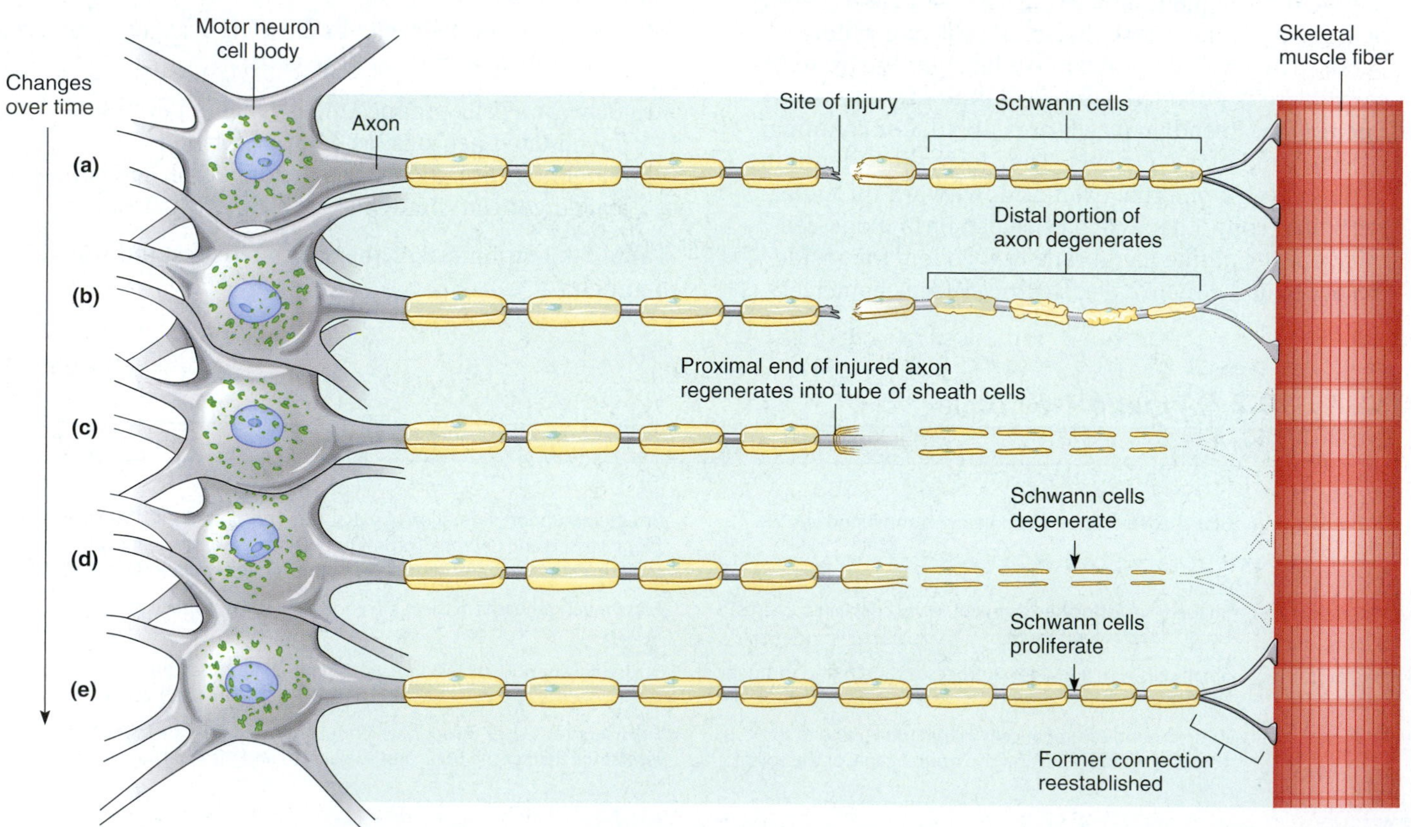

FIGURE 10.10 If a myelinated axon is injured, the following events may occur over several weeks to months: (*a*) The proximal portion of the axon may survive, but (*b*) the portion distal to the injury degenerates. (*c* and *d*) In time, the proximal portion may develop extensions that grow into the tube of basement membrane and connective tissue cells that the axon previously occupied and (*e*) possibly reestablish the former connection. Nerve growth factors that neuroglia secrete assist in the regeneration process.

Neurons do not divide. New neural tissue arises from neural stem cells, which give rise to neural progenitor cells that can give rise to neurons or neuroglia. In the adult brain, the rare neural stem cells are in a region called the dentate gyrus and near fluid-filled cavities called ventricles.

Neural stem cells were discovered in the 1980s, in songbirds—the cells were inferred to exist because the numbers of neurons waxed and waned with the seasons, peaking when the birds learned songs. Moving songbirds far from food, forcing them to sing longer, resulted in more brain neurons, thanks to the stem cells. In the 1990s, researchers identified the cells in brain slices from marmosets and tree shrews given a drug that marks dividing cells. Then they were discovered in humans when a researcher learned that patients with tongue and larynx cancer were taking the drug to mark their cancer cells. Five patients donated their brains after their deaths, and researchers identified the cells. Today, human neural stem and progenitor cells are being used to screen drugs and are being delivered as implants to experimentally treat a variety of brain disorders. One day, a person's neural stem cells may be coaxed to help heal from within.

10.5 THE SYNAPSE

Nerve impulses pass from neuron to neuron (or to other cells) at synapses (fig. 10.11). A **presynaptic neuron** brings the impulse to the synapse and, as a result, stimulates or inhibits a **postsynaptic neuron** (or a muscle or gland). A *synaptic cleft,* or gap, separates the two cells, which are connected functionally, not physically (fig. 10.12). The process by which the impulse in the presynaptic neuron signals the postsynaptic cell is called **synaptic transmission.**

A nerve impulse travels along the axon to the axon terminal. Axons usually have several rounded synaptic knobs at their terminals, which dendrites lack. These knobs have arrays of membranous sacs, called synaptic vesicles, that contain neurotransmitter molecules. When a nerve impulse reaches a synaptic knob, voltage-sensitive calcium channels open and calcium diffuses inward from the extracellular fluid. The increased calcium concentration inside the cell initiates a series of events that fuses the synaptic vesicles with the cell membrane, where they release their neurotransmitter by exocytosis.

Once the neurotransmitter binds to receptors on a postsynaptic cell, the action of neurotransmitter on the postsynaptic cell is either excitatory (turning a process on) or inhibitory (turning a process off). The net effect on the postsynaptic cell depends on the combined effect of the excitatory and inhibitory inputs from as few as 1 to 100,000 or more presynaptic neurons.

10.6 CELL MEMBRANE POTENTIAL

A cell membrane is usually electrically charged, or *polarized,* so that the inside is negatively charged with respect to the outside. This polarization is due to an unequal distribution of positive and negative ions on either side of the membrane. It is important in the conduction of muscle and nerve impulses.

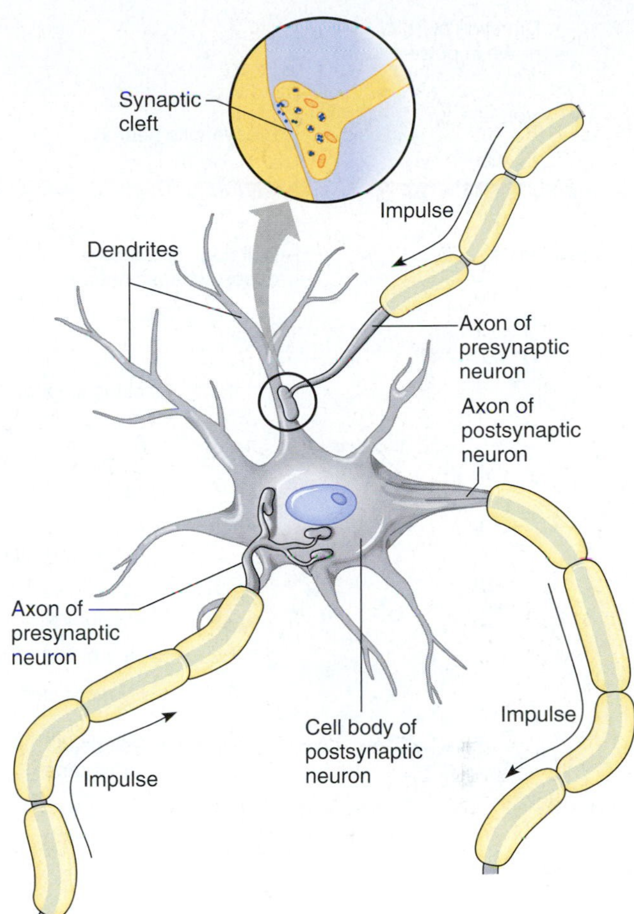

FIGURE 10.11 For an impulse to continue from one neuron to another, it must cross the synaptic cleft. A synapse usually separates an axon and a dendrite or an axon and a cell body.

Distribution of Ions

Potassium ions (K^+) are the major intracellular positive ion (cation), and sodium ions (Na^+) are the major extracellular cation. The distribution is created largely by the sodium–potassium pump (Na^+/K^+pump), which actively transports sodium ions out of the cell and potassium ions into the cell. It is also in part due to channels in the cell membrane that determine membrane permeability to these ions. These channels, formed by membrane proteins, can be selective; that is, a particular channel may allow only one type of ion to pass through and exclude all other ions of different size and charge. Thus, even though concentration gradients are present for sodium and potassium, the ability of these ions to diffuse across the cell membrane depends on the presence of channels.

RECONNECT
To Chapter 3, Cell Membrane, page 80.

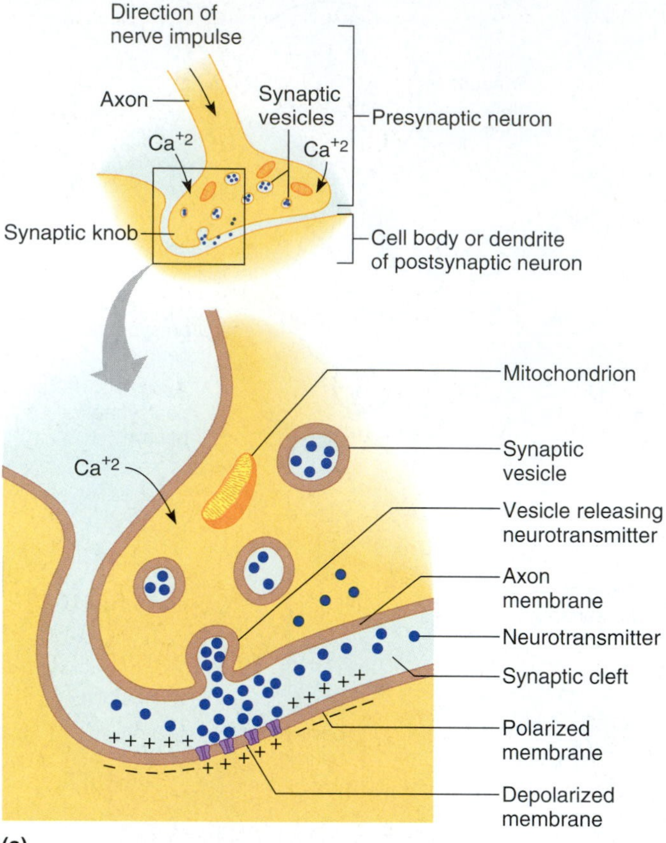

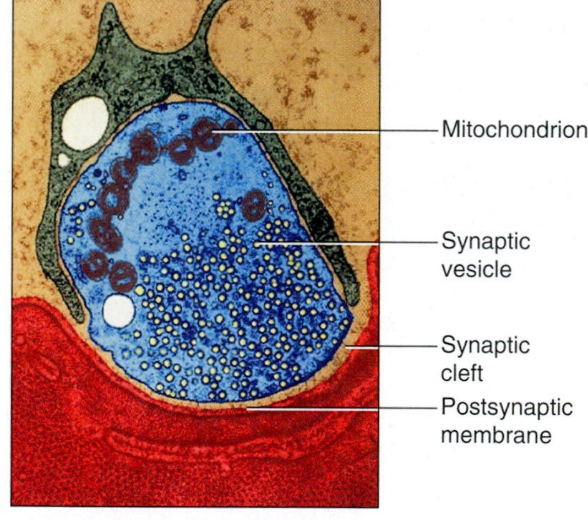

FIGURE 10.12 The synapse. (*a*) When a nerve impulse reaches the synaptic knob at the end of an axon, synaptic vesicles release a neurotransmitter that diffuses across the synaptic cleft. In this case the neurotransmitter is excitatory. (*b*) A transmission electron micrograph of a synaptic knob filled with synaptic vesicles (37,500×).

Some channels are always open, whereas others may be either open or closed, somewhat like a gate. Both chemical and electrical factors can affect the opening and closing of these *gated channels* (fig. 10.13).

Resting Potential

A resting nerve cell is not being stimulated to send a nerve impulse. Under resting conditions, nongated (always open) channels determine the membrane permeability to sodium and potassium ions.

Sodium and potassium ions follow the laws of diffusion described in chapter 3 (pp. 90 and 92) and show a net movement from areas of high concentration to areas of low concentration across a membrane as their permeabilities permit. The resting cell membrane is only slightly permeable to these ions, but the membrane is more permeable to potassium ions than to sodium ions (fig. 10.14*a*). Also, the cytoplasm of these cells has many negatively charged ions (anions) which include phosphate (PO_4^{-2}), sulfate (SO_4^{-2}), and proteins, that are synthesized inside the cell and cannot diffuse through cell membranes.

If we consider a hypothetical neuron, before a membrane potential has been established, we would expect potassium to diffuse out of the cell more rapidly than sodium could diffuse in. This means that every millisecond (as the membrane potential is being established in our hypothetical cell), a few more positive ions leave the cell than enter it (fig. 10.14*a*). As a result, the outside of the membrane gains a slight surplus of positive charges, and the inside reflects a surplus of the impermeant negatively charged ions. This creates a separation of positive and negative electrical charges between the inside and outside surfaces of the cell membrane (fig. 10.14*b*). All this time, the cell continues to expend metabolic energy in the form of ATP to actively transport sodium and potassium ions in opposite directions, thus maintaining the concentration gradients for those ions responsible for their diffusion in the first place.

The difference in electrical charge between two points is measured in units called volts. It is called a potential difference because it represents stored electrical energy that can be used to do work at some future time. The potential difference across the cell membrane is called the **membrane potential** (transmembrane potential) and is measured in millivolts.

In the case of a resting neuron, one that is not sending impulses or responding to other neurons, the membrane potential is termed the **resting potential** (resting membrane potential) and has a value of −70 millivolts (fig. 10.14*b*). The negative sign is relative to the inside of the cell and is due to the excess negative charges on the inside of the cell membrane. To understand how the resting potential provides the energy for sending a nerve impulse down the axon, we must first understand how neurons respond to signals called stimuli.

With the resting membrane potential established, a few sodium ions and potassium ions continue to diffuse across the cell membrane. The negative membrane potential helps sodium ions enter the cell despite sodium's low permeability, but it hinders potassium ions from leaving the cell despite potassium's higher permeability. The net effect is that three sodium ions "leak" into the cell for every two potassium ions that "leak" out. The Na^+/K^+ pump exactly balances these leaks by pumping three sodium ions out for every two potassium ions it pumps in (fig. 10.14*c*).

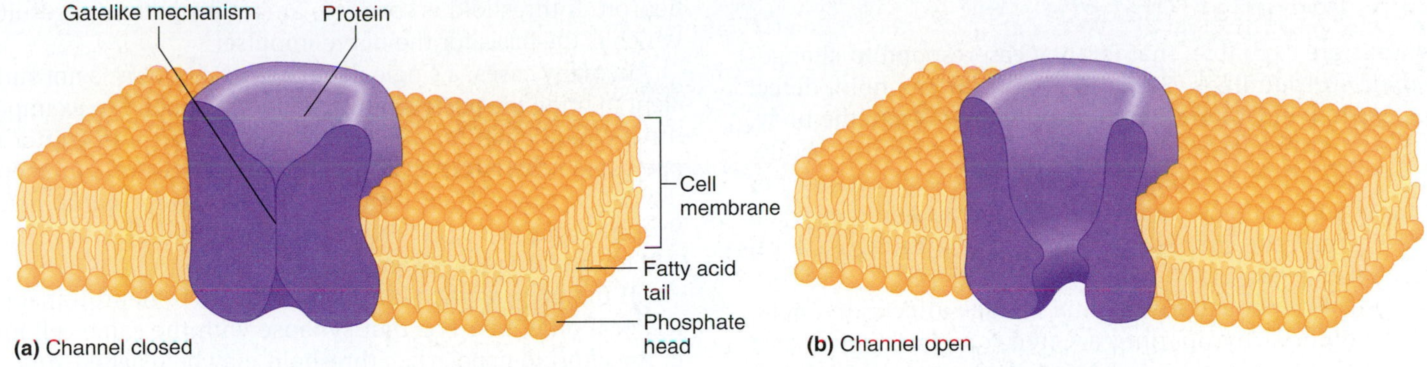

FIGURE 10.13 A gatelike mechanism can (*a*) close or (*b*) open some of the channels in cell membranes through which ions pass.

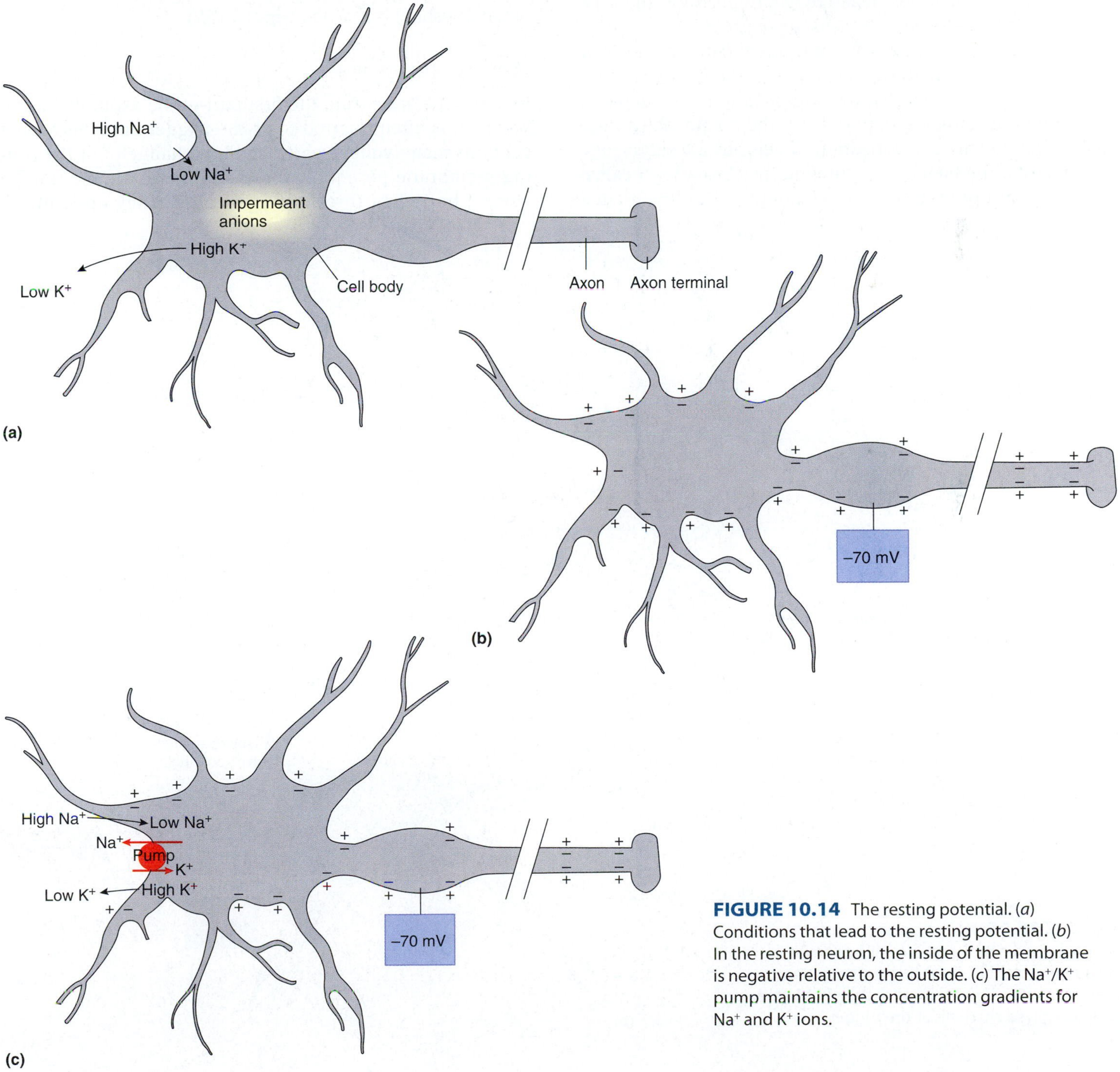

FIGURE 10.14 The resting potential. (*a*) Conditions that lead to the resting potential. (*b*) In the resting neuron, the inside of the membrane is negative relative to the outside. (*c*) The Na^+/K^+ pump maintains the concentration gradients for Na^+ and K^+ ions.

Local Potential Changes

Neurons are excitable; that is, they can respond to changes in their surroundings. Some neurons, for example, detect changes in temperature, light, or pressure outside the body, whereas others respond to signals from inside the body, often from other neurons. In either case, such changes or stimuli usually affect the membrane potential in the region of the membrane exposed to the stimulus, causing a local potential change.

Typically, the environmental change affects the membrane potential by opening a gated ion channel. If, as a result, the membrane potential becomes more negative than the resting potential, the membrane is *hyperpolarized.* If the membrane becomes less negative (more positive) than the resting potential, the membrane is *depolarized.*

Local potential changes are graded. This means that the degree of change in the resting potential is directly proportional to the intensity of the stimulation. For example, if the membrane is being depolarized, the greater the stimulus, the greater the depolarization. If neurons are sufficiently depolarized, the membrane potential reaches a level called the **threshold potential,** approximately −55 millivolts in a neuron. If threshold is reached, an **action potential** results, which is the basis for the nerve impulse.

In many cases, a single depolarizing stimulus is not sufficient to bring the postsynaptic cell to threshold. For example, if presynaptic neurons release enough neurotransmitter to open some chemically-gated sodium channels for a moment, the depolarization that results might be insufficient to reach threshold (fig. 10.15*a*). Such a subthreshold depolarization will not result in an action potential.

If the presynaptic neurons release more neurotransmitter, or if other neurons that synapse with the same cell join in the effort to depolarize, threshold may be reached, and an action potential results. The mechanism uses another type of ion channel, a voltage-gated sodium channel that opens when threshold is reached (fig. 10.15*b*).

Action Potentials

In a multipolar neuron, the first part of the axon, the *initial segment,* is often referred to as the **trigger zone** because it contains many voltage-gated sodium channels. At the resting membrane potential, these sodium channels remain closed, but when threshold is reached, they open for an

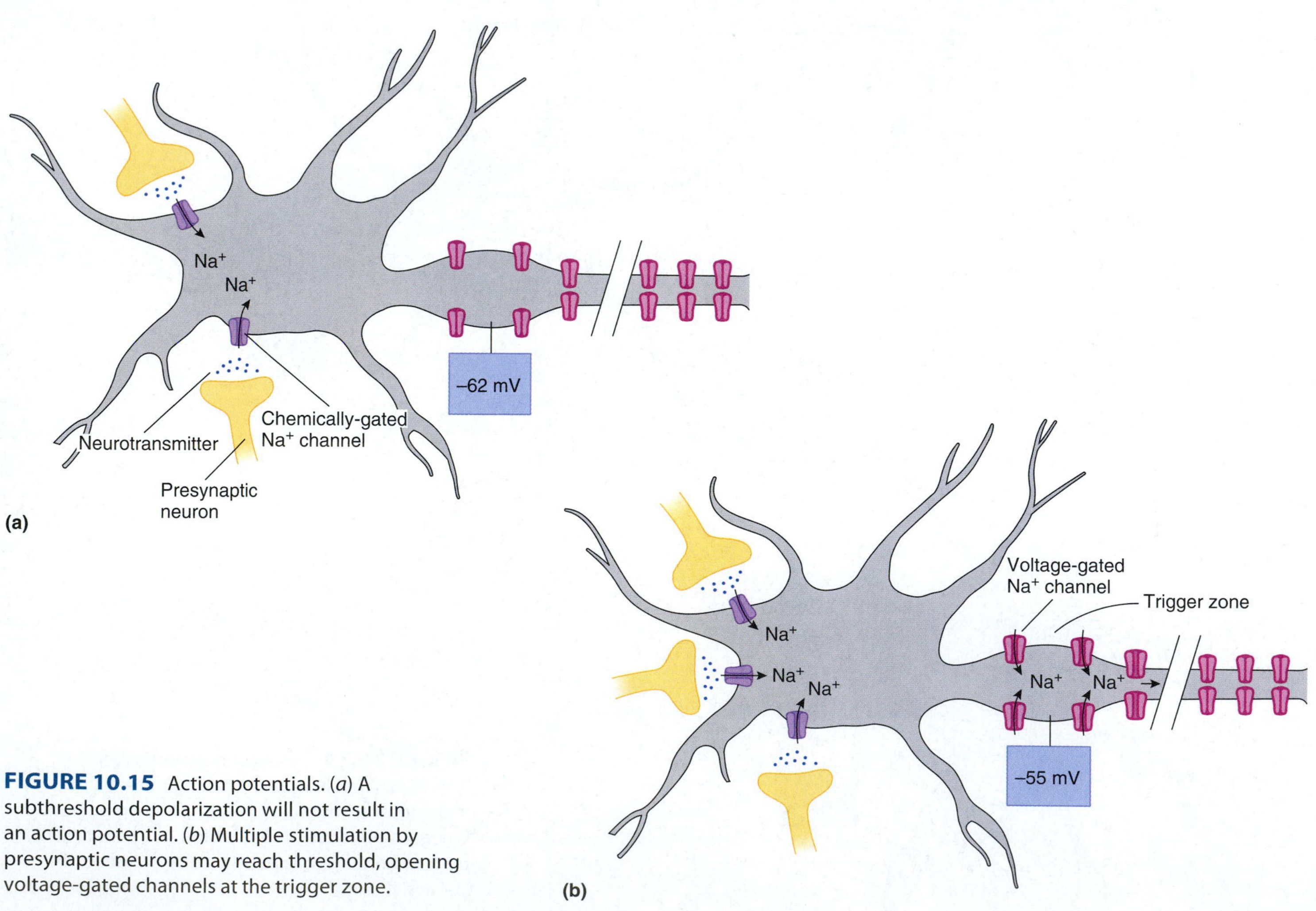

FIGURE 10.15 Action potentials. (*a*) A subthreshold depolarization will not result in an action potential. (*b*) Multiple stimulation by presynaptic neurons may reach threshold, opening voltage-gated channels at the trigger zone.

instant, briefly increasing sodium permeability. Sodium ions diffuse inward across that part of the cell membrane, down their concentration gradient, aided by the attraction of the sodium ions to the negative electrical condition on the inside of the membrane.

As the sodium ions diffuse inward, the membrane potential changes from its resting value (fig. 10.16*a*) and momentarily becomes positive on the inside (still considered depolarization). At the peak of the action potential, the membrane potential may reach +30mV (fig. 10.16*b*).

The voltage-gated sodium channels quickly close, but at almost the same time, slower voltage-gated potassium channels open and briefly increase potassium permeability. As potassium ions diffuse outward across that part of the membrane, the inside of the membrane becomes negatively charged once more. The membrane is thus repolarized (note in fig. 10.16*c* that it hyperpolarizes for an instant). The voltage-gated potassium channels then close as well. In this way, the resting potential is quickly reestablished, and it remains in the resting state until it is stimulated again (fig. 10.17). The active transport mechanism in the membrane works to maintain the original concentrations of sodium and potassium ions.

Axons are capable of action potentials, but the cell body and dendrites are not. An action potential at the trigger zone causes an electric current to flow a short distance down the axon, which stimulates the adjacent membrane to reach its threshold level, triggering another action potential. The second action potential causes another electric current to flow farther down the axon. This sequence of events results in a series of action potentials sequentially occurring all the way to the end of the axon without decreasing in amplitude, even if the axon branches. The propagation of action potentials along an axon is the nerve impulse (fig. 10.18).

A nerve impulse is similar to the muscle impulse mentioned in chapter 9, page 290. In the muscle fiber, stimulation at the motor end plate triggers an impulse to travel over

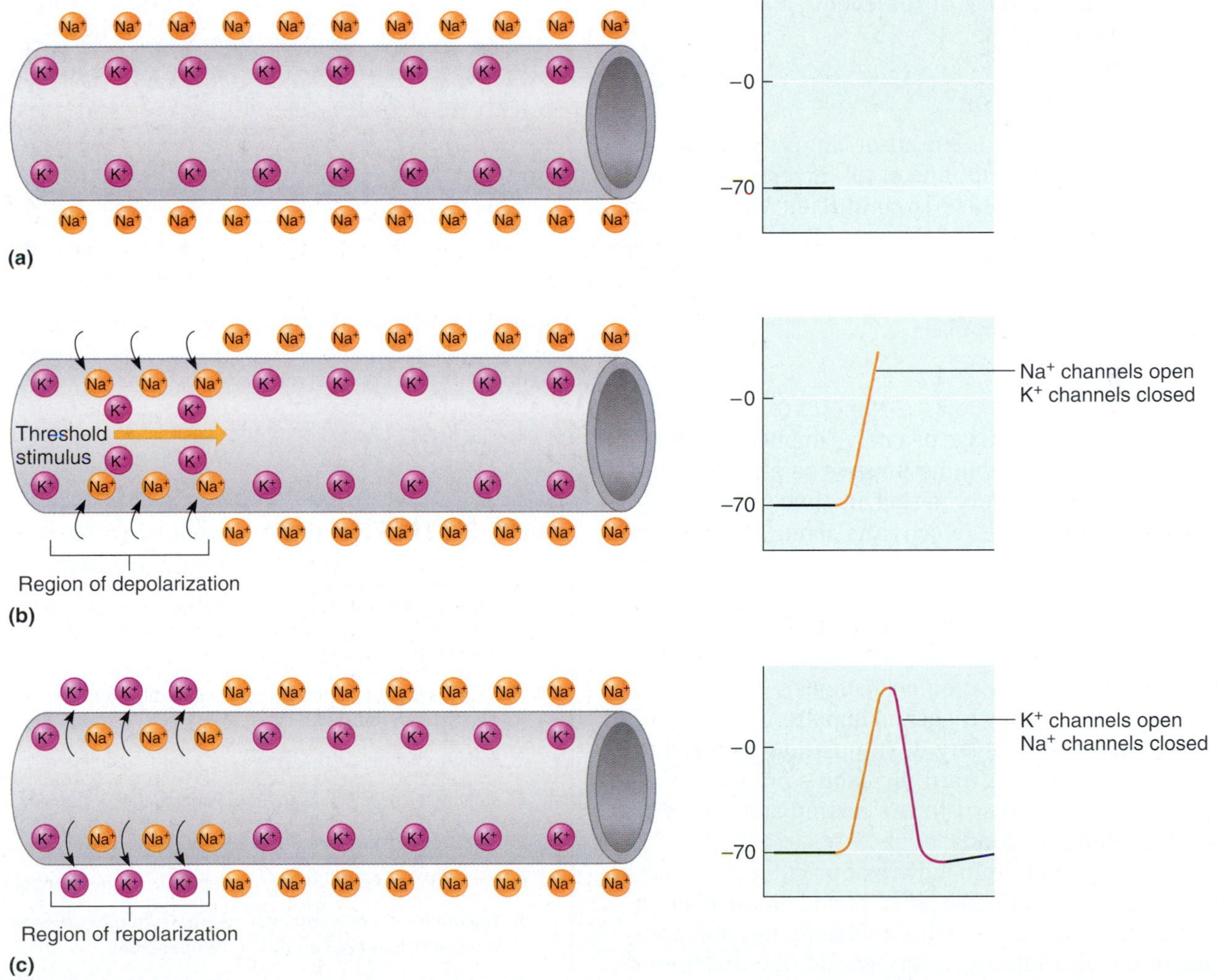

FIGURE 10.16 At rest (*a*), the membrane potential is about –70 millivolts. When the membrane reaches threshold (*b*), voltage-sensitive sodium channels open, some Na^+ diffuses inward, and the membrane is depolarized. Soon afterward (*c*), voltage-sensitive potassium channels open, K^+ diffuses out, and the membrane is repolarized. (Negative ions not shown.)

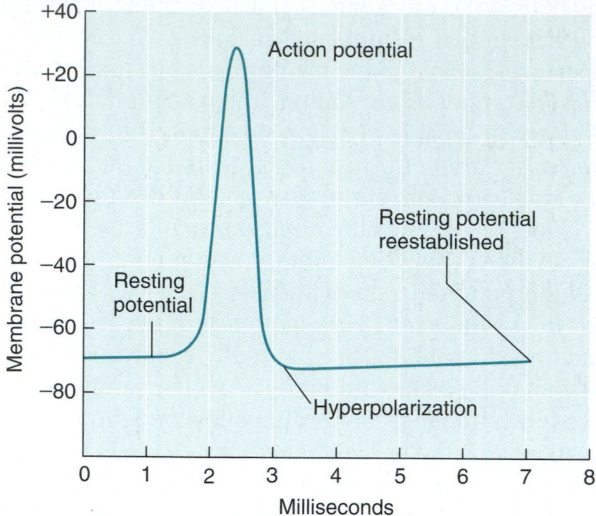

FIGURE 10.17 An oscilloscope records an action potential.

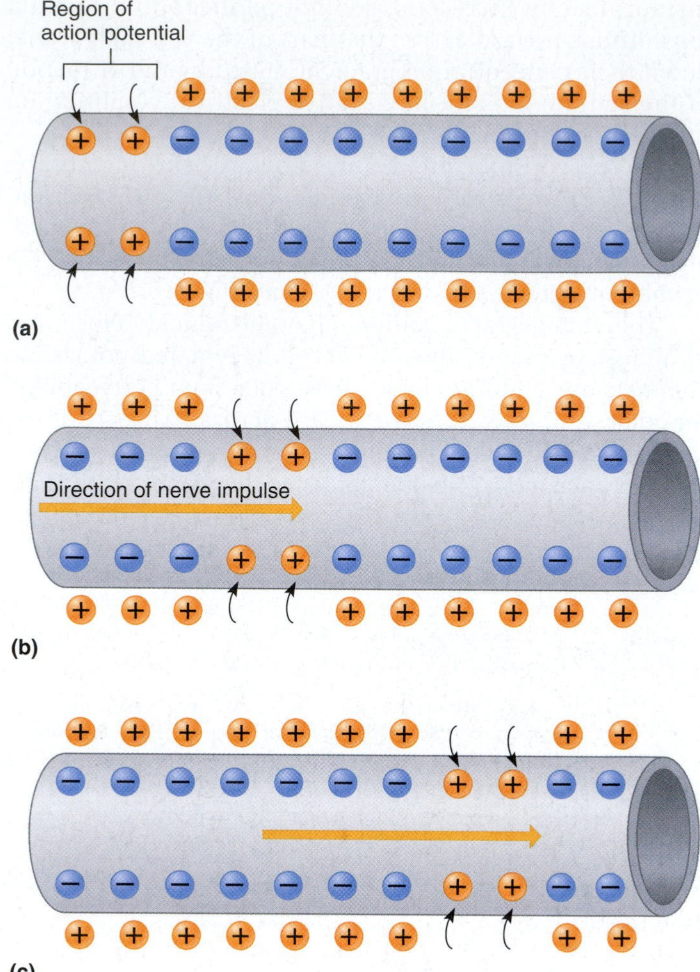

FIGURE 10.18 Nerve impulse. (*a*) An action potential in one region stimulates the adjacent region, and (*b* and *c*) a wave of action potentials (a nerve impulse) moves along the axon.

the surface of the fiber and down into its transverse tubules. See table 10.3 for a summary of the events leading to the conduction of a nerve impulse.

All-or-None Response

Nerve impulse conduction is an all-or-none response. In other words, if a neuron responds at all, it responds completely. Thus, a nerve impulse is conducted whenever a stimulus of threshold intensity or above is applied to an axon and all impulses carried on that axon are the same strength. A greater intensity of stimulation produces more impulses per second, not a stronger impulse.

Refractory Period

For a short time following passage of a nerve impulse, a threshold stimulus will not trigger another impulse on an axon. This brief period, called the **refractory period,** has two parts. During the *absolute refractory period,* which lasts about 1/2,500 of a second, the axon's membrane is changing in sodium permeability and cannot be stimulated. This is followed by a *relative refractory period,* when the membrane reestablishes its resting potential. While the membrane is in the relative refractory period, even though repolarization is incomplete, a threshold stimulus of high intensity may trigger an impulse.

As time passes, the intensity of stimulation required to trigger an impulse decreases until the axon's original excitability is restored. This return to the resting state usually takes from 10 to 30 milliseconds.

The refractory period limits how many action potentials may be generated in a neuron in a given period. Remembering that the action potential takes about a millisecond, and adding the time of the absolute refractory period, the maximum theoretical frequency of impulses in a neuron is about 700 per second. In the body, this limit is rarely achieved—frequencies of about 100 impulses per second are common.

TABLE 10.3 | Events Leading to Nerve Impulse Conduction

Events
1. Nerve cell membrane maintains resting potential by diffusion of Na^+ and K^+ down their concentration gradients as the cell pumps them up the gradients.
2. Neurons receive stimulation, causing local potentials, which may sum to reach threshold.
3. Sodium channels in the trigger zone of the axon open.
4. Sodium ions diffuse inward, depolarizing the membrane.
5. Potassium channels in the membrane open.
6. Potassium ions diffuse outward, repolarizing the membrane.
7. The resulting action potential causes an electric current that stimulates adjacent portions of the membrane.
8. Action potentials occur sequentially along the length of the axon as a nerve impulse.

Impulse Conduction

An unmyelinated axon conducts an impulse over its entire surface. A myelinated axon functions differently. Myelin contains a high proportion of lipid that excludes water and water-soluble substances. Thus, myelin serves as an electrical insulator and prevents almost all flow of ions through the membrane that it encloses.

It might seem that the myelin sheath would prevent conduction of a nerve impulse, and this would be true if the sheath were continuous. However, nodes of Ranvier between Schwann cells or oligodendrocytes interrupt the sheath (see fig. 10.3). At these nodes, the axon membrane has channels for sodium and potassium ions that open during a threshold depolarization.

When a myelinated axon is stimulated to threshold, an action potential occurs at the trigger zone. This causes an electric current to flow away from the trigger zone through the cytoplasm of the axon. As this local current reaches the first node, it stimulates the membrane to its threshold level, and an action potential occurs there, sending an electric current to the next node. Consequently, in a nerve impulse traveling along a myelinated axon, action potentials occur only at the nodes. The action potentials appear to jump from node to node, so this type of impulse conduction is called **saltatory conduction.** Conduction on myelinated axons is many times faster than conduction on unmyelinated axons (fig. 10.19).

The diameter of the axon also affects the speed of nerve impulse conduction—the greater the diameter, the faster the impulse. An impulse on a thick, myelinated axon, such as that of a motor neuron associated with a skeletal muscle, might travel 120 meters per second, whereas an impulse on a thin, unmyelinated axon, such as that of a sensory neuron associated with the skin, might move only 0.5 meter per second. Clinical Application 10.3 discusses factors that influence nerve impulse conduction.

PRACTICE 

9 Summarize how a resting potential is achieved.

10 Explain how a polarized axon responds to stimulation.

11 List the major events of an action potential.

12 Define *refractory* period.

13 Explain how impulse conduction differs in myelinated and unmyelinated axons.

10.7 SYNAPTIC TRANSMISSION

Released neurotransmitter molecules diffuse across the synaptic cleft and react with specific molecules called *receptors* in the postsynaptic neuron membrane. Effects of neurotransmitters vary. Some open ion channels, and others close them. These ion channels respond to neurotransmitter molecules, so they are called *chemically-gated,* in contrast to the voltage-gated ion channels that participate in action potentials. Changes in chemically-gated ion channels create local potentials, called **synaptic potentials,** which enable one neuron to affect another.

Synaptic Potentials

Synaptic potentials can depolarize or hyperpolarize the receiving cell membrane. For example, if a neurotransmitter binds to a postsynaptic receptor and opens sodium ion channels, the ions diffuse inward, depolarizing the membrane, possibly triggering an action potential. This type of membrane change is called an **excitatory postsynaptic potential** (EPSP), and it lasts for about 15 milliseconds.

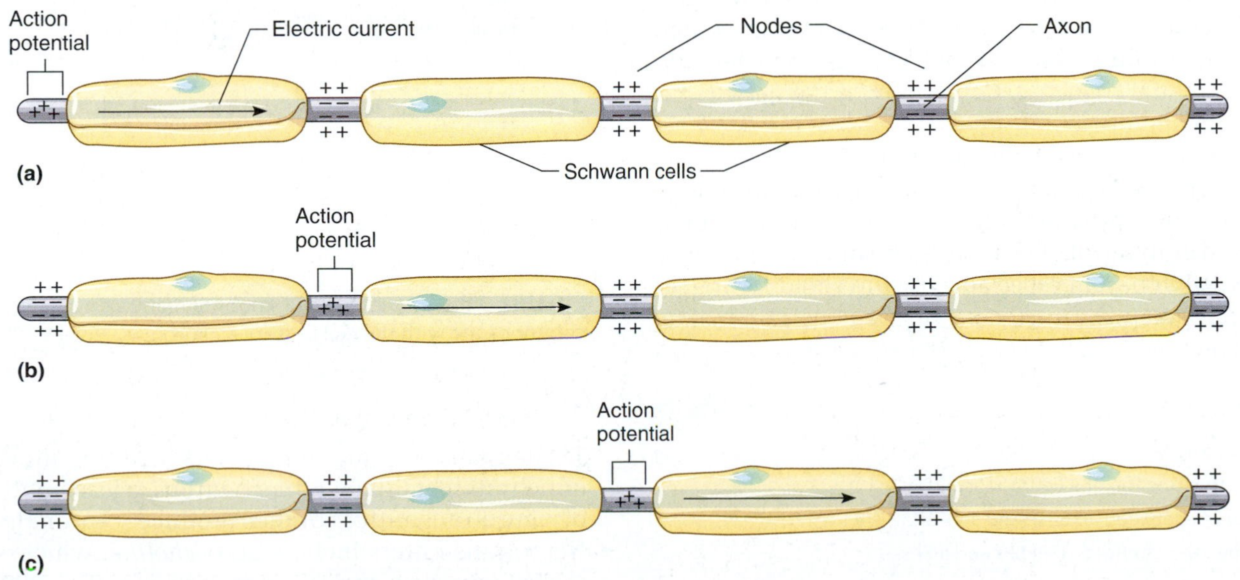

FIGURE 10.19 On a myelinated axon, a nerve impulse appears to jump from node to node.

10.3 CLINICAL APPLICATION

Factors Affecting Impulse Conduction

Painful muscle cramps, convulsions, paralysis, and anesthesia can each result from changes in the permeability of axons to particular ions. A number of substances alter axon membrane permeability to ions.

Calcium ions are required to close sodium channels in axon membranes during an action potential. If calcium is deficient, then sodium channels remain open, and sodium ions diffuse through the membrane continually so that impulses are transmitted repeatedly. If these spontaneous impulses travel along axons to skeletal muscle fibers, the muscles continuously spasm (tetanus or tetany). This can happen to women during pregnancy as the developing fetus uses maternal calcium. Tetanic contraction may also occur when the diet lacks calcium or vitamin D or when prolonged diarrhea depletes the body of calcium.

A small increase in the concentration of extracellular potassium ions causes the resting potential of nerve fibers to be less negative (partially depolarized). As a result, the threshold potential is reached with a less intense stimulus than usual. The affected fibers are excitable, and the person may experience convulsions.

If the extracellular potassium ion concentration is greatly decreased, the resting potentials of the nerve fibers may become so negative that action potentials are not generated. In this case, impulses are not triggered, and muscles become paralyzed.

Certain anesthetic drugs, such as procaine, decrease membrane permeability to sodium ions. In the tissue fluids surrounding an axon, these drugs prevent impulses from passing through the affected region. Consequently, the drugs keep impulses from reaching the brain, preventing perception of touch and pain. ■

If a different neurotransmitter binds other receptors and increases membrane permeability to potassium ions, these ions diffuse outward, hyperpolarizing the membrane. An action potential is now less likely to occur, so this change is called an **inhibitory postsynaptic potential** (IPSP). Some inhibitory neurotransmitters open chloride ion channels. In this case, if sodium ions enter the cell, negative chloride ions are free to follow, opposing the depolarization.

In the brain and spinal cord, each neuron may receive the synaptic knobs of a thousand or more axons on its dendrites and cell body (fig. 10.20). Furthermore, at any moment, some of the postsynaptic potentials are excitatory on a particular neuron, while others are inhibitory.

The integrated sum of the EPSPs and IPSPs determines whether an action potential results. If the net effect is more excitatory than inhibitory, threshold may be reached and an action potential triggered. Conversely, if the net effect is inhibitory, no impulse is transmitted.

Summation of the excitatory and inhibitory effects of the postsynaptic potentials commonly takes place at the trigger zone, usually in a proximal region of the axon, but found also in the distal peripheral process of some sensory neurons. This region has an especially low threshold for triggering an action potential; thus, it serves as a decision-making part of the neuron.

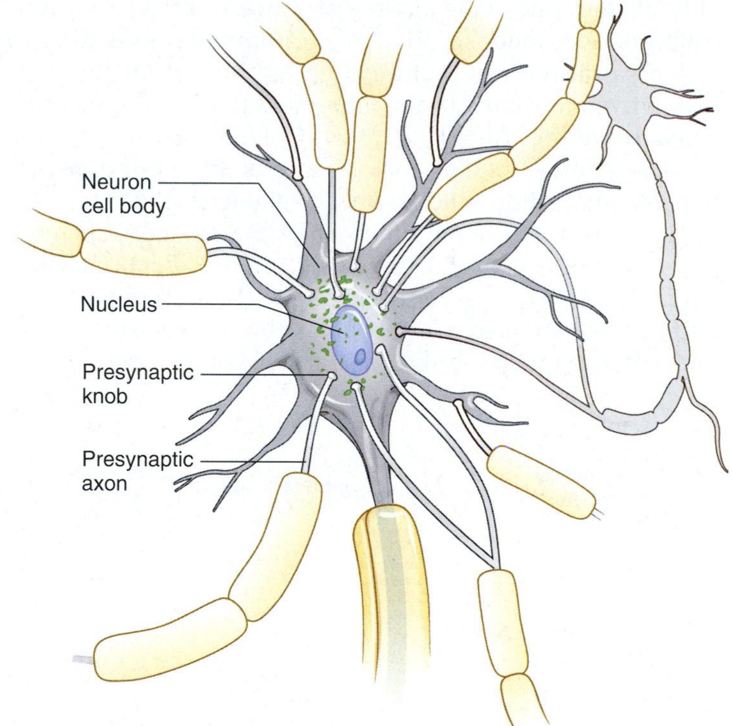

FIGURE 10.20 The synaptic knobs of many axons may communicate with the cell body of a neuron.

PRACTICE

14 Describe a synapse.
15 Explain the function of a neurotransmitter.
16 Distinguish between an EPSP and an IPSP.
17 Describe the net effects of EPSPs and IPSPs.

Neurotransmitters

The nervous system produces at least thirty different types of neurotransmitters. Some neurons release only one type of neurotransmitter; others produce two or three types. Neurotransmitters include *acetylcholine*, which stimulates skeletal muscle contractions (see chapter 9, p. 290); a group of compounds called *monoamines* (such as epinephrine, norepinephrine, dopamine, and serotonin), which are modifications of amino acids; a group of unmodified *amino acids* (such as glycine, glutamic acid, aspartic acid, and gamma-aminobutyric acid—GABA); and a large group of *peptides* (such as enkephalins and substance P), which are short chains of amino acids.

The peptide neurotransmitters are synthesized in the rough endoplasmic reticulum of the neuron cell bodies and transported in vesicles down the axon to the nerve terminal. Other neurotransmitters are synthesized in the cytoplasm of the nerve terminals and stored in vesicles. When an action potential passes along the membrane of a synaptic knob, it increases the membrane's permeability to calcium ions by opening its calcium ion channels. Calcium ions diffuse inward, and in response, some of the synaptic vesicles fuse with the presynaptic membrane and release their contents by exocytosis into the synaptic cleft. The more calcium that enters the synaptic knob, the more vesicles release neurotransmitter. Table 10.4 lists the major neurotransmitters and their actions. Tables 10.5 and 10.6 list disorders and drugs that alter neurotransmitter levels, respectively.

TABLE 10.4 | Some Neurotransmitters and Representative Actions

Neurotransmitter	Location	Major Actions
Acetylcholine	CNS	Controls skeletal muscle actions
	PNS	Stimulates skeletal muscle contraction at neuromuscular junctions. May excite or inhibit at autonomic nervous system synapses
Biogenic amines		
Norepinephrine	CNS	Creates a sense of well-being; low levels may lead to depression
	PNS	May excite or inhibit autonomic nervous system actions, depending on receptors
Dopamine	CNS	Creates a sense of well-being; deficiency in some brain areas associated with Parkinson disease
	PNS	Limited actions in autonomic nervous system; may excite or inhibit, depending on receptors
Serotonin	CNS	Primarily inhibitory; leads to sleepiness; action is blocked by LSD, enhanced by selective serotonin reuptake inhibitor antidepressant drugs
Histamine	CNS	Release in hypothalamus promotes alertness
Amino acids		
GABA	CNS	Generally inhibitory
Glutamate	CNS	Generally excitatory
Neuropeptides		
Enkephalins, endorphins	CNS	Generally inhibitory; reduce pain by inhibiting substance P release
Substance P	PNS	Excitatory; pain perception
Gases		
Nitric oxide	CNS	May play a role in memory
	PNS	Vasodilation

TABLE 10.5 | Disorders Associated with Neurotransmitter Imbalances

Condition	Symptoms	Imbalance of Neurotransmitter in Brain
Alzheimer disease	Memory loss, depression, disorientation, dementia, hallucinations, death	Deficient acetylcholine
Clinical depression	Debilitating, inexplicable sadness	Deficient norepinephrine and/or serotonin
Epilepsy	Seizures, loss of consciousness	Excess GABA leads to excess norepinephrine and dopamine
Huntington disease	Cognitive and behavioral changes, loss of coordination, uncontrollable dancelike movements, death	Deficient GABA
Hypersomnia	Excessive sleeping	Excess serotonin
Insomnia	Inability to sleep	Deficient serotonin
Mania	Elation, irritability, overtalkativeness, increased movements	Excess norepinephrine
Parkinson disease	Tremors of hands, slowed movements, muscle rigidity	Deficient dopamine
Schizophrenia	Inappropriate emotional responses, hallucinations	Deficient GABA leads to excess dopamine
Tardive dyskinesia	Uncontrollable movements of facial muscles	Deficient dopamine

TABLE 10.6 | Drugs That Alter Neurotransmitter Levels

Drug	Neurotransmitter Affected*	Mechanism of Action	Effect
Tryptophan	Serotonin	Stimulates neurotransmitter synthesis	Sleepiness
Reserpine	Norepinephrine	Decreases packaging of neurotransmitter into vesicles	Decreases blood pressure
Curare	Acetylcholine	Blocks receptor binding	Muscle paralysis
Valium	GABA	Enhances receptor binding	Decreases anxiety
Nicotine	Acetylcholine	Activates receptors	Increases alertness
	Dopamine	Elevates levels	Sense of pleasure
Cocaine	Dopamine	Blocks reuptake	Euphoria
Tricyclic antidepressants	Norepinephrine	Blocks reuptake	Antidepressant
	Serotonin	Blocks reuptake	Antidepressant
Monoamine oxidase inhibitors	Norepinephrine	Blocks enzymatic degradation of neurotransmitter in presynaptic cell	Antidepressant
Selective serotonin reuptake inhibitors	Serotonin	Blocks reuptake	Antidepressant, Anti-anxiety agent
Dual reuptake inhibitors	Serotonin, norepinephrine	Blocks reuptake	Mood elevation

*Others may be affected as well.

RECONNECT

To Chapter 3, Exocytosis, page 97.

After a vesicle releases its neurotransmitter, it becomes part of the cell membrane. Endocytosis eventually returns it to the cytoplasm, where it can provide material to form new secretory vesicles. Table 10.7 summarizes this process, called vesicle trafficking.

To keep signal duration short, enzymes in synaptic clefts and on postsynaptic membranes rapidly decompose some neurotransmitters. The enzyme **acetylcholinesterase,** for example, decomposes acetylcholine on postsynaptic membranes. Other neurotransmitters are transported back into the synaptic knob of the presynaptic neuron or into nearby neurons or neuroglia, a process called *reuptake.* The enzyme **monoamine oxidase** inactivates the monoamine neurotransmitters epinephrine and norepinephrine after reuptake. This enzyme is found in mitochondria in the synaptic knob. Destruction or removal of neurotransmitter prevents continuous stimulation of the postsynaptic neuron.

Neuropeptides

Neurons in the brain or spinal cord synthesize **neuropeptides.** These peptides act as neurotransmitters or as *neuromodulators*—substances that alter a neuron's response to a neurotransmitter or block the release of a neurotransmitter.

Among the neuropeptides are the *enkephalins,* present throughout the brain and spinal cord. Each enkephalin molecule is a chain of five amino acids. Synthesis of enkephalins increases during periods of painful stress, and they bind to the same receptors in the brain (opiate receptors) as the narcotic morphine. Enkephalins relieve pain sensations and probably have other functions. Another morphinelike peptide, *beta endorphin,* is found in the brain and cerebrospinal fluid. It acts longer than enkephalins and is a much more potent pain reliever (Clinical Application 10.4).

Substance P is a neuropeptide that consists of eleven amino acids and is widely distributed. It functions as a neurotransmitter (or perhaps as a neuromodulator) in the neurons that transmit pain impulses into the spinal cord and on to the brain. Enkephalins and endorphins may relieve pain by inhibiting the release of substance P from pain-transmitting neurons.

TABLE 10.7 | Events Leading to Neurotransmitter Release

1. Action potential passes along an axon and over the surface of its synaptic knob.
2. Synaptic knob membrane becomes more permeable to calcium ions, and they diffuse inward.
3. In the presence of calcium ions, synaptic vesicles fuse to synaptic knob membrane.
4. Synaptic vesicles release their neurotransmitter by exocytosis into the synaptic cleft.
5. Synaptic vesicles become part of the membrane.
6. The added membrane provides material for endocytotic vesicles.

10.8 IMPULSE PROCESSING

The way the nervous system processes and affects nerve impulses reflects, in part, the organization of neurons and axons in the brain and spinal cord.

Neuronal Pools

Interneurons, the neurons completely in the CNS, are organized into **neuronal pools.** These are groups of neurons that

10.4 CLINICAL APPLICATION

Opiates in the Human Body

Opiate drugs, such as morphine, heroin, codeine, and opium, are potent painkillers derived from the poppy plant. These drugs alter pain perception, making it easier to tolerate, and elevate mood.

The human body produces opiates, called endorphins (for "endogenous morphine"), that are peptides. Like the poppy-derived opiates that they structurally resemble, endorphins influence mood and perception of pain.

The discovery of endorphins began in 1971 in research laboratories at Stanford University and the Johns Hopkins School of Medicine, where researchers exposed pieces of brain tissue from experimental mammals to morphine. The morphine was radioactively labeled (some of the atoms were radioactive isotopes) so researchers could follow its destination in the brain.

The morphine bound receptors on neurons that transmit pain. Why, the investigators wondered, would an animal's brain cells have receptors for a plant chemical? One explanation was that a mammal's body could manufacture opiates. The opiate receptors, then, would normally bind the body's opiates (the endorphins) but would also bind the chemically similar compounds from poppies. Researchers have since identified several types of endorphins in the human brain and associated their release with situations involving pain relief, such as acupuncture and analgesia to mother and child during childbirth. Endorphin release is also associated with "runner's high." PET scans reveal endorphins binding opiate receptors after conditioned athletes run for two hours.

Endorphins explain why some people addicted to opiate drugs such as heroin experience withdrawal pain when they stop taking the drug. Initially, the body interprets the frequent binding of heroin to its endorphin receptors as an excess of endorphins. To bring the level down, the body slows its own production of endorphins. Then, when the person stops taking the heroin, the body becomes short of opiates (heroin and endorphins). The result is pain.

Opiate drugs can be powerfully addicting when abused—that is, taken repeatedly by a person who is not in pain. These same drugs, however, are extremely useful in dulling severe pain, particularly in terminal illnesses. ■

synapse with each other and perform a common function, even though their cell bodies may be in different parts of the CNS. Each neuronal pool receives input from neurons (which may be part of other pools), and each pool generates output. Neuronal pools may have excitatory or inhibitory effects on other pools or on peripheral effectors.

As a result of incoming impulses and neurotransmitter release, a particular neuron of a neuronal pool may be excited by some presynaptic neurons and inhibited by others. If the net effect is excitatory, threshold may be reached, and an outgoing impulse triggered. If the net effect is excitatory but subthreshold, an impulse will not be triggered, but because the neuron is close to threshold, it will be much more responsive to any further excitatory stimulation. This condition is called **facilitation** (fah-sil″ĭ-tā′shun).

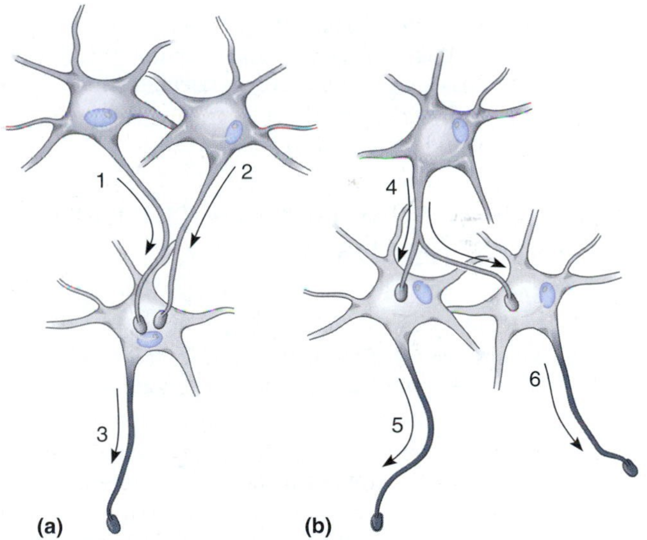

FIGURE 10.21 Impulse processing in neuronal pools. (*a*) Axons of neurons 1 and 2 converge to the cell body of neuron 3. (*b*) The axon of neuron 4 diverges to the cell bodies of neurons 5 and 6.

Convergence

Any single neuron in a neuronal pool may receive impulses from two or more other neurons. Axons originating from different parts of the nervous system leading to the same neuron exhibit **convergence** (kon-ver′jens).

Incoming impulses often represent information from various sensory receptors that detect changes. Convergence allows the nervous system to collect, process, and respond to information.

Convergence makes it possible for a neuron to sum impulses from different sources. For example, if a neuron receives subthreshold stimulation from one input neuron, it may reach threshold if it receives additional stimulation from a second input neuron. Thus, an output impulse triggered from this neuron reflects summation of impulses from two sources (fig. 10.21*a*). Such an output impulse may travel to a particular effector and evoke a response.

Divergence

A neuron has a single axon, but axons may branch at several points. Thus, impulses leaving a neuron of a neuronal pool may exhibit **divergence** (di-ver′jens) by reaching several other neurons. For example, one neuron may stimulate two others; each of these, in turn, may stimulate several others, and so forth. Such a pattern of diverging axons can amplify an impulse—that is, spread it to increasing numbers of neurons within the pool (fig. 10.21*b*).

As a result of divergence, an impulse originating from a single neuron in the CNS may be amplified so that sufficient impulses reach the motor units in a skeletal muscle to cause forceful contraction. Similarly, an impulse originating from a sensory receptor may diverge and reach several different regions of the CNS, where the resulting impulses can be processed and acted upon.

The nervous system enables us to experience the world and to think and feel emotion. This organ system is also sensitive to outside influences. Clinical Application 10.5 discusses one way that an outside influence can affect the nervous system—drug addiction.

PRACTICE 

18 Define *neuropeptide.*

19 What is a neuronal pool?

20 Define *facilitation.*

21 What is convergence?

22 What is the relationship between divergence and amplification?

CHAPTER SUMMARY

10.1 INTRODUCTION (PAGE 354)

1. The nervous system is a network of cells that sense and respond to stimuli in ways that maintain homeostasis.
2. The nervous system is composed of neural tissue, including neurons and neuroglia, blood vessels and connective tissue.
3. Neurons have processes that receive (dendrites) and send (axons) bioelectric signals (neurotransmitters) that cross spaces (synapses) between them.
4. Organs of the nervous system are divided into the central and peripheral nervous systems.

10.2 GENERAL FUNCTIONS OF THE NERVOUS SYSTEM (PAGE 355)

1. Sensory receptors detect changes in internal and external body conditions.
2. Integrative functions gather sensory information and make decisions that affect motor functions.
3. Motor impulses stimulate effectors to respond.
   a. The motor portion of the PNS that carries out voluntary activities is the somatic nervous system.
   b. The motor portion of the PNS that carries out involuntary activities is the autonomic nervous system.

10.3 DESCRIPTION OF CELLS OF THE NERVOUS SYSTEM (PAGE 356)

1. Neurons vary in size, shape, sizes and lengths of axons and dendrites, and number of dendrites.
2. A neuron includes a cell body, cell processes, and the organelles usually found in cells.
3. Neurofibrils support axons.
4. Chromatophilic substance is mostly rough ER and is scattered throughout the cytoplasm of neurons.
5. Dendrites and the cell body provide receptive surfaces.
6. A single axon arises from the cell body and may be enclosed in a myelin sheath and a neurilemma.
7. White matter consists of myelinated axons, and gray matter consists of unmyelinated axons and cell bodies.

10.4 CLASSIFICATION OF CELLS OF THE NERVOUS SYSTEM (PAGE 359)

1. Classification of neurons
   a. Neurons are structurally classified as multipolar, bipolar, or unipolar.
   b. Neurons are functionally classified as sensory neurons, interneurons, or motor neurons.
2. Classification of neuroglia
   a. Neuroglia are abundant and have several functions.
   b. They fill spaces, support neurons, hold nervous tissue together, help metabolize glucose, help regulate potassium ion concentration, produce myelin, carry on phagocytosis, rid synapses of excess ions and neurotransmitters, nourish neurons, and stimulate synapse formation.
   c. They include astrocytes, oligodendrocytes, microglia, and ependymal cells in the CNS and Schwann cells and satellite cells in the PNS.
   d. Malfunctioning neuroglia can cause disease.
   e. Neuroglia are involved in axonal regeneration.
      (1) If a neuron cell body is injured, the neuron is likely to die; neural stem cells may proliferate and produce replacements.

10.5 CLINICAL APPLICATION

Drug Addiction

Drug abuse and addiction are long-standing problems. A 3,500-year-old Egyptian document decries reliance on opium. In the 1600s, a smokable form of opium enslaved many Chinese, and the Japanese and Europeans discovered the addictive nature of nicotine. During the American Civil War, morphine was a widely used painkiller; cocaine was introduced a short time later to relieve veterans addicted to morphine. Today, abuse of drugs intended for medical use continues. LSD was originally used in psychotherapy but was abused in the 1960s as a hallucinogen. PCP was an anesthetic before being abused in the 1980s.

Why do certain drugs compel a person to repeatedly use them, even when knowing that doing so is dangerous? Eating hot fudge sundaes is highly enjoyable, but we usually don't feel driven to consume them repeatedly. The biology of neurotransmission helps to explain drug addiction.

When a drug alters the activity of a neurotransmitter on a postsynaptic neuron, it either halts or enhances synaptic transmission. A drug that binds to a receptor, blocking a neurotransmitter from binding, is called an *antagonist*. A drug that activates the receptor, triggering an action potential, or that helps a neurotransmitter to bind, is called an *agonist*. The effect of a drug depends upon whether it is an antagonist or an agonist; on the particular behaviors the affected neurotransmitter normally regulates; and in which parts of the brain drugs affect neurotransmitters and their binding to receptors. Many addictive substances bind to receptors for the neurotransmitter dopamine, in a brain region called the nucleus accumbens.

With repeated use of an addictive substance, the number of receptors it targets can decline. When this happens, the person must use more of the drug to feel the same effect. For example, neural pathways that use the neurotransmitter norepinephrine control arousal, dreaming, and mood. Amphetamine enhances norepinephrine activity, thereby heightening alertness and mood. Amphetamine's structure is so similar to that of norepinephrine that it binds to norepinephrine receptors and triggers the same changes in the postsynaptic membrane.

Cocaine has a complex mechanism of action, both blocking reuptake of norepinephrine and binding to molecules that transport dopamine to postsynaptic cells. The drug valium causes relaxation and inhibits seizures and anxiety by helping GABA, an inhibitory neurotransmitter used in a third of the brain's synapses, bind to receptors on postsynaptic neurons. Valium is therefore a GABA agonist.

Nicotine causes addiction, which supplies enough of the other chemicals in cigarette smoke to destroy health. An activated form of nicotine binds postsynaptic nicotinic receptors that normally receive acetylcholine. When sufficient nicotine binds, a receptor channel opens, allowing positive ions in (fig. 10A). When a certain number of positive ions enter, the neuron releases dopamine from its other end, which provides the pleasurable feelings associated with smoking.

When a smoker increases the number of cigarettes smoked, the number of nicotinic receptors increases. This happens because of the way that the nicotine binding impairs the recycling of receptor proteins, so receptors are produced faster than they are taken apart. After a period of steady nicotine exposure, many of the receptors malfunction and no longer admit the positive ions that trigger the nerve impulse. This may be why as time goes on it takes more nicotine to produce the same effects—a hallmark of addiction. ■

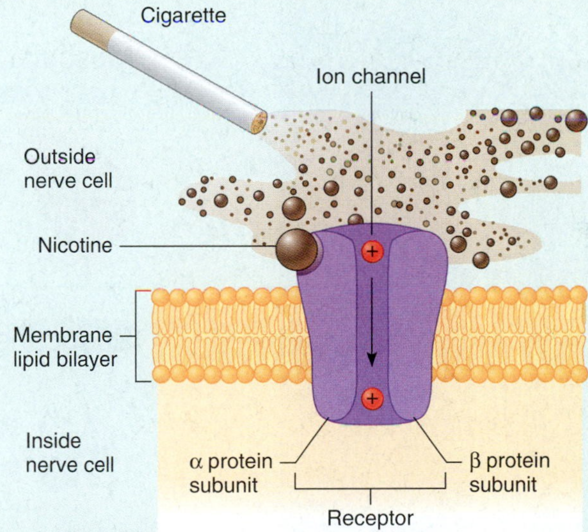

FIGURE 10A Nicotine binds and transiently alters postsynaptic receptors that normally bind the neurotransmitter acetylcholine. As a result, positive ions enter the cell, triggering dopamine release. With frequent smoking, receptors accumulate and soon become nonfunctional. Nicotine's effects on the nervous system are complex.

INNERCONNECTIONS | *Nervous System*

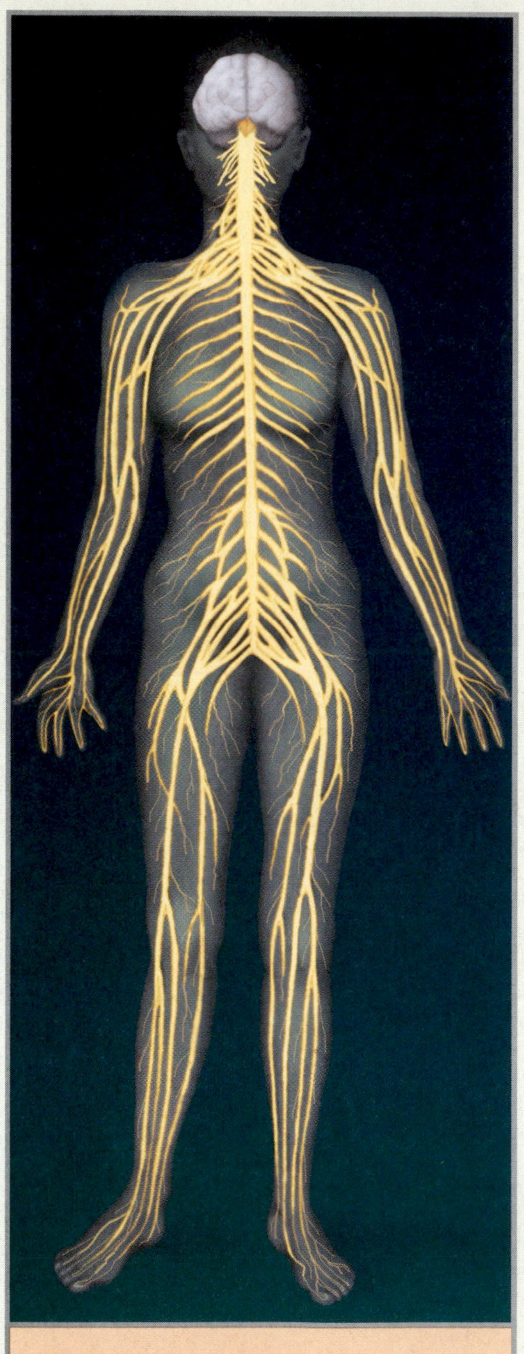

Nervous System

Nerves carry impulses that allow body systems to communicate.

Integumentary System

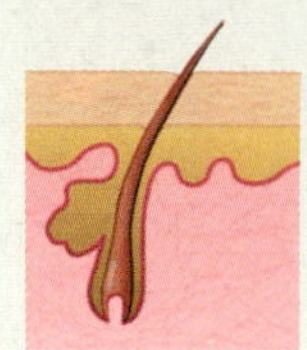

Sensory receptors provide the nervous system with information about the outside world.

Skeletal System

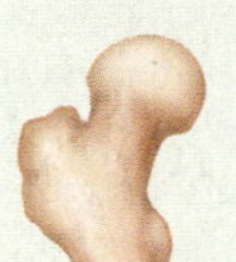

Bones protect the brain and spinal cord and help maintain plasma calcium, important to neuron function.

Muscular System

Nerve impulses control movement and carry information about the position of body parts.

Endocrine System

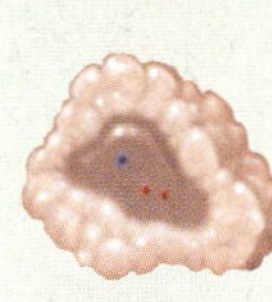

The hypothalamus controls secretion of many hormones.

Cardiovascular System

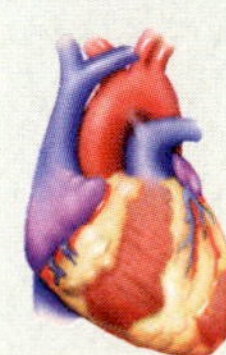

Nerve impulses help control blood flow and blood pressure.

Lymphatic System

Stress may impair the immune response.

Digestive System

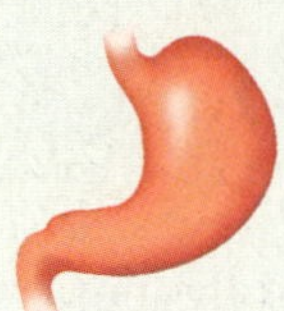

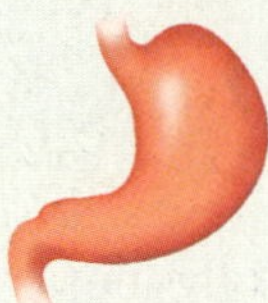

The nervous system can influence digestive function.

Respiratory System

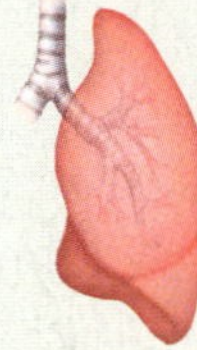

The nervous system alters respiratory activity to control oxygen levels and blood pH.

Urinary System

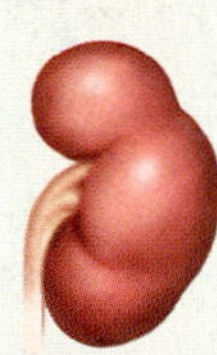

Nerve impulses affect urine production and elimination.

Reproductive System

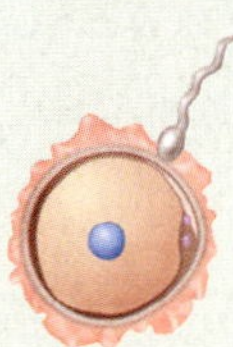

The nervous system plays a role in egg and sperm formation, sexual pleasure, childbirth, and nursing.

(2) If a peripheral axon is severed, its distal portion will die, but under the influence of nerve growth factors, the proximal portion may regenerate and reestablish connections, if a tube of connective tissue guides it.
(3) Significant regeneration is not likely in the CNS.

10.5 THE SYNAPSE (PAGE 365)

A synapse is a junction between two cells. A synaptic cleft is the gap between parts of two cells at a synapse. Synaptic transmission is the process by which the impulse in the presynaptic neuron signals the postsynaptic cell.

1. A nerve impulse travels along the axon to a synapse.
2. Axons have synaptic knobs at their distal ends that secrete neurotransmitters.
3. The neurotransmitter is released when a nerve impulse reaches the end of an axon, and the neurotransmitter diffuses across the synaptic cleft.
4. A neurotransmitter reaching a postsynaptic neuron or other cell may be excitatory or inhibitory.

10.6 CELL MEMBRANE POTENTIAL (PAGE 365)

A cell membrane is usually polarized as a result of an unequal distribution of ions on either side. Channels in membranes that allow passage of some ions but not others control ion distribution.

1. Distribution of ions
   a. Membrane ion channels, formed by proteins, may be always open or sometimes open and sometimes closed.
   b. Potassium ions pass more readily through resting neuron cell membranes than do sodium and calcium ions.
   c. A high concentration of sodium ions is on the outside of the membrane, and a high concentration of potassium ions is on the inside.
2. Resting potential
   a. Large numbers of negatively charged ions, which cannot diffuse through the cell membrane, are inside the cell.
   b. In a resting cell, more positive ions leave the cell than enter it, so the inside of the cell membrane develops a negative charge with respect to the outside.
3. Local potential changes
   a. Stimulation of a membrane affects its resting potential in a local region.
   b. The membrane is depolarized if it becomes less negative; it is hyperpolarized if it becomes more negative.
   c. Local potential changes are graded and subject to summation.
   d. Reaching threshold potential triggers an action potential.
4. Action potentials
   a. At threshold, sodium channels open and sodium ions diffuse inward, depolarizing the membrane.
   b. Slightly later, potassium channels open and potassium ions diffuse outward, repolarizing the membrane.
   c. This rapid change in potential is an action potential.
   d. Many action potentials can occur before active transport reestablishes the original resting potential.
   e. The propagation of action potentials along a nerve fiber is an impulse.
5. All-or-none response
   a. A nerve impulse is an all-or-none response. If a stimulus of threshold intensity is not applied to an axon, an action potential is not generated.
   b. All the impulses conducted on an axon are the same.
6. Refractory period
   a. The refractory period is a brief time following passage of a nerve impulse when the membrane is unresponsive to an ordinary stimulus.
   b. During the absolute refractory period, the membrane cannot be stimulated; during the relative refractory period, the membrane can be stimulated with a high-intensity stimulus.
7. Impulse conduction
   a. An unmyelinated axon conducts impulses that travel over its entire surface.
   b. A myelinated axon conducts impulses that travel from node to node.
   c. Impulse conduction is more rapid on myelinated axons with large diameters.

10.7 SYNAPTIC TRANSMISSION (PAGE 371)

Neurotransmitter molecules diffuse across the synaptic cleft and react with receptors in the postsynaptic neuron membrane.

1. Synaptic potentials
   a. Some neurotransmitters can depolarize the postsynaptic membrane, possibly triggering an action potential. This is an excitatory postsynaptic potential (EPSP).
   b. Others hyperpolarize the membrane, inhibiting an action potential. This is an inhibitory postsynaptic potential (IPSP).
   c. EPSPs and IPSPs are summed in a trigger zone of the neuron.
2. Neurotransmitters
   a. The nervous system produces at least thirty types of neurotransmitters.
   b. Calcium ions diffuse into synaptic knobs in response to action potentials, releasing neurotransmitters.
   c. Neurotransmitters are quickly decomposed or removed from synaptic clefts.
3. Neuropeptides
   a. Neuropeptides are chains of amino acids.
   b. Some neuropeptides are neurotransmitters or neuromodulators.
   c. They include enkephalins, endorphins, and substance P.

10.8 IMPULSE PROCESSING (PAGE 374)

The way impulses are processed reflects the organization of neurons in the brain and spinal cord.

1. Neuronal pools
   a. Neurons are organized into pools in the CNS.
   b. Each pool receives, processes, and may conduct impulses away.
   c. Each neuron in a pool may receive excitatory and inhibitory stimuli.
   d. A neuron is facilitated when it receives subthreshold stimuli and becomes more excitable.
2. Convergence
   a. Impulses from two or more axons may converge on a single postsynaptic neuron.
   b. Convergence enables a neuron to sum impulses from different sources.
3. Divergence
   a. Impulses from a presynaptic neuron may reach several postsynaptic neurons.
   b. Divergence amplifies impulses.

CHAPTER ASSESSMENTS

10.1 Introduction

1 Describe how the nervous system detects change associated with the body and reacts to that change to maintain homeostasis. (p. 354)

2 Distinguish between neurons and neuroglia. (p. 354)

3 Which of the following descriptions is accurate? (p. 354)
- a. A neuron has a single dendrite, which sends information.
- b. A neuron has a single axon, which sends information.
- c. A neuron has many axons, which receive information.
- d. A neuron has many dendrites, which send information.

4 Explain the difference between the central nervous system (CNS) and the peripheral nervous system (PNS). (p. 354)

10.2 General Functions of the Nervous System

5 List three general functions of the nervous system. (p. 355)

6 Distinguish a sensory receptor from an effector. (p. 355)

7 Distinguish between the types of activities that the somatic and autonomic nervous systems control. (p. 356)

10.3 Description of Cells of the Nervous System

8 Match the part of a neuron on the left with the description on the right (p. 356):

(1) dendrites	A. fine threads in an axon
(2) chromatophilic substance	B. part of neuron from which axon and dendrites extend
(3) axon	C. highly branched, multiple processes that may have spines
(4) cell body	D. sends nerve impulses
(5) neurofibrils	E. rough endoplasmic reticulum

9 Explain how Schwann cells encase large axons including the formation of myelin, the neurilemma, and the nodes of Ranvier. (p. 358)

10 What do Schwann cells and oligodendrocytes have in common, and how do they differ? (p. 358)

11 Distinguish between myelinated and unmyelinated axons. (p. 358)

10.4 Classification of Cells of the Nervous System

12 Describe the three types of neurons classified on the basis of structure. (p. 360)

13 Describe the three types of neurons classified on the basis of function (p. 360)

14 List six functions of neuroglia. (p. 361)

15 Describe the neuroglia of the CNS. (p. 361)

16 Explain how malfunctioning neuroglia can harm health. (p. 363)

17 Describe the neuroglia of the PNS. (p. 363)

18 Explain how an injured neuron may regenerate. (p. 364)

10.5 The Synapse

19 The _________ _________ brings the impulse to the synapse, whereas the _________ _________ on the other side of the synapse is stimulated or inhibited as a result of the synaptic transmission. (p. 365)

20 Explain how information is passed from a presynaptic neuron to a postsynaptic cell. (p. 365)

21 Diffusion of which of the following ions into the synaptic knob triggers the release of neurotransmitter? (p. 365)
- a. Na^{+}
- b. Ca^{+2}
- c. Cl^{-}
- d. K^{+}

10.6 Cell Membrane Potential

22 Define *resting potential.* (p. 366)

23 Distinguish among polarized, hyperpolarized, and depolarized. (p. 368)

24 Explain why the "trigger zone" of a neuron is named as such. (p. 368)

25 List in correct order the changes that occur during an action potential. (p. 368)

26 Explain the relationship between an action potential and a nerve impulse. (p. 369)

27 Define *refractory period.* (p. 370)

28 Explain the importance of the nodes of Ranvier and conduction in myelinated fibers as opposed to conduction in unmyelinated fibers. (p. 371)

10.7 Synaptic Transmission

29 Distinguish between excitatory and inhibitory postsynaptic potentials. (p. 371)

30 Explain how enzymes within synaptic clefts and reuptake of neurotransmitter prevents continuous stimulation of the postsynaptic cell. (p. 374):

10.8 Impulse Processing

31 Explain what determines the output of a neuronal pool in terms of input neurons, excitation, and inhibition. (p. 374)

32 Define *facilitation.* (p. 375)

33 Distinguish between convergence and divergence. (p. 375)

INTEGRATIVE ASSESSMENTS/CRITICAL THINKING

OUTCOMES 10.3, 10.4

1. Why are rapidly growing cancers that originate in nervous tissue more likely to be composed of neuroglia than of neurons?

OUTCOMES 10.3, 10.4, 10.6

2. In Tay-Sachs disease, an infant rapidly loses nervous system functions as neurons in the brain become covered in too much myelin. In multiple sclerosis, cells in the CNS have too little myelin. Identify the type of neuroglia implicated in each of these conditions.

OUTCOMES 10.4, 10.5, 10.7

3. How would you explain the following observations?
   a. When motor nerve fibers in the leg are severed, the muscles they innervate become paralyzed; however, in time, control over the muscles often returns.
   b. When motor nerve fibers in the spinal cord are severed, the muscles they control become permanently paralyzed.

OUTCOMES 10.5, 10.6, 10.7

4. Drugs that improve early symptoms of Alzheimer disease do so by slowing the breakdown of acetylcholine in synaptic clefts in certain parts of the brain. From this information, suggest a neurotransmitter imbalance that lies behind Alzheimer disease.

OUTCOME 10.6

5. What might be deficient in the diet of a pregnant woman complaining of leg muscle cramping? How would you explain this to her?

OUTCOME 10.6

6. People who inherit familial periodic paralysis often develop very low blood potassium concentrations. How would you explain that the paralysis may disappear quickly when potassium ions are administered intravenously?

WEB CONNECTIONS

Be sure to visit the text website at **www.mhhe.com/shier12** for answers to chapter assessments, additional quizzes, and interactive learning exercises.

ANATOMY & PHYSIOLOGY REVEALED

Anatomy & Physiology Revealed® (APR) includes cadaver photos that allow you to peel away layers of the human body to reveal structures beneath the surface. This program also includes animations, radiologic imaging, audio pronunciations, and practice quizzing. Check out **www.aprevealed.com**. APR has been proven to help improve student grades!

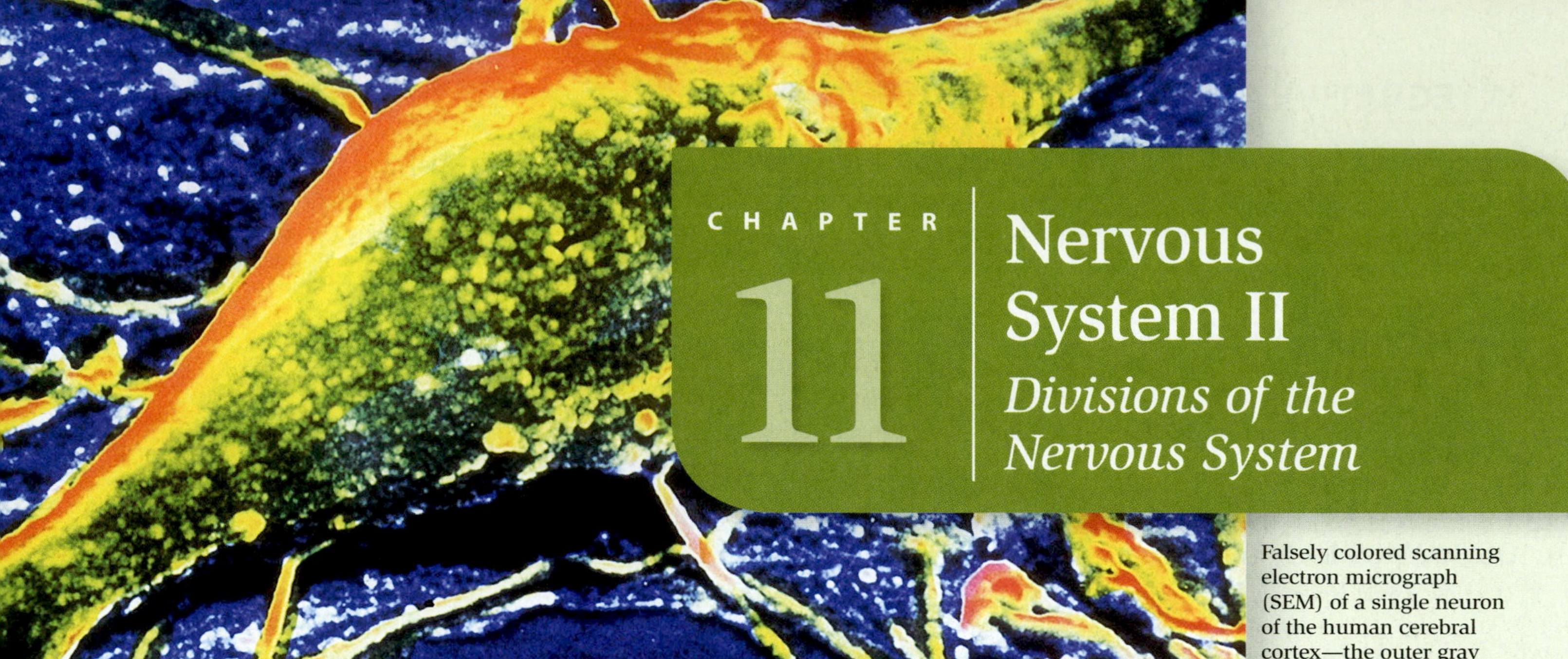

CHAPTER 11

Nervous System II

Divisions of the Nervous System

Falsely colored scanning electron micrograph (SEM) of a single neuron of the human cerebral cortex—the outer gray matter of the brain (7,200×).

UNDERSTANDING WORDS

cephal-, head: en*cephal*itis—inflammation of the brain.

chiasm-, cross: optic *chiasma*—X-shaped structure produced by the crossing over of optic nerve fibers.

flacc-, flabby: *flacc*id paralysis—paralysis characterized by loss of tone in muscles innervated by damaged axons.

funi-, small cord or fiber: *funi*culus—major nerve tract or bundle of myelinated axons within the spinal cord.

gangli-, swelling: *gangli*on—mass of neuron cell bodies.

mening-, membrane: *mening*es—membranous coverings of the brain and spinal cord.

plex-, interweaving: choroid *plexus*—mass of specialized capillaries associated with spaces in the brain.

LEARNING OUTCOMES

After you have studied this chapter, you should be able to:

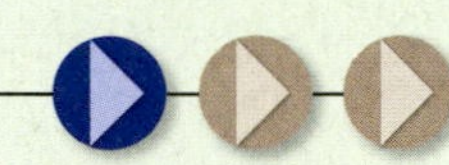

11.1 Introduction

1. Describe the relationship among the brain, brainstem, and spinal cord. (p. 384)

11.2 Meninges

2. Describe the coverings of the brain and spinal cord. (p. 384)

11.3 Ventricles and Cerebrospinal Fluid

3. Discuss the formation and function of cerebrospinal fluid. (p. 386)

11.4 Spinal Cord

4. Describe the structure of the spinal cord and its major functions. (p. 389)
5. Describe a reflex arc and reflex behavior. (p. 389)

11.5 Brain

6. Describe the development of the major parts of the brain and explain the functions of each part. (p. 398)
7. Distinguish among motor, sensory, and association areas of the cerebral cortex. (p. 401)
8. Discuss hemisphere dominance. (p. 403)
9. Explain the stages in memory storage. (p. 404)
10. Explain the functions of the limbic system and the reticular formation. (p. 407)

11.6 Peripheral Nervous System

11. Distinguish between the major parts of the peripheral nervous system. (p. 411)
12. Describe the structure of a peripheral nerve and how its fibers are classified. (p. 412)
13. Identify the cranial nerves and list their major functions. (p. 414)
14. Explain how spinal nerves are named and their functions. (p. 418)

11.7 Autonomic Nervous System

15. Characterize the autonomic nervous system. (p. 424)
16. Distinguish between the sympathetic and the parasympathetic divisions of the autonomic nervous system. (p. 425)
17. Describe a sympathetic and a parasympathetic nerve pathway. (p. 426)
18. Explain how the autonomic neurotransmitters differently affect visceral effectors. (p. 428)

11.8 Life-Span Changes

19. Describe aging-associated changes in the nervous system. (p. 431)

 LEARN  PRACTICE  ASSESS

FROM PHINEAS GAGE TO TERRY SCHIAVO

September 13, 1848, was a momentous day for Phineas Gage, a young man who worked in Vermont smoothing out terrain for railroad tracks. To blast away rock, he would drill a hole, fill it with gunpowder, cover that with sand, insert a fuse, and then press down with an iron rod called a tamping iron. The ensuing explosion would shatter the rock.

On that fateful September day, Gage began pounding on the tamping iron before his co-worker had put down the sand. The gunpowder exploded outward, slamming the inch-thick, 40-inch-long iron rod straight through Gage's skull. It pierced his brain like an arrow propelled through a soft melon, shooting out the other side of his head. Remarkably, Gage stood up just a few moments later, fully conscious and apparently unharmed.

Gage *was* harmed in the freak accident, but in ways so subtle that they were not at first evident. His friends reported that "Gage was no longer Gage." Although retaining his intellect and abilities to move, speak, learn, and remember, Gage's personality dramatically changed. Once a trusted, honest, and dedicated worker, the 25-year-old became irresponsible, shirking work, cursing, and pursuing what his doctor termed "animal propensities." Researchers as long ago as 1868 hypothesized that the tamping iron had ripped out a part of Gage's brain controlling personality. In 1994, computer analysis more precisely pinpointed the damage to the famous Gage brain, which, along with the tamping iron, went to a museum at Harvard University. Reconstruction of the trajectory of the tamping iron localized two small areas in the front of the brain that control rational decision-making and processing of emotion.

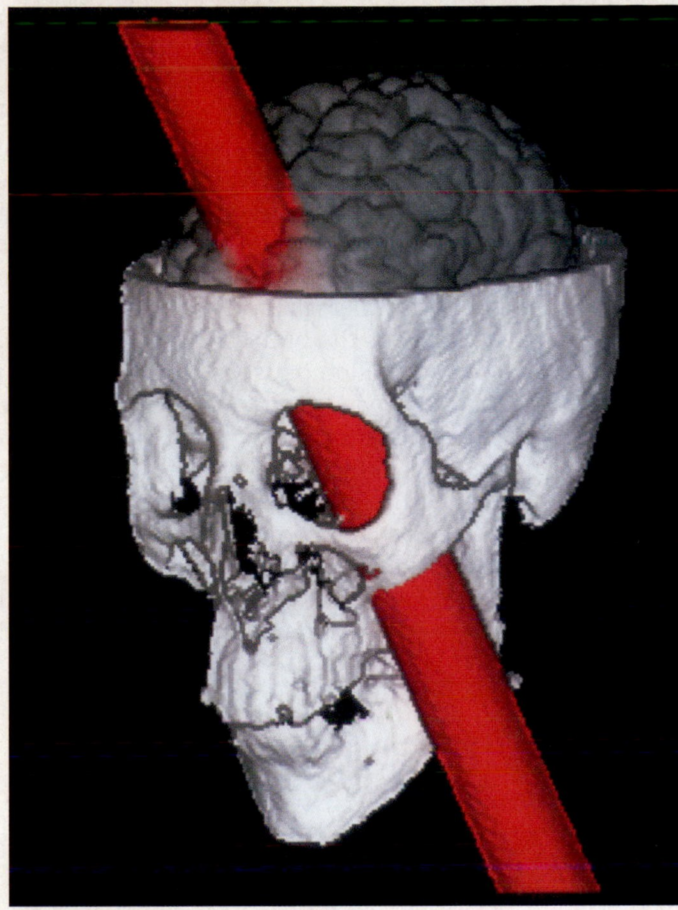

A rod that blasted through the head of a young railway worker has taught us much about the biology of personality.

More than a hundred years after Gage's accident, in 1975, 21-year-old Karen Ann Quinlan drank an alcoholic beverage after taking a prescription sedative, and her heart and lungs stopped functioning. When found, Quinlan had no pulse, was not breathing, had dilated pupils, and was unresponsive. Cardiopulmonary resuscitation restored her pulse, but at the hospital, she was placed on a ventilator. Within twelve hours, some functions returned—her pupils constricted, she moved, gagged, grimaced, and even opened her eyes. Within a few months, she could breathe unaided for short periods.

Quinlan's responses were random and not purposeful, and she was apparently unaware of herself and her environment, so she was said to be in a *persistent vegetative state.* Her basic life functions were intact, but she had to be fed and given water intravenously. Fourteen months after Quinlan took the pills and alcohol, her parents made a request that launched the right-to-die movement. They asked that Quinlan be taken off of life support. Doctors removed Quinlan's ventilator, and she lived for nine more years in a nursing home before dying of infection. She never regained awareness.

Throughout the Quinlan family's ordeal, researchers tried to fathom what had happened. A CAT scan performed five years after the accident showed atrophy in two major brain regions, the cerebrum and the cerebellum. But when researchers analyzed Karen Ann Quinlan's brain in 1993, they were surprised. The most severely damaged part of her brain was the thalamus, an area thought to function merely as a relay station to higher brain structures. Quinlan's tragic case revealed that the thalamus is also important in processing thoughts, in providing the awareness and responsiveness that makes a person a conscious being. In 2005, a similar case arose concerning a young woman named Terry Schiavo. Many people found the images of her facial movements on television disturbing. After much debate and discussion, she was permitted to die. On autopsy, her brain was found to be grossly degenerated.

The cases of Gage, Quinlan, and Schiavo dramatically illustrate the function of the human brain by revealing what can happen when it is damaged. Nearly every aspect of our existence depends upon the brain and other parts of the nervous system, from thinking and feeling; to sensing, perceiving, and responding to the environment; to carrying out vital functions such as breathing and heartbeat. This chapter describes how the billions of neurons and neuroglia comprising the nervous system interact in ways that enable us to survive and to enjoy the world around us. ■

11.1 INTRODUCTION

The central nervous system (CNS) consists of the brain and the spinal cord. The **brain** is the largest and most complex part of the nervous system. It oversees many aspects of physiology, such as sensation and perception, movement, and thinking. The brain includes the two cerebral hemispheres, the diencephalon, the brainstem (which attaches the brain to the spinal cord), and the cerebellum, all described in detail in the section 11.5 Brain. The brain includes about one hundred billion (10^{11}) multipolar neurons and countless branches of the axons (nerve fibers) by which these neurons communicate with each other and with neurons elsewhere in the nervous system.

The brainstem connects the brain and spinal cord and allows two-way communication between them. The spinal cord, in turn, provides two-way communication between the CNS and the peripheral nervous system (PNS).

Bones, membranes, and fluid surround the organs of the CNS. More specifically, the brain lies in the cranial cavity of the skull, whereas the spinal cord occupies the vertebral canal in the vertebral column. Beneath these bony coverings, membranes called **meninges,** located between the bone and the soft tissues of the nervous system, protect the brain and spinal cord (fig. 11.1*a*).

11.2 MENINGES

The meninges (sing., *meninx*) have three layers—dura mater, arachnoid mater, and pia mater (fig. 11.1*b*). The **dura mater** is the outermost layer. It is primarily composed of tough, white, dense connective tissue and contains many blood vessels and nerves. It attaches to the inside of the cranial cavity and forms the internal periosteum of the surrounding skull bones (see reference plate 13).

TABLE 11.1 | Partitions of the Dura Mater

Partition	Location
Falx cerebelli	Separates the right and left cerebellar hemispheres
Falx cerebri	Extends downward into the longitudinal fissure, and separates the right and left cerebral hemispheres (fig. 11.1*b*)
Tentorium cerebelli	Separates the occipital lobes of the cerebrum from the cerebellum (fig. 11.1 *a*)

In some regions, the dura mater extends inward between lobes of the brain and forms supportive and protective partitions (table 11.1). In other areas, the dura mater splits into two layers, forming channels called *dural sinuses,* shown in figure 11.1*b*. Venous blood flows through these channels as it returns from the brain to vessels leading to the heart.

The dura mater continues into the vertebral canal as a strong, tubular sheath that surrounds the spinal cord. It is attached to the cord at regular intervals by a band of pia mater (denticulate ligaments) that extends the length of the spinal cord on either side. The dural sheath terminates as a blind sac at the level of the second sacral vertebra, below the end of the spinal cord. The sheath around the spinal cord is not attached directly to the vertebrae but is separated by an *epidural space,* which lies between the dural sheath and the bony walls (fig. 11.2). This space contains blood vessels, loose connective tissue, and adipose tissue that pad the spinal cord.

A blow to the head may rupture some blood vessels associated with the brain, and the escaping blood may collect beneath the dura mater. This condition, called *subdural hematoma,* can increase pressure between the rigid bones of the skull and the soft tissues of the brain. Unless the accumulating blood is promptly evacuated, compression of the brain may lead to functional losses or even death.

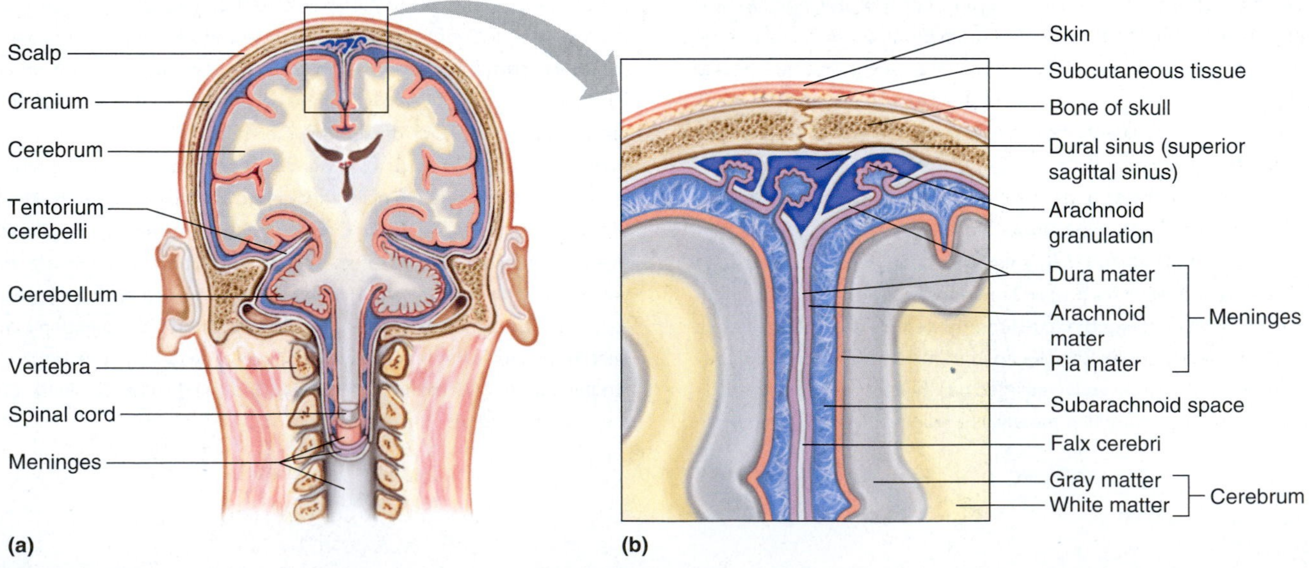

FIGURE 11.1 Meninges. (*a*) Membranes called meninges enclose the brain and spinal cord. (*b*) The meninges include three layers: dura mater, arachnoid mater, and pia mater.

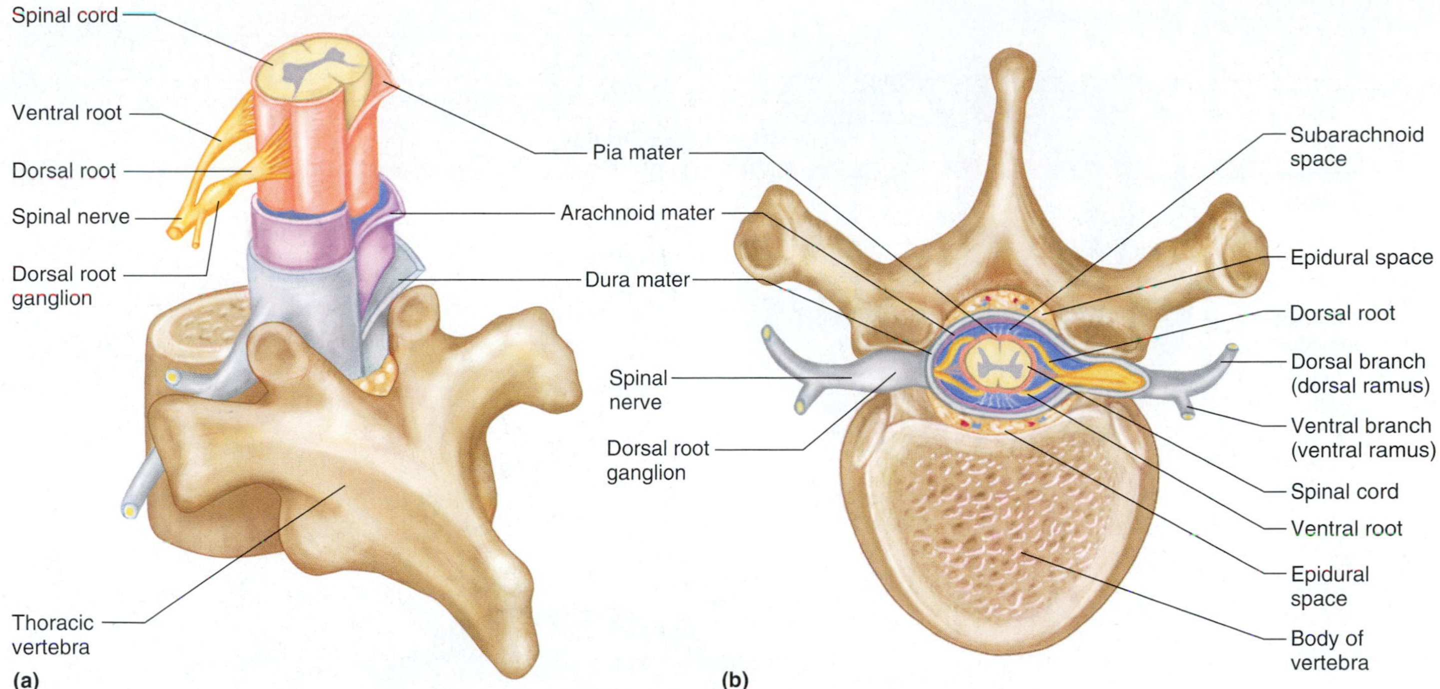

FIGURE 11.2 Meninges of the spinal cord. (*a*) The dura mater ensheaths the spinal cord. (*b*) Tissues forming a protective pad around the cord fill the epidural space between the dural sheath and the bone of the vertebra.

The **arachnoid mater** is a thin, weblike membrane that lacks blood vessels and is located between the dura and pia maters. It spreads over the brain and spinal cord but generally does not dip into the grooves and depressions on their surfaces. Many thin strands extend from its undersurface and are attached to the pia mater. Between the arachnoid and pia maters is a *subarachnoid space,* which contains the clear, watery **cerebrospinal fluid** (ser″ĕ-bro-spi′nal floo′id), or **CSF.**

The **pia mater** is thin and contains many nerves, as well as blood vessels that nourish the underlying cells of the brain and spinal cord. The pia mater is attached to the surfaces of these organs and follows their irregular contours, passing over the high areas and dipping into the depressions.

Meningitis is an inflammation of the meninges. Bacteria or viruses that infect the cerebrospinal fluid are typical causes of this condition. Meningitis may affect the dura mater, but it is more commonly limited to the arachnoid and pia maters. Meningitis most often affects infants and children and is serious. Complications include loss of vision, loss of hearing, paralysis, and mental retardation. It may be fatal.

PRACTICE

1 Describe the meninges.
2 Name the layers of the meninges.
3 Explain the location of cerebrospinal fluid.

11.3 VENTRICLES AND CEREBROSPINAL FLUID

Interconnected cavities called **ventricles** (ven′trĭ-klz) lie in the cerebral hemispheres and brainstem (fig. 11.3 and reference plates 13 and 14). These spaces are continuous with the central canal of the spinal cord and are filled with CSF.

The largest ventricles are the two *lateral ventricles.* The first ventricle is in the left cerebral hemisphere and the second ventricle is in the right cerebral hemisphere. They extend anteriorly and posteriorly into the cerebral hemispheres.

A narrow space that constitutes the *third ventricle* is in the midline of the brain beneath the corpus callosum, which is a bridge of axons that links the two cerebral hemispheres. This ventricle communicates with the lateral ventricles through openings (*interventricular foramina*) in its anterior end.

The *fourth ventricle* is in the brainstem, just anterior to the cerebellum. A narrow canal, the *cerebral aqueduct* (aqueduct of Sylvius), connects it to the third ventricle and passes lengthwise through the brainstem. This ventricle is continuous with the central canal of the spinal cord and has openings in its roof that lead into the subarachnoid space of the meninges.

Tiny, reddish cauliflowerlike masses of specialized capillaries from the pia mater, called **choroid plexuses** (ko′roid plek′sus-ez), secrete CSF. These structures project into the cavities of the ventricles (fig. 11.4). A single layer of specialized ependymal cells (see chapter 10, p. 361) joined closely by tight junctions covers the choroid plexuses. In much the same way that astrocytes provide a barrier between the blood and

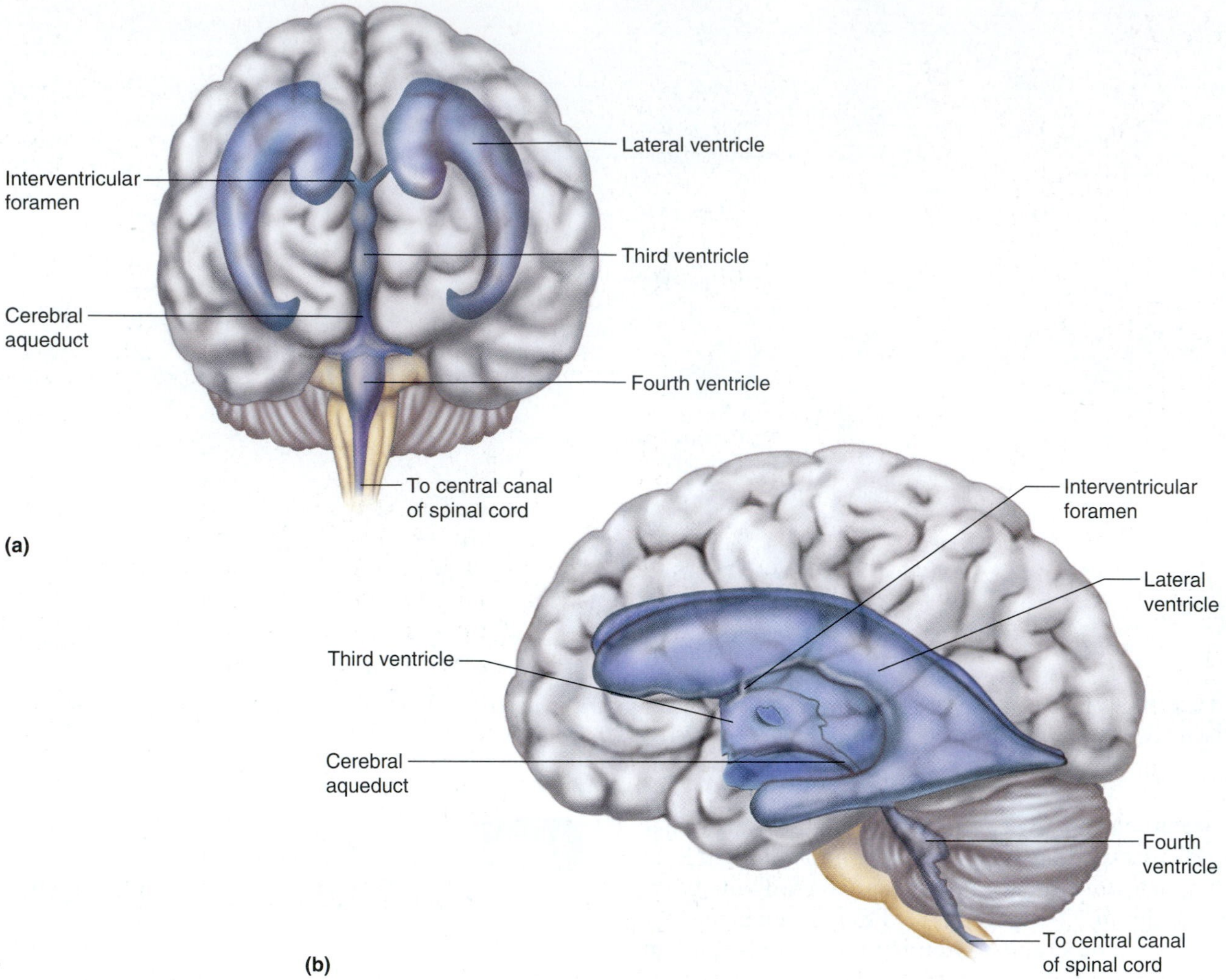

FIGURE 11.3 Ventricles in the brain. (*a*) Anterior view of the ventricles in the cerebral hemispheres and brainstem. (*b*) Lateral view.

the brain interstitial fluid (blood-brain barrier), ependymal cells block passage of water-soluble substances between the blood and the CSF (blood-CSF barrier). At the same time, the cells selectively transfer certain substances from the blood into the CSF by facilitated diffusion and transfer other substances by active transport (see chapter 3, pp. 93 and 95), regulating CSF composition.

Most CSF forms in the lateral ventricles, from where it slowly circulates into the third and fourth ventricles and into the central canal of the spinal cord. It also enters the subarachnoid space of the meninges by passing through the wall of the fourth ventricle near the cerebellum.

Humans secrete nearly 500 milliliters of CSF daily. However, only about 140 milliliters are in the nervous system at any time, because CSF is continuously reabsorbed into the blood through tiny, fingerlike structures called *arachnoid granulations* that project from the subarachnoid space into the blood-filled dural sinuses (see fig. 11.4).

CSF is a clear, somewhat viscid liquid that differs in composition from the fluid that leaves the capillaries in other parts of the body. Specifically, it contains a greater concentration of sodium and lesser concentrations of glucose and potassium than do other extracellular fluids. Its function is nutritive as well as protective. CSF helps maintain a stable ionic concentration in the CNS and provides a pathway to the blood for waste. The CSF may also supply information about the internal environment to autonomic centers in the hypothalamus and brainstem, because the fluid forms from blood plasma and therefore its composition reflects changes in body fluids. Clinical Application 11.1 discusses the pressure that CSF generates.

CSF occupies the subarachnoid space of the meninges, so it completely surrounds the brain and spinal cord. In effect, these organs float in the fluid. The CSF protects them by absorbing forces that might otherwise jar and damage their delicate tissues.

PRACTICE

4 Where are the ventricles of the brain located?

5 How does CSF form?

6 Describe the pattern of CSF circulation.

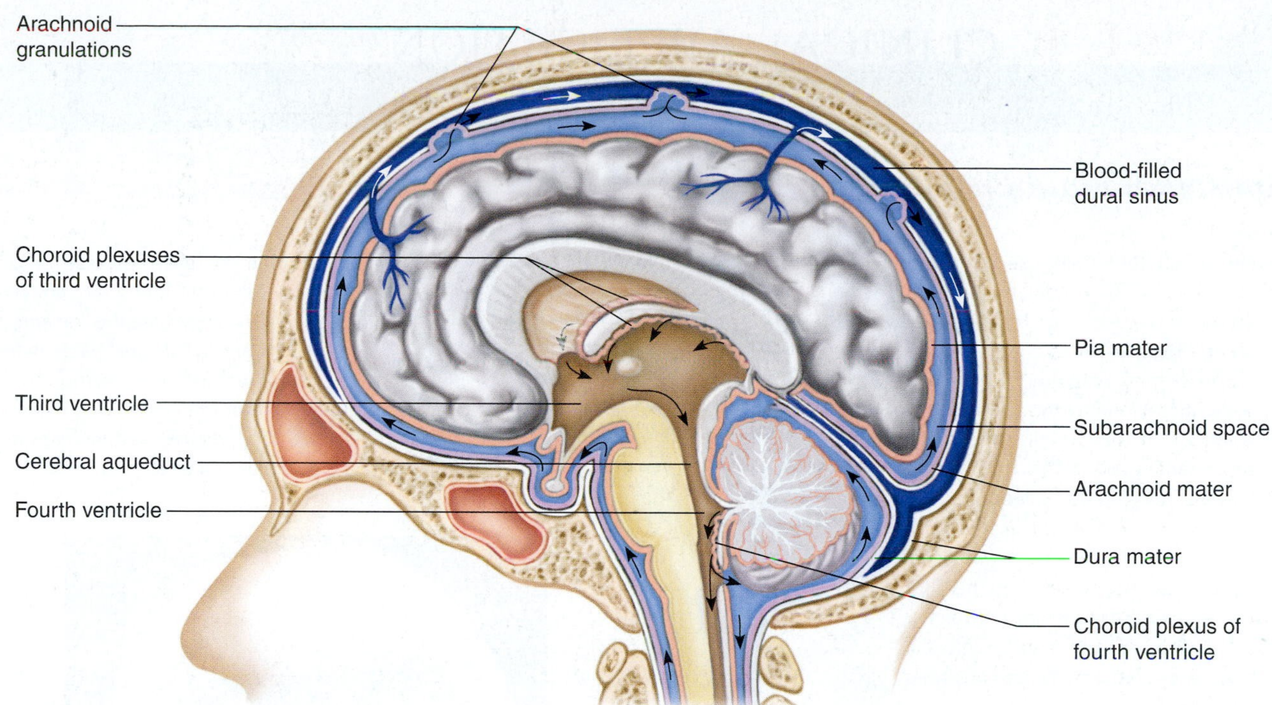

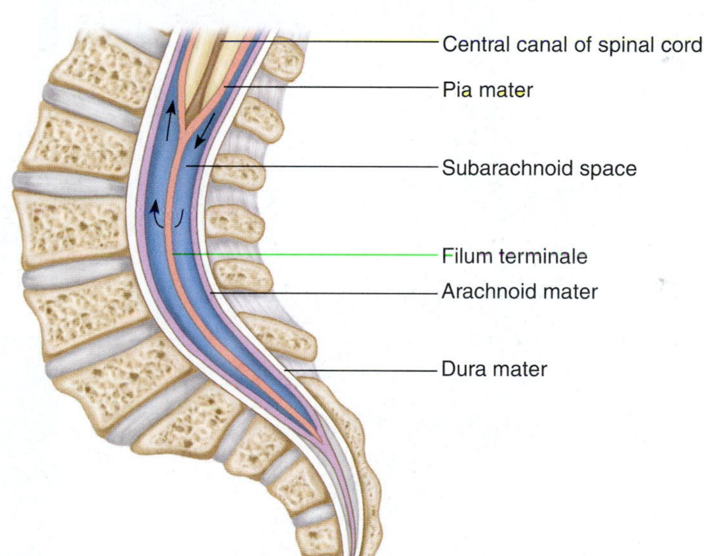

FIGURE 11.4 Choroid plexuses in ventricle walls secrete cerebrospinal fluid. The fluid circulates through the ventricles and central canal, enters the subarachnoid space, and is reabsorbed into the blood of the dural sinuses through arachnoid granulations. (Spinal nerves are not shown.)

> Chapters 9 and 10 distinguished between the term *nerve fiber*, which is part of a nerve cell, and *muscle fiber*, which refers to the entire muscle cell. "Nerve fiber" in the subsequent text is synonymous with axon.

11.4 SPINAL CORD

The **spinal cord** is a slender column of nervous tissue that is continuous with the brain and extends downward through the vertebral canal. The spinal cord originates where nervous tissue leaves the cranial cavity at the level of the foramen magnum (see reference plate 15). The cord tapers to a point and terminates near the intervertebral disc that separates the first and second lumbar vertebrae (fig. 11.5*a*).

Structure of the Spinal Cord

The spinal cord consists of thirty-one segments, each of which gives rise to a pair of **spinal nerves.** These nerves branch to various body parts and connect them with the CNS.

In the neck region, a thickening in the spinal cord, called the *cervical enlargement*, supplies nerves to the upper limbs. A similar thickening in the lower back, the *lumbar enlargement*, gives off nerves to the lower limbs. Just inferior to the

11.1 CLINICAL APPLICATION

Cerebrospinal Fluid Pressure

Cerebrospinal fluid (CSF) is secreted and reabsorbed continuously, so the fluid pressure in the ventricles remains relatively constant. However, infection, a tumor, or a blood clot can interfere with the fluid's circulation, increasing pressure in the ventricles (intracranial pressure or ICP). This can collapse cerebral blood vessels, retarding blood flow. Brain tissues forced against the skull may be injured.

A *lumbar puncture* (spinal tap) measures CSF pressure. A physician inserts a fine, hollow needle into the subarachnoid space between the third and fourth or between the fourth and fifth lumbar vertebrae—below the end of the spinal cord (fig. 11A). An instrument called a *manometer* measures the pressure of the fluid, usually about 130 millimeters of water (10 millimeters of mercury). At the same time, samples of CSF may be withdrawn and tested for abnormal constituents. Red blood cells in the CSF, for example, may indicate a hemorrhage in the central nervous system (CNS).

A temporary drain inserted into the subarachnoid space between the fourth and fifth lumbar vertebrae can relieve pressure. In a fetus or infant whose cranial sutures have not yet united, increasing ICP may enlarge the cranium, a condition called *hydrocephalus,* or "water on the brain" (fig. 11B). A shunt to relieve hydrocephalus drains fluid away from the cranial cavity and into the digestive tract, where it is either reabsorbed into the blood or excreted. ■

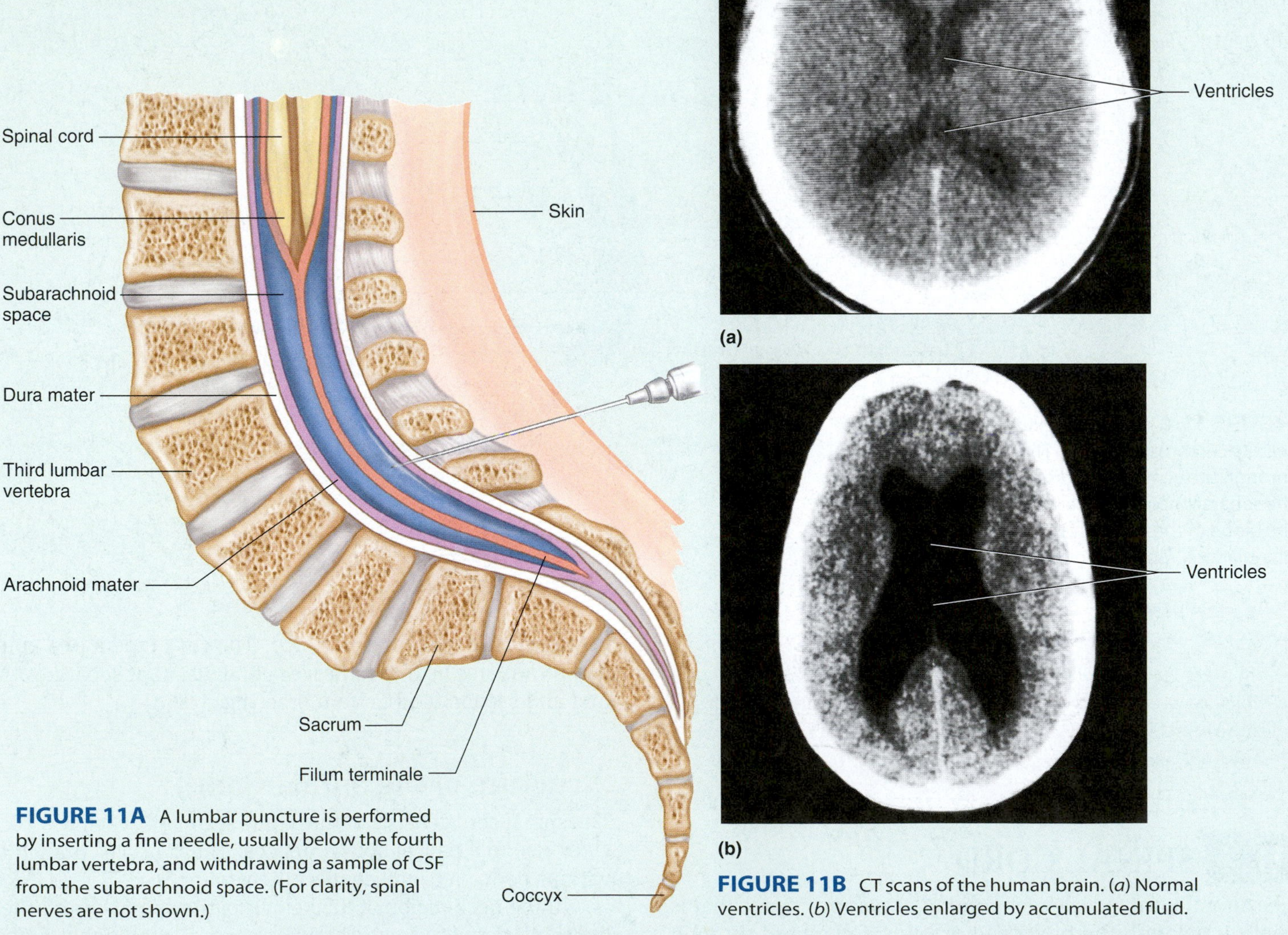

FIGURE 11A A lumbar puncture is performed by inserting a fine needle, usually below the fourth lumbar vertebra, and withdrawing a sample of CSF from the subarachnoid space. (For clarity, spinal nerves are not shown.)

FIGURE 11B CT scans of the human brain. (*a*) Normal ventricles. (*b*) Ventricles enlarged by accumulated fluid.

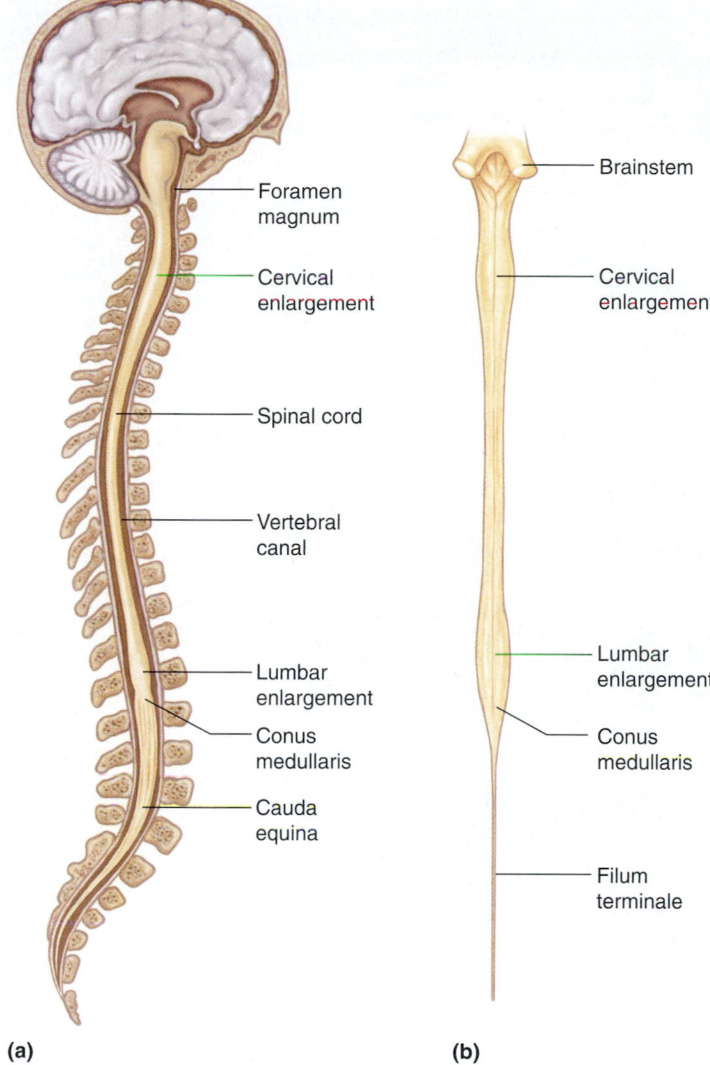

FIGURE 11.5 Spinal cord. (*a*) The spinal cord begins at the level of the foramen magnum. (*b*) Posterior view of the spinal cord with the spinal nerves removed.

lumbar enlargement, the spinal cord tapers to a structure called the *conus medullaris.* From this tip, nervous tissue, including axons of both motor and sensory neurons, extends downward to become spinal nerves at the remaining lumbar and sacral levels. Originating from among them, a thin cord of connective tissue descends to the upper surface of the coccyx. This cord is called the *filum terminale* (fig. 11.5*b*). The filum terminale and the spinal nerves below the conus medullaris form a structure that resembles a horse's tail, the *cauda equina.*

Two grooves, a deep *anterior median fissure* and a shallow *posterior median sulcus,* extend the length of the spinal cord, dividing it into right and left halves. A cross section of the cord (fig. 11.6) reveals that it consists of white matter surrounding a core of gray matter. The pattern the gray matter produces roughly resembles a butterfly with its wings outspread. The upper and lower wings of gray matter are called the *posterior horns* and the *anterior horns,* respectively. Between them on either side in some regions is a protrusion of gray matter called the *lateral horn.* Motor neurons with relatively large cell bodies in the anterior horns (anterior horn cells) give rise to axons that pass out through spinal nerves to various skeletal muscles. However, the majority of neurons in the gray matter are interneurons (see chapter 10, p. 358).

A horizontal bar of gray matter in the middle of the spinal cord, the *gray commissure,* connects the wings of the gray matter on the right and left sides. This bar surrounds the **central canal,** which is continuous with the ventricles of the brain and contains CSF. The central canal is prominent during embryonic development, but it becomes almost microscopic in adulthood.

The gray matter divides the white matter of the spinal cord into three regions on each side—the *anterior, lateral,* and *posterior funiculi.* Each column consists of longitudinal bundles of myelinated nerve fibers that comprise major nerve pathways called **nerve tracts.**

Functions of the Spinal Cord

The spinal cord has two main functions. First, it is a center for spinal reflexes. Second, it is a conduit for nerve impulses to and from the brain.

Reflex Arcs

Nerve impulses follow nerve pathways as they travel through the nervous system. The simplest of these pathways, including only a few neurons, constitutes a **reflex** (re′fleks) **arc.** Reflex arcs carry out the simplest responses—**reflexes.**

Recall that the nervous system receives sensory information, processes it, and initiates appropriate responses by activating effector organs. For example, as you read this book, your eyes send sensory information to your brain, where it processes the information, interprets its meaning, and even stores much of it in memory. For reading to continue, motor commands to muscles point the eyes at what you are reading and allow you to turn the pages. Some functions continue without your awareness, such as breathing and heartbeat.

To begin to understand how the nervous system does all of this, we will examine the simplest of the nervous system functions that reflect these processes—the reflexes. All reflexes share the basic reflex arc, as shown in figure 11.7*a*. A reflex arc begins with a **sensory receptor** at the dendritic end of a sensory neuron. Nerve impulses on these sensory neurons enter the CNS and constitute a sensory or afferent limb of the reflex. The CNS is a processing center. Afferent neurons may synapse with interneurons, which may in turn connect with other parts of the CNS. Afferent neurons or interneurons ultimately connect with motor neurons, whose fibers pass outward from the CNS to effectors. (It may help to remember that **eff**erent neurons control **eff**ector organs.)

Reflexes occur throughout the CNS. Those that involve the spinal cord are called spinal reflexes and reflect the simplest level of CNS function. Figure 11.7*b* shows the general components of a spinal reflex.

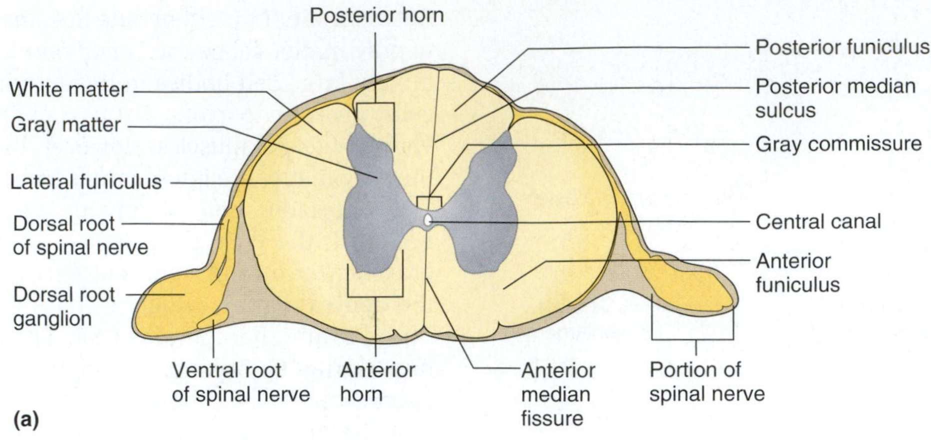

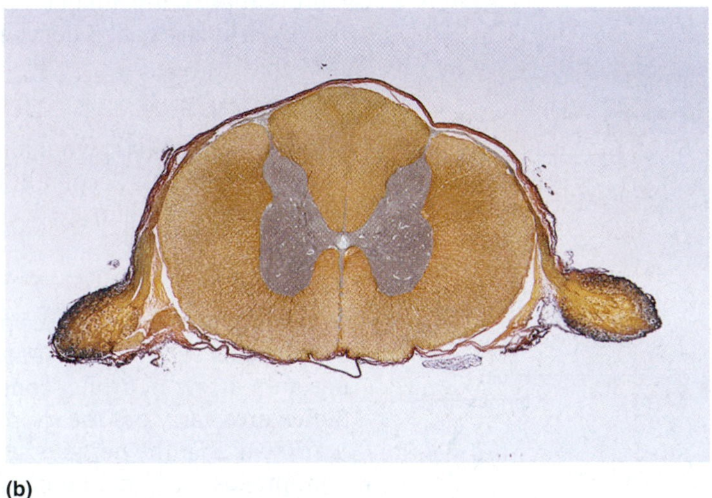

FIGURE 11.6 Spinal cord. (*a*) A cross section of the spinal cord. (*b*) Identify the parts of the spinal cord in this micrograph (7.5×).

Reflex Behavior

Reflexes are automatic responses to changes (stimuli) inside or outside the body. They help maintain homeostasis by controlling many involuntary processes such as heart rate, breathing rate, blood pressure, and digestion. Reflexes also carry out the automatic actions of swallowing, sneezing, coughing, and vomiting.

The *patellar reflex* (knee-jerk reflex) is an example of a simple monosynaptic reflex, so-called because it uses only two neurons—a sensory neuron communicating directly to a motor neuron. Striking the patellar ligament just below the patella initiates this reflex. The quadriceps femoris muscle group, attached to the patella by a tendon, is pulled slightly, stimulating stretch receptors in the muscle group. These receptors, in turn, trigger impulses that pass along the peripheral process (see fig. 10.7) of the axon of a unipolar sensory neuron, continuing along the central process of the axon into the lumbar region of the spinal cord. In the spinal cord, the sensory axon synapses with a motor neuron. An impulse is then triggered along the axon of the motor neuron and travels back to the quadriceps femoris. The muscles respond by contracting, and the reflex is completed as the leg extends (fig. 11.8).

The patellar reflex helps maintain an upright posture. For example, if a person is standing still and the knee begins to bend in response to gravity, the quadriceps femoris is stretched, the reflex is triggered, and the leg straightens again. Adjustments in the stretch receptors keep the reflex responsive at different muscle lengths.

Another type of reflex, called a *withdrawal reflex* (fig. 11.9), happens when a person touches something painful, as in stepping on a tack. Activated skin receptors send sensory impulses to the spinal cord. There the impulses pass on to interneurons of a reflex center and are directed to motor neurons. The motor neurons transmit signals to the flexor muscles of the leg and thigh, which contract in response, pulling the foot away from the painful stimulus. At the same time, some of the incoming impulses stimulate interneurons that

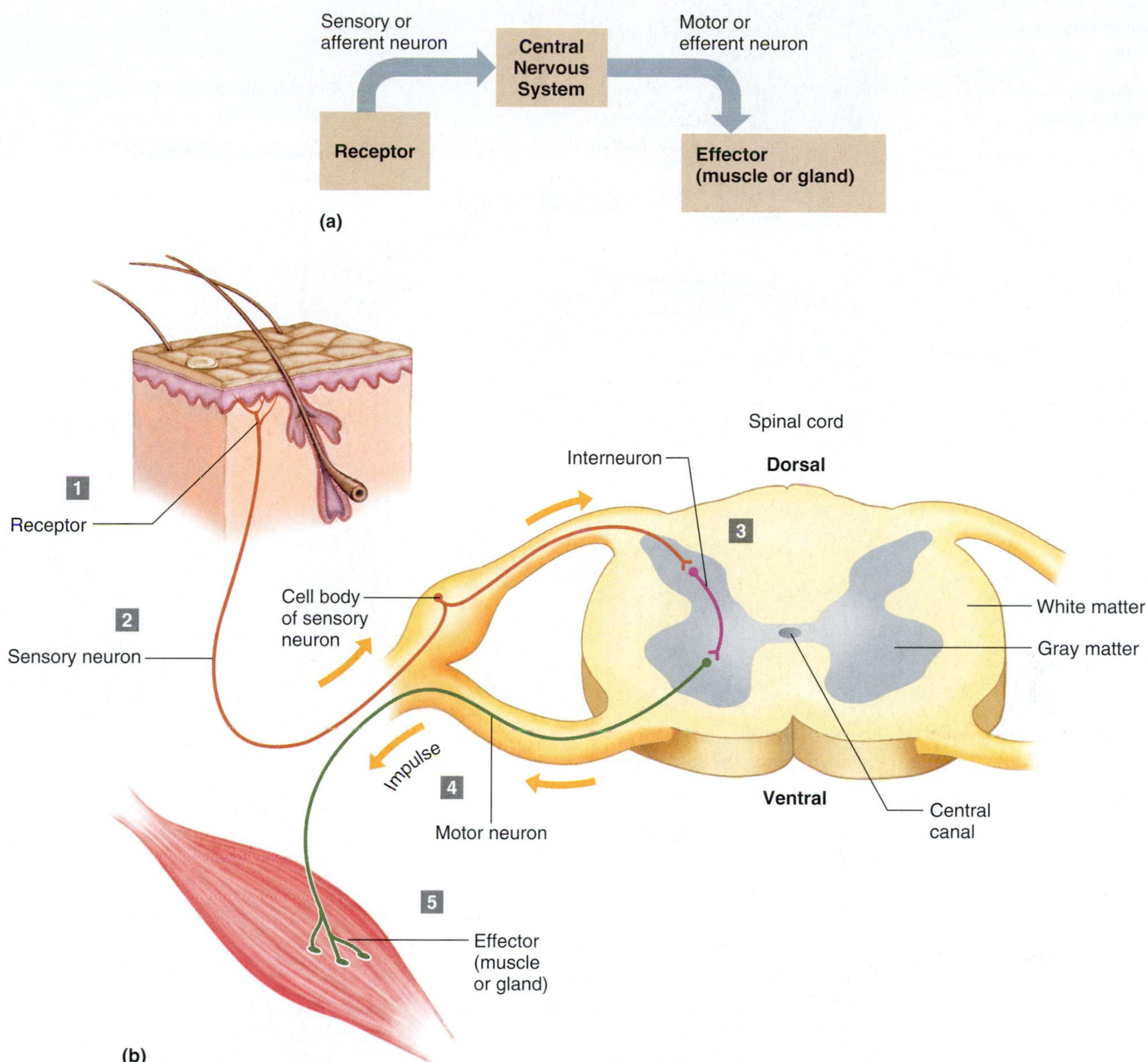

FIGURE 11.7 Reflex arc. (*a*) Schematic of a reflex arc. (*b*) A reflex arc usually includes a receptor (1), a sensory neuron (2), integration within the CNS involving at least one synapse (3), a motor neuron (4), and an effector (5). In this example of a spinal reflex, the integration center is in the spinal cord.

inhibit the action of the antagonistic extensor muscles (reciprocal innervation). This inhibition allows the flexor muscles to effectively withdraw the affected part (fig. 11.10).

While flexor muscles on the affected side (ipsilateral side) contract, the flexor muscles of the other limb (contralateral side) are inhibited. Furthermore, the extensor muscles on the contralateral side contract, helping to support the body weight shifted to that side. This phenomenon, called a *crossed extensor reflex,* is due to interneuron pathways in the reflex center of the spinal cord that allow sensory impulses arriving on one side of the cord to pass across to the other side and produce an opposite effect (fig. 11.10).

Concurrent with the withdrawal reflex, other interneurons in the spinal cord carry sensory impulses upward to the brain. The person becomes aware of the experience and may feel pain.

A withdrawal reflex protects because it prevents or limits tissue damage when a body part touches something potentially harmful. Table 11.2 summarizes the components of a reflex arc. Clinical Application 11.2 discusses some familiar reflexes.

PRACTICE

7 What is a nerve pathway?

8 Describe a reflex arc.

9 Define reflex.

10 Describe the actions that are part of a withdrawal reflex.

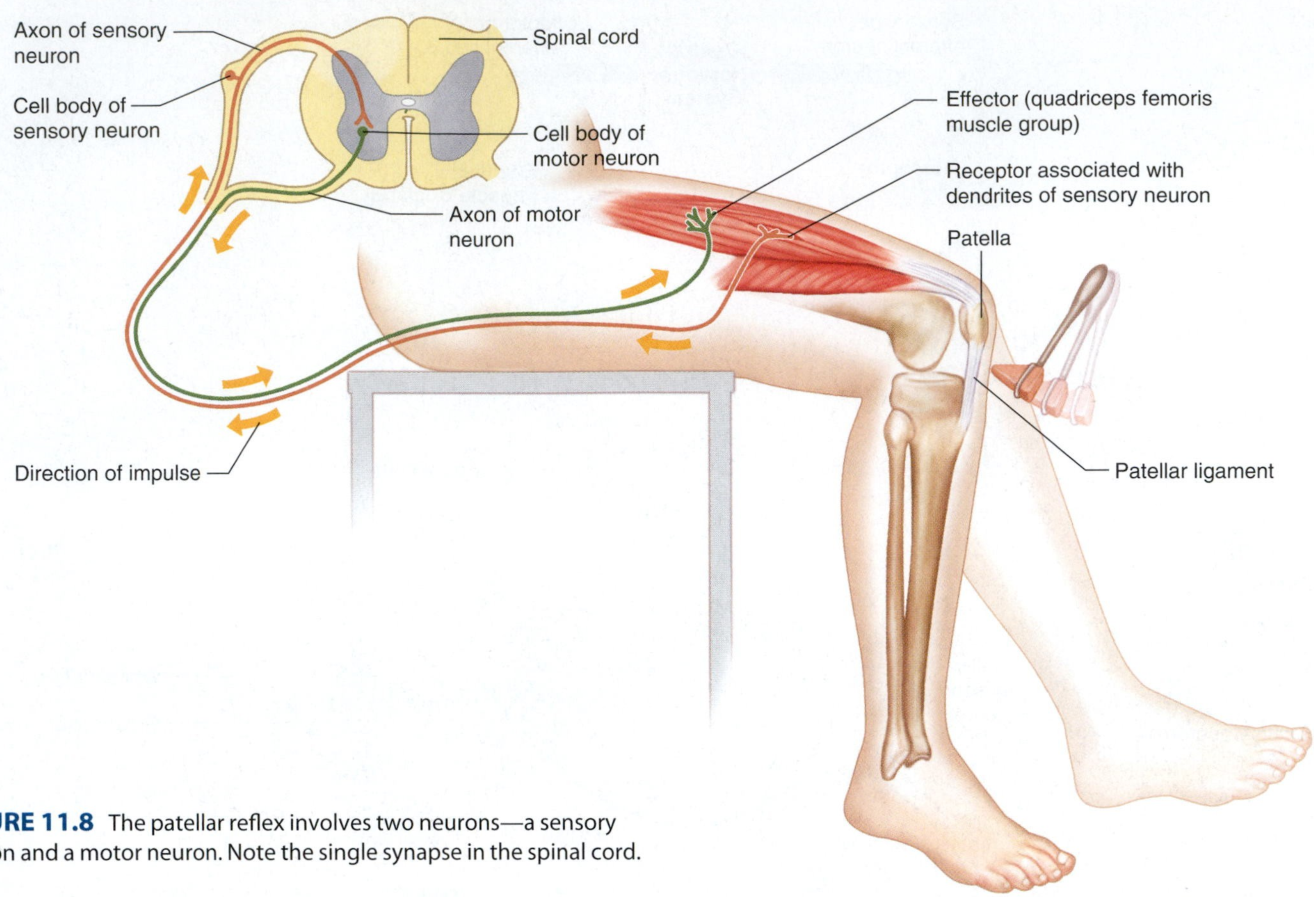

Axon of sensory neuron
Cell body of sensory neuron
Spinal cord
Cell body of motor neuron
Axon of motor neuron
Effector (quadriceps femoris muscle group)
Receptor associated with dendrites of sensory neuron
Patella
Patellar ligament
Direction of impulse

FIGURE 11.8 The patellar reflex involves two neurons—a sensory neuron and a motor neuron. Note the single synapse in the spinal cord.

Cell body of sensory neuron
Axon of sensory neuron
Interneuron
Direction of impulse
Spinal cord
Cell body of motor neuron
Axon of motor neuron
Effector (flexor muscle contracts and withdraws part being stimulated)
Dendrite of sensory neuron
Pain receptor in skin
Tack

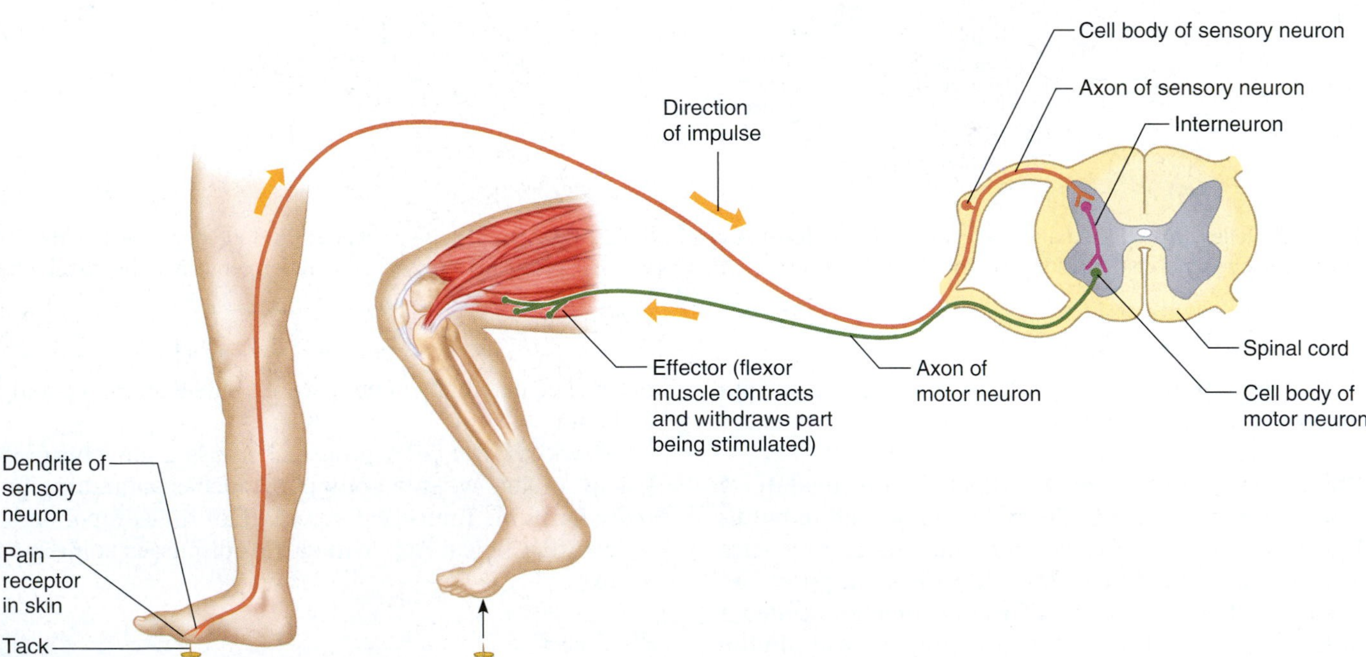

FIGURE 11.9 A withdrawal reflex involves a sensory neuron, an interneuron, and a motor neuron.

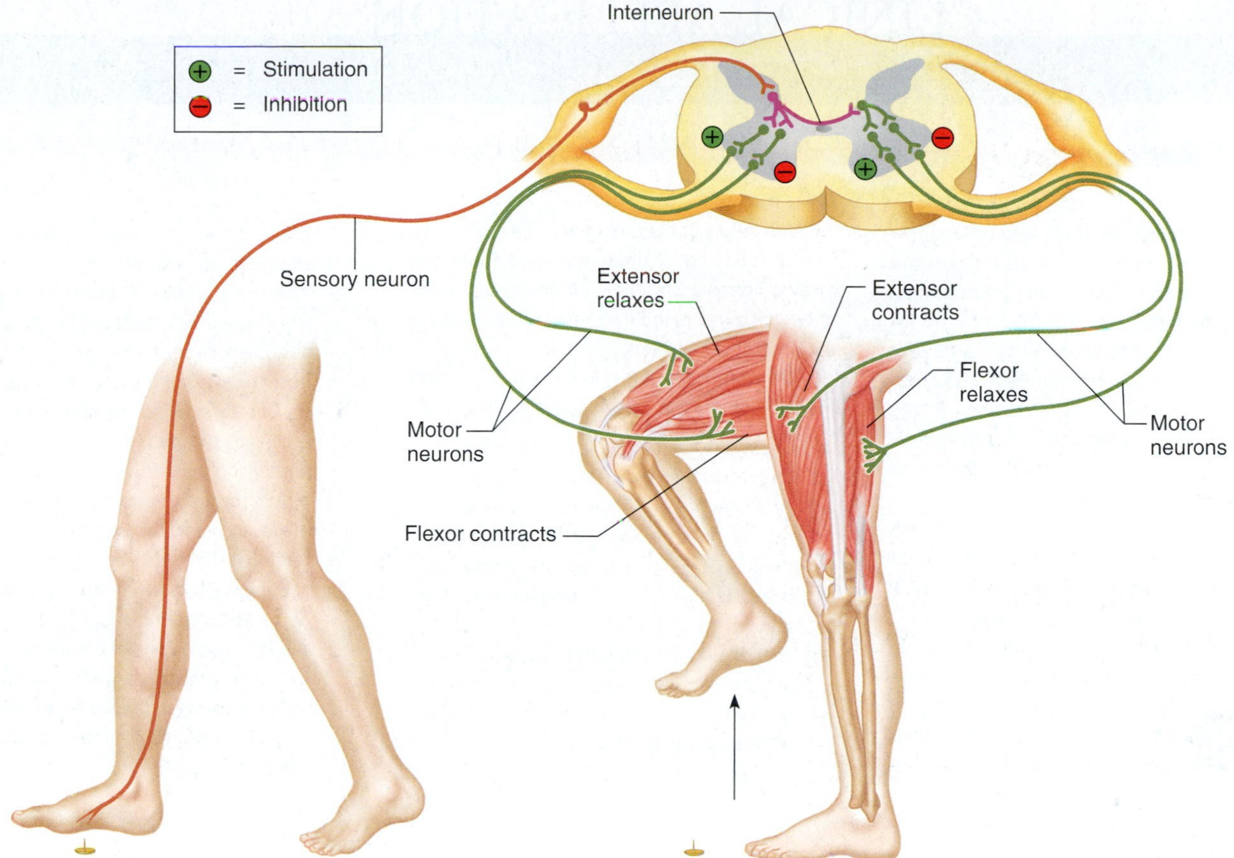

FIGURE 11.10 When the flexor muscle on one side is stimulated to contract in a withdrawal reflex, the extensor muscle on the opposite side also contracts. This helps to maintain balance.

TABLE 11.2 | Parts of a Reflex Arc

Part	Description	Function
Receptor	The receptor end of a dendrite or a specialized receptor cell in a sensory organ	Sensitive to a specific type of internal or external change
Sensory neuron	Dendrite, cell body, and axon of a sensory neuron	Transmits nerve impulse from the receptor into the brain or spinal cord
Interneuron	Dendrite, cell body, and axon of a neuron within the brain or spinal cord	Serves as processing center; conducts nerve impulse from the sensory neuron to a motor neuron
Motor neuron	Dendrite, cell body, and axon of a motor neuron	Transmits nerve impulse from the brain or spinal cord out to an effector
Effector	A muscle or gland	Responds to stimulation by the motor neuron and produces the reflex or behavioral action

Ascending and Descending Tracts

The nerve tracts of the spinal cord together with the spinal nerves provide a two-way communication system between the brain and body parts outside the nervous system. The tracts that conduct sensory impulses to the brain are called **ascending tracts;** those that conduct motor impulses from the brain to motor neurons reaching muscles and glands are **descending tracts.**

The ascending and descending tracts are comprised of axons. Typically, all the axons in a given tract originate from neuron cell bodies in the same part of the nervous system and end together in some other part. The names that identify nerve tracts often reflect these common origins and terminations. For example, a *spinothalamic tract* begins in the spinal cord and carries sensory impulses associated with the sensations of pain and touch to the thalamus of the brain (part of the diencephalon). A *corticospinal tract* originates in the cortex of the brain (the outer portion of the cerebrum) and carries motor impulses on upper motor neurons downward through the spinal cord. These impulses control lower motor neurons whose axons lead to skeletal muscles.

11.2 CLINICAL APPLICATION

Uses of Reflexes

Normal reflexes require and reflect normal neuron functions, so reflexes are commonly used to assess the condition of the nervous system. An anesthesiologist, for instance, may try to initiate a reflex in a patient being anesthetized to determine how the anesthetic drug is affecting nerve functions. In the case of injury to some part of the nervous system, observing reflexes may reveal the location and extent of damage.

Injury to any component of a reflex arc alters its function. For example, stroking the sole of the foot normally initiates a *plantar reflex*, which flexes the foot and toes. Damage to certain nerve pathways (corticospinal tract) may trigger an abnormal response called the *Babinski reflex*, a dorsiflexion, extending the great toe upward and fanning apart the smaller toes. If the injury is minor, the response may consist of plantar flexion with failure of the great toe to flex, or plantar flexion followed by dorsiflexion. The Babinski reflex is normally present in infants up to the age of twelve months and may reflect immaturity in their corticospinal tracts.

Other reflexes that may be tested during a neurological examination include the following:

1. *Biceps-jerk reflex.* Extending a person's forearm at the elbow elicits this reflex. The examiner places a finger on the inside of the extended elbow over the tendon of the biceps muscle and taps the finger. The biceps contracts in response, flexing at the elbow.
2. *Triceps-jerk reflex.* Flexing a person's forearm at the elbow and tapping the short tendon of the triceps muscle close to its insertion near the tip of the elbow elicits this reflex. The muscle contracts in response, extending the elbow.
3. *Abdominal reflexes.* These reflexes are a response to stroking the skin of the abdomen. For example, a dull pin drawn from the sides of the abdomen upward toward the midline and above the umbilicus contracts the abdominal muscles underlying the skin, and the umbilicus moves toward the stimulated region.
4. *Ankle-jerk reflex.* Tapping the calcaneal tendon just above its insertion on the calcaneus elicits this reflex. Contraction of the gastrocnemius and soleus muscles causes the response of plantar flexion.
5. *Cremasteric reflex.* This reflex is elicited in males by stroking the upper inside of the thigh. In response, contracting muscles elevate the testis on the same side. ■

Ascending Tracts Among the major ascending tracts of the spinal cord are the following:

1. **Fasciculus gracilis** (fah-sik′u-lus gras′il-is) and **fasciculus cuneatus** (ku′ne-at-us). These tracts are in the posterior funiculi of the spinal cord (fig. 11.11). Their fibers conduct sensory impulses from the skin, muscles, tendons, and joints to the brain, where they are interpreted as sensations of touch, pressure, and body movement.

At the base of the brain in an area called the medulla oblongata most of the fasciculus gracilis and fasciculus cuneatus fibers cross (decussate) from one side to the other—that is, those ascending on the left side of the spinal cord pass across to the right side, and vice versa. As a result, the impulses originating from sensory receptors on the left side of the body reach the right side of the brain, and those originating on the right side of the body reach the left side of the brain (fig. 11.12).

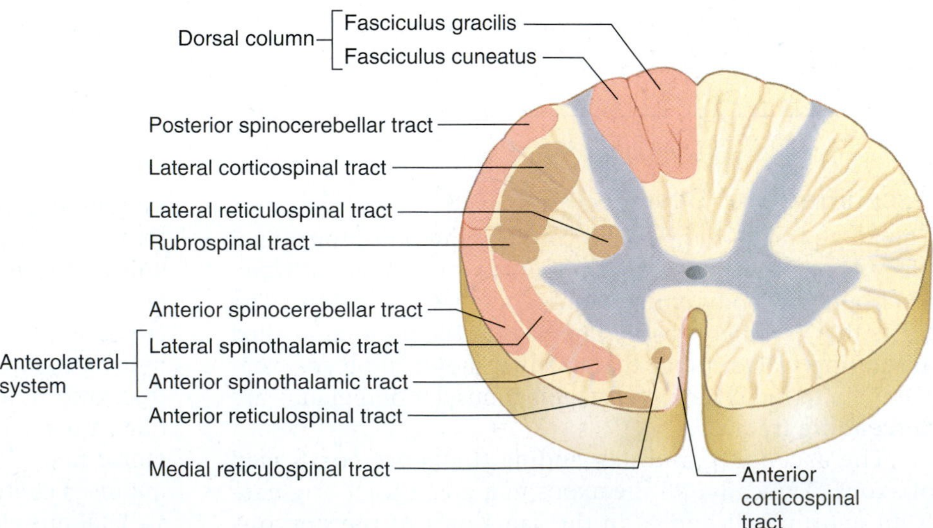

FIGURE 11.11 Major ascending and descending tracts in a cross section of the spinal cord. Ascending tracts are in pink, descending tracts in light brown. (Tracts are shown only on one side.) The pattern varies with the level of the spinal cord. This pattern is representative of the midcervical region.

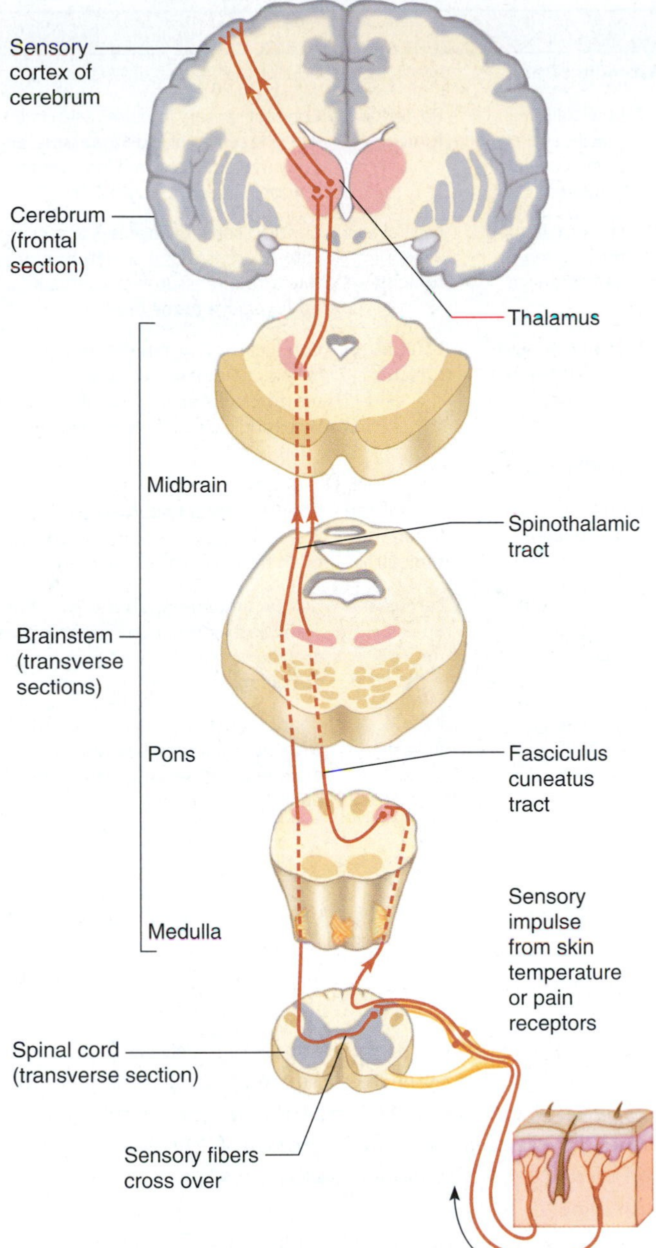

FIGURE 11.12 Sensory impulses originating in skin touch receptors ascend in the fasciculus cuneatus tract and cross over in the medulla of the brain. Pain and temperature information ascends in the lateral spinothalamic tract, which crosses over in the spinal cord.

2. **Spinothalamic** (spi″no-thah-lam′ik) **tracts.** The *lateral* and *anterior spinothalamic tracts* are in the lateral and anterior funiculi, respectively (see fig. 11.11). The lateral tracts conduct impulses from various body regions to the brain and give rise to sensations of pain and temperature. Impulses carried on fibers of the anterior tracts are interpreted as touch and pressure. Impulses in these tracts cross over in the spinal cord (fig. 11.12).
3. **Spinocerebellar** (spi″no-ser″ĕ-bel′ar) **tracts.** The *posterior* and *anterior spinocerebellar tracts* lie near the surface in the lateral funiculi of the spinal cord (see fig. 11.11). Fibers in the posterior tracts remain uncrossed, whereas those in the anterior tracts cross over in the medulla. Impulses conducted on their fibers originate in the muscles of the lower limbs and trunk and then travel to the cerebellum. These impulses coordinate muscular movements.

Descending Tracts The major descending tracts of the spinal cord include the following:

1. **Corticospinal** (kor″tĭ-ko-spi′nal) **tracts.** The *lateral* and *anterior corticospinal tracts* occupy the lateral and anterior funiculi, respectively (see fig. 11.11). Most of the fibers of the lateral tracts cross over in the lower medulla oblongata. Some fibers of the anterior tracts cross over at various levels of the spinal cord (fig. 11.13). The corticospinal tracts conduct motor impulses from the brain to spinal nerves and outward to various skeletal muscles. Thus, they help control voluntary movements.

   The corticospinal tracts are sometimes called *pyramidal tracts* after the pyramid-shaped regions in the medulla oblongata through which they pass. Other descending tracts are called *extrapyramidal tracts,* and they include the reticulospinal and rubrospinal tracts.
2. **Reticulospinal** (rĕ-tik″u-lo-spi′nal) **tracts.** The *lateral reticulospinal tracts* are in the lateral funiculi, whereas the *anterior* and *medial reticulospinal tracts* are in the anterior funiculi (see fig. 11.11). Some fibers in the lateral tracts cross over, whereas others remain uncrossed. Those of the anterior and medial tracts remain uncrossed. Motor impulses transmitted on the reticulospinal tracts originate in the brain and control muscular tone and activity of sweat glands.
3. **Rubrospinal** (roo″bro-spi′nal) **tracts.** The fibers of the rubrospinal tracts cross over in the brain and pass through the lateral funiculi (see fig. 11.11). They carry motor impulses from the brain to skeletal muscles, and they coordinate muscles and control posture.

Table 11.3 summarizes the nerve tracts of the spinal cord. Clinical Application 11.3 describes injuries to the spinal cord.

> A hemi-lesion of the spinal cord (severed on one side) affecting the corticospinal and spinothalamic tracts can cause Brown-Séquard syndrome. Ascending tracts cross over at different levels, so the injured side of the body becomes paralyzed and loses touch sensation. The other side of the body retains movement but loses sensations of pain and temperature.

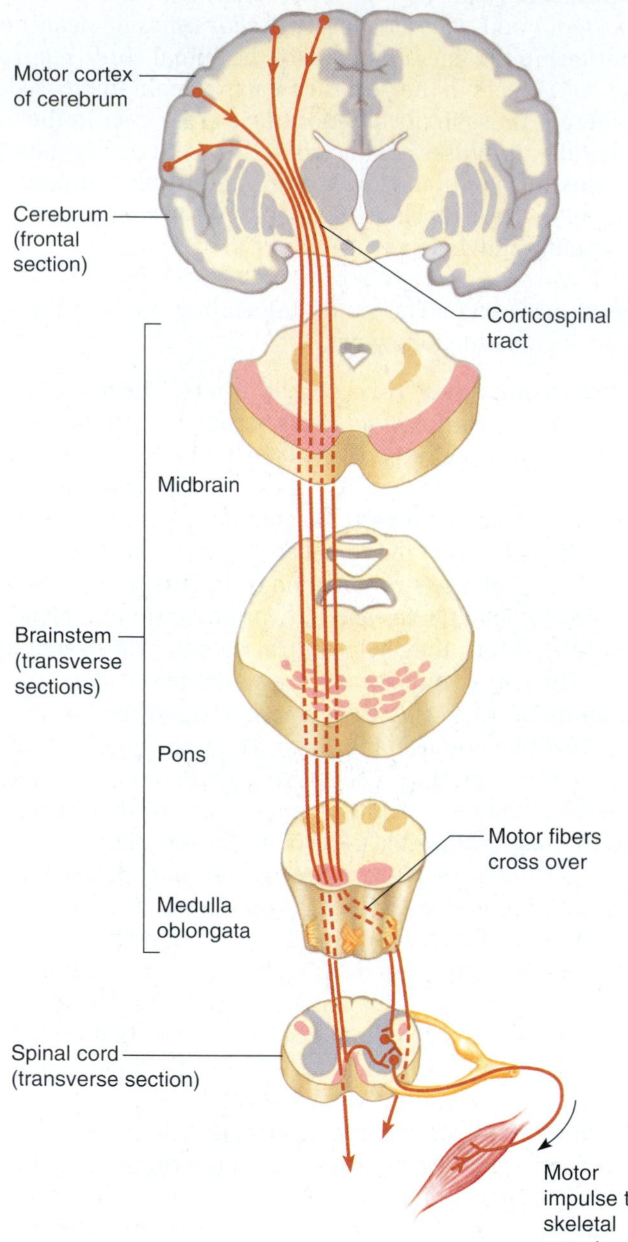

FIGURE 11.13 Most fibers of the corticospinal tract originate in the cerebral cortex, cross over in the medulla, and descend in the spinal cord, where they synapse with neurons whose fibers lead to spinal nerves supplying skeletal muscles. Some fibers cross over in the spinal cord.

PRACTICE

11 Describe the structure of the spinal cord.

12 What are ascending and descending tracts?

13 What is the consequence of fibers crossing over?

14 Name the major tracts of the spinal cord, and list the types of impulses each conducts.

TABLE 11.3 | Nerve Tracts of the Spinal Cord

Tract	Location	Function
Ascending Tracts		
1. Fasciculus gracilis and fasciculus cuneatus	Posterior funiculi	Conduct sensory impulses associated with the senses of touch, pressure, and body movement from skin, muscles, tendons, and joints to the brain
2. Spinothalamic tracts (lateral and anterior)	Lateral and anterior funiculi	Conduct sensory impulses associated with the senses of pain, temperature, touch, and pressure from various body regions to the brain
3. Spinocerebellar tracts (posterior and anterior)	Lateral funiculi	Conduct sensory impulses required for the coordination of muscle movements from muscles of the lower limbs and trunk to the cerebellum
Descending Tracts		
1. Corticospinal tracts (lateral and anterior)	Lateral and anterior funiculi	Conduct motor impulses associated with voluntary movements from the brain to skeletal muscles
2. Reticulospinal tracts (lateral, anterior, and medial)	Lateral and anterior funiculi	Conduct motor impulses associated with the maintenance of muscle tone and the activity of sweat glands from the brain
3. Rubrospinal tracts	Lateral funiculi	Conduct motor impulses associated with muscular coordination and the maintenance of posture from the brain

Amyotrophic lateral sclerosis (ALS or Lou Gehrig's disease) may begin with garbled speech, clumsiness, sudden fatigue, or limb weakness. Fasciculations (muscle twitches) that resemble moving ropes beneath the skin may prompt the person to seek medical attention. But because ALS is a diagnosis of exclusion, identifying it may take a year or more.

ALS affects the upper and lower parts of the body, and progresses faster if symptoms begin in the face or neck. Usually the battle is lost two to five years following diagnosis, typically from respiratory failure. Using assistive breathing devices and a drug, Riluzole, can extend life. The mind is often spared—one patient wrote a novel in his last months, and another remained a brilliant songwriter.

In ALS, motor neurons degenerate in the spinal cord, brainstem, and the cerebral cortex. ALS may be due to an inability of the motor neurons or associated astrocytes to counter buildup of oxidative free radicals, or the astrocytes may release a neurotoxin. Researchers have provided conditions that induced stem cells from ALS patients. In culture these cells give rise to motor neurons and astrocytes, providing peeks into the pathogenesis of this devastating disease. If researchers can understand how the disease begins, perhaps treatments will follow.

11.3 CLINICAL APPLICATION

Spinal Cord Injuries

On a bright May morning in 1995, actor Christopher Reeve sustained a devastating spinal cord injury when the horse that he was riding in a competition failed to clear a hurdle. Reeve rocketed forward, striking his head on the fence. He landed on the grass—unconscious, not moving or breathing.

Reeve had broken the first and second cervical vertebrae, between the neck and the brainstem. Someone performed mouth-to-mouth resuscitation until paramedics inserted a breathing tube and then stabilized him on a board. At a nearby hospital, Reeve received methylprednisolone, a drug that can save a fifth of the damaged neurons by reducing inflammation. Reeve was then flown to a larger medical center for further treatment.

Reeve's rehabilitation was slow, yet inspiring. Despite discouraging words from physicians, he persisted in trying to exercise. Suspended from a harness, he moved his feet over a treadmill. He moved other muscles in a swimming pool and rode a special recumbent bicycle, with electrical stimulation to his legs enabling him to pedal an hour a day. Five years after the accident, Reeve gradually started to move his fingers, and then his hips and legs, although he still required a wheelchair and a respirator. Following his example, hundreds of others with spinal cord injuries improved with exercise, too. Reeve's motto gave hope to many: "Nothing is impossible." He passed away in 2004. Most people with his level of injury—between the first and second cervical vertebrae—do not live more than seven years.

Thousands of people sustain spinal cord injuries each year. During the first few days the vertebrae are compressed and may break, which sets off action potentials in neurons, killing many of them. Dying neurons release calcium ions, which activate tissue-degrading enzymes. Then white blood cells arrive and produce inflammation that can destroy healthy as well as damaged neurons. Axons tear, myelin coatings are stripped off, and vital connections between nerves and muscles are cut. The tissue cannot regenerate.

The severity of a spinal cord injury depends on the extent and location of damage. Normal spinal reflexes require two-way communication between the spinal cord and the brain. Injuring nerve pathways depresses the cord's reflex activities in sites below the injury. At the same time, sensations and muscle tone in the parts the affected fibers innervate lessen. This condition, spinal shock, may last for days or weeks, although normal reflex activity may eventually return. However, if nerve fibers are severed, some of the cord's functions may be permanently lost.

Less severe injuries to the spinal cord, as from a blow to the head, whiplash, or rupture of an intervertebral disc, compress or distort the cord. Pain, weakness, and muscular atrophy in the regions the damaged nerve fibers supply may occur.

The most common cause of severe direct injury to the spinal cord is vehicular accidents (fig. 11C). Regardless of the cause, if nerve fibers in ascending tracts are cut, sensations arising from receptors below the level of the injury are lost. Damage to descending tracts results in loss of motor functions. For example, if the right lateral corticospinal tract is severed in the neck near the first cervical vertebra, control of the voluntary muscles in the right upper and lower limbs is lost, paralyzing them (hemiplegia). Problems of this type in fibers of the descending tracts produce *upper motor neuron syndrome,* characterized by *spastic paralysis* in which muscle tone increases, with little atrophy of the muscles. However, uncoordinated reflex activity (hyperreflexia) usually occurs, when the flexor and extensor muscles of affected limbs alternately spasm.

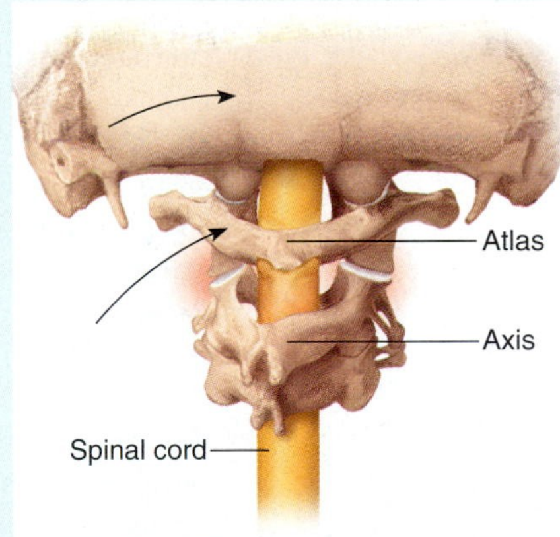

FIGURE 11C A dislocation of the atlas may cause a compression injury to the spinal cord.

Injury to motor neurons or their fibers in the horns of the spinal cord results in *lower motor neuron syndrome.* It produces *flaccid paralysis,* a total loss of muscle tone and reflex activity, and the muscles atrophy.

Several new treatments are on the horizon for spinal cord injuries. They work in three ways:

1. *Limiting damage during the acute phase.* An experimental drug called GM1 ganglioside is a carbohydrate normally found on neuron cell membranes. It blocks the actions of amino acids that function as excitatory neurotransmitters, which cuts the deadly calcium ion influx into cells. It also blocks apoptosis (programmed cell death) and stimulates synthesis of nerve growth factor.
2. *Restoring or compensating for function.* A new drug called 4-aminopyridine blocks potassium channels on neurons. This boosts electrical transmission and compensates for the myelin-stripping effects of the injury. Being developed for patients injured at least eighteen months previously, this drug can restore some sexual, bowel, and bladder function.
3. *Regeneration.* Many experiments have shown that paralyzed rodents given implants of human neural stem cells regain some ability to walk. Several research groups are working on creating tissue implants from embryonic or neural stem cells that can help damaged spinal cords to heal. ■

11.5 BRAIN

The brain contains nerve centers associated with sensory functions and is responsible for sensations and perceptions. It issues motor commands to skeletal muscles and carries on higher mental functions, such as memory and reasoning. It also contains centers that coordinate muscular movements, as well as centers and nerve pathways that regulate visceral activities. In addition to overseeing the function of the entire body, the brain also provides characteristics such as personality.

Brain Development

The basic structure of the brain reflects the way it forms during early (embryonic) development. It begins as the neural tube that gives rise to the CNS. The portion that becomes the brain has three major cavities, or vesicles, at one end—the *forebrain* (prosencephalon), *midbrain* (mesencephalon), and *hindbrain* (rhombencephalon) (fig. 11.14). Later, the forebrain divides into anterior and posterior portions (telencephalon and diencephalon, respectively), and the hindbrain partially divides into two parts (metencephalon and myelencephalon). The resulting five cavities persist in the mature brain as the fluid-filled *ventricles* and the tubes that connect them. Cells of the tissue surrounding the spaces differentiate into the structural and functional regions of the brain.

The wall of the anterior portion of the forebrain gives rise to the *cerebrum* and *basal nuclei,* whereas the posterior portion forms a section of the brain called the *diencephalon.*

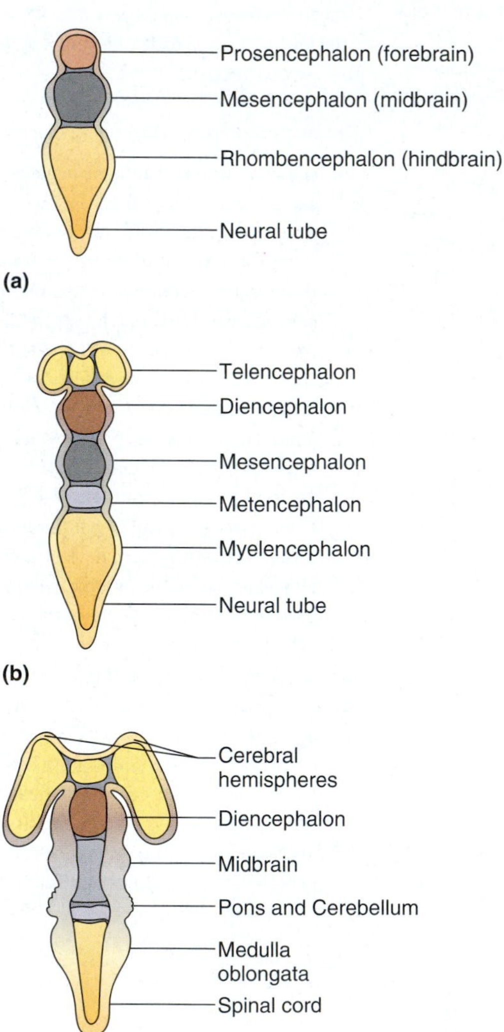

FIGURE 11.14 Brain development. (*a*) The brain develops from a tubular structure with three cavities. (*b*) The cavities persist as the ventricles and their interconnections. (*c*) The wall of the tube gives rise to various regions of the brain, brainstem, and spinal cord.

The region the midbrain produces continues to be called the *midbrain* in the adult structure, and the hindbrain gives rise to the *cerebellum, pons,* and *medulla oblongata* (fig. 11.15 and table 11.4). Together, the midbrain, pons, and medulla oblongata comprise the **brainstem** (brān′stem), which attaches the brain to the spinal cord.

On a cellular level, the brain develops as specific neurons attract others by secreting growth hormones. In the embryo and fetus, the brain overgrows, and then apoptosis (programmed cell death) destroys excess cells.

Structure of the Cerebrum

The **cerebrum** (ser′ē-brum), which develops from the anterior portion of the forebrain, is the largest part of the mature brain. It consists of two large masses, or **cerebral hemispheres** (ser′ĕ-bral hem′i-sfērz), which are essentially mirror images of each other (fig. 11.16 and reference plate 9). A deep bridge of nerve fibers called the **corpus callosum** connects the cerebral hemispheres. A layer of dura mater called the *falx cerebri* separates them (see fig. 11.1*b*).

Many ridges or convolutions, called **gyri** (ji′ri) (sing., *gyrus*), separated by grooves, mark the cerebrum's surface. Generally, a shallow to somewhat deep groove is called a **sulcus** (sul′kus; pl. *sulci,* sul′si), and a very deep groove is called a **fissure.** The pattern of these elevations and depressions is complex, and it is distinct in all normal brains. For example, a *longitudinal fissure* separates the right and left cerebral hemispheres; a *transverse fissure* separates the cerebrum from the cerebellum; and sulci divide each hemisphere into lobes (see figs. 11.15 and 11.16).

> A fetus or newborn with *anencephaly* has a face and lower brain structures but lacks most higher brain structures. A newborn with this anomaly survives only a day or two. Sometimes the parents donate the organs.
>
> Anencephaly is a type of neural tube defect (NTD). It occurs at about the twenty-eighth day of prenatal development, when a sheet of tissue that normally folds to form the neural tube, which develops into the CNS, remains open at the top. In *spina bifida,* an opening is farther down the neural tube, causing a lesion in the spine. Paralysis may occur from that point downward. Sometimes surgery can partially correct spina bifida. Taking folic acid supplements just before and during pregnancy can lower the risk of a neural tube defect.
>
> In a disorder called *lissencephaly* ("smooth brain"), a newborn has a smooth cerebral cortex, completely lacking convolutions. Absence of a protein early in prenatal development prevents certain neurons from migrating in the brain, which blocks formation of convolutions. The child is profoundly mentally retarded, with frequent seizures and other neurological problems.

The lobes of the cerebral hemispheres (fig. 11.16) are named after the skull bones that they underlie. The lobes include the following:

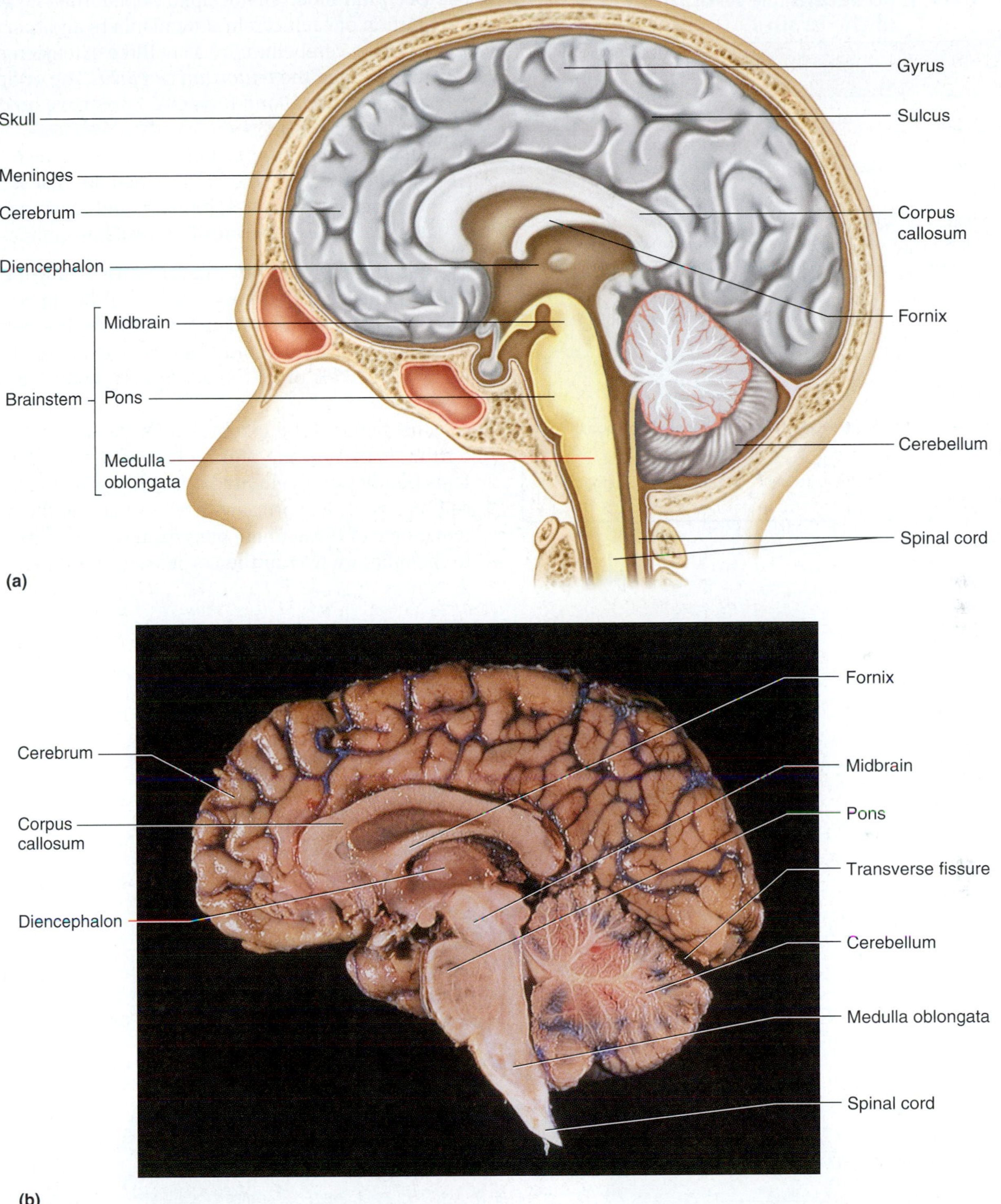

FIGURE 11.15 Sagittal section of brain. (*a*) The major portions of the brain include the cerebrum, the diencephalon, the cerebellum, and the brainstem. (*b*) Photo of human brain.

1. **Frontal lobe.** The frontal lobe forms the anterior portion of each cerebral hemisphere. It is bordered posteriorly by a *central sulcus* (fissure of Rolando), which passes out from the longitudinal fissure at a right angle, and inferiorly by a *lateral sulcus* (fissure of Sylvius), which exits the undersurface of the brain along its sides.
2. **Parietal lobe.** The parietal lobe is posterior to the frontal lobe and is separated from it by the central sulcus.
3. **Temporal lobe.** The temporal lobe lies inferior to the frontal and parietal lobes and is separated from them by the lateral sulcus.

TABLE 11.4 | Structural Development of the Brain

Embryonic Vesicle	Spaces Produced	Regions of the Brain Produced
Forebrain (prosencephalon)		
Anterior portion (telencephalon)	Lateral ventricles	Cerebrum Basal nuclei
Posterior portion (diencephalon)	Third ventricle	Thalamus Hypothalamus Posterior pituitary gland Pineal gland
Midbrain (mesencephalon)	Cerebral aqueduct	Midbrain
Hindbrain (rhombencephalon)		
Anterior portion (metencephalon)	Fourth ventricle	Cerebellum, pons
Posterior portion (myelencephalon)	Fourth ventricle	Medulla oblongata

4. **Occipital lobe.** The occipital lobe forms the posterior portion of each cerebral hemisphere and is separated from the cerebellum by a shelflike extension of dura mater called the *tentorium cerebelli.* The occipital lobe and the parietal and temporal lobes have no distinct boundary.
5. **Insula.** The insula (island of Reil) is a lobe deep within the lateral sulcus and is so named because it is covered by parts of the frontal, parietal, and temporal lobes. A *circular sulcus* separates the insula from the other lobes.

A thin layer of gray matter (2 to 5 millimeters thick) called the **cerebral cortex** (ser′ĕ-bral kor′teks) constitutes the outermost portion of the cerebrum. It covers the gyri, dipping into the sulci and fissures. The cerebral cortex contains nearly 75% of all the neuron cell bodies in the nervous system.

Just beneath the cerebral cortex is a mass of white matter that makes up the bulk of the cerebrum. This mass contains bundles of myelinated nerve fibers that connect neuron cell bodies of the cortex with other parts of the nervous system. Some of these fibers pass from one cerebral hemisphere to the other by way of the corpus callosum, and others carry

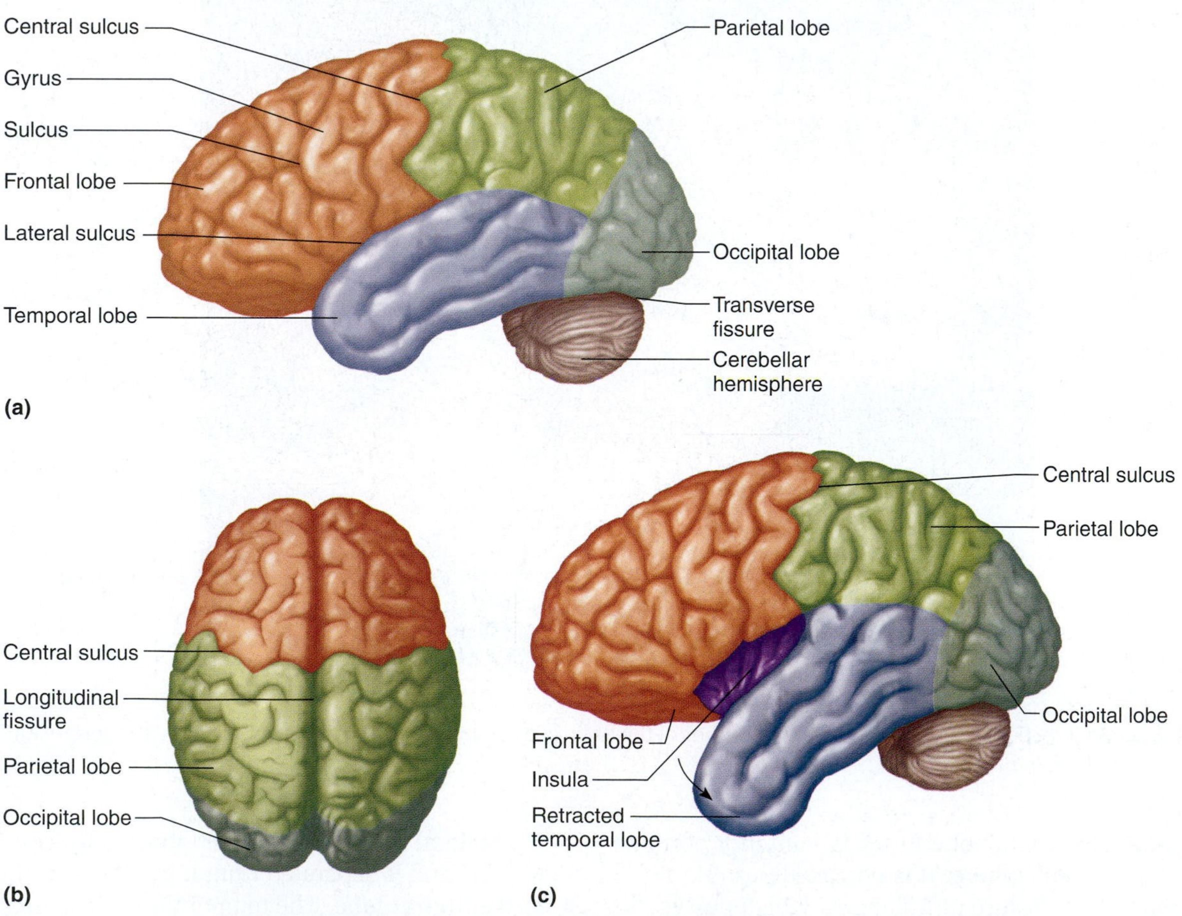

FIGURE 11.16 Colors in this figure distinguish the lobes of the cerebral hemispheres. (*a*) Lateral view of the left hemisphere. (*b*) Hemispheres viewed from above. (*c*) Lateral view of the left hemisphere with the insula exposed.

sensory or motor impulses from the cortex to nerve centers in the brain or spinal cord.

In a "stroke," or *cerebrovascular accident* (CVA), a sudden interruption in blood flow in a vessel supplying brain tissues damages the cerebrum. The affected blood vessel may rupture, bleeding into the brain, or be blocked by a clot. In either case, brain tissues downstream from the vascular accident die or permanently lose function. Temporary interruption in cerebral blood flow, perhaps by a clot that quickly breaks apart, produces a much less serious *transient ischemic attack* (TIA).

Functions of the Cerebrum

The cerebrum provides higher brain functions: interpreting impulses from sense organs, initiating voluntary muscular movements, storing information as memory, and retrieving this information in reasoning. The cerebrum is also the seat of intelligence and personality.

Functional Regions of the Cortex

The regions of the cerebral cortex that perform specific functions have been located using a variety of techniques. From Science to Technology 2.3, figure 2E (p. 71), shows how PET scanning is used to localize particular functions to specific areas of the cerebral cortex. Clues to cerebral functioning also come from people who have suffered brain disease or injury.

In other studies, areas of cortices have been exposed surgically and stimulated mechanically or electrically. Researchers observe the responses in certain muscles or the specific sensations that result. Based on such investigations, researchers have divided the cerebral cortex into sensory, association, and motor areas that overlap somewhat.

Sensory Areas

Sensory areas in several lobes of the cerebrum interpret impulses from sensory receptors, producing feelings or sensations. For example, the sensations of temperature, touch, pressure, and pain in the skin arise in the postcentral gyri of the anterior portions of the parietal lobes along the central sulcus and in the posterior wall of this sulcus (fig. 11.17). The posterior parts of the occipital lobes provide vision, whereas the superior posterior portions of the temporal lobes contain the centers for hearing. The sensory areas for taste are near the bases of the central sulci along the lateral sulci, and the sense of smell arises from centers deep in the cerebrum.

Like motor fibers, sensory fibers, such as those in the *fasciculus cuneatus tract,* cross over in the spinal cord or the brainstem (see fig. 11.12). Thus, the centers in the right central hemisphere interpret impulses originating from the left side of the body, and vice versa. However, the sensory areas concerned with vision receive impulses from both eyes, and those concerned with hearing receive impulses from both ears.

Not all sensory areas are bilateral. The *sensory speech area,* also called *Wernicke's area,* is in the parietal lobe near the temporal lobe, just posterior to the lateral sulcus, usually in the left hemisphere (fig. 11.17). This area receives and relays input from both the visual cortex and auditory cortex and is important for understanding written and spoken language.

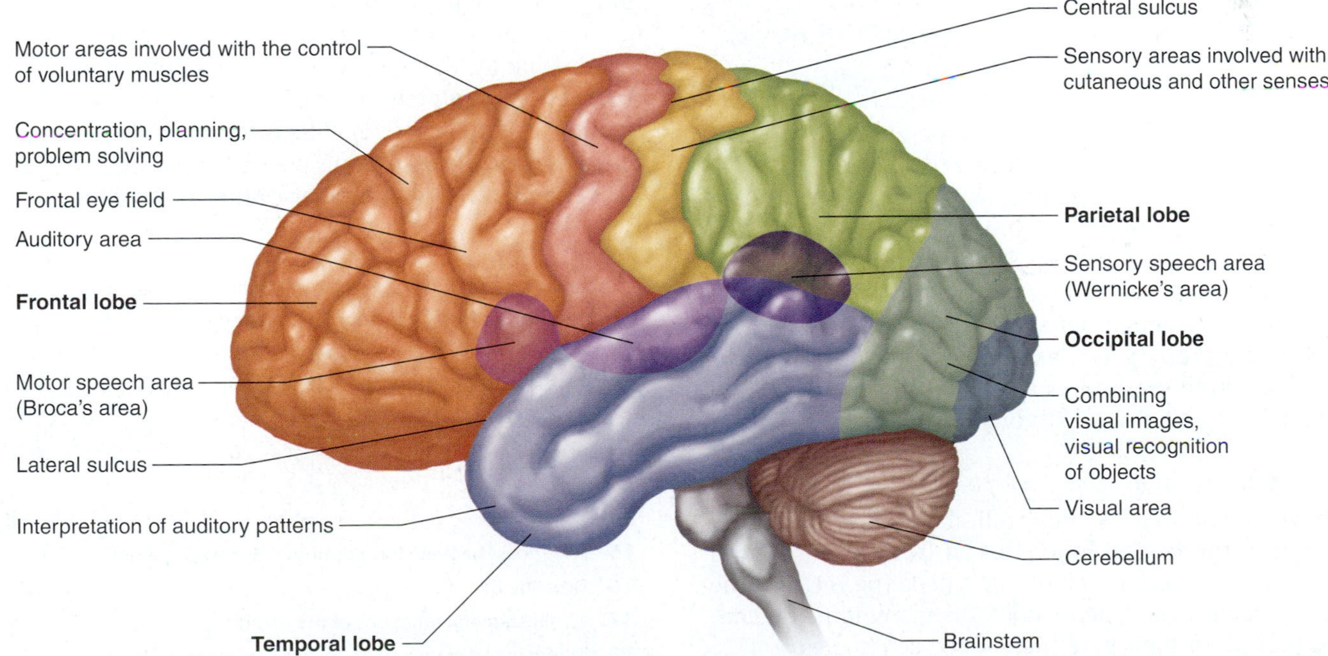

FIGURE 11.17 Some sensory association and motor areas of the left cerebral cortex.

Association Areas

Association areas are neither primarily sensory nor motor. They connect with each other and other brain structures. These areas occupy the anterior portions of the frontal lobes and are widespread in the lateral portions of the parietal, temporal, and occipital lobes. They analyze and interpret sensory experiences and help provide memory, reasoning, verbalizing, judgment, and emotions (fig. 11.17).

The association areas of the frontal lobes provide higher intellectual processes, such as concentrating, planning, and complex problem solving. The anterior and inferior portions of these lobes (prefrontal areas) control emotional behavior and produce awareness of the possible consequences of behavior.

The parietal lobes have association areas that help interpret sensory information and aid in understanding speech and choosing words to express thoughts and feelings. Awareness of the form of objects, including one's own body parts, stems from the posterior regions of these lobes.

The association areas of the temporal lobes and the regions at the posterior ends of the lateral sulci interpret complex sensory experiences, such as those needed to understand speech and to read. These regions also store memories of visual scenes, music, and other complex sensory patterns.

The occipital lobes have association areas adjacent to the visual centers. These are important in analyzing visual patterns and combining visual images with other sensory experiences—as when one recognizes another person.

> A person with *dyslexia* sees letters separately and must learn to read differently than people whose nervous systems can group letters into words. Three to 10% of people have dyslexia. The condition probably has several causes, with inborn visual and perceptual skills interacting with the way the child learns to read. Dyslexia has nothing to do with intelligence—many brilliant thinkers were "slow" in school because educators did not know how to help them.

Wernicke's area corresonds closely to a brain region that has been referred to as a "*general interpretive area,*" near where the occipital, parietal, and temporal lobes meet. The general interpretive area processes sensory information from all three of these association areas. It plays a role in integrating visual, auditory, and other sensory information and then interpreting a situation. For example, you hear a familiar voice, look up from your notes, see a friend from class, and realize that it is time for your study group.

Motor Areas

The *primary motor areas* of the cerebral cortex lie in the precentral gyri of the frontal lobes just in front of the central sulcus and in the anterior wall of this sulcus (fig. 11.17). The nervous tissue in these regions contains many large *pyramidal cells,* named for their pyramid-shaped cell bodies.

Impulses from the pyramidal cells move downward through the brainstem and into the spinal cord on the *corticospinal tracts.* Most of the nerve fibers in these tracts cross over from one side of the brain to the other within the brainstem and descend as lateral corticospinal tracts. Other fibers, in the anterior corticospinal tracts, cross over at various levels of the spinal cord (see fig. 11.13).

In the spinal cord, the corticospinal fibers synapse with motor neurons in the gray matter of the anterior horns. Axons of the motor neurons lead outward through peripheral nerves to voluntary muscles. Impulses transmitted on these pathways in special patterns and frequencies are responsible for fine movements in skeletal muscles. More specifically, as figure 11.18 shows, cells in the upper portions of the motor areas send impulses to muscles in the thighs and legs; those in the middle portions control muscles in the arms and forearms; and those in lower portions activate muscles of the head, face, and tongue.

The *reticulospinal* and *rubrospinal tracts* coordinate and control motor functions that maintain balance and posture. Many of these fibers pass into the basal nuclei on the way to the spinal cord. Some of the impulses conducted on these pathways normally inhibit muscular actions.

In addition to the primary motor areas, certain other regions of the frontal lobe control motor functions. For example, a region called the *motor speech area,* also known as *Broca's area,* is in the frontal lobe, usually in the left hemisphere, just anterior to the primary motor cortex and superior to the lateral sulcus. The motor speech area is important in generating the complex muscular actions of the mouth, tongue, and larynx, which make speech possible (see fig. 11.17). Bundles of axons directly and indirectly connect the motor speech area to the sensory speech area. A person with an injury to this area may be able to understand spoken words but may be unable to speak.

Above the motor speech area is a region called the *frontal eye field.* The motor cortex in this area controls voluntary movements of the eyes and eyelids. Nearby is the cortex responsible for movements of the head that direct the eyes. Another region just in front of the primary motor area controls the muscular movements of the hands and fingers that make such skills as writing possible (see fig. 11.17). Table 11.5 summarizes the functions of the cerebral lobes.

> An injury to the motor system may impair the ability to produce purposeful muscular movements. Such a condition that affects use of the upper and lower limbs, head, or eyes is called *apraxia* . When apraxia affects the speech muscles, disrupting speaking ability, it is called *aphasia.*

PRACTICE

15 How does the brain form during early development?
16 Describe the cerebrum.
17 List the general functions of the cerebrum.
18 Where in the brain are the sensory areas located?
19 Explain the functions of association areas.
20 Where in the brain are the motor areas located?

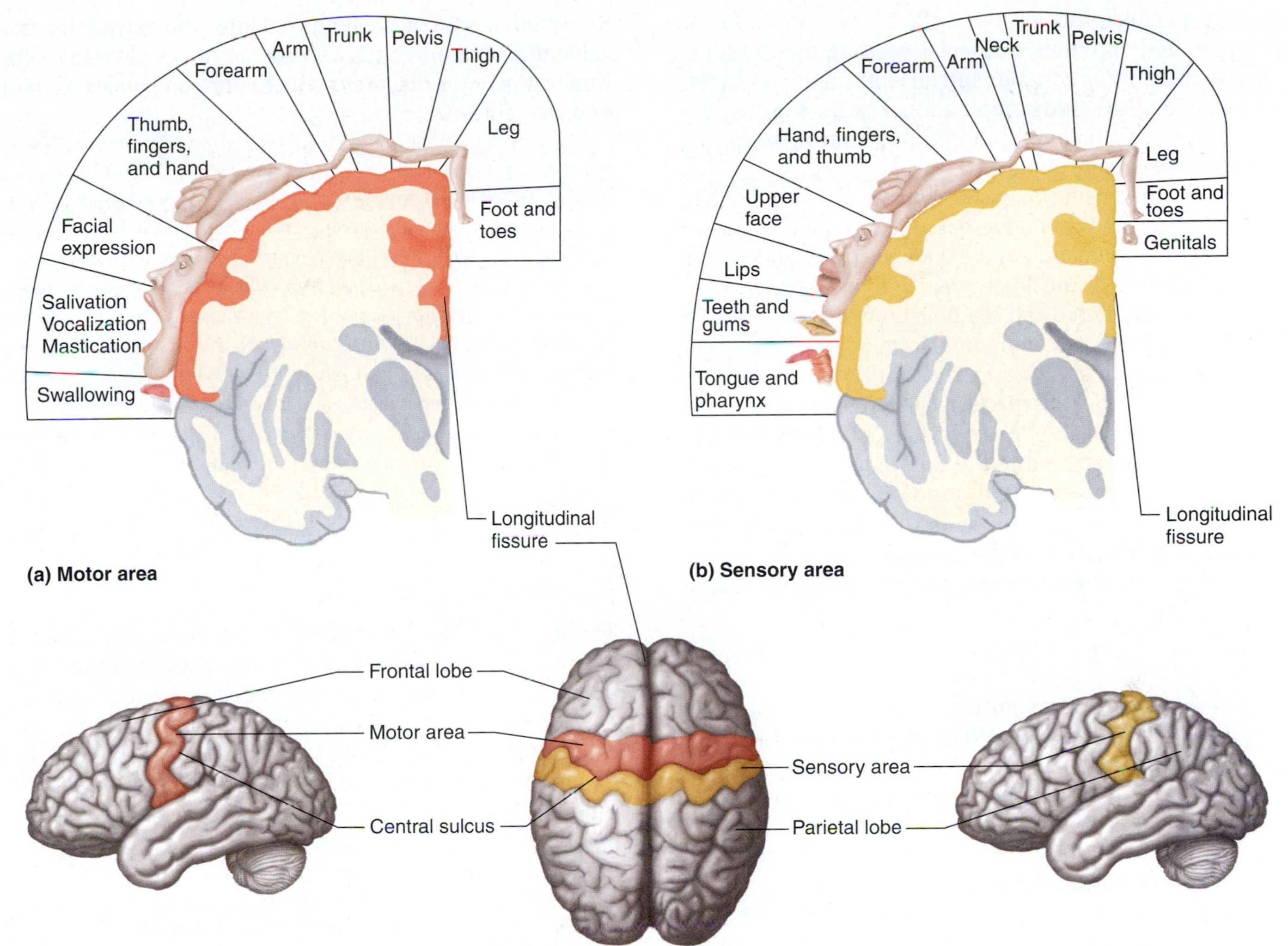

FIGURE 11.18 Functional regions of the cerebral cortex. (*a*) Motor areas that control voluntary muscles (only left hemisphere shown). (*b*) Sensory areas involved with cutaneous and other senses (only left hemisphere shown).

TABLE 11.5 | Functions of the Cerebral Lobes

Lobe	Functions
Frontal lobes	Association areas carry on higher intellectual processes for concentrating, planning, complex problem solving, and judging the consequences of behavior. Motor areas control movements of voluntary skeletal muscles.
Parietal lobes	Sensory areas provide sensations of temperature, touch, pressure, and pain involving the skin. Association areas function in understanding speech and in using words to express thoughts and feelings.
Temporal lobes	Sensory areas are responsible for hearing. Association areas interpret sensory experiences and remember visual scenes, music, and other complex sensory patterns.
Occipital lobes	Sensory areas are responsible for vision. Association areas combine visual images with other sensory experiences.

Hemisphere Dominance

Both cerebral hemispheres participate in basic functions, such as receiving and analyzing sensory impulses, controlling skeletal muscles on opposite sides of the body, and storing memory. However, one side usually acts as a *dominant hemisphere* for certain other functions.

Tests indicate that the left hemisphere is dominant in 90% of right-handed adults and in 64% of left-handed ones. The right hemisphere is dominant in 10% of right-handed adults and in 20% of left-handed ones. The hemispheres are equally dominant in the remaining 16% of left-handed persons. As a consequence of hemisphere dominance, the motor speech area on one side almost completely controls the motor activities associated with speech. For this reason, over 90% of patients with language impairment stemming from problems in the cerebrum have disorders in the left hemisphere.

In most persons, the left hemisphere is dominant for the language-related activities of speech, writing, and reading. It is also dominant for complex intellectual functions requiring verbal, analytical, and computational skills. In other persons, the right hemisphere is dominant, and in some, the hemispheres are equally dominant.

In addition to carrying on basic functions, the nondominant hemisphere specializes in nonverbal functions, such as motor tasks that require orientation of the body in space, understanding and interpreting musical patterns, and visual experiences. It also provides emotional and intuitive thought processes. For example, the region in the nondominant hemisphere that corresponds to the motor speech area does not control speech, but it influences the emotional aspects of spoken language.

Nerve fibers of the *corpus callosum,* which connect the cerebral hemispheres, enable the dominant hemisphere to control the motor cortex of the nondominant hemisphere. These fibers also transfer sensory information reaching the nondominant hemisphere to the general interpretative area of the dominant one, where the information can be used in decision making.

Memory

Memory, one of the most astonishing capabilities of the brain, is the consequence of learning. Whereas learning is the acquisition of new knowledge, memory is the persistence of that learning, with the ability to access it at a later time. Two types of memory, short term and long term, have been recognized for many years, and researchers are now beginning to realize that they differ in characteristics other than duration.

Short-term, or "working" memories are thought to be electrical. Neurons may be connected in a circuit so that the last in the series stimulates the first. As long as the pattern of stimulation continues, the thought is remembered. When the electrical events cease, so does the memory—unless it enters long-term memory.

Long-term memory probably changes the structure or function of neurons in ways that enhance synaptic transmission, perhaps by establishing certain patterns of synaptic connections. Synaptic patterns fulfill two requirements of long-term memory. First, there are enough synapses to encode an almost limitless number of memories—each of the 10 billion neurons in the cortex can make tens of thousands of synaptic connections to other neurons, forming 60 trillion synapses. Second, a certain pattern of synapses can remain unchanged for years.

Understanding how neurons in different parts of the brain encode memories and how short-term memories are converted to long-term memories, a process called **memory consolidation,** is at the forefront of research into the functioning of the human brain. According to one theory, **long-term synaptic potentiation,** near simultaneous repeated stimulation of the same neurons strengthens their synaptic connections. In response, in an area of the temporal lobe called the **hippocampus,** more frequent action potentials are triggered in postsynaptic cells.

Another area of the temporal lobe, the amygdala, assigns value to a memory, such as whether it was pleasant. Clinical Application 11.4 discusses some common causes of damage to the cerebrum.

Unusual behaviors and skills of people who have damaged the hippocampus have taught researchers much about this intriguing part of the brain. In 1953, a surgeon removed parts of the hippocampus and the amygdala of a young man called H. M., to relieve his severe epilepsy. His seizures became less frequent, but H. M. suffered a profound loss in the ability to consolidate short-term memories into long-term ones. As a result, events in H. M.'s life faded from memory as quickly as they occurred. He was unable to recall any events that took place since surgery, living always as if it was the 1950s. He would read the same magazine article repeatedly with renewed interest each time.

Basal Nuclei

The **basal nuclei** (basal ganglia) are masses of gray matter deep within the cerebral hemispheres. They are called the *caudate nucleus,* the *putamen,* and the *globus pallidus,* and they develop from the anterior portion of the forebrain (fig. 11.19). The basal nuclei produce the inhibitory neurotransmitter *dopamine.* The neurons of the basal nuclei interact with other brain areas, including the motor cortex, thalamus, and cerebellum. These interactions, through a combination of stimulation and inhibition, facilitate voluntary movement. Clinical Application 11.5 discusses Parkinson disease, in which neurons in the basal nuclei degenerate.

The functioning of the basal nuclei may be very specific, such as controlling the movements necessary to speak. A family in London with many members who have unintelligible speech led to the discovery of a single gene, *FoxP2,* that controls the ability to combine words into meaningful speech. The gene, which controls several other genes, also enables a person to understand and use grammar. Songbirds have the gene, too. Impairing *FoxP2* function in birds prevents them from learning their songs. In both humans and songbirds, *FoxP2* acts on a specific part of the basal nuclei called "area X."

PRACTICE

21 What is hemisphere dominance?

22 What are the functions of the nondominant hemisphere?

23 Distinguish between short-term and long-term memory.

24 What is the function of the basal nuclei?

Diencephalon

The **diencephalon** (di″en-sef′ah-lon) develops from the posterior forebrain and is located between the cerebral hemispheres and superior to the brainstem (see figs. 11.15 and 11.19). It

11.4 CLINICAL APPLICATION

Traumatic Brain Injury

A traumatic brain injury (TBI) is defined by what it is not: it is not a birth defect or degenerative, but instead happens from mechanical force. In the United States, more than 5 million people have such injuries.

TBI may result from a fall, accident, attack, or sports-related injury. It is on the rise in combat situations, where the cause and pattern of damage is so distinct that it has been designated "blast-related brain injury." The damage results from a change in atmospheric pressure, violent release of energy (sound, heat, pressure, or electromagnetic waves), and sometimes exposure to a neurotoxin released from the blast. Rocket-propelled grenades, improvised incendiary devices, and landmines create the situations that cause the injury. The brain is initially jolted forward at a force exceeding 1,600 feet per second, and then is hit by a second wave as air in the brain rushes forward.

Unlike a "conventional" TBI in which shrapnel penetrates the brain and surgery is obviously warranted, blast-related brain injury may not initially produce symptoms. A soldier who the day before could easily have run five miles and chat with friends may suddenly be unable to move or speak or may become blind or deaf. Yet others may have no initial effects, while the soft tissue of the brain has sustained severe damage. Blast-related brain injury is also on the rise in combat because of improved ability to treat other types of injuries, enabling soldiers to survive who in wars past would have perished. Effects are lasting—studies on veterans of the Vietnam War indicate cognitive decline years after the injury.

The exact mechanism of severe blast-related brain injury is not well understood, nor do military physicians have a precise definition for it or means of assessment. The presentation overlaps that of mild TBI, which also occurs in combat situations. Mild TBI, also known as a concussion, produces loss of consciousness or altered mental status. Its effects are more psychological than neurological; blast-related brain injury is the opposite.

Mild TBI does not appear to cause lasting damage. Symptoms include disturbed sleep, ringing in the ears, memory lapse, balance problems, irritability, and sensitivity to light and sounds. These physical symptoms are more severe if the person also suffers from depression or post-traumatic stress disorder (PTSD). Mild TBI may *cause* PTSD: the injury generates a shearing force as the brain hits the skull that impairs the prefrontal cortex's control of the amygdala. With an overactive amygdala, the person cannot let go of the psychological trauma—the essence of PTSD. ■

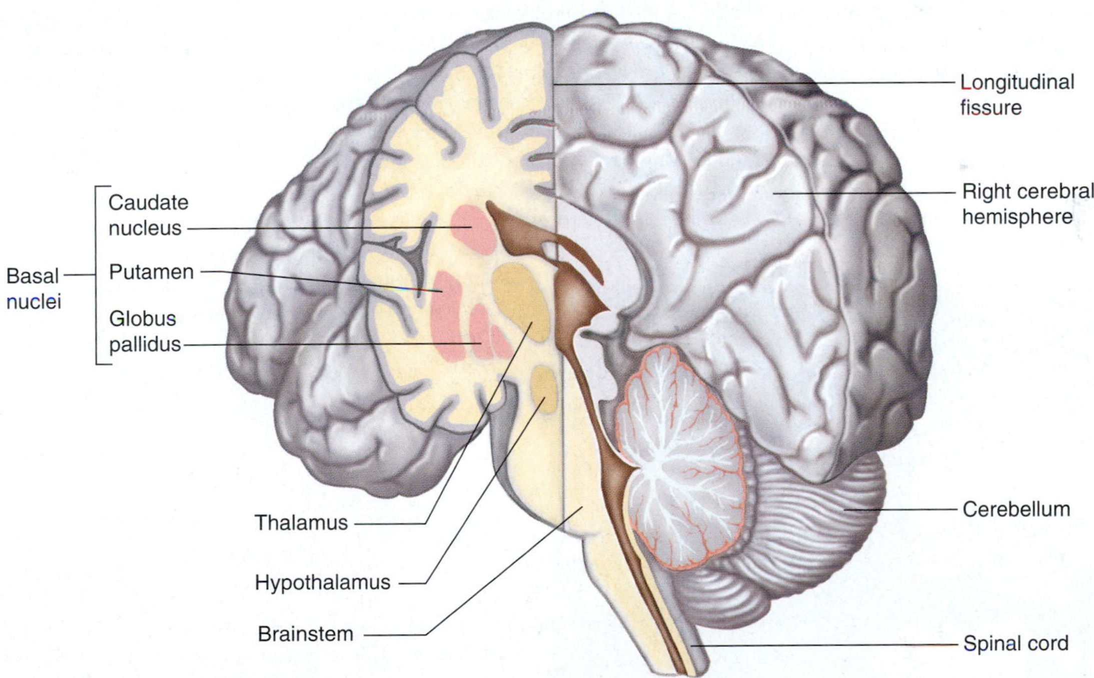

FIGURE 11.19 A coronal section of the left cerebral hemisphere reveals some of the basal nuclei.

surrounds the third ventricle and is largely composed of gray matter. In the diencephalon, a dense mass called the **thalamus** (thal′ah-mus) bulges into the third ventricle from each side. Another region of the diencephalon that includes many nuclei is the **hypothalamus** (hi″po-thal′ah-mus). It lies inferior to the thalamic nuclei and forms the lower walls and floor of the third ventricle (see reference plates 9 and 13).

Other parts of the diencephalon include (1) the **optic tracts** and the **optic chiasma,** formed by the optic nerve fibers crossing over; (2) the **infundibulum,** a conical process

11.5 CLINICAL APPLICATION

Parkinson Disease

Actor Michael J. Fox was in his late twenties when his wife, Tracy Pollan, noticed the first sign of Parkinson disease (PD)—he leaned when walking. When a finger began twitching, Fox consulted a physician, and so began the journey that would lead to his diagnosis with the neurological disorder. Of the approximately six million people worldwide who know that they have PD, only 10% develop symptoms before the age of 40. Michael J. Fox was one of them.

Fox kept his diagnosis private, but by the late 1990s, his co-workers began to notice symptoms that emerged when medication wore off—rigidity, a shuffling and off-balance gait, and poor small motor control. It was difficult to ignore Fox's expressionless, masklike face, a characteristic of PD called hypomimia. Fox had difficulty communicating; it took a huge effort to speak, a symptom called hypophia. Even though his brain could string thoughts into coherent sentences, the muscles of his jaw, lips, and tongue could not utter them. Oddest of all was micrographia, the tendency of his handwriting to become extremely small. PD also causes the sensation of not being able to stay in one spot.

In 1998 Fox publicly disclosed his condition. In 2000, he founded the Michael J. Fox Foundation for Parkinson's Research, and continues to act occasionally in television programs.

In PD, neurons in an area of the brainstem called the substantia nigra degenerate. *Substantia nigra* means "large black area," for the dark pigment that the neurons release as a by-product of synthesizing the neurotransmitter dopamine. When these neurons degenerate, less dopamine reaches synapses with neurons in the striatum of the basal nuclei. The decrease in dopamine causes the motor symptoms of PD. Some patients also develop nonmotor symptoms, including depression, dementia, constipation, incontinence, sleep problems, and orthostatic hypotension (dizziness upon standing).

So far, no treatments can cure or slow the course of PD, but replacing or enhancing use of dopamine can temporarily alleviate symptoms. The standard treatment for many years has been levodopa, a precursor to dopamine that can cross the blood-brain barrier. Once in the brain, levodopa is converted to dopamine. Levodopa provides temporary relief from the twitching and rigidity.

Drug treatment for PD becomes less effective over time. By a feedback mechanism the brain senses the external supply of dopamine and decreases its own production, so that eventually higher doses of levodopa are needed to achieve the effect. Taking too much levodopa leads to another condition, tardive dyskinesia, that produces uncontrollable facial tics and spastic extensions of the limbs. Tardive dyskinesia may result from effects of excess dopamine in brain areas other than those affected in PD.

Surgery can alleviate Parkinson's symptoms. Fox underwent thalamotomy, in which an electrode caused a lesion in his thalamus that calmed violent shaking in his left arm. Another surgical procedure, pallidotomy, causes lesions in the globus pallidus internus, a part of the basal nuclei, and the approach is also used on an area posterior to the thalamus. Deep brain implants of electrodes may also control some symptoms.

Researchers are turning to cells in a patient's own body as sources of the dopamine needed in PD. For example, cells at the back of the eye called retinal pigment epithelium can be cultured with biochemicals that stimulate them to produce dopamine or levodopa. Researchers hope that cells can be sampled from the patient without impairing vision and implanted in the substantia nigra. Neural stem and progenitor cells may also be useful. Implants of fetal dopamine-producing cells performed in the late 1990s had unexpected effects—by 2008, a few patients had died, and the implants showed signs of PD. Still, the implants alleviated symptoms in some patients for several years.

PD may have several causes. Symptoms have been attributed to use of certain designer drugs, exposure to pesticides, and frequent violent blows to the head (fig. 11D). Several genes may increase risk of developing PD, but in most cases it is not inherited. One gene that causes PD when mutant encodes a protein called alpha-synuclein. The abnormal protein folds improperly, forming deposits in the brain (fig. 11E). Understanding how the disease occurs in rare familial forms may provide clues to helping the many others who have this debilitating and common illness. ■

FIGURE 11D Professional boxers are at higher risk of developing Parkinson disease (PD) from repeated blows to the head. Muhammed Ali has PD from many years of head injuries. Michael J. Fox, an actor, not a boxer, first experienced symptoms of PD at age 29, which is unusual.

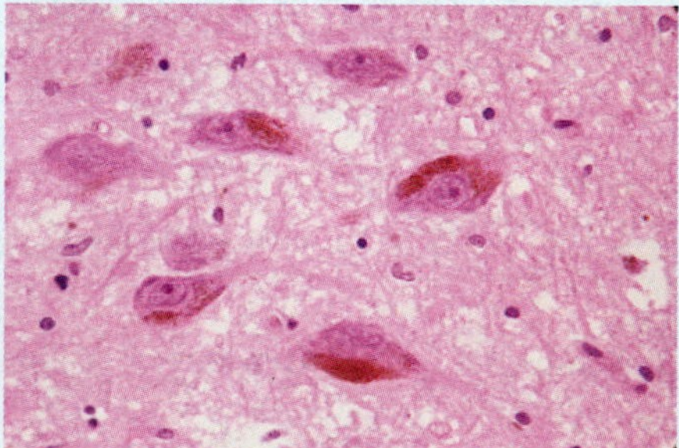

FIGURE 11E The chemical composition of Lewy bodies, characteristic of the brains of people with Parkinson disease, may provide clues to the cause of the condition. Lewy bodies include alpha-synuclein, cytoskeletal elements, and other components.

behind the optic chiasma to which the pituitary gland is attached; (3) the **posterior pituitary gland,** which hangs from the floor of the hypothalamus; (4) the **mammillary** (mam′ĭ-lar″e) **bodies,** two rounded structures behind the infundibulum; and (5) the **pineal gland,** which forms as a cone-shaped evagination from the roof of the diencephalon (see chapter 13, p. 511).

The thalamus is a selective gateway for sensory impulses ascending from other parts of the nervous system to the cerebral cortex. It receives all sensory impulses (except those associated with the sense of smell) and channels them to appropriate regions of the cortex for interpretation. In addition, all regions of the cerebral cortex can communicate with the thalamus by means of descending fibers.

The thalamus transmits sensory information by synchronizing action potentials. Consider vision. An image on the retina stimulates the *lateral geniculate nucleus* (LGN) region of the thalamus, which then sends action potentials to a part of the visual cortex. Those action potentials are synchronized—fired simultaneously—by the LGN's neurons only if the stimuli come from a single object, such as a bar. If the stimulus is two black dots, the resulting thalamic action potentials are not synchronized. The synchronicity of action potentials, therefore, may be a way that the thalamus selects which stimuli to relay to higher brain structures. Therefore, the thalamus is not only a messenger but also an editor.

Nerve fibers connect the hypothalamus to the cerebral cortex, thalamus, and parts of the brainstem so that it can receive impulses from them and send impulses to them. The hypothalamus maintains homeostasis by regulating a variety of visceral activities and by linking the nervous and endocrine systems.

The hypothalamus regulates:

1. Heart rate and arterial blood pressure.
2. Body temperature.
3. Water and electrolyte balance.
4. Control of hunger and body weight.
5. Control of movements and glandular secretions of the stomach and intestines.
6. Production of neurosecretory substances that stimulate the pituitary gland to release hormones that help regulate growth, control various glands, and influence reproductive physiology.
7. Sleep and wakefulness.

Structures in the region of the diencephalon also are important in controlling emotional responses. Parts of the cerebral cortex in the medial parts of the frontal and temporal lobes connect with the hypothalamus, thalamus, basal nuclei, and other deep nuclei. These structures form a complex called the **limbic system.** It controls emotional experience and expression and can modify the way a person acts, producing such feelings as fear, anger, pleasure, and sorrow. The limbic system reacts to potentially life-threatening upsets in a person's physical or psychological condition. By causing pleasant or unpleasant feelings about experiences, the limbic system guides behavior that may increase the chance of survival. In addition, parts of the limbic system interpret sensory impulses from the receptors associated with the sense of smell (olfactory receptors).

Brainstem

The **brainstem** connects the brain to the spinal cord. It consists of the midbrain, pons, and medulla oblongata. These structures include many tracts of nerve fibers and masses of gray matter called *nuclei* (see figs. 11.15, 11.19, and 11.20).

Midbrain

The **midbrain** (mesencephalon) is a short section of the brainstem between the diencephalon and the pons. It contains bundles of myelinated nerve fibers that join lower parts of the brainstem and spinal cord with higher parts of the brain. The midbrain includes several masses of gray matter that serve as reflex centers. It also contains the *cerebral aqueduct* that connects the third and fourth ventricles (fig. 11.21).

Two prominent bundles of nerve fibers on the underside of the midbrain comprise the *cerebral peduncles.* These fibers include the corticospinal tracts and are the main motor pathways between the cerebrum and lower parts of the nervous system (see fig. 11.20). Beneath the cerebral peduncles are large bundles of sensory fibers that carry impulses upward to the thalamus.

Two pairs of rounded knobs on the superior surface of the midbrain mark the location of four nuclei, known collectively as *corpora quadrigemina.* The upper masses (superior colliculi) contain the centers for certain visual reflexes, such as those responsible for moving the eyes to view something as the head turns. The lower ones (inferior colliculi) contain the auditory reflex centers that operate when it is necessary to move the head to hear sounds more distinctly (see fig. 11.20).

Near the center of the midbrain is a mass of gray matter called the *red nucleus.* This nucleus communicates with the cerebellum and with centers of the spinal cord, and it provides reflexes that maintain posture. It appears red because it is richly supplied with blood vessels.

Pons

The **pons** appears as a rounded bulge on the underside of the brainstem where it separates the midbrain from the medulla oblongata (see fig. 11.20). The dorsal portion of the pons largely consists of longitudinal nerve fibers, which relay impulses to and from the medulla oblongata and the cerebrum. Its ventral portion contains large bundles of transverse nerve fibers, which transmit impulses from the cerebrum to centers within the cerebellum.

Several nuclei of the pons relay sensory impulses from peripheral nerves to higher brain centers. Other nuclei function with centers of the medulla oblongata to maintain the basic rhythm of breathing.

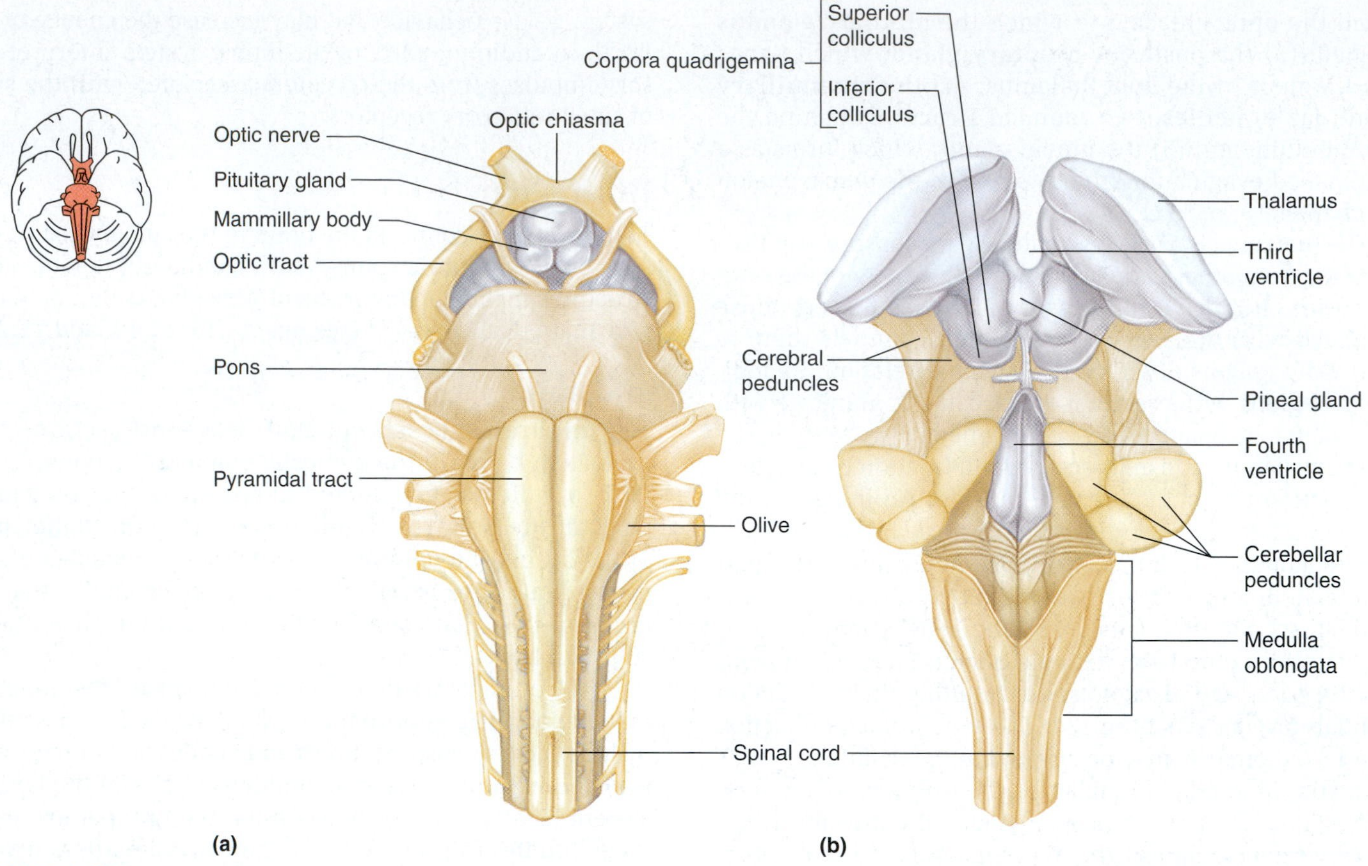

FIGURE 11.20 Brainstem. (*a*) Ventral view of the brainstem. (*b*) Dorsal view of the brainstem with the cerebellum removed, exposing the fourth ventricle.

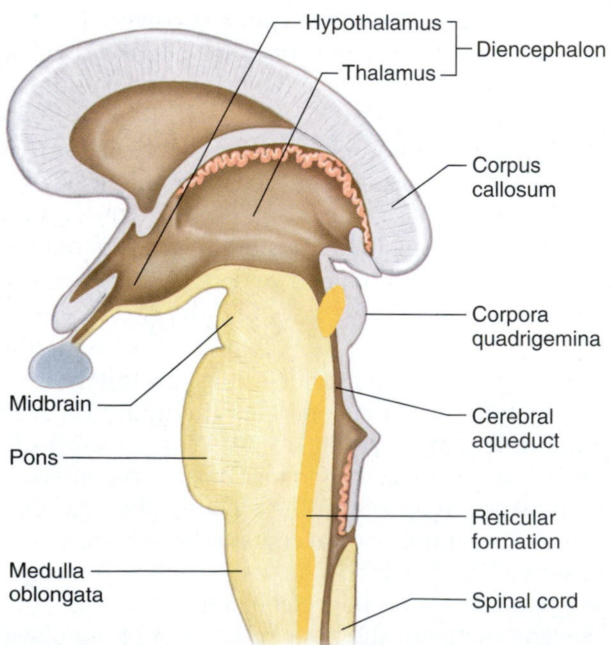

FIGURE 11.21 The reticular formation (shown in gold) extends from the superior portion of the spinal cord into the diencephalon.

Medulla Oblongata

The **medulla oblongata** (mĕ-dul′ah ob″long-ga′tah) is an enlarged continuation of the spinal cord, extending from the level of the foramen magnum to the pons (see fig. 11.20). Its dorsal surface flattens to form the floor of the fourth ventricle, and its ventral surface is marked by the corticospinal tracts, most of whose fibers cross over at this level. On each side of the medulla oblongata is an oval swelling called the *olive,* from which a large bundle of nerve fibers arises and passes to the cerebellum.

The ascending and descending nerve fibers connecting the brain and spinal cord must pass through the medulla oblongata because of its location. As in the spinal cord, the white matter of the medulla surrounds a central mass of gray matter. Here, however, the gray matter breaks up into nuclei separated by nerve fibers. Some of these nuclei relay ascending impulses to the other side of the brainstem and then on to higher brain centers. The *nucleus gracilis* and the *nucleus cuneatus,* for example, receive sensory impulses from fibers of the fasciculus gracilis and the fasciculus cuneatus and pass them on to the thalamus or the cerebellum.

Other nuclei in the medulla oblongata control vital visceral activities. These nuclei include the following:

1. **Cardiac center.** Peripheral nerves transmit impulses originating in the cardiac center to the heart, where they increase or decrease heart rate.

2. **Vasomotor center.** Certain cells of the vasomotor center initiate impulses that travel to smooth muscles in the walls of blood vessels and stimulate them to contract, constricting the vessels (vasoconstriction) and thereby increasing blood pressure. A decrease in the activity of these cells can produce the opposite effect—dilation of the blood vessels (vasodilation) and a consequent drop in blood pressure.
3. **Respiratory center.** The respiratory center adjusts the rate and depth of breathing and acts with the pons to maintain the basic rhythm of breathing.

Some nuclei in the medulla oblongata are centers for certain nonvital reflexes, such as those associated with coughing, sneezing, swallowing, and vomiting. However, because the medulla also contains vital reflex centers, injuries to this part of the brainstem are often fatal.

Reticular Formation

Scattered throughout the medulla oblongata, pons, and midbrain is a complex network of nerve fibers associated with tiny islands of gray matter. This network, the **reticular formation** (rĕ-tik′u-lar fōr-ma′shun), or reticular activating system, extends from the superior portion of the spinal cord into the diencephalon (fig. 11.21). Its intricate system of nerve fibers connects centers of the hypothalamus, basal nuclei, cerebellum, and cerebrum with fibers in all the major ascending and descending tracts.

When sensory impulses reach the reticular formation, it responds by activating the cerebral cortex into a state of wakefulness. Without this arousal, the cortex remains unaware of stimulation and cannot interpret sensory information or carry on thought processes. Decreased activity in the reticular formation results in sleep. If the reticular formation is injured and ceases to function, the person remains unconscious, even with strong stimulation. This is called a comatose state.

The reticular formation filters incoming sensory impulses. Impulses judged to be important, such as those originating in pain receptors, are passed on to the cerebral cortex, while others are disregarded. This selective action of the reticular formation frees the cortex from what would otherwise be a continual bombardment of sensory stimulation and allows it to concentrate on more significant information. The cerebral cortex can also activate the reticular system, so intense cerebral activity keeps a person awake. In addition, the reticular formation regulates motor activities so that various skeletal muscles move together evenly, and it inhibits or enhances certain spinal reflexes.

A person in a persistent vegetative state is occasionally awake, but not aware; a person in a coma is not awake or aware. Sometimes following a severe injury, a person will become comatose and then gradually enter a persistent vegetative state. Coma and persistent vegetative state are also seen in the end stage of neurodegenerative disorders such as Alzheimer disease; when there is an untreatable mass in the brain, such as a blood clot or tumor; or in anencephaly, when a newborn lacks higher brain structures.

Types of Sleep

The two types of normal sleep are *slow-wave* and *rapid eye movement* (REM). Slow-wave sleep (also called non-REM sleep) occurs when a person is very tired, and it reflects decreasing activity of the reticular formation. It is restful, dreamless, and accompanied by reduced blood pressure and respiratory rate. Slow-wave sleep may range from light to heavy and is usually described in four stages. It may last from seventy to ninety minutes. Slow-wave and REM sleep alternate.

REM sleep is also called "paradoxical sleep" because some areas of the brain are active. As its name implies, the eyes can be seen rapidly moving beneath the eyelids. Cats and dogs in REM sleep sometimes twitch their limbs. In humans, REM sleep usually lasts from five to fifteen minutes. This "dream sleep" is apparently important. If a person lacks REM sleep for just one night, sleep on the next night makes up for it. During REM sleep, heart and respiratory rates are irregular. Certain drugs, such as marijuana and alcohol, interfere with REM sleep. Table 11.6 describes several disorders of sleep.

TABLE 11.6 | Sleep Disorders

Disorder	Symptoms	Percent of Population
Fatal familial insomnia	Inability to sleep, emotional instability, hallucinations, stupor, coma, death within thirteen months of onset around age fifty, both slow-wave and REM sleep abolished.	Very rare
Insomnia	Inability to fall or remain asleep.	10%
Narcolepsy	Abnormal REM sleep causes extreme daytime sleepiness, begins between ages of fifteen and twenty-five.	0.02–0.06%
Obstructive sleep apnea syndrome	Upper airway collapses repeatedly during sleep, blocking breathing. Snoring and daytime sleepiness.	4–5%
Parasomnias	Sleepwalking, sleeptalking, and night terrors.	<5% of children
REM-sleep behavior disorder	Excessive motor activity during REM sleep, which disturbs continuous sleep.	Very rare
Restless legs syndrome	Brief, repetitive leg jerks during sleep. Leg pain forces person to get up several times a night.	5.5%
Sleep paralysis	Inability to move for up to a few minutes after awakening or when falling asleep.	Very rare

PRACTICE

25 What are the major functions of the thalamus? Of the hypothalamus?

26 How may the limbic system influence a person's behavior?

27 Which vital reflex centers are located in the brainstem?

28 What is the function of the reticular formation?

29 Describe two types of sleep.

Cerebellum

The **cerebellum** (ser″ĕ-bel′um) is a large mass of tissue located inferior to the occipital lobes of the cerebrum and posterior to the pons and medulla oblongata (see fig. 11.15). It consists of two lateral hemispheres partially separated by a layer of dura mater called the *falx cerebelli.* A structure called the *vermis* connects the cerebellar hemispheres at the midline.

Like the cerebrum, the cerebellum is primarily composed of white matter with a thin layer of gray matter, the **cerebellar cortex,** on its surface. This cortex doubles over on itself in a series of complex folds that have myelinated nerve fibers branching into them. A cut into the cerebellum reveals a treelike pattern of white matter, called the *arbor vitae,* surrounded by gray matter. A number of nuclei lie deep within each cerebellar hemisphere. The largest and most important is the *dentate nucleus.*

The cerebellum communicates with other parts of the CNS by means of three pairs of nerve tracts called **cerebellar peduncles** (ser″ĕ-bel′ar pe-dung′kls) (fig. 11.22). One pair, the *inferior peduncles,* brings sensory information concerning the position of body parts such as limbs and joints to the cerebellum via the spinal cord and medulla oblongata. The *middle peduncles* transmit information from the cerebral cortex about the desired position of these body parts. After integrating and analyzing the information from these two sources, the cerebellum sends correcting impulses from the dentate nucleus via the *superior peduncles* to the midbrain (fig. 11.22). These corrections are incorporated into motor impulses that travel downward through the pons, medulla oblongata, and spinal cord in the appropriate patterns to move the body in the desired way.

Overall, the cerebellum integrates sensory information concerning the position of body parts and coordinates skeletal muscle activity and maintains posture. It receives sensory impulses from receptors in muscles, tendons, and joints (proprioceptors) and from special sense organs, such as the eyes and ears. For example, the cerebellum uses sensory information from the semicircular canals of the inner ears concerning the motion and position of the head to help maintain equilibrium (see chapter 12, pp. 459–461). Damage to the cerebellum can cause tremors, inaccurate movements of voluntary muscles, loss of muscle tone, a reeling walk, and loss of equilibrium.

Table 11.7 summarizes the characteristics and functions of the major parts of the brain. Clinical Application 11.6 discusses how brain waves reflect brain activity.

PRACTICE 

30 Where is the cerebellum located?

31 What are the major functions of the cerebellum?

32 What types of receptors provide information to the cerebellum?

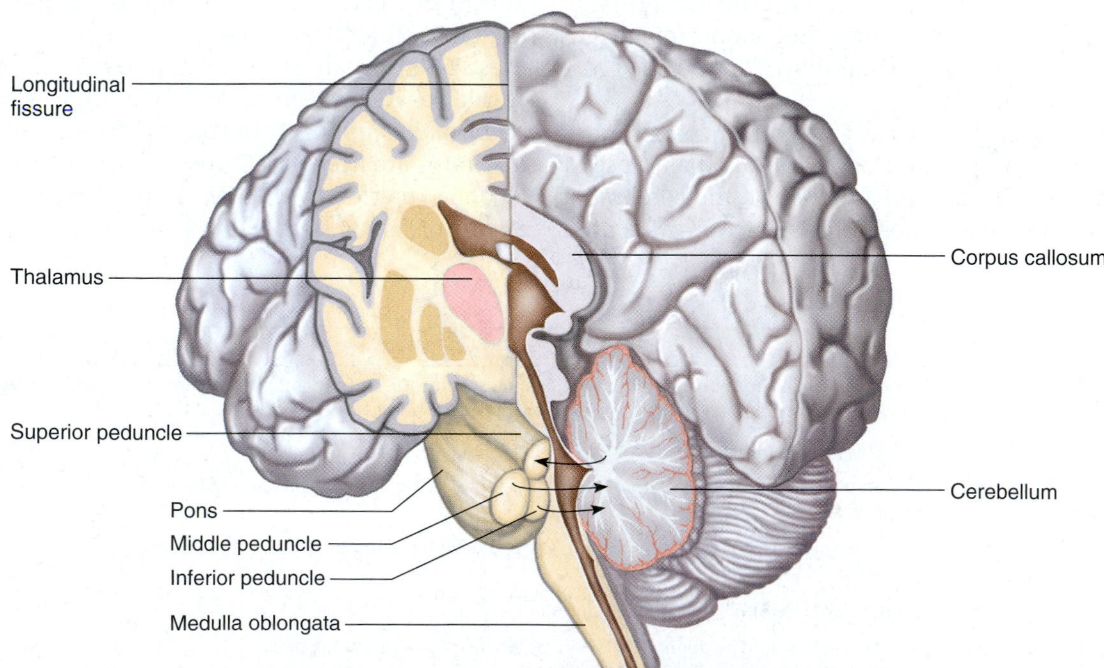

FIGURE 11.22 The cerebellum, located inferior to the occipital lobes of the cerebrum, communicates with other parts of the nervous system by means of the cerebellar peduncles.

11.6 CLINICAL APPLICATION

Brain Waves

Brain waves are recordings of fluctuating electrical changes in the brain. To obtain such a recording, electrodes are positioned on the surface of a surgically exposed brain (an electrocorticogram, ECoG) or on the outer surface of the head (an electroencephalogram, EEG). These electrodes detect electrical changes in the extracellular fluid of the brain in response to changes in potential among large groups of neurons. The resulting signals from the electrodes are amplified and recorded. Brain waves originate from the cerebral cortex but also reflect activities in other parts of the brain that influence the cortex, such as the reticular formation. The intensity of electrical changes is proportional to the degree of neuronal activity, so brain waves vary markedly in amplitude and frequency between sleep and wakefulness.

Brain waves are classified as alpha, beta, theta, and delta waves (fig. 11F). *Alpha waves* are recorded most easily from the posterior regions of the head and have a frequency of 8–13 cycles per second. They occur when a person is awake but resting, with the eyes closed. These waves disappear during sleep, and if the wakeful person's eyes open, higher-frequency beta waves replace the alpha waves.

Beta waves have a frequency of more than 13 cycles per second and are usually recorded in the anterior region of the head. They occur when a person is actively engaged in mental activity or is under tension.

Theta waves have a frequency of 4–7 cycles per second and occur mainly in the parietal and temporal regions of the cerebrum. They are normal in children but do not usually occur in adults. However, some adults produce theta waves in early stages of sleep or at times of emotional stress.

Delta waves have a frequency below 4 cycles per second and happen during sleep. They originate from the cerebral cortex when it is not being activated by the reticular formation.

Brain wave patterns can be useful for diagnosing disease conditions, such as distinguishing types of seizure disorders (epilepsy) and locating brain tumors. Brain waves are also used to detect *brain death,* in which neuronal activity ceases. An EEG that lacks waves (isoelectric EEG) verifies brain death. However, drugs that greatly depress brain functions must be excluded as the cause of the flat EEG pattern before confirming brain death. ■

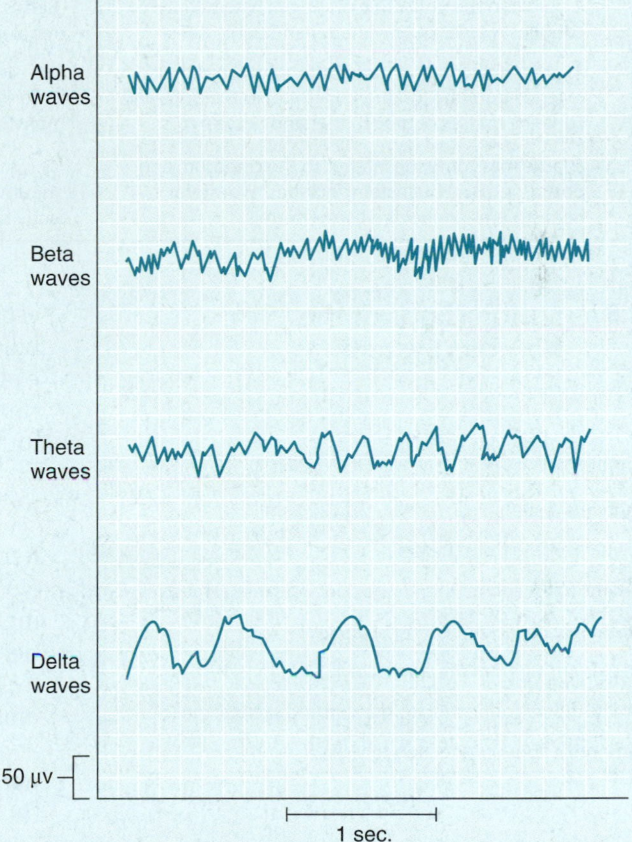

FIGURE 11F Brain waves record fluctuating electrical changes in the brain.

11.6 PERIPHERAL NERVOUS SYSTEM

The **peripheral nervous system** consists of the nerves that branch from the CNS, connecting it to other body parts. The PNS includes the *cranial nerves* that arise from the brain and the *spinal nerves* that arise from the spinal cord.

The PNS can also be subdivided into somatic and autonomic nervous systems. Generally, the **somatic nervous system** consists of the cranial and spinal nerve fibers that connect the CNS to the skin and skeletal muscles, so it oversees conscious activities. The **autonomic nervous system** (aw″to nom′ik ner′vus sis′tem) includes fibers that connect the CNS to viscera such as the heart, stomach, intestines, and various glands. The autonomic nervous system controls subconscious actions. Table 11.8 outlines the subdivisions of the nervous system.

TABLE 11.7 | Major Parts of the Brain

Part	Characteristics	Functions
1. Cerebrum	Largest part of the brain; two hemispheres connected by the corpus callosum	Controls higher brain functions, including interpreting sensory impulses, initiating muscular movements, storing memory, reasoning, and determining intelligence
2. Basal nuclei (ganglia)	Masses of gray matter deep within the cerebral hemispheres	Relay stations for motor impulses originating in the cerebral cortex and passing into the brainstem and spinal cord
3. Diencephalon	Includes masses of gray matter (thalamus and hypothalamus)	The thalamus is a relay station for sensory impulses ascending from other parts of the nervous system to the cerebral cortex; the hypothalamus helps maintain homeostasis by regulating visceral activities and by linking the nervous and endocrine systems
4. Brainstem	Connects the cerebrum to the spinal cord	
a. Midbrain	Contains masses of gray matter and bundles of nerve fibers that join the spinal cord to higher regions of the brain	Contains reflex centers that move the eyes and head, and maintains posture
b. Pons	A bulge on the underside of the brainstem that contains masses of gray matter and nerve fibers	Relays nerve impulses to and from the medulla oblongata and cerebrum; helps regulate rate and depth of breathing
c. Medulla oblongata	An enlarged continuation of the spinal cord that extends from the foramen magnum to the pons and contains masses of gray matter and nerve fibers	Conducts ascending and descending impulses between the brain and spinal cord; contains cardiac, vasomotor, and respiratory control centers and various nonvital reflex control centers
5. Cerebellum	A large mass of tissue inferior to the cerebrum and posterior to the brainstem; includes two lateral hemispheres connected by the vermis	Communicates with other parts of the CNS by nerve tracts; integrates sensory information concerning the position of body parts; and coordinates muscle activities and maintains posture

TABLE 11.8 | Subdivisions of the Nervous System

Subdivisions
1. Central nervous system (CNS)
a. Brain
b. Spinal cord
2. Peripheral nervous system (PNS)
a. Cranial nerves arising from the brain
(1) Somatic fibers connecting to the skin and skeletal muscles
(2) Autonomic fibers connecting to viscera
b. Spinal nerves arising from the spinal cord
(1) Somatic fibers connecting to the skin and skeletal muscles
(2) Autonomic fibers connecting to viscera

Structure of Peripheral Nerves

A peripheral nerve consists of connective tissue surrounding bundles of nerve fibers. The outermost layer of the connective tissue, called the *epineurium,* is dense and includes many collagenous fibers. Each bundle of nerve fibers (fascicle) is, in turn, enclosed in a sleeve of looser connective tissue called the *perineurium.* A small amount of loose connective tissue called *endoneurium* surrounds individual nerve fibers (figs. 11.23 and 11.24). Blood vessels in the epineurium and perineurium give rise to a network of capillaries in the endoneurium that provides oxygen and nutrients to the neurons.

The term "muscle fiber" refers to an entire muscle cell, whereas the term "nerve fiber" refers to a cellular process, especially an axon. The terminology for the connective tissue holding them together, however, is similar. In both cases, for example, fibers are bundled into fascicles, whereas epineurium in nerves corresponds to epimysium in muscles, and so forth (see figs. 11.23, 11.24, 9.2, and 9.3).

Nerve and Nerve Fiber Classification

Recall that nerves are bundles of nerve fibers, or axons. Nerves that have only fibers of sensory neurons, conducting impulses into the brain or spinal cord, are called **sensory nerves.** Nerves that have only fibers involved in motor control are **motor nerves.** Most nerves include both sensory and motor fibers and are called **mixed nerves.**

Nerves originating from the brain that communicate with other body parts are called **cranial nerves,** whereas those originating from the spinal cord that communicate with other body parts are called **spinal nerves.** The nerve fibers in these structures can be subdivided further into four groups as follows:

1. **General somatic efferent fibers** carry motor impulses outward from the brain or spinal cord to skeletal muscles and stimulate them to contract.
2. **General visceral efferent fibers** carry motor impulses outward from the brain or spinal cord to various smooth muscles and glands associated with internal organs, causing certain muscles to contract or glands to secrete.
3. **General somatic afferent fibers** carry sensory impulses inward to the brain or spinal cord from receptors in the skin and skeletal muscles.
4. **General visceral afferent fibers** carry sensory impulses to the CNS from blood vessels and internal organs.

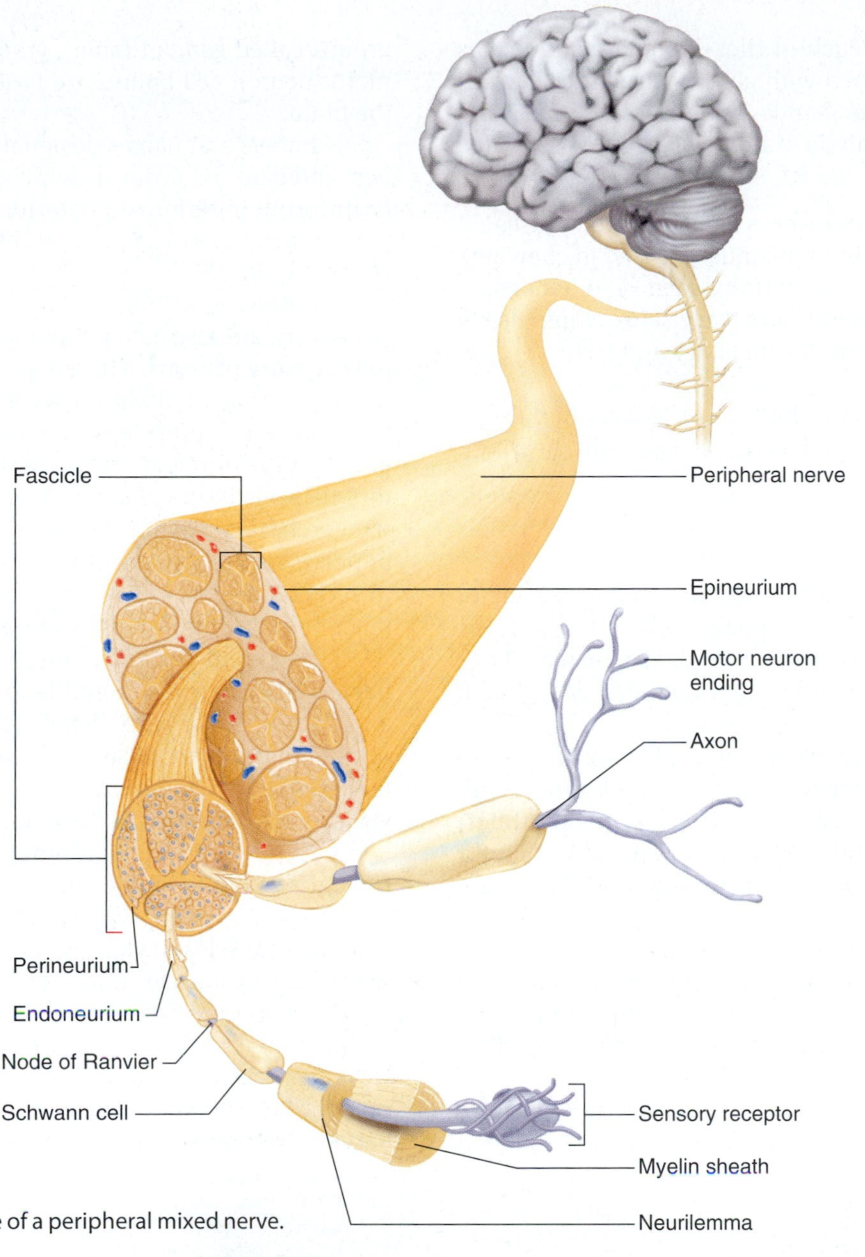

FIGURE 11.23 The structure of a peripheral mixed nerve.

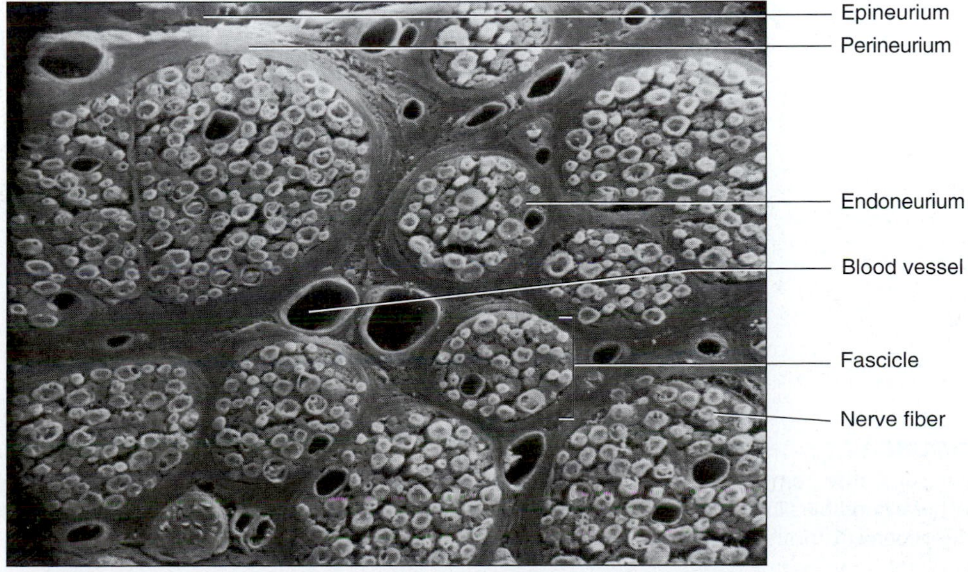

FIGURE 11.24 Scanning electron micrograph of a peripheral nerve in cross section (350×). Note the bundles or fascicles of nerve fibers. Fibers include axons of motor neurons as well as peripheral processes of sensory neurons. Copyright by R.G. Kessel and R.H. Kardon, *Tissues and Organs: A Text-Atlas of Scanning Electron Microscopy,* 1979 (W.H. Freeman & Co.).

The term *general* in each of these categories indicates that the fibers are associated with general structures such as the skin, skeletal muscles, glands, and viscera. Three other groups of fibers, found only in cranial nerves, are associated with more specialized, or *special,* structures:

1. **Special somatic efferent fibers** carry motor impulses outward from the brain to the muscles used in chewing, swallowing, speaking, and forming facial expressions.
2. **Special visceral afferent fibers** carry sensory impulses inward to the brain from the olfactory and taste receptors.
3. **Special somatic afferent fibers** carry sensory impulses inward to the brain from the receptors of sight, hearing, and equilibrium.

Cranial Nerves

Twelve pairs of **cranial nerves** arise from the underside of the brain. Except for the first two pairs, which begin in the cerebrum, these nerves originate from the brainstem. They pass from their sites of origin through foramina of the skull and lead to areas of the head, neck, and trunk.

Most cranial nerves are mixed nerves, but some of those associated with special senses, such as smell and vision, have only sensory fibers. Other cranial nerves that innervate muscles and glands are primarily composed of motor fibers and have only limited sensory functions. These are neurons associated with certain receptors (*proprioceptors*) that respond to the rate or degree of contraction of skeletal muscles. These fibers contribute directly to motor control, so cranial nerves whose only sensory component is from such proprioceptors are usually considered motor nerves. This pertains to cranial nerves III, IV, VI, XI, and XII.

Neuron cell bodies to which the sensory fibers in the cranial nerves attach are outside the brain and are usually in groups called *ganglia* (sing., *ganglion*). On the other hand, motor neuron cell bodies are typically in the gray matter of the brain.

Numbers and names designate cranial nerves. The numbers indicate the order in which the nerves arise from the brain, from anterior to posterior. The names describe primary functions or the general distribution of cranial nerve fibers (fig. 11.25).

The first pair of cranial nerves, the **olfactory** (ol-fak′to-re) **nerves** (I), are associated with the sense of smell and include only sensory neurons. These bipolar neurons, in the lining of the upper nasal cavity, serve as *olfactory receptor cells.* Axons from these receptors pass upward through the cribriform plates of the ethmoid bone, carrying impulses to the olfactory neurons in the *olfactory bulbs,* which are extensions of the cerebral cortex just beneath the frontal lobes. Sensory impulses move from the olfactory bulbs along *olfactory tracts* to cerebral centers where they are interpreted. The result of this interpretation is the sensation of smell.

The second pair of cranial nerves, the **optic** (op′tik) **nerves** (II), are sensory and lead from the eyes to the brain and are associated with vision. The cell bodies of these neurons form ganglion cell layers in the eyes, and their axons pass through the *optic foramina* of the orbits and continue into the visual nerve pathways of the brain (see chapter 12, p. 475).

The third pair, the **oculomotor** (ok″u-lo-mo′tor) **nerves** (III), arise from the midbrain and pass into the orbits of the eyes. One component of each nerve connects to a number of voluntary muscles, including those that raise the eyelids and four of the six muscles that move the eye.

A second portion of each oculomotor nerve is part of the autonomic nervous system, supplying involuntary muscles inside the eyes. These muscles help adjust the amount of light that enters the eyes and help focus the lenses. This nerve is considered motor, with some proprioceptive fibers.

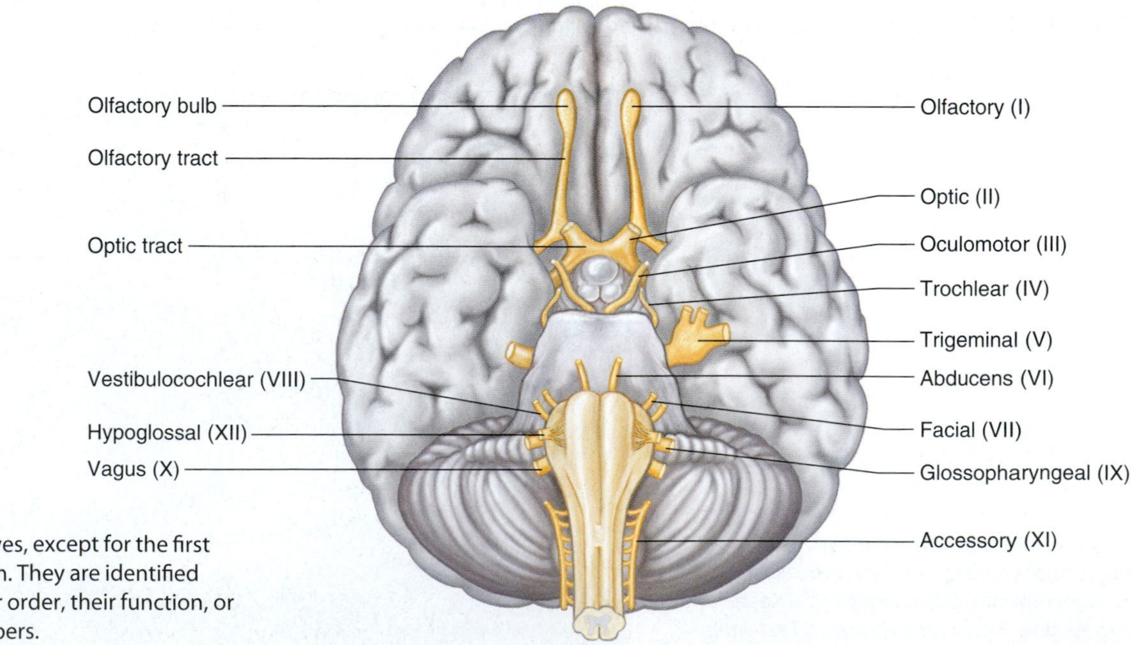

FIGURE 11.25 The cranial nerves, except for the first two pairs, arise from the brainstem. They are identified either by numbers indicating their order, their function, or the general distribution of their fibers.

The fourth pair, the **trochlear** (trok′le-ar) **nerves** (IV), are the smallest cranial nerves. They arise from the midbrain and carry motor impulses to a fifth pair of external eye muscles, the *superior oblique muscles,* which are not supplied by the oculomotor nerves. This nerve is considered motor, with some proprioceptive fibers.

The fifth pair, the **trigeminal** (tri-jem′i-nal) nerves (V), are the largest of the cranial nerves and arise from the pons. They are mixed nerves, with the sensory portions more extensive than the motor portions. Each sensory component includes three large branches, called the ophthalmic, maxillary, and mandibular divisions (fig. 11.26).

The *ophthalmic division* consists of sensory fibers that bring impulses to the brain from the surface of the eye; the tear gland; and the skin of the anterior scalp, forehead, and upper eyelid. The fibers of the *maxillary division* carry sensory impulses from the upper teeth, upper gum, and upper lip, as well as from the mucous lining of the palate and facial skin. The *mandibular division* includes both motor and sensory fibers. The sensory branches transmit impulses from the scalp behind the ears, the skin of the jaw, the lower teeth, the lower gum, and the lower lip. The motor branches supply the muscles of mastication and certain muscles in the floor of the mouth.

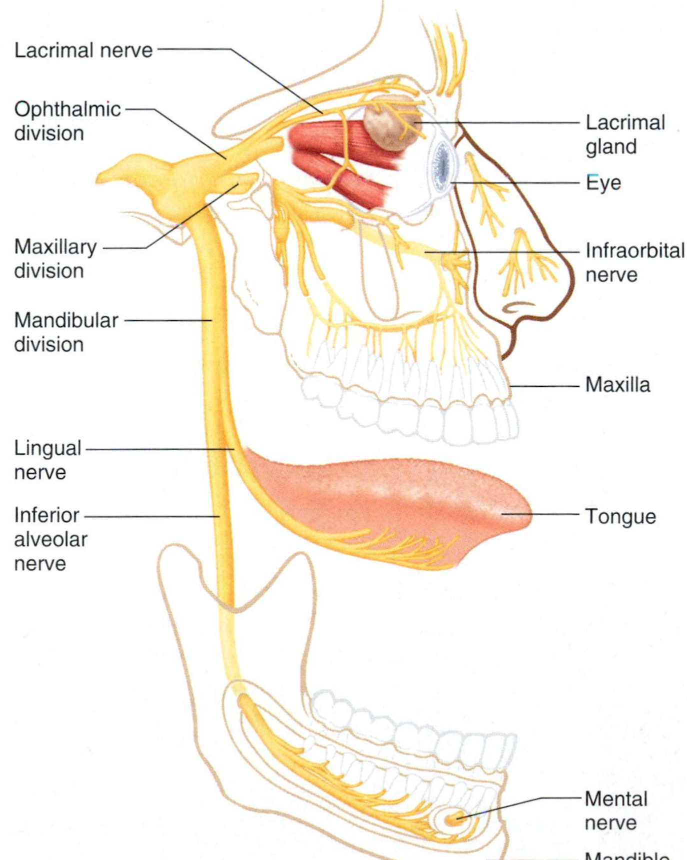

FIGURE 11.26 Each trigeminal nerve has three large branches that supply various regions of the head and face: the ophthalmic division (sensory), the maxillary division (sensory), and the mandibular division (sensory and motor to muscles of mastication).

A disorder of the trigeminal nerve called *trigeminal neuralgia* (tic douloureux) causes severe recurring pain in the face and forehead on the affected side. If drugs cannot control the pain, surgery may be used to sever the sensory part of the nerve. However, the patient loses sensations in other body regions that the sensory branch supplies. Consequently, after such surgery, care must be taken when eating or drinking hot foods or liquids, and the mouth must be inspected daily for food particles or damage to the cheeks from biting.

The sixth pair, the **abducens** (ab-du′senz) **nerves** (VI), are small and originate from the pons near the medulla oblongata. They enter the orbits of the eyes and supply motor impulses to the remaining pair of external eye muscles, the *lateral rectus muscles.* This nerve is considered motor, with some proprioceptive fibers.

The seventh pair of cranial nerves, the **facial** (fa′shal) **nerves** (VII), are mixed nerves that arise from the lower part of the pons and emerge on the sides of the face. Their sensory branches are associated with taste receptors on the anterior two-thirds of the tongue, and some of their motor fibers transmit impulses to muscles of facial expression (fig. 11.27). Still other motor fibers of these nerves function in the autonomic nervous system by stimulating secretions from tear glands and certain salivary glands (submandibular and sublingual glands).

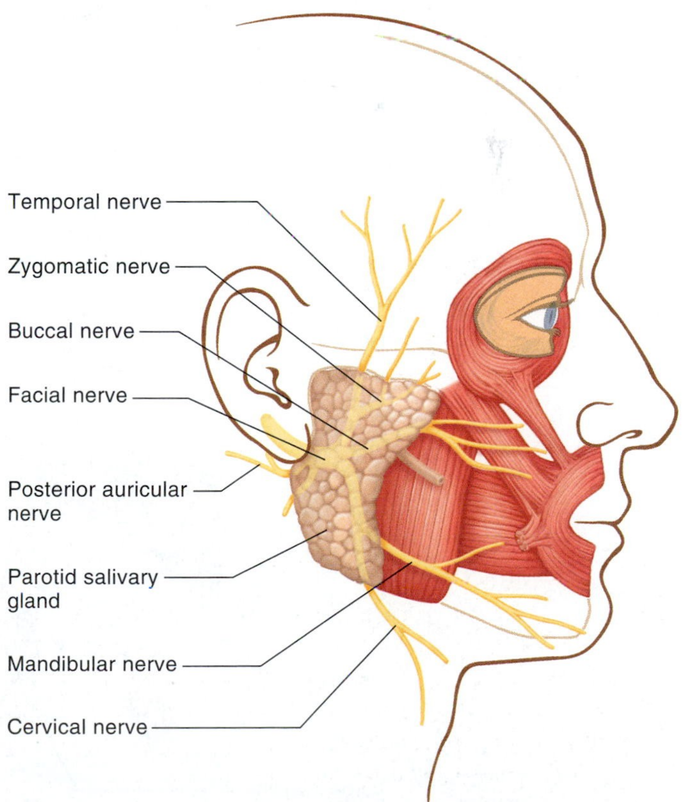

FIGURE 11.27 The facial nerves are associated with taste receptors on the tongue and with muscles of facial expression.

The eighth pair, the **vestibulocochlear** (ves-tib″u-lo-kok′le-ar) **nerves** (VIII, acoustic, or auditory, nerves), are sensory nerves that arise from the medulla oblongata. Each of these nerves has two distinct parts—a vestibular branch and a cochlear branch.

The neuron cell bodies of the *vestibular branch* fibers are located in ganglia near the vestibule and semicircular canals of the inner ear. These structures contain receptors that sense changes in the position of the head and, in response, initiate and send impulses to the cerebellum, where they are used in reflexes that maintain equilibrium.

The neuron cell bodies of the *cochlear branch* fibers are located in a ganglion of the cochlea, a part of the inner ear that houses the hearing receptors. Impulses from this branch pass through the medulla oblongata and midbrain on their way to the temporal lobe, where they are interpreted.

The ninth pair, the **glossopharyngeal** (glos″o-fah-rin′je-al) **nerves** (IX), are associated with the tongue and pharynx. These nerves arise from the medulla oblongata, and, although they are mixed nerves, their predominant fibers are sensory. These fibers carry impulses from the lining of the pharynx, tonsils, and posterior third of the tongue to the brain. Fibers in the motor component of the glossopharyngeal nerves innervate certain salivary glands and a constrictor muscle in the wall of the pharynx that functions in swallowing.

The tenth pair, the **vagus** (va′gus) **nerves** (X), originate in the medulla oblongata and extend downward through the neck into the chest and abdomen. These nerves are mixed, including both somatic and autonomic branches, with the autonomic fibers predominant.

Among the somatic components of the vagus nerves are motor fibers that carry impulses to muscles of the larynx. These fibers are associated with speech and swallowing reflexes that employ muscles in the soft palate and pharynx. Vagal sensory fibers carry impulses from the linings of the pharynx, larynx, and esophagus and from the viscera of the thorax and abdomen to the brain. Autonomic motor fibers of the vagus nerves supply the heart and many smooth muscles and glands in the viscera of the thorax and abdomen (fig. 11.28).

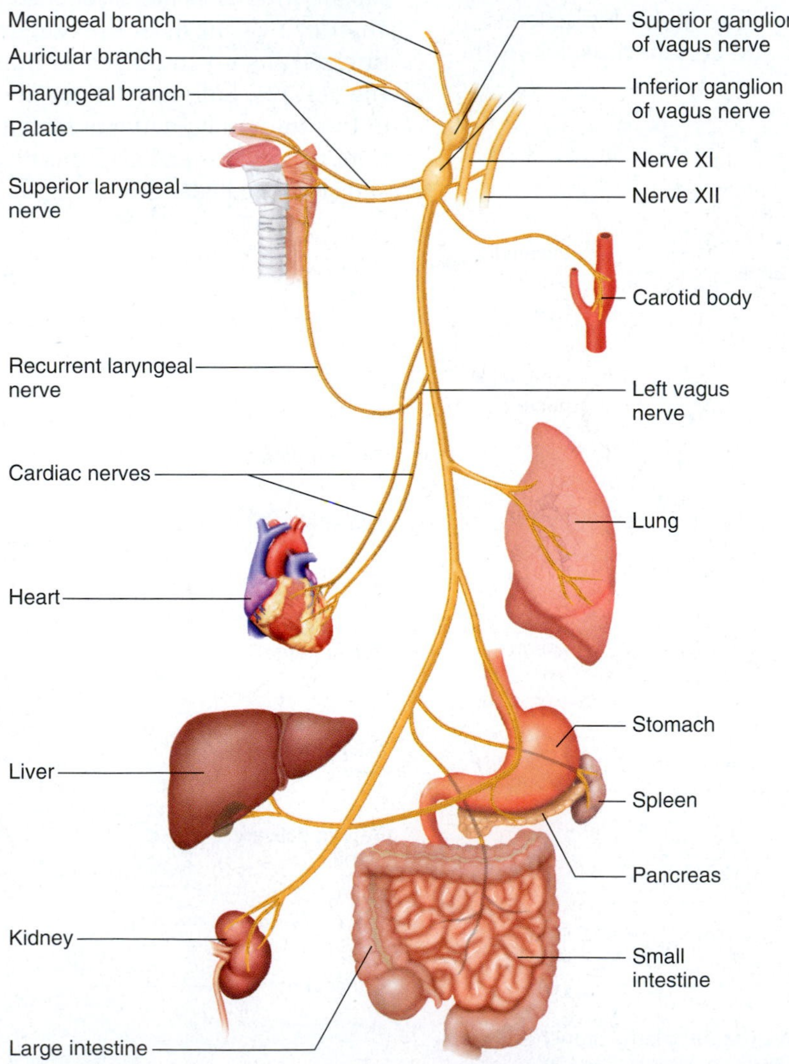

FIGURE 11.28 The vagus nerves (only the left vagus is shown) extend from the medulla oblongata downward into the chest and abdomen to supply many organs.

The eleventh pair, the **accessory** (ak-ses′o-re) **nerves** (XI, spinal accessory), originate in the medulla oblongata and the spinal cord; thus, they have both cranial and spinal branches. Each *cranial branch* of an accessory nerve joins a vagus nerve and carries impulses to muscles of the soft palate, pharynx, and larynx. The *spinal branch* descends into the neck and supplies motor fibers to the trapezius and sternocleidomastoid muscles. This nerve is considered motor, with some proprioceptive fibers.

The twelfth pair, the **hypoglossal** (hi″po-glos′al) **nerves** (XII), arise from the medulla oblongata and pass into the tongue. They primarily consist of fibers that carry impulses to muscles that move the tongue in speaking, chewing, and swallowing. This nerve is considered motor, with some proprioceptive fibers. Table 11.9 summarizes the functions of the cranial nerves.

PRACTICE 

33 Define *peripheral nervous system.*

34 Distinguish between somatic and autonomic nerve fibers.

35 Distinguish among sensory, motor, and mixed nerves.

36 Describe the structure of a peripheral nerve.

37 Name the cranial nerves, and list the major functions of each.

Spinal Nerves

Thirty-one pairs of spinal nerves originate from the spinal cord. All but the first pair are mixed nerves, and they provide two-way communication between the spinal cord and parts of the upper and lower limbs, neck, and trunk.

TABLE 11.9 | Functions of Cranial Nerves

Nerve		Type	Function
I	Olfactory	Sensory	Sensory fibers transmit impulses associated with the sense of smell.
II	Optic	Sensory	Sensory fibers transmit impulses associated with the sense of vision.
III	Oculomotor	Primarily motor	Motor fibers transmit impulses to muscles that raise the eyelids, move the eyes, adjust the amount of light entering the eyes, and focus the lenses.
			Some sensory fibers transmit impulses associated with proprioceptors.
IV	Trochlear	Primarily motor	Motor fibers transmit impulses to muscles that move the eyes.
			Some sensory fibers transmit impulses associated with proprioceptors.
V	Trigeminal	Mixed	
	Ophthalmic division		Sensory fibers transmit impulses from the surface of the eyes, tear glands, scalp, forehead, and upper eyelids.
	Maxillary division		Sensory fibers transmit impulses from the upper teeth, upper gum, upper lip, lining of the palate, and skin of the face.
	Mandibular division		Sensory fibers transmit impulses from the scalp, skin of the jaw, lower teeth, lower gum, and lower lip.
			Motor fibers transmit impulses to muscles of mastication and to muscles in the floor of the mouth.
VI	Abducens	Primarily motor	Motor fibers transmit impulses to muscles that move the eyes.
			Some sensory fibers transmit impulses associated with proprioceptors.
VII	Facial	Mixed	Sensory fibers transmit impulses associated with taste receptors of the anterior tongue.
			Motor fibers transmit impulses to muscles of facial expression, tear glands, and salivary glands.
VIII	Vestibulocochlear	Sensory	
	Vestibular branch		Sensory fibers transmit impulses associated with the sense of equilibrium.
	Cochlear branch		Sensory fibers transmit impulses associated with the sense of hearing.
IX	Glossopharyngeal	Mixed	Sensory fibers transmit impulses from the pharynx, tonsils, posterior tongue, and carotid arteries.
			Motor fibers transmit impulses to salivary glands and to muscles of the pharynx used in swallowing.
X	Vagus	Mixed	Somatic motor fibers transmit impulses to muscles associated with speech and swallowing; autonomic motor fibers transmit impulses to the viscera of the thorax and abdomen.
			Sensory fibers transmit impulses from the pharynx, larynx, esophagus, and viscera of the thorax and abdomen.
XI	Accessory	Primarily motor	
	Cranial branch		Motor fibers transmit impulses to muscles of the soft palate, pharynx, and larynx.
	Spinal branch		Motor fibers transmit impulses to muscles of the neck and back; some proprioceptor input.
XII	Hypoglossal	Primarily motor	Motor fibers transmit impulses to muscles that move the tongue; some proprioceptor input.

Spinal nerves are not named individually but are grouped by the level from which they arise. Each nerve is numbered in sequence (fig. 11.29). On each vertebra the vertebral notches, the major parts of the intervertebral foramina, are associated with the inferior portion of their respective vertebrae. For this reason, each spinal nerve, as it passes through the intervertebral foramen, is associated with the vertebra above it. The cervical spinal nerves are an exception because spinal nerve C1 passes superior to the vertebra C1. Thus, though there are seven cervical vertebrae, there are eight pairs of *cervical nerves* (numbered C1 to C8). There are twelve pairs of *thoracic nerves* (numbered T1 to T12), five

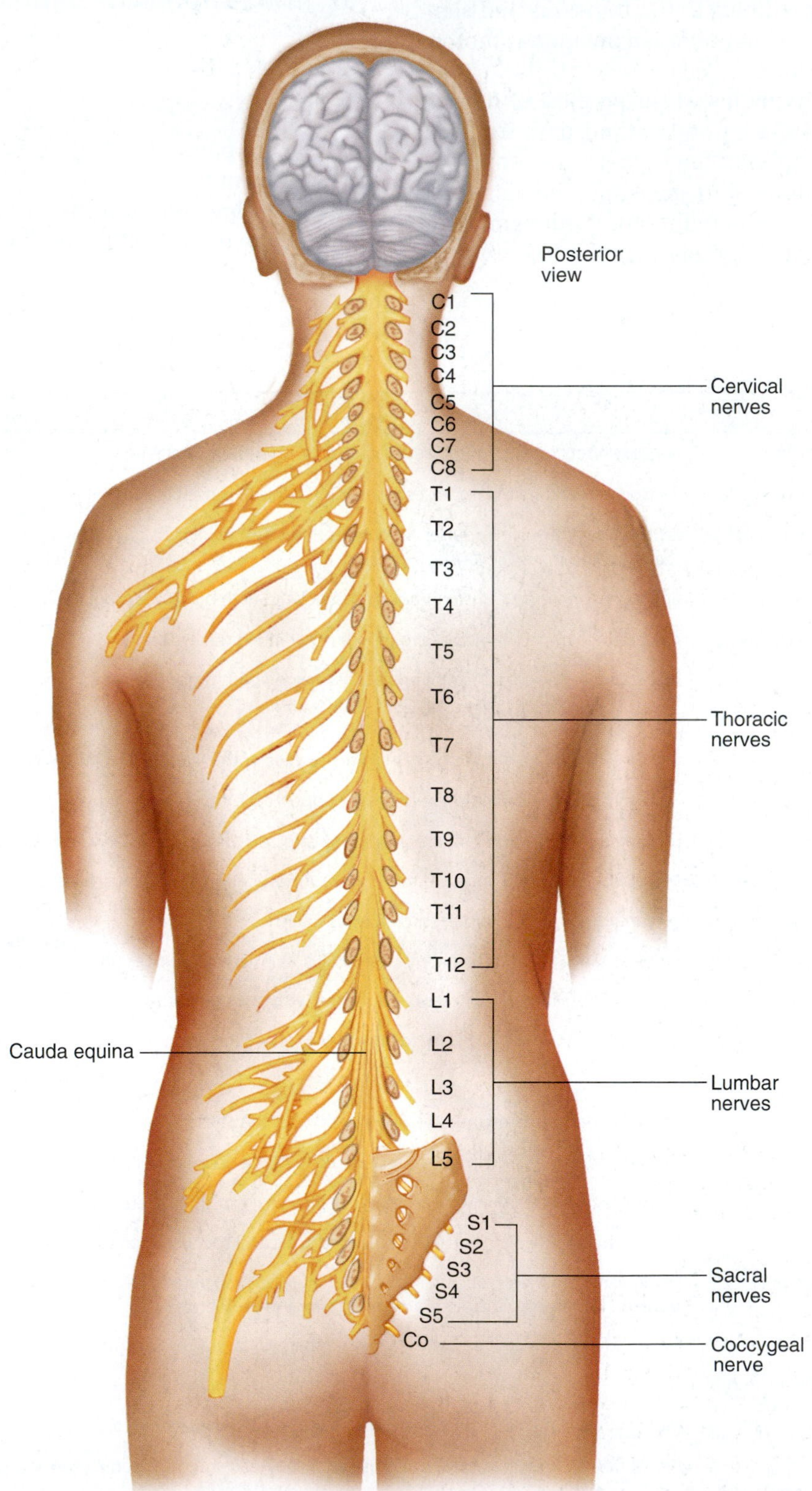

FIGURE 11.29 The thirty-one pairs of spinal nerves are grouped according to the level from which they arise and are numbered in sequence.

pairs of *lumbar nerves* (numbered L1 to L5), five pairs of *sacral nerves* (numbered S1 to S5), and one pair of *coccygeal nerves* (Co).

The nerves arising from the superior part of the spinal cord pass outward almost horizontally, whereas those from the inferior portions of the spinal cord descend at sharp angles. This anatomical organization is a consequence of growth. In early life, the spinal cord extends the entire length of the vertebral column, but with age, the column grows more rapidly than the cord. Thus, the adult spinal cord ends at the level between the first and second lumbar vertebrae, so the lumbar, sacral, and coccygeal nerves descend to their exits beyond the end of the cord. These descending nerves form a structure called the *cauda equina* (horse's tail) (fig. 11.29).

Each spinal nerve emerges from the cord by two short branches, or roots, which lie within the vertebral column. The **dorsal root** (posterior, or sensory, root) can be identified by an enlargement called the *dorsal root ganglion.* This ganglion contains the cell bodies of the sensory neurons whose axons (peripheral process) conduct impulses inward from the peripheral body parts. The axons of these neurons extend through the dorsal root and into the spinal cord (central process), where they form synapses with dendrites of other neurons (see fig. 10.7).

An area of skin that the sensory nerve fibers of a particular spinal nerve innervate is called a *dermatome.* Dermatomes are highly organized, but they vary considerably in size and shape, as figure 11.30 indicates. A map of the dermatomes is

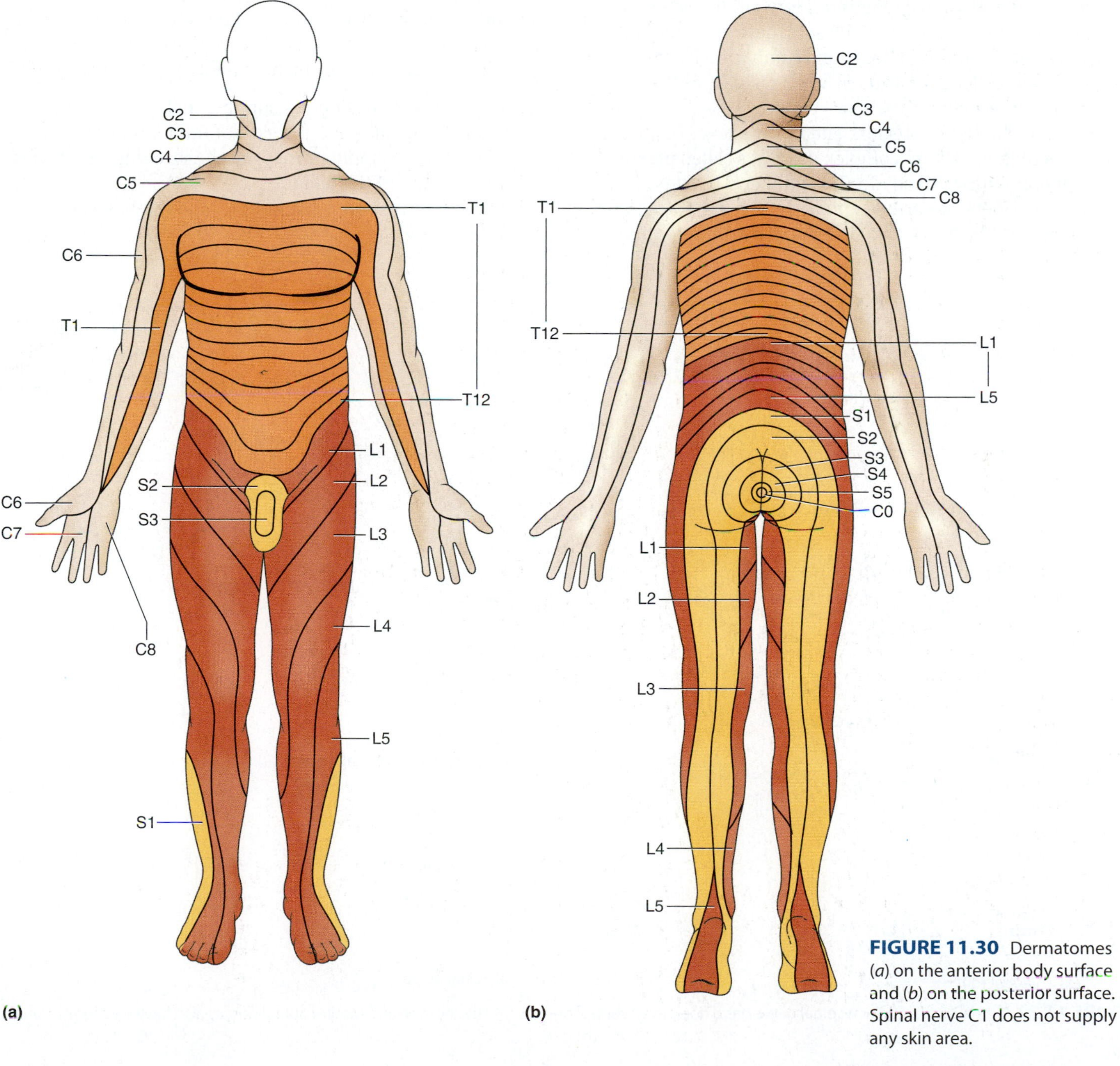

FIGURE 11.30 Dermatomes (*a*) on the anterior body surface and (*b*) on the posterior surface. Spinal nerve C1 does not supply any skin area.

often useful in localizing the sites of injuries to dorsal roots or to the spinal cord.

The **ventral root** (anterior, or motor, root) of each spinal nerve consists of axons from the motor neurons whose cell bodies lie within the gray matter of the cord. A ventral root and a dorsal root unite to form a spinal nerve, which extends outward from the vertebral canal through an *intervertebral foramen*. Just beyond its foramen, each spinal nerve branches. One of these parts, the small *meningeal branch*, reenters the vertebral canal through the intervertebral foramen and supplies the meninges and blood vessels of the cord, as well as the intervertebral ligaments and the vertebrae.

As figure 11.31 shows, a *dorsal branch* (dorsal ramus) of each spinal nerve turns posteriorly and innervates the muscles and skin of the back. The main portion of the nerve, the *ventral branch* (ventral ramus), continues forward to supply muscles and skin on the front and sides of the trunk and limbs. The spinal nerves in the thoracic and lumbar regions have a fourth branch, or *visceral branch*, which is part of the autonomic nervous system.

Except in the thoracic region, anterior branches of the spinal nerves combine to form complex networks called **plexuses** instead of continuing directly to the peripheral body parts. In a plexus, the fibers of various spinal nerves are sorted and recombined, so fibers associated with a particular peripheral body part reach it in the same nerve, even though the fibers originate from different spinal nerves (fig. 11.32).

Cervical Plexuses

The **cervical plexuses** lie deep in the neck on either side. They are formed by the anterior branches of the first four cervical nerves. Fibers from these plexuses supply the muscles and skin of the neck. In addition, fibers from the third, fourth, and fifth cervical nerves pass into the right and left **phrenic** (fren′ik) **nerves,** which conduct motor impulses to the muscle fibers of the diaphragm.

Brachial Plexuses

The anterior branches of the lower four cervical nerves and the first thoracic nerve give rise to **brachial plexuses.** These networks of nerve fibers are deep in the shoulders between the neck and the axillae (armpits). The major branches emerging from the brachial plexuses include the following (fig. 11.33):

1. *Musculocutaneous nerves* supply muscles of the arms on the anterior sides and the skin of the forearms.
2. *Ulnar nerves* supply muscles of the forearms and hands and the skin of the hands.
3. *Median nerves* supply muscles of the forearms and muscles and skin of the hands.

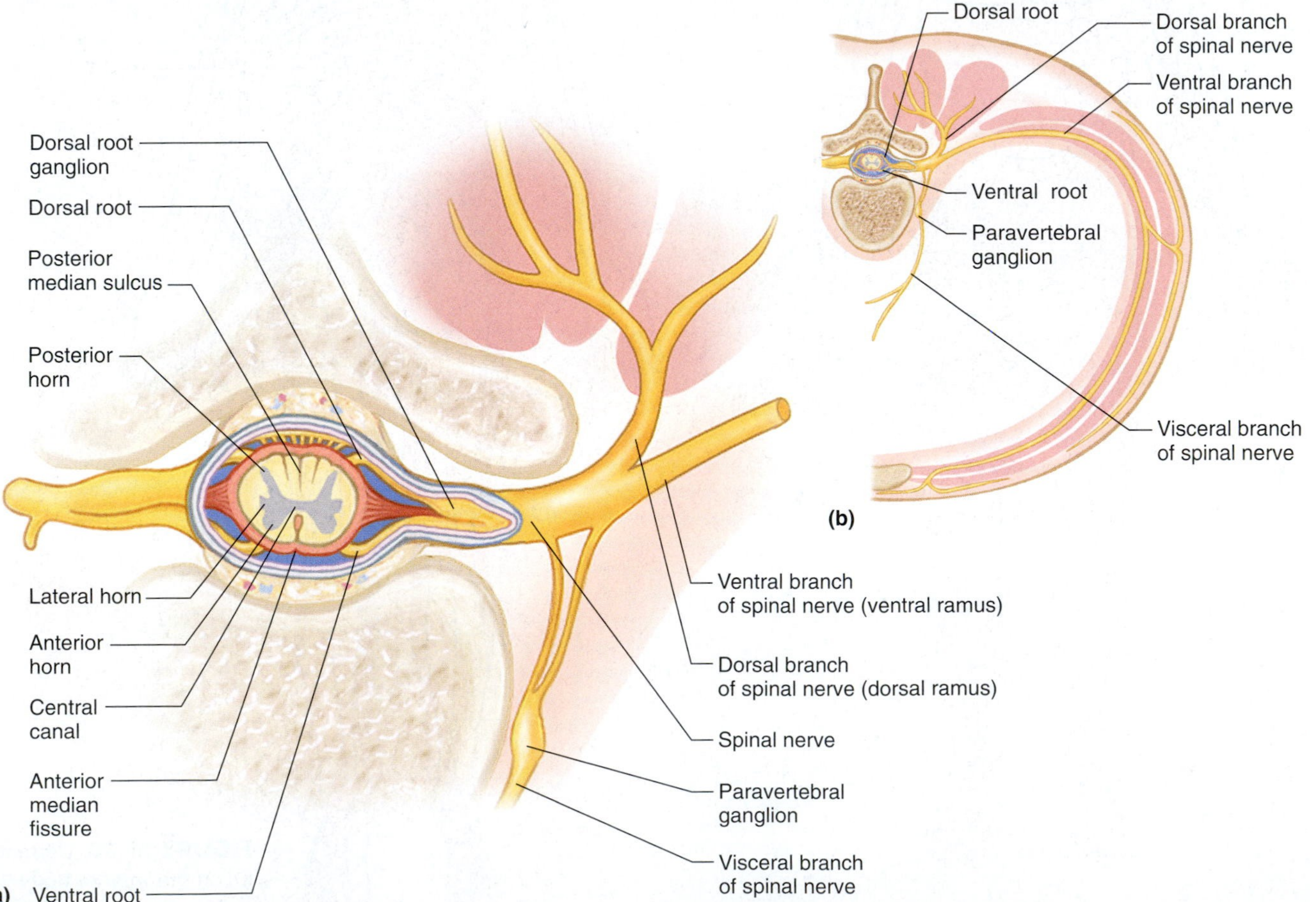

FIGURE 11.31 Spinal nerve. (*a*) Each spinal nerve has a dorsal and ventral branch. (*b*) The thoracic and lumbar spinal nerves also have a visceral branch.

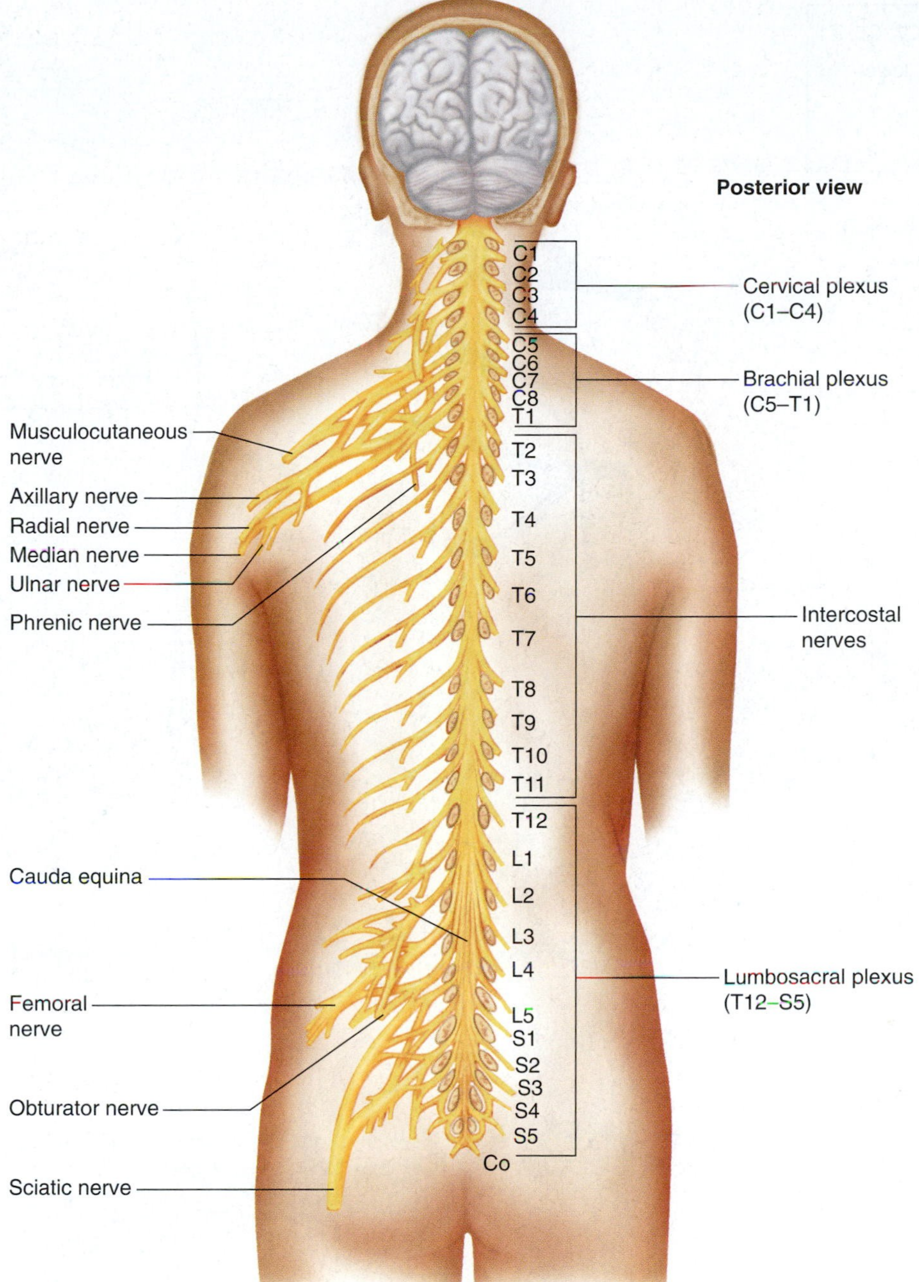

FIGURE 11.32 The ventral branches of the spinal nerves in the thoracic region give rise to intercostal nerves. Those in other regions combine to form complex networks called plexuses.

4. *Radial nerves* supply muscles of the arms on the posterior sides and the skin of the forearms and hands.
5. *Axillary nerves* supply muscles and skin of the anterior, lateral, and posterior regions of the arm.

Other nerves associated with the brachial plexus that innervate various skeletal muscles include the following:

1. The *lateral* and *medial pectoral nerves* supply the pectoralis major and pectoralis minor muscles.
2. The *dorsal scapular nerve* supplies the rhomboid major and levator scapulae muscles.
3. The *lower subscapular nerve* supplies the subscapularis and teres major muscles.
4. The *thoracodorsal nerve* supplies the latissimus dorsi muscle.
5. The *suprascapular nerve* supplies the supraspinatus and infraspinatus muscles.

Lumbosacral Plexuses

The **lumbosacral** (lum″bo-sa′kral) **plexuses** are formed by the last thoracic nerve and the lumbar, sacral, and coccygeal nerves. These networks of nerve fibers extend from the

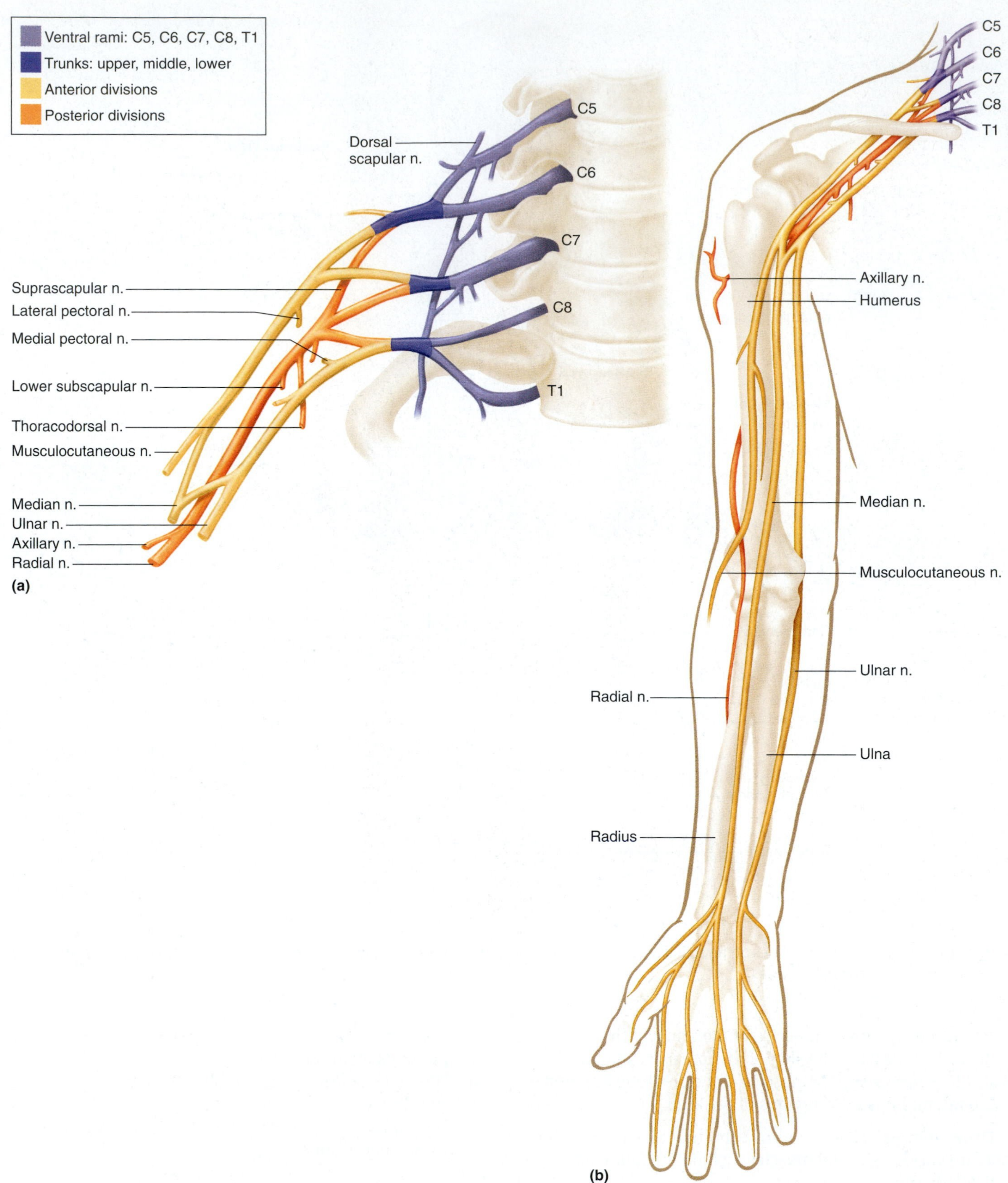

FIGURE 11.33 Nerves of the brachial plexus. (*a*) Close up. (*b*) Anterior view.

lumbar region of the back into the pelvic cavity, giving rise to a number of motor and sensory fibers associated with the lower abdominal wall, external genitalia, buttocks, thighs, legs, and feet. The major branches of these plexuses include the following (fig. 11.34):

1. The *obturator nerves* supply the adductor muscles of the thighs.
2. The *femoral nerves* divide into many branches, supplying motor impulses to muscles of the anterior thighs and receiving sensory impulses from the skin of the thighs and legs.
3. The *sciatic nerves* are the largest and longest nerves in the body. They pass downward into the buttocks and descend into the thighs, where they divide into *tibial* and *common fibular nerves.* The many branches of these nerves supply muscles and skin in the thighs, legs, and feet.

Other nerves associated with the lumbosacral plexus that innervate various skeletal muscles include the following:

1. The *pudendal nerve* supplies the muscles of the perineum.
2. The *inferior* and *superior gluteal nerves* supply the gluteal muscles and the tensor fasciae latae muscle.

The ventral branches of the thoracic spinal nerves do not enter a plexus. Instead, they travel into spaces between the ribs and become **intercostal** (in″ter-kos′tal) **nerves.** These nerves supply motor impulses to the intercostal muscles and the upper abdominal wall muscles. They also receive sensory impulses from the skin of the thorax and abdomen. Clinical Application 11.7 discusses injuries to the spinal nerves.

PRACTICE

38 How are spinal nerves grouped?

39 Describe how a spinal nerve emerges from the spinal cord.

40 Name and locate the major nerve plexuses.

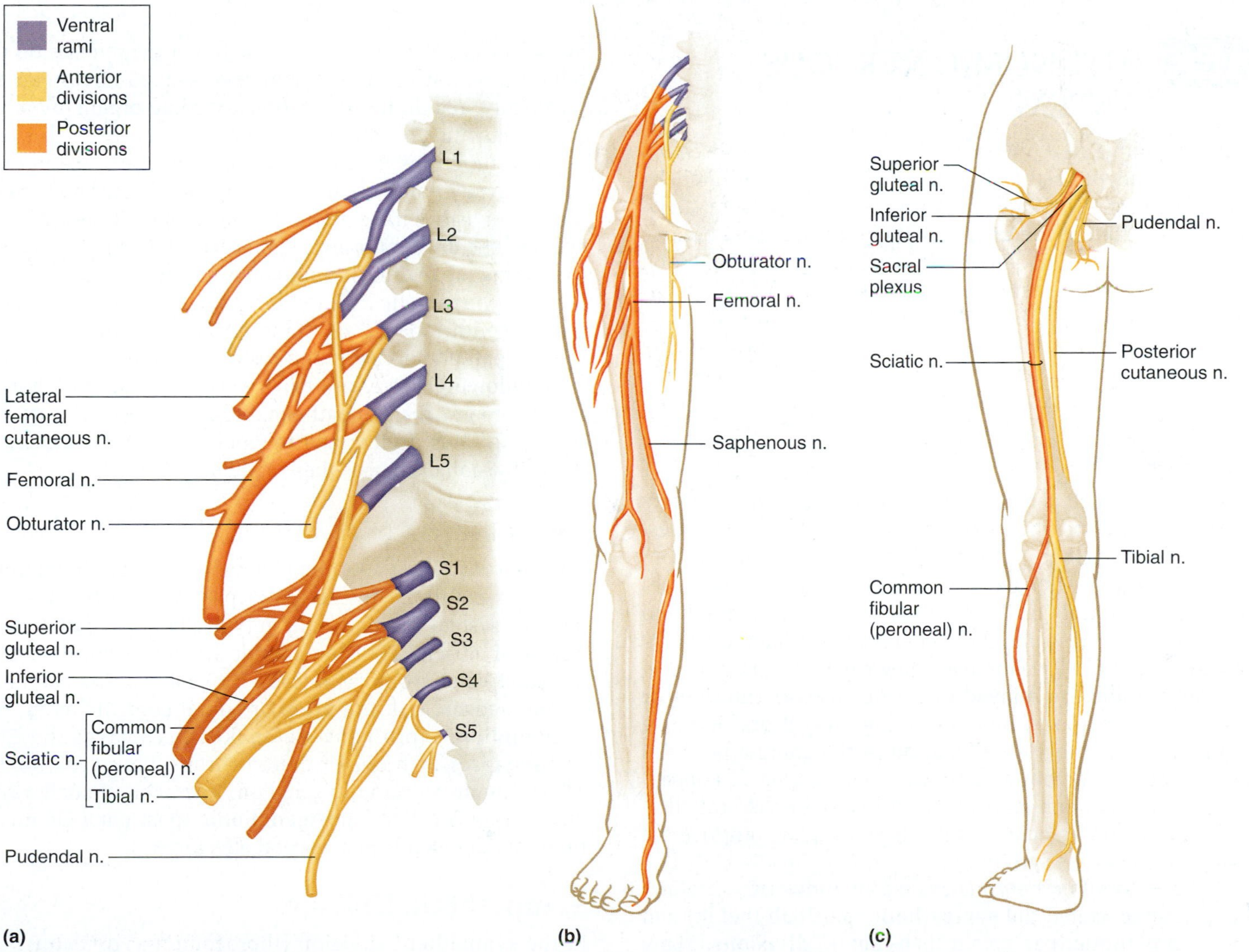

FIGURE 11.34 Nerves of the lumbosacral plexus. (*a*) Close up. (*b*)Anterior view. (*c*) Posterior view.

11.7 CLINICAL APPLICATION

Spinal Nerve Injuries

Birth injuries, dislocations, vertebral fractures, stabs, gunshot wounds, and pressure from tumors can all injure spinal nerves. Suddenly bending the neck, called whiplash, can compress the nerves of the cervical plexuses, causing persistent headache and pain in the neck and skin, which the cervical nerves supply. If a broken or dislocated vertebra severs or damages the phrenic nerves associated with the cervical plexuses, partial or complete paralysis of the diaphragm may result.

Intermittent or constant pain in the neck, shoulder, or upper limb may result from prolonged abduction of the upper limb, as in painting or typing. This is due to too much pressure on the brachial plexus. Called *thoracic outlet syndrome,* this condition may also result from a congenital skeletal malformation that compresses the plexus during upper limb and shoulder movements.

Degenerative changes may compress an intervertebral disc in the lumbar region, producing *sciatica,* which causes pain in the lower back and gluteal region that can radiate to the thigh, calf, ankle, and foot. Sciatica is most common in middle-aged people, particularly distance runners. It usually compresses spinal nerve roots between L2 and S1, some of which contain fibers of the sciatic nerve. Rest, drugs, or surgery are used to treat sciatica.

In *carpal tunnel syndrome,* repeated hand movements, such as typing, or weeding, inflame the tendons that pass through the carpal tunnel, a space between bones in the wrist. The swollen tendons compress the median nerve in the wrist, sending pain up the upper limb. Surgery or avoiding repetitive hand movements can relieve symptoms. ■

11.7 AUTONOMIC NERVOUS SYSTEM

The autonomic nervous system is the part of the PNS that functions independently (autonomously) and continuously, without conscious effort. This system controls visceral activities by regulating the actions of smooth muscles, cardiac muscles, and various glands. It oversees heart rate, blood pressure, breathing rate, body temperature, and other visceral activities that aid in maintaining homeostasis. Portions of the autonomic nervous system also respond during times of emotional stress and prepare the body to meet the demands of strenuous physical activity.

General Characteristics

Reflexes in which sensory signals originate from receptors in the viscera and the skin regulate autonomic activities. Afferent nerve fibers transmit these signals to nerve centers in the brain or spinal cord. In response, motor impulses leave these centers on efferent nerve fibers in cranial and spinal nerves. Typically, these efferent fibers lead to ganglia outside the CNS. The impulses they carry are integrated in the ganglia and are relayed to various organs (muscles or glands) that respond by contracting, secreting, or being inhibited. The integrative function of the ganglia provides the autonomic nervous system with some degree of independence from the brain and spinal cord, and the visceral efferent nerve fibers associated with these ganglia comprise the autonomic nervous system.

The autonomic nervous system includes two interacting divisions, called the **sympathetic** (sim″pah-thet′ik) and **parasympathetic** (par″ah-sim″pah-thet′ik) **divisions.** Many organs have nerve fibers from each division. Impulses on one set of fibers may activate an organ, and impulses on the other set inhibit it. Thus, the divisions may function antagonistically, regulating the actions of some organs by alternately activating or inhibiting them.

The functions of the autonomic divisions are varied; each activates some organs and inhibits others. This reveals that the divisions have important functional differences. The sympathetic division primarily prepares the body for energy-expending, stressful, or emergency situations. Conversely, the parasympathetic division is most active under ordinary, restful conditions. It also counterbalances the effects of the sympathetic division and restores the body to a resting state following a stressful experience. For example, during an emergency, the sympathetic division increases heart and breathing rates; following the emergency, the parasympathetic division decreases these activities.

Autonomic Nerve Fibers

The neurons of the autonomic nervous system are efferent, or motor, neurons. In the motor pathways of the somatic nervous system, a single neuron typically links the CNS and a skeletal muscle. In the autonomic system, motor pathways include two neurons, as figure 11.35 shows. The cell body of one neuron is in the brain or spinal cord. Its axon, the **preganglionic** (pre″gang-gle-on′ik) **fiber,** leaves the CNS and synapses with one or more neurons whose cell bodies are within an autonomic ganglion. The axon of such a second neuron is called a **postganglionic** (pōst″gang-gle-on′ik) **fiber,** and it extends to a visceral effector.

Sympathetic Division

In the sympathetic division (thoracolumbar division), the preganglionic fibers originate from neurons in the lateral

(a) Autonomic pathway

(b) Somatic pathway

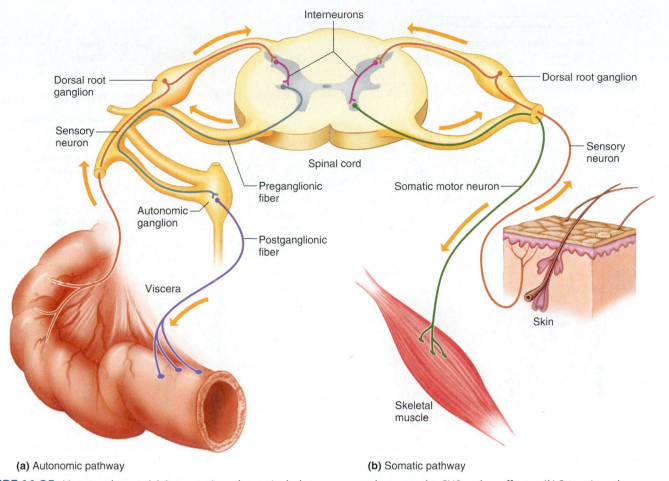

FIGURE 11.35 Motor pathways. (*a*) Autonomic pathways include two neurons between the CNS and an effector. (*b*) Somatic pathways usually have a single neuron between the CNS and an effector. In both cases the motor fibers pass through the ventral root of the spinal cord.

horn of the spinal cord. These neurons are found in all of the thoracic segments and in the upper two lumbar segments of the cord (T1–L2). Their axons exit through the ventral roots of spinal nerves along with various somatic motor fibers.

After traveling a short distance, preganglionic fibers leave the spinal nerves through branches called *white rami* (sing., *ramus*) and enter sympathetic ganglia. Two groups of such ganglia, called **sympathetic chain ganglia** (paravertebral ganglia) are located in chains along the sides of the vertebral column. These ganglia, with the fibers that connect them, comprise the **sympathetic trunks** (fig. 11.36).

The paravertebral ganglia lie just beneath the parietal pleura in the thorax and beneath the parietal peritoneum in the abdomen (see chapter 1, pp. 12 and 14). Although these ganglia are some distance from the viscera they help control, other sympathetic ganglia are nearer to the viscera. The *collateral ganglia,* for example, are in the abdomen, closely associated with certain large blood vessels (fig. 11.37).

Some of the preganglionic fibers that enter paravertebral ganglia synapse with neurons in these ganglia. Other fibers extend through the ganglia and pass up or down the sympathetic trunk and synapse with neurons in ganglia at higher or lower levels in the chain. Still other fibers pass through to

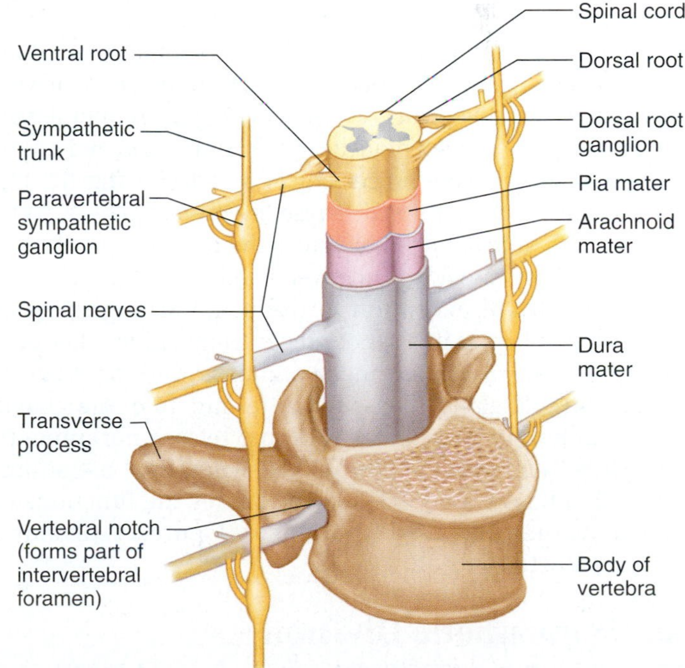

FIGURE 11.36 A chain of paravertebral ganglia extends along each side of the vertebral column.

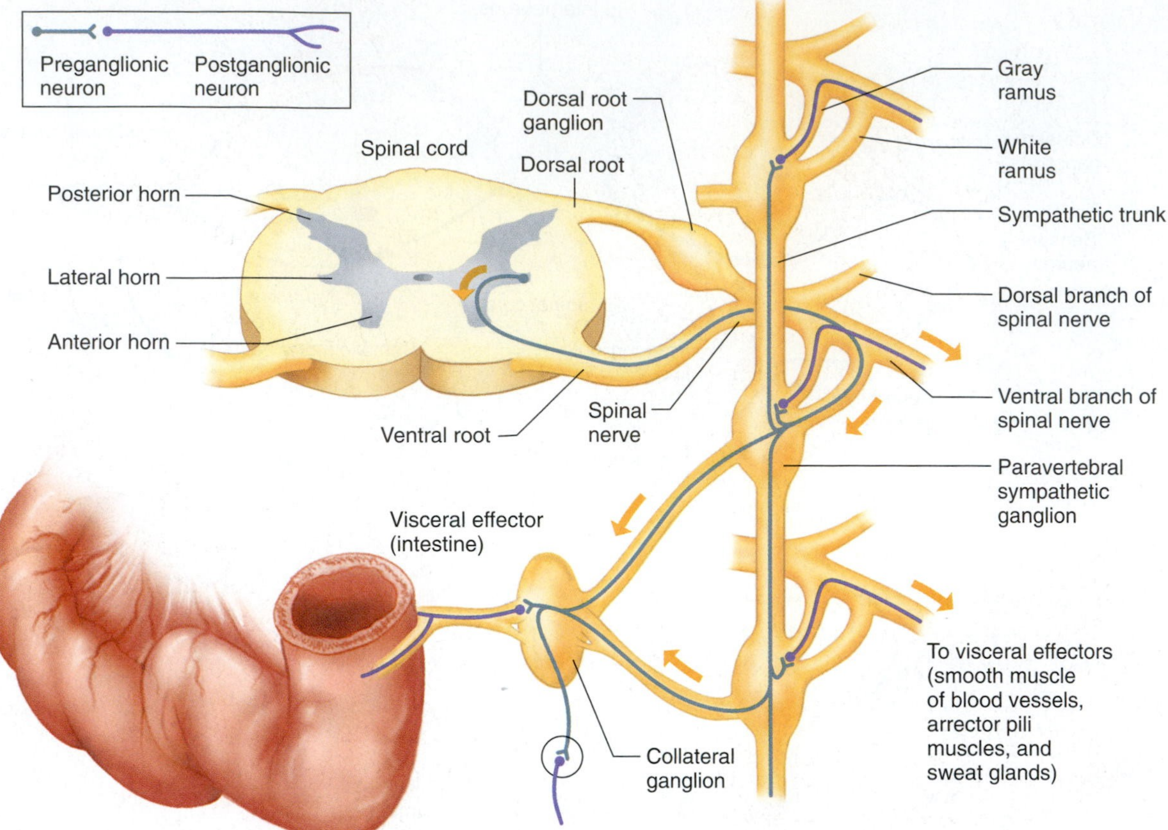

FIGURE 11.37 Sympathetic fibers leave the spinal cord in the ventral roots of spinal nerves, enter paravertebral ganglia, and synapse with other neurons that extend to visceral effectors.

collateral ganglia before they synapse. Typically, a preganglionic axon will synapse with several other neurons in a sympathetic ganglion (an example of divergence).

The axons of the second neurons in sympathetic pathways, the postganglionic fibers, extend from the sympathetic ganglia to visceral effectors. Those leaving paravertebral ganglia usually pass through branches called **gray rami** and return to a spinal nerve before proceeding to an effector (fig. 11.37). These branches appear gray because the postganglionic axons generally are unmyelinated, whereas the preganglionic axons in the white rami are nearly all myelinated.

An important exception to the usual arrangement of sympathetic fibers is in a set of preganglionic fibers that pass through the sympathetic ganglia and extend to the medulla of each adrenal gland. These fibers terminate in the glands on special hormone-secreting cells that release **norepinephrine** (20%) and **epinephrine** (80%) when they are stimulated. Chapter 13 (pp. 504–506) discusses the functions of the adrenal medulla and its hormones. Figure 11.38 shows the sympathetic division.

Parasympathetic Division

The preganglionic fibers of the parasympathetic division (craniosacral division) arise from neurons in the midbrain, pons, and medulla oblongata of the brainstem and from part of the sacral region (S2–4) of the spinal cord (fig. 11.39). From there, they lead outward on cranial or sacral nerves to ganglia near or in various organs (*terminal ganglia*). The short postganglionic fibers continue from the ganglia to specific muscles or glands in these organs (fig. 11.40). Parasympathetic preganglionic axons are usually myelinated, and the parasympathetic postganglionic fibers are unmyelinated.

The parasympathetic preganglionic fibers associated with parts of the head are included in the oculomotor, facial, and glossopharyngeal nerves. Those fibers that innervate organs of the thorax and upper abdomen are parts of the vagus nerves. (The vagus nerves carry about 75% of all parasympathetic fibers.) Preganglionic fibers arising from the sacral region of the spinal cord lie in the branches of the second through the fourth sacral spinal nerves, and they carry impulses to viscera in the pelvic cavity (fig. 11.39).

PRACTICE

41 What is the general function of the autonomic nervous system?

42 How are the divisions of the autonomic system distinguished?

43 Describe a sympathetic nerve pathway and a parasympathetic nerve pathway.

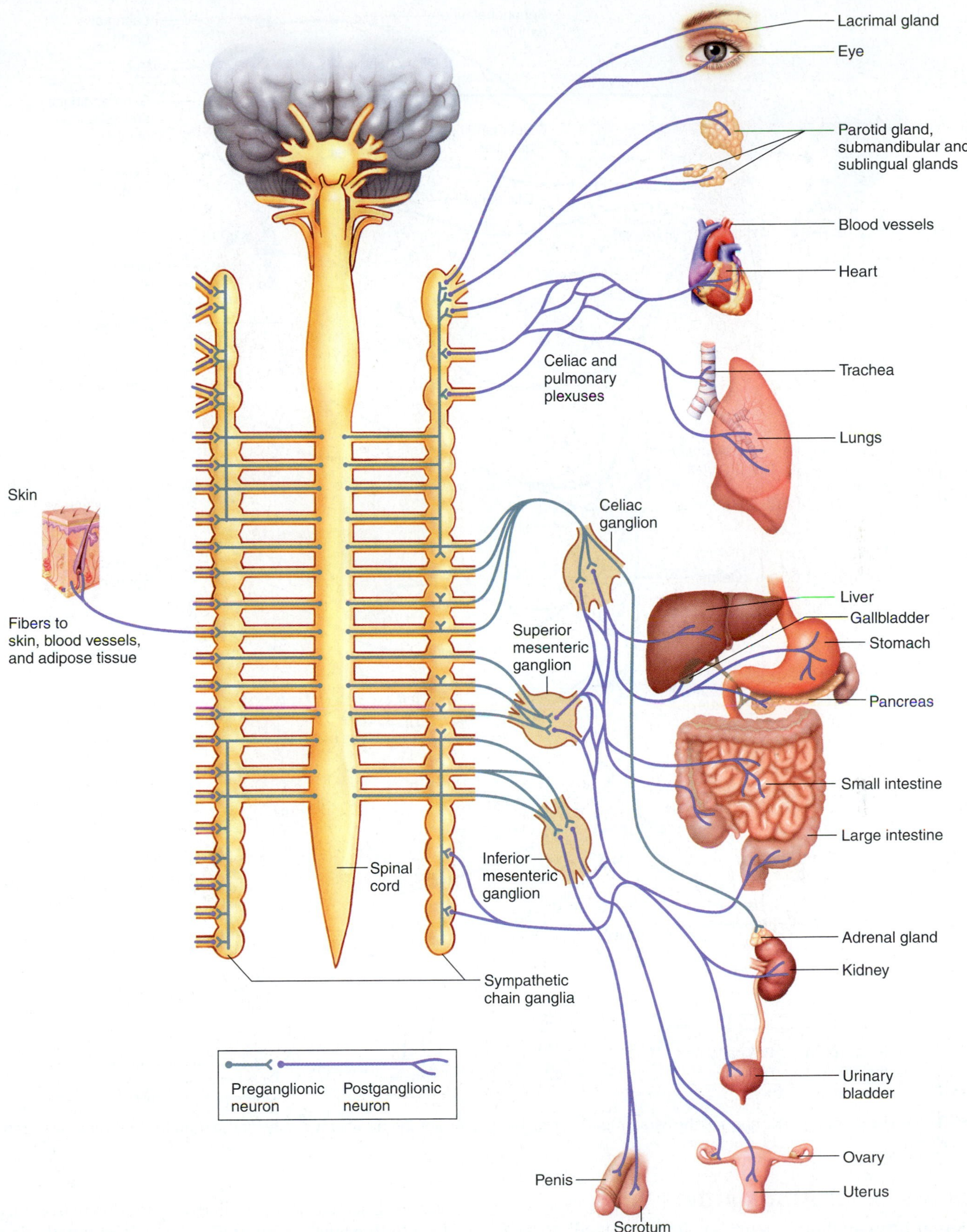

FIGURE 11.38 The preganglionic fibers of the sympathetic division of the autonomic nervous system arise from the thoracic and lumbar regions of the spinal cord. A preganglionic fiber directly innervates the adrenal medulla.

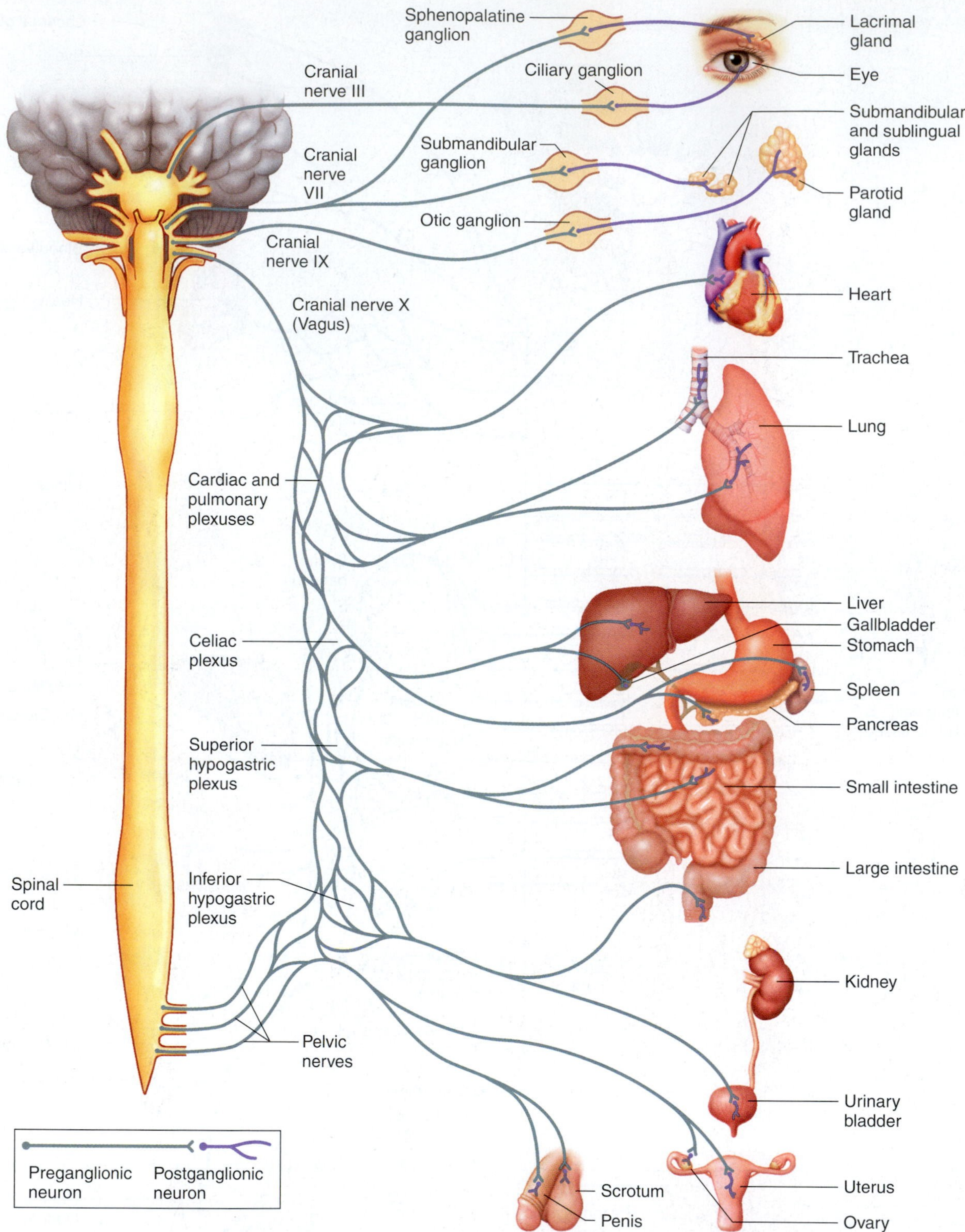

FIGURE 11.39 The preganglionic fibers of the parasympathetic division of the autonomic nervous system arise from the brain and sacral region of the spinal cord.

Autonomic Neurotransmitters

The different postganglionic neurotransmitters (mediators) are responsible for the different effects that the sympathetic and parasympathetic divisions have on organs. The preganglionic neurons of the sympathetic and parasympathetic divisions all secrete acetylcholine, and for this reason they are called **cholinergic** (ko″lin-er′jik). The parasympathetic postganglionic fibers are also cholinergic (One exception, parasympathetic neurons that secrete nitric oxide, is described in chapter 22, p. 844.) Most sympathetic postganglionic neu-

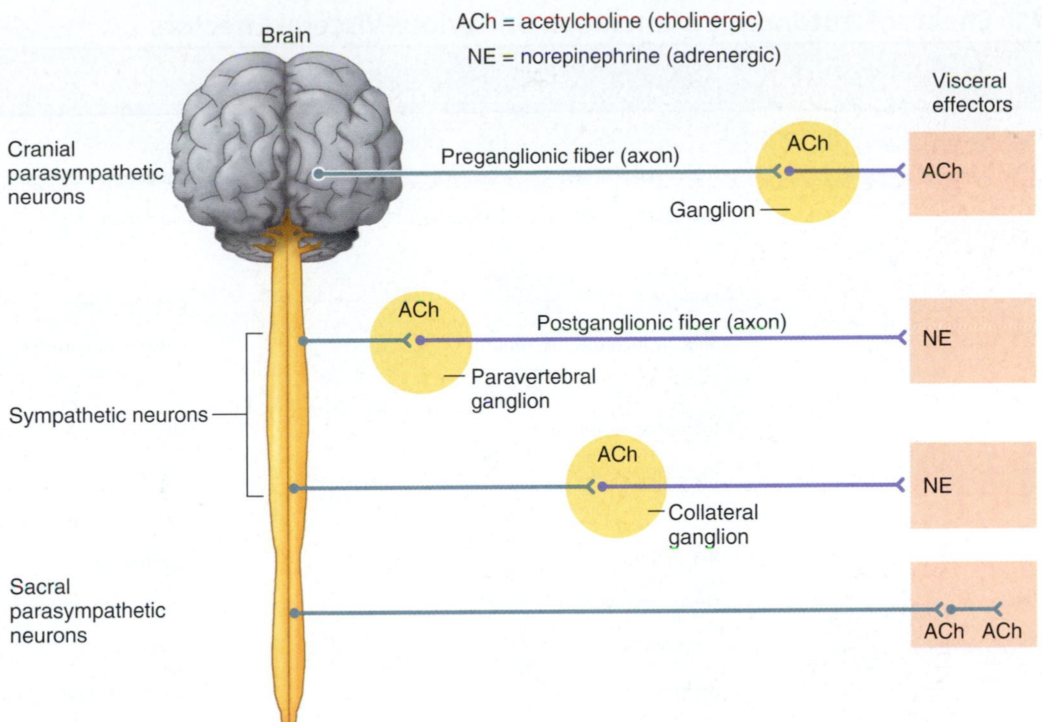

FIGURE 11.40 Most sympathetic fibers are adrenergic and secrete norepinephrine at the ends of the postganglionic fiber; parasympathetic fibers are cholinergic and secrete acetylcholine at the ends of the postganglionic fibers. Two arrangements of parasympathetic postganglionic fibers are seen in both cranial and sacral portions. Similarly, sympathetic paravertebral and collateral ganglia are seen in both the thoracic and lumbar portions of the nervous system. (Dendrites not shown.)

rons, however, secrete norepinephrine (noradrenalin) and are called **adrenergic** (ad″ren-er′jik) (fig. 11.40). Exceptions to this include the sympathetic postganglionic neurons that stimulate sweat glands and a few sympathetic neurons to blood vessels in skin (which cause vasodilation); these neurons secrete acetylcholine and therefore are cholinergic (adrenergic sympathetic fibers to blood vessels cause vasoconstriction).

Most organs receive innervation from both sympathetic and parasympathetic divisions, usually with opposing actions. For example, the sympathetic nervous system increases heart rate and dilates pupils, whereas parasympathetic stimulation decreases heart rate and constricts pupils. However, this is not always the case. For example, the diameters of most blood vessels lack parasympathetic innervation and are thus regulated by the sympathetic division. Smooth muscles in the walls of these vessels are continuously stimulated by sympathetic impulses; they are thereby maintained in a state of partial contraction called *sympathetic tone.* Decreasing sympathetic stimulation allows the muscular walls of such blood vessels to relax, increasing their diameters (vasodilation). Conversely, increasing sympathetic stimulation vasoconstricts vessels. Table 11.10 summarizes the effects of autonomic stimulation on various visceral effectors.

Actions of Autonomic Neurotransmitters

The actions of autonomic neurotransmitters result from their binding to protein receptors in the membranes of effector cells, as in the case of stimulation at neuromuscular junctions (see chapter 9, p. 290) and synapses (see chapter 10, p. 365). Receptor binding alters the membrane. For example, the membrane's permeability to certain ions may increase, and in smooth muscle cells, an action potential followed by muscular contraction may result. Similarly, a gland cell may respond to a change in its membrane by secreting a product.

Acetylcholine can combine with two types of cholinergic receptors, called *muscarinic receptors* and *nicotinic receptors.* These receptor names come from *muscarine,* a toxin from a fungus that can activate muscarinic receptors, and *nicotine,* a toxin of tobacco that can activate nicotinic receptors. The muscarinic receptors are in the membranes of effector cells at the ends of all postganglionic parasympathetic nerve fibers and at the ends of the cholinergic sympathetic fibers. Responses from these receptors are excitatory and relatively slow. The nicotinic receptors are in the synapses between the preganglionic and postganglionic neurons of the parasympathetic and sympathetic pathways. They produce rapid, excitatory responses (see Table 11.10). (Receptors at neuromuscular junctions of skeletal muscles are also nicotinic.)

Epinephrine and norepinephrine are the two chemical mediators of the sympathetic nervous system. The adrenal gland releases both as hormones, but only norepinephrine is released as a neurotransmitter by the sympathetic nervous system. These biochemicals can then bind adrenergic receptors of effector cells.

TABLE 11.10 | **Effects of Autonomic Stimulation on Various Visceral Effectors**

Effector Location	Response to Sympathetic Stimulation	Response to Parasympathetic Stimulation
Integumentary system		
Apocrine glands	Increased secretion	No action
Eccrine glands	Increased secretion (cholinergic effect)	No action
Special senses		
Iris of eye	Dilation	Constriction
Tear gland	Slightly increased secretion	Greatly increased secretion
Endocrine system		
Adrenal cortex	Increased secretion	No action
Adrenal medulla	Increased secretion	No action
Digestive system		
Muscle of gallbladder wall	Relaxation	Contraction
Muscle of intestinal wall	Decreased peristaltic action	Increased peristaltic action
Muscle of internal anal sphincter	Contraction	Relaxation
Pancreatic glands	Reduced secretion	Greatly increased secretion
Salivary glands	Reduced secretion	Greatly increased secretion
Respiratory system		
Muscles in walls of bronchioles	Dilation	Constriction
Cardiovascular system		
Blood vessels supplying muscles	Constriction (alpha adrenergic) Dilation (beta adrenergic)	No action
Blood vessels supplying skin	Constriction	No action
Blood vessels supplying heart (coronary arteries)	Constriction (alpha adrenergic) Dilation (beta adrenergic)	No action
Muscles in wall of heart	Increased contraction rate	Decreased contraction rate
Urinary system		
Muscle of bladder wall	Relaxation	Contraction
Muscle of internal urethral sphincter	Contraction	Relaxation
Reproductive systems		
Blood vessels to penis and clitoris	No action	Dilation leading to erection of penis and clitoris
Muscles associated with internal reproductive organs	Male ejaculation, female orgasm	

The two major types of adrenergic receptors are *alpha* and *beta*. Exciting them elicits different responses in the effector organs. For example, stimulation of the alpha receptors in vascular smooth muscle causes vasoconstriction, whereas stimulation of the beta receptors in bronchial smooth muscle causes relaxation leading to bronchodilation. Furthermore, although norepinephrine has a somewhat stronger effect on alpha receptors, both of these mediators can stimulate both types of receptors. Consequently, the way each of these adrenergic substances influences effector cells depends on the relative numbers of alpha and beta receptors in the cell membranes.

Terminating Autonomic Neurotransmitter Actions

The enzyme acetylcholinesterase rapidly decomposes the acetylcholine that cholinergic fibers release. (Recall that this decomposition also occurs at the neuromuscular junctions of skeletal muscle, chapter 9, p. 293.) Thus, acetylcholine usually affects the postsynaptic membrane for only a fraction of a second.

Much of the norepinephrine released from adrenergic fibers is removed from the synapse by active transport back into the nerve endings. The enzyme monoamine oxidase, in mitochondria, then inactivates norepinephrine. This may take a few seconds, during which some molecules may diffuse into nearby tissues or the bloodstream, where other enzymes decompose them. On the other hand, some norepinephrine molecules may escape decomposition and remain active for awhile. For these reasons, norepinephrine is likely to produce a more prolonged effect than acetylcholine. In fact, when the adrenal medulla releases norepinephrine and epinephrine into the blood in response to sympathetic stimulation, these substances may trigger sympathetic responses in organs throughout the body that last up to thirty seconds.

Many drugs influence autonomic functions. Some, like ephedrine, enhance sympathetic effects by stimulating release of norepinephrine from postganglionic sympathetic nerve endings. Others, like reserpine, inhibit sympathetic activity by preventing norepinephrine synthesis. Another group of drugs, which includes pilocarpine, produces parasympathetic effects, and some, like atropine, block the action of acetylcholine on visceral effectors.

Control of Autonomic Activity

The brain and spinal cord largely control the autonomic nervous system, although it has some independence resulting from impulse integration in its ganglia. For example, reflex centers in the medulla oblongata for cardiac, vasomotor, and respiratory activities receive sensory impulses from viscera on vagus nerve fibers and use autonomic nerve pathways to stimulate motor responses in various muscles and glands. Thus, these reflex centers control the autonomic nervous system. Similarly, the hypothalamus helps regulate body temperature, hunger, thirst, and water and electrolyte balance by influencing autonomic pathways.

Still higher levels in the brain, including the limbic system and the cerebral cortex, control the autonomic nervous system during emotional stress. In this way, the autonomic pathways can affect emotional expression and behavior. Subsequent chapters that deal with individual organs and organ systems discuss regulation of particular organs.

PRACTICE

44 Distinguish between cholinergic and adrenergic fibers.

45 Explain how the fibers of one autonomic division can control the actions of a particular organ.

46 Which neurotransmitters are used in the autonomic nervous system?

47 Describe two types of cholinergic receptors and two types of adrenergic receptors.

11.8 LIFE-SPAN CHANGES

The redundancies and overlap of function in our nervous systems ensure that we can perceive and interact with the environment for many decades. In a sense, aging of this organ system begins before birth, as apoptosis, a form of programmed cell death, carves out the structures that will remain in the brain. This normal dying off of neurons continues throughout life. When brain apoptosis fails, disease results. For example, the brains of individuals who die of schizophrenia as young adults contain the same numbers of neurons as do newborns. These extra neurons produce the extra dopamine that can lead to hallucinations, the hallmark of this illness.

By age thirty, the die-off of neurons accelerates somewhat, although pockets of neural stem cells lining the ventricles retain the capacity to give rise to cells that differentiate as neurons and neuroglia. Over an average lifetime, the brain shrinks by about 10%, with more loss in gray matter than white. Neuron loss is uneven—many cells die in the temporal lobe, but very few die in the brainstem. By age ninety, the frontal cortex has lost about half its neurons, but this deficit doesn't necessarily hamper function.

The nervous system changes over time in several ways. The number of dendritic branches in the cerebral cortex falls. Signs of slowing neurotransmission include decreasing levels of neurotransmitters, the enzymes necessary to synthesize them, and the numbers of postsynaptic receptors. The rate of action potential propagation may decrease by 5% to 10%. Nervous system disorders that may begin to cause symptoms in older adulthood include stroke, depression, Alzheimer disease, Parkinson disease, and multi-infarct dementia.

Noticeable signs of a normally aging nervous system include fading memory and slowed responses and reflexes. Decline in function of the sympathetic nervous system may cause transient drops in blood pressure, which, in turn, may cause fainting. By the seventh decade, waning ability of nerves in the ankles to respond to vibrations from walking may affect balance, raising the risk of falling. Poor eyesight, anemia, inner ear malfunction, and effects of drugs also contribute to poor balance in the later years. Because of these factors, nearly a third of individuals over age sixty-five have at least one serious fall a year.

Changes in sleep patterns accompany aging, reflecting the functioning of the reticular activating system. Older individuals generally sleep fewer hours per night than they once did, experiencing transient difficulty in getting to sleep and staying asleep, with more frequent movements when they are sleeping. Many have bouts with insomnia, sometimes not sleeping more than an hour or two a night. Changing electroencephalogram patterns indicate that stage IV slow-wave sleep as well as REM sleep diminish. All of these changes may result in daytime sleepiness.

PRACTICE

48 How does aging of the nervous system begin even before birth?

49 What are some diseases that affect the aging nervous system?

50 What are some of the physical and functional signs of an aging nervous system?

CHAPTER SUMMARY

11.1 INTRODUCTION (PAGE 384)

The central nervous sytem (CNS) consists of the brain and spinal cord.

1. The brain oversees sensation and perception, movement, and thinking.
2. The brainstem connects the brain and spinal cord, allowing communication between the two.
3. The spinal cord provides communication between the CNS and the peripheral nervous sytem (PNS).

11.2 MENINGES (PAGE 384)

Bone and protective membranes called meninges surround the brain and spinal cord.

1. The meninges consist of a dura mater, arachnoid mater, and pia mater.
2. Cerebrospinal fluid occupies the space between the arachnoid and pia maters.

11.3 VENTRICLES AND CEREBROSPINAL FLUID (PAGE 385)

Ventricles, filled with cerebrospinal fluid (CSF), are connected cavities in the cerebral hemispheres and brainstem.

1. Choroid plexuses in the walls of the ventricles secrete CSF.
2. Ependymal cells of the choroid plexus regulate the composition of CSF.
3. CSF circulates through the ventricles and is reabsorbed into the blood of the dural sinuses.
4. CSF helps maintain a stable ion concentration in the CNS and provides a pathway to the blood for waste.

11.4 SPINAL CORD (PAGE 387)

The spinal cord is a nerve column that extends from the brain into the vertebral canal. It terminates at the level between the first and second lumbar vertebrae.

1. Structure of the spinal cord
   a. The spinal cord is composed of thirty-one segments, each of which gives rise to a pair of spinal nerves.
   b. It is characterized by a cervical enlargement, a lumbar enlargement, and two deep longitudinal grooves that divide it into right and left halves.
   c. White matter surrounds a central core of gray matter.
   d. The white matter is composed of bundles of myelinated nerve fibers.
2. Functions of the spinal cord
   a. The spinal cord is the center for spinal reflexes.
      (1) Reflexes are automatic, subconscious responses to changes.
      (2) They help maintain homeostasis.
      (3) The knee-jerk reflex employs only two neurons. Other reflexes involve more neurons.
      (4) Withdrawal reflexes are protective actions.
   b. The cord provides a two-way communication system between the brain and structures outside the nervous system.
      (1) Ascending tracts carry sensory impulses to the brain; descending tracts carry motor impulses to muscles and glands.
      (2) Many of the fibers in the ascending and descending tracts cross over in the spinal cord or brain.

11.5 BRAIN (PAGE 397)

The brain is the largest and most complex part of the nervous system. It contains nerve centers associated with sensations. The brain issues motor commands and carries on higher mental functions.

1. Brain development
   a. Brain structure reflects the way it forms.
   b. The brain develops from a neural tube with three cavities—the forebrain, midbrain, and hindbrain.
   c. The cavities persist as ventricles, and the walls give rise to structural and functional regions.
2. Structure of the cerebrum
   a. The cerebrum consists of two cerebral hemispheres connected by the corpus callosum.
   b. Its surface is marked by ridges and grooves; sulci divide each hemisphere into lobes.
   c. The cerebral cortex is a thin layer of gray matter near the surface.
   d. White matter consists of myelinated nerve fibers that connect neurons and communicate with other body parts.
3. Functions of the cerebrum
   a. The cerebrum provides higher brain functions, such as thought, reasoning, interpretation of sensory impulses, control of voluntary muscles, and memory storage.
   b. The cerebral cortex has sensory, association, and motor areas.
   c. Areas that interpret sensory impulses from the skin are in the parietal lobes near the central sulcus; other specialized sensory areas are in the temporal and occipital lobes.
   d. Association areas analyze and interpret sensory impulses and provide memory, reasoning, verbalizing, judgment, and emotions.
   e. The primary motor regions lie in the frontal lobes near the central sulcus. Other areas of the frontal lobes control special motor functions.
   f. One cerebral hemisphere usually dominates for certain intellectual functions.
   g. Short-term memory is probably bioelectrical. Long-term memory is thought to be encoded in patterns of synaptic connections.
4. Basal nuclei
   a. Basal nuclei are masses of gray matter deep within the cerebral hemispheres.
   b. The neurons of the basal nuclei interact with other brain areas to facilitate voluntary movement.

5. Diencephalon
    a. The diencephalon includes the thalamus and hypothalamus.
    b. The thalamus selects incoming sensory impulses and relays them to the cerebral cortex.
    c. The hypothalamus is important in maintaining homeostasis.
    d. The limbic system produces emotional feelings and modifies behavior.
6. Brainstem
    a. The brainstem extends from the base of the brain to the spinal cord.
    b. The brainstem consists of the midbrain, pons, and medulla oblongata.
    c. The midbrain contains reflex centers associated with eye and head movements.
    d. The pons transmits impulses between the cerebrum and other parts of the nervous system and contains centers that help regulate rate and depth of breathing.
    e. The medulla oblongata transmits all ascending and descending impulses and contains several vital and nonvital reflex centers.
    f. The reticular formation filters incoming sensory impulses, arousing the cerebral cortex into wakefulness in response to meaningful impulses.
    g. Normal sleep results from decreasing activity of the reticular formation, and paradoxical sleep occurs when activating impulses are received by some parts of the brain, but not by others.
7. Cerebellum
    a. The cerebellum consists of two hemispheres connected by the vermis.
    b. A thin cortex of gray matter surrounds the white matter of the cerebellum.
    c. The cerebellum functions primarily as a reflex center, coordinating skeletal muscle movements and maintaining equilibrium.

11.6 PERIPHERAL NERVOUS SYSTEM (PAGE 411)

The peripheral nervous system consists of cranial and spinal nerves that branch out from the brain and spinal cord to all body parts. It can be subdivided into somatic and autonomic portions.

1. Structure of peripheral nerves
    a. A nerve consists of a bundle of nerve fibers surrounded by connective tissues.
    b. The connective tissues form an outer epineurium, a perineurium enclosing bundles of nerve fibers, and an endoneurium surrounding each fiber.
2. Nerve and nerve fiber classification
    a. Nerves are cordlike bundles of nerve fibers. Nerves can be classified as sensory nerves, motor nerves, or mixed nerves, depending on which type of fibers they contain.
    b. Nerve fibers in the CNS are subdivided into groups with general and special functions.
3. Cranial nerves
    a. Twelve pairs of cranial nerves connect the brain to parts in the head, neck, and trunk.
    b. Although most cranial nerves are mixed, some are pure sensory, and others are primarily motor.
    c. The names of cranial nerves indicate their primary functions or the general distributions of their fibers.
    d. Some cranial nerve fibers are somatic, and others are autonomic.
4. Spinal nerves
    a. Thirty-one pairs of spinal nerves originate from the spinal cord. All but the first pair are mixed nerves.
    b. These mixed nerves provide a two-way communication system between the spinal cord and the upper limbs, lower limbs, neck, and trunk.
    c. Spinal nerves are grouped according to the levels from which they arise, and they are numbered sequentially.
    d. Each nerve emerges by a dorsal and a ventral root.
        (1) A dorsal root contains sensory fibers and has a dorsal root ganglion.
        (2) A ventral root contains motor fibers.
    e. Just beyond its foramen, each spinal nerve divides into several branches.
    f. Most spinal nerves combine to form plexuses that direct nerve fibers to a particular body part.

11.7 AUTONOMIC NERVOUS SYSTEM (PAGE 424)

The autonomic nervous system functions without conscious effort. It regulates visceral activities that maintain homeostasis.

1. General characteristics
    a. Autonomic functions are reflexes controlled from centers in the hypothalamus, brainstem, and spinal cord.
    b. Autonomic nerve fibers are associated with ganglia where impulses are integrated before distribution to effectors.
    c. The integrative function of the ganglia provides a degree of independence from the CNS.
    d. The autonomic nervous system consists of the visceral efferent fibers associated with these ganglia.
    e. The autonomic nervous system is subdivided into two divisions—sympathetic and parasympathetic.
    f. The sympathetic division prepares the body for stressful and emergency conditions.
    g. The parasympathetic division is most active under ordinary conditions.
2. Autonomic nerve fibers

    The autonomic fibers are efferent, or motor.
3. Sympathetic division
    a. Sympathetic fibers leave the spinal cord and synapse in ganglia.
    b. Preganglionic fibers pass through white rami to reach paravertebral ganglia.
    c. Paravertebral ganglia and interconnecting fibers comprise the sympathetic trunks.
    d. Preganglionic fibers synapse with paravertebral or collateral ganglia.
    e. Postganglionic fibers usually pass through gray rami to reach spinal nerves before passing to effectors.
    f. A special set of sympathetic preganglionic fibers passes through ganglia and extends to the adrenal medulla.

4. Parasympathetic division includes the parasympathetic fibers that begin in the brainstem and sacral region of the spinal cord and synapse in ganglia near various organs or in the organs themselves.
5. Autonomic neurotransmitters
   a. Sympathetic and parasympathetic preganglionic fibers secrete acetylcholine.
   b. Most sympathetic postganglionic fibers secrete norepinephrine and are adrenergic; postganglionic parasympathetic fibers secrete acetylcholine and are cholinergic.
   c. The different effects of the autonomic divisions are due to the different neurotransmitters the postganglionic fibers release.
6. Actions of autonomic neurotransmitters
   a. Neurotransmitters combine with receptors and alter cell membranes.
   b. There are two types of cholinergic receptors and two types of adrenergic receptors.
   c. How cells respond to neurotransmitters depends upon the number and type of receptors in their membranes.
   d. Acetylcholine acts very briefly; norepinephrine and epinephrine may have more prolonged effects.
7. Terminating actions of autonomic neurotransmitters
   a. Acetylcholinesterase breaks down Ach.
   b. Norepinephrine is transported back into presynaptic neurons.
8. Control of autonomic activity
   a. The central nervous system largely controls the autonomic nervous system.
   b. The medulla oblongata uses autonomic fibers to regulate cardiac, vasomotor, and respiratory activities.
   c. The hypothalamus uses autonomic fibers in regulating visceral functions.
   d. The limbic system and cerebral cortex control emotional responses through the autonomic nervous system.

11.8 LIFE-SPAN CHANGES (PAGE 431)

Aging of the nervous system is a gradual elimination of cells and, eventually, slowed functioning.

1. Apoptosis of brain neurons begins before birth.
2. Neuron loss among brain regions is uneven.
3. In adulthood, numbers of dendrites in the cerebral cortex fall, as more generally neurotransmission slows.
4. Nervous system changes in older persons increase the risk of falling.
5. Sleep problems are common in the later years.

CHAPTER ASSESSMENTS

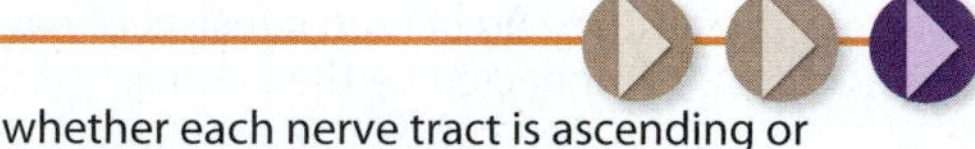

11.1 Introduction

1 Explain the general functions of the brain, spinal cord, and brainstem, and their interrelationship. (p. 384)

11.2 Meninges

2 Name the layers of the meninges, and explain their functions. (p. 384)

11.3 Ventricles and Cerebrospinal Fluid

3 Describe the relationship among the cerebrospinal fluid, the ventricles, the choroid plexuses, and arachnoid granulations. (p. 385)

4 List the functions of cerebrospinal fluid. (p. 386)

11.4 Spinal Cord

5 Describe the structure of the spinal cord. (p. 387)

6 Explain the two main functions of the spinal cord. (p. 389)

7 Distinguish between a reflex arc and a reflex. (p. 389)

8 Which of the choices is the correct sequence of events in a reflex arc? (p. 389)

a. effectors to motor neurons to interneurons to CNS to sensory receptor
b. sensory receptor to CNS to interneurons to motor neurons to effectors
c. effectors to CNS to interneurons to motor neurons to sensory receptor
d. sensory receptor to motor neurons to CNS to interneurons to effectors

9 Describe a withdrawal reflex. (p. 390)

10 Indicate whether each nerve tract is ascending or descending: (p. 393)

a. rubrospinal
b. corticospinal
c. spinothalamic
d. fasciculus gracilis
e. reticulospinal
f. spinocerebellar

11 Explain the consequences of nerve fibers crossing over. (p. 394)

11.5 Brain

12 Describe the events of brain development. (p. 398)

13 Which choice lists the parts of the brainstem? (p. 398)

a. midbrain, pons, and medulla oblongata
b. forebrain, midbrain, and hindbrain
c. sulci and fissures
d. frontal, parietal, and temporal lobes

14 Describe the structure of the cerebrum. (p. 398)

15 Define *cerebral cortex.* (p. 400)

16 Describe the location and function of the sensory areas of the cortex. (p. 401)

17 Explain the function of the association areas of the lobes of the cerebrum. (p. 402)

18 Describe the location and function of the motor areas of the cortex. (p. 402)

19 Broca's area controls ____________. (p. 402)

a. memory
b. defecation
c. understanding grammar
d. movements used in speaking

20 Explain *hemisphere dominance.* (p. 403)

21 Distinguish between short-term and long-term memory. (p. 404)

22 Explain the conversion of short-term to long-term memory. (p. 404)

23 The ___________ transmits sensory information from other parts of the nervous sytem to the cerebral cortex. (p. 407)

a. pineal gland c. thalamus
b. hypothalamus d. basal nuclei

24 List the parts of the limbic system, and explain its functions. (p. 407)

25 Name the functions of the midbrain, pons, and medulla oblongata. (p. 407)

26 Describe the location and function of the reticular formation. (p. 409)

27 Distinguish between normal and paradoxical sleep. (p. 409)

28 The cerebellum ____________. (p. 410)

a. communicates with the rest of the CNS
b. creates awareness of the body's location in space
c. maintains posture
d. all of the above

11.6 Peripheral Nervous System

29 Distinguish between the somatic and autonomic nervous systems. (p. 411)

30 Describe the connective tissue and nervous tissue making up a peripheral nerve. (p. 412)

31 Which of the following carry sensory impulses to the CNS from receptors in muscle or skin? (p. 412)

a. general somatic efferent fibers
b. general somatic afferent fibers
c. general visceral afferent fibers
d. general visceral efferent fibers

32 Draw the underside of a brain and label the cranial nerves. (p. 414)

33 Match the cranial nerve with its function. Functions may be used more than once. (p. 414)

(1) olfactory nerve	A. transmits impulses to muscles used in swallowing
(2) optic nerve	B. transmits impulses to muscles that move the tongue
(3) oculomotor nerve	C. transmits impulses to muscles that move the eyes
(4) trochlear nerve	D. transmits impulses to viscera
(5) trigeminal nerve	E. transmits impulses to muscles of facial expression
(6) abducens nerve	F. transmits impulses to muscles of neck
(7) facial nerve	G. transmits impulses associated with hearing
(8) vestibulocochlear nerve	H. transmits impulses to muscles that raise eyelids
(9) glossopharyngeal nerve	I. transmits impulses associated with sense of smell
(10) vagus nerve	J. transmits impulses from upper and lower teeth
(11) accessory nerve	K. transmits impulses associated with vision
(12) hypoglossal nerve	

34 Explain how the spinal nerves are grouped and numbered. (p. 417)

35 Define *cauda equina.* (p. 419)

36 Describe the parts and their functions of a spinal nerve. (p. 419)

37 Define *plexus,* and locate the major plexuses of the spinal nerves. (p. 420)

11.7 Autonomic Nervous System

38 The autonomic portion of the PNS functions __________. (p. 424)

a. consciously
b. voluntarily
c. without conscious effort
d. dependently

39 Contrast the sympathetic and parasympathetic divisions of the autonomic nervous system. (p. 424)

40 Distinguish between a preganglionic fiber and a postganglionic fiber. (p. 424)

41 Define *paravertebral ganglion.* (p. 425)

42 Trace a sympathetic nerve pathway through a ganglion to an effector. (p. 425)

43 Trace a parasympathetic nerve pathway. (p. 426)

44 Distinguish between cholinergic and adrenergic nerve fibers. (p. 428)

45 Define *sympathetic tone.* (p. 429)

46 Explain how autonomic neurotransmitters influence the actions of effector cells. (p. 429)

47 Distinguish between alpha adrenergic and beta adrenergic receptors. (p. 430)

11.8 Life-Span Changes

48 Explain the effects of apoptosis on the developing brain. (p. 431)

49 List three ways that the nervous system changes as we age. (p. 431)

50 Describe sleep problems that may accompany aging. (p. 431)

INTEGRATIVE ASSESSMENTS/CRITICAL THINKING

OUTCOMES 4.5, 11.5

1. In planning treatment for a patient who has had a cerebrovascular accident (CVA), why would it be important to know whether the CVA was caused by a ruptured or obstructed blood vessel?

OUTCOMES 7.6, 7.7, 11.2, 11.3

2. If a physician plans to obtain a sample of spinal fluid from a patient, what anatomical site can be safely used, and how should the patient be positioned to facilitate this procedure?

OUTCOMES 11.4, 11.5

3. What functional losses would you expect to observe in a patient who has suffered injury to the right occipital lobe of the cerebral cortex? To the right temporal lobe?

OUTCOMES 11.4, 11.5

4. The Brown-Seguard syndrome is due to an injury on one side of the spinal cord. It is characterized by paralysis below the injury and on the same side as the injury, and by loss of sensations of temperature and pain on the opposite side. How would you explain these symptoms?

OUTCOMES 11.4, 11.6

5. The biceps-jerk reflex employs motor neurons that exit from the spinal cord in the 5th spinal nerve (C5), that is, fifth from the top of the cord. The triceps-jerk reflex involves motor neurons in the 7th spinal nerve (C7). How might these reflexes be used to help locate the site of damage in a patient with a neck injury?

OUTCOME 11.7

6. What symptoms might the sympathetic division of the autonomic nervous system produce in a patient experiencing stress?

WEB CONNECTIONS

Be sure to visit the text website at **www.mhhe.com/shier12** for answers to chapter assessments, additional quizzes, and interactive learning exercises.

ANATOMY & PHYSIOLOGY REVEALED

Anatomy & Physiology Revealed® (APR) includes cadaver photos that allow you to peel away layers of the human body to reveal structures beneath the surface. This program also includes animations, radiologic imaging, audio pronunciations, and practice quizzing. Check out **www.aprevealed.com**. APR has been proven to help improve student grades!

CHAPTER 12

Nervous System III

Senses

The spiral organ, in the inner ear, has rows of hair cells, each bearing up to 100 hairs, which translate sound into neural messages that travel to the brain (2,700×).

UNDERSTANDING WORDS

aud-, to hear: *aud*itory—pertaining to hearing.

choroid, skinlike: *choroid* coat—middle, vascular layer of the eye.

cochlea, snail: *cochlea*—coiled tube in the inner ear.

corn-, horn: *corn*ea—transparent outer layer in the anterior portion of the eye.

iris, rainbow: *iris*—colored, muscular part of the eye.

labyrinth, maze: *labyrinth*—complex system of connecting chambers and tubes of the inner ear.

lacri-, tears: *lacri*mal gland—tear gland.

lut-, yellow: macula *lut*ea—yellowish spot on the retina.

macula, spot: *macula* lutea—yellowish spot on the retina.

malle-, hammer: *malle*us—one of the three bones in the middle ear.

ocul-, eye: orbicularis *ocul*i—muscle associated with the eyelid.

olfact-, to smell: *olfact*ory—pertaining to the sense of smell.

palpebra, eyelid: levator *palpebra*e superioris—muscle associated with the eyelid.

photo-, light: *photo*receptors—specialized structures in the eye responsive to light.

scler-, hard: *scler*a—tough, outer protective layer of the eye.

therm-, heat: *therm*oreceptor—receptor sensitive to changes in temperature.

tympan-, drum: *tympan*ic membrane—eardrum.

vitre-, glass: *vitre*ous humor—clear, jellylike substance within the eye.

LEARN PRACTICE ASSESS

LEARNING OUTCOMES

After you have studied this chapter, you should be able to:

12.1 Introduction

1 Differentiate between general senses and special senses. (p. 438)

12.2 Receptors, Sensation, and Perception

2 Name the five types of receptors and state the function of each. (p. 439)

3 Explain how receptors stimulate sensory impulses. (p. 439)

4 Explain sensation production and adaptation. (p. 439)

12.3 General Senses

5 Describe the differences among receptors associated with the senses of touch, pressure, temperature, and pain. (p. 440)

6 Describe how the sensation of pain is produced. (p. 440)

7 Explain the importance of stretch receptors in muscles and tendons. (p. 444)

12.4 Special Senses

8 Explain the relationship between the senses of smell and taste. (p. 446)

9 Describe how the sensations of smell and taste are produced and interpreted. (p. 446)

10 Name the parts of the ear and explain the function of each part. (p. 450)

11 Distinguish between static and dynamic equilibrium. (p. 459)

12 Describe the roles of the accessory organs to the eye. (p. 462)

13 Name the parts of the eye and explain the function of each part. (p. 465)

14 Explain how the eye refracts light. (p. 471)

15 Explain how the brain perceives depth and distance. (p. 475)

16 Describe the visual nerve pathways. (p. 475)

12.5 Life-Span Changes

17 Describe aging-associated changes that diminish the senses. (p. 476)

THE WORLD WITHOUT COLOR

John Dalton, a famous English chemist, saw things differently than most people. In a 1794 lecture, he described his visual world. Sealing wax that appeared red to other people was as green as a leaf to Dalton and his brother. Pink wildflowers were blue, and Dalton perceived the cranesbill plant as "sky blue" in daylight, but "very near yellow, but with a tincture of red" in candlelight. He concluded, ". . . that part of the image which others call red, appears to me little more than a shade, or defect of light." The Dalton brothers, like 7% of males and 0.4% of females today, had the inherited trait of colorblindness.

Dalton was very curious about the cause of his colorblindness, so he made arrangements with his personal physician, Joseph Ransome, to dissect his eyes after he died. Ransome snipped off the back of one eye, removing the retina, where the cone cells that provide color vision are nestled among the more abundant rod cells that impart black-and-white vision. Ransome could see red and green normally when he peered through the back of his friend's eyeball, so he concluded that it was not an abnormal filter in front of the eye that altered color vision.

Fortunately, Ransome stored the eyes in dry air, where they remained relatively undamaged. In 1994, Dalton's eyes underwent DNA analysis at London's Institute of Ophthalmology. The research showed that Dalton's remaining retina lacked one of three types of pigments, called photopigments, that enable cone cells to capture certain incoming wavelengths of light.

Although people have studied colorblindness for centuries, we are still learning more about it. Recently, researchers investigated why colorblind men lacking cones that capture green light are affected to different degrees. They discovered that colorblind men who can discern a few shades of green have red cone cells that can detect some wavelengths of light that fall within the green region of the spectrum. Color vision may be more complex than we had thought. ■

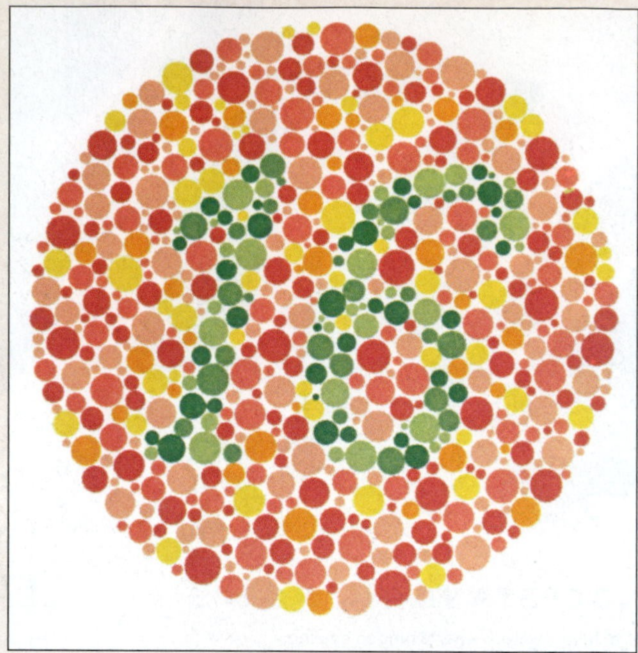

People who are colorblind must function in a multicolored world. To help them overcome the disadvantage of not seeing important color differences, researchers have developed computer algorithms that convert colored video pictures into shades those with colorblindness can see. This circle of dots is a test to determine whether someone is colorblind. Affected individuals cannot see a different color in certain of the dots in such a drawing. As a result, their brains cannot perceive the embedded pattern that forms the number 16 that others can see. The above has been reproduced from *Ishihara's Tests for Colour Blindness* published by Kanehara & Co. Ltd. Tokyo, Japan, but tests for colour blindness cannot be conducted with this material. For accurate testing, the original plates should be used.

12.1 INTRODUCTION

Our senses not only make our lives meaningful, connecting us to the sights, sounds, smells, tastes, and textures of the outside world, but also help our bodies maintain homeostasis by providing information about what is happening on the inside. Sensory receptors are the portals that link our nervous systems to all of these events. The **general senses** are those with receptors widely distributed throughout the body, including the skin, various organs, and joints. The **special senses** have more specialized receptors and are confined to structures in the head, such as the eyes and ears.

All senses work in basically the same way. Sensory receptors are specialized cells or multicellular structures that collect information from the environment and stimulate neurons to send impulses along sensory fibers to the brain. There the cerebral cortex forms a perception, a person's particular view of the stimulus. Table 12.1 outlines the pathways from sensation to perception that describe an apple. These are special senses.

Recall from chapter 11 (p. 412) that the terms "axon" and "nerve fiber" are used synonymously. Also recall that unipolar neurons, which include most sensory neurons, have an unusual structure in which the portion of the neuron associated with the dendrites, called a peripheral process, is considered to function like an axon (see fig. 10.7). Because of this, and for simplicity, the neuron processes that bring sensory information into the CNS will be called sensory fibers or afferent fibers, no matter what type of neuron is involved.

12.2 RECEPTORS, SENSATION, AND PERCEPTION

Sensory receptors are diverse but share certain features. Each type of receptor is particularly sensitive to a distinct type of environmental change and is much less sensitive to other forms of stimulation. The raw form in which these receptors send information to the brain is called **sensation.** The way our brains interpret this information is called **perception.**

TABLE 12.1 | Information Flow from the Environment Through the Nervous System

Information Flow	Smell	Taste	Sight	Hearing
Sensory receptors	Olfactory cells in nose	Taste bud receptor cells	Rods and cones in retina	Hair cells in cochlea
↓	↓	↓	↓	↓
Impulse in sensory fibers	Olfactory nerve fibers	Sensory fibers in various cranial nerves	Optic nerve fibers	Auditory nerve fibers
↓	↓	↓	↓	↓
Impulse reaches CNS	Cerebral cortex	Cerebral cortex	Midbrain and cerebral cortex	Midbrain and cerebral cortex
↓	↓	↓	↓	↓
Sensation (new experience, recalled memory)	A pleasant smell	A sweet taste	A small, round, red object	A crunching sound
↓	↓	↓	↓	↓
Perception	The smell of an apple	The taste of an apple	The sight of an apple	The sound of biting into an apple

Receptor Types

Five types of sensory receptors are recognized, based on their sensitivities to specific stimuli:

1. **Chemoreceptors** (ke″mo-re-sep′torz) respond to changes in the concentration of chemicals. Receptors associated with the senses of smell and taste are of this type. Chemoreceptors in internal organs detect changes in the blood concentrations of oxygen, hydrogen ions, glucose, and other chemicals.
2. **Pain receptors,** also called nociceptors (no″se-sep′torz), respond to tissue damage. Triggering stimuli include exposure to excess mechanical, electrical, thermal, or chemical energy.
3. **Thermoreceptors** (ther″mo-re-sep′torz) are sensitive to temperature change.
4. **Mechanoreceptors** (mek″ah-no re-sep′torz) are of several types and sense mechanical forces by detecting changes that deform the receptors. **Proprioceptors** (pro″pre-o-sep′torz) sense changes in the tensions of muscles and tendons, **baroreceptors** (bar″o-re-sep′torz), also called pressoreceptors, in certain blood vessels detect changes in blood pressure, and **stretch receptors** in the lungs sense degree of inflation.
5. **Photoreceptors** (fo″to-re-sep′torz) in the eyes respond to light energy of sufficient intensity.

Sensory Impulses

Sensory receptors can be ends of neurons or other types of cells close to them. In either case, stimulation causes local changes in their membrane potentials (receptor potentials), generating a graded electric current that reflects the intensity of stimulation (see chapter 10, p. 368).

If a receptor is a neuron and the change in membrane potential reaches threshold, an action potential is generated, and a sensory impulse is propagated along the afferent fiber. However, if the receptor is another type of cell, its receptor potential must be transferred to a neuron to trigger an action potential. Peripheral nerves transmit sensory impulses to the central nervous system (CNS), where they are analyzed and interpreted in the brain.

Sensation and Perception

A sensation occurs when the brain becomes aware of sensory impulses. A perception occurs when the brain interprets those sensory impulses. All the nerve impulses that travel away from sensory receptors into the CNS are alike, so the resulting sensation depends on which region of the cerebral cortex receives the impulse. For example, impulses reaching one region are always interpreted as sounds, and those reaching another are always sensed as touch. (Some receptors, such as those that measure oxygen levels in the blood, do not trigger sensations.)

Sensory receptors are specialized to respond to specific stimuli, but they may respond to other stimuli that are strong enough, in which case the sensations will be the same. Pain receptors, for example, can be stimulated by heat, cold, or pressure, but the sensation is always the same because, in each case, the same part of the brain interprets the resulting nerve impulses as pain. Similarly, stimuli other than light, such as a sharp blow to the head, may trigger nerve impulses in visual receptors. When this happens, the person may "see stars," even though no light is entering the eye, because any impulses reaching the visual cortex are interpreted as light. Normally receptors only respond to specific stimuli, so the brain creates the correct sensation for that particular stimulus.

At the same time that a sensation forms, the cerebral cortex interprets it to seem to come from the receptors being stimulated. This process is called **projection** because the brain projects the sensation back to its apparent source. Projection allows a person to pinpoint the region of stimulation. Thus, we perceive that the eyes see an apple, the nose smells it, and the ears hear the teeth crunch into it.

Sensory Adaptation

The brain must prioritize the sensory input it receives, or it would be overwhelmed by unimportant information. For example, until this sentence prompts you to think about it, you are probably unaware of the pressure of your clothing against your skin, or the background noise in the room. This ability to ignore unimportant stimuli is called **sensory adaptation** (sen′so-re ad″ap-ta′shun). It may reflect a decreased response to a particular stimulus from the receptors (peripheral adaptation) or along the CNS pathways leading to the sensory regions of the cerebral cortex (central adaptation). As adaptation occurs, sensory impulses become less frequent until they may cease. Once adaptation occurs, impulses are triggered only if the strength of the stimulus changes.

PRACTICE

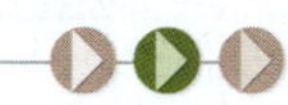

1. Distinguish between general and special senses.
2. List the five general types of sensory receptors.
3. What do all types of receptors have in common?
4. Explain how a sensation occurs.
5. What is sensory adaptation?

12.3 GENERAL SENSES

General senses are those whose sensory receptors are widespread, associated with the skin, muscles, joints, and viscera. These senses can be divided into three groups:

1. **Exteroreceptive senses** are associated with changes at the body surface. They include the senses of touch, pressure, temperature, and pain.
2. **Visceroreceptive** (interoceptive) **senses** are associated with changes in viscera (blood pressure stretching blood vessels, an ingested meal stimulating pH receptors in the small intestine, and so on).
3. **Proprioceptive senses** are associated with changes in muscles and tendons and in body position.

Touch and Pressure Senses

The senses of touch and pressure derive from three types of receptors (fig. 12.1). As a group, these receptors sense mechanical forces that deform or displace tissues. The touch and pressure receptors include the following:

1. **Free nerve endings.** These simplest receptors are common in epithelial tissues, where they lie between epithelial cells. They are responsible for the sensation of itching (fig. 12.1*a*).
2. **Tactile (Meissner's) corpuscles.** These are small, oval masses of flattened connective tissue cells in connective tissue sheaths. Two or more sensory nerve fibers branch into each corpuscle and end within it as tiny knobs.

   Tactile corpuscles are abundant in the hairless portions of the skin, such as the lips, fingertips, palms, soles, nipples, and external genital organs. They provide fine touch, such as distinguishing two points on the skin where an object touches, to judge its texture (fig. 12.1*b*).
3. **Lamellated (Pacinian) corpuscles.** These sensory bodies are relatively large, ellipsoidal structures composed of connective tissue fibers and cells. They are common in the deeper dermal tissues of the hands, feet, penis, clitoris, urethra, and breasts and also in tendons of muscles and ligaments of joints (fig. 12.1*c*). Heavier pressure and stretch stimulate lamellated corpuscles. They also detect vibrations in tissues.

Temperature Senses

Temperature receptors (thermoreceptors) include two groups of free nerve endings in the skin. Those that respond to warmer temperatures are *warm receptors,* and those that respond to colder temperatures are *cold receptors.*

The warm receptors are most sensitive to temperatures above 25°C (77°F) and become unresponsive at temperatures above 45°C (113°F). As 45°C is approached, pain receptors are also triggered, producing a burning sensation.

Cold receptors are most sensitive to temperatures between 10°C (50°F) and 20°C (68°F). If the temperature drops below 10°C, pain receptors are stimulated, and the person feels a freezing sensation.

At intermediate temperatures, the brain interprets sensory input from different combinations of these receptors as a particular temperature sensation. Both warm and cold receptors rapidly adapt, so within about a minute of continuous stimulation, the sensation of warm or cold begins to fade. This is why we quickly become comfortable after jumping into a cold swimming pool or submerging into a steaming hot tub.

Sense of Pain

Pain receptors (nociceptors) consist of free nerve endings. These receptors are widely distributed throughout the skin and internal tissues, except in the nervous tissue of the brain, which lacks pain receptors. Pain receptors protect in that they are stimulated when tissues are damaged. Pain sensation is usually perceived as unpleasant, signaling that action be taken to remove the source of the stimulation.

Most pain receptors can be stimulated by more than one type of change. However, some pain receptors are most sensitive to mechanical damage. Others are particularly sensitive to extremes in temperature. Some pain receptors are most responsive to chemicals, such as hydrogen ions; potassium ions; or specific breakdown products of proteins, histamine, and acetylcholine. A deficiency of blood (ischemia) and thus a deficiency of oxygen (hypoxia) in a tissue also triggers pain sensation. For example, pain elicited during a muscle cramp results from interruption of blood flow that occurs as the sustained contraction squeezes capillaries, as well as from the stimulation of mechanoreceptors. Also, when blood flow is interrupted, pain-stimulating chemicals accumulate.

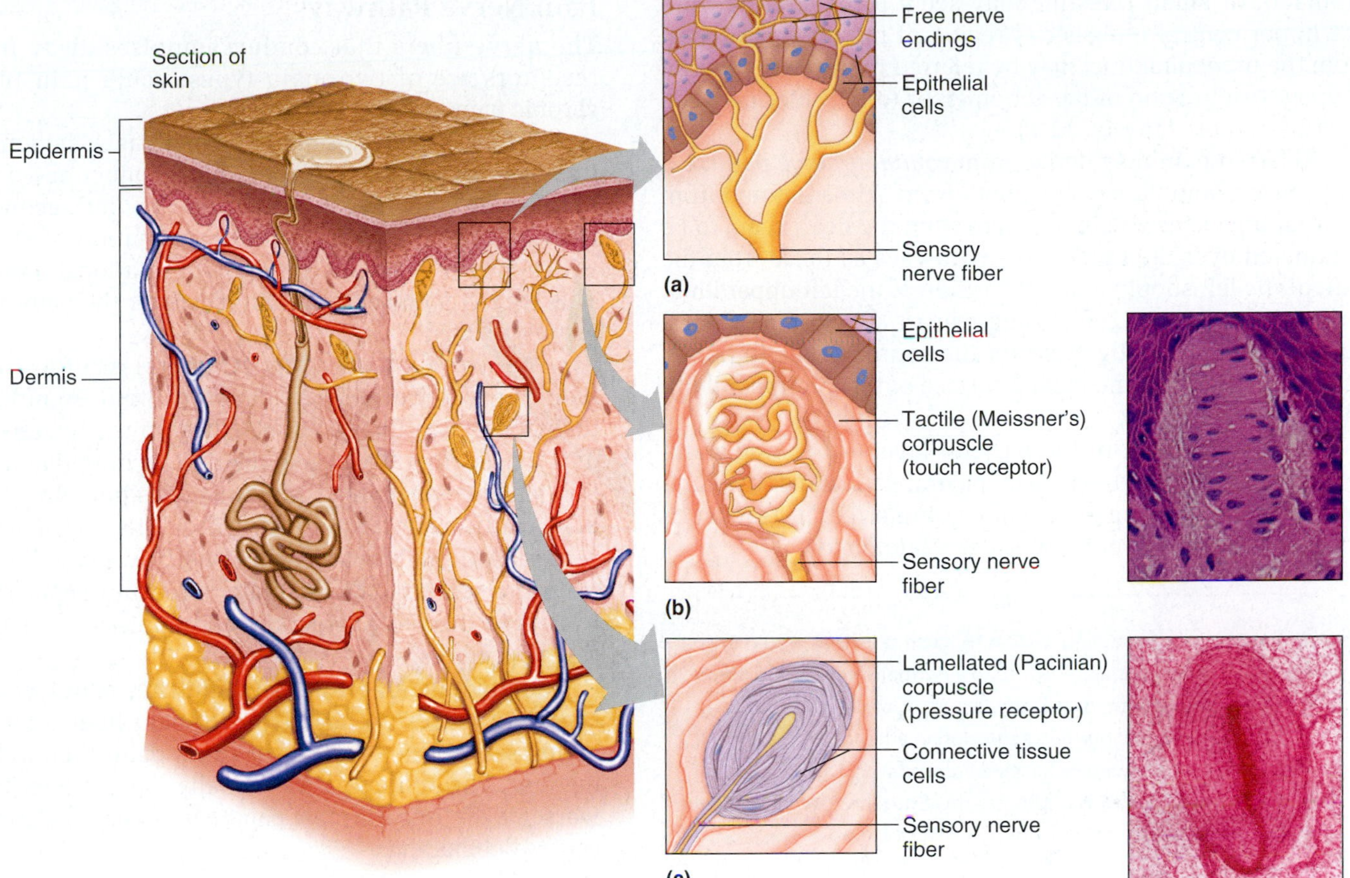

FIGURE 12.1 Touch and pressure receptors include (*a*) free ends of sensory nerve fibers, (*b*) tactile corpuscle (with 225× micrograph), and (*c*) lamellated corpuscle (with 50× micrograph).

Increasing blood flow through the sore tissue may relieve the resulting pain, and this is why heat is sometimes applied to reduce muscle soreness. The heat dilates blood vessels and thus promotes blood flow, which helps reduce the concentration of the pain-stimulating substances. In some conditions, accumulating chemicals lower the thresholds of pain receptors, making inflamed tissues more sensitive to heat or pressure.

The importance of the ability to feel pain is seen in people who cannot do so. Children who have hereditary sensory and autonomic neuropathy cannot feel pain, and as a result they inadvertently injure and mutilate themselves. More common is peripheral neuropathy, in which the hands and/or feet become numb due to decreasing numbers of tactile corpuscles. Skin biopsies are used to track the telltale loss of the corpuscles, but biopsies are painful, time-consuming, and give inconsistent results. An experimental approach uses reflectance confocal microscopy to image the number of corpuscles. In one pilot study, people with normal pain sensation had twelve corpuscles per square millimeter of skin, whereas people with peripheral neuropathy had 2.8. The most common causes of peripheral neuropathy are diabetes mellitus, cancer treatment, vitamin deficiency, and HIV infection.

Pain receptors adapt very little, if at all. Once such a receptor is activated, even by a single stimulus, it may continue to send impulses into the CNS for some time.

Visceral Pain

As a rule, pain receptors are the only receptors in viscera whose stimulation produces sensations. Pain receptors in these organs respond differently to stimulation than those associated with surface tissues. For example, localized damage to intestinal tissue during surgical procedures may not elicit any pain sensations, even in a conscious person. However, when visceral tissues are subjected to more widespread stimulation, as when intestinal tissues are stretched or when the smooth muscles in the intestinal walls undergo spasms, a strong pain sensation may follow. Once again, the resulting pain results from stimulation of mechanoreceptors and from decreased blood flow accompanied by lower tissue oxygen levels and accumulation of pain-stimulating chemicals.

Visceral pain may feel as if it is coming from some part of the body other than the part being stimulated—a phenomenon called **referred pain.** For example, pain originating in the heart may be referred to the left shoulder or the medial surface of the left upper limb. Pain from the lower esophagus, stomach, or small intestine may seem to be coming from the upper central (epigastric) region of the abdomen. Pain from the urogenital tract may be referred to the lower central (hypogastric) region of the abdomen or to the sides between the ribs and the hip (fig. 12.2).

Referred pain may derive from *common nerve pathways* that sensory impulses coming both from skin areas and from internal organs use. Pain impulses from the heart seem to be conducted over the same nerve pathways as those from the skin of the left shoulder and the inside of the left upper limb, as shown in figure 12.3. During a heart attack, the cerebral cortex may incorrectly interpret the source of the impulses as the shoulder and the medial surface of the left upper limb, rather than the heart.

Pain originating in the parietal layers of thoracic and abdominal membranes—parietal pleura, parietal pericardium, or parietal peritoneum—is usually not referred; instead, such pain is felt directly over the area being stimulated.

Neuropathic pain is an overreaction to a stimulus that would ordinarily cause pain or a pain response to a normally innocuous stimulus. Reflex sympathetic dystrophy is a form of neuropathic pain that causes an intense burning sensation in a hand or foot, even if the extremity is paralyzed or has been amputated. During the Civil War, it was called "causalgia." Union Army Surgeon S. Weir Mitchell described causalgia as "the most terrible of all tortures."

Pain Nerve Pathways

The nerve fibers that conduct impulses away from pain receptors are of two main types: acute pain fibers and chronic pain fibers.

The *acute pain fibers* (also known as A-delta fibers) are thin, myelinated nerve fibers. They conduct nerve impulses rapidly, at velocities up to 30 meters per second. These impulses are associated with the sensation of sharp pain, which typically seems to originate in a local area of skin. This type of pain seldom continues after the pain-producing stimulus stops.

The *chronic pain fibers* (C fibers) are thin, unmyelinated nerve fibers. They conduct impulses more slowly than acute pain fibers, at velocities up to 2 meters per second. These impulses cause the dull, aching pain sensation that may be widespread and difficult to pinpoint. Such pain may continue for some time after the original stimulus ceases. Although acute pain is usually sensed as coming from the surface, chronic pain is felt in deeper tissues as well as in the skin. Visceral pain impulses are usually carried on C fibers.

Commonly, an event that stimulates pain receptors will trigger impulses on both types of pain fibers. This causes a dual sensation—a sharp, pricking pain, then a dull, aching one. The aching pain is usually more intense and may worsen over time. Chronic pain that resists relief and control can be debilitating.

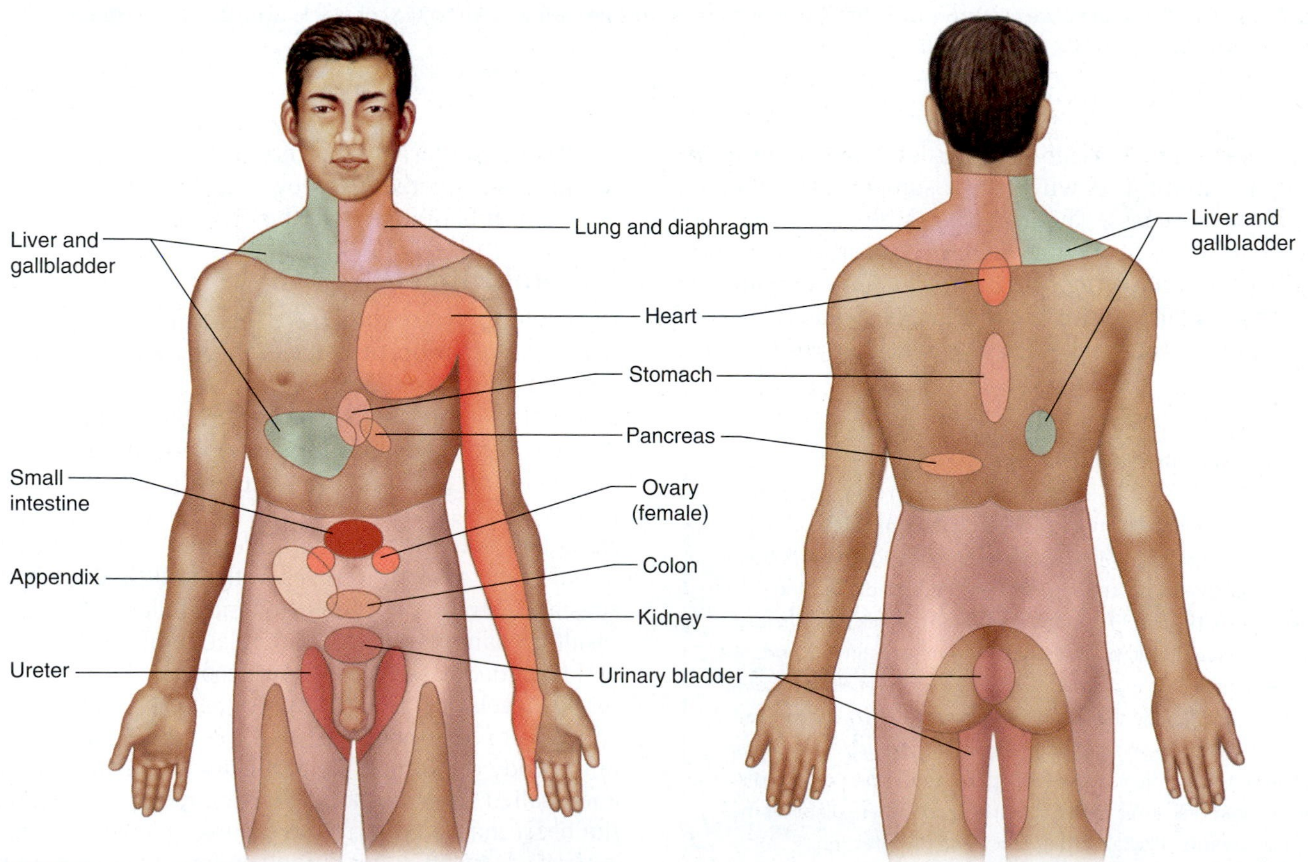

FIGURE 12.2 Surface regions to which visceral pain may be referred.

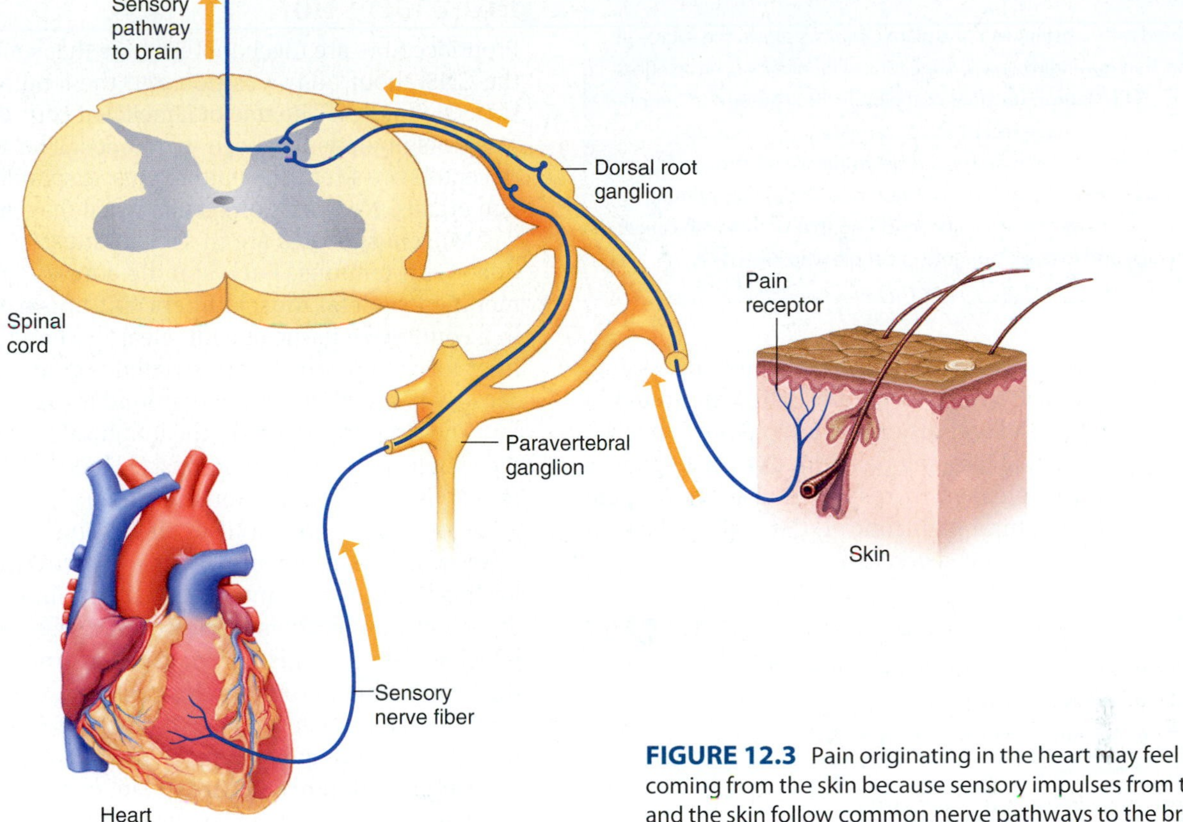

FIGURE 12.3 Pain originating in the heart may feel as if it is coming from the skin because sensory impulses from the heart and the skin follow common nerve pathways to the brain.

Pain impulses that originate from tissues of the head reach the brain on sensory fibers of the fifth, seventh, ninth, and tenth cranial nerves. All other pain impulses travel on sensory fibers of spinal nerves, and they pass into the spinal cord by way of the dorsal roots of these spinal nerves.

Upon reaching the spinal cord, pain impulses enter the gray matter of the posterior horn, where they are processed. The fast-conducting fibers synapse with long nerve fibers that cross over to the opposite side of the spinal cord in the anterior and lateral spinothalamic tracts (anterolateral system). The impulses carried on the slow-conducting fibers pass through one or more interneurons before reaching the long fibers that cross over and ascend to the brain.

In the brain, most of the pain fibers terminate in the reticular formation (see chapter 11, p. 409), and from there are conducted on fibers of still other neurons to the thalamus, hypothalamus, and cerebral cortex. Fibers of the spinothalamic tracts transmit pain and temperature information directly to the thalamus.

Regulation of Pain Impulses

Awareness of pain occurs when pain impulses reach the level of the thalamus—that is, even before they reach the cerebral cortex. However, the cerebral cortex must judge the intensity of pain and locate its source, and it is also responsible for emotional and motor responses to pain.

RECONNECT

To Chapter 10, Postsynaptic Potentials, pages 371–372.

Still other parts of the brain, including areas of gray matter in the midbrain, pons, and medulla oblongata, regulate the flow of pain impulses from the spinal cord (see chapter 11, pp. 407–408). Impulses from special neurons in these areas descend in the lateral funiculus to various levels within the spinal cord. The impulses stimulate the ends of certain nerve fibers to release biochemicals that can block pain signals by inhibiting presynaptic nerve fibers in the posterior horn of the spinal cord.

Among the inhibiting substances released in the posterior horn are neuropeptides called *enkephalins* and the monoamine *serotonin* (see chapter 10, p. 374). Enkephalins can suppress both acute and chronic pain impulses; thus, they can relieve strong pain sensations, much as morphine and other opiate drugs do. In fact, enkephalins were discovered because they bind to the same receptors on neuron membranes as does morphine. Serotonin stimulates other neurons to release enkephalins.

Cannabinoids are substances in the plant *Cannibus sativa,* the source of marijuana, that may relieve pain. Anecdotal evidence for such an effect dates to A.D. 315. Neurons in areas of the brain, brainstem, and peripheral nervous system have receptors for cannabinoids. A synthetic version of the compound in marijuana responsible for most of its effects (delta-9-tetra-hydrocannabinol), marketed as the prescription drug Marinol, is used to treat nausea and vomiting that occurs with cancer chemotherapy and to boost appetite in people who have AIDS.

Another group of neuropeptides with pain-suppressing, morphinelike actions are the *endorphins.* They are found in the pituitary gland and in regions of the nervous system, such as the hypothalamus, that transmit pain impulses. Enkephalins and endorphins are released in response to extreme pain impulses, providing natural pain control. Clinical Application 12.1 discusses treatments for severe pain.

PRACTICE

6 Describe three types of touch and pressure receptors.
7 Describe thermoreceptors.
8 What types of stimuli excite pain receptors?
9 What is referred pain?
10 Explain how neuropeptides control pain.

Proprioception

Proprioceptors are mechanoreceptors that send information to the CNS about body position and the length and tension of skeletal muscles. The role of lamellated corpuscles as pressure receptors in joints has been mentioned. The other main proprioceptors are stretch receptors: muscle spindles and Golgi tendon organs. No sensation results when they are stimulated.

Muscle spindles are in skeletal muscles near their junctions with tendons. Each spindle consists of several small, modified skeletal muscle fibers (intrafusal fibers) enclosed in a connective tissue sheath. Near its center, each intrafusal fiber has a specialized nonstriated region with the end of a sensory nerve fiber wrapped around it (fig. 12.4*a*).

The striated portions of the intrafusal fiber contract to keep the spindle taut at different muscle lengths. Thus, if the whole muscle is stretched, the muscle spindle is also stretched, triggering sensory nerve impulses on its nerve fiber. These sensory fibers synapse in the spinal cord with lower motor neurons leading back to the same muscle. Thus, impulses triggered by stretch of the muscle spindle contract the skeletal muscle of which it is a part. This action, called a **stretch reflex,** opposes the lengthening of the muscle and helps maintain the desired position of a limb despite gravitational or other forces tending to move it (see chapter 11, pp. 390–391).

Golgi tendon organs are in tendons close to their attachments to muscles. Each is connected to a set of skeletal mus-

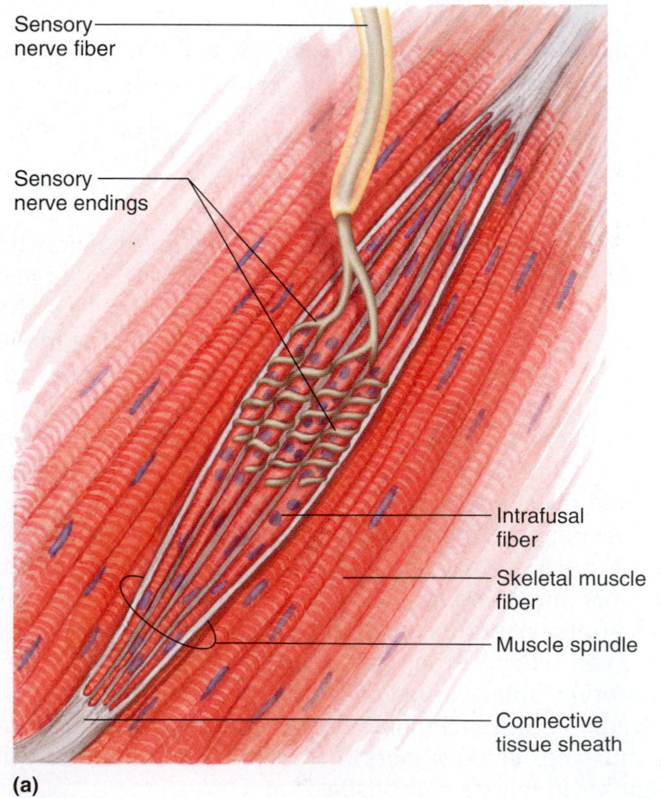

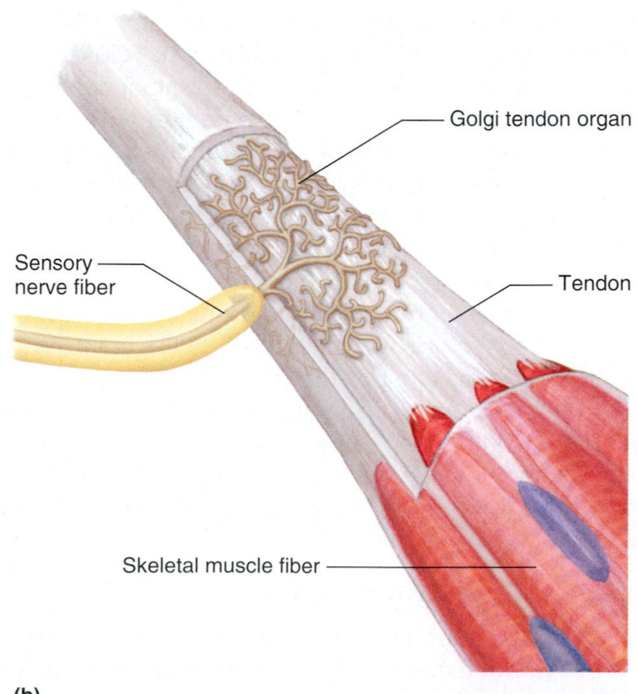

FIGURE 12.4 Stretch receptors maintain posture. (*a*) Increased muscle length stimulates muscle spindles, which stimulate muscle contraction. (*b*) Golgi tendon organs occupy tendons, where they inhibit muscle contraction.

12.1 CLINICAL APPLICATION

Treating Pain

Too many people are in pain. Several studies estimate that at any given time, one in four individuals worldwide is in moderate to severe chronic pain. A quarter of them may be undertreated.

Most pain remedies are nonsteroidal anti-inflammatory drugs (NSAIDs) such as aspirin, ibuprofen, and COX-2 inhibitors (fig. 12A), or opiates (fig. 12B). Many recently approved painkillers are variations on these traditional themes, with molecules tweaked or formulations changed, such as from injectable to oral delivery. The topselling painkiller is acetaminophen, sold under the brand name Tylenol. It has an interesting history.

Acetaminophen was first described in the chemical literature in 1878. In 1886, its analgesic and fever-lowering effects were noted, by accident, but not recognized. Two doctors in France were treating a patient with intestinal parasites with the chemical naphthalene. One day a pharmacist gave them the compound acetanilide in error. The mistake was fortuitous—fever dropped and pain faded. In 1899, a chemist discovered that acetanilide is metabolized to acetaminophen. By 1909, the compound had been further studied and produced as a drug, used first in England. It became available by prescription in the United States in 1953 and over-the-counter in 1960. The success of acetaminophen is attributed to its lack of side effects seen with the NSAIDs. It does not irritate the gastrointestinal lining, and does not thin the blood.

In 2005 a new painkiller became available. Ziconotide is a synthetic version of a peptide that the marine cone snail *Conus magus* releases to paralyze its fish prey (fig. 12C). When researchers noticed that the natural peptide binds to a type of calcium receptor on spinal cord neurons that receive pain impulses, the effort began to turn the snail's weapon into a pain reliever. Ziconotide is delivered by catheter to the affected body part and is prescribed to relieve intractable chronic pain. Similar drugs are in development.

The types of patients in greatest need of pain relief are people with cancer or chronic pain syndromes. More than half of people nearing the end of their battle against cancer suffer pain that can be treated, or at least dulled, but is not. The fear is unfounded that giving opiate drugs to ease their suffering will lead to addiction. Narcotics are much more likely to be addicting when they are abused to induce euphoria than when they are taken to relieve severe pain. Cancer patients take NSAIDs, weak narcotics such as hydrocodone, strong narcotics such as morphine, and opiates delivered directly to the spine via an implanted reservoir. Patients may use devices to control the delivery of pain medications. Anti-anxiety medications can ease the perception of pain.

Chronic pain is of three types: lower back pain, migraine, and myofascial syndrome (inflammation of muscles and their fascia). Treatment approaches include NSAIDs, stretching exercises, injection of local anesthetic drugs into cramping muscles, and antidepressants to raise serotonin levels in the CNS. Chronic pain may also be treated with electrodes implanted near the spinal cord (a dorsal column stimulator); transcutaneous electrical nerve stimulation (TENS), which also places electrodes on pain-conducting nerves; and an invasive nerve block, which interrupts a pain signal by freezing or introducing an anesthetic drug. ■

FIGURE 12A Painkillers come from nature. Aspirin derives from bark of the willow tree.

FIGURE 12B Poppies are the source of opiate drugs.

FIGURE 12C A newer analgesic for extreme and unrelenting chronic pain is based on a peptide from the marine cone snail *Conus magus.*

cle fibers and is innervated by a sensory neuron (fig. 12.4*b*). These receptors have high thresholds and are stimulated by increased tension. Sensory impulses from them produce a reflex that inhibits contraction of the muscle whose tendon they occupy. Thus, the Golgi tendon organs stimulate a reflex with an effect opposite that of a stretch reflex. The Golgi tendon reflex also helps maintain posture, and it protects muscle attachments from being pulled away from their insertions by excessive tension. Table 12.2 summarizes some of the receptors of the general senses and their functions.

TABLE 12.2 | Receptors Associated with General Senses

Type	Function	Sensation
Free nerve endings (mechanoreceptors)	Detect changes in pressure	Touch, pressure
Tactile corpuscles (mechanoreceptors)	Detect objects moving over the skin	Touch, texture
Lamellated corpuscles (mechanoreceptors)	Detect changes in pressure	Deep pressure, vibrations, fullness in viscera
Free nerve endings (thermoreceptors)	Detect changes in temperature	Heat, cold
Free nerve endings (pain receptors)	Detect tissue damage	Pain
Free nerve endings (mechanoreceptors)	Detect stretching of tissues, tissue spasms	Visceral pain
Muscle spindles (mechanoreceptors)	Detect changes in muscle length	None
Golgi tendon organs (mechanoreceptors)	Detect changes in muscle tension	None

Visceral Senses

Receptors in internal organs include lamellated corpuscles and free nerve endings. The information these receptors convey includes the sense of fullness after eating a meal as well as the discomfort of intestinal gas and the pain that signals a heart attack.

PRACTICE 

11 Describe a muscle spindle.
12 Explain how muscle spindles help maintain posture.
13 Where are Golgi tendon organs located?
14 What is the function of Golgi tendon organs?

12.4 SPECIAL SENSES

Special senses are those whose sensory receptors are part of large, complex sensory organs in the head. These senses and their respective organs include the following:

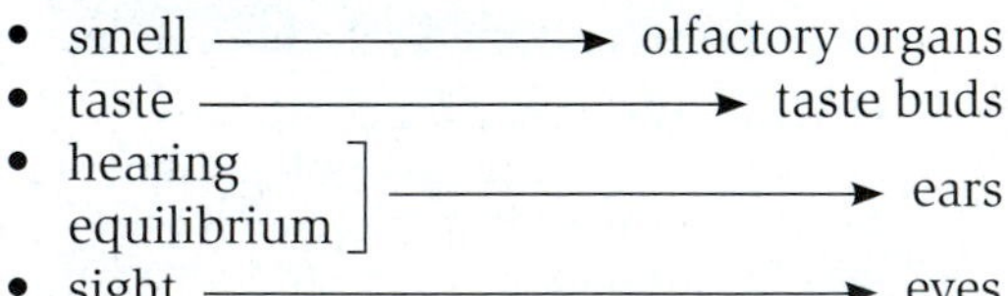

Clinical Application 12.2 discusses an unusual type of sensory abnormality.

Sense of Smell

The ability to detect the strong scent of a fish market, the antiseptic odor of a hospital, the aroma of a ripe melon—and thousands of other smells—is possible thanks to a yellowish patch of tissue the size of a quarter high up in the nasal cavity. This fabric of sensation is a layer of 12 million specialized cells.

Olfactory Receptors

Olfactory receptors, used to sense odors, are similar to those for taste in that they are chemoreceptors sensitive to chemicals dissolved in liquids. The two chemical senses function closely together and aid in food selection, because we smell food at the same time we taste it. It is often difficult to tell what part of a food sensation is due to smell and what part is due to taste. For this reason, an onion tastes different when sampled with the nostrils closed, because much of the usual onion sensation is due to odor. Similarly, if copious mucous secretions from an upper respiratory infection cover the olfactory receptors, food may seem tasteless. About 75% to 80% of flavor derives from the sense of smell.

Olfactory Organs

The olfactory organs, which contain the olfactory receptors, also include epithelial supporting cells. These organs appear as yellowish brown masses within pinkish mucous membrane. They cover the upper parts of the nasal cavity, the superior nasal conchae, and a portion of the nasal septum (fig. 12.5).

The **olfactory receptor cells** are bipolar neurons surrounded by columnar epithelial cells. These neurons have knobs at the distal ends of their dendrites covered with hairlike cilia. The cilia project into the nasal cavity and are the sensitive portions of the receptor cells (fig. 12.6). A person's 12 million olfactory receptor cells each have ten to twenty cilia.

> Humans smell the world using about 12 million olfactory receptor cells. Bloodhounds have more olfactory receptor cells than any other mammal, and therefore a keener sense of smell. The excellent sense of smell in canines is the basis of using service dogs to detect impending health problems in their owners. The dogs sense subtle odors that people emit when becoming ill with certain conditions. Service dogs are used to sense imminent seizures, drops in blood glucose, and accelerated heart rate. There is some evidence that dogs can sense cancer.

Chemicals that stimulate olfactory receptor cells, called odorant molecules, enter the nasal cavity as gases, where they must dissolve at least partially in the watery fluids that surround the cilia before they can bond to receptor proteins on the cilia and be detected. Odorant molecules bind to about 400 types of olfactory receptors that are part of the cell membranes of the olfactory receptor cells, depolarizing them and thereby generating nerve impulses. In addition, signaling proteins inside the receptor cell translate the chemical signal (binding of the odorant molecule to the receptor protein) into the electrochemical language of the nervous system.

Sensory receptors are not the same as membrane receptors. Sensory receptors may be as small as individual cells or as large as complex organs such as the eye or ear. They respond

12.2 CLINICAL APPLICATION

Mixed-Up Senses—Synesthesia

"The song was full of glittering orange diamonds."

"The paint smelled blue."

"The sunset was salty."

"The pickle tasted like a rectangle."

One in 1,000 people has a condition called synesthesia, in which the brain interprets a stimulus to one sense as coming from another. Most common is *grapheme-color type synesthesia,* in which letters, numbers, or time evoke specific colors, or music may evoke perception of a strong taste or odor (*lexical-gustatory synesthesia*). These associations are involuntary, are specific, and persist over a lifetime. For example, a person might report that three is always mustard yellow, Thursday brown, or a symphony bittersweet.

Most synesthesia is present for as long as a person can remember, but it may develop following brain damage. One woman began to feel touch sensations in response to certain sounds after suffering a stroke. Brain imaging has traced the formation of unusual connections between her midbrain and the tactile part of her cerebral cortex.

Synesthesia seems to be inherited and is more common in women. One of the authors of this book (R. L.) has it—to her, days of the week and months are specific colors. Synesthesia has been attributed to an immature nervous system that cannot sort out sensory stimuli or to altered brain circuitry that routes stimuli to the wrong part of the cerebral cortex. It may be a common inability in infants, but tends to be lost over time.

Positron emission tomography (PET) scanning reveals a physical basis to synesthesia. Brain scans of six nonsynesthetes were compared with those of six synesthetes who reported associating words with colors. Cortical blood flow was monitored while a list of words was read aloud to both groups. Interestingly, cortical blood flow was greatly elevated in the synesthetes compared with the nonsynesthetes. Furthermore, while blood flow was increased in word-processing areas for both groups, the scans revealed that areas important in vision and color processing were also lit up in those with synesthesia. ■

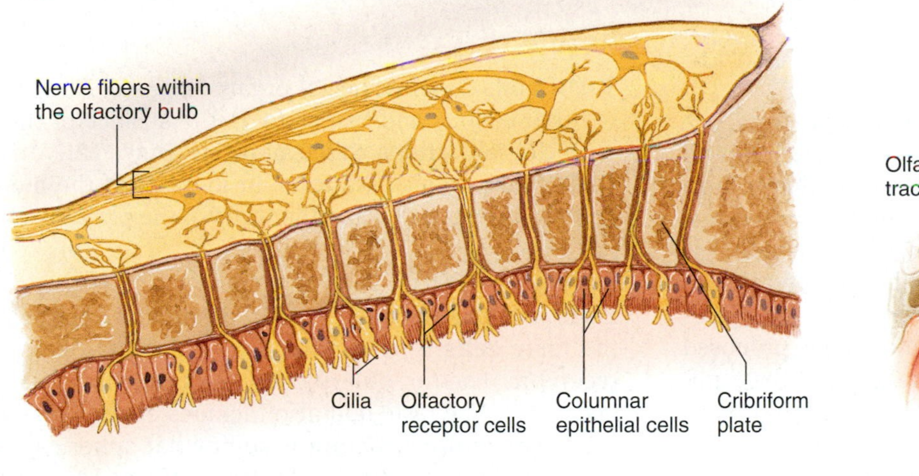

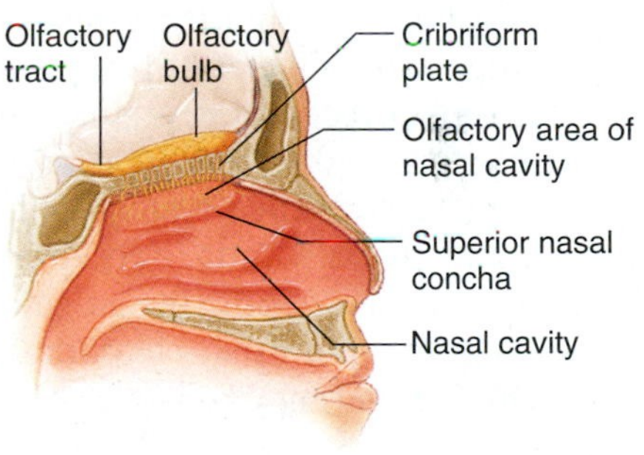

FIGURE 12.5 Olfactory receptors. (*a*) Columnar epithelial cells support olfactory receptor cells, which have cilia at their distal ends. (*b*) The olfactory area is associated with the superior nasal concha.

to sensory stimuli. Membrane receptors are molecules such as proteins and glycoproteins on the cell membranes. They allow cells, such as neurons and olfactory receptor cells, to respond to specific molecules. Thus, the olfactory receptors are cells that respond to chemical stimuli, but they require cell membrane receptors to do so.

Olfactory Nerve Pathways

Once olfactory receptor cells are stimulated, nerve impulses travel along their axons through tiny openings in the cribriform plates of the ethmoid bone. These fibers (which form the first cranial nerves) synapse with neurons located in the enlargements of the **olfactory bulbs,** structures that lie on either side of the crista galli of the ethmoid bone (see figs. 7.24 and 12.5).

In the olfactory bulbs, the sensory impulses are analyzed, and as a result, additional impulses travel along the **olfactory tracts** to portions of the limbic system (see chapter 11, p. 407), a brain center for memory and emotions. This is why we may become nostalgic over a scent from the past. A whiff of the perfume that grandma used to wear may bring back a flood of memories. The input to the limbic system also

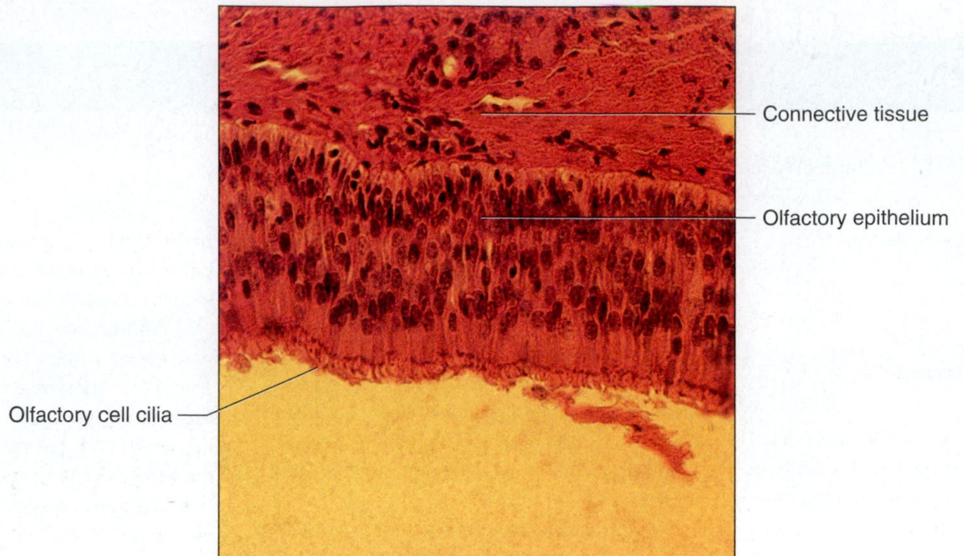

FIGURE 12.6 Light micrograph of the olfactory epithelium (250×).

explains why odors can alter mood so easily. For example, the scent of new-mown hay or rain on a summer's morning generally makes us feel good. The main interpreting areas for the olfactory impulses (olfactory cortex) are deep within the temporal lobes and at the bases of the frontal lobes, anterior to the hypothalamus.

Olfactory Stimulation

Biologists are not certain how stimulated receptors encode specific smells, but a leading hypothesis is that each odor likely stimulates a distinct set of receptor cells that in turn have distinct sets of receptor proteins. The brain then recognizes the particular combination as an *olfactory code.* For example, imagine a simplified system with ten types of odor receptors. Banana might stimulate receptors 2, 4, and 7; garlic, receptors 1, 5, and 9. Some investigators have proposed seven primary odors, but others hypothesize that the number is much higher and may reflect the functioning of hundreds of genes.

The olfactory organs are high in the nasal cavity above the usual pathway of inhaled air, so sniffing and forcing air over the receptor areas may be necessary to smell a faint odor. Olfactory receptors undergo sensory adaptation rather rapidly, so the intensity of a smell drops about 50% within a second following the stimulation. Within a minute, the receptors may become almost insensitive to a given odor, but even though they have adapted to one scent, their sensitivity to other odors persists.

The olfactory receptor neurons are the only nerve cells in direct contact with the outside environment. These neurons are subject to damage because of their exposed positions. Fortunately, basal cells along the basement membrane of the olfactory epithelium regularly divide and yield differentiated cells that replace lost olfactory receptor neurons. These are the only damaged neurons that are regularly replaced.

PRACTICE

15 Where are the olfactory receptors located?

16 Trace the pathway of an olfactory impulse from a receptor to the cerebrum.

Sense of Taste

Taste buds are the special organs of taste. They resemble orange sections and associate on the surface of the tongue with tiny elevations called **papillae** (figs. 12.7 and 12.8). Taste buds are also scattered in the roof of the mouth, the linings of the cheeks, and the walls of the pharynx.

Taste Receptors

Each taste bud includes a group of modified epithelial cells, the **taste cells** (gustatory cells), that function as sensory receptors. Each of our 10,000 taste buds houses 50 to 150 taste cells. The taste bud also includes epithelial supporting cells. The entire structure is somewhat spherical, with an opening, the **taste pore,** on its free surface. Tiny projections (microvilli), called **taste hairs,** protrude from the outer ends of the taste cells and jut out through the taste pore. These taste hairs are the sensitive parts of the receptor cells.

Interwoven among and wrapped around the taste cells is a network of nerve fibers. The ends of these fibers closely contact the receptor cell membranes. A stimulated receptor cell triggers an impulse on a nearby nerve fiber, which travels into the brain.

A chemical to be tasted must dissolve in saliva, the watery fluid surrounding the taste buds. The salivary glands supply this fluid. To demonstrate the importance of saliva, blot your tongue and try to taste some dry food; then repeat the test after moistening your tongue with saliva.

As is the case for smell, the sense of taste derives from combinations of chemicals binding specific receptor proteins

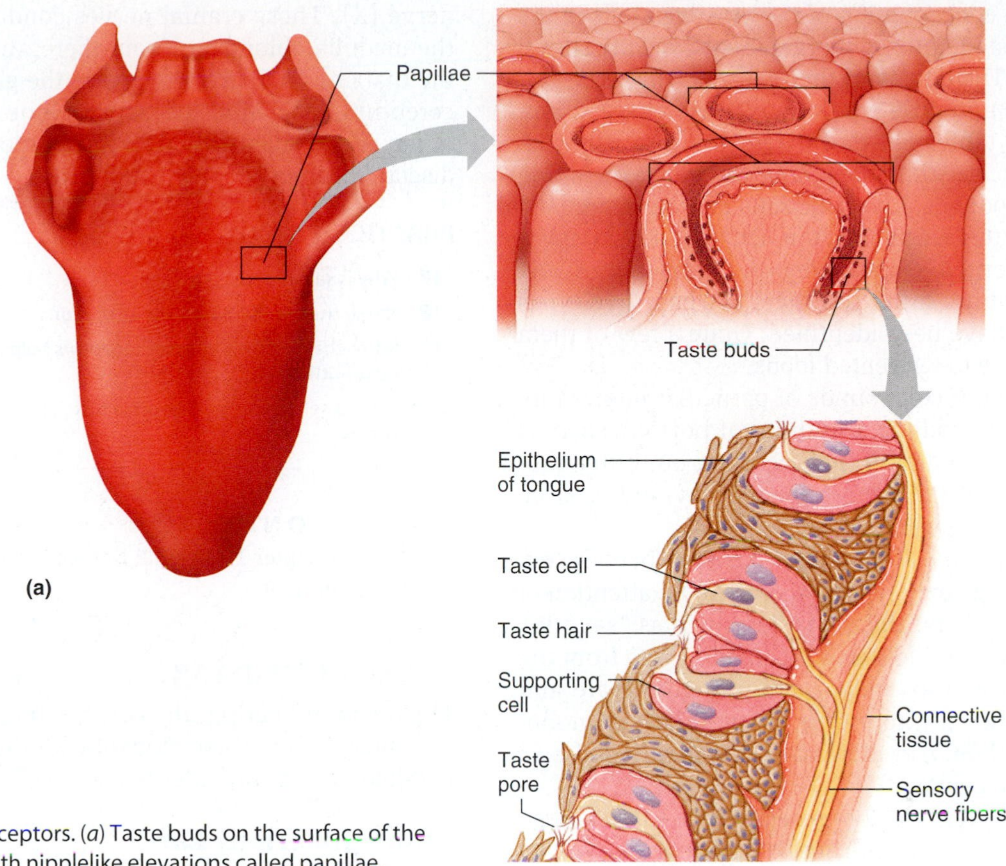

FIGURE 12.7 Taste receptors. (*a*) Taste buds on the surface of the tongue are associated with nipplelike elevations called papillae. (*b*) A taste bud contains taste cells and has an opening, the taste pore, at its free surface.

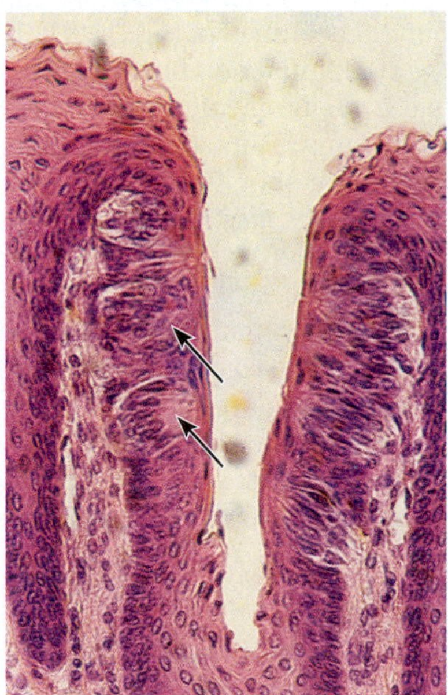

FIGURE 12.8 A light micrograph of some taste buds (arrows) (225×).

on taste hair surfaces. This binding alters membrane polarization, generating sensory impulses on nearby nerve fibers. The degree of change is directly proportional to the concentration of the stimulating substance.

Taste Sensations

The five primary taste sensations are sweet, sour, salty, bitter, and umami (oo-mom′ee). Each of the many flavors we experience results from one of the primary sensations or from a combination of them. The way we experience flavors may also reflect the concentration of chemicals as well as the sensations of smell, texture (touch), and temperature. Furthermore, chemicals in some foods—such as capsaisin in chili peppers—may stimulate pain receptors that cause a burning sensation.

Experiments indicate that each taste cell responds to one taste sensation only, with distinct receptors. Taste cells for each of the five taste sensations are in all areas of the tongue, but are distributed such that each sensation seems to arise most strongly from a particular region. Due to the distribution of taste cells, responsiveness to particular sensations varies from one region of the tongue to another. Sensitivity to a sweet stimulus peaks at the tip of the tongue, whereas responsiveness to sour is greatest at the margins of the tongue, and to bitter at the back. Receptors particularly responsive to salt are widely distributed.

Sweet receptors are usually stimulated by carbohydrates, but a few inorganic substances, including some salts of lead and beryllium, also elicit sweet sensations. Acids stimulate *sour receptors*. The intensity of a sour sensation is roughly proportional to the concentration of the hydrogen ions in the substance being tasted. Ionized inorganic salts mainly stimulate *salt receptors.* The quality of the sensation that each salt produces depends upon the type of positively charged ion, such as Na^+ from table salt, that it releases into solution. A variety of chemicals stimulates *bitter receptors,* including many organic compounds. Inorganic salts of magnesium and calcium produce bitter sensations, too. Extreme sensitivity to bitter tastes is inherited—this is why diet colas taste sweet to some people but are bitter to others. Twenty-five types of bitter receptors have been identified. Quite a few of them detect flavors unique to fermented foods.

One group of bitter compounds of particular interest are the *alkaloids,* which include a number of poisons such as strychnine, nicotine, and morphine. Spitting out bitter substances may be a protective mechanism to avoid ingesting poisonous alkaloids in foods.

The taste sensation called *umami* has long been recognized in Japan but has only recently come to the attention of western taste researchers. Umami is described as "savory," "pungent," "meaty," or "delicious." Umami arises from the binding of certain amino acids, including glutamate and aspartate, to specific receptors. The flavor enhancer monosodium glutamate (MSG), used in many prepared foods, also stimulates umami receptors.

Taste receptors, like olfactory receptors, rapidly undergo sensory adaptation. The resulting loss of taste can be avoided by moving bits of food over the surface of the tongue to stimulate different receptors at different moments.

Although taste cells are close to the surface of the tongue and are therefore exposed to environmental wear and tear, the sense of taste is not as likely to diminish with age as is the sense of smell. This is because taste cells are modified epithelial cells and divide continually. A taste cell functions for only about three days before it is replaced.

The sense of taste reflects what happens to food as it is chewed. Most foods are chemically complex, so they stimulate different receptors. In an experiment to track the act of tasting, chemists collected samples of air from participants' nostrils as they bit into juicy red tomatoes. An analytical technique called mass spectrometry revealed that chewing activates a sequence of chemical reactions in the tomato as its tissues are torn, releasing first aromatic hydrocarbons; then after a thirty-second delay, products of fatty acid breakdown; and, finally, several alcohols. This gradual release of stimulating molecules is why we experience a series of flavors as we savor a food.

Taste Nerve Pathways

Sensory impulses from taste receptor cells in the anterior two-thirds of the tongue travel on fibers of the facial nerve (VII); impulses from receptors in the posterior one-third of the tongue and the back of the mouth pass along the glossopharyngeal nerve (IX); and impulses from receptors at the base of the tongue and the pharynx travel on the vagus nerve (X). These cranial nerves conduct the impulses into the medulla oblongata. From there, the impulses ascend to the thalamus and are directed to the gustatory cortex of the cerebrum, located in the parietal lobe along a deep portion of the lateral sulcus. Clinical Application 12.3 and table 12.3 discuss disorders of smell and taste.

PRACTICE

17 Why is saliva necessary to taste?

18 Name the five primary taste sensations.

19 What characteristic of taste receptors helps maintain a sense of taste with age?

20 Trace a sensory impulse from a taste receptor to the cerebral cortex.

RECONNECT

To Chapter 11, Cranial Nerves, pages 414–417 and Table 11.9.

Sense of Hearing

The organ of hearing, the *ear,* has outer (external), middle, and inner (internal) sections. In addition to making hearing possible, the ear provides the sense of equilibrium.

Outer (External) Ear

The outer ear consists of all of the structures that face the outside. These include an outer, funnel-like structure called the **auricle** (pinna) and an S-shaped tube, the *external acoustic* (ah-kōōs′tik) *meatus* (external auditory canal) that leads inward for about 2.5 centimeters (fig. 12.9). The meatus terminates with the **tympanic membrane** (eardrum).

The external acoustic meatus passes into the temporal bone. Near this opening, hairs guard the tube. The opening and tube are lined with skin that has many modified sweat glands called *ceruminous glands,* which secrete wax (cerumen). The hairs and wax help keep large foreign objects, such as insects, out of the ear.

The transmission of vibrations through matter produces sound. Just as the sounds of some musical instruments are produced by vibrating strings or reeds, the sounds of the human voice are caused by vibrating vocal folds in the larynx. The auricle of the ear helps collect sound waves traveling through air and directs them into the external acoustic meatus.

TABLE 12.3 | Types of Smell and Taste Disorders

	Smell	Taste
Loss of sensation	Anosmia	Ageusia
Diminished sensation	Hyposmia	Hypogeusia
Heightened sensation	Hyperosmia	Hypergeusia
Distorted sensation	Dysosmia	Dysgeusia

12.3 CLINICAL APPLICATION

Smell and Taste Disorders

Imagine a spicy slice of pizza or freshly brewed coffee, and your mouth waters in anticipation. But for millions of people, the senses of smell and taste are dulled, distorted, or gone. Many more of us get some idea of their plight when a cold temporarily stifles these senses.

Compared to the loss of hearing or sight, being unable to taste or smell normally may seem more an oddity than an illness. People with such ailments would probably disagree. In some situations, a poor or absent sense of smell can be dangerous, such as in a house on fire.

The direct connection between the outside environment and the brain makes the sense of smell vulnerable to damage. Smell and taste disorders can be triggered by colds and flu, allergies, nasal polyps, swollen mucous membranes inside the nose, a head injury, chemical exposure, a nutritional or metabolic problem, or a disease. In many cases, a cause cannot be identified.

Drugs can alter taste and smell in many ways, affecting cell turnover, the neural conduction system, the status of receptors, and changes in nutritional status. Consider what happened to twelve hikers touring Peru and Bolivia. A day before a long hike, three of them had begun taking acetazolamide (Diamox), a drug that prevents acute mountain sickness. The night after the climb, the group went out for beer. To three of the people, the brew tasted unbearably bitter, and a drink of cola to wash away the taste was equally offensive. At fault: acetazolamide.

Drugs containing sulfur atoms squelch taste. They include the anti-inflammatory drug penicillamine, the antihypertensive drug captopril (Capoten), and transdermal (patch) nitroglycerin to treat chest pain. The antibiotic tetracycline and the antiprotozoan metronidazole (Flagyl) impart a metallic taste. Cancer chemotherapy and radiation treatment often alter taste and smell.

Exposure to toxic chemicals can affect taste and smell, too. One woman, suddenly found that once-pleasant smells had become offensive. Her doctor traced her problem to inhaling a paint stripper. Hydrocarbon solvents in the product—toluene, methanol, and methylene chloride—were responsible for her *cacosmia,* the association of an odor of decay with normally inoffensive stimuli. ■

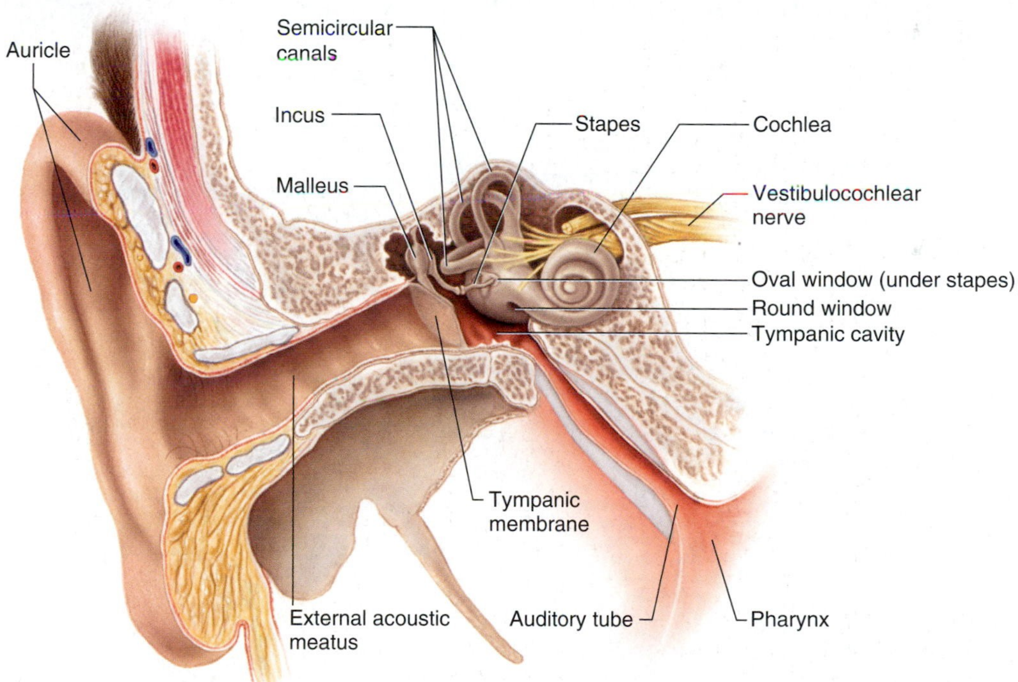

FIGURE 12.9 Major parts of the ear.

After entering the meatus, the sound waves pass to the end of the tube and alter the pressure on the tympanic membrane. The tympanic membrane is a semitransparent membrane covered by a thin layer of skin on its outer surface and by mucous membrane on the inside. It has an oval margin and is cone-shaped, with the apex of the cone directed inward. The tympanic membrane moves back and forth in response to sound waves, reproducing the vibrations of the sound-wave source.

Middle Ear

The **middle ear,** or the **tympanic cavity,** is an air-filled space in the temporal bone that separates the outer and inner ears. It is bounded by the tympanic membrane laterally and the inner ear medially and houses three small bones called **auditory ossicles** (aw′di-to″re os′i-klz).

The three auditory ossicles, called the *malleus,* the *incus,* and the *stapes,* are attached to the wall of the tympanic cavity by tiny ligaments and are covered by mucous membrane. These bones bridge the tympanic membrane and the inner ear, transmitting vibrations between these parts. Specifically, the malleus is attached to the tympanic membrane, helping to maintain its conical shape. When the tympanic membrane vibrates, the malleus vibrates in unison with it. The malleus vibrates the incus, and the incus passes the movement on to the stapes. Ligaments hold the stapes to an opening in the wall of the tympanic cavity called the **oval window** (fig. 12.9). Vibration of the stapes, which acts like a piston at the oval window, moves a fluid within the inner ear. These vibrations of the fluid stimulate the hearing receptors.

In addition to transmitting vibrations, the auditory ossicles form a lever system that helps increase (amplify) the force of the vibrations as they pass from the tympanic membrane to the oval window. Also, because the ossicles transmit vibrations from the large surface of the tympanic membrane to a much smaller area at the oval window, the vibrational force strengthens as it travels from the outer to the inner ear. As a result, the pressure (per square millimeter) that the stapes applies at the oval window is about twenty-two times greater than that which sound waves exert on the tympanic membrane.

The middle ear also has two small skeletal muscles attached to the auditory ossicles and controlled involuntarily. One of them, the *tensor tympani,* is inserted on the medial surface of the malleus and is anchored to the cartilaginous wall of the auditory tube. When it contracts, it pulls the malleus inward. The other muscle, the *stapedius,* is attached to the posterior side of the stapes and the inner wall of the tympanic cavity. It pulls the stapes outward (fig. 12.10). These muscles are the effectors in the **tympanic reflex,** elicited in about one-tenth second following a loud, external sound. When the reflex occurs, the muscles contract, and the malleus and stapes move. As a result, the bridge of ossicles in the middle ear becomes more rigid, reducing its effectiveness in transmitting vibrations to the inner ear.

The tympanic reflex reduces pressure from loud sounds that might otherwise damage the hearing receptors. Ordinary vocal sounds also elicit the tympanic reflex, as when a person speaks or sings. This action muffles the lower frequencies of such sounds, improving the hearing of higher frequencies, common in human vocal sounds. In addition, the tensor

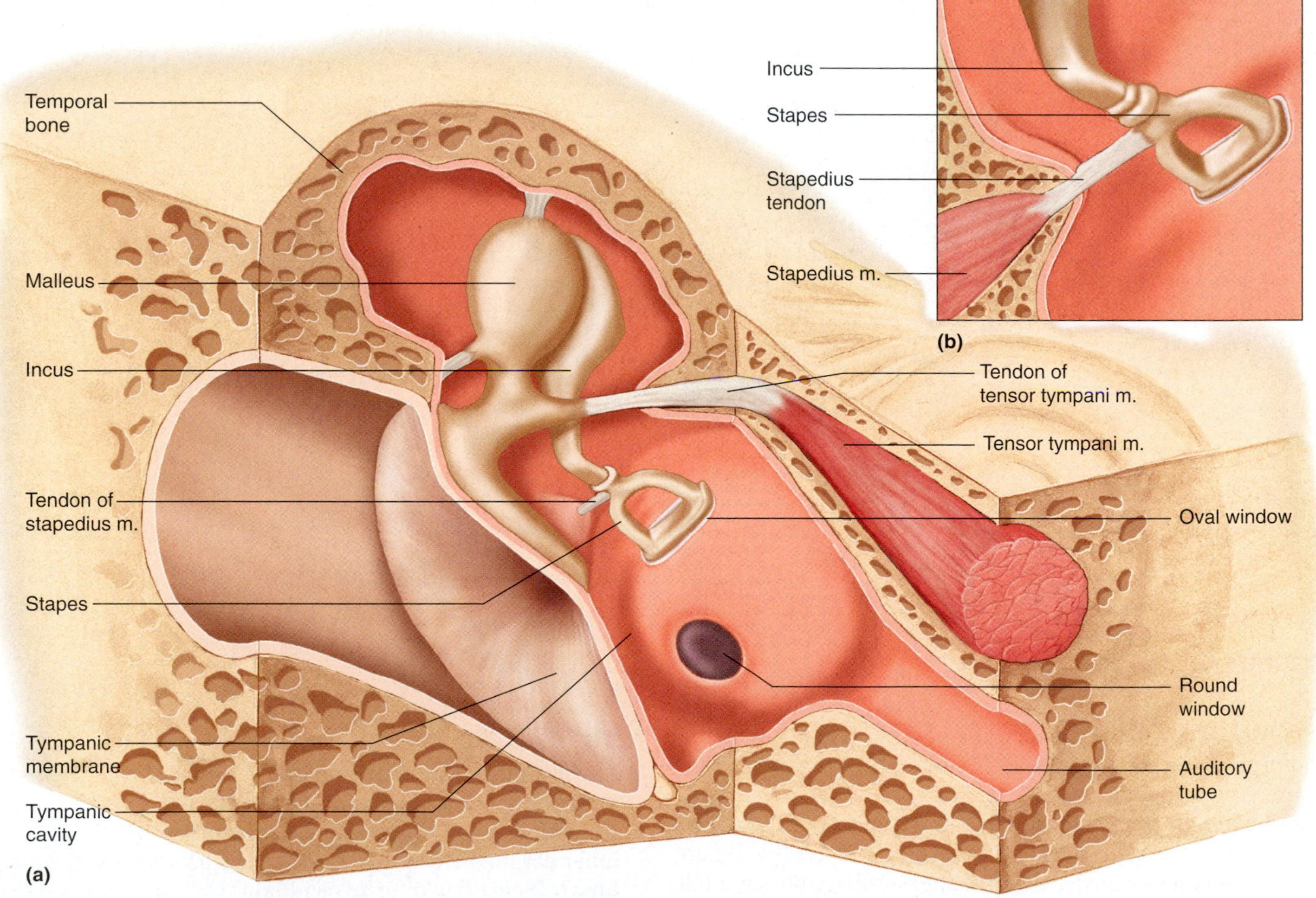

FIGURE 12.10 Two small muscles attached to the (*a*) malleus and (*b*) stapes, the tensor tympani and the stapedius, are effectors in the tympanic reflex. Figure 12.9 does not show these muscles. (*m.* stands for muscle.)

tympani muscle steadily pulls on the tympanic membrane. This is important because a loose tympanic membrane would not be able to effectively transmit vibrations to the auditory ossicles.

The muscles of the middle ear take 100 to 200 milliseconds to contract. For this reason, the tympanic reflex cannot protect the hearing receptors from the effects of very rapidly repeated loud sounds, such as from an explosion or a gunshot. On the other hand, this protective mechanism can reduce the effects of intense sounds that arise slowly, such as the roar of thunder.

Auditory Tube

An **auditory tube** (aw′di-to″re tub) (eustachian tube) connects each middle ear to the throat. This tube allows air to pass between the tympanic cavity and the outside of the body by way of the throat (nasopharynx) and mouth. It helps maintain equal air pressure on both sides of the tympanic membrane. This is necessary for normal hearing (see fig. 12.10).

The function of the auditory tube becomes noticeable during rapid change in altitude. For example, as a person moves from a high altitude to a lower one, the air pressure on the outside of the tympanic membrane steadily increases. As a result, the tympanic membrane may be pushed inward, out of its normal position, impairing hearing.

When the air pressure difference is great enough, some air may force its way up through the auditory tube into the middle ear. This equalizes the pressure on both sides of the tympanic membrane, which moves back into its regular position, causing a popping sound as normal hearing returns. A reverse movement of air ordinarily occurs when a person moves from a low altitude to a higher one.

The auditory tube is usually closed by valvelike flaps in the throat, which may inhibit air movements into the middle ear. Swallowing, yawning, or chewing aid in opening the flaps and can hasten equalization of air pressure.

Inner (Internal) Ear

The inner ear is a complex system of intercommunicating chambers and tubes called a **labyrinth** (lab′i-rinth). Each ear has two such regions—the osseous labyrinth and the membranous labyrinth.

The *osseous labyrinth* is a bony canal in the temporal bone; the *membranous labyrinth* is a tube that lies within the osseous labyrinth and has a similar shape (fig. 12.11*a*). Between the osseous and membranous labyrinths is a fluid called *perilymph,* secreted by cells in the wall of the bony canal. In the membranous labyrinth is a slightly different fluid called *endolymph.*

The parts of the labyrinths include a **cochlea** (kok′le-ah) that functions in hearing and three **semicircular canals** that provide a sense of equilibrium. A bony chamber called the **vestibule,** between the cochlea and the semicircular canals, houses membranous structures that serve both hearing and equilibrium.

The cochlea is shaped like a snail shell, coiled around a bony core (modiolus) with a thin, bony shelf (spiral lamina) that wraps around the core like a spiral staircase (fig. 12.11*b*). The shelf divides the bony labyrinth of the cochlea into upper and lower compartments. The upper compartment, called the *scala vestibuli,* leads from the oval window to the apex of the spiral. The lower compartment, the *scala tympani,* extends from the apex of the cochlea to a membrane-covered opening in the wall of the inner ear called the **round window.** These compartments constitute the bony labyrinth of the cochlea, and they are filled with perilymph. At the apex of the cochlea, a small opening (helicotrema) connects the fluids in the chambers (figs. 12.11*b* and 12.12).

A portion of the membranous labyrinth in the cochlea, called the *cochlear duct* (scala media), lies between the two bony compartments and is filled with endolymph. The cochlear duct ends as a closed sac at the apex of the cochlea. The duct is separated from the scala vestibuli by a *vestibular membrane* (Reissner's membrane) and from the scala tympani by a *basilar membrane* (see fig. 12.12). Clinical Application 12.4 describes an effective treatment for hearing loss called a cochlear implant.

The basilar membrane extends from the bony shelf of the cochlea and forms the floor of the cochlear duct. It has many thousands of stiff, elastic fibers that lengthen from the base of the cochlea to its apex. Vibrations entering the perilymph at the oval window travel along the scala vestibuli and pass through the vestibular membrane to enter the endolymph of the cochlear duct, where they move the basilar membrane. After passing through the basilar membrane, the vibrations enter the perilymph of the scala tympani, and their forces

Signs of a middle ear infection (otitis media) in an infant or toddler are hard to miss—irritability, screaming, fever, or tugging on the affected ear. Viewing the painful ear with an instrument called an otoscope reveals a red and bulging tympanic membrane.

Ear infections occur because the mucous membranes that line the auditory tubes are continuous with the linings of the middle ears, creating a conduit for bacteria infecting the throat or nasal passages to travel to the ear. The bacteria that most commonly cause middle ear infection are *Streptococcus pneumoniae, Haemophilus influenzae,* and *Moraxella catarrhalis.* This route to infection is greater in young children because their auditory tubes are shorter than in adults.

Physicians treat acute otitis media with antibiotics. Recurrent infections may cause hearing loss and interfere with learning, so children with recurrent otitis media may be fitted with tympanostomy tubes, inserted into affected ears during a brief surgical procedure. The tubes form a small tunnel through the tympanic membrane so the ears can drain. By the time the tubes fall out, the child has usually outgrown the susceptibility to ear infections.

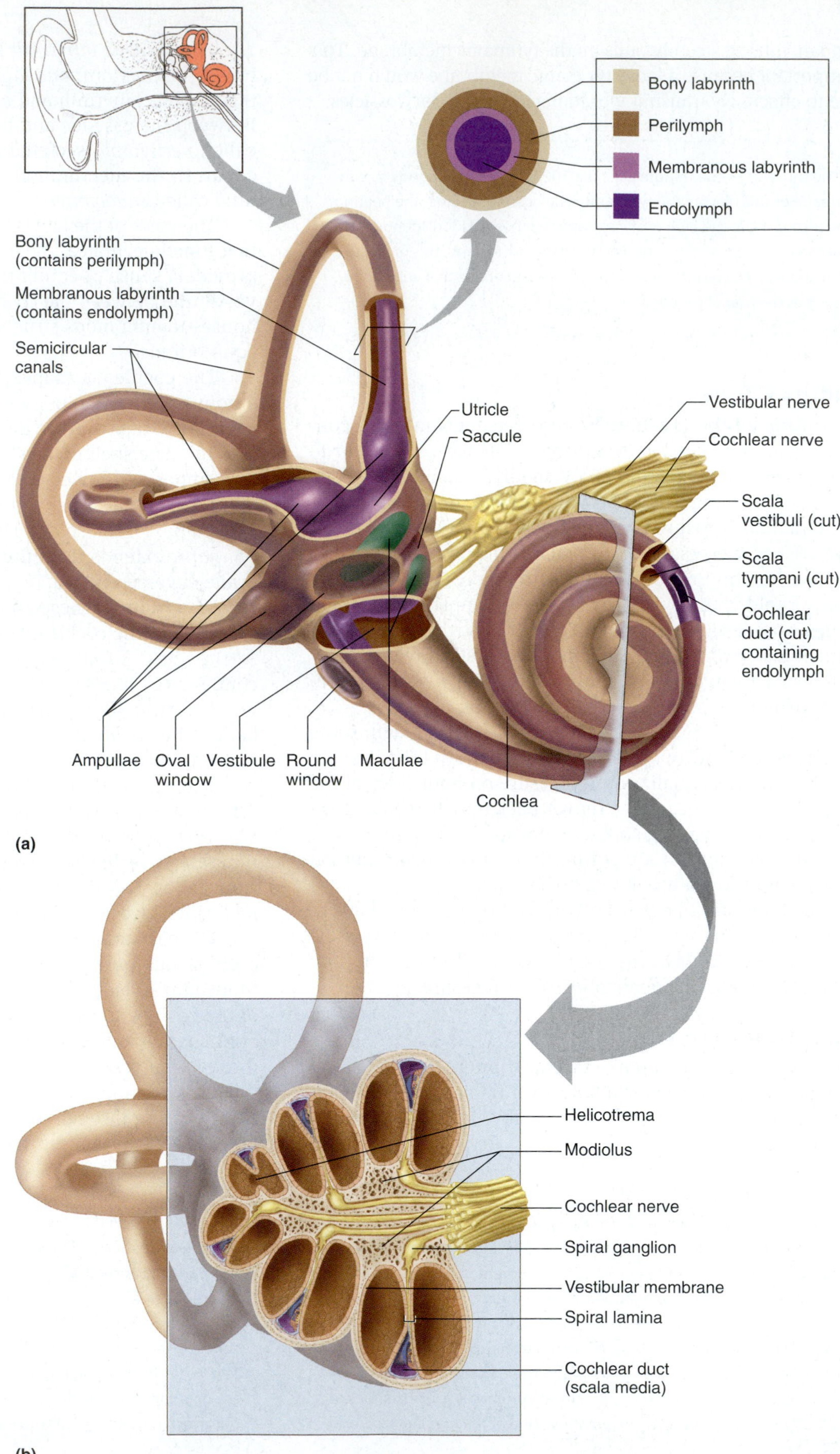

FIGURE 12.11 In the inner ear (*a*) perilymph separates the osseous labyrinth from the membranous labyrinth, which contains endolymph. (*b*) The spiral lamina coils around a bony core, the modiolus.

12.4 CLINICAL APPLICATION

Getting a Cochlear Implant

Yolanda Santana, of Rochester, New York, probably lost her hearing when she was only eight weeks old and suffered a high fever. But it wasn't until she was nine months old, when Yolanda didn't babble like her age-mates, that her parents suspected she might be deaf. With hearing aids she did well at a preschool for the deaf. Then Yolanda's parents read about the cochlear implant—it does not magically restore hearing, but enables a person to hear certain sounds. Teamed with speech therapy and use of sign language, the cochlear implant enables a person to make enough sense of sounds to speak.

When Carlos and Beth Santana read about the implants, Yolanda was already approaching three years of age. Before age three is the best time to receive a cochlear implant because this is when the brain is rapidly processing speech and hearing as a person masters language. Of the thousands of people in the United States who have received cochlear implants since they became available in 1984, about half have had them since early childhood.

The implant consists of a part inserted under the skin above the ear that leads to two dozen electrodes placed near the auditory nerve in the cochlea, the snail-shaped part of the inner ear. Yolanda wears a headset that includes a microphone lodged at the back of her ear to pick up incoming sounds and a fanny pack containing a speech processor that digitizes the sounds into coded signals. A transmitter on the headset sends the coded signals, as FM radio waves, to the implant, which changes them to electrical signals and delivers them to the cochlea. Here, the auditory nerve is stimulated and sends neural messages to the brain's cerebral cortex, which interprets the input as sound.

Yolanda's audiologist turned on the speech processor a month after the surgery. At first, the youngster heard low sounds and sometimes responded with a low hum. She grabbed at the processor, realizing it was the source of the sound. Gradually, the little girl learned from context what certain sounds meant. One day when Carlos signed "father" and said "poppy," Yolanda signed back and tried to say the word! Able to connect mouth movements to sounds to concepts, Yolanda was well on her way to hearing. ■

FIGURE 12D Yolanda Santana received a cochlear implant when she was three years old. The device enables her to detect enough sounds to effectively communicate.

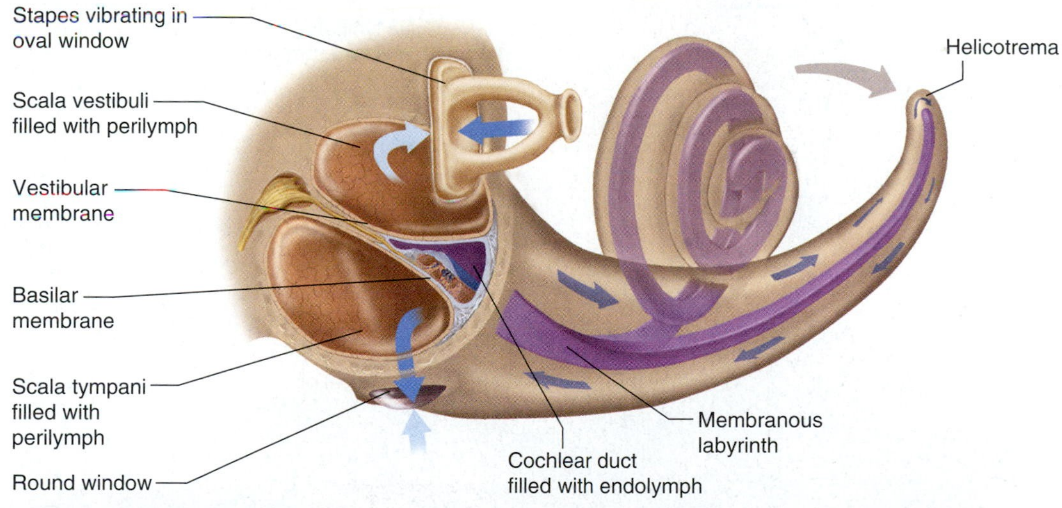

FIGURE 12.12 The cochlea is a coiled, bony canal with a membranous tube (labyrinth) inside. If the cochlea could be unwound, the membranous labyrinth would be seen ending as a closed sac at the apex where the bony canal makes a u-turn.

are dissipated into the air in the tympanic cavity by movement of the membrane covering the round window.

The **spiral organ** (organ of Corti), which contains about 16,000 hearing receptor cells, is on the superior surface of the basilar membrane and stretches from the apex to the base of the cochlea. The receptor cells, called **hair cells,** are in four parallel rows, with many hairlike processes (stereocilia) that extend into the endolymph of the cochlear duct. Above these hair cells is a *tectorial membrane,* attached to the bony shelf of the cochlea. It passes like a roof over the receptor cells, contacting the tips of their hairs (figs. 12.13 and 12.14).

Different frequencies of vibration move different parts of the basilar membrane. A particular sound frequency bends the hairs of a specific group of receptor cells against the tectorial membrane. Other frequencies deflect other sets of receptor cells.

Hearing receptor cells are epithelial cells, but they respond to stimuli somewhat like neurons (see chapter 10, pp. 365–370). For example, when a receptor cell is at rest, its membrane is polarized. When its hairs bend, selective ion channels open and its cell membrane depolarizes. The membrane then becomes more permeable, specifically to calcium ions. The receptor cell has no axon or dendrites, but it does have neurotransmitter-containing vesicles in the cytoplasm near its base. In the presence of calcium ions, some of these vesicles fuse with the cell membrane

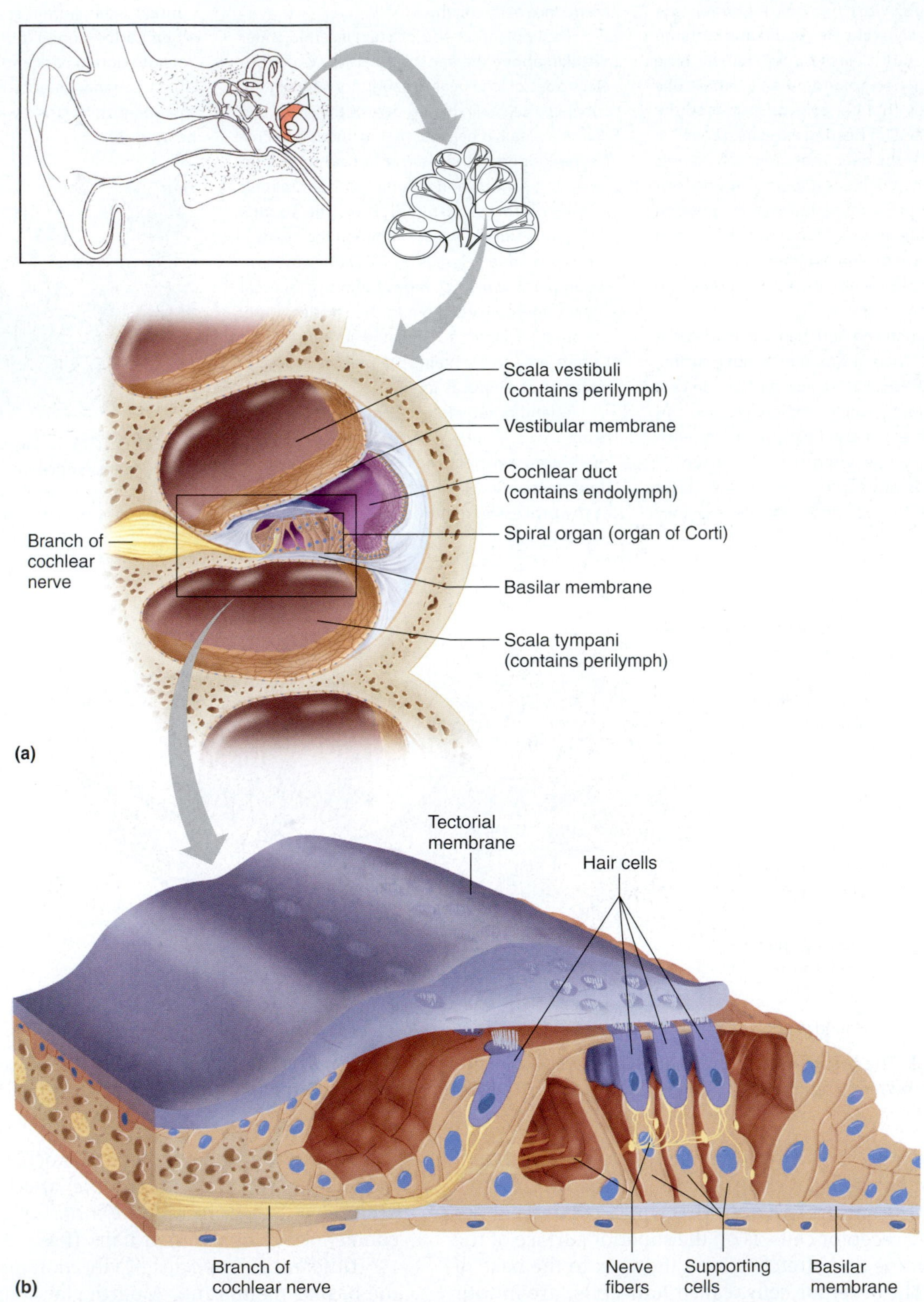

FIGURE 12.13 Cochlea. (*a*) Cross section of the cochlea. (*b*) Spiral organ and the tectorial membrane.

and release neurotransmitter to the outside. The neurotransmitter stimulates the ends of nearby sensory nerve fibers, and in response, they transmit nerve impulses along the cochlear branch of the vestibulocochlear nerve (cranial nerve VIII) to the brain.

The ear of a young person with normal hearing can detect sound waves with frequencies varying from about 20 to 20,000 or more vibrations per second. The range of greatest sensitivity is between 2,000 and 3,000 vibrations per second (fig. 12.15).

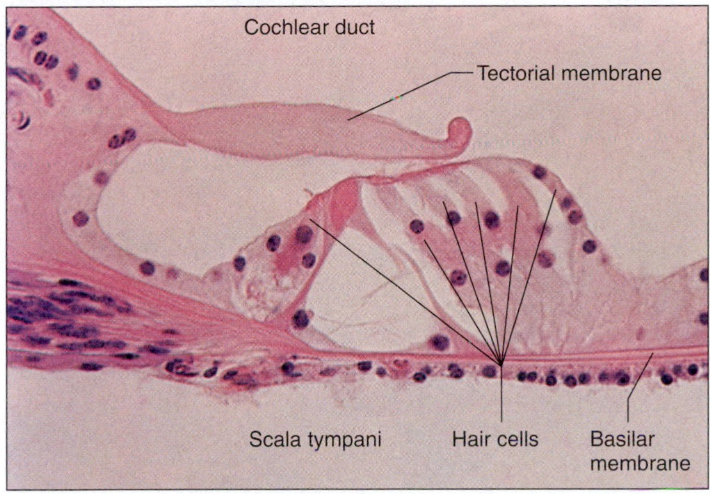

(a)

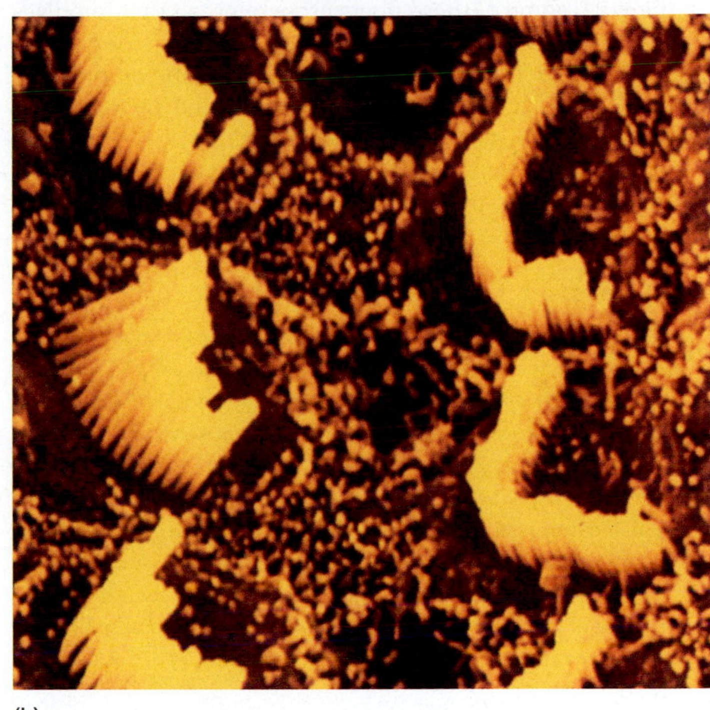

(b)

FIGURE 12.14 Spiral organ. (*a*) A micrograph of the spiral organ and the tectorial membrane (300×). (*b*) A scanning electron micrograph of hair cells in the spiral organ, looking down on the "hairs" (bright yellow) (6,700×).

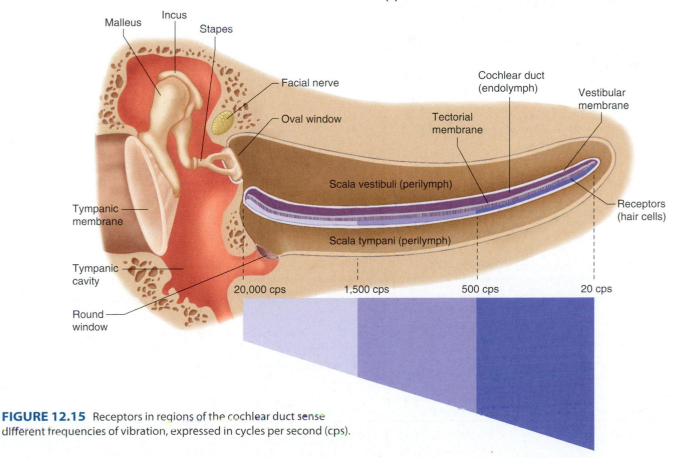

FIGURE 12.15 Receptors in regions of the cochlear duct sense different frequencies of vibration, expressed in cycles per second (cps).

Auditory Nerve Pathways

The cochlear branches of the vestibulocochlear nerves enter the auditory nerve pathways that extend into the medulla oblongata and proceed through the midbrain to the thalamus. From there they pass into the auditory cortices of the temporal lobes of the cerebrum, where they are interpreted. On the way, some of these fibers cross over, so that impulses arising from each ear are interpreted on both sides of the brain. Consequently, damage to a temporal lobe on one side of the brain is not necessarily accompanied by complete hearing loss in the ear on that side (fig. 12.16).

Table 12.4 summarizes the pathway of vibrations through the parts of the middle and inner ears. Clinical Application 12.5 examines types of hearing loss.

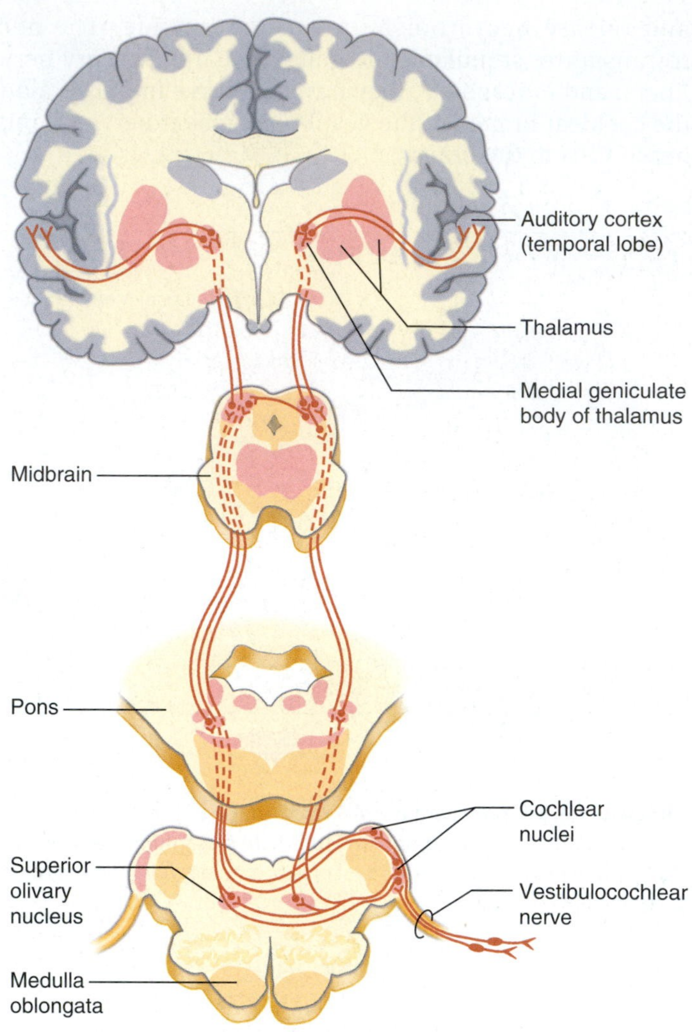

FIGURE 12.16 The auditory nerve pathway extends into the medulla oblongata, proceeds through the midbrain to the thalamus, and passes into the auditory cortex of the cerebrum.

> Units called *decibels* (dB) measure sound intensity as a logarithmic scale. The decibel scale begins at 0 dB, the intensity of the sound least perceptible by a normal human ear. A sound of 10 dB is 10 times as intense as the least perceptible sound; a sound of 20 dB is 100 times as intense; and a sound of 30 dB is 1,000 times as intense. A whisper has an intensity of about 40 dB, normal conversation measures 60–70 dB, and heavy traffic produces about 80 dB. A sound of 120 dB, such as a rock concert, produces discomfort; and a sound of 140 dB, such as a jet plane at takeoff, causes pain. Frequent or prolonged exposure to sounds with intensities above 90 dB can cause permanent hearing loss—as many aging rock stars have discovered.

PRACTICE

21 Describe the outer, middle, and inner ears.

22 Explain how sound waves are transmitted through the parts of the ear.

23 Describe the tympanic reflex.

24 Distinguish between the osseous and membranous labyrinths.

25 Explain the function of the spiral organ.

TABLE 12.4 | Steps in the Generation of Sensory Impulses from the Ear

1. Sound waves enter the external acoustic meatus.	7. A receptor cell depolarizes; its membrane becomes more permeable to calcium ions.
2. Waves of changing pressures cause the tympanic membrane to reproduce the vibrations coming from the sound-wave source.	8. In the presence of calcium ions, vesicles at the base of the receptor cell release neurotransmitter.
3. Auditory ossicles amplify and transmit vibrations to the end of the stapes.	9. Neurotransmitter stimulates the ends of nearby sensory neurons.
4. Movement of the stapes at the oval window transmits vibrations to the perilymph in the scala vestibuli.	10. Sensory impulses are triggered on fibers of the cochlear branch of the vestibulocochlear nerve.
5. Vibrations pass through the vestibular membrane and enter the endolymph of the cochlear duct.	11. The auditory cortex of the temporal lobe interprets the sensory impulses.
6. Different frequencies of vibration in endolymph move specific regions of the basilar membrane, stimulating specific sets of receptor cells.	

12.5 CLINICAL APPLICATION

Hearing Loss

Several factors can impair hearing, including interference with transmission of vibrations to the inner ear (*conductive deafness*) or damage to the cochlea or the auditory nerve and its pathways (*sensorineural deafness*). Disease, injury, and heredity all can impair hearing. There are more than 100 forms of inherited deafness, and many are part of syndromes. About 8% of people have some degree of hearing loss.

About 95% of cases of hearing loss are conductive. One cause is accumulated dry wax or a foreign object in the ear, which plugs the acoustic meatus. Changes in the tympanic membrane or auditory ossicles can also block hearing. The tympanic membrane may harden as a result of disease, becoming less responsive to sound waves, or an injury may tear or perforate it.

A common disorder of the auditory ossicles is *otosclerosis,* in which new bone is deposited abnormally around the base of the stapes. This interferes with the movement of the ossicles that is necessary to transmit vibrations to the inner ear. Surgery often can restore some hearing to a person with otosclerosis by chipping away the bone that holds the stapes in position or replacing the stapes with a wire or plastic substitute.

Two tests used to diagnose conductive deafness are the Weber test and the Rinne test. In the Weber test, the handle of a vibrating tuning fork is pressed against the forehead. A person with normal hearing perceives the sound coming from directly in front, whereas a person with sound conduction blockage in one middle ear hears the sound coming from the impaired side.

In the Rinne test, a vibrating tuning fork is held against the bone behind the ear. After the sound is no longer heard by conduction through the bones of the skull, the fork is moved to just in front of the external acoustic meatus. In middle ear conductive deafness, the vibrating fork can no longer be heard, but a normal ear will continue to hear its tone.

Very loud sounds can cause sensorineural deafness. If exposure is brief, hearing loss may be temporary, but when exposure is repeated and prolonged, such as occurs in foundries, near jackhammers, or on a firing range, impairment may be permanent. Such hearing loss begins as the hair cells develop blisterlike bulges that eventually pop. The tissue beneath the hair cells swells and softens until the hair cells, and sometimes the neurons, leaving the cochlea become blanketed with scar tissue and degenerate. Other causes of sensorineural deafness include tumors in the CNS, brain damage as a result of vascular accidents, and the use of certain drugs.

Hearing loss and other ear problems can begin gradually, so be aware of their signs, which may include the following:

- difficulty hearing people talking softly
- inability to understand speech when there is background noise
- ringing in the ears
- dizziness
- loss of balance

New parents should notice whether their infant responds to sounds in a way that indicates normal hearing. Hearing exams are part of a well-baby visit to a doctor. If the baby's responses indicate a possible problem, the next step is to see an audiologist, who identifies and measures hearing loss.

Often a hearing aid can help people with conductive hearing loss. A hearing aid has a tiny microphone that picks up sound waves and converts them to electrical signals, which are then amplified. An ear mold holds the device in place, either behind the outer ear, in the outer ear, or in the ear canal. ■

Sense of Equilibrium

The feeling of equilibrium derives from two senses—**static equilibrium** (stat′ik e′kwĭ-lib′re-um) and **dynamic equilibrium** (di-nam′ik e′kwĭ-lib′re-um). Different sensory organs provide these two components of equilibrium. The organs associated with static equilibrium sense the position of the head, maintaining stability and posture when the head and body are still. When the head and body suddenly move or rotate, the organs of dynamic equilibrium detect the motion and aid in maintaining balance.

Static Equilibrium

The organs of static equilibrium are in the vestibule, a bony chamber between the semicircular canals and the cochlea. More specifically, the membranous labyrinth inside the vestibule consists of two expanded chambers—a **utricle** (u′trĭ-kl) and a **saccule** (sak′ūl). The larger utricle communicates with the saccule and the membranous portions of the semicircular canals; the saccule, in turn, communicates with the cochlear duct (fig. 12.17).

The utricle and saccule each has a small patch of hair cells and supporting cells called a **macula** (mak′u-lah) on its wall. When the head is upright, the hairs of the macula in the utricle project vertically, while those in the saccule project horizontally. In both the utricle and saccule, the hairs contact a sheet of gelatinous material (otolithic membrane) that has crystals of calcium carbonate (otoliths) embedded on its surface. These particles add weight to the gelatinous sheet, making it more responsive to changes in position. The hair cells, which are sensory receptors, have nerve fibers wrapped around their bases. These fibers are associated with the vestibular portion of the vestibulocochlear nerve.

Gravity stimulates hair cells to respond. This usually happens when the head bends forward, backward, or to one side. Such movements tilt the gelatinous mass of one or more maculae, and as the material sags in response to gravity, the hairs projecting into it bend. This action stimulates the hair cells, and they signal their associated nerve fibers (figs. 12.18 and 12.19). The resulting nerve impulses travel into the CNS by means of the vestibular branch of the vestibulocochlear nerve, informing the brain of the head's

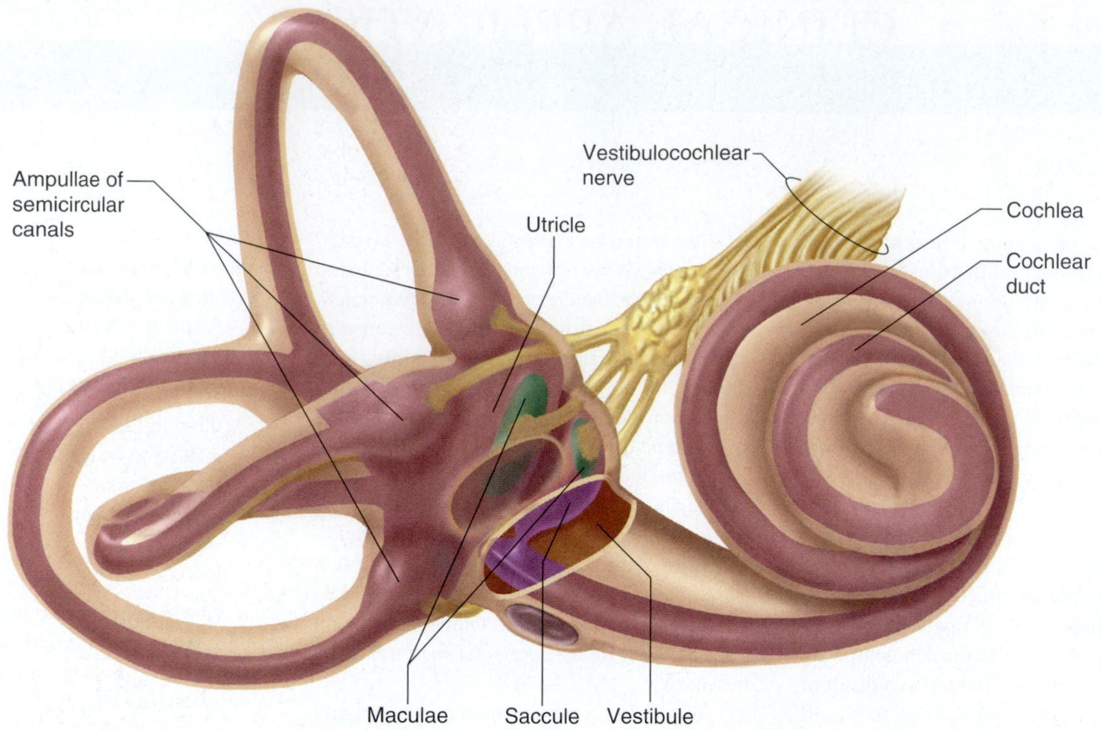

FIGURE 12.17 The saccule and utricle, expanded portions of the membranous labyrinth, are in the bony chamber of the vestibule. (Compare with figure 12.11.)

position. The brain responds by sending motor impulses to skeletal muscles, which may contract or relax appropriately to maintain balance.

The maculae also participate in the sense of dynamic equilibrium. For example, if the head or body is thrust forward or backward abruptly, the gelatinous mass of the maculae lags slightly behind, and the hair cells are stimulated. In this way, the maculae aid the brain in detecting movements such as falling and in maintaining posture while walking.

Dynamic Equilibrium

Each semicircular canal follows a circular path about 6 millimeters in diameter. The three bony semicircular canals lie at right angles to each other and occupy three different planes in space. Two of them, the *anterior canal* and the *posterior canal,* stand vertically, whereas the third, the *lateral canal,* is horizontal. Their orientations closely approximate the three body planes (see chapter 1, p. 22).

Suspended in the perilymph of each bony canal is a membranous semicircular canal that ends in a swelling called an **ampulla** (am-pul′ah). The ampullae communicate with the utricle of the vestibule.

An ampulla contains a septum that crosses the tube and houses a sensory organ. Each of these organs, called a **crista ampullaris,** has a number of sensory hair cells and supporting cells. As in the maculae, the hairs of the hair cells extend upward into a dome-shaped gelatinous mass called the *cupula.* Also, the hair cells are connected at their bases to nerve fibers that make up part of the vestibular branch of the vestibulocochlear nerve (fig. 12.20).

When the head or torso moves, the semicircular canals move as well, but initially the fluid inside the membranous canals tends to remain stationary because of inertia. This bends the cupula in one or more of the canals in a direction opposite that of the head and torso movement, and the hairs embedded in it also bend. The moving of the hairs stimulates the hair cells to signal their associated nerve fibers, and as a result, impulses travel to the brain (fig. 12.21). The orientation of the semicircular canals approximates the three anatomical planes, so movements in different directions affect different combinations of semicircular canals. The brain interprets impulses originating from these different combinations as movements in different directions.

Parts of the cerebellum are particularly important in interpreting impulses from the semicircular canals. Analysis of such information allows the brain to predict the consequences of rapid body movements, and by modifying signals to appropriate skeletal muscles, the cerebellum can maintain balance.

Other sensory structures aid in maintaining equilibrium. Various proprioceptors, particularly those associated with the joints of the neck, inform the brain about the position of body parts. The eyes detect changes in posture that result from body movements. Such visual information is so important that even if the organs of equilibrium are damaged, keeping the eyes open and moving slowly is sufficient to maintain normal balance.

FIGURE 12.18 The maculae respond to changes in head position. (*a*) Macula of the utricle with the head in an upright position. (*b*) Macula of the utricle with the head bent forward.

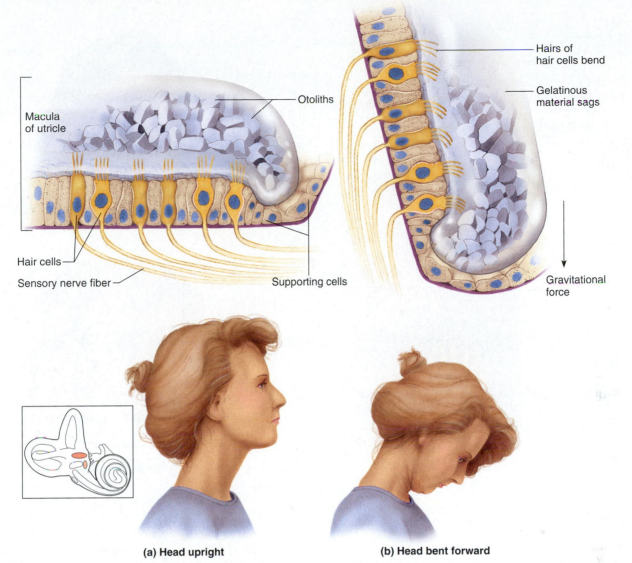

FIGURE 12.19 Scanning electron micrograph of hairs of hair cells, such as those in the utricle and saccule (8,000×).

Motion sickness is a disturbance of the inner ear's sensation of balance. Nine out of ten people have experienced this nausea and vomiting, usually when riding in a car or on a boat. Astronauts suffer a form of motion sickness called space adaptation syndrome.

Motion sickness is thought to result when visual information contradicts the inner ear's sensation that one is motionless. Consider a woman riding in a car. Her inner ears tell her that she is not moving, but the passing scenery tells her eyes that she is moving. The problem is compounded if she tries to read. The brain reacts to these seemingly contradictory sensations by signaling a "vomiting center" in the medulla oblongata.

PRACTICE

26 Distinguish between the senses of static and dynamic equilibrium.

27 Which structures provide the sense of static equilibrium? Of dynamic equilibrium?

28 How does sensory information from other receptors help maintain equilibrium?

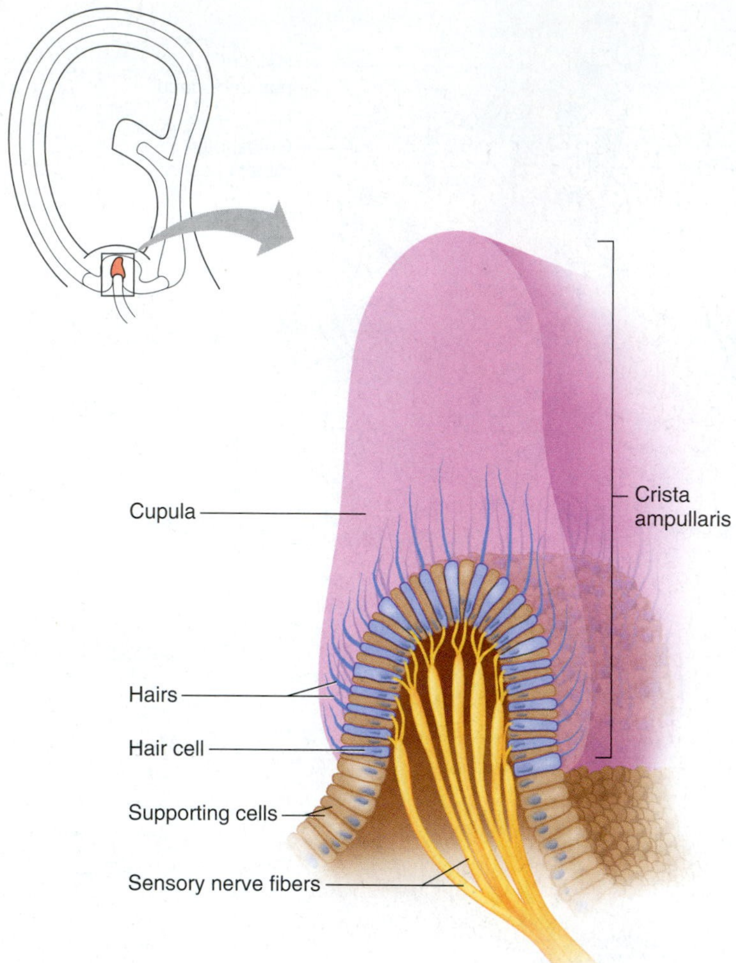

FIGURE 12.20 A crista ampullaris is in the ampulla of each semicircular canal.

(a) Head in still position

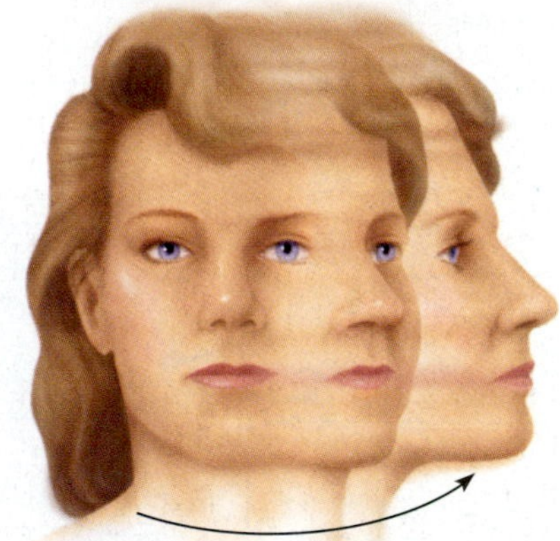

(b) Head rotating

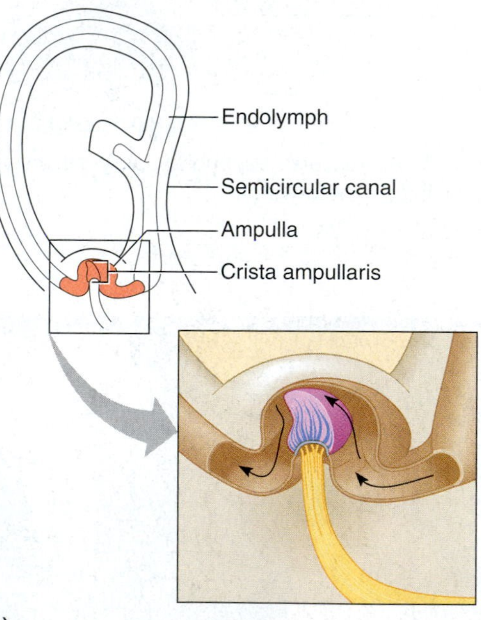

(c)

FIGURE 12.21 Equilibrium. (*a*) When the head is stationary, the cupula of the crista ampullaris remains upright. (*b*) When the head is moving rapidly, (*c*) the cupula bends opposite the motion of the head, stimulating sensory receptors.

Sense of Sight

A number of accessory organs assist the visual receptors in the eyes. These include the eyelids and lacrimal apparatus that help protect the eyes and a set of extrinsic muscles that move them.

Visual Accessory Organs

Each eye, lacrimal gland, and associated extrinsic muscles are housed in the orbital cavity of the skull. The orbit, lined with the periosteums of various bones, also contains fat, blood vessels, nerves, and connective tissues.

Each **eyelid** (palpebra) is composed of four layers—skin, muscle, connective tissue, and conjunctiva. The skin of the eyelid, the thinnest of the body, covers the lid's outer surface and fuses with its inner lining near the margin of the lid (fig. 12.22).

The muscles that move the eyelids include the *orbicularis oculi* and the *levator palpebrae superioris*. Fibers of the orbicularis oculi encircle the opening between the lids and spread out onto the cheek and forehead. This muscle acts as a sphincter that closes the lids when it contracts.

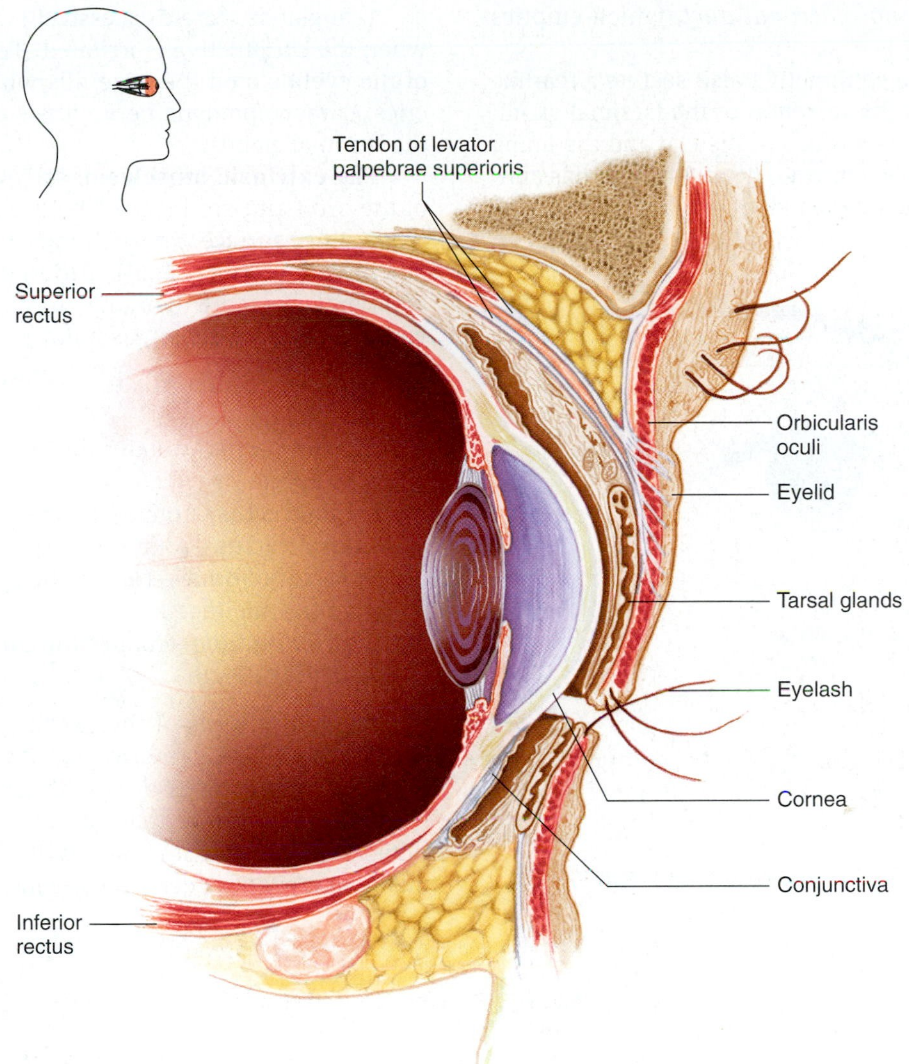

FIGURE 12.22 Sagittal section of the closed eyelids and the anterior portion of the eye.

Fibers of the levator palpebrae superioris muscle arise from the roof of the orbit and are inserted in the connective tissue of the upper lid. When these fibers contract, the upper lids are raised, and the eye opens.

The connective tissue layer of the eyelid, which helps give it form, contains many modified sebaceous glands (tarsal glands). Ducts carry the oily secretions of these glands to openings along the borders of the lids. This secretion helps keep the lids from sticking together.

The **conjunctiva** is a mucous membrane that lines the inner surfaces of the eyelids and folds back to cover the anterior surface of the eyeball, except for its central portion (cornea). Although the tissue that lines the eyelids is relatively thick, the conjunctiva that covers the eyeball is very thin. It is also freely movable and transparent, so that blood vessels are clearly visible beneath it.

A child in school with "pinkeye" is usually sent straight home. Bacteria cause this highly contagious form of inflammation of the conjunctiva, or *conjunctivitis*. Viral conjunctivitis is not usually contagious. Allergy or exposure to an irritating chemical may also cause conjunctivitis.

The **lacrimal apparatus** consists of the *lacrimal gland*, which secretes tears, and a series of *ducts*, which carry the tears into the nasal cavity (fig. 12.23). The gland is in the orbit, superior and lateral to the eye. It secretes tears continuously, which pass through tiny tubules and flow downward and medially across the eye.

Two small ducts (superior and inferior canaliculi) collect tears, and their openings (puncta) can be seen on the medial borders of the eyelids. From these ducts, the fluid moves into the *lacrimal sac*, which lies in a deep groove of the lacrimal bone, and then into the *nasolacrimal duct,* which empties into the nasal cavity.

Glandular cells of the conjunctiva also secrete a tearlike liquid that, together with the secretion of the lacrimal gland, moistens and lubricates the surface of the eye and the lining of the lids. Tears contain an enzyme, *lysozyme,* that has antibacterial properties, reducing the risk of eye infections.

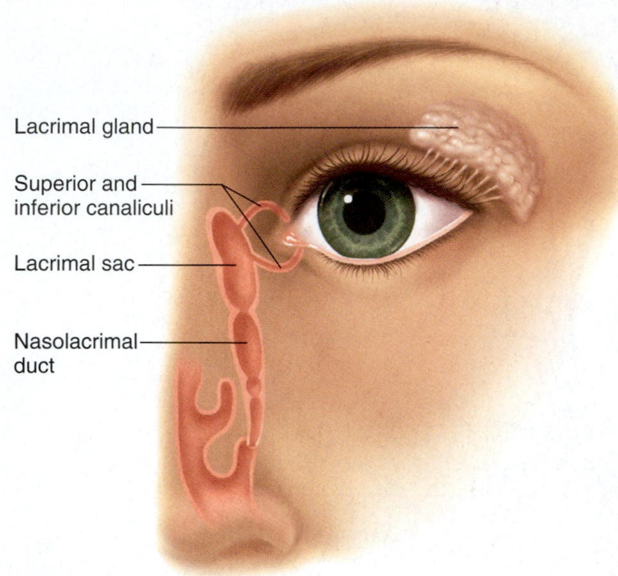

FIGURE 12.23 The lacrimal apparatus consists of a tear secreting gland and a series of ducts.

Tear glands secrete excessively when a person is upset or when the conjunctiva is irritated. Tears spill over the edges of the eyelids, and the nose fills with fluid. When a person cries, parasympathetic nerve fibers carry motor impulses to the lacrimal glands.

The **extrinsic muscles** of the eye arise from the bones of the orbit and are inserted by broad tendons on the eye's tough outer surface. Six such muscles move the eye in various directions (fig. 12.24). Although any given eye movement may use more than one muscle, each one is associated with one primary action, as follows:

1. **Superior rectus**—rotates the eye upward and toward the midline.
2. **Inferior rectus**—rotates the eye downward and toward the midline.
3. **Medial rectus**—rotates the eye toward the midline.
4. **Lateral rectus**—rotates the eye away from the midline.
5. **Superior oblique**—rotates the eye downward and away from the midline.
6. **Inferior oblique**—rotates the eye upward and away from the midline.

The motor units of the extrinsic eye muscles have the fewest muscle fibers (five to ten) of any muscles in the body, so they can move the eyes with great precision. Also, the eyes move together so that they align when looking at something. Such alignment is the result of complex motor adjustments that contract certain eye muscles while relaxing their

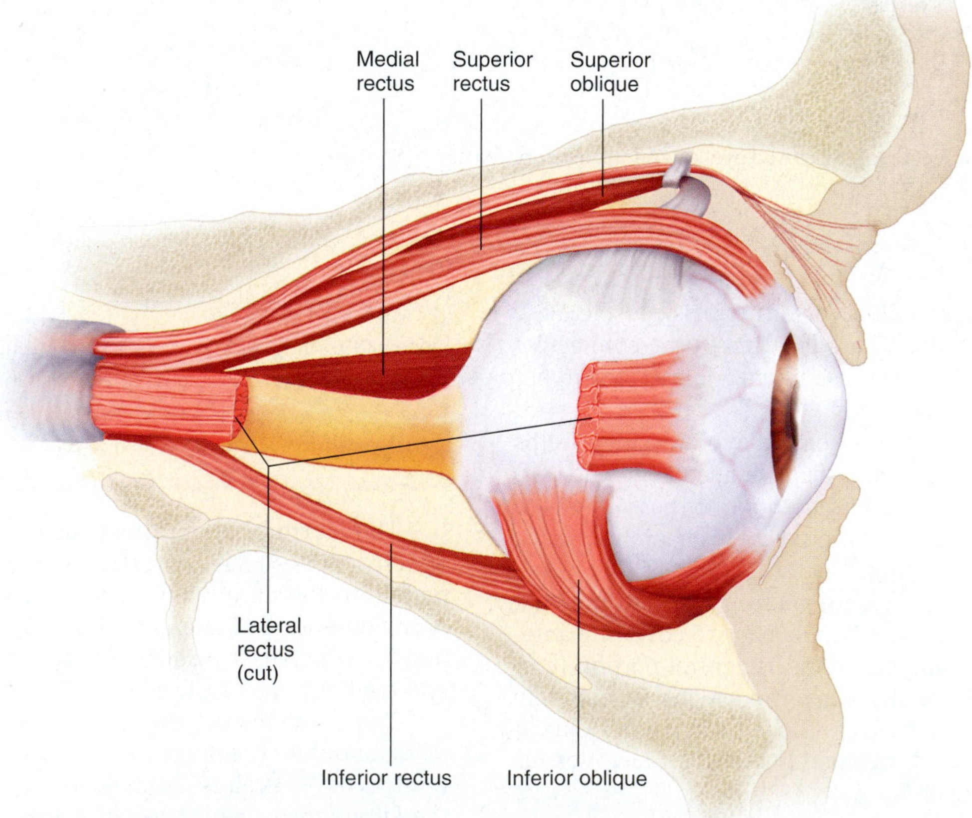

FIGURE 12.24 Extrinsic muscles of the right eye (lateral view).

antagonists. For example, when the eyes move to the right, the lateral rectus of the right eye and the medial rectus of the left eye must contract. At the same time, the medial rectus of the right eye and the lateral rectus of the left eye must relax. A person whose eyes are not coordinated well enough to align has *strabismus*. Table 12.5 summarizes the muscles associated with the eyelids and eye.

When one eye deviates from the line of vision, the person has double vision (diplopia). If this condition persists, the brain may eventually suppress the image from the deviated eye. As a result, the turning eye may become blind (suppression amblyopia). Treating the eye deviation early in life with exercises, eyeglasses, and surgery can prevent such monocular blindness. For this reason, vision screening programs for preschool children are important.

PRACTICE

29 Explain how the eyelid is moved.

30 Describe the conjunctiva.

31 What is the function of the lacrimal apparatus?

32 Describe the function of each extrinsic eye muscle.

Structure of the Eye

The eye is a hollow, spherical structure about 2.5 centimeters in diameter. Its wall has three distinct layers—an outer *fibrous tunic,* a middle *vascular tunic,* and an inner *nervous tunic.* The spaces in the eye are filled with fluids that support its wall and internal structures and help maintain its shape. Figure 12.25 shows the major parts of the eye.

The Outer Tunic

The anterior sixth of the outer tunic bulges forward as the transparent **cornea** (kor′ne-ah), the window of the eye that helps focus entering light rays. It is largely composed of connective tissue with a thin surface layer of epithelium. The cornea is transparent because it contains no blood vessels and the collagenous fibers form unusually regular patterns.

The cornea is well supplied with nerve fibers that enter its margin and radiate toward its center. These fibers are associated with many pain receptors that have very low thresholds. Cold receptors are also abundant in the cornea, but heat and touch receptors are not.

In 1905, doctors transplanted the cornea of an eleven-year-old boy who lost his eye in an accident into a man whose cornea had been destroyed by a splash of a caustic chemical, marking one of the first successful human organ transplants. Today, corneal transplants are commonly used to treat corneal disease, the most common cause of blindness worldwide. In this procedure, called a *penetrating keratoplasty,* a piece of donor cornea replaces the central two-thirds of the defective cornea. These transplants are highly successful because the cornea lacks blood vessels, and therefore, the immune system does not have direct access to the new, "foreign" tissue.

Along its circumference, the cornea is continuous with the **sclera** (skle′rah), the white portion of the eye. The sclera makes up the posterior five-sixths of the outer tunic and is opaque due to many large, seemingly disorganized collagenous and elastic fibers. The sclera protects the eye and is an attachment for the extrinsic muscles.

TABLE 12.5 | Muscles Associated with the Eyelids and Eyes

Skeletal Muscles Name	Innervation	Function
Muscles of the eyelids		
Orbicularis oculi	Facial nerve (VII)	Closes eye
Levator palpebrae superioris	Oculomotor nerve (III)	Opens eye
Extrinsic muscles of the eyes		
Superior rectus	Oculomotor nerve (III)	Rotates eye upward and toward midline
Inferior rectus	Oculomotor nerve (III)	Rotates eye downward and toward midline
Medial rectus	Oculomotor nerve (III)	Rotates eye toward midline
Lateral rectus	Abducens nerve (VI)	Rotates eye away from midline
Superior oblique	Trochlear nerve (IV)	Rotates eye downward and away from midline
Inferior oblique	Oculomotor nerve (III)	Rotates eye upward and away from midline

Smooth Muscles Name	Innervation	Function
Ciliary muscles	Oculomotor nerve (III) parasympathetic fibers	Relax suspensory ligaments
Iris, circular muscles	Oculomotor nerve (III) parasympathetic fibers	Constrict pupil
Iris, radial muscles	Sympathetic fibers	Dilate pupil

In the back of the eye, the **optic** (op′tik) **nerve** and blood vessels pierce the sclera. The dura mater that encloses these structures is continuous with the sclera.

The Middle Tunic

The middle, or vascular, tunic of the eyeball (uveal layer) includes the **choroid coat,** the ciliary body, and the iris. The choroid coat, in the posterior five-sixths of the globe of the eye, loosely joins the sclera. Blood vessels pervade the choroid coat and nourish surrounding tissues. The choroid coat also contains abundant pigment-producing melanocytes that give it a brownish-black appearance. The melanin of these cells absorbs excess light and helps keep the inside of the eye dark.

The **ciliary body,** which is the thickest part of the middle tunic, extends forward from the choroid coat and forms an internal ring around the front of the eye. In the ciliary body are many radiating folds called *ciliary processes* and two distinct groups of muscle fibers that constitute the *ciliary muscles.* Figure 12.26 shows these structures.

Many strong but delicate fibers, called *suspensory ligaments* (zonular fibers), extend inward from the ciliary processes and hold the transparent **lens** in position. The distal ends of these fibers are attached along the margin of a thin

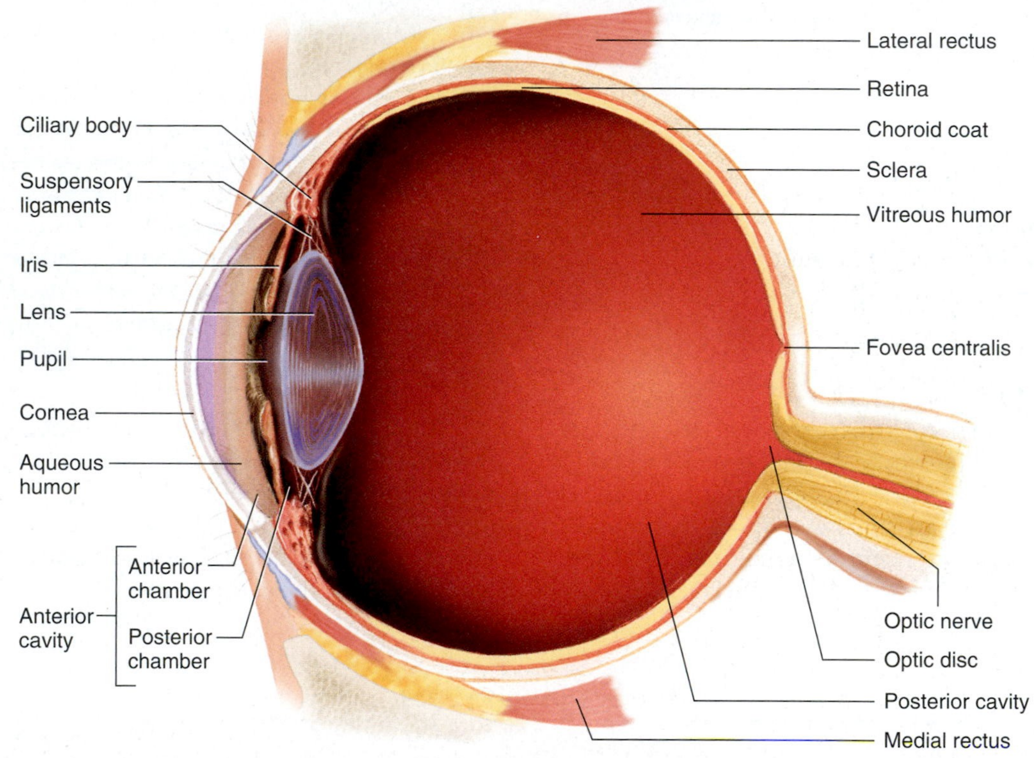

FIGURE 12.25 Transverse section of the right eye (superior view).

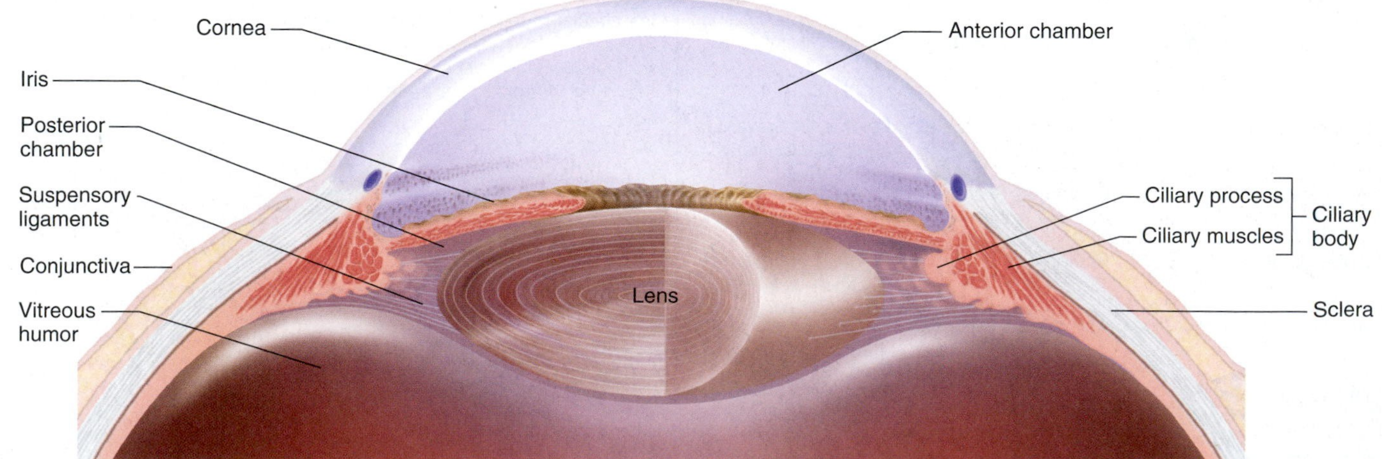

FIGURE 12.26 Anterior portion of the eye.

capsule that surrounds the lens. The body of the lens, which lacks blood vessels, lies directly behind the iris and pupil and is composed of specialized epithelial cells.

The cells of the lens originate from a single layer of epithelium beneath the anterior portion of the lens capsule. The cells divide, and the new cells on the surface of the lens capsule differentiate into columnar cells called *lens fibers,* which constitute the substance of the lens. Lens fiber production continues slowly throughout life, thickening the lens from front to back. Simultaneously, the deeper lens fibers are compressed toward the center of the structure (fig. 12.27).

The lens capsule is a clear, membranelike structure largely composed of intercellular material. It is quite elastic, a quality that keeps it under constant tension. As a result, the lens can assume a globular shape. However, the suspensory ligaments attached to the margin of the capsule are also under tension, and they pull outward, flattening the capsule and the lens (fig. 12.28).

If the tension on the suspensory ligaments relaxes, the elastic capsule rebounds, and the lens surface becomes more convex. This change, called **accommodation** (ah-kom″o-da′shun), occurs in the lens when the eye focuses to view a close object.

The ciliary muscles relax the suspensory ligaments during accommodation. One set of these muscle fibers forms a circular sphincterlike structure around the ciliary processes. The fibers of the other set extend back from fixed points in the sclera to the choroid coat. When the circular muscle fibers contract, the diameter of the ring formed by the ciliary processes decreases; when the other fibers contract, the choroid coat is pulled forward, and the ciliary body shortens. Both of these actions relax the suspensory ligaments, thickening the lens. In this thickened state, the lens is focused for viewing objects closer than before (fig. 12.29*a*).

To focus on a distant object, the ciliary muscles relax, increasing tension on the suspensory ligaments. The lens thins again (fig. 12.29*b*).

PRACTICE

33 Describe the outer and middle tunics of the eye.

34 What factors contribute to the transparency of the cornea?

35 How does the shape of the lens change during accommodation?

36 Why would reading for a long time lead to "eye fatigue," while looking at something distant is restful?

The **iris** is a thin diaphragm mostly composed of connective tissue and smooth muscle fibers. Seen from the outside, it is the colored portion of the eye. The iris extends forward from the periphery of the ciliary body and lies between the cornea and the lens. It divides the space separating these parts, called the *anterior cavity,* into an *anterior chamber* (between the cornea and the iris) and a *posterior chamber* (between the iris and the vitreous humor, occupied by the lens).

The epithelium on the inner surface of the ciliary body continuously secretes a watery fluid called **aqueous humor** into the posterior chamber. The fluid circulates from this chamber through the **pupil,** a circular opening in the center of the iris, and into the anterior chamber (fig. 12.30). Aqueous humor fills the space between the cornea and the

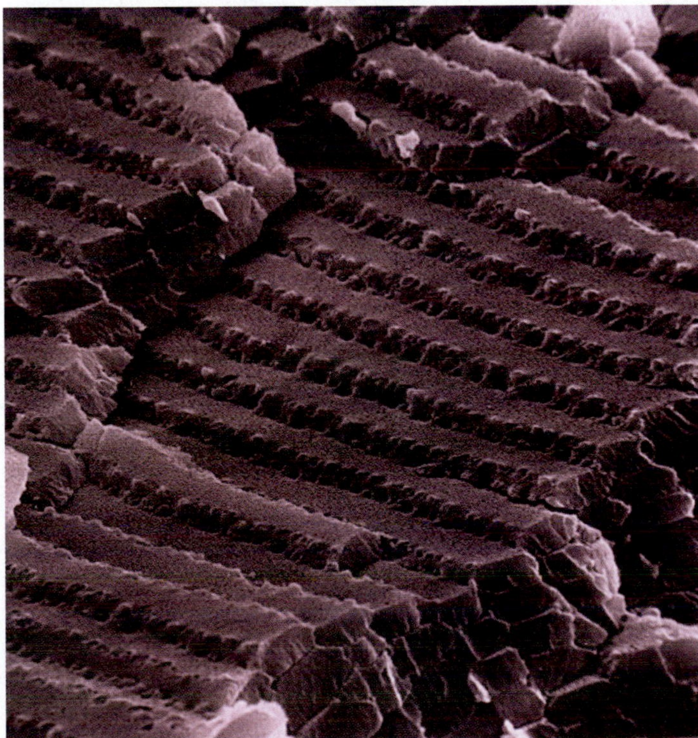

FIGURE 12.27 A scanning electron micrograph of the long, flattened lens fibers (2,650×). Note the fingerlike junctions where one fiber joins another.

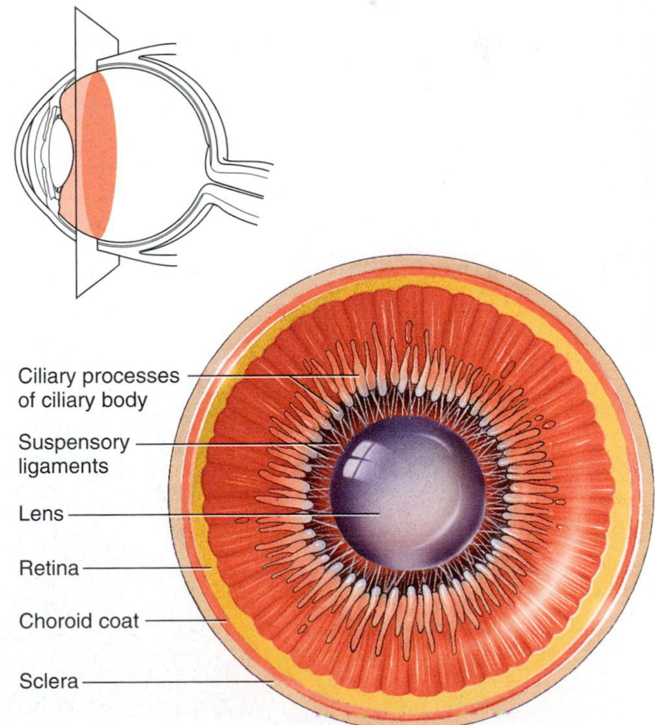

FIGURE 12.28 Lens and ciliary body viewed from behind.

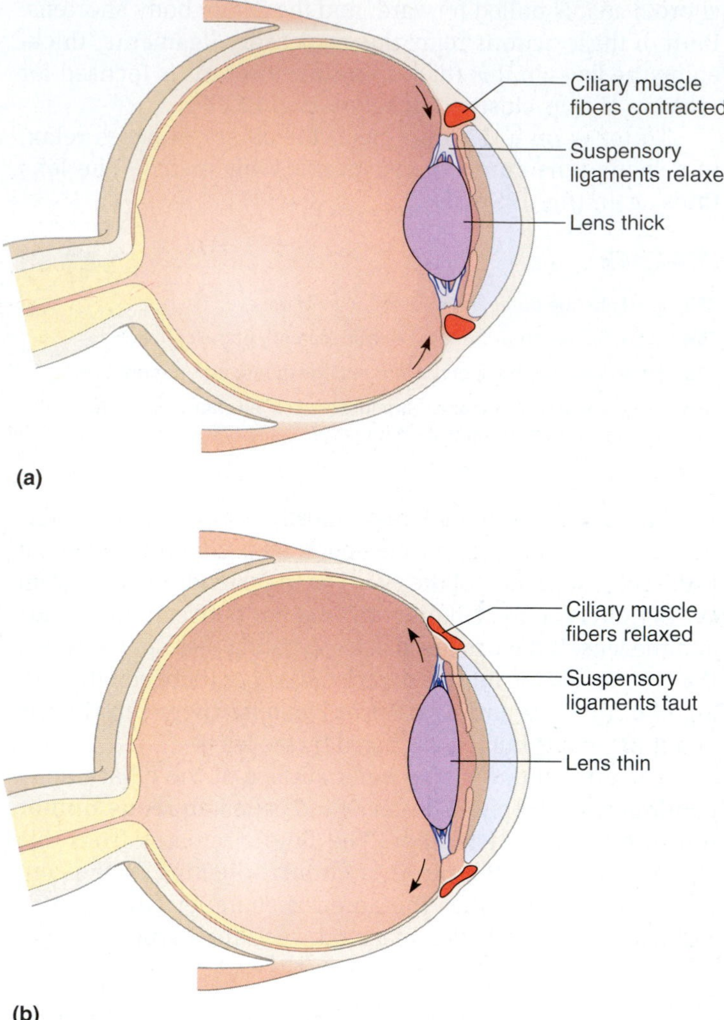

FIGURE 12.29 In accommodation, (*a*) the lens thickens as the ciliary muscle fibers contract. (*b*) The lens thins as ciliary muscle fibers relax.

lens, providing nutrients and maintaining the shape of the front of the eye. It subsequently leaves the anterior chamber through veins and a special drainage canal, the scleral venous sinus (canal of Schlemm), in its wall at the junction of the cornea and the sclera.

The smooth muscle fibers of the iris form two groups, a *circular set* and a *radial set.* These muscles control the size of the pupil, through which light passes. The circular set of muscle fibers acts as a sphincter, and when it contracts, the pupil gets smaller (constricts) and the intensity of the light entering decreases. When the radial muscle fibers contract, the diameter of the pupil increases (dilates) and the intensity of the light entering increases.

The sizes of the pupils change constantly in response to pupillary reflexes triggered by such factors as light intensity, gaze, accommodation, and variations in emotional state. For example, bright light elicits a reflex, and impulses travel along parasympathetic nerve fibers to the *circular muscles* of the irises. The pupils constrict in response. Conversely, in dim light, impulses travel on sympathetic nerve fibers to the *radial muscles* of the irises, and the pupils dilate (fig. 12.31).

RECONNECT

To Chapter 11, Autonomic Nervous System, pages 424–428.

The amount and distribution of melanin in the irises and the density of the tissue in the body of the iris determine eye color. If melanin is present only in the epithelial cells on the iris's posterior surface, the iris reflects more wavelengths of light, and appears blue or green. When the same distribution of melanin is denser in the body of the iris, eye color is gray. When melanin is within the body of the iris as well as in the posterior epithelial covering, the iris appears brown.

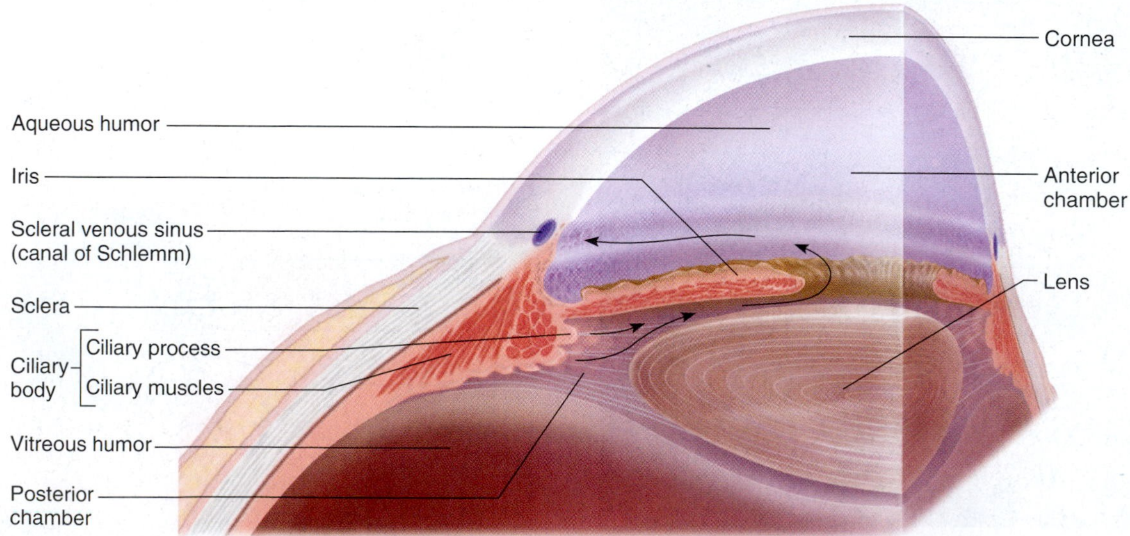

FIGURE 12.30 Aqueous humor (arrows), secreted into the posterior chamber, circulates into the anterior chamber and leaves it through the scleral venous sinus (canal of Schlemm).

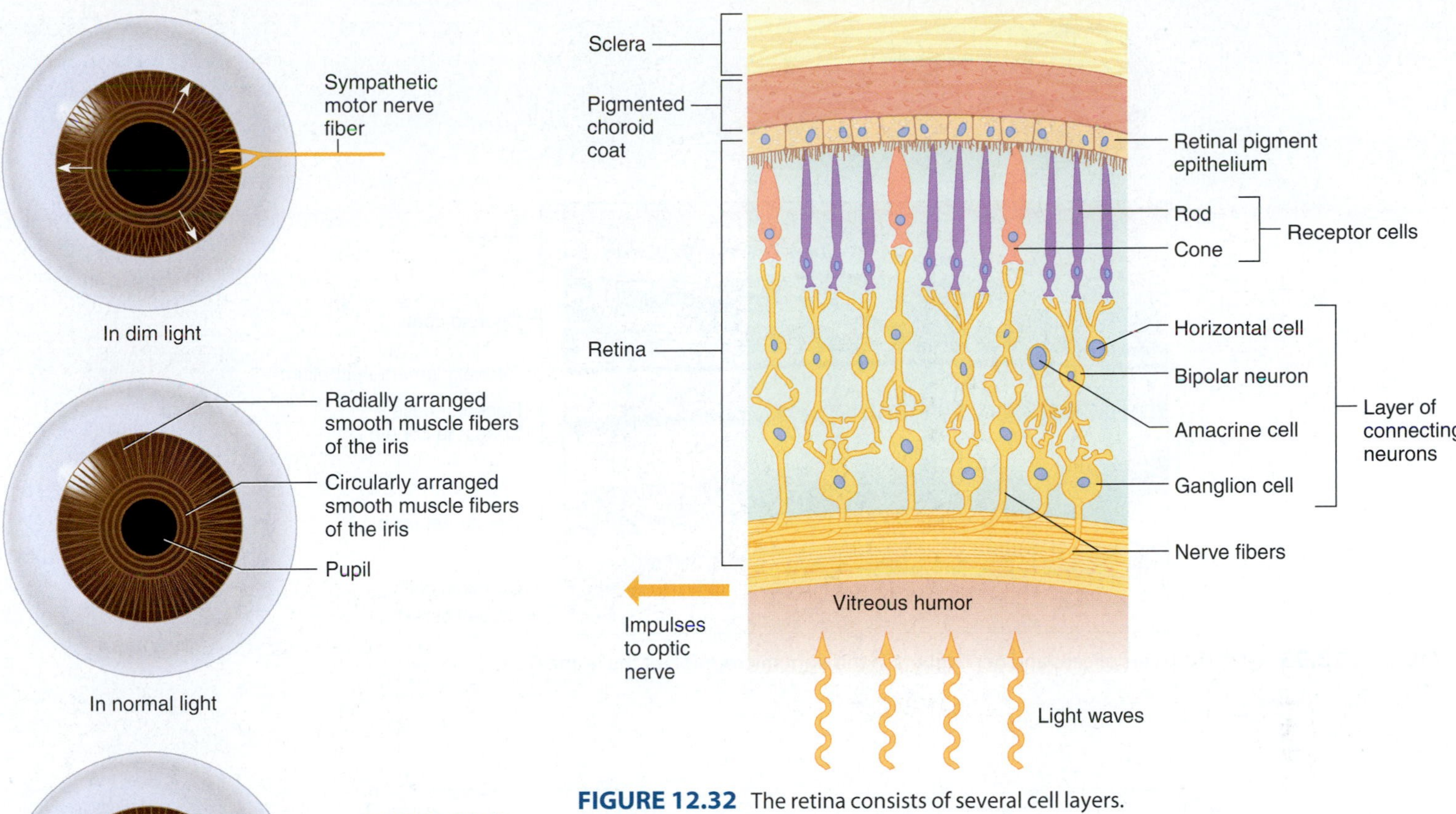

FIGURE 12.32 The retina consists of several cell layers.

FIGURE 12.31 Dim light stimulates the radial muscles of the iris to contract, and the pupil dilates. Bright light stimulates the circular muscles of the iris to contract, and the pupil constricts.

The Inner Tunic

The inner tunic of the eye consists of the **retina** (ret′ĭ-nah), which contains the visual receptor cells (photoreceptors). This nearly transparent sheet of tissue is continuous with the optic nerve in the back of the eye and extends forward as the inner lining of the eyeball. It ends just behind the margin of the ciliary body.

The retina is thin and delicate, but its structure is complex. It has distinct layers, including retinal pigment epithelium, neurons, nerve fibers, and limiting membranes (figs. 12.32 and 12.33).

There are five major groups of retinal neurons. The nerve fibers of three of these groups—the *receptor cells, bipolar neurons,* and *ganglion cells*—provide a direct pathway for impulses triggered in the receptors to the optic nerve and brain. The nerve fibers of the other two groups of retinal cells, called *horizontal cells* and *amacrine cells,* pass laterally between retinal cells (see fig. 12.32). The horizontal and amacrine cells modify the impulses transmitted on the fibers of the direct pathway.

In the central region of the retina is a yellowish spot called the **macula lutea** that occupies about 1 square millimeter. A depression in its center, called the **fovea centralis,** is in the region of the retina that produces the sharpest vision.

Just medial to the fovea centralis is an area called the **optic disc** (fig. 12.34). Here the nerve fibers from the retina leave the eye and become parts of the optic nerve. A central artery and vein also pass through at the optic disc. These vessels are continuous with capillary networks of the retina, and together with vessels in the underlying choroid coat, they supply blood to the cells of the inner tunic. The optic disc lacks receptor cells, so it is commonly referred to as the *blind spot* of the eye.

The space enclosed by the lens, ciliary body, and retina is the largest compartment of the eye and is called the *posterior cavity.* It is filled with a transparent, jellylike fluid called **vitreous humor,** which with some collagenous fibers comprise the **vitreous body.** The vitreous body supports the internal structures of the eye and helps maintain its shape.

In summary, light waves entering the eye must pass through the cornea, aqueous humor, lens, vitreous humor, and several layers of the retina before they reach the photoreceptors (see fig. 12.32). Table 12.6 summarizes the layers of the eye.

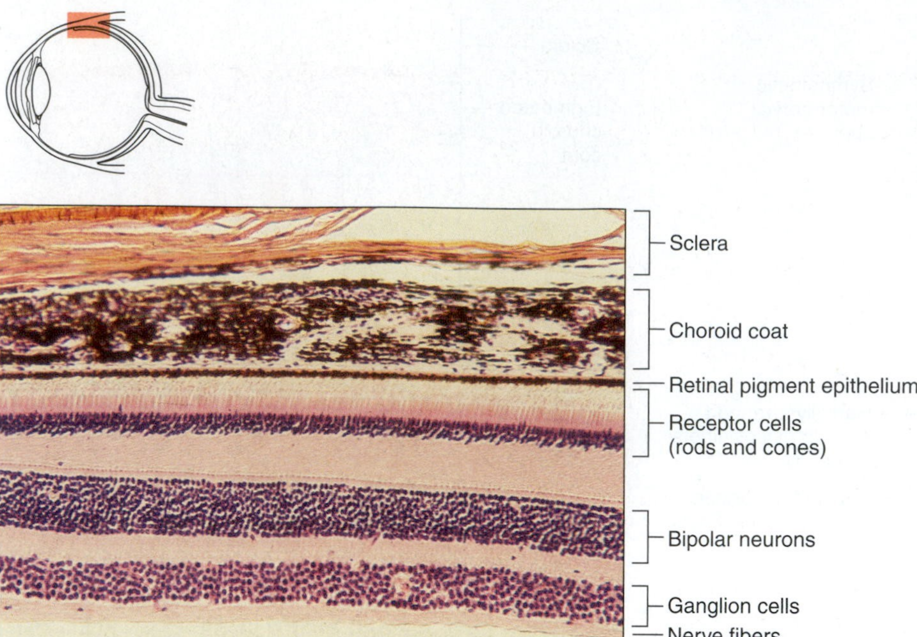

FIGURE 12.33 Note the layers of cells and nerve fibers in this light micrograph of the retina (75×).

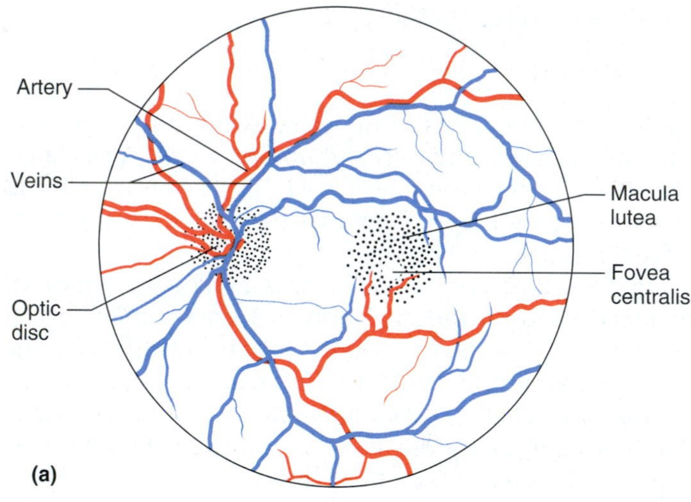

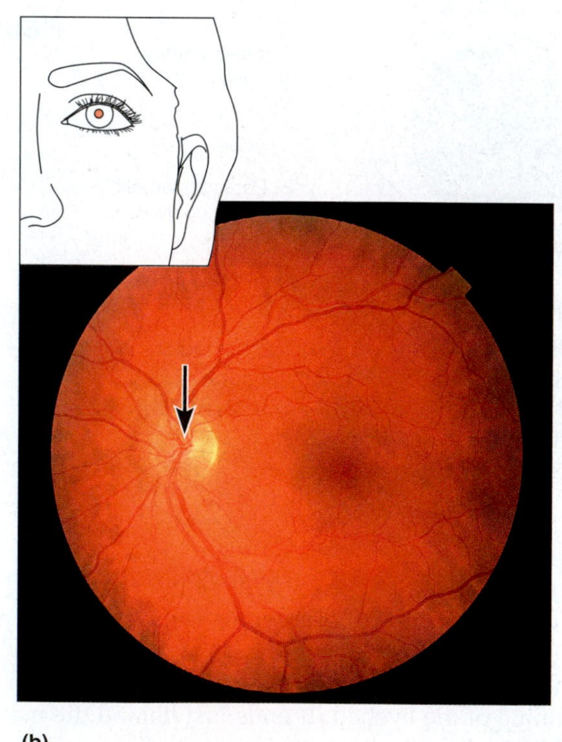

FIGURE 12.34 The retina. (*a*) Major features of the retina. (*b*) Nerve fibers leave the retina in the area of the optic disc (arrow) to form the optic nerve in this magnified view (53×).

TABLE 12.6 | Layers of the Eye

Layer/Tunic	Posterior Portion	Function	Anterior Portion	Function
Outer layer	Sclera	Protection	Cornea	Light transmission and refraction
Middle layer	Choroid coat	Blood supply, pigment prevents reflection	Ciliary body, iris	Accommodation; controls light intensity
Inner layer	Retina	Photoreception, impulse transmission	None	

PRACTICE

37 Explain the origin of aqueous humor and trace its path through the eye.

38 How is the size of the pupil regulated?

39 Describe the structure of the retina.

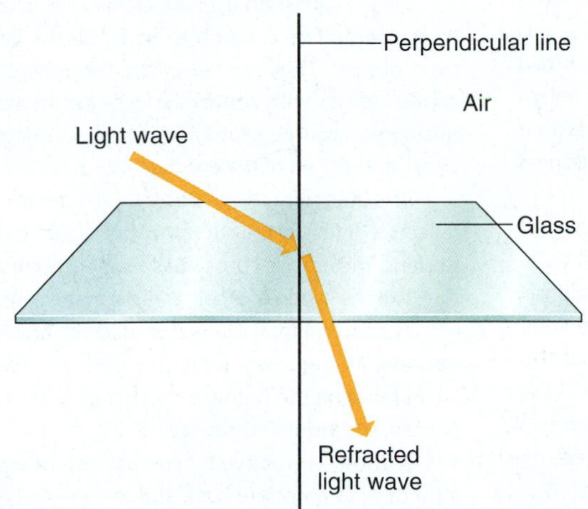

FIGURE 12.35 When light passes at an oblique angle from air into glass, the light waves bend toward a line perpendicular to the surface of the glass.

Light Refraction

When a person sees an object, either the object is giving off light, or light waves from another source are reflected from it. These light waves enter the eye, and an image of what is seen focuses upon the retina. The light rays must bend to be focused, a phenomenon called **refraction** (re-frak′shun).

Refraction occurs when light waves pass at an oblique angle from a medium of one optical density into a medium of a different optical density. For example, as figure 12.35 shows, when light passes obliquely from a less-dense medium such as air into a denser medium such as glass, or from air into the cornea of the eye, the light is bent toward a line perpendicular to the surface between these substances. When the surface between such refracting media is curved, a lens is formed. A lens with a *convex* surface causes light waves to converge, and a lens with a *concave* surface causes light waves to diverge (fig. 12.36). Clinical Application 12.6 discusses some familiar problems with refraction.

The convex surface of the cornea refracts light waves from objects outside the eye, providing about 75% of the total refractive power of the eye. The light is refracted again by the convex surface of the lens and to a lesser extent by the surfaces of the fluids in the eye chambers.

If the shape of the eye is normal, light waves are focused sharply upon the retina, much as a motion-picture image is focused on a screen for viewing. Unlike the motion-picture image, however, the one formed on the retina is upside down and reversed from left to right (fig. 12.37). When the visual

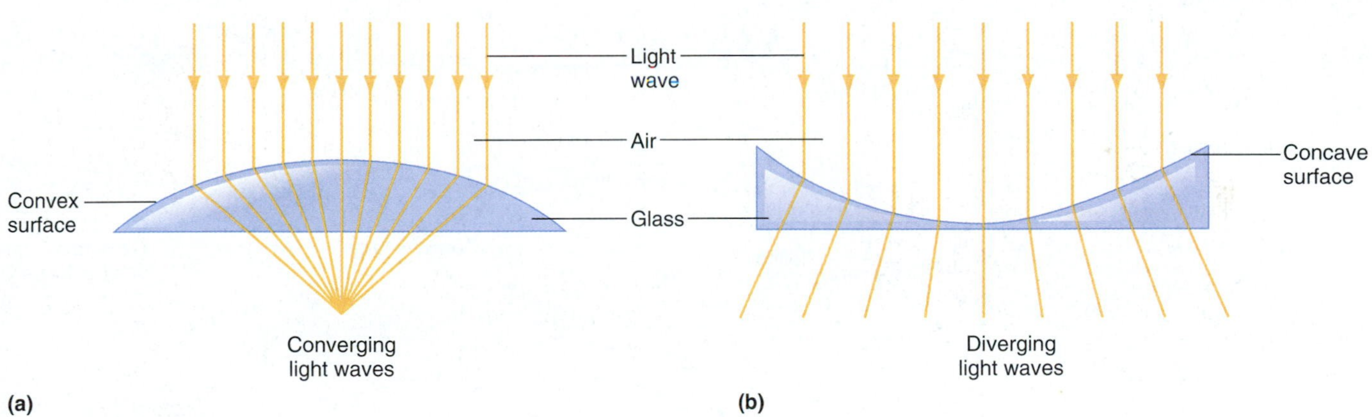

FIGURE 12.36 Light waves passing through a lens. (*a*) A lens with a convex surface causes light waves to converge. (*b*) A lens with a concave surface causes them to diverge.

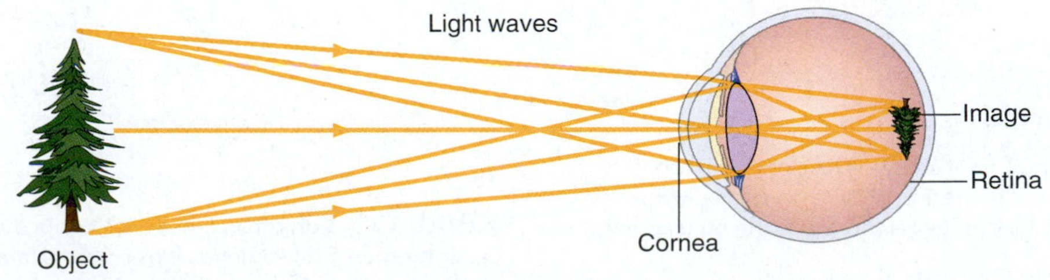

FIGURE 12.37 The image of an object forms upside down on the retina.

12.6 CLINICAL APPLICATION

Refraction Disorders

For many people, after the age of forty-five, it seems as if the print in magazines and on medicine bottles suddenly becomes too small to read. The problem is not in the print, but in a lessening of the elastic quality of the lens capsule. In the condition presbyopia, or farsightedness of age, eyes remain focused for distant vision. Eyeglasses or contact lenses can usually make up for the eye's loss of refracting power.

Other visual problems result from eyeballs that are too long or too short for sharp focusing. If an eyeball is too long, light waves focus in front of the retina, blurring the image. In other words, the refracting power of the eye, even when the lens is flattened, is too great. Although a person with this problem may be able to focus on close objects by accommodation, distance vision is invariably poor. For this reason, the person is said to be *nearsighted*. Eyeglasses or contact lenses with concave surfaces that focus images farther from the front of the eye treat nearsightedness (myopia).

If an eye is too short, light waves are not focused sharply on the retina because their point of focus lies behind it. A person with this condition may be able to bring the image of distant objects into focus by accommodation, but this requires contraction of the ciliary muscles at times when these muscles are at rest in a normal eye. Still more accommodation is necessary to view closer objects, and the person may suffer from ciliary muscle fatigue, pain, and headache when doing close work.

People with short eyeballs are usually unable to accommodate enough to focus on the very close objects. They are *farsighted*. Eyeglasses or contact lenses with *convex* surfaces can remedy this condition (hyperopia) by focusing images closer to the front of the eye (figs. 12E and 12F).

Another refraction problem, *astigmatism*, reflects a defect in the curvature of the cornea or the lens. The normal cornea has a spherical curvature, like the inside of a ball; an astigmatic cornea usually has an elliptical curvature, like the bowl of a spoon. As a result, some portions of an image are in focus on the retina, but other portions are blurred, and vision is distorted.

Without corrective lenses, astigmatic eyes tend to accommodate back and forth reflexly in an attempt to sharpen focus. The consequence of this continual action is often ciliary muscle fatigue and headache. ■

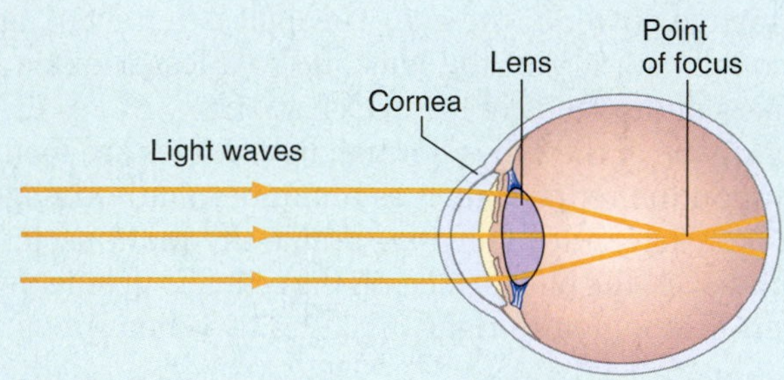

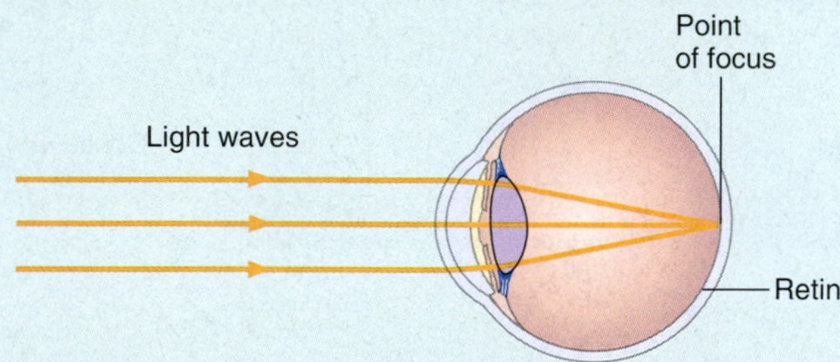

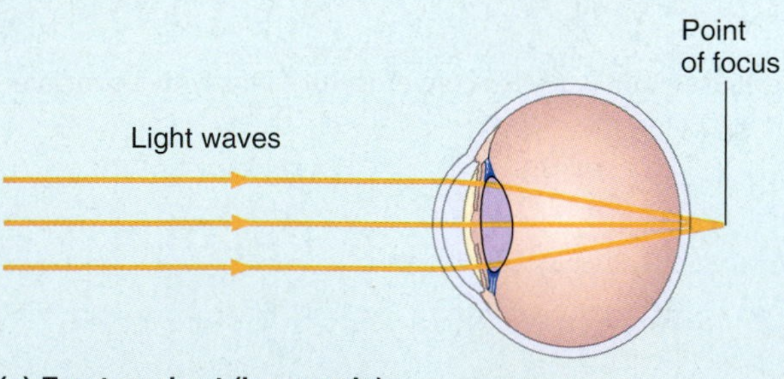

FIGURE 12E Point of focus. (*a*) If an eye is too long, the focus point of images lies in front of the retina. (*b*) In a normal eye, the focus point is on the retina. (*c*) If an eye is too short, the focus point lies behind the retina.

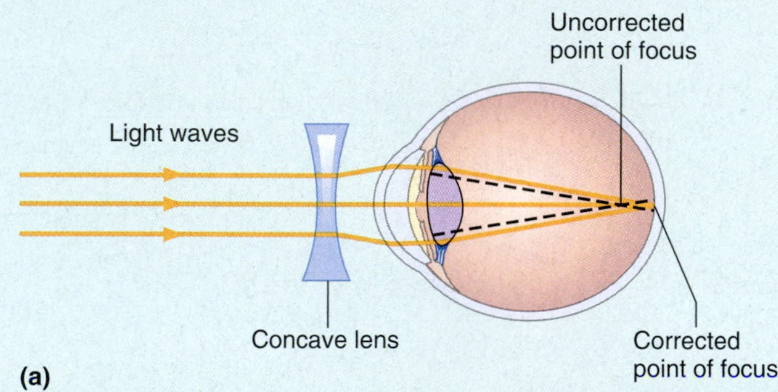

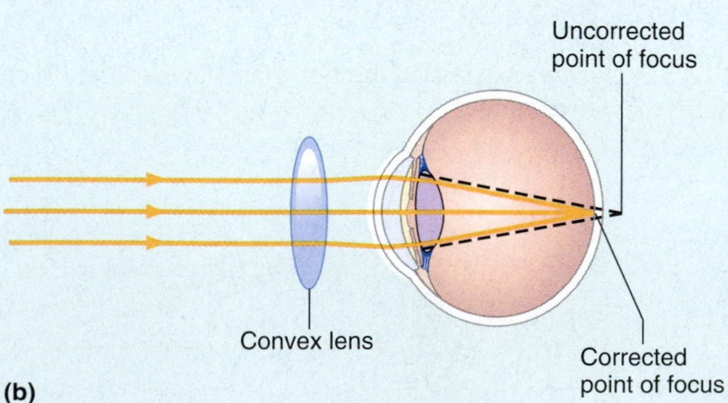

FIGURE 12F Corrective lenses. (*a*) A concave lens corrects nearsightedness. (*b*) A convex lens corrects farsightedness.

cortex of the cerebrum interprets such an image, it corrects this, and objects are seen in their real positions.

Light waves coming from objects more than 20 feet away are traveling in nearly parallel lines, and the cornea and the lens in its more flattened or "at-rest" condition focuses the light waves on the retina. Light waves arriving from objects less than 20 feet away, however, reach the eye along more divergent lines—the closer the object, the more divergent the lines.

Divergent light waves focus behind the retina unless something increases the refracting power of the eye. Accommodation accomplishes this increase, thickening the lens. As the lens thickens, light waves converge more strongly so that diverging light waves coming from close objects focus on the retina.

PRACTICE 

40 What is refraction?

41 What parts of the eye provide refracting surfaces?

42 Why is it necessary to accommodate for viewing close objects?

Visual Receptors

The photoreceptors of the eye are modified neurons of two distinct types. One group of receptor cells, called **rods,** have long, thin projections at their terminal ends. The cells of the other group, called **cones,** have short, blunt projections. The retina has about 100 million rods and 3 million cones.

Rods and cones occupy a deep layer of the retina, closely associated with a layer of retinal pigment epithelium (see figs. 12.32 and 12.33). The projections from the receptors extend into the pigmented layer and contain light-sensitive visual pigments.

The retinal pigment epithelium absorbs light waves that the receptor cells do not, and with the pigment of the choroid coat, keeps light from reflecting off the surfaces inside the eye. The retinal pigment epithelium also stores vitamin A, which the receptor cells use to synthesize visual pigments.

Researchers can grow retinal pigment epithelium cells in laboratory cultures, and the cells retain their pigment. This means that someday scientists may be able to grow tissue that can be implanted into a person's eye to treat some forms of blindness.

The visual receptors are stimulated only when light reaches them. When a light image is focused on an area of the retina, some receptors are stimulated and send impulses to the brain. However, the impulse leaving each activated receptor provides only a small portion of the information required for the brain to interpret a total scene.

Rods and cones function differently. Rods are hundreds of times more sensitive to light than are cones, and as a result, rods provide vision in dim light. In addition, rods produce colorless vision, whereas cones can detect colors.

Albinism is an inherited condition in which an enzyme necessary to produce pigment is missing, causing very pale, highly sun-sensitive skin. More severe forms of albinism also affect the eyes, making vision blurry and intolerant to light. A person may squint even in very faint light. This separate extrasensitivity is because light reflects inside the lenses, over-stimulating visual receptors. The eyes of many people with albinism also dart about uncontrollably, a condition called *nystagmus.*

Cones provide sharp images, whereas rods produce more general outlines of objects. This is because nerve fibers from many rods may converge, and their impulses may be transmitted to the brain on the same nerve fiber (see chapter 10, p. 375). Thus, if light stimulates a rod, the brain cannot tell which one of many receptors has been stimulated. Such a convergence of impulses occurs to a much lesser degree among cones, so when a cone is stimulated, the brain is able to pinpoint the stimulation more accurately (fig. 12.38).

The area of sharpest vision, the fovea centralis in the macula lutea, lacks rods but has densely packed cones with few or no converging fibers. Also, the overlying layers of the retina, as well as the retinal blood vessels, are displaced to the sides in the fovea, which more fully exposes the receptors to incoming light. Consequently, to view something in detail, a person moves the eyes so that the important part of an image falls upon the fovea centralis.

The concentration of cones decreases in areas farther away from the macula lutea, whereas the concentration of rods increases in these areas. Also, the degree of convergence among the rods and cones increases toward the periphery of the retina. As a result, the visual sensations from images focused on the sides of the retina are blurred compared with those focused on the central portion of the retina.

The bony orbit usually protects the eye, but a forceful blow can displace structures in and around the eye. The suspensory ligaments may tear and the lens become dislocated into the posterior cavity, or the retina may pull away from the underlying vascular choroid coat. Once the retina is detached, photoreceptor cells may die because of lack of oxygen and nutrients. Unless such a *detached retina* is repaired surgically, this injury may cause visual loss or blindness.

Visual Pigments

Rods and cones contain light-sensitive pigments that decompose when they absorb light energy. The light-sensitive pigment in rods is **rhodopsin** (ro-dop′sin), or visual purple, and it is embedded in membranous discs stacked in these receptor cells (fig. 12.39). A single rod cell may have 2,000 interconnected discs, derived from the cell membrane. In the presence of light, rhodopsin molecules break down into molecules of a colorless protein called *opsin* and a yellowish organic molecule called *retinal* (retinene) synthesized from vitamin A.

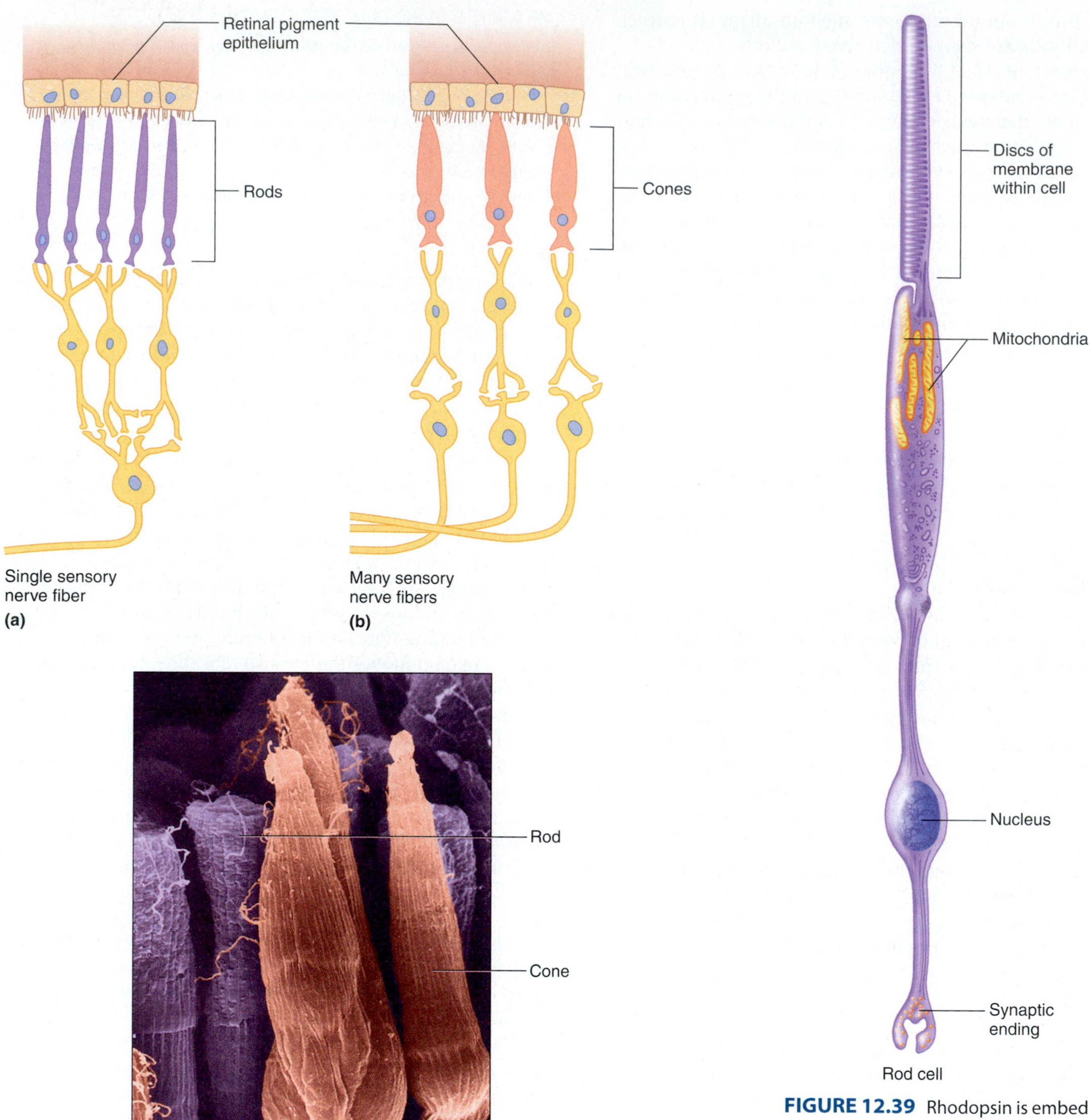

FIGURE 12.38 Rods and cones. (*a*) A single sensory nerve fiber transmits impulses from several rods to the brain. (*b*) Separate sensory nerve fibers transmit impulses from cones to the brain. (*c*) Scanning electron micrograph of rods and cones (1,350×).

FIGURE 12.39 Rhodopsin is embedded in discs of membrane stacked in the rod cells.

In darkness, sodium channels in portions of the receptor cell membranes are kept open by a nucleotide called *cyclic guanosine monophosphate* (cGMP). When rhodopsin molecules absorb light, they change shape and release opsin, in trillionths of a second. The released opsin then becomes an active enzyme, which activates a second enzyme (transducin), which, in turn, activates another enzyme (phosphodiesterase). The third enzyme of this series breaks down cGMP, and as the concentration of cGMP decreases, sodium channels close, and the receptor cell membrane hyperpolarizes (see chapter 10, p. 368). The degree of hyperpolarization is directly proportional to the intensity of the light stimulating the receptor cells.

The hyperpolarization reaches the synaptic end of the cell, inhibiting release of neurotransmitter. Through a complex mechanism, decreased release of neurotransmitter by photoreceptor cells either stimulates or inhibits nerve impulses (action potentials) in nearby retinal neurons. Consequently, complex patterns of nerve impulses travel away from the retina, through the optic nerve, and into the brain, where they are interpreted as vision.

In bright light, nearly all of the rhodopsin in the rods decomposes, sharply reducing the sensitivity of these receptors (the rhodopsin loses its purplish color as a result, and is said to have "bleached"). The cones continue to function, however, and in bright light, we therefore see in color. In dim light, rhodopsin can be regenerated from opsin and retinal faster than it is broken down. This regeneration requires cellular energy, which ATP provides (see chapter 4, p. 119). Under these conditions, the rods continue to function and the cones remain unstimulated. Hence, we see only shades of gray in dim light.

The light sensitivity of an eye whose rods have converted the available opsin and retinal to rhodopsin increases about 100,000 times, and the eye is said to be *dark adapted.* A person needs a dark-adapted eye to see in dim light. For example, when going from daylight into a darkened theater, it may be difficult to see well enough to locate a seat, but soon the eyes adapt to the dim light, and vision improves. Later, leaving the theater and entering the sunlight may cause discomfort or even pain. This occurs at the moment that most of the rhodopsin decomposes in response to the bright light. At the same time, the light sensitivity of the eyes decreases greatly, and they become *light adapted.*

Too little vitamin A in the diet reduces the amount of retinal, impairing rhodopsin production and sensitivity of the rods. The result is poor vision in dim light, called nightblindness.

The light-sensitive pigments of cones, called *iodopsins,* are similar to rhodopsin in that they are composed of retinal combined with a protein; the protein, however, differs from the protein in the rods. The three sets of cones in the retina all contain an abundance of one of three different visual pigments.

The wavelength of a particular type of light determines the color perceived from it. For example, the shortest wavelengths of visible light are perceived as violet, whereas the longest wavelengths of visible light are seen as red. One type of cone pigment (erythrolabe) is most sensitive to red light waves, another (chlorolabe) to green light waves, and a third (cyanolabe) to blue light waves. The sensitivities of these pigments overlap somewhat. For example, both red and green light pigments are sensitive to orange light waves. On the other hand, red pigment absorbs orange light waves more effectively.

The color perceived depends upon which sets of cones the light in a given image stimulates. If all three types of sets of cones are stimulated, the light is perceived as white, and if none are stimulated, it is seen as black.

Examination of the retinas of different people reveals that individuals have unique patterns of cone types, all apparently able to provide color vision. Some parts of the retina are even normally devoid of one particular type, yet the brain integrates information from all over to "fill in the gaps," creating a continuous overall image. People who lack a cone type due to a mutation are colorblind.

As primates, we humans enjoy a more multicolored world than many other mammals. This is because the visual systems of nonprimate mammals funnel input from groups of photoreceptor cells into the CNS. That is, several photoreceptors signal the same bipolar neurons, which, in turn, pool their input to ganglion cells. Primates are the only mammals to have three types of cones (others have two), and it appears that primates excel in color vision because the cones connect individually to neural pathways to the brain.

Stereoscopic Vision

Stereoscopic vision (stereopsis) simultaneously perceives distance, depth, height, and width of objects. Such vision is possible because the pupils are 6–7 centimeters apart. Consequently, close objects (less than 20 feet away) produce slightly different retinal images. That is, the right eye sees a little more of one side of an object, while the left eye sees a little more of the other side. The visual cortex superimposes and interprets the two images. The result is the perception of a single object in three dimensions (fig. 12.40).

Stereoscopic vision requires vision with two eyes (binocular vision), so a one-eyed person is less able to judge distance and depth accurately. To compensate, a person with one eye can use the relative sizes and positions of familiar objects as visual clues.

A woman had a stroke that damaged part of her visual cortex, so that she could no longer integrate images to perceive motion. She saw movement as a series of separate, static images. Her deficit had profound effects on her life. She could not pour a drink, because she could not tell when the cup would overflow. She could not cross a street because she could not detect cars moving toward her.

Visual Nerve Pathways

As mentioned in chapter 11 (p. 414), the axons of the ganglion cells in the retina leave the eyes to form the *optic nerves.* Just anterior to the pituitary gland, these nerves give rise to the X-shaped *optic chiasma,* and in the chiasma, some

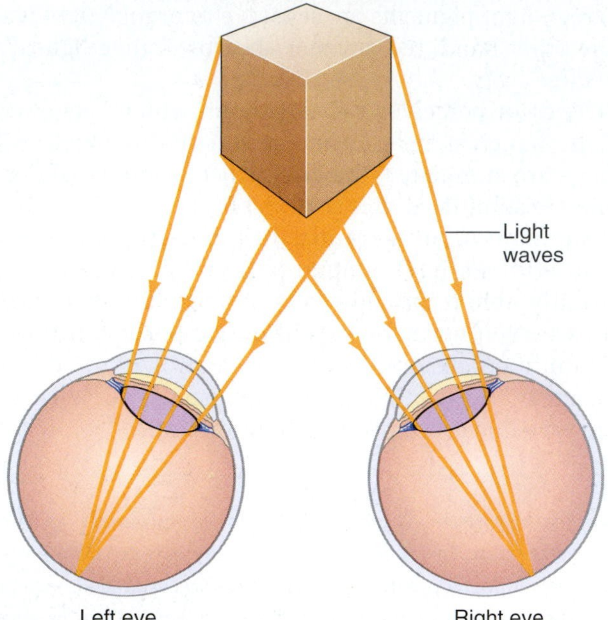

FIGURE 12.40 Stereoscopic vision results from formation of two slightly different retinal images.

of the fibers cross over. More specifically, the fibers from the nasal (medial) half of each retina cross over, whereas those from the temporal (lateral) sides do not. Thus, fibers from the nasal half of the left eye and the temporal half of the right eye form the right *optic tract;* fibers from the nasal half of the right eye and the temporal half of the left eye form the left optic tract.

The nerve fibers continue in the optic tracts, and before they reach the thalamus, a few of them leave to enter nuclei that function in various visual reflexes. Most of the fibers, however, enter the thalamus and synapse in its posterior portion (lateral geniculate body). From this region, the visual impulses enter nerve pathways called *optic radiations,* and the pathways lead to the visual cortex of the occipital lobes (fig. 12.41).

> Each visual cortex receives impulses from each eye, so a person may develop partial blindness in both eyes if either visual cortex is injured. For example, if the right visual cortex (or the right optic tract) is injured, sight may be lost in the temporal side of the right eye and the nasal side of the left eye. Similarly, damage to the central portion of the optic chiasma, where fibers from the nasal sides of the eyes cross over, blinds the nasal sides of both eyes.

Fibers not leading to the thalamus conduct visual impulses downward into the brainstem. These impulses are important for controlling head and eye movements associated with visually tracking an object; for controlling the simultaneous movements of both eyes; and for controlling certain visual reflexes, such as those that move the muscles of the iris.

PRACTICE 

43 Distinguish between the rods and the cones of the retina.

44 Explain the roles of visual pigments.

45 What factors make stereoscopic vision possible?

46 Trace the pathway of visual impulses from the retina to the occipital cortex.

12.5 LIFE-SPAN CHANGES

We often first become aware of aging-associated changes through diminished senses. By age forty, a book may need to be held farther away from the eyes. By the fifties, the senses of smell and taste may begin to diminish, which usually reflects anosmia, a loss of olfactory receptors.

By age sixty, a quarter of the population experiences noticeable hearing loss, and from ages sixty-five to seventy-four, the percentage reaches a third. Half of all people over age eighty-five cannot hear adequately. Age-related hearing loss may be the result of decades of cumulative damage to the sensitive hair cells of the spiral organ in the inner ear. It becomes more difficult to hear high pitches, as well as par-

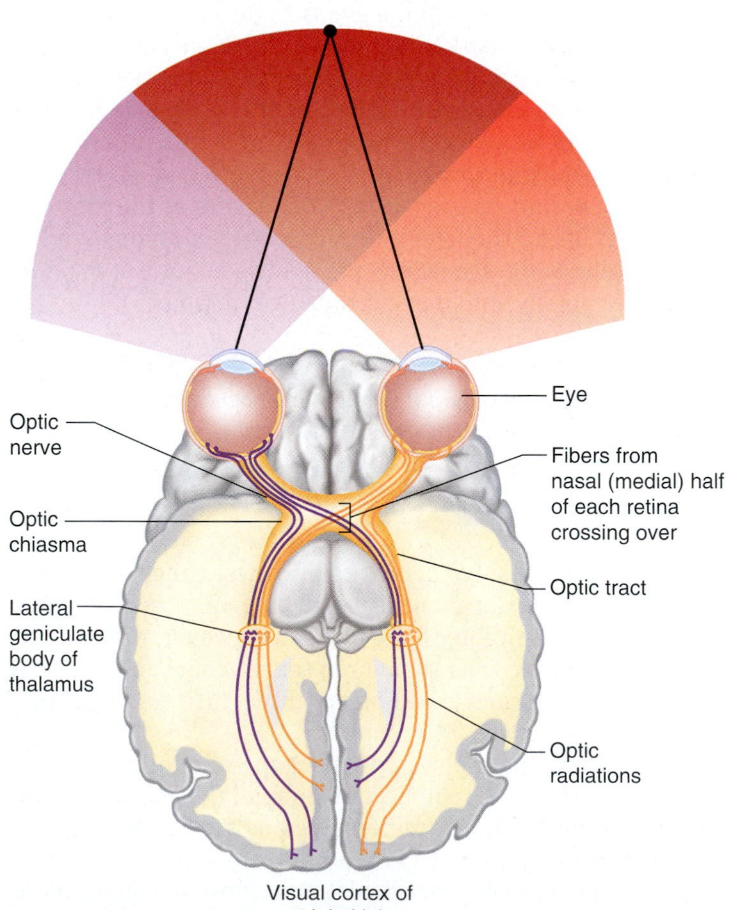

FIGURE 12.41 The visual pathway includes the optic nerve, optic chiasma, optic tract, and optic radiations.

ticular sounds, such as *f, g, s, t, sh, th, z* and *ch.* Hearing loss may also be due to a degeneration or failure of nerve pathways to the brain. This condition, called presbycusis, may affect the ability to understand speech. It gradually worsens. Tinnitus, a ringing or roaring in the ears, is also more common among older adults. Hearing aids can often restore some hearing. A person to whom the ordinary sounds of life are hopelessly garbled may show, understandably, symptoms of paranoia, depression, or social withdrawal.

Vision may decline with age for several reasons. "Dry eyes" are common. Too few tears, or poor quality tears, lead to itching and burning eyes, and diminished vision. In some cases, too many tears result from oversensitivity to environmental effects, such as wind, intense light, or a change in temperature.

With age, tiny dense clumps of gel or crystal-like deposits form in the vitreous humor. When these clumps cast shadows on the retina, the person sees small moving specks in the field of vision. These specks, or *floaters,* are most apparent when looking at a plain background, such as the sky or a blank wall. Also with age, the vitreous humor may shrink and pull away from the retina. This may mechanically stimulate receptor cells of the retina, and the person may see flashes of light.

The inability to read small print up close, called presbyopia, results from a loss of elasticity in the lens, preventing it from changing shape easily. After age seventy, the iris cannot dilate as well as it once did, halving the amount of light that can enter the eye. Brighter lights can counter this effect.

Glaucoma develops in the eyes as a person ages when the rate of aqueous humor formation exceeds the rate of its removal. Fluid accumulates in the anterior chamber of the eye, and the fluid pressure rises. As this pressure is transmitted to all parts of the eye, in time, the blood vessels that supply the receptor cells of the retina may squeeze shut, cutting off the nutrient and oxygen supply. The result may eventually be permanent blindness.

Drugs, or traditional or laser surgery to promote the outflow of aqueous humor, can treat glaucoma if it is diagnosed early. However, because glaucoma in its early stages typically produces no symptoms, discovery of the condition usually depends on measuring the intraocular pressure using an instrument called a *tonometer.*

A common eye disorder particularly in older people is *cataract.* The lens or its capsule slowly becomes cloudy, opaque, and discolored, adding a yellowish tinge to a person's view of the world. Clear images cannot focus on the retina, and in time, the person may become blind. Removing the lens with a laser and replacing it with an artificial implant can treat cataract. Afterward, patients report that their surroundings are no longer yellow.

Several conditions affect the retinas of an older person. Age-related macular degeneration is an impairment of the macula, the most sensitive part of the retina, which blurs images. Retinal detachment becomes more common. People with diabetes are at high risk of developing diabetic retinopathy, an interference with retinal function because of damaged blood vessels or growth of new ones that block vision.

Despite these various problems, many older individuals continue to enjoy sharp, functional senses well into the upper decades of life.

PRACTICE 

47 Why do smell and taste diminish with age?

48 What are some causes of age-related hearing loss?

49 Describe visual problems likely to arise with age.

CHAPTER SUMMARY

12.1 INTRODUCTION (PAGE 438)

Sensory receptors are sensitive to internal and external environmental changes and initiate impulses to the brain and spinal cord.

12.2 RECEPTORS, SENSATION, AND PERCEPTION (PAGE 438)

1. Receptor types
   a. Each type of receptor is sensitive to a distinct type of stimulus.
   b. The major types of receptors include:
      (1) Chemoreceptors, sensitive to changes in chemical concentration.
      (2) Pain receptors (nociceptors), sensitive to tissue damage.
      (3) Thermoreceptors, sensitive to temperature changes.
      (4) Mechanoreceptors, sensitive to mechanical forces.
      (5) Photoreceptors, sensitive to light.
2. Sensory impulses
   a. When receptors are stimulated, membrane potentials change.
   b. Receptor potentials are transferred to nerve fibers, triggering action potentials.
3. Sensation and perception
   a. Sensation is a feeling resulting from sensory stimulation.
   b. Perception is when a particular part of the sensory cortex interprets the sensory stimulation.
   c. The cerebral cortex projects a sensation back to the region of stimulation.
4. Sensory adaptations are adjustments of sensory receptors to continuous stimulation. Impulses are triggered at slower rates.

12.3 GENERAL SENSES (PAGE 440)

General senses receive information from receptors in skin, muscles, joints, and viscera. They can be grouped as exteroceptive, visceroceptive, and proprioceptive senses.

1. Touch and pressure senses
   a. Free ends of sensory nerve fibers are the receptors for the sensations of touch and pressure.
   b. Tactile corpuscles are the receptors for the sensations of light touch.
   c. Lamellated corpuscles are the receptors for the sensations of heavy pressure and vibrations.
2. Temperature senses
   a. Thermoreceptors include two sets of free nerve endings that are heat and cold receptors.
   b. Combinations of input from both receptor types are interpreted as intermediate temperatures.
3. Sense of pain
   a. Pain receptors
      (1) Pain receptors are free nerve endings that tissue damage stimulates.
      (2) Pain receptors provide protection; do not adapt rapidly; and can be stimulated by changes in temperature, mechanical force, and chemical concentration.
   b. The only receptors in viscera that provide sensations are pain receptors.
      (1) These receptors are most sensitive to certain chemicals and lack of blood flow.
      (2) The sensations they produce may feel as if they come from some other part of the body (referred pain).
   c. Pain nerve pathways
      (1) The two main types of pain fibers are acute pain fibers and chronic pain fibers.
      (2) Acute pain fibers are fast conducting; chronic pain fibers are slower conducting.
      (3) Pain impulses are processed in the dorsal horn of the spinal cord, and they ascend in the spinothalamic tracts.
      (4) Within the brain, pain impulses pass through the reticular formation before being conducted to the cerebral cortex.
   d. Regulation of pain impulses
      (1) Awareness of pain occurs when impulses reach the thalamus.
      (2) The cerebral cortex judges the intensity of pain and locates its source.
      (3) Impulses descending from the brain cause neurons to release pain-relieving substances, such as enkephalins and serotonin.
      (4) Endorphin is a pain-relieving biochemical produced in the brain.
   e. Certain neuropeptides synthesized in the CNS inhibit pain impulses.
4. Proprioception
   a. Stretch receptors provide information about the condition of muscles and tendons.
   b. Muscle spindles are stimulated when a muscle is relaxed, and they initiate a reflex that contracts the muscle.
   c. Golgi tendon organs are stimulated when muscle tension increases, and they initiate a reflex that relaxes the muscle.
5. Visceral receptors include lamellated corpuscles and free nerve endings.

12.4 SPECIAL SENSES (PAGE 446)

Special senses have receptors in complex sensory organs of the head.

1. Sense of smell
   a. Olfactory receptors
      (1) Olfactory receptors are chemoreceptors that chemicals dissolved in nasal secretions stimulate.
      (2) Olfactory receptors function with taste receptors and aid in food selection.
   b. Olfactory organs
      (1) The olfactory organs consist of receptors and supporting cells in the nasal cavity.
      (2) Olfactory receptors are neurons with cilia.
   c. Olfactory nerve pathways.
      (1) Nerve impulses travel from the olfactory receptors through the olfactory nerves, olfactory bulbs, and olfactory tracts.
      (2) They go to interpreting centers in the limbic system.
   d. Olfactory stimulation
      (1) Olfactory impulses may result when an odorant molecule stimulates a distinct set of receptor cells.
      (2) Olfactory receptors adapt rapidly.
      (3) Olfactory receptors are often damaged by environmental factors and are replaced from a pool of stem cells.
2. Sense of taste
   a. Taste receptors
      (1) Taste buds consist of receptor cells and supporting cells.
      (2) Taste cells have taste hairs that are sensitive to particular chemicals dissolved in water.
      (3) Taste hair surfaces have receptor sites to which chemicals combine and trigger impulses to the brain.
   b. Taste sensations
      (1) The five primary taste sensations are sweet, sour, salty, bitter, and umami.
      (2) Various taste sensations result from the stimulation of one or more sets of taste receptors.
      (3) Each of the five primary types of taste cells is particularly sensitive to a certain group of chemicals.
   c. Taste nerve pathways
      (1) Sensory impulses from taste receptors travel on fibers of the facial, glossopharyngeal, and vagus nerves.
      (2) These impulses are carried to the medulla and ascend to the thalamus and then to the gustatory cortex in the parietal lobes.

3. Sense of hearing
   a. The outer ear includes the auricle, the external acoustic meatus, and the tympanic membrane. It collects sound waves created by vibrating objects.
   b. Middle ear
      (1) Auditory ossicles of the middle ear conduct sound waves from the tympanic membrane to the oval window of the inner ear. They also increase the force of these waves.
      (2) Skeletal muscles attached to the auditory ossicles provide the tympanic reflex, which protects the inner ear from the effects of loud sounds.
   c. Auditory tubes connect the middle ears to the throat and help maintain equal air pressure on both sides of the tympanic membranes.
   d. Inner ear
      (1) The inner ear consists of a complex system of connected tubes and chambers—the osseous and membranous labyrinths. It includes the cochlea, which houses the spiral organ.
      (2) The spiral organ includes the hearing receptors that vibrations in the fluids of the inner ear stimulate.
      (3) Different frequencies of vibrations stimulate different sets of receptor cells; the human ear can detect sound frequencies from about 20 to 20,000 vibrations per second.
   e. Auditory nerve pathways
      (1) The nerve fibers from hearing receptors travel in the cochlear branch of the vestibulocochlear nerves.
      (2) Auditory impulses travel into the medulla oblongata, midbrain, and thalamus and are interpreted in the temporal lobes of the cerebrum.
4. Sense of equilibrium
   a. Static equilibrium maintains the stability of the head and body when they are motionless. The organs of static equilibrium are in the vestibule.
   b. Dynamic equilibrium balances the head and body when they are suddenly moved or rotated. The organs of this sense are in the ampullae of the semicircular canals.
   c. Other structures that help maintain equilibrium include the eyes and the proprioceptors associated with certain joints.
5. Sense of sight
   a. Visual accessory organs include the eyelids and lacrimal apparatus that protect the eye and the extrinsic muscles that move the eye.
   b. Structure of the eye
      (1) The wall of the eye has an outer, a middle, and an inner tunic that function as follows:
         (a) The outer layer (sclera) is protective, and its transparent anterior portion (cornea) refracts light entering the eye.
         (b) The middle layer (choroid coat) is vascular and has pigments that help keep the inside of the eye dark.
         (c) The inner layer (retina) includes visual receptor cells.
      (2) The lens is a transparent, elastic structure. The ciliary muscles control its shape.
      (3) The iris is a muscular diaphragm that controls the amount of light entering the eye; the pupil is an opening in the iris.
      (4) Spaces in the eye are filled with fluids (aqueous and vitreous humors) that help maintain its shape.
   c. Light refraction
      (1) Light waves are primarily refracted by the cornea and lens to focus an image on the retina.
      (2) The lens must thicken to focus on close objects.
   d. Visual receptors
      (1) The visual receptors are rods and cones.
      (2) Rods are responsible for colorless vision in dim light, and cones provide color vision.
   e. Visual pigments
      (1) A light-sensitive pigment in rods (rhodopsin) decomposes in the presence of light and triggers a complex series of reactions that initiate nerve impulses on the optic nerve.
      (2) Three sets of cones provide color vision. Each set has a different light-sensitive pigment, and each set is sensitive to a different wavelength of light; the color perceived depends on which set or sets of cones are stimulated.
   f. Stereoscopic vision
      (1) Stereoscopic vision provides perception of distance and depth.
      (2) Stereoscopic vision occurs because of the formation of two slightly different retinal images that the brain superimposes and interprets as one image in three dimensions.
      (3) A one-eyed person uses relative sizes and positions of familiar objects to judge distance and depth.
   g. Visual nerve pathways
      (1) Nerve fibers from the retina form the optic nerves.
      (2) Some fibers cross over in the optic chiasma.
      (3) Most of the fibers enter the thalamus and synapse with others that continue to the visual cortex of the occipital lobes.
      (4) Other impulses pass into the brainstem and function in various visual reflexes.

12.5 LIFE-SPAN CHANGES (PAGE 476)

Diminished senses are often one of the first noticeable signs of aging.

1. Age-related hearing loss may reflect damage to hair cells of the spiral organ, degeneration of nerve pathways to the brain, or tinnitus.
2. Age-related visual problems include dry eyes, floaters and light flashes, presbyopia, glaucoma, cataracts, macular degeneration, and retinal detachment.

CHAPTER ASSESSMENTS

12.1 Introduction

1 Explain the difference between a general sense and a special sense. (p. 438)

12.2 Receptors, Sensation, and Perception

2 Match each sensory receptor to the type of stimulus to which it is likely to respond: (p. 439)

(1) Chemoreceptor	A. approaching headlights
(2) Pain receptor	B. a change in blood pressure
(3) Thermoreceptor	C. the smell of roses
(4) Mechanoreceptor	D. an infected tooth
(5) Photoreceptor	E. a cool breeze

3 Explain how sensory receptors stimulate sensory impulses. (p. 439)

4 Explain the difference between a sensation and a perception. (p. 439)

5 Explain the projection of a sensation. (p. 439)

6 Define *sensory adaptation.* (p. 440)

7 You fill up the tub to take a hot bath, but the water is too hot. You test it a second and third time within a few seconds, and it feels OK. Which of the following is the most likely explanation? (p. 440)

a. The water has cooled down unusually quickly.

b. Your ability to sense heat has adapted.

c. Your nervous system is suddenly not functioning properly.

12.3 General Senses

8 Explain how general senses can be grouped. (p. 440)

9 Describe the functions of free nerve endings, tactile corpuscles, and lamellated corpuscles. (p. 440)

10 Describe the functions of the two classes of thermoreceptors. (p. 440)

11 Compare pain receptors with the other types of somatic receptors. (p. 440)

12 List the conditions likely to stimulate visceral pain receptors. (p. 441)

13 Define *referred pain,* and provide an example. (p. 441)

14 Contrast the nerve pathways involved in the production of acute and chronic pain. (p. 442)

15 Explain how neuropeptides relieve pain. (p. 443)

16 Distinguish between muscle spindles and Golgi tendon organs. (p. 444)

12.4 Special Senses

17 Explain how the senses of smell and taste function together to create the perception of the flavors of foods. (p. 446)

18 Which two of the following are part of the olfactory organs? (p. 446)

a. olfactory receptors

b. columnar epithelial cells in the nasal mucosa

c. the nose

d. the brain

19 Trace each step in the pathway from an olfactory receptor to the interpreting center of the cerebrum. (p. 447)

20 Salivary glands are important in taste because ________. (p. 448)

a. they provide the fluid in which food molecules dissolve

b. the taste receptors are located in salivary glands

c. salivary glands are part of the brain

d. lamellar corpuscles are activated

21 Name the five primary taste sensations and indicate a specific stimulus for each. (p. 449)

22 Explain why taste sensation is less likely to diminish with age than olfactory sensation. (p. 450)

23 Trace each step in the pathway from a taste receptor to the interpreting center of the cerebrum. (p. 450)

24 Match the ear area with the associated structure: (p. 450)

(1) outer ear	A. cochlea
(2) middle ear	B. eardrum
(3) inner ear	C. auditory ossicles

25 Trace each step in the pathway from the external acoustic meatus to hearing receptors. (p. 450)

26 Describe the functions of the auditory ossicles. (p. 452)

27 Identify the parts of the tympanic reflex, explain how they work, and explain the importance of this reflex. (p. 452)

28 The function of the auditory tube is to ________. (p. 453)

a. equalize air pressure on both sides of the eardrum

b. transmit sound vibrations to the eardrum

c. contain the hearing receptors

d. none of the above

29 Distinguish between the osseous and membranous labyrinths. (p. 453)

30 Describe the cochlea and its function. (p. 453)

31 Which of the following best describes hearing receptor "hair cells"? (p. 456)

a. They are neurons.

b. They lack ion channels.

c. They are epithelial, but function like neurons.

d. They are made of keratin.

32 Explain how a hearing receptor stimulates a sensory neuron. (p. 456)

33 Trace each step in the pathway from the spiral organ to the interpreting centers of the cerebrum. (p. 456)

34 Describe the organs of static and dynamic equilibrium and their functions. (p. 459)

35 Explain how the sense of vision helps maintain equilibrium. (p. 461)

36 Match the visual accessory organ with its function: (p. 462)

(1) Eyelid	A. moves the eye
(2) Conjunctiva	B. covers the eye
(3) Lacrimal gland	C. lines the eyelids
(4) Extrinsic muscle	D. produces tears

37 Name the three layers of the eye wall and describe the functions of each layer. (p. 465)

38 Explain why looking at a close object causes fatigue in terms of how accommodation is accomplished. (p. 467)

39 Explain the mechanisms of pupil constriction and pupil dilation. (p. 468)

40 Distinguish between the fovea centralis and the optic disc. (p. 469)

41 The following are compartments in the eye. In which one is vitreous humor found? (p. 469)
   a. anterior chamber
   b. posterior chamber
   c. anterior cavity
   d. posterior cavity

42 Explain how light is focused on the retina. (p. 471)

43 Distinguish between rods and cones. (p. 473)

44 Explain why cone vision is generally more acute than rod vision. (p. 473)

45 Describe the function of rhodopsin. (p. 473)

46 Explain why rod vision may be more important under dim light conditions. (p. 475)

47 Describe the relationship between light wavelength and color vision. (p. 475)

48 Define *stereoscopic vision.* (p. 475)

49 Explain why a person with normal binocular vision is able to judge distance and depth of close objects more accurately than a person who has lost one eye. (p. 475)

50 Trace each step in the pathway from the retina to the visual cortex. (p. 475)

12.5 Life-Span Changes

51 Explain the basis of fading senses of smell and taste with aging. (p. 476)

52 List three causes of hearing loss associated with aging. (p. 476)

53 Explain five problems that can interfere with vision as a person ages. (p. 477)

INTEGRATED ASSESSMENTS/CRITICAL THINKING

OUTCOMES 2.2, 11.5, 12.2, 12.3, 12.4

1. Positron emission tomography (PET) scans of the brains of people who have been blind since birth reveal high neural activity in the visual centers of the cerebral cortex when these people read Braille. When sighted individuals run their fingers over the raised letters of Braille, their visual centers do not show increased activity. Explain these findings.

OUTCOMES 6.5, 11.6, 12.2

2. Why are some serious injuries, like a bullet entering the abdomen, relatively painless, but others, such as a burn, considerably more painful?

OUTCOMES 11.5, 12.2, 12.4

3. Loss of the sense of smell often precedes the major symptoms of Alzheimer disease and Parkinson disease. What additional information is needed to use this association to prevent or treat these diseases?

OUTCOMES 12.2, 12.3

4. A patient with heart disease experiences pain at the base of the neck and in the left shoulder and upper limb during exercise. How would you explain the likely origin of this pain to the patient?

OUTCOMES 12.2, 12.4

5. People who are deaf due to cochlear damage do not suffer motion sickness. Why not?

OUTCOMES 12.2, 12.4

6. Labyrinthitis is an inflammation of the inner ear. What symptoms would you expect in a patient with this disorder?

WEB CONNECTIONS

Be sure to visit the text website at **www.mhhe.com/shier12** for answers to chapter assessments, additional quizzes, and interactive learning exercises.

ANATOMY & PHYSIOLOGY REVEALED

Anatomy & Physiology Revealed® (APR) includes cadaver photos that allow you to peel away layers of the human body to reveal structures beneath the surface. This program also includes animations, radiologic imaging, audio pronunciations, and practice quizzing. Check out **www.aprevealed.com**. APR has been proven to help improve student grades!

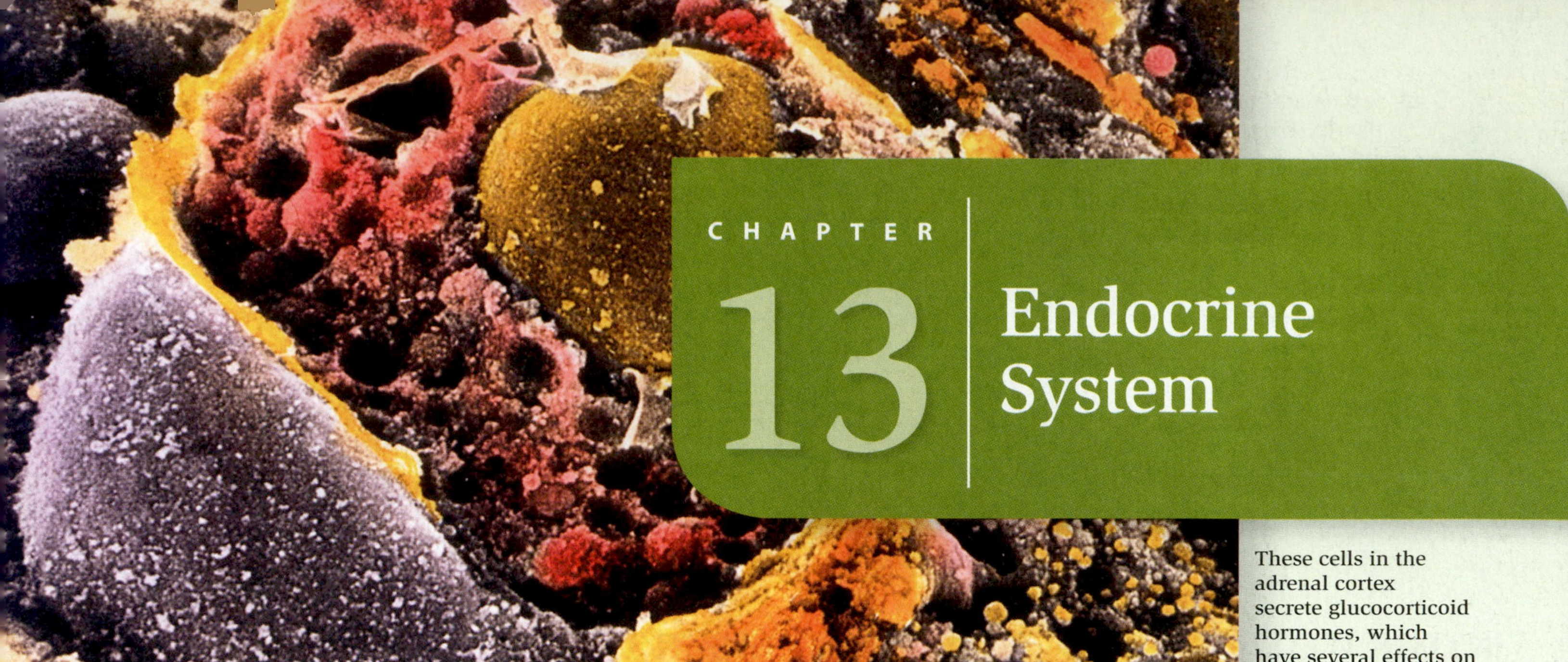

CHAPTER

13 Endocrine System

These cells in the adrenal cortex secrete glucocorticoid hormones, which have several effects on metabolism (14,400×).

UNDERSTANDING WORDS

cort-, bark, rind: adrenal *cortex*—outer portion of an adrenal gland.

-crin, to secrete: endo*crine*—internal secretions.

diuret-, to pass urine: *diuretic*—substance that promotes urine production.

endo-, inside: *endo*crine gland—gland that internally secretes into a body fluid.

exo-, outside: *exo*crine gland—gland that secretes to the outside through a duct.

horm-, impetus, impulse: *horm*one—substance that a cell secretes that affects another cell.

hyper-, above: *hyper*thyroidism—condition resulting from an above-normal secretion of thyroid hormone.

hypo-, below: *hypo*thyroidism—condition resulting from a below-normal secretion of thyroid hormone.

lact-, milk: pro*lact*in—hormone that promotes milk production.

med-, middle: adrenal *med*ulla—middle section of an adrenal gland.

para-, beside: *para*thyroid glands—set of glands near the surface of the thyroid gland.

toc-, birth: oxy*toc*in—hormone that stimulates the uterine muscles to contract during childbirth.

-tropic, influencing: adrenocortico*tropic* hormone—a hormone secreted by the anterior pituitary gland that stimulates the adrenal cortex.

vas-, vessel: *vas*opressin—substance that causes blood vessel walls to contract.

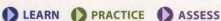

LEARNING OUTCOMES

After you have studied this chapter, you should be able to:

13.1 Introduction

1 Distinguish between endocrine and exocrine glands. (p. 483)

13.2 General Characteristics of the Endocrine System

2 Explain what makes a cell a target cell for a hormone. (p. 483)

3 List some important functions of hormones. (p. 484)

13.3 Hormone Action

4 Describe how hormones can be classified according to their chemical composition (p. 485)

5 Explain how steroid and nonsteroid hormones affect their target cells (p. 485)

13.4 Control of Hormonal Secretions

6 Discuss how negative feedback mechanisms regulate hormone secretion. (p. 492)

7 Explain how the nervous system controls hormone secretion. (p. 492)

13.5–13.10 Pituitary Gland–Other Endocrine Glands

8 Name and describe the locations of the major endocrine glands, and list the hormones that they secrete. (p. 492)

9 Describe the actions of the various hormones and their contributions to homeostasis. (p. 494)

10 Explain how the secretion of each hormone is regulated. (p. 494)

13.11 Stress and Its Effects

11 Distinguish between physical and psychological stress. (p. 514)

12 Describe the general stress response. (p. 514)

13.12 Life-Span Changes

13 Describe some of the changes associated with aging of the endocrine system. (p. 515)

ASSESS

ON THE ORIGIN OF MALE BODY ODOR

The endocrine system produces hormones, which are biochemicals that spread messages in an individual. Less well understood are pheromones, which are chemical signals sent between members of the same species. In insects and rodents, pheromones stimulate social behavior, including mating. Experiments suggest that sensing pheromones, or at least certain odors, may influence human sexual attraction, too.

Mice and rats choose mates dissimilar to themselves in a group of genes that provide immunity. Their sense of smell helps them discern appropriate mates. Biologists think this mate choice based on smell may protect offspring in two ways: it prevents close relatives from mating and it may team immune systems with different strengths.

A mouse's mating behavior stems from receptors in the olfactory epithelium, a collection of sensory neurons high in the nasal cavity. These receptors—called trace amine-associated receptors—are attuned to molecules in mouse urine that direct social behavior. The genes that encode the receptors are also found in the human genome.

To test if heterosexual humans use the sense of smell to respond to pheromones in mate selection, as rodents do, researchers recruited forty-nine young women and forty-four young men. Each donated DNA, which was typed for the human counterparts of genes that affect mating in rodents. The women used a nasal spray for two weeks to clear their nasal passages. The men wore the same tee shirt on two consecutive days without using deodorant or soap. Each woman was then given three tee shirts from men genetically similar to her and three tee shirts from men genetically dissimilar to her, not knowing which shirts came from which men. The women sniffed the shirts—preferable to smelling urine, as mice do—and rated the sweat stains on intensity, pleasantness, and sexiness. Like female rodents, female humans preferred the sweaty tees from the men least like them.

The tee shirt study provided indirect evidence of the existence of human pheromones. Analysis of the human genome has taken the study of human pheromones a step further by revealing a possible basis for the female response to male sweat. The gene that encodes a particular odorant receptor that affects social behavior in other animals comes in two variants in humans. Women of genotype RT/WM either do not associate male sweat with a scent at all, or detect a faint, pleasant, vanilla-like odor. But for women of genotype RT/RT, male sweat smells like stale urine! In an experiment on 400 women asked to sniff sixty-six odors in low or high concentrations, those who responded adversely to androstenone, a metabolite of testosterone found in sweat, shared the RT/RT genotype. Therefore, perception of male body odor may lie in the nose of the beholder! ■

13.1 INTRODUCTION

Regulating the functions of the human body to maintain homeostasis is an enormous job. Two organ systems function coordinately to enable body parts to communicate with each other and to adjust constantly to changing incoming signals. The nervous system is one biological communication system: it uses nerve impulses and chemicals called neurotransmitters. The other is the endocrine system.

The **endocrine system** is so named because the cells, tissues, and organs that comprise it, collectively called endocrine glands, secrete substances into the internal environment. ("Endocrine" means "internal secretion.") The secreted substances, called **hormones,** diffuse from the interstitial fluid into the bloodstream and eventually act on cells, called **target cells,** some distance away.

Other glands secrete substances into the internal environment that are not hormones by the traditional definition, but they function in similar fashion as messenger molecules and are sometimes referred to as "local hormones." These include **paracrine** secretions, which enter the interstitial fluid but affect only neighboring cells, and **autocrine** secretions, which affect only the secreting cell.

Another category of substances, secreted by **exocrine glands,** enter tubes or ducts that lead to body surfaces. In contrast to endocrine secretions, exocrine secretions are released externally. Two examples are stomach acid reaching the lumen of the digestive tract and sweat released at the skin's surface (fig. 13.1).

The interrelationships of the glands of the endocrine system are obvious in families that have an inherited cancer syndrome called *multiple endocrine neoplasia* (MEN). Different glands are affected in different family members, although the genetic cause is the same. One family member might have a tumor of the adrenal glands called pheochromocytoma; another might have thyroid cancer; yet a third relative might have parathyroid hyperplasia, a precancerous condition.

13.2 GENERAL CHARACTERISTICS OF THE ENDOCRINE SYSTEM

Cells of the endocrine system and the nervous system communicate using chemical signals that bind to receptor molecules. Table 13.1 summarizes some similarities and differences between the two systems. In contrast to the nervous system, which releases neurotransmitter molecules into synapses, the endocrine system releases hormones into the bloodstream, which carries these messenger molecules everywhere. However, the endocrine system is also precise, because only target cells can respond to a hormone (fig. 13.2). A hormone's target cells have specific receptors that other cells lack. These receptors are proteins or glycoproteins with binding sites for a specific hormone. The other chemical messengers, paracrine and autocrine substances, also bind to specific receptors, and some examples of these are included in the chapter.

Endocrine glands and their hormones help regulate metabolic processes. They control the rates of certain chemical reactions; aid in transporting substances through membranes; and help regulate water balance, electrolyte balance, and blood pressure. Endocrine hormones also play vital roles in reproduction, development, and growth.

Small groups of specialized cells produce some hormones. However, the larger endocrine glands—the pituitary gland, thyroid gland, parathyroid glands, adrenal glands, and pancreas—are the subject of this chapter (fig. 13.3). Subsequent chapters discuss several other hormone-secreting glands and tissues.

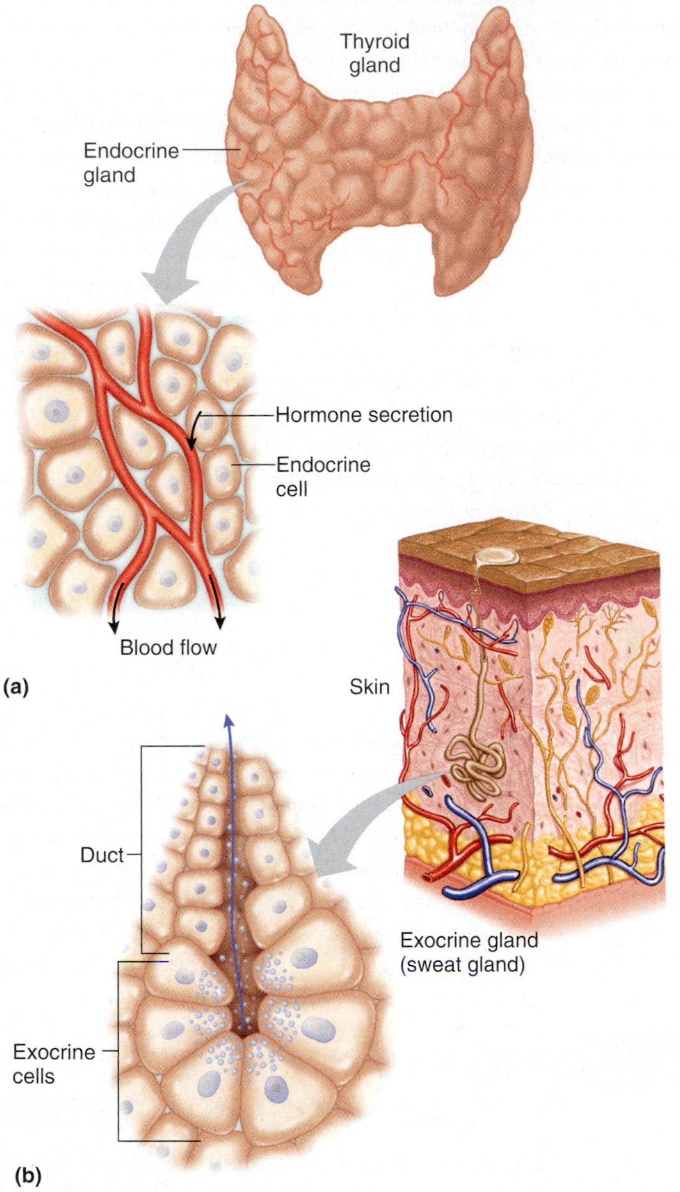

FIGURE 13.1 Types of glands. (*a*) Endocrine glands release hormones into the internal environment (body fluids). (*b*) Exocrine glands secrete to the outside environment through ducts that lead to body surfaces.

13.3 HORMONE ACTION

Hormones are released into the extracellular spaces surrounding endocrine cells. From there, they diffuse into the bloodstream and are carried to all parts of the body.

Chemistry of Hormones

Chemically, most hormones are either steroids or steroidlike substances—or they are nonsteroids, including amines, peptides, proteins, or glycoproteins. Thus, hormones are organic compounds. Hormones can stimulate changes in target cells even in extremely low concentrations.

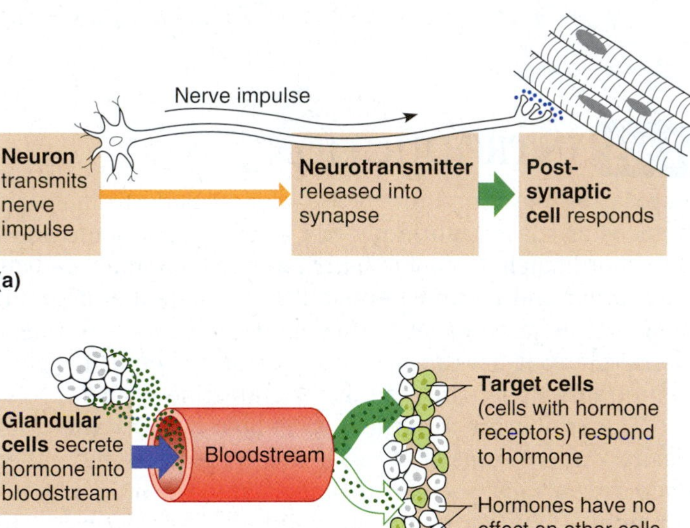

FIGURE 13.2 Chemical communication. (*a*) Neurons release neurotransmitters into synapses, affecting postsynaptic cells. (*b*) Glands release hormones into the bloodstream. Blood carries hormone molecules throughout the body, but only target cells respond.

TABLE 13.1 | A Comparison Between the Nervous System and the Endocrine System

	Nervous System	Endocrine System
Cells	Neurons	Glandular epithelium
Chemical signal	Neurotransmitter	Hormone
Specificity of action	Receptors on postsynaptic cell	Receptors on target cell
Speed of onset	Seconds	Seconds to hours
Duration of action	Very brief unless neuronal activity continues	May be brief or may last for days even if secretion ceases

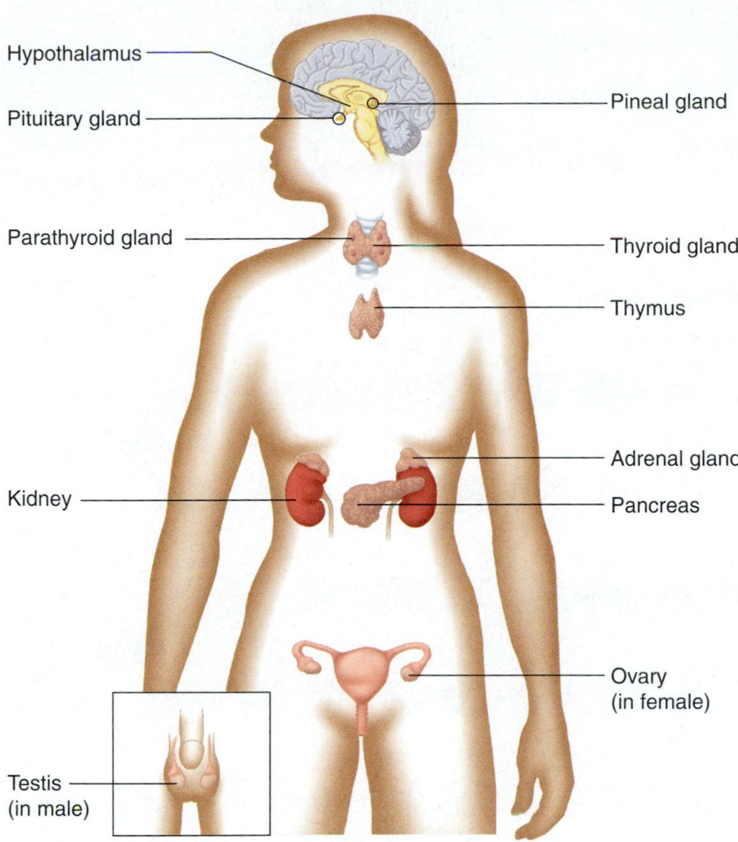

FIGURE 13.3 Locations of major endocrine glands.

Steroid Hormones

Steroids (ste′roidz) are lipids that include complex rings of carbon and hydrogen atoms (fig. 13.4*a*). Steroids differ by the types and numbers of atoms attached to these rings and the ways they are joined (see fig. 2.16). All steroid hormones are derived from cholesterol (see chapter 2, p. 64). They include sex hormones such as testosterone and the estrogens, and secretions of the adrenal cortex (the outer portion of the adrenal gland), including aldosterone and cortisol. Vitamin D is a modified steroid and can be converted into a hormone, as is discussed later in this chapter, in the section entitled "Parathyroid Hormone" (see also chapter 18, p. 711).

Nonsteroid Hormones

Hormones called *amines,* including norepinephrine and epinephrine, are derived from the amino acid tyrosine. These hormones are also synthesized in the adrenal medulla (the inner portion of the adrenal gland) (fig. 13.4*b*).

Protein hormones, like all proteins, are composed of long chains of amino acids, linked to form intricate molecular structures (see chapter 2, pp. 65–67 and fig. 13.4*c*). They include the hormone secreted by the parathyroid gland and some of those secreted by the anterior pituitary gland. Certain other hormones secreted from the anterior pituitary gland are *glycoproteins,* which consist of proteins joined to carbohydrates.

The *peptide* hormones are short chains of amino acids (fig. 13.4*d*). This group includes hormones associated with the posterior pituitary gland and some produced in the hypothalamus.

Another group of compounds, called *prostaglandins* (pros″tah-glan′dinz), are paracrine substances. They regulate neighboring cells. Prostaglandins are lipids (20-carbon fatty acids that include 5-carbon rings) and are synthesized from a type of fatty acid (arachidonic acid) in cell membranes (fig. 13.4*e*). Prostaglandins are produced in a wide variety of cells, including those of the liver, kidneys, heart, lungs, thymus gland, pancreas, brain, and reproductive organs.

Table 13.2 lists the names and abbreviations of some of the hormones discussed in this chapter. Table 13.3 and figure 13.4 summarize the chemical composition of hormones. Other hormones related to specific organ systems are discussed in their appropriate chapters.

PRACTICE 

1 What is a hormone?

2 How do endocrine glands and exocrine glands differ?

3 How are hormones chemically classified?

Actions of Hormones

Hormones exert their effects by altering metabolic processes. A hormone might change the activity of an enzyme necessary for synthesizing a particular substance or alter the rate at which particular chemicals are transported through cell membranes. A hormone delivers its message to a cell by uniting with the binding site of its receptor. The more receptors the hormone binds on its target cells, the greater the response.

The number of receptors on target cells may change. *Up-regulation* is an increase in the number of receptors on a target cell, often in response to a prolonged decrease in the level of a hormone. *Down-regulation* is the opposite, a decrease in the number of receptors due to a prolonged increase in hormone levels. Therefore, the number of receptors changes in ways that maintain an appropriate response to hormone level.

Steroid Hormones and Thyroid Hormones

Steroid hormones and thyroid hormones are insoluble in water. They are carried in the bloodstream weakly bound to plasma proteins in a way that they are released in sufficient quantity to affect their target cells. However, unlike amine, peptide, and protein hormones, steroid and thyroid hormones are soluble in the lipids that make up the bulk of cell membranes. For this reason, these hormones can diffuse into cells relatively easily and may enter any cell in the body.

Once inside a target cell, steroid and thyroid hormones combine (usually in the nucleus) with specific protein receptors. The resulting *hormone-receptor complex* binds to particular DNA sequences, either activating or repressing specific genes. Activated genes are transcribed into messenger RNA (mRNA), which enters the cytoplasm

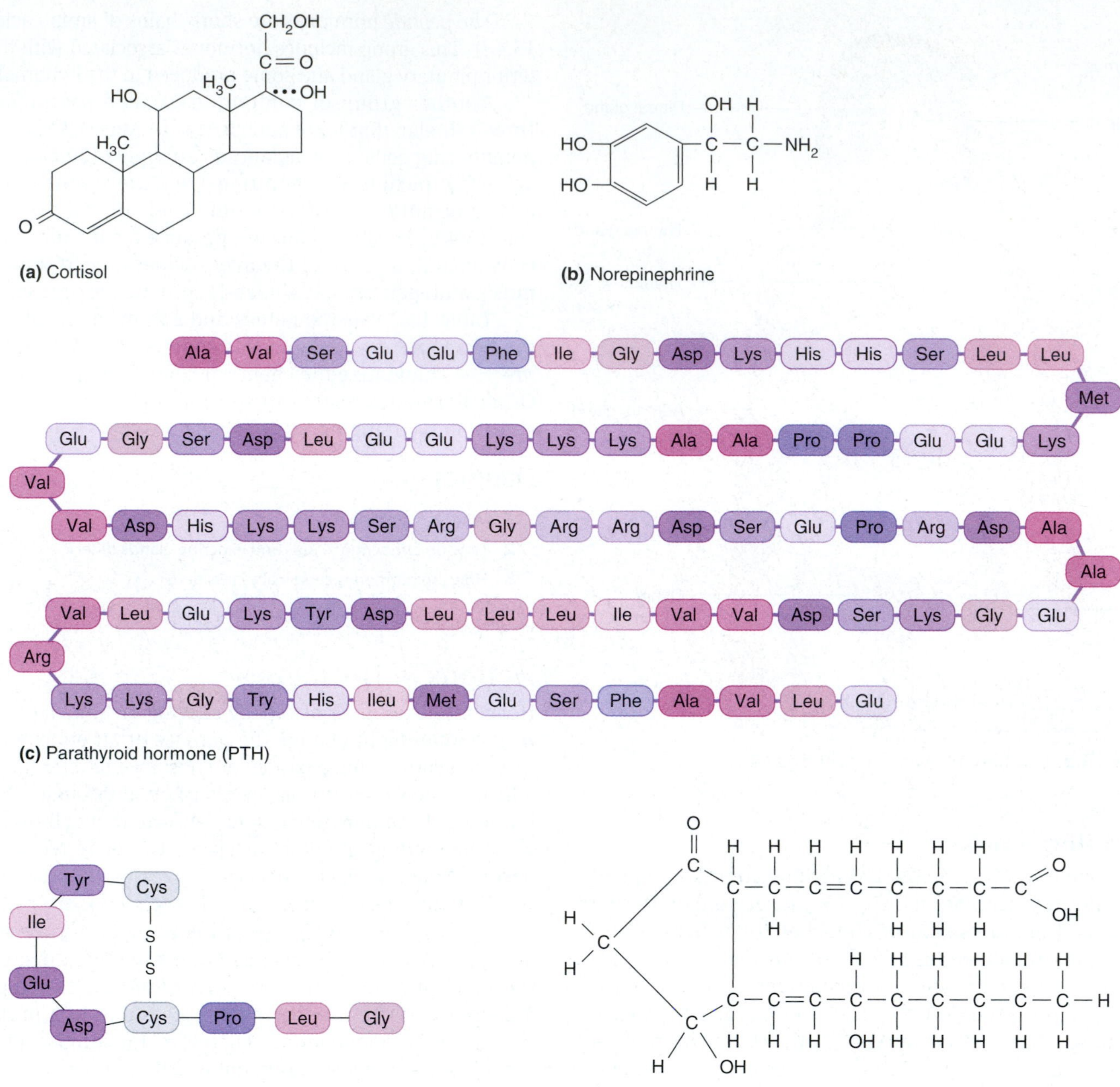

FIGURE 13.4 Structural formulas of (*a*) a steroid hormone (cortisol) and (*b*) an amine hormone (norepinephrine). Amino acid sequences of (*c*) a protein hormone (PTH) and (*d*) a peptide hormone (oxytocin). Structural formula of (*e*) a prostaglandin (PGE_2).

where it directs synthesis of specific proteins, which may be enzymes, transport proteins, or even hormone receptors. The activities of these hormones produce the cellular changes associated with the particular hormone (fig. 13.5, table 13.4, and Clinical Application 13.1). An example is the steroid hormone **aldosterone** (al′do-ster-ōn″, al-dos′ter-ōn), from the adrenal gland, whose action is to stimulate the kidneys to retain sodium. In response to aldosterone, cells that form tubules in the kidney begin to synthesize more Na^+/K^+ pumps. These are the proteins that actively transport these ions across the cell membrane, retaining sodium.

In some cases, steroid hormones may repress a particular gene, so transcription of it does not occur. The cellular response results from decreased levels of a particular protein.

Nonsteroid Hormones

A nonsteroid hormone, such as an amine, peptide, or protein, usually combines with specific receptor molecules on the target cell membrane. Each receptor molecule is a protein that has a *binding site* and an *activity site*. The hormone combines with the binding site, which causes the receptor's activity site to interact with other membrane proteins. Receptor binding

TABLE 13.2 | Hormone Names and Abbreviations

Source	Name	Abbreviation	Synonym
Hypothalamus	Corticotropin-releasing hormone	CRH	
	Gonadotropin-releasing hormone	GnRH	Luteinizing hormone releasing hormone (LHRH)
	Somatostatin	SS	Growth hormone release-inhibiting hormone (GRIH)
	Growth hormone-releasing hormone	GHRH	
	Prolactin release-inhibiting hormone	PIH	Dopamine
	Prolactin-releasing factor*	PRF*	
	Thyrotropin-releasing hormone	TRH	
Anterior pituitary gland	Adrenocorticotropic hormone	ACTH	Corticotropin
	Follicle-stimulating hormone	FSH	Follitropin
	Growth hormone	GH	Somatotropin (STH)
	Luteinizing hormone	LH	Lutropin, interstitial cell-stimulating hormone (ICSH)
	Prolactin	PRL	
	Thyroid-stimulating hormone	TSH	Thyrotropin
Posterior pituitary gland	Antidiuretic hormone	ADH	Vasopressin
	Oxytocin	OT	
Thyroid gland	Calcitonin		
	Thyroxine	T_4	Tetraiodothyronine
	Triiodothyronine	T_3	
Parathyroid gland	Parathyroid hormone	PTH	Parathormone
Adrenal medulla	Epinephrine	EPI	Adrenalin
	Norepinephrine	NE	Noradrenalin
Adrenal cortex	Aldosterone		
	Cortisol		Hydrocortisone
Pancreas	Glucagon		
	Insulin		
	Somatostatin	SS	

*"Factor" is used because a specific prolactin-releasing hormone has not yet been identified.

TABLE 13.3 | Types of Hormones

Type of Compound	Formed from	Examples
Amines	Amino acids	Norepinephrine, epinephrine
Peptides	Amino acids	ADH, OT, TRH, SS, GnRH
Proteins	Amino acids	PTH, GH, PRL
Glycoproteins	Protein and carbohydrate	FSH, LH, TSH
Steroids	Cholesterol	Estrogens, testosterone, aldosterone, cortisol

TABLE 13.4 | Sequence of Steroid Hormone Action

1. Endocrine gland secretes steroid hormone.
2. Steroid hormone diffuses through target cell membrane and enters cytoplasm or nucleus.
3. Hormone combines with a receptor molecule in the cytoplasm or nucleus.
4. Steroid hormone-receptor complex binds to DNA and promotes transcription of messenger RNA.
5. Messenger RNA enters the cytoplasm and directs protein synthesis.
6. Newly synthesized proteins produce hormone's specific effects.

may alter the function of enzymes or membrane transport mechanisms, changing the concentrations of still other cellular components. The hormone that triggers this cascade of biochemical activity is considered a *first messenger*. The biochemicals in the cell that induce the changes recognized as responses to the hormone are called *second messengers*.

Many hormones use **cyclic adenosine monophosphate** (cyclic AMP, or cAMP) as a second messenger. In this mechanism, a hormone binds to its receptor, and the resulting hormone-receptor complex activates a protein called a

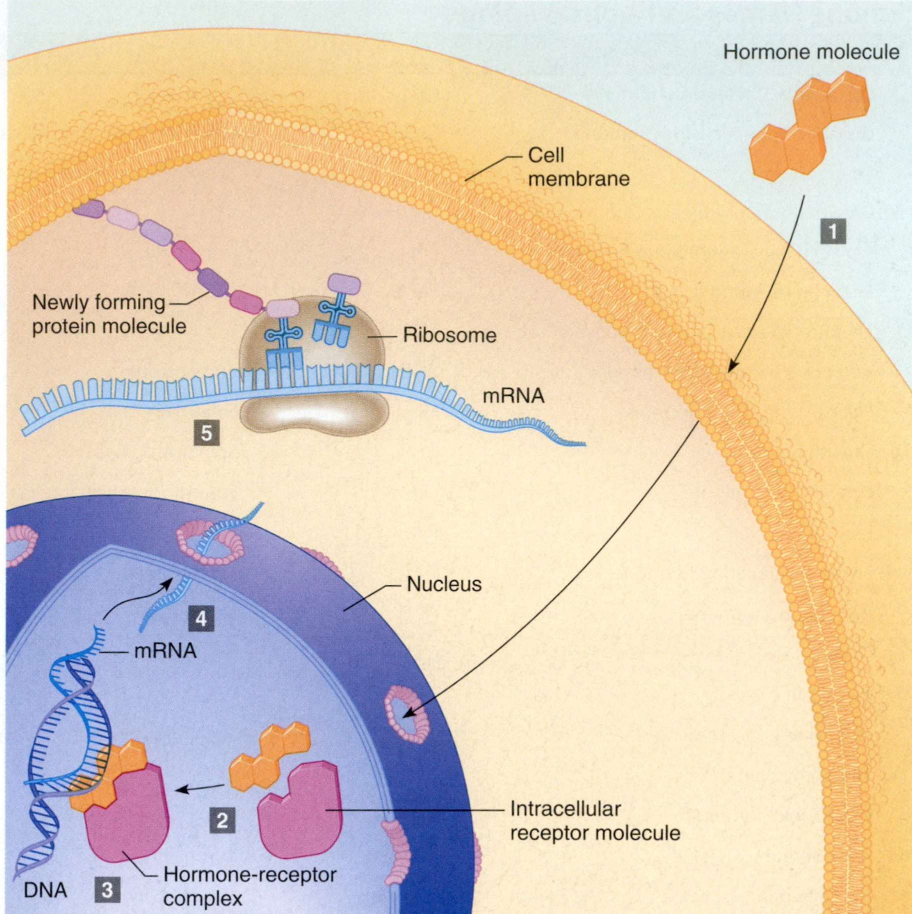

FIGURE 13.5 Steroid hormones. (*1*) A steroid hormone crosses a cell membrane and (*2*) combines with a protein receptor, usually in the nucleus. (*3*) The hormone-receptor complex activates transcription of specific messenger RNA (mRNA) molecules from DNA. (*4*) The mRNA molecules leave the nucleus and enter the cytoplasm (*5*) where they guide synthesis of their encoded proteins. In the bloodstream, most molecules of a particular steroid are bound to proteins. Only the few that are not bound are free to enter cells, as shown here.

G protein, which then activates an enzyme called **adenylate cyclase** (ah-den′ĭ-lāt si′klās), an integral membrane protein with its active site facing the inside of the cell. The activated enzyme removes two phosphates from ATP and circularizes it, forming *cyclic AMP* (fig. 13.6). Cyclic AMP, in turn, activates another set of enzymes called **protein kinases** (ki′nās-ez). Protein kinases transfer phosphate groups from ATP molecules to protein substrate molecules. This phosphorylation alters the shapes of the substrate molecules, in some cases activating inactive forms.

The activated proteins then alter various cellular processes, bringing about the effect of that particular hormone (fig. 13.7). The response of a particular cell to such a hormone is determined by the type of membrane receptors present and by the types of protein substrate molecules in the cell. Table 13.5 summarizes these actions.

Cellular responses to second messenger activation include altering membrane permeabilities, activating enzymes, promoting synthesis of certain proteins, stimulating or inhibiting specific metabolic pathways, promoting cellular movements, and initiating secretion of hormones and other substances. A specific example is the action of epinephrine to raise blood sugar during periods of physical stress. Epinephrine acts through the second messenger cAMP to increase the activity of the enzyme that breaks down liver

TABLE 13.5 | Sequence of Actions of Nonsteroid Hormone Using Cyclic AMP

1. Endocrine gland secretes nonsteroid hormone.
2. Body fluid carries hormone to its target cell.
3. Hormone combines with receptor site on membrane of its target cell, activating G protein.
4. Adenylate cyclase molecules are activated in target cell's membrane.
5. Adenylate cyclase circularizes ATP into cyclic AMP.
6. Cyclic AMP activates protein kinases.
7. Protein kinases activate protein substrates in the cell that change metabolic processes.
8. Cellular changes produce the hormone's effects.

13.1 CLINICAL APPLICATION

Using Hormones to Improve Athletic Performance

It is difficult to keep up with the numbers of athletes, amateur and professional, caught using performance-enhancing drugs. In late 1999 the International Olympic Committee founded the World Anti-Doping Agency to set rules and regulations, but it has not stopped the problem. At the summer games in Sydney, Australia, in 2000, the first time the agency participated, thirty-six athletes and coaches were dismissed for breaking the new rules. Four years later, at the summer games in Athens, six medals were revoked. By 2008, the agency released an updated list of banned drugs (table 13A). Famous athletes are often caught too, as baseball hero Barry Bonds and sprinter Marion Jones discovered.

Athletes have used drugs to aid performance since the earliest Olympics, when cocaine, heroin, morphine, and strychnine were the drugs of choice. During World War II, soldiers took amphetamines to mask the fatigue that accompanies great exertion. Shortly after the war, use of amphetamines spread to the sports world. Today, the general focus of performance enhancement is misuse of certain powerful hormones of the endocrine system. Three types of approaches are described here.

TABLE 13A | List of Prohibited Drugs

Anabolic agents
Anabolic androgenic steroids
Other anabolic agents
Hormones and related substances
EPO
Growth factors
Gonadotropins (LH, hCG in males)
Insulin
Corticotropins
Beta-2 agonists
Hormone antagonists and modulators
Aromatase inhibitors
Selective estrogen receptor modulators
Other antiestrogens
Myostatin inhibitors
Diuretics

Source: World Anti-Doping Agency

Steroids

Athletes who abuse steroids seek the hormone's ability to increase muscular strength. They are caught when the steroids or their metabolites are detected in urine or when natural testosterone levels plummet in a negative feedback response to the outside supply of the hormone.

Steroids are powerful drugs, and abusing them carries serious risks to health. Steroids hasten adulthood, stunting height and causing early hair loss. In males, excess steroid hormones lead to breast development, and in females to a deepened voice, hairiness, and a male physique. The kidneys, liver, and heart may be damaged, and atherosclerosis may develop because steroids raise LDL and lower HDL—the opposite of a healthy cholesterol profile. Steroids can also cause psychiatric symptoms, including delusions, depression, and violence.

For some athletes, illicit steroid use cannot be detected because of a natural mutation that deletes part of a gene. The result is a block in the conversion of testosterone from fat soluble to water soluble, and the hormone is not excreted into the urine as it normally is. This mutation is responsible for variations seen in different populations in the amount of urinary testosterone excreted. In one telling experiment, researchers in Sweden injected fifty-five male volunteers with a high dose of testosterone—and seventeen of the men showed no traces of the steroid in their urine! The ratio of testosterone to epitestosterone (synthesized together) serves as a test of taking exogenous testosterone. The body produces testosterone and an inactive form of it, called epitestosterone, in about equal amounts. An amount of testosterone in great excess of the amount of epitestosterone therefore indicates "doping." In the study, the men without the mutation had a T:E of 100, reflecting their massive doses. Men missing one copy of the gene had a T:E of 50, and those missing both copies of the gene had a T:E of 0 to 4.

Growth Hormone

Some athletes take human growth hormone (HGH) preparations to supplement the effects of steroids, because HGH enlarges muscles, as steroids strengthen them. HGH has been available as a drug since 1985, and it is prescribed to treat children with certain forms of inherited dwarfism. However, HGH is available from other nations and can be obtained illegally to enhance athletic performance. Unlike steroids, HGH has a half-life of only seventeen to forty-five minutes, which means that it becomes so scant that it is undetectable in body fluids within an hour. (Half-life is the time that it takes half of a given number of particles to break down into another substance.)

Erythropoietin

Increasing the number of red blood cells can increase oxygen delivery to muscles and thereby enhance endurance. Swedish athletes introduced "blood doping" in 1972. The athletes would have blood removed a month or more prior to performance, then have the blood reinfused shortly before a competition, boosting the red blood cell supply. Easier than blood doping is to take erythropoietin (EPO), a hormone secreted from the kidneys that signals the bone marrow to produce more red blood cells. EPO is used to treat certain forms of anemia. Using it to improve athletic performance is ill advised. In 1987, EPO abuse led to heart attacks and death in twenty-six cyclists from the Netherlands. Runners and swimmers also use EPO.

Testing for drugs or for mutations that enable athletes to mask abuse is expensive. While geneticists and biochemists sort out natural nuances in the metabolism of these substances, sports officials are using an "athlete's passport" to try to expose use of performance-enhancing drugs. The passport is a record of drug test results taken over several years, against which new drug activity is detectable. ■

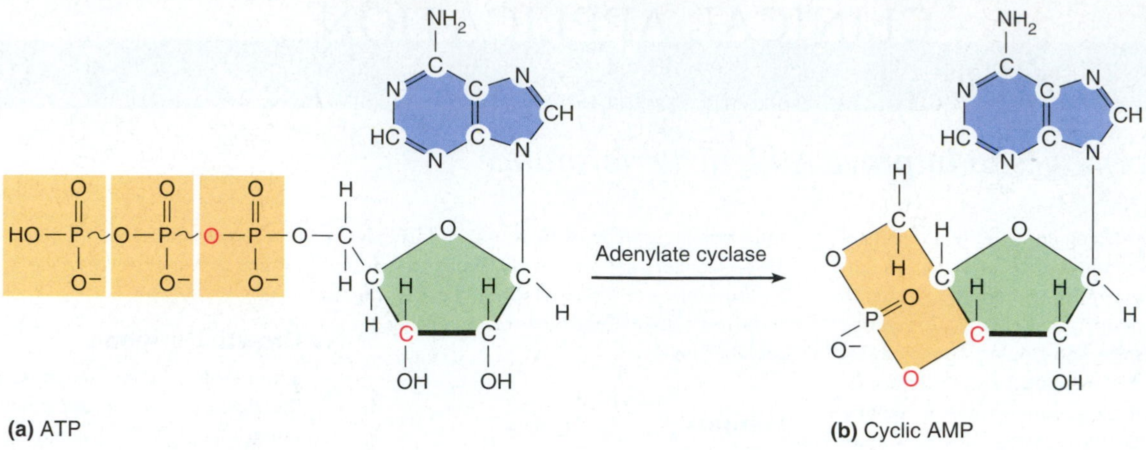

FIGURE 13.6 Adenylate cyclase catalyzes conversion of (*a*) ATP molecules into cyclic AMP (*b*). The atoms forming the new bond are shown in red.

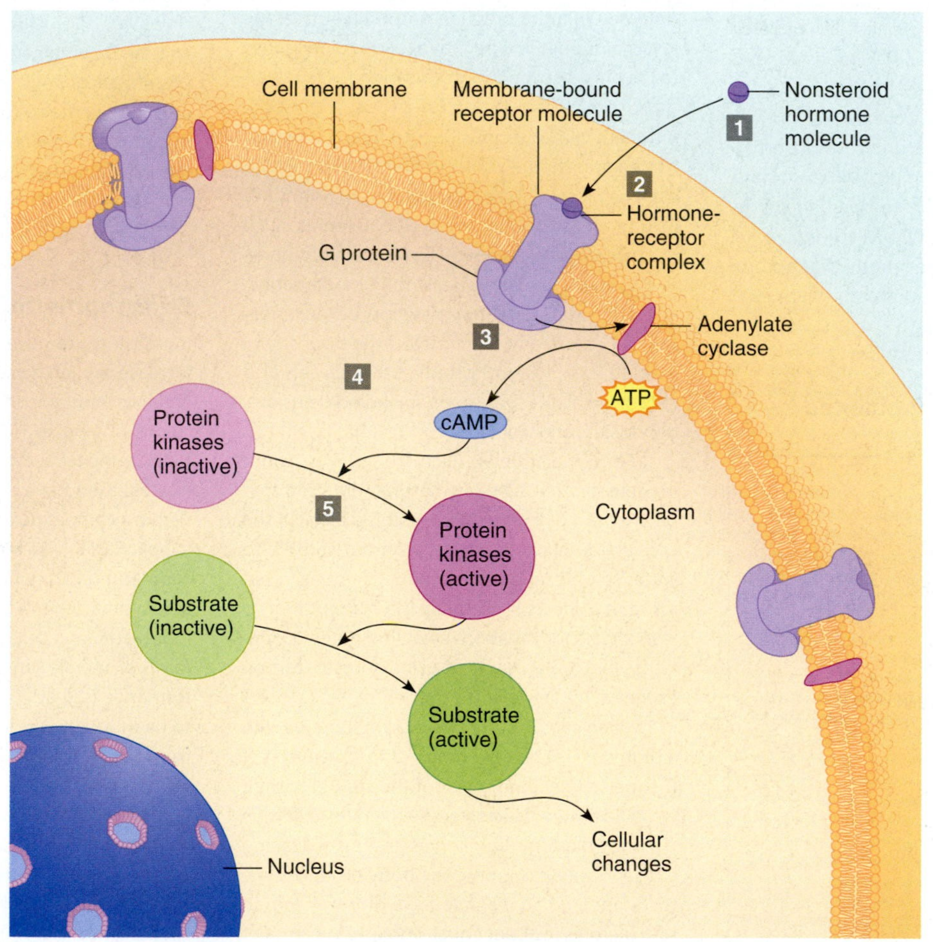

FIGURE 13.7 Nonsteroid hormone action. (*1*) Body fluids carry nonsteroid hormone molecules to the target cell, where (*2*) they bind receptor molecules on the cell membrane. (*3*) This activates molecules of adenylate cyclase, which (*4*) catalyze conversion of ATP into cyclic adenosine monophosphate (cAMP). (*5*) The cAMP promotes a series of reactions leading to the cellular changes associated with the hormone's action.

glycogen, increasing the number of glucose molecules that can diffuse out of liver cells and enter the bloodstream.

Another enzyme, phosphodiesterase, quickly and continuously inactivates cAMP, so its action is short-lived. For this reason, a continuing response in a target cell requires a continuing signal from hormone molecules binding receptors in the target cell membrane.

Hormones whose actions require cyclic AMP include releasing hormones from the hypothalamus; thyroid-stimulating hormone (TSH), adrenocorticotropic hormone (ACTH), follicle-stimulating hormone (FSH), and luteinizing hormone (LH) from the anterior pituitary gland; antidiuretic hormone (ADH) from the posterior pituitary gland; parathyroid hormone (PTH) from the parathyroid glands; norepinephrine and epinephrine from the adrenal glands; calcitonin from the thyroid gland; and glucagon from the pancreas.

An abnormality in cAMP-mediated signaling can lead to symptoms from many endocrine glands. In McCune-Albright syndrome, for example, a defect in the G protein that activates adenylate cyclase results in conversion of ATP to cAMP even without hormonal stimulation. As a result, cells in the pituitary, thyroid, gonads, and adrenal glands secrete hormones in excess. One symptom is precocious puberty. Infant girls menstruate, and boys as young as six years produce mature sperm. Symptoms vary widely because the syndrome results from a mutation that is not present in all cells of an individual.

Certain nonsteroid hormones use second messengers other than cAMP. For example, a second messenger called diacylglycerol (DAG), like cAMP, activates a protein kinase leading to a cellular response.

In another mechanism, a hormone binding its receptor increases calcium ion concentration in the cell. Such a hormone may stimulate transport of calcium ions inward through the cell membrane or induce release of calcium ions from cellular storage sites via a second messenger called inositol triphosphate (IP3). The calcium ions combine with the protein *calmodulin* (see chapter 9, p. 301), altering its molecular structure in a way that activates the molecule. Activated calmodulin can then interact with enzymes, altering their activities and thus eliciting diverse responses.

Still another hormonal mechanism uses *cyclic guanosine monophosphate* (cyclic GMP, or cGMP). Like cAMP, cGMP is a nucleotide derivative and functions in much the same manner as a second messenger.

Cellular response to a steroid hormone (and thyroid hormone) is directly proportional to the number of hormone-receptor complexes that form. In contrast, response to a hormone operating through a second messenger is greatly amplified. This is possible because many second messenger molecules can be activated in response to just a few hormone-receptor complexes. Cells are highly sensitive to changes in the concentrations of nonsteroid hormones because of such amplification.

PRACTICE 

4 How does a steroid hormone act on its target cells?

5 How does a nonsteroid hormone act on its target cells?

6 What is a second messenger?

Prostaglandins

Prostaglandins are paracrine substances, acting locally, that are potent and present in small amounts. They are not stored in cells but are synthesized just before they are released. They are rapidly inactivated.

Some prostaglandins regulate cellular responses to hormones. For example, different prostaglandins can either activate or inactivate adenylate cyclase in cell membranes, thereby controlling production of cAMP and altering the cell's response to a hormone.

Prostaglandins produce a variety of effects. Some prostaglandins can relax smooth muscle in the airways of the lungs and in the blood vessels, dilating these passageways. Yet other prostaglandins can contract smooth muscle in the walls of the uterus, causing menstrual cramps and labor contractions. They stimulate secretion of hormones from the adrenal cortex and inhibit secretion of hydrochloric acid from the wall of the stomach. Prostaglandins also influence movements of sodium ions and water in the kidneys, help regulate blood pressure, and have powerful effects on both male and female reproductive physiology. When tissues are injured, prostaglandins promote inflammation (see chapter 16, p. 627).

Understanding prostaglandin function has medical applications. Drugs such as aspirin and certain steroids that relieve the joint pain of rheumatoid arthritis inhibit production of prostaglandins in the synovial fluid of affected joints. Daily doses of aspirin may reduce the risk of heart attack by altering prostaglandin activity. Prostaglandins may be used as drugs to dilate constricted blood vessels to relieve hypertension.

PRACTICE 

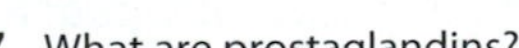

7 What are prostaglandins?

8 Describe one possible function of prostaglandins.

9 What are effects of prostaglandins?

13.4 CONTROL OF HORMONAL SECRETIONS

The body must be able to turn processes on and off. (Recall how acetylcholine is removed from the neuromuscular junction to end muscle contraction.) In the case of hormones, half-life measures removal. Half-life is the time for half of the hormone molecules to be removed from the plasma. For example, a hormone with a half-life of ten minutes would start out at 100% of its blood concentration, and if secretion were to stop, it would drop to 50% in ten minutes, 25% in another ten minutes, 12.5% in another ten minutes, and so on. Hormones with short half-lives (a few minutes) control body functions that turn on and off quickly, whereas the effects of hormones with longer half-lives, such as thyroid hormone and steroids, may last for days.

Hormones are continually excreted in the urine and broken down by enzymes, primarily in the liver. Therefore, increasing or decreasing blood levels of a hormone requires increased or decreased secretion. Hormone secretion is precisely regulated.

Control Sources

Control of hormone secretion is essential to maintaining the internal environment. In a few cases, primarily in the reproductive systems, positive feedback affects this control.

Generally, hormone secretion is controlled in three ways, all of which employ **negative feedback** (see chapter 1, p. 9). In each case, an endocrine gland or the system controlling it senses the concentration of the hormone the gland secretes, a process the hormone controls, or an action the hormone has on the internal environment (fig. 13.8).

1. The hypothalamus controls the anterior pituitary gland's release of **tropic hormones,** which stimulate other endocrine glands to release hormones (fig. 13.8*a*). The hypothalamus constantly receives information about the internal environment from neural connections and cerebrospinal fluid, made possible by its location near the thalamus and the third ventricle (fig. 13.9).
2. The nervous system directly stimulates some glands. The adrenal medulla, for example, secretes its hormones (epinephrine and norepinephrine) in response to preganglionic sympathetic nerve impulses. The secretory cells replace the postganglionic sympathetic neurons, which would normally secrete norepinephrine alone as a neurotransmitter (see fig. 13.8*b*).
3. Another group of glands responds directly to changes in the composition of the internal environment. For example, when the blood glucose level rises, the pancreas secretes insulin, and when the blood glucose level falls, it secretes glucagon, as covered in the section entitled "Hormones of the Pancreatic Islets" (see fig. 13.8*c*).

In each of these cases, as hormone levels rise in the blood and the hormone exerts its effects, negative feedback inhibits the system and hormone secretion decreases. Then, as hormone levels in the blood decrease and the hormone's effects wane, inhibition of the system ceases, and secretion of that hormone increases again (fig. 13.10). As a result of negative feedback, hormone levels in the bloodstream remain relatively stable, fluctuating slightly around an average value (fig. 13.11).

PRACTICE 

10 How does the nervous system help regulate hormonal secretions?

11 How does a negative feedback system control hormonal secretion?

13.5 PITUITARY GLAND

The **pituitary** (pĭ-tu′ĭ-tār″e) **gland** (hypophysis), at the base of the brain, is about one centimeter in diameter. It is attached to the hypothalamus by the pituitary stalk, or *infundibulum*, and lies in the sella turcica of the sphenoid bone, as figure 13.9 shows.

The pituitary gland consists of two distinct portions: an *anterior lobe* (adenohypophysis) and a *posterior lobe* (neurohypophysis). The anterior lobe secretes a number of hormones, including growth hormone (GH), thyroid-stimulating hormone (TSH), adrenocorticotropic hormone (ACTH), follicle-stimulating hormone (FSH), luteinizing hormone (LH), and prolactin (PRL). Although the cells of the posterior lobe (pituicytes) do not synthesize any hormones, specialized

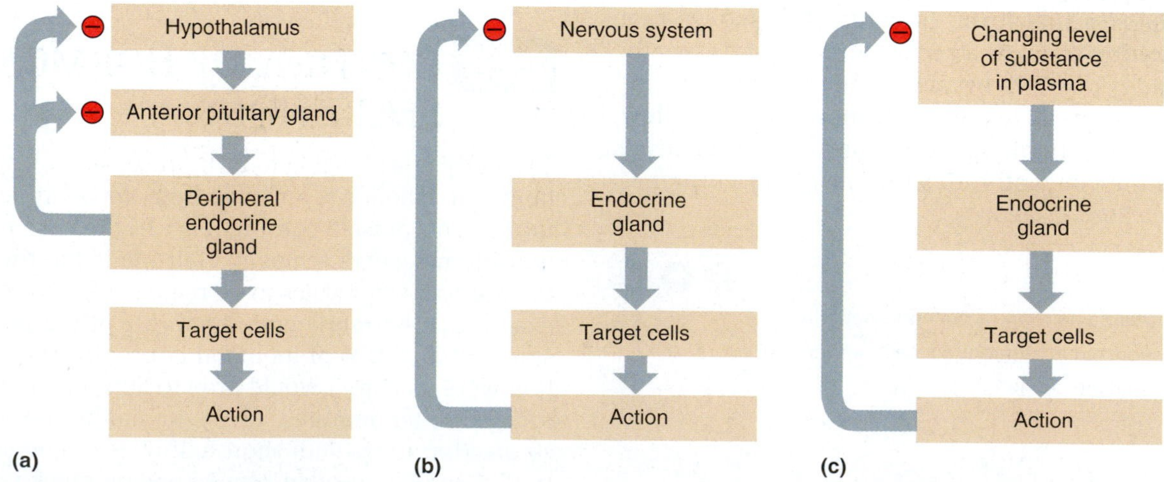

FIGURE 13.8 Examples of endocrine system control: (*a*) one way the hypothalamus controls the anterior pituitary, (*b*) the nervous system controls some glands directly, and (*c*) some glands respond directly to changes in the internal environment. ⊖ indicates negative feedback inhibition.

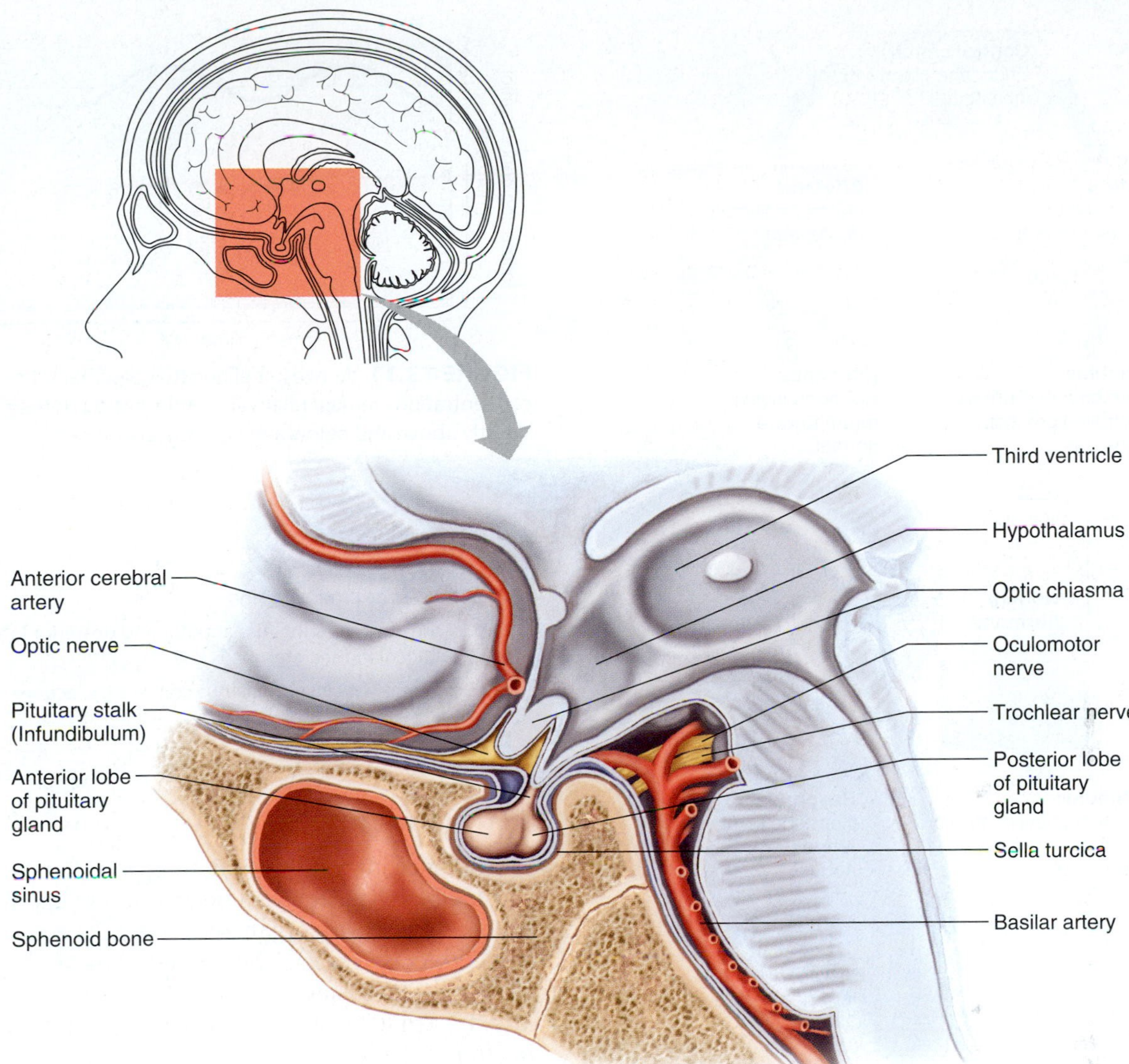

FIGURE 13.9 The pituitary gland is attached to the hypothalamus and lies in the sella turcica of the sphenoid bone.

neurons called neurosecretory cells secrete two important hormones, antidiuretic hormone (ADH) and oxytocin (OT), into the bloodstream from their nerve endings in the posterior lobe. The cell bodies of these neurosecretory cells are in the hypothalamus.

During fetal development, a narrow region between the anterior and posterior lobes of the pituitary gland, the *intermediate lobe* (pars intermedia), produces melanocyte-stimulating hormone (MSH). This hormone regulates the formation of melanin—the pigment in the skin and in parts of the eyes and brain. This intermediate lobe seems to disappear during fetal development, but its secretory cells persist and become part of the two remaining lobes.

The brain controls most of the pituitary gland's activities (fig. 13.12). The pituitary gland's posterior lobe releases hormones into the bloodstream in response to nerve impulses from the hypothalamus. A different mechanism controls the anterior lobe. Here, releasing hormones from the hypothalamus primarily control secretion. These releasing hormones are carried in the blood via a capillary bed associated with the hypothalamus. The vessels merge to form the **hypophyseal** (hi″po-fiz′e-al) **portal veins** that pass downward along the pituitary stalk and give rise to a capillary bed in the anterior lobe. Thus, substances released into the blood from the hypothalamus are carried directly to the anterior lobe. The hypothalamus, therefore, is an endocrine gland, yet it also controls other endocrine glands. This is also true of the anterior pituitary.

The arrangement of two capillaries in series is unusual and is called a *portal system*. It exists in only three places in the body: the hepatic portal vein connects intestinal capillaries to special liver capillaries called sinusoids, the efferent arteriole of kidney nephrons connects two sets of capillaries, and the hypophyseal portal vein gives rise to a capillary net in the anterior lobe of the pituitary gland.

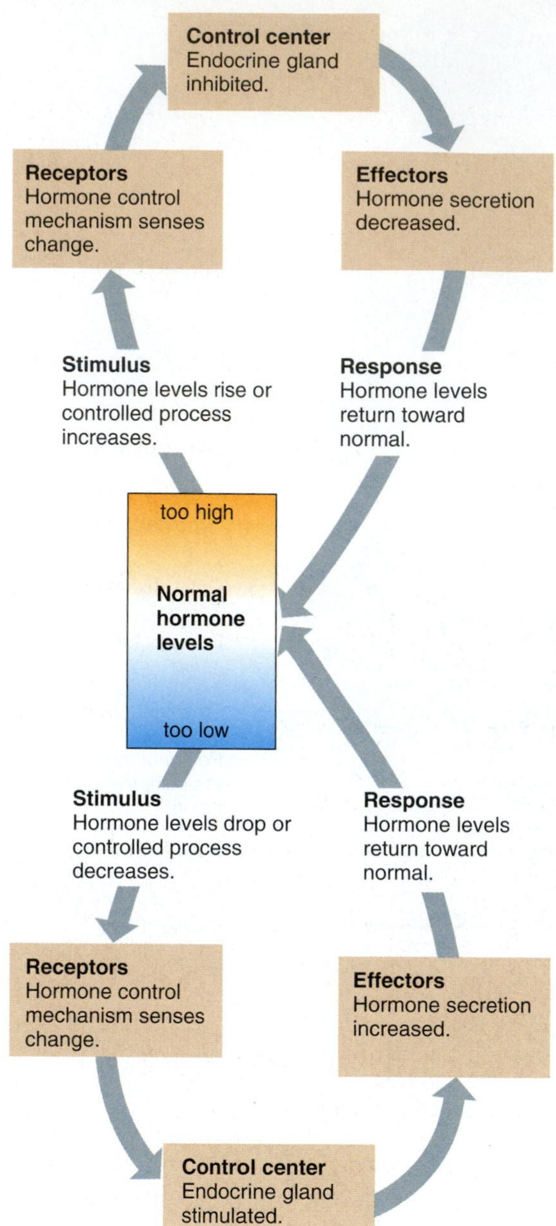

FIGURE 13.10 Hormone secretion is under negative feedback control.

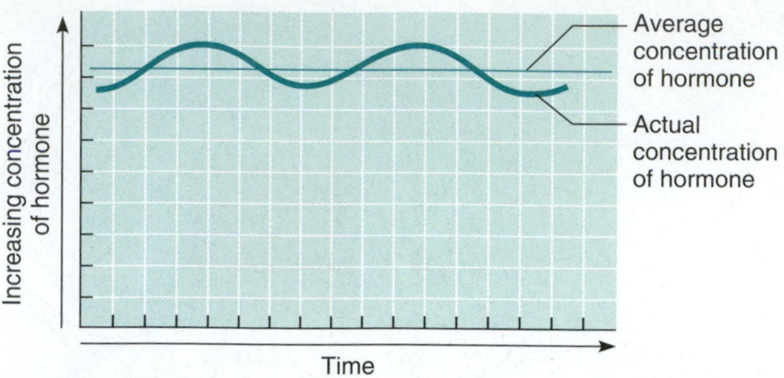

FIGURE 13.11 As a result of negative feedback, hormone concentrations remain relatively stable, although they may fluctuate slightly above and below average concentrations.

Upon reaching the anterior lobe of the pituitary, each of the hypothalamic releasing hormones acts on a specific population of cells. Some of the resulting actions are inhibitory (prolactin release-inhibiting hormone and somatostatin), but most stimulate the anterior pituitary to release hormones that stimulate the secretions of peripheral endocrine glands. In many of these cases, important negative feedback relationships regulate hormone levels in the bloodstream. Figure 13.13 shows this general relationship.

PRACTICE  

12 Where is the pituitary gland?

13 List the hormones that the anterior and posterior lobes of the pituitary gland secrete.

14 Explain how the hypothalamus controls the actions of the pituitary gland.

Anterior Pituitary Hormones

The anterior lobe of the pituitary gland is enclosed in a dense capsule of collagenous connective tissue and largely consists of epithelial tissue organized in blocks around many thin-walled blood vessels. The epithelial tissue has five types of secretory cells. They are *somatotropes* that secrete GH, *mammatropes* that secrete PRL, *thyrotropes* that secrete TSH, *corticotropes* that secrete ACTH, and *gonadotropes* that secrete FSH and LH (figs. 13.14 and 13.15). In males, LH (luteinizing hormone) is known as ICSH (interstitial cell-stimulating hormone) because it affects the interstitial cells of the testes (see chapter 22, p. 845).

Growth hormone, also called *somatotropin* (STH), is a protein that stimulates cells to enlarge and more rapidly divide. It enhances the movement of amino acids through cell membranes and increases the rate of protein synthesis. GH also decreases the rate at which cells use carbohydrates and increases the rate at which they use fats.

Growth hormone secretion varies during the day, peaking during sleep. Two biochemicals from the hypothalamus control its secretion. They are released alternately, exerting opposite effects. *Growth hormone-releasing hormone* (GHRH) stimulates secretion of GH, and *somatostatin* (SS) inhibits secretion.

Nutritional state can affect control of GH. More GH is released during periods of protein deficiency and abnormally low blood glucose concentration. Conversely, when blood protein and glucose concentrations increase, growth hormone secretion decreases. Apparently the hypothalamus can sense changes in the concentrations of certain blood nutrients and it releases GHRH in response to some of them.

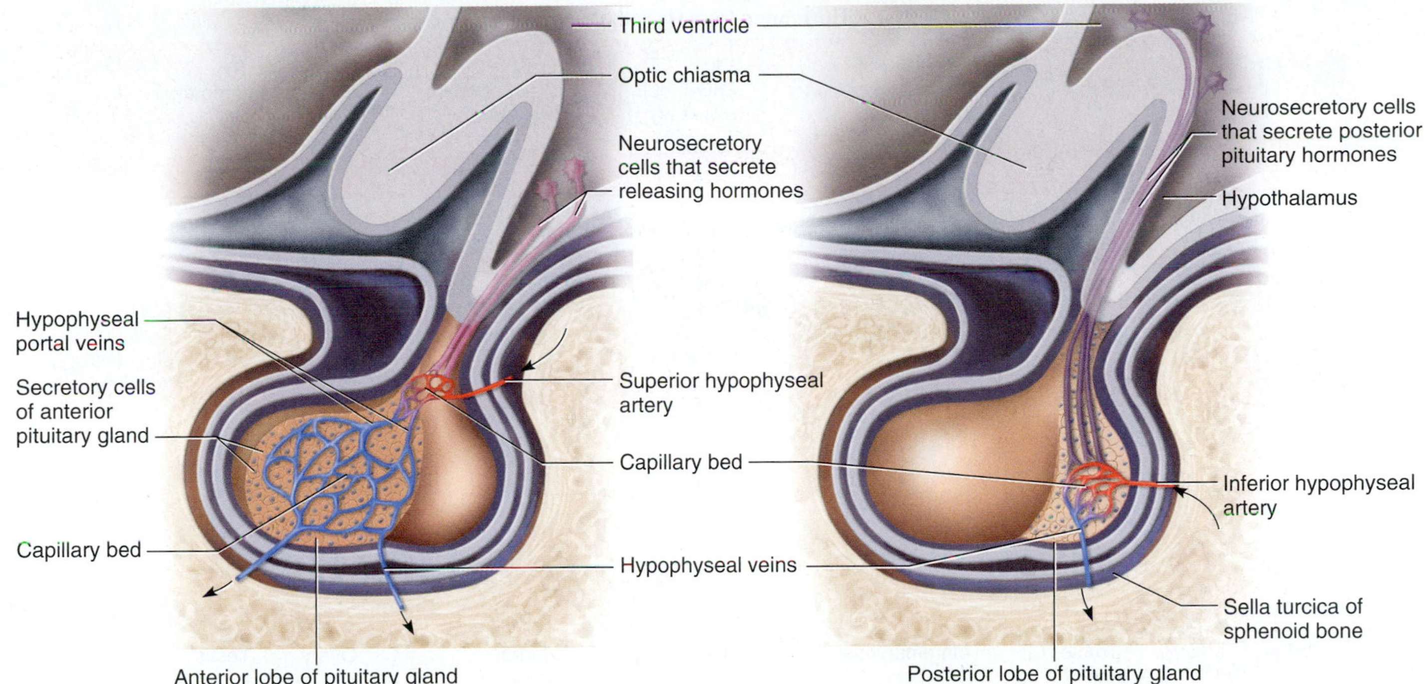

FIGURE 13.12 Hypothalamic releasing hormones stimulate cells of the anterior lobe to secrete hormones. Nerve impulses originating in the hypothalamus stimulate nerve endings in the posterior lobe of the pituitary gland to release hormones.

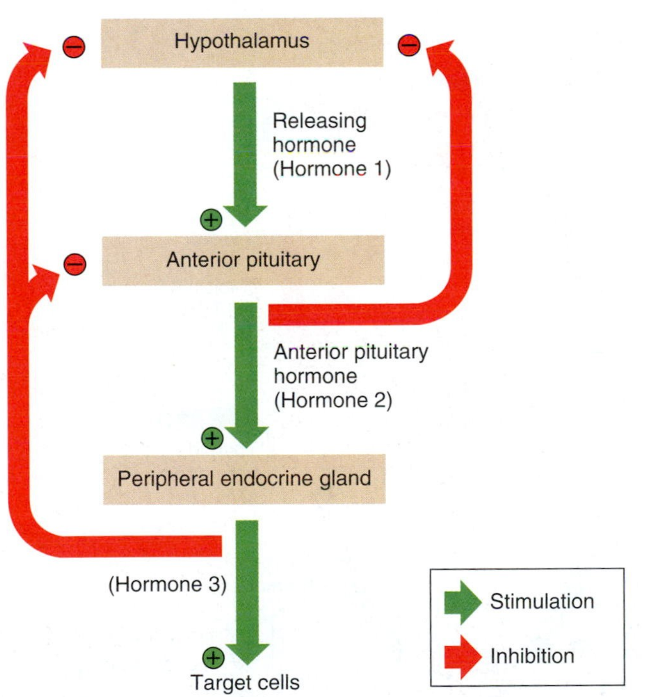

FIGURE 13.13 Hypothalamic control of the peripheral endocrine glands may use as many as three types of hormones, with multiple negative feedback controls. (⊕ = stimulation; ⊖ = inhibition)

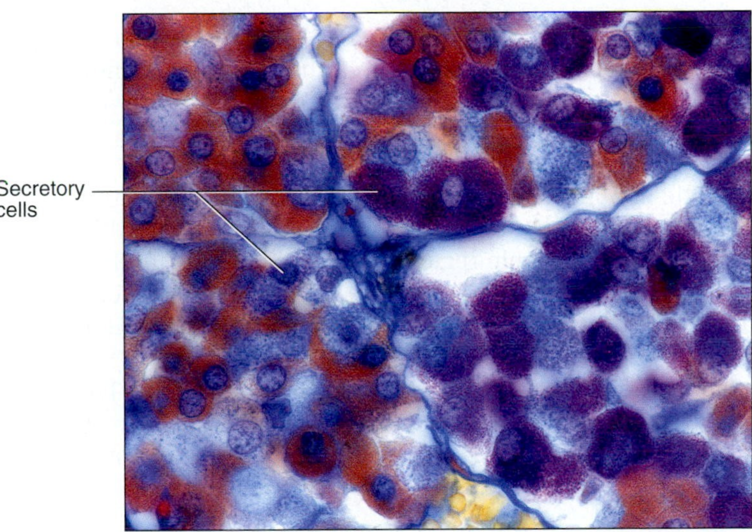

FIGURE 13.14 Light micrograph of the anterior pituitary gland (240×).

Growth hormone can stimulate elongation of bone tissue directly, but its effect on cartilage requires a mediator substance, insulin-like growth factor-1 (IGF-1). Growth hormone releases IGF-1 from the liver and other tissues. Clinical Application 13.2 discusses some clinical uses of growth hormone.

Prolactin is a protein, and as its name suggests, it promotes milk production. No normal physiological role in human males has been firmly established, although prolactin may help maintain normal sperm production. In contrast, abnormally elevated levels of the hormone can disrupt sexual function in both sexes.

Prolactin secretion is mostly under inhibitory control by dopamine from the hypothalamus, also called *prolactin-release inhibiting hormone* (PIH). The hypothalamus likely releases more than one *prolactin-releasing factor* (PRF).

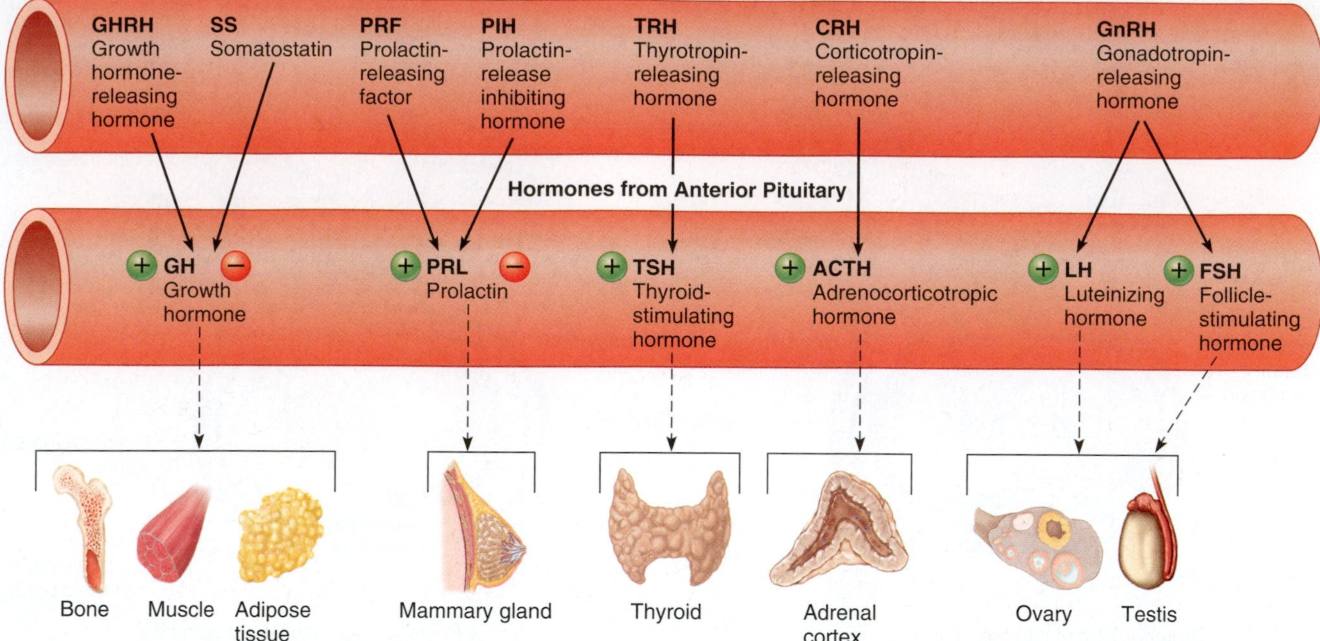

FIGURE 13.15 Hormones released from the hypothalamus, the corresponding hormones released from the anterior lobe of the pituitary gland, and their target organs.

Thyroid-stimulating hormone, also called *thyrotropin*, is a glycoprotein. It controls secretion of certain hormones from the thyroid gland. TSH can also stimulate growth of the gland, and abnormally high TSH levels may lead to an enlarged thyroid gland, or *goiter*.

The hypothalamus partially regulates TSH secretion by producing thyrotropin-releasing hormone (TRH). Circulating thyroid hormones help regulate TSH secretion by inhibiting release of TRH and TSH; therefore, as the blood concentration of thyroid hormones increases, secretion of TRH and TSH decline (fig. 13.16).

External factors influence release of TRH and TSH. These include exposure to extreme cold, accompanied by increased hormonal secretion and emotional stress, which can either increase or decrease hormonal secretion, depending upon circumstances.

PRACTICE

15 How does growth hormone affect the cellular metabolism of carbohydrates, fats, and proteins?

16 What are the functions of prolactin?

17 How is TSH secretion regulated?

Adrenocorticotropic (ah-dre″no-ka″te-ko-trōp′ik) **hormone** is a peptide that controls the manufacture and secretion of certain hormones from the outer layer (cortex) of the adrenal gland. The secretion of ACTH is regulated in part by *corticotropin-releasing hormone* (CRH), which the hypothalamus releases in response to decreased concentrations of

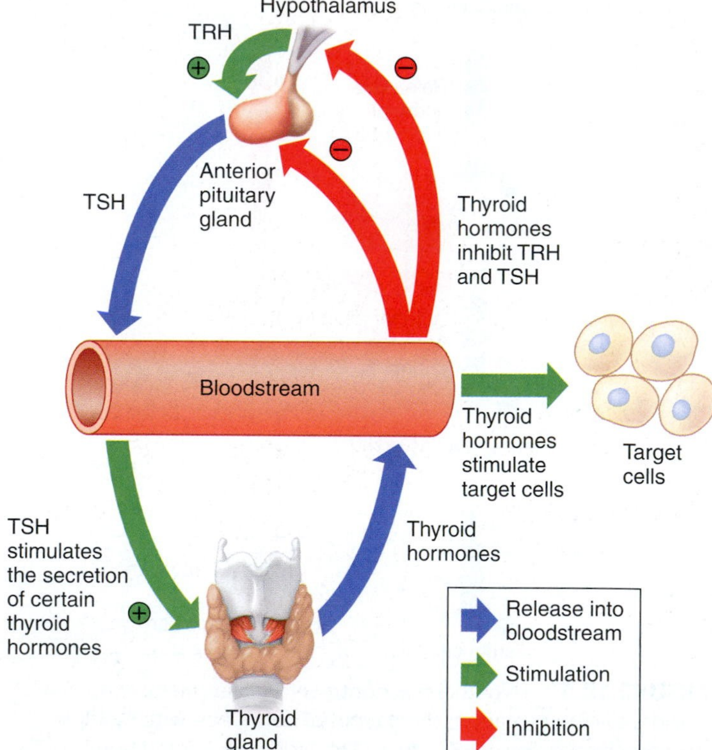

FIGURE 13.16 Thyrotropin-releasing hormone (TRH) from the hypothalamus stimulates the anterior pituitary gland to release thyroid-stimulating hormone (TSH), which stimulates the thyroid gland to release hormones. These thyroid hormones reduce the secretion of TSH and TRH by negative feedback. (+ = stimulation; − = inhibition)

13.2 CLINICAL APPLICATION

Growth Hormone Ups and Downs

Insufficient secretion of human growth hormone (HGH) during childhood produces *hypopituitary dwarfism.* Body proportions and mental development are normal, but because secretion of other anterior pituitary hormones is also below normal, additional hormone deficiency symptoms may appear. For example, a child with this condition often fails to develop adult sexual features unless he or she receives hormone therapy.

HGH can treat hypopituitary dwarfism if administration begins before the bones completely ossify. It is also used to treat various conditions that include very short stature but in which HGH is not deficient. These include chronic renal failure, Turner syndrome, intrauterine growth retardation, and Prader-Willi syndrome, as well as some cases of extreme short stature for which no cause is known. HGH may also delay muscle wasting in people who have AIDS.

The reputation of HGH as an antiaging agent stems from a 1990 study in which a dozen men over age sixty who received the hormone showed slight improvements in muscle mass and bone mineral density. The media interpreted this tiny study to indicate that HGH reverses aging, fueling the many websites that attempt to sell the hormone, in various guises, as a treatment against the ravages of time. A more recent meta-analysis—a study of studies—affirmed that HGH in older individuals can both increase muscle mass and decrease fat. It did not, however, improve strength, so may have merely replaced some fat with water. Adverse effects that can occur with excess HGH include joint pain and swelling and increased diabetes mellitus risk.

Oversecretion of growth hormone in childhood may result in *gigantism,* in which height may eventually exceed 8 feet. Gigantism is usually caused by a tumor of the pituitary gland, which secretes excess pituitary hormones including GH. As a result, a person with gigantism may have other metabolic disturbances.

If growth hormone is oversecreted in an adult after the epiphyses of the long bones have ossified, the person does not grow taller. The soft tissues, however, continue to enlarge and the bones thicken, producing a large tongue, nose, hands and feet, and a protruding jaw. This condition, acromegaly, is also associated with a pituitary tumor (fig. 13A). ■

(a)

(b)

(c)

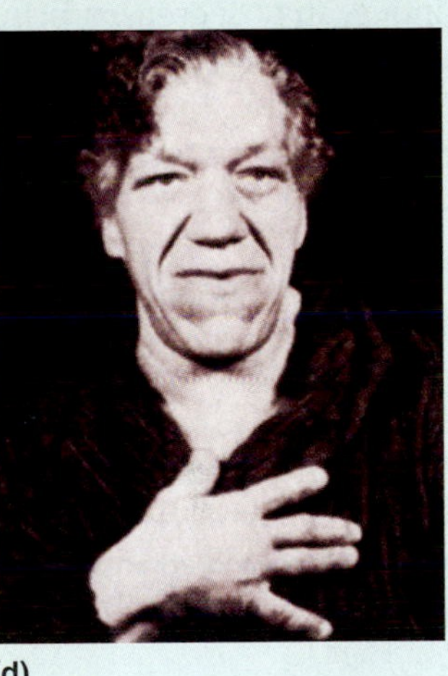
(d)

FIGURE 13A Natural oversecretion of growth hormone in adulthood causes acromegaly. Note the changes in this woman's facial features at ages (*a*) nine, (*b*) sixteen, (*c*) thirty-three, and (*d*) fifty-two. Abuse of growth hormone (as a drug) can cause acromegaly.

adrenal cortical hormones. Stress can increase secretion of ACTH by stimulating release of CRH.

Both **follicle-stimulating hormone** and **luteinizing** (lu′te′in-īz″ing) **hormone** are glycoproteins and are called *gonadotropins,* which means they act on the gonads or reproductive organs. FSH controls growth and development of follicles that house egg cells in the ovaries. It also stimulates the follicular cells to secrete a group of female sex hormones, collectively called *estrogen* (or estrogens).

In males, FSH stimulates the production of sperm cells in the testes. LH promotes secretion of sex hormones in both males and females and is essential for release of egg cells from the ovaries. Other functions of the gonadotropins and their interactions are discussed in chapter 22.

The mechanism that regulates secretion of gonadotropins is not well understood. However, starting at puberty, the hypothalamus secretes a *gonadotropin-releasing hormone* (GnRH). Gonadotropins are absent in the body fluids of infants and children.

PRACTICE 

18 What is the function of ACTH?

19 Describe the functions of FSH and LH in a male and in a female.

20 What is a gonadotropin?

Posterior Pituitary Hormones

Unlike the anterior lobe of the pituitary gland, which is primarily composed of glandular epithelial cells, the posterior lobe largely consists of nerve fibers and neuroglia (*pituicytes*). The neuroglia support the nerve fibers that originate in the hypothalamus. The hypothalamic cells that give rise to these fibers are called neurosecretory cells because their secretions function not as neurotransmitters but as hormones.

Specialized neurons in the hypothalamus produce the two hormones associated with the posterior pituitary—**antidiuretic** (an″tĭ-di″u-ret′ik) **hormone** (also known as *vasopressin*) and **oxytocin** (ok″sĭ-to′sin) (see fig. 13.12). These hormones travel down axons through the pituitary stalk to the posterior pituitary and are stored in vesicles (secretory granules) near the ends of the axons. The hormones are released into the blood in response to nerve impulses coming from the hypothalamus. Thus, posterior pituitary hormones are synthesized in the hypothalamus, but they are named for where they enter the bloodstream.

Antidiuretic hormone and oxytocin are short polypeptides with similar sequences (fig. 13.17). A *diuretic* is a chemical that increases urine production. An *antidiuretic*, then, is a chemical that decreases urine formation. ADH produces its antidiuretic effect by reducing the volume of water that the kidneys excrete. In this way, ADH plays an important role in regulating the concentration of body fluids (see chapter 20, pp. 794–795).

Frequent and copious urination often follows drinking alcoholic beverages. This is because alcohol (ethyl alcohol) inhibits ADH secretion. A person must replace the lost body fluid to maintain normal water balance. Although it seems counterintuitive, drinking too much beer can lead to dehydration because the body loses more fluid than it replaces.

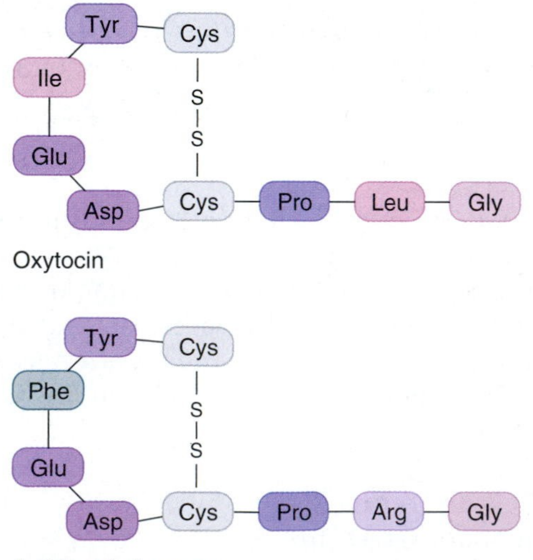

FIGURE 13.17 The structure of oxytocin differs from that of ADH by only two amino acids, yet they function differently.

Sufficient concentration of ADH contracts certain smooth muscles, including those in the walls of blood vessels. As a result, vascular resistance and blood pressure may increase. (This is why ADH is also called vasopressin.) Although ADH is seldom abundant enough to cause high blood pressure, its secretion increases following severe blood loss. In this situation, ADH's vasopressor effect may help to minimize the drop in blood pressure that results from profuse bleeding and return blood pressure toward normal.

ADH's two effects—vasoconstriction and water retention—are possible because the hormone binds two different receptors on target cells. The binding of ADH to V1 receptors increases the concentration of the second messenger inositol triphosphate, which increases intracellular calcium ion concentration, leading to vasoconstriction. The second receptor, V2, is on parts of the kidneys' microscopic tubules called collecting ducts. ADH binding there activates the cAMP second messenger system, which ultimately causes collecting duct cells to reabsorb water that would otherwise be excreted as urine.

The hypothalamus regulates secretion of ADH. Certain neurons in this part of the brain, called *osmoreceptors*, sense changes in the concentration of body fluids. For example, if a person is dehydrating due to a lack of water intake, the solutes in blood become more concentrated. The osmoreceptors, sensing the resulting increase in osmotic pressure, signal the posterior pituitary to release ADH, which causes the kidneys to retain water.

On the other hand, if a person drinks a large volume of water, body fluids become more dilute, which inhibits the release of ADH. Consequently, the kidneys excrete more dilute urine until the concentration of body fluids returns to normal.

Blood volume also affects ADH secretion. Increased blood volume stretches the walls of certain blood vessels, stimulating volume receptors that signal the hypothalamus to inhibit release of ADH. However, if hemorrhage decreases blood volume, these receptors are stretched less and therefore send fewer inhibiting impulses. As a result, ADH secretion increases, and as before, ADH causes the kidneys to conserve water. This helps prevent further volume loss.

The baby first displayed symptoms at five months of age—he drank huge volumes of water. By thirteen months, he had become severely dehydrated, although he drank nearly continuously. His parents were constantly changing his wet diapers. Doctors finally diagnosed a form of *diabetes insipidus,* which impairs ADH regulation of water balance. The boy was drinking sufficient fluids, but his kidneys could not retain the water. ADH V2 receptors on the kidney collecting ducts were defective. The hormone could bind, but the receptor failed to trigger cAMP formation. The boy's ADH was still able to constrict blood vessels because the V1 receptors were unaffected. A high-calorie diet and providing lots of water preserved the boy's mental abilities, but he remained small for his age. Tumors and injury affecting the hypothalamus and posterior pituitary can also cause diabetes insipidus.

Oxytocin also has an antidiuretic action, but less so than ADH. In addition, oxytocin can contract smooth muscles in the uterine wall, playing a role in the later stages of childbirth. The uterus becomes more sensitive to oxytocin's effects during pregnancy. Stretching of uterine and vaginal tissues late in pregnancy, caused by the growing fetus, initiates nerve impulses to the hypothalamus, which then signals the posterior pituitary to release oxytocin, which, in turn, stimulates the uterine contractions of labor.

In the breasts, oxytocin contracts certain cells near the milk-producing glands and their ducts. In lactating breasts, this action forces liquid from the milk glands into the milk ducts and ejects the milk.

The mechanical stimulation of suckling initiates nerve impulses that travel to the mother's hypothalamus, which responds by signaling the posterior pituitary to release oxytocin, which, in turn, stimulates milk release. Thus, milk is normally not ejected from the milk glands and ducts until the baby suckles. The fact that milk is ejected from both breasts in response to suckling is a reminder that all target cells respond to a hormone.

> If the uterus is not sufficiently contracting to expel a fully developed fetus, oxytocin is sometimes given intravenously to stimulate uterine contractions, inducing labor. Oxytocin is also administered to the mother following childbirth to ensure that the uterine muscles contract enough to squeeze broken blood vessels closed, minimizing the danger of hemorrhage.

Oxytocin has no established function in males, although it is present in the male posterior pituitary. There is evidence that it may stimulate the movement of certain fluids in the male reproductive tract during sexual activity. Table 13.6 reviews the hormones of the pituitary gland.

PRACTICE

21 What is the function of ADH?

22 How is the secretion of ADH controlled?

23 What effects does oxytocin produce in females?

13.6 THYROID GLAND

The **thyroid gland** (thi'roid gland), as figure 13.18 shows, is a vascular structure that consists of two large lateral lobes connected by a broad **isthmus** (is'mus). The thyroid lies just below the larynx on either side and anterior to the trachea. The gland is specialized to remove iodine from the blood.

Structure of the Gland

A capsule of connective tissue covers the thyroid gland, which is made up of many secretory parts called *follicles*. The cavities in these follicles are lined with a single layer of cuboidal epithelial cells and are filled with a clear viscous *colloid*, which consists primarily of a glycoprotein called *thyroglobulin*. The follicular cells produce and secrete hormones that may either be stored in the colloid or released into nearby capillaries (fig. 13.19). Other hormone-secreting cells, called *extrafollicular cells* (C cells), lie outside the follicles.

TABLE 13.6 | Hormones of the Pituitary Gland

Anterior Lobe Hormone	Action	Source of Control
Growth hormone (GH)	Stimulates increase in size and rate of division of body cells; enhances movement of amino acids through membranes; promotes growth of long bones	Secretion inhibited by somatostatin (SS) and stimulated by growth hormone-releasing hormone (GHRH) from the hypothalamus
Prolactin (PRL)	Sustains milk production after birth; amplifies effect of LH in males	Secretion inhibited by prolactin-release inhibiting hormone (PIH) and may be stimulated by yet to be identified prolactin-releasing factor (PRF) from the hypothalamus
Thyroid-stimulating hormone (TSH)	Controls secretion of hormones from the thyroid gland	Thyrotropin-releasing hormone (TRH) from the hypothalamus
Adrenocorticotropic hormone (ACTH)	Controls secretion of certain hormones from the adrenal cortex	Corticotropin-releasing hormone (CRH) from the hypothalamus
Follicle-stimulating hormone (FSH)	Development of egg-containing follicles in ovaries; stimulates follicular cells to secrete estrogen; in males, stimulates production of sperm cells	Gonadotropin-releasing hormone (GnRH) from the hypothalamus
Luteinizing hormone (LH)	Promotes secretion of sex hormones; releases egg cell in females	Gonadotropin-releasing hormone (GnRH) from the hypothalamus
Posterior Lobe Hormone	**Action**	**Source of Control**
Antidiuretic hormone (ADH)	Causes kidneys to reduce water excretion; in high concentration, raises blood pressure	Hypothalamus in response to changes in blood water concentration and blood volume
Oxytocin (OT)	Contracts muscles in uterine wall and those associated with milk-secreting glands	Hypothalamus in response to stretch in uterine and vaginal walls and stimulation of breasts

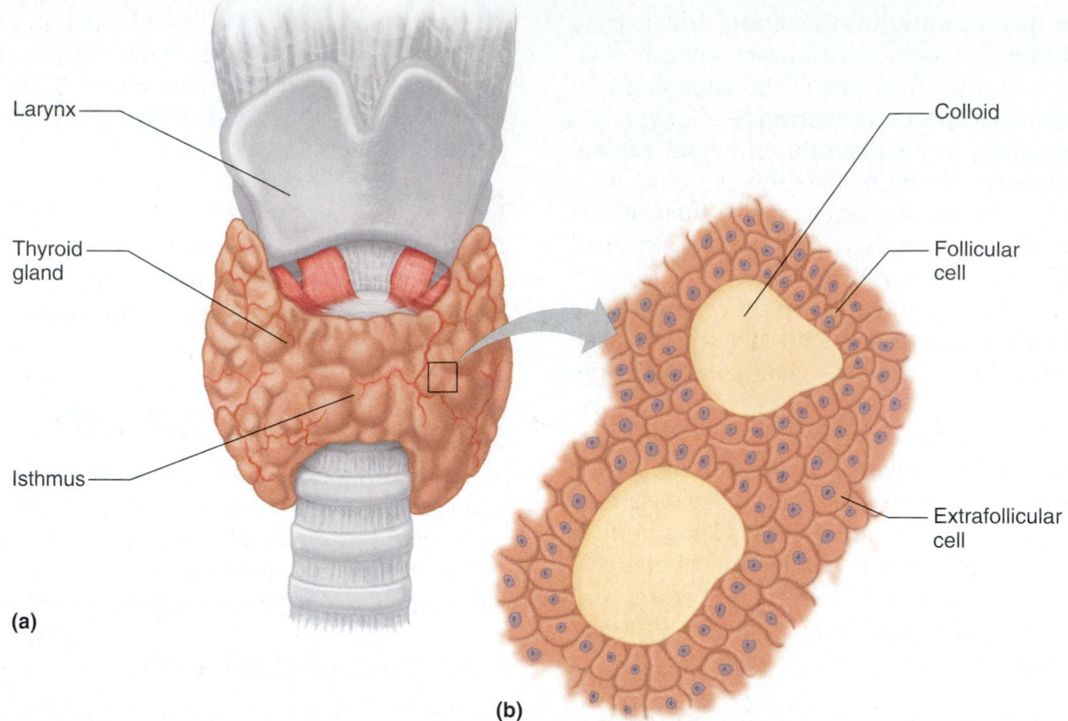

FIGURE 13.18 Thyroid gland. (*a*) The thyroid gland consists of two lobes connected anteriorly by an isthmus. (*b*) Follicular cells secrete thyroid hormones.

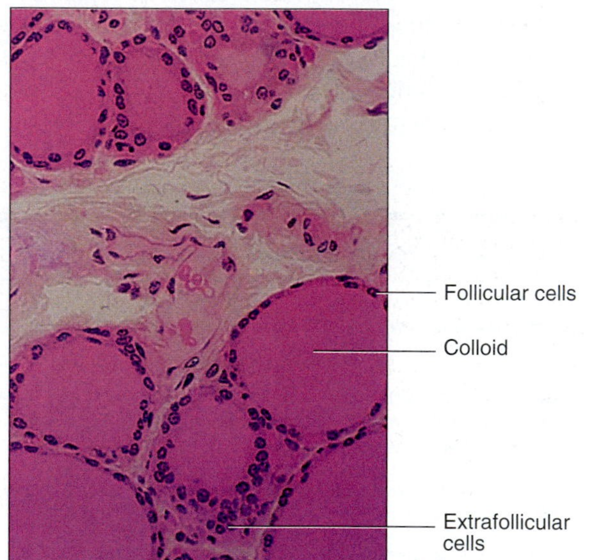

FIGURE 13.19 A light micrograph of thyroid gland tissue (240×). The open spaces that follicular cells surround are filled with colloid.

Thyroid Hormones

The thyroid gland produces three important hormones. The follicular cells synthesize two of these, which have marked effects on the metabolic rates of body cells. The extrafollicular cells produce the third type of hormone, which influences blood concentrations of calcium and phosphate ions.

The two important thyroid hormones that affect cellular metabolic rates are **thyroxine** (thi-rok′sin), or tetraiodothyronine (also called T_4 because it includes four atoms of iodine), and **triiodothyronine** (tri″i-o″do-thi′ro-nēn), also called T_3 because it includes three atoms of iodine. These hormones help regulate the metabolism of carbohydrates, lipids, and proteins. Specifically, thyroxine and triiodothyronine increase the rate at which cells release energy from carbohydrates, enhance the rate of protein synthesis, and stimulate breakdown and mobilization of lipids. These hormones are the major factors determining how many calories the body must consume at rest to maintain life, measured as the *basal metabolic rate* (BMR). They are essential for normal growth and development and for maturation of the nervous system (fig. 13.20). TSH from the anterior pituitary gland controls levels of thyroid hormones.

Follicular cells require iodine salts (iodides) to produce thyroxine and triiodothyronine. Such salts are normally obtained from foods, and after they have been absorbed from the intestine, the blood carries some of them in the form of iodide (I^-) to the thyroid gland. An efficient active transport protein called the *iodide pump* moves the iodides into the follicular cells, where they are converted to iodine and concentrated. The iodine, with the amino acid tyrosine, is used to synthesize these thyroid hormones.

Follicular cells synthesize thyroglobulin, whose protein portion includes molecules of tyrosine, many of which have already had iodine attached by an enzymatic reaction. As the thyroglobulin protein twists and coils into its tertiary structure, bonds form between some of the tyrosine molecules, creating potential thyroid hormones waiting to be released. The follicular cells take up molecules of thyroglobulin by

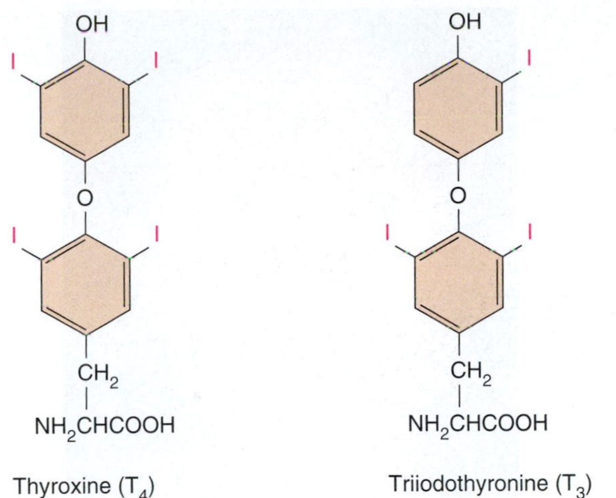

FIGURE 13.20 The hormones thyroxine and triiodothyronine have very similar molecular structures.

endocytosis, break down the protein, and release the individual thyroid hormones into the bloodstream. When the thyroid hormone levels in the bloodstream drop below a certain level, this process occurs more rapidly, returning thyroid hormone levels to normal.

Once in the blood, thyroid hormones combine with blood proteins (alpha globulins) and are transported to body cells. About a third of T_4 is converted to T_3 in peripheral tissues. Triiodothyronine is nearly five times more potent, but thyroxine accounts for at least 95% of circulating thyroid hormones.

The thyroid gland produces **calcitonin,** which is usually not referred to as a thyroid hormone because it is synthesized by the C cells, distinct from the gland's follicles. Calcitonin plays a role in the control of blood calcium and phosphate ion concentrations. It helps lower concentrations of calcium and phosphate ions by decreasing the rate at which they leave the bones and enter extracellular fluids by inhibiting the bone-destroying activity of osteoclasts. At the same time, calcitonin increases the rate at which calcium and phosphate ions are deposited in bone matrix by stimulating activity of osteoblasts (see chapter 7, p. 201). Calcitonin also increases the excretion of calcium ions and phosphate ions by the kidneys.

Calcitonin secretion is stimulated by a high blood calcium ion concentration, as may occur following absorption of calcium ions from a recent meal. Certain hormones also prompt its secretion, such as gastrin, released from active digestive organs. Calcitonin helps prevent prolonged elevation of blood calcium ion concentration after eating.

Research suggests that calcitonin may be most important during early growth and physiological stress. For example, in the young, calcitonin stimulates the increase in bone deposition associated with growth. In females, its actions help protect bones from resorption during pregnancy and lactation, when calcium is needed for growth of the fetus and synthesis of breast milk.

Table 13.7 summarizes the actions and sources of control of the thyroid hormones. From Science to Technology 2.1 (p. 54), table 13.8, and figures 13.21, 13.22, and 13.23 discuss disorders of the thyroid gland.

TABLE 13.7 | Hormones of the Thyroid Gland

Hormone	Action	Source of Control
Thyroxine (T_4)	Increases rate of energy release from carbohydrates; increases rate of protein synthesis; accelerates growth; stimulates activity in the nervous system	TSH from the anterior pituitary gland
Triiodothyronine (T_3)	Same as above, but five times more potent than thyroxine	Same as above
Calcitonin	Lowers blood calcium and phosphate ion concentrations by inhibiting release of calcium and phosphate ions from bones and by increasing the rate at which calcium and phosphate ions are deposited in bones; increases excretion of calcium by the kidneys	Elevated blood calcium ion concentration, digestive hormones

TABLE 13.8 | Disorders of the Thyroid Gland

Condition	Mechanism/Symptoms
Hyperthyroid	
Hyperthyroidism	High metabolic rate, sensitivity to heat, restlessness, hyperactivity, weight loss, protruding eyes, goiter
Graves disease	Autoantibodies (against self) bind TSH receptors on thyroid cell membranes, mimicking action of TSH, overstimulating gland (hyperthyroidism); exopthalmia (protrusion of the eyes) and goiter
Hypothyroid	
Hashimoto disease	Autoantibodies (against self) attack thyroid cells, resulting in hypothyroidism
Hypothyroidism (infantile)	Cretinism—stunted growth, abnormal bone formation, mental retardation, low body temperature, sluggishness
Hypothyroidism (adult)	Myxedema—low metabolic rate, sensitivity to cold, sluggishness, poor appetite, swollen tissues, mental dullness
Simple goiter	Deficiency of thyroid hormones due to iodine deficiency; because no thyroid hormones inhibit pituitary release of TSH, thyroid is overstimulated and enlarges but functions below normal (hypothyroidism)

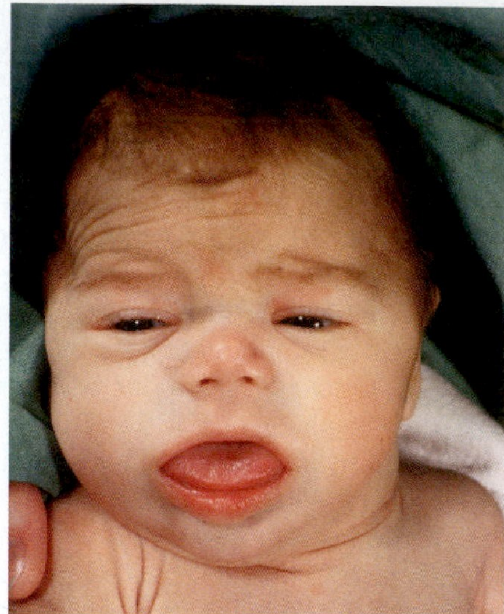

FIGURE 13.21 Cretinism is due to an underactive thyroid gland during infancy and childhood.

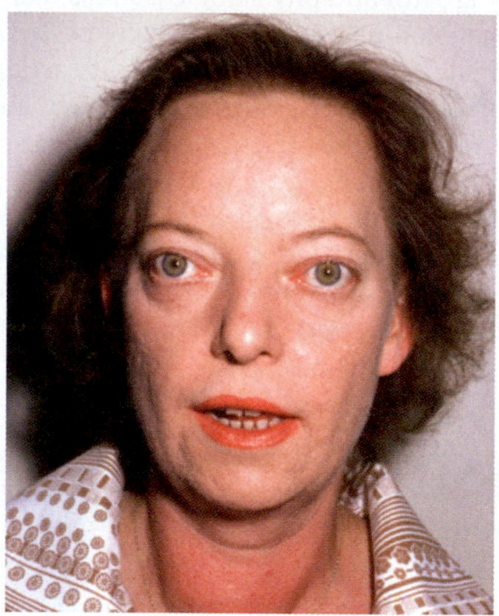

FIGURE 13.22 Graves disease may cause the eyes to protrude (exopthalmia).

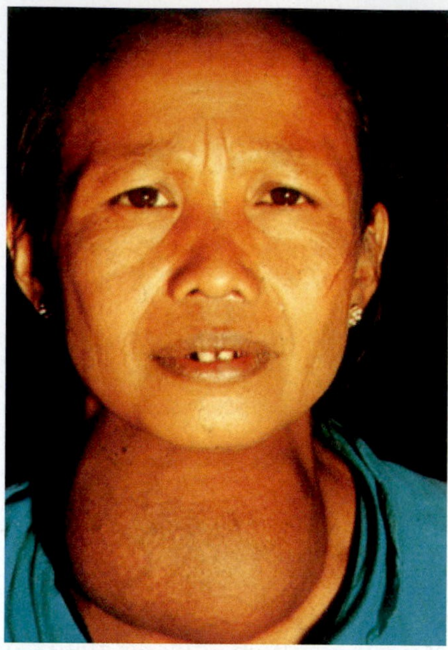

FIGURE 13.23 An iodine deficiency causes simple (endemic) goiter and results in high levels of TSH.

PRACTICE

24 Where is the thyroid gland located?

25 Which hormones of the thyroid gland affect carbohydrate metabolism, the mobilization of lipids, and protein synthesis?

26 What substance is essential for the production of thyroxine and triiodothyronine?

27 How does calcitonin influence the concentrations of blood calcium and phosphate ions?

13.7 PARATHYROID GLANDS

The **parathyroid glands** (par″ah-thi′roid glandz) are on the posterior surface of the thyroid gland, as figure 13.24 shows. Usually there are four of them—a superior and an inferior gland associated with each of the thyroid's lateral lobes. The parathyroid glands secrete a hormone that regulates the concentrations of calcium and phosphate ions in the blood.

Structure of the Glands

Each parathyroid gland is a small, yellowish brown structure covered by a thin capsule of connective tissue. The body of the gland consists of many tightly packed secretory cells closely associated with capillary networks (fig. 13.25).

Parathyroid Hormone

The parathyroid glands secrete a protein, **parathyroid hormone** (PTH), or *parathormone* (see fig. 13.4*c*). This hormone increases blood calcium ion concentration and decreases blood phosphate ion concentration through actions in the bones, kidneys, and intestines.

The extracellular matrix of bone tissue contains a considerable amount of calcium phosphate and calcium carbonate. PTH stimulates bone resorption by osteoclasts and inhibits the activity of osteoblasts (see chapter 7, p. 201). As bone resorption increases, calcium and phosphate ions are released into the blood. At the same time, PTH causes the kidneys to conserve blood calcium ions and to excrete more phosphate ions in the urine. PTH also indirectly stimulates absorption of calcium ions from food in the intestine by influencing metabolism of vitamin D.

Vitamin D (cholecalciferol) is synthesized from dietary cholesterol, which intestinal enzymes convert into provitamin D (7-dehydrocholesterol). This provitamin is largely stored in the skin, and exposure to the ultraviolet wavelengths of sunlight changes it to vitamin D. Some vitamin D also comes from foods.

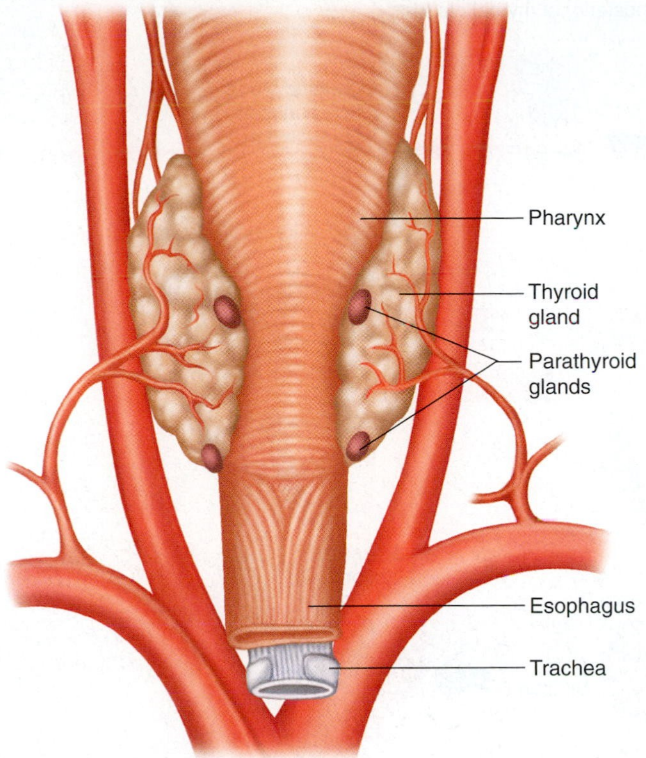

FIGURE 13.24 The parathyroid glands are embedded in the posterior surface of the thyroid gland.

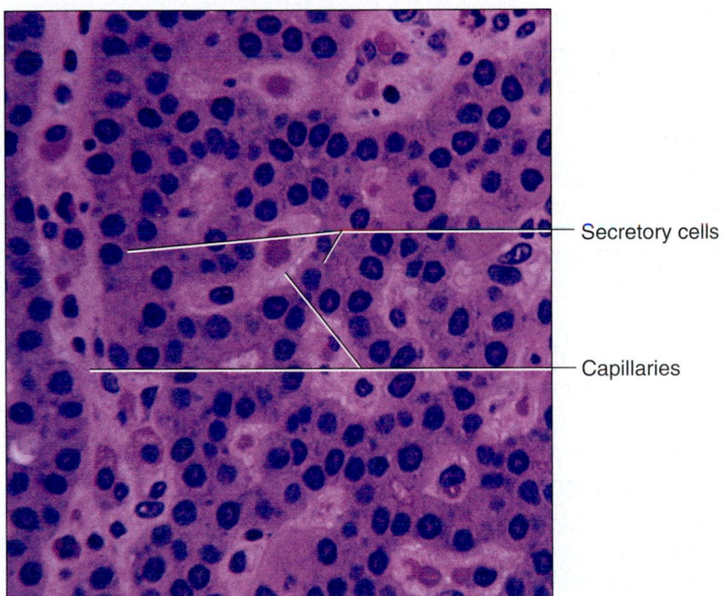

FIGURE 13.25 Light micrograph of the parathyroid gland (540×).

The liver changes vitamin D to hydroxycholecalciferol, which is carried in the bloodstream or stored in tissues. When PTH is present, hydroxycholecalciferol can be changed in the kidneys into an active form of vitamin D (dihydroxycholecalciferol), which controls absorption of calcium ions from the intestine (fig. 13.26).

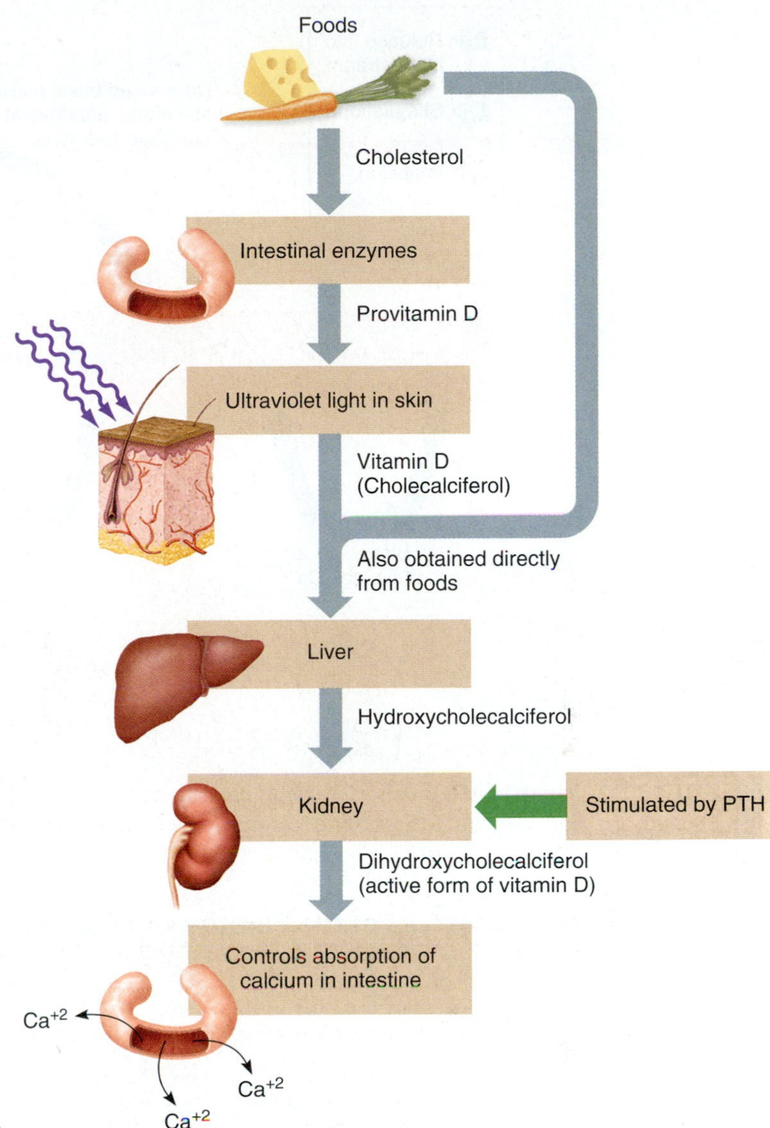

FIGURE 13.26 Mechanism by which PTH promotes calcium absorption in the intestine.

A negative feedback mechanism operating between the parathyroid glands and the blood calcium ion concentration regulates secretion of PTH (fig. 13.27). As the concentration of blood calcium ions rises, less PTH is secreted; as the concentration of blood calcium ions drops, more PTH is released.

The opposite effects of calcitonin and PTH maintain calcium ion homeostasis. This is important in a number of physiological processes. For example, as the blood calcium ion concentration drops (hypocalcemia), the nervous system becomes abnormally excitable, and impulses may be triggered spontaneously. As a result, muscles, including the respiratory muscles, may undergo tetanic contractions, and the person may suffocate. In contrast, an abnormally high concentration of blood calcium ions (hypercalcemia) depresses the nervous system. Consequently, muscle contractions are weak, and reflexes are sluggish. Table 13.9 lists parathyroid disorders.

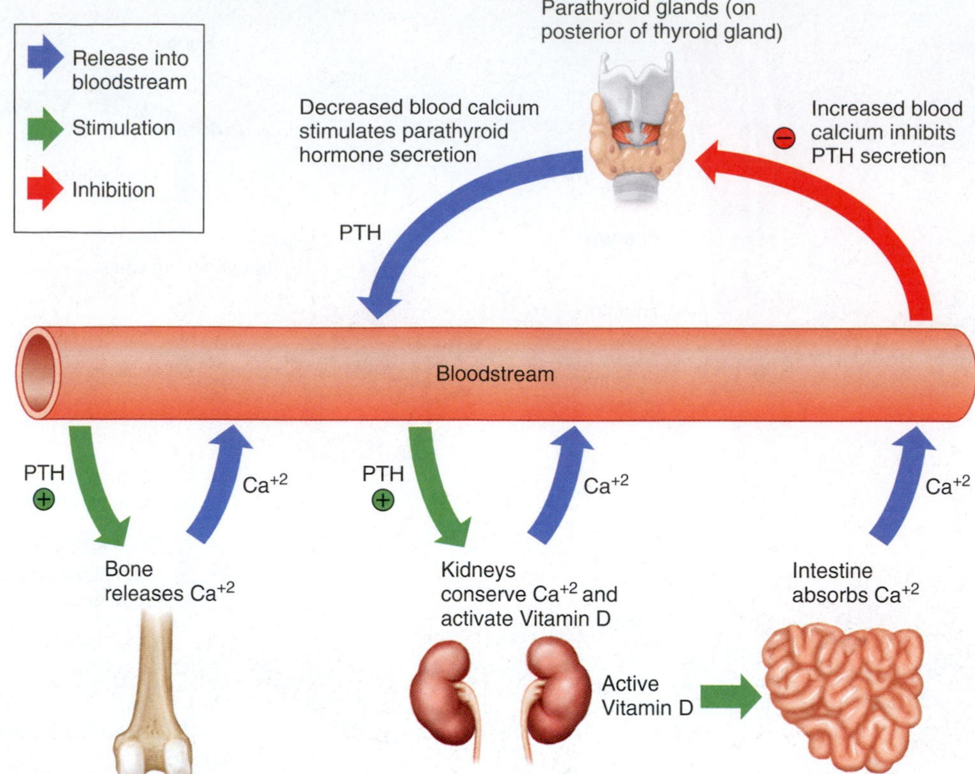

FIGURE 13.27 Parathyroid hormone (PTH) stimulates bone to release calcium (Ca^{+2}) and the kidneys to conserve calcium. It indirectly stimulates the intestine to absorb calcium. The resulting increase in blood calcium concentration inhibits secretion of PTH by negative feedback. (⊕ = stimulation; ⊖ = inhibition)

TABLE 13.9 | Disorders of the Parathyroid Glands

Condition	Symptoms/Mechanism	Cause	Treatment
Hyperparathyroidism	Fatigue, muscular weakness, painful joints, altered mental functions, depression, weight loss, bone weakening. Increased PTH secretion overstimulates osteoclasts.	Tumor	Remove tumor, correct bone deformities
Hypoparathyroidism	Muscle cramps and seizures. Decreased PTH secretion reduces osteoclast activity, diminishing blood calcium ion concentration.	Inadvertent surgical removal; injury	Calcium salt injections, massive doses of vitamin D

PRACTICE

28 Where are the parathyroid glands located?

29 How does parathyroid hormone help regulate the concentrations of blood calcium and phosphate ions?

30 How does the negative feedback system of the parathyroid glands differ from that of the thyroid gland?

13.8 ADRENAL GLANDS

The **adrenal glands** (suprarenal glands) are closely associated with the kidneys. A gland sits atop each kidney like a cap and is embedded in the mass of adipose tissue that encloses the kidney.

Structure of the Glands

The adrenal glands are shaped like pyramids. Each adrenal gland is vascular and consists of two parts. The central portion is the adrenal medulla, and the outer part is the adrenal cortex (fig. 13.28). These regions are not sharply divided, but they are distinct glands that secrete different hormones.

The **adrenal medulla** (ah-dre′nal me-dul′ah) consists of irregularly shaped cells grouped around blood vessels. These cells are intimately connected with the sympathetic division of the autonomic nervous system. The adrenal medullary cells are modified postganglionic neurons, and preganglionic autonomic nerve fibers lead to them directly from the central nervous system (see chapter 11, p. 424).

The **adrenal cortex** (ah-dre′nal kor′teks) makes up the bulk of the adrenal gland. It is composed of closely packed masses of epithelial layers that form an outer, a middle, and an inner zone of the cortex—the zona glomerulosa, the zona fasciculata, and the zona reticularis, respectively (figs. 13.28 and 13.29).

Hormones of the Adrenal Medulla

The cells of the adrenal medulla (chromaffin cells) produce, store, and secrete two closely related hormones, **epinephrine**

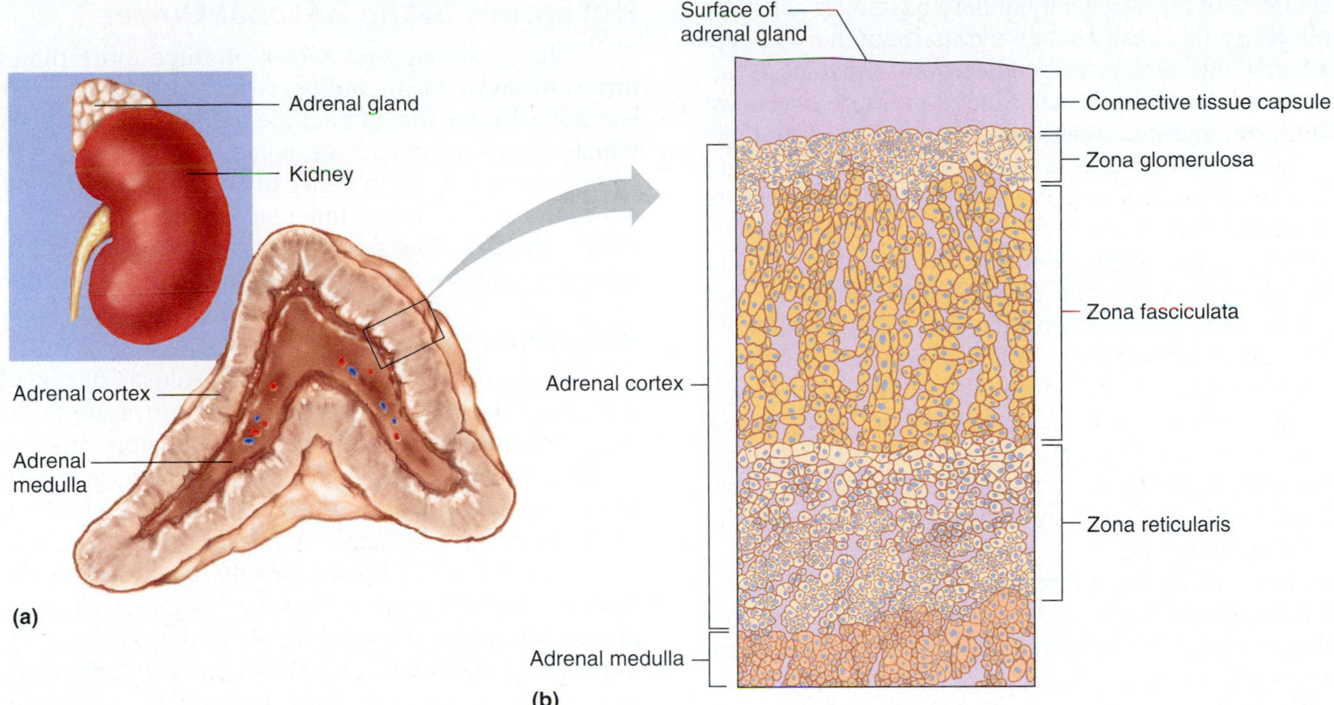

FIGURE 13.28 Adrenal glands. (*a*) An adrenal gland consists of an outer cortex and an inner medulla. (*b*) The cortex consists of three layers, or zones, of cells.

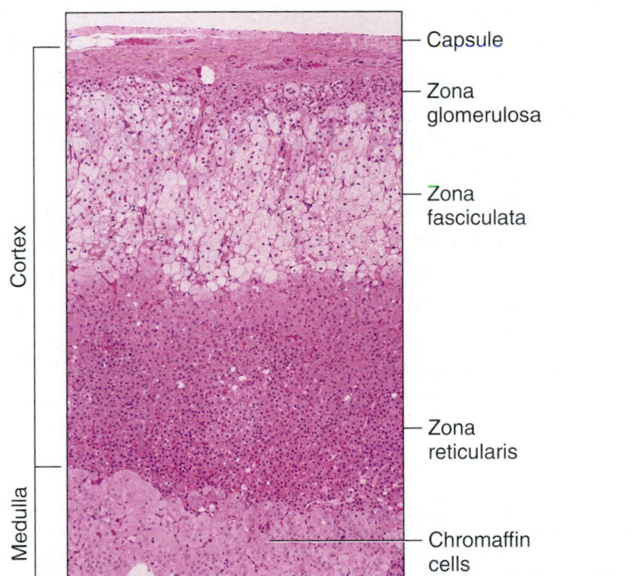

FIGURE 13.29 Light micrograph of the adrenal medulla and the adrenal cortex (75×).

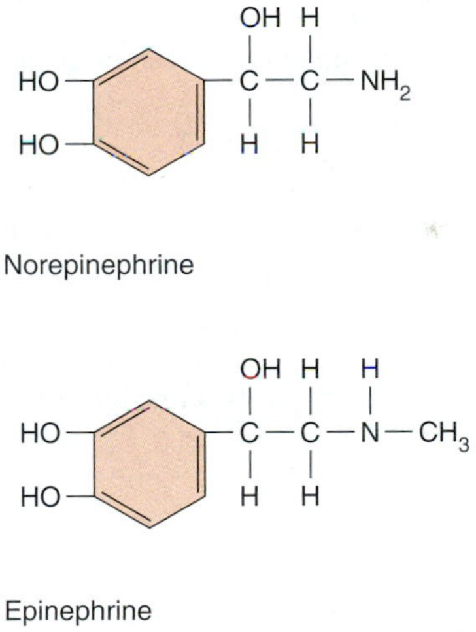

Norepinephrine

Epinephrine

FIGURE 13.30 Epinephrine and norepinephrine have similar molecular structures and similar functions.

(ep″ĭ-nef′rin), also called adrenalin and **norepinephrine** (nor″ep-ĭ-nef′rin), also called noradrenalin. Both of these substances are a type of amine called a *catecholamine,* and they have similar molecular structures and physiological functions (fig. 13.30). Epinephrine is synthesized from norepinephrine.

The synthesis of catecholamines begins with the amino acid tyrosine. In the first step of the pathway, an enzyme (tyrosine hydroxylase) in the secretory cells catalyzes a reaction that converts tyrosine into a substance called *dopa.* A second enzyme (dopa decarboxylase) catalyzes a reaction that modifies dopa into dopamine, and a third enzyme (dopamine betahydroxylase) catalyzes a reaction that alters dopamine to form norepinephrine. Still another enzyme (phenylethanolamine N-methyltransferase) then catalyzes conversion of norepinephrine to epinephrine. About 15% of the norepinephrine is stored unchanged. The hormones occupy tiny vesicles (chromaffin granules), much like neurotransmitters are stored in vesicles in neurons.

The effects of the adrenal medullary hormones generally resemble those that result when sympathetic nerve fibers stimulate their effectors: increased heart rate and force of cardiac muscle contraction, elevated blood pressure, increased breathing rate, and decreased digestive activity (see table 11.10). The hormonal effects last up to ten times longer than the neurotransmitter effects because the hormones are slowly removed from the tissues.

The ratio of the two hormones in the adrenal medullary secretion varies with different physiological conditions, but usually it is about 80% epinephrine and 20% norepinephrine. Although these hormones' effects are generally similar, certain effector cells respond differently, due to the relative numbers of alpha and beta receptors in their membranes. Both hormones can stimulate both classes of receptors, although norepinephrine has a greater effect on alpha receptors. Table 13.10 compares some of the differences in the effects of these hormones.

Impulses arriving on sympathetic nerve fibers stimulate the adrenal medulla to release its hormones at the same time sympathetic impulses stimulate other effectors. As a rule, these impulses originate in the hypothalamus in response to stress. Thus, the medullary secretions function together with the sympathetic division of the autonomic nervous system in preparing the body for energy-expending action—"fight or flight."

RECONNECT

To Chapter 11, Sympathetic Division, pages 424–426.

PRACTICE 

31 Describe the location and structure of the adrenal glands.

32 Name the hormones the adrenal medulla secretes.

33 What general effects do hormones secreted by the adrenal medulla produce?

34 What usually stimulates release of hormones from the adrenal medulla?

Hormones of the Adrenal Cortex

The cells of the adrenal cortex produce more than thirty different steroids, including several hormones (corticosteroids). Unlike the adrenal medullary hormones, without which a person can survive, some of those released by the cortex are vital. In the absence of adrenal cortical secretions, a person usually dies within a week without extensive electrolyte therapy. The most important adrenal cortical hormones are aldosterone, cortisol, and certain sex hormones.

Aldosterone

Cells in the outer zone (zona glomerulosa) of the adrenal cortex synthesize **aldosterone.** This hormone is called a *mineralocorticoid* because it helps regulate the concentration of mineral electrolytes, such as sodium and potassium ions. More specifically, aldosterone causes the kidney to conserve sodium ions and to excrete potassium ions. The cells that secrete aldosterone respond directly to changes in the composition of blood plasma. However, whereas an increase in plasma potassium strongly stimulates these cells, a decrease in plasma sodium only slightly stimulates them. Control of aldosterone secretion is indirectly linked to plasma sodium level by the **renin-angiotensin system.**

Groups of specialized kidney cells (juxtaglomerular cells) are able to respond to changes in blood pressure and the plasma sodium ion concentration. If the level of either of these factors decreases, the cells release an enzyme called **renin** (re′nin). Renin reacts with a blood protein called **angiotensinogen** (an″je-o-ten-sin′o-jen) to release a peptide called **angiotensin I.** Another enzyme (angiotensin-converting enzyme, or ACE) primarily in lung blood vessels catalyzes a reaction that converts angiotensin I into another form, **angiotensin II,** carried in the bloodstream (fig. 13.31). When angiotensin II reaches the adrenal cortex, it stimulates the release of aldosterone. ACTH is necessary for aldosterone secretion to respond to this and other stimuli.

Aldosterone, in conserving sodium ions, indirectly retains water by osmosis. This helps maintain blood sodium

TABLE 13.10 | Comparative Effects of Epinephrine and Norepinephrine

Structure or Function Affected	Epinephrine	Norepinephrine
Heart	Rate increases Force of contraction increases	Rate increases Force of contraction increases
Blood vessels	Vessels in skeletal muscle vasodilate, decreasing resistance to blood flow	Blood flow to skeletal muscles increases, resulting from constriction of blood vessels in skin and viscera
Systemic blood pressure	Some increase due to increased cardiac output	Great increase due to vasoconstriction, counteracted in muscle blood vessels during exercise
Airways	Dilated	Some dilation
Reticular formation of brain	Activated	Little effect
Liver	Promotes breakdown of glycogen to glucose, increasing blood sugar level	Little effect on blood glucose level
Metabolic rate	Increases	Increases

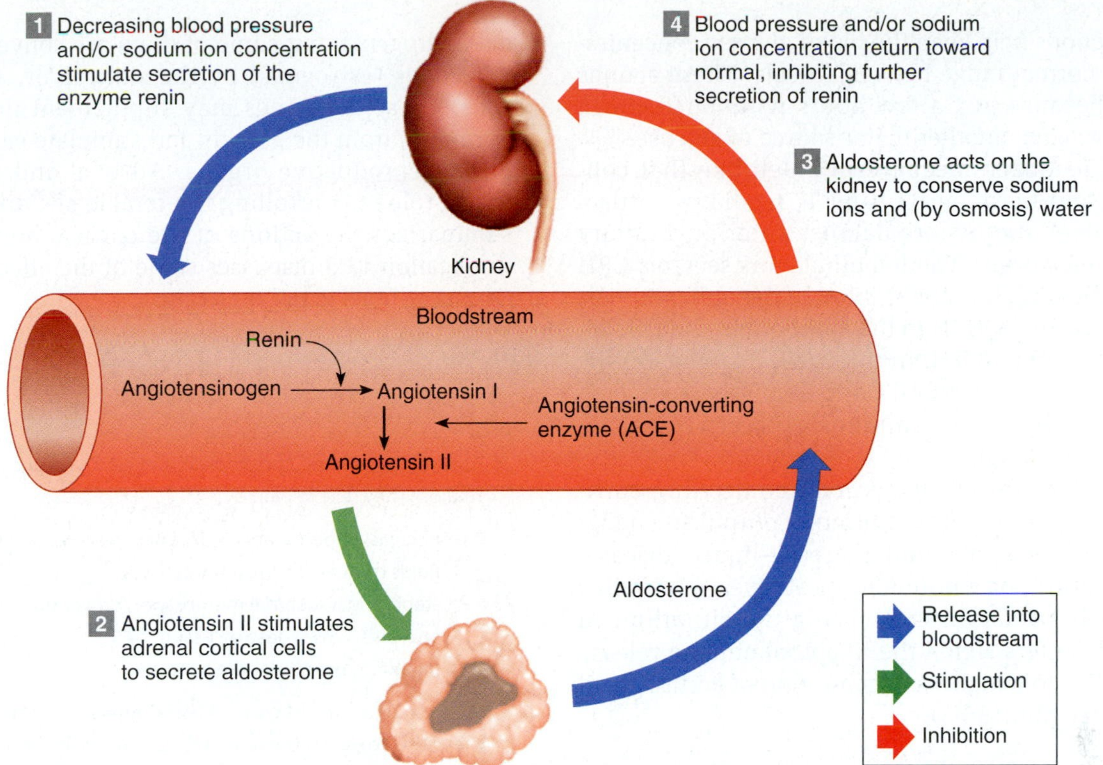
1 Decreasing blood pressure and/or sodium ion concentration stimulate secretion of the enzyme renin
4 Blood pressure and/or sodium ion concentration return toward normal, inhibiting further secretion of renin
3 Aldosterone acts on the kidney to conserve sodium ions and (by osmosis) water
Kidney
Bloodstream
Renin
Angiotensinogen
Angiotensin I
Angiotensin-converting enzyme (ACE)
Angiotensin II
Aldosterone
2 Angiotensin II stimulates adrenal cortical cells to secrete aldosterone
Release into bloodstream
Stimulation
Inhibition

FIGURE 13.31 Aldosterone increases blood volume and pressure by promoting conservation of sodium ions and water (steps *1–4*).

ion concentration and blood volume (fig. 13.31). Angiotensin II is also a powerful vasoconstrictor and thereby helps maintain systemic blood pressure.

ACE inhibitors are a class of drugs used to treat some forms of high blood pressure (hypertension). They work by competing with angiotensin-converting enzyme, blocking formation of angiotensin II and preventing inactivation of bradykinin, a vasodilator. Both effects dilate blood vessels, lowering blood pressure.

Cortisol

Cortisol (hydrocortisone) is a *glucocorticoid,* which means it affects glucose metabolism. It is produced in the middle zone (zona fasciculata) of the adrenal cortex and has a molecular structure similar to aldosterone (fig. 13.32). In addition to affecting glucose, cortisol influences protein and fat metabolism. Among the more important actions of cortisol are the following:

1. It inhibits the synthesis of protein in various tissues, increasing blood concentration of amino acids.
2. It promotes the release of fatty acids from adipose tissue, increasing the use of fatty acids and decreasing the use of glucose as energy sources.
3. It stimulates liver cells to synthesize glucose from noncarbohydrates (gluconeogenesis), such as circulating amino acids and glycerol, thus increasing blood glucose concentration.

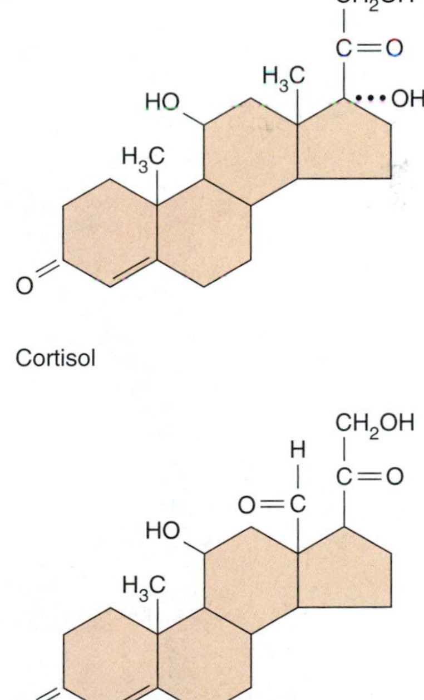
CH2OH
C=O
H3C
HO
OH
H3C
O
Cortisol
CH2OH
H
C=O
O=C
HO
H3C
O
Aldosterone

FIGURE 13.32 Cortisol and aldosterone are steroids with similar molecular structures.

Cortisol's actions help keep the blood glucose concentration within the normal range between meals. These actions are important because just a few hours without food can exhaust liver glycogen, another major source of glucose.

A negative feedback mechanism much like that controlling the thyroid hormones T_3 and T_4 regulates cortisol release. It involves the hypothalamus, anterior pituitary gland, and adrenal cortex. The hypothalamus secretes CRH (corticotropin-releasing hormone) into the hypophyseal portal veins, which carry the CRH to the anterior pituitary gland, stimulating it to secrete ACTH. In turn, ACTH stimulates the adrenal cortex to release cortisol. Cortisol inhibits release of both CRH and ACTH. As concentration of these substances falls, cortisol production drops.

The set point of the feedback loop controlling cortisol secretion changes, adapting hormone output to changing conditions. For example, under stress—injury, disease, extreme temperature, or emotional upset—nerve impulses send the brain information concerning the situation. In response, brain centers signal the hypothalamus to release more CRH, leading to a higher concentration of cortisol until the stress subsides (fig. 13.33).

Sex Hormones

Cells in the inner zone (zona reticularis) of the adrenal cortex produce sex hormones. These hormones are male (adrenal androgens), but some of them are converted into female hormones (estrogens) by the skin, liver, and adipose tissues. These hormones may supplement the supply of sex hormones from the gonads and stimulate early development of the reproductive organs. Adrenal androgens may also play a role in controlling the female sex drive. Table 13.11 summarizes the actions of the cortical hormones. Clinical Application 13.3 discusses some of the effects of a malfunctioning adrenal gland on health.

Cortisol and related compounds are used as drugs to reduce inflammation. They relieve pain by

- decreasing permeability of capillaries, preventing leakage of fluids that swell surrounding tissues
- stabilizing lysosomal membranes, preventing release of their enzymes, which destroy tissue
- inhibiting prostaglandin synthesis

The concentration of cortisol compounds necessary to stifle inflammation is toxic, so these drugs can be used for only a short time. They are used to treat autoimmune disorders, allergies, asthma, and patients who have received organ transplants or tissue grafts.

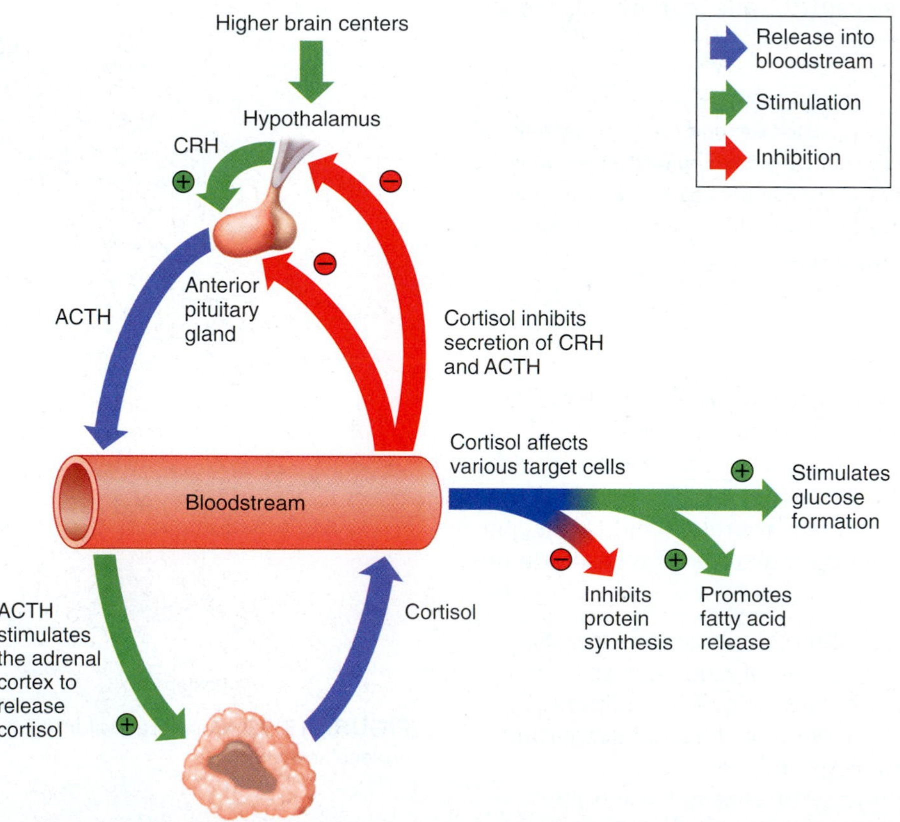

FIGURE 13.33 Negative feedback regulates cortisol secretion, similar to the regulation of thyroid hormone secretion (see fig. 13.16). (⊕ = stimulation; ⊖ = inhibition)

13.3 CLINICAL APPLICATION

Disorders of the Adrenal Cortex

John F. Kennedy's beautiful bronze complexion may have resulted not from sunbathing, but from a disorder of the adrenal glands. When he ran for president in 1960, Kennedy knew he had *Addison disease,* but his staff kept his secret, for fear it would affect his career. Kennedy had almost no adrenal tissue but was able to function by receiving mineralocorticoids and glucocorticoids, the standard treatment.

In Addison disease, the adrenal cortex does not secrete hormones sufficiently due to immune system attack (autoimmunity) or an infection such as tuberculosis. Signs and symptoms include decreased blood sodium, increased blood potassium, low blood glucose level (hypoglycemia), dehydration, low blood pressure, frequent infections, fatigue, nausea and vomiting, loss of appetite, and increased skin pigmentation. Some sufferers experience salt cravings—one woman reported eating many bowls of salty chicken noodle soup, with pickles and briny pickle juice added! Without treatment, death comes within days from severe disturbances in electrolyte balance.

An adrenal tumor or oversecretion of ACTH by the anterior pituitary causes hypersecretion of glucocorticoids (primarily cortisol), resulting in *Cushing syndrome.* It may also result from taking corticosteroid drugs for many years, such as to treat asthma or rheumatoid arthritis. Tissue protein level plummets, due to muscle wasting and loss of bone tissue. Blood glucose level remains elevated, and excess sodium is retained. As a result, tissue fluid increases, blood pressure rises, and the skin appears puffy. The skin may appear thin due to inhibition of collagen synthesis by the excess cortisol. Adipose tissue deposited in the face and back produce a characteristic "moon face" and "buffalo hump." Increase in adrenal sex hormone secretion may masculinize a female, causing growth of facial hair and a deepening voice. Other symptoms include extreme fatigue, sleep disturbances, skin rashes, headache, and leg muscle cramps.

Treatment of Cushing syndrome attempts to reduce ACTH secretion. This may entail removing a tumor in the pituitary gland or partially or completely removing the adrenal glands.

Both Addison disease and Cushing syndrome are rare, and for this reason, they are often misdiagnosed, or, in early stages, the patient's report of symptoms is not taken seriously. Addison disease affects thirty-nine to sixty people of every million, and Cushing syndrome affects five to twenty-five people per million. ■

PRACTICE

35 Name the important hormones of the adrenal cortex.

36 What is the function of aldosterone?

37 What does cortisol do?

38 How are blood concentrations of aldosterone and cortisol regulated?

13.9 PANCREAS

The **pancreas** (pan′kre-as) consists of two major types of secretory tissues. This organization of cell types reflects its dual function as an exocrine gland that secretes digestive juice through a duct and an endocrine gland that releases hormones into body fluids.

Structure of the Gland

The pancreas is an elongated, somewhat flattened organ posterior to the stomach and behind the parietal peritoneum (fig. 13.34). It is attached to the first section of the small intestine (duodenum) by a duct, which transports its digestive juice into the intestine. The digestive functions of the pancreas are discussed in chapter 17 (pp. 671–672).

The endocrine portion of the pancreas consists of cells grouped around blood vessels. These groups, called *pancreatic islets* (islets of Langerhans), include three distinct types of hormone-secreting cells—*alpha cells,* which secrete glucagon; *beta cells,* which secrete insulin; and *delta cells,* which secrete somatostatin (figs. 13.34 and 13.35).

Hormones of the Pancreatic Islets

Glucagon is a protein that stimulates the liver to break down glycogen into glucose (glycogenolysis) and to convert noncarbohydrates, such as amino acids, into glucose (gluconeogenesis). Glucagon also stimulates breakdown of fats into fatty acids and glycerol.

In a negative feedback system, a low concentration of blood glucose stimulates release of glucagon from the alpha

TABLE 13.11 | Hormones of the Adrenal Cortex

Hormone	Action	Factors Regulating Secretion
Aldosterone	Helps regulate the concentration of extracellular electrolytes by conserving sodium ions and excreting potassium ions	Electrolyte concentrations in body fluids and renin-angiotensin mechanism
Cortisol	Decreases protein synthesis, increases fatty acid release, and stimulates glucose synthesis from noncarbohydrates	CRH from the hypothalamus and ACTH from the anterior pituitary gland
Adrenal androgens	Supplement sex hormones from the gonads; may be converted into estrogens	

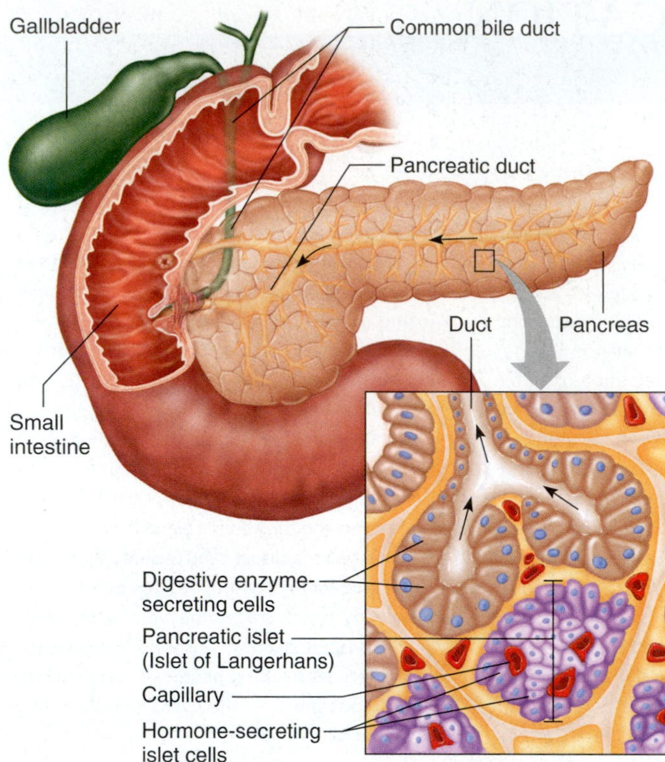

FIGURE 13.34 The hormone-secreting cells of the pancreas are grouped in clusters, or islets, closely associated with blood vessels. Other pancreatic cells secrete digestive enzymes into ducts.

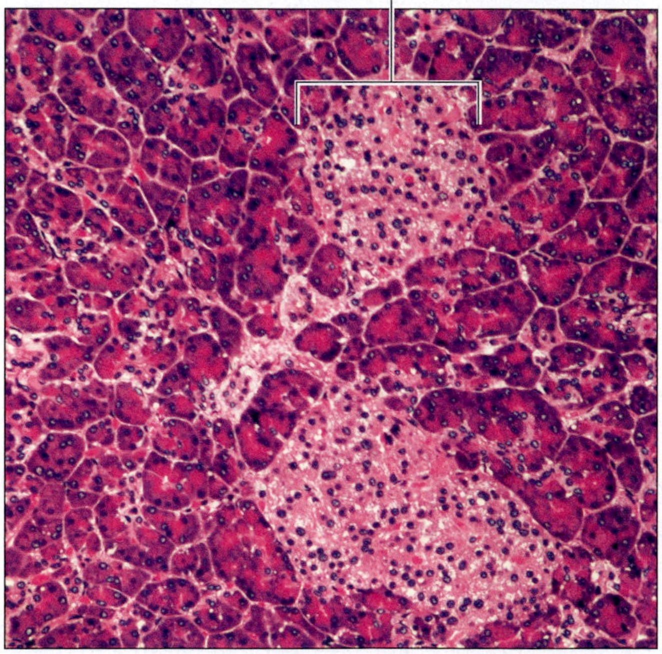

Pancreatic islet (Islet of Langerhans)

FIGURE 13.35 Light micrograph of pancreatic islets (200×).

cells. When blood glucose concentration returns toward normal, glucagon secretion decreases (fig. 13.36). This mechanism prevents hypoglycemia from occurring at times when glucose concentration is relatively low, such as between meals, or when glucose is being used rapidly—during periods of exercise, for example.

The hormone **insulin** is also a protein, and its main effect is exactly opposite that of glucagon. Insulin stimulates the liver to form glycogen from glucose and inhibits conversion of noncarbohydrates into glucose. Insulin also has the special effect of promoting the facilitated diffusion (see chapter 3, p. 93) of glucose through the membranes of cells bearing insulin receptors. These cells include those of cardiac muscle, adipose tissues, and resting skeletal muscles (glucose uptake by exercising skeletal muscles is not dependent on insulin). Insulin action decreases the concentration of blood glucose, promotes transport of amino acids into cells, and increases protein synthesis. It also stimulates adipose cells to synthesize and store fat.

An enzyme called glucokinase enables pancreatic cells to "sense" glucose level, important information in determining rates of synthesis of glucagon and insulin. One form of a rare type of diabetes mellitus, maturity-onset diabetes of the young (MODY), is caused by a mutation in a gene encoding glucokinase—the beta cells cannot accurately assess when they must produce insulin. Other mutations that cause MODY alter insulin's structure, secretion, or cell surface receptors or the ability of liver cells to form glycogen in response to insulin. MODY is treated with drugs or dietary modification.

A negative feedback system sensitive to the concentration of blood glucose regulates insulin secretion. When glucose concentration is relatively high, such as after a meal, the beta cells release insulin. By promoting formation of glycogen in the liver and entrance of glucose into adipose and muscle cells, insulin helps prevent excessive rise in blood glucose concentration (hyperglycemia). Then, when the glucose concentration falls, between meals or during the night, insulin secretion decreases (fig. 13.36).

As insulin concentration falls, less glucose enters the adipose and muscle cells, and the glucose remaining in the blood is available for cells that lack insulin receptors, such as nerve cells. Neurons readily tap the energy in a continuous supply of glucose to produce ATP.

Hypoglycemia, or low blood glucose level due to excess insulin in the bloodstream, causes episodes of shakiness, weakness, and anxiety. Following a diet of frequent, small meals low in carbohydrates and high in protein can often control symptoms by preventing the surges of insulin that lower the blood glucose level. Hypoglycemia is most often seen when a person with diabetes injects too much insulin, but it can also reflect a tumor of the insulin-producing cells of the pancreas, or it may occur transiently following strenuous exercise.

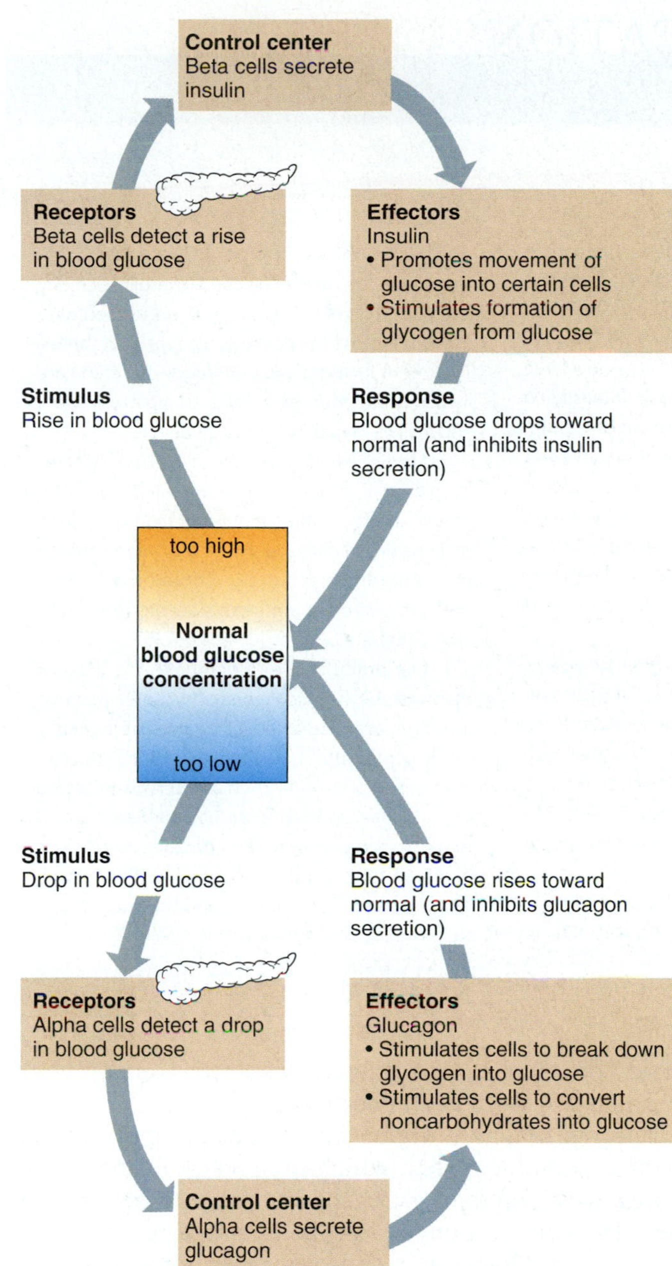

FIGURE 13.36 Insulin and glucagon function together to stabilize blood glucose concentration. Negative feedback responding to blood glucose concentration controls the levels of both hormones.

Neurons, including those of the brain, obtain glucose by a facilitated diffusion mechanism not dependent on insulin, but rather only on blood glucose concentration. For this reason, neurons are particularly sensitive to changes in blood glucose concentration. Conditions that cause such changes—excess insulin secretion, for example—are likely to affect brain functions.

At the same time that insulin concentration is decreasing, glucagon secretion is increasing. Therefore, these hormones function together to maintain a relatively constant blood glucose concentration, despite great variations in the amounts of ingested carbohydrates.

Somatostatin (similar to the hypothalamic hormone), which the delta cells release, helps regulate glucose metabolism by inhibiting secretion of glucagon and insulin. Table 13.12 summarizes the hormones of the pancreatic islets, and Clinical Application 13.4 and From Science to Technology 13.1 discuss diabetes mellitus, a derangement of the control of glucose metabolism that affects millions of people.

PRACTICE 

39 Name the endocrine portion of the pancreas.

40 What is the function of glucagon?

41 What is the function of insulin?

42 How are the secretions of glucagon and insulin controlled?

43 Why are nerve cells particularly sensitive to changes in blood glucose concentration?

13.10 OTHER ENDOCRINE GLANDS

Additional organs produce hormones. These are part of the endocrine system too. They include the pineal gland; the thymus; reproductive glands; and certain cells of the digestive tract, the heart, and the kidneys.

Pineal Gland

The **pineal gland** (pin′e-al gland) is a small, oval structure deep between the cerebral hemispheres, where it attaches to the upper portion of the thalamus near the roof of the third ventricle. It largely consists of specialized *pineal cells* and supportive neuroglia (see fig. 11.20*b*).

TABLE 13.12 | Hormones of the Pancreatic Islets

Hormone	Action	Source of Control
Glucagon	Stimulates the liver to break down glycogen and convert noncarbohydrates into glucose; stimulates breakdown of fats	Blood glucose concentration
Insulin	Promotes formation of glycogen from glucose, inhibits conversion of noncarbohydrates into glucose, and enhances movement of glucose through adipose and muscle cell membranes, decreasing blood glucose concentration; promotes transport of amino acids into cells; enhances synthesis of proteins and fats	Blood glucose concentration
Somatostatin	Helps regulate carbohydrates	Not determined

13.4 CLINICAL APPLICATION

Diabetes Mellitus

About 7% of the population of the United States has some form of diabetes mellitus. In Latin, *diabetes* means "increased urine output," and *mellitus* means "honey," referring to urine's sugar content.

In type 1 diabetes mellitus (juvenile or insulin-dependent diabetes mellitus), the pancreas cannot produce insulin. Symptoms usually begin before age twenty. About 15% of people with diabetes mellitus have this form. It is an autoimmune disorder in which the immune system attacks pancreatic beta cells, ultimately destroying them and halting insulin secretion. Lack of insulin decreases movement of glucose into skeletal muscle and adipose cells, inhibiting glycogen formation. As a result, blood glucose concentration rises (hyperglycemia), and when it reaches a certain level, the kidneys begin to excrete the excess. Glucose appears in the urine (glycosuria). Water follows the glucose by osmosis, causing dehydration and intense thirst.

Untreated type 1 diabetes decreases protein synthesis, shrinking tissues as glucose-starved cells use protein for energy. Weight falls, and wounds cannot heal. Fatty acids accumulate in the blood as a result of decreased fat synthesis and storage. Ketone bodies, a by-product of fat metabolism, also build up in the blood. They are excreted in the urine as sodium salts, and a large volume of water follows by osmosis, intensifying dehydration and lowering sodium ion concentration in the blood. Accumulation of ketones and loss of sodium ions lead to metabolic acidosis, a condition that lowers the pH of body fluids. Acidosis and dehydration adversely affect brain neurons. Without treatment (insulin replacement), the person becomes disoriented and may enter a diabetic coma and die.

Daily life for a person with type 1 diabetes mellitus means constant awareness of the illness—insulin delivery; frequent finger punctures to monitor blood glucose level; a restrictive diet; and concern over complications, which include loss of vision, leg ulcers, and kidney damage. The many symptoms reflect disturbances in carbohydrate, protein, and fat metabolism.

In another form of diabetes, type 2 or noninsulin-dependent diabetes mellitus, begins gradually, usually in people over forty. Cells lose insulin receptors and are less able to respond to insulin. Heredity and a lifestyle of overeating and underexercising are risk factors for developing type 2 diabetes. Treatment includes avoiding foods that stimulate insulin production, exercising, taking medication, and maintaining desirable body weight.

About 54 million people in the United States have prediabetes, which is an elevated blood glucose level after eating, that is not high enough to be considered diabetes. However, organ damage may already be occurring. Adopting a healthy diet and exercise habits can slow progression to type 2 diabetes.

The oral glucose tolerance test is used to diagnose both major types of diabetes mellitus and indicate prediabetes. The patient ingests a known quantity of glucose, and blood glucose concentration is measured at intervals to assess glucose use. If the person has diabetes, blood glucose concentration rises greatly and remains elevated for several hours. In a healthy person, glucose rise is less dramatic, and the level returns to normal in about an hour and a half. ■

The pineal gland secretes a hormone, **melatonin,** that is synthesized from serotonin. Varying patterns of light and dark outside the body control the gland's activities. In the presence of light, nerve impulses from the eyes travel to the hypothalamus, then to the reticular formation, and then downward into the spinal cord. From here, the impulses travel along sympathetic nerve fibers back into the brain, and finally they reach the pineal gland, where they decrease melatonin secretion. In the absence of light, nerve impulses from the eyes decrease and secretion of melatonin increases.

Melatonin secretion is part of the regulation of **circadian rhythms,** which are patterns of repeated activity associated with cycles of day and night, such as sleep/wake rhythms. Melatonin binds to two types of receptors on brain neurons, one that is abundant and one that is scarce. The major receptors are on cells of the suprachiasmatic nucleus, a region that regulates the circadian clock. Binding to the second, less abundant receptors, however, induces sleepiness.

The **thymus** (thi′mus), which lies in the mediastinum posterior to the sternum and between the lungs, is large in young children but shrinks with age. This gland secretes a group of hormones, called **thymosins,** that affect production and differentiation of certain white blood cells (T lymphocytes). The thymus gland plays an important role in immunity and is discussed in chapter 16 (p. 623).

The reproductive organs that secrete important hormones include the **testes,** which produce testosterone; the **ovaries,** which produce estrogens and progesterone; and the **placenta,** which produces estrogens, progesterone, and a gonadotropin. Chapters 22 and 23 discuss these glands and their secretions (pp. 846, 857, and 884).

The digestive glands that secrete hormones are generally associated with the linings of the stomach and small intestine. The small intestine alone produces dozens of hormones, many of which have not been well studied. Chapter 17 (pp. 668 and 672) describes these structures and their secretions.

The fact that melatonin secretion is a response to the relative lengths of day and night explains why traveling across several time zones produces the temporary insomnia of jet lag. Melatonin supplements are advertised as preventing jet lag, based on anecdotal reports and small studies. However, the first large study, conducted on 257 doctors traveling from Norway to New York, testing three nightly doses of melatonin supplement versus placebo, showed no effect at all from melatonin in preventing or alleviating jet lag.

13.1 FROM SCIENCE TO TECHNOLOGY

Treating Diabetes

The sweet-smelling urine that is the hallmark of type 1 diabetes mellitus was noted as far back as an Egyptian papyrus from 1500 B.C. In A.D. 96 in Greece, Aretaeus of Cappadocia described the condition as a "melting down of limbs and flesh into urine." One of the first people to receive insulin as a drug was a three-year-old boy whose body could not produce the hormone. In December 1922, before treatment, he weighed only fifteen pounds. The boy rapidly improved after beginning insulin treatment, doubling his weight in just two months.

In 1921, Canadian physiologists Sir Frederick Grant Banting and Charles Herbert Best discovered the link between lack of insulin and diabetes. They induced diabetes symptoms in a dog by removing its pancreas, then cured it by administering insulin from another dog's healthy pancreas. A year later, people with diabetes began to receive insulin extracted from pigs or cattle.

In 1982, pure human insulin became available by genetically altering bacteria to produce the human protein (recombinant DNA technology). Human insulin helps people with diabetes who are allergic to the product from pigs or cows. Today, people receive insulin in several daily injections, from an implanted insulin pump, and/or in aerosol form (fig. 13B).

Providing new pancreatic islets is a longer-lasting treatment for type 1 diabetes. Islet cell transplantation was first attempted in 1893, when an English surgeon transplanted bits of a sheep's pancreas into an adolescent near death. He died a few days later. After insulin was discovered, interest in islet transplantation revived once researchers realized that the more frequent the daily doses of insulin, the healthier the patient. Transplants succeeded, in rats, in 1972. Difficulties arose, however, in treating humans. It was challenging to separate islets from cadaver pancreases, and then collect enough beta cells, which account for only 2% of pancreas cells. Many patients' immune systems rejected transplants. By the 1990s, automated islet isolation and new anti-rejection drugs helped. In 1996 in Germany, and then in 1999 in Edmonton, Canada, islet transplantation began.

Since 2000, several hundred people have received islet transplants in a procedure called the Edmonton protocol, which introduces islets into a vein in the liver. By a year after transplant, from 50% to 68% of patients do not need to receive additional insulin, but by five years after the procedure, fewer than 10% of total patients are free of daily insulin supplementation. The procedure is risky—12% of patients hemorrhage, and 4% develop blood clots in the liver vein. These risks, plus the apparent short-term improvement, have prompted physicians to carefully evaluate which patients are most likely to benefit from the few years of insulin independence that the procedure may offer.

Researchers are currently investigating various stem and progenitor cells to see if they can be implanted to treat diabetes, using a patient's own cells. Reprogramming cells back to a stem-like state might not be necessary—one research group has succeeded in culturing pancreatic alpha cells (from the digestive part of the pancreas) in a way that stimulates them to give rise to beta cells.

A possible treatment for type 2 diabetes has come from an unexpected place—gastric bypass surgery, which removes parts of the stomach and small intestine to help people lose weight. Isolated reports since the 1950s noted cases of morbidly obese people with diabetes having gastric bypass surgery, and then, within days and with lasting effect, not needing to inject insulin. By the 1980s, doctors noticed that some patients who had normal regulation of blood glucose before the surgery, a few months after began to experience confusion, altered behavior, seizures, and unconsciousness—signs of low blood glucose. Today some surgeons are performing gastric bypass surgery on people who do not weigh as much as typical candidates for the surgery, but who have severe diabetes. Treating diabetes with weight-loss surgery is controversial, because the mechanism of how it corrects blood glucose regulation is not understood. ■

FIGURE 13B A person with either common form of diabetes mellitus must monitor his or her blood glucose level and be very diligent about proper diet and exercise.

Other organs that produce hormones include the heart, which secretes *atrial natriuretic peptide* (chapter 15, p. 556), and the kidneys, which secrete *erythropoietin* that stimulates red blood cell production (chapter 14, p. 527). Clinical Application 13.1 discusses abuse of EPO to improve athletic performance.

PRACTICE

44 Where is the pineal gland located?
45 What is the function of the pineal gland?
46 Where is the thymus gland located?

13.11 STRESS AND ITS EFFECTS

Because survival depends upon maintaining homeostasis, factors that change the body's internal environment are potentially life threatening. Sensing such dangers directs nerve impulses to the hypothalamus, triggering physiological responses that resist a loss of homeostasis. These responses include increased activity in the sympathetic division of the autonomic nervous system and increased secretion of adrenal hormones. A factor capable of stimulating such a response is called a **stressor,** and the condition it produces in the body is called **stress.**

Types of Stress

Stressors may be physical or psychological. They may also be a combination.

Physical stress threatens tissues. Extreme heat or cold, decreased oxygen concentration, infections, injuries, prolonged heavy exercise, and loud sounds inflict physical stress. Unpleasant or painful sensations often accompany physical stress.

Psychological stress results from thoughts about real or imagined dangers, personal losses, unpleasant social interactions (or lack of social interactions), or any threatening factors. Feelings of anger, fear, grief, anxiety, depression, and guilt cause psychological stress. Psychological stress may also stem from pleasant stimuli, such as friendly social contact, feelings of joy or happiness, or sexual arousal. The factors that produce psychological stress vary greatly from person to person. A situation that is stressful to one person may not affect another, and what is stressful at one time may not be at another time.

Responses to Stress

The hypothalamus controls response to stress, termed the *general stress* (or *general adaptation*) *syndrome.* This response, evoked to stress of any type, maintains homeostasis.

Inflammation is the immune system's generalized response to limit the effects of injury or infection. However, inflammation is painful and possibly destructive. The endocrine system keeps the immune system in check by increasing secretion by the pituitary and adrenal glands to temper inflammation. This is an example of how homeostasis operates between organ systems as well as within them.

Recall that the hypothalamus receives information from nearly all body parts, including visceral receptors, the cerebral cortex, the reticular formation, and limbic system. At times of stress, the hypothalamus responds to incoming impulses by activating the "fight-or-flight" response. Specifically, sympathetic impulses from the hypothalamus raise blood glucose concentration, the level of blood glycerol and fatty acids, heart rate, blood pressure and breathing rate, and dilate the air passages. The response also shunts blood from the skin and digestive organs into the skeletal muscles and increases secretion of epinephrine from the adrenal medulla. The epinephrine, in turn, intensifies these sympathetic responses and prolongs their effects (fig. 13.37).

At the same time that sympathetic activity increases, the hypothalamus's release of corticotropin-releasing hormone (CRH) stimulates the anterior pituitary gland to secrete

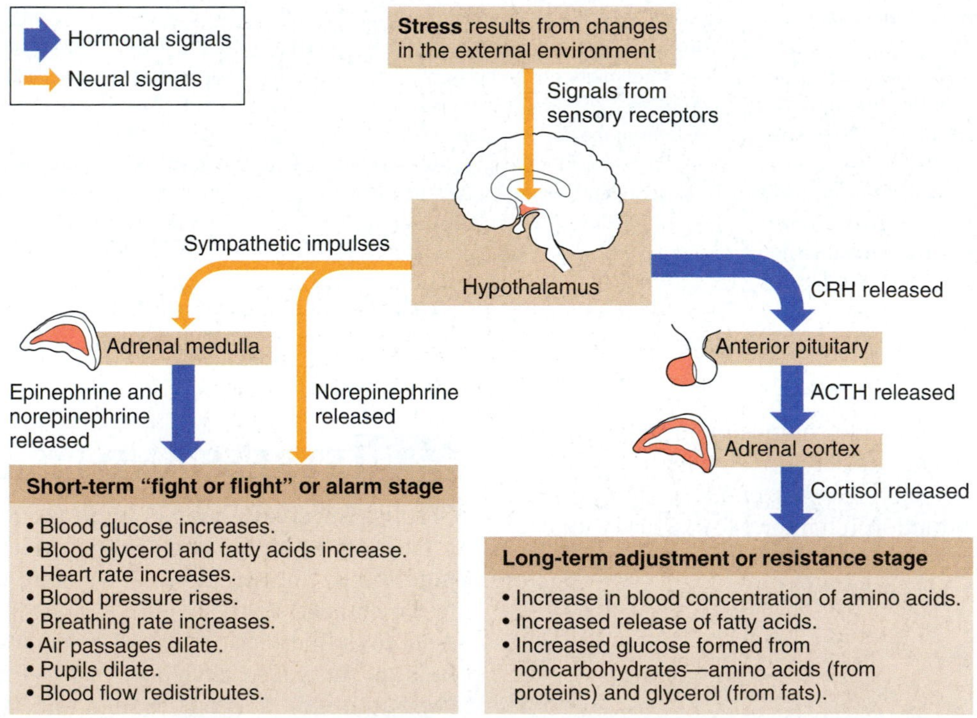

FIGURE 13.37 During stress, the hypothalamus helps prepare the body for "fight or flight" by triggering sympathetic impulses to various organs. It also stimulates epinephrine release, intensifying the sympathetic responses. The hypothalamus also stimulates the adrenal cortex to release cortisol, which promotes longer-term responses that resist the effects of stress.

ACTH, which increases the adrenal cortex's secretion of cortisol. Cortisol supplies cells with amino acids and extra energy sources and diverts glucose from skeletal muscles to brain tissue (fig. 13.37). Stress can also stimulate release of glucagon from the pancreas, growth hormone (GH) from the anterior pituitary, antidiuretic hormone (ADH) from the posterior pituitary gland, and renin from the kidney.

Glucagon and growth hormone help mobilize energy sources, such as glucose, glycerol, and fatty acids, and stimulate cells to take up amino acids, facilitating repair of injured tissues. ADH stimulates the kidneys to retain water. This action decreases urine output and helps to maintain blood volume—particularly important if a person is bleeding or sweating heavily. Renin, by increasing angiotensin II levels, helps stimulate the kidneys to retain sodium (through aldosterone), and through the vasoconstrictor action of angiotensin II contributes to maintaining blood pressure. Table 13.13 summarizes the body's reactions to stress.

PRACTICE 

47 What is stress?

48 Distinguish between physical stress and psychological stress.

49 Describe the general stress syndrome.

TABLE 13.13 | Major Events in the General Stress Syndrome

1. In response to stress, nerve impulses are transmitted to the hypothalamus.
2. Sympathetic impulses arising from the hypothalamus increase blood glucose concentration, blood glycerol concentration, blood fatty acid concentration, heart rate, blood pressure, and breathing rate. They dilate air passages, shunt blood into skeletal muscles, and increase secretion of epinephrine from the adrenal medulla.
3. Epinephrine intensifies and prolongs sympathetic actions.
4. The hypothalamus secretes CRH, which stimulates secretion of ACTH by the anterior pituitary gland.
5. ACTH stimulates release of cortisol by the adrenal cortex.
6. Cortisol increases the concentration of blood amino acids, releases fatty acids, and forms glucose from noncarbohydrate sources.
7. Secretion of glucagon from the pancreas and growth hormone from the anterior pituitary increase.
8. Glucagon and growth hormone aid mobilization of energy sources and stimulate uptake of amino acids by cells.
9. Secretion of ADH from the posterior pituitary increases.
10. ADH promotes the retention of water by the kidneys, which increases blood volume.
11. Renin increases blood levels of angiotensin II, which acts as a vasoconstrictor and also stimulates aldosterone secretion by the adrenal cortex.
12. Aldosterone stimulates sodium retention by the kidneys.

13.12 LIFE-SPAN CHANGES

With age, the glands of the endocrine system generally decrease in size and increase in the proportion of each gland that is fibrous in nature. At the cellular level, lipofuscin pigment accumulates as glands age. Functionally, hormone levels may change with advancing years. Treatments for endocrine disorders associated with aging supplement deficient hormones, remove part of an overactive gland or use drugs to block the action of an overabundant hormone.

Aging affects different hormones in characteristic ways. For growth hormone, the surge in secretion that typically occurs at night lessens somewhat with age. Lower levels of GH are associated with declining strength in the skeleton and muscles with advancing age. However, supplementing older people with GH in an attempt to duplicate the effects of exercise can dangerously raise blood pressure and blood glucose levels and enlarge the spleen, liver, and kidneys.

Levels of antidiuretic hormone increase with age, but this is due to slowed breakdown in the liver and kidneys, rather than increased synthesis. As a result, the kidneys are stimulated to reabsorb more water.

The thyroid gland shrinks with age, as individual follicles shrink and increasing amounts of fibrous connective tissue separate them. Thyroid nodules, which may be benign or cancerous, become more common with age, and are often first detected upon autopsy. Although blood levels of T_3 and T_4 may diminish with age, in general, the thyroid gland's control over the metabolism of various cell types is maintained throughout life. Calcitonin levels decline with age, which raises the risk of osteoporosis.

Parathyroid function differs between the sexes with age. Secretion peaks in males at about age fifty, whereas in women, the level of parathyroid hormone decreases until about age forty, after which it rises and contributes to osteoporosis risk. Fat accumulates between the cells of the parathyroid glands.

The adrenal glands illustrate the common theme of aging-related physical changes, yet continued function. Fibrous connective tissue, lipofuscin pigment, and increased numbers of abnormal cells characterize the aging adrenal glands. However, thanks to the fine-tuning of negative feedback systems, blood levels of glucocorticoids and mineralocorticoids usually remain within the normal range, although the ability to maintain homeostasis of osmotic pressure, blood pressure, acid/base balance and sodium and potassium ion distributions may falter with age.

The most obvious changes in endocrine function that occur with age involve blood glucose regulation. The pancreas may be able to maintain secretion of insulin and glucagon, but lifestyle changes, such as increase in fat intake and less exercise, may increase the blood insulin level. The development of insulin resistance—the decreased ability of muscle, liver, and fat cells to take in glucose even in the presence of insulin—reflects impaired ability of these target cells to respond to the hormone, rather than compromised pancreatic function. Blood glucose buildup may signal the pancreas to secrete more insulin, setting the stage for type 2 diabetes mellitus.

INNERCONNECTIONS | *Endocrine System*

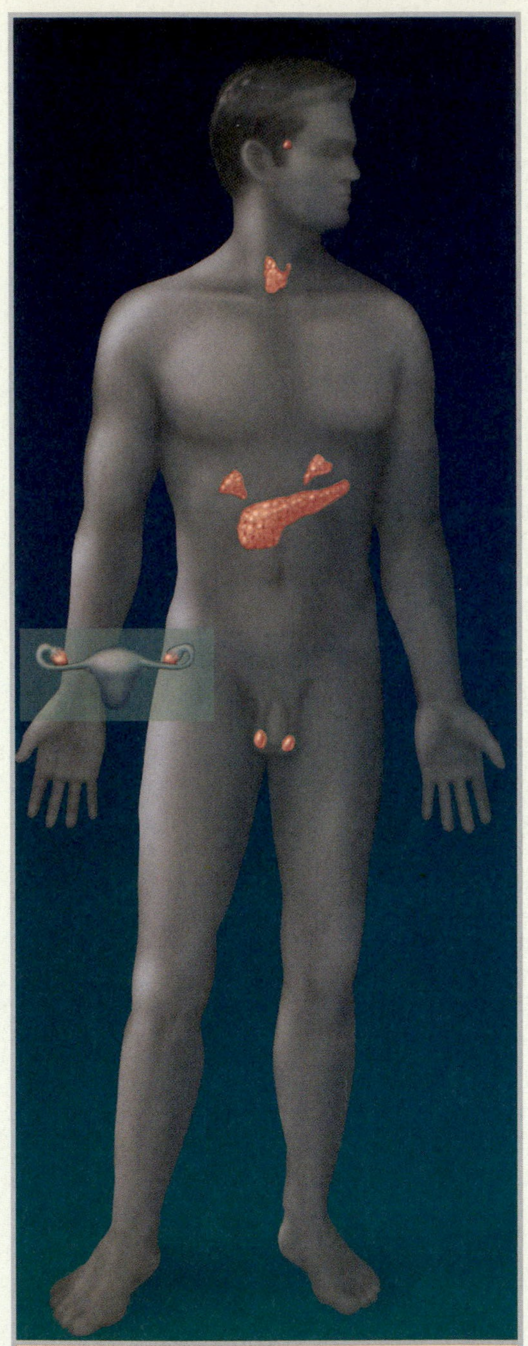

Endocrine System

Glands secrete hormones that have a variety of effects on cells, tissues, organs, and organ systems.

Integumentary System

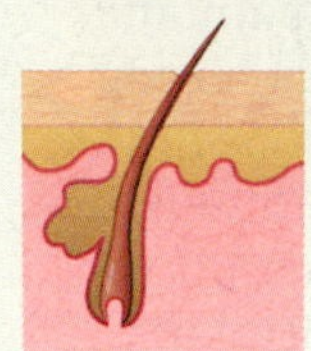

Melanocytes produce skin pigment in response to hormonal stimulation.

Lymphatic System

Hormones stimulate lymphocyte production.

Skeletal System

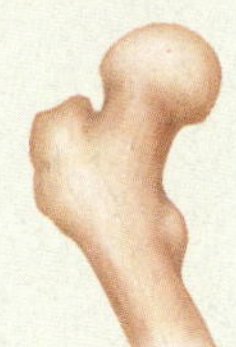

Hormones act on bones to control calcium balance.

Digestive System

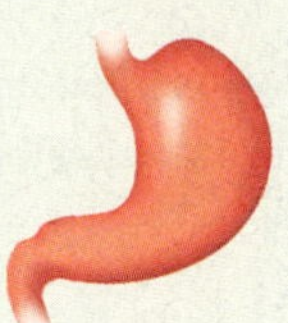

Hormones help control digestive system activity.

Muscular System

Hormones help increase blood flow to exercising muscles.

Respiratory System

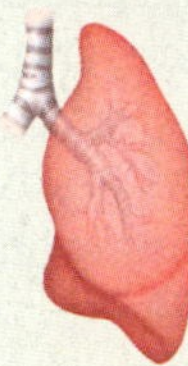

Decreased oxygen causes hormonal stimulation of red blood cell production; red blood cells transport oxygen and carbon dioxide.

Nervous System

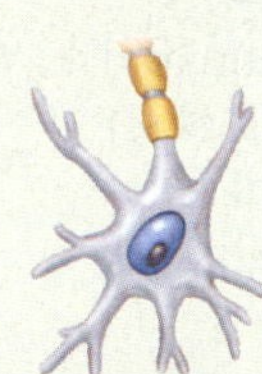

Neurons control the secretions of the anterior and posterior pituitary glands and the adrenal medulla.

Urinary System

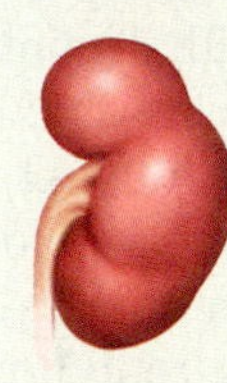

Hormones act on the kidneys to help control water and electrolyte balance.

Cardiovascular System

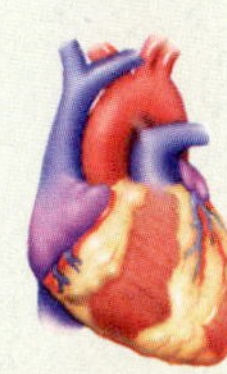

Hormones are carried in the bloodstream; some have direct actions on the heart and blood vessels.

Reproductive System

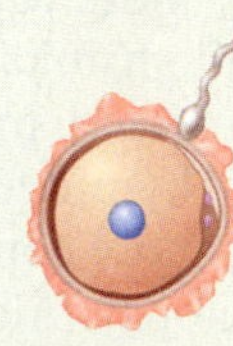

Sex hormones play a major role in development of secondary sex characteristics, egg, and sperm.

The daily fall and rise of melatonin levels may even out somewhat with age, which may alter control of the sleep/wake cycle. People usually require less sleep as they age. Changes to the tempo of the body clock may, in turn, affect secretion of other hormones.

The thymus gland begins to noticeably shrink before age twenty, with accompanying declining levels of thymosins. By age sixty, thymosin secretion is nil. The result is a slowing of the maturation of B and T cells, which increases susceptibility to infections as a person ages.

PRACTICE 

50 What general types of changes occur in the glands of the endocrine system with aging?

51 How do the structures and functions of particular endocrine glands change over a lifetime?

CHAPTER SUMMARY

13.1 INTRODUCTION (PAGE 483)

The nervous sytem and the endocrine system work together to control body functions. Endocrine glands secrete their products into body fluids (the internal environment); exocrine glands secrete their products into ducts that lead to the outside of the body.

13.2 GENERAL CHARACTERISTICS OF THE ENDOCRINE SYSTEM (PAGE 483)

A hormone's target cells have specific receptors. Hormones from endocrine glands regulate metabolic processes.

13.3 HORMONE ACTION (PAGE 484)

Endocrine glands secrete hormones into the bloodstream which carries them to all parts of the body.

1. Chemistry of hormones
   a. Steroid hormones are lipids that include complex rings of carbon and hydrogen atoms.
   b. Nonsteroid hormones are amines, peptides, and proteins.
2. Actions of hormones
   a. Steroid hormones and thyroid hormones
      (1) Steroid hormones enter target cells and combine with receptors to form complexes.
      (2) These complexes activate specific genes in the nucleus, which direct synthesis of specific proteins.
      (3) The degree of cellular response is proportional to the number of hormone-receptor complexes formed.
   b. Nonsteroid hormones
      (1) Nonsteroid hormones combine with receptors in the target cell membrane.
      (2) A hormone-receptor complex stimulates membrane proteins, such as adenylate cyclase, to induce the formation of second messenger molecules.
      (3) A second messenger, such as cAMP, activates protein kinases.
      (4) Protein kinases activate certain protein substrate molecules, which, in turn, change cellular processes.
      (5) The cellular response to a nonsteroid hormone is amplified because the enzymes induced by a small number of hormone-receptor complexes can catalyze formation of a large number of second messenger molecules.
3. Prostaglandins
   a. Prostaglandins are paracrine substances that have powerful hormonelike effects, even in small amounts.
   b. Prostaglandins modulate hormones that regulate formation of cyclic AMP.

13.4 CONTROL OF HORMONAL SECRETIONS (PAGE 491)

The concentration of each hormone in the body fluids is precisely regulated.

1. Control sources
   a. Some endocrine glands secrete hormones in response to releasing hormones the hypothalamus secretes.
   b. Some endocrine glands secrete in response to nerve impulses.
   c. Some endocrine glands secrete in response to changes in the plasma concentration of a substance.
2. Negative feedback systems
   a. In a negative feedback system, a gland is sensitive to the concentration of a substance it regulates.
   b. When the concentration of the regulated substance reaches a certain concentration, it inhibits the gland.
   c. As the gland secretes less hormone, the controlled substance also decreases.

13.5 PITUITARY GLAND (PAGE 492)

The pituitary gland, attached to the base of the brain, has an anterior lobe and a posterior lobe. Releasing hormones from the hypothalamus control most pituitary secretions.

1. Anterior pituitary hormones
   a. The anterior pituitary consists largely of epithelial cells, and it secretes GH, PRL, TSH, ACTH, FSH, and LH.
   b. Growth hormone (GH)
      (1) Growth hormone stimulates body cells to grow and divide.
      (2) Growth hormone-releasing hormone and somatostatin from the hypothalamus control GH secretion.
   c. Prolactin (PRL)
      (1) PRL promotes breast development and stimulates milk production.
      (2) In males, prolactin decreases secretion of LH (ICSH).

(3) Prolactin release-inhibiting hormone from the hypothalamus restrains secretion of prolactin, whereas the yet to be identified prolactin-releasing factor is thought to promote its secretion.

d. Thyroid-stimulating hormone (TSH)
   (1) TSH controls secretion of hormones from the thyroid gland.
   (2) The hypothalamus, by secreting thyrotropin-releasing hormone, regulates TSH secretion.

e. Adrenocorticotropic hormone (ACTH)
   (1) ACTH controls the secretion of certain hormones from the adrenal cortex.
   (2) The hypothalamus, by secreting corticotropin-releasing hormone, regulates ACTH secretion.

f. Follicle-stimulating hormone (FSH) and luteinizing hormone (LH) are gonadotropins that affect the reproductive organs.

2. Posterior pituitary hormones
   a. The posterior lobe of the pituitary gland largely consists of neuroglia and nerve fibers that originate in the hypothalamus.
   b. The two hormones of the posterior pituitary are produced in the hypothalamus.
   c. Antidiuretic hormone (ADH)
      (1) ADH causes the kidneys to excrete less water.
      (2) In high concentration, ADH constricts blood vessel walls, raising blood pressure.
      (3) The hypothalamus regulates ADH secretion.
   d. Oxytocin (OT)
      (1) OT has an antidiuretic effect and can contract muscles in the uterine wall.
      (2) OT also contracts certain cells associated with production and ejection of milk from the milk glands of the breasts.

13.6 THYROID GLAND (PAGE 499)

The thyroid gland is located in the neck and consists of two lateral lobes.

1. Structure of the gland
   a. The thyroid gland consists of many hollow secretory parts called follicles.
   b. The follicles are fluid filled and store the hormones the follicle cells secrete.
   c. Extrafollicular cells secrete calcitonin.
2. Thyroid hormones
   a. Thyroxine and triiodothyronine
      (1) These hormones increase the rate of metabolism, enhance protein synthesis, and stimulate lipid breakdown.
      (2) These hormones are needed for normal growth and development and for maturation of the nervous system.
   b. Calcitonin
      (1) Calcitonin lowers blood calcium and phosphate ion concentrations.
      (2) This hormone prevents prolonged elevation of calcium after a meal.

13.7 PARATHYROID GLANDS (PAGE 502)

The parathyroid glands are on the posterior surface of the thyroid.

1. Structure of the glands
   a. Each gland is small and yellow-brown, within a thin connective tissue capsule.
   b. Each gland consists of secretory cells well supplied with capillaries.
2. Parathyroid hormone (PTH)
   a. PTH increases blood calcium ion concentration and decreases blood phosphate ion concentration.
   b. PTH stimulates resorption of bone tissue, causes the kidneys to conserve calcium ions and excrete phosphate ions, and indirectly stimulates absorption of calcium ions from the intestine.
   c. A negative feedback mechanism operating between the parathyroid glands and the blood regulates these glands.

13.8 ADRENAL GLANDS (PAGE 504)

The adrenal glands are located atop the kidneys.

1. Structure of the glands
   a. Each adrenal gland consists of a medulla and a cortex.
   b. The adrenal medulla and adrenal cortex are distinct glands that secrete different hormones.
2. Hormones of the adrenal medulla
   a. The adrenal medulla secretes epinephrine and norepinephrine.
   b. These hormones are synthesized from tyrosine and are chemically similar.
   c. These hormones produce effects similar to those of the sympathetic nervous system.
   d. Sympathetic impulses originating from the hypothalamus stimulate secretion of these hormones.
3. Hormones of the adrenal cortex
   a. The cortex produces several types of steroids that include hormones.
   b. Aldosterone
      (1) It causes the kidneys to conserve sodium ions and water and to excrete potassium ions.
      (2) It is secreted in response to increased potassium ion concentration or presence of angiotensin II.
      (3) By conserving sodium ions and water, it helps maintain blood volume and pressure.
   c. Cortisol
      (1) It inhibits protein synthesis, releases fatty acids, and stimulates glucose formation from noncarbohydrates.
      (2) A negative feedback mechanism involving secretion of CRH from the hypothalamus and ACTH from the anterior pituitary gland controls its level.
   d. Adrenal sex hormones
      (1) These hormones are of the male type although some can be converted into female hormones.
      (2) They supplement the sex hormones produced by the gonads.

13.9 PANCREAS (PAGE 509)

The pancreas secretes digestive juices as well as hormones.

1. Structure of the gland
   a. The pancreas is posterior to the stomach and is attached to the small intestine.

b. The endocrine portion, called the pancreatic islets (islets of Langerhans), secretes glucagon, insulin, and somatostatin.

2. Hormones of the pancreatic islets
   a. Glucagon stimulates the liver to produce glucose, increasing concentration of blood glucose. It also breaks down fat.
   b. Insulin activates facilitated diffusion of glucose through cell membranes, stimulates its storage, promotes protein synthesis, and stimulates fat storage.
   c. Facilitated diffusion of glucose into nerve cells does not depend on insulin.
   d. Somatostatin inhibits insulin and glucagon release.

13.10 OTHER ENDOCRINE GLANDS (PAGE 511)

1. Pineal gland
   a. The pineal gland is attached to the thalamus near the roof of the third ventricle.
   b. Postganglionic sympathetic nerve fibers innervate it.
   c. It secretes melatonin, part of the regulation of circadian rhythms.
2. Thymus gland
   a. The thymus gland lies posterior to the sternum and between the lungs.
   b. It shrinks with age.
   c. It secretes thymosin, which affects the production of certain lymphocytes that, in turn, provide immunity.
3. Reproductive glands
   a. The testes secrete testosterone.
   b. The ovaries secrete estrogens and progesterone.
   c. The placenta secretes estrogens, progesterone, and a gonadotropin.
4. The digestive glands include certain glands of the stomach and small intestine that secrete hormones.
5. Other hormone-producing organs include the heart and kidneys.

13.11 STRESS AND ITS EFFECTS (PAGE 513)

Stress occurs when the body responds to stressors that threaten the maintenance of homeostasis. Stress responses include increased activity of the sympathetic nervous system and increased secretion of adrenal hormones.

1. Types of stress
   a. Physical stress results from environmental factors that are harmful or potentially harmful to tissues.
   b. Psychological stress results from thoughts about real or imagined dangers.
   c. Factors that produce psychological stress vary with the individual and the situation.
2. Responses to stress
   a. Responses to stress maintain homeostasis.
   b. The hypothalamus controls a general stress syndrome.

13.12 LIFE-SPAN CHANGES (PAGE 515)

With age, endocrine glands shrink and accumulate fibrous connective tissue, fat, and lipofuscin, but hormonal activities usually remain within the normal range.

1. GH levels even out, as muscular strength declines.
2. ADH levels increase due to slowed breakdown.
3. The thyroid shrinks but control of metabolism continues.
4. Decreasing levels of calcitonin and increasing levels of parathyroid hormone increase osteoporosis risk.
5. The adrenal glands show aging-related changes, but negative feedback maintains functions.
6. Muscle, liver, and fat cells may develop insulin resistance.
7. Changes in melatonin secretion affect the body clock.
8. Thymosin production declines, hampering infectious disease resistance.

CHAPTER ASSESSMENTS

13.1 Introduction

1 Contrast the definitions of *endocrine gland* and *exocrine gland.* (p. 483)

13.2 General Characteristics of the Endocrine System

2 Explain the specificity of a hormone for its target cell. (p. 483)

3 List six general functions of hormones. (p. 484)

13.3 Hormone Action

4 Explain how hormones can be grouped on the basis of their chemical composition. (p. 484)

5 List the steps of steroid hormone action. (p. 485)

6 List the steps of the action of most nonsteroid hormones. (p. 486)

7 Explain how prostaglandins are similar to hormones and how they are different. (p. 491)

13.4 Control of Hormonal Secretions

8 Diagram the three mechanisms that control hormone secretion, including negative feedback. (p. 492)

13.5 Pituitary Gland

9 Describe the location and structure of the pituitary gland. (p. 492)

10 List the hormones that the anterior pituitary secretes. (p. 492)

11 Explain two ways that the brain controls pituitary gland activity. (p. 494)

12 Releasing hormones come from which one of the following? (p. 494)

a. thyroid gland
b. anterior pituitary gland
c. posterior pituitary gland
d. hypothalamus

13 Match the following hormones with their actions: (p. 494)

(1) growth hormone	A. milk synthesis
(2) thyroid stimulating hormone	B. cell division
(3) prolactin	C. metabolic rate
(4) adrenocorticotropic hormone	D. acts on gonads
(5) follicle-stimulating hormone	E. controls secretion of adrenal cortical hormones
(6) luteinizing hormone	

14 Explain how growth hormone produces its effects. (p. 494)

15 Describe the control of growth hormone secretion. (p. 494)

16 Describe the anatomical differences between the anterior and posterior lobes of the pituitary gland. (p. 498)

17 Name and describe the functions of the posterior pituitary hormones. (p. 498)

18 Under which of the following conditions would you expect an increase in antidiuretic hormone secretion? (p. 498)

a. An individual ingests excess water.

b. The posterior pituitary is removed because it has a tumor.

c. An individual is rescued after three days in the desert without food or water.

d. An individual receives an injection of synthetic antidiuretic hormone.

13.6 Thyroid Gland

19 Describe the location and structure of the thyroid gland. (p. 499)

20 Match the hormones from the thyroid gland with their descriptions. (p. 500)

(1) thyroxine	A. most potent at controlling metabolism
(2) triiodothyronine	B. regulates blood calcium
(3) calcitonin	C. has four iodine atoms

21 Define *iodide pump.* (p. 500)

22 Diagram the control of thyroid hormone secretion. (p. 501)

13.7 Parathyroid Glands

23 Describe the location and structure of the parathyroid glands. (p. 502)

24 Explain the general function of parathyroid hormone. (p. 502)

25 Diagram the regulation of parathyroid hormone secretion. (p. 503)

13.8 Adrenal Glands

26 Distinguish between the adrenal medulla and the adrenal cortex. (p. 504)

27 Match the adrenal hormones with their source and actions: (p. 505)

(1) cortisol	A. cortex; sodium retention
(2) aldosterone	B. cortex; female sex hormones
(3) epinephrine	C. cortex; male sex hormones
(4) androgens	D. medulla; fight-or-flight response
(5) estrogens	E. cortex; gluconeogenesis

28 Diagram control of aldosterone secretion. (p. 506)

29 Diagram control of cortisol secretion. (p. 508)

13.9 Pancreas

30 Describe the location and structure of the pancreas. (p. 509)

31 List the hormones the pancreatic islets secrete and their general functions. (p. 509)

32 Diagram the control of pancreatic hormone secretion. (p. 510)

13.10 Other Endocrine Glands

33 Describe the location and general function of the pineal gland. (p. 511)

34 Describe the location and general function of the thymus. (p. 512)

35 Name five additional hormone-secreting organs. (p. 512)

13.11 Stress and Its Effects

36 Distinguish between a stressor and stress. (p. 513)

37 List several factors that cause physical and/or psychological stress. (p. 514)

38 Describe hormonal and nervous responses to stress. (p. 514)

13.12 Life-Span Changes

39 Levels of which hormones decrease with age? Which increase? (p. 515)

INTEGRATIVE ASSESSMENTS/CRITICAL THINKING

OUTCOMES 2.2, 13.3, 13.4, 13.5, 13.6

1. When a reactor of the Chernobyl Nuclear Power Station in the Ukraine exploded at 1:23 P.M. on April 26, 1986, a great plume of radioactive isotopes erupted into the air and spread for thousands of miles. Most of the isotopes emitted immediately following the blast were of the element iodine. Which of the glands of the endocrine system would be most seriously—and immediately—affected by the blast, and how do you think this would become evident in the nearby population?

OUTCOMES 4.5, 13.4, 13.9

2. Why might oversecretion of insulin reduce glucose uptake by nerve cells?

OUTCOMES 13.3, 13.5, 13.11, 13.12

3. A young mother feels shaky, distracted, and generally ill. She lives with her mother, who is dying. A friend tells the young woman, "It's just stress, it's all in your head." Is it?

OUTCOMES 13.4, 13.5

4. Growth hormone is administered to people who have pituitary dwarfism. Parents wanting their normal children to be taller have requested the treatment for them. Do you think that this is a wise request? Why or why not?

OUTCOMES 13.4, 13.5, 13.6, 13.8, 13.10

5. An adult has had her anterior pituitary removed. Which hormone supplements will she require?

OUTCOMES 13.5, 13.8

6. The adrenal cortex of a patient who has lost a large volume of blood will increase secretion of aldosterone. What effect will this increased secretion have on the patient's blood concentrations of sodium and potassium ions?

OUTCOMES 13.5, 13.8, 13.11

7. What problems might result from the prolonged administration of cortisol to a person with severe inflammatory disease?

WEB CONNECTIONS

Be sure to visit the text website at **www.mhhe.com/shier12** for answers to chapter assessments, additional quizzes, and interactive learning exercises.

ANATOMY & PHYSIOLOGY REVEALED

Anatomy & Physiology Revealed® (APR) includes cadaver photos that allow you to peel away layers of the human body to reveal structures beneath the surface. This program also includes animations, radiologic imaging, audio pronunciations, and practice quizzing. Check out **www.aprevealed.com**. APR has been proven to help improve student grades!

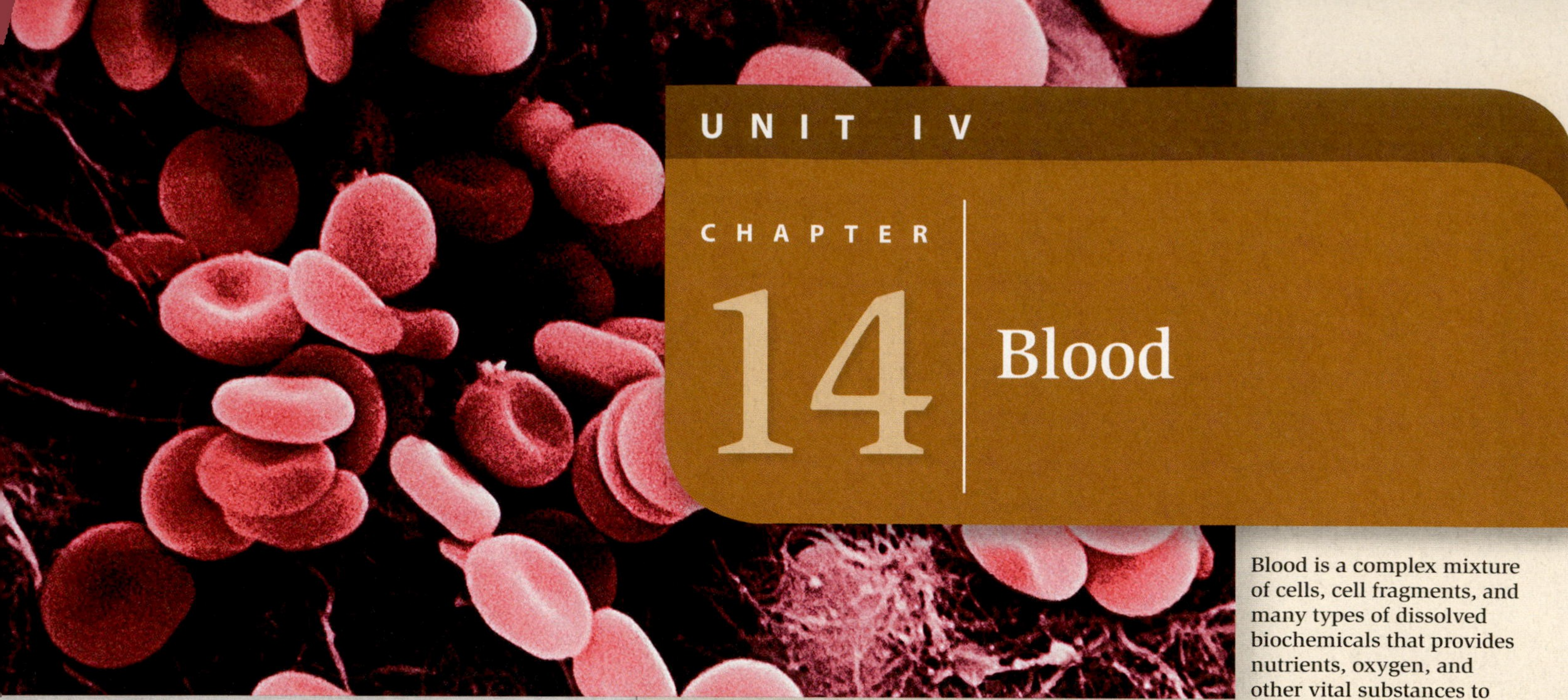

Blood is a complex mixture of cells, cell fragments, and many types of dissolved biochemicals that provides nutrients, oxygen, and other vital substances to our cells (1,800×).

UNDERSTANDING WORDS

agglutin-, to glue together: *agglutin*ation—clumping of red blood cells.

bil-, bile: *bil*irubin—pigment excreted in the bile.

-crit, to separate: hemato*crit*—percentage by volume of red blood cells in a blood sample, determined by separating the red blood cells from the plasma.

embol-, stopper: *embol*ism—obstruction of a blood vessel.

erythr-, red: *erythr*ocyte—red blood cell.

hema-, blood: *hema*tocrit—percentage of red blood cells in a given volume of blood.

hemo-, blood: *hemo*globin—red pigment responsible for the color of blood.

hepa-, liver: *hepa*rin—anticoagulant secreted by liver cells.

leuko-, white: *leuko*cyte—white blood cell.

-lys, to break up: fibrino*lys*in—protein-splitting enzyme that can digest fibrin.

macro-, large: *macro*phage—large phagocytic cell.

-osis, abnormal condition: leukocyt*osis*—overproduction of white blood cells.

-poie, make, produce: erythro*poie*tin—hormone that stimulates the production of red blood cells.

poly-, many: *poly*cythemia—overproduction of red blood cells.

-stasis, halt, make stand: hemo*stasis*—arrest of bleeding from damaged blood vessels.

thromb-, clot: *thromb*ocyte—blood platelet involved in the formation of a blood clot.

 LEARN  PRACTICE ASSESS

LEARNING OUTCOMES

After you have studied this chapter, you should be able to:

14.1 Introduction

1. Describe the general characteristics of blood, and discuss its major functions. (p. 523)
2. Distinguish among the formed elements of blood and the liquid portion of blood. (p. 523)

14.2 Blood Cells

3. Describe the origin of blood cells. (p. 524)
4. Explain the significance of red blood cell counts and how they are used to diagnose disease. (p. 527)
5. Discuss the life cycle of a red blood cell. (p. 527)
6. Summarize the control of red blood cell production. (p. 527)
7. Distinguish among the five types of white blood cells, and give the function(s) of each type. (p. 530)
8. Describe a blood platelet, and explain its functions. (p. 534)

14.3 Blood Plasma

9. Describe the functions of each of the major components of plasma. (p. 536)

14.4 Hemostasis

10. Define *hemostasis,* and explain the mechanisms that help to achieve it. (p. 538)
11. Review the major steps in coagulation. (p. 538)
12. Explain how to prevent coagulation. (p. 542)

14.5 Blood Groups and Transfusions

13. Explain blood typing and how it is used to avoid adverse reactions following blood transfusions. (p. 544)
14. Describe how blood reactions may occur between fetal and maternal tissues. (p. 547)

UNIVERSAL PRECAUTIONS

Blood can contain more than cells, nutrients, proteins, and water—a single drop from an infected individual can harbor billions of viruses. In the wake of the AIDS epidemic, in 1988 the U.S. Centers for Disease Control and Prevention (CDC) devised "universal precautions," which are specific measures that health-care workers should take to prevent transmission of bloodborne infectious agents in the workplace. The CDC singled out HIV and the hepatitis B virus. The guidelines grew out of earlier suggestions for handling patients suspected to have been exposed to viruses. The term *universal* refers to the assumption that *any* patient may have been exposed to a pathogen that can be transmitted in a body fluid.

Attention to safety in the health-care setting can prevent transmission of infectious diseases. The World Health Organization estimates that 4% to 7% of new infections worldwide are transmitted via unsafe injections. Specific recommendations include:

- Use of personal protection equipment, such as gloves, goggles, and masks.
- Engineering controls, such as fume hoods and sharps containers.
- Work-practice controls, such as enforcing hand washing before and after performing procedures.

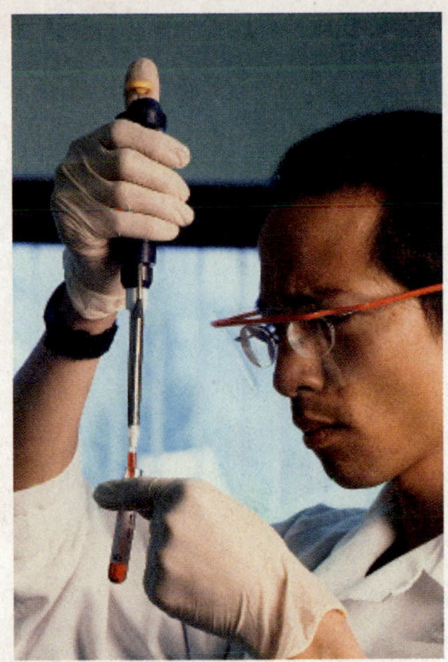

This health-care worker is practicing safe handling of a patient's blood sample.

Universal precautions were designed for, and work well in, preventing transmission of viral illnesses in settings already relatively safe, such as clinics. This isn't the case for outbreaks, natural disasters, and combat zones. For example, several pediatric nurses who aided neighbors infected with the Marburg virus in the isolated town of Uige in Angola, South Africa, in 2005 and hundreds of others died from this hemorrhagic fever. Headache, fever, vomiting, and diarrhea begin three to nine days after exposure to the virus. Then the person bleeds from all body openings, internally and under the skin. Plummeting blood pressure kills most infected individuals within a week, and anyone contacting their blood is in danger of infection. Victims must be isolated and not touched, but the scourge spreads because many family members become infected while tending their loved ones. In the 2005 outbreak, contaminated medical equipment caused the rapid and deadly spread of the infection. Nontrained clinic workers reused needles, and some people reused needles and intravenous equipment in their homes. Universal precautions could not contain the virus, which spreads in vomit, sweat, and saliva, as well as the huge volumes of blood. Another situation in which circumstances may overwhelm attempts at universal precautions is in treating patients with combat-related blast injuries. ■

14.1 INTRODUCTION

Blood signifies life, and for good reason—it has many vital functions. This complex mixture of cells, cell fragments, and dissolved biochemicals transports nutrients, oxygen, wastes, and hormones; helps maintain the stability of the interstitial fluid; and distributes heat. The blood, heart, and blood vessels form the cardiovascular system and link the body's internal and external environments.

Blood is a type of connective tissue whose cells are suspended in a liquid extracellular matrix. Blood is vital in transporting substances between body cells and the external environment, thereby promoting homeostasis.

Whole blood is slightly heavier and three to four times more viscous than water. Its cells, which form mostly in red bone marrow, include red blood cells and white blood cells. Blood also contains cellular fragments called blood platelets (fig. 14.1). The cells and platelets are termed "formed elements" of the blood, in contrast to the liquid portion.

Blood volume varies with body size, changes in fluid and electrolyte concentrations, and the amount of adipose tissue. Blood volume is typically about 8% of body weight. An average-sized adult has a blood volume of about 5 liters.

If a blood sample stands in a tube for awhile and is prevented from clotting, the cells separate from the liquid portion of the blood and settle to the bottom. Centrifuging the sample quickly packs the cells into the lower part of the centrifuge tube, as figure 14.2 shows. The percentage of cells and liquid in the blood sample can then be calculated.

A blood sample is usually about 45% red blood cells by volume. This percentage is called the **hematocrit** (HCT), or **packed cell volume** (PCV). The white blood cells and platelets account for less than 1%. The remaining blood sample, about 55%, is the clear, straw-colored **plasma** (plaz′mah). Plasma is a complex mixture that includes water, amino acids, proteins, carbohydrates, lipids, vitamins, hormones, electrolytes, and cellular wastes (fig. 14.3). Appendix B, Laboratory Tests of Clinical Significance (pp. 940–942), lists values for the hematocrit and other blood tests commonly performed on healthy individuals.

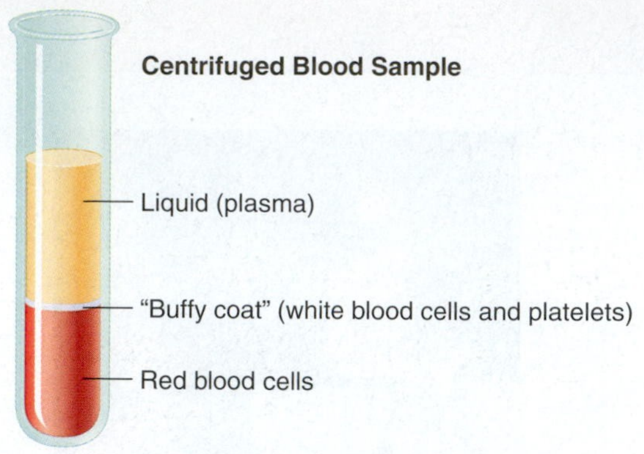

FIGURE 14.1 Blood consists of a liquid portion called plasma and a solid portion (the formed elements) that includes red blood cells, white blood cells, and platelets. (Note: When blood components are separated, the white blood cells and platelets form a thin layer, called the "buffy coat," between the plasma and the red blood cells.) Blood cells and platelets can be seen under a light microscope when a blood sample is smeared onto a glass slide.

PRACTICE

1. What are the major components of blood?
2. What factors affect blood volume?
3. How is hematocrit determined?

14.2 BLOOD CELLS

The Origin of Blood Cells

Blood cells originate in red bone marrow from **hematopoietic** (he″mat-o-poi-et′ik) **stem cells** or *hemocytoblasts* (fig. 14.4). A stem cell can divide to give rise to specialized (more differentiated) cells as well as more stem cells.

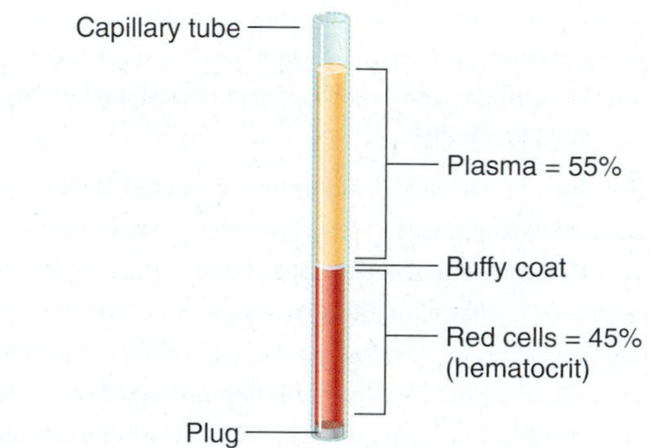

FIGURE 14.2 If a blood-filled capillary tube is centrifuged, the red cells pack in the lower portion and the percentage of red cells (hematocrit) can be determined. Values shown are within the normal range for healthy humans.

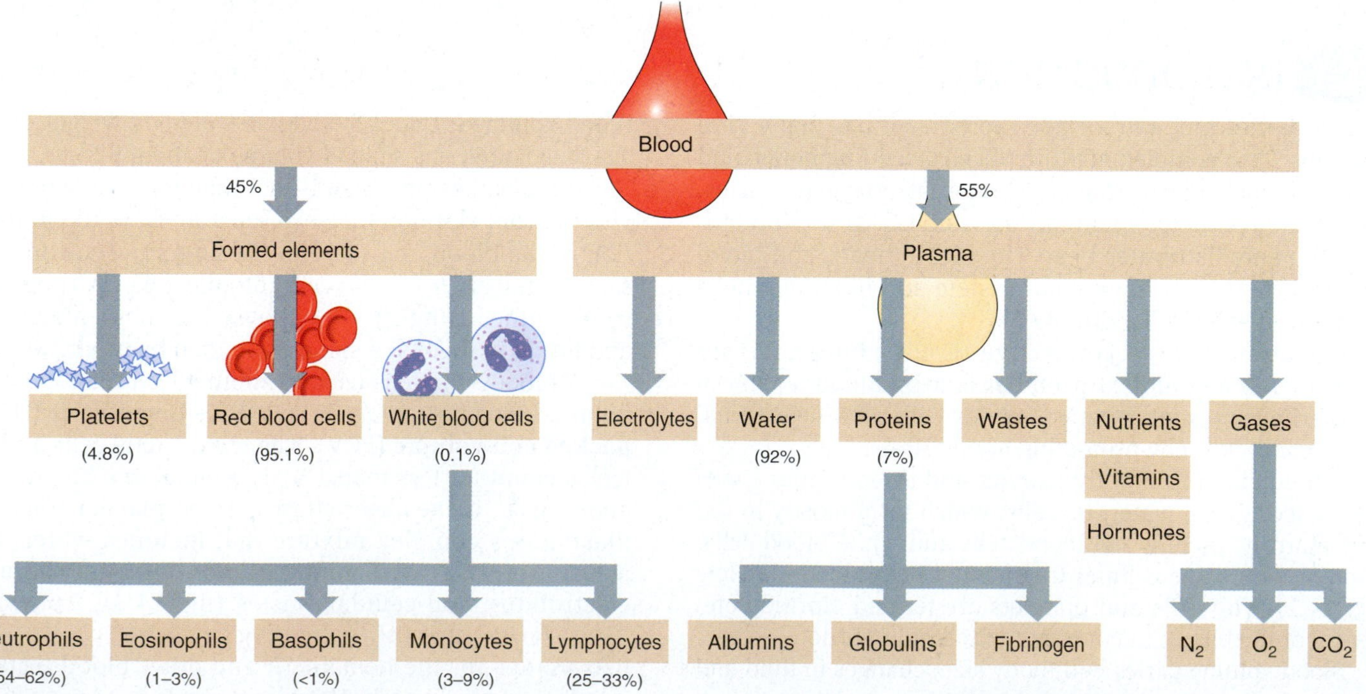

FIGURE 14.3 Blood composition. Blood is a complex mixture of formed elements in a liquid extracellular matrix, plasma.

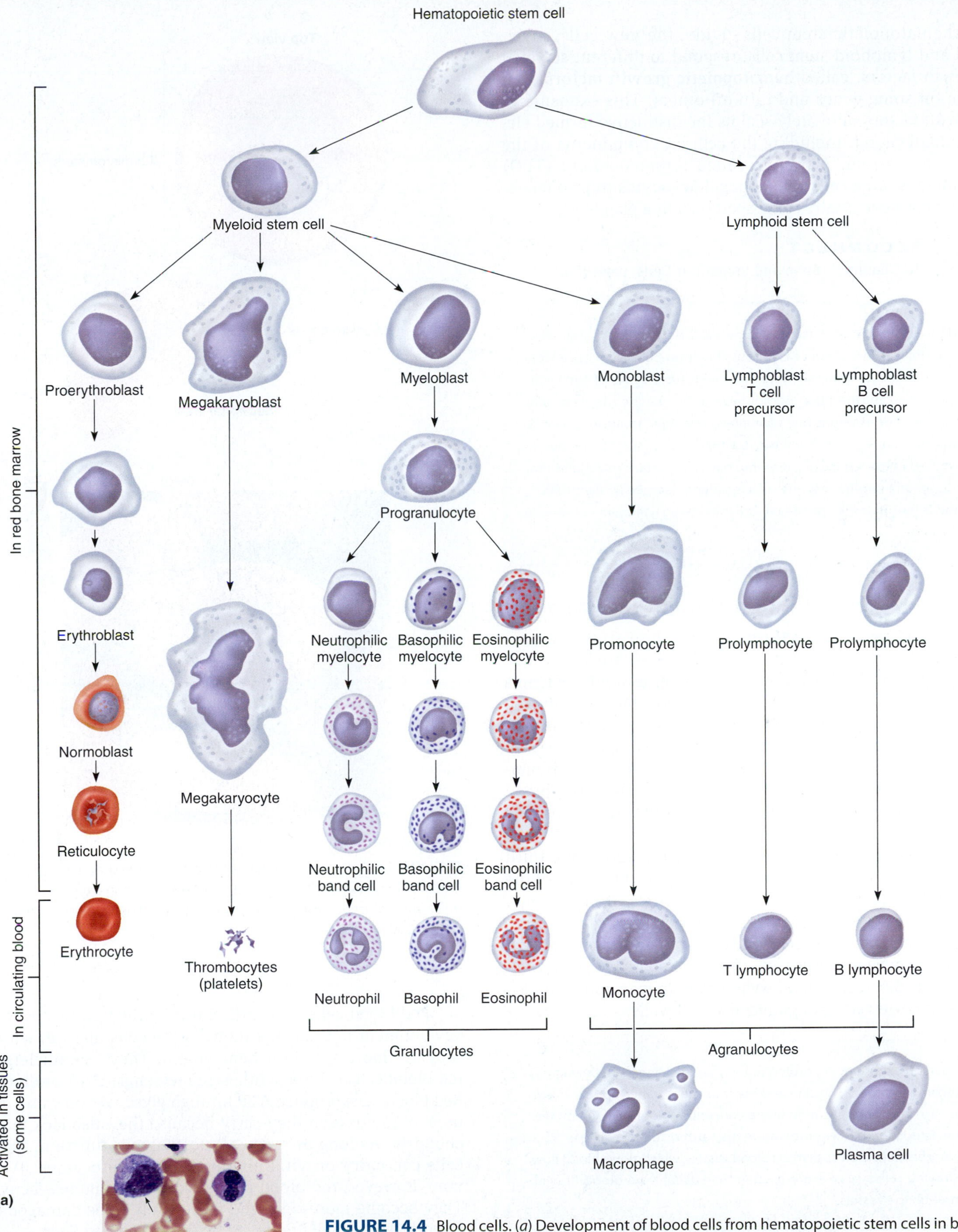

FIGURE 14.4 Blood cells. (*a*) Development of blood cells from hematopoietic stem cells in bone marrow. (*b*) Light micrograph of a hematopoietic stem cell (arrow) in red bone marrow (500×).

As hematopoietic stem cells divide, the new cells, myeloid and lymphoid stem cells, respond to different secreted growth factors, called **hematopoietic growth factors,** that turn on some genes and turn off others. This exposure to growth factors ultimately sculpts the distinctive formed elements of blood, including the cellular components of the immune system. A protein called *thrombopoietin* (TPO) stimulates large cells called **megakaryocytes** to proliferate. These cells eventually come apart, yielding platelets.

RECONNECT
To Chapter 3, Stem and Progenitor Cells, page 104.

Many companies offer to store a person's bone marrow stem cells or umbilical cord stem cells for possible future use to treat cancers of the blood or immune system. In the future, banks of such cells may be established that do not have to be matched to particular patients. For example, one company is testing a "multipotent adult progenitor cell" from bone marrow that can be easily expanded in number in the laboratory, does not have cell-surface molecules that trigger an immune response in a recipient, dampens inflammation, and is pluripotent—can be coaxed into giving rise to any of several cell types.

Characteristics of Red Blood Cells

Red blood cells, or **erythrocytes** (ĕ-rith′ro-sītz), are tiny, approximately 7.5 μm in diameter. They are biconcave discs, thin near their centers and thicker around their rims (fig. 14.5). This distinctive shape is an adaptation for the red blood cell's function of transporting gases; it increases the surface area through which gases can diffuse. The shape also places the cell membrane closer to oxygen-carrying **hemoglobin** (he″mo-glo′bin) molecules in the cell. A red blood cell's shape enables it to readily squeeze through the narrow capillaries.

The volume of each red blood cell is about one-third hemoglobin. This protein is responsible for the color of the blood. The rest of the cell mainly consists of membrane, water, electrolytes, and enzymes. When hemoglobin combines with oxygen, the resulting *oxyhemoglobin* is bright red; when the oxygen is released, the resulting *deoxyhemoglobin* is darker. Blood rich in deoxyhemoglobin may appear bluish when it is viewed through blood vessel walls.

A person experiencing prolonged oxygen deficiency (hypoxia) may become *cyanotic.* The skin and mucous membranes appear bluish due to an abnormally high blood concentration of deoxyhemoglobin. Exposure to low temperature may also result in cyanosis. Such exposure constricts superficial blood vessels, which slows blood flow, allowing removal of more oxygen than usual from blood flowing through the vessels.

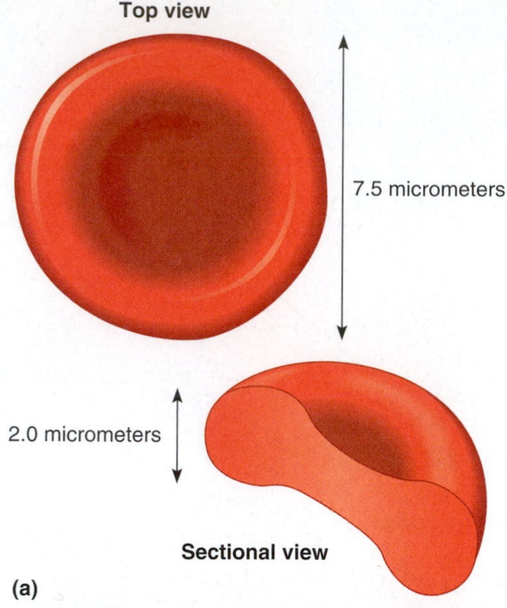

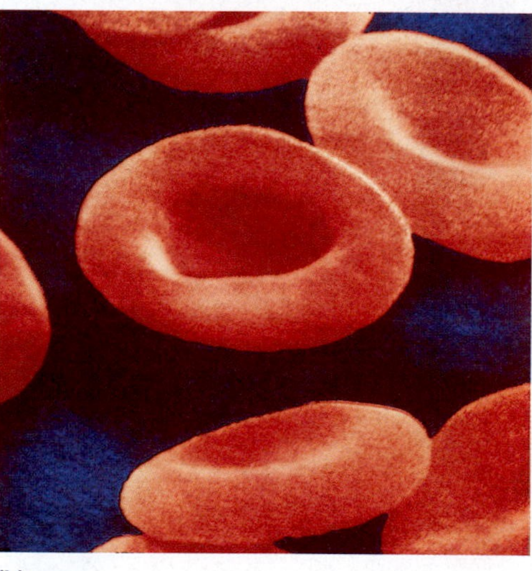

FIGURE 14.5 Red blood cells. (*a*) The biconcave shape of a red blood cell makes possible its function of transporting oxygen. (*b*) Falsely colored scanning electron micrograph of human red blood cells (5,000×).

Red blood cells have nuclei during their early stages of development but extrude them as the cells mature, which provides more space for hemoglobin. They lack nuclei, so red blood cells cannot synthesize messenger RNA or divide. Red blood cells produce ATP through glycolysis only and use none of the oxygen they carry because they also lack mitochondria. As long as cytoplasmic enzymes function, these cells can carry on vital energy-releasing processes. With time, however, red blood cells become less and less active. They become more rigid and more likely to be damaged or worn and eventually removed by the spleen and liver.

In *sickle cell disease*, a single DNA base mutation changes one amino acid in the protein portion of hemoglobin, causing hemoglobin to crystallize in a low oxygen environment. This bends the red blood cells containing the hemoglobin into a sickle shape, which blocks circulation in small vessels, causing excruciating joint pain and damaging many organs. As the spleen works harder to recycle the too short-lived red blood cells, infection becomes likely.

Children with sickle cell disease are typically diagnosed at birth and receive antibiotics daily for years to prevent infection. Hospitalization for blood transfusions may be necessary if the person experiences painful sickling "crises" of blocked circulation.

A cancer drug called hydroxyurea is used to reactivate production of a form of hemoglobin normally produced only in a fetus. The fetal hemoglobin slows the crystallization of the sickle hemoglobin. Sickling is delayed, which enables the red blood cells to reach the lungs—where fresh oxygen restores the cells' normal shapes. A bone marrow transplant or umbilical cord stem cell transplant can completely cure sickle cell disease, but has a 15% risk of fatality. Thanks to pain management, transfusions, drugs, and transplants, half of all sickle cell patients now live beyond their fiftieth birthdays. In the 1960s affected individuals rarely survived childhood.

PRACTICE 

4 How do blood cells form?

5 Describe a red blood cell.

6 How does the biconcave shape of a red blood cell make possible its function?

7 What is the function of hemoglobin?

Red Blood Cell Counts

The number of red blood cells in a microliter (μL or mcL or 1 mm^3) of blood is called the *red blood cell count* (RBCC or RCC). Although this number varies from time to time even in healthy individuals, the typical range for adult males is 4,600,000–6,200,000 cells per microliter, and that for adult females is 4,200,000–5,400,000 cells per microliter. For children, the average range is 4,500,000–5,100,000 cells per microliter. These values may vary slightly with the hospital, physician, and type of equipment used to make blood cell counts. The number of red blood cells generally increases after several days following strenuous exercise or an increase in altitude.

An increasing number of circulating red blood cells increases the blood's *oxygen-carrying capacity*, so changes in this number may affect health. For this reason, red blood cell counts are routinely consulted to help diagnose and evaluate the courses of various diseases.

Red Blood Cell Production and Its Control

Red blood cell formation (erythropoiesis) initially occurs in the yolk sac, liver, and spleen. After birth, these cells are produced almost exclusively by tissue lining the spaces in bones, filled with red bone marrow.

RECONNECT

To Chapter 7, Blood Cell Formation, pages 202–203.

In the red bone marrow, hematopoietic stem cells divide and give rise to **erythroblasts** (ĕ-rith′ro-blastz) that can synthesize 2 million to 3 million hemoglobin molecules per second. The erythroblasts also divide and give rise to many new cells. The nuclei of these newly formed cells soon shrink and are extruded by being pinched off in thin coverings of cytoplasm and cell membrane. The resulting cells are erythrocytes. Some of these young red cells may contain a netlike structure (reticulum) for a day or two. This network is the remainder of the endoplasmic reticulum, and such cells are called **reticulocytes** (rĕ-tik′u-lo-sitz). This is the stage that exits the bone marrow to enter the blood. When the reticulum degenerates, the cells are fully mature.

The average life span of a red blood cell is 120 days. During that time, it travels through the body about 75,000 times. Many red blood cells are removed from the circulation each day, yet the number in the circulating blood remains relatively stable. These numbers suggest a homeostatic control of the rate of red blood cell production.

A *negative feedback mechanism* using the hormone **erythropoietin** (e-rith″ro-poi′ĕ-tin) (EPO) controls the rate of red blood cell formation. In response to prolonged oxygen deficiency, EPO is released, from the kidneys and to a lesser extent from the liver. (In a fetus, the liver is the main site of EPO production.) At high altitudes, for example, although the percentage of oxygen in the air remains the same, the atmospheric pressure decreases, reducing availability of oxygen. The amount of oxygen delivered to the tissues initially decreases. As figure 14.6 shows, this drop in oxygen triggers release of EPO, which travels via the blood to the red bone marrow and stimulates increased erythrocyte production.

After a few days, many new red blood cells begin to appear in the circulating blood. The increased rate of production continues until the number of erythrocytes in the circulation supplies sufficient oxygen to tissues. When the availability of oxygen returns to normal, EPO release decreases and the rate of red blood cell production returns to normal.

RECONNECT

To Chapter 13, Using Hormones to Improve Athletic Performance, page 489.

Other conditions can lower oxygen levels and stimulate EPO release. These include loss of blood, which decreases the oxygen-carrying capacity of the cardiovascular system, and chronic lung diseases, which decrease the respiratory surface area available for gas exchange.

PRACTICE 

8 What is the typical red blood cell count for an adult male? For an adult female?

9 Where are red blood cells produced?

10 How does a red blood cell change as it matures?

11 How is red blood cell production controlled?

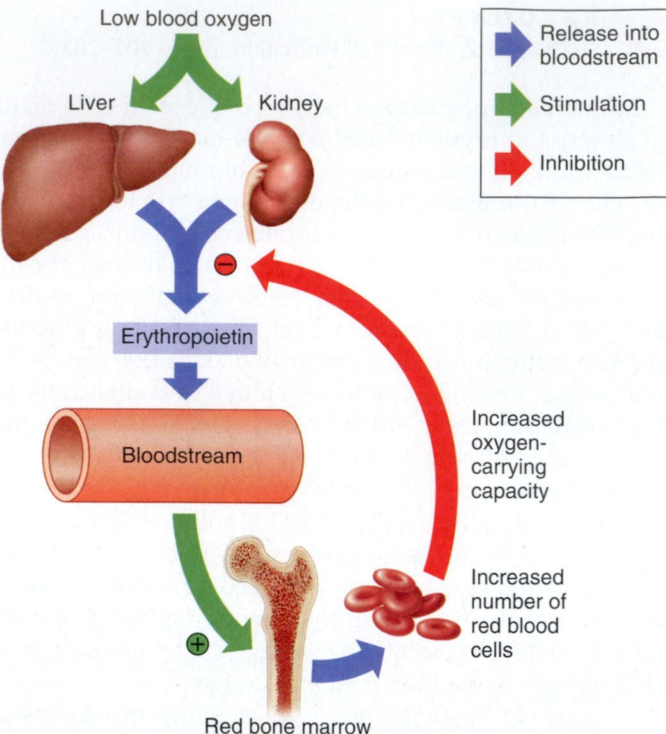

FIGURE 14.6 Low blood oxygen causes the kidneys and liver to release erythropoietin. Erythropoietin travels to red bone marrow and stimulates the production of red blood cells that carry oxygen to tissues.

Dietary Factors Affecting Red Blood Cell Production

The availability of two B-complex vitamins—vitamin B_{12} and folic acid—significantly influences red blood cell production. These vitamins are required for DNA synthesis, so they are necessary for the growth and division of all cells. Cell division is rapid in hematopoietic tissue, so this tissue is especially vulnerable to deficiency of either of these vitamins. Lack of vitamin B_{12} is usually due to a disorder in the stomach lining rather than to a dietary deficiency, because parietal cells in the stomach secrete a substance called *intrinsic factor* required to absorb vitamin B_{12}.

Iron is required for hemoglobin synthesis. Although much of the iron released during the decomposition of hemoglobin is available for reuse, some iron is lost each day and must be replaced. Only a small fraction of ingested iron is absorbed. Iron absorption is slow, although the rate varies with the total amount of iron in the body. When iron stores are low, absorption rate increases, and when the tissues are becoming saturated with iron, the rate greatly decreases. Figure 14.7 summarizes the life cycle of a red blood cell. Table 14.1 summarizes the dietary factors that affect red blood cell production.

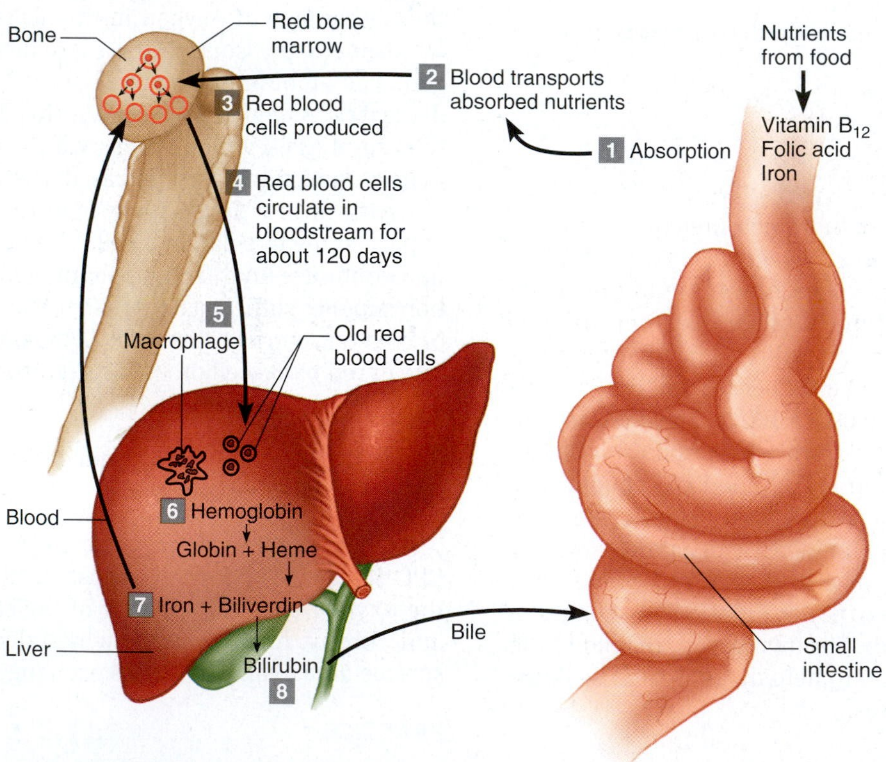

FIGURE 14.7 Life cycle of a red blood cell. (*1*) The small intestine absorbs essential nutrients. (*2*) Blood transports nutrients to red bone marrow. (*3*) In the red bone marrow, red blood cells arise from the division of less-specialized progenitor cells. (*4*) Mature red blood cells are released into the bloodstream, where they circulate for about 120 days. (*5*) Macrophages destroy old red blood cells in the liver and spleen. (*6*) Hemoglobin liberated from red blood cells is broken down into heme and globin. (*7*) Iron from heme returns to red bone marrow and is reused. (*8*) Biliverdin and bilirubin are excreted in bile.

TABLE 14.1 | Dietary Factors Affecting Red Blood Cell Production

Substance	Source	Function
Vitamin B_{12} (requires intrinsic factor for absorption via small intestine)	Absorbed from small intestine	DNA synthesis
Iron	Absorbed from small intestine; conserved during red blood cell destruction and made available for reuse	Hemoglobin synthesis
Folic acid	Absorbed from small intestine	DNA synthesis

Vitamin C increases absorption of iron in the digestive tract. Drinking orange juice with a meal is a good way to boost iron intake. Drinking tea with a meal reduces absorption of iron because tannic acid in tea binds the iron and prevents its absorption.

A deficiency of red blood cells or a reduction in the amount of hemoglobin they contain results in a condition called **anemia.** This reduces the oxygen-carrying capacity of the blood, and the affected person may appear pale and lack energy. Table 14.2 describes types of anemia, and figure 14.8 shows normal red blood cells and those of someone who has anemia. A pregnant woman may become anemic if she doesn't eat iron-rich foods, because her blood volume increases due to fluid retention to accommodate the requirements of the fetus. This increased blood volume decreases the hematocrit. Clinical Application 14.1 discusses another disorder of red blood cells that affected British royalty.

In the absence of intrinsic factor, vitamin B_{12} absorption decreases, causing the red bone marrow to form abnormally large, irregularly shaped, thin-membraned fragile cells. This condition, called *pernicious anemia,* can cause permanent brain damage if not treated promptly with vitamin B_{12} injections. Taking excess folic acid can mask a vitamin B_{12} deficiency.

PRACTICE

12 Which vitamins are necessary for red blood cell production?

13 Why is iron required for the formation of red blood cells?

Destruction of Red Blood Cells

Red blood cells are elastic and flexible, and they readily bend as they pass through small blood vessels. With age, however, these cells become more fragile, and may be damaged by passing through capillaries, particularly those in active muscles.

Damaged or worn red blood cells rupture as they pass through the spleen or liver. In these organs, macrophages (see chapter 5, p. 155) phagocytize and destroy damaged red blood cells and their contents. Hemoglobin molecules liberated from the red blood cells break down into their four component polypeptide "globin" chains, each surrounding a heme group.

TABLE 14.2 | Types of Anemia

Type	Cause	Defect
Aplastic anemia	Toxic chemicals, radiation	Damaged bone marrow
Hemolytic anemia	Toxic chemicals	Red blood cells destroyed
Iron deficiency anemia	Dietary lack of iron	Hemoglobin deficient
Pernicious anemia	Inability to absorb vitamin B_{12}	Excess of immature cells
Sickle cell disease	Defective gene	Red blood cells abnormally shaped
Thalassemia	Defective gene	Hemoglobin deficient; red blood cells short-lived

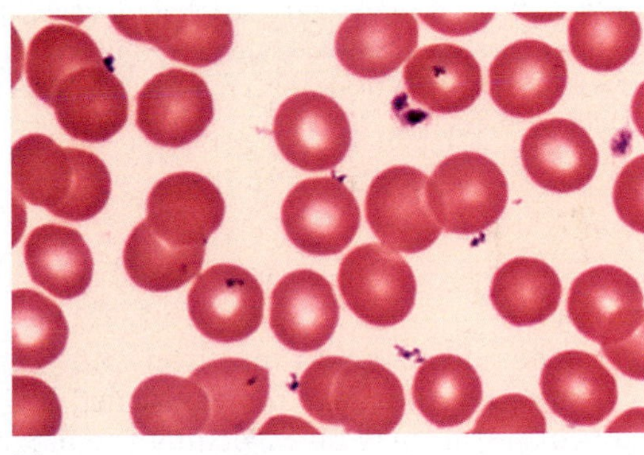

(a)

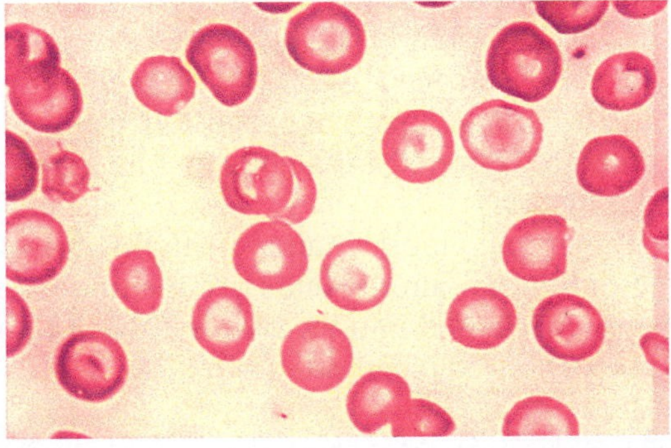

(b)

FIGURE 14.8 Red blood cells, normal and abnormal. (*a*) Light micrograph of normal human erythrocytes (1,000×). (*b*) Light micrograph of erythrocytes from a person with hypochromic anemia (1,000×).

14.1 CLINICAL APPLICATION

King George III and Porphyria Variegata

King George III, who ruled England at the time of the American revolution, inherited an abnormality of hemoglobin synthesis that, combined with arsenic poisoning, caused a strange sequence of signs and symptoms.

At age fifty, the king first experienced abdominal pain and constipation, followed by weak limbs, fever, a rapid pulse, hoarseness, and dark red urine. Next, nervous system symptoms began, including insomnia, headaches, visual problems, restlessness, delirium, convulsions, and stupor. His confused and racing thoughts, combined with his ripping off his wig and running about naked while at the peak of a fever, convinced court observers that the king was mad. Just as Parliament was debating his ability to rule, he mysteriously recovered.

But George III's plight was far from over. He suffered a relapse thirteen years later, then again three years after that. Always the symptoms appeared in the same order—abdominal pain, fever, and weakness progressing to the nervous system symptoms. When an attack in 1811 caused permanent stupor, the Prince of Wales dethroned him. George III lived for several more years, experiencing further episodes.

In George III's time, physicians were permitted to do little to the royal body, basing diagnoses on what the patient told them. Twentieth-century researchers found that George III's red urine was caused by an inborn error of metabolism. In porphyria variegata, because of the absence of an enzyme, part of the blood pigment hemoglobin, called a porphyrin ring, is routed into the urine instead of being broken down and metabolized by cells. Porphyrin builds up, causing many of the nervous system symptoms. Examination of the medical records of King George III's descendants reveals several of them also had porphyria variegata. The underlying defect in red blood cell recycling had appeared in its various guises as different problems.

King George III's porphyria, however, was highly unusual—it appeared later in life, with frequent, severe episodes. In 2005, British and Australian researchers sought another causative factor—heavy metal poisoning. Arsenic disturbs hemoglobin synthesis and in combination with lead found in certain alcoholic beverages of the time, triggers severe porphyria symptoms.

Clues to arsenic poisoning came from the king's hair, which had been on exhibit at the Science Museum in London since 1928. The hairs had failed to yield DNA, but they did have whopping levels of arsenic—17 parts-per-million (ppm), compared to 0.05 ppm in control hairs. A level above 1 ppm is considered evidence of poisoning!

The presence of arsenic throughout the royal hairs indicated a slow, steady exposure, rather than contamination. Researchers identified the source of the poison in the king's medical records. While at a "provincial madhouse," he had been given "emetic tartar," a concoction of potassium antimony tartrate. When mined, the antimony is often contaminated with arsenic. The king was forced to take the medication, which, in combination with his mutant genes, caused the episodes of porphyria.

People with various porphyria-related symptoms, including reddish teeth, pink urine, excess hair, and photosensitivity (avoidance of daylight), may have inspired the vampire and werewolf legends (see fig. 4.26). Today, porphyria variegata remains rare, and people who have it are often misdiagnosed with a seizure disorder. Unfortunately, some seizure medications and anesthetics worsen symptoms. ■

Heme further decomposes into iron and a greenish pigment called **biliverdin.** The iron, combined with a protein called *transferrin,* may be carried by the blood to the hematopoietic (red blood cell-forming) tissue in the red bone marrow and reused in synthesizing new hemoglobin. About 80% of the iron is stored in the liver cells in the form of an iron-protein complex called *ferritin.* In time, the biliverdin is converted to an orange pigment called **bilirubin.** Biliverdin and bilirubin are excreted in the bile as bile pigments (see fig. 14.7 and fig. 14.9).

The polypeptide globin chains break down into amino acids. The individual amino acids are metabolized by the macrophages or released into the blood. Table 14.3 summarizes the process of red blood cell destruction.

PRACTICE

14 What happens to damaged red blood cells?

15 What are the products of hemoglobin breakdown?

Types of White Blood Cells

White blood cells, or **leukocytes** (lu′ko-sītz), protect against disease. Leukocytes develop from hematopoietic stem cells in the red bone marrow in response to hormones, much as red cells form from precursors upon stimulation from EPO. These hormones fall into two groups—**interleukins** (in″ter-lu-kinz) and **colony-stimulating factors** (CSFs). Interleukins are numbered, while most colony-stimulating factors are named for the cell population they stimulate. Blood transports white blood cells to sites of infection. White blood cells may then leave the bloodstream, as described later in this chapter.

Normally, five types of white cells are in circulating blood. They differ in size, the composition of cytoplasm, the shape of the nucleus, and their staining characteristics, and are named for these distinctions. For example, leukocytes with granular cytoplasm are called **granulocytes** (gran′u-lo-sītz″), whereas those without cytoplasmic granules are called **agranulocytes** (a-gran′u-lo-sītz).

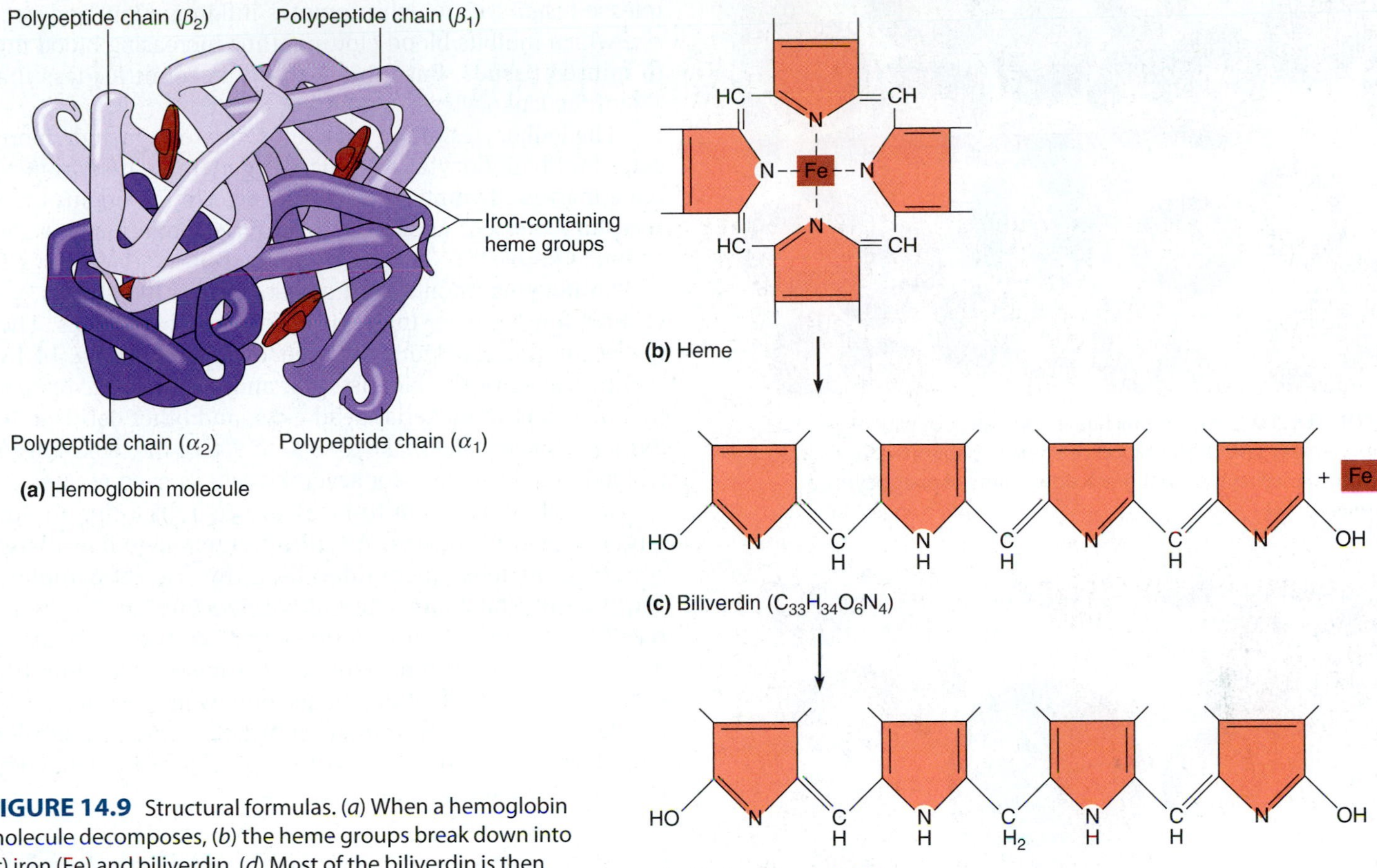

FIGURE 14.9 Structural formulas. (*a*) When a hemoglobin molecule decomposes, (*b*) the heme groups break down into (*c*) iron (Fe) and biliverdin. (*d*) Most of the biliverdin is then converted to bilirubin.

TABLE 14.3 | Major Events in Red Blood Cell Destruction

1. Squeezing through the capillaries of active tissues damages red blood cells.
2. Macrophages in the spleen and liver phagocytize damaged red blood cells.
3. Hemoglobin from the red blood cells is decomposed into heme and globin.
4. Heme is decomposed into iron and biliverdin.
5. Iron is made available for reuse in the synthesis of new hemoglobin or is stored in the liver as ferritin.
6. Some biliverdin is converted into bilirubin.
7. Biliverdin and bilirubin are excreted in bile as bile pigments.
8. The globin is broken down into amino acids metabolized by macrophages or released into the blood.

A typical granulocyte is about twice the size of a red blood cell. The members of this group include neutrophils, eosinophils, and basophils. These cells develop in the red bone marrow in much the same manner as red blood cells. However, they have a short life span, averaging about twelve hours.

Neutrophils (nu′tro-filz) have fine cytoplasmic granules that appear light purple with a combination of acid and base stains. The nucleus of an older neutrophil is lobed and consists of two to five sections (segments, so these cells are sometimes called *segs*) connected by thin strands of chromatin (fig. 14.10). They are also called *polymorpho-nuclear leukocytes* (PMNs) due to the variation of nucleus shape from cell to cell. Younger neutrophils are also called *bands* because their nuclei are C-shaped. Neutrophils are the first white blood cells to arrive at an infection site. These cells phagocytize bacteria, fungi, and some viruses. Neutrophils account for 54% to 62% of the leukocytes in a typical blood sample from an adult.

Eosinophils (e″o-sin′o-filz) contain coarse, uniformly sized cytoplasmic granules that stain deep red in acid stain (fig. 14.11). The nucleus usually has only two lobes (bilobed). Eosinophils moderate allergic reactions and defend against parasitic worm infestation. These cells make up 1% to 3% of the total number of circulating leukocytes.

Basophils (ba′so-filz) are similar to eosinophils in size and in the shape of their nuclei. However, they have fewer, more irregularly shaped cytoplasmic granules than eosinophils, and these granules appear deep blue in basic stain (fig. 14.12). A basophil's granules can obscure a view of the nucleus. Basophils migrate to damaged tissues where they release *histamine,* which promotes inflammation, and *heparin,* which inhibits blood clotting, thus increasing blood flow to injured tissues. Basophils usually account for less than 1% of the leukocytes.

The leukocytes of the agranulocyte group include monocytes and lymphocytes. Monocytes generally arise from red bone marrow. Lymphocytes are formed in the organs of the lymphatic system as well as in the red bone marrow (see chapter 16, p. 628).

Monocytes (mon′o-sītz) are the largest blood cells, two to three times greater in diameter than red blood cells. Their nuclei are spherical, kidney-shaped, oval, or lobed (fig. 14.13). Monocytes leave the bloodstream and become *macrophages* that phagocytize bacteria, dead cells, and other debris in the tissues. They usually make up 3% to 9% of the leukocytes in a blood sample and live for several weeks or even months.

Lymphocytes (lim′fo-sītz) are usually only slightly larger than erythrocytes. A typical lymphocyte has a large, spherical nucleus surrounded by a thin rim of cytoplasm (fig. 14.14). The major types of lymphocytes are *T cells* and *B cells,* both important in *immunity.* T cells directly attack microorganisms, tumor cells, and transplanted cells (see chapter 16, p. 630). B cells produce *antibodies* (see chapter 16, p. 632), which are proteins that attack foreign molecules. Lymphocytes account for 25% to 33% of the circulating leukocytes. They may live for years.

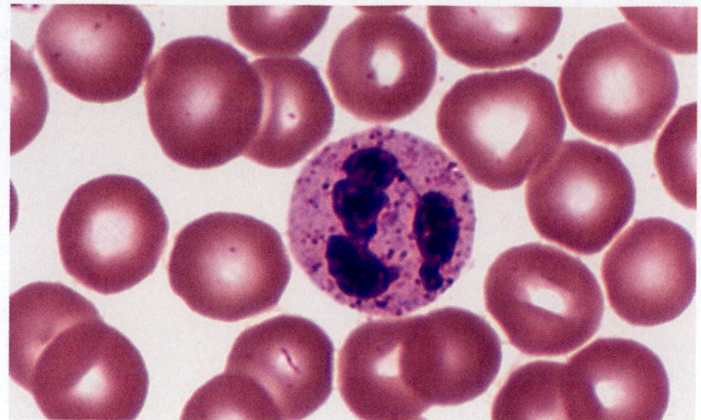

FIGURE 14.10 A neutrophil has a lobed nucleus with two to five components (2,000×). This blood cell type has abundant lysosomes, which contain enzymes that break down parts of phagocytized bacteria.

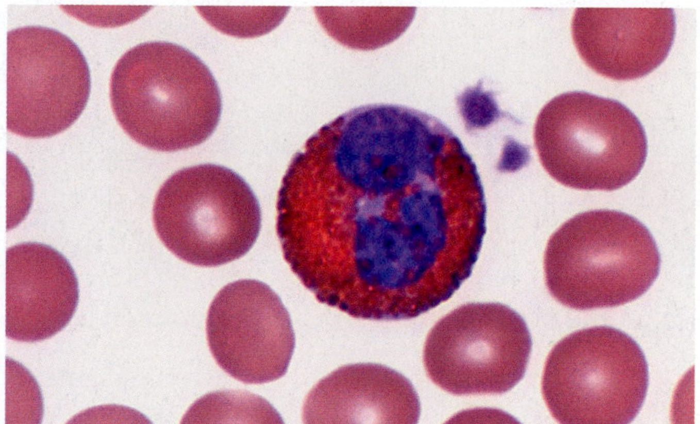

FIGURE 14.11 An eosinophil has red-staining cytoplasmic granules (2,000×). This type of white blood cell kills certain parasites and helps to control inflammation and allergic reactions.

PRACTICE

16 Which hormones are necessary for the differentiation of white blood cells from hematopoietic stem cells in red bone marrow?

17 Distinguish between granulocytes and agranulocytes.

18 List five types of white blood cells, and explain how they differ from one another.

19 Describe the function of each type of white blood cell.

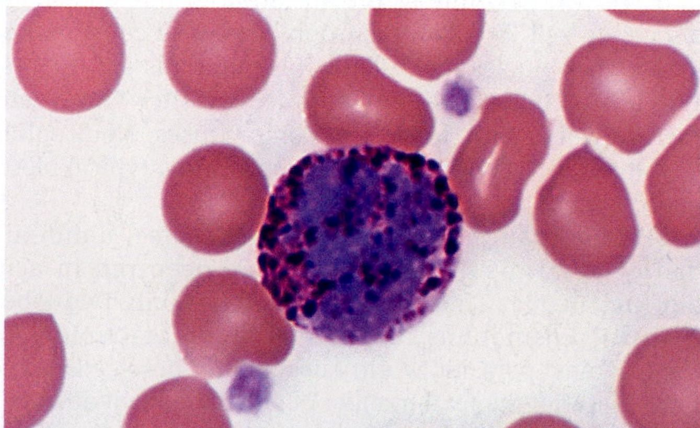

FIGURE 14.12 A basophil has cytoplasmic granules that stain deep blue (2,000×). This type of white blood cell produces heparin, which prevents inappropriate blood clotting, and histamines, which increase circulation to injured tissues. Basophils also take part in certain allergic reactions.

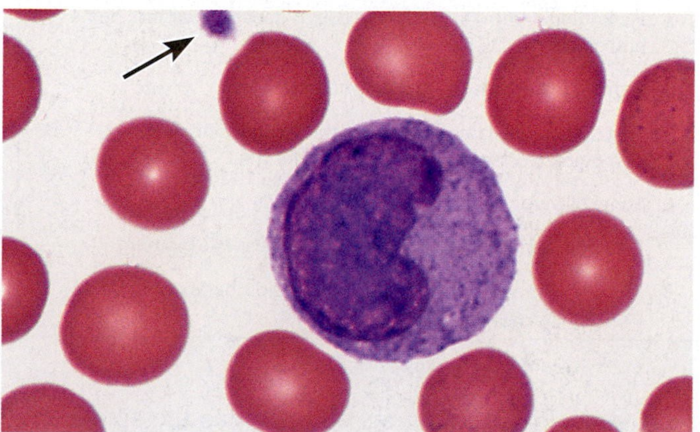

FIGURE 14.13 A monocyte is the largest of the blood cells (2,000×). It may leave the bloodstream and become a macrophage, which is a wandering phagocytic cell that destroys damaged red blood cells. Note the platelet indicated by the arrow.

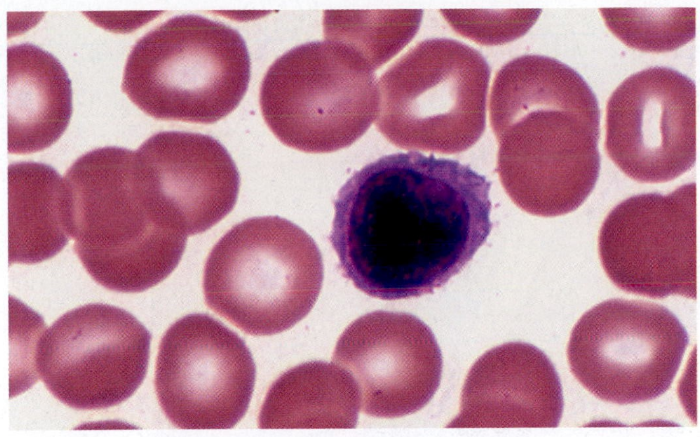

FIGURE 14.14 The lymphocyte, the smallest of the white blood cells, has a large, round nucleus (2,000×). Lymphocytes carry out the immune response and are discussed further in chapter 16.

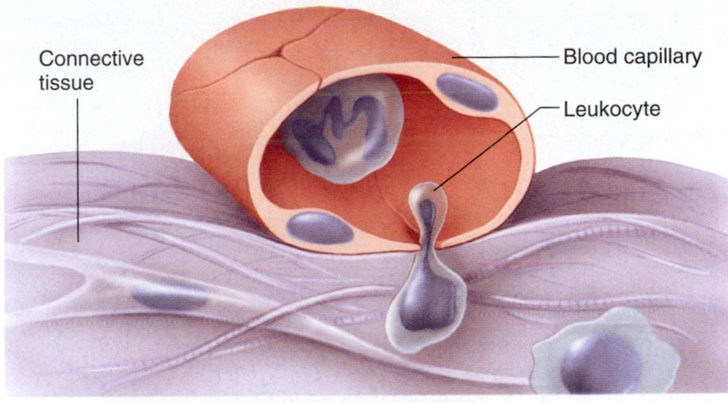

FIGURE 14.15 In a type of movement called diapedesis, leukocytes squeeze between the cells of a capillary wall and enter the tissue space outside the blood vessel.

Functions of White Blood Cells

White blood cells protect against infection in various ways. Some leukocytes phagocytize bacterial cells in the body, and others produce antibodies.

Leukocytes can squeeze between the cells that form the walls of the smallest blood vessels. This movement, called **diapedesis** (di″ah-pĕ-de′sis), allows the white blood cells to leave the circulation (fig. 14.15). A series of proteins called cellular adhesion molecules help guide leukocytes to the site of injury. Once outside the blood, leukocytes move through interstitial spaces using a form of self-propulsion called *ameboid motion.*

RECONNECT
To Chapter 3, Cellular Adhesion Molecules, page 81.

The most mobile and active phagocytic leukocytes are neutrophils and monocytes. Although neutrophils are unable to ingest particles much larger than bacterial cells, monocytes can engulf larger structures. Monocytes contain numerous lysosomes, filled with digestive enzymes that break down organic molecules in captured bacteria. Neutrophils and monocytes often become so engorged with digestive products and bacterial toxins that they also die.

When microorganisms invade human tissues, basophils respond by releasing biochemicals that dilate local blood vessels. For example, histamine dilates smaller blood vessels and makes the smallest vessels leaky. As more blood flows through the smallest vessels, the tissues redden and copious fluids leak into the interstitial spaces. The swelling that this inflammatory reaction produces delays the spread of invading microorganisms into other regions (see chapter 16, p. 627). At the same time, damaged cells release chemicals that attract leukocytes. This phenomenon is called **positive chemotaxis** (poz′ĭ-tiv ke″mo-tak′sis) and, when combined with diapedesis, brings many white blood cells into inflamed areas quickly (fig. 14.16).

As bacteria, leukocytes, and damaged cells accumulate in an inflamed area, a thick fluid called *pus* often forms and remains while the invading microorganisms are active. If the pus is not moved to the outside of the body or into a body cavity, it may remain trapped in the tissues for some time. Eventually, surrounding cells absorb it.

PRACTICE 

20 How do white blood cells fight infection?

21 Which white blood cells are the most active phagocytes?

22 How do white blood cells reach microorganisms outside blood vessels?

White Blood Cell Counts

The procedure used to count white blood cells is similar to that used for counting red blood cells. However, before a *white blood cell count* (WBCC or WCC) is made, the red blood cells in the blood sample are destroyed so they will not be mistaken for white blood cells. Normally, a microliter of blood includes 4,500 to 10,000 white blood cells.

The total number and percentages of different white blood cell types are of clinical interest. A rise in the number of circulating white blood cells may indicate infection. A total number of white blood cells exceeding 10,000 per microliter of blood constitutes **leukocytosis** (lu″ko-si-to′sis), indicating acute infection, such as appendicitis. Leukocytosis may also follow vigorous exercise, emotional disturbances, or great loss of body fluids.

A total white blood cell count below 4,500 per microliter of blood is called **leukopenia** (lu″ko-pe′ne-ah). Such a deficiency may accompany typhoid fever, influenza, measles, mumps, chickenpox, AIDS, or poliomyelitis. Leukopenia may also result from anemia or from lead, arsenic, or mercury poisoning.

A *differential white blood cell count* (DIFF) lists percentages of the types of leukocytes in a blood sample. This test is useful because the relative proportions of white blood cells

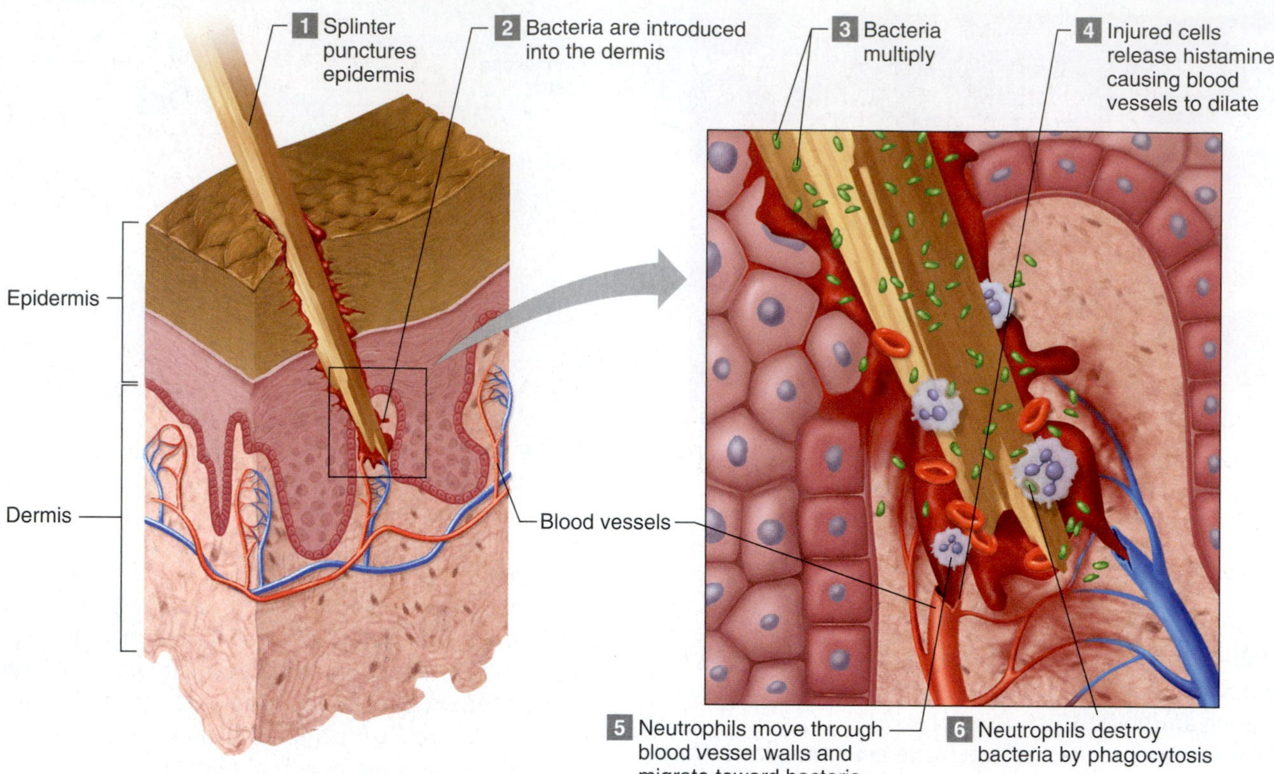
1 Splinter punctures epidermis
2 Bacteria are introduced into the dermis
3 Bacteria multiply
4 Injured cells release histamine, causing blood vessels to dilate
Epidermis
Dermis
Blood vessels
5 Neutrophils move through blood vessel walls and migrate toward bacteria
6 Neutrophils destroy bacteria by phagocytosis

FIGURE 14.16 When bacteria invade the tissues, leukocytes migrate into the region and destroy the microbes by phagocytosis.

may change in particular diseases. The number of neutrophils, for instance, usually increases during bacterial infections, and eosinophils may become more abundant during certain parasitic infections and allergic reactions. In HIV infection and AIDS, the numbers of a type of lymphocyte called helper T cells plummet.

Table 14.4 lists some disorders that alter the numbers of particular types of white blood cells. Clinical Application 14.2 examines leukemia, cancer of white blood cells.

PRACTICE

23 What is the normal human white blood cell count?

24 Distinguish between leukocytosis and leukopenia.

25 What is a differential white blood cell count?

TABLE 14.4 | Abnormal White Blood Cell Numbers

White Blood Cell Population Change	Illness
Elevated lymphocytes	Hairy cell leukemia, whooping cough, mononucleosis
Elevated eosinophils	Tapeworm infestation, hookworm infestation, allergic reactions
Elevated monocytes	Typhoid fever, malaria, tuberculosis
Elevated neutrophils	Bacterial infections
Too few helper T cells (lymphocytes)	AIDS

Blood Platelets

Platelets (plāt′letz), also called **thrombocytes** (throm′bo-sītz), are not complete cells. They arise from very large cells in the red bone marrow, called *megakaryocytes* (meg″ah-kar′o-sītz), that fragment like a shattered plate, releasing small sections of cytoplasm—platelets—into the circulation. The larger fragments of the megakaryocytes shrink and become platelets as they pass through the blood vessels of the lungs (see fig. 14.13). Megakaryocytes, and therefore platelets, develop from hematopoietic stem cells (see fig. 14.4) in response to the hormone **thrombopoietin** (throm″bo-poi′ĕ-tin).

Each platelet lacks a nucleus and is less than half the size of a red blood cell. It is capable of ameboid movement and may live for about ten days. In normal blood, the *platelet count* varies from 130,000 to 360,000 platelets per microliter.

Platelets help repair damaged blood vessels by sticking to broken surfaces. They release **serotonin,** which contracts smooth muscles in the vessel walls, reducing blood flow. Table 14.5 summarizes the characteristics of blood cells and platelets.

PRACTICE

26 What is the normal human blood platelet count?

27 What is the function of blood platelets?

TABLE 14.5 | Cellular Components of Blood

Component	Description	Number Present	Function
Red blood cell (erythrocyte)	Biconcave disc without a nucleus, about one-third hemoglobin	4,200,000 to 6,200,000 per microliter	Transports oxygen and carbon dioxide
White blood cell (leukocyte)		4,500 to 10,000 per microliter	Destroys pathogenic microorganisms and parasites and removes worn cells
Granulocytes	About twice the size of red blood cells; cytoplasmic granules are present		
Neutrophil	Nucleus with two to five lobes; cytoplasmic granules stain light purple in combined acid and base stains	54%–62% of white blood cells present	Phagocytizes small particles
Eosinophil	Nucleus bilobed; cytoplasmic granules stain red in acid stain	1%–3% of white blood cells present	Kills parasites and moderates allergic reactions
Basophil	Nucleus lobed; cytoplasmic granules stain blue in basic stain	Less than 1% of white blood cells present	Releases heparin and histamine
Agranulocytes	Cytoplasmic granules are absent		
Monocyte	Two to three times larger than a red blood cell; nuclear shape varies from spherical to lobed	3%–9% of white blood cells present	Phagocytizes large particles
Lymphocyte	Only slightly larger than a red blood cell; its nucleus nearly fills cell	25%–33% of white blood cells present	Provides immunity
Platelet (thrombocyte)	Cytoplasmic fragment	130,000 to 360,000 per microliter	Helps control blood loss from broken vessels

14.3 BLOOD PLASMA

Plasma is the clear, straw-colored, liquid part of the blood in which the cells and platelets are suspended. It is approximately 92% water and contains a complex mixture of organic and inorganic biochemicals. Functions of plasma constituents include transporting nutrients, gases, and vitamins; helping to regulate fluid and electrolyte balance; and maintaining a favorable pH.

Plasma Proteins

By weight, **plasma proteins** are the most abundant dissolved substances (solutes) in plasma. These proteins remain in the blood and interstitial fluids and ordinarily are not used as energy sources. The three main types of plasma proteins are albumins, globulins, and fibrinogen. The groups differ in composition and function.

Albumins (al-bu′minz) are the smallest of the plasma proteins, yet account for 60% of these proteins by weight. They are synthesized in the liver, and because they are so plentiful, albumins are an important determinant of the *osmotic pressure* of the plasma.

Recall from chapter 3 (pp. 93–94) that the presence of an impermeant solute on one side of a selectively permeable membrane creates an osmotic pressure and that water always diffuses toward a greater osmotic pressure. Plasma proteins are too large to pass through the capillary walls, they are impermeant, and they create an osmotic pressure that holds water in the capillaries despite blood pressure forcing water out of capillaries by filtration (see chapter 3, p. 94). The term *colloid osmotic pressure* is used to describe this osmotic effect due to the plasma proteins.

By maintaining the colloid osmotic pressure of plasma, albumins and other plasma proteins help regulate water movement between the blood and the tissues. In doing so, they help control blood volume, which, in turn, directly affects blood pressure (see chapter 15, p. 582). For this reason, it is important that the concentration of plasma proteins remains relatively stable. Albumins also bind and transport certain molecules, such as bilirubin, free fatty acids, many hormones, and certain drugs.

> If the concentration of plasma proteins falls, tissues swell, a condition called *edema*. This may result from starvation or a protein-deficient diet, either of which requires the body to use protein for energy, or from an impaired liver that cannot synthesize plasma proteins. As the concentration of plasma proteins drops, so does the colloid osmotic pressure, sending fluids into the interstitial spaces.

Globulins (glob′u-linz), which make up about 36% of the plasma proteins, can be further subdivided into *alpha, beta,* and *gamma globulins.* The liver synthesizes alpha and beta globulins, which have a variety of functions, including transport of lipids and fat-soluble vitamins. Lymphatic tissues produce the gamma globulins, which are a type of antibody (see chapter 16, p. 633).

Fibrinogen (fi-brin′o-jen), which constitutes about 4% of the plasma proteins, plays a primary role in blood coagulation. Synthesized in the liver, it is the largest of the plasma proteins. The function of fibrinogen is discussed later in this chapter under the section "Blood Coagulation" on page 538. Table 14.6 summarizes the characteristics of the plasma proteins.

14.2 CLINICAL APPLICATION

Leukemia

When twenty-three-year-old magazine editor Erin Zammett Ruddy had a routine physical examination in late 2001, she expected reassurance that her healthy lifestyle had indeed been keeping her healthy (figure 14A). After all, she felt great. What she got, a few days later, was a shock. Instead of having 4,500 to 10,000 white blood cells per microliter of blood, she had more than ten times that number—and many of the cells were cancerous. Erin had chronic myeloid leukemia (CML). Her red bone marrow was flooding her circulation with too many granulocytes, most of them poorly differentiated.

FIGURE 14A "My third bone marrow biopsy—you never get used to the pain," said Erin Zammett Ruddy. The drug Gleevec has treated her leukemia. Bone marrow biopsies are required at regular intervals, even after successful treatment, to be certain that the disease has not returned.

Another type of leukemia is lymphoid, in which the cancer cells are lymphocytes, produced in lymph nodes. Both myeloid and lymphoid leukemia can cause fatigue, headaches, nosebleeds and other bleeding, frequent respiratory infections, fever, bone pain, bruising, and other signs of slow blood clotting. The symptoms arise from the disrupted proportions of the blood's formed elements and their malfunction (figure 14B).

Immature white blood cells increase the risk of infection. Leukemic cells crowd out red blood cells and their precursors in the red marrow, causing anemia and resulting fatigue. Platelet deficiency (thrombocytopenia) slows clotting time, causing bruises and bleeding. Finally, spread of the cancer cells outside the marrow painfully weakens surrounding bone. Eventually, without treatment, cancer cells spread outside the cardiovascular system, causing other tissues that would normally not produce white blood cells to do so.

Leukemia is also classified as acute or chronic. An acute condition appears suddenly, symptoms progress rapidly, and without treatment, death occurs in a few months. Chronic forms begin more slowly and may remain undetected for months or even years or, in rare cases, decades. Without treatment, life expectancy after symptoms develop is about three years.

Erin was diagnosed just as a new drug was flying through clinical trials. She became one of the first patients to take Gleevec, now standard treatment for CML and several other cancers. Gleevec specifically targets cancer cells by nestling into ATP-binding sites on a type of enzyme called a tyrosine kinase, which blocks the message to divide. The drug worked so well that it was approved in just ten weeks. People with leukemia have other options too, such as standard chemotherapies, and bone marrow and stem cell transplants.

TABLE 14.6 | Plasma Proteins

Protein	Percentage of Total	Origin	Function
Albumin	60%	Liver	Helps maintain colloid osmotic pressure
Globulin	36%		
Alpha globulins		Liver	Transport lipids and fat-soluble vitamins
Beta globulins		Liver	Transport lipids and fat-soluble vitamins
Gamma globulins		Lymphatic tissues	Constitute the antibodies of immunity
Fibrinogen	4%	Liver	Plays a key role in blood coagulation

PRACTICE

28 List three types of plasma proteins.

29 How do albumins help maintain water balance between the blood and the tissues?

30 Which of the globulins functions in immunity?

31 What is the role of fibrinogen?

Gases and Nutrients

The most important *blood gases* are oxygen and carbon dioxide. Plasma also contains a considerable amount of dissolved nitrogen, which ordinarily has no physiological function. Chapter 19 (pp. 762–766) discusses blood gases and their transport.

The *plasma nutrients* include amino acids, simple sugars, nucleotides, and lipids absorbed from the digestive tract. For example, plasma transports glucose from the small intestine to the liver, where it may be stored as glycogen or con-

Another way that leukemia treatment is improving is refining diagnosis by identifying the proteins that leukemia cells produce. This information is used to predict which drugs are most likely to be effective and which will cause intolerable side effects or not work in particular individuals. For example, some people with acute lymphoblastic leukemia (ALL), diagnosed on the basis of the appearance of the cancer cells in a blood smear, do not respond to standard chemotherapy. However, DNA microarray (also called DNA chip) technology revealed that the cells of patients who do not improve produce different proteins than the cancer cells of patients who do respond to the drugs used to treat ALL—the nonresponders have a different form of leukemia, called mixed-lineage leukemia. These people respond to different drugs.

As for Erin, she went off Gleevec to have a baby and is doing well. If her leukemia should return, and if Gleevec becomes ineffective, she will have newer drugs to try that bind more strongly to cancer cells or that fit slightly altered resistant cancer cells. ■

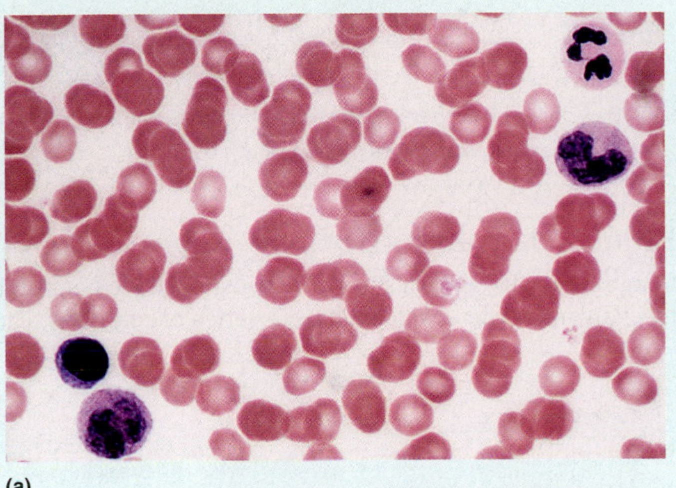

(a)

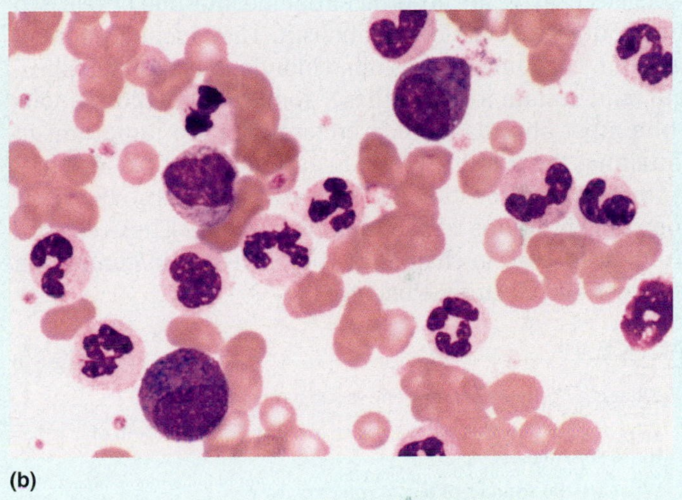

(b)

FIGURE 14B Leukemia and blood cells. (*a*) Normal blood cells (700×). (*b*) Blood cells from a person with granulocytic leukemia, a type of myeloid leukemia (700×). Note the increased number of leukocytes.

As a rule, blood gases are evaluated using a fresh sample of whole blood obtained from an artery. This blood is cooled to decrease the rates of metabolic reactions, and an anticoagulant is added to prevent clotting. In the laboratory, the levels of oxygen and carbon dioxide of the blood are determined, the blood pH is measured, and the plasma bicarbonate concentration is calculated. Such information is used to diagnose and treat disorders of circulation, respiration, and electrolyte balance. Appendix B (p. 941) lists average values for these laboratory tests.

verted to fat. If blood glucose concentration drops below the normal range, glycogen may be broken down into glucose, as described in chapter 13 (p. 509).

Recently absorbed amino acids are also carried to the liver. Here they may be used to manufacture proteins or deaminated and used as an energy source (see chapter 18, p. 704).

Plasma lipids include fats (triglycerides), phospholipids, and cholesterol. Lipids are not water soluble and plasma is almost 92% water, so these lipids are carried in the plasma by joining with proteins, forming lipoprotein complexes.

PRACTICE

32 Which gases are in plasma?

33 Which nutrients are in plasma?

Nonprotein Nitrogenous Substances

Molecules that contain nitrogen atoms but are not proteins comprise a group called **nonprotein nitrogenous substances** (NPNs). In plasma, this group includes amino acids, urea, uric acid, creatine (kre′ah-tin), and creatinine (kre-at′ĭ-nin). Amino acids come from protein digestion and amino acid absorption. Urea and uric acid are products of protein and nucleic acid catabolism, respectively, and creatinine results from the metabolism of creatine. As discussed in chapter 9 (p. 294), creatine is present as **creatine phosphate** in muscle and brain tissues as well as in the blood, where it stores energy in phosphate bonds, much like those of ATP molecules.

Normally, the concentration of nonprotein nitrogenous substances in plasma remains relatively stable because protein intake and use are balanced with excretion of nitrogenous wastes. Because about half of the NPN substances is urea, which the kidneys ordinarily excrete, a rise in the blood urea nitrogen (BUN) may suggest a kidney disorder. Excess protein catabolism or infection may also elevate BUN.

Plasma Electrolytes

Recall that electrolytes release ions when dissolved in water. Plasma contains a variety of these ions, also called *electrolytes.* Plasma electrolytes are absorbed from the intestine or released as by-products of cellular metabolism. They include sodium, potassium, calcium, magnesium, chloride, bicarbonate, phosphate, and sulfate ions. Of these, sodium and chloride ions are the most abundant. Bicarbonate ions are important in maintaining the osmotic pressure and the pH of plasma, and like other plasma constituents, they are regulated so that their blood concentrations remain relatively stable. These electrolytes are discussed in chapter 21 (pp. 815–818) in connection with water and electrolyte balance.

PRACTICE 

34 What is a nonprotein nitrogenous substance?

35 Why does kidney disease increase the blood concentration of these substances?

36 What are the sources of plasma electrolytes?

14.4 HEMOSTASIS

Hemostasis (he″mo-sta′sis) refers to the stoppage of bleeding, which is vitally important when blood vessels are damaged. Following an injury to the blood vessels, several actions may help to limit or prevent blood loss, including blood vessel spasm, platelet plug formation, and blood coagulation. These mechanisms are most effective in minimizing blood losses from small vessels. Injury to a larger vessel may result in a severe hemorrhage that requires special treatment.

Blood Vessel Spasm

Cutting or breaking a smaller blood vessel stimulates the smooth muscles in its wall to contract, called *vasospasm.* Blood loss lessens almost immediately, and the ends of the severed vessel may close completely. This effect results from direct stimulation of the vessel wall as well as from reflexes elicited by pain receptors in the injured tissues.

Although the reflex response may last only a few minutes, the effect of the direct stimulation usually continues for about thirty minutes. By then, a blockage called a *platelet plug* has formed, and blood is coagulating. Also, platelets release serotonin, which contracts smooth muscles in the blood vessel walls. This vasoconstriction further helps to reduce blood loss.

Platelet Plug Formation

Platelets adhere to exposed ends of injured blood vessels. They adhere to any rough surface, particularly to the collagen in connective tissue underlying the endothelial lining of blood vessels.

When platelets contact collagen, their shapes change drastically, and many spiny processes begin to protrude from their membranes. At the same time, platelets adhere to each other, forming a platelet plug in the vascular break. A plug may control blood loss from a small break, but a larger break may require a blood clot to halt bleeding. Figure 14.17 shows the steps in platelet plug formation.

PRACTICE 

37 What is hemostasis?

38 How does a blood vessel spasm help control bleeding?

39 Describe the formation of a platelet plug.

Blood Coagulation

Coagulation (ko-ag″u-la′shun), the most effective hemostatic mechanism, forms a *blood clot* in a series of reactions, each

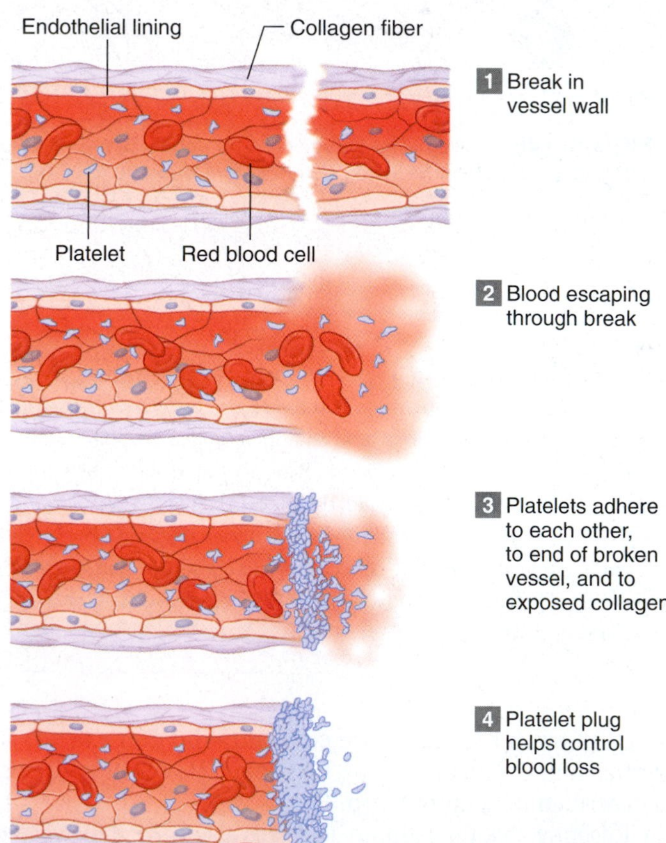
Endothelial lining
Collagen fiber
1 Break in vessel wall
Platelet
Red blood cell
2 Blood escaping through break
3 Platelets adhere to each other, to end of broken vessel, and to exposed collagen
4 Platelet plug helps control blood loss

FIGURE 14.17 Steps in platelet plug formation.

one activating the next in a *cascade.* Coagulation may occur extrinsically or intrinsically. Release of biochemicals from broken blood vessels or damaged tissues triggers the **extrinsic clotting mechanism.** Blood contact with foreign surfaces in the absence of tissue damage stimulates the **intrinsic clotting mechanism.** The following sections describe these responses.

Blood coagulation is complex and uses many biochemicals called *clotting factors.* They are designated by Roman numerals indicating the order of their discovery. Vitamin K is necessary for some clotting factors to function. Whether the blood coagulates depends on the balance between factors that promote coagulation (procoagulants) and others that inhibit it (anticoagulants). Normally, the anticoagulants prevail, and the blood does not clot. However, as a result of injury (trauma), biochemicals that favor coagulation may increase in concentration, and the blood may coagulate.

The major event in blood clot formation is conversion of the soluble plasma protein *fibrinogen* (factor I) into insoluble threads of the protein **fibrin** (fig. 14.18). Activation of certain plasma proteins by still other protein factors triggers conversion of fibrinogen to fibrin. Table 14.7 summarizes the three primary hemostatic mechanisms: blood vessel spasm, platelet plug formation, and blood coagulation.

Extrinsic Clotting Mechanism

The extrinsic clotting mechanism is triggered when blood contacts damaged blood vessel walls or tissues outside blood vessels. Such damaged tissues release a complex of substances called *tissue thromboplastin* (factor III), that is associated with disrupted cell membranes. Tissue thromboplastin activates factor VII, which combines with and activates factor X. Further, factor X combines with and activates factor V. These reactions, which also require calcium ions (factor IV), lead to production and release of *prothrombin activator* by the platelets.

Prothrombin (factor II) is an alpha globulin that the liver continually produces and is thus a normal constituent of plasma. In the presence of calcium ions, prothrombin activator converts prothrombin into **thrombin** (factor IIa). Thrombin, in turn, catalyzes a reaction that fragments fibrinogen (factor I). The fibrinogen fragments join, forming long threads of fibrin. Fibrinogen is a soluble plasma protein, but fibrin is insoluble. Thrombin also activates factor XIII, which strengthens and stabilizes fibrin threads.

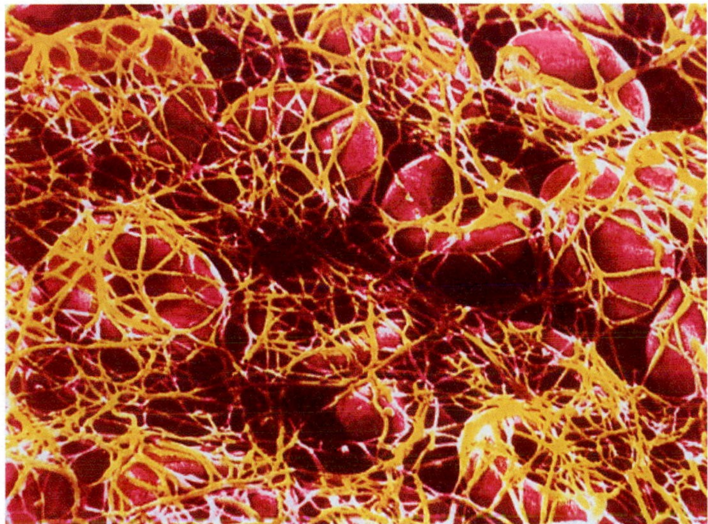

FIGURE 14.18 A scanning electron micrograph of fibrin threads (2,800×).

Once fibrin threads form, they stick to exposed surfaces of damaged blood vessels, creating a meshwork that entraps blood cells and platelets. The resulting mass is a blood clot, which may block a vascular break and prevent further blood loss.

The amount of prothrombin activator in the blood is directly proportional to the degree of tissue damage. Once a blood clot begins to form, it promotes additional clotting, because thrombin also acts directly on blood clotting factors other than fibrinogen, causing prothrombin to form still more thrombin. This type of self-initiating action is an example of a **positive feedback system,** in which the original action stimulates more of the same type of action. Such a mechanism produces unstable conditions and can operate for only a short time in a living system, because life requires the maintenance of a stable internal environment (see chapter 1, p. 9).

Normally, blood flow throughout the body prevents formation of a massive clot in the cardiovascular system by rapidly carrying excess thrombin away and keeping its concentration too low to enhance further clotting. In addition, a substance called *antithrombin,* in the blood and on the surfaces of endothelial cells that line blood vessels, limits thrombin formation. Consequently, blood coagulation usually happens only in blood that is standing still or moving slowly, and clotting ceases where a clot contacts circulating blood.

In *disseminated intravascular clotting,* coagulation is abnormally activated in several regions of the cardiovascular system. This condition is usually associated with bacterial infection or bacterial toxins in the blood or with a disorder causing widespread tissue damage. Many small clots form and obstruct blood flow into various tissues and organs, particularly the kidneys. As plasma clotting factors and platelets are depleted, severe bleeding occurs.

TABLE 14.7 | Hemostatic Mechanisms

Mechanism	Stimulus	Effect
Blood vessel spasm	Direct stimulus to vessel wall or to pain receptors; platelets release serotonin, a vasoconstrictor	Smooth muscles in vessel wall contract reflexly; vasoconstriction helps maintain prolonged vessel spasm
Platelet plug formation	Exposure of platelets to rough surfaces or to collagen of connective tissue	Platelets adhere to rough surfaces and to each other, forming a plug
Blood coagulation	Cellular damage and blood contact with foreign surfaces activate factors that favor coagulation	Blood clot forms as a result of a series of reactions, terminating in the conversion of fibrinogen into fibrin

Intrinsic Clotting Mechanism

Unlike extrinsic clotting, all of the components necessary for intrinsic clotting are in the blood. Activation of a substance called the *Hageman factor* (factor XII) initiates intrinsic clotting. This happens when blood is exposed to a foreign surface such as collagen in connective tissue instead of the smooth endothelial lining of intact blood vessels or when blood is stored in a glass container. Activated factor XII activates factor XI, which activates factor IX. Factor IX then joins with factor VIII and platelet phospholipids to activate factor X. These reactions, which also require calcium ions, lead to the production of prothrombin activator. The subsequent steps of blood clot formation are the same as those described for the extrinsic mechanism (fig. 14.19). Table 14.8 compares

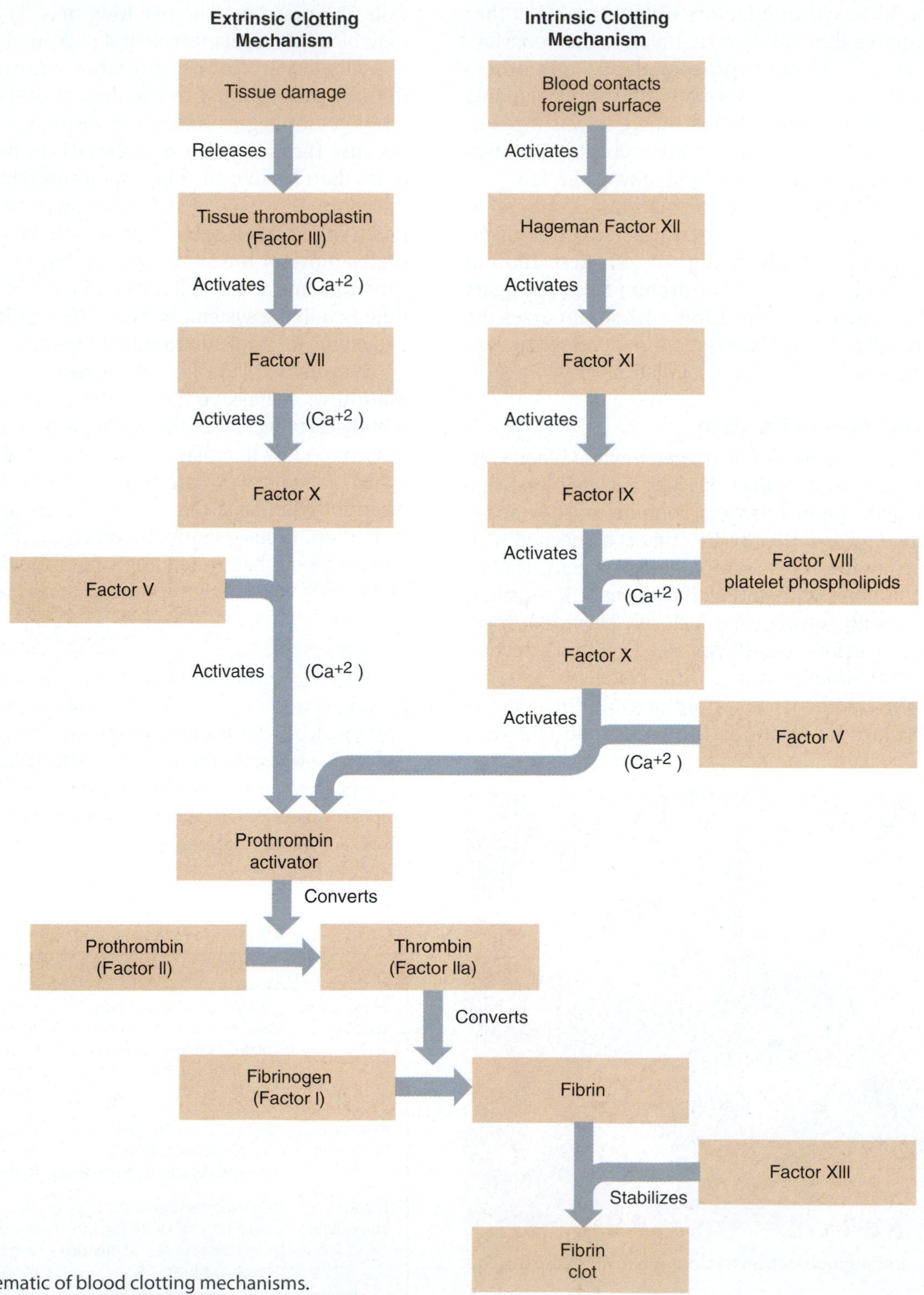

FIGURE 14.19 Schematic of blood clotting mechanisms.

extrinsic and intrinsic clotting mechanisms. Table 14.9 lists the clotting factors, their sources, and clotting mechanisms.

Laboratory tests commonly used to evaluate blood coagulation mechanisms include *prothrombin time* (PT) and *partial thromboplastin time* (PTT). These tests measure the time it takes for fibrin threads to form in a sample of blood plasma. The prothrombin time test checks the extrinsic clotting mechanism, whereas the partial thromboplastin test evaluates intrinsic clotting.

Fate of Blood Clots

After a blood clot forms, it soon begins to retract as the tiny processes extending from the platelet membranes adhere to strands of fibrin within the clot and contract. The blood clot shrinks, pulling the edges of the broken vessel closer together and squeezing a fluid called **serum** from the clot. Serum is essentially plasma minus all of its fibrinogen and most other clotting factors. Platelets associated with a blood clot also release *platelet-derived growth factor* (PDGF), which stimulates smooth muscle cells and fibroblasts to repair damaged blood vessel walls.

Fibroblasts (see chapter 5, p. 153) invade blood clots that form in ruptured vessels, producing connective tissue with many fibers throughout the clots, which helps strengthen and seal vascular breaks. Many clots, including those that form in tissues as a result of blood leakage (hematomas), disappear in time. In clot dissolution, fibrin threads absorb a plasma protein called *plasminogen* (profibrinolysin). Then a substance called plasminogen activator released from the lysosomes of damaged tissue cells converts plasminogen to *plasmin*. Plasmin is a protein-splitting enzyme that can digest fibrin threads and other proteins associated with blood clots. Plasmin formation may dissolve a whole clot; however, clots that fill large blood vessels are seldom removed naturally.

A blood clot abnormally forming in a vessel is a **thrombus** (throm′bus). A clot that dislodges, or a fragment of a clot that breaks loose and is carried away by the blood flow, is called an **embolus** (em′bo-lus). Generally, emboli continue to move until they reach narrow places in vessels where they may lodge and block blood flow, causing an **embolism.**

A blood clot forming in a vessel that supplies a vital organ, such as the heart (coronary thrombosis) or the brain (cerebral thrombosis), blocks blood flow and kills tissues the vessel serves (*infarction*) and may be fatal. A blood clot that

TABLE 14.8 | Blood Coagulation

Steps	Extrinsic Clotting Mechanism	Intrinsic Clotting Mechanism
Trigger	Damage to vessel or tissue	Blood contacts foreign surface
Initiation	Tissue thromboplastin	Hageman factor
Series of reactions involving several clotting factors and calcium ions (Ca^{+2}) lead to the production of:	Prothrombin activator	Prothrombin activator
Prothrombin activator and calcium ions cause the conversion of:	Prothrombin to thrombin	Prothrombin to thrombin
Thrombin causes fragmentation, then joining of:	Fibrinogen to fibrin	Fibrinogen to fibrin

TABLE 14.9 | Clotting Factors

Clotting Factor		Source	Mechanism(s)
I	(fibrinogen)	Synthesized in liver	Extrinsic and intrinsic
II	(prothrombin)	Synthesized in liver, requires vitamin K	Extrinsic and intrinsic
III	(tissue thromboplastin)	Damaged tissue	Extrinsic
IV	(calcium ions)	Diet, bone	Extrinsic and intrinsic
V	(proaccelerin)	Synthesized in liver, released by platelets	Extrinsic and intrinsic
VII	(serum prothrombin conversion accelerator)	Synthesized in liver, requires vitamin K	Extrinsic
VIII	(antihemophilic factor)	Released by platelets and endothelial cells	Intrinsic
IX	(plasma thromboplastin component)	Synthesized in liver, requires vitamin K	Intrinsic
X	(Stuart-Prower factor)	Synthesized in liver, requires vitamin K	Extrinsic and intrinsic
XI	(plasma thromboplastin antecedent)	Synthesized in liver	Intrinsic
XII	(Hageman factor)	Synthesized in liver	Intrinsic
XIII	(fibrin-stabilizing factor)	Synthesized in liver, released by platelets	Extrinsic and intrinsic

There is no clotting factor VI. The chemical once thought to be factor VI is apparently a combination of activated factors V and X.

travels and then blocks a vessel that supplies a vital organ, such as the lungs (pulmonary embolism), affects the portion of the organ that the blocked blood vessel supplies.

Drugs based on "clot-busting" biochemicals can be lifesavers. *Tissue plasminogen activator* (tPA) may restore blocked coronary or cerebral circulation if given within an hour of a heart attack or within three hours of a stroke.

Abnormal clot formations are often associated with conditions that change the endothelial linings of vessels. For example, in *atherosclerosis* (ath″er-o″skle-ro′sis), accumulations of fatty deposits change arterial linings, sometimes initiating inappropriate clotting. This is the most common cause of thrombosis in medium-sized arteries (fig. 14.20). Clinical Application 14.3 discusses deep vein thrombosis.

Coagulation may also occur in blood flowing too slowly. The concentration of clot-promoting substances may increase to a critical level instead of being carried away by more rapidly moving blood, and a clot may form. This imbalance is the usual cause of thrombosis in veins.

PRACTICE 

40 Distinguish between extrinsic and intrinsic clotting mechanisms.

41 What is the major event in blood clot formation?

42 What factors initiate the formation of fibrin?

43 What prevents the formation of massive clots throughout the cardiovascular system?

44 Distinguish between a thrombus and an embolus.

45 How might atherosclerosis promote the formation of blood clots?

Prevention of Coagulation

In a healthy cardiovascular system, the endothelium of the blood vessels partly prevents spontaneous blood clot formation. This smooth lining lowers the risk of platelets and clotting factors accumulating. Endothelial cells also produce a prostaglandin (see chapter 13, p. 491) called *prostacyclin* (PGI_2), which inhibits the adherence of platelets to the inner surface of healthy blood vessel walls.

When a clot is forming, fibrin threads latch onto or *adsorb* thrombin, thus helping prevent the spread of the clotting reaction. A plasma alpha globulin, *antithrombin,* inactivates additional thrombin by binding to it and blocking its action on fibrinogen. In addition, basophils and mast cells in the connective tissue surrounding capillaries secrete the anticoagulant *heparin.* This substance interferes with formation of prothrombin activator, prevents the action of thrombin on fibrinogen, and promotes removal of thrombin by antithrombin and fibrin adsorption. Heparin and another compound, coumadin (warfarin), are used to prevent abnormal clotting.

Thrombocytopenia (throm″bo-si″to-pe′ne-ah) occurs when the platelet count drops below 100,000 platelets per microliter of blood. Symptoms include bleeding easily; capillary hemorrhages throughout the body; and small, bruiselike spots on the skin called petechiae. Thrombocytopenia is a common side effect of cancer chemotherapy and radiation treatments and can be a complication of pregnancy, leukemia, bone marrow transplantation, infectious disease, cardiac surgery, or anemia. Conventional treatment is transfusion of platelets. Treatment with thrombopoietin (TPO) stimulates formation and maturation of megakaryocytes and thereby boosts platelet levels.

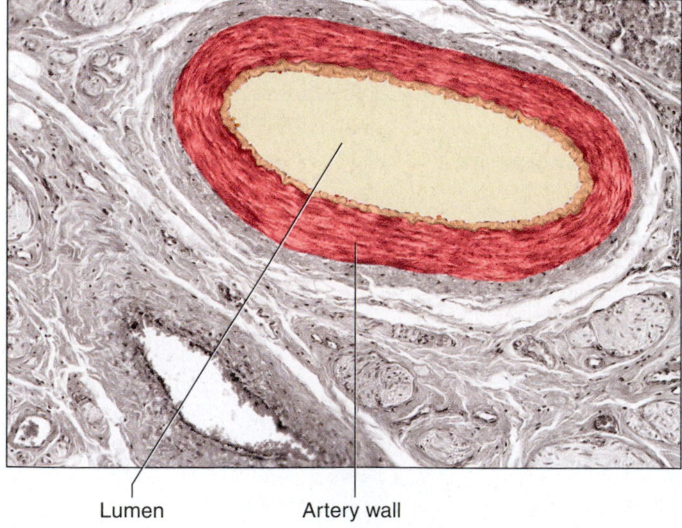

(a)

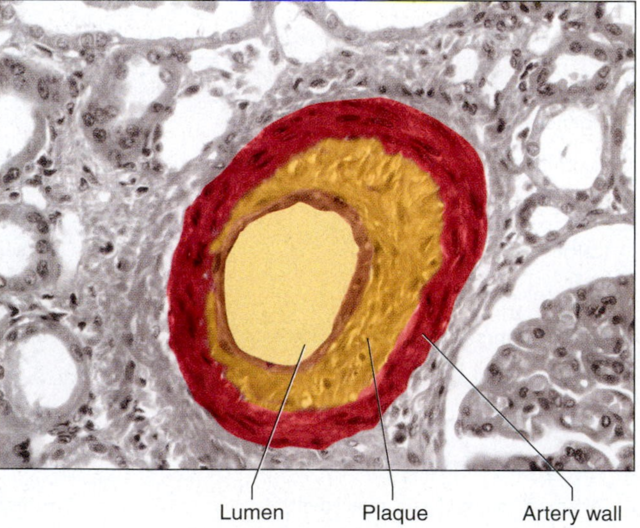

(b)

FIGURE 14.20 Artery cross sections. (*a*) Light micrograph of a normal artery (50×). (*b*) The inner wall of an artery changed as a result of atherosclerosis (100×).

14.3 CLINICAL APPLICATION

Deep Vein Thrombosis

On April 2, 2003, NBC news correspondent David Bloom, traveling with the United States infantry near Baghdad, mentioned on the phone to his wife Melanie that he felt cramps behind his knees. He also reported the symptom to physicians in the United States, who advised that he seek immediate medical attention. But he didn't. Instead, Bloom continued his reporting from the "Bloom-mobile," a tank turned into a traveling, cramped newsroom. Three days later, as the military unit prepared to leave for the final approach to Baghdad, Bloom suddenly collapsed and died.

David Bloom died from a pulmonary embolism, the result of a condition called deep vein thrombosis (DVT). In the United States, approximately 2 million people a year develop DVT, and 200,000 die from pulmonary embolism.

In DVT, prolonged immobility causes blood to pool, leading to clot formation, typically in the femoral or popliteal veins or in the deep veins of the pelvis. Symptoms occur in half of all affected individuals. These include deep muscle pain, redness, swelling, and possibly discoloration and dilation of surface veins (phlebitis). Part of the clot may break off hours or days after it forms and follow the path of circulation, lodging in the pulmonary arteries. This is pulmonary embolism, and it is rapidly fatal. Symptoms include chest pain, anxiety, racing pulse, sweating, cough with bloody sputum, and loss of consciousness.

Bloom had several risk factors for DVT: prolonged periods of immobility in the Bloom-mobile and on flights; dehydration; and an inherited clotting disorder that he had not known about called factor V Leiden. Other risk factors for DVT are prolonged immobility due to surgery; oral contraceptive use; hormone replacement therapy (estrogen); surgery (of the abdomen, pelvis, or limbs); and cancer (of the ovaries, pancreas, colon, liver, stomach, or lymphoma). Measures to prevent DVT include taking anti-coagulants if immobilization is expected; wearing compression stockings that keep blood flowing in the legs; and doing exercises while immobilized during travel. Some airlines advise passengers on how to exercise on cramped flights, such as by curling the toes up and down (figure 14C). DVT is sometimes called "traveler's thrombosis" and "economy-class syndrome" for good reason. ■

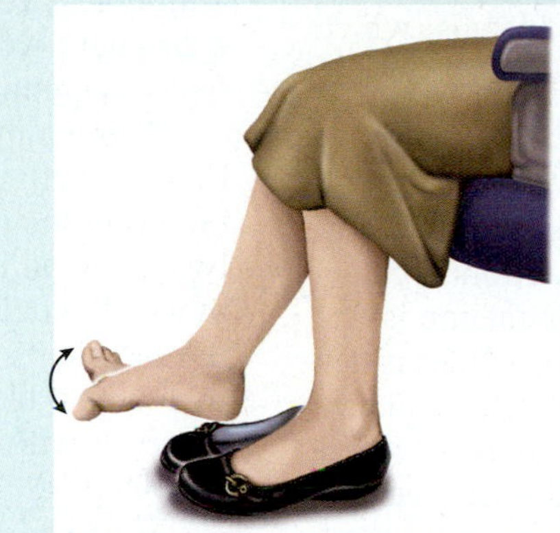

FIGURE 14C Exercising the toes and ankles on a long flight can lower the risk of deep vein thrombosis.

TABLE 14.10 | Factors That Inhibit Blood Clot Formation

Factor	Action	Factor	Action
Smooth lining of blood vessel	Prevents activation of intrinsic blood clotting mechanism	Antithrombin in plasma	Interferes with the action of thrombin
Prostacyclin	Inhibits adherence of platelets to blood vessel wall	Heparin from mast cells and basophils	Interferes with the formation of prothrombin activator
Fibrin threads	Adsorbs thrombin		

Heparin-secreting cells are particularly abundant in the liver and lungs, where capillaries trap small blood clots that commonly form in the slow-moving blood of veins. These cells continually secrete heparin, preventing additional clotting in the cardiovascular system. Table 14.10 summarizes clot-inhibiting factors.

Leeches have been used since the time of the ancient Egyptians to break up blood clots. The saliva of the medicinal leech *Hirudo medicinalis* contains a potent and long-acting anticoagulant, hirudin. Most people do not like three-inch long, slimy green-gray invertebrates dining on a wound, so a biotech company developed a drug based on the leech protein. However, the drug was *too* long acting—it caused bleeding problems, could not be easily monitored, and could damage the kidneys. Synthetic anticoagulants have taken its place.

PRACTICE

46 How does the lining of a blood vessel help prevent blood clot formation?

47 What is the function of antithrombin?

48 How does heparin help prevent blood clot formation?

14.5 BLOOD GROUPS AND TRANSFUSIONS

Early blood transfusion experiments, which date from the late 1600s, used lamb blood. By the 1800s, human blood was being used with unpredictable results—some recipients were cured, but some were killed when their kidneys failed under the strain of handling clumping red blood cells when blood types were incompatible. So poor was the success rate that, by the late 1800s, many nations banned transfusions.

Around this time, Austrian physician Karl Landsteiner began investigating why transfusions sometimes worked and sometimes did not. In 1900, he determined that blood was of differing types and that only certain combinations of them were compatible. In 1910, identification of the ABO blood antigen gene explained the observed blood type incompatibilities. Today, thirty-one different genes are known to contribute to the surface features of red blood cells, which determine compatibility between blood types.

Antigens and Antibodies

The clumping of red blood cells when testing blood compatibility or resulting from a transfusion reaction is called **agglutination** (ah-gloo″tĭna′shun). This phenomenon is due to a reaction between red blood cell surface molecules called **antigens** (an′ti-jenz), formerly called *agglutinogens,* and protein **antibodies** (an′ti-bod″ez), formerly called *agglutinins,* carried in the plasma. Antibodies are called *anti-* because they are "against" specific antigens. Although many different antigens are associated with human erythrocytes, only a few of them are likely to produce serious transfusion reactions. These include the antigens of the ABO group and those of the Rh group. Avoiding the mixture of certain types of antigens and antibodies prevents adverse transfusion reactions. From Science to Technology 14.1 discusses a DNA chip that prevents transfusion mismatches.

> A mismatched blood transfusion quickly produces telltale signs of agglutination—anxiety; breathing difficulty; facial flushing; headache; and severe pain in the neck, chest, and lumbar area. Red blood cells burst, releasing free hemoglobin. Macrophages phagocytize the hemoglobin, breaking it down into heme and globin. Some of the heme is recycled. The rest of the heme is converted to bilirubin, which may sufficiently accumulate to cause the yellow skin of jaundice. Free hemoglobin in the kidneys may ultimately cause them to fail.

ABO Blood Group

The *ABO blood group* is based on the presence (or absence) of two major antigens on red blood cell membranes—*antigen A* and *antigen B.* A person's erythrocytes have one of four antigen combinations: only A, only B, both A and B, or neither A nor B. An individual with only antigen A has *type A blood;* a person with only antigen B has *type B blood;* one with both antigen A and antigen B has *type AB blood;* and one with neither antigen A nor antigen B has *type O blood.* Table 14.11 indicates some of the frequencies of ABO blood types in the diverse population of the United States.

ABO blood group antibodies are synthesized in the plasma about two to eight months following birth. The stimulus for their synthesis has not clearly been established. However, we know that whenever antigen A is absent in the red blood cells, an antibody called *anti-A* is produced, and whenever antigen B is absent, an antibody called *anti-B* is manufactured. Therefore, persons with type A blood also have anti-B antibody in their plasma; those with type B blood have anti-A antibody; those with type AB blood have neither antibody; and those with type O blood have both anti-A and anti-B antibodies (fig. 14.21 and table 14.12). The anti-A and anti-B antibodies are large and do not cross the placenta. Thus, if a pregnant woman and her fetus are of different ABO blood types, agglutination in the fetus will not occur.

The major concern in blood transfusion procedures is that the cells in the donated blood not clump due to antibodies in the recipient's plasma. For example, a person with type A blood must not receive blood of type B or AB, either of which would clump in the presence of anti-B antibodies in the recipient's type A blood (fig. 14.22). Likewise, a person with type B blood must not be given type A or AB blood, and a person with type O blood must not be given type A, B, or AB blood.

TABLE 14.11 | Some ABO Blood Type Frequencies (%) in the United States

Population	Type O	Type A	Type B	Type AB
Caucasian	45	40	11	4
African American	49	27	20	4
Native American	79	16	4	1
Hispanic	63	14	20	3
Chinese American	42	27	25	6
Japanese American	31	38	21	10
Korean American	32	28	30	10

TABLE 14.12 | Antigens and Antibodies of the ABO Blood Group

Blood Type	Antigen	Antibody
A	A	anti-B
B	B	anti-A
AB	A and B	Neither anti-A nor anti-B
O	Neither A nor B	Both anti-A and anti-B

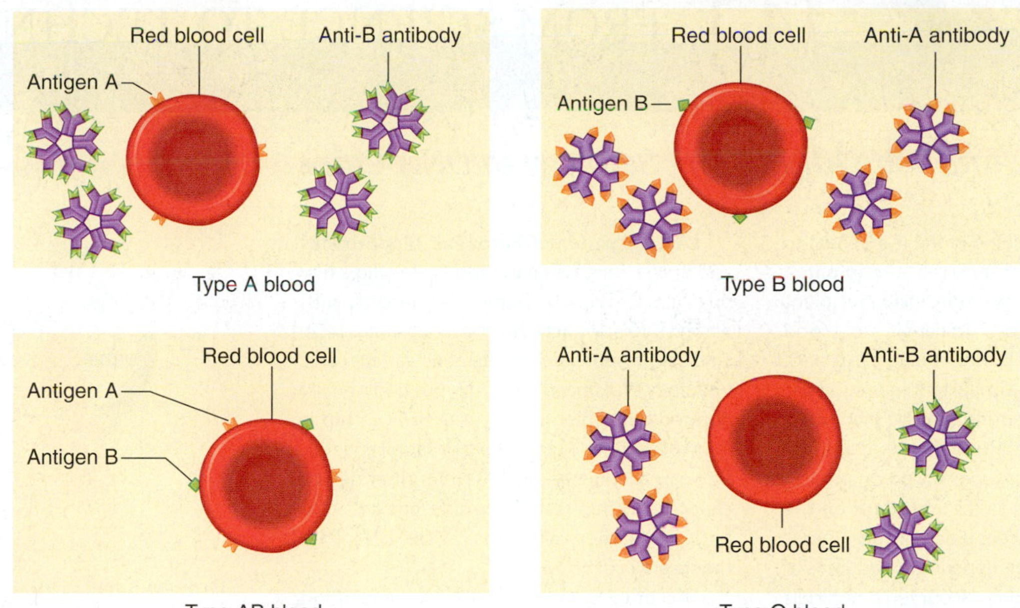

FIGURE 14.21 Different combinations of antigens and antibodies distinguish blood types. (Cells and antibodies not drawn to scale.)

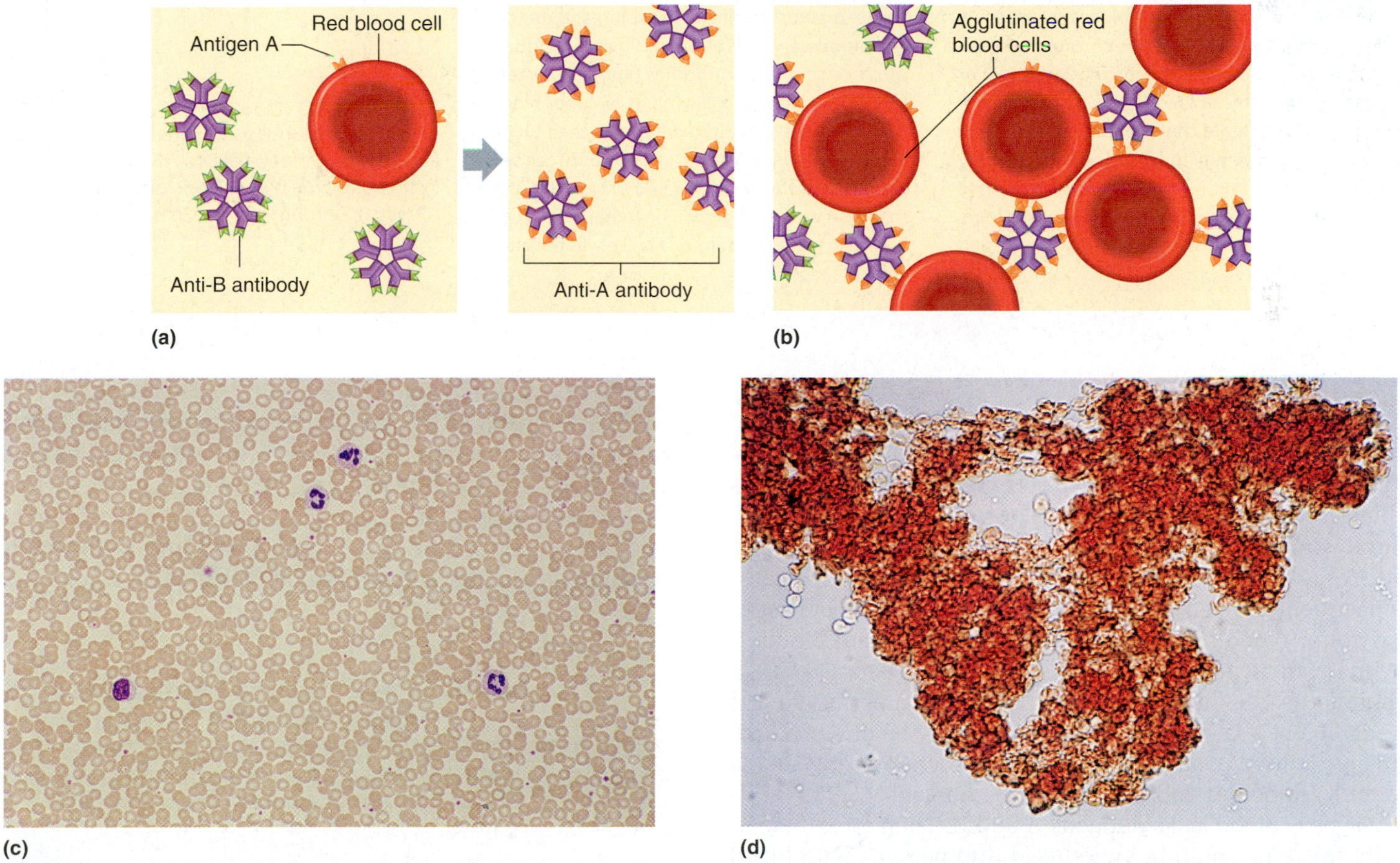

Antigen A
Red blood cell
Anti-B antibody
Anti-A antibody
Agglutinated red blood cells

(a) (b)

(c) (d)

FIGURE 14.22 Agglutination. (*a*) If red blood cells with antigen A are added to blood containing anti-A antibody, (*b*) the antigens react with the antibodies, causing clumping (agglutination). (*c*) Nonagglutinated blood (210×). (*d*) Agglutinated blood (220×). (Cells and antibodies in *a* and *b* not drawn to scale.)

14.1 FROM SCIENCE TO TECHNOLOGY

Blood Typing and Matching: From Serology to DNA Chips

Typing and matching blood is essential to minimize the chance that the immune response of a blood transfusion recipient will reject the blood. The popular view of blood types is that it is a simple matter of A, B, AB, or O. Rh blood types become important in pregnancy, when woman and offspring may be incompatible. However, human blood can be classified into twenty-nine major types based on protein and carbohydrate molecules (antigens) on the surfaces of red blood cells. Each of these types includes many subtypes, generating hundreds of ways that the topographies of our red blood cells differ from individual to individual.

For more than a century, an approach called serology has been used to type blood into major groups by identifying red blood cell antigens. Different reagents are required to detect different antigens, and so typing by serology beyond a few blood groups is costly and time consuming. It can also miss variants of antigens that are small or hidden among the embedded proteins and emanating carbohydrate chains of the cell surface. A more informative way to type blood is to identify the *instructions* for the cell-surface antigens—the genes that encode these proteins. This approach, termed genotyping, is being tested in Europe and Canada.

A consortium of European blood banks and universities, called Bloodgen, is using a tiny device called a BLOODchip that detects, with one test, 100 distinct DNA "signatures" (fig. 14D). These represent some of the gene variant combinations that underlie blood types traditionally defined by serology. So far the BLOODchip is able to accurately type 99.8% of samples; serology can accurately type 97%. Another device called BeadChip tests for a different group of antigens. Other variations on the DNA chip theme are in development.

Accurate blood matching for many blood groups may not be of great importance for the average person who may require just one or two transfusions in a lifetime. But for a person who has a chronic disorder that requires multiple transfusions—such as leukemia or sickle cell disease—greater specificity in typing and matching blood can be lifesaving. Such individuals produce so many antibodies (immune system proteins) against so many types of donor blood that it is often difficult to determine their blood types by serology. DNA typing, however, can tell. The head of the American Red Cross estimates that DNA blood typing will be routine within fifteen years, but that serology is likely to remain a valuable tool. ■

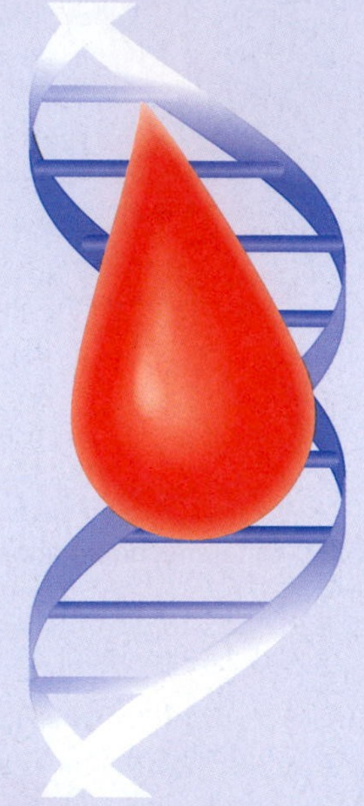

FIGURE 14D Bloodgen is a consortium of European blood banks and universities investigating the use of DNA microarrays ("chips") to type blood. Their logo is a drop of blood superimposed on a DNA double helix.

Type AB blood lacks both anti-A and anti-B antibodies, so an AB person can receive a transfusion of blood of any other type. For this reason, type AB persons are sometimes called *universal recipients*. However, type A blood, type B blood, and type O blood still contain antibodies (either anti-A and/or anti-B) that could agglutinate type AB cells. Consequently, even for AB individuals, it is always best to use donor blood of the same type as the recipient blood. If the matching type is not available and type A, B, or O is used, it should be transfused slowly and in limited amounts so that the recipient's larger blood volume dilutes the donor blood. This precaution usually avoids serious reactions between the donor's antibodies and the recipient's antigens.

Type O blood lacks antigens A and B. Therefore, theoretically this type could be transfused into persons with blood of any other type. Individuals with type O blood are sometimes called *universal donors*. Type O blood, however, does contain both anti-A and anti-B antibodies, and if it is given to a person with blood type A, B, or AB, it too should be transfused slowly and in limited amounts to minimize the chance of an adverse reaction. When type O blood is given to blood types A, B, or AB, it is generally transfused as "packed cells," meaning the plasma has been removed. This also minimizes adverse reactions due to the anti-A and anti-B antibodies found in the plasma of type O blood. Table 14.13 summarizes

TABLE 14.13 | Preferred and Permissible Blood Types for Transfusions

Blood Type of Recipient	Preferred Blood Type of Donor	Permissible Blood Type of Donor (In an Extreme Emergency)
A	A	O
B	B	O
AB	AB	A, B, O
O	O	No alternate types

preferred blood types for normal transfusions and permissible blood types for emergency transfusions.

PRACTICE

49 Distinguish between antigens and antibodies.

50 What is the main concern when blood is transfused from one individual to another?

51 Why is a type AB person called a universal recipient?

52 Why is a type O person called a universal donor?

When is type O blood not really type O blood? This occurs when a certain mutation is present. A person with a rare genetic condition called the Bombay phenotype lacks an enzyme that inserts a particular molecule onto red blood cell surfaces. Without that molecule, the A and B antigens cannot bind. The result is blood that tests as O (because it lacks A and B antigens) but can genetically be of any ABO type—A, B, AB, or O. Although the Bombay phenotype does not affect health, it can sometimes explain a child's ABO type that cannot be derived from those of the parents.

Rh Blood Group

The *Rh blood group* was named after the rhesus monkey in which it was first studied. In humans, this group includes several Rh antigens (factors). The most prevalent of these is *antigen D:* however, if any of the antigen D and other Rh antigens are present on the red blood cell membranes, the blood is said to be *Rh-positive.* Conversely, if the red blood cells lack the Rh antigens, the blood is called *Rh-negative.* About 15% of Caucasians and 5% of African Americans in the U.S. population are Rh-negative. The remaining ethnic groups are all Rh-positive.

As in the case of antigens A and B, the presence (or absence) of Rh antigens is an inherited trait. Antibodies for Rh (*anti-Rh*) form only in Rh-negative persons in response to the presence of red blood cells with Rh antigens.

If an Rh-negative person receives a transfusion of Rh-positive blood, the recipient's antibody-producing cells are stimulated by the presence of the Rh antigens and will begin producing *anti-Rh antibodies.* Generally, no serious consequences result from this initial transfusion, but if the Rh-negative person—now sensitized to Rh-positive blood—receives another transfusion of Rh-positive blood some months later, the donated red blood cells are likely to agglutinate.

A related condition may occur when an Rh-negative woman is pregnant with an Rh-positive fetus. If this is her first pregnancy with an Rh-positive fetus, it may be uneventful; however, at the time of this infant's birth (or if a miscarriage occurs), the placental membranes that separated the maternal blood from the fetal blood during the pregnancy tear, and some of the infant's Rh-positive blood cells may enter the maternal circulation. These Rh-positive cells may then stimulate the maternal tissues to begin producing anti-Rh antibodies (fig. 14.23).

If a woman who has already developed anti-Rh antibodies becomes pregnant with a second Rh-positive fetus, these anti-Rh antibodies, called hemolysins, cross the placental membrane and destroy the fetal red cells. The fetus then develops a condition called *erythroblastosis fetalis* (ĕrith″ro-blas-to′sis fe′tal-iz), or hemolytic disease of the

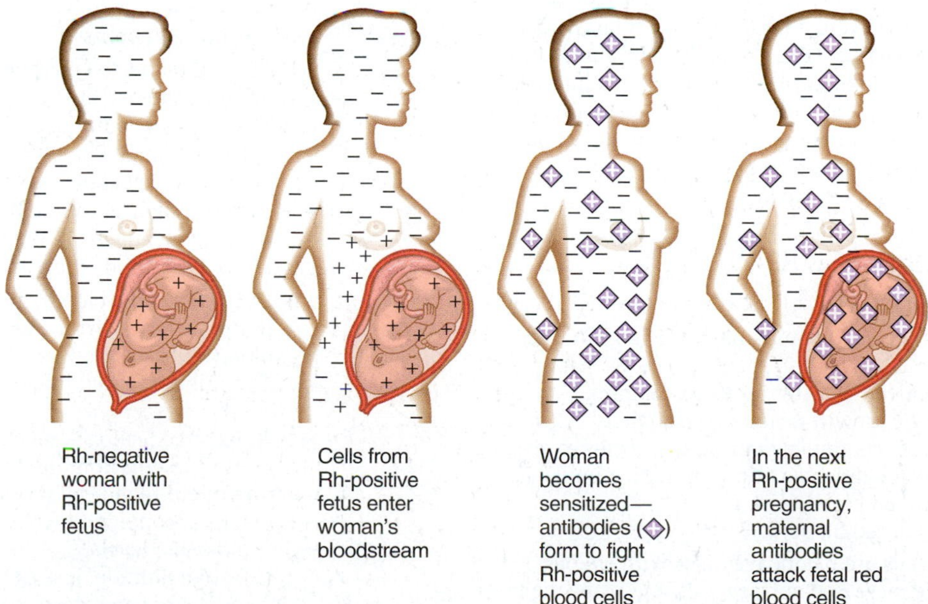
Rh-negative woman with Rh-positive fetus
Cells from Rh-positive fetus enter woman's bloodstream
Woman becomes sensitized—antibodies (◇) form to fight Rh-positive blood cells
In the next Rh-positive pregnancy, maternal antibodies attack fetal red blood cells

FIGURE 14.23 Rh incompatibility. If a man who is Rh-positive and a woman who is Rh-negative conceive a child who is Rh-positive, the woman's body may manufacture antibodies that attack future Rh-positive offspring.

TABLE 14.14 | Some Inherited Disorders of Blood

Disorder	Abnormality
Chronic granulomatous disease	Granulocytes cannot produce superoxide, which kills pathogenic bacteria
Erythrocytosis	Reticulocytes have extra EPO receptors, enhancing stamina
Factor V Leiden	Increases risk of abnormal clotting; elevates risk of deep vein thrombosis
Hemophilia (several types)	Lack of specific clotting factor causes bleeding
Hereditary hemochromatosis	Excess absorption of dietary iron into bloodstream deposits iron in various organs
Porphyria variegata	Enzyme deficiency excretes porphyrin ring of hemoglobin into urine; metabolic blockage causes sequence of varied symptoms
Sickle cell disease	Abnormal hemoglobin crystallizes under low-oxygen conditions, sickling red blood cells, which block circulation causing anemia, pain, and other symptoms
Von Willebrand disease	Lack of clotting factor (von Willebrand factor), which stabilizes factor VIII, causes bleeding; less severe than hemophilia

newborn. Transfusions with Rh-negative blood are given to the fetus before birth and to the neonate after birth to avoid brain damage or possibly death.

Table 14.14 describes a few of the many and diverse inherited disorders that affect the blood.

PRACTICE

53 What is the Rh blood group?

54 What are two ways that Rh incompatibility can arise?

Erythroblastosis fetalis is extremely rare today because obstetricians carefully track Rh status. An Rh^- woman who might carry an Rh^+ fetus is given an injection of a drug called RhoGAM. This is anti-Rh antibodies, which bind to and shield any Rh^+ fetal cells that might contact the woman's cells and sensitize her immune system. RhoGAM must be given within seventy-two hours of possible contact with Rh^+ cells—including giving birth, terminating a pregnancy, miscarrying, or undergoing amniocentesis (a prenatal test in which a needle is inserted into the uterus).

CHAPTER SUMMARY

14.1 INTRODUCTION (PAGE 523)

Blood is often considered a type of connective tissue in which cells are suspended in a liquid extracellular matrix. It transports substances between the body cells and the external environment and helps maintain a stable internal environment. Blood volume varies with body size, fluid and electrolyte balance, and adipose tissue content. Blood can be separated into formed elements (mostly red blood cells) and liquid portions—plasma.

14.2 BLOOD CELLS (PAGE 524)

1. The origin of blood cells
   a. Blood cells develop from hematopoietic stem cells in red bone marrow.
   b. Cells descended from stem cells respond to hematopoietic growth factors to specialize.
   c. Thrombopoietin stimulates megakaryocytes to give rise to platelets, much as erythropoietin stimulates formation of red blood cells.
2. Characteristics of red blood cells
   a. Red blood cells are biconcave discs with shapes that provide increased surface area and place their cell membranes close to internal structures.
   b. Red blood cells contain hemoglobin, which combines loosely with oxygen.
   c. The mature form lacks nuclei and other organelles, but contains enzymes needed for energy-releasing processes.
3. Red blood cell counts
   a. The red blood cell count equals the number of cells per microliter of blood.
   b. The average count may range from approximately 4,000,000 to 6,000,000 cells per microliter.
   c. Red blood cell count is related to the oxygen-carrying capacity of the blood and is used in diagnosing and evaluating the courses of diseases.
4. Red blood cell production and its control
   a. During fetal development, red blood cells form in the yolk sac, liver, and spleen; after birth, red blood cells are produced by the red bone marrow.
   b. The number of red blood cells remains relatively stable.
   c. A negative feedback mechanism involving erythropoietin from the kidneys and liver controls rate of red blood cell production.
      (1) Erythropoietin is released in response to low oxygen levels.
      (2) High altitude, loss of blood, or chronic lung disease can lower oxygen concentration in the blood.

5. Dietary factors affecting red blood cell production
   a. The availability of vitamin B_{12}, iron, and folic acid affects red blood cell production.
   b. The rate of iron absorption varies with the amount of iron in the body.
6. Destruction of red blood cells
   a. Red blood cells are fragile and are damaged while moving through capillaries.
   b. Macrophages in the spleen and liver phagocytize damaged red blood cells.
   c. Hemoglobin molecules are decomposed, and the iron from the heme portion is recycled.
   d. Biliverdin and bilirubin are pigments, released from the heme portion, excreted in bile.
   e. The globin portion is broken down into amino acids metabolized by macrophages or released into the blood.
7. Types of white blood cells
   a. Granulocytes include neutrophils, eosinophils, and basophils.
   b. Agranulocytes include monocytes and lymphocytes.
8. Functions of white blood cells
   a. Neutrophils and monocytes phagocytize foreign particles.
   b. Chemicals released by damaged cells attract and stimulate leukocytes.
9. White blood cell counts
   a. Normal total white blood cell counts vary from 4,500 to 10,000 cells per microliter of blood.
   b. The number of white blood cells may change in abnormal conditions such as infections, emotional disturbances, or excessive loss of body fluids.
   c. A differential white blood cell count indicates the percentages of various types of leukocytes.
10. Blood platelets
   a. Blood platelets are fragments of megakaryocytes that enter the circulation.
   b. The normal count varies from 130,000 to 360,000 platelets per microliter.
   c. Platelets help close breaks in blood vessels.

14.3 BLOOD PLASMA (PAGE 535)

Plasma is the liquid part of the blood that is composed of water and a mixture of organic and inorganic substances. It transports nutrients and gases, helps regulate fluid and electrolyte balance, and helps maintain stable pH.

1. Plasma proteins
   a. Plasma proteins remain in the blood and interstitial fluids and are not normally used as energy sources.
   b. Three major types exist.
      (1) Albumins help maintain the osmotic pressure of plasma.
      (2) Globulins provide immunity (antibodies) and transport lipids and fat-soluble vitamins.
      (3) Fibrinogen functions in blood clotting.
2. Gases and nutrients
   a. Gases in plasma include oxygen, carbon dioxide, and nitrogen.
   b. Plasma nutrients include simple sugars, amino acids, and lipids.
      (1) Glucose is stored in the liver as glycogen and is released whenever the blood glucose concentration falls.
      (2) Amino acids are used to synthesize proteins and are deaminated for use as energy sources.
      (3) Lipoproteins function in the transport of lipids.
3. Nonprotein nitrogenous substances
   a. Nonprotein nitrogenous substances are composed of molecules that contain nitrogen atoms but are not proteins.
   b. They include amino acids, urea, uric acid, creatine, and creatinine.
      (1) Urea and uric acid are products of catabolism.
      (2) Creatinine results from the metabolism of creatine.
   c. Levels of these substances usually remain stable; an increase may indicate a kidney disorder.
4. Plasma electrolytes
   a. Plasma electrolytes are absorbed from the intestines and are released as by-products of cellular metabolism.
   b. They include ions of sodium, potassium, calcium, magnesium, chloride, bicarbonate, phosphate, and sulfate.
   c. They are important in the maintenance of osmotic pressure and pH.

14.4 HEMOSTASIS (PAGE 538)

Hemostasis refers to the stoppage of bleeding. Hemostatic mechanisms are most effective in controlling blood loss from small vessels.

1. Blood vessel spasm (vasospasm)
   a. Smooth muscles in the walls of smaller blood vessels reflexly contract following injury.
   b. Platelets release serotonin that stimulates vasoconstriction and helps maintain blood vessel spasm.
2. Platelet plug formation
   a. Platelets adhere to rough surfaces and exposed collagen.
   b. Platelets adhere together at the sites of injuries and form platelet plugs in broken vessels.
3. Blood coagulation
   a. Blood clotting, the most effective means of hemostasis, is a series of reactions wherein each reaction stimulates the next (cascade), which may be initiated by extrinsic or intrinsic mechanisms.
   b. The extrinsic clotting mechanism is triggered when blood contacts damaged tissue.
   c. The intrinsic clotting mechanism is triggered when blood contacts a foreign surface.
   d. Clot formation reflects balance between clotting factors that promote or inhibit clotting.
   e. The major event of coagulation is the conversion of soluble fibrinogen into insoluble fibrin.
   f. A formed clot retracts and pulls the edges of a broken blood vessel closer together.
   g. A thrombus is an abnormal blood clot in a blood vessel; an embolus is a clot or fragment of a clot that moves in a blood vessel.

h. Fibroblasts invade a clot, forming connective tissue throughout.
i. Protein-splitting enzymes may eventually destroy a clot.

4. Prevention of coagulation
   a. The smooth inner lining of blood vessels discourages the accumulation of platelets.
   b. As a clot forms, fibrin adsorbs thrombin and prevents the reaction from spreading.
   c. Antithrombin interferes with the action of excess thrombin.
   d. Some cells secrete heparin, an anticoagulant.

14.5 BLOOD GROUPS AND TRANSFUSIONS (PAGE 544)

Blood can be typed on the basis of the surface structures of its cells.

1. Antigens and antibodies
   a. Red blood cell membranes may display specific antigens.
   b. Blood plasma may contain antibodies against certain of these antigens.
2. ABO blood group
   a. Blood can be grouped according to the presence or absence of antigens A and B.
   b. Wherever antigen A is absent, anti-A antibody is present; wherever antigen B is absent, anti-B antibody is present.
   c. Preventing the mixing of red blood cells that have an antigen with plasma that contains the corresponding antibody avoids a transfusion reaction.
   d. Adverse reactions are due to agglutination (clumping) of the red blood cells.
3. Rh blood group
   a. Rh antigens are present on the red blood cell membranes of Rh-positive blood; they are absent in Rh-negative blood.
   b. An Rh-negative person exposed to Rh-positive blood produces anti-Rh antibodies in response.
   c. Mixing Rh-positive red cells with plasma that contains anti-Rh antibodies agglutinates the positive cells.
   d. If an Rh-negative female is pregnant with an Rh-positive fetus, some of the positive cells may enter the maternal blood at the time of birth and stimulate the maternal tissues to produce anti-Rh antibodies.
   e. Anti-Rh antibodies in maternal blood may pass through the placental tissues and react with the red blood cells of an Rh-positive fetus.

CHAPTER ASSESSMENTS

14.1 Introduction

1 Major functions of blood include ___________. (p. 523)
   a. nutrient, hormone, oxygen, and waste transport
   b. helping maintain the stability of interstitial fluid
   c. heat distribution
   d. all of the above

2 Formed elements in blood are ___________, ___________, and ___________. (p. 523)

3 Define *hematocrit,* and explain how it is determined. (p. 523)

4 The liquid portion of the blood is called ___________. (p. 523)

14.2 Blood Cells

5 Indicate where blood cells differentiate, and explain the process. (p. 524)

6 Describe a red blood cell. (p. 526)

7 Contrast oxyhemoglobin and deoxyhemoglobin. (p. 526)

8 Explain the significance of red blood cell counts. (p. 527)

9 Describe the life cycle of a red blood cell from differentiation through destruction. (p. 527)

10 Define *erythropoietin,* and explain its function. (p. 527)

11 Explain how vitamin B_{12} and folic acid deficiencies affect red blood cell production. (p. 528)

12 List two sources of iron that can be used for the synthesis of hemoglobin. (p. 528)

13 Distinguish between biliverdin and bilirubin. (p. 530)

14 Distinguish between granulocytes and agranulocytes. (p. 530)

15 Name five types of leukocytes, and list the major functions of each type. (p. 531)

16 Explain the significance of white blood cell counts as aids to diagnosing disease. (p. 533)

17 ___________ are fragments of megakaryocytes that function in ___________. (p. 534)

14.3 Blood Plasma

18 The most abundant component of plasma is ___________. (p. 535)
   a. vitamins
   b. oxygen
   c. proteins
   d. water
   e. electrolytes

19 Name three types of plasma proteins, and indicate the function of each type. (p. 535)

20 Name the gases and nutrients in plasma. (p. 536)

21 Define *nonprotein nitrogenous substances,* and name those commonly present in plasma. (p. 537)

22 The most abundant plasma electrolytes are ___________ and ___________. (p. 538)

23 Name several plasma electrolytes. (p. 538)

14.4 Hemostasis

24 ___________ is the term for stoppage of bleeding. (p. 538)

25 Explain how blood vessel spasm is stimulated following an injury. (p. 538)

26 Platelets adhering to form a plug may control blood loss from a ___________ break, but a larger break may require a ___________ to halt bleeding. (p. 538)

27 Name a vitamin required for blood clotting. (p. 539)

28 Distinguish between fibrinogen and fibrin. (p. 539)

29 Describe the major steps leading to the formation of a blood clot. (p. 539)

30 Indicate the trigger and outline the steps for extrinsic clotting and for intrinsic clotting. (p. 539)

31 Describe a positive feedback system that operates during blood clotting. (p. 539)

32 Define *serum.* (p. 541)

33 Distinguish between a thrombus and an embolus. (p. 541)

34 Explain how a blood clot may be removed naturally from a blood vessel. (p. 541)

35 Describe how blood coagulation may be prevented. (p. 542)

14.5 Blood Groups and Transfusions

36 Distinguish between an antigen and an antibody. (p. 544)

37 Explain the basis of ABO blood types. (p. 544)

38 Indicate the blood type (both ABO and Rh) of this individual: (p. 544)

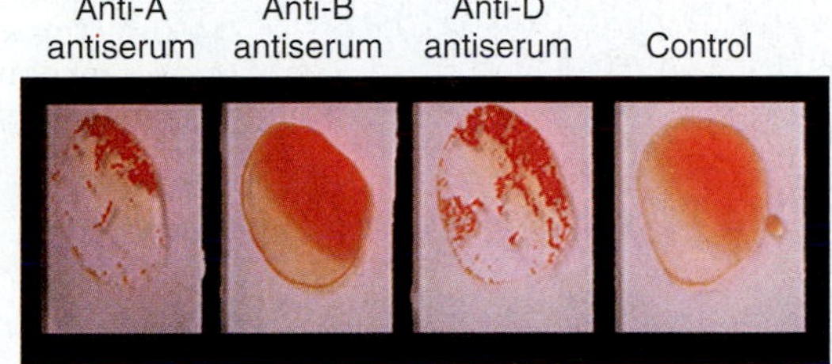

39 What blood type(s) could safely receive blood from this individual (#38)? (p. 546)

40 What blood type(s) could this individual (#38) safely receive? (p. 546)

41 Explain why a person with blood type AB is sometimes called a universal recipient. (p. 546)

42 Explain why a person with blood type O is sometimes called a universal donor. (p. 546)

43 Distinguish between Rh-positive and Rh-negative blood. (p. 547)

44 Describe how a person may become sensitized to Rh-positive blood. (p. 547)

45 Describe *erythroblastosis fetalis,* and explain how this condition may develop. (p. 547)

INTEGRATIVE ASSESSMENTS/CRITICAL THINKING

OUTCOMES 3.4, 14.2

1. If a patient with inoperable cancer is treated using a drug that reduces the rate of cell division, how might the patient's white blood cell count change? How might the patient's environment be modified to compensate for the effects of these changes?

OUTCOMES 9.3, 14.2

2. Erythropoietin is available as a drug. Why would athletes abuse it?

OUTCOMES 14.2, 14.4

3. In the United States, between 1977 and 1985, more than 10,000 men contracted the human immunodeficiency virus (HIV) from contaminated factor VIII that they received to treat hemophilia. What are two abnormalities in the blood of these men?

OUTCOME 14.2

4. How would you explain to a patient with leukemia, who has a greatly elevated white blood cell count, the importance of avoiding bacterial infections?

OUTCOME 14.2

5. Hypochromic (iron-deficiency) anemia is common among aging persons admitted to hospitals for other conditions. What environmental and sociological factors might promote this form of anemia?

OUTCOME 14.4

6. Why do patients with liver diseases commonly develop blood clotting disorders?

WEB CONNECTIONS

Be sure to visit the text website at **www.mhhe.com/shier12** for answers to chapter assessments, additional quizzes, and interactive learning exercises.

ANATOMY & PHYSIOLOGY REVEALED

Anatomy & Physiology Revealed® (APR) includes cadaver photos that allow you to peel away layers of the human body to reveal structures beneath the surface. This program also includes animations, radiologic imaging, audio pronunciations, and practice quizzing. Check out **www.aprevealed.com**. APR has been proven to help improve student grades!

CHAPTER 15

Cardiovascular System

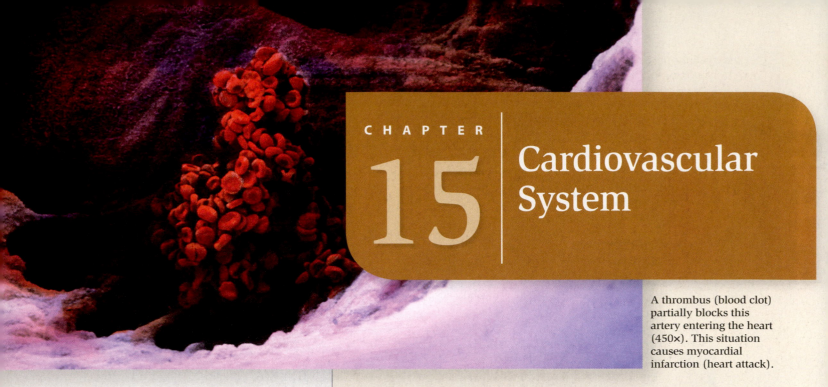

A thrombus (blood clot) partially blocks this artery entering the heart (450×). This situation causes myocardial infarction (heart attack).

UNDERSTANDING WORDS

angio-, vessel: *angio*tensin—substance that constricts blood vessels.

ather-, porridge: *athero*sclerosis—deposits of plaque in arteries.

brady-, slow: *brady*cardia—abnormally slow heartbeat.

diastol-, dilation: *diastol*ic pressure—blood pressure when the ventricle of the heart is relaxed.

edem-, swelling: *edema*—accumulation of fluids in the tissues that causes them to swell.

-gram, something written: electrocardio*gram*—recording of the electrical changes in the myocardium during a cardiac cycle.

lun-, moon: semi*lun*ar valve—valve with crescent-shaped flaps.

myo-, muscle: *myo*cardium—muscle tissue within the wall of the heart.

papill-, nipple: *papill*ary muscle—small mound of muscle projecting into a ventricle of the heart.

phleb-, vein: *phleb*itis—inflammation of a vein.

scler-, hard: arterio*scler*osis—loss of elasticity and hardening of a blood vessel wall.

syn-, together: *syn*cytium—mass of merging cells that act together.

systol-, contraction: *systol*ic pressure—blood pressure resulting from a single ventricular contraction.

tachy-, rapid: *tachy*cardia—abnormally fast heartbeat.

LEARNING OUTCOMES

After you have studied this chapter, you should be able to:

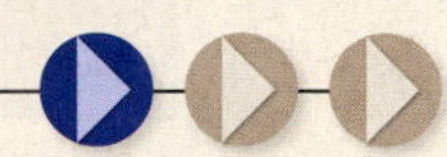

15.1 Introduction

1 Discuss the functions of the organs of the cardiovascular system. (p. 553)

15.2 Structure of the Heart

2 Distinguish between the various coverings of the heart and the layers that compose the wall of the heart. (p. 554)

3 Identify and locate the major parts of the heart and discuss the function of each part. (p. 555)

4 Trace the pathway of the blood through the heart and the vessels of coronary circulation. (p. 560)

15.3 Heart Actions

5 Describe the cardiac cycle and explain how heart sounds are produced. (p. 564)

6 Identify the parts of a normal ECG pattern and discuss the significance of this pattern. (p. 568)

7 Explain control of the cardiac cycle. (p. 568)

15.4 Blood Vessels

8 Compare the structures and functions of the major types of blood vessels. (p. 574)

9 Describe how substances are exchanged between blood in capillaries and the tissue fluid surrounding body cells. (p. 578)

15.5 Blood Pressure

10 Explain how blood pressure is produced and controlled. (p. 580)

11 Describe the mechanisms that aid in returning venous blood to the heart. (p. 589)

15.6 Paths of Circulation

12 Compare the pulmonary and systemic circuits of the cardiovascular system. (p. 590)

15.7–15.8 Arterial System–Venous System

13 Identify and locate the major arteries and veins. (p. 592)

15.9 Life-Span Changes

14 Describe life-span changes in the cardiovascular system. (p. 611)

ASSESS

DEFIBRILLATORS SAVE LIVES

A man rushing to catch a flight at a busy airport stops suddenly, looks about in confusion, and collapses. People congregate around him, as a woman runs to a device mounted on a nearby wall. It is an automated external defibrillator (AED) and looks like a laptop computer. The woman learned how to use it in a cardiopulmonary resuscitation class. She brings it over to the man, opens it, and places electrode pads over the man's chest, as indicated in a drawing on the inner cover of the defibrillator. Then the device speaks. "Analyzing heart rhythm," it declares as a computer assesses the heart rhythm. After a short pause, the device says, "charging, stand clear," and then "push button." The woman does so, and the device delivers a shock to the man's chest. It assesses the heart rhythm again, and instructs the woman to deliver a second shock. Soon, the man recovers, just as emergency technicians arrive.

The AEDs found in airports, malls, schools, health clubs, and other public places can save the life of a person suffering sudden cardiac arrest. One study conducted at Chicago's O'Hare and Midway airports found that over a ten-month period, AEDs saved 64% of the people on which they were used. Without defibrillation, only 5% to 7% of people survive sudden cardiac arrest. Each minute, the odds of survival shrink by 10%, and after six minutes, brain damage is irreversible. Sudden cardiac arrest can result from an accelerated heartbeat (ventricular tachycardia) or a chaotic and irregular heartbeat (ventricular fibrillation). The electrical malfunction that usually causes these conditions may result from an artery blocked with plaque or from build up of scar tissue from a previous myocardial infarction (heart attack).

For people who know that they have an inherited disorder that causes sudden cardiac arrest (by having suffered an event and then had genetic tests), a device called an implantable cardioverter defibrillator (ICD) can be placed under the skin of the chest in a one-hour procedure. Like the AED, the ICD monitors heart rhythm, and when the telltale deviations of ventricular tachycardia or ventricular fibrillation begin, it delivers a shock.

ICDs have been so successful in preventing subsequent cardiac arrests that they may soon be offered to people at high risk for the condition. The two major risk factors are having had a previous myocardial infarction and a low ejection fraction, which is the volume of blood pumped with each heartbeat. Normal ejection fraction is 50% to 60%; low is below 30% to 40%. Scarring lowers the ejection fraction. An echocardiogram, which is an ultrasound scan of the heart, can reveal the ejection fraction.

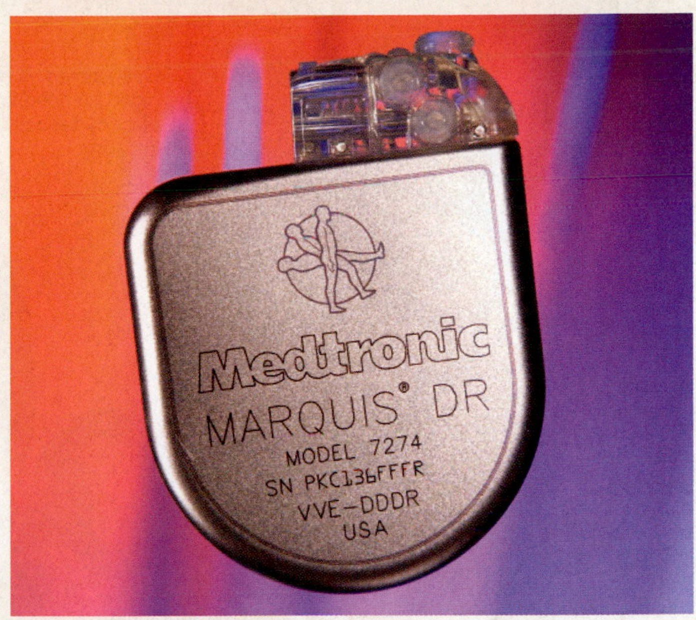

An implantable cardioverter defibrillator delivers a shock to a heart whose ventricles are wildly contracting, restoring a normal heartbeat.

In a four-year study at seventy-six medical centers involving 1,232 patients with the two risk factors, use of an ICD decreased death rate from sudden cardiac arrest by 31%. Many nations use the device as a preventative. In the United States, wider use of ICDs could save thousands of the 300,000 people who die each year of sudden cardiac arrest. ■

15.1 INTRODUCTION

The heart pumps 7,000 liters of blood through the body each day, contracting some 2.5 billion times in an average lifetime. This muscular pump forces blood through arteries, which connect to smaller-diameter vessels, called arterioles. Arterioles branch into the tiniest tubes, the capillaries, the sites of nutrient, electrolyte, gas, and waste exchange. Capillaries converge into venules, which in turn converge into veins that return blood to the heart, completing the closed system of blood circulation. These structures—the pump and its vessels—form the **cardiovascular system.**

The blood vessels form two circuits. The **pulmonary** (pul′mo-ner″e) **circuit** sends oxygen-depleted (deoxygenated) blood to the lungs to pick up oxygen and unload carbon dioxide. The **systemic** (sis-tem′ik) **circuit** sends oxygen-rich (oxygenated) blood and nutrients to all body cells and removes wastes. Without circulation, tissues would lack a supply of oxygen and nutrients, and wastes would accumulate. Such deprived cells soon begin irreversible change, which quickly leads to death. Figure 15.1 shows the general pattern of blood transport in the cardiovascular system.

15.2 STRUCTURE OF THE HEART

The heart is a hollow, cone-shaped, muscular pump. It is in the mediastinum of the thorax and rests upon the diaphragm.

Size and Location of the Heart

Heart size varies with body size. An average adult's heart is generally about 14 centimeters long and 9 centimeters wide (fig. 15.2).

The heart is bordered laterally by the lungs, posteriorly by the vertebral column, and anteriorly by the sternum (fig. 15.3 and reference plates 10, 16, 21, and 22). The *base* of the heart, which attaches to several large blood vessels, lies beneath the second rib. The heart's distal end extends

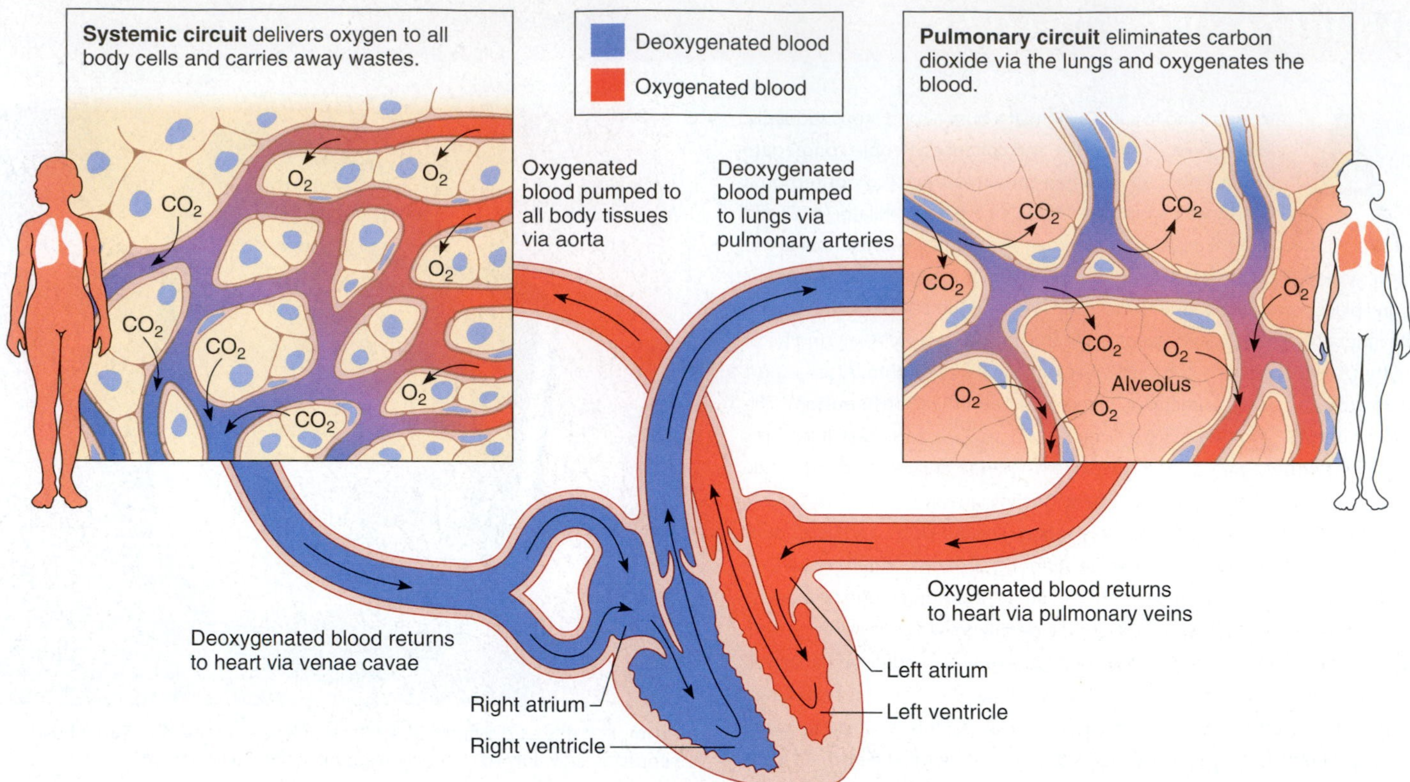
Systemic circuit delivers oxygen to all body cells and carries away wastes.
Deoxygenated blood
Oxygenated blood
Pulmonary circuit eliminates carbon dioxide via the lungs and oxygenates the blood.
O_2
CO_2
Oxygenated blood pumped to all body tissues via aorta
Deoxygenated blood pumped to lungs via pulmonary arteries
Alveolus
Oxygenated blood returns to heart via pulmonary veins
Deoxygenated blood returns to heart via venae cavae
Left atrium
Left ventricle
Right atrium
Right ventricle

FIGURE 15.1 The cardiovascular system transports blood between the body cells and organs such as the lungs, intestines, and kidneys that communicate with the external environment. Vessels in the pulmonary circuit carry blood from the heart to the lungs and back to the heart, replenishing oxygen (O_2) and releasing the metabolic waste carbon dioxide (CO_2). Vessels of the systemic circuit supply all of the other cells.

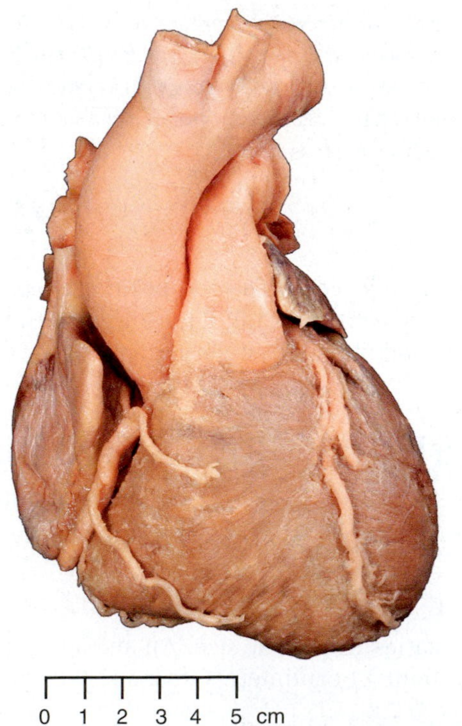
0 1 2 3 4 5 cm

FIGURE 15.2 Anterior view of a human heart. This photo is not life-size, so a proportionately reduced ruler has been included to help the student grasp the true size of the organ.

downward and to the left, terminating as a bluntly pointed *apex* at the level of the fifth intercostal space. For this reason, it is possible to sense the *apical heartbeat* by feeling or listening to the chest wall between the fifth and sixth ribs, about 7.5 centimeters to the left of the midline.

Coverings of the Heart

The **pericardium** (per″i-kar′de-um), or pericardial sac, is a covering that encloses the heart and the proximal ends of the large blood vessels to which it attaches. The pericardium consists of an outer fibrous bag, the *fibrous pericardium,* that surrounds a more delicate, double-layered serous membrane. The innermost layer of this serous membrane, the *visceral pericardium* (epicardium), covers the heart. At the base of the heart, the visceral pericardium turns back upon itself to become the *parietal pericardium,* which forms the inner lining of the fibrous pericardium.

The fibrous pericardium is a tough, protective sac largely composed of dense connective tissue. It is attached to the central portion of the diaphragm, the posterior of the sternum, the vertebral column, and the large blood vessels emerging from the heart (see figs. 1.9*b*, 15.4 and reference plates 16 and 17). Between the parietal and visceral layers of the pericardium is a space, the *pericardial cavity,* that contains a small volume of serous fluid that the pericardial membranes secrete. This fluid reduces friction between the pericardial membranes as the heart moves within them.

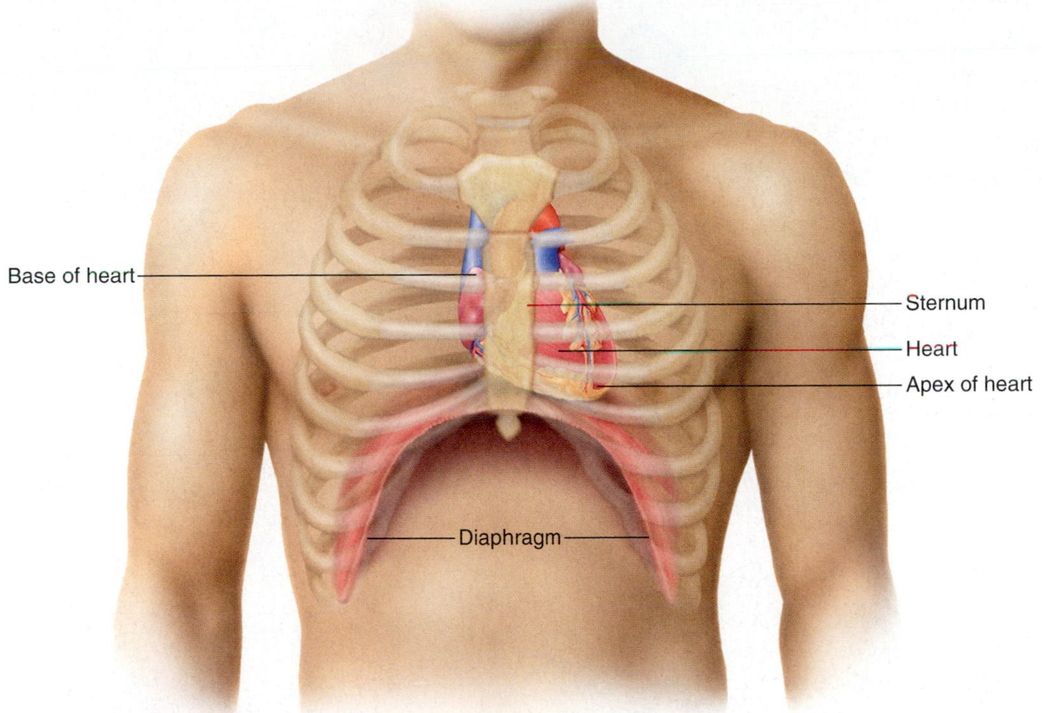

FIGURE 15.3 The heart is posterior to the sternum, where it rests upon the diaphragm.

In *pericarditis,* inflammation of the pericardium due to viral or bacterial infection produces adhesions that attach the layers of the pericardium to each other. This condition is painful and interferes with heart movements.

PRACTICE

1. Where is the heart located?
2. Where would you listen to hear the apical heartbeat?
3. Distinguish between the visceral pericardium and the parietal pericardium.
4. What is the function of the fluid in the pericardial cavity?

Wall of the Heart

The wall of the heart is composed of three distinct layers: an outer epicardium, a middle myocardium, and an inner endocardium (fig. 15.5).

The **epicardium** (ep″ĭ-kar′de-um), which corresponds to the visceral pericardium, protects the heart by reducing friction. It is a serous membrane that consists of connective tissue covered by epithelium, and it includes capillaries and nerve fibers. The deeper portion of the epicardium often contains fat, particularly along the paths of coronary arteries and cardiac veins that provide blood flow through the myocardium.

The middle layer, or **myocardium** (mi″o-kar′de-um), is thick and largely consists of the cardiac muscle tissue that pumps blood out of the heart chambers. The muscle fibers are arranged in planes, separated by connective tissues richly supplied with blood capillaries, lymph capillaries, and nerve fibers.

The inner layer, or **endocardium** (en″do-kar′de-um), consists of epithelium and underlying connective tissue. The endocardium also contains blood vessels and some specialized cardiac muscle fibers called *Purkinje fibers,* described later in this chapter in the section entitled "Cardiac Conduction System."

The endocardium lines all of the heart chambers and covers the structures, such as the heart valves, that project into them. This inner lining is also continuous with the inner linings (endothelium) of the blood vessels attached to the heart and throughout the cardiovascular system. Table 15.1 summarizes the characteristics of the three layers of the heart wall.

Heart Chambers and Valves

Internally the heart is divided into four hollow chambers, two on the left and two on the right. The upper chambers, called **atria** (a′tre-ah) (sing., *atrium*), have thin walls and receive blood returning to the heart. Small, earlike projections called **auricles** (aw′ri-klz) extend anteriorly from the atria, slightly increasing atrial volume (see fig. 15.4). The lower chambers, the **ventricles** (ven′tri-klz), force the blood out of the heart into arteries.

A structure called the *interatrial septum* separates the right from the left atrium. An *interventricular septum* separates the two ventricles. The atrium on each side communicates with its corresponding ventricle through an opening called the **atrioventricular orifice** (a″tre-o-ven-trik′u-lar ori-fis), guarded by an *atrioventricular valve* (AV valve).

Grooves on the surface of the heart mark the divisions between its chambers, and they also contain major blood vessels that supply the heart tissues. The deepest of

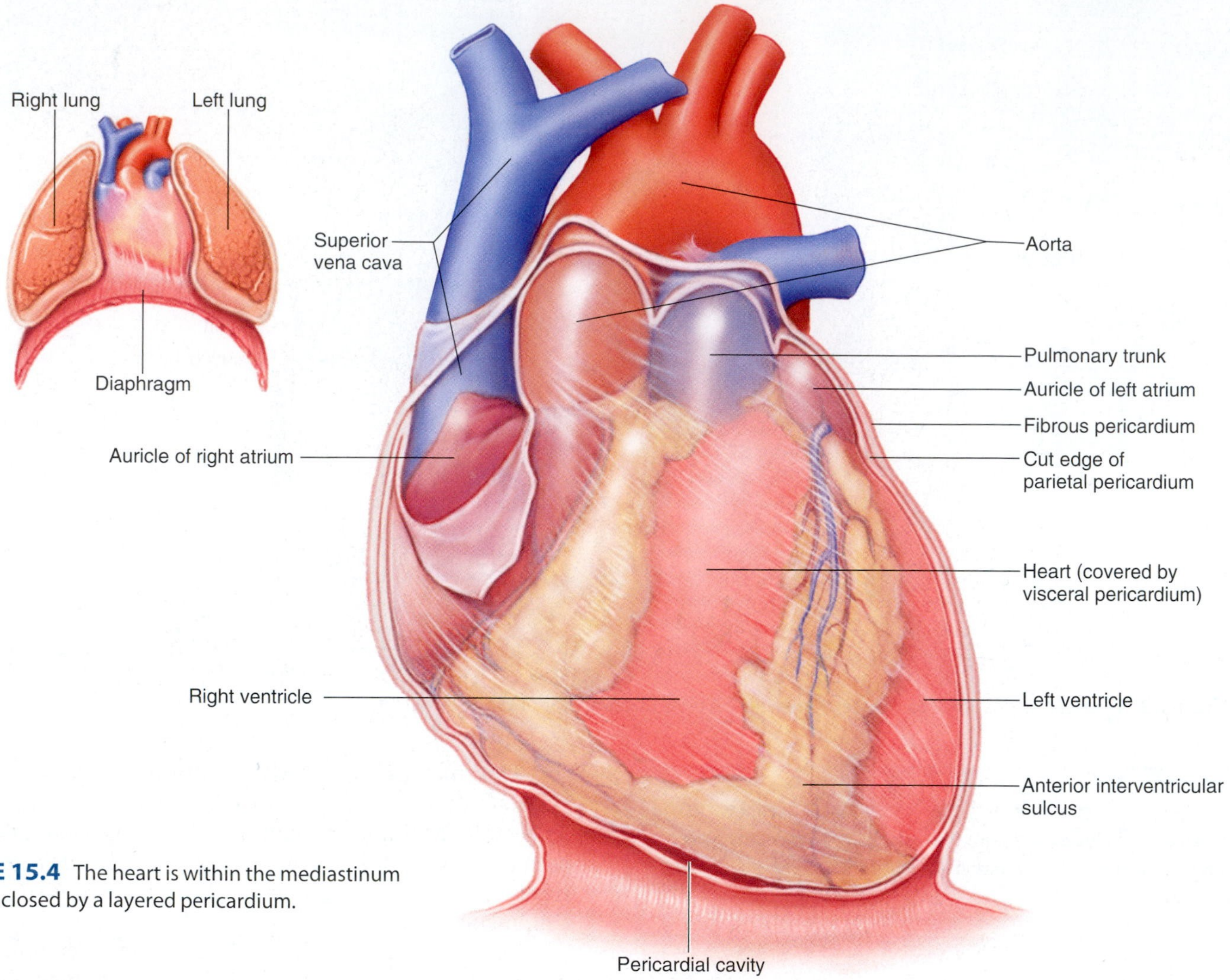

FIGURE 15.4 The heart is within the mediastinum and is enclosed by a layered pericardium.

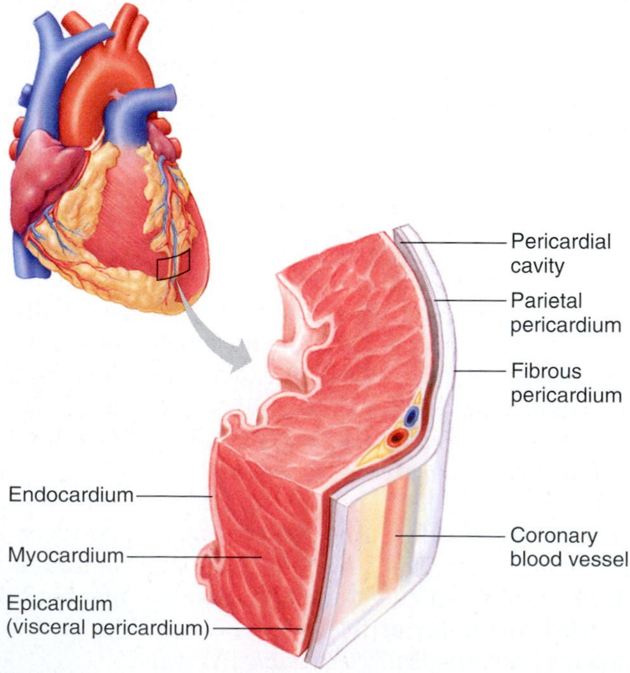

FIGURE 15.5 The heart wall has three layers: an endocardium, a myocardium, and an epicardium.

these grooves is the **atrioventricular** (coronary) **sulcus** (a″tre-o-ven-trik′u-lar sul′kus), which encircles the heart between the atria and ventricles. Two **interventricular** (anterior and posterior) **sulci** mark the septum that separates the right and left ventricles (see fig. 15.4).

When increasing blood volume stretches muscle cells associated with the atria, the cells secrete a peptide hormone called *atrial natriuretic peptide* (ANP). ANP inhibits release of renin from the kidneys and of aldosterone from the adrenal cortex. The result is increased excretion of sodium ions and water from the kidneys and lowered blood volume and blood pressure. Researchers are investigating use of ANP to treat high blood pressure.

PRACTICE

5 Describe the layers of the heart wall.

6 Name and locate the four chambers of the heart.

7 Name the orifices between the upper and the lower chambers of the heart.

TABLE 15.1 | Wall of the Heart

Layer	Composition	Function
Epicardium (visceral pericardium)	Serous membrane of connective tissue covered with epithelium and including blood capillaries, lymph capillaries, and nerve fibers	Forms a protective outer covering; secretes serous fluid
Myocardium	Cardiac muscle tissue separated by connective tissues and including blood capillaries, lymph capillaries, and nerve fibers	Contracts to pump blood from the heart chambers
Endocardium	Membrane of epithelium and underlying connective tissue, including blood vessels and specialized muscle fibers	Forms a protective inner lining of the chambers and valves

The right atrium receives blood from two large veins: the *superior vena cava* and the *inferior vena cava*. These veins return blood, low in oxygen, from tissues. A smaller vein, the *coronary sinus*, also drains blood into the right atrium from the myocardium of the heart.

A large **tricuspid valve** (right atrioventricular valve) guards the atrioventricular orifice between the right atrium and the right ventricle. It is composed of three leaflets, or cusps, as its name implies. This valve permits the blood to move from the right atrium into the right ventricle and prevents it from moving in the opposite direction. The cusps fold passively out of the way against the ventricular wall when the blood pressure is greater on the atrial side, and they close passively when the pressure is greater on the ventricular side (figs. 15.6, 15.7, 15.8, and 15.9).

Strong, fibrous strings, called *chordae tendineae* (kor′de ten′dĭ-ne), attach to the cusps of the tricuspid valve on the ventricular side. These strings originate from small mounds of cardiac muscle tissue, the **papillary muscles** (pap′ĭ-ler″e mus′elz), that project inward from the walls of the ventricle. The papillary muscles contract when the right ventricle contracts. As the tricuspid valve closes, these muscles pull on the chordae tendineae and prevent the cusps from swinging backwards into the right atrium.

The right ventricle has a thinner muscular wall than the left ventricle (see fig. 15.6). This right chamber pumps the blood a fairly short distance to the lungs against a relatively low resistance to blood flow. The left ventricle, on the other hand, must force the blood to all the other parts of the body against a much greater resistance to flow.

When the muscular wall of the right ventricle contracts, the blood inside its chamber is put under increasing pressure, and the tricuspid valve closes passively. As a result, the only exit for the blood is through the *pulmonary trunk*, which divides to form the left and right *pulmonary arteries* that lead to the lungs. At the base of this trunk is a **pulmonary valve** (pulmonary semilunar valve), which consists of three cusps (see figs. 15.8 and 15.9). This valve opens as the right ventricle contracts. However, when the ventricular muscles relax, the blood begins to back up in the pulmonary trunk. This closes the pulmonary valve, preventing a return flow into the right ventricle. Unlike the tricuspid valve, the pulmonary valve does not have chordae tendineae or papillary muscles attached to its cusps.

The left atrium receives the blood from the lungs through four *pulmonary veins*—two from the right lung and two from the left lung. The blood passes from the left atrium into the left ventricle through the atrioventricular orifice, which a valve guards. This valve consists of two leaflets and is named the **mitral valve** (shaped like a miter, a type of headpiece) or bicuspid valve or left atrioventricular valve. It prevents the blood from flowing back into the left atrium from the left ventricle when the ventricle contracts. As with the tricuspid valve, the papillary muscles and the chordae tendineae prevent the cusps of the mitral valve from swinging backwards into the left atrium.

When the left ventricle contracts, the mitral valve closes passively, and the only exit is through a large artery called the *aorta*. Its branches distribute blood to all parts of the body.

At the base of the aorta is an **aortic valve** (aortic semilunar valve) that consists of three cusps (see figs. 15.8 and 15.9). It opens and allows blood to leave the left ventricle as it contracts. When the ventricular muscles relax, this valve closes and prevents blood from backing up into the left ventricle.

The mitral and tricuspid valves are also called atrioventricular valves because they are between atria and ventricles. The pulmonary and aortic valves are also called semilunar because of the half-moon shapes of their cusps. Table 15.2 summarizes the locations and functions of the heart valves.

Mitral valve prolapse (MVP) affects up to 6% of the U.S. population. In this condition, one (or both) of the cusps of the mitral valve stretches and bulges into the left atrium during ventricular contraction. The valve usually continues to function adequately, but sometimes, blood regurgitates into the left atrium. Through a stethoscope, a regurgitating MVP sounds like a click at the end of ventricular contraction, then a murmur as blood goes back through the valve into the left atrium. Symptoms of MVP include chest pain, palpitations, fatigue, and anxiety.

Certain species of *Streptococcus* bacteria can damage the mitral valve. Endocarditis, an inflammation of the endocardium due to an infection, appears as a plantlike growth on the valve. People with MVP are particularly susceptible to endocarditis. They take antibiotics before undergoing dental work to prevent *Streptococcus* bacteria in the mouth from migrating through the blood to the heart and causing infection.

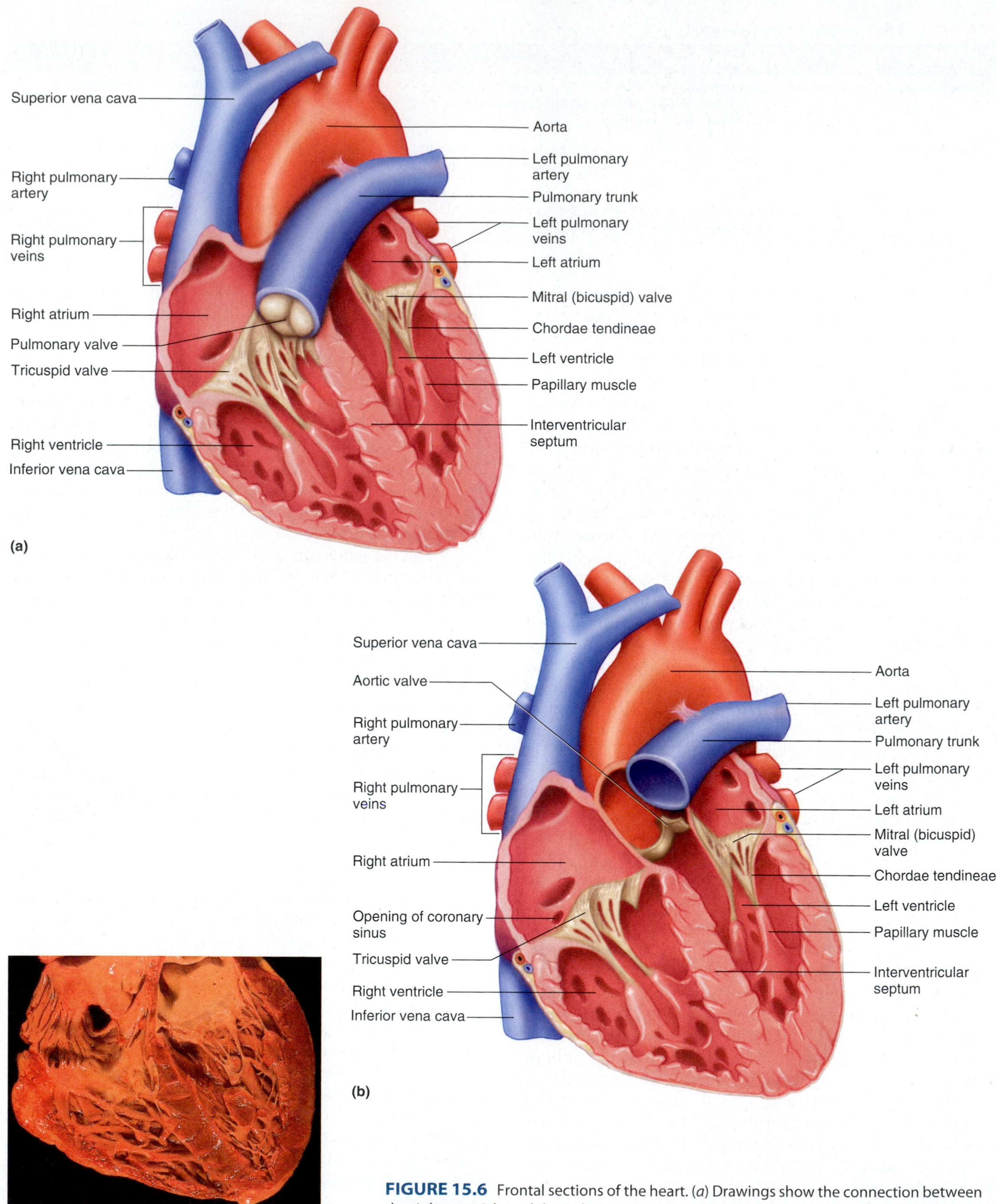

FIGURE 15.6 Frontal sections of the heart. (*a*) Drawings show the connection between the right ventricle and the pulmonary trunk and (*b*) the connection between the left ventricle and the aorta, as well as the four hollow chambers. (*c*) A cadaver heart.

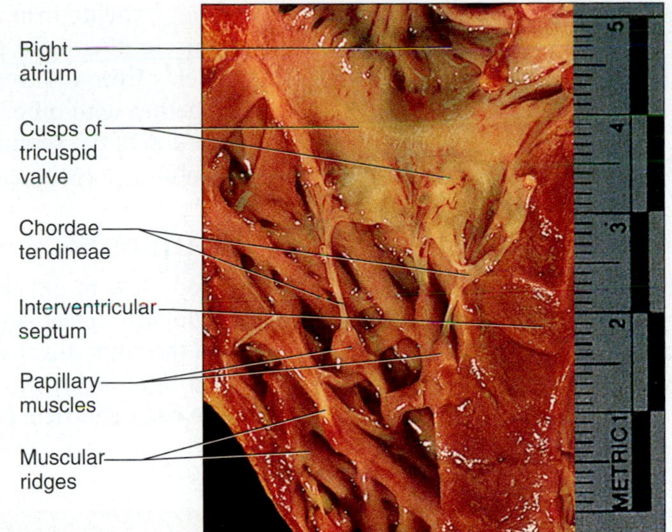

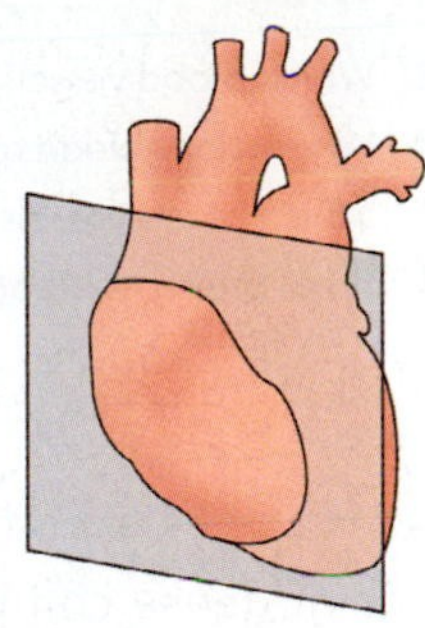

FIGURE 15.7 Photograph of a human tricuspid valve.

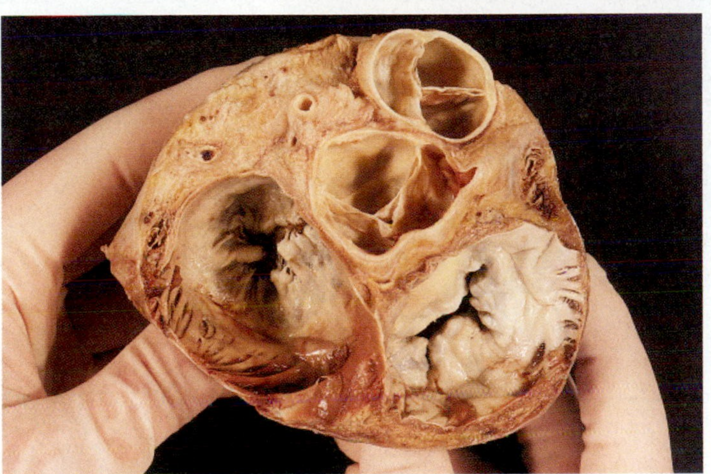

FIGURE 15.8 Photograph of the heart valves (superior view). Figure 15.9 labels the valves as seen in this photograph.

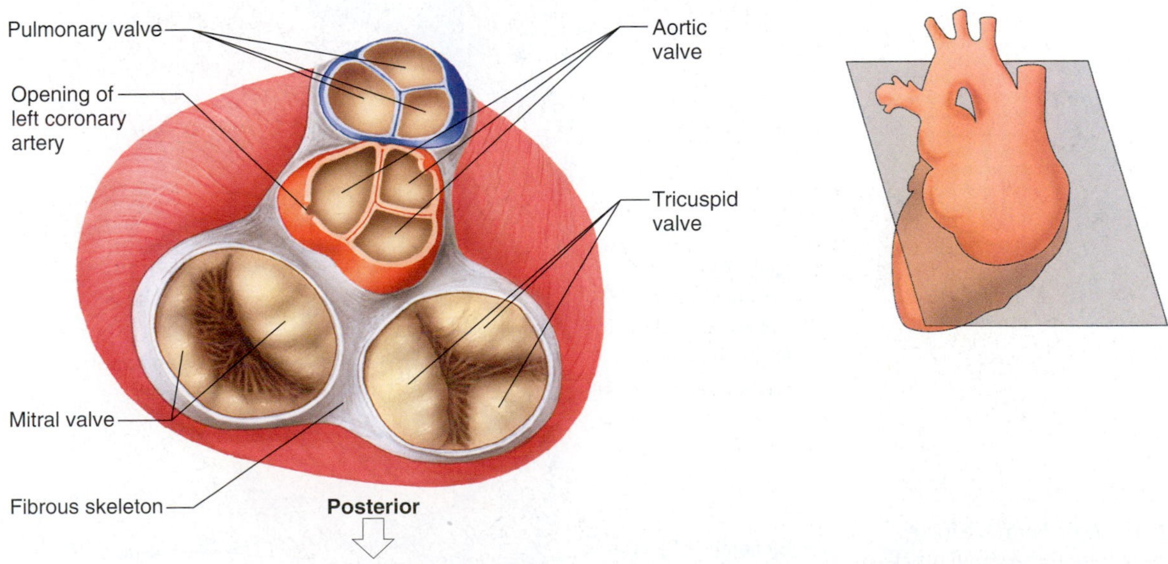

FIGURE 15.9 The skeleton of the heart consists of fibrous rings to which the heart valves are attached (superior view).

PRACTICE

8 Which blood vessels carry blood into the right atrium?

9 Where does blood go after it leaves the right ventricle?

10 Which blood vessels carry blood into the left atrium?

11 What prevents blood from flowing back into the ventricles when they relax?

Skeleton of the Heart

Rings of dense connective tissue surround the pulmonary trunk and aorta at their proximal ends. These rings are continuous with others that encircle the atrioventricular orifices. They provide firm attachments for the heart valves and for muscle fibers and prevent the outlets of the atria and ventricles from dilating during contraction. The fibrous rings, together with other masses of dense connective tissue in the portion of the septum between the ventricles (interventricular septum), constitute the *skeleton of the heart* (fig. 15.9).

Path of Blood Through the Heart

Blood low in oxygen and high in carbon dioxide enters the right atrium through the venae cavae and the coronary sinus. As the right atrial wall contracts, the blood passes through the right atrioventricular orifice and enters the chamber of the right ventricle (fig. 15.10).

TABLE 15.2 | Valves of the Heart

Valve	Location	Function
Tricuspid valve	Right atrioventricular orifice	Prevents blood from moving from right ventricle into right atrium during ventricular contraction
Pulmonary valve	Entrance to pulmonary trunk	Prevents blood from moving from pulmonary trunk into right ventricle during ventricular relaxation
Mitral valve	Left atrioventricular orifice	Prevents blood from moving from left ventricle into left atrium during ventricular contraction
Aortic valve	Entrance to aorta	Prevents blood from moving from aorta into left ventricle during ventricular relaxation

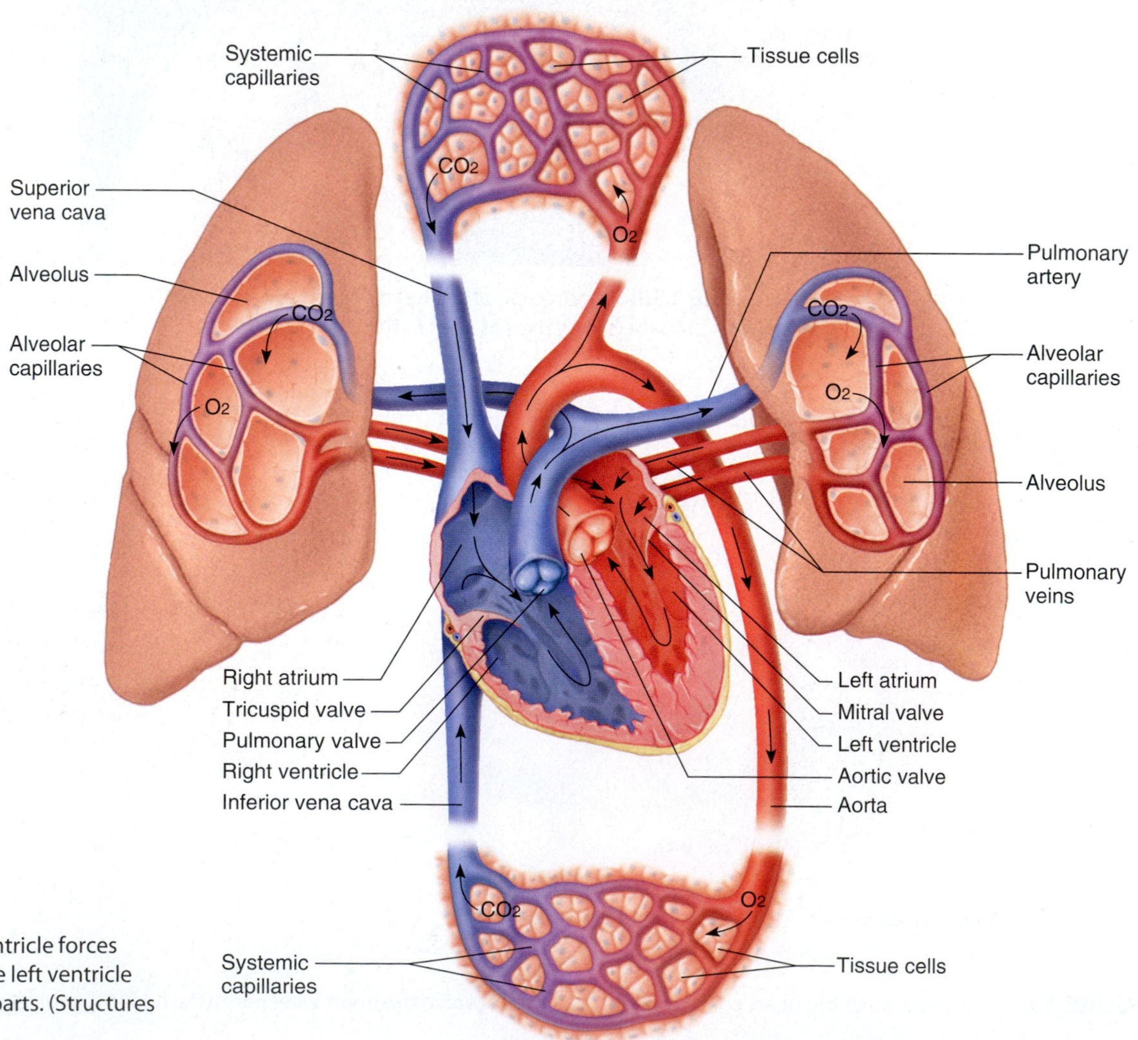

FIGURE 15.10 The right ventricle forces blood to the lungs, whereas the left ventricle forces blood to all other body parts. (Structures are not drawn to scale.)

When the right ventricular wall contracts, the tricuspid valve closes the right atrioventricular orifice, and the blood moves through the pulmonary valve into the pulmonary trunk and its branches (pulmonary arteries). From these vessels, blood enters the capillaries associated with the alveoli (microscopic air sacs) of the lungs. Gas exchange occurs between the blood in the capillaries and the air in the alveoli. The freshly oxygenated blood, now relatively low in carbon dioxide, returns to the heart through the pulmonary veins that lead to the left atrium.

The left atrial wall contracts, and the blood moves through the left atrioventricular orifice and into the chamber of the left ventricle. When the left ventricular wall contracts, the mitral valve closes the left atrioventricular orifice, and the blood passes through the aortic valve into the aorta and its branches. Figure 15.11 summarizes the path the blood takes as it moves through the heart to the alveolar capillaries and systemic capillaries, then back to the heart.

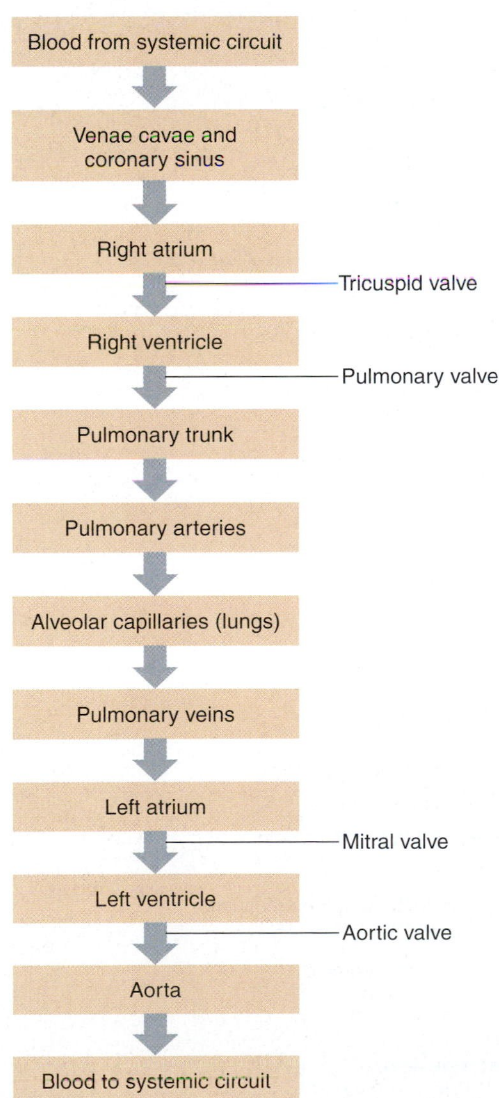

FIGURE 15.11 Path of blood through the heart and pulmonary circuit.

Blood Supply to the Heart

The first two branches of the aorta, called the right and left **coronary arteries,** supply blood to the tissues of the heart. Their openings lie just beyond the aortic valve (fig. 15.12).

The right coronary artery passes along the atrioventricular sulcus between the right atrium and the right ventricle. It gives off two major branches—a *posterior interventricular artery,* which travels along the posterior interventricular sulcus and supplies the walls of both ventricles, and a *marginal artery,* which passes along the lower border of the heart. Branches of the marginal artery supply the walls of the right atrium and the right ventricle (figs. 15.13 and 15.14).

One branch of the left coronary artery, the *circumflex artery,* follows the atrioventricular sulcus between the left atrium and the left ventricle. Its branches supply blood to the walls of the left atrium and the left ventricle. Another branch of the left coronary artery, the *anterior interventricular artery* (or *left anterior descending artery,*) lies in the anterior interventricular sulcus. Its branches supply the walls of both ventricles (figs. 15.13 and 15.14).

Magnetic resonance imaging (MRI) can image coronary arteries. Blood flow appears as a bright signal, and areas of diminished or absent blood flow, or blood turbulence, appear as blank areas. This approach is less invasive than the standard procedure of *coronary angiography,* in which a catheter is snaked through a blood vessel into the heart and a contrast agent is used to show heart structure.

The heart must beat continually to supply blood to the tissues. To do this, myocardial cells require a constant supply of freshly oxygenated blood. The myocardium contains many capillaries fed by branches of the coronary arteries.

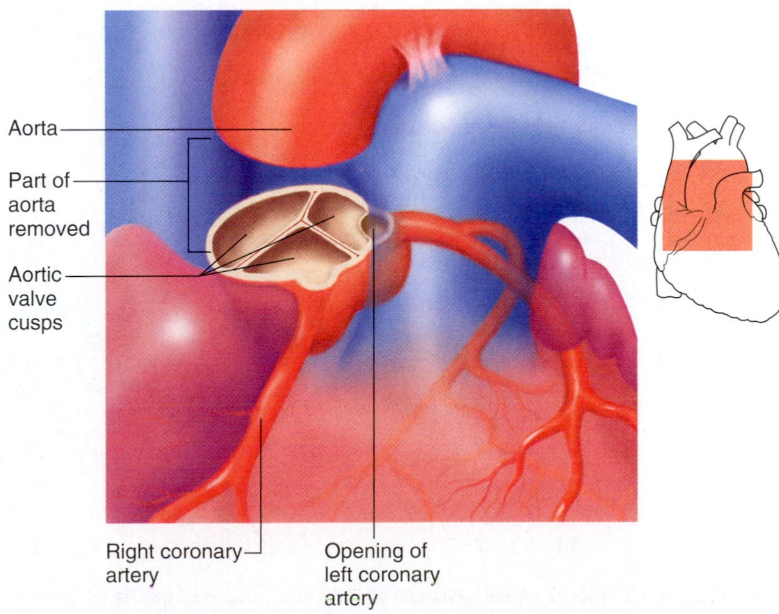

FIGURE 15.12 The openings of the coronary arteries lie just beyond the aortic valve.

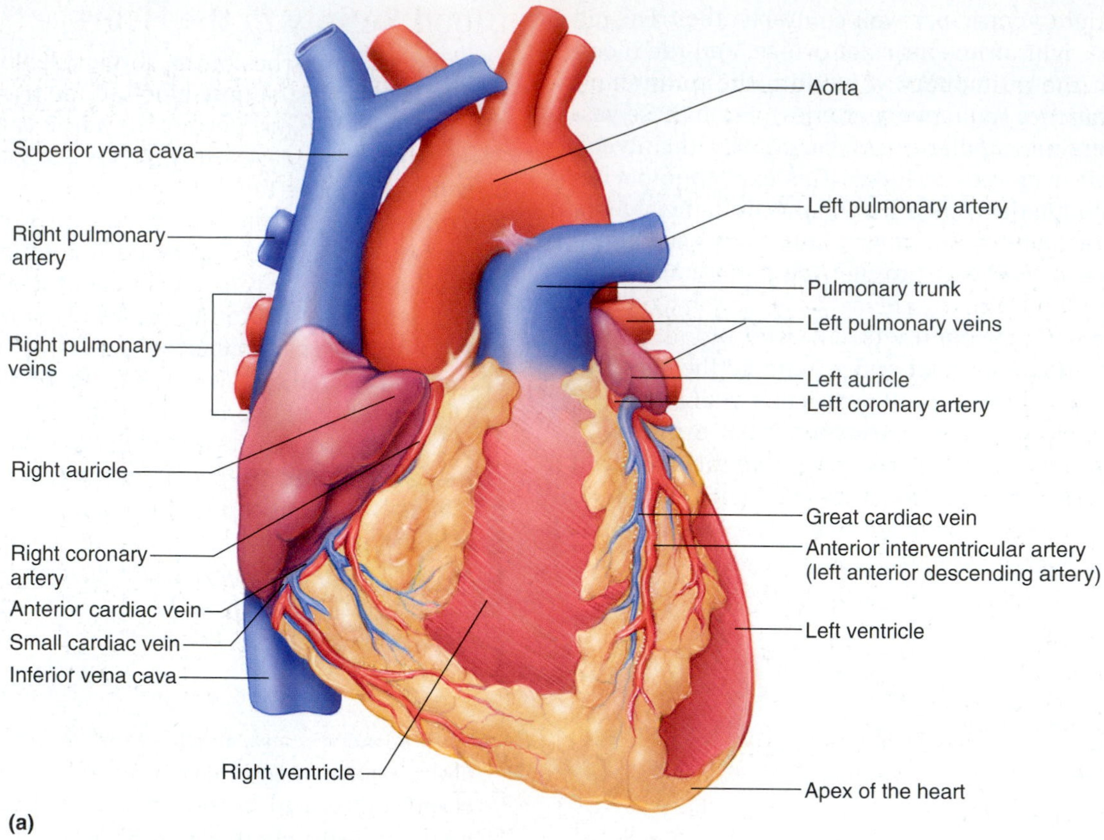

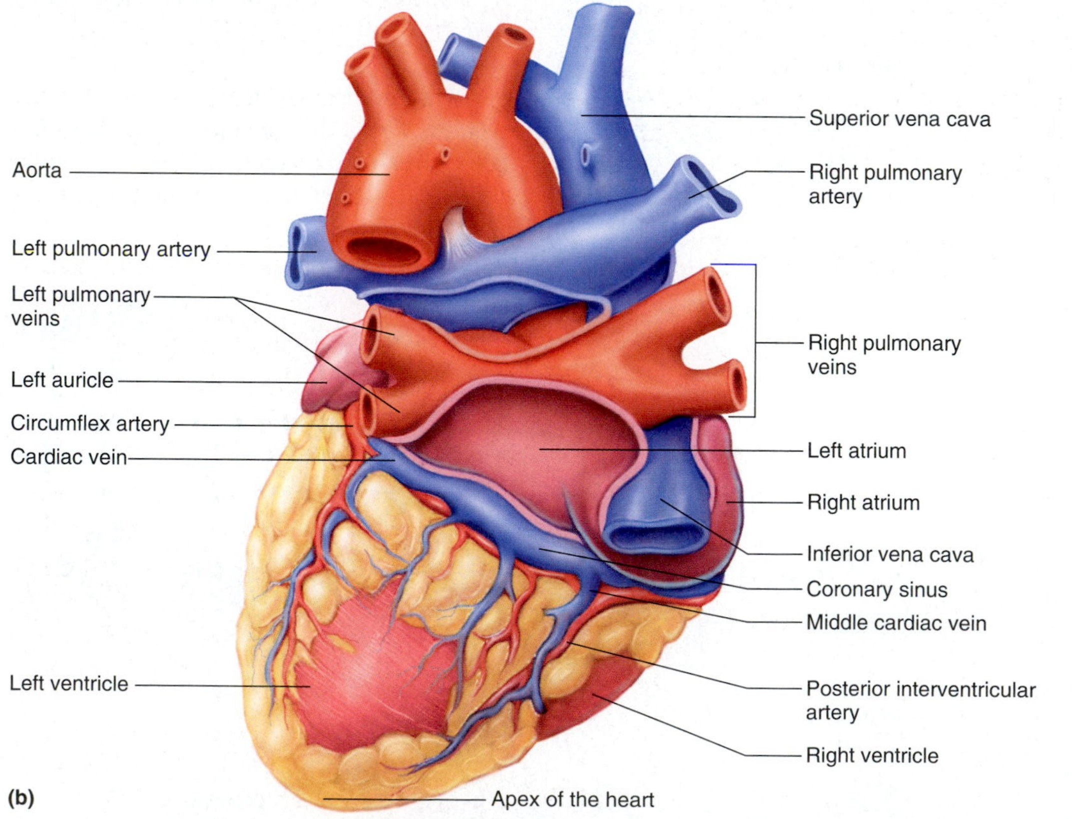

FIGURE 15.13 Blood vessels associated with the surface of the heart. (*a*) Anterior view. (*b*) Posterior view.

The smaller branches of these arteries usually have connections (anastomoses) between vessels that provide alternate pathways for blood, called collateral circulation. These detours in circulation may supply oxygen and nutrients to the myocardium when a coronary artery is blocked.

A thrombus or embolus that blocks or narrows a coronary artery branch deprives myocardial cells of oxygen, producing ischemia and painful *angina pectoris*. The pain usually occurs during physical activity, when oxygen requirements exceed supply. Pain lessens with rest. Emotional stress may also trigger angina pectoris.

Angina pectoris may cause a heavy pressure, tightening, or squeezing sensation in the chest. The pain is usually felt behind the sternum or in the anterior portion of the upper thorax, but may radiate to the neck, jaw, throat, shoulder, upper limb, back, or upper abdomen. Other symptoms include profuse perspiration (diaphoresis), difficulty breathing (dyspnea), nausea, or vomiting.

A blood clot completely obstructing a coronary artery or one of its branches (coronary thrombosis) kills part of the heart. This is a *myocardial infarction* (MI), more commonly known as a heart attack.

In most body parts, blood flow in arteries peaks during ventricular contraction. However, blood flow in the vessels of the myocardium is poorest during ventricular contraction. This is because the muscle fibers of the myocardium compress nearby vessels as they contract, interfering with blood flow. Also, the openings into the coronary arteries are partially blocked as the flaps of the aortic valve open. Conversely, during ventricular relaxation, the myocardial vessels are no longer compressed, and the aortic valve does not block the orifices of the coronary arteries. This increases blood flow into the myocardium.

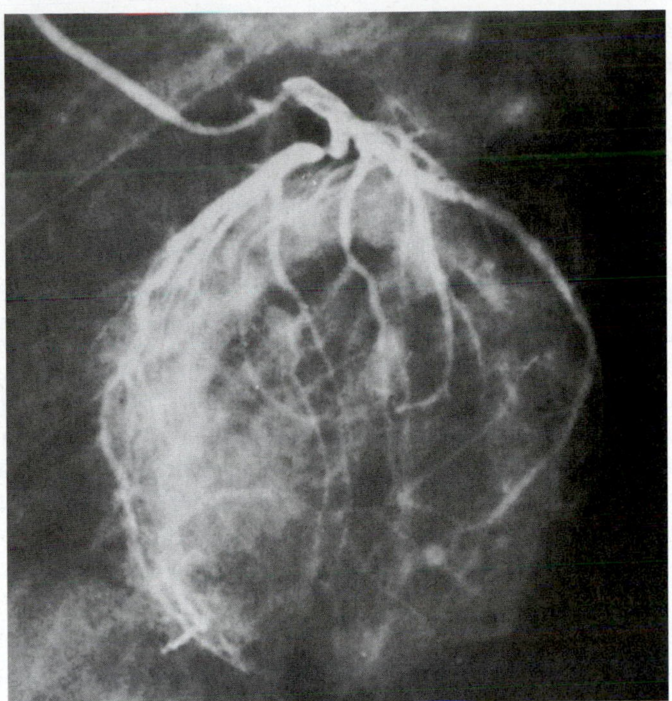

FIGURE 15.14 An angiogram (radiograph) of the coronary arteries is a diagnostic procedure used to examine specific blood vessels.

Branches of the **cardiac veins** drain blood that has passed through the capillaries of the myocardium. Their paths roughly parallel those of the coronary arteries. As figure 15.13*b* shows, these veins join the **coronary sinus,** an enlarged vein on the posterior surface of the heart in the atrioventricular sulcus. The coronary sinus empties into the right atrium. Figure 15.15 summarizes the path of blood that supplies the tissues of the heart.

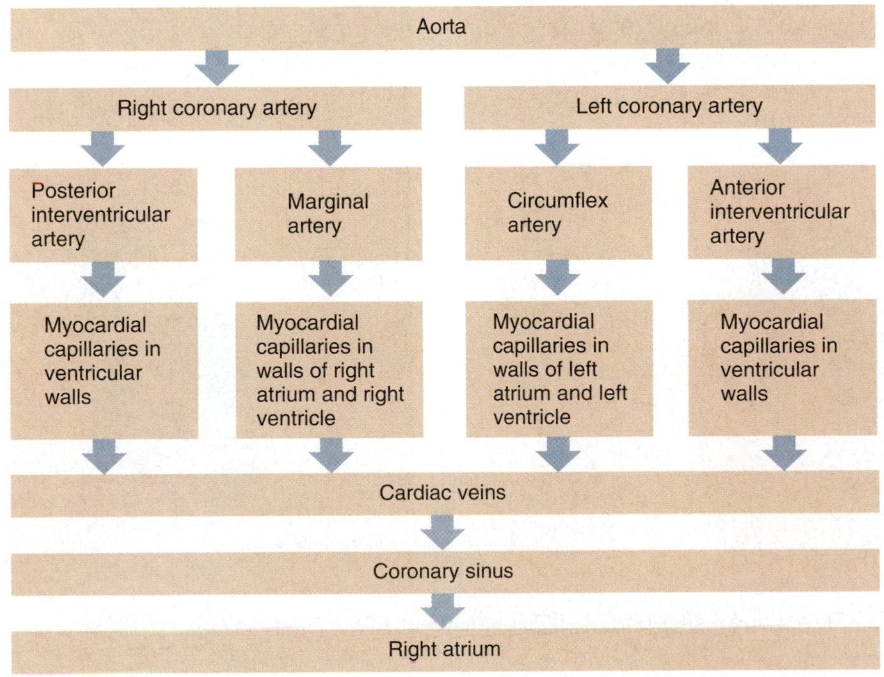

FIGURE 15.15 Path of blood through the coronary circulation.

PRACTICE 

12 Which structures make up the skeleton of the heart?

13 Review the path of blood through the heart.

14 How does blood composition differ in the right and left ventricle?

15 Which vessels supply blood to the myocardium?

16 How does blood return from the cardiac tissues to the right atrium?

15.3 HEART ACTIONS

The heart chambers function in coordinated fashion. Their actions are regulated so that atria contract, called atrial **systole** (sis′to-le), while ventricles relax, called ventricular **diastole** (di-as′to-le); then ventricles contract (ventricular systole) while atria relax (atrial diastole). Then the atria and ventricles both relax for a brief interval. This series of events constitutes a complete heartbeat, or **cardiac cycle** (kar′de-ak si′kl).

Cardiac Cycle

During a cardiac cycle, the pressure in the heart chambers rises and falls. These changes open and close the valves, much like a door being blown open or closed by the wind. Pressure in the ventricles is low early in diastole, and the pressure difference between atria and ventricles opens the AV valves. The ventricles fill. About 70% of the returning blood enters the ventricles prior to contraction, and ventricular pressure gradually increases. During atrial systole, the remaining 30% of returning blood is pushed into the ventricles, and ventricular pressure increases. Then, as the ventricles contract, ventricular pressure rises sharply. As soon as the ventricular pressure exceeds atrial pressure, the AV valves close. At the same time, the papillary muscles contract. By pulling on the chordae tendineae, they prevent the cusps of the AV valves from bulging too far into the atria.

During ventricular systole, the AV valves remain closed. The atria are now relaxed, and pressure in the atria is low, even lower than venous pressure. As a result, blood flows into the atria from the large, attached veins. That is, as the ventricles are contracting, the atria are filling, already preparing for the next cardiac cycle (fig. 15.16).

As ventricular systole progresses, ventricular pressure continues to increase until it exceeds the pressure in the pulmonary trunk (right side) and aorta (left side). At this point, the pressure differences across the semilunar valves open the pulmonary and aortic valves, and blood is ejected from each valve's ventricle into these arteries.

As blood flows out of the ventricles, ventricular pressure begins to drop, and it falls even farther as the ventricles relax. When ventricular pressure is lower than the blood pressure in the aorta and pulmonary trunk, the pressure difference is reversed, and the semilunar valves close. The ventricles continue to relax, and as soon as ventricular pressure is less than atrial pressure, the AV valves open, and the ventricles begin to fill once more. Atria and ventricles relax for a brief interval.

Heart Sounds

A heartbeat heard through a stethoscope sounds like "lubb-dupp." These sounds are due to vibrations in the heart tissues produced as the blood flow is suddenly slowed with the contraction and relaxation of the heart chambers and with the closing of the valves.

The first part of a heart sound (*lubb*) occurs during the ventricular systole, when the AV valves are closing. The second part (*dupp*) occurs during ventricular diastole, when the pulmonary and aortic valves are closing.

Sometimes during inspiration, the interval between the closure of the pulmonary and the aortic valves is long enough that a sound comes from each of these events. In this case, the second heart sound is said to be *split*.

Heart sounds are of particular interest because they can indicate the condition of the heart valves. For example, inflammation of the endocardium (endocarditis) may

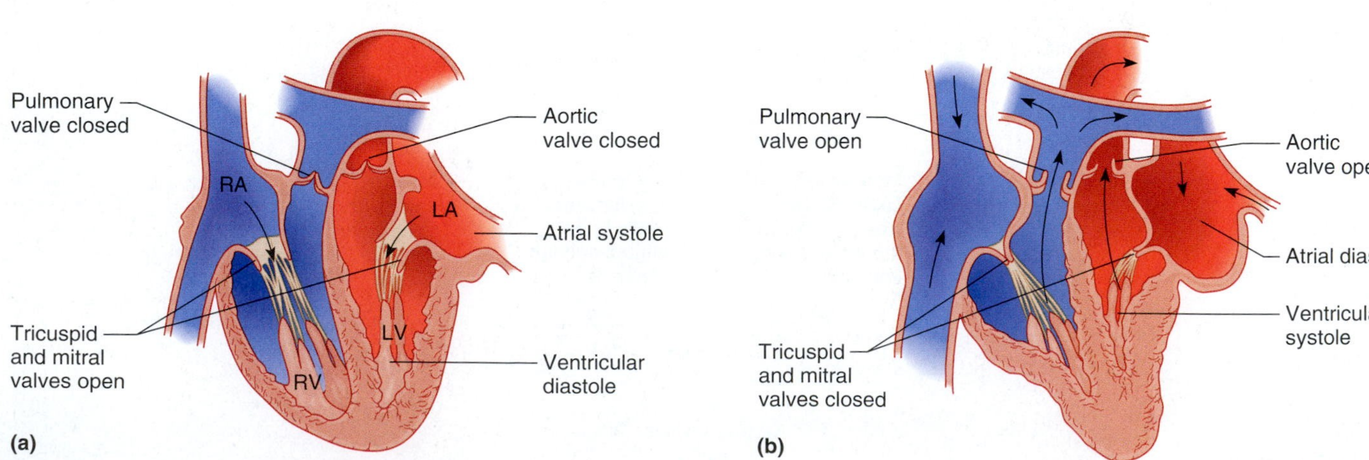

FIGURE 15.16 The atria (*a*) empty during atrial systole and (*b*) fill with blood during atrial diastole.

erode the edges of the valvular cusps. As a result, the cusps may not close completely, and some blood may leak back through the valve, producing an abnormal sound called a *murmur.* The seriousness of a murmur depends on the degree of valvular damage. Many heart murmurs are harmless. It is often possible to repair damaged valves or to replace them. From Science to Technology 15.1 describes treatments for a failing heart.

Using a stethoscope, it is possible to hear sounds associated with the aortic and pulmonary valves by listening from the second intercostal space on either side of the sternum. The *aortic sound* comes from the right, and the *pulmonic sound* from the left. The sound associated with the mitral valve can be heard from the fifth intercostal space at the nipple line on the left. The sound of the tricuspid valve can be heard at the fifth intercostal space just to the right of the sternum (fig. 15.17).

Cardiac Muscle Fibers

Recall that cardiac muscle fibers function like those of skeletal muscles, but the fibers connect in branching networks (chapter 9, p. 301). Stimulation to any part of the network sends impulses throughout the heart, which contracts as a unit.

A mass of merging cells that act as a unit is called a **functional syncytium** (funk′shun-al sin-sish′e-um). Two such structures are in the heart—in the atrial walls and in the ventricular walls. Portions of the heart's fibrous skeleton separate these masses of cardiac muscle fibers, except for a small area in the right atrial floor. In this region, fibers of the cardiac conduction system connect the *atrial syncytium* and the *ventricular syncytium.*

PRACTICE  

17 Describe the pressure changes in the atria and ventricles during a cardiac cycle.

18 What causes heart sounds?

19 What is a functional syncytium?

20 Where are the functional syncytia of the heart?

Cardiac Conduction System

Throughout the heart are clumps and strands of specialized cardiac muscle tissue whose fibers contain only a few myofibrils. Instead of contracting, these areas initiate and distribute impulses (cardiac impulses) throughout the myocardium. They comprise the **cardiac conduction system,** which coordinates the events of the cardiac cycle.

A key portion of this conduction system is the **SA node (sinoatrial node** or **sinuatrial node)**, a small, elongated mass of specialized cardiac muscle tissue just beneath the epicardium. It is in the right atrium near the opening of the superior

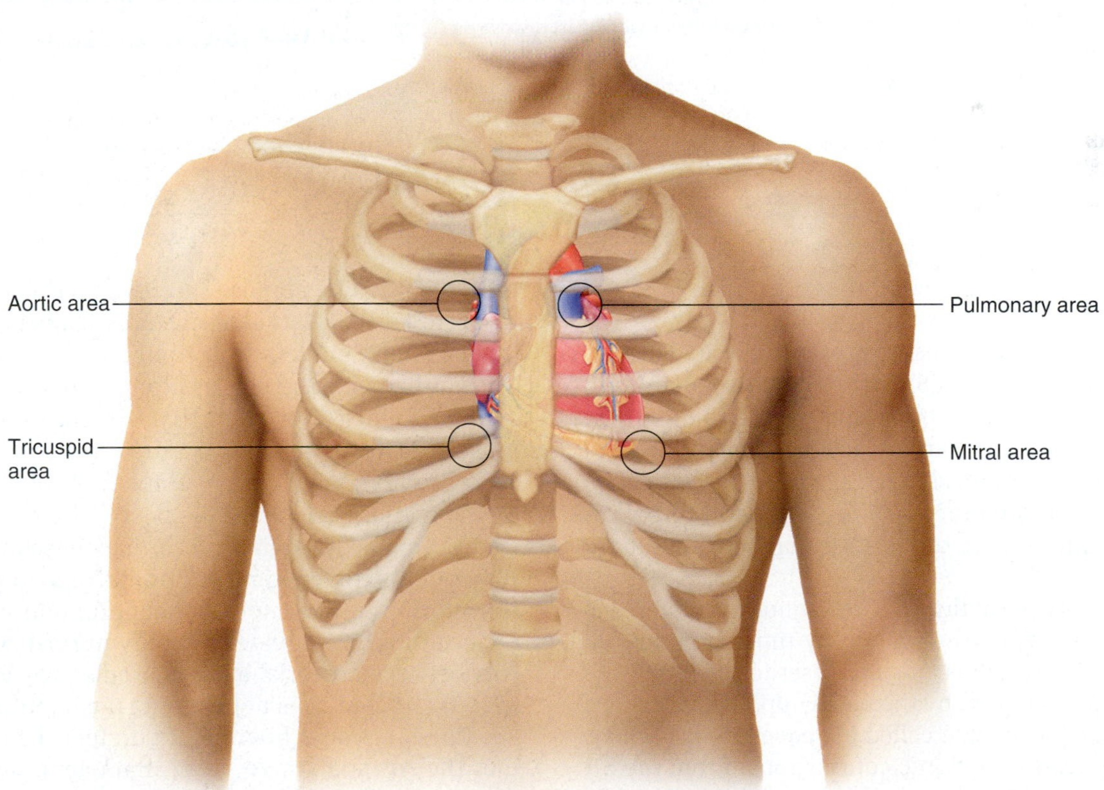

FIGURE 15.17 Thoracic regions where the sounds of each heart valve are most easily heard.

15.1 FROM SCIENCE TO TECHNOLOGY

Replacing the Heart—From Transplants To Stem Cell Implants

Medical science offers several ways to aid or even replace a failing heart. In a *heart transplant,* the recipient's failing heart is removed, except for the posterior walls of the right and left atria and their connections to the venae cavae and pulmonary veins. The donor heart is similarly prepared and is attached to the atrial cuffs remaining in the recipient's thorax. Finally, the recipient's aorta and pulmonary arteries are connected to those of the donor heart (fig. 15A).

Donor hearts are scarce. A mechanical half-heart, called a *left ventricular assist device* (LVAD), can often maintain cardiac function long enough for a heart to become available. An LVAD allows a patient to resume some activities and to increase physical fitness, which can increase the chance of success of an eventual heart transplant. A few patients in England too ill to receive transplants are surviving with permanently implanted LVADs.

An implantable replacement heart became available in 2006 for people who are not candidates for heart transplantation and have less than a month to live. The two-pound, titanium and plastic cardiac stand-in consists of an internal motor-driven hydraulic pump, a battery and electronics package, and an external battery pack. The electronics component manages the rate and force of the pump's actions, tailoring them to the patient's condition. Newer implantable replacement hearts are smaller and can provide up to five years of life.

Stem cell technology may allow researchers to patch failing hearts with new cardiac muscle. Human cardiac muscle tissue can be cultured from stem cells. In laboratory dishes, the tissue contracts, as cardiac muscle would in the body. These cells are combined with a synthetic, compatible biomaterial elastic enough to stretch as the cells divide and degrade in a controlled way. The idea is that "stem cell heart patches" would consist of scaffolding made of such a biomaterial that would support the cells as they nestle into a damaged heart. As the cells contract, the synthetic portion degrades, leaving the pulsating patch. If the cells originate from the patient, the patch would not be rejected. Experiments in mice use stem cells that divide to give rise to cardiac muscle, endothelium and vascular smooth muscle. Mice with heart disease given mixtures of these cells showed improvement in heart function, suggesting that it might be possible to engineer customized heart parts. ■

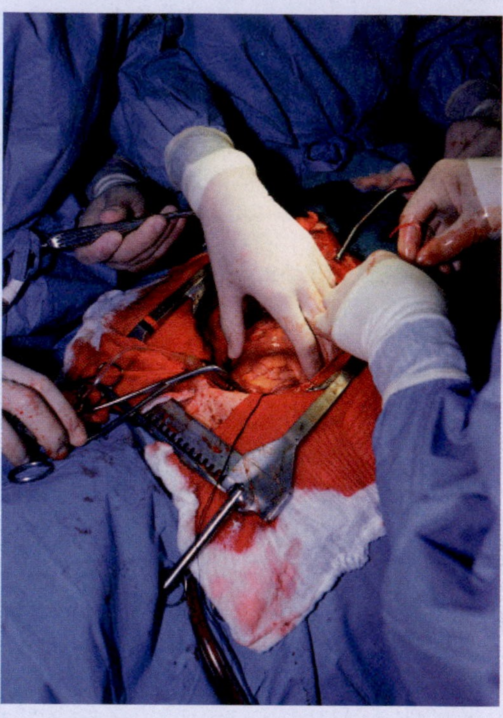

FIGURE 15A A heart transplant can save a life. A heart that might have died with its donor can provide a new lease on life for a recipient, thanks to our understanding of the immune system—and a well-trained medical team.

vena cava, and its fibers are continuous with those of the atrial syncytium.

The cells of the SA node reach threshold spontaneously. Recall from chapter 10 (p. 368) that an action potential in a neuron is triggered by a depolarizing input from presynaptic neurons. In contrast, an SA node reaches threshold on its own. A number of mechanisms appear to play a role, including a progresssive increase in permeability to calcium ions and sodium ions, and decreasing permeability to potassium ions.

SA node activity is rhythmic. The SA node initiates one impulse after another, more than eighty times a minute in an adult (resting heart rate is usually closer to seventy beats per minute due to inhibition by the parasympathetic nervous system). The SA node is often called the **pacemaker** because it generates the heart's rhythmic contractions. From the SA node, bundles of atrial muscle, called *internodal atrial muscle,* preferentially conduct impulses to more distant regions of the atria. Then, because gap junctions connect cardiac muscle cells, the resulting impulse spreads into the surrounding atrial myocardium and stimulates the muscle fibers to contract.

The right and left atria contract almost simultaneously. The cardiac impulse passes along fibers (junctional fibers) of the conduction system that are continuous with atrial muscle fibers, instead of passing directly into the ventricular syncytium, which is separated from the atrial syncytium by the fibrous skeleton of the heart. These conducting fibers lead to a mass of specialized cardiac muscle tissue called the **AV node (atrioventricular node).** This node is in the inferior portion of the interatrial septum and just beneath the endocardium. It provides the only normal conduction pathway between the atrial and ventricular syncytia, because the fibrous skeleton does not conduct the impulse.

The junctional fibers that conduct the cardiac impulse into the AV node have small diameters. Small fibers conduct impulses slowly because they delay transmission of the impulse. The impulse is delayed further as it moves through the AV node. This allows time for the atria to completely contract so they empty all their blood into the ventricles prior to ventricular systole.

Once the cardiac impulse reaches the distal side of the AV node, it passes into a group of large fibers that make up the **AV bundle** (atrioventricular bundle or bundle of His), and the impulse moves rapidly through them. The AV bundle enters the upper part of the interventricular septum and divides into right and left bundle branches that lie just beneath the endocardium. About halfway down the septum, the branches give rise to enlarged **Purkinje fibers** (pur-kin′je fi′berz). These larger fibers carry the impulse to distant regions of the ventricular myocardium much faster than cell-to-cell conduction could. As a result, the massive ventricular myocardium contracts as a functioning unit.

The base of the aorta, which includes the aortic valve, is enlarged and protrudes somewhat into the interatrial septum close to the AV bundle. Consequently, inflammatory conditions, such as bacterial endocarditis affecting the aortic valve (aortic valvulitis), may also affect the AV bundle.

If a portion of the bundle is damaged, it may no longer conduct impulses normally. As a result, cardiac impulses may reach the two ventricles at different times so that they fail to contract together. This condition is called a *bundle branch block.*

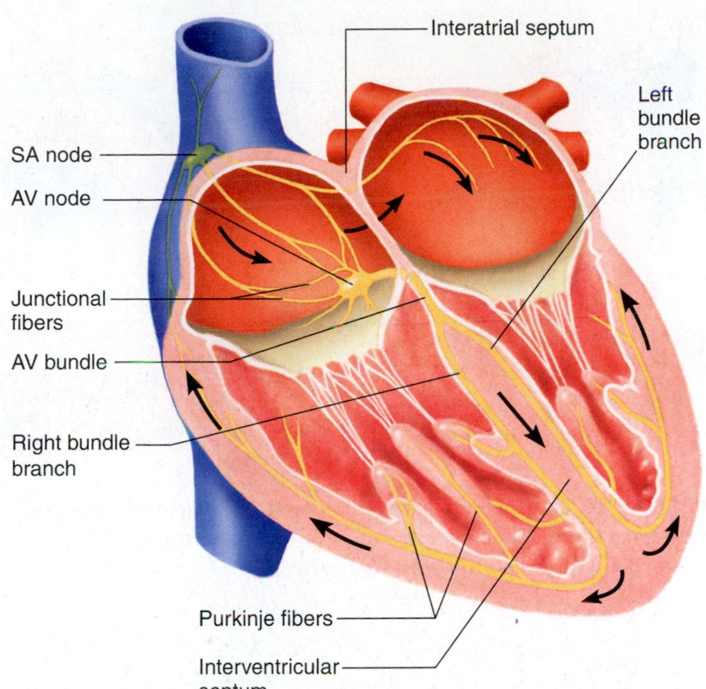

FIGURE 15.18 The cardiac conduction system coordinates the cardiac cycle.

The Purkinje fibers spread from the interventricular septum into the papillary muscles, which project inward from the ventricular walls, and then continue downward to the apex of the heart. There they curve around the tips of the ventricles and pass upward over the lateral walls of these chambers. Along the way, the Purkinje fibers give off many small branches, which become continuous with cardiac muscle fibers. Figure 15.18 shows the parts of the conduction system and figure 15.19 summarizes them.

The muscle fibers in the ventricular walls form irregular whorls. When impulses on the Purkinje fibers stimulate these muscle fibers, the ventricular walls contract with a twisting motion (fig. 15.20). This action squeezes blood out of the ventricular chambers and forces it into the aorta and pulmonary trunk.

Another property of the conduction system is that the Purkinje fibers transmit the impulse to the apex of the heart first. As a result, contraction begins at the apex and pushes the blood superiorly toward the aortic and pulmonary semilunar valves, rather than originating the impulse superiorly and pushing blood toward the apex, as it would if the impulse traveled from cell to cell.

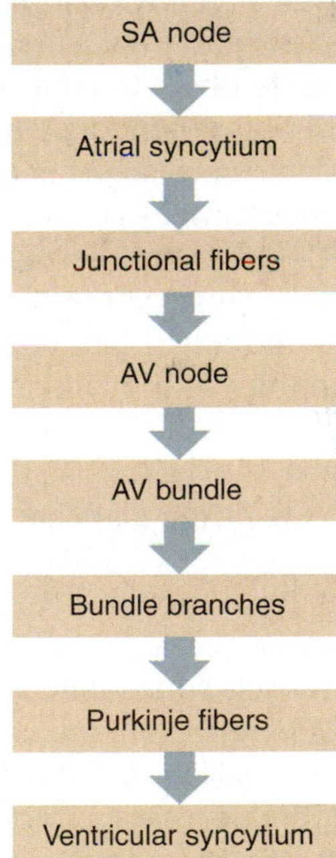

FIGURE 15.19 Components of the cardiac conduction system.

PRACTICE

21 What is the function of the cardiac conduction system?

22 What types of tissues make up the cardiac conduction system?

23 How is a cardiac impulse initiated?

24 How is a cardiac impulse transmitted from the right atrium to the other heart chambers?

A significant percentage of cases of heart failure in adults of African descent may be due to an inherited condition called familial amyloidosis. A protein called amyloid forms deposits in the heart, causing angina (chest pain), failure of cardiac muscle function (cardiomyopathy), blockage of conduction of electrical impulses, and disturbed heart rhythm (arrhythmia). Echocardiography can detect the amyloid deposits that thicken the ventricular walls. It is important to distinguish amyloidosis from other forms of arrhythmias, because drug treatments are different.

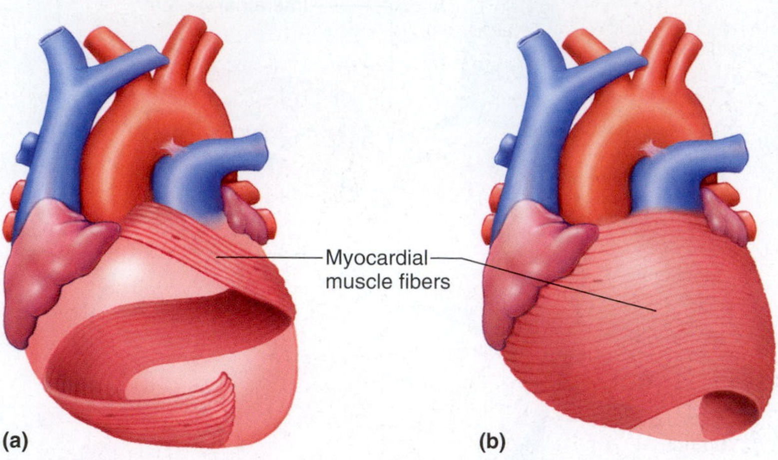

FIGURE 15.20 The muscle fibers within the ventricular walls form whorled patterns. The fibers of groups (*a*) and (*b*) surround both ventricles in these anterior views of the heart.

Electrocardiogram

An **electrocardiogram** (e-lek″tro-kar′de-o-gram″) (ECG) is a recording of the electrical changes in the myocardium during a cardiac cycle. (This pattern occurs as action potentials stimulate cardiac muscle fibers to contract, but it is not the same as individual action potentials.) Because body fluids can conduct electrical currents, such changes can be detected on the surface of the body.

To record an ECG, electrodes are placed on the skin and connected by wires to an instrument that responds to weak electrical changes by moving a pen or stylus on a moving strip of paper. Up-and-down movements of the pen correspond to electrical changes in the myocardium. The paper moves past the pen at a known rate, so the distance between pen deflections indicates time elapsing between phases of the cardiac cycle.

A normal ECG pattern includes several deflections, or *waves,* during each cardiac cycle, as figure 15.21*a* illustrates. Between cycles, the muscle fibers remain polarized, with no detectable electrical changes. The pen does not move and marks along the baseline. When the SA node triggers a cardiac impulse, the atrial fibers depolarize, producing an electrical change. The pen moves, and at the end of the electrical change, returns to the base position. This first pen movement produces a *P wave,* corresponding to depolarization of the atrial fibers that will lead to contraction of the atria (fig. 15.21*b–d*).

When the cardiac impulse reaches the ventricular fibers, they rapidly depolarize. The ventricular walls are thicker than those of the atria, so the electrical change is greater, and the pen deflects more. When the electrical change ends, the pen returns to the baseline. This leaves a mark called the *QRS complex,* which usually consists of a *Q wave,* an *R wave,* and an *S wave.* The complex appears due to depolarization of the ventricular fibers just prior to the contraction of the ventricular walls (fig. 15.21*e* and *f*).

The electrical changes that accompany ventricular muscle fiber repolarization slowly produce a *T wave* as the pen deflects again, ending the ECG pattern (fig. 15.21*g* and *h*). The record of the atrial repolarization seems to be missing from the pattern because the atrial fibers repolarize at the same time that the ventricular fibers depolarize. Thus, the QRS complex obscures the recording of the atrial repolarization. The graph in figure 15.22 summarizes some of the changes that occur during a cardiac cycle with corresponding ECG patterns and heart sounds.

In addition to the waves that comprise the classic electrocardiogram are repeating subpatterns of other waves that occur at different timescales and in an irregular pattern. Although it may seem counterintuitive, this complex, varying backdrop to the cardiac cycle seems to be necessary for health. It is disrupted in congestive heart failure.

Physicians use ECG patterns to assess the heart's ability to conduct impulses. For example, the period between the beginning of a P wave and the beginning of a QRS complex called the *PQ interval* (or if the initial portion of the QRS wave is upright, the *PR interval*) indicates the time for the cardiac impulse to travel from the SA node through the AV node. Ischemia or other problems affecting the fibers of the AV conduction pathways can increase this PQ interval. Similarly, injury to the AV bundle can extend the QRS complex, because it may take longer for an impulse to spread throughout the ventricular walls (fig. 15.23).

PRACTICE

25 What is an electrocardiogram?

26 Which cardiac events do the P wave, QRS complex, and T wave represent?

Regulation of the Cardiac Cycle

The volume of blood pumped changes to accommodate cellular requirements. For example, during strenuous exercise, skeletal muscles require more blood, and heart rate increases in response. The SA node normally controls heart rate, so changes in this rate often involve factors that affect the pacemaker, such as the motor impulses carried on the parasympathetic and sympathetic nerve fibers (see figs. 11.38, 11.39, 15.24, 15.37, and 15.38).

The parasympathetic fibers that innervate the heart arise from neurons in the medulla oblongata and make up parts of the *vagus nerves.* Most of these fibers branch to the SA and AV nodes. When the nerve impulses reach nerve fiber endings, they secrete acetylcholine, which decreases SA and AV nodal activity. As a result, heart rate decreases.

The vagus nerves continually carry impulses to the SA and AV nodes, "braking" heart action. Consequently, parasympathetic activity can change heart rate in either direction. An increase in the impulses slows the heart rate, and a decrease in the impulses releases the parasympathetic "brake" and increases heart rate.

Sympathetic fibers reach the heart by means of the *accelerator nerves,* whose branches join the SA and AV nodes as well as other areas of the atrial and ventricular myocardium. The endings of these fibers secrete norepinephrine in response to nerve impulses. Norepinephrine increases the rate and force of myocardial contractions.

Reflexes called *baroreceptor reflexes* arising from the *cardiac control center* of the medulla oblongata maintain

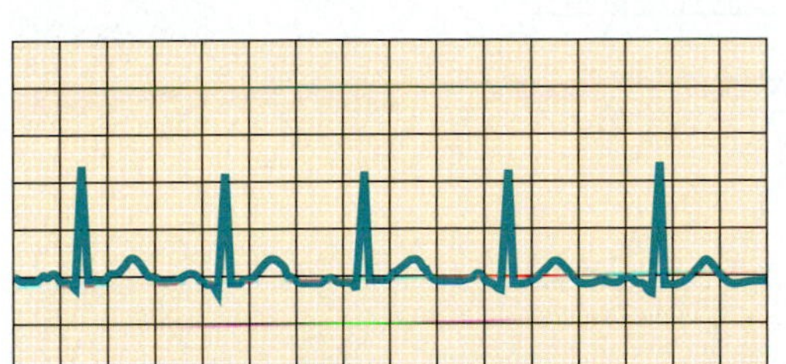

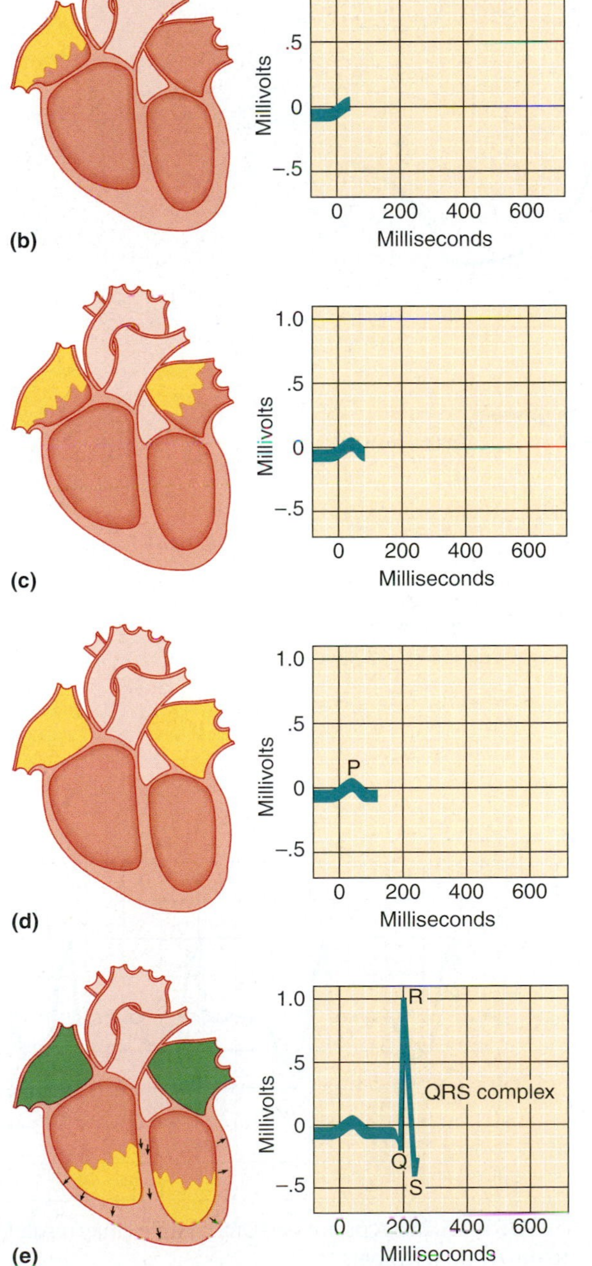

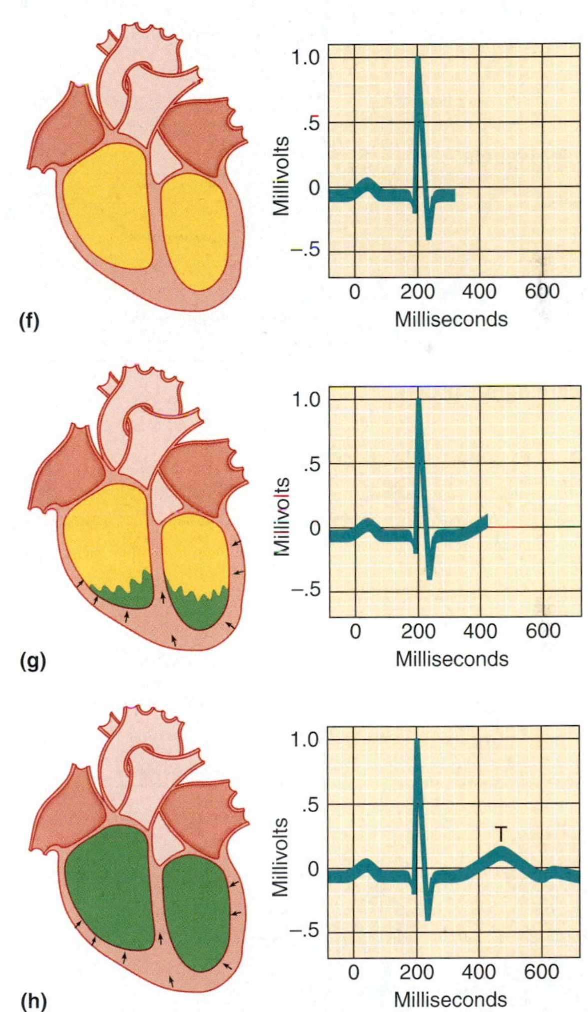

FIGURE 15.21 ECG pattern. (*a*) A normal ECG. In this set of drawings (*b–h*), the yellow areas of the hearts indicate where depolarization is occurring, and the green areas indicate where tissues are repolarizing; the portion of the ECG pattern produced at each step is shown by the continuation of the line on the graph paper.

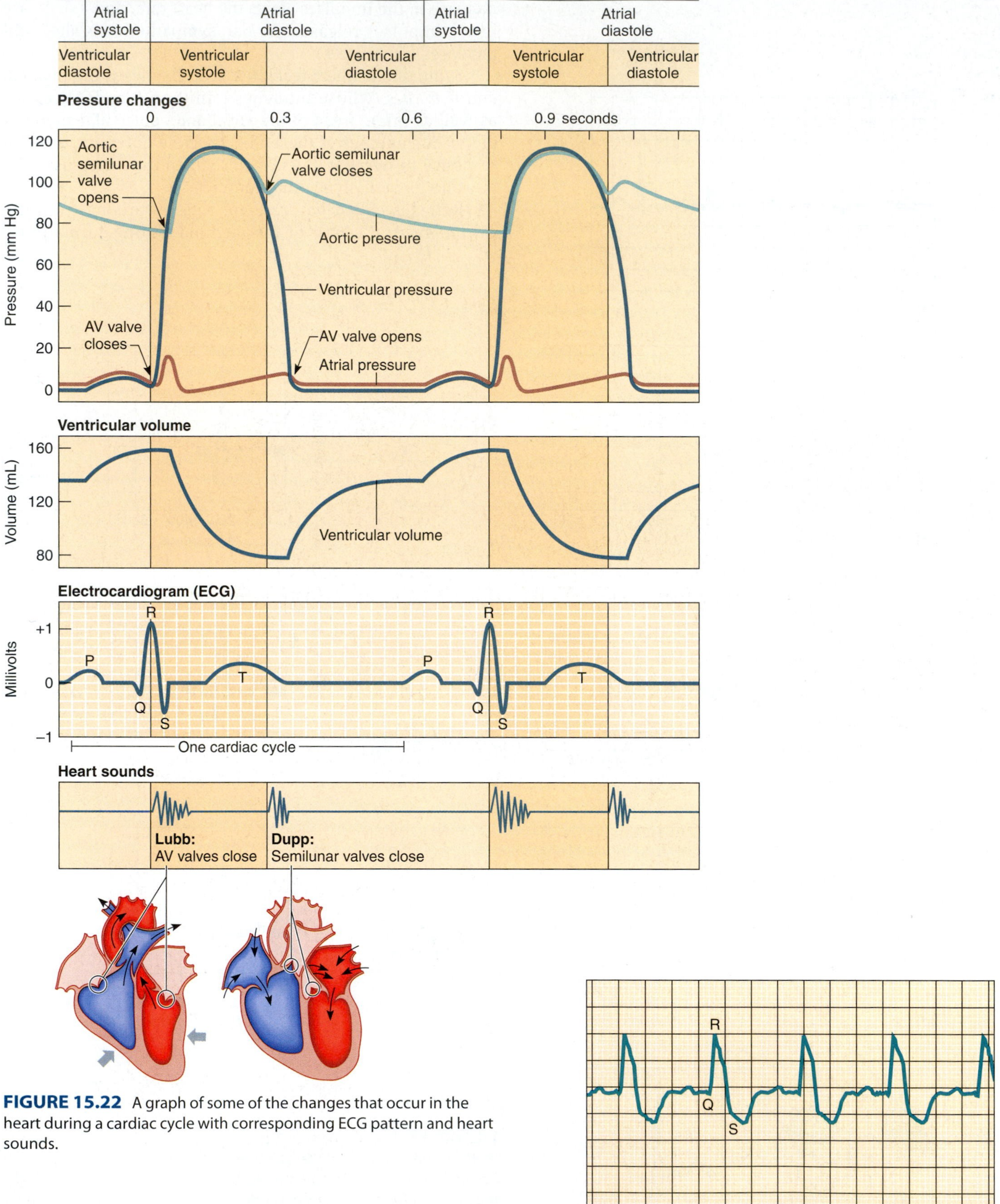

FIGURE 15.22 A graph of some of the changes that occur in the heart during a cardiac cycle with corresponding ECG pattern and heart sounds.

FIGURE 15.23 A prolonged QRS complex may result from damage to the AV bundle fibers.

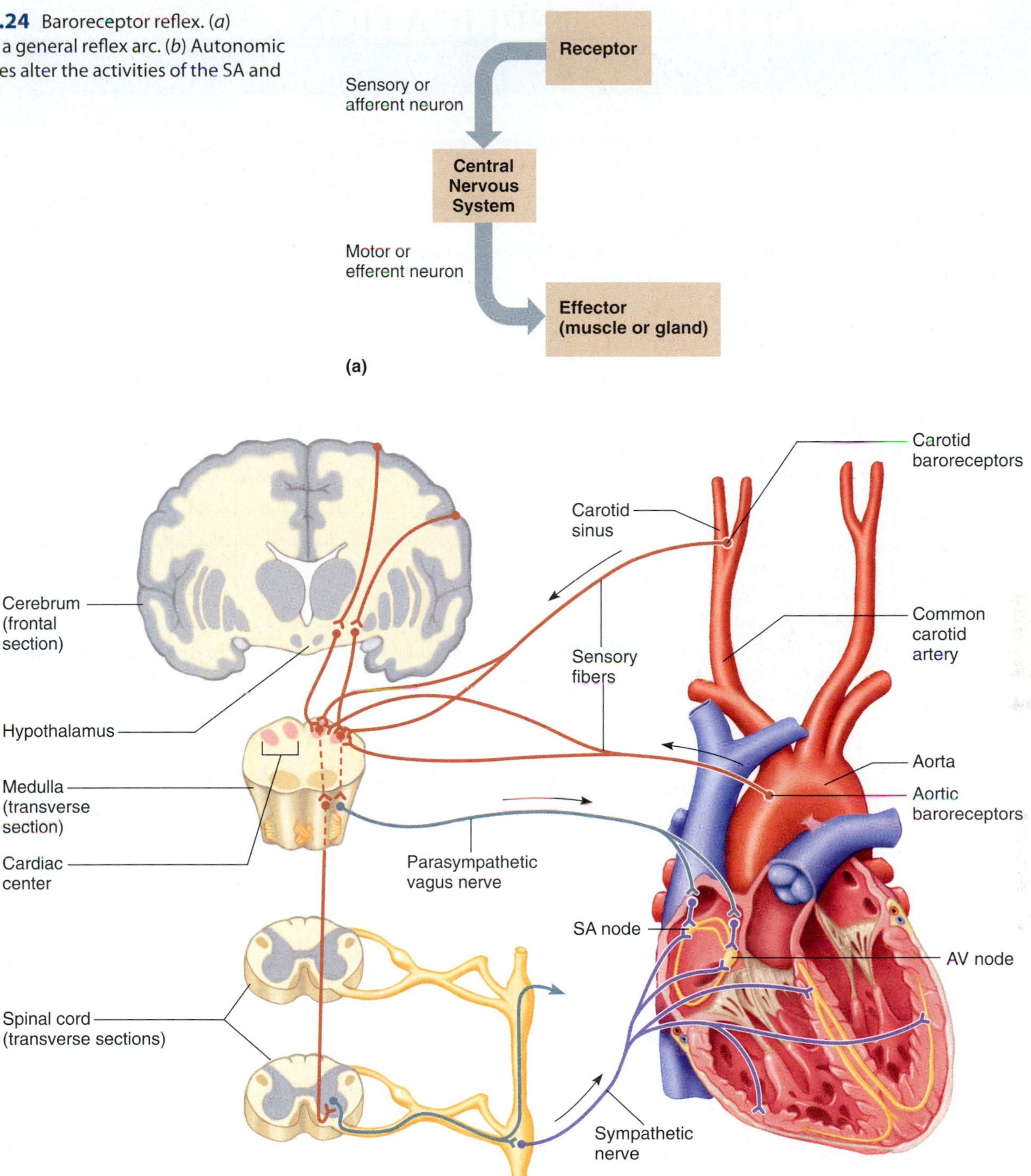

FIGURE 15.24 Baroreceptor reflex. (*a*) Schematic of a general reflex arc. (*b*) Autonomic nerve impulses alter the activities of the SA and AV nodes.

balance between inhibitory effects of the parasympathetic fibers and excitatory effects of the sympathetic fibers. In this region of the brain, masses of neurons function as *cardioinhibitor* and *cardioaccelerator reflex centers.* These centers receive sensory impulses from throughout the cardiovascular system and relay motor impulses to the heart in response. For example, receptors sensitive to stretch are located in certain regions of the aorta (aortic arch) and in the carotid arteries (carotid sinuses). These receptors, called *baroreceptors* (pressoreceptors), can detect changes in blood pressure (fig. 15.24). Rising pressure stretches the receptors, and they signal the cardioinhibitor center in the medulla. In response, the medulla sends parasympathetic motor impulses to the heart via the vagus nerve, decreasing the heart rate. This action helps lower blood pressure toward normal.

Another regulatory reflex uses stretch receptors in the venae cavae near the entrances to the right atrium. If venous blood pressure abnormally increases in these vessels, the

15.1 CLINICAL APPLICATION

Arrhythmias

Each year, thousands of people die from a fast or irregular heartbeat. These altered heart rhythms are called *arrhythmias.*

In *fibrillation,* small areas of the myocardium contract in an uncoordinated, chaotic fashion (fig. 15B). As a result, the myocardium fails to contract as a whole, and blood is no longer pumped. Atrial fibrillation is not life threatening, because the ventricles still pump blood, but ventricular fibrillation is often deadly. Ventricular fibrillation can be caused by an obstructed coronary artery, toxic drug exposure, electric shock, or traumatic injury to the heart or chest wall. A defibrillator device can deliver a shock to restore a normal heartbeat, as described in the vignette that opens this chapter.

An abnormally fast heartbeat, usually more than 100 beats per minute, is called *tachycardia.* Increase in body temperature, nodal stimulation by sympathetic fibers, certain drugs or hormones, heart disease, excitement, exercise, anemia, or shock can cause tachycardia. Figure 15C shows the ECG of a tachycardic heart.

Bradycardia means a slow heart rate, usually fewer than sixty beats per minute. Decreased body temperature, nodal stimulation by parasympathetic impulses, or certain drugs may cause bradycardia. It also may occur during sleep. Figure 15D shows the ECG of a bradycardic heart. Athletes sometimes have unusually slow heartbeats because their hearts pump a greater-than-normal volume of blood with each beat. The slowest heartbeat recorded in a healthy athlete was twenty-five beats per minute!

A *premature beat* occurs before it is expected in a normal series of cardiac cycles. Cardiac impulses originating from unusual (ectopic) regions of the heart probably cause a premature beat. That is, the impulse originates from a site other than the SA node. Cardiac impulses may arise from ischemic tissues or from muscle fibers irritated by disease or drugs.

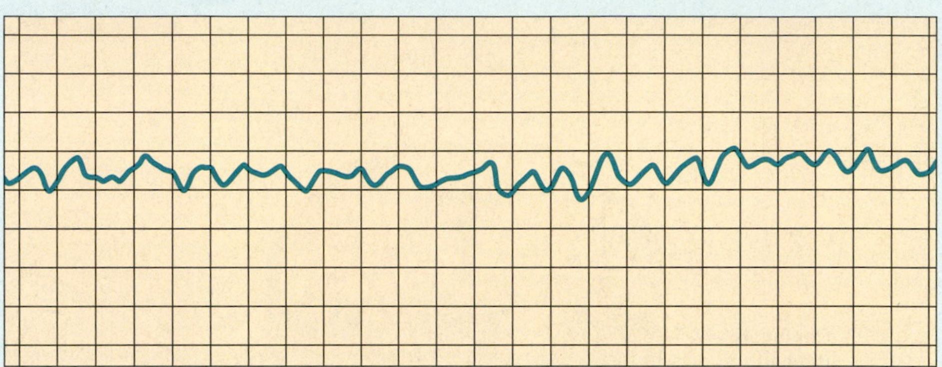

FIGURE 15B Ventricular fibrillation is rapid, uncoordinated depolarization of the ventricles.

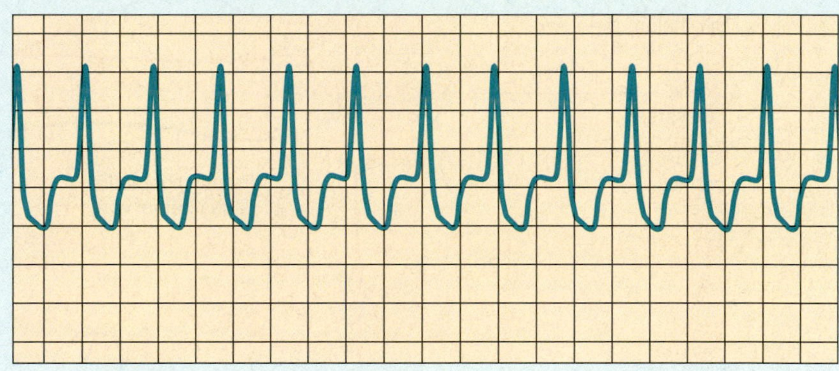

FIGURE 15C Tachycardia is a rapid heartbeat.

receptors signal the cardioaccelerator center, and sympathetic impulses reach the heart. As a result, heart rate and force of contraction increase, and the venous pressure is reduced.

Impulses from the cerebrum or hypothalamus also influence the cardiac control center. Such impulses may decrease heart rate, as occurs when a person faints following an emotional upset, or they may increase heart rate during a period of anxiety.

Two other factors that influence heart rate are temperature change and certain ions. Rising body temperature increases heart action, which is why heart rate usually increases during fever. On the other hand, abnormally low body temperature decreases heart action.

The most important ions that influence heart action are potassium (K^+) and calcium (Ca^{+2}). Potassium affects the electrical potential of the cell membrane, altering its ability to reach the threshold for conducting an impulse (see chapter 10, p. 368). The sarcoplasmic reticula of cardiac muscle fibers have less calcium than do the sarcoplasmic reticula of skeletal muscle fibers. Therefore, cardiac muscle depends more on extracellular (blood-borne) calcium. Although homeostatic mechanisms normally maintain the concentrations of these ions within narrow ranges, these mechanisms sometimes fail, and the consequences can be serious or even fatal. Clinical Application 15.1 examines abnormal heart rhythms.

A heart chamber *flutters* when it contracts regularly, but very rapidly, such as 250–350 times per minute. Although normal hearts may flutter occasionally, this condition is more likely to be due to damage to the myocardium (fig. 15E).

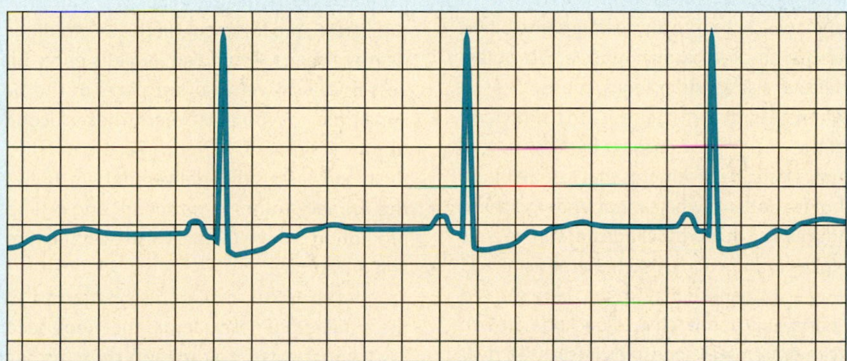

FIGURE 15D Bradycardia is a slow heartbeat.

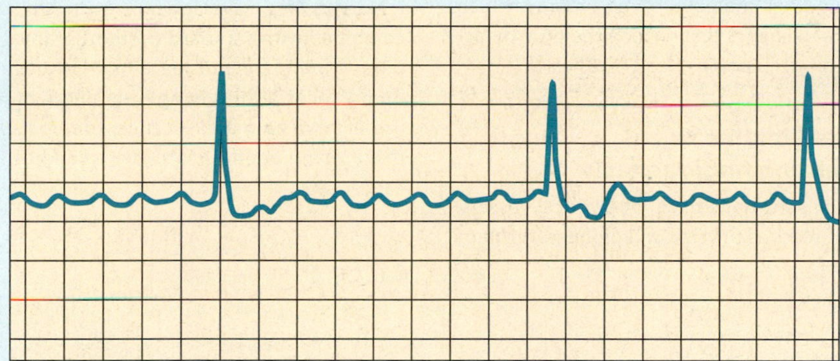

FIGURE 15E Atrial flutter is an abnormally rapid rate of atrial depolarization.

Any interference or block in cardiac impulse conduction may cause arrhythmia, the type varying with the location and extent of the block. Such arrhythmias arise because certain cardiac tissues other than the SA node can function as pacemakers.

The SA node usually initiates seventy to eighty heartbeats per minute, called a sinus rhythm. If the SA node is damaged, impulses originating in the AV node may travel upward into the atrial myocardium and downward into the ventricular walls, stimulating them to contract. Under the influence of the AV node acting as a *secondary pacemaker,* the heart may continue to pump blood, but at a rate of forty to sixty beats per minute, called a nodal rhythm. Similarly, the Purkinje fibers can initiate cardiac impulses, contracting the heart fifteen to forty times per minute.

An *artificial pacemaker* can treat a disorder of the cardiac conduction system. This device includes an electrical pulse generator and a lead wire that communicates with a portion of the myocardium. The pulse generator contains a permanent battery that provides energy and a microprocessor that can sense the cardiac rhythm and signal the heart to alter its contraction rate.

An artificial pacemaker is surgically implanted beneath the patient's skin in the shoulder. An external programmer adjusts its functions from the outside. The first pacemakers, made in 1958, were crude. Today, thanks to telecommunications advances, a physician can check a patient's pacemaker over the phone. A device called a pacemaker-cardioverter-defibrillator can correct both abnormal heart rhythm and cardiac arrest. ■

Excess potassium ions (hyperkalemia) alter the usual polarized state of the cardiac muscle fibers, decreasing the rate and force of contractions. High potassium ion concentration may block conduction of cardiac impulses, and heart action may suddenly stop (cardiac arrest). Conversely, if the potassium concentration drops below normal (hypokalemia), the heart may develop a potentially life-threatening abnormal rhythm (arrhythmia).

Excess calcium ions (hypercalcemia) increases heart action, risking that the heart will undergo a prolonged contraction. Conversely, low calcium ion concentration (hypocalcemia) depresses heart action because these ions help initiate muscle contraction.

PRACTICE

27 Which nerves supply parasympathetic fibers to the heart? Which nerves supply sympathetic fibers?

28 How do parasympathetic and sympathetic impulses help control heart rate?

29 How do changes in body temperature affect heart rate?

15.4 BLOOD VESSELS

The blood vessels are organs of the cardiovascular system. They form a closed circuit of tubes that carries blood from the heart to the body cells and back again. These vessels

15.2 FROM SCIENCE TO TECHNOLOGY

Altering Angiogenesis

Angiogenesis is the formation of new blood vessels. Under the influence of specific growth factors, endothelial cells divide and assemble into the tubules that form capillaries as well as the innermost linings of larger blood vessels. In normal development, angiogenesis is crucial to build a blood supply to serve a growing body. New blood vessels deliver nutrients, hormones, and growth factors to tissues and remove wastes. Angiogenesis is also essential for healing. After a heart attack, for example, new vessels form in the remaining healthy cardiac muscle to supply blood.

As with most biological processes, angiogenesis must be highly controlled. Excess, deficient, or inappropriate angiogenesis can cause, or worsen, a variety of illnesses. By understanding how angiogenesis proceeds, researchers are developing ways to direct new blood vessel formation, with two specific applications—healing hearts and starving cancerous tumors.

Heart Attacks: Promoting Angiogenesis

An errant clot blocks a coronary artery. Within seconds, the localized lack of oxygen stimulates muscle cells to release hypoxia-inducible factor (HIF-1). This protein activates several genes whose products restore homeostasis by stimulating glycolysis (anaerobic respiration); signalling the kidneys to produce erythropoietin, which boosts the red blood cell supply; and triggering angiogenesis by turning on production of vascular endothelial growth factor (VEGF). The growth factor stimulates certain cells to proliferate and aggregate to form capillaries, which, eventually, restore some blood flow to the blocked cardiac muscle. Fibroblast growth factor also assists in angiogenesis.

When natural angiogenesis isn't sufficient, part of the heart dies. Coronary bypass surgery and angioplasty are treatments that restore blood flow, but for patients who cannot undergo these procedures or whose blockages are in vessels too narrow or difficult to reach, harnessing and targeting angiogenesis may help to save starved heart parts. One approach is to package growth factors in time-release capsules implanted near small vessels while large ones are being surgically bypassed. In one clinical trial, this technique increased blood flow to the area and halted chest pain. Another strategy is gene therapy, which delivers the genes that encode the growth factors to oxygen-starved areas of the heart.

Cancer Treatment: Preventing Angiogenesis

A tumor surrounds itself with blood vessels. Once it reaches the size of a pinhead, a tumor secretes growth factors that stimulate nearby capillaries to sprout new branches that extend toward it. Endothelial cells that are part of the tumor assemble into sheets, roll into tubules, and, eventually, snake out of the tumor as new capillaries. Other cancer cells wrap around the capillaries, spreading out on this scaffolding into nearby tissues. Some cancer cells enter blood vessels and travel to other parts of the body. For a time, maybe even years, these secondary tumors stay small, adhering to the outsides of the blood vessels that delivered them. But when the primary tumor is removed, angiogenesis-promoting growth factors wash over the tumors, and they grow.

In the 1970s, researchers began to study the antiangiogenesis factors that keep secondary tumors small, to develop them as cancer treatments. The first antiangiogenesis drug to treat cancer became available in 2004, for colorectal cancer that has spread to other organs. It extends life an average of five months, when combined with standard chemotherapy, compared to chemotherapy alone. The drug, a monoclonal antibody against VEGF, may be useful in treating breast cancer too. Today several antiangiogenesis drugs are used to treat cancer. ■

include arteries, arterioles, capillaries, venules, and veins. The arteries and arterioles conduct blood away from the ventricles of the heart and lead to the capillaries, where substances are exchanged between blood and the body cells. Venules and veins return blood from the capillaries to the atria. From Science to Technology 15.2 describes angiogenesis, the formation of new blood vessels in the body.

An intriguing new way to study angiogenesis is to grow human capillaries in mice. Human endothelial cells cultured in a laboratory dish aggregate into tiny spheres. Researchers embedded the spheres in a gel and subcutaneously injected the material into mice that lack immune systems, along with growth factors. Human capillaries formed under the mouse skin and joined the blood vessels of the mice. The method is being used to study angiogenesis as well as engineering replacement vascular tissue.

Arteries and Arterioles

Arteries (ar′-terēz) are strong, elastic vessels adapted for carrying the blood away from the heart under high pressure. These vessels subdivide into progressively thinner tubes and eventually give rise to the finer branched **arterioles** (ar-te′re-olz).

The wall of an artery consists of three distinct layers, or *tunics*, shown in figure 15.25*a*. The innermost tunic, tunica interna (intima), is composed of a layer of simple squamous epithelium, called *endothelium*, that rests on a connective tissue membrane rich in elastic and collagenous fibers.

The endothelial lining of an artery provides a smooth surface that allows blood cells and platelets to flow through without being damaged. Additionally, endothelium helps prevent blood clotting by secreting biochemicals that inhibit platelet aggregation (see chapter 14, p. 542). Endothelium also may help regulate local blood flow by secreting substances that either dilate or constrict blood vessels. For

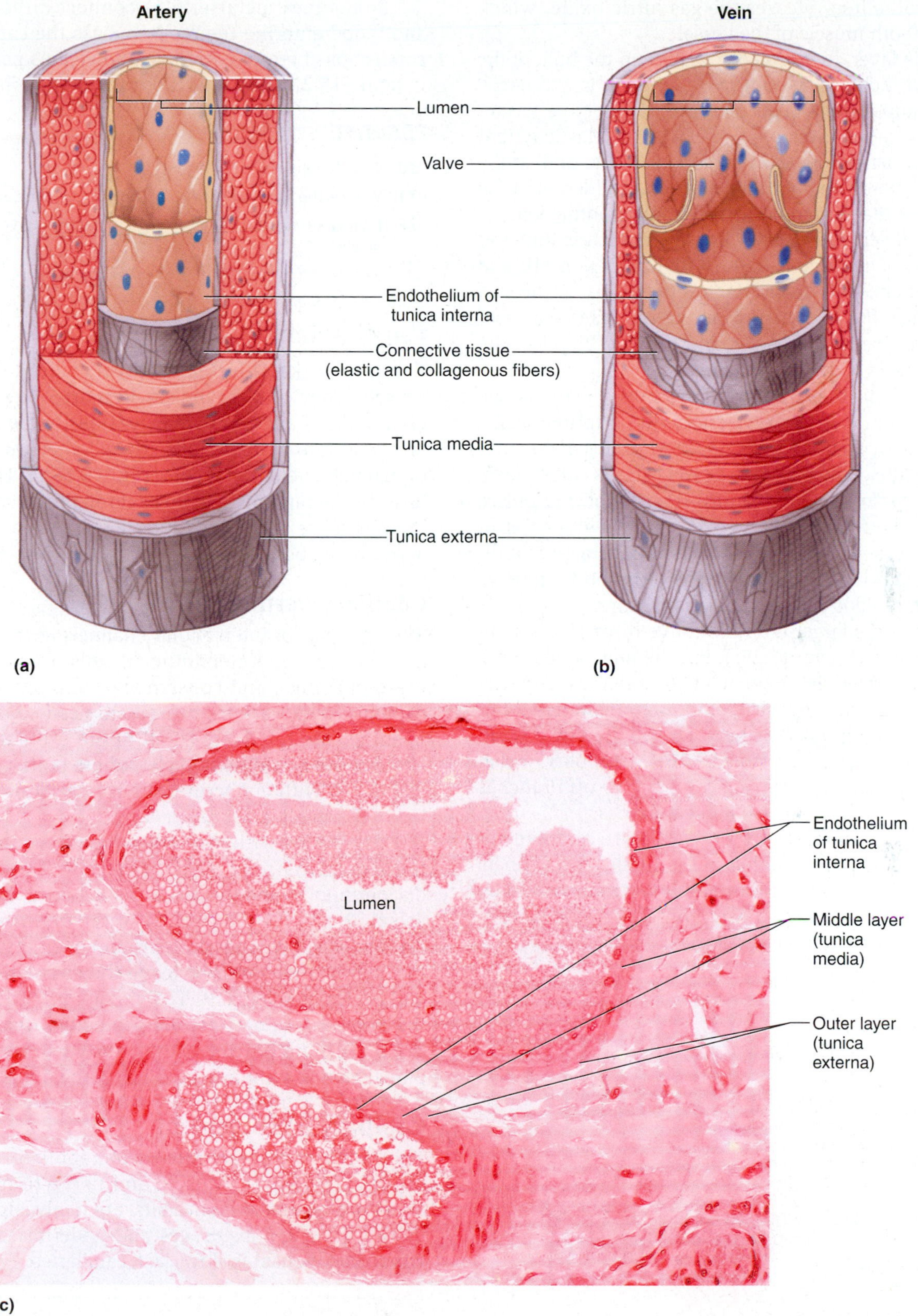

FIGURE 15.25 Blood vessels. (*a*) The wall of an artery. (*b*) The wall of a vein. (*c*) Note the structural differences in these cross sections of an arteriole (bottom) and a venule (top) (200×).

example, endothelium releases the gas nitric oxide, which relaxes the smooth muscle of the vessel.

The middle layer, tunica media, makes up the bulk of the arterial wall. It includes smooth muscle fibers, which encircle the tube, and a thick layer of elastic connective tissue. The connective tissue gives the vessel a tough elasticity that enables it to withstand the force of blood pressure and, at the same time, to stretch and accommodate the sudden increase in blood volume that accompanies ventricular contraction.

The outer layer, tunica externa (adventitia), is thin and chiefly consists of connective tissue with irregular elastic and collagenous fibers. This layer attaches the artery to the surrounding tissues. It also contains minute vessels (vasa vasorum) that give rise to capillaries and provide blood to the more external cells of the artery wall.

The sympathetic branches of the autonomic nervous system innervate smooth muscle in artery and arteriole walls. *Vasomotor fibers* stimulate the smooth muscle cells to contract, reducing the diameter of the vessel. This is called **vasoconstriction** (vas″o-kon-strik′-shun). If vasomotor impulses are inhibited, the muscle fibers relax, and the diameter of the vessel increases. This is called **vasodilation** (vas″o-di-la′shun). Changes in the diameters of arteries and arterioles greatly influence blood flow and blood pressure.

The walls of the larger arterioles have three layers similar to those of arteries (fig. 15.25*c*), but the middle and outer layers thin as the arterioles approach the capillaries. The wall of a very small arteriole consists only of an endothelial lining and some smooth muscle fibers, surrounded by a small amount of connective tissue (fig. 15.26). Arterioles, which are microscopic continuations of arteries, give off branches called *metarterioles* that, in turn, join capillaries.

The arteriole and metarteriole walls are adapted for vasoconstriction and vasodilation in that their muscle fibers respond to impulses from the autonomic nervous system by contracting or relaxing. In this way, these vessels help control the flow of blood into the capillaries.

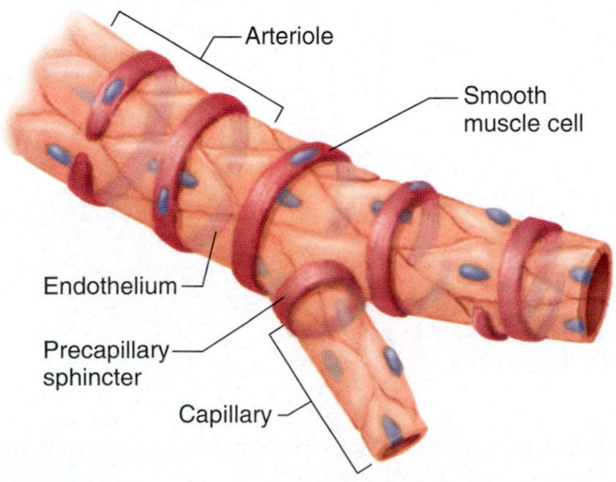

FIGURE 15.26 The smallest arterioles have only a few smooth muscle fibers in their walls. Capillaries lack these fibers.

Sometimes metarterioles connect directly to venules, and blood entering them can bypass the capillaries. These connections between arteriole and venous pathways, shown in figure 15.27, are called *arteriovenous shunts*.

PRACTICE

30 Describe the wall of an artery.

31 What is the function of the smooth muscle in the arterial wall?

32 How is the structure of an arteriole different from that of an artery?

Capillaries

Capillaries (kap′ĭ-ler″ēz) are the smallest diameter blood vessels. They connect the smallest arterioles and the smallest venules. Capillaries are extensions of the inner linings of arterioles in that their walls are endothelium—a single layer of squamous epithelial cells (fig. 15.28*a*). These thin walls form the semipermeable layer through which substances in the blood are exchanged for substances in the tissue fluid surrounding body cells.

Capillary Permeability

The openings or intercellular channels in the capillary walls are thin slits where endothelial cells overlap. The sizes of these openings, and consequently the permeability of the capillary wall, vary from tissue to tissue. For example, the openings are relatively small in the capillaries of smooth, skeletal, and cardiac muscle, whereas those in capillaries associated with endocrine glands, the kidneys, and the lining of the small intestine are larger.

Capillaries with the largest openings include those of the liver, spleen, and red bone marrow. These capillaries are discontinuous, and the distances between their cells appear as small cavities (sinusoids) in the organ. Discontinuous capillaries allow large proteins and even intact cells to pass through as they enter or leave the circulation (fig. 15.28*b* and *c*). From Science to Technology 5.1 (p. 145) discusses the blood-brain barrier, the protective tight capillaries in the brain. The barrier is not present in the pituitary and pineal glands and parts of the hypothalamus.

Capillary Arrangement

The higher a tissue's rate of metabolism, the denser its capillary networks. Muscle and nerve tissues, which use abundant oxygen and nutrients, are richly supplied with capillaries; cartilaginous tissues, the epidermis, and the cornea, where metabolism is slow, lack capillaries.

If the capillaries of an adult were unwound and spread end to end, they would cover from 25,000 to 60,000 miles.

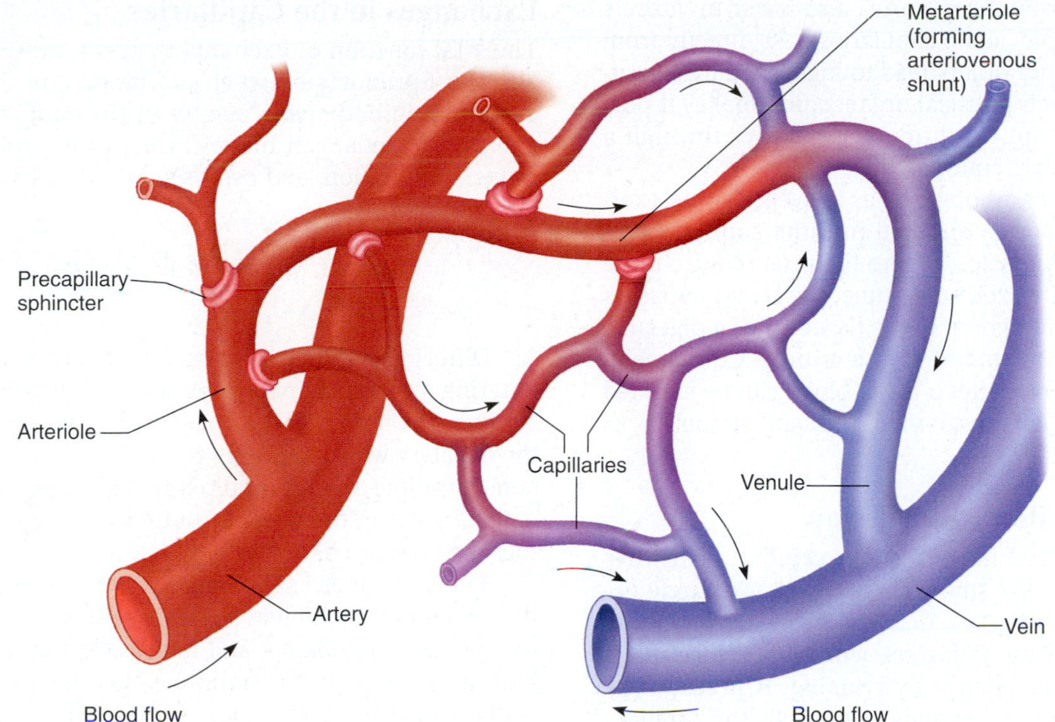

FIGURE 15.27 Some metarterioles provide arteriovenous shunts by connecting arterioles directly to venules.

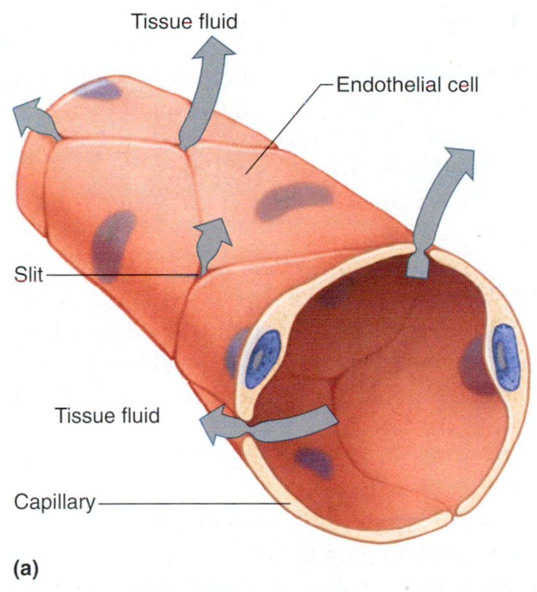

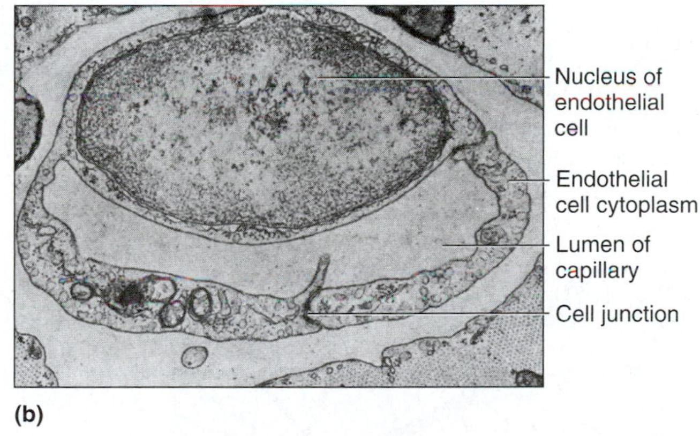

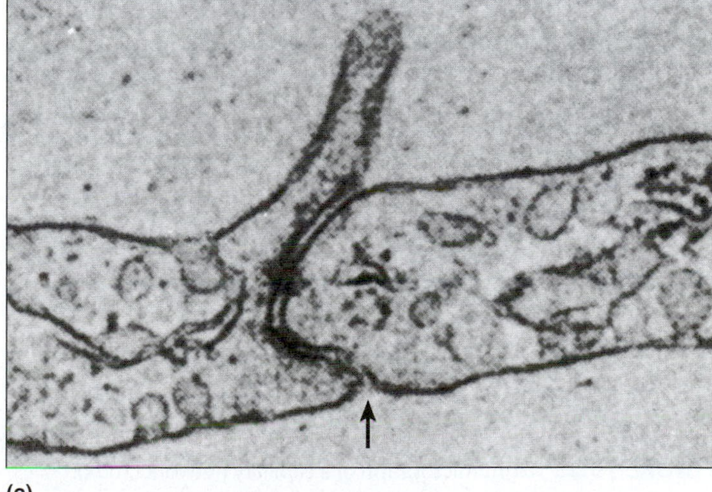

FIGURE 15.28 Capillary structure. (*a*) Substances are exchanged between the blood and tissue fluid through openings (slits) separating endothelial cells. (*b*) Transmission electron micrograph of a capillary cross section (11,500×). (*c*) Note the narrow slitlike openings at the cell junctions (arrow) (micrograph *b* enlarged to 62,500×).

The spatial patterns of capillaries also differ in various body parts. For example, some capillaries pass directly from arterioles to venules, but others lead to highly branched networks (fig. 15.29). Such physical organization makes it possible for the blood to follow different pathways through a tissue attuned to cellular requirements.

Blood flow can vary among tissues as well. During exercise, for example, blood is directed into the capillary networks of the skeletal muscles, where the cells require more oxygen and nutrients. At the same time, the blood bypasses some of the capillary networks in the tissues of the digestive tract, where demand for blood is less critical. Conversely, when a person is relaxing after a meal, blood can be shunted from the inactive skeletal muscles into the capillary networks of the digestive organs.

Regulation of Capillary Blood Flow

The distribution of blood in the various capillary pathways is mainly regulated by the smooth muscles that encircle the capillary entrances. As figures 15.26 and 15.27 show, these muscles form *precapillary sphincters*, which may close a capillary by contracting or open it by relaxing. A precapillary sphincter responds to the demands of the cells the capillary supplies. When these cells have low concentrations of oxygen and nutrients, the precapillary sphincter relaxes, and blood flow increases; when cellular requirements have been met, the precapillary sphincter may contract again.

PRACTICE 

33 Describe a capillary wall.

34 What is the function of a capillary?

35 What controls blood flow into capillaries?

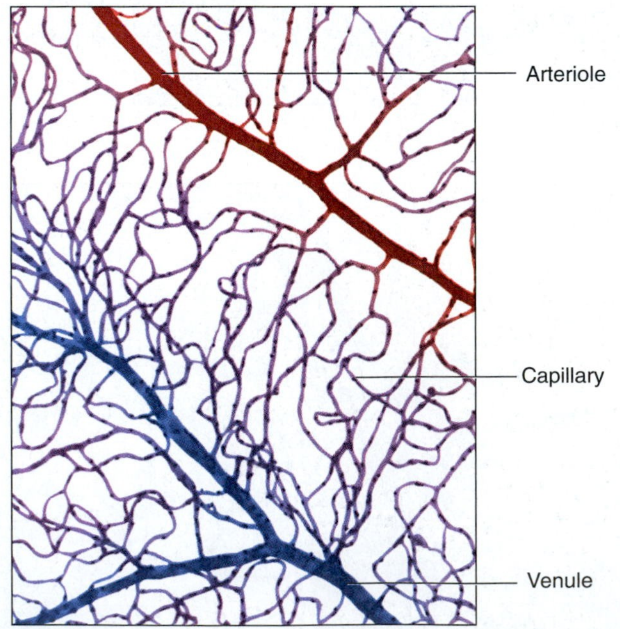

FIGURE 15.29 Light micrograph of a capillary network (100×).

Exchanges in the Capillaries

The vital function of exchanging gases, nutrients, and metabolic by-products between the blood and the tissue fluid surrounding body cells occurs in the capillaries. The biochemicals exchanged move through the capillary walls by diffusion, filtration, and osmosis.

RECONNECT

To Chapter 3, Movements Into and Out of the Cell, pages 90–94.

Diffusion is the most important means of transfer. Blood entering systemic capillaries carries high concentrations of oxygen and nutrients, so these substances diffuse through the capillary walls and enter the tissue fluid. Conversely, the concentrations of carbon dioxide and other wastes are generally greater in the tissues, and such wastes tend to diffuse into the capillary blood.

The paths these substances follow depend primarily on their solubilities in lipids. Substances soluble in lipid, such as oxygen, carbon dioxide, and fatty acids, can diffuse through most areas of the cell membranes that make up the capillary wall because the membranes are largely lipid. Lipid-insoluble substances, such as water, sodium ions, and chloride ions, diffuse through pores in the cell membranes and through the slitlike openings between the endothelial cells that form the capillary wall (see fig. 15.28). Plasma proteins generally remain in the blood because they are not soluble in the lipid of the endothelial cell membranes, and they are too large to diffuse through the membrane pores or slitlike openings between the endothelial cells of most capillaries.

In *filtration, hydrostatic pressure* forces molecules through a membrane. In the capillaries, the blood pressure generated when ventricle walls contract provides the force for filtration.

Blood pressure also moves blood through the arteries and arterioles. This pressure decreases as the distance from the heart increases because of friction (peripheral resistance) between the blood and the vessel walls. For this reason, blood pressure is greater in the arteries than in the arterioles and greater in the arterioles than in the capillaries. It is similarly greater at the arteriolar end of a capillary than at the venular end.

The walls of arteries and arterioles are too thick to allow blood components to pass through. However, the hydrostatic pressure of the blood pushes small molecules through capillary walls by filtration primarily at the arteriolar ends of capillaries, whereas diffusion takes place along their entire lengths.

The presence of an impermeant solute on one side of a cell membrane creates an osmotic pressure. Plasma proteins trapped in the capillaries create an osmotic pressure that draws water into the capillaries. The term *colloid osmotic pressure* describes this osmotic effect due solely to the plasma proteins.

The effect of capillary blood pressure, which favors filtration, opposes the actions of the plasma colloid osmotic pressure, which favors reabsorption. At the arteriolar end of capillaries, the blood pressure is higher (35 mm Hg outward) than the colloid osmotic pressure (24 mm Hg inward), so at the arteriolar end of the capillary, filtration predominates. At the venular end, the colloid osmotic pressure is essentially unchanged (24 mm Hg inward), but the blood pressure has decreased due to resistance through the capillary (16 mm Hg outward). Thus, at the venular end, reabsorption predominates (fig. 15.30). (The interstitial fluid also has hydrostatic pressure and osmotic pressure, but the values are low and tend to cancel each other; as such, they are omitted from this discussion.)

Normally, more fluid leaves the capillaries than returns to them because the net inward pressure at the venular ends of the capillaries is less than the net outward pressure at the arteriolar ends of the capillaries. Closed-ended vessels called lymphatic capillaries collect the excess fluid and return it through lymphatic vessels to the venous circulation. Chapter 16 (pp. 617–618) discusses this mechanism.

Sometimes unusual events increase blood flow to capillaries, and excess fluid enters spaces between tissue cells (interstitial spaces). This may occur, for instance, in response to certain chemicals such as *histamine* that vasodilate the metarterioles and increase capillary permeability. Enough fluid may leak out of the capillaries to overwhelm lymphatic drainage, and affected tissues become swollen (edematous) and painful.

PRACTICE

36 What forces affect the exchange of substances between blood and the tissue fluid?

37 Why is the fluid movement out of a capillary greater at its arteriolar end than at its venular end?

38 More fluid leaves the capillary than returns to it, so how is the remainder returned to the vascular system?

If the right ventricle of the heart is unable to pump blood out as rapidly as it enters, other parts of the body may develop edema because the blood backs up into the veins, venules, and capillaries, increasing blood pressure in these vessels. As a result of this increased *back pressure,* osmotic pressure of the blood in the venular ends of the capillaries is less effective in attracting water from tissue fluid, and the tissues swell. This is true particularly in the lower extremities if the person is upright, or in the back if the person is supine. In the terminal stages of heart failure, edema is widespread, and fluid accumulates in the peritoneal cavity of the abdomen. This condition is called *ascites.*

Venules and Veins

Venules (ven′ūlz) are the microscopic vessels that continue from the capillaries and merge to form **veins** (vānz). The veins, which carry blood back to the atria, follow pathways that roughly parallel those of the arteries.

The walls of veins are similar to those of arteries in that they are composed of three distinct layers. However, the middle layer of the venous wall is poorly developed. Consequently, veins have thinner walls that contain less smooth muscle and less elastic tissue than those of comparable arteries, but their lumens have a greater diameter (see fig. 15.25*b*).

Many veins, particularly those in the upper and lower limbs, contain flaplike *valves* (called semilunar valves), which project inward from their linings. Valves, shown in figure 15.31, are usually composed of two leaflets pushed closed if the blood begins to back up in a vein. These valves aid in returning blood to the heart because they are open as long as the flow is toward the heart but close if it is in the opposite direction.

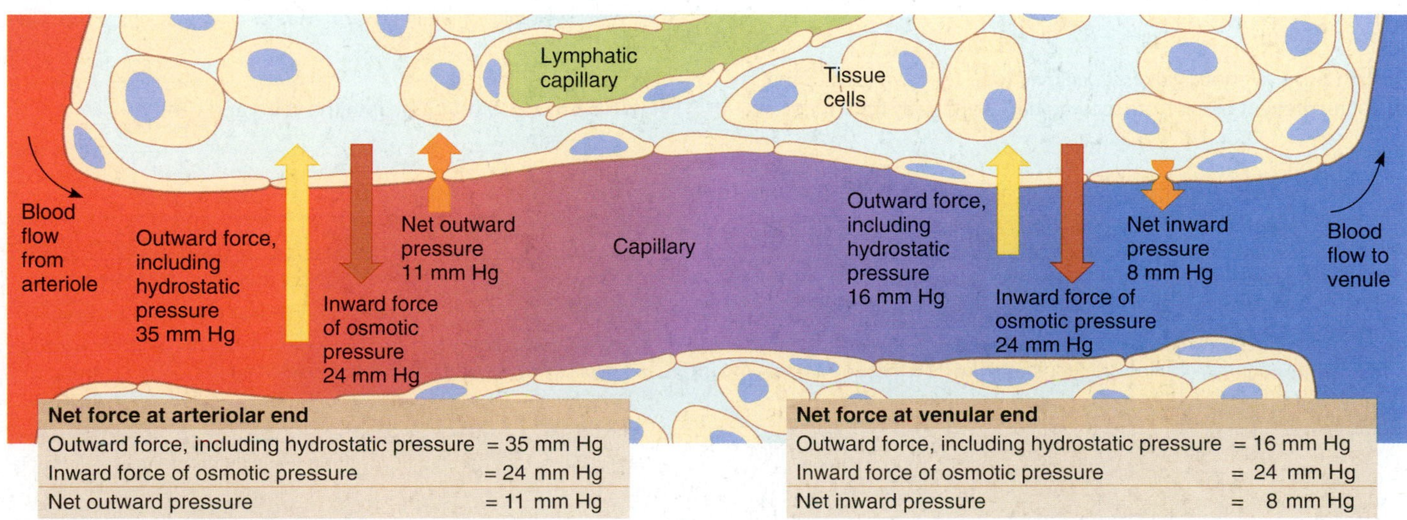

Net force at arteriolar end	
Outward force, including hydrostatic pressure	= 35 mm Hg
Inward force of osmotic pressure	= 24 mm Hg
Net outward pressure	= 11 mm Hg

Net force at venular end	
Outward force, including hydrostatic pressure	= 16 mm Hg
Inward force of osmotic pressure	= 24 mm Hg
Net inward pressure	= 8 mm Hg

FIGURE 15.30 Water and other substances leave capillaries because of a net outward pressure at the capillaries' arteriolar ends. Water enters at the capillaries' venular ends because of a net inward pressure. Substances move in and out along the length of the capillaries according to their respective concentration gradients.

Veins also function as *blood reservoirs*, useful in times of blood loss. For example, in hemorrhage accompanied by a drop in arterial blood pressure, sympathetic nerve impulses reflexly stimulate the muscular walls of the veins. The resulting venous constrictions help maintain blood pressure by returning more blood to the heart. This mechanism ensures a nearly normal blood flow even when as much as 25% of the blood volume is lost. Figure 15.32 illustrates the relative volumes of blood in the veins and other blood vessels.

Table 15.3 summarizes the characteristics of blood vessels. Clinical Application 15.2 examines disorders of blood vessels.

PRACTICE

39 How does the structure of a vein differ from that of an artery?

40 What are the functions of veins and venules?

41 How does venous circulation help to maintain blood pressure when hemorrhaging causes blood loss?

15.5 BLOOD PRESSURE

Blood pressure is the force the blood exerts against the inner walls of the blood vessels. Although this force is present throughout the vascular system, the term *blood pressure* most commonly refers to pressure in arteries supplied by branches of the aorta (systemic arteries).

Arterial Blood Pressure

The arterial blood pressure rises and falls in a pattern corresponding to the phases of the cardiac cycle. That is, when the ventricles contract (ventricular systole), their walls squeeze the blood inside their chambers and force it into the pulmonary trunk and aorta. As a result, the pressures in these arteries sharply increase. The maximum pressure achieved during ventricular contraction is called the **systolic pressure.** When the ventricles relax (ventricular diastole), the arterial pressure drops, and the lowest pressure that remains in the arteries before the next ventricular contraction is termed the **diastolic pressure.**

The surge of blood entering the arterial system during ventricular systole distends the elastic walls of the arteries,

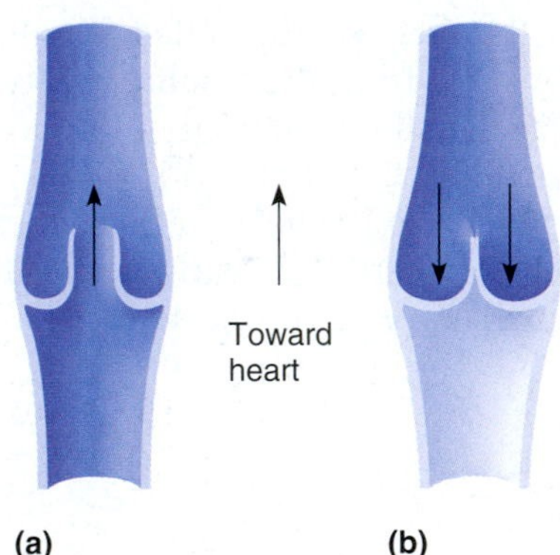

FIGURE 15.31 Venous valves. (*a*) allow blood to move toward the heart, but (*b*) prevent blood from moving backward away from the heart.

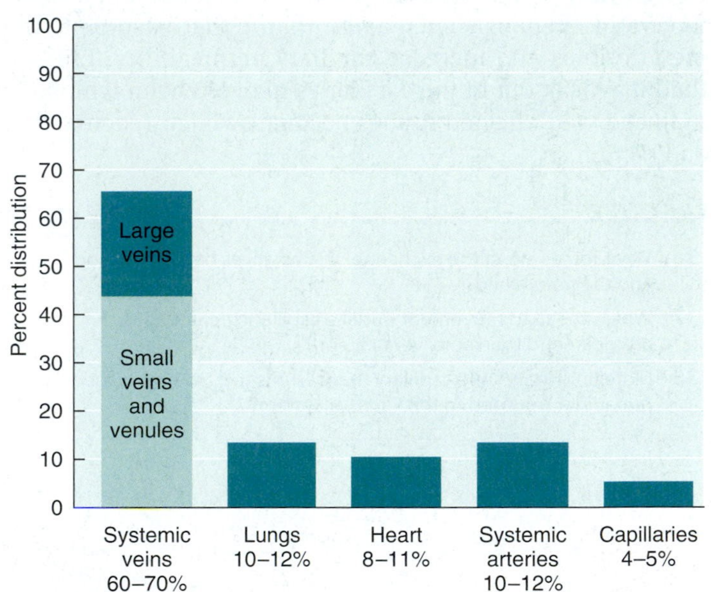

FIGURE 15.32 Most blood is in the veins and venules.

TABLE 15.3 | Characteristics of Blood Vessels

Vessel	Type of Wall	Function
Artery	Thick, strong wall with three layers—an endothelial lining, a middle layer of smooth muscle and elastic tissue, and an outer layer of connective tissue	Carries blood under relatively high pressure from the heart to arterioles
Arteriole	Thinner wall than an artery but with three layers; smaller arterioles have an endothelial lining, some smooth muscle tissue, and a small amount of connective tissue	Connects an artery to a capillary, helps control the blood flow into a capillary by vasoconstricting or vasodilating
Capillary	Single layer of squamous epithelium	Provides a membrane through which nutrients, gases, and wastes are exchanged between the blood and tissue fluid; connects an arteriole to a venule
Venule	Thinner wall than an arteriole, less smooth muscle and elastic tissue	Connects a capillary to a vein
Vein	Thinner wall than an artery but with similar layers; the middle layer is more poorly developed; some have flaplike valves	Carries blood under relatively low pressure from a venule to the heart; valves prevent a backflow of blood; serves as blood reservoir

15.2 CLINICAL APPLICATION

Blood Vessel Disorders

In the arterial disease *atherosclerosis* (ath″er-o-sklĕ-ro′sis), deposits of fatty materials, particularly cholesterol, form within the intima and inner lining of the arterial walls. Such deposits, called *plaque,* protrude into the lumens of the vessels and interfere with blood flow (fig. 15F). Furthermore, plaque often forms a surface texture that can initiate formation of a blood clot, increasing the risk of developing thrombi or emboli that cause blood deficiency (*ischemia*) or tissue death (*necrosis*) downstream from the obstruction.

The walls of affected arteries may degenerate, losing their elasticity and becoming hardened or *sclerotic.* In this stage of the disease, called *arteriosclerosis,* a sclerotic vessel may rupture under the force of blood pressure.

Risk factors for developing atherosclerosis include a fatty diet, elevated blood pressure, tobacco smoking, obesity, and lack of physical exercise (see chapter 18, pp. 703–704). Genetic factors may also increase the risk of developing atherosclerosis.

If atherosclerosis so weakens the wall of an artery that blood pressure dilates a region of it, a pulsating sac called an *aneurysm* may form. Aneurysms tend to grow. If the resulting sac develops by a longitudinal splitting of the middle layer of the arterial wall, it is called a *dissecting aneurysm.* An aneurysm may cause symptoms by pressing on nearby organs, or it may rupture, producing great blood loss.

Aneurysms may also result from trauma, high blood pressure, infections, inherited disorders such as Marfan syndrome, or congenital defects in blood vessels. Common sites of aneurysms include the thoracic and abdominal aorta and an arterial circle at the base of the brain (circle of Willis).

Phlebitis, or inflammation of a vein, is relatively common. It may occur in association with an injury or infection or after surgery, or it may develop for no apparent reason.

If inflammation is restricted to a superficial vein, such as the greater or lesser saphenous veins, blood flow may be rechanneled through other vessels. But if it occurs in a deep vein, such as the tibial, peroneal, popliteal, or femoral veins, the consequences can be serious, particularly if the blood in the affected vessel clots and blocks normal circulation. This condition, called *thrombophlebitis,* introduces a risk that a blood clot in a vein will detach, move with the venous blood, pass through the heart, and lodge in the pulmonary arterial system in a lung. Such an obstruction is called a *pulmonary embolism.*

Varicose veins are abnormal and irregular dilations in superficial veins, particularly in the legs. This condition is usually associated with prolonged, increased back pressure in the affected vessels due to gravity, such as when a person stands. Crossing the legs or sitting in a chair so that its edge presses against the area behind the knee can obstruct venous blood flow and aggravate varicose veins.

Increased venous back pressure stretches and widens the veins. The valves in these vessels do not change size, so they soon lose their abilities to block the backward flow of blood, and blood accumulates in the enlarged regions.

Increased venous pressure is also accompanied by rising pressure in the venules and capillaries that supply the veins. Consequently, tissues in affected regions typically become edematous and painful.

Genetics, pregnancy, obesity, and standing for long periods raise the risk of developing varicose veins. Elevating the legs above the level of the heart or putting on support hosiery before arising in the morning can relieve discomfort. Intravenous injection of a substance that destroys veins (a sclerosing agent) or surgical removal of the affected veins may be necessary. ■

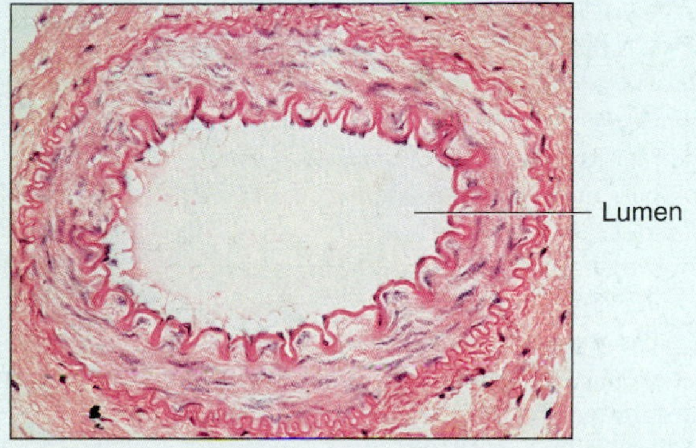

(a)

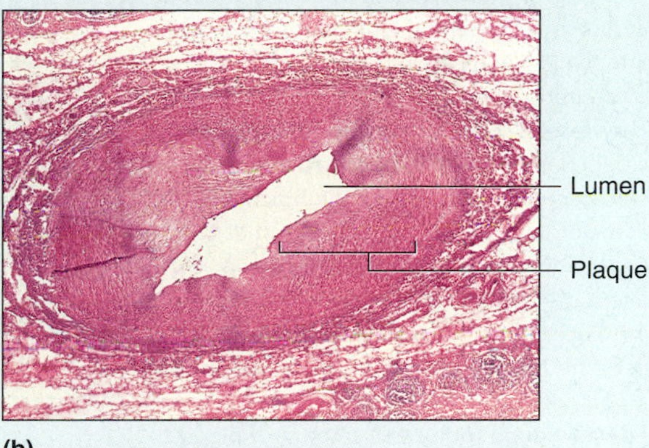

(b)

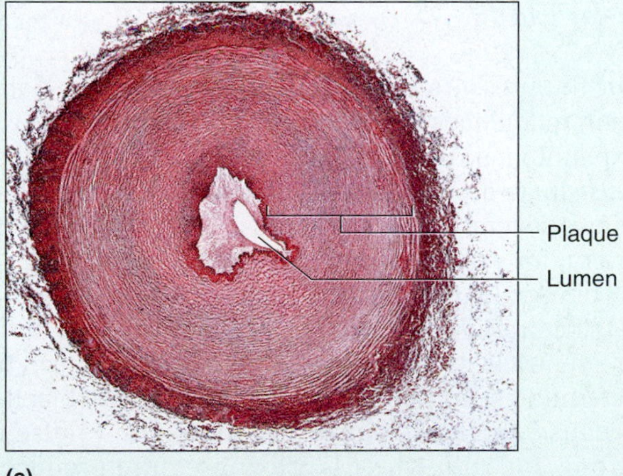

(c)

FIGURE 15F Development of atherosclerosis. (*a*) Normal arteriole (100×). (*b, c*) Accumulation of plaque on the inner wall of the arteriole (*b* and *c* 100×).

15.3 CLINICAL APPLICATION

Measurement of Arterial Blood Pressure

Systemic arterial blood pressure usually is measured using an instrument called a sphygmomanometer (sfig″mo-mah-nom′ĕ-ter) (fig. 15G). This device consists of an inflatable cuff connected by tubing to a compressible bulb and a pressure gauge. The bulb is used to pump air into the cuff, and a rise in pressure is indicated on the pressure gauge. The pressure in the cuff is expressed in millimeters of mercury (mm Hg) based on older equipment that used a glass tube containing a column of mercury in place of a pressure gauge. The older devices have been discontinued because of the danger of mercury.

To measure arterial blood pressure, the cuff of the sphygmomanometer is usually wrapped around the arm so that it surrounds the brachial artery. Air is pumped into the cuff until the cuff pressure exceeds the pressure in that artery. The vessel is squeezed closed and its blood flow stopped. At this moment, if the diaphragm of a stethoscope is placed over the brachial artery at the distal border of the cuff, no sounds can be heard from the vessel because the blood flow is interrupted. As air is slowly released from the cuff, the air pressure inside it decreases. When the cuff pressure is approximately equal to the systolic blood pressure in the brachial artery, the artery opens enough for a small amount of blood to spurt through. This movement produces a sharp

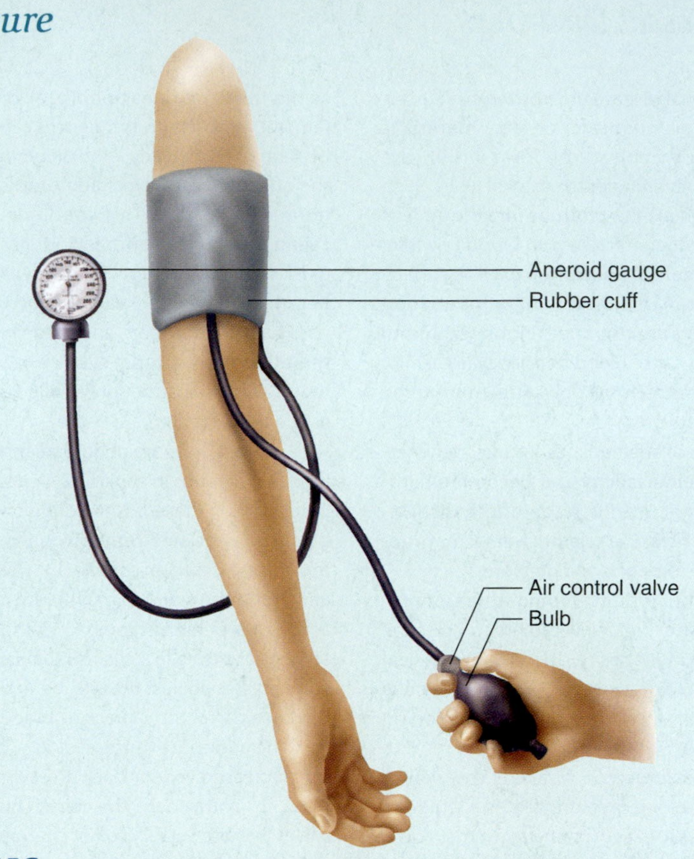

FIGURE 15G A sphygmomanometer is used to measure arterial blood pressure. The use of the column of mercury is the most accurate measurement, but due to environmental concerns, it has been replaced by alternative gauges and digital readouts.

but the pressure begins to drop almost immediately as the contraction ends, and the arterial walls recoil. This alternate expanding and recoiling of the arterial wall can be felt as a *pulse* in an artery that runs close to the surface. Figure 15.33 shows several sites where a pulse can be detected. The radial artery, for example, courses near the surface at the wrist and is commonly used to sense a person's radial pulse.

The radial pulse rate is equal to the rate at which the left ventricle contracts, and for this reason, it can be used to determine heart rate. A pulse can also reflect blood pressure, because an elevated pressure produces a pulse that feels strong and full, whereas a low pressure produces a pulse that is weak and easily compressed. Clinical Application 15.3 describes how to measure arterial blood pressure.

PRACTICE 

42 Distinguish between systolic and diastolic blood pressure.

43 Which cardiac event causes systolic pressure? Diastolic pressure?

44 What causes a pulse in an artery?

Factors That Influence Arterial Blood Pressure

Arterial pressure depends on a variety of factors. These include heart action which includes stroke volume and heart rate; blood volume; resistance to flow; and blood viscosity (fig. 15.34).

Heart Action

Each ventricular contraction determines the volume of blood that enters the arterial system, called the **stroke volume.** It equals about 70 milliliters in an average-weight male at rest. The volume discharged from the ventricle per minute is called the **cardiac output.** It is calculated by multiplying the stroke volume by the heart rate in beats per minute. (Cardiac output = stroke volume × heart rate.) For example, if the stroke volume is 70 milliliters and the heart rate is 72 beats per minute, the cardiac output is 5,040 milliliters per minute.

sound (Korotkoff's sound) heard through the stethoscope. The sound results from turbulence in the narrowed artery. The pressure indicated on the pressure gauge when this first tapping sound is heard represents the *arterial systolic pressure* (SP).

As the cuff pressure continues to drop, increasingly louder sounds are heard. Then, when the cuff pressure is approximately equal to that within the fully opened artery, the sounds become abruptly muffled and disappear. The pressure indicated on the pressure gauge when this happens represents the *arterial diastolic pressure* (DP).

The results of a blood pressure measurement are reported as a fraction, such as 120/80. The upper number indicates the systolic pressure in mm Hg (SP), and the lower number indicates the diastolic pressure in mm Hg (DP). Figure 15H shows how these pressures decrease as distance from the left ventricle increases. The difference between the systolic and diastolic pressures (SP-DP), called the *pulse pressure* (PP), is about 40 mm Hg.

The average pressure in the arterial system is also of interest because it represents the force effective throughout the cardiac cycle for driving blood to the tissues. This force, called the *mean arterial pressure*, is approximated by adding the diastolic pressure and one-third of the pulse pressure (DP + 1/3PP). ■

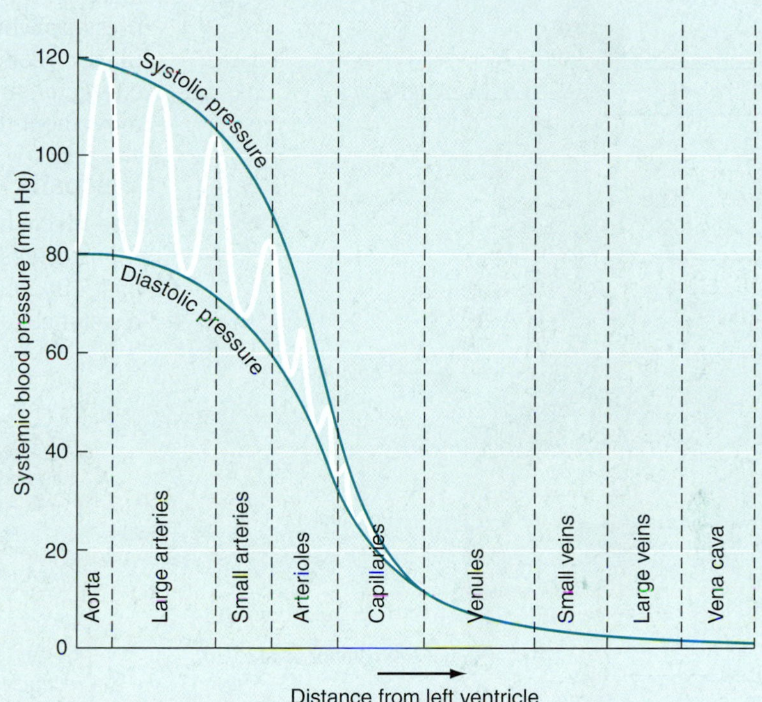

FIGURE 15H Blood pressure decreases as the distance from the left ventricle increases. Systolic pressure occurs during maximal ventricular contraction. Diastolic pressure occurs when the ventricles relax.

Blood pressure varies with the cardiac output. If either the stroke volume or the heart rate increases, so does the cardiac output, and, blood pressure initially rises. Conversely, if the stroke volume or the heart rate decreases, the cardiac output decreases, and blood pressure also initially decreases.

Blood Volume

Blood volume equals the sum of the formed elements and plasma volumes in the vascular system. Although the blood volume varies somewhat with age, body size, and sex, it is usually about 5 liters for adults or 8% of body weight in kilograms (1 kilogram of water equals 1 liter).

Blood volume is determined by injecting a known volume of an indicator, such as radioactive iodine, into the blood. After thorough mixing, a blood sample is withdrawn, and the concentration of the indicator measured. The total blood volume is calculated using this formula: blood volume = amount of indicator injected/concentration of indicator in blood sample.

Blood pressure is normally directly proportional to the volume of the blood in the cardiovascular system. Thus, any changes in the blood volume can initially alter the blood pressure. For example, if a hemorrhage reduces blood volume, blood pressure at first drops. A transfusion that restores normal blood volume may reestablish normal pressure. Blood volume can also fall if the fluid balance is upset, as happens in dehydration. Fluid replacement can reestablish normal blood volume and pressure. Clinical Application 15.4 describes how the unusual conditions of microgravity in outer space affect the distribution of blood volume and control of blood pressure.

Peripheral Resistance

Friction between blood and the walls of the blood vessels produces a force called **peripheral resistance** (pĕ-rif′er-al re-zis′tans), which hinders blood flow. Blood pressure must overcome this force if the blood is to continue flowing. Therefore, factors that alter the peripheral resistance change blood pressure. For example, contraction of smooth muscles in the walls of contracting arterioles increases the peripheral resistance by constricting these vessels. Blood tends to back up into the arteries supplying the arterioles, and the arterial pressure rises. Dilation of the arterioles has the opposite effect—peripheral resistance lessens, and the arterial blood pressure drops in response (fig. 15.35).

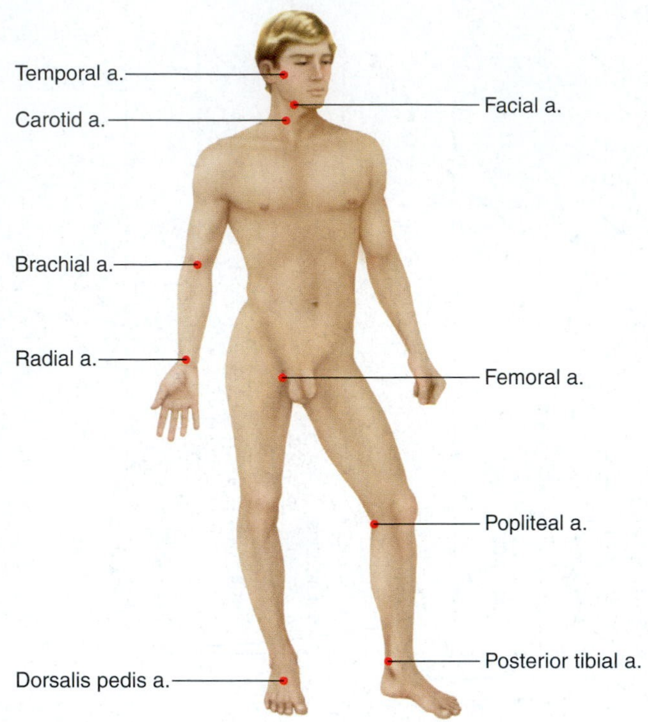

FIGURE 15.33 Sites where an arterial pulse is most easily detected. (*a.* stands for artery.)

Arterial walls are elastic, so when the ventricles discharge a surge of blood, arteries swell. Almost immediately, the elastic tissues recoil, and the vessel walls press against the blood inside. This action helps force the blood onward against the peripheral resistance in arterioles and capillaries. Recoiling of the arteries maintains blood pressure during diastole. If there were no elasticity in the arterial walls, blood pressure would fall to zero between ventricular contractions. Elastic recoil also converts the intermittent flow of blood, characteristic of the arterial system, into a more continuous movement through the capillaries.

Viscosity

The **viscosity** (vis-kos′ĭ-te) of a fluid is a physical property that derives from the ease with which its molecules flow past one another. The greater the viscosity, the greater the resistance to flow.

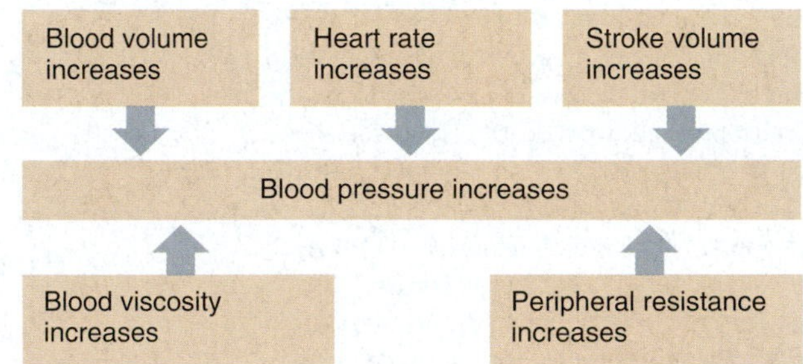

FIGURE 15.34 Some of the factors that influence arterial blood pressure.

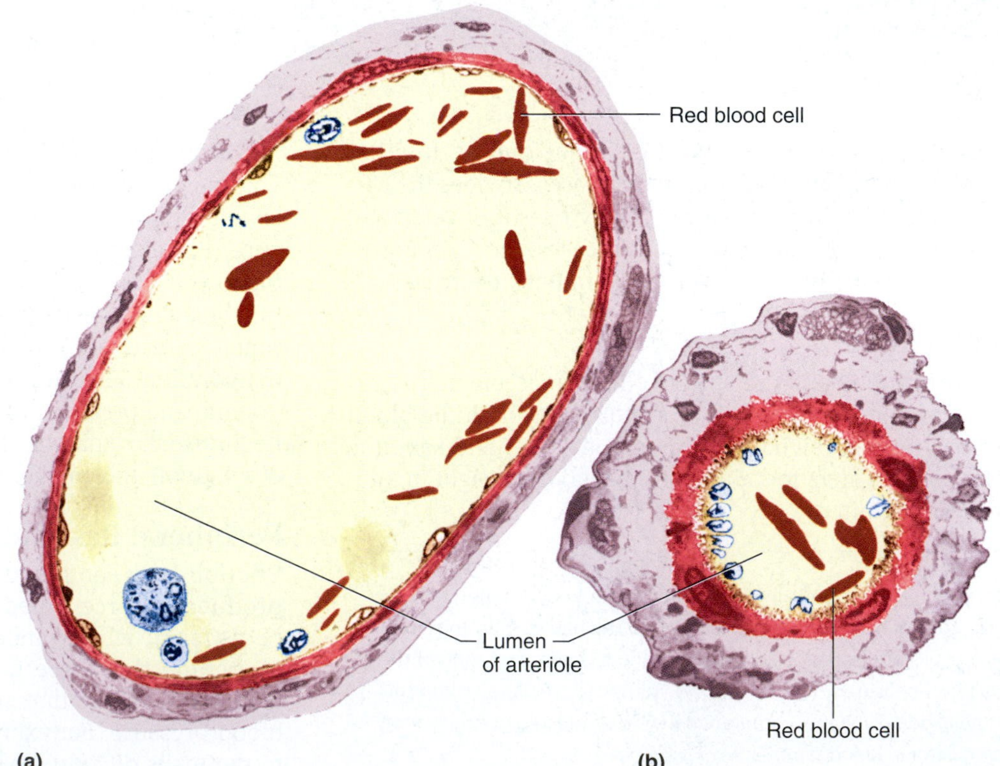

FIGURE 15.35 Vasodilation and vasoconstriction. (*a*) Relaxation of smooth muscle in the arteriole wall produces dilation, whereas (*b*) contraction of the smooth muscle causes constriction (*a* and *b* 1,500×).

15.4 CLINICAL APPLICATION

Space Medicine

When the rescue team approached the space shuttle *Atlantis* just after it landed on September 26, 1996, they brought a stretcher, expecting to carry off Mission Specialist Shannon Lucid, Ph.D. The fifty-three-year-old biochemist had just spent 188 days aboard the Russian *Mir* space station (fig. 15I). About 70% of astronauts cannot stand at all upon reencountering gravity, but Lucid walked, albeit a little wobbly, the 25 feet to the crew transporter.

The human body evolved under conditions of constant gravity. When a body is exposed to microgravity (very low gravity) or weightlessness for extended periods, changes occur. Lucid was expected to require the stretcher because of decreased muscle mass, mineral-depleted bones, and low blood volume. The 400 hours that she logged on the *Mir*'s treadmill and stationary bicycle may have helped her stay in shape.

Lucid was poked and prodded, monitored and tested, as medical researchers attempted to learn how six months in space affects cardiovascular functioning, respiratory capacity, mood, blood chemistry, circadian rhythms, muscular strength, body fluid composition, and many other aspects of anatomy and physiology.

Feeling unsteady upon returning to earth is one of the better-studied physiologic responses to low-gravity conditions. It is called orthostatic intolerance. Normally, gravity helps blood circulate in the lower limbs. In microgravity or no gravity, blood pools in blood vessels in the center of the body, registering on receptors there. The body interprets this as excess blood, and in response, signals the kidneys to excrete more fluid. But there really isn't an increased blood volume. On return to earth, the body has a pint to a quart less blood than it should, up to a 10% to 20% decrease in total blood volume. If blood vessels cannot sufficiently constrict to counter the plummeting blood pressure, orthostatic intolerance results. To minimize the effect, astronauts wear lower-body suction suits, which apply a vacuum force that helps draw blood into the blood vessels of the lower limbs. Maintaining fluid intake helps prevent dehydration. ■

FIGURE 15I Shannon Lucid's 188-day stay in space revealed to researchers much about the body's responses to microgravity conditions. While aboard the space station *Mir,* Lucid conducted experiments on quail embryos and growth of protein crystals.

Blood cells and some plasma proteins increase blood viscosity. The greater the blood's resistance to flowing, the greater the force needed to move it through the vascular system, so blood pressure rises as blood viscosity increases and drops as blood viscosity decreases.

The viscosity of blood normally remains stable. However, any condition that alters the concentrations of blood cells or specific plasma proteins may alter blood viscosity. For example, anemia may decrease viscosity and consequently lower blood pressure. Excess red blood cells increase viscosity and blood pressure.

PRACTICE 

45 How is cardiac output calculated?

46 How are cardiac output and blood pressure related?

47 How does blood volume affect blood pressure?

48 What is the relationship between peripheral resistance and blood pressure? Between blood viscosity and blood pressure?

Control of Blood Pressure

Blood pressure (BP) is determined by cardiac output (CO) and peripheral resistance (PR) according to this relationship: $BP = CO \times PR$. Maintenance of normal blood pressure therefore requires regulation of these two factors (fig. 15.36).

Cardiac output depends on the stroke volume and heart rate. Stroke volume, the amount of blood pumped in a single beat, is reflected by the difference between **end-diastolic volume** (EDV), the volume of blood in each ventricle at the end of ventricular diastole, and **end-systolic volume** (ESV), the volume of blood in each ventricle at the end of ventricular systole. Mechanical, neural, and chemical factors affect stroke volume and heart rate.

Cardiac output is limited by the amount of blood returning to the ventricles, called the *venous return.* Usually, however, stroke volume can be increased by sympathetic stimulation, which increases the force of ventricular contraction. Only about 60% of the end-diastolic volume is pumped out in a

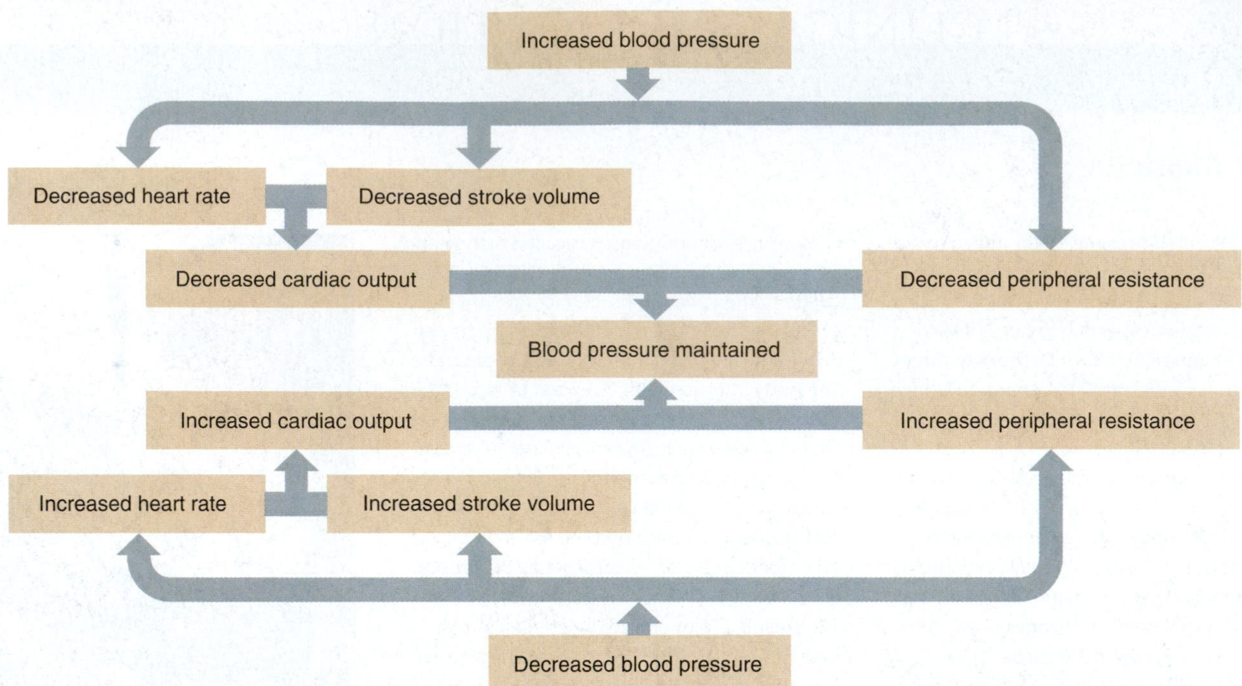

FIGURE 15.36 Controlling cardiac output and peripheral resistance regulates blood pressure.

normal contraction, so increasing the force of ventricular contraction may increase that fraction and help maintain stroke volume if venous return should decrease.

RECONNECT

To Chapter 9, Recording a Muscle Contraction, pages 296–297.

Another mechanism increases stroke volume independently of sympathetic stimulation. As blood enters the ventricles, myocardial fibers are mechanically stretched. This constitutes the **preload.** The greater the EDV, the greater the preload. Within limits, the longer these fibers, the greater the force with which they contract. This relationship between fiber length (due to stretching of the cardiac muscle cell just before contraction) and force of contraction is called the **Frank-Starling law of the heart,** or Starling's law of the heart. This becomes important, for example, during exercise, when venous return increases. The more blood that enters the heart from the veins, the greater the ventricular distension, the stronger the contraction, the greater the stroke volume, and the greater the cardiac output.

Conversely, the less blood that returns from the veins, the less the ventricle distends, the weaker the ventricular contraction, and the lesser the stroke volume and cardiac output. This mechanism ensures that the volume of blood discharged from the heart is equal to the volume entering its chambers.

Some blood remains in the ventricles after contraction and stroke volume ejection. This ESV is influenced by preload, contractility of the ventricle, and afterload. **Contractility,** the amount of force produced during a contraction at a given preload, is influenced by autonomic innervation and hormones (epinephrine, norepinephrine, thyroid hormones). Sympathetic stimulation contracts the ventricles more forcefully, increasing the volume ejected and decreasing the ESV. Decreased sympathetic stimulation produces the opposite effect. The amount of force the ventricle must produce to open the semilunar valves to eject blood is the **afterload.** Increased arterial pressure (hypertension) increases afterload. As the afterload increases, stroke volume decreases and ESV increases.

Recall that baroreceptors in the walls of the aortic arch and carotid sinuses sense changes in blood pressure. If arterial pressure increases, nerve impulses travel from the receptors to the *cardiac center* of the medulla oblongata. This center relays parasympathetic impulses to the SA node in the heart, and heart rate decreases in response. As a result of this *cardioinhibitor reflex,* cardiac output falls, and blood pressure decreases toward the normal level. Figure 15.37 summarizes this mechanism.

Conversely, decreasing arterial blood pressure initiates the *cardioaccelerator reflex,* in which sympathetic impulses go to the SA node. The heart then beats faster. This response increases cardiac output, increasing arterial pressure.

Recall that epinephrine increases heart rate (chapter 13, p. 506) and consequently alters cardiac output and blood pressure. Other factors that increase heart rate and blood pressure include emotional responses, such as fear and anger; physical exercise; and a rise in body temperature.

Changes in arteriole diameters regulate peripheral resistance. Blood vessels with smaller diameters offer a greater resistance to blood flow, so factors that cause arteriole

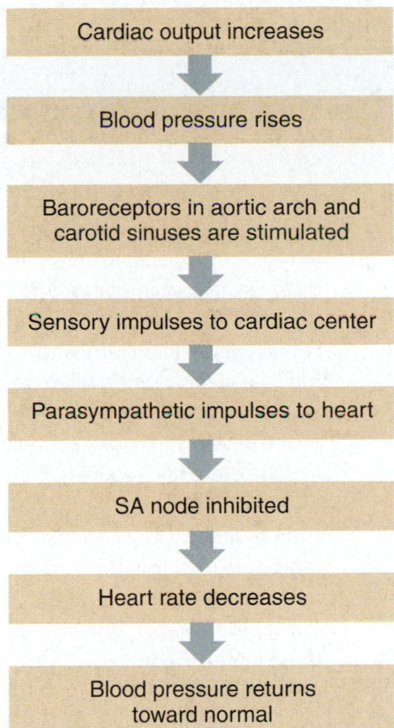

FIGURE 15.37 If blood pressure rises, baroreceptors initiate the cardioinhibitor reflex, which lowers the blood pressure.

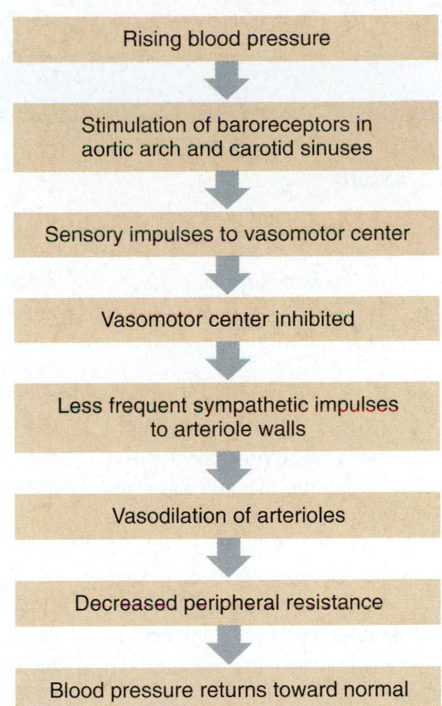

FIGURE 15.38 Dilating arterioles helps regulate blood pressure.

vasoconstriction increase peripheral resistance, increasing blood pressure; factors causing vasodilation decrease peripheral resistance, decreasing blood pressure.

The *vasomotor center* of the medulla oblongata continually sends sympathetic impulses to the smooth muscles in the arteriole walls, keeping them in a state of tonic contraction, which helps maintain the peripheral resistance associated with normal blood pressure. The vasomotor center responds to changes in blood pressure, so it can increase peripheral resistance by increasing its outflow of sympathetic impulses, or it can decrease such resistance by decreasing its sympathetic outflow. In the latter case, the vessels vasodilate as sympathetic stimulation decreases.

Whenever arterial blood pressure suddenly increases, baroreceptors in the aortic arch and carotid sinuses signal the vasomotor center, and the sympathetic outflow to the arteriole walls falls (fig. 15.38). The resulting vasodilation decreases peripheral resistance, and blood pressure decreases toward the normal level.

Similarly, if blood pressure drops, as following a hemorrhage, the vasomotor center increases sympathetic outflow. The resulting release of epinephrine and norepinephrine vasoconstricts most systemic vessels, increasing peripheral resistance. This helps return blood pressure toward normal.

The vasomotor center's control of vasoconstriction and vasodilation is especially important in the arterioles of the *abdominal viscera* (splanchnic region). These vessels, if fully dilated, could accept nearly all the blood of the body and send the arterial pressure toward zero. Thus, control of their diameters is essential in regulating normal peripheral resistance.

Certain chemicals, including carbon dioxide, oxygen, and hydrogen ions, influence peripheral resistance by affecting precapillary sphincters and smooth muscles in arteriole and metarteriole walls. For example, increasing blood carbon dioxide, decreasing blood oxygen, and lowering of the blood's pH relaxes smooth muscle in the systemic circulation. This increases local blood flow to tissues with high metabolic rates, such as exercising skeletal muscles.

Other chemicals also influence peripheral resistance and thus blood pressure. Nitric oxide, produced by endothelial cells, and bradykinin, formed in the blood, are both vasodilators. Angiotensin plays a role in vasoconstriction; and endothelin, released by cells of the endothelium, is a powerful vasoconstrictor. Clinical Application 15.5 discusses high blood pressure.

PRACTICE 

49 What factors affect cardiac output?

50 Explain the Frank-Starling law of the heart.

51 What is the function of the baroreceptors in the walls of the aortic arch and carotid sinuses?

52 How does the vasomotor center control peripheral resistance?

Venous Blood Flow

Blood pressure decreases as the blood moves through the arterial system and into the capillary networks, so little pressure remains at the venular ends of capillaries (see fig. 15G). Instead, blood flow through the venous system is only partly

15.5 CLINICAL APPLICATION

Hypertension

Hypertension, or high blood pressure, is persistently elevated arterial pressure. It is one of the more common diseases of the cardiovascular system in industrialized nations.

High blood pressure with unknown cause is called *essential* (also primary or idiopathic) *hypertension.* Elevated blood pressure that is a consequence of another problem, such as arteriosclerosis or kidney disease, is called secondary *hypertension.*

Arteriosclerosis is accompanied by decreased elasticity of the arterial walls and narrowed vessel lumens, which raise blood pressure. Kidney diseases often produce changes that interfere with blood flow to kidney cells. In response, the affected tissues may release an enzyme called *renin* that leads to the production of *angiotensin II,* a powerful vasoconstrictor that increases peripheral resistance in the arterial system, raising arterial pressure (fig. 15J). Angiotensin II also stimulates the adrenal cortex to release *aldosterone,* which stimulates the kidneys to retain sodium ions and water. The resulting increase in blood volume contributes to increased blood pressure. Normally, this mechanism ensures that a decrease in blood flow to the kidneys is followed by an increase in arterial pressure, which, in turn, restores blood flow to the kidneys. If the decreased blood flow is the result of disease, such as atherosclerosis, the mechanism may cause high blood pressure and promote further deterioration of the arterial system.

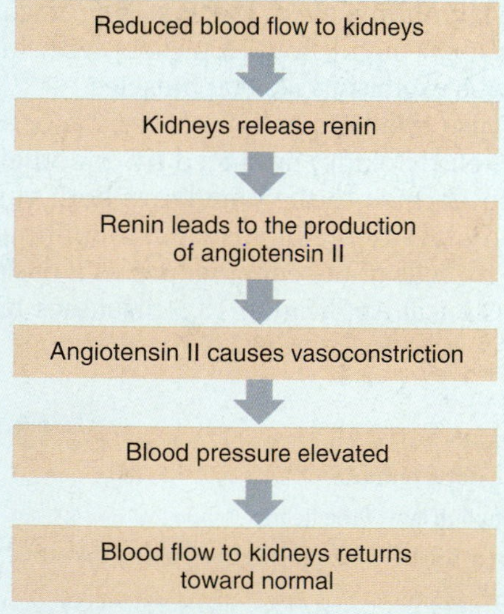

Reduced blood flow to kidneys
Kidneys release renin
Renin leads to the production of angiotensin II
Angiotensin II causes vasoconstriction
Blood pressure elevated
Blood flow to kidneys returns toward normal

FIGURE 15J Renin stimulates production of angiotensin II, which elevates blood pressure.

In some individuals, high sodium intake leads to vasoconstriction, raising blood pressure. Obesity also is a risk factor for hypertension because it increases peripheral resistance. Psychological stress, which activates sympathetic nerve impulses that cause generalized vasoconstriction, may also lead to hypertension. Yet another cause of hypertension may be an inability of endothelium to respond to a relaxing factor, leading to vasoconstriction.

Hypertension is called a "silent killer" because it may not have direct symptoms, yet can raise the risk for serious cardiovascular complications. For example, as the left ventricle works harder to pump blood at a higher pressure, the myocardium thickens, enlarging the heart. If the coronary blood vessels cannot support this overgrowth, parts of the heart muscle die and fibrous tissue replaces them. Eventually, the enlarged and weakened heart dies.

Hypertension also contributes to the development of atherosclerosis. As arteries accumulate plaque, a *coronary thrombosis* or a *coronary embolism* may occur. Similar changes in the arteries of the brain increase the chances of a *cerebral vascular accident* (CVA), or stroke, due to a cerebral thrombosis, embolism, or hemorrhage.

When an embolus or hemorrhage causes a stroke, paralysis and other functional losses suddenly appear. A thrombus-caused stroke is slower. It may begin with clumsiness, progress to partial visual loss, then affect speech. One arm becomes paralyzed, then a day later, perhaps an entire side of the body is affected. Table 15A lists risk factors for a stroke.

A *transient ischemic attack* (TIA, or "ministroke") is a temporary block in a small artery. Symptoms include difficulty in speaking or understanding speech; numbness or weakness in the face, upper limb, lower limb, or one side; dizziness; falling; an unsteady gait; blurred vision; or blindness. These symptoms typically resolve within twenty-four hours with no lasting effects, but may be a warning of an impending, more serious stroke.

Treatment of hypertension varies and may include exercising regularly, controlling weight, reducing stress, and limiting the diet to low-sodium foods. Drugs, such as diuretics and/or inhibitors of sympathetic nerve activity, may help control blood pressure. Diuretics increase urinary excretion of sodium and water, reducing the volume of body fluids. Sympathetic inhibitors block the synthesis of neurotransmitters, such as norepinephrine, or block receptor sites of effector cells. Table 15B describes how drugs that treat hypertension work. ■

TABLE 15A | Risk Factors for Stroke

Risk Factors for Stroke
Alcohol consumption
Diabetes
Elevated serum cholesterol
Family history of cardiovascular disease
Hypertension
Smoking
Transient ischemic attacks

TABLE 15B | Drugs to Treat Hypertension

Type of Drug	Mechanism of Action
Angiotensin-converting enzyme (ACE) inhibitors	Block formation of angiotensin II, preventing vasoconstriction
Beta blockers	Lower heart rate
Calcium channel blockers	Dilate blood vessels by keeping calcium ions out of muscle cells in vessel walls
Diuretics	Increase urine output, lowering blood volume

the direct result of heart action and depends on other factors, such as skeletal muscle contraction, breathing movements, and vasoconstriction of veins. For example, contracting skeletal muscles press on veins, moving blood from one valve section to another. This massaging action of contracting skeletal muscles helps push the blood through the venous system toward the heart (fig. 15.39).

Respiratory movements also move venous blood. During inspiration, the pressure in the thoracic cavity is reduced as the diaphragm contracts and the rib cage moves upward and outward. At the same time, the pressure in the abdominal cavity is increased as the diaphragm presses downward on the abdominal viscera. Consequently, blood is squeezed out of the abdominal veins and forced into thoracic veins. During exercise, these respiratory movements act with skeletal muscle contractions to increase return of venous blood to the heart.

Venoconstriction also returns venous blood to the heart. When venous pressure is low, sympathetic reflexes stimulate smooth muscles in the walls of veins to contract. The veins also provide a blood reservoir that can adapt its capacity to changes in blood volume (see fig. 15.32). If some blood is lost and blood pressure falls, venoconstriction can force blood out of this reservoir, returning venous blood to the heart. By maintaining venous return, venoconstriction helps to maintain blood pressure.

Central Venous Pressure

All veins, except those returning to the heart from the lungs, drain into the right atrium. Therefore, the pressure in this heart chamber is called *central venous pressure*. It affects the pressure in the peripheral veins. For example, if the heart is beating weakly, the central venous pressure increases, and blood backs up in the venous network, raising its pressure too. However, if the heart is beating forcefully, the central venous pressure and the pressure in the venous network decrease.

> Blood or tissue fluid accumulating in the pericardial cavity increases pressure, causing a condition called *acute cardiac tamponade*. It can be life threatening. As the pressure around the heart increases, it may compress the heart, interfere with the flow of blood into its chambers, and prevent pumping action. An early symptom of acute cardiac tamponade may be increased central venous pressure, with visible engorgement of the veins in the neck. Other symptoms include anxiety, rapid or difficulty breathing, light-headedness, palpitations, pallor, and chest pain. Acute cardiac tamponade has several causes, including bacterial or viral infection, injury, acute myocardial infarction, advanced lung cancer, and dissecting aortic aneurysm.

Increase in blood volume or widespread venoconstriction also increase blood flow into the right atrium, elevating the central venous pressure. An increase in central venous

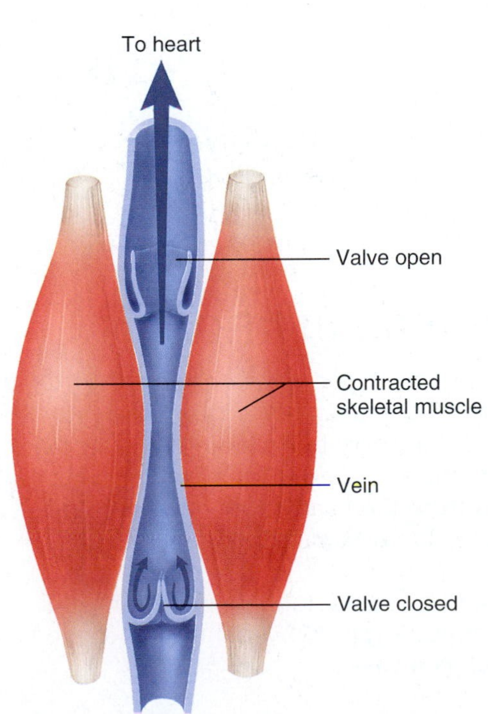

FIGURE 15.39 The massaging action of skeletal muscles helps move blood through the venous system toward the heart.

15.6 CLINICAL APPLICATION

Exercise and the Cardiovascular System

We know that exercise is good for the heart. Yet each year, a few individuals die of sudden cardiac arrest while shoveling snow, running, or engaging in some other strenuous activity. The explanation for this apparent paradox is that exercise is good for the heart—but only if it is a regular part of life.

Physiological responses to intense aerobic exercise generally increase blood flow, and therefore oxygen delivery, to active muscles. In muscles, vasodilation opens more capillaries. At the same time, vasoconstriction diminishes blood flow where it is not immediately needed, such as to the digestive tract. Blood flow, however, is maintained in the brain and kidneys, which require a steady stream of oxygen and nutrients to function. Respiratory movements and skeletal muscle activity increase venous return to the heart. As venous return to the heart increases, ventricular walls stretch, stimulating them to contract with greater force. Heart rate increases as well.

The cardiovascular system adapts to exercise. The conditioned athlete experiences increases in heart pumping efficiency, blood volume, blood hemoglobin concentration, and the number of mitochondria in muscle fibers. These adaptations improve oxygen delivery to, and use by, muscle tissue.

An athlete's heart typically changes in response to these increased demands and may enlarge 40% or more. Myocardial mass increases, the ventricular cavities expand, and the ventricle walls thicken. Stroke volume increases, and heart rate decreases, as does blood pressure. To a physician unfamiliar with a conditioned cardiovascular system, a trained athlete may appear to be abnormal.

The cardiovascular system responds beautifully to a slow, steady buildup in exercise frequency and intensity. It does not react well to sudden demands—such as when a person who never exercises suddenly shovels snow or runs 3 miles.

For exercise to benefit the cardiovascular system, the heart rate must be elevated to 70% to 85% of its "theoretical maximum" for at least half an hour three times a week. You can calculate your theoretical maximum by subtracting your age from 220. If you are eighteen years old, your theoretical maximum is 202 beats per minute. One hundred forty-one to 172 beats per minute is 70% to 85% of this value. Some good activities for raising the heart rate are tennis, skating, skiing, handball, vigorous dancing, hockey, basketball, biking, and fast walking.

It is wise to consult a physician before starting an exercise program. People over the age of thirty are advised to have a stress test, which is an electrocardiogram taken while exercising. (The standard electrocardiogram is taken at rest.) An arrhythmia that appears only during exercise may indicate heart disease that has not yet produced symptoms.

The American Heart Association suggests that after a physical exam, a sedentary person wishing to start an exercise program begin with 30 minutes of activity (perhaps broken into two 15-minute sessions at first) at least five times per week. ■

pressure can lead to peripheral edema because the resulting higher capillary hydrostatic pressure favors movement of fluid into the tissues. Clinical Application 15.6 discusses the effects of exercise on the heart and blood vessels.

PRACTICE

53 What is the function of the venous valves?

54 How do skeletal muscles affect venous blood flow?

55 How do respiratory movements affect venous blood flow?

56 What factors stimulate venoconstriction?

15.6 PATHS OF CIRCULATION

Recall from figure 15.1 that the blood vessels can be divided into two major pathways. The *pulmonary circuit* (or circulation) consists of vessels that carry blood from the heart to the lungs and back to the heart. The *systemic circuit* (or circulation) carries blood from the heart to all other parts of the body and back again. The systemic circuit includes the coronary circulation.

The pathways described in the following sections are those of an adult. Chapter 23 (pp. 897–898) describes the somewhat different fetal pathways.

Pulmonary Circuit

Blood enters the pulmonary circuit as it leaves the right ventricle through the pulmonary trunk. The pulmonary trunk extends upward and posteriorly from the heart, and about 5 centimeters above its origin, it divides into the right and left pulmonary arteries. These branches penetrate the right and left lungs, respectively. In the lungs, they diverge into *lobar branches* (three on the right side and two on the left) that accompany the main divisions of the bronchi (airways) into the lobes of the lungs. After repeated divisions, the lobar branches give rise to arterioles that continue into the capillary networks associated with the walls of the alveoli (air sacs) (fig. 15.40).

The blood in the arteries and arterioles of the pulmonary circuit is low in oxygen and high in carbon dioxide. Gases are exchanged between the blood and the air as the blood moves through the *alveolar capillaries*, discussed in chapter 19 (pp. 760–761).

The right ventricle contracts with less force than the left ventricle, so the arterial pressure in the pulmonary circuit is less than that in the systemic circuit. Therefore, the alveolar capillary pressure is low.

The force that moves fluid out of an alveolar capillary is 23 mm Hg; the force pulling fluid into it is 22 mm Hg. Thus, such a capillary has a net filtration pressure of 1 mm Hg. This pressure propels a slight, continuous flow of fluid into the narrow interstitial space between the alveolar capillary and the alveolus.

The epithelial cells of the alveoli are so tightly joined that sodium, chloride, and potassium ions, as well as glucose and urea, enter the interstitial space but usually fail to enter the alveoli. This helps maintain a high osmotic pressure in the interstitial fluid. Consequently, osmosis rapidly moves any water that gets into the alveoli back into the interstitial space. Although the alveolar surface must be moist to allow diffusion of oxygen and carbon dioxide, this mechanism keeps excess water out of the alveoli, preventing them from filling with fluid (fig. 15.41).

Fluid in the interstitial space may be drawn back into the alveolar capillaries by the somewhat higher osmotic pressure

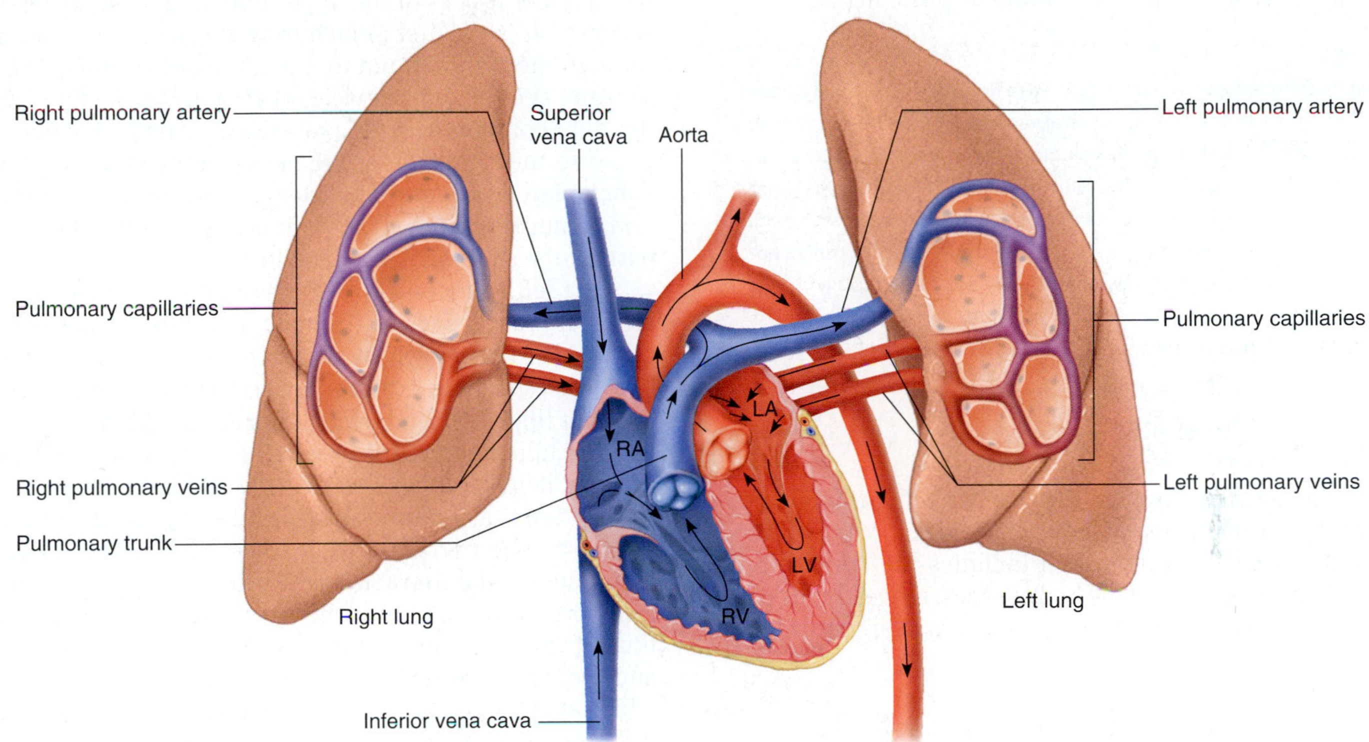

FIGURE 15.40 Blood reaches the lungs through branches of the pulmonary arteries, and it returns to the heart through pulmonary veins. (Structures are not drawn to scale.)

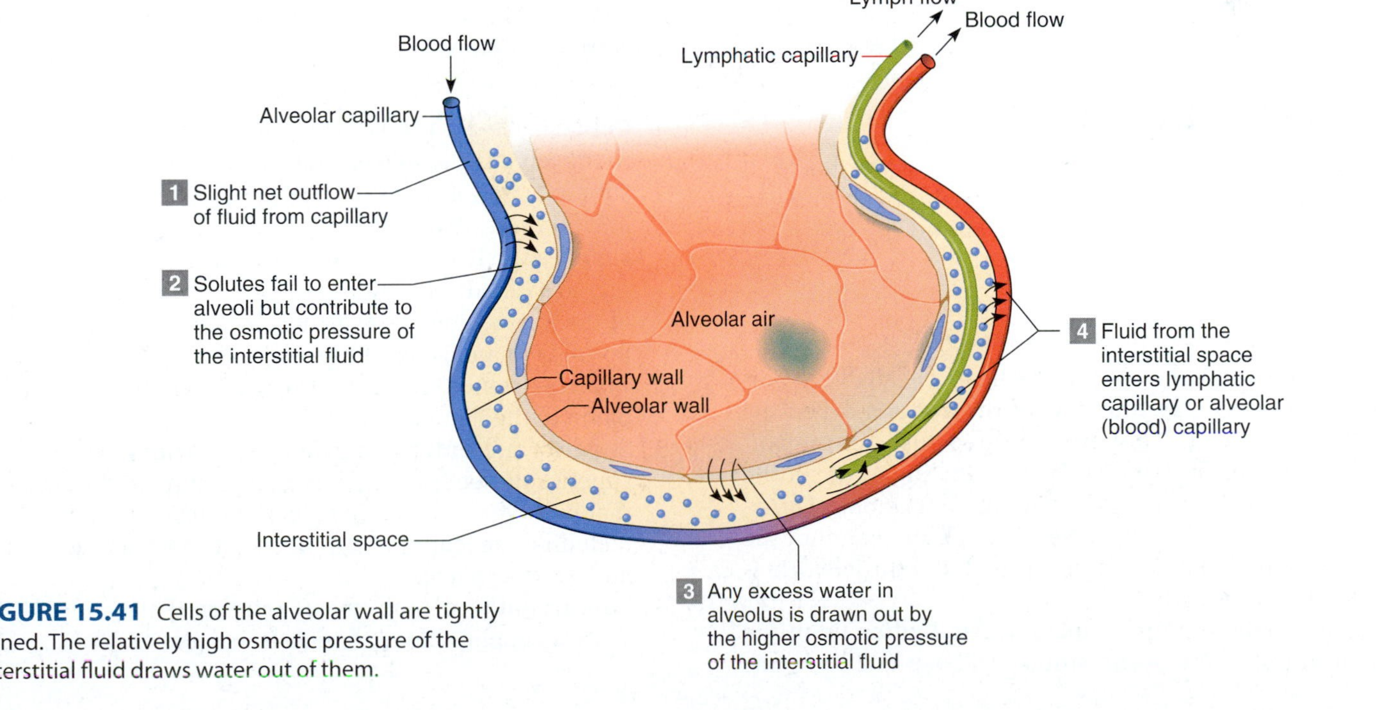

FIGURE 15.41 Cells of the alveolar wall are tightly joined. The relatively high osmotic pressure of the interstitial fluid draws water out of them.

of the blood. Alternatively, lymphatic vessels (see chapter 16, pp. 617–618) may return it to the circulation.

As a result of the gas exchanges between the blood and the alveolar air, blood entering the venules of the pulmonary circuit is rich in oxygen and low in carbon dioxide. These venules merge to form small veins, and these veins in turn converge to form larger veins. Four *pulmonary veins,* two from each lung, return blood to the left atrium, and this completes the vascular loop of the pulmonary circuit.

> *Pulmonary edema,* in which lungs fill with fluid, can accompany a failing left ventricle or a damaged mitral valve. A weak left ventricle may be unable to move the normal volume of blood into the systemic circuit. Blood backing up into the pulmonary circuit increases pressure in the alveolar capillaries, flooding the interstitial spaces with fluid. Increasing pressure in the interstitial fluid may rupture the alveolar membranes, and fluid may enter the alveoli more rapidly than it can be removed. This reduces the alveolar surface available for gas exchange, and the person may suffocate.

Systemic Circuit

Freshly oxygenated blood moves from the left atrium into the left ventricle. Contraction of the left ventricle forces this blood into the systemic circuit, which includes the aorta and its branches that lead to all of the body tissues, as well as the companion system of veins that returns blood to the right atrium.

PRACTICE

57 Distinguish between the pulmonary and systemic circuits of the cardiovascular system.

58 Trace the path of blood through the pulmonary circuit from the right ventricle.

59 Explain why the alveoli normally do not fill with fluid.

15.7 ARTERIAL SYSTEM

The **aorta** is the largest diameter artery in the body. It extends upward from the left ventricle, arches over the heart to the left, and descends just anterior and to the left of the vertebral column.

Principal Branches of the Aorta

The part of the aorta attached to the heart is called the aortic root. From there, the first part of the aorta is called the *ascending aorta.* At the root are the three cusps of the aortic valve, and opposite each cusp is a swelling in the aortic wall called an **aortic sinus.** The right and left *coronary arteries* arise from two of these sinuses. The elastic recoil of the aortic wall following contraction of the left ventricle drives blood flow into these arteries.

Several small structures called **aortic bodies** lie in the epithelial lining of the aortic sinuses. These bodies house chemoreceptors that sense blood concentrations of oxygen and carbon dioxide.

Three major arteries originate from the *arch of the aorta* (aortic arch). They are the brachiocephalic artery, the left common carotid artery, and the left subclavian artery. The aortic arch has baroreceptors that detect changes in blood pressure.

The **brachiocephalic** (brak″e-o-sĕ-fal′ik) **artery** supplies blood to the tissues of the upper limb and head, as its name suggests. It is the first branch from the aortic arch and rises through the mediastinum to a point near the junction of the sternum and the right clavicle. There it divides, giving rise to the right **common carotid** (kah-rot′id) **artery,** which carries blood to the right side of the neck and head, and the right **subclavian** (sub-kla′ve-an) **artery,** which leads into the right arm. Branches of the subclavian artery also supply blood to parts of the shoulder, neck, and head.

The left *common carotid artery* and the left *subclavian artery* are respectively the second and third branches of the aortic arch. They supply blood to regions on the left side of the body corresponding to those supplied by their counterparts on the right (fig. 15.42 and reference plates 21, 22, and 23).

The upper part of the *descending aorta* is left of the midline, but it gradually moves medially and lies directly in front of the vertebral column at the level of the twelfth thoracic vertebra. The portion of the descending aorta above the diaphragm is the **thoracic aorta** (tho-ras′ik a-or′tah), and it gives off many small branches to the thoracic wall and the thoracic viscera. These branches, the *bronchial, pericardial,* and *esophageal arteries,* supply blood to the structures for which they were named. Other branches become *mediastinal arteries,* supplying various tissues in the mediastinum, and *posterior intercostal arteries,* which pass into the thoracic wall.

Below the diaphragm, the descending aorta becomes the **abdominal aorta,** and it branches to the abdominal wall and various abdominal organs. These branches include the following:

1. **Celiac** (se′le-ak) **artery.** This single vessel gives rise to the left *gastric, splenic,* and *hepatic arteries,* which supply upper portions of the digestive tract, the spleen, and the liver, respectively. (*Note:* The hepatic artery supplies the liver with about one-third of its blood flow, and this blood is oxygen-rich. The remaining two-thirds of the liver's blood flow arrives by means of the hepatic portal vein and is oxygen-poor.)
2. **Phrenic** (fren′ik) **arteries.** These paired arteries supply blood to the diaphragm.
3. **Superior mesenteric** (mes″en-ter′ik) **artery.** The superior mesenteric artery is a large, unpaired vessel that branches to many parts of the intestinal tract, including the jejunum, ileum, cecum, ascending colon, and transverse colon.
4. **Suprarenal** (soo″prah-re′nal) **arteries.** This pair of vessels supplies blood to the adrenal glands.

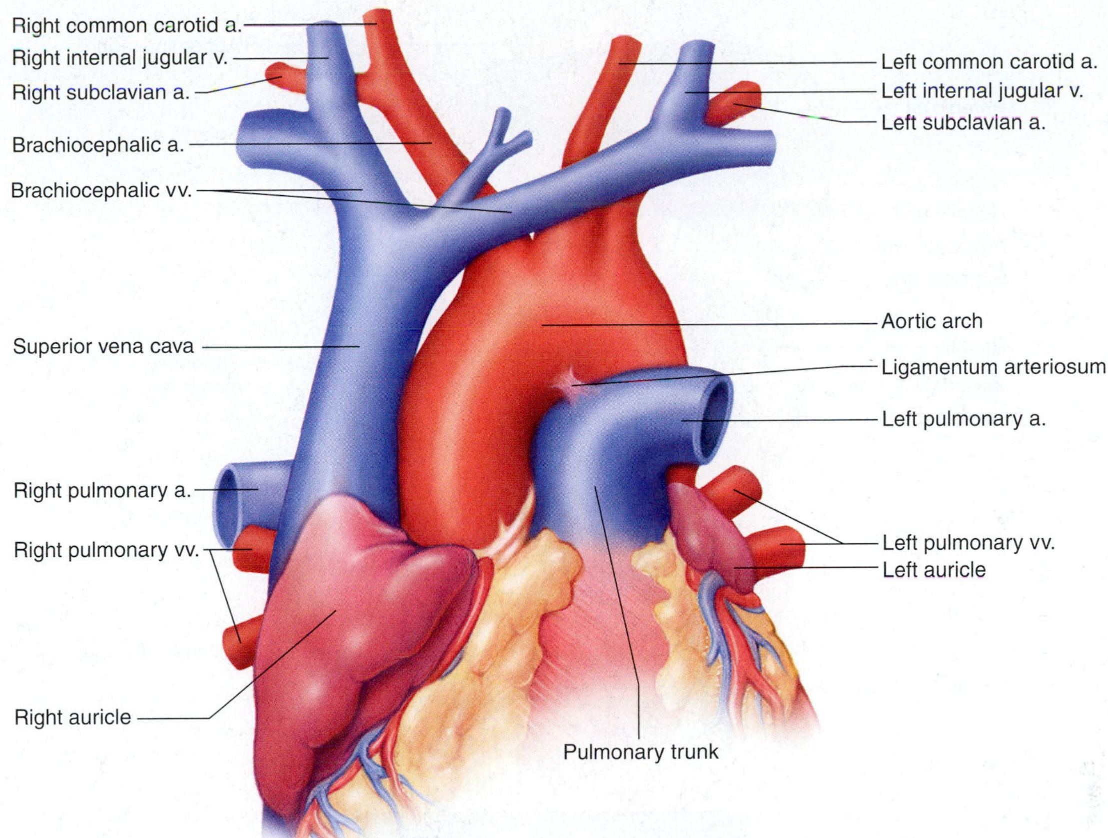

FIGURE 15.42 The major blood vessels associated with the heart. (*a.* stands for artery, *v.* stands for vein, *vv.* stands for veins.)

5. **Renal** (re′nal) **arteries.** The renal arteries pass laterally from the aorta into the kidneys. Each artery then divides into several lobar branches in the kidney tissues.
6. **Gonadal** (go′nad-al) **arteries.** In a female, paired ovarian arteries arise from the aorta and pass into the pelvis to supply the ovaries. In a male, *spermatic arteries* originate in similar locations. They course downward and pass through the body wall by way of the *inguinal canal* to supply the testes.
7. **Inferior mesenteric artery.** Branches of this single artery lead to the descending colon, the sigmoid colon, and the rectum.
8. **Lumbar arteries.** Three or four pairs of lumbar arteries arise from the posterior surface of the aorta in the region of the lumbar vertebrae. These arteries supply muscles of the skin and the posterior abdominal wall.
9. **Middle sacral artery.** This small, single vessel descends medially from the aorta along the anterior surfaces of the lower lumbar vertebrae. It carries blood to the sacrum and coccyx.

The abdominal aorta terminates near the brim of the pelvis, where it divides into right and left *common iliac arteries.* These vessels supply blood to lower regions of the abdominal wall, the pelvic organs, and the lower extremities (fig. 15.43). Table 15.4 summarizes the major branches of the aorta.

Arteries to the Brain, Head, and Neck

Branches of the subclavian and common carotid arteries supply blood to structures in the brain, head, and neck (figs. 15.44 and 15.45). The main divisions of the subclavian artery to these regions are the vertebral, thyrocervical, and costocervical arteries. The common carotid artery communicates with these regions by means of the internal and external carotid arteries.

The **vertebral arteries** arise from the subclavian arteries in the base of the neck near the tips of the lungs. They pass upward through the foramina of the transverse processes of the cervical vertebrae and enter the skull by way of the foramen magnum. Along their paths, these vessels supply blood to vertebrae and to their associated ligaments and muscles.

In the cranial cavity, the vertebral arteries unite to form a single *basilar artery.* This vessel passes along the ventral brainstem and gives rise to branches leading to the pons, midbrain, and cerebellum. The basilar artery terminates by dividing into two *posterior cerebral arteries* that supply portions of the occipital and temporal lobes of the cerebrum. The posterior cerebral arteries also help form the **cerebral arterial circle** (*circle of Willis*) at the base of the brain, which connects the vertebral artery and internal carotid artery systems (fig. 15.46). The union of these systems provides alternate pathways for blood to circumvent blockages and reach brain tissues. It also equalizes blood pressure in the brain's

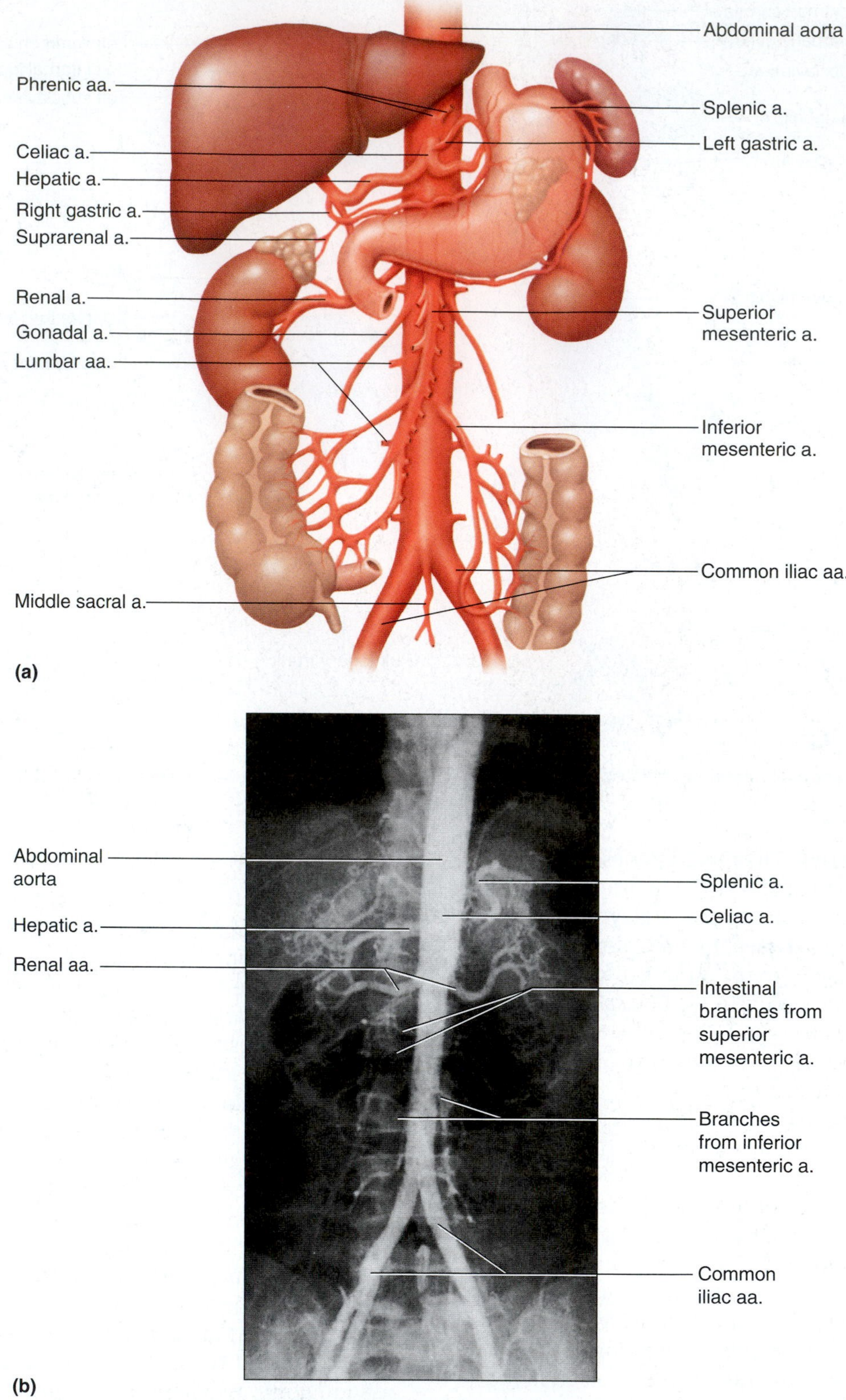

FIGURE 15.43 Abdominal aorta. (*a*) Its major branches. (*b*) Angiogram (radiograph). (*a.* stands for artery, *aa.* stands for arteries.)

TABLE 15.4 | The Aorta and Its Major Branches

Portion of Aorta	Branch	General Regions or Organs Supplied
Ascending aorta	Right and left coronary arteries	Heart
Arch of aorta	Brachiocephalic artery	Right upper limb, right side of head
	Left common carotid artery	Left side of head
	Left subclavian artery	Left upper limb
Descending aorta		
Thoracic aorta	Bronchial artery	Bronchi
	Pericardial artery	Pericardium
	Esophageal artery	Esophagus
	Mediastinal artery	Mediastinum
	Posterior intercostal artery	Thoracic wall
Abdominal aorta	Celiac artery	Organs of upper digestive tract
	Phrenic artery	Diaphragm
	Superior mesenteric artery	Portions of small and large intestines
	Suprarenal artery	Adrenal gland
	Renal artery	Kidney
	Gonadal artery	Ovary or testis
	Inferior mesenteric artery	Lower portions of large intestine
	Lumbar artery	Posterior abdominal wall
	Middle sacral artery	Sacrum and coccyx
	Common iliac artery	Lower abdominal wall, pelvic organs, and lower limb

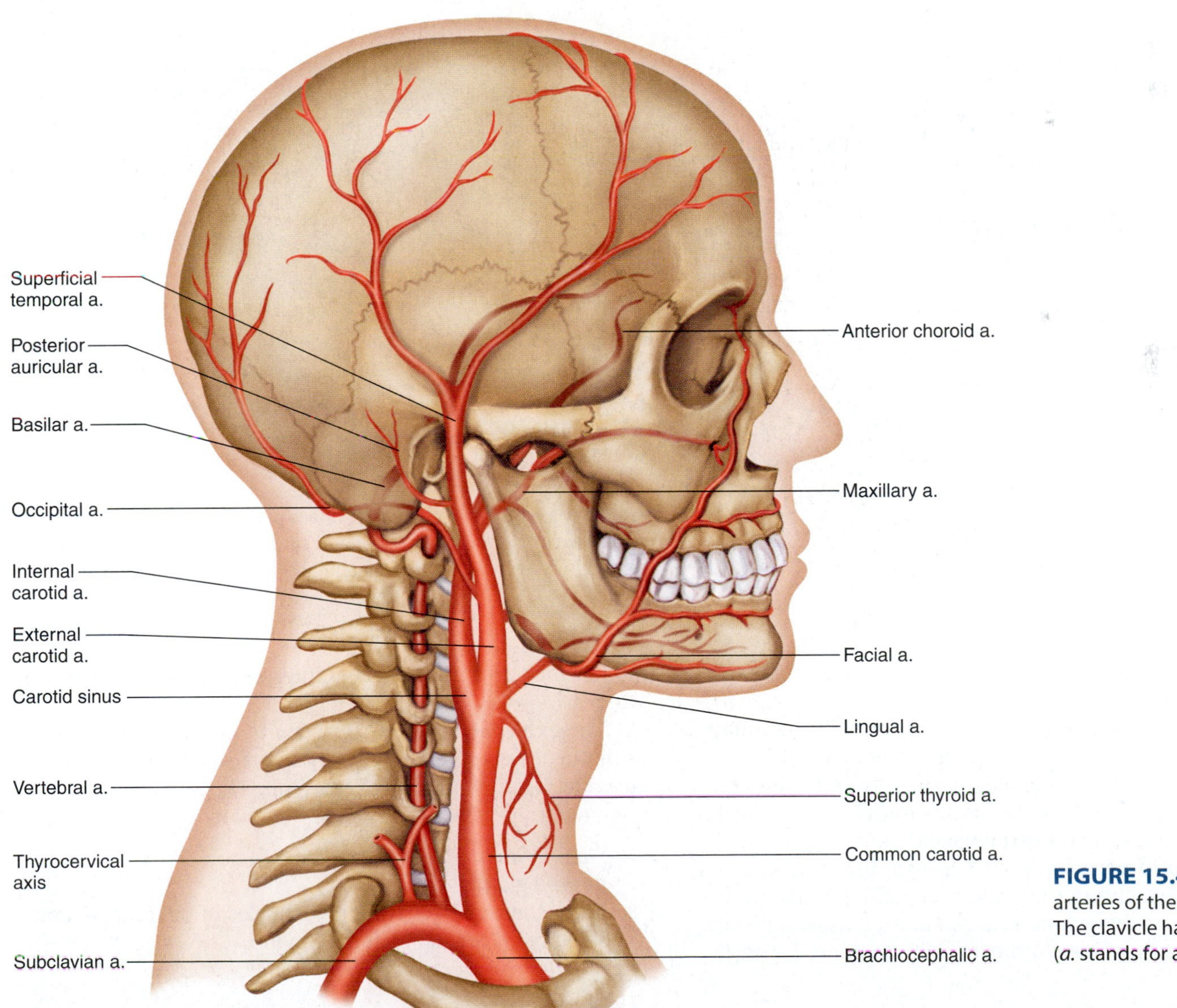

FIGURE 15.44 The major arteries of the head and neck. The clavicle has been removed. (*a.* stands for artery.)

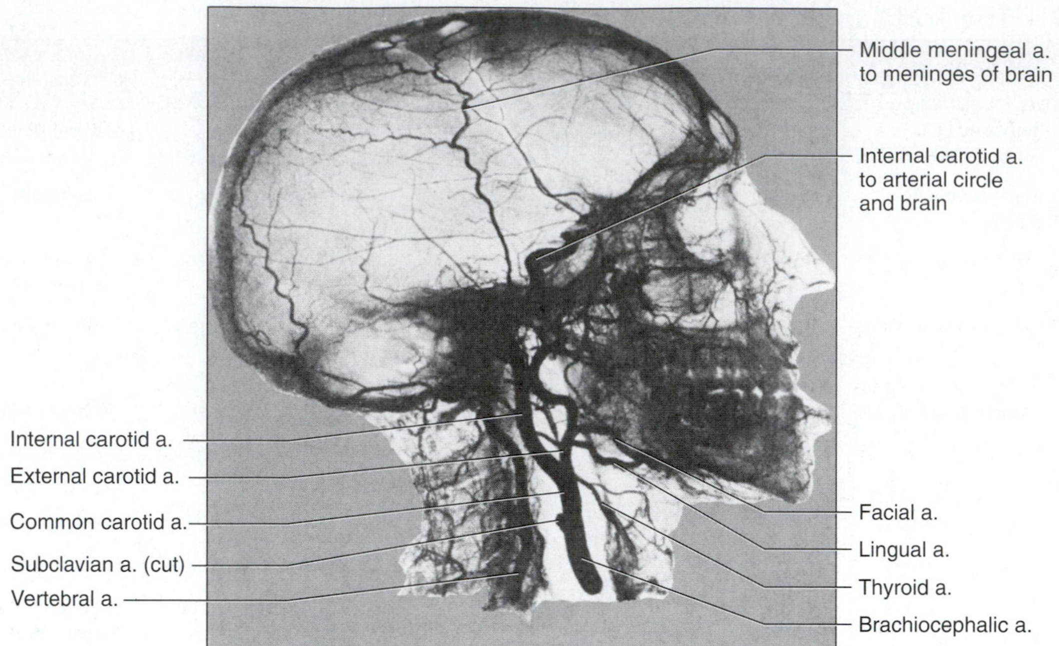

FIGURE 15.45 An angiogram of the arteries associated with the head. (*a.* stands for artery.)

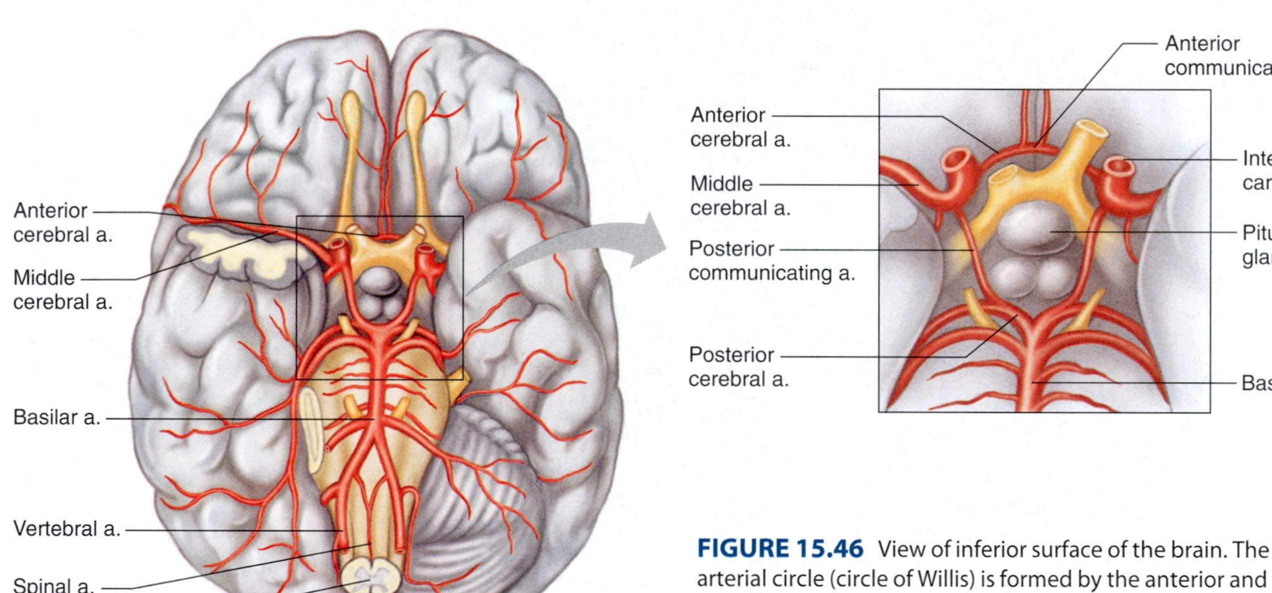

FIGURE 15.46 View of inferior surface of the brain. The cerebral arterial circle (circle of Willis) is formed by the anterior and posterior cerebral arteries, which join the internal carotid arteries. (*a.* stands for artery.)

blood supply. The circle is complete in only 20% to 30% of the population.

The **thyrocervical** (thi″ro-ser′vĭ-kal) **arteries** are short vessels that give off branches at the thyrocervical axis to the thyroid gland, parathyroid glands, larynx, trachea, esophagus, and pharynx, as well as to various muscles in the neck, shoulder, and back. The **costocervical** (kos″to-ser′vĭ-kal) **arteries,** the third vessels to branch from the subclavians, carry blood to muscles in the neck, back, and thoracic wall.

The left and right *common carotid arteries* ascend deeply in the neck on either side. At the level of the upper laryngeal border, they divide to form the internal and external carotid arteries.

The **external carotid artery** courses upward on the side of the head, giving off branches to structures in the neck, face, jaw, scalp, and base of the skull. The main vessels that originate from this artery include the following:

1. *Superior thyroid artery* to the hyoid bone, larynx, and thyroid gland.
2. *Lingual artery* to the tongue, muscles of the tongue, and salivary glands beneath the tongue.

3. *Facial artery* to the pharynx, palate, chin, lips, and nose.
4. *Occipital artery* to the scalp on the back of the skull, the meninges, the mastoid process, and various muscles in the neck.
5. *Posterior auricular artery* to the ear and the scalp over the ear.

The external carotid artery terminates by dividing into *maxillary* and *superficial temporal arteries.* The maxillary artery supplies blood to the teeth, gums, jaws, cheek, nasal cavity, eyelids, and meninges. The temporal artery extends to the parotid salivary gland and to various surface regions of the face and scalp.

The **internal carotid artery** follows a deep course upward along the pharynx to the base of the skull. Entering the cranial cavity, it provides the major blood supply to the brain. The major branches of the internal carotid artery include the following:

1. *Ophthalmic artery* to the eyeball and to various muscles and accessory organs within the orbit.
2. *Posterior communicating artery* that forms part of the cerebral arterial circle.
3. *Anterior choroid artery* to the choroid plexus within the lateral ventricle of the brain and to nerve structures in the brain.

The internal carotid artery terminates by dividing into *anterior* and *middle cerebral arteries.* The middle cerebral artery passes through the lateral tissue and supplies the lateral surface of the cerebrum, including the primary motor and sensory areas of the face and upper limbs, the optic radiations, and the speech area (see chapter 11, pp. 401–402). The anterior cerebral artery extends anteriorly between the cerebral hemispheres and supplies the medial surface of the brain.

Near the base of each internal carotid artery is an enlargement called a **carotid sinus.** Like the aortic sinuses, these structures contain baroreceptors that control blood pressure. A number of small epithelial masses, called **carotid bodies,** are also in the wall of the carotid sinus. These bodies are vascular and have chemoreceptors that act with those of the aortic bodies to regulate circulation and respiration.

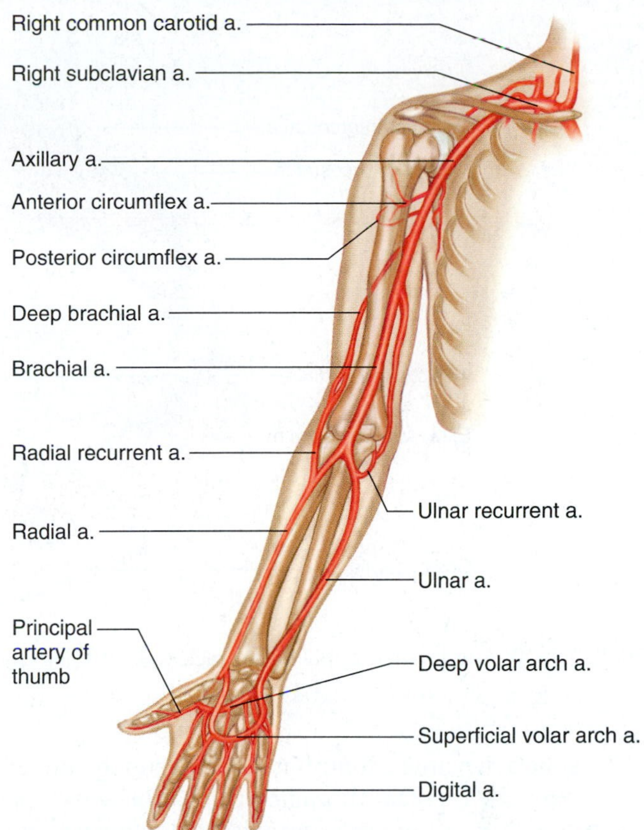

FIGURE 15.47 The major arteries to the shoulder and upper limb. (*a.* stands for artery.)

Arteries to the Shoulder and Upper Limb

The subclavian artery, after giving off branches to the neck, continues into the arm (fig. 15.47). It passes between the clavicle and the first rib and becomes the axillary artery.

The **axillary artery** supplies branches to structures in the axilla and the chest wall, including the skin of the shoulder; part of the mammary gland; the upper end of the humerus; the shoulder joint; and muscles in the back, shoulder, and chest. As this vessel leaves the axilla, it becomes the brachial artery.

The **brachial artery** courses along the humerus to the elbow. It gives rise to a *deep brachial artery* that curves posteriorly around the humerus and supplies the triceps muscle. Shorter branches pass into the muscles on the anterior side of the arm, whereas others descend on each side to the elbow and connect with arteries in the forearm. The resulting arterial network allows blood to reach the forearm even if a portion of the distal brachial artery becomes obstructed.

In the elbow, the brachial artery divides into an ulnar artery and a radial artery. The **ulnar artery** leads downward on the ulnar side of the forearm to the wrist. Some of its branches join the anastomosis around the elbow joint, whereas others supply blood to flexor and extensor muscles in the forearm.

The **radial artery,** a continuation of the brachial artery, extends along the radial side of the forearm to the wrist. As it nears the wrist, it comes close to the surface and provides a convenient vessel for taking the pulse (radial pulse). Branches of the radial artery join the anastomosis of the elbow and supply the lateral muscles of the forearm.

At the wrist, the branches of the ulnar and radial arteries join to form a network of vessels. Arteries arising from this network supply blood to structures in the hand.

Arteries to the Thoracic and Abdominal Walls

Blood reaches the thoracic wall through several vessels. These include branches from the subclavian artery and the thoracic aorta (fig. 15.48).

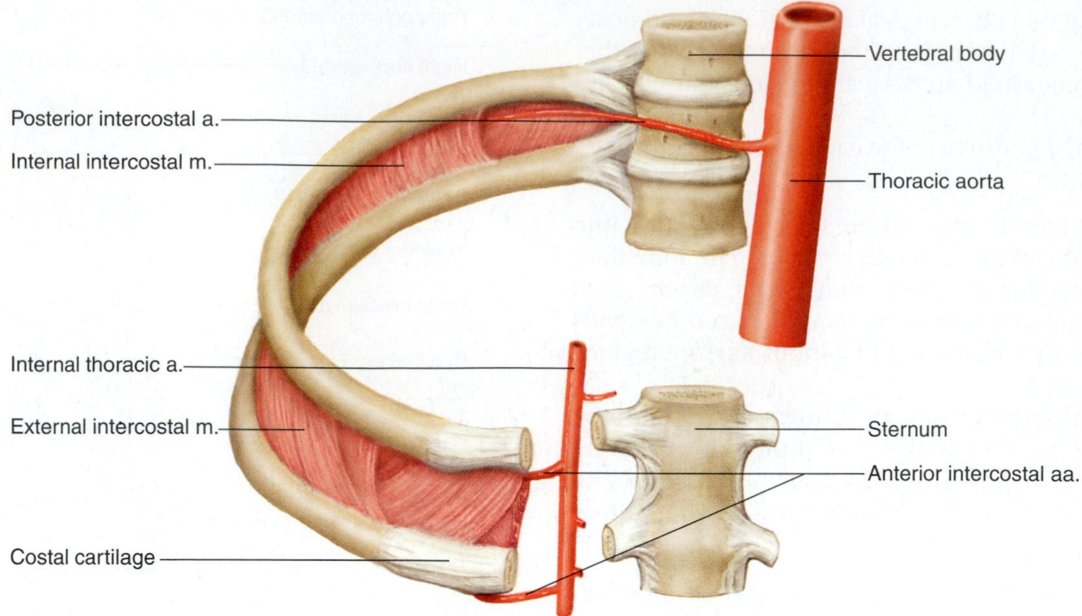

FIGURE 15.48 Arteries that supply the thoracic wall. (*a.* stands for artery, *aa.* stands for arteries , *m.* stands for muscle.)

The subclavian artery contributes to this supply through a branch called the **internal thoracic artery.** This vessel originates in the base of the neck and passes downward on the pleura and behind the cartilages of the upper six ribs. It gives off two *anterior intercostal arteries* to each of the upper six intercostal spaces; these two arteries supply the intercostal muscles, other intercostal tissues, and the mammary glands.

The *posterior intercostal arteries* arise from the thoracic aorta and enter the intercostal spaces between the third through the eleventh ribs. These arteries branch to supply the intercostal muscles, the vertebrae, the spinal cord, and deep muscles of the back.

Branches of the *internal thoracic* and *external iliac arteries* provide blood to the anterior abdominal wall. Paired vessels originating from the abdominal aorta, including the *phrenic* and *lumbar arteries,* supply blood to structures in the posterior and lateral abdominal wall.

Arteries to the Pelvis and Lower Limb

The abdominal aorta divides to form the **common iliac** (il′e-ak) **arteries** at the level of the pelvic brim. These vessels provide blood to the pelvic organs, gluteal region, and lower limbs.

Each common iliac artery descends a short distance and divides into an internal (hypogastric) branch and an external branch. The **internal iliac artery** gives off many branches to various pelvic muscles and visceral structures, as well as to the gluteal muscles and the external genitalia. Parts of figure 15.49 show important branches of this vessel, including the following:

1. *Iliolumbar artery* to the ilium and muscles of the back.
2. *Superior and inferior gluteal arteries* to the gluteal muscles, pelvic muscles, and skin of the buttocks.
3. *Internal pudendal artery* to muscles in the distal portion of the alimentary canal, the external genitalia, and the hip joint.
4. *Superior* and *inferior vesical arteries* to the urinary bladder. In males, these vessels also supply the seminal vesicles and the prostate gland.
5. *Middle rectal artery* to the rectum.
6. *Uterine artery* to the uterus and vagina.

The **external iliac artery** provides the main blood supply to the lower limbs (fig. 15.50). It passes downward along the brim of the pelvis and gives off two large branches—an *inferior epigastric artery* and a *deep circumflex iliac artery.* These vessels supply the muscles and skin in the lower abdominal wall. Midway between the symphysis pubis and the anterior superior iliac spine of the ilium, the external iliac artery becomes the femoral artery.

The **femoral** (fem′or-al) **artery,** which passes fairly close to the anterior surface of the upper thigh, gives off many branches to muscles and superficial tissues of the thigh. These branches also supply the skin of the groin and the lower abdominal wall. Important subdivisions of the femoral artery include the following:

1. *Superficial circumflex iliac artery* to the lymph nodes and skin of the groin.
2. *Superficial epigastric artery* to the skin of the lower abdominal wall.
3. *Superficial* and *deep external pudendal arteries* to the skin of the lower abdomen and external genitalia.
4. *Deep femoral artery* (the largest branch of the femoral artery) to the hip joint and muscles of the thigh.
5. *Deep genicular artery* to distal ends of thigh muscles and to an anastomosis around the knee joint.

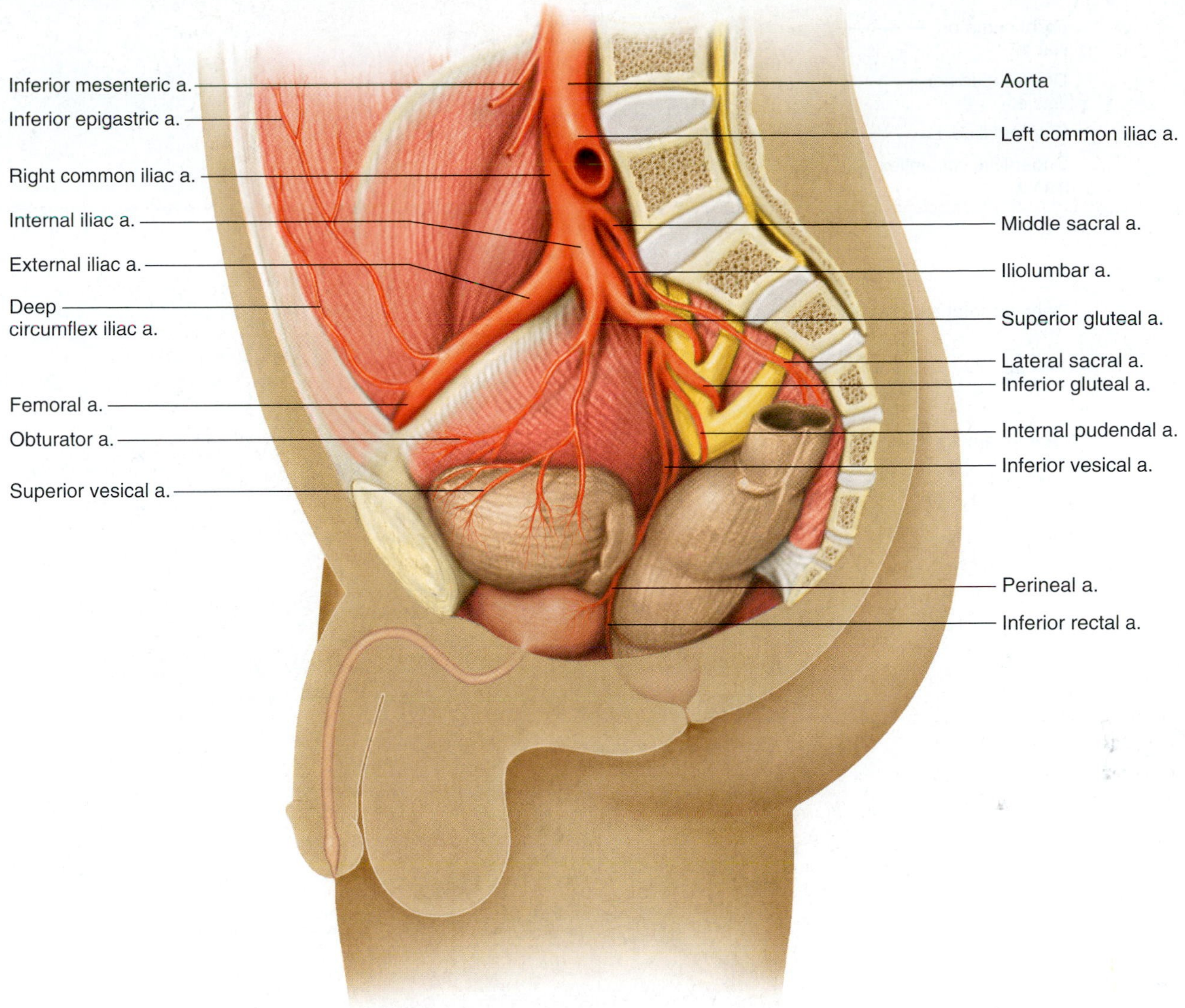

FIGURE 15.49 Arteries that supply the pelvic region. (*a.* stands for artery.)

As the femoral artery reaches the proximal border of the space behind the knee (popliteal fossa), it becomes the **popliteal** (pop″lĭ-te′al) **artery.** Branches of this artery supply blood to the knee joint and to certain muscles in the thigh and calf. Also, many of its branches join the anastomosis of the knee and help provide alternate pathways for blood in the case of arterial obstructions. At the lower border of the popliteal fossa, the popliteal artery divides into the anterior and posterior tibial arteries.

The **anterior tibial** (tib′e-al) **artery** passes downward between the tibia and the fibula, giving off branches to the skin and muscles in the anterior and lateral regions of the leg. It also communicates with the anastomosis of the knee and with a network of arteries around the ankle. This vessel continues into the foot as the *dorsalis pedis artery,* which supplies blood to the instep and toes.

The **posterior tibial artery,** the larger of the two popliteal branches, descends beneath the calf muscles, giving off branches to the skin, muscles, and other tissues of the leg along the way. Some of these vessels join the anastomoses of the knee and ankle. As it passes between the medial malleolus and the heel, the posterior tibial artery divides into the *medial* and *lateral plantar arteries.* Branches from these arteries supply blood to tissues of the heel, instep, and toes.

The largest branch of the posterior tibial artery is the *fibular artery,* which extends downward along the fibula and contributes to the anastomosis of the ankle. Figure 15.51 shows the major vessels of the arterial system.

PRACTICE

60 Name the portions of the aorta.

61 Name the vessels that arise from the aortic arch.

62 Name the branches of the thoracic and abdominal aorta.

63 Which vessels supply blood to the head? To the upper limb? To the abdominal wall? To the lower limb?

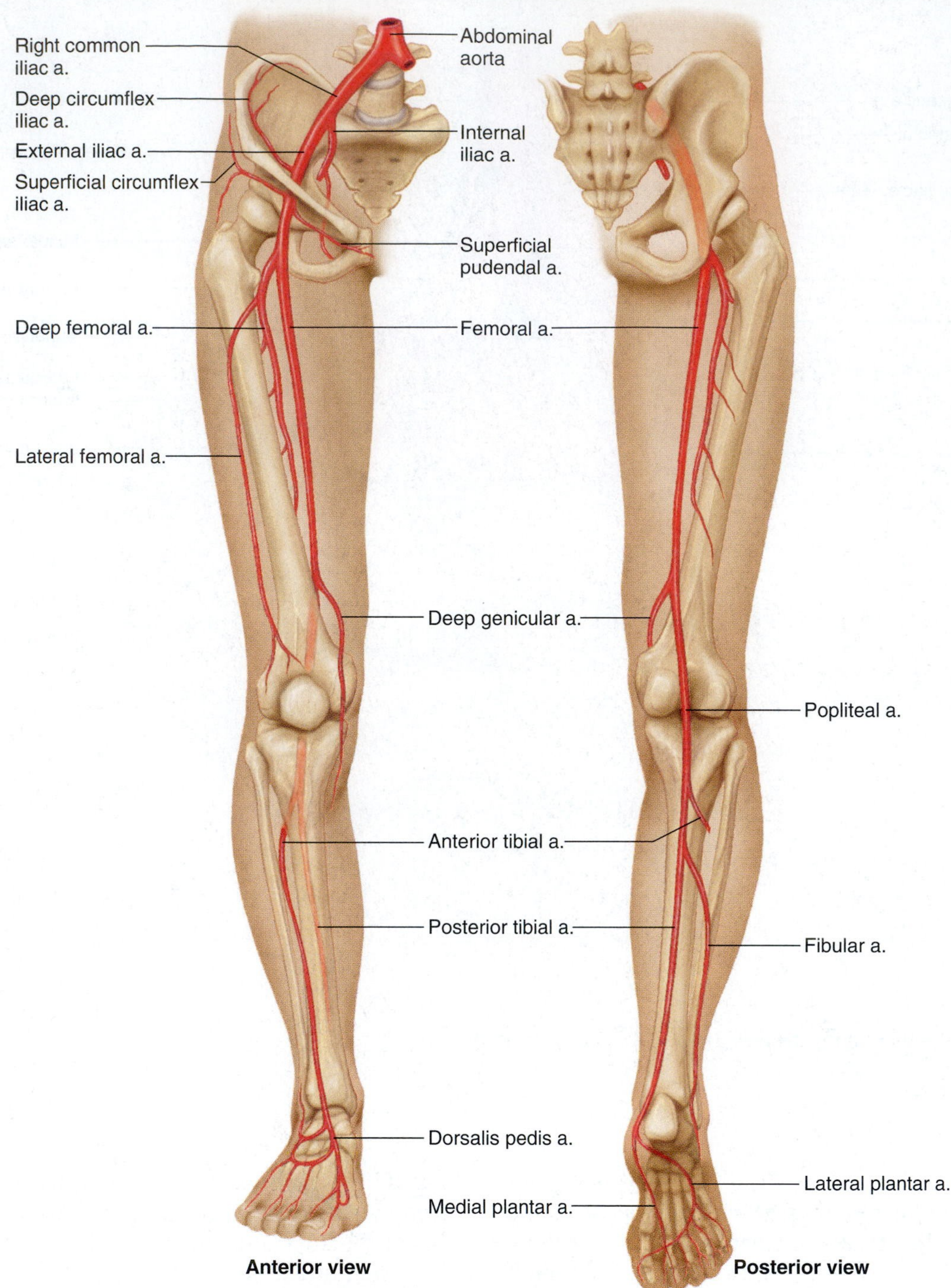

FIGURE 15.50 Major branches of the external iliac artery. (*a.* stands for artery.)

15.8 VENOUS SYSTEM

Venous circulation returns blood to the heart after gases, nutrients, and wastes are exchanged between the blood and body cells.

Characteristics of Venous Pathways

The vessels of the venous system originate with the merging of capillaries into venules, venules into small veins, and small veins into larger ones. Unlike the arterial pathways, those of the venous system are difficult to follow. This is because the vessels commonly connect in irregular networks, so many unnamed tributaries may join to form a large vein. On the other hand, larger veins typically parallel the courses of named arteries, and these veins often have the same names as their counterparts in the arterial system. For example, the renal vein parallels the renal artery, and the common iliac vein accompanies the common iliac artery.

The veins that carry the blood from the lungs and myocardium back to the heart have already been described. The

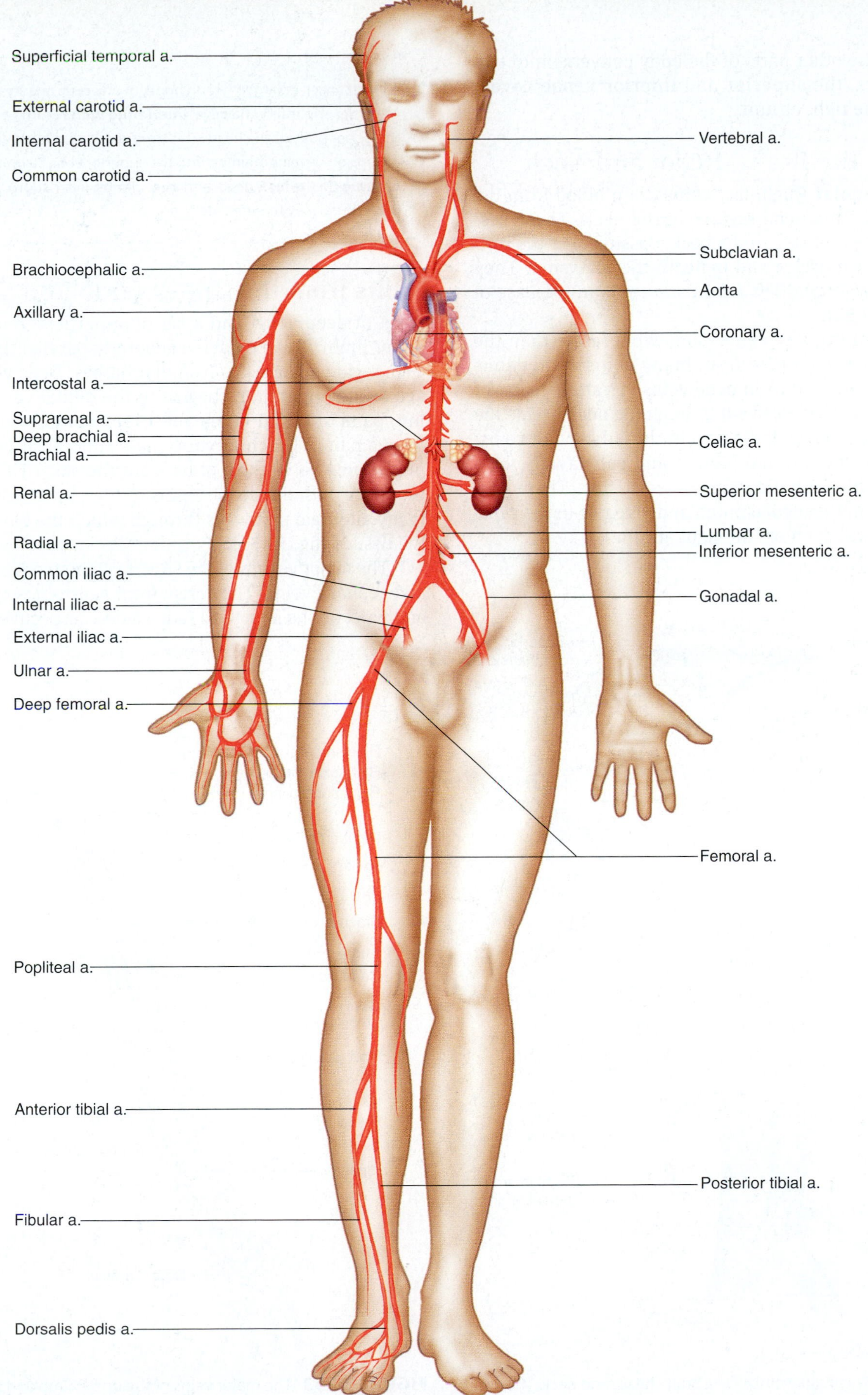

FIGURE 15.51 Major vessels of the arterial system. (*a.* stands for artery.)

veins from all the other parts of the body converge into two major pathways, the **superior** and **inferior venae cavae,** which lead to the right atrium.

Veins from the Brain, Head, and Neck

The **external jugular** (jug′u-lar) **veins** drain blood from the face, scalp, and superficial regions of the neck. These vessels descend on either side of the neck, passing over the sternocleidomastoid muscles and beneath the platysma. They empty into the *right* and *left subclavian veins* in the base of the neck (fig. 15.52).

The **internal jugular veins,** somewhat larger than the external jugular veins, arise from many veins and venous sinuses of the brain and from deep veins in various parts of the face and neck. They descend through the neck beside the common carotid arteries and also join the subclavian veins. These unions of the internal jugular and subclavian veins form large **brachiocephalic veins** on each side. These vessels then merge in the mediastinum and give rise to the *superior vena cava,* which enters the right atrium.

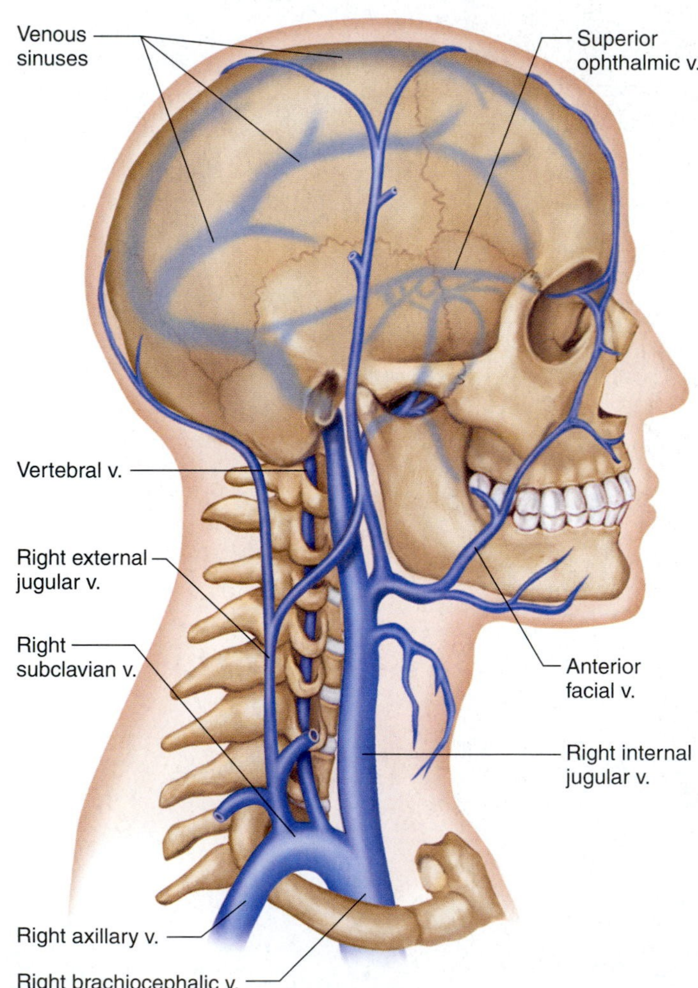

FIGURE 15.52 The major veins of the brain, head, and neck. The clavicle has been removed. (*v.* stands for vein.)

A lung cancer, enlarged lymph node, or an aortic aneurysm can compress the superior vena cava, interfering with return of blood from the upper body to the heart. This produces pain; shortness of breath; distension of veins draining into the superior vena cava; and swelling of tissues in the face, head, and lower limbs. Restriction of blood flow to the brain may threaten life.

Veins from the Upper Limb and Shoulder

A set of deep veins and a set of superficial ones drain the upper limb. The deep veins generally parallel the arteries in each region and are given similar names. Deep venous drainage of the upper limbs begins in the digital veins that drain into pairs of **radial veins** and **ulnar veins,** which merge to form a pair of **brachial veins.** The superficial veins connect in complex networks just beneath the skin. They also communicate with the deep vessels of the upper limb, providing many alternate pathways through which the blood can leave the tissues (fig. 15.53).

The major vessels of the superficial network are the basilic and cephalic veins. They arise from anastomoses in the palm and wrist on the ulnar and radial sides, respectively.

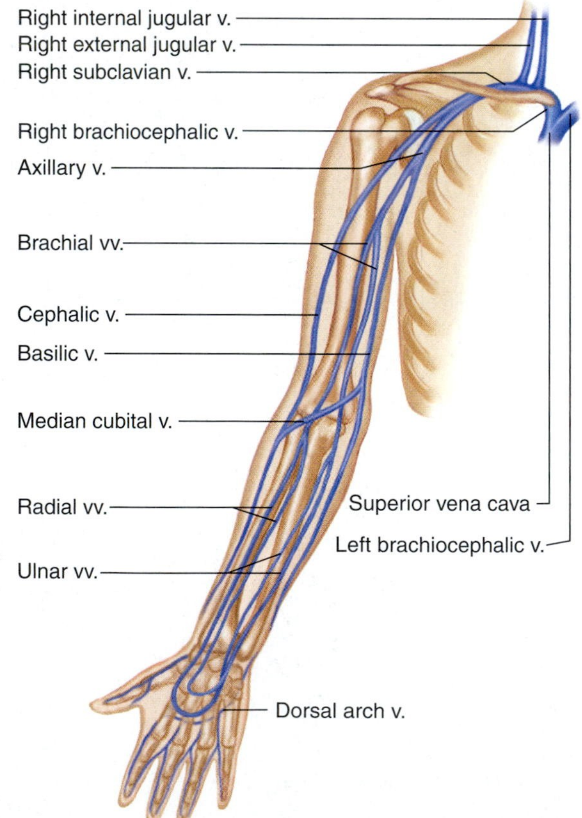

FIGURE 15.53 The major veins of the upper limb and shoulder. (*v.* stands for vein, *vv.* stands for veins.)

The **basilic** (bah-sil′ik) **vein** passes along the back of the forearm on the ulnar side for a distance and then curves forward to the anterior surface below the elbow. It continues ascending on the medial side until it reaches the middle of the arm. There it deeply penetrates the tissues and joins the *brachial vein.* As the basilic and brachial veins merge, they form the *axillary vein.*

The **cephalic** (sĕ-fal′ik) **vein** courses upward on the lateral side of the upper limb from the hand to the shoulder. In the shoulder, it pierces the tissues and joins the axillary vein, which beyond the axilla becomes the *subclavian vein.*

In the bend of the elbow, a *median cubital vein* ascends from the cephalic vein on the lateral side of the forearm to the basilic vein on the medial side. This large vein is usually visible. It is often used as a site for *venipuncture,* when it is necessary to remove a sample of blood for examination or to add fluids to the blood.

Veins from the Abdominal and Thoracic Walls

Tributaries of the brachiocephalic and azygos veins drain the abdominal and thoracic walls. For example, the *brachiocephalic vein* receives blood from the *internal thoracic vein,* which generally drains the tissues the internal thoracic artery supplies. Some *intercostal veins* also empty into the brachiocephalic vein (fig. 15.54).

The **azygos** (az′ĭ-gos) **vein** originates in the dorsal abdominal wall and ascends through the mediastinum on the right side of the vertebral column to join the superior vena cava. It drains most of the muscular tissue in the abdominal and thoracic walls.

Tributaries of the azygos vein include the *posterior intercostal veins* on the right side, which drain the intercostal spaces, and the *superior* and *inferior hemiazygos veins,* which receive blood from the posterior intercostal veins on the left. The right and left *ascending lumbar veins,* with tributaries that include vessels from the lumbar and sacral regions, also connect to the azygos system.

Veins from the Abdominal Viscera

Veins carry blood directly to the atria of the heart, except those that drain the abdominal viscera (fig. 15.55). They originate in the capillary networks of the stomach, intestines, pancreas, and spleen and carry blood from these organs through a **hepatic portal** (por′tal) **vein** to the liver (fig. 15.56). There the blood enters capillary-like **hepatic sinusoids** (hĕ-pat′ik si′nŭ-soidz). This unique venous pathway is called the **hepatic portal system.**

The tributaries of the hepatic portal vein include the following vessels:

1. Right and left *gastric veins* from the stomach.
2. *Superior mesenteric vein* from the small intestine, ascending colon, and transverse colon.

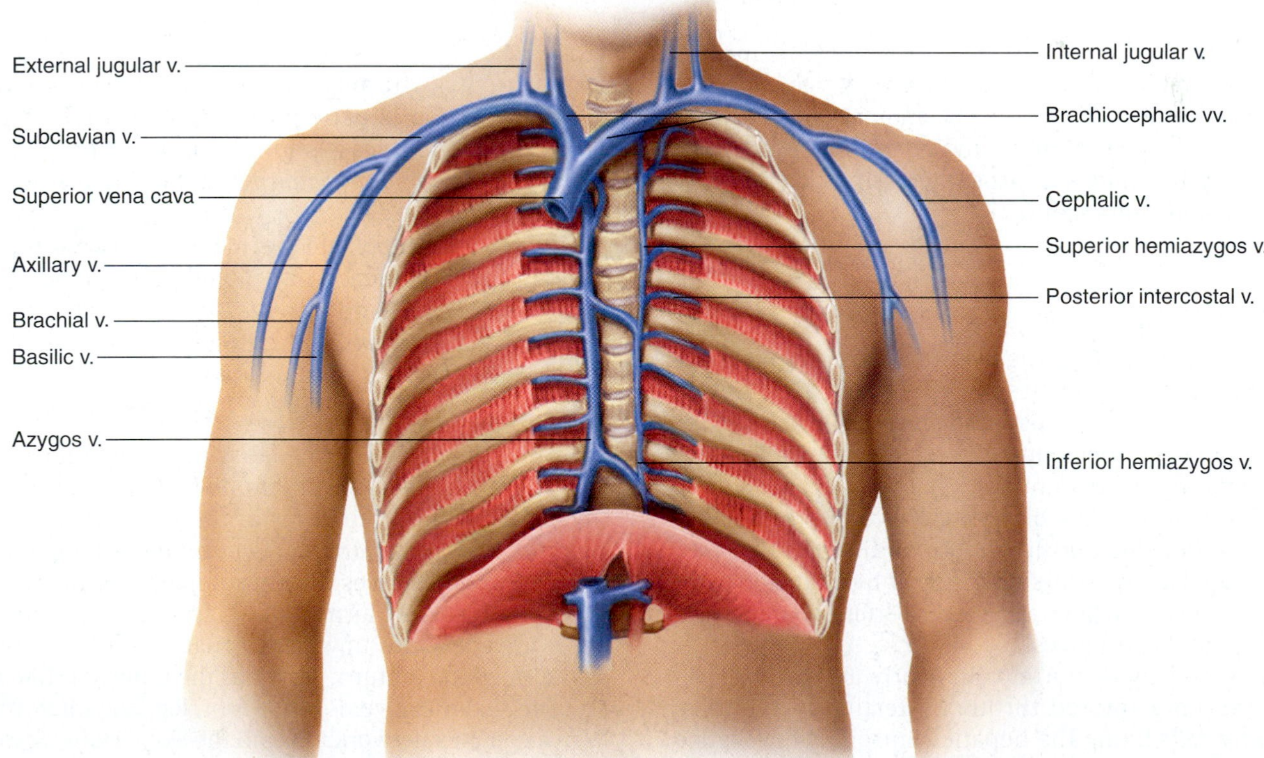

FIGURE 15.54 Veins that drain the thoracic wall. (*v.* stands for vein, *vv.* stands for veins.)

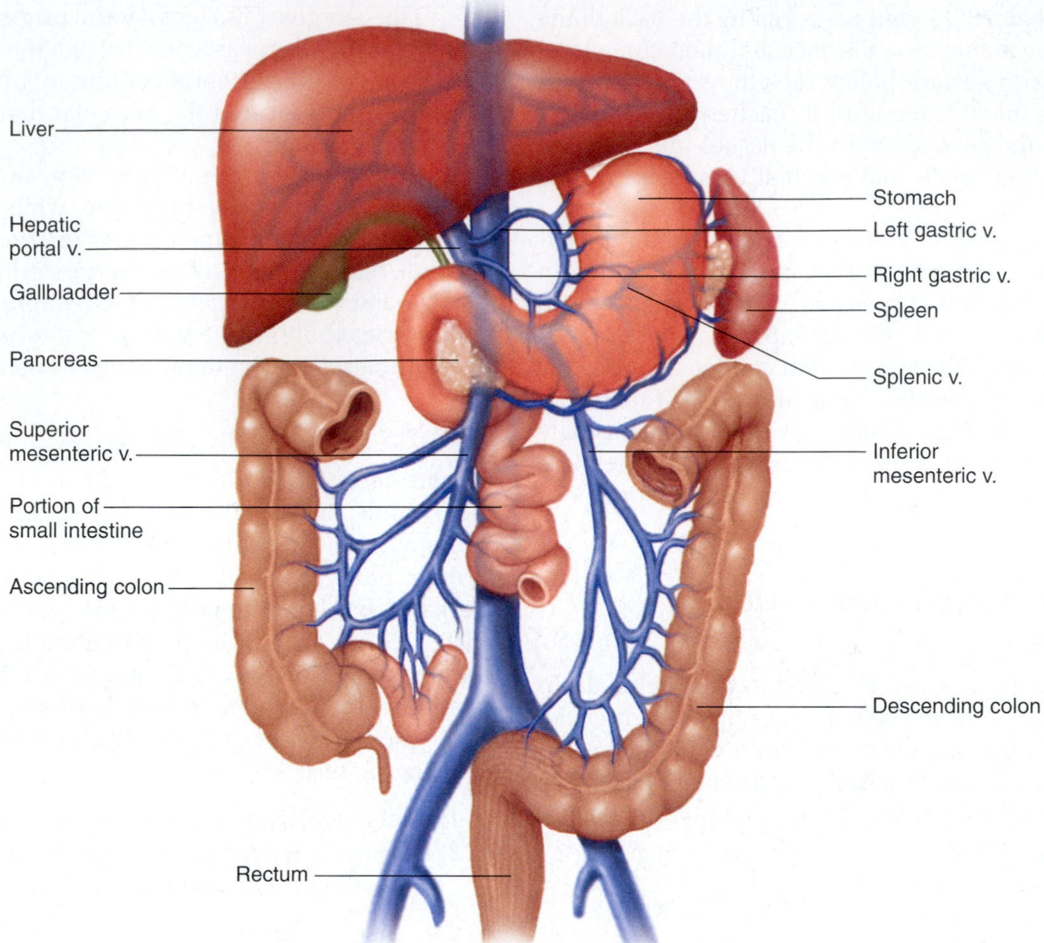

FIGURE 15.55 Veins that drain the abdominal viscera. (*v.* stands for vein.)

3. *Splenic vein* from a convergence of several veins draining the spleen, the pancreas, and a portion of the stomach. Its largest tributary, the *inferior mesenteric vein,* brings blood upward from the descending colon, sigmoid colon, and rectum.

About 80% of the blood flowing to the liver in the hepatic portal system comes from the capillaries in the stomach and intestines and is oxygen-poor, but nutrient-rich. As discussed in chapter 17 (pp. 674–675), the liver handles these nutrients in a variety of ways. It regulates blood glucose concentration by polymerizing excess glucose into glycogen for storage or by breaking down glycogen into glucose when blood glucose concentration drops below normal.

The liver helps regulate blood concentrations of recently absorbed amino acids and lipids by modifying them into forms cells can use, by oxidizing them, or by changing them into storage forms. The liver also stores certain vitamins and detoxifies harmful substances.

Blood in the hepatic portal vein nearly always contains bacteria that have entered through intestinal capillaries. Large *Kupffer cells* lining the hepatic sinusoids phagocytize these microorganisms, removing them from the portal blood before it leaves the liver.

After passing through the hepatic sinusoids of the liver, the blood in the hepatic portal system travels through a series of merging vessels into **hepatic veins.** These veins empty into the *inferior vena cava,* returning the blood to the general circulation.

Other veins empty into the inferior vena cava as it ascends through the abdomen. They include the *lumbar, gonadal, renal, suprarenal,* and *phrenic veins.* These vessels drain regions that arteries with corresponding names supply.

Veins from the Lower Limb and Pelvis

As in the upper limb, veins that drain the blood from the lower limb can be divided into deep and superficial groups (fig. 15.57). The deep veins of the leg, such as the paired *anterior* and *posterior tibial veins,* have names that correspond to the arteries they accompany. At the level of the knee, these vessels form a single trunk, the **popliteal vein.** This vein continues upward through the thigh as the **femoral vein,** which, in turn, becomes the **external iliac vein.**

The superficial veins of the foot, leg, and thigh connect to form a complex network beneath the skin. These vessels drain into two major trunks: the small and great saphenous veins.

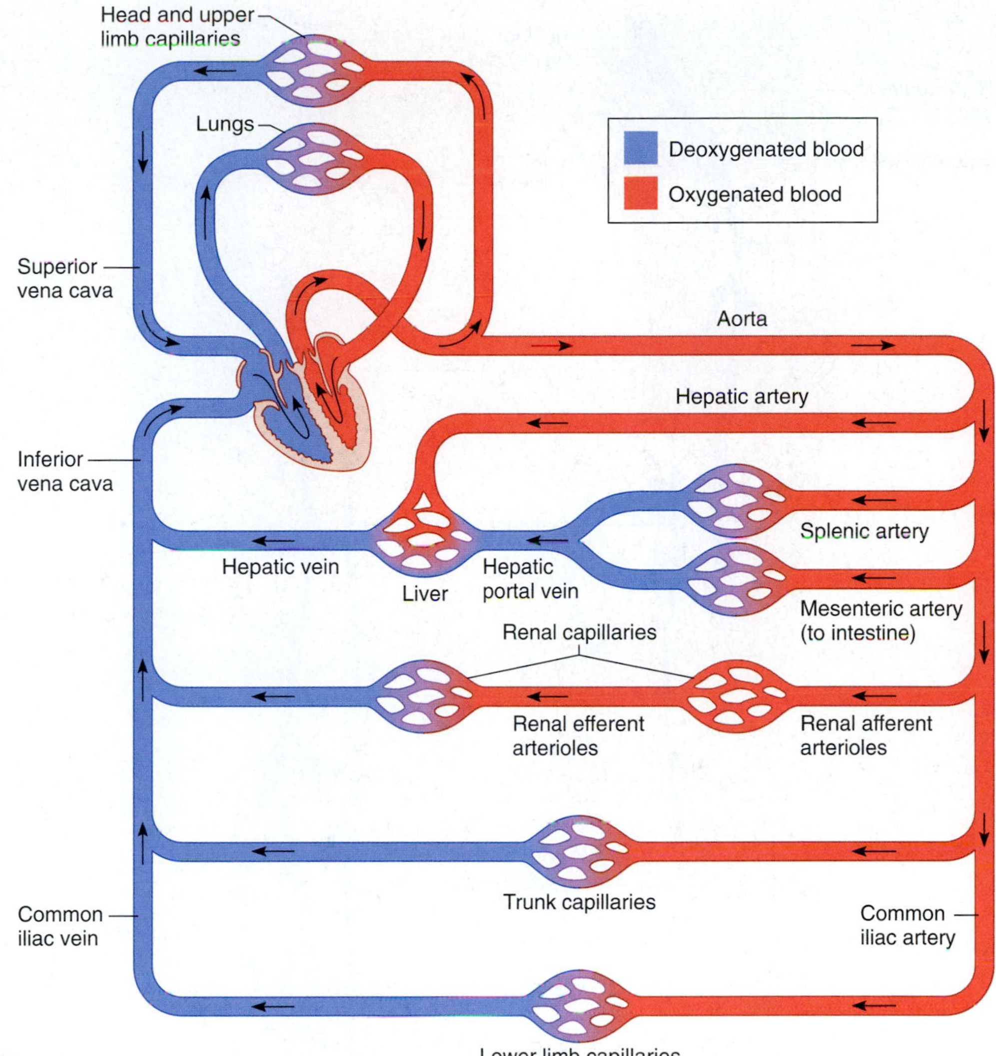

FIGURE 15.56 In this schematic drawing of the cardiovascular system, note how the hepatic portal vein drains one set of capillaries and leads to another set. A similar relationship exists in the kidneys.

The **small saphenous** (sah-fe′nus) **vein** begins in the lateral portion of the foot and passes upward behind the lateral malleolus. It ascends along the back of the calf, enters the popliteal fossa, and joins the popliteal vein.

The **great saphenous vein,** the longest vein in the body, originates on the medial side of the foot. It ascends in front of the medial malleolus and extends upward along the medial side of the leg and thigh. In the thigh just below the inguinal ligament, it deeply penetrates and joins the femoral vein. Near its termination, the great saphenous vein receives tributaries from a number of vessels that drain the upper thigh, groin, and lower abdominal wall.

In addition to communicating freely with each other, the saphenous veins communicate extensively with the deep veins of the leg and thigh. Blood can thus return to the heart from the lower extremities by several routes.

In the pelvic region, vessels leading to the **internal iliac vein** carry blood away from organs of the reproductive, urinary, and digestive systems. This vein is formed by tributaries corresponding to the branches of the internal iliac artery, such as the *gluteal, pudendal, vesical, rectal, uterine,* and *vaginal veins.* Typically, these veins have many connections and form complex networks (plexuses) in the regions of the rectum, urinary bladder, and prostate gland (in the male) or uterus and vagina (in the female).

The internal iliac veins originate deep within the pelvis and ascend to the pelvic brim. There they unite with the right and left external iliac veins to form the **common iliac veins.** These vessels, in turn, merge to produce the *inferior vena cava* at the level of the fifth lumbar vertebra. Figure 15.58 shows the major vessels of the venous system.

Clinical Application 15.7 looks at molecular explanations of certain cardiovascular disorders. Clinical Application 15.8 discusses coronary artery disease.

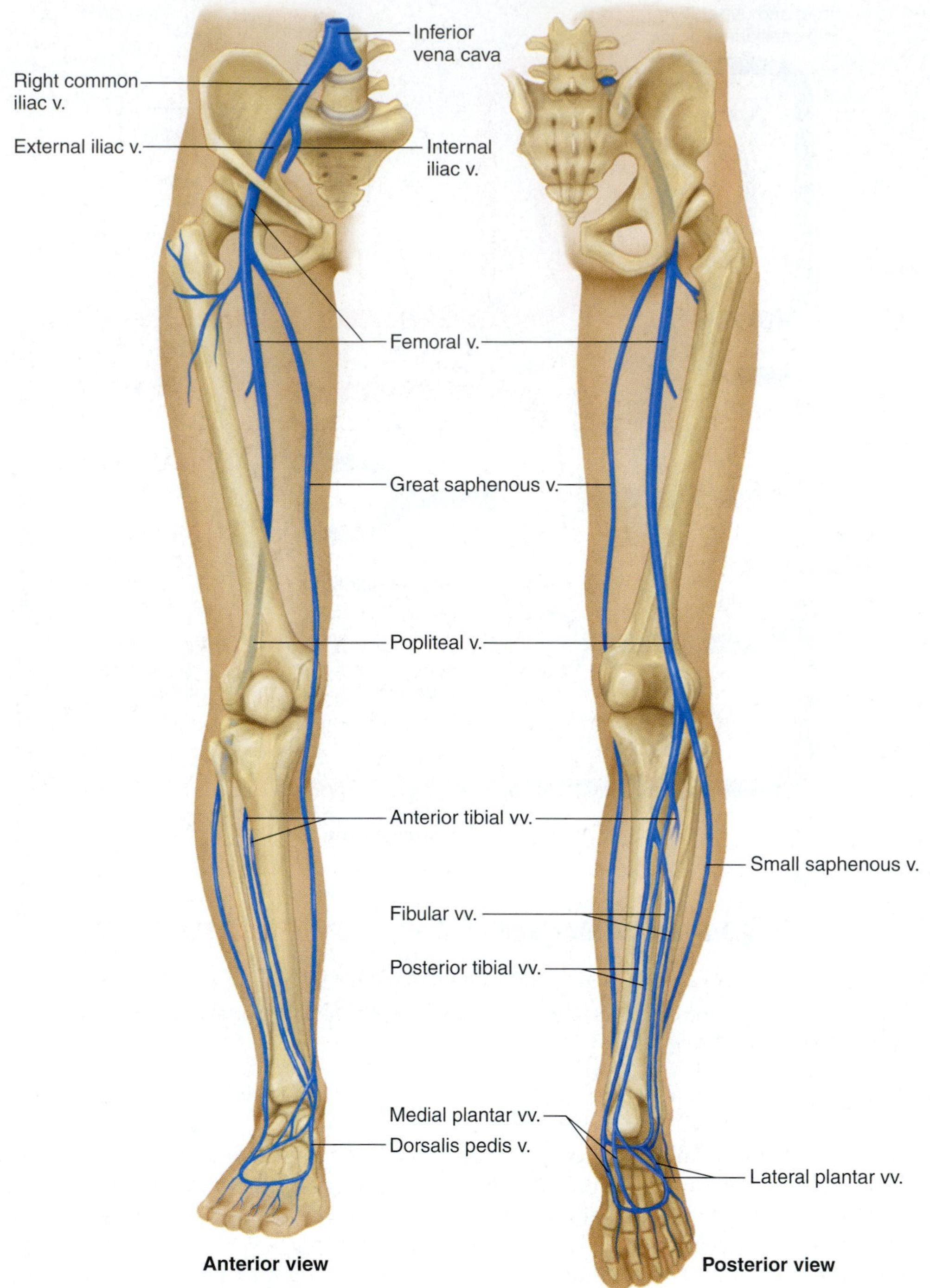

FIGURE 15.57 The major veins of the lower limb and pelvis. (*v.* stands for vein, *vv.* stands for veins.)

PRACTICE

64 Name the veins that return blood to the right atrium.

65 Which major veins drain blood from the head? From the upper limbs? From the abdominal viscera? From the lower limbs?

15.9 LIFE-SPAN CHANGES

The years take a toll on the cardiovascular system. Signs of cardiovascular disease may appear long before symptoms arise. Autopsies of soldiers killed in the Korean and Vietnam Wars, for example, revealed significant plaque buildup in the arterial walls of otherwise healthy young men. Incidence

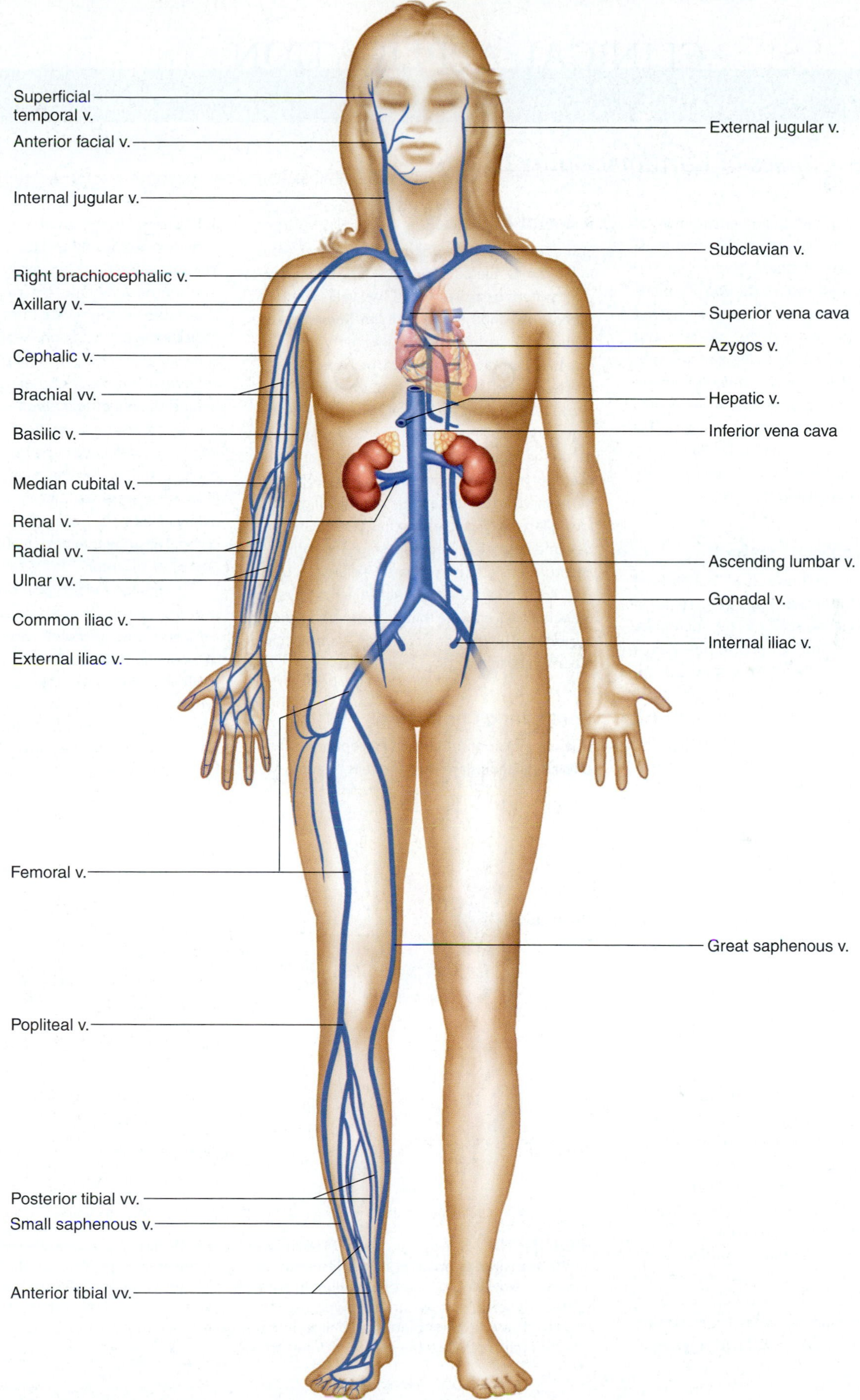

FIGURE 15.58 Major vessels of the venous system. (*v.* stands for vein, *vv.* stands for veins.)

15.7 CLINICAL APPLICATION

Molecular Causes of Cardiovascular Disease

A variety of inherited and environmental factors contribute to causing cardiovascular disease, including poor diet and sedentary lifestyle, against a backdrop of genetic predisposition. Disorders of the heart and blood vessels caused by single genes are rare, but understanding how they arise can provide insights useful in developing treatments for more prevalent forms of disease. For example, widely used cholesterol-lowering drugs called statins were developed based on understanding familial hypercholesterolemia, an inherited condition that affects one-in-a-million children.

A Connective Tissue Defect

Just after midnight on June 8, 2004, Florida State University basketball player Ronalda Pierce lay in bed watching TV. At about 2 A.M., her roommate heard sounds of disturbed breathing—Pierce could not be roused. The nineteen-year-old athlete died within an hour of reaching the hospital. Her aorta had burst. Pierce had *Marfan syndrome,* an inherited condition that also caused the characteristics that led her to excel in her sport—her great height and long fingers. In Marfan syndrome, an abnormal form of a connective tissue protein called fibrillin weakens the aorta wall, dilating the aortic root (figure 15K).

Ronalda Pierce had a strong family history of heart disease, but Marfan syndrome had never been diagnosed. Identifying the mutation can make it possible to locate and surgically repair a weakened aorta, or take a drug that can slow the effect of the disease on the aorta. This can prevent a tragic first symptom—sudden death.

A Myosin Defect

Each year, one or two seemingly healthy young people die suddenly during a sports event, usually basketball. The cause of death is often *familial hypertrophic cardiomyopathy,* an inherited overgrowth of the heart muscle. The defect in this disorder is different from that behind Marfan syndrome. It is an abnormality in one of the myosin chains that comprise cardiac muscle. Again, detecting the responsible gene can alert affected individuals to their increased risk of sudden death. They can adjust the type of exercise they do to avoid stressing the cardiovascular system.

A Metabolic Block

Sometimes inherited heart disease strikes early in life. Jim D. died at four days of age, two days after suffering cardiac arrest. Two years later, his parents had another son. Like Jim, Kerry seemed normal at birth, but when he was thirty-six hours old, his heart rate plummeted, he had a seizure, and he stopped breathing. He was resuscitated. A blood test revealed excess long-chain fatty acids, indicating inability to use fatty acids. Hunger triggered the symptoms because the boys could not use fatty acids for energy, as healthy people do. Kerry survived for three years by following a diet low in fatty acids and eating frequently. Once he became comatose because he missed a meal. Eventually, he died of respiratory failure.

Jim and Kerry had inherited a deficiency of a mitochondrial enzyme that processes long-chain fatty acids. This is a primary energy source for cardiac muscle, so their tiny hearts failed.

Controlling Cholesterol

Low-density lipoprotein (LDL) receptors on liver cells admit cholesterol into the cells, keeping the lipid from building up in the bloodstream and occluding arteries. When LDL receptors bind cholesterol, they activate a negative feedback system that temporarily halts the cell's production of cholesterol. In the severe form of *familial hypercholesterolemia,* a person inherits two defective copies of the gene encoding the LDL receptors. Yellowish lumps of cholesterol can be seen behind the knees and elbows, and heart failure usually causes death in childhood. People who inherit one defective gene have a milder form of the illness. They develop coronary artery disease in young adulthood, but can delay symptoms by following a heart-healthy diet and regularly exercising. These people have half the normal number of LDL receptors.

In Niemann-Pick type C disease, a defective protein disturbs the fate of cholesterol inside cells. Normally, the protein escorts cholesterol out of a cell's lysosomes, which triggers the negative feedback mechanism that shuts off cholesterol synthesis. When the protein is absent or malfunctions, the cell keeps producing cholesterol and LDL receptors. Coronary artery disease develops, and is typically fatal in childhood. ■

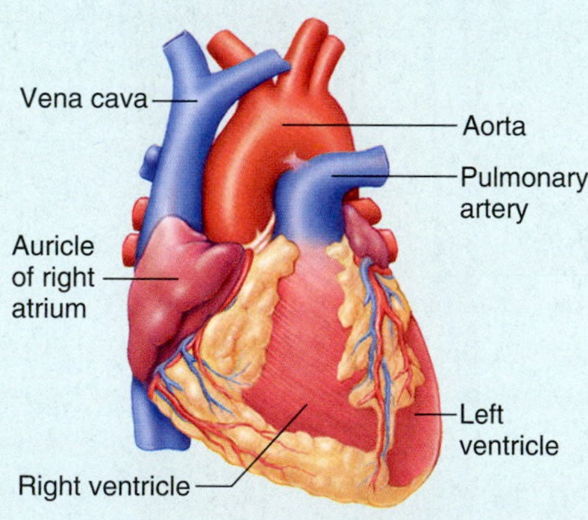

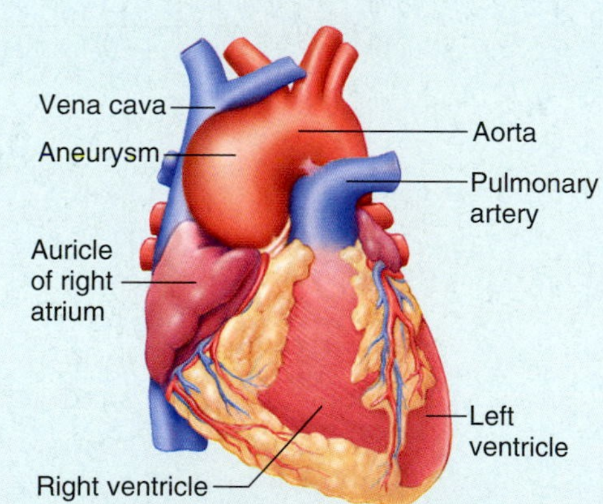

FIGURE 15K The symptoms of Marfan syndrome, including a progressive dilation of the aortic root, arise from an abnormal form of the connective tissue protein fibrillin that in turn increases signaling by a protein called transforming growth factor B (TGF-B). A drug usually used to treat hypertension, losartan, lowers levels of TGF-B and has been shown in preliminary clinical trials to slow the life-threatening ballooning out of the ascending aorta. Surgery can replace the affected part of the aorta with a synthetic graft.

15.8 CLINICAL APPLICATION

Coronary Artery Disease

Dave R., a fifty-two-year-old overweight accountant, had been having occasional chest pains for several months. The mild pain occurred during his usual weekend tennis match, and he attributed it to indigestion. The discomfort almost always diminished after the game, but recently, the pain seemed more severe and prolonged. Dave asked his physician about the problem.

The physician explained that Dave was probably experiencing *angina pectoris,* a symptom of *coronary artery disease* (CAD), and suggested that he undergo an *exercise stress test.* Dave walked on a treadmill, increasing speed and incline while he exercised. An ECG was recorded and his blood pressure monitored. Near the end of the test, when Dave's heart reached the desired rate, a small amount of radioactive thallium-201 was injected into a vein. A *scintillation counter* scanned Dave's heart to determine if branches of his coronary arteries carried the blood marked with the thallium uniformly throughout the myocardium (see fig. 15.14).

The test revealed that Dave was developing CAD. In addition, he had hypertension and high serum cholesterol. The physician advised Dave to stop smoking; to reduce his intake of foods high in saturated fats, cholesterol, refined carbohydrates, and sodium; and to exercise regularly. He was given medications to lower his blood pressure and to relieve the pain of angina. The doctor also cautioned Dave to avoid stressful situations and to lose weight.

Six months later, despite following medical advice, Dave suffered a heart attack—a sign that blood flow to part of his myocardium had been obstructed, producing oxygen deficiency. The attack began as severe, crushing chest pain, shortness of breath, and sweating. Paramedics stabilized Dave's condition and transported him to a hospital. There, a cardiologist concluded from an ECG that Dave's heart attack was caused by a blood clot obstructing a coronary artery (occlusive coronary thrombosis). The cardiologist intravenously administered a thrombolytic ("clot-busting") drug.

A repeat ECG a few hours later showed that the blood vessel remained partially obstructed, so the cardiologist ordered a *coronary angiogram.* In this X-ray procedure, conducted in a cardiac catheterization laboratory, a thin plastic catheter was passed through a guiding sheath inserted into the femoral artery of Dave's right inguinal area. From there, the catheter was pushed into the aorta until it reached the region of the opening to the left coronary artery, and then near the opening to the right coronary artery.

X-ray fluoroscopy monitored the progress of the catheter. Each time the catheter was in proper position, a radiopaque dye (contrast medium) was released from its distal end into the blood. X-ray images that revealed the path of the dye as it entered a coronary artery and its branches were recorded on videotape and on motion-picture film, later analyzed frame by frame. A single severe narrowing was discovered near the origin of Dave's left anterior descending artery. The cardiologist decided to perform *percutaneous transluminal coronary angioplasty* (PTCA) to enlarge the opening (lumen) of that vessel.

The PTCA was performed by passing another plastic catheter through the guiding sheath used for the angiogram. This second tube had a tiny deflated balloon at its tip. The balloon was placed in the region of the arterial narrowing and inflated for a short time under high pressure. The inflating balloon compressed the atherosclerotic plaque (atheroma) obstructing the arterial wall and stretched the blood vessel wall, widening its lumen (recanalization). Blood flow to the myocardial tissue downstream from the obstruction immediately improved.

About 50% of the time, a vessel opened with PTCA becomes occluded again, because the underlying disease persists. To prevent this restenosis, the doctor inserted a *coronary stent,* which is an expandable tube or coil that holds the vessel wall open. The cardiologist had two other options that have a slightly higher risk of causing damage. She might have vaporized the plaque obstructing the vessel with an excimer laser pulse delivered along optical fibers threaded through the catheter. Or, she could have performed atherectomy, in which a cutting device attached to the balloon inserted into the catheter spins, removing plaque by withdrawing it on the catheter tip.

Should the coronary stent fail, or an obstruction block another heart vessel, Dave might benefit from *coronary bypass surgery.* A portion of his internal mammary artery inside his chest wall or his great saphenous vein would be removed and sutured (with the vein reversed to allow blood flow through the valves) between the aorta and the blocked coronary artery at a point beyond the obstruction, restoring circulation through the heart. ■

of disease of the heart and blood vessels increases exponentially with age. About 60% of men over age sixty have at least one narrowed coronary artery; the same is true for women over age eighty.

Assessing cardiac output over a lifetime vividly illustrates how cardiovascular disease prevalence can interfere with studying the changes associated with normal aging. Recall that cardiac output is the ability of the heart to meet the body's oxygen requirements and is calculated as the heart rate in beats per minute multiplied by the stroke volume in milliliters per beat. For many years, studies indicated that cardiac output declines with age, but when researchers began to screen participants for hidden heart disease with treadmill stress tests, then evaluated only individuals with completely healthy cardiovascular systems, they discovered that cardiac output at rest is maintained as a person ages. It does decline during exercise for some people, however.

The heart may normally shrink slightly with age, but disease may enlarge it. The proportion of the heart that is cardiac muscle declines with age, even in a healthy person, because cardiac muscle cells do not divide. Lipofuscin pigments become especially prominent in these cells. Fibrous connective tissue and adipose tissue fill in the spaces left by the waning population of cardiac muscle cells, thickening the endocardium. Adipose cells may also accumulate in the ventricle walls and the septum between them. The left ventricular wall may be up to 25% thicker at age eighty than it was at age thirty.

INNERCONNECTIONS | *Cardiovascular System*

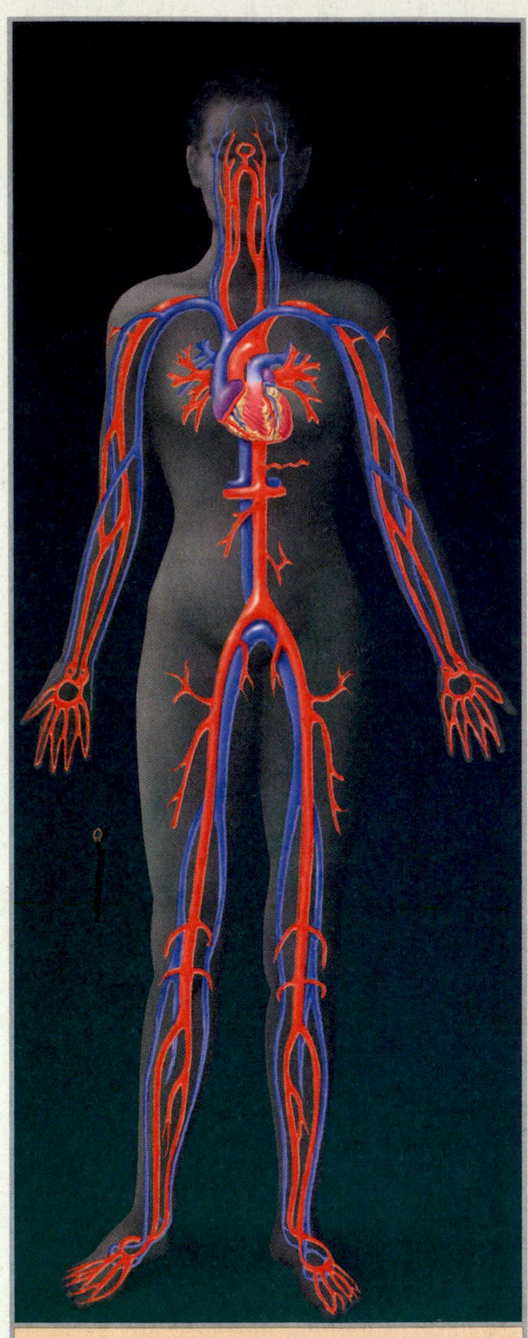

Cardiovascular System

The heart pumps blood through as many as 60,000 miles of blood vessels, delivering nutrients to, and removing wastes from, all body cells.

Integumentary System

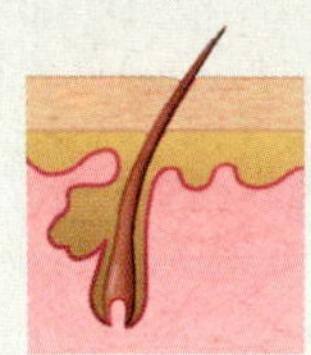

Changes in skin blood flow are important in temperature control.

Lymphatic System

The lymphatic system returns tissue fluids to the bloodstream.

Skeletal System

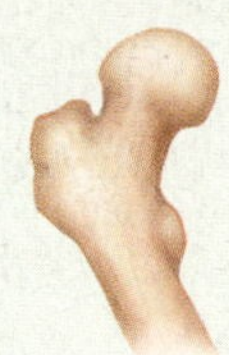

Bones help control plasma levels of calcium ions, which influence heart action.

Digestive System

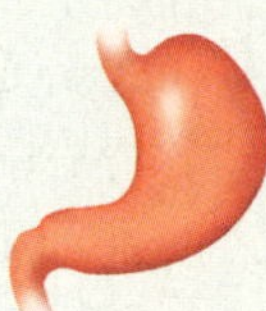

The digestive system breaks down nutrients into forms readily absorbed by the bloodstream.

Muscular System

Blood flow increases to exercising skeletal muscle, delivering oxygen and nutrients and removing wastes. Muscle actions help the blood circulate.

Respiratory System

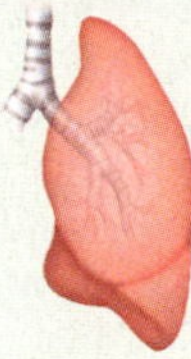

The respiratory system oxygenates the blood and removes carbon dioxide. Respiratory movements help the blood circulate.

Nervous System

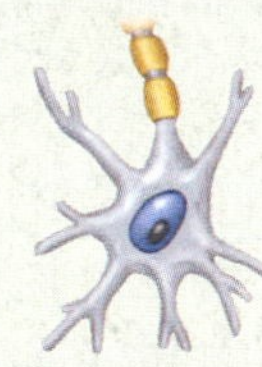

The brain depends on blood flow for survival. The nervous system helps control blood flow and blood pressure.

Urinary System

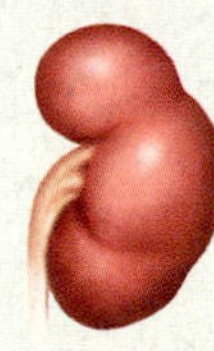

The kidneys clear the blood of wastes and substances present in the body. The kidneys help control blood pressure and blood volume.

Endocrine System

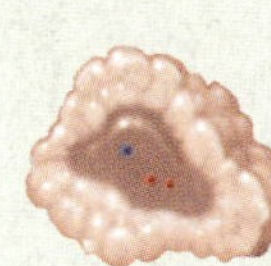

Hormones are carried in the bloodstream. Some hormones directly affect the heart and blood vessels.

Reproductive System

Blood pressure is important in normal function of the sex organs.

The heart slows down, ever so slightly, with age. A study of 5,000 healthy people aged forty-five to eighty-four showed that the cardiac cycle lengthens by 2% to 5% per year. Past studies that showed a slight increase in ejection fraction per year were flawed because this measure represents the ratio of blood volume pumped with each heartbeat to total blood volume, and it is the end-diastolic volume that diminishes over time, due to shrinking cardiac muscle and thickening heart walls. In actuality, the aging heart pumps about 8 milliliters less per year.

The heart valves thicken and become more rigid after age sixty, changes that may begin as early as the third decade. The valves may calcify.

Just as the heart need not falter with age, the cardiac conduction system may remain functional despite change. The sinoatrial and atrioventricular nodes and the atrioventricular bundle become more elastic. However, these changes may alter the ECG pattern.

Systolic blood pressure increases with age; a blood pressure reading of 140/90 is not abnormal in an older person. In about 40% of the elderly, the systolic pressure exceeds 160. The increase may be due to the decreasing diameters and elasticity of arteries. Regular exercise can slow these changes. Resting heart rate declines from 145 or more beats per minute in a fetus to 140 beats per minute in a newborn, then levels out in an adult to about 70 (range of 60–99) beats per minute.

In the vascular system, age-related changes are most apparent in the arteries. The tunica interna thickens. Dividing smooth muscle cells in the tunica media may push up the endothelium in places, and over time, the lumens of the larger arteries narrow. Rigidity increases as collagen, calcium, and fat are deposited as elastin production declines. Arterial elasticity at age seventy is only about half of what it was at age twenty. The arterioles have diminished ability to contract in response to cold temperatures and to dilate in response to heat, contributing to the loss of temperature control common among the elderly. The extent of change in arteries may reflect stress—that is, not all arteries "age" at the same rate.

Veins may accumulate collagen and calcify but, in general, do not change as much with age as do arteries. Thickened patches may appear in the inner layer, and fibers in the valves, but venous diameters are large enough that these changes have little impact on function. The venous supply to many areas is so redundant that alternate vessels can often take over for damaged ones.

The once-sleek endothelium changes as the cells become less uniform in size and shape. The endothelial inner linings of blood vessels are important to health because these cells release nitric oxide, which signals the vessels to dilate to increase blood flow, which counters atherosclerosis and thrombosis. In addition to the changes in arteries and veins, the number of capillaries declines with age.

Exercise can help maintain a "young" vascular system. One study compared the vascular endothelial linings of athletic and sedentary individuals of various ages and found that the status of the vessels of the exercising elderly were very similar to those of either athletic or sedentary people in their twenties. Many studies have correlated regular exercise to lowered heart disease risk in older people.

Overall, aging-related changes affect many components of the cardiovascular system. But in the absence of disease, the system is so fine-tuned and redundant that effective oxygen delivery can continue well into the later decades of life.

PRACTICE

66 Explain why the heart may enlarge with age.

67 Describe what happens to resting heart rate with age.

CHAPTER SUMMARY

15.1 INTRODUCTION (PAGE 553)

The cardiovascular system is composed of the heart and blood vessels, which circulate blood to supply oxygen to, and remove wastes from, body cells.

15.2 STRUCTURE OF THE HEART (PAGE 553)

1. Size and location of the heart
    a. The heart is about 14 centimeters long and 9 centimeters wide.
    b. It is located in the mediastinum and rests on the diaphragm.
2. Coverings of the heart
    a. A layered pericardium encloses the heart.
    b. The pericardial cavity is a space between the visceral and parietal layers of the pericardium.
3. Wall of the heart
    a. The wall of the heart has three layers.
    b. These layers include an epicardium, a myocardium, and an endocardium.
4. Heart chambers and valves
    a. The heart is divided into four chambers—two atria and two ventricles—that communicate through atrioventricular orifices on each side.
    b. Right chambers and valves
        (1) The right atrium receives blood from the venae cavae and coronary sinus.
        (2) The tricuspid valve guards the right atrioventricular orifice.
        (3) The right ventricle pumps blood into the pulmonary trunk.
        (4) A pulmonary valve guards the base of the pulmonary trunk.
    c. Left chambers and valves
        (1) The left atrium receives blood from the pulmonary veins.
        (2) The mitral valve guards the left atrioventricular orifice.
        (3) The left ventricle pumps blood into the aorta.
        (4) An aortic valve guards the base of the aorta.

5. Skeleton of the heart
   a. The skeleton of the heart consists of fibrous rings that enclose the bases of the pulmonary artery, aorta, and atrioventricular orifices.
   b. The fibrous rings provide attachments for valves and muscle fibers and prevent the orifices from excessively dilating during ventricular contractions.
6. Path of blood through the heart
   a. Blood low in oxygen and high in carbon dioxide enters the right side of the heart from the venae cavae and coronary sinus and then is pumped into the pulmonary circulation.
   b. After the blood is oxygenated in the lungs and some of its carbon dioxide is removed, it returns to the left side of the heart through the pulmonary veins.
   c. From the left ventricle, it moves into the aorta.
7. Blood supply to the heart
   a. The coronary arteries supply blood to the myocardium.
   b. It is returned to the right atrium through the cardiac veins and coronary sinus.

15.3 HEART ACTIONS (PAGE 564)

1. Cardiac cycle
   a. The atria contract (atrial systole) while the ventricles relax (ventricular diastole); the ventricles contract (ventricular systole) while the atria relax (atrial diastole).
   b. Pressure in the chambers rises and falls in cycles.
2. Heart sounds
   a. Heart sounds can be described as *lubb-dupp*.
   b. Heart sounds are due to the vibrations that the valve movements produce.
   c. The first part of the sound occurs as AV valves close, and the second part is associated with the closing of pulmonary and aortic valves.
3. Cardiac muscle fibers
   a. Cardiac muscle fibers connect to form a functional syncytium.
   b. If any part of the syncytium is stimulated, the whole structure contracts as a unit.
   c. Except for a small region in the floor of the right atrium, the fibrous skeleton separates the atrial syncytium from the ventricular syncytium.
4. Cardiac conduction system
   a. This system, composed of specialized cardiac muscle tissue, initiates and conducts depolarization waves through the myocardium.
   b. Impulses from the SA node pass slowly to the AV node; impulses travel rapidly along the AV bundle and Purkinje fibers.
   c. Muscle fibers in the ventricular walls form whorls that squeeze blood out of the contracting ventricles.
5. Electrocardiogram
   a. An electrocardiogram (ECG) records electrical changes in the myocardium during a cardiac cycle.
   b. The pattern contains several waves.
      (1) The P wave represents atrial depolarization.
      (2) The QRS complex represents ventricular depolarization.
      (3) The T wave represents ventricular repolarization.
6. Regulation of the cardiac cycle
   a. Physical exercise, body temperature, and concentration of various ions affect heartbeat.
   b. Branches of sympathetic and parasympathetic nerve fibers innervate the SA and AV nodes.
      (1) Parasympathetic impulses decrease heart action; sympathetic impulses increase heart action.
      (2) The cardiac center in the medulla oblongata regulates autonomic impulses to the heart.

15.4 BLOOD VESSELS (PAGE 573)

The blood vessels form a closed circuit of tubes that transports blood between the heart and body cells. The tubes include arteries, arterioles, capillaries, venules, and veins.

1. Arteries and arterioles
   a. The arteries are adapted to carry blood under relatively high pressure away from the heart.
   b. The arterioles are branches of arteries.
   c. The walls of arteries and arterioles consist of layers of endothelium, smooth muscle, and connective tissue.
   d. Autonomic fibers innervate smooth muscles in vessel walls.
2. Capillaries

Capillaries connect arterioles and venules. The capillary wall is a single layer of cells that forms a semipermeable membrane.

   a. Capillary permeability
      (1) Openings in the capillary walls are thin slits between endothelial cells.
      (2) The sizes of the openings vary from tissue to tissue.
      (3) Endothelial cells of brain capillaries are tightly fused, forming a blood-brain barrier through which substances move by facilitated diffusion.
   b. Capillary arrangement
      Capillary density varies directly with tissue metabolic rates.
   c. Regulation of capillary blood flow
      (1) Precapillary sphincters regulate capillary blood flow.
      (2) Precapillary sphincters open when cells are low in oxygen and nutrients and close when cellular needs are met.
   d. Exchanges in the capillaries
      (1) Gases, nutrients, and metabolic by-products are exchanged between the capillary blood and the tissue fluid.
      (2) Diffusion provides the most important means of transport.
      (3) Diffusion pathways depend on lipid solubilities.
      (4) Plasma proteins generally remain in the blood.
      (5) Filtration, due to the hydrostatic pressure of blood, causes a net outward movement of fluid at the arteriolar end of a capillary.
      (6) Osmosis due to colloid osmotic pressure causes a net inward movement of fluid at the venular end of a capillary.
      (7) Some factors cause fluids to accumulate in the tissues.

3. Venules and veins
   a. Venules continue from capillaries and merge to form veins.
   b. Veins carry blood to the heart.
   c. Venous walls are similar to arterial walls but are thinner and contain less muscle and elastic tissue.

15.5 BLOOD PRESSURE (PAGE 580)

Blood pressure is the force blood exerts against the insides of blood vessels.

1. Arterial blood pressure
   a. The arterial blood pressure is produced primarily by heart action; it rises and falls with phases of the cardiac cycle.
   b. Systolic pressure occurs when the ventricle contracts; diastolic pressure occurs when the ventricle relaxes.
2. Factors that influence arterial blood pressure
   a. Heart action, blood volume, resistance to flow, and blood viscosity influence arterial blood pressure.
   b. Arterial pressure increases as cardiac output, blood volume, peripheral resistance, or blood viscosity increases.
3. Control of blood pressure
   a. Blood pressure is controlled in part by the mechanisms that regulate cardiac output and peripheral resistance.
   b. Cardiac output depends on the volume of blood discharged from the ventricle with each beat (stroke volume) and on the heart rate.
      (1) The more blood that enters the heart, the stronger the ventricular contraction, the greater the stroke volume, and the greater the cardiac output.
      (2) The cardiac center of the medulla oblongata regulates heart rate.
   c. Changes in the diameter of arterioles, controlled by the vasomotor center of the medulla oblongata, regulate peripheral resistance.
4. Venous blood flow
   a. Venous blood flow is not a direct result of heart action; it depends on skeletal muscle contraction, breathing movements, and venoconstriction.
   b. Many veins contain flaplike valves that prevent blood from backing up.
   c. Venous constriction can increase venous pressure and blood flow.
5. Central venous pressure
   a. Central venous pressure is the pressure in the right atrium.
   b. Factors that influence it alter the flow of blood into the right atrium.
   c. It affects pressure in the peripheral veins.

15.6 PATHS OF CIRCULATION (PAGE 590)

1. Pulmonary circuit
   a. The pulmonary circuit consists of vessels that carry blood from the right ventricle to the alveolar capillaries in the lungs, and vessels that lead back to the left atrium.
   b. Alveolar capillaries exert less pressure than those of the systemic circuit.
   c. Tightly joined epithelial cells of alveoli walls prevent most substances from entering the alveoli.
   d. Osmotic pressure rapidly draws water out of alveoli into the interstitial fluid, so alveoli do not fill with fluid.
2. Systemic circuit
   a. The systemic circuit is composed of vessels that lead from the left ventricle to all body parts (including vessels supplying the heart itself) and back to the heart.
   b. It includes the aorta and its branches as well as the system of veins that return blood to the right atrium.

15.7 ARTERIAL SYSTEM (PAGE 592)

1. Principal branches of the aorta
   a. The branches of the ascending aorta include the right and left coronary arteries.
   b. The branches of the aortic arch include the brachiocephalic, left common carotid, and left subclavian arteries.
   c. The branches of the descending aorta include the thoracic and abdominal groups.
   d. The abdominal aorta terminates by dividing into right and left common iliac arteries.
2. Arteries to the brain, head, and neck include branches of the subclavian and common carotid arteries.
3. Arteries to the shoulder and upper limb
   a. The subclavian artery passes into the arm, and in various regions, it is called the axillary and brachial artery.
   b. Branches of the brachial artery include the ulnar and radial arteries.
4. Arteries to the thoracic and abdominal walls
   a. Branches of the subclavian artery and thoracic aorta supply the thoracic wall.
   b. Branches of the abdominal aorta and other arteries supply the abdominal wall.
5. Arteries to the pelvis and lower limb

   The common iliac artery supplies the pelvic organs, gluteal region, and lower limb.

15.8 VENOUS SYSTEM (PAGE 600)

1. Characteristics of venous pathways
   a. The veins return blood to the heart.
   b. Larger veins usually parallel the paths of major arteries.
2. Veins from the brain, head, and neck
   a. The jugular veins drain these regions.
   b. Jugular veins unite with subclavian veins to form the brachiocephalic veins.
3. Veins from the upper limb and shoulder
   a. Sets of superficial and deep veins drain the upper limb.
   b. Digital veins drain into pairs of radial veins and ulnar veins, which merge to form a pair of brachial veins.
   c. The major superficial veins are the basilic and cephalic veins.
   d. Basilic and brachial veins merge to form the axillary vein.
   e. The median cubital vein in the bend of the elbow is often used as a site for venipuncture.

4. Tributaries of the brachiocephalic and azygos veins drain the abdominal and thoracic walls.
5. Veins from the abdominal viscera
   a. The blood from the abdominal viscera generally enters the hepatic portal system and is carried to the liver.
   b. The blood in the hepatic portal system is rich in nutrients.
   c. The liver helps regulate the blood concentrations of glucose, amino acids, and lipids.
   d. Phagocytic cells in the liver remove bacteria from the portal blood.
   e. From the liver, hepatic veins carry blood to the inferior vena cava.
6. Veins from the lower limb and pelvis
   a. Sets of deep and superficial veins drain these regions.
   b. The deep veins include the tibial veins, and the superficial veins include the saphenous veins.

15.9 LIFE-SPAN CHANGES (PAGE 606)

1. Plaque build-up may begin early.
2. Fibrous connective tissue and adipose tissue enlarge the heart by filling in when the number and size of cardiac muscle cells fall.
3. Heart rate and output decline slightly with age.
4. Blood pressure increases with age, while resting heart rate decreases with age.
5. Moderate exercise correlates to lowered risk of heart disease in older people.

CHAPTER ASSESSMENTS

15.1 Introduction

1 Match the structure and its function. (p. 553)

(1) Arteries	A. sites of nutrient, electrolyte, gas, and waste exchange
(2) Arterioles	B. muscular pump that forces blood through arteries
(3) Capillaries	C. branch into capillaries
(4) Venules	D. capillaries converge into these
(5) Veins	E. transport blood away from the heart
(6) Heart	F. transport blood to the heart

15.2 Structure of the Heart

2 Describe the pericardium. (p. 554)

3 Compare the layers of the heart wall. (p. 555)

4 Draw a heart and label the chambers and valves. (p. 555)

5 Blood flows through the vena cavae and coronary sinus into the right atrium, through the ______________ to the right ventricle, through the pulmonary valve to the pulmonary trunk into the right and left ______________ to the lungs, then leaves the lungs through the pulmonary veins and flows into the ____________, through the mitral valve to the ____________, and through the ____________ to the aorta. (p. 561)

6 List the vessels through which blood flows from the aorta to the myocardium and back to the right atrium. (p. 561)

15.3 Heart Actions

7 Describe the pressure changes in the atria and ventricles during a cardiac cycle. (p. 564)

8 Explain the origins of heart sounds. (p. 564)

9 Describe the arrangement of cardiac muscle fibers. (p. 565)

10 Distinguish between the roles of the SA node and AV node. (p. 565)

11 Explain how the cardiac conduction system controls the cardiac cycle. (p. 566)

12 Describe and explain the normal ECG pattern. (p. 568)

13 Discuss how the nervous system regulates the cardiac cycle. (p. 568)

14 Describe two factors other than the nervous system that affect the cardiac cycle. (p. 572)

15.4 Blood Vessels

15 Distinguish between an artery and an arteriole. (p. 574)

16 Explain control of vasoconstriction and vasodilation. (p. 576)

17 Describe the structure and function of a capillary. (p. 576)

18 Describe the function of the blood-brain barrier. (p. 576)

19 Explain control of blood flow through a capillary. (p. 578)

20 Relate how diffusion functions in the exchange of substances between blood plasma and tissue fluid. (p. 578)

21 Explain why water and dissolved substances leave the arteriolar end of a capillary and enter the venular end. (p. 578)

22 Describe the effect of histamine on a capillary. (p. 579)

23 Distinguish between a venule and a vein. (p. 579)

24 Explain how veins function as blood reservoirs. (p. 580)

15.5 Blood Pressure

25 Arterial blood pressure peaks when the ventricles contract. This maximum pressure achieved is called the ______________. (p. 580)

26 Name several factors that influence blood pressure, and explain how each produces its effect. (p. 582)

27 Describe the control of blood pressure. (p. 585)

28 List the major factors that promote venous blood flow. (p. 589)

29 Define *central venous pressure.* (p. 589)

15.6 Paths of Circulation

30 Distinguish between the pulmonary and systemic circuits of the cardiovascular system. (p. 590)

31 Trace the path of blood through the pulmonary circuit. (p. 590)

32 Explain why the alveoli normally do not fill with fluid. (p. 590)

15.7–15.8 Arterial System–Venous System

33 Describe the aorta, and name its principal branches. (p. 592)

34 Discuss the relationship between the major venous pathways and the major arterial pathways to the head, upper limbs, abdominal viscera, and lower limbs. (p. 600)

15.9 Life-Span Changes

35 List and discuss changes in the aging cardiovascular system. (p. 606)

INTEGRATIVE ASSESSMENTS/CRITICAL THINKING

OUTCOMES 15.2, 15.3

1. What structures and properties should an artificial heart have?

OUTCOMES 15.2, 15.3

2. Why is ventricular fibrillation more likely to be life threatening than an atrial fibrillation?

OUTCOMES 15.2, 15.4, 15.6, 15.7

3. If a cardiologist inserts a catheter into a patient's right femoral artery, which arteries will the tube have to pass through to reach the entrance of the left coronary artery?

OUTCOMES 15.3, 15.4, 15.5, 15.6

4. How might the results of a cardiovascular exam differ for an athlete in top condition and a sedentary, overweight individual?

OUTCOMES 15.3, 15.4, 15.5, 15.6

5. Cigarette smoke contains thousands of chemicals, including nicotine and carbon monoxide. Nicotine constricts blood vessels. Carbon monoxide prevents oxygen binding to hemoglobin. How do these two components of smoke affect the cardiovascular system?

OUTCOME 15.4

6. Given the way capillary blood flow is regulated, do you think it is wiser to rest or to exercise following a heavy meal? Cite a reason for your answer.

OUTCOMES 15.4, 15.5, 15.6, 15.8

7. Cirrhosis of the liver, a disease commonly associated with alcoholism, obstructs blood flow through the hepatic blood vessels. As a result the blood backs up, and the capillary pressure greatly increases in the organs drained by the hepatic portal system. What effects might this increasing capillary pressure produce, and which organs would it affect?

OUTCOMES 15.6, 15.7, 15.8

8. If a patient develops a blood clot in the femoral vein of the left lower limb and a portion of the clot breaks loose, where is the blood flow likely to carry the embolus? What symptoms are likely?

WEB CONNECTIONS

Be sure to visit the text website at **www.mhhe.com/shier12** for answers to chapter assessments, additional quizzes, and interactive learning exercises.

ANATOMY & PHYSIOLOGY REVEALED

Anatomy & Physiology Revealed® (APR) includes cadaver photos that allow you to peel away layers of the human body to reveal structures beneath the surface. This program also includes animations, radiologic imaging, audio pronunciations, and practice quizzing. Check out **www.aprevealed.com**. APR has been proven to help improve student grades!

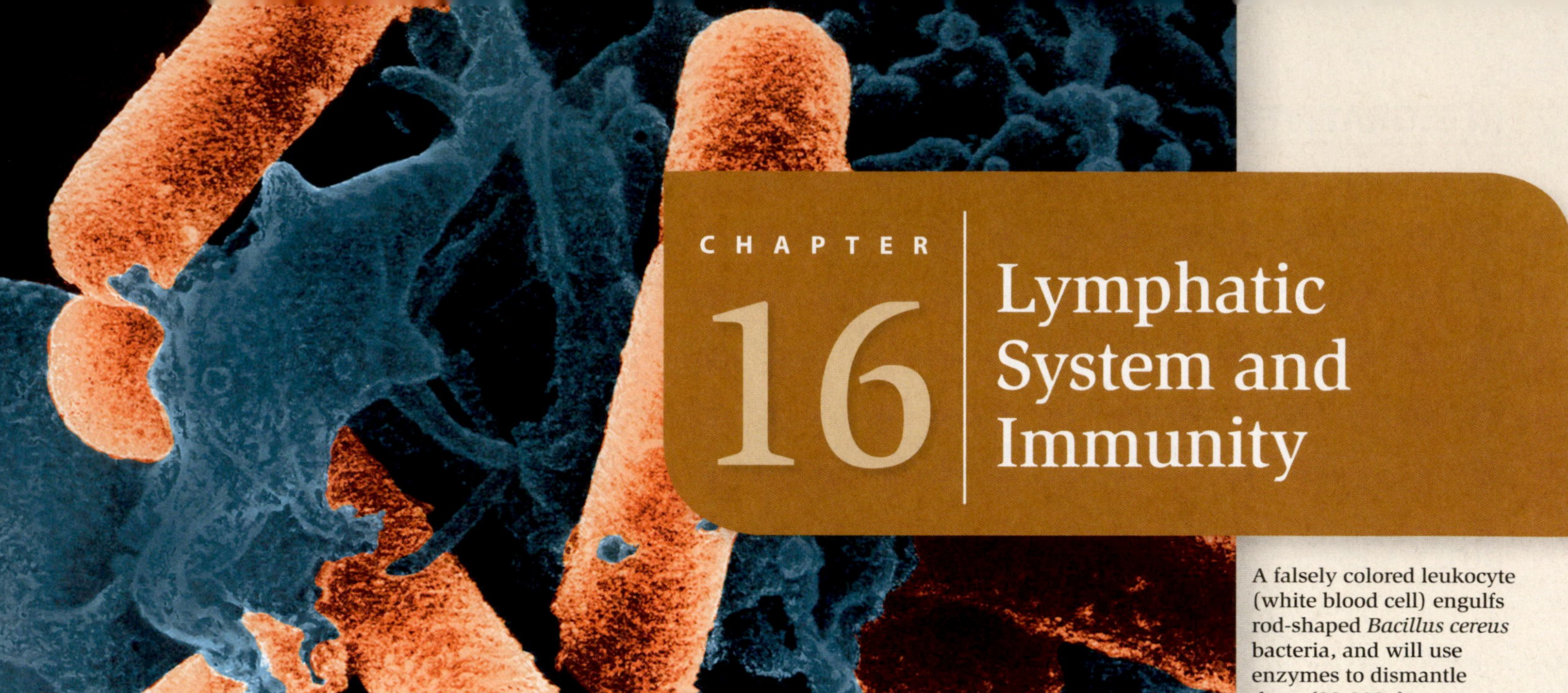

CHAPTER 16

Lymphatic System and Immunity

A falsely colored leukocyte (white blood cell) engulfs rod-shaped *Bacillus cereus* bacteria, and will use enzymes to dismantle them (20,000×).

UNDERSTANDING WORDS

auto-, self: *auto*immune disease—the immune system attacking the body's tissues.

-gen, become, be produced: aller*gen*—substance that evokes an allergic response.

humor-, moisture, fluid: *humor*al immunity—immunity resulting from antibodies in body fluids.

immun-, free, exempt: *immun*ity—resistance to (freedom from) a specific disease.

inflamm-, to set on fire: *inflamm*ation—localized redness, heat, swelling, and pain in the tissues.

nod-, knot: *nod*ule—small mass of lymphocytes surrounded by connective tissue.

patho-, disease, sickness: *patho*gen—disease-causing agent.

LEARNING OUTCOMES

After you studied this chapter, you should be able to:

16.1 Introduction
- **1** Describe the general functions of the lymphatic system. (p. 617)

16.2 Lymphatic Pathways
- **2** Identify and describe the parts of the major lymphatic pathways. (p. 617)

16.3 Tissue Fluid and Lymph
- **3** Describe how tissue fluid and lymph form, and explain the function of lymph. (p. 619)

16.4 Lymph Movement
- **4** Explain how lymphatic circulation is maintained, and describe the consequence of lymphatic obstruction. (p. 621)

16.5 Lymph Nodes
- **5** Describe a lymph node and its major functions. (p. 621)
- **6** Identify the locations of the major chains of lymph nodes. (p. 622)

16.6 Thymus and Spleen
- **7** Discuss the locations and functions of the thymus and spleen. (p. 623)

16.7 Body Defenses Against Infection
- **8** Distinguish between innate (nonspecific) and adaptive (specific) defenses. (p. 626)

16.8 Innate (Nonspecific) Defenses
- **9** List seven innate body defense mechanisms, and describe the action of each mechanism. (p. 626)

16.9 Adaptive (Specific) Defenses or Immunity
- **10** Explain how two major types of lymphocytes are formed and activated and how they function in immune mechanisms. (p. 628)
- **11** Identify the parts of an antibody molecule. (p. 634)
- **12** Discuss the actions of the five types of antibodies. (p. 635)
- **13** Distinguish between primary and secondary immune responses. (p. 637)
- **14** Distinguish between active and passive immunity. (p. 638)
- **15** Explain how allergic reactions, tissue rejection reactions, and autoimmunity arise from immune mechanisms. (p. 639)

16.10 Life-Span Changes
- **16** Describe life-span changes in immunity. (p. 644)

 LEARN  PRACTICE  ASSESS

PEANUT ALLERGY

The young woman went to the emergency department for sudden onset of difficulty breathing. She was also flushed and had vomited. An astute medical student taking a quick history from the woman's roommates discovered that she had just eaten cookies from a vending machine in their dorm. Suspecting that the cookies may have contained peanuts, the student alerted the attending physician, who treated the woman for suspected peanut allergy—giving oxygen, an antihistamine, a steroid drug, and epinephrine. She recovered.

Peanut allergy is common and on the rise in certain westernized countries. In the United States, about 1% of children under the age of five and 2% of the population over ten years of age have had allergic reactions to peanuts. About 30,000 people react each year, and about 200 die.

Peculiarities of peanuts and our fondness for them may explain why peanut allergy prevalence is increasing. Three glycoproteins in peanuts are allergens, causing the misdirected immune response that is an allergy. These glycoproteins are highly concentrated in the peanut, and when ingested, they disturb the intestinal lining in such a way that they enter the circulation rapidly, without being digested. Many allergens confront cells of the immune system beneath the intestinal lining.

Compounding the rapidity with which peanut allergens flood the bloodstream is that people in the United States eat many peanuts—virtually everyone has eaten one by two years of age. This is sufficient exposure to set the stage for later allergy in genetically predisposed individuals. That the average age of first allergic reaction to peanuts is fourteen months suggests that the initial exposure—necessary to "prime" the immune system for future response—happens through breast milk or in the uterus. Countries where peanuts are rarely eaten, such as Denmark and Norway, have very low incidence of peanut allergy.

The dry roasting of peanuts in the United States may make the three glycoproteins that evoke the allergic response more active. In China, where peanuts are equally popular but are boiled or fried, allergy is rare. However, children of Chinese immigrants in the United States have the same incidence of peanut allergy as other children, supporting the idea that method of preparation contributes to allergenicity.

Peanut allergy is probably impossible to eradicate. Creating a genetically modified peanut that lacks allergens is difficult because at least three of its proteins would have to be removed—it would no longer be a peanut. The best approach is for people who know they are allergic to avoid peanuts. Always ask if it is an ingredient in a food. Parents of small children with peanut allergy should read all food labels meticulously and carry an injectible "pen" of epinephrine in case the child reacts. A person in the midst of an allergic response to peanuts should remain in the emergency department for a few hours, because a dangerous late-phase response can occur three to ten hours after the initial symptoms of rash, difficulty breathing, and/or gastrointestinal upset. ■

16.1 INTRODUCTION

The **lymphatic** (lim-fat′ik) **system** is a vast collection of cells and biochemicals that travel in lymphatic vessels, and the organs and glands that produce them. The lymphatic system includes a network of vessels that assist in circulating body fluids, so it is closely associated with the cardiovascular system. Lymphatic vessels transport excess fluid away from the interstitial spaces in most tissues and return it to the bloodstream (fig. 16.1). Without the lymphatic system, this fluid would accumulate in tissue spaces. The organs of the lymphatic system also help defend the body against infection by disease-causing agents, or **pathogens** (path′o-jenz).

16.2 LYMPHATIC PATHWAYS

The **lymphatic pathways** begin as lymphatic capillaries that merge to form larger lymphatic vessels. These, in turn, lead to larger vessels that unite with the veins in the thorax.

Lymphatic Capillaries

Lymphatic capillaries are microscopic, closed-ended tubes. They extend into the interstitial spaces, forming complex networks that parallel the networks of the blood capillaries (fig. 16.2). The walls of lymphatic capillaries are similar to those of blood capillaries. Each consists of a single layer of squamous epithelial cells called endothelium. These thin walls allow tissue fluid (interstitial fluid) from the interstitial space to enter the lymphatic capillaries. Fluid inside a lymphatic capillary is called **lymph** (limf). Special lymphatic capillaries (*lacteals*) in the lining of the small intestine absorb digested fats, then transport the fats to the venous circulation.

Lymphatic Vessels

The walls of **lymphatic vessels** are similar to those of veins, but thinner. Each is composed of three layers: an endothelial lining, a middle layer of smooth muscle and elastic fibers, and an outer layer of connective tissue. Also like those veins below the heart, the lymphatic vessels have semilunar valves, which help prevent backflow of lymph. Figure 16.3 shows one of these valves.

The larger lymphatic vessels lead to specialized organs called **lymph nodes** (limf nōdz). After leaving the nodes, the vessels merge into larger lymphatic trunks.

Lymphatic Trunks and Collecting Ducts

The **lymphatic trunks,** which drain lymph from the lymphatic vessels, are named for the regions they serve. For example, the *lumbar trunk* drains lymph from the lower limbs, lower abdominal wall, and pelvic organs; the *intestinal trunk* drains the abdominal viscera; the *intercostal* and *bronchomediastinal trunks* drain lymph from portions of the thorax; the *subclavian trunk* drains the upper limb; and the

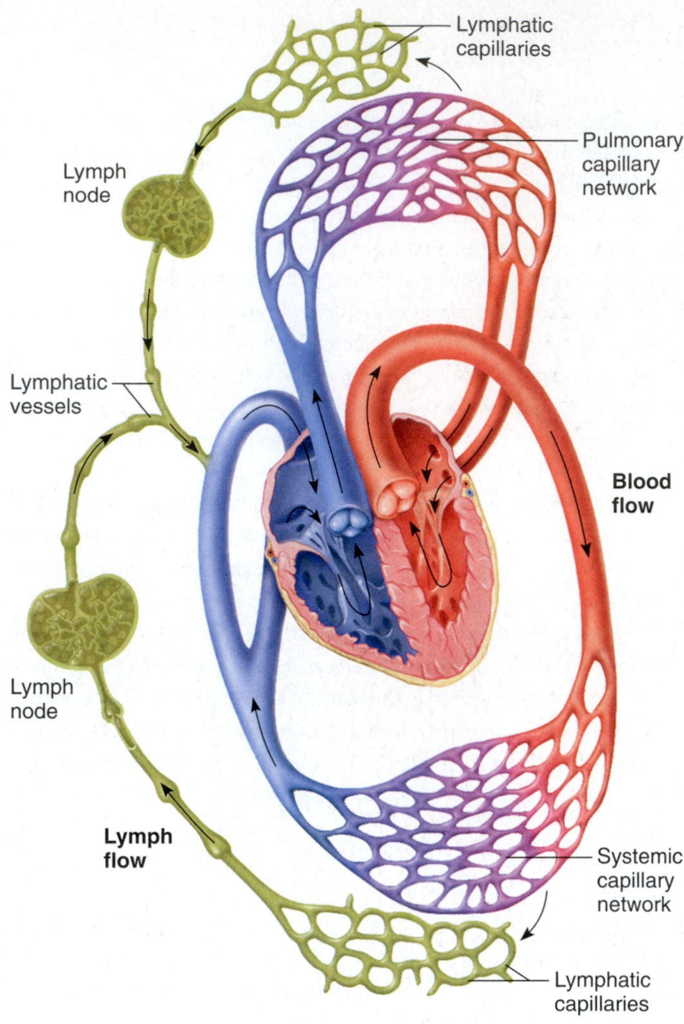

FIGURE 16.1 Schematic representation of lymphatic vessels transporting fluid from interstitial spaces to the bloodstream.

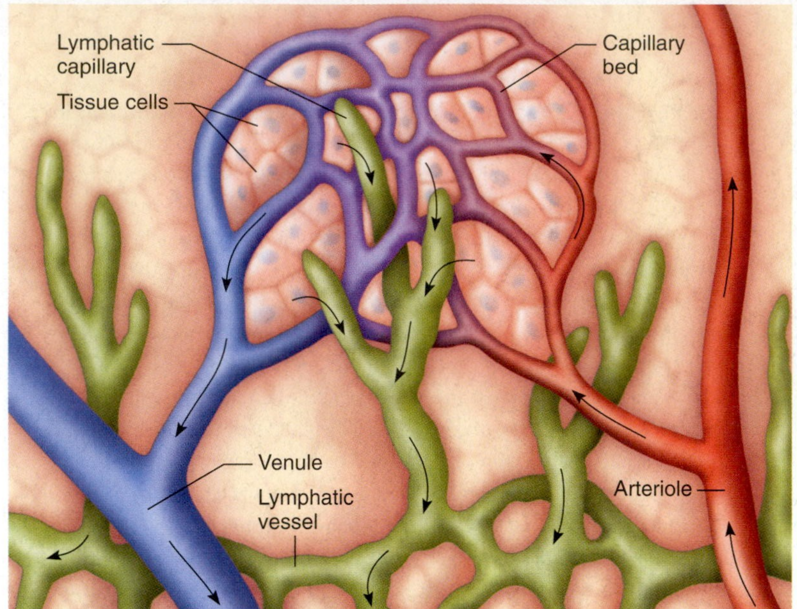

FIGURE 16.2 Lymphatic capillaries are microscopic, closed-ended tubes that originate in the interstitial spaces of most tissues.

jugular trunk drains portions of the neck and head. These lymphatic trunks then join one of two **collecting ducts**—the thoracic duct or the right lymphatic duct. Figure 16.4 shows the location of the major lymphatic trunks and collecting ducts, and figure 16.5 shows a lymphangiogram, or radiograph, of some lymphatic vessels and lymph nodes.

The **thoracic duct** is the larger and longer of the two collecting ducts. It originates in the abdomen, passes upward through the diaphragm beside the aorta, ascends anterior to the vertebral column through the mediastinum, and empties into the left subclavian vein near the junction of the left jugular vein. This duct drains lymph from the intestinal, lumbar, and intercostal trunks, as well as from the left subclavian, left jugular, and left bronchomediastinal trunks.

The **right lymphatic duct** originates in the right thorax at the union of the right jugular, right subclavian, and right bronchomediastinal trunks. It empties into the right subclavian vein near the junction of the right jugular vein.

Lymph leaves the two collecting ducts, then enters the venous system and becomes part of the plasma prior to the

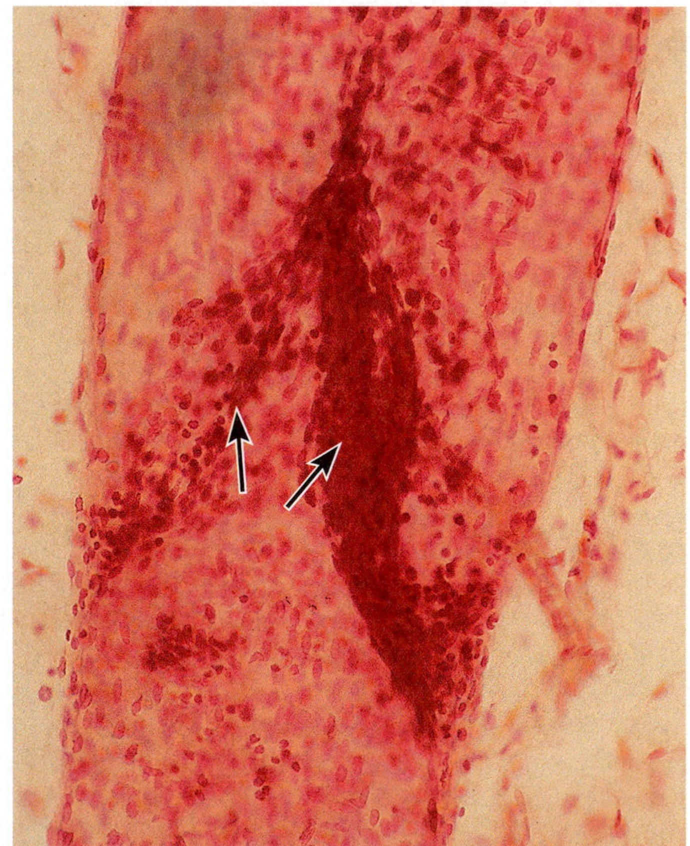

FIGURE 16.3 Light micrograph of the flaplike valve (arrows) within a lymphatic vessel (60×).

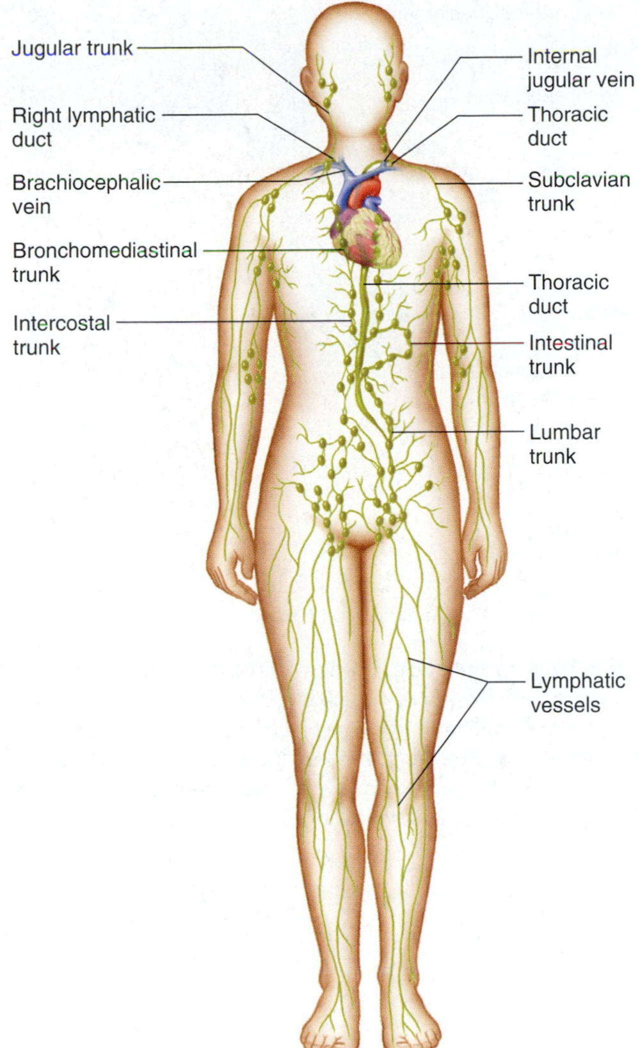

FIGURE 16.4 Lymphatic vessels merge into larger lymphatic trunks, which, in turn, drain into collecting ducts.

blood's return to the right atrium. Thus, lymph from the lower body regions, the left upper limb, and the left side of the head and neck enters the thoracic duct; lymph from the right side of the head and neck, the right upper limb, and the right thorax enters the right lymphatic duct (fig. 16.6). Figure 16.7 summarizes the lymphatic pathway.

> The skin has many lymphatic capillaries. Consequently, if the skin is broken, or if something is injected into it (such as venom from a stinging insect), foreign substances rapidly enter the lymphatic system.

PRACTICE

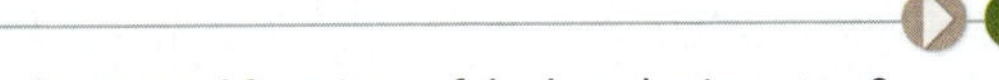

1. What are the general functions of the lymphatic system?
2. Distinguish between the thoracic duct and the right lymphatic duct.
3. Through which lymphatic structures would lymph pass in traveling from a lower limb back to the bloodstream?

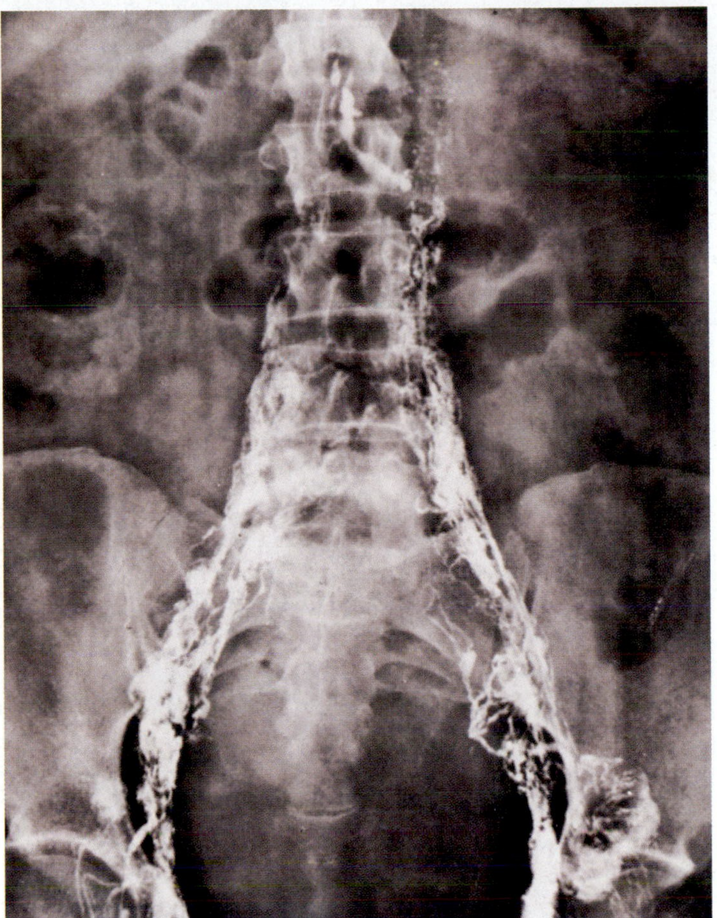

FIGURE 16.5 A lymphangiogram (radiograph) of the lymphatic vessels and lymph nodes of the pelvic region.

16.3 TISSUE FLUID AND LYMPH

Lymph is essentially tissue fluid that has entered a lymphatic capillary. Thus, lymph formation depends upon tissue fluid formation.

Tissue Fluid Formation

Capillary blood pressure filters water and small molecules from the plasma. The resulting fluid has much the same composition as the plasma (including nutrients, gases, and hormones), with the important exception of the plasma proteins, which are generally too large to pass through the capillary walls. The osmotic effect of these proteins (called the *plasma colloid osmotic pressure*) helps to draw fluid back into the capillaries by osmosis.

RECONNECT
To Chapter 15, Exchanges in the Capillaries, pages 578–579.

Lymph Formation

Filtration from the plasma normally exceeds reabsorption, leading to the net formation of tissue fluid. This increases

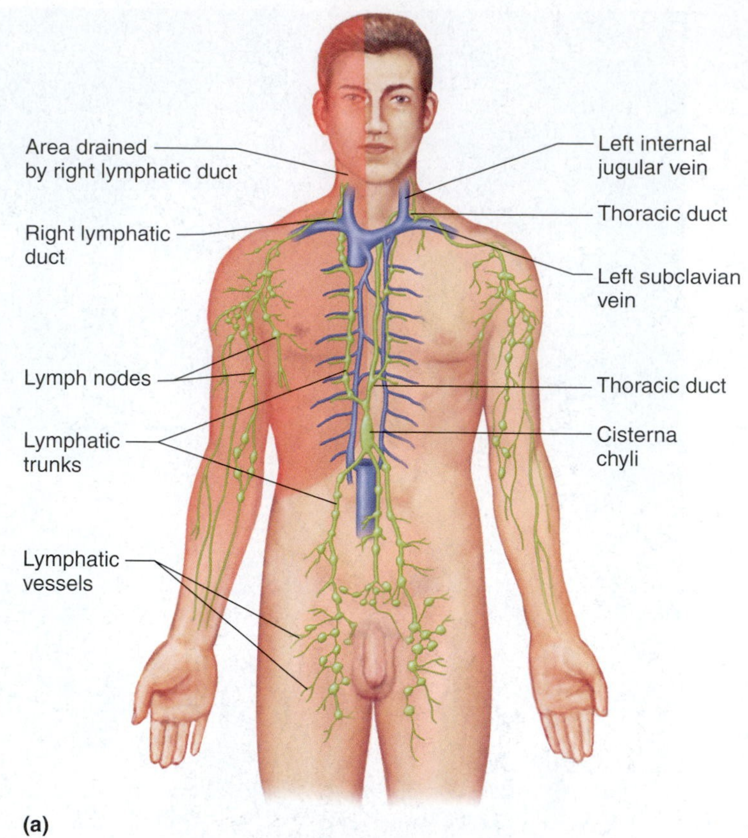

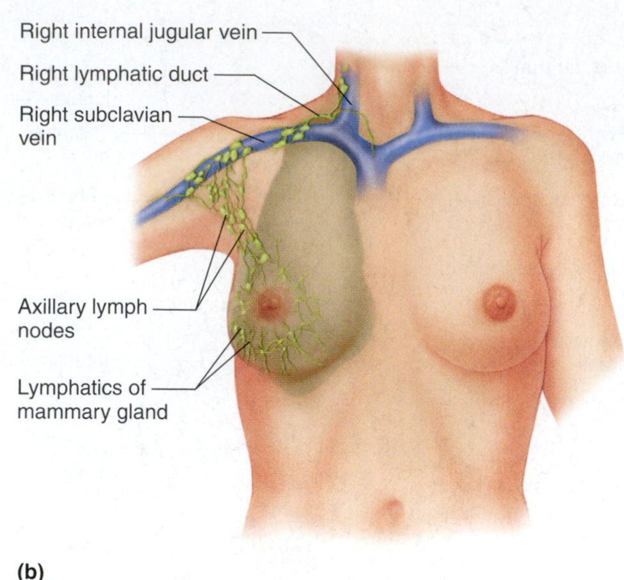

FIGURE 16.6 Lymphatic pathways. (*a*) The right lymphatic duct drains lymph from the upper right side of the body, whereas the thoracic duct drains lymph from the rest of the body. (*b*) Lymph drainage of the right breast illustrates a localized function of the lymphatic system. Surgery to remove a cancerous breast can disrupt this drainage, causing painful swelling.

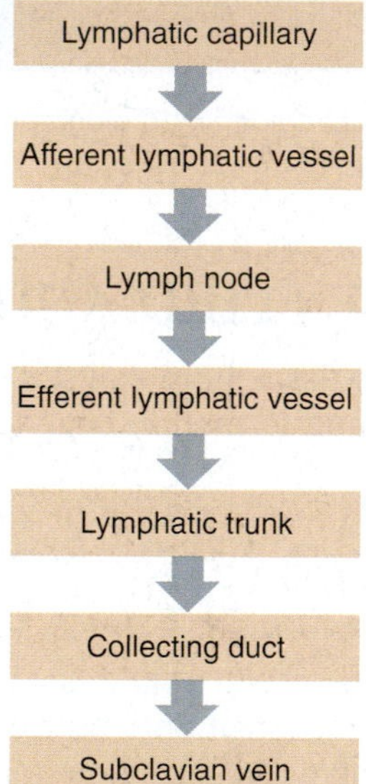

FIGURE 16.7 The lymphatic pathway.

the tissue fluid hydrostatic pressure moving tissue fluid into lymphatic capillaries, forming lymph. Thus, lymph formation prevents the accumulation of excess tissue fluid, or *edema* (ĕ′de-mah).

Lymph Function

Lymphatic vessels in the small intestine play a major role in the absorption of dietary fats (chapter 17, pp. 684–685). Lymph also returns to the bloodstream most of the small proteins that the blood capillaries filtered. At the same time, lymph transports foreign particles, such as bacteria and viruses, to lymph nodes.

Lymphatic capillaries are adapted to receive proteins and foreign particles in a way that blood capillaries are not. The epithelial cells that form the walls of lymphatic vessels overlap but are not attached to each other. This configuration, shown in figure 16.8, creates flaplike valves in the lymphatic capillary wall. The valves are pushed inward when the pressure is greater on the outside of the capillary but close when the pressure is greater on the inside.

The epithelial cells of the lymphatic capillary wall are also attached to surrounding connective tissue cells by thin protein filaments. As a result, the lumen of a lymphatic capillary remains open even when the outside pressure is greater than the pressure inside the lymph capillary.

PRACTICE 

4 What is the relationship between tissue fluid and lymph?
5 How do plasma proteins in tissue fluid affect lymph formation?
6 What are the major functions of lymph?

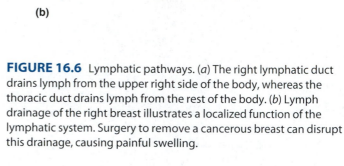

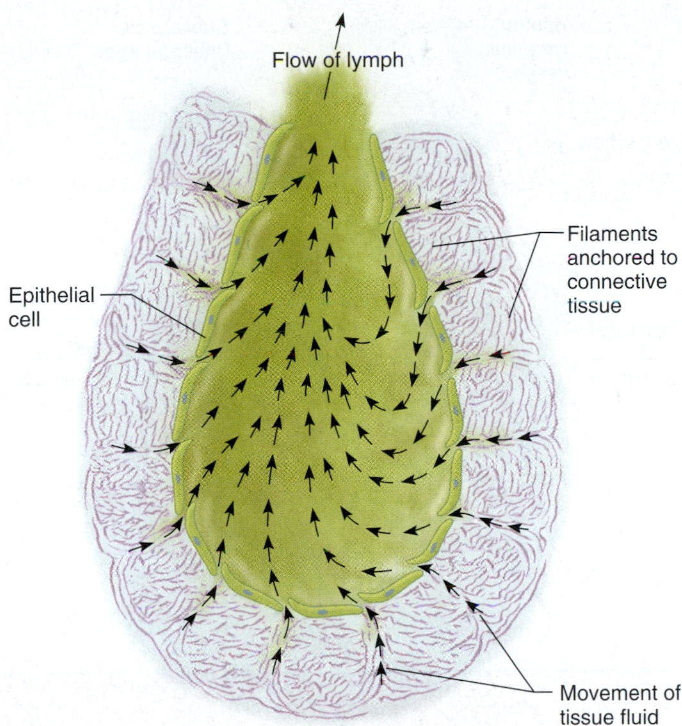

FIGURE 16.8 Tissue fluid enters lymphatic capillaries through flaplike valves between epithelial cells.

of the abdominal vessels and forced into the thoracic vessels. Once again, the valves of the lymphatic vessels prevent lymph backflow.

Obstruction of Lymph Movement

The continuous movement of fluid from interstitial spaces into blood capillaries and lymphatic capillaries stabilizes the volume of fluid in these spaces. Conditions that interfere with lymph movement cause tissue fluid to accumulate in interstitial spaces, producing edema. For example, a surgeon removing a cancerous breast tumor also usually removes nearby axillary lymph nodes to prevent associated lymphatic vessels from transporting cancer cells to other sites (metastasis). Removing the lymphatic tissue can obstruct drainage from the upper limb, causing edema (see fig. 16.6*b*).

Edema can swell the lower limbs of a person with congestive heart failure to enormous proportions. Elevating the lower limbs relieves edema somewhat.

PRACTICE

7 What factors promote lymph flow?

8 What is the consequence of lymphatic obstruction?

16.4 LYMPH MOVEMENT

The hydrostatic pressure of tissue fluid drives lymph into lymphatic capillaries. However, muscular activity largely influences the movement of lymph through the lymphatic vessels.

Lymph Flow

Lymph, like venous blood, is under relatively low hydrostatic pressure. It may not flow readily through the lymphatic vessels without help from contracting skeletal muscles in the limbs, pressure changes from the action of skeletal muscles used in breathing, and contraction of smooth muscles in the walls of the larger lymphatic trunks. Lymph flow peaks during physical exercise, due to the actions of skeletal muscles and pressure changes associated with breathing.

Contracting skeletal muscles compress lymphatic vessels. This squeezing action moves the lymph inside a vessel, but because the lymphatic vessels have valves that prevent backflow, the lymph can move only toward a collecting duct. Additionally, the smooth muscles in the walls of the larger lymphatic trunks may contract and compress the lymph inside, forcing the fluid onward.

Breathing aids lymph circulation by creating a relatively low pressure in the thorax during inhalation. At the same time, the contracting diaphragm increases the pressure in the abdominal cavity. Consequently, lymph is squeezed out

16.5 LYMPH NODES

Lymph nodes (lymph glands) are located along the lymphatic pathways. They contain many *lymphocytes* and *macrophages* (histiocytes) that fight invading pathogens.

Structure of a Lymph Node

Lymph nodes vary in size and shape but are usually less than 2.5 centimeters long and are somewhat bean-shaped (fig. 16.9). Figure 16.10 illustrates a section of a typical lymph node.

Blood vessels and nerves join a lymph node through the indented region of the node, called the **hilum.** The lymphatic vessels leading to a node (afferent vessels) enter separately at various points on its convex surface, but the lymphatic vessels leaving the node (efferent vessels) exit from the hilum.

A *capsule* of connective tissue with many fibers encloses each lymph node. The capsule extends into the node and partially subdivides it into compartments. Masses of lymphocytes (B cells) and macrophages in the cortex, called **lymph nodules,** (lymph follicles) are the functional units of the lymph node.

Lymph nodules are found singly or in groups associated with the mucous membranes of the respiratory and digestive tracts. The *tonsils,* described in chapter 17 (p. 657), are partially encapsulated lymph nodules. Aggregates of nodules called *Peyer's patches* pervade the mucosal lining of the distal small intestine. Within the Peyer's patches are scattered

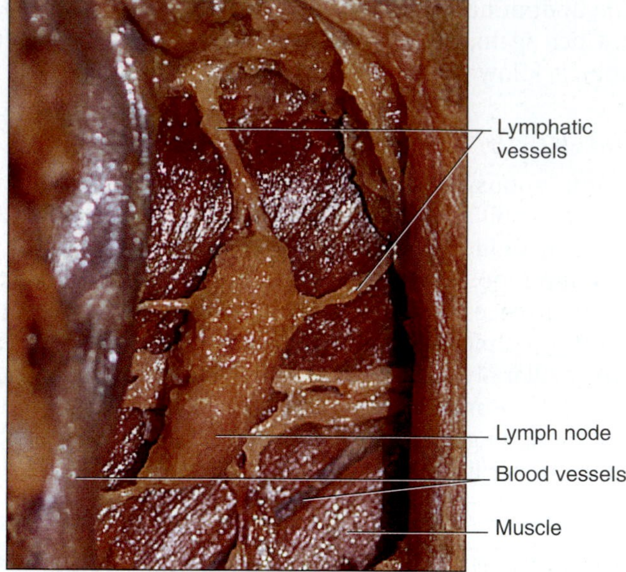

FIGURE 16.9 Lymph enters and leaves a lymph node through lymphatic vessels.

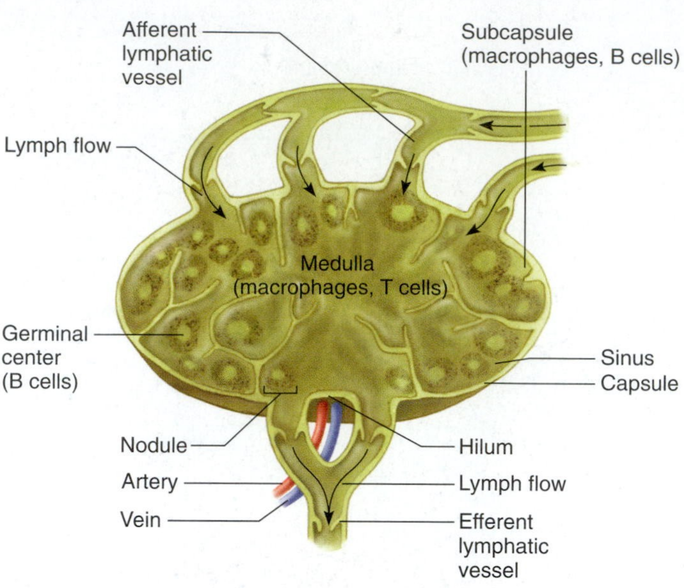

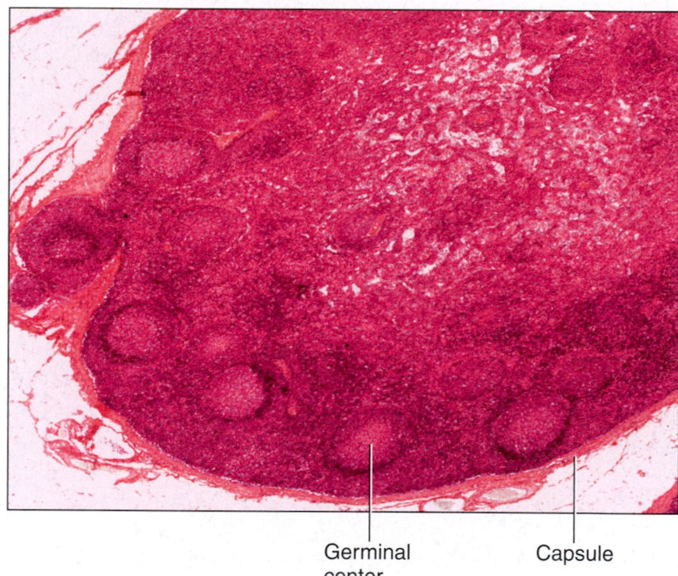

FIGURE 16.10 Lymph node. (*a*) A section of a lymph node. (*b*) Light micrograph of a lymph node (20×).

M cells, through which certain ingested molecules pass by transcytosis, then face lymphocytes and other immune system cells that then may initiate an immune response. The lymphoid tissues in the appendix, Peyer's patches, tonsils, adenoids, and mesenteric lymph nodes are collectively termed mucosa-associated lymphoid tissues (MALT).

Spaces in a lymph node, called **lymph sinuses,** provide a complex network of chambers and channels through which lymph circulates. Lymph enters a lymph node through *afferent lymphatic vessels,* moves slowly through the lymph sinuses, and leaves through *efferent lymphatic vessels* (fig. 16.10*a*).

Superficial lymphatic vessels inflamed by bacterial infection appear as red streaks beneath the skin, a condition called *lymphangitis.* Inflammation of the lymph nodes, called *lymphadenitis,* often follows. Affected nodes enlarge and may be painful.

Locations of Lymph Nodes

Lymph nodes are found in groups or chains along the paths of the larger lymphatic vessels throughout the body, but they are not in the central nervous system. The major locations of the lymph nodes, shown in figure 16.11, are as follows:

1. **Cervical region.** These lymph nodes follow the lower border of the mandible, anterior to and posterior to the ears, and deep in the neck along the paths of the larger blood vessels. These nodes are associated with the lymphatic vessels that drain the skin of the scalp and face, as well as the tissues of the nasal cavity and pharynx.
2. **Axillary region.** Lymph nodes in the underarm region receive lymph from vessels that drain the upper limbs, the wall of the thorax, the mammary glands (breasts), and the upper wall of the abdomen.
3. **Supratrochlear region.** These lymph nodes are located superficially on the medial side of the elbow. They often enlarge in children in response to infections acquired through cuts and scrapes on the hands.
4. **Inguinal region.** Lymph nodes in the inguinal region receive lymph from the lower limbs, the external genitalia, and the lower abdominal wall.
5. **Pelvic cavity.** Here lymph nodes primarily follow the iliac blood vessels. They receive lymph from the lymphatic vessels of the pelvic viscera.

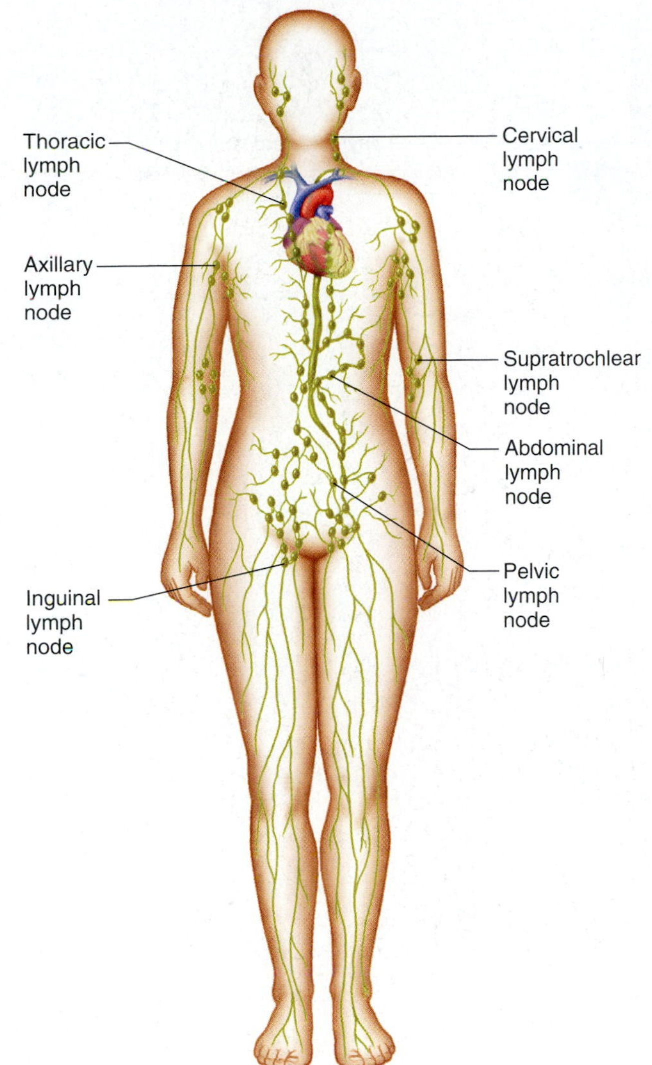

FIGURE 16.11 Major locations of lymph nodes.

6. **Abdominal cavity.** These lymph nodes form chains along the main branches of the mesenteric arteries and the abdominal aorta. These lymph nodes receive lymph from the abdominal viscera.
7. **Thoracic cavity.** These lymph nodes are in the mediastinum and along the trachea and bronchi. They receive lymph from the thoracic viscera and from the internal wall of the thorax.

> The illness described as "swollen glands" refers to enlarged cervical lymph nodes associated with throat or respiratory infection.

Functions of Lymph Nodes

Lymph nodes have two primary functions: filtering potentially harmful particles from lymph before returning it to the bloodstream, and monitoring body fluids (immune surveillance) provided by lymphocytes and macrophages. Along with the red bone marrow, the lymph nodes are centers for lymphocyte production. These cells attack viruses, bacteria, and other parasitic cells that lymphatic vessels bring to the lymph nodes. Macrophages in the lymph nodes engulf and destroy foreign substances, damaged cells, and cellular debris.

PRACTICE 

9 Distinguish between a lymph node and a lymph nodule.

10 What factors promote the flow of lymph through a node?

11 In what body regions are lymph nodes most abundant?

12 What are the major functions of lymph nodes?

16.6 THYMUS AND SPLEEN

Two other lymphatic organs, whose functions are similar to those of the lymph nodes, are the thymus and the spleen.

Thymus

The **thymus** (thi′mus) is a soft, bilobed structure enclosed in a connective tissue capsule. It is in the mediastinum, anterior to the aortic arch and posterior to the upper part of the body of the sternum, and extends from the root of the neck to the pericardium (fig. 16.12*a*). The thymus varies in size and is usually proportionately larger during infancy and early childhood. After puberty, the thymus shrinks, and in an adult, it may be small (fig. 16.13). In elderly persons, adipose and connective tissues replace lymphatic tissue in the thymus.

Connective tissues extend inward from the surface of the thymus, subdividing it into lobules (see fig. 16.12*b*). The lobules house many lymphocytes that developed from progenitor cells in the bone marrow. Most of these cells (thymocytes) are inactivated; however, some mature into **T lymphocytes,** or (T cells)which leave the thymus and provide immunity. Epithelial cells in the thymus secrete protein hormones called *thymosins,* which stimulate maturation of T lymphocytes.

Spleen

The **spleen** (splēn) is the largest lymphatic organ. It is in the upper left portion of the abdominal cavity, just inferior to the diaphragm, posterior and lateral to the stomach (see fig. 16.12*a* and reference plates 4, 5, and 6).

The spleen resembles a large lymph node in that it is enclosed in connective tissue that extends inward from the surface and partially subdivides the organ into chambers, or lobules. The organ also has a hilum on one surface through which blood vessels and nerves enter. However, unlike the sinuses of a lymph node that are filled with lymph, the spaces (venous sinuses) in the chambers of the spleen are filled with blood.

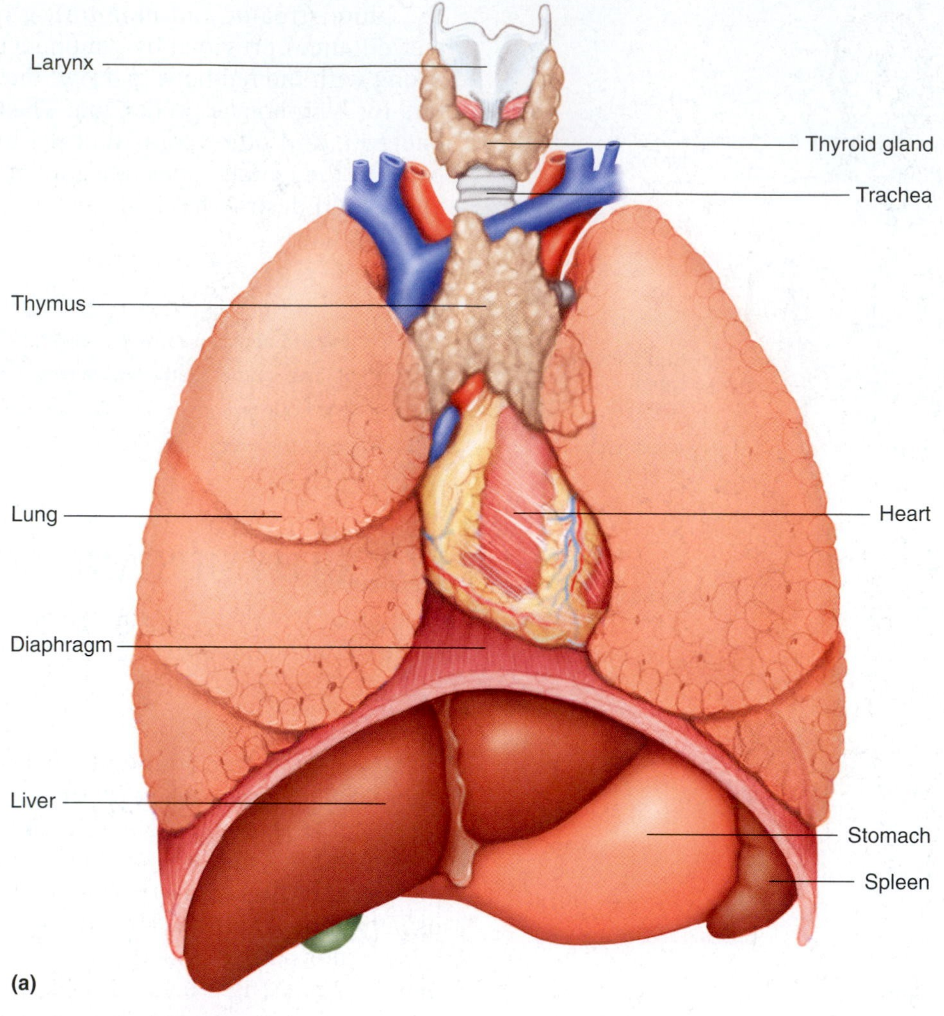

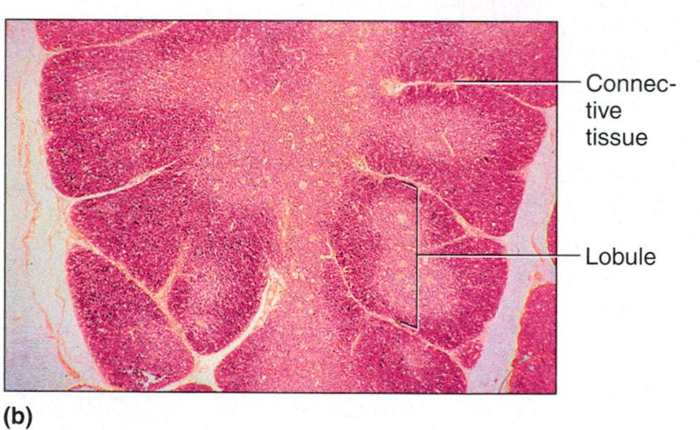

FIGURE 16.12 Thymus and spleen. (*a*) The thymus is bilobed, between the lungs, and superior to the heart. The spleen is inferior to the diaphragm, posterior and lateral to the stomach. (*b*) A cross section of the thymus (15×). Note how the thymus is subdivided into lobules.

The tissues in the lobules of the spleen are of two types. *White pulp* is distributed throughout the spleen in tiny islands. This tissue is composed of nodules (splenic nodules), similar to those in lymph nodes and are packed with lymphocytes. The *red pulp,* which fills the remaining spaces of the lobules, surrounds the venous sinuses. This pulp contains abundant red blood cells, which impart its color, plus many lymphocytes and macrophages (fig. 16.14).

During fetal development, pulp cells of the spleen produce blood cells, much as red bone marrow cells do after birth. As the time of birth approaches, this splenic function ceases. However, in certain diseases, such as *erythroblastosis fetalis* (see chapter 14, page 547), in which many red blood cells are destroyed, the splenic pulp cells may resume their hematopoietic activity.

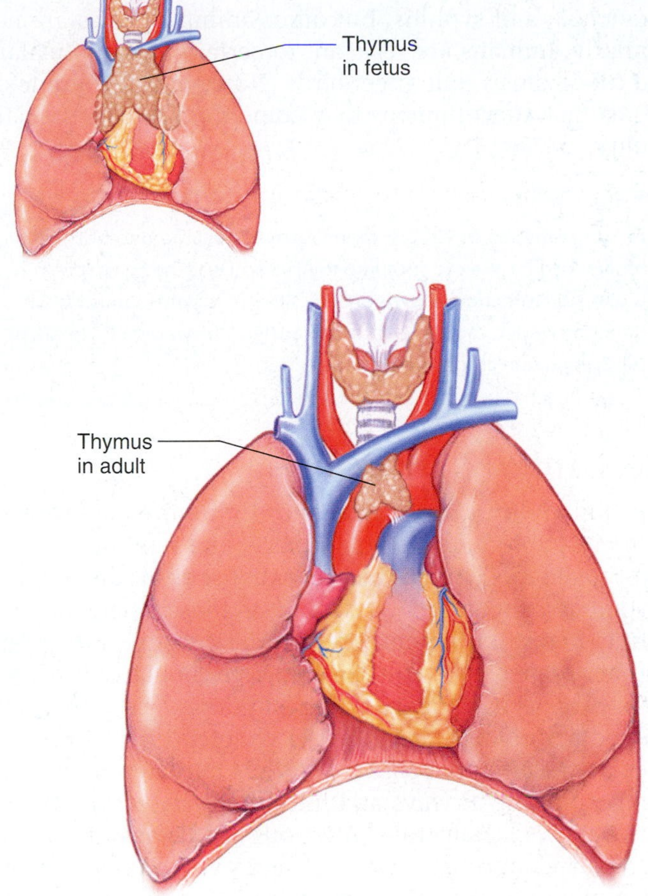

FIGURE 16.13 Compared to other thoracic organs, the thymus in the fetus is large, but in the adult is small. Figure is not to scale.

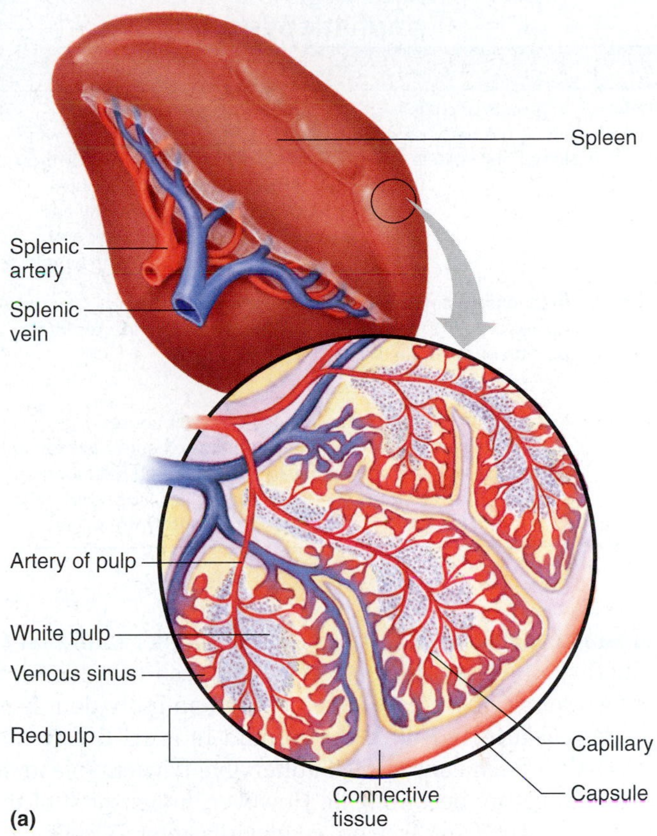

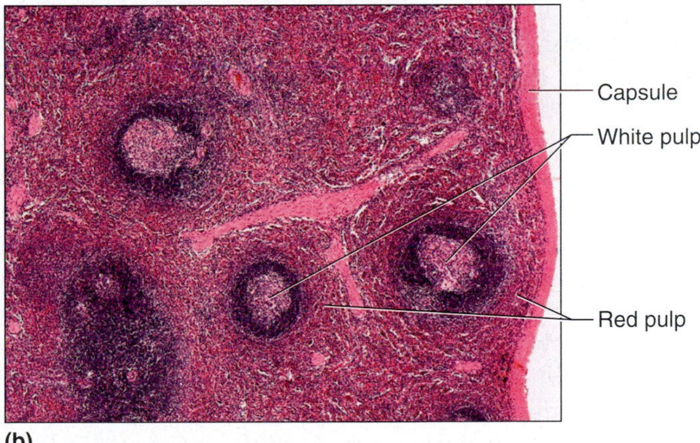

FIGURE 16.14 Spleen. *(a)* The spleen resembles a large lymph node. *(b)* Light micrograph of the spleen (40×).

Blood capillaries in the red pulp are permeable. Red blood cells can squeeze through the pores in these capillary walls and enter the venous sinuses. The older, more fragile red blood cells may rupture during this passage. Macrophages in the splenic sinuses remove the resulting cellular debris.

Macrophages engulf and destroy foreign particles, such as bacteria, that may be carried in the blood as it flows through the splenic sinuses. Lymphocytes of the spleen, like those of the thymus, lymph nodes, and nodules, also help defend the body against infections. Thus, the spleen filters blood much as the lymph nodes filter lymph. Table 16.1 summarizes the characteristics of the major organs of the lymphatic system.

PRACTICE

13 Why are the thymus and spleen considered organs of the lymphatic system?

14 What are the major functions of the thymus and the spleen?

16.7 BODY DEFENSES AGAINST INFECTION

The presence and multiplication of a pathogen in the body may cause an infection. Pathogens include simple microorganisms such as bacteria, complex microorganisms such as protozoa, and even spores of multicellular organisms such as fungi. Viruses are pathogens, but they are not considered

TABLE 16.1 | Major Organs of the Lymphatic System

Organ	Location	Function
Lymph nodes	In groups or chains along the paths of larger lymphatic vessels	Filter foreign particles and debris from lymph; produce and house lymphocytes that destroy foreign particles in lymph; house macrophages that engulf and destroy foreign particles and cellular debris carried in lymph
Thymus	In the mediastinum posterior to the upper portion of the body of the sternum	Houses lymphocytes; differentiates thymocytes into T lymphocytes
Spleen	In upper-left portion of the abdominal cavity, inferior to the diaphragm, posterior and lateral to the stomach	Blood reservoir houses macrophages that remove foreign particles, damaged red blood cells, and cellular debris from the blood; contains lymphocytes

organisms because their structure is far simpler than that of a living cell and they must infect a living cell to reproduce. An infection may be present even though an individual feels well. People infected with the human immunodeficiency virus (HIV), which causes AIDS, often live for many years in good health before becoming ill. However, impairment of the immune system begins as soon as the virus enters T cells.

The human body can prevent entry of pathogens or destroy them if they enter. Some mechanisms are general and protect against many types of pathogens, providing **innate** (*nonspecific*) **defense.** They function the same way regardless of the pathogen or the number of exposures. These mechanisms include species resistance, mechanical barriers, chemical barriers (enzyme action, interferon, and complement), natural killer cells, inflammation, phagocytosis, and fever. Other defense mechanisms are very precise, targeting specific pathogens with an **adaptive** (*specific*) **defense,** or immunity. These more directed responses are carried out by specialized lymphocytes that recognize foreign molecules (nonself antigens) in the body and act against them. Innate and adaptive defense mechanisms work together to protect the body against infection. While the innate defenses respond rapidly, slower-to-respond adaptive defenses begin as well.

16.8 INNATE (NONSPECIFIC) DEFENSES

Species Resistance

Species resistance refers to the fact that a species may be resistant to diseases that affect other species because its cells do not have receptors for the pathogen or its tissues do not provide the temperature or chemical environment that a particular pathogen requires. For example, humans are infected by the infectious agents that cause measles, mumps, gonorrhea, and syphilis, but other animal species are not. Similarly, humans are resistant to certain forms of malaria and tuberculosis that affect birds. However, new influenza strains that affect humans may come from birds, especially poultry.

In San Francisco in 1982, Simon Guzman became one of the first recorded individuals to succumb to AIDS and was the first to die from a parasitic infection known only in sheep—*cryptosporidiosis.* This infection, which causes relentless diarrhea, illustrates a hallmark of AIDS: alteration of species resistance.

Mechanical Barriers

The skin and mucous membranes lining the passageways of the respiratory, digestive, urinary, and reproductive systems create **mechanical barriers** that prevent the entrance of some infectious agents. As long as these barriers remain intact, many pathogens are unable to penetrate them. Another protection is that the epidermis sloughs off, removing superficial bacteria with it. In addition, the mucus-coated ciliated epithelium, described in chapter 19 (p. 742), that lines the respiratory passages entraps particles and sweeps them out of the airways and into the pharynx, where they are swallowed. Hair traps infectious agents associated with the skin and mucous membranes and sweat and mucus rinse away microorganisms. Tears, saliva, and urine also wash away organisms before they become firmly attached. These barriers provide a *first line of defense.* The rest of the nonspecific defenses discussed in this section are part of the *second line of defense.*

Chemical Barriers

Enzymes in body fluids provide a **chemical barrier** to pathogens. Gastric juice, for example, contains the protein-splitting enzyme pepsin and has a low pH due to hydrochloric acid in the stomach. The combined effect of pepsin and hydrochloric acid kills many pathogens that enter the stomach. Similarly, tears contain the enzyme lysozyme, which destroys certain bacteria on the eyes. The accumulation of salt from perspiration also kills certain bacteria on the skin.

Interferons (in″ter-fēr′onz) are hormonelike peptides that lymphocytes and fibroblasts produce in response to viruses or tumor cells. Once released from a virus-infected cell, interferon binds to receptors on uninfected cells, stimulating them to synthesize proteins that block replication of a variety of viruses. Thus, interferon's effect is nonspecific. Interferons also stimulate phagocytosis and enhance the activity of other cells that help to resist infections and the growth of tumors.

Other antimicrobial biochemicals are defensins and collectins. **Defensins** are peptides produced by neutrophils and other types of granular white blood cells in the intestinal epithelium, the urogenital tract, the kidneys, and the skin. Recognition of a nonself cell surface or viral particle triggers the expression of genes that encode defensins. Some defensins make holes in bacterial cell walls and membranes, crippling the microbes. **Collectins** are proteins that provide broad protection against bacteria, yeasts, and some viruses. These proteins home in on slight differences in the structures and arrangements of sugars that protrude from the surfaces of pathogens. Collectins detect not only the sugar molecules, but the pattern in which they are clustered, grabbing on much like velcro clings to fabric, thus making the pathogen more easily phagocytized.

Complement (kom′plĕ-ment) is a group of proteins (complement system), in plasma and other body fluids, that interact in a series of reactions or cascade. Complement activation can rapidly occur by the *classical pathway* when a complement protein binds to an antibody attached to its specific antigen (discussed later in this chapter, p. 636), or more slowly by the *alternative pathway* triggered by exposure to foreign antigens, in the absence of antibodies. Activation of complement stimulates inflammation, attracts phagocytes, and enhances phagocytosis.

Natural Killer (NK) Cells

Natural killer (NK) cells are a small population of lymphocytes that are distinctly different from the lymphocytes that provide adaptive defense mechanisms. NK cells defend the body against various viruses and cancer cells by secreting cytolytic ("cell-cutting") substances called **perforins** that lyse the cell membrane, destroying the infected cell. NK cells also secrete chemicals that enhance inflammation.

Inflammation

Inflammation produces localized redness, swelling, heat, and pain. The redness is a result of blood vessel dilation that increases blood flow and volume in affected tissues (hyperemia). This effect, coupled with an increase in permeability of nearby capillaries and subsequent leakage of protein-rich fluid into tissue spaces, swells tissues (edema). The heat comes as blood enters from deeper body parts, which are warmer than the surface. Pain results from stimulation of nearby pain receptors. Most inflammation is a tissue response to pathogen invasion, but physical factors (heat, ultraviolet light) or chemical factors (acids, bases) can also cause it.

White blood cells accumulate at the sites of inflammation, where some of them help control pathogens by phagocytosis. Neutrophils are the first to arrive at the site, followed by monocytes. Monocytes pass through capillary walls (diapedesis), becoming macrophages that remove pathogens from surrounding tissues. In bacterial infections, the resulting mass of white blood cells, bacterial cells, and damaged tissue may form a thick fluid called *pus*.

Tissue fluids (exudate) also collect in inflamed tissues. These fluids contain fibrinogen and other clotting factors that may stimulate formation of a network of fibrin threads in the affected region. Later, fibroblasts arrive and secrete fibers around the area, enclosing it in a sac of connective tissue. This walling off of the infected area helps inhibit the spread of pathogens and toxins to adjacent tissues.

Once an infection is controlled, phagocytic cells remove dead cells and other debris from the site of inflammation. Cell division replaces lost cells. Table 16.2 summarizes the process of inflammation.

Phagocytosis

Phagocytosis (fag″o-si-to′sis) removes foreign particles from the lymph as it moves from the interstitial spaces to the bloodstream. Phagocytes in the blood vessels and in the tissues of the spleen, liver, or bone marrow usually remove particles that reach the blood. Recall from chapter 14 (p. 533) that the most active phagocytic cells of the blood are *neutrophils* and *monocytes*. Chemicals released from injured tissues attract these cells (chemotaxis). Neutrophils engulf and digest smaller particles; monocytes phagocytize larger ones.

Monocytes that leave the blood become macrophages, which may be *free* or *fixed* in various tissues including lymph nodes, the spleen, the liver, and the lungs, or attached to the inner walls of blood and lymphatic vessels. A macrophage can engulf up to 100 bacteria, compared to the twenty or so bacteria that a neutrophil can engulf. Monocytes, macrophages, and neutrophils constitute the **mononuclear phagocytic system** (reticuloendothelial system).

TABLE 16.2 | Major Actions of an Inflammation Response

Action	Result
Blood vessels dilate. Capillary permeability increases and fluid leaks into tissue spaces.	Tissues become red, swollen, warm, and painful.
White blood cells invade the region.	Pus may form as white blood cells, bacterial cells, and cellular debris accumulate.
Tissue fluids containing clotting factors seep into the area.	A clot containing threads of fibrin may form.
Fibroblasts arrive.	A connective tissue sac may form around the injured tissues.
Phagocytes are active.	Bacteria, dead cells, and other debris are removed.
Cells divide.	Newly formed cells replace injured ones.

Fever

A **fever** is a nonspecific defense that offers powerful protection. A fever begins as a viral or bacterial infection stimulates lymphocytes to proliferate, producing cells that secrete a substance called *interleukin-1* (IL-1), more colorfully known as endogenous pyrogen ("fire maker from within"). IL-1 raises the thermoregulatory set point in the brain's hypothalamus to maintain a higher body temperature.

Fever indirectly counters microbial growth because higher body temperature causes the liver and spleen to sequester iron, which reduces the level of iron in the blood. Bacteria and fungi require iron for normal metabolism, so their growth and reproduction in a fever-ridden body slows and may cease. Also, phagocytic cells attack more vigorously when the temperature rises. For these reasons, low-grade fever of short duration may be a desired natural response, not a symptom to be treated aggressively with medications. Table 16.3 summarizes the types of innate (nonspecific) defenses.

RECONNECT

To Chapter 6, Clinical Application 6.4, page 184.

PRACTICE 

15 What may cause an infection?

16 Explain seven innate (nonspecific) defense mechanisms.

16.9 ADAPTIVE (SPECIFIC) DEFENSES OR IMMUNITY

The *third line of defense,* **immunity** (ĭ-mu′nĭ-te), is resistance to specific pathogens or to their toxins or metabolic by-products. An immune response is based upon the ability to distinguish molecules that are part of the body ("self") from those that are not ("nonself," or foreign). Such molecules that can elicit an immune response are called **antigens** (an′tĭ-jenz). Lymphocytes and macrophages that recognize specific nonself antigens carry out adaptive immune responses.

Antigens

Before birth, cells inventory the proteins and other large molecules in the body, learning to identify these as self. The lymphatic system responds to nonself, or foreign, antigens, but not normally to self antigens. Receptors on lymphocyte surfaces enable these cells to recognize foreign antigens.

Antigens may be proteins, polysaccharides, glycoproteins, or glycolipids. The antigens most effective in eliciting an immune response are large and complex, with few repeating parts. Sometimes, a smaller molecule that cannot by itself stimulate an immune response combines with a larger one, which makes it able to do so (antigenic). Such a small molecule is called a **hapten** (hap′ten). Stimulated lymphocytes react either to the hapten or to the larger molecule of the combination. Hapten molecules are in drugs, such as penicillin; in household and industrial chemicals; in dust particles; and in products of animal skins (dander).

Lymphocyte Origins

During fetal development, red bone marrow releases unspecialized precursors of lymphocytes into the circulation. About half of these cells reach the thymus, where they remain for a time. Here, these thymocytes specialize into T cells. ("T" refers to *thymus-derived* lymphocytes.) Later, some of these T cells constitute 70% to 80% of the circulating lymphocytes in blood (fig. 16.15). Other T cells reside in lymphatic organs and are particularly abundant in the lymph nodes, the thoracic duct, and the white pulp of the spleen.

Other lymphocytes remain in the red bone marrow until they differentiate fully into **B lymphocytes,** or **B cells.** (Historically, the "B" stands for *bursa of Fabricius,* an organ in the chicken where these cells were discovered.) The blood distributes B cells, which constitute 20% to 30% of circulating lymphocytes. B cells settle in lymphatic organs along

TABLE 16.3 | Types of Innate (Nonspecific) Defenses

Type	Description
Species resistance	A species is resistant to certain diseases to which other species are susceptible.
Mechanical barriers	Unbroken skin and mucous membranes prevent the entrance of some infectious agents. Fluids wash away microorganisms before they can firmly attach to tissues.
Chemical barriers	Enzymes in various body fluids kill pathogens. pH extremes and high salt concentration also harm pathogens. Interferons induce production of other proteins that block reproduction of viruses, stimulate phagocytosis, and enhance the activity of cells such that they resist infection and the growth of tumors. Defensins damage bacterial cell walls and membranes. Collectins grab onto microbes. Complement stimulates inflammation, attracts phagocytes, and enhances phagocytosis.
Natural killer cells	Distinct type of lymphocyte that secretes perforins that lyse virus-infected cells and cancer cells.
Inflammation	A tissue response to injury that helps prevent the spread of infectious agents into nearby tissues.
Phagocytosis	Neutrophils, monocytes, and macrophages engulf and destroy foreign particles and cells.
Fever	Elevated body temperature inhibits microbial growth and increases phagocytic activity.

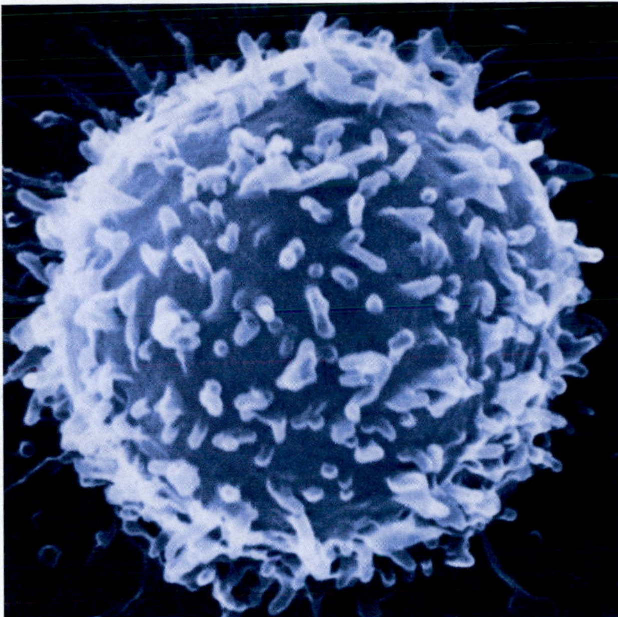

FIGURE 16.15 Falsely colored scanning electron micrograph of a circulating lymphocyte (8,600×).

with T cells and are abundant in lymph nodes, the spleen, bone marrow, and the intestinal lining (fig. 16.16).

Each person has millions of varieties of T and B cells. The members of each variety originate from a single early cell, so they are all alike, forming a **clone** (klōn) of cells (genetically identical cells originating from division of a single cell). The members of each variety have a particular type of antigen receptor on their cell membranes that can respond only to a specific antigen. Table 16.4 compares the characteristics of T cells and B cells.

PRACTICE  

17 What is immunity?

18 What is the difference between an antigen and a hapten?

19 How do T cells and B cells originate?

T Cells and the Cellular Immune Response

A lymphocyte must be activated before it can respond to an antigen. T cell activation requires processed fragments of antigen attached to the surface of another type of cell, called

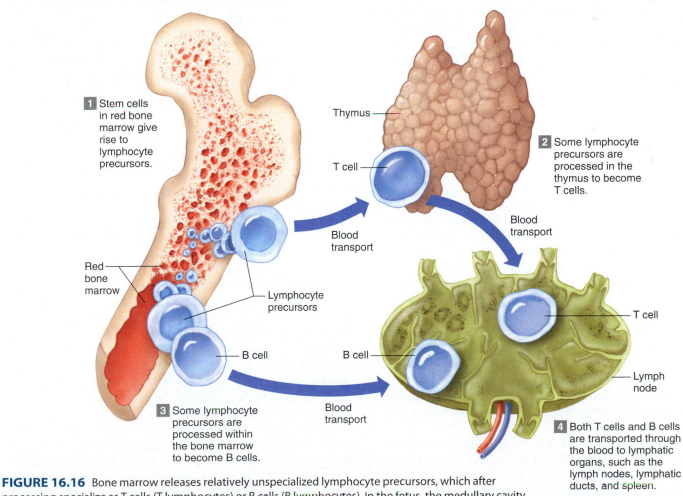

FIGURE 16.16 Bone marrow releases relatively unspecialized lymphocyte precursors, which after processing specialize as T cells (T lymphocytes) or B cells (B lymphocytes). In the fetus, the medullary cavity contains red marrow.

TABLE 16.4 | A Comparison of T Cells and B Cells

Characteristic	T Cells	B Cells
Origin of undifferentiated cell	Red bone marrow	Red bone marrow
Site of differentiation	Thymus	Red bone marrow
Primary locations	Lymphatic tissues, 70% to 80% of the circulating lymphocytes in blood	Lymphatic tissues, 20% to 30% of the circulating lymphocytes in blood
Primary functions	Provide cellular immune response in which T cells interact directly with the antigens or antigen-bearing agents to destroy them	Provide humoral immune response in which B cells interact indirectly, producing antibodies that destroy the antigens or antigen-bearing agents

an **antigen-presenting cell** (accessory cell). Macrophages, B cells, and several other cell types can be antigen-presenting cells.

T cell activation begins when a macrophage phagocytizes a bacterium, digesting it in its lysosomes. Some bacterial antigens exit the lysosomes and move to the macrophage's surface. Here, they are displayed on the cell membrane near certain protein molecules that are part of a group of proteins called the **major histocompatibility complex** (MHC) or *human leukocyte antigens* (HLA) because they were first identified on white blood cells. MHC antigens help T cells recognize that an antigen is foreign, not self. Class I MHC antigens are in cell membranes of all body cells except red blood cells. Class II MHC antigens are on the surfaces of antigen-presenting cells, thymus cells, and activated T cells.

Activated T cells interact directly with the antigen-presenting cell. Such cell-to-cell contact is called the **cellular immune response,** or cell-mediated immunity.

T cells (and some macrophages) also synthesize and secrete polypeptides called **cytokines** that enhance certain cellular responses to antigens. For example, *interleukin-1* and *interleukin-2* stimulate synthesis of several cytokines from other T cells. In addition, interleukin-1 helps activate T cells, whereas interleukin-2 causes T cells to proliferate. Other cytokines called *colony-stimulating factors* (CSFs) stimulate production of leukocytes in the red bone marrow, cause B cells to grow and mature, and activate macrophages. Certain cytokine combinations shut off the immune response. Table 16.5 summarizes several cytokine types.

T cells may also secrete toxins that kill their antigen-bearing target cells, growth-inhibiting factors that prevent target cell growth, or interferon that inhibits the proliferation of viruses and tumor cells. Several types of T cells have distinct functions.

A specialized type of T cell, called a **helper T cell,** becomes activated when its antigen receptor combines with displayed foreign antigen (fig. 16.17). Once activated, the helper T cell stimulates the B cell to produce antibodies specific for the displayed antigen.

A type of helper T cell called a CD4 cell is the prime target of HIV, the virus that causes AIDS. (CD4 stands for the "cluster-of-differentiation" antigen it bears that enables it to recognize a macrophage displaying a foreign antigen.) Considering the role of CD4 helper T cells as key players in establishing immunity—they stimulate B cells and secrete cytokines—it is no wonder that harming them destroys immunity.

TABLE 16.5 | Types of Cytokines

Cytokine	Function
Colony-stimulating factors	Stimulate bone marrow to produce lymphocytes
Interferons	Block viral replication, stimulate macrophages to engulf viruses, stimulate B cells to produce antibodies, attack cancer cells
Interleukins	Control lymphocyte differentiation and growth
Tumor necrosis factor	Stops tumor growth, releases growth factors, causes fever that accompanies bacterial infection, stimulates lymphocyte differentiation

Another type of T cell is a **cytotoxic T cell,** which recognizes and combines with nonself antigens that cancerous cells or virally infected cells display on their surfaces near certain MHC proteins. Cytokines from helper T cells activate the cytotoxic T cell (fig. 16.17*a*). Next the cytotoxic T cell proliferates, enlarging its clone of cells. Cytotoxic T cells then bind to the surfaces of antigen-bearing cells, where they release *perforin* protein that cuts porelike openings, destroying these cells. In this way, cytotoxic T cells continually monitor the body's cells, recognizing and eliminating tumor cells and cells infected with viruses.

Certain cytotoxic T cells, called CD8 T cells, give rise to **memory T cells** that provide for future immune protection. When a CD8 T cell contacts an antigen-presenting cell, it contorts into a dumbbell shape. The side of the dumbbell that contacts the antigen-presenting cell accumulates different receptors and other proteins from the side facing farthest from the provoking antigen. When the CD8 T cell divides, the daughter cell that was the part of the original cell closest to the antigen becomes an active cytotoxic T cell. The daughter cell farther from the antigen becomes a memory T cell. As its name implies, a memory T cell does not respond to an initial exposure to an antigen, but upon subsequent exposure immediately divides and differentiates into a cytotoxic T cell. This response usually vanquishes the pathogen before it can cause the body to produce signs and symptoms of disease.

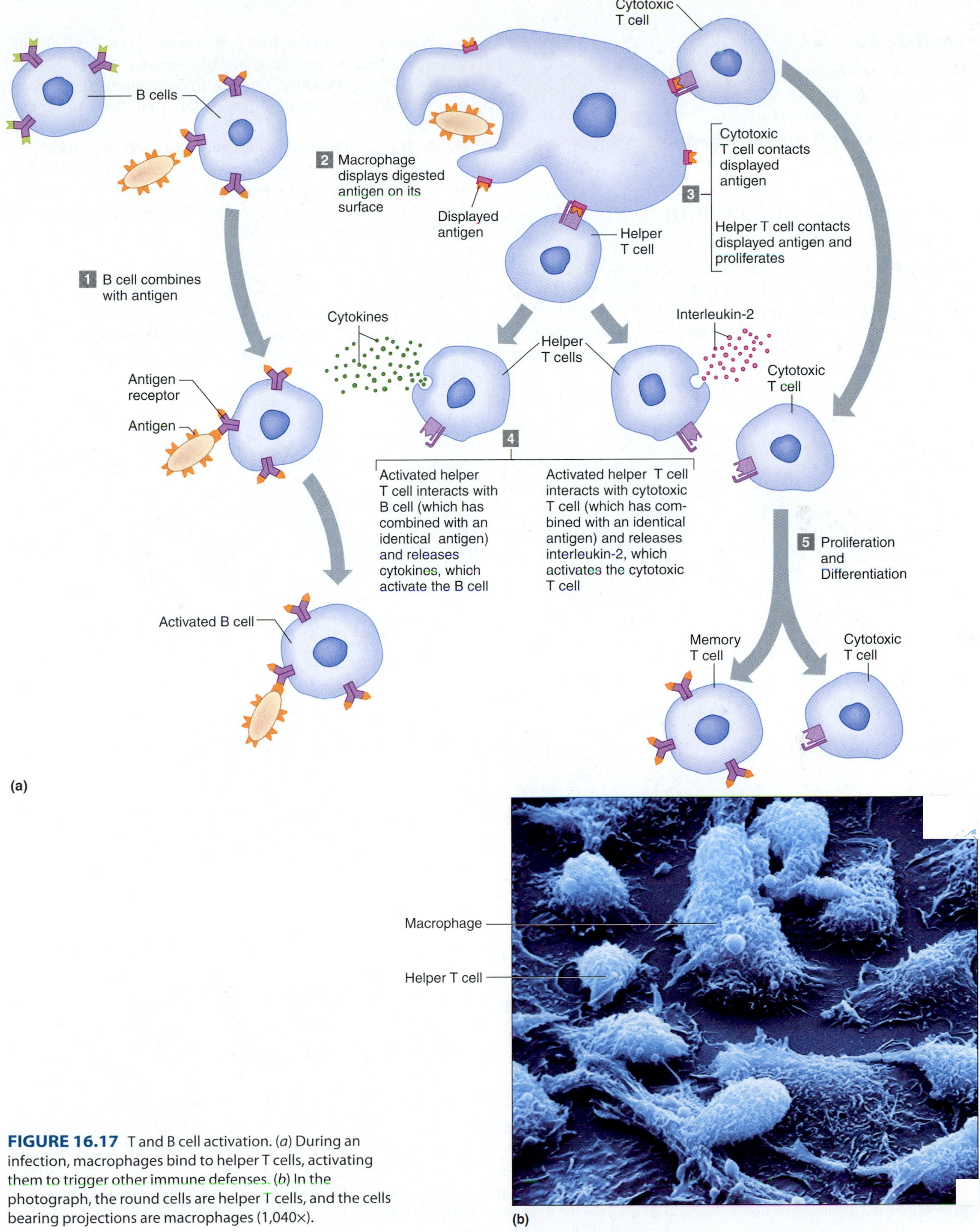
Cytotoxic T cell
B cells
2 Macrophage displays digested antigen on its surface
Displayed antigen
Cytotoxic T cell contacts displayed antigen
3
Helper T cell contacts displayed antigen and proliferates
Helper T cell
1 B cell combines with antigen
Cytokines
Helper T cells
Interleukin-2
Cytotoxic T cell
Antigen receptor
Antigen
4
Activated helper T cell interacts with B cell (which has combined with an identical antigen) and releases cytokines, which activate the B cell
Activated helper T cell interacts with cytotoxic T cell (which has combined with an identical antigen) and releases interleukin-2, which activates the cytotoxic T cell
5 Proliferation and Differentiation
Activated B cell
Memory T cell
Cytotoxic T cell
(a)
Macrophage
Helper T cell
(b)

FIGURE 16.17 T and B cell activation. (*a*) During an infection, macrophages bind to helper T cells, activating them to trigger other immune defenses. (*b*) In the photograph, the round cells are helper T cells, and the cells bearing projections are macrophages (1,040×).

PRACTICE 

20 How do T cells become activated?

21 What are some functions of cytokines?

22 Name three types of T cells.

23 How do cytotoxic T cells destroy antigen-bearing cells?

B Cells and the Humoral Immune Response

Sometimes a B cell may become activated when it encounters an antigen whose molecular shape fits the shape of the B cell's antigen receptors. In response to the receptor-antigen combination, the B cell divides repeatedly, expanding its clone. However, most of the time B cell activation requires T cell "help."

When an activated helper T cell encounters a B cell already combined with an identical foreign antigen, the helper cell releases certain cytokines. These cytokines stimulate the B cell to proliferate, thus enlarging its clone of antibody-producing cells (fig. 16.18). The cytokines also attract macrophages and leukocytes into inflamed tissues and help keep them there.

Some members of the activated B cell's clone differentiate further into **memory cells** (fig. 16.19). Like memory T cells, these memory B cells respond rapidly to subsequent exposure to a specific antigen.

In 2008, researchers discovered that memory B cells are incredibly long-lived. B cells taken from living survivors of the 1918 flu pandemic produced antibodies in the laboratory that, when injected into mice, fully protected them against any form of influenza.

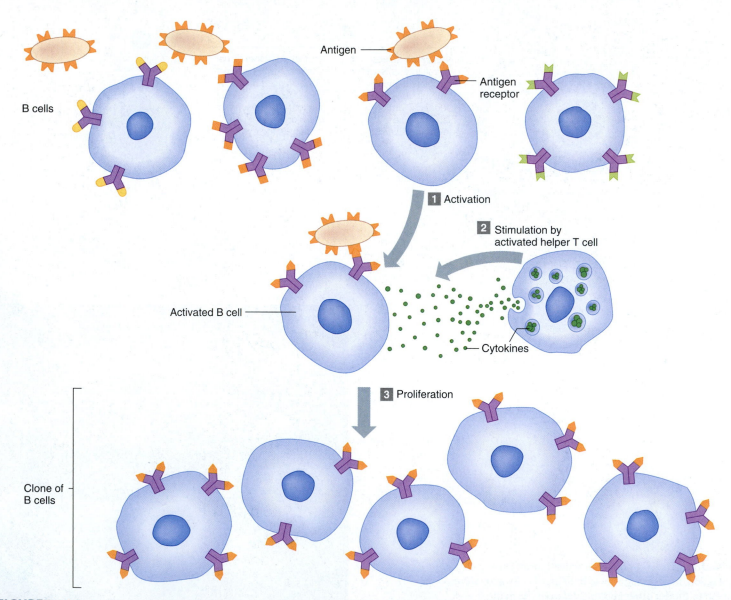

FIGURE 16.18 B cell clone. 1. When a B cell encounters an antigen that fits its antigen receptor, it becomes activated. 2. Stimulated by helper T cell cytokines, the B cell proliferates. 3. Proliferation enlarges the B cell clone. All cells in the clone have the same antigen receptor.

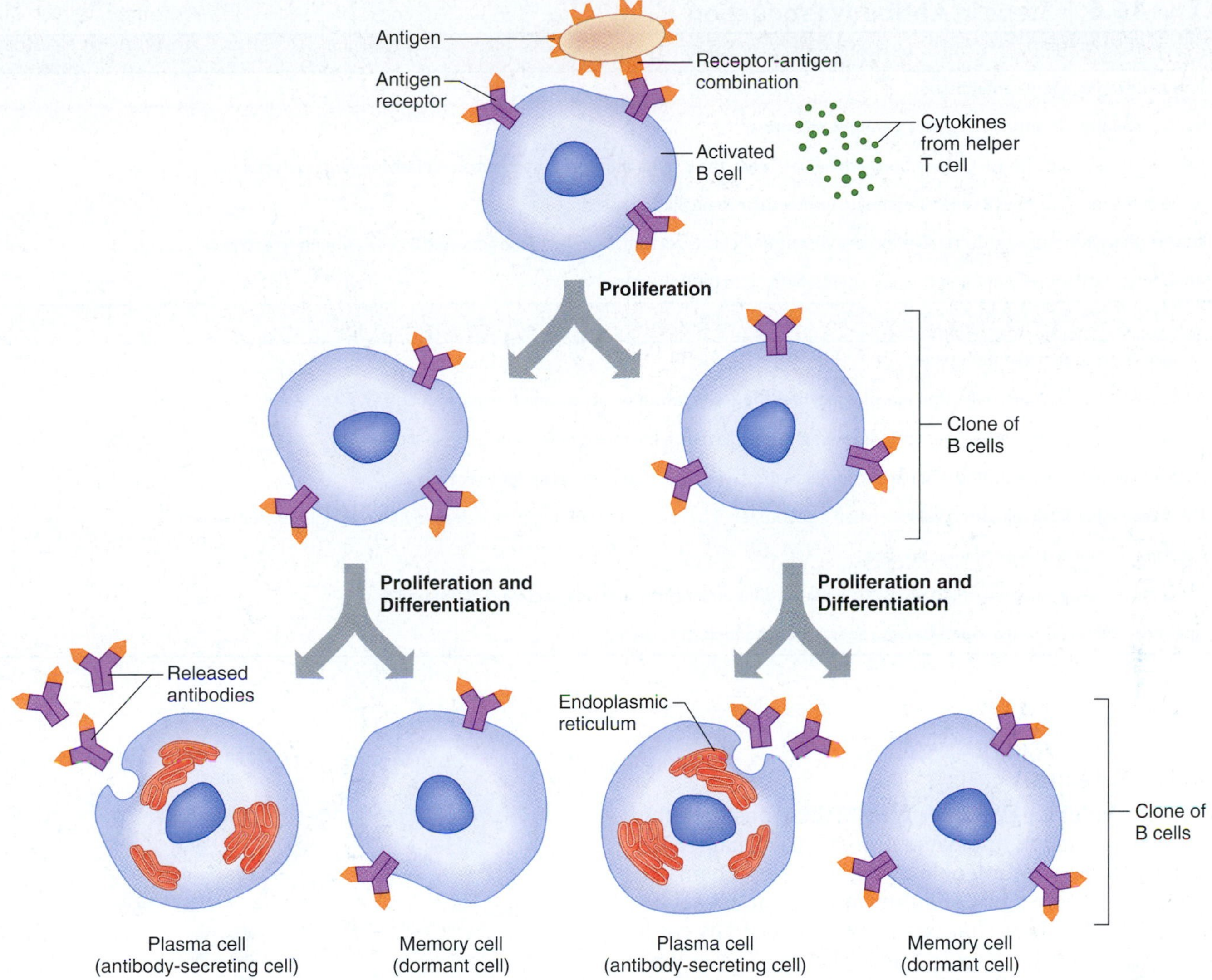

FIGURE 16.19 An activated B cell proliferates after stimulation by cytokines released by helper T cells. The B cell's clone enlarges. Some cells of the clone give rise to antibody-secreting plasma cells and others to dormant memory cells.

Other members of the activated B cell's clone differentiate further into **plasma cells,** which produce and secrete large globular proteins called **antibodies** (an′tĭ-bod″ēz) or **immunoglobulins** (im″u-no-glob′u-linz) similar in structure to the antigen-receptor molecules on the original B cell's surface (fig. 16.19). These antibodies can combine with the antigen on the pathogen and react against it. A plasma cell is an antibody factory, as evidenced by its characteristically huge Golgi apparatus. At the peak of an infection, a plasma cell may produce and secrete 2,000 antibody molecules per second! Body fluids carry antibodies, which then react in various ways to destroy specific antigens or antigen-bearing particles. This antibody-mediated immune response is called the **humoral immune response** ("humoral" refers to fluid). Table 16.6 summarizes the steps leading to antibody production as a result of B and T cell activities. T cells can suppress antibody formation by releasing cytokines that inhibit B cell function.

A single type of B cell carries information to produce a single type of antibody. However, different B cells respond to different antigens on a pathogen's surface. Therefore, an immune response may include several types of antibodies manufactured against a single microbe or virus. This is called a **polyclonal response.** From Science to Technology 16.1 discusses how researchers use clones of single B cells to produce single, or monoclonal, antibodies.

The human body can manufacture an apparently limitless number of different antibodies, but we have a limited number of antibody genes. During the early development of B cells, sections of their antibody genes move to other chromosomal locations, creating new genetic instructions for antibodies. The great diversity of antibody types is increased further because different antibody protein subunits combine. Antibody diversity is like using the limited number of words in a language to compose an infinite variety of stories.

TABLE 16.6 | Steps in Antibody Production

B Cell Activities
1. Antigen-bearing agents enter tissues.
2. B cell encounters an antigen that fits its antigen receptors.
3. Either alone or more often in conjunction with helper T cells, the B cell is activated. The B cell proliferates, enlarging its clone.
4. Some of the newly formed B cells differentiate further to become plasma cells.
5. Plasma cells synthesize and secrete antibodies whose molecular structure is similar to the activated B cell's antigen receptors.
6. Antibodies combine with antigen-bearing agents, helping to destroy them.
T Cell Activities
1. Antigen-bearing agents enter tissues.
2. An accessory cell, such as a macrophage, phagocytizes the antigen-bearing agent, and the macrophage's lysosomes digest the agent.
3. Antigens from the digested antigen-bearing agents are displayed on the membrane of the accessory cell.
4. Helper T cell becomes activated when it encounters a displayed antigen that fits its antigen receptors.
5. Activated helper T cell releases cytokines when it encounters a B cell previously combined with an identical antigen-bearing agent.
6. Cytokines stimulate the B cell to proliferate.
7. Some of the newly formed B cells give rise to cells that differentiate into antibody-secreting plasma cells.
8. Antibodies combine with antigen-bearing agents, helping to destroy them.

Antibody Molecules

Antibodies are soluble, globular proteins that constitute the *gamma globulin* fraction of plasma proteins (see chapter 14, p. 537). Each antibody molecule consists of four chains of amino acids linked by pairs of sulfur atoms that attract by disulfide bonds. The four chains form a Y-shaped structure (fig. 16.20). Two of these amino acid chains are identical **light chains** (L-chains), and two are identical **heavy chains** (H-chains). The heavy chains have about twice as many amino acids as the light chains. The five major types of antibody molecules are distinguished by a particular type of heavy chain. Most of the types of antibody molecules consist of a single Y-shaped structure, but some have as many as five (see fig. 14.21).

As with other proteins, the sequences of amino acids of the heavy and light chains confer the unique, three-dimensional structure (conformation) of each antibody. This special conformation, in turn, imparts the physiological properties of the molecule. For example, one end of each of the heavy and light chains consists of variable sequences of amino acids (variable regions). These regions are specialized to fit the shape of a specific antigen molecule.

Antibodies can bind to certain antigens because of the conformation of the variable regions. The antibody contorts to form a pocket around the antigen. These specialized ends of the antibody molecule are called **antigen-binding sites,** and the parts that bind the antigen are called **idiotypes** (id′e-o-tīpz′).

The remaining parts of the chains are termed constant regions because their amino acid sequences are similar.

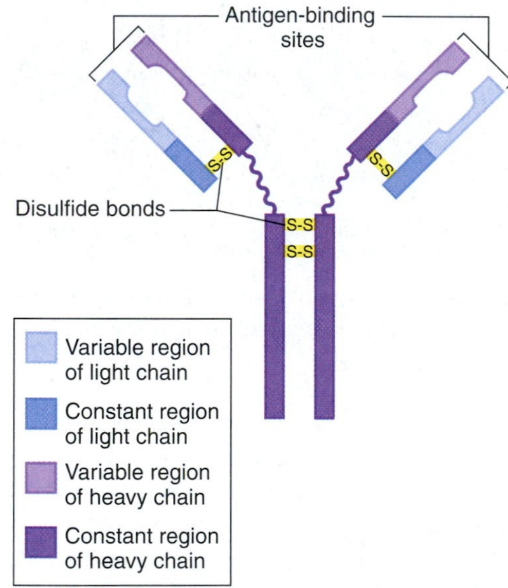

FIGURE 16.20 An antibody (immunoglobulin) molecule consists basically of two identical light chains of amino acids and two identical heavy chains of amino acids.

Constant regions impart other properties of the antibody molecule, such as its ability to bond to cellular structures or to combine with certain chemicals (fig. 16.20).

Types of Immunoglobulins

Of the five major types of immunoglobulins, three constitute the bulk of the circulating antibodies. They are immunoglobulin G, which accounts for about 80% of the antibodies;

immunoglobulin A, which makes up about 13%; and immunoglobulin M, responsible for about 6%. The remainder of the antibodies are immunoglobulin D or immunoglobulin E.

Immunoglobulin G (IgG) is in plasma and tissue fluids and is effective against bacteria, viruses, and toxins. IgG also activates complement proteins introduced in section 16.7 and described further in the following section "Antibody Actions." Anti-Rh antibodies are examples of IgG and, as described in chapter 14 (p. 547), can cross the placenta.

Immunoglobulin A (IgA) is in exocrine gland secretions. It is in breast milk, tears, nasal fluid, gastric juice, intestinal juice, bile, and urine.

A newborn does not yet have its own antibodies but has IgG that passed through the placenta from the mother. These maternal antibodies protect the infant against some illnesses to which the mother is immune. The maternal antibody supply begins to fall just about when the infant begins to make its own antibodies. The newborn also receives IgA from colostrum, which is secreted from the mother's breasts for the first few days after birth. Antibodies in colostrum protect against certain digestive and respiratory infections.

Immunoglobulin M (IgM) is a type of antibody produced in plasma in response to contact with certain antigens in foods or bacteria. Examples of IgM are the anti-A and anti-B antibodies, described in chapter 14 (p. 544). IgM also activates complement.

Immunoglobulin D (IgD) is on the surfaces of most B cells, especially those of infants. IgD acts as an antigen receptor and is important in activating B cells (see fig. 16.18).

Immunoglobulin E (IgE) appears in exocrine secretions with IgA. It is associated with allergic reactions, described later in this chapter in the section "Allergic Reactions." Table 16.7 summarizes the major immunoglobulins and their functions.

PRACTICE

24 How are B cells activated?

25 How does the antibody response protect against diverse infections?

26 What is an immunoglobulin?

27 Describe the structure of an immunoglobulin molecule, and name the five major types of immunoglobulins.

Antibody Actions

In general, antibodies react to antigens in three ways. Antibodies directly attack antigens, activate complement, or stimulate localized changes (inflammation) that help prevent spread of the pathogen.

In a direct attack, antibodies combine with antigens and cause them to clump (agglutinate) or to form insoluble substances (precipitation). Such actions make it easier for phagocytic cells to engulf the antigen-bearing pathogens and eliminate them. In other instances, antibodies cover the toxic parts of antigen molecules and neutralize their effects (neutralization). However, under normal conditions, complement activation is more important in protecting against infection than is direct antibody attack.

When certain IgG or IgM antibodies combine with antigens, they expose reactive sites on the antibody constant regions. This triggers a series of reactions leading to activation of complement proteins, which, in turn, produce a variety of effects, including coating the antigen-antibody complexes (opsonization), making the complexes more susceptible to phagocytosis; attracting macrophages and neutrophils into the region (chemotaxis); clumping antigen-bearing cells; rupturing membranes of foreign cells (lysis); and altering the molecular structure of viruses, making them harmless. Other proteins promote inflammation, which helps prevent the spread of infectious agents.

Immunoglobulin E promotes inflammation that may be so intense that it damages tissues. This antibody is usually attached to the membranes of widely distributed *mast cells* (see chapter 5, p. 155). When antigens combine with the antibodies, the resulting antigen-antibody complexes stimulate mast cells to release biochemicals, such as histamine, that cause the changes associated with inflammation, such as vasodilation and edema. Table 16.8 summarizes the actions of antibodies.

PRACTICE

28 In what general ways do antibodies function?

29 How is complement activated?

30 What is the function of complement?

TABLE 16.7 | Characteristics of Major Immunoglobulins

Type	Occurrence	Major Function
IgG	Plasma and tissue fluid	Defends against bacteria, viruses, and toxins; activates complement
IgA	Exocrine gland secretions	Defends against bacteria and viruses
IgM	Plasma	Reacts with antigens on some red blood cell membranes following mismatched blood transfusions; activates complement
IgD	Surface of most B lymphocytes	B cell activation
IgE	Exocrine gland secretions	Promotes inflammation and allergic reactions

16.1 FROM SCIENCE TO TECHNOLOGY

Immunotherapy

At the turn of the last century, German bacteriologist Paul Ehrlich developed the concept of the "magic bullet"—a substance that could enter the body and destroy diseased cells, yet spare the healthy ones. The biochemicals and cells of the immune system, with their great specificity for attacking foreign tissue, would be ideal magic bullets. *Immunotherapy* uses immune system components to fight disease—both the humoral immune response (antibodies) and the cellular immune response (cytokines).

Monoclonal Antibodies

Tapping the specificity of a single B cell and using its single type, or *monoclonal,* antibody to target a specific antigen (such as on a cancer or bacterial cell) awaited finding a way to entice the normally short-lived mature B cells into persisting in culture. In 1975, British researchers Cesar Milstein and Georges Köhler devised *monoclonal antibody* (MAb) technology to capture the antibody-making capacity of a single B cell.

Milstein and Köhler injected a mouse with antigen-laden red blood cells from a sheep. They then isolated a single B cell from the mouse's spleen and fused it with a cancerous white blood cell from a mouse. The result was a fused cell, or *hybridoma,* with a valuable pair of talents: Like the B cell, it produces large amounts of a single antibody type; like the cancer cell, it divides continuously (fig. 16A). Human versions of MAbs are now used.

MAbs are important in basic research, veterinary and human health care, and agriculture. Cell biologists use pure antibodies to localize and isolate proteins. Diagnostic MAb "kits" detect tiny amounts of a single molecule. Most kits consist of a paper strip impregnated with a MAb, to which the user adds a body fluid. For example, a woman who suspects she is pregnant places drops of her urine onto the paper. A color change ensues if the MAb binds to human

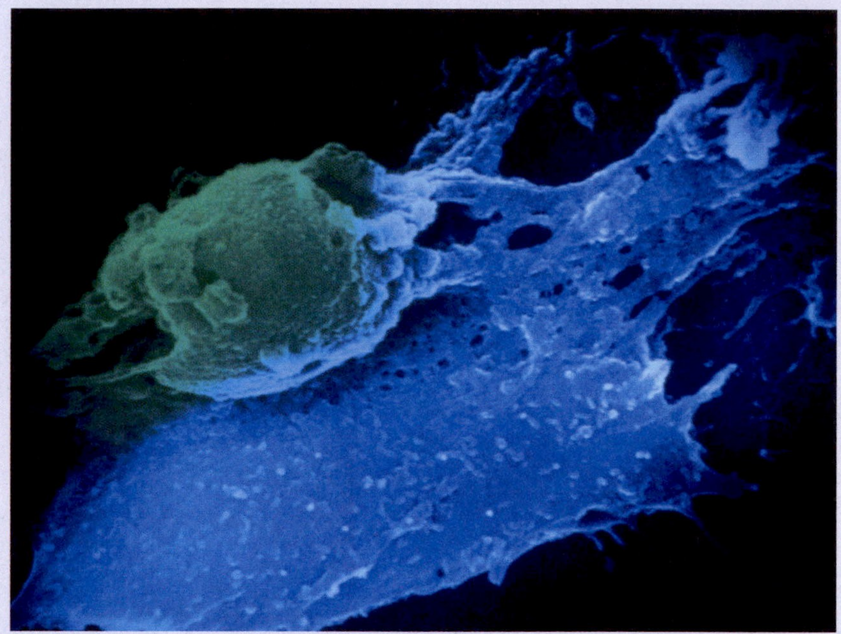

FIGURE 16A Monoclonal antibodies are produced by a type of artificial cell combination called a hybridoma. It consists of a cancer cell (the flat blue cell) fused with a B cell (the round green cell). The cancer cell contributes rapid and continuous division; the B cell secretes a single antibody type (7,000×).

TABLE 16.8 | Actions of Antibodies

General Action	Type of Effect	Description
Direct Attack		
	Agglutination	Antigens clump
	Precipitation	Antigens become insoluble
	Neutralization	Antigens lose toxic properties
Activation of Complement (Antibodies combined with antigens)	Opsonization	Alters antigen cell membranes so cells are more susceptible to phagocytosis
	Chemotaxis	Attracts macrophages and neutrophils into the region
	Agglutination	Clumping of antigen-bearing cells
	Lysis	Allows rapid movement of water and ions into the foreign cell causing osmotic rupture of the foreign cell
	Neutralization	Altering the molecular structure of viruses, making them harmless
Localized Changes		
	Inflammation	Helps prevent the spread of antigens

chorionic gonadotropin (see chapter 23, p. 882), indicating pregnancy.

MAbs can highlight a new cancer or detect recurrence. The MAb is attached to a radioactive chemical, which emits a signal when the MAb binds an antigen unique to the cancer cell surface.

MAbs can ferry conventional cancer treatments to where they are needed and spare healthy tissue. Drugs or radioactive chemicals are attached to MAbs that deliver them to antigens on cancer cells. When injected into a patient, the MAb and its cargo are engulfed by the cancer cells, which are destroyed. MAbs can also combat cancer directly by blocking growth factors that otherwise bind cancer cells, stimulating them to divide. Table 16A lists some MAb-based drugs and how they work. They have rare adverse effects, including anaphylactic shock and severe anemia.

Cytokines

Immunotherapy experiments were difficult to do in the late 1960s because cytokines and antibodies could be obtained only in small amounts from cadavers. In the 1970s, recombinant DNA and monoclonal antibody technologies yielded unlimited amounts of pure proteins—just as the AIDS epidemic was making it essential to find a purer source of biochemicals than cadavers.

Interferon was the first cytokine tested on a large scale. It is used to treat a dozen or so conditions, including a type of leukemia, multiple sclerosis, hepatitis, and genital warts. Interleukin-2 is used to treat kidney cancer, and colony-stimulating factors boost the white blood cell supply in people whose immune systems are temporarily suppressed, such as those receiving drugs to treat cancer or AIDS, or transplant recipients.

Increasingly, cancer treatment consists of combinations of immune system cells or biochemicals, plus standard therapies. Immunotherapy can enable a patient to withstand higher doses of a conventional drug or destroy cancer cells remaining after standard treatment. ■

TABLE 16A | **Drugs Based on Monoclonal Antibodies**

Drug	Condition	Mechanism
Tysabri	Multiple sclerosis	Binds cells that produce autoantibodies against brain and/or spinal cord tissue
Erbitux	Colorectal cancer	Blocks growth factor receptor overabundant on cancer cells, preventing cell division
Herceptin	Breast cancer	Blocks growth factor receptor overabundant on cancer cells, preventing cell division
Iressa	Lung cancer	Blocks growth factor receptor overabundant on cancer cells, preventing cell division
Avastin	Colorectal cancer Lung cancer	Blocks growth factor from stimulating formation of blood vessels to tumor
Raptiva	Psoriasis	Blocks T cell functions
Xolair	Asthma	Blocks IgE, preventing allergic response
Synagis	Respiratory syncytial virus infection	Binds virus part required to enter cells
Remicade	Crohn disease Rheumatoid arthritis	Blocks excess tumor necrosis factor
Rituxan	B cell lymphoma	Blocks CD20 protein receptor on all B cells—bone marrow stem cells then restore healthy B cells
Zenapax	Rejection of transplant	Blocks IL-2 from rejecting transplant

Immune Responses

When B cells or T cells become activated after first encountering the antigens for which they are specialized to react, their actions constitute a **primary immune response.** During such a response, plasma cells release antibodies (IgM, followed by IgG) into the lymph. The antibodies are transported to the blood and then throughout the body, where they help destroy antigen-bearing agents. Production and release of antibodies continues for several weeks.

After a primary immune response, some of the B cells produced during proliferation of the clone remain dormant and serve as *memory cells* (see fig. 16.19). If the identical antigen is encountered in the future, the clones of these memory cells enlarge, and they can respond rapidly with IgG to the antigen to which they were previously sensitized. These memory B cells with memory T cells produce a **secondary immune response.** In lymph nodes, *follicular dendritic cells* may help memory by harboring and slowly releasing viral antigens after an initial infection. This constantly stimulates memory B cells, which present the antigens to memory T cells, maintaining immunity.

As a result of a primary immune response, detectable concentrations of antibodies usually appear in the plasma within five to ten days after exposure to antigens. If the identical antigen is encountered later, a secondary immune response may produce additional antibodies within a day or two (fig. 16.21). Although newly formed antibodies may persist in the body for only a few months or years, memory cells live much longer. A secondary immune response may be very long-lasting.

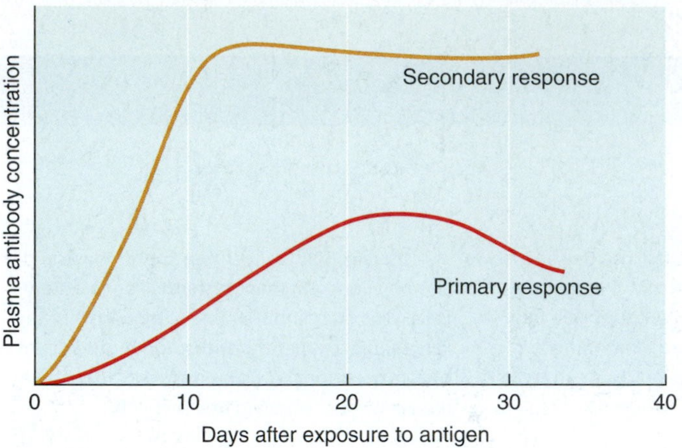

FIGURE 16.21 A primary immune response causes less-vigorous antibody production than does a secondary immune response.

Practical Classification of Immunity

Before vaccines against "childhood diseases" began to be developed in the 1960s, suffering through measles, mumps, rubella, and chickenpox was part of attending elementary school. However, each child usually had each illness only once, thanks to *naturally acquired active immunity.* This form of immunity develops after a primary immune response and is a response to exposure to a live pathogen.

Today, most children in developed countries do not contract measles, mumps, rubella, or chickenpox because they develop another type of active immunity, produced in response to receiving a **vaccine** (vak′sēn). A vaccine is a preparation that includes an antigen that can stimulate a primary immune response against a particular pathogen but does not produce symptoms of that disease.

A vaccine might include bacteria or viruses that have been killed or attenuated (weakened) so that they cannot cause a serious infection, or a toxoid, a toxin from an infectious organism that has been chemically altered to destroy its dangerous effects. A "subunit" vaccine consists of a single glycoprotein or similar large molecule from the pathogen's surface, which provides enough of a foreign antigen to alert the immune system. A vaccine causes a person to develop *artificially acquired active immunity.*

Although an individual receives a vaccine, the ultimate effect is at the population level. That is, if a critical number of people are vaccinated, becoming immune, the infectious agent can no longer easily pass from person to person. This protection that results from widespread vaccination is called *herd immunity.*

Viruses whose genetic material rapidly mutates present a great challenge to vaccine development because their surfaces, which serve as antigens, change. It is a little like fighting an enemy who is constantly changing disguises. For this reason, pharmaceutical companies must develop a new vaccine against influenza each year. HIV is particularly changeable, which has severely hampered efforts to produce a vaccine.

Vaccines stimulate active immunity against a variety of diseases, including typhoid fever, cholera, whooping cough, diphtheria, tetanus, polio, influenza, hepatitis A and B, and bacterial pneumonia. Vaccines have virtually eliminated natural smallpox from the world. Vaccine distribution is not equitable worldwide. Many thousands of people in underdeveloped countries die of infectious diseases for which vaccines are available in other nations.

For some infections, science makes developing a vaccine difficult. This is the case for norovirus infection, also known as "winter vomiting disease" or erroneously as "stomach flu" (influenza is respiratory). The virus and receptors for it on small intestine lining cells are diverse, so the virus always finds a place to infect. Norovirus passes readily in feces and vomit, especially in crowded areas such as cruise ships and schools. Some people do not develop symptoms, yet can pass the virus, and immunity does not last until the next winter. Practically, a vaccine is not a priority because for most people, the illness resolves quickly and is not severe.

Sometimes a person who has been exposed to infection needs protection against a pathogen but lacks time to develop active immunity from a vaccine. This happens with hepatitis A, a viral infection of the liver. In such a case, it may be possible to inject the person with antiserum, which has ready-made antibodies from gamma globulin separated from the blood plasma of persons who have already developed immunity against the disease.

An injection of antibodies or antitoxin (antibodies against a toxin) provides *artificially acquired passive immunity.* It is called passive because the recipient's cells do not produce the antibodies. Such immunity is short-term, seldom lasting more than a few weeks. Furthermore, because the recipient's lymphocytes might not have time to react to the pathogens for which protection was needed, susceptibility to infection may persist.

During pregnancy, certain antibodies (IgG) pass from the maternal blood into the fetal bloodstream. Receptor-mediated endocytosis (see chapter 3, p. 96) using receptor sites on cells of the fetal yolk sac accomplishes the transfer. These receptor sites bind to a region common to the structure of IgG molecules. After entering the fetal cells, the antibodies are secreted into the fetal blood. The fetus acquires limited immunity against the pathogens for which the pregnant woman has developed active immunities. Thus, the fetus has *naturally acquired passive immunity,* which may persist for six months to a year after birth. The newborn may naturally acquire passive immunity through breast milk as well. Table 16.9 summarizes the types of immunity.

PRACTICE 

31 Distinguish between a primary and a secondary immune response.

32 Explain the difference between active and passive immunities.

TABLE 16.9 | Practical Classification of Immunity

Type	Mechanism	Result
Naturally acquired active immunity	Exposure to live pathogens	Stimulation of an immune response with symptoms of a disease
Artificially acquired active immunity	Exposure to a vaccine containing weakened or dead pathogens or their components	Stimulation of an immune response without the symptoms of a disease
Artificially acquired passive immunity	Injection of gamma globulin containing antibodies or antitoxin	Short-term immunity without stimulating an immune response
Naturally acquired passive immunity	Antibodies passed to fetus from pregnant woman with active immunity or to newborn through breast milk from a woman with active immunity	Short-term immunity for newborn without stimulating an immune response

Allergic Reactions

Both allergic reactions and immune responses entail the sensitizing of lymphocytes or the combining of antigens with antibodies. An allergic reaction, however, is an immune response to a nonharmful substance and can damage tissues. An allergy is also called a hypersensitivity reaction. One form of allergic reaction can occur in almost anyone, but another affects only people with an inherited tendency toward exaggerated immune responses. The antigens that trigger allergic responses are called **allergens** (al′er-jenz).

An *immediate-reaction* (type I or anaphylactic) *allergy* occurs within minutes after contact with an allergen. Persons with this type of allergy have inherited the tendency to overproduce IgE antibodies in response to certain antigens. IgE normally comprises a tiny fraction of plasma proteins.

An immediate-reaction allergy activates B cells, which become sensitized when the allergen is first encountered. Subsequent exposures to the allergen trigger allergic reactions. In the initial exposure, IgE attaches to the membranes of widely distributed mast cells and basophils. When a subsequent allergen-antibody reaction occurs, these cells release allergy mediators such as *histamine, prostaglandin* D_2, and leukotrienes (fig. 16.22). These substances affect physiology by dilating arterioles and increasing vascular permeability, both of which cause edema. They also cause contraction of bronchial and intestinal smooth muscles, and increased mucus production. The result is a severe inflammation reaction responsible for the symptoms of the allergy, such as hives, hay fever, asthma, eczema, or gastric disturbances.

Anaphylactic shock is a severe form of immediate-reaction allergy in which mast cells release allergy mediators throughout the body. The person may at first feel an inexplicable apprehension, and then suddenly, the entire body itches and breaks out in red hives. Vomiting and diarrhea may follow. The face, tongue, and larynx begin to swell, and breathing becomes difficult. Unless the person receives an injection of epinephrine (adrenalin) and sometimes a tracheotomy (an incision into the windpipe to restore breathing), he or she will lose consciousness and may die within five minutes. Anaphylactic shock most often results from an allergy to penicillin or insect stings. Fortunately, thanks to prompt medical attention and avoidance of allergens by people who know they have allergies, fewer than 100 people a year die of anaphylactic shock. The peanut allergy described in the chapter opening vignette causes many of the symptoms of anaphylactic shock, but usually not the sensation of the throat closing.

> One theory of the origin of allergies, particularly anaphylactic shock, is that they evolved at a time when insect bites and the natural substances from which antibiotics such as penicillin are made threatened human survival. Today, that once-protective response is an overreaction. The observation that IgE protects against roundworm and flatworm infections, in addition to taking part in allergic reactions, supports the idea that this antibody class is a holdover from times past, when challenges to the immune system differed from what they are today.

Hypersensitivities that take one to three hours to develop include *antibody-dependent cytotoxic reactions* (type II) and *immune complex reactions* (type III). In an antibody-dependent cytotoxic reaction, an antigen binds to a specific cell, stimulating phagocytosis and complement-mediated lysis of the antigen. A transfusion reaction to mismatched blood is a type II hypersensitivity reaction. In an immune complex or type III reaction, phagocytosis and lysis cannot clear widespread antigen-antibody complexes from the circulation. The complexes may block small vessels, which damages the tissues that they reach. **Autoimmunity,** the loss of the ability to tolerate self-antigens, illustrates this type of hypersensitivity reaction. It is discussed later in the chapter in the section "Autoimmunity."

> The *tuberculin skin test* is used to detect individuals who have tuberculosis (TB) or who have had it (or a closely related infection) or been exposed to it. The test introduces a tuberculin preparation called *purified protein derivative* (PPD) into the superficial layers of the skin (Mantoux test). If the person's T cells have been sensitized to the antigens of the mycobacteria that cause tuberculosis, an allergic reaction occurs within forty-eight to seventy-two hours, and a localized region of the skin and subcutaneous tissue hardens (indurates). The absence of this reaction (negative result) signifies that the person's T cells have not previously encountered the mycobacterial antigens.

Allergic
reaction

Allergen

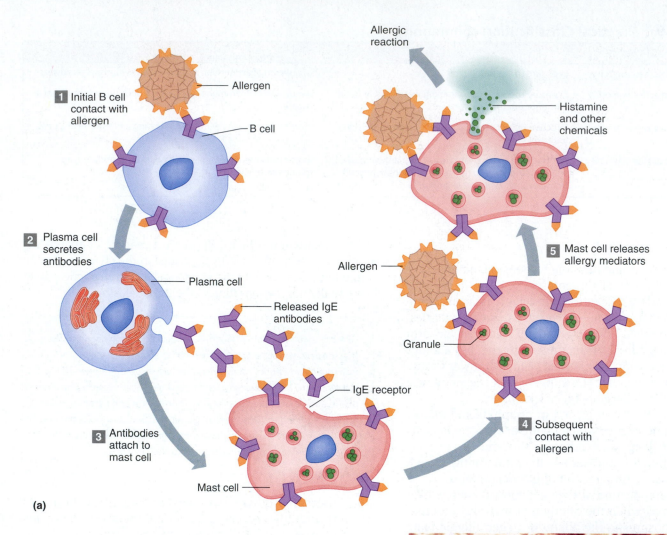

1 Initial B cell contact with allergen

B cell

Histamine and other chemicals

2 Plasma cell secretes antibodies

Plasma cell

5 Mast cell releases allergy mediators

Allergen

Released IgE antibodies

Granule

IgE receptor

3 Antibodies attach to mast cell

4 Subsequent contact with allergen

Mast cell

(a)

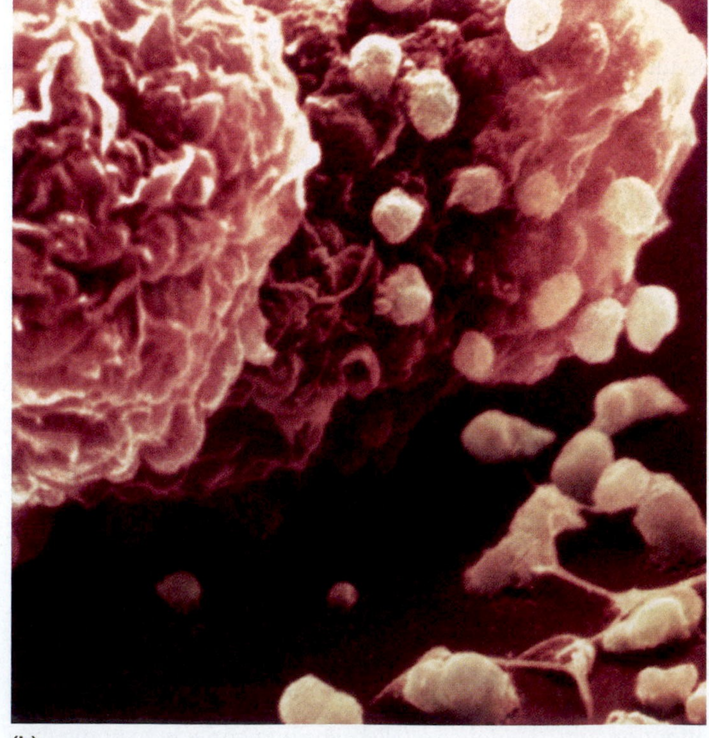

(b)

FIGURE 16.22 Immediate-reaction allergy. (*a*) 1. B cells are activated when they contact an allergen. 2. An activated B cell differentiates further into an antibody-secreting plasma cell. 3. Antibodies attach to mast cells. 4. When allergens are encountered, they combine with the antibodies on the mast cells. 5. The mast cells release allergy mediators, which cause the symptoms of the allergy attack. (*b*) A mast cell releases histamine granules (3,000×).

A *delayed-reaction allergy* (type IV) may affect anyone. It results from repeated exposure of the skin to certain chemicals—commonly, household or industrial chemicals or some cosmetics. Eventually the foreign substance activates T cells, many of which collect in the skin. The T cells and the macrophages they attract release chemical factors, which, in turn, cause eruptions and inflammation of the skin (dermatitis). This reaction is called *delayed* because it usually takes about forty-eight hours to occur.

Transplantation and Tissue Rejection

When a car breaks down, replacing the damaged or malfunctioning part often fixes the trouble. The same is sometimes true for the human body. Transplanted tissues and organs include corneas, kidneys, lungs, pancreases, bone marrow, pieces of skin, livers, and hearts. A transplant is risky. The recipient's cells may recognize the donor's tissues as foreign and attempt to destroy the transplanted tissue in a **tissue rejection reaction.** The transplanted tissue may also produce molecules that harm the recipient's tissue, a response called graft-versus-host disease (GVHD).

Tissue rejection resembles the cellular immune response against a foreign antigen. The greater the antigenic difference between the cell surface molecules (MHC antigens, discussed earlier in this chapter on page 630) of the recipient tissues and the donor tissues, the more rapid and severe the rejection reaction. Matching the cell surface molecules of donor and recipient tissues can minimize the rejection reaction. This means locating a donor whose tissues are antigenically similar to those of the person needing a transplant.

The four major types of grafts (transplant tissue) include

- *Isograft.* Tissue is from an identical twin.
- *Autograft.* Tissue is taken from elsewhere in a person's body. (Technically, this is not a transplant because it is within an individual.)
- *Allograft.* Tissue comes from another person who is not an identical twin.
- *Xenograft.* Tissue comes from a different species, such as pigs and baboons.

Table 16.10 presents examples of transplants.

Transplanting a body part, such as a hand or face, is much more difficult than transplanting tissue or an organ, because a greater diversity of cell and tissue types must be replaced, and each has different effects on the recipient's immune system. A hand transplant, for example, requires skin, muscle, nerve, and bone.

Immunosuppressive drugs are used to reduce the recipient immune system's rejection of transplanted tissue. These drugs suppress T cell or antibody production, thereby dampening the cellular and humoral immune responses, but often have severe side effects, including infection, kidney damage, and cancer. Since their use began in the 1980s, immunosuppressive drugs were typically given for the rest of the recipient's life. That view is changing, as cases accumulate of patients who received well-matched transplants and are surviving *without* continued immunosuppressive therapy. These patients received, along with standard transplants such as kidneys, bone marrow stem cells from the donors. Apparently the influx of donor cells establishes a "stable coexistence" with the recipient's cells—they accept each other. When this balance is disturbed, tissue rejection or graft-versus-host disease is the outcome. Researchers hypothesize that bombarding the recipient's body with immunosuppressive drugs immediately after the transplant can disturb this process. The emerging new view is to give the drugs *before* the transplant, and in some cases minimize their use after.

TABLE 16.10 | Transplant Types

Type	Donor	Example
Isograft	Identical twin	Bone marrow transplant from a healthy twin to a twin who has leukemia
Autograft	Self	Skin graft from one part of the body to replace burned skin
Allograft	Same species	Kidney transplant from a relative or closely matched donor
Xenograft	Different species	Heart valves from a pig

Autoimmunity

The immune system can fail to distinguish self from nonself, producing antibodies, called **autoantibodies,** and cytotoxic T cells that attack and damage the body's tissues and organs. This attack against self is called *autoimmunity.* The signs and symptoms of autoimmune disorders reflect the affected cell types. In autoimmune hemolytic anemia, autoantibodies destroy red blood cells. Autoimmune ulcerative colitis harms colon cells and severe abdominal pain results. Table 16.11 lists some autoimmune disorders.

Why might the immune system attack body tissues? Perhaps a virus, while replicating in a human cell, "borrows" proteins from the host cell's surface and incorporates them onto its own surface. When the immune system "learns" the surface of the virus to destroy it, it also learns to attack the human cells that normally bear the particular protein. Another explanation of autoimmunity is that T cells never learn in the thymus to distinguish self from nonself.

A third possible route of autoimmunity is when a nonself antigen coincidentally resembles a self antigen. For example, damage to heart valve cells in acute rheumatic fever is due to attack by antibodies present from a recent throat infection with group A streptococcus bacteria. The surfaces of the cells that make up the heart valve resemble those of the bacteria.

Some disorders thought to be autoimmune may have a stranger cause—fetal cells persisting in a woman's circulation for decades. In response to an as yet unknown trigger, the fetal cells, perhaps "hiding" in a tissue such as skin, emerge,

16.1 CLINICAL APPLICATION

Immunity Breakdown: AIDS

Natural History of a Modern Plague

In late 1981 and early 1982, physicians from large cities began reporting to the United States Centers for Disease Control and Prevention cases of formerly rare infections in otherwise healthy young men. Infections prevalent in the general population, such as herpes simplex and cytomegalovirus, were unusually severe in these young men. Some infections were caused by organisms known to infect only nonhuman animals. Other infections, particularly pneumonia caused by the microorganism *Pneumocystis jirovecii,* and a cancer, Kaposi sarcoma, were known only in individuals whose immune systems were suppressed (fig. 16B). The bodies of the sick young men had become nesting places for all types of infectious agents, including viruses, bacteria, protozoa, and fungi. The infections were *opportunistic,* which means that they took advantage of a weakened immune system.

As the unusual infections spread, a portrait of a lethal infectious disease emerged. Table 16B lists how it is, and isn't, spread. *Acquired immune deficiency syndrome,* or AIDS, starts with recurrent fever, weakness, and weight loss. This may begin years after infection with the human immunodeficiency virus (HIV) that causes AIDS. Then, usually after another relatively healthy period, infections begin.

How HIV Ravages the Immune System

HIV infection gradually shuts down the immune system. First, HIV crosses a mucosal barrier, such as in the anus or vagina (see figure 3.34). Then, the virus enters macrophages, impairing this first line of defense. In these cells and later in helper T cells, the virus adheres with a surface protein, called gp120, to coreceptors on the host cell surface, called CD4 and CCR5, discussed in the vignette to Chapter 3. Once the virus enters the cell, a viral enzyme, reverse transcriptase, catalyzes the construction of a DNA strand complementary to the viral RNA sequence (the virus has RNA as its genetic material). The initial viral DNA strand replicates to form a DNA double helix, which enters the cell's nucleus and inserts into a chromosome. The viral DNA sequences are then transcribed and translated. The cell fills with pieces of HIV, which are then assembled into new viral particles that eventually burst from the cell.

Once infected helper T cells start to rapidly die, bacterial infections begin, because B cells aren't activated to produce antibodies. Much later in infection, HIV variants arise that bind to receptors called CXCR4 on cytotoxic T cells, killing them too. Loss of these cells renders the body vulnerable to other infections and to cancers.

Viral Variabilty

HIV replicates quickly, and can hide, twisting and altering its surface features in ways that evade recognition and attack by antibodies or cytotoxic T cells. The virus is especially prone to mutation, because it cannot repair DNA replication errors. The immune system cannot keep up; antibodies against one viral variant are useless against another. For several years, the bone marrow produces 2 billion new T and B cells a day, counter-

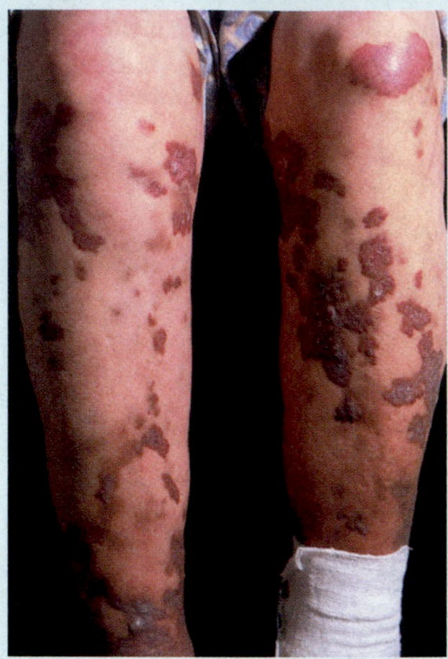

FIGURE 16B Prior to the appearance of AIDS, Kaposi sarcoma was a rare cancer seen only in elderly Jewish and Italian men and in people with suppressed immune systems. In these groups, it produces purplish patches on the lower limbs, but in AIDS patients, Kaposi sarcoma patches appear all over the body and sometimes internally too. These lower limbs display characteristic lesions.

TABLE 16.11 | Autoimmune Disorders

Disorder	Symptoms	Antibodies Against
Glomerulonephritis	Lower back pain	Kidney cell antigens that resemble streptococcal bacteria antigens
Graves disease	Restlessness, weight loss, irritability, increased heart rate and blood pressure	Thyroid gland antigens near thyroid-stimulating hormone receptor, causing overactivity
Type I diabetes mellitus	Thirst, hunger, weakness, emaciation	Pancreatic beta cells
Hemolytic anemia	Fatigue and weakness	Red blood cells
Multiple sclerosis	Weakness, incoordination, speech disturbances, visual complaints	Myelin in the white matter of the central nervous system
Myasthenia gravis	Muscle weakness	Receptors for neurotransmitters on skeletal muscle
Pernicious anemia	Fatigue and weakness	Binding site for vitamin B on cells lining stomach
Rheumatic fever	Weakness, shortness of breath	Heart valve cell antigens that resemble streptococcal bacteria antigens
Rheumatoid arthritis	Joint pain and deformity	Cells lining joints
Systemic lupus erythematosus	Red rash on face, prolonged fever, weakness, kidney damage, joint pain	Connective tissue
Ulcerative colitis	Lower abdominal pain	Colon cells

ing the million to billion new HIV particles that infected cells release daily.

So genetically diverse is the population of HIV in a human host that within days of initial infection, viral variants can arise that resist drugs. Combining drugs that act in different ways minimizes the number of viruses ("viral load") and delays symptom onset and progression. Classes of HIV/AIDS drugs target different points of viral vulnerability. The mechanisms of action parallel viral infection. They block HIV from binding to T cells, fusing with the cell membrane, entering the cell, replicating its genetic material once in the cell, or processing its proteins to a functional size. More than 200 drugs are also used to treat AIDS-associated infections and cancers.

Viral variability has also stymied vaccine development. Many candidates have failed. Reasons vary. Some trials did not go on for long enough to see an effect. Others did not accurately consider the expected incidence of infection in the studied population, critical to predicting the number of expected cases in a trial and following enough patients to see an effect. Many people dropped out of clinical trials or failed to comply with complex drug regimens. Other reasons for the failure to develop a vaccine are scientific—a live vaccine is too dangerous; a killed vaccine does not evoke a sufficient immune response; and insertion of viral genetic material into human chromosomes happens too fast to raise an antibody response, let alone enable memory cells to form. Entry of a single virus into a single cell can cause infection, and so any vaccine must be extremely powerful and fast-acting.

It may not be possible to devise an HIV vaccine. There are other ways to prevent HIV infection, such as education about reducing risk factors, circumcising men, and giving pregnant women antiretroviral drugs to prevent transmission to offspring. For people already infected, drugs work so well that for some individuals who have access to them, HIV infection and AIDS may become chronic diseases.

The epidemic may have peaked. The Joint UN Programme on HIV/AIDS estimates that incidence worldwide is declining by 2.5 percent per year—but this still projects 20 to 60 million new infections by the year 2028. Perhaps like other infectious diseases, AIDS will become less harmful as time goes on. ■

TABLE 16B | HIV Transmission

How HIV Is Transmitted
Sexual contact, particularly anal intercourse, but also vaginal intercourse and oral sex
Contaminated needles (intravenous drug use, injection of anabolic steroids, accidental needle stick in medical setting)
During birth from infected mother
Breast milk from an infected mother
Receiving infected blood or other tissue (precautions usually prevent this)
How HIV Is Not Transmitted
Casual contact (social kissing, hugging, handshakes)
Objects (toilet seats, deodorant sticks, doorknobs)
Mosquitoes
Sneezing and coughing
Sharing food
Swimming in the same water
Donating blood

stimulating antibody production. The resulting antibodies and symptoms appear to be an autoimmune disorder. The presence of more than one genetically distinct cell population in an individual is called microchimerism ("small mosaic"). Microchimerism that reflects the retention of cells from a fetus may explain the higher prevalence of autoimmune disorders among women. It is seen in a disorder called scleroderma, which means "hard skin."

Scleroderma, which typically begins between ages forty-five and fifty-five, is described as "the body turning to stone." Symptoms include fatigue, swollen joints, stiff fingers, and a masklike face. The hardening may affect blood vessels, the lungs, and the esophagus. Clues that scleroderma is a delayed response to persisting fetal cells include the following observations:

- It is much more common among women.
- Symptoms resemble those of graft-versus-host disease (GVHD), in which transplanted tissue produces chemicals that destroy the recipient's tissues. Antigens on cells in scleroderma lesions match those that cause GVHD.
- Mothers who have scleroderma and their sons have cell surfaces more similar than those of unaffected mothers and their sons. Perhaps the similarity of cell surfaces enabled the fetal cells to escape destruction by the woman's immune system. Female fetal cells probably have the same effect, but these cells cannot be distinguished from maternal cells by the presence of a Y chromosome.

Perhaps other disorders considered autoimmune reflect an immune system response to lingering fetal cells.

PRACTICE

33 How are allergic reactions and immune reactions similar yet different?

34 How does a tissue rejection reaction involve an immune response?

35 How is autoimmunity an abnormal functioning of the immune response?

16.10 LIFE-SPAN CHANGES

In a sense, aging of the immune system begins before birth, when nonself T cells are selected for destruction, via programmed cell death (apoptosis), in the thymus. The immune system begins to decline early in life. The thymus reaches its maximal size in adolescence and then slowly shrinks. By age seventy, the thymus is one-tenth the size it was at the age of ten, and the immune system is only 25% as powerful.

The declining strength of the immune response is why elderly people have a higher risk of developing cancer and succumb more easily to infections that they easily fought off at an earlier age, such as influenza, tuberculosis, and pneumonia. Encephalitis due to infection by the West Nile virus may cause very minor symptoms in young people, but it can kill the elderly. HIV infection progresses to AIDS faster in people older than forty. AIDS is more difficult to diagnose in older people, sometimes because physicians do not initially suspect the condition, instead attributing the fatigue, confusion, loss of appetite, and swollen glands to other causes. However, 11% of new cases of AIDS occur in those over age fifty.

Interestingly, numbers of T cells diminish only slightly with increasing age, and numbers of B cells not at all. However, activity levels change for both types of lymphocytes. T cell function controls production of B cells, so effects on B cells are secondary. The antibody response to antigens is slower, and as a result, vaccines that would ordinarily be effective in one dose may require an extra dose. The proportions of the different antibody classes shift, with IgA and IgG increasing, and IgM and IgE decreasing. A person may produce more autoantibodies than at a younger age, increasing the risk of developing an autoimmune disorder.

Elderly people may not be candidates for certain medical treatments that suppress immunity, such as cancer chemotherapy and steroids to treat inflammatory disorders, because of the declining function of the immune system. Overall, the immune system enables us to survive in a world that is home to many microorganisms. Clinical Application 16.1 looks at the devastation of immunity that is AIDS.

PRACTICE 

36 When is maximum size of the thymus reached?

37 Explain the decline in strength of the immune response in elderly people.

CHAPTER SUMMARY

16.1 INTRODUCTION (PAGE 617)

The lymphatic system is closely associated with the cardiovascular system. It transports excess fluid to the bloodstream, absorbs fats, and helps defend the body against disease-causing agents.

16.2 LYMPHATIC PATHWAYS (PAGE 617)

1. Lymphatic capillaries
   a. Lymphatic capillaries are microscopic, closed-ended tubes that extend into interstitial spaces.
   b. They receive tissue fluid through their thin walls.
   c. Lacteals are lymphatic capillaries in the villi of the small intestine.
2. Lymphatic vessels
   a. Lymphatic vessels are formed by the merging of lymphatic capillaries.
   b. They have walls similar to veins, but thinner, and valves that prevent backflow of lymph.
   c. Larger lymphatic vessels lead to lymph nodes and then merge into lymphatic trunks.
3. Lymphatic trunks and collecting ducts
   a. Lymphatic trunks drain lymph from large body regions.
   b. Trunks lead to two collecting ducts—the thoracic duct and the right lymphatic duct.
   c. Collecting ducts join the subclavian veins.

16.3 TISSUE FLUID AND LYMPH (PAGE 619)

1. Tissue fluid formation
   a. Tissue fluid originates from plasma and includes water and dissolved substances that have passed through the capillary wall.
   b. Tissue fluid generally lacks large proteins, but some smaller proteins are filtered out of blood capillaries into interstitial spaces.
   c. As the protein concentration of tissue fluid increases, colloid osmotic pressure increases.
2. Lymph formation
   a. Increasing hydrostatic pressure in interstitial spaces forces some tissue fluid into lymphatic capillaries. This fluid becomes lymph.
   b. Lymph formation prevents accumulation of excess tissue fluid (edema).
3. Lymph function
   a. Lymph returns the smaller protein molecules and fluid to the bloodstream.
   b. It transports foreign particles to the lymph nodes.

16.4 LYMPH MOVEMENT (PAGE 621)

1. Lymph flow
   a. Lymph is under low pressure and may not flow readily without external aid.
   b. Contraction of skeletal muscles and low pressure in the thorax created by breathing movements move lymph.

INNERCONNECTIONS | *Lymphatic System*

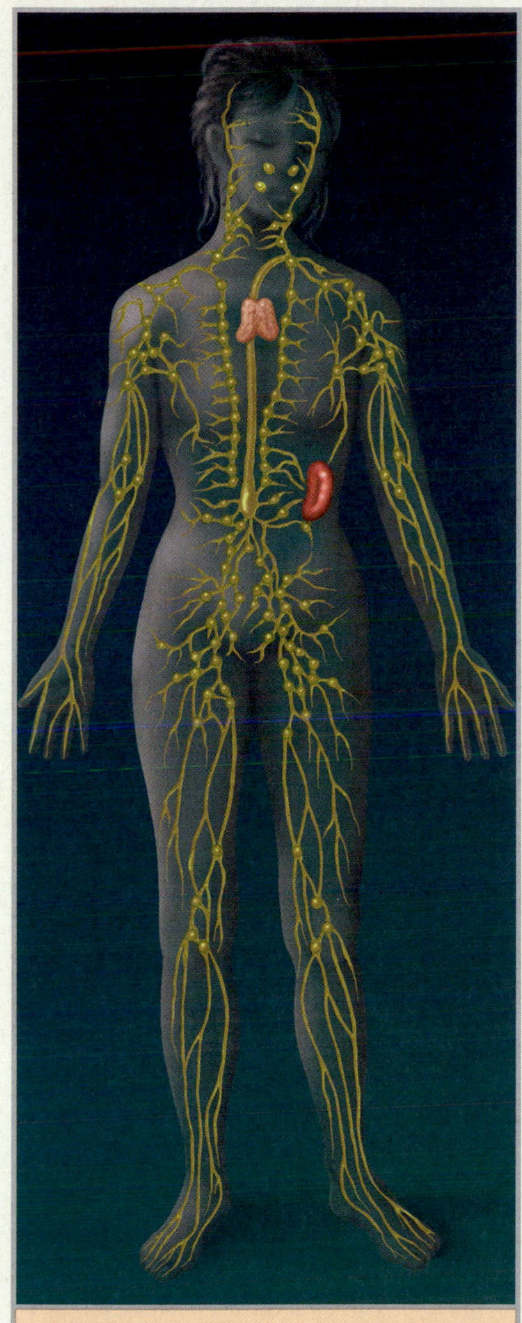

Lymphatic System

The lymphatic system is an important link between tissue fluid and the plasma; it also plays a major role in the response to infection.

Integumentary System

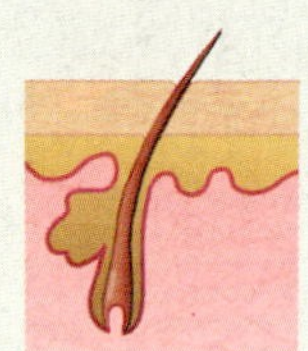

The skin is a first line of defense against infection.

Skeletal System

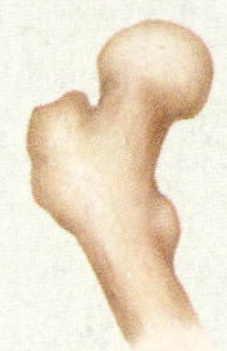

Cells of the immune system originate in the bone marrow.

Muscular System

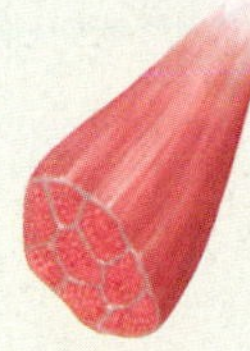

Muscle action helps pump lymph through the lymphatic vessels.

Nervous System

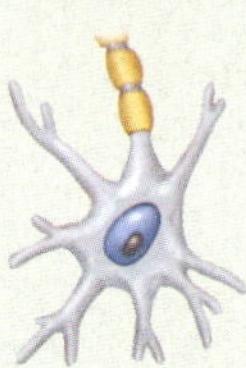

Stress may impair the immune response.

Endocrine System

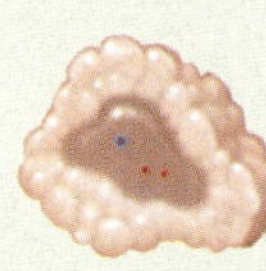

Hormones stimulate lymphocyte production.

Cardiovascular System

The lymphatic system returns tissue fluid to the bloodstream. Lymph originates as tissue fluid, formed by the action of blood pressure.

Digestive System

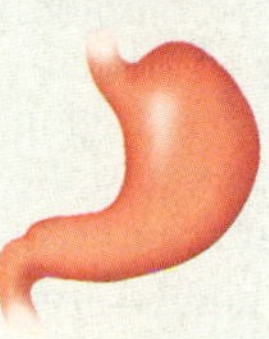

Lymph plays a major role in the absorption of fats.

Respiratory System

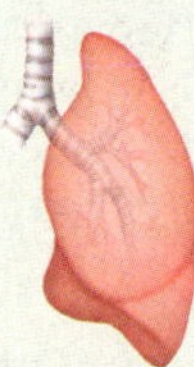

Cells of the immune system patrol the respiratory system to defend against infection.

Urinary System

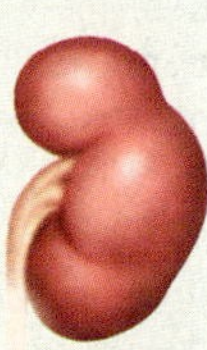

The kidneys control the volume of extracellular fluid, including lymph.

Reproductive System

Special mechanisms inhibit the female immune system in its attack of sperm as foreign invaders.

2. Obstruction of lymph movement
    a. Any condition that interferes with the flow of lymph results in edema.
    b. Obstruction of lymphatic vessels due to surgical removal of lymph nodes causes edema in the affected area.

16.5 LYMPH NODES (PAGE 621)

1. Structure of a lymph node
    a. Lymph nodes are usually bean-shaped, with blood vessels, nerves, and efferent lymphatic vessels attached to the indented region; afferent lymphatic vessels enter at points on the convex surface.
    b. Lymph nodes are enclosed in connective tissue that extends into the nodes and subdivides them into nodules.
    c. Nodules contain masses of lymphocytes and macrophages and spaces through which lymph flows.
2. Locations of lymph nodes
    a. Lymph nodes aggregate in groups or chains along the paths of larger lymphatic vessels.
    b. They are in the cervical, axillary, supratrochlear, and inguinal regions and in the pelvic, abdominal, and thoracic cavities.
3. Functions of lymph nodes
    a. Lymph nodes filter potentially harmful foreign particles from the lymph before it is returned to the bloodstream.
    b. Lymph nodes are centers for the production of lymphocytes that act against foreign particles.
    c. They contain macrophages that remove foreign particles from lymph.

16.6 THYMUS AND SPLEEN (PAGE 623)

1. Thymus
    a. The thymus is a soft, bilobed organ within the mediastinum.
    b. It slowly shrinks after puberty.
    c. It is composed of lymphatic tissue subdivided into lobules.
    d. Lobules contain lymphocytes.
    e. T lymphocytes leave the thymus and provide immunity.
    f. The thymus secretes thymosins, which stimulate maturation of T lymphocytes.
2. Spleen
    a. The spleen is in the upper left portion of the abdominal cavity.
    b. It resembles a large lymph node encapsulated and subdivided into lobules by connective tissue.
    c. Spaces in splenic lobules are filled with blood.
    d. The spleen, which filters foreign particles and damaged red blood cells from the blood, contains many macrophages and lymphocytes.

16.7 BODY DEFENSES AGAINST INFECTION (PAGE 625)

The presence and reproduction of pathogens may cause an infection. Pathogens include bacteria, complex single-celled organisms, fungi, and viruses. An infection may not immediately cause symptoms. The body has innate (nonspecific) and adaptive (specific) defenses against infection.

16.8 INNATE (NONSPECIFIC) DEFENSES (PAGE 626)

1. Species resistance

    Each species is resistant to certain diseases that may affect other species but is susceptible to diseases other species may resist.
2. Mechanical barriers
    a. Mechanical barriers include the skin and mucous membranes.
    b. Intact mechanical barriers prevent entrance of some pathogens.
    c. Hair traps infectious agents; and fluids such as tears, sweat, saliva, mucus, and urine wash away organisms before they can firmly attach.
3. Chemical barriers
    a. Enzymes in gastric juice and tears kill some pathogens.
    b. Low pH in the stomach prevents growth of some bacteria.
    c. High salt concentration in perspiration kills some bacteria.
    d. Interferons stimulate uninfected cells to synthesize antiviral proteins that block proliferation of viruses, stimulate phagocytosis, and enhance activity of cells that help resist infections and stifle tumor growth.
    e. Defensins make holes in bacterial cell walls and membranes.
    f. Collectins provide broad protection against a wide variety of microbes by grabbing onto them, easing phagocytosis.
    g. Activation of complement proteins in plasma stimulates inflammation, attracts phagocytes, and enhances phagocytosis.
4. Natural killer (NK) cells

    Natural killer cells secrete perforins, which destroy cancer cells and cells infected with viruses.
5. Inflammation
    a. Inflammation is a tissue response to damage, injury, or infection.
    b. The response includes localized redness, swelling, heat, and pain.
    c. Chemicals released by damaged tissues attract white blood cells to the site.
    d. Clotting may occur in body fluids that accumulate in affected tissues.
    e. Connective tissue may form a sac around the injured tissue and thus aid in preventing the spread of pathogens.

6. Phagocytosis
    a. The most active phagocytes in blood are neutrophils and monocytes; monocytes give rise to macrophages, which remain fixed in tissues.
    b. Phagocytic cells associated with the linings of blood vessels, in the red bone marrow, liver, spleen, and lymph nodes constitute the mononuclear phagocytic system.
    c. Phagocytes remove foreign particles from tissues and body fluids.
7. Fever
    a. Viral or bacterial infection stimulates certain lymphocytes to secrete IL-1, which temporarily raises body temperature.
    b. Physical factors, such as heat or ultraviolet light, or chemical factors, such as acids or bases, can cause fever.
    c. Elevated body temperature and the resulting decrease in blood iron level and increased phagocytic activity hamper infection.

16.9 ADAPTIVE (SPECIFIC) DEFENSES OR IMMUNITY (PAGE 628)

1. Antigens
    a. Before birth, body cells inventory "self" proteins and other large molecules.
    b. After inventory, lymphocytes develop receptors that allow them to differentiate between nonself (foreign) and self antigens.
    c. Nonself antigens combine with T cell and B cell surface receptors and stimulate these cells to cause an immune reaction.
    d. Haptens are small molecules that can combine with larger ones, becoming antigenic.
2. Lymphocyte origins
    a. Lymphocytes originate in red bone marrow and are released into the blood.
    b. Some reach the thymus where they mature into T cells.
    c. Others, the B cells, mature in the red bone marrow.
    d. Both T cells and B cells reside in lymphatic tissues and organs.
    e. Varieties of T cells and B cells number in the millions.
    f. The members of each variety respond only to a specific antigen.
    g. As a group, the members of each variety form a clone.
3. T cells and the cellular immune response
    a. T cells are activated when an antigen-presenting cell displays a foreign antigen.
    b. When a macrophage acts as an accessory cell, it phagocytizes an antigen-bearing agent, digests the agent, and displays the antigens on its cell membrane in association with certain MHC proteins.
    c. T cells respond to antigens by cell-to-cell contact (cellular immune response).
    d. T cells secrete cytokines, such as interleukins, that enhance cellular responses to antigens.
    e. T cells may also secrete substances that are toxic to their target cells.
    f. A helper T cell becomes activated when it encounters displayed antigens for which it is specialized to react.
    g. Once activated, helper T cells stimulate B cells to produce antibodies.
    h. Cytotoxic T cells recognize foreign antigens on tumor cells and cells whose surfaces indicate that they are infected by viruses. Stimulated cytotoxic T cells secrete perforin to destroy these cells.
    i. Memory T cells allow for immediate response to second and subsequent exposure to the same antigen.
4. B cells and the humoral immune response
    a. B cell activation
        (1) Sometimes a B cell is activated when it encounters an antigen that fits its antigen receptors or more often a B cell is activated when stimulated by a helper T cell.
        (2) An activated B cell proliferates (especially when stimulated by a T cell), enlarging its clone.
        (3) Some activated B cells differentiate further into memory cells.
        (4) Other activated B cells differentiate into antibody-producing plasma cells.
        (5) Antibodies react against the antigen-bearing agent that stimulated their production (humoral immune response).
        (6) An individual's diverse B cells defend against many pathogens.
    b. Antibody molecules
        (1) Antibodies are soluble proteins called immunoglobulins.
        (2) They constitute the gamma globulin fraction of plasma.
        (3) Each immunoglobulin molecule consists of four linked chains of amino acids.
        (4) Variable regions at the ends of these chains are specialized into antigen binding sites to react with different antigens.
    c. Types of immunoglobulins
        (1) The five major types of immunoglobulins are IgG, IgA, IgM, IgD, and IgE.
        (2) IgG, IgA, and IgM make up most of the circulating antibodies.
    d. Antibody actions
        (1) Antibodies directly attach to antigens, activate complement, or stimulate local tissue changes that are unfavorable to antigen-bearing agents.
        (2) Direct attachment results in agglutination, precipitation, or neutralization.
        (3) Activated complement proteins alter infected cells so they become more susceptible to phagocytosis, attract phagocytes, and lyse foreign cell membranes.
5. Immune responses
    a. B cells or T cells first encountering an antigen for which they are specialized to react constitutes a primary immune response.
        (1) During this response, antibodies are produced for several weeks.
        (2) Some T cells and B cells remain dormant as memory cells.

b. A secondary immune response occurs rapidly as memory cells respond to subsequent exposure to an antigen.

6. Practical classification of immunity
   a. A person who encounters a pathogen and has a primary immune response develops naturally acquired active immunity.
   b. A person who receives a vaccine containing a dead or weakened pathogen, or part of it, develops artificially acquired active immunity. Herd immunity protects populations.
   c. A person who receives an injection of antibodies or antitoxin has artificially acquired passive immunity.
   d. When antibodies pass through a placental membrane from a pregnant woman to her fetus, the fetus develops naturally acquired passive immunity.
   e. Active immunity lasts much longer than passive immunity.
7. Allergic reactions
   a. Allergic or hypersensitivity reactions are excessive misdirected immune responses that may damage tissues.
   b. Immediate-reaction allergy is an inborn ability to overproduce IgE.
      (1) Allergic reactions result from mast cells bursting and releasing allergy mediators such as histamine and serotonin.
      (2) The released chemicals cause allergy symptoms such as hives, hay fever, asthma, eczema, or gastric disturbances.
      (3) In anaphylactic shock, allergy mediators flood throughout the body, causing severe symptoms, including decreased blood pressure and difficulty breathing.
   c. Antibody-dependent cytotoxic allergic reactions occur when blood transfusions are mismatched.
   d. Immune complex allergic reactions involve autoimmunity, an immune reaction against self antigens.
   e. Delayed-reaction allergy, which can occur in anyone and inflame the skin, results from repeated exposure to allergens.
8. Transplantation and tissue rejection
   a. A transplant recipient's immune system may react against the donated tissue in a tissue rejection reaction.
   b. Matching cell surface molecules (MHC antigens) of donor and recipient tissues, transplanting stem cells from the donor, and discontinuing immunosuppressive drugs after the procedure can help the body accept the foreign tissue.
   c. Transplants may take place between genetically identical twins, from one body part to another, between unrelated individuals of the same species, or between individuals of different species.
9. Autoimmunity
   a. In autoimmune disorders, autoantibodies attack the body's tissues.
   b. Autoimmune disorders may result from a previous viral infection, faulty T cell development, or reaction to a nonself antigen that resembles a self antigen.
   c. Retained fetal cells can cause a condition that resembles an autoimmune disorder.

16.10 LIFE-SPAN CHANGES (PAGE 644)

1. The immune system begins to decline early in life, in part due to the shrinking thymus.
2. Numbers of T cells and B cells do not significantly change, but activity levels do.
3. Proportions of the different antibody classes shift.

CHAPTER ASSESSMENTS

16.1 Introduction

1 Explain the functions of the lymphatic system. (p. 617)

16.2 Lymphatic Pathways

2 Trace the general pathway of lymph from the interstitial spaces to the bloodstream. (p. 617)

16.3 Tissue Fluid and Lymph

3 Distinguish between tissue fluid and lymph. (p. 619)

4 Describe the primary functions of lymph. (p. 620)

16.4 Lymph Movement

5 Explain why physical exercise promotes lymphatic circulation. (p. 621)

6 Explain how a lymphatic obstruction leads to edema. (p. 621)

16.5 Lymph Nodes

7 Draw a lymph node, and label its parts. (p. 621)

8 On a drawing of the body locate the major body regions containing lymph nodes. (p. 622)

9 Explain the functions of a lymph node. (p. 623)

16.6 Thymus and Spleen

10 Indicate the locations of the thymus and spleen. (p. 623)

11 Compare and contrast the functions of the thymus and spleen. (p. 623)

16.7 Body Defenses Against Infection

12 Defense mechanisms that prevent the entry of many types of pathogens and destroy them if they enter provide ______________ (nonspecific) defense. Precise mechanisms targeting specific pathogens provide ________________ (specific) defense. (p. 626)

16.8 Innate (Nonspecific) Defenses

13 Define *species resistance.* (p. 626)

14 Identify the barriers that provide the body's first line of defense against infectious agents. (p. 626)

15 Describe how enzymatic actions function as defense mechanisms against pathogens. (p. 626)

16 Distinguish among the chemical barriers (interferons, defensins, collectins, and complement proteins), and give examples of their different actions. (p. 626)

17 Describe natural killer cells and their actions. (p. 627)

18 List the major effects of inflammation. (p. 627)

19 Identify the major phagocytic cells in the blood and other tissues. (p. 627)

20 List possible causes of fever, and explain the benefits of fever. (p. 628)

16.9 Adaptive (Specific) Defenses or Immunity

21 Distinguish between an antigen and a hapten. (p. 628)

22 Review the origin of T cells and B cells. (p. 628)

23 Define *clone of lymphocytes.* (p. 629)

24 Explain the cellular immune response including the activation of T cells. (p. 630)

25 Define *cytokine.* (p. 630)

26 List three types of T cells, and describe the function of each in the immune response. (p. 630)

27 Explain the humoral immune response, including the activation of B cells. (p. 632)

28 Explain the function of plasma cells. (p. 632)

29 Draw and label the parts of an antibody molecule. (p. 634)

30 Distinguish between the variable region and the constant region of an antibody molecule. (p. 634)

31 Match the types of antibodies with their function and/or where each is found. (p. 635)

(1) associated with allergic reactions	A. IgA
(2) important in B cell activation, on surfaces of most B cells	B. IgM
(3) activates complement, anti-A and anti-B in blood	C. IgG
(4) effective against bacteria, viruses, toxins in plasma and tissue fluids	D. IgD
(5) in exocrine secretions, including breast milk	E. IgE

32 Describe three ways in which an antibody's direct attack on an antigen helps remove that antigen. (p. 635)

33 Explain the functions of complement. (p. 635)

34 Contrast a primary and a secondary immune response. (p. 635)

35 Contrast active and passive immunity. (p. 638)

36 Define *vaccine.* (p. 638)

37 Explain how a vaccine produces its effect. (p. 638)

38 Describe how a fetus may obtain antibodies from maternal blood. (p. 638)

39 Explain the relationship between an allergic reaction and an immune response. (p. 639)

40 Distinguish between an antigen and an allergen. (p. 639)

41 Describe how an immediate-reaction allergic response may occur. (p. 639)

42 List the major events leading to a delayed-reaction allergic response. (p. 641)

43 Explain the relationship between tissue rejection and an immune response. (p. 641)

44 Describe two methods used to reduce the severity of a tissue rejection reaction. (p. 641)

45 Explain the goal of using immunosuppressive drugs before a transplant. (p. 641)

46 Explain the relationship between autoimmunity and an immune response. (p. 641)

16.10 Life-Span Changes

47 Explain the causes for a decline in the strength of the immune response in the elderly. (p. 644)

INTEGRATIVE ASSESSMENTS/CRITICAL THINKING

OUTCOMES 16.2, 16.3, 16.4, 16.5

1. How can removal of enlarged lymph nodes for microscopic examination aid in diagnosing certain diseases?

OUTCOMES 16.6, 16.9

2. What functions of the lymphatic system would be affected in a person born without a thymus?

OUTCOME 16.9

3. The immune response is specific, diverse, and has memory. Give examples of each of these characteristics.

OUTCOME 16.9

4. Some parents keep their preschoolers away from other children to prevent them from catching illnesses. How might these well-meaning parents be harming their children?

OUTCOME 16.9

5. Why does vaccination provide long-lasting protection against disease, whereas gamma globulin (IgG) provides only short-term protection?

OUTCOME 16.9

6. Why is a transplant consisting of fetal tissue less likely to provoke an immune rejection response than tissue from an adult?

OUTCOME 16.9

7. An eighteen-year-old female athlete received a kidney transplant from her brother, who was an HLA match, along with stem cells from his blood. The surgery was so successful that she was able to return to competitive swimming within a year and made the Olympic team. However, she was disqualified due to blood test results that found a male Y chromosome in her blood. Explain one way that this could have happened.

WEB CONNECTIONS

Be sure to visit the text website at **www.mhhe.com/shier12** for answers to chapter assessments, additional quizzes, and interactive learning exercises.

ANATOMY & PHYSIOLOGY REVEALED

Anatomy & Physiology Revealed® (APR) includes cadaver photos that allow you to peel away layers of the human body to reveal structures beneath the surface. This program also includes animations, radiologic imaging, audio pronunciations, and practice quizzing. Check out **www.aprevealed.com**. APR has been proven to help improve student grades!

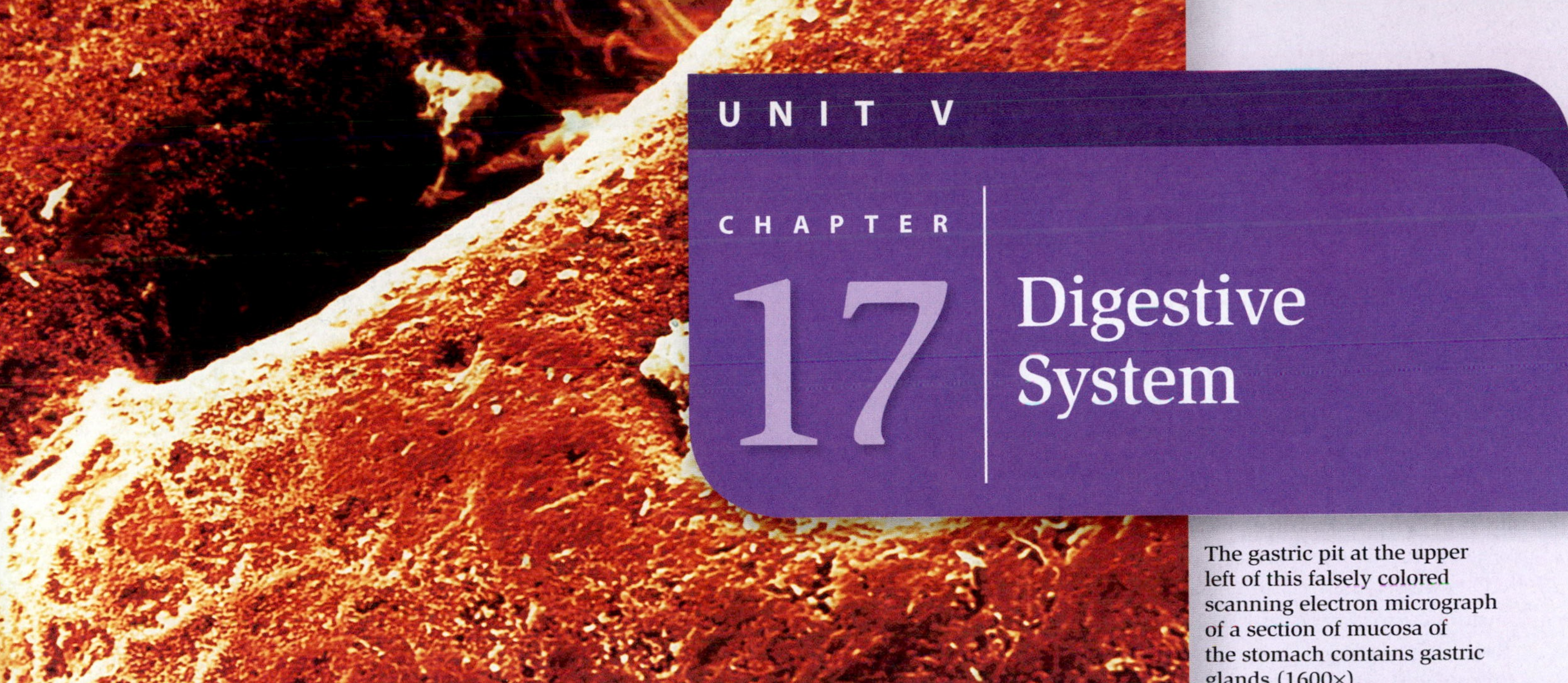

The gastric pit at the upper left of this falsely colored scanning electron micrograph of a section of mucosa of the stomach contains gastric glands (1600×).

UNDERSTANDING WORDS

aliment-, food: *aliment*ary canal—tubelike part of the digestive system.

cari-, decay: dental *caries*—tooth decay.

cec-, blindness: *cec*um—blind-ended sac at the origin of the large intestine.

chym-, juice: *chyme*—semifluid paste of food particles and gastric juice formed in the stomach.

decidu-, falling off: *decidu*ous teeth—teeth shed during childhood.

frenul-, bridle, restraint: *frenul*um—membranous fold that anchors the tongue to the floor of the mouth.

gastr-, stomach: *gastr*ic gland—part of the stomach that secretes gastric juice.

hepat-, liver: *hepat*ic duct—duct that carries bile from the liver to the bile duct.

hiat-, opening: esophageal *hiat*us—opening through which the esophagus penetrates the diaphragm.

lingu-, tongue: *lingu*al tonsil—mass of lymphatic tissue at the root of the tongue.

peri-, around: *peri*stalsis—wavelike ring of contraction that moves material along the alimentary canal.

pyl-, gatekeeper, door: *pyl*oric sphincter—muscle that serves as a valve between the stomach and small intestine.

rect-, straight: *rect*um—distal part of the large intestine.

sorpt-, to soak up: ab*sorpt*ion—uptake of substances.

vill-, hairy: *vill*i—tiny projections of mucous membrane in the small intestine.

LEARN PRACTICE ASSESS

LEARNING OUTCOMES

After you have studied this chapter, you should be able to:

17.1 Introduction

1. Describe the general functions of the digestive system. (p. 652)
2. Name the major organs of the digestive system. (p. 652)

17.2 General Characteristics of the Alimentary Canal

3. Describe the structure of the wall of the alimentary canal. (p. 653)
4. Explain how the contents of the alimentary canal are mixed and moved. (p. 654)

17.3 Mouth

5. Describe the functions of the structures associated with the mouth. (p. 656)
6. Describe how different types of teeth are adapted for different functions, and list the parts of a tooth. (p. 658)

17.4–17.10 Salivary Glands—Large Intestine

7. Locate each of the organs and glands; then describe the general function of each. (p. 660)
8. Identify the function of each enzyme secreted by the digestive organs and glands. (p. 660)
9. Describe how digestive secretions are regulated. (p. 660)
10. Explain control of movement of material through the alimentary canal. (p. 663)
11. Describe the mechanisms of swallowing, vomiting, and defecating. (p. 664)
12. Explain how the products of digestion are absorbed. (p. 669)

17.11 Life-Span Changes

13. Describe aging-related changes in the digestive system. (p. 690)

THE GUT MICROBIOME

Ninety percent of the cells in an adult body, if the digestive system is included, are not human—they are microorganisms collectively called microflora. Terms from ecology are used to describe the microbes within us, including community, ecosystem, and biome. The Human Oral Microbiome Database (www.homd.org), for example, lists more than 600 species that live in the mouth. Each person has about 200 of the oral bacterial types.

To assess the microbial biome at the other end of the digestive tract, researchers analyzed DNA fragments in stool samples from a twenty-eight-year-old woman and a thirty-seven-year-old man, both of whom had not taken medications that could have affected the microflora. By comparing the DNA pieces to those of known microorganisms, the researchers discovered that the "distal gut microbial community" includes more than 6,800 species.

Researchers also tracked the formation and changing nature of the human gut microflora by classifying bacterial DNA in a year's worth of stool collected daily from soiled diapers. Bacteria in the stool varied greatly from baby to baby at the outset, but by their first birthdays, the gut communities were more alike and more closely resembled the microbial communities in adults. Each person's large intestine comes to support 500 to 1,000 bacterial species.

The microorganisms that live in our large intestines are crucial to our health. They produce more than eighty types of enzymes that digest plant polysaccharides that our bodies cannot break down, as well as easing processing of certain sugars. Our "gut" residents also synthesize essential vitamins and amino acids and break down certain toxins and drugs.

We can use knowledge of our gut microbiome to improve health. For example, unusual bacterial communities can reflect disease. Specific microfloral profiles are associated with colorectal cancer, diarrhea, inflammatory bowel disease, and peptic ulcers. A new focus of drug development is to target our microbial residents. Also, we can add bacteria to foods to prevent certain infections, an approach called probiotics. For example, certain *Lactobacillus* strains added to yogurt can protect against *Salmonella* foodborne infection.

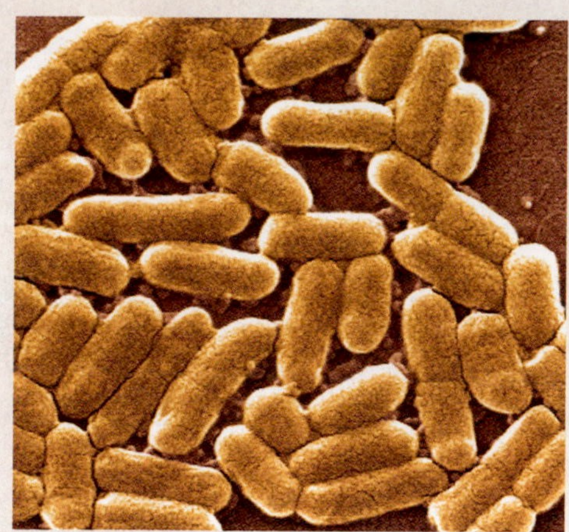

Several million microorganisms are normal residents of our digestive tracts. *Escherichia coli*, pictured here (6,800×), produce vitamin K and if present in low numbers, will not cause diarrhea.

The numbers and types of microorganisms that live in our intestines vary somewhat from person to person, and these differences may be one reason why some people can eat a great deal and not gain weight, yet others gain weight easily. Studies show that an item of food may yield different numbers of calories when eaten by different people. One investigation of the energy in a cookie found that even though the package listed 110 calories, it yielded anywhere from 90 to 110 calories, depending upon who ate it. ■

17.1 INTRODUCTION

Digestion (di-jest′yun) is the mechanical and chemical breakdown of foods into forms that cell membranes can absorb. *Mechanical digestion* breaks large pieces into smaller ones without altering their chemical composition. *Chemical digestion* breaks food into simpler chemicals. The organs of the **digestive system** carry out these processes, as well as ingestion, propulsion, absorption, and defecation.

The digestive system consists of the **alimentary canal** (al″i-men′tar-e kah-nal′), extending from the mouth to the anus, and several accessory organs, which release secretions into the canal. The alimentary canal includes the mouth, pharynx, esophagus, stomach, small intestine, large intestine, and anal canal. The accessory organs include the salivary glands, liver, gallbladder, and pancreas. Figure 17.1 and reference plates 4, 5, and 6 show the major organs of the digestive system.

The digestive system originates from the inner layer (endoderm) of the embryo, which folds to form the tube of the alimentary canal. The accessory organs develop as buds from the tube.

PRACTICE

1 What are the general functions of the digestive system?
2 Which organs constitute the digestive system?

17.2 GENERAL CHARACTERISTICS OF THE ALIMENTARY CANAL

The alimentary canal is a muscular tube about 8 meters long that passes through the body's thoracic and abdominopelvic cavities (fig. 17.2). The structure of its wall, how it moves food, and its innervation are similar throughout its length.

Structure of the Wall

The wall of the alimentary canal consists of four distinct layers developed to different degrees from region to region.

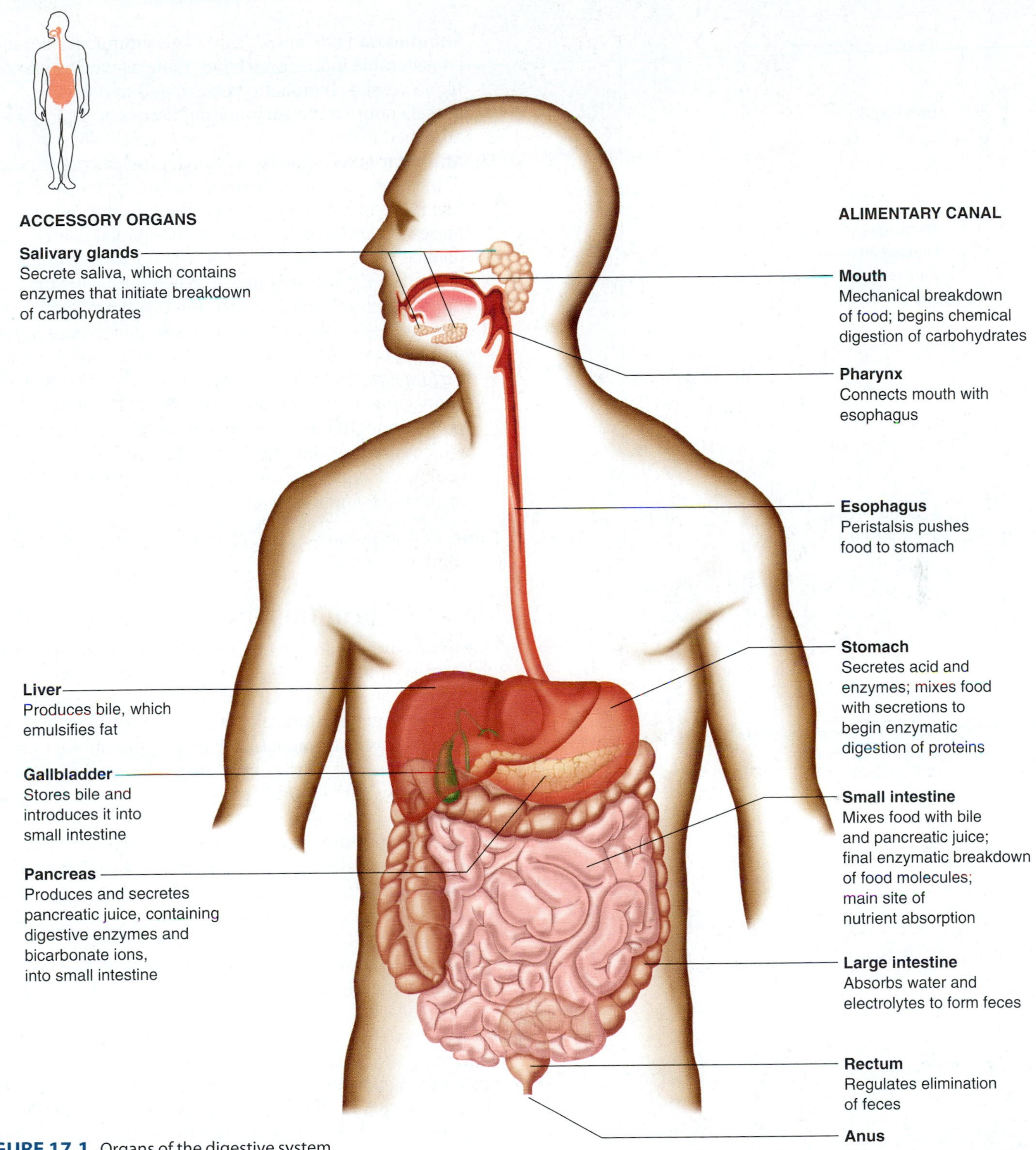
ACCESSORY ORGANS
Salivary glands
Secrete saliva, which contains enzymes that initiate breakdown of carbohydrates
Liver
Produces bile, which emulsifies fat
Gallbladder
Stores bile and introduces it into small intestine
Pancreas
Produces and secretes pancreatic juice, containing digestive enzymes and bicarbonate ions, into small intestine
ALIMENTARY CANAL
Mouth
Mechanical breakdown of food; begins chemical digestion of carbohydrates
Pharynx
Connects mouth with esophagus
Esophagus
Peristalsis pushes food to stomach
Stomach
Secretes acid and enzymes; mixes food with secretions to begin enzymatic digestion of proteins
Small intestine
Mixes food with bile and pancreatic juice; final enzymatic breakdown of food molecules; main site of nutrient absorption
Large intestine
Absorbs water and electrolytes to form feces
Rectum
Regulates elimination of feces
Anus

FIGURE 17.1 Organs of the digestive system.

Although the four-layered structure persists throughout the alimentary canal, certain regions are specialized for particular functions. Beginning with the innermost tissues, these layers, shown in figure 17.3, include the following:

1. **Mucosa** (mu-ko′sah), or **mucous membrane.** This layer is formed of surface epithelium, underlying connective tissue (lamina propria), and a small amount of smooth muscle (muscularis mucosae). In some regions, the mucosa is folded with tiny projections that extend into the passageway, or **lumen,** of the digestive tube; the folds increase the absorptive surface area. The mucosa also has glands that are tubular invaginations into which the lining cells secrete mucus and digestive enzymes. The mucosa protects the tissues beneath it and carries on secretion and absorption.

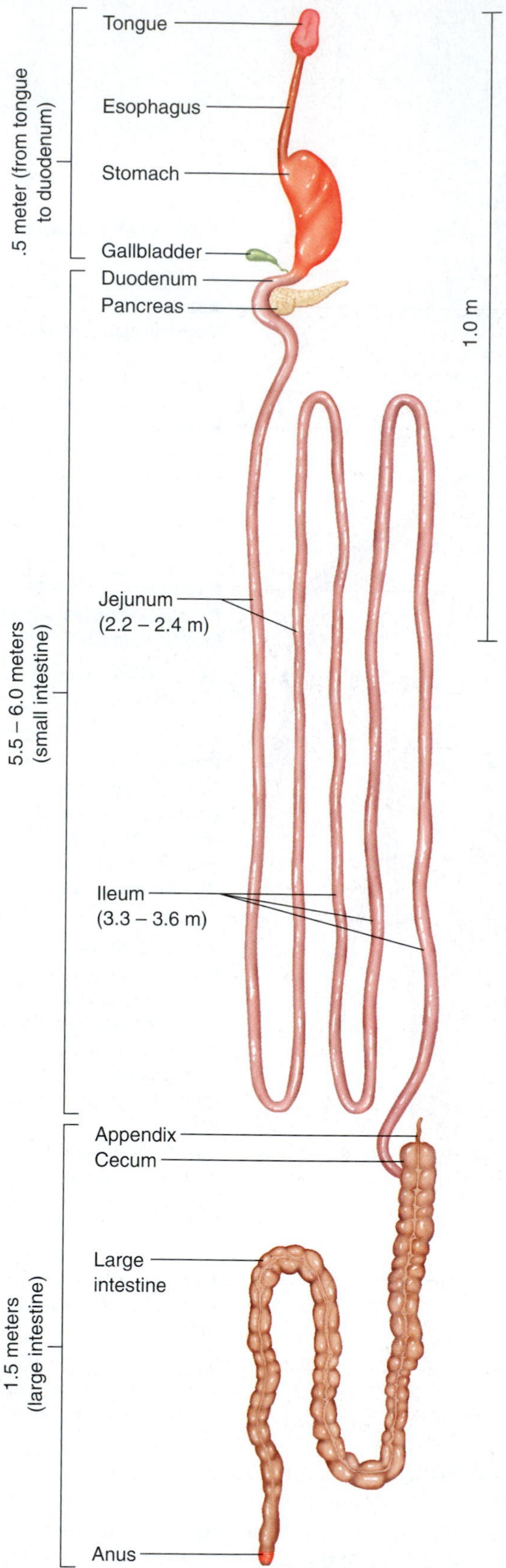

FIGURE 17.2 The alimentary canal is a muscular tube about 8 meters long.

2. **Submucosa** (sub″mu-ko′sah). The submucosa contains considerable loose connective tissue as well as glands, blood vessels, lymphatic vessels, and nerves. Its vessels nourish the surrounding tissues and carry away absorbed materials.
3. **Muscular layer.** This layer, which provides movements of the tube, consists of two coats of smooth muscle tissue. The fibers of the inner coat encircle the tube. When these *circular fibers* (they are closed spirals) contract, the diameter of the tube decreases. The fibers of the outer muscular coat run lengthwise. When these *longitudinal fibers* (open spirals) contract, the tube shortens.
4. **Serosa** (se-r-o′sah), or **serous layer.** The serous layer, or outer covering of the tube, is composed of the *visceral peritoneum,* formed of epithelium on the outside and connective tissue beneath. The cells of the serosa protect underlying tissues and secrete serous fluid, which moistens and lubricates the tube's outer surface so that the organs in the abdominal cavity slide freely against one another.

Table 17.1 summarizes the characteristics of the layers of the alimentary canal.

Movements of the Tube

The motor functions of the alimentary canal are of two basic types—*mixing movements* and *propelling movements.* Mixing occurs when smooth muscles in small segments of the tube contract rhythmically. For example, when the stomach is full, waves of muscular contractions move along its wall from one end to the other (fig. 17.4*a*). These waves occur every twenty seconds or so, and they mix foods with the digestive juices that the mucosa secretes. In the small intestine, **segmentation** is a type of movement that aids mixing by alternately contracting and relaxing the smooth muscle in nonadjacent segments of the organ (fig. 17.4*b*). Segmentation does not follow a set pattern, so materials are not moved along the tract in one direction.

Propelling movements include a wavelike motion called **peristalsis** (per″ĭ-stal′sis), in which a ring of contraction occurs in the wall of the tube (fig. 17.4*c*). At the same time, the muscular wall just ahead of the ring relaxes—a phenomenon called *receptive relaxation.* As the wave moves along the tube, it pushes the contents of the tube ahead of it. Peristalsis begins when food expands the tube. It causes the sounds that can be heard through a stethoscope applied to the abdominal wall.

A "GI camera" the size of a large vitamin pill can image the alimentary canal, revealing blockages and sites of bleeding. The patient swallows the capsule, which contains a camera, a light source, radio transmitter, and batteries. Peristalsis moves it along, and about six hours after swallowing, it transmits images from the small intestine to a device worn on the physician's belt. The information goes to a computer, and still or video images are downloaded. The device, which is disposable, leaves the body in the feces within a day or two.

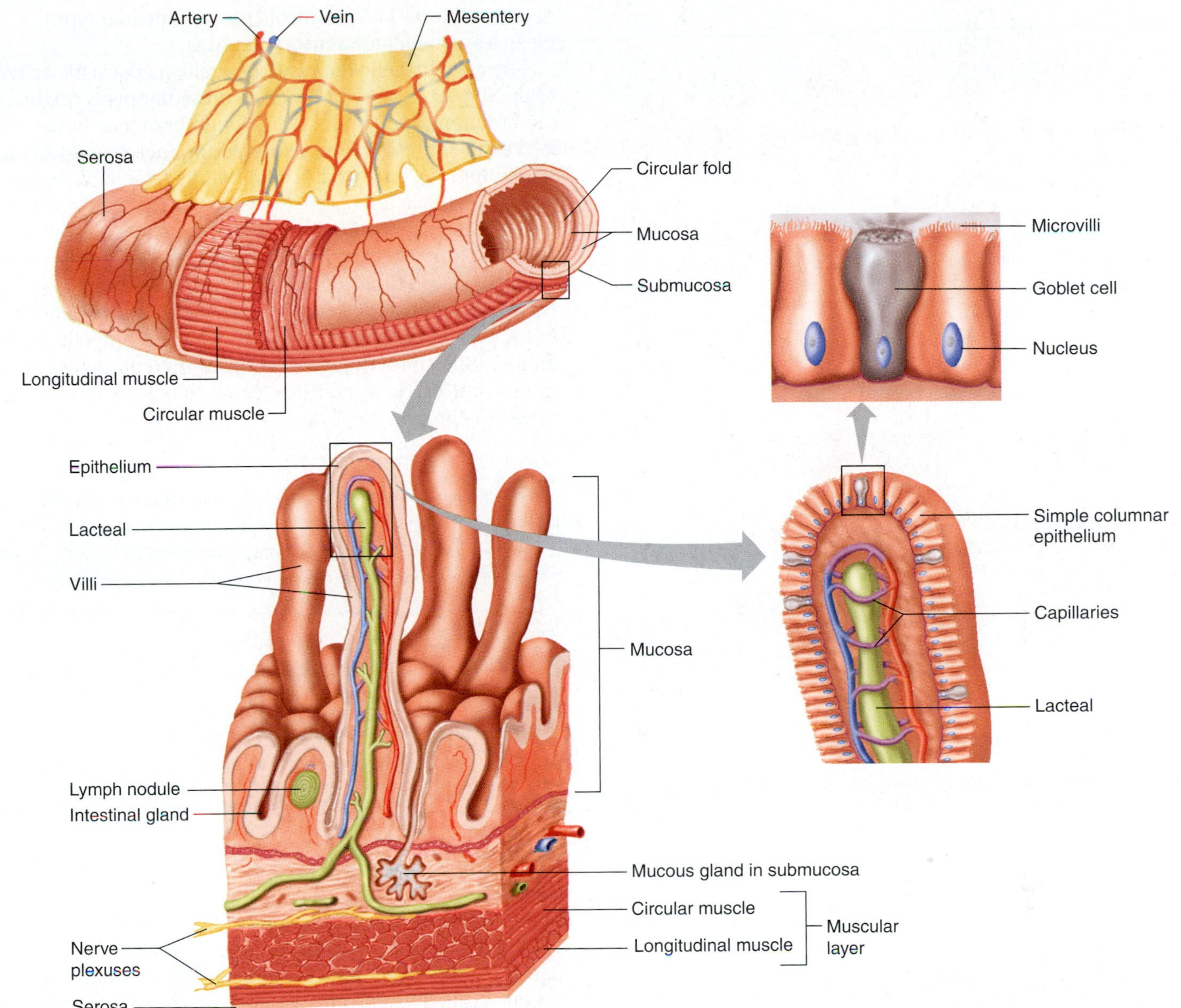

FIGURE 17.3 The wall of the small intestine, as in other portions of the alimentary canal, consists of four layers: an inner mucosa, a submucosa, a muscular layer, and an outer serosa.

TABLE 17.1 | Layers of the Wall of the Alimentary Canal

Layer	Composition	Function
Mucosa	Epithelium, connective tissue, smooth muscle	Protection, secretion, absorption
Submucosa	Loose connective tissue, blood vessels, lymphatic vessels, nerves	Nourishes surrounding tissues, transports absorbed materials
Muscular layer	Smooth muscle fibers in circular and longitudinal groups	Movements of the tube and its contents
Serosa	Epithelium, connective tissue	Protection, lubrication

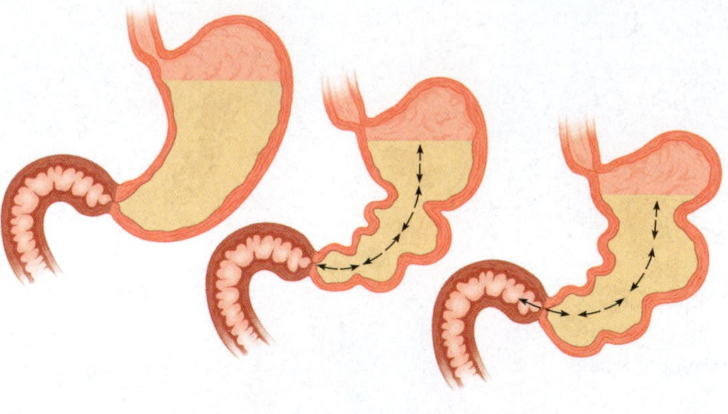
(a)

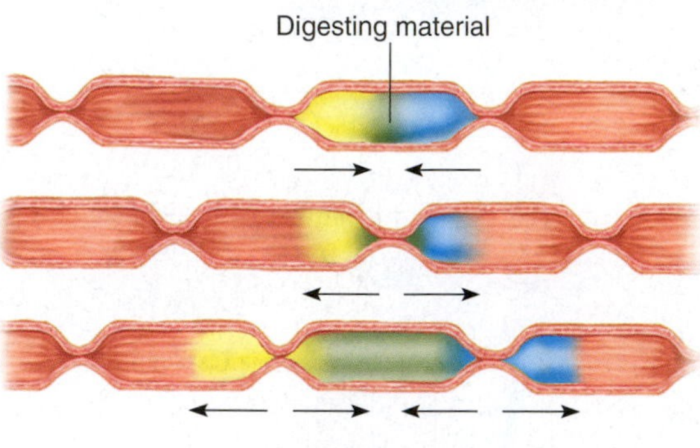

(b)

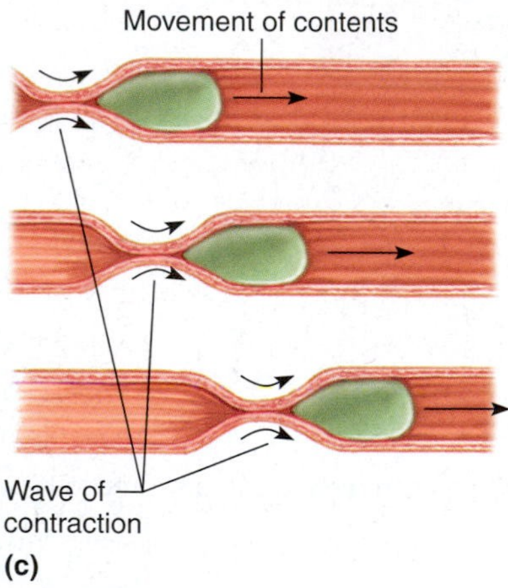

(c)

FIGURE 17.4 Movements through the alimentary canal. (*a*) Mixing movements occur when small segments of the muscular wall of the stomach rhythmically contract. (*b*) Segmentation mixes the contents of the small intestine. (*c*) Peristaltic waves move the contents along the canal.

Innervation of the Tube

Branches of the sympathetic and parasympathetic divisions of the autonomic nervous system extensively innervate the alimentary canal. These nerve fibers, mainly associated with the tube's muscular layer, maintain muscle tone and regulate the strength, rate, and velocity of muscular contractions. Many of the postganglionic fibers are organized into a network or nerve plexus within the wall of the canal (see fig. 17.3). The *submucosal plexus* is important in controlling secretions by the gastrointestinal tract. The *myenteric plexus* of the muscular layer controls gastrointestinal motility. The plexuses also include sensory neurons.

Parasympathetic impulses generally increase the activities of the digestive system. Some of these impulses originate in the brain and are conducted through branches of the vagus nerves to the esophagus, stomach, pancreas, gallbladder, small intestine, and proximal half of the large intestine. Other parasympathetic impulses arise in the sacral region of the spinal cord and supply the distal half of the large intestine.

Sympathetic nerve impulses' effects on digestive actions usually oppose those of the parasympathetic division. That is, sympathetic impulses inhibit certain digestive actions. For example, such impulses inhibit mixing and propelling movements, but stimulate contraction of the sphincter muscles in the wall of the alimentary canal, blocking movement of materials through the tube.

So extensive are the nerve plexuses of the gastrointestinal tract that it is sometimes said to have a "second brain." The small intestine, for example, has at least 100 million neurons, many neuroglia, and abundant and diverse neurotransmitters, neuropeptides, and growth factors.

PRACTICE 

3 Describe the wall of the alimentary canal.
4 Name the types of movements in the alimentary canal.
5 How do parasympathetic nerve impulses affect digestive actions? What effect do sympathetic nerve impulses have?

17.3 MOUTH

The **mouth,** the first portion of the alimentary canal, receives food and begins digestion by mechanically breaking up solid particles into smaller pieces and mixing them with saliva. This action is called *mastication.* (mas″tĭ-ka′shun) The mouth also functions as an organ of speech and sensory reception. It is surrounded by the lips, cheeks, tongue, and palate and includes a chamber between the palate and tongue called the *oral cavity,* as well as a narrow space between the teeth, cheeks, and lips called the *vestibule* (fig. 17.5 and reference plate 9).

Cheeks and Lips

The **cheeks** form the lateral walls of the mouth. They consist of outer layers of skin, pads of subcutaneous fat, muscles associated with expression and chewing, and inner linings of moist, stratified squamous epithelium.

Cells lining the cheek are commonly used as a source of DNA for genetic testing. A person scrapes the inside of the cheek with a cotton swab or swishes with mouthwash and expectorates into a small tube. The sample is sent to a lab, and specific genetic variants are identified in the DNA.

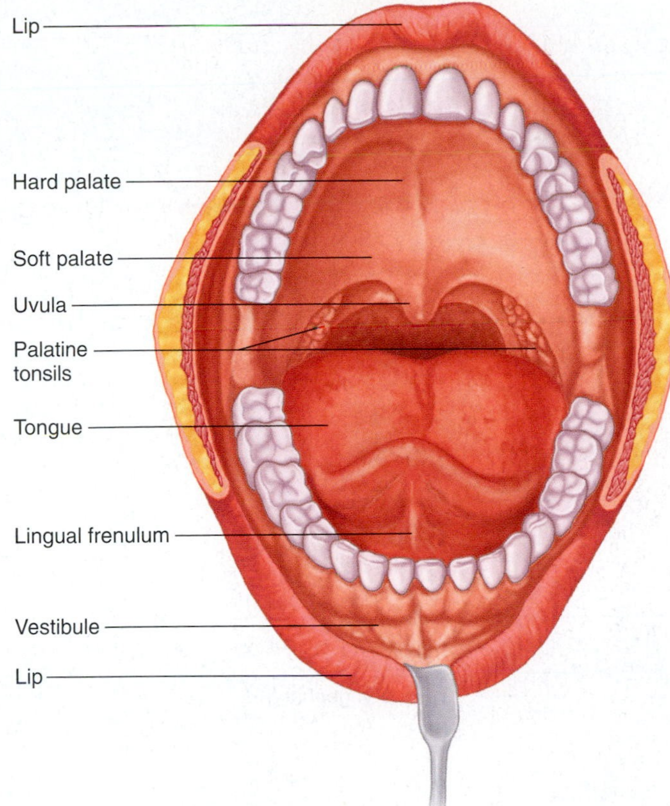

FIGURE 17.5 The mouth is adapted for ingesting food and preparing it for digestion, both mechanically and chemically.

The **lips** are highly mobile structures that surround the mouth opening. They contain skeletal muscles and sensory receptors useful in judging the temperature and texture of foods. Their normal reddish color is due to the many blood vessels near their surfaces. The external borders of the lips mark the boundaries between the skin of the face and the mucous membrane that lines the alimentary canal.

Tongue

The **tongue** (tung) is a thick, muscular organ that occupies the floor of the mouth and nearly fills the oral cavity when the mouth is closed. Mucous membrane covers the tongue, and a membranous fold called the **lingual frenulum** (ling′gwahl fren′u-lum) connects the midline of the tongue to the floor of the mouth.

The *body* of the tongue is largely composed of skeletal muscle fibers that run in several directions. These muscles mix food particles with saliva during chewing and move food toward the pharynx during swallowing. The tongue also helps move food underneath the teeth for chewing. The surface of the tongue has rough projections, called **papillae** (pah-pil′a) (fig. 17.6). Some of these provide friction, which helps handle food. Other papillae contain most of the taste buds (see chapter 12, p. 448). Some taste buds are scattered elsewhere in the mouth, particularly in children.

The posterior region, or *root,* of the tongue is anchored to the hyoid bone. It is covered with rounded masses of lymphatic tissue called **lingual tonsils** (ton′silz) (fig. 17.7).

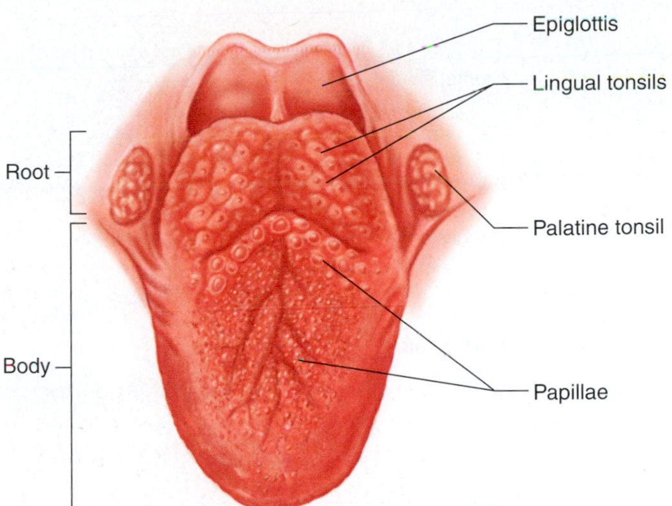

FIGURE 17.6 The surface of the tongue, superior view.

Palate

The **palate** (pal′at) forms the roof of the oral cavity and consists of a hard anterior part and a soft posterior part. The *hard palate* is formed by the palatine processes of the maxillary bones in front and the horizontal portions of the palatine bones in back. The *soft palate* forms a muscular arch, which extends posteriorly and downward as a cone-shaped projection called the **uvula** (u′vu-lah).

During swallowing, muscles draw the soft palate and the uvula upward. This action closes the opening between the nasal cavity and the pharynx, preventing food from entering the nasal cavity.

In the back of the mouth, on either side of the tongue and closely associated with the palate, are masses of lymphatic tissue called **palatine** (pal′ah-tīn) **tonsils** (fig. 17.7). These structures lie beneath the epithelial lining of the mouth and, like other lymphatic tissues, help protect the body against infections (see chapter 16, p. 621).

Other masses of lymphatic tissue, called **pharyngeal** (fah-rin′je-al) **tonsils,** or *adenoids,* are on the posterior wall of the pharynx, above the border of the soft palate (fig. 17.7). If the adenoids enlarge and block the passage between the nasal cavity and pharynx, they may be surgically removed.

> The palatine tonsils are common sites of infection and when inflamed, produce *tonsillitis.* Infected tonsils may swell so greatly that they block the passageways of the pharynx and interfere with breathing and swallowing. The mucous membranes of the pharynx, auditory tubes, and middle ears are continuous, so such an infection can spread from the throat into the middle ears (otitis media).
>
> When tonsillitis occurs repeatedly and does not respond to antibiotic treatment, the tonsils may be surgically removed. Such tonsillectomies are done less often today than they were a generation ago because the tonsils′ role in immunity is now recognized.

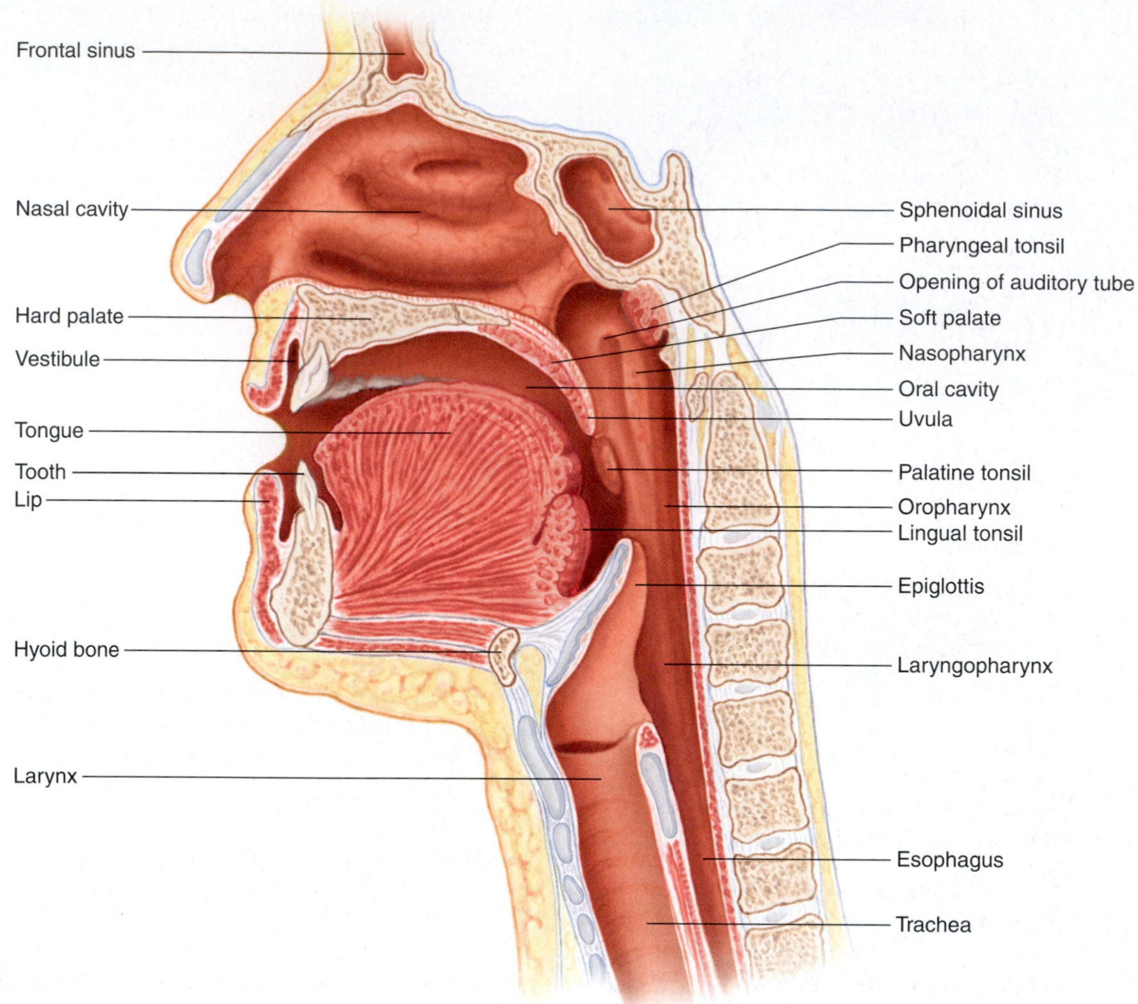

FIGURE 17.7 Sagittal section of the mouth, nasal cavity, and pharynx.

PRACTICE 

6 What are the functions of the mouth?

7 How does the tongue function as part of the digestive system?

8 What is the role of the soft palate in swallowing?

9 Where are the tonsils located?

Teeth

The **teeth** are the hardest structures in the body. They are not considered part of the skeletal system because they have at least two types of proteins that are not also found in bone, and their structure is different.

Teeth develop in sockets in the alveolar processes of the mandibular and maxillary bones. Teeth are unique structures in that two sets form during development (fig. 17.8). The first set, the *primary teeth* (deciduous teeth), usually erupt through the gums (gingiva) at regular intervals between the ages of six months and two to four years. The ten primary teeth are anchored in each jaw from the midline toward the sides in the following sequence: central incisor, lateral incisor, canine (cuspid), first molar, and second molar.

The primary teeth are usually shed in the same order they erupted, after their roots are resorbed. Then, the *secondary* (permanent) *teeth* push the primary teeth out of their sockets. This secondary set consists of thirty-two teeth—sixteen in each jaw—and they are arranged from the midline as follows: central incisor, lateral incisor, canine (cuspid), first premolar (bicuspid), second premolar (bicuspid), first molar, second molar, and third molar (fig. 17.9). Table 17.2 summarizes the types and numbers of primary and secondary teeth.

The secondary teeth usually begin to erupt at six years, but the set may not be completed until the third molars emerge between seventeen and twenty-five years. Sometimes these third molars, also called wisdom teeth, become wedged in abnormal positions in the jaws and fail to erupt. Such *impacted* wisdom teeth must be removed to alleviate pain.

The teeth break food into smaller pieces, which begins mechanical digestion. Chewing increases the surface area of the food particles, enabling digestive enzymes to interact more effectively with nutrient molecules.

Different teeth are adapted to handle food in different ways. The *incisors* are chisel-shaped, and their sharp edges bite off large pieces of food. The *canines* are cone-shaped, and they grasp and tear food. The *premolars* and *molars* have flattened surfaces and are specialized for grinding food particles.

Each tooth consists of two main portions—the *crown,* which projects beyond the gum, and the *root,* anchored to the alveolar process of the jaw. The region where these portions meet is called the *neck* of the tooth. Glossy, white *enamel* covers the crown. Enamel mainly consists of calcium salts and is the hardest substance in the body. If abrasive action or injury damages enamel, it is not replaced. Enamel may wear away with age.

The bulk of a tooth beneath the enamel is composed of a living cellular tissue called *dentin,* a substance much like bone, but somewhat harder. Dentin, in turn, surrounds the tooth's central cavity (pulp cavity), which contains a combination of blood vessels, nerves, and connective tissue, collectively called pulp. Blood vessels and nerves reach this cavity through tubular *root canals,* extending into the root. Tooth loss is most often associated with diseases of the gums (gingivitis) and the dental pulp (endodontitis).

FIGURE 17.8 This partially dissected child's skull reveals primary and developing secondary teeth in the maxilla and mandible.

TABLE 17.2 | Primary and Secondary Teeth

Primary Teeth (Deciduous)		Secondary Teeth (Permanent)	
Type	*Number*	*Type*	*Number*
Incisor		Incisor	
Central	4	Central	4
Lateral	4	Lateral	4
Canine (cuspid)	4	Canine (cuspid)	4
		Premolar (bicuspid)	
		First	4
		Second	4
Molar		Molar	
First	4	First	4
Second	4	Second	4
		Third	4
Total	20	**Total**	32

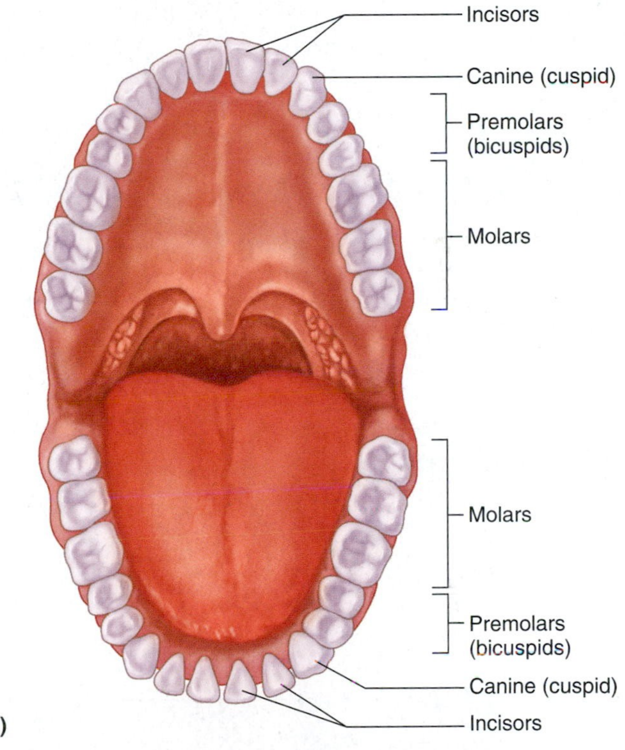

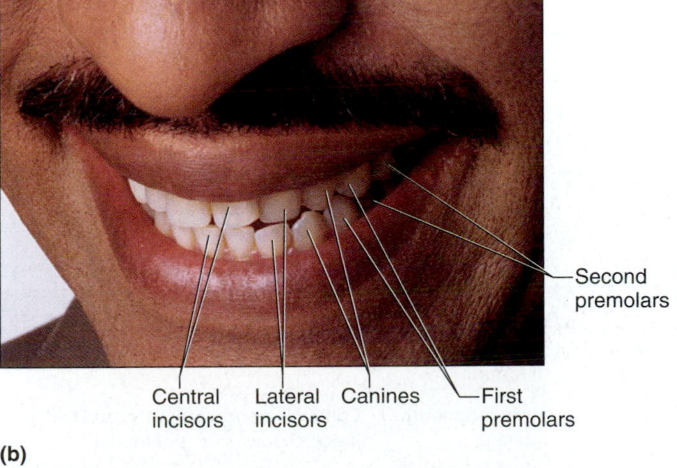

FIGURE 17.9 Permanent teeth. (*a*) The secondary teeth of the upper and lower jaws. (*b*) Anterior view of the secondary teeth.

A thin layer of bonelike material called *cementum,* surrounded by a *periodontal ligament* (periodontal membrane), encloses the root. This ligament, composed of collagen, passes between the cementum and the bone of the alveolar process, firmly attaching the tooth to the jaw. The ligament also contains blood vessels and nerves near the surface of the cementum-covered root (fig. 17.10). Clinical Application 17.1 describes the effect of bacteria on teeth. Table 17.3 summarizes the mouth parts and their functions.

Extracted primary and wisdom teeth may one day provide stem cells that can be used to regenerate tooth roots and supporting periodontal ligaments. The stem cells are in the pulp and a region called the apical papilla. Dental researchers hope that these stem cells may be cultured to yield replacement teeth for people who do not have enough jawbone to support dental implants.

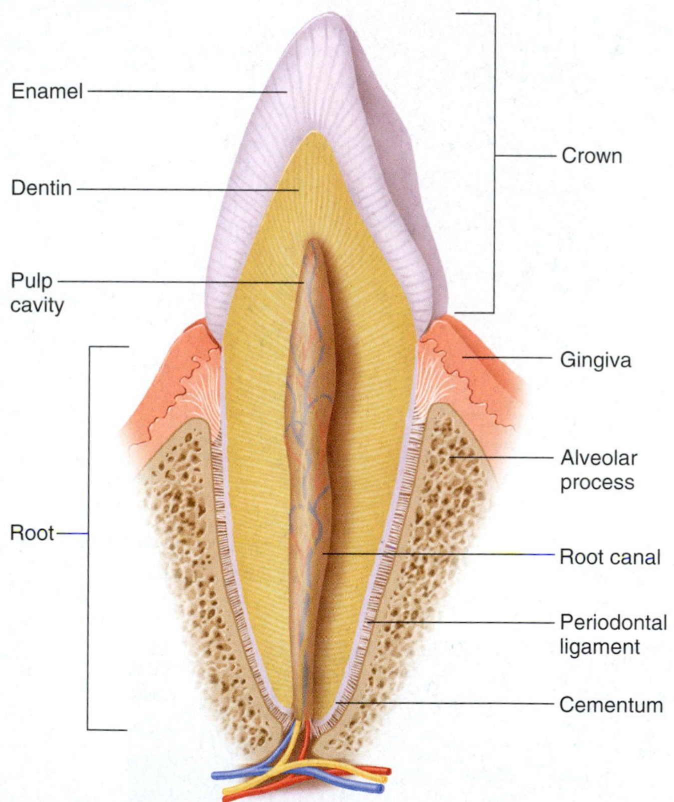

FIGURE 17.10 A section of a tooth.

PRACTICE 

10 How do primary teeth differ from secondary teeth?

11 How are types of teeth adapted to provide specialized functions?

12 Describe the structure of a tooth.

13 Explain how a tooth is attached to the bone of the jaw.

17.4 SALIVARY GLANDS

The **salivary** (sal′ĭ-ver-e) **glands** secrete saliva. This fluid moistens food particles, helps bind them, and begins the chemical digestion of carbohydrates. Saliva is also a solvent, dissolving foods so that they can be tasted, and it helps cleanse the mouth and teeth. Bicarbonate ions (HCO_3^-) in saliva help buffer the acid concentration so that the pH of saliva usually remains near neutral, between 6.5 and 7.5. This is a favorable range for the action of the salivary enzyme and protects the teeth from exposure to acids in foods.

Many minor salivary glands are scattered throughout the mucosa of the tongue, palate, and cheeks. They continuously secrete fluid, keeping the lining of the mouth moist. The three pairs of major salivary glands are the parotid glands, the submandibular glands, and the sublingual glands.

Salivary Secretions

The different salivary glands have varying proportions of two types of secretory cells, *serous cells* and *mucous cells.* Serous cells produce a watery fluid that contains a digestive enzyme called **salivary amylase** (am′ĭ-lās). This enzyme splits starch and glycogen molecules into disaccharides—the first step in the chemical digestion of carbohydrates. Mucous cells secrete a thick liquid called **mucus,** which binds food particles and acts as a lubricant during swallowing.

Like other digestive structures, the salivary glands are innervated by branches of both sympathetic and parasympathetic nerves. Impulses arriving on sympathetic fibers stimulate the gland cells to secrete a small volume of viscous saliva. Parasympathetic impulses, on the other hand, elicit the secretion of a large volume of watery saliva. Such parasympathetic impulses are activated reflexly when a person sees, smells, tastes, or even thinks about pleasant foods. Conversely, if food looks, smells, or tastes unpleasant, parasympathetic activity is inhibited, so less saliva is produced, and swallowing may become difficult.

TABLE 17.3 | Mouth Parts and Their Functions in Digestion

Part	Location	Function	Part	Location	Function
Cheeks	Form lateral walls of mouth	Hold food in mouth; muscles chew food	Tongue	Floor of mouth	Mixes food with saliva; moves food toward pharynx; contains taste receptors
Lips	Surround mouth opening	Contain sensory receptors used to judge characteristics of foods	Palate	Forms roof of mouth	Holds food in mouth; directs food to pharynx
			Teeth	In sockets of mandibular and maxillary bones	Break food particles into smaller pieces; help mix food with saliva during chewing

17.1 CLINICAL APPLICATION

Dental Caries

Sticky foods, such as caramel, lodge between the teeth and in the crevices of molars, feeding bacteria such as *Actinomyces*, *Streptococcus mutans*, and *Lactobacillus*. These microbes metabolize carbohydrates in the food, producing acid by-products that destroy tooth enamel and dentin (fig. 17A). The bacteria also produce sticky substances that hold them in place.

If a person eats a candy bar, for example, but does not brush the teeth soon afterward, the actions of the acid-forming bacteria will produce decay, called dental caries. Unless a dentist cleans and fills the resulting cavity that forms where enamel is destroyed, the damage will spread to the underlying dentin. As a result, the tooth becomes sensitive.

Dental caries can be prevented in several ways:

1. Brush and floss teeth regularly.
2. Have regular dental exams and cleanings.
3. Drink fluoridated water or receive a fluoride treatment. Fluoride is incorporated into the enamel's chemical structure, strengthening it.
4. Have a dentist apply a sealant to children's and adolescents' teeth where crevices might hold onto decay-causing bacteria. The sealant is a coating that keeps acids from eating away at tooth enamel. ■

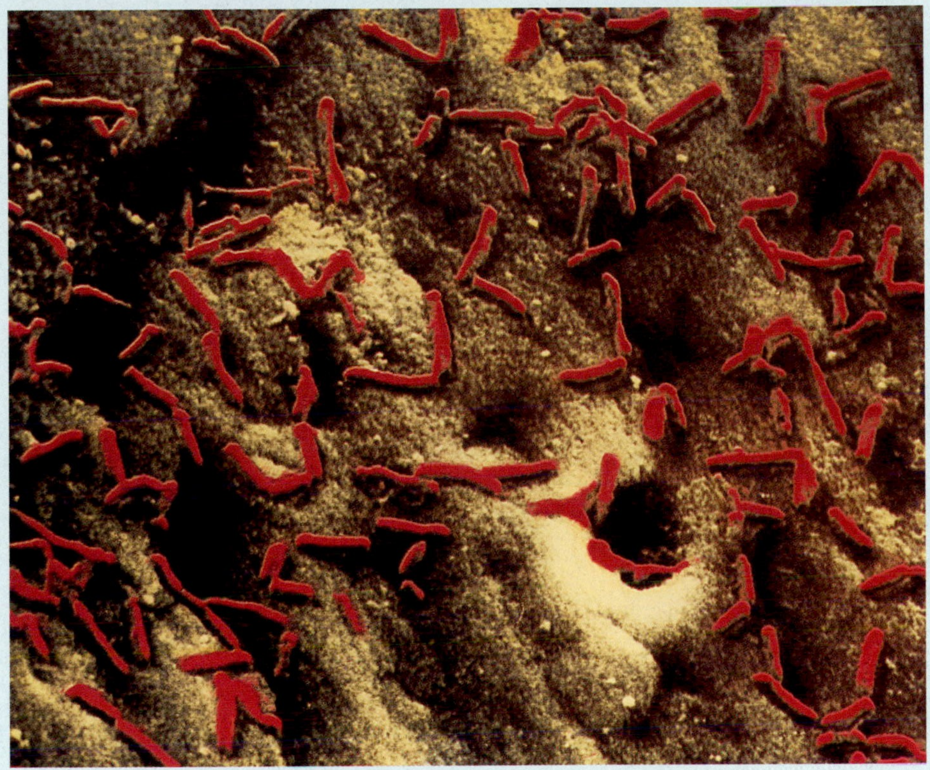

FIGURE 17A *Actinomyces* bacteria (falsely colored) clinging to teeth release acids that decay tooth enamel (1,250×).

Major Salivary Glands

The **parotid** (pah-rot′id) **glands** are the largest of the major salivary glands. Each gland lies anterior to and somewhat inferior to each ear, between the skin of the cheek and the masseter muscle. A *parotid duct* (Stensen's duct) passes from the gland inward through the buccinator muscle, entering the mouth just opposite the upper second molar on either side of the jaw. The parotid glands secrete a clear, watery fluid rich in salivary amylase (figs. 17.11 and 17.12*a*).

The **submandibular** (sub″man-dib′u-lar) **glands** are in the floor of the mouth on the inside surface of the lower jaw. The secretory cells of these glands are about equally serous and mucous. Consequently, the submandibular glands secrete a more viscous fluid than the parotid glands (see figs. 17.11 and 17.12*b*). The ducts of the submandibular glands (Wharton's ducts) open inferior to the tongue, near the lingual frenulum.

The **sublingual** (sub-ling′gwal) **glands** are the smallest of the major salivary glands. They are on the floor of the mouth inferior to the tongue. Their cells are primarily the mucous type, so their secretions, which enter the mouth through many separate ducts (Rivinus's ducts), are thick and stringy (see figs. 17.11 and 17.12*c*). Table 17.4 summarizes the characteristics of the major salivary glands.

PRACTICE  

14 What is the function of saliva?

15 What stimulates the salivary glands to secrete saliva?

16 Where are the major salivary glands?

17.5 PHARYNX AND ESOPHAGUS

The pharynx is a cavity posterior to the mouth from which the tubular esophagus leads to the stomach. The pharynx and the esophagus do not digest food, but both are important passageways, and their muscular walls function in swallowing.

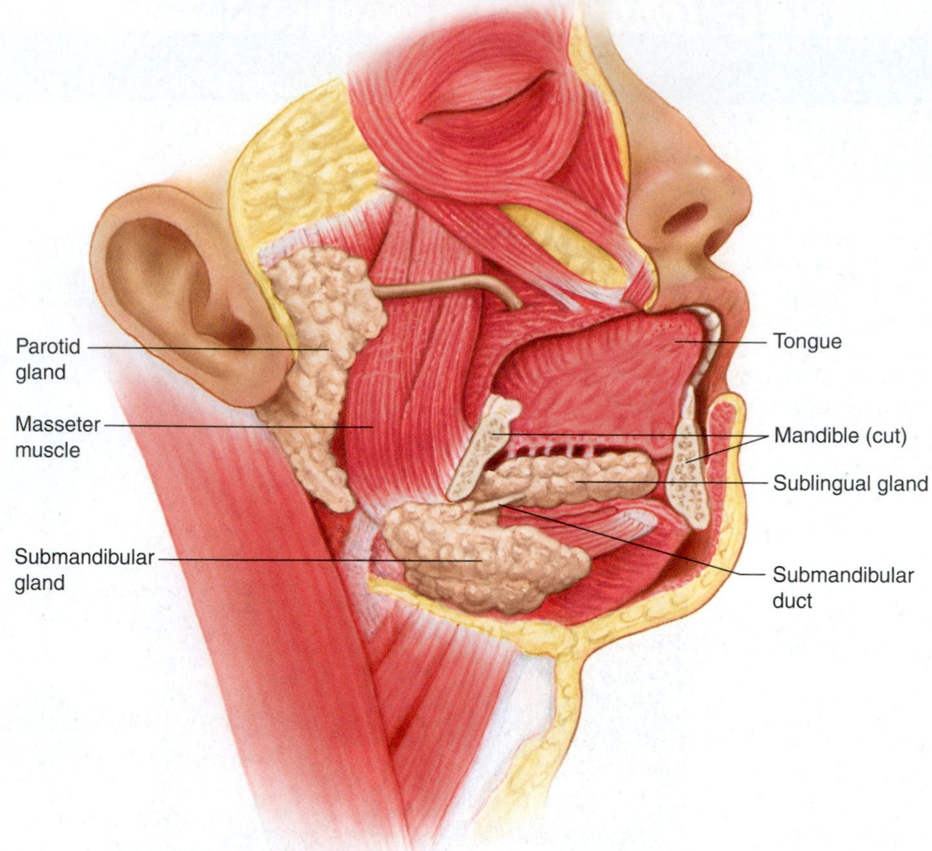

FIGURE 17.11 Locations of the major salivary glands.

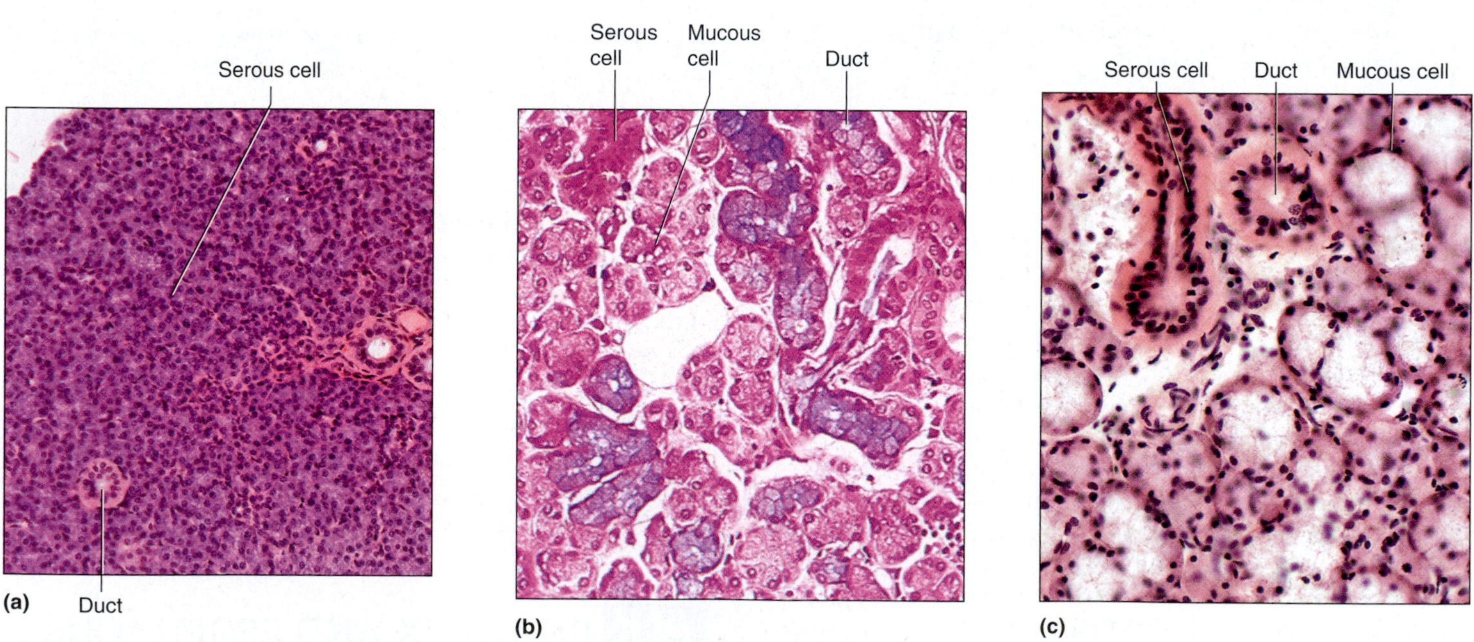

FIGURE 17.12 Light micrographs of (*a*) the parotid salivary gland (100×), (*b*) the submandibular salivary gland (180×), and (*c*) the sublingual salivary gland (200×).

TABLE 17.4 | The Major Salivary Glands

Gland	Location	Duct	Type of Secretion
Parotid glands	Anterior to and somewhat inferior to the ears between the skin of the cheeks and the masseter muscles	Parotid ducts pass through the buccinator muscles and enter the mouth opposite the upper second molars	Clear, watery serous fluid, rich in salivary amylase
Submandibular glands	In the floor of the mouth on the inside surface of the mandible	Ducts open inferior to the tongue near the frenulum	Some serous fluid with some mucus; more viscous than parotid secretion
Sublingual glands	In the floor of the mouth inferior to the tongue	Many separate ducts	Primarily thick, stringy mucus

Structure of the Pharynx

The **pharynx** (far′ingks) connects the nasal and oral cavities with the larynx and esophagus (see fig. 17.7). It can be divided into the following parts:

1. The **nasopharynx** (na″zo-far′ingks) is superior to the soft palate. It communicates with the nasal cavity and provides a passageway for air during breathing. The auditory tubes, which connect the pharynx with the middle ears, open through the walls of the nasopharynx (see chapter 12, p. 453).
2. The **oropharynx** (o″ro-far′ingks) is posterior to the mouth. It is posterior to the soft palate and inferior to the nasopharynx, projecting downward to the upper border of the epiglottis. This portion is a passageway for food moving downward from the mouth and for air moving to and from the nasal cavity.
3. The **laryngopharynx** (lah-ring″go-far′inks) is just inferior to the oropharynx. It extends from the upper border of the epiglottis downward to the lower border of the cricoid cartilage of the larynx and is a passageway to the esophagus.

The muscles in the walls of the pharynx form inner circular and outer longitudinal groups (fig. 17.13). The circular muscles, called *constrictor muscles,* pull the walls inward during swallowing. The *superior constrictor muscles,* attached to bony processes of the skull and mandible, curve around the upper part of the pharynx. The *middle constrictor muscles* arise from projections on the hyoid bone and fan around the middle of the pharynx. The *inferior constrictor muscles* originate from cartilage of the larynx and pass around the lower portion of the pharyngeal cavity. Some of the lower inferior constrictor muscle fibers contract most of the time, which prevents air from entering the esophagus during breathing.

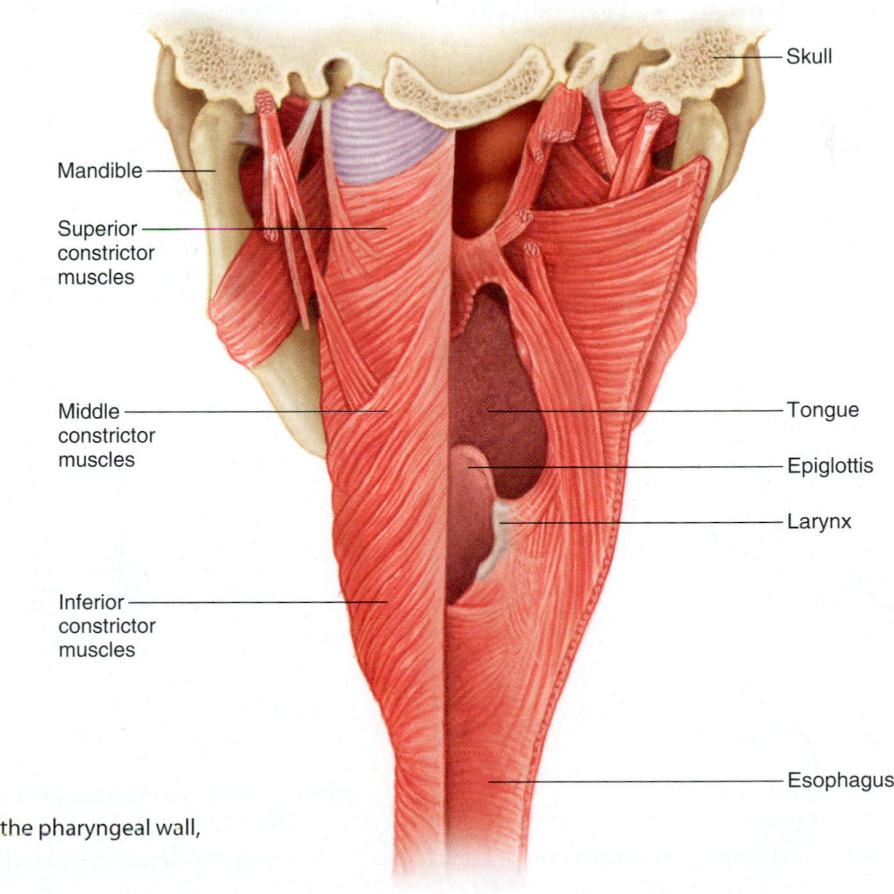

FIGURE 17.13 Muscles of the pharyngeal wall, posterior view.

Although the pharyngeal muscles are skeletal muscles, they are under voluntary control only in the sense that swallowing (deglutition) can be voluntarily initiated. Complex reflexes control the precise actions of these muscles during swallowing.

Swallowing Mechanism

Swallowing can be divided into three stages. In the first stage, which is voluntary, food is chewed and mixed with saliva. Then, the tongue rolls this mixture into a mass, or **bolus,** and forces it into the pharynx.

The second stage of swallowing begins as food reaches the pharynx and stimulates sensory receptors around the pharyngeal opening. This triggers the swallowing reflex, illustrated in figure 17.14, which includes the following actions:

1. The soft palate (including the uvula) raises, preventing food from entering the nasal cavity.
2. The hyoid bone and the larynx are elevated. A flaplike structure attached to the larynx, called the *epiglottis,* (ep″ĭ-glot′is) closes off the top of the trachea so that food is less likely to enter the trachea.
3. The tongue is pressed against the soft palate and uvula, sealing off the oral cavity from the nasal cavity.
4. The longitudinal muscles in the pharyngeal wall contract, pulling the pharynx upward toward the food.
5. The lower portion of the inferior constrictor muscles relaxes, opening the esophagus.
6. The superior constrictor muscles contract, stimulating a peristaltic wave to begin in other pharyngeal muscles. This wave forces the food into the esophagus.

The swallowing reflex momentarily inhibits breathing. Then, during the third stage of swallowing, peristalsis transports the food in the esophagus to the stomach.

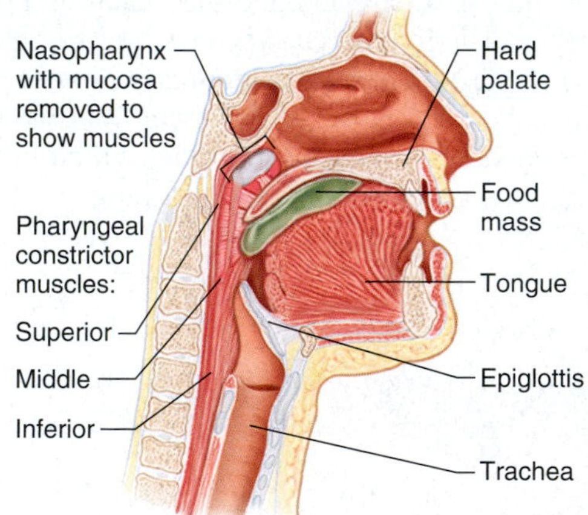

(a) The tongue forces food into the pharynx.

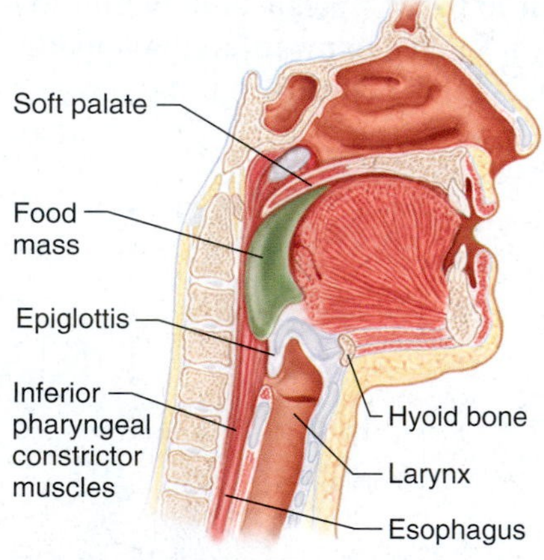

(b) The soft palate, hyoid bone, and larynx are raised, the tongue is pressed against the palate, the epiglottis closes, and the inferior constrictor muscles relax so that the esophagus opens.

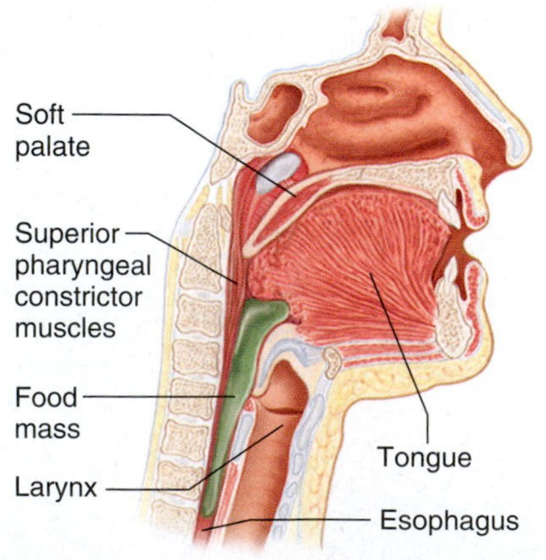

(c) Superior constrictor muscles contract and force food into the esophagus.

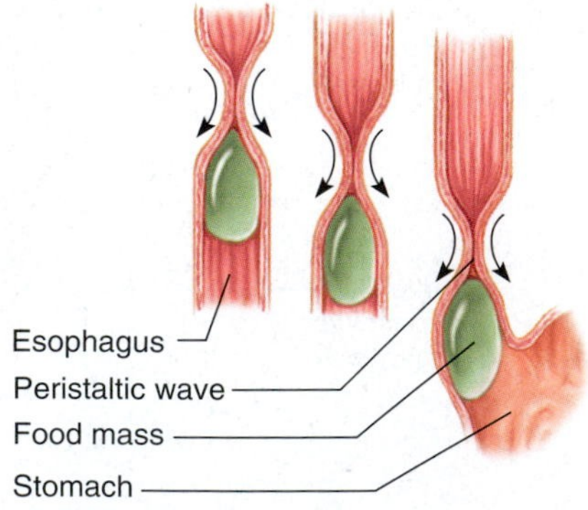

(d) Peristaltic waves move food through the esophagus to the stomach.

FIGURE 17.14 Steps in the swallowing reflex. (The mucosa in (*a*), (*b*), and (*c*) has been removed to reveal the underlying muscles.)

Esophagus

The **esophagus** (ĕ-sof′ah-gus) is a straight, collapsible tube about 25 centimeters long. It provides a passageway for food, and its muscular wall propels food from the pharynx to the stomach. The esophagus descends through the thorax posterior to the trachea, passing through the mediastinum. It penetrates the diaphragm through an opening, the *esophageal hiatus* (ĕ-sof″ah-je′al hi-a′tus), and is continuous with the stomach on the abdominal side of the diaphragm (fig. 17.15 and reference plates 17, 23).

In a *hiatal hernia,* part of the stomach protrudes through a weakened area of the diaphragm, through the esophageal hiatus and into the thorax. Regurgitation (reflux) of gastric juice into the esophagus may inflame the esophageal mucosa, causing heartburn, difficulty in swallowing, or ulceration and blood loss. In response to the destructive action of gastric juice, columnar epithelium may replace the squamous epithelium that normally lines the esophagus (see chapter 5, pages 147–148). This condition, called *Barrett's esophagus,* increases the risk of developing esophageal cancer.

Mucous glands are scattered throughout the submucosa of the esophagus (fig. 17.16). Their secretions moisten and lubricate the inner lining of the tube.

Just superior to the point where the esophagus joins the stomach, some of the circular muscle fibers have increased sympathetic muscle tone, forming the **lower esophageal sphincter** (loh′er ĕ-sof″ah-je′al sfingk′ter), or cardiac sphincter (fig. 17.17). These fibers usually remain contracted, and they close the entrance to the stomach. In this way, they help prevent regurgitation of the stomach contents into the esophagus. When peristaltic waves reach the stomach, the muscle fibers that guard its entrance relax briefly, and the swallowed food enters.

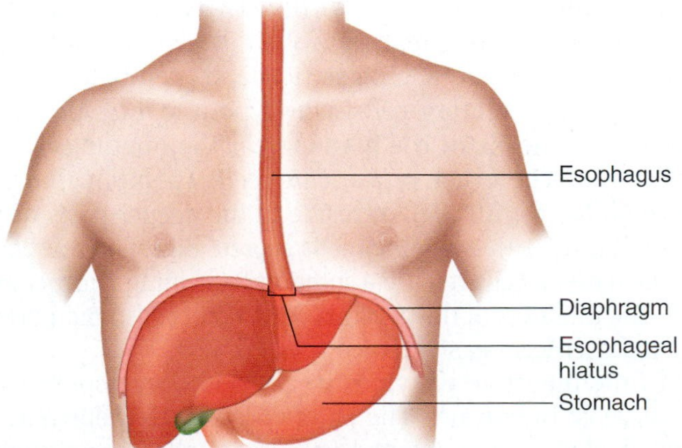

FIGURE 17.15 The esophagus is a passageway between the pharynx and the stomach.

PRACTICE 

17 Describe the regions of the pharynx.

18 List the major events of swallowing.

19 What is the function of the esophagus?

17.6 STOMACH

The **stomach** (stum′ak) is a J-shaped, pouchlike organ, about 25–30 centimeters long, which hangs inferior to the diaphragm in the upper-left portion of the abdominal cavity (see figs. 17.1 and 17.15; reference plates 4 and 5). The stomach has a capacity of about one liter or more. Its inner lining has thick gastric folds (rugae) of the mucosal and submucosal layers that disappear when the wall is distended. The stomach receives food from the esophagus, mixes it with gastric juice, initiates the digestion of proteins, carries on limited absorption, and moves food into the small intestine.

In addition to the two layers of smooth muscle—an inner circular layer and an outer longitudinal layer—also in other regions of the alimentary canal, some parts of the stomach have an additional inner layer of oblique fibers, which strengthen the stomach wall and help the mixing and churning. This third innermost muscular layer is most highly developed near the opening of the esophagus and in the body of the stomach (fig. 17.17).

Parts of the Stomach

The stomach, shown in figures 17.17 and 17.18 and reference plate 11, can be divided into the cardia, fundus, body, and pylorus. The *cardia* is a small area near the esophageal opening. The *fundus,* which balloons superior to the cardia, is a temporary storage area and sometimes fills with swallowed

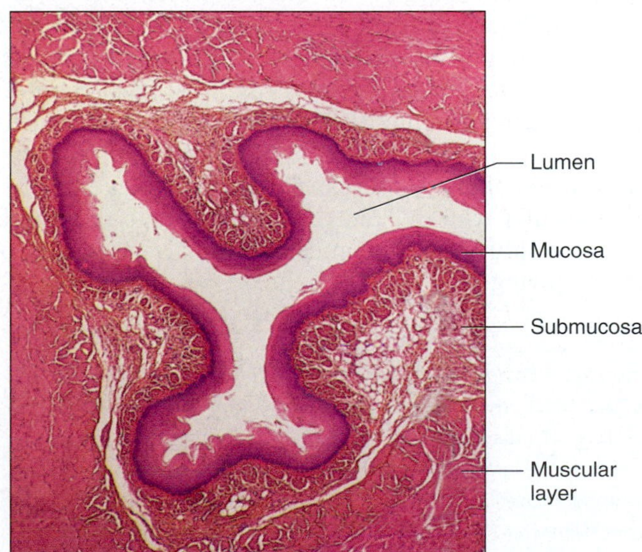

FIGURE 17.16 This cross section of the esophagus shows its muscular wall (10×).

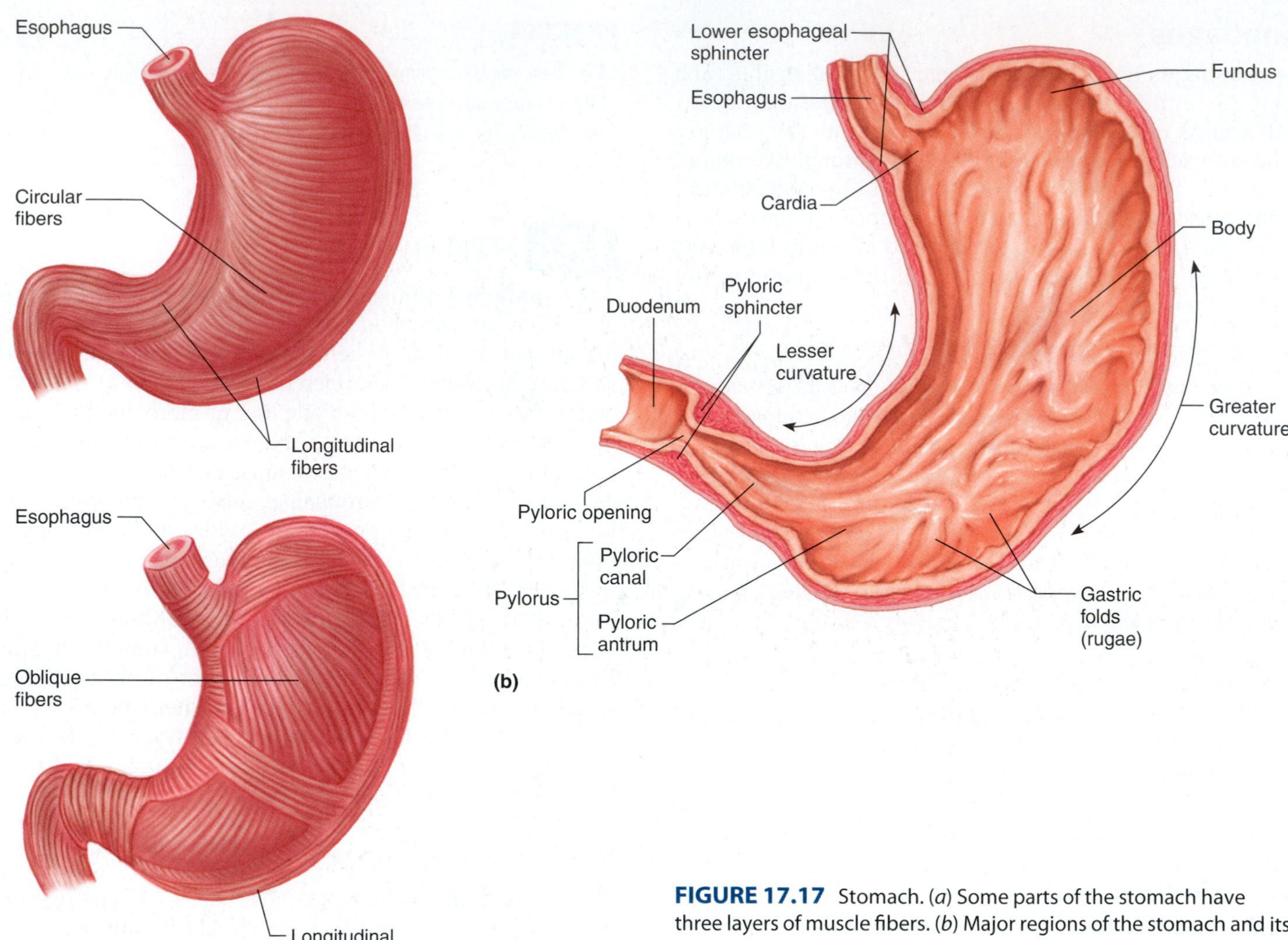

FIGURE 17.17 Stomach. (*a*) Some parts of the stomach have three layers of muscle fibers. (*b*) Major regions of the stomach and its associated structures.

air. This produces a gastric air bubble, which may be used as a landmark on a radiograph of the abdomen. The dilated *body,* the main part of the stomach, is between the fundus and pylorus. The *pyloric antrum* is a funnel-shaped portion that narrows and becomes the *pyloric canal* as it approaches the small intestine.

At the end of the pyloric canal, the circular layer of fibers in its muscular wall thickens, forming a powerful muscle, the **pyloric sphincter.** This muscle is a valve that controls gastric emptying.

> *Hypertrophic pyloric stenosis* is a birth defect in which muscle overgrowth blocks the pyloric canal. The newborn vomits, with increasing force. To diagnose the condition, a radiograph is taken of the area after the infant drinks formula containing a radiopaque barium compound. Surgical splitting of the muscle blocking the passageway from stomach to small intestine is necessary to enable the infant to eat normally. Pyloric stenosis can occur later in life as a result of ulcers or cancer.

Gastric Secretions

The mucous membrane that forms the inner lining of the stomach is thick. Its surface is studded with many small openings, called *gastric pits,* located at the ends of tubular **gastric glands** (oxyntic glands) (fig. 17.19). Although their structure and the composition of their secretions vary in different parts of the stomach, gastric glands generally contain three types of secretory cells. One type, the *mucous cell,* is found in the necks of the glands near the openings of the gastric pits. The other types, *chief cells* (peptic cells) and *parietal cells* (oxyntic cells), reside in the deeper parts of the glands (fig. 17.19). The chief cells secrete digestive enzymes, and the parietal cells release a solution containing hydrochloric acid. The products of the mucous cells, chief cells, and parietal cells together form **gastric juice** (gas′trik jo͞os).

Of the digestive enzymes in gastric juice, **pepsin** is by far the most important. The chief cells secrete pepsin as an inactive, nonerosive enzyme precursor called **pepsinogen.** When pepsinogen contacts the hydrochloric acid from the parietal cells, however, it breaks down rapidly, forming

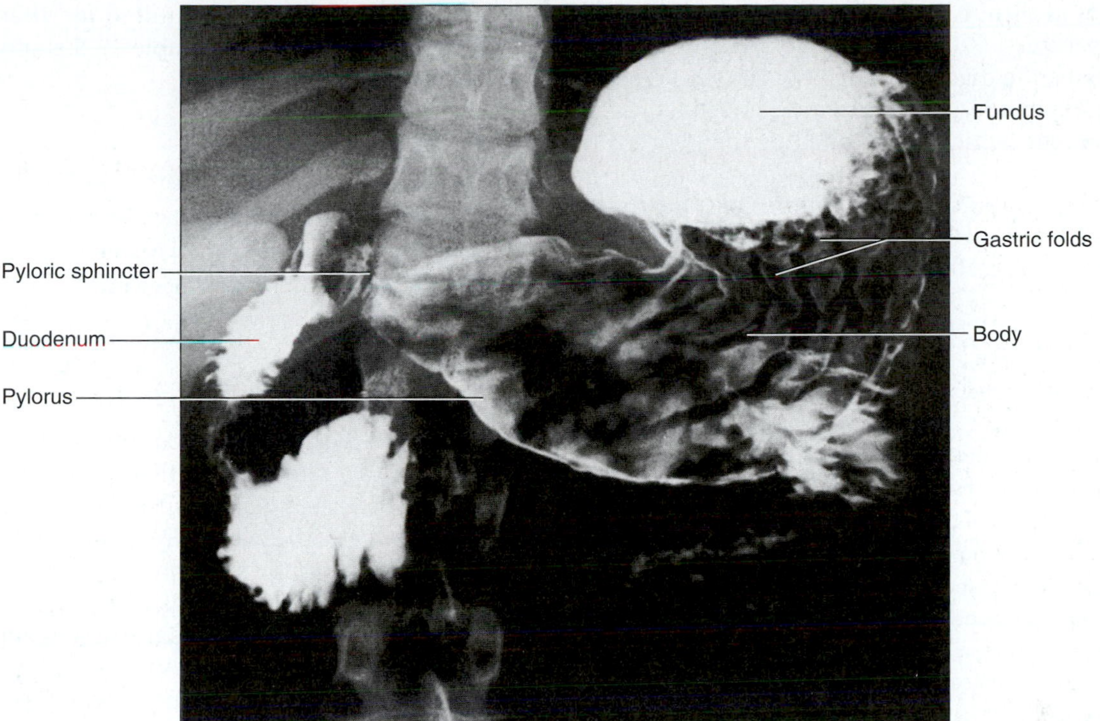

FIGURE 17.18 Radiograph of a stomach. (*Note:* A radiopaque compound the patient swallowed appears white in the radiograph.)

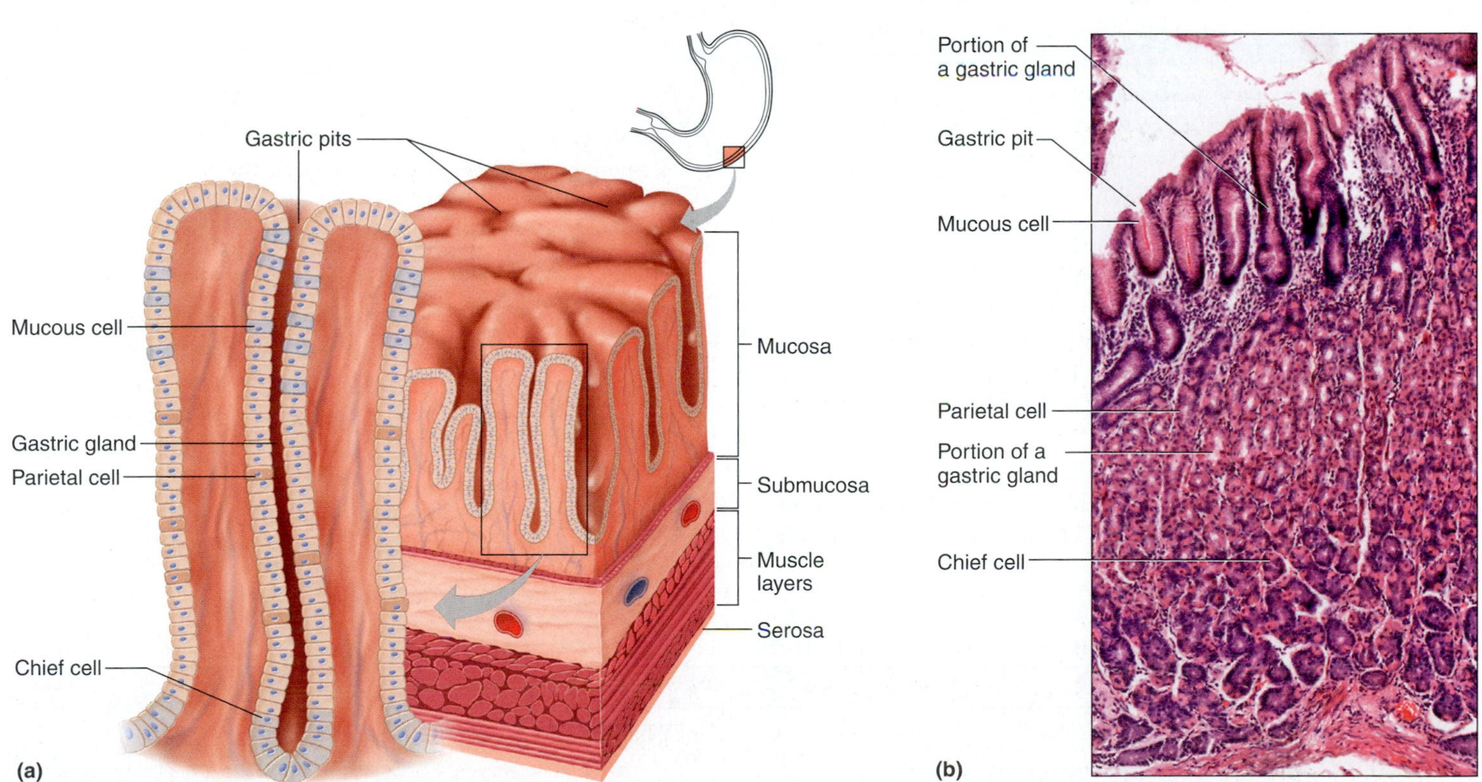

FIGURE 17.19 Lining of the stomach. (*a*) Gastric glands include mucous cells, parietal cells, and chief cells. The mucosa of the stomach is studded with gastric pits that are the openings of the gastric glands. (*b*) A light micrograph of cells associated with the gastric glands (60×).

pepsin. Pepsin, in turn, can also break down pepsinogen to release more pepsin.

Pepsin begins the digestion of nearly all types of dietary protein into polypeptides. This enzyme is most active in an acidic environment, which is provided by the hydrochloric acid in gastric juice.

Gastric juice contains small quantities of a fat-splitting enzyme, *gastric lipase.* However, its action is weak due in part to the low pH of gastric juice. Gastric lipase acts mainly on butterfat.

Much of what we know about the stomach's functioning comes from a French-Canadian explorer, Alexis St. Martin, who in 1822 accidentally shot himself in the abdomen. His extensive injuries eventually healed, but a hole, called a fistula, was left, allowing observers to look at his stomach in action. A U.S. Army surgeon, William Beaumont, spent eight years watching food digesting in the stomach, noting how the stomach lining changed in response to stress.

In 1984, our knowledge of digestive function expanded when medical resident Barry Marshall at Royal Perth Hospital in western Australia performed a daring experiment. His mentor, J. Robin Warren, had hypothesized that a bacterial infection causes gastritis (inflammation of the stomach lining) and peptic ulcers (sores in the lining of the esophagus, stomach, or small intestine). At the time, these conditions were attributed to poor diet and stress. Marshall drank "swamp water"—billions of bacteria. He developed gastritis, which, fortunately, cleared up. A colleague who repeated the experiment developed an ulcer and required antibiotics. After a decade of debate, the medical community finally concurred that the bacterium *Helicobacter pylori,* which thrives under acidic conditions, causes many cases of gastritis and peptic ulcers. A short course of antibiotics and acid-lowering drugs has replaced lifelong treatments. Marshall and Warren were awarded a Nobel Prize in 2005 for their discovery.

The mucous cells of the gastric glands (*mucous neck cells*) and the mucous cells, associated with the stomach's inner surface, release a viscous, alkaline secretion that coats the inside of the stomach wall. This coating is especially important because pepsin can digest the proteins of stomach tissues, as well as those in foods. The coating normally prevents the stomach from digesting itself.

Another component of gastric juice is **intrinsic factor** (in-trin′sik fak′tor). The parietal cells of the gastric glands secrete intrinsic factor, required for vitamin B_{12} absorption from the small intestine. Table 17.5 summarizes the components of gastric juice.

PRACTICE 

20 Where is the stomach located?

21 What are the secretions of the chief cells and parietal cells?

22 Which is the most important digestive enzyme in gastric juice?

23 Why doesn't the stomach digest itself?

Regulation of Gastric Secretions

Gastric juice is produced continuously, but the rate varies considerably and is controlled both neurally and hormonally. In the gastric glands, specialized cells closely associated with the parietal cells secrete the hormone *somatostatin,* which inhibits acid secretion. However, acetylcholine (ACh) released from nerve endings in response to parasympathetic impulses arriving on the vagus nerves suppresses the secretion of somatostatin and stimulates the gastric glands to secrete abundant gastric juice, which is rich in hydrochloric acid and pepsinogen. These parasympathetic impulses also stimulate certain stomach cells, mainly in the pyloric region, to release a peptide hormone called **gastrin,** which increases the secretory activity of gastric glands (fig. 17.20). Furthermore, parasympathetic impulses and gastrin promote release of *histamine* from gastric mucosal cells, which, in turn, stimulates additional gastric secretion.

Gastrin stimulates cell growth in the mucosa of the stomach and intestines, except where gastrin is produced. This effect helps replace mucosal cells damaged by normal stomach function, disease, or medical treatments.

Gastric secretion occurs in three stages—the cephalic, gastric, and intestinal phases. The *cephalic phase* begins before food reaches the stomach and possibly even before eating. In this stage, parasympathetic reflexes operating through the vagus nerves stimulate gastric secretion at the taste, smell, sight, or thought of food. The greater the hunger, the greater the gastric secretion. The cephalic phase is responsible for 30% to 50% of the secretory response to a meal.

TABLE 17.5 | Major Components of Gastric Juice

Component	Source	Function
Pepsinogen	Chief cells of the gastric glands	Inactive form of pepsin
Pepsin	Formed from pepsinogen in the presence of hydrochloric acid	A protein-splitting enzyme that digests nearly all types of dietary protein
Hydrochloric acid	Parietal cells of the gastric glands	Provides the acid environment needed for production and action of pepsin
Mucus	Mucous cells	Provides a viscous, alkaline protective layer on the stomach's inner surface
Intrinsic factor	Parietal cells of the gastric glands	Aids in vitamin B_{12} absorption

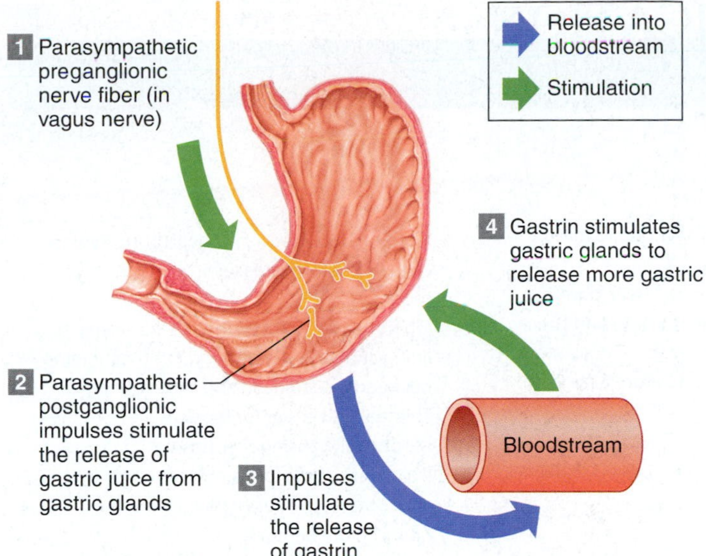
Release into bloodstream
Stimulation
1 Parasympathetic preganglionic nerve fiber (in vagus nerve)
4 Gastrin stimulates gastric glands to release more gastric juice
2 Parasympathetic postganglionic impulses stimulate the release of gastric juice from gastric glands
Bloodstream
3 Impulses stimulate the release of gastrin

FIGURE 17.20 The secretion of gastric juice is regulated in part by parasympathetic nerve impulses that stimulate the release of gastric juice and gastrin.

The *gastric phase* of gastric secretion, which accounts for 40% to 50% of the secretory activity, starts when food enters the stomach. The presence of food and the distension of the stomach wall trigger the stomach to release gastrin, which stimulates production of more gastric juice.

As food enters the stomach and mixes with gastric juice, the pH of the contents rises, which enhances gastrin secretion. Consequently, the pH of the stomach contents drops. As the pH approaches 3.0, secretion of gastrin is inhibited. When the pH reaches 1.5, gastrin secretion ceases.

For the stomach to secrete hydrochloric acid, hydrogen ions are removed from the blood, and an equivalent number of alkaline bicarbonate ions are released into the blood. Following a meal, the blood concentration of bicarbonate ions increases, and the urine excretes excess bicarbonate ions. This phenomenon is called the *alkaline tide.*

The *intestinal phase* of gastric secretion, which accounts for about 5% of the total secretory response to a meal, begins when food leaves the stomach and enters the small intestine. When food first contacts the intestinal wall, it stimulates intestinal cells to release a hormone, *intestinal gastrin,* that again enhances gastric gland secretion.

As more food moves into the small intestine, a sympathetic reflex triggered by acid in the upper part of the small intestine inhibits secretion of gastric juice from the stomach wall. At the same time, proteins and fats in this region of the intestine stimulate release of the peptide hormone **cholecystokinin** (ko″le-sis″to-ki′nin) from the intestinal wall, which decreases gastric motility. Similarly, fats in the small intestine stimulate intestinal cells to release *intestinal somatostatin,* which inhibits release of gastric juice. Overall, these actions decrease gastric secretion and motility as the small intestine fills with food. Table 17.6 summarizes the phases of gastric secretion.

PRACTICE 

24 What controls gastric juice secretion?

25 Distinguish among the cephalic, gastric, and intestinal phases of gastric secretion.

26 What is the function of cholecystokinin?

Gastric Absorption

Gastric enzymes begin breaking down proteins, but the stomach wall is not well-adapted to absorb digestive products. The stomach absorbs only some water and certain salts, as well as certain lipid-soluble drugs. Most nutrients are absorbed in the small intestine. Alcohol, which is not a nutrient, is absorbed both in the small intestine and stomach. This is why the intoxicating effects of alcohol are felt soon after consuming alcoholic beverages.

Mixing and Emptying Actions

Food stretches the smooth muscles of the stomach wall. The stomach may enlarge, but its muscles maintain their tone, and internal pressure of the stomach normally is unchanged. When a person eats more than the stomach can comfortably hold, the internal pressure may rise enough to stimulate pain receptors. The result is a stomachache. Clinical Application 17.2 discusses this common problem along with its associated indigestion.

Following a meal, the mixing movements of the stomach wall aid in producing a semifluid paste of food particles and gastric juice called **chyme** (kīm). Peristaltic waves push the chyme toward the pylorus of the stomach, and as chyme accumulates near the pyloric sphincter, this muscle begins to relax. Stomach contractions push chyme a little (5–15 milliliters) at a time into the small intestine. These stomach contraction waves push most of the chyme backward into the stomach, mixing it further. The lower esophageal sphincter prevents reflux of stomach contents into the esophagus. Figure 17.21 illustrates this process.

The rate at which the stomach empties depends on the fluidity of the chyme and the type of food. Liquids usually pass through the stomach rapidly, but solids remain until

TABLE 17.6 | Phases of Gastric Secretion

Phase	Action
Cephalic phase	The sight, taste, smell, or thought of food triggers parasympathetic reflexes. Gastric juice is secreted in response.
Gastric phase	Food in stomach chemically and mechanically stimulates release of gastrin, which, in turn, stimulates secretion of gastric juice; reflex responses also stimulate gastric juice secretion.
Intestinal phase	As food enters the small intestine, it stimulates intestinal cells to release intestinal gastrin, which, in turn, promotes the secretion of gastric juice from the stomach wall.

17.2 CLINICAL APPLICATION

Oh, My Aching Stomach!

At the barbecue, Perry W. consumed two burgers, three hot dogs, beans in a spicy sauce, loads of chips, several beers, and ice cream. Later, a feeling of fullness became abdominal pain, then heartburn, as his stomach contents backed up into his esophagus.

Perry found temporary relief with an over-the-counter antacid product, which quickly raised the pH of the stomach. These products usually include a compound containing either sodium, calcium, magnesium, or aluminum. Another ingredient in some products is simethicone, which breaks up gas bubbles in the digestive tract. If antacids do not help within a few minutes or they are used for longer than two weeks, a doctor should be consulted. The problem may be more serious than overeating.

Avoiding acid indigestion and heartburn is a more healthful approach than gorging and then reaching for medication—or even taking products that lower acid production before a large or spicy meal. Some tips:

- Avoid large meals. The more food, the more acid the stomach produces.
- Eat slowly so that stomach acid secretion is more gradual.
- Do not lie down immediately after eating. Being upright enables gravity to help food move along the alimentary canal.
- If prone to indigestion or heartburn, avoid caffeine, which increases stomach acid secretion.
- Cigarettes and alcohol irritate the stomach lining and relax the sphincter at the junction between the stomach and the esophagus. This makes it easier for food to return to the esophagus, causing heartburn.
- Do not eat acidic foods, such as citrus fruits and tomatoes, unless it is at least three hours before bedtime.
- Use a pillow that elevates the head six to eight inches above the stomach. ■

they are well mixed with gastric juice. Fatty foods may remain in the stomach three to six hours; foods high in proteins move through more quickly; carbohydrates usually pass through more rapidly than either fats or proteins.

As chyme fills the duodenum, internal pressure on the organ increases, stretching the intestinal wall. These actions stimulate sensory receptors in the wall, triggering an **enterogastric reflex** (en-ter-o-gas′trik re′fleks). The name of this reflex, like those of other digestive reflexes, describes the origin and termination of reflex impulses. That is, the enterogastric reflex begins in the small intestine (*entero*) and ends in the stomach (*gastro*). As a result of the enterogastric reflex, fewer parasympathetic impulses arrive at the stomach, inhibiting peristalsis, and intestinal filling slows (fig. 17.22). If chyme entering the intestine is fatty, the intestinal wall releases the hormone cholecystokinin, which further inhibits peristalsis.

Vomiting results from a complex reflex that empties the stomach in the reverse of the normal direction. Irritation or distension in the stomach or intestines can trigger vomiting. Sensory impulses travel from the site of stimulation to the *vomiting center* in the medulla oblongata, and motor responses follow. These include taking a deep breath, raising the soft palate and thus closing the nasal cavity, closing the opening to the trachea (glottis), relaxing the circular muscle fibers at the base of the esophagus, contracting the

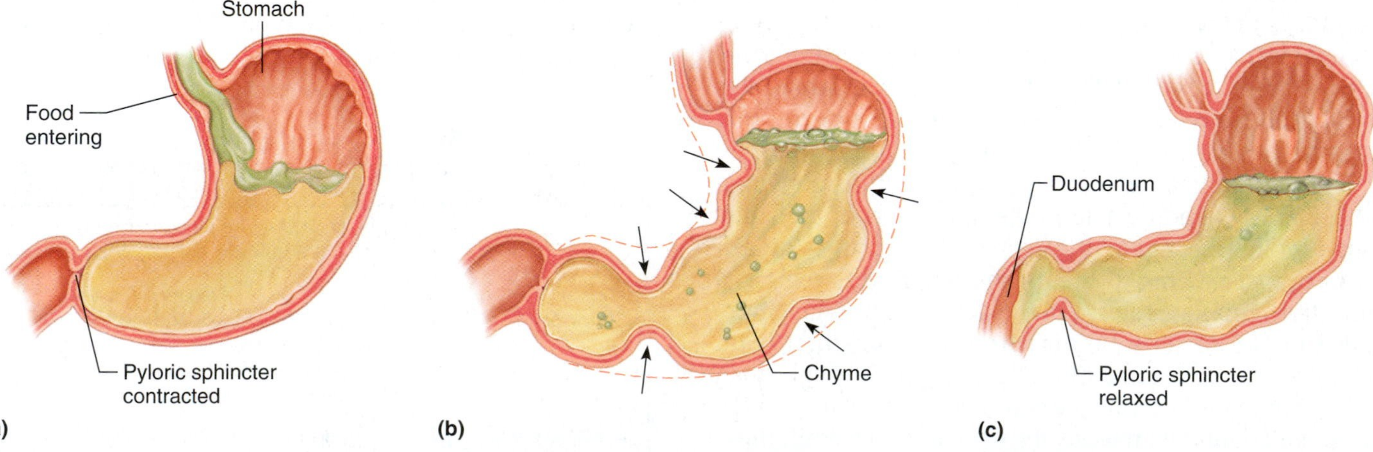

FIGURE 17.21 Stomach movements. (*a*) As the stomach fills, its muscular wall stretches, but the pyloric sphincter remains closed. (*b*) Mixing movements combine food and gastric juice, creating chyme. (*c*) Peristaltic waves move the chyme toward the pyloric sphincter, which relaxes and admits some chyme into the duodenum.

diaphragm so it presses downward over the stomach, and contracting the abdominal wall muscles to increase pressure inside the abdominal cavity. As a result, the stomach is squeezed from all sides, forcing its contents upward and out through the esophagus, pharynx, and mouth.

Drugs (emetics), toxins in contaminated foods, and rapid changes in body motion stimulate activity in the vomiting center. With changes in motion, sensory impulses from the labyrinths of the inner ears reach the vomiting center and can produce motion sickness. The vomiting center can also be activated by stimulation of higher brain centers through sights, sounds, odors, tastes, emotions, or mechanical stimulation of the back of the pharynx.

Nausea emanates from activity in the vomiting center or in nerve centers near it. During nausea, stomach movements usually are diminished or absent, and duodenal contents may move back into the stomach.

PRACTICE

27 How is chyme produced?

28 What factors influence how quickly chyme leaves the stomach?

29 Describe the enterogastric reflex.

30 Describe the vomiting reflex.

31 Which factors may stimulate the vomiting reflex?

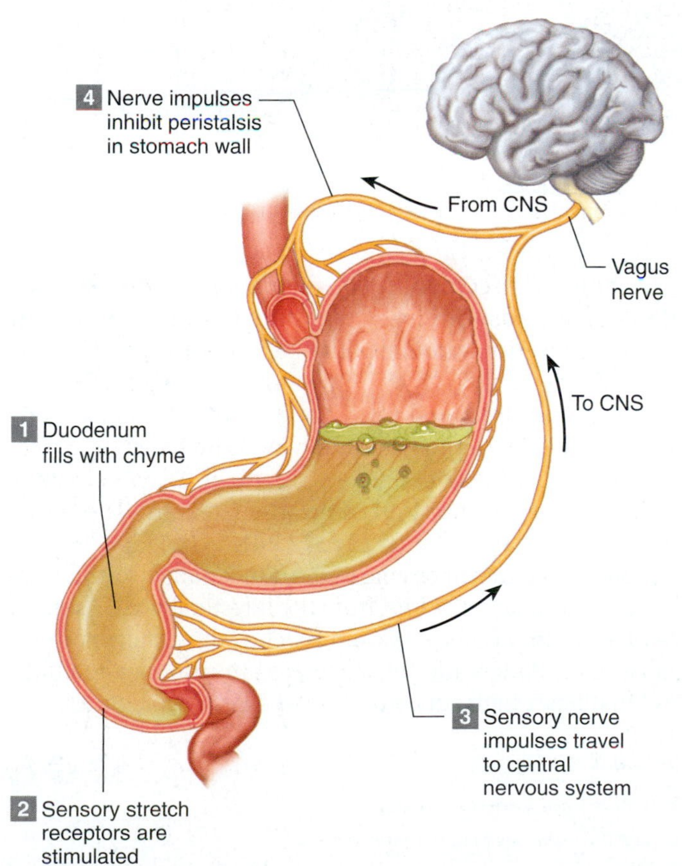
4 Nerve impulses inhibit peristalsis in stomach wall
From CNS
Vagus nerve
To CNS
1 Duodenum fills with chyme
3 Sensory nerve impulses travel to central nervous system
2 Sensory stretch receptors are stimulated

FIGURE 17.22 The enterogastric reflex partially regulates the rate at which chyme leaves the stomach.

17.7 PANCREAS

The **pancreas** was discussed as an endocrine gland in chapter 13 (p. 509). It also has an exocrine function—secretion of a digestive fluid called **pancreatic juice** (pan″kre-at′ik jo͞os).

Structure of the Pancreas

The pancreas is closely associated with the small intestine and is posterior to the parietal peritoneum. It extends horizontally across the posterior abdominal wall, with its head in the C-shaped curve of the duodenum (portion of the small intestine) and its tail against the spleen (fig. 17.23 and reference plate 19).

The cells that produce pancreatic juice, called *pancreatic acinar cells,* make up the bulk of the pancreas. These cells form clusters called *acini* (acinus, singular) around tiny tubes into which they release their secretions. The smaller tubes unite to form larger ones, which, in turn, give rise to a *pancreatic duct* extending the length of the pancreas and transporting pancreatic juice to the small intestine. The pancreatic duct usually connects with the duodenum at the same place where the bile duct from the liver and gallbladder joins the duodenum, although other connections may be present (see figs. 13.34 and 17.23).

The pancreatic and bile ducts join at a short, dilated tube called the *hepatopancreatic ampulla* (ampulla of Vater). A band of smooth muscle, called the *hepatopancreatic sphincter* (sphincter of Oddi), surrounds this ampulla.

Pancreatic Juice

Pancreatic juice contains enzymes that digest carbohydrates, fats, proteins, and nucleic acids. The carbohydrate-digesting enzyme, **pancreatic amylase,** splits molecules of starch or glycogen into disaccharides. The fat-digesting enzyme, **pancreatic lipase,** breaks triglyceride molecules into fatty acids and monoglycerides. (A monoglyceride molecule consists of one fatty acid bound to glycerol.)

The protein-splitting (proteolytic) enzymes are **trypsin, chymotrypsin,** and **carboxypeptidase.** Each of these enzymes splits the bonds between particular combinations of amino acids in proteins. No single enzyme can split all possible amino acid combinations, so several enzymes are necessary to completely digest protein molecules.

The proteolytic enzymes are stored in tiny cellular structures called *zymogen granules.* These enzymes, like gastric pepsin, are secreted in inactive forms and must be activated by other enzymes after they reach the small intestine. For example, the pancreatic cells release inactive **trypsinogen,** activated to trypsin when it contacts the enzyme *enterokinase,* which the mucosa of the small intestine secretes. Chymotrypsin and carboxypeptidase are activated, in turn, by trypsin. This mechanism prevents enzymatic digestion of proteins in the secreting cells and the pancreatic ducts.

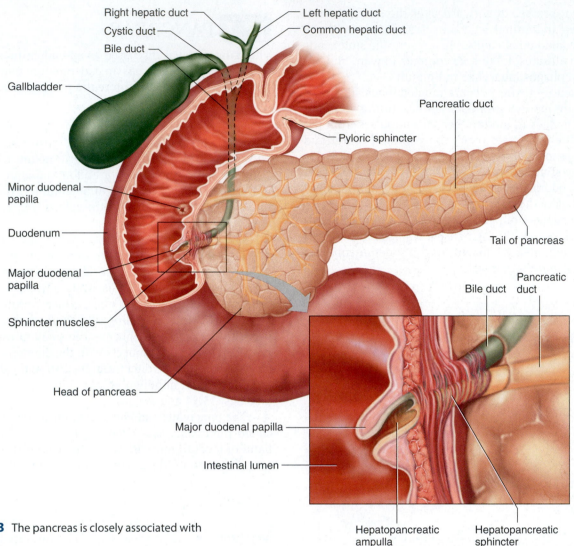

FIGURE 17.23 The pancreas is closely associated with the duodenum.

> Painful *acute pancreatitis* results from blockage in the release of pancreatic juice. Trypsin, activated as pancreatic juice builds up, digests parts of the pancreas. Alcoholism, gallstones, certain infections, traumatic injuries, or the side effects of some drugs can cause pancreatitis.

Pancreatic juice contains two types of **nucleases,** which are enzymes that break down nucleic acid molecules into nucleotides. A high concentration of bicarbonate ions makes the juice alkaline, which provides a favorable environment for the actions of the digestive enzymes and helps neutralize acidic chyme as it arrives from the stomach. The alkaline environment in the small intestine also blocks the action of pepsin, which might otherwise damage the duodenal wall.

Regulation of Pancreatic Secretion

The nervous and endocrine systems regulate release of pancreatic juice, much as they regulate gastric and small intestinal secretions. For example, during the cephalic and gastric phases of gastric secretion, parasympathetic impulses stimulate the pancreas to release digestive enzymes. A peptide hormone, **secretin,** stimulates the pancreas to secrete abundant fluid when acidic chyme enters the duodenum. Secretin is released into the blood from the duodenal mucous membrane in response to the acid in chyme. The pancreatic juice secreted at this time contains few, if any, digestive enzymes but has a high concentration of bicarbonate ions. These ions neutralize the acid in chyme (fig. 17.24).

Proteins and fats in chyme in the duodenum also stimulate the release of *cholecystokinin* from the intestinal wall. As in the case of secretin, cholecystokinin reaches the pancreas in the bloodstream. Pancreatic juice secreted in response to cholecystokinin has a high concentration of digestive enzymes.

PRACTICE

32 Where is the pancreas located?

33 List the enzymes in pancreatic juice.

34 What are the functions of the enzymes in pancreatic juice?

35 What regulates secretion of pancreatic juice?

17.8 LIVER

The **liver,** the largest internal organ, is located in the upper right quadrant of the abdominal cavity, just inferior to the diaphragm. It is partially surrounded by the ribs and extends from the level of the fifth intercostal space to the lower margin of the ribs. It is reddish brown in color and well supplied with blood vessels (figs. 17.25, 17.26, 17.27 and reference plates 8, 17, 24).

Liver Structure

A fibrous capsule encloses the liver, and connective tissue divides the organ into a large *right lobe* and a smaller *left lobe.* The *falciform ligament* is a fold of visceral peritoneum that separates the lobes and fastens the liver to the abdominal wall anteriorly. The liver also has two minor lobes, the *quadrate lobe,* near the gallbladder, and the *caudate lobe,* close to the vena cava (see fig. 17.26). The area where the four lobes meet and blood vessels and ducts enter or exit the liver is the *porta hepatis.*

A fold of visceral peritoneum called the *coronary ligament* attaches the liver to the diaphragm on its superior surface. Each lobe is separated into many tiny **hepatic lobules,** the liver's functional units (fig. 17.27). A lobule consists of many *hepatic cells* radiating outward from a *central vein.* Vascular channels called **hepatic sinusoids** separate platelike groups of these cells from each other. Blood from the digestive tract, carried in the *hepatic portal vein* (see chapter 15, pp. 603–604), brings newly absorbed nutrients into the sinusoids (fig. 17.28). At the same time, oxygenated blood from the hepatic artery mixes freely with the blood containing nutrients, then flows through the liver sinusoids and nourishes the hepatic cells.

Often blood in the portal vein contains some bacteria that have entered through the intestinal wall. However, large **Kupffer cells,** fixed to the inner lining (endothelium) of the hepatic sinusoids, remove most of the bacteria from the blood by phagocytosis. Then the blood passes into the *central veins* of the hepatic lobules and exits the liver via the hepatic vein.

Within the hepatic lobules are many fine *bile canaliculi,* which carry secretions from hepatic cells to *bile ductules.* The ductules of neighboring lobules unite to form larger bile ducts, which then converge to become the **hepatic ducts.** These ducts merge, in turn, to form the **common hepatic duct.**

Liver Functions

The liver carries on many important metabolic activities. From Science to Technology 17.1 discusses a bioengineered liver. Recall from chapter 13 (pp. 509–511) that the liver plays a key

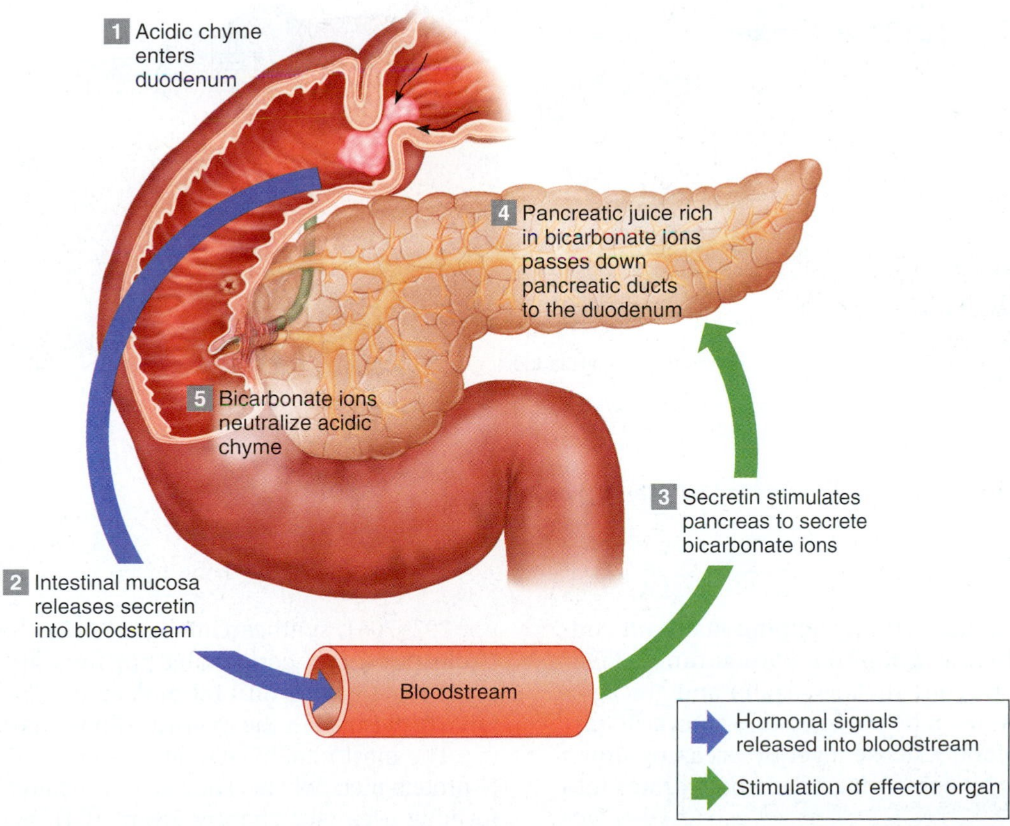

1 Acidic chyme enters duodenum
4 Pancreatic juice rich in bicarbonate ions passes down pancreatic ducts to the duodenum
5 Bicarbonate ions neutralize acidic chyme
3 Secretin stimulates pancreas to secrete bicarbonate ions
2 Intestinal mucosa releases secretin into bloodstream
Bloodstream
Hormonal signals released into bloodstream
Stimulation of effector organ

FIGURE 17.24 Acidic chyme entering the duodenum from the stomach stimulates the release of secretin, which, in turn, stimulates the release of pancreatic juice.

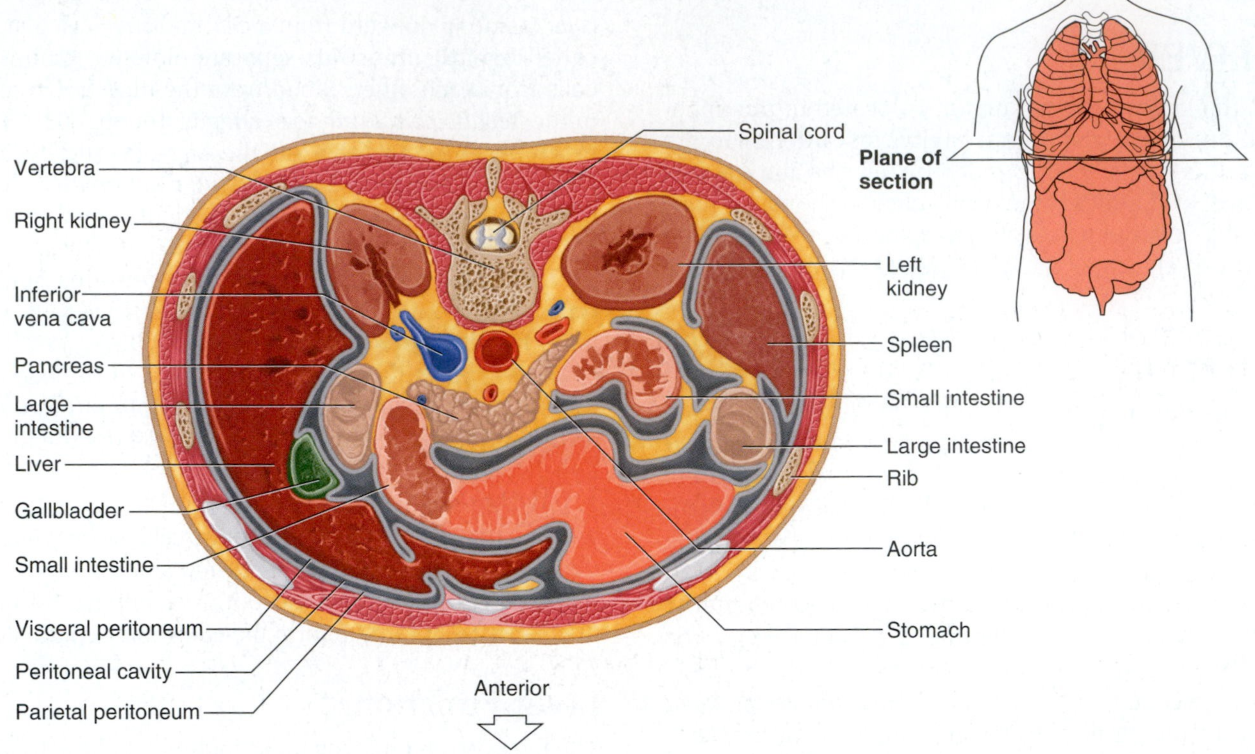

FIGURE 17.25 This transverse section of the abdomen reveals the liver and other organs in the upper part of the abdominal cavity.

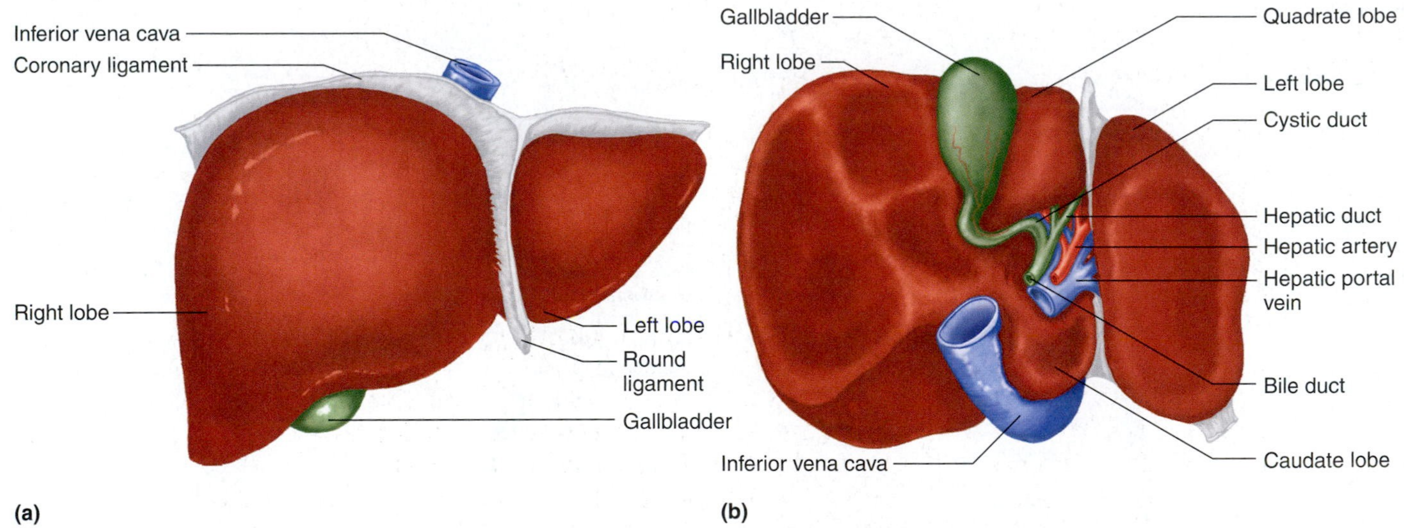

FIGURE 17.26 Lobes of the liver, viewed (*a*) anteriorly and (*b*) inferiorly.

role in carbohydrate metabolism by helping maintain concentration of blood glucose within the normal range. Liver cells responding to hormones such as insulin and glucagon lower the blood glucose level by polymerizing glucose to glycogen and raise the blood glucose level by breaking down glycogen to glucose or by converting noncarbohydrates into glucose.

The liver's effects on lipid metabolism include oxidizing fatty acids at an especially high rate (see chapter 18, pp. 702–703); synthesizing lipoproteins, phospholipids, and cholesterol; and converting portions of carbohydrate and protein molecules into fat molecules. The blood transports fats synthesized in the liver to adipose tissue for storage.

The most vital liver functions are probably those related to protein metabolism. They include deaminating amino acids; forming urea (see chapter 18, p. 704); synthesizing plasma proteins such as clotting factors (see chapter 14, p. 535); and converting certain amino acids to other amino acids.

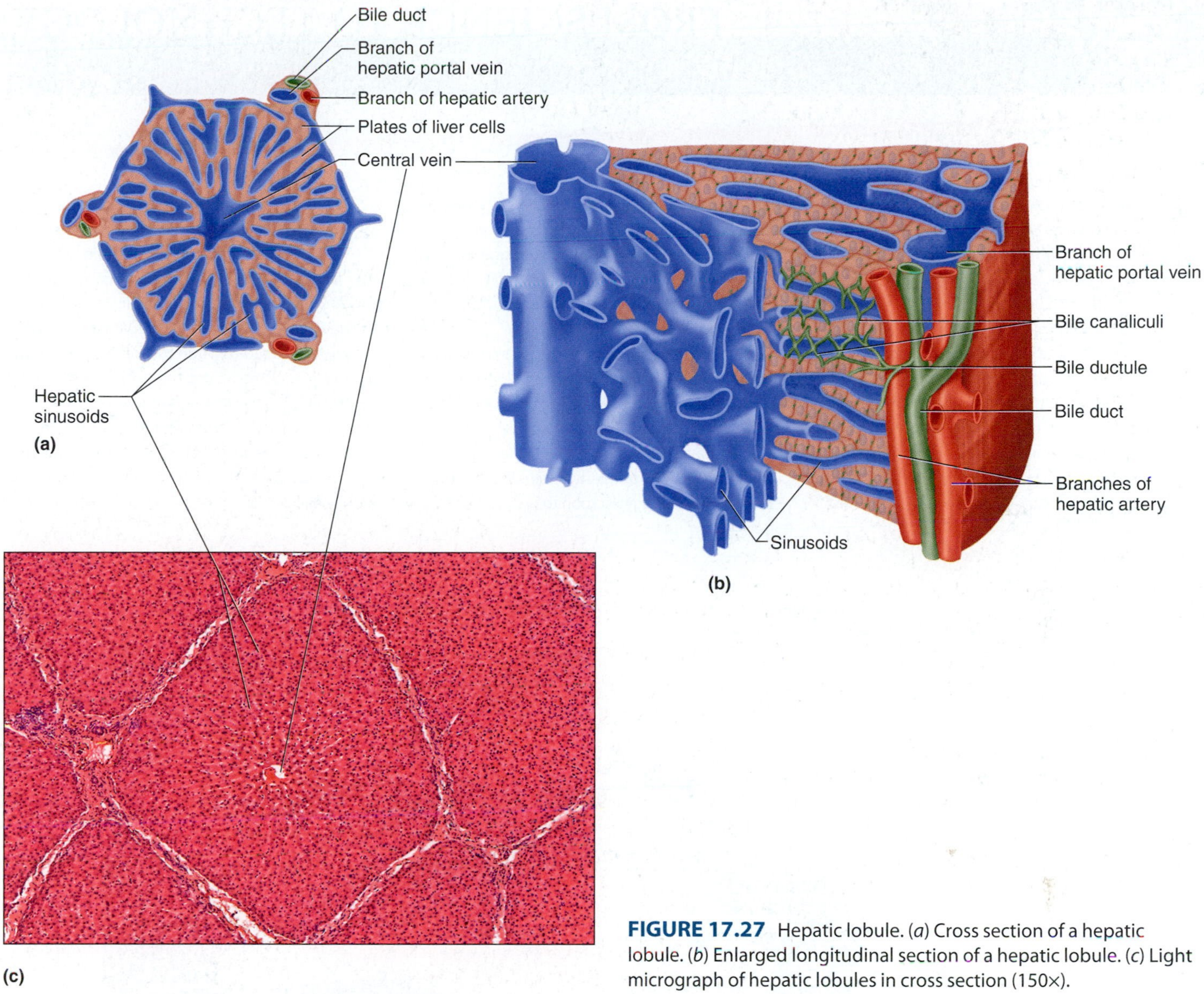

FIGURE 17.27 Hepatic lobule. (*a*) Cross section of a hepatic lobule. (*b*) Enlarged longitudinal section of a hepatic lobule. (*c*) Light micrograph of hepatic lobules in cross section (150×).

Bacteria in the intestine produce ammonia, which is carried in the blood to the liver, where it reacts to yield urea. When this liver function fails, concentration of blood ammonia sharply rises, causing *hepatic coma,* a condition that can lead to death.

The liver also stores many substances, including glycogen, iron, and vitamins A, D, and B_{12}. Extra iron from the blood combines with a protein (apoferritin) in liver cells, forming *ferritin.* The iron is stored in this form until blood iron concentration falls, when some of the iron is released. Thus, the liver is important in iron homeostasis.

Liver cells help destroy damaged red blood cells and phagocytize foreign antigens. The liver removes toxic substances such as alcohol and certain other drugs from the blood (detoxification). The liver can also serve as a blood reservoir, storing 200 to 400 milliliters of blood. The liver's role in digestion is to secrete bile. Table 17.7 summarizes the major functions of the liver. Clinical Application 17.3 discusses hepatitis, an inflammation of the liver.

PRACTICE

36 Locate the liver.

37 Describe a hepatic lobule.

38 Review liver functions.

39 How does the liver aid in digestion?

Composition of Bile

Bile (bīl) is a yellowish-green liquid that hepatic cells continuously secrete. In addition to water, it contains *bile salts, bile pigments, cholesterol,* and *electrolytes.* Of these, bile salts are the most abundant. They are the only bile substances that have a digestive function.

17.1 FROM SCIENCE TO TECHNOLOGY

Replacing the Liver

Life without a liver is not possible. A person can survive only a few days once the liver stops functioning. In fulminant hepatic failure, for example, an otherwise healthy, young person suddenly experiences liver failure, caused by exposure to a toxin, reaction to a drug, or a viral infection. Jaundice and fatigue progress rapidly to coma and death. Once cancer spreads to the liver, survival is generally only weeks or a few months.

Livers are in great demand for transplant, but as is the case for other organs, are scarce. Each year in the United States, only about 4,500 of the 12,000 or so individuals requiring livers survive long enough to undergo a transplant. Sometimes a person can receive part of a liver donated by a living relative or other close match. The donor continues to live because only 15% of the liver need be functional for survival.

A promising solution to the problem of liver failure in light of the organ shortage is an "extracorporeal liver assist device" (ELAD), which can take over the liver's blood-cleansing function until a cadaver organ becomes available, or it can enable remaining functional liver tissue to sufficiently stimulate regeneration to restore health.

ELAD is a "bioartificial" liver because it has synthetic as well as biological components. The device consists of four cartridges filled with hollow fibers that house millions of continuously dividing ("immortalized") human liver cells (hepatocytes). A patient's plasma is separated from the blood and passed through the device, where the liver cells remove toxins and add liver-secreted products, such as clotting factors. The plasma is then filtered, the formed elements of the blood added back, and the blood reinfused into the patient. ELAD is essentially a hepatic version of kidney dialysis. ■

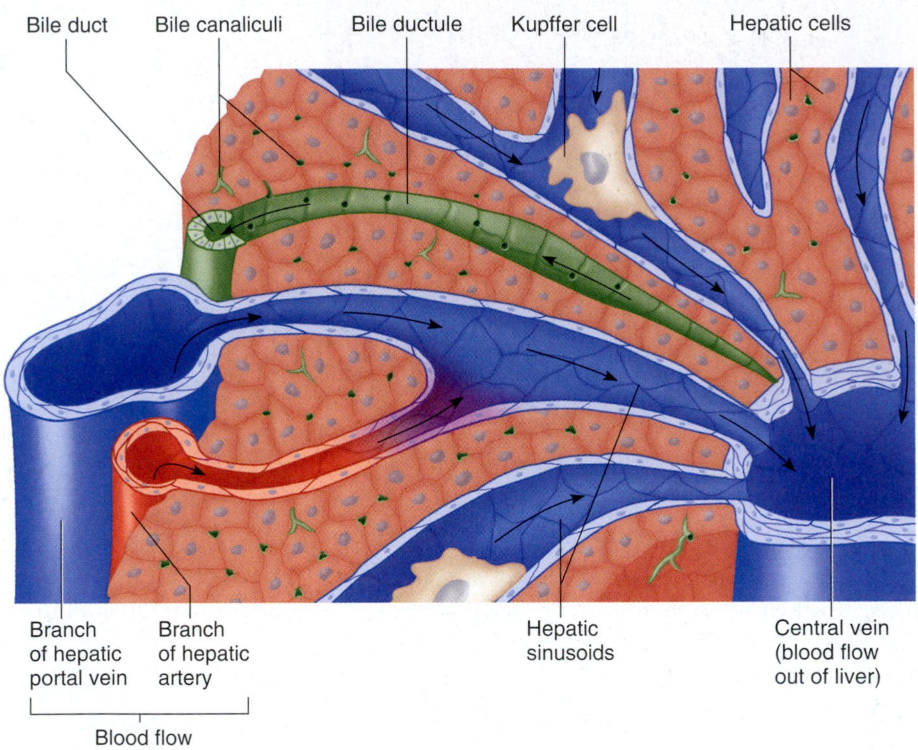

FIGURE 17.28 The paths of blood and bile in a hepatic lobule.

Hepatic cells use cholesterol to produce bile salts, and in secreting these salts, they release some cholesterol into the bile. Cholesterol by itself has no special function in bile or in the alimentary canal.

Bile pigments (bilirubin and biliverdin) are breakdown products of hemoglobin from red blood cells (see chapter 14, p. 530). These pigments are normally excreted in the bile. The yellowish skin, sclerae, and mucous membranes of jaundice result from excess deposition of bile pigments.

Jaundice can have several causes. In *obstructive jaundice,* bile ducts are blocked (as with gallstones or tumors). In hepatocellular jaundice, the liver is diseased (as in cirrhosis or hepatitis). In *hemolytic jaundice,* red blood cells are destroyed too rapidly (as with a blood transfusion from an incompatible blood group or a blood infection such as malaria).

17.3 CLINICAL APPLICATION

Hepatitis

Hepatitis is an inflammation of the liver. It has several causes, but the various types have similar symptoms.

For the first few days, hepatitis may resemble the flu, producing mild headache, low fever, fatigue, lack of appetite, nausea and vomiting, and sometimes stiff joints. By the end of the first week, more distinctive symptoms arise: a rash, pain in the upper right quadrant of the abdomen, dark and foamy urine, and pale feces. The skin and sclera of the eyes begin to turn yellow due to accumulating bile pigments (jaundice). Great fatigue may continue for two or three weeks, and then gradually the person begins to feel better.

This is hepatitis in its most common, least dangerous acute guise. About half a million people develop hepatitis in the United States each year, and 6,000 die. In a rare form called *fulminant hepatitis,* symptoms are sudden and severe, along with altered behavior and personality. Medical attention is necessary to prevent kidney or liver failure or coma. Hepatitis that persists for more than six months is termed chronic. As many as 300 million people worldwide are hepatitis carriers. They do not have symptoms but can infect others. Five percent of carriers eventually develop liver cancer.

Only rarely does hepatitis result from alcoholism, autoimmunity, or the use of certain drugs. Usually, one of several types of viruses causes hepatitis. Viral types are distinguished by the route of infection, surface features, and whether the viral genetic material is DNA or RNA. Hepatitis B virus has DNA; the others have RNA. The viral types are classified as follows:

Hepatitis A spreads by contact with food or objects contaminated with virus-containing feces, including diapers. The course of hepatitis A is short and mild.

Hepatitis B spreads by contact with virus-containing body fluids, such as blood, saliva, or semen. It may be transmitted by blood transfusions, hypodermic needles, or sexual activity.

Hepatitis C accounts for about half of all known cases of hepatitis. This virus is primarily transmitted in blood—by sharing razors or needles, from pregnant woman to fetus, or through blood transfusions or use of blood products. As many as 60% of individuals infected with the hepatitis C virus suffer chronic symptoms.

Hepatitis D occurs in people already infected with the hepatitis B virus. It is blood-borne and associated with blood transfusions and intravenous drug use. About 20% of individuals infected with this virus die from the infection.

Hepatitis E virus is usually transmitted in water contaminated with feces. It most often affects visitors to developing nations.

Hepatitis G is rare but seems to account for a significant percentage of cases of fulminant hepatitis. In people with healthy immune systems, it produces symptoms so mild that they may not even be noticed.

A virus usually causes hepatitis, so antibiotic drugs, effective against bacteria, are not helpful. Usually, the person must just wait out the symptoms. Hepatitis C, however, sometimes responds to a form of interferon, an immune system biochemical given as a drug. ■

TABLE 17.7 | Major Functions of the Liver

General Function	Specific Function
Carbohydrate metabolism	Polymerizes glucose to glycogen; breaks down glycogen to glucose; converts noncarbohydrates to glucose
Lipid metabolism	Oxidizes fatty acids; synthesizes lipoproteins, phospholipids, and cholesterol; converts portions of carbohydrate and protein molecules into fats
Protein metabolism	Deaminates amino acids; forms urea; synthesizes plasma proteins; converts certain amino acids to other amino acids
Storage	Stores glycogen, vitamins A, D, and B_{12}, iron, and blood
Blood filtering	Removes damaged red blood cells and foreign substances by phagocytosis
Detoxification	Removes toxins from the blood
Secretion	Secretes bile

Gallbladder

The **gallbladder** is a pear-shaped sac in a depression on the inferior surface of the liver. It is connected to the **cystic duct,** which, in turn, joins the common hepatic duct (see fig. 17.23 and reference plate 19). The gallbladder has a capacity of 30–50 milliliters, is lined with columnar epithelial cells, and has a strong muscular layer in its wall. It stores bile between meals, concentrates bile by reabsorbing water, and contracts to release bile into the duodenum when stimulated by cholecystokinin from the small intestine.

The **bile duct** is formed by the union of the common hepatic and cystic ducts. It leads to the duodenum, where the hepatopancreatic sphincter muscle guards its exit (see fig. 17.23). This sphincter normally remains contracted, so as bile collects in the common bile duct it backs up into the cystic duct. When this happens, the bile flows into the gallbladder, where it is stored.

Bile salts, bile pigments, and cholesterol become increasingly concentrated as the gallbladder lining reabsorbs water and electrolytes. The cholesterol normally remains in solution, but under certain conditions it may precipitate and form solid crystals. If cholesterol continues to come out of solution, the crystals enlarge, forming *gallstones* (fig. 17.29). This may happen if the bile is too concentrated, hepatic cells secrete too much cholesterol, or the gallbladder is inflamed (cholecystitis). Gallstones in the bile duct may block the flow of bile, causing obstructive jaundice and considerable pain. Clinical Application 17.4 discusses disorders of the gallbladder.

Regulation of Bile Release

Normally, bile does not enter the duodenum until cholecystokinin stimulates the gallbladder to contract. The intestinal mucosa releases this hormone in response to proteins and fats in the small intestine. (Recall from earlier in this chapter, page 672, that cholecystokinin also stimulates pancreatic enzyme secretion.) The hepatopancreatic sphincter usually remains contracted until a peristaltic wave in the duodenal wall approaches. Just before the wave hits, the sphincter relaxes, and bile squirts into the duodenum (fig. 17.30). Table 17.8 summarizes the hormones that control digestion.

Functions of Bile Salts

Bile salts aid digestive enzymes. Molecules of fats clump into *fat globules*. Bile salts reduce surface tension and break fat globules into droplets, much like the action of soap or detergent. This process is called **emulsification.** Monoglycerides that form from the action of pancreatic lipase on triglyceride molecules aid emulsification. Overall, emulsification greatly increases the total surface area of the fatty substance, and the resulting droplets mix with water. Lipases can then digest the fat molecules more effectively.

Bile salts enhance absorption of fatty acids and cholesterol by forming complexes (micelles) that are very soluble in chyme and that epithelial cells can more easily absorb. The fat-soluble vitamins A, D, E, and K are also absorbed in the presence of bile salts. Lack of bile salts results in poor lipid absorption and vitamin deficiencies.

The mucous membrane of the small intestine reabsorbs nearly all of the bile salts, along with fatty acids. The blood carries bile salts to the liver, where hepatic cells resecrete them into the bile ducts. Liver cells synthesize bile salts, which replace the small amounts lost in the feces.

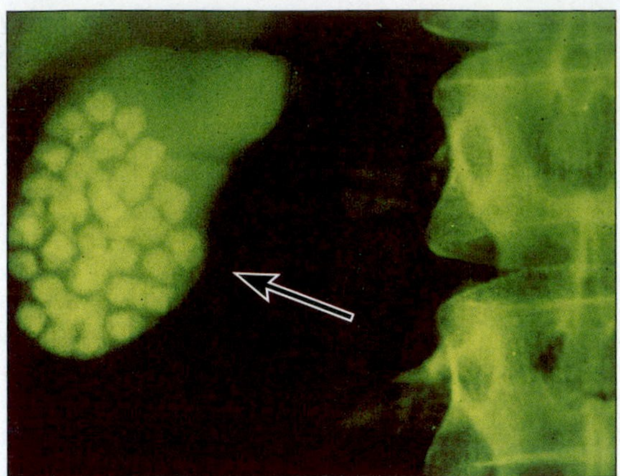

FIGURE 17.29 Falsely colored radiograph of a gallbladder that contains gallstones (arrow).

PRACTICE  

40 Explain how bile forms.

41 Describe the function of the gallbladder.

42 How is secretion of bile regulated?

43 How do bile salts function in digestion?

17.9 SMALL INTESTINE

The **small intestine** is a tubular organ that extends from the pyloric sphincter to the beginning of the large intestine. With its many loops and coils, it fills much of the abdominal cavity (see fig. 17.1 and reference plates 4 and 5). The small intestine is 5.5–6.0 meters (18–20 feet) long in a cadaver when the muscular wall lacks tone, but it may be only half this long in a living person.

The small intestine receives secretions from the pancreas and liver. It also completes digestion of the nutrients in chyme, absorbs the products of digestion, and transports the remaining residue to the large intestine.

TABLE 17.8 | Hormones of the Digestive Tract

Hormone	Source	Function
Gastrin	Gastric cells, in response to food	Increases secretory activity of gastric glands
Intestinal gastrin	Cells of small intestine, in response to chyme	Increases secretory activity of gastric glands
Somatostatin	Gastric cells	Inhibits secretion of acid by parietal cells
Intestinal somatostatin	Intestinal wall cells, in response to fats	Inhibits secretion of acid by parietal cells
Cholecystokinin	Intestinal wall cells, in response to proteins and fats in the small intestine	Decreases secretory activity of gastric glands and inhibits gastric motility; stimulates pancreas to secrete fluid with a high digestive enzyme concentration; stimulates gallbladder to contract and release bile
Secretin	Cells in the duodenal wall, in response to acidic chyme entering the small intestine	Stimulates pancreas to secrete fluid with a high bicarbonate ion concentration

17.4 CLINICAL APPLICATION

Gallbladder Disease

Molly G., an overweight, forty-seven-year-old college administrator and mother of four, had pain in the upper-right quadrant of her abdomen (see fig. 1.24b). Sometimes the discomfort seemed to radiate around to her back and move upward into her right shoulder. She usually felt this pain after her evening meal, but it occasionally happened at night, awakening her. After an episode of severe pain accompanied by sweating (diaphoresis) and nausea, Molly visited her physician, who discovered tenderness in the epigastric region (see fig. 1.24a). She decided that Molly's symptoms symptoms could indicate *acute cholecystitis*—an inflammation of the gallbladder. Molly needed a *cholecystogram*—an X ray of the gallbladder.

Molly took tablets containing a contrast medium the night before the X-ray procedure, which allowed time for the small intestine to absorb the substance and it to reach the liver and be excreted into the bile. Later, the bile and contrast medium would be stored and concentrated in the gallbladder and would make the contents of the gallbladder opaque to X rays.

Molly's cholecystogram revealed several stones (calculi) in her gallbladder, a condition called *cholelithiasis* (see fig. 17.29). Molly's symptoms were worsening, so her physician recommended *cholecystectomy*—surgical removal of the gallbladder.

An incision was made in Molly's right subcostal region and her gallbladder excised from the liver. Fortunately, the cystic duct and hepatic ducts did not have stones (see fig. 17.26).

Unfortunately, Molly's symptoms persisted following her recovery from surgery. So her surgeon ordered a *cholangiogram*—an X-ray series of the bile ducts. This study showed a residual stone at the distal end of Molly's bile duct (see fig. 17.23).

The surgeon extracted the residual stone using a *fiber-optic endoscope,* a long, flexible tube passed through the esophagus and stomach and into the duodenum. This instrument enables a surgeon to observe features of the gastrointestinal tract directly through the eyepiece of the endoscope or on a monitor. A surgeon can also perform manipulations using specialized tools passed through the endoscope to its distal end.

In Molly's case, the surgeon performed an *endoscopic papillotomy*—-an incision of the hepatopancreatic sphincter by applying an electric current to a wire extending from the end of the endoscope (see fig. 17.23). She then removed the exposed stone by manipulating a tiny basket at the tip of the endoscope. Many patients undergo only the endoscopic procedure to remove the gallbladder, performed on an outpatient basis. ■

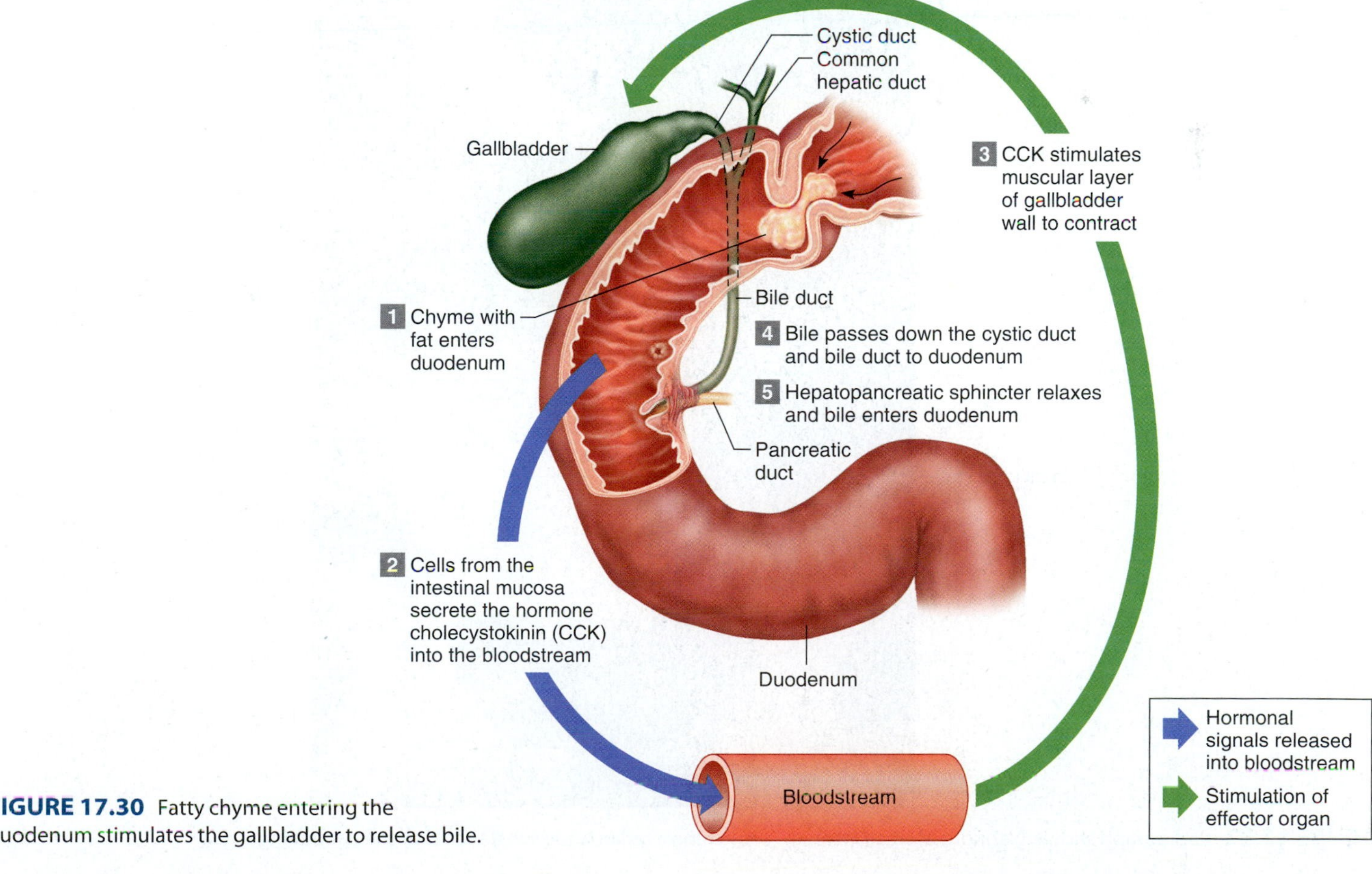

FIGURE 17.30 Fatty chyme entering the duodenum stimulates the gallbladder to release bile.

Parts of the Small Intestine

The small intestine, shown in figures 17.31 and 17.32 and in reference plates 12 and 18, consists of three portions: the duodenum, the jejunum, and the ileum.

The **duodenum** (du″o-de′num), about 25 centimeters long and 5 centimeters in diameter, lies posterior to the parietal peritoneum (retroperitoneal). It is the shortest and most fixed portion of the small intestine. The duodenum follows a C-shaped path as it passes anterior to the right kidney and the upper three lumbar vertebrae.

The remainder of the small intestine is mobile and lies free in the peritoneal cavity. The proximal two-fifths of this portion is the **jejunum** (jĕ-joo′num), and the remainder is the **ileum** (il′e-um). There is no distinct separation between the jejunum and ileum, but the diameter of the jejunum is usually greater, and its wall is thicker, more vascular, and more active than that of the ileum. The ileum has more lymph nodules (Peyer's patches) and a higher bacterial population.

The jejunum and ileum are suspended from the posterior abdominal wall by a double-layered fold of peritoneum called **mesentery** (mes′en-ter″e) (fig. 17.33). The mesentery supports the blood vessels, nerves, and lymphatic vessels that supply the intestinal wall.

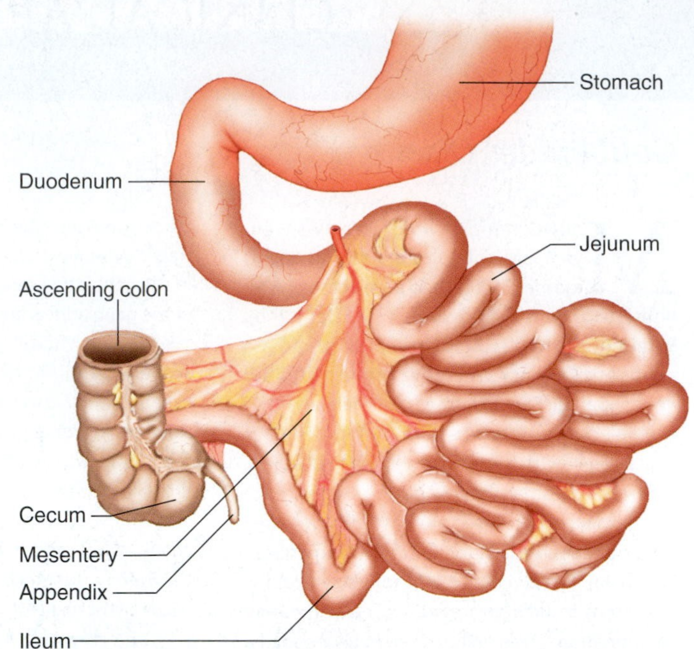

FIGURE 17.31 The three parts of the small intestine are the duodenum, the jejunum, and the ileum.

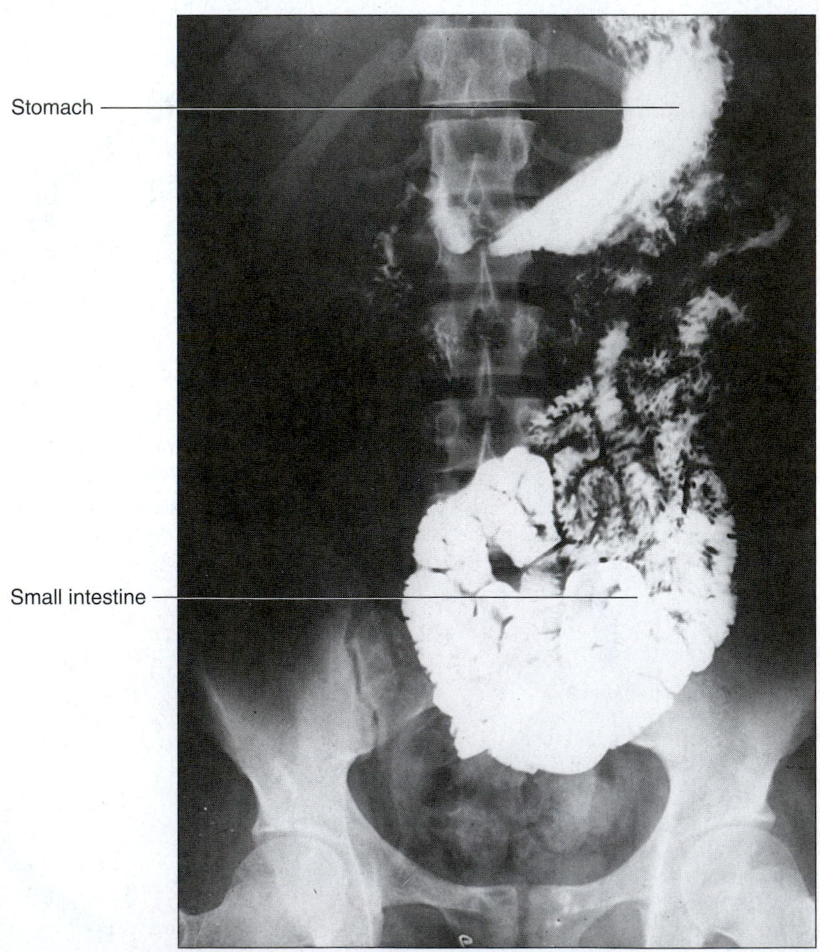

FIGURE 17.32 Radiograph showing abnormal small intestine containing a radiopaque substance that the patient ingested.

A filmy, double fold of peritoneal membrane called the *greater omentum* drapes like an apron from the stomach over the transverse colon and the folds of the small intestine (fig. 17.34). If the wall of the alimentary canal becomes infected, cells from the omentum may adhere to the inflamed region and help seal it off, lowering the risk that the infection will spread to the peritoneal cavity.

Structure of the Small Intestinal Wall

The inner wall of the small intestine has a velvety appearance throughout its length, due to many tiny projections of mucous membrane called **intestinal villi** (figs. 17.35 and 17.36; see fig. 17.3). These structures are most numerous in the duodenum and the proximal jejunum. They project into the lumen of the alimentary canal, contacting the intestinal contents. Villi greatly increase the surface area of the intestinal lining, aiding absorption of digestive products.

Each villus consists of a layer of simple columnar epithelium and a core of connective tissue containing blood capillaries, a lymphatic capillary called a **lacteal,** and nerve fibers (see fig. 17.35). At their free surfaces, the epithelial cells have many fine extensions called *microvilli* that form a brushlike border and greatly increase the surface area of the intestinal cells, further enhancing absorption (see figs. 17.3 and 17.37). The blood capillaries and lacteals carry away absorbed nutrients, and impulses transmitted by the nerve fibers can stimulate or inhibit activities of the villus.

Between the bases of adjacent villi are tubular **intestinal glands** (crypts of Lieberkühn) that dip into the mucous

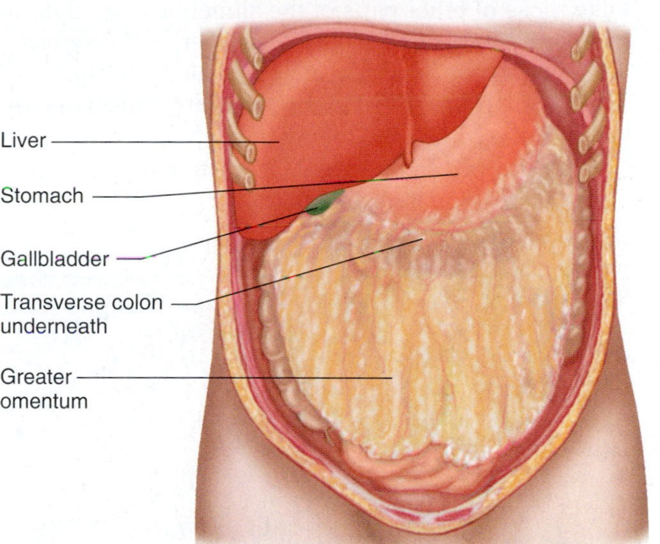

FIGURE 17.34 The greater omentum hangs like an apron over the abdominal organs.

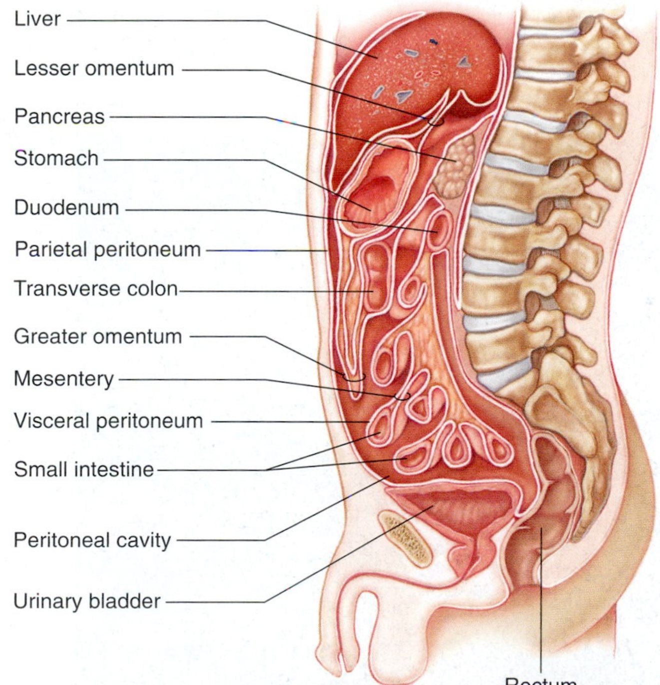

FIGURE 17.33 Mesentery formed by folds of the peritoneal membrane suspends parts of the small intestine from the posterior abdominal wall.

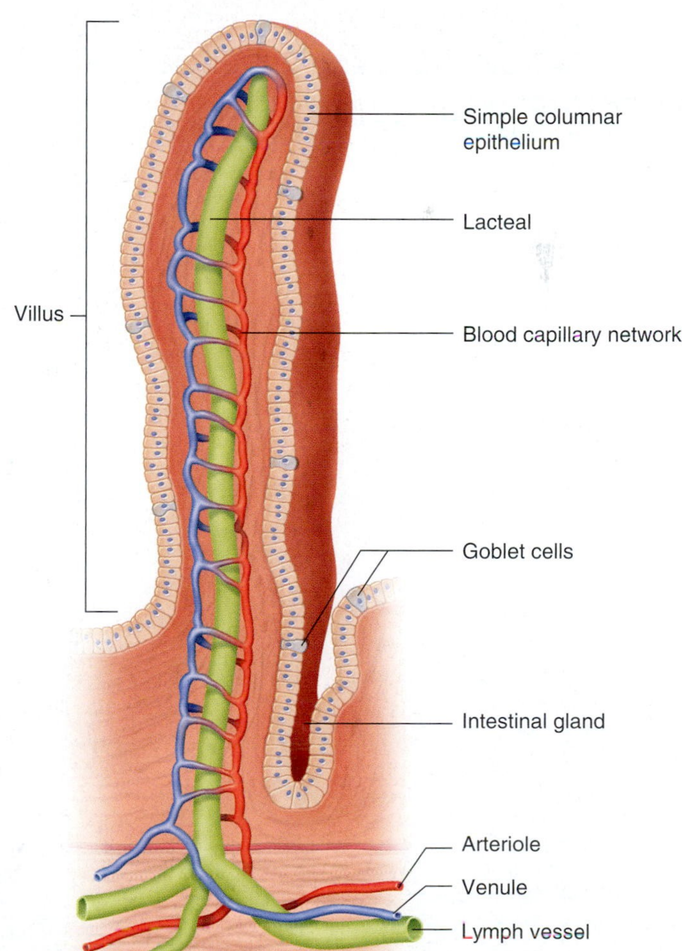

FIGURE 17.35 Structure of a single intestinal villus.

membrane. The deeper layers of the small intestinal wall are much like those of other parts of the alimentary canal in that they include a submucosa, a muscular layer, and a serosa.

The lining of the small intestine has many circular folds of mucosa, called *plicae circulares,* that are especially well developed in the lower duodenum and upper jejunum (fig. 17.38). With the villi and microvilli, these folds help increase the surface area of the intestinal lining.

The epithelial cells that form the lining of the small intestine are continually replaced. New cells form in the intestinal glands by mitosis and migrate outward onto the villus surface. When the migrating cells reach the tip of the villus, they are shed. This *cellular turnover* renews the small intestine's epithelial lining every three to six days. As a result, nearly one-quarter of the bulk of feces consists of dead epithelial cells from the small intestine.

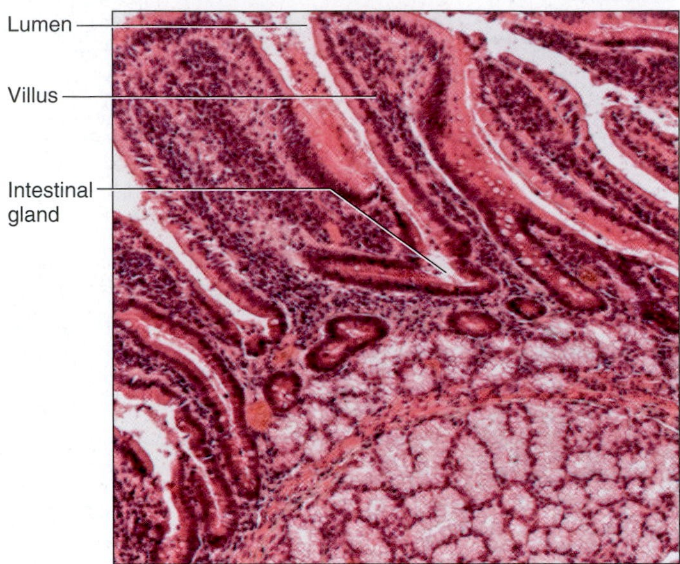

FIGURE 17.36 Light micrograph of intestinal villi from the wall of the duodenum (50×).

Secretions of the Small Intestine

In addition to the mucous-secreting goblet cells, abundant throughout the mucosa of the small intestine, many specialized *mucous-secreting glands* (Brunner's glands) are in the submucosa in the proximal portion of the duodenum. These glands secrete a thick, alkaline mucus in response to certain stimuli.

The intestinal glands at the bases of the villi secrete large volumes of a watery fluid (see fig. 17.35). The villi rapidly reabsorb this fluid, which carries digestive products into the villi. The fluid the intestinal glands secrete has a nearly neutral pH (6.5–7.5), and it lacks digestive enzymes. However, the epithelial cells of the intestinal mucosa have digestive enzymes embedded in the membranes of the microvilli on their luminal surfaces. These enzymes break down food molecules just before absorption takes place. The enzymes include **peptidases,** which split peptides into their constituent amino acids; **sucrase, maltase,** and **lactase,** which split the disaccha-

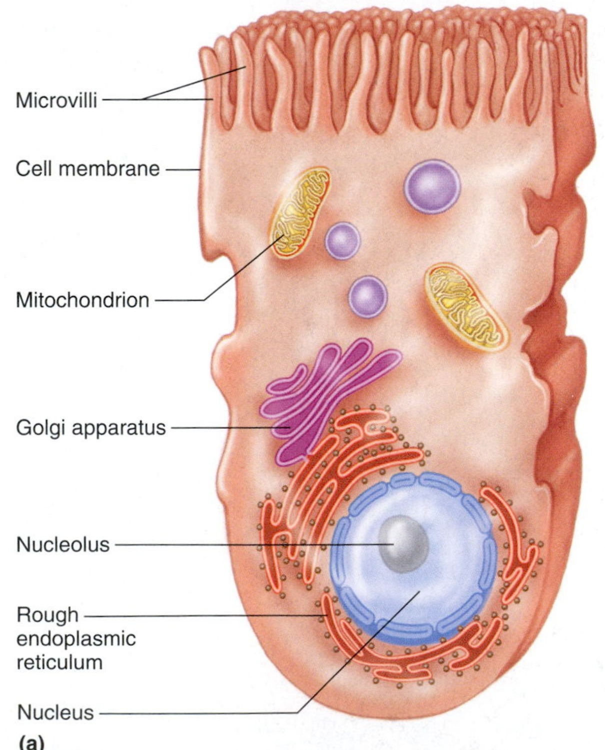

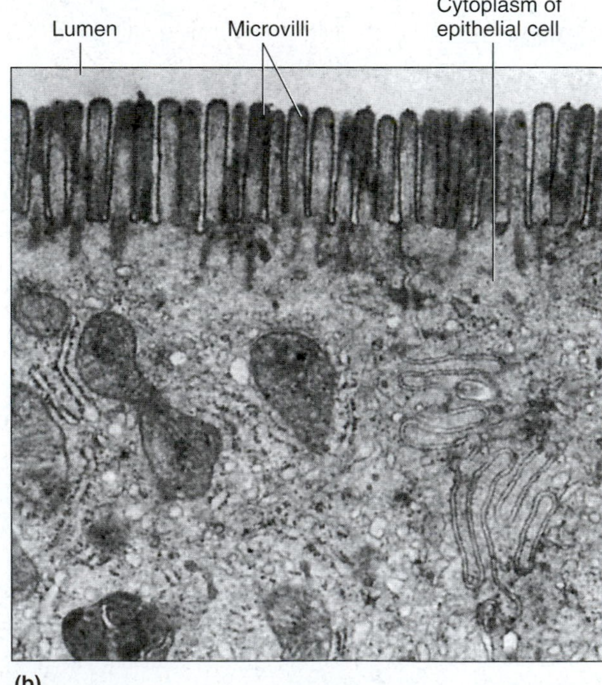

FIGURE 17.37 Intestinal epithelium. (*a*) Microvilli increase the surface area of intestinal epithelial cells. (*b*) Transmission electron micrograph of microvilli (16,000×).

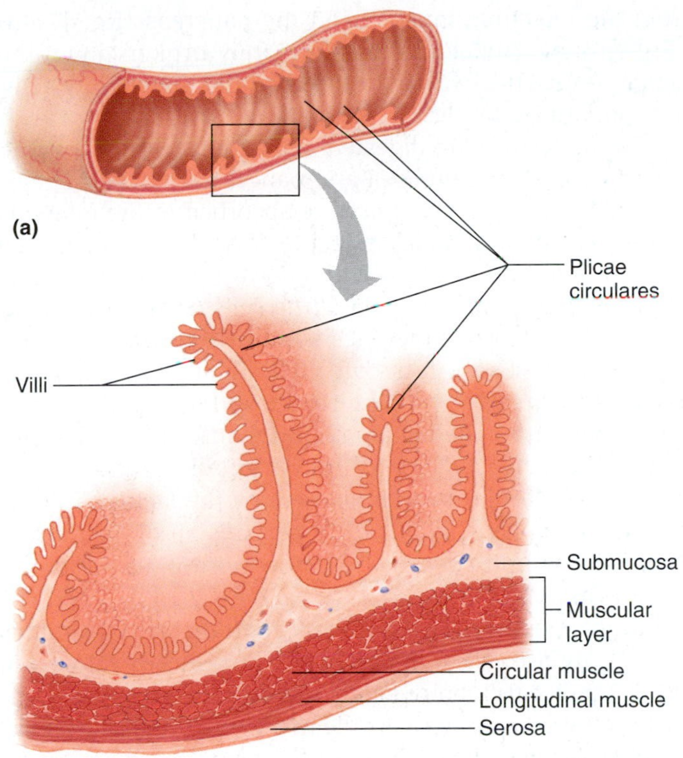

FIGURE 17.38 Section of small intestine. (*a*) The inner lining of the small intestine contains many circular folds, the plicae circulares. (*b*) A longitudinal section through some of these folds.

rides sucrose, maltose, and lactose into the monosaccharides glucose, fructose, and galactose; and **intestinal lipase,** which splits fats into fatty acids and glycerol. Table 17.9 summarizes the sources and actions of the major digestive enzymes.

Regulation of Small Intestinal Secretions

Mucus protects the intestinal wall in the same way it protects the stomach lining, it is not surprising that mucous secretion increases in response to mechanical stimulation and the presence of irritants, such as gastric juice. Stomach contents entering the small intestine stimulate the duodenal mucous glands to release mucus.

Direct contact with chyme chemically and mechanically stimulates goblet cells and intestinal glands to secrete their products. Distension of the intestinal wall activates the nerve plexuses therein and stimulates parasympathetic reflexes that also trigger release of small intestine secretions.

PRACTICE

44 Describe the parts of the small intestine.
45 What is the function of an intestinal villus?
46 Distinguish between intestinal villi and microvilli.
47 How is surface area maximized in the small intestine?
48 What is the function of the intestinal glands?
49 List intestinal digestive enzymes.

TABLE 17.9 | Summary of the Major Digestive Enzymes

Enzyme	Source	Digestive Action
Salivary Enzyme		
Salivary amylase	Salivary glands	Begins carbohydrate digestion by breaking down starch and glycogen to disaccharides
Gastric Enzymes		
Pepsin	Gastric glands	Begins protein digestion
Gastric lipase	Gastric glands	Begins butterfat digestion
Pancreatic Enzymes		
Pancreatic amylase	Pancreas	Breaks down starch and glycogen into disaccharides
Pancreatic lipase	Pancreas	Breaks down fats into fatty acids and glycerol
Trypsin, chymotrypsin	Pancreas	Breaks down proteins or partially digested proteins into peptides
Carboxypeptidase	Pancreas	Breaks down peptides into amino acids
Nucleases	Pancreas	Breaks down nucleic acids into nucleotides
Intestinal Enzymes		
Peptidase	Mucosal cells	Breaks down peptides into amino acids
Sucrase, maltase, lactase	Mucosal cells	Breaks down disaccharides into monosaccharides
Intestinal lipase	Mucosal cells	Breaks down fats into fatty acids and glycerol
Enterokinase	Mucosal cells	Shortens trypsinogen into trypsin

Many adults do not produce sufficient lactase to adequately digest lactose, or milk sugar. In this *lactose intolerance,* the sugar remains undigested, increasing osmotic pressure of the intestinal contents and drawing water into the intestines. At the same time, intestinal bacteria metabolize undigested sugar, producing organic acids and gases. The overall result is bloating, intestinal cramps, and diarrhea. To avoid these symptoms, people with lactose intolerance can take lactase pills before eating dairy products, and infants can drink formula based on soybeans. Genetic evidence suggests that lactose intolerance may be the "normal" condition, with the ability to digest lactose the result of a mutation that occurred recently in our evolutionary past and became advantageous when the advent of agriculture brought dairy foods to human populations. The trait of ability to digest lactose has increased in parallel to increased use of dairy foods at least three times in history, in different populations.

Absorption in the Small Intestine

Villi greatly increase the surface area of the intestinal mucosa, making the small intestine the most important absorbing organ of the alimentary canal. The small intestine is so effective in absorbing digestive products, water, and electrolytes, that very little absorbable material reaches the organ's distal end.

Carbohydrate digestion begins in the mouth with the activity of salivary amylase and is completed in the small intestine by enzymes from the intestinal mucosa and pancreas (fig. 17.39). The resulting monosaccharides are absorbed by facilitated diffusion or active transport into the villi and enter blood capillaries (see chapter 3, pp. 93 and 95).

Protein digestion begins in the stomach as a result of pepsin activity and is completed in the small intestine by enzymes from the intestinal mucosa and the pancreas (fig. 17.40). Large protein molecules are ultimately broken down into amino acids, which are then absorbed into the villi by active transport and enter the circulation.

Fat molecules are digested almost entirely by enzymes from the intestinal mucosa and pancreas (fig. 17.41). The resulting fatty acid molecules are absorbed in the following steps: (1) The fatty acid molecules dissolve in the epithelial cell membranes of the villi and diffuse through them. (2) The endoplasmic reticula of the cells use the fatty acids to resynthesize fat molecules similar to those previously digested. (3) These fats collect in clusters that become encased in protein. (4) The resulting large molecules of lipoprotein are called *chylomicrons,* and they make their way to the lacteals of the villi. (5) Periodic contractions of smooth muscles in the villi help empty the lacteals into the cysterna chyli (see fig. 16.6*a*), an expansion of the thoracic duct. The lymph carries the chylomicrons to the bloodstream (fig. 17.42).

Chylomicrons in the blood transport dietary fats to muscle and adipose cells. Similarly, very-low-density lipoprotein (VLDL) molecules, produced in the liver, transport triglycerides synthesized from excess dietary carbohydrates. As VLDL molecules reach adipose cells, an enzyme, *lipoprotein lipase,* catalyzes reactions that unload their triglycerides, converting VLDL to low-density lipoprotein (LDL) molecules. Most of the triglycerides have been removed, so LDL molecules have a higher cholesterol content than the original VLDL molecules. Cells in the peripheral tissues remove LDL from plasma by receptor-mediated endocytosis, thus obtaining a supply of cholesterol (see chapter 3, p. 96).

While LDL delivers cholesterol to tissues, high-density lipoprotein (HDL) removes cholesterol from tissues and delivers it to the liver. The liver produces the basic HDL

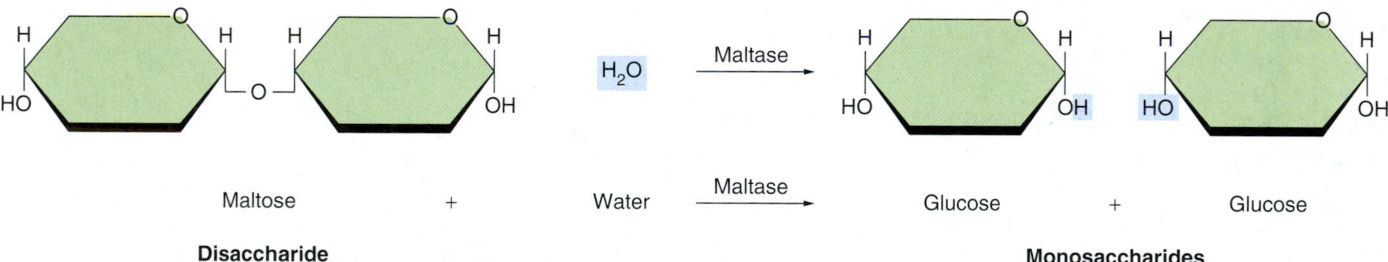

FIGURE 17.39 Digestion breaks down complex carbohydrates into disaccharides, which are then broken down into monosaccharides, which are small enough for intestinal villi to absorb. The monosaccharides then enter the bloodstream.

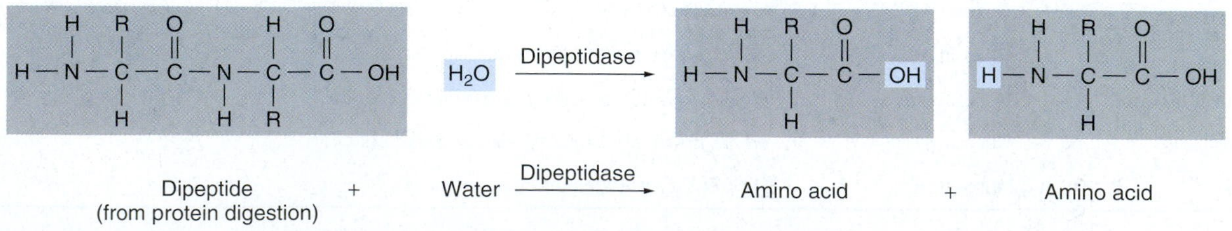

H_2O Dipeptidase

Dipeptide (from protein digestion) + Water —Dipeptidase→ Amino acid + Amino acid

FIGURE 17.40 The amino acids that result from dipeptide digestion are absorbed by intestinal villi and enter the blood.

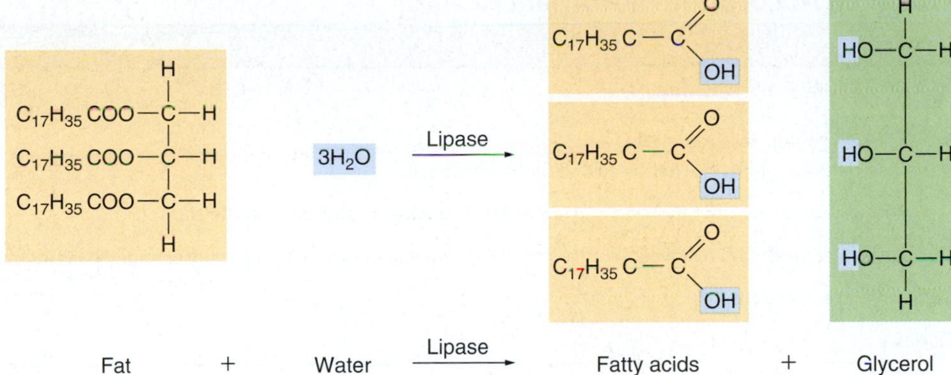

FIGURE 17.41 Fatty acids and glycerol result from fat digestion. Intestinal villi absorb them, and most are resynthesized into fat molecules before they enter the lacteals.

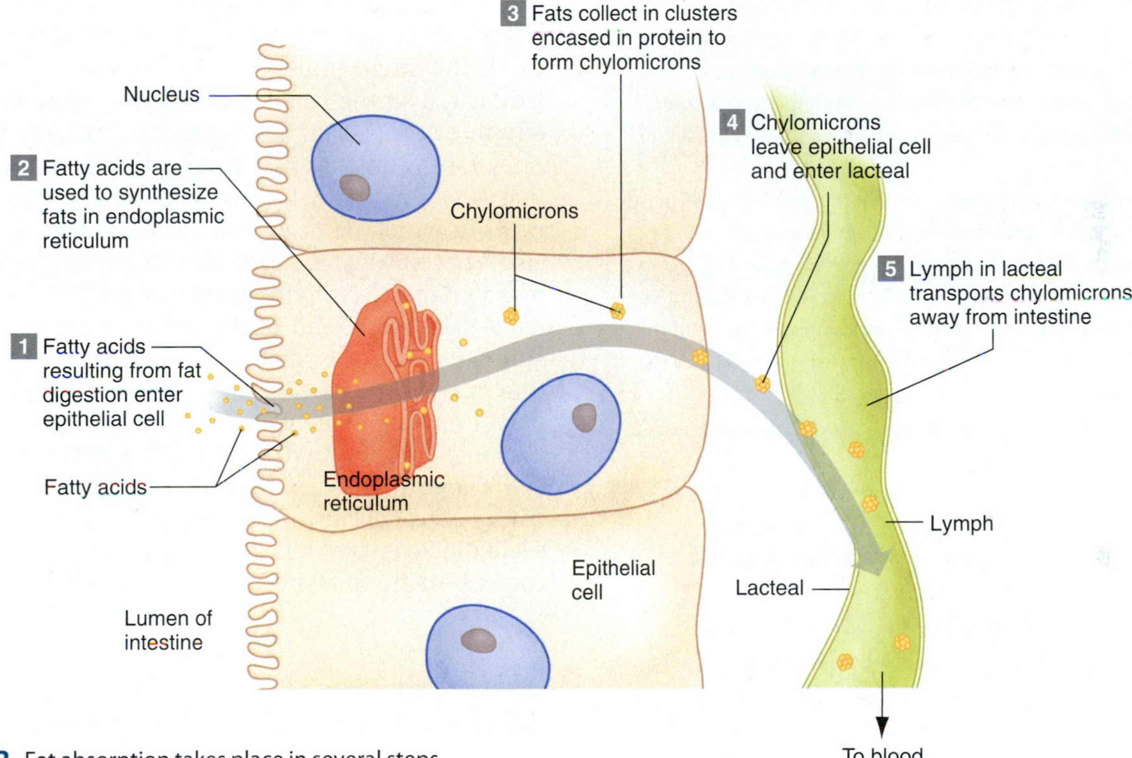

FIGURE 17.42 Fat absorption takes place in several steps.

framework and secretes HDL molecules into the bloodstream. As it circulates, the HDL picks up cholesterol from peripheral tissues and returns the cholesterol to the liver. Here, the cholesterol molecules enter liver cells by receptor-mediated endocytosis. The liver disposes of the cholesterol it obtains in this manner by secreting it into bile or by using it to synthesize bile salts.

The intestine reabsorbs much of the cholesterol and bile salts in bile, which are then transported back to the liver, and the secretion-reabsorption cycle repeats. During each cycle, some of the cholesterol and bile salts escape reabsorption, reach the large intestine, and are eliminated with the feces.

In addition to absorbing the products of carbohydrate, protein, and fat digestion, the intestinal villi absorb electrolytes and water. Certain ions, such as those of sodium, potassium, chloride, nitrate, and bicarbonate, are readily absorbed; but others, including ions of calcium, magnesium, and sulfate, are poorly absorbed.

Electrolytes are usually absorbed by active transport, and water by osmosis. Thus, even though the intestinal contents may be hypertonic to the epithelial cells at first, as nutrients and electrolytes are absorbed, the contents become slightly hypotonic to the cells. Then, water follows the nutrients and electrolytes into the villi by osmosis. Table 17.10 summarizes the absorption process.

TABLE 17.10 | Intestinal Absorption of Nutrients

Nutrient	Absorption Mechanism	Means of Transport
Monosaccharides	Facilitated diffusion and active transport	Blood in capillaries
Amino acids	Active transport	Blood in capillaries
Fatty acids and glycerol	Facilitated diffusion of glycerol; diffusion of fatty acids into cells	
	(a) Most fatty acids are resynthesized into fats and incorporated in chylomicrons for transport.	Lymph in lacteals
	(b) Some fatty acids with relatively short carbon chains are transported without being changed back into fats.	Blood in capillaries
Electrolytes	Diffusion and active transport	Blood in capillaries
Water	Osmosis	Blood in capillaries

In *malabsorption,* the small intestine digests, but does not absorb, some nutrients. Symptoms of malabsorption include diarrhea, weight loss, weakness, vitamin deficiencies, anemia, and bone demineralization. Causes of malabsorption include surgical removal of a portion of the small intestine, obstruction of lymphatic vessels due to a tumor, or interference with the production and release of bile as a result of liver disease.

Another cause of malabsorption is a reaction to *gluten,* which is a protein found in certain grains, especially wheat and rye. This condition is called *celiac disease.* Microvilli are damaged, and in severe cases, villi are destroyed. Both of these effects reduce the absorptive surface of the small intestine, preventing absorption of some nutrients. Health-food grocery stores sell gluten-free products, including pasta and baked goods.

PRACTICE

50 Which substances resulting from digestion of carbohydrate, protein, and fat molecules does the small intestine absorb?

51 Which ions does the small intestine absorb?

52 What transport mechanisms do intestinal villi use?

53 Describe how fatty acids are absorbed and transported.

Movements of the Small Intestine

Like the stomach, the small intestine carries on mixing movements and peristalsis. The major mixing movement is segmentation, in which small, ringlike, periodic contractions cut the chyme into segments and move it back and forth. Segmentation also slows the movement of chyme through the small intestine.

Peristaltic waves propel chyme through the small intestine. These waves are usually weak, and they stop after pushing the chyme a short distance. Consequently, chyme moves slowly through the small intestine, taking from three to ten hours to travel its length.

As might be expected, parasympathetic impulses enhance both mixing and peristaltic movements, and sympathetic impulses inhibit them. Reflexes involving parasympathetic impulses to the small intestine sometimes originate in the stomach. For example, food distends the stomach wall, triggering the gastroenteric reflex, which greatly increases peristaltic activity in the small intestine. Another reflex begins when the duodenum fills with chyme, stretching its wall. This reflex speeds movement through the small intestine.

If the small intestine wall becomes overdistended or irritated, a strong *peristaltic rush* may pass along the entire length of the organ, sweeping the contents into the large intestine so quickly that water, nutrients, and electrolytes that would normally be absorbed are not. The result is *diarrhea,* in which defecation becomes more frequent and the stools become watery. Prolonged diarrhea causes imbalances in water and electrolyte concentrations.

At the distal end of the small intestine, the **ileocecal sphincter** joins the small intestine's ileum to the large intestine's cecum. Normally, this sphincter remains constricted, preventing the contents of the small intestine from entering the large intestine, and at the same time keeping the contents of the large intestine from backing up into the ileum. However, eating a meal elicits a gastroileal reflex that increases peristalsis in the ileum and relaxes the sphincter, forcing some of the contents of the small intestine into the cecum.

PRACTICE

54 Describe the movements of the small intestine.

55 How are the movements of the small intestine initiated?

56 What is a peristaltic rush?

57 What stimulus relaxes the ileocecal sphincter?

17.10 LARGE INTESTINE

The **large intestine** is so named because its diameter is greater than that of the small intestine. This part of the alimentary canal is about 1.5 meters long, and it begins in the lower right side of the abdominal cavity where the ileum joins the cecum. From there, the large intestine ascends on the right side, crosses obliquely to the left, and descends into the pelvis. At its distal end, it opens to the outside of the body as the anus.

The large intestine absorbs ingested water and electrolytes remaining in the alimentary canal. Additionally it reabsorbs and recycles water and remnants of digestive secretions. The large intestine also forms and stores feces.

Parts of the Large Intestine

The large intestine consists of the cecum, the colon, the rectum, and the anal canal. Figures 17.43 and 17.44 and reference plates 11, 12, 18, and 25 depict the large intestine.

The **cecum,** at the beginning of the large intestine, is a dilated, pouchlike structure that hangs slightly inferior to the ileocecal opening. Projecting downward from it is a narrow tube with a closed end called the **appendix.** The human appendix has no known digestive function. However, it contains lymphatic tissue. One suggested function of the appendix is that it sequesters useful bacteria when a person has diarrhea and vomits due to an infection.

> In *appendicitis,* the appendix becomes inflamed and infected. Surgery is required to prevent the appendix from rupturing. If it breaks open, the contents of the large intestine may enter the abdominal cavity and cause a serious infection of the peritoneum called *peritonitis.*

The **colon** is divided into four parts—the ascending, transverse, descending, and sigmoid colons. The **ascending colon** begins at the cecum and extends upward against the posterior abdominal wall to a point just inferior to the liver. There it turns sharply to the left (as the right colic, or hepatic, flexure) and becomes the **transverse colon.** The transverse colon is the longest and most movable part of the large intestine. It is suspended by a fold of peritoneum and sags in the middle below the stomach. As the transverse colon approaches the spleen, it turns abruptly downward (as the left colic, or splenic, flexure) and becomes the **descending colon.** At the brim of the pelvis, the descending colon makes an S-shaped curve, called the **sigmoid colon,** and then becomes the rectum.

The **rectum** lies next to the sacrum and generally follows its curvature. The peritoneum firmly attaches it to the sacrum, and it ends about 5 centimeters inferior to the tip of the coccyx, where it becomes the anal canal (fig. 17.45).

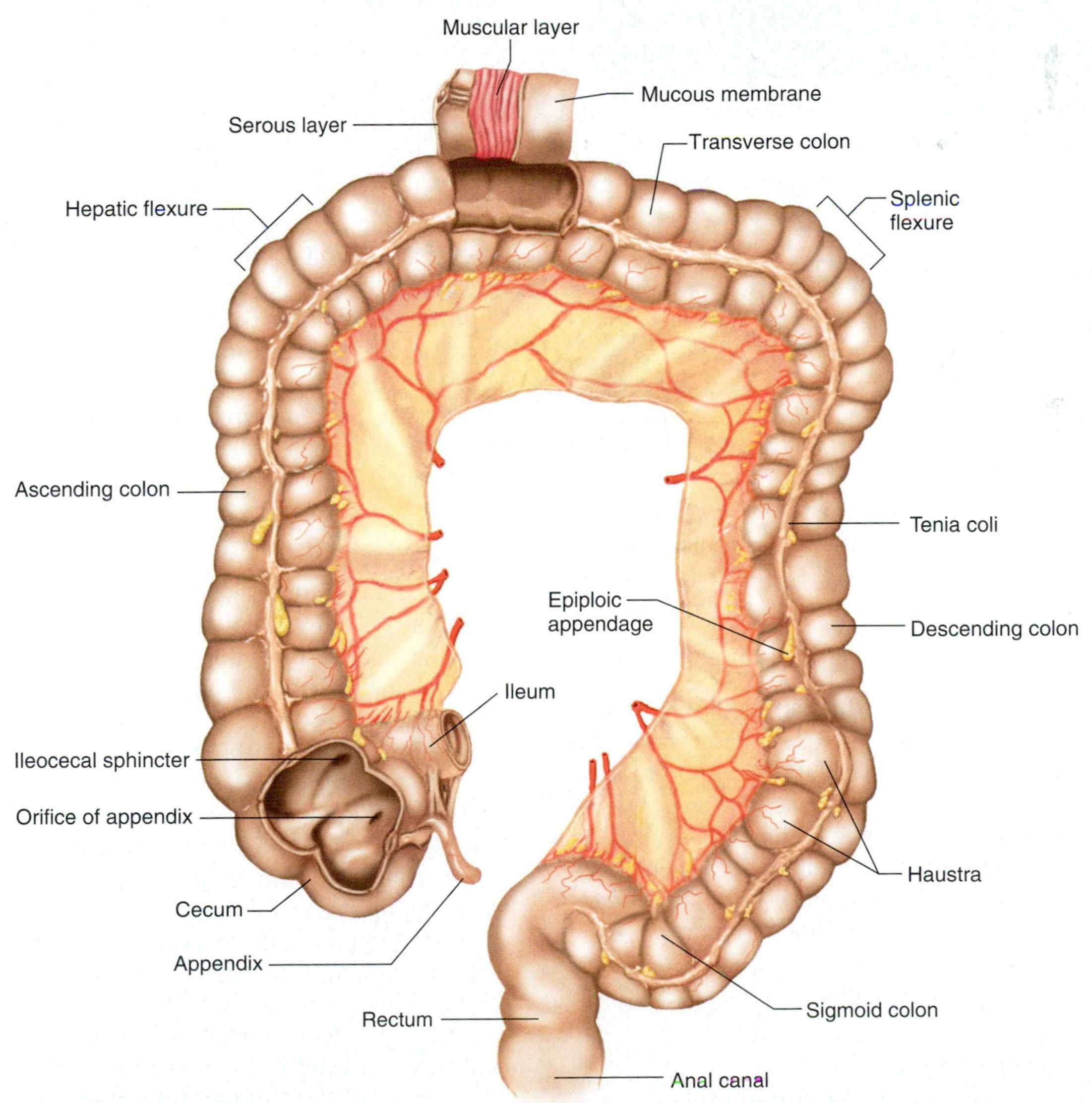

FIGURE 17.43 Parts of the large intestine (anterior view).

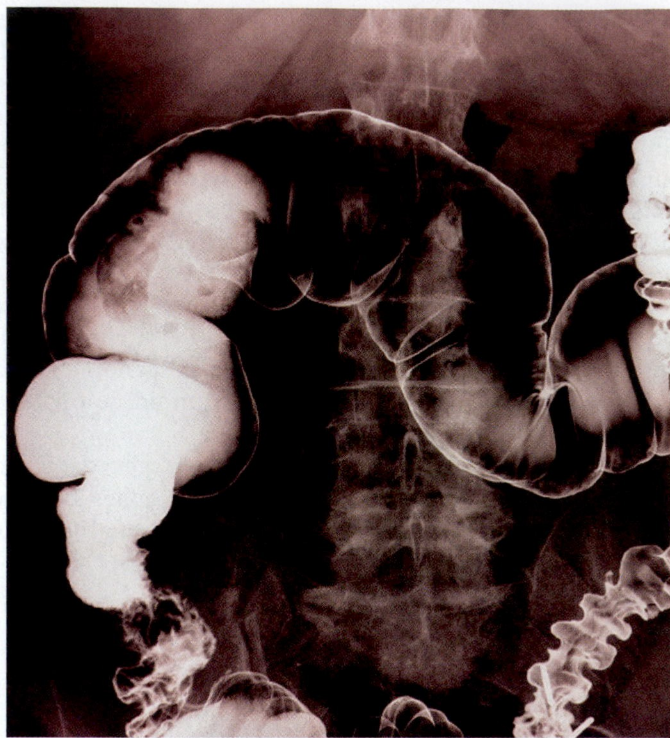

FIGURE 17.44 Radiograph of the large intestine containing a radiopaque substance that the patient ingested.

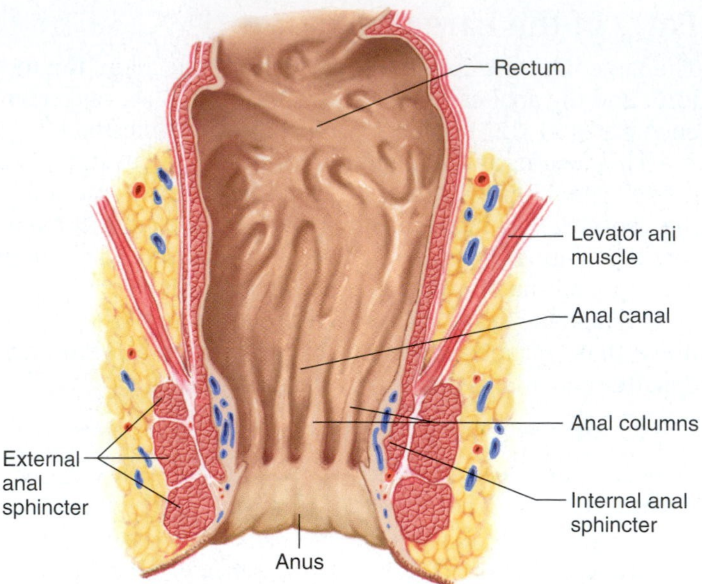

FIGURE 17.45 The rectum and the anal canal are at the distal end of the alimentary canal.

Hemorrhoids are enlarged and inflamed branches of the rectal vein in the anal columns that cause intense itching, sharp pain, and sometimes bright red bleeding. The hemorrhoids may be internal (which do not produce symptoms) or bulge out of the anus. Causes of hemorrhoids include anything that puts prolonged pressure on the delicate rectal tissue, including obesity, pregnancy, constipation, diarrhea, and liver disease.

Eating more fiber-rich foods and drinking lots of water can usually prevent or cure hemorrhoids. Warm soaks in the tub, cold packs, and careful wiping of painful areas also help, as do external creams and ointments. Surgery—with a scalpel or a laser—can remove severe hemorrhoids.

The **anal canal** is formed by the last 2.5 to 4.0 centimeters of the large intestine. The mucous membrane in the canal is folded into a series of six to eight longitudinal *anal columns.* At its distal end, the canal opens to the outside as the **anus.** Two sphincter muscles guard the anus—an *internal anal sphincter muscle,* composed of smooth muscle under involuntary control, and an *external anal sphincter muscle,* composed of skeletal muscle under voluntary control.

PRACTICE

58 What is the general function of the large intestine?

59 Describe the parts of the large intestine.

60 Distinguish between the internal sphincter muscle and the external sphincter muscle of the anus.

Structure of the Large Intestinal Wall

The wall of the large intestine includes the same types of tissues found in other parts of the alimentary canal but also has some unique features (fig. 17.46). The large intestinal wall lacks the villi and plicae circularis characteristic of the small intestine. Also, the layer of longitudinal muscle fibers is not uniformly distributed throughout the large intestine wall; the fibers are in three distinct bands (teniae coli) that extend the entire length of the colon. These bands exert tension lengthwise on the wall, creating a series of pouches (haustra). The large intestinal wall also has small collections of fat (epiploic appendages) in the serosa on its outer surface (see fig. 17.43).

Functions of the Large Intestine

The large intestine has little or no digestive function. This is in contrast to the small intestine, which secretes digestive enzymes and absorbs the products of digestion. However, the mucous membrane that forms the inner lining of the large intestine includes many tubular glands. Structurally, these glands are similar to those of the small intestine, but they are composed almost entirely of goblet cells (fig. 17.47). Consequently, mucus is the only significant secretion of this part of the alimentary canal.

Mechanical stimulation from chyme and parasympathetic impulses control the rate of mucous secretion. In both cases, the goblet cells respond by increasing mucous production, which, in turn, protects the intestinal wall against the abrasive action of materials passing through it. Mucus also holds particles of fecal matter together, and, because it is alkaline, mucus helps control the pH of the large intestinal contents. This is important because acids are sometimes released from the feces as a result of bacterial activity.

Chyme entering the large intestine usually has few nutrients remaining in it and mostly consists of materials not digested or absorbed in the small intestine. It also contains water, electrolytes, mucus, and bacteria.

Absorption in the large intestine is normally limited to water and electrolytes, and this usually occurs in the proximal half of the tube. Electrolytes such as sodium ions can be absorbed by active transport, while the water follows passively, entering the mucosa by osmosis. About 90% of the water that enters the large intestine is absorbed, and little sodium or water is lost in the feces.

The many bacteria that normally inhabit the large intestine, called *intestinal flora,* break down some of the molecules that escape the actions of human digestive enzymes. For instance, cellulose, a complex carbohydrate in food of plant origin, passes through the alimentary canal almost unchanged, but colon bacteria can break down cellulose and use it as an energy source. These bacteria, in turn, synthesize vitamins, such as K, B_{12}, thiamine, and riboflavin, which the intestinal mucosa absorbs. Bacterial actions in the large intestine may produce intestinal gas (flatus).

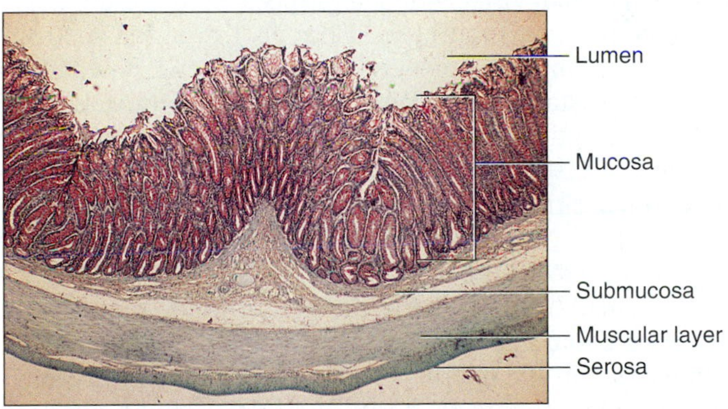

FIGURE 17.46 Light micrograph of the large intestinal wall (64×).

Intestinal gas is 99% nitrogen and oxygen taken in while breathing and eating, plus methane (CH_4), carbon dioxide (CO_2), and hydrogen contributed from the bacterial fermentation of undigested food. The characteristic odor comes from bacterial action on the nitrogen and sulfur in proteins, which yields pungent-smelling ammonia (NH_3) and foul hydrogen sulfide (H_2S). Most people release a half liter of intestinal gas a day. Foods rich in sulfur-containing amino acids make intestinal gas more foul. These include beans, broccoli, bran, brussels sprouts, cabbage, cauliflower, and onions.

PRACTICE  

61 How does the structure of the large intestine differ from that of the small intestine?

62 What substances does the large intestine absorb?

63 What useful substances do bacteria inhabiting the large intestine produce?

Movements of the Large Intestine

The movements of the large intestine—mixing and peristalsis—are similar to those of the small intestine, although usually slower. The mixing movements break the fecal matter into segments and turn it so that all portions are exposed to the intestinal mucosa. This helps absorb water and electrolytes.

The peristaltic waves of the large intestine are different from those of the small intestine. Instead of occurring frequently, they happen only two or three times each day. These waves produce *mass movements* in which a large section of the intestinal wall constricts vigorously, forcing the intestinal contents to move toward the rectum. Typically, mass movements follow a meal, as a result of the gastrocolic reflex initiated in the small intestine. Irritation of the intestinal mucosa can also trigger such movements. For instance, a person suffering from an inflamed colon (colitis) may experience frequent mass movements. Clinical Application 17.5 examines conditions affecting the large intestine.

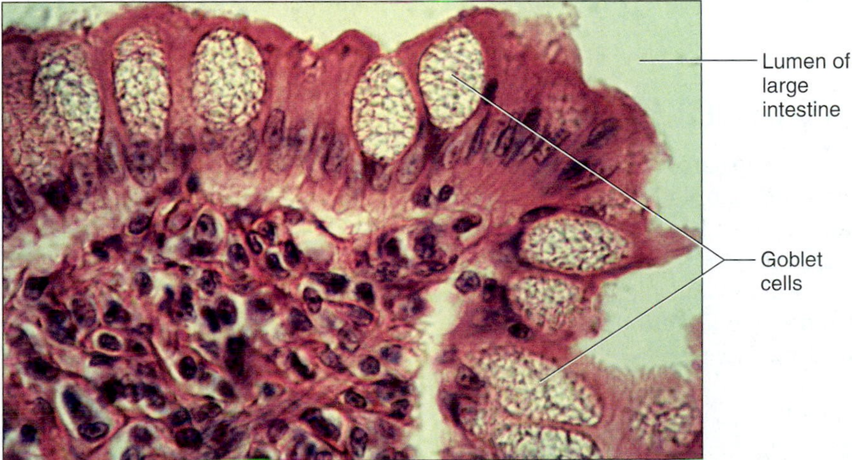

FIGURE 17.47 Light micrograph of the large intestinal mucosa (560×).

When it is appropriate to defecate, a person usually can initiate a *defecation reflex* by holding a deep breath and contracting the abdominal wall muscles. This action increases the internal abdominal pressure and forces feces into the rectum. As the rectum fills its wall distends, triggering the defecation reflex, which stimulates peristaltic waves in the descending colon. The internal anal sphincter relaxes. At the same time, other reflexes involving the sacral region of the spinal cord strengthen the peristaltic waves, lower the diaphragm, close the glottis, and contract the abdominal wall muscles. These actions further increase the internal abdominal pressure and squeeze the rectum. The external anal sphincter is signaled to relax, and the feces are forced to the outside. A person can voluntarily inhibit defecation by contracting the external anal sphincter.

Hirschsprung disease causes extreme, chronic constipation and abdominal distension. The part of the large intestine distal to the distension lacks innervation. The person does not feel the urge to defecate. The problem begins in the embryo, when a mutant gene prevents neurons from migrating to this portion of the gastrointestinal tract. Surgery may be used to treat the once lethal Hirschsprung disease.

Feces

Feces (fe′sēz) are composed of materials not digested or absorbed, along with water, electrolytes, mucus, and bacteria. Usually the feces are about 75% water, and their color derives from bile pigments altered by bacterial action.

The pungent odor of the feces results from a variety of compounds that bacteria produce. These compounds include phenol, hydrogen sulfide, indole, skatole, and ammonia.

PRACTICE

64 How does peristalsis in the large intestine differ from peristalsis in the small intestine?

65 List the major events of defecation.

66 Describe the composition of feces.

17.11 LIFE-SPAN CHANGES

Changes to the digestive system associated with the passing years are slow and slight, so most people can enjoy eating a variety of foods as they grow older. Maintaining healthy teeth is vital to obtaining adequate nutrition. This requires frequent dental checkups, cleanings, and plaque removal, plus care of the gums. Tooth loss due to periodontal disease becomes more likely after age thirty-five.

Despite regular dental care, some signs of aging may affect the teeth. The enamel often thins from years of brushing, teeth grinding, and eating acidic foods. Thinning enamel may make the teeth more sensitive to hot and cold foods. At the same time, the cementum may thicken. The dentin heals more slowly and enlarges as the pulp shrinks. Loss of neurons in the pulp may make it more difficult to notice tooth decay. The gums recede, creating more pockets to harbor the bacteria whose activity contributes to periodontal disease. The teeth may loosen as the bones of the jaw weaken. On a functional level, older people sometimes do not chew their food thoroughly, swallowing larger chunks of food that may present a choking hazard.

A common complaint of older individuals is "dry mouth," or xerostomia. This condition is not a normal part of aging—studies have shown that the oldest healthy people make just as much saliva as healthy younger people. Dry mouth is common, however, because it is a side effect of more than 400 medications, many of which are more likely to be taken by older persons. These include antidepressants, antihistamines, and drugs that treat cancer or hypertension. In addition, radiation and chemotherapy used to treat cancer can cause mouth sores and tooth decay. It is a good idea for cancer patients to coordinate dental visits with other aspects of their care.

Once past the mouth, food travels through a gastrointestinal tract that declines gradually in efficiency with age. Slowing peristalsis may cause frequent heartburn as food backs up into the esophagus. The stomach lining thins with age, and secretion of hydrochloric acid, pepsin, and intrinsic factor decline. Exit of chyme from the stomach slows. Overall, these changes may affect the rate at which certain medications are absorbed.

The small intestine is the site of absorption of nutrients, so it is here that noticeable signs of aging on digestion arise. Subtle shifts in the microbial species that inhabit the small intestine alter the rates of absorption of particular nutrients. With age, the small intestine becomes less efficient at absorbing vitamins A, D, and K and the mineral zinc. This raises the risk of deficiency symptoms—effects on skin and vision due to a lack of vitamin A; weakened bones from inadequate vitamin D; impaired blood clotting seen in vitamin K deficiency; and slowed healing, decreased immunity, and altered taste evidenced in zinc deficiency.

Many people who have inherited lactose intolerance begin to notice the telltale cramping after eating dairy foods in the middle years. They must be careful that by avoiding dairy products, they do not also lower their calcium intake. Less hydrochloric acid also adversely affects the absorption of calcium, as well as iron. Too little intrinsic factor may lead to vitamin B_{12} deficiency anemia.

The lining of the large intestine changes too, thinning and containing less smooth muscle and mucus. A dampening of the responsiveness of the smooth muscle to neural

17.5 CLINICAL APPLICATION

Disorders of the Large Intestine

The large intestine (colon) is the source of familiar digestive discomforts as well as more serious disorders.

Diverticulosis and Inflammatory Bowel Disease

In diverticulosis, parts of the intestinal wall weaken, and the inner mucous membrane protrudes through. If chyme accumulates in the outpouching and becomes infected (diverticulitis), antibiotics or surgical removal of the area may become necessary. Lack of dietary fiber may set the stage for diverticulosis. The condition does not occur in populations that eat high fiber diets, and began to appear in the United States only after refined foods were introduced in the middle of the twentieth century.

Inflammatory bowel disease is a group of disorders that includes ulcerative colitis and Crohn disease. They differ by the site and extent of inflammation and ulceration of the intestines. In the United States, about 100,000 people suffer the abdominal cramps and diarrhea of ulcerative colitis, and 500,000 individuals have similar symptoms of Crohn disease.

Ulcerative colitis affects the mucosa and submucosa of the distal large intestine and the rectum. In about 25% of cases, the disease extends no farther than the rectum. Bouts of bloody diarrhea and cramps may last for days or weeks and may recur frequently or only rarely. The severe diarrhea leads to weight loss and electrolyte imbalances and may develop into colon cancer or affect other organs, including the skin, eyes, or liver. The inflamed and ulcerous tissue is continuous.

Crohn disease is more extensive than ulcerative colitis, extending into the small and large intestines and penetrating all tissue layers. In contrast to the uniformity of ulcerative colitis, affected portions of intestine in Crohn disease are interspersed with unaffected areas, producing a "cobblestone" effect after many years. The ileum and cecum are affected in about 40% of affected individuals, only the small intestine is involved in 30% of cases, and only the large intestine is involved in 25% of cases. Rarely, the disease affects more proximal structures of the gastrointestinal tract. The diarrhea is often not bloody, and complications such as cancer are rare.

Overall, about 20% of people with symptoms of irritable bowel disease seem to fall between the descriptions of ulcerative colitis and Crohn disease and are classified as having "indeterminate colitis." Surgery is sometimes used to treat inflammatory bowel disease. Autoimmunity, infection, or a genetic predisposition may contribute to causing inflammatory bowel disease.

Colorectal Cancer

Cancer of the large intestine or rectum, known as *colorectal cancer,* is the fourth most prevalent cancer in the United States and the second most common cause of cancer death. More than 30,000 new cases are diagnosed each year, and more than 56,000 people die of the condition. It tends to run in families.

Symptoms of colorectal cancer include

- a change in frequency or consistency of bowel movements
- blood in the feces
- a narrowing of feces
- abdominal discomfort or pain
- weight loss
- fatigue
- unexplained vomiting

Diagnostic tests can detect colorectal cancer, described in table 17A. These tests are of two general types—the fecal occult blood test performed on a stool sample and imaging the large intestine wall. In development is a test that screens the DNA from cells in feces for mutations associated with colorectal cancer.

A rite of passage for those over fifty is fiberoptic colonoscopy. Under sedation, a flexible lit tube is inserted into the rectum, and polyps and tumors are identified and removed. Those with a family history of colon cancer should be screened at an earlier age. Fiberoptic colonoscopy takes less than an hour. A newer procedure, computed tomographic colonography (popularly called a virtual colonoscopy), requires the same preparatory bowel cleansing, but does not require sedation and is faster. However, if a lesion is detected, the more invasive approach must be used to remove the suspicious tissue.

Treatment for colorectal cancer is to remove the affected tissue. If a large portion of the intestine is removed, surgery is used to construct a new opening to release feces. The free end of the intestine is attached to an opening created through the skin of the abdomen, and a bag is attached to the opening to collect the fecal matter. This procedure is called a colostomy.

For people who have a certain inherited form of colon cancer (familial adenomatous polyposis), nonsteroidal anti-inflammatory drugs (NSAIDs) called *Cox-2 inhibitors* are used to treat the disease and may even help to prevent it in those identified by a genetic test to be at high risk. However, these drugs increase the risk of heart disease in certain individuals. ■

TABLE 17A | Diagnostic Tests for Colorectal Cancer

Diagnostic Test	Description
Digital rectal exam	Physician palpates large intestine and rectum
Double-contrast barium enema	X-ray exam following ingestion of contrast agent highlights blockages in large intestine
Fecal occult blood test	Blood detected in feces sample
Colorectal cancer gene test (experimental)	Mutations associated with colorectal cancer detected in DNA of cells shed with feces
Sigmoidoscopy	Endoscope views rectum and lower colon
Colonoscopy	Endoscope views rectum and entire colon

INNERCONNECTIONS | *Digestive System*

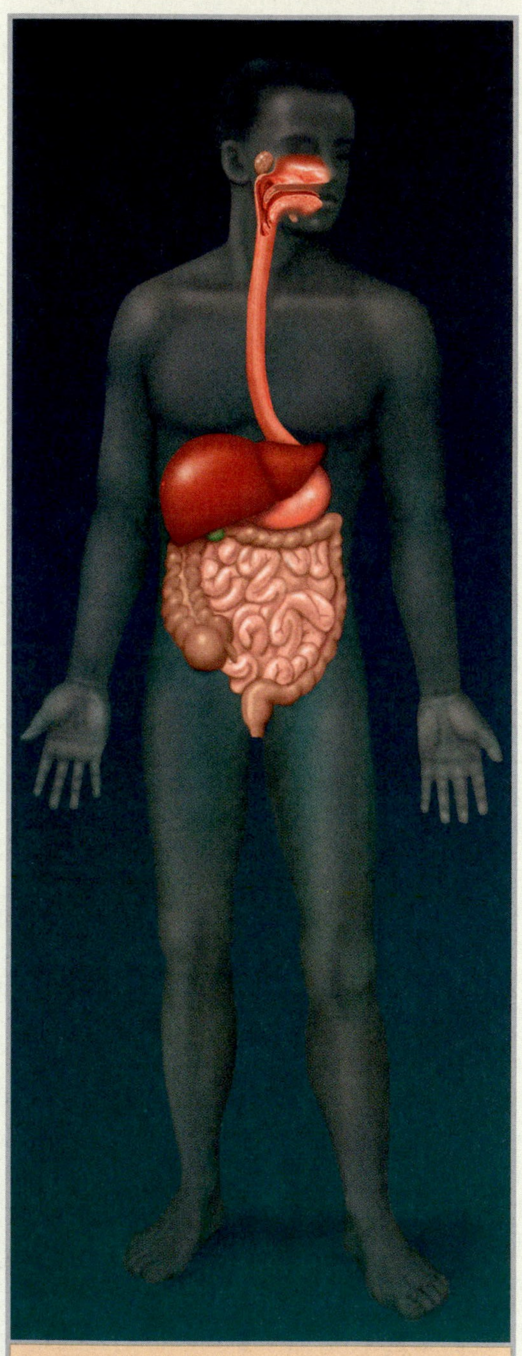

Digestive System

The digestive system ingests, digests, and absorbs nutrients for use by all body cells.

Integumentary System

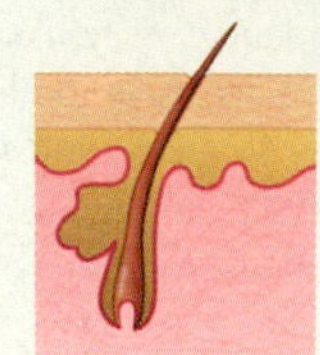

Vitamin D activated in the skin plays a role in absorption of calcium from the digestive tract.

Cardiovascular System

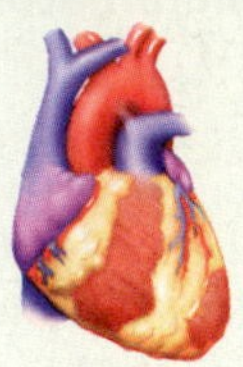

The bloodstream carries absorbed nutrients to all body cells.

Skeletal System

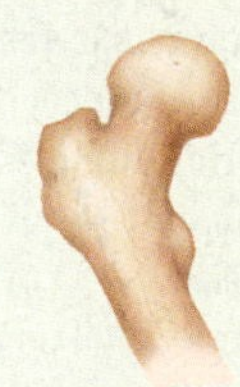

Bones are important in mastication. Calcium absorption is necessary to maintain bone matrix.

Lymphatic System

The lymphatic system plays a major role in the absorption of fats.

Muscular System

Muscles are important in mastication, swallowing, and the mixing and moving of digestion products through the gastrointestinal tract.

Respiratory System

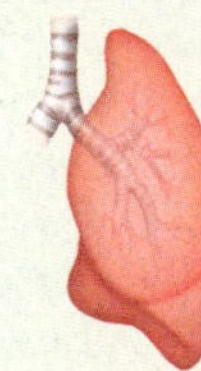

The digestive system and the respiratory system share common anatomical structures.

Nervous System

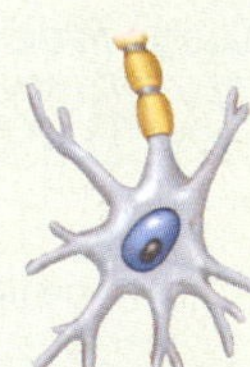

The nervous system can influence digestive system activity.

Urinary System

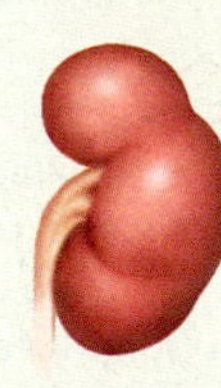

The kidneys and liver work together to activate vitamin D.

Endocrine System

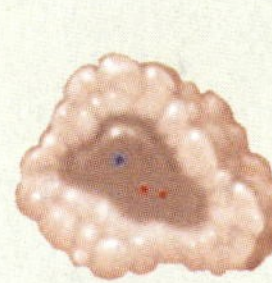

Hormones can influence digestive system activity.

Reproductive System

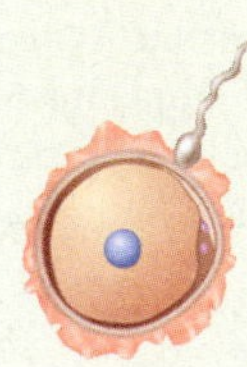

In a woman, nutrition is essential for conception and normal development of an embryo and fetus.

stimulation slows peristalsis, ultimately causing constipation. Compounding this common problem is a loss of elasticity in the walls of the rectum and declining strength and responsiveness of the internal and external sphincters.

The accessory organs to digestion age too, but not necessarily in ways that affect health. Both the pancreas and the liver are large organs with cells to spare, so a decline in their secretion abilities does not usually hamper digestion. Only 10% of the pancreas and 20% of the liver are required to digest foods. However, the liver may not be able to detoxify certain medications as quickly as it once did. The gallbladder becomes less sensitive to cholecystokinin, but in a classic feedback response, cells of the intestinal mucosa secrete more of it into the bloodstream, and the gallbladder continues to be able to contract. The bile ducts widen in some areas, but the end of the bile duct narrows as it approaches the small intestine. As long as gallstones do not become entrapped in the ducts, the gallbladder generally functions well into the later years.

PRACTICE 

67 Describe the effects of aging on the teeth.

68 What conditions might be caused by the slowing of peristalsis in the digestive tract that occurs with aging?

CHAPTER SUMMARY

17.1 INTRODUCTION (PAGE 652)

Digestion is the process of mechanically and chemically breaking down foods so that they can be absorbed. The digestive system consists of an alimentary canal and several accessory organs that carry out the processes of ingestion, propulsion, digestion, absorption, and defecation.

17.2 GENERAL CHARACTERISTICS OF THE ALIMENTARY CANAL (PAGE 652)

Regions of the alimentary canal perform specific functions.

1. Structure of the wall
   a. The wall consists of four layers.
   b. These layers include the mucosa, submucosa, muscular layer, and serosa.
2. Movements of the tube
   a. Motor functions include mixing and propelling movements.
   b. Peristalsis is responsible for propelling movements.
   c. The wall of the tube undergoes receptive relaxation just ahead of a peristaltic wave.
3. Innervation of the tube
   a. The tube is innervated by branches of the sympathetic and parasympathetic divisions of the autonomic nervous system.
   b. Parasympathetic impulses generally increase digestive activities; sympathetic impulses generally inhibit digestive activities.
   c. Sympathetic impulses contract certain sphincter muscles, controlling movement through the alimentary canal.

17.3 MOUTH (PAGE 656)

The mouth is adapted to receive food and begin preparing it for digestion. It also serves as an organ of speech and sensory perception.

1. Cheeks and lips
   a. Cheeks form the lateral walls of the mouth.
   b. Lips are highly mobile and have a variety of sensory receptors useful in judging the characteristics of food.
2. Tongue
   a. The tongue is a thick, muscular organ that mixes food with saliva and moves it toward the pharynx.
   b. The rough surface of the tongue handles food and has taste buds.
   c. Lingual tonsils are located on the root of the tongue.
3. Palate
   a. The palate comprises the roof of the mouth and includes hard and soft portions.
   b. The soft palate, including the uvula, closes the opening to the nasal cavity during swallowing.
   c. Palatine tonsils are located on either side of the tongue in the back of the mouth.
   d. Tonsils consist of lymphatic tissues.
4. Teeth
   a. Two sets of teeth develop in sockets of the mandibular and maxillary bones.
   b. There are twenty primary and thirty-two secondary teeth.
   c. Teeth mechanically break food into smaller pieces, increasing the surface area exposed to digestive actions.
   d. Different types of teeth are adapted to handle foods in different ways, such as biting, grasping, or grinding.
   e. Each tooth consists of a crown and root and is composed of enamel, dentin, pulp, nerves, and blood vessels.
   f. A tooth is attached to the alveolar process by the periodontal ligament.

17.4 SALIVARY GLANDS (PAGE 660)

Salivary glands secrete saliva, which moistens food, helps bind food particles, begins chemical digestion of carbohydrates, makes taste possible, helps cleanse the mouth, and regulates pH in the mouth.

1. Salivary secretions
   a. Salivary glands include serous cells that secrete digestive enzymes and mucous cells that secrete mucus.
   b. Parasympathetic impulses stimulate the secretion of serous fluid.

2. Major salivary glands
   a. The parotid glands are the largest, and they secrete saliva rich in amylase.
   b. The submandibular glands in the floor of the mouth produce viscous saliva.
   c. The sublingual glands in the floor of the mouth primarily secrete mucus.

17.5 PHARYNX AND ESOPHAGUS (PAGE 661)

The pharynx and esophagus serve as passageways.

1. Structure of the pharynx
   a. The pharynx is divided into a nasopharynx, oropharynx, and laryngopharynx.
   b. The muscular walls of the pharynx contain fibers in circular and longitudinal groups.
2. Swallowing mechanism
   a. Swallowing occurs in three stages.
      (1) Food is mixed with saliva and forced into the pharynx.
      (2) Involuntary reflex actions move the food into the esophagus.
      (3) Peristalsis transports food to the stomach.
   b. Swallowing reflexes momentarily inhibit breathing.
3. Esophagus
   a. The esophagus passes through the mediastinum and penetrates the diaphragm.
   b. Circular muscle fibers at the distal end of the esophagus help prevent regurgitation of food from the stomach.

17.6 STOMACH (PAGE 665)

The stomach receives food, mixes it with gastric juice, carries on a limited amount of absorption, and moves food into the small intestine.

1. Parts of the stomach
   a. The stomach is divided into the cardia, fundus, body, and pylorus.
   b. The lower esophageal sphincter serves as a valve between the esophagus and the stomach.
   c. The pyloric sphincter serves as a valve between the stomach and the small intestine.
2. Gastric secretions
   a. Gastric glands secrete gastric juice.
   b. Gastric juice contains pepsin(begins digestion of proteins), hydrochloric acid, lipase, and intrinsic factor.
3. Regulation of gastric secretions
   a. Parasympathetic impulses and the hormone gastrin enhance gastric secretion.
   b. The three stages of gastric secretion are the cephalic, gastric, and intestinal phases.
   c. The presence of food in the small intestine reflexly inhibits gastric secretions.
4. Gastric absorption
   a. The stomach is not well adapted for absorption.
   b. A few substances such as water and other small molecules are absorbed through the stomach wall.
5. Mixing and emptying actions
   a. As the stomach fills, its wall stretches, but its internal pressure remains unchanged.
   b. Mixing movements aid in producing chyme; peristaltic waves move chyme into the pylorus.
   c. The muscular wall of the pylorus regulates chyme movement into the small intestine.
   d. The rate of emptying depends on the fluidity of the chyme and the type of food present.
   e. The upper part of the small intestine fills, and an enterogastric reflex inhibits peristalsis in the stomach.
   f. Vomiting results from a complex reflex that has many stimuli.

17.7 PANCREAS (PAGE 671)

The pancreas is closely associated with the duodenum.

1. Structure of the pancreas
   a. It produces pancreatic juice secreted into a pancreatic duct.
   b. The pancreatic duct leads to the duodenum.
2. Pancreatic juice
   a. Pancreatic juice contains enzymes that can split carbohydrates, proteins, fats, and nucleic acids.
   b. Pancreatic juice has a high bicarbonate ion concentration that helps neutralize chyme and causes the intestinal contents to be alkaline.
3. Regulation of pancreatic secretion
   a. Secretin from the duodenum stimulates the release of pancreatic juice that contains few digestive enzymes but has a high bicarbonate ion concentration.
   b. Cholecystokinin from the intestinal wall stimulates the release of pancreatic juice that has a high concentration of digestive enzymes.

17.8 LIVER (PAGE 673)

The liver is located in the upper-right quadrant of the abdominal cavity.

1. Liver structure
   a. The liver is a highly vascular organ, enclosed in a fibrous capsule, and divided into lobes.
   b. Each lobe consists of hepatic lobules, the functional units of the liver.
   c. Bile from the lobules is carried by bile ductules to hepatic ducts that unite to form the common hepatic duct.
2. Liver functions
   a. The liver has many functions. It metabolizes carbohydrates, lipids, and proteins; stores some substances; filters blood; destroys toxins; and secretes bile.
   b. Bile is the only liver secretion that directly affects digestion.
3. Composition of bile
   a. Bile contains bile salts, bile pigments, cholesterol, and electrolytes.
   b. Only the bile salts have digestive functions.
   c. Bile pigments are products of red blood cell breakdown.

4. Gallbladder
   a. The gallbladder stores bile between meals.
   b. A sphincter muscle controls release of bile from the bile duct.
   c. Gallstones may form within the gallbladder.
5. Regulation of bile release
   a. Cholecystokinin from the small intestine stimulates bile release.
   b. The sphincter muscle at the base of the bile duct relaxes as a peristaltic wave in the duodenal wall approaches.
6. Functions of bile salts
   a. Bile salts emulsify fats and aid in the absorption of fatty acids, cholesterol, and certain vitamins.
   b. Bile salts are reabsorbed in the small intestine.

17.9 SMALL INTESTINE (PAGE 678)

The small intestine extends from the pyloric sphincter to the large intestine. It receives secretions from the pancreas and liver, completes digestion of nutrients, absorbs the products of digestion, and transports the residues to the large intestine.

1. Parts of the small intestine
   a. The small intestine consists of the duodenum, jejunum, and ileum.
   b. The small intestine is suspended from the posterior abdominal wall by mesentery.
2. Structure of the small intestinal wall
   a. The wall is lined with villi that greatly increase the surface area and aid in mixing and absorption.
   b. Microvilli on the free ends of epithelial cells increase the surface area even more.
   c. Intestinal glands are located between the villi.
   d. Circular folds in the lining of the intestinal wall also increase its surface area.
3. Secretions of the small intestine
   a. Intestinal glands secrete a watery fluid that lacks digestive enzymes but provides a vehicle for moving chyme to the villi.
   b. Digestive enzymes embedded in the surfaces of microvilli split molecules of sugars, proteins, and fats.
4. Regulation of small intestinal secretions
   a. Secretion is stimulated by gastric juice, chyme, and reflexes stimulated by distension of the small intestinal wall.
5. Absorption in the small intestine
   a. Blood capillaries in the villi absorb monosaccharides, amino acids, fatty acids, and glycerol.
   b. Blood capillaries in the villi also absorb water and electrolytes.
   c. Fat molecules with longer chains of carbon atoms enter the lacteals of the villi; fatty acids with short carbon chains enter the blood capillaries of the villi.
6. Movements of the small intestine
   a. Movements include mixing by segmentation and peristalsis.
   b. Overdistension or irritation may stimulate a peristaltic rush and result in diarrhea.
   c. The ileocecal sphincter controls movement of the intestinal contents from the small intestine into the large intestine.

17.10 LARGE INTESTINE (PAGE 686)

The large intestine absorbs water and electrolytes and forms and stores feces.

1. Parts of the large intestine
   a. The large intestine consists of the cecum, colon, rectum, and anal canal.
   b. The colon is divided into ascending, transverse, descending, and sigmoid portions.
2. Structure of the large intestinal wall
   a. The large intestinal wall resembles the wall in other parts of the alimentary canal.
   b. The large intestinal wall has a unique layer of longitudinal muscle fibers, arranged in distinct bands, that extend the entire length of the colon.
3. Functions of the large intestine
   a. The large intestine has little or no digestive function, although it secretes mucus.
   b. Mechanical stimulation and parasympathetic impulses control the rate of mucous secretion.
   c. The large intestine absorbs water and electrolytes.
   d. Many bacteria inhabit the large intestine, where they break down some undigestible substances, such as cellulose, and synthesize vitamins K, B_{12}, thiamine, and riboflavin.
4. Movements of the large intestine
   a. Movements are similar to those in the small intestine.
   b. Mass movements occur two to three times each day.
   c. A reflex stimulates defecation.
5. Feces
   a. The large intestine forms and stores feces.
   b. Feces consist of water, undigested material, mucus, and bacteria.
   c. The color of feces is due to bile pigments that have been altered by bacterial actions.

17.11 LIFE-SPAN CHANGES (PAGE 690)

1. Older people sometimes do not chew food thoroughly because thinning enamel makes teeth more sensitive to hot and cold foods, gums recede, and teeth may loosen.
2. Slowing peristalsis in the digestive tract may cause heartburn and constipation.
3. Aging affects nutrient absorption in the small intestine.
4. Accessory organs to digestion also age, but not necessarily in ways that affect health.

CHAPTER ASSESSMENTS

17.1 Introduction

1 Functions of the digestive system include ______. (p. 652)
   a. ingestion
   b. propulsion
   c. mechanical and chemical digestion
   d. absorption
   e. all of the above

2 List the major parts of the alimentary canal, then separately list the accessory organs of the digestive system. (p. 652)

17.2 General Characteristics of the Alimentary Canal

3 Contrast the composition of the four layers in the wall of the alimentary canal. (p. 653)

4 Distinguish between mixing movements and propelling movements. (p. 654)

5 Define *peristalsis.* (p. 654)

6 Explain the relationship between peristalsis and receptive relaxation. (p. 654)

7 Describe the effects of parasympathetic and sympathetic impulses on the alimentary canal. (p. 656)

17.3 Mouth

8 Discuss the functions of the mouth and its parts. (p. 656)

9 Distinguish among lingual, palatine, and pharyngeal tonsils. (p. 657)

10 Compare the primary and secondary teeth. (p. 658)

11 The teeth best adapted for grasping and tearing food are the ____________. (p. 659)
   a. incisors
   b. canines
   c. premolars
   d. molars
   e. bicuspids

12 Describe the structure of a tooth. (p. 659)

13 Explain how a tooth is anchored in its socket. (p. 660)

17.4–17.10 Salivary Glands–Large Intestine

14 Discuss the digestive functions of saliva. (p. 660)

15 Describe the locations of the major salivary glands. (p. 661)

16 Explain how secretions of the salivary glands differ. (p. 661)

17 Name and locate the three major regions of the pharynx. (p. 663)

18 Discuss the mechanism of swallowing. (p. 664)

19 Explain the function of the esophagus. (p. 665)

20 Describe the structure of the stomach. (p. 665)

21 Explain how gastric secretions are regulated. (p. 668)

22 Describe the mechanism the controls the emptying of the stomach. (p. 669)

23 Describe the enterogastric reflex. (p. 670)

24 Explain the mechanism of vomiting. (p. 670)

25 Describe the location of the pancreas and the pancreatic duct. (p. 671)

26 Explain how pancreatic secretions are regulated. (p. 672)

27 Describe the structure of the liver. (p. 673)

28 Trace the path of bile from a bile canaliculus to the small intestine by filling in the blanks: bile canaliculi carry secretions from liver cells to ____________, which unite to form large bile ducts, which converge to form ____________ that merge to form the hepatic duct, which joins the cystic duct to form the ____________, which empties into the small intestine (duodenum). (p. 673)

29 List the major functions of the liver. (p. 673)

30 Describe the composition of bile. (p. 675)

31 Explain how gallstones form. (p. 677)

32 Explain the regulation of bile release. (p. 678)

33 Discuss the functions of bile salts. (p. 678)

34 Describe the locations of the parts of the small intestine. (p. 680)

35 Match the organ or gland with the enzyme(s) it secretes. Enzymes may be used more than once. An organ or gland may secrete more than one enzyme. (pp. 661–683)

(1) salivary glands (serous cells)	A. peptidase
(2) stomach (chief cells)	B. amylase
(3) pancreas (acinar cells)	C. nuclease
(4) small intestine (mucosal cells)	D. lipase
	E. pepsin
	F. trypsin, chymotrypsin, carboxypeptidase
	G. sucrase, maltase, lactase

36 Match the enzyme(s) with its (their) function(s). (pp. 661–683)

(1) peptidase	A. begins protein digestion
(2) amylase	B. breaks fats into fatty acids and glycerol
(3) nuclease	C. breaks down proteins into peptides
(4) lipase	D. breaks down starch and glycogen into disaccharides
(5) pepsin	E. breaks down peptides into amino acids
(6) trypsin, chymotrypsin, carboxypeptidase	F. breaks down nucleic acids into nucleotides
(7) sucrase, maltase, lactase	G. breaks down disaccharides into monosaccharides

37 Explain the regulation of the secretions of the small intestine. (p. 683)

38 Describe the functions of the intestinal villi. (p. 684)

39 Discuss absorption of amino acids, monosaccharides, glycerol, fatty acids, electrolytes, and water from substances in the small intestine. (p. 684)

40 Explain control of the movement of the intestinal contents. (p. 686)

41 Describe the locations of the parts of the large intestine. (p. 687)

42 Explain the general functions of the large intestine. (p. 688)

43 Explain the defecation reflex. (p. 690)

17.11 Life-Span Changes

44 How does digestive function change with age? (p. 690)

45 What are the effects of altered rates of absorption, due to aging, in the small intestine? (p. 690)

INTEGRATIVE ASSESSMENTS/CRITICAL THINKING

OUTCOMES 17.4, 17.7, 17.9, 17.10

1. Several years ago, an extract from kidney beans was sold in health-food stores as a "starch blocker." Advertisements claimed that one could eat a plate of spaghetti, yet absorb none of it, because starch-digesting enzyme function would be blocked. The kidney bean product indeed kept salivary amylase from functioning. However, people who took the starch blocker developed abdominal pain, bloating, and gas. Suggest a reason for these ill effects.

OUTCOMES 17.6, 17.9

2. How would removal of 95% of the stomach (subtotal gastrectomy) to treat severe ulcers or cancer affect digestion and absorption? How would the patients have to alter eating habits? Why? Do you think that people should have this type of surgery to treat life-threatening obesity?

OUTCOMES 17.6, 17.9

3. What effect is a before-dinner alcoholic cocktail likely to have on digestion? Why are such beverages inadvisable for persons with ulcers?

OUTCOMES 17.6, 17.9

4. What type of acid-base imbalance is likely to develop if the stomach contents are repeatedly lost by vomiting over a prolonged period? Which acid-base imbalance may develop as a result of prolonged diarrhea?

OUTCOMES 17.7, 17.8

5. Why might a person with inflammation of the gallbladder (cholecystitis) also develop inflammation of the pancreas (pancreatitis)?

WEB CONNECTIONS

Be sure to visit the text website at **www.mhhe.com/shier12** for answers to chapter assessments, additional quizzes, and interactive learning exercises.

ANATOMY & PHYSIOLOGY REVEALED

Anatomy & Physiology Revealed® (APR) includes cadaver photos that allow you to peel away layers of the human body to reveal structures beneath the surface. This program also includes animations, radiologic imaging, audio pronunciations, and practice quizzing. Check out **www.aprevealed.com**. APR has been proven to help improve student grades!

CHAPTER 18

Nutrition and Metabolism

People can satisfy nutritional requirements in a wide variety of ways.

UNDERSTANDING WORDS

bas-, base: *bas*al metabolic rate—metabolic rate of body under resting (basal) conditions.

calor-, heat: *calor*ie—unit used to measure heat or energy content of foods.

carot-, carrot: *carot*ene—yellowish plant pigment that imparts the color of carrots and other yellowish plant tissues.

lip-, fat: *lip*ids—fat or fatlike substance insoluble in water.

mal-, bad, abnormal: *mal*nutrition—poor nutrition resulting from lack of food or failure to adequately use available foods.

-meter, measure: calori*meter*—instrument used to measure the caloric content of food.

nutri-, nourish: *nutri*ent—substance needed to nourish cells.

obes-, fat: *obes*ity—condition in which the body has excess fat.

pell-, skin: *pell*agra—vitamin deficiency condition characterized by inflammation of the skin and other symptoms.

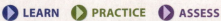

LEARN

PRACTICE ASSESS

LEARNING OUTCOMES

After you have studied this chapter, you should be able to:

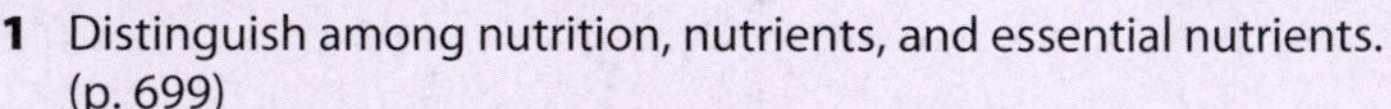

18.1 Introduction

1. Distinguish among nutrition, nutrients, and essential nutrients. (p. 699)
2. Explain appetite control.(p. 699)

18.2–18.4 Carbohydrates–Proteins

3. List the major sources of carbohydrates, lipids, and proteins. (p. 700)
4. Describe how cells use carbohydrates, lipids, and proteins. (p. 701)
5. Identify examples of positive and negative nitrogen balance. (p. 706)

18.5 Energy Expenditures

6. Explain how energy values of foods are determined. (p. 706)
7. Explain the factors that affect an individual's energy requirements. (p. 707)
8. Contrast the physiological impact of positive and negative energy balance. (p. 708)
9. Explain what is meant by desirable weight. (p. 708)

18.6 Vitamins

10. List the fat-soluble and water-soluble vitamins and summarize the general functions of each vitamin. (p. 709)

18.7 Minerals

11. Distinguish between a vitamin and a mineral. (p. 717)
12. List the major minerals and trace elements and summarize the general functions of each. (p. 717)

18.8 Healthy Eating

13. Describe an adequate diet. (p. 722)
14. Distinguish between primary and secondary malnutrition. (p. 723)

18.9 Life-Span Changes

15. List the factors that may lead to inadequate nutrition later in life. (p. 728)

LANCE ARMSTRONG: SECRETS OF HIS SUCCESS

For an elite athlete such as seven-time Tour de France winner Lance Armstrong, carbohydrates are essential for powering muscles. His diet of 70% carbohydrates and 15% each of protein and fat equals 6,000 to 7,000 calories per day and rises to 9,000 calories on the most grueling days of competition.

Diet is crucial to the success of endurance athletes like Armstrong. The goals are simple: fuel and recovery. At most levels of effort, fuel comes from both carbohydrates and fat, but as the effort intensifies, most energy comes from carbohydrates. The body can store only about 1,800 calories worth of carbohydrates, which an endurance athlete's body will burn through in a few hours, so the athlete needs to eat often to fuel the muscles and to maintain the supply of glucose to the brain. Protein is required for recovery, to repair microscopic muscle tears.

Coaches have the diet of Tour de France participants down to a precise science. The athletes get their food in three meals and many snacks. Meals include rice, pasta, cereal, potatoes, fruits, vegetables, and whole grains for carbohydrates, and protein from lean meats, eggs, and yogurt. Fats come from cheese, butter, and olive oil. While biking, the riders carry bags, called "musettes," that contain power bars, meat sandwiches, and potatoes, which they eat whole for a rapid infusion of carbohydrates. Adding drinks that replace electrolytes and provide sugar brings the intake to 300 to 400 calories per hour on the bike. A rider nibbles or sips about every fifteen minutes. After riding, bikers get new musettes that provide drinks that are one-fifth protein and four-fifths carbohydrate to aid in recovery, because the muscles soak up the carbohydrates in the hour after riding. In the bus, they continue their recovery by drinking fruit smoothies.

The 7,000-calorie diet plan works for Lance Armstrong, but he is not average. In fact, University of Texas at Austin researcher Ed Coyle probed Armstrong's physiology from ages 21 to 28 for clues to his prowess—even in the face of testicular cancer during that period. Armstrong has an exceptionally strong heart and extensive vascular system, but he also works hard. For example, he reduced his weight by 10 pounds before each Tour de France victory, which increased power per kilogram of body weight by 18%. Armstrong uses a machine in the Human Performance Laboratory at the university to measure his muscle power at given oxygen intakes. His training and diet enable him to ride twice as fast as the average cyclist. Armstrong's athletic abilities are eclectic: he ran the New York City and Boston marathons in under three hours. ■

18.1 INTRODUCTION

The human body requires fuel as well as materials to develop, grow, and heal. Nutrients from food fulfill these requirements. **Nutrients** (nu′tre-ents) are chemicals supplied from the environment that an organism requires for survival. There are two major classes of nutrients. The **macronutrients,** needed in bulk, are the carbohydrates, proteins, and fats. **Micronutrients** are essential in small daily doses and include vitamins and minerals. The body also requires water.

In countries with adequate food supplies, most healthy individuals can obtain nourishment by eating a variety of foods and limiting fat intake. People who do not eat meat products can also receive adequate nutrition but must pay more attention to food choices to avoid developing nutrient deficiencies. For example, eliminating red meat also means eliminating an excellent source of iron, copper, zinc, and vitamin B_{12}. The fiber that often makes up much of a vegetarian's diet, although very healthful in many ways, also decreases absorption of iron. Therefore, a vegetarian must obtain sufficient iron from nonmeat sources. This is easily done, providing proper **nutrition** (adequate nutrients) when sources, actions, and interactions of nutrients are considered. Fortified foods, green leafy vegetables, and especially whole grains provide many of the nutrients also in meat. Table 18.1 lists types of vegetarian diets.

TABLE 18.1 | Types of Vegetarian Diets

Type	Food Restrictions
Vegan	No animal foods
Ovo-vegetarian	Eggs allowed; no dairy or meat
Lacto-vegetarian	Dairy allowed; no eggs or meat
Lacto-ovo-vegetarian	Dairy and eggs allowed; no meat
Pesco-vegetarian	Dairy, eggs, and fish allowed; no other meat
Semivegetarian	Dairy, eggs, chicken, and fish allowed; no other meat

Digestion breaks down nutrients to sizes that can be absorbed and transported in the bloodstream. **Metabolism** refers to the ways that nutrients are altered chemically and used in anabolism (building up or synthesis) and catabolism (breaking down) of chemical compounds to support the activities of life. (Chapter 4, pp. 120–124, introduced metabolism of carbohydrates.) Nutrients that human cells cannot synthesize, such as certain amino acids, are particularly important and are therefore called **essential nutrients.**

We eat to obtain the nutrients that power the activities of life. Eating is a complex, finely tuned homeostatic mechanism that balances nutrient intake with nutrient use. Too few nutrients, and disorders associated with malnutrition result. Too many nutrients, and obesity is the consequence. Several factors influence food intake, including smell, taste, and texture of food; neural signals triggered by stretch receptors in the stomach; stress; and hormones. Several types of interacting hormones control appetite by affecting part of the hypothalamus called the *arcuate nucleus* (table 18.2).

Insulin, secreted from the pancreas, regulates fat stores by stimulating adipocytes to take up glucose and store fat, and by stimulating certain other cells to take up glucose and form

TABLE 18.2 | Substances That Control Appetite

Substance	Site of Secretion	Function
Insulin	Pancreas	Stimulates adipocytes to admit glucose and store fat; glycogen synthesis
Leptin	Adipocytes	Suppresses appetite and increases metabolic rate after eating
Neuropeptide Y	Hypothalamus	Enhances appetite
Ghrelin	Stomach	Enhances appetite

glycogen, a storage carbohydrate. Eating stimulates adipocytes to secrete **leptin,** which acts on target cells in the hypothalamus. The effect suppresses appetite by inhibiting release of the hypothalamic neurotransmitter **neuropeptide Y** that stimulates eating. This produces a negative feedback response to ingested calories. Leptin also stimulates metabolic rate (fig. 18.1). Low leptin levels indicate depleted fat stores, a condition in which metabolism slows to conserve energy and appetite increases. Inherited leptin deficiency is rare, but the resultant loss of this appetite "brake" results in obesity.

Conversely, **ghrelin,** another recently discovered hormone that the stomach secretes, enhances appetite by stimulating the release of neuropeptide Y. Therefore, a compound that blocks ghrelin production or activity might help people lose weight. Fasting and a low-calorie diet stimulate ghrelin production. The success of gastric bypass surgery may be due in part to decreased ghrelin secretion resulting from loss of stomach tissue.

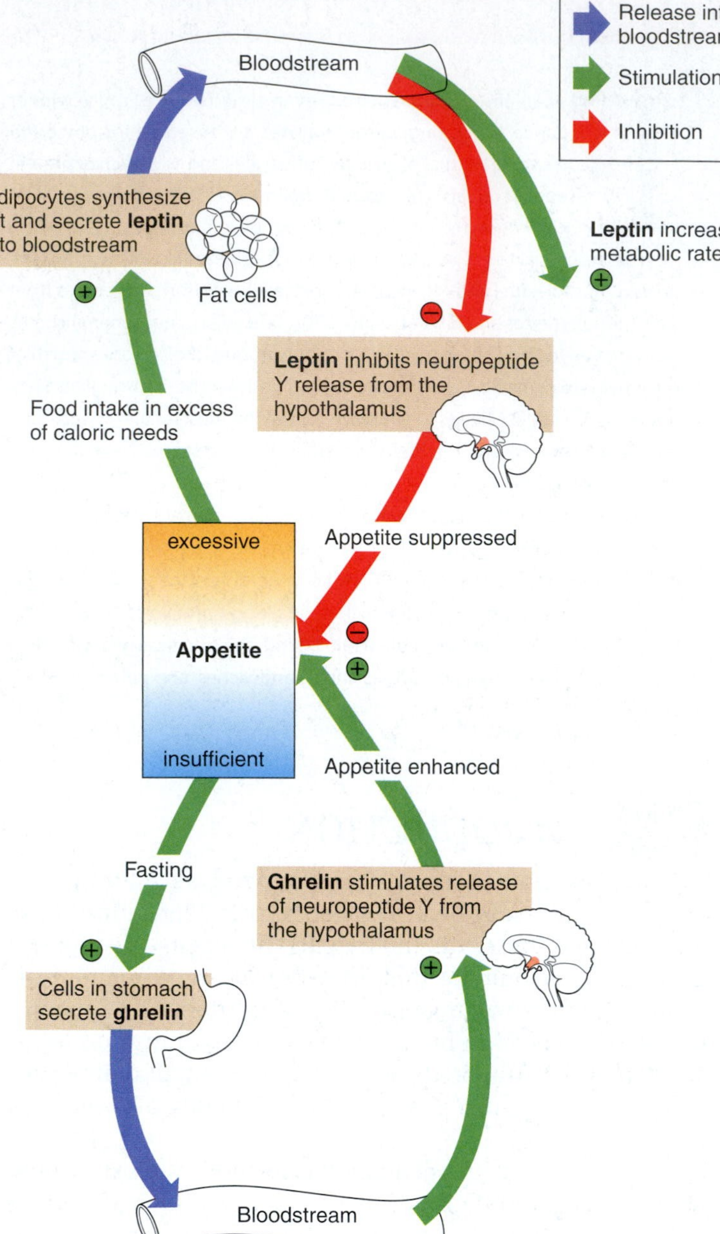

FIGURE 18.1 Appetite control is complex. Illustrated here are the effects of leptin and ghrelin on appetite.

Adiponectin is another hormone that affects body weight. It is a protein hormone synthesized in adipose cells and secreted into the bloodstream that mediates response to insulin and regulates fatty acid catabolism. Adiponectin also has an anti-inflammatory effect. The hormone boosts basal metabolic rate (calories burned at rest) without affecting appetite. Levels of adiponectin are higher in people who do not have diabetes and have healthy body weights, so boosting its levels may one day help people to lose weight. Most information about adiponectin, however, is still from work on nonhuman animals.

PRACTICE

1 Identify and distinguish among macronutrients and micronutrients.

2 Describe how hormones control appetite.

18.2 CARBOHYDRATES

Carbohydrates are organic compounds and include the sugars and starches. The energy held in their chemical bonds is used to power cellular processes.

Carbohydrate Sources

Carbohydrates are ingested in a variety of forms. Complex carbohydrates include the *polysaccharides,* such as starch from plant foods and glycogen from meats. Foods containing starch and glycogen usually have many other nutrients, including valuable vitamins and minerals. The simple carbohydrates include *disaccharides* from milk sugar, cane sugar, beet sugar, and molasses and *monosaccharides* from honey and fruits. Digestion breaks complex carbohydrates down to monosaccharides, which are small enough to be absorbed into the bloodstream.

Sugar substitutes provide concentrated sweetness, so fewer calories are needed to sweeten a food. Aspartame, a dipeptide (two joined amino acids), is 200 times as sweet as table sugar (sucrose); saccharin is 300 times as sweet as sugar; and sucralose (Splenda) is 600 times as sweet as sucrose. Sucralose is derived from sucrose, with chloride attached. The body does not recognize it as a carbohydrate, and so it has zero calories.

Cellulose is a complex carbohydrate abundant in our food—it provides the crunch to celery and the crispness to lettuce. We cannot digest cellulose, and most of it passes through the alimentary canal largely unchanged. However, cellulose provides bulk (also called fiber or roughage) against which the muscular wall of the digestive system can push, facilitating the movement of food. *Hemicellulose, pectin,* and *lignin* are other plant carbohydrates that provide fiber.

Carbohydrate Use

The monosaccharides absorbed from the digestive tract include *fructose, galactose,* and *glucose.* Liver enzymes catalyze reactions that convert fructose and galactose into glucose (fig. 18.2). Recall that glucose is the carbohydrate most commonly oxidized in glycolysis for cellular fuel.

RECONNECT

To Chapter 4, Cellular Respiration, pages 120–124.

Some excess glucose is polymerized to form *glycogen* (glycogenesis), which the liver and muscles store as a glucose reserve. Glycogen can be rapidly broken down to yield glucose (glycogenolysis) when it is required to supply energy. However, only a certain amount of glycogen can be stored. Excess glucose beyond what is stored as glycogen usually reacts to form fat and is stored in adipose tissue (fig. 18.3).

Many cells can also oxidize fatty acids to obtain energy. However, some cells, such as neurons, normally require continuous glucose for survival. (Under some conditions, such as prolonged starvation, other fuel sources may become available for neurons.) Even a temporary decrease in the glucose supply may seriously impair nervous system function. Consequently, the body requires a minimum amount of carbohydrate. If adequate carbohydrates are not obtained from foods, the liver may convert some noncarbohydrates, such as amino acids from proteins or glycerol from fats, into glucose—a process called *gluconeogenesis.* The requirement for glucose has physiological priority over the need to synthesize certain other substances, such as proteins, from available amino acids.

Cells also use carbohydrates as starting materials for synthesizing such vital biochemicals as the 5-carbon sugars *ribose* and *deoxyribose.* These sugars are required for the production of the nucleic acids RNA and DNA and the disaccharide *lactose* (milk sugar), synthesized when the breasts are actively secreting milk.

PRACTICE

3 List several common sources of carbohydrates.
4 In what form are carbohydrates used as a cellular fuel?
5 Explain what happens to excess glucose in the body.
6 How does the body obtain glucose when its food supply of carbohydrates is insufficient?
7 Name two uses of carbohydrates other than supplying energy.

An adult's liver stores about 100 grams of glycogen, and muscle tissue stores another 200 grams, providing enough reserve to meet energy demands for about twelve hours when the person is resting. Whether these stores are filled depends on diet. People consume widely varying amounts of carbohydrates, often reflecting economic conditions. In the United States, a typical adult's diet supplies about 50% of total body energy from carbohydrates. In Asian countries where rice is a staple, carbohydrates contribute even more to the diet.

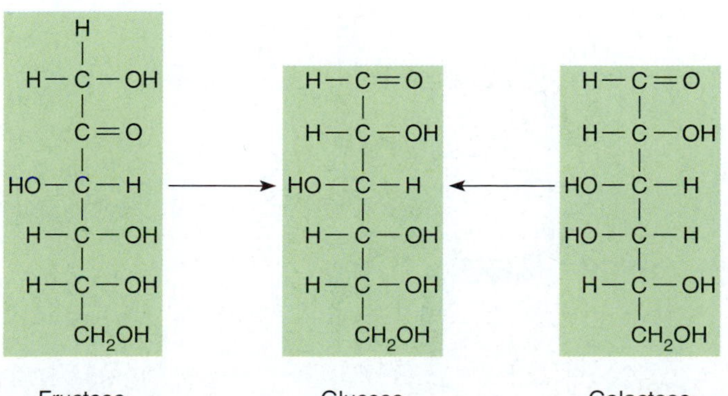

FIGURE 18.2 Liver enzymes catalyze reactions that convert the monosaccharides fructose and galactose into glucose.

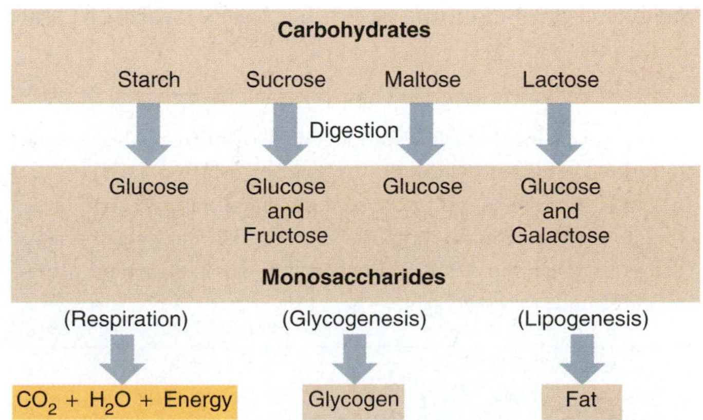

FIGURE 18.3 Monosaccharides from foods are used for energy, stored as glycogen, or reacted to produce fat.

Carbohydrate Requirements

Carbohydrates provide the primary source of fuel for cellular processes, so the need for carbohydrates varies with individual energy requirements. Physically active individuals require more carbohydrates than those who are sedentary. The minimal requirement for carbohydrates in the human diet is unknown. It is estimated, however, that an intake of at least 125 to 175 grams daily is necessary to spare protein (to avoid protein breakdown) and to avoid metabolic disorders resulting from excess fat use. An average diet includes 200 to 300 grams of carbohydrates daily.

PRACTICE 

8 Why do daily requirements for carbohydrates vary from person to person?

9 What is the daily minimum requirement for carbohydrates?

18.3 LIPIDS

Lipids are organic compounds that include fats, oils, and fat-like substances such as phospholipids and cholesterol (see chapter 2, pp. 62–64). They supply energy for cellular processes and help build structures, such as cell membranes. The most common dietary lipids are the fats called *triglycerides* (tri-glis′er-īdz) (see fig. 2.14).

Lipid Sources

Triglycerides are found in plant- and animal-based foods. Saturated fats (which should comprise no more than 10% of the diet) are mainly found in foods of animal origin, such as meat, eggs, milk, and lard, as well as in palm and coconut oil. Unsaturated fats are in seeds, nuts, and plant oils. Monounsaturated fats, such as those in olive, peanut, and canola oils, are the healthiest. Saturated fats in excess are a risk factor for cardiovascular disease.

Cholesterol is abundant in liver and egg yolk and, to a lesser extent, in whole milk, butter, cheese, and meats. Foods of plant origin do not contain cholesterol. A label on a plant-based food claiming that it is "cholesterol-free" states the obvious.

> Be wary of claims that a food product is "99% fat-free." This usually refers to percentage by weight—not calories, which is what counts. A 99% fat-free creamy concoction may be largely air and water, and therefore in that form, fat comprises very little of it. But when the air is compressed and the water absorbed, as happens in the stomach, the fat percentage may skyrocket.

Lipid Use

The lipids in foods are phospholipids, cholesterol, and, most commonly, fats (triglycerides). A triglyceride consists of one glycerol and three fatty acids.

Lipids provide a variety of physiological functions; however, fats mainly supply energy. Gram for gram, fats contain more than twice as much chemical energy as carbohydrates or proteins.

Before a triglyceride molecule can release energy, it must undergo hydrolysis. This happens when digestion breaks triglycerides down into fatty acids and glycerol. After being absorbed, these products are transported in the lymph to the blood, then on to tissues. As figure 18.4 shows, some of the resulting fatty acid portions can then form molecules of acetyl coenzyme A (acetyl CoA) in a series of reactions called **beta oxidation,** which occurs in the mitochondria.

In the first phase of beta oxidation, fatty acids are activated. This change requires energy from ATP and a special group of enzymes called thiokinases. Each of these enzymes can act upon a fatty acid that has a particular carbon chain length.

Once fatty acid molecules have been activated, other enzymes called **fatty acid oxidases** in mitochondria break them down. This phase of the reactions removes successive two-carbon segments of fatty acid chains. In the liver, some of these segments react to produce acetyl coenzyme A

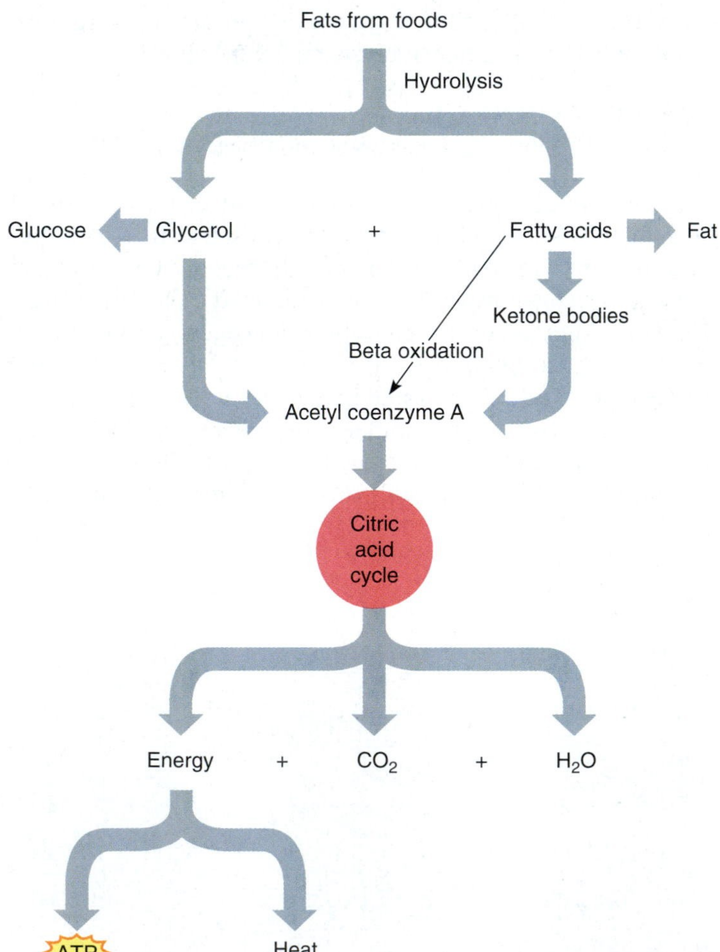

FIGURE 18.4 The body digests fat from foods into glycerol and fatty acids, which may enter catabolic pathways and provide energy.

molecules. Excess acetyl CoA molecules react to form compounds called **ketone bodies,** such as acetone, which later may react to form acetyl coenzyme A once again. In either case, the citric acid cycle can oxidize the acetyl coenzyme A molecules. The glycerol parts of the triglyceride molecules can also enter metabolic pathways leading to the citric acid cycle, or they can be used to synthesize glucose.

> When ketone bodies form faster than they can be decomposed, some of them are eliminated through the lungs and kidneys. When this happens, the ketone acetone may impart a fruity odor to the breath and urine. This can happen when a person fasts, forcing body cells to metabolize a large amount of fat, and in persons suffering from diabetes mellitus who develop a serious imbalance in pH called acidosis, which results when acetone and other acidic ketones accumulate.

Glycerol and fatty acid molecules resulting from the hydrolysis of fats can also combine to form fat molecules in anabolic reactions and be stored in fat tissue. Additional fat molecules can be synthesized from excess glucose or amino acids.

The liver can convert fatty acids from one form to another. However, the liver cannot synthesize certain fatty acids, called **essential fatty acids.** *Linoleic acid,* for example, is an essential fatty acid required to synthesize phospholipids, which, in turn, are necessary for constructing cell membranes and myelin sheaths, and for transporting circulating lipids. Good sources of linoleic acid include corn oil, cottonseed oil, and soy oil. *Linolenic acid* is also an essential fatty acid.

The liver uses free fatty acids to synthesize triglycerides, phospholipids, and lipoproteins that may then be released into the blood (fig. 18.5). These lipoproteins are large and consist of a surface layer of phospholipid, cholesterol, and protein surrounding a triglyceride core. The protein constituents of lipoproteins in the outer layer, called *apoproteins* or apolipoproteins, can combine with receptors on the membranes of specific target cells. Lipoprotein molecules vary in the proportions of the lipids they contain.

Lipids are less dense than proteins. As the proportion of lipids in a lipoprotein increases, the density of the particle decreases. Conversely, as the proportion of lipids decreases, the density increases. Lipoproteins are classified on the basis of their densities, which reflect their composition. *Very-low-density lipoproteins* (VLDL) have a high concentration of triglycerides. *Low-density lipoproteins* (LDL) have a high concentration of cholesterol and are the major cholesterol-carrying lipoproteins. *High-density lipoproteins* (HDL) have a relatively high concentration of protein and a lower concentration of lipids.

In addition to regulating circulating lipids, the liver controls the total amount of cholesterol in the body by synthesizing and releasing it into the blood or by removing cholesterol from the blood and excreting it into the bile. The liver uses cholesterol to produce bile salts. Cholesterol is not used as an energy source, but it does provide structural material for cell and organelle membranes, and it furnishes starting materials for the synthesis of certain sex hormones and hormones produced by the adrenal cortex.

Adipose tissue stores excess triglycerides. If the blood lipid concentration drops (in response to fasting, for example), some of these stored triglycerides are hydrolyzed into free fatty acids and glycerol and released into the bloodstream.

Lipid Requirements

The lipid content of human diets varies widely. A person who eats mostly burgers, fries, and shakes may consume 50% or more of total daily calories from fat. For a vegetarian, the percentage may be far lower. The American Heart Association advises that the diet not exceed 30% of total daily calories from fat.

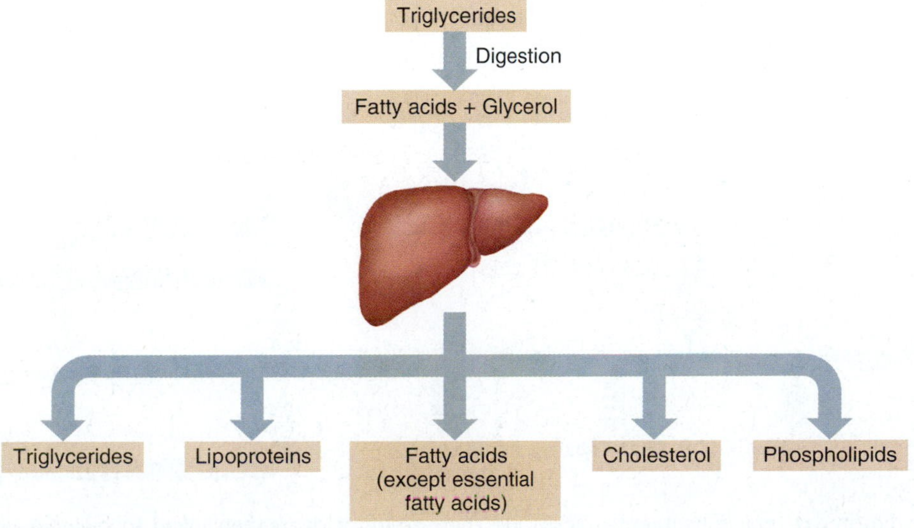

FIGURE 18.5 The liver uses fatty acids to synthesize a variety of lipids.

The types and locations of the chemical bonds between carbon atoms of fatty acid molecules affect how healthful the fat is. For example, monounsaturated fats (such as from avocado and olives), promote cardiovascular health, whereas saturated fats, such as those in butter or lard, contribute to heart disease. The site of the double bond that contributes to a fat's degree of unsaturation is also important. Omega-3 fatty acids, which have double bonds between the third and fourth carbons, are more healthful than omega-6 fatty acids, with double bonds between the sixth and seventh carbons. Omega-3 fatty acids are found in fish; omega-6 fatty acids are in red meat.

The amounts and types of fats required for health vary with individuals' habits and goals. Linoleic acid is an essential fatty acid, so nutritionists recommend that infants fed formula receive 3% of the energy intake in the form of linoleic acid to prevent deficiency conditions. Fat intake must be sufficient to supply and carry fat-soluble vitamins. A typical adult diet consisting of a variety of foods usually provides adequate fats.

PRACTICE 

10 Which foods commonly supply lipids?

11 Which fatty acids are essential nutrients?

12 What is the role of the liver in the use of lipids?

13 What are the functions of cholesterol?

18.4 PROTEINS

Proteins are polymers of amino acids. They have a wide variety of functions. When dietary proteins are digested, the resulting amino acids are absorbed and transported by the blood to cells. Many of these amino acids are used to form new protein molecules, as specified by DNA base sequences. These new proteins include enzymes that control the rates of metabolic reactions; clotting factors; the keratins of skin and hair; elastin and collagen of connective tissue; plasma proteins that regulate water balance; the muscle components actin and myosin; certain hormones; and the antibodies that protect against infection (fig. 18.6).

Protein molecules may also supply energy. To do this, they must first be broken down into amino acids. The amino acids then undergo **deamination,** a process in the liver that removes the nitrogen-containing (—NH_2) groups from the amino acids (see fig. 2.17). These —NH_2 groups subsequently react to form a waste called **urea** (u-re′ah). The liver therefore produces urea from amino groups released by deamination of amino acids. The blood carries urea to the kidneys, where it is excreted in urine.

Certain kidney disorders impair the removal of urea from the blood, raising the blood urea concentration. A blood test called blood urea nitrogen (BUN) determines the blood urea concentration and is often used to evaluate kidney function.

Several pathways decompose the remaining deaminated parts of amino acids. The specific pathways that are followed depend upon the particular type of amino acid being dismantled. Some pathways form acetyl coenzyme A, and others more directly lead to steps of the citric acid cycle. As energy is released from the cycle, some of it is captured in molecules of ATP (fig. 18.7). If energy is not required immediately, the deaminated parts of the amino acids may react to form glucose or fat molecules in other metabolic pathways (see fig. 18.6).

A few hours after a meal, protein catabolism, through the process of gluconeogenesis (see chapter 13, p. 509), becomes a major source of blood glucose. However, metabolism in most tissues soon shifts away from glucose and toward fat catabolism as a source of ATP. Thus, energy needs are met in a way that spares proteins for tissue building and repair, rather than being broken down and reassembled into carbohydrates to supply energy. Using structural proteins to generate energy causes the tissue-wasting of starvation.

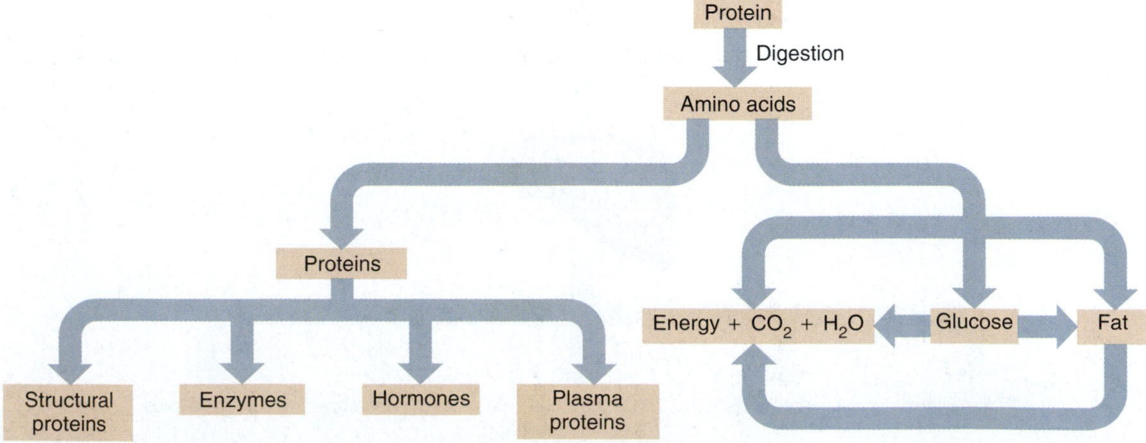

FIGURE 18.6 Proteins are digested to their constituent amino acids. These amino acids are then linked, following genetic instructions, to build new proteins. Free amino acids are also used to supply energy under certain conditions.

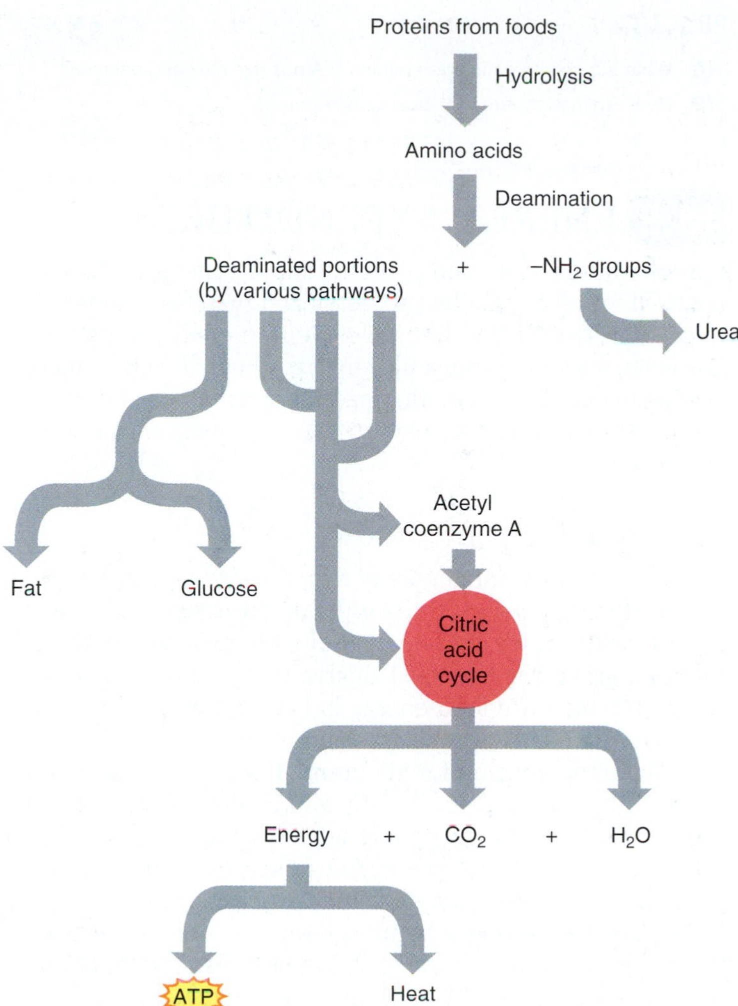

FIGURE 18.7 The body digests proteins from foods into amino acids, but must deaminate these smaller molecules before they can be used as energy sources.

Protein Sources

Foods rich in proteins include meats, fish, poultry, cheese, nuts, milk, eggs, and cereals. Legumes, including beans and peas, contain less protein.

The human body can synthesize many amino acids (nonessential amino acids). However, eight amino acids the adult body needs (ten required for growing children) cannot be synthesized sufficiently or at all, and they are called **essential amino acids.** This term refers only to dietary intake, because all amino acids are required for normal protein synthesis. Table 18.3 lists the amino acids in foods and indicates those that are essential.

All twenty types of amino acids must be in the body at the same time for growth and tissue repair to occur. In other words, if the diet lacks one essential amino acid, the cells cannot synthesize protein. Essential amino acids are not stored, so those not used in protein synthesis are oxidized as energy sources or are converted into carbohydrates or fats.

TABLE 18.3 | Amino Acids in Foods

Alanine	Glycine	Proline
Arginine (ch)	Histidine (ch)	Serine
Asparagine	Isoleucine (e)	Threonine (e)
Aspartic acid	Leucine (e)	Tryptophan (e)
Cysteine	Lysine (e)	Tyrosine
Glutamic acid	Methionine (e)	Valine (e)
Glutamine	Phenylalanine (e)	

Eight essential amino acids (e) cannot be synthesized by human cells and must be provided in the diet. Two additional amino acids (ch) are essential in growing children.

Proteins are classified as complete or incomplete based on the amino acid types they provide. **Complete proteins** have adequate amounts of the essential amino acids to maintain human body tissues and promote normal growth and development. Certain proteins in milk, meat and eggs are complete. Incomplete proteins cannot by themselves maintain human tissues or support normal growth and development. Zein in corn, for example, has too little of the essential amino acids tryptophan and lysine to be complete. A *partially complete protein* does not have enough amino acid variety to promote growth, but it has enough to maintain life. A protein in wheat called gliadin is partially complete because it has very little of the amino acid lysine.

Many proteins in plants do not provide enough of one or more essential amino acids for the human diet. However, combining appropriate plant foods can supply an adequate diversity of dietary amino acids. For example, beans are low in methionine but have enough lysine. Rice lacks lysine but has enough methionine. A meal of beans and rice offers enough of both types of amino acids.

Plants can be genetically modified to make their protein more "complete." For example, genetic instructions for producing the amino acid tryptophan inserted into corn cells can compensate for the low levels of this nutrient normally found in corn.

PRACTICE 

14 How do cells use proteins?
15 Which foods are rich sources of protein?
16 Why are some amino acids called essential?
17 Distinguish between a complete protein and an incomplete protein.

Nitrogen Balance

In a healthy adult, proteins are continuously built up and broken down. This occurs at different rates in different tissues, but the overall gain of body proteins equals the loss, producing a state of **dynamic equilibrium** (di-nam′ik e″kwĭ-lib′re-um). Proteins have a high percentage of nitrogen, so dynamic equilibrium also brings **nitrogen balance** (ni′tro-jen bal'ans)—a condition in which the amount of nitrogen taken in equals the amount excreted.

A person who is starving has a *negative nitrogen balance* because the amount of nitrogen excreted as a result of amino acid oxidation exceeds the amount the diet replaces. Conversely, a growing child, a pregnant woman, or an athlete in training is likely to have a *positive nitrogen balance* because more protein is being built into new tissue and less is being used for energy or excreted.

Protein Requirements

In addition to supplying essential amino acids, proteins provide nitrogen and other elements for the synthesis of nonessential amino acids and certain nonprotein nitrogenous substances. The amount of dietary protein individuals require varies according to body size, metabolic rate, and nitrogen balance condition.

For an average adult, nutritionists recommend a daily protein intake of about 0.8 gram per kilogram (0.4 gram per pound) of body weight or 10% of a person's diet. Another way to estimate desirable protein intake is to divide weight in pounds by 2. Most people should consume 60–150 grams of protein a day. For a pregnant woman, who needs to maintain a positive nitrogen balance, the recommendation adds 30 grams of protein per day. Similarly, a nursing mother requires an additional 20 grams of protein per day to maintain a high level of milk production.

Protein deficiency causes tissue wasting and also decreases the level of plasma proteins, which decreases the colloid osmotic pressure of the plasma. As a result, fluids collect in the tissues, producing a condition called *nutritional edema*. Table 18.4 summarizes the sources, requirements, and uses for carbohydrate, lipid, and protein nutrients.

RECONNECT

To Chapter 15, Exchanges in the Capillaries, pages 578–579.

PRACTICE 

18 What is a negative nitrogen balance? A positive nitrogen balance?

19 How can inadequate nutrition cause edema?

18.5 ENERGY EXPENDITURES

Carbohydrates, fats, and proteins supply energy, which is required for all metabolic processes and therefore important to cell survival. If the diet is deficient in energy-supplying nutrients, structural molecules may gradually be consumed, leading to death. On the other hand, excess intake of energy-supplying nutrients may lead to obesity, which also threatens health.

Energy Values of Foods

The amount of potential energy a food contains is expressed as **calories** (kal'o-rēz), units of heat. Although a calorie is defined as the amount of heat required to raise the temperature of a gram of water by 1 degree Celsius (°C), the calorie used to measure food energy is 1,000 times greater. This *large calorie* (Cal.) equals the amount of heat required to raise the temperature of a kilogram (1,000 grams) of water by 1°C (from 14.5°C to 15.5°C) and is also equal to 4.184 joules. A joule is the international unit of heat and energy. A large calorie is sometimes called a *kilocalorie*, but it is customary in nutritional studies to refer to it as a calorie.

Figure 18.8 shows a bomb calorimeter, used to measure the caloric contents of foods. It consists of a metal chamber submerged in a known volume of water. A food sample is dried, weighed, and placed in a nonreactive dish inside the chamber. The chamber is filled with oxygen and submerged in the water. Then, the food is ignited and allowed to completely oxidize. Heat released from the food raises the

TABLE 18.4 | Carbohydrate, Lipid, and Protein Nutrients

Nutrient	Sources and RDA* for Adults	Calories per Gram	Use	Conditions Associated with	
				Excesses	Deficiencies
Carbohydrate	Primarily from starch and sugars in foods of plant origin and from glycogen in meats 125–175 g	4.1	Oxidized for energy; used in production of ribose, deoxyribose, and lactose; stored in liver and muscles as glycogen; converted to fats and stored in adipose tissue	Obesity, dental caries, nutritional deficits	Metabolic acidosis
Lipid	Meats, eggs, milk, lard, plant oils 80–100 g	9.5	Oxidized for energy; production of triglycerides, phospholipids, lipoproteins, and cholesterol, stored in adipose tissue; glycerol portions of fat molecules may be used to synthesize glucose	Obesity, increased serum cholesterol, increased risk of heart disease	Weight loss, skin lesions
Protein	Meats, cheese, nuts, milk, eggs, cereals, legumes 0.8 g/kg body weight	4.1	Production of protein molecules used to build cell structure and to function as enzymes or hormones; used in the transport of oxygen, regulation of water balance, control of pH, formation of antibodies; amino acids may be broken down and oxidized for energy or converted to carbohydrates or fats for storage	Obesity	Extreme weight loss, wasting, anemia, growth retardation

*RDA = recommended dietary allowance.

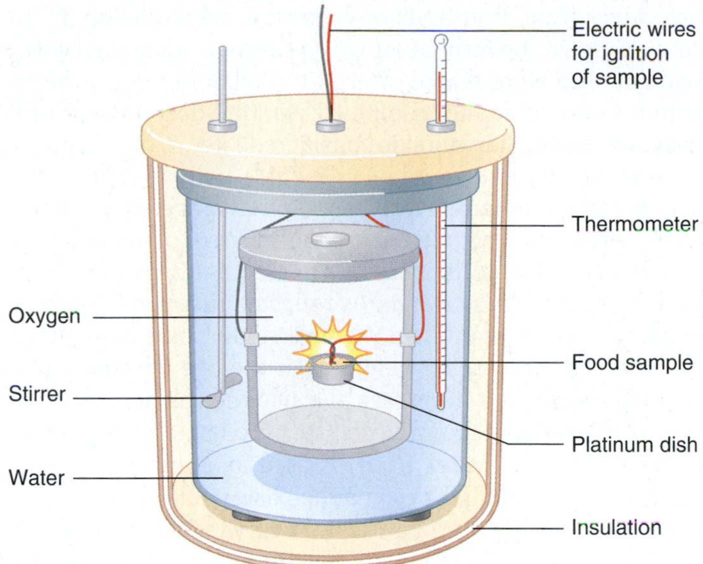

FIGURE 18.8 A bomb calorimeter measures the caloric content of a food sample.

temperature of the surrounding water, and the change in temperature is measured. The volume of the water is known, so the amount of heat released from the food can be calculated in calories.

Caloric values determined in a bomb calorimeter are somewhat higher than the amount of energy that metabolic oxidation releases, because nutrients generally are not completely absorbed from the digestive tract. Also, the body does not completely oxidize amino acids, but excretes parts of them in urea or uses them to synthesize other nitrogenous substances. When such losses are considered, cellular oxidation yields on the average about 4.1 calories from 1 gram of carbohydrate, 4.1 calories from 1 gram of protein, and 9.5 calories from 1 gram of fat. More than twice as much energy is derived from equal amounts by weight of fats as from either proteins or carbohydrates. This is one reason why avoiding fatty foods helps weight loss, if intake of other nutrients does not substantially increase. Fats encourage weight gain because they add flavor to food, which can cause overeating. However, fatty foods satisfy hunger longer than carbohydrate-rich foods.

PRACTICE

20 What term designates the potential energy in a food?

21 How is the energy value of a food determined?

22 What is the energy value of a gram of carbohydrate? A gram of protein? A gram of fat?

Energy Requirements

The amount of energy required to support metabolic activities for twenty-four hours varies from person to person. The factors that influence individual energy needs include a measurement called the basal metabolic rate, the degree of muscular activity, body temperature, and rate of growth.

The **basal metabolic rate** (ba′sal met″ah-bol′ik rāt), or BMR, measures the rate at which the body expends energy under *basal conditions*—when a person is awake and at rest; after an overnight fast; and in a comfortable, controlled environment. Tests of thyroid function can be used to estimate a person's BMR.

The amount of oxygen the body consumes is directly proportional to the amount of energy released by cellular respiration. The BMR indicates the total amount of energy expended in a given time to support the activities of such organs as the brain, heart, lungs, liver, and kidneys.

The BMR for an average adult indicates a requirement for approximately 1 calorie of energy per hour for each kilogram of body weight. However, this requirement varies with sex, body size, body temperature, and level of endocrine gland activity. For example, because heat loss is directly proportional to the body surface area, and a smaller person has a greater surface area relative to body mass, he or she will have a higher BMR. Males typically have higher metabolic rates than females. As body temperature, blood level of thyroxine or blood level of epinephrine increase, so does the BMR. The BMR can also increase when the level of physical activity increases during the day.

Maintaining the BMR usually requires the body's greatest expenditure of energy. The energy required to support voluntary muscular activity comes next, though this amount varies greatly with the type of activity (table 18.5). For example, the energy to maintain posture while sitting at a desk might require 100 calories per hour above the basal need, whereas running or swimming might require 500–600 calories per hour.

Maintenance of body temperature may require additional energy expenditure, particularly in cold weather. In this case, extra energy is expended in involuntary muscular contractions, such as shivering, or through voluntary muscular actions, such as walking. Growing children and pregnant women, because their bodies are actively producing new tissues, also require more calories.

TABLE 18.5 | **Calories Used During Various Activities**

Activity	Calories (per Hour)
Walking up stairs	1,100
Running (jogging)	570
Swimming	500
Vigorous exercise	450
Slow walking	200
Dressing and undressing	118
Sitting at rest	100

Energy Balance

A state of **energy balance** exists when caloric intake in the form of foods equals caloric output from the basal metabolic rate and muscular activities. Under these conditions, body weight remains constant, except perhaps for slight variations due to changes in water content.

If, however, caloric intake exceeds output, a *positive energy balance* occurs, and tissues store excess nutrients. This increases body weight because 3,500 excess calories is stored as a pound of fat. Conversely, if caloric output exceeds input, the energy balance is negative, and stored materials are mobilized from the tissues for oxidation, causing weight loss. All diet plans, no matter which foods they stress, boil down to this fact: to maintain weight, calories in must equal calories out.

PRACTICE

23 What is basal metabolic rate?

24 What factors influence the BMR?

25 What is energy balance?

Desirable Weight

The most obvious and common nutritional disorders reflect calorie imbalances, which may result from societal and geographic factors. Obesity is prevalent in nations where food is plentiful and diverse. The tendency to become obese may be a holdover from thousands of years ago, when the ability to store energy in the form of fat was a survival advantage when food supplies were scarce or erratic. Today in many African nations, natural famines combined with political unrest cause mass starvation. Starvation is considered later in the chapter.

It is difficult to determine a desirable body weight. In the past, weight standards were based on average weights and heights in a certain population, and the degrees of underweight and overweight were expressed as percentage deviations from these averages. These standards reflected the gradual gain in weight as people age. Then medical researchers recognized that such an increase in weight after the age of twenty-five to thirty years is not necessary and may not be healthy. This lead to standards of *desirable weights.* Today a measurement termed **body mass index** (BMI) is used to assess weight considering height, and has become the basis of classifying a person as underweight, normal weight, overweight, or obese.

Overweight is defined as exceeding desirable weight by 10% to 20%, or a BMI between 25 and 30. A person more than 20% above the desired weight or with a BMI over 30, is *obese,* although **obesity** (o-bēs′ĭ-te) is more correctly defined as excess adipose tissue. Therefore, overweight and obesity are not the same. For example, as figure 18.9 shows, an athlete or a person whose work requires heavy muscular activity may be overweight, but not obese. Clinical Application 18.1 discusses obesity.

When a person needs to gain weight, diet can be altered to include more calories and to emphasize particular macronutrients. For example, a person recovering

(a)

(b)

FIGURE 18.9 Weight. (*a*) An obese person is overweight and has excess adipose tissue. (*b*) An athlete may be overweight due to muscle overgrowth but is not considered obese. Many athletes have very low percentages of body fat.

from a debilitating illness might consume more carbohydrates, whereas a bodybuilder might eat extra protein to hasten muscle development. An infant also needs to gain weight rapidly, best accomplished by drinking human milk, which has more total carbohydrate than prepared formulas. The high fat content of human milk is important for the rapid growth of the infant's brain, where many neurons are ensheathed in lipids.

PRACTICE 

26 What is desirable weight?

27 Distinguish between being overweight and obese.

28 Under what conditions is weight gain desirable?

18.6 VITAMINS

Vitamins (vi'tah-minz) are organic compounds (other than carbohydrates, lipids, and proteins) required in small amounts for normal metabolism, that body cells cannot synthesize in adequate amounts. Vitamins are essential nutrients that must come directly from foods or indirectly from **provitamins,** which are precursor substances.

Vitamins are classified on the basis of solubility, because some are soluble in fats (or fat solvents) and others are soluble in water. *Fat-soluble* vitamins are A, D, E, and K; the *water-soluble* group includes the B vitamins and vitamin C. Table 18.6 lists, and corrects, some common misconceptions about vitamins.

Different species have different vitamin requirements. For example, ascorbic acid is a required vitamin (C) in humans, guinea pigs, and Indian fruit bats, but not in other animals, which can manufacture their own.

TABLE 18.6 | Vitamin Fallacies and Facts

Fallacy	Fact
The more vitamins, the better	Too much of a water-soluble vitamin results in excretion of the vitamin through urination; too much of a fat-soluble vitamin can harm health
A varied diet provides all needed vitamins	Many people do need vitamin supplements, particularly pregnant and breast-feeding women
Vitamins provide energy	Vitamins do not directly supply energy; they aid in the release of energy from carbohydrates, fats, and proteins

Fat-Soluble Vitamins

Fat-soluble vitamins dissolve in fats, so they associate with lipids and are influenced by the same factors that affect lipid absorption. For example, bile salts in the intestine promote absorption of fat-soluble vitamins. As a group, the fat-soluble vitamins are stored in moderate quantities in various tissues, which is why excess intake can lead to overdose conditions. Fat-soluble vitamins resist the effects of heat, so cooking and food processing do not usually destroy them.

PRACTICE 

29 What are vitamins?

30 How are vitamins classified?

31 How do bile salts affect the absorption of fat-soluble vitamins?

Vitamin A exists in several forms, including retinol and retinal (retinene). Body cells synthesize this vitamin from a group of yellowish plant pigments, provitamins called *carotenes* (fig. 18.10). Excess vitamin A or its precursors are mainly stored in the liver, which regulates their concentration in the body. An adult's liver stores enough vitamin A to supply body requirements for a year. Infants and children usually lack such reserves and are therefore more likely to develop vitamin A deficiencies if their diets are inadequate.

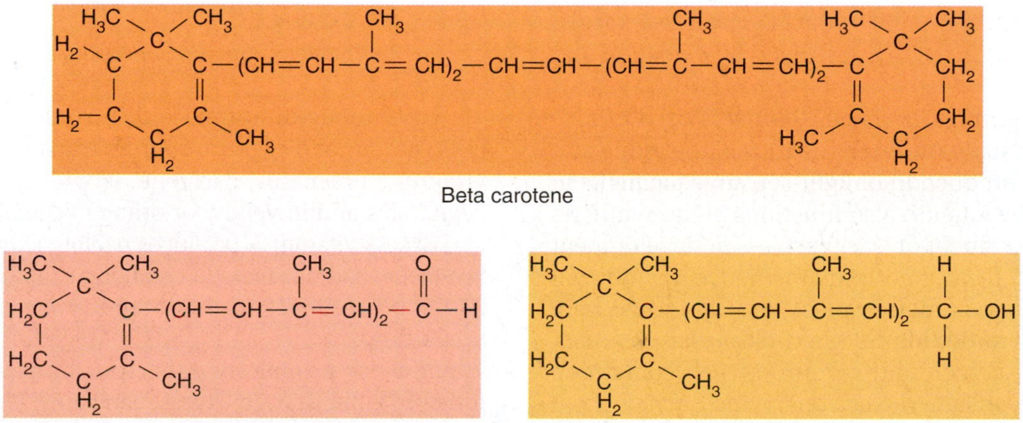

Beta carotene

Retinal (retinene)

Retinol

FIGURE 18.10 A molecule of beta carotene can react to form two molecules of retinal, which, in turn, can react to form retinol.

18.1 CLINICAL APPLICATION

Obesity

In the United States, obesity is epidemic. Nearly a third of all adults are obese, defined as 20% above "ideal" weight based on population statistics considering age, sex, and build, or a body mass index above 30. Obesity raises risks for type 2 diabetes, digestive disorders, heart disease, kidney failure, hypertension, stroke, and cancers of the female reproductive organs and the gallbladder. The body has to support the extra weight—miles of extra blood vessels are needed to nourish the additional pounds. Another third of the adult population of the United States is overweight. Obesity is the second leading cause of preventable death, following cigarette smoking. People in the United States are overweight because of overeating and underexercising. The average person today consumes 3,700 calories daily, compared to 3,100 in the 1960s.

Obesity refers to extra pounds of fat. The proportion of fat in a human body ranges from 5% to more than 50%, with "normal" for males falling between 12% and 23% and for females between 16% and 28%. An elite athlete may have a body fat level as low as 4%. Fat distribution also affects health. Excess poundage above the waist is linked to increased risk of heart disease, type 2 diabetes, hypertension, and lipid disorders. The body mass index (BMI) accounts somewhat for a person's build (fig. 18A). A person who weighs 170 pounds and is 6 feet tall is slim, whereas a person of the same weight who is 5 feet tall is obese. The tall person's BMI is 23; the short person's is 33.5.

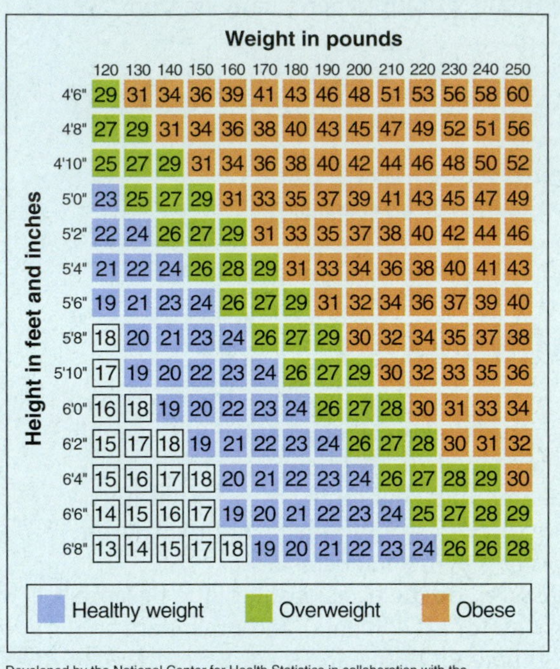

Weight in pounds

Height in feet and inches	120	130	140	150	160	170	180	190	200	210	220	230	240	250
4'6"	29	31	34	36	39	41	43	46	48	51	53	56	58	60
4'8"	27	29	31	34	36	38	40	43	45	47	49	52	51	56
4'10"	25	27	29	31	34	36	38	40	42	44	46	48	50	52
5'0"	23	25	27	29	31	33	35	37	39	41	43	45	47	49
5'2"	22	24	26	27	29	31	33	35	37	38	40	42	44	46
5'4"	21	22	24	26	28	29	31	33	34	36	38	40	41	43
5'6"	19	21	23	24	26	27	29	31	32	34	36	37	39	40
5'8"	18	20	21	23	24	26	27	29	30	32	34	35	37	38
5'10"	17	19	20	22	23	24	26	27	29	30	32	33	35	36
6'0"	16	18	19	20	22	23	24	26	27	28	30	31	33	34
6'2"	15	17	18	19	21	22	23	24	26	27	28	30	31	32
6'4"	15	16	17	18	20	21	22	23	24	26	27	28	29	30
6'6"	14	15	16	17	19	20	21	22	23	24	25	27	28	29
6'8"	13	14	15	17	18	19	20	21	22	23	24	26	26	28

Developed by the National Center for Health Statistics in collaboration with the National Center for Chronic Disease Prevention and Health Promotion

FIGURE 18A Body mass index (BMI). BMI equals weight/height^2, with weight measured in kilograms and height measured in meters. This chart provides a shortcut—the calculations have been done and converted to the English system of measurement. The uncolored squares indicate lower than healthy weight according to this index.

Both heredity and the environment contribute to obesity. Dozens of genes interact to control energy balance and therefore body weight. The observation that identical twins reared in different households can grow into adults of vastly different weights indicates that environment influences weight too. Even the environment before birth can affect body weight later. Individuals born at full

Vitamin A is relatively stable to the effects of heat, acids, and bases. However, it is readily destroyed by oxidation and is unstable in light.

Vitamin A is important in vision. Retinal is used to synthesize *rhodopsin* (visual purple) in the rods of the retina and may be required for production of light-sensitive pigments in the cones as well. The vitamin also functions in the synthesis of mucoproteins and mucopolysaccharides, in development of normal bones and teeth, and in maintenance of epithelial cells in skin and mucous membranes. Vitamin A and beta carotenes also act as **antioxidants** (an′tĭ-ok′sĭ-dant) by readily combining with oxygen and certain oxygen-containing molecules that have unshared electrons, which makes them highly reactive and damaging to cellular structures. These unstable molecules are called oxygen free radicals, and they accumulate in certain diseases and with age.

RECONNECT

To Chapter 2, Bonding of Atoms, pages 55–57.

Only foods of animal origin such as liver, fish, whole milk, butter, and eggs are sources of vitamin A. However, the vitamin's precursor, carotene, is widespread in leafy green vegetables and in yellow or orange vegetables and fruits.

Excess vitamin A produces peeling skin, hair loss, nausea, headache, and dizziness, a condition called *hypervitaminosis A.* Chronic overdoses of the vitamin may inhibit growth and break down bones and joints. "Megadosing" on fat-soluble vitamins is particularly dangerous during pregnancy. Some forms of vitamin A, in excess, can cause birth defects.

A deficiency of vitamin A causes *night blindness,* in which a person cannot see normally in dim light. Xeropthalmia, a dryness of the conjunctiva and cornea, is due to vitamin A

term, but undernourished as fetuses, are at high risk of obesity. Physiological changes that countered starvation in the uterus cause obesity when they persist.

Certain genes encode proteins that connect sensations in the gastrointestinal tract with centers in the hypothalamus that control hunger and satiety. It is how we satisfy those signals—what we eat—that provides the environmental component to body weight. A certain set of gene variants may have led to a trim figure in a human many thousands of years ago, when food had to be hunted or gathered—and meat was leaner. Today those same gene variants do not foster slimness in a person who takes in many more calories than he or she expends.

Treatments for Obesity

Diet and Exercise

A safe goal for weight loss using dietary restriction and exercise is 1 pound of fat per week. A pound of fat contains 3,500 calories of energy, so that pound can be shed by an appropriate combination of calorie cutting and exercise. This might mean eating 500 fewer calories per day or exercising off 500 calories each day. More than a pound of weight will drop because water is lost as well as fat.

Dieting should apply to the energy-providing nutrients (carbohydrates, proteins, and fats) but never to the vitamins and minerals. Choose foods that you like and distribute them into three or four balanced meals of 250 to 500 calories each.

Appetite is an important consideration in dieting to control weight. Many people in the 1990s, following advice from the U.S. government, followed low-fat diets, which caused weight gain if the dieters compensated by eating more highly refined carbohydrates. These foods escalate the rise and fall of blood glucose following a meal, which stimulates hunger sooner than if the meal contained more protein and fat. Substituting whole grains for "white" carbohydrates slows the rate of entry of glucose into the bloodstream (the glycemic index), and this can better control the urge to eat.

Ideally, weight loss can be accomplished by changing diet and exercise habits. However, realistically, two-thirds of those who lose weight regain it within five years. Physicians are increasingly regarding obesity as a chronic illness that for some people may require more drastic measures than dieting and exercising.

Drug Therapy

Some physicians recommend drug therapy if the BMI exceeds 30 or if it exceeds 27 and the person also has hypertension, type 2 diabetes mellitus, or hyperlipidemia. Several types of "diet drugs" are no longer in use because they are dangerous. Amphetamines, for example, carried the risk of addiction, and the combination of fenfluramine and phentermine damaged heart valves.

Newer antiobesity drugs target fat in diverse ways. Tetrahydrolipostatin, marketed as Orlistat and Xenical, inhibits the function of pancreatic lipase, preventing the digestion and absorption of about a third of dietary fat, which is eliminated in loose feces. This effect is not disruptive as long as the person follows a low-fat diet. Future weight control drugs may manipulate appetite-control hormones, such as ghrelin and leptin.

Surgery

For people with BMIs above 40, or above 35 in addition to an obesity-related disorder, bariatric (weight loss) surgery can lead to great weight loss. Two major types of procedures are done. In laparoscopic adjustable gastric banding, a silicone band ties off part of the stomach, limiting the capacity of the organ to hold food. The band can be inflated or deflated in a doctor's office by adding or removing saline. The second type of bariatric surgery is gastric bypass, in which part of the stomach is stapled shut, forming a pouch surgically connected to the jejunum, bypassing the duodenum. Both procedures lead to decreased hunger, drastically reduced food intake, and some decrease in the absorption of nutrients. A special diet, liquid at first, must be followed. Many patients who have had bariatric surgery report improvement in or disappearance of type 2 diabetes, back pain, arthritis, varicose veins, sleep apnea, and hypertension. ■

deficiency. Vitamin A deficiency also causes degenerative changes in certain epithelial tissues, and the body becomes more susceptible to infection.

PRACTICE

32 What chemical in the body is the precursor to vitamin A?

33 What conditions destroy vitamin A?

34 Which foods are good sources of vitamin A?

Vitamin D is a group of steroids that have similar properties. One of these substances, vitamin D_3 (cholecalciferol), is found in foods such as milk, egg yolk, and fish liver oils. Vitamin D_2 (ergocalciferol) is commercially produced by exposing a steroid obtained from yeasts (ergosterol) to ultraviolet light. Vitamin D can also be synthesized from dietary cholesterol that has been metabolized to provitamin D by intestinal enzymes, then stored in the skin and exposed to ultraviolet light (see chapter 13, p. 485).

Like other fat-soluble vitamins, vitamin D resists the effects of heat, oxidation, acids, and bases. It is primarily stored in the liver and is less abundant in the skin, brain, spleen, and bones.

Vitamin D stored in the form of hydroxycholecalciferol is released as needed into the blood. When parathyroid hormone is present, this form of vitamin D is converted in the kidneys into an active form of the vitamin (dihydroxycholecalciferol). This substance, in turn, is carried as a hormone in the blood to the intestines where it stimulates production of calcium-binding protein. Here, it promotes absorption of calcium and phosphorus, ensuring that adequate amounts of these minerals are available in the blood for tooth and bone formation and metabolic processes.

Natural foods are often poor sources of vitamin D, so it is often added to food during processing. For example, homogenized, nonfat, and evaporated milk are typically fortified with vitamin D. *Fortified* means essential nutrients have been added to a food where they originally were absent or scarce. *Enriched* means essential nutrients have been partially replaced in a food that has lost nutrients during processing.

Excess vitamin D, or *hypervitaminosis D,* produces diarrhea, nausea, and weight loss. Over time it may also calcify certain soft tissues and irreversibly damage the kidneys.

In children, vitamin D deficiency results in *rickets,* in which the bones and teeth fail to develop normally (fig. 18.11). In adults or in the elderly who have little exposure to sunlight, such a deficiency may lead to *osteomalacia,* in which the bones decalcify and weaken due to disturbances in calcium and phosphorus metabolism. Risk of developing vitamin deficiency increases in people who stay out of the sun or liberally use sun block to prevent skin cancer. However, just five minutes of sun exposure two to three times a week can maintain skeletal health without elevating skin cancer risk. Older people are usually outdoors less than younger individuals, so the Institute of Medicine suggests that daily vitamin D intake increase with age (table 18.7).

PRACTICE

35 Where is vitamin D stored?

36 What are the functions of vitamin D?

37 Which foods are good sources of vitamin D?

38 What are symptoms of vitamin D excess and deficiency?

FIGURE 18.11 Vitamin D deficiency causes rickets, in which the bones and teeth do not develop normally. Note this boy's bowed legs.

TABLE 18.7 | Vitamin D Requirements Increase with Age

Age Range (Years)	International Units of Vitamin D
<50	200
50–70	400
70+	600

Vitamin E includes a group of compounds, the most active of which is *alpha-tocopherol.* This vitamin is resistant to the effects of heat, acids, and visible light but is unstable in bases and in the presence of ultraviolet light or oxygen. Vitamin E is a strong antioxidant.

Vitamin E is found in all tissues but is primarily stored in the muscles and adipose tissue. It is also highly concentrated in the pituitary and adrenal glands.

The precise functions of vitamin E are unknown, but it is thought to act as an antioxidant by preventing oxidation of vitamin A and polyunsaturated fatty acids in the tissues. It may also help maintain the stability of cell membranes.

Vitamin E is widely distributed among foods. Its richest sources are oils from cereal seeds such as wheat germ. Other good sources are salad oils, margarine, shortenings, fruits, nuts, and vegetables. Excess vitamin E may cause nausea, headache, fatigue, easy bruising and bleeding, and muscle weakness. This vitamin is so easily obtained that deficiency conditions are rare.

PRACTICE

39 Where is vitamin E stored?

40 What are the functions of vitamin E?

41 Which foods are good sources of vitamin E?

Vitamin K, like the other fat-soluble vitamins, is in several chemical forms. One of these, vitamin K_1 (phylloquinone), is found in foods, whereas another, vitamin K_2, is produced by bacteria (*Escherichia coli*) that normally inhabit the human intestinal tract. These vitamins resist the effects of heat but are destroyed by oxidation or by exposure to acids, bases, or light. The liver stores them to a limited degree.

Vitamin K primarily functions in the liver, where it is necessary for the formation of several proteins needed for blood clotting, including *prothrombin* (see chapter 14, p. 539). Consequently, deficiency of vitamin K prolongs blood clotting time and may increase risk of hemorrhage. Excess vitamin K may occur in formula-fed infants, causing jaundice, hemolytic anemia, and hyperbilirubinemia.

The richest sources of vitamin K are leafy green vegetables. Other good sources are egg yolk, pork liver, soy oil, tomatoes, and cauliflower. Table 18.8 summarizes the fat-soluble vitamins and their properties.

TABLE 18.8 | Fat-Soluble Vitamins

Vitamin	Characteristics	Functions	Sources and RDA* for Adults	Conditions Associated with	
				Excesses	Deficiencies
Vitamin A	Exists in several forms; synthesized from carotenes; stored in liver; stable in heat, acids, and bases; unstable in light	An antioxidant necessary for synthesis of visual pigments, mucoproteins, and mucopolysaccharides; for normal development of bones and teeth; and for maintenance of epithelial cells	Liver, fish, whole milk, butter, eggs, leafy green vegetables, yellow and orange vegetables and fruits 4,000–5,000 IU**	Nausea, headache, dizziness, hair loss, birth defects	Night blindness, degeneration of epithelial tissues
Vitamin D	A group of steroids; resistant to heat, oxidation, acids, and bases; stored in liver, skin, brain, spleen, and bones	Promotes absorption of calcium and phosphorus; promotes development of teeth and bones	Produced in skin exposed to ultraviolet light; in milk, egg yolk, fish liver oils, fortified foods 400 IU	Diarrhea, calcification of soft tissues, renal damage	Rickets, bone decalcification and weakening
Vitamin E	A group of compounds; resistant to heat and visible light; unstable in presence of oxygen and ultraviolet light; stored in muscles and adipose tissue	An antioxidant; prevents oxidation of vitamin A and polyunsaturated fatty acids; may help maintain stability of cell membranes	Oils from cereal seeds, salad oils, margarine, shortenings, fruits, nuts, and vegetables 30 IU	Nausea, headache, fatigue, easy bruising and bleeding	Rare, uncertain effects
Vitamin K	Exists in several forms; resistant to heat but destroyed by acids, bases, and light; stored in liver	Required for synthesis of prothrombin, which functions in blood clotting	Leafy green vegetables, egg yolk, pork liver, soy oil, tomatoes, cauliflower 55–70 μg	Jaundice in formula-fed newborns	Prolonged clotting time

*RDA = recommended daily allowance.
**IU = international unit.

About 1 in every 200 to 400 newborns develops vitamin K deficiency because of an immature liver, poor transfer of vitamin K through the placenta, or lack of intestinal bacteria that can synthesize this vitamin. The deficiency causes "hemorrhagic disease of the newborn," which is associated with abnormal bleeding that typically appears during the second to fifth day of life. Injections of vitamin K shortly after birth prevent this condition. Adults may develop vitamin K deficiency if they are treated with antibiotic drugs that kill the intestinal bacteria that manufacture the vitamin. People with cystic fibrosis develop vitamin K deficiency, and/or deficiency of other fat-soluble vitamins, because they cannot digest fats well.

PRACTICE

42 Where in the body is vitamin K synthesized?

43 What is the function of vitamin K?

44 Which foods are good sources of vitamin K?

Water-Soluble Vitamins

The water-soluble vitamins include the B vitamins and vitamin C. Cooking and food processing destroy some of them. The **B vitamins** are several compounds essential for normal cellular metabolism. They help oxidize (remove electrons from) carbohydrates, lipids, and proteins during cellular respiration. The B vitamins are usually in the same foods, so they are called the *vitamin B complex*. Members of this group differ chemically and functionally.

The B-complex vitamins include the following:

1. **Thiamine,** or **vitamin B_1**. In its pure form, thiamine is a crystalline compound called thiamine hydrochloride. Exposure to heat and oxygen destroys it, especially in alkaline environments. (See fig. 18.18 for its molecular structure.)

   Thiamine is part of a coenzyme called *cocarboxylase*, which oxidizes carbohydrates. Specifically, thiamine is required for pyruvic acid to enter the citric acid cycle (see chapter 4, p. 122); in the absence of this vitamin, pyruvic acid accumulates in the blood. Thiamine also functions as a coenzyme in the synthesis of the sugar ribose, which is part of the nucleic acid RNA.

   Thiamine is primarily absorbed through the wall of the duodenum and is transported by the blood to body cells. Only small amounts are stored in the tissues, and excess is excreted in the urine.

   Vitamin B_1 oxidizes carbohydrates, so cellular requirements vary with caloric intake. It is recommended that an adult diet contain 0.5 milligram (mg) of thiamine for every 1,000 calories ingested daily. Good sources of thiamine are lean meats, liver, eggs, whole-grain cereals, leafy green vegetables, and legumes.

   Excess thiamine is not as common as excesses of fat-soluble vitamins, due to the excretion of thiamine in urine. Toxicity effects include vasodilation, cardiac dysrhythmias, headache, weakness, and convulsions.

   A mild deficiency of thiamine produces loss of appetite, fatigue, and nausea. Prolonged deficiency leads to a disease called *beriberi*, which causes gastrointestinal disturbances, mental confusion, muscular weakness and paralysis, and heart enlargement. In severe cases, the heart may fail.

In developed nations, beriberi affects mostly people with chronic alcoholism who have substituted alcohol for foods. Moreover, because thiamine is required for the metabolic oxidation of alcohol, people with alcoholism are particularly likely to develop a thiamine deficiency.

2. **Riboflavin,** or **vitamin B_2.** Riboflavin is a yellowish brown crystalline substance that is relatively stable to the effects of heat, acids, and oxidation but is destroyed by exposure to bases and ultraviolet light. This vitamin is part of several enzymes and coenzymes known as *flavoproteins*. One such coenzyme, FAD, is an electron carrier in the citric acid cycle and electron transport chain of aerobic respiration. Flavoproteins are essential for the oxidation of glucose and fatty acids and for cellular growth. An active transport system controls the amount of riboflavin entering the intestinal mucosa. Riboflavin is carried in the blood combined with proteins called *albumins*. Excess riboflavin in the blood is excreted in the urine, turning it yellow-orange, and any that remains unabsorbed in the intestine is lost in the feces.

   The amount of riboflavin the body requires varies with caloric intake. About 0.6 mg of riboflavin per 1,000 calories is sufficient to meet daily cellular requirements.

   Riboflavin is widely distributed in foods, and rich sources include meats and dairy products. Leafy green vegetables, whole-grain cereals, and enriched cereals provide lesser amounts. Vitamin B_2 deficiency produces dermatitis and blurred vision.
3. **Niacin** or **vitamin B_3.** Niacin, also known as *nicotinic acid,* is in plant tissues and is stable in the presence of heat, acids, and bases. After ingestion, it is converted to a physiologically active form called *niacinamide* (fig. 18.12). Niacinamide is the form of niacin in foods of animal origin.

   Niacin functions as part of two coenzymes (coenzyme I, also called NAD [fig. 18.13], and coenzyme II, called NADP) essential in glucose oxidation. These coenzymes are electron carriers in glycolysis, the citric acid cycle, and the electron transport chain, as well as in the synthesis of proteins and fats. They are also required for the synthesis of the sugars (ribose and deoxyribose) that are part of nucleic acids.

   Niacin is readily absorbed from foods, and human cells synthesize it from the essential amino acid *tryptophan*. Consequently, the daily requirement for niacin varies with tryptophan intake. Nutritionists recommend a daily niacin (or niacin equivalent) intake of 6.6 mg per 1,000 calories.

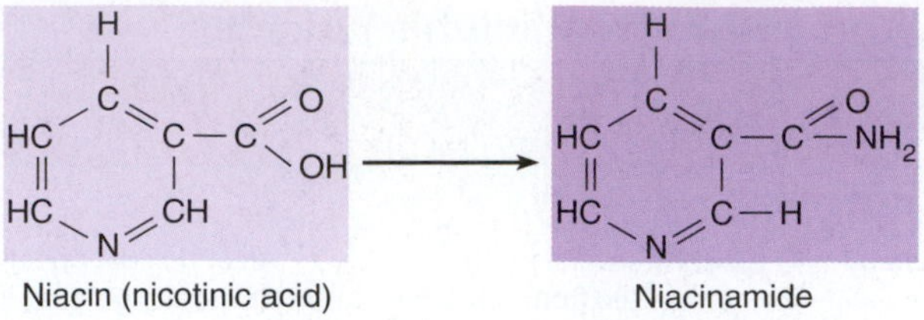

Niacin (nicotinic acid) → Niacinamide

FIGURE 18.12 Enzymes catalyze reactions that convert niacin from foods into physiologically active niacinamide.

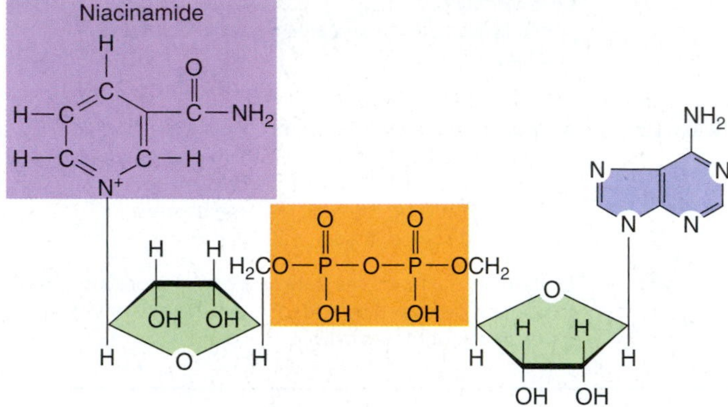

FIGURE 18.13 Niacinamide is incorporated into molecules of NAD.

   Rich sources of niacin (and tryptophan) include liver, lean meats, peanut butter, and legumes. Milk is a poor source of niacin but a good source of tryptophan.

   Excess niacin can cause acute toxicity with effects such as flushing, wheezing, vasodilation, headache, diarrhea, and vomiting. Chronic toxicity effects include liver problems.

   Historically, niacin deficiencies have been associated with diets largely consisting of corn and corn products, which are very low in niacin and lack adequate tryptophan. Such a deficiency causes a disease called *pellagra* that produces dermatitis, inflammation of the digestive tract, diarrhea, and mental disorders.

   Pellagra is rare today, but it was a serious problem in the rural South of the United States in the early 1900s. Pellagra is less common in cultures that extensively treat corn with lime ($CaCO_3$) to release niacin bound to protein. It sometimes affects people with chronic alcoholism who drink alcohol rather than eat.
4. **Pantothenic acid,** or **vitamin B_5.** Pantothenic acid is a yellowish oil destroyed by heat, acids, and bases. It functions as part of a complex molecule called *coenzyme A,* which, in turn, reacts with intermediate products of carbohydrate and fat metabolism to yield *acetyl coenzyme A,* which enters the citric acid cycle. Pantothenic acid is therefore essential to cellular energy release.

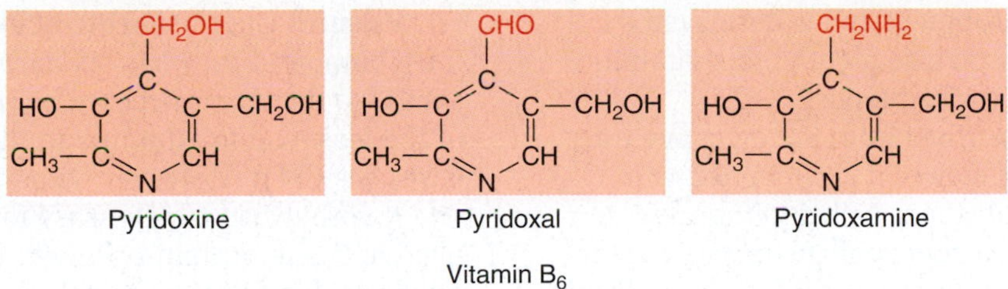

FIGURE 18.14 Vitamin B_6 includes three similar chemical compounds.

A daily adult intake of 4–7 mg of pantothenic acid is adequate. Most diets provide sufficient amounts, so deficiencies are rare. Good sources of pantothenic acid include meats, whole-grain cereals, legumes, milk, fruits, and vegetables.

5. **Vitamin B_6.** Vitamin B_6 is a group of three compounds, *pyridoxine, pyridoxal,* and *pyridoxamine,* which are chemically similar (fig. 18.14). These compounds have similar actions and are fairly stable in the presence of heat and acids. Oxidation or exposure to bases or ultraviolet light destroys them. The vitamin B_6 compounds function as coenzymes essential in several metabolic pathways, including those that synthesize proteins, amino acids, antibodies, and nucleic acids, as well as the reaction of tryptophan to produce niacin.

   Vitamin B_6 functions in the metabolism of nitrogen-containing substances, so the requirement for this vitamin varies with the protein content of the diet rather than with caloric intake. The recommended daily allowance of vitamin B_6 is 2.0 mg, but because it is so widespread in foods, deficiency conditions are rare. Good sources of vitamin B_6 include liver, meats, bananas, avocados, beans, peanuts, whole-grain cereals, and egg yolk. Excess vitamin B_6 produces burning pains, numbness, clumsiness, diminished reflexes, and paralysis.

6. **Cyanocobalamin,** or **vitamin B_{12}.** Cyanocobalamin has a complex molecular structure, including a single atom of the element *cobalt* (fig. 18.15). In its pure form, this vitamin is red. It is stable to the effects of heat but is inactivated by light or strong acids or strong bases.

   Secretion of *intrinsic factor* from the parietal cells of the gastric glands regulates cyanocobalamin absorption. Intrinsic factor combines with cyanocobalamin and facilitates its transfer through the epithelial lining of the small intestine and into the blood. Calcium ions must be present for the process to take place.

> When the gastric glands of some individuals fail to secrete adequate amounts of intrinsic factor, cyanocobalamin is poorly absorbed. This leads to *pernicious anemia,* in which abnormally large red blood cells called macrocytes are produced when bone marrow cells do not divide properly because of defective DNA synthesis.

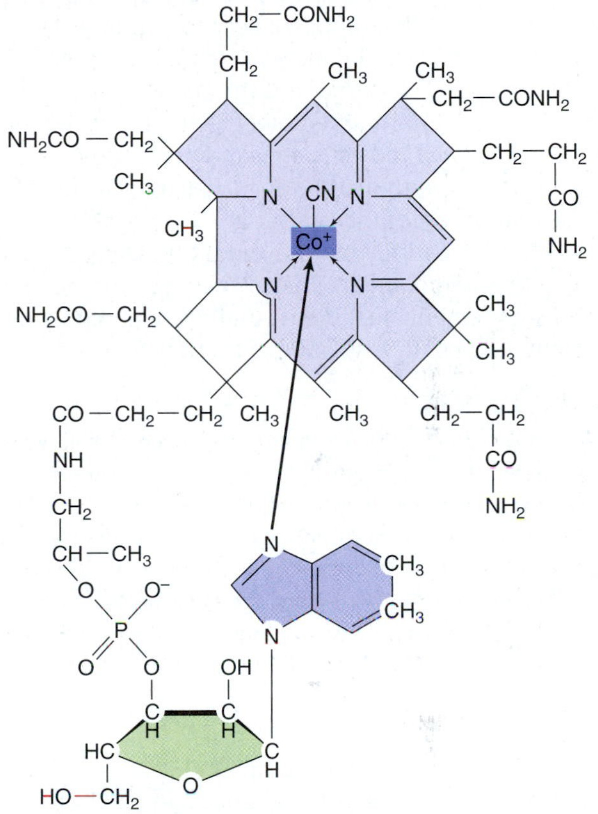

Vitamin B_{12} (cyanocobalamin)

FIGURE 18.15 Vitamin B_{12}, which has the most complex molecular structure of the vitamins, contains cobalt (Co).

Various tissues store cyanocobalamin, particularly those of the liver. An average adult has a reserve sufficient to supply cells for three to five years. This vitamin is essential for the functions of all cells. It is part of coenzymes required for the synthesis of nucleic acids and the metabolism of carbohydrates and fats. Vitamin B_{12} is important to erythrocyte production. Cyanocobalamin also helps form myelin in the central nervous system.

Only foods of animal origin contain cyanocobalamin. Good sources include liver, meats, milk, cheese, and eggs. Excessive intake does not appear to be toxic. In most countries, dietary lack of this vitamin is rare, although strict vegetarians may develop a deficiency.

7. **Folacin,** or **folic acid.** Folacin is a yellow crystalline compound that exists in several forms. It is easily

oxidized in an acid environment and is destroyed by heat in alkaline solutions; consequently, this vitamin may be lost in stored or cooked foods.

Folacin is readily absorbed from the digestive tract and is stored in the liver, where it is converted to a physiologically active substance called *folinic acid.* Folinic acid functions as a coenzyme necessary for the metabolism of certain amino acids and for the synthesis of DNA. It also acts with cyanocobalamin in producing normal red blood cells.

Good sources of folacin include liver, leafy green vegetables, whole-grain cereals, and legumes. Due to excess excretion in the urine, toxicity is rare. Folacin deficiency leads to *megaloblastic anemia,* in which the number of normal red blood cells is reduced and large, nucleated red cells appear. Folacin deficiency has been linked to neural tube defects, in which the tube that becomes the central nervous system in a fetus fails to entirely close. Neural tube defects include spina bifida and anencephaly. Taking synthetic folic acid supplements just before and during pregnancy can greatly reduce the risk of a neural tube defect.

> Naturally occurring folate is a mixture of compounds that collectively have the same activity as synthetic folic acid, also called pteroylmonoglutamic acid. However, synthetic folic acid is much more stable and enters the bloodstream much more readily. This difference has led to confusion. For example, 200 micrograms of synthetic folic acid is prescribed to treat anemia, but the same effect requires 400 micrograms of folate from foods. Synthetic folic acid used to enrich grain foods has a greater effect on health than folate.

8. **Biotin.** Biotin is a simple compound, that is stable to the effects of heat, acids, and light but may be destroyed by oxidation or bases. (See fig. 18.18 for the molecular structure of biotin.) It is a coenzyme in metabolic pathways for amino acids and fatty acids. It also plays a role in the synthesis of the purine nitrogenous bases of nucleic acids.

   Metabolically active organs such as the brain, liver, and kidneys store some biotin. Bacteria that inhabit the intestinal tract synthesize biotin. The vitamin is widely distributed in foods, and dietary deficiencies are rare. Good sources include liver, egg yolk, nuts, legumes, and mushrooms. Excess biotin does not produce toxic effects.

PRACTICE

45 Which biochemicals comprise the vitamin B complex?

46 Which foods are good sources of vitamin B complex?

47 Which of the B-complex vitamins can be synthesized from tryptophan?

48 What is the general function of each member of the vitamin B complex?

Vitamin C, or **ascorbic acid.** Ascorbic acid is a crystalline compound that has six carbon atoms. Chemically, it is similar to the monosaccharides (fig. 18.16). Vitamin C is one of the least stable vitamins in that oxidation, heat, light, or bases destroy it. However, vitamin C is fairly stable in acids.

Ascorbic acid is necessary for the production of the connective tissue protein *collagen,* for conversion of folacin to folinic acid, and in the metabolism of certain amino acids. It also promotes iron absorption and synthesis of certain hormones from cholesterol.

Overall, vitamin C is not stored in any great amount, but the adrenal cortex, pituitary gland, and intestinal glands contain high concentrations of it. Excess vitamin C is excreted in the urine or oxidized.

Individual requirements for ascorbic acid may vary. Ten mg per day is sufficient to prevent deficiency symptoms, and 80 mg per day saturate the tissues within a few weeks. Many nutritionists recommend a daily adult intake of 60 mg, which is enough to replenish normal losses and satisfy cellular requirements.

Ascorbic acid is fairly widespread in plant foods, with high concentrations in citrus fruits and tomatoes. Leafy green vegetables are also good sources.

Prolonged deficiency of ascorbic acid leads to *scurvy,* which is more likely to affect infants and children. Scurvy impairs bone development and causes swollen, painful joints. The gums may swell and bleed easily, resistance to infection is lowered, and wounds heal slowly (fig. 18.17). If a woman takes large doses of ascorbic acid during pregnancy, the newborn may develop symptoms of scurvy when the daily dose of the vitamin drops after birth because it is no longer delivered through the placenta.

Table 18.9 summarizes the water-soluble vitamins and their characteristics.

PRACTICE 

49 What factors destroy vitamin C?

50 What are functions of vitamin C?

51 Which foods are good sources of vitamin C?

52 What are symptoms of vitamin C deficiency?

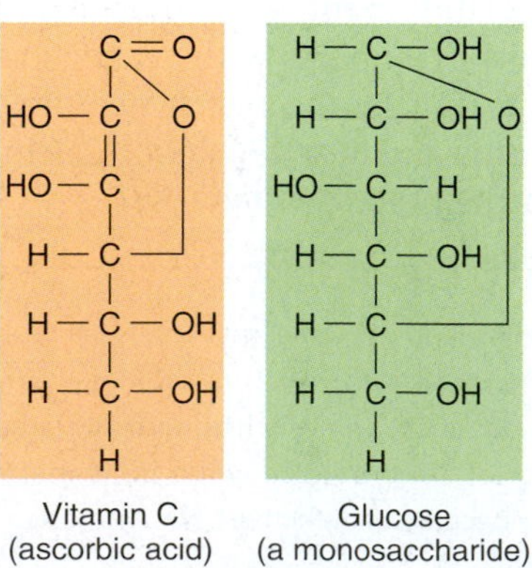

FIGURE 18.16 Vitamin C is chemically similar to some 6-carbon monosaccharides.

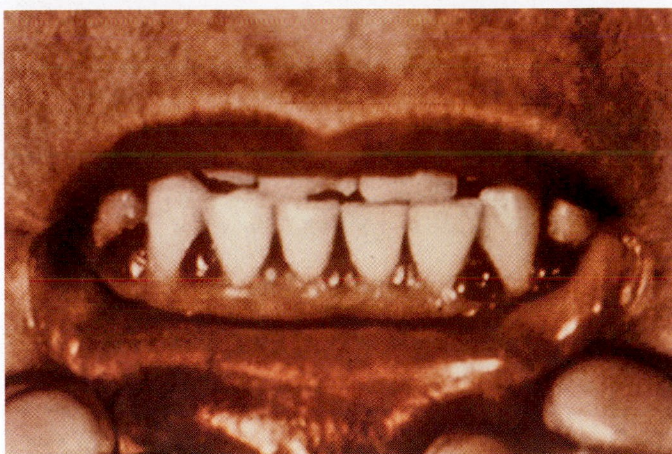

FIGURE 18.17 Vitamin C deficiency causes bleeding gums and other symptoms of scurvy.

Millions of Americans regularly take vitamin supplements. Consumer spending on vitamins and minerals is well into the billions of dollars annually. This practice has led to clinical signs of excess vitamin and mineral toxicity. Iron-containing vitamins are the most toxic, especially in acute pediatric ingestions.

18.7 MINERALS

Carbohydrates, lipids, proteins, and vitamins are organic compounds. Dietary **minerals** are inorganic elements essential in human metabolism. Plants usually extract these elements from the soil, and humans obtain them from plant foods or from animals that have eaten plants.

Characteristics of Minerals

Minerals contribute about 4% of body weight and are most concentrated in the bones and teeth. The minerals *calcium* and *phosphorus* are very abundant in these tissues.

Minerals are usually incorporated into organic molecules. Examples include phosphorus in phospholipids, iron in hemoglobin, and iodine in thyroxine. However, some minerals are part of inorganic compounds, such as the calcium phosphate of bone. Other minerals are free ions, such as sodium, chloride, and calcium ions in the blood.

Minerals compose parts of the structural materials of all cells. They also constitute portions of enzyme molecules, contribute to the osmotic pressure of body fluids, and play vital roles in nerve impulse conduction, muscle fiber contraction, blood coagulation, and maintenance of the pH of body fluids. The physiologically active form of minerals is the ionized form, such as Ca^{+2}.

Homeostatic mechanisms regulate the concentrations of minerals in body fluids. This ensures that excretion of minerals matches intake. Mineral toxicity may result not only from consumption of too much of a mineral, but also from overexposure to industrial pollutants, household chemicals, or certain drugs. Certain diseases, lead to mineral toxicity. Hemochromatosis, for example, is an inherited form of "iron overload." Injuries may also lead to mineral toxicity, such as severe trauma leading to hyperkalemia (high potassium).

PRACTICE 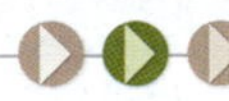

53 How do minerals differ from other nutrients?

54 What are the major functions of minerals?

Major Minerals

Calcium and phosphorus account for nearly 75% by weight of the mineral elements in the body; thus, they are **major minerals** (macrominerals). Other major minerals, each of which accounts for 0.05% or more of body weight, include potassium, sulfur, sodium, chlorine, and magnesium. Descriptions of the major minerals follow:

1. **Calcium.** Calcium (Ca) is widely distributed in cells and body fluids, even though 99% of the body's supply is in the inorganic salts of the bones and teeth. It is essential for nerve impulse conduction, muscle fiber contraction, and blood coagulation. Calcium also decreases the permeability of cell membranes and activates certain enzymes.

   The amount of calcium absorbed varies with a number of factors. For example, the proportion of calcium absorbed increases as the body's need for calcium increases. Vitamin D and high protein intake promote calcium absorption; increased motility of the digestive tract or an excess intake of fats decreases absorption. Daily intake of 800 mg is sufficient to cover adult requirements even with variations in absorption.

   Only a few foods contain significant amounts of calcium. Milk and milk products and fish with bones, such as salmon or sardines, are the richest sources. Leafy green vegetables, such as mustard greens, turnip greens, and kale, are good sources, but because one must consume large amounts of these vegetables to obtain sufficient minerals, most people must regularly consume milk or milk products to get enough calcium.

   Calcium toxicity is rare, but overconsumption of calcium supplements can deposit calcium phosphate in soft tissues. Calcium deficiency in children causes stunted growth, misshapen bones, and enlarged wrists and ankles. In adults, such a deficiency may remove calcium from the bones, thinning them and raising risk of fracture. Calcium is required for normal closing of the sodium channels in nerve cell membranes, so too little calcium (hypocalcemia) can cause tetany. Extra calcium demands in pregnancy can cause cramps.
2. **Phosphorus.** Phosphorus (P) accounts for about 1% of total body weight, most of it in the calcium phosphate of bones and teeth. The remainder serves as structural components and plays important roles in nearly all metabolic reactions. Phosphorus is a constituent of

TABLE 18.9 | Water-Soluble Vitamins

Vitamin	Characteristics	Functions	Sources and RDA* for Adults	Conditions Associated with	
				Excesses	Deficiencies
Thiamine (Vitamin B_1)	Destroyed by heat and oxygen, especially in alkaline environment	Part of coenzyme required for oxidation of carbohydrates; coenzyme required for ribose synthesis	Lean meats, liver, eggs, whole-grain cereals, leafy green vegetables, legumes 1.5 mg	Uncommon, vasodilation, cardiac dysrhythmias	Beriberi, muscular weakness, enlarged heart
Riboflavin (Vitamin B_2)	Stable to heat, acids, and oxidation; destroyed by bases and ultraviolet light	Part of enzymes and coenzymes, such as FAD, required for oxidation of glucose and fatty acids and for cellular growth	Meats, dairy products, leafy green vegetables, whole-grain cereals 1.7 mg	None known	Dermatitis, blurred vision
Niacin (Nicotinic acid, Vitamin B_3)	Stable to heat, acids, and bases; converted to niacinamide by cells; synthesized from tryptophan	Part of coenzymes NAD and NADP required for oxidation of glucose and synthesis of proteins, fats, and nucleic acids	Liver, lean meats, peanuts, legumes 20 mg	Flushing, vasodilation, wheezing, liver problems	Pellagra, photosensitive dermatitis, diarrhea, mental disorders
Pantothenic acid (Vitamin B_5)	Destroyed by heat, acids, and bases	Part of coenzyme A required for oxidation of carbohydrates and fats	Meats, whole-grain cereals, legumes, milk, fruits, vegetables 10 mg	None known	Rare, loss of appetite, mental depression, muscle spasms
Vitamin B_6	Group of three compounds; stable to heat and acids; destroyed by oxidation, bases, and ultraviolet light	Coenzyme required for synthesis of proteins and various amino acids, for conversion of tryptophan to niacin, for production of antibodies, and for nucleic acid synthesis	Liver, meats, bananas, avocados, beans, peanuts, whole-grain cereals, egg yolk 2 mg	Numbness, clumsiness, paralysis	Rare, convulsions, vomiting, seborrhea lesions
Cyanocobalamin (Vitamin B_{12})	Complex, cobalt-containing compound; stable to heat; inactivated by light, strong acids, and strong bases; absorption regulated by intrinsic factor from gastric glands; stored in liver	Part of coenzyme required for synthesis of nucleic acids and for metabolism of carbohydrates; plays role in myelin synthesis; required for normal red blood cell production	Liver, meats, milk, cheese, eggs 3–6 μg	None known	Pernicious anemia
Folacin (Folic acid)	Occurs in several forms; destroyed by oxidation in acid environment or by heat in alkaline environment; stored in liver where it is converted into folinic acid	Coenzyme required for metabolism of certain amino acids and for DNA synthesis; promotes production of normal red blood cells	Liver, leafy green vegetables, whole-grain cereals, legumes 0.4 mg	None known	Megaloblastic anemia
Biotin	Stable to heat, acids, and light; destroyed by oxidation and bases	Coenzyme required for metabolism of amino acids and fatty acids and for nucleic acid synthesis	Liver, egg yolk, nuts, legumes, mushrooms 0.3 mg	None known	Rare, elevated blood cholesterol, nausea, fatigue, anorexia
Ascorbic acid (Vitamin C)	Chemically similar to monosaccharides; stable in acids but destroyed by oxidation, heat, light, and bases	Required for collagen production, conversion of folacin to folinic acid, and metabolism of certain amino acids; promotes absorption of iron and synthesis of hormones from cholesterol	Citrus fruits, tomatoes, potatoes, leafy green vegetables 60 mg	Exacerbates gout and kidney stone formation	Scurvy, lowered resistance to infection, wounds heal slowly

*RDA = recommended daily allowance.

nucleic acids, many proteins, some enzymes, and some vitamins. It is also in the phospholipids of cell membranes, in the energy-carrying molecule ATP, and in the phosphates of body fluids that regulate pH. (Review the molecular structure of ATP in fig. 4.7.)

The recommended daily adult intake of phosphorus is 800 mg, and because this mineral is abundant in protein foods, diets adequate in proteins are also adequate in phosphorus. Phosphorus-rich foods include meats, cheese, nuts, whole-grain cereals, milk, and legumes.

PRACTICE

55 Which are the most abundant minerals in the body?

56 What are the functions of calcium?

57 What are the functions of phosphorus?

58 Which foods are good sources of calcium and phosphorus?

3. **Potassium.** Potassium (K) is widely distributed throughout the body and is concentrated inside cells rather than in extracellular fluids. On the other hand, sodium, which has similar chemical properties, is concentrated

outside cells. The ratio of potassium to sodium in a cell is 10:1, whereas the ratio outside the cell is 1:28.

Potassium helps maintain intracellular osmotic pressure and pH. It is a cofactor for enzymes that catalyze reactions of carbohydrate and protein metabolism and is vital in establishing the membrane potential in nerve impulse conduction.

Nutritionists recommend a daily adult intake of 2.5 grams (2,500 mg) of potassium. This mineral is in many foods, so a typical adult diet provides between 2 and 6 grams each day. Excess potassium in the blood is uncommon because of the uptake of potassium by body cells and the excretion of potassium in urine. Potassium deficiency due to diet is rare, but it may occur for other reasons. For example, when a person has diarrhea, the intestinal contents may pass through the digestive tract so rapidly that potassium absorption is greatly reduced. Vomiting or using diuretic drugs may also deplete potassium. Such losses may cause muscular weakness, cardiac abnormalities, and edema.

Foods rich in potassium are avocados, dried apricots, meats, milk, peanut butter, potatoes, and bananas. Citrus fruits, apples, carrots, and tomatoes provide lesser amounts.

PRACTICE

59 How is potassium distributed in the body?

60 What is the function of potassium?

61 Which foods are good sources of potassium?

4. **Sulfur.** Sulfur (S) is responsible for about 0.25% of body weight and is widely distributed throughout tissues. It is abundant in skin, hair, and nails. Most sulfur is part of the amino acids *methionine* and *cysteine.* Other sulfur-containing compounds include thiamine, insulin, and biotin (fig. 18.18). In addition, sulfur is a constituent of mucopolysaccharides in cartilage, tendons, and bones and of sulfolipids in the liver, kidneys, salivary glands, and brain.

   No daily requirement for sulfur has been established. It is thought, however, that a diet providing adequate amounts of protein will also meet the body's sulfur requirement. Good food sources of this mineral include meats, milk, eggs, and legumes.

5. **Sodium.** About 0.15% of adult body weight is sodium (Na), which is widely distributed throughout the body. Only about 10% of this mineral is inside cells, and about 40% is in the extracellular fluids. The remainder is bound to the inorganic salts of bone.

   Active transport readily absorbs sodium from foods. The kidneys regulate the blood concentration of sodium under the influence of the adrenal cortical hormone *aldosterone,* which causes the kidneys to reabsorb sodium while expelling potassium.

   Sodium makes a major contribution to the solute concentration of extracellular fluids and thus helps regulate water movement between cells and their surroundings. It is necessary for nerve impulse conduction and helps to move substances, such as chloride ions, through cell membranes (see chapter 21, p. 818).

   The usual human diet probably provides more than enough sodium to meet the body's requirements. Sodium toxicity, which shrinks cells, including those of the brain, requires unusual ingestion of additional sodium, such as drinking ocean water or accidentally using table salt instead of sugar for feeding infants. Sodium may be lost as a result of diarrhea, vomiting, kidney disorders, sweating, or using diuretics. Sodium loss may cause a variety of symptoms, including nausea, cramps, and convulsions.

   The amount of sodium naturally present in foods varies greatly, and it is commonly added to foods in the form of table salt (sodium chloride). In some geographic regions, drinking water contains significant concentrations of sodium. Foods high in sodium include cured ham, sauerkraut, and cheese.

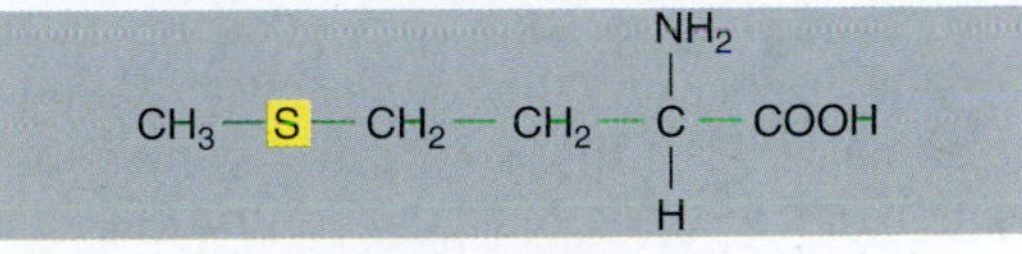

Methionine

Thiamine hydrochloride (vitamin B_1)

Biotin

FIGURE 18.18 Three essential sulfur-containing nutrients.

PRACTICE 

62 In which compounds and tissues of the body is sulfur found?

63 Which hormone regulates the blood concentration of sodium?

64 What are the functions of sodium?

6. **Chlorine.** Chlorine (Cl) in the form of chloride ions is found throughout the body and is most highly concentrated in cerebrospinal fluid and in gastric juice.

With sodium, chlorine helps to regulate pH and maintain electrolyte balance and the solute concentration of extracellular fluids. Chlorine is also essential for the formation of hydrochloric acid in gastric juice and in the transport of carbon dioxide by red blood cells.

Chlorine and sodium are usually ingested in table salt (sodium chloride), and as in the case for sodium, an ordinary diet usually provides considerably more chlorine than the body requires. Vomiting, diarrhea, kidney disorders, sweating, or using diuretics can deplete chlorine in the body.

7. **Magnesium.** Magnesium (Mg) is responsible for about 0.05% of body weight and is found in all cells. It is particularly abundant in bones in the form of phosphates and carbonates.

   Magnesium is important in ATP-forming reactions in mitochondria, as well as in breaking down ATP to ADP. Therefore, it is important in providing energy for cellular processes.

   Magnesium absorption in the intestinal tract adapts to dietary intake of the mineral. When the intake of magnesium is high, a smaller percentage is absorbed from the intestinal tract, and when the intake is low, a larger percentage is absorbed. Absorption increases as protein intake increases, and decreases as calcium and vitamin D intake increase. Bone tissue stores a reserve supply of magnesium, and excess is excreted in the urine.

   The recommended daily allowance of magnesium is 300 mg for females and 350 mg for males. A typical diet usually provides only about 120 mg of magnesium for every 1,000 calories, barely meeting the body's needs. Good sources of magnesium include milk and dairy products (except butter), legumes, nuts, and leafy green vegetables.

PRACTICE 

65 Where are chloride ions most highly concentrated in the body?

66 Where is magnesium stored?

67 What factors influence the absorption of magnesium from the intestinal tract?

Table 18.10 summarizes the major minerals.

Trace Elements

Trace elements (microminerals) are essential minerals found in minute amounts, each making up less than 0.005% of adult body weight. They include iron, manganese, copper, iodine, cobalt, zinc, fluorine, selenium, and chromium.

Iron (Fe) is most abundant in the blood, but is stored in the liver, spleen, and bone marrow and is found to some extent in all cells. Iron enables *hemoglobin* molecules in red blood cells to carry oxygen (fig. 18.19). Iron is also part of *myoglobin*, which stores oxygen in muscle cells. In addition,

CH_2 = CH; CH_3; CH_3; N; N--Fe--N; N; CH_3; CH = CH_2; CH_2–CH_2–COOH; CH_2–CH_2–COOH; CH_3

(a) Heme group

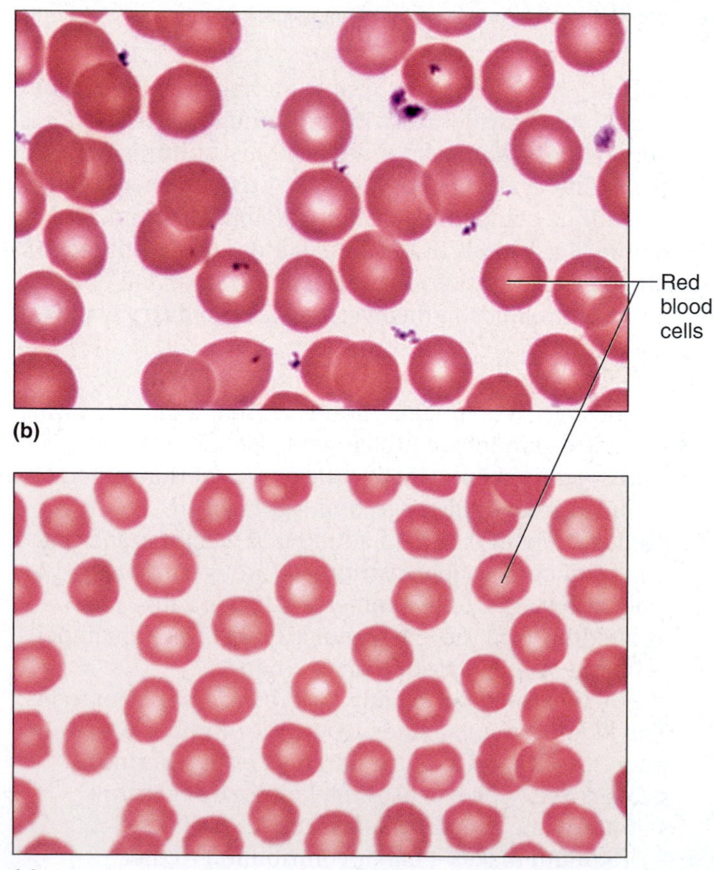

FIGURE 18.19 Iron in hemoglobin. (*a*) A hemoglobin molecule contains four heme groups, each of which houses a single iron atom (Fe) that can combine with oxygen. Iron deficiency anemia can result from a diet poor in iron-containing foods. The red blood cells in (*b*) are normal (1250x), but many of those in (*c*) are small and pale (1250x). They contain too little hemoglobin, because iron is lacking in the diet. Vegetarians must be especially careful to consume sufficient iron.

TABLE 18.10 | Major Minerals

Mineral	Distribution	Functions	Sources and RDA* for Adults	Conditions Associated with Excesses	Conditions Associated with Deficiencies
Calcium (Ca)	Mostly in the inorganic salts of bones and teeth	Structure of bones and teeth; essential for neurotransmitter release, muscle fiber contraction, and blood coagulation; increases permeability of cell membranes; activates certain enzymes	Milk, milk products, leafy green vegetables 800 mg	Kidney stones, deposition of calcium phosphate in soft tissues	Stunted growth, misshapen bones, fragile bones, tetany
Phosphorus (P)	Mostly in the inorganic salts of bones and teeth	Structure of bones and teeth; in nearly all metabolic reactions; in nucleic acids, many proteins, some enzymes, and some vitamins; in cell membrane, ATP, and phosphates of body fluids	Meats, cheese, nuts, whole-grain cereals, milk, legumes 800 mg	None known	Stunted growth
Potassium (K)	Widely distributed; tends to be concentrated inside cells	Helps maintain intracellular osmotic pressure and regulate pH; required for nerve impulse conduction	Avocados, dried apricots, meats, nuts, potatoes, bananas 2,500 mg	Uncommon	Muscular weakness, cardiac abnormalities, edema
Sulfur (S)	Widely distributed; abundant in skin, hair, and nails	Essential part of certain amino acids, thiamine, insulin, biotin, and mucopolysaccharides	Meats, milk, eggs, legumes No RDA established	None known	None known
Sodium (Na)	Widely distributed; mostly in extracellular fluids and bound to inorganic salts of bone	Helps maintain osmotic pressure of extracellular fluids; regulates water movement; plays a role in nerve impulse conduction; regulates pH and transport of substances across cell membranes	Table salt, cured ham, sauerkraut, cheese, graham crackers 2,500 mg	Hypertension, edema, body cells shrink	Nausea, cramps, convulsions
Chlorine (Cl)	Closely associated with sodium; most highly concentrated in cerebrospinal fluid and gastric juice	Helps maintain osmotic pressure of extracellular fluids, regulates pH; maintains electrolyte balance; forms hydrochloric acid; aids transport of carbon dioxide by red blood cells	Same as for sodium No RDA established	Vomiting	Cramps
Magnesium (Mg)	Abundant in bones	Required in metabolic reactions in mitochondria that produce ATP; plays a role in the breakdown of ATP to ADP	Milk, dairy products, legumes, nuts, leafy green vegetables 300–350 mg	Diarrhea	Neuromuscular disturbances

*RDA = recommended daily allowance.

iron assists in vitamin A synthesis, is incorporated into a number of enzymes, and is included in the cytochrome molecules that participate in ATP-generating reactions.

An adult male requires from 0.7 to 1 mg of iron daily, and a female needs 1.2 to 2 mg. A typical diet supplies about 10 to 18 mg of iron each day, but only 2% to 10% of the iron is absorbed. For some people, this may not be enough iron. Eating foods rich in vitamin C along with iron-containing foods can increase absorption of this important mineral.

Liver is the only rich source of dietary iron, and because liver is not a popular food, iron is one of the more difficult nutrients to obtain from natural sources in adequate amounts. Foods that contain some iron include lean meats; dried apricots, raisins, and prunes; enriched whole-grain cereals; legumes; and molasses.

> Pregnant women require extra iron to support the formation of a placenta and the growth and development of a fetus. Iron is required for the synthesis of hemoglobin in a fetus as well as in a pregnant woman, whose blood volume increases by a third.

Manganese (Mn) is most concentrated in the liver, kidneys, and pancreas. It is necessary for normal growth and development of skeletal structures and other connective tissues. Manganese is part of enzymes essential for the synthesis of fatty acids and cholesterol, for urea formation, and for the normal functions of the nervous system.

The daily requirement for manganese is 2.5–5 mg. The richest sources include nuts, legumes, and whole-grain cereals; leafy green vegetables and fruits are good sources.

PRACTICE

68 What is the primary function of iron?

69 Why does the usual diet provide only a narrow margin of safety in supplying iron?

70 How is manganese used?

71 Which foods are good sources of manganese?

Copper (Cu) is in all body tissues but is most highly concentrated in the liver, heart, and brain. It is essential for hemoglobin synthesis, bone development, melanin production, and formation of myelin in the nervous system.

A daily intake of 2 mg of copper is sufficient to supply cells. A typical adult diet has about 2–5 mg of this mineral, so adults seldom develop copper deficiencies. Foods rich in copper include liver, oysters, crabmeat, nuts, whole-grain cereals, and legumes.

Iodine (I) is found in minute quantities in all tissues but is highly concentrated in the thyroid gland. Its only known function is as an essential component of thyroid hormones. (Figure 13.20 shows the molecular structures of two of these hormones, thyroxine and triiodothyronine.)

A daily intake of 1 microgram (0.001 mg) of iodine per kilogram of body weight is adequate for most adults. The iodine content of foods varies with the iodine content of soils in different geographic regions, so many people use *iodized* table salt to season foods to prevent deficiencies.

Cobalt (Co) is widely distributed in the body because it is an essential part of cyanocobalamin (vitamin B_{12}). It is also necessary for the synthesis of several important enzymes.

The amount of cobalt required in the daily diet is unknown. This mineral is found in a great variety of foods, and the quantity in the average diet is apparently sufficient. Good sources of cobalt include liver, lean meats, and milk.

Zinc (Zn) is most concentrated in the liver, kidneys, and brain. It is part of many enzymes involved in digestion, respiration, and bone and liver metabolism. It is also necessary for normal wound healing and for maintaining the integrity of the skin.

The daily requirement for zinc is about 15 mg, and most diets provide 10–15 mg. Only some may be absorbed, so zinc deficiencies may occur. The richest sources of zinc are meats; cereals, legumes, nuts, and vegetables provide lesser amounts.

Fluorine (F), as part of the compound fluoroapatite, replaces hydroxyapatite in teeth, strengthening the enamel and preventing dental caries. **Selenium** (Se) is stored in the liver and kidneys. It is a constituent of certain enzymes and participates in heart function. This mineral is found in lean meats, whole-grain cereals, and onions. **Chromium** (Cr) is widely distributed throughout the body and regulates glucose use. It is found in liver, lean meats, yeast, and pork kidneys. Table 18.11 summarizes the characteristics of trace elements.

The term "dietary supplement" traditionally refers to minerals, vitamins, carbohydrates, proteins, and fats—the micronutrients and macronutrients. Clinical Application 18.2 discusses the more commercial meaning of "dietary supplement."

PRACTICE 

72 How is copper used?

73 What is the function of iodine?

74 Why might zinc deficiencies be common?

A compulsive disorder that may result from mineral deficiency is *pica*, in which people consume huge amounts of nondietary substances such as ice chips, soil, sand, laundry starch, clay and plaster, and even hair, toilet paper, matchheads, inner tubes, mothballs, and charcoal. The condition is named for the magpie bird, *Pica pica*, which eats a range of odd things.

Pica affects people of all cultures and was noted as early as 40 B.C. The connection to dietary deficiency stems from the observation that slaves suffering from pica in colonial America recovered when their diets improved, particularly when given iron supplements. Another clue comes from a variation on pica called geophagy—"eating dirt"—that affects many types of animals, including humans. Researchers discovered that when parrots eat a certain claylike soil in their native Peru, soil particles bind alkaloid toxins in their seed food and carry the toxins out of the body. Perhaps pica in humans is protective in some way, too.

18.8 HEALTHY EATING

An *adequate diet* provides sufficient energy (calories), essential fatty acids, essential amino acids, vitamins, and minerals to support optimal growth and to maintain and repair body tissues. Individual nutrient requirements vary greatly with age, sex, growth rate, level of physical activity and stress, as well as with genetic and environmental factors, so it is not possible to design a diet adequate for everyone. However, nutrients are so widely found in foods that satisfactory amounts and combinations can usually be obtained despite individual food preferences, assuming that foods are available.

It is difficult to keep track of the different nutrients in a diet and be certain that an adequate amount of each is consumed daily. Nutritionists have devised several ways to help consumers make healthy food choices, recognizing that people can meet dietary requirements in many and diverse ways. Most familiar is the RDA guideline that has appeared on several tables in the chapter. *RDA* stands for United States Recommended Daily Allowance. An RDA is the upper limit of another measurement, the Recommended Dietary Allowance, which lists optimal calorie intake for each sex at various ages, and the amounts of vitamins and minerals needed to avoid deficiency or excess conditions. The RDA values on food packages are set high, ensuring that most people who follow them receive sufficient amounts of each nutrient. Government panels meet every five years to evaluate the RDAs in light of new data.

Placing foods into groups is a simpler way to follow a healthy diet. Diagrams called **food pyramids** organize foods according to suggested proportions of the diet, often in serving sizes. One food pyramid, developed by the U.S. Department of Agriculture, dominated for years, but new ones offer more specific suggestions geared to age, health, ethnicity, food preferences such as vegetarianism, or

TABLE 18.11 | Trace Elements

Trace Element	Distribution	Functions	Sources and RDA* for Adults	Conditions Associated with	
				Excesses	Deficiencies
Iron (Fe)	Primarily in blood; stored in liver, spleen, and bone marrow	Part of hemoglobin molecule; catalyzes formation of vitamin A; incorporated into a number of enzymes	Liver, lean meats, dried apricots, raisins, enriched whole-grain cereals, legumes, molasses 10–18 mg	Liver damage	Anemia
Manganese (Mn)	Most concentrated in liver, kidneys, and pancreas	Activates enzymes required for fatty acids and cholesterol synthesis, formation of urea, and normal functioning of the nervous system	Nuts, legumes, whole-grain cereals, leafy green vegetables, fruits 2.5–5 mg	None known	None known
Copper (Cu)	Most highly concentrated in liver, heart, and brain	Essential for hemoglobin synthesis, bone development, melanin production, and myelin formation	Liver, oysters, crabmeat, nuts, whole-grain cereals, legumes 2–3 mg	Rare	Rare
Iodine (I)	Concentrated in thyroid gland	Essential component for synthesis of thyroid hormones	Food content varies with soil content in different geographic regions; iodized table salt 0.15 mg	Decreased uptake by the thyroid gland	Decreased synthesis of thyroid hormones
Cobalt (Co)	Widely distributed	Component of cyanocobalamin; required for synthesis of several enzymes	Liver, lean meats, milk No RDA established	Heart disease	Pernicious anemia
Zinc (Zn)	Most concentrated in liver, kidneys, and brain	Component of enzymes involved in digestion, respiration, bone metabolism, liver metabolism; necessary for normal wound healing and maintaining integrity of the skin	Meats, cereals, legumes, nuts, vegetables 15 mg	Slurred speech, problems walking	Depressed immunity, loss of taste and smell, learning difficulties
Fluorine (F)	Primarily in bones and teeth	Component of tooth structure	Fluoridated water 1.5–4 mg	Mottled teeth	None known
Selenium (Se)	Concentrated in liver and kidneys	Components of certain enzymes	Lean meats, fish, cereals 0.05–2 mg	Vomiting, fatigue	None known
Chromium (Cr)	Widely distributed	Essential for use of carbohydrates	Liver, lean meats, wine 0.05–2 mg	None known	None known

RDA = recommended daily allowance.

weight loss goals. Past pyramids can seem strange in light of today's individualized goals, depicted in figure 18.20. A plan from the 1940s had eight categories, including separate groups for butter and margarine, and for eggs—foods now associated with the development of heart disease. In the 1920s, an entire food group was devoted to sweets! Although many types of diets work well for many people, a good general approach is to emphasize fresh fruits and vegetables and whole grains and to minimize processed foods, which are often high in fat and/or sugar.

When making individual food choices, it helps to read and understand food labels. Disregard claims such as "light" and "low fat" and skip right to the calories or grams of different ingredients. "Light" can mean many things: that the nutritionally altered form contains one-third fewer calories or half the fat of the reference food, that the sodium content has been reduced by 50%, or it may describe the texture and color of the food. "Low fat" indicates 3 grams of fat or less per serving. Many times when fat is removed, sugar is added, so the calories of a low-fat product may be more, not less—just compare the ingredients lists on various ice cream cartons to see this. Ingredients are listed in descending order by weight. Clinical Application 18.3 discusses some ways that understanding nutrition can help athletic performance.

PRACTICE

75 What is an adequate diet?

76 What factors influence individual needs for nutrients?

77 Describe ways that consumers can make wise food choices.

Malnutrition

Malnutrition (mal″nu-trish′un) is poor nutrition that results from a lack of essential nutrients or a failure to utilize them. It may result from *undernutrition* and produce the symptoms of deficiency diseases, or it may be due to *overnutrition* arising from excess nutrient intake.

A variety of factors can lead to malnutrition. For example, a deficiency condition may stem from lack of availability or poor quality of food. On the other hand, malnutrition may result from overeating or taking too many vitamin supplements. Malnutrition from diet alone is called *primary malnutrition.*

Secondary malnutrition occurs when an individual's characteristics make a normally adequate diet insufficient.

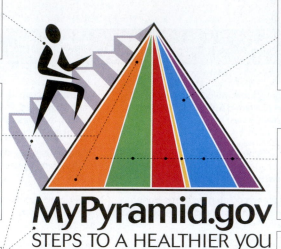

(a)

Activity
Activity is represented by the steps and the person climbing them, as a reminder of the importance of daily physical activity.

Moderation
Moderation is represented by the narrowing of each food group from bottom to top. The wider base stands for foods with little or no solid fats or added sugars. These should be selected more often. The narrower top area stands for foods containing more added sugars and solid fats. The more active you are, the more of these foods can fit into your diet.

Personalization
Personalization is shown by the person on the steps and the slogan.

Proportionality
Proportionality is shown by the different widths of the food group bands. The widths suggest how much food a person should choose from each group. The widths are just a general guide, not exact proportions.

Variety
Variety is symbolized by the 6 color bands representing the 5 food groups of the Pyramid and oils. This illustrates that foods from all groups are needed each day for good health.

Gradual Improvement
Gradual improvement is encouraged by the slogan. It suggests that individuals can benefit from taking small steps to improve their diet and lifestyle each day.

MyPyramid.gov
STEPS TO A HEALTHIER YOU

USDA U.S. Department of Agriculture Center for Nutrition Policy and Promotion April 2005 CNPP-16

USDA is an equal opportunity provider and employer.

GRAINS | VEGETABLES | FRUITS | OILS | MILK | MEAT & BEANS

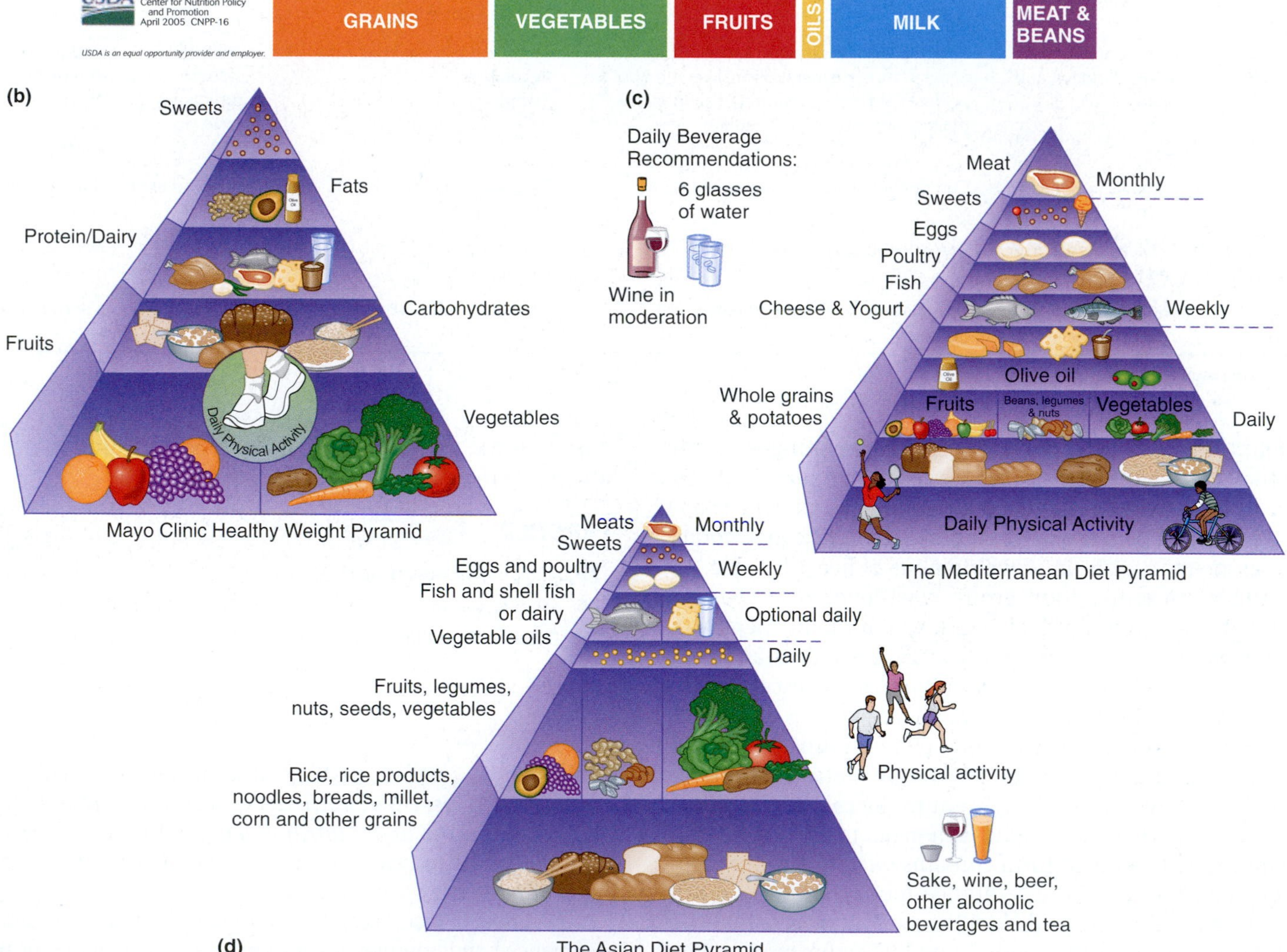

FIGURE 18.20 Food guide pyramids show, at a glance, the relative amounts of food that should compose a particular type of diet. (*a*) The USDA food pyramid symbolizes an individual approach to healthy eating and physical exercise. It requires entering personal information into the website http://mypyramid.gov. (*b*) The Mayo Clinic Healthy Weight Pyramid rates sweets as less healthy than fats. (*c*) The Mediterranean Diet Pyramid includes olive oil and other foods appropriate to the geographical region. (*d*) The Asian food guide pyramid suits the cuisine of a particular ethnic group, stressing its reliance on vegetables.

18.2 CLINICAL APPLICATION

Dietary Supplements—Proceed with Caution

Displayed prominently among the standard vitamin and mineral preparations in the pharmacy or health food store is a dizzying collection of products (fig. 18B). Some obviously come from organisms (bee pollen and shark cartilage), some have chemical names (glucosamine with chondroitin, see chapter 8 vignette on page 261) and some names are historical or cultural (St. John's Wort). These "dietary supplements" are neither food nor drug, but they contain active compounds that may function as pharmaceuticals in the human body.

FIGURE 18B Some dietary supplements are natural substances that function as drugs in the human body. Proceed with caution!

By law dietary supplements include: "a product (other than tobacco) that is intended to supplement the diet that bears or contains one or more of the following dietary ingredients: a vitamin, a mineral, an herb or other botanical, an amino acid, a dietary substance for use by man to supplement the diet by increasing the total daily intake, or a concentrate, metabolite, constituent, extract, or combinations of these ingredients."

Labels cannot claim that a dietary supplement diagnoses, prevents, mitigates, treats, or cures any specific disease. Instead, the language is positive. For example, Valerian root "promotes restful sleep" and echinacea and goldenseal "may help support the immune system." In 2004, the U.S. Food and Drug Administration relaxed requirements that evidence back up claims of cure or prevention. It allows the following specific food and health claims:

- Dietary calcium decreases the risk of osteoporosis (see Clinical Application 7.2 on page 204).
- A low-fat diet lowers the risk of some cancers.
- A diet low in saturated fat and cholesterol lowers the risk of coronary heart disease.
- Fiber, fruits and vegetables, and whole grains reduce the risk of some cancers and coronary heart disease.
- Lowering sodium intake lowers blood pressure.
- Folic acid lowers the risk of neural tube defects.

Many dietary supplements contain pharmaceutical agents, so a physician should be consulted before using these products, particularly if a person has a serious illness or is taking medication. The active ingredients in supplements may interact with other drugs. For example, the active ingredient in St. John's Wort, hypericin, lowers blood levels of nearly half of all prescription drugs by interfering with liver enzymes that metabolize many drugs. Some patients have experienced intracranial hemorrhage after taking ginkgo biloba, a tree extract reported to enhance memory.

Certain dietary supplements are of dubious value. For example, the marketing of shark cartilage followed initial studies that suggested sharks do not get cancer. Sharks have cartilaginous skeletons, so the idea arose that their cartilage somehow protects against cancer. It turned out that sharks indeed get cancer, and shark cartilage has no magical properties. Similarly, pyruvic acid and ATP are not necessary to boost energy levels—these biochemicals are abundant in the cellular respiration pathways. Some health-food stores sell DNA, merely expensive brewer's yeast, and unnecessary, because any food consisting of cells is packed with DNA. The list is long of supplements with little scientific evidence of value, yet dietary supplements are a multi-billion dollar industry. Buyer beware! ■

For example, a person who secretes insufficient bile salts is likely to develop a deficiency of fat-soluble vitamins because bile salts promote absorption of fats. Likewise, severe and prolonged emotional stress may lead to secondary malnutrition, because stress can change hormonal concentrations in ways that break down amino acids or excrete nutrients.

Starvation

A healthy human can stay alive for 50 to 70 days without food. In prehistoric times, this margin allowed survival during seasonal famines. In some areas of Africa today, famine is not a seasonal event but a constant condition, and millions of people have starved to death. Starvation is also seen in hunger strikers, in prisoners of concentration camps, and in sufferers of psychological eating disorders such as *anorexia nervosa* and *bulimia.*

Whatever the cause, the starving human body begins to digest itself. After only one day without eating, the body's reserves of sugar and starch are gone. Next, the body extracts energy from fat and then from muscle protein. By the third day, hunger ceases as the body uses energy from fat reserves. Gradually, metabolism slows to conserve energy, blood pressure drops, the pulse slows, and chills set in. Skin becomes dry and hair falls out as the proteins in these structures are

18.3 CLINICAL APPLICATION

Nutrition and the Athlete

An endurance athlete and a sedentary individual have different nutritional requirements, as the description of cyclist Lance Armstrong's diet in the chapter opener indicates. A diet that is predominantly complex carbohydrate supports a lifestyle that includes frequent strenuous activity. A distribution of 60% or more carbohydrate, 18% protein, and 22% fat supports frequent, strenuous activity.

Macronutrients

Athletes should get the bulk of their carbohydrates from vegetables and grains to avoid cholesterol, and they should eat frequently, because the muscles can store only 1,800 calories worth of glycogen.

Athletes need to consume only slightly more protein than less-active individuals. The American Dietetic Association suggests that athletes eat 1 gram of protein per kilogram of weight per day, compared to 0.8 gram for nonathletes. Athletes should not rely solely on meat for protein, because these foods can be high in fat. Protein supplements may be necessary for only young athletes at the start of training, under a doctor's supervision. Too little protein in an athlete is linked to "sports anemia," in which hemoglobin levels decline and blood may appear in the urine.

Water

A sedentary person loses a quart of water a day as sweat; an athlete may lose 2 to 4 quarts of water an hour! To stay hydrated, athletes should drink 3 cups of cold water two hours before an event, then 2 more cups fifteen minutes before the event, and small amounts every fifteen minutes during the event. They should drink afterward too. Another way to determine water needs is to weigh in before and after training. For each pound lost, athletes should drink a pint of water. They should also avoid sugary fluids, which slow water's trip through the digestive system, and alcohol, which increases fluid loss. However, athletes should also avoid drinking too much water during competition, which can cause hyponatremia (too little sodium in the bloodstream, see Clinical Applications 21.1 and 21.2 on pages 816–817 and 820 respectively).

Vitamins and Minerals

If an athlete eats an adequate, balanced diet, vitamin supplements are not needed. Supplements of sodium and potassium are usually not needed either, because the active body naturally conserves these nutrients. To be certain of enough sodium, athletes may want to salt their food; to get enough potassium, they can eat bananas, dates, apricots, oranges, or raisins.

A healthy pregame meal should be eaten two to five hours before the game, provide 500 to 1,500 calories, and include 4 or 5 cups of fluid. The pregame meal should also be high in carbohydrates, which taste good, provide energy, and are easy to digest.

Creatine

Creatine is advertised to increase energy stores and provide a safe alternative to steroids for bulking up muscles. This is deceptive.

Creatine may be obtained from foods, through supplements or by synthesis from the amino acids arginine, glycine, and methionine. Creatine, in the form of creatine phosphate, provides energy to muscle cells by phosphorylating ADP to generate ATP. Creatine is converted to its metabolite, creatinine, at such a constant rate that the excretion of creatinine in the urine is used as a marker for normal kidney function.

Do creatine supplements enhance performance? The emerging picture suggests that during peak exertion, especially repetitive peak exertion (such as multiple sprints), conditions in which creatine levels may become depleted, supplemental creatine may be advantageous. Muscle mass may appear to increase in athletes taking creatine supplements because creatine draws water into muscle cells by osmosis. However, the disturbance in water distribution that creatine supplementation can cause may create problems if the athlete encounters extreme heat—sweating becomes inadequate to effectively cool the body. Swelled muscle cells may burst, causing a potentially fatal condition called rhabdomyolysis. The Food and Drug Administration has received many adverse event reports of muscle cramps, seizures, diarrhea, loss of appetite, muscle strains, dehydration and even deaths, associated with creatine use among athletes. ■

broken down to release amino acids used for the more vital functioning of the brain, heart, and lungs. When the immune system's antibody proteins are dismantled for their amino acids, protection against infection declines. Mouth sores and anemia develop, the heart beats irregularly, and bone begins to degenerate. After several weeks without food, coordination is gradually lost. Near the end, the starving human is blind, deaf, and emaciated.

Marasmus and Kwashiorkor

Lack of nutrients is called **marasmus** (mah-raz′mus), and it causes people to resemble living skeletons (fig. 18.21*a*). Children under the age of two with marasmus often die of measles or other infections. Their immune systems become too weakened to fight off normally mild viral illnesses.

Some starving children have protruding bellies. These youngsters suffer from a form of protein starvation called **kwashiorkor** (kwash″-e-or′kor), which in the language of Ghana means "the evil spirit which infects the first child when the second child is born" (fig. 18.21*b*). Kwashiorkor typically appears in a child who has recently been weaned from the breast, usually because of the birth of a sibling. The switch from protein-rich breast milk to the protein-poor gruel that is the staple of many developing nations is the source of this protein deficiency. The children's bellies swell with filtered fluid, lost from capillaries in greater than normal volume due to a lack of plasma proteins. This condition is called **ascites** (ah-si′tēz). Their skin may develop lesions. Infections overwhelm the body as the immune system becomes depleted of its protective antibodies.

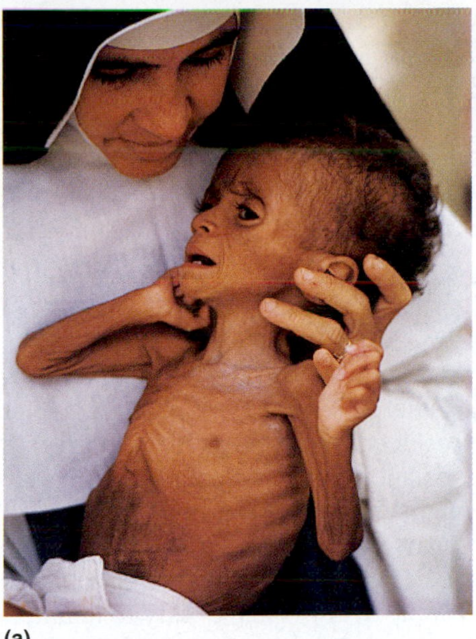

(a)

(b)

FIGURE 18.21 Two types of starvation in the young. (*a*) This child, suffering from marasmus, did not have adequate nutrition as an infant. (*b*) These children suffer from kwashiorkor. Although they may have received adequate nourishment from breast milk early in life, they became malnourished when their diet switched to a watery, white extract from cassava that looks like milk but has very little protein. The lack of protein in the diet causes edema and the ascites that swells their bellies.

Anorexia Nervosa

Anorexia nervosa (an″o-rek′se-ah ner′vo-sah) is self-imposed starvation. The condition is reported to affect 1 out of 250 adolescents, most of them female, although the true number among males is not known and may be higher than has been thought. The sufferer, typically a well-behaved adolescent girl from an affluent family, perceives herself to be overweight and eats barely enough to survive (fig. 18.22). She is terrified of gaining weight and usually loses 25% of her original body weight. In addition to eating only small amounts of low-calorie foods, she further loses weight by vomiting, taking laxatives and diuretics, or exercising intensely. Her eating behavior is often ritualized. She may meticulously arrange her meager meal on her plate or consume only a few foods. She develops low blood pressure, a slowed or irregular heartbeat, constipation, and constant chilliness. She stops menstruating as her body fat level plunges. Like any starving person, the hair becomes brittle and the skin dries out. To conserve body heat, she may develop soft, pale, fine body hair called lanugo, normally seen only on a fetus.

When the person with anorexia reaches an obviously emaciated state, her parents usually have her hospitalized, where she is fed intravenously so that she does not starve to death or die suddenly of heart failure due to an electrolyte imbalance. She also receives psychotherapy and nutritional counseling. Despite these efforts, 15% to 21% of people with anorexia die.

Anorexia nervosa has no known physical cause. One hypothesis is that the person is rebelling against approaching womanhood. Indeed, her body is astonishingly childlike, and she has often ceased to menstruate. She typically has low self esteem and believes that others, particularly her parents, are controlling her life. Her weight is something that she can control. Anorexia can be a one-time, short-term experience or a lifelong obsession.

FIGURE 18.22 Perceiving herself overweight, this young woman is tying the measuring tape extraordinarily tight around her waist to have a waist measurement as small as possible.

Bulimia

A person suffering from **bulimia** (bu-lim′e-ah) is often of normal weight. She eats whatever she wants, often in huge amounts, but she then rids her body of the thousands of extra calories by vomiting, taking laxatives, or exercising frantically. For an estimated one in five college students, the majority of them female, "bingeing and purging" appears to be a way of coping with stress.

Sometimes a dentist is the first to spot bulimia by observing a patient with teeth decayed from frequent vomiting. The backs of her hands may bear telltale scratches from efforts to induce vomiting. Her throat is raw and her esophageal lining ulcerated from the stomach acid forced forward by vomiting. The binge and purge cycle is hard to break, even with psychotherapy and nutritional counseling.

Underweight and overweight are societal problems complexly connected to economics. A phenomenon called a "dual-burden household" affects nations that were recently considered to be developing, but where the gross national product is on the rise. The problem is that in poor countries, the poorest people lack food and toil at physically taxing jobs, and so are severely underweight. But as economic resources come to a country, poor people who find jobs in newly urbanized areas begin to eat the cheapest food, of low nutritional quality, and sometimes too much of it. Office jobs rather than working in fields, and wider availability of television, promote a sedentary lifestyle. The result: weight gain. At the same time, underweight in children under five is increasing in urban areas of countries whose socioeconomic status is changing, according to the World Bank. This may reflect lack of home-grown foods, previously a major dietary staple, as both parents work in nonagricultural jobs. One researcher calls the coexistence of underweight and overweight in transitional countries "a nutritional paradox."

PRACTICE

78 What is primary malnutrition? Secondary malnutrition?

79 What happens to the body during starvation?

80 How do marasmus and kwashiorkor differ?

81 How do anorexia nervosa and bulimia differ?

18.9 LIFE-SPAN CHANGES

Dietary requirements remain generally the same throughout life, but the ability to acquire those nutrients may change drastically. The basal metabolic rate (BMR) changes with age. It rises from birth to about age five and then declines until adolescence, when it peaks again. During adulthood, the BMR drops in parallel to decreasing activity levels and shrinking muscle mass. In women, it may spike during pregnancy and breastfeeding, when caloric requirements likewise increase. Table 18.12 shows changes in energy requirements for adults who are healthy and engage in regular, light exercise.

TABLE 18.12 | Energy Requirements Decline with Age

	CAL/DAY	
Age	Female	Male
23–50	2,000	2,700
51–74	1,800	2,400
75+	1,600	2,050

For all ages, weight gain occurs when energy in exceeds energy out, and weight loss happens when energy out exceeds energy in. Age fifty seems to be a key point in energy balance. For most people, energy balance is positive, and weight is maintained before this age, but afterwards, weight may creep up. However, being aware of a decrease in activity, and curbing food consumption accordingly, enables many people over the age of fifty to maintain their weight.

Changing nutrition with age often reflects effects of medical conditions, many of which are more common among older people, and social and economic circumstances. Medications can dampen appetite directly through side effects such as nausea or altered taste perception or affect a person's mood in a way that prevents eating. Poverty may take a greater nutritional toll on older people who either cannot get out to obtain food or who give whatever is available to younger people.

Medical conditions that affect the ability to obtain adequate nutrition include depression, tooth decay and periodontal disease, diabetes mellitus, lactose intolerance, and alcoholism. These conditions may lead to deficiencies that are not immediately obvious. Vitamin A deficiency, for example, may take months or years to become noticeable because the liver stores this fat-soluble vitamin. Calcium depletion may not produce symptoms, even as the mineral is taken from bones. The earliest symptom of malnutrition, fatigue, may easily be attributed to other conditions or ignored.

Evidence for vitamin D deficiency related to sun avoidance has a long history. The link between lack of sunlight and development of rickets was noted in 1822, and a century later, researchers realized that sun exposure helps reverse the disease in children. Other evidence comes from diverse sources, such as women who wear veils and naval personnel serving three-month tours of duty on submarines.

It is important to obtain a good balance of nutrients and enough energy throughout life. Many studies link caloric restriction to increased longevity in species such as mice and fruit flies. However, these observations cannot be extrapolated to humans because the experimental laboratory animals were kept extremely healthy. Human starvation is usually the consequence of many other problems and is more likely to lead to malnutrition than increased longevity.

PRACTICE

82 List factors that affect nutrient acquisition.

83 Describe changes in BMR throughout life.

CHAPTER SUMMARY

18.1 INTRODUCTION (PAGE 699)

Nutrients include carbohydrates, lipids, proteins, vitamins, and minerals. The ways nutrients are used to support life processes constitute metabolism. Essential nutrients are required for health, and body cells cannot synthesize them. Hormones communicate from the gastrointestinal tract to the hypothalamus to control appetite, and monitor fat stores.

18.2 CARBOHYDRATES (PAGE 700)

Carbohydrates are organic compounds primarily used to supply cellular energy.

1. Carbohydrate sources
   a. Carbohydrates are ingested in a variety of forms.
   b. Polysaccharides, disaccharides, and monosaccharides are carbohydrates.
   c. Cellulose is a polysaccharide that human enzymes cannot digest, but it provides bulk that facilitates movement of intestinal contents.
2. Carbohydrate use
   a. Carbohydrates are absorbed as monosaccharides.
   b. Enzymes in the liver catalyze reactions that convert fructose and galactose into glucose.
   c. Oxidation releases energy from glucose.
   d. Excess glucose is stored as glycogen or combined to produce fat.
3. Carbohydrate requirements
   a. Most carbohydrates supply energy; some are used to produce sugars.
   b. Some cells require a continuous supply of glucose to survive.
   c. If glucose is scarce, amino acids may react to produce glucose.
   d. Humans survive with a wide range of carbohydrate intakes.
   e. Poor nutritional status is usually related to low intake of nutrients other than carbohydrates.

18.3 LIPIDS (PAGE 702)

Lipids are organic compounds that supply energy and are used to build cell structures. They include fats, phospholipids, and cholesterol.

1. Lipid sources
   a. Triglycerides are obtained from foods of plant and animal origins.
   b. Cholesterol is mostly obtained in foods of animal origin.
2. Lipid use
   a. Before fats can be used as an energy source, they must be broken down into glycerol and fatty acids.
   b. Beta oxidation decomposes fatty acids.
      (1) Beta oxidation activates fatty acids and breaks them down into segments of two carbon atoms each.
      (2) Fatty acid segments are converted into acetyl coenzyme A, which can then be oxidized in the citric acid cycle.
   c. The liver and adipose tissue control triglyceride metabolism.
   d. Liver enzymes can alter the molecular structures of fatty acids.
   e. Linoleic acid and linolenic acid are essential fatty acids.
   f. The liver regulates cholesterol level by synthesizing or excreting it.
3. Lipid requirements
   a. Humans survive with a wide range of lipid intakes.
   b. The amounts and types of lipids needed for health are unknown.
   c. Fat intake must be sufficient to supply fat-soluble vitamins.

18.4 PROTEINS (PAGE 704)

Proteins are organic compounds that serve as structural materials, act as enzymes, and provide energy. Amino acids are incorporated into various structural and functional proteins, including enzymes. During starvation, tissue proteins may be used as energy sources; thus, the tissues waste away.

1. Protein sources
   a. Proteins are mainly obtained from meats, dairy products, cereals, and legumes.
   b. During digestion, proteins are broken down into amino acids.
   c. The resulting amino acids can be used to form new protein molecules such as enzymes, clotting factors, keratin, elastin, collagen, actin, myosin, hormones, and antibodies, or can be used as energy sources.
   d. Before amino acids can be used as energy sources, they must be deaminated.
   e. The deaminated portions of amino acids can be broken down into carbon dioxide and water or used to produce glucose or fat.
   f. Eight amino acids are essential for adults, whereas ten are essential for growing children.
   g. All essential amino acids must be present at the same time for growth and repair of tissues to take place.
   h. Complete proteins contain adequate amounts of all the essential amino acids needed to maintain the tissues and promote growth.
   i. Incomplete proteins lack adequate amounts of one or more essential amino acids.
2. Nitrogen balance
   a. In healthy adults, the gain of protein equals the loss of protein, and a nitrogen balance exists.
   b. A starving person has a negative nitrogen balance; a growing child, a pregnant woman, or an athlete in training usually has a positive nitrogen balance.
3. Protein requirements
   a. Proteins and amino acids are needed to supply essential amino acids and nitrogen for the synthesis of nitrogen-containing molecules.
   b. The consequences of protein deficiencies are particularly severe among growing children.

18.5 ENERGY EXPENDITURES (PAGE 706)

Energy is of prime importance to survival and may be obtained from carbohydrates, fats, or proteins.

1. Energy values of foods
   a. The potential energy values of foods are expressed in calories.
   b. When energy losses due to incomplete absorption and incomplete oxidation are taken into account, 1 gram of carbohydrate or 1 gram of protein yields about 4 calories, whereas 1 gram of fat yields about 9 calories.
2. Energy requirements
   a. The amount of energy required varies from person to person.
   b. Factors that influence energy requirements include basal metabolic rate, muscular activity, body temperature, and nitrogen balance.
3. Energy balance
   a. Energy balance exists when caloric intake equals caloric output.
   b. If energy balance is positive, body weight increases; if energy balance is negative, body weight decreases.
4. Desirable weight
   a. The most common nutritional disorders involve caloric imbalances.
   b. Average weights of persons 25–30 years of age are desirable for older persons as well.
   c. Body mass index assesses weight taking height into account.
   d. A BMI between 25 and 30 indicates overweight, and above 30, obesity.

18.6 VITAMINS (PAGE 709)

Vitamins are organic compounds (other than carbohydrates, lipids, and proteins), that cannot be synthesized by body cells in adequate amounts and are essential for normal metabolic processes.

1. Fat-soluble vitamins
   a. General characteristics
      (1) Fat-soluble vitamins are carried in lipids and are influenced by the same factors that affect lipid absorption.
      (2) They resist the effects of heat; thus, they are not destroyed by cooking or food processing.
   b. Vitamin A
      (1) Vitamin A exists in several forms, is synthesized from carotenes, and is stored in the liver.
      (2) It is an antioxidant required for production of visual pigments.
   c. Vitamin D
      (1) Vitamin D is a group of related steroids.
      (2) It is found in certain foods and is produced commercially; it can also be synthesized in the skin.
      (3) When needed, vitamin D is converted by the kidneys to an active form that functions as a hormone and promotes the intestine's absorption of calcium and phosphorus.
   d. Vitamin E
      (1) Vitamin E is an antioxidant.
      (2) It is stored in muscles and adipose tissue.
      (3) It prevents breakdown of polyunsaturated fatty acids and stabilizes cell membranes.
   e. Vitamin K
      (1) Vitamin K is in foods and is produced by intestinal bacteria.
      (2) Some vitamin K is stored in the liver.
      (3) It is used to produce prothrombin, required for blood clotting.
2. Water-soluble vitamins
   a. General characteristics
      (1) Water-soluble vitamins include the B vitamins and vitamin C.
      (2) Cooking or processing food destroys some water-soluble vitamins.
      (3) B vitamins make up a group called the vitamin B complex and oxidize carbohydrates, lipids, and proteins.
   b. Vitamin B complex
      (1) Thiamine (vitamin B_1)
         (a) Thiamine functions as part of coenzymes that oxidize carbohydrates and synthesize essential sugars.
         (b) Small amounts are stored in the tissues; excess is excreted in the urine.
         (c) Quantities needed vary with caloric intake.
      (2) Riboflavin (vitamin B_2)
         (a) Riboflavin functions as part of several enzymes and coenzymes essential to the oxidation of glucose and fatty acids.
         (b) Its absorption is regulated by an active transport system; excess is excreted in the urine.
         (c) Quantities required vary with caloric intake.
      (3) Niacin (nicotinic acid or vitamin B_3)
         (a) Niacin functions as part of coenzymes required for the oxidation of glucose and for the synthesis of proteins and fats.
         (b) It can be synthesized from tryptophan; daily requirement varies with the tryptophan intake.
      (4) Pantothenic acid (vitamin B_5)
         (a) Pantothenic acid functions as part of coenzyme A; thus, it is essential for energy-releasing mechanisms.
         (b) Most diets provide sufficient amounts; deficiencies are rare.
      (5) Vitamin B_6
         (a) Vitamin B_6 is a group of compounds that function as coenzymes in metabolic pathways that synthesize proteins, certain amino acids, antibodies, and nucleic acids.
         (b) Its requirement varies with protein intake.
      (6) Cyanocobalamin (vitamin B_{12})
         (a) The cyanocobalamin molecule contains cobalt.
         (b) Its absorption is regulated by the secretion of intrinsic factor from the gastric glands.
         (c) It functions as part of coenzymes needed for nucleic acid synthesis and for the metabolism of carbohydrates and fats.

(7) Folacin (folic acid)
    (a) Liver enzymes catalyze reactions that convert folacin to physiologically active folinic acid.
    (b) It is a coenzyme needed for the metabolism of certain amino acids, DNA synthesis, and the normal production of red blood cells.
(8) Biotin
    (a) Biotin is a coenzyme required for the metabolism of amino acids and fatty acids, and for nucleic acid synthesis.
    (b) It is stored in metabolically active organs, including the brain, liver, and kidneys.

c. Ascorbic acid (vitamin C)
    (1) Vitamin C is similar chemically to monosaccharides.
    (2) It is required for collagen production, the metabolism of certain amino acids, and iron absorption.
    (3) It is not stored in large amounts; excess is excreted in the urine.

18.7 MINERALS (PAGE 717)

1. Characteristics of minerals
    a. Minerals account for about 4% of body weight.
    b. About 75% by weight of the minerals are found in bones and teeth as calcium and phosphorus.
    c. Minerals are usually incorporated into organic molecules, although some are in inorganic compounds or are free ions.
    d. They compose structural materials, function in enzymes, and play vital roles in various metabolic processes.
    e. Homeostatic mechanisms regulate mineral concentrations.
    f. The physiologically active form of minerals is the ionized form.
2. Major minerals
    a. Calcium
        (1) Calcium is essential for forming bones and teeth, neurotransmitter release, contracting muscle fibers, clotting blood, and activating various enzymes.
        (2) Existing calcium concentration, vitamin D, protein intake, and motility of the digestive tract affect calcium absorption.
    b. Phosphorus
        (1) Phosphorus is incorporated into the salts of bones and teeth.
        (2) It participates in nearly all metabolic reactions as a constituent of nucleic acids, proteins, and some vitamins.
        (3) It also is in the phospholipids of cell membranes, in ATP, and in phosphates of body fluids.
    c. Potassium
        (1) Potassium is concentrated inside cells.
        (2) It maintains osmotic pressure, regulates pH, and plays a role in nerve impulse conduction.
    d. Sulfur
        (1) Sulfur is incorporated into two of the twenty amino acids.
        (2) It is also in thiamine, insulin, biotin, and mucopolysaccharides.
    e. Sodium
        (1) Most sodium is in extracellular fluids or is bound to the inorganic salts of bone.
        (2) The kidneys, under the influence of aldosterone, regulate the blood concentration of sodium.
        (3) Sodium helps maintain solute concentration and regulates water balance.
        (4) It is essential for nerve impulse conduction and moving substances through cell membranes.
    f. Chlorine
        (1) Chlorine is closely associated with sodium as chloride ions.
        (2) It acts with sodium to help maintain osmotic pressure, regulate pH, and maintain electrolyte balance.
        (3) Chlorine is essential for hydrochloric acid formation and for carbon dioxide transport by red blood cells.
    g. Magnesium
        (1) Magnesium is abundant in the bones as phosphates and carbonates.
        (2) It functions in ATP production and in the breakdown of ATP to ADP.
        (3) A reserve supply of magnesium is stored in the bones; excesses are excreted in the urine.
3. Trace elements
    a. Iron
        (1) Iron is part of hemoglobin in red blood cells and myoglobin in muscles.
        (2) A reserve supply of iron is stored in the liver, spleen, and bone marrow.
        (3) It is required to catalyze vitamin A formation; it is also incorporated into various enzymes and the cytochrome molecules.
    b. Manganese
        (1) Most manganese is concentrated in the liver, kidneys, and pancreas.
        (2) It is necessary for normal growth and development of skeletal structures and other connective tissues; it is essential for the synthesis of fatty acids, cholesterol, and urea.
    c. Copper
        (1) Most copper is concentrated in the liver, heart, and brain.
        (2) It is required for hemoglobin synthesis, bone development, melanin production, and myelin formation.
    d. Iodine
        (1) Iodine is most highly concentrated in the thyroid gland.
        (2) It is an essential component of thyroid hormones.
        (3) It is often added to foods as iodized table salt.
    e. Cobalt
        (1) Cobalt is widely distributed throughout the body.
        (2) It is an essential part of cyanocobalamin and is required for the synthesis of several enzymes.

f. Zinc
   (1) Zinc is most concentrated in the liver, kidneys, and brain.
   (2) It is a component of several enzymes that take part in digestion, respiration, and metabolism.
g. Fluorine
   (1) The teeth concentrate fluorine.
   (2) It is incorporated into enamel and prevents dental caries.
h. Selenium
   (1) The liver and kidneys store selenium.
   (2) It is a component of certain enzymes.
i. Chromium
   (1) Chromium is widely distributed throughout the body.
   (2) It regulates glucose use.

18.8 HEALTHY EATING (PAGE 722)

1. An adequate diet provides sufficient energy and essential nutrients to support optimal growth, as well as maintenance and repair, of tissues.
2. Individual needs vary so greatly that it is not possible to design a diet adequate for everyone.
3. Devices to help consumers make healthy food choices include Recommended Daily Allowances, Recommended Dietary Allowances, food group plans, food pyramids, and food labels.
4. Malnutrition
   a. Poor nutrition is due to lack of foods or failure to wisely use available foods.
   b. Primary malnutrition is due to poor diet.
   c. Secondary malnutrition is due to an individual characteristic that makes a normal diet inadequate.
5. Starvation
   a. A person can survive fifty to seventy days without food.
   b. A starving body digests itself, starting with carbohydrates, then fats, then proteins.
   c. Symptoms include low blood pressure, slow pulse, chills, dry skin, hair loss, and poor immunity. Finally, vital organs cease to function.
   d. Marasmus is lack of all nutrients.
   e. Kwashiorkor is protein starvation.
   f. Anorexia nervosa is a self-starvation eating disorder.
   g. Bulimia is an eating disorder characterized by bingeing and purging.

18.9 LIFE-SPAN CHANGES (PAGE 728)

1. Changing nutrition with age reflects medical conditions and social and economic circumstances.
2. Basal metabolic rate rises in early childhood, declines, then peaks again in adolescence, with decreasing activity during adulthood.
3. Weight gain, at any age, occurs when energy in exceeds energy out, and weight loss occurs when energy out exceeds energy in.

CHAPTER ASSESSMENTS

18.1 Introduction

1 Define *nutrition.* (p. 699)

2 Contrast nutrients and essential nutrients. (p. 699)

3 Contrast leptin and ghrelin as they affect appetite control. (p. 700)

18.2–18.4 Carbohydrates–Proteins

4 Identify dietary sources of carbohydrates. (p. 700)

5 Summarize the importance of cellulose in the diet. (p. 701)

6 Explain what happens to excess glucose in the body. (p. 701)

7 Explain why a temporary drop in blood glucose concentration may impair nervous system functioning. (p. 701)

8 List some factors that affect an individual's need for carbohydrates. (p. 702)

9 Identify dietary sources of lipids. (p. 702)

10 Define *triglyceride.* (p. 702)

11 Define *beta oxidation.* (p. 702)

12 Explain how fats may provide energy. (p. 702)

13 Describe the liver's role in fat metabolism. (p. 703)

14 Review the major functions of cholesterol. (p. 703)

15 Define *deamination,* and explain its importance. (p. 704)

16 Identify dietary sources of proteins. (p. 705)

17 Distinguish between essential and nonessential amino acids. (p. 705)

18 Explain why all of the essential amino acids must be present for growth. (p. 705)

19 Distinguish between complete and incomplete proteins. (p. 705)

20 ____________________ is when the amount of nitrogen taken in is equal to the amount excreted. (p. 705)

21 Explain why a protein deficiency may accompany edema. (p. 706)

18.5 Energy Expenditures

22 Define *calorie.* (p. 706)

23 Explain how the caloric values of foods are determined. (p. 706)

24 Define *basal metabolic rate.* (p. 707)

25 List some of the factors that effect the BMR. (p. 707)

26 ____________________ exists when caloric intake in the form of foods equals caloric output from basal metabolic rate and muscular activities. (p. 708)

27 Distinguish among desirable weight, overweight, and obesity. (p. 708)

18.6 Vitamins

28 Match the vitamins with their general functions, and indicate if the vitamin is fat-soluble or water-soluble. Functions may be used more than once, and more than one function may be applied to a vitamin. (pp. 709–716)

(1) vitamin A
(2) vitamin B_1 (thiamine)
(3) vitamin B_2 (riboflavin)
(4) vitamin B_3 (niacin)
(5) vitamin B_5 (pantothenic acid)
(6) vitamin B_6
(7) vitamin B_{12} (cyanocobalamin)
(8) folacin
(9) biotin
(10) vitamin C (ascorbic acid)
(11) vitamin D
(12) vitamin E
(13) vitamin K

A. Part of coenzyme A in oxidation of carbohydrates
B. Required for ribose synthesis
C. Necessary for synthesis of visual pigments
D. Required for synthesis of prothrombin
E. Required to produce collagen
F. Required to synthesize nucleic acids
G. Promotes red blood cell production
H. Plays a role in myelin synthesis
I. Antioxidant, helps stabilize cell membranes
J. Promotes development of teeth and bones
K. Required to produce anti-bodies
L. Required for cellular reproduction
M. Part of coenzymes to synthesize proteins, fats, and nucleic acids

18.7 Minerals

29 Match the minerals/elements with their functions, and indicate whether each is a major mineral or a trace element required for nutrition. Functions may be used more than once, and more than one function may be applied to a mineral or trace element. (pp. 717–722)

(1) calcium
(2) chlorine
(3) chromium
(4) cobalt
(5) copper
(6) fluorine
(7) iodine
(8) iron
(9) magnesium
(10) manganese
(11) phosphorus
(12) potassium
(13) selenium
(14) sodium
(15) sulfur
(16) zinc

A. Essential for the use of carbohydrates
B. Component of certain enzymes
C. Component of tooth enamel
D. Component of teeth and bones
E. Helps maintain intracellular osmotic pressure
F. Essential part of certain amino acids
G. Helps maintain extracellular fluid osmotic pressure
H. Necessary for normal wound healing
I. Component of cyanocobalamin
J. Essential for synthesis of thyroid hormones
K. Required in metabolic reactions associated with ATP production
L. Component of hemoglobin
M. Essential for hemoglobin synthesis and melanin production
N. Required for cholesterol synthesis and urea formation

18.8 Healthy Eating

30 Define *adequate diet.* (p. 722)

31 Explain various methods to eat an adequate diet. (p. 722)

32 Define *malnutrition.* (p. 723)

33 Contrast primary and secondary malnutrition. (p. 723)

34 Discuss bodily changes during starvation. (p. 723)

35 Distinguish among marasmus, kwashiorkor, anorexia nervosa, and bulimia. (p. 726)

18.9 Life-Span Changes

36 Factors that may lead to inadequate nutrition later in life include __________. (p. 728)

a. medical conditions
b. social circumstances
c. economic circumstances
d. medication
e. all of the above

37 Name some medical conditions that affect the ability to obtain adequate nutrition as a person ages. (p. 728)

INTEGRATIVE ASSESSMENTS/CRITICAL THINKING

OUTCOMES 4.3, 9.4, 18.2, 18.3, 18.4, 18.5, 18.8

1. Which of the diets described in the following chart would be most appropriate for an athlete training for a triathlon (biking, swimming, running event)? What is a problem with all of these diets for such a person?

Diet	Total Calories/Day	% Fat	% Carbohydrate	% Protein
Dean Ornish	1450	10–20	70	17
Jenny Craig	1450	25	60	15
Zone	1400	60	10	30

OUTCOMES 4.3, 18.2, 18.3, 18.4, 18.5, 18.6

2. A young man takes several vitamin supplements each day, claiming that they give him energy. Is he correct? Why or why not?

OUTCOMES 13.9, 18.1, 18.2, 18.3, 18.4

3. Why does the blood sugar concentration of a person whose diet is low in carbohydrates remain stable?

OUTCOMES 18.1, 18.2, 18.3, 18.4, 18.5, 18.6, 18.7, 18.8, 18.9

4. How do you think the nutritional requirements of a healthy twelve-year-old boy, a twenty-four-year-old pregnant woman, and a healthy sixty-year-old man differ?

OUTCOMES 18.2, 18.3, 18.4, 18.5

5. Using nutrient tables, calculate the number of grams of carbohydrate, lipid, and protein that you eat in a typical day, and the total calories in these foods. Suggest ways to improve your diet.

OUTCOMES 18.2, 18.3, 18.4, 18.5, 18.6, 18.7, 18.8

6. For each of the following diets, indicate how the diet is nutritionally unsound (if it is) and why it would be easy or difficult to follow.
   a. For the first ten days of the Beverly Hills diet, only fruit is eaten. On day ten, you can eat a bagel and butter, and then only fruit until day nineteen, when you can eat steak or lobster. The cycle repeats, adding more meat. This diet is based on "conscious combining"—the idea that eating certain combinations of foods leads to weight loss.
   b. The Weight Loss Clinic diet consists of 800 calories per day, with 46.1% protein, 35.2% carbohydrate, and 18.7% fat.
   c. The macrobiotic diet includes 10% to 20% protein, 70% carbohydrate, and 10% fat, with a half hour of walking each day. Most familiar foods are forbidden, but you can eat many unusual foods—such as rice cakes, seaweed, barley stew, pumpkin soup, rice gruel, kasha and onions, millet balls, wheat berries, and parsnip chips.
   d. The No Aging diet maintains that eating food rich in nucleic acids (RNA and DNA) can prolong life, because these are the genetic materials. Recommended foods include sardines, salmon, calves' liver, lentils, and beets.

OUTCOMES 18.2, 18.3, 18.4, 18.6, 18.7

7. Examine the label information on the packages of a variety of breakfast cereals. Which types of cereals provide the best sources of carbohydrates, lipids, proteins, vitamins, and minerals? Which major nutrients are lacking in these cereals?

OUTCOME 18.4

8. If a person decided to avoid eating meat and other animal products, such as milk, cheese, and eggs, what foods might be included in the diet to provide essential amino acids?

WEB CONNECTIONS

Be sure to visit the text website at **www.mhhe.com/shier12** for answers to chapter assessments, additional quizzes, and interactive learning exercises.

ANATOMY & PHYSIOLOGY REVEALED

Anatomy & Physiology Revealed® (APR) includes cadaver photos that allow you to peel away layers of the human body to reveal structures beneath the surface. This program also includes animations, radiologic imaging, audio pronunciations, and practice quizzing. Check out **www.aprevealed.com**. APR has been proven to help improve student grades!

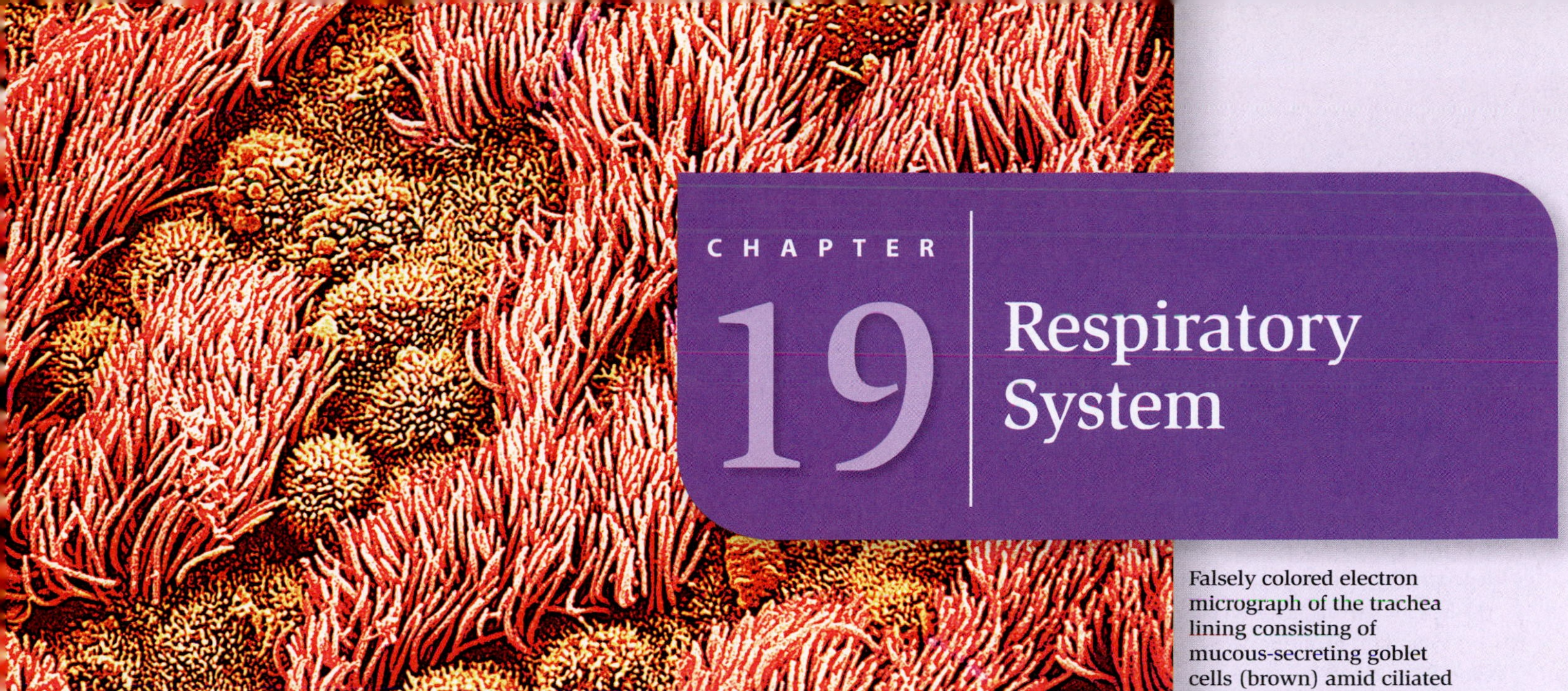

CHAPTER

19 Respiratory System

Falsely colored electron micrograph of the trachea lining consisting of mucous-secreting goblet cells (brown) amid ciliated epithelium (red) (3,700×).

UNDERSTANDING WORDS

alveol-, small cavity: *alveol*us—microscopic air sac in a lung.

bronch-, windpipe: *bronch*us—primary branch of the trachea.

carcin-, spreading sore: *carcin*oma—type of cancer.

carin-, keel-like: *carina*—ridge of cartilage betweeen the right and left bronchi.

cric-, ring: *cric*oid cartilage—ring-shaped mass of cartilage at the base of the larynx.

epi-, upon: *epi*glottis—flaplike structure that partially covers the opening into the larynx during swallowing.

hem-, blood: *hem*oglobin—pigment in red blood cells.

inhal-, to breathe in: *inhal*ation—to take air into the lungs.

phren-, diaphragm: *phren*ic nerve—nerve associated with the cervical plexuses that stimulates the muscle fibers of the diaphragm to contract.

tuber-, swelling: *tuber*culosis—disease in which fibrous masses form in the lungs.

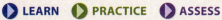

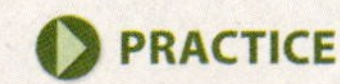

LEARN PRACTICE ASSESS

LEARNING OUTCOMES

After you have studied this chapter, you should be able to:

19.1 Introduction
- **1** Identify the general functions of the respiratory system. (p. 736)

19.2 Why We Breathe
- **2** Explain why respiration is necessary for cellular survival. (p. 736)

19.3 Organs of the Respiratory System
- **3** Name and describe the locations of the organs of the respiratory system. (p. 737)
- **4** Describe the functions of each organ of the respiratory system. (p. 737)

19.4 Breathing Mechanism
- **5** Explain how inspiration and expiration are accomplished. (p. 750)
- **6** Describe each of the respiratory air volumes and capacities. (p. 752)
- **7** Show how alveolar ventilation rate is calculated. (p. 754)
- **8** List several nonrespiratory air movements, and explain how each occurs. (p. 754)

19.5 Control of Breathing
- **9** Locate the respiratory areas, and explain control of normal breathing. (p. 755)
- **10** Discuss how various factors affect breathing. (p. 757)

19.6 Alveolar Gas Exchanges
- **11** Describe the structure and function of the respiratory membrane. (p. 760)
- **12** Explain the importance of partial pressure in diffusion of gases. (p. 760)

19.7 Gas Transport
- **13** Explain how the blood transports oxygen and carbon dioxide. (p. 762)
- **14** Describe gas exchange in the pulmonary and systemic circuits. (p. 763)

19.8 Life-Span Changes
- **15** Describe the effects of aging on the respiratory system. (p. 767)

DANGER! SECONDHAND SMOKE

Exposure to environmental tobacco smoke (ETS)—also called secondhand smoke—may be as dangerous as smoking. ETS has two sources: *sidestream* smoke comes from lit cigarettes, cigars, or pipes and *mainstream* smoke is exhaled by smokers. The smoke contains more than 4,000 chemical compounds, including irritants; carcinogens; mutagens; and many toxins, such as benzene, formaldehyde, vinyl chloride, ammonia, arsenic, and cyanide.

An estimated 53,000 nonsmokers in the United States die each year from ETS. That includes 46,000 heart disease deaths, 3,400 lung cancer deaths, and 1,900 to 2,700 infant deaths. In children, ETS is also responsible for hundreds of thousands of cases of lower respiratory infections (bronchitis or pneumonia), worsening asthma and ear infections, and low birth weight. One study found that two-to-five-year-olds absorb six times as much nicotine as older children and have higher levels of biomarkers of vascular injury.

Even short exposures are dangerous, to anyone. Experiments that exposed healthy nonsmokers to a room simulating a smoky bar showed that just thirty minutes of breathing secondhand smoke activates platelets, decreases coronary artery blood flow, and produces heart rate variability—all changes that raise the risk for cardiovascular disease. The normally smooth endothelial inner lining of blood vessels becomes damaged, and endothelial progenitor cells that typically travel to such injury sites from the circulation are impaired. So not only are blood vessel linings damaged, they cannot readily heal. The effects on progenitor cells persist for at least twenty-four hours.

The only way to decrease exposure to ETS is to eliminate it. Dividing a space into smoking and nonsmoking areas is futile, because smoke lingers for hours. This is why airplanes, restaurants, and many workplaces have been made smoke-free. In areas where new laws ban smoking in bars, restaurants, and workplaces, hospital admissions for heart attacks have dropped 20%. ■

The danger of secondhand smoke, long debated, is now widely accepted, and new rules and regulations attempt to limit exposure.

19.1 INTRODUCTION

The respiratory system consists of passages that filter incoming air and transport it into the body, into the lungs, and to the many microscopic air sacs where gases are exchanged. The entire process of exchanging gases between the atmosphere and body cells is called **respiration** (res″pĭ-ra′shun). It consists of several events:

- Movement of air in and out of the lungs, commonly called breathing, or *ventilation.*
- Exchange of gases between the air in the lungs and the blood, sometimes called *external respiration.*
- Transport of gases by the blood between the lungs and body cells.
- Exchange of gases between the blood and the body cells, sometimes called *internal respiration.*
- Oxygen (O_2) use and production of carbon dioxide (CO_2) by body cells as part of the process of *cellular respiration.*

19.2 WHY WE BREATHE

Respiration occurs on a macroscopic level—it is a function of an organ system. However, the reason that body cells must exchange gases—take up oxygen and release carbon dioxide—is apparent at the cellular and molecular levels.

RECONNECT

To Chapter 4, Aerobic Reactions, pages 122–124.

Cellular respiration enables cells to harness energy held in the chemical bonds of nutrient molecules. In aerobic reactions, cells liberate energy from these molecules by removing electrons and channeling them through a series of carriers called the electron transport chain. At the end of this chain, electrons bind oxygen atoms and hydrogen ions to produce water molecules. Without oxygen, these reactions cease.

Besides producing ATP, the aerobic reactions produce CO_2, a metabolic waste that combines with water to form carbonic acid, helping to maintain blood pH. Too much CO_2, however, lowers blood pH, compromising homeostasis. The respiratory system both provides oxygen for aerobic reaction, and eliminates CO_2 rapidly enough to maintain the pH of the internal environment.

19.3 ORGANS OF THE RESPIRATORY SYSTEM

The organs of the respiratory system can be divided into two groups, or tracts. Those in the *upper respiratory tract* include the nose, nasal cavity, sinuses, and pharynx. Those in the *lower respiratory tract* include the larynx, trachea, bronchial tree, and lungs (fig. 19.1).

Nose

The nose is covered with skin and is supported internally by muscle, bone, and cartilage. Its two *nostrils* (external nares) provide openings through which air can enter and leave the nasal cavity. Many internal hairs in these openings prevent entry of large particles carried in the air.

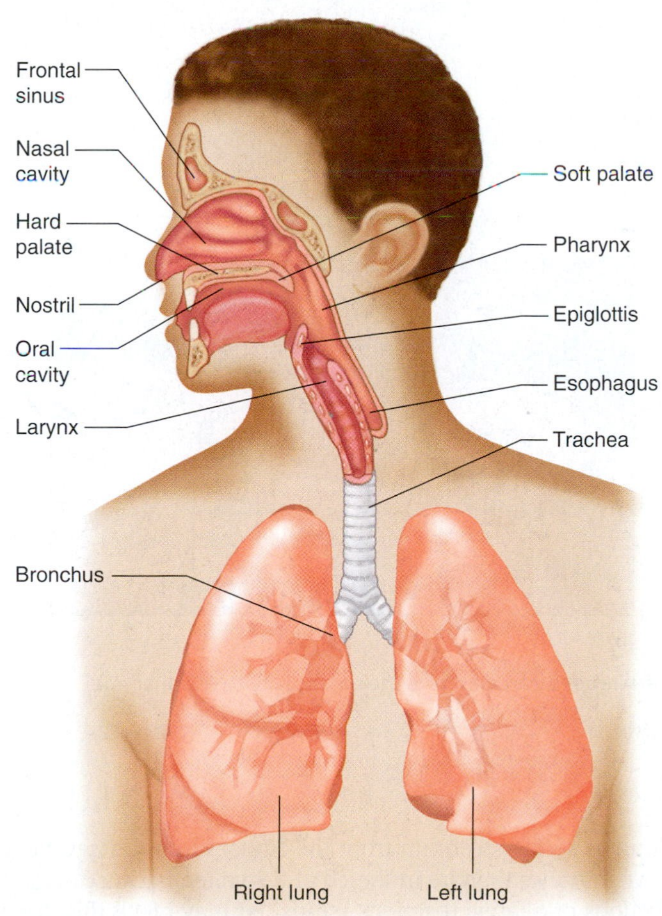

FIGURE 19.1 Organs of the respiratory system.

Nasal Cavity

The **nasal cavity,** a hollow space behind the nose, is divided medially into right and left portions by the **nasal septum.** This cavity is separated from the cranial cavity by the cribriform plate of the ethmoid bone and from the oral cavity by the hard palate.

> The nasal septum may bend during birth or shortly before adolescence. Such a *deviated septum* may obstruct the nasal cavity, making breathing difficult.

As figure 19.2 shows, **nasal conchae** (turbinate bones) curl out from the lateral walls of the nasal cavity on each side, forming passageways called the *superior, middle,* and *inferior meatuses* (see chapter 7, pp. 212 and 214). The nasal chonchae support the mucous membrane that lines the nasal cavity and help increase its surface area.

The upper posterior portion of the nasal cavity, below the cribriform plate, is slitlike, and its lining contains the olfactory receptors that provide the sense of smell. The remainder of the cavity conducts air to and from the nasopharynx.

The mucous membrane lining the nasal cavity has pseudostratified ciliated epithelium rich in mucous-secreting goblet cells (see chapter 5, p. 146). It also includes an extensive network of blood vessels and normally appears pinkish. As air passes over the membrane, heat radiates from the blood and warms the air, adjusting its temperature to that of the body, as evaporation of water from the mucous lining moistens the air. The sticky mucus the mucous membrane secretes entraps dust and other small particles entering with the air.

As the cilia of the epithelial cells move, a thin layer of mucus and any entrapped particles are pushed toward the pharynx (fig. 19.3). When the mucus reaches the pharynx, it is swallowed. In the stomach, gastric juice destroys microorganisms in the mucus, including pathogens. Thus, the filtering from the mucous membrane prevents particles from reaching the lower air passages and consequently prevents respiratory infections. Clinical Application 19.1 discusses how cigarette smoking impairs the respiratory system, beginning with the cleansing mucus and cilia.

PRACTICE

1. What is respiration?
2. Which organs constitute the respiratory system?
3. What is the function of the mucous membrane that lines the nasal cavity?
4. What is the function of the cilia on the cells that line the nasal cavity?

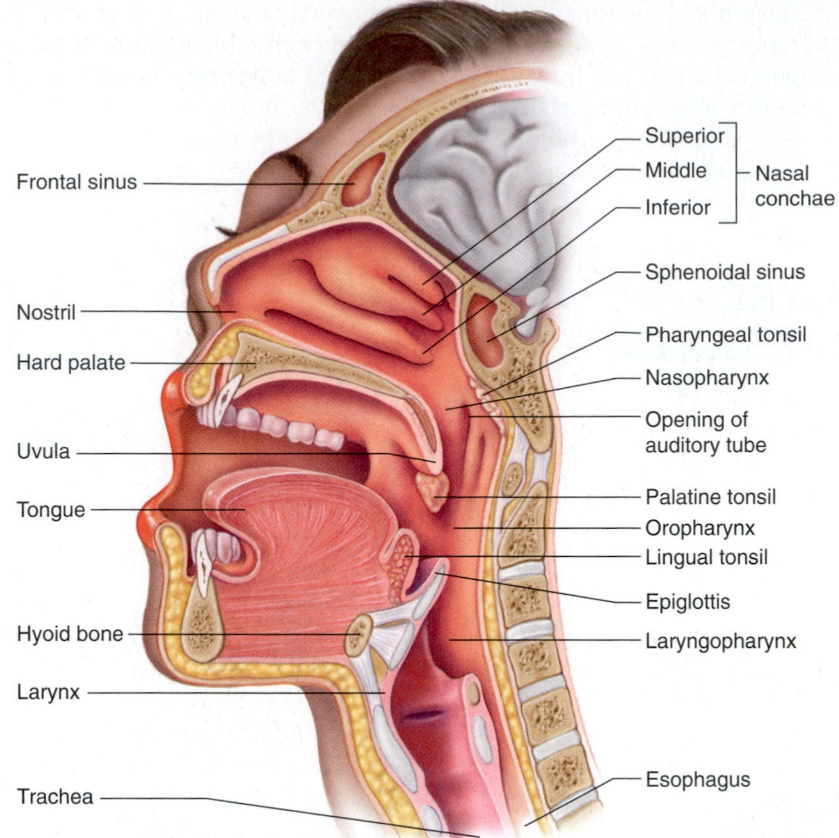

FIGURE 19.2 Major structures associated with the respiratory tract in the head and neck.

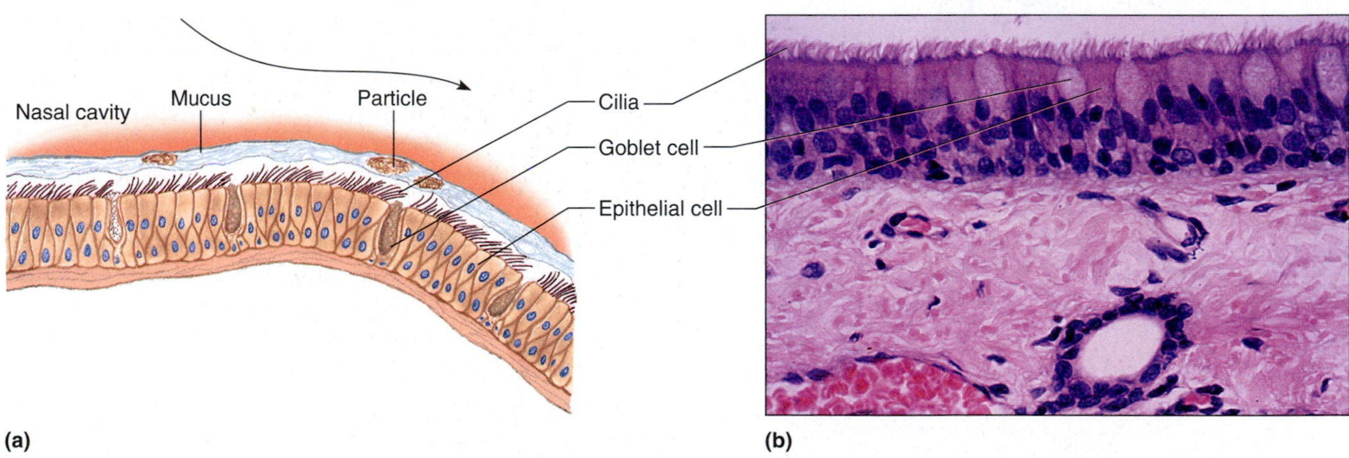

FIGURE 19.3 Mucus movement in the respiratory tract. (*a*) Cilia move mucus and trapped particles from the nasal cavity to the pharynx. (*b*) Micrograph of ciliated epithelium in the respiratory tract (275×).

Sinuses

Recall from chapter 7 (pp. 208 and 212) that the *sinuses* (paranasal sinuses) are air-filled spaces in the *frontal, sphenoid, ethmoid,* and *maxillary bones* of the skull (fig. 19.4). These spaces open into the nasal cavity and are lined with mucous membranes that are continuous with the lining of the nasal cavity, so mucus secretions drain from the sinuses into the nasal cavity. Membranes that are inflamed and swollen because of nasal infections or allergic reactions (sinusitis) may block this drainage, increasing pressure in a sinus and causing headache.

The sinuses reduce the weight of the skull. They also serve as resonant chambers that affect the quality of the voice.

19.1 CLINICAL APPLICATION

The Effects of Cigarette Smoking on the Respiratory System

Damage to the respiratory system from cigarette smoking is slow, progressive, and deadly. A healthy respiratory system is continuously cleansed. The mucus produced by the respiratory tubules traps dirt and pathogens, which cilia sweep toward the mouth, where they can be eliminated. Smoking greatly impairs this housekeeping. With the first inhalation of smoke, the beating of cilia slows. With time, the cilia become paralyzed and, eventually, disappear. The loss of cilia leads to "smoker's cough." The cilia no longer effectively remove mucus, which must be coughed up. Coughing is usually worse in the morning because mucus has accumulated during sleep.

To make matters worse, smokers produce excess mucus and it accumulates, clogging the air passageways. Pathogens normally removed now have easier access to the respiratory surfaces, and the resulting lung congestion favors their growth. This is why smokers have respiratory infections more often than nonsmokers. A lethal chain reaction begins. Cough leads to chronic bronchitis, and increased mucus production and bronchial lining thickening compromises breathing. The smaller airways lose elasticity and are no longer able to absorb the pressure changes of coughing. A cough can increase the air pressure in the alveoli (microscopic air sacs) enough to rupture their delicate walls. This is the beginning of smoking-induced *emphysema*. The burst alveoli worsen the cough, fatigue, wheezing, and impaired breathing. Emphysema is fifteen times more common among individuals who smoke a pack of cigarettes a day than among nonsmokers.

Simultaneous with the structural changes progressing to emphysema may be cellular changes leading to lung cancer. First, cells in the outer border of the bronchial lining begin to divide more rapidly than usual. Eventually, these cells displace the ciliated cells. Their nuclei begin to resemble those of cancerous cells—large and distorted with abnormal numbers of chromosomes. Up to this point, the damage can be repaired if smoking ceases. If smoking continues, these cells may eventually break through the basement membrane and begin dividing within the lung tissue, forming a tumor with the potential of spreading throughout lung tissue (figs. 19A and 19B) and beyond, such as to the brain or bones. While more than 80% of lung cancer cases are due to cigarette smoking, only 20% of smokers develop the cancer. Genetics may explain this seeming discrepancy. A DNA sequence on chromosome 15 affects part of the receptor for nicotine on neurons. A certain variant of that sequence increases the risk of developing the most common type of lung cancer—but only in smokers. Inheriting this DNA sequence does not affect cancer risk unless a person smokes.

It pays to quit. Much of the damage to the respiratory system can be repaired. Cilia are restored, and the thickening of alveolar walls due to emphysema can be reversed. But ruptured alveoli are gone forever. The nicotine in tobacco smoke causes a powerful dependency by binding to certain receptors on brain cells. ■

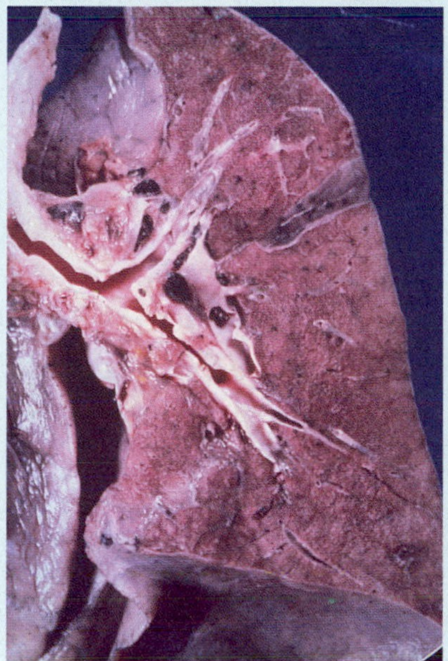

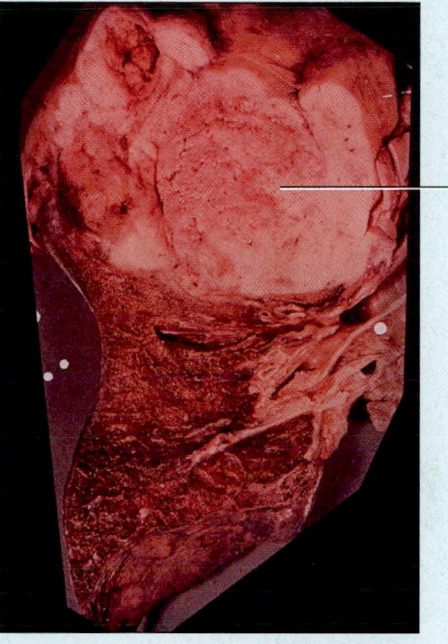

FIGURE 19A The lung on the left is healthy. A cancerous tumor has invaded the lung on the right, taking up nearly half of the lung space.

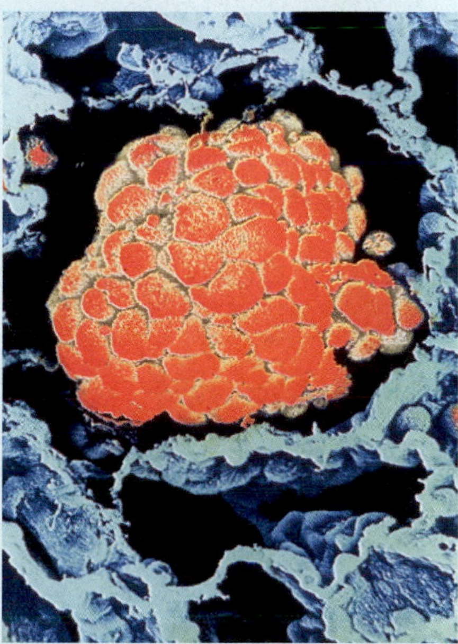

FIGURE 19B Lung cancer may begin as a tiny tumor growing in an alveolus, a microscopic air sac (125×). This image is falsely colored.

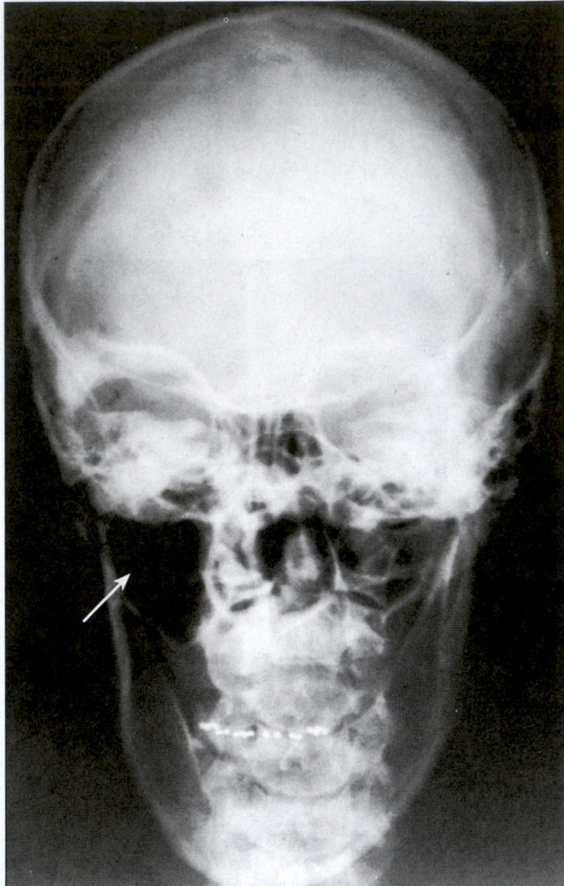

(a)

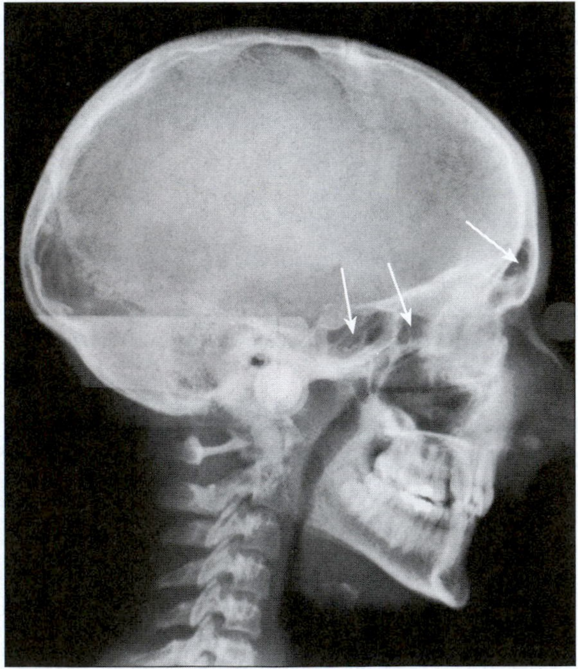

(b)

FIGURE 19.4 Radiograph of a skull (*a*) from the anterior view and (*b*) from the lateral view, showing air-filled sinuses (arrows) within the bones.

It is possible to illuminate a person's frontal sinus in a darkened room by holding a small flashlight just beneath the eyebrow. Similarly, holding the flashlight in the mouth illuminates the maxillary sinuses.

PRACTICE 

5 Where are the sinuses located?

6 What are the functions of the sinuses?

Pharynx

The **pharynx** (throat) is posterior to the nasal cavity, oral cavity, and larynx. It is a passageway for food moving from the oral cavity to the esophagus and for air passing between the nasal cavity and the larynx (see fig. 19.2). It also aids in producing the sounds of speech. The subdivisions of the pharynx—the nasopharynx, oropharynx, and laryngopharynx—are described in chapter 17 (p. 663).

Larynx

The **larynx** is an enlargement in the airway superior to the trachea (see reference plates 9 and 21). It is a passageway for air moving in and out of the trachea and prevents foreign objects from entering the trachea. The larynx also houses the *vocal cords*.

The larynx is composed of a framework of muscles and cartilages bound by elastic tissue. The largest of the cartilages are the thyroid, cricoid, and epiglottic cartilages (fig. 19.5). These structures are single. The other laryngeal cartilages—the arytenoid, corniculate, and cuneiform cartilages—are paired.

The **thyroid cartilage** was named for the thyroid gland that covers its lower area. This cartilage is the shieldlike structure that protrudes in the front of the neck and is sometimes called the Adam's apple. The protrusion typically is more prominent in males than in females because of an effect of male sex hormones on the development of the larynx.

The **cricoid cartilage** lies inferior to the thyroid cartilage. It marks the lowermost portion of the larynx.

The **epiglottic cartilage,** the only one of the laryngeal cartilages that is elastic, not hyaline, cartilage, is attached to the upper border of the thyroid cartilage and supports a flaplike structure called the **epiglottis.** The epiglottis usually stands upright and allows air to enter the larynx. During swallowing, however, muscular contractions raise the larynx, and the base of the tongue presses the epiglottis downward. The epiglottis partially covers the opening into the larynx, helping prevent foods and liquids from entering the air passages.

Posteriorly on the larynx, the pyramid-shaped **arytenoid cartilages** are superior to and on either side of the cricoid cartilage. Attached to the tips of the arytenoid cartilages are the tiny, conelike **corniculate cartilages.** These cartilages are attachments for muscles that help regulate tension on the

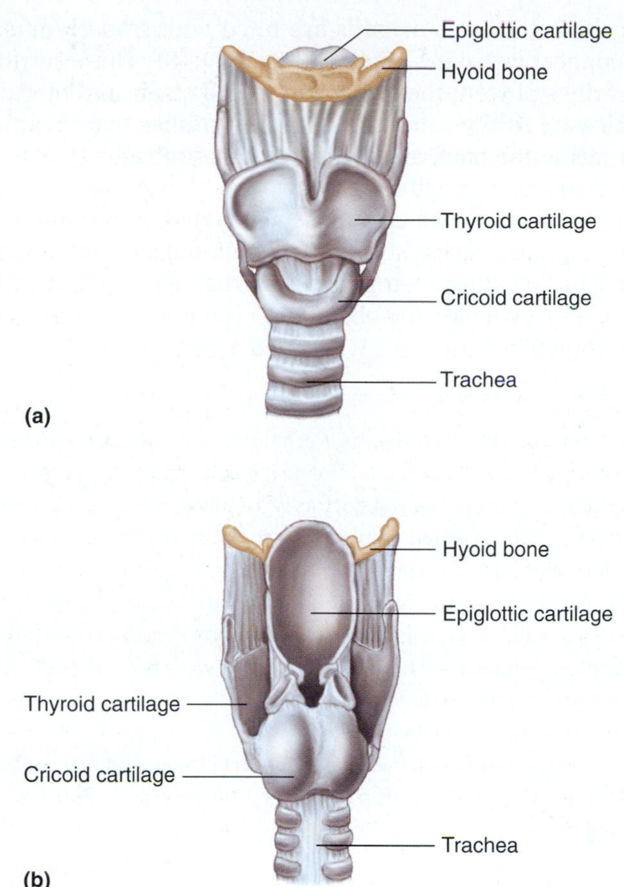

FIGURE 19.5 Larynx. (*a*) Anterior and (*b*) posterior views of the larynx.

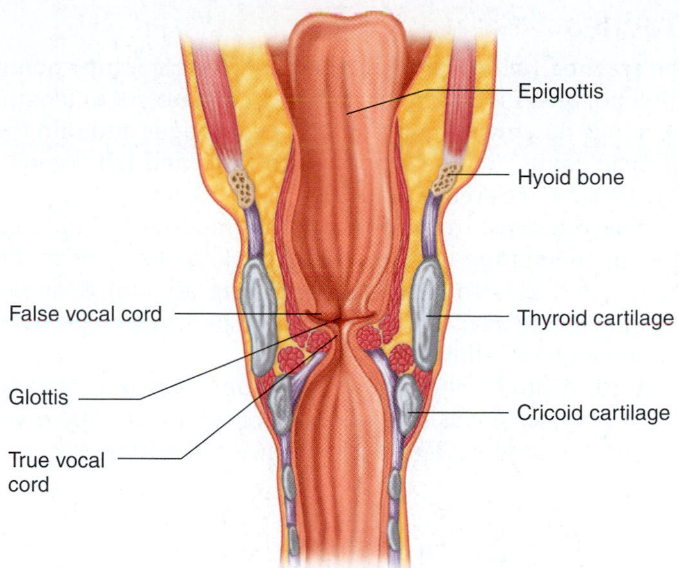

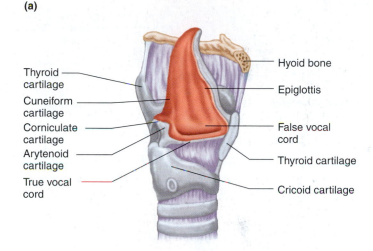

FIGURE 19.6 Larynx. (*a*) Frontal section and (*b*) sagittal section of the larynx.

vocal cords during speech and aid in closing the larynx during swallowing.

The **cuneiform cartilages** are small, cylindrical structures in the mucous membrane between the epiglottic and the arytenoid cartilages. They stiffen the soft tissues in this region.

Inside the larynx, two pairs of horizontal folds composed of muscle tissue and connective tissue with a covering of mucous membrane extend inward from the lateral walls. The upper folds (vestibular folds) are called *false vocal cords* because they do not produce sounds. Muscle fibers within these folds help close the larynx during swallowing.

The lower folds are the *true vocal cords*. They have elastic fibers and are responsible for vocal sounds, which are created when air is forced between these folds, vibrating them. This action generates sound waves, formed into words by changing the shapes of the pharynx and oral cavity and by using the tongue and lips. Figure 19.6 shows both pairs of cords.

Changing tension on the vocal cords, by contracting or relaxing laryngeal muscles, controls *pitch* (musical tone) of the voice. Increasing the tension produces a higher pitch, and decreasing the tension creates a lower pitch.

The *intensity* (loudness) of a vocal sound depends upon the force of the air passing over the vocal cords. Stronger blasts of air result in greater vibration of the vocal cords and louder sound.

During normal breathing the vocal cords remain relaxed, and the opening between them, called the **glottis** (glot′is), is a triangular slit. However, when food or liquid is swallowed, muscles close the glottis in the false vocal folds. Along with closing of the epiglottis, this action helps prevent food or liquid from entering the trachea (fig. 19.7). The mucous membrane that lines the larynx continues to filter incoming air by entrapping particles and moving them toward the pharynx by ciliary action.

PRACTICE

7 What part of the respiratory tract is shared with the alimentary canal?

8 Describe the structure of the larynx.

9 How do the vocal cords produce sounds?

10 What is the function of the epiglottis? Of the glottis?

Trachea

The **trachea** (windpipe) is a flexible cylindrical tube about 2.5 centimeters in diameter and 12.5 centimeters in length. It extends downward anterior to the esophagus and into the thoracic cavity, where it splits into right and left bronchi (fig. 19.8 and reference plate 9).

The inner wall of the trachea is lined with a ciliated mucous membrane that has many goblet cells. This membrane continues to filter the incoming air and to move entrapped particles upward into the pharynx where the mucus can be swallowed.

Within the tracheal wall are about twenty C-shaped pieces of hyaline cartilage, one above the other. The open ends of these incomplete rings are directed posteriorly, and the gaps between their ends are filled with smooth muscle and connective tissues (figs. 19.9 and 19.10). These cartilaginous rings prevent the trachea from collapsing and blocking the airway. At the same time, the soft tissues that complete the rings in the back allow the nearby esophagus to expand as food moves through it on the way to the stomach.

A blocked trachea can cause asphyxiation in minutes. If swollen tissues, excess secretions, or a foreign object obstruct the trachea, making a temporary, external opening in the tube so that air can bypass the obstruction is lifesaving. This procedure, shown in figure 19.11, is called a *tracheostomy.*

> On December 13, 1799, George Washington spent the day walking on his estate in a freezing rain. The next day, he had trouble breathing and swallowing. Several doctors were called in. One suggested a tracheostomy, cutting a hole in the throat so that the president could breathe. He was voted down. The other physicians suggested bleeding the patient, plastering his throat with bran and honey, and placing beetles on his legs to produce blisters. No treatment was provided, and within a few hours, Washington's voice became muffled, breathing was more labored, and he was restless. For a short time he seemed euphoric, and then he died.
>
> George Washington had *epiglottitis,* an inflammation that swells the epiglottis to ten times its normal size. A tracheostomy might have saved his life.

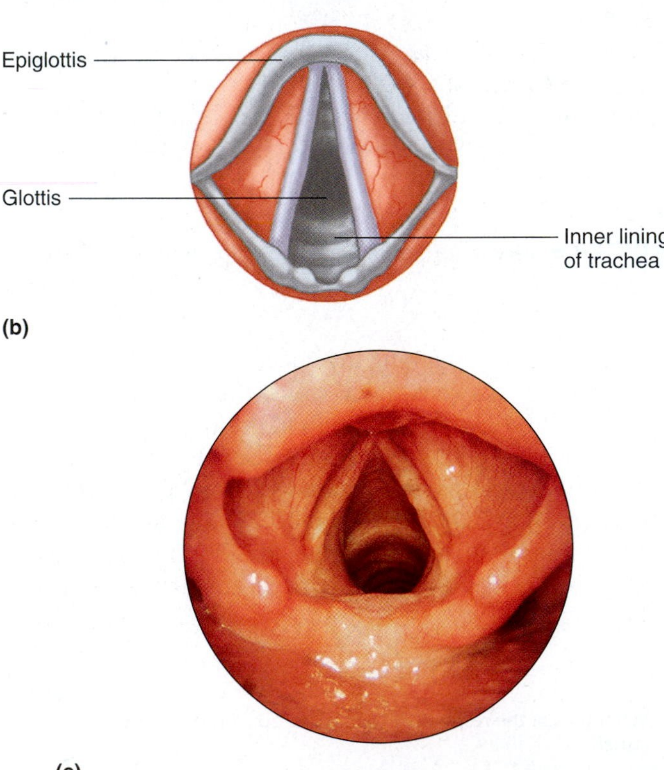

FIGURE 19.7 The vocal cords as viewed from above with the glottis (*a*) closed and (*b*) open. (*c*) Photograph of the glottis and vocal cords.

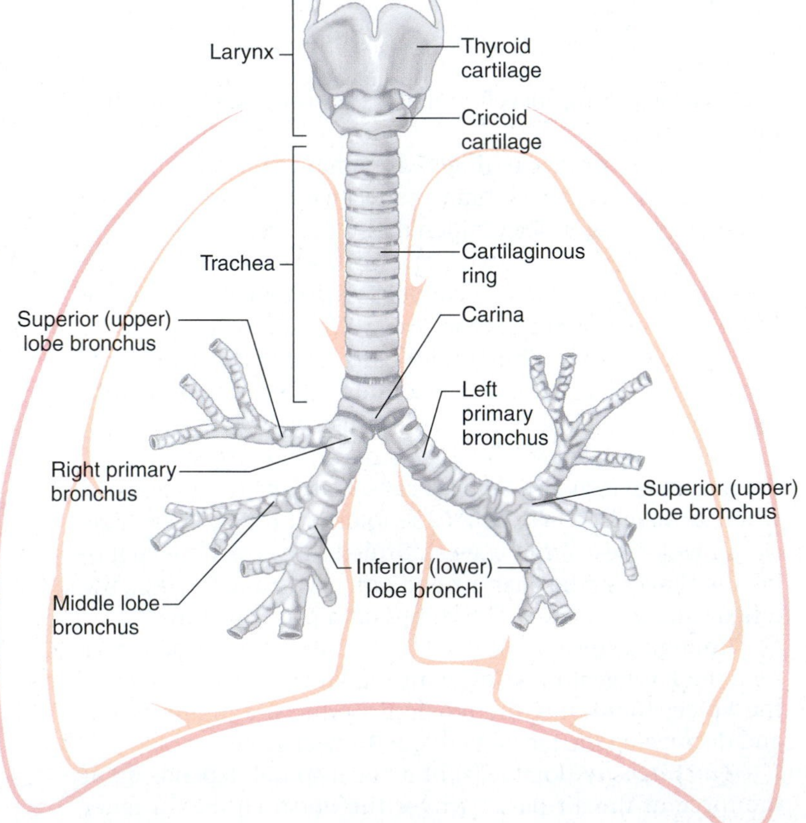

FIGURE 19.8 The trachea transports air between the larynx and the bronchi.

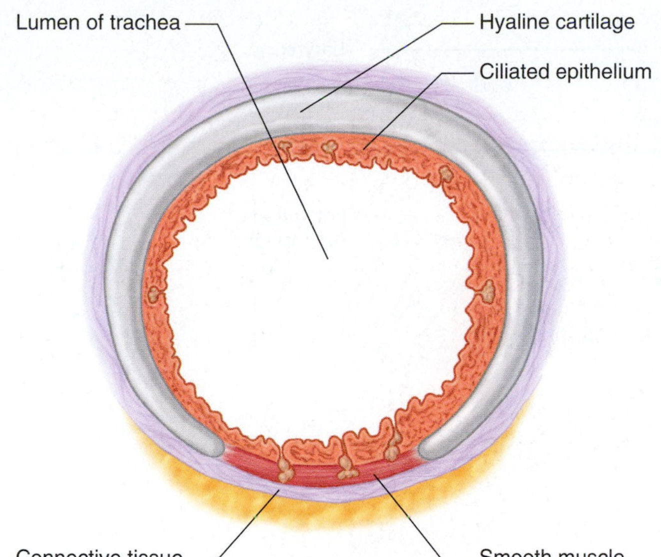

FIGURE 19.9 Cross section of the trachea. Note the C-shaped ring of hyaline cartilage in the wall.

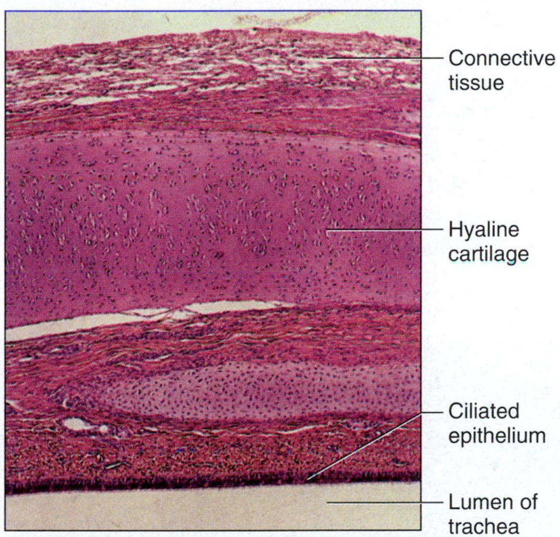

FIGURE 19.10 Light micrograph of a section of the tracheal wall (63×).

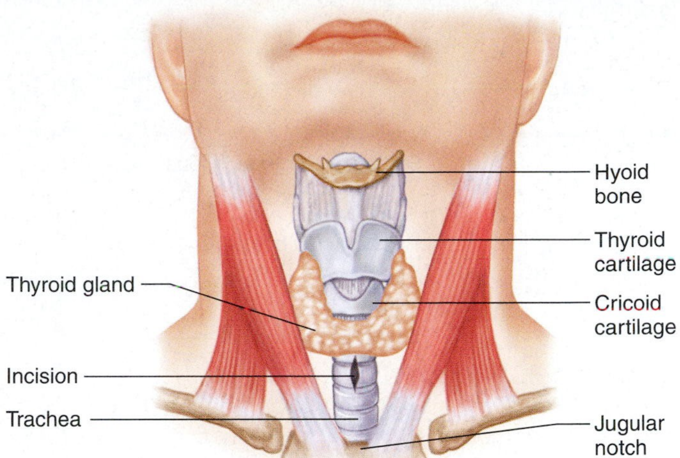

FIGURE 19.11 A tracheostomy may be performed to allow air to bypass an obstruction in the larynx.

Bronchial Tree

The **bronchial tree** (brong′ke-al trē) consists of branched airways leading from the trachea to the microscopic air sacs in the lungs. Its branches begin with the right and left **primary bronchi,** which arise from the trachea at the level of the fifth thoracic vertebrae. The openings of the primary bronchi are separated by a ridge of cartilage called the *carina* (see fig. 19.8). Each bronchus, accompanied by large blood vessels, enters its respective lung.

Branches of the Bronchial Tree

A short distance from its origin, each primary bronchus divides into **secondary,** or **lobar, bronchi** (two on the left and three on the right) that, in turn, branch repeatedly (figs. 19.12 and 19.13). When stripped of their associated blood vessels and tissues, the airways appear as an upside down tree. The successive divisions of these branches from the trachea to the microscopic air sacs follow:

1. **Right** and **left primary bronchi.**
2. **Secondary,** or **lobar, bronchi.** Three branch from the right primary bronchus, and two branch from the left.
3. **Tertiary,** or **segmental, bronchi.** Each of these branches supplies a portion of the lung called a *bronchopulmonary segment.* Usually ten such segments are in the right lung and eight are in the left lung.
4. **Intralobular bronchioles.** These small branches of the segmental bronchi enter the basic units of the lung—the *lobules.*
5. **Terminal bronchioles.** These tubes branch from an intralobular bronchiole. Fifty to eighty terminal bronchioles occupy a lobule of the lung.
6. **Respiratory bronchioles.** Two or more respiratory bronchioles branch from each terminal bronchiole. Short and about 0.5 millimeter in diameter, these structures are called "respiratory" because a few air sacs bud from their sides, enabling them to take part in gas exchange.
7. **Alveolar ducts.** Alveolar ducts branch from each respiratory bronchiole (fig. 19.14).
8. **Alveolar sacs.** Alveolar sacs are thin-walled, closely packed outpouchings of the alveolar ducts.
9. **Alveoli** (al-ve′o-li). Alveoli are thin-walled, microscopic air sacs that open to an alveolar sac. Air can diffuse freely from the alveolar ducts, through the alveolar sacs, and into the alveoli (fig. 19.15).

Dust particles, asbestos fibers, and other pollutants travel at speeds of 200 centimeters per second in the trachea but slow to 1 centimeter per second when deep in the lungs. Gravity deposits such particles, particularly at branchpoints

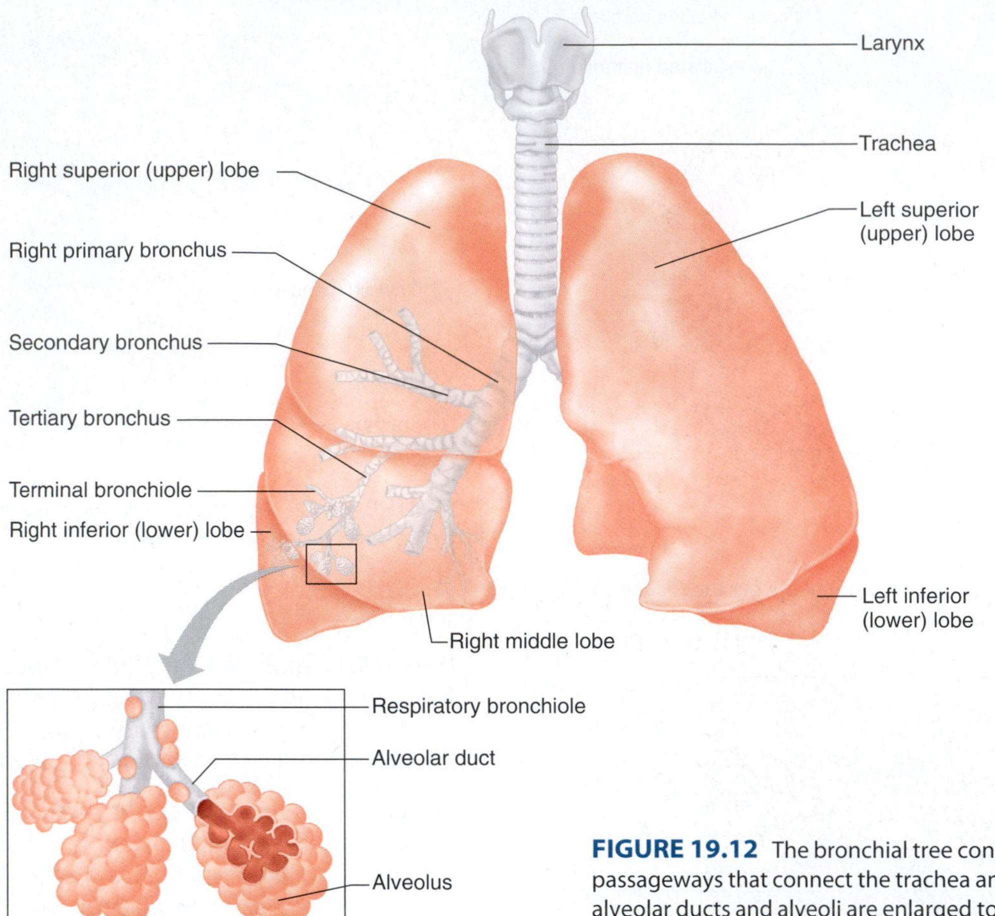

FIGURE 19.12 The bronchial tree consists of the passageways that connect the trachea and the alveoli. The alveolar ducts and alveoli are enlarged to show their locations.

in the respiratory tree. It is a little like traffic backing up at an exit from a highway.

> In severe cases of the inherited illness *cystic fibrosis,* airways become clogged with thick, sticky mucus, which attracts bacteria. As damaged white blood cells accumulate at the infection site, their DNA may leak out and further clog the area. A treatment that moderately eases breathing is deoxyribonuclease (DNase), an enzyme that normally degrades accumulating extracellular DNA.

Structure of the Respiratory Tubes

The structure of a bronchus is similar to that of the trachea, but the C-shaped cartilaginous rings are replaced with cartilaginous plates where the bronchus enters the lung. These plates are irregularly shaped and completely surround the tube. However, as the branching tubes become thinner, the amount of cartilage decreases. It finally disappears in the bronchioles, which have diameters of about 1 millimeter. Also as the tubes thin, layers of surrounding smooth muscle just beneath the mucosa become more prominent. This muscular layer persists in the walls to the ends of the respiratory bronchioles, and only a few muscle fibers are in the walls of the alveolar ducts.

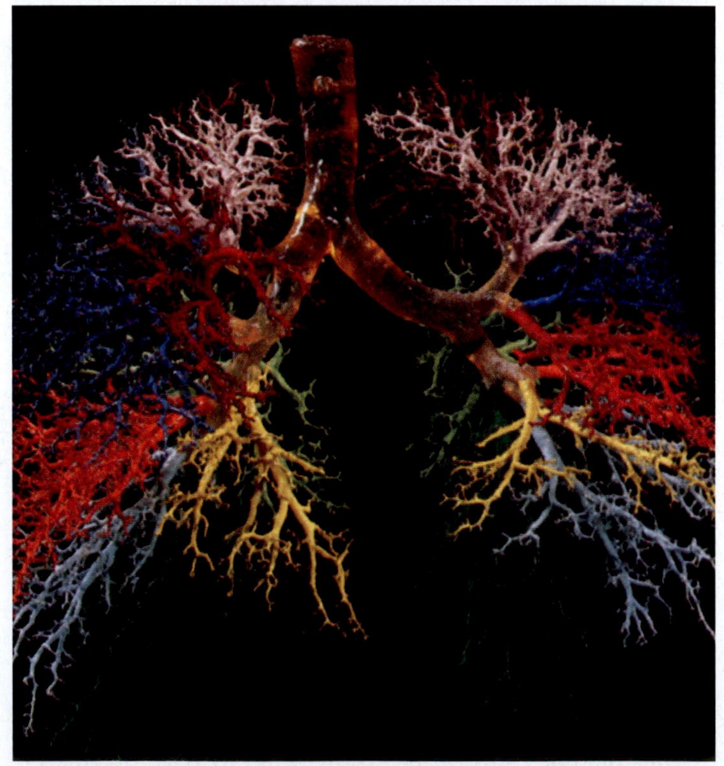

FIGURE 19.13 A plastic cast of the bronchial tree.

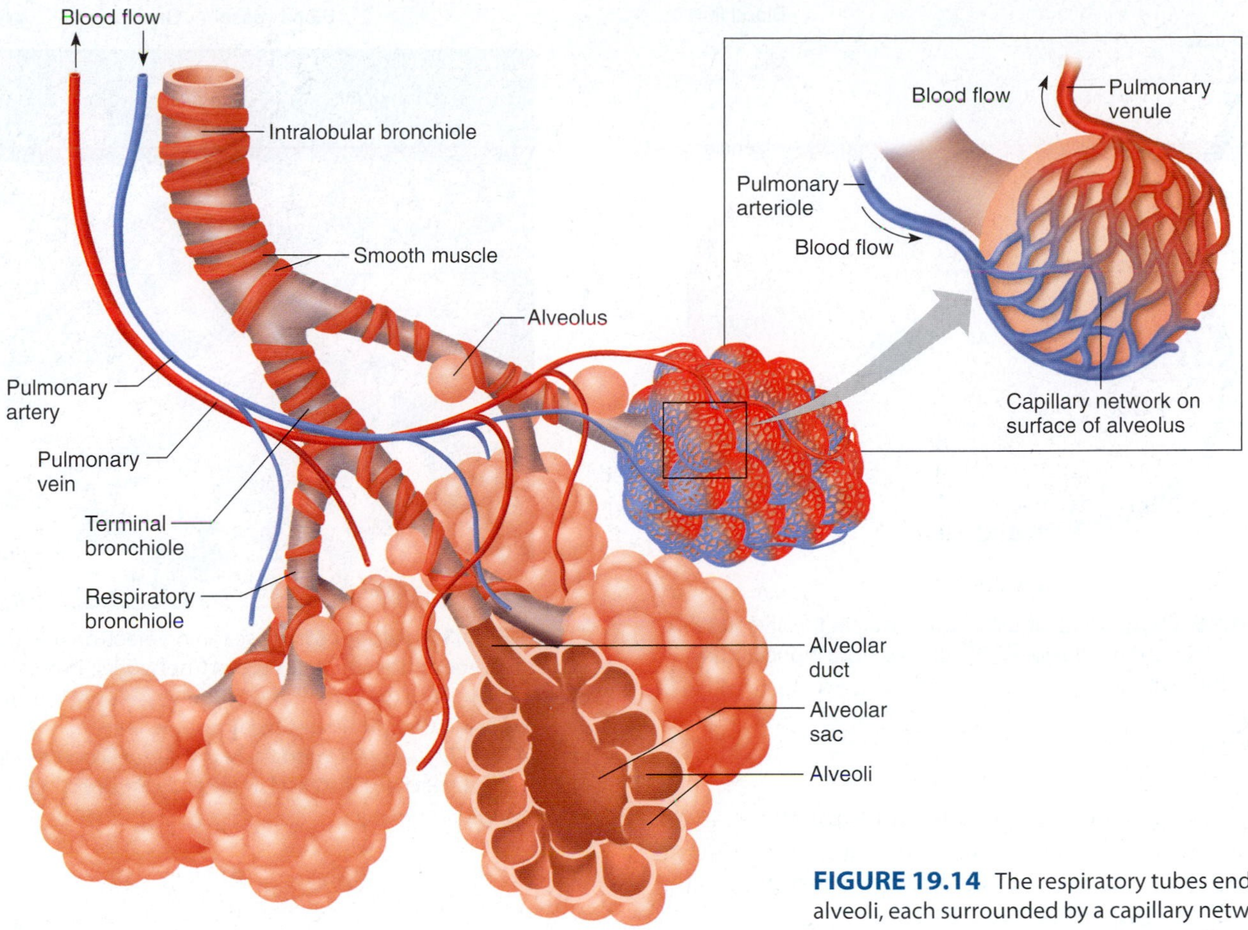

FIGURE 19.14 The respiratory tubes end in tiny alveoli, each surrounded by a capillary network.

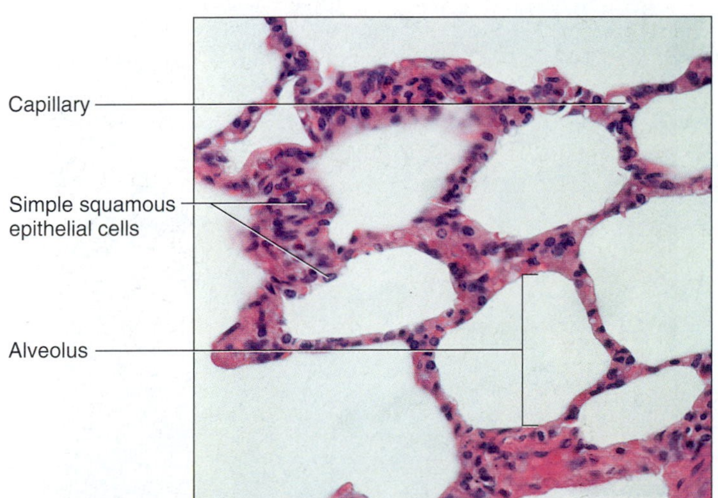

FIGURE 19.15 Light micrograph of alveoli (250×).

Elastic fibers are scattered among the smooth muscle cells and are abundant in the connective tissue that surrounds the respiratory tubes. These fibers play an important role in breathing, as is explained later in this chapter.

As the tubes become smaller in diameter, the type of cells that line them changes. The lining of the larger tubes consists of pseudostratified, ciliated columnar epithelium and mucous-secreting goblet cells. However, along the way, the number of goblet cells and the height of the other epithelial cells decline, and cilia become scarcer. In the finer tubes, beginning with the respiratory bronchioles, the lining is cuboidal epithelium; in the alveoli, it is simple squamous epithelium closely associated with a dense network of capillaries. The mucous lining gradually thins, until none appears in the alveoli.

> A flexible optical instrument called a *fiberoptic bronchoscope* is used to examine the trachea and bronchial tree. This procedure (bronchoscopy) is used in diagnosing tumors or other pulmonary diseases and to locate and remove aspirated foreign bodies in the air passages.

Functions of the Respiratory Tubes and Alveoli

The branches of the bronchial tree are air passages, which continue to filter incoming air and distribute it to the alveoli in all parts of the lungs. The alveoli, in turn, provide a large surface area of thin epithelial cells through which gas exchanges can occur (fig. 19.16). If the 300 million alveoli in the human lung were spread out, they would cover an area of between 70 and 80 square meters—nearly half the size of a tennis court.

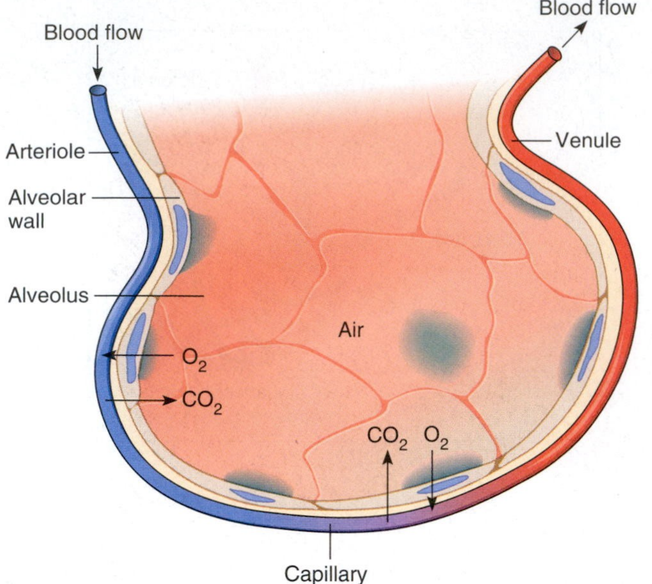

FIGURE 19.16 Oxygen (O_2) diffuses from the air in the alveolus into the capillary, while carbon dioxide (CO_2) diffuses from blood in the capillary into the alveolus.

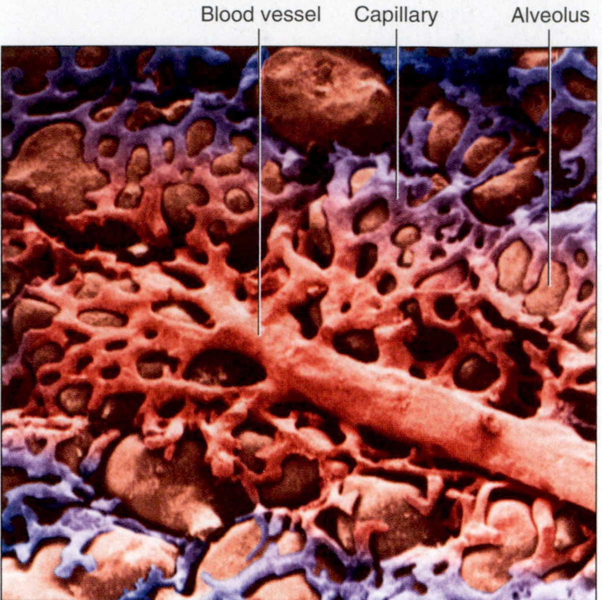

FIGURE 19.17 Falsely colored scanning electron micrograph of casts of alveoli and associated capillary networks. These casts were prepared by filling the alveoli and blood vessels with resin and later removing the soft tissues by digestion, leaving only the resin casts (420×). *Tissues and Organs: A Text-Atlas of Scanning Electron Microscopy,* by Richard D. Kessel and Randy Kardon.© 1979 W. H. Freeman and Company.

During gas exchange, oxygen diffuses through the alveolar walls and enters the blood in nearby capillaries. Carbon dioxide diffuses from the blood through these walls and enters the alveoli (figs. 19.17 and 19.18).

Several techniques enable a person who has stopped breathing to survive. In *artificial respiration,* a person blows into the mouth of a person who has stopped breathing. The oxygen in the rescuer's exhaled breath can keep the victim alive.

In *extracorporeal membrane oxygenation,* blood is pumped out of the body and across a gas-permeable membrane that adds oxygen and removes CO_2, simulating lung function. Such a device can keep a person alive until he or she recovers from other problems, but is too costly and cumbersome to maintain life indefinitely.

A lung assist device, called an *intravascular oxygenator,* consists of hundreds of tiny porous hair-thin fibers surgically implanted in the inferior vena cava. Here, deoxygenated blood returning to the heart receives oxygen and is rid of CO_2—but only at about 30% the capacity of a healthy respiratory system.

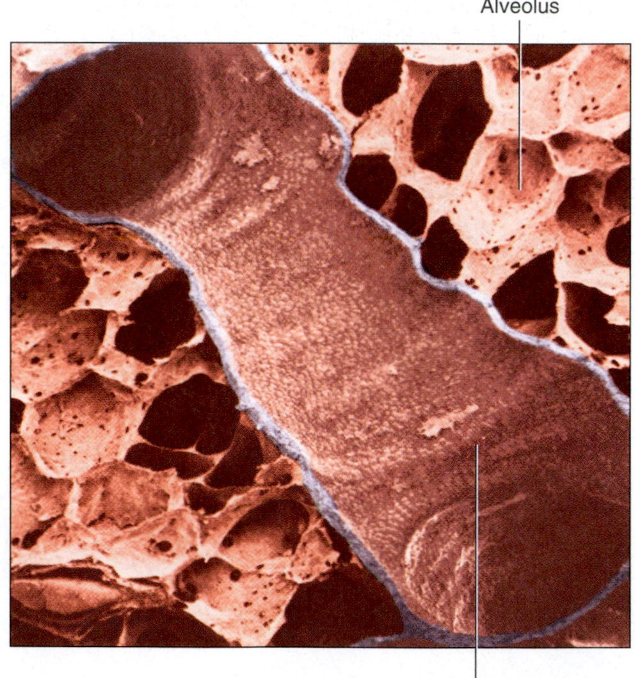

FIGURE 19.18 Falsely colored scanning electron micrograph of lung alveoli and a bronchiole (70×).

PRACTICE

11 What is the function of the cartilaginous rings in the tracheal wall?

12 How do the right and left bronchi differ in structure?

13 List the branches of the bronchial tree.

14 Describe structural changes in the respiratory tubes as their diameters decrease.

15 How are gases exchanged in the alveoli?

Lungs

The lungs are soft, spongy, cone-shaped organs in the thoracic cavity. The right and left lungs are separated medially by the heart and the mediastinum, and they are enclosed by the diaphragm and the thoracic cage (see figs. 1.9, 19.19 and reference plates 16, 17, and 21).

Each lung occupies most of the thoracic space on its side and is suspended in the cavity by a bronchus and some large blood vessels. These tubular structures enter the lung on its medial surface through a region called the **hilum.** A layer of serous membrane, the *visceral pleura,* is firmly attached to the surface of each lung, and this membrane folds back at the hilus to become the *parietal pleura.* The parietal pleura, in turn, forms part of the mediastinum and lines the inner wall of the thoracic cavity (fig. 19.20).

No significant space exists between the visceral and parietal pleurae, because they are essentially in contact with each other. The potential space between them, the **pleural cavity,** contains only a thin film of serous fluid that lubricates the adjacent pleural surfaces, reducing friction as they move against one another during breathing. This fluid also helps hold the pleural membranes together.

The right lung is larger than the left lung, and fissures divide it into three parts, called the superior, middle, and inferior lobes. The left lung is similarly divided and consists of two parts, a superior and an inferior lobe.

A lobar bronchus of the bronchial tree supplies each lobe. A lobe also has connections to blood and lymphatic vessels and is enclosed by connective tissues. Connective tissue further subdivides a lobe into **lobules,** each of which contains terminal bronchioles together with their alveolar ducts, alveolar sacs, alveoli, nerves, and associated blood and lymphatic vessels.

Table 19.1 summarizes the characteristics of the major parts of the respiratory system. Clinical Application 19.2 considers substances that irritate the lungs.

PRACTICE 

16 Where are the lungs located?

17 What is the function of the serous fluid in the pleural cavity?

18 How does the structure of the right lung differ from that of the left lung?

19 What types of structures make up a lung?

19.4 BREATHING MECHANISM

Breathing, also called ventilation, is the movement of air from outside the body into the bronchial tree and alveoli, followed by a reversal of this air movement. The actions responsible for these air movements are termed **inspiration** (in″spĭ-ra′shun), or inhalation, and **expiration** (ek″spi-ra′shun), or exhalation.

Inspiration

Atmospheric pressure due to the weight of the air is the force that moves air into the lungs. At sea level, this pressure is sufficient to support a column of mercury about 760 millimeters high in a tube. Thus, normal air pressure equals 760 millimeters (mm) of mercury (Hg). (Other units are in common usage: 760 mm Hg = 760 Torr = 1 atmosphere.)

Air pressure is exerted on all surfaces in contact with the air, and because people breathe air, the inside surfaces of their lungs are also subjected to pressure. In other words, when the respiratory muscles are at rest, the pressures on the inside of the lungs and alveoli and on the outside of the thoracic wall are about the same (fig. 19.21).

Pressure and volume are related in an opposite, or inverse, way (this is known as Boyle's law). For example, pulling back on the plunger of a syringe increases the volume inside the barrel, lowering the air pressure inside. Atmospheric pressure then pushes outside air into the syringe (fig. 19.22*a*). In contrast, pushing on the plunger of a syringe reduces the volume inside the syringe, but the

TABLE 19.1 | Parts of the Respiratory System

Part	Description	Function
Nose	Part of face centered above the mouth and inferior to the space between the eyes	Nostrils provide entrance to nasal cavity; internal hairs begin to filter incoming air
Nasal cavity	Hollow space behind nose	Conducts air to pharynx; mucous lining filters, warms, and moistens incoming air
Sinuses	Hollow spaces in various bones of the skull	Reduce weight of the skull; serve as resonant chambers
Pharynx	Chamber posterior to the nasal cavity, oral cavity, and larynx	Passageway for air moving from nasal cavity to larynx and for food moving from oral cavity to esophagus
Larynx	Enlargement at the top of the trachea	Passageway for air; prevents foreign objects from entering trachea; houses vocal cords
Trachea	Flexible tube that connects larynx with bronchial tree	Passageway for air; mucous lining continues to filter air
Bronchial tree	Branched tubes that lead from the trachea to the alveoli	Conducts air to the alveoli; mucous lining continues to filter incoming air
Lungs	Soft, cone-shaped organs that occupy a large portion of the thoracic cavity	Contain the air passages, alveoli, blood vessels, connective tissues, lymphatic vessels, and nerves of the lower respiratory tract

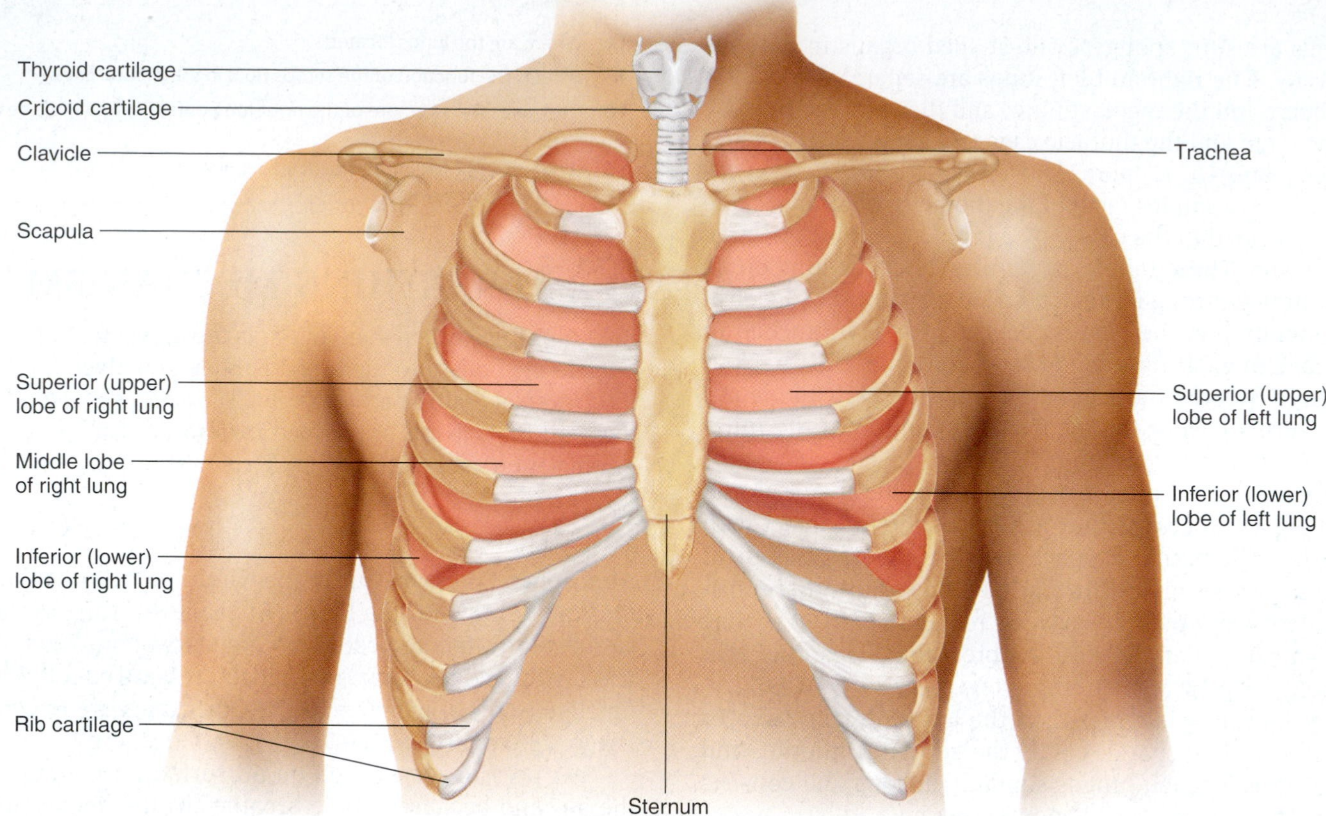

FIGURE 19.19 Location of the lungs in the thoracic cavity.

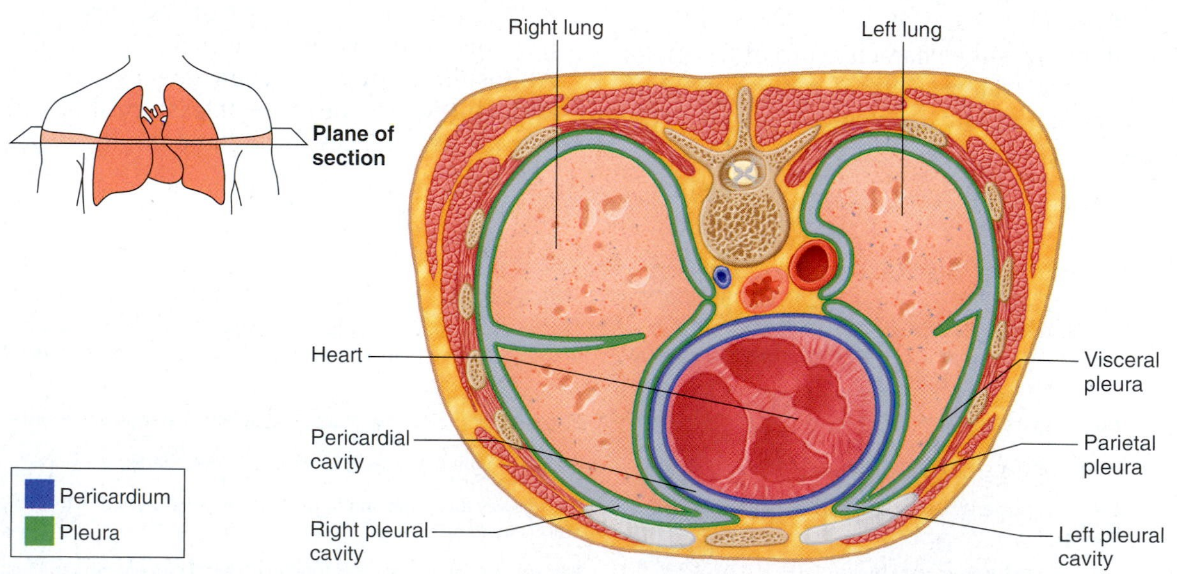

FIGURE 19.20 The potential spaces between the pleural membranes, called the left and right pleural cavities, are shown here as actual spaces.

19.2 CLINICAL APPLICATION

Lung Irritants

The lungs are exquisitely sensitive to the presence of inhaled particles. Such exposures can cause a variety of symptoms, both acute and chronic, that range from a persistent cough to cancer.

Asbestos

Asbestos, a naturally occurring mineral, was once widely used in buildings and on various products because it resists burning and chemical damage. Asbestos easily crumbles into fibers, which, when airborne, can enter human respiratory passages. Asbestos-related problems include asbestosis (shortness of breath resulting from scars in lungs), lung cancer, and mesothelioma (a rare cancer of the pleural membrane).

Asbestos fibers longer than 5 micrometers (0.0002 inch) and thinner than 2 micrometers (0.00008 inch) can cause illness when inhaled, but asbestos clearly causes respiratory illness only if the fibers are airborne. Experts must determine whether it is safer to encapsulate asbestos in a building and leave it in place or remove it. Table 19A indicates how risk of becoming ill rises with duration of exposure to asbestos.

Berylliosis

Beryllium is an element used in fluorescent powders, metal alloys, and in the nuclear power industry. A small percentage of workers exposed to beryllium dust or vapor develop an immune response, which damages the lungs. Symptoms include cough, shortness of breath, fatigue, loss of appetite, fevers and night sweats, and weight loss. Radiographs show granuloma scars in the lungs, and pulmonary function tests and listening to breath sounds with a stethoscope reveal impaired breathing.

Symptoms of berylliosis typically begin about a decade after the first exposure. It is distinguished from other lung ailments with a blood test that detects antibodies to beryllium. Affected individuals and those who do not have symptoms but know that they were exposed to beryllium can have periodic blood tests and chest radiographs to detect the condition early. The steroid drug prednisone is used to control symptoms.

A Disorder with Many Names

Repeatedly inhaling dust of organic origin can cause a lung irritation called extrinsic allergic alveolitis. An acute form of this reaction impairs breathing and causes a fever a few hours after encountering dust. In the chronic form, lung changes occur gradually over several years. The condition is associated with several occupations and has a variety of colorful names, including bathtub refinisher's lung, maple bark stripper disease, popcorn worker's lung, and wheat weevil disease.

9/11-Associated Air Pollution

When more than a million tons of dust and debris fell on lower Manhattan as the World Trade Center collapsed on September 11, 2001, the dust and debris presented a mixture of chemicals that no human respiratory system had ever encountered (fig. 19C). Particles spewed into the air from paint, plaster, foam, glass, ceramics, concrete, vermiculite, wood, soot, and textiles. Particle size was important. Although the human respiratory tract easily ejects particles greater than 10 micrometers in diameter in coughs or sneezes, particles with diameters between 2.5 and 10 micrometers can enter the upper airways, and the finest particulates, with diameters from 0.09 to 0.24 micrometers, pass the respiratory system's initial barriers and may make it as far as the alveoli. These finest particles included sulfur compounds, tiny bits of silicon and metals, including vanadium and nickel from fuel oil; titanium from concrete; and iron, copper, and zinc.

Fortunately, asbestos and dangerous organic compounds were very scant in the debris, and more than 95% of the particulates were large enough to sneeze out. However, some of the workers in the area developed "World Trade Center cough" from inhaling alkaline large particles, mostly from fiberglass. We will be learning about the effects of this disaster on air quality and the human respiratory system for many years.

FIGURE 19C Urban aerosols compromise lung function (temporarily, we hope). This image highlights the sources of some of the debris that covered lower Manhattan after the September 11, 2001 attacks. Red and yellow represent concrete dust, and purple indicates a mineral, gypsum, found in wallboard.

TABLE 19A | Asbestos-Related Respiratory Illness

Situation	Level of Exposure (fibers/cubic centimeter)	Cancer Cases per Million Exposed People
Asbestos workers with twenty years' exposure	10 fiber/cc	200,000
Permissible upper limit in buildings today	0.1 fiber/cc	2,000
Child in school with asbestos six hours/day	0.0005 fiber/cc	6
Most modern buildings	0.0002 fiber/cc	4

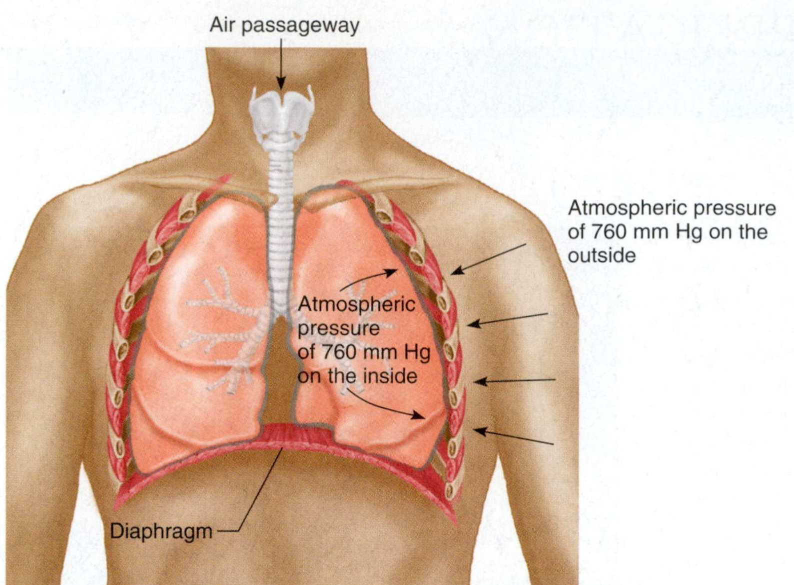

FIGURE 19.21 When the lungs are at rest, the pressure on the inside of the lungs is equal to the pressure on the outside of the thorax.

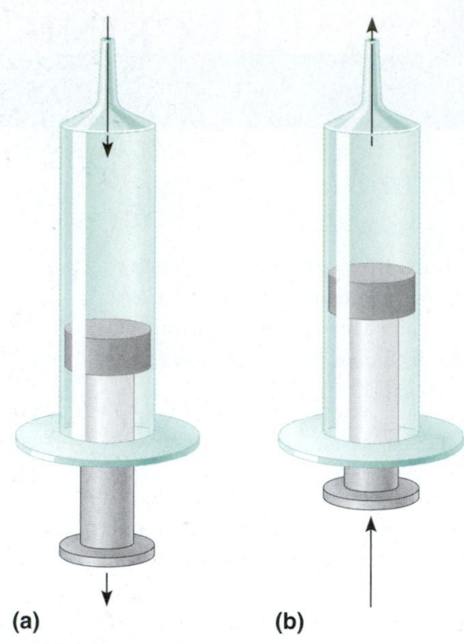

FIGURE 19.22 Moving the plunger of a syringe causes air to move (*a*) in or (*b*) out of the syringe. Air movements in and out of the lungs occur in much the same way.

pressure inside increases, forcing air out into the atmosphere (fig. 19.22*b*). The movement of air into and out of the lungs occurs in much the same way.

If the pressure inside the lungs and alveoli (intra-alveolar pressure) decreases, outside air will then be pushed into the airways by atmospheric pressure. This is what happens during normal inspiration, and it uses muscle fibers in the dome-shaped *diaphragm.*

The diaphragm is just inferior to the lungs. It consists of an anterior group of skeletal muscle fibers (costal fibers) that originate from the ribs and sternum, and a posterior group (crural fibers) that originate from the vertebrae. Both groups of muscle fibers are inserted on a tendinous central portion of the diaphragm (reference plate 21).

The muscle fibers of the diaphragm are stimulated to contract by impulses carried by the *phrenic nerves,* which are associated with the cervical plexuses. When this happens, the diaphragm moves downward, the thoracic cavity enlarges, and the intra-alveolar pressure falls about 2 mm Hg below atmospheric pressure. In response to this decreased pressure, air is forced into the airways by atmospheric pressure (fig. 19.23).

While the diaphragm is contracting and moving downward, the *external (inspiratory) intercostal muscles* and certain thoracic muscles may be stimulated to contract. This action raises the ribs and elevates the sternum, increasing the size of the thoracic cavity even more. The intra-alveolar pressure falls farther, and atmospheric pressure forces more air into the airways.

Lung expansion in response to movements of the diaphragm and chest wall depends on movements of the pleural membranes. Any separation of the pleural membranes decreases pressure in the intrapleural space, holding these membranes together. In addition, only a thin film of serous fluid separates the parietal pleura on the inner wall of the thoracic cavity from the visceral pleura attached to the surface of the lungs. The water molecules in this fluid greatly attract the pleural membranes and each other, helping to hold the moist surfaces of the pleural membranes tightly together, much as a wet coverslip sticks to a microscope slide. As a result of these factors, when the intercostal muscles move the thoracic wall upward and outward, the parietal pleura moves too, and the visceral pleura follows it. This helps expand the lung in all directions.

Although the moist pleural membranes help expand the lungs, the moist inner surfaces of the alveoli have the opposite effect. Here the attraction of water molecules to each other creates a force called **surface tension** that makes it difficult to inflate the alveoli and may collapse them. Certain alveolar cells, however, synthesize a mixture of lipoproteins called **surfactant,** which is secreted continuously into alveolar air spaces. Surfactant reduces the alveoli's tendency to collapse, especially when lung volumes are low, and makes it easier for inspiratory efforts to inflate the alveoli. Table 19.2 summarizes the steps of inspiration.

> Surfactant is particularly important in the minutes after birth, when the newborn's lungs inflate for the first time. Premature infants often suffer respiratory distress syndrome because they do not produce sufficient surfactant. To help many of these newborns survive, physicians inject synthetic surfactant into the tiny lungs through an endotracheal tube. A ventilator machine especially geared to an infant's size assists breathing.

If a person needs to take a deeper than normal breath, the diaphragm and external intercostal muscles contract

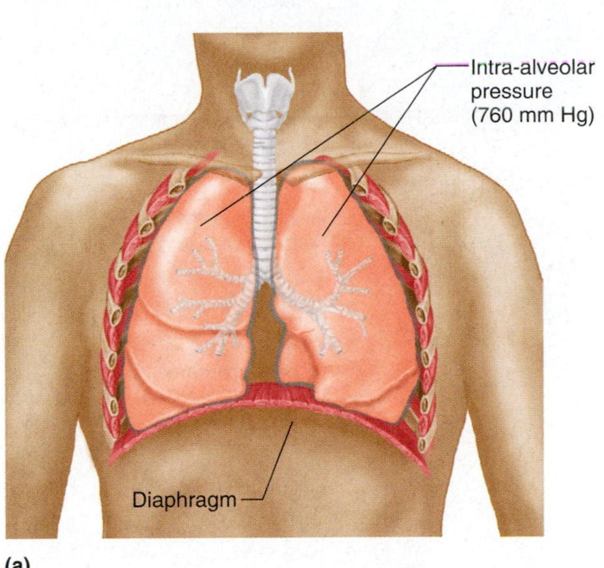

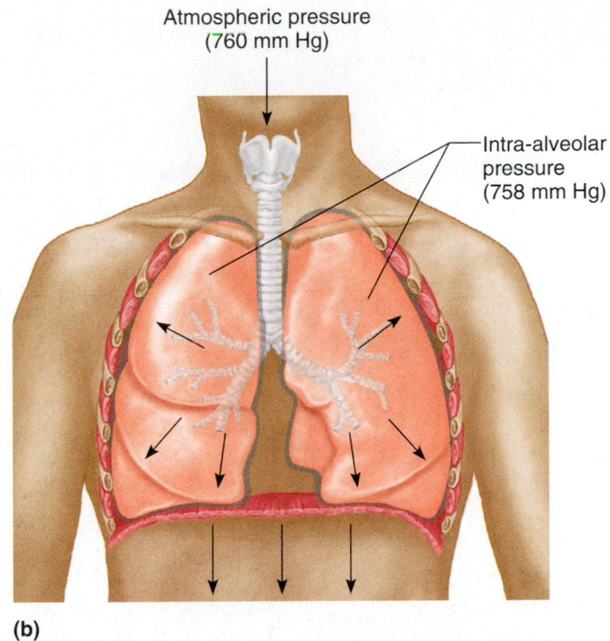

FIGURE 19.23 Normal inspiration. (*a*) Prior to inspiration, the intra-alveolar pressure is 760 mm Hg. (*b*) The intra-alveolar pressure decreases to about 758 mm Hg as the thoracic cavity enlarges, and atmospheric pressure forces air into the airways.

TABLE 19.2 | Major Events in Inspiration

1. Nerve impulses travel on phrenic nerves to muscle fibers in the diaphragm, contracting them.
2. As the dome-shaped diaphragm moves downward, the thoracic cavity expands.
3. At the same time, the external intercostal muscles may contract, raising the ribs and expanding the thoracic cavity further.
4. The intra-alveolar pressure decreases.
5. Atmospheric pressure, greater on the outside, forces air into the respiratory tract through the air passages.
6. The lungs fill with air.

more forcefully. Additional muscles, such as the pectoralis minors and sternocleidomastoids, can also be used to pull the thoracic cage farther upward and outward, enlarging the thoracic cavity and decreasing intra-alveolar pressure even more (fig. 19.24).

The ease with which the lungs can expand as a result of pressure changes during breathing is called *compliance* (distensibility). In a normal lung, compliance decreases as lung volume increases, because an inflated lung is more difficult to expand than a lung at rest. Conditions that obstruct air passages, destroy lung tissue, or impede lung expansion in other ways also decrease compliance.

Expiration

The forces responsible for normal resting expiration come from *elastic recoil* of lung tissues and from surface tension. The lungs contain a considerable amount of elastic tissue, which stretches as the lungs expand during inspiration. When the diaphragm lowers, the abdominal organs inferior to it are compressed. As the diaphragm and the external intercostal muscles relax following inspiration, the elastic tissues cause the lungs to recoil, and they return to their original shapes. Similarly, elastic tissues cause abdominal organs to spring back into their previous shapes, pushing the diaphragm upward. At the same time, surface tension that develops between the moist surfaces of the alveolar linings shrinks alveoli. Each of these factors increases the intra-alveolar pressure about 1 mm Hg above atmospheric pressure, so the air inside the lungs is forced out through the respiratory passages. Normal resting expiration occurs without the contraction of muscles. Thus, it is a passive process.

The recoil of elastic fibers in lung tissues reduces pressure in the pleural cavity. Consequently, the pressure between the pleural membranes (intrapleural pressure) is usually about 4 mm Hg less than atmospheric pressure.

The visceral and parietal pleural membranes are held closely together because of the low intrapleural pressure, and no significant space normally separates them in the pleural cavity. However, if the thoracic wall is punctured, atmospheric air may enter the pleural cavity and create a substantial space between the membranes. This condition, called *pneumothorax,* can collapse the lung on the affected side because of its elasticity.

Pneumothorax may be treated by covering the chest wound with an impermeable bandage, passing a tube (chest tube) through the thoracic wall into the pleural cavity, and applying suction to the tube. The suction reestablishes negative pressure in the cavity and the collapsed lung expands.

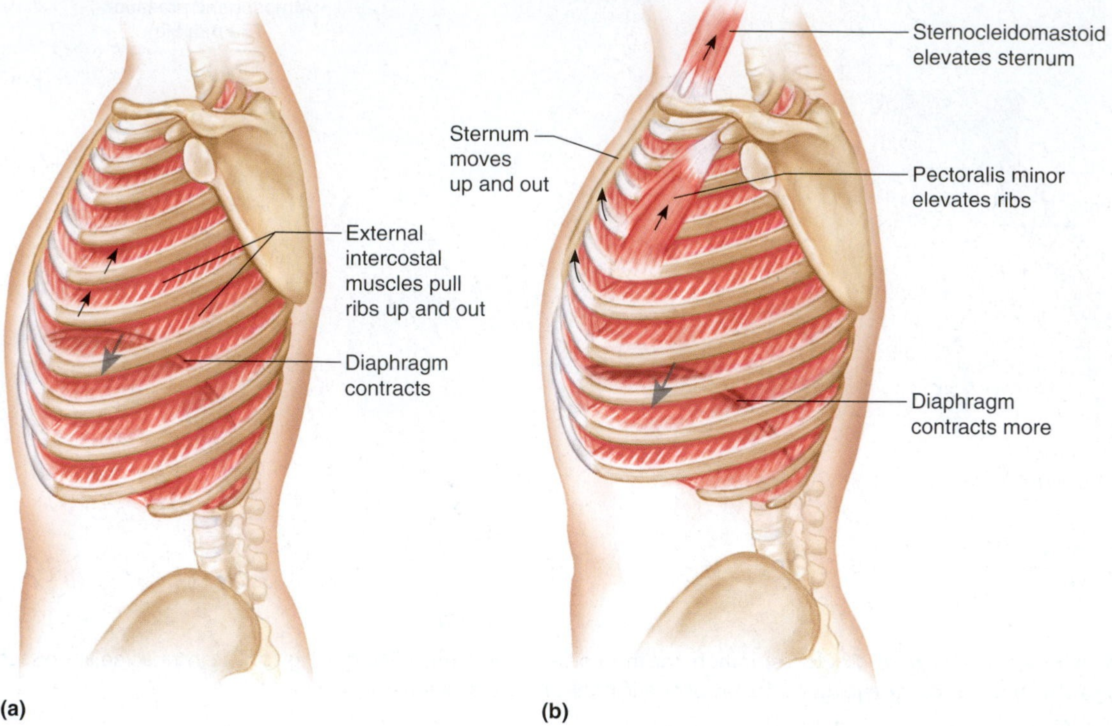

FIGURE 19.24 Maximal inspiration. (*a*) Shape of the thorax at the end of normal inspiration. (*b*) Shape of the thorax at the end of maximal inspiration, aided by contraction of the sternocleidomastoid and pectoralis minor muscles.

A person can exhale more air than normal by contracting the posterior *internal* (*expiratory*) *intercostal muscles*. These muscles pull the ribs and sternum downward and inward, increasing the pressure in the lungs. Also, the *abdominal wall muscles*, including the external and internal obliques, the transversus abdominis, and the rectus abdominis, squeeze the abdominal organs inward. Thus, the abdominal wall muscles can increase pressure in the abdominal cavity and force the diaphragm still higher against the lungs, pushing additional air out of the lungs (fig. 19.25). Table 19.3 summarizes the steps in expiration.

PRACTICE

20 Describe the events in inspiration.

21 How does surface tension aid in expanding the lungs during inspiration?

22 What forces are responsible for normal expiration?

Respiratory Air Volumes and Capacities

Different degrees of effort in breathing move different volumes of air in or out of the lungs. The measurement of such air volumes is called *spirometry*, and it describes four distinct **respiratory volumes.**

One inspiration plus the following expiration is called a **respiratory cycle.** The volume of air that enters or leaves during a respiratory cycle is termed the **tidal volume.** About 500 milliliters (mL) of air enter during a normal, resting inspiration. On the average, the same volume leaves during a normal, resting expiration. Thus, the **resting tidal volume** is about 500 mL (fig. 19.26).

During forced maximal inspiration, a volume of air in addition to the resting tidal volume enters the lungs. This additional volume is called the **inspiratory reserve volume** (complemental air), and it equals about 3,000 mL.

During a maximal forced expiration, about 1,100 mL of air in addition to the resting tidal volume can be expelled from the lungs. This volume is called the **expiratory reserve volume** (supplemental air). However, even after the most forceful expiration, about 1,200 mL of air remains in the lungs. This is the **residual volume.**

Residual air remains in the lungs at all times, and consequently, newly inhaled air always mixes with air already in the lungs. This prevents the oxygen and carbon dioxide concentrations in the lungs from fluctuating greatly with each breath.

Once the respiratory volumes are known, four *respiratory capacities* can be calculated by combining two or more of

TABLE 19.3 | Major Events in Expiration

1. The diaphragm and external respiratory muscles relax.
2. Elastic tissues of the lungs and thoracic cage, stretched during inspiration, suddenly recoil, and surface tension collapses alveolar walls.
3. Tissues recoiling around the lungs increase the intra-alveolar pressure.
4. Air is squeezed out of the lungs.

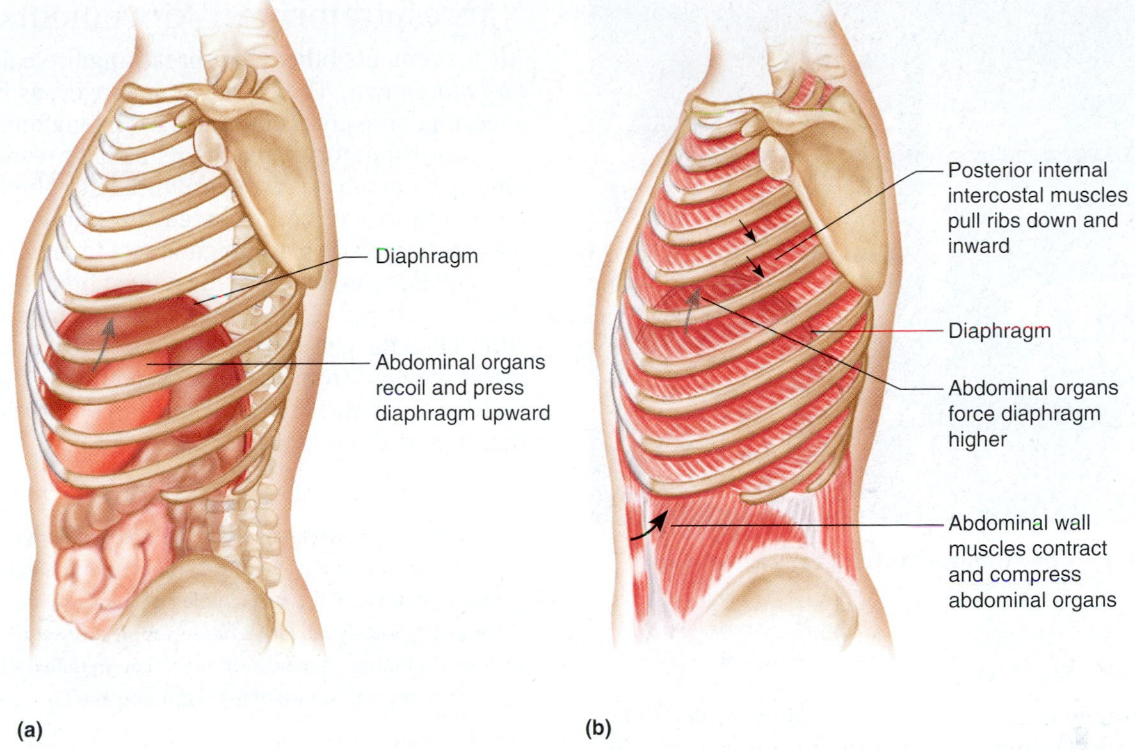

FIGURE 19.25 Expiration. (*a*) Normal resting expiration is due to elastic recoil of the lung tissues and the abdominal organs. (*b*) Contraction of the abdominal wall muscles and posterior internal intercostal muscles aids maximal expiration.

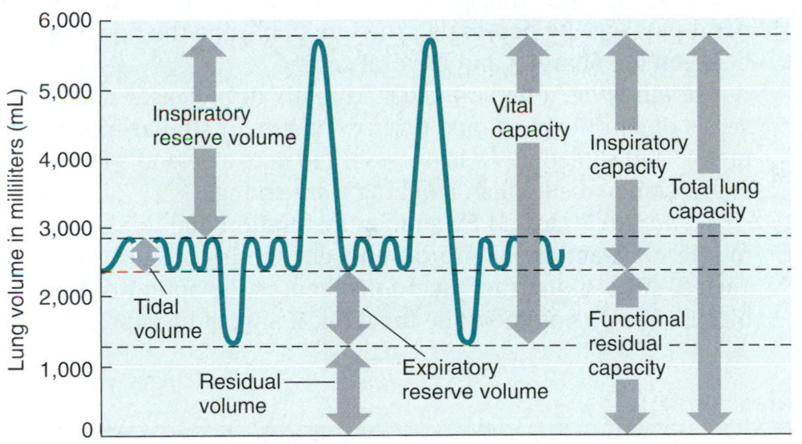

FIGURE 19.26 Respiratory volumes and capacities.

the volumes. If the inspiratory reserve volume (3,000 mL) is combined with the tidal volume (500 mL) and the expiratory reserve volume (1,100 mL), the total is termed the **vital capacity** (4,600 mL). This capacity is the maximum volume of air a person can exhale after taking the deepest breath possible.

The tidal volume (500 mL) plus the inspiratory reserve volume (3,000 mL) gives the **inspiratory capacity** (3,500 mL), which is the maximum volume of air a person can inhale following a resting expiration. Similarly, the expiratory reserve volume (1,100 mL) plus the residual volume (1,200 mL) equals the **functional residual capacity** (2,300 mL), which is the volume of air that remains in the lungs following a resting expiration.

The vital capacity plus the residual volume equals the **total lung capacity** (about 5,800 mL) (see fig. 19.26). This total varies with age, sex, and body size.

Some of the air that enters the respiratory tract during breathing fails to reach the alveoli. This volume (about 150 mL) remains in the passageways of the trachea, bronchi, and bronchioles. Gases are not exchanged through the walls of these passages, so this air is said to occupy *anatomic dead space.*

Occasionally, alveoli in some regions of the lungs are nonfunctional due to poor blood flow in the adjacent capillaries. This creates *alveolar dead space.* The anatomic and alveolar dead space volumes combined equal *physiologic dead space.* In a normal lung, the anatomic and physiologic dead spaces are essentially the same (about 150 mL).

A spirometer (fig. 19.27) is used to measure respiratory air volumes (except the residual volume) to evaluate the course of respiratory illnesses, such as emphysema, pneumonia, lung cancer, and bronchial asthma. Table 19.4 summarizes respiratory air volumes and capacities.

PRACTICE

23 What is tidal volume?

24 Distinguish between inspiratory and expiratory reserve volumes.

25 How is vital capacity measured?

26 How is the total lung capacity calculated?

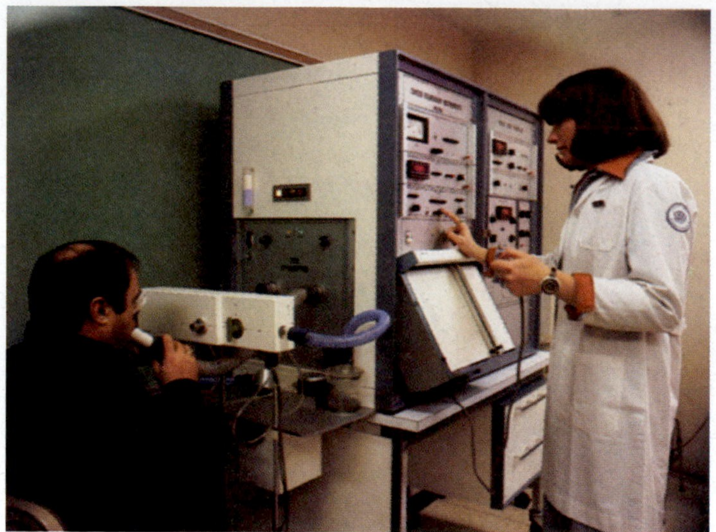

FIGURE 19.27 A spirometer measures respiratory air volumes.

Alveolar Ventilation

The volume of new atmospheric air moved into the respiratory passages each minute is called the *minute ventilation.* It equals the tidal volume multiplied by the breathing rate. Thus, if the tidal volume is 500 mL and the breathing rate is 12 breaths per minute, the minute ventilation is 500 mL × 12, or 6,000 mL per minute. However, for each breath much of the new air remains in the physiologic dead space.

For each respiratory cycle, the volume of new air that does reach the alveoli and is available for gas exchange is calculated by subtracting the physiologic dead space (150 mL) from the tidal volume (500 mL). The resulting volume (350 mL) multiplied by the breathing rate (12 breaths per minute) is the *alveolar ventilation rate* (4,200 mL per minute). This is the more important value physiologically because it affects the concentrations of oxygen and carbon dioxide in the alveoli and thus in the blood.

Nonrespiratory Air Movements

Air movements other than breathing are called *nonrespiratory movements.* They clear air passages, as in coughing and sneezing, or express emotions, as in laughing and crying.

Nonrespiratory movements usually result from *reflexes,* although sometimes they are initiated voluntarily. A cough, for example, can be produced through conscious effort or may be triggered by a foreign object in an air passage.

Coughing involves taking a deep breath, closing the glottis, and forcing air upward from the lungs against the closure. Then the glottis is suddenly opened, and a blast of air is forced upward from the lower respiratory tract. Usually this rapid rush of air is of sufficient force to dislodge the object that triggered the reflex.

> The most sensitive areas of the air passages are in the larynx, the carina, and in regions near the branches of the major bronchi. The distal portions of the bronchioles (respiratory bronchioles), alveolar ducts, and alveoli lack a nerve supply. Consequently, before any material in these parts can trigger a cough reflex, it must be moved into the larger passages of the respiratory tract.

A *sneeze* is much like a cough, but it clears the upper respiratory passages rather than the lower ones. Usually a mild irritation in the lining of the nasal cavity forces a blast of air up through the glottis. The air is directed into the nasal passages by depressing the uvula, closing the opening between the pharynx and the oral cavity.

In *laughing,* a person takes a breath and releases it in a series of short expirations. *Crying* consists of similar movements, and sometimes it is necessary to note a person's facial expression to distinguish laughing from crying.

A *hiccup* is caused by sudden inspiration due to a spasmodic contraction of the diaphragm while the glottis is closed. Air striking the vocal folds causes the sound of the hiccup. We do not know the function, if any, of hiccups.

TABLE 19.4 | Respiratory Air Volumes and Capacities

Name	Volume*	Description
Tidal volume (TV)	500 mL	Volume moved in or out of the lungs during a respiratory cycle
Inspiratory reserve volume (IRV)	3,000 mL	Volume that can be inhaled during forced breathing in addition to resting tidal volume
Expiratory reserve volume (ERV)	1,100 mL	Volume that can be exhaled during forced breathing in addition to resting tidal volume
Residual volume (RV)	1,200 mL	Volume that remains in the lungs at all times
Inspiratory capacity (IC)	3,500 mL	Maximum volume of air that can be inhaled following exhalation of resting tidal volume: IC = TV + IRV
Functional residual capacity (FRC)	2,300 mL	Volume of air that remains in the lungs following exhalation of resting tidal volume: FRC = ERV + RV
Vital capacity (VC)	4,600 mL	Maximum volume of air that can be exhaled after taking the deepest breath possible: VC = TV + IRV + ERV
Total lung capacity (TLC)	5,800 mL	Total volume of air that the lungs can hold: TLC = VC + RV

*Values are typical for a tall, young adult.

Yawning is familiar to everyone, yet its significance and the mechanism by which yawning is contagious remain poorly understood. Evidence points away from a role in increasing oxygen intake. Yawning, and its effect of getting others yawning, may be rooted in primitive brainstem mechanisms that maintain alertness.

Table 19.5 summarizes the characteristics of nonrespiratory air movements. Clinical Application 19.3 discusses respiratory problems that affect ventilation.

PRACTICE

27 How is the minute ventilation calculated? The alveolar ventilation rate?

28 Which nonrespiratory air movements help clear the air passages?

29 Which nonrespiratory air movements are used to express emotions?

30 What seems to be the function of a yawn?

19.5 CONTROL OF BREATHING

Normal breathing is a rhythmic, involuntary act that continues when a person is unconscious. However, the respiratory muscles can be voluntarily controlled too (Take a deep breath and consider this!).

Respiratory Areas

Groups of neurons in the brainstem comprise the **respiratory areas,** which control breathing. These areas periodically initiate impulses that travel on cranial and spinal nerves to breathing muscles, causing inspiration and expiration. The respiratory areas also adjust the rate and depth of breathing to meet cellular requirements for supply of oxygen and removal of CO_2, even during strenuous physical exercise.

The components of the respiratory areas are widely scattered throughout the pons and medulla oblongata. However, two parts of the respiratory areas are of special interest. They are the respiratory center of the medulla and the respiratory group of the pons (fig. 19.28).

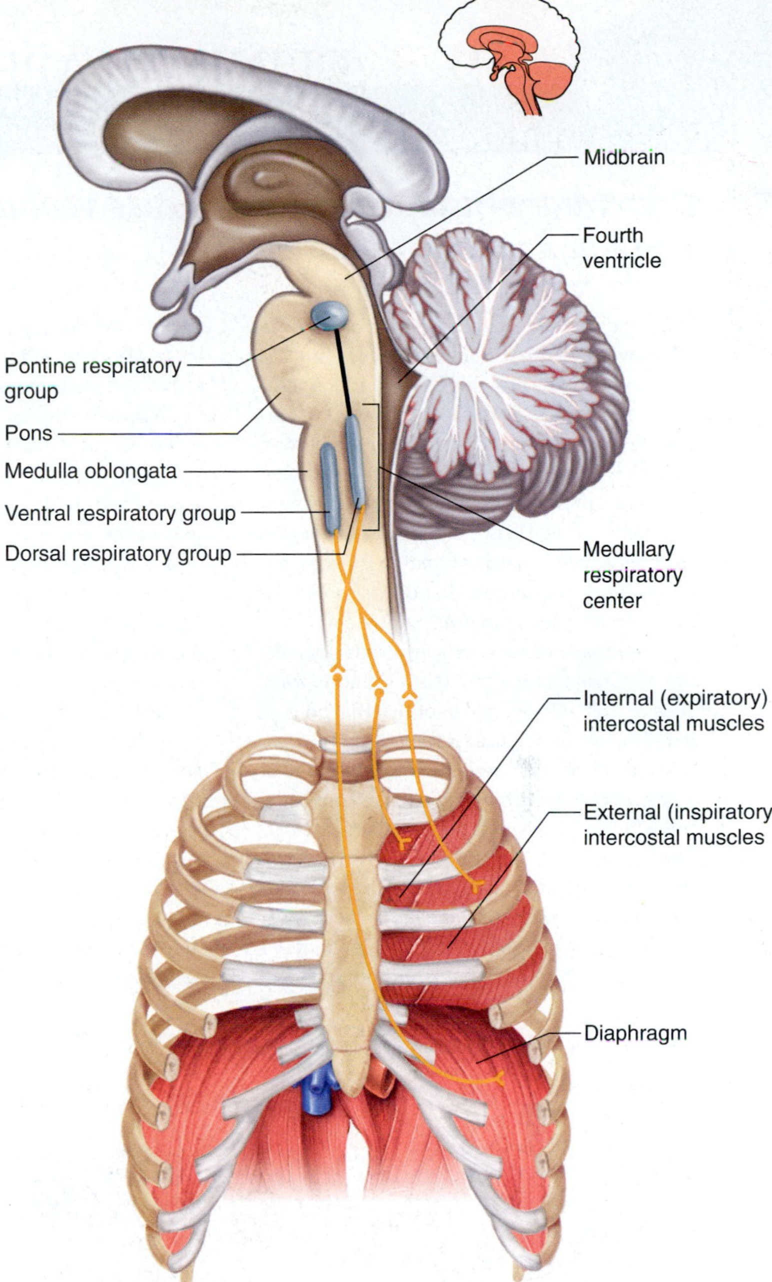

FIGURE 19.28 The respiratory areas are located in the pons and the medulla oblongata.

TABLE 19.5 | Nonrespiratory Air Movements

Air Movement	Mechanism	Function
Coughing	Deep breath is taken, glottis is closed, and air is forced against the closure; suddenly the glottis is opened, and a blast of air passes upward	Clears lower respiratory passages
Sneezing	Same as coughing, except air moving upward is directed into the nasal cavity by depressing the uvula	Clears upper respiratory passages
Laughing	Deep breath is released in a series of short expirations	Expresses happiness
Crying	Same as laughing	Expresses sadness
Hiccuping	Diaphragm contracts spasmodically while glottis is closed	No useful function known
Yawning	Deep breath is taken	Some hypotheses, but no established function
Speech	Air is forced through the larynx, causing vocal cords to vibrate; actions of lips, tongue, and soft palate form words	Vocal communication

19.3 CLINICAL APPLICATION

Respiratory Disorders That Decrease Ventilation: Bronchial Asthma and Emphysema

Injuries to the respiratory center or to spinal nerve tracts that transmit motor impulses may paralyze breathing muscles. Paralysis may also be due to a disease that affects the central nervous system and injures motor neurons, such as *poliomyelitis*. Sometimes, other muscles, by increasing their responses, can compensate for functional losses of a paralyzed muscle. Otherwise, mechanical ventilation is necessary. More common disorders that decrease ventilation are bronchial asthma and emphysema.

Bronchial asthma is usually an allergic reaction to foreign antigens in the respiratory tract, such as from inhaled pollen or material on dust mites. Cells of the larger airways secrete abundant mucus, which traps allergens. Ciliated columnar epithelial cells move the mucus up and out of the bronchi, then up and out of the trachea, clearing the upper respiratory structures. However, in the lower respiratory areas, mucus drainage plus edematous secretions accumulate because fewer cells are ciliated. The allergens and secretions irritate smooth muscles, stimulating bronchoconstriction. Breathing becomes increasingly difficult, and inhalation produces a characteristic wheezing sound as air moves through narrowed passages.

A person with asthma usually finds it harder to force air out of the lungs than to bring it in. This is because inspiration uses powerful breathing muscles, and, as they contract, the lungs expand, opening the air passages. Expiration, on the other hand, is a passive process due to elastic recoil of stretched tissues. Expiration also compresses the tissues and constricts the bronchioles, further impairing air movement through the narrowed air passages.

Increase in the prevalence of asthma in the United States may be due to a too-clean environment, especially for children. Many studies have shown that children who are with others and contract minor respiratory infections, as well as children raised with cats or dogs, are less likely to develop asthma than are children who do not have these exposures. This association of a primed immune system with lower risk of developing asthma is called the hygiene hypothesis.

Emphysema is a progressive, degenerative disease that destroys many alveolar walls. As a result, clusters of small air sacs merge into larger chambers, which decreases the total surface area of the alveolar walls. At the same time, the alveolar walls lose their elasticity, and the capillary networks associated with the alveoli diminish (fig. 19D).

A person with emphysema finds it increasingly difficult to force air out of the lungs because of the loss of tissue elasticity. Abnormal muscular efforts are required to compensate for the lack of elastic recoil that normally contributes to expiration. Only 3% of the 2 million people in the United States who have emphysema inherit the condition; the majority of the other cases are due to smoking or exposure to other respiratory irritants.

An experimental treatment for severe emphysema is lung volume reduction surgery. As its name suggests, the procedure reduces lung volume, which opens collapsed airways and eases breathing. So far, it seems to noticeably improve lung function (as measured by distance walked in six minutes) and quality of life. ■

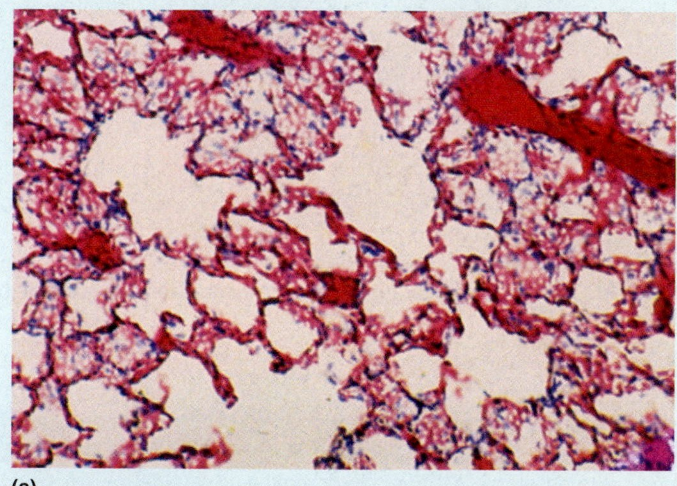
(a)

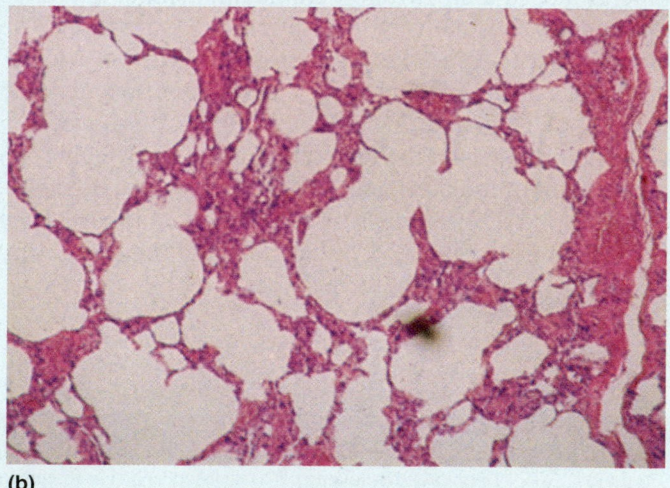
(b)

FIGURE 19D Comparison of lung tissues. (*a*) Normal lung tissue (100×). (*b*) As emphysema develops, alveoli coalesce, forming larger chambers (100×).

The **medullary respiratory center** includes two bilateral groups of neurons that extend throughout the length of the medulla oblongata. They are called the ventral respiratory group and the dorsal respiratory group.

Current evidence suggests that the basic rhythm of breathing arises from the *ventral respiratory group*. Two different populations of neurons in this area have been implicated in maintaining inspiration and expiration, depending on the experimental conditions. Indeed, it has been suggested that such an important physiological process might have more than one way of occurring.

The *dorsal respiratory group* stimulates the inspiratory muscles, primarily the diaphragm, and may contribute to more forceful breathing. The dorsal respiratory goup also helps process sensory information regarding the respiratory system.

Neurons in another part of the brainstem, the pons, compose the *pontine respiratory group* (formerly the *pneumotaxic center*). They may contribute to the rhythm of breathing by limiting inspiration (fig. 19.29).

A condition called *sleep apnea* is responsible for some cases of sudden infant death and for snoring. Infant apneas are almost always *central,* due to a problem with respiratory control centers. In adults, these apneas are usually *obstructive,* involving airway blockage,

Babies who have difficulty breathing just after birth are often sent home with monitoring devices, which sound an alarm when the child stops breathing, alerting parents to resuscitate the infant. The position in which the baby sleeps seems to affect the risk of sleep apnea—sleeping on the back or side is safest during the first year of life.

Adults with sleep apnea may cease breathing for ten to twenty seconds, hundreds of times a night. Bedmates may be aware of the problem because the frequent cessation in breathing causes snoring. The greatest danger of adult sleep apnea is the fatigue, headache, depression, and drowsiness that follows during waking hours.

Sleep apnea is diagnosed in a sleep lab, which monitors breathing during slumber. One treatment for obstructive sleep apnea is *nasal continuous positive airway pressure.* A device is strapped onto the nose at night that maintains air flow into the respiratory system.

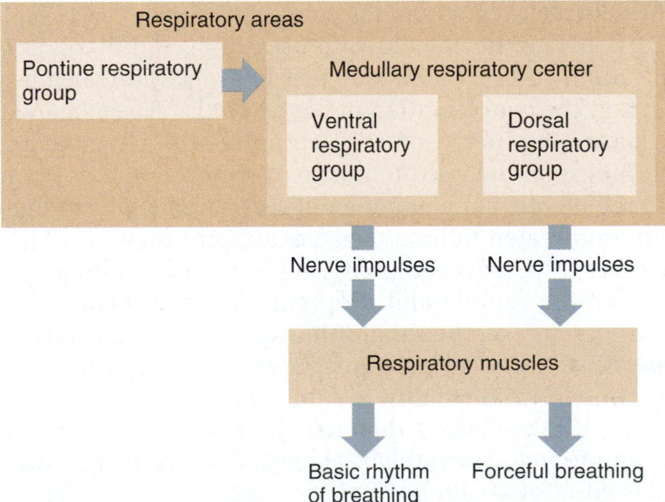

FIGURE 19.29 The medullary respiratory center and the pontine respiratory group control breathing.

PRACTICE

31 Where are the respiratory areas?

32 Describe how the respiratory areas maintain a normal breathing pattern.

33 Explain how the breathing pattern may be changed.

Factors Affecting Breathing

In a mixture of gases such as air, each gas accounts for a portion of the total pressure the mixture produces. The amount of pressure each gas contributes is called the **partial pressure** of that gas and is proportional to its concentration. For example, because air is 21% oxygen, oxygen accounts for 21% of the atmospheric pressure (21% of 760 Hg), or 160 mm Hg (.21 × 760 = 160). Thus, the partial pressure of oxygen, symbolized P_{O_2}, in atmospheric air is 160 mm Hg. Similarly, the partial pressure of carbon dioxide (P_{CO_2}) in air is 0.3 mm Hg.

Gas molecules from the air may enter, or dissolve, in liquid. This is what happens when CO_2 is added to a carbonated beverage or when inspired gases dissolve in the blood in the alveolar capillaries. Using partial pressures simplifies the calculation of the concentration of a dissolved gas. The partial pressure of a gas dissolved in a liquid equals the partial pressure of that gas in the air the liquid has equilibrated with. For example, the P_{O_2} in a glass of water that has been on your desk for awhile must be 160mm Hg, the same as in the air around it. Thus, instead of concentrations of oxygen and carbon dioxide in the body fluids, we will refer to P_{O_2} and P_{CO_2}.

A number of factors influence breathing rate and depth. These include P_{O_2} and P_{CO_2} in body fluids, the degree to which lung tissues are stretched, emotional state, and level of physical activity. The receptors involved include mechanoreceptors that sense stretch as well as central and peripheral chemoreceptors.

Central chemoreceptors are in chemosensitive areas in the ventral portion of the medulla oblongata near the origin of the vagus nerve. These chemoreceptors respond to changes in blood pH, but only indirectly, because hydrogen ions do not easily cross the blood-brain barrier. However, if plasma P_{CO_2} rises, the CO_2 easily diffuses into the brain, where it combines with water in the cerebrospinal fluid to form carbonic acid (H_2CO_3):

$$CO_2 + H_2O \rightarrow H_2CO_3$$

The carbonic acid soon ionizes, releasing hydrogen ions (H^+) and bicarbonate ions (HCO_3^-):

$$H_2CO_3 \rightarrow H^+ + HCO_3^-$$

Hydrogen ions rather than carbon dioxide influence the central chemoreceptors. Breathing rate and tidal volume increase when a person inhales air rich in CO_2 or when body cells produce excess CO_2. These changes increase alveolar ventilation. More CO_2 is exhaled, and the blood P_{CO_2} and hydrogen ion concentration return toward normal.

Adding CO_2 to air can stimulate the rate and depth of breathing. Ordinary air is about 0.04% CO_2. If a patient inhales air containing 4% CO_2, breathing rate usually doubles.

Low blood P_{O_2} has little direct effect on the central chemoreceptors associated with the medulla oblongata. Instead, changes in the blood P_{O_2} are primarily sensed by **peripheral chemoreceptors** in specialized structures called the *carotid bodies* and *aortic bodies,* located in the walls of the carotid sinuses and aortic arch (fig. 19.30). When decreased P_{O_2} stimulates these peripheral receptors, impulses are transmitted to the respiratory center, and the breathing rate and tidal volume increase, increasing alveolar ventilation. This mechanism does not usually play a major role until the P_{O_2} decreases to about 50% of normal. Therefore, oxygen plays only a minor role in the control of normal respiration.

The limited role of P_{O_2} may be surprising, considering the importance of oxygen for sustaining life. Because most blood oxygen is carried on the hemoglobin in red blood cells, deoxygenated systemic venous blood still has 75% of the oxygen it had when it was fully oxygenated. This large excess of oxygen "frees up" respiratory control from paying attention to blood oxygen levels under most circumstances. Thus, the respiratory system can focus on blood P_{CO_2} and hydrogen ion concentration, which are important in maintaining the pH of the internal environment.

The peripheral chemoreceptors of the carotid and aortic bodies are also stimulated by changes in the blood P_{CO_2} and pH. However, CO_2 and hydrogen ions have a much greater effect on the central chemoreceptors of the respiratory center than they do on the carotid and aortic bodies, although this relationship may change with intense exercise.

Patients who have chronic obstructive pulmonary diseases (COPD), such as asthma, bronchitis, and emphysema, gradually adapt to high concentrations of carbon dioxide. For them, low oxygen concentrations may serve as a necessary respiratory stimulus. When such a patient is placed on 100% oxygen, the low arterial P_{O_2} may be corrected, the stimulus removed, and breathing may stop.

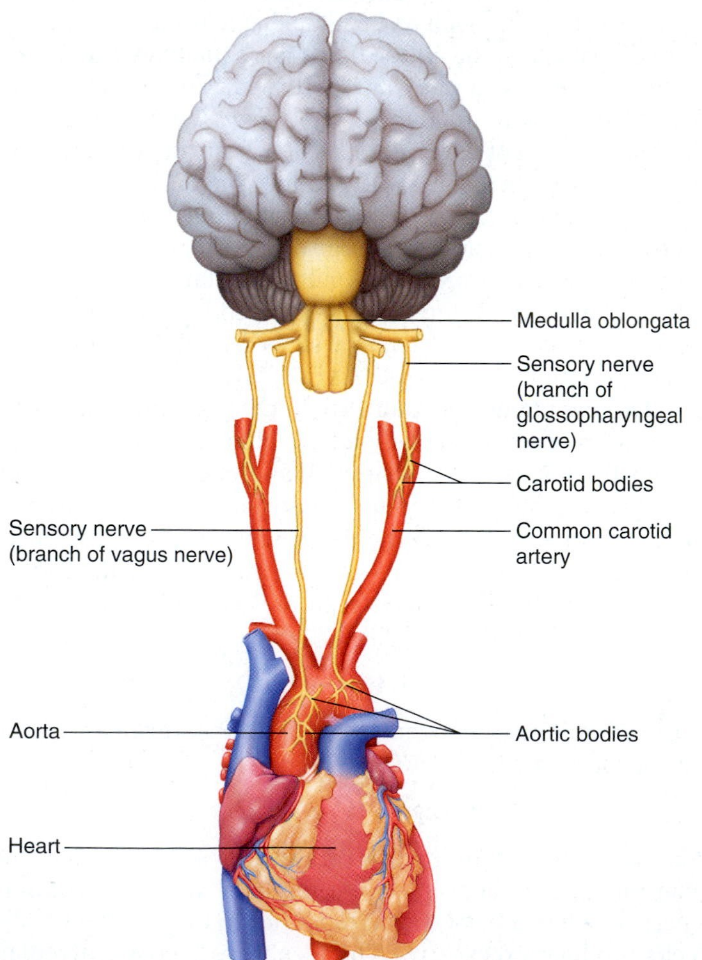

FIGURE 19.30 Decreased P_{O_2} stimulates peripheral chemoreceptors in the carotid and aortic bodies.

An *inflation reflex* (Hering-Breuer reflex) helps regulate the depth of breathing. This reflex occurs when stretch receptors in the visceral pleura, bronchioles, and alveoli are stimulated as lung tissues are stretched. The sensory impulses of the reflex travel via the vagus nerves to the pontine respiratory group and shorten the duration of inspiratory movements. This action prevents overinflation of the lungs during forceful breathing (fig. 19.31).

Emotional upset or strong sensory stimulation may alter the normal breathing pattern. Gasping and rapid breathing are familiar responses to fear, anger, shock, excitement, horror, surprise, sexual stimulation, or even the chill of stepping into a cold shower. Control of the respiratory muscles is voluntary, so we can alter breathing pattern consciously or stop it altogether for a short time. During childbirth, for example, women often concentrate on controlling their breathing, which distracts them from the pain.

If a person decides to stop breathing, the blood concentrations of carbon dioxide and hydrogen ions begin to rise, and the concentration of oxygen falls. These changes (primarily the increased CO_2) stimulate the chemoreceptors, and soon the need to inhale overpowers the desire to hold the breath—much to the relief of parents when young children threaten to hold their breath until they turn blue! However, a person can increase the breath-holding time by breathing rapidly and deeply in advance. (This could be dangerous, see box that follows.) This action, termed **hyperventilation** (hi″per-ven″tĭ-la′shun), lowers the blood CO_2 concentration below normal. Following hyperventilation, it takes longer than usual for the carbon dioxide concentration to reach the level needed to override the conscious effort of breath holding.

Table 19.6 discusses factors affecting breathing. Clinical Application 19.4 focuses on one influence on breathing—exercise.

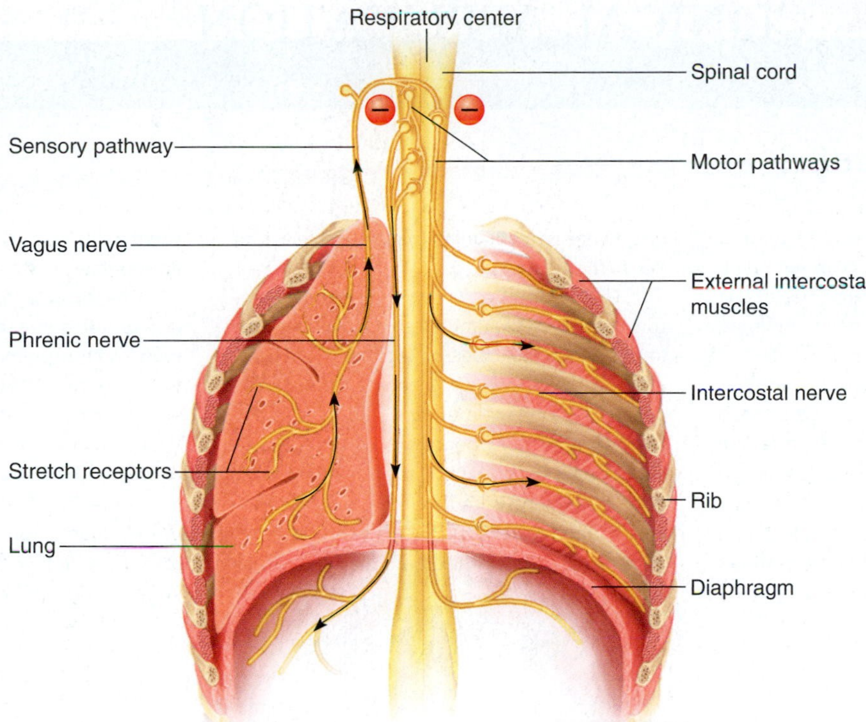

FIGURE 19.31 In the process of inspiration, motor impulses travel from the respiratory center to the diaphragm and external intercostal muscles, which contract and cause the lungs to expand. This expansion stimulates stretch receptors in the lungs to send inhibiting impulses to the respiratory center, preventing overinflation.

TABLE 19.6 | Factors Affecting Breathing

Factors	Receptors Stimulated	Response	Effect
Stretch of tissues	Stretch receptors in visceral pleura, bronchioles, and alveoli	Inhibits inspiration	Prevents overinflation of lungs during forceful breathing
Low plasma Po_2	Chemoreceptors in carotid and aortic bodies	Increases alveolar ventilation	Increases plasma Po_2
High plasma Pco_2	Chemosensitive areas of the respiratory center	Increases alveolar ventilation	Decreases plasma Pco_2
High cerebrospinal fluid hydrogen ion concentration	Chemosensitive areas of the respiratory center	Increases alveolar ventilation	Decreases plasma Pco_2

Sometimes a person who is emotionally upset may hyperventilate, become dizzy, and lose consciousness. This is due to a lowered CO_2 concentration followed by a rise in pH (respiratory alkalosis) and a localized vasoconstriction of cerebral arterioles, decreasing blood flow to nearby brain cells. Hampered oxygen supply to the brain causes fainting. A person should never hyperventilate to help hold the breath while swimming, because the person may lose consciousness under water and drown.

PRACTICE

34 Which chemical factors affect breathing?

35 Describe the inflation reflex.

36 How does hyperventilation decrease respiratory rate?

19.6 ALVEOLAR GAS EXCHANGES

The tubelike parts of the respiratory system move air in and out of the air passages. The alveoli are the sites of the vital process of gas exchange between the air and the blood.

Alveoli

Alveoli are microscopic air sacs clustered at the distal ends of the finest respiratory tubes—the alveolar ducts. Each alveolus is a tiny space within a thin wall that separates it from adjacent alveoli. Tiny openings, called **alveolar pores,** in the walls of some alveoli may permit air to pass from one alveolus to another (fig. 19.32). The pores provide alternate air pathways if the passages in some parts of the lung become obstructed.

Phagocytic cells called *alveolar macrophages* are in alveoli and in the pores. These macrophages phagocytize

19.4 CLINICAL APPLICATION

Exercise and Breathing

Moderate to heavy exercise greatly increases the amount of oxygen skeletal muscles use. A young man at rest uses about 250 milliliters of oxygen per minute but may require 3,600 milliliters per minute during maximal exercise. While oxygen use is increasing, CO_2 production increases also. Decreased blood oxygen and increased blood CO_2 stimulate the respiratory center, so exercise is accompanied by increased breathing rate. However, blood oxygen and CO_2 levels usually do not change during exercise—this reflects the respiratory system's effectiveness in obtaining oxygen and releasing CO_2 to the outside.

The cerebral cortex and the proprioceptors associated with muscles and joints are also implicated in the increased breathing rate associated with exercise (see chapter 12, p. 444). The cortex transmits stimulating impulses to the respiratory center whenever it signals the skeletal muscles to contract. At the same time, muscular movements stimulate the proprioceptors, triggering a *joint reflex.* In this reflex, sensory impulses are transmitted from the proprioceptors to the respiratory center, and breathing accelerates.

The increase in breathing rate during exercise requires increased blood flow to skeletal muscles. Thus, exercise increases demand on both the respiratory and the cardiovascular systems. If either of these systems fails to keep pace with cellular demands, the person will begin to feel out of breath. This sensation, however, is usually due to the inability of the cardiovascular system to move enough blood between the lungs and the cells, rather than to the inability of the respiratory system to provide enough air. ■

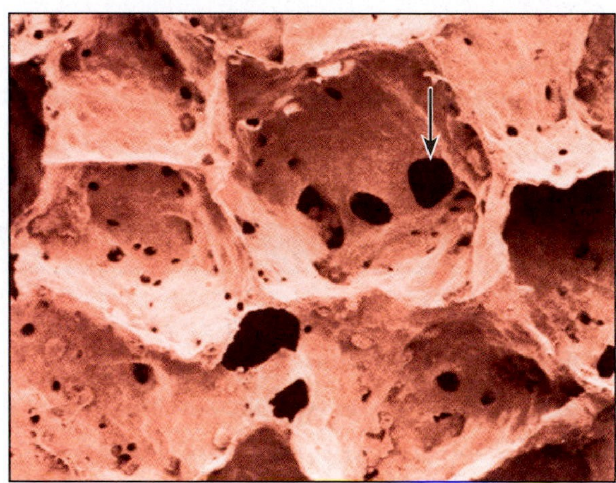

FIGURE 19.32 Alveolar pores (arrow) allow air to pass from one alveolus to another (300×).

airborne agents, including bacteria, thereby cleaning the alveoli (fig. 19.33).

Respiratory Membrane

Part of the wall of an alveolus is made up of cells (type II cells) that secrete pulmonary surfactant, described earlier, but the bulk of the wall of an alveolus consists of a layer of simple squamous epithelium (type I cells). Each alveolus is associated with a dense network of capillaries lined with simple squamous epithelial cells (fig. 19.33). Thin basement membranes separate the layers of these flattened cells, and in the spaces between them are elastic and collagenous fibers that help support the alveolar wall. Thus, two thicknesses of epithelial cells and basement membranes separate the air in an alveolus and the blood in a capillary. These layers make up the **respiratory membrane** (alveolar-capillary membrane), through which gas exchange occurs between the alveolar air and the blood (figs. 19.34 and 19.35).

Exposure to high oxygen concentration (hyperoxia) for a prolonged time may damage lung tissues, particularly capillary walls. Excess fluid may escape the capillaries and flood the alveolar air spaces, interfering with gas exchange, which can be lethal. Similarly, hyperoxia can damage the retinal capillaries of premature infants, causing retrolental fibroplasia (RLF), a condition that may lead to blindness.

Diffusion Through the Respiratory Membrane

Molecules diffuse from regions where they are in higher concentration toward regions where they are in lower concentration. Thus, in determining the direction of diffusion of a solute, we must know the concentration gradient. In the case of gases, it is more convenient to think in terms of a partial pressure gradient, such that a gas will diffuse from an area of higher partial pressure to an area of lower partial pressure.

RECONNECT
To Chapter 3, Diffusion, pages 90–92.

When a mixture of gases dissolves in blood, the resulting concentration of each dissolved gas is proportional to its partial pressure. Each gas diffuses between blood and its surroundings from areas of higher partial pressure to areas of lower partial pressure until the partial pressures in the two regions reach equilibrium. For example, the P_{CO_2} of blood entering the pulmonary capillaries is 45 mm Hg, but

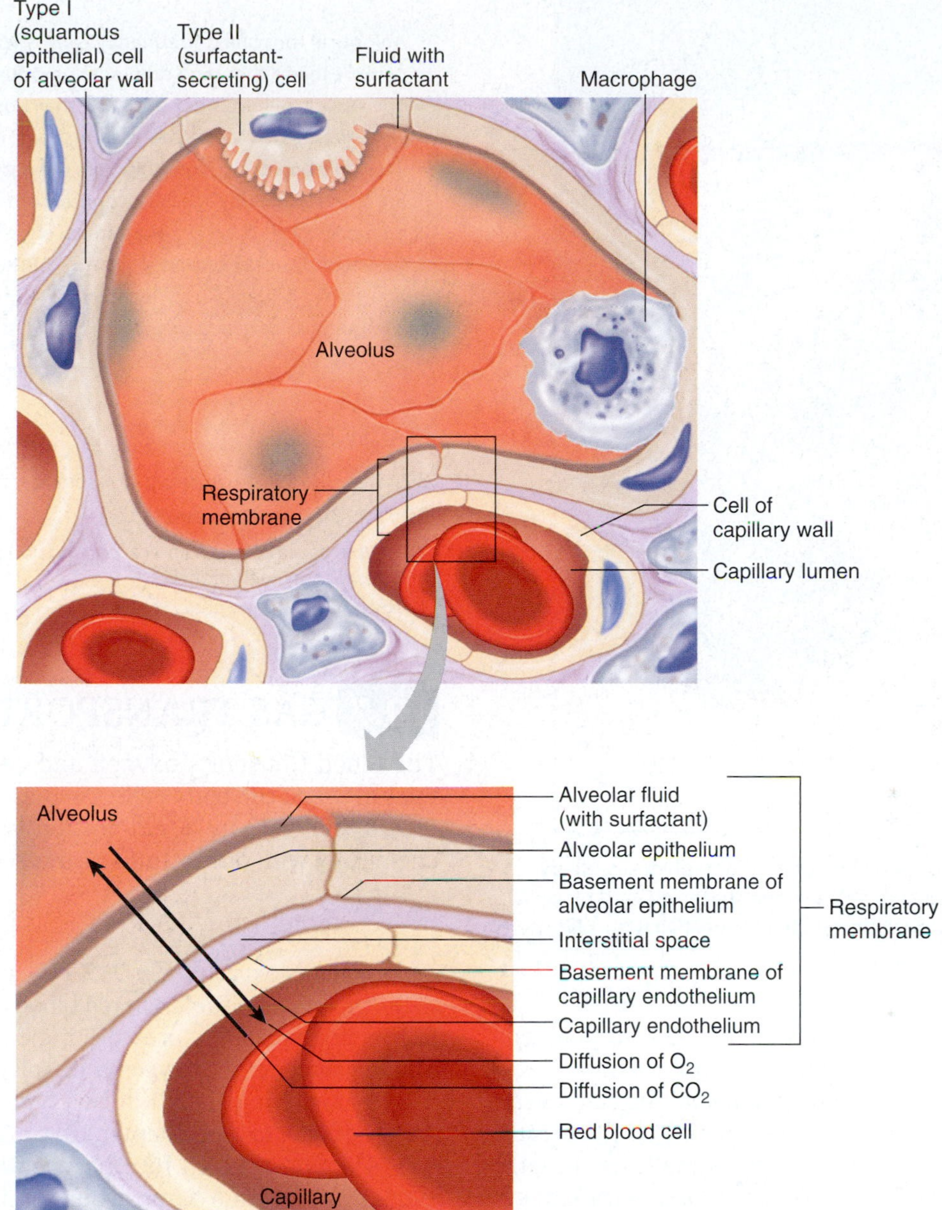

FIGURE 19.33 The respiratory membrane consists of the walls of the alveolus and the capillary.

the P_{CO_2} in alveolar air is 40 mm Hg. Because of the difference in these partial pressures, carbon dioxide diffuses from blood, where its partial pressure is higher, across the respiratory membrane and into alveolar air. When blood leaves the lungs, its P_{CO_2} is 40 mm Hg, the same as the P_{CO_2} of alveolar air. Similarly, the P_{O_2} of blood entering the pulmonary capillaries is 40 mm Hg, but reaches 104 mm Hg as oxygen diffuses from alveolar air into the blood. Thus, because equilibrium is reached, blood leaves the alveolar capillaries with a P_{O_2} of 104 mm Hg. (Some venous blood draining the bronchi and bronchioles mixes with this blood before returning to the heart, so the P_{O_2} of left atrial, left ventricular, and systemic arterial blood is 95 mm Hg.) Because of the large volume of air always in the lungs as long as breathing continues, alveolar P_{O_2} stays relatively constant at 104 mm Hg. Clinical Application 19.5 looks at a respiratory effect that occurs under specific conditions—high altitude.

The respiratory membrane is normally thin (about 1 micrometer thick), and gas exchange is rapid. However, a number of factors may affect diffusion across the respiratory membrane. More surface area, shorter distance, greater solubility of gases, and a steeper partial pressure gradient all favor increased diffusion. Diseases that harm the respiratory membrane, such as pneumonia, or reduce the surface area for diffusion, such as emphysema, impair gas exchange. These conditions may require increased P_{O_2} for treatment. Clinical Application 19.6 examines illnesses that result from impaired gas exchange.

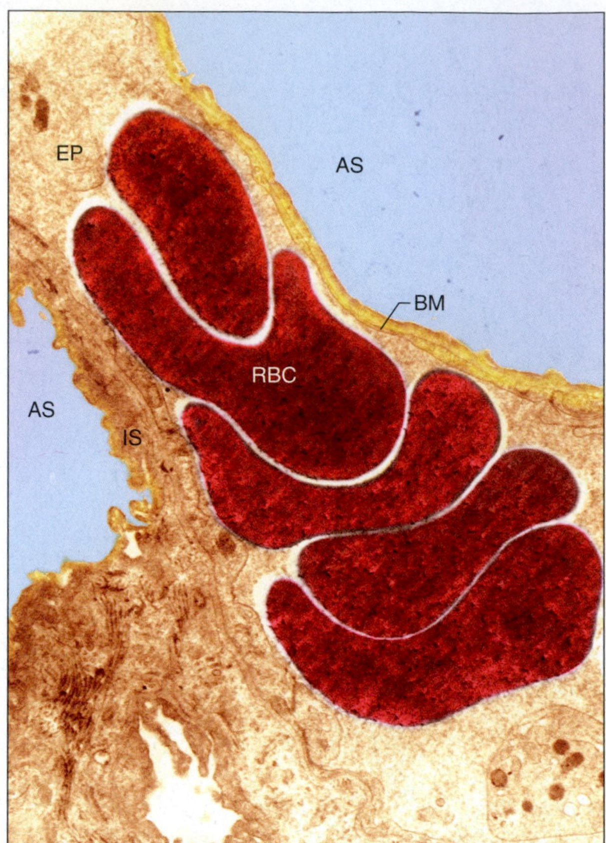

FIGURE 19.34 Falsely colored electron micrograph of a capillary located between alveoli (7,000×). (*AS,* alveolar space; *RBC,* red blood cell; *BM,* basement membrane; *IS,* interstitial connective tissue; *EP,* epithelial nucleus.)

The respiratory membrane is normally so thin that certain soluble chemicals other than CO_2 may diffuse into alveolar air and be exhaled. This is why breath analysis can reveal alcohol in the blood or acetone can be smelled on the breath of a person who has untreated diabetes mellitus. Breath analysis may also detect substances associated with kidney failure, certain digestive disturbances, and liver disease.

An optical tool called "frequency combs" can detect trace amounts of more than a thousand different compounds in an exhaled breath. The sets of chemicals detected and their concentrations can provide clues to health. A special laser sends short pulses of light into a device holding a breath. An algorithm uses the pattern of light wavelengths absorbed to identify specific chemical compounds that may serve as biomarkers (see chapter 2 vignette on page 51). For example, a signature of nitric oxide, carbon monoxide, nitrites, nitrates, pentane, ethane, and hydrogen peroxide indicates asthma, but a different set of compounds in exhaled breath indicates cystic fibrosis.

PRACTICE

37 Describe the structure of the respiratory membrane.

38 What is partial pressure?

39 What causes oxygen and carbon dioxide to move across the respiratory membrane?

19.7 GAS TRANSPORT

The blood transports oxygen and carbon dioxide between the lungs and the body cells. As these gases enter the blood, they dissolve in the liquid portion, the plasma, or combine chemically with other atoms or molecules.

Oxygen Transport

Almost all the oxygen (over 98%) is carried in the blood bound to the protein *hemoglobin* in red blood cells. The oxygen bound to iron in hemoglobin provides the color of these blood cells. The remainder of the oxygen is dissolved in the blood plasma.

Hemoglobin consists of two types of components called *heme* and *globin* (see chapter 18, p. 720). Globin is a protein of 574 amino acids in four polypeptide chains. Each chain is associated with a heme group, and each heme group surrounds an atom of iron. Each iron atom can loosely bind an

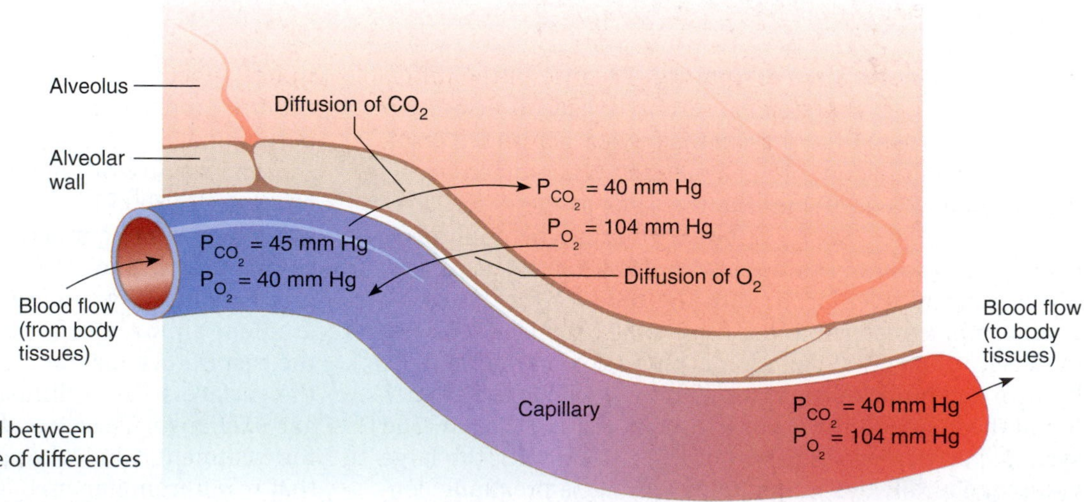

FIGURE 19.35 Gases are exchanged between alveolar air and capillary blood because of differences in partial pressures.

19.5 CLINICAL APPLICATION

Effects of High Altitude

Every year, about 100,000 mountain climbers experience varying degrees of altitude sickness, because at high elevations, the proportion of oxygen in air remains the same (about 21%), but the P_{O_2} decreases. When a person ascends rapidly, oxygen diffuses more slowly from the alveoli into the blood, and the hemoglobin becomes less saturated with oxygen. In some individuals, the body's efforts to get more oxygen—increased breathing and heart rate and enhanced red blood cell and hemoglobin production—cannot keep pace with the plummeting oxygen supply.

Severe altitude sickness includes a condition called high-altitude pulmonary edema (HAPE). Symptoms are sudden severe headache, nausea and vomiting, rapid heart rate and breathing, and a cyanotic (blue) cast to the skin, often first apparent under the fingernails.

The hypoxia associated with high altitude can cause vasoconstriction of pulmonary blood vessels. In some persons, this shunts blood under high pressure through less constricted vessels in the pulmonary circuit, raising capillary pressure and filtering fluid from the blood vessels into the alveoli. Persons with severe HAPE commonly develop high-altitude cerebral edema (HACE).

HAPE is treated by giving oxygen and coming down from the mountain. Delay may prove fatal. Exertion may worsen the symptoms, and victims often need to be carried. Some prescription vasodilators, such as nifedipine, may help reduce the pulmonary hypertension, but they can be dangerous without proper medical attention.

Mountain climbing is an extreme activity that endangers the respiratory system. Regularly exercising at moderately high altitude, however, can strengthen the system. Analysis of the results of 1,460 international football competitions, covering 100 years, revealed that teams accustomed to altitudes greater than 2,500 meters (about 1.5 miles) above sea level scored significantly more touchdowns, at any altitude, than did teams that practiced at much lower altitudes. In response to the study, the Federation of International Football Associations banned matches above this level because the teams that trained at higher elevations had an unfair advantage. Distance runners, too, often train at high altitudes because of beneficial effects on fitness. ■

oxygen molecule. As oxygen dissolves in blood, it rapidly combines with hemoglobin, forming a new compound called **oxyhemoglobin** (ok″sĭ-he″mo-glo′bin). Each hemoglobin molecule can bind up to four oxygen molecules.

The P_{O_2} determines the amount of oxygen that hemoglobin binds. The greater the P_{O_2}, the more oxygen binds until the hemoglobin molecules are saturated (fig. 19.36). At normal arterial P_{O_2} (95 mm Hg), hemoglobin is essentially completely saturated.

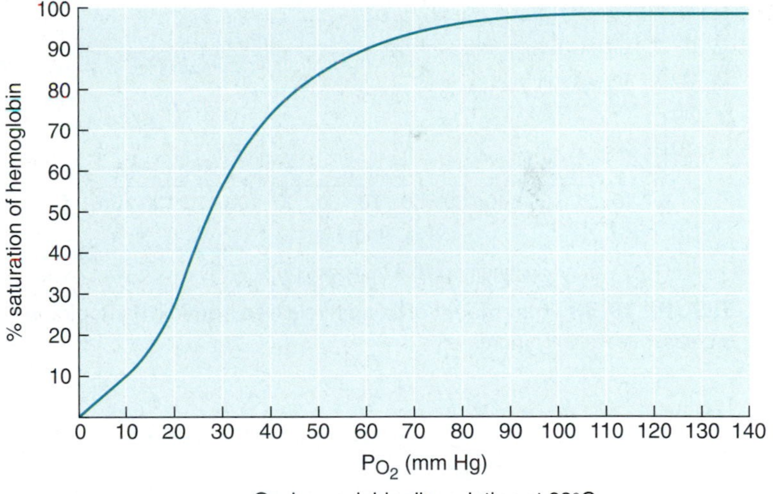

FIGURE 19.36 Hemoglobin is completely saturated at normal systemic arterial P_{O_2} but readily releases oxygen at the P_{O_2} of the body tissues.

The chemical bonds between oxygen and hemoglobin molecules are relatively unstable, and as the P_{O_2} decreases, oxyhemoglobin releases oxygen molecules (fig. 19.36). This happens in tissues in which cells have used oxygen in respiration. The free oxygen diffuses from the blood into nearby cells, as figure 19.37 shows.

Increasing blood concentration of carbon dioxide (P_{CO_2}), acidity, and temperature all increase the amount of oxygen that oxyhemoglobin releases (figs. 19.38, 19.39, and 19.40). These influences explain why more oxygen is released from the blood to the skeletal muscles during exercise. The increased muscular activity accompanied by increased oxygen use increases the P_{CO_2}, decreases the pH, and raises the local temperature. At the same time, less-active cells receive less oxygen.

As described earlier, respiratory control under most circumstances is responding to plasma P_{CO_2} and pH, not P_{O_2}, despite the central role of oxygen in cellular metabolism. Notice, however, in figures 19.36 and 19.37, that the deoxygenated systemic venous blood retains 75% of the oxygen it had when it was fully oxygenated. This safety margin makes it possible for the respiratory system to safely maintain CO_2 levels, and thereby the pH, of the internal environment.

PRACTICE

40 How is oxygen transported from the lungs to body cells?

41 What factors affect the release of oxygen from oxyhemoglobin?

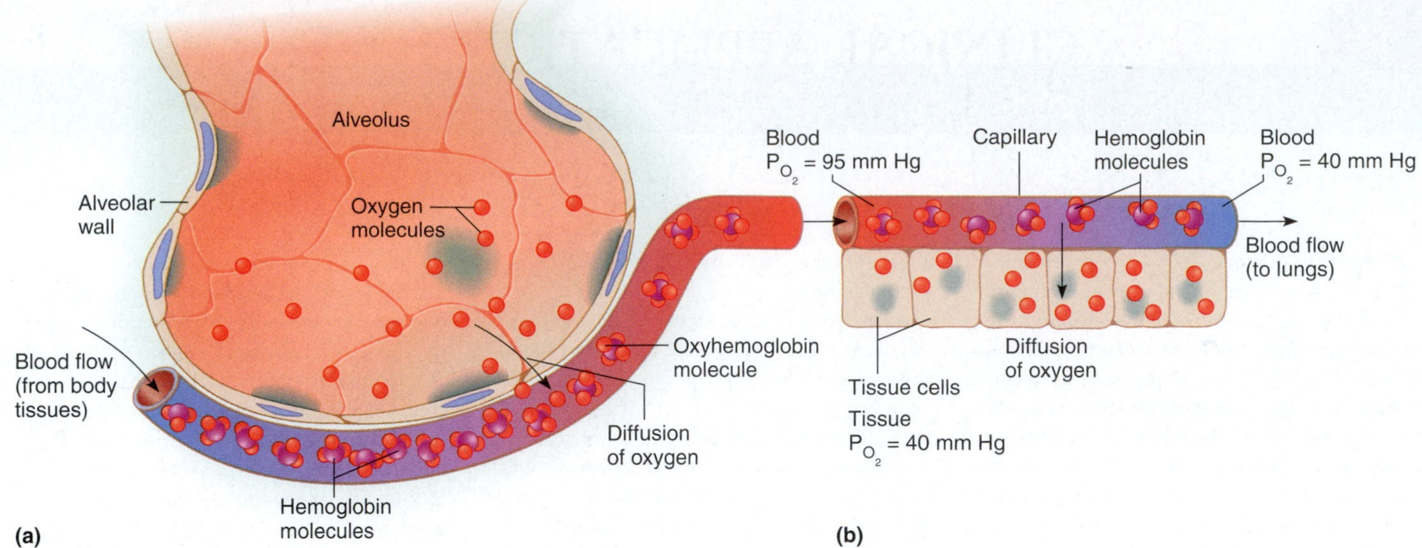

FIGURE 19.37 Blood transports oxygen. (*a*) Oxygen molecules, entering the blood from the alveolus, bond to hemoglobin, forming oxyhemoglobin. (*b*) In the regions of the body cells, oxyhemoglobin releases oxygen. Much oxygen is still bound to hemoglobin at the P_{O_2} of systemic venous blood.

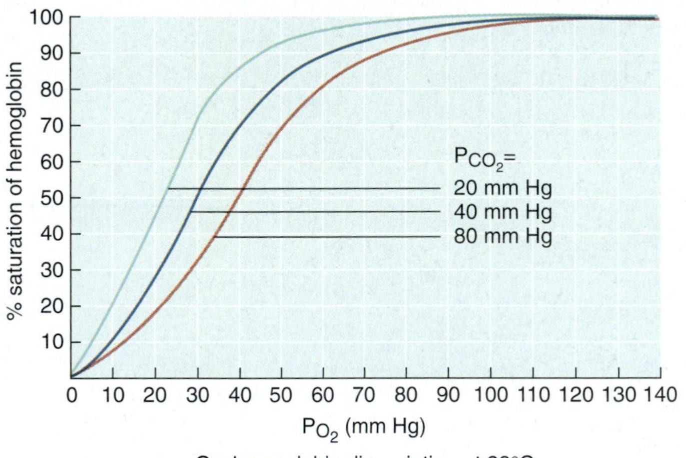

FIGURE 19.38 The amount of oxygen released from oxyhemoglobin increases as the P_{CO_2} increases.

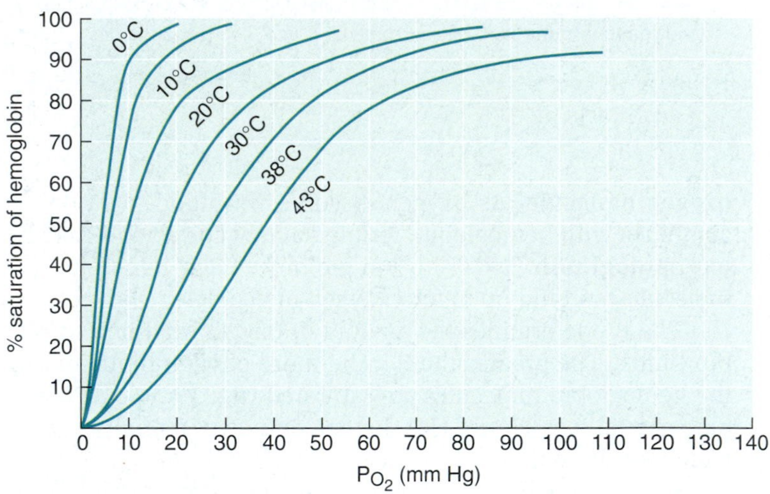

FIGURE 19.40 The amount of oxygen released from oxyhemoglobin increases as the blood temperature increases.

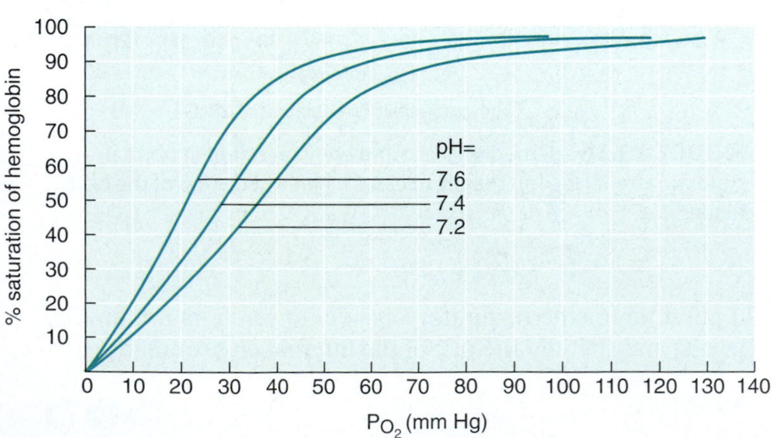

FIGURE 19.39 The amount of oxygen released from oxyhemoglobin increases as the blood pH decreases.

Carbon monoxide (CO) is a toxic gas produced in gasoline engines and some stoves as a result of incomplete combustion of fuels. It is also a component of tobacco smoke. Carbon monoxide is toxic because it binds hemoglobin many times more effectively than does oxygen and therefore does not readily dissociate from hemoglobin. Thus, when a person breathes carbon monoxide, less hemoglobin is available for oxygen transport, depriving cells of oxygen. Treatment for carbon monoxide poisoning is to administer oxygen in high concentration to replace some of the carbon monoxide bound to hemoglobin molecules. Carbon dioxide (CO_2) is usually given simultaneously to stimulate the respiratory center, which, in turn, increases breathing rate. Rapid breathing helps reduce the concentration of carbon monoxide in the alveoli.

19.6 CLINICAL APPLICATION

Disorders That Impair Gas Exchange: Pneumonia, Tuberculosis, and Adult Respiratory Distress Syndrome

Five-year-old Carly became ill very quickly. After twelve hours of flulike symptoms, her temperature soared, her chest began to hurt, and her breathing became rapid and shallow. A chest radiograph indicated that she had *pneumonia*. The bacteria that had caused a mild upper respiratory infection in her sisters had infected her lower respiratory structures.

Antibiotics treated Carly's pneumonia. Many of the 50 million people who died in the 1918 influenza pandemic were not so lucky, in that era before antibiotics. Although viruses can cause pneumonia, most of the flu deaths were actually caused by secondary bacterial infections. Bacteria that normally inhabit upper respiratory structures easily moved downward to the lungs made vulnerable by the viral infection. Fungi can also cause pneumonia, such as the fungus *Pneumocystis jirovecii* that affects people with AIDS.

For all types of pneumonia, events in the infected lung are similar: alveolar linings swell with edema and become abnormally permeable, allowing fluids and white blood cells to accumulate in the air sacs. As the alveoli fill, the surface area available for gas exchange diminishes. Breathing becomes difficult. Untreated, pneumonia can kill.

Tuberculosis is a different type of lung infection, caused by the bacterium *Mycobacterium tuberculosis*. Fibrous connective tissue develops around the sites of infection, forming structures called *tubercles* (fig. 19E). By walling off the bacteria, the tubercles help stop their spread. Sometimes this protective mechanism fails, and the bacteria flourish throughout the lungs and may spread to other organs. In the later stages of infection, other types of bacteria may cause secondary infections. As lung tissue is destroyed, the surface area for gas exchange decreases. In addition, the widespread fibrous tissue thickens the respiratory membrane, further restricting gas exchange. A variety of drugs are used to treat tuberculosis, but in recent years, strains resistant to drugs have arisen, and these can be swiftly deadly.

Another type of condition that impairs gas exchange is *atelectasis*. This is the collapse of a lung, or some part of it, with the collapse of the blood vessels that supply the affected region. Obstruction of a respiratory tube, such as by an inhaled foreign object or excess mucus secretion, may cause atelectasis. The air in the alveoli beyond the obstruction is absorbed, and as the air pressure in the alveoli decreases, their elastic walls collapse, and they can no longer function. Fortunately, after a portion of a lung collapses, the functional regions that remain are often able to carry on enough gas exchange to sustain the body cells.

In *acute respiratory distress syndrome* (ARDS), which is a special form of atelectasis, alveoli collapse. Causes include pneumonia and other infections, near drowning, shock, sepsis, aspiration of stomach acid into the respiratory system, or physical trauma to the lungs from an injury or surgical procedure. Anesthetic drugs can cause ARDS by suppressing surfactant production. Until surfactant levels return to normal—in a day or two—the respiratory membrane that separates the air in the alveoli from the blood in the pulmonary capillaries is damaged enough to allow protein-rich fluid to escape from the capillaries and flood the alveoli. In response the tiny air sacs collapse. Blood vessels and airways narrow, greatly elevating blood pressure in the lungs. Delivery of oxygen to tissues is seriously impaired. ARDS is fatal about 60% of the time. (It was until recently called adult respiratory distress syndrome.) ■

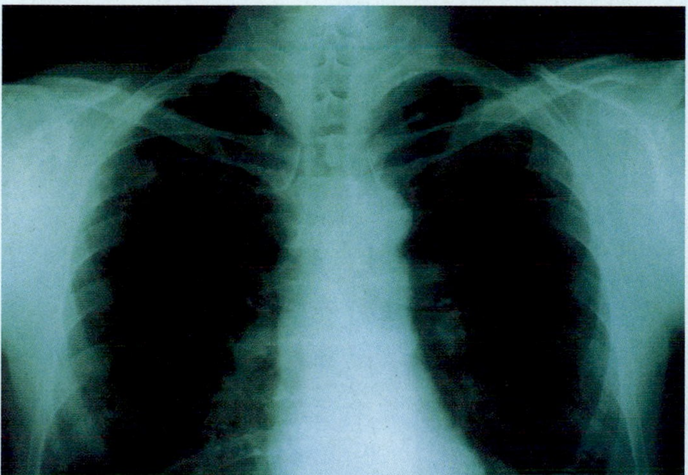

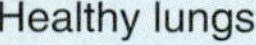

Healthy lungs

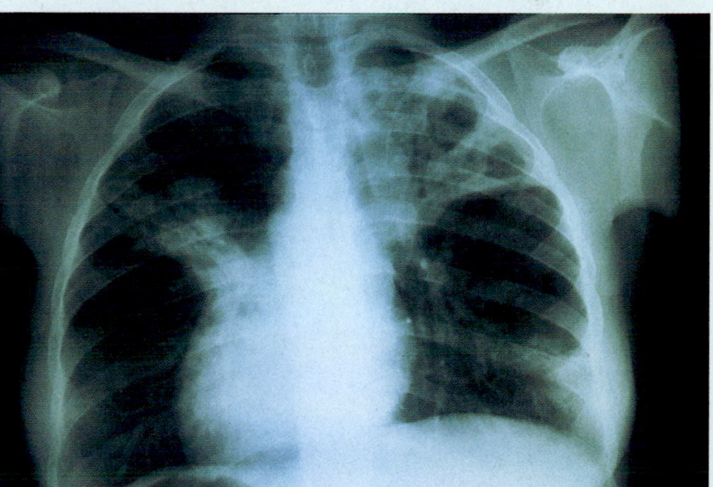

Tuberculosis

FIGURE 19E Healthy lungs appear dark and clear on a radiograph. Lungs with tuberculosis have cloudy areas where fibrous tissue grows, walling off infected areas.

Carbon Dioxide Transport

Blood flowing through capillaries gains CO_2 because the tissues have a high P_{CO_2}. This CO_2 is transported to the lungs in one of three forms: as CO_2 dissolved in plasma, as part of a compound formed by bonding to hemoglobin, or as part of a bicarbonate ion (fig. 19.41).

The amount of carbon dioxide that dissolves in plasma is determined by its partial pressure. The higher the P_{CO_2} of the tissues, the more carbon dioxide will go into solution. However, only about 7% of the carbon dioxide is transported in this form.

Unlike oxygen, which binds the iron atoms of hemoglobin molecules, carbon dioxide bonds with the amino groups ($—NH_2$) of these molecules. Consequently, oxygen and carbon dioxide do not directly compete for binding sites—a hemoglobin molecule can transport both gases at the same time.

Carbon dioxide binding hemoglobin forms a loosely bound compound called **carbaminohemoglobin** (kar-bam″ĭ-no-he″mo-globin). This molecule readily decomposes in regions where the P_{CO_2} is low, releasing its carbon dioxide. Although this method of transporting carbon dioxide is theoretically quite effective, carbaminohemoglobin forms relatively slowly. Only about 15% to 25% of the total CO_2 is carried this way.

In the most important CO_2 transport mechanism **bicarbonate ions** (HCO_3^-) form. Recall that carbon dioxide reacts with water to form carbonic acid (H_2CO_3). This reaction occurs slowly in the blood plasma, but much of the CO_2 diffuses into the red blood cells. These cells contain an enzyme, **carbonic anhydrase** (kar-bon′ik an-hi′drās), which speeds the reaction between CO_2 and water.

The resulting carbonic acid dissociates almost immediately, releasing hydrogen ions (H^+) and bicarbonate ions (HCO_3^-):

$$H_2CO_3 \rightarrow H^+ + HCO_3^-$$

These new hydrogen ions might be expected to lower blood pH, but this reaction occurs in the systemic capillaries, where deoxyhemoglobin is generated. Deoxyhemoglobin is an excellent buffer because hydrogen ions readily bind it. The bicarbonate ions diffuse out of the red blood cells and enter the blood plasma. As much as 70% of the carbon dioxide transported in the blood is carried in this form.

As the bicarbonate ions leave the red blood cells and enter the plasma, *chloride ions,* which also have negative charges, are electrically repelled, and they move from the plasma into the red blood cells. This exchange in position of the two negatively charged ions, shown in figure 19.42, maintains the ionic balance between the red blood cells and the plasma. It is termed the **chloride shift.**

As blood passes through the capillaries of the lungs, the dissolved carbon dioxide diffuses into the alveoli, in response to the relatively low P_{CO_2} of the alveolar air. As the plasma P_{CO_2} drops, hydrogen ions and bicarbonate ions in the red blood cells recombine to form carbonic acid, and under the influence of carbonic anhydrase, the carbonic acid quickly yields new molecules of CO_2 and water:

$$H^+ + HCO_3^- \rightarrow H_2CO_3 \rightarrow CO_2 + H_2O$$

Carbaminohemoglobin also releases its CO_2, and both of these events contribute to the P_{CO_2} of the alveolar capillary blood. CO_2 diffuses out of the blood until an equilibrium is established between the P_{CO_2} of the blood and the P_{CO_2} of the

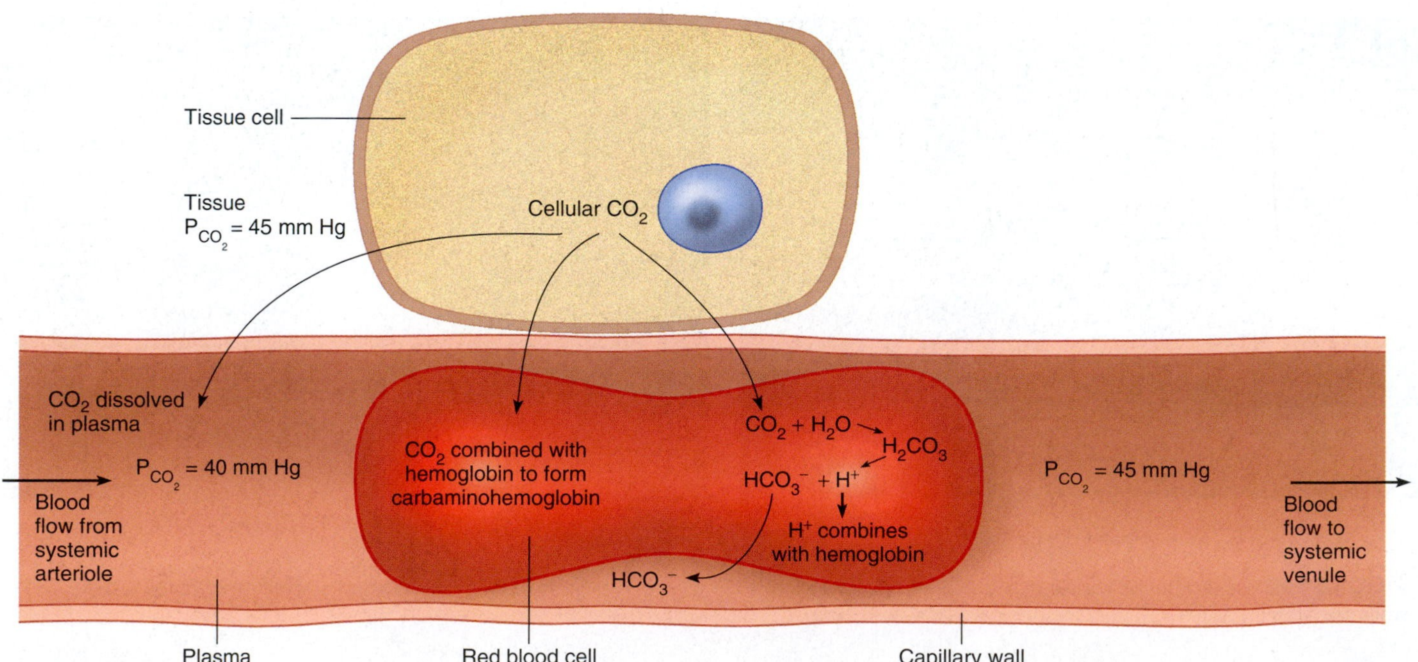

FIGURE 19.41 Carbon dioxide produced by cells is transported in the blood plasma in a dissolved state, bound to hemoglobin, or in the form of bicarbonate ions (HCO_3^-).

alveolar air. Figure 19.43 summarizes this process, and table 19.7 summarizes the transport of blood gases.

PRACTICE

42 Describe three ways carbon dioxide can be transported from cells to the lungs.

43 How can hemoglobin carry oxygen and carbon dioxide at the same time?

44 How do bicarbonate ions help buffer the blood (maintain its pH)?

45 What is the chloride shift?

46 How is carbon dioxide released from the blood into the lungs?

19.8 LIFE-SPAN CHANGES

Changes in the respiratory system over a lifetime reflect both the accumulation of environmental influences and the effects of aging in other organ systems. The lungs and respiratory passageways of a person who has breathed only clean air are pinker and can exchange gases much more efficiently as the years pass than can the respiratory system of a person who has breathed polluted air and smoked for many years. Those who have been exposed to foul air are more likely to develop chronic bronchitis, emphysema, and/or lung cancer. Long-term exposure to particulates in the workplace can also raise the risk of developing these conditions. Still, many age-associated changes in the respiratory system are unavoidable.

With age, protection of the lungs and airways falters, as ciliated epithelial cells become fewer, and their cilia less active or gone. At the same time, mucus thickens; the swallowing, gagging, and coughing reflexes slow; and macrophages lose their efficiency in phagocytizing bacteria. These changes combine to slow the clearance of pathogens

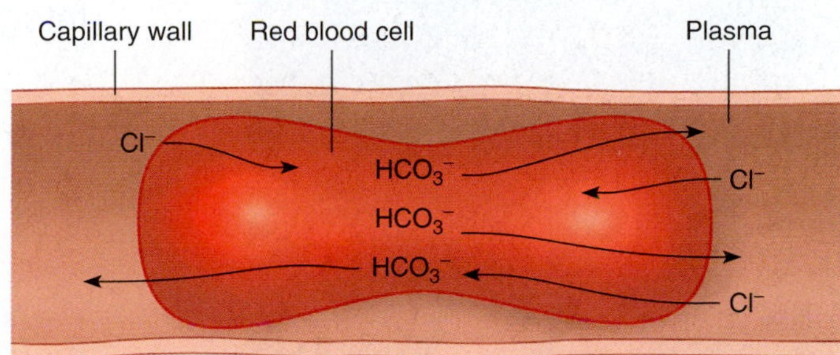

FIGURE 19.42 As bicarbonate ions (HCO_3^-) diffuse out of the red blood cell, chloride ions (Cl⁻) from the plasma diffuse into the cell, maintaining the electrical balance between ions. This exchange of ions is called the chloride shift.

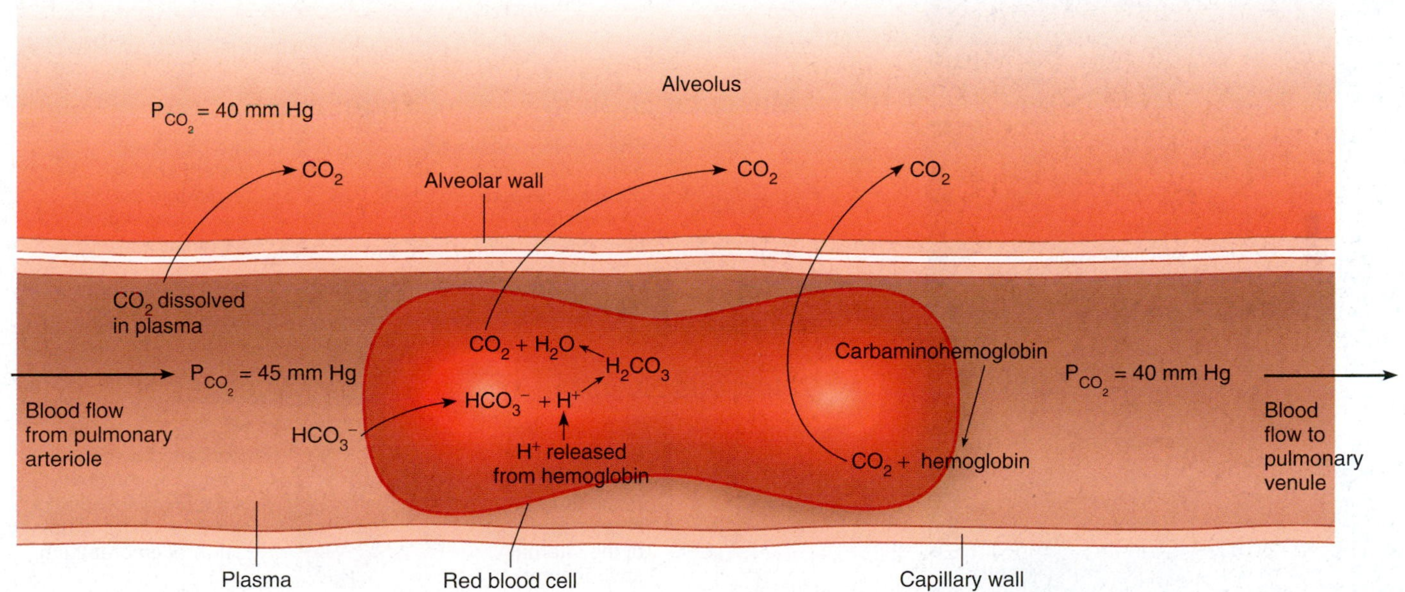

FIGURE 19.43 In the lungs, carbon dioxide diffuses from the blood into the alveoli.

TABLE 19.7 | Gases Transported in Blood

Gas	Reaction	Substance Transported
Oxygen	Combines with iron atoms of hemoglobin molecules	Oxyhemoglobin
Carbon dioxide	About 7% dissolves in plasma	Carbon dioxide
	About 23% combines with the amino groups of hemoglobin molecules	Carbaminohemoglobin
	About 70% reacts with water to form carbonic acid; the carbonic acid then dissociates to release hydrogen ions and bicarbonate ions	Bicarbonate ions

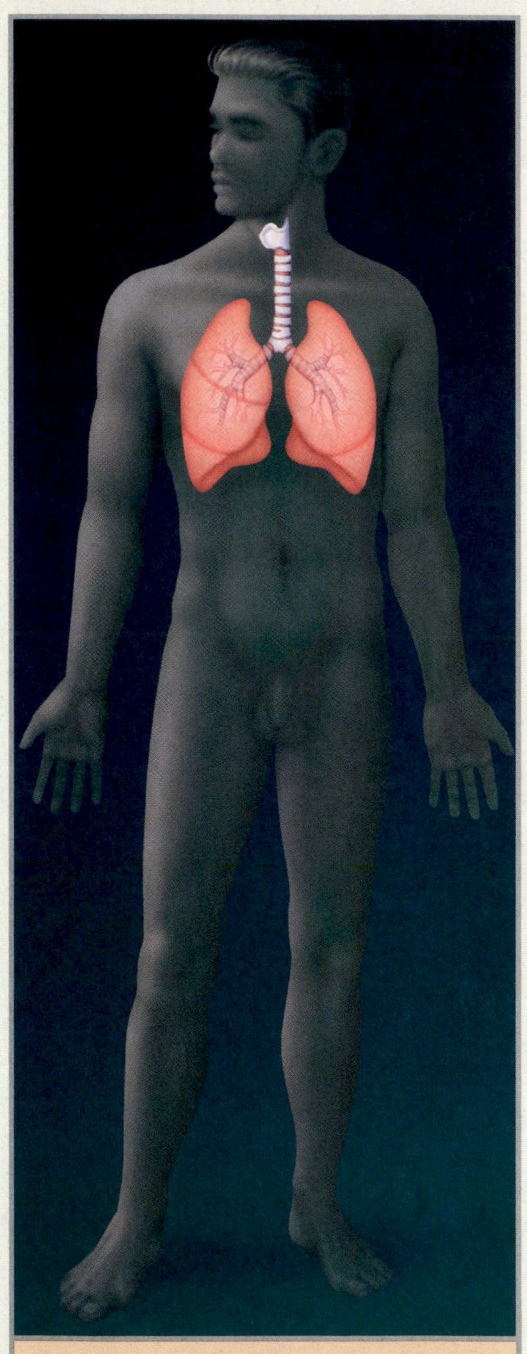

Respiratory System

The respiratory system provides oxygen for the internal environment and excretes carbon dioxide.

Integumentary System

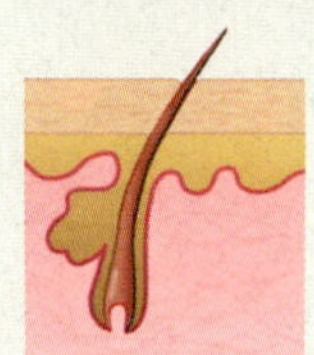

Stimulation of skin receptors may alter respiratory rate.

Cardiovascular System

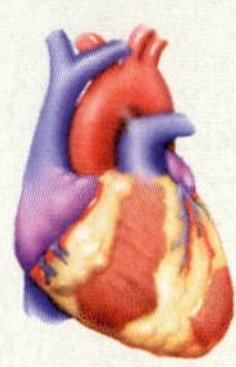

As the heart pumps blood through the lungs, the lungs oxygenate the blood and excrete carbon dioxide.

Skeletal System

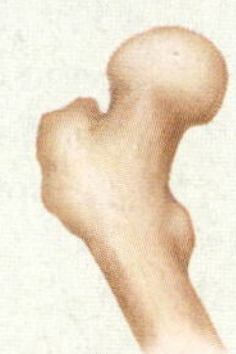

Bones provide attachments for muscles involved in breathing.

Lymphatic System

Cells of the immune system patrol the lungs and defend against infection.

Muscular System

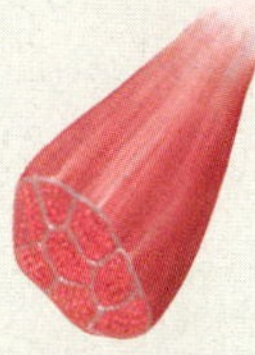

The respiratory system eliminates carbon dioxide produced by exercising muscles.

Digestive System

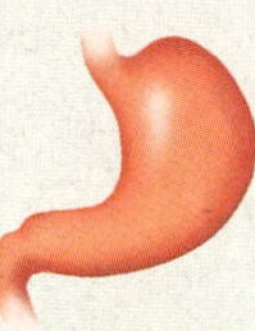

The digestive system and respiratory system share openings to the outside.

Nervous System

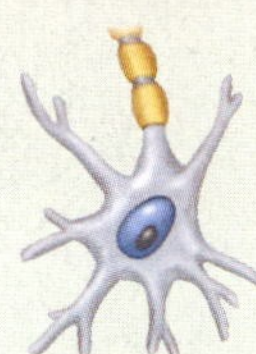

The brain controls the respiratory system. The respiratory system helps control pH of the internal environment.

Urinary System

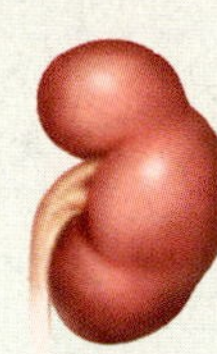

The kidneys and the respiratory system work together to maintain blood pH. The kidneys compensate for water lost through breathing.

Endocrine System

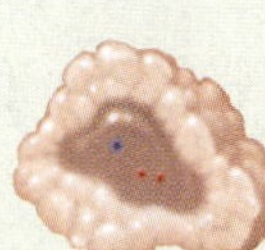

Hormonelike substances control the production of red blood cells that transport oxygen and carbon dioxide.

Reproductive System

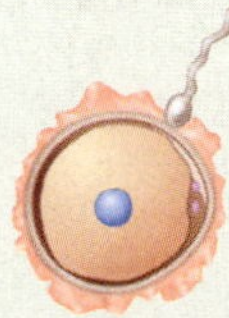

Respiration increases during sexual activity. Fetal "respiration" begins before birth.

from the lungs and respiratory passages, which increases susceptibility to and severity of respiratory infections.

Several changes contribute to an overall increase in effort required to breathe that accompanies aging. Cartilage between the sternum and ribs calcifies and stiffens, and skeletal shifts change the shape of the thoracic cavity into a "barrel chest" as posture too changes with age. In the bronchioles, fibrous connective tissue replaces some smooth muscle, decreasing contractility. As muscles lose strength, breathing comes to depend more upon the diaphragm. The vital capacity, which reaches a maximum by age forty, may drop by a third by the age of seventy years.

Keeping fresh air in the lungs becomes more difficult with age. As the farthest reaches of the bronchiole walls thin, perhaps in response to years of gravity, they do not stay as open as they once did, trapping residual air in the lower portions of the lungs. Widening of the bronchi and alveolar ducts increases dead space. The lungs can still handle the same volume of air, but a greater proportion of that air is "stale," reflecting lessened ability to move air in and out. The maximum minute ventilation drops by 50% from age twenty to age eighty.

Aging-associated changes occur at the microscopic level too. The number of alveoli is about 24 million at birth, peaking at 300 million by age eight years. The number remains constant throughout life, but the alveoli expand. Alveolar walls thin and may coalesce, and the depth of alveoli begins to diminish by age forty, decreasing the surface area available for gas exchange—about three square feet per year. In addition, an increase in the proportion of collagen to elastin and a tendency of the collagen to cross-link impair the ability of alveoli to expand fully. Oxygen transport from the alveoli to the blood, as well as oxygen loading onto hemoglobin in red blood cells, becomes less efficient. Diffusion of CO_2 out of the blood and through the alveolar walls slows too.

As with other organ systems, the respiratory system undergoes specific changes, but these may be unnoticeable at the whole-body level. A person who is sedentary or engages only in light activity would probably not be aware of the slowing of air flow in and out of the respiratory system. Unaccustomed exercise, however, would quickly reveal how difficult breathing has become with age.

PRACTICE 

47 How does the environment influence the effects of aging on the respiratory system?

48 Which aging-related changes raise the risk of respiratory infection?

49 How do alveoli change with age?

CHAPTER SUMMARY

19.1 INTRODUCTION (PAGE 736)

The respiratory system includes the passages that transport air to and from the lungs and the air sacs in which gas exchanges occur. Respiration is the entire process by which gases are exchanged between the atmosphere and the body cells.

19.2 WHY WE BREATHE (PAGE 736)

Respiration is necessary because of cellular respiration. Cells require oxygen to extract maximal energy from nutrient molecules and release carbon dioxide, a metabolic waste.

19.3 ORGANS OF THE RESPIRATORY SYSTEM (PAGE 737)

The respiratory system includes the nose, nasal cavity, sinuses, pharynx, larynx, trachea, bronchial tree, and lungs. The upper respiratory tract includes the nose, nasal cavity, sinuses, and pharynx; the lower respiratory tract includes the larynx, trachea, bronchial tree, and lungs.

1. Nose
    a. Bone and cartilage support the nose.
    b. Nostrils provide entrances for air.
2. Nasal cavity
    a. The nasal cavity is a space posterior to the nose.
    b. The nasal septum divides it medially.
    c. Nasal conchae divide the cavity into passageways and help increase the surface area of the mucous membrane.
    d. Mucous membrane filters, warms, and moistens incoming air.
    e. Particles trapped in the mucus are carried to the pharynx by ciliary action and are swallowed.
3. Sinuses
    a. Sinuses are spaces in the bones of the skull that open into the nasal cavity.
    b. They are lined with mucous membrane that is continuous with the lining of the nasal cavity.
4. Pharynx
    a. The pharynx is posterior to the mouth, between the nasal cavity and the larynx.
    b. It provides a common passage for air and food.
    c. It aids in creating vocal sounds.
5. Larynx
    a. The larynx is an enlargement at the top of the trachea.
    b. It is a passageway for air and helps prevent foreign objects from entering the trachea.
    c. It is composed of muscles and cartilages; some of these cartilages are single, whereas others are paired.
    d. It contains the vocal cords, which produce sounds by vibrating as air passes over them.
        (1) The pitch of a sound is related to the tension on the cords.
        (2) The intensity of a sound is related to the force of the air passing over the cords.
    e. The glottis and epiglottis help prevent food and liquid from entering the trachea.

6. Trachea
    a. The trachea extends into the thoracic cavity in front of the esophagus.
    b. It divides into the right and left bronchi.
    c. The mucous lining continues to filter incoming air.
    d. Incomplete cartilaginous rings support the wall.
7. Bronchial tree
    a. The bronchial tree consists of branched air passages that connect the trachea to the air sacs.
    b. The branches of the bronchial tree include primary bronchi, lobar bronchi, segmental bronchi, intralobular bronchioles, terminal bronchioles, respiratory bronchioles, alveolar ducts, alveolar sacs, and alveoli.
    c. Structure of the respiratory tubes
        (1) As tubes branch, the amount of cartilage in the walls decreases, and the muscular layer becomes more prominent.
        (2) Elastic fibers in the walls aid breathing.
        (3) The epithelial lining changes from pseudostratified and ciliated to cuboidal and simple squamous as the tubes become progressively smaller.
    d. Functions of the respiratory tubes include distribution of air and exchange of gases between the alveolar air and the blood.
8. Lungs
    a. The left and right lungs are separated by the mediastinum and are enclosed by the diaphragm and the thoracic cage.
    b. The visceral pleura is attached to the surface of the lungs; parietal pleura lines the thoracic cavity.
    c. The right lung has three lobes, and the left lung has two.
    d. Each lobe is composed of lobules that contain alveolar ducts, alveolar sacs, alveoli, nerves, blood vessels, lymphatic vessels, and connective tissues.

19.4 BREATHING MECHANISM (PAGE 747)

Inspiration and expiration movements are accompanied by changes in the size of the thoracic cavity.

1. Inspiration
    a. Atmospheric pressure forces air into the lungs.
    b. Inspiration occurs when the intra-alveolar pressure is reduced.
    c. The intra-alveolar pressure is reduced when the diaphragm moves downward and the thoracic cage moves upward and outward.
    d. Surface tension holding the pleural membranes together aids lung expansion.
    e. Surfactant reduces surface tension in the alveoli.
2. Expiration
    a. The forces of expiration come from the elastic recoil of tissues and from surface tension in the alveoli.
    b. Expiration can be aided by thoracic and abdominal wall muscles that pull the thoracic cage downward and inward and compress the abdominal organs inward and upward.
3. Respiratory air volumes and capacities
    a. One inspiration followed by one expiration is a respiratory cycle.
    b. The amount of air that moves in or out during a respiratory cycle is the tidal volume.
    c. Additional air that can be inhaled is the inspiratory reserve volume; additional air that can be exhaled is the expiratory reserve volume.
    d. Residual air remains in the lungs and is mixed with newly inhaled air.
    e. The total lung capacity is equal to the vital capacity plus the residual air volume.
    f. The inspiratory capacity is the maximum volume of air a person can inhale following exhalation of the tidal volume.
    g. The functional residual capacity is the volume of air that remains in the lungs following the exhalation of the tidal volume.
    h. The vital capacity is the maximum amount of air a person can exhale after taking the deepest breath possible.
    i. Air in the anatomic and alveolar dead spaces is not available for gas exchange.
4. Alveolar ventilation
    a. Minute ventilation is tidal volume multiplied by breathing rate.
    b. Alveolar ventilation rate is the physiologic dead space subtracted from the tidal volume multiplied by breathing rate.
    c. The alveolar ventilation rate is a major factor affecting gas exchange between the alveolar air and the blood.
5. Nonrespiratory air movements
    a. Nonrespiratory air movements are air movements other than breathing.
    b. They include coughing, sneezing, laughing, crying, hiccupping, and yawning.

19.5 CONTROL OF BREATHING (PAGE 755)

Normal breathing is rhythmic and involuntary, although the respiratory muscles can be controlled voluntarily.

1. Respiratory areas
    a. The respiratory areas are in the brainstem and include parts of the medulla oblongata and pons.
    b. The medullary respiratory center includes two groups of neurons.
        (1) The dorsal respiratory group is most important in inspiration.
        (2) The ventral respiratory group increases inspiratory and expiratory movements during forceful breathing.
    c. The pontine respiratory group may contribute to the rhythm of breathing.
2. Factors affecting breathing
    a. The partial pressure of a gas is determined by the concentration of that gas in a mixture of gases or the concentration of gas dissolved in a liquid.
    b. Chemicals, lung tissue stretching, and emotional state affect breathing.
    c. Chemosensitive areas (central chemoreceptors) are associated with the respiratory center.
        (1) CO_2 combines with water to form carbonic acid, which, in turn, releases hydrogen ions in the CSF.
        (2) Stimulation of these areas increases alveolar ventilation.

d. Peripheral chemoreceptors are in the carotid bodies and aortic bodies of certain arteries.
   (1) These chemoreceptors sense low oxygen levels.
   (2) When oxygen levels are low, alveolar ventilation increases.
e. Stretching the lung tissues triggers an inflation reflex.
   (1) This reflex reduces the duration of inspiratory movements.
   (2) This prevents overinflation of the lungs during forceful breathing.
f. Hyperventilation decreases CO_2 levels, but *this is dangerous when associated with breath holding during underwater swimming.*

19.6 ALVEOLAR GAS EXCHANGES (PAGE 759)

Gas exchange between the air and the blood occurs in the alveoli.

1. Alveoli
   a. The alveoli are tiny sacs clustered at the distal ends of the alveolar ducts.
   b. Some alveoli open into adjacent air sacs that provide alternate pathways for air when passages are obstructed.
2. Respiratory membrane
   a. The respiratory membrane consists of the alveolar and capillary walls.
   b. Gas exchange takes place through these walls.
3. Diffusion through the respiratory membrane
   a. Gases diffuse from regions of higher partial pressure toward regions of lower partial pressure.
   b. Oxygen diffuses from the alveolar air into the blood; CO_2 diffuses from the blood into the alveolar air.

19.7 GAS TRANSPORT (PAGE 762)

Blood transports gases between the lungs and the body cells.

1. Oxygen transport
   a. Oxygen is mainly transported with hemoglobin molecules.
   b. The resulting oxyhemoglobin is unstable and releases its oxygen in regions where the P_{O_2} is low.
   c. More oxygen is released as the plasma P_{CO_2} increases, as the blood becomes more acidic, and as the blood temperature increases.
2. Carbon dioxide transport
   a. CO_2 may be carried in solution, either as dissolved CO_2, CO_2 bound to hemoglobin, or as a bicarbonate ion.
   b. Most CO_2 is transported in the form of bicarbonate ions.
   c. Carbonic anhydrase speeds the reaction between CO_2 and water to form carbonic acid.
   d. Carbonic acid dissociates to release hydrogen ions and bicarbonate ions.

19.8 LIFE-SPAN CHANGES (PAGE 767)

The lungs, respiratory passageways, and alveoli undergo aging-associated changes exacerbated by exposure to polluted air. However, the increased work required to breathe with age is typically not noticeable unless one engages in vigorous exercise.

1. Exposure to pollutants, smoke, and other particulates raises risk of developing diseases of the respiratory system.
2. Loss of cilia, thickening of mucus, and impaired macrophages raise the risk of infection.
3. Calcified cartilage, skeletal changes, altered posture, and replacement of smooth muscle with fibrous connective tissue in bronchioles make breathing more difficult. Vital capacity diminishes.
4. The lungs contain a greater proportion of "stale" air.
5. Alveoli coalesce and become shallower, slowing gas exchange.

CHAPTER ASSESSMENTS

19.1 Introduction

1 List the general functions of the respiratory system. (p. 736)

19.2 Why We Breathe

2 Explain why oxygen is required at the cellular level. (p. 736)

19.3 Organs of the Respiratory System

3 Distinguish between the upper and lower respiratory tracts. (p. 737)

4 Explain how the nose and nasal cavity filter incoming air. (p. 737)

5 Name and describe the locations of the major sinuses. (p. 738)

6 Explain how a sinus headache may occur. (p. 738)

7 The pharynx is also known as the: (p. 740)
a. nasal cavity
b. oral cavity
c. voice box
d. throat

8 Name and describe the functions of the cartilages of the larynx. (p. 740)

9 Match the following structures with their descriptions: (pp. 740–746)

(1) true vocal cords	A. serous membrane on lungs
(2) false vocal cords	B. contains the vocal cords
(3) larynx	C. vibrate to make sound
(4) visceral pleura	D. air sacs
(5) alveoli	E. muscular folds that close the glottis

10 Name the successive branches of the bronchial tree, from the primary bronchi to the alveoli, and identify their functions. (p. 743)

11 Describe how the structure of the respiratory tubes changes as the branches become finer. (p. 744)

12 Distinguish between the visceral pleura and the parietal pleura. (p. 746)

13 Name the lobes of the lungs and identify their locations. (p. 747)

19.4 Breathing Mechanism

14 Compare the muscles used in a resting inspiration with those used in a forced inspiration. (p. 747)

15 Define *surface tension,* and explain how it works against the breathing mechanism. (p. 750)

16 Define *surfactant,* and explain its function. (p. 750)

17 Define *compliance.* (p. 751)

18 Compare the muscles used (if any) in a resting expiration with those used in a forced expiration. (p. 751)

19 Match the air volumes with their descriptions: (p. 752)

(1) tidal volume	A. air that remains after most forceful expiration
(2) inspiratory reserve volume	B. volume of air, in addition to resting tidal volume, that can enter lungs
(3) expiratory reserve volume	C. amount of air that enters and leaves lungs during a respiratory cycle
(4) residual volume	D. volume of air, in addition to resting tidal volume, that can be expelled from the lungs

20 Distinguish between vital capacity and total lung capacity. (p. 753)

21 Physiologic dead space is equal to ____________. (p. 753)

a. anatomic dead space

b. anatomic dead space plus alveolar dead space

c. alveolar dead space

22 Calculate both minute ventilation and alveolar ventilation given the following: (p. 754)

respiratory rate = 12 breaths per minute

tidal volume = 500 mL per breath

physiologic dead space = 150 mL per breath

23 Explain the mechanisms of coughing and sneezing, and give the functions of each. (p. 754)

24 Describe a possible function of yawning. (p. 755)

19.5 Control of Breathing

25 Locate the respiratory areas and name their major components. (p. 756)

26 Explain control of the basic rhythm of breathing. (p. 757)

27 Which one of the following is most important in forceful breathing? (p. 757)

a. dorsal respiratory group

b. ventral respiratory group

c. pontine respiratory group

28 Explain the effect increasing CO_2 levels have on the central chemoreceptors. (p. 757)

29 Describe the function of the peripheral chemoreceptors in the carotid and aortic bodies. (p. 758)

30 Describe the inflation reflex. (p. 758)

31 Describe the effects of emotions on breathing. (p. 758)

32 Hyperventilation is which one of the following? (p. 758)

a. any increase in breathing

b. an increase in breathing that brings in oxygen too quickly

c. an increase in breathing that eliminates CO_2 too quickly

d. an increase in breathing that has no effect on blood gases

19.6 Alveolar Gas Exchanges

33 Describe the respiratory membrane. (p. 760)

34 Explain the relationship between the partial pressure of a gas and its rate of diffusion. (p. 760)

35 Summarize the exchange of oxygen and CO_2 across the respiratory membrane. (p. 760)

19.7 Gas Transport

36 Describe how the blood transports oxygen. (p. 762)

37 List three factors that increase the release of oxygen from hemoglobin. (p. 763)

38 Explain why carbon monoxide is toxic. (p. 764)

39 Give the percentages of the three ways CO_2 is transported in blood. (p. 766)

40 Explain the function of carbonic anhydrase. (p. 766)

41 Define chloride shift. (p. 766)

19.8 Life-Span Changes

42 Describe the changes that make it harder to breathe with advancing years. (p. 769)

INTEGRATIVE ASSESSMENT/CRITICAL THINKING

OUTCOMES 3.2, 5.2, 5.3, 5.5, 19.3

1. Describe the following structures that are part of the respiratory tubes and state their locations.
   a. pseudostratified epithelium
   b. cuboidal epithelium
   c. simple squamous epithelium
   d. goblet cells
   e. cartilage
   f. smooth muscle
   g. elastic fibers
   h. cilia

OUTCOMES 19.3, 19.4

2. Patients experiencing asthma attacks are often advised to breathe through pursed (puckered) lips. How might this help reduce the symptoms of asthma?

OUTCOMES 19.3, 19.4, 19.5, 19.6, 19.7

3. What changes would you expect to occur in the levels of blood oxygen and carbon dioxide in a patient who breathes rapidly and deeply for a prolonged time?

OUTCOMES 19.3, 19.4, 19.6

4. If a tracheostomy bypasses the upper respiratory passages, how might the air entering the trachea differ from air normally passing through this tube? What problems might this cause for the patient?

OUTCOMES 19.3, 19.4, 19.6

5. Certain respiratory disorders, such as emphysema, reduce the capacity of the lungs to recoil elastically. Which respiratory volumes will this condition affect? Explain the impact on gas exchange.

OUTCOMES 19.5, 19.6, 19.7

6. If a person is receiving supplemental oxygen to restore blood oxygen levels, why might it be better to administer a combination of oxygen and carbon dioxide rather than pure oxygen?

OUTCOMES 19.5, 19.7

7. Why is it impossible, under normal circumstances, for a person to hold the breath long enough to pass out?

OUTCOMES 19.6, 19.7

8. What problem might a person with a serious respiratory disorder encounter flying in the passenger compartment of a commercial aircraft that has an air pressure equivalent to an altitude of 8,000 feet?

WEB CONNECTIONS

Be sure to visit the text website at **www.mhhe.com/shier12** for answers to chapter assessments, additional quizzes, and interactive learning exercises.

ANATOMY & PHYSIOLOGY REVEALED

Anatomy & Physiology Revealed® (APR) includes cadaver photos that allow you to peel away layers of the human body to reveal structures beneath the surface. This program also includes animations, radiologic imaging, audio pronunciations, and practice quizzing. Check out **www.aprevealed.com**. APR has been proven to help improve student grades!

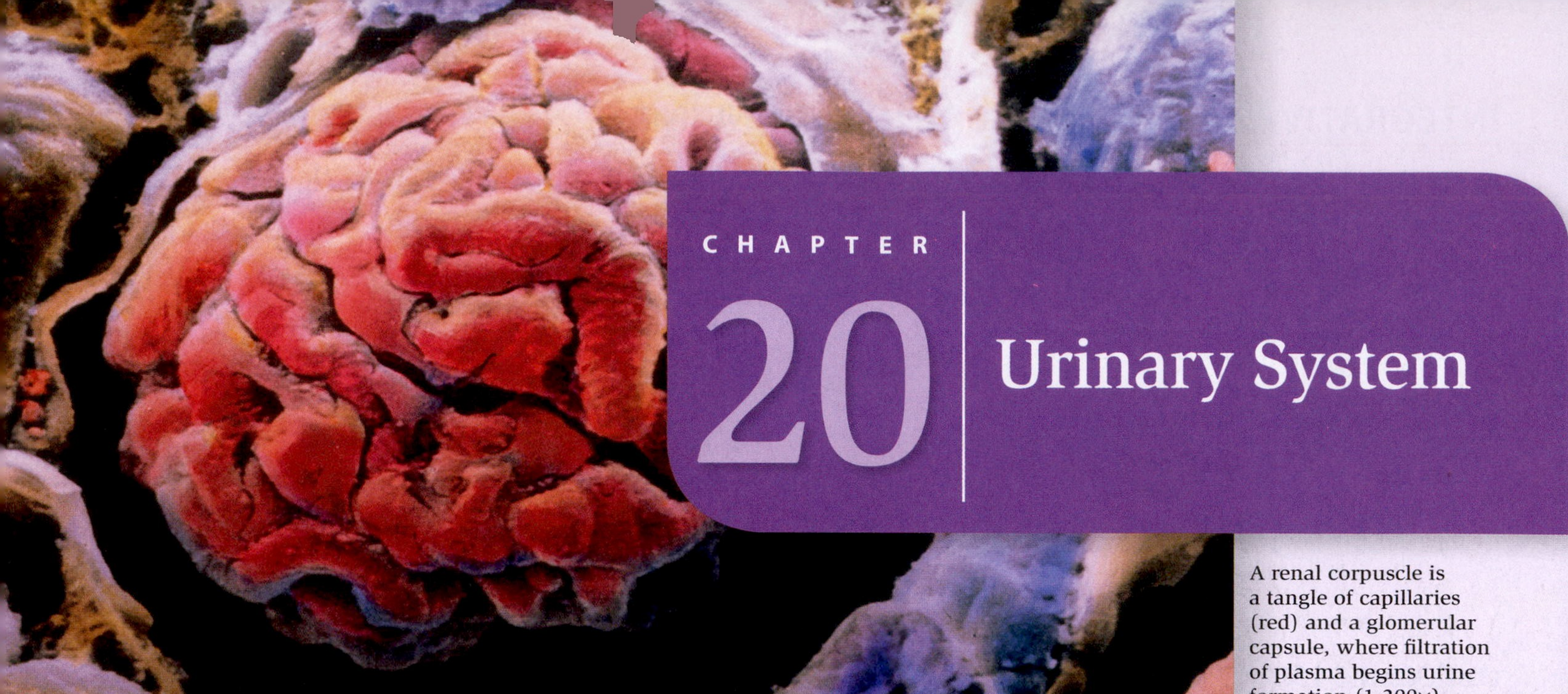

CHAPTER 20 Urinary System

A renal corpuscle is a tangle of capillaries (red) and a glomerular capsule, where filtration of plasma begins urine formation (1,200×).

UNDERSTANDING WORDS

af-, to: *af*ferent arteriole—arteriole that leads to a nephron.

calyc-, small cup: major *calyc*es—cuplike subdivisions of the renal pelvis.

cort-, covering: renal *cort*ex—shell of tissue surrounding the inner region of a kidney.

cyst-, bladder: *cyst*itis—inflammation of the bladder.

detrus-, to force away: *detrus*or muscle—muscle in the bladder wall that causes urine to be expelled.

glom-, little ball: *glom*erulus—cluster of capillaries in a renal corpuscle.

juxta-, near to: *juxta*medullary nephron—nephron located near the renal medulla.

mict-, to pass urine: *mict*urition—expelling urine from the bladder.

nephr-, pertaining to the kidney: *nephr*on—functional unit of a kidney.

papill-, nipple: renal *papill*ae—small elevations that project into a renal calyx.

prox-, nearest: *prox*imal tubule—coiled portion of the renal tubule leading from the glomerular capsule.

ren-, kidney: *ren*al cortex—outer region of a kidney.

trigon-, triangular shape: *trigone*—triangular area on the internal floor of the bladder.

LEARNING OUTCOMES

After you have studied this chapter, you should be able to:

20.1 Introduction

1. Name the organs of the urinary system and list their general functions. (p. 775)

20.2 Kidneys

2. Describe the locations of the kidneys and the structure of the kidney. (p. 776)
3. List the functions of the kidneys. (p. 777)
4. Trace the pathway of blood flow through the major vessels within a kidney. (p. 779)
5. Describe a nephron and explain the functions of its major parts. (p. 779)

20.3 Urine Formation

6. Explain how glomerular filtrate is produced and describe its composition. (p. 785)
7. Explain how various factors affect the rate of glomerular filtration and identify ways that this rate is regulated. (p. 788)
8. Explain tubular reabsorption, and its role in urine formation. (p. 790)
9. Identify the changes in the osmotic concentration of the glomerular filtrate as it passes through the renal tubule. (p. 792)
10. Explain tubular secretion, and its role in urine formation. (p. 793)
11. Identify the characteristics of a countercurrent mechanism, and explain its role in concentrating the urine. (p. 795)
12. Explain how the final composition of urine contributes to homeostasis. (p. 796)

20.4 Elimination of Urine

13. Describe the structures of the ureters, urinary bladder, and urethra. (p. 798)
14. Explain how micturition occurs, and how it is controlled. (p. 802)

20.5 Life-Span Changes

15. Describe how the components of the urinary system change with age. (p. 803)

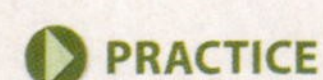  LEARN PRACTICE ASSESS

A MEDICAL MYSTERY

The abandoned buildings stained with creeping dark mold are called "black houses" in the Croatian villages. They once housed farm families wiped out by a mysterious illness. Typically a mother would sicken first, perhaps not noticing the initial fatigue, insomnia, and frequent urination, eventually dying of renal (kidney) failure. Her husband would soon follow, and then the children would leave, but within months or years would become ill too. People fortunate enough to receive medical care discovered that they had anemia, hypertension, and protein in the urine, indicators of kidney damage. Some villages had many sick people; others didn't. The pattern did not seem to indicate inheritance. A family would typically vanish in three to five years as the disease took its toll.

Doctors noticed the disease in the 1950s, calling it Balkan endemic nephropathy, in honor of its apparent origin in the northern Balkan peninsula. Romania, Bulgaria, Serbia, and Bosnia also had cases. Investigators ruled out various causes: viruses, heavy metals, tainted water, fungus, weeds. Local doctors preserved the damaged organs of their patients, in the hope that someday the specimens might be useful in developing a treatment. The illness was devastating. At death, the kidneys were about a third the size of normal kidneys and fibrous, and often the tubes leading from them to the bladder were lined with tumors.

The clue that solved the mystery of the black houses came in 2000, when a prominent medical journal reported that a Chinese herb (*Aristolochia clematitis*) taken for weight loss caused symptoms remarkably similar to those of the Balkan disease. Long-term use of the herb damaged delicate kidney tissues (interstitial fibrosis) leading to renal failure, as well as causing cancer of the urinary tract lining. A biochemist at Stony Brook University, Arthur Grollman, alerted Croatian colleagues to the possible link between the disease and the herb, and soon additional clues turned up. Horses that ate hay on which the herb grew developed renal failure, and the herb fed to rats and rabbits turned their kidneys fibrous.

By this time, people with the illness in the Balkans were being treated with dialysis, which slowed disease progression. When researchers showed them photographs of the herb, the patients recognized it immediately—a weed, called Wolf's Paw, that grew on their wheat. Further analysis revealed that the herb produces a pair of powerful toxins that home in on the kidneys, causing fibrosis as well as entering cells and binding to a cancer-causing gene called p53.

Kidney failure from a popular herb. A mysterious illness that cripples the urinary system was first described among villagers in Croatia who grew wheat. *Aristolochia clematitis,* which grew on the wheat, is also used as an herbal supplement.

"Balkan endemic nephropathy" is now called aristolochic acid nephropathy, in recognition of the poisonous herb. The discovery of the botanical connection has explained many cases of renal failure in other cultures—the mysterious illness was not unique to the black houses of Croatia. ■

20.1 INTRODUCTION

A major part of homeostasis is maintaining the composition, pH, and volume of body fluids within normal ranges. The urinary system accomplishes this task. It removes metabolic wastes and chemicals in excess, yet at the same time is sensitive enough so that the body is not depleted of essential substances. The urinary system also excretes foreign substances, such as drugs and their metabolites that can be toxic if they remain in the body fluids after they have exerted their therapeutic effects.

The urinary system consists of a pair of glandular kidneys, which remove substances from the blood, form urine, and help regulate certain metabolic processes; a pair of tubular ureters, which transport urine from the kidneys; a saclike urinary bladder, which collects urine from the ureters and serves as a urine reservoir; and a tubular urethra, which conveys urine to the outside of the body. Figures 20.1 and 20.2 show these organs.

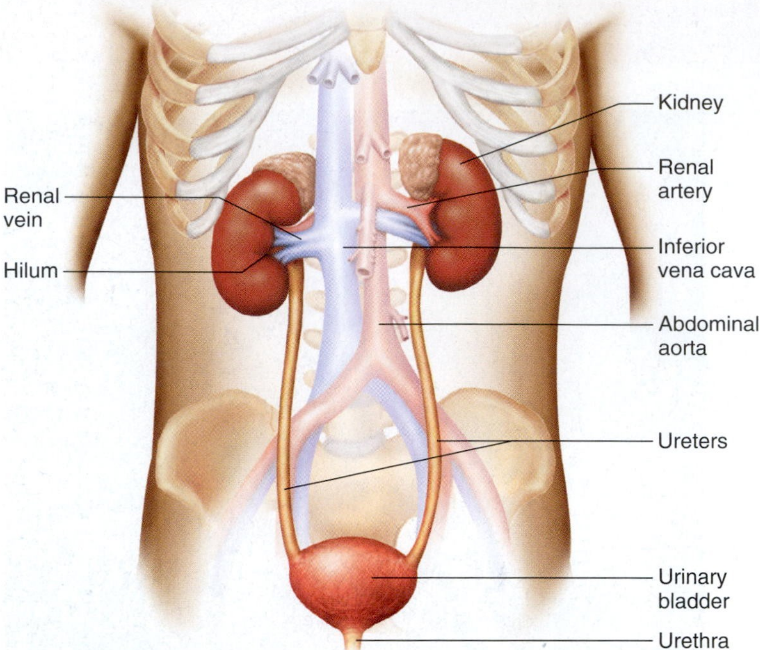

FIGURE 20.1 The urinary system includes the kidneys, ureters, urinary bladder, and urethra. Notice the relationship of these structures to the major blood vessels.

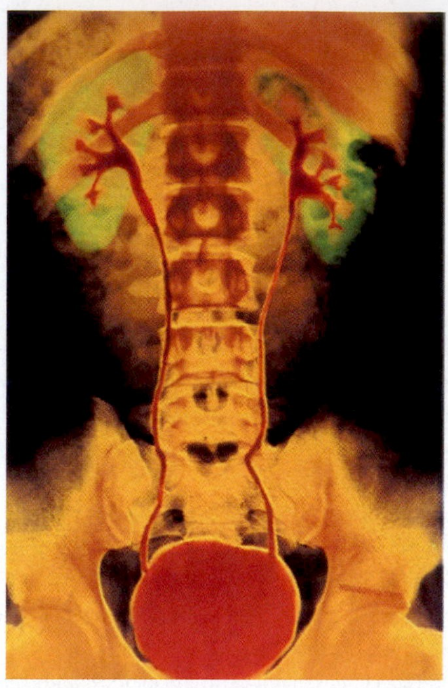

FIGURE 20.2 Structures of the urinary system are visible in this falsely colored radiograph (anterior view).

20.2 KIDNEYS

A **kidney** is a reddish brown, bean-shaped organ with a smooth surface. It is about 12 centimeters long, 6 centimeters wide, and 3 centimeters thick in an adult, and it is enclosed in a tough, fibrous capsule (tunic fibrosa).

Location of the Kidneys

The kidneys lie on either side of the vertebral column in a depression high on the posterior wall of the abdominal cavity. The upper and lower borders of the kidneys are generally at the levels of the twelfth thoracic and third lumbar vertebrae, respectively, although the positions of the kidneys may vary slightly with changes in posture and with breathing movements. The left kidney is usually about 1.5 to 2 centimeters higher than the right one.

The kidneys are positioned *retroperitoneally* (re″tro-per″ĭ-to-ne′al-le); they are behind the parietal peritoneum and against the deep muscles of the back. Connective tissue (renal fascia) and masses of adipose tissue (renal fat) surrounding the kidneys hold them in place (fig. 20.3 and reference plates 18, 19).

Kidney Structure

The lateral surface of each kidney is convex, but its medial side is deeply concave. The resulting medial depression leads into a hollow chamber called the **renal sinus.** Through the entrance to this sinus, termed the *hilum,* pass blood vessels, nerves, lymphatic vessels, and the ureter (see fig. 20.1).

The superior end of the ureter expands to form a funnel-shaped sac called the **renal pelvis,** located mostly inside the renal sinus. The pelvis is formed by the convergence of two or three tubes, called *major calyces* (sing., *calyx*), and they, in turn, are formed by the convergence of seven to twenty *minor calyces* (fig. 20.4). At least one small projection called a *renal papilla* extends into each minor calyx.

The kidney includes two distinct regions: an inner medulla and an outer cortex. The **renal medulla** (re′nal mĕ-dul′ah) is composed of conical masses of tissue called *renal pyramids.* Their bases orient toward the convex surface of the kidney, and their apexes form the renal papillae. The tissue of the medulla appears striated because it consists of microscopic tubules that lead from the cortex to the renal papillae.

The **renal cortex** (re′nal kor′teks), which appears somewhat granular, forms a shell around the medulla. Its tissue dips into the medulla between the renal pyramids, forming *renal columns.* The **renal capsule** is a fibrous membrane that surrounds the cortex and helps maintain the shape of the kidney. It also is protective (figs. 20.4 and 20.5).

Functions of the Kidneys

The main function of the kidneys is to regulate the volume, composition, and pH of body fluids. In the process, the kidneys remove metabolic wastes from the blood and excrete

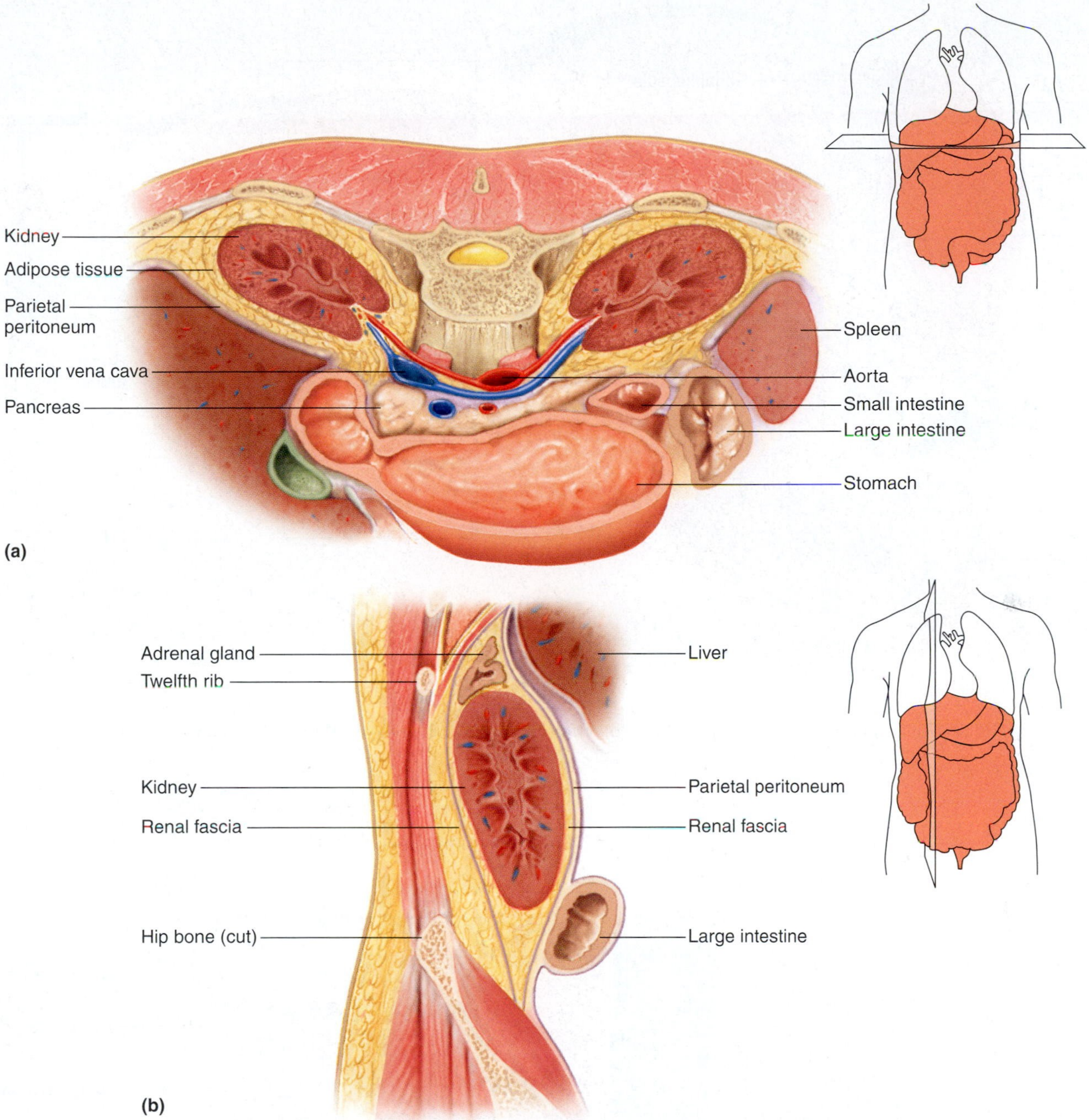

Kidney
Adipose tissue
Parietal peritoneum
Inferior vena cava
Pancreas
Spleen
Aorta
Small intestine
Large intestine
Stomach
(a)

Adrenal gland
Twelfth rib
Kidney
Renal fascia
Hip bone (cut)
Liver
Parietal peritoneum
Renal fascia
Large intestine
(b)

FIGURE 20.3 The kidneys are located retroperitoneally. (*a*) Transverse section through the posterior abdominal cavity including the kidneys, behind the parietal peritoneum. Adipose and other connective tissues surround and support the kidneys. (*b*) Sagittal section through the posterior abdominal cavity showing the kidney.

them to the outside. These wastes include nitrogenous and sulfur-containing products of protein metabolism. The kidneys also help control the rate of red blood cell formation by secreting the hormone *erythropoietin* (see chapter 14, p. 527), regulate blood pressure by secreting the enzyme *renin* (see chapter 13, p. 506), and regulate absorption of calcium ions by activating *vitamin D* (see chapter 18, p. 717).

Medical technology can take over some of the roles of a kidney. In *hemodialysis,* a person's blood is rerouted across an artificial membrane that "cleanses" it, removing substances that would normally be excreted in the urine. A patient usually must use this artificial kidney three times a week, for several hours each time. Clinical Application 20.1 further discusses hemodialysis.

PRACTICE

1 Where are the kidneys located?

2 Describe the structure of a kidney.

3 What are the general functions of the kidneys?

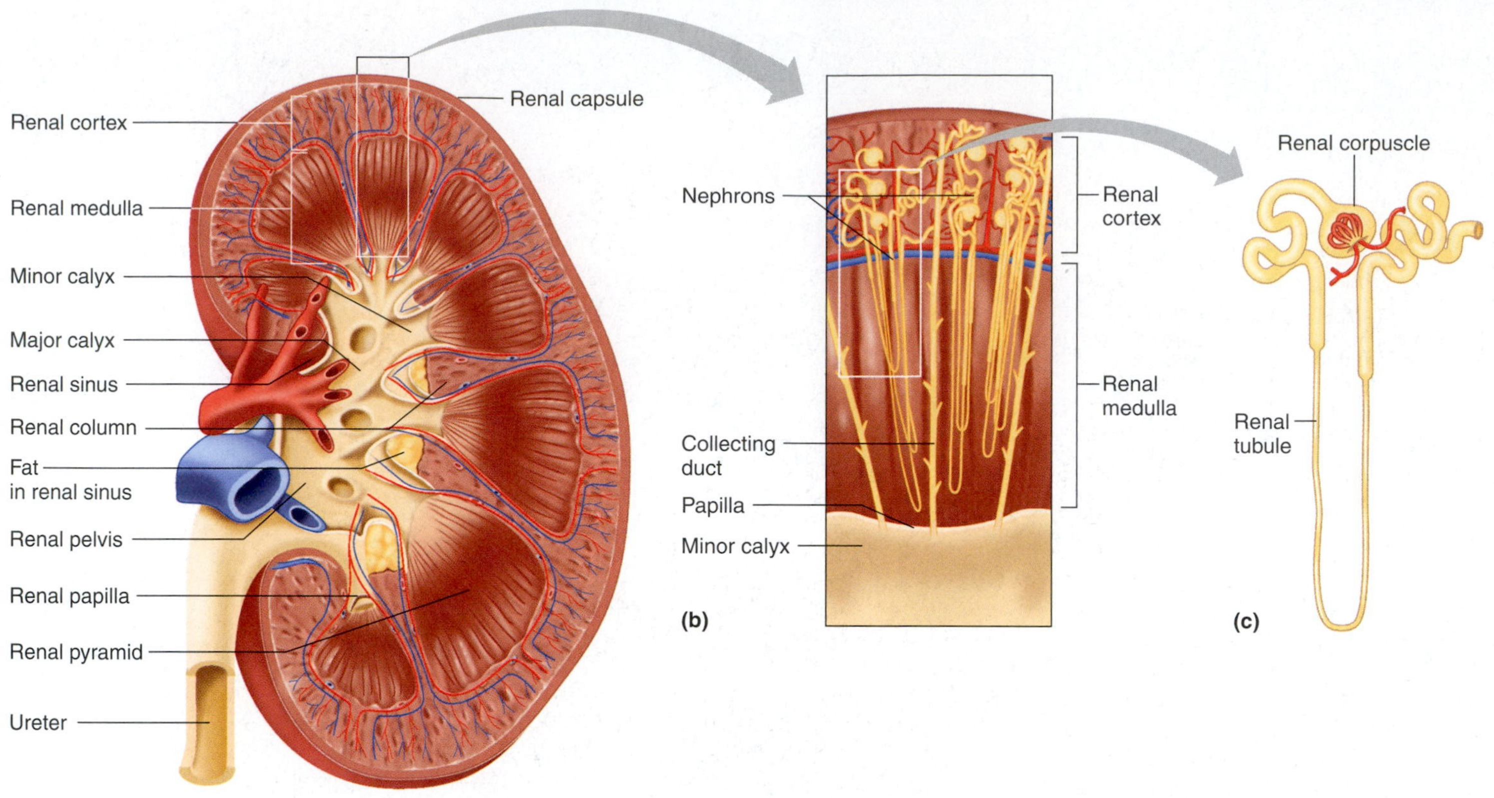

FIGURE 20.4 The kidney. (*a*) Longitudinal section of a kidney. (*b*) A renal pyramid containing nephrons. (*c*) A single nephron.

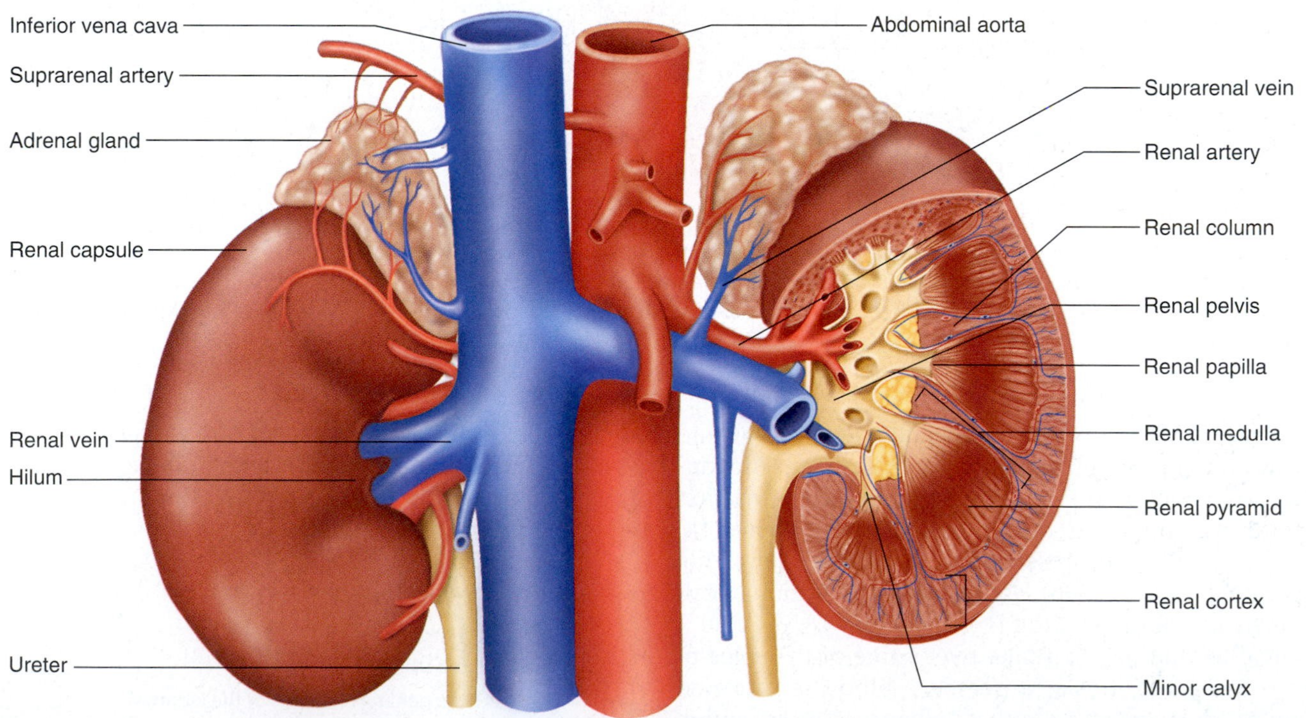

FIGURE 20.5 Blood vessels are associated with the kidneys and adrenal glands. Note their relationship with the renal pelvis and ureters.

20.1 CLINICAL APPLICATION

Chronic Kidney Failure

Charles B., a forty-three-year-old construction worker, had been feeling unusually tired for several weeks, with occasional dizziness and difficulty sleeping. More recently he had noticed a burning pain in his lower back, just below his rib cage, and his urine had darkened. In addition, his feet, ankles, and face were swollen. His wife suggested that he consult their family physician.

Charles had elevated blood pressure (hypertension) and the regions of his kidneys were sensitive to pressure. His urine had excess protein (proteinuria) and blood (hematuria), and his blood had elevated blood urea nitrogen (BUN), elevated serum creatinine, and decreased serum protein (hypoproteinemia) concentrations.

Charles likely had *chronic glomerulonephritis,* an inflammation of the capillaries in the glomeruli of the nephrons. It is a progressive degeneration with no direct treatment, although drugs can help to control blood pressure. Microscopic examination of a small sample of kidney tissue (biopsy) confirmed the diagnosis.

Despite antihypertensive drugs and careful attention to his diet, Charles's condition deteriorated rapidly. When it appeared that most of his kidney function had been lost (end-stage renal disease, or ESRD), he was offered artificial kidney treatments (hemodialysis).

To prepare Charles for hemodialysis, a vascular surgeon created a fistula in his left forearm by surgically connecting an artery to a vein. The greater pressure of the blood in the artery that now flowed directly into the vein swelled the vein, making it more accessible.

During hemodialysis treatment, a hollow needle was inserted into the vein of the fistula near its arterial connection. This allowed the blood to flow, with the aid of a blood pump, through a tube leading to the blood compartment of a dialysis machine. In this compartment, the blood passed over a selectively permeable membrane. On the opposite side of the membrane was a dialysate solution with a controlled composition. Negative pressure on the dialysate side of the membrane, created by a vacuum pump, increased the movement of fluid through the membrane. At the same time, waste and excess electrolytes diffused from the blood through the membrane and entered the dialysate solution. The blood was then returned through a tube to the vein of the fistula.

To maintain favorable blood concentrations of waste, electrolytes, and water, Charles had to undergo hemodialysis three times per week, with each treatment lasting three to four hours. During the treatments, he was given an anticoagulant to prevent blood clotting, an antibiotic drug to control infections, and an antihypertensive drug.

Charles was advised to carefully control his intake of water, sodium, potassium, proteins, and total calories between treatments. He was also asked to consider a kidney transplant, which could free him from dependence on hemodialysis.

In a transplant, a kidney from a living donor or a cadaver, whose tissues are antigenically similar (histocompatible) to those of the recipient, is placed in the depression on the medial surface of the right or left ilium (iliac fossa). The renal artery and vein of the donor kidney are connected to the recipient's iliac artery and vein, respectively, and the kidney's ureter is attached to the dome of the recipient's urinary bladder. ■

About two in every ten patients with renal failure can use a procedure that can be done at home called *continuous ambulatory peritoneal dialysis* instead of hemodialysis. The patient infuses a solution into the abdominal cavity through a permanently implanted tube. The solution stays in for four to eight hours, while it takes up substances that would normally be excreted into urine. Then the patient drains the waste-laden solution out of the tube, replacing it with clean fluid. Infection is a risk associated with this procedure.

Renal Blood Vessels

The **renal arteries,** which arise from the abdominal aorta, transport a large volume of blood to the kidneys (fig. 20.5). When a person is at rest, the renal arteries usually carry from 15% to 30% of the total cardiac output into the kidneys, although the kidneys account for only 1% of body weight.

A renal artery enters a kidney through the hilum and gives off several branches, called the *interlobar arteries,* which pass between the renal pyramids. At the junction between the medulla and the cortex, the interlobar arteries branch to form a series of incomplete arches, the *arcuate arteries* (arciform arteries), which, in turn, give rise to *cortical radiate arteries* (interlobular arteries). The final branches of the cortical radiate arteries, called **afferent arterioles** (af′er-ent ar-te′re-ōlz), lead to the nephrons, the functional units of the kidneys.

Venous blood returns through a series of vessels that generally correspond to the arterial pathways. For example, the venous blood passes through cortical radiate, arcuate, interlobar, and renal veins. The **renal vein** then joins the inferior vena cava as it courses through the abdominal cavity. (Figs. 20.6 and 20.7 show branches of the renal arteries and veins.)

Nephrons

Structure of a Nephron

Some organs can be broken down into subunits, each of which performs the functions of the organ as a whole. Each kidney contains about 1 million such functional units, the **nephrons** (nef′ronz). Each nephron in turn consists of a **renal corpuscle** (re′nal kor′pusl) and a **renal tubule** (re′nal tu′būl) (see fig. 20.4).

A renal corpuscle consists of a filtering unit composed of a tangled cluster of blood capillaries called a **glomerulus** (glo-mer′u-lus) and a surrounding thin-walled, saclike structure

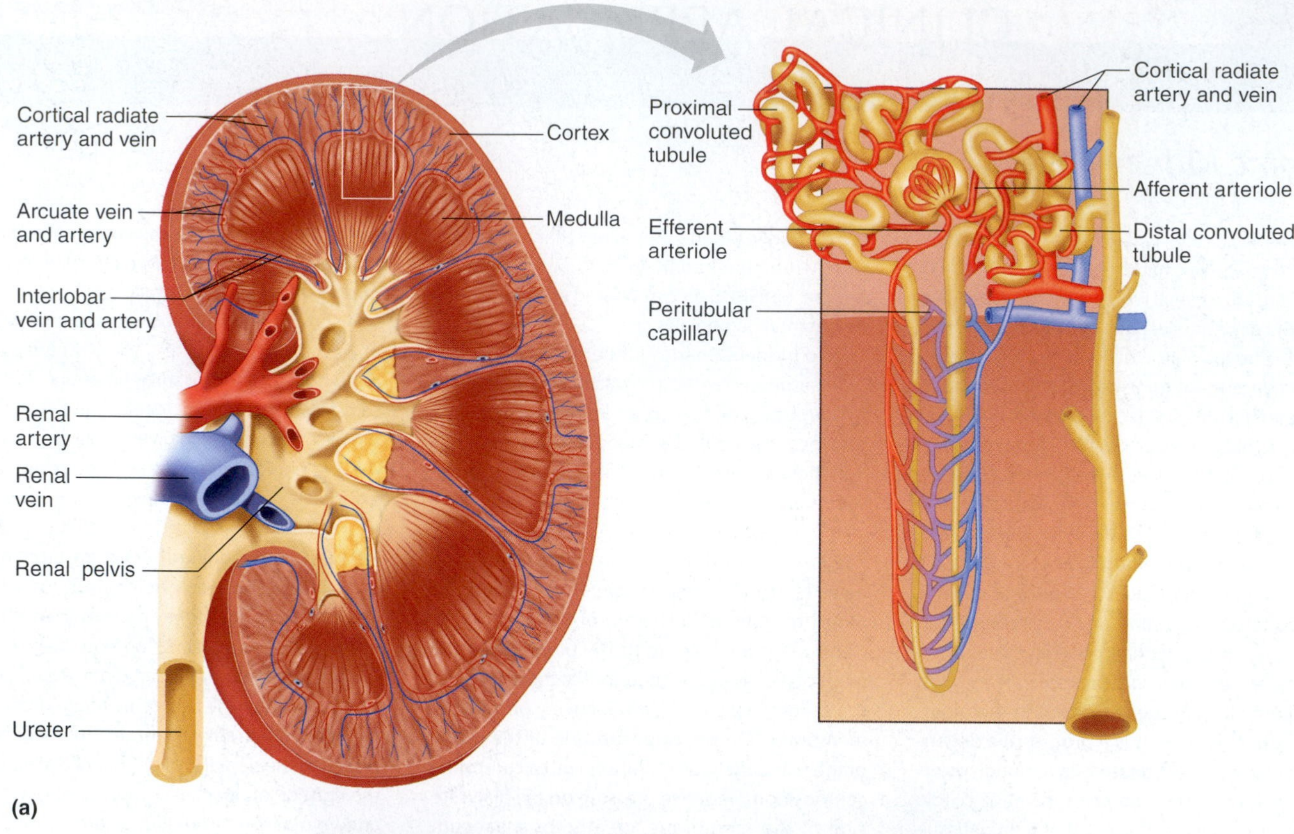

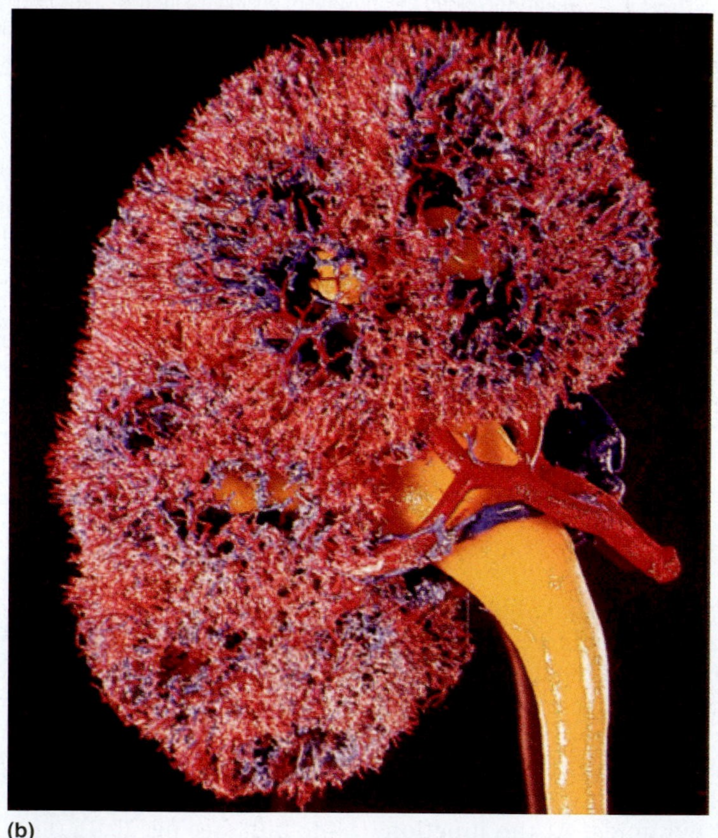

FIGURE 20.6 Renal blood vessels. (*a*) Main branches of the renal artery and vein. (*b*) Corrosion cast of the renal arterial system. Not all blood vessels associated with the nephron are shown.

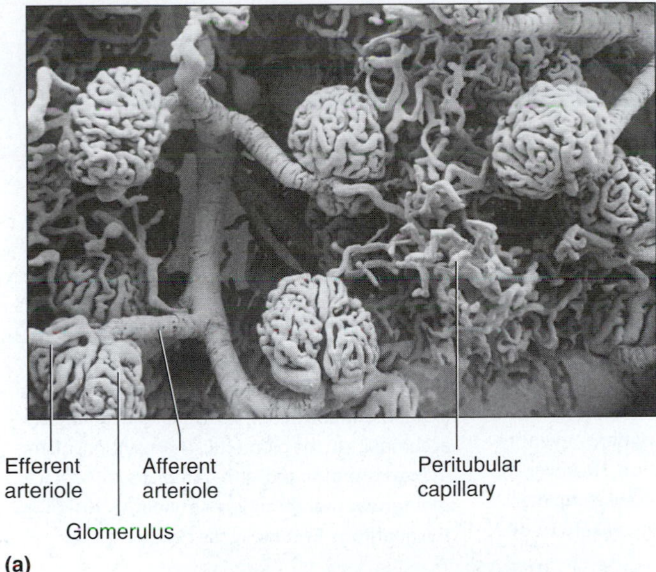

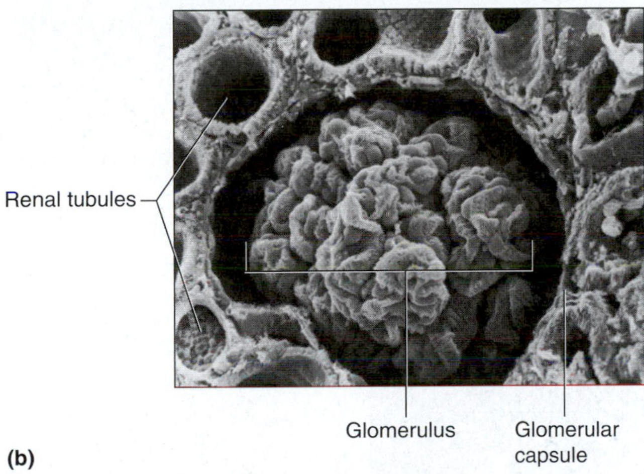

FIGURE 20.7 Blood vessels associated wtih nephrons. (*a*) A scanning electron micrograph of a cast of the renal blood vessels associated with the glomeruli (200×). (*b*) A scanning electron micrograph of a glomerular capsule surrounding a glomerulus (480×). *Tissues and Organs: A Text-Atlas of Scanning Electron Microscopy,* by R. G. Kessel and R. H. Kardon. ©1979 W. H. Freeman and Company.

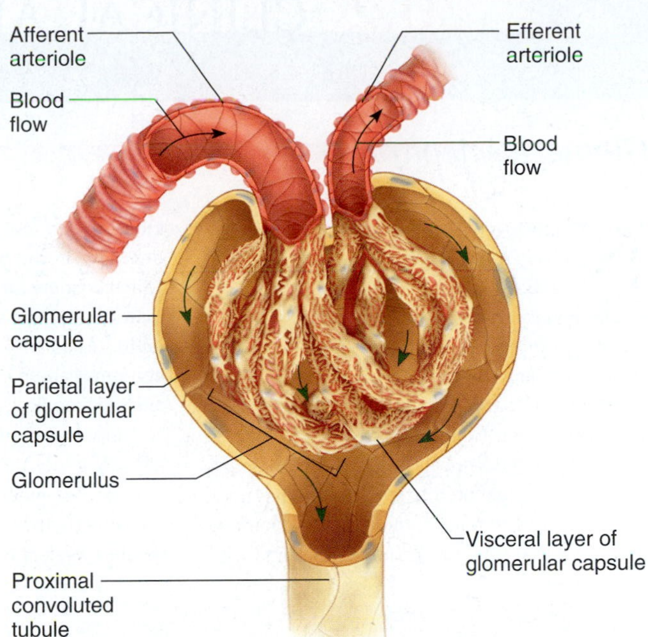

FIGURE 20.8 The glomerular capsule has a visceral layer and a parietal layer.

called a **glomerular** (Bowman's) **capsule.** Afferent arterioles give rise to these capillaries, which lead to **efferent arterioles** (ef′er-ent ar-te′re-ōlz) (fig. 20.7). Filtration of fluid from the glomerular capillaries is the first step in urine formation.

The glomerular capsule is an expansion at the end of a renal tubule that receives the fluid filtered at the glomerulus. The capsule is composed of two layers of squamous epithelial cells: a visceral layer that closely covers the glomerulus and an outer parietal layer continuous with the visceral layer and with the wall of the renal tubule (fig. 20.8).

The cells of the parietal layer are typical squamous epithelial cells; however, those of the visceral layer are highly modified epithelial cells called *podocytes.* Each podocyte has several primary processes extending from its cell body, and these processes, in turn, bear numerous secondary processes, or *pedicels.* The pedicels of each cell interdigitate with those of adjacent podocytes, and the clefts between them form a complicated system of *slit pores* (figs. 20.8 and 20.9).

A cascade of cell-to-cell communication forms glomeruli in the embryo. Podocyte precursor cells give rise to podocytes, but also secrete vascular endothelial growth factor (VEGF), which attracts the squamous epithelium that forms the parietal layer. These epithelial cells then produce platelet-derived growth factor (PDGF) and extracellular matrix proteins, which in turn signal certain cells to specialize as mesangial cells. These cells are closely associated with the capillary tuft, and can contract, which decreases the rate at which the glomerulus filters plasma. Mesangial cells also provide structural support to the glomerulus and phagocytize debris.

The renal tubule leads away from the glomerular capsule and becomes highly coiled. This coiled portion is the *proximal convoluted tubule.* Following it is the **nephron loop** (loop of Henle). The proximal convoluted tubule dips toward the renal pelvis to become the *descending limb* of the nephron loop. The tubule then curves back toward its renal corpuscle and forms the *ascending limb* of the nephron loop. The ascending limb returns to its renal corpuscle of origin, where it becomes the *distal convoluted tubule* and tightly coils again. This distal portion is shorter and straighter than the proximal tubule.

Several distal convoluted tubules merge in the renal cortex to form a *collecting duct* (collecting tubule), technically not part of any one nephron. The collecting duct passes into the renal medulla, widening as it joins other collecting ducts. The resulting tube empties into a minor calyx through an

20.2 CLINICAL APPLICATION

Glomerulonephritis

Nephritis is an inflammation of the kidney. *Glomerulonephritis* is an inflammation of the glomeruli, and it may be acute or chronic and can lead to renal failure.

Acute glomerulonephritis (AGN) usually results from an abnormal immune reaction that develops one to three weeks following bacterial infection by beta-hemolytic *Streptococcus.* As a rule, the infection is not in the kidneys, but bacterial antigens trigger production of antibodies that form insoluble immune complexes (see chapter 16, p. 639) that travel in the bloodstream to the kidneys. The antigen-antibody complexes are deposited in and block the glomerular capillaries, which become further obstructed as the inflammatory response sends white blood cells to the region. Capillaries remaining open may become abnormally permeable, allowing plasma proteins and red blood cells to enter the urine.

Most glomerulonephritis patients eventually regain normal kidney function. However, in severe cases, renal functions may fail completely. Without treatment, the person is likely to die within a week or so.

Chronic glomerulonephritis is a progressive disease in which increasing numbers of nephrons are slowly damaged until the kidneys are unable to function. This condition is usually associated with certain diseases other than streptococcal infections, and it also involves formation of antigen-antibody complexes that precipitate and accumulate in the glomeruli. The resulting inflammation is prolonged, and fibrous tissue replaces glomerular membranes, permanently disabling the nephrons. Eventually the kidneys fail. ■

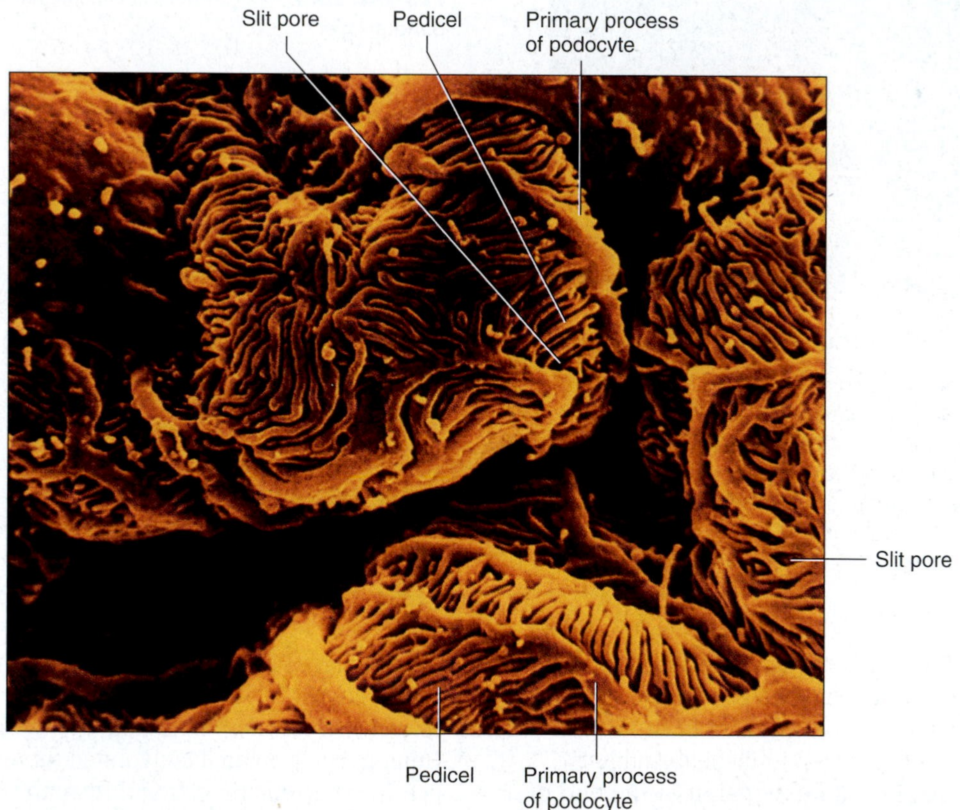

FIGURE 20.9 Scanning electron micrograph of a portion of a glomerulus (8,000×). Note the slit pores between the pedicels.

opening in a renal papilla. Figures 20.10 and 20.11 show the parts of a nephron. Clinical Application 20.2 examines glomerulonephritis, an inflammation of the glomeruli.

Juxtaglomerular Apparatus

At the end of the nephron loop, before it becomes the distal convoluted tubule, the ascending limb passes between the afferent and efferent arterioles of its own glomerulus and makes contact with them. At the point of contact, the epithelial cells of the ascending limb are tall and densely packed. These cells comprise a structure called the *macula densa.*

Close by, in the wall of the afferent arteriole near its attachment to the glomerulus, are large, vascular smooth muscle cells called *juxtaglomerular cells.* Together with the

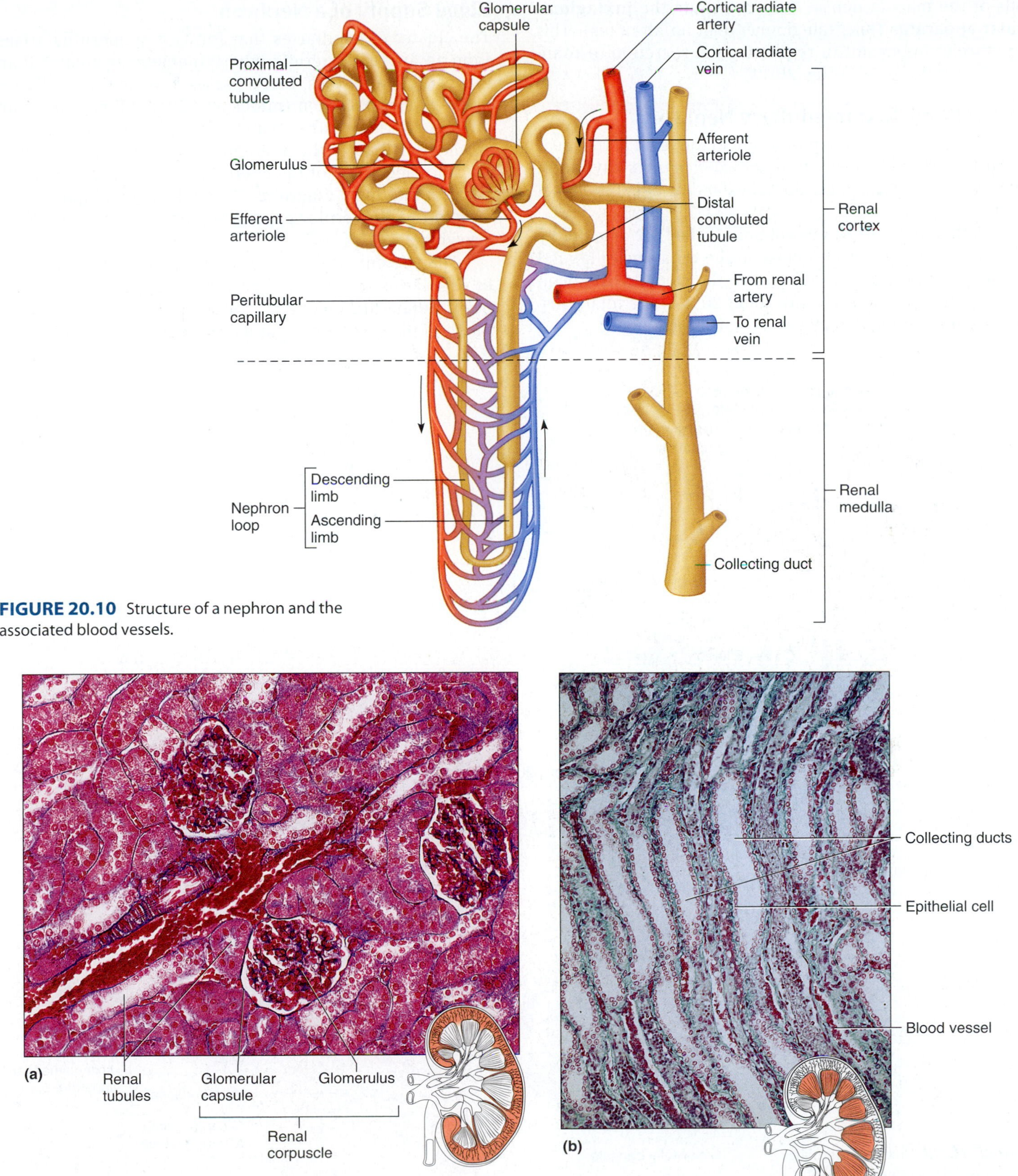

FIGURE 20.10 Structure of a nephron and the associated blood vessels.

FIGURE 20.11 Microscopic view of the kidney. (*a*) Light micrograph of a section of the human renal cortex (220×). (*b*) Light micrograph of the renal medulla (80×).

cells of the macula densa, they constitute the **juxtaglomerular apparatus** (juks″tah-glo-mer′u-lar ap″ah-ra′tus). This structure is important in regulating the secretion of renin (see chapter 13, p. 506) (fig. 20.12).

Cortical and Juxtamedullary Nephrons

The corpuscles of most nephrons are in the renal cortex near the surface of the kidney. These *cortical nephrons* have short nephron loops that usually do not extend as far as the renal medulla.

Another group of kidney subunits, called *juxtamedullary nephrons,* have corpuscles close to the renal medulla, and their nephron loops extend deep into the medulla. Although these nephrons represent only about 20% of the total, they are important in regulating water balance (fig. 20.13).

Blood Supply of a Nephron

The cluster of capillaries that forms a glomerulus arises from an afferent arteriole whose diameter is greater than that of other arterioles. Blood passes through the capillaries of the glomerulus, then (minus any filtered fluid) enters an efferent arteriole (rather than a venule), whose diameter is smaller than that of the afferent arteriole. The greater resistance to blood flow of the efferent arteriole causes blood to back up into the glomerulus. This results in a higher pressure in the glomerular capillaries compared to capillaries elsewhere.

The efferent arteriole branches into a complex network of capillaries that surrounds the renal tubule called the **peritubular capillary** (per″ĭ-tu′bu-lar kap′ĭ-ler″e) **system.** Blood in the system has passed through two arterioles and

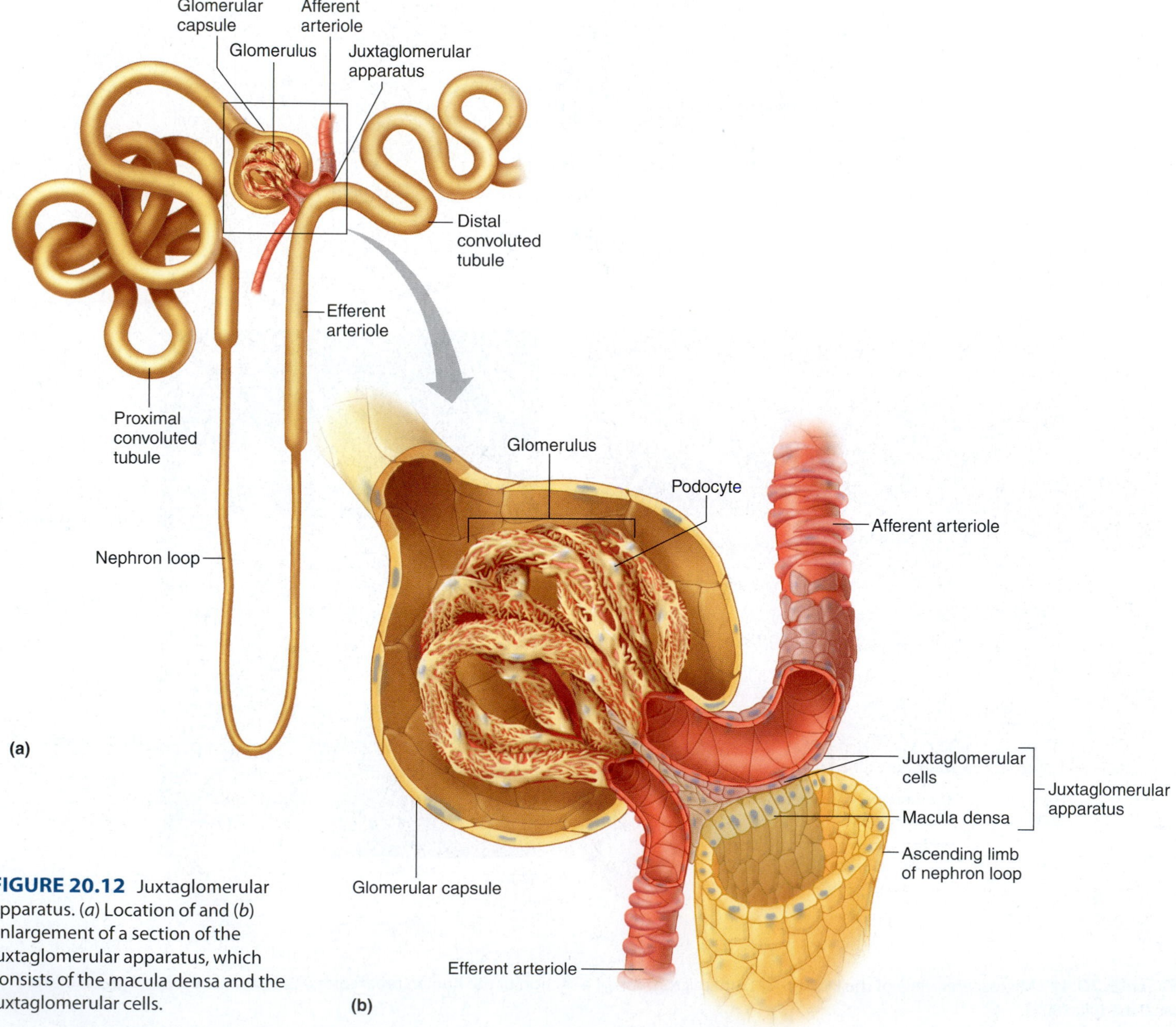

FIGURE 20.12 Juxtaglomerular apparatus. (*a*) Location of and (*b*) enlargement of a section of the juxtaglomerular apparatus, which consists of the macula densa and the juxtaglomerular cells.

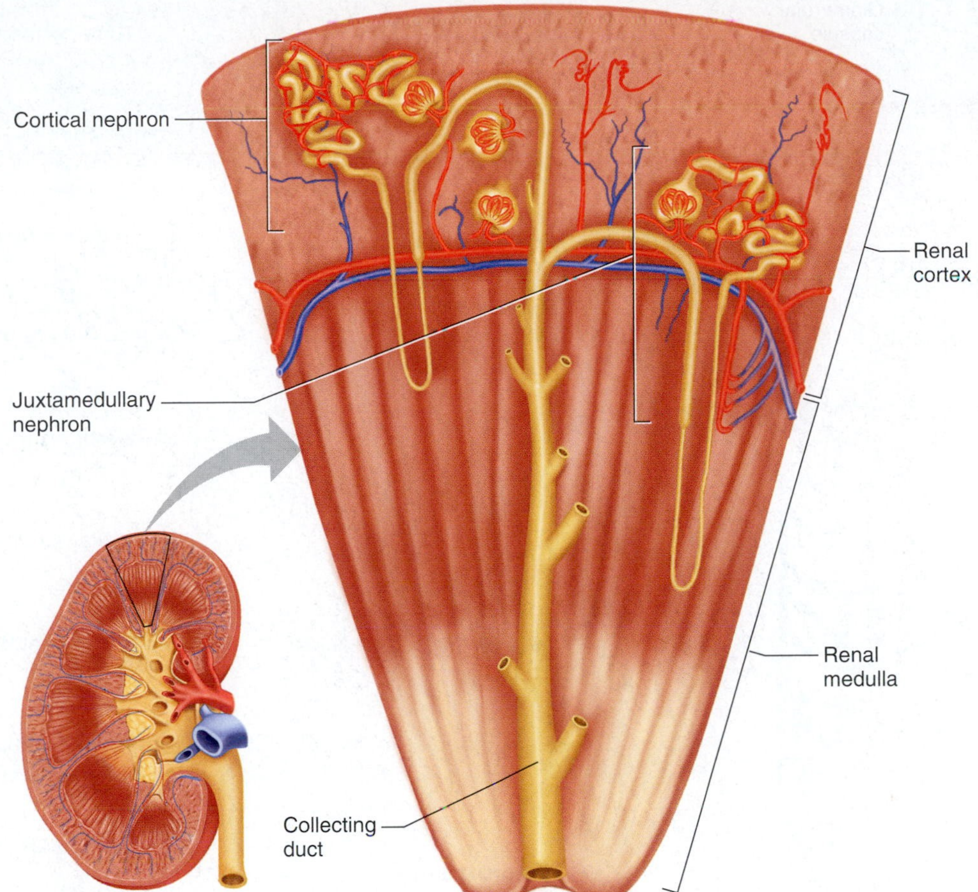

FIGURE 20.13 Cortical nephrons are close to the surface of a kidney; juxtamedullary nephrons are near the renal medulla.

is under relatively low pressure (see fig. 20.10). Branches of this system that primarily receive blood from the efferent arterioles of the juxtamedullary nephrons form capillary loops called *vasa recta*. These loops dip into the renal medulla and are closely associated with the loops of the juxtamedullary nephrons (fig. 20.14). Blood flows through the vasa recta and returns to the renal cortex, where it joins blood from other branches of the peritubular capillary system and enters the venous system of the kidney. Figure 20.15 summarizes the pathway that blood follows as it passes through the blood vessels of the kidney and nephron.

PRACTICE

4 Explain why nephrons are considered functional units.
5 Describe the system of vessels that supplies blood to the kidney.
6 Name the parts of a nephron.
7 Which structures comprise the juxtaglomerular apparatus?
8 Distinguish a cortical nephron from a juxtamedullary nephron.
9 Describe the blood supply of a nephron.

20.3 URINE FORMATION

The main function of the nephrons and collecting ducts is to control the composition of body fluids and remove wastes from the blood. The product is **urine,** which is excreted from the body. Urine contains wastes, excess water, and electrolytes.

Urine formation begins when the glomerular capillaries filter plasma, a process called **glomerular filtration** (glo-mer′u-lar fil-tra′shun). Recall from chapter 15 (pp. 578–579) that the force of blood pressure drives filtration at capillaries throughout the body. Most of this fluid is reabsorbed into the bloodstream by the colloid osmotic pressure of the plasma, leaving only a small volume of interstitial fluid (fig. 20.16*a*).

Nephrons take filtration to another level, using two capillaries working in series. The first capillary bed is specialized only to filter. However, instead of forming interstitial fluid, the filtered fluid (filtrate) moves into the renal tubule, where some of it is destined to become urine (fig. 20.16*b*).

Glomerular filtration produces 180 liters of fluid, more than four times the total body water, every 24 hours. Obviously this could not continue for long unless most of this filtered fluid were returned to the internal environment.

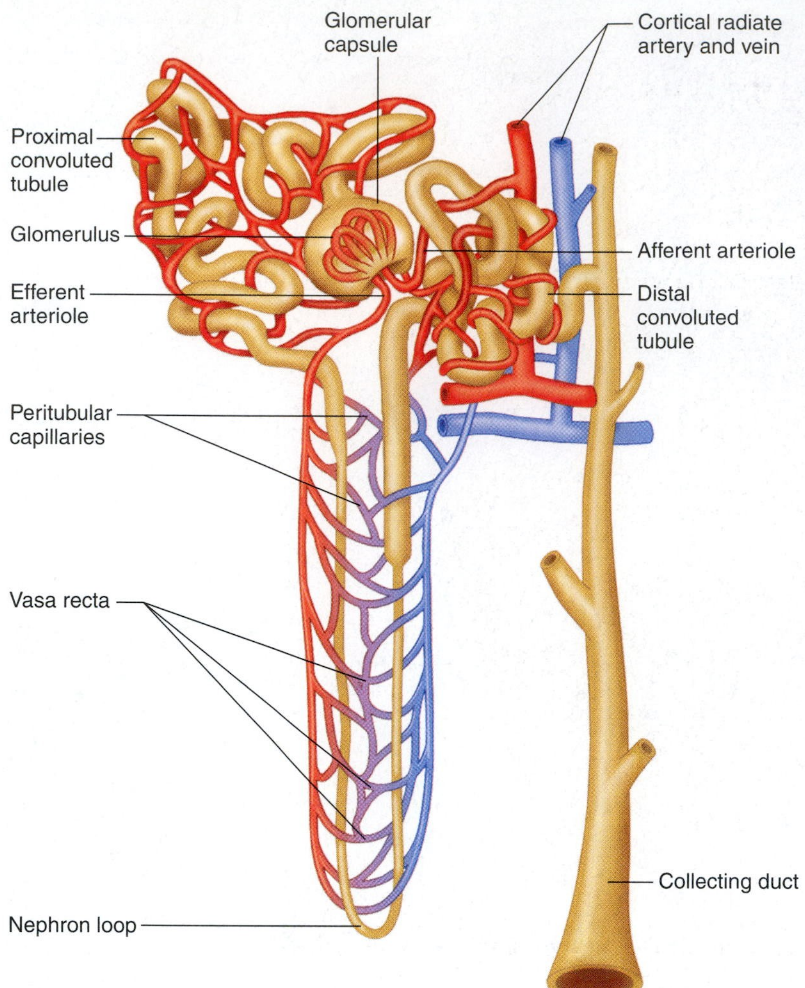

FIGURE 20.14 The capillary loop of the vasa recta is closely associated with the nephron loop of a juxtamedullary nephron.

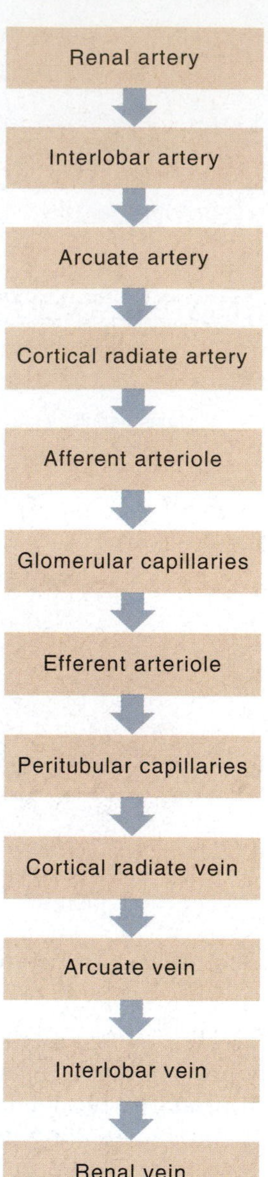

FIGURE 20.15 Pathway of blood through the blood vessels of the kidney and nephron.

The kidney accomplishes this by the process of **tubular reabsorption** (too′bu-lar re-ab-sorp′shun), selectively reclaiming just the right amounts of substances, such as water, electrolytes, and glucose, that the body requires. Waste products and substances in excess are allowed out of the body. Some substances that the body must eliminate, such as hydrogen ions and certain toxins, are removed even faster than through filtration alone by the process of **tubular secretion** (too′bu-lar se-kre′shun).

Overall, the following relationship determines the volume of substances excreted in the urine:

urinary excretion = **glomerular filtration**
+ tubular secretion
− tubular reabsorption

The final product of these processes is *urine.*

Glomerular Filtration

Urine formation begins when *glomerular filtration* filters water and other small dissolved molecules and ions out of the glomerular capillary plasma and into the glomerular capsules. Large molecules, such as proteins, are restricted primarily because of their size. The filtration of these materials through the capillary walls is much like the filtration at the arteriole ends of other capillaries throughout the body. The glomerular capillaries, however, are many times more permeable to small molecules than are the capillaries in other tissues, due to the many tiny openings (fenestrae) in their walls (fig. 20.17).

RECONNECT

To Chapter 15, Exchanges in the Capillaries, pages 578–579.

The glomerular capsule receives the resulting **glomerular filtrate,** which has about the same composition as

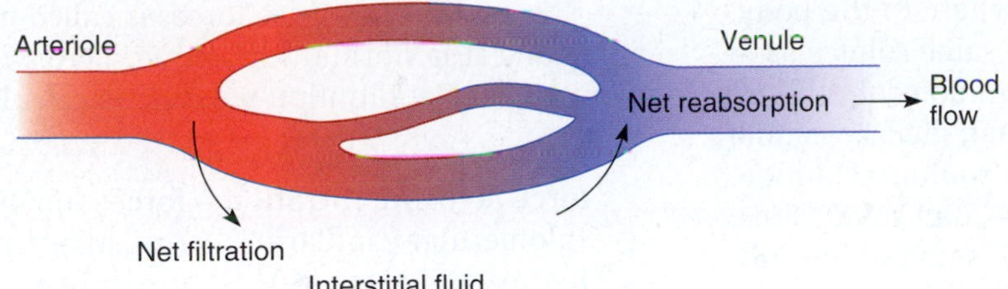

(a) In most systemic capillaries, filtration predominates at the arteriolar end and osmotic reabsorption predominates at the venular end.

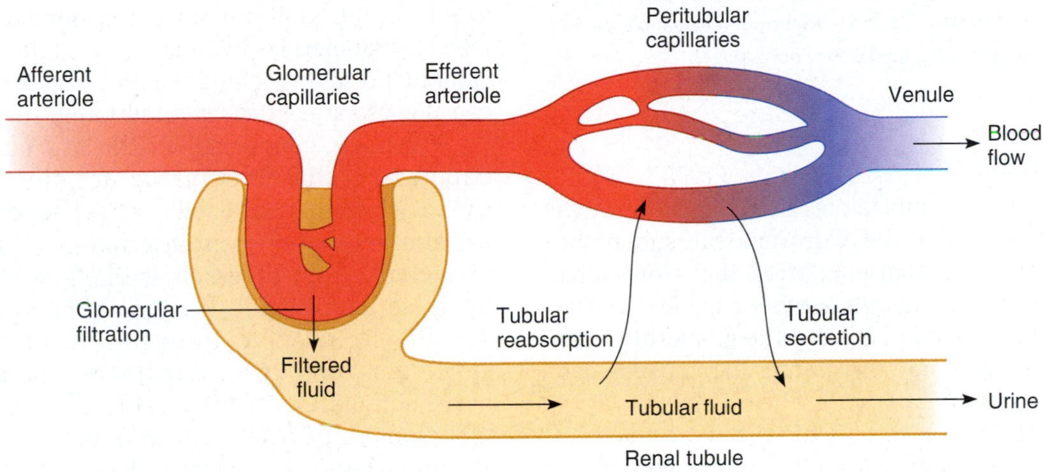

(b) In the kidneys, the glomerular capillaries are specialized for filtration. The renal tubule is specialized to control movements of substances back into the blood of the peritubular capillaries (tubular reabsorption) or from the blood into the renal tubule (tubular secretion).

FIGURE 20.16 Compared to most capillaries in the (*a*) systemic circulation, those in the (*b*) kidneys are highly specialized for filtration.

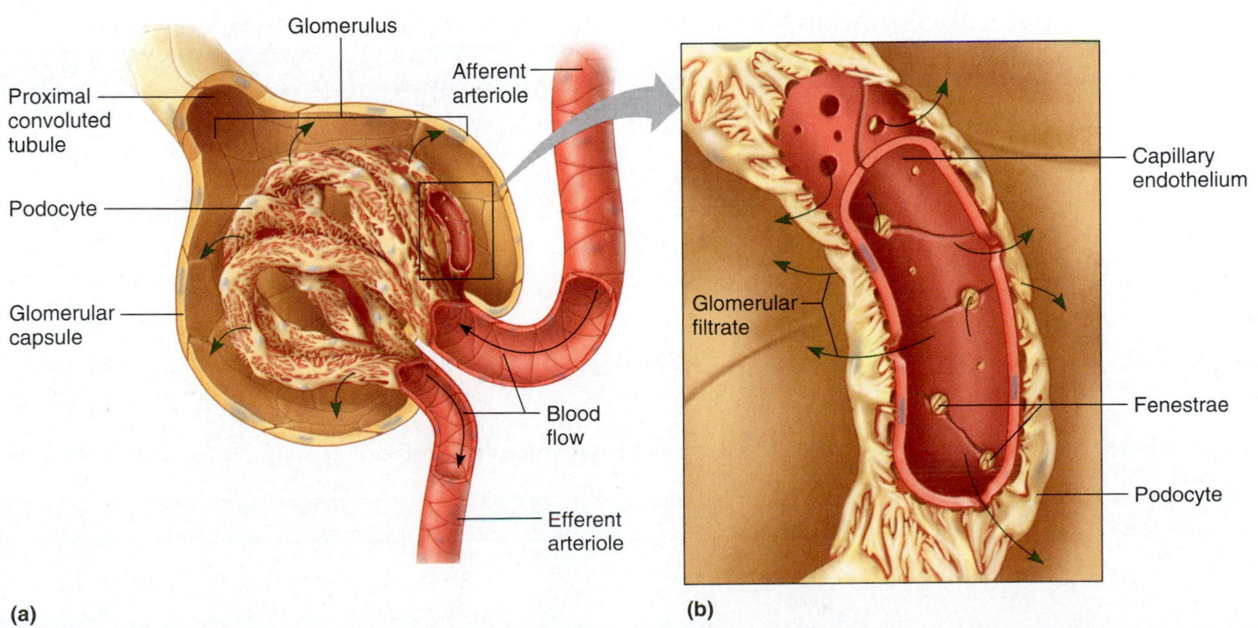

FIGURE 20.17 Glomerular filtration. (*a*) The first step in urine formation is filtration of substances out of glomerular capillaries and into the glomerular capsule. (*b*) Glomerular filtrate passes through the fenestrae of the capillary endothelium.

the filtrate that becomes tissue fluid elsewhere in the body. Glomerular filtrate is mostly water and the same solutes as in blood plasma, except for the larger protein molecules. More specifically, glomerular filtrate includes water, glucose, amino acids, urea, uric acid, creatine, creatinine, sodium, chloride, potassium, calcium, bicarbonate, phosphate, and sulfate ions. Table 20.1 compares the concentrations of some of the substances in the blood plasma, glomerular filtrate, and urine.

> The concentrations of certain components of the blood plasma can be used to evaluate kidney functions. For example, if the kidneys are functioning inadequately, the plasma concentrations of urea (a nitrogenous waste) indicated by a blood urea nitrogen test and creatinine may increase as much as tenfold above normal.

Filtration Pressure

The main force that moves substances by filtration through the glomerular capillary wall is the hydrostatic pressure of the blood inside, as in other capillaries. (Recall that glomerular capillary pressure is high compared to other capillaries.) The osmotic pressure of the blood plasma in the glomerulus and the hydrostatic pressure inside the glomerular capsule also influence glomerular filtration.

The colloid osmotic pressure of the plasma caused by plasma proteins is always higher than that of the glomerular filtrate (except in some types of kidney disease). This draws water back into the glomerular capillaries, opposing filtration. Any increase in glomerular capsule hydrostatic pressure also opposes filtration (fig. 20.18).

The net effect of these forces is called **net filtration pressure,** and it is normally positive, favoring filtration at the glomerulus. Net filtration pressure is calculated as follows:

Net filtration pressure =
force favoring filtration – forces opposing filtration
(glomerular capillary hydrostatic pressure) (capsular hydrostatic pressure and glomerular capillary osmotic pressure)

Filtration Rate

The glomerular filtration rate (GFR) is directly proportional to the net filtration pressure. Consequently, the factors that affect the glomerular hydrostatic pressure, glomerular plasma osmotic pressure, or hydrostatic pressure in the glomerular capsule also affect the rate of filtration (fig. 20.18).

Normally, glomerular hydrostatic pressure is the most important factor determining net filtration pressure and GFR. Each glomerular capillary lies between two arterioles—the afferent and efferent arterioles—so any change in the diameters of these vessels is likely to change glomerular hydrostatic pressure, affecting glomerular filtration rate. The afferent arteriole, through which the blood enters the glomerulus, may vasoconstrict in response to sympathetic nerve stimulation. If this occurs, net filtration pressure in that glomerulus decreases, and filtration rate drops. If, on the other hand, the efferent arteriole (through which the blood leaves the glomerulus) vasoconstricts, blood backs up into the glomerulus, net filtration pressure increases, and filtration rate rises. Vasodilation of these vessels produces opposite effects.

TABLE 20.1 | Relative Concentrations of Plasma, Glomerular Filtrate, and Urine Components

Substance	Plasma	Glomerular Filtrate Concentrations (mEq/L)	Urine
Sodium (Na^+)	142	142	128
Potassium (K^+)	5	5	60
Calcium (Ca^{+2})	4	4	5
Magnesium (Mg^{+2})	3	3	15
Chloride (Cl^-)	103	103	134
Bicarbonate (HCO_3^-)	27	27	14
Sulfate (SO_4^{-2})	1	1	33
Phosphate (PO_4^{-3})	2	2	40
(mEq/L [milliequivalents per liter] is a commonly used measure of concentration based on how many charges an ion carries. For a substance with a charge of 1, such as Cl^-, a mEq is equal to a millimole.)			
Substance	**Plasma**	**Glomerular Filtrate Concentrations (mEq/L)**	**Urine**
Glucose	100	100	0
Urea	26	26	1,820
Uric acid	4	4	53
Creatinine	1	1	196

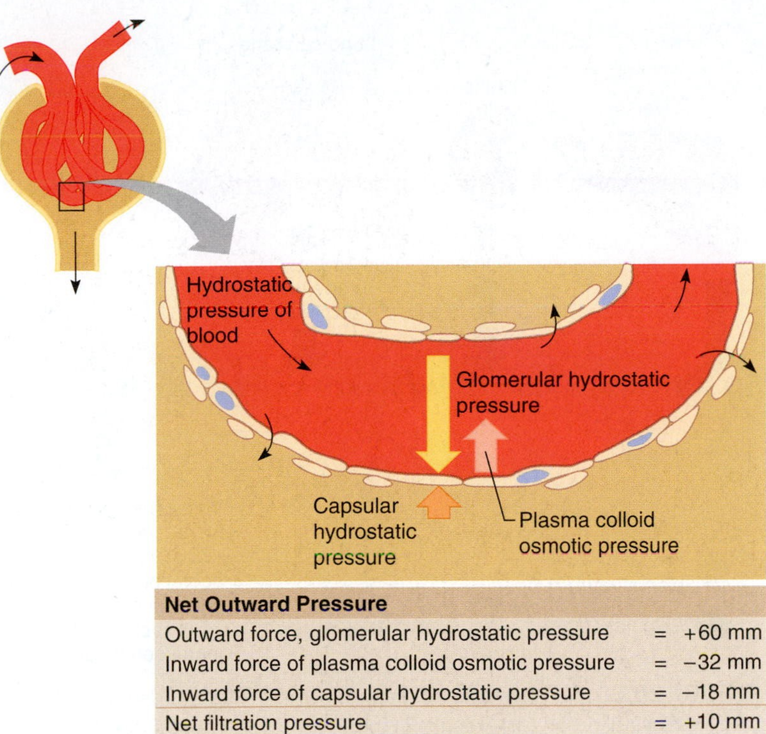

FIGURE 20.18 Normally the glomerular net filtration pressure is positive, causing filtration. The responsible forces include the hydrostatic and osmotic pressure of the plasma and the hydrostatic pressure of the fluid in the glomerular capsule.

> If arterial blood pressure drops drastically, such as during *shock*, the glomerular hydrostatic pressure may fall below the level required for filtration, leading to acute renal failure. At the same time, the epithelial cells of the renal tubules may not receive sufficient nutrients to maintain their high rates of metabolism. Cells may die (tubular necrosis), and renal functions may be lost permanently, resulting in chronic renal failure.

The colloid osmotic pressure of the glomerular plasma also influences net filtration pressure and the rate of filtration. In other systemic capillaries, filtration occurs at the beginning of the capillary, but the osmotic effect of the plasma proteins predominates at the end of the capillary, and most filtered fluid is thus reabsorbed. The small excess remaining eventually becomes lymph.

Much more fluid is filtered by glomerular capillaries than by capillaries elsewhere because of the relatively high hydrostatic pressure in the glomerular capillaries. In fact, as filtration occurs through the capillary wall, proteins remaining in the plasma raise the colloid osmotic pressure within the glomerular capillaries. Despite this, the glomerular capillary hydrostatic pressure is sufficiently great that the net filtration pressure is normally positive. That is, the forces favoring filtration in the glomerular capillaries always predominate. Of course, conditions that lower plasma colloid osmotic pressure, such as a decrease in plasma protein concentration, would increase filtration rate.

The hydrostatic pressure in the glomerular capsule is another factor that may affect net filtration pressure and rate. This capsular pressure can change as a result of an obstruction, such as a stone in a ureter or an enlarged prostate gland pressing on the urethra. If this occurs, fluids back up into the renal tubules and raise the hydrostatic pressure in the glomerular capsules. Any increase in capsular pressure opposes glomerular filtration, so filtration rate may significantly decrease.

At rest, the kidneys receive approximately 25% of the cardiac output, and about 20% of the blood plasma is filtered as it flows through the glomerular capillaries. This means that in an average adult, the glomerular filtration rate for the nephrons of both kidneys is about 125 milliliters per minute, or 180,000 milliliters (180 liters) in twenty-four hours. Assuming that the blood plasma volume is about 3 liters, the production of 180 liters of filtrate in twenty-four hours means that all of the plasma must be filtered through the glomeruli about sixty times each day (fig. 20.19). This twenty-four-hour volume is nearly 45 gallons, so it is obvious that not all of it is excreted as urine. Instead, most of the fluid that passes through the renal tubules is reabsorbed and reenters the plasma.

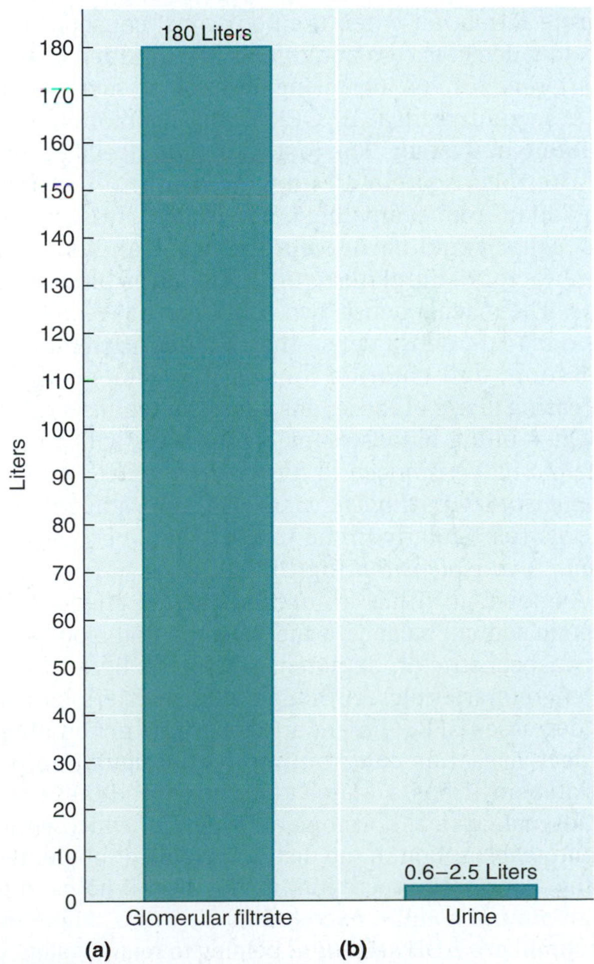

FIGURE 20.19 Relative volumes of (*a*) glomerular filtrate and (*b*) urine formed in twenty-four hours.

The volume of plasma the kidneys filter also depends on the *surface area* of the glomerular capillaries. This surface area is estimated to be about 2 square meters—approximately equal to the surface area of an adult's skin.

PRACTICE 

10 What processes occur as urine forms?

11 How is filtration pressure calculated?

12 What factors influence the rate of glomerular filtration?

Control of Filtration Rate

In general, glomerular filtration rate remains relatively constant through a process called **autoregulation.** However, certain conditions override autoregulation. GFR may increase, for example, when body fluids are in excess and decrease when the body must conserve fluid.

Recall from chapter 15 (p. 587) that sympathetic nervous system fibers synapse with the vascular smooth muscle of arterioles. Reflexes responding to changes in blood pressure and volume control the activity of these sympathetic fibers. If blood pressure and volume drop, vasoconstriction of the afferent arterioles results, decreasing filtration pressure and thus GFR. The result is an appropriate decrease in the rate of urine formation when the body must conserve water. If receptors detect excess body fluids, vasodilation of the afferent arteriole results, increasing filtration pressure and GFR.

A second control of GFR is the hormonelike **renin-angiotensin system.** The juxtaglomerular cells of the afferent arterioles secrete an enzyme, **renin,** in response to stimulation from sympathetic nerves and pressure-sensitive cells, called **renal baroreceptors,** in the afferent arteriole. These factors stimulate renin secretion if blood pressure drops. The macula densa also controls renin secretion. Cells of the macula densa sense the concentrations of sodium, potassium, and chloride ions in the distal renal tubule. Decreasing levels of these ions stimulate renin secretion.

Once in the bloodstream, *renin* reacts with the plasma protein *angiotensinogen* to form *angiotensin I.* An enzyme, *angiotensin-converting enzyme* (ACE), on capillary endothelial cells (particularly in the lungs), rapidly converts angiotensin I to *angiotensin II* (fig. 20.20).

Angiotensin II has a number of renal effects that help maintain sodium balance, water balance, and blood pressure. As a vasoconstrictor, angiotensin II affects both the afferent and efferent arterioles. Although afferent arteriolar constriction decreases GFR, efferent arteriolar constriction minimizes the decrease, thus contributing to autoregulation of GFR. Angiotensin II has a major effect on the kidneys through the adrenal cortical hormone aldosterone, which stimulates sodium reabsorption in the distal convoluted tubule. By stimulating aldosterone secretion, angiotensin II helps to reduce the amount of sodium excreted in the urine. Angiotensin II also stimulates ADH secretion, helping to retain water.

The hormone **atrial natriuretic peptide** (ANP) also affects sodium excretion. ANP secretion increases when the atria of the heart stretch due to increased blood volume. ANP stimulates sodium excretion through a number of mechanisms, including increasing GFR.

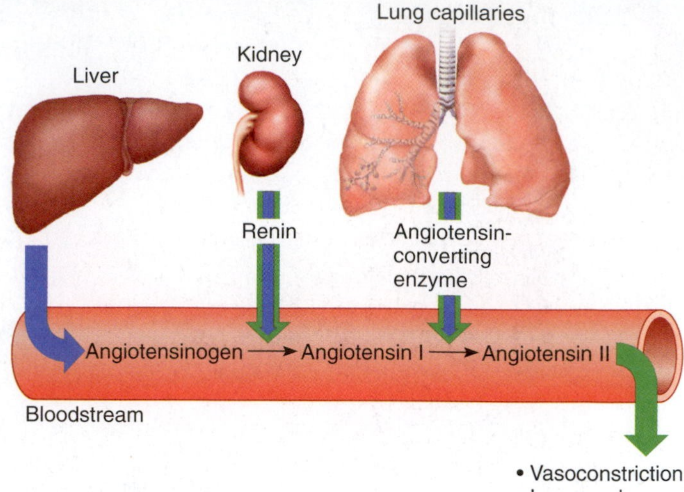

FIGURE 20.20 The formation of angiotensin II in the bloodstream involves several organs and results in multiple actions that conserve sodium and water.

Eating raw spinach or undercooked hamburger, drinking unprocessed apple cider, and petting animals at country fairs have all led to hemolytic uremic syndrome (HUS). The direct cause is a poison called shigatoxin that a certain strain of *E. coli* (bacteria) produces. The toxin finds its way to humans through excrement—in the water used to grow spinach, in beef mixed from many exposed animals, from apples that dropped into droppings, and from tiny hands petting animals. Food poisoning from toxin-producing *E. coli* begins with sharp abdominal pain and bloody diarrhea, and about 16% of the time HUS develops. About 5% of HUS cases are fatal.

Shigatoxin causes fibrin to form thrombi that obstruct the narrow glomerular capillaries. Platelets join the clumping, depleting the circulation of these cell fragments (thrombocytopenia). Red blood cells in the blocked glomerular capillaries break apart, causing hemolytic anemia. As GFR plummets, the person, typically a child, goes into acute renal failure.

Tubular Reabsorption

If the composition of the glomerular filtrate entering the renal tubule is compared with that of the urine leaving the tubule, it is apparent that the fluid changes as it passes through the tubule (see table 20.1). For example, glucose is present in the filtrate but absent in the urine. In contrast, urea and uric acid are considerably more concentrated in urine than they are in the glomerular filtrate. Such changes in fluid composition are largely the result of *tubular reabsorption*, the process by which substances are transported out of the tubular fluid, through the epithelium of the renal tubule, and into the interstitial fluid. These substances then diffuse into the peritubular capillaries (fig. 20.21).

Tubular reabsorption returns substances to the internal environment. The term *tubular* is used because this process is controlled by the epithelial cells that make up the renal tubules and collecting ducts. In tubular reabsorption, substances must first cross the cell membrane facing the inside of the tubule (mucosal surface) and then the cell membrane facing the interstitial fluid (serosal surface).

The basic rules for movements across cell membranes apply to tubular reabsorption. Substances moving down a concentration gradient must either be lipid soluble or there must be a carrier or channel for that substance. Active transport, requiring ATP, may move substances uphill against a concentration gradient. If active transport is involved at any step of the way, the process is considered active tubular reabsorption. In all other cases, the process is considered passive.

RECONNECT

To Chapter 3, Movements Into and Out of the Cell, pages 90–99.

Peritubular capillary blood is under relatively low pressure because it has already passed through two arterioles. Also, the wall of the peritubular capillary is more permeable than that of other capillaries. Finally, the relatively high rate of glomerular filtration has increased the protein concentration and, thus, the colloid osmotic pressure of the peritubular capillary plasma. These factors enhance the rate of fluid reabsorption from the renal tubule.

Tubular reabsorption occurs throughout the renal tubule. However, most of it is in the proximal convoluted portion. The epithelial cells in this portion have many *microvilli* that form a "brush border" on their free surfaces facing the tubular lumen. These tiny extensions greatly increase the surface area exposed to the glomerular filtrate and enhance reabsorption.

Segments of the renal tubule are adapted to reabsorb specific substances, using particular modes of transport. Glucose reabsorption, for example, takes place through the walls of the proximal convoluted tubule by active transport. Water also is rapidly reabsorbed through the epithelium of the proximal convoluted tubule by osmosis; however, portions of the distal convoluted tubule and collecting duct may be almost impermeable to water. This characteristic of the distal convoluted tubule and collecting duct is important in the regulation of urine concentration and volume, as described in a subsequent section.

Recall that active transport requires carrier proteins in a cell membrane. The molecule to be transported binds to the carrier; the carrier changes shape, releases the transported molecule on the other side of the cell membrane, and then returns to its original shape and repeats the process. Such a mechanism has a *limited transport capacity;* it can transport only a certain number of molecules in a given time because the number of carriers is limited.

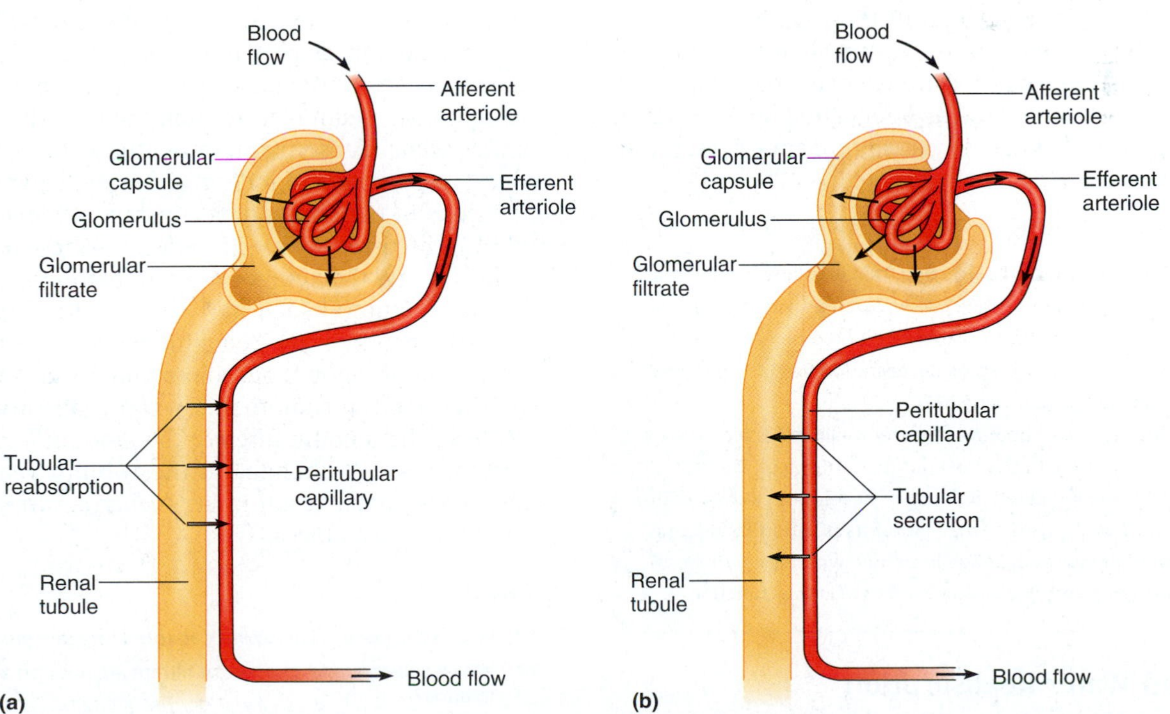

FIGURE 20.21 Two processes in addition to glomerular filtration help to form urine. (*a*) Tubular reabsorption transports substances from the glomerular filtrate into the blood in the peritubular capillary. (*b*) Tubular secretion transports substances from the blood in the peritubular capillary into the renal tubule.

Usually all of the glucose in the glomerular filtrate is reabsorbed because there are enough carrier molecules to transport it. When the plasma glucose concentration increases to a critical level, called the **renal plasma threshold,** more glucose molecules are in the filtrate than the active transport mechanism can handle. As a result, some glucose remains in the filtrate and is excreted in the urine. This explains why the elevated blood glucose of diabetes mellitus results in glucose in the urine.

Any increase in urine volume is called *diuresis.* Nonreabsorbed glucose in the tubular fluid increases the osmotic concentration of the tubular fluid, which reduces the volume of water reabsorbed by osmosis from the proximal tubule and collecting duct. The resultant increase in urine volume is called an *osmotic diuresis.*

Amino acids enter the glomerular filtrate and are reabsorbed in the proximal convoluted tubule. Three different active transport mechanisms reabsorb different groups of amino acids, whose members have similar structures. Normally only a trace of amino acids remains in the urine.

The glomerular filtrate is nearly free of protein, but a number of smaller protein molecules, such as albumin, may squeeze through the glomerular capillaries. These proteins are transported by *endocytosis* through the brush border of epithelial cells lining the proximal convoluted tubule. Once inside an epithelial cell, the proteins are degraded to amino acids, which are moved into the blood of the peritubular capillary.

The epithelium of the proximal convoluted tubule also reabsorbs creatine; lactic, citric, uric, and ascorbic (vitamin C) acids; and phosphate, sulfate, calcium, potassium, and sodium ions. Active transport mechanisms with limited transport capacities reabsorb all of these chemicals. Such substances begin to appear in the urine when their concentrations in the glomerular filtrate exceed their respective renal plasma thresholds. Clinical Application 20.3 discusses how the nephrotic syndrome causes plasma proteins to appear in the urine.

Glucose in the urine is called *glucosuria.* It may follow intravenous administration of glucose, or eating candy, or it may occur in a person with diabetes mellitus. In type 1 diabetes, blood glucose concentration rises because of insufficient insulin secretion from the pancreas (see Clinical Application 13.4, page 512).

One in three people who have diabetes mellitus sustains kidney damage (nephropathy). In the past, large amounts of the protein albumin in the urine indicated high risk of developing kidney damage. Recent studies show that small amounts of albumin (microalbuminuria) predict kidney damage in people with type 1 diabetes. Following a low-protein diet can slow the loss of kidney function.

Sodium and Water Reabsorption

Water reabsorption occurs passively by osmosis, primarily in the proximal convoluted tubule, and is closely associated with the active reabsorption of sodium ions. In the proximal convoluted tubule, if sodium reabsorption increases, water reabsorption increases; if sodium reabsorption decreases, water reabsorption decreases also.

Much of the sodium ion reabsorption occurs in the proximal segment of the renal tubule by active transport (sodium pump mechanism). When the positively charged sodium ions (Na^+) are moved through the tubular wall, negatively charged ions, including chloride ions (Cl^-), phosphate ions (PO_4^{-3}), and bicarbonate ions (HCO_3^-), accompany them. This movement of negatively charged ions is due to the electrochemical attraction between particles of opposite electrical charge. Although this movement of negatively charged ions depends on active transport of sodium, it is considered a passive process because it does not require a direct expenditure of cellular energy. Active transport also reabsorbs some of these ions directly, such as HCO_3^- and PO_4^{-3}.

As more sodium ions are reabsorbed into the peritubular capillary along with negatively charged ions, the concentration of solutes in the peritubular blood might be expected to increase. However, because water moves by osmosis through cell membranes from regions of lesser solute concentration (hypotonic) toward regions of greater solute concentration (hypertonic), water is also reabsorbed, following the ions from the renal tubule into the peritubular capillary.

The proximal convoluted tubule reabsorbs about 70% of the filtered sodium, other ions, and water. By the end of the proximal convoluted tubule, osmotic equilibrium is reached, and the remaining tubular fluid is isotonic (fig. 20.22).

Active transport continues to reabsorb sodium ions as the tubular fluid moves through the nephron loop, the distal convoluted tubule, and the collecting duct. Consequently, almost all of the sodium and water (97% to 99%) that enters the renal tubules as part of the glomerular filtrate may be reabsorbed before the urine is excreted. However, aldosterone controls sodium reabsorption, and antidiuretic hormone controls water reabsorption. Under the influence of these hormones, reabsorption of sodium and water can change to keep conditions in the body fluids constant. Chapter 21 (pp. 818 and 815) discusses the specific effects of these hormones.

Recall that the kidneys filter an extremely large volume of fluid (180 liters) each day. If 99% of the glomerular filtrate is reabsorbed, the remaining 1% excreted includes a relatively large amount of sodium and water (table 20.2). On the other hand, if sodium and water reabsorption decrease to 97% of the amount filtered, the amount excreted triples! Therefore, small changes in the tubular reabsorption of sodium and water result in large changes in urinary excretion of these substances.

PRACTICE 

13 How is the peritubular capillary adapted for reabsorption?

14 Which substances in glomerular filtrate are not normally present in urine?

15 Which mechanisms reabsorb solutes from glomerular filtrate?

16 Define *renal plasma threshold.*

17 Describe the role of passive transport in urine formation.

20.3 CLINICAL APPLICATION

The Nephrotic Syndrome

The *nephrotic syndrome* is a set of symptoms that often appears in patients with renal diseases. Considerable loss of plasma proteins into the urine (proteinuria), results in widespread edema, and increased susceptibility to infection.

Plasma proteins enter the urine because of increased permeability of the glomerular membranes, which accompanies renal disorders such as glomerulonephritis. As a consequence of a decreasing plasma protein concentration (hypoproteinemia), the plasma colloid osmotic pressure falls, increasing net filtration pressure in capillaries throughout the body. This may lead to widespread, severe edema as a large volume of fluid accumulates in interstitial spaces in tissues and in body spaces such as the abdominal cavity, pleural cavity, pericardial cavity, and joint cavities.

Also, as edema develops, blood volume decreases and blood pressure drops. These changes may activate the renin-angiotensin system, leading to the release of aldosterone from the adrenal cortex (see chapter 13, p. 506), which, in turn, stimulates the kidneys to conserve sodium ions and water. This action reduces the urine output and may aggravate the edema.

The nephrotic syndrome sometimes appears in young children who have *lipoid nephrosis*. The cause of this condition is unknown, but it alters the epithelial cells of the glomeruli so that the glomerular membranes enlarge and distort, allowing proteins through. ■

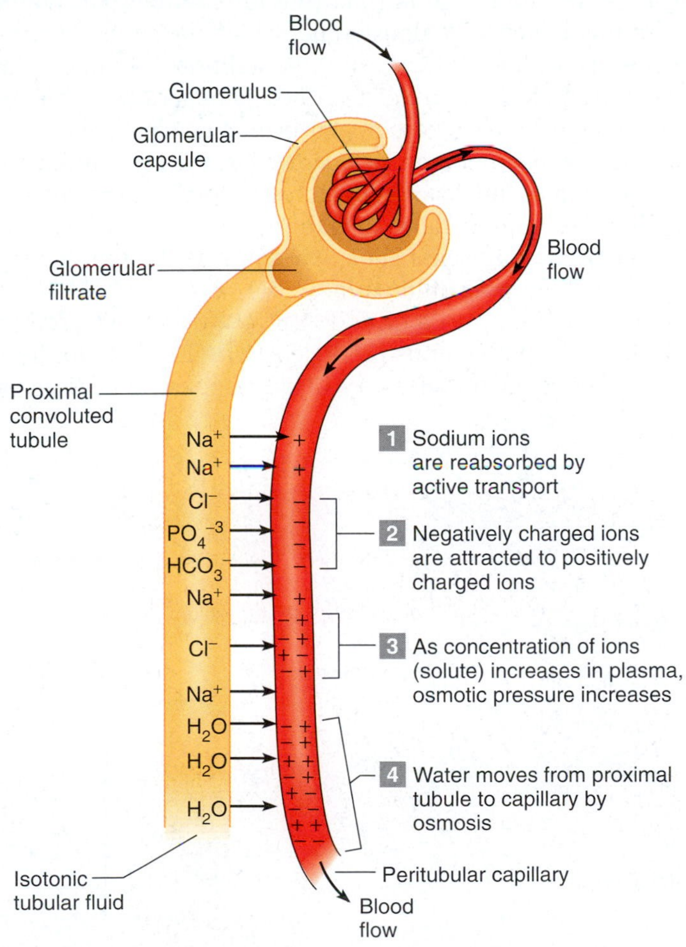

FIGURE 20.22 In the proximal part of the renal tubule, osmosis reabsorbs water in response to active transport reabsorbing sodium and other solutes.

Tubular Secretion

In *tubular secretion*, certain substances move from the plasma of the peritubular capillary into the fluid of the renal tubule. The amount of a particular chemical excreted in the urine may exceed the amount filtered from the plasma in the glomerulus (see fig. 20.21). As in the case of tubular reabsorption, the term *tubular* is used because the epithelial cells of the renal tubules control the process.

Active transport mechanisms similar to those that function in reabsorption secrete some substances. However, the secretory mechanisms transport substances in the opposite direction. For example, the epithelium of the proximal convoluted tubules actively secretes certain organic compounds, including penicillin and histamine, into the tubular fluid.

Hydrogen ions are actively secreted throughout the entire renal tubule. Urine is usually acidic by the time it is excreted, although the urinary pH can vary considerably. The secretion of hydrogen ions is important in regulating the pH of body fluids, as chapter 21 (p. 823) explains.

The kidneys respond to certain chemicals, including antibiotics, by eliminating them from the body as rapidly as possible. In some cases, such as eliminaiton of toxins, this is beneficial. In the case of helpful drugs, however, it is undesirable. The drug probenecid is given with certain antibiotics to block their tubular secretion, increasing their levels in the plasma. For example, probenecid increases plasma concentration of penicillin two to four fold. It also acts on ampicillin, methicillin, and certain other drugs in the penicillin family, as well as certain cephalosporin antibiotics.

TABLE 20.2 | Average Values for Sodium and Water Filtration, Reabsorption, and Excretion

	Amount Filtered per Day	Amount Reabsorbed per Day (%)	Amount Excreted per Day
Water (L)	180	178.2 (99%)	1.8 (1%)
Na^+ (g)	630	626.8 (99.5%)	3.2 (0.5%)

Most of the potassium ions in the glomerular filtrate are actively reabsorbed in the proximal convoluted tubule, but some may be secreted in the distal convoluted tubule and collecting duct. During this process, the active reabsorption of sodium ions out of the tubular fluid under the influence of aldosterone produces a negative electrical charge in the tube. Positively charged potassium ions (K^+) are attracted to negatively charged regions, so these ions move passively through the tubular epithelium and enter the tubular fluid. Potassium ions are also actively secreted (fig. 20.23).

To summarize, urine forms as a result of the following:

- Glomerular filtration of materials from blood plasma.
- Tubular reabsorption of substances, including glucose; water; urea; proteins; creatine; amino, lactic, citric, and uric acids; and phosphate, sulfate, calcium, potassium, and sodium ions.
- Tubular secretion of substances, including penicillin, histamine, phenobarbital, hydrogen ions, ammonia, and potassium ions.

PRACTICE

18 Define *tubular secretion.*

19 Which substances are actively secreted? Passively secreted?

20 How does the reabsorption of sodium affect the secretion of potassium?

Regulation of Urine Concentration and Volume

Hormones such as aldosterone and ANP affect the solute concentration of urine, particularly sodium. However, the ability of the kidneys to maintain the internal environment rests in large part on their ability to concentrate urine by reabsorbing large volumes of water.

In contrast to conditions in the proximal convoluted tubule, the tubular fluid reaching the distal convoluted tubule is hypotonic because of changes that occur through the loop segment of each nephron. The cells lining the distal convoluted tubule and the collecting duct that follows continue to reabsorb sodium ions (chloride ions follow passively) under the influence of aldosterone, which the adrenal cortex secretes (see chapter 13, p. 506). In addition, the interstitial fluid surrounding the collecting ducts is hypertonic, particularly in the medulla. These might seem to be ideal conditions for water reabsorption as well. However, the cells lining the later portion of the distal convoluted tubule and the collecting duct are impermeable to water unless antidiuretic hormone (ADH) is present. Thus, water inside the tubule may be excreted, forming dilute urine.

As discussed in chapter 13 (p. 498), neurosecretory cells in the hypothalamus produce ADH. The posterior lobe of the pituitary gland releases ADH in response to decreasing

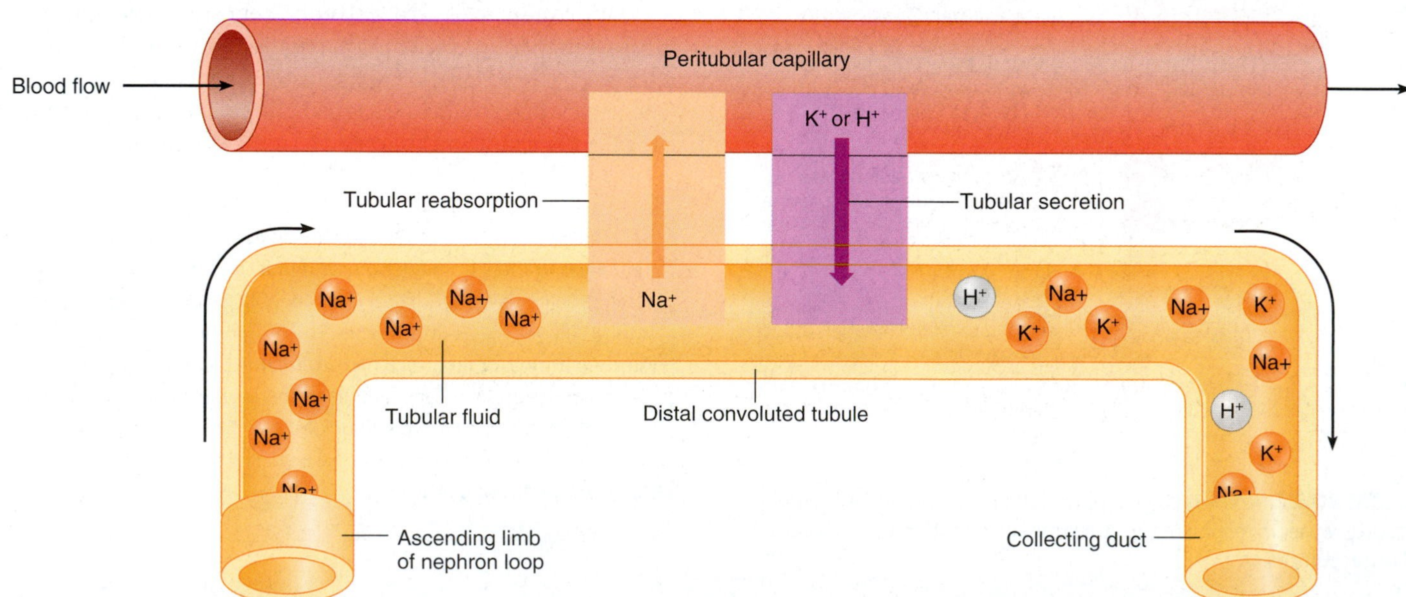

FIGURE 20.23 In the distal convoluted tubule, potassium ions (or hydrogen ions) may be passively secreted in response to the active reabsorption of sodium ions.

concentration of water in the body fluids or to decreasing blood volume and blood pressure. When ADH reaches the kidney, it stimulates cells in the distal convoluted tubules and collecting ducts to insert proteins called aquaporins into their cell membranes, which form water channels. These channels greatly increase permeability to water; consequently, water rapidly moves out of these structures by osmosis, especially where the distal tubules and collecting ducts pass through the extremely hypertonic medulla. The urine becomes more concentrated, and water is retained in the internal environment (fig. 20.24).

A **countercurrent mechanism** in the nephron loops, particularly of the juxtamedullary nephrons, ensures that the medullary interstitial fluid becomes hypertonic. This mechanism is possible because the descending and ascending limbs of the nephron loops lie parallel and close to one another. The mechanism is named partly for the fact that fluid moving down the descending limb creates a current that is counter to that of the fluid moving up in the ascending limb.

The different parts of the nephron loop have important functional distinctions. For example, the epithelial lining in the thick upper portion of the ascending limb (thick segment) is relatively impermeable to water. However, the epithelium does actively reabsorb sodium and chloride ions (some potassium is actively reabsorbed as well). As these solutes accumulate in the interstitial fluid outside the ascending limb, it becomes hypertonic, while the tubular fluid inside becomes hypotonic because it is losing its solute.

In contrast to the ascending limb, the epithelium of the descending limb (thin segment) is permeable to water, but relatively impermeable to solutes. This segment is surrounded by hypertonic fluid created by the ascending limb, so water tends to leave the descending limb by osmosis. The contents of the descending limb become more concentrated, or hypertonic (fig. 20.25).

The very concentrated tubular fluid now moves into the ascending limb, and sodium chloride (NaCl) is again actively reabsorbed into the medullary interstitial fluid, raising the interstitial NaCl concentration further. With the increased interstitial fluid solute concentration, even more water diffuses out of the descending limb, further increasing the salt concentration of the tubular fluid. Each time this circuit is completed, the concentration of NaCl increases, or multiplies. For this reason, the mechanism is called a *countercurrent multiplier*. In humans, this mechanism creates a tubular fluid solute concentration near the tip of the loop more than four times the solute concentration of plasma (fig. 20.25).

The descending limb of the loop is permeable to water, so the interstitial fluid at any level of the loop is essentially in equilibrium with the fluid in the tubule. Thus, the concentration gradient in the loop is also found in the interstitial fluid. The solute concentration of the tubular fluid progressively decreases as the fluid moves toward the renal cortex (fig. 20.25).

The vasa recta is another countercurrent mechanism that maintains the NaCl concentration gradient in the renal medulla. Blood flows slowly down the descending portion of the vasa recta, and NaCl enters it by diffusion. Then, as the blood moves back up toward the renal cortex, most of the NaCl diffuses from the blood and reenters the medullary interstitial fluid. Consequently, the bloodstream carries little NaCl away from the renal medulla, preserving the gradient (fig. 20.26).

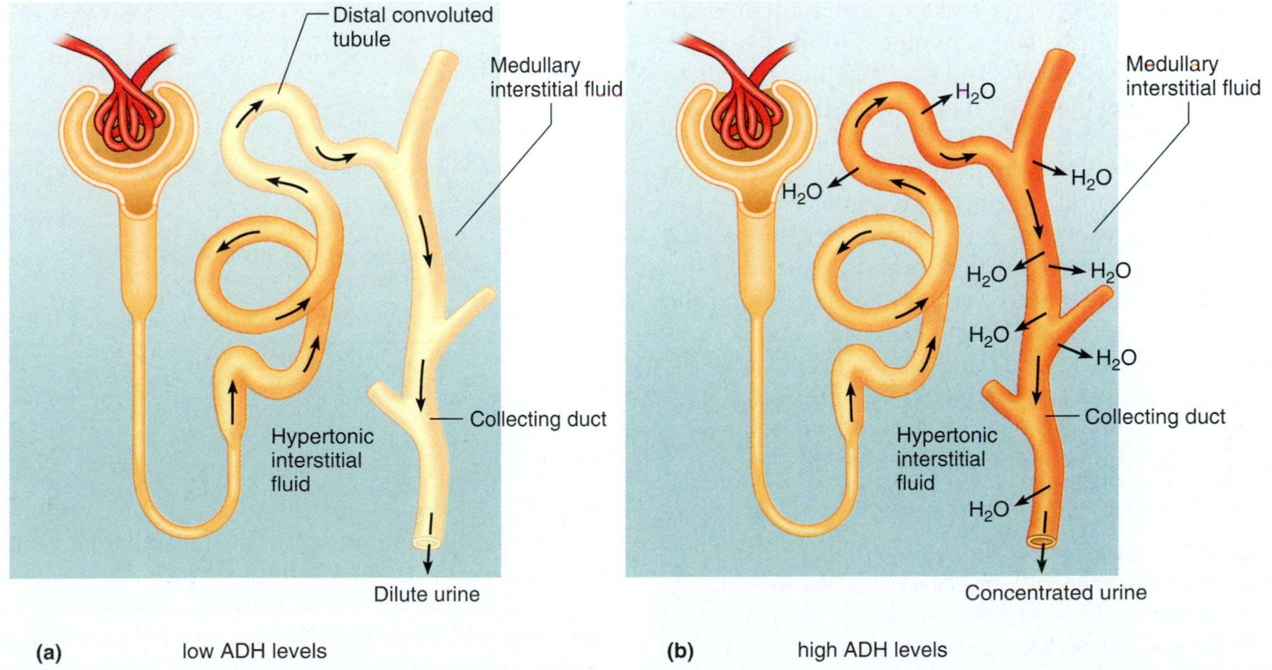

FIGURE 20.24 Urine concentrating mechanism. (*a*) The distal convoluted tubule and collecting duct are impermeable to water, so water may be excreted as dilute urine. (*b*) If ADH is present, however, these segments become permeable, and water is reabsorbed by osmosis into the hypertonic medullary interstitial fluid.

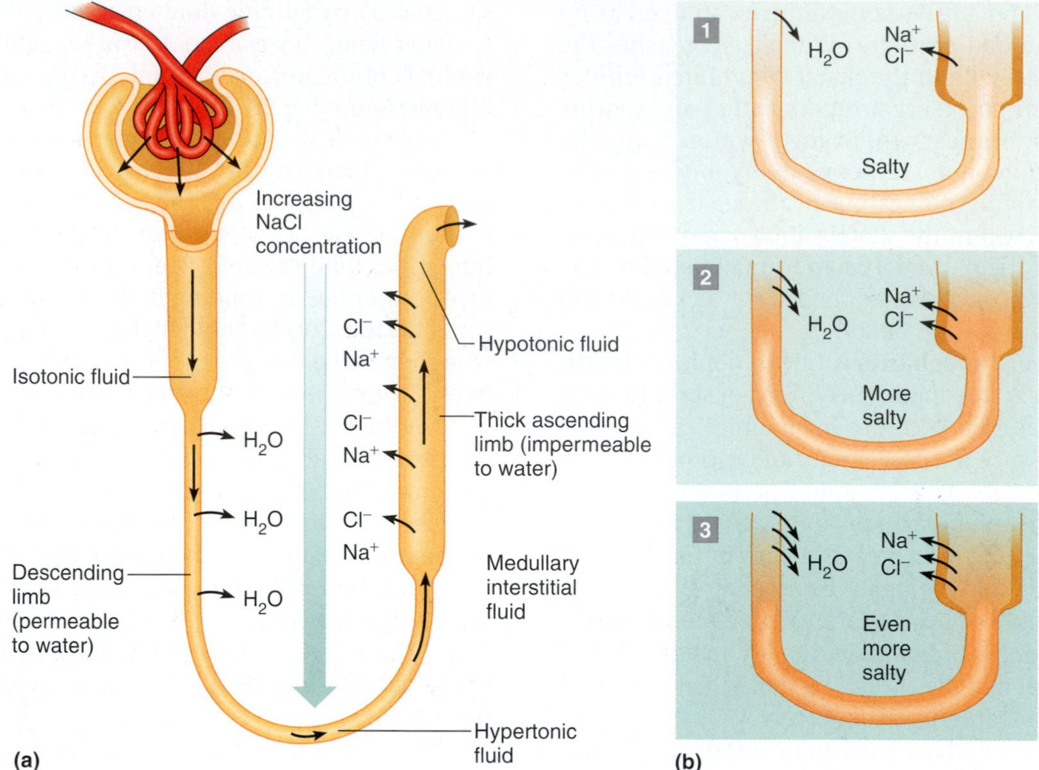

FIGURE 20.25 The countercurrent multiplier. (*a*) The solute concentration of interstitial fluid in the medulla equilibrates with tubular fluid, which loses water in the descending limb, and thus becomes hypertonic by the tip of the nephron loop. The ascending limb of the loop actively reabsorbs solute. (*b*) Active solute reabsorption from the ascending limb of the loop causes even more water loss from the descending limb as tubular fluid continues to flow. The countercurrent multiplier progressively increases the solute concentration of the interstitial fluid, up to a maximum near the tip of the loop more than four times that of plasma.

To summarize, the countercurrent multiplier creates a large osmotic gradient for water reabsorption in the interstitial fluid surrounding the distal convoluted tubules and the collecting ducts of the nephron. The epithelial lining of these structures is impermeable to water, unless ADH is present. The higher the blood levels of ADH, the more permeable the epithelial lining becomes, increasing water reabsorption and concentrating the urine. In this way, soluble wastes and other substances are excreted in minimal water, preserving body water when dehydration is a threat. If the body fluids contain excess water, ADH secretion decreases and the epithelial linings of the distal convoluted tubule and the collecting duct become less permeable to water. Less water is reabsorbed, and the urine becomes more dilute. Table 20.3 summarizes the role of ADH in urine production. Table 20.4 summarizes the functions of different parts of the nephron.

A substance that causes diuresis is called a *diuretic.* Medically, a "water pill" may help patients abnormally retaining water. However, other diuretics are encountered more commonly. Caffeine inhibits proximal tubular sodium reabsorption, leading to an osmotic diuresis. Alcohol inhibits secretion of ADH from the posterior pituitary gland, directly decreasing water reabsorption.

TABLE 20.3 | Role of ADH in Regulating Urine Concentration and Volume

1. Concentration of water in the blood decreases.
2. Increase in the osmotic pressure of body fluids stimulates osmoreceptors in the hypothalamus.
3. Hypothalamus signals the posterior pituitary gland to release ADH.
4. Blood carries ADH to the kidneys.
5. ADH causes the distal convoluted tubules and collecting ducts to increase water reabsorption by osmosis.
6. Urine becomes more concentrated, and urine volume decreases.

Urea and Uric Acid Excretion

Urea is a by-product of amino acid catabolism in the liver. Therefore, the amount of urea that must be eliminated in the urine reflects the amount of protein in the diet. Urea enters the renal tubule by filtration and is both reabsorbed and secreted by different portions of the renal tubule. The pattern of these processes effectively recycles up to 80% of the filtered urea, which provides much of the osmotic concentration of the medullary interstitial fluid. As a result,

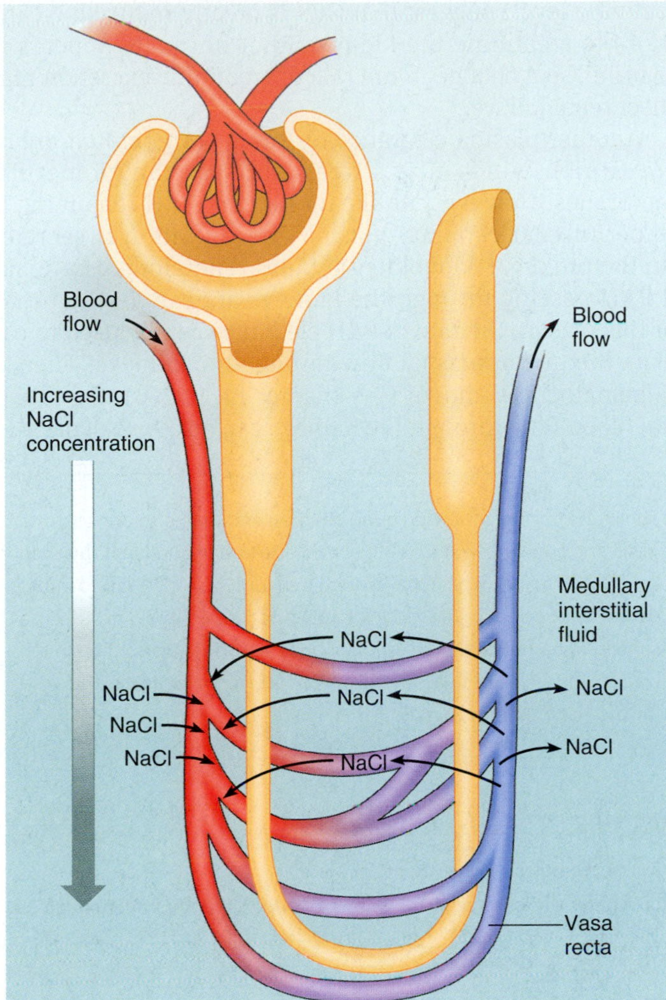

FIGURE 20.26 A countercurrent mechanism in the vasa recta helps maintain the NaCl concentration gradient in the medullary interstitial fluid (see fig. 20.14).

urea contributes to the reabsorption of water from the collecting duct.

Uric acid is a product of the metabolism of certain nucleic acid bases (the purines adenine and guanine). Active transport completely reabsorbs the filtered uric acid. Some uric acid, equal to approximately 10% of the amount filtered, is excreted in the urine. This reflects uric acid secretion into the renal tubule.

PRACTICE 

21 Describe a countercurrent mechanism.

22 How does the hypothalamus regulate urine concentration and volume?

23 Explain how urea and uric acid are excreted.

TABLE 20.4 | Functions of Nephron Components

Part	Function
Renal Corpuscle	
Glomerulus	Filtration of water and dissolved substances from the plasma
Glomerular capsule	Receives the glomerular filtrate
Renal Tubule	
Proximal convoluted tubule	Reabsorption of glucose; amino acids; creatine; lactic, citric, uric, and ascorbic acids; phosphate, sulfate, calcium, potassium, and sodium ions by active transport
	Reabsorption of proteins by endocytosis
	Reabsorption of water by osmosis
	Reabsorption of chloride ions and other negatively charged ions by electrochemical attraction
	Active secretion of substances such as penicillin, histamine, creatinine, and hydrogen ions
Descending limb of nephron loop	Reabsorption of water by osmosis
Ascending limb of nephron loop	Reabsorption of sodium, potassium, and chloride ions by active transport
Distal convoluted tubule	Reabsorption of sodium ions by active transport
	Reabsorption of water by osmosis
	Active secretion of hydrogen ions
	Secretion of potassium ions both actively and by electrochemical attraction
Collecting Duct*	Reabsorption of water by osmosis

*Although the collecting duct is not anatomically part of the nephron, it is functionally linked in the process of urine formation.

Excess uric acid may precipitate in the plasma and be deposited as crystals in joints, causing the inflammation and extreme pain of gout, particularly in the digits and especially in the great toe. Gout has had an interesting history. Hippocrates mentioned it. King Charles I of Spain gave up his vast empire in 1556 due to the painful condition. In 2006, Spanish researchers confirmed the diagnosis by detecting uric acid deposits in the terminal joint of a finger that, for reasons unknown, had been preserved in a small box apart from the rest of the king. Today, gout is treated with drugs that inhibit uric acid reabsorption or block an enzyme in the biosynthetic pathway for uric acid. Limiting foods rich in uric acid, such as organ meats and seafood, and drinking more to dilute urine can help. Gout is inherited, but an attack may not occur until the person eats the offending foods. The condition was once attributed to gluttony because only wealthy people could afford to eat meat.

Urine Composition

Urine composition reflects the volumes of water and solutes that the kidneys must eliminate from the body or retain in the internal environment to maintain homeostasis. It varies considerably from time to time because of differences in dietary intake and physical activity. Urine is about 95% water and usually also contains urea and uric acid from the catabolism of amino acids and nucleic acids, and creatinine from metabolism of creatine. Urine may also have a trace of amino acids, as well as electrolytes whose concentrations reflect diet (see table 20.1). Appendix B (p. 943) lists the normal concentrations of urine components.

Not all abnormal constituents of urine indicate illness. Glucose may enter urine after a sugary meal or toward the end of pregnancy; protein may appear in the urine following vigorous physical exercise; ketones are in urine during a prolonged fast or when a person follows a very low-calorie or low-carbohydrate diet.

The volume of urine produced usually varies between 0.6 and 2.5 liters per day. Such factors as fluid intake; environmental temperature; relative humidity of the surrounding air; and a person's emotional condition, respiratory rate, and body temperature influence the exact urine volume. An output of 50–60 milliliters of urine per hour is considered normal, and an output of less than 30 milliliters per hour may indicate kidney failure.

Renal Clearance

The rate at which a particular chemical is removed from the plasma indicates kidney efficiency. This rate is called *renal clearance.*

Tests of renal clearance detect glomerular damage or monitor the progression of renal disease. One such test, the *inulin clearance test,* uses *inulin* (not to be confused with insulin), a complex polysaccharide found in certain plant roots. In the test, a known amount of inulin is infused into the blood at a constant rate. The inulin passes freely through the glomerular membranes, so its concentration in the glomerular filtrate equals that in the plasma. In the renal tubule, inulin is not reabsorbed to any significant degree, nor is it secreted. Consequently, the rate at which it appears in the urine can be used to calculate the rate of glomerular filtration.

Similarly, the kidneys remove creatinine from the blood. Creatinine is produced at a constant rate during muscle metabolism. Like inulin, creatinine is filtered but neither reabsorbed nor secreted by the kidneys. The *creatinine clearance test,* which compares a patient's blood and urine creatinine concentrations, can also be used to calculate the GFR. A significant advantage is that the bloodstream normally has a constant level of creatinine. Therefore, a single measurement of plasma creatinine levels provides a rough index of kidney function. For example, significantly elevated plasma creatinine levels suggest that GFR is greatly reduced. Nearly all of the creatinine the kidneys filter normally appears in the urine, so a change in the rate of creatinine excretion may reflect renal failure.

Another plasma clearance test uses *para-aminohippuric acid* (PAH), which filters freely through the glomerular membranes. However, unlike inulin, any PAH remaining in the peritubular capillary plasma after filtration is secreted into the proximal convoluted tubules. Therefore, essentially all PAH passing through the kidneys appears in the urine. For this reason, the rate of PAH clearance can be used to calculate the rate of plasma flow through the kidneys. Then, if the hematocrit is known (see chapter 14, p. 523), the rate of total blood flow through the kidneys can also be calculated.

The kidneys of infants and young children are unable to concentrate urine and conserve water as effectively as those of adults, so they can lose water rapidly, which may lead to dehydration. A 20-pound infant can lose a pound in just a day of an acute viral illness, and this is a sufficiently significant proportion of body weight to warrant hospitalization. Intravenous fluids are given to restore water and electrolyte balance (see chapter 21, p. 811).

PRACTICE

24 List the normal constituents of urine.

25 What is the normal hourly output of urine? The minimal hourly output?

20.4 ELIMINATION OF URINE

After forming along the nephrons, urine passes from the collecting ducts through openings in the renal papillae and enters the minor and major calyces of the kidney. From there it passes through the renal pelvis, into a ureter, and into the urinary bladder. The urethra delivers urine to the outside.

Ureters

Each **ureter** is a tubular organ about 25 centimeters long, which begins as the funnel-shaped renal pelvis. It extends downward posterior to the parietal peritoneum and parallel to the vertebral column. In the pelvic cavity, each ureter courses forward and medially to join the urinary bladder from underneath.

The wall of a ureter is composed of three layers. The inner layer, or *mucous coat,* includes several thicknesses of transitional epithelial cells and is continuous with the linings of the renal tubules and the urinary bladder. The middle layer, or *muscular coat,* largely consists of smooth muscle fibers in circular and longitudinal bundles. The outer layer, or *fibrous coat,* is composed of connective tissue (fig. 20.27).

Muscular peristaltic waves, originating in the renal pelvis, help move the urine along the length of the ureter. The presence of urine in the renal pelvis initiates these waves, whose

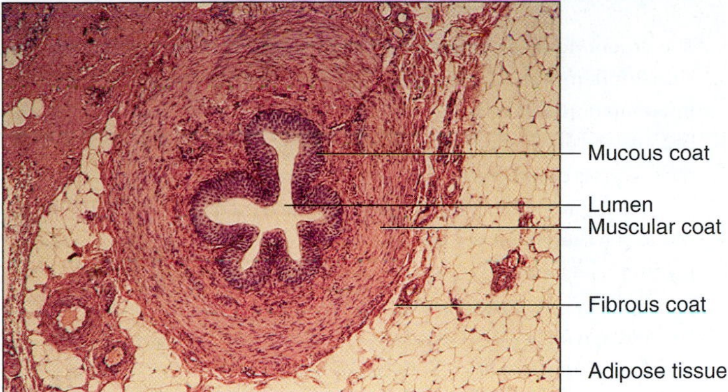

FIGURE 20.27 Cross section of a ureter (75×).

Kidney stones, usually composed of calcium oxalate, calcium phosphate, uric acid, or magnesium phosphate, can form in the renal pelvis. A stone passing into a ureter produces severe pain in the region of the kidney and radiating into the abdomen, pelvis, and lower limbs. Nausea and vomiting may also occur.

About 60% of kidney stones pass spontaneously; the others must be removed. Shock waves applied from outside the body are used to fragment kidney stones. This procedure, called *extracorporeal shock-wave lithotripsy* (ESWL), focuses high-energy shock waves through water (either in a tub or in a water-filled sack placed against the patient). The stones shatter into fragments small enough to be eliminated with urine.

frequency keeps pace with the rate of urine formation. If urine forms quickly, a peristaltic wave may pass every few seconds; if it forms more slowly, a wave may pass every few minutes.

A peristaltic wave reaching the urinary bladder spurts urine into it. A flaplike fold of mucous membrane covers the opening where the urine enters. This fold acts as a valve, allowing urine to enter the bladder from the ureter but preventing it from backing up from the bladder into the ureter.

If a ureter becomes obstructed, such as by a small kidney stone (renal calculus) in its lumen, strong peristaltic waves begin in the proximal part of the tube, which may help move the stone into the bladder. The presence of a stone usually also stimulates a sympathetic reflex (ureterorenal reflex) that constricts the renal arterioles and reduces urine production in the affected kidney.

PRACTICE

26 Describe the structure of a ureter.

27 How is urine moved from the renal pelvis to the urinary bladder?

28 What prevents urine from backing up from the urinary bladder into the ureters?

29 How does an obstruction in a ureter affect urine production?

Urinary Bladder

The **urinary bladder** is a hollow, distensible, muscular organ in the pelvic cavity, posterior to the symphysis pubis and inferior to the parietal peritoneum (fig. 20.28 and reference plate 8). In a female, the bladder contacts the anterior walls of the uterus and vagina, and in a male, the bladder lies posteriorly against the rectum.

The pressure of surrounding organs alters the spherical shape of the bladder. When the bladder is empty, its inner wall forms many folds, but as it fills with urine, the wall becomes smoother. At the same time, the superior surface of the bladder expands upward into a dome.

When greatly distended, the bladder pushes above the pubic crest and into the region between the abdominal wall and the parietal peritoneum. The dome can reach the level of the umbilicus and press against the coils of the small intestine.

The internal floor of the bladder includes a triangular area called the *trigone*, which has an opening at each of its three angles (fig. 20.29). Posteriorly, at the base of the trigone, the openings are those of the ureters. Anteriorly, at the apex of the trigone, is a short, funnel-shaped extension called the *neck* of the bladder, which contains the opening into the

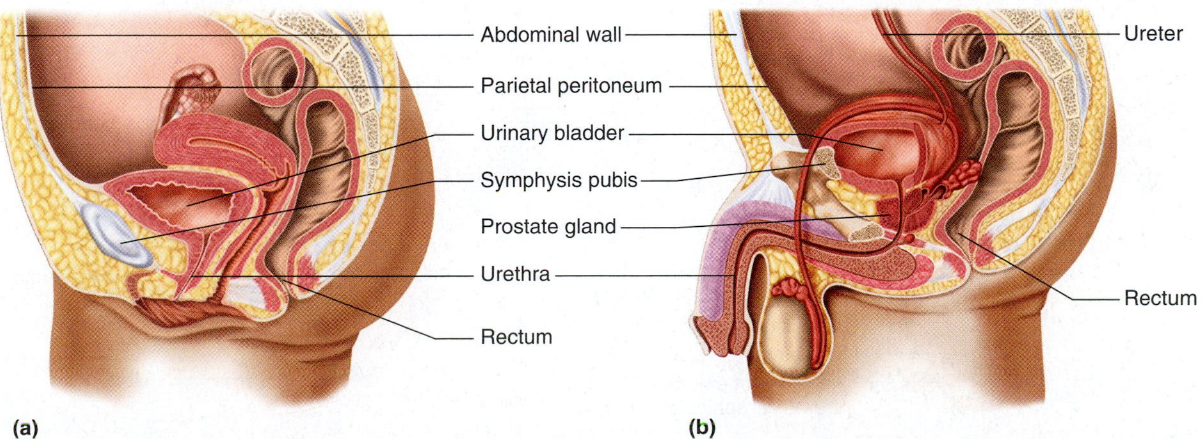

FIGURE 20.28 The urinary bladder is in the pelvic cavity and behind the symphysis pubis. (*a*) In a female, it contacts the uterus and vagina. (*b*) In a male, it lies against the rectum.

urethra. The trigone generally remains in a fixed position, even though the rest of the bladder distends and contracts.

The wall of the urinary bladder consists of four layers. The inner layer, or *mucous coat*, includes several thicknesses of transitional epithelial cells similar to those lining the ureters and the upper portion of the urethra. The thickness of this tissue changes as the bladder expands and contracts. During distension, the tissue appears to be only two or three cells thick, but during contraction, it appears to be five or six cells thick (see fig. 5.9).

The second layer of the bladder wall is the *submucous coat*. It consists of connective tissue and has many elastic fibers.

The third layer of the bladder wall, the *muscular coat*, is primarily composed of coarse bundles of smooth muscle fibers. These bundles are interlaced in all directions and at all depths, and together they comprise the **detrusor muscle** (de-truz′or mus′l). The portion of the detrusor muscle that surrounds the neck of the bladder forms an *internal urethral sphincter*. Sustained contraction of this sphincter muscle prevents the bladder from emptying until the pressure within it increases to a certain level. The detrusor muscle has parasympathetic nerve fibers that function in the reflex that passes urine.

The outer layer of the wall, the *serous coat*, consists of the parietal peritoneum. It is found only on the upper surface of the bladder. Elsewhere, the outer coat is composed of fibrous connective tissue (fig. 20.30). From Science to Technology 5.2 on page 166 discusses a tissue-engineered replacement bladder.

The urgent need to urinate and the burning pain of urination may signal inflammation or infection of the bladder and/or ureters. Inflammation of the bladder (cystitis) is more common in women because the urethral pathway is shorter than in men. Inflammation of the ureters is called ureteritis.

In a urinary tract infection (UTI), bacteria ascend from the bladder to the ureters on the continuous linings. Urination is frequent, painful, scant, and may be bloody, with accompanying abdominal pain. Usually pathogenic bacteria in the urinary tract remain outside the cells and are easily killed with antibiotic drugs or prevented from attaching to ureter lining cells by exposure to compounds in cranberry and blueberry juices that the person drinks. However, certain bacteria, such as *Escherichia coli*, enter the lining cells, forming "intracellular bacterial communities." The microbes can emerge and start new infections, explaining why some people experience frequent UTIs.

PRACTICE

30 Describe the trigone of the urinary bladder.

31 Describe the structure of the bladder wall.

32 What type of nerve fibers supply the detrusor muscle?

Urethra

The **urethra** is a tube that conveys urine from the urinary bladder to the outside of the body. Its wall is lined with mucous membrane and has a thick layer of longitudinal smooth muscle fibers. The urethral wall also has many

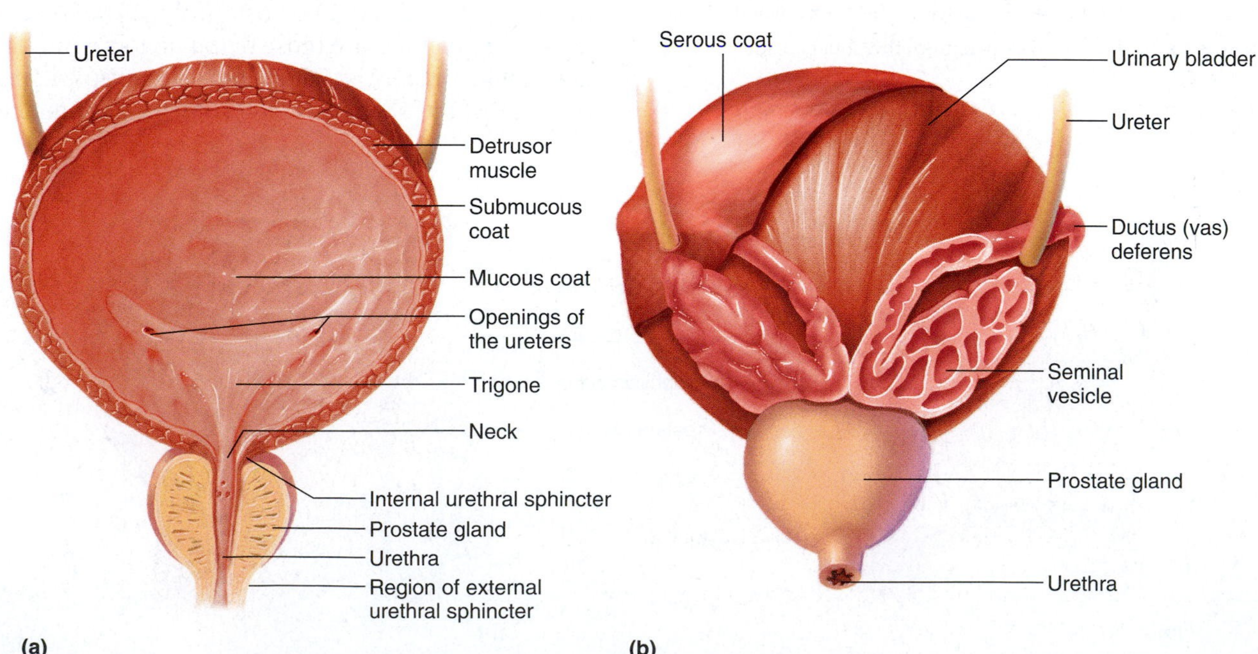

FIGURE 20.29 A male urinary bladder. (*a*) Longitudinal section. (*b*) Posterior view.

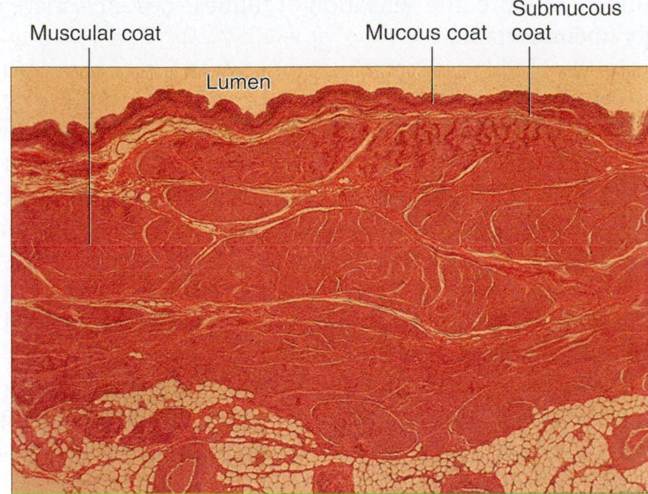

FIGURE 20.30 Light micrograph of the human urinary bladder wall (6×).

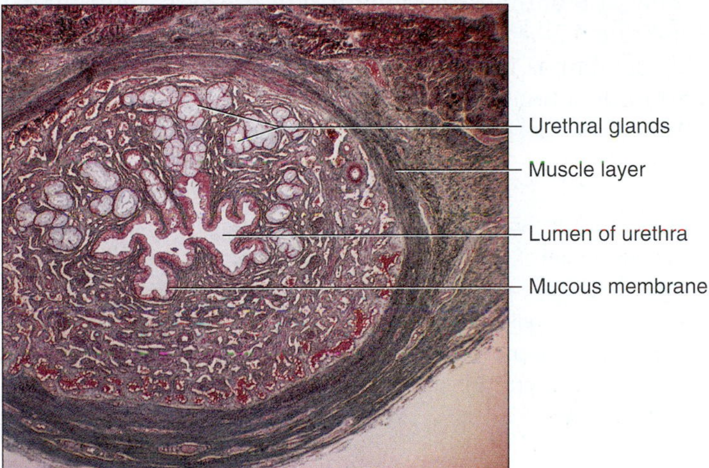

FIGURE 20.31 Cross section through the urethra (10×).

mucous glands, called *urethral glands,* which secrete mucus into the urethral canal (fig. 20.31).

In a female, the urethra is about 4 centimeters long. It passes forward from the bladder, courses below the symphysis pubis, and empties between the labia minora. Its opening, the *external urethral orifice* (urinary meatus), is anterior to the vaginal opening and is about 2.5 centimeters posterior to the clitoris (fig. 20.32*a*).

In a male, the urethra, which functions both as a urinary canal and a passageway for cells and secretions from the reproductive organs, can be divided into three sections: the prostatic urethra, the membranous urethra, and the penile urethra (see fig. 20.32*b* and reference plate 20).

The **prostatic urethra** is about 2.5 centimeters long and passes from the urinary bladder through the *prostate gland,* located just below the bladder. Ducts from reproductive structures join the urethra in this region.

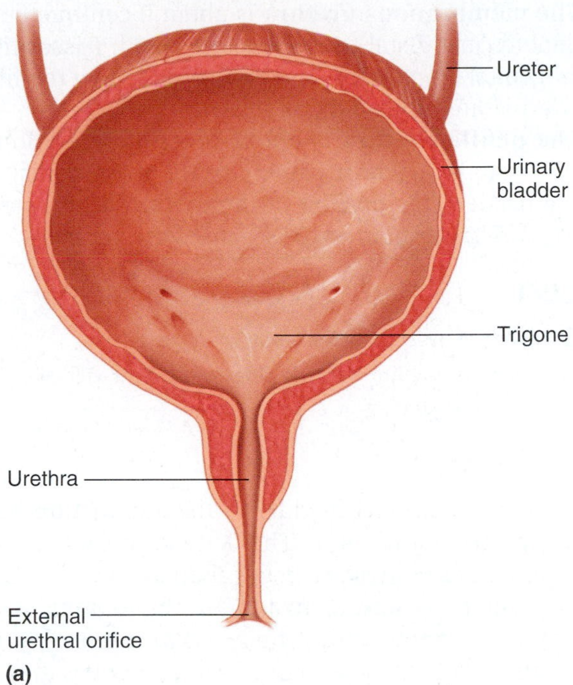

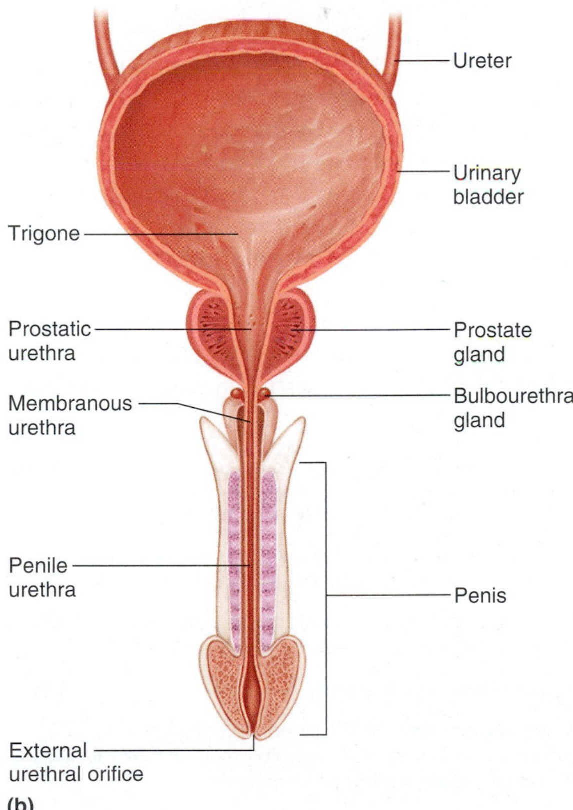

FIGURE 20.32 Urinary bladder and urethra (*a*) of the female (longitudinal section) and (*b*) of the male (longitudinal section).

The **membranous urethra** is about 2 centimeters long. It originates just distal to the prostate gland, passes through the urogenital diaphragm, and is surrounded by the fibers of the external urethral sphincter muscle.

The **penile urethra** is about 15 centimeters long and passes through the corpus spongiosum of the penis, within erectile tissue. This portion of the urethra terminates with the *external urethral* orifice at the tip of the penis.

PRACTICE 

33 Describe the structure of the urethra.

34 How does the urethra of a male differ from that of a female?

Micturition

Urine leaves the urinary bladder by the **micturition** (mik″tu-rish′un) or urination reflex. The detrusor muscle contracts, and contractions of muscles in the abdominal wall and pelvic floor may help, as well as fixation of the thoracic wall and diaphragm. In micturition, the *external urethral sphincter* also relaxes. This muscle, part of the urogenital diaphragm (see chapter 9, pp. 322–323), surrounds the urethra about 3 centimeters from the bladder and is composed of voluntary skeletal muscle tissue.

Distension of the bladder wall as it fills with urine stimulates the urge to urinate. The wall expands, stimulating stretch receptors, which triggers the micturition reflex.

The *micturition reflex center* is in the sacral portion of the spinal cord. When sensory impulses from the stretch receptors signal the reflex center, parasympathetic motor impulses travel out to the detrusor muscle, which contracts rhythmically in response. A sensation of urgency accompanies this action.

The urinary bladder may hold up to 600 milliliters of urine. The desire to urinate usually begins when it contains about 150 milliliters. Then, as urine volume increases to 300 milliliters or more, the sensation of fullness becomes increasingly uncomfortable.

As the bladder fills with urine and its internal pressure increases, contractions of its wall intensify. When these contractions become strong enough to force the internal urethral sphincter open, another reflex signals the external urethral sphincter to relax, and the bladder may empty. However, because the external urethral sphincter is composed of skeletal muscle, it is under conscious control, and therefore usually remains contracted until a person decides to urinate. Nerve centers in the brainstem and cerebral cortex that inhibit the micturition reflex aid this control.

When a person decides to urinate, the external urethral sphincter relaxes and inhibition of the micturition reflex lifts. Nerve centers in the pons and the hypothalamus heighten the micturition reflex. The detrusor muscle contracts and urine is excreted through the urethra. Within a few moments, the neurons of the micturition reflex tire, the detrusor muscle relaxes, and the bladder begins to fill with urine again. Table 20.5 outlines the micturition process, and Clinical Application 20.4 discusses urinalysis and health. Table 20.6 lists conditions that result from abnormal development of part of the urinary system. About 1 in 500 newborns has a birth defect affecting the urinary system.

Damage to the spinal cord above the sacral region may abolish voluntary control of urination. However, if the micturition reflex center and its sensory and motor fibers are uninjured, micturition may continue to occur reflexively. In this case, the bladder collects urine until its walls stretch enough to trigger a micturition reflex, and the detrusor muscle contracts in response. This condition is called an *automatic bladder.*

TABLE 20.5 | Major Events of Micturition

1. Urinary bladder distends as it fills with urine.
2. Stretch receptors in the bladder wall are stimulated, and they signal the micturition center in the sacral spinal cord.
3. Parasympathetic nerve impulses travel to the detrusor muscle, which responds by contracting rhythmically.
4. The need to urinate is urgent.
5. Voluntary contraction of the external urethral sphincter and inhibition of the micturition reflex by impulses from the brainstem and the cerebral cortex prevent urination.
6. Following the decision to urinate, the external urethral sphincter is relaxed, and impulses from the pons and the hypothalamus facilitate the micturition reflex.
7. The detrusor muscle contracts, and urine is expelled through the urethra.
8. Neurons of the micturition reflex center fatigue, the detrusor muscle relaxes, and the bladder begins to fill with urine again.

TABLE 20.6 | Developmental Abnormalities of the Urinary System

Condition	Description
Crossed fused ectopia	Fused kidneys that lie on one side of the midline
Horseshoe kidney	Fusion of kidneys at one pole, usually lower, with most of each kidney on opposing side of midline
Nephrotic syndrome	Proteinuria (protein in urine) due to abnormal glomeruli
Oligomeganephronia	Reduced number of nephrons that are abnormally large
Polycystic kidney disease	Cysts form in renal tubules and/or collecting ducts
Renal agenesis	Absence of a kidney
Renal dysplasia	Abnormal kidney structure
Renal hypoplasia	Small kidney with fewer nephrons, but development normal
Tubular dysgenesis	Abnormal formation of proximal tubules
Vesicoureteral reflux	Urine backs up from bladder to ureter or kidney

20.4 CLINICAL APPLICATION

Urinalysis: Clues to Health

Urine has long fascinated medical minds. As a folk remedy, urine has been used as a mouthwash, toothache treatment, and a cure for sore eyes. Hippocrates (460–377 B.C.) was the first to observe that the condition of the urine can reflect health, noting that frothy urine denoted kidney disease. During the Middle Ages, health practitioners consulted charts that matched urine colors to diseases. In the seventeenth century, British physicians diagnosed diabetes by having their medical students taste sugar in patients' urine. Today, urine composition is still used as a window on health and also to check for illicit drug use.

Certain inherited disorders can noticeably alter urine. The name *maple syrup urine disease* vividly describes what this inborn error of metabolism does to the urine. This condition, which causes mental retardation, results from a block in the breakdown pathways for certain amino acids. In *alkaptonuria,* urine turns black when it is left to stand. This condition also produces painful arthritis and blackened ear tips. People with *Wilson disease* have an inherited inability to excrete copper. If they are properly diagnosed and given the drug penicillamine, their urine becomes the color of copper.

Other genetic conditions alter urine without causing health problems. People with beeturia excrete dark pink urine after they eat beets. The problem for people with *urinary excretion of odoriferous component* of asparagus is obvious. Parents of newborns who have inherited blue diaper syndrome are shocked when they change their child's first diaper. Due to a defect in transport of the amino acid tryptophan in the small intestine, bacteria degrade the partially digested tryptophan, producing a compound that turns blue on contact with oxygen. ■

PRACTICE

35 Describe micturition.

36 How is it possible to consciously inhibit the micturition reflex?

Incontinence is the loss of control of micturition. Stress incontinence, caused by pressure on the bladder, is particularly common among women who have had children, especially if they have gained weight. An effective treatment is at least two months of doing Kegel exercises, in which a woman contracts the muscles that support the bladder, several times daily. Treatments for severe cases include a tamponlike cone inserted into the vagina to raise the pelvic floor; a small foam pad placed over the urethra to catch small amounts of urine; collagen injections around the urethra to tighten it; and surgery. Many people use absorbent pads.

Nighttime bedwetting was noted as long ago as 1500 B.C. Treatments have ranged from drinking the broth from boiled hens' combs, to blocking the urethra at night, to punishment and ridicule. In many cases, this *nocturnal enuresis* is inherited. Drug treatment and pads to absorb urine help to manage the problem in children, who usually outgrow the condition.

20.5 LIFE-SPAN CHANGES

The urinary system is sufficiently redundant, in both structure and function, to mask aging-related changes. However, the kidneys become slower to remove nitrogenous wastes and toxins and to compensate for changes that might alter homeostasis.

From the outside, the kidneys change with age, appearing scarred and grainy as arterioles serving the cortex constrict and fibrous connective tissue accumulates around the capsules. On the inside, kidney cells begin to die as early as age twenty years, but the gradual shrinkage is not generally noticeable until after age forty. By eighty years, the kidneys have lost about a third of their mass.

Kidney shrinkage is largely due to the gradual loss of glomeruli—they may atrophy, cease functioning, become blocked with fibrous connective tissue, or untwist. About 5% of glomeruli are abnormal by age forty; 37% are abnormal by age ninety. The progressive shut down of glomeruli decreases the surface area for filtration, and as a result, glomerular filtration rate (GFR) begins to drop in the fourth decade. By age seventy-five, GFR is about half that in a young adult, falling from about 125 milliliters/minute to about 60. With this decline, proteins are more likely to enter the urine. About a third of the elderly have proteinuria.

Further along the nephron, renal tubules thicken, accumulating fatty coats. They may shorten, forming small outpouches as cell death disrupts their sleek symmetry. Urine may become more dilute as reabsorption of sodium and glucose and other molecules becomes less efficient. The renal tubules also slow in processing certain drugs, which remain in the circulation longer. It becomes harder to clear nonsteroidal anti-inflammatory drugs such as aspirin, as well as opiates, antibiotics, urea, uric acid, creatinine, and various toxins. Therefore, a person's age should be considered when prescribing drugs.

Cardiovascular changes slow the journey of blood through the kidneys. A college student's kidneys may process about a fourth of the cardiac output, or about 1,200 milliliters, per minute. Her eighty-year-old grandfather's kidneys can handle about half that volume. Starting at about age twenty, renal blood flow rate diminishes by about 1% per year. The blood vessels that serve the kidneys become

INNERCONNECTIONS | *Urinary System*

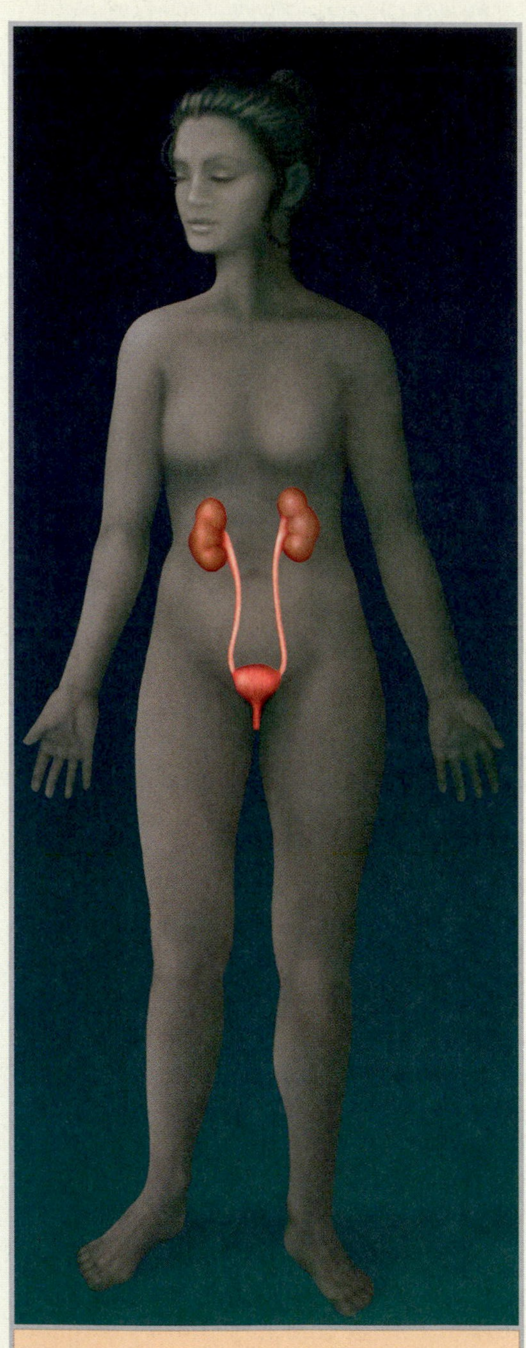

Urinary System

The urinary system controls the composition of the internal environment.

Integumentary System

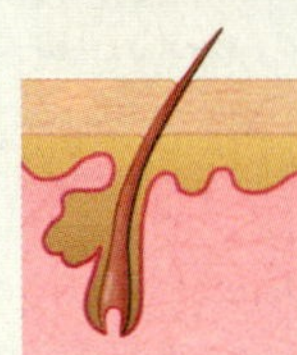

The urinary system compensates for water loss due to sweating. The kidneys and skin both play a role in vitamin D production.

Skeletal System

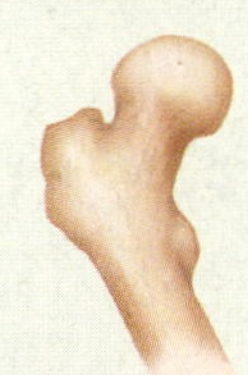

The kidneys and bone tissue work together to control plasma calcium levels.

Muscular System

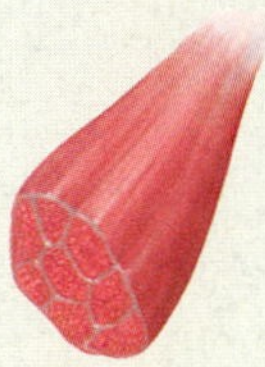

Muscle tissue controls urine elimination from the bladder. Kidneys excrete creatinine, produced by muscle metabolism.

Nervous System

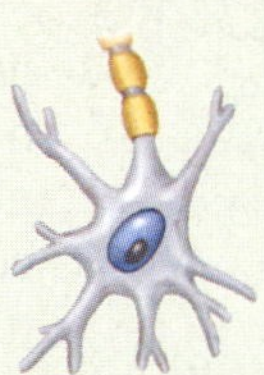

The nervous system influences urine production and elimination.

Endocrine System

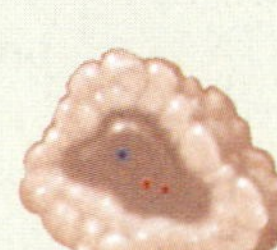

The endocrine system influences urine production.

Cardiovascular System

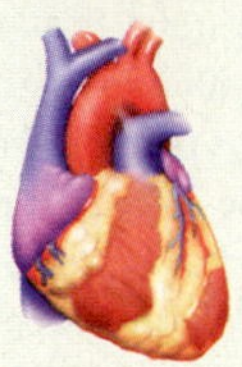

The urinary system controls blood volume. Blood volume and blood pressure play a role in determining water and solute excretion.

Lymphatic System

The kidneys control extracellular fluid volume and composition (including lymph).

Digestive System

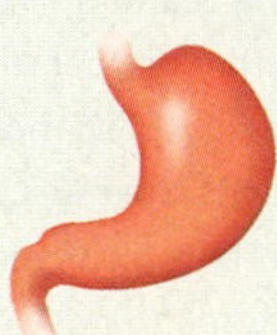

The kidneys compensate for fluids lost by the digestive system.

Respiratory System

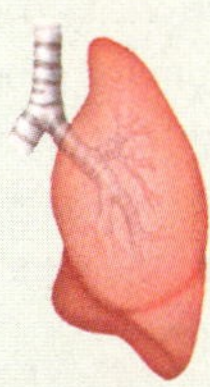

The kidneys and the lungs work together to control the pH of the internal environment.

Reproductive System

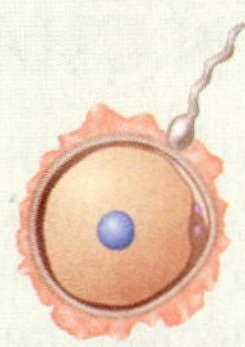

The urinary system in males shares organs with the reproductive system. The kidneys compensate for fluids lost from the male and female reproductive systems.

slower to dilate or constrict in response to body conditions. The kidneys' release of renin declines, hampering control of osmotic pressure, blood pressure, and sodium and potassium ion concentrations in the blood. The kidneys are also less able to activate vitamin D, which may contribute to the higher prevalence of osteoporosis among the elderly.

The urinary bladder, ureters, and urethra lose elasticity and recoil with age. In the later years, the bladder holds less than half of what it did in young adulthood, and may retain more urine after urination. In the elderly, the urge to urinate may become delayed, so when it does happen, it is sudden. Older individuals have to urinate at night more than younger people.

Controlling bladder function is a challenge at the beginning of life and much later too. A child usually learns to control urination by about age two or three years. Incontinence becomes more common in advanced years, although it is not considered a normal part of aging. It results from loss of muscle tone in the bladder, urethra and ureters. Incontinence affects 15% to 20% of women over sixty-five and half of all men. In women, incontinence reflects the stresses of childbirth and the effects of less estrogen during menopause. Bladder sphincter muscles atrophy, muscles in the pelvic floor weaken, and muscle tone of the urethra wanes. In males, incontinence usually is a response to an enlarged prostate gland pressing on the bladder.

PRACTICE

37 How do the kidneys change in appearance with advancing years?

38 What happens to glomeruli as a person ages?

39 How does kidney function change with age?

40 How do aging-related changes in the cardiovascular system affect the kidneys?

41 How do the urinary bladder, ureters, and urethra change with age?

CHAPTER SUMMARY

20.1 INTRODUCTION (PAGE 775)

The urinary system maintains homeostasis by regulating the composition, pH, and volume of body fluids. It does this by producing urine. The urinary system includes the kidneys, ureters, urinary bladder, and urethra.

20.2 KIDNEYS (PAGE 776)

1. Location of the kidneys
   a. The kidneys are bean-shaped organs on either side of the vertebral column, high on the posterior wall of the abdominal cavity.
   b. They are posterior to the parietal peritoneum and anchored by adipose and connective tissues.
2. Kidney structure
   a. A kidney has a hollow renal sinus.
   b. The ureter expands into the renal pelvis, which, in turn, leads to the major and minor calyces.
   c. Renal papillae project into the minor calyces.
   d. Kidney tissue is divided into a medulla and a cortex.
3. Functions of the kidneys
   a. The kidneys remove metabolic wastes from the blood and excrete them.
   b. They also help regulate red blood cell production, blood pressure, calcium ion absorption, and the volume, composition, and pH of the blood.
4. Renal blood vessels
   a. Arterial blood flows through the renal artery, interlobar arteries, arcuate arteries, cortical radiate arteries, afferent arterioles, glomerular capillaries, efferent arterioles, and peritubular capillaries.
   b. Venous blood returns through a series of vessels that correspond to those of the arterial pathways.
5. Nephrons
   a. Structure of a nephron
      (1) A nephron is the functional unit of the kidney.
      (2) It consists of a renal corpuscle and a renal tubule.
         (a) The corpuscle consists of a glomerulus and a glomerular capsule.
         (b) Parts of the renal tubule include the proximal convoluted tubule, the nephron loop (ascending and descending limbs), and the distal convoluted tubule.
      (3) The nephron joins a collecting duct, which empties into a minor calyx.
   b. Juxtaglomerular apparatus
      (1) The juxtaglomerular apparatus is between the ascending limb of the nephron loop and the afferent and efferent arterioles.
      (2) It consists of the macula densa and the juxtaglomerular cells.
   c. Cortical and juxtamedullary nephrons
      (1) Cortical nephrons are the most numerous and have corpuscles near the surface of the kidney.
      (2) Juxtamedullary nephrons have corpuscles near the medulla.
   d. Blood supply of a nephron
      (1) The glomerular capillary receives blood from the afferent arteriole and passes it to the efferent arteriole.
      (2) The efferent arteriole gives rise to the peritubular capillary system, which surrounds the renal tubule.
      (3) Capillary loops, called vasa recta, dip down into the medulla.

20.3 URINE FORMATION (PAGE 785)

Nephrons remove wastes from the blood and regulate water and electrolyte concentrations. Urine is the product.

1. Glomerular filtration
   a. Urine formation begins when water and dissolved materials are filtered out of the glomerular capillary.
   b. The glomerular capillaries are much more permeable than the capillaries in other tissues.
2. Filtration pressure
   a. Filtration is mainly due to hydrostatic pressure inside the glomerular capillaries.
   b. The osmotic pressure of the blood plasma and hydrostatic pressure in the glomerular capsule also affect filtration.
   c. Filtration pressure is the net force moving material out of the glomerulus and into the glomerular capsule.
   d. The composition of the filtrate is similar to that of tissue fluid.
3. Filtration rate
   a. The rate of filtration varies with the filtration pressure.
   b. Filtration pressure changes with the diameters of the afferent and efferent arterioles.
   c. As the osmotic pressure in the glomerulus increases, filtration decreases.
   d. As the hydrostatic pressure in a glomerular capsule increases, the filtration rate decreases.
   e. The kidneys produce about 125 milliliters of glomerular fluid per minute; most is reabsorbed.
   f. The volume of filtrate varies with the surface area of the glomerular capillary.
4. Control of filtration rate
   a. Glomerular filtration rate (GFR) remains relatively constant but may increase or decrease with need. Increased sympathetic nerve activity can decrease GFR.
   b. When tubular fluid NaCl concentration decreases, the macula densa causes the juxtaglomerular cells to release renin. This triggers a series of changes leading to vasoconstriction, which may affect GFR, and secretion of aldosterone, which stimulates tubular sodium reabsorption.
   c. Autoregulation is the ability of an organ or tissue to maintain a constant blood flow under certain conditions when the arterial blood pressure is changing.
5. Tubular reabsorption
   a. Substances are selectively reabsorbed from the glomerular filtrate.
   b. The peritubular capillary is adapted for reabsorption.
      (1) It carries low-pressure blood.
      (2) It is permeable.
   c. Most reabsorption is in the proximal tubule, where epithelial cells have microvilli.
   d. Different modes of transport reabsorb various substances in particular segments of the renal tubule.
      (1) Glucose and amino acids are reabsorbed by active transport.
      (2) Water is reabsorbed by osmosis.
      (3) Proteins are reabsorbed by endocytosis.
   e. Active transport mechanisms have limited capacities.
   f. If the concentration of a substance in the filtrate exceeds its renal plasma threshold, the excess is excreted in the urine.
   g. Substances that remain in the filtrate are concentrated as water is reabsorbed.
   h. Sodium ions are reabsorbed by active transport.
      (1) Negatively charged ions accompany positively charged sodium ions out of the filtrate.
      (2) Water is passively reabsorbed by osmosis as sodium ions are actively reabsorbed.
6. Tubular secretion
   a. Tubular secretion transports certain substances from the plasma to the tubular fluid.
   b. Some substances are actively secreted.
      (1) These include various organic compounds and hydrogen ions.
      (2) The proximal and distal convoluted tubules secrete hydrogen ions.
   c. Potassium ions are secreted actively and passively in the distal convoluted tubule and collecting duct.
7. Regulation of urine concentration and volume
   a. Most of the sodium ions are reabsorbed before the urine is excreted.
   b. The countercurrent mechanism concentrates sodium ions in the renal medulla.
      (1) Sodium and chloride ions are actively reabsorbed in the ascending limb.
      (2) Tubular fluid in the ascending limb becomes hypotonic as it loses solutes.
      (3) Water leaves the descending limb by osmosis, and NaCl enters this limb by diffusion.
      (4) Tubular fluid in the descending limb becomes hypertonic as it loses water and gains NaCl.
      (5) As NaCl repeats this circuit, its concentration in the medulla increases.
   c. The vasa recta countercurrent mechanism helps maintain NaCl concentration in the medulla.
   d. The distal convoluted tubule and collecting duct are impermeable to water, which therefore is excreted in urine.
   e. ADH from the posterior pituitary gland increases the permeability of the distal convoluted tubule and collecting duct, promoting water reabsorption.
8. Urea and uric acid excretion
   a. Urea is a by-product of amino acid metabolism.
      (1) It is passively reabsorbed by diffusion.
      (2) About 50% of the urea is excreted in urine.
      (3) A countercurrent mechanism involving urea helps reabsorb water.
   b. Uric acid results from the metabolism of nucleic acids.
      (1) Most is reabsorbed by active transport.
      (2) Some is secreted into the renal tubule.
9. Urine composition
   a. Urine is about 95% water, and it usually contains urea, uric acid, and creatinine.
   b. It may contain a trace of amino acids and varying amounts of electrolytes, depending upon diet.
   c. The volume of urine varies with the fluid intake and with certain environmental factors.

10. Renal clearance
    a. Renal clearance is the rate at which a chemical is removed from the plasma.
    b. The inulin clearance test, creatinine clearance test, and para-aminohippuric acid test can be used to calculate GFR.

20.4 ELIMINATION OF URINE (PAGE 798)

1. Ureters
    a. The ureter is a tubular organ that extends from each kidney to the urinary bladder.
    b. Its wall has mucous, muscular, and fibrous layers.
    c. Peristaltic waves in the ureter force urine to the urinary bladder.
    d. Obstruction in the ureter stimulates strong peristaltic waves and a reflex that decreases urine production.
2. Urinary bladder
    a. The urinary bladder is a distensible organ that stores urine and forces it into the urethra.
    b. The ureters and urethra open at the three angles of the trigone in the floor of the urinary bladder.
    c. Muscle fibers in the wall form the detrusor muscle.
    d. A portion of the detrusor muscle forms an internal urethral sphincter.
3. Urethra
    a. The urethra conveys urine from the urinary bladder to the outside.
    b. In females, it empties between the labia minora.
    c. In males, it conveys products of reproductive organs as well as urine.
        (1) Three portions of the male urethra are prostatic, membranous, and penile.
        (2) The urethra empties at the tip of the penis.
4. Micturition
    a. Micturition is the process of expelling urine.
    b. In micturition, the detrusor muscle contracts and the external urethral sphincter relaxes.
    c. Micturition reflex
        (1) Distension stimulates stretch receptors in the urinary bladder wall.
        (2) The micturition reflex center in the sacral portion of the spinal cord sends parasympathetic motor impulses to the detrusor muscle.
        (3) As the urinary bladder fills, its internal pressure increases, forcing the internal urethral sphincter open.
        (4) A second reflex relaxes the external urethral sphincter, unless its contraction is voluntarily controlled.
        (5) Nerve centers in the brainstem and cerebral cortex aid control of urination.

20.5 LIFE-SPAN CHANGES (PAGE 803)

The kidneys, ureters, and urethra change with age, but nephrons are so numerous that a healthy person is usually unaware of kidney shrinkage and slowed cleansing of the blood.

1. With age, the kidneys appear grainy and scarred.
2. GFR drops significantly with age as glomeruli atrophy, fill with connective tissue, or unwind.
3. Renal tubules accumulate fat on their outsides and become asymmetric. Reabsorption and secretion may slow or become impaired. Drugs remain longer in the circulation as a person ages.
4. Changes in the cardiovascular system slow the rate of processing through the urinary system. The kidneys slow in their response to changes, and are less efficient at activating vitamin D.
5. The urinary bladder, ureters, and urethra lose elasticity, with effects on the urge and timing of urination.

CHAPTER ASSESSMENTS

20.1 Introduction

1 Explain why the urinary system is necessary for survival. (p. 775)

2 Identify the organs of the urinary system and list their general functions. (p. 775)

20.2 Kidneys

3 Describe the external and internal structure of a kidney. (p. 776)

4 List the functions of the kidneys. (p. 777)

5 List in correct order the vessels through which blood passes as it travels from the renal artery to the renal vein. (p. 779)

6 Distinguish between a renal corpuscle and a renal tubule. (p. 781)

7 Name in correct order the structures through which fluid passes from the glomerulus to the collecting duct. (p. 781)

8 Describe the location and structure of the juxtaglomerular apparatus. (p. 782)

9 Distinguish between cortical and juxtamedullary nephrons. (p. 784)

20.3 Urine Formation

10 Distinguish among filtration, tubular reabsorption, and tubular secretion as they relate to urine formation. (p. 785)

11 Which one of the following is abundant in blood plasma, but present only in small amounts in glomerular filtrate? (p. 788)

a. sodium ions
b. water
c. glucose
d. protein
e. potassium ions

12 Define *filtration pressure.* (p. 788)

13 Explain how the diameters of the afferent and efferent arterioles affect the rate of glomerular filtration. (p. 788)

14 Explain how changes in the osmotic pressure of blood plasma affect the glomerular filtration rate. (p. 789)

15 Explain how the hydrostatic pressure of a glomerular capsule affects the rate of glomerular filtration. (p. 789)

16 Define *autoregulation.* p. 790)

17 Describe the two mechanisms by which the body regulates glomerular filtration rate. (p. 790)

18 Discuss how tubular reabsorption is selective. (p. 791)

19 Explain how the peritubular capillary is adapted for tubular reabsorption. (p. 791)

20 Explain how epithelial cells of the proximal convoluted tubule are adapted for tubular reabsorption. (p. 791)

21 Explain why active transport mechanisms have limited transport capacities. (p. 791)

22 Define *renal plasma threshold,* and explain its significance in tubular reabsorption. (p. 792)

23 Explain how amino acids and proteins are reabsorbed. (p. 792)

24 Describe the effect of sodium reabsorption on the reabsorption of negatively charged ions. (p. 792)

25 Explain how sodium reabsorption affects water reabsorption. (p. 792)

26 Explain how the renal tubule is adapted to secrete hydrogen ions. (p. 793)

27 Explain how potassium ions may be passively secreted. (p. 794)

28 Explain how hypotonic fluid is produced in the ascending limb of the nephron loop. (p. 795)

29 Explain why fluid in the descending limb of the nephron loop is hypertonic. (p. 795)

30 The major action of ADH on the kidneys is to increase ____________. (p. 796)

a. water reabsorption by the proximal convoluted tubule
b. glomerular filtration rate
c. water reabsorption by the collecting duct
d. potassium excretion

31 Explain how urine may become concentrated as it moves through the collecting duct. (p. 796)

32 Compare the processes by which urea and uric acid are absorbed. (p. 796)

33 List the common constituents of urine and their sources. (p. 798)

34 List some of the factors that affect the daily urine volume. (p. 798)

20.4 Elimination of Urine

35 Describe the structure and function of a ureter. (p. 798)

36 Explain how the muscular wall of the ureter helps move urine. (p. 798)

37 Describe what happens if a ureter becomes obstructed. (p. 799)

38 Describe the location and structure of the urinary bladder. (p. 799)

39 Define *detrusor muscle.* (p. 800)

40 Distinguish between the internal and external urethral sphincters. (p. 800)

41 Compare the urethra of a female with that of a male. (p. 800)

42 Describe the micturition reflex. (p. 802)

43 Which movement involves skeletal muscle? (p. 802)

a. contraction of the internal urethral sphincter
b. contraction of the external urethral sphincter
c. ureteral peristalsis
d. detrusor muscle contraction

20.5 Life-Span Changes

44 Describe changes in the urinary system with age. (p. 803)

INTEGRATIVE ASSESSMENTS/CRITICAL THINKING

OUTCOMES 15.5, 20.2, 20.3

1. If the blood pressure of a patient in shock as a result of a severe injury decreases greatly, how would you expect the volume of urine to change? Why?

OUTCOMES 15.7, 15.8, 20.2, 20.3, 20.4

2. A physician prescribes oral pencillin therapy for a patient with an infection of the urinary bladder. How would you describe for the patient the route the drug follows to reach the bladder?

OUTCOMES 16.8, 20.2, 20.3

3. Why are people with the nephrotic syndrome, in which plasma proteins are lost into the urine, more susceptible to infection?

OUTCOMES 16.8, 20.4

4. Inflammation of the urinary bladder is more common in women than in men. What anatomical differences between the female and male urethra explain this observation?

OUTCOMES 20.2, 20.3

5. What effect would being born with narrowed renal arteries have on the volume and composition of urine?

OUTCOMES 20.2, 20.3, 20.4

6. If a patient who has had major abdominal surgery receives intravenous fluids equal to the volume of blood lost during surgery, would you expect the volume of urine produced to be greater than or less than normal? Why?

WEB CONNECTIONS

Be sure to visit the text website at **www.mhhe.com/shier12** for answers to chapter assessments, additional quizzes, and interactive learning exercises.

ANATOMY & PHYSIOLOGY REVEALED

Anatomy & Physiology Revealed® (APR) includes cadaver photos that allow you to peel away layers of the human body to reveal structures beneath the surface. This program also includes animations, radiologic imaging, audio pronunciations, and practice quizzing. Check out **www.aprevealed.com**. APR has been proven to help improve student grades!

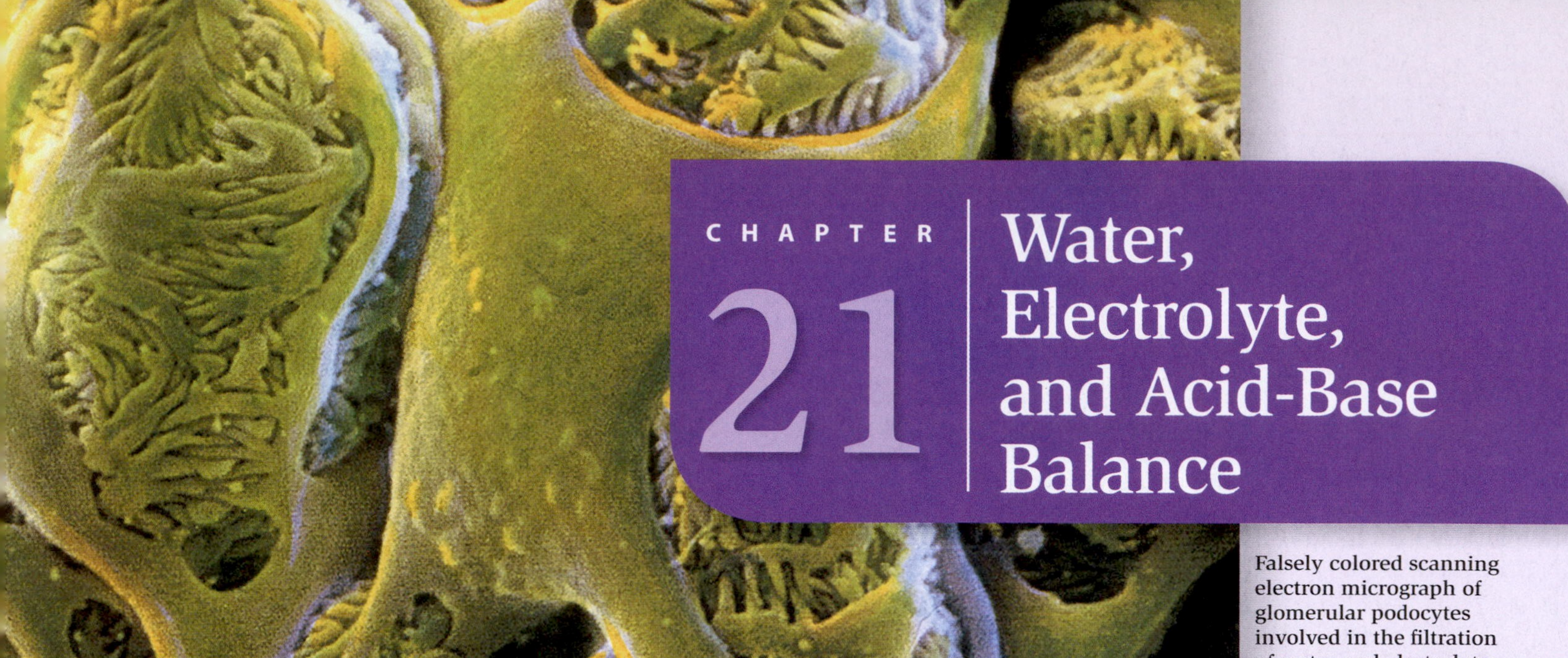

CHAPTER 21 Water, Electrolyte, and Acid-Base Balance

Falsely colored scanning electron micrograph of glomerular podocytes involved in the filtration of water and electrolytes (5,700×).

UNDERSTANDING WORDS

de-, separation from: *de*hydration—removal of water from the cells or body fluids.

edem-, swelling: *edema*—swelling due to abnormal accumulation of extracellular fluid.

-emia, a blood condition: hypoprotein*emia*—abnormally low concentration of blood plasma proteins.

extra-, outside: *extra*cellular fluid—fluid outside of the body cells.

im- (or in-), not: *im*balance—condition in which factors are not in equilibrium.

intra-, within: *intra*cellular fluid—fluid in body cells.

neutr-, neither one nor the other: *neutr*al—solution that is neither acidic nor basic.

-osis, a state of: acid*osis*—condition in which hydrogen ion concentration is abnormally high.

-uria, a urine condition: keto*uria*—ketone bodies in the urine.

LEARN PRACTICE ASSESS

LEARNING OUTCOMES

After you have studied this chapter, you should be able to:

21.1 Introduction

1 Explain the balance concept. (p. 811)

2 Explain the importance of water and electrolyte balance. (p. 811)

21.2 Distribution of Body Fluids

3 Describe how body fluids are distributed in compartments. (p. 811)

4 Explain how fluid composition varies among compartments and how fluids move from one compartment to another. (p. 812)

21.3 Water Balance

5 List the routes by which water enters and leaves the body. (p. 813)

6 Explain the regulation of water input and water output. (p. 815)

21.4 Electrolyte Balance

7 List the routes by which electrolytes enter and leave the body. (p. 815)

8 Explain the regulation of the input and output of electrolytes. (p. 818)

21.5 Acid-Base Balance

9 Explain acid-base balance. (p. 819)

10 Identify how pH number describes the acidity and alkalinity of a body fluid. (p. 819)

11 List the major sources of hydrogen ions in the body. (p. 819)

12 Distinguish between strong acids and weak acids. (p. 820)

13 Explain how chemical buffer systems, the respiratory center, and the kidneys keep the pH of body fluids relatively constant. (p. 821)

21.6 Acid-Base Imbalances

14 Describe the causes and consequences of increase or decrease in body fluid pH. (p. 824)

DISCOVERING THE UNDERPINNINGS OF HEATSTROKE

Heatstroke is a response to extreme environmental heat that can be quickly fatal. It occurs when the body is exposed to a heat index (heat considering humidity) of more than 105°F and body temperature exceeds 106°F. Under these conditions, evaporation of sweat becomes less efficient at cooling the body, and organs begin to fail.

The symptoms of heatstroke happen in a sequence. First come headache, dizziness, and exhaustion. Sweating is profuse, then stops, as the skin becomes dry, hot, and red. Respiratory rate rises and the pulse may race up to 180 beats per minute. If the person isn't cooled with fluids, water applied to the skin, fanning, and removal of clothing, neurological symptoms may begin, including disorientation, hallucinations, and odd behavior. Kidney failure and/or heart arrhythmia prove fatal.

During heat waves, the very young and the very old are more susceptible to heatstroke because their temperature control mechanisms may be poor. However, heatstroke also affects two groups of young, otherwise healthy individuals—athletes who work out in extreme heat and soldiers deployed to hot climates. In the Persian Gulf in July, the temperature may soar to 122°F at midday, dipping down only to about 100° at night. Given these conditions it isn't surprising that some cases of unexpected sudden death in soldiers in Iraq and Afghanistan are due to heatstroke. For this reason, military officials insist that soldiers carry drinking water with them at all times and drink throughout the day, whether they feel thirsty or not.

Clues to the cause of heatstroke have come from an unexpected source—mutant mice bred to have a version of a disease that also affects humans, called malignant hyperthermia. In this condition, a patient receiving general anesthesia experiences increased heart rate, whole-body muscle rigidity, and a spike in body temperature up to 112°F. Affected mice die upon exposure to anesthesia—and also when subjected to 105° heat. The mutant gene encodes a receptor that admits calcium ions into skeletal muscle cells. Under intense heat, the receptors allow too many calcium ions to enter, causing uncontrolled contraction. In addition, calcium ions leak from the receptor area and free radicals are released, which causes oxidative damage. Adding an antioxidant drug to the animals' water dampened their reaction to heat.

The experiments done with mice that were genetically susceptible to heatstroke suggest that people, particularly those with malignant hyperthermia, might be predisposed to react dangerously to heat, too. Researchers are using the mice as models to test treatments for heatstroke—but it is still wise to avoid extreme heat and get medical attention swiftly if the initial symptoms of heatstroke appear. ■

21.1 INTRODUCTION

The term *balance* suggests a state of equilibrium. For water and electrolytes, balance means that the quantities entering the body equal the quantities leaving. Mechanisms that replace lost water and electrolytes and excrete excesses maintain this balance. As a result, the levels of water and electrolytes in the body remain relatively stable at all times.

Water balance and electrolyte balance are interdependent, because electrolytes are dissolved in the water of body fluids. Consequently, anything that alters the concentrations of the electrolytes will alter the concentration of the water by adding solutes to it or by removing solutes from it. Likewise, anything that changes the concentration of the water will change the concentrations of the electrolytes by concentrating or diluting them.

PRACTICE 

1 How are fluid balance and electrolyte balance interdependent?

21.2 DISTRIBUTION OF BODY FLUIDS

Body fluids are not uniformly distributed. Instead, they occupy regions, or *compartments,* of different volumes that contain fluids of varying compositions. The movement of water and electrolytes between these compartments is regulated to stabilize their distribution and the composition of body fluids.

Fluid Compartments

The body of an average adult female is about 52% water by weight, and that of an average male is about 63% water. This difference between the sexes is because females generally have more adipose tissue, which has little water. Males have more muscle tissue, which contains a great deal of water. Water in the body (about 40 liters), with its dissolved electrolytes, is distributed into two major compartments: an intracellular fluid compartment and an extracellular fluid compartment (fig. 21.1).

The **intracellular** (in″trah-sel′u-lar) **fluid compartment** includes all the water and electrolytes that cell membranes enclose. In other words, intracellular fluid is the fluid inside cells, and, in an adult, it accounts for about 63% by volume of total body water.

RECONNECT
To Chapter 1, Homeostasis, page 9.

The **extracellular** (ek″strah-sel′u-lar) **fluid compartment** includes all the fluid outside cells—in tissue spaces (interstitial fluid), blood vessels (plasma), and lymphatic vessels (lymph). Epithelial layers separate a specialized fraction of the extracellular fluid from other extracellular fluids. This *transcellular* (trans-sel′ular) *fluid* includes cerebrospinal fluid of the central nervous system, aqueous and vitreous humors of the eyes, synovial fluid of the joints, serous fluid in the body cavities, and fluid secretions of the exocrine glands. The fluids of the extracellular compartment constitute about 37% by volume of the total body water (fig. 21.2).

Body Fluid Composition

Extracellular fluids generally are similar in composition, including high concentrations of sodium, chloride, calcium, and bicarbonate ions and lesser concentrations of potassium, magnesium, phosphate, and sulfate ions. The blood plasma fraction of extracellular fluid has considerably more protein than do either interstitial fluid or lymph.

Intracellular fluid has high concentrations of potassium, phosphate, and magnesium ions. It includes a greater concentration of sulfate ions and lesser concentrations of sodium, chloride, and bicarbonate ions than does extracellular fluid. Intracellular fluid also has a greater concentration of protein than plasma. Figure 21.3 shows these relative concentrations.

PRACTICE 

2 Describe the normal distribution of water in the body.

3 Which electrolytes are in higher concentrations in extracellular fluids? In intracellular fluid?

4 How does the concentration of protein vary in the various body fluids?

FIGURE 21.1 Of the 40 liters of water in the body of an average adult male, about two-thirds is intracellular, and one-third is extracellular.

Movement of Fluid Between Compartments

Two major factors regulate the movement of water and electrolytes from one fluid compartment to another: hydrostatic pressure and osmotic pressure. For example, as explained in chapter 15 (p. 579), fluid leaves the plasma at the arteriolar ends of capillaries and enters the interstitial spaces because of the net outward force of *hydrostatic pressure* (blood pressure). Fluid returns to the plasma from the interstitial spaces at the venular ends of capillaries because of the net inward force of *colloid osmotic pressure.* Likewise, as mentioned in chapter 16 (p. 620), fluid leaves the interstitial spaces and enters the lymph capillaries due to the hydrostatic pressure of the interstitial fluid. The circulation of lymph returns interstitial fluid to the plasma.

Hydrostatic pressure in the cells and surrounding interstitial fluid is ordinarily equal and remains stable, so any net fluid movement is likely to be the result of changes in osmotic pressure (fig. 21.4). Recall that osmotic pressure is due to impermeant solutes on one side of a cell membrane. Sodium (extracellular) and potassium (intracellular) ions function as impermeant solutes and create an osmotic pressure because of the Na^+/K^+ pump. For example, because most cell membranes in the body are freely permeable to water, a decrease in extracellular sodium ion concentration causes a net movement of water from the extracellular compartment into the intracellular compartment by osmosis. The cell swells. Conversely, if the extracellular sodium

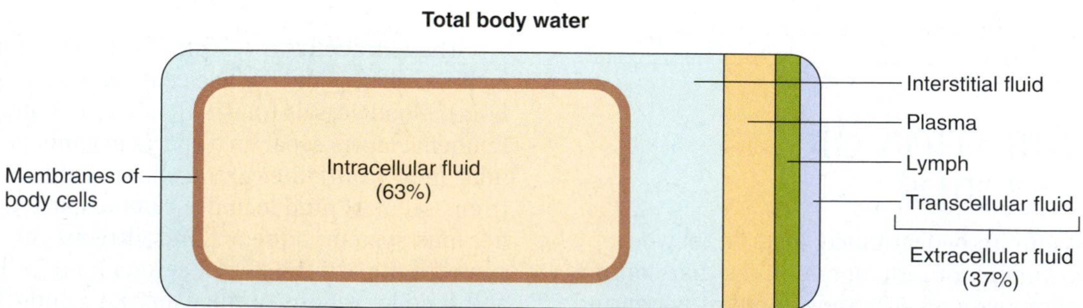

FIGURE 21.2 Cell membranes separate fluid in the intracellular compartment from fluid in the extracellular compartment. Approximately two-thirds of the water in the body is inside cells.

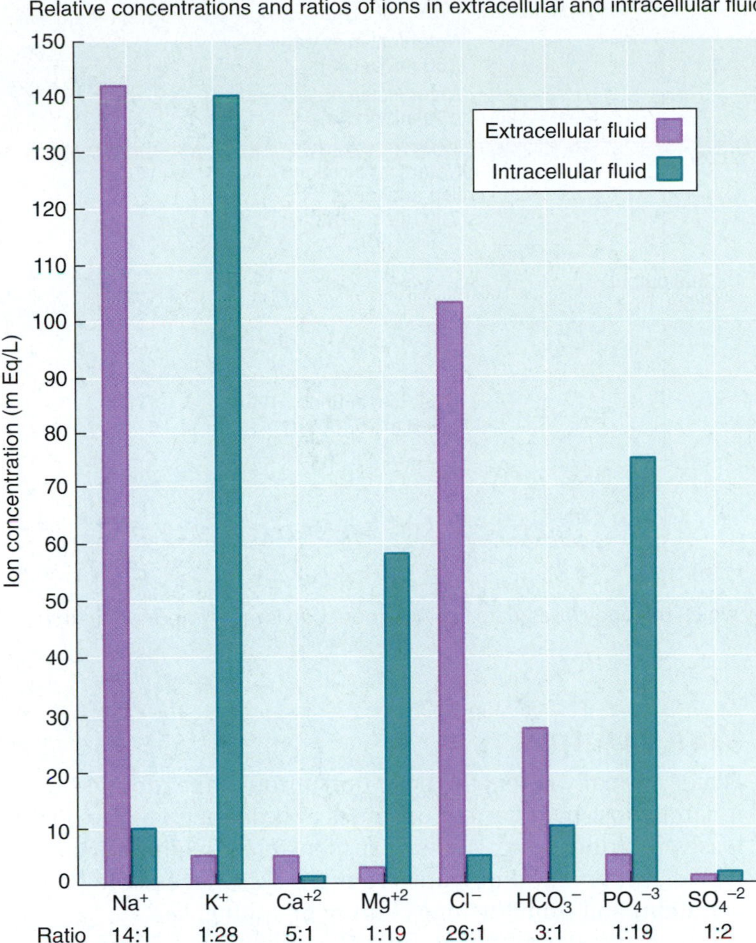

FIGURE 21.3 Extracellular fluids have relatively high concentrations of sodium (Na^+), calcium (Ca^{+2}), chloride (Cl^-), and bicarbonate (HCO_3^-) ions. Intracellular fluid has relatively high concentrations of potassium (K^+), magnesium (Mg^{+2}), phosphate (PO_4^{-3}), and sulfate (SO_4^{-2}) ions.

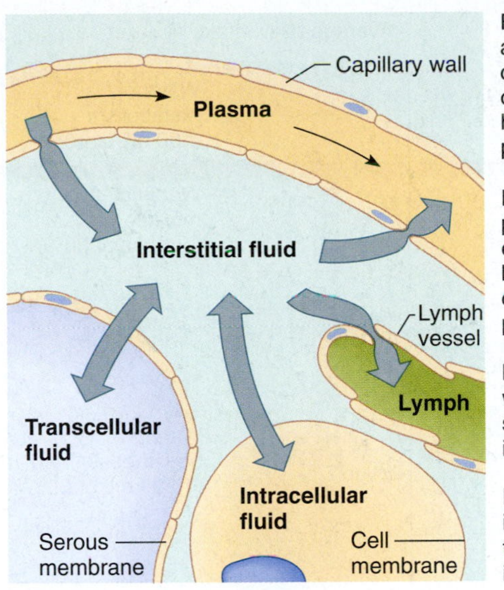

FIGURE 21.4 Net movements of fluid between compartments result from differences in hydrostatic and osmotic pressures.

PRACTICE

5 Which factors control the movement of water and electrolytes from one fluid compartment to another?

6 How does the sodium ion concentration in body fluids affect the net movement of water between the compartments?

ion concentration increases, cells shrink as they lose water. Although the solute composition of body fluids varies between intracellular and extracellular compartments, water will "follow salt" and distribute by osmosis such that the water concentration (and total solute concentration) is essentially equal inside and outside cells.

Different substances may be distributed to different compartments. For example, an infusion of 1 liter of isotonic sodium chloride solution is restricted largely to the extracellular fluid because of the active transport sodium pumps in cell membranes. In contrast, a liter of isotonic glucose solution may be given intravenously without damaging red blood cells, but as the glucose is metabolized aerobically, it reacts to release carbon dioxide and water. Thus, the liter of isotonic glucose yields a liter of water that can be distributed throughout intracellular and extracellular compartments.

21.3 WATER BALANCE

Water balance exists when water intake equals water output. Homeostasis requires control of both water intake and water output. Ultimately, maintenance of the internal environment depends on thirst centers in the brain to vary water intake and on the kidneys' ability to vary water output.

Water Intake

The volume of water gained each day varies among individuals. An average adult living in a moderate environment takes in about 2,500 milliliters daily. Probably 60% is obtained from drinking water or beverages, and another 30% comes from moist foods. The remaining 10% is a by-product of the oxidative metabolism of nutrients, called **water of metabolism** (fig. 21.5*a*).

Regulation of Water Intake

The primary regulator of water intake is thirst. The intense feeling of thirst derives from the osmotic pressure of extracellular fluids and a *thirst center* in the hypothalamus of the brain.

As the body loses water, the osmotic pressure of the extracellular fluids increases. Such a change stimulates *osmoreceptors* (oz″mo-re-sep′torz) in the thirst center, and

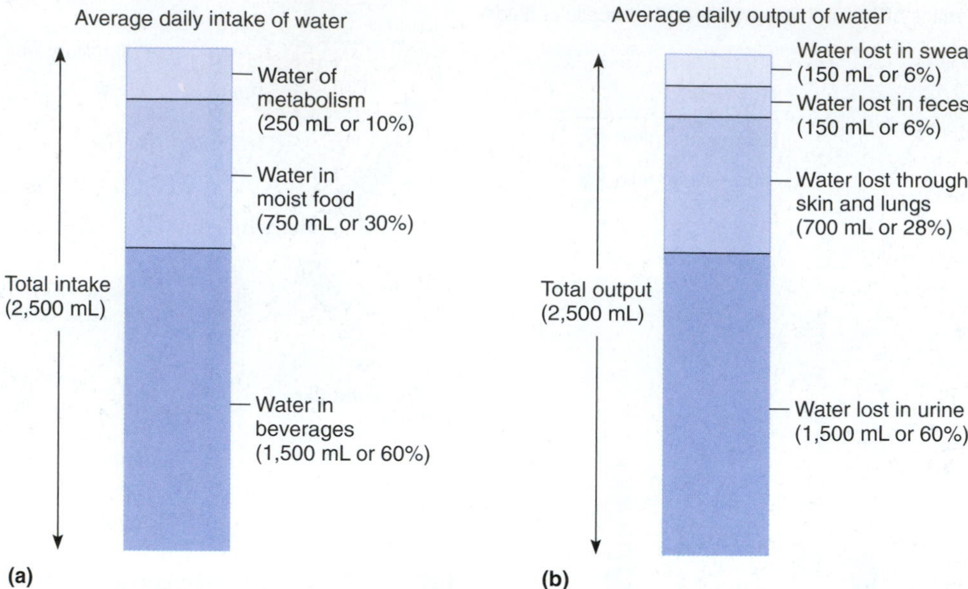

FIGURE 21.5 Water balance. (*a*) Major sources of body water. (*b*) Routes by which the body loses water. Urine production is most important in the regulation of water balance.

in response, the hypothalamus causes the person to feel thirsty and to seek water. A thirsty person usually has a dry mouth, caused by loss of extracellular water and the resulting decreased flow of saliva.

The thirst mechanism is normally triggered whenever the total body water decreases by as little as 1%. The act of drinking and the resulting distension of the stomach wall trigger nerve impulses that inhibit the thirst mechanism. Thus, drinking stops long before the swallowed water is absorbed. This inhibition helps prevent the person from drinking more than is required to replace the volume lost, avoiding development of an imbalance. Table 21.1 summarizes this mechanism.

PRACTICE

7 What is water balance?

8 Where is the thirst center?

9 What stimulates fluid intake? What inhibits it?

TABLE 21.1 | Regulation of Water Intake

1. The body loses as little as 1% of its water.
2. An increase in the osmotic pressure of extracellular fluid due to water loss stimulates osmoreceptors in the thirst center.
3. Activity in the hypothalamus causes the person to feel thirsty and to seek water.
4. Drinking and the resulting distension of the stomach by water stimulate nerve impulses that inhibit the thirst center.
5. Water is absorbed through the walls of the stomach and small intestine.
6. The osmotic pressure of extracellular fluid returns to normal.

Water Output

Water normally enters the body only through the mouth, but it can be lost by a variety of routes. These include obvious losses in urine, feces, and sweat (sensible perspiration), as well as evaporation of water from the skin (insensible perspiration) and from the lungs during breathing.

If an average adult takes in 2,500 milliliters of water each day, then 2,500 milliliters must be eliminated to maintain water balance. Of this volume, perhaps 60% will be lost in urine, 6% in feces, and 6% in sweat. About 28% will be lost by evaporation from the skin and lungs (fig. 21.5*b*). These percentages vary with such environmental factors as temperature and relative humidity and with physical exercise.

If water intake is insufficient, water output must be reduced to maintain balance. Water lost in sweat is a necessary part of the body's temperature control mechanism; water lost in feces accompanies the elimination of undigested food materials; and water lost by evaporation is largely unavoidable. Therefore, the primary means of regulating water output is control of urine production.

Proteins called *aquaporins* form water-selective membrane channels that enable certain cells, including red blood cells and cells in the proximal convoluted tubules and descending limbs of the nephron loops, to admit water. A mutation in one aquaporin gene (which instructs cells to manufacture a type of aquaporin protein) causes a form of *diabetes insipidus,* in which the renal tubules fail to reabsorb water. Rare individuals who lack other aquaporin genes apparently have no symptoms. This suggests that cells have more than one way to admit water.

Regulation of Water Output

The distal convoluted tubules and collecting ducts of the nephrons regulate the volume of water excreted in the urine. The epithelial linings of these segments of the renal tubule remain relatively impermeable to water unless antidiuretic hormone (ADH) is present.

Recall from chapter 13 (p. 498) that osmoreceptors in the hypothalamus help control release of ADH. If the blood plasma becomes more concentrated because of excessive water loss, the osmoreceptors lose water by osmosis and shrink. This change triggers impulses that signal the posterior pituitary gland to release ADH. The ADH released into the bloodstream reaches the kidneys, where it increases the permeability of the distal convoluted tubules and collecting ducts. Consequently, water reabsorption increases, conserving water. This action resists further osmotic change in the plasma. The *osmoreceptor-ADH mechanism* can reduce a normal urine production of 1,500 milliliters per day to about 500 milliliters per day when the body is dehydrated.

If a person drinks too much water, the plasma becomes less concentrated, and the osmoreceptors swell as they receive extra water by osmosis. In this instance, ADH release is inhibited, and the distal tubules and collecting ducts remain impermeable to water. Consequently, less water is reabsorbed and more urine produced. Table 21.2 summarizes this mechanism. Clinical Application 21.1 discusses disorders resulting from water imbalance.

> *Diuretics* are chemicals that promote urine production. They act in different ways. Alcohol and certain narcotic drugs promote urine formation by inhibiting ADH release. Caffeine inhibits the reabsorption of sodium ions or other solutes in parts of the renal tubules. As a consequence, the osmotic pressure of the tubular fluid increases, reducing osmotic reabsorption of water and increasing urine volume.

PRACTICE

10 By what routes does the body lose water?

11 What is the primary regulator of water loss?

12 What types of water loss are unavoidable?

13 How does the hypothalamus regulate water balance?

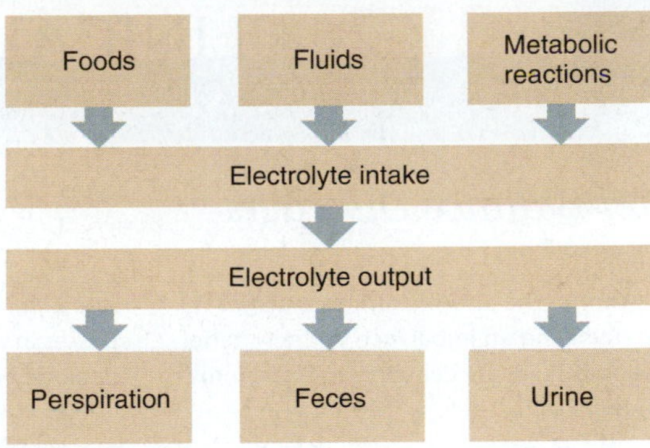

FIGURE 21.6 Electrolyte balance exists when the intake of electrolytes from all sources equals the output of electrolytes.

21.4 ELECTROLYTE BALANCE

An **electrolyte balance** (e-lek′tro-līt bal′ans) exists when the quantities of electrolytes (molecules that release ions in water) the body gains equal those lost (fig. 21.6).

Electrolyte Intake

The electrolytes of greatest importance to cellular functions release sodium, potassium, calcium, magnesium, chloride, sulfate, phosphate, bicarbonate, and hydrogen ions. These electrolytes are primarily obtained from foods, but they may also be found in drinking water and other beverages. In addition, some electrolytes are by-products of metabolic reactions.

Regulation of Electrolyte Intake

Ordinarily, a person obtains sufficient electrolytes by responding to hunger and thirst. However, a severe electrolyte deficiency may cause *salt craving*, a strong desire to eat salty foods.

Electrolyte Output

The body loses some electrolytes by perspiring (sweat has about half the solute concentration of plasma). The quantities of electrolytes leaving vary with the amount of

TABLE 21.2 | Events in Regulation of Water Output

Dehydration	Excess Water Intake
1. Extracellular fluid becomes osmotically more concentrated.	1. Extracellular fluid becomes osmotically less concentrated.
2. Osmoreceptors in the hypothalamus are stimulated by the increase in the osmotic pressure of body fluids.	2. This change stimulates osmoreceptors in the hypothalamus.
3. The hypothalamus signals the posterior pituitary gland to release ADH into the blood.	3. The posterior pituitary gland decreases ADH release.
4. Blood carries ADH to the kidneys.	4. Renal tubules decrease water reabsorption.
5. ADH causes the distal convoluted tubules and collecting ducts to increase water reabsorption.	5. Urine output increases, and excess water is excreted.
6. Urine output decreases, and further water loss is minimized.	

21.1 CLINICAL APPLICATION

Water Balance Disorders

Among the more common disorders involving an imbalance in the water of body fluids are dehydration, water intoxication, and edema.

Dehydration

In 1994, thousands of starving people died in the African nation of Rwanda. It wasn't lack of food that killed most of these people, but cholera, a bacterial infection that cripples the ability of intestinal mucosal cells to reabsorb water. The severe diarrhea that develops can kill in days, sometimes even hours. The dehydration it causes is deadly.

Dehydration is a deficiency condition that occurs when output of water exceeds intake. It is a great problem for athletes, military personnel, and certain industrial workers. Dehydration may develop following excessive sweating or as a result of prolonged water deprivation accompanied by continued water output. In either case, as water is lost, the extracellular fluid becomes more concentrated, and water leaves cells by osmosis (fig. 21A). Dehydration may also accompany illnesses in which prolonged vomiting or diarrhea depletes body fluids.

During dehydration, the skin and mucous membranes of the mouth feel dry, and body weight drops. Severe hyperthermia may develop as the body temperature regulating mechanism falters due to lack of water for sweat. In severe dehydration, as waste products accumulate in the extracellular fluid, symptoms of cerebral disturbances, including mental confusion, delirium, and coma, may develop. Dehydration is responsible for some of the symptoms of heatstroke, discussed in this chapter's opening vignette (p. 811).

Infants are more likely to become dehydrated because their kidneys are less efficient at conserving water than those of adults. Elderly people are also especially susceptible to developing water imbalances because the sensitivity of their thirst mechanisms decreases with age, and physical disabilities may make it difficult to obtain adequate fluids.

The treatment for dehydration is to replace the lost water and electrolytes. If only water is replaced, the extracellular fluid will become more dilute than normal. This may produce a condition called water intoxication.

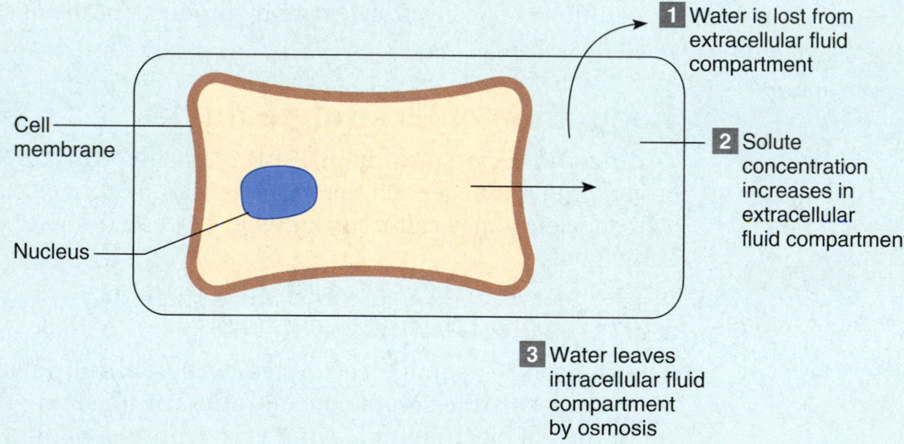
1 Water is lost from extracellular fluid compartment
Cell membrane
2 Solute concentration increases in extracellular fluid compartment
Nucleus
3 Water leaves intracellular fluid compartment by osmosis

FIGURE 21A If excess extracellular fluids are lost, cells dehydrate by osmosis.

Water Intoxication

Until recently, runners were advised to drink as much fluid as they could, particularly in long events. But the death of a young woman in the 2002 Boston Marathon from low blood sodium (*hyponatremia,* "water intoxication") due to excessive fluid intake, inspired further study and a reevaluation of this advice. Researchers from Harvard Medical School studied 488 runners from the race, and found that 13% of them developed hyponatremia. Tendency to develop the condition was associated with longer race time, high or low body mass index, and significant weight gain during the race. Drinking sports drinks instead of water does not make a difference because these beverages are mostly water.

In recognition of the possibility of hyponatremia, USA Track and Field, the national governing body for the sport, offers on their website (www.usatf.org) instructions for runners to determine exactly how much to consume during a one-hour training run. The goal is to replace exactly what is lost.

Edema

Edema is an abnormal accumulation of extracellular fluid in the interstitial spaces (fig. 21B). Causes include a decrease in the plasma protein concentration (*hypoproteinemia*), obstructions in lymphatic vessels, increased venous pressure, and increased capillary permeability.

Hypoproteinemia may result from failure of the liver to synthesize plasma proteins; kidney

perspiration. More electrolytes are lost in sweat on warmer days and during strenuous exercise. Varying amounts of electrolytes are lost in the feces. The greatest electrolyte output occurs as a result of kidney function and urine production. The kidneys alter renal electrolyte losses to maintain the proper composition of body fluids, thereby promoting homeostasis.

Recall from chapter 2 (p. 57) that water molecules are polar. Molecules that have polar regions, such as carbohydrates and proteins, dissolve in water but remain intact. In contrast, ionically bonded molecules, such as electrolytes, dissociate in water to release ions.

The total solute concentration of a body fluid determines its *osmolarity*. One molecule of glucose yields one dissolved

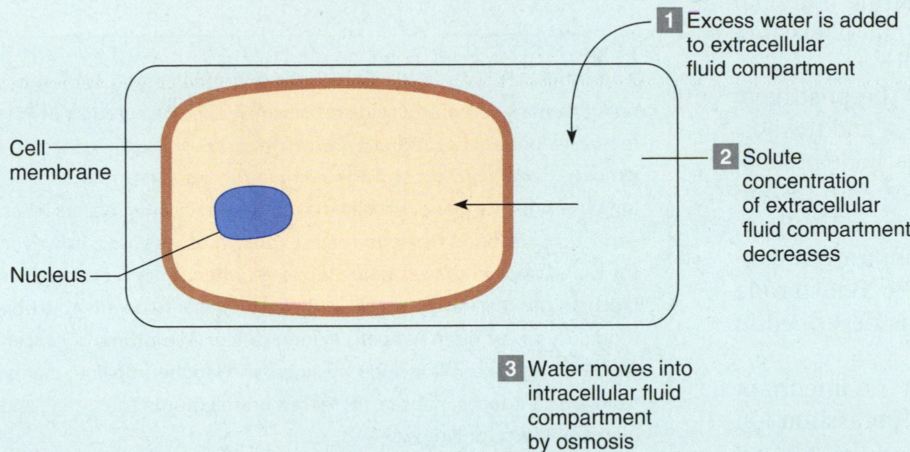
1 Excess water is added to extracellular fluid compartment
Cell membrane
2 Solute concentration of extracellular fluid compartment decreases
Nucleus
3 Water moves into intracellular fluid compartment by osmosis

FIGURE 21B If excess water is added to the extracellular fluid compartment, cells gain water by osmosis.

disease (glomerulonephritis) that damages glomerular capillaries, allowing proteins to enter the urine; or starvation, in which amino acid intake is insufficient to support synthesis of plasma proteins. In each of these instances, the plasma protein concentration is decreased, which decreases plasma osmotic pressure, reducing the normal return of tissue fluid to the venular ends of capillaries. Consequently, tissue fluid accumulates in the interstitial spaces.

As discussed in chapter 16 (p. 621), *lymphatic* obstructions may result from surgery or from parasitic infections of lymphatic vessels. Back pressure develops in the lymphatic vessels, which interferes with the normal movement of tissue fluid into them. At the same time, proteins that the lymphatic circulation ordinarily removes accumulate in the interstitial spaces, raising osmotic pressure of the interstitial fluid. This effect attracts still more fluid into the interstitial spaces.

If the outflow of blood from the liver into the inferior vena cava is blocked, the venous pressure in the liver and portal blood vessels greatly increases. This, in turn, raises pressure in liver sinusoids and intestinal capillaries. Fluid with a high concentration of protein is exuded from the surfaces of the liver and intestine into the peritoneal cavity. This elevates the osmotic pressure of the abdominal fluid, which, in turn, attracts more water into the peritoneal cavity by osmosis. This condition, called *ascites*, distends the abdomen. It is painful.

Edema may also result from increased capillary permeability accompanying *inflammation*. Recall that inflammation is a response to tissue damage and usually releases chemicals such as histamine from damaged cells. Histamine causes vasodilation and increased capillary permeability, so excess fluid leaks out of the capillary and enters the interstitial spaces. Table 21A summarizes the factors that result in edema. ■

TABLE 21A | Factors Associated with Edema

Factor	Cause	Effect
Low plasma protein concentration	Liver disease and failure to synthesize proteins; kidney disease and loss of proteins in urine; lack of proteins in diet due to starvation	Plasma osmotic pressure decreases; less fluid enters venular ends of capillaries by osmosis
Obstruction of lymph vessels	Surgical removal of portions of lymphatic pathways; certain parasitic infections	Back pressure in lymph vessels interferes with movement of fluid from interstitial spaces into lymph capillaries
Increased venous pressure	Venous obstructions or faulty venous valves	Back pressure in veins increases capillary filtration and interferes with return of fluid from interstitial spaces into venular ends of capillaries
Inflammation	Tissue damage	Capillaries become abnormally permeable; fluid leaks from plasma into interstitial spaces

particle, and one molecule of sodium chloride yields two, a sodium ion and a chloride ion. The total number of dissolved particles determines the osmolarity of body solutions, irrespective of the source, so the term *osmoles* is used. Thus, one mole of glucose yields one osmole of dissolved particles, and one mole of sodium chloride yields two osmoles. The total number of osmoles per liter gives the osmolarity of the solution.

PRACTICE

14 Which electrolytes are most important to cellular functions?
15 Which mechanisms ordinarily regulate electrolyte intake?
16 By what routes does the body lose electrolytes?

Regulation of Electrolyte Output

The concentrations of positively charged ions, such as sodium (Na^+), potassium (K^+), and calcium (Ca^{+2}), are particularly important. Certain concentrations of these ions are vital for nerve impulse conduction, muscle fiber contraction, and maintenance of cell membrane permeability. Potassium is especially important in maintaining the resting potential of nerve and cardiac muscle cells, and abnormal potassium levels may disrupt the functioning of these cells.

Sodium ions account for nearly 90% of the positively charged ions in extracellular fluids. The kidneys and the hormone aldosterone primarily regulate these ions. Aldosterone, which the adrenal cortex secretes, increases sodium ion reabsorption in the distal convoluted tubules and collecting ducts of the nephrons. A decrease in sodium ion concentration in the extracellular fluid stimulates aldosterone secretion via the renin-angiotensin system, as described in chapter 20 (p. 790 and fig. 20.20).

Aldosterone also regulates *potassium ions.* An important stimulus for aldosterone secretion is a rising potassium ion concentration, which directly stimulates cells of the adrenal cortex. This hormone enhances the renal tubular reabsorption of sodium ions and, at the same time, stimulates renal tubular secretion of potassium ions (fig. 21.7).

Recall from chapter 13 (p. 502) that the calcium ion concentration dropping below normal directly stimulates the parathyroid glands to secrete parathyroid hormone. Parathyroid hormone increases activity in bone-resorbing cells (osteoclasts), which increases the concentrations of both calcium and phosphate ions in the extracellular fluids. Parathyroid hormone also indirectly stimulates calcium absorption from the intestine. Concurrently, this hormone causes the kidneys to conserve calcium ions (through increased tubular reabsorption) and increases the urinary excretion of phosphate ions. The increased phosphate excretion offsets the increased plasma phosphate. Thus, the net effect of the hormone is to return the *calcium ion* concentration of the extracellular fluid to normal levels but to maintain a normal *phosphate ion* concentration (fig. 21.8).

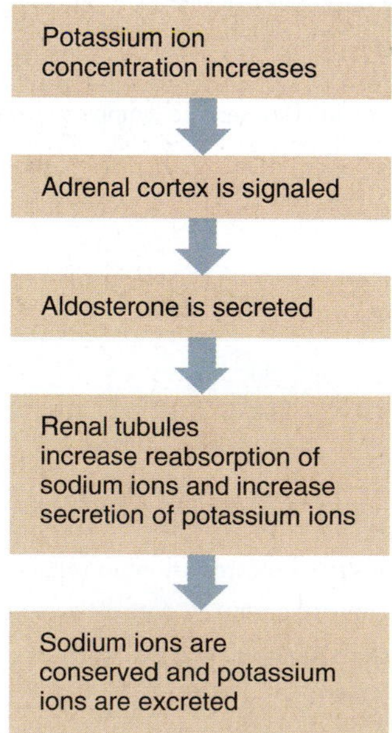

FIGURE 21.7 If the potassium ion concentration increases, the kidneys conserve sodium ions and excrete potassium ions.

> Abnormal increases in blood calcium (hypercalcemia) sometimes result from hyperparathyroidism, in which excess secretion of PTH increases bone resorption. Hypercalcemia may also be caused by cancers, particularly those originating in the bone marrow, breasts, lungs, or prostate gland. Usually the increase in calcium occurs when cancer causes bone tissue to release ions. In other cases, however, the blood calcium concentration increases when cancer cells produce biochemicals that have physiological effects similar to parathyroid hormone. This most often happens in lung cancer. Symptoms of cancer-induced hypercalcemia include weakness and fatigue, impaired mental function, headache, nausea, increased urine volume (polyuria), and increased thirst (polydipsia).
>
> Abnormal decreases in blood calcium (hypocalcemia) may result from reduced availability of PTH following removal of the parathyroid glands, or from vitamin D deficiency, which may result from decreased absorption following gastrointestinal surgery or excess excretion due to kidney disease. Hypocalcemia may be life threatening if it causes muscle spasms in the airways and cardiac arrhythmias. Administering calcium salts and high doses of vitamin D to promote calcium absorption can correct this condition.

Generally, the regulatory mechanisms that control positively charged ions secondarily control the concentrations of negatively charged ions. For example, chloride ions (Cl^-), the most abundant negatively charged ions in the extracellular fluids, are passively reabsorbed from the renal tubules in response to the active reabsorption of sodium ions. That is, the negatively charged chloride ions are electrically attracted to the positively charged sodium ions and accompany them as they are reabsorbed.

Some negatively charged ions, such as phosphate ions (PO_4^{-3}) and sulfate ions (SO_4^{-2}), also are partially regulated by active transport mechanisms that have limited transport capacities. Thus, if the extracellular phosphate ion concentration is low, the phosphate ions in the renal tubules are conserved. On the other hand, if the renal plasma threshold is exceeded, the excess phosphate will be excreted in the urine. Clinical Application 21.2 discusses symptoms associated with sodium and potassium imbalances.

PRACTICE

17 How does aldosterone regulate sodium and potassium ion concentration?

18 How is calcium regulated?

19 What mechanism regulates the concentrations of most negatively charged ions?

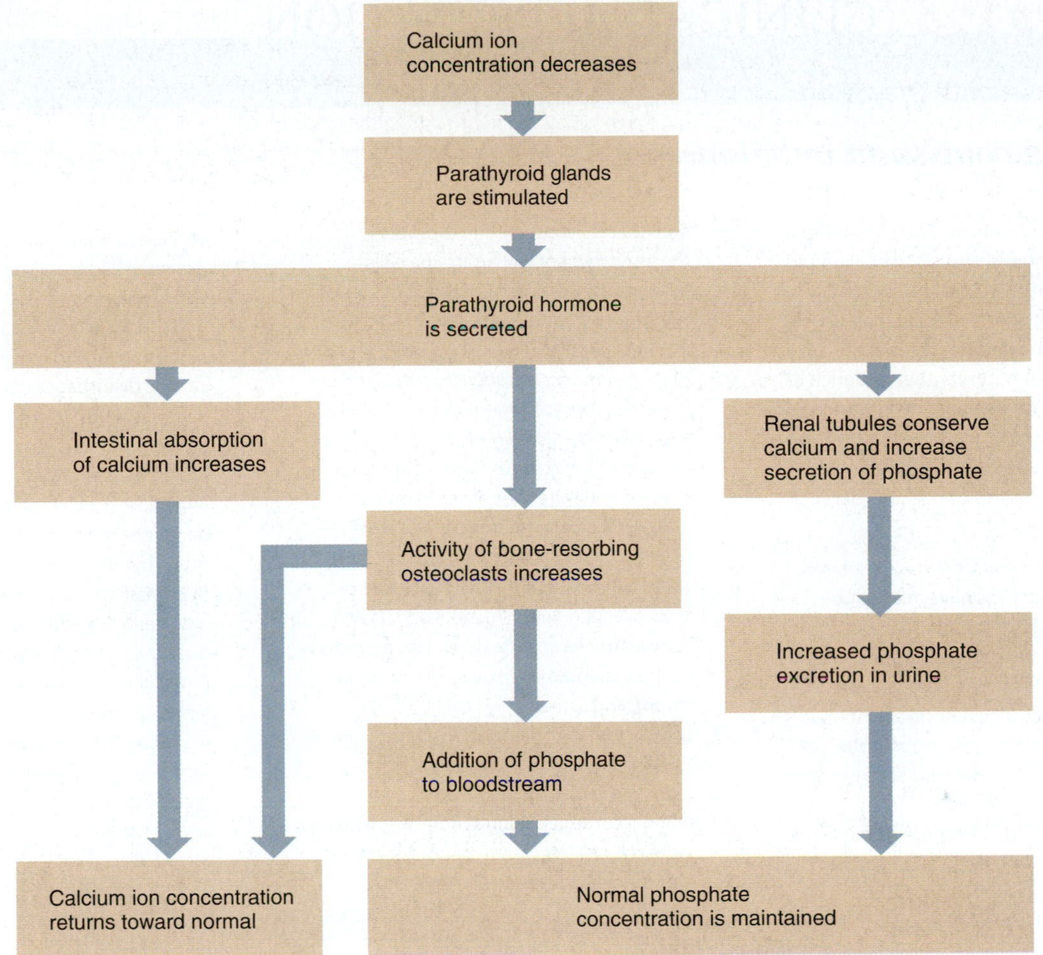

FIGURE 21.8 If the concentration of calcium ions decreases, parathyroid hormone increases calcium ion concentration. Increased urinary phosphate excretion offsets bone resorption (which increases blood phosphate levels) to maintain a normal concentration of phosphate ions.

21.5 ACID-BASE BALANCE

As discussed in chapter 2 (p. 59), electrolytes that ionize in water and release hydrogen ions are **acids.** Substances that combine with hydrogen ions are **bases.** Acid-base balance entails regulation of the hydrogen ion concentration of body fluids. This is important because slight changes in hydrogen ion concentrations can alter the rates of enzyme-controlled metabolic reactions, shift the distribution of other ions, or modify hormone actions. Recall from chapter 2 that pH number indicates the degree to which a solution is acidic or basic (alkaline). The more acid the solution, the lower its pH, and vice versa. The internal environment is normally maintained between pH 7.35 and 7.45.

Sources of Hydrogen Ions

Most of the hydrogen ions in body fluids originate as by-products of metabolic processes, although the digestive tract may directly absorb some hydrogen ions. The major metabolic sources of hydrogen ions include the following. (These are reversible reactions but, for clarity, are presented as the net reaction only. Remember, it is the concentration of H^+ at equilibrium that determines the pH.)

1. **Aerobic respiration of glucose.** This process produces carbon dioxide and water. Carbon dioxide diffuses out of the cells and reacts with water in the extracellular fluids to form *carbonic acid:*

$$CO_2 + H_2O \rightarrow H_2CO_3$$

   The resulting carbonic acid then ionizes to release hydrogen ions and bicarbonate ions:

$$H_2CO_3 \rightarrow H^+ + HCO_3^-$$

2. **Anaerobic respiration of glucose.** Glucose metabolized anaerobically produces *lactic acid,* which adds hydrogen ions to body fluids.
3. **Incomplete oxidation of fatty acids.** The incomplete oxidation of fatty acids produces *acidic ketone bodies,* which increase hydrogen ion concentration.
4. **Oxidation of amino acids containing sulfur.** The oxidation of sulfur-containing amino acids yields *sulfuric acid* (H_2SO_4), which ionizes to release hydrogen ions.

21.2 CLINICAL APPLICATION

Sodium and Potassium Imbalances

Extracellular fluids usually have high sodium ion concentrations, and intracellular fluid usually has high potassium ion concentration. The renal regulation of sodium is closely related to that of potassium because active reabsorption of sodium (under the influence of aldosterone) is accompanied by secretion (and excretion) of potassium. Thus, it is not surprising that conditions that alter sodium ion balance also affect potassium ion balance. Such disorders can be summarized as follows:

1. *Low sodium concentration* (*hyponatremia*). Possible causes of hyponatremia include prolonged sweating, vomiting, or diarrhea; renal disease in which sodium is inadequately reabsorbed; adrenal cortex disorders in which aldosterone secretion is insufficient to promote the reabsorption of sodium (Addison disease); and drinking too much water.

   Possible effects of hyponatremia include the development of extracellular fluid that is hypotonic and promotes the movement of water into the cells by osmosis. This is accompanied by the symptoms of water intoxication described in Clinical Application 21.1 on page 816.
2. *High sodium concentration* (*hypernatremia*). Possible causes of hypernatremia include excessive water loss by evaporation and diffusion, as may occur during high fever, or increased water loss accompanying diabetes insipidus, in one form of which ADH secretion is insufficient to maintain water conservation by the renal tubules and collecting ducts. Possible effects of hypernatremia include disturbances of the central nervous system, such as confusion, stupor, and coma.
3. *Low potassium concentration* (*hypokalemia*). Possible causes of hypokalemia include excessive release of aldosterone by the adrenal cortex (Cushing syndrome), which increases renal excretion of potassium; use of diuretic drugs that promote potassium excretion; kidney disease; and prolonged vomiting or diarrhea. Possible effects of hypokalemia include muscular weakness or paralysis, respiratory difficulty, and severe cardiac disturbances, such as atrial or ventricular arrhythmias.
4. *High potassium concentration* (*hyperkalemia*). Possible causes of hyperkalemia include renal disease, which decreases potassium excretion; use of drugs that promote renal conservation of potassium; insufficient secretion of aldosterone by the adrenal cortex (Addison disease); or a shift of potassium from the intracellular fluid to the extracellular fluid, a change that accompanies an increase in plasma hydrogen ion concentration (acidosis). Possible effects of hyperkalemia include paralysis of the skeletal muscles and severe cardiac disturbances, such as cardiac arrest. ■

5. **Breakdown (hydrolysis) of phosphoproteins and nucleic acids.** Phosphoproteins and nucleic acids contain phosphorus. Their oxidation produces *phosphoric acid* (H_3PO_4), which ionizes to release hydrogen ions.

The acids resulting from metabolism vary in strength. Thus, their effects on the hydrogen ion concentration of body fluids vary (fig. 21.9).

Strengths of Acids and Bases

Acids that ionize more completely (release more H^+) are strong acids, and those that ionize less completely are weak acids. For example, the hydrochloric acid (HCl) of gastric juice is a strong acid and dissociates completely to release a lot of H^+, but the carbonic acid (H_2CO_3) produced when carbon dioxide reacts with water is weak and dissociates less completely to release less H^+.

Bases release ions, such as hydroxide ions (OH^-), which can combine with hydrogen ions, thereby lowering their concentration. Thus, sodium hydroxide (NaOH), which releases hydroxide ions, and sodium bicarbonate ($NaHCO_3$), which releases bicarbonate ions ($HCO3^-$), are bases. Strong bases dissociate to release more OH^- or its equivalent than do weak bases. Often, the negative ions themselves are called bases. For example, HCO_3^- acting as a base combines with H^+ from the strong acid HCl to form the weak acid carbonic acid (H_2CO_3).

Regulation of Hydrogen Ion Concentration

Either an acid shift or an alkaline (basic) shift in the body fluids could threaten the internal environment. However, normal metabolic reactions generally produce more acid than base. These reactions include cellular metabolism of glucose, fatty acids, and amino acids. Consequently, the maintenance of acid-base balance usually eliminates acid. This is accomplished in three ways: acid-base buffer systems; respiratory excretion of carbon dioxide; and renal excretion of hydrogen ions.

PRACTICE 

20 Explain why the regulation of hydrogen ion concentration is so important.

21 What are the major sources of hydrogen ions in the body?

Acid-Base Buffer Systems

Acid-base buffer systems are in all body fluids and are based on chemicals that combine with excess acids or bases. Buffers are substances that stabilize the pH of a solution,

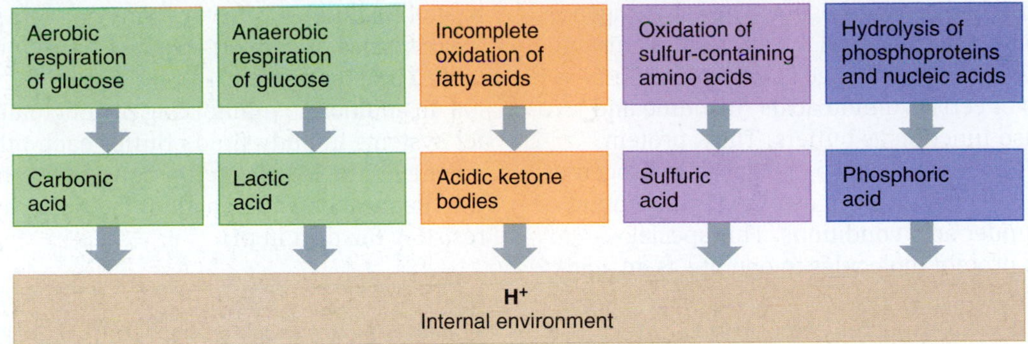

FIGURE 21.9 Some of the metabolic processes that provide hydrogen ions.

despite the addition of an acid or a base. More specifically, the chemical components of a buffer system can combine with strong acids to convert them into weak acids. Likewise, these buffers can combine with strong bases to convert them into weak bases. Such activity helps minimize pH changes in the body fluids. The three most important buffer systems in body fluids are the bicarbonate buffer system, the phosphate buffer system, and the protein buffer system.

In the following discussion, associated anions and cations have been omitted for clarity. For example, the weak base sodium bicarbonate ($NaHCO_3$) is represented by bicarbonate (HCO_3^-). Sodium is also the cation associated with the phosphate ions.

1. **Bicarbonate buffer system.** In the bicarbonate buffer system, present in both intracellular and extracellular fluids, the bicarbonate ion (HCO_3^-) acts as a weak base, and carbonic acid (H_2CO_3) acts as a weak acid. In the presence of excess hydrogen ions, bicarbonate ions combine with hydrogen ions to form carbonic acid, minimizing any increase in the hydrogen ion concentration of body fluids:

$$H^+ + HCO_3^- \rightarrow H_2CO_3$$

On the other hand, if conditions are basic or alkaline, carbonic acid dissociates to release bicarbonate ion and hydrogen ion:

$$H_2CO_3 \rightarrow H^+ + HCO_3^-$$

Although this reaction releases bicarbonate ion, it is the increase of free hydrogen ions at equilibrium that is important in minimizing the shift toward a more alkaline pH.

2. **Phosphate buffer system.** The phosphate buffer system is also present in both intracellular and extracellular fluids. However, it is particularly important in the control of hydrogen ion concentration in the intracellular fluid and in renal tubular fluid and urine. This buffer system consists of two phosphate ions, dihydrogen phosphate ($H_2PO_4^-$) and monohydrogen phosphate (HPO_4^{2-}).

In the presence of excess hydrogen ions, monohydrogen phosphate ions act as a weak base, combining with hydrogen ions to form dihydrogen phosphate, minimizing increase in the hydrogen ion concentration of body fluids.

$$H^+ + HPO_4^{-2} \rightarrow H_2PO_4^-$$

On the other hand, if conditions are basic or alkaline, dihydrogen phosphate, acting as a weak acid, dissociates to release hydrogen ion:

$$H_2PO_4^- \rightarrow H^+ + HPO_4^{-2}$$

3. **Protein buffer system.** The protein acid-base buffer system consists of the plasma proteins, such as albumins, and certain proteins in cells, including hemoglobin in red blood cells.

As described in chapter 2 (p. 64), proteins are chains of amino acids. Some of these amino acids have freely exposed groups of atoms, called carboxyl groups. If the H^+ concentration drops, a carboxyl group (—COOH) can become ionized, releasing a hydrogen ion, thus resisting the pH change:

$$\text{—COOH} \leftrightarrow \text{—COO}^- + H^+$$

This is a reversible reaction. (However, the degree to which it is reversible depends on the particular amino acids.) In the presence of excess hydrogen ions, the —COO⁻ parts of the protein molecules accept hydrogen ions and become —COOH groups again. This action decreases the number of free hydrogen ions in the body fluids and again minimizes the pH change.

Some of the amino acids of a protein molecule also have freely exposed amino groups (—NH_2). If the H^+ concentration rises, these amino groups can accept hydrogen ions in another reversible reaction (Once again, the degree to which it is reversible depends on the particular amino acids.):

$$\text{—}NH_2 + H^+ \leftrightarrow \text{—}NH_3^+$$

In the presence of excess hydroxyl ions (OH^-), the —NH_3^+ groups of protein molecules give up hydrogen ions and become —NH_2 groups again. These hydrogen ions then combine with hydroxyl ions to form water molecules. Once again, pH change is minimized. The "R"-groups of certain amino acids (histidine and cysteine) can also function as buffers. Thus, protein molecules can function as acids by releasing hydrogen ions under alkaline conditions or as bases by accepting hydrogen ions under acid conditions. This special property allows protein molecules to operate as an acid-base buffer system.

Hemoglobin is an especially important protein that buffers hydrogen ions. As explained in chapter 19 (p. 766), carbon dioxide, produced by cellular oxidation of glucose, diffuses through the capillary wall and enters the plasma and then the red blood cells. The red blood cells contain an enzyme, *carbonic anhydrase,* that speeds the reaction between carbon dioxide and water, producing carbonic acid:

$$CO_2 + H_2O \rightarrow H_2CO_3$$

The carbonic acid quickly dissociates, releasing hydrogen ions and bicarbonate ions:

$$H_2CO_3 \rightarrow H^+ + HCO_3^-$$

In the peripheral tissues, where CO_2 is generated, oxygen is used in the metabolism of glucose. Hemoglobin gives up much of its oxygen and is in the form of deoxyhemoglobin. In this form, hemoglobin can bind the hydrogen ions generated in red blood cells, thus acting as a buffer to minimize the pH change that would otherwise occur.

The above two reactions can be written as a single reversible reaction:

$$CO_2 + H_2O \leftrightarrow H_2CO_3 \leftrightarrow H^+ + HCO_3^-$$

Thus, in the peripheral tissues, where CO_2 levels are high, the reaction equilibrium shifts to the right, generating H^+, buffered by hemoglobin, and HCO_3^-, which becomes a plasma electrolyte. In the lungs, where oxygen levels are high, hemoglobin is no longer a good buffer, and it releases its H^+. However, the released H^+ combines with plasma HCO_3, shifting the reaction equilibrium to the left, generating carbonic acid, which quickly dissociates to form CO_2 and water. The water is added to the body fluids, and the CO_2 is exhaled. Carbonic acid is sometimes called a *volatile acid* because of this relationship to CO_2 (see figs. 19.41 and 19.43).

Individual amino acids in body fluids can also function as acid-base buffers by accepting or releasing hydrogen ions. This is possible because every amino acid has an amino group (—NH_2) and a carboxyl group (—COOH).

To summarize, acid-base buffer systems take up hydrogen ions when body fluids are becoming more acidic and give up hydrogen ions when the fluids are becoming more basic (alkaline). Buffer systems convert stronger acids into weaker acids or convert stronger bases into weaker bases, as table 21.3 summarizes.

In addition to minimizing pH fluctuations, acid-base buffer systems in body fluids buffer each other. Consequently, whenever the hydrogen ion concentration begins to change, the chemical balances in all of the buffer systems change too, resisting the drift in pH.

Neurons are sensitive to changes in the pH of body fluids. If the interstitial fluid becomes more alkaline than normal (alkalosis), neurons become more excitable, and seizures may result. Conversely, acidic conditions (acidosis) depress neuron activity and level of consciousness may decrease.

PRACTICE

22 What is the difference between a strong acid or base and a weak acid or base?

23 How does a chemical buffer system help regulate pH of body fluids?

24 List the major buffer systems of the body.

Chemical buffer systems only temporarily solve the problem of acid-base balance. Ultimately, the body must eliminate excess acid or base. The lungs (controlled by the respiratory center) and the kidneys accomplish this task.

Respiratory Excretion of Carbon Dioxide

The **respiratory center** in the brainstem helps regulate hydrogen ion concentrations in the body fluids by controlling the rate and depth of breathing. Specifically, if body cells increase their production of carbon dioxide, as occurs during periods of physical exercise, carbonic acid production increases. As the carbonic acid dissociates, the concentra-

TABLE 21.3 | Chemical Acid-Base Buffer Systems

Buffer System	Constituents	Actions
Bicarbonate system	Bicarbonate ion (HCO_3^-)	Converts a strong acid into a weak acid
	Carbonic acid (H_2CO_3)	Converts a strong base into a weak base
Phosphate system	Monohydrogen phosphate ion (HPO_4^{-2})	Converts a strong acid into a weak acid
	Dihydrogen phosphate ($H_2PO_4^-$)	Converts a strong base into a weak base
Protein system (and amino acids)	—NH_3^+ group of an amino acid or protein	Releases a hydrogen ion in the presence of excess base
	—COO^- group of an amino acid or protein	Accepts a hydrogen ion in the presence of excess acid

tion of hydrogen ions increases, and the pH of the internal environment drops (see chapter 19, pp. 757–758). Such an increasing concentration of carbon dioxide in the central nervous system and the subsequent increase in hydrogen ion concentration in the cerebrospinal fluid stimulate chemosensitive areas in the respiratory center.

In response, the respiratory center increases the depth and rate of breathing so that the lungs excrete more carbon dioxide. Hydrogen ion concentration in body fluids returns toward normal, because the released carbon dioxide is in equilibrium with carbonic acid and CO_2 (fig. 21.10):

$$CO_2 + H_2O \leftrightarrow H_2CO_3 \leftrightarrow H^+ + HCO_3^-$$

Conversely, if body cells are less active, concentrations of carbon dioxide and hydrogen ions in body fluids remain low. Breathing rate and depth fall. This increases the carbon dioxide level in the body fluids, returning pH to normal. If the pH drops below normal, the respiratory center is stimulated to increase the rate and depth of breathing.

Activity of the respiratory center, therefore, changes in response to shifts in the pH of the body fluids, reducing these shifts to a minimum. Most of the hydrogen ions in the body fluids originate from carbonic acid produced when carbon dioxide reacts with water, so the respiratory regulation of hydrogen ion concentration is important.

Renal Excretion of Hydrogen Ions

Nephrons help regulate the hydrogen ion concentration of body fluids by excreting hydrogen ions in the urine. Recall from chapter 20 (p. 793) that the epithelial cells lining the proximal and distal convoluted tubules and the collecting ducts secrete these ions into the tubular fluid. The tubular secretion of hydrogen ions is linked to tubular reabsorption of bicarbonate ions. In this way, the kidneys also regulate the concentration of bicarbonate ions in body fluids. These mechanisms also help balance the sulfuric acid, phosphoric acid, and various organic acids that appear in body fluids as by-products of metabolic processes.

The metabolism of certain amino acids, for example, produces sulfuric and phosphoric acids. Consequently, a diet high in proteins may form excess acid. The kidneys compensate by altering the tubular secretion of hydrogen ions, thus resisting a shift in the pH of body fluids (fig. 21.11). Once hydrogen ions are secreted, phosphates filtered into the fluid of the renal tubule buffer them, aided by ammonia (NH_3).

Through deamination of certain amino acids, the cells of the renal tubules produce ammonia, which diffuses readily through cell membranes and enters the renal tubule. When increase in the hydrogen ion concentration of body fluids is prolonged, the renal tubules increase ammonia production. Ammonia is a weak base, so it can accept hydrogen ions to form *ammonium ions* (NH_4^+):

$$H^+ + NH_3 \rightarrow NH_4^+$$

Cell membranes are impermeable to ammonium ions, which are trapped in the renal tubule as they form and are excreted with the urine. This mechanism helps to transport excess hydrogen ions to the outside and helps prevent the urine from becoming too acidic.

Time Course of Hydrogen Ion Regulation

The various regulators of hydrogen ion concentration operate at different rates. Acid-base buffers function rapidly and can convert strong acids or bases into weak acids or bases almost immediately. For this reason, these chemical buffer systems are called the body's *first line of defense* against shifts in pH.

Physiological buffer systems, such as the respiratory and renal mechanisms, function more slowly and constitute the *second line of defense*. The respiratory mechanism may require several minutes to begin resisting a change in pH, and the renal mechanism may require one to three days to regulate a changing hydrogen ion concentration (fig. 21.12).

Cells increase production of CO_2
CO_2 reacts with H_2O to produce H_2CO_3
H_2CO_3 releases H^+
Respiratory center is stimulated
Rate and depth of breathing increase
More CO_2 is eliminated through lungs

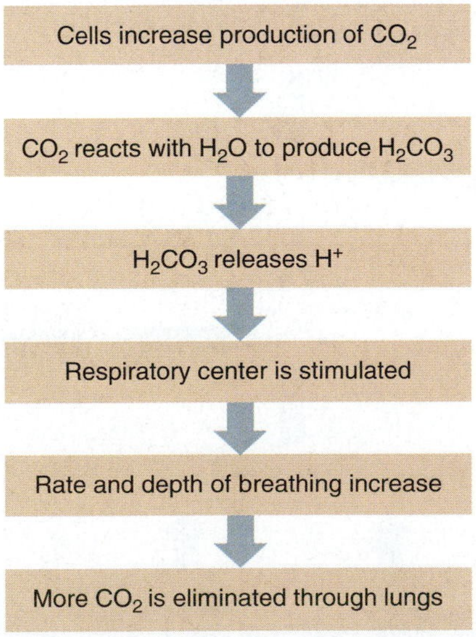

FIGURE 21.10 An increase in carbon dioxide elimination follows an increase in carbon dioxide production.

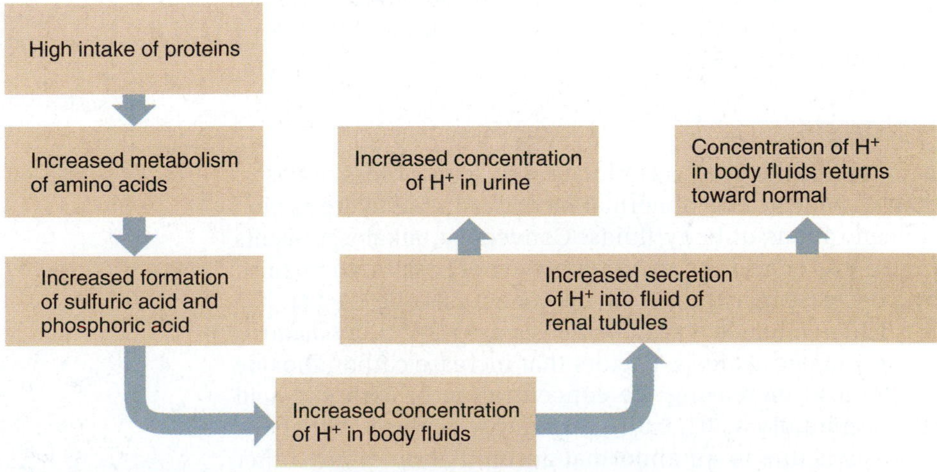

High intake of proteins
Increased metabolism of amino acids
Increased formation of sulfuric acid and phosphoric acid
Increased concentration of H^+ in body fluids
Increased secretion of H^+ into fluid of renal tubules
Increased concentration of H^+ in urine
Concentration of H^+ in body fluids returns toward normal

FIGURE 21.11 If the concentration of hydrogen ions in body fluids increases, the renal tubules increase their secretion of hydrogen ions into the urine.

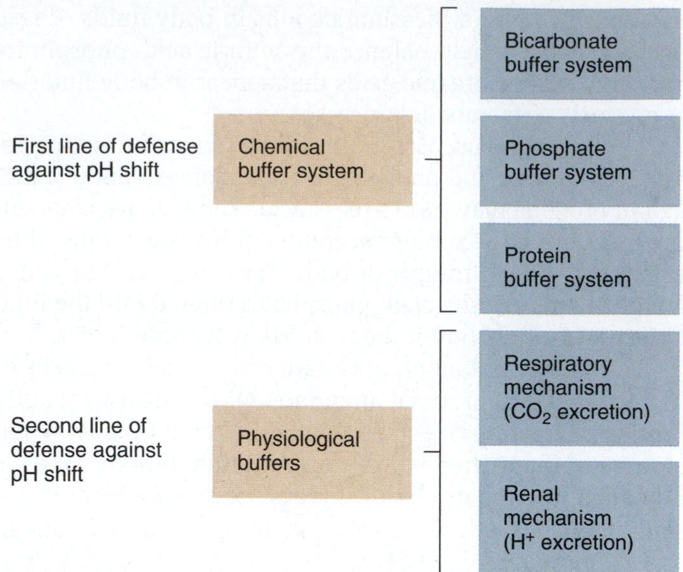

FIGURE 21.12 Chemical buffers act rapidly, while physiological buffers may require several minutes to several days to begin resisting a change in pH.

PRACTICE 

25 How does the respiratory system help regulate acid-base balance?

26 How do the kidneys respond to excess hydrogen ions?

27 How do the rates at which chemical and physiological buffer systems act differ?

21.6 ACID-BASE IMBALANCES

Chemical and physiological buffer systems ordinarily maintain the hydrogen ion concentration of body fluids within very narrow pH ranges. Abnormal conditions may disturb the acid-base balance. For example, the pH of arterial blood is normally 7.35–7.45. A value below 7.35 produces *acidosis.* A pH above 7.45 produces *alkalosis.* Such shifts in the pH of body fluids may be life threatening. A person usually cannot survive if the pH drops to 6.8 or rises to 8.0 for more than a few hours (fig. 21.13).

Acidosis

Acidosis results from accumulation of acids or loss of bases, both of which cause abnormal increases in the hydrogen ion concentrations of body fluids. Conversely, alkalosis results from a loss of acids or an accumulation of bases accompanied by a decrease in hydrogen ion concentrations (fig. 21.14).

The two major types of acidosis are *respiratory acidosis* and *metabolic acidosis.* Factors that increase carbon dioxide levels, also increasing the concentration of carbonic acid (the respiratory acid), cause respiratory acidosis. Metabolic acidosis is due to an abnormal accumulation of any other acids in the body fluids or to a loss of bases, including bicarbonate ions. Similarly, the two major types of alkalosis are *respiratory alkalosis* and *metabolic alkalosis.* Excessive loss of carbon dioxide and consequent loss of carbonic acid cause respiratory alkalosis. Metabolic alkalosis is due to excessive loss of hydrogen ions or gain of bases.

In respiratory acidosis carbon dioxide accumulates, so this can result from factors that hinder pulmonary ventilation (fig. 21.15). These include the following:

1. Injury to the respiratory center of the brainstem, decreasing rate and depth of breathing.
2. Obstructions in air passages that interfere with air movement into the alveoli.
3. Diseases that decrease gas exchange, such as pneumonia, or those that reduce surface area of the respiratory membrane, such as emphysema.

Any of these conditions can increase the level of carbonic acid and hydrogen ions in body fluids, lowering pH.

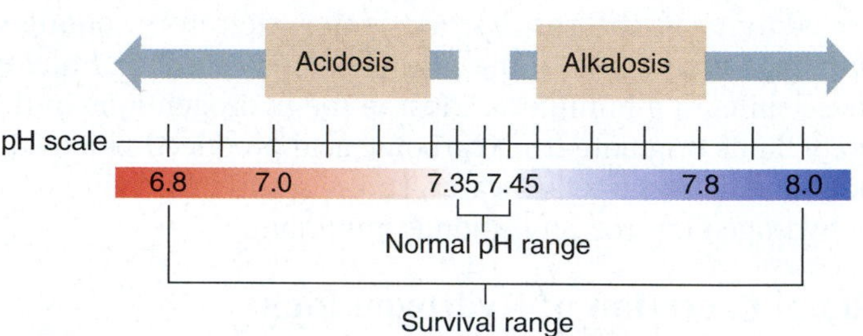

FIGURE 21.13 If the pH of arterial blood drops to 6.8 or rises to 8.0 for more than a few hours, the person usually cannot survive.

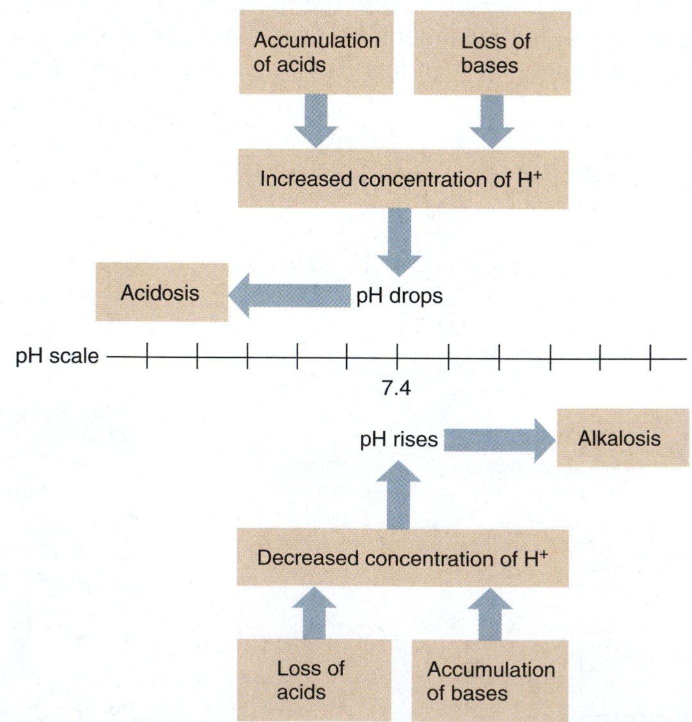

FIGURE 21.14 Acidosis results from accumulation of acids or loss of bases. Alkalosis results from loss of acids or accumulation of bases.

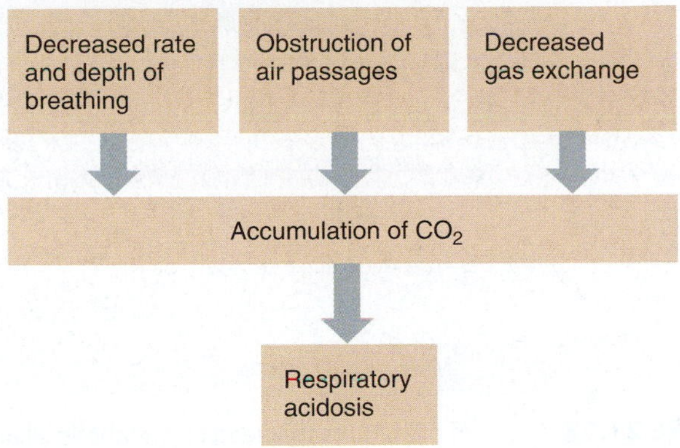

FIGURE 21.15 Some of the factors that lead to respiratory acidosis.

Chemical buffers, such as hemoglobin, may resist this shift in pH. At the same time, increasing levels of carbon dioxide and hydrogen ions stimulate the respiratory center, increasing breathing rate and depth and thereby lowering carbon dioxide levels. Also, the kidneys may begin to excrete more hydrogen ions.

Eventually, thanks to these chemical and physiological buffers, the pH of the body fluids may return to normal. When this happens, the acidosis is said to be *compensated*.

The symptoms of respiratory acidosis result from depression of central nervous system function and include drowsiness, disorientation, and stupor. Evidence of respiratory insufficiency, such as labored breathing and cyanosis, is usually also evident. In *uncompensated acidosis*, the person may become comatose and die.

Metabolic acidosis is due to either accumulation of nonrespiratory acids or loss of bases (fig. 21.16). Factors that may lead to this condition include

1. Kidney disease that reduces glomerular filtration and fails to excrete the acids produced in metabolism (uremic acidosis).
2. Prolonged vomiting that loses the alkaline contents of the upper intestine and stomach contents. (Losing only the stomach contents produces metabolic alkalosis.)
3. Prolonged diarrhea, in which excess alkaline intestinal secretions are lost (especially in infants).
4. Diabetes mellitus, in which some fatty acids react to produce ketone bodies, such as *acetoacetic acid, beta-hydroxybutyric acid,* and *acetone.* Normally, these molecules are scarce, and cells oxidize them as energy sources. However, if fats are being used at an abnormally high rate, as may occur in diabetes mellitus, ketone bodies may accumulate faster than they can be oxidized, and spill over into the urine (ketonuria); in addition, the lungs may release acetone, which is volatile and imparts a fruity odor to the breath. More seriously, the accumulation of acetoacetic acid and beta-hydroxybutyric acid may lower pH (ketonemic acidosis).

These acids may also combine with bicarbonate ions in the urine. Excess bicarbonate ions are excreted, interfering with the function of the bicarbonate acid-base buffer system.

Whatever the cause, metabolic acidosis shifts pH downward. However, the following factors resist this shift: chemical buffer systems, which accept excess hydrogen ions; the respiratory center, which increases breathing rate and depth; and the kidneys, which excrete more hydrogen ions.

Alkalosis

Respiratory alkalosis develops as a result of *hyperventilation,* described in chapter 19 (p. 758). Hyperventilation is accompanied by too great a loss of carbon dioxide and consequent decreases in carbonic acid and hydrogen ion concentrations (fig. 21.17).

Hyperventilation may happen during periods of anxiety. It may also accompany fever or poisoning from salicylates, such as aspirin. At high altitudes, hyperventilation may be a response to low oxygen pressure. Also, musicians, such as bass tuba players, who must provide a large volume of air when playing sustained passages, sometimes hyperventilate. In each case, rapid, deep breathing depletes carbon dioxide, and the pH of body fluids increases.

Chemical buffers, such as hemoglobin, that release hydrogen ions resist this pH change. Also, the lower levels of carbon dioxide and hydrogen ions stimulate the respiratory center to a lesser degree. This inhibits hyperventilation, thus reducing further carbon dioxide loss. At the same time, the kidneys decrease their secretion of hydrogen ions, and the urine becomes alkaline as bases are excreted.

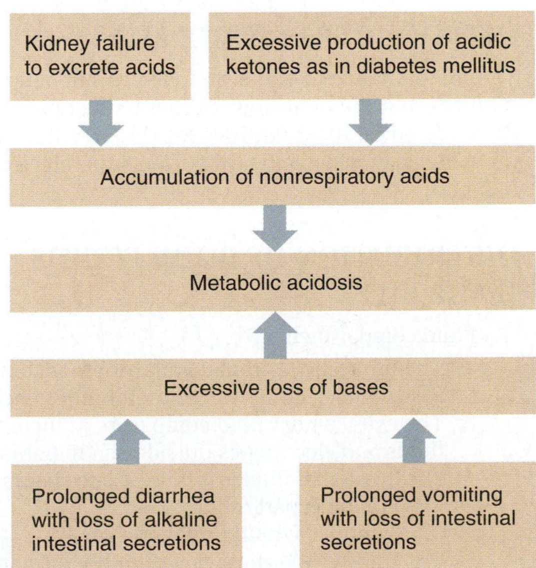

FIGURE 21.16 Some of the factors that lead to metabolic acidosis.

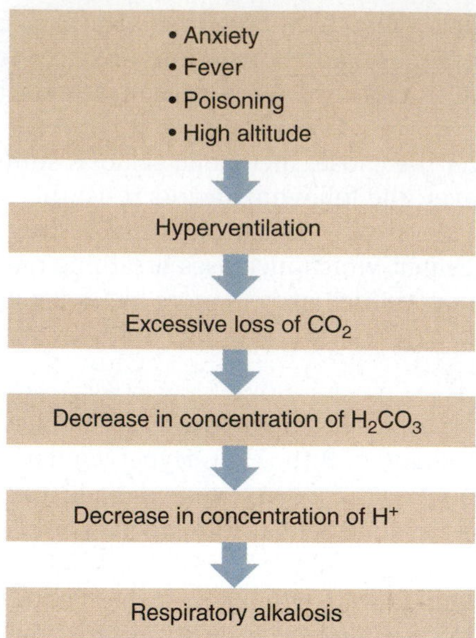

FIGURE 21.17 Some of the factors that lead to respiratory alkalosis.

The symptoms of respiratory alkalosis include lightheadedness, agitation, dizziness, and tingling sensations. In severe cases, impulses may be triggered spontaneously on peripheral nerves, and muscles may respond with tetanic contractions (see chapter 9, p. 297).

Metabolic alkalosis results from a great loss of hydrogen ions or from a gain in bases, both accompanied by a rise in the pH of the blood (alkalemia) (fig. 21.18). This condition may occur following gastric drainage (lavage), prolonged vomiting in which only the stomach contents are lost, or the use of certain diuretic drugs. Gastric juice is acidic, so its loss leaves the body fluids with a net increase of basic substances and a pH shift toward alkaline values. Metabolic alkalosis may also develop as a result of ingesting too much antacid, such as sodium bicarbonate, to relieve the symptoms of indigestion. The symptoms of metabolic alkalosis include a decrease in the breathing rate and depth, which, in turn, results in an increased concentration of carbon dioxide in the blood.

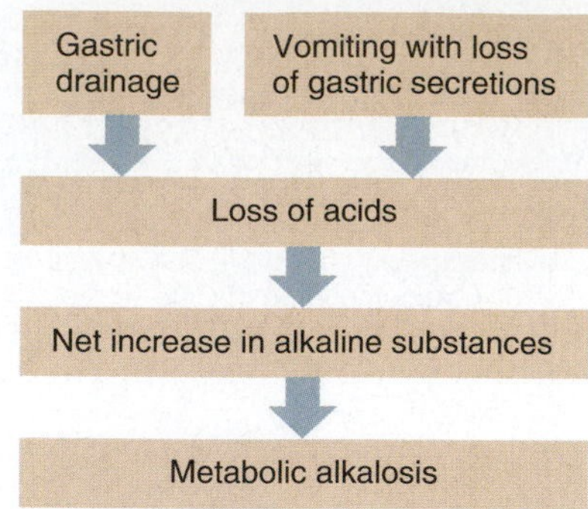

FIGURE 21.18 Some of the factors that lead to metabolic alkalosis.

PRACTICE 

28 What is the difference between a respiratory acid-base disturbance and a metabolic disturbance?

29 How do the symptoms of alkalosis compare with those of acidosis?

CHAPTER SUMMARY

21.1 INTRODUCTION (PAGE 811)

The maintenance of water and electrolyte balance requires that the quantities of these substances entering the body equal the quantities leaving it. Altering the water balance necessarily affects the electrolyte balance.

21.2 DISTRIBUTION OF BODY FLUIDS (PAGE 811)

1. Fluid compartments
   a. The intracellular fluid compartment includes the fluids and electrolytes cell membranes enclose.
   b. The extracellular fluid compartment includes all fluids and electrolytes outside cell membranes.
      (1) Interstitial fluid is within tissue spaces.
      (2) Plasma is part of the blood.
      (3) Lymph is within lymphatic vessels.
      (4) Transcellular fluid is within body cavities.
2. Body fluid composition
   a. Extracellular fluids have high concentrations of sodium, chloride, calcium, and bicarbonate ions, with less potassium, calcium, magnesium, phosphate, and sulfate ions. Plasma contains more protein than does either interstitial fluid or lymph.
   b. Intracellular fluid contains relatively high concentrations of potassium, magnesium, and phosphate ions; it also contains a greater concentration of sulfate ions and lesser concentrations of sodium, chloride, calcium, and bicarbonate ions than does extracellular fluid.
3. Movement of fluid between compartments
   a. Hydrostatic and osmotic pressure regulate fluid movements.
      (1) Fluid leaves plasma because of hydrostatic pressure and returns to plasma because of osmotic pressure.
      (2) Hydrostatic pressure drives fluid into lymph vessels.

(3) Osmotic pressure regulates fluid movement in and out of cells.
b. Sodium ion concentrations are especially important in fluid movement regulation.

21.3 WATER BALANCE (PAGE 813)

1. Water intake
   a. The volume of water taken in varies from person to person.
   b. Most water comes from consuming liquid or moist foods.
   c. Oxidative metabolism produces some water.
2. Regulation of water intake
   a. The thirst mechanism is the primary regulator of water intake.
   b. Drinking and the resulting stomach distension inhibit the thirst mechanism.
3. Water output
   a. Water is lost in a variety of ways.
      (1) It is excreted in the urine, feces, and sweat.
      (2) Insensible loss occurs through evaporation from the skin and lungs.
   b. Urine production regulates water output.
4. Regulation of water output
   a. The distal convoluted tubules and collecting ducts of the nephrons regulate water output.
      (1) ADH from the hypothalamus and posterior pituitary gland stimulates water reabsorption in these segments.
      (2) The mechanism involving ADH can reduce normal output of 1,500 milliliters to 500 milliliters per day.
   b. If excess water is taken in, the ADH mechanism is inhibited.

21.4 ELECTROLYTE BALANCE (PAGE 815)

1. Electrolyte intake
   a. The most important electrolytes in the body fluids are those that release ions of sodium, potassium, calcium, magnesium, chloride, sulfate, phosphate, and bicarbonate.
   b. These ions are obtained in foods and beverages or as by-products of metabolic processes.
2. Regulation of electrolyte intake
   a. Electrolytes are usually obtained in sufficient quantities in response to hunger and thirst mechanisms.
   b. In a severe electrolyte deficiency, a person may experience a salt craving.
3. Electrolyte output
   a. Electrolytes are lost through perspiration, feces, and urine.
   b. Quantities lost vary with temperature and physical exercise.
   c. The greatest electrolyte loss occurs as a result of kidney functions.
4. Regulation of electrolyte output
   a. Concentrations of sodium, potassium, and calcium ions in the body fluids are particularly important.
   b. The regulation of sodium ions involves the secretion of aldosterone from the adrenal glands.
   c. The regulation of potassium ions also involves aldosterone.
   d. Parathyroid hormone from the parathyroid glands regulate calcium ion concentration.
   e. The mechanisms that control positively charged ions secondarily regulate negatively charged ions.
      (1) Chloride ions are passively reabsorbed in renal tubules as sodium ions are actively reabsorbed.
      (2) Some negatively charged ions, such as phosphate ions, are reabsorbed partially by limited-capacity active transport mechanisms.

21.5 ACID-BASE BALANCE (PAGE 819)

Acids are electrolytes that release hydrogen ions. Bases combine with hydrogen ions.

1. Sources of hydrogen ions.
   a. Aerobic respiration of glucose produces carbon dioxide, which reacts with water to form carbonic acid. Carbonic acid dissociates to release hydrogen and bicarbonate ions.
   b. Anaerobic respiration of glucose produces lactic acid.
   c. Incomplete oxidation of fatty acids releases acidic ketone bodies.
   d. Oxidation of sulfur-containing amino acids produces sulfuric acid.
   e. Hydrolysis of phosphoproteins and nucleic acids gives rise to phosphoric acid.
2. Strengths of acids and bases
   a. Acids vary in the extent to which they ionize.
      (1) Strong acids, such as hydrochloric acid, ionize more completely.
      (2) Weak acids, such as carbonic acid, ionize less completely.
   b. Bases vary in strength also.
      (1) Strong bases, such as hydroxide ions, combine readily with hydrogen ions.
      (2) Weak bases, such as bicarbonate ions, combine with hydrogen ions less readily.
3. Regulation of hydrogen ion concentration
   a. Acid-base buffer systems minimize pH changes.
      (1) Buffer systems are composed of sets of two or more chemicals.
      (2) They convert strong acids into weaker acids or strong bases into weaker bases.
      (3) They include the bicarbonate buffer system, phosphate buffer system, and protein buffer system.
   b. The respiratory system excretes carbon dioxide.
      (1) The respiratory center is located in the brainstem.
      (2) It helps regulate pH by controlling the rate and depth of breathing.
      (3) Increasing carbon dioxide and hydrogen ion concentrations stimulates chemoreceptors associated with the respiratory center; breathing rate and depth increase, and carbon dioxide concentration decreases.

(4) If the carbon dioxide and hydrogen ion concentrations are low, the respiratory center inhibits breathing.

c. The kidneys excrete hydrogen ions.
   (1) Nephrons secrete hydrogen ions to regulate pH.
   (2) Phosphates buffer hydrogen ions in urine.
   (3) Ammonia produced by renal cells helps transport hydrogen ions to the outside of the body.

d. Chemical buffers act rapidly; physiological buffers act more slowly.

21.6 ACID-BASE IMBALANCES (PAGE 824)

1. Acidosis
   a. Respiratory acidosis results from increased levels of carbon dioxide and carbonic acid.
   b. Metabolic acidosis results from accumulation of other acids or loss of bases.
2. Alkalosis
   a. Respiratory alkalosis results from loss of carbon dioxide and carbonic acid.
   b. Metabolic alkalosis results from loss of other acids or gain of bases.

CHAPTER ASSESSMENTS

21.1 Introduction

1 Explain how water balance and electrolyte balance are interdependent. (p. 811)

21.2 Distribution of Body Fluids

2 The water and electrolytes enclosed by cell membranes constitute the ______________. (p. 811)

a. transcellular fluid
b. intracellular fluid
c. extracellular fluid
d. lymph
e. plasma

3 Explain how the fluids in the compartments differ in composition. (p. 812)

4 Describe how fluid movements between the compartments are controlled. (p. 812)

21.3 Water Balance

5 Prepare a list of sources of normal water gain and loss to illustrate how the input of water equals the output of water. (p. 813)

6 Define *water of metabolism.* (p. 813)

7 Explain how water intake is regulated. (p. 813)

8 Explain how the kidneys regulate water output. (p. 815)

21.4 Electrolyte Balance

9 Electrolytes in body fluids of importance to cellular functions include ______________. (p. 815)

a. sodium
b. potassium
c. calcium
d. chloride
e. all of the above

10 Explain how electrolyte intake is regulated. (p. 815)

11 List the routes by which electrolytes leave the body. (p. 816)

12 Explain how the adrenal cortex functions to regulate electrolyte balance. (p. 818)

13 Describe the role of the parathyroid glands in regulating electrolyte balance. (p. 818)

14 Describe the mechanisms by which the renal tubules regulate electrolyte balance. (p. 818)

21.5 Acid-Base Balance

15 Define an *acid* and a *base.* (p. 819)

16 List five sources of hydrogen ions in the body fluids, and name an acid that originates from each source. (p. 819)

17 ______________ dissociate to release hydrogen ions more completely. An example is hydrochloric acid. (p. 820)

18 ______________ dissociate to release fewer hydroxide ions. (p. 820)

19 Explain how an acid-base buffer system functions. (p. 820)

20 Explain how the bicarbonate buffer system resists changes in pH. (p. 821)

21 Explain why a protein has both acidic and basic properties. (p. 821)

22 Describe how a protein functions as a buffer system. (p. 821)

23 Describe the role of hemoglobin as a buffer. (p. 822)

24 Explain how the respiratory system functions in the regulation of acid-base balance. (p. 822)

25 Explain how the kidneys function in the regulation of acid-base balance. (p. 823)

26 Describe the role of ammonia in the transport of hydrogen ions to the outside of the body. (p. 823)

27 Distinguish between a chemical buffer system and a physiological buffer system. (p. 823)

21.6 Acid-Base Imbalances

28 Distinguish between respiratory and metabolic acid-base imbalances. (p. 824)

29 Explain how the body compensates for acid-base imbalances. (p. 825)

INTEGRATIVE ASSESSMENTS/CRITICAL THINKING

OUTCOMES 14.3, 19.6, 19.7, 21.5, 21.6

1. An elderly, semiconscious patient is tentatively diagnosed as having acidosis. What components of the arterial blood will be most valuable in determining if the acidosis is of respiratory origin?

OUTCOMES 15.4, 15.5, 16.3, 21.2

2. If the right ventricle of a patient's heart is failing, increasing the systemic venous pressure, what changes might occur in the patient's extracellular fluid compartments?

OUTCOMES 17.2, 17.6, 17.9, 21.4, 21.6

3. Radiation therapy may damage the mucosa of the stomach and intestines. What effect might this have on the patient's electrolyte balance?

OUTCOMES 17.6, 17.9, 19.5, 21.5, 21.6

4. Describe what might happen to the plasma pH of a patient as a result of:
   a. prolonged diarrhea.
   b. suction of the gastric contents.
   c. hyperventilation.
   d. hypoventilation.

OUTCOMES 19.5, 21.5, 21.6

5. A student hyperventilates and is disoriented just before an exam. Is this student likely to be experiencing acidosis or alkalosis? How will the body compensate in an effort to maintain homeostasis?

WEB CONNECTIONS

Be sure to visit the text website at **www.mhhe.com/shier12** for answers to chapter assessments, additional quizzes, and interactive learning exercises.

ANATOMY & PHYSIOLOGY REVEALED

Anatomy & Physiology Revealed® (APR) includes cadaver photos that allow you to peel away layers of the human body to reveal structures beneath the surface. This program also includes animations, radiologic imaging, audio pronunciations, and practice quizzing. Check out **www.aprevealed.com**. APR has been proven to help improve student grades!

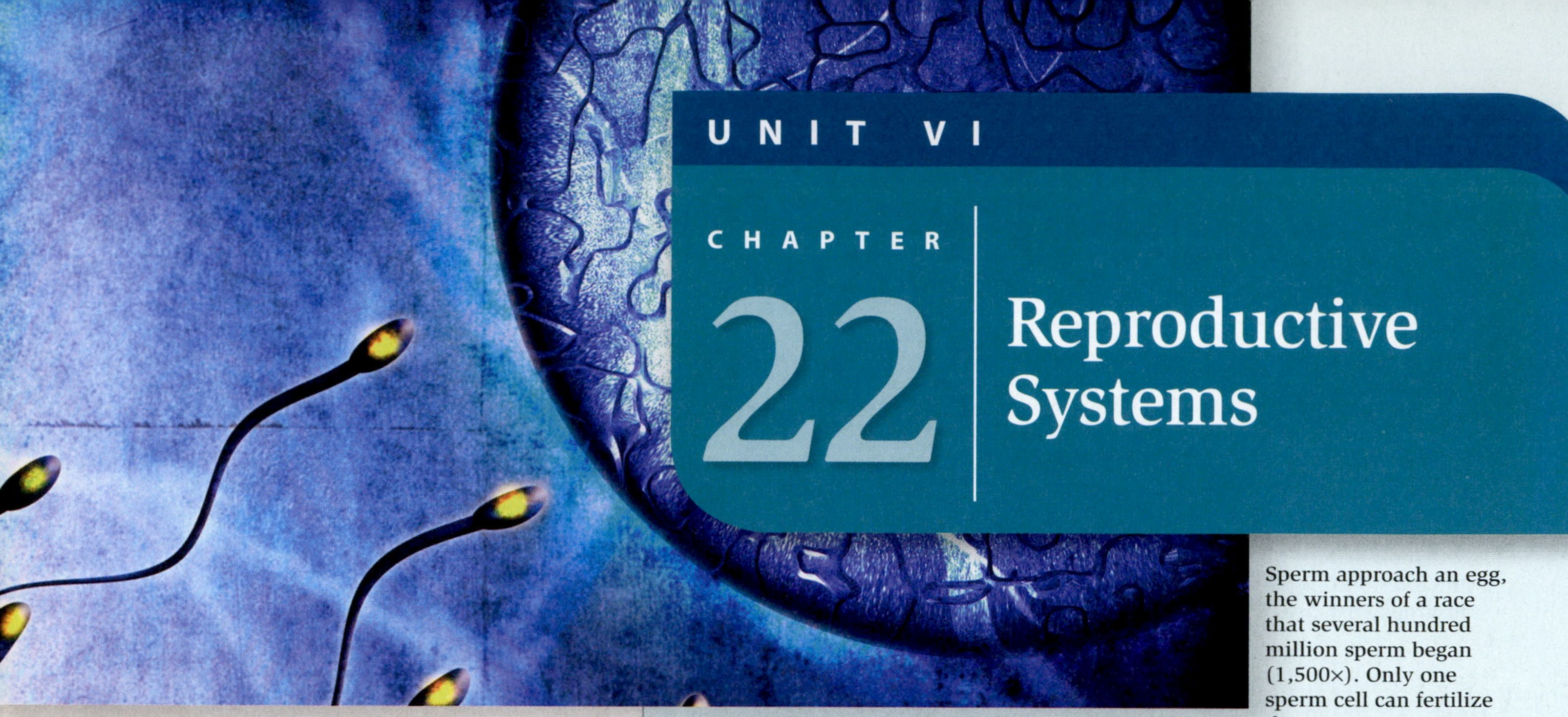

UNIT VI
CHAPTER 22 Reproductive Systems

Sperm approach an egg, the winners of a race that several hundred million sperm began (1,500×). Only one sperm cell can fertilize the egg.

UNDERSTANDING WORDS

andr-, man: *andr*ogens—male sex hormones.

contra-, against, counter: *contra*ception—prevention of fertilization.

crur-, lower part: *crura*—diverging parts at the base of the penis by which it attaches to the pelvic arch.

ejacul-, to shoot forth: *ejacul*ation—expulsion of semen from the male reproductive tract.

fimb-, fringe: *fimb*riae—irregular extensions on the margin of the infundibulum of the uterine tube.

follic-, small bag: *follic*le—ovarian structure that contains an egg.

-genesis, origin: spermato*genesis*—formation of sperm cells.

germ-, to bud, to sprout: *germ*inal epithelium-tissue that gives rise to sex cells by special cell division.

gubern-, to steer, to guide: *gubern*aculum—fibromuscular cord that guides the descent of a testis.

labi-, lip: *labi*a minora—flattened, longitudinal folds that extend along the margins of the female vestibule.

mamm-, breast: *mamm*ary gland—female accessory gland that secretes milk.

mast-, breast: *mast*itis—inflammation of the mammary gland.

mens-, month: *mens*es—monthly flow of blood from the female reproductive tract.

mons-, an eminence: *mons* pubis—rounded elevation of fatty tissue overlying the symphysis pubis in a female.

oo-, egg: *oo*genesis—formation of an egg.

prim-, first: *prim*ordial follicle—ovarian follicle composed of an oocyte surrounded by a single layer of cells.

puber-, adult: *puber*ty—time when a person becomes able to reproduce.

zon-, belt: *zon*a pellucida—transparent layer surrounding an oocyte.

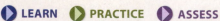

LEARN

PRACTICE
ASSESS

LEARNING OUTCOMES

After you have studied this chapter, you should be able to:

22.1 Introduction

1. State the general functions of the male and female reproductive systems. (p. 831)
2. Outline the process of meiosis, and explain how it mixes up parental genes. (p. 831)

22.2 Organs of the Male Reproductive System

3. Describe the function(s) of each part of the male reproductive system. (p. 833)
4. Outline the process of spermatogenesis. (p. 836)
5. Describe semen production and exit from the body. (p. 840)
6. Describe the structure of the penis, and explain how its parts produce an erection. (p. 843)

22.3 Hormonal Control of Male Reproductive Functions

7. Explain how hormones control the activities of the male reproductive organs and the development of male secondary sex characteristics. (p. 845)

22.4 Organs of the Female Reproductive System

8. Describe the function(s) of each part of the female reproductive system. (p. 846)
9. Outline the process of oogenesis. (p. 849)

22.5 Hormonal Control of Female Reproductive Functions

10. Explain how hormones control the activities of the female reproductive organs and the development of female secondary sex characteristics. (p. 857)
11. Describe the major events during a female reproductive cycle. (p. 857)

22.6 Mammary Glands

12. Review the structure of the mammary glands. (p. 861)

22.7 Birth Control

13. Describe several methods of birth control, including the relative effectiveness of each method. (p. 862)

22.8 Sexually Transmitted Infections

14. List the general symptoms of sexually transmitted infections. (p. 867)

EGGS FOR SALE

The ad in the student newspaper seemed too good to be true—the fee for donating a few eggs would pay nearly a semester's tuition. Intrigued, the young woman submitted a health history, had a checkup, and a month later received a call. A young couple struggling with infertility sought an egg donor. They'd chosen Sherrie because, with her strawberry-blond hair, she looked like Linda, the woman whose cancer had left her unable to conceive. The donor eggs would be fertilized in a laboratory dish (*in vitro*) with sperm from Linda's partner, and then implanted in Linda's uterus.

For two weeks Sherrie injected herself in the thigh with a drug that acts like gonadotropin-releasing hormone, suppressing release of an egg from an ovary (ovulation). When daily hormone checks indicated that her endocrine system was in sync with Linda's, Sherrie began giving herself shots twice a day at the back of the hip. This second drug mimicked follicle-stimulating hormone, and it caused several ovarian follicles to mature. Finally, injections of luteinizing hormone brought the eggs to full maturity. Then, at a health-care facility, Sherrie was given pain medication and light sedation. A needle was inserted through her vaginal wall to remove a dozen of the most mature eggs as they swelled to the surface of her ovary. This is the least invasive way to retrieve eggs.

Two *in vitro* fertilized ova divided a few times, forming an early embryo, and were then implanted into Linda's uterus. The rest were frozen, for possible later use. The preparation and procedure weren't too painful. Sherrie had felt a dull aching the last day, and felt bloated for a few days after the egg retrieval, but she did not experience bleeding, infection, cramping, or mood swings. Nor did she develop a complication in which too many eggs mature, causing fluid to leak from blood vessels and accumulate in the abdomen. About 6% of egg donors develop the syndrome, which can cause infertility, kidney failure, and even death. Future risks, however, are uncertain, because eggs haven't yet been collected long enough to know the consequences.

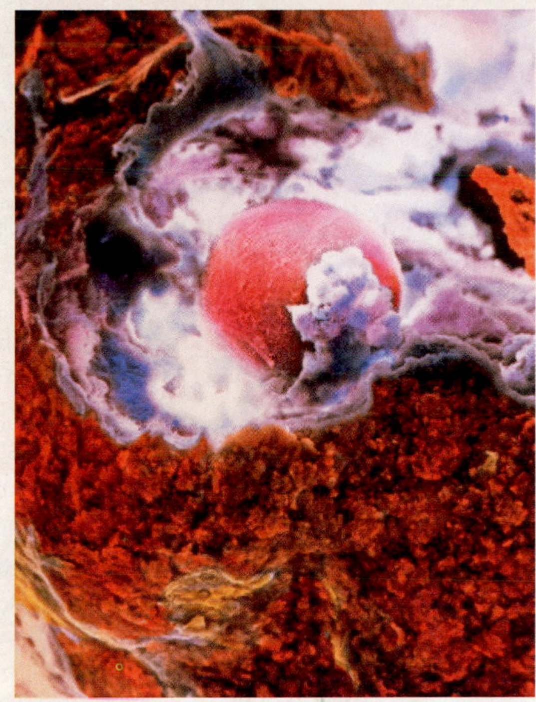

Ads in student newspapers seek egg donors for infertile couples. The protocol is uncomfortable and takes several weeks. Pay is high, and is often based on SAT scores, college grades, and even the reputation of a particular university.

Case reports point to ovary scarring and possibly even cancer. Another side effect that Sherrie had not fully considered was how she would feel afterward. Although she was happy to have helped the couple and to have paid her tuition, she feared she would always wonder about the twins her eggs had become. ■

22.1 INTRODUCTION

The male and female reproductive systems are connected sets of organs and glands. Some of the reproductive organs and glands secrete hormones vital to the development and maintenance of secondary sex characteristics and the regulation of reproductive functions. Reproductive organs produce and nurture sex cells and transport them to sites of fertilization. Male sex cells are **sperm.** Female sex cells are eggs, or **oocytes** (o′o-sītz), which in Latin means "egg cells." These cells are produced by a special type of cell division called **meiosis** (mi-o′sis).

Meiosis includes two successive divisions, called the *first* and *second meiotic divisions.* The first meiotic division (meiosis I) separates homologous chromosome pairs. Homologous pairs are the same, gene for gene. They may not be identical, however, because a gene may have variants, and the chromosome that comes from the person's mother may carry a different variant for the corresponding gene from the father's homologous chromosome. Before meiosis I, each homologous chromosome is replicated, so it consists of two DNA strands called *chromatids.* Each chromatid has the complete genetic information associated with that chromosome. The chromatids of a replicated chromosome attach at regions called *centromeres.*

Each of the cells that undergoes the second meiotic division (*meiosis II*) emerges with one member of each homologous pair, a condition termed **haploid.** That is, a haploid cell has one set of chromosomes. This second division separates the chromatids, producing cells that are still haploid, but whose chromosomes are no longer in the replicated form. After meiosis II, each of the chromatids is an independent chromosome.

The steps of meiosis are clearer when considered in a time sequence (fig. 22.1). However, keep in mind that, like mitosis, meiosis is a continuous process. Considering it in steps makes it easier to follow.

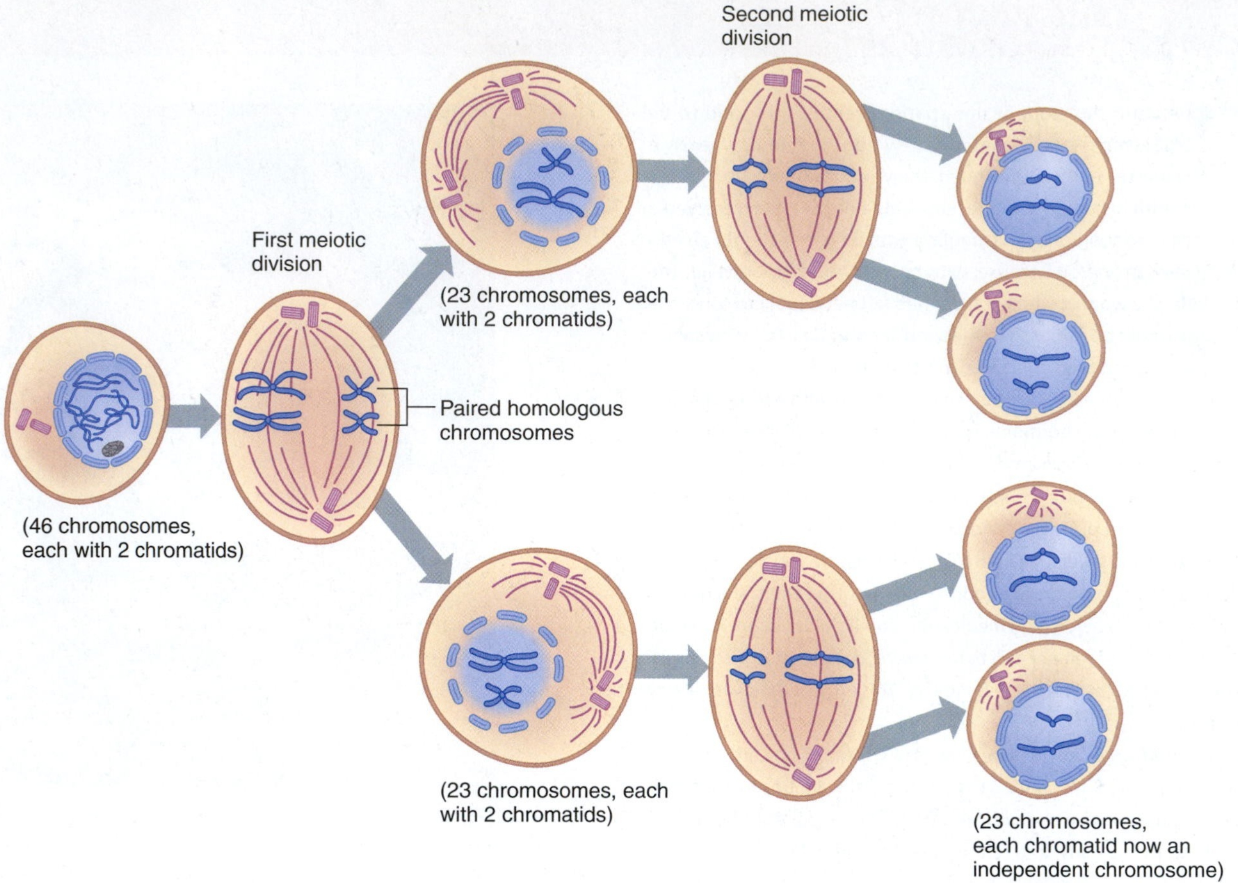

FIGURE 22.1 Sex cells are formed by a special type of cell division, meiosis. This illustration follows two representative pairs of homologous chromosomes.

First Meiotic Division

Prophase I. Individual chromosomes appear as thin threads in the nucleus, then shorten and thicken. Nucleoli disappear, the nuclear membrane temporarily disassembles, and microtubules begin to build the spindle that will separate the chromosomes. The DNA of the chromosomes has already been replicated.

As prophase I continues, homologous chromosomes pair up side by side and tightly intertwine. During this pairing, called *synapsis,* the chromatids of the homologous chromosomes contact one another at various points along their lengths. Often, the chromatids break in one or more places and exchange parts, forming chromatids with new combinations of genetic information (fig. 22.2). One chromosome of a homologous pair is from a person's mother and the other is from the father, so an exchange, or **cross over,** between homologous chromosomes produces chromatids that contain genetic information from both parents.

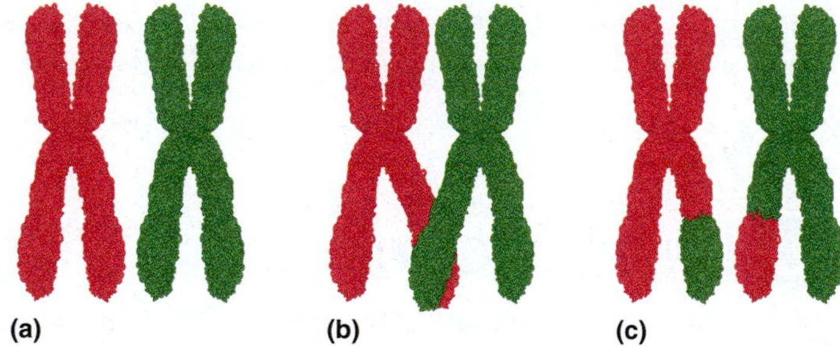

FIGURE 22.2 Crossing over mixes up genetic traits. (*a*) Homologous chromosome pair, (*b*) chromatids cross over, (*c*) crossing over recombines genes. The different colors indicate that one of the homologous chromosomes comes from the individual's father and one from the mother.

Metaphase I. During the first metaphase, chromosome pairs line up about midway between the poles of the developing spindle, and they are held under great tension, like two groups of people playing tug-of-war. Each chromosome pair consists of two chromosomes, which equals four chromatids. Each chromosome attaches to spindle fibers from one pole. The chromosome alignment is random with respect to maternal and paternal origin of the chromosomes. Each of the 23 chromosomes contributed from the mother may be on the left or the right, and the same is true for the paternal chromosomes—it is similar to the number of ways that 23 pairs of children could line up, while maintaining the pairs. Chromosomes can line up with respect to each other in many combinations.

Anaphase I. Homologous chromosome pairs separate, and each replicated member moves to one end of the spindle. Each new, or daughter, cell receives only one replicated member of a homologous pair of chromosomes, overall halving the chromosome number.

Telophase I. The original cell divides in two. Nuclear membranes form around the chromosomes, nucleoli reappear, and the spindle fibers disassemble into their constituent microtubules.

Second Meiotic Division

After telophase I, the second meiotic division begins. Meiosis II is similar to a mitotic division (see fig. 22.1). During *prophase II,* chromosomes condense and reappear, still replicated. They move into positions midway between the poles of the developing spindle. In *metaphase II,* the replicated chromosomes attach to spindle fibers. In *anaphase II,* centromeres separate, freeing the chromatids to move to opposite poles of the spindles. The former chromatids are now considered to be chromosomes. In *telophase II,* each of the two cells resulting from meiosis I divides to form two cells. Therefore, each cell undergoing meiosis has the potential to produce four gametes. In males, the gametes mature into four sperm cells. In females, three of the products of meiosis are "cast aside" as polar bodies, and one cell becomes the egg.

Meiosis generates astounding genetic variety. Any one of a person's more than 8 million possible combinations of 23 chromosomes can combine with any one of the more than 8 million combinations of his or her mate, raising the potential variability to more than 70 trillion genetically unique individuals! Crossing over contributes even more genetic variability. Figure 22.3 illustrates in a simplified manner how maternal and paternal traits reassort during meiosis.

Sex cells have one set of genetic instructions, carried on 23 chromosomes, compared to two sets on 46 chromosomes in other cells. When sex cells join at fertilization, the amount of genetic information held in 46 chromosomes is restored.

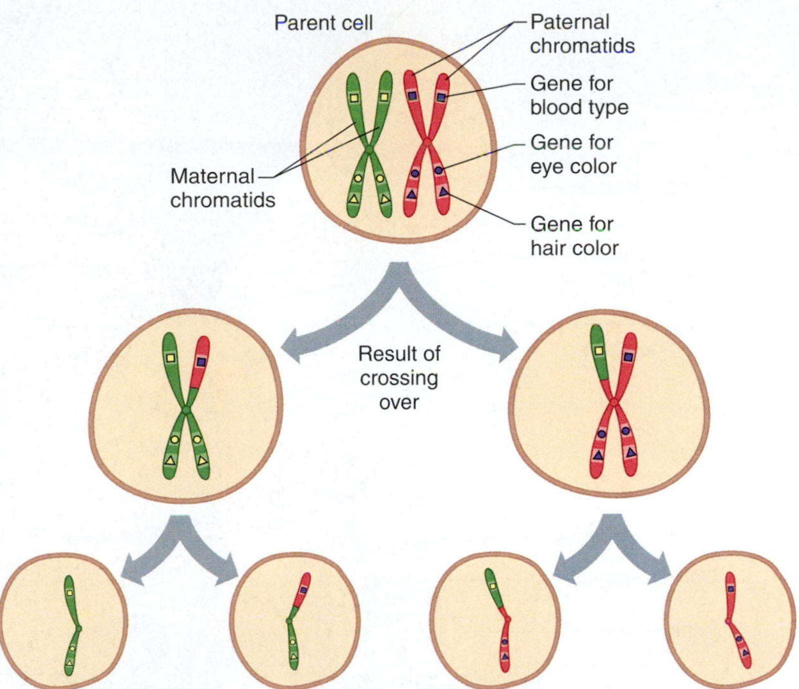

FIGURE 22.3 As a result of crossing over, the genetic information in sex cells varies from cell to cell. Colors represent parent of origin. Although only one eye color gene pair is illustrated, eye color is polygenic (involving more than one gene pair).

PRACTICE

1. What are male and female sex cells called?
2. Describe the major events of meiosis.
3. How does meiosis provide genetic variability?

22.2 ORGANS OF THE MALE REPRODUCTIVE SYSTEM

Organs of the male reproductive system are specialized to produce and maintain the male sex cells, or *sperm cells;* transport these cells and supporting fluids to the outside; and secrete male sex hormones.

The *primary sex organs* (gonads) of this system are the two testes in which the sperm cells (spermatozoa) and the male sex hormones are formed. The other structures of the male reproductive system are termed *accessory sex organs* (secondary sex organs). They include the internal reproductive organs and the external reproductive organs (fig. 22.4; reference plates 3 and 4).

Testes

The **testes** (tes′tēz; sing., *testis*) are ovoid structures about 5 centimeters in length and 3 centimeters in diameter. Both testes, each suspended by a spermatic cord, are within the cavity of the saclike *scrotum* (fig. 22.4 and reference plate 12).

Descent of the Testes

In a male fetus, the testes originate from masses of tissue posterior to the parietal peritoneum, near the developing kidneys. Usually a month or two before birth, the testes descend to the lower abdominal cavity and pass through the abdominal wall into the scrotum.

The male sex hormone *testosterone,* which the developing testes secrete, stimulates the testes to descend. A fibromuscular cord called the **gubernaculum** (goo″ber′-nak′u-lum) aids movement of the testes. This cord is attached to each developing testis and extends into the inguinal region of the abdominal cavity. It passes through the abdominal wall and is fastened to the skin on the outside of the scrotum. The testis descends, guided by the gubernaculum, passing through the **inguinal canal** (ing′gwĭ-nal kah-nal′) of the abdominal

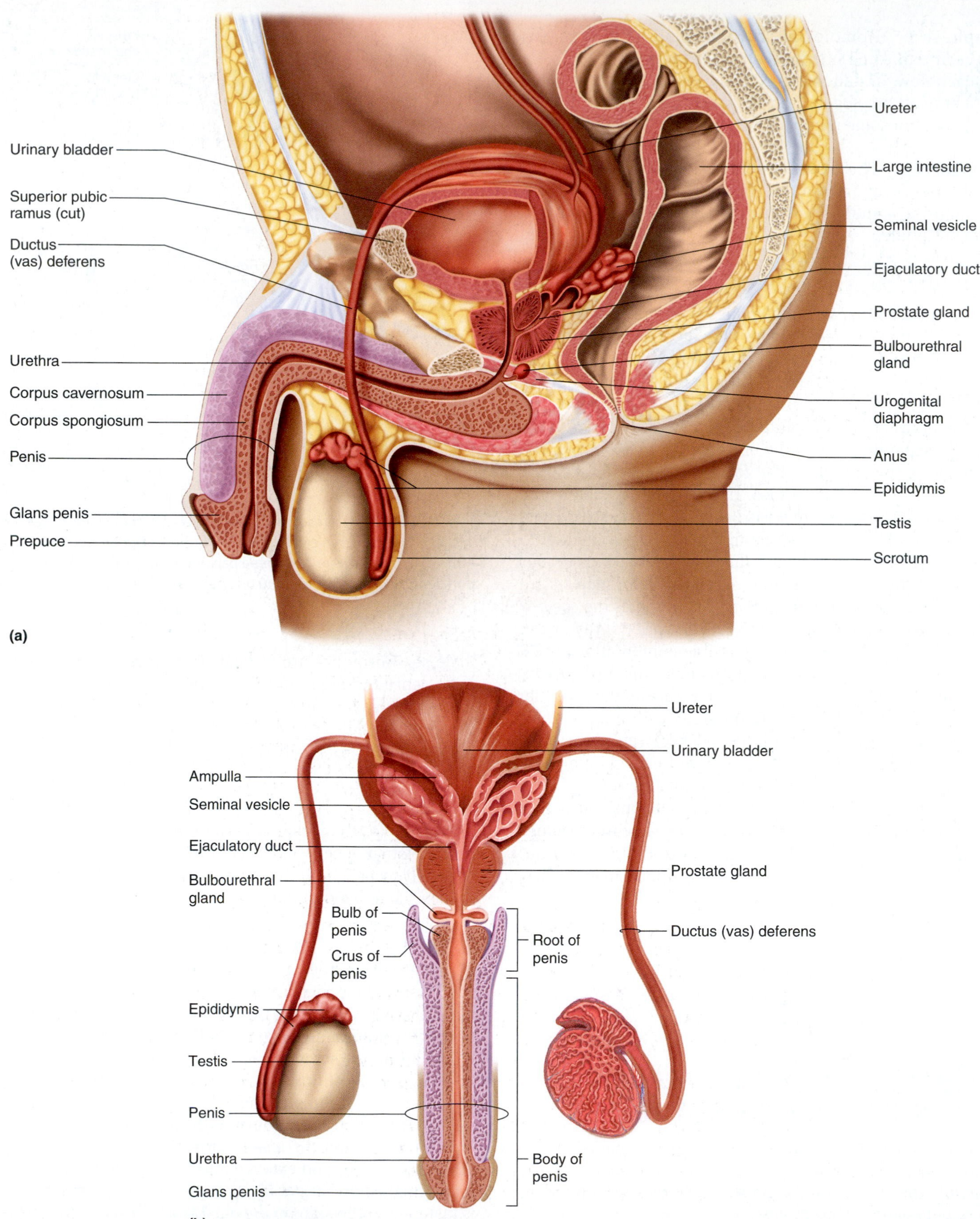

FIGURE 22.4 Male reproductive organs. (*a*) Sagittal view and (*b*) posterior view. The paired testes are the primary sex organs, and the other reproductive structures, both internal and external, are accessory sex organs.

wall and entering the scrotum, where it remains anchored by the gubernaculum. Each testis carries a developing *ductus (vas) deferens,* blood vessels, and nerves. These structures later form parts of the **spermatic cord** by which the testis is suspended in the scrotum (fig. 22.5).

If the testes fail to descend into the scrotum, they will not produce sperm cells because the temperature in the abdominal cavity is too high. If this condition, called *cryptoorchidism,* is left untreated, the cells that normally produce sperm cells degenerate, and the male is infertile.

PRACTICE 

4 What are the primary sex organs of the male reproductive system?

5 Describe the descent of the testes.

6 What happens if the testes fail to descend into the scrotum?

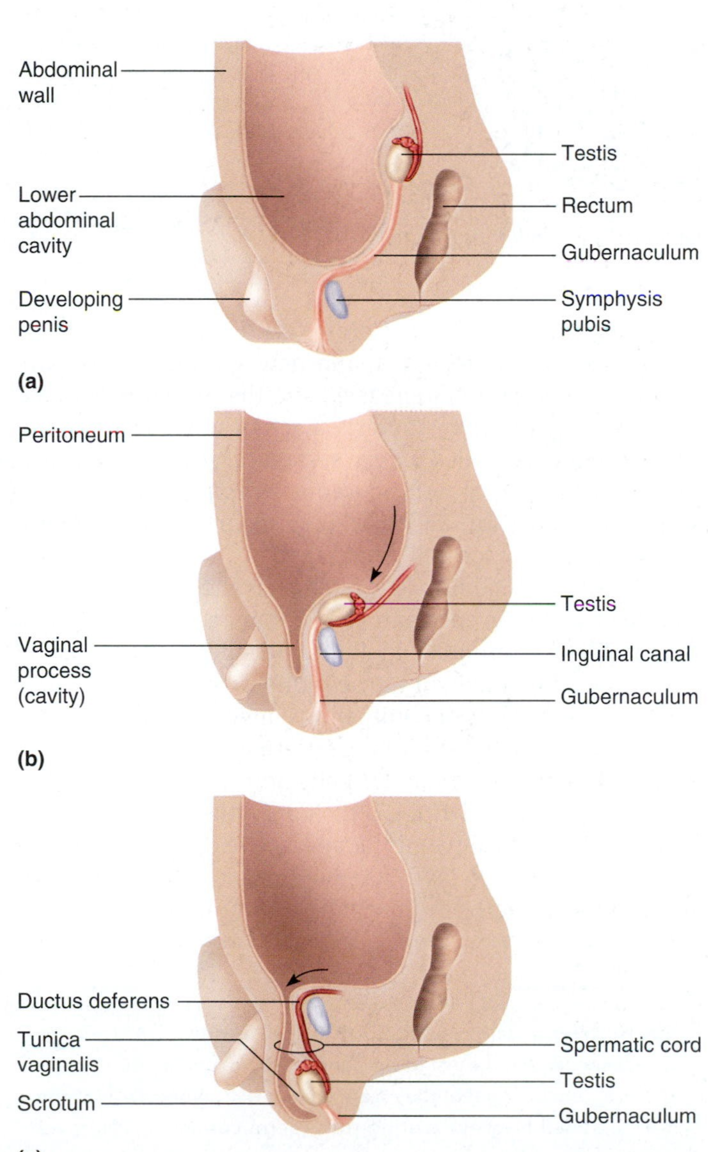

FIGURE 22.5 During fetal development, each testis develops near a kidney and then descends through an inguinal canal and enters the scrotum, completing the journey by the eighth gestational month (*a–c*).

During the descent of a testis, a pouch of peritoneum, called the *vaginal process,* moves through the inguinal canal and into the scrotum. In about one-quarter of males, this pouch remains open, providing a potential passageway through which a loop of intestine may be forced by great abdominal pressure, producing an *indirect inguinal hernia.* If the protruding intestinal loop is so tightly constricted within the inguinal canal that its blood supply stops, the condition is called a *strangulated hernia.* Without prompt treatment, the strangulated tissues may die.

Structure of the Testes

A tough, white, fibrous capsule called the *tunica albuginea* encloses each testis. Along its posterior border, the connective tissue thickens and extends into the organ, forming a mass called the *mediastinum testis.* From this structure, thin layers of connective tissue, called *septa,* pass into the testis and subdivide it into about 250 *lobules.*

A lobule contains one to four highly coiled, convoluted **seminiferous tubules** (sem″ĭ-nif′er-us tu′būlz), each approximately 70 centimeters long when uncoiled. These tubules course posteriorly and unite to form a complex network of channels called the *rete testis* (re′te tes′tis). The rete testis is in the mediastinum testis and gives rise to several ducts that join a tube called the *epididymis.* The epididymis, in turn, is coiled on the outer surface of the testis and continues to become the *ductus deferens.*

The seminiferous tubules are lined with a specialized stratified epithelium, which includes the **spermatogenic cells** that give rise to the sperm cells. Other specialized cells, called **interstitial** (in″ter-stish′al) **cells** (cells of Leydig), lie between the seminiferous tubules. Interstitial cells produce and secrete male sex hormones (figs. 22.6 and 22.7).

The epithelial cells of the seminiferous tubules can give rise to *testicular cancer,* a common cancer in young men. In most cases, the first sign is a painless testis enlargement or a scrotal mass attached to a testis. If a biopsy (tissue sample) reveals cancer cells, surgery is performed to remove the affected testis (orchiectomy). Radiation and/or chemotherapy often prevents the cancer from recurring.

PRACTICE 

7 Describe the structure of a testis.

8 Where in the testes are sperm cells produced?

9 Which cells produce male sex hormones?

Formation of Sperm Cells

The epithelium of the seminiferous tubules consists of supporting cells called *sustentacular cells* (Sertoli cells) and spermatogenic cells. The sustentacular cells are columnar and extend the full thickness of the epithelium from its base

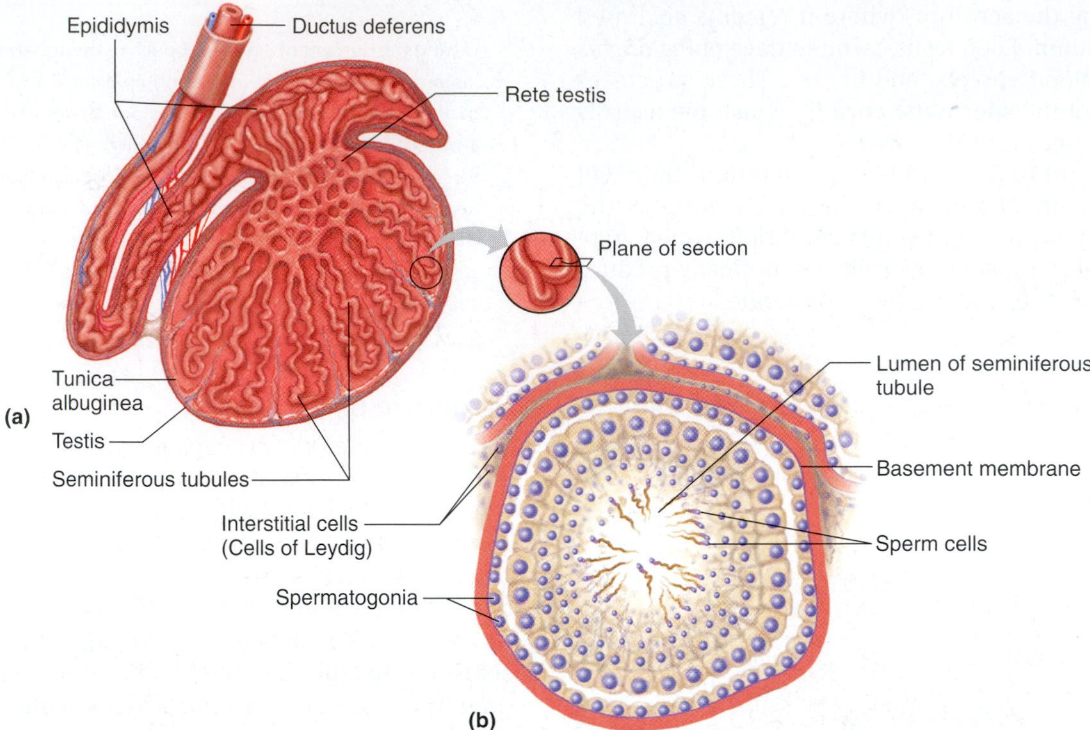

FIGURE 22.6 Structure of the testis. (*a*) Sagittal section of a testis. (*b*) Cross section of a seminiferous tubule.

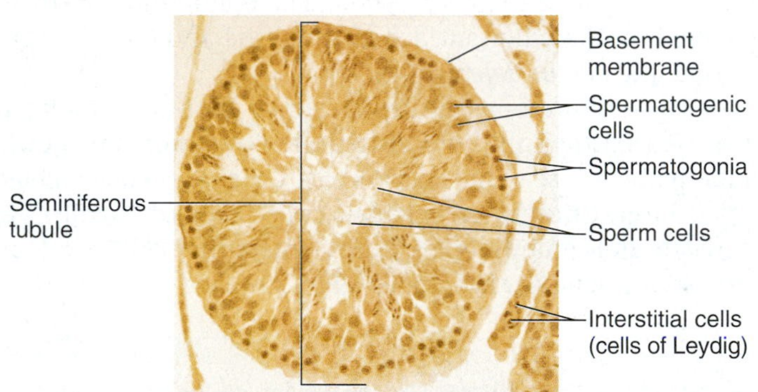

FIGURE 22.7 Light micrograph of a seminiferous tubule (250×).

to the lumen of the seminiferous tubule. The sustentacular cells support, nourish, and regulate the spermatogenic cells, which give rise to sperm cells (spermatozoa).

In the male embryo, undifferentiated spermatogenic cells are called *spermatogonia.* Each spermatogonium has 46 chromosomes (23 pairs) in its nucleus, the usual number for human body cells. Spermatogonia are located within the seminiferous tubules, adjacent to the inside surface of the basement membrane surrounding each seminiferous tubule.

Hormones stimulate the spermatogonia to become active. Some of the cells undergo mitosis (see chapter 3, pp. 99–100). Each cell division gives rise to two new cells, one (type A) of which maintains the supply of undifferentiated cells, the other (type B) of which differentiates to become a *primary spermatocyte.* Sperm production or **spermatogenesis** (sper″mah-to-jen′ĕ-sis) is arrested at this stage (fig. 22.8). At puberty, mitosis resumes, and new spermatogonia form. Testosterone secretion increases, and the primary spermatocytes then reproduce by meiosis. Each primary spermatocyte divides to form two *secondary spermatocytes.* Each of these cells, in turn, divides to form two *spermatids,* which mature into sperm cells (fig. 22.9). Meiosis reduces the number of chromosomes in each cell by one half.

The spermatogonia are located near the wall of the seminiferous tubule. As spermatogenesis proceeds, cells in more advanced stages are pushed along the sides of sustentacular cells toward the lumen of the seminiferous tubule.

Near the base of the epithelium, membranous processes from adjacent sustentacular cells fuse by tight junctions (fig. 22.9). The sustentacular cells and their tight junctions form the *blood-testis barrier,* which prevents some substances from reaching the developing sperm. The blood-testis barrier helps maintain a favorable environment by isolating the developing sperm from the male's immune system, which might otherwise view the sperm as abnormal cells.

Sperm have fascinated biologists for centuries. Anton van Leeuwenhoek was the first to view human sperm under a microscope in 1678, concluding that they were parasites in semen. By 1685, he had modified his view, writing that sperm contain a preformed human being and are seeds requiring nurturing in a female to start a new life.

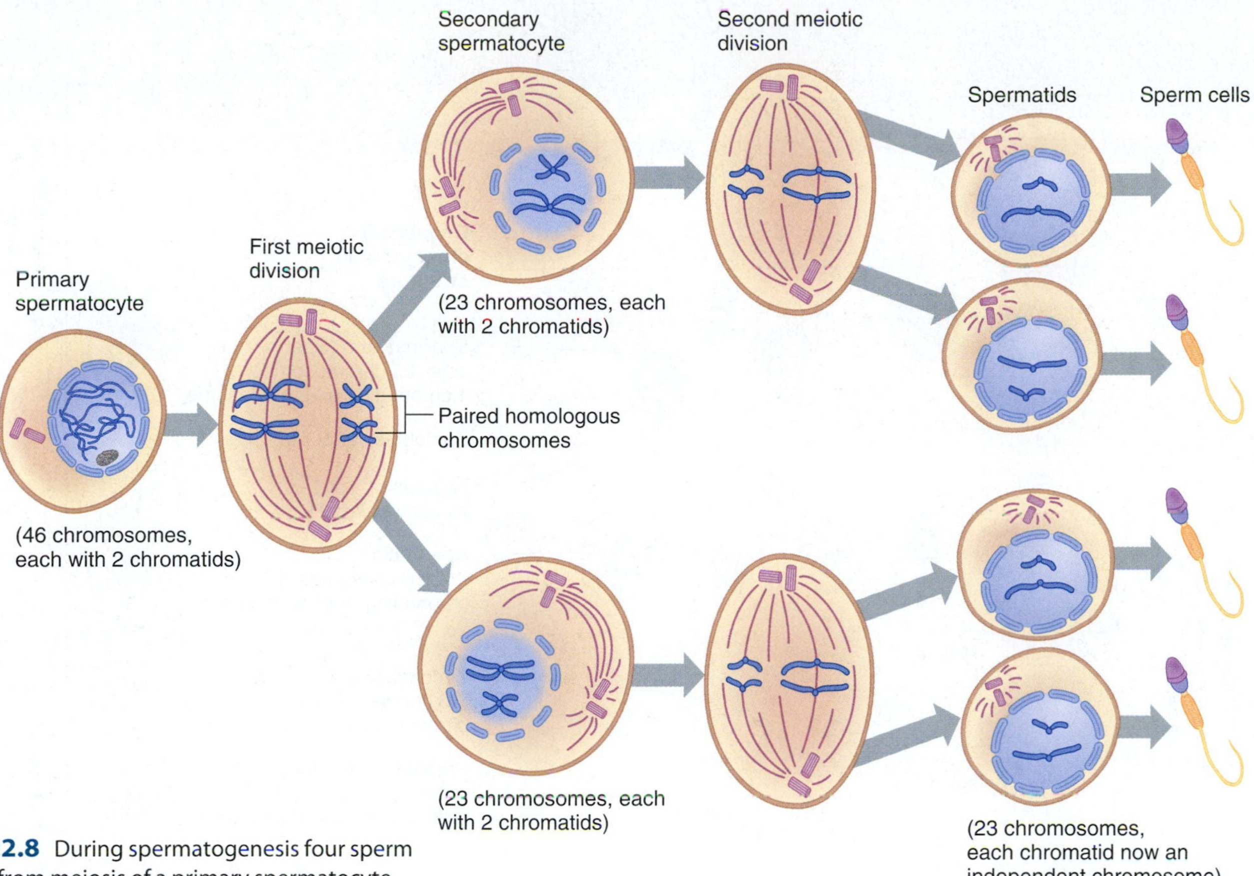

FIGURE 22.8 During spermatogenesis four sperm cells result from meiosis of a primary spermatocyte. Two representative homologous chromosome pairs are shown.

Spermatogenesis occurs continually in a male, starting at puberty. The resulting sperm cells collect in the lumen of each seminiferous tubule, then pass through the rete testis to the epididymis, where they accumulate and mature.

Structure of a Sperm Cell

Spermiogenesis is the process of sperm cell maturation. A mature sperm cell is a tiny, tadpole-shaped structure about 0.06 millimeter long. It consists of a flattened head, a cylindrical midpiece (body), and an elongated tail.

The oval *head* of a sperm cell is primarily composed of a nucleus and contains highly compacted chromatin consisting of 23 chromosomes. A small protrusion at its anterior end, called the *acrosome,* contains enzymes, including hyaluronidase, that aid the sperm cell in penetrating the cell layers surrounding the oocyte during fertilization (fig. 22.10).

The *midpiece* of a sperm has a central, filamentous core and many mitochondria organized in a spiral. The *tail* (flagellum) consists of several microtubules enclosed in an extension of the cell membrane. The mitochondria provide ATP for the lashing movement of the tail that propels the sperm cell through fluid. The micrograph in figure 22.11 shows a few mature sperm cells.

Many toxic chemicals that affect sperm hamper their ability to swim, so the cells cannot transmit the toxin to an egg. An exception is cocaine, which attaches to thousands of binding sites on human sperm cells, without apparently harming the cells or impeding their movements. Sperm can ferry cocaine to an egg, but it is not known what harm, if any, the drug causes. We do know that fetuses exposed to cocaine in the uterus may suffer a stroke, or, as infants, be unable to react normally to their surroundings.

PRACTICE 

10 Review the events of spermatogenesis.

11 Explain the function of the sustentacular cells in the seminiferous tubules.

12 Describe the structure of a sperm cell.

Male Internal Accessory Organs

The internal accessory organs of the male reproductive system are specialized to nurture and transport sperm cells. These structures include the two epididymides, two ductus deferentia, two ejaculatory ducts, and urethra, as well as

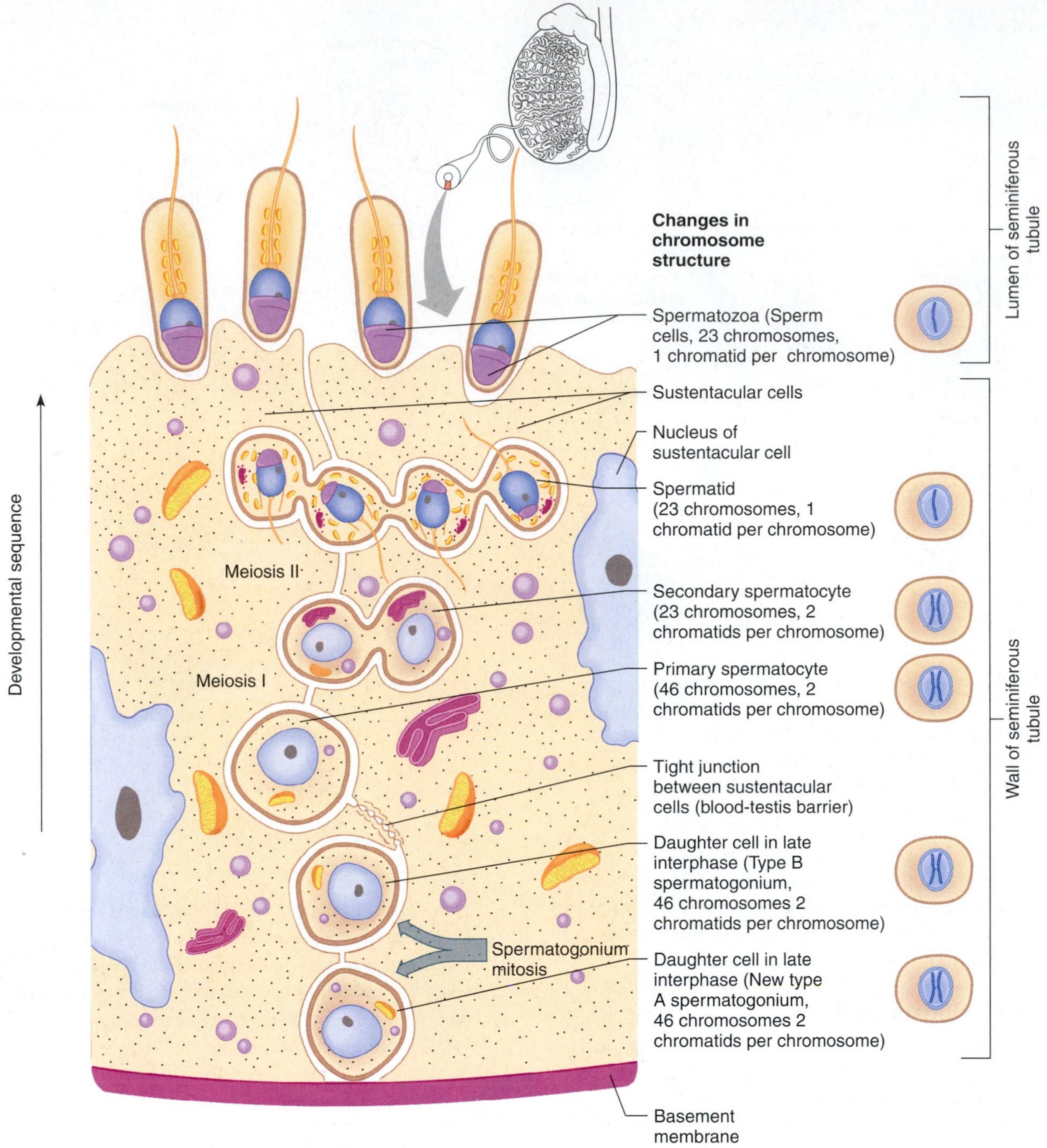

FIGURE 22.9 Spermatogonia (type B) give rise to primary spermatocytes by mitosis; the spermatocytes, in turn, give rise to sperm cells by meiosis. Type A spermatogonia continue the germ cell line.

the two seminal vesicles, prostate gland, and two bulbourethral glands.

Epididymides

The **epididymides** (ep″ĭ-di-dy′mides; sing., *epididymis*) are tightly coiled, threadlike tubes about 6 meters long (see figs. 22.4, 22.12, and reference plate 12). Each epididymis is connected to ducts in a testis. It emerges from the top of the testis, descends along its posterior surface, and then courses upward to become the ductus deferens.

The inner lining of the epididymis is composed of pseudostratified columnar cells that bear nonmotile cilia. These cells secrete glycogen and other substances that support stored sperm cells and promote their maturation.

When immature sperm cells reach the epididymis, they are nonmotile. However, as they travel through the epididymis as a result of rhythmic peristaltic contractions, they mature. Following this aging process, the sperm cells can move independently and fertilize egg cells (ova). However, they usually do not "swim" until after ejaculation.

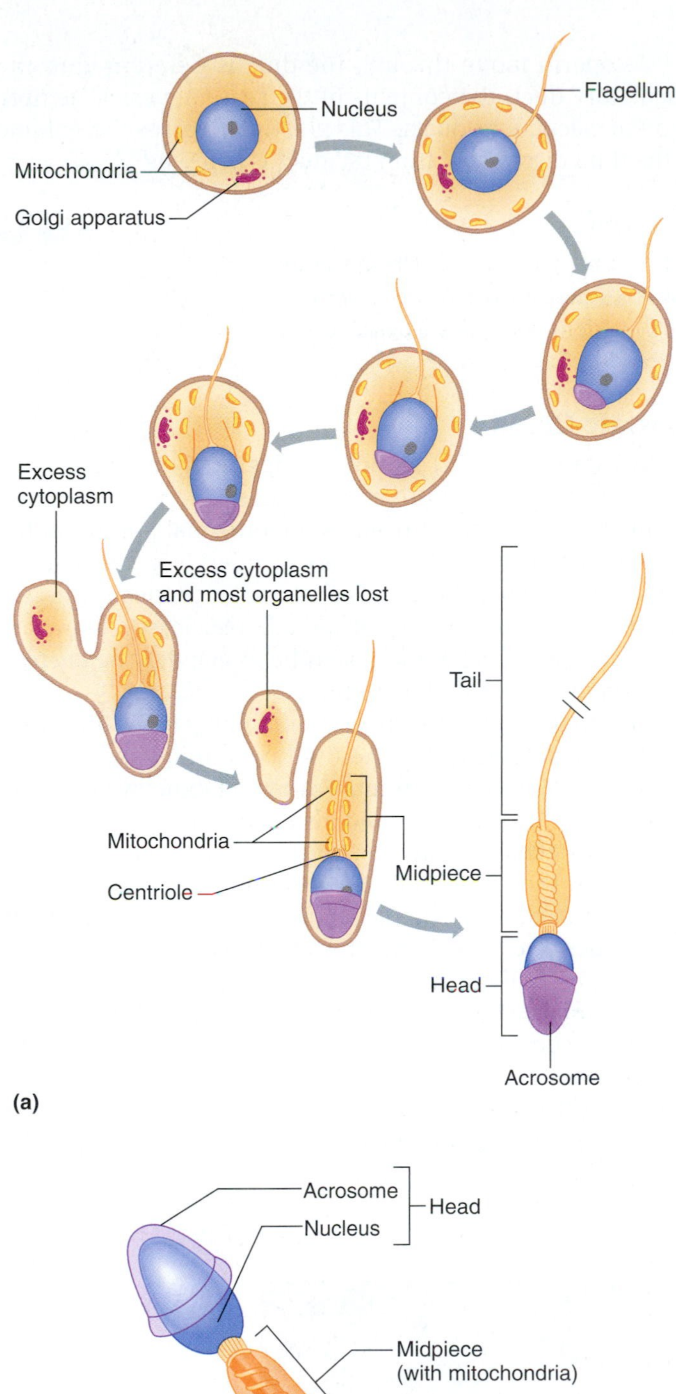

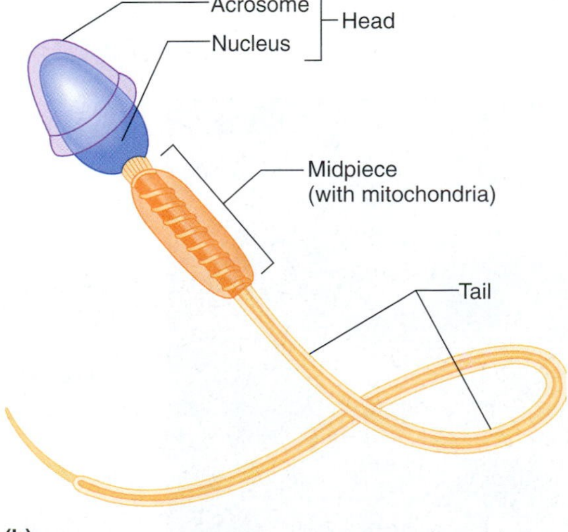

FIGURE 22.10 Spermiogenesis—sperm cell maturation. (*a*) The head of the sperm develops largely from the nucleus of the formative cell. (*b*) Parts of a mature sperm cell.

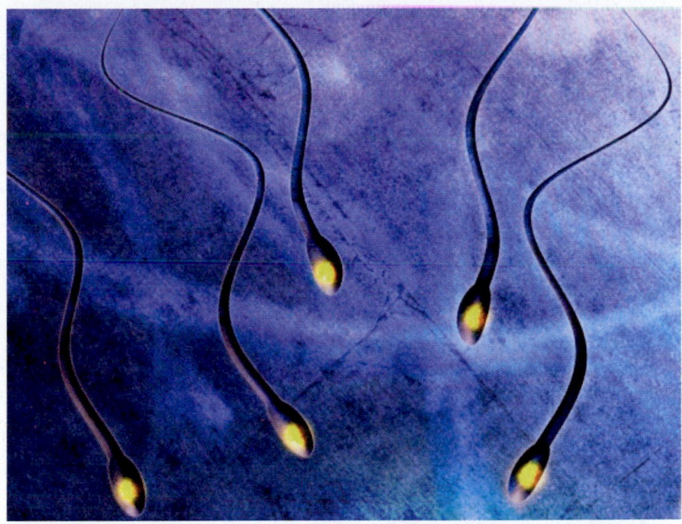

FIGURE 22.11 Falsely colored scanning electron micrograph of human sperm cells (1,400×).

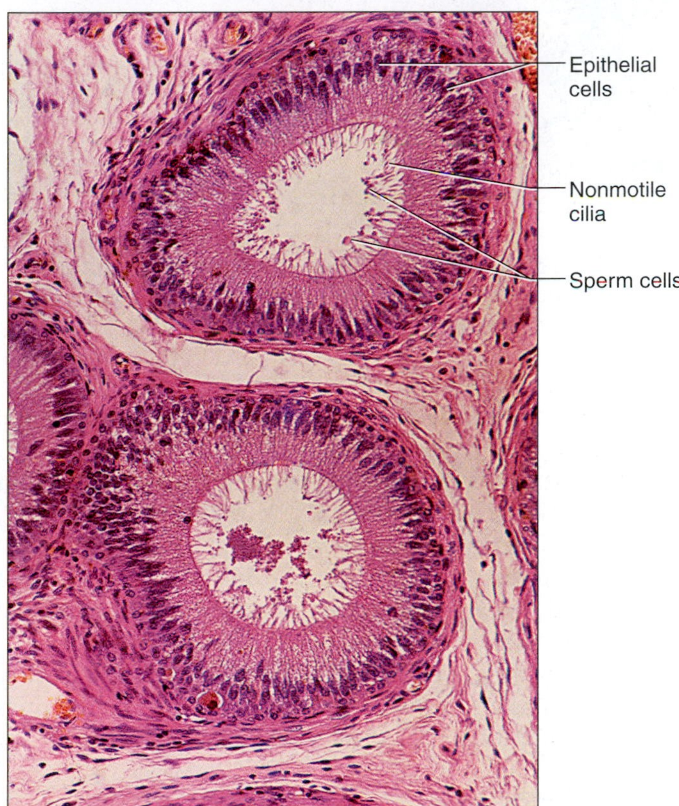

FIGURE 22.12 Cross section of a human epididymis (200×).

Ductus Deferentia

The **ductus deferentia** (duk′tus def′er-en′sha; sing., *ductus deferens*), also called *vasa deferentia,* are muscular tubes about 45 centimeters long lined with pseudostratified columnar epithelium (fig. 22.13). Each ductus deferens originates at the lower end of the epididymis and passes upward along the medial side of a testis to become

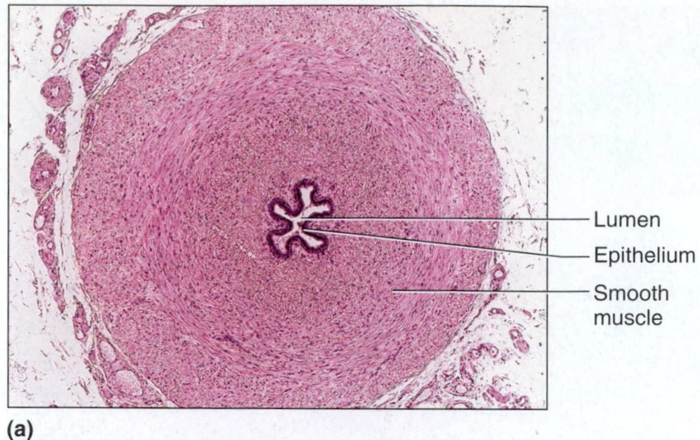

(a)

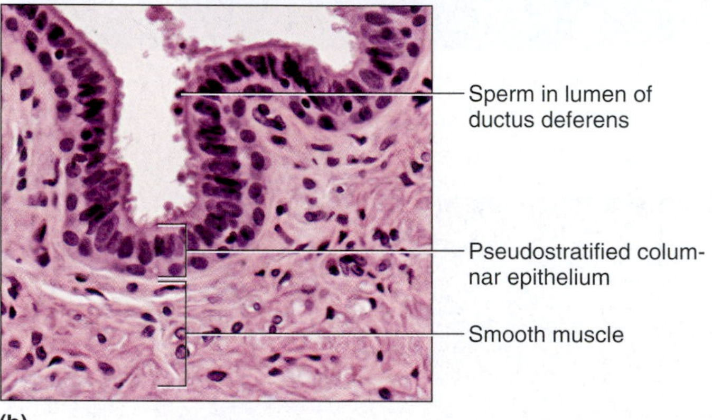

(b)

FIGURE 22.13 Ductus (vas) deferens. (*a*) Micrograph of a cross section of the ductus deferens (40×). (*b*) Light micrograph of the wall of the ductus deferens (400×).

part of the spermatic cord. It passes through the inguinal canal, enters the abdominal cavity outside the parietal peritoneum, and courses over the pelvic brim. From there, it extends backward and medially into the pelvic cavity, where it ends behind the urinary bladder.

Near its termination, the ductus deferens dilates into a portion called the *ampulla.* Just outside the prostate gland, the tube becomes slender again and unites with the duct of a seminal vesicle. The fusion of these two ducts forms an **ejaculatory duct,** which passes through the prostate gland and empties into the urethra through a slitlike opening (see fig. 22.4).

Seminal Vesicles

The **seminal vesicles** (see fig. 22.4) are convoluted, saclike structures about 5 centimeters long each attached to the ductus deferens near the base of the urinary bladder. The glandular tissue lining the inner wall of the seminal vesicle secretes a slightly alkaline fluid. This fluid helps regulate the pH of the tubular contents as sperm cells travel to the outside. The secretion of the seminal vesicle also contains *fructose,* a monosaccharide that provides energy to the sperm cells, and *prostaglandins,* which stimulate muscular contractions of the female reproductive organs, aiding the movement of sperm cells toward the egg cell.

As sperm move through the ductus deferens into the ejaculatory duct, the contents of the seminal vesicles empty into the ejaculatory ducts. This greatly increases the volume of the fluid discharged from the ductus deferens.

PRACTICE

13 Describe the structure of the epididymis.

14 Trace the path of the ductus deferens.

15 What is the function of a seminal vesicle?

Prostate Gland

The **prostate** (pros′tāt) **gland** (see figs. 22.4 and 22.14) is a chestnut-shaped structure about 4 centimeters across and 3 centimeters thick that surrounds the proximal portion of the urethra, just inferior to the urinary bladder. It is composed of many branched tubular glands enclosed in connective tissue. Septa of connective tissue and smooth muscle extend inward from the capsule, separating the tubular glands. The ducts of these glands open into the urethra.

The prostate gland secretes a thin, milky fluid. This alkaline secretion neutralizes the fluid containing sperm cells, which is acidic from accumulation of metabolic wastes from the sperm cells. Prostatic fluid also enhances the motility of sperm cells, which remain relatively nonmotile in the acidic contents of the epididymis. In addition, the prostatic fluid helps neutralize the acidic secretions of the vagina, helping to sustain sperm cells that enter the female reproductive tract.

The prostate gland releases its secretions into the urethra as smooth muscles contract in its capsular wall. As this release occurs, the contents of the ductus deferens and the seminal vesicles enter the urethra, which increases the volume of the fluid. Clinical Application 22.1 discusses the effects of prostate enlargement.

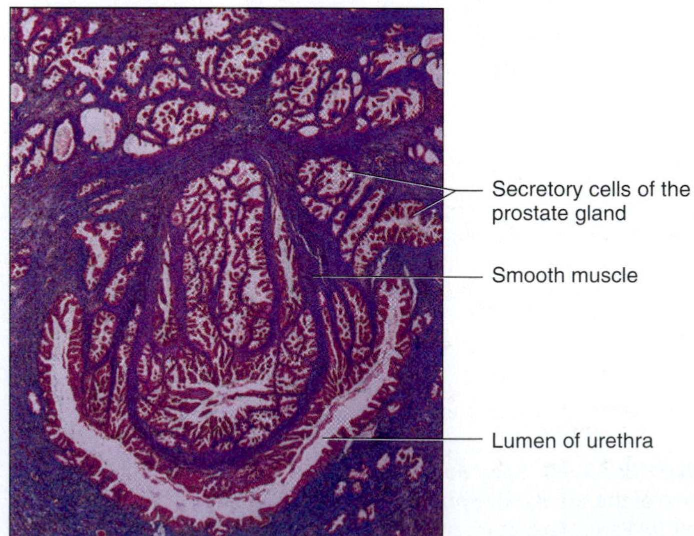

FIGURE 22.14 Light micrograph of the prostate gland (10×).

22.1 CLINICAL APPLICATION

Prostate Enlargement

The prostate gland is small in boys, begins to grow in early adolescence, and reaches adult size several years later. An adult's prostate gland is about the size of a walnut. Usually, the gland does not grow again until age fifty, when in half of all men, it enlarges enough to press on the urethra. This condition is called benign prostatic hypertrophy (BPH). As many as 90% of men over age seventy may have BPH. It produces a feeling of pressure on the bladder because it cannot empty completely, and the man feels the urge to frequently urinate. An early sign may be dribbling after urination. Retained urine can lead to infection and inflammation, bladder stones, or kidney disease.

Medical researchers do not know what causes prostate enlargement. Risk factors include a fatty diet, having had a vasectomy, possible occupational exposure to batteries or the metal cadmium, and inheriting a particular gene that also causes breast cancer. The enlargement may be benign or cancerous. Prostate cancer is highly treatable if detected early, so men should have their prostates examined regularly. Four out of five men who have prostate cancer are over age sixty-five.

Diagnostic tests for prostate cancer include a rectal exam; visualization of the prostate, urethra, and urinary bladder with a device inserted through the penis, called a cytoscope; as well as a blood test to detect elevated prostate specific antigen (PSA), a cell surface protein normally found on prostate cells. Elevated PSA levels indicate an enlarged prostate, possibly from a benign or cancerous growth. Ultrasound may provide further information on whether a benign or cancerous growth is present.

Table 22A summarizes treatments for an enlarged prostate. The components of treatment vary greatly from individual to individual. In some men, the recommended course is "watchful waiting," continuing to have frequent checkups to monitor the enlargement, but not acting until symptoms arise. Surgery to treat prostate cancer is highly effective. It once commonly left a man incontinent and with erectile dysfunction. However, control of urination often returns within a few weeks, and newer surgical methods preserve the nerves necessary for erection. ■

TABLE 22A | Some Medical Treatments for an Enlarged Prostate Gland

Surgical removal of prostate
Radiation
Drug (Proscar, or finasteride) to block testosterone's growth-stimulating effect on the prostate
Alpha blocker drugs, which relax muscles near the prostate, relieving pressure
Microwave energy delivered through a probe inserted into the urethra or rectum
Balloon inserted into the urethra and inflated with liquid
Tumor frozen with liquid nitrogen delivered by a probe through the skin
Device (stent) inserted between lobes of prostate to relieve pressure on the urethra

Bulbourethral Glands

The **bulbourethral** (bul″bo-u-re′thral) **glands** (Cowper's glands) are two small structures, each about a centimeter in diameter. They are inferior to the prostate gland lateral to the membranous urethra and are enclosed by muscle fibers of the urogenital diaphragm (see fig. 22.4).

The bulbourethral glands are composed of many tubes whose epithelial linings secrete a mucuslike fluid. This fluid is released in response to sexual stimulation and lubricates the end of the penis in preparation for sexual intercourse (coitus). However, females secrete most of the lubricating fluid for intercourse.

Semen

The fluid the urethra conveys to the outside during ejaculation is called **semen** (se′men). It consists of sperm cells from the testes and secretions of the seminal vesicles, prostate gland, and bulbourethral glands. Semen is slightly alkaline (pH about 7.5), and it includes prostaglandins and nutrients.

The volume of semen released at one time varies from 2 to 5 milliliters. The average number of sperm cells in the fluid is about 120 million per milliliter.

Sperm cells remain nonmotile while they are in the ducts of the testis and epididymis, but begin to swim as they mix with the secretions of accessory glands. However, sperm cells cannot fertilize an egg cell until they enter the female reproductive tract. Here, they undergo *capacitation,* which weakens the acrosomal membranes of the sperm cells. When sperm cells are placed with egg cells in a laboratory dish to achieve fertilization—a technique called *in vitro* fertilization, discussed in From Science to Technology 23.1 (p. 878)—chemicals are added to simulate capacitation.

Although sperm cells can live for many weeks in the ducts of the male reproductive tract, they usually survive only up to six days after being expelled to the outside, even when they are maintained at body temperature. (The ability of a sperm cell to fertilize an oocyte generally lasts only twenty-four to forty-eight hours after the sperm enter the femal reproductive tract.) On the other hand, sperm cells can be stored and kept viable for years if they are frozen at a temperature below −100°C. Clinical Application 22.2 describes some causes of male infertility.

22.2 CLINICAL APPLICATION

Male Infertility

Male infertility—the inability of sperm cells to fertilize an egg cell—has several causes. If, during fetal development, the testes do not descend into the scrotum, the higher temperature of the abdominal cavity or inguinal canal causes the developing sperm cells in the seminiferous tubules to degenerate. Certain diseases, such as mumps, may inflame the testes (orchitis), impairing fertility by destroying cells in the seminiferous tubules.

The quality and quantity of sperm cells are essential factors in the ability of a man to father a child. If a sperm head is misshapen, if a sperm cannot swim, or if there are too few sperm cells, completing the arduous journey to the well-protected egg may be impossible. Sometimes even a sperm cell that enters an egg is unsuccessful because it lacks the microtubules necessary to attract and merge the nuclei of the two cells.

In the past, sperm analysis was based on microscopic examination, which was not very accurate because the human eye cannot easily follow multiple moving targets. More objective is computer-aided sperm analysis (CASA), which has been in use since the late 1980s. CASA can analyze up to 200 moving sperm in a few seconds, assessing the number of cells per milliliter of seminal fluid (density), sperm movement (motility), and the size and shape of sperm cell parts (morphology).

In a sperm analysis, a man abstains from intercourse for two to three days, then provides a sperm sample. This may be done either in a

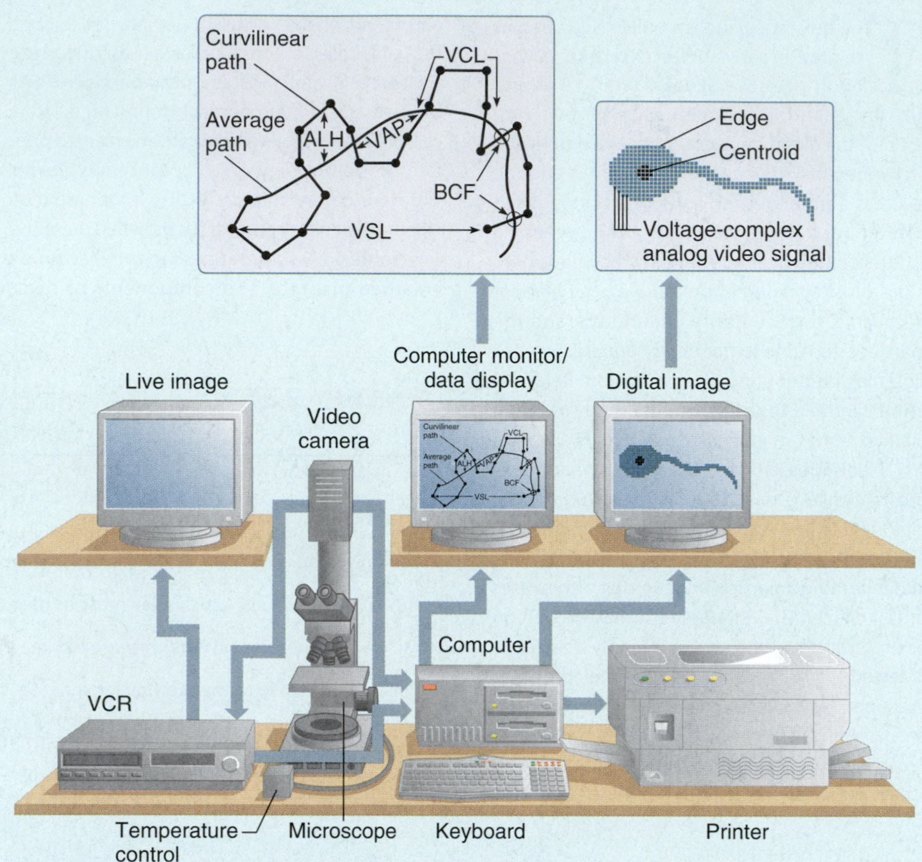

FIGURE 22A Computer analysis improves the consistency and accuracy of describing sperm density, motility, and morphology, each important in diagnosing male infertility.

PRACTICE 

16 Where is the prostate gland located?

17 What are the functions of the prostate gland's secretion?

18 What is the function of the bulbourethral glands?

19 What are the components of semen?

Male External Reproductive Organs

The male external reproductive organs are the scrotum, which encloses two testes, and the penis. The urethra passes through the penis.

Scrotum

The **scrotum** is a pouch of skin and subcutaneous tissue that hangs from the lower abdominal region posterior to the penis. The subcutaneous tissue of the scrotal wall lacks fat but contains a layer of smooth muscle fibers that constitute the *dartos muscle.* Exposure to cold stimulates these muscles to contract, the scrotal skin to wrinkle, and the testes to move closer to the pelvic cavity, where they can absorb heat. Exposure to warmth stimulates the fibers to relax and the scrotum to hang loosely and provides an environment 3°C (about 5°F) below body temperature, more conducive to sperm production and survival.

A medial septum divides the scrotum into two chambers, each of which encloses a testis. Each chamber also contains a serous membrane, which covers the front and sides of the testis and the epididymis, helping to ensure that the testis and epididymis move smoothly within the scrotum (see fig. 22.4).

clinical setting or at home using a kit ordered from the Internet, with the sample mailed to a lab. The CASA system captures images with a digital camera and analyzes and integrates information on sperm density, motility, and morphology with details of the patient's health and reproductive history (fig. 22A). The result is a "spermiogram." Figure 22B shows the results for a normal sample.

The first sperm classification system, in the 1950s, considered only sperm head shape. Then World Health Organization classification evaluated dimensions of the sperm head, midpiece, and tail, adding the requirement of analyzing at least 200 moving sperm to establish fertility. The National Institutes of Health have more recently revised the WHO standards, and these are presented in Table 22B. ■

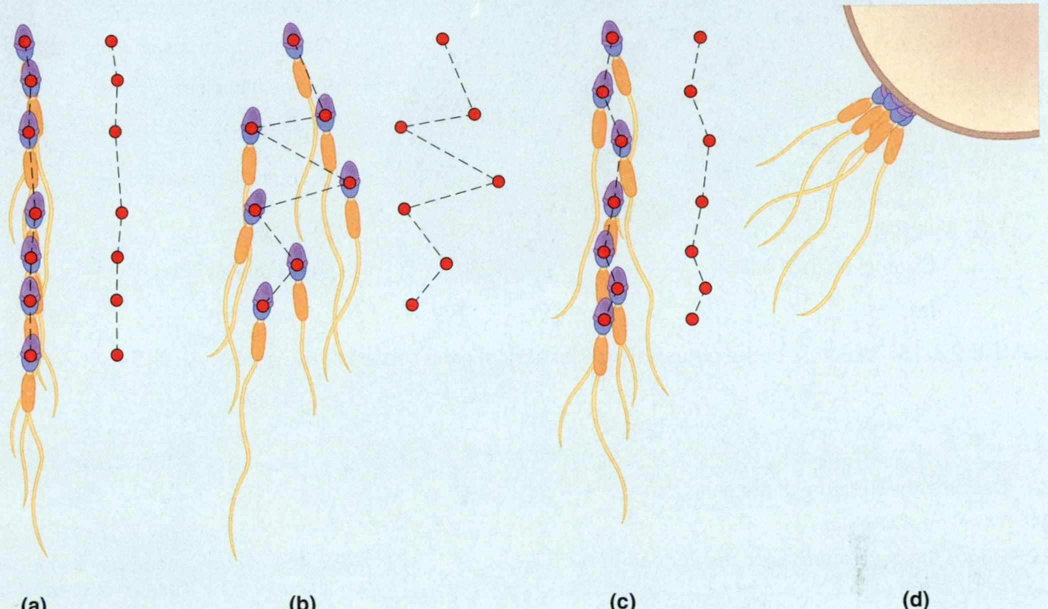

FIGURE 22B A computer tracks sperm cell movements. In semen, sperm cells swim in a straight line (*a*), but as they are activated by biochemicals in the woman's body, their trajectories widen (*b*). The sperm cells in (*c*) are in the mucus of a woman's cervix, and the sperm cells in (*d*) are attempting to digest through the structures surrounding an egg cell.

TABLE 22B | Sperm Cell Analysis

	Density (million cells/milliliter)	% Motile	% Normal morphology
Fertile	>48	>63%	>12%
Borderline fertile	13.5–48	>63%	>12%
Infertile	<13.5	<32%	<9%

Penis

The **penis** is a cylindrical organ that conveys urine and semen through the urethra to the outside. It is also specialized to enlarge and stiffen, which enables it to enter the vagina during sexual intercourse.

The *body*, or shaft, of the penis is composed of three columns of erectile tissue, which include a pair of dorsally located *corpora cavernosa* and a single, ventral *corpus spongiosum*. A tough capsule of white dense connective tissue called a *tunica albuginea* surrounds each column. Skin, a thin layer of subcutaneous tissue, and a layer of connective tissue enclose the penis (fig. 22.15).

The corpus spongiosum, through which the urethra extends, enlarges at its distal end to form a sensitive, cone-shaped **glans penis**. The glans covers the ends of the corpora cavernosa and bears the urethral opening—the *external urethral orifice*. The skin of the glans is very thin, hairless, and contains sensory receptors for sexual stimulation. A loose fold of skin called the *prepuce* (foreskin) begins just posterior to the glans and extends anteriorly to cover it as a sheath. The prepuce is sometimes removed by a surgical procedure called *circumcision*.

At the *root* of the penis, the columns of erectile tissue separate. The corpora cavernosa diverge laterally in the perineum and are firmly attached to the inferior surface of the pubic arch by connective tissue. These diverging parts form the *crura* (sing., *crus*) of the penis. The single corpus spongiosum is enlarged between the crura as the *bulb* of the penis, attached to membranes of the perineum (see fig. 22.4*b*).

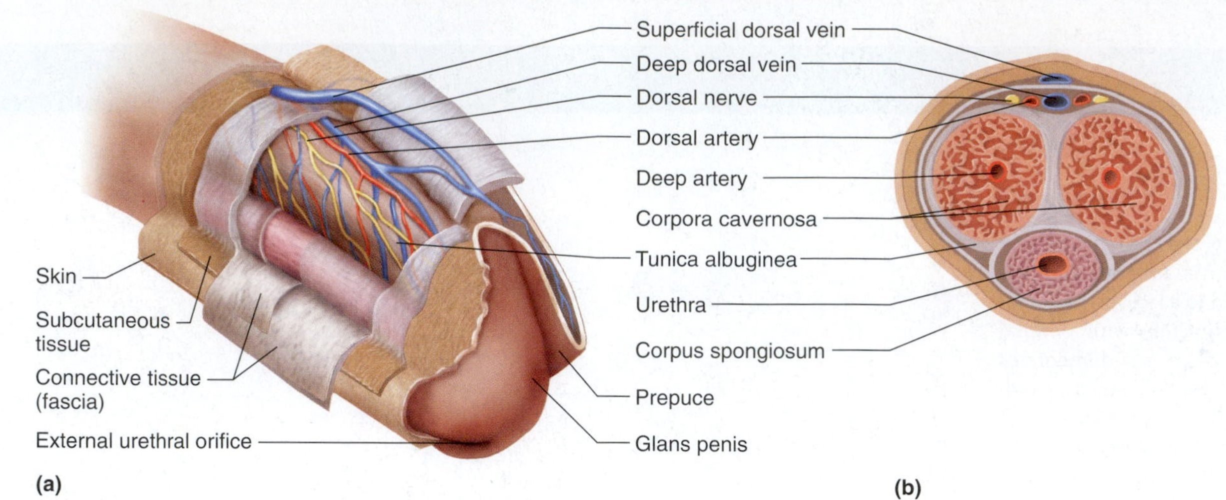

FIGURE 22.15 Structure of the penis. (*a*) Interior and (*b*) cross section of the penis.

PRACTICE

20 Describe the structure of the penis.

21 What is circumcision?

22 How is the penis attached to the perineum?

Erection, Orgasm, and Ejaculation

During sexual stimulation, parasympathetic nerve impulses from the sacral portion of the spinal cord release the vasodilator nitric oxide, which dilates the arteries leading into the penis, increasing blood flow into erectile tissues. At the same time, the increasing pressure of arterial blood entering the vascular spaces of the erectile tissue compresses the veins of the penis, reducing flow of venous blood away from the penis. Consequently, blood accumulates in the erectile tissues, and the penis swells and elongates, producing an **erection** (fig. 22.16).

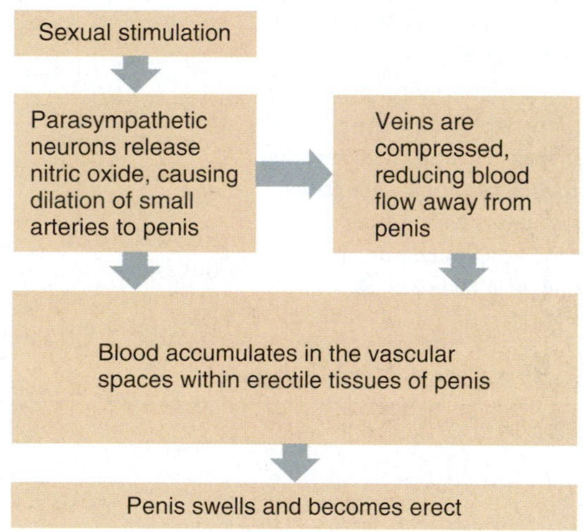

FIGURE 22.16 Mechanism of penile erection in the male.

In erectile dysfunction (impotence), the penis cannot become erect or sustain an erection. Causes of erectile dysfunction include underlying disease such as diabetes mellitus; paralysis; treatments such as prostate surgery or certain drugs; and excess smoking or drinking alcohol. Development of drugs to treat erectile dysfunction grew out of understanding the physiology of erection. The first drug, Viagra (sildenafil), blocks the enzyme that breaks down cyclic monophosphate, which is necessary for an erection to persist.

The culmination of sexual stimulation is **orgasm** (or′gazm), a pleasurable feeling of physiological and psychological release. Orgasm in the male is accompanied by emission and ejaculation.

Emission (e-mish′un) is the movement of sperm cells from the testes and secretions from the prostate gland and seminal vesicles into the urethra, where they mix to form semen. Emission is a response to sympathetic nerve impulses from the spinal cord, which stimulate peristaltic contractions in smooth muscles in the walls of the testicular ducts, epididymides, ductus deferentia, and ejaculatory ducts. Other sympathetic impulses stimulate rhythmic contractions of the seminal vesicles and prostate gland.

As the urethra fills with semen, sensory impulses are stimulated and pass into the sacral part of the spinal cord. In response, motor impulses are transmitted from the spinal cord to certain skeletal muscles at the base of the erectile columns of the penis, rhythmically contracting them. This increases the pressure in the erectile tissues and aids in forcing the semen through the urethra to the outside—a process called **ejaculation** (e-jak″u-la′shun).

The sequence of events during emission and ejaculation is coordinated so that the fluid from the bulbourethral glands is expelled first. This is followed by the release of fluid from the prostate gland, the passage of the sperm cells, and finally, the ejection of fluid from the seminal vesicles (fig. 22.17).

Immediately after ejaculation, sympathetic impulses constrict the arteries that supply the erectile tissue, reducing the inflow of blood. Smooth muscles in the walls of the vascular

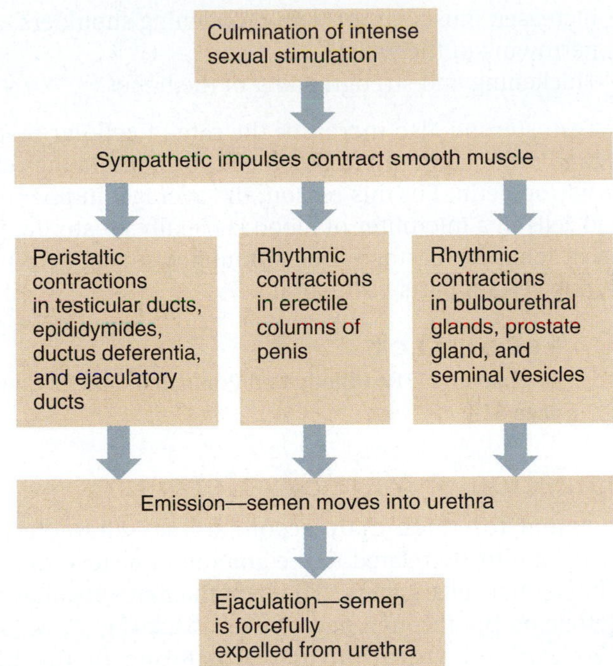

FIGURE 22.17 Mechanism of emission and ejaculation in the male.

spaces partially contract again, and the veins of the penis carry the excess blood out of these spaces. The penis gradually returns to its flaccid state, and usually another erection and ejaculation cannot be triggered for a period of ten to thirty minutes or longer. Table 22.1 summarizes the functions of the male reproductive organs.

> Spontaneous emission and ejaculation commonly occur in adolescent males during sleep and thus are called *nocturnal emissions.* Changes in hormonal concentrations that accompany adolescent development and sexual maturation cause these emissions.

PRACTICE

23 What controls blood flow into penile erectile tissues?

24 Distinguish among orgasm, emission, and ejaculation.

25 Review the events associated with emission and ejaculation.

TABLE 22.1 | Functions of the Male Reproductive Organs

Organ	Function
Testis	
Seminiferous tubules	Produce sperm cells
Interstitial cells	Produce and secrete male sex hormones
Epididymis	Promotes sperm cell maturation; stores sperm cells; conveys sperm cells to ductus deferens
Ductus deferens	Conveys sperm cells to ejaculatory duct
Seminal vesicle	Secretes an alkaline fluid containing nutrients and prostaglandins that helps neutralize the acidic components of semen
Prostate gland	Secretes an alkaline fluid that helps neutralize the acidic components of semen and enhances sperm cell motility
Bulbourethral gland	Secretes fluid that lubricates end of the penis
Scrotum	Encloses, protects, and regulates temperature of testes
Penis	Conveys urine and semen to outside of body; inserted into the vagina during sexual intercourse; the glans penis is richly supplied with sensory nerve endings associated with feelings of pleasure during sexual stimulation

22.3 HORMONAL CONTROL OF MALE REPRODUCTIVE FUNCTIONS

Hormones secreted by the *hypothalamus,* the *anterior pituitary gland,* and the testes control male reproductive functions. These hormones initiate and maintain sperm cell production and oversee the development and maintenance of male sex characteristics.

Hypothalamic and Pituitary Hormones

Prior to ten years of age, the male body is reproductively immature. It is childlike, and the spermatogenic cells of the testes are undifferentiated. Then a series of changes leads to development of a reproductively functional adult. The hypothalamus controls many of these changes.

Recall from chapter 13 (p. 497) that the hypothalamus secretes gonadotropin-releasing hormone (GnRH), which enters the blood vessels leading to the anterior pituitary gland. In response, the anterior pituitary gland secretes the **gonadotropins** (go-nad″o-trōp′inz) called *luteinizing hormone* (LH) and *follicle-stimulating hormone* (FSH). LH, which in males has been referred to as interstitial cell-stimulating hormone (ICSH), promotes development of the interstitial cells of the testes, and they, in turn, secrete male sex hormones. FSH stimulates the sustentacular cells of the seminiferous tubules to proliferate, grow, mature, and respond to the effects of the male sex hormone testosterone. Then, in the presence of FSH and testosterone, these cells stimulate the spermatogenic cells to undergo spermatogenesis, giving rise to sperm cells (fig. 22.18). The sustentacular cells also secrete a hormone called *inhibin,* which inhibits the anterior pituitary gland by negative feedback and thus prevents oversecretion of FSH.

Male Sex Hormones

Male sex hormones are termed **androgens** (an′dro-jenz). The interstitial cells of the testes produce most of them, but small amounts are synthesized in the adrenal cortex (see chapter 13, p. 508).

The hormone **testosterone** (tes-tos′tĕ-rōn) is the most important androgen. It is secreted and transported in the blood, loosely attached to plasma proteins. Like other steroid hormones, testosterone binds receptor molecules which are usually in the nuclei of its target cells (see chapter 13, p. 485). However, in many target cells, such as those in the prostate gland, seminal vesicles, and male external accessory organs, testosterone is first converted to another androgen called **dihydrotestosterone** (di-hi″dro-tes-tos′ter-ōn), which stimulates the cells of these organs. Androgen molecules that do not reach receptors in target cells are usually changed by the liver into forms that can be excreted in bile or urine.

Testosterone secretion begins during fetal development and continues for several weeks following birth; then it nearly ceases during childhood. Between the ages of thirteen and fifteen, a young man's androgen production usually increases rapidly. This phase in development, when an individual becomes reproductively functional, is **puberty** (pu′ber-te). After puberty, testosterone secretion continues throughout the life of a male.

> In a group of disorders called male pseudohermaphroditism, testes are usually present, but a block in testosterone synthesis prevents the genetically male fetus from developing male structures, and as a result, later, the child appears to be a girl. But at puberty, the adrenal glands begin to produce testosterone, as they normally do in any male. This leads to masculinization. The voice deepens, and muscles build up into a masculine physique. Breasts do not develop, nor does menstruation occur. The clitoris may enlarge so greatly under the adrenal testosterone surge that it looks like a penis. Individuals with a form of this condition prevalent in the Dominican Republic are called *guevedoces,* which means "penis at age twelve."

Actions of Testosterone

Cells of the embryonic testes first produce testosterone after about eight weeks of development. This hormone stimulates the formation of the male reproductive organs, including the penis, scrotum, prostate gland, seminal vesicles, and ducts. Later in development, testosterone causes the testes to descend into the scrotum.

During puberty, testosterone stimulates enlargement of the testes (the primary male sex characteristic) and accessory organs of the reproductive system, as well as development of male *secondary sex characteristics,* which are special features associated with the adult male body. Secondary sex characteristics in the male include:

1. Increased growth of body hair, particularly on the face, chest, axillary region, and pubic region. Sometimes growth of hair on the scalp slows.
2. Enlargement of the larynx and thickening of the vocal folds, with lowering of the pitch of the voice.
3. Thickening of the skin.
4. Increased muscular growth, broadening shoulders, and narrowing of the waist.
5. Thickening and strengthening of the bones.

Testosterone also increases the rate of cellular metabolism and production of red blood cells by stimulating release of erythropoietin. For this reason, the average number of red blood cells in a microliter of blood is usually greater in males than in females. Testosterone stimulates sexual activity by affecting certain parts of the brain.

RECONNECT

To Chapter 14, Red Blood Cell Production and Its Control, page 527.

Regulation of Male Sex Hormones

The extent to which male secondary sex characteristics develop is directly related to the amount of testosterone that the interstitial cells secrete. The hypothalamus regulates testosterone output through negative feedback (fig. 22.18).

As the concentration of testosterone in the blood increases, the hypothalamus becomes inhibited, decreasing its stimulation of the anterior pituitary gland by GnRH. As the pituitary's secretion of LH falls in response, the amount of testosterone the interstitial cells release decreases.

As the blood testosterone concentration drops, the hypothalamus becomes less inhibited, and it once again stimulates the anterior pituitary gland to release LH. The increasing secretion of LH causes the interstitial cells to release more testosterone, and blood testosterone concentration increases. Testosterone level decreases somewhat during and after the *male climacteric,* which is a decline in sexual function that accompanies aging. At any given age, the testosterone concentration in the male body is regulated to remain relatively constant.

PRACTICE

26 Which hormone initiates the changes associated with male sexual maturity?

27 Describe several male secondary sex characteristics.

28 Explain how the secretion of male sex hormones is regulated.

22.4 ORGANS OF THE FEMALE REPRODUCTIVE SYSTEM

The organs of the female reproductive system are specialized to produce and maintain the female sex cells, the *egg cells* (or oocytes); transport these cells to the site of fertilization; provide a favorable environment for a developing offspring; move the offspring to the outside; and produce female sex hormones.

The *primary sex organs* (gonads) of this system are the two ovaries, which produce the female sex cells and sex hormones. The *accessory sex organs* of the female reproductive system are the internal and external reproductive organs (fig. 22.19; reference plates 5 and 6).

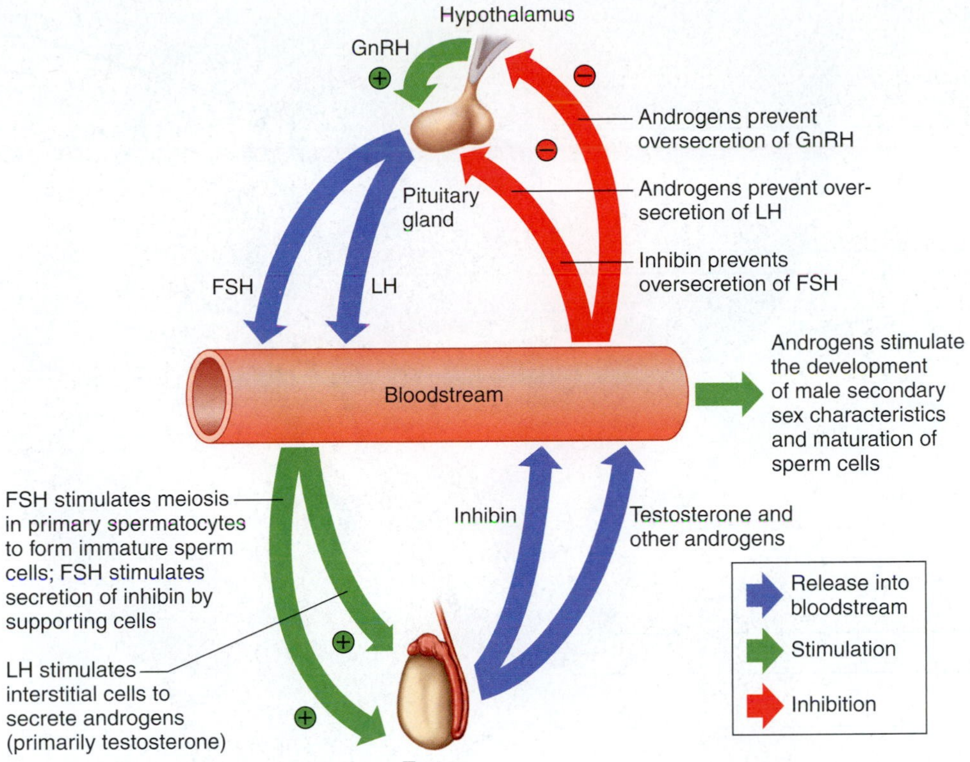

FIGURE 22.18 The hypothalamus controls maturation of sperm cells and development of male secondary sex characteristics. Negative feedback among the hypothalamus, the anterior lobe of the pituitary gland, and the testes controls the concentration of testosterone in the male body.

Ovaries

The two **ovaries** are solid, ovoid structures measuring about 3.5 centimeters in length, 2 centimeters in width, and 1 centimeter in thickness. The ovaries lie in shallow depressions (ovarian fossae) on each side in the lateral wall of the pelvic cavity (fig. 22.20).

Ovary Attachments

Several ligaments help hold each ovary in position. The largest of these, formed by a fold of peritoneum, is called the *broad ligament.* It is also attached to the uterine tubes and the uterus.

A small fold of peritoneum, called the *suspensory ligament,* holds the ovary at its upper end. This ligament also contains the ovarian blood vessels and nerves. At its lower end, the ovary is attached to the uterus by a rounded, cordlike thickening of the broad ligament called the *ovarian ligament* (fig. 22.20).

Ovary Descent

Like the testes in a male fetus, the ovaries in a female fetus originate from masses of tissue posterior to the parietal peritoneum, near the developing kidneys. During development, these structures descend to locations just inferior to the pelvic brim, where they remain attached to the lateral pelvic wall.

Ovary Structure

The tissues of an ovary can be subdivided into two rather indistinct regions, an inner *medulla* and an outer *cortex.* The ovarian medulla is mostly composed of loose connective tissue and contains many blood vessels, lymphatic vessels, and nerve fibers. The ovarian cortex consists of more compact tissue and has a granular appearance due to tiny masses of cells called *ovarian follicles.*

A layer of cuboidal epithelial cells (germinal epithelium) covers the free surface of the ovary. Just beneath this epithelium is a layer of dense connective tissue called the *tunica albuginea* (too′nĭ-kah al″bu-jin′e-ah).

PRACTICE 

29 What are the primary sex organs of the female?

30 Describe the descent of the ovary.

31 Describe the structure of an ovary.

Primordial Follicles

During prenatal (before birth) development of a female, oogonia divide by mitosis to produce more oogonia. The oogonia develop into *primary oocytes.* Each primary oocyte is closely surrounded by a layer of flattened epithelial cells called *follicular cells,* forming a **primordial follicle.**

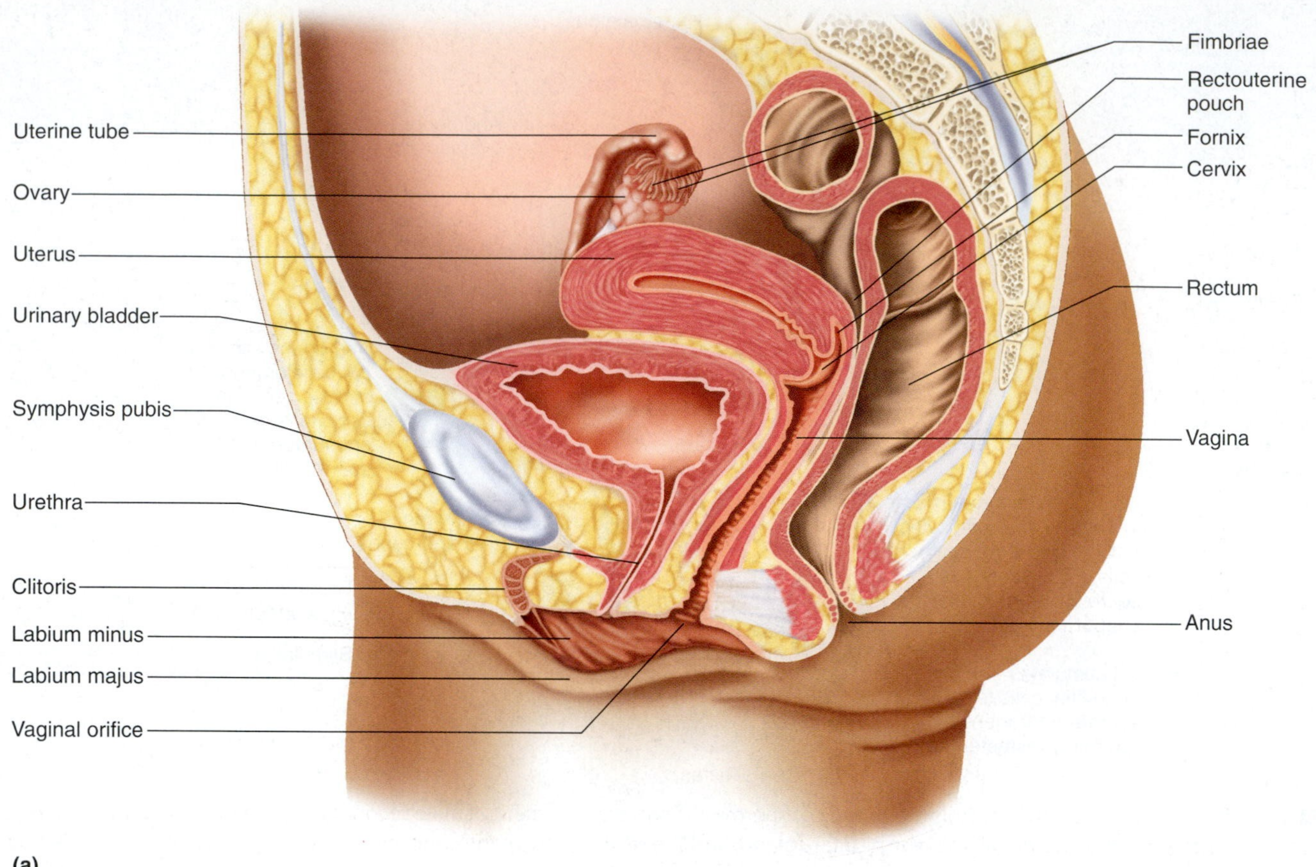

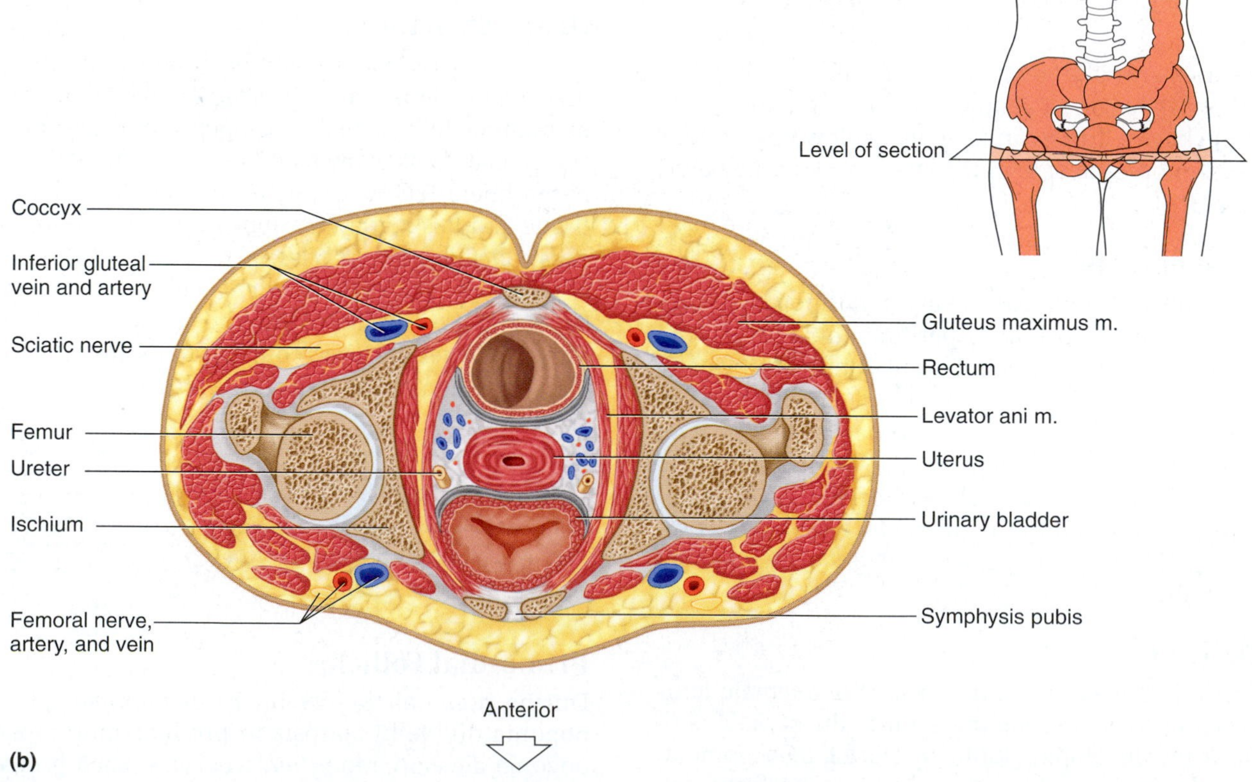

FIGURE 22.19 The paired ovaries are the primary female sex organs, and the other structures, both internal and external, are accessory sex organs. (*a*) Sagittal view. (*b*) Transverse section of the female pelvic cavity. (*m.* stands for muscle.)

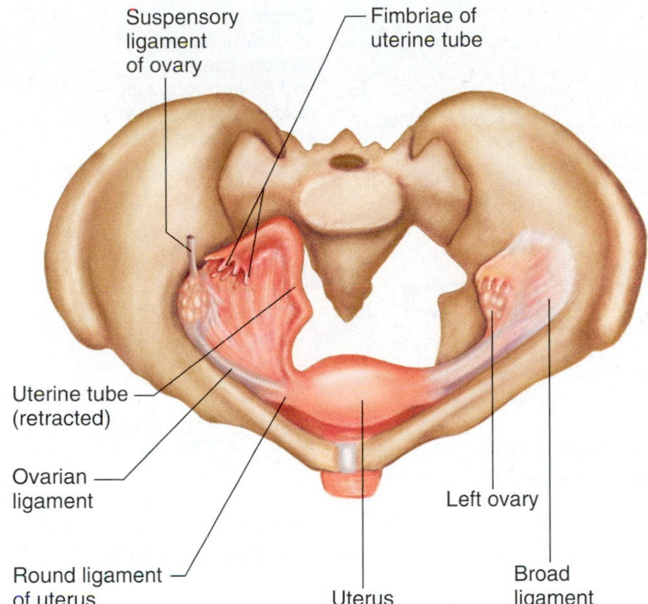

FIGURE 22.20 The ovaries are located on each side against the lateral walls of the pelvic cavity. The right uterine tube is retracted to reveal the ovarian ligament.

Early in development, the primary oocytes begin to undergo meiosis, but the process soon halts and does not continue until the individual reaches puberty. Once the primordial follicles appear, no new ones form. Instead, the number of oocytes in the ovary steadily declines, as many of the oocytes degenerate. Of the several million oocytes that formed in the embryo, only a million or so remain at the time of birth, and perhaps 400,000 are present at puberty. Of these, probably fewer than 400 or 500 will be released from the ovary during the reproductive life of a female. Probably fewer than ten will go on to form a new individual.

A possible explanation for the increased incidence of chromosome defects in children of older mothers is that the primary oocytes, having been present for several decades, had time to be extensively exposed to damaging agents, such as radiation, viruses, and toxins.

Oogenesis

Oogenesis (o′o-jen′ĕ-sis) is the process of egg cell formation. Beginning at puberty, some primary oocytes are stimulated to continue meiosis. As in the case of sperm cells, the resulting cells have one-half as many chromosomes (23) in their nuclei as their parent cells, constituting one chromosome set.

Unlike a primary spermatocyte, when a primary oocyte divides, the cytoplasm is distributed unequally. One of the resulting cells, called a *secondary oocyte,* is large, and the other, called the *first polar body,* is small (fig. 22.21).

The large secondary oocyte represents a future *egg cell* (ovum) that can be fertilized by uniting with a sperm cell. If this happens, the oocyte divides unequally to produce a tiny *second polar body* and a large fertilized egg cell, or **zygote** (zi′gōt), that can divide and develop into an **embryo** (em′bre-o). An embryo is the stage of prenatal development when the rudiments of all organs form. The polar bodies have no further function, and they begin to degenerate fifteen hours post fertilization.

Formation of polar bodies may appear wasteful, but it has an important biological function. It allows for production of an egg cell that has the massive amounts of cytoplasm and abundant organelles required to carry a zygote through the first few cell divisions, yet the right number of chromosomes.

PRACTICE 

32 Describe the major events of oogenesis.

33 What is the function of polar body formation?

Follicle Maturation

At puberty, the anterior pituitary gland secretes increased amounts of FSH, and the ovaries enlarge in response. With each reproductive cycle, some of the primordial follicles mature (fig. 22.22). Within each maturing primordial follicle, the oocyte enlarges and the surrounding follicular cells divide mitotically, giving rise to a stratified epithelium composed of *granulosa cells.* A layer of glycoprotein, called the **zona pellucida** (zo′nah pel-u′cĭ-dah), gradually separates the primary oocyte from the granulosa cells; at this stage, the structure is called a *primary follicle.*

Meanwhile, the ovarian cells outside the follicle organize into layers. The *inner vascular layer* (theca interna) is largely composed of steroid-secreting cells, plus some loose connective tissue and blood vessels. The *outer fibrous layer* (theca externa) consists of tightly packed connective tissue cells.

The follicular cells continue to proliferate, and when there are six to twelve layers of cells, irregular, fluid-filled spaces appear among them. These spaces soon join to form a single cavity (antrum), and the primary oocyte is pressed to one side of the follicle. At this stage, the follicle is about 0.2 millimeter in diameter and is called a *secondary follicle.*

About one week into the cycle, one follicle becomes the *dominant follicle.* Maturation of this follicle takes ten to fourteen days. The *mature follicle* (preovulatory, or Graafian, follicle) is about 10 millimeters or more in diameter, and its fluid-filled cavity bulges outward on the surface of the ovary, like a blister. The secondary oocyte within the mature follicle is a large, spherical cell, surrounded by a thick zona pellucida, attached to a mantle of follicular cells called the *corona radiata.* Processes from these follicular cells extend through the zona pellucida and supply nutrients to the oocyte (fig. 22.23).

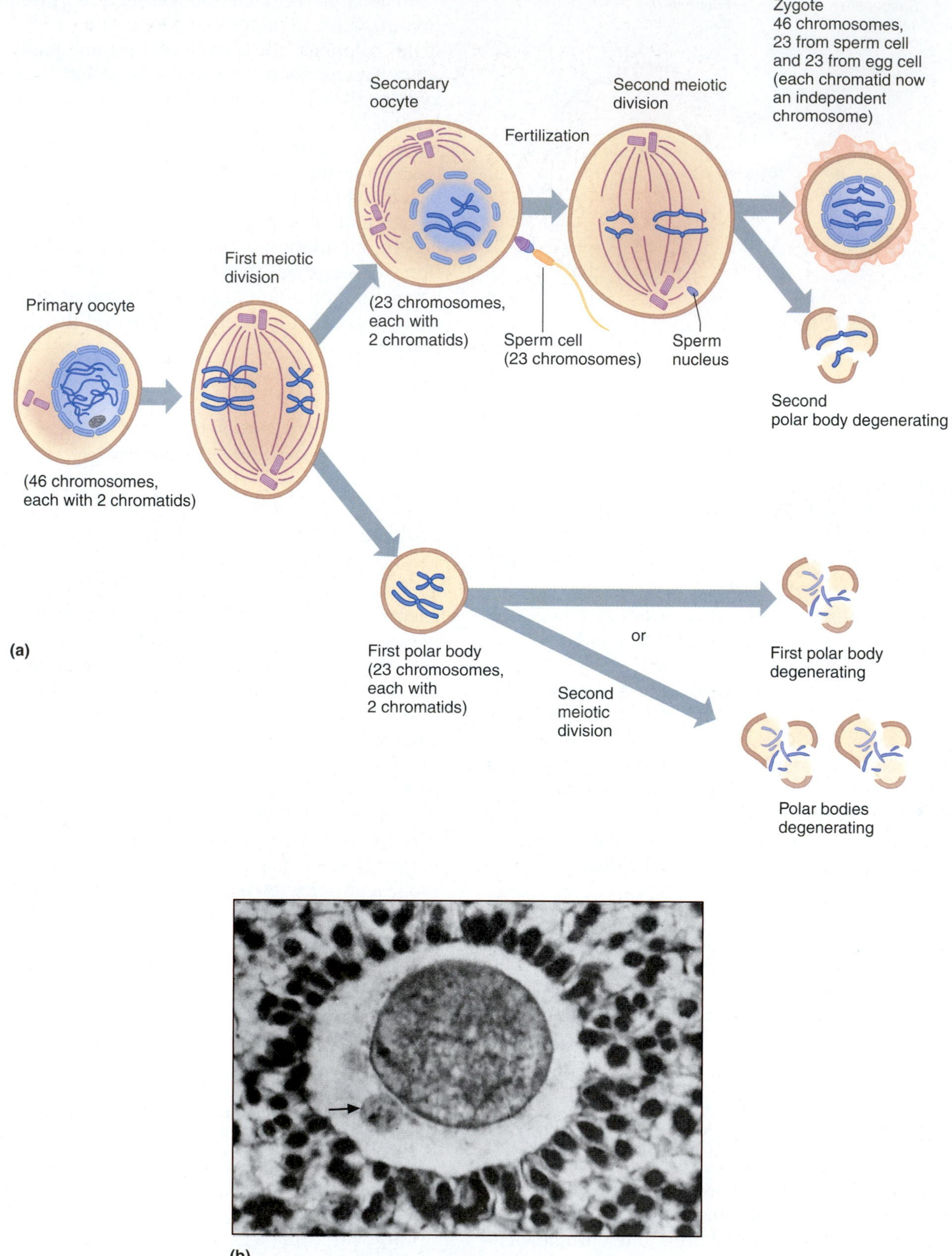

FIGURE 22.21 During oogenesis, (*a*) a single egg cell (secondary oocyte) results from meiosis of a primary oocyte. If the egg cell is fertilized, it generates a second polar body and becomes a zygote. (Note: The second meiotic division does not occur in the egg cell if it is not fertilized.) (*b*) Light micrograph of a secondary oocyte and a polar body (arrow) (700×).

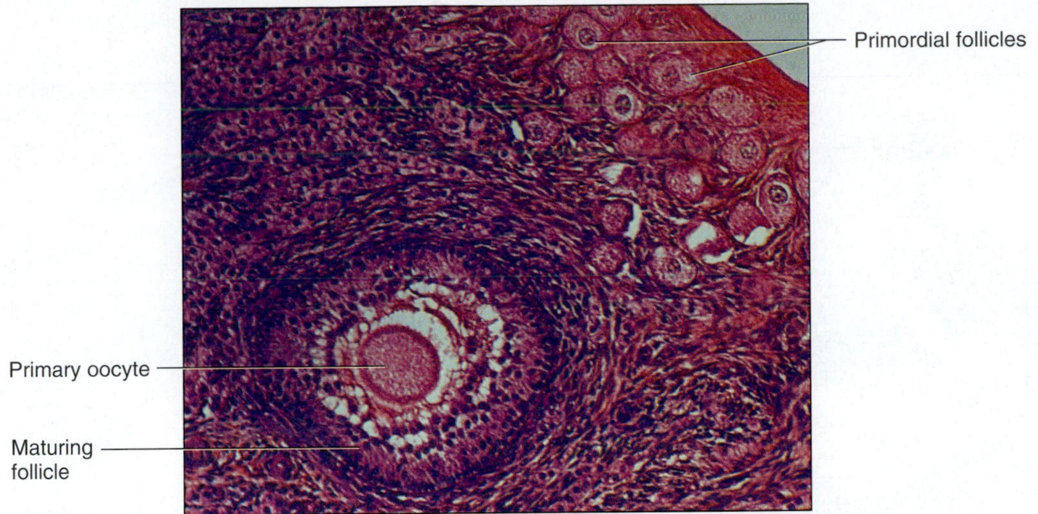

FIGURE 22.22 Light micrograph of the surface of a mammalian ovary (200×).

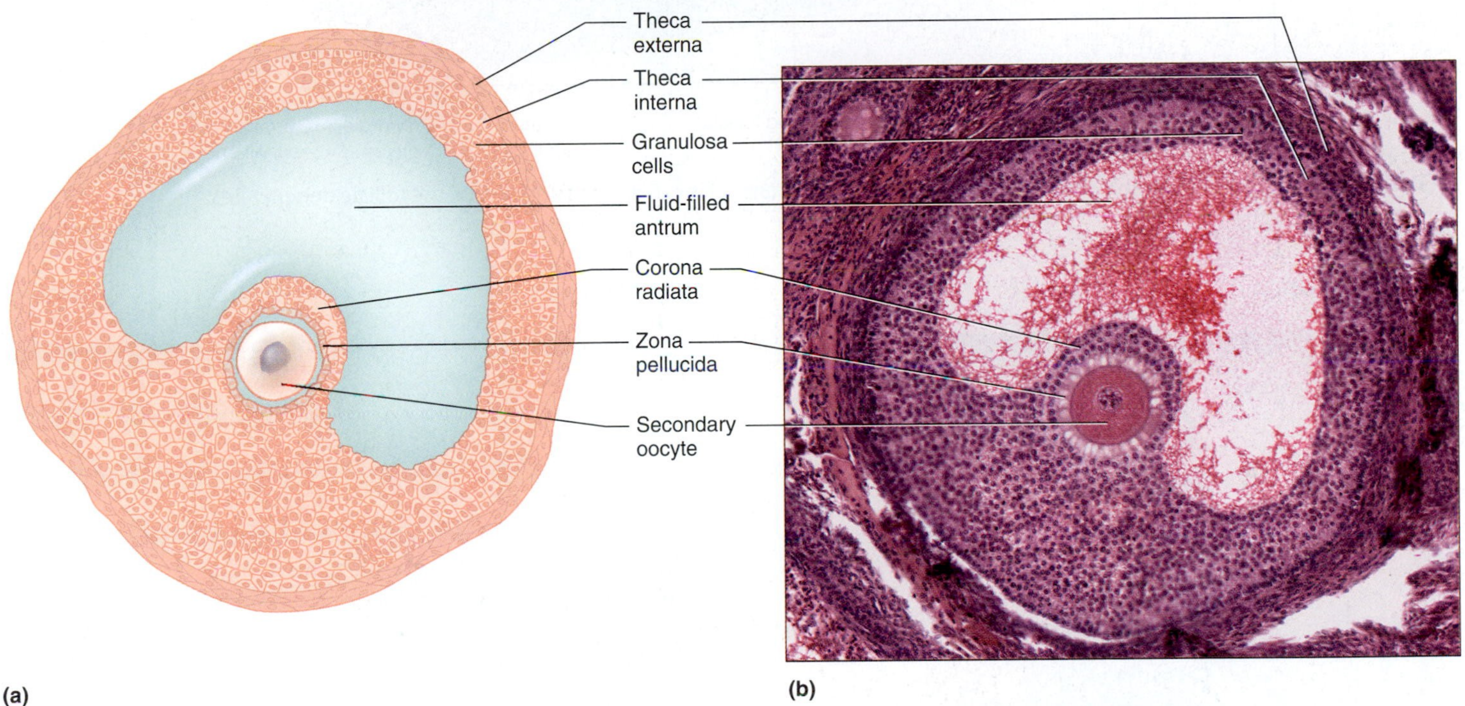

FIGURE 22.23 Ovarian follicle. (*a*) Structure of a mature (Graafian) follicle. (*b*) Light micrograph of a mature follicle (250×).

Although as many as twenty primary follicles may begin maturing at any one time, one dominant follicle usually outgrows the others. Typically, only the dominant follicle fully develops, and the other follicles degenerate (fig. 22.24).

Certain drugs used to treat female infertility, such as Clomid (clomiphene), may cause a woman to "superovulate." More than one follicle grows, more than one secondary oocyte is released, and if all of these secondary oocytes are fertilized and complete prenatal development, multiple births may result.

Ovulation

As a follicle matures, its primary oocyte undergoes meiosis I, giving rise to a secondary oocyte and a first polar body. A process called **ovulation** (o″vu-la′shun) releases these cells from the follicle.

Release of LH from the anterior pituitary gland triggers ovulation, which rapidly swells the mature follicle and weakens its wall. Eventually the wall ruptures, and the follicular fluid, accompanied by the secondary oocyte, oozes outward from the surface of the ovary. Figure 22.25 shows expulsion of a mammalian oocyte.

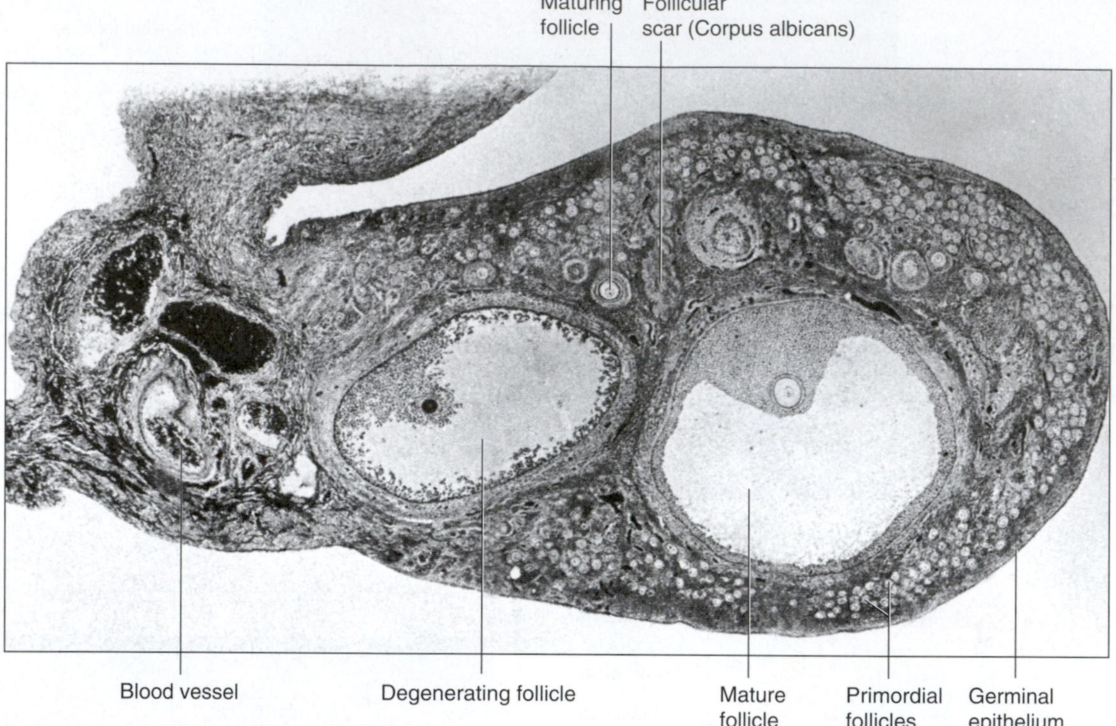

FIGURE 22.24 Light micrograph of a mammalian (monkey) ovary (30×). If ovulation does not occur, the follicle degenerates.

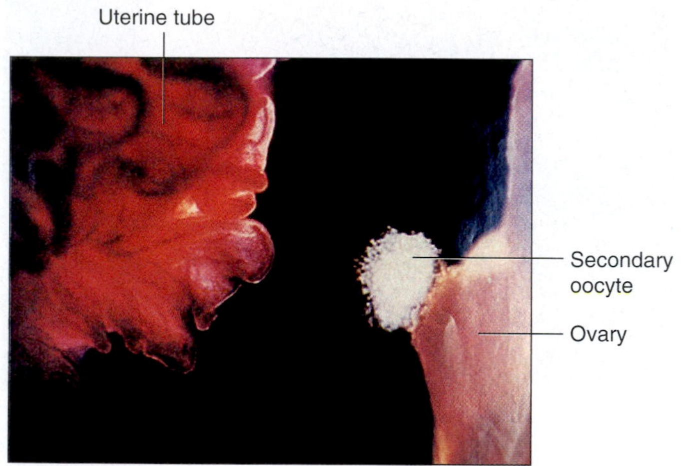

FIGURE 22.25 Light micrograph of a follicle during ovulation (75×).

After ovulation, the secondary oocyte and one or two layers of follicular cells surrounding it are usually propelled to the opening of a nearby uterine tube. If the secondary oocyte is not fertilized within hours, it degenerates. Figure 22.26 illustrates a conceptual progression of maturation of a follicle over time and the release of an oocyte. In reality, the secondary oocyte could be released from any external region of the ovary and not directly into the uterine tube.

PRACTICE

34 What changes occur in a follicle and its oocyte during maturation?

35 What causes ovulation?

36 What happens to an oocyte following ovulation?

Female Internal Accessory Organs

The internal accessory organs of the female reproductive system include a pair of uterine tubes, a uterus, and a vagina.

Uterine Tubes

The **uterine tubes** (fallopian tubes, or oviducts) are suspended by portions of the broad ligament and open near the ovaries. Each tube, about 10 centimeters long and 0.7 centimeters in diameter, passes medially to the uterus, penetrates its wall, and opens into the uterine cavity.

Near each ovary, a uterine tube expands to form a funnel-shaped **infundibulum** (in″fun-dib′u-lum), which partially encircles the ovary medially. On its margin, the infundibulum bears a number of irregular, branched extensions called **fimbriae** (fim′bre) (fig. 22.27). Although the infundibulum generally does not touch the ovary, one of the larger extensions (ovarian fimbria) connects directly to the ovary.

The wall of a uterine tube consists of an inner mucosal layer, a middle muscular layer, and an outer covering of peritoneum. The mucosal layer is drawn into many longitudinal

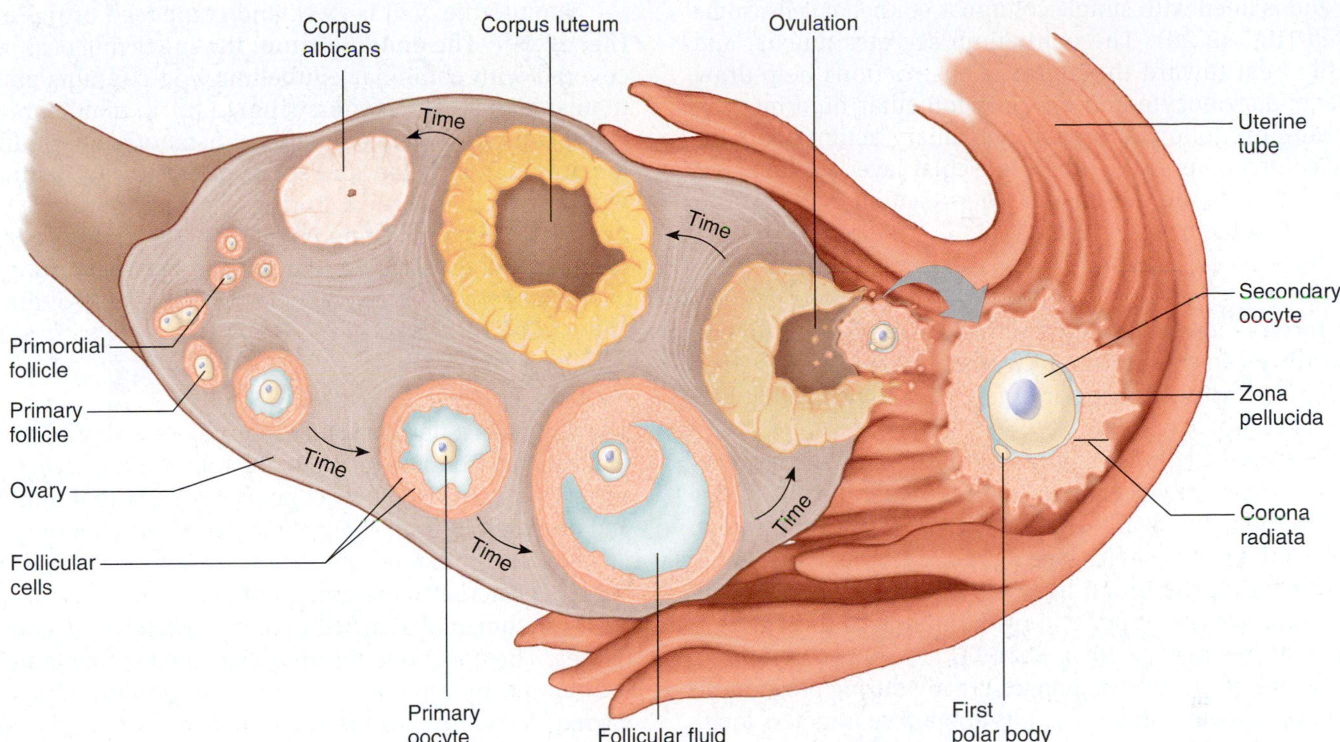

FIGURE 22.26 In an ovary, as a follicle matures, a developing oocyte enlarges and becomes surrounded by follicular cells and fluid. Eventually, the mature follicle ruptures, releasing the secondary oocyte and layers of surrounding follicular cells.

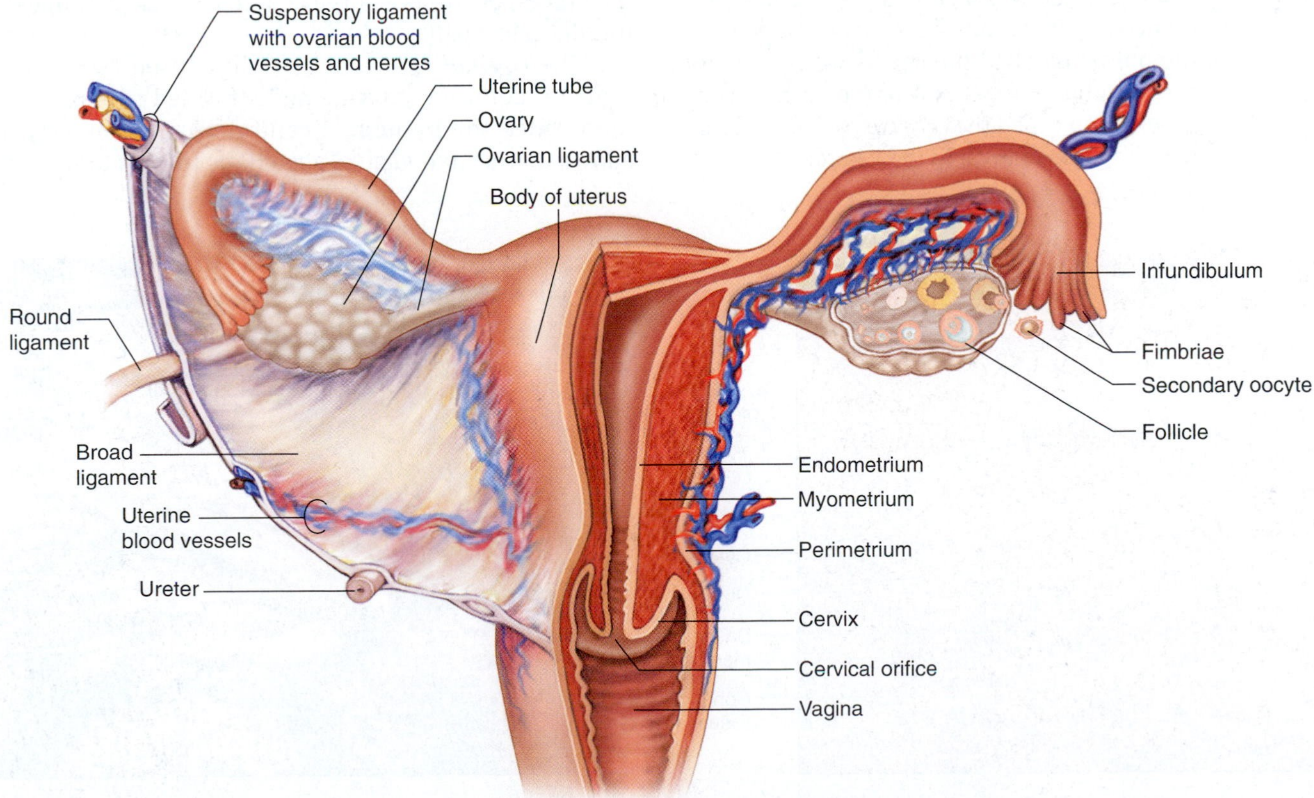

FIGURE 22.27 The funnel-shaped infundibulum of the uterine tube partially encircles the ovary (posterior view).

folds and is lined with simple columnar epithelial cells, some *ciliated* (fig. 22.28). The epithelium secretes mucus, and the cilia beat toward the uterus. These actions help draw the secondary oocyte and expelled follicular fluid into the infundibulum following ovulation. Ciliary action and peristaltic contractions of the tube's muscular layer aid transport of the egg down the uterine tube. Fertilization may occur in the uterine tube.

Uterus

The **uterus** receives the embryo that develops from an egg cell fertilized in the uterine tube and sustains its development. It is a hollow, muscular organ, shaped somewhat like an inverted pear.

The *broad ligament,* which also attaches to the ovaries and uterine tubes, extends from the lateral walls of the uterus to the pelvic walls and floor, creating a drape across the top of the pelvic cavity (see fig. 22.27). A flattened band of tissue within the broad ligament, called the *round ligament,* connects the upper end of the uterus to the anterior pelvic wall (see figs. 22.20 and 22.27).

The size of the uterus changes greatly during pregnancy. In its nonpregnant, adult state, it is about 7 centimeters long, 5 centimeters wide (at its broadest point), and 2.5 centimeters in diameter. The uterus is located medially in the anterior part of the pelvic cavity, superior to the vagina, and usually bends forward over the urinary bladder.

The upper two-thirds, or *body,* of the uterus has a dome-shaped top, called the *fundus,* and is joined by the uterine tubes, which enter its wall at its broadest part. The lower one-third, or neck, of the uterus is called the **cervix.** This tubular part extends downward into the upper part of the vagina. The cervix surrounds the opening called the *cervical orifice* (ostium uteri), through which the uterus opens to the vagina.

The uterine wall is thick and composed of three layers (fig. 22.29). The **endometrium,** the inner mucosal layer, is covered with columnar epithelium and contains abundant tubular glands. The **myometrium,** a thick, middle, muscular layer, consists largely of bundles of smooth muscle fibers in longitudinal, circular, and spiral patterns and is interlaced with connective tissues. During the monthly female reproductive cycles and during pregnancy, the endometrium and myometrium extensively change. The **perimetrium** consists of an outer serosal layer, which covers the body of the uterus and part of the cervix.

Vagina

The **vagina** is a fibromuscular tube, about 9 centimeters long, that extends from the uterus to the outside. It conveys uterine secretions, receives the erect penis during sexual intercourse, and provides an open channel for the offspring during birth.

The vagina extends upward and back into the pelvic cavity. It is posterior to the urinary bladder and urethra, anterior to the rectum, and attached to these structures by connective tissues. The upper one-fourth of the vagina is separated from the rectum by a pouch (rectouterine pouch). The tubular vagina also surrounds the end of the cervix, and the recesses between the vaginal wall and the cervix are termed *fornices* (sing., *fornix*). The fornices are clinically important because they are thin-walled and allow the physician to palpate the internal abdominal organs during a physical examination. Also, the posterior fornix, which is somewhat longer than the others, provides a surgical access to the peritoneal cavity through the vagina.

The *vaginal orifice* is partially closed by a thin membrane of connective tissue and stratified squamous epithelium called the **hymen.** A central opening of varying size allows uterine and vaginal secretions to pass to the outside.

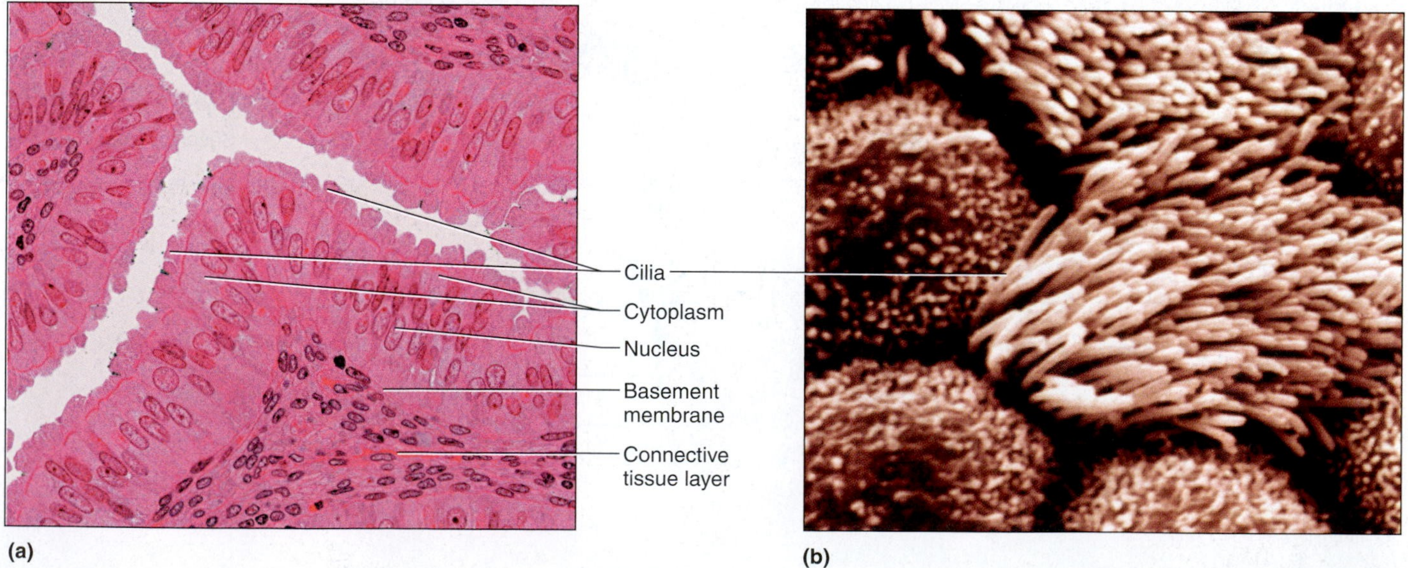

FIGURE 22.28 Uterine tube. (*a*) Light micrograph of a uterine tube (800×). (*b*) Falsely colored scanning electron micrograph of ciliated cells that line the uterine tube (4,000×).

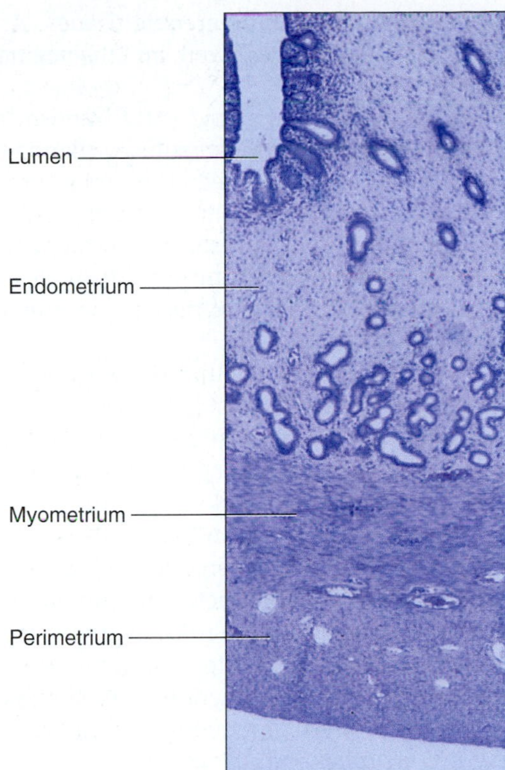

FIGURE 22.29 Light micrograph of the uterine wall (10×).

The vaginal wall has three layers. The inner *mucosal layer* is stratified squamous epithelium and is drawn into many longitudinal and transverse ridges (vaginal rugae). This layer lacks mucous glands; the mucus in the lumen of the vagina comes from the glands of the cervix and the vestibular glands at the mouth of the vagina.

The middle *muscular layer* of the vagina mainly consists of smooth muscle fibers in longitudinal and circular patterns. At the lower end of the vagina is a thin band of striated muscle. This band helps close the vaginal opening; however, a voluntary muscle (bulbospongiosus) is primarily responsible for closing this orifice.

The outer *fibrous layer* consists of dense connective tissue interlaced with elastic fibers. It attaches the vagina to surrounding organs.

PRACTICE

37 How does a secondary oocyte move into the infundibulum following ovulation?

38 How is a secondary oocyte moved along a uterine tube?

39 Describe the structure of the uterus.

40 What is the function of the uterus?

41 Describe the structure of the vagina.

Female External Reproductive Organs

The *external accessory organs* of the female reproductive system include the labia majora, the labia minora, the clitoris, and the vestibular glands. These structures that surround the openings of the urethra and vagina compose the **vulva** (fig. 22.30).

Labia Majora

The **labia majora** (sing., *labium majus*) enclose and protect the other external reproductive organs. They correspond to the scrotum of the male and are composed of rounded folds of adipose tissue and a thin layer of smooth muscle, covered by skin. On the outside, this skin includes hairs, sweat glands, and sebaceous glands, whereas on the inside, it is thinner and hairless.

The labia majora lie close together and are separated longitudinally by a cleft (pudendal cleft), which includes the urethral and vaginal openings. At their anterior ends, the labia merge to form a medial, rounded elevation of adipose tissue called the *mons pubis,* which overlies the symphysis pubis. At their posterior ends, the labia taper and merge into the perineum near the anus.

Labia Minora

The labia minora (sing., *labium minus*) are flattened longitudinal folds between the labia majora. They are composed of connective tissue richly supplied with blood vessels, giving a pinkish appearance. Stratified squamous epithelium covers this tissue. Posteriorly, the labia minora merge with the labia majora, whereas anteriorly, they converge to form a hood-like covering around the clitoris.

Clitoris

The **clitoris** (kli′to-ris) is a small projection at the anterior end of the vulva between the labia minora. It is usually about 2 centimeters long and 0.5 centimeter in diameter,

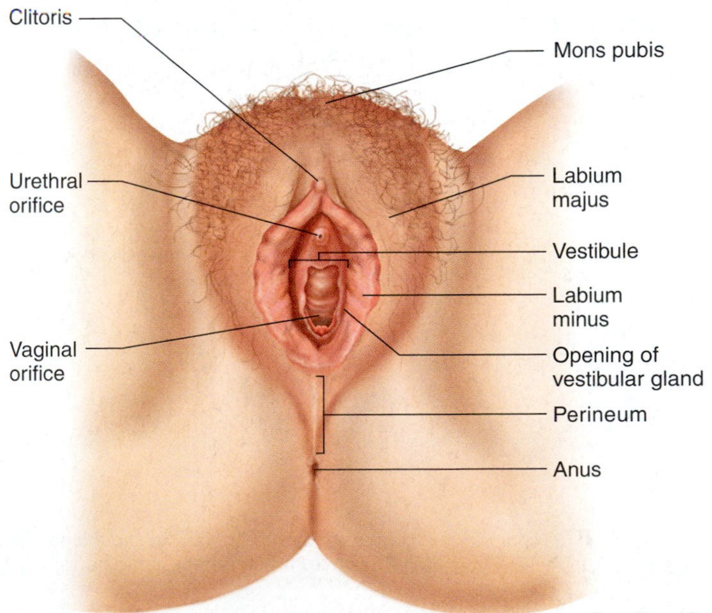

FIGURE 22.30 Female external reproductive organs and associated structures.

including a portion embedded in surrounding tissues. The clitoris corresponds to the penis and has a similar structure. It is composed of two columns of erectile tissue called *corpora cavernosa*. A septum separates these columns, which are covered with dense connective tissue.

At the root of the clitoris, the corpora cavernosa diverge to form *crura*, which, in turn, attach to the sides of the pubic arch. At its anterior end, a small mass of erectile tissue forms a glans, which is richly supplied with sensory nerve fibers.

Vestibule

The labia minora enclose the space called the **vestibule.** The vagina opens into the posterior portion of the vestibule, and the urethra opens in the midline, just anterior to the vagina and about 2.5 centimeters posterior to the glans of the clitoris.

A pair of **vestibular glands** (Bartholin's glands), corresponding to the bulbourethral glands in the male, lie on either side of the vaginal opening. Their ducts open into the vestibule near the lateral margins of the vaginal orifice.

Beneath the mucosa of the vestibule on either side is a mass of vascular erectile tissue. These structures are called the *vestibular bulbs*. They are separated from each other by the vagina and the urethra, and they extend forward from the level of the vaginal opening to the clitoris.

PRACTICE 

42 What is the male counterpart of the labia majora? Of the clitoris?

43 Which structures are within the vestibule?

Erection, Lubrication, and Orgasm

Erectile tissues in the clitoris and around the vaginal entrance respond to sexual stimulation. Following such stimulation, parasympathetic nerve impulses from the sacral portion of the spinal cord release the vasodilator nitric oxide, dilating the arteries associated with the erectile tissues. As a result, blood inflow increases, tissues swell, and the vagina expands and elongates.

If sexual stimulation is sufficiently intense, parasympathetic impulses stimulate the vestibular glands to secrete mucus into the vestibule. This secretion moistens and lubricates the tissues surrounding the vestibule and the lower end of the vagina, facilitating insertion of the penis into the vagina. Mucus secretion continuing during sexual intercourse helps prevent irritation of tissues that might occur if the vagina remained dry.

The clitoris is abundantly supplied with sensory nerve fibers, which are especially sensitive to local stimulation. The culmination of such stimulation is orgasm, the pleasurable sensation of physiological and psychological release.

Just prior to orgasm, the tissues of the outer third of the vagina engorge with blood and swell. This increases the friction on the penis during intercourse. Orgasm initiates a series of reflexes involving the sacral and lumbar parts of the spinal cord. In response to these reflexes, the muscles of the perineum and the walls of the uterus and uterine tubes contract rhythmically. These contractions help transport sperm cells through the female reproductive tract toward the upper ends of the uterine tubes (fig. 22.31).

Following orgasm, the flow of blood into the erectile tissues slackens, and the muscles of the perineum and reproductive tract relax. Consequently, the organs return to a state similar to that prior to sexual stimulation. Table 22.2 summarizes the functions of the female reproductive organs.

PRACTICE 

44 What events result from parasympathetic stimulation of the female reproductive organs?

45 What changes occur in the vagina just prior to and during female orgasm?

46 How do the uterus and the uterine tubes respond to orgasm?

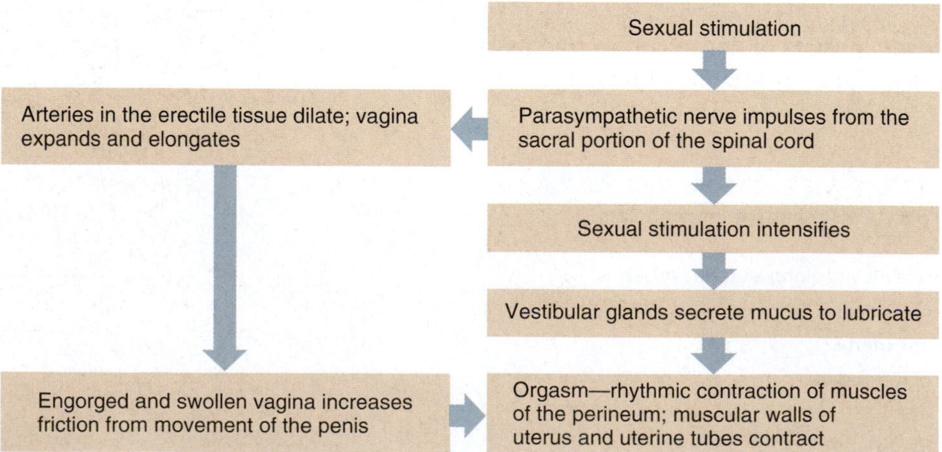

FIGURE 22.31 Mechanism of erection, lubrication, and orgasm in the female.

TABLE 22.2 | Functions of the Female Reproductive Organs

Organ	Function
Ovary	Produces oocytes and female sex hormones
Uterine tube	Conveys secondary oocyte toward uterus; site of fertilization; conveys developing embryo to uterus
Uterus	Protects and sustains embryo during pregnancy
Vagina	Conveys uterine secretions to outside of body; receives erect penis during sexual intercourse; provides open channel for offspring during birth process
Labia majora	Enclose and protect other external reproductive organs
Labia minora	Form margins of vestibule; protect openings of vagina and urethra
Clitoris	Produces feelings of pleasure during sexual stimulation due to abundant sensory nerve endings in glans
Vestibule	Space between labia minora that contains vaginal and urethral openings
Vestibular glands	Secrete fluid that moistens and lubricates the vestibule

22.5 HORMONAL CONTROL OF FEMALE REPRODUCTIVE FUNCTIONS

The *hypothalamus,* the *anterior pituitary gland,* and the *ovaries* secrete hormones that control development and maintenance of female secondary sex characteristics, maturation of female sex cells, and changes during the monthly reproductive cycle.

Female Sex Hormones

A female body is reproductively immature until about ten years of age. Then, the hypothalamus begins to secrete increasing amounts of GnRH, which, in turn, stimulate the anterior pituitary gland to release the gonadotropins FSH and LH. These hormones play primary roles in controlling female sex cell maturation and in producing female sex hormones.

Several tissues, including the ovaries, the adrenal cortices, and the placenta (during pregnancy), secrete female sex hormones. These hormones include the group of **estrogens** (es′tro-jenz) and **progesterone** (pro-jes′te-rōn). *Estradiol* is the most abundant of the estrogens, which also include *estrone* and *estriol.*

The primary source of estrogens in a nonpregnant female is the ovaries, although some estrogens are also synthesized in adipose tissue from adrenal androgens. At puberty, under the influence of the anterior pituitary gland, the ovaries secrete increasing amounts of estrogens. Estrogens stimulate enlargement of accessory organs, including the vagina, uterus, uterine tubes, and ovaries, as well as the external structures; stimulate the endometrium to thicken; and are also responsible for the development and maintenance of female *secondary sex characteristics.* These are listed in figure 22.32 and include the following:

1. Development of the breasts and the ductile system of the mammary glands in the breasts.
2. Increased deposition of adipose tissue in the subcutaneous layer generally and in the breasts, thighs, and buttocks particularly.
3. Increased vascularization of the skin.

The ovaries are also the primary source of progesterone in a nonpregnant female. This hormone promotes changes in the uterus during the female reproductive cycle, affects the mammary glands, and helps regulate secretion of gonadotropins from the anterior pituitary gland.

Certain other changes in females at puberty are related to *androgen* (male sex hormone) concentrations. For example, increased growth of hair in the pubic and axillary regions is due to androgen secreted by the adrenal cortices. Conversely, development of the female skeletal configuration, which includes narrow shoulders and broad hips, is a response to a low concentration of androgen.

Female athletes who train for endurance events, such as a marathon, typically maintain about 6% body fat. Male endurance athletes usually have about 4% body fat. This difference of 50% in proportion of body fat reflects the actions of sex hormones in males and females. Testosterone promotes deposition of protein throughout the body and especially in skeletal muscles, whereas estrogens deposit adipose tissue in the breasts, thighs, buttocks, and the subcutaneous layer of the skin.

PRACTICE

47 What stimulates sexual maturation in a female?

48 Name the major female sex hormones.

49 What is the function of estrogens?

50 What is the function of androgen in a female?

Female Reproductive Cycle

The female reproductive cycle is characterized by regular, recurring changes in the endometrium, which culminate in menstrual bleeding (menses). Such cycles usually begin near the thirteenth year of life and continue into middle age, then cease.

A female's first reproductive cycle, called **menarche** (mĕ-nar′ke), occurs after the ovaries and other organs of the female reproductive control system mature and begin responding to certain hormones. Then, the hypothalamic secretion of GnRH stimulates the anterior pituitary gland to release threshold levels of FSH and LH. As its name implies, FSH stimulates maturation of an ovarian follicle. The granulosa cells of the

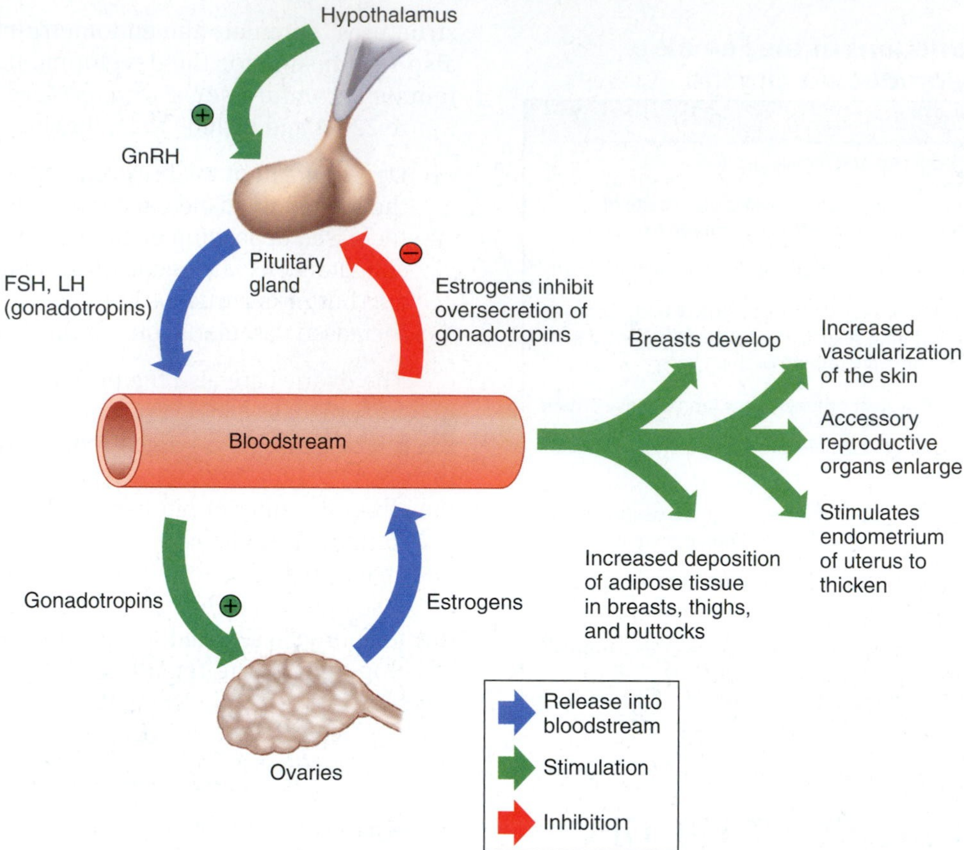

FIGURE 22.32 Control of female secondary sex development. Estrogens inhibit LH and FSH during most of the reproductive cycle except during egg cell production and ovulation.

follicle produce increasing amounts of estrogens and some progesterone. LH stimulates certain ovarian cells (theca interna) to secrete precursor molecules (such as testosterone), also used to produce estrogens.

In a young female, estrogens stimulate development of various secondary sex characteristics. Estrogens secreted during subsequent reproductive cycles continue development of these traits and maintain them. Table 22.3 summarizes the hormonal control of female secondary sex characteristics.

Increasing concentration of estrogens during the first week or so of a reproductive cycle changes the uterine lining, thickening the glandular endometrium (proliferative phase). Meanwhile, the developing follicle fully matures, and by around the fourteenth day of the cycle, the follicle appears on the surface of the ovary as a blisterlike bulge. Within the follicle, the granulosa cells, which surround the secondary oocyte and connect it to the inner wall, loosen. Follicular fluid rapidly accumulates.

While the follicle matures, it secretes estrogens that inhibit the release of LH from the anterior pituitary gland but allow LH to be stored in the gland. Estrogens also make the anterior pituitary cells more sensitive to the action of GnRH, which is released from the hypothalamus in rhythmic pulses about ninety minutes apart.

Near the fourteenth day of follicular development, the anterior pituitary cells finally respond to the pulses of GnRH and release stored LH. The resulting surge in LH concentration, which lasts for about thirty-six hours, weakens and ruptures the bulging follicular wall. This event sends the secondary oocyte and follicular fluid out of the ovary (ovulation).

Following ovulation, the remnants of the follicle and the theca interna within the ovary rapidly change. The space containing the follicular fluid fills with blood, which soon clots. Under the influence of LH, the follicular and thecal cells expand, forming a temporary glandular structure in the ovary called a **corpus luteum** ("yellow body") (see fig. 22.26).

TABLE 22.3 | Hormonal Control of Female Secondary Sex Characteristics

1. The hypothalamus releases GnRH, which stimulates the anterior pituitary gland.
2. The anterior pituitary gland secretes FSH and LH.
3. FSH stimulates the maturation of a follicle.
4. Granulosa cells of the follicle produce and secrete estrogens; LH stimulates certain cells to secrete estrogen precursor molecules.
5. Estrogens are responsible for the development and maintenance of most of the female secondary sexcharacteristics.
6. Concentrations of androgen affect other secondary sex characteristics, including skeletal growth and growth of hair.
7. Progesterone, secreted by the ovaries, affects cyclical changes in the uterus and mammary glands.

Follicular cells secrete some progesterone during the first part of the reproductive cycle. However, corpus luteum cells secrete abundant progesterone and estrogens during the second half of the cycle. Consequently, as a corpus luteum is established, the blood concentration of progesterone sharply increases.

Progesterone causes the endometrium to become more vascular and glandular. It also stimulates the uterine glands to secrete more glycogen and lipids (secretory phase). The endometrial tissues fill with fluids containing nutrients and electrolytes, which provide a favorable environment for embryo development.

The endometrium contains stem cells that enable this tissue to proliferate each month. A company sells a device to collect a menstrual sample and send it in to their laboratory, where the stem cells are extracted and stored. These menstrual stem cells in a lab dish give rise to a variety of specialized cell types. Storing them could provide a resource for future medical treaments.

High levels of estrogens and progesterone inhibit the release of LH and FSH from the anterior pituitary gland. Consequently, no other follicles are stimulated to develop when the corpus luteum is active. However, if the secondary oocyte released at ovulation is not fertilized, the corpus luteum begins to degenerate (regress) on about the twenty-fourth day of the cycle. Eventually, connective tissue replaces it. The remnant of such a corpus luteum is called a *corpus albicans* (see fig. 22.26).

When the corpus luteum ceases to function, concentrations of estrogens and progesterone rapidly decline, and in response, blood vessels in the endometrium constrict. This reduces the supply of oxygen and nutrients to the thickened endometrium, and these lining tissues (decidua) soon disintegrate and slough off. At the same time, blood leaves damaged capillaries, creating a flow of blood and cellular debris, which passes through the vagina as the *menstrual flow* (menses). This flow usually begins about the twenty-eighth day of the cycle and continues for three to five days, while the concentrations of estrogens are relatively low.

About 30% of women suffer from premenstrual syndrome (PMS), and 3 to 8% of them have severe enough symptoms for the condition to be considered premenstrual dysphoric disorder (PMDD). PMS is associated with water retention (bloating), fatigue, headache, depression, crying jags, and difficulty concentrating. For women with PMDD, the symptoms are much more pronounced, and interfere with daily functioning. Some women with PMDD may sink into such despair that they have suicidal thoughts. Diagnosis of either disorder is based mostly on timing—the symptoms appear only during the days leading up to the start of menstrual flow and perhaps a day or two after. Selective serotonin reuptake inhibitors, a class of anti-depressant, and certain oral contraceptives, improve symptoms in some women with PMDD.

The beginning of the menstrual flow marks the end of a reproductive cycle and the beginning of a new cycle. This cycle is summarized in table 22.4 and diagrammed in figure 22.33.

Low blood concentrations of estrogens and progesterone at the beginning of the reproductive cycle mean that the hypothalamus and anterior pituitary gland are no longer inhibited. Consequently, the concentrations of FSH and LH soon increase, and a new follicle is stimulated to mature. As this follicle secretes estrogens, the uterine lining undergoes repair, and the endometrium begins to thicken again. Clinical Application 22.3 addresses some causes of infertility in the female.

Elite female athletes may have disturbed reproductive cycles, ranging from diminished menstrual flow (oligomenorrhea) to complete stoppage (amenorrhea). The more active an athlete, the more likely it is that she will have menstrual irregularities, and this may impair her ability to conceive. The culprit in infertility appears to be too little body fat. The diminished fat reserves results in decreased secretion of the hormone leptin, which lowers secretion of gonadotropin releasing hormone from the hypothalamus, which in turn lowers estrogen levels. The infertility apparently results from too little estrogen. Adipose tissue also contains some estrogen, a small supply made even smaller in the trim elite athlete.

TABLE 22.4 | Major Events in a Reproductive Cycle

1. The anterior pituitary gland secretes FSH and LH.
2. FSH stimulates maturation of a follicle.
3. Granulosa cells of the follicle produce and secrete estrogens.
    a. Estrogens maintain secondary sex traits.
    b. Estrogens cause the endometrium to thicken.
4. The anterior pituitary gland releases a surge of LH, which stimulates ovulation.
5. Follicular and thecal cells become corpus luteum cells, which secrete estrogens and progesterone.
    a. Estrogens continue to stimulate uterine wall development.
    b. Progesterone stimulates the endometrium to become more glandular and vascular.
    c. Estrogens and progesterone inhibit secretion of FSH and LH from the anterior pituitary gland.
6. If the secondary oocyte is not fertilized, the corpus luteum degenerates and no longer secretes estrogens and progesterone.
7. As the concentrations of luteal hormones decline, blood vessels in the endometrium constrict.
8. The uterine lining disintegrates and sloughs off, producing a menstrual flow.
9. The anterior pituitary gland is no longer inhibited and again secretes FSH and LH.
10. The reproductive cycle repeats.

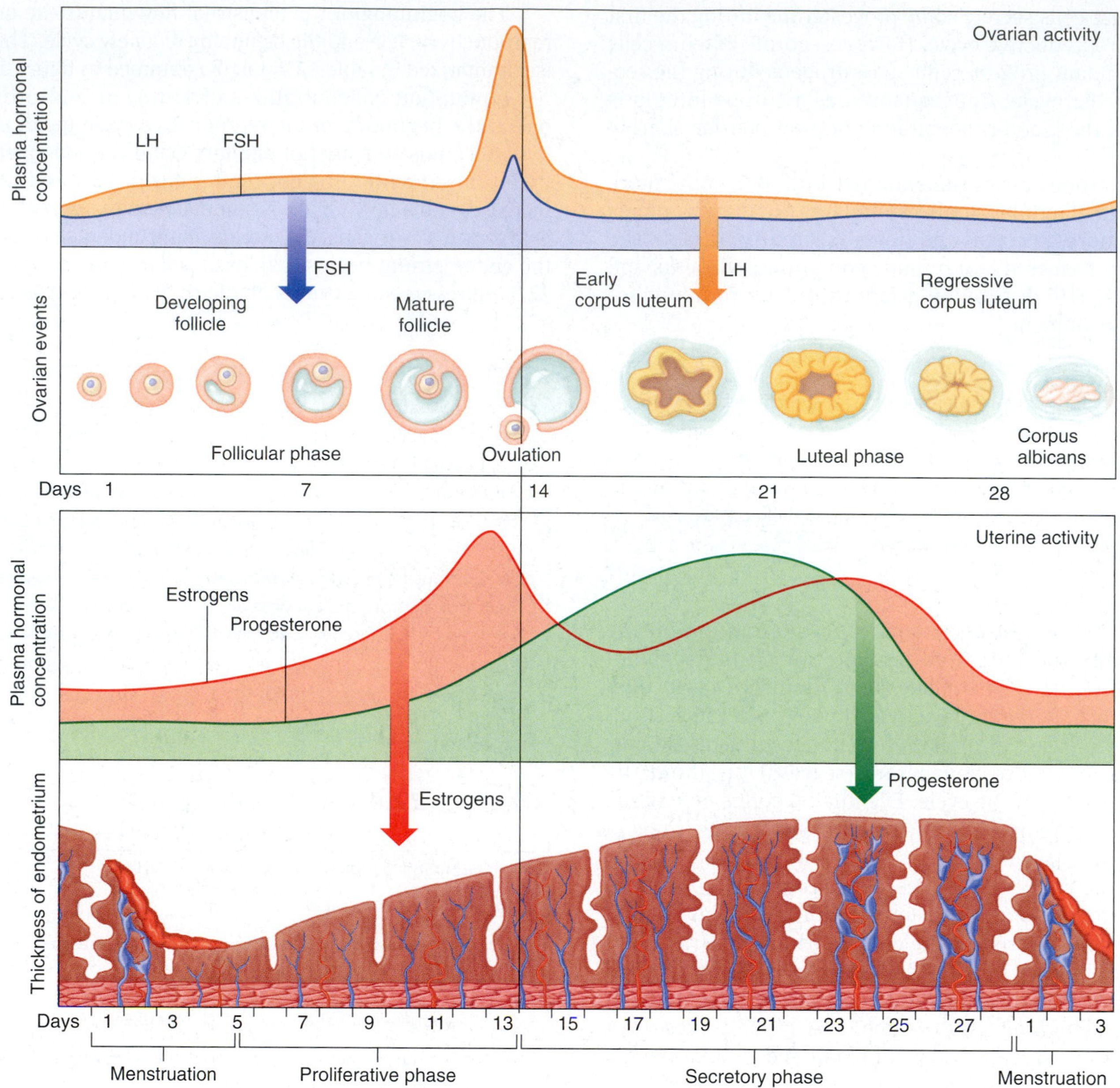

FIGURE 22.33 Major events in the female reproductive cycle.

Menopause

After puberty, reproductive cycles continue at regular intervals into the late forties or early fifties, when they usually become increasingly irregular. Then within a few months or years, the cycles cease. This period in life is called **menopause** (men′o-pawz), or female climacteric.

Aging of the ovaries causes menopause. After about thirty-five years of cycling, few primary follicles remain to respond to pituitary gonadotropins. The follicles no longer mature, ovulation does not occur, and the blood concentration of estrogens plummets, although many women continue to synthesize some estrogens from adrenal androgens.

As a result of reduced concentrations of estrogens and lack of progesterone, the female secondary sex characteristics may change. The breasts, vagina, uterus, and uterine tubes may shrink, and the pubic and axillary hair may thin. The epithelial linings associated with urinary and reproductive organs may thin. There may be increased loss of bone matrix (osteoporosis) and thinning of the skin. The pituitary secretions of FSH and LH are no longer inhibited, so these hormones may be continuously released for some time.

About 50% of women reach menopause by age fifty, and 85% reach it by age fifty-two. Of these, perhaps 20% have no unusual health effects—they simply stop menstruating. However, about 50% of women experience unpleasant vasomotor signs during menopause, including sensations of heat in the face, neck, and upper body called "hot flashes." Such a sensation may last for thirty seconds to five minutes and may

22.3 CLINICAL APPLICATION

Female Infertility

For one out of six couples, trying for parenthood is a time of increasing concern, as pregnancy remains elusive. Infertility is the inability to conceive after a year of trying. A physical cause is found in 90% of cases, and 60% of the time, the abnormality lies in the female's reproductive system. Some medical specialists (reproductive endocrinologists) use the term "subfertility" to distinguish individuals and couples who can conceive unaided, but for whom this may take longer than usual.

One of the more common causes of female infertility is hyposecretion of gonadotropic hormones from the anterior pituitary gland, followed by failure to ovulate (anovulation). This type of anovulatory cycle can sometimes be detected by testing the woman's urine for *pregnanediol,* a product of progesterone metabolism. The concentration of progesterone normally rises following ovulation, so no increase in pregnanediol in the urine during the latter part of the reproductive cycle suggests lack of ovulation.

Fertility specialists can treat absence of ovulation due to too little secretion of gonadotropic hormones by administering hCG (obtained from human placentas) or another ovulation-stimulating biochemical, human menopausal gonadotropin (hMG), which contains LH and FSH and is obtained from urine of women past menopause. However, either hCG or hMG may overstimulate the ovaries and cause many follicles to release egg cells simultaneously, resulting in multiple births if fertilization occurs.

Another cause of female infertility is *endometriosis,* in which tissue resembling the inner lining of the uterus (endometrium) grows in the abdominal cavity. This may happen if small pieces of the endometrium move up through the uterine tubes during menses and implant in the abdominal cavity. Here the tissue changes as it would in the uterine lining during the reproductive cycle. However, when the tissue begins to break down at the end of the cycle, it cannot be expelled to the outside. Instead, material remains in the abdominal cavity where it may irritate the lining (peritoneum) and cause considerable abdominal pain. These breakdown products also stimulate formation of fibrous tissue (fibrosis), which may encase the ovary and prevent ovulation or obstruct the uterine tubes. Conception becomes impossible.

Some women become infertile as a result of infections, such as gonorrhea. Infections can inflame and obstruct the uterine tubes or stimulate production of viscous mucus that can plug the cervix and prevent entry of sperm.

The first step in finding the right treatment for a particular patient is to determine the cause of the infertility. Table 22C describes diagnostic tests that a woman having difficulty conceiving may undergo. ■

TABLE 22C | Tests to Assess Female Infertility

Test	What It Checks
Hormone levels	If ovulation occurs
Ultrasound	Placement and appearance of reproductive organs and structures
Postcoital test	Cervix examined soon after unprotected intercourse to see if mucus is thin enough to allow sperm through
Endometrial biopsy	Small piece of uterine lining sampled and viewed under microscope to see if it can support an embryo
Hysterosalpingogram	Dye injected into uterine tube and followed with scanner shows if tube is clear or blocked
Laparoscopy	Small, lit optical device inserted near navel to detect scar tissue blocking tubes, which ultrasound may miss

be accompanied by chills and sweating. Women may also experience migraine headache, backache, and fatigue during menopause. These vasomotor symptoms may result from changes in the rhythmic secretion of GnRH by the hypothalamus in response to declining concentrations of sex hormones.

To minimize menopause symptoms, some women take hormone replacement therapy (HRT), which consists of estrogen plus progestin to lower the risk of developing endometrial cancer. A woman whose uterus has been removed may take estrogen alone, called estrogen replacement therapy (ERT). A doctor prescribes the therapy in any of several forms, including rings, patches, pills, creams, and gels. The lowest effective dose is taken for the shortest possible time. HRT is not advised for women who have a history of or high risk of abnormal blood clotting, heart disease, stroke, breast cancer, or gallbladder disease.

PRACTICE

51 Trace the events of the female reproductive cycle.
52 What effect does progesterone have on the endometrium?
53 What causes menstrual flow?
54 What are some changes that may occur at menopause?

22.6 MAMMARY GLANDS

The **mammary glands** are accessory organs of the female reproductive system specialized to secrete milk following pregnancy.

Location of the Glands

The mammary glands are located in the subcutaneous tissue of the anterior thorax within the hemispherical elevations called *breasts.* The breasts overlie the *pectoralis major* muscles and extend from the second to the sixth ribs and from the sternum to the axillae (fig. 22.34*a*).

A *nipple* is located near the tip of each breast at about the level of the fourth intercostal space. It is surrounded by a circular area of pigmented skin called the *areola* (fig. 22.34*b*).

Structure of the Glands

A mammary gland is composed of fifteen to twenty irregularly shaped lobes. Each lobe contains glands (alveolar glands), drained by alveolar ducts, which drain into a lactiferous duct that leads to the nipple and opens to the outside. Dense connective and adipose tissues separate the lobes. These tissues also support the glands and attach them to the fascia of the underlying pectoral muscles. Other connective tissue, which forms dense strands called *suspensory ligaments,* extends inward from the dermis of the breast to the fascia, helping support the breast. Clinical Application 22.4 discusses breast cancer.

Development of the Breasts

The mammary glands of males and females are similar. As children reach *puberty,* the glands in males do not develop, whereas ovarian hormones stimulate development of the glands in females. The alveolar glands and ducts enlarge, and fat is deposited so that each breast becomes surrounded by adipose tissue, except for the region of the areola. Chapter 23 (pp. 902–903) describes the hormonal mechanism that stimulates mammary glands to produce and secrete milk.

PRACTICE 

55 Describe the structure of a mammary gland.

56 How does ovarian hormone secretion change the mammary glands?

22.7 BIRTH CONTROL

Birth control is the voluntary regulation of the number of offspring produced and the time they are conceived. This control requires a method of **contraception** (kon″trah-sep′shun) designed to avoid fertilization of an egg cell following sexual intercourse or to prevent implantation of a hollow ball of cells (a *blastocyst*) that will develop into an embryo.

Coitus Interruptus

Coitus interruptus is the practice of withdrawing the penis from the vagina before ejaculation, preventing entry of sperm cells into the female reproductive tract. This method of contraception often proves unsatisfactory and may result in pregnancy, because a male may find it difficult to withdraw just prior to ejaculation. Also, some semen containing sperm cells may reach the vagina before ejaculation occurs.

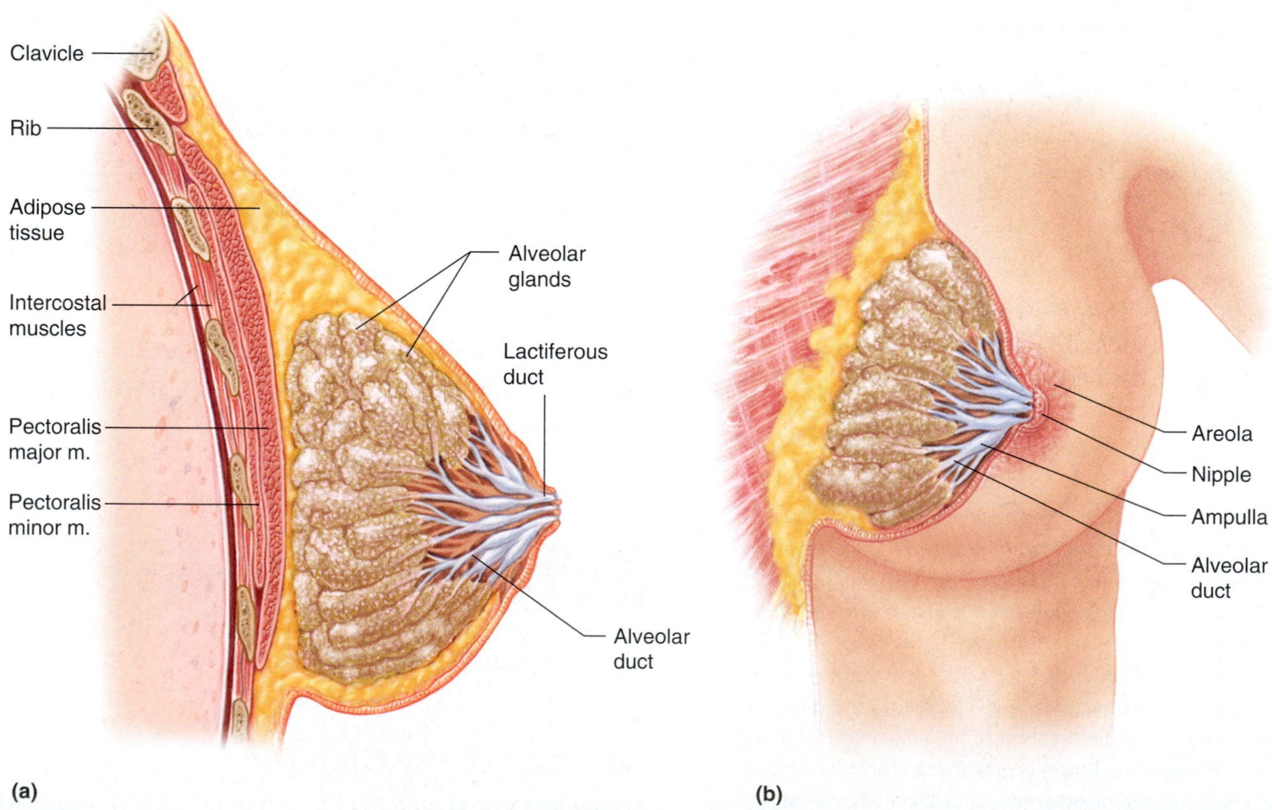

FIGURE 22.34 Structure of the female breast and mammary glands. (*a*) Sagittal section. (*b*) Anterior view. (*m.* stands for *muscle.*)

Rhythm Method

The *rhythm method* (also called timed coitus or natural family planning) requires abstinence from sexual intercourse two days before and one day after ovulation. The rhythm method results in a relatively high rate of pregnancy because accurately identifying infertile times to have intercourse is difficult. Another disadvantage of the rhythm method is that it requires adherence to a particular pattern of behavior and restricts spontaneity in sexual activity.

> The effectiveness of the rhythm method can increase by measuring and recording the woman's body temperature when she awakens each morning for several months. Body temperature typically rises about 0.6°F immediately following ovulation. However, this technique does not work for all women. More helpful may be an "ovulation predictor kit" that detects the surge in LH preceding ovulation.

PRACTICE

57 Why is coitus interruptus unreliable?

58 Describe the idea behind the rhythm method of contraception.

59 What factors make the rhythm method less reliable than some other methods of contraception?

Mechanical Barriers

Mechanical barriers prevent sperm cells from entering the female reproductive tract during sexual intercourse. The *male condom* is a thin latex or natural membrane sheath placed over the erect penis before intercourse to prevent semen from entering the vagina upon ejaculation. A *female condom* resembles a small plastic bag (fig. 22.35*a*). A woman inserts it into her vagina prior to intercourse. The device blocks sperm from reaching the cervix. A condom is inexpensive, and it may also help protect the user against contracting sexually transmitted infections and prevent the user from spreading them. However, some men often feel that a condom decreases the sensitivity of the penis during intercourse. Also, its use interrupts the sex act.

Another mechanical barrier is the *diaphragm*. It is a cup-shaped structure with a flexible ring forming the rim. The diaphragm is inserted into the vagina so that it covers the cervix, preventing entry of sperm cells into the uterus (fig. 22.35*b*). To be effective, a diaphragm must be fitted for size by a physician, inserted properly, and used in conjunction with a spermicide applied to the diaphragm surface adjacent to the cervix and to the rim of the diaphragm. The device must be left in position for several hours following sexual intercourse, and can be inserted into the vagina up to six hours before sexual contact.

Similar to but smaller than the diaphragm is the *cervical cap*, which adheres to the cervix by suction. A woman inserts it with her fingers before intercourse. Cervical caps have been used for centuries in different cultures and have been made of such varied substances as beeswax, lemon halves, paper, and opium poppy fibers.

Chemical Barriers

Chemical barrier contraceptives include creams, foams, and jellies that have spermicidal properties. These chemicals create an environment in the vagina that is unfavorable for sperm cells (fig. 22.35*c*).

Chemical barrier contraceptives are easy to use but have a high failure rate when used alone. They are more effective when used with a condom or diaphragm.

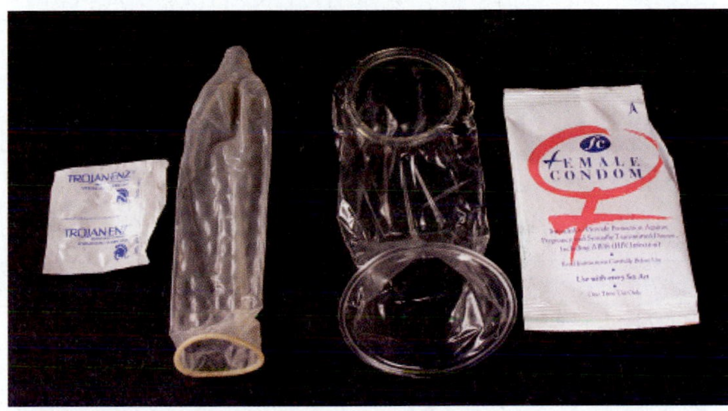

(a)

(b)

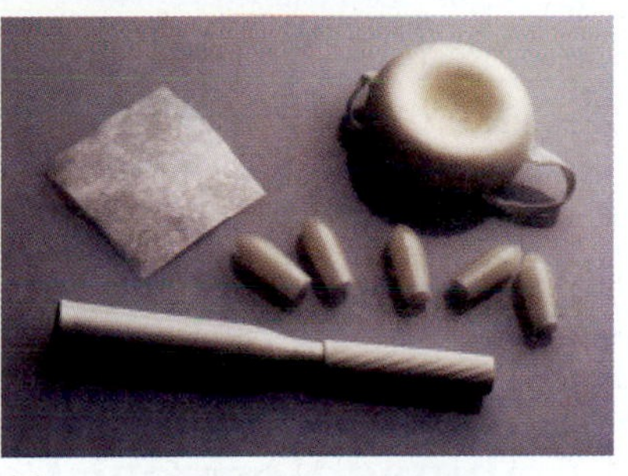

(c)

(d)

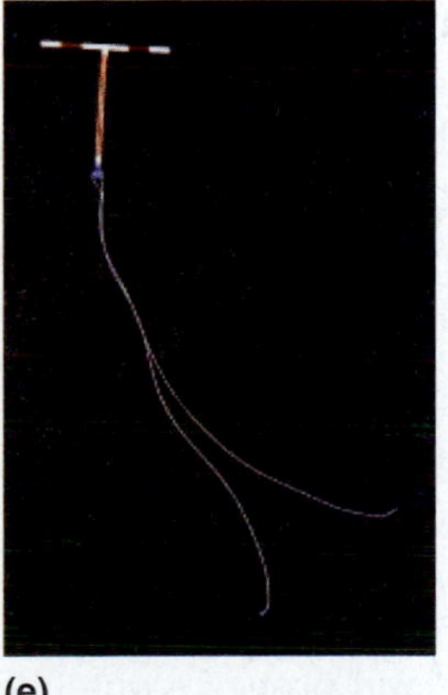

(e)

FIGURE 22.35 Devices and substances used for birth control include (*a*) male and female condoms, (*b*) a diaphragm, (*c*) spermicide in film, sponge, suppositories, gel (*d*) oral contraceptive, and (*e*) an IUD.

22.4 CLINICAL APPLICATION

Treating Breast Cancer

One in eight women will develop breast cancer at some point in her life (table 22D). About 1% of breast cancer cases are in men. Breast cancer is several illnesses. As research reveals the cellular and molecular characteristics that distinguish subtypes of the disease, treatments old and new are being increasingly tailored to individuals, at the time of diagnosis. This approach may delay disease progression and increase survival rate, while enabling patients to avoid drug treatments that will not work.

Warning Signs

Changes that could signal breast cancer include a small area of thickened tissue; a dimple; a change in contour; or a nipple that is flattened, points in an unusual direction, or produces a discharge. A woman can note these changes by performing a monthly "breast self-exam," in which she lies flat on her back with the arm raised behind her head and systematically feels all parts of each breast. But sometimes breast cancer gives no warning—fatigue and feeling ill may not occur until the disease has spread beyond the breast.

After finding a lump, the next step is a physical exam, where a health-care provider palpates the breast and does a mammogram, an X-ray scan that can pinpoint the location and approximate extent of abnormal tissue (fig. 22C). An ultrasound scan can distinguish between a cyst (a fluid-filled sac of glandular tissue) and a tumor (a solid mass). If an area is suspicious, a thin needle is used to take a biopsy (sample) of the tissue, whose cells are scrutinized under a microscope for the telltale characteristics of cancer. Further tests can identify estrogen and progesterone receptors on the cancer cells, which is information used to guide treatment choices.

TABLE 22D | Breast Cancer Risk

By Age	Odds	By Age	Odds
25	1 in 19,608	60	1 in 24
30	1 in 2,525	65	1 in 17
35	1 in 622	70	1 in 14
40	1 in 217	75	1 in 11
45	1 in 93	80	1 in 10
50	1 in 50	85	1 in 9
55	1 in 33	95 or older	1 in 8

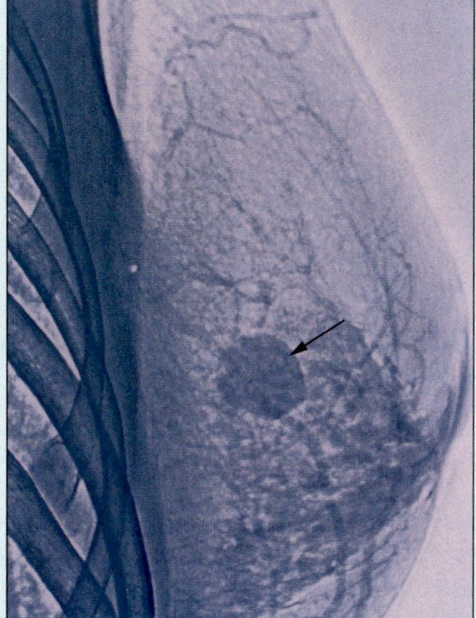

FIGURE 22C Mammogram of a breast with a tumor (arrow).

Combined Hormone Contraceptives

Combined hormone contraceptives deliver estrogen and progestin to prevent pregnancy. Various methods are used to administer the hormones, but all work on the same principle with about the same efficacy, although the amounts of the component hormones may vary. One such method is a small flexible chemical ring (Nuvaring®) inserted deep into the vagina once a month, remaining in place three out of four weeks. A plastic patch (Ortho Evra®) impregnated with the hormones may be applied to the skin on the buttocks, stomach, arm, or upper torso once a week for three out of four weeks. The most commonly used method to deliver the hormones is orally, in pill form (fig. 22.35*d*).

Combined hormone contraceptives contain synthetic estrogen-like and progesterone-like chemicals. These drugs disrupt the normal pattern of gonadotropin (FSH and LH) secretion, preventing follicle maturation and the LH surge that triggers ovulation. They also interfere with buildup of the uterine lining necessary for implantation of a blastocyst. One oral contraceptive does not include placebo days when no hormones are taken, and as a result, menstruation does not occur.

If used correctly, combined hormone contraceptives prevent pregnancy nearly 100% of the time. However, they may cause nausea, retention of body fluids, increased skin pigmentation, and breast tenderness. Some women, particularly those over thirty-five years of age who smoke, may develop intravascular blood clots, liver disorders, or high blood pressure when using certain types of these contraceptives. The patch, for example, has a higher amount of estrogen than other products and bears a warning label about the association with blood clots.

Similar to, but different from the combined hormone contraceptives is the "minipill," which contains only progestin. The progestin thickens the cervical mucus so the sperm have difficulty reaching the egg. The minipill must be taken every day at approximately the same time for maximum

Eighty percent of the time, a breast lump is a sign of fibrocystic breast disease, which is benign (noncancerous). The lump may be a cyst or a solid, fibrous mass of connective tissue called a fibroadenoma. Treatment for fibrocystic breast disease includes vitamin E, synthetic androgens, and lowering caffeine intake.

Surgery, Radiation, and Chemotherapies

If biopsied breast cells are cancerous, treatment usually begins with surgery. A lumpectomy removes a small tumor and some surrounding tissue; a simple mastectomy removes a breast; and a modified radical mastectomy removes the breast and surrounding lymph nodes but preserves the pectoral muscles. Radical mastectomies, which remove the muscles too, are rarely done anymore. In addition, a few lymph nodes are typically examined, which allows a physician to identify the ones that are affected and must be removed.

Most breast cancers are then treated with radiation and combinations of chemotherapeutic drugs, plus sometimes newer drugs targeted to certain types of breast cancer. Standard chemotherapies kill all rapidly dividing cells, and those used for breast cancer include fluorouracil, doxorubicin, cyclophosphamide, methotrexate, and paclitaxol. Protocols that provide more frequent, lower doses can temper some of the side effects of these powerful drugs.

Newer treatments developed specifically for breast cancer are easier to tolerate and can be extremely effective. They are usually given after the standard therapies, but in the future may become a first line of attack. Three types of drugs keep signals (estrogen and growth factors) from stimulating cancer cells to divide:

1. Selective estrogen receptor modulators (SERMs), such as tamoxifen and raloxifene, block estrogen receptors. About half of people with breast cancer have receptors for estrogen on their cancer cells.
2. Aromatase inhibitors block an enzyme required for tissues other than those of the ovaries to synthesize estrogens. These drugs are used in women past menopause, whose ovaries no longer synthesize estrogen. They are prescribed after a five-year course of a SERM.
3. Trastuzumab can help people whose cancer cells bear too many receptors that bind a particular growth factor. It is a monoclonal antibody, based on an immune system protein. Trastuzumab blocks the growth factor from signaling cell division. Marketed as Herceptin, this drug treats a particularly aggressive form of the disease that strikes younger women.

Prevention Strategies

Health-care providers advise women to have baseline mammograms by the age of forty, and yearly mammograms after that, or beginning at age fifty, depending upon medical and family histories. Although a mammogram can detect a tumor up to two years before it can be felt, it can also miss some tumors. Thus, breast self-exam is also important in early detection.

Genetic tests can identify women who have inherited certain variants of genes—such as *BRCA1*, *BRCA2*, *p53*, and *HER-2/neu*—that place them at high risk for developing breast cancer. Some of these women have their breasts removed to prevent the disease. Only 5% to 10% of all breast cancers arise from an inherited tendency. Much research seeks to identify the environmental triggers that contribute to causing the majority of cases. Gene expression profiling is beginning to be used to identify which drugs are most likely to help particular patients, such as those with leukemia, as described in Clinical Application 14.2 (pp. 536–537). Gene expression profiling for breast cancer can also predict risk of recurrence after surgery. This type of information is important in choosing follow-up treatment. ■

effectiveness. It is slightly less effective than combined hormone contraceptives.

Injectable Contraception

An intramuscular injection of Depo-Provera (medroxyprogesterone acetate) protects against pregnancy for three months by preventing maturation and release of a secondary oocyte. It also alters the uterine lining, making it less hospitable for a developing embryo. Depo-Provera is long-acting; it takes ten to eighteen months after the last injection for the effects to wear off.

Use of Depo-Provera requires a doctor's care, because of potential side effects and risks. The most common side effect is weight gain. Women with a history of breast cancer, depression, kidney disease, high blood pressure, migraine headaches, asthma, epilepsy, or diabetes, or strong family histories of these conditions, should probably not use this form of birth control.

A large dose of high-potency estrogens can prevent implantation of a developing embryo in the uterus. Such a "morning-after pill," taken shortly after unprotected intercourse, promotes powerful contractions of smooth muscle in a woman's reproductive tract. This may dislodge and expel a fertilized egg or early embryo. However, if the embryo has already implanted, this treatment may harm it.

PRACTICE  

60 Describe two methods of contraception that use mechanical barriers.

61 What action can increase the effectiveness of chemical contraceptives?

62 What substances are contained in oral contraceptives?

63 How do combined hormone contraceptives, including oral contraceptives, and injectable contraceptives prevent pregnancy?

Intrauterine Devices

An *intrauterine device,* or *IUD,* is a small, solid object that a physician places in the uterine cavity. An IUD interferes with implantation of a blastocyst, perhaps by inflaming the uterine tissues (fig. 22.35*e*).

An IUD may be spontaneously expelled from the uterus or produce abdominal pain or excessive menstrual bleeding. It may also harm the uterus or produce other serious health problems and should be checked regularly by a physician. A few babies have been born with IUDs attached to them.

Surgical Methods

Surgical methods of contraception sterilize the male or female. In the male, a physician removes a small section of each ductus (vas) deferens near the epididymis and ties the cut ends of the ducts. This is a *vasectomy,* and it is an operation that produces few side effects, although it may cause some pain for a week or two.

After a vasectomy, sperm cells cannot leave the epididymis, thus they are excluded from the semen. However, sperm cells may already be present in the ducts distal to the cuts. Consequently, the sperm count may not reach zero for several weeks.

The corresponding procedure in the female is called *tubal ligation.* The uterine tubes are cut and tied so that sperm cells cannot reach an egg cell.

Neither a vasectomy nor a tubal ligation changes hormonal concentrations or sex drives. These procedures, shown in figure 22.36, provide the most reliable forms of contraception. Reversing them requires microsurgery. Table 22.5 describes several contraceptive approaches and devices and indicates their effectiveness.

PRACTICE 

64 How does an IUD prevent pregnancy?

65 Describe the surgical methods of contraception for a male and for a female.

TABLE 22.5 | Birth Control Methods

	Method	Mechanism	Advantages	Disadvantages	Pregnancies per Year per 100 Women*
	None				85
Barrier and Spermicidal	Condom	Worn over penis or within vagina, keeps sperm out of vagina or from entering cervix	Protection against sexually transmitted diseases (latex only)	Disrupts spontaneity, can break, reduces sensation in male	2–12
Barrier and Spermicidal	Condom and spermicide	Worn over penis or within vagina, keeps sperm out of vagina, and kills sperm that escape	Protection against sexually transmitted diseases (latex only)	Disrupts spontaneity, can break, reduces sensation in male	2–5
Barrier and Spermicidal	Diaphragm and spermicide	Kills sperm and blocks uterus	Inexpensive	Disrupts spontaneity, messy, needs to be fitted by doctor	6–18
Barrier and Spermicidal	Cervical cap and spermicide	Kills sperm and blocks uterus	Inexpensive, can be left in 24 hours	May slip out of place, messy, needs to be fitted by doctor	6–18
Barrier and Spermicidal	Spermicidal film, sponge, suppository, foam, or gel	Kills sperm and blocks vagina	Inexpensive, easy to use and carry	Messy, irritates 25% of users, male and female	3–21
Hormonal	Combination estrogen and progestin (pill, patch, ring, or injection)	Prevents follicle maturation, ovulation, and implantation	Does not interrupt spontaneity, lowers risk of some cancers, decreases menstrual flow (one pill eliminates menstruation)	Raises risk of cardiovascular disease in some women, causes weight gain and breast tenderness	3
Hormonal	Minipill	Thickens cervical mucus	Does not interrupt spontaneity	Menstrual changes	5
Hormonal	Medroxyprogesterone acetate (Depo-Provera)	Prevents ovulation, alters uterine lining	Easy to use	Menstrual changes, weight gain	0.3
Behavioral	Rhythm method	No intercourse during fertile times	No cost	Difficult to do, hard to predict timing	20
Behavioral	Withdrawal (coitus interruptus)	Removal of penis from vagina before ejaculation	No cost	Difficult to do	4–18
Surgical	Vasectomy	Sperm cells never reach penis	Permanent, does not interrupt spontaneity	Requires surgery	0.15
Surgical	Tubal ligation	Egg cells never reach uterus	Permanent, does not interrupt spontaneity	Requires surgery, entails some risk of infection	0.4
Other	Intrauterine device	Prevents implantation	Does not interrupt spontaneity	Severe menstrual cramps, increases risk of infection	3

*The lower figures apply when the contraceptive device is used correctly. The higher figures reflect human error in using birth control.

22.8 SEXUALLY TRANSMITTED INFECTIONS

Sexually transmitted infection (STI) is replacing the term *sexually transmitted disease (STD)* because a person can be infected with a pathogen and transmit the pathogen to others, but not develop symptoms of the disease. By the time symptoms appear, it is often too late to prevent complications or the spread of the infection to sexual partners. Many STIs have similar symptoms, and some of the symptoms are also seen in diseases or allergies not sexually related, so it is wise to consult a physician if one or a combination of these symptoms appears:

1. Burning sensation during urination
2. Pain in the lower abdomen
3. Fever or swollen glands in the neck
4. Discharge from the vagina or penis
5. Pain, itching, or inflammation in the genital or anal area
6. Pain during intercourse
7. Sores, blisters, bumps, or a rash anywhere on the body, particularly the mouth or genitals
8. Itchy, runny eyes

Table 22.6 describes some prevalent sexually transmitted infections.

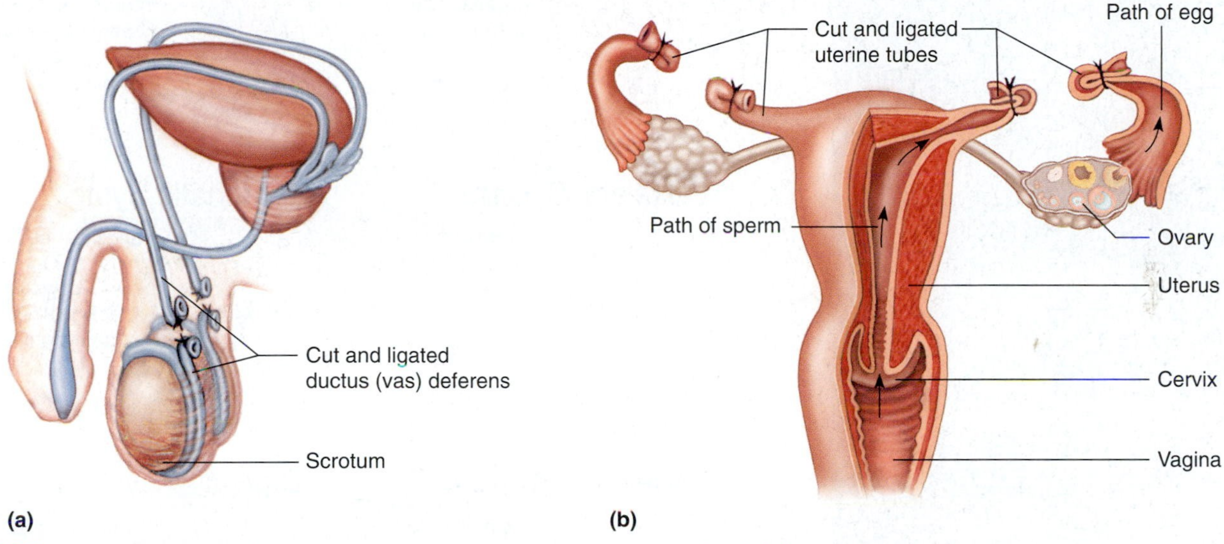

FIGURE 22.36 Surgical methods of birth control. (*a*) In a vasectomy, each ductus (vas) deferens is cut and ligated (tied). (*b*) In a tubal ligation, each uterine tube is cut and ligated.

TABLE 22.6 | Some Sexually Transmitted Infections

Disease	Cause	Symptoms	Number of Reported Cases (U.S.)*	Effects on Fetus	Treatment	Complications
Acquired immune deficiency syndrome	Human immunodeficiency virus	Fever, weakness, infections, cancer	450,000	Exposure to HIV and other infections	Drugs to treat or delay symptoms; no cure	Body overrun by infection and cancer
Chlamydia infection	*Chlamydia trachomatis* bacteria	Painful urination and intercourse, mucous discharge from penis or vagina	1 million (most cases go undiagnosed)	Premature birth, blindness, pneumonia	Antibiotics	Pelvic inflammatory disease, infertility, arthritis, ectopic pregnancy
Genital herpes	Herpes simplex 2 virus	Genital sores, fever	310,000	Brain damage, stillbirth	Antiviral drug (acyclovir)	Increased risk of cervical cancer
Genital warts	Human papilloma virus	Warts on genitals	422,000	None known	Chemical or surgical removal	Increased risk of cervical cancer
Gonorrhea	*Neisseria gonorrhoeae* bacteria	In women, usually none; in men, painful urination	350,000	Blindness, stillbirth	Antibiotics	Arthritis, rash, infertility, pelvic inflammatory disease
Syphilis	*Treponema pallidum* bacteria	Initial chancre sore usually on genitals or mouth; rash 6 months later; several years with no symptoms as infection spreads; finally damage to heart, liver, nerves, brain	37,000	Miscarriage, premature birth, birth defects, stillbirth	Antibiotics	Dementia

*2006 CDC statistics, Centers for Disease Control and Prevention

INNERCONNECTIONS | *Reproductive System*

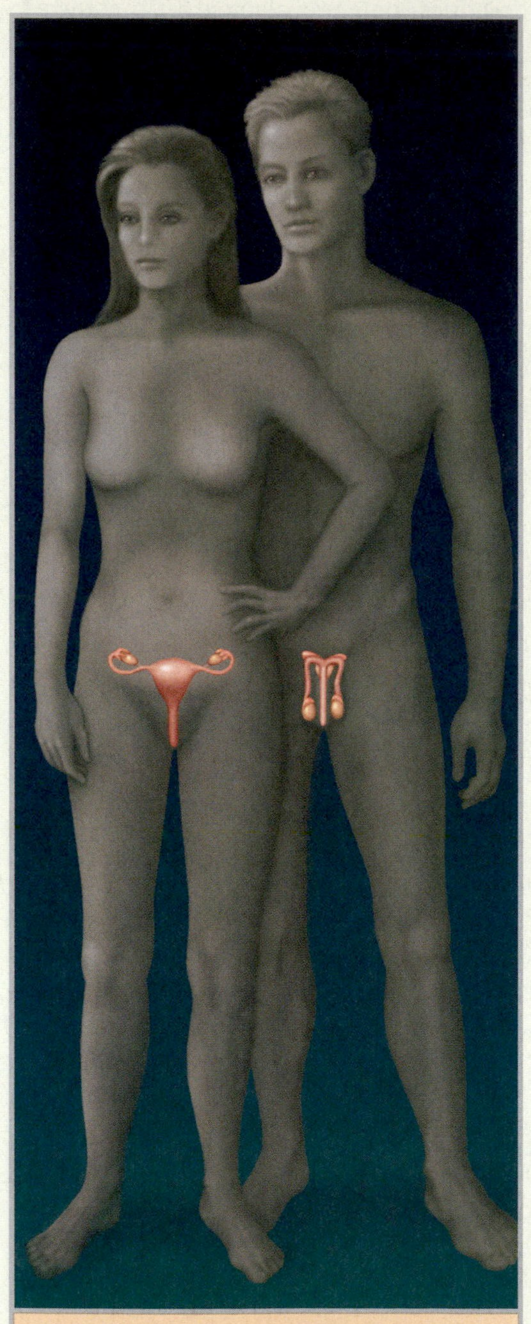

Reproductive System

Gamete production, fertilization, fetal development, and childbirth are essential for survival of the species.

Integumentary System

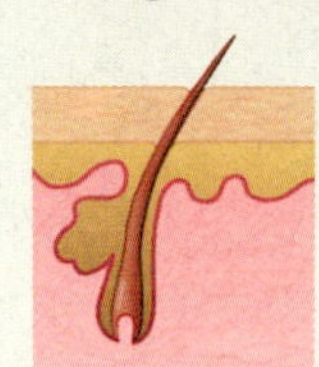

Skin sensory receptors play a role in sexual pleasure.

Skeletal System

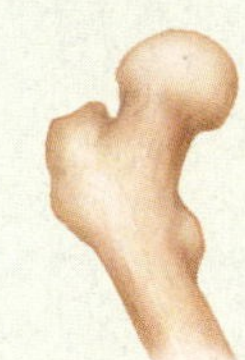

Bones can be a temporary source of calcium during lactation.

Muscular System

Skeletal, cardiac, and smooth muscles all play a role in reproductive processes and sexual activity.

Nervous System

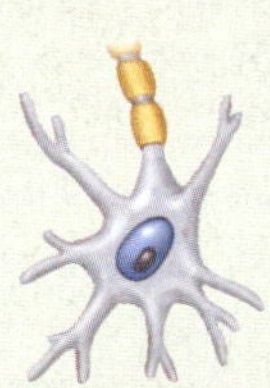

The nervous system plays a major role in sexual activity and sexual pleasure.

Endocrine System

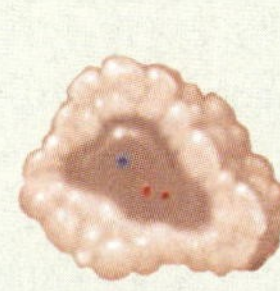

Hormones control the production of eggs in the female and sperm in the male.

Cardiovascular System

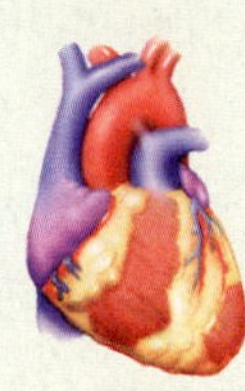

Blood pressure is necessary for the normal function of erectile tissue in the male and female.

Lymphatic System

Special mechanisms inhibit the female immune system from attacking sperm as foreign invaders.

Digestive System

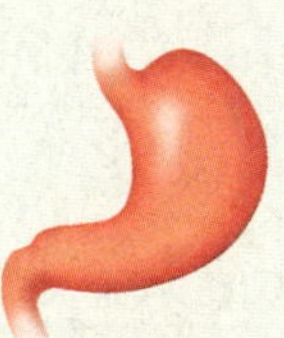

Proper nutrition is essential for the formation of normal gametes.

Respiratory System

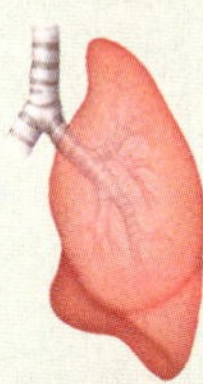

Breathing provides oxygen that assists in the production of ATP needed for egg and sperm development.

Urinary System

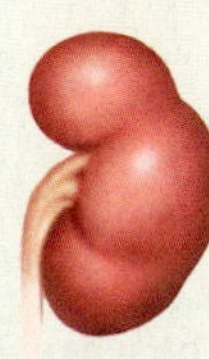

Male urinary and reproductive systems share common structures. Kidneys help compensate for fluid loss from the reproductive systems.

One possible complication of the STDs gonorrhea and chlamydia is **pelvic inflammatory disease,** in which bacteria enter the vagina and spread throughout the reproductive organs. The disease begins with intermittent cramps, followed by sudden fever, chills, weakness, and severe cramps. Hospitalization and intravenous antibiotics can stop the infection. The uterus and uterine tubes are often scarred, resulting in infertility and increased risk of ectopic pregnancy, in which the embryo develops in a uterine tube.

Acquired immune deficiency syndrome (AIDS) is a steady deterioration of the body's immune defenses in which the body becomes overrun by infection and often cancer. The human immunodeficiency virus (HIV) that causes AIDS is transmitted in body fluids such as semen, blood, and milk. It is most frequently passed during unprotected intercourse or by using a needle containing contaminated blood. Clinical Application 16.1 (pp. 636–637) explores HIV infection.

> Certain subtypes of human papilloma virus (HPV) are responsible for genital warts and nearly all cases of cervical cancer. A vaccine that became available in 2006 consists of proteins from HPV types 6, 11, 16, and 18. The vaccine stimulates the immune system to manufacture antibodies against the viruses, and it cannot cause cancer because it does not contain viral DNA. The HPV vaccine, marketed as Gardasil, is administered in three doses over a six-month period, and is most effective if given to 9 to 15 year old girls. Women aged 16 to 26 can also benefit from the vaccine.

PRACTICE

66 Why are sexually transmitted diseases also called sexually transmitted infections?

67 Why are sexually transmitted diseases sometimes difficult to diagnose?

68 What are some common symptoms of sexually transmitted diseases?

CHAPTER SUMMARY

22.1 INTRODUCTION (PAGE 831)

1. Reproductive organs produce sex cells and sex hormones, nurture these cells and transport them.
2. Sex cells are produced by meiosis consisting of two divisions, each progressing through prophase, metaphase, anaphase, and telophase.
    a. In the first meiotic division, homologous, replicated chromosomes (each consisting of two chromatids held together by a centromere) separate, and their number is halved.
    b. In the second meiotic division, the chromatids part, producing four haploid cells.
    c. Meiosis leads to genetic variability because of the random alignment of maternally and paternally derived chromosomes in metaphase I and crossing over.

22.2 ORGANS OF THE MALE REPRODUCTIVE SYSTEM (PAGE 833)

The primary male sex organs are the two testes, which produce sperm cells and male sex hormones. Accessory organs are internal and external.

1. Testes
    a. Descent of the testes
        (1) Testes originate posterior to the parietal peritoneum near the level of the developing kidneys.
        (2) The gubernaculum guides the descent of the testes into the lower abdominal cavity and through the inguinal canal.
        (3) Undescended testes cannot produce sperm cells because of the high abdominal temperature.
    b. Structure of the testes
        (1) The testes are lobules separated by connective tissue and filled with seminiferous tubules.
        (2) The seminiferous tubules unite to form the rete testis that joins the epididymis.
        (3) The seminiferous tubules contain undifferentiated cells that give rise to sperm cells.
        (4) The interstitial cells that produce male sex hormones are between the seminiferous tubules.
    c. Formation of sperm cells
        (1) The epithelium lining the seminiferous tubules includes sustentacular cells and spermatogenic cells.
            (a) The sustentacular cells support and nourish the spermatogenic cells.
            (b) The spermatogenic cells give rise to spermatogonia.
        (2) Spermatogenesis produces sperm cells from spermatogonia that have differentiated to become primary spermatocytes.
            (a) Spermatogensis produces four sperm cells from each primary spermatocyte.
            (b) Meiosis halves the number of chromosomes in sperm cells (46 to 23).
        (3) Membranous processes of adjacent sustentacular cells form a barrier in the epithelium.
            (a) The barrier separates early and late stages of spermatogenesis.
            (b) It helps provide a favorable environment for differentiating cells.
    d. Structure of a sperm cell
        (1) The sperm head contains a nucleus with 23 chromosomes.
        (2) The sperm body has many mitochondria.
        (3) The sperm tail propels the cell.

2. Male internal accessory organs
   a. Epididymides
      (1) Each epididymis is a tightly coiled tube on the outside of the testis that leads into the ductus deferens.
      (2) It stores and nourishes immature sperm cells and promotes their maturation.
   b. Ductus deferentia
      (1) Each ductus deferens is a muscular tube that forms part of the spermatic cord.
      (2) It passes through the inguinal canal, enters the abdominal cavity, courses medially into the pelvic cavity, and ends behind the urinary bladder.
      (3) It fuses with the duct from the seminal vesicle to form the ejaculatory duct.
   c. Seminal vesicles
      (1) Each seminal vesicle is a saclike structure attached to the ductus deferens.
      (2) It secretes an alkaline fluid that contains nutrients, such as fructose, and prostaglandins.
      (3) This secretion is added to sperm cells entering the ejaculatory duct.
   d. Prostate gland
      (1) This gland surrounds the urethra just below the urinary bladder.
      (2) It secretes a thin, milky fluid, which enhances sperm motility and neutralizes the fluid containing the sperm cells and acidic secretions of the vagina.
   e. Bulbourethral glands
      (1) These glands are two small structures inferior to the prostate gland.
      (2) They secrete a fluid that lubricates the penis in preparation for sexual intercourse.
   f. Semen
      (1) Semen consists of sperm cells and secretions of the seminal vesicles, prostate gland, and bulbourethral glands.
      (2) This fluid is slightly alkaline and contains nutrients and prostaglandins.
      (3) Sperm cells in semen swim, but cannot fertilize egg cells until they enter the female reproductive tract.
3. Male external reproductive organs
   a. Scrotum
      (1) The scrotum is a pouch of skin and subcutaneous tissue that encloses the testes.
      (2) The dartos muscle in the scrotal wall causes the scrotal skin to wrinkle and the testes to move closer to the pelvic cavity or the fibers relax and the scrotum hangs loosely, regulating the temperature for sperm production and survival.
   b. Penis
      (1) The penis conveys urine and semen.
      (2) It becomes erect for insertion into the vagina during sexual intercourse.
      (3) Its body is composed of three columns of erectile tissue surrounded by connective tissue.
      (4) The root of the penis is attached to the pelvic arch and membranes of the perineum.
   c. Erection, orgasm, and ejaculation
      (1) During erection, vascular spaces in the erectile tissue become engorged with blood as arteries dilate and veins are compressed.
      (2) Orgasm is the culmination of sexual stimulation and is accompanied by emission and ejaculation.
      (3) Semen moves along the reproductive tract as smooth muscle in the walls of the tubular structures contract, stimulated by a reflex.
      (4) Following ejaculation, the penis becomes flaccid.

22.3 HORMONAL CONTROL OF MALE REPRODUCTIVE FUNCTIONS (PAGE 845)

1. Hypothalamic and pituitary hormones

   The male body remains reproductively immature until the hypothalamus releases GnRH, which stimulates the anterior pituitary gland to release gonadotropins.
   a. FSH stimulates spermatogenesis.
   b. LH (ICSH) stimulates the interstitial cells to produce male sex hormones.
   c. Inhibin prevents oversecretion of FSH.
2. Male sex hormones
   a. Male sex hormones are called androgens.
   b. Testosterone is the most important androgen.
   c. Testosterone is converted into dihydrotestosterone in some organs.
   d. Androgens that fail to become fixed in tissues are metabolized in the liver and excreted.
   e. Androgen production increases rapidly at puberty.
3. Actions of testosterone
   a. Testosterone stimulates the development of the male reproductive organs and causes the testes to descend.
   b. It is responsible for the development and maintenance of male secondary sex characteristics.
4. Regulation of male sex hormones
   a. A negative feedback mechanism regulates testosterone concentration.
      (1) As the concentration of testosterone rises, the hypothalamus is inhibited, and the anterior pituitary secretion of gonadotropins is reduced.
      (2) As the concentration of testosterone falls, the hypothalamus signals the anterior pituitary to secrete gonadotropins.
   b. The concentration of testosterone remains relatively stable from day to day.

22.4 ORGANS OF THE FEMALE REPRODUCTIVE SYSTEM (PAGE 846)

The primary female sex organs are the two ovaries, which produce female sex cells and sex hormones. Accessory organs are internal and external.

1. Ovaries
   a. Ovary attachments
      (1) Several ligaments hold the ovaries in position.
      (2) These ligaments include broad, suspensory, and ovarian ligaments.

b. Ovary descent
   (1) The ovaries descend from posterior to the parietal peritoneum near the developing kidneys.
   (2) They are attached to the pelvic wall just inferior to the pelvic brim.

c. Ovary structure
   (1) The ovaries are subdivided into a medulla and a cortex.
   (2) The medulla is composed of connective tissue, blood vessels, lymphatic vessels, and nerves.
   (3) The cortex contains ovarian follicles and is covered by cuboidal epithelium.

d. Primordial follicles
   (1) During prenatal development, oogonia divide by mitosis to produce more oogonia that develop into primary oocytes.
   (2) Each primordial follicle contains a primary oocyte and a layer of flattened epithelial cells.
   (3) The primary oocyte begins to undergo meiosis, but the process soon halts and does not resume until puberty.
   (4) The number of oocytes steadily declines throughout the life of a female.

e. Oogenesis
   (1) Beginning at puberty, some oocytes are stimulated to continue meiosis.
   (2) When a primary oocyte undergoes oogenesis, it gives rise to a secondary oocyte in which the original chromosome number is halved (from 46 to 23).
   (3) A secondary oocyte may be fertilized to produce a zygote.

f. Follicle maturation
   (1) At puberty, FSH initiates follicle maturation.
   (2) During maturation of the follicle, the primary oocyte enlarges, the follicular cells proliferate, and a fluid-filled cavity appears and produces a secondary follicle.
   (3) Ovarian cells surrounding the follicle form two layers.
   (4) A mature follicle contains a secondary oocyte surrounded by a zona pellucida and corona radiata.
   (5) Usually only one follicle per reproductive cycle reaches full development.

g. Ovulation
   (1) Ovulation is the release of a secondary oocyte from an ovary.
   (2) The secondary oocyte is released when its follicle ruptures.
   (3) After ovulation, the secondary oocyte is drawn into the opening of the uterine tube.

2. Female internal accessory organs
   a. Uterine tubes
      (1) These tubes convey egg cells toward the uterus.
      (2) The end of each uterine tube is expanded, and its margin bears irregular extensions.
      (3) Ciliated cells that line the tube and peristaltic contractions in the wall of the tube move an egg cell into the tube's opening.
   b. Uterus
      (1) The uterus receives the embryo and sustains it during development.
      (2) The uterine wall includes the endometrium, myometrium, and perimetrium.
   c. Vagina
      (1) The vagina connects the uterus to the vestibule.
      (2) It receives the erect penis, conveys uterine secretions to the outside, and provides an open channel for the fetus during birth.
      (3) A thin membrane, the hymen, partially closes the vaginal orifice.
      (4) The vaginal wall consists of a mucosa, muscularis, and outer fibrous coat.

3. Female external reproductive organs
   a. Labia majora
      (1) The labia majora are rounded folds of adipose tissue and skin that enclose and protect the other external reproductive parts.
      (2) The anterior ends form a rounded elevation over the symphysis pubis.
   b. Labia minora
      (1) The labia minora are flattened, longitudinal folds between the labia majora.
      (2) They are well supplied with blood vessels.
   c. Clitoris
      (1) The clitoris is a small projection at the anterior end of the vulva; it corresponds to the male penis.
      (2) It is composed of two columns of erectile tissue.
      (3) Its root is attached to the sides of the pubic arch.
   d. Vestibule
      (1) The vestibule is the space between the labia minora that encloses the vaginal and urethral openings.
      (2) The vestibular glands secrete mucus into the vestibule during sexual stimulation.
   e. Erection, lubrication, and orgasm
      (1) During periods of sexual stimulation, the erectile tissues of the clitoris and vestibular bulbs become engorged with blood and swollen.
      (2) The vestibular glands secrete mucus into the vestibule and vagina.
      (3) During orgasm, the muscles of the perineum, uterine wall, and uterine tubes contract rhythmically.

22.5 HORMONAL CONTROL OF FEMALE REPRODUCTIVE FUNCTIONS (PAGE 857)

Hormones from the hypothalamus, anterior pituitary gland, and ovaries play important roles in the control of sex cell maturation, the development and maintenance of female secondary sex characteristics, and changes that occur during the monthly reproductive cycle.

1. Female sex hormones
   a. A female body remains reproductively immature until about ten years of age, when gonadotropin secretion increases.

b. The most important female sex hormones are estrogens and progesterone.
   (1) Estrogens are responsible for the development and maintenance of most female secondary sex characteristics.
   (2) Progesterone prepares the uterus for pregnancy.

2. Female reproductive cycle
   a. The reproductive cycle is characterized by regularly recurring changes in the uterine lining culminating in menstrual flow.
   b. Menarche is the female's first reproductive cycle.
   c. A reproductive cycle is initiated by FSH, which stimulates maturation of a follicle.
   d. Granulosa cells of a maturing follicle secrete estrogens, responsible for maintaining the secondary sex traits and thickening the uterine lining.
   e. Ovulation is triggered when the anterior pituitary gland releases a relatively large amount of LH.
   f. Following ovulation, the follicular cells and thecal cells give rise to the corpus luteum.
      (1) The corpus luteum secretes estrogens and progesterone, which cause the uterine lining to become more vascular and glandular.
      (2) If a secondary oocyte is not fertilized, the corpus luteum begins to degenerate.
      (3) As the concentrations of estrogens and progesterone decline, the uterine lining disintegrates, causing menstrual flow.
   g. During this cycle, estrogens and progesterone inhibit the release of LH and FSH; as the concentrations of these hormones decline, the anterior pituitary secretes FSH and LH again, stimulating a new reproductive cycle.
3. Menopause
   a. Eventually the ovaries cease responding to FSH, and cycling ceases.
   b. Menopause is characterized by a low concentration of estrogens and a continuous secretion of FSH and LH.
   c. The female reproductive organs regress.

22.6 MAMMARY GLANDS (PAGE 861)

1. Location of the glands
   a. The mammary glands are located in the subcutaneous tissue of the anterior thorax within the breasts.
   b. The breasts extend between the second and sixth ribs and from sternum to axillae.
2. Structure of the glands
   a. The mammary glands are composed of lobes that contain tubular glands.
   b. Dense connective and adipose tissues separate the lobes.
   c. Ducts connect the mammary glands to the nipple.
3. Development of the breasts
   a. Breasts of males remain nonfunctional.
   b. Estrogens stimulate breast development in females.
      (1) Alveolar glands and ducts enlarge.
      (2) Fat is deposited around and in the breasts.

22.7 BIRTH CONTROL (PAGE 862)

Birth control is the voluntary regulation of the number of children produced and the time they are conceived. This usually involves some method of contraception.

1. Coitus interruptus
   a. Coitus interruptus is withdrawal of the penis from the vagina before ejaculation.
   b. Some semen may be expelled from the penis before ejaculation.
2. Rhythm method
   a. Abstinence from sexual intercourse two days before and one day after ovulation is the rhythm method.
   b. It is difficult to predict the time of ovulation.
3. Mechanical barriers
   a. Males and females can use condoms.
   b. Females use diaphragms and cervical caps.
4. Chemical barriers
   a. Spermicidal film, sponges, suppositories, foams, and gels are chemical barriers to conception.
   b. These provide an unfavorable environment in the vagina for sperm survival.
5. Combined hormone contraceptives
   a. A flexible ring inserted deep into the vagina, a plastic patch, or pill can deliver estrogen and progestin to prevent pregnancy.
   b. They disrupt the normal pattern of gonadotropin secretion preventing follicle maturation, ovulation, and the normal buildup of the uterine lining.
   c. When used correctly, combined hormone contraceptives are almost 100% effective.
   d. Some women develop undesirable side effects.
   e. A minipill contains only progestin and must be taken at the same time daily.
6. Injectable contraceptives
   a. Intramuscular injection with medroxyprogesterone acetate every three months prevents pregnancy.
   b. High levels of hormone prevent maturation and release of a secondary oocyte.
   c. Very effective if administered promptly at the end of the three months.
   d. Women may experience side effects; in some women, use is contraindicated.
7. Intrauterine devices
   a. An IUD is a solid object inserted in the uterine cavity.
   b. It prevents pregnancy by interfering with implantation.
   c. It may be expelled spontaneously or produce undesirable side effects.
8. Surgical methods
   a. These are sterilization procedures.
      (1) Vasectomy is performed in males.
      (2) Tubal ligation is performed in females.
   b. Surgical methods are the most reliable contraception.

22.8 SEXUALLY TRANSMITTED INFECTIONS (PAGE 867)

1. Sexually transmitted infections or sexually transmitted diseases are passed during sexual contact and may go undetected for years.
2. Many of the sexually transmitted diseases share symptoms.

CHAPTER ASSESSMENTS

22.1 Introduction

1 General functions of the male and female reproductive system include ___________. (p. 831)

a. producing sex cells
b. nurturing sex cells
c. transporting sex cells to sites of fertilization
d. secreting hormones
e. all of the above

2 Construct a table of the steps in meiosis including the major event(s). (p. 831)

3 Define *cross over,* and explain how it produces genetic variation. (p. 832)

22.2 Organs of the Male Reproductive System

4 Distinguish between the primary and accessory male reproductive organs, then explain how each organ's structure affects the organ's function. (p. 833)

5 Describe the descent of the testes. (p. 833)

6 Define *cryptoorchidism.* (p. 835)

7 List the major steps in spermatogenesis. (p. 836)

8 Explain the function of the sustentacular cells in the testis. (p. 836)

9 Describe a sperm cell. (p. 836)

10 Trace the path of the ductus deferens from the epididymis to the ejaculatory duct. (p. 838)

11 On a diagram, locate the seminal vesicles, prostate gland, and bulbourethral glands, and describe the composition of their secretions. (p. 840)

12 Describe the composition of semen. (p. 841)

13 Define *capacitation.* (p. 841)

14 Explain the mechanism that produces an erection of the penis. (p. 844)

15 Define *orgasm.* (p. 844)

16 Distinguish between emission and ejaculation. (p. 844)

17 Explain the mechanism of ejaculation. (p. 844)

22.3 Hormonal Control of Male Reproductive Functions

18 Describe the role of GnRH in the control of male reproductive functions. (p. 845)

19 Distinguish between androgen and testosterone. (p. 845)

20 Define *puberty.* (p. 846)

21 Discuss the actions of testosterone. (p. 846)

22 List several male secondary sex characteristics. (p. 846)

23 Explain the regulation of testosterone concentration. (p. 846)

22.4 Organs of the Female Reproductive System

24 Distinguish between the primary and accessory female reproductive organs, then explain how each organ's structure affects the organ's function. (p. 846)

25 Describe how the ovaries are held in position. (p. 847)

26 Describe the descent of the ovaries. (p. 847)

27 Define *primordial follicle.* (p. 847)

28 List the major steps in oogenesis. (p. 849)

29 Distinguish between a primary and a secondary follicle. (p. 849)

30 Describe how a follicle matures. (p. 849)

31 Define *ovulation.* (p. 851)

32 Define *vestibule.* (p. 856)

33 Describe the process of an erection in the female reproductive organs. (p. 856)

22.5 Hormonal Control of Female Reproductive Functions

34 Describe the role of GnRH in regulating female reproductive functions. (p. 857)

35 List several female secondary sex characteristics. (p. 857)

36 Explain how a reproductive cycle is initiated. (p. 857)

37 Summarize the major events in a reproductive cycle. (p. 857)

38 A female's first reproductive cycle is ______________, whereas ______________ is when the female's reproductive cycles cease. (p. 857)

22.6 Mammary Glands

39 Describe the structure of a mammary gland. (p. 862)

22.7 Birth Control

40 Match the birth control method with its description. (pp. 862–866)

(1) withdrawal
(2) rhythm method
(3) condom
(4) spermicide (gel, foam)
(5) estrogen/ progesterone
(6) IUD
(7) vasectomy
(8) tubal ligation

A. Kills sperm (not very effective when used alone)
B. Keeps sperm out of vagina or from entering cervix (may help prevent disease)
C. Prevents implantation
D. No intercourse during fertile times (ineffective)
E. Penis removed from vagina before ejaculation
F. Sperm cells never reach penis (very effective)
G. Prevents follicle maturation and ovulation
H. Oocytes never reach uterus (very effective)

22.8 Sexually Transmitted Infections

41 Common symptoms of sexually transmitted infections include ___________. (p. 867)

a. a burning sensation during urination
b. discharge from penis or vagina
c. pain during intercourse
d. sores, blisters, or rash on genitals
e. all of the above

42 If left untreated, a complication of the sexually transmitted infections gonorrhea and chlamydia is ____________________. (p. 869)

INTEGRATIVE ASSESSMENTS/CRITICAL THINKING

OUTCOMES 13.5, 13.8, 22.2, 22.3

1. What changes, if any, might occur in the secondary sex characteristics of an adult male following removal of one testis? Following removal of both testes? Following removal of the prostate gland?

OUTCOMES 13.5, 13.8, 22.2, 22.3, 22.4, 22.5, 22.7

2. Understanding the causes of infertility can be valuable in developing new birth control methods. Cite a type of contraceptive based on each of the following causes of infertility: (a) failure to ovulate due to a hormonal imbalance; (b) a large fibroid tumor that disturbs the uterine lining; (c) endometrial tissue blocking uterine tubes; (d) low sperm count (too few sperm per ejaculate).

OUTCOMES 13.5, 13.8, 22.4, 22.5

3. What effect would removal of an ovary have on a woman's reproductive cycles? What effect would removal of both ovaries have?

OUTCOMES 13.5, 13.8, 22.4, 22.5, 22.7

4. Does a tubal ligation cause a woman to enter menopause prematurely? Why or why not?

OUTCOME 22.2

5. Some men are unable to become fathers because their spermatids do not mature into sperm. Injection of their spermatids into their partner's secondary oocytes sometimes results in conception. A few men have fathered healthy babies this way. Why would this procedure work with spermatids but not with primary spermatocytes?

OUTCOMES 22.2, 22.3

6. As a male reaches adulthood, what will be the consequences if his testes have remained undescended since birth? Why?

OUTCOMES 22.4, 22.5

7. Sometimes a sperm cell fertilizes a polar body rather than a secondary oocyte. An embryo does not develop, and the fertilized polar body degenerates. Why is the polar body unable to support development of an embryo?

WEB CONNECTIONS

Be sure to visit the text website at **www.mhhe.com/shier12** for answers to chapter assessments, additional quizzes, and interactive learning exercises.

ANATOMY & PHYSIOLOGY REVEALED

Anatomy & Physiology Revealed® (APR) includes cadaver photos that allow you to peel away layers of the human body to reveal structures beneath the surface. This program also includes animations, radiologic imaging, audio pronunciations, and practice quizzing. Check out **www.aprevealed.com**. APR has been proven to help improve student grades!

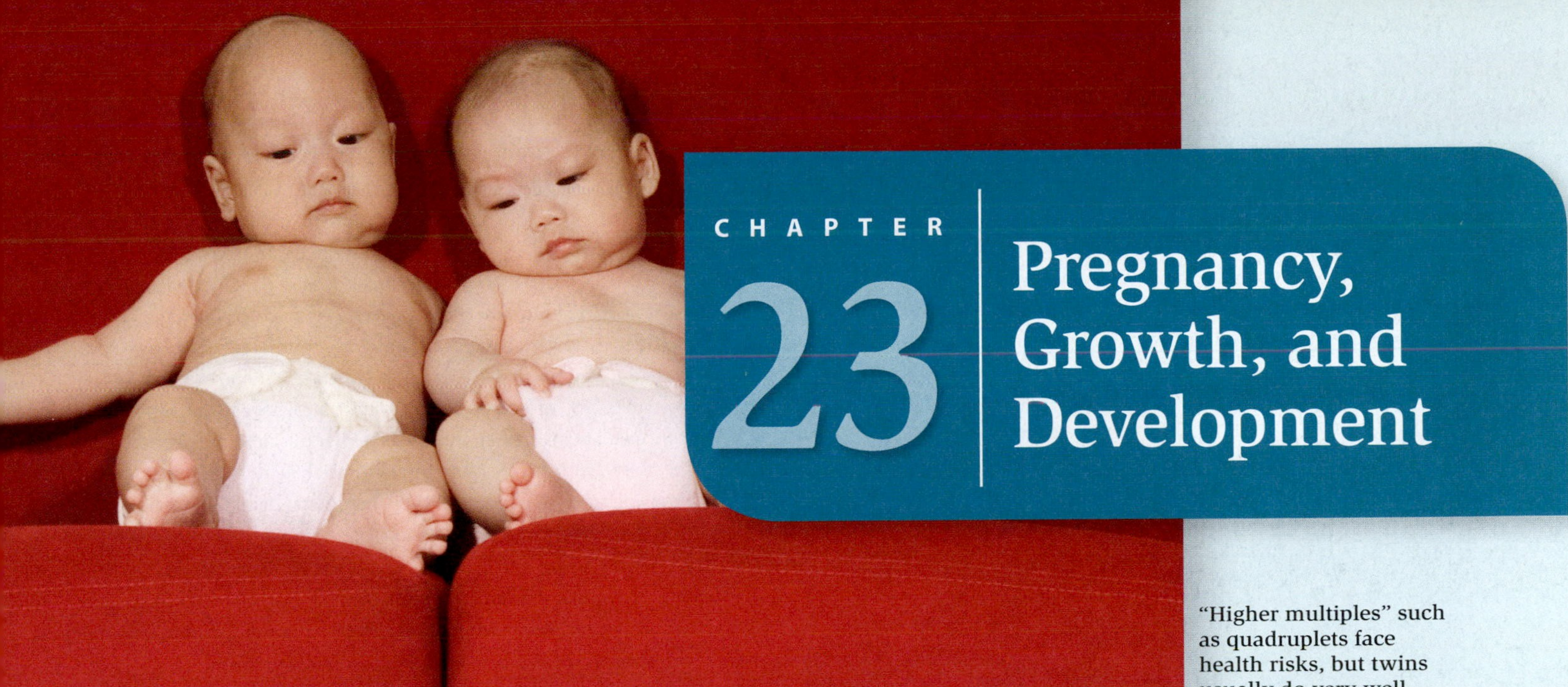

CHAPTER 23 Pregnancy, Growth, and Development

"Higher multiples" such as quadruplets face health risks, but twins usually do very well.

UNDERSTANDING WORDS

allant-, sausage: *allant*ois—tubelike structure extending from the yolk sac into the connecting stalk of an embryo.

chorio-, skin: *chorion*—outermost membrane surrounding the fetus and its membranes.

cleav-, to divide: *cleav*age—period of development when a zygote divides, producing increasingly smaller cells.

ect-, outside: *ect*oderm—outermost germ layer of embryo.

lacun-, pool: *lacuna*—space between the chorionic villi that fills with maternal blood.

lanug-, down: *lanugo*—fine hair covering the fetus.

mes-, middle: *mes*oderm—middle germ layer of embryo.

morul-, mulberry: *morula*—embryonic structure consisting of a solid ball of about sixteen cells that resembles a mulberry.

nat-, to be born: pre*nat*al—period of development before birth.

ne-, new, young: *ne*onatal—period of development including the first four weeks after birth.

post-, after: *post*natal—period of development after birth.

pre-, before: *pre*natal—period of development before birth.

sen-, old: *sen*escence—process of growing old.

troph-, well fed: *troph*oblast—cellular layer that surrounds the inner cell mass and helps nourish it.

umbil-, navel: *umbil*ical cord—structure attached to the fetal navel (umbilicus) that connects the fetus to the placenta.

LEARN

PRACTICE ASSESS

LEARNING OUTCOMES

After you have studied this chapter, you should be able to:

23.1 Introduction

1. Distinguish between growth and development. (p. 876)
2. Distinguish between prenatal and postnatal. (p. 876)

23.2 Pregnancy

3. Describe fertilization. (p. 876)

23.3 Prenatal Period

4. List and provide details of the major events of cleavage. (p. 879)
5. Describe implantation. (p. 882)
6. Discuss the hormonal and other changes in the maternal body during pregnancy. (p. 883)
7. Explain how the primary germ layers originate, and list the structures each layer produces. (p. 885)
8. Describe the major events of the embryonic stage of development. (p. 885)
9. Describe the formation and function of the placenta. (p. 888)
10. Define *fetus,* and describe the major events of the fetal stage of development. (p. 892)
11. Trace the path of blood through the fetal cardiovascular system. (p. 897)
12. Explain the role of hormones in the birth process and milk production. (p. 899)

23.4 Postnatal Period

13. Name the postnatal stages of development of a human, and indicate the general characteristics of each stage. (p. 904)
14. Describe the major cardiovascular and physiological adjustments in the newborn. (p. 905)

23.5 Aging

15. Distinguish between passive and active aging. (p. 909)
16. Contrast life span and life expectancy. (p. 911)

POSTMORTEM SPERM RETRIEVAL

Bruce and Gaby Vernoff, in their early thirties, had delayed becoming parents, confident that their good health would make pregnancy possible. But Bruce suddenly died of an allergic reaction to a medication. Gaby knew how much he had wanted to be a father, so she requested that physicians take some of Bruce's sperm after his death. Thirty hours after Bruce died, the medical examiner collected a sperm sample and sent it to California Cryobank (a sperm bank), where it lay deeply frozen for more than a year. In the summer of 1978, Dr. Cappy Rothman, medical director of the bank, used the defrosted sperm to fertilize one of Gaby's eggs. On March 17, their daughter was born. It was the first case of "postmortem sperm retrieval" and use in which the father had not actively participated in the decision. In other cases, the dying men had had time to state, in writing, their wishes to be fathers posthumously.

Sperm for the future. Some servicemen left frozen samples of sperm at sperm banks before reporting for active duty, along with written statements expressing their wishes for their wives to use the sperm to conceive children should they not return.

Postmortem sperm retrieval raises legal and ethical issues. In another case, a woman conceived twins sixteen months after her husband had died of leukemia at age thirty, with his consent. But the Social Security Administration refused to provide survivor benefits to their daughters, claiming that the man was not a father, but a sperm donor. The Massachusetts Superior Court reversed this decision.

Postmortem sperm retrieval, like other assisted reproductive technologies, is not regulated at the federal level in the United States, so bioethicists suggest that men document their wishes, and have identified situations to avoid:

- Someone other than a spouse wishing to use the sperm
- A too-hasty decision based on grief
- Use of the sperm for monetary gain

Some nations, including Germany, Australia, Canada, and Sweden, ban the procedure. ■

23.1 INTRODUCTION

A sperm cell and a secondary oocyte unite, forming a zygote, and the journey of prenatal development begins. Following thirty-eight weeks of cell division, growth, and specialization into distinctive tissues and organs, a new human being enters the world.

Humans grow, develop, and age. **Growth** is an increase in size. In humans and other many-celled organisms, growth entails an increase in cell numbers as a result of mitosis, followed by enlargement of the newly formed cells and of the body.

Development, which includes growth, is the continuous process by which an individual changes from one life phase to another. These life phases include a **prenatal period** (pre-na′tal pe′re-od), which begins with the fertilization of an egg cell and ends at birth, and a **postnatal period** (pōst-na′tal pe′re-od), which begins at birth and ends with death.

PRACTICE

1 Distinguish between growth and development.
2 Distinguish between prenatal and postnatal.

23.2 PREGNANCY

The union of a secondary oocyte and a sperm cell is called **fertilization** (fer″tĭ-lĭ-za′shun), or conception, which typically takes place in a uterine tube. **Pregnancy** (preg′nan-se) is the presence of a developing offspring in the uterus. Pregnancy consists of three periods called trimesters, each about three months long.

Transport of Sex Cells

Ordinarily, before fertilization can occur, a secondary oocyte must be ovulated and enter a uterine tube. During sexual intercourse, the male deposits semen containing sperm cells in the vagina near the cervix. To reach the secondary oocyte, the sperm cells must then move upward through the uterus and uterine tube. Prostaglandins in the semen stimulate lashing of sperm tails and muscular contractions within the walls of the uterus and uterine tube aid the sperm cells' journey. Also, under the influence of high concentrations of estrogens during the first part of the reproductive cycle, the uterus and cervix contain a thin, watery secretion that promotes sperm transport and survival. Conversely, during the latter part of the cycle, when the progesterone concentration is high, the female reproductive tract secretes a viscous fluid that hampers sperm transport and survival (fig. 23.1).

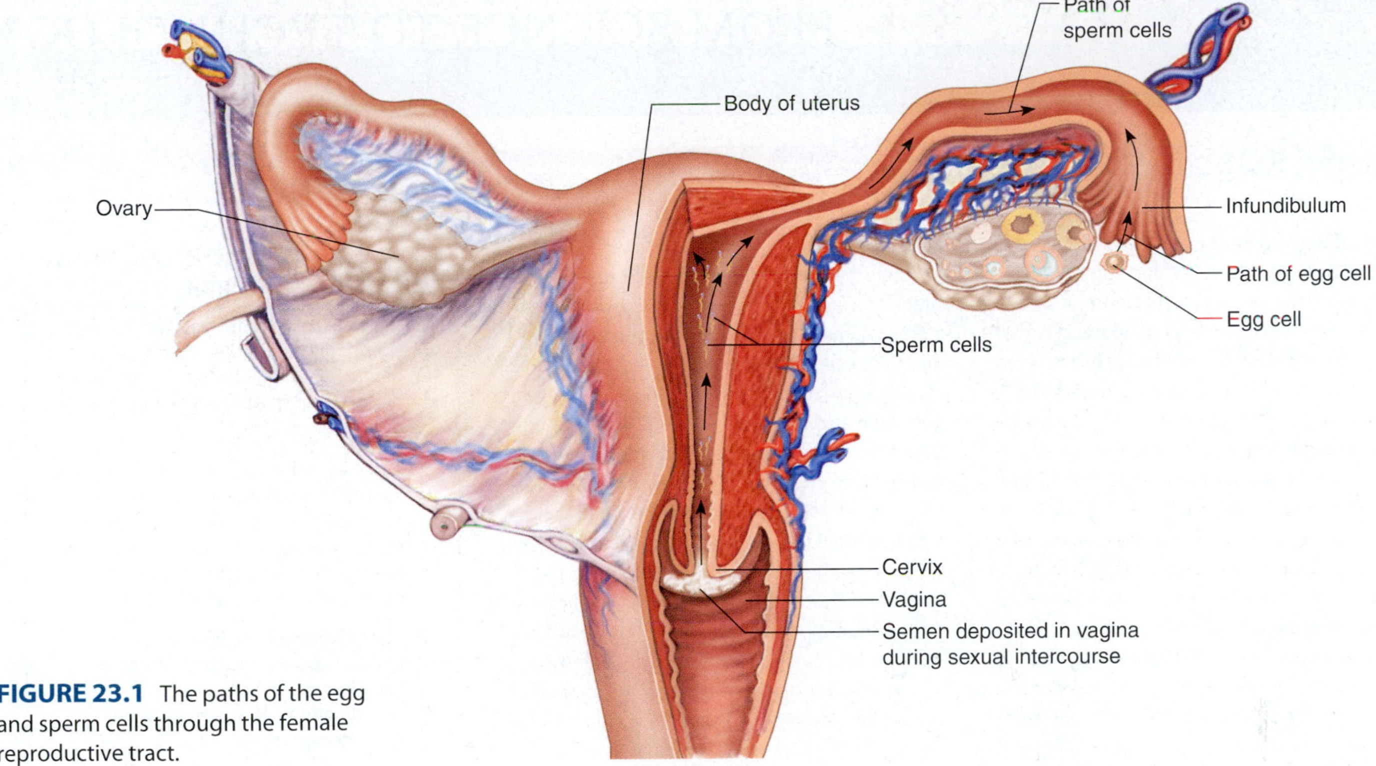

FIGURE 23.1 The paths of the egg and sperm cells through the female reproductive tract.

Sperm movement is inefficient. Even though as many as 200 million to 600 million sperm cells may be deposited in the vagina by a single ejaculation, only a few hundred ever reach a secondary oocyte. The journey to the upper part of the uterine tube takes less than an hour following sexual intercourse. Many sperm cells may reach a secondary oocyte, but usually only one sperm cell fertilizes it (fig. 23.2). If a second sperm were to enter, the fertilized ovum would have three sets of chromosomes and be very unlikely to develop. About one in a million births produces a severely deformed child who has inherited three sets of chromosomes.

A secondary oocyte may survive for only twelve to twenty-four hours following ovulation. In contrast, sperm cells may survive up to six days, but only have the power to fertilize an oocyte within twenty-four to forty-eight hours after entering the female reproductive tract. Consequently, sexual intercourse probably should occur not earlier than forty-eight hours before ovulation, or within twenty-four hours following ovulation if fertilization is to take place. From Science to Technology 23.1 describes assisted routes to conception.

Fertilization

When a sperm reaches a secondary oocyte, it invades the follicular cells that adhere to the oocyte's surface (corona radiata) and binds to the *zona pellucida* that surrounds the oocyte's cell membrane. The acrosome of a sperm cell releases enzymes (including hyaluronidase) that aid penetration of the sperm head by digesting proteins in the zona pellucida (fig. 23.3).

The head portion of one sperm cell enters the secondary oocyte, leaving the mitochondria-rich middle section and tail outside. This action triggers lysosome-like vesicles just beneath the oocyte cell membrane to release enzymes that harden the zona pellucida. This reduces the chance that other sperm cells will penetrate.

The sperm nucleus enters the secondary oocyte's cytoplasm and swells. The secondary oocyte then divides unequally to form a large cell, whose nucleus contains the female's genetic contribution, and a tiny second polar body,

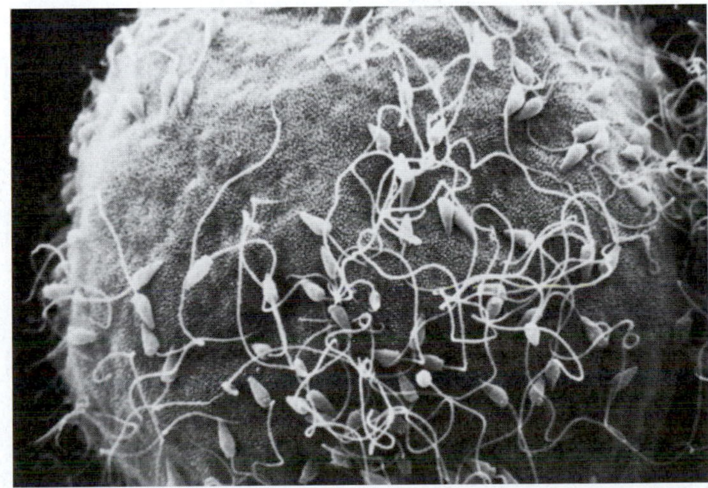

FIGURE 23.2 Scanning electron micrograph of sperm cells on the surface of an egg cell (1,200×). For normal development to proceed, only one sperm cell can fertilize the egg cell.

23.1 FROM SCIENCE TO TECHNOLOGY

Assisted Reproductive Technologies

Conception requires the meeting and merging of sperm cell and egg cell, which naturally occurs in the woman's uterine tube. Abnormal gametes or blockages that impede this meeting of cells can result in infertility (see Clinical Applications 22.2 and 22.3, pp. 842–843, 861). Assisted reproductive technologies (ART) can help couples conceive. The procedures usually involve a laboratory technique and sometimes participation of a third individual. These techniques are often costly and may take several attempts, and some have very low success rates.

Most ARTs were developed in nonhuman animals. For example, the first intrauterine inseminations were performed in dogs in 1782, and the first successful *in vitro* fertilization was accomplished in 1959, in a rabbit. ARTs are now commonplace. In the United States about 1% of the 4 million or so births each year used an ART. Here is a look at some of the procedures.

Donated Sperm—Intrauterine Insemination

In intrauterine ("artificial") insemination, a doctor places donated sperm in a woman's reproductive tract. A woman might seek intrauterine insemination if her partner is infertile or carries a gene for an inherited illness that could affect a child's health, or if she desires to be a single parent. Millions of babies have been born worldwide as a result of this procedure. The first human intrauterine inseminations by donor were done in the 1890s. For many years, physicians donated sperm, and this became a way for male medical students to earn a few extra dollars. By 1953, sperm could be frozen and stored for later use. Today, sperm banks freeze and store donated sperm and then provide it to physicians who perform the procedure. Since 1983, sperm banks have asked donors if they wished to be contacted by their children years later. The first such meetings occurred in 2002, and thousands of others have followed. Today several websites offer DNA tests that enable people to find their sperm donor fathers.

A woman or couple choosing intrauterine insemination can select sperm from a catalog that lists the personal characteristics of the donors, including blood type; hair, skin, and eye color; build; and even educational level and interests. Not all of these traits are inherited. Rarely, intrauterine insemination has led to dilemmas (table 23A).

In Vitro Fertilization

In *in vitro* fertilization (IVF), which means "fertilization in glass," sperm cell meets egg cell outside the woman's body. The fertilized egg cell divides two or three times and is then introduced into

TABLE 23A | Assisted Reproductive Dilemmas

1. A physician in California used his own sperm for intrauterine insemination of fifteen patients and told them that he had used sperm from anonymous donors.
2. A plane crash killed the wealthy parents of two early embryos stored at −320°F (−195°C) in a hospital in Melbourne, Australia. Adult children of the couple were asked to share their estate with two eight-celled balls.
3. Several couples in Chicago planning to marry discovered that they were half-siblings. Their mothers had been inseminated with sperm from the same donor.
4. Two Rhode Island couples sued a fertility clinic for misplacing embryos.
5. A man sued his ex-wife for possession of their frozen embryos as part of the divorce settlement.
6. A man who donated sperm when he was healthy later developed a late-onset genetic disease, cerebellar ataxia. Each of the 18 children conceived using his sperm faces a 1 in 2 chance of having inherited the condition.

which is later expelled. Meiosis is completed. The approaching nuclei from the two sex cells are called pronuclei, until they meet and merge. Next, the pronuclei unite. Their nuclear membranes disassemble, and their chromosomes mingle, completing the process of fertilization.

Each sex cell provides 23 chromosomes, so the product of fertilization is a cell with 46 chromosomes—the usual number in a human body cell. This cell, called a **zygote** (zi′gōt), is the first cell of the future offspring.

PRACTICE

3 What factors aid the movements of the secondary oocyte and sperm cells through the female reproductive tract?

4 Where in the female reproductive system does fertilization normally take place?

5 List the events of fertilization.

A couple expecting a child can estimate the approximate time of fertilization (conception) by adding fourteen days to the date of the onset of the last menstrual period. They can predict the time of birth by adding 266 days to the fertilization date. Most babies are born within ten to fifteen days of this calculated time.

Tracking a pregnancy's progress can be confusing, because some health-care providers measure 40 weeks from the last menstrual period, rather than the more accurate 38 weeks from fertilization. Obstetricians can, however, estimate the date of conception by scanning the embryo with ultrasound and comparing the crown-to-rump length to known values that are the average for each day of gestation. This approach is inaccurate if an embryo is smaller or larger than usual due to a medical problem.

the egg donor's (or another woman's) uterus. If all goes well, a pregnancy begins.

A woman might undergo IVF if her ovaries and uterus function but her uterine tubes are blocked. To begin, she takes a hormone that hastens maturity of several secondary oocytes. Using a laparoscope to view the ovaries and uterine tubes, a physician removes a few of the largest egg cells and transfers them to a dish, then adds chemicals similar to those in the female reproductive tract, and sperm cells.

If a sperm cell cannot penetrate the egg cell *in vitro,* it may be sucked up into a tiny syringe and injected using a tiny needle into the female cell (fig. 23A). This variant of IVF, called intracytoplasmic sperm injection (ICSI), has a 68% fertilization rate. It can help men with very low sperm counts, high numbers of abnormal sperm, or injuries or illnesses that prevent ejaculation. However, if ICSI is done to help a man become a father who has an inherited form of infertility, he may be passing the problem on to a son. In ICSI, minor surgery removes testicular tissue, from which viable sperm are isolated and injected into eggs. A day or so later, a physician transfers some of the resulting balls of eight or sixteen cells to the woman's uterus. The birth rate following IVF for a woman implanted with her own egg is about 26%, compared with 31% for natural conceptions.

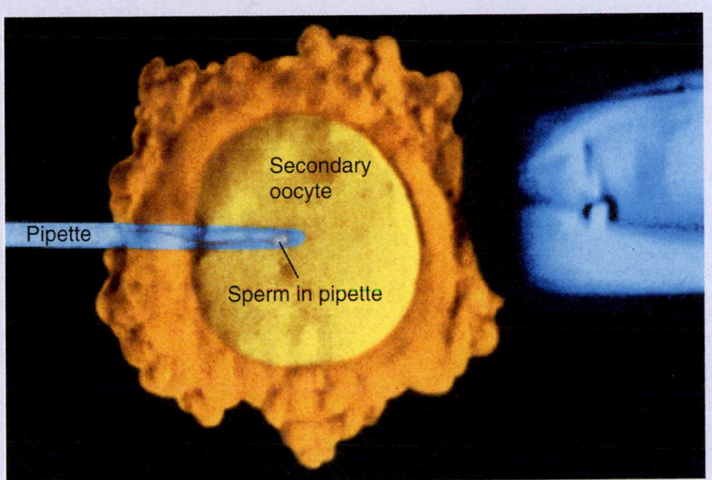

FIGURE 23A Intracytoplasmic sperm injection (ICSI) enables some infertile men and men with spinal cord injuries and other illnesses to become fathers. A single sperm cell is injected into the cytoplasm of an egg.

Gamete Intrafallopian Transfer

One reason that IVF rarely works is the artificial fertilization environment. A procedure called GIFT, which stands for gamete intrafallopian transfer, moves fertilization to the woman's body. A woman takes a superovulation drug for a week and then has several of her largest eggs removed. A man donates a sperm sample, and a physician separates the most active cells. The collected eggs and sperm are deposited together in the woman's uterine tube, at a site past any obstruction so that implantation can occur. GIFT is 22% successful.

In another variation called zygote intrafallopian transfer (ZIFT), a physician places an *in vitro* fertilized ovum in a woman's uterine tube. ■

23.3 PRENATAL PERIOD

The prenatal period of development usually lasts for thirty-eight weeks from conception. It can be divided into a period of cleavage, an embryonic stage, and a fetal stage.

Period of Cleavage

Conception occurs when the pronuclei of sperm cell and secondary oocyte merge, forming a zygote. Thirty hours later, the zygote undergoes mitosis, giving rise to two new cells. These cells, in turn, divide to form four cells, which then divide into eight cells, and so forth. The divisions occur rapidly with little time for the cells to grow (fig. 23.4). Thus, with each subsequent early division, the resulting cells are smaller and smaller. This rapid cell division and distribution of the zygote's cytoplasm into progressively smaller cells is called **cleavage** (klēv′ij), and the cells produced in this way are called *blastomeres.* The ball of cells that results from these initial cell divisions is also called a cleavage embryo. From Science to Technology 23.2 describes genetic tests of blastomeres.

The tiny mass of cells moves through the uterine tube to the uterine cavity, aided by the beating of cilia of the tubular epithelium and by weak peristaltic contractions of smooth muscles in the tubular wall. Secretions from the epithelial lining bring nutrients to the developing organism.

The trip to the uterus takes about three days, and by then, the structure consists of a solid ball, called a **morula** (mor′u-lah), of about sixteen cells (fig. 23.5). The morula remains free within the uterine cavity for about three days. Cell division continues, and the solid ball of cells gradually hollows out. During this stage, the zona pellucida of the original secondary oocyte degenerates, and the structure, now hollow and called a **blastocyst** (blas′to-sist), drops into one of the tubules in the endometrium. Figure 23.5 diagrams this process.

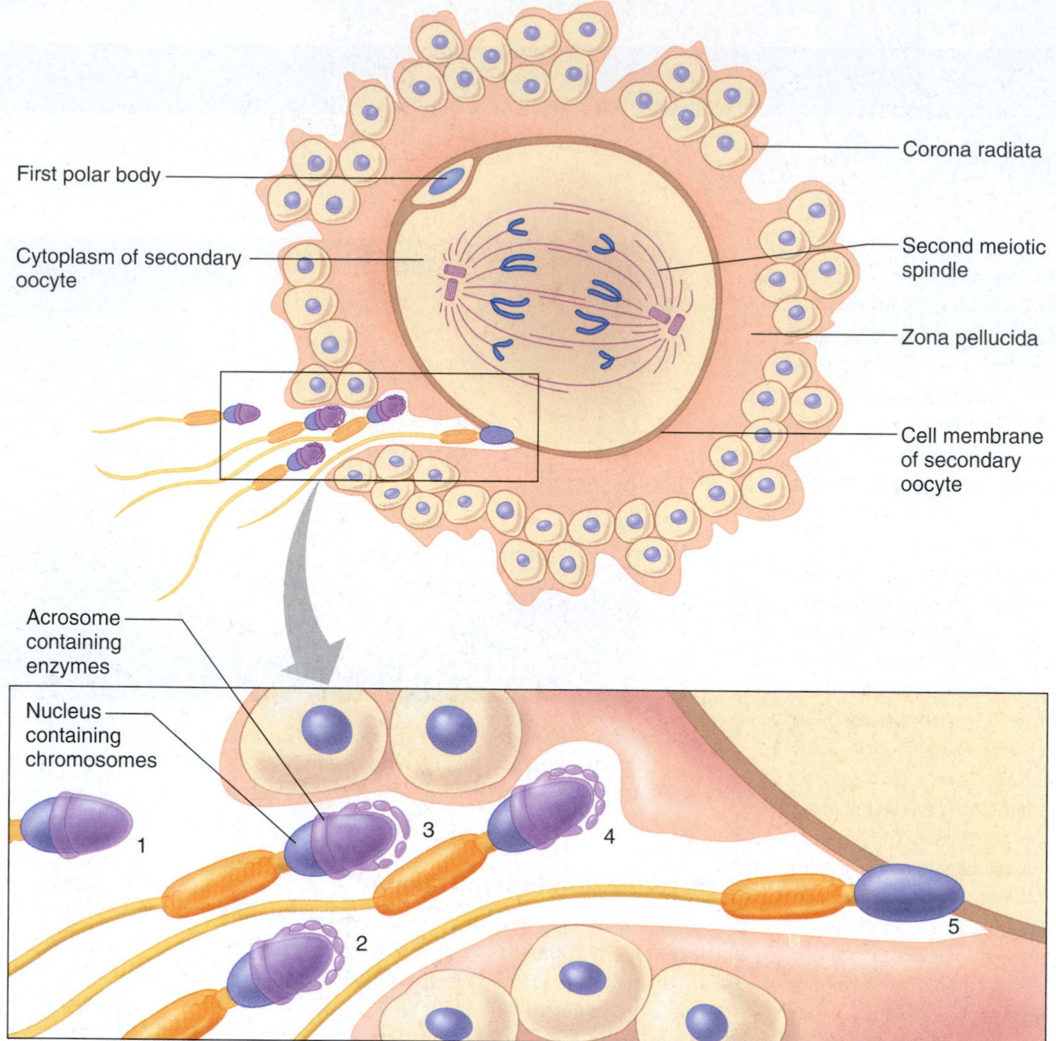

FIGURE 23.3 Steps in fertilization: (*1*) The sperm cell reaches the corona radiata surrounding the oocyte. (*2*) and (*3*) The acrosome of the sperm cell releases a protein-digesting enzyme. (*4*) The sperm cell penetrates the zona pellucida surrounding the oocyte. (*5*) The sperm cell's membrane fuses with the oocyte's membrane.

By the end of the first week of development, the blastocyst superficially implants in the endometrium (fig. 23.6*a*). Up until this point, the cells that will become developing offspring are pluripotent stem cells, which means they can give rise to several specialized types of cells, as well as yield additional stem cells.

Within the blastocyst, cells in one region group to form an *inner cell mass* that eventually gives rise to the **embryo proper** (em′bre-o prop′er)—the body of the developing offspring. The cells that form the wall of the blastocyst make up the *trophoblast,* which develops into structures that assist the embryo.

Inner cell mass cells can be cultured in a laboratory dish and, in the presence of a specific "cocktail" of growth factors and other biochemicals, become human embryonic stem cells. These cells are useful in research to study normal development as well as the beginnings of diseases.

If two ovarian follicles release secondary oocytes simultaneously, and both are fertilized, the resulting zygotes develop into fraternal (dizygotic) twins, which are no more alike genetically than any nontwin siblings. Twins may develop from a single fertilized oocyte (monozygotic twins) if two inner cell masses form within a blastocyst and each produces an embryo. Twins of this type usually share a single placenta, and they are genetically identical. They are always the same sex and are similar in appearance.

The blastocyst attaches to the uterine lining, aided by its secretion of proteolytic enzymes that digest part of the endometrium (fig. 23.6*b, c*). The blastocyst sinks slowly into the resulting depression, becoming completely buried in the uterine lining. At the same time, the uterine lining is stimulated to thicken below the implanting blastocyst, and cells of the trophoblast begin to produce tiny, fingerlike extensions

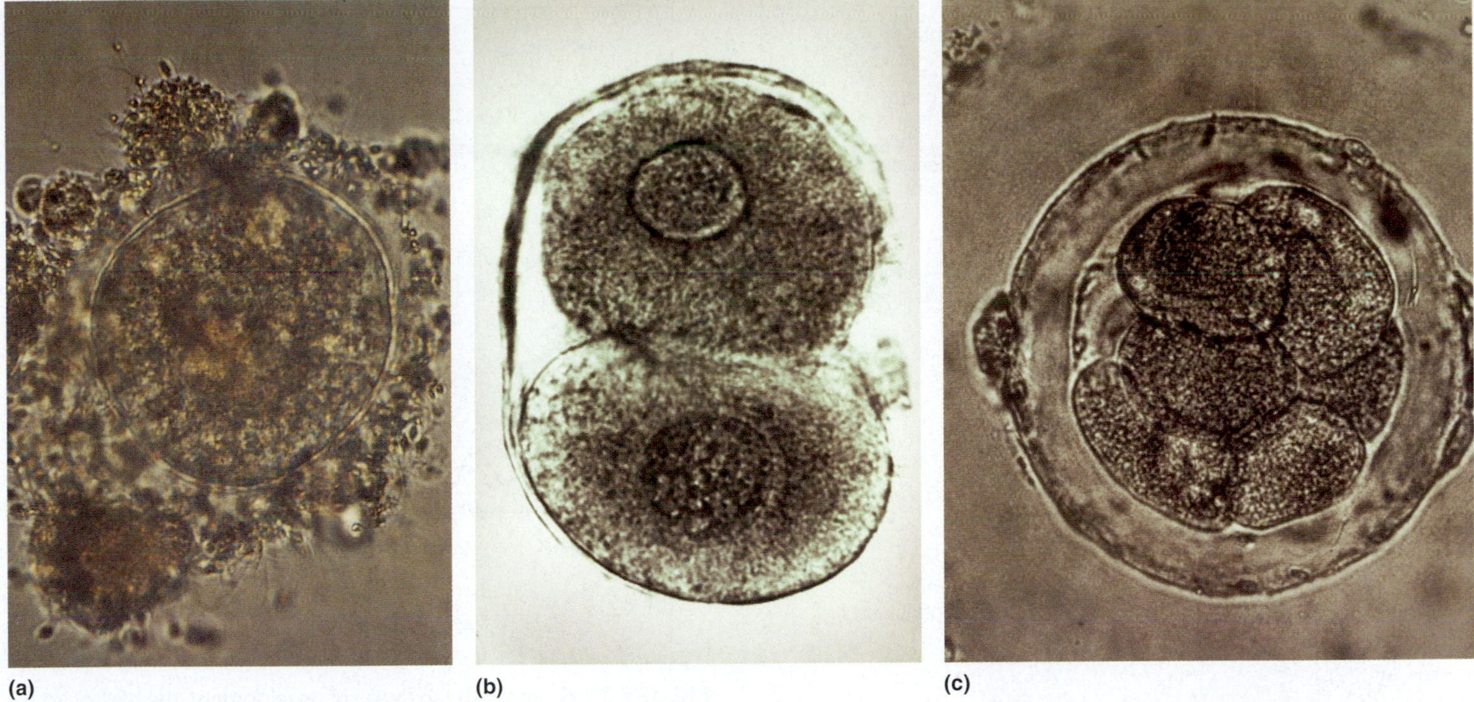

FIGURE 23.4 Light micrographs of (*a*) a human secondary oocyte surrounded by follicular cells and sperm cells (250×), (*b*) the two-cell stage (600×), and (*c*) a morula (500×).

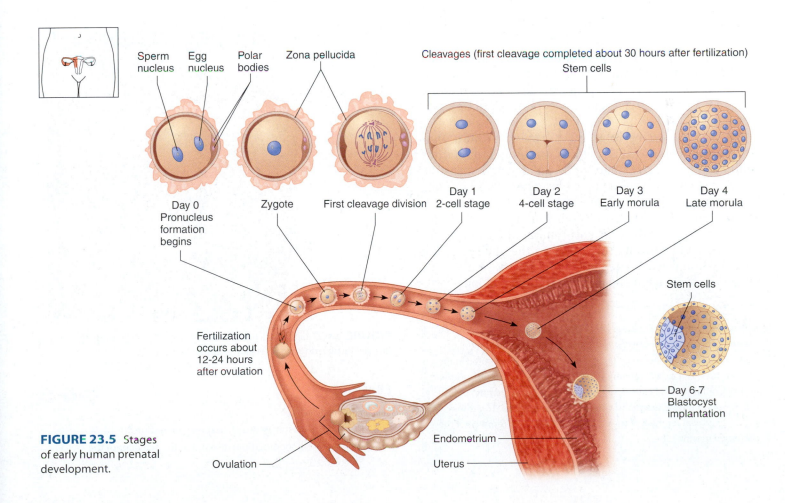

FIGURE 23.5 Stages of early human prenatal development.

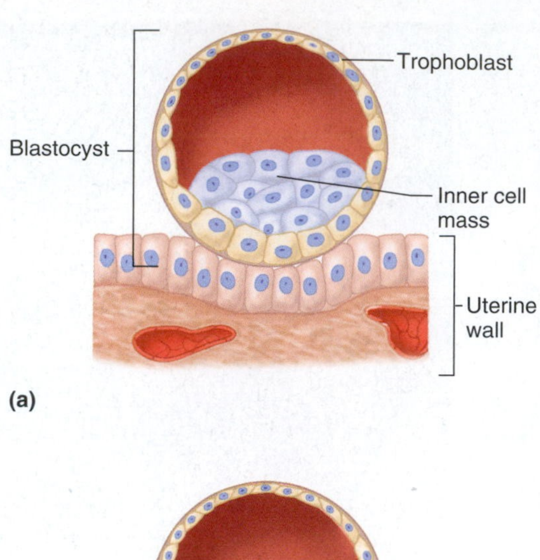

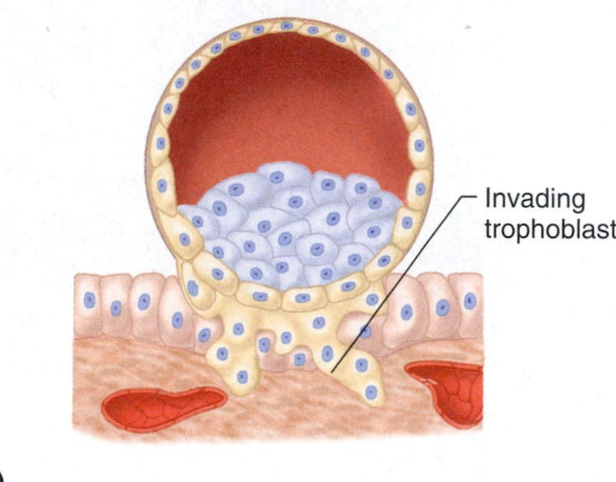

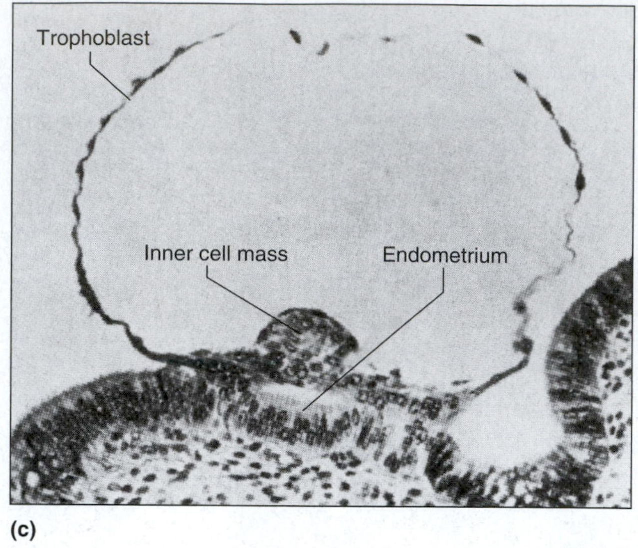

FIGURE 23.6 About the sixth day of development, the blastocyst (*a*) contacts the uterine wall and (*b*) begins to implant. The trophoblast, which will help form the placenta, secretes hCG, a hormone that maintains the pregnancy. (*c*) Light micrograph of a blastocyst from a monkey in contact with the endometrium of the uterine wall (150×).

(microvilli) that grow into the endometrium. This nestling of the blastocyst into the uterine lining is called **implantation** (im″-plan-ta′shun). It begins near the end of the first week and is completed during the second week of development (fig. 23.7).

The trophoblast secretes the hormone **human chorionic gonadotropin** (hCG), which maintains the corpus luteum during the early stages of pregnancy and keeps the immune system from rejecting the blastocyst. This hormone also stimulates synthesis of other hormones from the developing placenta. The **placenta** (plah-sen′tah) is a vascular structure, formed by the cells surrounding the embryo and cells of the endometrium, that attaches the embryo to the uterine wall and exchanges nutrients, gases, and wastes between the maternal blood and the embryo's blood.

If an embryo implants in tissues outside the uterus, such as those of a uterine tube, an ovary, the cervix, or an organ in the abdominal cavity, the result is an *ectopic pregnancy*. A fertilized egg implanted in the uterine tube is a *tubal pregnancy*. The tube usually ruptures as the embryo enlarges and causes severe pain and heavy vaginal bleeding, threatening the pregnant woman and the embryo. Treatment is prompt surgical removal of the embryo and repair or removal of the damaged uterine tube.

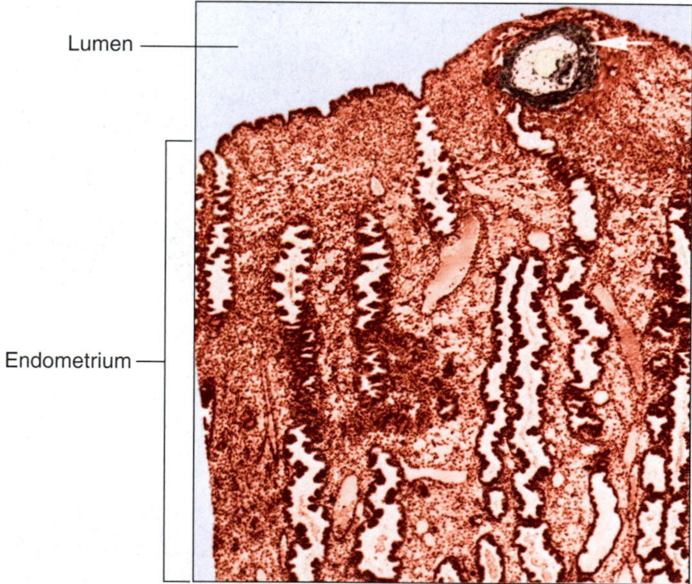

FIGURE 23.7 Light micrograph of a human cleavage embryo (arrow) implanting in the endometrium (18×).

PRACTICE

6 What is cleavage?

7 How does a blastocyst attach to the endometrium?

8 How does the endometrium respond to the activities of the blastocyst?

23.2 FROM SCIENCE TO TECHNOLOGY

Preimplantation Genetic Diagnosis

Six-year-old Molly Nash would probably have died within a year or two of Fanconi anemia had she not received a very special gift from her baby brother Adam—his umbilical cord stem cells. Adam was not only free of the gene that causes the anemia, but his cell surfaces matched those of his sister, making a transplant likely to succeed. But the parents didn't have to wait until Adam's birth in August 2000 to know that his cells could save Molly—they knew when he was a mere eight-celled cleavage embryo (fig. 23B).

When the Nashs could not find a compatible bone marrow donor for Molly, they turned to preimplantation genetic diagnosis (PGD). Following *in vitro* fertilization, described in From Science to Technology 23.1, researchers at the Reproductive Genetics Institute at Illinois Masonic Medical Center removed a single cell from each of several eight-celled cleavage embryos and probed the cells to detect those free of the disease-causing mutation. They also scrutinized the HLA genes, which control rejection of a transplanted organ, and chose the ball of cells that would be Adam. The cleavage embryo divided in the laboratory until it was about 120 cells, and then it was implanted into Lisa Nash's uterus. Adam was born, and a month later physicians infused the umbilical cord stem cells into his sister. Today, Molly is healthy.

PGD works because of a feature of many animal species called indeterminate cleavage. Up until a certain point in early development, a cell or two can be removed, yet the remainder of the embryo can continue to develop normally if implanted into a uterus. PGD, invented in 1989, at first helped a few families avoid devastating inherited illnesses in their sons. Then, in 1992, Chloe O'Brien was born free of the cystic fibrosis that made her brother very ill, thanks to PGD. In 1994 came another milestone, when a girl was conceived and selected to provide umbilical cord stem cells that cured her teenage sister's leukemia. This case became the basis of a popular novel.

Thousands of children have been born worldwide following PGD, free of the disorders that run in their families. In addition to enabling families to circumvent many inherited conditions, PGD enables couples who repeatedly lose early embryos due to chromosome abnormalities—they suffer repeat miscarriages—to select chromosomally normal embryos. PGD is becoming routine with IVF, because it ensures that only embryos with normal-appearing chromosomes or lacking a family's mutation are implanted. ■

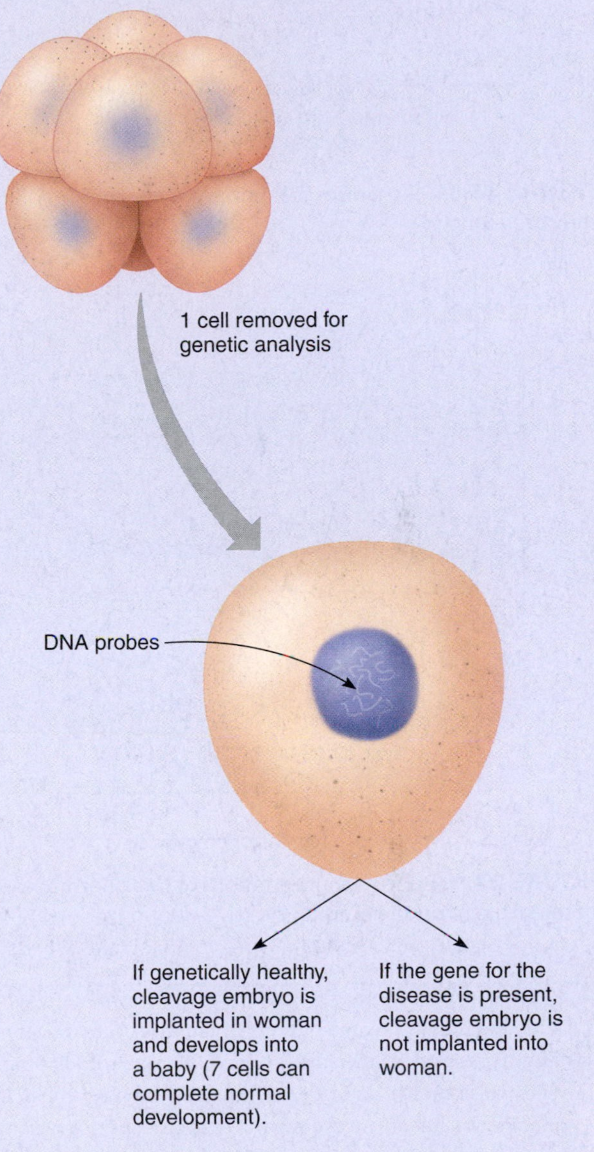
1 cell removed for genetic analysis
DNA probes
If genetically healthy, cleavage embryo is implanted in woman and develops into a baby (7 cells can complete normal development).
If the gene for the disease is present, cleavage embryo is not implanted into woman.

FIGURE 23B Preimplantation genetic diagnosis probes disease-causing genes in an eight-celled cleavage embryo.

Hormonal Changes During Pregnancy

During a typical reproductive cycle, the corpus luteum degenerates about two weeks after ovulation. Consequently, concentrations of estrogens and progesterone decline rapidly, the uterine lining is no longer maintained, and the endometrium sloughs off as menstrual flow. If this occurs following implantation, the embryo is lost in a spontaneous abortion.

The hormone hCG normally helps prevent spontaneous abortion. It functions similarly to LH, and it maintains the corpus luteum, which continues secreting estrogens and progesterone. Thus, the uterine wall continues to grow and develop (fig. 23.8). At the same time, release of FSH and LH from the anterior pituitary gland is inhibited, so normal reproductive cycles cease.

Secretion of hCG continues at a high level for about two months, then declines to a low level by the end of four months. Although the corpus luteum persists throughout pregnancy, its function as a source of hormones becomes less important after the first three months (first trimester), when the placenta secretes sufficient estrogens and progesterone (fig. 23.9).

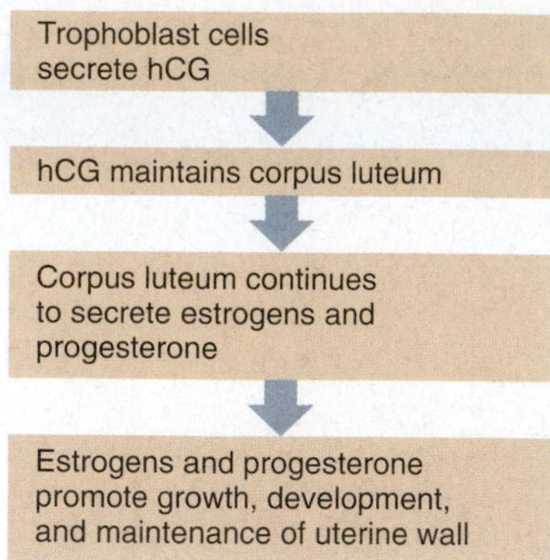

FIGURE 23.8 Mechanism that preserves the uterine lining during early pregnancy.

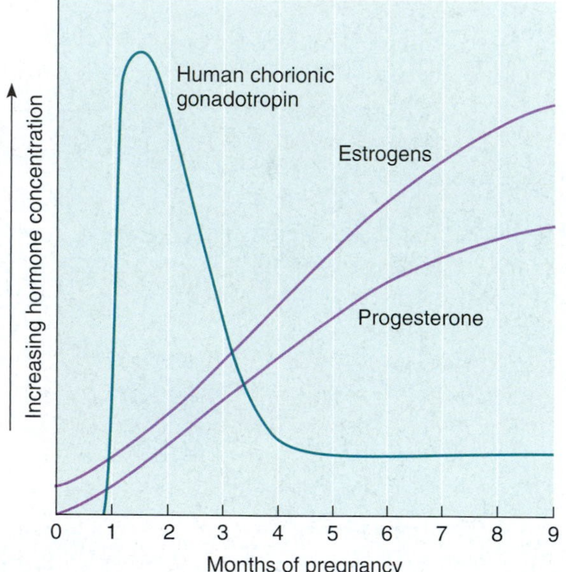

FIGURE 23.9 Relative concentrations of three hormones in maternal blood during pregnancy.

> Detecting hCG in a woman's urine or blood is used to confirm pregnancy. The level of hCG in a pregnant woman's body fluids peaks at fifty to sixty days of gestation, then falls for the remainder of her pregnancy. Later on, measuring hCG has other uses. If a woman miscarries but her blood still shows hCG, fetal tissue may remain in her uterus, and this material must be removed. At the fifteenth week of pregnancy, most women have a blood test that measures levels of four substances produced by the fetus—alpha fetoprotein (AFP), estriol (an estrogen), pregnancy-associated plasma protein A (PAPP-A), and hCG. If estriol and PAPP-A are low but hCG is elevated, the fetus is at risk of having an extra chromosome 21 (Down syndrome). Further tests are advisable for a definitive diagnosis.

For the remainder of the pregnancy, *placental estrogens* and *placental progesterone* maintain the uterine wall. The placenta also secretes a hormone called **placental lactogen** that may stimulate breast development and prepare the mammary glands to secrete milk, with the aid of placental estrogens and progesterone. Placental progesterone and a polypeptide hormone called *relaxin* from the corpus luteum inhibit the smooth muscles in the myometrium, suppressing uterine contractions until the birth process begins.

The high concentration of placental estrogens during pregnancy enlarges the vagina and the external reproductive organs. Also, relaxin relaxes the connective tissue of the symphysis pubis and sacroiliac joints. This action, which usually occurs during the last week of pregnancy, allows for greater movement at these joints, aiding passage of the fetus through the birth canal.

Other hormonal changes during pregnancy include increased secretion of aldosterone from the adrenal cortex and of parathyroid hormone from the parathyroid glands. Aldosterone promotes renal reabsorption of sodium, leading to fluid retention. Parathyroid hormone helps to maintain a high concentration of maternal blood calcium, because fetal demand for calcium can cause hypocalcemia, which promotes cramps. Table 23.1 summarizes the hormonal changes of pregnancy.

RECONNECT

To Chapter 13, Parathyroid Glands, pages 502–503.

PRACTICE

9 What are the sources of the hormones that sustain the uterine wall during pregnancy?

10 What other hormonal changes occur during pregnancy?

Other Changes During Pregnancy

Other changes in a woman's body respond to the increased requirements of a growing offspring. As the offspring grows, the uterus enlarges greatly, and instead of being confined to its normal location in the pelvic cavity, it extends upward and may eventually reach the level of the ribs. The abdominal organs are displaced upward and compressed against the diaphragm. The enlarging uterus also presses on the urinary bladder. As a result, a pregnant woman may be unable to eat large meals and may develop heartburn and have to urinate often.

The growing and developing placenta requires more blood, and as the offspring enlarges, it needs more oxygen and produces more waste that must be excreted. The pregnant woman's blood volume, cardiac output, breathing rate, and urine production all increase to handle offspring growth.

The pregnant woman must eat more to obtain adequate nutrition for the offspring. Her intake must supply sufficient vitamins, minerals, and proteins for herself and the offspring. The offspring tissues have a greater capacity to capture available nutrients than do the maternal tissues. Consequently, if the pregnant woman's diet is inadequate, her body will usually show symptoms of a deficiency condition before offspring growth is adversely affected.

TABLE 23.1 | Hormonal Changes During Pregnancy

1. Following implantation, cells of the trophoblast begin to secrete hCG.
2. hCG maintains the corpus luteum, which continues to secrete estrogens and progesterone.
3. As the placenta develops, it secretes abundant estrogens and progesterone.
4. Placental estrogens and progesterone
a. stimulate the uterine lining to continue development.
b. maintain the uterine lining.
c. inhibit secretion of FSH and LH from the anterior pituitary gland.
d. stimulate development of the mammary glands.
e. inhibit uterine contractions (progesterone).
f. enlarge the reproductive organs (estrogens).
5. Relaxin from the corpus luteum also inhibits uterine contractions and relaxes the pelvic ligaments.
6. The placenta secretes placental lactogen that stimulates breast development.
7. Aldosterone from the adrenal cortex promotes reabsorption of sodium.
8. Parathyroid hormone from the parathyroid glands helps maintain a high concentration of maternal blood calcium.

Nausea and vomiting in pregnancy may shield a fetus from foods that might contain toxins or pathogens. The condition affects two in three pregnancies and coincides with the time in gestation when a woman's immune system is at its weakest. An analysis of more than 80,000 pregnant women found that they avoid foods that spoil easily, such as eggs and meats, as well as coffee and alcohol. Yet many pregnant women eat more fruits and vegetables than usual. In societies where the diet is mostly grains with little if any meat, incidence of morning sickness is much lower than in groups with more varied, and possibly dangerous, diets. Rates of morning sickness are highest in Japan, where raw fish is a dietary staple, and European countries, where undercooked meat is often eaten. Evolution has likely selected for "morning sickness" where it correlates to, and possibly contributes to, better birth outcomes.

Embryonic Stage

The **embryonic stage** extends from the beginning of the second week through the eighth week of prenatal development. During this time, the placenta forms, the main internal organs develop, and the major external body structures appear.

During the second week of prenatal development, the blastocyst completes implantation, and the inner cell mass changes. A space, the *amniotic cavity,* forms between the inner cell mass and the part of the trophoblast that "invades" the endometrium. The inner cell mass then flattens and is called the **embryonic disc.** By the end of the second week, layers form.

The embryonic disc initially consists of two distinct layers: an outer *ectoderm* and an inner *endoderm.* A short time later, through a process called gastrulation, a third layer of cells, the *mesoderm,* forms between the ectoderm and endoderm. These three layers of cells are called the **primary germ layers** (pri′ma-re jerm la′erz) of the primordial embryo. They are the primitive tissues from which all organs form. At this point, the embryo is termed a **gastrula** (gas′troo-lah). Also during this time, a structure called a connecting stalk appears. It attaches the embryo to the developing placenta (fig. 23.10). Table 23.2 summarizes the stages of early human prenatal development.

Gastrulation is an important process in prenatal development because a cell's fate is determined by which layer it is in. The cells of the ectoderm and endoderm are epithelia. The mesoderm is loosely organized connective tissue. *Ectodermal cells* give rise to the nervous system, parts of special sensory organs, the epidermis, hair, nails, glands of the skin, and linings of the mouth and anal canal. *Mesodermal cells* form all types of muscle tissue, bone tissue, bone marrow, blood, blood vessels, lymphatic vessels, internal reproductive organs, kidneys, and the mesothelium of the body cavities. *Endodermal cells* produce the epithelial linings of the digestive tract, respiratory tract, urinary bladder, and urethra (fig. 23.11). The primary germ layers retain stem cells, a few of which persist in the adult, enabling tissues to grow and repair damage.

As the embryo implants in the uterus, proteolytic enzymes from the trophoblast break down endometrial tissue, providing nutrients for the developing embryo. A second layer of cells begins to line the trophoblast, and together these two layers form a structure called the **chorion** (ko′re-on). Soon, slender projections grow out from the trophoblast, including the new cell layer, eroding their way into the surrounding endometrium by continuing to secrete proteolytic enzymes. These projections become increasingly complex and form the highly branched **chorionic villi,** which are well established by the end of the fourth week. Figure 24.13 depicts prenatal genetic tests performed on chorionic villi or shed fetal skin cells.

Continued secretion of proteolytic enzymes forms irregular spaces called **lacunae** in the endometrium around and between the chorionic villi. These spaces fill with maternal blood that escapes from endometrial blood vessels eroded by enzyme action. At the same time, embryonic blood vessels carrying blood to and from the embryo extend through the connecting stalk and establish capillary networks in the developing chorionic villi. These embryonic vessels allow nutrient exchange with blood in the lacunae, meeting the increased nutrient demands of the growing embryo.

During the fourth week of development, the flat embryonic disc becomes cylindrical, and the precursor of the central nervous system, called the neural tube, forms. By the end of week four, the head and jaws appear, the heart beats and

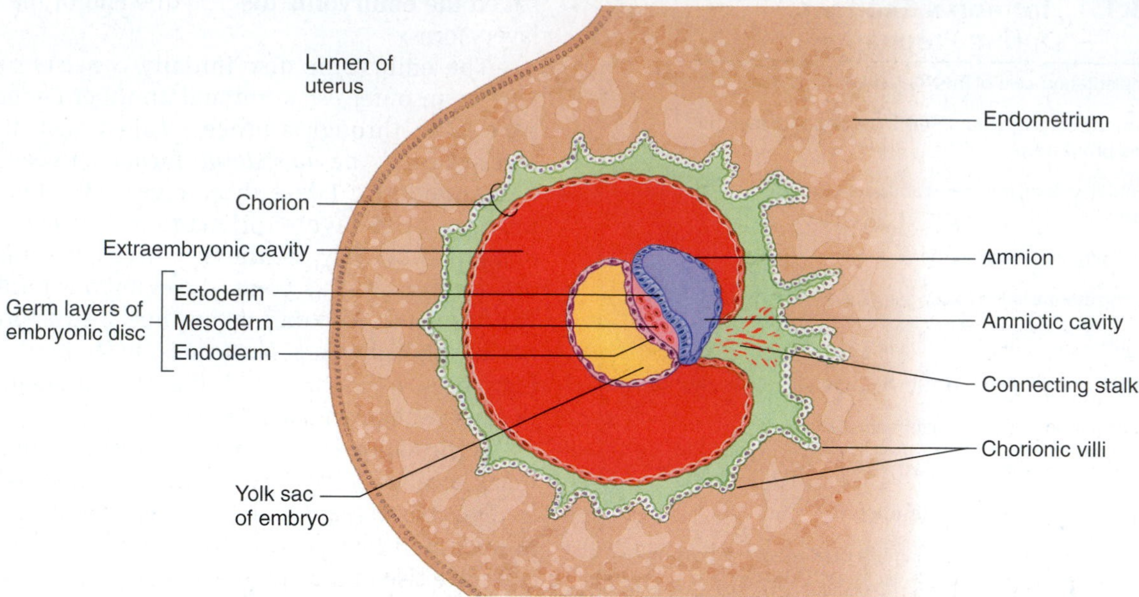

FIGURE 23.10 Early in the embryonic stage of development, the three primary germ layers form.

TABLE 23.2 | Stages and Events of Early Human Prenatal Development

Stage	Time Period	Principal Events
Zygote	12-24 hours following ovulation	Secondary oocyte fertilized, meiosis is completed; zygote has 46 chromosomes and is genetically distinct
Cleavage	30 hours to third day	Mitosis increases cell number
Morula	Third to fourth day	Solid ball of cells
Blastocyst	Fifth day through second week	Hollowed ball forms trophoblast (outside) and inner cell mass, which implants and flattens to form embryonic disc
Gastrula	End of second week	Primary germ layers form

forces blood through blood vessels, and tiny buds form, which will give rise to the upper and lower limbs (fig. 23.12).

During the fifth through the seventh weeks, as figure 23.13 shows, the head grows rapidly and becomes rounded and erect. The face develops eyes, nose, and mouth. The upper and lower limbs elongate, and fingers and toes form (fig. 23.14). By the end of the seventh week, all the main internal organs are established, and as these structures enlarge and elaborate, the body takes on a humanlike appearance.

Until about the end of the eighth week, the chorionic villi cover the entire surface of the former trophoblast. However, as the embryo and the surrounding chorion enlarge, only villi that contact the endometrium endure. The others degenerate, and the areas of the chorion where they were attached become smooth. The region of the chorion still in contact with the uterine wall is restricted to a disc-shaped area that becomes the placenta (fig. 23.15).

A thin **placental membrane** separates embryonic blood in the capillary of a chorionic villus from maternal blood in a lacuna. This membrane is composed of the epithelium of the chorionic villus and the endothelium of the capillary inside the villus (fig. 23.16). Through this membrane, substances are exchanged between the maternal blood and the embryo's blood. Oxygen and nutrients diffuse from the maternal blood into the embryo's blood, and carbon dioxide and other wastes diffuse from the embryo's blood into the maternal blood. Active transport and pinocytosis also move substances through the placental membrane.

If a pregnant woman repeatedly ingests an addictive substance that crosses the placenta, her newborn may suffer from withdrawal symptoms when amounts of the chemical the fetus was accustomed to receiving suddenly plummet. Newborn addiction occurs with certain addictive drugs of abuse, such as heroin and with certain prescription drugs used to treat anxiety.

PRACTICE

11 Describe the major events of the embryonic stage of development.

12 Which tissues and structures develop from ectoderm? From mesoderm? From endoderm?

13 Describe the structure of a chorionic villus.

14 What is the function of the placental membrane?

15 How are substances exchanged between the embryo's blood and the maternal blood?

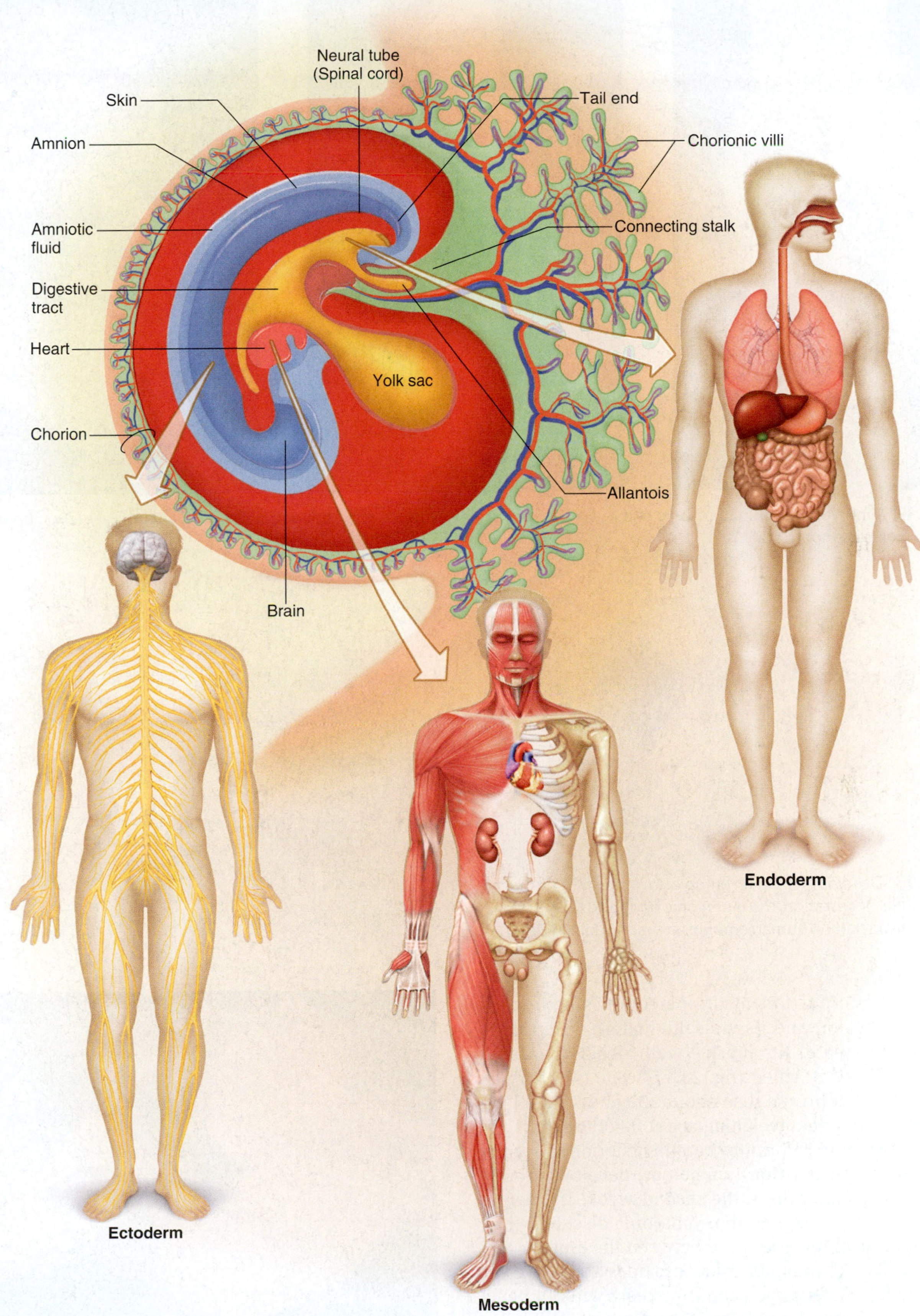

FIGURE 23.11 Each of the primary germ layers collectively serve as primitive tissues from which all organs form.

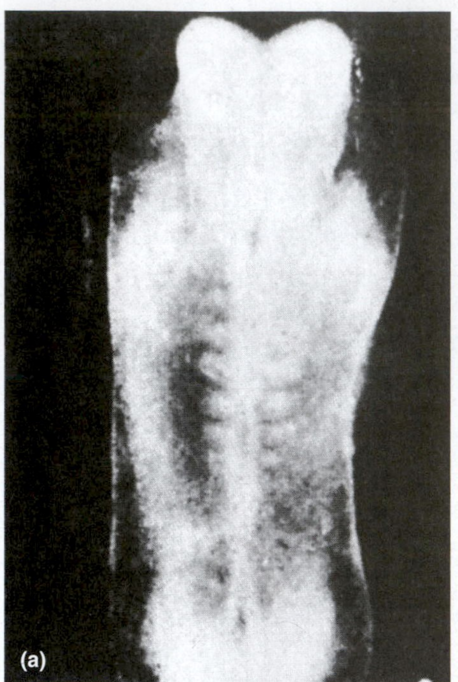

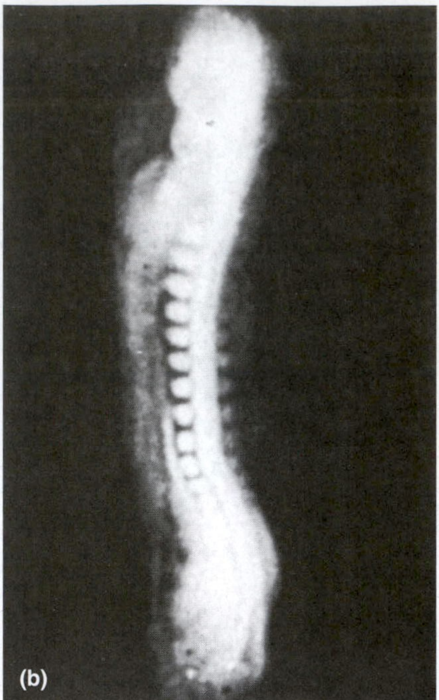

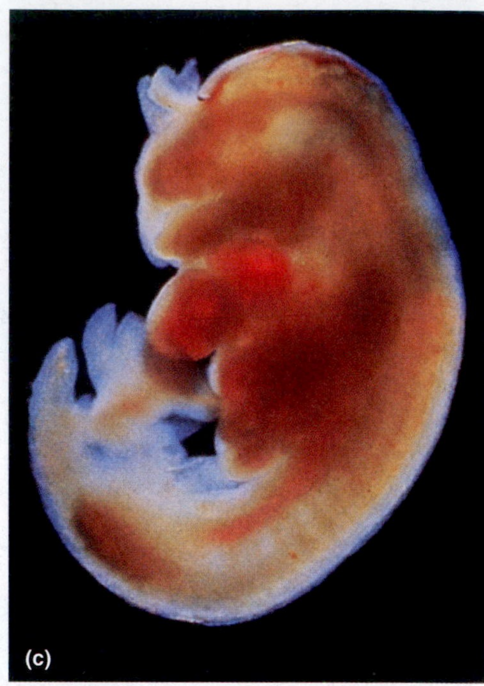

FIGURE 23.12 Embryo. (*a*) A human embryo at three weeks, posterior view; (*b*) at three and one-half weeks, lateral view; (*c*) at about four weeks, lateral view. (Figures are not to scale.)

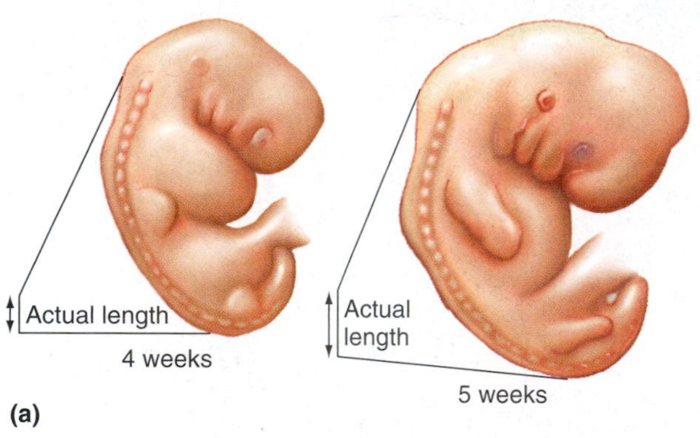

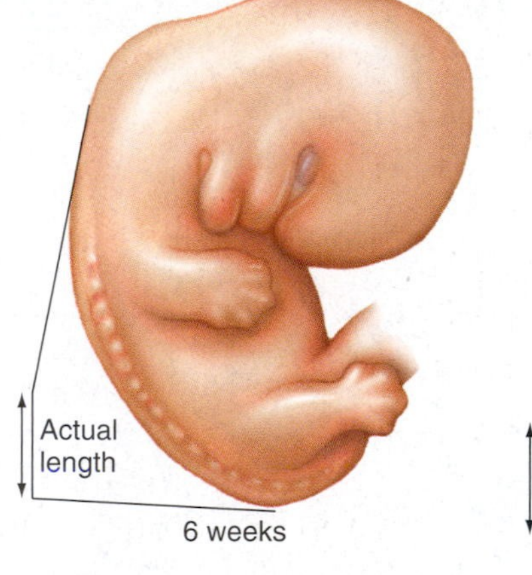

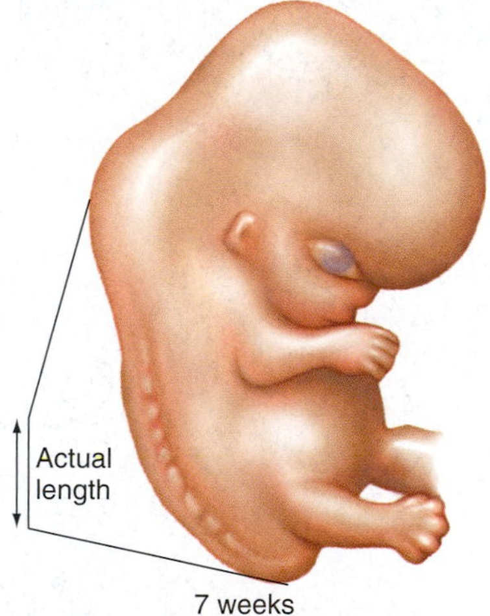

FIGURE 23.13 Development of an embryo. (*a*) In the fifth through the seventh weeks of gestation, the embryonic body and face become more distinct. (*b*) A human embryo after about six weeks of development.

The embryonic portion of the placenta is composed of parts of the chorion and its villi; the maternal portion is composed of the area of the uterine wall (decidua basalis) to which the villi are attached (fig. 23.17). The fully formed placenta is a reddish brown disc about 20 centimeters long and 2.5 centimeters thick, weighing about 0.5 kilogram.

While the placenta is forming from the chorion, a second membrane, called the **amnion** (am′ne-on), develops around the embryo. It appears during the second week. Its margin is attached around the edge of the embryonic disc, and fluid called **amniotic fluid** fills the space between the amnion and the embryonic disc. The amniotic fluid provides a watery environment in which the embryo can grow freely without being compressed by surrounding tissues. The amniotic fluid also protects the embryo from being jarred by the movements of

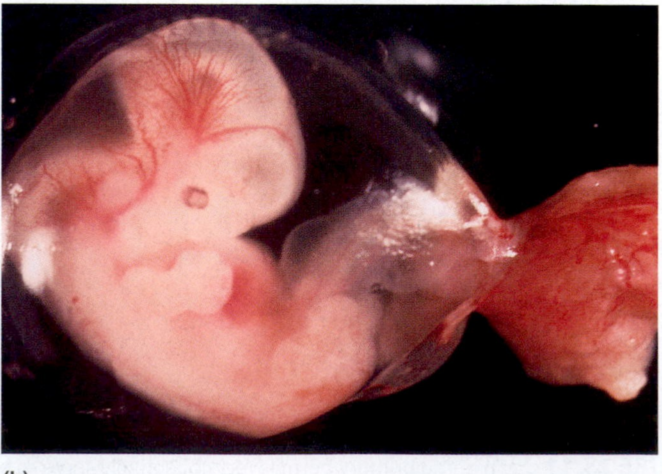

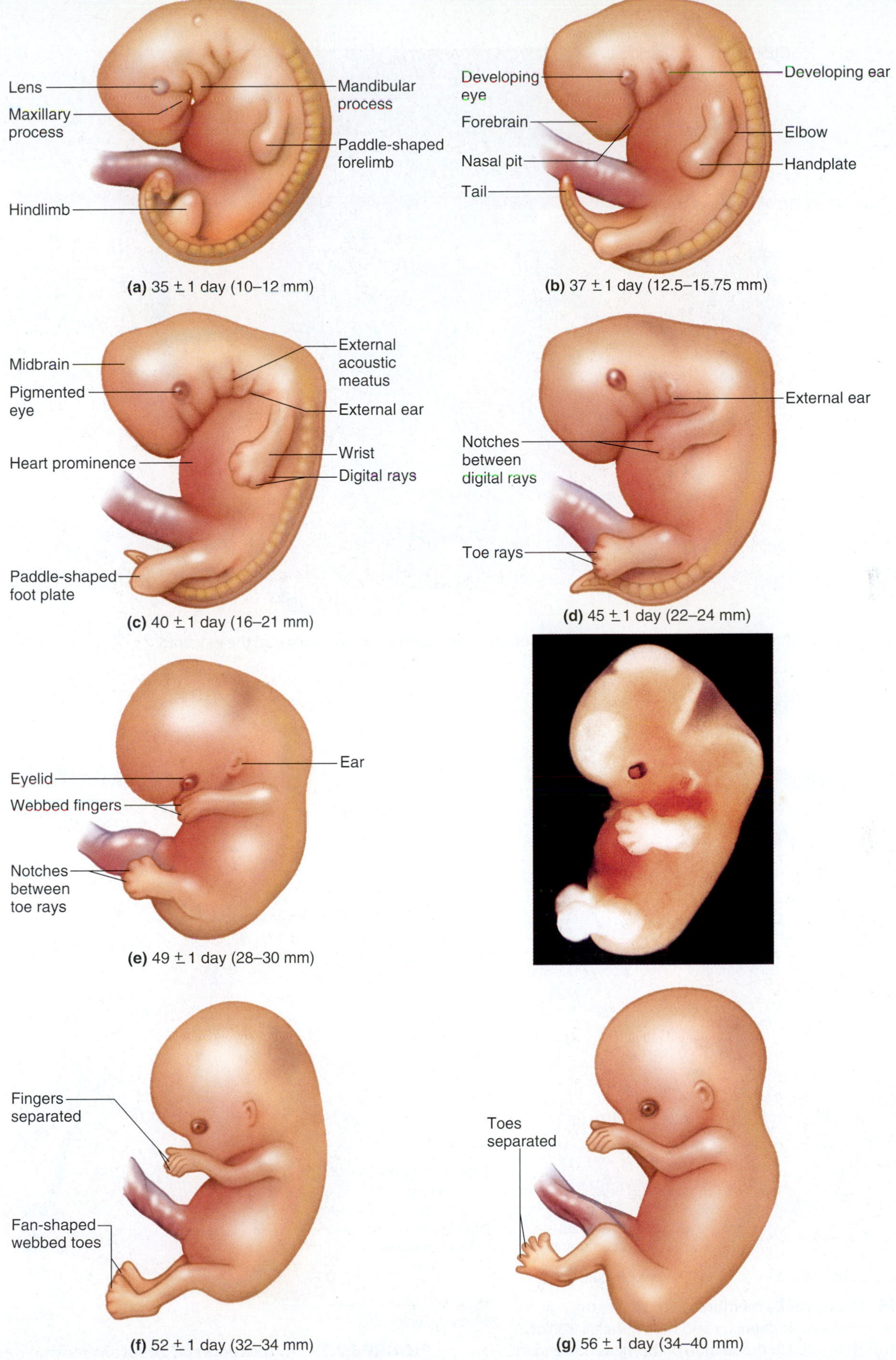

FIGURE 23.14 Changes occurring during the fifth (*a–c*), sixth (*d*), and seventh (*e–g*) weeks of development. The photo corresponds to forty-nine days of development.

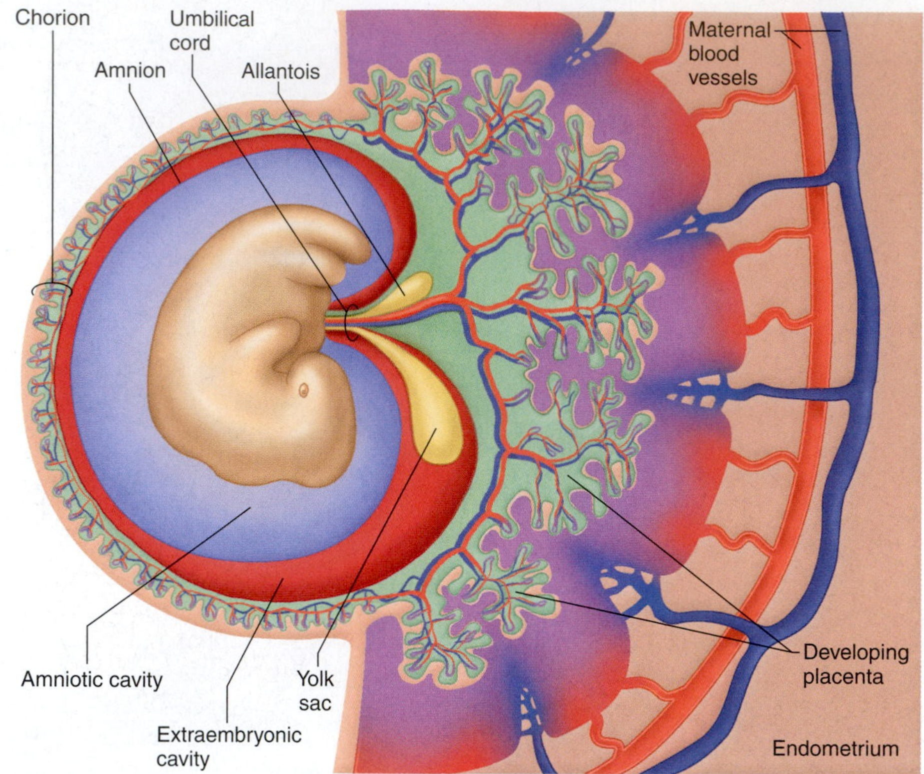

FIGURE 23.15 The disc-shaped area where the chorion still contacts the uterine wall develops into the placenta.

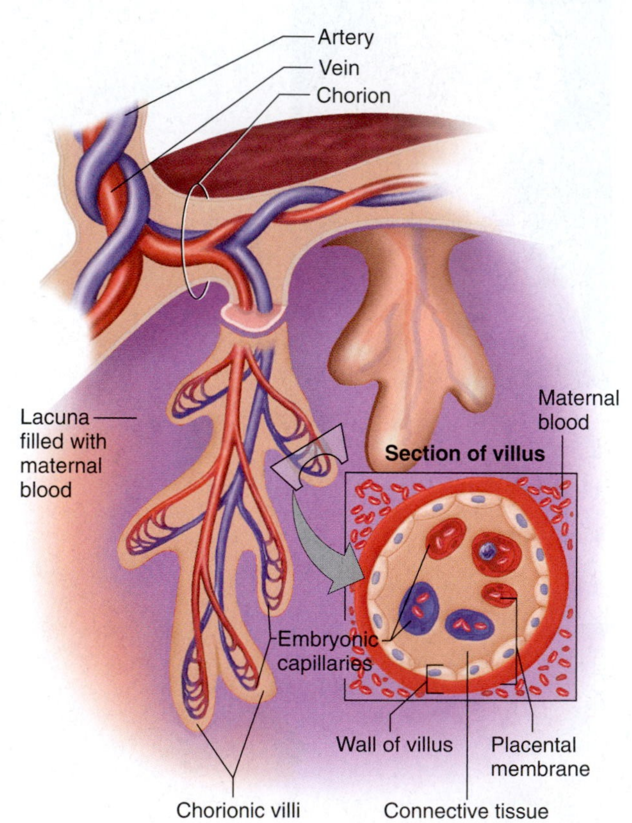

FIGURE 23.16 The placental membrane consists of the endothelium of an embryonic capillary and the epithelial wall of a chorionic villus, as illustrated in the section of the villus (lower part of the figure).

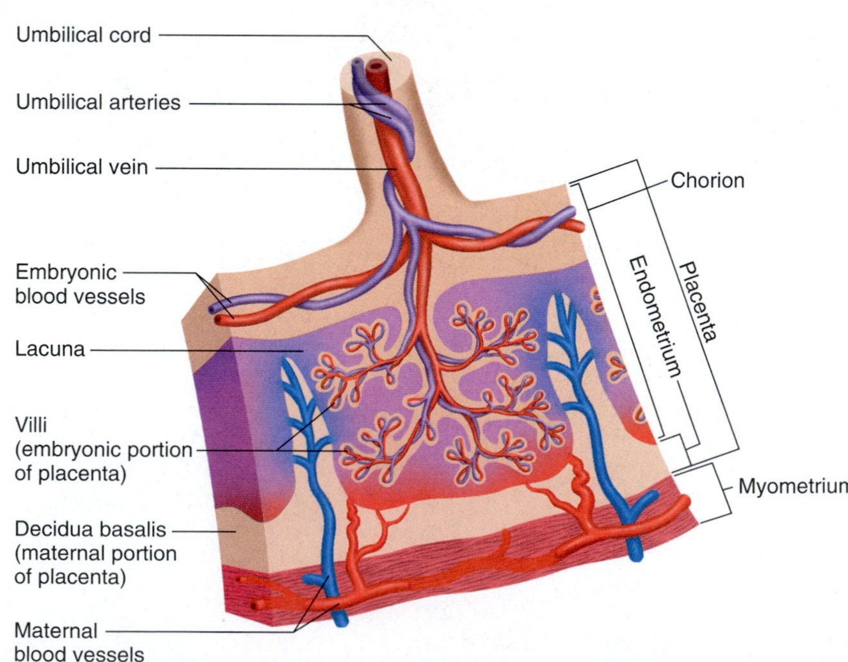

FIGURE 23.17 The placenta consists of an embryonic portion and a maternal portion.

the woman's body, and maintains a stable temperature for proper embryonic and fetal development.

> Certain epithelial cells of the amnion can be cultured in laboratory dishes with biochemicals that coax them to become pluripotent stem cells. Changing the culture medium stimulates these cells to differentiate as cartilage, bone, fat, muscle, pancreas, liver, or neuron—representatives of all three primary germ layers. These amniotic cells are just one example of stem cell sources from what would otherwise be medical waste.

The developing placenta synthesizes progesterone from cholesterol in the maternal blood. Cells associated with the developing fetal adrenal glands use the placental progesterone to synthesize estrogens. The estrogens, in turn, promote changes in the maternal uterus and breasts and influence maternal metabolism and the development of fetal organs.

As the embryo becomes more cylindrical, the margins of the amnion fold, enclosing the embryo in the amnion and amniotic fluid. The amnion envelopes the tissues on the underside of the embryo, particularly the connecting stalk, by which it is attached to the chorion and the developing placenta. In this manner, the **umbilical cord** (um-bil′ĭ-kal kord) forms (see fig. 23.15).

The fully developed umbilical cord is about 1 centimeter in diameter and about 55 centimeters in length. It originates at the umbilicus of the embryo and inserts into the center of the placenta. The cord contains three blood vessels—two *umbilical arteries* and one *umbilical vein*—that transport blood between the embryo and the placenta (fig. 23.17). The umbilical cord also suspends the embryo in the *amniotic cavity.*

In addition to the amnion and chorion, two other extraembryonic membranes form during development. They are the yolk sac and the allantois.

The **yolk sac** forms during the second week, and it is attached to the underside of the embryonic disc (see fig. 23.15). This structure forms blood cells in the early stages of development and gives rise to the cells that later become sex cells. The yolk sac also produces stem cells of the bone marrow, which are precursors to many cell types, but predominantly to blood cells. Parts of the yolk sac form the embryonic digestive tube as well. Part of the membrane derived from the yolk sac becomes incorporated into the umbilical cord, and the remainder lies in the cavity between the chorion and the amnion near the placenta.

The **allantois** (ah-lan′to-is) forms during the third week as a tube extending from the early yolk sac into the connecting stalk of the embryo. It, too, forms blood cells and gives rise to the umbilical arteries and vein (see figs. 23.15 and 23.17).

Eventually, the amniotic cavity becomes so enlarged that the membrane of the amnion contacts the thicker chorion around it. The two membranes fuse into an *amniochorionic membrane* (fig. 23.18).

By the beginning of the eighth week, the embryo is usually 30 millimeters long and weighs less than 5 grams. Although its body is unfinished, it looks human (fig. 23.19).

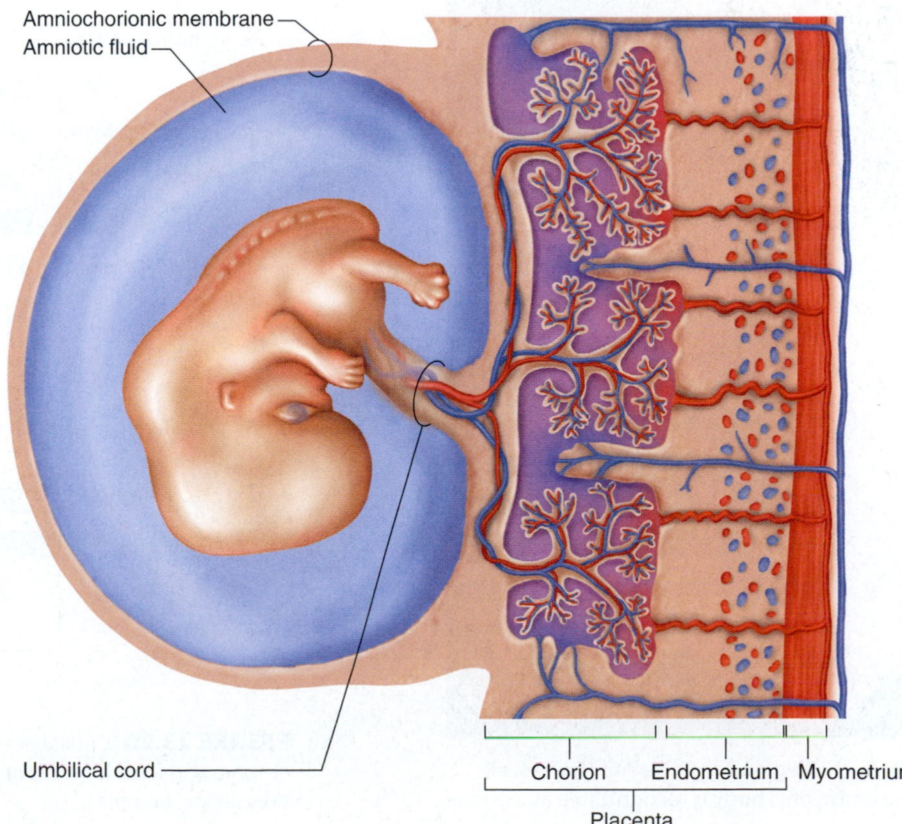

FIGURE 23.18 The amniochorionic membrane forms from fusion of the amnion and the chorion.

The *embryonic stage* concludes at the end of the eighth week. It is the most critical period of development, because during it, the embryo implants in the uterine wall, and all the essential external and internal body parts form. Disturbances to development during the embryonic stage can cause major malformations or malfunctions. This is why early prenatal care is important.

Factors that cause congenital malformations by affecting an embryo during its period of rapid growth and development are called **teratogens.** Such agents include drugs, viruses, radiation, and even large amounts of otherwise healthful substances, such as fat-soluble vitamins. Each prenatal structure has a time in development, called its *critical period,* when it is sensitive to teratogens (fig. 23.20).

A critical period may extend over many months or be just a day or two. Neural tube defects, for example, are traced to day twenty-eight in development, when a sheet of ectoderm folds into the neural tube. When this process is disrupted, an opening may remain in the spine (spina bifida). In contrast, the critical period for the developing brain begins when the anterior neural tube begins to swell into a brain, and continues throughout gestation. This is why so many teratogens affect the brain. Clinical Application 23.1 discusses some teratogens and their effects.

RECONNECT

To Chapter 11, Brain Development, page 398.

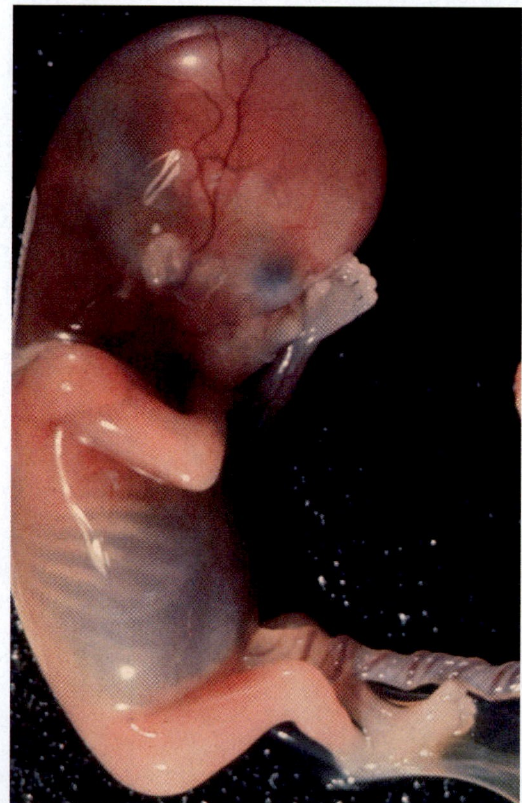

FIGURE 23.19 By the beginning of the eighth week of development, the embryonic body is recognizable as a human.

PRACTICE

16 Describe the development of the amnion.

17 What is the function of amniotic fluid?

18 Which blood vessels are in the umbilical cord?

19 What types of cells and other structures are derived from the yolk sac?

20 How do teratogens cause birth defects?

Fetal Stage

The **fetal stage** begins at the end of the eighth week of prenatal development and lasts until birth. During this period, growth is rapid, and body proportions change considerably. At the beginning of the fetal stage, the head is disproportionately large, and the lower limbs are relatively short. Gradually, proportions come to more closely resemble those of a child (fig. 23.21).

During the third month, body lengthening accelerates, but growth of the head slows. The upper limbs of the **fetus** (fe′tus) achieve the relative length they will maintain throughout development, and ossification centers appear in most bones. By the twelfth week, the external reproductive organs are distinguishable as male or female. Figure 23.22 illustrates how these external reproductive organs of the male and female differentiate from precursor structures.

(a) When physical structures develop

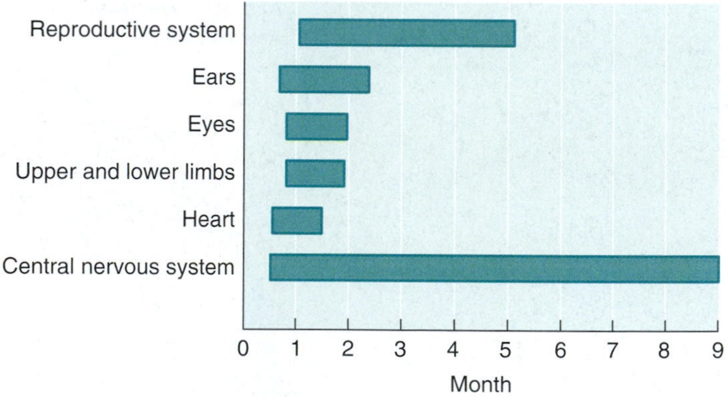

(b) When different teratogens disrupt development

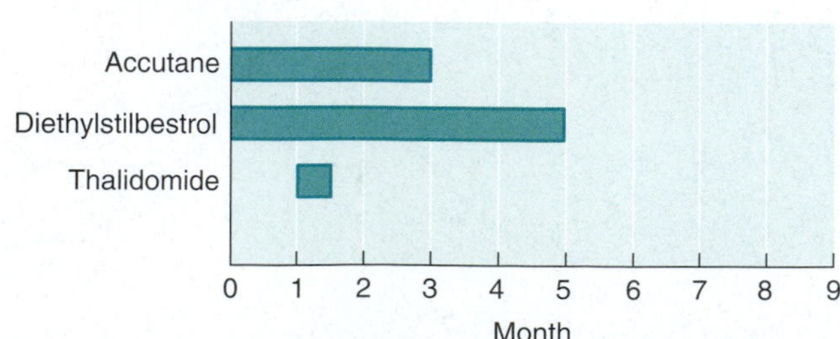

FIGURE 23.20 Critical periods. *(a)* Structures in the developing embryo and fetus *(b)* are sensitive to specific teratogens at different times in gestation.

In the fourth month, the body grows rapidly and reaches a length of up to 20 centimeters and weighs about 170 grams. The lower limbs considerably lengthen, and the skeleton continues to ossify. The fetus has hair, nipples, and nails, and may even scratch itself.

In the fifth month, growth slows. The lower limbs achieve their final relative proportions. Skeletal muscles contract, and the pregnant woman may feel fetal movements for the first time. Some hair grows on the fetal head, and fine, downy hair called lanugo covers the skin. A cheesy mixture of sebum from the sebaceous glands and dead epidermal cells (vernix caseosa) also coats the skin. The fetus, weighing about 450 grams and about 30 centimeters long, curls into the fetal position.

During the sixth month, the fetus gains substantial weight. Eyebrows and eyelashes appear. The skin is wrinkled and translucent. Blood vessels in the skin cause a reddish appearance.

In the seventh month, the skin becomes smoother as fat is deposited in the subcutaneous tissues. The eyelids, which fused during the third month, reopen. At the end of this month, the fetus is about 40 centimeters long.

In the final trimester, fetal brain cells rapidly form networks, as organs specialize and grow. A layer of fat is laid down beneath the skin. The testes of males descend from regions near the developing kidneys, through the inguinal canal, and into the scrotum (see chapter 22, pp. 833–835). The digestive and respiratory systems mature last, which is why premature infants may have difficulty digesting milk and breathing.

> Premature infants' survival chances increase directly with age and weight, and parallel increasing maturity of the lungs. The ability of the alveoli to exchange gases and the presence of surfactant to reduce alveolar surface tension are important. A baby born at 25 weeks has a 50% chance of survival; at 24 weeks, 39% survival; and at 23 weeks, 17% survival. One of the smallest and earliest premature babies is Amillia Taylor, born at 21 weeks, 6 days and weighing slightly under 10 ounces (about 280 grams).

Approximately 266 days after a single sperm burrowed its way into a secondary oocyte, a baby is ready to be born, *full-term*. It is about 50 centimeters long and weighs 2.7 to 3.6 kilograms. The skin has lost its downy hair but is still coated with sebum and dead epidermal cells. The scalp is usually covered with hair; the fingers and toes have well-developed

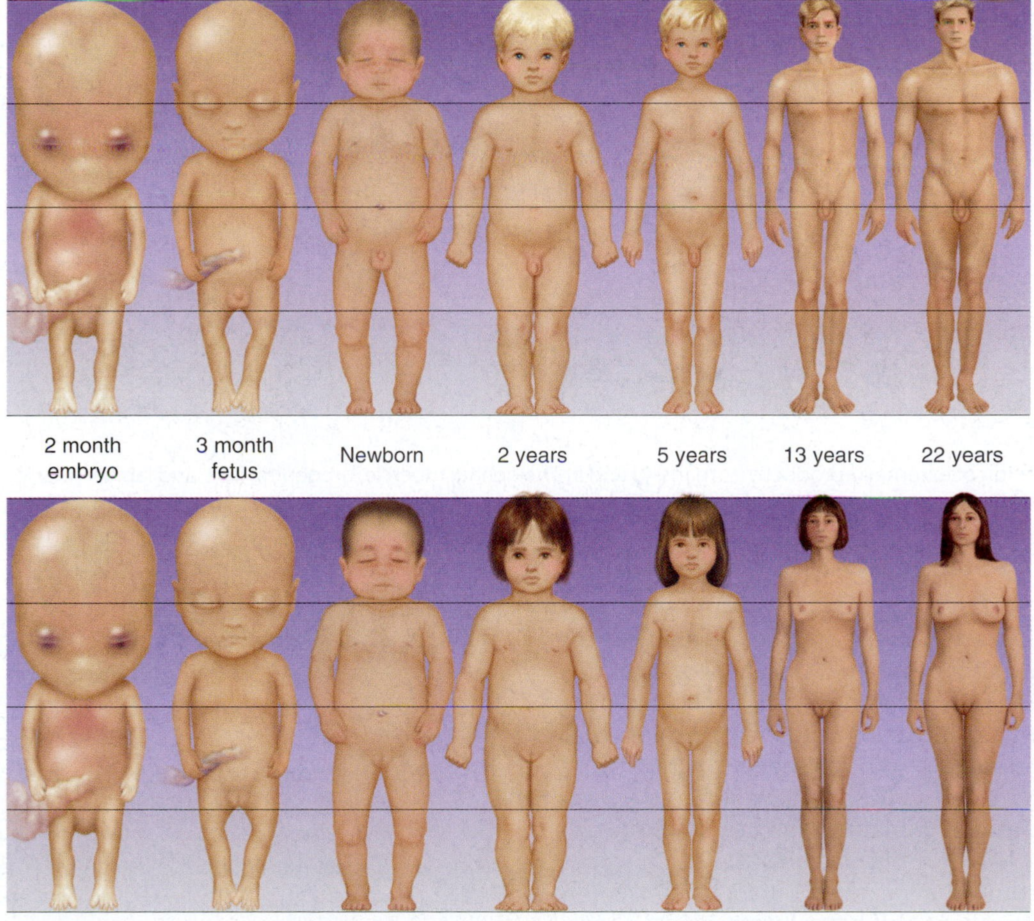

FIGURE 23.21 During development, body proportions change considerably.

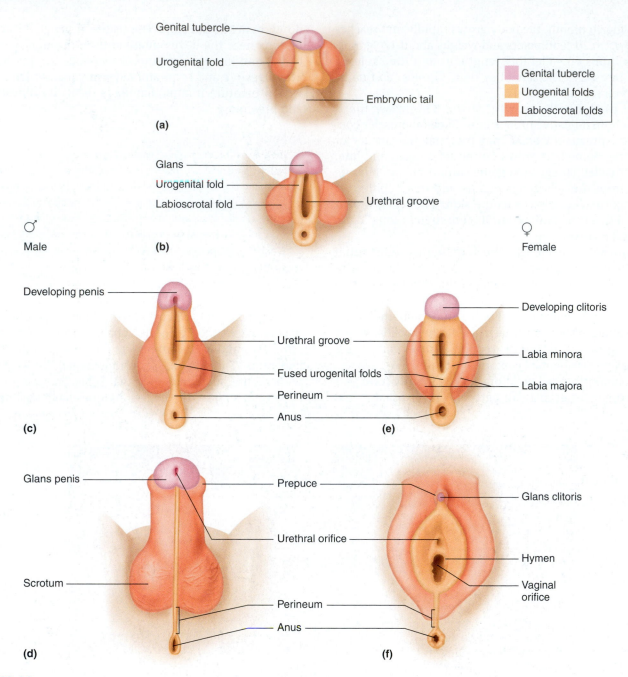

FIGURE 23.22 Formation of external reproductive organs. (*a* and *b*) The genital tubercle, urogenital fold, and labioscrotal folds that appear during the fourth week of development may differentiate into (*c* and *d*) male external reproductive organs or (*e* and *f*) female external reproductive organs.

nails; and the skull bones are largely ossified. As figure 23.23 shows, the fetus is usually positioned upside down with its head toward the cervix (*vertex position*).

The birth of a live, healthy baby is against the odds, considering human development from the beginning. Of every 100 secondary oocytes exposed to sperm, eighty-four are fertilized. Of these, sixty-nine implant in the uterus, forty-two survive one week or longer, thirty-seven survive six weeks or longer, and only thirty-one are born alive. Of those that do not survive to birth, about half have chromosomal abnormalities too severe to maintain life. Table 23.3 summarizes the stages of prenatal development.

PRACTICE

21 What major changes happen during the fetal stage of development?

22 When can the sex of a fetus be determined visually?

23 How is a fetus usually positioned in the uterus as birth nears?

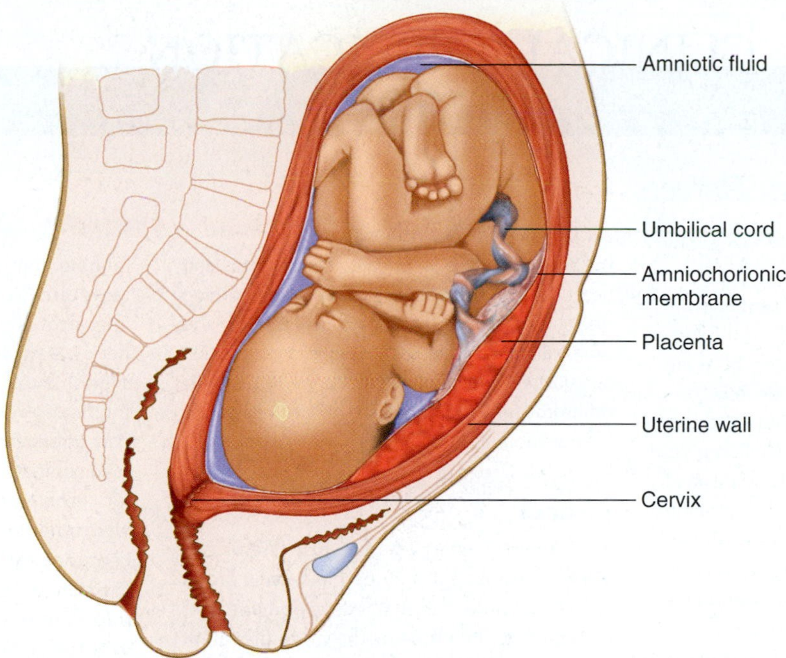

FIGURE 23.23 A full-term fetus is usually positioned with its head near the cervix.

TABLE 23.3 | Stages of Prenatal Development

Stage	Time Period	Major Events
Preembryonic stage	First week	Cells undergo mitosis, blastocyst forms; inner cell mass appears; blastocyst implants in uterine wall Size: 1/4 inch (0.63 centimeter), weight: 1/120 ounce (0.21 gram)
Embryonic stage	Second through eighth week	Inner cell mass becomes embryonic disc; primary germ layers form, embryo proper becomes cylindrical; main internal organs and external body structures appear; placenta and umbilical cord form, embryo proper is suspended in amniotic fluid Size: 1 inch (2.5 centimeters), weight: 1/30 ounce (0.8 gram)
Fetal stage	Ninth through twelfth week	Ossification centers appear in bones, sex organs differentiate, nerves and muscles coordinate so that the fetus can move its limbs Size: 4 inches (10 centimeters), weight: 1 ounce (28 grams)
	Thirteenth through sixteenth week	Body grows rapidly; ossification continues Size: 8 inches (20 centimeters), weight: 6 ounces (170 grams)
	Seventeenth through twentieth week	Muscle movements are stronger, and woman may be aware of slight flutterings; skin is covered with fine downy hair (lanugo) and coated with sebum mixed with dead epidermal cells (vernix caseosa) Size: 12 inches (30.5 centimeters), weight: 1 pound (454 grams)
	Twenty-first through thirty-eighth week	Body gains weight, subcutaneous fat deposited; eyebrows and lashes appear; eyelids reopen; testes descend Size: 21 inches (53 centimeters), weight: 6 to 10 pounds (2.7 to 4.5 kilograms)

A fetus can receive medical treatments. Tubes can be inserted to remove abnormal fluid accumulations, such as in the bladder or in the brain, and the problems corrected after birth. One young lady, Macie Hope McCartney, had a tumor the size of a grapefruit at the base of her spine removed 10 weeks before her birth. She was gently lifted from the uterus, remained attached during surgery, and was then tucked back into the uterus to finish development.

Fetal Blood and Circulation

Throughout fetal development, the maternal blood supplies oxygen and nutrients and carries away wastes. These substances diffuse between the maternal and fetal blood through the placental membrane, and the umbilical blood vessels carry them to and from the fetus (fig. 23.24). Consequently, the fetal blood and cardiovascular system are adapted to intrauterine existence. For example, the concentration of

23.1 CLINICAL APPLICATION

Some Causes of Birth Defects

Thalidomide

The idea that the placenta protects the embryo and fetus from harmful substances was tragically disproven between 1957 and 1961, when 10,000 children in Europe were born with flippers in place of limbs. Doctors soon identified a mild tranquilizer, thalidomide, which all of the mothers of deformed infants had taken early in pregnancy, during the time of limb formation. Although some women in the United States did use thalidomide and had affected children, the United States was spared a thalidomide disaster because an astute government physician noted adverse effects of the drug on monkeys in experiments, and she halted use of the drug. However, thalidomide is used today to treat leprosy and certain blood disorders.

Rubella

The virus that causes rubella (German measles) is a powerful teratogen. Australian physicians first noted its effects in 1941, and a rubella epidemic in the United States in the early 1960s caused 20,000 birth defects and 30,000 stillbirths. Exposure in the first trimester leads to cataracts, deafness, and heart defects, and later exposure causes learning disabilities, speech and hearing problems, and type 1 diabetes mellitus. Widespread vaccination has slashed the incidence of "congenital rubella syndrome," and today it occurs only where people are not vaccinated.

Alcohol

A pregnant woman who has just one or two alcoholic drinks a day, or perhaps many drinks at a crucial time in prenatal development, risks *fetal alcohol syndrome* or the more prevalent *fetal alcohol effects* in her unborn child. The effects of small amounts of alcohol at different stages of pregnancy are not yet well understood and because each woman metabolizes alcohol slightly differently, it is best to avoid drinking alcohol entirely when pregnant or when trying to become pregnant.

A child with fetal alcohol syndrome has a characteristic small head, misshapen eyes, and a flat face and nose (fig. 23C). Growth is slow before and after birth. Intellect is impaired, ranging from minor learning disabilities to mental retardation. Teens and young adults with fetal alcohol syndrome are short and have small heads. Many individuals remain at early grade-school level. They often lack social and communication skills, such as understanding the consequences of actions, forming friendships, taking initiative, and interpreting social cues.

Problems in children of alcoholic mothers were noted by Aristotle more than twenty-three centuries ago. Today, fetal alcohol syndrome is the third most common cause of mental retardation in newborns. One to 3 in every 1,000 infants has the syndrome, and more than 40,000 affected children are born each year.

Cigarettes

Chemicals in cigarette smoke stress a fetus. Carbon monoxide crosses the placenta and plugs up the sites on the fetus's hemoglobin molecules that would normally bind oxygen. Other chemicals in smoke prevent nutrients from reaching the fetus. Studies comparing placentas of smokers and nonsmokers show that smoke-exposed

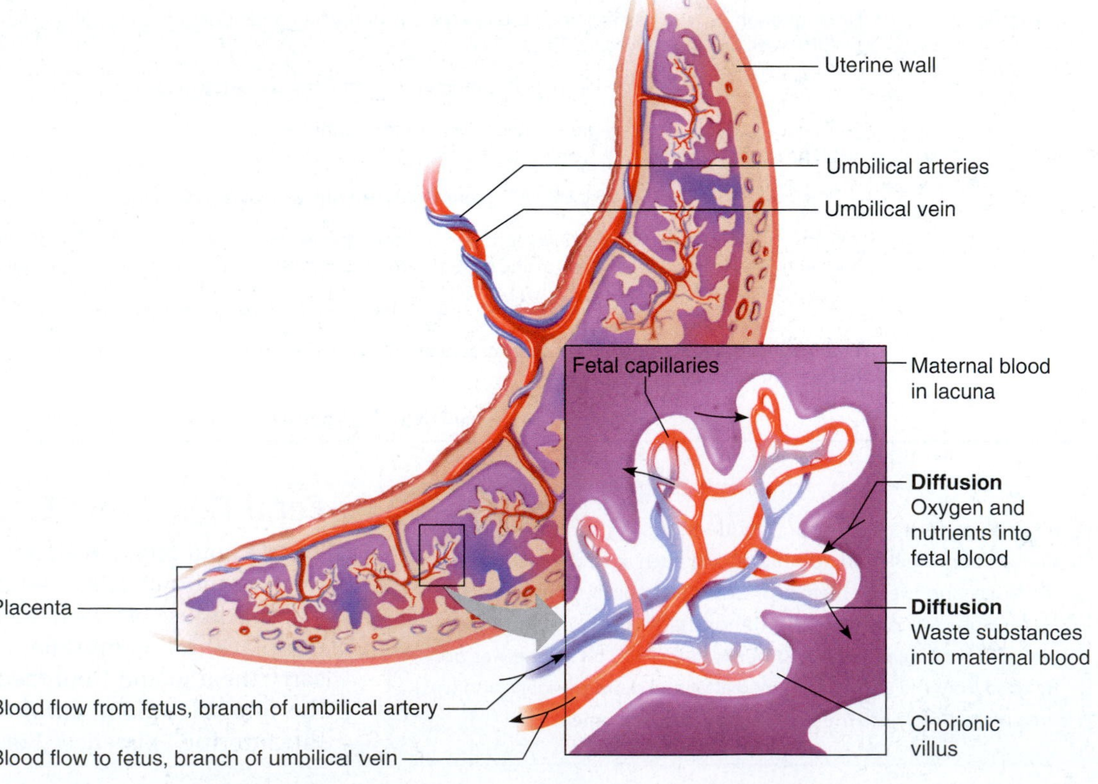

FIGURE 23.24 Oxygen and nutrients diffuse into the fetal blood from the maternal blood. Waste diffuses into the maternal blood from the fetal blood.

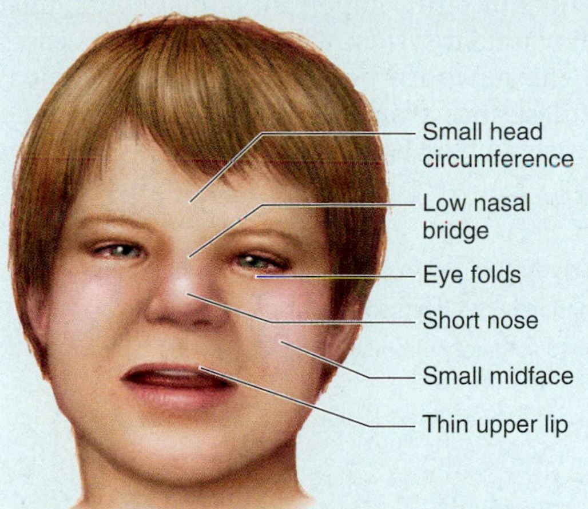

FIGURE 23C Fetal alcohol syndrome. Some children whose mothers drank alcohol during pregnancy have characteristic flat faces. Women who drink excessively while pregnant have a 30% to 45% chance of having a child affected to some degree by prenatal exposure to alcohol. Two mixed drinks per day seems to be the level above which damage is likely to occur.

placentas lack important growth factors. The result of these assaults is poor growth before and after birth. Cigarette smoking during pregnancy raises the risk of spontaneous abortion, stillbirth, prematurity, and low birth weight.

Nutrients and Malnutrition

Certain nutrients in large amounts, particularly vitamins, act in the body as drugs. The acne medication *isotretinoin* (Accutane) is a derivative of vitamin A that causes spontaneous abortions and defects of the heart, nervous system, and face in the fetus. A vitamin A-based drug used to treat psoriasis, as well as excesses of vitamin A itself, also cause birth defects. This is because some forms of vitamin A are stored in body fat for up to three years after ingestion.

Malnutrition during pregnancy causes intrauterine growth retardation (IUGR), which may have delayed health effects. Fetal physiology adapts to starvation to best use available nutrients. Insulin resistance changes to compensate for lack of muscle tissue. Cardiovascular changes shunt blood to vital organs. Starvation also raises stress hormone levels, stiffens arteries, and too few kidney tubules form. These changes set the stage for type 2 diabetes mellitus, hypertension, stroke, and coronary artery disease years later. Paradoxically, the infant is scrawny, but the older child tends to be obese, and difficulty losing weight may persist.

Occupational Hazards

Some teratogens are encountered in the workplace. Increased rates of spontaneous abortion and birth defects have been noted among women who work with textile dyes, lead, certain photographic chemicals, semiconductor materials, mercury, and cadmium. We do not know much about the role of the male in environmentally caused birth defects. Men whose jobs expose them to sustained heat, such as smelter workers, glass manufacturers, and bakers, may produce sperm that can fertilize an egg and possibly lead to spontaneous abortion or a birth defect. A virus or a toxic chemical carried in semen may also cause a birth defect. ■

oxygen-carrying hemoglobin in the fetal blood is about 50% greater than in the maternal blood. Also, fetal hemoglobin has a greater attraction for oxygen than does adult hemoglobin. Thus, at the oxygen partial pressure of the placental capillaries, fetal hemoglobin can carry 20% to 30% more oxygen than adult hemoglobin. Different genes encode the protein subunits of hemoglobin in embryos, fetuses, and individuals after birth.

In the fetal cardiovascular system, the *umbilical vein* transports blood rich in oxygen and nutrients from the placenta to the fetal body. This vein enters the body through the umbilical ring and travels along the anterior abdominal wall to the liver. About half the blood it carries passes into the liver, and the rest enters a vessel called the **ductus venosus** (duk′tus ven-o′sus), which bypasses the liver.

The ductus venosus extends a short distance and joins the inferior vena cava. There, oxygenated blood from the placenta mixes with deoxygenated blood from the lower parts of the fetal body. This mixture continues through the inferior vena cava to the right atrium.

RECONNECT

To Chapter 15, Path of Blood Through the Heart, pages 560–561.

In an adult heart, the blood from the right atrium enters the right ventricle and is pumped through the pulmonary trunk and pulmonary arteries to the lungs. In the fetus, however, the lungs are nonfunctional, and the blood largely bypasses them. As blood from the inferior vena cava enters the fetal right atrium, much of it is shunted directly into the left atrium through an opening in the atrial septum. This opening is called the **foramen ovale** (fo-ra′men o-val′e), and the blood passes through it because the blood pressure in the right atrium is somewhat greater than that in the left atrium. Furthermore, a small valvelike structure (septum primum) located on the left side of the atrial septum overlies the foramen ovale and helps prevent blood from moving in the reverse direction.

The rest of the fetal blood entering the right atrium, including a large proportion of the deoxygenated blood entering from the superior vena cava, passes into the right ventricle and out through the pulmonary trunk. Only a small volume of blood enters the pulmonary circuit because the lungs are collapsed and their blood vessels have a high resistance to blood flow. However, enough blood reaches the lung tissues to sustain them.

Most of the blood in the pulmonary trunk bypasses the lungs by entering a fetal vessel called the **ductus arteriosus** (duk′tus ar-te″re-o′sus), which connects the pulmonary trunk to the descending portion of the aortic arch. As a result of this connection, the blood with a relatively low oxygen concentration, returning to the heart through the superior vena cava, bypasses the lungs and does not enter the portion of the aorta that branches to the heart and brain.

The more highly oxygenated blood that enters the left atrium through the foramen ovale mixes with a small amount of deoxygenated blood returning from the pulmonary veins. This mixture moves into the left ventricle and is pumped into the aorta. Some of it reaches the myocardium through the coronary arteries, and some reaches the brain tissues through the carotid arteries.

Blood carried by the descending aorta includes the less oxygenated blood from the ductus arteriosus. Some of the blood is carried into the branches of the aorta that lead to the lower regions of the body. The rest passes into the *umbilical arteries,* which branch from the internal iliac arteries and lead to the placenta. There the blood is reoxygenated (figs. 23.25 and 23.26).

The umbilical cord usually contains two arteries and one vein. Rarely, newborns have only one umbilical artery. This condition is often associated with other cardiovascular, urogenital, or gastrointestinal disorders. The vessels within the severed cord are routinely counted following a birth because of the possibility of these conditions.

Table 23.4 summarizes the major features of fetal circulation. At the time of birth, important adjustments must occur in the cardiovascular system when the placenta ceases to function and the newborn begins to breathe. Clinical Application 23.2 describes a case in which fetal ultrasound revealed two hearts and bloodstreams, yet a single body.

PRACTICE  

24 Which umbilical vessel carries oxygenated blood to the fetus?

25 What is the function of the ductus venosus?

26 What characteristic of fetal pulmonary circulation shunts blood away from the lungs?

27 How does fetal circulation allow blood to bypass the lungs?

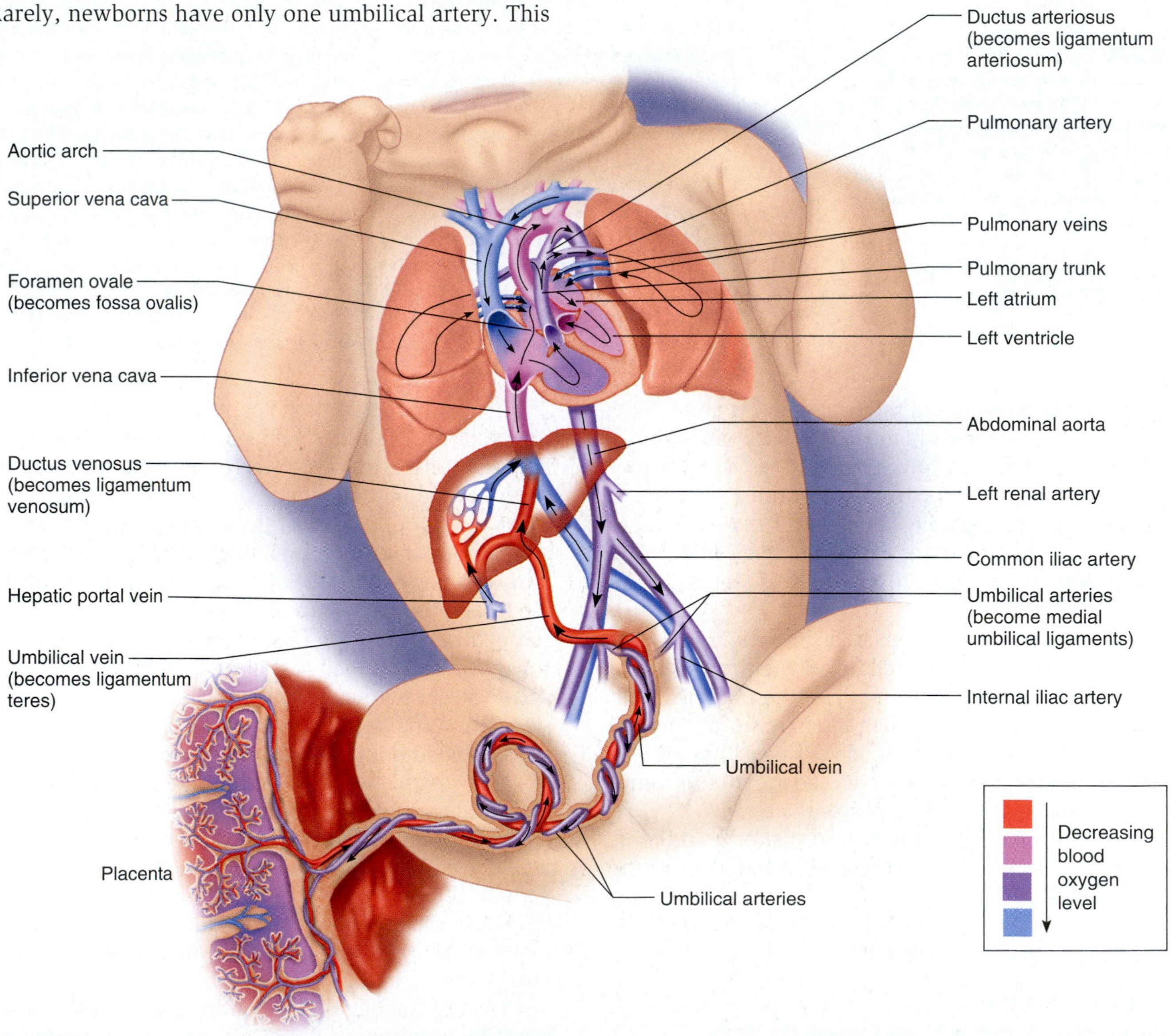

FIGURE 23.25 The general pattern of fetal circulation is shown anatomically.

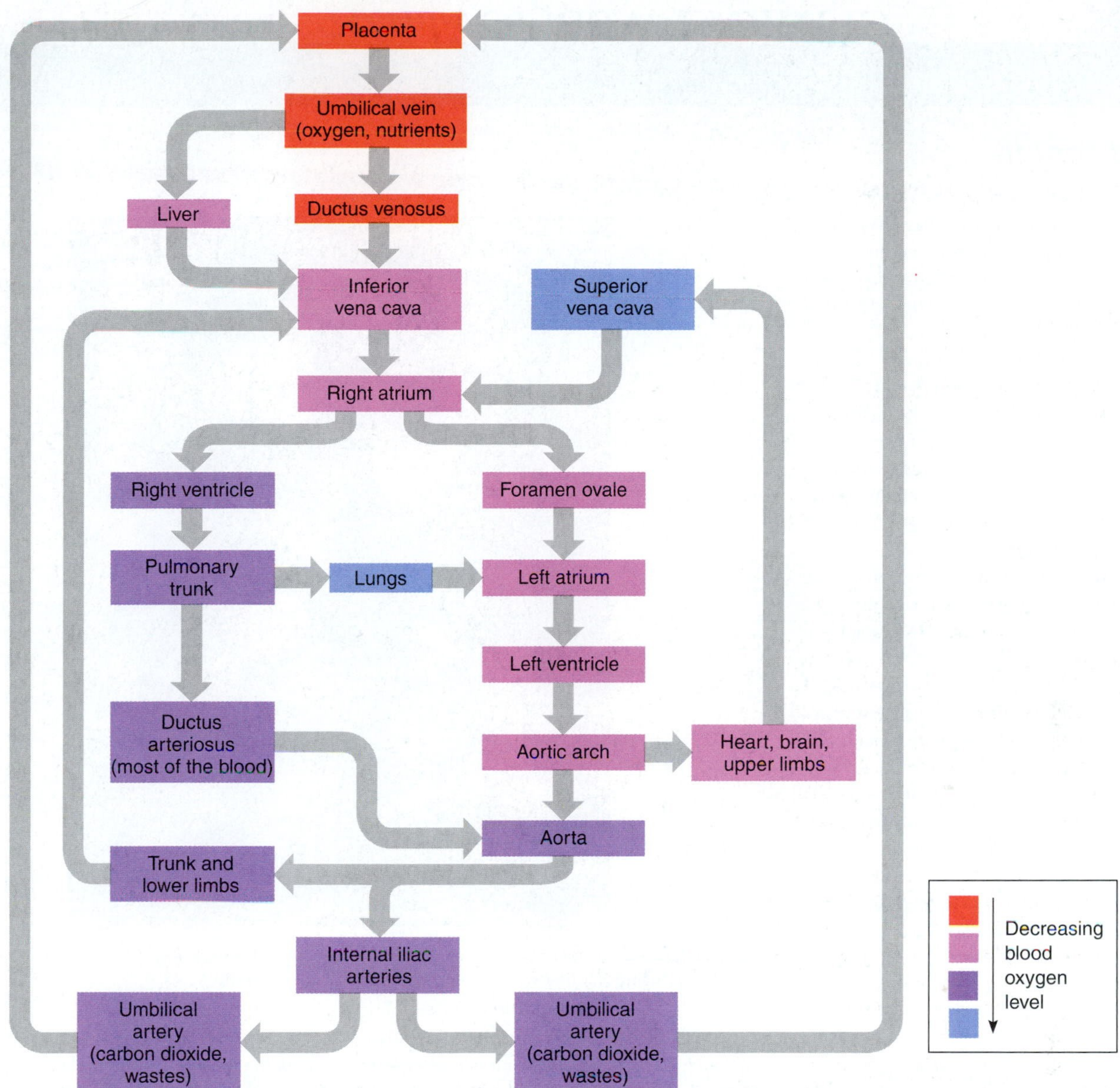

FIGURE 23.26 The general pattern of fetal circulation is shown schematically. Flow arrows are shown in gray so as not to be confused with blood oxygenation.

TABLE 23.4 | Fetal Cardiovascular Adaptations

Adaptation	Function
Fetal blood	Has greater oxygen-carrying capacity than adult blood
Umbilical vein	Carries nutrient-rich, oxygenated blood from the placenta to the fetus
Ductus venosus	Conducts about half the blood from the umbilical vein directly to the inferior vena cava, bypassing the liver
Foramen ovale	Conveys a large proportion of the blood entering the right atrium from the inferior vena cava, through the atrial septum, and into the left atrium, bypassing the lungs
Ductus arteriosus	Conducts some blood from the pulmonary trunk to the aorta, bypassing the lungs
Umbilical arteries	Carry the blood containing carbon dioxide and wastes from the internal iliac arteries to the placenta

Birth Process

Pregnancy terminates with the *birth process* (parturition). It is complex. Progesterone plays a major role in its start. During pregnancy, this hormone suppresses uterine contractions. As the placenta ages, progesterone concentration in the uterus declines, which stimulates synthesis of a prostaglandin that promotes uterine contractions. At the same time, the cervix begins to thin and then open. Changes in the cervix may begin a week or two before other signs of labor.

Stretching of the uterine and vaginal tissues late in pregnancy also stimulates the birth process. This initiates nerve impulses to the hypothalamus, which, in turn, signals the posterior pituitary gland to release the hormone **oxytocin** (see chapter 13, p. 499), which stimulates powerful uterine contractions. Combined with the greater excitability of the

23.2 CLINICAL APPLICATION

Joined For Life

Patty Hensel's pregnancy was uneventful. An ultrasound scan revealed an apparently normal fetus, although at one medical exam, Mike Hensel thought he heard two heartbeats.

A cesarean section was necessary because the baby was positioned bottom-first. To everyone's amazement, the baby had two heads and two necks, yet appeared to share the rest of the body, with two legs and two arms in the correct places, and a third arm between the heads. The ultrasound had probably imaged the twins from an angle that superimposed one head on the other. Patty, dopey from medication, recalls hearing the word "Siamese" and thinking she had given birth to cats. She had delivered conjoined, or Siamese, twins.

The baby was two individuals, named Abigail and Brittany. Each twin had her own neck, head, heart, stomach, and gallbladder. Remarkably, each also had her own nervous system. The twins shared a large liver, a single bloodstream, and all organs below the navel, including the reproductive tract. They had three lungs and three kidneys.

Abby and Brittany were strong and healthy. Doctors suggested surgery to separate the twins. Aware that only one child would likely survive surgery, Mike and Patty chose to let their daughters be. As teens, Abby and Brittany are glad their parents did not choose to separate them, because they would have been unable to walk or run, as they can today. They enjoy kickball, volleyball, basketball, and cycling. Like any teen girls, they have distinctive tastes in clothing and in food (fig. 23D).

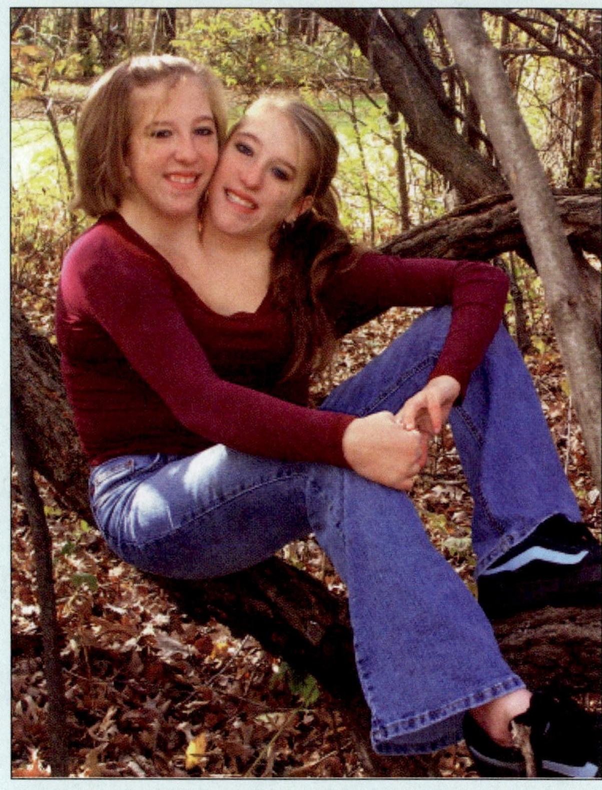

FIGURE 23D Abby and Brittany Hensel are conjoined twins, the result of incomplete twinning during the first two weeks of prenatal development.

Conjoined twins occur in 1 in 50,000 births, and about 40% are stillborn. They can be attached in any of several ways. The Hensel twins were joined in a manner seen only four times before. They are the result of incomplete twinning, which occurred during the first two weeks of gestation. The girls have shared tissue derived from ectoderm, mesoderm, and endoderm, so the partial twinning event must have occurred before the three germ layers were established, at day fourteen.

The term "Siamese twins" comes from Chang and Eng, born in Thailand, then called Siam, in 1811. They were joined by a ligament from the navel to the breastbone, which surgeons could easily correct today. Chang and Eng lived for sixty-three years, and each married. ■

myometrium due to the decline in progesterone secretion, stimulation by oxytocin aids *labor* in its later stages.

During labor, muscular contractions force the fetus through the birth canal. Rhythmic contractions that begin at the top of the uterus and travel down its length force the contents of the uterus toward the cervix.

The fetus is usually positioned head downward, so labor contractions force the head against the cervix. This action stretches the cervix, which elicits a reflex that stimulates still stronger labor contractions. Thus, a *positive feedback system* operates in which uterine contractions produce more intense uterine contractions until effort is maximal (fig. 23.27). At the same time, dilation of the cervix reflexly stimulates an increased release of oxytocin from the posterior pituitary gland.

As labor continues, positive feedback stimulates abdominal wall muscles to contract, helping to propel the fetus through the cervix and vagina to the outside. Table 23.5 summarizes some of the factors promoting labor. Figure 23.28 illustrates the steps of the birth process.

Following birth of the fetus, the placenta, which initially remains inside the uterus, separates from the uterine wall and is pushed by uterine contractions through the birth canal. This expelled placenta, called the *afterbirth*, is

TABLE 23.5 | Factors Contributing to the Labor Process

1. As the time of birth approaches, secretion of progesterone declines, and its inhibiting effect on uterine contractions lessens.
2. Decreasing progesterone concentration stimulates synthesis of prostaglandins, which initiate labor.
3. Stretching uterine tissues stimulates release of oxytocin from the posterior pituitary gland.
4. Oxytocin stimulates uterine contractions and aids labor in its later stages.
5. As the fetal head stretches the cervix, a positive feedback mechanism results in stronger and stronger uterine contractions and a greater release of oxytocin.
6. Positive feedback stimulates abdominal wall muscles to contract with greater and greater force.
7. The fetus is forced through the birth canal to the outside.

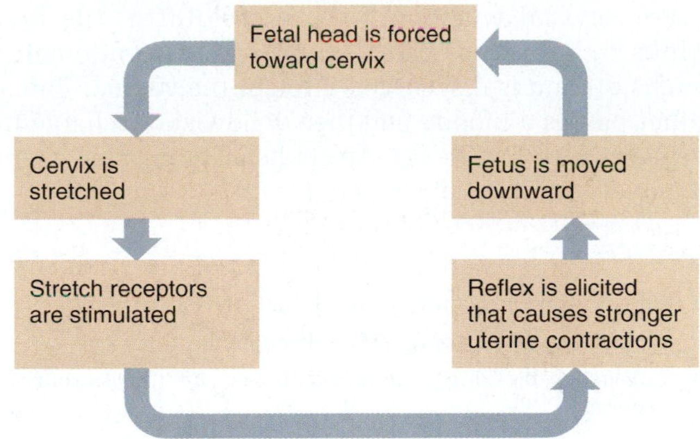

FIGURE 23.27 A positive feedback mechanism propels the birth process.

accompanied by bleeding, because vascular tissues are damaged in the process. However, continued contraction of the uterus compresses the bleeding vessels, minimizing blood loss. Oxytocin stimulates this contraction. Breastfeeding also contributes to returning the uterus to its original, prepregnancy size, because suckling by the newborn stimulates the mother's posterior pituitary to release oxytocin.

An infant passing through the birth canal can stretch and tear the perineum (the tissues between the vulva and anus). Before the birth is complete, a physician may make an incision along the midline of the perineum from the vestibule to within 1.5 centimeters of the anus. This procedure, called an *episiotomy,* ensures that the perineal tissues are cut cleanly rather than torn, which aids healing.

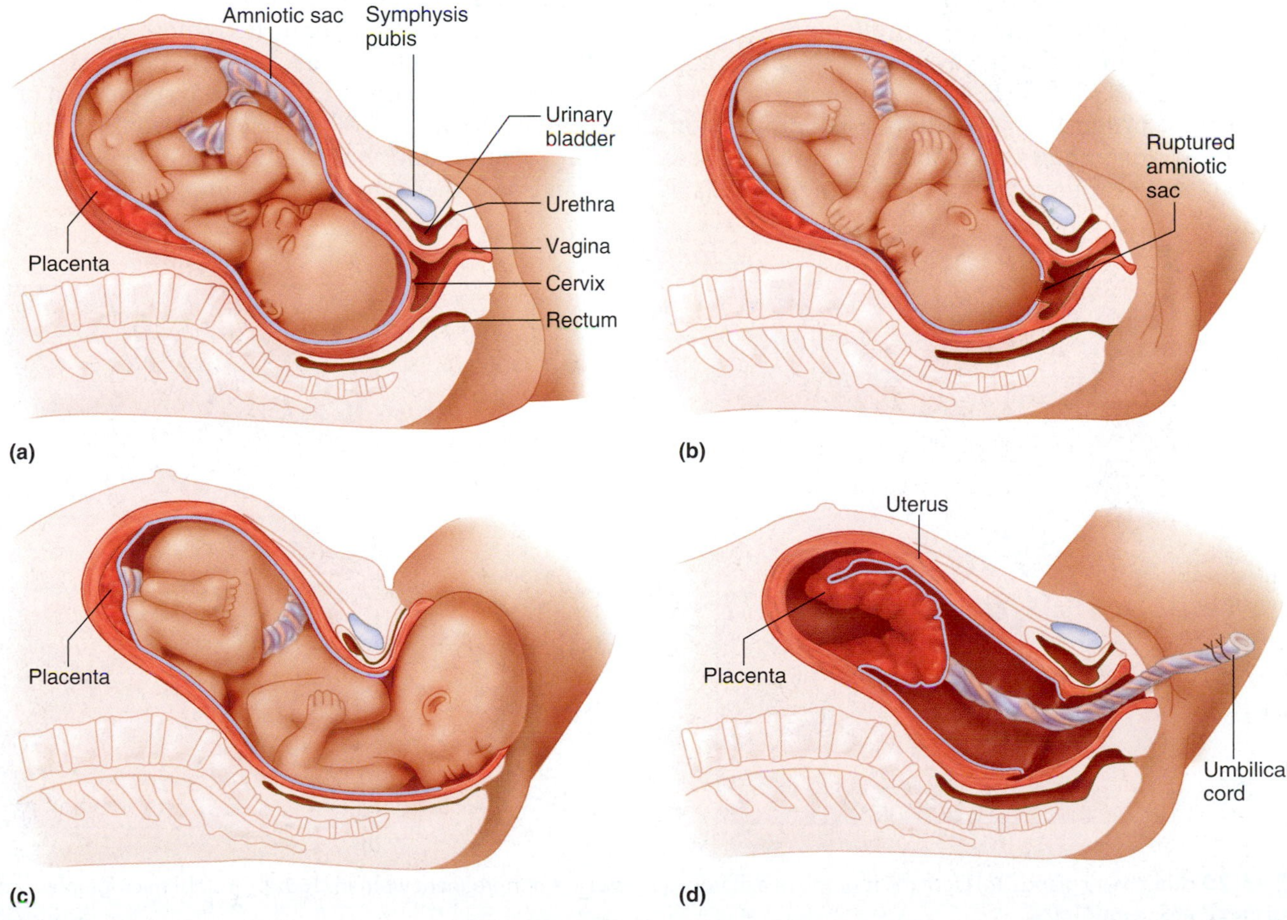

FIGURE 23.28 Stages in birth. (*a*) Fetal position before labor, (*b*) dilation of the cervix, (*c*) expulsion of the fetus, (*d*) expulsion of the placenta.

For several weeks following childbirth, the uterus shrinks by a process called *involution*. Its endometrium sloughs off and is discharged through the vagina. The new mother passes a bloody and then yellowish discharge from the vagina for a few weeks. An epithelial lining characteristic of a nonpregnant female then returns.

PRACTICE

28 Describe the role of progesterone in initiating labor.

29 Explain how dilation of the cervix affects labor.

30 Explain how bleeding is naturally controlled after the placenta is expelled.

Milk Production and Secretion

During pregnancy, placental estrogens and progesterone stimulate further development of the mammary glands. Estrogens cause the ductile systems to grow and branch, and deposit abundant fat around them. Progesterone stimulates the development of the alveolar glands at the ends of the ducts. Placental lactogen also promotes these changes.

The breasts may double in size during pregnancy because of hormonal activity. At the same time, glandular tissue replaces the adipose tissue of the breasts. Beginning about the fifth week of pregnancy, the anterior pituitary gland releases increasing amounts of *prolactin*. Prolactin is synthesized from early pregnancy throughout gestation, peaking at the time of birth. However, milk secretion does not begin until after birth. This is because during pregnancy, placental progesterone inhibits milk production, and placental lactogen blocks the action of prolactin (see chapter 13, p. 495). Consequently, even though the mammary glands can secrete milk, none is produced. The micrographs in figure 23.29 compare the mammary gland tissues of a nonpregnant woman with those of a lactating woman.

Following childbirth and the expulsion of the placenta, the maternal blood concentrations of placental hormones rapidly decline. The action of prolactin is no longer inhibited. Prolactin stimulates the mammary glands to secrete abundant milk. This hormonal effect does not occur until two or three days following birth. In the meantime, the glands secrete a thin, watery fluid called *colostrum*. It is rich in proteins, particularly antibodies from the mother's immune system that protect the newborn from certain infections, but has lower concentrations of carbohydrates and fats than milk.

Milk does not flow readily through the ductile system of the mammary gland but must be actively ejected as specialized *myoepithelial cells* surrounding the alveolar glands contract (fig. 23.30). A reflex action controls this process and is elicited when the breast is suckled or the nipple or areola is otherwise mechanically stimulated. Then, impulses from sensory receptors in the breasts travel to the hypothalamus, which signals the posterior pituitary gland to release oxytocin. The oxytocin reaches the breasts by means of the blood and stimulates the myoepithelial cells in both breasts to contract (fig. 23.31). Within about thirty seconds, milk squirts into a suckling infant's mouth.

Sensory impulses triggered by mechanical stimulation of the nipples also signal the hypothalamus to continue secreting prolactin. Thus, prolactin is released as long as breastfeeding continues. However, if stimulation of the nipple does not occur regularly, the hypothalamus inhibits secretion of prolactin, and within about one week, the mammary glands lose their capacity to produce milk.

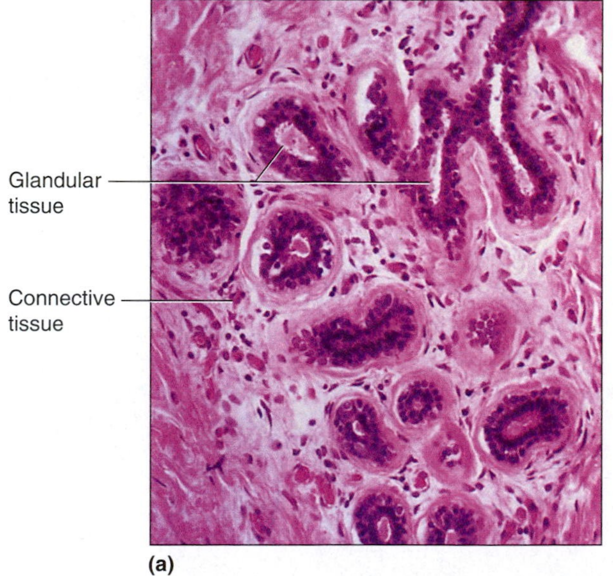

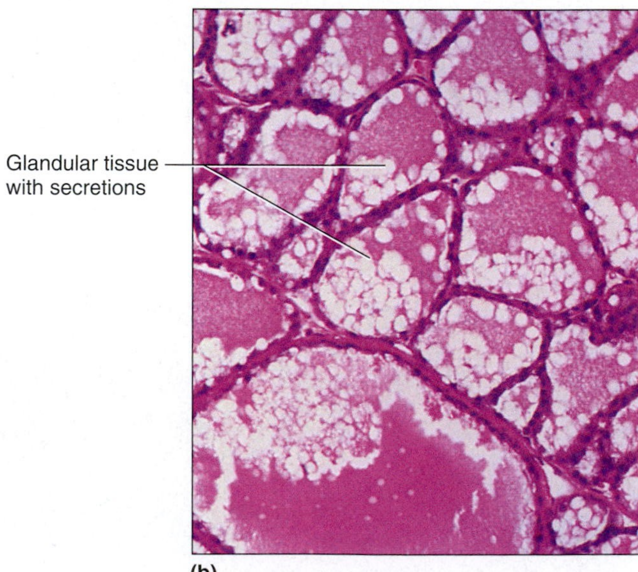

FIGURE 23.29 Mammary glands. *(a)* Light micrograph of a mammary gland in a nonpregnant woman (160×). *(b)* Light micrograph of an active (lactating) mammary gland (160×).

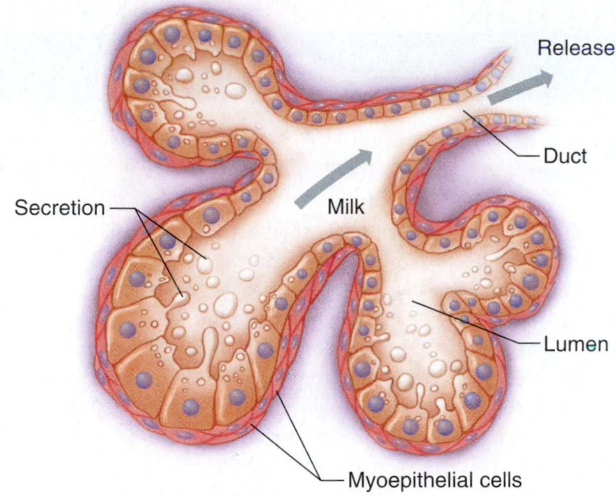

Release
Duct
Secretion
Milk
Lumen
Myoepithelial cells

FIGURE 23.30 Myoepithelial cells contract to release milk from an alveolar gland.

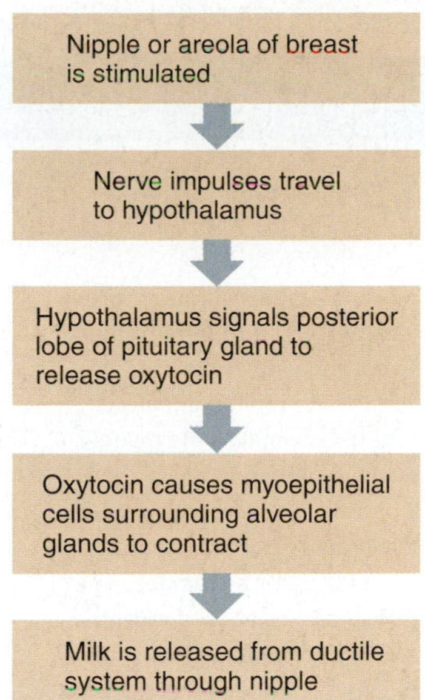

Nipple or areola of breast is stimulated
Nerve impulses travel to hypothalamus
Hypothalamus signals posterior lobe of pituitary gland to release oxytocin
Oxytocin causes myoepithelial cells surrounding alveolar glands to contract
Milk is released from ductile system through nipple

FIGURE 23.31 Mechanism that releases milk from the breasts.

A woman who is breastfeeding feels her milk "let down," or flood her breasts, when her infant suckles. If the baby nurses on a regular schedule, the mother may feel the letdown shortly before the baby is due to nurse. The connection between mind and hormonal control of lactation is so strong that if a nursing mother hears a baby cry, her milk may flow. If this occurs in public, she can keep from wetting her shirt by pressing her arms strongly against her chest.

To wean a nursing child, it is best to stop breastfeeding gradually, by eliminating one feeding per day each week, for example. If a woman stops nursing abruptly, her breasts will become painfully engorged for several days.

A woman who is breastfeeding usually does not ovulate for several months. This may be because prolactin suppresses release of gonadotropins from the anterior pituitary gland. When a woman discontinues breastfeeding, the anterior pituitary no longer secretes prolactin. Then, FSH is released, and the reproductive cycle is activated. If a new mother does not wish to have another child soon, she or her partner should practice contraception, because she will be fertile approximately two weeks prior to the return of her menstrual period.

Table 23.6 summarizes the hormonal control of milk production, and table 23.7 lists some agents that adversely affect lactation or harm the child. Clinical Application 23.3 explains the benefits of breastfeeding.

PRACTICE

31 How does pregnancy affect the mammary glands?

32 What stimulates the mammary glands to produce milk?

33 What causes milk ejection?

34 What happens to milk production if milk is not regularly removed from the breast?

TABLE 23.6 | Hormonal Control of the Mammary Glands

Before Pregnancy (Beginning of Puberty)	Following Childbirth
Ovarian hormones secreted during reproductive cycles stimulate alveolar glands and ducts of mammary glands to develop.	1. Placental hormonal concentrations decline, so the action of prolactin is no longer inhibited.
During Pregnancy	2. The breasts begin producing milk.
1. Estrogens cause the ductile system to grow and branch.	3. Mechanical stimulation of the breasts releases oxytocin from the posterior pituitary gland.
2. Progesterone stimulates development of alveolar glands.	4. Oxytocin stimulates release of milk from ducts.
3. Placental lactogen promotes development of the breasts.	5. As long as breastfeeding continues, more prolactin is released; if the nipple is not stimulated regularly, milk production ceases.
4. Prolactin is secreted throughout pregnancy, but placental progesterone inhibits milk production and placental lactogen blocks the action of prolactin.	

23.3 CLINICAL APPLICATION

Human Milk—The Perfect Food for Human Babies

The female human body manufactures milk that is a perfect food for a human newborn in several ways. Human milk is rich in the lipids required for rapid brain growth, and it is low in protein. Cow milk is the reverse, with three times as much protein as human milk. Much of cow milk protein is casein, which spurs a calf's rapid muscle growth, but forms hard-to-digest curds in a human baby's stomach. The protein in human milk has a balance of essential amino acids more suited to human growth and development than does the protein in cow's milk.

Human milk protects a newborn from many infections. For the first few days after giving birth, a new mother's breasts produce colostrum, which has less sugar and fat than mature milk but more protein, and is rich in antibodies. The antibodies protect the baby from such infections as *Salmonella* poisoning and polio. When the milk matures by a week to ten days, it has antibodies, enzymes, and white blood cells from the mother that continue infection protection. A milk protein called lactoferrin binds iron, making it unavailable to microorganisms that might use it to thrive in the newborn's digestive tract. Another biochemical in human milk, bifidus factor, encourages the growth of the bacteria *Lactobacillus bifidus,* which manufacture acids in the baby's digestive system that kill harmful bacteria.

A breastfed baby typically nurses until he or she is full, not until a certain number of ounces have been drunk, which may explain why breastfed babies are less likely to be obese than bottlefed infants. Babies nurtured on human milk are also less likely to develop allergies to cow's milk. A nursing mother must eat about 500 calories per day more than usual to meet the energy requirements of milk production—but she also loses weight faster than a mother who bottlefeeds, because the fat reserves set aside during pregnancy are used to manufacture milk.

Breastfeeding is not the best choice for all women. It may be impossible to be present for each feeding or to provide milk. Also, many drugs a mother takes may enter breast milk and can affect the baby. Another disadvantage of breastfeeding is that the father cannot do it.

An alternative to breastfeeding is infant formula, which is usually cow milk plus fats, proteins, carbohydrates, vitamins, and minerals added to make it as much like human milk as possible. Although infant formula is nutritionally sound, the foul-smelling and bulkier bowel movements of the bottlefed baby compared to the odorless, loose, more frequent and less abundant feces of a breastfed baby indicate that breast milk is a more digestible first food than infant formula. ■

TABLE 23.7 | Agents Contraindicated During Breastfeeding

Agent	Use	Effect on Lactation or Baby
Doxorubicin, methotrexate	Cancer chemotherapy, psoriasis, rheumatoid arthritis	Immune suppression
Cyclosporine	Immune suppression in transplant patients	Immune suppression
Radioactive isotopes	Cancer diagnosis and therapy	Radioactivity in milk
Phenobarbitol	Anticonvulsant	Sedation, spasms on weaning
Oral contraceptives	Birth control	Decreased milk production
Caffeine (large amounts)	Food additive	Irritability, poor sleeping
Cocaine	Drug of abuse	Intoxication, seizures, vomiting, diarrhea
Ethanol (alcohol) (large amounts)	Drug of abuse	Weak, drowsy; infant decreases in length but gains weight; decreased milk ejection reflex
Heroin	Drug of abuse	Tremors, restlessness, vomiting, poor feeding
Nicotine	Drug of abuse	Diarrhea, shock, increased heart rate; lowered milk production
Phencyclidine	Drug of abuse	Hallucinations

23.4 POSTNATAL PERIOD

Following birth, both mother and newborn experience physiological and anatomical changes. The postnatal period of development lasts from birth until death. It can be divided into the neonatal period, infancy, childhood, adolescence, adulthood, and senescence. Dying is also part of the life cycle.

Neonatal Period

The **neonatal** (ne″o-na′tal) **period,** which extends from birth to the end of the first four weeks, begins very abruptly at birth. At that moment, physiological adjustments must occur quickly because the newborn (neonate) must suddenly do for itself what the mother's body had been doing for it. The newborn must respire, obtain and digest nutrients, excrete wastes, and regulate body temperature. However, a newborn's most immediate need is to obtain oxygen and excrete carbon dioxide, so the first breath is critical.

The first breath must be particularly forceful because the newborn's lungs are collapsed and the airways are small, offering considerable resistance to air movement. Also, surface tension tends to hold the moist membranes of the lungs together. However, the lungs of a full-term fetus continuously secrete *surfactant* (see chapter 19, p. 750), which reduces surface tension. After the first powerful breath begins to expand the lungs, breathing eases.

A newborn's first breath is stimulated by increasing concentration of carbon dioxide, decreasing pH, low oxygen concentration, drop in body temperature, and mechanical stimulation during and after birth. Also, in response to the stress the fetus experiences during birth, blood concentrations of epinephrine and norepinephrine rise significantly (see chapter 13, p. 514). These hormones promote normal breathing by increasing the secretion of surfactant and dilating the airways.

For energy, the fetus primarily uses glucose and fatty acids in the pregnant woman's blood. The newborn, on the other hand, is suddenly without an external source of nutrients. The mother will not produce mature milk for two to three days, by which time the infant's gastrointestinal tract will be able to digest it. The early milk, *colostrum,* is an adaptation to the state of the newborn's digestive physiology. The newborn has a high metabolic rate, and its liver, not fully mature, may be unable to supply enough glucose to support metabolism. Instead, the newborn uses stored fat for energy.

A newborn's kidneys are usually unable to produce concentrated urine, so they excrete a dilute fluid. For this reason, the newborn may become dehydrated and develop a water and electrolyte imbalance. Also, certain homeostatic control mechanisms may not function adequately. For example, during the first few days of life, body temperature may respond to slight stimuli by fluctuating above or below the normal level.

When the placenta ceases to function and breathing begins, the newborn's cardiovascular system changes. Following birth, the umbilical vessels constrict. The umbilical arteries close first, and if the umbilical cord is not clamped or severed for a minute or so, blood continues to flow from the placenta to the newborn through the umbilical vein, adding to the newborn's blood volume.

The proximal portions of the umbilical arteries persist in the adult as the *superior vesical arteries* that supply blood to the urinary bladder. The more distal portions become solid cords (lateral umbilical ligaments). The umbilical vein becomes the cordlike *ligamentum teres* that extends from the umbilicus to the liver in an adult. The ductus venosus constricts shortly after birth and appears in the adult as a fibrous cord (ligamentum venosum) superficially embedded in the wall of the liver.

The foramen ovale closes as a result of blood pressure changes in the right and left atria. As blood ceases to flow from the umbilical vein into the inferior vena cava, the blood pressure in the right atrium falls. Also, as the lungs expand with the first breathing movements, resistance to blood flow through the pulmonary circuit decreases, more blood enters the left atrium through the pulmonary veins, and blood pressure in the left atrium increases.

As the blood pressure in the left atrium rises and that in the right atrium falls, the valvelike septum primum on the left side of the atrial septum closes the foramen ovale. In most individuals, this valve gradually fuses with the tissues along the margin of the foramen. In an adult, a depression called the *fossa ovalis* marks the site of the past opening.

The ductus arteriosus, like other fetal vessels, constricts after birth. After this, blood can no longer bypass the lungs by moving from the pulmonary trunk directly into the aorta. In an adult, a cord called the *ligamentum arteriosum* represents the ductus arteriosus.

In *patent ductus arteriosus* (PDA), the ductus arteriosus fails to close completely. After birth, the metabolic rate and oxygen consumption in neonatal tissues increase, in large part to maintain body temperature. If the ductus arteriosus remains open, the neonate's blood oxygen concentration may be too low to adequately supply body tissues, including the myocardium. If PDA is not corrected surgically, the heart may fail, even though the myocardium is normal.

Changes in the newborn's cardiovascular system are gradual. Although constriction of the ductus arteriosus may be functionally complete within fifteen minutes, the permanent closure of the foramen ovale may take up to a year. These cardiovascular changes are illustrated in figure 23.32 and summarized in table 23.8.

Fetal hemoglobin production falls after birth. By the time an infant is four months old, most of the circulating hemoglobin is the adult type.

PRACTICE

35 Define *neonatal period.*

36 What factors stimulate the first breath?

37 What does a newborn use for energy during its first few days?

38 How do the kidneys of a newborn differ from those of an adult?

39 What is the fate of the foramen ovale? Of the ductus arteriosus?

Infancy

The period of continual development extending from the end of the first four weeks to one year is called **infancy.** During this time, the infant grows rapidly and may triple its birth weight. Its teeth begin to erupt through the gums, and its muscular and nervous systems mature so that coordinated muscular activities become possible. The infant is soon able to visually follow objects; reach for and grasp objects; and sit, creep, and stand.

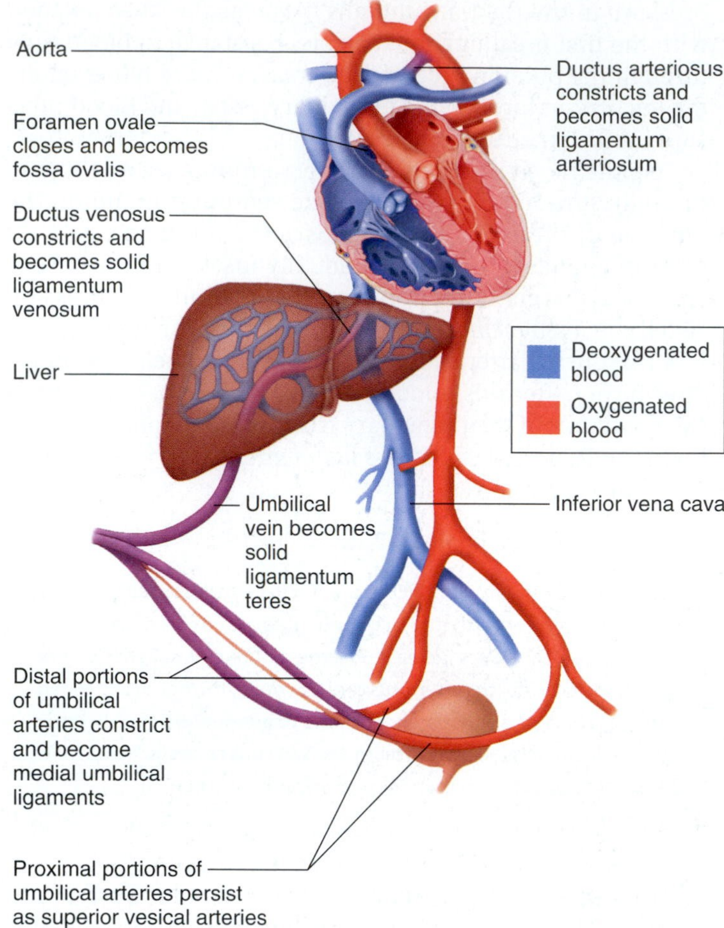
Aorta
Ductus arteriosus constricts and becomes solid ligamentum arteriosum
Foramen ovale closes and becomes fossa ovalis
Ductus venosus constricts and becomes solid ligamentum venosum
Liver
Deoxygenated blood
Oxygenated blood
Umbilical vein becomes solid ligamentum teres
Inferior vena cava
Distal portions of umbilical arteries constrict and become medial umbilical ligaments
Proximal portions of umbilical arteries persist as superior vesical arteries

FIGURE 23.32 Major changes occur in the newborn's cardiovascular system.

Infancy also brings the beginning of the ability to communicate. The infant learns to smile, laugh, and respond to some sounds. By the end of the first year, the infant may be able to say two or three words. Often one of a child's first words is the name of a beloved pet.

Infancy (as well as childhood) is a period of rapid growth, so the infant has particular nutritional requirements. In addition to an energy source, the body requires proteins to provide the amino acids necessary to form new tissues; calcium and vitamin D to promote the development and ossification of skeletal structures (see chapter 7, p. 200); iron to support blood cell formation; and vitamin C for production of structural tissues such as cartilage and bone.

Childhood

Childhood begins at the end of the first year and ends at puberty. During this period, growth continues at a rapid rate. The primary teeth appear, and then secondary teeth replace them (see chapter 17, pp. 658–659). The child develops voluntary muscular control and learns to walk, run, and climb. Bladder and bowel controls are established. The child learns to communicate effectively by speaking, and later, usually learns to read, write, and reason objectively. At the same time, the child is maturing emotionally.

PRACTICE

40 Define *infancy.*

41 What developmental changes characterize infancy?

42 Define *childhood.*

43 What developmental changes characterize childhood?

Adolescence

Adolescence is the period of development between puberty and adulthood. It is a time of anatomical and physiological changes that result in reproductively functional individuals (see chapter 22, pp. 846 and 857). Most of these changes are hormonally controlled, and they include the appearance of secondary sex characteristics as well as growth spurts in the muscular and skeletal systems.

Females usually experience these changes somewhat earlier than males, so early in adolescence, females may be taller and stronger than their male peers. However, females attain full growth at earlier ages, and in late adolescence, the average male is taller and stronger than the average female.

The periods of rapid growth in adolescence, which usually begin between the ages of eleven and thirteen in females and between thirteen and fifteen in males, increase demands for certain nutrients. It is not uncommon for a teenager to consume a huge plate of food, go back for more—and still

TABLE 23.8 | Cardiovascular Adjustments in the Newborn

Structure	Adjustment	In the Adult
Umbilical vein	Constricts	Becomes ligamentum teres that extends from the umbilicus to the liver
Ductus venosus	Constricts	Becomes ligamentum venosum superficially embedded in the wall of the liver
Foramen ovale	Closes by valvelike septum primum as blood pressure in right atrium decreases and blood pressure in left atrium increases	Valve fuses along margin of foramen ovale and is marked by a depression called the fossa ovalis
Ductus arteriosus	Constricts	Becomes ligamentum arteriosum that extends from the pulmonary trunk to the aorta
Umbilical arteries	Distal portions constrict	Distal portions become lateral umbilical ligaments; proximal portions function as superior vesical arteries

remain thin. In addition to energy sources, foods must provide ample amounts of proteins, vitamins, and minerals to support growth of new tissues. Adolescence also brings increasing levels of motor skills, intellectual ability, and emotional maturity.

Adulthood

Adulthood (maturity) extends from adolescence to old age. As we age, we become gradually aware of certain declining functions—yet other abilities remain adequate throughout life. The "Life-Span Changes" sections in previous chapters have chronicled the effects of aging on particular organ systems. It is interesting to note the peaks of particular structures or functions throughout an average human life.

By age eighteen, the human male is producing the most testosterone that he will ever have, and as a result his sex drive is strong. In the twenties, muscle strength peaks in both sexes. Hair is at its fullest, each hair its thickest. By the end of the third decade of life, obvious signs of aging may first appear as a loss in the elasticity of facial skin, producing small wrinkles around the mouth and eyes. Height is already starting to decrease, but not yet at a detectable level.

The age of thirty seems to be a developmental turning point. After this, hearing often becomes less acute. Heart muscle begins to thicken. The elasticity of the ligaments between the small bones in the back lessens, setting the stage for the slumping posture that becomes apparent in later years. Some researchers estimate that beginning roughly at age thirty, the human body becomes functionally less efficient by about 0.8% every year.

During their forties, many people weigh 10 to 20 pounds (4.5 to 9 kilograms) more than they did at the age of twenty, thanks to a slowing of metabolism and decrease in activity level. They may be 1/8 inch (0.3 centimeter) shorter, too. Hair may be graying as melanin-producing biochemical pathways lose efficiency, and some hair may fall out. Vision may become farsighted. The immune system is less efficient, making the body more prone to infection and cancer. Skeletal muscles lose strength as connective tissue appears within them; the cardiovascular system is strained as the lumens of arterioles and arteries narrow with fatty deposits; skin loosens and wrinkles as elastic fibers in the dermis break down.

The early fifties bring further declines. Nail growth slows, taste buds die, and the skin continues to lose elasticity. For most people, the ability to see close objects becomes impaired, but for the nearsighted, vision improves. Women stop menstruating, although interest in sex continues (see chapter 22, pp. 860–861). Delayed or reduced insulin release by the pancreas, in response to a glucose load, may lead to diabetes. By the decade's end, muscle mass and weight begin to decrease. A male produces less semen but is still sexually active. His voice may become higher as his vocal cords degenerate. A man has half the strength in his upper limb muscles and half the lung function as he did at age twenty-five. He is about 3/4 inch (2 centimeters) shorter.

The sixty-year-old may experience minor memory losses. A few million of the person's billions of brain cells have been lost over his or her lifetime, but for the most part, intellect remains sharp. By age seventy, height decreases a full inch (2.5 centimeters). Sagging skin and loss of connective tissue, combined with continued growth of cartilage, make the nose, ears, and eyes more prominent.

Senescence

Senescence (se-nes′ens) is the process of growing old. It is a continuation of the degenerative changes that begin during adulthood. The body becomes less able to cope with the demands placed on it by the individual and by the environment.

Senescence is a result of the normal wear-and-tear of body parts over many years. For example, the cartilage covering the ends of bones at joints may wear away, leaving the joints stiff and painful. Other degenerative changes are caused by disease processes that interfere with vital functions, such as gas exchanges or blood circulation. Metabolic rate and distribution of body fluids may change. The rate of division of certain cell types declines, and immune responses weaken. The person becomes less able to repair damaged tissue and more susceptible to disease.

Decreasing efficiency of the central nervous system accompanies senescence. The person may lose some intellectual functions. Also, the physiological coordinating capacity of the nervous system may decrease, and homeostatic mechanisms may fail to operate effectively. Sensory functions decline with age also.

Death usually results, not from these degenerative changes, but from mechanical disturbances in the cardiovascular system, failure of the immune system, or disease processes that affect vital organs. Table 23.9 summarizes the major phases of postnatal life and their characteristics, and table 23.10 lists some aging-related changes.

> From 65% to 80% of all deaths in the United States take place in hospitals, often with painful and sometimes unwanted interventions to prolong life. One study found that about half of conscious patients suffer severe pain prior to death. In Oregon, which has pioneered education on caring for the dying patient and allows assisted suicide, a greater percentage of patients live out their last days at home; in nursing homes; or in hospices, facilities dedicated to providing comfort and support for the dying. The medical community is trying to remedy shortcomings in the treatment of the dying. Medical training is increasing emphasis on providing palliative care for the terminally ill. Such care seeks to make a patient comfortable, even if the treatment does not cure the disease or extend life.

The End of Life

Nearing the end of life is a personal process, influenced by belief as well as circumstance. However, if the person has

TABLE 23.9 | Stages in Postnatal Development

Stage	Time Period	Major Events
Neonatal period	Birth to end of fourth week	Newborn begins to carry on respiration, obtain nutrients, digest nutrients, excrete wastes, regulate body temperature, and make cardiovascular adjustments
Infancy	End of fourth week to one year	Growth rate is high; teeth begin to erupt; muscular and nervous systems mature so that coordinated activities are possible; communication begins
Childhood	One year to puberty	Growth rate is high; primary teeth erupt and are replaced by secondary teeth; high degree of muscular control is achieved; bladder and bowel controls are established; intellectual abilities mature
Adolescence	Puberty to adulthood	Person becomes reproductively functional and emotionally more mature; growth spurts occur in skeletal and muscular systems; high levels of motor skills are developed; intellectual abilities increase
Adulthood	Adolescence to old age	Person remains relatively unchanged anatomically and physiologically; degenerative changes begin
Senescence	Old age to death	Degenerative changes continue; body becomes less able to cope with demands; death usually results from mechanical disturbances in the cardiovascular system or from diseases that affect vital organs. Signs of approaching death may appear in a common sequence

TABLE 23.10 | Aging-Related Changes

Organ System	Aging-Related Changes
Integumentary system	Degenerative loss of collagenous and elastic fibers in dermis; decreased production of pigment in hair follicles; reduced activity of sweat and sebaceous glands; skin thins, wrinkles, and dries out; hair turns gray and then white
Skeletal system	Degenerative loss of bone matrix; bones become thinner, less dense, and more likely to fracture; stature may shorten due to compression of intervertebral discs and vertebrae
Muscular system	Loss of skeletal muscle fibers; degenerative changes in neuromuscular junctions; loss of muscular strength
Nervous system	Degenerative changes in neurons; loss of dendrites and synaptic connections; accumulation of lipofuscin in neurons; decreases in sensation; decreasing efficiency in processing and recalling information; decreasing ability to communicate; diminished senses of smell and taste; loss of elasticity of lenses and consequent loss of ability to accommodate for close vision
Endocrine system	Reduced hormonal secretions; decreased metabolic rate; reduced ability to cope with stress; reduced ability to maintain homeostasis
Cardiovascular system	Degenerative changes in cardiac muscle; decrease in lumen diameters of arteries and arterioles; decreased cardiac output; increased resistance to blood flow; increased blood pressure
Lymphatic system	Decrease in efficiency of immune system; increased incidence of infections and neoplastic diseases; increased incidence of autoimmune diseases
Digestive system	Decreased motility in gastrointestinal tract; reduced secretion of digestive juices; reduced efficiency of digestion
Respiratory system	Degenerative loss of elastic fibers in lungs; fewer alveoli; reduced vital capacity; increase in dead air space; reduced ability to clear airways by coughing
Urinary system	Degenerative changes in kidneys; fewer functional nephrons; reductions in filtration rate, tubular secretion, and tubular reabsorption
Reproductive systems	
Male	Reduced secretion of sex hormones; enlargement of prostate gland; decrease in sexual energy
Female	Degenerative changes in ovaries; decrease in secretion of sex hormones; menopause; regression of secondary sex characteristics

been chronically ill and is receiving comfort care, certain signs of impending death may appear, often in a sequence. A person may exhibit some or all of these signs. Health-care professionals view the dying process in two stages—preactive dying and active dying.

Preactive dying may take up to three months. During this time, some people are aware of what is happening and begin the psychological process of coming to terms with their mortality. A month or more before death, the person starts to withdraw, losing interest in news from the outside world and possibly requesting that visits from friends and relatives cease or shorten. He or she sleeps more, and might not even get out of bed on some days. Conversation lags. Gradually, the loss of interest in everyday activities extends to eating. This parallels physical changes, such as difficulty swallowing, that make eating increasingly difficult. The person might first give up eating meats, then fibrous vegetables, until it is clear that softer foods are preferred. The person might eat and drink astonishingly little, and the family might feel the need to try to force eating—which could cause the dying person discomfort. Dry mouth is common. The caregiver can provide ice chips or popsicles, or frequently wet the mouth with a swab.

Active dying presents a distinct set of signs, which might appear only on the day before death, or might begin up to two weeks earlier. During this phase the person sleeps often, but can easily be awakened. Even if sleep is deep, the person can hear—this is the last sense to fade. He or she may confuse time, place, and identities. A nurse might be mistaken for a relative, or an adult child might not be recognized. An actively dying person may go back in time, talking to a deceased spouse, for example. Signs of agitation appear, such as picking lint on the blanket or thrashing the arms about. Appetite may be nil.

In active dying, the organ systems slowly shut down. Cardiovascular signs include falling blood pressure (systolic below 70, diastolic below 50). The pulse may race or slow, or alternate. Poor circulation, which redirects the blood supply to the body's core, ushers in peripheral changes. The limbs feel cool to the touch, and the person may note numbness. The extremities become pale, then take on a bluish tinge. Skin areas under pressure, such as the undersides of the limbs, become mottled. Sensitivity to touch and pain declines.

The slowing circulation affects muscles. Poor ability to cough and swallow causes secretions to build up in the lungs. Secretions aren't suctioned, because this increases their rate of accumulation. The person can be repositioned to provide some relief. Congestion is intermittent. One day it may be so severe that eating is impossible; the next day breathing may ease. Cheyne-Stokes breathing—shallow mouth-breathing interspersed with increasingly long periods of apnea—is common. The normal rate of 16 to 20 breaths per minute may speed to more than 50, slow precipitously, perhaps pausing for 10 to 30 seconds, and then the person gasps and breathes rapidly again. As the throat muscles relax, exhalation over the vocal cords causes a passive moaning sound—this does not indicate that the person is in pain. A day or two before death, breathing may become loud—a sound called the "death rattle."

Body temperature control changes, and the person may have elevated body temperature or feel cold. The skin may be alternately flushed, then blue with an internal chill. Sweating is common, and as death nears, the skin takes on a yellowish pallor.

In the day or two before death, the signs intensify, although a last burst of energy may occur. A bedridden person may suddenly wish to be propped up in the living room and see people, or, after weeks of barely eating, suddenly request an ice cream sundae. Meanwhile, respiration and circulation are slowing, and decreased oxygen delivery may cause restlessness and agitation. The pulse becomes thready. Often right before death the person loses control of the bladder or bowels. Breathing becomes more irregular, with longer periods between breaths, and the lung rattling becomes louder. Consciousness seems to fade in and out—often the eyes do not focus and appear glassy, or may be only partially open. The eyes may tear frequently. Finally, the person can no longer respond. After one, two, or three long, last breaths, the eyes become fixed and open, the pupils dilate, the jaw relaxes, and the mouth may slightly open. The journey of life has ended.

PRACTICE 

44 How does the body change during adolescence?

45 Define *adulthood*.

46 What changes occur during adulthood?

47 What changes accompany senescence?

48 What are the signs of preactive and active dying?

23.5 AGING

The aging process is difficult to analyze because of the intricate interactions of the body's organ systems. Breakdown of one structure ultimately affects the functioning of others. The medical field of gerontology examines the biological changes of aging at the molecular, cellular, organismal, and population levels. Aging is both passive and active.

Passive Aging

Aging as a passive process is a breakdown of structures and slowing of functions. At the molecular level, passive aging is seen in the degeneration of the elastin and collagen proteins of connective tissues, causing skin to sag and muscle to lose its firmness.

During a long lifetime, biochemical abnormalities accumulate. Mistakes occur throughout life when DNA replicates in dividing cells. Usually, repair enzymes correct this damage immediately. But over many years, exposure to chemicals, viruses, and radiation disrupts DNA repair mechanisms so that the error burden becomes too great to be fixed. The cell may die as a result of faulty genetic instructions.

Another sign of passive aging at the biochemical level is the breakdown of lipids. As aging membranes leak during lipid degeneration, a fatty, brown pigment called lipofuscin accumulates. Mitochondria also begin to break down in older cells, decreasing the supply of chemical energy to power the cell's functions.

The cellular degradation associated with aging may be set into action by highly reactive chemicals called **free radicals.** A molecule that is a free radical has an unpaired electron in its outermost valence shell. This causes the molecule to grab electrons from other molecules, destabilizing them, and a chain reaction of chemical instability begins that could kill the cell. Free radicals are a by-product of normal metabolism and also form by exposure to radiation or toxic chemicals. The bile pigment bilirubin protects against free radicals. Enzymes that usually inactivate free radicals diminish in number and activity in the later years. One such enzyme is *superoxide dismutase* (SOD).

Active Aging

Aging also entails new activities or the appearance of new substances. Lipofuscin granules, for example, may be considered an active sign of aging, but they result from the passive breakdown of lipids. Another example of active

23.4 CLINICAL APPLICATION

Living to 100—And Beyond

People who live past 100 years are called centenarians, and they account for 1 in about 10,000 people. Usually they enjoy excellent health, remaining active and engaged in activities, then succumb rapidly to diseases that typically claim people decades earlier. Some never get these disorders at all. Researchers hope that learning which gene variants and environmental factors centenarians share will lead to a better understanding of the common disorders of later adulthood—heart disease, cancers, stroke, type 2 diabetes mellitus, and dementias.

While the environment seems to play an important role in the deaths of people ages 60 to 85, past that age, genes predominate. Someone who dies at age 65 of lung cancer can probably blame decades of smoking, but a smoker who dies at age 101 of the same type of cancer probably had protective gene variants. Centenarians have higher levels of HDL cholesterol than other people, which researchers estimate adds 20 years of life. Factors that vary widely among centenarians include religion, education level, socioeconomic status, ethnicity, and diet. They share normal or low weight, not smoking, handling stress well, and not having dementia. Women who gave birth late in life are overrepresented, perhaps because their reproductive systems, and the rest of their bodies, age slowly.

Children and siblings of centenarians tend to be long-lived too, suggesting a large inherited component. The saying of the New England Centenarian Study, one of the longest-running programs, is "The older you get, the healthier you've been," rather than associating extreme old age with sickness. Their website includes a tool to calculate your chances of joining this elite group living to 100.com.

Centenarians have luckily inherited two general types of gene variants: those that are directly protective, and normal variants of genes that, when mutant, cause disease. Specifically, single genes important in aging affect:

- Control of glucose metabolism and insulin secretion
- Immune system functioning
- Cell cycle control
- Lipid (cholesterol) metabolism
- Response to stress
- Production of antioxidant enzymes

Several studies are identifying the gene variants and lifestyle practices that contribute to living long and well. The New England study is amassing the "healthy standard genome." Investigators at the Coriell Institute in New Jersey are probing the genomes of the oldest old in nursing homes. So far, what these people share is never having had heart disease and never having smoked. Several have had cancer, indicating that these disorders are survivable. Researchers at the University of Pittsburgh are pursuing gene variants that preserve cognition. The National Institute on Aging is currently seeking families with more than two long-lived members living near Boston, New York, or Pittsburgh to participate in the Long Life Family Study which will collect and parse genetic and health information for clues to longevity. Perhaps these studies will provide information that will help the majority of us, who have not inherited longevity gene variants, discover how to control our environments in ways that promote long and healthy lives. ■

FIGURE 23E This woman has enjoyed living for more than a century. Researchers are discovering clues to good health by probing the genomes of centenarians.

aging is autoimmunity, in which the immune system turns against the body, attacking its cells as if they were invading organisms.

Active aging begins before birth, as certain cells die as part of the developmental program encoded in the genes. This process of programmed cell death, called **apoptosis** (ap″o-tō′sis), occurs regularly in the embryo, degrading certain structures to pave the way for new ones. The number of neurons in the fetal brain, for example, is halved as those that make certain synaptic connections are spared from death. In the fetal thymus, T cells that do not recognize "self" cell surfaces die, thereby building the immune system. Throughout life, apoptosis enables organs to maintain their characteristic shapes.

Mitosis and apoptosis are opposite, but complementary, processes. As organs grow, the number of cells in some regions increases, but in others, it decreases. Cell death is not a phenomenon only of the aged. It is a normal part of life. Clinical Application 23.4 discusses characteristics that people who live past 100 share.

The Human Life Span

In the age-old quest for longer life, people have sampled everything from turtle soup to owl meat to human blood. A Russian-French microbiologist, Ilya Mechnikov, believed that a life span of 150 years could be achieved with the help of a steady diet of milk cultured with bacteria. He thought that the bacteria would live in the large intestine and somehow increase the human life span. (He died at age 71.) Ironically, many people have died in pursuit of a literal "fountain of youth."

The human *life span*—the length of time that a human can theoretically live—is 120 years. Although most people succumb to disease or injury long before that point, in many countries the fastest growing age group is those over age eighty. These "oldest old," having passed the age when cancer and cardiovascular disease typically strike, are often quite healthy.

Life expectancy is a realistic projection of how long an individual will live, based on epidemiological information. In the United States, life expectancy is 75.4 years for men and 83.2 years for women. Yet in some African nations being decimated by the AIDS epidemic, life expectancy is in the thirties.

Medical advances have greatly contributed to improved life expectancy. Antibiotics have tamed some once-lethal infections, drugs enable many people with cancer to survive, and such advances as beta-blocker drugs and coronary bypass surgery have extended the lives of people with heart disease. However, the rise of new or renewed infectious diseases, such as AIDS, polio, and measles, also indicates that we cannot yet conquer all illnesses. Although we can alter our environment more than other species can, some forces of nature remain beyond our control. Table 23.11 lists the top 10 causes of death in the United States.

PRACTICE 

49 Why is it difficult to sort out the causes of aging?

50 How is aging a passive process?

51 How is aging an active process?

52 Distinguish between life span and life expectancy.

TABLE 23.11 | The Ten Leading Causes of Death in the United States, 2004

Cause	% of total 2,397,615 deaths
1. Heart disease	27.2
2. Cancer	23.1
3. Stroke	6.2
4. Chronic obstructive pulmonary disease (COPD)	0.5
5. Injuries	0.4
6. Diabetes mellitus	0.3
7. Alzheimer disease	0.3
8. Influenza and pneumonia	0.2
9. Kidney disease	0.2
10. Infection (septicemia)	0.1

Source: U.S. National Center for Health Statistics, *Health, United States, 2007.*

CHAPTER SUMMARY

23.1 INTRODUCTION (PAGE 876)

Growth refers to an increase in size; development is the process of changing from one phase of life to another. Development includes both prenatal (ending at birth) and postnatal (beginning at birth) life phases.

23.2 PREGNANCY (PAGE 876)

Pregnancy is the presence of a developing offspring in the uterus.

1. Transport of sex cells
   a. Ciliary action aids movement of the secondary oocyte into the uterine tube.
   b. A sperm cell moves, by its tail lashing and muscular contraction in the female reproductive tract, into the uterine tube.
2. Fertilization
   a. With the aid of an enzyme, a sperm cell penetrates the zona pellucida.
   b. When a sperm cell penetrates a secondary oocyte membrane, changes in the oocyte cell membrane and the zona pellucida prevent entry of additional sperm.
   c. Completion of meiosis II forms the second polar body.
   d. Fusion of the nuclei of a sperm and a secondary oocyte completes fertilization.
   e. The product of fertilization is a zygote with 46 chromosomes.

23.3 PRENATAL PERIOD (PAGE 879)

1. Period of cleavage
   a. The zygote undergoes mitosis, and the newly formed cells divide mitotically.
   b. Each subsequent division produces smaller and smaller cells.
   c. A solid ball of cells (morula) forms, and it becomes a hollow ball called a blastocyst.
   d. The inner cell mass that gives rise to the embryo proper forms within the blastocyst.
   e. The blastocyst implants in the uterine wall.
      (1) Enzymes digest the endometrium around the blastocyst.
      (2) Fingerlike processes from the blastocyst penetrate the endometrium.
   f. Cleavage lasts through the first week of development.
   g. The trophoblast secretes hCG, which helps maintain the corpus luteum, helps protect the blastocyst against being rejected, and stimulates the developing placenta to secrete hormones.
2. Hormonal changes during pregnancy
   a. Embryonic cells produce hCG that maintains the corpus luteum, which continues to secrete estrogens and progesterone.
   b. Placental tissue produces high concentrations of estrogens and progesterone.
      (1) Estrogens and progesterone maintain the uterine wall and inhibit secretion of FSH and LH.
      (2) Progesterone and relaxin inhibit contractions of uterine muscles.
      (3) Estrogens enlarge the vagina.
      (4) Relaxin helps relax the ligaments of the pelvic joints.
   c. The placenta secretes placental lactogen that stimulates the development of the breasts and mammary glands.
   d. During pregnancy, increasing secretion of aldosterone promotes retention of sodium and body fluid, and increasing secretion of parathyroid hormone helps maintain a high concentration of maternal blood calcium.
3. Other changes during pregnancy
   a. The uterus greatly enlarges.
   b. The woman's blood volume, cardiac output, breathing rate, and urine production increase.
   c. The woman's dietary needs increase, but if intake is inadequate, fetal tissues have priority for use of available nutrients.
4. Embryonic stage
   a. The embryonic stage extends from the second through the eighth weeks.
   b. The placenta and main internal and external body structures develop.
   c. The embryonic disc becomes cylindrical and attaches to the developing placenta by the connecting stalk.
   d. The cells of the inner cell mass fold inward, forming a gastrula that has two and then three primary germ layers.
      (1) Ectoderm gives rise to the nervous system, portions of the skin, the lining of the mouth, and the lining of the anal canal.
      (2) Mesoderm gives rise to muscles, bones, blood vessels, lymphatic vessels, reproductive organs, kidneys, and linings of body cavities.
      (3) Endoderm gives rise to linings of the digestive tract, respiratory tract, urinary bladder, and urethra.
   e. Chorionic villi develop and are surrounded by spaces filled with maternal blood.
   f. The embryo develops a head, face, upper limbs, lower limbs, and mouth, and appears more humanlike.
   g. The placental membrane consists of the epithelium of the chorionic villi and the endothelium of the capillaries inside the chorionic villi.
      (1) Oxygen and nutrients diffuse from the maternal blood through the placental membrane and into the fetal blood.
      (2) Carbon dioxide and other wastes diffuse from the fetal blood through the placental membrane and into the maternal blood.
   h. The placenta develops in the disc-shaped area where the chorion contacts the uterine wall.
      (1) The embryonic portion consists of the chorion and its villi.
      (2) The maternal portion consists of the endometrium.
   i. A fluid-filled amnion develops around the embryo.
   j. The umbilical cord is formed as the amnion envelopes the tissues attached to the underside of the embryo.
      (1) The umbilical cord includes two arteries and a vein.
      (2) It suspends the embryo in the amniotic cavity.
   k. The chorion and amnion fuse.
   l. The yolk sac forms on the underside of the embryonic disc.
      (1) It gives rise to blood cells and cells that later form sex cells.
      (2) It helps form the digestive tube.
   m. The allantois extends from the yolk sac into the connecting stalk.
      (1) It forms blood cells.
      (2) It gives rise to the umbilical vessels.
   n. By the beginning of the eighth week, the embryo is recognizable as a human.
5. Fetal stage
   a. This stage extends from the end of the eighth week and continues until birth.
   b. Existing structures grow and mature; only a few new parts appear.
   c. The body enlarges, upper and lower limbs reach final relative proportions, the skin is covered with sebum and dead epidermal cells, the skeleton continues to ossify, muscles contract, and fat is deposited in subcutaneous tissue.
   d. The fetus is full term at the end of the ninth month, which equals approximately 266 days.
      (1) It is about 50 centimeters long and weighs 2.7–3.6 kilograms.
      (2) It is positioned with its head toward the cervix.
6. Fetal blood and circulation
   a. Umbilical vessels carry blood between the placenta and the fetus.
   b. Fetal blood carries a greater concentration of oxygen than does maternal blood because the

concentration of oxygen-carrying hemoglobin is greater in fetal blood, and fetal hemoglobin has greater affinity for oxygen.
   c. Blood enters the fetus through the umbilical vein and partially bypasses the liver by means of the ductus venosus.
   d. Blood enters the right atrium and partially bypasses the lungs by means of the foramen ovale.
   e. Blood entering the pulmonary trunk partially bypasses the lungs by means of the ductus arteriosus.
   f. Blood enters the umbilical arteries from the internal iliac arteries.
7. Birth process
   a. Pregnancy usually lasts thirty-eight weeks after fertilization.
   b. During pregnancy, placental progesterone inhibits uterine contractions.
   c. Several events occur at birth.
      (1) A decreasing concentration of progesterone and the release of prostaglandins may initiate the birth process.
      (2) The posterior pituitary gland releases oxytocin.
      (3) Uterine muscles are stimulated to contract, and labor begins.
      (4) Positive feedback causes stronger contractions and greater release of oxytocin.
   d. Following the birth of the infant, placental tissues are expelled.
8. Milk production and secretion
   a. During pregnancy, the breasts change.
      (1) Estrogens cause the ductile system to grow.
      (2) Progesterone causes development of alveolar glands.
      (3) Prolactin is released during pregnancy, but progesterone inhibits milk production.
   b. Following childbirth, the concentrations of placental hormones decline.
      (1) The action of prolactin is no longer blocked.
      (2) The mammary glands begin to secrete milk.
   c. Reflex response to mechanical stimulation of the nipple causes the posterior pituitary to release oxytocin, which releases milk from the alveolar ducts.
   d. As long as milk is removed from glands, more milk is produced; if milk is not removed, production ceases.
   e. During the period of milk production, the reproductive cycle is partially inhibited.

23.4 POSTNATAL PERIOD (PAGE 904)

1. Neonatal period
   a. This period extends from birth to the end of the fourth week.
   b. The newborn must begin to respire, obtain nutrients, excrete wastes, and regulate its body temperature.
   c. The first breath must be powerful to expand the lungs.
      (1) Surfactant reduces surface tension.
      (2) A variety of factors stimulate the first breath.
   d. The liver is immature and unable to supply sufficient glucose, so the newborn depends primarily on stored fat for energy.
   e. Immature kidneys cannot concentrate urine very well.
      (1) The newborn may become dehydrated.
      (2) Water and electrolyte imbalances may develop.
   f. Homeostatic mechanisms may function imperfectly, and body temperature may be unstable.
   g. The cardiovascular system changes when placental circulation ceases.
      (1) Umbilical vessels constrict.
      (2) The ductus venosus constricts.
      (3) The foramen ovale is closed by a valve as blood pressure in the right atrium falls and blood pressure in the left atrium rises.
      (4) The ductus arteriosus constricts.
2. Infancy
   a. Infancy extends from the end of the fourth week to one year of age.
   b. Infancy is a period of rapid growth.
      (1) The muscular and nervous systems mature, and coordinated activities become possible.
      (2) Communication begins.
   c. Rapid growth depends on an adequate intake of proteins, vitamins, and minerals in addition to energy sources.
3. Childhood
   a. Childhood extends from the end of the first year to puberty.
   b. Primary teeth erupt and are replaced by secondary teeth.
   c. It is a period of rapid growth, development of muscular control, and establishment of bladder and bowel control.
4. Adolescence
   a. Adolescence extends from puberty to adulthood.
   b. Physiological and anatomical changes result in a reproductively functional individual.
   c. Females may be taller and stronger than males in early adolescence, but the situation reverses in late adolescence.
   d. Adolescents develop high levels of motor skills, their intellectual abilities increase, and they continue to mature emotionally.
5. Adulthood
   a. Adulthood extends from adolescence to old age.
   b. The adult remains relatively unchanged physiologically and anatomically for many years.
   c. After age thirty, degenerative changes begin.
      (1) Skeletal muscles lose strength.
      (2) The cardiovascular system becomes less efficient.
      (3) The skin loses elasticity.
      (4) The capacity to produce sex cells declines.
6. Senescence
   a. Senescence is growing old.
   b. Degenerative changes continue, and the body becomes less able to cope with demands.
   c. Changes occur because of prolonged use, effects of disease, and cellular alterations.
   d. An aging person usually loses some intellectual functions, sensory functions, and physiological coordinating capacities.
   e. Death usually results from mechanical disturbances in the cardiovascular system or from disease processes that affect vital organs.

7. The end of life
    a. Certain signs may appear in sequence when a person dies of a chronic illness.
    b. Preactive dying takes up to three months. The person withdraws socially and appetite wanes.
    c. Active dying takes up to two weeks. The person rests, may become confused or agitated, and eats very little. Gradually the organ systems shut down. The skin becomes mottled as circulation slows and congestion and loud breathing occur.

23.5 AGING (PAGE 909)

1. Passive aging
    a. Passive aging entails breakdown of structures and slowing or failure of functions.
    b. Connective tissue breaks down.
    c. DNA errors accumulate.
    d. Lipid breakdown in aging membranes releases lipofuscin.
    e. Free radical damage escalates.
2. Active aging
    a. In autoimmunity, the immune system attacks the body.
    b. Apoptosis is a form of programmed cell death. It occurs throughout life, shaping organs.
3. The human life span
    a. The theoretical maximum life span is 120 years.
    b. Life expectancy, based on real populations, is 75.4 years for men and 83.2 years for women in the United States, and may be lower in poorer nations and those ravaged by AIDS.
    c. Medical technology makes life expectancy more closely approach life span.

CHAPTER ASSESSMENTS

23.1 Introduction

1 ______________ is an increase in the size of the individual, whereas ______________ is the continuous process by which an individual changes from one phase to another. (p. 876)

2 ______________ is the period of development from fertilization to birth, whereas ________________ is the period of development from birth to death. (p. 876)

23.2 Pregnancy

3 Define *pregnancy.* (p. 876)

4 Describe how sperm cells move in the female reproductive tract. (p. 876)

5 Summarize the events occurring after the sperm cell head enters the oocyte's cytoplasm. (p. 878)

23.3 Prenatal Period

6 Describe the process of cleavage. (p. 879)

7 Distinguish between a morula and a blastocyst. (p. 880)

8 Describe the formation of the inner cell mass, and explain its significance. (p. 880)

9 Explain what happens when the blastocyst nestles into the endometrium. (p. 882)

10 List the functions of hGG. (p. 882)

11 Describe the formation of the placenta, and explain its functions. (p. 882)

12 Explain the hormonal changes that occur in the maternal body during pregnancy. (p. 884)

13 Describe the nonhormonal changes that occur in the maternal body during pregnancy. (p. 885)

14 Explain how the primary germ layers form. (p. 885)

15 List the structures derived from the primitive tissues of the ectoderm, mesoderm, and endoderm. (p. 885)

16 Define *placental membrane.* (p. 886)

17 Distinguish between the chorion and amnion. (p. 888)

18 Explain the function of the amniotic fluid. (p. 888)

19 Describe the formation of the umbilical cord. (p. 891)

20 Explain how the yolk sac and the allantois are related, and list the functions of each. (p. 891)

21 Explain why the embryonic period is so critical. (p. 892)

22 Give the time frame for the fetus, listing the major changes that occur during fetal development. (p. 892)

23 Describe a full-term fetus. (p. 893)

24 Explain how the fetal cardiovascular system is adapted to intrauterine life. (p. 895)

25 Compare the properties of fetal hemoglobin with those of adult hemoglobin. (p. 897)

26 Trace the pathway of blood from the placenta to the fetus and back to the placenta. (p. 897)

27 Describe the role of progesterone in initiating the birth process. (p. 899)

28 Discuss the events that occur during the birth process. (p. 899)

29 Explain positive feedback and the role of hormones in the expulsion of the fetus and the afterbirth. (p. 900)

30 Detail the roles of prolactin and oxytocin in milk production and secretion. (p. 902)

23.4 Postnatal Period

31 Distinguish between a newborn and an infant. (p. 904)

32 Explain why a newborn's first breath must be particularly forceful. (p. 905)

33 List some of the factors that stimulate the first breath. (p. 905)

34 Explain why newborns tend to develop water and electrolyte imbalances. (p. 905)

35 Discuss the cardiovascular changes in the newborn. (p. 905)

36 Describe the characteristics of an infant. (p. 905)

37 Distinguish between a child and an adolescent. (p. 906)

38 Define *adulthood.* (p. 907)

39 List some of the degenerative changes that begin during adulthood. (p. 907)

40 Define *senescence.* (p. 907)

41 List some of the factors that promote senescence. (p. 907)

42 Contrast preactive dying and active dying. (p. 908)

23.5 Aging

42 Discuss the signs of passive and active aging and the physiological causes of these signs. (p. 909)

43 ________________ is the length of time a human can theoretically live, whereas ______________ is the realistic projection of how long an individual will live. (p. 911)

INTEGRATIVE ASSESSMENTS/CRITICAL THINKING

OUTCOMES 6.6, 7.13, 8.7, 9.9, 11.8, 12.5, 13.12, 15.9, 16.10, 17.11, 18.9, 19.8, 20.5, 23.4, 23.5

1. If an aged relative came to live with you, what special provisions could you make in your household environment and routines that would demonstrate your understanding of the changes brought on by aging?

OUTCOMES 15.2, 15.6, 23.2, 23.3

2. One of the more common congenital cardiac disorders is a ventricular septal defect in which an opening remains between the right and left ventricles. What problem would such a defect create as blood moves through the heart?

OUTCOMES 15.2, 15.6, 23.2, 23.3, 23.4

3. What symptoms may appear in a newborn if its ductus arteriosus fails to close?

OUTCOMES 15.5, 23.4, 23.5

4. Why is it important for a middle-aged adult who has neglected physical activity for many years to have a physical examination before beginning an exercise program?

OUTCOMES 16.9, 23.2

5. Why can twins resulting from a single fertilized secondary oocyte exchange blood or receive organ transplants from each other without rejection, while twins resulting from two fertilized secondary oocytes sometimes cannot?

OUTCOMES 23.2, 23.3

6. Toxins usually cause more severe medical problems if exposure is during the first eight weeks of pregnancy rather than during the later weeks. Why?

OUTCOMES 23.2, 23.3, 23.4

7. What technology would enable a fetus born in the fourth month to survive in a laboratory setting? (This is not yet possible.)

WEB CONNECTIONS

Be sure to visit the text website at **www.mhhe.com/shier12** for answers to chapter assessments, additional quizzes, and interactive learning exercises.

ANATOMY & PHYSIOLOGY REVEALED

Anatomy & Physiology Revealed® (APR) includes cadaver photos that allow you to peel away layers of the human body to reveal structures beneath the surface. This program also includes animations, radiologic imaging, audio pronunciations, and practice quizzing. Check out **www.aprevealed.com**. APR has been proven to help improve student grades!

CHAPTER

24 Genetics and Genomics

Note the similar profiles of the four generations of this family. We do not always know how information in DNA sequences manifests as a trait such as facial structure.

UNDERSTANDING WORDS

chromo-, color: *chromo*some—a "colored body" in a cell's nucleus that includes the genes.

gen-, born: *gen*etics—the study of inheritance of characteristics.

hetero-, other, different: *hetero*zygous—different members of a gene pair.

hom-, same, common: *hom*ologous chromosomes—pair of chromosomes that have similar genetic information.

karyo-, nucleus: *karyo*type—a chart that displays chromosomes in size order.

mono-, one: *mono*somy—one type of chromosome present in only one copy.

phen-, show, be seen: *pheno*type—physical appearance or health condition that results from the way genes are expressed in an individual.

tri-, three: *tri*somy—three copies of a chromosome.

LEARNING OUTCOMES

After you have studied this chapter, you should be able to:

24.1 Introduction

1. Distinguish among genome, gene, and chromosome (p. 917)
2. Define *genetics.* (p. 917)
3. Explain how the human genome is an economical storehouse of information. (p. 917)
4. Explain how the environment influences how genes are expressed. (p. 918)

24.2 Modes of Inheritance

5. Describe a karyotype, and explain what it represents. (p. 918)
6. Explain the basis of multiple alleles of a gene. (p. 919)
7. Distinguish between heterozygous and homozygous; genotype and phenotype; dominant and recessive. (p. 919)
8. Distinguish between autosomal recessive and autosomal dominant inheritance. (p. 920)

24.3 Factors That Affect Expression of Single Genes

9. Explain how and why the same genotype can have different phenotypes among individuals. (p. 924)

24.4 Multifactorial Traits

10. Describe and give examples of how traits may be determined by genes and the environment. (p. 924)

24.5 Matters of Sex

11. Describe how and when sex is determined. (p. 927)
12. Explain how X-linked inheritance differs from inheritance of autosomal traits. (p. 928)
13. Discuss factors that affect how phenotypes may differ between the sexes. (p. 928)

24.6 Chromosome Disorders

14. Describe three ways that chromosomes can be abnormal. (p. 929)
15. Explain how prenatal tests provide information about chromosomes. (p. 931)

24.7 Gene Expression Explains Aspects of Anatomy and Physiology

16. Describe the type of information gene expression profiling provides. (p. 933)
17. Explain how a DNA microarray works. (p. 933)

 LEARN 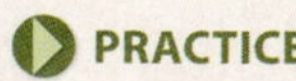 PRACTICE  ASSESS

DIRECT-TO-CONSUMER GENETIC TESTING

Some websites offer genetic tests. Submit a DNA sample on a cheek swab or by spitting into a tube, and for a fee, receive information about traits influenced by genetics. The tests are either for mutations of well-studied single genes or "associations" of patterns of variation throughout the genome statistically linked to certain illnesses or susceptibilities.

DNA tests may detect benign traits such as ear wax consistency or ability to taste bitter substances. For health matters, however, without genetic counseling consumers can make decisions based on partial or inappropriate information. Passage of the Genetic Information Nondiscrimination Act in 2008, which prevents employers and insurers from using results of genetic tests to deny opportunities, may encourage people to have their DNA probed.

A study from 2006, however, provided information about DNA testing companies that offer nutritional supplements supposedly matched to personal genetic profiles. After the media spread the word of these services, the U.S. government's General Accounting Office researched the tests. Investigators sent to four "nutrigenetics" companies two DNA samples—one from a nine-month-old female and the other from a forty-eight-year-old man. But the samples were sent along with different invented lifestyle dietary profiles, creating fourteen "fictitious consumers"—twelve for the female, two for the male.

Here is an example of the information sent as part of the study:

- The DNA from the baby girl was submitted as a thirty-three-year-old woman, 185 pounds, 5′9″, who smokes, drinks a lot of coffee, doesn't exercise, and eats a lot of dairy, grains, and fats.
- The DNA from the baby girl was also submitted as a fifty-nine-year-old man, 140 pounds, 5′7″, who exercises, never smoked, takes vitamins, hates coffee, and eats a lot of protein and fried foods.
- The DNA from the man was submitted as being from a thirty-two-year-old male, 150 pounds, 5′9″, who smokes, rarely exercises, drinks coffee, doesn't exercise, and eats a lot of dairy, grains, and fats.

The disorders with elevated risks found for the three fake people were exactly the same: osteoporosis, hypertension, type 2 diabetes, and heart disease. One company offered the appropriate multivitamin supplements for $1,200, which the investigation found to be worth about $35. Recommendations tended to state the obvious, such as advising a smoker to quit. The advice tracked with the fictional lifestyle/diet information and not genetics. Concluded the study: "Although these recommendations may be beneficial to consumers in that they constitute common sense health and dietary guidance, DNA analysis is not needed to generate this advice." Some of the suggestions could even be dangerous, such as recommending vitamin excesses in people with certain medical conditions. ■

24.1 INTRODUCTION

Packaged into our cells are instruction manuals for our bodies. The manual—the human genome—is written in the language of DNA molecules. Recall from chapter 4 that DNA consists of sequences of the nucleotide building blocks *A* (adenine), *G* (guanine), *C* (cytosine), and *T* (thymine). Sequences of DNA that encode particular proteins are called genes. A gene has different forms, because its sequence can vary from individual to individual. In somatic cells, two copies of the human genome of 3 billion building blocks each are dispersed among the forty-six chromosomes (figure 24.1).

Genetics (jĕ-net′iks) is the study of inheritance of characteristics. Genes provide our diversity, including eye, skin, and hair color; many aspects of health; athletic ability and talents; and hard-to-define characteristics such as personality traits. We often equate the study of genetics with disease, but it is more accurately described as the study of inherited variation. Our genomes are more than 99.9% alike in DNA sequence; within that less than 0.1% of genetic variation lies our individuality.

The transfer of genetic information from one generation to the next occurs through the processes of meiosis and fertilization, when one copy of the genome from each parent join. The cells that give rise to eggs and sperm, like all somatic cells, are diploid, with two copies of each of the 23 chromosomes. Eggs and sperm are haploid, with one set of chromosomes. Joining of eggs and sperm reconstitutes diploidy.

RECONNECT

To Chapter 22, Sperm Cell Formation, pages 835–836, and Oogenesis, page 849

The human genome is an economical information store. It includes about 24,000 protein-encoding genes. Different cell types access different subsets of the genome, using the information to produce particular proteins and in this way sculpt the hundreds of types of specialized cells in the body. Yet at the same time, the genome encodes much more than 24,000 bits of information. RNA molecules can represent parts of different genes, so that the 24,000 genes encode 100,000 to 200,000 different proteins. It is a little like having a wardrobe of twenty items, but wearing them in different combinations to create many dozens of different outfits.

Genetic information functions at several levels. It is encoded in DNA and expressed in RNA and protein; affects cells and tissues; affects the individual; and is also passed to the next generation. Figure 24.2 depicts the inherited disease cystic fibrosis (CF) viewed at several levels.

RECONNECT

To Chapter 4, Nucleic Acids and Protein Synthesis, pages 131–133.

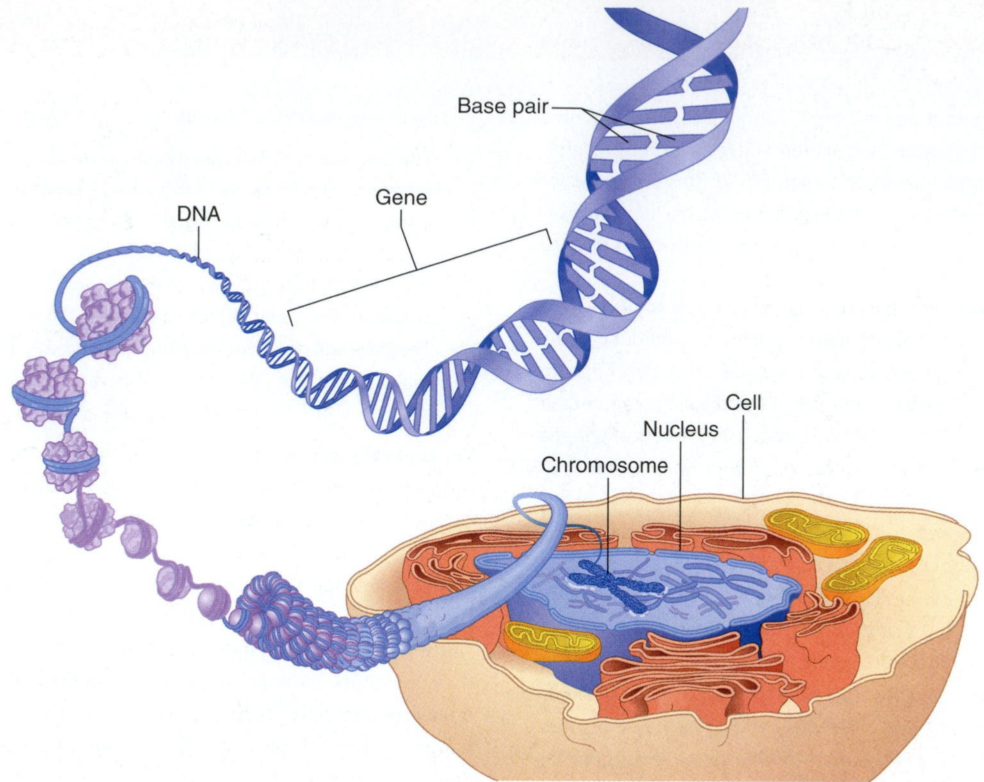

FIGURE 24.1 From DNA to gene to chromosome. Chromosomes consist of a continuous DNA double helix and associated proteins. Chromosomes condense to become visible under a microscope just prior to cell division.

Until recently, the field of medical genetics dealt mostly with rare disorders that can be traced to the malfunction or absence of single genes. However, information from the human genome sequence is providing a new view of physiology as a complex interplay of gene functions.

> Looking at the human body in terms of multiple, interacting genes is termed **genomics.** A related field, proteomics, focuses on the proteins that specific cell types produce. A proteomics approach to studying the function of the breast as a gland, for example, compares the proteins in a healthy epithelial cell in a milk duct of a woman who is not pregnant or breastfeeding to the proteins in the same type of cell from a new mother who is breastfeeding.

As important as genes are, they do not determine who we are, nor do they act alone. The environment influences how genes are expressed—under which circumstances they are transcribed into mRNA and those mRNA molecules translated into protein. The environment includes the chemical, physical, social and biological factors surrounding an individual. A common medical condition may result from several inherited risk factors and exposure to certain environmental influences. For example, development of cardiovascular disease may reflect inheritance of specific variants of specific genes that control blood pressure, blood clotting, and cholesterol metabolism, but also lifestyle influences such as stress, smoking, poor diet, and lack of physical exercise that may affect the expression of those genes in negative ways.

PRACTICE 

1 What is the chemical composition of our genetic instructions?
2 Define *gene* and *genetics*.
3 Explain how the human genome encodes more proteins than it has genes.

24.2 MODES OF INHERITANCE

Genetics has the power of prediction. Knowing how genes are distributed in meiosis and the combinations in which they can join at fertilization makes it possible to calculate the probability that a certain trait will appear in the offspring of two particular individuals. The patterns in which genes are transmitted in families are termed *modes of inheritance.*

Chromosomes and Genes Come in Pairs

The basis of being able to make genetic predictions is the way in which genes are carried as parts of chromosomes. From the moment of conception, a human cell is diploid. Chromosome charts called **karyotypes** display the 23 chromosome pairs in size order (fig. 24.3). Pairs 1 through 22 are **autosomes** (aw′to-sōmz), chromosomes that do not carry genes that

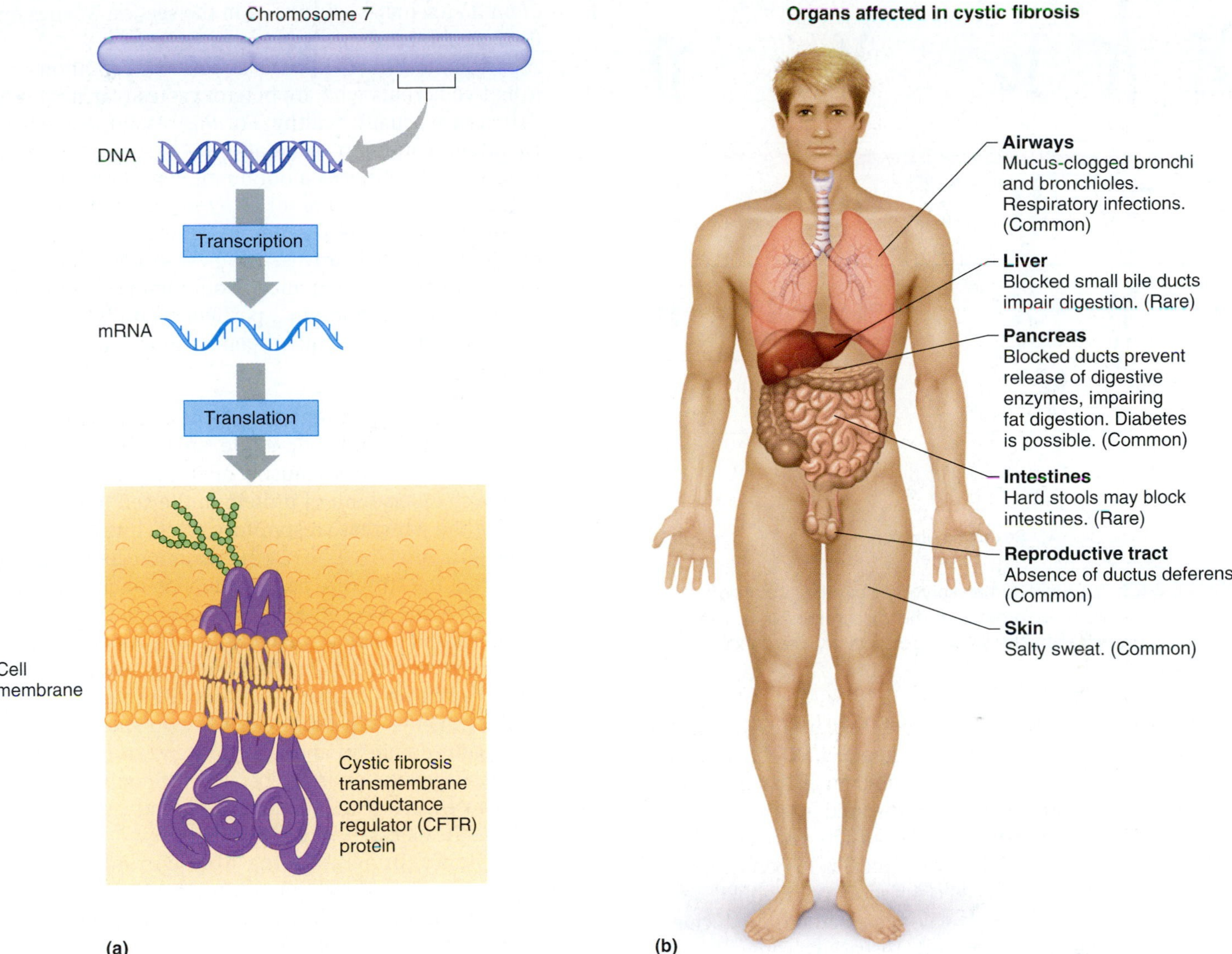
Chromosome 7
DNA
Transcription
mRNA
Translation
Cell membrane
Cystic fibrosis transmembrane conductance regulator (CFTR) protein
(a)
Organs affected in cystic fibrosis
Airways
Mucus-clogged bronchi and bronchioles. Respiratory infections. (Common)
Liver
Blocked small bile ducts impair digestion. (Rare)
Pancreas
Blocked ducts prevent release of digestive enzymes, impairing fat digestion. Diabetes is possible. (Common)
Intestines
Hard stools may block intestines. (Rare)
Reproductive tract
Absence of ductus deferens. (Common)
Skin
Salty sweat. (Common)
(b)

FIGURE 24.2 From gene to protein to person. *(a)* The gene encoding the CFTR protein, and causing cystic fibrosis when mutant, is on the seventh largest chromosome. CFTR protein folds into a channel that regulates the flow of chloride ions into and out of cells lining the respiratory tract, pancreas, intestines, and elsewhere. *(b)* In cystic fibrosis, the CFTR protein is abnormal, usually missing an amino acid. Its shape is altered, which entraps the chloride ions inside cells. Water entering these cells leaves behind very thick mucus and other secretions in the places highlighted in the illustration. The sticky secretions cause the symptoms of the illness. Source: Data from M.C. Iannuzi and F.S. Collins, "Reverse Genetics and Cystic Fibrosis" in *American Journal of Respiratory Cellular and Molecular Biology,* 2:309–316, 1990.

determine sex. The other two chromosomes, the X and the Y, include genes that determine sex and are called **sex chromosomes.** They are discussed later in the chapter in the section entitled "Matters of Sex."

Each chromosome includes hundreds or thousands of genes. We have two copies of each chromosome, so we also have two copies of each gene. The two members of a gene pair are located at the same position on homologous chromosomes. Sometimes the members of a gene pair are identical in DNA sequence. However, because a gene consists of hundreds of nucleotide building blocks, it exists in variant forms, called **alleles** (ah-lēlz′), that differ in DNA sequence. An individual who has two identical alleles of a particular gene is **homozygous** (ho″mo-zi′gus) for that gene. A person with two different alleles for a gene is **heterozygous** (het″er-o-zi′gus) for it. A gene may have many alleles, but an individual person can have a maximum of two alleles for a particular gene.

The allele that causes most cases of cystic fibrosis was discovered in 1989, and a test developed to detect it. However, other alleles were soon discovered. Today, more than 1,000 mutations (changes) in the cystic fibrosis gene are known. Different allele combinations produce different combinations and severities of symptoms. Tests for the disorder detect ninety-six or more of the most common alleles.

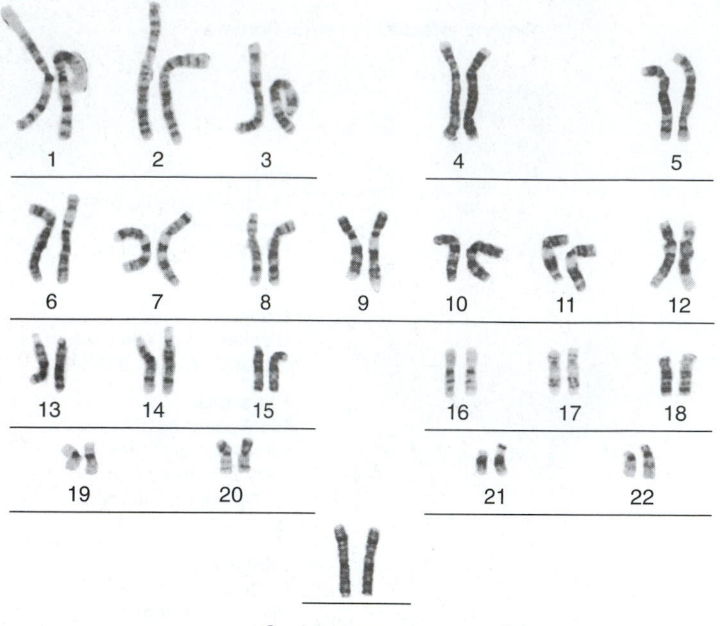

FIGURE 24.3 A normal human karyotype has the 22 pairs of autosomes aligned in size order, plus the sex chromosomes. The individual represented by this karyotype is female, with two *X* chromosomes.

The particular combination of gene variants (alleles) in a person's genome constitutes the **genotype** (je'no-tīp). The appearance or health condition of the individual that develops as a result of the ways the genes are expressed is termed the **phenotype** (fe'no-tīp). An allele is **wild type** if its associated phenotype is either normal function or the most common expression in a particular population. Wild type is indicated with a + sign. An allele that is a change from wild type, perhaps producing an uncommon phenotype, is **mutant.** Disease-causing alleles are mutant. Some mutations, however, are advantageous, as the vignette in chapter 3 on page 76 describes.

Dominant and Recessive Inheritance

For many genes, in heterozygotes, one allele determines the phenotype. Such an allele whose action masks that of another allele is termed **dominant.** The allele whose expression is masked is **recessive.** For genes with only two alleles, the dominant ones are usually indicated with a capital letter.

An allele that causes a disease can be recessive or dominant. It may also be *autosomal* (carried on a nonsex chromosome) or *X-linked* (carried on the X chromosome) or *Y-linked* (carried on the Y chromosome). The mode of inheritance refers to whether a trait is dominant or recessive, autosomal or carried on a sex chromosome. This designation has important consequences in predicting the chance that offspring will inherit an illness or trait. The following rules emerge:

1. An autosomal condition is equally likely to affect either sex. X-linked characteristics affect males much more often than females, discussed in the section "Genes on the Sex Chromosomes."
2. A person most likely inherits a recessive condition from two parents who are heterozygotes (carriers). The parents are usually healthy. For this reason, recessive conditions can "skip" generations.
3. A person who inherits a dominant condition has at least one affected parent. Therefore, dominant conditions do not skip generations. (An exception is if the dominant allele arises, as a new mutation, in the sperm or egg.) If, by chance, a dominant trait does not appear in a generation in a particular family, it does not reappear in subsequent generations, as a recessive trait might.

Cystic fibrosis (CF) is an example of an autosomal recessive disorder. The wild type allele for the gene, which is dominant over the disease-causing allele, specifies formation of chloride channels built of protein in the cell membrane of cells lining the pancreas, respiratory tract, intestine, testes, and other structures (see fig. 24.2). Certain recessive mutant alleles disrupt the structure and/or function of chloride channels. An individual who inherits two such mutant alleles has cystic fibrosis and is homozygous recessive. A person inheriting only one recessive mutant allele plus a dominant wild type allele is a carrier and transmits the disease-causing allele in half of the gametes. A person who has two wild type alleles is homozygous dominant and does not have or carry CF. The three possible genotypes are associated with only two phenotypes, because carriers and homozygous dominant individuals do not have the illness.

Using logic, understanding how chromosomes and genes are apportioned into gametes in meiosis, and knowing that mutant alleles that cause CF are autosomal recessive, we can predict genotypes and phenotypes of the next generation. Figure 24.4 illustrates two carriers of CF. Half of the man's sperm contain the mutant allele, as do half of the woman's eggs. Sperm and eggs combine at random, so each offspring has a

- 25% chance of inheriting two wild type alleles (homozygous dominant, healthy, and not a carrier)
- 50% chance of inheriting a mutant allele from either parent (heterozygous and a carrier, but healthy)
- 25% chance of inheriting a mutant allele from each parent (homozygous recessive, has CF)

Genetic counselors use two tools to explain inheritance to families: Punnett squares and pedigrees. A Punnett square is a table that symbolizes the logic used to deduce the probabilities of particular genotypes in offspring. The mother's alleles (for a particular gene) are listed atop the four boxes comprising the square, and the father's alleles are listed along the left side. Each box records an allele combination at fertilization. (When more than one gene is considered, the Punnett square has more boxes.)

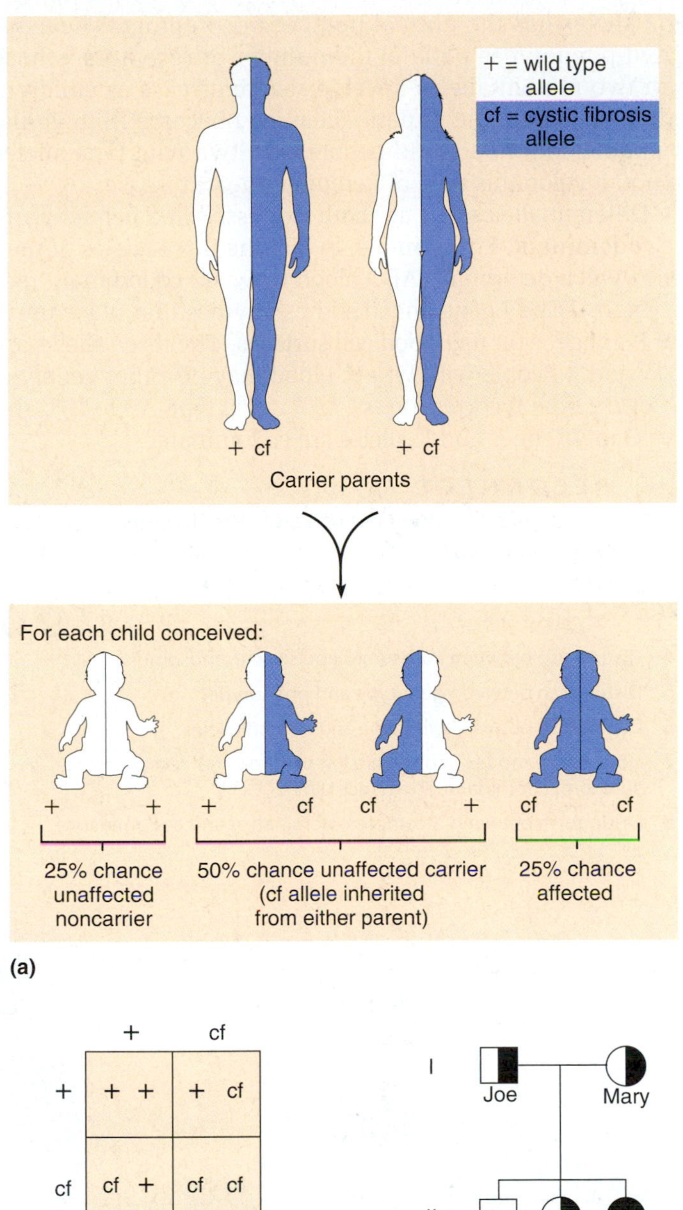

FIGURE 24.4 Inheritance of cystic fibrosis from carrier parents illustrates autosomal recessive inheritance. (*a*) Each child has a 25% chance of being unaffected and not a carrier, a 50% chance of being an unaffected carrier, and a 25% chance of being affected. Sexes are affected with equal frequency. A Punnett square (*b*) and a pedigree (*c*) are other ways of depicting this information. Symbols in the pedigree with both black and white indicate unaffected carriers (heterozygotes). The pedigree illustrates the makeup of one possible family. Each child is an independent event. The three siblings could have any combination of genotypes (allele combinations).

A **pedigree** is a diagram that depicts family relationships and known genotypes and phenotypes. Circles are females and squares are males; shaded-in symbols represent people who have a trait or condition; half-shaded symbols denote carriers. Roman numerals indicate generations. Figures 24.4 and 24.5 show Punnett squares and pedigrees.

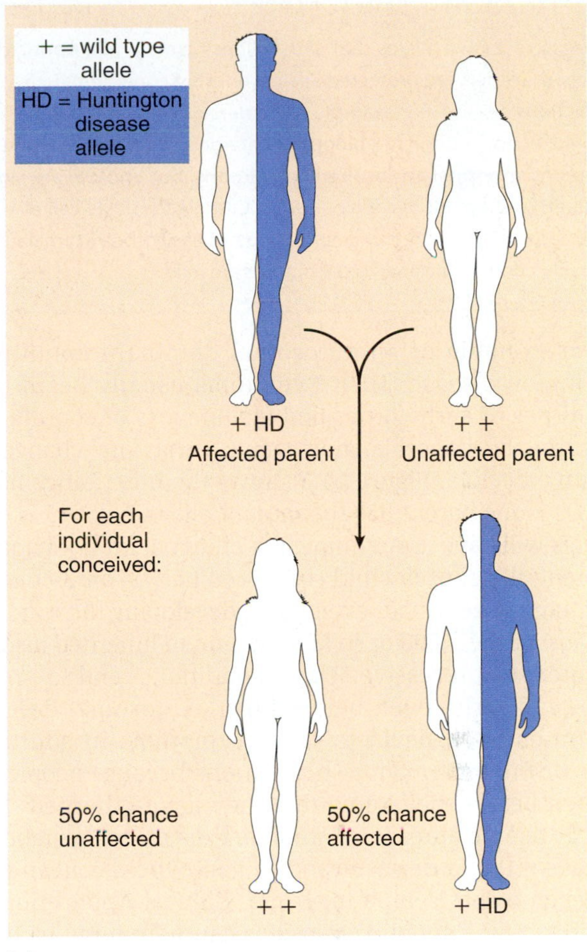

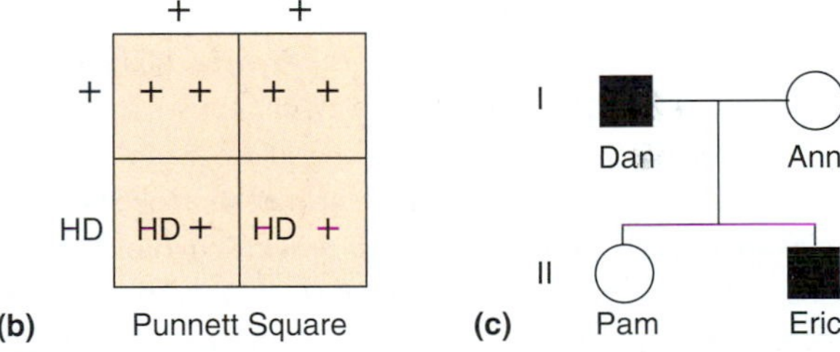

FIGURE 24.5 Inheritance of Huntington disease from a parent who will be affected in middle age illustrates autosomal dominant inheritance. (*a*) A person with just one HD allele develops the disease. A Punnett square (*b*) and pedigree (*c*) depict the inheritance of HD. The pedigree symbols for HD are completely filled in to indicate that the person is affected. Autosomal dominant conditions affect both sexes.

In an autosomal recessive illness, an affected person's parents are usually carriers—they do not have the illness. Or, if the phenotype is mild, a parent might be homozygous recessive and affected. In an autosomal dominant condition, an affected person typically has an affected parent. He or she need inherit only one copy of the mutant allele to have the associated phenotype; in contrast, expression of an autosomal recessive condition requires two copies of the mutant allele.

Certain recessive alleles that cause illness remain in a population because carriers are protected against another condition, such as an infectious disease. For example, in carriers of sickle cell disease (see fig. 4.24), not enough red blood cells are deformed to block circulation under normal atmospheric conditions, but enough are sickle shaped to keep out malaria parasites. Carriers for sickle-cell disease and certain other inherited anemias do not easily contract malaria. If they do contract malaria, the symptoms are mild.

An example of an autosomal dominant condition is Huntington disease (HD). Symptoms usually begin in the late thirties or early forties and include loss of coordination, uncontrollable dancelike movements, behavioral changes and cognitive decline. Figure 24.5 shows the inheritance pattern for HD. If one parent has the mutant allele, half of his or her gametes will have it. Assuming the other parent does not have a mutant allele, each child conceived has a 1 in 2 chance of inheriting the gene and, eventually, developing the condition.

Most of the 3,000 or so known human inherited disorders are autosomal recessive. These conditions tend to produce symptoms early, even before birth. Autosomal dominant conditions often begin to cause symptoms in adulthood. These disorders remain in populations because people have children before they know that they have inherited the illness. For some autosomal dominant disorders, genetic tests can reveal that a disease-causing genotype has been inherited, even before symptoms begin. Clinical Application 24.1 explores how a genetic counselor can help families understand the implications of different modes of inheritance.

Gene therapy, still experimental, alters, replaces, silences, or augments a gene's function to prevent, delay, or improve specific symptoms. In humans gene therapy targets only affected somatic cells and therefore cannot be transmitted to the next generation. A gene therapy for an inherited bleeding disorder provides genes that encode deficient, inactive, or missing clotting factors. Delivering gene therapy is challenging. To treat CF, for example, a wild type copy of the CFTR gene is inhaled; for muscular dystrophy, a wild type dystrophin gene is implanted in muscle. Many gene therapies deliver healing genes in viruses, which themselves can cause problems.

Different Dominance Relationships

Most genes exhibit complete dominance or recessiveness. Interesting exceptions are incomplete dominance and codominance.

In **incomplete dominance,** the heterozygous phenotype is intermediate between that of either homozygote. For example, in familial hypercholesterolemia (FH), a person with two disease-causing alleles completely lacks LDL (low-density lipoprotein) receptors on liver cells that take up cholesterol from the bloodstream (fig. 24.6). A person with one disease-causing allele (a heterozygote) has half the normal number of cholesterol receptors. Someone with two wild type alleles has the normal number of receptors. The associated phenotypes parallel the number of receptors—those with two mutant alleles develop severe disease as children, individuals with one mutant allele may become ill in young or middle adulthood, and people with two wild type alleles do not develop this type of hereditary heart disease.

Different alleles that are both expressed in a heterozygote are **codominant.** For example, two of the three alleles of the *I* gene, which determines ABO blood type, are codominant (see fig. 14.21). The *I* gene encodes the enzymes that place the A and B antigens on red blood cell surfaces. The three alleles are I^A, I^B, and i. People with type A blood may be either genotype I^AI^A or I^Ai; type B corresponds to I^BI^B or I^Bi; type AB to I^AI^B; and type O to ii. The I^A and I^B alleles are codominant.

RECONNECT

To Chapter 14, Blood Groups and Transfusions, pages 544–546.

PRACTICE

4 Distinguish between autosomes and sex chromosomes.
5 Distinguish between genotype and phenotype.
6 Distinguish between wild type and mutant alleles.
7 How do the modes of transmission of autosomal recessive and autosomal dominant inheritance differ?
8 Distinguish between incomplete dominance and codominance.

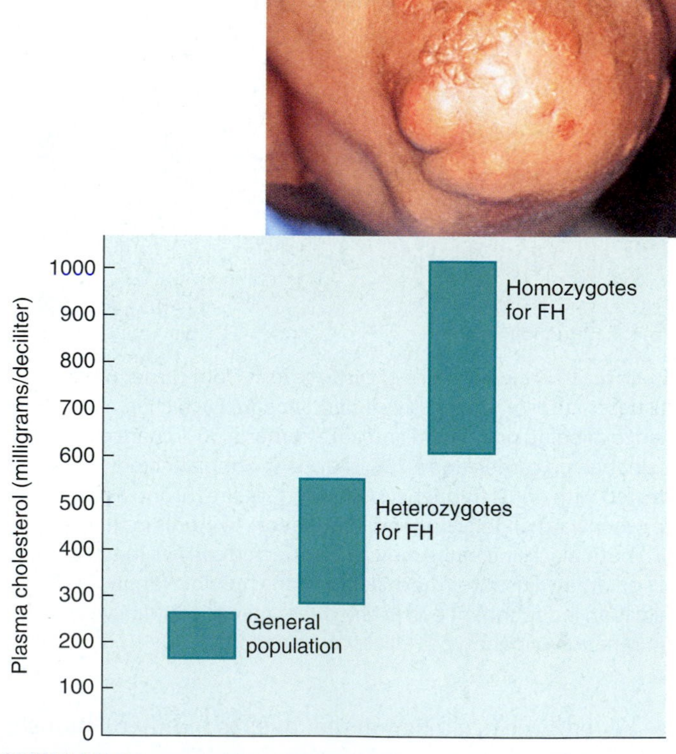

FIGURE 24.6 Incomplete dominance appears in the plasma cholesterol levels of heterozygotes and homozygotes for familial hypercholesterolemia (FH). This condition is one of many that increase the cholesterol level in the blood, raising the risk of developing cardiovascular disease. The photograph shows cholesterol deposits on the elbow of a young man who is a homozygote for the disease-causing allele.

24.1 CLINICAL APPLICATION

Genetic Counselors Communicate Modes of Inheritance

Genetic counselors are medical professionals who help families understand the implications of an inherited illness and guide them in obtaining tests, treatments, and services (fig. 24A). The tools of the genetic counselor include pedigree charts, family and medical histories, communication skills, and compassion. For disorders caused by a single gene, the counselor must explain the mode of inheritance—as the following real cases demonstrate.

Autosomal Recessive Inheritance: Cystic Fibrosis

Molly and Russell received unexpected news during her fourth month of pregnancy—a routine blood test had revealed that she was a carrier of CF. She had one wild type copy of the gene and one mutant copy. That allele was rare and associated with mild disease, the genetic counselor explained. Did Molly or any of her blood relatives have frequent respiratory problems? Molly thought a moment. Her mother and maternal grandmother seemed to have bronchitis nearly every winter, and just last year her mother had been hospitalized with pneumonia. One of her brothers had frequent colds, and she herself had always suffered from sinusitis.

Molly's family history was consistent with CF. The next step to better determine risk to the fetus was to have Russell tested. If he was a carrier, too, then the fetus faced a one in four chance of having inherited CF. The severity would depend upon which allele Russell had. Russell was homozygous for the wild type allele. Had he carried a mutation too, further testing could have diagnosed the fetus. CF testing is routine for pregnant women and newborns are screened for the disease.

Autosomal Dominant Inheritance: Huntington Disease

Stuart had completed several sessions with his genetic counselor and felt he was ready to take the predictive test for Huntington Disease (HD). It hadn't been an easy decision. At age thirty, he was able to see the inheritance pattern of his family's illness with tragic clarity—his grandmother had recently died after a fifteen-year battle with the disease, his father lived in a nursing home where his uncontrollable movements could be controlled, and his older brother Mike was just starting to show symptoms. Inherited in an autosomal dominant manner, HD strikes every generation and affects males and females. If Stuart had the mutation—a "triplet repeat" in which a sequence of DNA, CAG, is present in many extra copies—he, too, would develop the behavioral and cognitive changes and motor symptoms of HD. He was about to get married and would use the information in his decision about having children.

X-Linked Recessive Inheritance: Duchenne Muscular Dystrophy

Malcolm had always been a clumsy child, but when he started tripping more often at the age of 4, his mother Emelda took him to a pediatrician. The doctor observed Malcolm walking, noticing his enlarged calf muscles, and also had the boy rise from a seated position. Rather than standing directly, he pushed off with his hands. He'd been standing that way for so long that Emelda had thought it normal. The doctor thought otherwise and referred the entire family to a muscular dystrophy clinic at a nearby medical center.

Genetic testing revealed that the gene that encodes a protein called dystrophin was absent—Malcolm had Duchenne muscular dystrophy. His muscle cells lack a protein required to withstand the force of contraction. His skeletal muscles were already affected, and he would be in a wheelchair by age ten. Then his cardiac and smooth muscle cells might be affected too.

The family met with a genetic counselor. She asked questions about Malcolm's cousins on his mother's side, but they were all girls. The counselor explained that Duchenne muscular dystrophy is caused by a mutation in a gene on the X chromosome. It passes from carrier mothers, such as Emelda, to affected sons. Malcolm's father George could not have transmitted the gene, because he passed his Y chromosome to his son, not his X. The counselor told Emelda and George that they could have their younger son Thad tested. He, too, tripped often, but the parents had attributed his clumsiness to imitating his brother. Thad did not have the mutation. ■

Reasons to seek genetic counseling:

- Family history of abnormal chromosomes
- Elevated risk of single gene disorder
- Family history of multifactorial disorder
- Family history of cancer

↓

Genetic counseling sessions:

- Family history
- Pedigree construction
- Information provided on specific disorders, modes of inheritance, tests to identify at-risk family members
- Testing arranged, discussion of results
- Links to support groups, appropriate services
- Follow-up contact

FIGURE 24A The genetic counseling process.

24.3 FACTORS THAT AFFECT EXPRESSION OF SINGLE GENES

Most genotypes vary somewhat from person to person, due to the effects of the environment and other genes. Even identical twins may not exhibit the symptoms of an inherited illness in the exact same way. The terms penetrance, expressivity, and pleiotropy are used to describe some of these distinctions of genotype.

Penetrance and Expressivity

Some disease-causing allele combinations are **completely penetrant,** which means that everyone who inherits a particular genotype has some symptoms. A genotype is **incompletely penetrant** if some individuals do not express the associated phenotype. Polydactyly, having extra fingers or toes, is incompletely penetrant (see fig. 7.46). Some people who inherit the autosomal allele have more than five digits on a hand or foot, yet others known to have the allele (because they have an affected parent and child) have ten fingers and ten toes.

The penetrance of a gene is described numerically. If 80 of 100 people who have inherited the dominant polydactyly allele have extra digits, the allele is 80% penetrant. However, incomplete penetrance is difficult to detect because it describes the absence of a phenotype.

A phenotype is **variably expressive** if the symptoms vary in intensity in different people—which is nearly always the case. One person with polydactyly might have an extra digit on both hands and a foot; another might have two extra digits on both hands and both feet; a third person might have just one extra fingertip. Penetrance refers to the all-or-none expression of a genotype in an individual; expressivity refers to the severity of a phenotype. Polydactyly is both incompletely penetrant and variably expressive.

Pleiotropy

A single genetic disorder can produce several symptoms, a phenomenon called **pleiotropy** (ple′o-tro-pe). Family members who have different symptoms can appear to have different illnesses.

Pleiotropy is seen in genetic diseases that affect a single protein found in different parts of the body. This is the case for Marfan syndrome, an autosomal dominant defect in an elastic connective tissue protein called fibrillin. The protein's abundance in the lens of the eye, in the bones of the limbs, fingers, and ribs, and in the aorta explains the symptoms of lens dislocation, long limbs, spindly fingers, and a caved-in chest. The most serious symptom is a life-threatening weakening in the aorta wall, which sometimes causes the vessel to suddenly burst. If the weakening is found early, a synthetic graft can be used to patch that part of the vessel wall, saving the person's life. Clinical Application 14.1 (p. 530) discusses a pleiotropic disorder that left its mark on American history, porphyria variegata.

Genetic Heterogeneity

The same phenotype resulting from the actions of different genes is called **genetic heterogeneity** (jĕ-net′ik het″er-o-je-ne′ĭ-te). For example, the nearly 200 forms of hereditary deafness are each due to impaired actions of a different gene. Different genes affect different aspects of hearing.

Genetic heterogeneity occurs when genes encode different enzymes that catalyze the same biochemical pathway, or encode different proteins that are part of the pathway. For example, eleven biochemical reactions lead to blood clot formation. Clotting disorders may result from mutations in the genes that specify any of the enzymes that catalyze these reactions, leading to several types of bleeding disorders.

> A couple was arrested on suspicion of child abuse because their toddler suffered repeated broken bones. Tests for the gene known to cause osteogenesis imperfecta, "brittle bone disease," which can break bones even in a fetus, revealed that the parents were wild type. They were found guilty. However, discovery of a second gene that causes osteogenesis imperfecta enabled them to be tested again, for this gene, and they were exonerated. Their child had inherited a rare form of the disease.

PRACTICE 

9 Distinguish between penetrance and expressivity.

10 What is pleiotropy?

11 What is genetic heterogeneity?

24.4 MULTIFACTORIAL TRAITS

Most if not all characteristics and disorders considered "inherited" reflect input from the environment as well as genes. Characteristics molded by one or more genes plus the environment are termed **multifactorial traits** or complex traits.

Traits determined by more than one gene are termed **polygenic.** Most polygenic traits are also influenced by the environment. Eye color is an example of a trait that is as close to purely polygenic as we know. Usually, several genes each contribute to differing degrees toward molding an overall phenotype. A polygenic trait, with many degrees of expression because of the input of several genes, is said to be continuously varying. Height, skin color, and eye color are polygenic traits (figs. 24.7, 24.8, and 24.9).

Although the expression of a polygenic trait is continuous, we can categorize individuals into classes and calculate the frequencies of the classes. When we do this and

(a)

(b)

FIGURE 24.7 Previous editions of this (and other) textbooks have used the photograph in (*a*) to illustrate the continuously varying nature of height. In the photo, taken around 1920, 175 cadets at the Connecticut Agricultural College lined up by height. In 1997, Professor Linda Strausbaugh asked her genetics students at the school, today the University of Connecticut at Storrs, to re-create the scene (*b*). They did, and confirmed the continuously varying nature of human height. But they also elegantly demonstrated how height increased during the twentieth century, largely due to improved nutrition. The tallest people in the old photograph (*a*) are 5′9″ tall, whereas the tallest people in the more recent photograph (*b*) are 6′5″ tall.

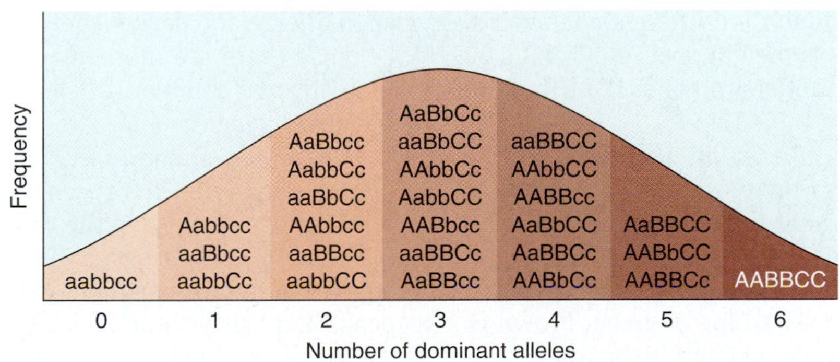

FIGURE 24.8 Variations in skin color. A model of three genes, with two alleles each, can explain some of the hues of human skin. In actuality, this trait likely involves many more than three genes. The mid-range colors are more common.

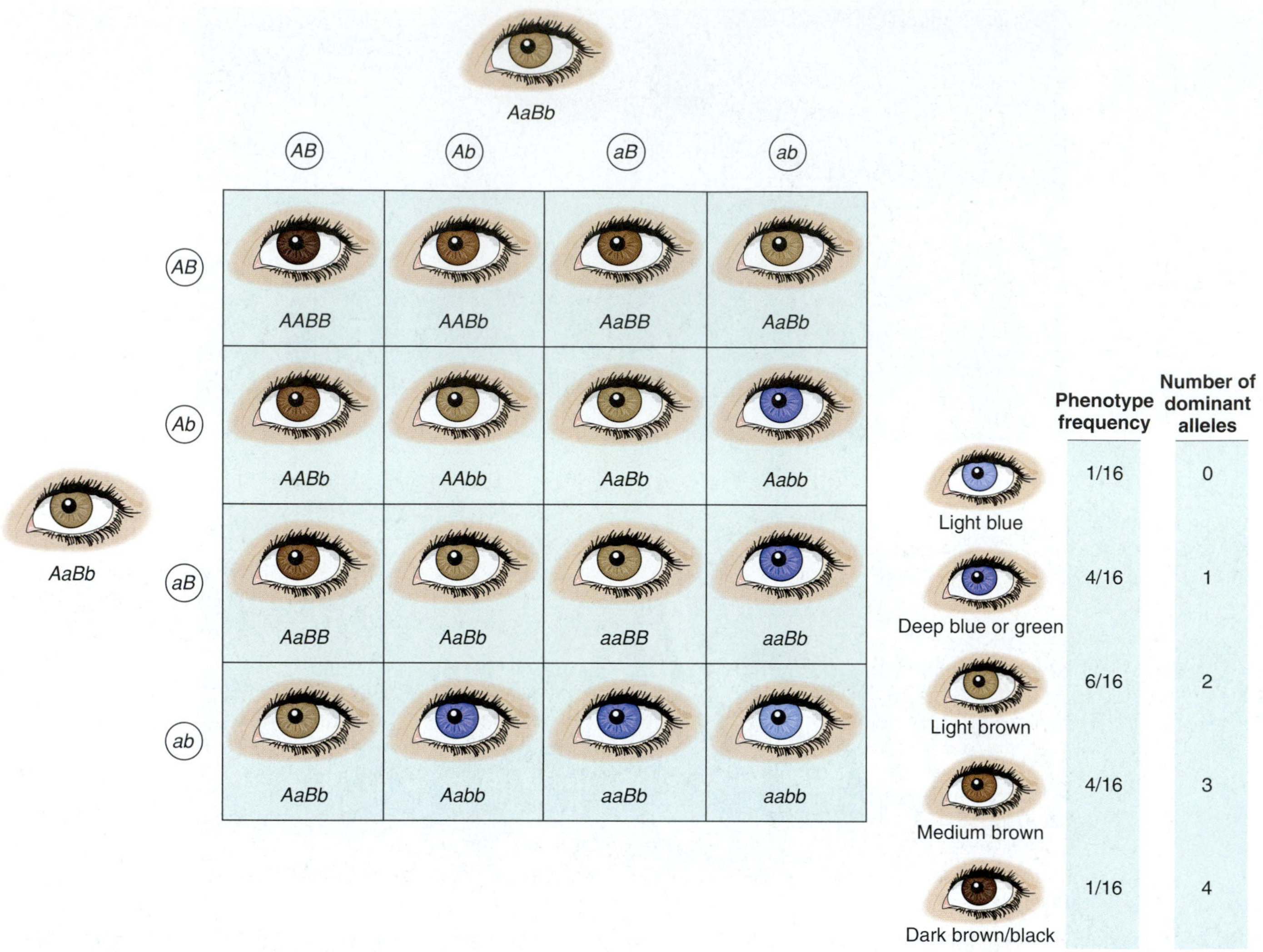

FIGURE 24.9 Variations in eye color. A model of two genes, with two alleles each, can explain five human eye colors. If eye color is controlled by two genes *A* and *B*, each of which comes in two allelic forms *A* and *a* and *B* and *b*, then the lightest color would be genotype *aabb;* the darkest, *AABB.*

plot the frequency for each phenotype class, a bell-shaped curve results. This curve indicating continuous variation of a polygenic trait is strikingly similar for different characteristics, such as fingerprint patterns, height, eye color, and skin color. Even when different numbers of genes contribute to the phenotype, the curve is the same shape.

Eye color illustrates how interacting genes can mold a single trait. The colored part of the eye, the iris, darkens as melanocytes produce the pigment melanin. Unlike melanin in skin melanocytes, the pigment in the eye tends to stay in the cell that produces it. Blue eyes have just enough melanin to make the color opaque, and dark blue or green, brown or black eyes have increasingly more melanin in the iris.

Two genes (*OCA2* and *HERC2*) control melanin synthesis and deposition. The alleles of the two genes interact additively, producing distinct eye colors. Figure 24.9 depicts how this might happen to account for five distinct eye colors—light blue, deep blue or green, light brown, medium brown, and dark brown/black. If each dominant allele contributes a certain amount of pigment, then the greater the number of such alleles, the darker the eye color. The bell curve arises because there are more ways to inherit light brown eyes, with any two dominant alleles, than there are ways to inherit the other colors.

Other genes modify the expression of the two melanin-controlling genes. Two genes add greenish colors, called lipochromes. Overlying the colors and tones are specks and flecks, streaks and rings, and regions of dark versus light that arise from the way pigment is laid down onto the distinctive peaks and valleys at the back of the iris.

Height and skin color are multifactorial as well as polygenic, because environmental factors influence them: good nutrition enables a person to reach the height dictated by genes, and sun exposure affects skin color. Most of the more common illnesses, including heart disease, diabetes mellitus, hypertension, and cancers, are multifactorial.

Studies on well-nourished populations indicate that about 90% of variation in height is due to genetics. In populations where nutrition is inadequate, environmental effects on height—stunting—are more pronounced.

PRACTICE

12 How does polygenic inheritance make possible many variations of a trait?

13 How can two genes specify five phenotypes?

14 How may the environment influence gene expression?

24.5 MATTERS OF SEX

Human somatic (nonsex) cells include an X and a Y chromosome in males and two X chromosomes in females. All eggs carry a single X chromosome, and sperm carry either an X or a Y chromosome. Sex is determined at conception: a Y-bearing sperm fertilizing an egg conceives a male, and an X-bearing sperm conceives a female (fig. 24.10). The female is termed the homogametic sex because she has two of the same type of sex chromosome, and the human male is called the heterogametic sex because his two sex chromosomes are different. This is not the case for all types of animals. In birds, for example, the female is the heterogametic sex.

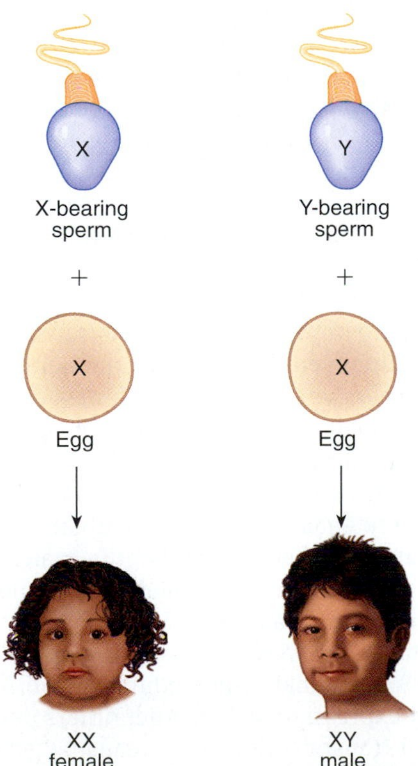

FIGURE 24.10 Sex determination. An egg contributes an X chromosome, and a sperm, either an X or a Y. If an X-bearing sperm fertilizes an egg, the zygote is female (XX). If a Y-bearing sperm fertilizes an egg, the zygote is male (XY). A gene on the Y chromosome, *SRY*, determines sex.

Sex Determination

Maleness derives from a *Y* chromosome gene called *SRY*, for sex-determining region of the *Y*. The *SRY* gene encodes a type of protein called a transcription factor, which switches on other genes. *SRY* activates transcription of genes that direct development of male structures in the embryo, while suppressing formation of female structures. Figure 24.11 shows the sex chromosomes. Absence of the *SRY* transcription factor, plus expression of a gene called *Wnt4*, triggers development of a female body.

Genes on the Sex Chromosomes

Genes that are part of the X and Y chromosomes are inherited in different patterns than are autosomal genes because of the different sex chromosome constitutions of males and females. Traits transmitted on the X chromosome are X-linked, and on the Y, Y-linked. The X chromosome has more than 1,500 genes; the Y chromosome has only 231 protein-encoding genes.

Y-linked genes are considered in three groups, based on their similarity to X-linked genes. One group consists of genes at the tips of the Y chromosome that have counterparts on the X chromosome. These genes encode a variety of proteins that function in both sexes, participating in or controlling such

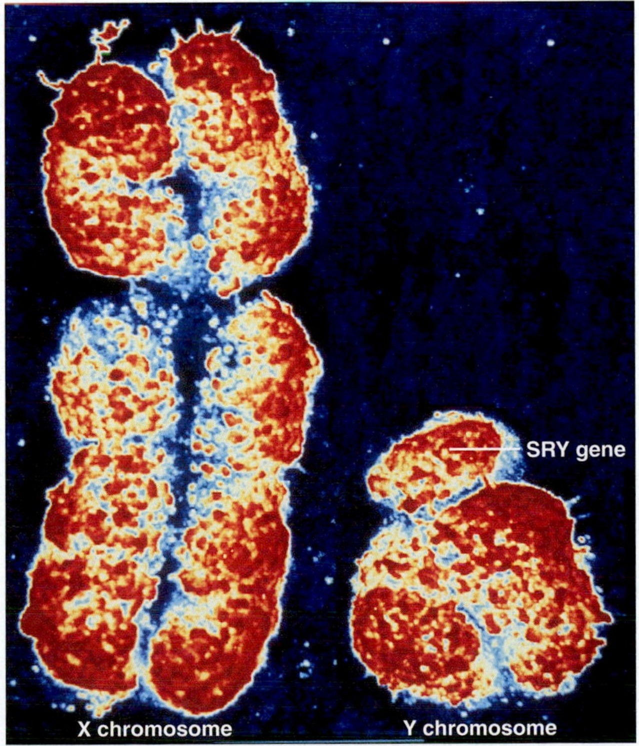

FIGURE 24.11 The X and Y chromosomes. The *SRY* gene, at one end of the short arm of the Y chromosome, starts the cascade of gene activity that directs development of a male (31,000×).

activities as bone growth, signal transduction, the synthesis of hormones and receptors, and energy metabolism. The members of the second functional group of Y chromosome genes are similar in DNA sequence to certain genes on the X chromosome, but they are not identical. These genes are expressed in nearly all tissues, including those found only in males. The third group of genes includes those unique to the Y chromosome. Many of them control male fertility, such as the *SRY* gene. Some cases of male infertility arise from tiny deletions of these parts of the Y chromosome. Other genes in this group encode proteins that participate in cell cycle control; proteins that regulate gene expression; enzymes; and protein receptors for immune system biochemicals.

Y-linked genes are transmitted only from fathers to sons, because only males have Y chromosomes. The differences in inheritance patterns of X-linked genes between females and males result from the fact that any gene on the X chromosome of a male is expressed in his phenotype, because he has no second allele on a second X chromosome to mask its expression. The human male is **hemizygous** for X-linked traits because he has only one copy of each X chromosome gene. *Red-green colorblindness* and the most common form of the clotting disorder *hemophilia* are recessive X-linked traits.

A male always inherits his Y chromosome from his father and his X chromosome from his mother. A female inherits one X chromosome from each parent. If a mother is heterozygous for a particular X-linked gene, then her son has a 50% chance of inheriting either allele from her. X-linked genes are therefore passed from mother to son. A male does not receive an X chromosome from his father (he inherits the Y chromosome from his father), so an X-linked trait is not passed from father to son.

Consider the inheritance of hemophilia A. It is passed from carrier mother to affected son with a risk of 50%, because he can inherit either her normal allele or the mutant one. A daughter has a 50% chance of inheriting the hemophilia allele and being a carrier like her mother and a 50% chance of not inheriting the allele.

The contribution of the X chromosome is equal in males and females, even though males have one X and females have two, because one X is silenced in every somatic cell of a female mammal. A female is a mosaic, with genes from her father's X chromosome expressed in some cells, and genes from her mother's in others. This X inactivation occurs at random and is detectable for some genes. A woman who is a carrier (a heterozygote) for *Duchenne muscular dystrophy*, for example, has a wild type allele for the dystrophin gene on one X chromosome and a mutant allele on the other. Cells in which the X chromosome bearing the wild type allele is inactivated do not produce the dystrophin protein. However, cells in which the mutant allele is inactivated produce dystrophin. When a stain for dystrophin is applied to a sample of her muscle tissue, only some cells may turn blue, revealing her carrier status. If by chance many wild type dystrophin alleles are turned off in her muscle cells, she may experience mild muscle weakness.

A daughter can inherit an X-linked recessive disorder or trait if her father is affected and her mother is a carrier. She inherits one affected X chromosome from each parent. Without a biochemical test, though, a woman would not know that she is a carrier of an X-linked recessive trait unless she has an affected son.

For X-linked recessive traits that seriously impair health, affected males may not feel well enough to have children. A female affected by an X-linked trait must inherit the mutant allele from a carrier mother and an affected father, so such traits that are nearly as common among females as males tend to be those associated with milder phenotypes. Colorblindness is a mild X-linked trait—men who are colorblind are as likely to have children as men with full color vision.

Dominant disease-causing alleles on the X chromosome are rarely seen. Males are usually much more severely affected than females, who have a second X to offer a protective effect. In a condition called incontinentia pigmenti, for example, an affected girl has swirls of pigment in her skin where melanin in the epidermis extends into the dermis. She may have abnormal teeth, sparse hair, visual problems, and seizures. However, males inheriting the dominant gene on their X chromosomes are so severely affected that they do not survive to be born.

Gender Effects on Phenotype

Certain autosomal traits are expressed differently in males and females, due to differences between the sexes.

A **sex-limited trait** affects a structure or function present in only males or only females. Such a gene may be X-linked or autosomal. Beard growth and breast size are sex-limited traits. A woman cannot grow a beard because she does not manufacture sufficient hormones required for facial hair growth, but she can pass to her sons the genes that specify heavy beard growth. In animal breeding, milk yield and horn development are important sex-limited traits.

In **sex-influenced inheritance,** an allele is dominant in one sex but recessive in the other. Again, such a gene may be X-linked or autosomal. This difference in expression reflects hormonal differences between the sexes. For example, a gene for hair growth pattern has two alleles, one that produces hair all over the head and another that causes pattern baldness. The baldness allele is dominant in males but recessive in females, which is why more men than women are bald. A heterozygous male is bald, but a heterozygous female is not. A bald woman would have two mutant alleles.

About 1% of human genes exhibit **genomic imprinting,** in which the expression of a disorder differs depending upon which parent transmits the disease-causing gene or chromosome. The phenotype may differ in degree of severity, in age of onset, or even in the nature of the symptoms. The physical basis of genomic imprinting is that methyl ($—CH_3$) groups are placed on the gene inherited from one parent, preventing it from being transcribed and translated.

PRACTICE

15 Which chromosomes and genes determine sex?

16 What are the three functional classes of genes on the Y chromosome?

17 Why do X-linked recessive conditions appear most commonly in males?

18 How can gender affect gene expression?

24.6 CHROMOSOME DISORDERS

Deviations from the normal human chromosome number of 46 produce syndromes because of the excess or deficit of genes. Rearrangement of chromosomes, such as an inversion of a section of a chromosome, or two nonhomologous chromosomes exchanging parts, may also cause symptoms. This may happen if the rearrangement disrupts a vital gene or if it results in "unbalanced" gametes that contain too little or too much genetic material. Chromosome number abnormalities may involve single chromosomes or entire sets of chromosomes.

Polyploidy

The most drastic upset in chromosome number is an entire extra set, a condition called **polyploidy** (pol′e-ploi″de). This results from formation of a diploid, rather than a normal haploid, gamete. For example, if a haploid sperm fertilizes a diploid egg, the fertilized egg is *triploid,* with three copies of each chromosome. Most human polyploids cease developing as embryos or fetuses, but occasionally an infant survives for a few days, with many anomalies. Eight cases of tetraploidy (4 copies of each chromosome) have been reported. One such child, at age 26 months, had severe delayed growth and development, a small head with tiny features, and a heart defect.

Some organs normally have a few polyploid cells, with no adverse effects on health. Liver cells, for example, may be tetraploid or even octaploid (8 chromosome sets). Polyploidy is common in flowering plants and is seen in some insects, but it is rare in vertebrates.

Aneuploidy

A normal chromosome number is termed **euploid** (u′ploid). Cells missing a chromosome or having an extra one are **aneuploid** (an′u-ploid). Aneuploidy results from a meiotic error called **nondisjunction** (non″dis-jungk′shun) (fig. 24.12). In normal meiosis, pairs of homologous chromosomes separate, and each of the resulting gametes contains only one member of each pair. In nondisjunction, a chromosome pair fails to separate, either at the first or at the second meiotic division, producing a sperm or egg that has two copies of a particular chromosome or none, rather than the normal one copy. When such a gamete fuses with its mate at fertilization, the resulting zygote has either 47 or 45 chromosomes, instead of the normal 46.

Symptoms that result from aneuploidy reflect which chromosome is missing or extra. Autosomal aneuploidy often results in mental retardation, indicating that many genes affect brain function. Sex chromosome aneuploidy is less severe. Extra genetic material is apparently less dangerous than missing material, and this is why most children born with the wrong number of chromosomes have an extra one, called a **trisomy** (tri′so-me), rather than a missing one, called a **monosomy** (mon′o-so-me).

Aneuploid conditions have historically been named for the researchers or clinicians who identified them, but today chromosome designations are preferred because they are more precise. Down syndrome, for example, refers to a distinct set of symptoms usually caused by trisomy 21. It is the most common autosomal aneuploid. However, the syndrome may also arise from one copy of chromosome 21 exchanging parts with a different chromosome in a gamete and the fertilized ovum receiving excess chromosome 21 material. This is a type of chromosomal aberration called a *translocation.* Knowing whether an individual has trisomy 21 or translocation Down syndrome is important, because the probability of trisomy 21 recurring in a sibling is about 1 in 100, but the chance of translocation Down syndrome recurring is considerably greater. Clinical Application 24.2 takes a closer look at trisomy 21.

Trisomies 13 and 18 are the next most common autosomal aneuploids and usually result in miscarriage. An infant with trisomy 13 has an underdeveloped face, extra and fused fingers and toes, heart defects, small adrenal glands, and a cleft lip or palate. An infant with trisomy 18 suffers many of the problems seen in trisomy 13, plus a peculiar positioning of the fingers and flaps of extra abdominal skin called a "prune belly."

Table 24.1 indicates the rarity of trisomies 13, 18, and 21 and that it is rarer still for an affected newborn to survive infancy. Trisomies of the other autosomes do not develop beyond the embryonic period.

Sex chromosome aneuploids are less severely affected than are autosomal aneuploids. XO syndrome (Turner syndrome) affects 1 in 2,000 newborn girls, but these represent only 1% of XO conceptions. Often the only symptom is a lag in sexual development, and with hormone supplements, life can be fairly normal, except for infertility.

About 1 in every 1,000 to 2,000 females has an extra X chromosome in each cell, a condition called triplo-X. Often the only associated characteristics are great height and menstrual irregularities. Males with an extra X chromosome have XXY syndrome (Klinefelter syndrome). Like XO females, many XXY males typically do not realize they have an unusual number of chromosomes until they encounter fertility problems and their chromosomes are checked. Associated characteristics are sexual underdevelopment (rudimentary testes and prostate glands and no pubic or facial hair), growth of breast tissue, long limbs, and large hands and feet. XXY syndrome affects 1 in every 500 to 2,000 male births.

One male in 1,000 has an extra Y chromosome, called XYY syndrome, or Jacobs syndrome. Until 1974, the extra chromosome was linked to criminal behavior, because the

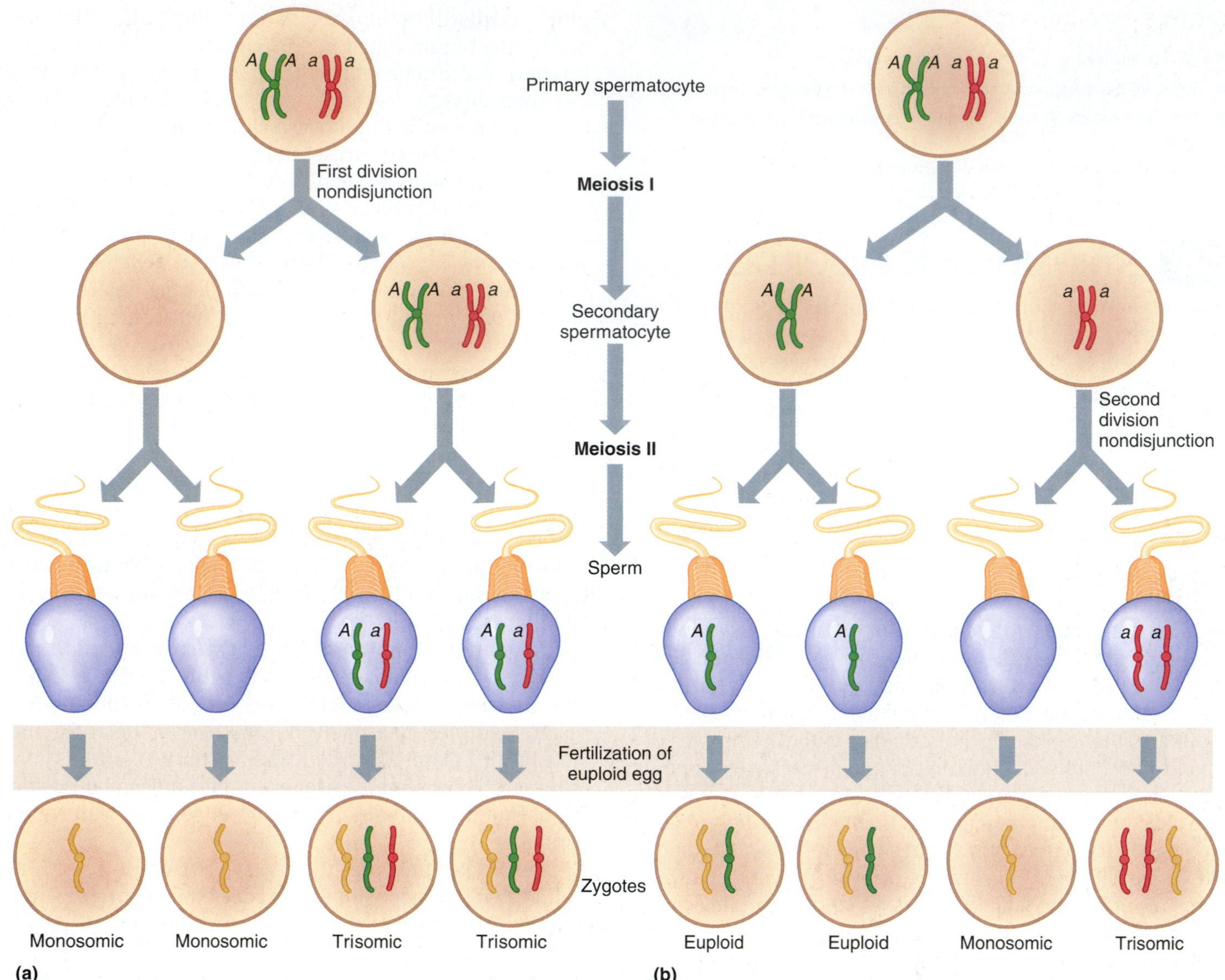

FIGURE 24.12 Extra or missing chromosomes constitute aneuploidy. Unequal division of chromosome pairs into sperm and egg cells can occur at either the first or the second meiotic division. (*a*) A single pair of chromosomes is unevenly partitioned into the two cells arising from the first division of meiosis in a male. The result: two sperm cells that have two copies of the chromosome and two sperm cells that have no copies of that chromosome. When a sperm cell with two copies of the chromosome fertilizes a normal egg cell, the zygote produced is trisomic for that chromosome; when a sperm cell lacking the chromosome fertilizes a normal egg cell, the zygote is monosomic for that chromosome. Symptoms depend upon which chromosome is involved. (*b*) This nondisjunction occurs at the second meiotic division. The two products of the first division are unaffected, so two of the mature sperm are normal, and two are aneuploid. Egg cells can undergo nondisjunction as well, leading to zygotes with extra or missing chromosomes when they are fertilized by normal sperm cells.

TABLE 24.1 | Comparing and Contrasting Trisomies 13, 18, and 21

Type of Trisomy	Incidence at Birth	Percent of Conceptions That Survive 1 Year After Birth
13 (Patau)	1/12,500–1/21,700	<5%
18 (Edward)	1/6,000–1/10,000	<5%
21 (Down)	1/800–1/826	85%

first studies to detect it were performed on inmates at a high-security mental facility. However, 96% of men with XYY syndrome share only great height, acne, and speech and reading problems. Perhaps teachers, employers, parents, and others may expect more of these physically large boys and men than of their peers, and a small percentage of them cope with this stress by becoming aggressive.

A fertilized ovum that has one Y chromosome and no X chromosome has never been observed. Apparently, when a zygote lacks an X chromosome, so much genetic material is missing that only a few, if any, cell divisions are possible.

24.2 CLINICAL APPLICATION

Down Syndrome

The most common autosomal aneuploid is *trisomy 21*, an extra chromosome 21. The characteristic slanted eyes and flat face of affected individuals prompted Sir John Langdon Haydon Down to coin the inaccurate term "mongolism" when he described the syndrome in 1886. As the medical superintendent of a facility for the profoundly mentally retarded, Down noted that about 10% of his patients resembled people of the Mongolian race. The resemblance is coincidental. Males and females of all races can have the syndrome.

A person with Down syndrome (either trisomy or translocation) is short and has straight, sparse hair and a tongue protruding through thick lips. The eyes slant and have upward "epicanthal" skin folds in the inner corners. Ears are abnormally shaped. The hands have an unusual pattern of creases, the joints are loose, and reflexes and muscle tone are poor. Developmental milestones (such as sitting, standing, and walking) are slow, and toilet training may take several years. Intelligence varies greatly, from profound mental retardation to being able to follow simple directions, read, and use a computer. At least two colleges specialize in educating people with Down syndrome (fig. 24B).

Down syndrome (either type) is associated with many physical problems, including heart or kidney defects, susceptibility to infections, and blockages in the digestive system. An affected child is fifteen times more likely to develop leukemia than a healthy child, but this is still a low figure. Prenatal testing cannot reveal how severely affected an individual with Down syndrome will be.

The likelihood of giving birth to a child with trisomy 21 Down syndrome increases dramatically with the age of the mother (table 24A). However, 80% of children with trisomy 21 are born to women under age thirty-five, because younger women are more likely to become pregnant and have been less likely to have prenatal testing. About 5% of cases of trisomy 21 can be traced to nondisjunction in the sperm.

The age factor in Down syndrome may be because meiosis in the female is completed after conception. The older a woman is, the longer her oocytes have been arrested on the brink of completing meiosis. During this time, the oocytes may have been exposed to chromosome-damaging chemicals or radiation. Other trisomies are more likely to occur among the offspring of older women, too. In the nineteenth century, when physicians noted that people with Down syndrome were often the youngest in their families, they attributed the condition to "maternal reproductive exhaustion."

Many of the medical problems that people with Down syndrome suffer are treatable, so life expectancy is now fifty-five years. In 1910, life expectancy was only to age nine. ■

FIGURE 24B Many people with Down syndrome can learn, go to school, and hold jobs. This young lady is learning from a chef.

TABLE 24A | Risk of Trisomy 21 Increases with Maternal Age

Maternal Age	Trisomy 21 Risk	Risk for Any Aneuploid
20	1/1,667	1/526
24	1/1,250	1/476
28	1/1,053	1/435
30	1/952	1/385
32	1/769	1/322
35	1/378	1/192
36	1/289	1/156
37	1/224	1/127
38	1/173	1/102
40	1/106	1/66
45	1/30	1/21
48	1/14	1/10

Prenatal Tests Detect Chromosome Abnormalities

Several types of tests performed on pregnant women can identify anatomical or physiological features of fetuses that can indicate a chromosomal problem or detect the abnormal chromosomes (fig. 24.13). An ultrasound scan, for example, can reveal the fusion of the eyes, cleft lip and/or palate, malformed nose, and extra fingers and toes that indicate trisomy 13 (fig. 24.14). A blood test performed on the woman during the fifteenth week of pregnancy detects levels of certain biochemicals in serum. Levels of these maternal serum markers, which include alpha fetoprotein, an estrogen, pregnancy-associated plasma protein A and human chorionic gonadotropin, can indicate an underdeveloped liver, a sign of trisomies 13, 18, and 21. Screening maternal serum markers is routine in the management of pregnancy, and it identifies fetuses at elevated risk.

After a maternal serum marker pattern indicates increased risk, the patient is offered **amniocentesis,** in which a needle is inserted into the amniotic sac and withdraws about 5 milliliters of fluid. Fetal fibroblasts in the

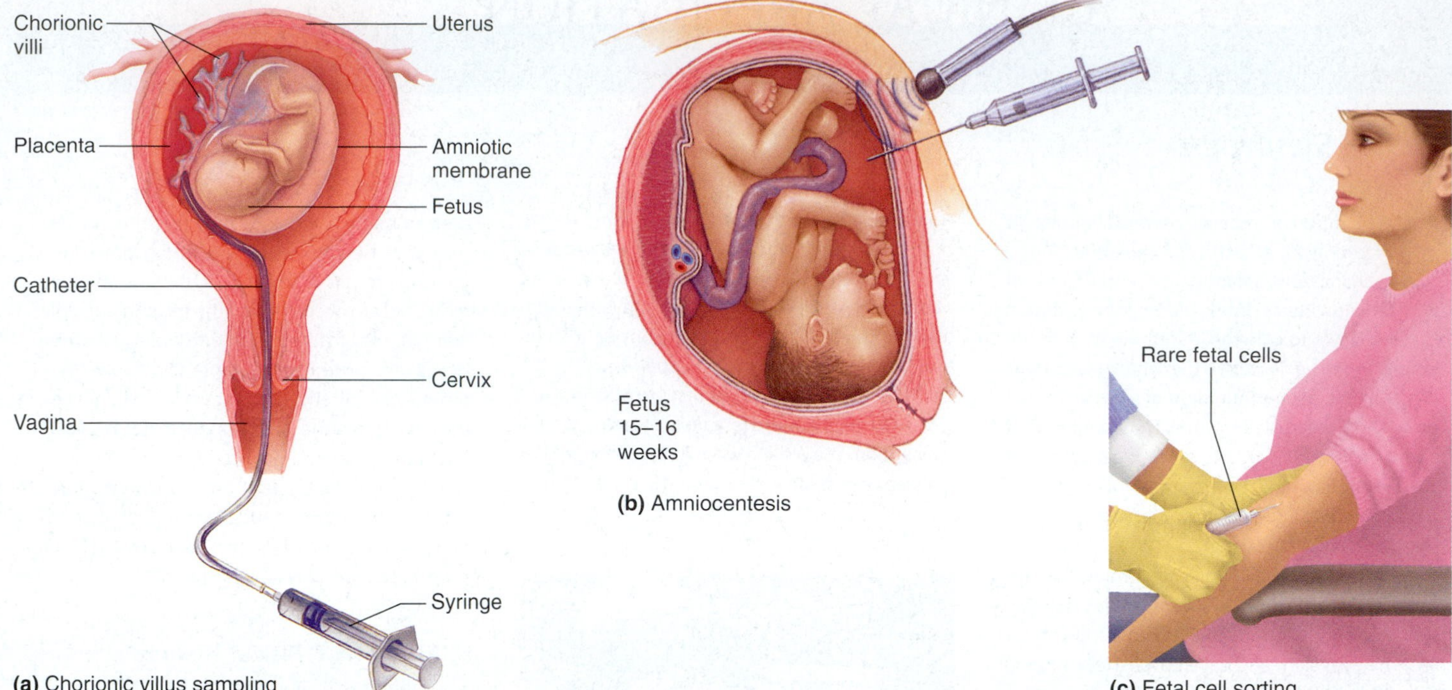

FIGURE 24.13 Three ways to check a fetus's chromosomes. (*a*) Chorionic villus sampling removes cells that would otherwise develop into the placenta. Since these cells descended from the fertilized ovum, they should have the same chromosomes as the fetus. (*b*) Amniocentesis draws out amniotic fluid. Fetal cells shed into the fluid are collected and their chromosomes examined. (*c*) Improved techniques for extracting and identifying specific cells allow researchers to detect fetal cells in a sample of blood from the woman.

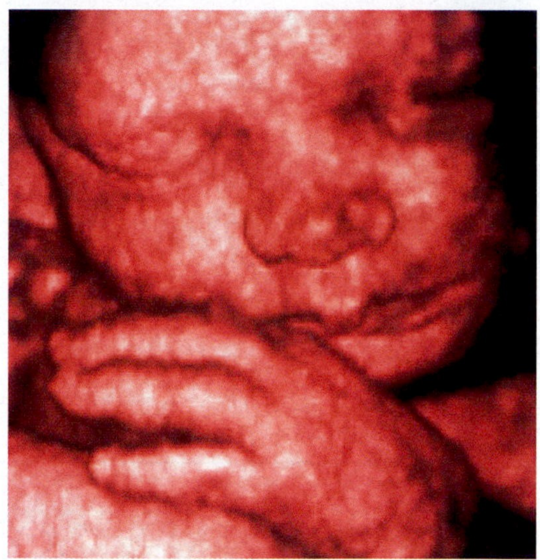

FIGURE 24.14 In an ultrasound exam, sound waves are bounced off the embryo or fetus, and the pattern of deflected sound waves is converted into an image. By thirteen weeks, the face can be discerned. Ultrasound can now provide images that look three-dimensional. Videos of embryos and fetuses are also possible.

sample are cultured and a karyotype constructed, which reveals extra, missing, or translocated chromosomes or smaller anomalies. However, additional tests on amniotic fluid or fibroblast DNA are necessary to detect mutations in individual genes.

The decision to have amniocentesis is based on risk. In the past, the procedure carried about the same risk of being followed by miscarriage as the age-related risk of miscarriage for a thirty-five-year-old without a family history of chromosome abnormalities. This is why the test was offered only to women of "advanced maternal age"—thirty-five—for many years. Recently, however, the safety of the procedure has greatly improved, so that it is now justified for younger pregnant women, too.

Couples who have already had a child with a chromosome abnormality can have **chorionic villus sampling** (CVS), which has the advantage of being performed as early as the tenth week from conception, but carries a higher risk of being followed by miscarriage than does amniocentesis. In CVS, a physician samples chorionic villus cells through the cervix. The basis of the test is that, theoretically, these cells are genetically identical to fetal cells because they too descend from the fertilized ovum. However, sometimes a mutation can occur in a villus cell only, or a fetal cell only, creating a false positive or false negative test result, respectively.

Fetal cell sorting is an experimental prenatal test that separates the rare fetal cells that normally cross the placenta and enter the woman's circulation; then a karyotype is constructed from the sampled cells. It can be performed early in pregnancy, but so far, it is too costly to be widely implemented. Fetal cell sorting is safer than amniocentesis or CVS because it samples only maternal blood, yet it provides the high accuracy of these tests. It is also more accurate than measuring maternal serum markers. Table 24.2 and figure 24.13 summarize the tests used to visualize fetal chromosomes as a window onto health.

PRACTICE

19 Why do deviations from the normal chromosome number of 46 affect health?

20 Distinguish between polyploidy and aneuploidy.

21 How do extra sets of chromosomes or extra individual chromosomes arise?

22 How are fetal chromosomes examined?

24.7 GENE EXPRESSION EXPLAINS ASPECTS OF ANATOMY AND PHYSIOLOGY

This book opened with a look at the deep roots of anatomy and physiology. Today much more recent fields—genetics and the even newer genomics—are adding to what we know about the structure and function of the human body. Specifically, identifying which genes are active and inactive in particular cell types, under particular conditions, can add to our understanding of physiology. This approach goes well beyond the rare, single-gene disorders on which the field of genetics focused for many years. Gene expression monitors the proteins that a cell produces, providing snapshots of physiology in action.

The technology that provides these glimpses of gene function is termed *gene expression profiling,* and identifying the sets of proteins in a cell is **proteomics.** Clinical Applications 3.1 and 3.2 (pp. 82 and 87) describe single gene disorders. In contrast, gene expression profiling considers sets of genes whose functioning underlies cell survival and specialization as well as how cells interact as they respond to the environment and form tissues.

Devices called DNA microarrays (or "DNA chips") are used to reveal the subset of genes expressed in a particular cell type. A microarray is a square of glass or nylon, smaller than a postage stamp, to which DNA molecules of known, short sequences are affixed at known points, creating a grid pattern. A DNA microarray might include selected genes expected to be present in a particular medical condition, or the entire human genome. For example, a DNA microarray used to study cardiovascular disease includes thousands of genes whose protein products control blood pressure; blood clotting; and synthesis, transport, and metabolism of cholesterol and other lipids. Using a whole genome "chip," however, can detect genes not expected to be active in a particular condition.

To track gene expression, a cell type of interest is sampled and separated from its tissue. Its messenger RNA molecules are collected and copied using a special enzyme (reverse transcriptase) into DNA, and during the copying process, a chemical "tag" is included that makes the DNA fluoresce under a laser scanner. Sometimes two samples of cells to be compared are each labeled with a different color. The DNA copies are then added to the DNA microarray. Genetic material from the sample binds complementary sequences embedded in the microarray.

The resulting pattern of fluorescent spots seen with a laser scanner reveals which genes are expressed in the sampled cells. Fluorescence intensity reflects degree of gene expression. Software analyzes the patterns and identifies which genes are turned on or off in a sample. The mRNAs in a differentiated cell type reflect instructions for "housekeeping" proteins essential for all cells, as well as the proteins that provide the cells' particular characteristics, such as contractile proteins in muscle cells or signaling proteins in cells of the nervous or endocrine systems.

RECONNECT
To Chapter 3, Stem and Progenitor Cells, pages 104–105.

RECONNECT
To Chapter 4, Nucleic Acids and Protein Synthesis, pages 131–133.

DNA microarrays that profile gene expression have varied uses in clinical medicine and in basic research. In cancer management, for example, microarray-based gene expression profiles on tissue samples can identify cancer cells very early, when treatment is more likely to work; estimate if

TABLE 24.2 | Prenatal Tests

Procedure	Time (Weeks)	Source	Information Provided
Maternal serum markers	15–16	Maternal blood	Small liver may indicate increased risk of trisomy
Amniocentesis	14–16	Fetal skin, urinary bladder, digestive system cells in amniotic fluid	Karyotype of cell from fetus
CVS	10–12	Chorionic villi	Karyotype of cell from chorionic villus
Fetal cell sorting	Not yet established	Maternal blood	Karyotype of cell from fetus
Ultrasound	Any time	Applied externally or through vagina	Growth rate, head size, size and location of organs

and how quickly the disease will progress; indicate which drugs are likely to be effective and which will likely produce intolerable side effects; and monitor response to treatment. Tests based on gene expression profiling enable physicians to prescribe the antidepressant or cholesterol-lowering drug most effective for a particular patient—a more personalized approach to medicine than trying one drug at a time, based on the fact that it is effective in some people.

The power of gene expression profiling in normal anatomy and physiology, as well as in pathology, comes from comparing sets of mRNAs. A muscle cell from a person with diabetes mellitus expresses different genes (makes different proteins) than a muscle cell from a person whose glucose metabolism lies within the normal range. These differences are expected. Analyzing gene expression can also *discover* new aspects of physiology. This was the case for spinal cord injury. Comparing gene expression profiles in cerebrospinal fluid from individuals who had suffered spinal cord injury to those of individuals who had not been injured revealed activation of the same set of genes in the spinal cord injury patients whose protein products heal injury to the dermis. Looking at the protein profiles in the fluid added to our knowledge of tissue repair.

Gene expression profiling is perhaps most valuable in anatomy and physiology when it enables medical researchers to see distinctions that their eyes cannot detect. Clinical Application 14.2 (p. 536) describe a subtype of leukemia lumped in with a different type for many years, because the cancerous white blood cells looked alike. On a biochemical level however, the cells were different. Once gene expression profiling revealed the distinctions, affected children were given different treatments, and survival increased (fig. 24.15).

Identifying proteins produced in abnormal or injured cells also suggests new drug targets. Consider rheumatoid arthritis (RA), an autoimmune disorder that causes great pain and deformity to joints (see the opening photo to chapter 8, p. 260). Researchers prepared DNA microarrays from cells in joint fluid from pairs of monozygous (identical) twins in which one had RA and the other didn't, indicating that it is not inherited. The affected twins had high levels of three types of mRNA. The encoded proteins made sense: one destroys bone and cartilage; one deactivates the hormone cortisol, whose levels are diminished in RA; and the other stimulates blood vessel formation, which enhances inflammation. Researchers can now test existing drugs and develop new ones against these new targets.

In the coming years, as the human genome continues to be analyzed, detailed portraits of the living chemistry within our cells, and the chemical crosstalk among cells, will continue to be painted of the human body. It is a new view of anatomy and physiology.

PRACTICE

23 State the three types of molecules that gene expression profiling describes.

24 Explain how gene expression profiling provides information about physiology.

25 Discuss how gene expression profiling can be helpful in medical practice.

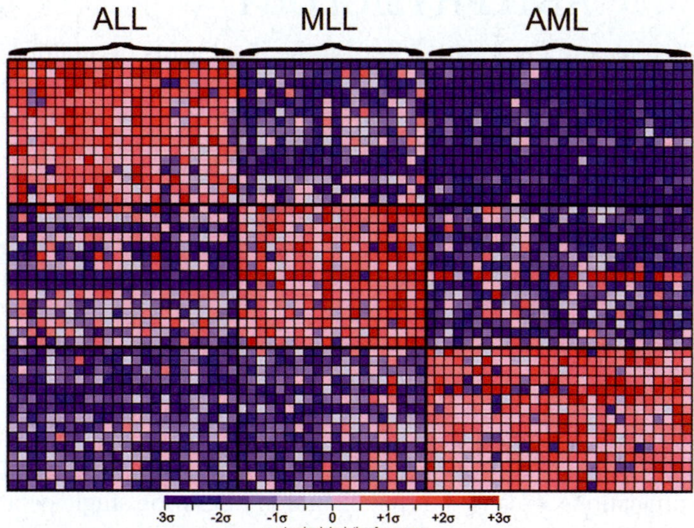

FIGURE 24.15 DNA microarrays reveal a "hidden" type of leukemia. It is easy to see that these three leukemias—ALL, MLL, and AML—differ in their gene expression patterns. The vertical columns of squares represent tumor samples, and the horizontal rows compare the activities of particular genes. Red tones indicate higher-than-normal expression and blue tones show lower-than-normal expression. The different patterns indicate very distinct cancers, although the cells may look alike . Once children with MLL were correctly diagnosed, they were given more effective treatments.

CHAPTER SUMMARY

24.1 INTRODUCTION (PAGE 917)

1. Genes are DNA sequences that encode proteins.
2. Genetics is the study of inheritance of characteristics and human variation.
3. Genetic information is passed from generation to generation through meiosis and fertilization, when the haploid genomes of the parents join.
4. The human genome assembles many more proteins than there are genes by combining gene parts.
5. Genetic information functions at the biochemical, cell, tissue, individual, family, and population levels.
6. Genes influence each other and their expression responds to environmental influences.

24.2 MODES OF INHERITANCE (PAGE 918)

1. Chromosomes and genes come in pairs
   a. Chromosome charts are called karyotypes.
   b. Chromosomes 1 through 22, numbered in decreasing size order, are autosomes. They do not have genes that determine sex.
   c. The X and Y chromosomes are sex chromosomes. They have genes that determine sex.
   d. Chromosomes and the genes they carry are paired.
   e. An allele is an alternate form of a gene. An individual can have two different alleles for a particular gene. The gene itself can have many alleles, because a gene consists of many building blocks, any of which may be altered.
   f. An individual with a pair of identical alleles for a particular gene is homozygous; if the alleles are different, the individual is heterozygous.
   g. The combination of genes present in an individual's cells constitutes a genotype; the appearance of the individual is its phenotype.
   h. A wild type allele provides normal or the most common function. A mutant allele causes disease or an unusual trait; it is a change from the wild type condition.
2. Dominant and recessive inheritance
   a. In the heterozygous condition, an allele expressed when the other is not is dominant. The masked allele is recessive.
   b. Recessive and dominant genes may be autosomal or X-linked or Y-linked.
   c. An autosomal recessive condition affects both sexes and may skip generations. The homozygous dominant and heterozygous individuals have normal phenotypes. The homozygous recessive individual has the condition. The heterozygote is a carrier. An affected individual inherits one mutant allele from each parent.
   d. An autosomal dominant condition affects both sexes and does not skip generations. A person inherits it from one affected parent.
   e. Pedigrees and Punnett squares are used to depict modes of inheritance.
3. Different dominance relationships
   a. In incomplete dominance, a heterozygote has a phenotype intermediate between those of both homozygotes.
   b. In codominance, each of the alleles in the heterozygote is expressed.

24.3 FACTORS THAT AFFECT EXPRESSION OF SINGLE GENES (PAGE 924)

1. Penetrance and expressivity
   a. A genotype is incompletely penetrant if not all individuals inheriting it express the phenotype.
   b. A genotype is variably expressive if it is expressed to different degrees in different individuals.
2. Pleiotropy
   a. A pleiotropic disorder has several symptoms, different subsets of which are expressed among individuals.
   b. Pleiotropy reflects a gene product that is part of more than one biochemical reaction or is in several organs or structures.
3. Genetic heterogeneity
   a. Genetic heterogeneity refers to a phenotype resulting from mutation in more than one gene.
   b. The same symptoms may result from mutations in genes whose products are enzymes in the same biochemical pathway.

24.4 MULTIFACTORIAL TRAITS (PAGE 924)

1. A trait caused by the action of a single gene is monogenic, and by the action of more than one gene, polygenic.
2. A trait caused by the action of one or more genes and the environment is multifactorial.
3. Height, skin color, eye color, and many common illnesses are multifactorial traits.
4. A frequency distribution for a polygenic trait forms a bell curve.

24.5 MATTERS OF SEX (PAGE 927)

A female has two X chromosomes; a male has one X and one Y chromosome. The X chromosome has many more genes than the Y.

1. Sex determination
   a. A male zygote forms when a Y-bearing sperm fertilizes an egg. A female zygote forms when an X-bearing sperm fertilizes an egg.
   b. A gene on the Y chromosome, called *SRY*, switches on genes in the embryo that promote development of male characteristics.
   c. Lack of *SRY* and activation of *Wnt4* directs development as a female.
2. Genes on the sex chromosomes
   a. Genes on the sex chromosomes follow different inheritance patterns than those on autosomes.

b. Y-linked genes are considered in three functional groups: those with counterparts on the X; those similar to genes on the X; and genes unique to the Y, many of which affect male fertility. Y-linked genes pass from fathers to sons.
c. Males are hemizygous for X-linked traits; they can have only one copy of an X-linked gene, because they have only one X chromosome.
d. Females can be heterozygous or homozygous for genes on the X chromosome, because they have two copies of it.
e. A male inherits an X-linked trait from a carrier mother. These traits are more common in males than in females.
f. A female inherits an X-linked mutant gene from her carrier mother and/or from her father if the associated trait does not impair his ability to have children.
g. Dominant X-linked traits are rarely seen because affected males typically die before birth.

3. Gender effects on phenotype
   a. Sex-limited traits affect structures or functions seen in only one sex and may be autosomal.
   b. Sex-influenced traits are dominant in one sex and recessive in the other.
   c. In genomic imprinting, the severity, age of onset, or nature of symptoms varies according to which parent transmits the causative gene.

24.6 CHROMOSOME DISORDERS (PAGE 929)

Extra, missing, or rearranged chromosomes or parts of them can cause syndromes, because they either cause an imbalance of genetic material or disrupt a vital gene.

1. Polyploidy
   a. Polyploidy is an extra chromosome set.
   b. Polyploidy results from fertilization in which one gamete is diploid.
   c. Human polyploids do not survive beyond a few days of birth.
2. Aneuploidy
   a. Cells with the normal chromosome number are euploid. Cells with an extra or missing chromosome are aneuploid.
   b. Aneuploidy results from nondisjunction, in which a chromosome pair does not separate, either in meiosis I or meiosis II, producing a gamete with a missing or extra chromosome. At fertilization, a monosomic or trisomic zygote results.
   c. A cell with an extra chromosome is trisomic. A cell with a missing chromosome is monosomic. Individuals with trisomies are more likely to survive to be born than those with monosomies.
   d. Autosomal aneuploids are more severe than sex chromosome aneuploids.
3. Prenatal tests detect chromosome abnormalities
   a. Ultrasound can detect large-scale structural abnormalities and assess growth.
   b. Maternal serum marker tests indirectly detect a small fetal liver, which can indicate a trisomy.
   c. Amniocentesis samples and examines fetal chromosomes in amniotic fluid.
   d. Chorionic villus sampling obtains and examines chorionic villus cells, which descend from the fertilized egg and therefore are presumed to be genetically identical to fetal cells.
   e. Fetal cell sorting obtains and analyzes rare fetal cells in the maternal circulation.

24.7 GENE EXPRESSION EXPLAINS ASPECTS OF ANATOMY AND PHYSIOLOGY (PAGE 933)

1. Gene expression patterns reveal the identities and activities of proteins in cells (proteomics), which can explain physiological processes.
2. DNA microarrays are used to monitor gene expression.
3. Comparing gene expression profiles for the same cell type under different conditions can provide information on pathology.

CHAPTER ASSESSMENTS

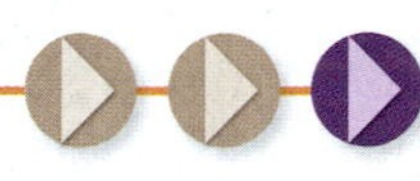

24.1 Introduction

1 Which choice places the structures in order of increasing size? (p. 917)
   a. genome, chromosome, gene, DNA base
   b. DNA base, gene, chromosome, genome
   c. gene, DNA base, genome, chromosome
   d. population, family, individual, organ, tissue, cell, DNA

2 Discuss the origin of the 23 chromosome pairs in a diploid human cell. (p. 917)

3 Explain how a certain number of genes hold enough information to encode a greater number of proteins. (p. 917)

4 Explain how genes can respond to environmental factors. (p. 918)

24.2 Modes of Inheritance

5 Which is a chromosome chart? (p. 918)
   a. karyotype
   b. pedigree
   c. Punnett square
   d. none of the above

6 Distinguish between autosome and sex chromosome, homozygous and heterozygous, phenotype and genotype, mutant and wild type, dominant and recessive, and incomplete dominance and codominance. (p. 919)

7 Explain how a gene might have hundreds of alleles. (p. 919)

8 Which of the following is a mode of inheritance? (p. 920)
   a. autosomal recessive
   b. autosomal dominant
   c. X-linked recessive
   d. all of the above

24.3 Factors That Affect Expression of Single Genes

9 Explain the distinction between penetrance and expressivity. (p. 924)

10 A single gene disorder that produces several symptoms is _______________. (p. 924)

11 A single syndrome that can have more than one genetic cause exhibits _______________. (p. 924)

a. pleiotropy
b. incomplete penetrance
c. genetic heterogeneity
d. variable expressivity

24.4 Multifactorial Traits

12 Define *multifactorial trait.* (p. 924)

13 List three multifactorial traits. (p. 924)

14 Explain why the frequency distributions of different multifactorial traits give very similar bell curves. (p. 926)

15 Give an example of how the environment can influence a multifactorial trait. (p. 926)

24.5 Matters of Sex

16 Explain how genes and chromosomes determine sex. (p. 927)

17 Explain why Y-linked traits are passed only from fathers to sons. (p. 928)

18 Explain why the inheritance pattern of X-linked traits differs in males and females. (p. 928)

19 Distinguish between a sex-limited and a sex-influenced trait. (p. 928)

20 Define *genomic imprinting.* (p. 929)

24.6 Chromosome Disorders

21 State whether trisomy 21 Down syndrome is euploid, aneuploid, or polyploid. (p. 929)

22 In nondisjunction _______________. (p. 929)

a. multiple sets of chromosomes are inherited
b. gametes (sperm and eggs) receive only even-numbered chromosomes or only odd-numbered chromosomes
c. a chromosome pair fails to separate during meiosis, and as a result a gamete has an extra or missing chromosome
d. two eggs fertilize each other

23 Describe three types of prenatal tests. (p. 931)

24.7 Gene Expression Explains Aspects of Anatomy and Physiology

24 Explain how gene expression profiling can add to our knowledge of anatomy and physiology. (p. 933)

25 Explain how gene expression profiling differs from studying mutations in a single gene. (p. 933)

26 List the steps in using gene expression profiling to compare the proteins in a skin cell from an oily part of a face to a cell from a dry part of the same face. (p. 933)

27 Explain how gene expression profiling can improve the efficacy of drugs to treat cancer. (p. 934)

INTEGRATIVE ASSESSMENTS/CRITICAL THINKING

OUTCOMES 4.7, 24.1, 24.2

1. Bob and Joan know from a blood test that they are each heterozygous (carriers) for the autosomal recessive gene that causes sickle cell disease. If their first three children are healthy, what is the probability that their fourth child will have the disease?

OUTCOMES 6.2, 24.4

2. Why are medium-brown skin colors more common than very white or very black skin?

OUTCOMES 6.3, 24.2, 24.3

3. A balding man undergoes a treatment that transfers some of the hair from the sides of his head, where it is still plentiful, to the top. Is he altering his phenotype or his genotype?

OUTCOMES 24.1, 24.2

4. A young couple is devastated when their second child is born and has CF. Their older child is healthy, and no one else in the family has CF. How is this possible?

OUTCOMES 24.1, 24.2, 24.3

5. CF is an autosomal recessive disorder that varies widely in phenotype, from great difficulty in breathing and digesting, to frequent bouts of sinusitis, bronchitis, and/or pneumonia. In the Maxwell family, the mother, Matilida, has a mild case—she often suffers from respiratory infections. Her husband Jake is a heterozygote, and testing reveals that he carries an allele associated with severe disease. Of the four Maxwell children, Katie and Jim have fairly severe disease, although Jim has to be hospitalized more frequently than Katie. The other two children, Emily and Rose, do not have CF.
   a. Explain how Katie and Jim can have severe cases of CF when one parent does not have the disease and the other has a mild case.
   b. Explain how Katie and Jim have different severities of the illness.
   c. What is the probability that Emily or Rose is a carrier, like their father?

OUTCOMES 24.4, 24.7

6. A DNA microarray can predict which individuals who smoke cigarettes are at increased risk of developing lung cancer. Explain how making such a prediction entails genetics and environmental influences.

OUTCOME 24.5

7. In Hunter syndrome, lack of an enzyme leads to build up of sticky carbohydrates in the liver, spleen, and heart. The individual is also deaf and has unusual facial features. Hunter syndrome is inherited as an X-linked recessive condition. Intellect is usually unimpaired, and life span can be normal. A man who has mild Hunter syndrome has a child with a woman who is a carrier (heterozygote).
   a. What is the probability that a son inherits the syndrome?
   b. What is the chance that a daughter inherits it?
   c. What is the chance that a girl would be a carrier?

WEB CONNECTIONS

Be sure to visit the text website at **www.mhhe.com/shier12** for answers to chapter assessments, additional quizzes, and interactive learning exercises.

ANATOMY & PHYSIOLOGY REVEALED

Anatomy & Physiology Revealed® (APR) includes cadaver photos that allow you to peel away layers of the human body to reveal structures beneath the surface. This program also includes animations, radiologic imaging, audio pronunciations, and practice quizzing. Check out **www.aprevealed.com**. APR has been proven to help improve student grades!

APPENDIX A

Periodic Table of Elements

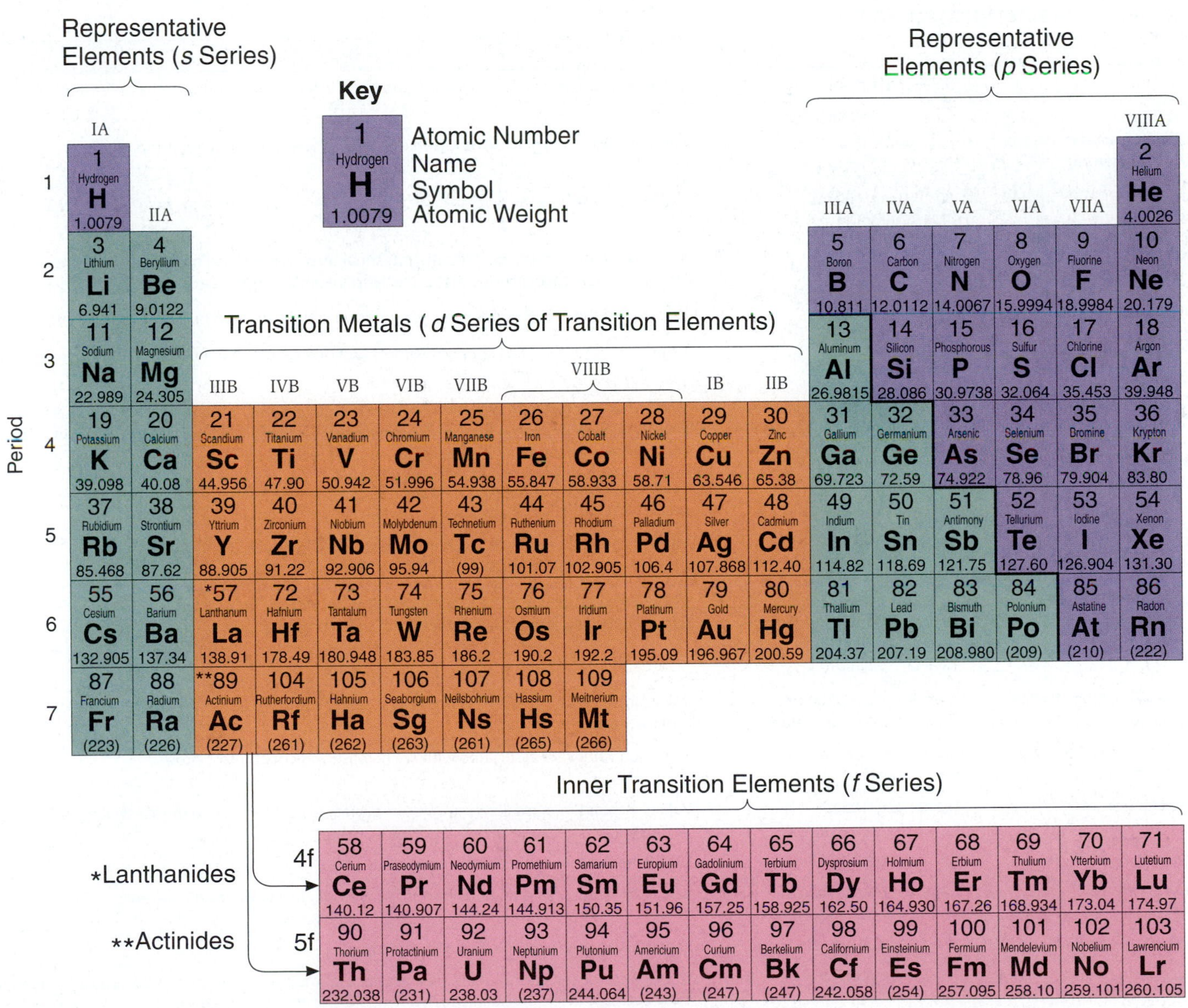

Elements 110 to 114 have been reported in experiments, but have not yet been confirmed.

APPENDIX B

Laboratory Tests of Clinical Importance

Common Tests Performed on Blood

Test	Normal Values* (Adult)	Clinical Significance
Albumin (serum)	3.2–5.5 g/100 mL	Values increase in multiple myeloma and decrease with proteinuria and as a result of severe burns.
Albumin-globulin ratio, or A/G ratio (serum)	1.5:1 to 2.5:1	Ratio of albumin to globulin is lowered in kidney diseases and malnutrition.
Ammonia	80–110 mg/100 mL (12–55 μ mol/L)	Values increase in severe liver disease, pneumonia, shock, and congestive heart failure.
Amylase (serum)	4–25 units/mL	Values increase in acute pancreatitis, intestinal obstructions, and mumps. They decrease in chronic pancreatitis, cirrhosis of the liver, and toxemia of pregnancy.
Bilirubin, total (serum)	0–1.0 mg/100 mL	Values increase in conditions causing red blood cell destruction or biliary obstruction.
Blood urea nitrogen, or BUN (plasma or serum)	8–25 mg/100 mL (2.5–9.3 mmol/L)	Values increase in various kidney disorders and decrease in liver failure and during pregnancy.
Calcium (serum)	8.5–10.5 mg/100 mL	Values increase in hyperparathyroidism, hypervitaminosis D, and respiratory conditions that cause a rise in CO_2 concentration. They decrease in hypoparathyroidism, malnutrition, and severe diarrhea.
Carbon dioxide (serum)	24–30 mEq/L	Values increase in respiratory diseases, intestinal obstruction, and vomiting. They decrease in acidosis, nephritis, and diarrhea.
Chloride (serum)	100–106 mEq/L	Values increase in nephritis, Cushing syndrome, dehydration, and hyperventilation. They decrease in metabolic acidosis, Addison disease, diarrhea, and following severe burns.
Cholesterol, total (serum)	120–220 mg/100 mL (below 200 mg/100 mL recommended by the American Heart Association)	Values increase in diabetes mellitus and hypothyroidism. They decrease in pernicious anemia, hyperthyroidism, and acute infections.
Cholesterol, high-density lipoprotein (HDL)	Men: 30–70 mg/100 mL Women: 30–80 mg/100 mL	Values increase in liver disease. Decreased values are associated with an increased risk of atherosclerosis.
Cholesterol, low-density lipoprotein (LDL)	62–185 mg/100 mL	Increased values are associated with an increased risk of atherosclerosis.
Creatine (serum)	0.2–0.8 mg/100 mL	Values increase in muscular dystrophy, nephritis, severe damage to muscle tissue, and during pregnancy.
Creatinine (serum)	0.6–1.5 mg/100 mL	Values increase in various kidney diseases.
Ferritin (serum)	Men: 10–270 mg/100 mL Women: 5–280 mg/100 mL	Values correlate with total body iron store. They decrease with iron deficiency.
Globulin (serum)	2.3–3.5 g/100 mL	Values increase as a result of chronic infections.
Glucose (plasma)	70–110 mg/100 mL	Values increase in diabetes mellitus, liver diseases, nephritis, hyperthyroidism, and pregnancy. They decrease in hyperinsulinism, hypothyroidism, and Addison disease.
Hematocrit (whole blood)	Men: 40–54% Women: 37–47% Children: 35–49% (varies with age)	Values increase in polycythemia due to dehydration or shock. They decrease in anemia and following severe hemorrhage.

*These values may vary with hospital, physician, and type of equipment used to make measurements.

Common Tests Performed on Blood—*continued*

Test	Normal Values* (Adult)	Clinical Significance
Hemoglobin (whole blood)	Men: 14–18 g/100 mL Women: 12–16 g/100 mL Children: 11.2–16.5 g/100 mL (varies with age)	Values increase in polycythemia, obstructive pulmonary diseases, congestive heart failure, and at high altitudes. They decrease in anemia, pregnancy, and as a result of severe hemorrhage or excessive fluid intake.
Iron (serum)	50–150 μg/100 mL	Values increase in various anemias and liver disease. They decrease in iron-deficiency anemia.
Iron-binding capacity (serum)	250–410 μg/100 mL	Values increase in iron-deficiency anemia and pregnancy. They decrease in pernicious anemia, liver disease, and chronic infections.
Lactic acid (whole blood)	0.6–1.8 mEq/L	Values increase with muscular activity and in congestive heart failure, severe hemorrhage, and shock.
Lactic dehydrogenase, or LDH (serum)	70–200 U/L	Values increase in pernicious anemia, myocardial infarction, liver disease, acute leukemia, and widespread carcinoma.
Lipids, total (serum)	450–850 mg/100 mL	Values increase in hypothyroidism, diabetes mellitus, and nephritis. They decrease in hyperthyroidism.
Magnesium	1.3–2.1 mEq/L	Values increase in renal failure, hypothyroidism, and Addison disease. They decrease in renal disease, liver disease, and pancreatitis.
Mean corpuscular hemoglobin (MCH)	26–32 pg/RBC	Values increase in macrocytic anemia. They decrease in microcytic anemia.
Mean corpuscular volume (MCV)	86–98 μ mm^3/RBC	Values increase in liver disease and pernicious anemia. They decrease in iron-deficiency anemia.
Osmolality	275–295 mOsm/kg	Values increase in dehydration, hypercalcemia, and diabetes mellitus. They decrease in hyponatremia, Addison disease, and water intoxication.
Oxygen saturation (whole blood)	Arterial: 96–100% Venous: 60–85%	Values increase in polycythemia and decrease in anemia and obstructive pulmonary diseases.
Pco_2	35–45 mmHg	Values increase in respiratory diseases, intestinal obstruction, and vomiting. They decrease in acidosis, nephritis, and diarrhea.
pH (whole blood)	7.35–7.45	Values increase due to mild vomiting, Cushing syndrome, and hyperventilation. They decrease as a result of hypoventilation, severe diarrhea, Addison disease, and diabetic acidosis.
Po_2	75–100 mmHg	Values increase in polycythemia. They decrease in anemia and obstructive pulmonary diseases.
Phosphatase acid (serum)	Men: 0.13–0.63 Sigma U/mL Women: 0.01–0.56 Sigma U/mL	Values increase in cancer of the prostate gland, hyperparathyroidism, certain liver diseases, myocardial infarction, and pulmonary embolism.
Phosphatase, alkaline (serum)	Adults: 13–39 U/L Children: up to 104 U/L	Values increase in hyperparathyroidism (and in other conditions that promote resorption of bone), liver diseases, and pregnancy.
Phosphorus (serum)	3.0–4.5 mg/100 mL	Values increase in kidney diseases, hypoparathyroidism, acromegaly, and hypervitaminosis D. They decrease in hyperparathyroidism.
Platelet count (whole blood)	150,000–350,000/μL	Values increase in polycythemia and certain anemias. They decrease in acute leukemia and aplastic anemia.
Potassium (serum)	3.5–5.0 mEq/L	Values increase in Addison disease, hypoventilation, and conditions that cause severe cellular destruction. They decrease in diarrhea, vomiting, diabetic acidosis, and chronic kidney disease.
Protein, total (serum)	6.0–8.4 g/100 mL	Values increase in severe dehydration and shock. They decrease in severe malnutrition and hemorrhage.
Prothrombin time (serum)	12–14 sec (one stage)	Values increase in certain hemorrhagic diseases, liver disease, vitamin K deficiency, and following the use of various drugs.
Red cell count (whole blood)	Men: 4,600,000–6,200,000/μL Women: 4,200,000–5,400,000/μL Children: 4,500,000–5,100,000/μL (varies with age)	Values increase as a result of severe dehydration or diarrhea, and decrease in anemia, leukemia, and following severe hemorrhage.
Red cell distribution width (RDW)	8.5–11.5 microns	Variation in cell width changes with pernicious anemia.

*These values may vary with hospital, physician, and type of equipment used to make measurements.

Common Tests Performed on Blood—*continued*

Test	Normal Values* (Adult)	Clinical Significance
Sedimentation rate, erythrocyte (whole blood)	Men: 1–13 mm/hr Women: 1–20 mm/hr	Values increase in infectious diseases, menstruation, pregnancy, and as a result of severe tissue damage.
Serum glutamic pyruvic transaminase (SGPT)	Men: 6–24 U/L Women: 4–17 U/L	Values increase in liver disease, pancreatitis, and acute myocardial infarction.
Sodium (serum)	135–145 mEq/L	Values increase in nephritis and severe dehydration. They decrease in Addison disease, myxedema, kidney disease, and diarrhea.
Thromboplastin time, partial (plasma)	35–45 sec	Values increase in deficiencies of blood factors VIII, IX, and X.
Thyroid-stimulating hormone (TSH)	0.5–5.0 μU/mL	Values increase in hypothyroidism and decrease in hyperthyroidism.
Thyroxine, or T_4 (serum)	4–12 μg/100 mL	Values increase in hyperthyroidism and pregnancy. They decrease in hypothyroidism.
Transaminases, or SGOT (serum)	7–27 units/mL	Values increase in myocardial infarction, liver disease, and diseases of skeletal muscles.
Triglycerides	40–150 mg/100 mL	Values increase in liver disease, nephrotic syndrome, hypothyroidism, and pancreatitis. They decrease in malnutrition and hyperthyroidism.
Triiodothyronine, or T_3 (serum)	75–195 ng/100 mL	Values increase in hyperthyroidism and decrease in hypothyroidism.
Uric acid (serum)	Men: 2.5–8.0 mg/100 mL Women: 1.5–6.0 mg/100 mL	Values increase in gout, leukemia, pneumonia, toxemia of pregnancy, and as a result of severe tissue damage.
White blood cell count, differential (whole blood)	Neutrophils 54–62% Eosinophils 1–3% Basophils <1% Lymphocytes 25–33% Monocytes 3–7%	Neutrophils increase in bacterial diseases; lymphocytes and monocytes increase in viral diseases; eosinophils increase in collagen diseases, allergies, and in the presence of intestinal parasites.
White blood cell count, total (whole blood)	4,500–10,000/μL	Values increase in acute infections, acute leukemia, and following menstruation. They decrease in aplastic anemia and as a result of drug toxicity.

*These values may vary with hospital, physician, and type of equipment used to make measurements.

Common Tests Performed on Urine

Test	Normal Values* (Adult)	Clinical Significance
Acetone and acetoacetate	0	Values increase in diabetic acidosis.
Albumin, qualitative	0 to trace	Values increase in kidney disease, hypertension, and heart failure.
Ammonia	20–70 mEq/L	Values increase in diabetes mellitus and liver diseases.
Bacterial count	Under 10,000/mL	Values increase in urinary tract infection.
Bile and bilirubin	0	Values increase in melanoma and biliary tract obstruction.
Calcium	Under 300 mg/24 hr	Values increase in hyperparathyroidism and decrease in hypoparathyroidism.
Creatinine (24 hours)	15–25 mg/kg body weight/day	Values increase in infections, and decrease in muscular atrophy, anemia, leukemia, and kidney diseases.
Creatinine clearance (24 hours)	100–140 mL/min	Values increase in renal diseases.
Glucose	0	Values increase in diabetes mellitus and various pituitary gland disorders.
Hemoglobin	0	Blood may occur in urine as a result of extensive burns, crushing injuries, hemolytic anemia, or blood transfusion reactions.
17-hydroxycorticosteroids	3–8 mg/24 hr	Values increase in Cushing syndrome and decrease in Addison disease.
Osmolality	850 mOsm/kg	Values increase in hepatic cirrhosis, congestive heart failure, and Addison disease. They decrease in hypokalemia, hypercalcemia, and diabetes insipidus.
pH	4.6–8.0	Values increase in urinary tract infections and chronic renal failure. They decrease in diabetes mellitus, emphysema, and starvation.
Phenylpyruvic acid	0	Values increase in phenylketonuria.
Specific gravity (SG)	1.003–1.035	Values increase in diabetes mellitus, nephrosis, and dehydration. They decrease in diabetes insipidus, glomerulonephritis, and severe renal injury.
Urea	25–35 g/24 hr	Values increase as a result of excessive protein breakdown. They decrease as a result of impaired renal function.
Urea clearance	Over 40 mL blood cleared of urea/min	Values increase in renal diseases.
Uric acid	0.6–1.0 g/24 hr as urate	Values increase in gout and decrease in various kidney diseases.
Urobilinogen	0–4 mg/24 hr	Values increase in liver diseases and hemolytic anemia. They decrease in complete biliary obstruction and severe diarrhea.

*These values may vary with hospital, physician, and type of equipment used to make measurements.

APPENDIX C

Cellular Respiration

GLYCOLYSIS

Figure C.1 illustrates the chemical reactions of glycolysis. In the early steps of this metabolic pathway, the original glucose molecule is altered by the addition of phosphate groups (*phosphorylation*) and by the rearrangement of its atoms. ATP supplies the phosphate groups and the energy to drive these reactions. The result is a molecule of fructose bound to two phosphate groups (fructose-1,6-bisphosphate). This molecule is split through two separate reactions into two 3-carbon molecules (glyceraldehyde-3-phosphate). Since each of these is converted to pyruvic acid, the following reactions, 1 through 5, must be counted twice to account for breakdown of a single glucose molecule.

1. An inorganic phosphate group is added to glyceraldehyde-3-phosphate to form 1,3-bisphosphoglyceric acid, releasing two hydrogen atoms, to be used in ATP synthesis, described later.
2. 1,3-bisphosphoglyceric acid is changed to 3-phosphoglyceric acid. As this occurs, some energy in the form of a high-energy phosphate is transferred from the 1,3-bisphosphoglyceric acid to an ADP molecule, phosphorylating the ADP to ATP.
3. A slight alteration of 3-phosphoglyceric acid forms 2-phosphoglyceric acid.
4. A change in 2-phosphoglyceric acid converts it into phosphoenolpyruvic acid.
5. Finally, a high-energy phosphate is transferred from the phosphoenolpyruvic acid to an ADP molecule, phosphorylating it to ATP. A molecule of pyruvic acid remains.

Overall, one molecule of glucose is ultimately broken down to two molecules of pyruvic acid. Also, a total of four hydrogen atoms are released (step *a*), and four ATP molecules form (two in step *b* and two in step *e*). However, because two molecules of ATP are used early in glycolysis, there is a net gain of only two ATP molecules during this phase of cellular respiration.

In the presence of oxygen, each pyruvic acid molecule is oxidized to an acetyl group, which then combines with a molecule of coenzyme A (obtained from the vitamin pantothenic acid) to form acetyl coenzyme A. As this occurs, two more hydrogen atoms and one carbon dioxide molecule are released for each molecule of acetyl coenzyme A formed. The acetyl coenzyme A is then broken down by means of the citric acid cycle, which figure C.2 illustrates.

Because obtaining energy for cellular metabolism is vital, disruptions in glycolysis or the reactions that follow it can devastate health. The opening vignette to chapter 4 (p. 115) describes how arsenic poisoning kills by blocking the products of glycolysis from entering the mitochondria.

CITRIC ACID CYCLE

An acetyl coenzyme A molecule enters the citric acid cycle by combining with a molecule of oxaloacetic acid to form citric acid. As citric acid is produced, coenzyme A is released and thus can be used again to form acetyl coenzyme A from pyruvic acid. The citric acid is then changed by a series of reactions back into oxaloacetic acid, and the cycle may repeat.

Steps in the citric acid cycle release carbon dioxide and hydrogen atoms. More specifically, for each glucose molecule metabolized in the presence of oxygen, two molecules of acetyl coenzyme A enter the citric acid cycle. The cycle releases four carbon dioxide molecules and sixteen hydrogen atoms. At the same time, two more molecules of ATP form.

The released carbon dioxide dissolves in the cytoplasm and leaves the cell, eventually entering the bloodstream. Most of the hydrogen atoms released from the citric acid cycle, and those released during glycolysis and during the formation of acetyl coenzyme A, supply electrons used to produce ATP.

ATP SYNTHESIS

Note that in figures C.1 and C.2 various metabolic reactions release hydrogen atoms. The electrons of these hydrogen atoms contain much of the energy associated with the chemical bonds of the original glucose molecule. To keep this energy in a usable form, these hydrogen atoms, with their high energy electrons, are passed in pairs to *hydrogen carriers*. One of these carriers is NAD^+ (nicotinamide adenine dinucleotide). When NAD^+ accepts a pair of hydrogen atoms, the two electrons and one hydrogen nucleus bind to NAD^+ to form NADH, and the remaining hydrogen nucleus (a hydrogen ion) is released as follows:

$$NAD^+ + 2H \rightarrow NADH + H^+$$

NAD^+ is a coenzyme obtained from a vitamin (niacin), and when it combines with the energized electrons it is said to be *reduced*. Reduction results from the addition of electrons, often as part of hydrogen atoms. Another electron acceptor, FAD (flavine adenine dinucleotide), acts in a similar manner, combining with two electrons and two hydrogen nuclei to form $FADH_2$

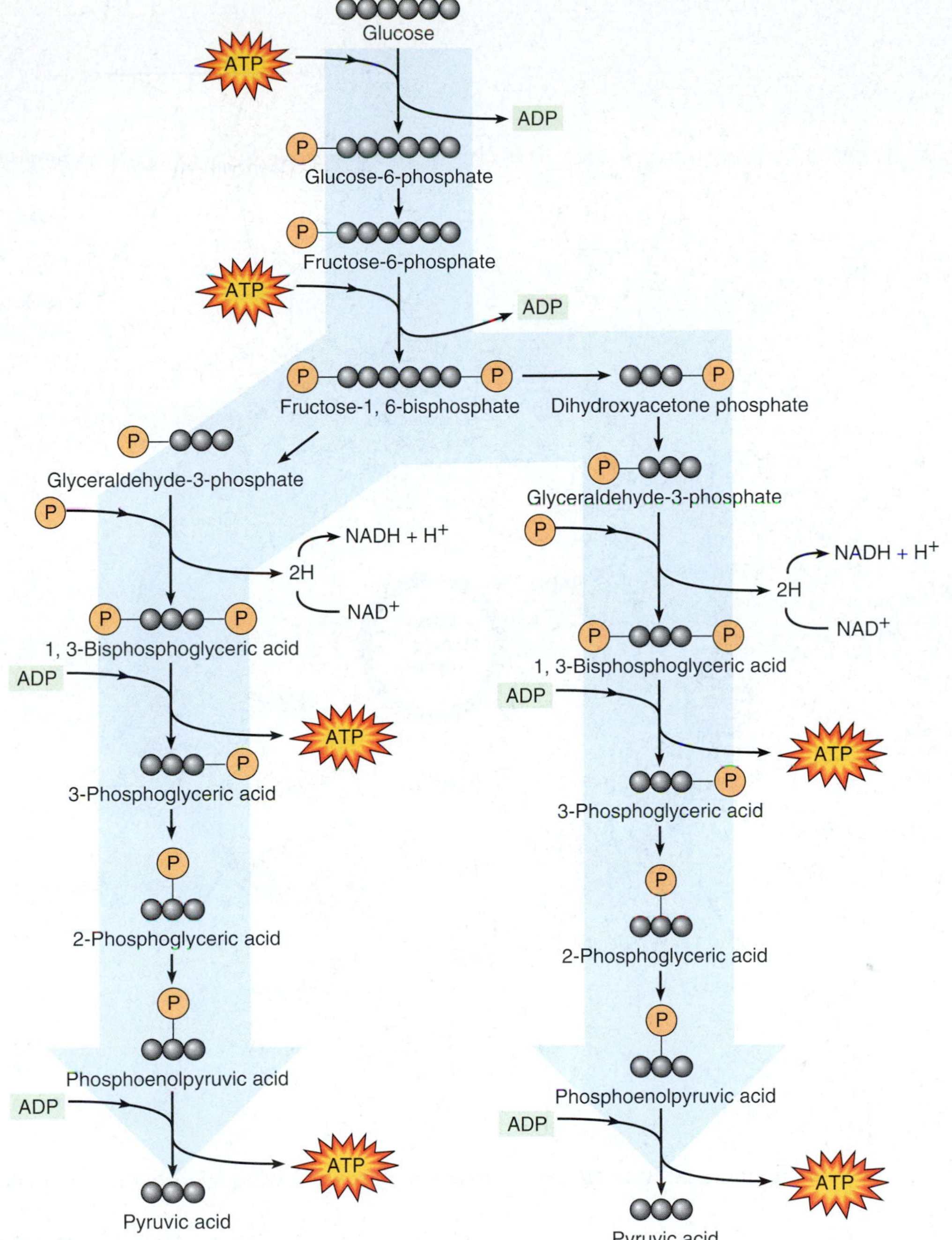

FIGURE C.1 Chemical reactions of glycolysis. There is a net production of 2 ATP molecules from each glucose molecule. The four hydrogen atoms released provide high-energy electrons that may be used to generate ATP in the electron transport chain, described later.

(fig. C.2). In their reduced states, the hydrogen carriers NADH and $FADH_2$ now hold most of the energy once held in the bonds of the original glucose molecule.

Figure 4.12 shows that NADH can release the electrons and hydrogen nucleus. Since this reaction removes electrons, the resulting NAD^+ is said to be *oxidized*. Oxidation results from the removal of electrons, often as part of hydrogen atoms; it is the opposite of reduction. The two electrons this reaction releases pass to a series of electron carriers. The regenerated NAD^+ can once again accept electrons, and is recycled.

The molecules that act as electron carriers comprise an **electron transport chain** described in Chapter 4 (pp. 122–123). As electrons are passed from one carrier to another, the carriers are alternately reduced and oxidized as they accept or release electrons. The transported electrons gradually lose energy as they proceed down the chain.

Among the members of the electron transport chain are several proteins, including a set of iron-containing molecules called **cytochromes.** The chain is located in the inner membranes of the mitochondria (fig. C.3; also see chapter 3, p. 84). The folds

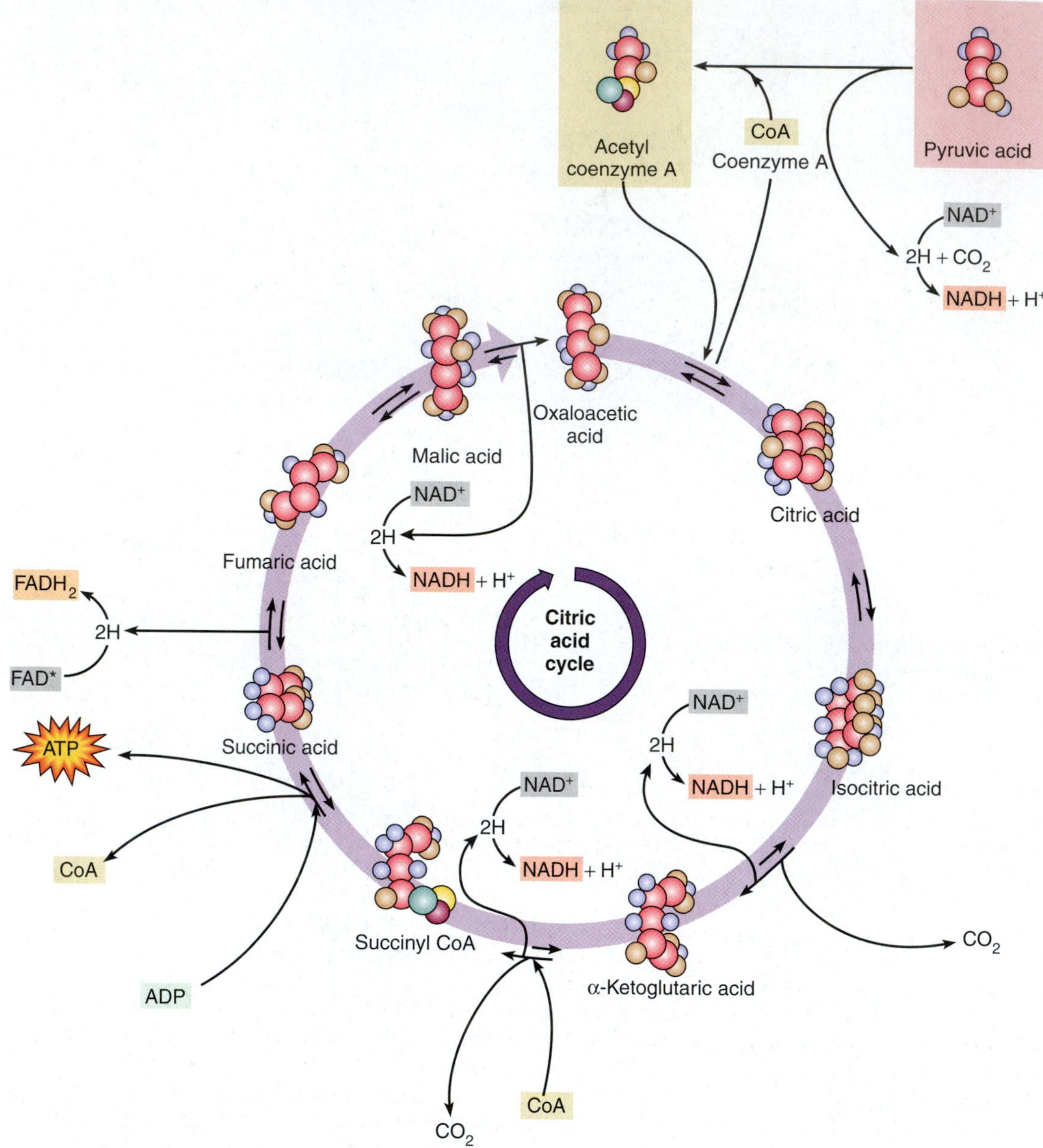

FIGURE C.2 Chemical reactions of the citric acid cycle. NADH and $FADH_2$ molecules carrying hydrogens are highlighted.

of the inner mitochondrial membrane provide surface area on which the energy reactions take place. In a muscle cell, the inner mitochondrial membrane, if stretched out, may be as much as forty-five times as long as the cell membrane!

Note in figure C.3, that as electrons pass through the electron transport chain, hydrogen ions are forced into the space between the inner and outer mitochondrial membranes. This sets up a concentration gradient for the hydrogen ions to diffuse back into the mitochondrial matrix via the enzyme complex *ATP synthase*. Much of their energy is used by the synthase to synthesize ATP.

The final cytochrome of the electron transport chain (cytochrome oxidase) gives up a pair of electrons and causes two hydrogen ions (formed at the beginning of the sequence) to combine with an atom of oxygen. This process produces a water molecule:

$$2e^- + 2H^+ + 1/2\ O_2 \rightarrow H_2O$$

Thus, oxygen is the final electron acceptor. In the absence of oxygen, electrons cannot pass through the electron transport chain, NAD+ cannot be regenerated, and aerobic respiration halts.

As electrons pass through the electron transport chain, energy is released. Some of this energy is used by a mechanism involving the ATP synthase to combine phosphate and ADP by a high-energy bond (phosphorylation), forming ATP.

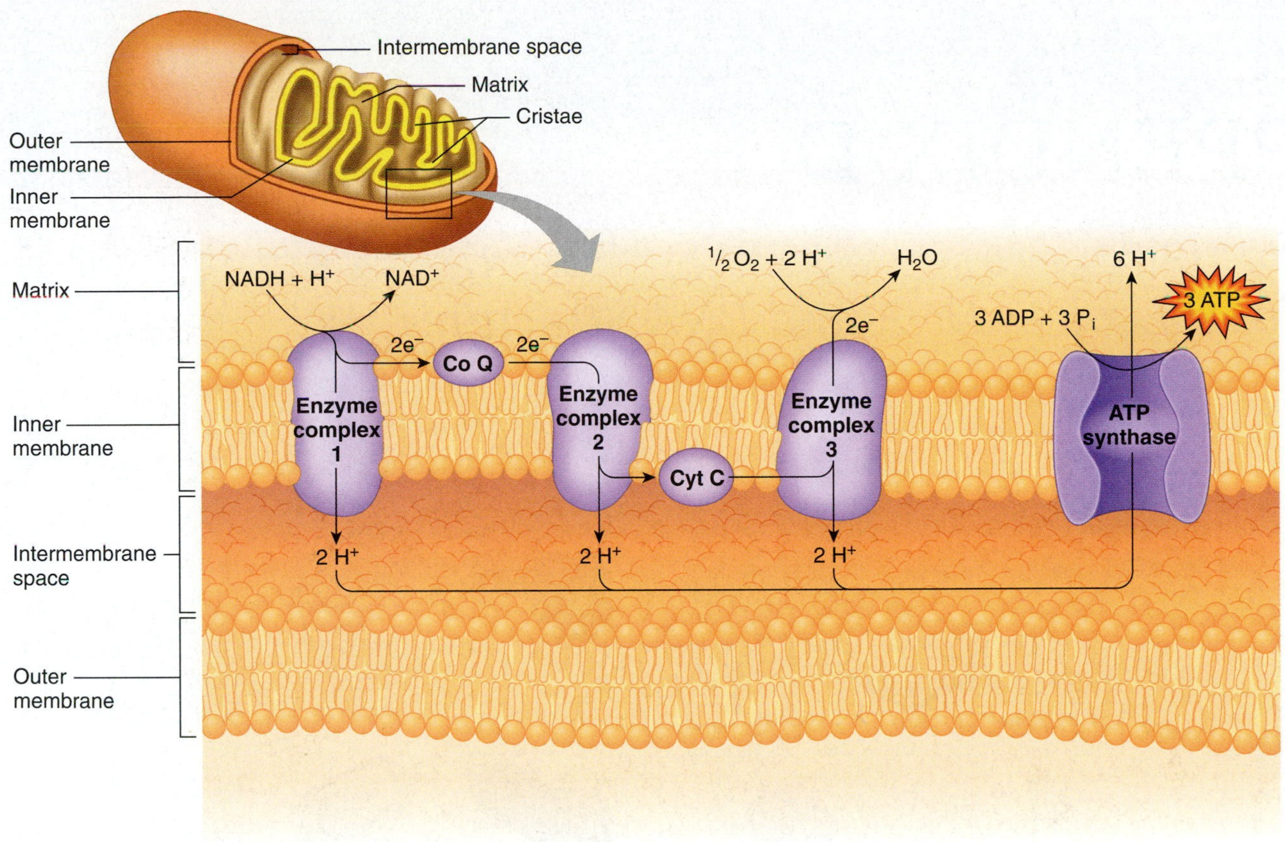

FIGURE C.3 High energy electrons moving down the electron transport chain force hydrogen ions into the space between the mitochondrial membranes. This sets up a gradient for hydrogen ions to diffuse back into the matrix by way of ATP synthase, which converts much of their energy into ATP.

Note in figures C.1 and C.2 that twelve pairs of hydrogen atoms are released during the complete breakdown of one glucose molecule—two pairs from glycolysis, two pairs from the conversion of pyruvic acid to acetyl coenzyme A (one pair from each of two pyruvic acid molecules), and eight pairs from the citric acid cycle (four pairs for each of two acetyl coenzyme A molecules).

High-energy electrons from ten pairs of these hydrogen atoms produce thirty ATP molecules in the electron transport chain. Two pairs enter the chain differently and form four ATP molecules. Because this process of forming ATP involves both the oxidation of hydrogen atoms and the bonding of phosphate to ADP, it is called oxidative phosphorylation. Also, there is a net gain of two ATP molecules during glycolysis, and two ATP molecules form by direct enzyme action in two turns of the citric acid cycle. Thus, a maximum of thirty-eight ATP molecules form for each glucose molecule metabolized.

APPENDIX D

A Closer Look at DNA and RNA Structures

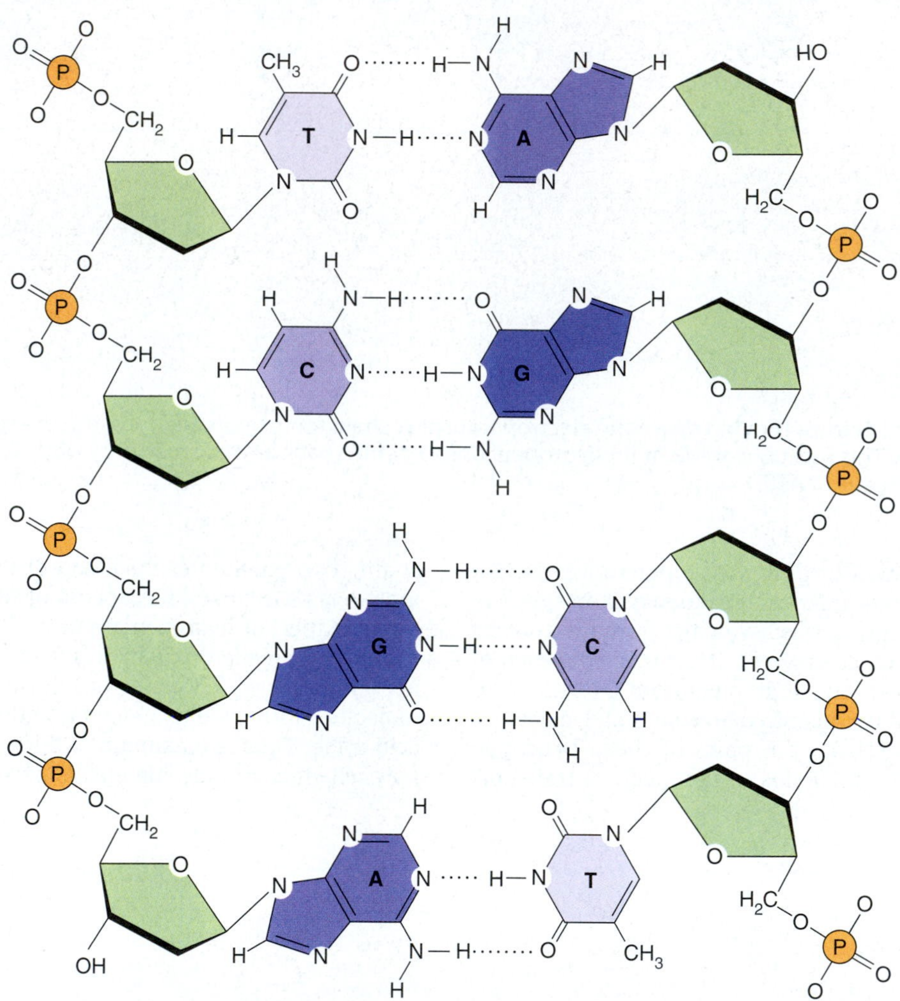

FIGURE D.1 The nucleotides of a double-stranded DNA molecule pair so that an adenine (A) of one strand hydrogen bonds to a thymine (T) of the other strand, and a guanine (G) of one strand hydrogen bonds to a cytosine (C) of the other. The dotted lines represent hydrogen bonds.

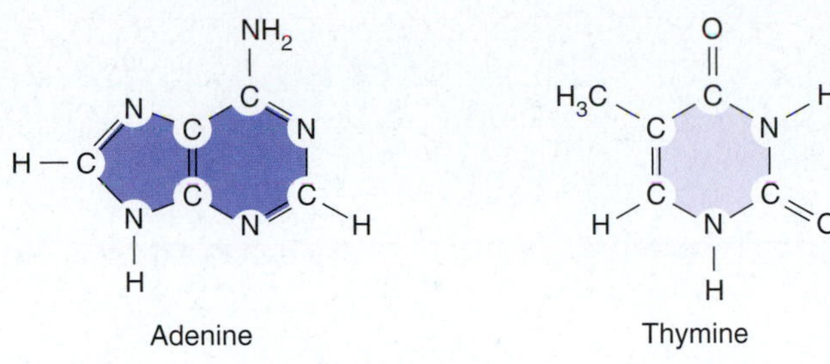

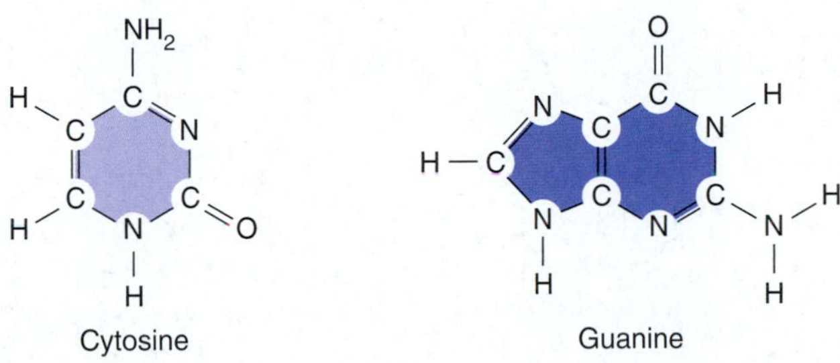

FIGURE D.2 The deoxyribonucleotides contain adenine, thymine, cytosine, or guanine.

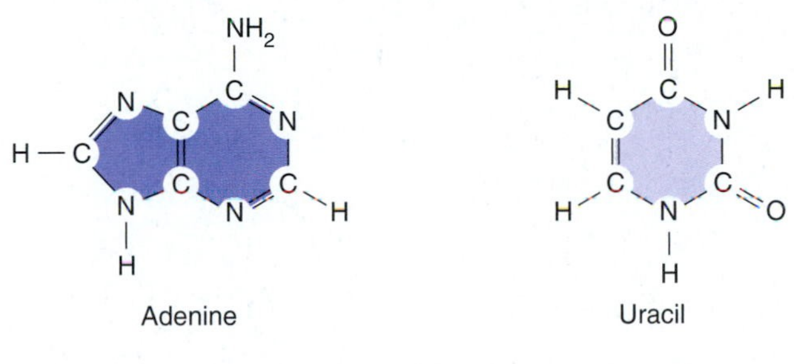

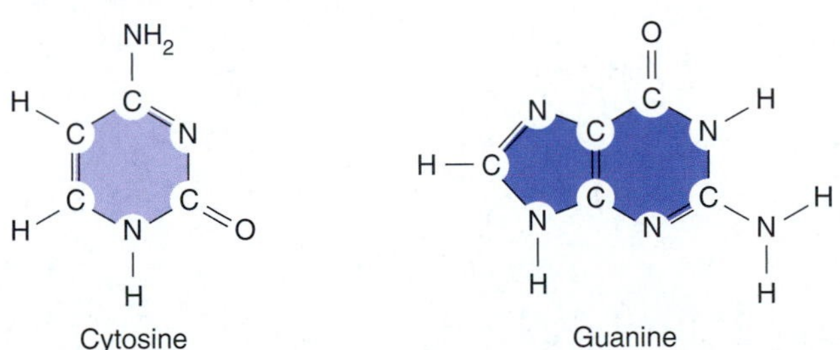

FIGURE D.3 The ribonucleotides contain adenine, uracil, cytosine, or guanine.

GLOSSARY

Each word in this glossary is followed by a phonetic guide to pronunciation. In this guide, any unmarked vowel that ends a syllable or stands alone as a syllable has the long sound. Thus, *play* would be spelled *pla.* Any unmarked vowel that is followed by a consonant has the short sound. *Tough,* for instance, is spelled *tuf.* If a long vowel appears in the middle of a syllable (followed by a consonant), it is marked with the macron (¯), the sign for a long vowel. Thus, the word *plate* would be phonetically spelled plāt. Similarly, if a vowel stands alone or ends a syllable, but has the short sound, it is marked with a breve (˘).

A

abdominal (ab-dom′ĭ-nal) Pertaining to the portion of the body between the diaphragm and the pelvis. p. 23

abdominal cavity (ab-dom′ĭ-nal kav′ĭ-tē) Space between the diaphragm and the pelvic inlet that contains the abdominal viscera. p. 12

abdominopelvic cavity (ab-dom″ĭ-no-pel′vik kav′ĭ-tē) Space between the diaphragm and the lower portion of the trunk of the body. p. 12

abduction (ab-duk′shun) Movement of a body part away from the midline. p. 269

absorption (ab-sorp′shun) The taking in of substances by cells or across membranes. p. 8

accessory organ (ak-ses′o-re or′gan) Organ that supplements the functions of other organs. p. 833

accommodation (ah-kom″o-da′shun) Adjustment of the lens of the eye for close or distant vision. p. 487

acetylcholine (as″ĕ-til-ko′lēn) Type of neurotransmitter, which is a biochemical secreted at axon ends of many neurons; transmits nerve impulses across synapses; ACH. p. 290

acetylcholinesterase (as″ĕ-til-ko″lin-es′ter-ās) Enzyme that catalyzes breakdown of acetylcholine. p. 293

acetyl coenzyme A (as′ĕ-til ko-en′zīm) Intermediate compound produced from the oxidation of carbohydrates and fats. p. 122

acid (as′id) Substance that ionizes in water to release hydrogen ions. p. 59

acid-base buffer system (as′id-bās buf′er sis′tem) Pair of chemicals, one a weak acid, the other a weak base, that resists pH changes. p. 820

acidosis (as″ĭ-do′sis) Increase in acidity of body fluids below pH 7.35. p. 60

acoustic (ah-koo-s′tik) Pertaining to sound. p. 450

acromial (ah-kro′me-al) Pertaining to the shoulder. p. 23

ACTH Adrenocorticotropic hormone. p. 496

actin (ak′tin) A protein in a muscle fiber that forms filaments that slide between filaments of the protein myosin, contracting muscle fibers. p. 287

action potential (ak′shun po-ten′shal) Sequence of electrical changes in part of a nerve cell membrane exposed to a stimulus that exceeds threshold. p. 368

activation energy (ak″tĭ-va′shun en′er-je) Energy required to initiate a chemical reaction. p. 117

active site (ak′tiv sīt) Part of an enzyme that temporarily binds a substrate. p. 117

active transport (ak′tiv trans′port) Process that requires energy and a carrier molecule to move a substance across a cell membrane, usually against the concentration gradient. p. 95

adaptive defense (a-dap′tiv dē-fenc) Specific defenses T and B lymphocytes carry out. p. 626

adduction (ah-duk′shun) Movement of a body part toward the midline. p. 269

adenoids (ad′ĕ-noids) The pharyngeal tonsils in the nasopharynx. p. 657

adenosine diphosphate (ah-den′o-sēn di-fos′fāt) Molecule produced when adenosine triphosphate loses a terminal phosphate; ADP. p. 119

adenosine triphosphate (ah-den′o-sēn tri-fos′fāt) Organic molecule that stores and releases energy, used in cellular processes; ATP. p. 84

adenylate cyclase (ah-den′ĭ-lāt si′klās) Enzyme activated when certain hormones bind receptors on cell membranes. It catalyzes the circularization of ATP to cyclic AMP. p. 488

ADH Antidiuretic hormone. p. 498

adipose tissue (ad′ĭ-pōs tish′u) Fat-storing tissue. p. 156

adolescence (ad″o-les′ens) Period of life between puberty and adulthood. p. 906

ADP Adenosine diphosphate. p. 119

adrenal cortex (ah-dre′nal kor′teks) Outer part of the adrenal gland. p. 504

adrenal gland (ah-dre′nal gland) Endocrine gland on the superior portion of each kidney. p. 504

adrenalin (ah-dren′ah-lin) Epinephrine. Adrenal gland hormone. p. 505

adrenal medulla (ah-dre′nal me-dul′ah) Inner part of the adrenal gland. p. 504

adrenergic fiber (ad″ren-er′jik fi′ber) Axon that secretes norepinephrine at its terminal. p. 429

adrenocorticotropic hormone (ah-dre″no-kor″te-ko-trōp′ik hor′mōn) Hormone that the anterior pituitary secretes to stimulate activity in the adrenal cortex; ACTH. p. 496

adulthood (ah-dult′hood) Period of life between adolescence and senescence. p. 907

aerobic (a″er-ōb′ik) Presence of molecular oxygen. p. 120

afferent (af′er-ent) Conducting toward a central part of a structure. p. 779

afterload (af′ter-lōd) The force required to open the semilunar valves to eject blood from the ventricles. p. 586

agglutination (ah-gloo″ti-na′shun) Clumping of blood cells in response to a reaction between an antibody and an antigen. p. 544

agonist (ag′o-nist) A prime mover. p. 304

agranulocyte (a-gran′u-lo-sīt) Nongranular leukocyte. p. 530

albumin (al-bu′min) Plasma protein that helps regulate the osmotic concentration of blood. p. 535

aldosterone (al-dos′ter-ōn) Adrenal cortical hormone that regulates sodium and potassium ion concentrations and fluid volume. p. 486

alimentary canal (al″i-men′tar-e kah-nal′) Tubular part of the digestive tract from the mouth to the anus. p. 652

alkaline (al′kah-līn) Pertaining to or having the properties of a base. p. 60

alkaloid (al′kah-loid) Group of organic compounds that are bitter and have toxic effects. p. 450

alkalosis (al″kah-lo′sis) Increase in the pH of body fluids above 7.45. p. 60

allantois (ah-lan′to-is) Structure in the embryo from which the umbilical cord blood vessels develop. p. 891

allele (ah-lēl) One of two or more different forms of a gene. p. 919

allergen (al′er-jen) A nonself substance that can provoke an allergic reaction. p. 639

all-or-none response (al′or-nun′ re-spons′) Complete response of a muscle fiber or neuron to a stimulus of threshold strength. p. 297

alpha receptor (al′fah re-sep′tor) Receptor on effector cell membrane that binds epinephrine or norepinephrine. p. 430

alveolar duct (al-ve′o-lar dukt) Fine tube that conducts inhaled air to an air sac of the lungs. p. 743

alveolar pore (al-ve′o-lar pōr) Small opening in the wall of an air sac that permits air to pass from one alveolus to another. p. 759

alveolar process (al-ve′o-lar pros′es) Projection on the border of the jaw that holds the bony sockets of the teeth. p. 213

alveolus (al-ve′o-lus) Air sac of a lung; a saclike structure (pl., *alveoli*). p. 743

amacrine cell (am′ah-krin sel) Retinal neuron whose fibers pass laterally between other retinal cells. p. 469

amine (am′in) Type of nitrogen-containing organic compound, including hormones from the adrenal medulla. p. 485

amino acid (ah-me′no as′id) Organic compound that includes an amino group ($-NH_2$) and a carboxyl group ($-COOH$); the structural unit of a protein molecule. p. 65

amniocentesis (am″ne-o-sen-te′sis) Removal of a sample of amniotic fluid through the abdominal wall of a pregnant woman. Fetal cells in it are cultured and checked for chromosome complement. p. 931

amnion (am′ne-on) Extraembryonic membrane that encircles a fetus and contains amniotic fluid. p. 888

amniotic cavity (am″ne-ot′ik kav′ĭ-te) Fluid-filled space within the amnion. p. 885

amniotic fluid (am″ne-ot′ik floo′id) Fluid in the amniotic cavity that surrounds the developing fetus. p. 888

ampulla (am-pul′ah) Expansion at the end of each semicircular canal that houses a crista ampullaris. p. 460

amylase (am′ĭ-lās) Enzyme that hydrolyzes polysaccharides. p. 116

anabolism (ah-nab′o-liz″em) Synthesis of larger molecules from smaller ones; anabolic metabolism. p. 115

anaerobic (an-a″er-ōb′ik) Absence of molecular oxygen. p. 120

anal canal (a′nal kah-nal′) The most distal two or three inches of the large intestine that open to the outside as the anus. p. 688

anaphase (an′ah-fāz) Stage in mitosis when replicated chromosomes separate and move to opposite poles of the cell. p. 101

anatomical position (an″ah-tom′ĭ-kal po-zish′un) Body posture with the body erect, face forward, arms at the sides with palms facing forward, and toes pointing straight ahead. p. 20

anatomy (ah-nat′o-me) Branch of science dealing with the form and structure of body parts. p. 4

androgen (an′dro-jen) Male sex hormone such as testosterone. p. 845

anemia (ah-ne′me-ah) Deficiency of red blood cells or hemoglobin. p. 529

aneuploid (an′u-ploid) Cell with one or more extra or missing chromosomes. p. 929

aneurysm (an′u-rizm) Saclike expansion of a blood vessel wall. p. 582

angiotensin I (an″je-o-ten′sin wun) A peptide released when angiotensinogen is decomposed. p. 506

angiotensin II (an″je-o-ten′sin too) Vasoconstricting biochemical released in response to reduced blood flow to the kidneys, elevating blood pressure. p. 506

angiotensinogen (an″je-o-ten-sin′o-jen) Liver serum globulin that renin converts to angiotensin I. p. 506

anion (an′i-on) Atom carrying a negative charge due to one or more extra electrons. p. 55

antagonist (an-tag′o-nist) A muscle that opposes a prime mover. p. 304

antebrachial (an″te-bra′ke-al) Pertaining to the forearm. p. 23

antecubital (an″te-ku′bi-tal) Region in front of the elbow joint. p. 23

anterior (an-te′re-or) Pertaining to the front. p. 21

anterior pituitary (an-te′re-or pi-tu′i-tār″e) Front lobe of the pituitary gland. p. 492

antibody (an′ti-bod″e) Protein (immunoglobulin) that B cells of the immune system produce in response to a nonself antigen; it reacts with the antigen. p. 532

anticoagulant (an″tĭ-ko-ag′u-lant) Biochemical that inhibits blood clotting. p. 543

anticodon (an″ti-ko′don) Three contiguous nucleotides of a transfer RNA molecule that are complementary to a specific mRNA codon. p. 131

antidiuretic hormone (an″tĭ-di″u-ret′ik hor′mōn) Hormone of the posterior pituitary gland that enhances water conservation in the kidneys; ADH, vasopressin. p. 498

antigen (an′tĭ-jen) Chemical that stimulates B lymphocytes to produce antibodies. p. 544

antigen-binding site (an′tĭ-jen-bīn′ding sīt) Specialized ends of antibodies that bind specific antigens. p. 634

antigen-presenting cell (an′tĭ-jen-pre-sen′ting cel) The cell that displays the antigen to the cells of the immune system so they can defend the body against that particular antigen. p. 630

antioxidant (an′tĭ-ok′sĭ-dant) Substance that inhibits oxidation of another substance. p. 710

antithrombin (an″tĭ-throm′bin) Substance that inhibits the action of thrombin and thus inhibits blood clotting. p. 542

anus (a′nus) Inferior outlet of the digestive tube. p. 688

aorta (a-or′tah) Major systemic artery that receives blood from the left ventricle. p. 592

aortic body (a-or′tik bod′e) Structure associated with the wall of the aorta that contains chemoreceptors. p. 592

aortic sinus (a-or′tik si′nus) Swelling in the aortic wall, behind each cusp of the semilunar valve, that contains baroreceptors. p. 592

aortic valve (a-or′tik valv) Flaplike structures in the wall of the aorta near its origin that prevent blood from returning to the left ventricle of the heart. p. 557

apocrine gland (ap′o-krin gland) Type of gland whose secretions have parts of secretory cells. p. 150

aponeurosis (ap″o-nu-ro′sis) Sheet of connective tissue that attaches muscles to bone and fascia. p. 285

apoptosis (ap″o-to′-sis) Programmed cell death. p. 910

appendicular (ap″en-dik′u-lar) Pertaining to the upper or lower limbs. p. 12

appendix (ah-pen′diks) Small, tubular appendage of lymphatic tissue that extends outward from the cecum; vermiform appendix. p. 687

aqueous humor (a′kwe-us hu′mor) Watery fluid that fills the anterior cavity of the eye. p. 467

arachnoid granulation (ah-rak′noid gran″u-la′shun) Fingerlike structure that projects from the subarachnoid space of the meninges into blood-filled dural sinuses and reabsorbs cerebrospinal fluid. p. 386

arachnoid mater (ah-rak′noid ma′ter) Delicate, weblike middle layer of the meninges. p. 385

arbor vitae (ar′bor vi′ta) Treelike pattern of white matter in a section of cerebellum. p. 410

areola (ah-re′o-lah) Pigmented region surrounding the nipple of the mammary gland or breast. p. 862

areolar tissue (ah-re′o-lar tish′u) Connective tissue composed mainly of fibers. p. 156

arrector pili muscle (ah-rek′tor pil′i mus′l) Smooth muscle in the skin associated with a hair follicle. p. 179

arrhythmia (ah-rith′me-ah) An irregular heartbeat. p. 572

arteriole (ar-te′re-ōl) Small branch of an artery that communicates with a capillary network. p. 574

arteriosclerosis (ar-te″re-o-sklĕ-ro′sis) Condition in which the walls of arteries thicken and lose their elasticity; hardening of the arteries. p. 582

artery (ar′ter-e) Vessel that transports blood from the heart. p. 574

arthritis (ar-thri′tis) Joint inflammation. p. 278

articular cartilage (ar-tik′u-lar kar′tĭ-lij) Hyaline cartilage that covers the ends of bones in synovial joints. p. 194

articulation (ar-tik″u-la′shun) The union of two or more bones; a joint. p. 261

ascending colon (ah-send′ing ko′lon) Portion of the large intestine that passes upward on the right side of the abdomen from the cecum to the lower edge of the liver. p. 687

ascending tract (ah-send′ing trakt) Group of nerve fibers in the spinal cord that transmits sensory impulses upward to the brain. p. 393

ascites (ah-si′tez) Serous fluid accumulation in the abdominal cavity. p. 726

ascorbic acid (as-kor′bik as′id) One of the water-soluble vitamins; vitamin C. p. 716

assimilation (ah-sim″ĭ-la′shun) Chemically changing absorbed substances in the body. p. 8

association area (ah-so″se-a′shun a′re-ah) Region of the cerebral cortex controlling memory, reasoning, judgment, and emotions. p. 402

astigmatism (ah-stig′mah-tizm) Visual defect due to errors in refraction caused by abnormal curvatures in the surface of the cornea or lens. p. 472

astrocyte (as′tro-sīt) Type of neuroglia that connects neurons to blood vessels and provides growth factors. p. 361

atherosclerosis (ath″er-o-sklĕ-ro′sis) Condition in which fatty substances accumulate on the inner linings of arteries. p. 542

atmospheric pressure (at″mos-fer′ik presh′ur) Pressure exerted by the weight of the air; about 760 mm of mercury at sea level. p. 8

atom (at′om) Smallest particle of an element that has the properties of that element. p. 4

atomic number (ah-tom′ik num′ber) Number of protons in an atom of an element. p. 52

atomic weight (ah-tom′ik wāt) Number of protons plus the number of neutrons in an atom of an element. p. 52

ATP Adenosine triphosphate. p. 84

ATPase Enzyme that releases the energy in terminal phosphate bonds of ATP. p. 291

atria (a′tre-ah) Upper chambers of the heart. p. 555

atrial natriuretic peptide (a′trē-al na″trē-u-ret′ik pep′tīd) Family of polypeptide hormones that increase sodium excretion. p. 513

atrioventricular bundle (a″tre-o-ven-trik′u-lar bun′dl) Group of specialized fibers that conducts impulses from the atrioventricular node to the Purkinje fibers in the ventricular muscle of the heart; AV bundle; bundle of His. p. 567

atrioventricular node (a″tre-o-ven-trik′u-lar nōd) Specialized mass of cardiac muscle fibers in the interatrial septum of the heart; transmits cardiac impulses from the sinoatrial node to the AV bundle; AV node. p. 566

atrioventricular orifice (a″tre-o-ven-trik′u-lar or′i-fis) Opening between the atrium and the ventricle on each side of the heart. p. 555

atrioventricular sulcus (a″tre-o-ven-trik′u-lar sul′kus) Groove on the surface of the heart that marks the division between an atrium and a ventricle. p. 556

atrioventricular valve (a″tre-o-ven-trik′u-lar valv) Cardiac valve between an atrium and a ventricle. p. 555

atrophy (at′ro-fe) Shrinking of an organ or tissue. p. 300

auditory (aw′di-to″re) Pertaining to the ear or the sense of hearing. p. 451

auditory ossicle (aw′di-to″re os′i-kl) A bone of the middle ear. p. 451

auditory tube (aw′di-to″re tūb) Tube that connects the middle ear cavity to the pharynx; eustachian tube. p. 453

auricles (aw′ri-klz) Earlike structures; the parts of the heart that form the walls of the atria. p. 450

autoantibody (aw″to-an′tĭ-bod″e) An antibody produced against oneself. p. 641

autocrine (aw′to-krin) Hormone that acts on the same cell that secreted it. p. 483

autoimmunity (aw″to-ĭ-mū′ni-tē) An immune response against a person′s own tissues. p. 639

autonomic nervous system (aw″to-nom′ik ner′vus sis′tem) Part of the nervous system that controls the viscera. p. 356

autoregulation (aw″to-reg″u-la′shun) Ability of an organ or tissue to maintain a constant blood flow despite changing arterial blood pressure. p. 790

autosome (aw′to-sōm) A chromosome that does not include a gene that determines sex. p. 918

AV bundle (bun′dl) Atrioventricular bundle. p. 567

AV node (nōd) Atrioventricular node. p. 566

axial (ak′se-al) Pertaining to the head, neck, and trunk. p. 12

axial skeleton (ak′se-al skel′ĕ-ton) Part of the skeleton that supports and protects the organs of the head, neck, and trunk. p. 206

axillary (ak′sĭ-ler″e) Pertaining to the armpit. p. 23

axon (ak′son) A nerve fiber; conducts a nerve impulse away from a neuron cell body. p. 354

axonal transport (ak′so-nal trans′port) Transport of substances from the neuron cell body to an axon terminal. p. 356

B

ball-and-socket joint (bawl-and-sok′et joint) A bone with a spherical mass on one end joined with a bone that has a complementary hollow depression. Also called a spheroidal joint. p. 265

baroreceptor (bar″o-re-sep′tor) Sensory receptor in the blood vessel wall stimulated by changes in pressure (pressoreceptor). p. 439

basal metabolic rate (ba′sal met″ah-bo′lic rāt) Rate of metabolic reactions when the body is at rest; BMR. p. 500

basal nucleus (bas′al nu′kle-us) Mass of gray matter deep within a cerebral hemisphere of the brain. p. 404

base (bās) Substance that ionizes in water, releasing hydroxide ions (OH^-) or other ions that combine with hydrogen ions. p. 59

basement membrane (bās′ment mem′brān) Layer of nonliving material that anchors epithelial tissue to underlying connective tissue. p. 144

basophil (ba′so-fil) White blood cell containing cytoplasmic granules that stain with basic dye. p. 531

beta oxidation (ba′tah ok″sĭ-da′shun) Chemical process that breaks fatty acids down into molecules of acetyl which bind coenzyme A, entering the citric acid cycle. p. 702

beta receptor (ba′tah re-sep′tor) Receptor on an effector cell membrane that combines mainly with epinephrine and only slightly with norepinephrine. p. 430

bicarbonate buffer system (bi-kar′bo-nāt buf′er sis′tem) Mixture of carbonic acid and sodium bicarbonate that weakens a strong base and a strong acid, respectively; resists a change in pH. p. 821

bicarbonate ion (bi-kar′bon-āt i′on) HCO_3^-. p. 766

bicuspid tooth (bi-kus′pid tooth) Premolar specialized for grinding hard particles of food. p. 658

bile (bīl) Fluid secreted by the liver and stored in the gallbladder. p. 675

bile duct (bīl dukt) Tube that transports bile from the cystic duct and common hepatic duct to the duodenum. p. 677

bilirubin (bil″ĭ-roo′bin) A bile pigment produced from hemoglobin breakdown. p. 530

biliverdin (bil″ĭ-ver′din) A bile pigment produced from hemoglobin breakdown. p. 530

biochemistry (bi″o-kem′is-tre) Branch of science dealing with the chemistry of living organisms. p. 51

biotin (bi′o-tin) A water-soluble vitamin; member of the vitamin B complex. p. 716

bipolar neuron (bi-po′lar nu′ron) A nerve cell whose cell body has only one axon and one dendrite. p. 360

blastocyst (blas′to-sist) Early stage of prenatal development when the embryo is a hollow ball of cells. p. 879

blood (blud) Cells in a liquid matrix that circulate through the heart and vessels carrying substances throughout the body. p. 162

B lymphocyte (B lim′fo-sīt) Lymphocyte that produces and secretes antibodies that bind and destroy nonself molecules; B cell. p. 628

BMI Body mass index. p. 708

BMR Basal metabolic rate. p. 500

body mass index (bŏdē mas in′-dex) A measure of relative weight (underweight, normal weight, overweight, obesity) calculated as weight in kilograms divided by the square of height in meters; BMI. p. 708

bolus (bo′lus) Mass of food passing through the gastrointestinal tract. p. 664

bond (bond) Connection between atoms in a compound. p. 55

bone (bōn) Part of the skeleton composed of cells and inorganic, mineral matrix; also a connective tissue. p. 160

brachial (bra′ke-al) Pertaining to the arm. p. 23

bradycardia (brad″e-kar′de-ah) An abnormally slow heart rate or pulse rate. p. 572

brainstem (brān′stem) Portion of the brain that includes the midbrain, pons, and medulla oblongata. p. 398

Broca's area (bro′kahz a′re-ah) Region of the frontal lobe that coordinates complex muscular actions of the mouth, tongue, and larynx, making speech possible. p. 402

bronchial tree (brong′ke-al trē) The bronchi and their branches that carry air from the trachea to the alveoli of the lungs. p. 743

bronchiole (brong′ke-ōl) Small branch of a bronchus in the lung. p. 743

bronchus (brong′kus) Branch of the trachea that leads to a lung (pl., *bronchi*). p. 743

buccal (buk′al) Pertaining to the mouth and inner lining of the cheeks. p. 23

buffer (buf′er) Substance that can react with a strong acid or base to form a weaker acid or base, and thus resist a change in pH. p. 60

bulbourethral gland (bul″bo-u-re′thral gland) Gland that secretes a viscous fluid into the male urethra during sexual excitement; Cowper's gland. p. 841

bulimia (bu-lim′e-ah) Disorder of binge eating followed by purging. p. 725

bulk element (bulk el′ĕ-ment) Basic chemical required in abundance. p. 52

bursa (bur′sah) Saclike, fluid-filled structure, lined with synovial membrane, near a joint. p. 265

bursitis (bur-si′tis) Inflammation of a bursa. p. 278

C

calcitonin (kal″sĭ-to′-nin) Thyroid hormone that regulates blood calcium concentration. p. 501

calorie (kal′o-re) Unit that measures heat energy and the energy contents of foods. p. 706

calorimeter (kal″o-rim′ĕ-ter) Device that measures the heat energy content of foods; bomb calorimeter. p. 706

canaliculus (kan″ah-lik′u-lus) Microscopic canal that connects lacunae of bone tissue (pl. *canaliculi*). p. 195

capacitation (kah-pas″i-ta′shun) Activation of a sperm cell to fertilize an egg cell. p. 841

capillaries (kap′ĭ-ler″ēz) A small blood vessel that connects an arteriole and a venule. p. 576

carbaminohemoglobin (kar-bam″ĭ-no-he″mo-glo′bin) Bonded carbon dioxide and hemoglobin. p. 766
carbohydrate (kar″bo-hi′drāt) Organic compound consisting of carbon, hydrogen, and oxygen, in a 1:2:1 ratio. p. 61
carbonic anhydrase (kar-bon′ik an-hi′drās) Enzyme that catalyzes the reaction between carbon dioxide and water to form carbonic acid. p. 766
carbon monoxide (kar′bon mon-ok′sīd) Toxic gas that binds hemoglobin, forming a relatively stable compound; CO. p. 764
carboxypeptidase (kar-bok″se-pep′ti-dās) Protein-splitting enzyme in pancreatic juice. p. 671
cardiac center (kar′de-ak sen′ter) Neurons in the medulla oblongata that control heart rate. p. 408
cardiac conduction system (kar′de-ak kon-duk′shun sis′tem) System of specialized cardiac muscle fibers that conduct cardiac impulses from the SA node into the myocardium. p. 565
cardiac cycle (kar′de-ak si′kl) Series of myocardial contractions and relaxations that constitute a complete heartbeat. p. 564
cardiac muscle tissue (kar′de-ak mus′el tish′u) Specialized muscle tissue found only in the heart. p. 164
cardiac output (kar′de-ak owt′poot) The volume of blood per minute that the heart pumps (multiply stroke volume in milliliters by the heart rate in beats per minute). p. 582
cardiac vein (kar′de-ak vān) Blood vessel that returns blood from the venules of the myocardium to the coronary sinus. p. 563
cardiovascular (kar″de-o-vas′ku-lar) Pertaining to the heart and blood vessels. p. 16
carina (kah-ri′nah) Cartilaginous ridge between the openings of the right and left bronchi. p. 743
carotene (kar′o-tēn) Yellow, orange, or reddish pigment in plants and a precursor of vitamin A. p. 176
carotid bodies (kah-rot′id bod′ēz) Masses of chemoreceptors in the wall of the internal carotid artery near the carotid sinus. p. 597
carpal (kar′pal) Wrist bone. p. 23
carpus (kar′pus) Wrist; the wrist bones as a group. p. 23
cartilage (kar′tĭ-lij) Type of connective tissue in which cells are in lacunae separated by a semisolid extracellular matrix. p. 158
cartilaginous joint (kar″tĭ-laj′ĭ-nus joint) Two or more bones joined by cartilage. p. 262
catabolism (ka-tab′o-lizm) Breakdown of large molecules; catabolic metabolism. p. 115
catalyst (kat′ah-list) Chemical that increases the rate of a chemical reaction, but is not permanently altered by the reaction. p. 59
catecholamine (kat″ĕ-kol′am-in) Type of organic compound that includes epinephrine and norepinephrine. p. 505
cation (kat′i-on) Atom carrying a positive charge due to a deficiency of electrons. p. 55
cecum (se′kum) Pouchlike part of the large intestine attached to the small intestine. p. 687
celiac (se′le-ak) Pertaining to the abdomen. p. 23
cell (sel) The structural and functional unit of an organism. p. 4
cell body (sel bod′e) Part of a nerve cell that includes a cytoplasmic mass and a nucleus, and from which the nerve fibers extend. p. 356
cell cycle (sel sī-kl) Life cycle of a cell consisting of G_1 (growth), S (DNA synthesis), G_2 (growth), and mitosis (division). p. 100
cell membrane (sel mem′brān) The selectively permeable outer boundary of a cell consisting of a phospholipid bilayer embedded with proteins. p. 76
cellular adhesion molecules (sel′u-lar ad-hee′zhon mol′ĕ-kūlz) Proteins that guide movement of cells; CAMs. p. 80
cellular immune response (sel′u-lar ĭ-mūn re-spons′) The body's attack of T cells and their secreted products on nonself cells. p. 630
cellular respiration (sel′u-lar res″pĭ-ra′shun) A biochemical pathway that releases energy from organic compounds. p. 119
cellulose (sel′u-lōs) Polysaccharide abundant in plant tissues that human digestive enzymes cannot break down. p. 701
cementum (se-men′tum) Bonelike material that fastens the root of a tooth into its bony socket. p. 660
central canal (sen′tral kah-nal′) Tiny channel in bone tissue that contains a blood vessel; Haversian canal; tube in the spinal cord continuous with the ventricles of the brain and contains cerebrospinal fluid. p. 363
central nervous system (sen′tral ner′vus sis′tem) The brain and spinal cord; CNS. p. 354
centriole (sen′tre-ōl) Cellular structure built of microtubules that organizes the mitotic spindle. p. 86
centromere (sen′tro-mēr) Region of a chromosome where spindle fibers attach during mitosis. p. 101
centrosome (sen′tro-sōm) Cellular organelle consisting of two centrioles. p. 86
cephalic (sĕ-fal′ik) Pertaining to the head. p. 23
cerebellar cortex (ser″ĕ-bel′ar kor′teks) Outer layer of the cerebellum. p. 410
cerebellar peduncles (ser″ĕ-bel′ar pe-dung′kl) Bundle of nerve fibers connecting the cerebellum and the brainstem. p. 410
cerebellum (ser″ĕ-bel′um) Part of the brain that coordinates skeletal muscle movement. p. 410
cerebral aqueduct (ser′ĕ-bral ak′wĕ-dukt″) Tube that connects the third and fourth ventricles of the brain. p. 385
cerebral cortex (ser′ĕ-bral kor′teks) Outer layer of the cerebrum. p. 400
cerebral hemisphere (ser′ĕ-bral hem′ĭ-sfēr) One of the large, paired structures that constitute the cerebrum. p. 398
cerebrospinal fluid (ser″ĕ-bro-spi′nal floo′id) Fluid in the ventricles of the brain, the subarachnoid space of the meninges, and the central canal of the spinal cord; CSF. p. 385
cerebrum (ser′ĕ-brum) Part of the brain in the upper part of the cranial cavity that provides higher mental functions. p. 398
cerumen (sĕ-roo′men) Waxlike substance produced by cells that line the external ear canal. p. 450
cervical (ser′vĭ-kal) Pertaining to the neck. p. 23
cervix (ser′viks) Narrow, inferior end of the uterus that leads into the vagina. p. 854
chemoreceptor (ke″mo-re-sep′tor) Receptor stimulated by the binding of certain chemicals. p. 439
chemotaxis (ke″mo-tak′sis) Attraction of leukocytes to chemicals released from damaged cells. p. 635
chief cell (chēf sel) Cell type in gastric gland that secretes various digestive enzymes. p. 666
childhood (child′hood) Period of life between infancy and adolescence. p. 906

chloride shift (klo′rīd shift) Movement of chloride ions from the blood plasma into red blood cells as bicarbonate ions diffuse out of the red blood cells into the plasma. p. 766

cholecystokinin (ko″le-sis″to-ki′nin) Hormone the small intestine secretes that stimulates release of pancreatic juice from the pancreas and bile from the gallbladder. p. 669

cholesterol (ko-les′ter-ol) A lipid that cells produce and use to synthesize steroid hormones. p. 64

cholinergic fiber (ko″lin-er′jik fi′ber) Axon that secretes acetylcholine at its terminal. p. 428

chondrocyte (kon′dro-sīt) A cartilage cell. p. 158

chorion (ko′re-on) Extraembryonic membrane that forms the outermost layer around a fetus and contributes to formation of the placenta. p. 885

chorionic villus (ko″re-on′ik vil′us) Projection that extends from the outer surface of the chorion and helps attach an embryo to the uterine wall. p. 885

choroid coat (ko′roid kōt) Vascular, pigmented middle layer of the wall of the eye. p. 466

choroid plexus (ko′roid plek′sus) Mass of specialized capillaries that secretes cerebrospinal fluid into a ventricle of the brain. p. 363

chromatid (kro′mah-tid) One longitudinal half of a replicated chromosome. p. 831

chromatin (kro′mah-tin) DNA and complexed protein that condenses to form chromosomes during mitosis. p. 90

chromatophilic substance (kro″mah-to-fil′ik sub′stans) Membranous sacs within the cytoplasm of nerve cells that have ribosomes attached to their surfaces; Nissl bodies. p. 356

chromosome (kro′mo-sōm) Rodlike structure that condenses from chromatin in a cell′s nucleus during mitosis. p. 90

chylomicron (kil″o-mi′kron) Microscopic droplet of fat in the blood that forms following fat digestion. p. 684

chyme (kīm) Semifluid mass of partially digested food that passes from the stomach to the small intestine. p. 669

chymotrypsin (ki″mo-trip′sin) Protein-splitting enzyme in pancreatic juice. p. 671

cilia (sil′e-ah) Microscopic, hairlike processes on the exposed surfaces of certain epithelial cells. p. 87

ciliary body (sil′e-er″e bod′e) Structure associated with the choroid layer of the eye that secretes aqueous humor and contains the ciliary muscle. p. 466

circadian rhythm (ser″kah-de′an rithm) Pattern of repeated activity associated with the cycles of night and day. p. 512

circular muscles (ser′ku-lar mus′lz) Muscles whose fibers are organized in circular patterns, usually around an opening or in the wall of a tube; sphincter muscles. p. 654

circumduction (ser″kum-duk′shun) Movement of a body part, such as a limb, so that the end follows a circular path. p. 269

cisterna (sis-ter′nah) Enlarged portion of the sarcoplasmic reticulum near the actin and myosin filaments of a muscle fiber. p. 288

citric acid cycle (sit′rik as′id si′kl) Series of chemical reactions that oxidizes certain molecules, releasing energy; Krebs cycle. p. 120

cleavage (klēv′ij) Early successive divisions of cells of the blastocyst, reducing cell size. p. 879

clitoris (kli′to-ris) Small erectile organ in the anterior vulva, corresponding to the penis. p. 855

clone (klōn) Group of cells that descend from a single cell and are therefore genetically identical to it. p. 629

CNS Central nervous system. p. 354

coagulation (ko-ag″u-la′shun) Blood clotting. p. 538

cochlea (kok′le-ah) Part of the inner ear that has hearing receptors. p. 453

codominant (ko-dom′ĭ-nant) Alleles that are each expressed in a heterozygote. p. 922

codon (ko′don) Set of three nucleotides of a messenger RNA molecule corresponding to one of the 20 types of amino acids. p. 131

coenzyme (ko-en′zīm) Nonprotein organic molecule required for the activity of a particular enzyme. p. 119

coenzyme A (ko-en′zīm) A molecule that reacts with acetyl, forming acetyl coenzyme A, which enters the citric acid cycle. p. 714

cofactor (ko′fak-tor) Small molecule or ion that must combine with an enzyme for activity. p. 118

collagen (kol′ah-jen) Protein in the white fibers of connective tissues and in bone matrix. p. 155

collateral (ko-lat′er-al) A branch of an axon or blood vessel. p. 356

collectin (ko-lek′tin) Protein that protects against bacteria, yeasts, and some viruses. p. 627

collecting duct (ko-lek′ting dukt) In the lymphatic system, ducts into which lymphatic trunks drain. In the kidney, tubule that receives fluid from several nephrons. pp. 618, 781

colon (kolon) Part of the large intestine. p. 687

colony stimulating factor (ko′le-ne stim′yu-lay″ting fak′tor) Protein that stimulates differentiation and maturation of white blood cells. p. 530

colorblindness (kul′erblīnd′nes) Inherited inability to distinguish certain colors. p. 928

colostrum (ko-los′trum) The first secretion of a woman′s mammary glands after she gives birth. p. 902

compact bone (kom′pakt bōn) Dense tissue in which cells are organized in osteons without apparent spaces. p. 194

complement (kom′plĕ-ment) Group of proteins activated when an antibody binds an antigen; enhances reaction against nonself substances. p. 627

complementary base pair (kom″plĕ-men′tă-re bās pār) Hydrogen bonded adenine and thymine or guanine and cytosine in DNA. Adenine bonds to uracil in RNA. p. 127

completely penetrant (kom-plēt′le pen′e-trent) In genetics, indicates that the frequency of expression of a genotype is 100%. p. 924

complete protein (kom-plēt pro′tēn) Protein that contains adequate amounts of the essential amino acids to maintain body tissues and to promote normal growth and development. p. 705

compound (kom′pownd) Substance composed of two or more chemically bonded elements. p. 72

concentric contraction (kon-sen′-trik kon-trak′-shun) A contraction that shortens a muscle. p. 298

condom (kon′dum) Latex sheath used to cover the penis in the male and line the vagina, preventing sperm from entering the uterus in the female, during sexual intercourse; used as a contraceptive and to minimize the risk of transmitting infection. p. 863

conduction (kon-duk′shun) Movement of body heat into the molecules of cooler objects in contact with the body surface. p. 182

condyle (kon′dīl) Rounded process of a bone, usually at the articular end. p. 208

condylar joint (kon′dĭ-lar joint) Bone with an ovoid projection at one end joined with a bone with a complementary elliptical cavity; ellipsoidal joint. p. 265

cone (kōn) Color receptor in the retina of the eye. p. 473

conformation (kon-for-ma′shun) Three-dimensional form of a protein, determined by its amino acid sequence and attractions and repul-sions between amino acids. p. 65

conjunctiva (kon″junk-ti′vah) Membranous covering on the anterior surface of the eye. p. 463

connective tissue (kŏ-nek′tiv tish′u) Basic type of tissue that consists of cells within an extracellular matrix, including bone, cartilage, blood, loose and dense connective tissue. p. 152

contraception (kon″trah-sep′shun) Behavior or device that prevents fertilization. p. 862

contractility (kon″trak-til′ĭ-te) Shortening of a muscle in response to stimulation. p. 586

contralateral (kon″trah-lat′er-al) Positioned on the opposite side of something else. p. 21

convection (kon-vek′shun) Transmission of heat from one substance to another through the circulation of heated air particles. p. 182

convergence (kon-ver′jens) Nerve impulses arriving at the same neuron. p. 375

cornea (kor′ne-ah) Transparent anterior portion of the outer layer of the eye wall. p. 465

corona radiata (ko-ro′nah ra-di-ă′ta) Follicular cells surrounding the zona pellucida of an ovum. p. 849

coronary artery (kor′o-na″re ar′ter-e) An artery that supplies blood to the wall of the heart. p. 561

coronary sinus (kor′o-na″re si′nus) Large vessel on the posterior surface of the heart into which the cardiac veins drain. p. 563

corpus callosum (kor′pus kah-lo′sum) Mass of white matter in the brain composed of nerve fibers connecting the right and left cerebral hemispheres. p. 398

corpus luteum (kor′pus lu′te-um) Structure that forms from the tissues of a ruptured ovarian follicle and secretes female hormones. p. 858

cortex (kor′teks) Outer layer of an organ such as the adrenal gland, cerebrum, or kidney. p. 504

cortical nephron (kor′tĭ-kl nef′ron) Nephron with a corpuscle in the renal cortex. p. 784

cortisol (kor′ti-sol) Glucocorticoid secreted by the adrenal cortex. p. 507

costal (kos′tal) Pertaining to the ribs. p. 23

countercurrent mechanism (kown′ter-kar″ent me′kĕ-nĭ″zm) Part of the process by which the kidneys concentrate urine. p. 795

covalent bond (ko′va-lent bond) Chemical bond formed by electron sharing between atoms. p. 56

coxal (kok′sel) Pertaining to the hip. p. 23

cranial (kra′ne-al) Pertaining to the cranium, the part of the skull that does not include the face. p. 12

cranial cavity (kran′e-al kav′i-te) Space in the cranium containing the brain. p. 12

cranial nerve (kra′ne-al nerv) Nerve that arises from the brain or brainstem. p. 411

cranium (kra′ne-um) Structure formed by the eight bones of the head. p. 208

creatine phosphate (kre′ah-tin fos′fāt) Muscle biochemical that stores energy. p. 294

crest (krest) Ridgelike projection of a bone. p. 208

cricoid cartilage (kri′koid kar′tĭ-lij) A ringlike cartilage that forms the lower end of the larynx. p. 740

crista ampullaris (kris′tah am-pul-lah′ris) Sensory organ in a semicircular canal that functions in the sense of dynamic equilibrium. p. 460

cross-over (kros o′ver) The exchange of genetic material between homologous chromosomes during meiosis. p. 832

crural (krur′al) Pertaining to the leg. p. 23

cubital (ku′bi-tal) Pertaining to the elbow. p. 23

cuspid (kus′pid) A canine tooth. p. 658

cutaneous (ku-ta′ne-us) Pertaining to the skin. p. 163

cyanosis (si″ah-no′sis) Bluish skin coloration due to decreased blood oxygen concentration. p. 176

cyclic adenosine monophosphate (sik′lik ah-den′o-sēn mon″o-fos′fāt) Circularized derivative of ATP that responds to messages entering a cell and triggers the cell's response; cyclic AMP or cAMP. p. 487

cystic duct (sis′tik dukt) Tube that connects the gallbladder to the bile duct. p. 677

cytokine (si′to-kīn) Type of protein secreted by a T lymphocyte that attacks viruses, virally infected cells, and cancer cells. p. 630

cytokinesis (si-to-kin-e′sis) Division of the cytoplasm. p. 101

cytoplasm (si′to-plazm) The contents of a cell, excluding the nucleus and cell membrane. p. 76

cytoskeleton (si′to-skel″e-ton) System of protein tubules and filaments that reinforces a cell's three-dimensional form and provides scaffolding and transport tracts for organelles. p. 82

cytosol (si′to-sol) Fluid matrix of the cytoplasm. p. 76

D

deamination (de-am″ĭ-na′shun) Removing amino groups (NH_2) from amino acids. p. 704

decomposition (de-kom″po-zish′un) The breakdown of molecules. p. 58

deep (dēp) Far beneath the surface. p. 21

defecation (def″ĕ-ka′shun) Discharge of feces from the rectum through the anus. p. 690

defensin (di-fen′sin) Antimicrobial peptide. p. 626

dehydration (de″hi-dra′shun) Excess water loss. p. 816

dehydration synthesis (de″hi-dra′shun sin′thĕ-sis) Anabolic process that joins small molecules by releasing the equivalent of a water molecule; synthesis. p. 115

dendrite (den′drīt) Process of a neuron that receives input from other neurons. p. 354

dental caries (den′tal kar′ēz) Decalcification and decay of teeth. p. 661

dentin (den′tin) Bonelike substance that forms the bulk of a tooth. p. 659

deoxyhemoglobin (de-ok″sĭ-he″mo-glo′bin) Hemoglobin that does not bind oxygen. p. 526

deoxyribonucleic acid (dē-ok′si-rī″bō-nu-klē″ik as′id) The genetic material; a double-stranded polymer of nucleotides, each containing a phosphate group, a nitrogenous base (adenine, thymine, guanine, or cytosine), and the sugar deoxyribose; DNA. p. 68

depolarization (de-po″lar-ĭ-za′shun) The loss of an electrical charge on the surface of a cell membrane. p. 368

depression (de-presh′un) Downward displacement. p. 270

dermatome (der′mah-tōm) Area of the body supplied by sensory nerve fibers associated with a particular dorsal root of a spinal nerve. p. 419

dermis (der′mis) The thick layer of the skin beneath the epidermis. p. 172

descending colon (de-send′ing ko′lon) Part of the large intestine that passes downward along the left side of the abdominal cavity to the brim of the pelvis. p. 687

descending tract (de-send′ing trakt) Group of nerve fibers that carries nerve impulses downward from the brain through the spinal cord. p. 393

desmosome (des′mo-sōm) Specialized junction between cells, which serves as a "spot weld." p. 144

detrusor muscle (de-trūz′or mus′l) Muscular wall of the urinary bladder. p. 800

dextrose (dek′strōs) Glucose. p. 62

diabetes insipidus (di″ah-be′tēz in-sip′ĭ-dus) Extremely copious urine produced due to a deficiency of antidiuretic hormone or lack of ADH response. p. 498

diabetes mellitus (di″ah-be′tēz mel-li′tus) Elevated glucose in the urine and blood due to a deficiency of insulin or poor response to it. p. 512

diapedesis (di″ah-pĕ-de′sis) Squeezing of leukocytes between the cells of blood vessel walls. p. 533

diaphragm (di′ah-fram) Sheetlike structure of skeletal muscle and connective tissue that separates the thoracic and abdominal cavities; also a caplike contraceptive device inserted in the vagina. pp. 12, 863

diaphysis (di-af′ĭ-sis) Shaft of a long bone. p. 194

diastole (di-as′to-le) Phase of the cardiac cycle when a heart chamber wall relaxes. p. 564

diastolic pressure (di-a-stol′ik presh′ur) Lowest arterial blood pressure during the diastolic phase of the cardiac cycle. p. 580

diencephalon (di″en-sef′ah-lon) Part of the brain in the region of the third ventricle that includes the thalamus and hypothalamus. p. 404

differentiation (dif′er-en″she-a′shun) Cell specialization. p. 76

diffusion (dĭ-fu′zhun) Random movement of molecules from a region of higher concentration toward one of lower concentration. p. 90

digestion (di-jest′yun) Breaking down of large nutrient molecules into smaller molecules that can be absorbed; hydrolysis. p. 652

digital (di′ji-tal) Pertaining to the finger or toe. p. 23

dihydrotestosterone (di-hi″dro-tes-tos′ter-ōn) Hormone produced from testosterone that stimulates certain cells of the male reproductive system. p. 846

dipeptide (di-pep′tīd) Molecule composed of two joined amino acids. p. 116

diploid (dip′loid) Body cell with two full sets of chromosomes, in humans 46. p. 917

disaccharide (di-sak′ah-rīd) Sugar produced by the union of two monosaccharides. p. 62

distal (dis′tal) Farther from the trunk or origin; opposite of proximal. p. 21

diuretic (di″u-ret′ik) Substance that increases urine production. p. 792

divergence (di-ver′jens) Spreading apart. p. 376

DNA Deoxyribonucleic acid. p. 68

DNA damage response A group of reactions that enable a cell to repair certain types of DNA changes that occur during replication. p. 137

dominant allele (dom′ĭ-nant ah-lēl) The form of a gene that is expressed. p. 920

dorsal root (dor′sal rōot) Sensory branch of a spinal nerve by which it joins the spinal cord. p. 419

dorsal root ganglion (dor′sal rōot gang′gle-on) Mass of sensory neuron cell bodies in the dorsal root of a spinal nerve. p. 419

dorsiflexion (dor″si-flek′shun) Ankle movement that brings foot closer to shin. p. 269

dorsum (dor′sum) Pertaining to the back surface of a body part. p. 23

ductus arteriosus (duk′tus ar-te″re-o′sus) Blood vessel that connects the pulmonary artery and the aorta in a fetus. p. 897

ductus deferens (duk′tus def′er-ens) Tube that leads from the epididymis to the urethra of the male reproductive tract (pl., *ductus deferentia; vas deferens*). p. 839

ductus venosus (duk′tus ven-o′sus) Blood vessel that connects the umbilical vein and the inferior vena cava in a fetus. p. 897

duodenum (du″o-de′num) First portion of the small intestine that leads from the stomach to the jejunum. p. 680

dural sinus (du′ral si′nus) Blood-filled channel formed by the splitting of the dura mater into two layers. p. 384

dura mater (du′rah ma′ter) Tough outer layer of the meninges. p. 384

dynamic equilibrium (di-nam′ik e″kwĭ-lib′re-um) Maintenance of balance when the head and body are suddenly moved or rotated. p. 459

E

eccentric contraction (ek-sen′trik kon-trak′shun) Force in a muscle less than that required to lift or move an object. p. 298

eccrine gland (ek′rin gland) Sweat gland that maintains body temperature. p. 180

ECG Electrocardiogram; EKG. p. 568

ectoderm (ek′to-derm) Outermost primary germ layer in the embryo. p. 885

edema (ĕ-de′mah) Fluid accumulation in tissue spaces. p. 620

effector (ĕ-fek′tor) A muscle or gland that effects change in the body. p. 9

efferent (ef′er-ent) Conducting away from the center. p. 781

ejaculation (e-jak″u-la′shun) Discharge of sperm-containing semen from the male urethra. p. 844

ejaculatory duct (e-jak′u-lah-to″re dukt) Tube, formed by the joining of the ductus deferens and the tube from the seminal vesicle, that transports sperm to the urethra. p. 844

elastic cartilage (e-las′tik kar′tĭ-lij) Opaque, flexible connective tissue with branching yellow fibers throughout the extracellular matrix. p. 160

elastic fiber (e-las′tik fi′ber) Yellow, stretchy, threadlike structure in connective tissue. p. 155

elastin (e-las′tin) Protein that comprises the yellow, elastic fibers of connective tissue. p. 155

electrocardiogram (e-lek″tro-kar′de-o-gram″) Recording of the electrical activity associated with the heartbeat; ECG or EKG. p. 568

electrolyte (e-lek′tro-līt) Substance that ionizes in a water solution. p. 59

electrolyte balance (e-lek′tro-līt bal′ans) Condition when the quantities of electrolytes entering the body equal those leaving it. p. 61

electron (e-lek′tron) A small, negatively charged particle that encircles the nucleus of an atom. p. 52

electron shell (e-lek′tron shel) The path formed by an electron or several electrons encircling the nucleus of an atom at a particular energy level. p. 55

electron transport chain (e-lek′tron trans′pohrt) Series of oxidation-reduction reactions that takes high-energy electrons from glycolysis and the citric acid cycle to form water and ATP. p. 120

element (el′ĕ-ment) Chemical substance with only one type of atom. p. 51

elevation (el-e-vā-shun) Upward movement of a body part. p. 270

embolus (em′bo-lus) Blood clot or gas bubble carried in the circulation that may obstruct a blood vessel. p. 541

embryo (em′bre-o) Prenatal stage of development after germ layers form and rudiments of all organs are present. p. 849

embryonic disc (em″brē-on′ik disk) Flattened area in the cleavage embryo from which the embryo arises. p. 885

emission (e-mish′un) Movement of sperm cells from the ductus deferens into the ejaculatory duct and urethra. p. 844

emulsification (e-mul″sĭ-fĭ′ka′shun) Breaking up of fat globules into smaller droplets by the action of bile salts. p. 678

enamel (e-nam′el) Hard covering on the exposed surface of a tooth. p. 659

end-diastolic volume (end di-a-stol′ik vol′ūm) Blood volume remaining in a ventricle at the end of ventricular diastole. p. 585

endocardium (en″do-kar′de-um) Inner lining of the heart chambers. p. 555

endochondral bone (en′do-kon′dral bōn) Bone that begins as hyaline cartilage that is subsequently replaced by bone tissue. p. 197

endocrine gland (en′do-krīn gland) Gland that secretes hormones directly into the blood. p. 150

endocytosis (en″do-si-to′sis) Process by which a cell membrane envelopes a substance and draws it into the cell in a vesicle. p. 96

endoderm (en′do-derm) Innermost primary germ layer in the embryo. p. 885

endolymph (en′do-limf) Fluid in the membranous labyrinth of the inner ear. p. 453

endometrium (en″do-me′tre-um) Inner lining of the uterus. p. 854

endomysium (en″do-mis′e-um) Sheath of connective tissue surrounding each skeletal muscle fiber. p. 286

endoneurium (en″do-nu′re-um) Layer of loose connective tissue that surrounds individual nerve fibers. p. 412

endoplasmic reticulum (en-do-plaz′mic rē-tik′u-lum) Organelle composed of a system of connected membranous tubules and vesicles. p. 82

endorphin (en-dor′fin) Neuropeptide synthesized in the pituitary gland that suppresses pain. p. 374

endosteum (en-dos′tē-um) Tissue lining the medullary cavity in a bone. p. 194

endothelium (en″do-the′le-um) Layer of epithelial cells that forms the inner lining of blood vessels and heart chambers. p. 574

end-systolic volume (end sis-tol′ik vol′ūm) Blood volume remaining in a ventricle at the end of ventricular systole. p. 585

energy (en′er-je) An ability to move something and thus do work. p. 119

energy balance (en′er-je bal′ans) When the caloric intake of the body equals its caloric output. p. 708

enkephalin (en-kef′ah-lin) Neuropeptide in the brain and spinal cord that inhibits pain impulses. p. 374

enterogastric reflex (en-ter-o-gas′trik re′fleks) Inhibition of gastric (stomach) peristalsis and secretions when food enters the small intestine. p. 670

enzyme (en′zīm) Protein that catalyzes a specific biochemical reaction. p. 64

eosinophil (e″o-sin′o-fil) White blood cell containing cytoplasmic granules that stain with acidic dye. p. 531

ependyma (ĕ-pen′dĭ-mah) Membrane composed of neuroglial cells that lines the ventricles of the brain. p. 363

epicardium (ep″ĭ-kar′de-um) Visceral part of the pericardium on the surface of the heart. p. 555

epicondyle (ep″ĭ-kon′dīl) Projection of a bone above a condyle. p. 208

epidermis (ep″ĭ-der′mis) Outer epithelial layer of the skin. p. 172

epididymis (ep″ĭ-did′ĭ-mis) Highly coiled tubule that leads from the seminiferous tubules of the testis to the ductus deferens (pl., *epididymides*). p. 838

epidural space (ep″ĭ-du′ral spās) Space between the dural sheath of the spinal cord and the bone of the vertebral canal. p. 384

epigastric region (ep″ĭ-gas′trik re′jun) Upper middle part of the abdomen. p. 22

epiglottis (ep″ĭ-glot′is) Flaplike, cartilaginous structure at the back of the tongue near the entrance to the trachea. p. 664

epimysium (ep″ĭ-mis′e-um) Outer sheath of connective tissue surrounding a skeletal muscle. p. 286

epinephrine (ep″ĭ-nef′rin) Hormone the adrenal medulla secretes during times of stress. p. 426

epineurium (ep″ĭ-nu′re-um) Outermost layer of connective tissue surrounding a nerve. p. 412

epiphyseal plate (ep″ĭ-fiz′e-al plāt) Cartilaginous layer in the long bone epiphysis that grows. p. 198

epiphysis (e-pif′ĭ-sis) End of a long bone. p. 194

epithelial membrane (ep″ĭ-the′le-al mem′brān) Thin layer of tissue lining a cavity or covering a surface. p. 163

epithelial tissue (ep″ĭ-the′le-al tish′u) One of the basic types of tissue that covers all free body surfaces. p. 144

equilibrium (e″kwĭ-lib′re-um) State of balance between two opposing forces. p. 92

erection (ĕ-rek′shun) The filling of penile tissues with blood, stiffening and elevating the structure. p. 844

erythroblast (ĕ-rith′ro-blast) A red blood cell at an immature stage. p. 527

erythroblastosis fetalis (ĕ-rith″ro-blas-to′sis fe-tal′is) Life-threatening massive agglutination in the fetus or neonate due to the mother's anti-Rh antibodies reacting with the baby's Rh-positive red blood cells. p. 547

erythrocyte (ĕ-rith′ro-sīt) Red blood cell. p. 526

erythropoietin (ĕ-rith″ro-poi′ĕ-tin) Kidney hormone that promotes red blood cell formation; EPO. p. 527

esophageal hiatus (ĕ-sof″ah-je′al hi-a′tus) Opening in the diaphragm through which the esophagus passes. p. 665

esophagus (ĕ-sof′ah-gus) Tubular part of the digestive tract between the pharynx and the stomach. p. 665

essential amino acid (ĕ-sen′shal ah-me′no as′id) Amino acid required for health that body cells cannot synthesize in adequate amounts. p. 705

essential fatty acid (ĕ-sen′shal fat′e as′id) Fatty acid required for health that body cells cannot synthesize in adequate amounts. p. 703

essential nutrient (ĕ-sen′shal nu′trē-ent) Nutrient necessary for growth, normal functioning, and maintaining life that the diet must supply because the body cannot synthesize it. p. 699

estrogens (es′tro-jenz) Group of hormones (including estradiol, estrone, and estriol) that stimulates the development of female secondary sex characteristics and produces an environment suitable for fertilization, implantation, and growth of an embryo. p. 857

euploid (u'ploid) Having a balanced set of chromosomes. p. 929
evaporation (e″vap′o-ra-shun) Changing a liquid into a gas. p. 182
eversion (e-ver′zhun) Turning the sole of the foot outward. p. 269
exchange reaction (eks-chānj re-ak′shun) Chemical reaction in which parts of two types of molecules trade positions. p. 58
excretion (ek-skre′shun) Elimination of metabolic wastes. p. 18
exocrine gland (ek′so-krin gland) Gland that secretes its products into a duct or onto a body surface. p. 150
exocytosis (eks″o-si-to′sis) Transport of a substance out of a cell in a membrane-bounded vesicle. p. 98
expiration (ek″spĭ-ra′shun) Expulsion of air from the lungs. p. 747
expiratory reserve volume (eks-pi′rah-to″re re-zerv′ vol′ūm) Volume of air that can be exhaled in addition to the tidal volume. p. 752
extension (ek-sten′shun) Movement increasing the angle between parts at a joint. p. 267
exteroreceptive sense (eks″ter-o-re-sep′tiv sens) Associated with changes at the body surface. p. 440
extracellular fluid (eks″trah-selu′-lar floo′id) Body fluids outside cells. p. 9
extracellular matrix (eks″trah-selu′-lar ma′trix) Fibers and ground substance among connective tissue cells. p. 153

F

facet (fas′et) Small, flattened surface of a bone. p. 22
facilitated diffusion (fah-sil′ĭ-tāt′id dĭ-fu′zhun) Diffusion in which a carrier molecule transports a substance across a cell membrane from a region of higher concentration to a region of lower concentration. p. 90
facilitation (fah-sil″ĭ-tā′shun) Subthreshold stimulation of a neuron that increases response to further stimulation. p. 375
fascia (fash′e-ah) Sheet of fibrous connective tissue that encloses a muscle. p. 285
fascicle (fas′ĭ-k′l) Small bundle of muscle fibers. p. 286
fat (fat) Adipose tissue; or an organic molecule that includes glycerol and fatty acids. p. 62
fatty acid (fat′e as′id) Building block of a fat molecule. p. 63
fatty acid oxidase (fat′e as′id ok′si-days″) An enzyme that catalyzes the removal of hydrogen or electrons from a fatty acid molecule. p. 702
feces (fe′sēz) Material expelled from the digestive tract during defecation. p. 688
femoral (fem′or-al) Pertaining to the thigh. p. 23
fertilization (fer″tĭ-lĭ-za′shun) Union of an egg cell and a sperm cell. p. 876
fetus (fe′tus) Prenatal human after eight weeks of development. p. 892
fever (fe′ver) Elevation of body temperature above normal. p. 628
fibrillation (fi″brĭ-la′shun) Uncoordinated contraction of cardiac muscle fibers. p. 572
fibrin (fi′brin) Insoluble, fibrous protein formed from fibrinogen during blood coagulation. p. 539
fibrinogen (fi-brin′o-jen) Plasma protein converted into fibrin during blood coagulation. p. 535
fibroblast (fi′bro-blast) Cell that produces fibers in connective tissues. p. 153
fibrocartilage (fi″bro-kar′ti-lij) Strongest and most durable cartilage; made up of cartilage cells and many collagenous fibers. p. 160
fibrous joint (fi′brus joint) Two or more bones joined by connective tissue containing many fibers. p. 261
filtration (fil-tra′shun) Movement of material through a membrane as a result of hydrostatic pressure. p. 58
fimbria (fim′bre-ah) Fringelike process at the distal end of the infundibulum of the uterine tube. p. 852
fissure (fish′ūr) Narrow cleft separating parts, such as the lobes of the cerebrum. p. 398
flagellum (flah-jel′um) Relatively long, motile process that extends from the surface of a sperm cell. p. 87
flexion (flek′shun) Bending at a joint to decrease the angle between bones. p. 267
folacin (fōl′ah-sin) B complex vitamin necessary for normal cellular synthetic processes; folic acid. p. 715
follicle (fol′ĭ-kl) Pouchlike depression or cavity. p. 499
follicle-stimulating hormone (fol′ĭ-kl stim′u-la″ting hor′mōn) Substance secreted by the anterior pituitary gland to stimulate development of an ovarian follicle in a female or production of sperm cells in a male; FSH. p. 497
follicular cells (fŏ-lik′u-lar selz) Ovarian cells that surround a developing egg cell and secrete female sex hormones. p. 847
fontanel (fon″tah-nel′) Membranous region between certain cranial bones in the skull of a fetus or infant. p. 216
food pyramid (food pēr′ah-mid) Triangular structure divided into sections representing different types and proportions of nutrients the body requires. p. 722
foramen (fo-ra′men) An opening, usually in a bone or membrane (pl., *foramina*). p. 208
foramen magnum (fo-ra′men mag′num) Opening in the occipital bone of the skull through which the spinal cord passes. p. 208
foramen ovale (fo-ra′men o-val′e) Opening in the interatrial septum of the fetal heart. p. 897
forebrain (fōr′brān) Anteriormost part of the developing brain that gives rise to the cerebrum and basal nuclei. p. 398
fossa (fos′ah) Depression in a bone or other part. p. 208
fovea centralis (fo′ve-ah sen-tral′is) Depressed region of the retina, consisting of densely packed cones, that provides the greatest visual acuity. p. 469
fracture (frak′chur) Break in a bone. p. 202
Frank-Starling law of the heart (frank star′ling law of thĕ hart) The relatively longer the cardiac muscle fibers, the greater the force of contraction. p. 586
free radical (frē rad′eh-kel) Highly reactive by-product of metabolism that can damage tissue. p. 909
frontal (frun′tal) Pertaining to the forehead. p. 22
FSH Follicle-stimulating hormone. p. 497
functional residual capacity (funk′shun-al re-zid′u-al kah-pas′i-te) Amount of air remaining after a normal quiet respiration. p. 753
functional syncytium (funk′shun-al sin-sish′e-um) Merging cells performing as a unit; those of the heart are joined electrically. A syncytium lacks cell boundaries, appearing as a multinucleated structure. p. 565

G

galactose (gah-lak′tōs) Monosaccharide component of the disaccharide lactose. p. 62
gallbladder (gawl′blad-er) Saclike organ associated with the liver that stores and concentrates bile. p. 677

ganglion (gang′gle-on) Mass of neuron cell bodies, usually outside the central nervous system (pl., *ganglia*). p. 360

gastric gland (gas′trik gland) Gland in the stomach wall that secretes gastric juice. p. 666

gastric juice (gas′trik joo-s) Secretion of the gastric glands in the stomach. p. 666

gastrin (gas′trin) Hormone secreted by the stomach lining that stimulates secretion of gastric juice. p. 668

gastrula (gas′troo-lah) Embryonic stage following the blastula; cells differentiate into endoderm, mesoderm, and ectoderm. p. 885

gene (jēn) Part of a DNA molecule that encodes the information to synthesize a protein, a control sequence, or tRNA or rRNA. The unit of inheritance. p. 125

general sense (jen′er-al sens) Involves receptors widely distributed throughout the body. p. 438

genetic code (jĕ-net′ik kōd) Information for synthesizing proteins encoded in the nucleotide sequence of DNA molecules. p. 124

genetic heterogeneity (jĕ-net′ik het″er-o-je-ne′ĭ-te) A trait or condition that is inherited in more than one way. p. 924

genetics (jĕ-net′iks) The study of the transmission of inherited traits. p. 917

genital (jen′i-tal) Pertaining to the genitalia (internal and external organs of reproduction). p. 23

genome (jeh′nōm) Complete set of genetic instructions for an organism. p. 125

genomics (je-nom′iks) Study of all of the genetic information in an individual organism. p. 918

genotype (je′no-tīp) The alleles of a particular gene in an individual. p. 920

glans penis (glanz pe′nis) Enlarged mass of corpus spongiosum at the end of the penis; may be covered by the foreskin. p. 843

gliding joint (glīd′eng joint) Two joined bones with nearly flat surfaces. p. 266

globin (glo′bin) Protein part of a hemoglobin molecule. p. 762

globulin (glob′u-lin) Type of protein in blood plasma. p. 535

glomerular capsule (glo-mer′u-lar kap′sŭl) Proximal part of a renal tubule that encloses the glomerulus; Bowman's capsule. p. 781

glomerular filtrate (glo-mer′u-lar fil′trāt) Liquid that passes out of the glomerular capillaries in the kidney into the glomerular capsule. p. 786

glomerular filtration (glo-mer′u-lar fil-tra′shun) Process in which blood pressure forces fluid through the glomerular capillaries in the kidney into the glomerular capsule. p. 785

glomerulus (glo-mer′u-lus) Capillary tuft in the glomerular capsule of a nephron. p. 779

glottis (glot′is) Slitlike opening between the true vocal cords. p. 741

glucagon (gloo′kah-gon) Hormone secreted by the pancreatic islets that releases glucose from glycogen. p. 509

glucocorticoid (gloo″ko-kor′tĭ-koid) Any one of a group of hormones secreted by the adrenal cortex that influences carbohydrate, fat, and protein metabolism. p. 507

gluconeogenesis (gloo″ko-ne″o-jen′ĕ-sis) Synthesis of glucose from noncarbohydrates such as amino acids. p. 701

glucose (gloo′kōs) Monosaccharide in blood that is the primary source of cellular energy. p. 62

gluteal (gloo′te-al) Pertaining to the buttocks. p. 23

glycerol (glis′er-ol) Organic compound that is a building block for fat molecules. p. 63

glycogen (gli′ko-jen) Polysaccharide that stores glucose in the liver and muscles. p. 62

glycolysis (gli-kol′ĭ-sis) The energy-releasing breakdown of glucose to pyruvic acid during cellular respiration. p. 120

glycoprotein (gli″ko-pro′te-in) Compound composed of a carbohydrate and a protein. p. 485

goblet cell (gob′let sel) Epithelial cell specialized to secrete mucus. p. 146

goiter (goi′ter) Enlarged thyroid gland. p. 496

Golgi apparatus (gol′jē ap″ah-ra′tus) Organelle that prepares cellular products for secretion. p. 83

Golgi tendon organ (gol″jē ten′dun or′gan) Sensory receptors in tendons close to muscle attachments that are involved in reflexes that help maintain posture. p. 444

gomphosis (gom-fo′sis) Type of joint in which a cone-shaped process is fastened in a bony socket. p. 262

gonad (go′nad) A sex cell-producing organ; an ovary or testis. p. 497

gonadotropin (go-nad″o-trōp′in) Hormone that stimulates activity in the gonads. p. 845

G protein (g pro′tēn) Organic compound which activates an enzyme bound to the inner surface of the cell membrane, eliciting a signal. p. 488

granulocyte (gran′u-lo-sīt) Leukocyte with granules in its cytoplasm. p. 530

gray matter (grā mat′er) Region of the central nervous system that lacks myelin and thus appears gray. p. 358

gray ramus (grā ra′mus) Short nerve containing postganglionic axons returning to a spinal nerve. p. 426

growth (grōth) Process by which a structure enlarges. p. 876

growth hormone (grōth hor′mōn) Hormone released by the anterior lobe of the pituitary gland that promotes the growth of the organism; GH. p. 201

gubernaculum (goo″ber′nak′u-lum) Structure that guides another structure. p. 833

gyrus (ji′rus) Elevation on the brain's surface caused by infolding. p. 398

H

hair cell (hār sel) Mechanoreceptor in the inner ear between the basilar membrane and the tectorial membrane that triggers action potentials in fibers of the auditory nerve. p. 455

hair follicle (hār fol′ĭ-kl) Tubelike depression in the skin where a hair develops. p. 178

haploid (hap′loid) Sex cell with a single set of chromosomes, in humans 23. p. 831

hapten (hap′ten) Small molecule that combines with a larger one, forming an antigen. p. 628

haustra (haws′trah) Pouches in the wall of the large intestine. p. 688

hematocrit (he-mat′o-krit) The percentage by volume of red blood cells in a sample of whole blood. p. 523

hematopoiesis (hem″ah-to-poi-e′sis) Production of blood and blood cells; hemopoiesis. p. 202

hematopoietic stem cell (hem″ah-to-poi-e′tik stem sel) Stem cell that gives rise to blood cells. p. 524

heme (hēm) Iron-containing part of a hemoglobin molecule. p. 762

hemizygous (hem″ĭ-zi′gus) A gene on the Y chromosome in humans. p. 928

hemoglobin (he″mo-glo′bin) Oxygen-carrying pigment in red blood cells. p. 203

hemostasis (he″mo-sta′sis) Stoppage of bleeding. p. 538

hepatic (hĕ-pat′ik) Pertaining to the liver. p. 673

hepatic lobule (hĕ-pat′ik lob′ūl) Functional unit of the liver. p. 673

hepatic sinusoid (hĕ-pat′ik si′nŭ-soid) Vascular channel in the liver. p. 673

heterozygous (het″er-o-zi′gus) Different alleles in a gene pair. p. 919

hilum (hi′lum) Depression where vessels, nerves, and other structures (bronchus, ureter, etc.) enter an organ. p. 621

hilus (hi′lus) Hilum. p. 621

hindbrain (hīnd′brān) Posteriormost part of the developing brain that gives rise to the cerebellum, pons, and medulla oblongata. p. 398

hinge joint (hinj joint) Two bones joined where the convex end of one bone fits into the complementary concave end of another. p. 266

hippocampus (hip″o-kam′pus) Part of the cerebral cortex where memories form. p. 404

histamine (his′tah-min) Substance released from stressed cells that promotes inflammation. p. 579

holocrine gland (ho′lo-krin gland) Gland whose secretion contains entire secretory cells. p. 150

homeostasis (ho″me-o-sta′sis) State of equilibrium in which the internal environment of the body remains in the normal range. p. 9

homeostatic mechanism (ho″me-o-stat′ik mek′ah-nizm) Process used to maintain a normal internal environment in the body. p. 9

homozygous (ho″mo-zi′gus) Identical alleles in a gene pair. p. 919

hormone (hor′mōn) Substance secreted by an endocrine gland and transported in the blood. p. 16

human chorionic gonadotropin (hu′man ko″re-on′ik gon″ah-do-tro′pin) Hormone, secreted by an embryo, that helps form the placenta; hCG. p. 882

humoral immune response (hu′mor-al i-mūn′ ri-spons′) Circulating antibodies′ destruction of cells bearing nonself antigens. p. 633

hyaline cartilage (hi′ah-līn kar′tĭ-lij) Semitransparent, flexible connective tissue with an ultra-fine fiber matrix. p. 160

hydrogen bond (hi′ dro-jen bond) Weak bond between a hydrogen atom and an atom of oxygen or nitrogen. p. 57

hydrolysis (hi-drol′ĭ-sis) Enzymatically adding a water molecule to split a molecule. p. 116

hydrostatic pressure (hi″dro-stat′ik presh′ur) Pressure exerted by fluids, such as blood pressure. p. 8

hymen (hi′men) Membranous fold of tissue that partially covers the vaginal opening. p. 854

hyperextension (hi″per-ek-sten′shun) Extreme extension; continuing extension beyond the anatomical position. p. 267

hyperglycemia (hi″per-gli-se′me-ah) Elevated blood glucose. p. 510

hyperkalemia (hi″per-kah-le′me-ah) Elevated blood potassium. p. 513

hypernatremia (hi″per-nah-tre′me-ah) Elevated blood sodium. p. 816

hyperparathyroidism (hi″per-par″ah-thi′roi-dizm) Oversecretion of parathyroid hormone. p. 504

hyperpolarization (hi″per-po″lar-i-za′shun) Increase in the negativity of the resting potential of a cell membrane. p. 368

hypertension (hi″per-ten′shun) Elevated blood pressure. p. 588

hyperthyroidism (hi″per-thi′roi-dizm) Oversecretion of thyroid hormones. p. 501

hypertonic (hi″per-ton′ik) Solution with a greater osmotic pressure than the solution with which it is compared (usually body fluids). p. 94

hypertrophy (hi-per′tro-fe) Enlargement of an organ or tissue. p. 300

hyperventilation (hi″per-ven″tĭ-la′shun) Abnormally deep and prolonged breathing. p. 758

hypochondriac region (hi″po-kon′dre-ak re′jun) Portion of the abdomen on either side of the epigastric region. p. 22

hypogastric region (hi″po-gas′trik re′jun) Lower middle portion of the abdomen. p. 22

hypoglycemia (hi″po-gli-se′me-ah) Low blood glucose. p. 510

hypokalemia (hi″po-kah-le′me-ah) Low blood potassium. p. 573

hyponatremia (hi″po-nah-tre′me-ah) Low blood sodium. p. 816

hypoparathyroidism (hi″po-par″ah-thi′roi-dizm) Undersecretion of parathyroid hormone. p. 504

hypophysis (hi-pof′i-sis) Pituitary gland. p. 493

hypoproteinemia (hi″po-pro″te-ĭ-ne′me-ah) Low blood proteins. p. 816

hypothalamus (hi″po-thal′ah-mus) Part of the brain located below the thalamus and forming the floor of the third ventricle. p. 405

hypothyroidism (hi″po-thi′roi-dizm) Undersecretion of thyroid hormones. p. 501

hypotonic (hi″po-ton′ik) Solution with a lower osmotic pressure than the solution with which it is compared (usually body fluids). p. 94

hypoxia (hi-pok′se-ah) Deficiency of oxygen in the tissues. p. 440

I

idiotype (id′e-o-tīp′) Parts of an antibody′s antigen binding site that are complementary in conformation to a particular antigen. p. 634

ileocecal sphincter (il″e-o-se′kal sfingk′ter) Ring of muscle fibers at the distal end of the ileum where it joins the cecum. p. 686

ileum (il′e-um) Part of the small intestine between the jejunum and the cecum. p. 680

iliac region (il′e-ak re′jun) Part of the abdomen on either side of the hypogastric region. p. 22

ilium (il′e-um) One of the bones of a hipbone. p. 231

immunity (ĭ-mu′nĭ-te) Resistance to the effects of specific disease-causing agents. p. 532

immunoglobulin (im″u-no-glob′u-lin) Globular plasma protein that functions as an antibody. p. 633

immunosuppressive drugs (im″u-no-sŭ-pres′iv drugz) Substances that suppress the immune response against transplanted tissue. p. 641

implantation (im″plan-ta′shun) Embedding of a cleavage embryo in the lining of the uterus. p. 882

impulse (im′puls) Wave of depolarization conducted along a nerve fiber or muscle fiber. p. 16

incisor (in-si′zor) One of the front teeth adapted for cutting food. p. 659

inclusion (in-kloo′zhun) Inert chemicals in the cytoplasm. p. 89

incomplete dominance (in″kom-plēt′ do′meh-nents) Heterozygote whose phenotype is intermediate between the phenotypes of the two homozygotes. p. 922

incompletely penetrant (in″kom-plēt′le pen′e-trent) When the frequency of genotype expression is less than 100%. p. 924

inert (in-ert′) Nonreactive with other elements. p. 55

infancy (in′fan-se) Period of life from the fifth week after birth through the end of the first year. p. 905

infection (in-fek′shun) Invasion and multiplication of microorganisms or infectious agents in body tissues. p. 626

inferior (in-fēr′e-or) Situated below something else; pertaining to the lower surface of a part. p. 21

inflammation (in″flah-ma′shun) Tissue response to stress that includes dilation of blood vessels and fluid accumulation in the affected region. p. 183

infundibulum (in″fun-dib′u-lum) Stalk attaching the pituitary gland to the base of the brain. p. 405

inguinal (ing′gwĭ-nal) Pertaining to the groin region. p. 23

inguinal canal (ing′gwĭ-nal kah-nal′) Passage in the lower abdominal wall through which a testis descends into the scrotum. p. 833

inhibin (in′hib′in) Hormone secreted by cells of the testes and ovaries that inhibits the secretion of FSH from the anterior pituitary gland. p. 845

innate defense (in-nāt′ de-fens′) Inborn, nonspecific defense that blocks entry of or destroys pathogens. p. 626

inorganic (in″or-gan′ik) Chemical that does not include carbon and hydrogen atoms. p. 60

insertion (in-ser′shun) End of a muscle attached to a movable part. p. 303

inspiration (in″spĭ-ra′shun) Breathing in; inhalation. p. 747

inspiratory capacity (in-spi′rah-to″re kah-pas′i-te) Volume of air equal to the tidal volume plus the inspiratory reserve volume. p. 753

inspiratory reserve volume (in-spi′rah-to″re re-zerv′ vol′ūm) Amount of air that can be inhaled in addition to the tidal volume. p. 752

insula (in′su-lah) Cerebral lobe deep within the lateral sulcus. p. 400

insulin (in′su-lin) Hormone the pancreatic islets secrete that stimulates cells to take up glucose. p. 93

integumentary (in-teg-u-men′tar-e) Pertaining to the skin and its accessory organs. p. 14

intercalated disc (in-ter″kăh-lāt′ed disk) Membranous boundary between adjacent cardiac muscle cells. p. 301

intercellular junction (in″ter-sel′u-lar junk′shun) Site of union between cells. p. 144

interferon (in″ter-fēr′on) Class of immune system chemicals (cytokines) that inhibit multiplication of viruses and growth of tumors. p. 626

interleukin (in″ter-lu′kin) Class of immune system chemicals (cytokines) with varied effects. p. 530

internal environment (in-ter′nal en-vi′ron-ment) Conditions and elements that make up the inside of the body, surrounding the cells. p. 9

interneuron (in″ter-nu′ron) Neuron between a sensory neuron and a motor neuron; intercalated; internuncial, or association neuron. p. 361

interphase (in′ter-fāz) Period between two cell divisions when a cell metabolizes and prepares for division. p. 100

interstitial cell (in″ter-stish′al sel) Hormone-secreting cell between the seminiferous tubules of the testis. p. 835

intervertebral disc (in″ter-ver′tĕ-bral disk) Layer of fibrocartilage between the bodies of adjacent vertebrae. p. 206

intestinal gland (in-tes′tĭ-nal gland) Tubular gland at the base of a villus in the intestinal wall. p. 681

intracellular fluid (in″trah-sel′u-lar floo′id) Fluid in cells. p. 812

intramembranous bone (in″trah-mem′brah-nus bōn) Bone that forms from membranelike layers of primitive connective tissue. p. 197

intrauterine device (in″trah-u′ter-in de-vīs) Solid object placed in the uterine cavity to prevent implantation of a fertilized ovum; IUD. p. 866

intrinsic factor (in-trin′sik fak′tor) Substance that gastric glands produce to promote absorption of vitamin B_{12}. p. 528

inversion (in-ver′zhun) Turning the sole of the foot inward. p. 269

involuntary (in-vol′un-tār″e) Not consciously controlled; functions automatically. p. 163

ion (i′on) Atom or molecule with an electrical charge. p. 55

ionic bond (i-on′ik bond) Chemical bond formed between two ions by transfer of electrons; electrovalent bond. p. 55

ipsilateral (ip″sī-lat′er-al) On the same side. p. 21

iris (i′ris) Colored, muscular part of the eye around the pupil that regulates its size. p. 467

ischemia (is-ke′me-ah) Deficiency of blood in a body part. p. 440

isometric contraction (i″so-met′rik kon-trak′shun) Muscular contraction that does not change the muscle length. p. 298

isotonic (i″so-ton′ik) Solution with the same osmotic pressure as the solution with which it is compared (usually body fluids). p. 94

isotonic contraction (i″so-ton′ik kon-trak′shun) Muscular contraction that changes the muscle length. p. 298

isotope (i′so-tōp) Atom that has the same number of protons as other atoms of an element but has a different number of neutrons in its nucleus. p. 53

isthmus (is′mus) Narrow connection between two larger parts. p. 499

J

jejunum (jĕ-joo′num) Part of the small intestine between the duodenum and the ileum. p. 680

joint capsule (joint kap′sul) An envelope, attached to the end of each bone at the joint, enclosing the cavity of a synovial joint. p. 264

juxtaglomerular apparatus (juks″tah-glo-mer′u-lār ap″ah-ra′tus) A group of cells in the wall of the afferent arteriole in the kidney that plays a role in the control of renin secretion. p. 784

juxtamedullary nephron (juks″tah-med′u- lār-e nef′ron) A nephron with its corpuscle near the renal medulla. p. 784

K

karyotype (kar′ē-o-tīp) A chart of the chromosomes arranged in size-ordered homologous pairs. The human karyotype has 23 chromosome pairs. p. 918

keratin (ker′ah-tin) Protein in epidermis, hair, and nails. p. 147

keratinization (ker″ah-tin″ĭ-za′shun) Process by which cells form fibrils of keratin and harden. p. 173

ketone body (ke′tōn bod′e) Type of compound produced during fat catabolism, including acetone, acetoacetic acid, and betahydroxybutyric acid. p. 703

Kupffer cell (koop′fer sel) Large, fixed phagocyte in the liver that removes bacterial cells from the blood. p. 673

kwashiorkor (kwash″e-or′kor) Starvation resulting from a switch from breast milk to food deficient in protein. p. 726

L

labor (la′bor) Process of childbirth. p. 900

labyrinth (lab′ĭ-rinth) System of connecting tubes in the inner ear, including the cochlea, vestibule, and semicircular canals. p. 453

lacrimal gland (lak′rĭ-mal gland) Tear-secreting gland. p. 463

lactase (lak′tās) Enzyme that catalyzes breakdown of lactose into glucose and galactose. p. 118

lacteal (lak′te-al) Lymphatic capillary associated with a villus of the small intestine. p. 681

lactic acid (lak′tik as′id) Organic compound formed from pyruvic acid during the anaerobic reactions of cellular respiration. p. 122

lactose (lak′tōs) A disaccharide in milk; milk sugar. p. 701

lacuna (lah-ku′nah) Hollow cavity. p. 158

lamella (lah-mel′ah) Layer of matrix in bone tissue. p. 160

lamellated corpuscle (lah-mel′a-ted kor′pusl) Nerve endings deep in the dermis providing perception of pressure; Pacinian corpuscle. p. 440

large intestine (lahrj in-tes′tin) Part of the gastrointestinal tract from the ileum to the anus, divided into the cecum, colon, rectum, and anal canal. p. 686

laryngopharynx (lah-ring″go-far′ingks) Lower part of the pharynx near the posterior to the larynx. p. 663

larynx (lar′ingks) Structure between the pharynx and trachea that houses the vocal cords. p. 740

latent period (la′tent pe′re-od) Time between the application of a stimulus and the beginning of a response in a muscle fiber. p. 296

lateral (lat′er-al) Pertaining to the side. p. 21

leptin (lep′tin) Hormone, produced by fat cells, that communicates with the hypothalamus to indicate the degree of hunger. p. 700

leukocyte (lu′ko-sīt) White blood cell. p. 530

leukocytosis (lu″ko-si-to′sis) Too many white blood cells in the blood. p. 533

leukopenia (lu″ko-pe′ne-ah) Too few white blood cells in the blood. p. 533

lever (lev′er) Simple mechanical device consisting of a rod, fulcrum, weight, and a source of energy that is applied to some point on the rod. p. 301

ligament (lig′ah-ment) Cord or sheet of connective tissue binding two or more bones at a joint. p. 155

limbic system (lim′bik sis′tem) Connected structures in the brain that produce emotional feelings. p. 407

linea alba (lin′e-ah al′bah) Narrow band of tendinous connective tissue in the midline of the anterior abdominal wall. p. 320

lingual (ling′gwal) Pertaining to the tongue. p. 657

lingual frenulum (ling′gwal fren′ u-lum) Fold of tissue that anchors the tongue to the floor of the mouth. p. 657

lipase (lī′pās) Fat-digesting enzyme. p. 118

lipid (lip′id) Fat, oil, or fatlike compound that usually has fatty acids in its molecular structure. p. 62

lipoprotein (lip″o-pro′te-in) A complex of lipid and protein. p. 684

liver (liv′er) Large, dark red organ in the upper part of the abdomen on the right side that detoxifies blood, stores glycogen and fat-soluble vitamins and synthesizes proteins, including clotting factors and enzymes. p. 673

lobule (lob′ul) Small, well-defined part of an organ. p. 743

long-term synaptic potentiation (long-term sĭ-nap′tik po-ten′she-a-shun) Theory that frequent, repeated stimulation of the same neurons in the hippocampus strengthens their synaptic connections. p. 404

lower esophageal sphincter (loh′er ĕ-sof″ah-je′al sfingk′ter) Ring of muscle, at the distal end of the esophagus where it joins the stomach, that prevents food from re-entering the esophagus when the stomach contracts; cardiac sphincter. p. 665

lower limb (loh′er lim) Inferior appendage consisting of the thigh, leg, and foot. p. 234

lumbar (lum′bar) Pertaining to the region of the loins, part of back between the thorax and pelvis. p. 23

lumen (lu′men) Space in a tubular structure such as a blood vessel or intestine. p. 653

luteinizing hormone (lu′te-in-īz″ing hor′mōn) A hormone that the anterior pituitary secretes that controls formation of the corpus luteum in females and testosterone secretion in males. p. 497

lymph (limf) Fluid that the lymphatic vessels carry. p. 617

lymph node (limf nōd) Mass of lymphoid tissue located along the course of a lymphatic vessel. p. 617

lymphocyte (lim′fo-sīt) Type of white blood cell that provides immunity; B cell or T cell. p. 532

lysosome (li′so-sōm) Organelle that contains digestive enzymes. p. 84

M

macromolecule (mak′ro-mol′ĕ-kūl) Very large molecule. p. 4

macronutrient (mak′ro-nu′tre-ent) Nutrient (carbohydrate, lipid, and protein) required in large amount. p. 699

macrophage (mak′ro-fāj) Large phagocytic cell. p. 155

macula (mak′u-lah) Hair cells and supporting cells associated with an organ of static equilibrium. p. 459

macula lutea (mak′u-lah lu′te-ah) Yellowish depression in the retina of the eye associated with acute vision. p. 469

major histocompatibility complex (ma′jŏr his″to-kom-pat″ĭ-bil′ĭ-te kom′pleks) Cluster of genes that code for cell surface proteins; MHC. p. 630

major mineral (ma′jor min″er-al) Inorganic substance necessary for metabolism that is part of a group that accounts for 75% of the mineral elements in the body; macro mineral. p. 717

malignant (mah-lig′nant) The power to threaten life; cancerous. p. 104

malnutrition (mal″nu-trish′un) Symptoms resulting from lack of specific nutrients. p. 723

maltase (mawl′tās) Enzyme that catalyzes breakdown of maltose into glucose. p. 118

maltose (mawl′tōs) Disaccharide composed of two glucose molecules. p. 683

mammary (mam′ar-e) Pertaining to the breast. p. 23

mammillary body (mam′ĭ-lar″e bod′e) One of two small, rounded bodies posterior to the hypothalamus involved with reflexes associated with the sense of smell. p. 407

marasmus (mah-raz′mus) Starvation due to profound nutrient deficiency. p. 726

marrow (mar′o) Connective tissue in bones that includes stem cells. p. 195

mast cell (mast sel) Cell to which antibodies, formed in reponse to allergens attach, bursting the cell and releasing allergy mediators. p. 155

mastication (mas′tĭ-ka′shun) Chewing movements. p. 656

matter (mat′er) Anything that has weight and occupies space. p. 51

meatus (me-a′tus) Passageway or channel, or the external opening of a passageway. p. 208

mechanoreceptor (mek″ah-no-re-sep′tor) Sensory receptor sensitive to mechanical stimulation, such as changes in pressure or tension. p. 439

medial (me′de-al) Toward or near the midline. p. 21

mediastinum (me″de-ah-sti′num) Tissues and organs of the thoracic cavity that form a septum between the lungs. p. 12

medulla (mĕ-dul′ah) Inner portion of an organ. p. 504

medulla oblongata (mĕ-dul′ah ob″long-gah′tah) Part of the brainstem between the pons and the spinal cord. p. 408

medullary cavity (med′u-lār″e kav′ĭ-te) Cavity containing marrow in the diaphysis of a long bone. p. 194

medullary respiratory center (med′u-lār″e re-spi′rah-to″re sen′ter) Area of the brainstem that controls the rate and depth of breathing. p. 756

megakaryocyte (meg″ah-kar′e-o-sīt) Large bone marrow cell that shatters to yield blood platelets. p. 526

meiosis (mi-o′sis) Cell division that halves the genetic material, resulting in egg and sperm cells (gametes). p. 100

melanin (mel′ah-nin) Dark pigment in skin and hair. p. 174

melanocyte (mel′ah-no-sīt) Melanin-producing cell. p. 174

melatonin (mel″ah-to′nin) Hormone that the pineal gland secretes. p. 512

membrane potential (mem′brān po-ten′shal) Unequal distribution of positive and negative ions on two sides of a membrane. p. 366

memory cell (mem′o-re sel) B lymphocyte or T lymphocyte produced in a primary immune response that can be activated rapidly if the same antigen is encountered in the future. p. 632

memory consolidation (mem′o-re kon-sol″ĭ-da′shun) Conversion of short-term memories to long-term memories. p. 404

menarche (mĕ-nar′ke) First menstrual period. p. 857

meninx (me′-ninks) Membrane that covers the brain and spinal cord (pl., *meninges*). p. 265

meniscus (men-is′kus) Fibrocartilage that separates the articulating surfaces of bones in the knee (pl., *menisci*). p. 265

menopause (men′o-pawz) Cessation of the female reproductive cycle. p. 860

menses (men′sēz) Shedding of blood and tissue from the uterine lining at the end of a female reproductive cycle. p. 859

mental (men′tal) Pertaining to the mind; pertaining to the chin body region. p. 23

merocrine gland (mer′o-krĭn gland) A structure whose cells remain intact while secreting. p. 150

mesentery (mes′en-ter″e) Fold of peritoneal membrane that attaches an abdominal organ to the abdominal wall. p. 680

mesoderm (mez′o-derm) Middle primary germ layer of the embryo. p. 885

messenger RNA (mes′in-jer RNA) RNA that transmits information for a protein′s amino acid sequence from the nucleus of a cell to the cytoplasm; mRNA. p. 130

metabolic pathway (met″ah-bol′ik path′wa) Series of linked, enzymatically controlled chemical reactions. p. 118

metabolism (mĕ-tab′o-lizm) The chemical reactions in cells that use or release energy. p. 6

metacarpal (met″ah-kar′pal) Bone of the hand between the wrist and finger bones. p. 229

metaphase (met′ah-fāz) Stage in mitosis when chromosomes align in the middle of the cell p. 101

metatarsal (met″ah-tar′sal) Foot bone between the ankle and toe bones. p. 236

microfilament (mi″kro-fil′ah-ment) Rod of the protein actin that provides structural support or motility in the cytoplasm. p. 88

microglia (mi-krog′le-a) Neuroglia that support neurons and phagocytize. p. 363

micronutrient (mi-kro-nu′tre-ent) Nutrient (vitamin or mineral) required in small amount. p. 699

microtubule (mi′kro-tu′būl) Hollow rod of the protein tubulin in the cytoplasm. p. 88

microvillus (mi″kro-vil′us) Cylindrical process that extends from some epithelial cell membranes and increases the membrane surface area (pl., *microvilli*). p. 681

micturition (mik″tu-rish′un) Urination. p. 802

midbrain (mid′brān) Small region of the brainstem between the diencephalon and the pons. p. 398

mineral (min′er-al) Inorganic element essential in human metabolism. p. 717

mineralocorticoid (min″er-al-o-kor′tĭ-koid) Hormone the adrenal cortex secretes that affects electrolyte concentrations in body fluids. p. 506

mitochondrion (mi″to-kon′dre-on) Organelle housing enzymes that catalyze aerobic reactions of cellular respiration (pl., *mitochondria*). p. 83

mitosis (mi-to′sis) Division of a somatic cell, forming two genetically identical cells. p. 100

mitral valve (mi′tral valv) Heart valve located between the left atrium and the left ventricle; bicuspid valve. p. 557

mixed nerve (mikst nerv) Nerve that includes both sensory and motor nerve fibers. p. 412

molar (mo′lar) Rear tooth with a flattened surface adapted for grinding food. p. 659

molecular formula (mo-lek′u-lar for′mu-lah) Abbreviation for the number of atoms of each element in a compound. p. 53

molecule (mol′ĕ-kūl) Particle composed of two or more joined atoms. p. 4

monoamine oxidase (mon″o-am′ēn ok′sĭ-dās) Enzyme that catalyzes the removal of an amine group from a compound. p. 374

monocyte (mon′o-sīt) Type of white blood cell that is a phagocyte. p. 532

monosaccharide (mon″o-sak′ah-rīd) Single sugar, such as glucose or fructose. p. 62

monosomy (mon′o-so″me) Cell missing one chromosome. p. 929

morula (mor′u-lah) Early stage in prenatal development; solid ball of cells. p. 879

motor area (mo′tor a′re-ah) Region of the brain that sends impulses to muscles or glands. p. 402

motor end plate (mo′tor end plāt) Specialized part of a muscle fiber membrane at a neuromuscular junction. p. 290

motor nerve (mo′tor nerv) Nerve that consists of motor nerve fibers. p. 412

motor neuron (mo′tor nu′ron) Neuron that transmits impulses from the central nervous system to an effector. p. 289

motor unit (mo′tor unit) A motor neuron and its associated muscle fibers. p. 290

mucosa (mu-ko′sah) Innermost layer of the alimentary canal. p. 653

mucous membrane (mu′kus mem′brān) Membrane that lines tubes and body cavities that open to the outside of the body. p. 163

mucus (mu′kus) Fluid secretion of mucous cells. p. 152

multiple motor unit summation (mul′tĭ-pl mo′tor u′nit sum-mā′shun) Sustained muscle contraction of increasing strength in response to input from many motor units. p. 298

multipolar neuron (mul″tĭ-po′lar nu′ron) Nerve cell that has many processes emanating from its cell body. p. 360

muscle fiber (mus′el fi′ber) Muscle cell. p. 163

muscle impulse (mus′el im′puls) Impulse propagated along the sarcolemma into the transverse tubules. p. 290

muscle spindle (mus′el spin′dul) Modified skeletal muscle fiber that can respond to changes in muscle length. p. 444

muscle tissue (mus′el tish′u) Contractile tissue of filaments of actin and myosin, which slide past each other, shortening cells. p. 163

muscle tone (mus′el tōn) Contraction of some fibers in skeletal muscle at any given time. p. 298

mutagen (mu′tah-jen) Agent that can cause mutations. p. 136

mutant (mu′tant) Allele for a certain gene that has been altered from the "normal" condition. p. 920

mutation (mu-ta′shun) Change in a gene. p. 135

myelin (mi′ĕ-lin) Fatty material that forms a sheathlike covering around certain nerve fibers. p. 356

myocardium (mi″o-kar′de-um) Muscle tissue of the heart. p. 555

myofibril (mi″o-fi′bril) Contractile fibers in muscle cells. p. 287

myoglobin (mi″o-glo′bin) Pigmented compound in muscle tissue that stores oxygen. p. 294

myogram (mi′o-gram) Recording of a muscular contraction. p. 296

myometrium (mi″o-me′tre-um) Layer of smooth muscle tissue within the uterine wall. p. 854

myopia (mi-o′pe-ah) Nearsightedness. p. 472

myosin (mi′o-sin) Protein that, with actin, contracts and relaxes muscle fibers. p. 287

N

nail (nāl) Horny plate at the distal end of a finger or toe. p. 177

nasal cavity (na′zal kav′ĭ-te) Space in the nose. p. 12

nasal concha (na′zal kong′kah) Shell-like bone extending outward from the wall of the nasal cavity; a turbinate bone. p. 737

nasal septum (na′zal sep′tum) Wall of bone and cartilage that separates the nasal cavity into two parts. p. 737

nasopharynx (na″zo-far″ingks) Part of the pharynx posterior to the nasal cavity. p. 663

natural killer cell (nat′u-ral kil′er sel) Lymphocyte that bursts an infected or cancerous cell. p. 627

negative feedback (neg′ah-tiv fēd′bak) A mechanism that restores a biochemical or other balance in which build up of a product suppresses its synthesis. p. 9

neonatal (ne″o-na′tal) The first four weeks after birth. p. 904

nephron (nef′ron) Functional unit of a kidney, consisting of a renal corpuscle and a renal tubule. p. 779

nerve (nerv) Bundle of nerve fibers. p. 354

nerve cell (nerv sel) Neuron. p. 354

nerve fiber (nerv fi′ber) Axon of a neuron. p. 354

nerve impulse (nerv im′puls) Depolarization and repolarization along a nerve fiber. p. 354

nerve tract (nerv trakt) Long bundle of nerve fibers in the CNS having the same origin, function, and termination. p. 389

nervous tissue (ner′vus tish′u) Neurons and neuroglia composing the brain, spinal cord and nerves. p. 164

net filtration pressure (fil-tra′shun presh′ur) Equal to the hydrostatic pressure of the blood entering the glomerulus minus the pressure of the opposing forces (the hydrostatic pressure in the glomerular capsule and the plasma osmotic pressure of the blood in the glomerulus). p. 788

neurilemma (nur″ĭ-lem′ah) Sheath on certain nerve fibers, formed from Schwann cells. p. 356

neurofibril (nu″ro-fi′bril) Fine cytoplasmic thread that extends from the cell body into the process of a neuron. p. 356

neuroglia (nu-ro′gle-ah) Specialized cells of the nervous system that produce myelin, communicate between cells, maintain the ionic environment, provide growth factors that support neurons, and provide structural support. p. 354

neuromodulator (nu″ro-mod′u-lā-tor) Substance that alters a neuron′s response to a neurotransmitter. p. 374

neuromuscular junction (nu″ro-mus′ku-lar jungk′shun) Synapse between a motor neuron and a skeletal muscle fiber; myoneural junction. p. 290

neuron (nu′ron) Nerve cell. p. 354

neuronal pool (nu′ro-nal po͞ol) Accumulation of nerve cells. p. 374

neuropeptide (nu″ro-pep′tīd) Peptide in the brain that functions as a neurotransmitter or neuromodulator. p. 374

neurosecretory cell (nu″ro-se-kre′to-re sel) Cell in the hypothalamus that functions as a neuron at one end but like an endocrine cell at the other, by receiving messages and secreting the hormones ADH and oxytocin. p. 493

neurotransmitter (nu″ro-trans-mit′er) Chemicals that an axon end secretes that stimulates a muscle fiber to contract or a neuron to fire an impulse. p. 289

neutral (nu′tral) Neither acidic nor alkaline; pH 7. p. 59

neutron (nu′tron) Electrically neutral subatomic particle. p. 52

neutrophil (nu′tro-fil) Type of phagocytic white blood cell. p. 531

niacin (ni′ah-sin) Vitamin of the B-complex group; nicotinic acid. p. 714

nitrogen balance (ni′tro-jen bal′ans) Condition in which the amount of nitrogen ingested equals the amount excreted. p. 705

node of Ranvier (nōd of Ron′vee-ay) Short region of exposed (unmyelinated) axon between Schwann cells on neurons of the peripheral nervous system. p. 356

nondisjunction (non″dis-jungk′shun) A pair of chromosomes that remains together rather than separating during meiosis. p. 929

nonprotein nitrogenous substance (non-pro′tēn ni-troj′ĕ-nus sub′stans) A nitrogen-containing molecule that is not a protein. p. 537

norepinephrine (nor″ep-ĭ-nef′rin) Neurotransmitter released from the axons of some nerve fibers. p. 301

normal range (nor′mal rānj) Measurements or values obtained from a statistical sample of the healthy population for reference or comparison. p. 9

nuclear envelope (nu′kle-ar en′vĕ-lōp) Membrane surrounding the cell nucleus and separating it from the cytoplasm. p. 89

nuclear pore (nu′kle-ar pōr) Protein-lined channel in the nuclear envelope. p. 89

nuclease (nu′kle-ās) Enzyme that catalyzes decomposition of nucleic acids. p. 672

nucleic acid (nu-kle′ik as′id) A molecule that is composed of bonded nucleotides; RNA or DNA. p. 68

nucleolus (nu-kle′o-lus) Small structure in the cell nucleus that contains RNA and proteins and is the site of synthesis of ribosome components (pl., *nucleoli*). p. 90

nucleoplasm (nu′kle-o-plazm″) Contents of the cell nucleus. p. 90

nucleotide (nu′kle-o-tīd″) Building block of a nucleic acid molecule, consisting of a sugar, nitrogenous base, and phosphate group. 68

nucleus (nu′kle-us) Cellular organelle enclosed by a double-layered, porous membrane and containing DNA; the dense core of an atom that is composed of protons and neutrons (pl., *nuclei*). pp. 52, 76

nutrient (nu′tre-ent) Chemical that the body requires from the environment. p. 699

nutrition (nu-trish′un) Study of the sources, actions, and interactions of nutrients. p. 699

O

obesity (o-bēs′ĭ-te) Excess adipose tissue; a body mass index greater than 30. p. 708

occipital (ok-sip′i-tal) Pertaining to the lower, back portion of the head. p. 24

olfactory (ol-fak′to-re) Pertaining to the sense of smell. p. 446

oligodendrocyte (ol″ĭ-go-den′dro-sīt) Type of neuroglia in the CNS that produces myelin. p. 358

oncogene (ong′ko-jēn) Gene that normally controls cell division but when overexpressed leads to cancer. p. 104

oocyte (o′o-sīt) Cell formed by oogenesis. An egg cell. p. 831

oogenesis (o′o-jen′ĕ-sis) Differentiation of an egg cell. p. 849

optic chiasma (op′tik ki-az′mah) X-shaped structure on the underside of the brain formed by a partial crossing of optic nerve fibers. p. 405

optic disc (op′tik disk) Region in the retina of the eye where nerve fibers exit, becoming part of the optic nerve. p. 469

oral (o′ral) Pertaining to the mouth. p. 24

orbital (or′bĭ-tal) Region in the atom containing electrons. p. 23

organ (or′gan) Structure consisting of a group of tissues with a specialized function. p. 4

organelle (or″gah-nel′) A structure or compartment in cells that houses the biochemical reactions that carry out a specific function. p. 4

organic (or-gan′ik) A molecule that contains carbon. p. 60

organism (or′gah-nizm) An individual living thing. p. 4

organ system (or′gan sis′tem) Group of organs coordinated to carry on a specialized function. p. 4

orgasm (or′gaz-em) An intense sensation that is the peak of sexual excitement. p. 844

origin (or′ĭ-jin) End of a muscle that attaches to a relatively immovable part. p. 303

oropharynx (o″ro-far′ingks) Part of the pharynx in the posterior to the oral cavity. p. 663

osmoreceptor (oz″mo-re-sep′tor) Receptor that senses changes in the osmotic pressure of body fluids. p. 498

osmosis (oz-mo′sis) Movement of water through a selectively permeable membrane in response to a concentration gradient created by an impermeant solute. p. 90

osmotic pressure (oz-mot′ik presh′ur) Pressure needed to stop osmosis; a solution's potential pressure caused by impermeant solute particles in the solution. p. 94

osseous tissue (os′e-us tish′u) Bone tissue. p. 453

ossification (os″ĭ-fĭ-ka′shun) Formation of bone tissue. p. 197

osteoblast (os′te-o-blast″) Bone-forming cell. p. 197

osteoclast (os′te-o-klast″) Cell that erodes bone. p. 198

osteocyte (os′te-o-sīt) Mature bone cell. p. 195

osteon (os′te-on) Cylinder-shaped unit containing bone cells that surround a central canal; Haversian system. p. 195

osteoporosis (os″te-o-po-ro′sis) Condition in which bones break easily because calcium is lost faster than it is replaced. p. 204

otic (o′tik) Pertaining to the ear. p. 24

otolith (o′to-lith) Small particle of calcium carbonate associated with the receptors of equilibrium. p. 459

oval window (o′val win′do) Opening between the stapes and the inner ear. p. 452

ovarian (o-va′re-an) Pertaining to the ovary. p. 512

ovary (o′var-e) Primary female reproductive organ; an egg cell-producing organ. p. 512

ovulation (o″vu-la′shun) Release of an egg cell from a mature ovarian follicle. p. 851

oxidation (ok″sĭ-da′shun) Process by which oxygen combines with another chemical; removal of hydrogen or loss of electrons; the opposite of reduction. p. 120

oxygen debt (ok′sĭ-jen det) Amount of oxygen required following physical exercise to react accumulated lactic acid to form glucose. p. 295

oxyhemoglobin (ok″sĭ-he″mo-glo′bin) Compound formed when oxygen binds hemoglobin. p. 763

oxytocin (ok″sĭ-to′sin) Hormone released by the posterior lobe of the pituitary gland that contracts smooth muscles in the uterus and mammary glands. p. 498

P

pacemaker (pās′māk-er) Mass of specialized cardiac muscle tissue that controls the rhythm of the heartbeat; the sinoatrial node. p. 566

packed cell volume (pakt sel vol′ūm) Number of red blood cells in milliliters per 100 mL of centrifuged blood. p. 523

pain receptor (pān re″sep′tor) Sensory nerve ending that transmits impulses interpreted as pain. p. 439

palate (pal′at) Roof of the mouth. p. 657

palatine (pal′ah-tīn) Pertaining to the palate. p. 657

palmar (pahl′mar) Pertaining to the palm of the hand. p. 23

pancreas (pan′kre-as) Glandular organ in the abdominal cavity that secretes hormones and digestive enzymes. p. 509

pancreatic (pan″kre-at′ik) Pertaining to the pancreas. p. 509

pantothenic acid (pan″to-the′nik as′id) Vitamin of the B-complex group; vitamin B_5. p. 714

papilla (pah-pil′ah) Tiny, nipplelike projection. p. 448

papillary muscles (pap′ĭ-ler″e mus′elz) Muscles that extend inward from the ventricular walls of the heart and to which the chordae tendineae attach. p. 557

paracrine (par′ah-krin) Type of endocrine secretion in which the hormone affects nearby cells. p. 483

paradoxical sleep (par″ah-dok′se-kal slēp) Sleep in which some areas of the brain are active, producing dreams and rapid eye movements. p. 409

paranasal sinus (par″ah-na′zal si-nus) Air-filled cavity in a cranial bone; lined with mucous membrane and connected to the nasal cavity. p. 208

parasympathetic division (par″ah-sim″pah-thet′ik dĭ-vizh′un) Part of the autonomic nervous system that arises from the brain and sacral region of the spinal cord. p. 424

parathyroid gland (par″ah-thi′roid gland) One of four small endocrine glands embedded in the posterior part of the thyroid gland. p. 502

parathyroid hormone (par″ah-thi′roid hor′mōn) Hormone secreted by the parathyroid glands that helps regulate the level of blood calcium and phosphate ions; PTH. p. 502

parietal (pah-ri′ĕ-tal) Pertaining to the wall of an organ or cavity. p. 12

parietal cell (pah-ri′ĕ-tal sel) Cell of a gastric gland that secretes hydrochloric acid and intrinsic factor. p. 666

parietal pleura (pah-ri′ĕ-tal ploo′rah) Membrane that lines the inner wall of the thoracic cavity. p. 12

parotid glands (pah-rot′id glandz) Large salivary glands on the sides of the face just in front and below the ears. p. 661

partial pressure (par′shal presh′ur) Pressure one gas produces in a mixture of gases. p. 757

patellar (pah-tel′ar) Pertaining to the kneecap. 23

pathogen (path′o-jen) Disease-causing agent. p. 617

pectoral (pek′tor-al) Pertaining to the chest. p. 24

pectoral girdle (pek′tor-al ger′dl) Part of the skeleton that supports and attaches the upper limbs. p. 206

pedal (ped′al) Pertaining to the foot. p. 24

pedigree (ped′ĭ-gre) Chart that displays relationships among family members and their inherited traits and disorders. p. 921

pelvic (pel′vik) Pertaining to the pelvis. p. 24

pelvic cavity (pel′vik kav′ĭ-te) Hollow place within the ring formed by the sacrum and coxae. p. 12

pelvic girdle (pel′vik ger′dl) Part of the skeleton to which the lower limbs attach. p. 206

pelvic inflammatory disease (pel′vik in-flam′ah-tore dĭ-zēz′) Ascending infection of the upper female genital tract. p. 869

pelvis (pel′vis) Bony ring formed by the sacrum and coxae. p. 206

penis (pe′nis) Male external reproductive organ through which the urethra passes. p. 843

pepsin (pep′sin) Protein-splitting enzyme that the gastric glands secrete. p. 666

pepsinogen (pep-sin′o-jen) Inactive form of pepsin. p. 666

peptidase (pep′tĭ-dās) Enzyme that catalyzes the breakdown of polypeptides. p. 682

peptide (pep′tīd) Compound composed of two or more amino acids. p. 373

peptide bond (pep′tīd bond) Bond between the carboxyl group of one amino acid and the amino group of another. p. 116

perception (per-sep′shun) Mental awareness of sensory stimulation. p. 438

perforating canal (per′fo-rāt″eng kah-nal′) Transverse channel that connects central canals in compact bone; Volkmann's canal. p. 195

perforin (per′fo-rin) Protein that cytotoxic T cells release that binds antigen. p. 627

pericardial (per″ĭ-kar′de-al) Pertaining to the pericardium. p. 12

pericardium (per″ĭ-kar′de-um) Serous membrane that surrounds the heart. p. 554

perichondrium (per″ĭ-kon′dre-um) Layer of fibrous connective tissue that encloses cartilaginous structures. p. 158

perilymph (per-ĭ-limf) Fluid in the space between the membranous and osseous labyrinths of the inner ear. p. 453

perimetrium (per-ĭ-me′tre-um) Outer serosal layer of the uterine wall. p. 854

perimysium (per″ĭ-mis′e-um) Sheath of connective tissue that encloses a bundle of skeletal muscle fibers or a fascicle. p. 286

perineal (per″ĭ-ne′al) Pertaining to the perineum. p. 24

perineum (per″ĭ-ne′um) Body region between the scrotum (male) or urethral opening (female) and the anus. p. 412

perineurium (per″ĭ-nu-re-um) Layer of connective tissue that encloses a bundle of nerve fibers in a nerve. p. 412

periodontal ligament (per″e-o-don′tal lig′ah-ment) Fibrous membrane that surrounds a tooth and attaches it to the jawbone. p. 660

periosteum (per″e-os′te-um) Fibrous connective tissue covering the surface of a bone. p. 194

peripheral (pĕ-rif′er-al) Pertaining to parts near the surface or toward the outside. p. 21

peripheral nervous system (pĕ-rif′er-al ner′vus sis′tem) Parts of the nervous system outside the central nervous system; PNS. p. 354

peripheral protein (pĕ-rif′er-al pro′tēn) Globular protein associated with the inner surface of the cell membrane. p. 80

peripheral resistance (pĕ-rif′er-al re-zis′tans) Resistance to blood flow due to friction between blood and blood vessel walls. p. 583

peristalsis (per″ĭ-stal′sis) Rhythmic waves of muscular contraction in the walls of certain tubular organs. p. 300

peritoneal (per″ĭ-to-ne′al) Pertaining to the peritoneum. p. 14

peritoneal cavity (per″ĭ-to-ne′al kav′ĭ-te) The potential space between the parietal and visceral peritoneal membranes. p. 14

peritoneum (per″ĭ-to-ne′um) Serous membrane that lines the abdominal cavity and encloses the abdominal viscera. p. 14

peritubular capillary (per″ĭ-tu′bu-lar kap′ĭ-ler″e) Capillary that surrounds a renal tubule and functions in tubular reabsorption and tubular secretion during urine formation. p. 784

permeable (per′me-ah-bl) Open to passage or penetration. p. 93

peroxisome (pĕ-roks′ĭ-sōm) Membranous cytoplasmic vesicle that contains enzymes that catalyze reactions that produce and decompose hydrogen peroxide. p. 86

phagocyte (fag′o-sīt) Cell that ingests particulate matter. p. 96

phagocytosis (fag″o-si-to′sis) Process by which a cell engulfs and digests solids. p. 96

phalanx (fa′langks) Bone of a finger or toe (pl., *phalanges*). p. 206

pharynx (far′ingks) Portion of the digestive tube posterior to the nasal and oral cavities, as well as the larynx. p. 661

phenotype (fe′no-tīp) The expression of a gene or genes. p. 920

phosphate buffer system (fos′fāt buf′er sis′tem) A mix of sodium monohydrogen phosphate and sodium dihydrogen phosphate that weakens a strong acid and a strong base, respectively; resists changes in pH. p. 821

phospholipid (fos″fo-lip′id) Molecule consisting of two fatty acids and a phosphate group bound to a glycerol molecule. p. 64

phosphorylation (fos″for-ĭ-la′shun) Metabolic process that adds a phosphate to an organic molecule. p. 119

photoreceptor (fo″to-re-sep′tor) Nerve ending sensitive to light energy. p. 439

pH scale (pH skāl) Shorthand notation for the hydrogen ion concentration used to indicate the acidic or alkaline condition of a solution; values range from 0 to 14. p. 59

physiology (fiz″e-ol′o-je) The study of body functions. p. 4

pia mater (pi′ah ma′ter) Inner layer of meninges that encloses the brain and spinal cord. p. 385

pineal gland (pin′e-al gland) Small structure in the central part of the brain that secretes the hormone melatonin, which controls certain biological rhythms. p. 407

pinocytosis (pin″o-si-to′sis) Process by which a cell engulfs droplets of fluid from its surroundings. p. 96

pituitary gland (pĭ-tu′ĭ-tār″e gland) Endocrine gland attached to the base of the brain that consists of anterior and posterior lobes; the hypophysis. p. 492

pivot joint (piv′ut joint) End of a bone moving within a ring formed by another bone and connective tissue. p. 266

placenta (plah-sen′tah) Structure that attaches the fetus to the uterine wall, delivering nutrients to and removing wastes from the fetus. p. 512

placental lactogen (plah-sen′tahl lak′to-jen) Hormone secreted by the placenta that inhibits maternal insulin activity during pregnancy. p. 884

plantar (plan′tar) Pertaining to the sole of the foot. p. 24

plantar flexion (plan′tar flek′shun) Ankle movement that brings the foot farther from the shin. p. 269

plasma (plaz′mah) Fluid portion of circulating blood. p. 523

plasma cell (plaz′mah sel) Type of antibody-producing cell that forms when activated B cells proliferate. p. 633

plasma protein (plaz′mah pro′tēn) Protein dissolved in blood plasma. p. 535

plasmin (plaz′min) Protein-splitting enzyme that can digest fibrin in a blood clot. p. 541

platelet (plāt′let) Cytoplasmic fragment formed in the bone marrow that helps blood clot. p. 162

pleiotropy (plē′o-tro-pē) Gene that has several expressions (phenotypes). p. 924

pleural (ploo′ral) Pertaining to the pleura or membranes surrounding the lungs. p. 12

pleural cavity (ploo′ral kav′ĭ-te) Potential space between pleural membranes. p. 12

pleural membrane (ploo′ral mem′brān) Serous membrane that encloses the lungs and lines the chest wall. p. 12

plexus (plek′sus) Network of interlaced nerves or blood vessels. p. 420

pluripotent (ploo-rip′o-tent) Cell able to differentiate to yield several specialized cell types. p. 106

PNS Peripheral nervous system. p. 354

polar body (po′lar bod′e) Small, nonfunctional cell that is a product of meiosis in the female. p. 849

polarization (po″lar-ĭ-za′shun) Electrical charge on a cell membrane surface due to an unequal distribution of positive and negative ions on either side of the membrane. p. 365

polar molecule (po′lar mol′ĕ-kūl) Combination of atoms in which the electrical charge is not distributed symmetrically. p. 57

polygenic (pol″ĕ-jĕn′ik) An inherited trait that results from the actions of more than one gene. p. 924

polymorphonuclear leukocyte (pol″e-mor″fo-nu′kle-ar lu′ko-sīt) White blood cell with an irregularly lobed nucleus; neutrophil. p. 531

polynucleotide (pol″e-noo′-kle-o-tīd) Compound formed by the union of many nucleotides; a nucleic acid. p. 68

polypeptide (pol″e-pep′tīd) Compound formed by the union of many amino acid molecules. p. 116

polyploidy (pol′e-ploi″de) A cell with one or more extra sets of chromosomes. p. 929

polysaccharide (pol″e-sak′ah-rīd) Carbohydrate composed of many joined monosaccharides. p. 62

pons (ponz) Part of the brainstem above the medulla oblongata and below the midbrain. p. 407

popliteal (pop″lĭ-te′al) Pertaining to the region behind the knee. p. 24

positive chemotaxis (poz′ĭ-tiv ke″mo-tak′sis) Movement of a cell toward the greater concentration of a substance. p. 533

positive feedback (poz′ĭ-tiv fēd′bak) Process by which changes cause additional similar changes, producing unstable conditions. p. 539

posterior (pos-tēr′e-or) Toward the back; opposite of anterior. p. 21

postganglionic fiber (pōst″gang-gle-on′ik fi′ber) Autonomic nerve fiber on the distal side of a ganglion. p. 424

postnatal (pōst-na′tal) After birth. p. 875

postsynaptic neuron (pōst″sĭ-nap′tik nu′ron) One of two adjacent neurons transmitting an impulse; cell on the "receiving" side of a synapse. p. 365

postsynaptic potential (pōst″sĭ-nap′tik po-ten′shal) Membrane polarization is increased (excitatory) or decreased (inhibitory) in the postsynaptic neuron with repeated stimulation over an excitatory or inhibitory pathway so that the neuron will either fire or not respond. p. 371

preganglionic fiber (pre″gang-gle-on′ik fi′ber) Autonomic nerve fiber located on the proximal side of a ganglion. p. 424

pregnancy (preg′nan-se) Condition in which a female has a developing offspring in her uterus. p. 876

preload (pre′lōd) Blood entering and filling the relaxed ventricles prior to their contraction p. 586

prenatal (pre-na′tal) Before birth. p. 876

presbyopia (pres″be-o′pe-ah) Loss of the eye's ability to accommodate due to declining elasticity in the lens; farsightedness of age. p. 472

presynaptic neuron (pre″sĭ-nap′tik nu′ron) One of two adjacent neurons transmitting an impulse; cell on the "sending" side of a synapse. p. 365

primary germ layers (pri′ma-re jerm lā′erz) Three layers (endoderm, mesoderm, and ectoderm) of cells in the embryo that divide and differentiate into specific tissues and organs. p. 885

primary immune response (pri′ma-re ĭ-mūn′ re-spons′) Immune system's response to an initial encounter with a nonself antigen. p. 637

primary sex organs (pri′ma-re seks or′ganz) Sex cell-producing parts; testes in males and ovaries in females. p. 833

prime mover (prīm moo-v′er) Muscle that provides a particular body movement. p. 304

primordial follicle (pri-mor′de-al fol′lĭ-kl) Egg enclosed by a single layer of cells in the ovary. p. 847

product (prod′ukt) The result of a chemical reaction. p. 58

progenitor cell (pro-jen′ĭ-tor sel) Daughter cell of a stem cell whose own daughter cells are restricted to follow specific lineages. p. 105

progesterone (pro-jes′tĕ-rōn) Female hormone secreted by the corpus luteum of the ovary and the placenta. p. 857

projection (pro-jek′shun) Process by which the brain causes a sensation to seem to come from the region of the body being stimulated. p. 439

prolactin (pro-lak′tin) Hormone secreted by the anterior pituitary gland that stimulates the production of milk in the mammary glands; PRL. p. 495

pronation (pro-na′shun) Downward or backward rotation of the palm. p. 269

prophase (pro′fāz) Stage of mitosis when chromosomes become visible when stained and viewed under a microscope. p. 101

proprioceptor (pro″pre-o-sep′tor) Nerve ending that senses changes in muscle or tendon tension. p. 439

prostaglandins (pros′tah-glan′dins) Group of compounds that have powerful, hormonelike effects. p. 491

prostate gland (pros′tāt gland) Gland surrounding the male urethra below the urinary bladder that secretes into semen prior to ejaculation. p. 801

protein (pro′tēn) Nitrogen-containing organic compound composed of bonded amino acid molecules. p. 64

protein buffer system (pro′tēn buf′er sis′tem) Amino acids of a protein accept or donate hydrogen ions to keep the concentration of hydrogen ions in solution constant; resists changes in pH. p. 821

protein kinase (pro′tēn ki′nās) Enzyme that catalyzes the reaction to form a phosphoprotein. p. 488

prothrombin (pro-throm′bin) Plasma protein that functions in blood clotting. p. 539

proton (pro′ton) Positively charged particle in an atomic nucleus. p. 52

protraction (pro-trak′shun) Forward movement of a body part. p. 269

provitamin (pro-vi′tah-min) Precursor of a vitamin. p. 709

proximal (prok′sĭ-mal) Closer to the trunk or origin; opposite of distal. p. 21

pseudostratified (soo″do-strat′ĭ-fīd) Single layer of epithelial cells that appear as more than one layer because the nuclei occupy different positions in the cells. p. 146

puberty (pu′ber-te) Stage of development in which the reproductive organs become functional. p. 846

pulmonary circuit (pul′mo-ner″e ser′kit) System of blood vessels that carries blood between the heart and the lungs. p. 553

pulmonary valve (pul′mo-ner″e valv) Valve leading from the right ventricle to the pulmonary trunk (artery); pulmonary semilunar valve. p. 557

pulse (puls) Surge of blood felt through the walls of arteries due to the contraction of the heart ventricles. p. 582

Punnett square (pun′it skwair) A grid diagram that displays possible progeny that can be predicted based on parental gametes. p. 920

pupil (pu′pil) Opening in the iris through which light enters the eye. p. 467

purine (pu rēn) Type of nitrogen-containing base that is part of DNA and RNA and has two organic rings; adenine and guanine. p. 126

Purkinje fibers (pur-kin′je fi′berz) Specialized muscle fibers that conduct the cardiac impulse from the AV bundle into the ventricular walls. p. 567

pyloric sphincter (pi-lor′ik sfingk′ter) Ring of muscle between the stomach and the duodenum; controls food entry into the duodenum. p. 666

pyridoxine (pir′ĭ-dok′sēn) A vitamin of the B-complex group; vitamin B_6. p. 715

pyrimidine (pe-ri′mĭ-dēn) Type of nitrogen-containing base that is part of DNA and RNA and has two organic rings; thymine, cytosine, and uracil. p. 126

pyruvic acid (pi-roo′vik as′id) Intermediate product of carbohydrate oxidation. p. 120

R

radiation (ra″de-a′shun) Form of energy that includes visible light, ultraviolet light, and X rays; the means by which body heat is lost as infrared rays. pp. 53, 182

radioactive (ra″de-o-ak′tiv) Property of an atom that releases energy at a constant rate. p. 53

rate-limiting enzyme (rāt lim′ĭ-ting en′zīm) Enzyme, usually present in small amounts, that controls the rate of a metabolic pathway by regulating one of its steps. p. 118

reactant (re-ak′tant) A chemical that takes part in a chemical reaction. A starting material. p. 58

receptor (re″sep′tor) Specialized cells that provide information about the environment. Also, cell surface structures that bind particular molecules, called ligands, thereby transmitting a signal to inside the cell. pp. 9, 80

receptor-mediated endocytosis (re″sep′tor-me′de-ā-tid en″do-si-to′sis) Selective uptake of molecules into a cell by binding to a specific receptor. p. 97

recessive allele (re-ses′iv ah-lēl) Form of a gene not expressed if the dominant form is present. p. 920

recruitment (re-krōōt′ment) Increase in number of motor units activated as stimulation intensity increases. p. 298

rectum (rek′tum) Terminal end of the digestive tube between the sigmoid colon and the anus. p. 687

red blood cell (red blud sel) Disc-shaped cell, lacking a nucleus, that is packed with the oxygen-carrying molecule hemoglobin; erythrocyte. p. 162

red marrow (red mar′o) Blood cell-forming tissue in bones. p. 203

red fiber (red fi′ber) Slow-contracting postural muscle fiber that contains abundant myoglobin. p. 299

referred pain (re-ferd′ pān) Pain that feels as if it is originating from a part other than the site being stimulated. p. 441

reflex (re′fleks) Rapid, automatic response to a stimulus. p. 389

reflex arc (re′fleks ark) Nerve pathway, consisting of a sensory neuron, interneuron, and motor neuron, that forms the structural and functional bases for a reflex. p. 389

refraction (re-frak′shun) Bending of light as it passes between media of different densities. p. 471

refractory period (re-frak′to-re pe′re-od) Period following stimulation during which a neuron or muscle fiber will not respond to a stimulus. p. 370

renal (re′nal) Pertaining to the kidney. p. 776

renal corpuscle (re′nal kor′pusl) Part of a nephron that consists of a glomerulus and a glomerular capsule; Malpighian corpuscle. p. 779

renal cortex (re′nal kor′teks) Outer part of a kidney. p. 776

renal medulla (re′nal mĕ-dul′ah) Inner part of a kidney. p. 776

renal pelvis (re′nal pel′vis) Cavity in a kidney that channels urine to the ureter. p. 776

renal plasma threshold (re′nal plaz′mah thresh′old) Concentration of a substance in blood at which it begins to be excreted in the urine. p. 792

renal tubule (re′nal tu′būl) Part of a nephron that extends from the renal corpuscle to the collecting duct. p. 779

renin (re′nin) Enzyme that kidneys release that maintains blood pressure, plasma sodium, and blood volume. p. 506

renin-angiotensin system (re′nin-an″je-o-ten′sin sis′tem) An enzyme, renin, that converts angiotensinogen to angiotensin, ultimately stimulating aldosterone secretion. p. 506

repair enzyme (re-pār′ en′zīm) Protein that removes mismatched nucleotides from a section of DNA and replaces them with complementary nucleotides. p. 137

replication (rep″lĭ-ka′shun) Copying of a DNA molecule. p. 127

repolarization (re-po″lar-ĭ-za′shun) Returning the cell membrane potential to resting potential. p. 369

reproduction (re″pro-duk′shun) Offspring formation. p. 18

residual volume (re-zid′u-al vol′ūm) Volume of air remaining in the lungs after the most forceful expiration. p. 752

resorption (re-sorp′shun) Decomposition of a structure as a result of physiological activity. p. 200

respiration (res″pĭ-ra′shun) Breathing. p. 736

respiratory center (re-spi′rah-to″re sen′ter) Part of the brainstem that controls breathing depth and rate. p. 409

respiratory capacity (re-spi′rah-to″re kah-pas′ĭ-te) The sum of any two or more respiratory volumes. p. 752

respiratory cycle (re-spi′rah-to″re si′kl) An inspiration followed by an expiration. p. 752

respiratory membrane (re-spi′rah-to″re mem′brān) Membrane composed of a capillary wall, an alveolar wall, and their respective basement membranes through which blood and inspired air exchange gases. p. 760

respiratory volume (re-spi′rah-to″re vol′ūm) Any one of several distinct volumes of air in the lungs. p. 752

resting potential (res′ting po-ten′shal) Difference in electrical charge between the inside and outside of an undisturbed nerve cell membrane. p. 366

resting tidal volume (res′ting tīd′al vol′ūm) Volume of air entering and leaving the lungs in a respiratory cycle at rest. p. 752

reticular fiber (rĕ-tik′u-lar fi′ber) Threadlike structure within a network of like structures in connective tissue. p. 156

reticular formation (rĕ-tik′u-lar fōr-ma′shun) Complex network of nerve fibers in the brainstem that arouses the cerebrum. p. 409

reticulocyte (rĕ-tik′u-lo-sīt) Immature red blood cell that has a network of fibrils in its cytoplasm. p. 527

retina (ret′ĭ-nah) Inner layer of the eye wall that contains the visual receptors. p. 469

retinal (ret′ĭ-nal) A form of vitamin A; retinene. p. 473

retinene (ret′ĭ-nēn) Chemical precursor of rhodopsin, a visual pigment. p. 473

retraction (rĕ-trak′shun) Movement of a part toward the back. p. 269

retroperitoneal (ret″ro-per″ĭ-to-ne′al) Behind the peritoneum. p. 776

reversible reaction (re-ver′sĭ-b′l re-ak′shun) Chemical reaction in which the end products can react again, yielding the original reactants. p. 59

rhodopsin (ro-dop′sin) Light-sensitive pigment in the rods of the retina; visual purple. p. 473

riboflavin (ri″bo-fla′vin) A vitamin of the B-complex group; vitamin B_2. p. 714

ribonucleic acid (ri″bo-nu-kle′ik as′id) Single stranded polymer of nucleotides, in which each nucleotide includes a phosphate group, a nitrogen-containing base (adenine, uracil, cytosine, or guanine), and the sugar ribose; RNA. p. 68

ribose (ri′bōs) 5-carbon sugar in RNA. p. 125

ribosomal RNA (ri-bo-sōm′al) Type of RNA that forms part of the ribosome; rRNA. p. 132

ribosome (ri′bo-sōm) Organelle composed of RNA and protein that is a structural support for protein synthesis. p. 82

RNA Ribonucleic acid. p. 68

rod (rod) Type of light receptor that provides colorless vision. p. 473

rotation (ro-ta′shun) Movement turning a body part on its longitudinal axis. p. 269

round window (rownd win′do) Membrane-covered opening between the inner ear and the middle ear. p. 453

S

saccule (sak′ūl) Saclike cavity that makes up part of the membranous labyrinth of the inner ear. p. 459

sacral (sa′kral) Pertaining to the five fused (pelvic) vertebrae at the distal end of the spinal column. p. 24

saddle joint (sad′l joint) Two bones joined each with a convex and concave surface that are complementary. p. 267

sagittal (saj′i-tal) Plane or section that divides a structure into right and left portions. p. 22

salivary gland (sal′ĭ-ver-e gland) Gland, associated with the mouth, that secretes saliva. p. 660

salt (sawlt) Compound produced by a reaction between an acid and a base. p. 59

saltatory conduction (sal′tah-tor-e kon-duk′shun) Nerve impulse conduction that seems to jump from one node to the next. p. 371

SA node (nōd) Sinoatrial node. p. 565

sarcolemma (sar″ko-lem′ah) Cell membrane of a muscle fiber. p. 287

sarcomere (sar′ko-mēr) Structural and functional unit of a myofibril. p. 287

sarcoplasm (sar′ko-plazm) Cytoplasm in a muscle fiber. p. 287

sarcoplasmic reticulum (sar″ko-plaz′mik rĕ-tik′u-lum) Membranous network of channels and tubules in a muscle fiber, corresponding to the endoplasmic reticulum of other cells. p. 288

satellite cells (sat′il-līte selz) Glia in the peripheral nervous system that support ganglia. p. 363

saturated fat (sat′u-rāt″ed fat) Fat molecule that contains only fatty acid molecules with as many hydrogen atoms as possible, and therefore no double-bonded carbon atoms. p. 63

Schwann cell (shwahn sel) Type of neuroglia that surrounds a fiber of a peripheral nerve, forming the neurilemma and myelin. p. 356

sclera (skle′rah) White fibrous outer layer of the eyeball. p. 465

scrotum (skro′tum) Pouch of skin that encloses the testes. p. 833

sebaceous gland (sĕ-ba′shus gland) Skin gland that secretes sebum. p. 179

sebum (se′bum) Oily secretion of the sebaceous glands. p. 180

secondary immune response (sek′un-der″e i-mun′ re-spons′) Immune system′s response to subsequent encounters with a nonself antigen. p. 637

secretin (se-kre′tin) Hormone from the small intestine that stimulates the pancreas to release pancreatic juice. p. 672

secretion (se-kre′shun) Substance produced in and released from a gland cell. p. 150

segmentation (seg″men-ta′shun) Alternating contraction and relaxation of circular muscle fibers that propels chyme through the intestines. p. 654

selectively permeable (se-lĕk′tiv-le per′me-ah-b′l) Membrane that allows some types of molecules through but not others. p. 79

semen (se′men) Fluid containing sperm cells and secretions discharged from the male reproductive tract at ejaculation. p. 841

semicircular canal (sem″ĭ-ser′ku-lar kah-nal′) Tubular structure in the inner ear that contains the receptors providing the sense of dynamic equilibrium. p. 453

seminal vesicle (sem″ĭ-nal ves′ĭ-kel) One of a pair of pouches that adds fructose and prostaglandins to sperm as semen forms. p. 840

seminiferous tubule (sem″ĭ-nif′er-us tu′būl) Tubule within the testes where sperm cells form. p. 835

senescence (sĕ-nes′ens) Aging. p. 907

sensation (sen-sa′shun) A feeling resulting from the brain's interpretation of sensory nerve impulses. p. 438

sensory adaptation (sen′so-re ad″ap-ta′shun) Sensory receptors becoming unresponsive or inhibition along the CNS pathways leading to sensory regions of the cerebral cortex after constant repeated stimulation. p. 440

sensory area (sen′so-re a′re-ah) Part of the cerebral cortex that receives and interprets sensory nerve impulses. p. 401

sensory nerve (sen′so-re nerv) Nerve composed of sensory nerve fibers. p. 412

sensory neuron (sen′so-re nu′ron) Neuron that transmits an impulse from a receptor to the central nervous system. p. 360

sensory receptor (sen′so-re re″sep′tor) Specialized structure associated with the peripheral end of a sensory neuron specific to detecting a particular sensation and triggering a nerve impulse in response. p. 355

serosa (sēr-o′sah) Outer covering of the alimentary canal. p. 654

serotonin (se″ro-to′nin) Vasoconstrictor that blood platelets release when blood vessels break, controlling bleeding. Also a neurotransmitter. p. 443

serous cell (se′rus sel) Glandular cell that secretes a watery fluid (serous fluid) with a high enzyme content. p. 152

serous fluid (se′rus floo′id) Secretion of a serous membrane. p. 152

serous membrane (se′rus mem′brān) Membrane that lines a cavity without an opening to the outside of the body. p. 12

serum (se′rum) Fluid portion of coagulated blood. p. 541

sesamoid bone (ses′ah-moid bōn) Round bone within tendons adjacent to joints. p. 193

set point (set point) Target value of a physiological measure maintained automatically in the body. p. 9

sex chromosome (seks kro′mo-sōm) Chromosome that carries genes responsible for the development of characteristics associated with maleness or femaleness; an X or Y chromosome. p. 919

sex-influenced inheritance (seks-in′floo-enst in-her′ĭ-tens) Transmission of a trait dominant in one sex but recessive in the other. p. 928

sex-limited inheritance (seks′-lim′it-ed in-her′ĭ-tens) Transmission of a trait expressed in one sex only. p. 928

sexually transmitted infection (sek′shoo-ah-le trans-mi′ted in-fek′shun) Infection transmitted from one individual to another by direct contact during sexual activity. p. 867

sigmoid colon (sig′moid ko′lon) S-shaped part of the large intestine between the descending colon and the rectum. p. 687

signal transduction (sig′nahl trans-duk′shun) Series of biochemical reactions that allows cells to receive and respond to messages coming in through the cell membrane. p. 79

simple sugar (sim′pl shoog′ar) Monosaccharide. p. 62

single nucleotide polymorphism (sing′el nu′kle-o-tīd pol″e-mor′fiz-em) A variant base at a particular place in the genome that occurs in at least 1% of a population. p. 135

sinoatrial node (si″no-a′tre-al nōd) Specialized tissue in the wall of the right atrium that initiates cardiac cycles; the pacemaker; SA node. p. 565

sinus (si′nus) Cavity or space in a bone or other body part. p. 738

skeletal muscle (skel′ĕ-tal mus′l) Type of voluntary muscle tissue in muscles attached to bones. p. 14

sliding filament model (slī′ding fil′eh-ment ma′dul) Muscles contract when the thin (actin) and thick (myosin) filaments move past each other, shortening the skeletal muscle cells. p. 291

small intestine (smawl in-tes′tin) Part of the digestive tract extending from the stomach to the cecum; consisting of the duodenum, jejunum, and ileum. p. 678

smooth muscle (smoo-th mus′l) Type of involuntary muscle tissue in the walls of hollow viscera; visceral muscle. p. 163

sodium pump (so′de-um pump) Active transport mechanism that concentrates sodium ions outside a cell membrane. p. 365

solute (sol′ūt) Chemical dissolved in a solution. p. 92

solution (so-lu′shun) Homogenous mixture of chemicals (solutes) in a dissolving medium (solvent). p. 92

solvent (sol′vent) Liquid portion of a solution in which a solute is dissolved. p. 92

somatic cell (so-mat′ik sel) Any cell of the body other than the sex cells. p. 100

somatic nervous system (so-mat′-ik ner′vus sis′tem) Motor pathways of the peripheral nervous system that lead to the skin and skeletal muscles. p. 411

somatostatin (so-mat′o-sta′tin) Hormone secreted by the pancreatic islets that inhibits the release of growth hormone. p. 511

somatotropin (so″mah-to-tro′pin) Growth hormone. p. 494

special sense (spesh′al sens) Sense that stems from receptors associated with specialized sensory organs, such as the eyes and ears. p. 438

species resistance (spe′sēz re-zis′tans) Natural ability of one type of organism to resist infection by pathogens that cause disease in another type of organism. p. 626

spermatic cord (sper-mat′ik kord) Structure consisting of blood vessels, nerves, the ductus deferens, and other vessels extending from the abdominal inguinal ring to the testis. p. 835

spermatid (sper′mah-tid) Intermediate stage in sperm cell formation. p. 836

spermatocyte (sper-mat′o-sīt) Early stage in sperm cell formation. p. 836

spermatogenesis (sper″mah-to-jen′ĕ-sis) Sperm cell production. p. 836
spermatogonium (sper″mah-to-go′ne-um) Undifferentiated spermatogenic cell in the outer part of a seminiferous tubule. p. 836
sphygmomanometer (sfĭg′mo-mah-nom′ĕ-ter) Instrument used for measuring blood pressure. p. 582
spinal cord (spi′nal kord) Part of the central nervous system extending from the brainstem through the vertebral canal. p. 387
spinal nerve (spi′nal nerv) Nerve that arises from the spinal cord. p. 387
spiral organ (spī rul or′gan) Organ in the cochlear duct containing the receptors for hearing. It consists of hair cells and supporting cells. p. 455
spleen (splēn) Large organ in the upper left region of the abdomen that processes old red blood cells. p. 623
spongy bone (spunj′e bōn) Bone that consists of bars and plates separated by irregular spaces; cancellous bone. p. 194
squamous (skwa′mus) Flat or platelike. p. 146
starch (starch) Polysaccharide common in foods of plant origin. p. 62
static equilibrium (stat′ik e′kwĭ-lib′re-um) Maintenance of balance when the head and body are motionless. p. 459
stem cell (stem sel) Undifferentiated cell that can divide to yield two daughter stem cells or a stem cell and a progenitor cell. p. 105
stereoscopic vision (ster″e-o-skop′ik vizh′un) The ability to perceive objects as three-dimensional; depth perception. p. 475
sternal (ster′nal) Pertaining to the sternum. p. 24
steroid (ste′roid) Type of organic molecule including complex rings of carbon and hydrogen atoms. p. 64
stomach (stum′ak) Digestive organ between the esophagus and the small intestine. p. 665
strabismus (strah-biz′mus) Lack of visual coordination. p. 465
stratified (strat′ĭ-fīd) Organized in layers. p. 147
stratum basale (stra′tum ba′sal) Deepest layer of the epidermis, where cells divide; stratum germinativum. p. 173
stratum corneum (stra′tum kor′ne-um) Outer, horny layer of the epidermis. p. 173
stress (stres) Response to factors perceived as life-threatening. p. 513
stressor (stres′or) Factor that can stimulate a stress response. p. 513
stretch receptor (strech re-sep′ter) Sensory nerve ending that responds to tension. p. 439
stretch reflex (strech re′fleks) Muscle contraction in response to stretching the muscle. p. 444
stroke volume (strōk vol′ūm) Volume of blood the ventricle discharges with each heartbeat. p. 582
structural formula (struk′cher-al fōr′mu-lah) Representation of the way atoms bond in a molecule, using symbols for each element and lines to indicate chemical bonds. p. 57
subarachnoid space (sub″ah-rak′noid spās) Space in the meninges between the arachnoid mater and the pia mater. p. 385
subatomic particles (sub″ah-tom′ik par′tĕ-kalz) Parts of an atom, including protons, electrons, and neutrons. p. 4
subcutaneous (sub″ku-ta′ne-us) Loose connective tissue layer that is mostly fat and beneath the skin; hypodermis. p. 172
sublingual (sub-ling′gwal) Beneath the tongue. p. 661
submucosa (sub″mu-ko′sah) Layer of the alimentary canal underneath the mucosa. p. 654
substrate (sub′strāt) Target of enzyme action. p. 117
sucrase (su′krās) Digestive enzyme that catalyzes the breakdown of sucrose. p. 118
sucrose (soo′krōs) Disaccharide; table sugar. p. 62
sugar (shoog′ar) Sweet carbohydrate. p. 62
sulcus (sul′kus) Shallow groove, such as that between convolutions on the surface of the brain (pl., *sulci*). p. 398
summation (sum-ma′shun) Increased force of contraction by a skeletal muscle fiber when twitches occur before the previous twitch relaxes. p. 297
superficial (soo′per-fish′al) Near the surface. p. 21
superior (su-pe′re-or) Structure higher than another structure. p. 21
supination (soo″pĭ-na′shun) Upward or forward rotation of palm of hand. p. 269
surface tension (ser′fas ten′shun) Force that holds moist membranes together due to the attraction of water molecules. p. 750
surfactant (ser-fak′tant) Substance produced by the lungs that reduces the surface tension in alveoli. p. 750
suture (soo′cher) Immovable joint, such as that between flat bones of the skull. p. 205
sweat gland (swet gland) Exocrine gland in skin that secretes a mixture of water, salt, urea, and other bodily wastes. p. 180
sympathetic nervous system (sim″pah-thet′ik ner′vus sis′tem) Part of the autonomic nervous system that arises from the thoracic and lumbar regions of the spinal cord. p. 424
symphysis (sim′fĭ-sis) Slightly movable joint between bones separated by a pad of fibrocartilage. p. 263
synapse (sin′aps) Functional connection between the axon of one neuron and the dendrite or cell body of another neuron or the membrane of another cell type. p. 289
synaptic cleft (sĭ-nap′tik kleft) A narrow extracellular space between the presynaptic and postsynaptic neurons. p. 290
synaptic knob (sĭ-nap′tik nob) Tiny enlargement at the end of an axon that secretes a neurotransmitter. p. 356
synaptic potential (sĭ-nap′tik po-ten′shal) Electrical activity generated in the space between two neurons. p. 371
synaptic transmission (sĭ-nap′tik trans-mish′un) Communication of an impulse from one neuron to the next. p. 365
synchondrosis (sin″kon-dro′sis) Type of joint in which bands of hyaline cartilage unite bones. p. 262
syndesmosis (sin″des-mo′sis) Type of joint in which long fibers of connective tissue unite bones. p. 261
synergist (sin′er-jist) Muscle that assists the action of a prime mover. p. 304
synovial fluid (sĭno′ve-al floo′id) Fluid that the synovial membrane secretes. p. 264
synovial joint (sĭ-no′ve-al joint) Freely movable joint. p. 263
synovial membrane (sĭ-no′ve-al mem′brān) Membrane that forms the inner lining of the capsule of a freely movable joint. p. 163
synthesis (sin′thĕ-sis) Building large molecules from bonding smaller ones. p. 58
systemic circuit (sis-tem′ik ser′kit) Vessels that conduct blood between the heart and all body tissues except the lungs. p. 553
systole (sis′to-le) Phase of the cardiac cycle when a heart chamber wall contracts. p. 564

systolic pressure (sis-tol′ik presh′ur) Arterial blood pressure reached during the systolic phase of the cardiac cycle. p. 580

T

tachycardia (tak″e-kar′de-ah) Abnormally rapid heartbeat. p. 572

tactile corpuscle (tak′til kor′pus-l) Sensory receptor close to the surface of the skin that is sensitive to light touch; Meissner's corpuscle. p. 440

target cell (tar′get sel) Cell with specific receptors on which a hormone exerts its effect. p. 16

tarsal (tahr′sal) Bone in the area between the foot and leg. p. 24

tarsus (tar′sus) Ankle bones. p. 206

taste bud (tāst bud) Organ containing receptors associated with the sense of taste. p. 448

telophase (tel′o-fāz) Stage in mitosis when newly formed cells separate. p. 101

tendon (ten′don) Cordlike or bandlike mass of white fibrous connective tissue that connects a muscle to a bone. p. 155

teratogen (ter′ah-to-jen) Chemical or other environmental agent that causes a birth defect. p. 892

testis (tes′tis) Primary male reproductive organ; sperm cell-producing organ (pl., *testes*). p. 512

testosterone (tes-tos′tĕ-rōn) Male sex hormone secreted by the interstitial cells of the testes. p. 833

tetanic contraction (tĕ-tan′ik kon-trak′shun) Continuous, forceful muscular contraction without relaxation. p. 297

thalamus (thal′ah-mus) Mass of gray matter at the base of the cerebrum in the wall of the third ventricle. p. 405

thermoreceptor (ther″mo-re-sep′tor) Sensory receptor sensitive to temperature changes; heat and cold receptors. p. 439

thiamine (thi′ah-min) Vitamin of the B-complex group; vitamin B_1. p. 713

thoracic (tho-ras′ik) Pertaining to the chest. p. 12

thoracic cavity (tho-ras′ik kav′i-te) Hollow place in the chest. p. 12

threshold potential (thresh′old po-ten′shal) Level of potential at which an action potential or nerve impulse is produced. p. 368

threshold stimulus (thresh′old stim′u-lus) Stimulation level that must be exceeded to elicit a nerve impulse or a muscle contraction. p. 296

thrombin (throm′bin) Blood-clotting enzyme that catalyzes formation of fibrin from fibrinogen. p. 39

thrombocyte (throm′bo-sīt) Blood platelet. p. 534

thrombocytopenia (throm″bo-si″to-pe′ne-ah) Low number of platelets in the circulating blood. p. 542

thromobopoietin (throm′bo-poi′ĕ-tin) A hormone that stimulates megakaryocytes to differentiate from hematopoietic stem cells. p. 534

thrombus (throm′bus) Blood clot that remains where it forms in a blood vessel. p. 541

thymosins (thi′mo-sins) Group of peptides the thymus gland secretes that increases production of certain types of white blood cells. p. 512

thymus (thi′mus) Glandular organ in the mediastinum, behind the sternum and between the lungs. p. 512

thyroid gland (thi′roid gland) Endocrine gland just below the larynx and in front of the trachea that secretes thyroid hormones. p. 499

thyroid-stimulating hormone (thi′roid-stim″u-lāt′eng hor′mōn) Hormone secreted from the anterior pituitary gland that controls secretion from the thyroid gland; TSH. p. 496

thyroxine (thi-rok′sin) Hormone secreted by the thyroid gland; T4. p. 500

tidal volume (tīd′al vol′ūm) Volume of air entering and leaving the lungs in a respiratory cycle. p. 752

tissue (tish′u) Group of similar cells that performs a specialized function. p. 4

titin (ti′tin) Protein that attaches myosin filaments to z lines in muscle. p. 287

T lymphocyte (T lim′fo-sīt) Type of white blood cell that interacts directly with antigen-bearing cells and particles and secretes cytokines, producing the cellular immune response; T cell. p. 623

tonsil (ton′sil) Collection of lymphatic tissue in the throat. p. 621

total lung capacity (toh′tal lung kah-pas′i-te) Vital capacity plus residual volume. p. 753

totipotent (to-tip′o-tent) Ability of a cell to differentiate into any type of cell. Only a fertilized ovum is totipotent. p. 105

trabecula (trah-bek′u-lah) Branching bony plate that separates irregular spaces within spongy bone. p. 194

trace element (trās el′ĕ-ment) Basic chemical substance needed in small quantity. p. 52

trachea (tra′ke-ah) Tubular organ that leads from the larynx to the bronchi. p. 742

transcellular fluid (trans″sel′u-lar floo′-id) Part of the extracellular fluid, including the fluid in special body cavities. p. 811

transcription (trans-krip′shun) Manufacturing a complementary RNA from DNA. p. 130

transcytosis (trans″si-to′sis) Combination of receptor-mediated endocytosis and exocytosis that moves particles through a cell. p. 98

transfer RNA (trans′fer RNA) RNA molecule that carries an amino acid to a ribosome in protein synthesis; tRNA. p. 131

translation (trans-la′ shun) Assembly of an amino acid chain according to the sequence of base triplets in an mRNA molecule. p. 131

transverse (trans-vers′) Plane that divides a structure into superior and inferior parts. p. 22

transverse colon (trans-vers′ ko′lon) Part of the large intestine that extends across the abdomen from right to left below the stomach. p. 687

transverse tubule (trans-vers′ tu′būl) Membranous channel that extends inward from a muscle fiber membrane and passes through the fiber. p. 288

triad (tri′ad) Group of three structures. p. 288

tricuspid valve (tri-kus′pid valv) Heart valve between the right atrium and the right ventricle. p. 557

trigger zone (trig′ger zōn) Sensitive part of an axon where a nerve impulse originates. p. 368

triglyceride (tri-glis′er-īd) Lipid composed of three fatty acids and a glycerol molecule. p. 63

triiodothyronine (tri″i-o″do-thi′ro-nēn) Type of thyroid hormone; T_3. p. 500

trisomy (tri′so-me) Condition in which a cell has three chromosomes of a particular type instead of two. p. 929

trochanter (tro-kan′ter) Broad process on a bone. p. 208

trophoblast (trof′o-blast) Outer cells of a blastocyst that give rise to cells that become part of the placenta and extraembryonic membranes. p. 880

tropic hormone (trōp′ik hor′mōn) Hormone that has an endocrine gland as its target tissue. p. 492

tropomyosin (tro″po-mi′o-sin) Protein that blocks muscle contraction until calcium ions are present. p. 287

troponin (tro′po-nin) Protein that functions with tropomyosin to block muscle contraction until calcium ions are present. p. 287

trypsin (trip′sin) Enzyme in pancreatic juice that breaks down protein molecules. p. 671

trypsinogen (trip-sin′o-jen) Substance pancreatic cells secrete that is enzymatically cleaved to yield trypsin. p. 671

tubercle (tu′ber-kl) Small, rounded process on a bone. p. 223

tuberosity (tu″bĕ-ros′ĭ-te) Elevation or protuberance on a bone. p. 208

tubular reabsorption (too′bu-lar re″ab-sorp′shun) Transport of substances out of the renal tubule into the interstitial fluid from which the substances diffuse into peritubular capillaries. p. 786

tubular secretion (too′bu-lar se-kre′shun) Movement of substances out of the peritubular capillaries into the renal tubule for excretion in the urine. p. 786

tumor (too′mor) Tissue mass formed when cells lose division control. p. 104

tumor suppressor gene (too′mor sŭ-pres′or jēn) A gene (DNA sequence) that encodes a protein that ordinarily inhibits cell division. p. 104

twitch (twich) Brief muscular contraction followed by relaxation. p. 296

tympanic membrane (tim-pan′ik mem′brān) Thin membrane that covers the auditory canal and separates the outer ear from the middle ear; the eardrum. p. 450

U

ultratrace element (ul′trah-trās el′ĕ-ment) Basic chemical substance needed in very small quantity. p. 52

umbilical (um-bil′ĭ-kal) Pertaining to the navel p. 24

umbilical cord (um-bil′ĭ-kal kord) Cordlike structure that connects the fetus to the placenta. p. 891

umbilical region (um-bil′ĭ-kal re′jun) Central portion of the abdomen. p. 22

umbilicus (um-bil′ĭ-kus) Region to which the umbilical cord was attached; the navel. p. 891

unipolar neuron (un″ĭ-po′lar nu′ron) Neuron that has a single nerve fiber extending from its cell body. p. 360

unsaturated fat (un-sat′u-rāt″ed fat) Fat molecule that includes unsaturated fatty acids with one or more double bonds between the atoms of the carbon chains. p. 64

upper limb (uh′per lim) Superior appendage consisting of the arm, forearm, and hand. p. 226

urea (u-re′ah) Nonprotein nitrogenous substance produced as a result of protein metabolism. p. 704

ureter (u-re′ter) Muscular tube that carries urine from the kidney to the urinary bladder. p. 798

urethra (u-re′thrah) Tube leading from the urinary bladder to the outside of the body. p. 800

uric acid (u′rik as′id) Product of nucleic acid metabolism in the body. p. 798

urine (u′rin) Wastes and excess water removed from the blood and excreted by the kidneys into the ureters to the urinary bladder and out of the body through the urethra. p. 785

uterine (u′ter-in) Pertaining to the uterus. p. 852

uterine tube (u′ter-in tūb) Tube that extends from the uterus on each side toward an ovary and transports sex cells; fallopian tube or oviduct. p. 852

uterus (u′ter-us) Hollow muscular organ in the pelvis of a female where a fetus develops. p. 854

utricle (u′trĭ-kl) Enlarged portion of the membranous labyrinth of the inner ear. p. 459

uvula (u′vu-lah) Fleshy portion of the soft palate that hangs down above the root of the tongue. p. 657

V

vaccine (vak′sēn) Preparation that includes antigens used to stimulate an immune response to prevent an infectious disease. p. 638

vagina (vah-ji′nah) Tubular organ that leads from the uterus to the vestibule of the female reproductive tract. p. 854

variably expressive (va′re-ah-ble eks″pres′iv) Phenotype whose symptoms vary in intensity in different people. p. 924

varicose veins (var′ĭ-kos vānz) Abnormally swollen and enlarged veins, especially in the legs. p. 582

vasa recta (va′sah rek′tah) Branch of the peritubular capillary that receives blood from the efferent arterioles of juxtamedullary nephrons. p. 785

vasoconstriction (vas″o-kon-strik′shun) Decrease in the diameter of a blood vessel. p. 576

vasodilation (vas″o-di-la′shun) Increase in the diameter of a blood vessel. p. 576

vasomotor center (vas″o-mo′tor sen′ter) Neurons in the brainstem that control the diameter of the arteries. p. 587

vasopressin (vas″o-pres′in) Antidiuretic hormone. p. 498

vein (vān) Vessel that carries blood toward the heart. p. 579

vena cava (vēn′ah kāv′ah) One of two large veins that convey deoxygenated blood to the right atrium of the heart. p. 557

ventral root (ven′tral roo-t) Motor branch of a spinal nerve by which it attaches to the spinal cord. p. 420

ventricles (ven′trĭ-klz) Cavities, such as brain ventricles filled with cerebrospinal fluid, or heart ventricles that contain blood. pp. 363, 555

venule (ven′ūl) Vessel that carries blood from capillaries to a vein. p. 579

vertebral (ver′te-bral) Pertaining to the bones of the spinal column. p. 24

vertebral canal (ver′te-bral kah-nal′) Hollow area in vertebrae containing the spinal cord. p. 12

vesicle (ves′ĭ-kal) Membranous, cytoplasmic sac formed by an infolding of the cell membrane. p. 83

vestibule (ves′tĭ-būl) Space at the opening to a canal. p. 453

villus (vil′us) Tiny, fingerlike projection that extends outward from the inner lining of the small intestine (pl., *villi*). p. 681

viscera (vis′er-ah) Organs in a body cavity. p. 12

visceral (vis′er-al) Pertaining to the contents of a body cavity. p. 12

visceral peritoneum (vis′er-al per″ĭ-to-ne′um) Membrane that covers organ surfaces in the abdominal cavity. p. 14

visceral pleura (vis′er-al ploo′rah) Membrane that covers the surfaces of the lungs. p. 12

visceroreceptive sense (vis″er-o-re′cep-tiv sens) Detecting changes in the viscera. p. 440

viscosity (vis-kos′ĭ-te) Tendency for a fluid to resist flowing due to the internal friction of its molecules. p. 584

vital capacity (vi′tal kah-pas′i-te) The maximum volume of air a person can exhale after taking the deepest breath possible. p. 753

vitamin (vi′tah-min) Organic compound other than a carbohydrate, lipid, or protein needed for normal metabolism but that the body cannot synthesize in adequate amounts and must therefore be obtained in the diet. p. 119

vitreous body (vit′re-us bod′e) Collagenous fibers and fluid in the posterior cavity of the eye. p. 469

vitreous humor (vit′re-us hu′mor) Fluid between the lens and the retina of the eye. p. 469

vocal cords (vo′kal kordz) Folds of tissue in the larynx that produce sounds when they vibrate. p. 740

voluntary (vol′un-tār″e) Consciously controlled. p. 163

W

water balance (wot′er bal′ans) When the volume of water entering the body is equal to the volume leaving it. p. 813

water of metabolism (wot′er uv mĕ-tab′o-lizm) Water produced as a by-product of metabolic processes. p. 813

white blood cell (whīt blud sel) Cell that helps fight infection; leukocyte. p. 162

white fiber (whīt fi′ber) Fast-contracting skeletal muscle fiber. p. 299

wild type (wīld tīp) Phenotype or allele that is the most common for a certain gene in a population. p. 920

X

X-linked trait (eks-linkt′trāt) Trait determined by a gene on an X chromosome. p. 920

x-ray (eks′ ray) Used as a verb: to photograph using radiation. p. 679

X ray (eks ray) Used as a noun: a photograph produced by radiation. May also be used as an adjective: X-ray. p. 202

Y

yellow marrow (yel′o mar′o) Fat storage tissue in the cavities in certain bones. p. 203

yolk sac (yōk sak) Extraembryonic membrane connected to the embryo by a long, narrow tube. p. 891

Z

zona pellucida (zo′nah pel-u′cĭ-dah) Thick, transparent, noncellular layer surrounding a secondary oocyte. p. 849

zygote (zi′gōt) Cell produced by the fusion of an egg and sperm; a fertilized egg cell. p. 849

zymogen granule (zi-mo′jen gran′-ūl) Cellular structure that stores inactive forms of protein-splitting enzymes in a pancreatic cell. p. 671

CREDITS

PHOTOGRAPHS

Preview Chapter/Foundations for Success

Opener: © Comstock Images/Jupiter Images

Chapter 1

Opener: © Brand X Pictures/Punchstock; **p. 2:** © Stockbyte/PunchStock; **figure 1.1:** © Andreas Vesalius, 2nd book, De Humani Corporis Fabrica, 1543; **figure 1A:** © Keith Brofsky/Getty Images; **figure 1B:** © Jim Wehtje/Getty Images; **figure 1C:** © SPL/Photo Researchers, Inc.; **figure 1.4:** © McGraw-Hill Higher Education, Inc./ Carol D. Jacobson, PhD, Dept. of Veterinary Anatomy, Iowa State University; **figure 1.21a–d:** © McGraw-Hill Higher Education, Inc./Joe De Grandis, photographer; **figure 1.22a:** © Patrick J. Lynch/Photo Researchers, Inc.; **figure 1.22b:** © Biophoto Associates/Photo Researchers, Inc.; **figure 1.22c:** © A. Glauberman/Photo Researchers, Inc.

Chapter 2

Opener: © Dr. Tim Evans/SPL/Photo Researchers, Inc.; **figure 2A:** © Mark Antman/ Image Works; **figure 2B(a):** © SIU/Peter Arnold; **figures 2.7, 2.23a–e:** Courtesy of John W. Hole, Jr.; **figure 2.23f:** Courtesy Ealing Corporation; **figure 2D(a):** © SPL/Photo Researchers, Inc.; **figure 2D(b):** © CNRI/SPL/Photo Researchers, Inc.; **figure 2E:** From M.I. Posner, 10/29/93, "Seeing the Mind," Science 262:673, Fig. 1, © AAAS/Photo, Marcus Raichle.

Chapter 3

Opener: © K.R. Porter/Photo Researchers, Inc.; **figure 3.4:** © Steve Allen/Getty Images; **figure 3.5a:** © Robert Becker/Custom Medical Stock Photo; **figure 3.5b:** © R. Roseman/Custom Medical Stock Photo; **figure 3.5c:** © M. Kalab/ Custom Medical Stock Photo; **figure 3.6a:** © Biophoto Associates/Photo Researchers, Inc.; **figure 3.9a:** © Don W. Fawcett/Photo Researchers, Inc.; **figure 3.10a:** © Gordon Leedale/Biophoto Associates; **figure 3.12a:** © Bill Longcore/Photo Researchers, Inc.; **figure 3.13:** © K.G. Murti/Visuals Unlimited; **figure 3.14a:** © Don W. Fawcett/Visuals Unlimited; **figure 3.15a:** © Oliver Meckes/Photo Researchers, Inc.; **figure 3.16:** © Colin Anderson/Brand X/ CORBIS; **figure 3.17:** © M. Schliwa/Visuals Unlimited; **figure 3.18b:** © K.G. Murti/Visuals Unlimited; **figure 3.19b:** © Stephen L. Wolfe; **figure 3.25a–c:** © David M. Phillips/Visuals Unlimited; **figure 3.36a–e:** © Ed Reschke; **figure 3.37a–c:** From Scanning Electron Microscopy in Biology, by R.G. Kessel and C.Y. Shih. © 1976 Springer-Verlag; **figure 3.38:** © Tony Brain/Photo Researchers, Inc.; **figure 3.42:** © Peter Skinner/ Photo Researchers, Inc.

Chapter 4

Opener: © Biophoto Associates, SPL/Photo Researchers, Inc.; **figure 4.15:** © Royalty Free/ CORBIS.

Chapter 5

Opener: © The McGraw-Hill Companies, Inc./ Al Telser, photographer; **figure 5.2b,d:** © Ed Reschke; **figures 5.3b, 5.4b:** © The McGraw-Hill Companies, Inc./Al Telser, photographer; **figure 5.5:** © Fawcett, Hirokawa, Heuser/ Photo Researchers, Inc.; **figure 5.6b:** © The McGraw-Hill Companies, Inc./Dennis Strete, photographer; **figures 5.7b, 5.8b, 5.9b:** © The McGraw-Hill Companies, Inc./Al Telser, photographer; **figure 5.10b,d:** © Ed Reschke; **figure 5.13:** © The McGraw-Hill Companies, Inc./Al Telser, photographer; **figure 5.14:** © David M. Phillips/Visuals Unlimited; **figure 5.15:** © Manfred Kage/Peter Arnold; **figure 5.16:** © Microworks Color/Phototake; **figure 5.17:** © Prof. P. Motta/Univ. "La Sapienza"/Photo Researchers, Inc.; **figures 5.18b, 5.19b:** © The McGraw-Hill Companies, Inc./Dennis Strete, photographer; **figure 5.20b:** © The McGraw-Hill Companies, Inc./Al Telser, photographer; **figure 5.21b:** © The McGraw-Hill Companies, Inc./Dennis Strete, photographer; **figures 5.22b, 5.23b, 5.24b, 5.25b:** © The McGraw-Hill Companies, Inc./Al Telser, photographer; **figure 5.26b:** © The McGraw-Hill Companies, Inc./Dennis Strete, photographer; **figure 5.26c:** © Prof. P. Motta/Univ."La Sapienza"/Photo Researchers, Inc.; **figures 5.27b, 5.28b:** © The McGraw-Hill Companies, Inc./Al Telser, photographer; **figure 5.29b:** © The McGraw-Hill Companies, Inc./ Dennis Strete, photographer; **figure 5.30b:** © The McGraw-Hill Companies, Inc./Al Telser, photographer; **figure 5.31b:** © Ed Reschke.

Chapter 6

Opener: © Dr. Jeremy Burgess/SPL/Photo Researchers, Inc.; **p. 171(top):** Courtesy of Achillean Franjakis, EMBL; **figure 6.1:** © The McGraw-Hill Companies, Inc./Al Telser, photographer; **figure 6.2b:** © Victor Eroschenko; **figure 6.3b:** © The McGraw-Hill Companies, Inc./Al Telser, photographer; **figure 6A(a),(b):** © Biophoto Associates/Photo Researchers, Inc.; **figure 6A©:** © SPL/Photo Researchers, Inc.; **figure 6.4a:** © M. Schliwa/Visuals Unlimited; **figure 6.5:** © Mediscan/Visuals Unlimited; **figure 6.7b:** © The McGraw-Hill Companies, Inc./ Al Telser, photographer; **figure 6B:** © Anthony Saint James/Getty Images; **figure 6.8:** © CNRI/ SPL/Photo Researchers, Inc.; **figure 6.9:** © Per H. Kjeldsen; **figure 6.11:** © Dr. John D. Cunningham/Visuals Unlimited; **figure 6C:** © Kenneth Greer/Visuals Unlimited; **figure 6.15:** © Kjell Sandvad/Visuals Unlimited.

Chapter 7

Opener: © David Scharf/Peter Arnold; **p. 193:** Reprinted by permission from Macmillan Publishers Ltd: Nature, "Pleistocene Homo sapiens from Middle Awash, Ethiopia", FIG 1, VOL 423, © 2003; **figure 7.3a:** © Ed Reschke; **figure 7.3b,c:** Courtesy of John W. Hole, Jr.; **figure 7.5:** Tissues and Organs: A Text-Atlas of Scanning Electron Microscopy, by R.G. Kessel and R.H. Kardon. © 1979 W.H. Freeman and Company; **figure 7.6a:** © Biophoto Associates/ Photo Researchers, Inc.; **figure 7.6b:** Courtesy of Dr. Francis Collins; **figure 7.7:** © Secchi, Lecaque, Roussel, Uclaf, CNRA/SPL/Photo Researchers, Inc.; **figure 7.9b:** © The McGraw-Hill Companies, Inc./Al Telser, photographer; **figure 7.10:** © Biophoto Associates/Photo Researchers, Inc.; **figure 7.11:** © James L. Shaffer; **figure 7.12:** Courtesy of John W. Hole, Jr.; **figure 7.30a:** © LumaGraphix/ Photo Researchers, Inc.; **figure 7.30b:** © David Roberts/SPL/Photo Researchers, Inc.; **figure**

7.35: © Royalty Free/CORBIS; **figure 7.38b:** © Victor B. Eichler, PhD; **figure 7.40b:** Courtesy of Eastman Kodak; **figure 7.42d:** © Martin Rotker; **figure 7.45c:** © Ed Reschke; **figure 7.46:** © Dr. Charles Eaton/e-hand Electronic Textbook of Hand Surgery (www.e-hand.com); **figures 7.47c, 7.50a, 7.53a:** © Martin Rotker; **figure 7.54b:** Courtesy of St. Joseph's Hospital Medical Media, Milwaukee, WI; **figure 7.55(left),(right):** Courtesy of Pat Hesse.

Chapter 8

Opener: © Keith Brofsky/Getty Images; **figure 8.3a,b:** Courtesy of John W. Hole, Jr.; **figures 8.10, 8.11, 8.12:** © McGraw-Hill Companies / Womack Photography Ltd.; **figures 8.13b, 8.15b:** © Paul Reimann; **figure 8.17a,b:** Courtesy of John W. Hole, Jr.; **figures 8.18b, 8.20b:** © Paul Reimann; **figure 8.22:** © 1966 C.R. Wheeless/ Wheeless' Textbook of Orthopaedics; **figure 8.23a,b:** © Science 00/Visuals Unlimited.

Chapter 9

Opener: © CNRI/SPL/Photo Researchers, Inc.; **figure 9.3:** © Ed Reschke; **figure 9.5a:** © H.E. Huxley; **figure 9.8b:** © McGraw-Hill Higher Education, Inc./Carol D. Jacobson Ph.D., Dept. Veterinary Anatomy, Iowa State University; **figure 9.11b:** © H.E. Huxley; **figure 9.19:** © Dr. Paul Heidger, University of Iowa College of Medicine; **p. 339(left),(right):** © Royalty-Free/ CORBIS.

Chapter 10

Opener: © Nancy Kedersha/SPL/Photo Researchers, Inc.; **p. 354:** © Dr. Dennis Steindler; **figure 10.1:** © Ed Reschke; **figure 10.4b:** © Biophoto Associates/Photo Researchers, Inc.; **figure 10.5:** © Dennis Emery; **figure 10.9:** Tissues and Organs: A Text-Atlas of Scanning Electron Microscopy, by R.G. Kessel and R.H. Kardon. ©1979 W.H. Freeman and Company; **figure 10.12b:** © Don W. Fawcett/Photo Researchers, Inc.

Chapter 11

Opener: © Secchi-Lecaque/Roussel-UCLAF/ CNRA/SPL/Photo Researchers, Inc.; **p. 383:** From: H. Damasio, T. Grabowski, R. Frank, A.M. Galaburda, A.R. Damasio (1994) The return of Phineas Gage: clues about the brain from a famous patient, *Science*, 264, 1102–1105. Dornsife Neuroscience Imaging Center and Brain and Creativity Institute, University of Southern California.; **figure 11B(a),(b):** © Dr. Kent M. Van De Graaff; **figure 11.6b:** © Ed Reschke; **figure 11.15b:** © Martin M. Rotker/Photo Researchers, Inc.; **figure 11D:** © AFP/Getty Images; **figure 11E:** © M. Abbey/Photo Researchers, Inc.; **figure 11.24:** Tissues and Organs: A Text-Atlas of Scanning Electron Microscopy, by R.G. Kessel and R.H. Kardon. ©1979 W.H. Freeman and Company.

Chapter 12

Opener: © SPL/Photo Researchers, Inc.; **p. 438:** © Kanehara & Co., Ltd.; **figure 12.1b,c:** © Ed Reschke; **figure 12A:** © Imagebroker/Alamy; **figure 12B:** © Pallava Bagla/CORBIS; **figure 12C:** © K.S. Matz; **figure 12.6:** © Dwight Kuhn; **figure 12.8:** © Victor B. Eichler, PhD; **figure 12D:** © Democrat and Chronicle, Rochester, N.Y.; **figure 12.14a:** © John D. Cunningham/Visuals Unlimited; **figure 12.14b:** © Fred Hossler/Visuals Unlimited; **figure 12.19:** Courtesy of Dean E. Hillman; **figure 12.27:** © SPL/Photo Researchers, Inc.; **figure 12.33:** © Per H. Kjeldsen; **figure 12.34b:** © Mediscan/CORBIS; **figure 12.38c:** © Frank S. Werblin, PhD.

Chapter 13

Opener: © Prof. P. Motta/Dept. of Anatomy/ University of "La Sapienza" Rome/SPL/Photo Researchers, Inc.; **figure 13.14:** © Michael Ross/ Photo Researchers, Inc.; **figure 13A(a)–(d):** Reprinted by permission of publisher from "Albert Mendeloff, "Acromegaly, diabetes, hypermetabolism, proteinurea and heart failure," American Journal of Medicine: 20:1, 01–'56, p. 135/© by Excerpta Medica, Inc.; **figure 13.19:** © Fred Hossler/Visuals Unlimited; **figure 13.21:** © Mediscan/Visuals Unlimited; **figure 13.22:** © Imagingbody.com; **figure 13.23:** © L.V. Bergman/ The Bergman Collection; **figure 13.25:** © R. Calentine/Visuals Unlimited; **figure 13.29:** © Ed Reschke; **figure 13.35:** From Kent M. Van De Graaff and Stuart Ira Fox, Concepts of Human Anatomy and Physiology, 2nd ed. ©1989 Wm. C. Brown Publishers, Dubuque, Iowa. All Rights Reserved. Reprinted with permission; **figure 13B:** © Imagezoo/PunchStock.

Chapter 14

Opener: © Image Source; **p. 523:** © Photodisc Collection/Getty Images; **figure 14.4b:** © The McGraw-Hill Companies, Inc./Al Telser, photographer; **figure 14.5b:** © Bill Longcore/ Photo Researchers, Inc.; **figure 14.8a:** © The McGraw-Hill Companies, Inc./Al Telser, photographer; **figures 14.8b, 14.10–14.12:** © Ed Reschke; **figure 14.13:** © R. Kessel/Visuals Unlimited; **figure 14.14:** © Ed Reschke; **figure 14A:** © 2005 Basil Childers; **figure 14B(a):** © The McGraw-Hill Companies, Inc./Al Telser, photographer; **figure 14B(b):** © Andrew Syred/ SPL/Photo Researchers, Inc.; **figure 14.18:** © SPL/Photo Researchers, Inc.; **figure 14.20a,b:** © The McGraw-Hill Companies, Inc./Al Telser, photographer; **figure 14.22c:** © G.W. Willis/ Visuals Unlimited; **figure 14.22d:** © George W. Wilder/Visuals Unlimited; **p. 551:** © Image Source/Getty Images.

Chapter 15

Opener: © Prof. P. Motta/G. Macchiarelli/ University "La Sapienza" Rome/SPL/Photo Researchers, Inc.; **p. 553:** © McGraw-Hill Higher Education, Inc./University of Michigan Biomedical Communications; **figure 15.2:** © McGraw-Hill Companies, Inc./Photo and dissection by Christine Eckel; **figure 15.6c:** © The McGraw-Hill Companies, Inc.; **figures 15.7,15.8:** © McGraw-Hill Higher Education, Inc./University of Michigan Biomedical Communications; **figure 15.14:** Courtesy of Eastman Kodak; **p. 566:** © F. Schussler/PhotoLink/Getty Images; **figure 15.25c:** © The McGraw-Hill Companies, Inc./Al Telser, photographer; **figures 15.28b,c, 15.29:** © Don. W. Fawcett/Visuals Unlimited; **figure 15E(a):** © Ed Reschke/Peter Arnold; **figure 15E(b):** © J & L Weber/Peter Arnold; **figure 15E:** © Alfred Pasieka/Peter Arnold; **figure 15.35a,b:** From Patricia Phelps and J. Luft, "Electron Study of Relaxation and Constriction in Frog Arterioles." American Journal of Anatomy, 125:409, fig. 2–3, ©1969, Wiley-Liss, a division of John Wiley and Sons, Inc.; **figure 15H:** NASA; **figure 15.43b:** © Dr. Kent M. Van De Graaff; **figure 15.45:** From J. Dankmeijer, H.J. Lammers en J.M.F. Landsmeer, Practische Ontleedkunde, Bohn, Scheltema & Holkema Publisher, Netherlands, 1955.

Chapter 16

Opener: © Dr. Kari Lounatmaa/SPL/ Photo Researchers, Inc.; **figure 16.3:** © The McGraw-Hill Companies, Inc./Dennis Strete, photographer; **figure 16.5:** Courtesy of Eastman Kodak; **figure 16.9:** © Dr. Kent M. Van De Graaff; **figure 16.10b:** © The McGraw-Hill Companies, Inc./Al Telser, photographer; **figure 16.12b:** © The McGraw-Hill Companies, Inc./Dennis Strete, photographer; **figure 16.14b:** © The McGraw-Hill Companies, Inc./Al Telser, photographer; **figure 16.15:** Courtesy of Dr. Etienne de Harven and Miss Nina Lampen, from the Sloan-Kettering Institute for Cancer Research; **figure 16.17b:** © Manfred Kage/Peter Arnold; **figure 16A:** Courtesy of Schering-Plough Corp./Photo by Phillip A. Harrington; **figure 16.22b:** © Lennart Nilsson/Bonnierforlagen AB; **figure 16B:** © Zeva Oelbaum/Peter Arnold.

Chapter 17

Opener: © CNRI/SPL/Photo Researchers, Inc.; **p. 652:** CDC; **figure 17.8:** © The McGraw-Hill Companies, Inc./Rebecca Gray, photographer; **figure 17.9b:** © Nick Koudis/Getty Images; **figure 17A:** © SPL/Photo Researchers, Inc.; **figure 17.12a:** © The McGraw-Hill Companies, Inc./Al Telser, photographer; **figure 17.12b:** © Biophoto Associates/Photo Researchers, Inc.; **figure 17.12c:** © The McGraw-Hill Companies, Inc./Al Telser, photographer; **figure 17.16:** © Ed Reschke; **figure 17.18:** ©Dr. Kent M. Van De Graaff; **figures 17.19b, 17.27c:** © The McGraw-Hill Companies, Inc./Al Telser, photographer; **figure 17.29:** © Carroll Weiss/Camera M.D. Studios; **figure 17.32:** © Armed Forces Institute of Pathology; **figures 17.36, 17.37b:** © The McGraw-Hill Companies, Inc./Al Telser, photographer; **figure 17.44:** © Jim Wehtje/ Getty Images; **figure 17.46:** © Ed Reschke/Peter Arnold; **figure 17.47:** © Ed Reschke.

Chapter 18

Opener: © Digital Vision/Getty Images; **figure 18.9a:** © The McGraw-Hill Companies, Inc./Lars A. Niki, photographer; **figure 18.9b:** © PhotoLink/Getty Images; **figure 18.11:** © Jeffrey L. Rotman/CORBIS; **figure 18.17:** CDC; **figure 18.19b:** © The McGraw-Hill Companies, Inc./Al Telser, photographer; **figure 18.19c:** © CNRI/PhotoTake; **figure 18B:** © Charles D. Winters/Photo Researchers, Inc.; **figure 18.21a:** © Reuters/CORBIS; **figure 18.21b:** © Charles O. Cecil/Visuals Unlimited; **figure 18.22:** © Brand X Pictures/Punchstock.

Chapter 19

Opener: © Susumu Nishinaga/SPL/Photo Researchers, Inc.; **p. 736:** © Brand X Pictures; **figure 19.3b:** © Biophoto Associates/Photo Researchers, Inc.; **figure 19A(left),(right):** Used with permission of the American Cancer Society; **figure 19B:** © Moredun Animal Health/SPL/Photo Researchers, Inc.; **figure 19.4a,b:** Courtesy of Eastman Kodak; **figure 19.7c:** © CNRI/PhotoTake; **figure 19.10:** © Ed Reschke; **figure 19.13:** © Ralph Hutchings/Visuals Unlimited; **figure 19.15:** © McGraw-Hill Higher Education, Inc./Bob Coyle; **figure 19.17:** Tissues and Organs: A Text-Atlas of Scanning Electron Microscopy, by R.G. Kessel and R.H. Kardon. © 1979 W.H. Freeman and Company; **figure 19.18:** Courtesy of the American Lung Association; **p. 749:** USGS; **figure 19.27:** © Edward Lettau/Photo Researchers, Inc.; **figure 19D(a),(b):** © Victor B. Eichler, PhD; **figure 19.32:** Courtesy of the American Lung Association; **figure 19.34:** © Imagingbody.com; **figure 19E(left),(right):** © CNRI/PhotoTake.

Chapter 20

Opener: © Prof. P. Motta/Dept. of Anatomy/University "La Sapienza" Rome/SPL/Photo Researchers, Inc.; **p. 775:** © Mirko Beovic; **figure 20.2:** © CNRI/SPL/Photo Researchers, Inc.; **figure 20.6b:** © L.V. Bergman/The Bergman Collection; **figure 20.7a:** Tissues and Organs: A Text-Atlas of Scanning Electron Microscopy, by R.G. Kessel and R.H. Kardon. ©1979 W.H. Freeman and Company; **figure 20.7b:** Courtesy of R.B. Wilson MD, Eppeley Institute for Research in Cancer, University of Nebraska Medical Center; **figure 20.9:** © David M. Phillips/Visuals Unlimited; **figure 20.11a:** © Biophoto Associates/Photo Researchers, Inc.; **figure 20.11b:** © Manfred Kage/Peter Arnold; **figure 20.27:** © Per H. Kjeldsen; **figure 20.30:** © John D. Cunningham/Visuals Unlimited; **figure 20.31:** © Ed Reschke.

Chapter 21

Opener: © Prof. P.M. Motta & M. Castellucci/SPL/Photo Researchers, Inc.

Chapter 22

Opener: © Brand X Pictures/CORBIS; **p. 831:** © Profs. P.M. Motta & J. Van Blerkom/SPL/Photo Researchers, Inc.; **figure 22.7:** © The McGraw-Hill Companies, Inc./Al Telser, photographer; **figure 22.11:** © Brand X Pictures/CORBIS; **figure 22.12:** © Image Source; **figure 22.13a,b:** © The McGraw-Hill Companies, Inc./Al Telser, photographer; **figure 22.14:** © Manfred Kage/Peter Arnold; **figure 22.21b:** Courtesy of R.J. Blandau; **figure 22.22:** © Ed Reschke; **figure 22.23b:** © The McGraw-Hill Companies, Inc./Al Telser, photographer; **figure 22.24:** © H. Mizoguchi, from W. Bloom & D.W. Fawcett, A Textbook of Histology, 10/e; **figure 22.25:** © 2007 Landrum B. Shettles; **figure 22.28a:** © The McGraw-Hill Companies, Inc./Al Telser, photographer; **figure 22.28b:** © Mediscan/Visuals Unlimited; **figure 22.29:** © McGraw-Hill Higher Education, Inc./Carol D. Jacobson, PhD, Dept. of Veterinary Anatomy, Iowa State University; **figure 22.35a,b:** © The McGraw-Hill Companies, Inc./Jill Braaten, photographer; **figure 22.35c:** © Photolink/Getty Images; **figure 22.35d:** © Don Farrall/Getty Images; **figure 22.35e:** © The McGraw-Hill Companies, Inc./Jill Braaten, photographer; **figure 22C:** Courtesy of Southern Illinois University School of Medicine/Biomedical.

Chapter 23

Opener: Courtesy of Michelle Watnick; **p. 876:** © J. Meyer/Custom Medical Stock Photo; **figure 23.2:** From M. Tegner and D. Epel. 16 February 1973. "Sea Urchin Sperm." Science, 179:685–688. © 1973 American Association for the Advancement of Science; **figure 23A:** Courtesy of Southern Illinois University School of Medicine/Biomedical; **figure 23.4a:** © A. Tsiara/Photo Researchers, Inc.; **figure 23.4b:** © Omikron/Photo Researchers, Inc.; **figure 23.4c:** © Petit Format/Nestle/Photo Researchers, Inc.; **figures 23.6c, 23.7:** Courtesy of Ronan O'Rahilly, M.D. Carnegie Institute of Washington; **figure 23.12a,b:** © 2007 Landrum B. Shettles; **figure 23.12c:** © Petit Format/Nestle/Photo Researchers, Inc.; **figure 23.13b:** © Carroll Weiss/Camera M.D. Studios; **figure 23.14e(bottom):** © 2007 Landrum B. Shettles; **figure 23.19:** © Donald Yaeger/Camera M.D. Studios; **figure 23D:** © Abigail and Brittany Hensel; **figure 23.29a,b:** © Biophoto Associates/Photo Researchers, Inc.; **figure 23G:** © Mitch Wojnarowicz/The Image Works.

Chapter 24

Opener: © Barros & Barros/Getty Images; **figure 24.3:** © SPL/Photo Researchers, Inc.; **figure 24.6:** From Genest, Jacques, Jr., Lavoie, Marc-Andre. August 12, 1999. "Images in Clinical Medicine." New England Journal of Medicine, p. 490. ©1999, Massachusetts Medical Society. All Rights Reserved; **figure 24A:** © Mediacolor's/Alamy; **figure 24.7a:** © Library of Congress; **figure 24.7b:** © Peter Morenus/University of Connecticut at Storrs; **figure 24.11:** © Biophoto Associates/Photo Researchers, Inc.; **figure 24B:** © Michael Greenlar/Image Works; **figure 24.14:** © Bernard Bennot/SPL/Photo Researchers, Inc.; **figure 24.15:** © S.A. Armstrong et al. MLL translocations specify a distinct gene expression profile that distinguishes a unique leukemia. Nature Genetics Vol. 30, Fig. 5, pp. 41–47, January 2002.

Reference Plates

Plates 8–25: © McGraw-Hill Higher Education, Inc./Karl Rubin; **Plates 26–54:** Courtesy of John W. Hole, Jr.; **Plates 55–65:** © Dr. Kent M. Van De Graaff; **Plates 66–75:** © McGraw-Hill Higher Education, Inc./Karl Rubin.

INDEX

Note: Page numbers followed by f refer to figures and by t to tables; page numbers in **boldface** refer to plates.

A

A bands, 287, 288f
Abdominal aorta, **37, 40,** 592, 594f, 600f, 778f, 898f
Abdominal cavity, 12, **37,** 623, 777f
Abdominal reflexes, 394
Abdominal region, 23, 24f, 25f
Abdominal viscera, **48–49,** 587, 603–604
Abdominal wall, 320, 321f, 322t, 597–98, 603, 752, 799f
Abdominopelvic cavity, 12, 13f, 15f
Abdominopelvic membranes, 12–14
Abducens nerves, 415, 417t
Abduction, 269
Abductor muscles, 315
ABO blood groups, 544, 546–47
Absolute refractory period, 370
Absorption
  characteristics of life, 8t
  definition, 651
  gastric, 669
  nutrients, 686t
  organ systems, 17–18
  small intestine, 684–85
Accelerator nerves, 569
Accessory nerves, 417
Accessory organs
  of digestive system, 653f
  of reproductive system, 833, 837–42, 846, 852–56
Accessory structures
  of joints, 264
  of skin, 177–81
Accommodation, of eye, 467, 468f
ACE inhibitors, 293, 507
Acetabulum, 192, 231, 232f, 234t
Acetaminophen, 445
Acetazolamide, 451
Acetoacetic acid, 825, 943
Acetone, 825, 943
Acetylcholine (Ach), 290, 301, 372, 373t, 429
Acetylcholinesterase, 293, 374, 430
Acetylcholinesterase inhibitors, 293, 507
Acetyl coenzyme A (acetyl CoA), 121f, 122, 714, 944, 946f
Achilles tendon, 330
Acid(s), 59–60
Acid–base balance, 819–26
Acid–base buffer systems, 820–22
Acidic ketone bodies, 819
Acid indigestion, 670
Acidosis, 60, 810, 822, 824–25
Acini, 671
Acne, 180, 181
Acoustic meatus, 210, 211f, **253,** 450
Acquired immune deficiency syndrome (AIDS), 76, 278, 534t, 630, 637, 642, 644, 862t, 869. *See also* Human immunodeficiency virus
Acromegaly, 201, 497
Acromial region, 23, 25f
Acromioclavicular joint, 270t
Acromion process, 226, 227f, 272f, **342–43**
Acrosome, 837
Actin, 287, 288f, 289
Action potential, 368–70
Activation energy, 117
Active aging, 909–10
Active dying, 909
Active sites, 117
Active transport, 95–96, 100t, 291
Activity site, of hormone, 486
Acute cardiac tamponade, 589
Acute cholecystitis, 679
Acute glomerulonephritis (AGN), 782
Acute intermittent porphyria, 138f
Acute lymphoblastic leukemia (ALL), 537
Acute pain fibers, 442
Acute pancreatitis, 672
Acute respiratory distress syndrome (ARDS), 765. *See also* Respiratory distress syndrome
Adaptive defense, 626
Addison disease, 509, 820
Adduction, 269
Adductor brevis muscle, **36,** 323, 324f, 327t
Adductor longus muscle, **34–37,** 305f, 323, 324f, 327t, **350**
Adductor magnus muscle, **37,** 306f, 323, 324f, 325, 326f, 327t
Adductor muscles, of thigh, 323, 325
Adenine, 128f, 948f, 949f
Adenoids, 657
Adenosine diphosphate (ADP), 119, 294
Adenosine triphosphate (ATP)
  cellular metabolism, 119, 120f
  cellular respiration, 944–47
  deamination of amino acids, 704
  magnesium, 720
  mitochondrion, 84, 720
  muscle contraction, 291–92, 293, 294
  phosphorus, 718
  protein synthesis, 132
Adenylate cyclase, 488, 490f
ADH. *See* Antidiuretic hormone
Adipocytes, 156
Adiponectin, 700
Adipose tissue, 143, 156, 158, 162t
Adolescence. *See also* Children
  development, 906–907, 908t
  skeleton, 236
ADP. *See* Adenosine diphosphate
Adrenal androgens, 508, 509t
Adrenal cortex, 430t, 487t, 504, 505f, 506–508, 509
Adrenal gland, **36,** 482, 504–509, 515, 777f, 778f
Adrenal medulla, 430t, 482, 487t, 492, 504–506
Adrenergic receptors, 429, 430
Adrenocorticotropic hormone (ACTH), 482, 487t, 490, 492, 496, 499t
Adrenoleukodystrophy (ALD), 87
Adulthood, and development, 907, 908t
Aerobic exercise, 334
Aerobic reactions, 120, 122
Aerobic respiration, 114, 125f, 294, 819
Aerobic training, 296
Afferent arteriole, 774, 779, 780f, 783f, 786f
Afferent lymphatic vessels, 622
Afferent neurons, 389, 391f
Afterbirth, 900–901
Afterload, and blood pressure, 586
Age. *See also* Adolescence; Adulthood; Aging; Children; Infants; Life-span changes
  aging-related changes in development, 908t
  breast cancer risk, 864t
  chromosome defects in children and maternal, 849, 931t
  falls by elderly, 240, 280
  incontinence in elderly, 805
  nutritional requirements, 712t, 728t
"Age spots," 186

Ageusia, 450t
Agglutination, 522, 544, 545f, 635, 636t
Aging, process of, 909–11. *See also* Age; Life-span changes
Agonists, 304, 377
Agranulocytes, 530, 535t
Agriculture, and human health, 3
AIDS. *See* Acquired immune deficiency syndrome
Air pollution, 749
ALA dehydratase deficiency, 138f
Alanine, 135t, 705t
Albinism, 170, 174, 176, 178, 473
Albumin(s), 535, 536t, 714, 792, 940
Albumin-globulin ratio, 940
Alcohol
ADH secretion and dehydration, 498
breath tests, 762
gastric absorption, 669
pregnancy and fetal alcohol syndrome, 896, 897
urinary system, 796
Alcoholism, 714
Aldosterone, 486, 487t, 506–507, 509t, 719, 818, 885t
Ali, Muhammed, 406f
Alimentary canal, 651, 652–56
Alkaline tide, 669
Alkaloids, 450
Alkalosis, 60, 822, 824, 825–26
Alkaptonuria, 803
Allantois, 875, 891
Alleles, 919
Allergen, 616
Allergies, 534t, 617, 639–41
Allograft, 186, 641
All-or-none response, and nerve impulse conduction, 370
Alopecia areata, 179
Alpha fetoprotein (AFP), 884
Alpha globulin, 535
Alpha radiation, 53, 56
Alpha receptors, 430
Alpha-tocopherol, 712
Alpha waves, 411
Alternative pathway, of complement, 627
Altitude sickness, 763
Alveolar arch, 213, **252**
Alveolar border, 214
Alveolar capillaries, 560f, 590–91
Alveolar dead space, 753
Alveolar ducts, 743, 744f, 745f, 862
Alveolar gas exchanges, 759–62
Alveolar macrophages, 759–60
Alveolar pores, 759
Alveolar process, 213, 215f
Alveolar sacs, 743, 745f
Alveolar ventilation, 754–55
Alveolar ventilation rate, 754
Alveolar wall, 591f
Alveolus (alveoli), 560f, 735, 743, 744f, 745–46, 759–60
Alzheimer disease, 20, 373t, 409, 911t
Amacrine cells, 469
Ameboid motion, 533
Amenorrhea, 859
American Dietetic Association, 726
American Heart Association, 590, 703
Amines, 485, 487t
Amino acids. *See also* Proteins
digestion, 684f, 686t
glomerular filtration, 792
neurotransmitters, 373
nutrition, 704, 705
oxidation, 819
peptide bonds, 117f
structure, 65
Ammonia, 940, 943
Ammonium, 823
Amniocentesis, 931–32, 933t
Amniochorionic membrane, 891
Amnion, 888, 891
Amniotic cavity, 885, 891
Amniotic fluid, 888, 891
Amoeba, 9
Amphetamines, 377, 489, 711
Amphiarthrotic joints, 261, 267t
Amplification, of nucleic acids, 132
Ampulla, 460, 834f, 840
Amygdala, 404
Amylase, 118
Amyloid, 567
Amyotrophic lateral sclerosis (ALS), 363, 396
Anabolic steroids, 489t
Anabolism, 114, 115–16
Anaerobic reactions, 120, 121–22
Anaerobic respiration, 114, 294, 819
Anal canal, 688
Anal columns, 688
Anal sphincter, 320, 322f, 688
Anaphase, 101, 102f, 103f, 833
Anaphylactic shock, 639
Anatomical neck, 226, 229f
Anatomical position, 20
Anatomic dead space, 753
Anatomy. *See also* Cardiovascular system; Digestive system; Endocrine system; Integumentary system; Lymphatic system; Muscular system; Nervous system; Reproductive system; Respiratory system; Skeletal system; Urinary system
characteristics of life, 6, 8t
definition, 1, 4
gene expression, 933–34
history, 3
levels of organization, 4–5
life-span changes, 20
maintenance of life, 7–12
organization of human body, 12–19
terminology, 20–25
Andreas Vesalius, 3f
Androgen(s), 508, 830, 845–46, 857
Androgenic alopecia, 179
Anemia, 529. *See also* Megaloblastic anemia; Pernicious anemia; Sickle cell disease
Anencephaly, 398, 409
Anesthetics and anesthesia, 82t, 394
Aneuploidy, 929–30
Aneurysm, 581, 608f
Angina pectoris, 563, 609
Angiogenesis, 574
Angiogram, 563f
Angioplasty, 574
Angiotensin(s), 506, 552, 587, 588, 790
Angiotensin-converting enzyme (ACE) inhibitors, 588t, 790
Angiotensinogen, 506
Anions, 55
Ankle, 270t
Ankle-jerk reflex, 394
Annulus fibrosus, 263
Anorexia nervosa, 725, 727
Anosmia, 450t
Antacids, 670
Antagonists, 304, 377
Antebrachial region, 23, 25f
Antecubital region, 23, 25f
Anterior, as term of relative position, 21
Anterior canal, 460
Anterior cavity, 467
Anterior chamber, 467
Anterior crest, 235, 236f
Anterior fontanel, 218f, **258**
Anterior funiculus, 390f
Anterior group, of thigh muscles, 323
Anterior horns, 389, 390f
Anthropology, and human evolution, 84, 193
Antiangina drugs, 82t
Anti-angiogenesis drugs, 104
Antiarrhythmia drugs, 82t
Antibiotics, 181t, 793
Antibodies, 532, 544, 545f, 633, 634, 636t
Antibody-dependent cytotoxic reactions, 639
Anticodon, 131
Anticonvulsant drugs, 82t
Antidiabetic drugs, 82t
Antidiuretic hormone (ADH)
actions and control of, 499t
aging and levels of, 515
mechanisms of nonsteroid hormones, 487t, 491
pituitary gland and secretion of, 493, 498
stress response, 515
urine concentration and volume, 498, 794–95, 796t, 815
Antigen(s), 544, 545f, 628, 635
Antigen-binding sites, 634
Antigen-presenting cell, 630
Antihemophilic factor, 541t
Antihistamines, 145, 155
Anti-HIV drugs, 76
Antihypertensive drugs, 82t
Antioxidants, 710
Antiparallel strands, of DNA, 125
Anti-Rh antibodies, 547
Antithrombin, 539, 542, 543t
Anular ligament, 260, 272, **273**
Anus, 653f, 688, 834f, 848f
Aorta, **38, 40, 43–46,** 558f, 560f, 561f, 562f, 567, 592–93, 595t, 599f, 601f. *See also* Abdominal aorta; Arch of aorta; Descending aorta; Thoracic aorta
Aortic bodies, 592, 758
Aortic sinus, 592
Aortic sound, 565
Aortic valve, 557, 558f, 559f, 560t, 567
Aortic valvulitis, 567
Apex, of heart, 562f
Apical surface, 144

Aplastic anemia, 529t
Apocrine glands, 150, 152f, 153t, 180, 182t, 430t
Aponeuroses, 285, 312f
Apoproteins, 97, 703
Apoptosis, 75, 106, 109, 184, 910
Appendicitis, 687
Appendicular portion, 1, 12, 207f
Appendicular skeleton, 205t, 206
Appendix, **34–35,** 687
Appetite control, 700t, 711
Apraxia, 402
Aquaporins, 814
Aqueous humor, 467–68
Arachnoid granulations, 386, 387f
Arachnoid mater, 385, 387f, 388f
Arbor vitae, 410
Arch of aorta, **34–37, 47,** 593f, 595t, 898f
Arcuate arteries, 779, 780f
Arcuate nucleus, 699
Arcuate popliteal ligament, 276, 277f
Areola, **31,** 862
Areolar tissue, 156, 157f, 162t
Aretaeus of Cappadocia, 513
Arginine, 135t, 705t
Aristolochia clematitis, 775
Aristotle, 896
Arm, and muscles, 312, 314t, 317f, **343, 349**
Armstrong, Lance, 699, 726
Aromatase inhibitors, 489t, 865
Arrector pili muscle, 179
Arrhythmias, 572–73
Arsenic poisoning, 115, 530
Arterial blood pressure, 580, 582–85
Arteries, 574–76, 580t, 592–99, 600–601f. *See also* Cardiovascular system; *specific arteries*
Arterioles, 574–76, 580t
Arteriosclerosis, 552, 581, 588
Arthritis, 261, 265, 278, 279t. *See also* Rheumatoid arthritis
Arthrology, 260
Arthroscopy, 274
Articular capsule, 272f
Articular cartilage, 194, 263, 265, **273,** 275f
Articulations, 261
Artificial insemination, 878
Artificially acquired active immunity, 638, 639t
Artificially acquired passive immunity, 638, 639t
Artificial pacemaker, 573
Artificial respiration, 746
Arytenoid cartilages, 740–41
Asbestos, 749
Ascending aorta, 592
Ascending colon, **34–35, 49,** 604f, 687
Ascending lumbar veins, 603, 607f
Ascending tract, 393–96
Ascites, 579, 726, 817
Ascorbic acid, 709, 716, 717t
Asian food guide pyramid, 724f
Asparagine, 135t, 705t
Aspartame, 701
Aspartic acid, 135t, 705t
Aspirin, 491
Assimilation, as characteristic of life, 8t
Assisted reproductive technologies (ART), 878–79
Association areas, of brain, 402
Asthma, 637t, 756
Astigmatism, 472
Astrocytes, 353, 361, 362f, 363t
Atelectasis, 765
Atherosclerosis, 63, 154, 542, 552, 581, 588
Atlantis (space shuttle), 585
Atlantoaxial joint, 270t
Atlas, 221
Athletes and athletic performance. *See also* Exercise
  cardiovascular system, 590
  female reproductive system, 857, 859
  hormones, 489
  nutrition, 699, 726
  weight lifting, 300
Atmospheric pressure, 8, 747, 750, 757
Atom(s), 4, 5f, 51t, 52–53, 55–58
Atomic number, 52
Atomic radiation, 53
Atomic weight, 52
ATP. *See* Adenosine triphosphate
ATPase, 291–92
ATP synthase, 946
Atria, 555
Atrial natriuretic peptide (ANP), 513, 556, 790
Atrial syncytium, 565
Atrioventricular orifice, 555
Atrioventricular sulcus, 556
Atrioventricular valve (AV valve), 555, 566–67, 571f, 573
Atrium, **40**
Atrophy, 300
Auditory, 437
Auditory meatus, 192
Auditory nerve pathways, 458
Auditory ossicles, 451
Auditory tube, 453
Auricles, 450, 555, 593f, 608f
Auricular artery, posterior, 595f, 597
Auricular surfaces, 222, 223f
Autoantibodies, 641
Autocrine secretions, 483
Autograft, 186, 641
Autoimmune disease, 616
Autoimmunity, 639, 641, 643
Automated external defibrillator (AED), 553
Automatic bladder, 802
Autonomic nervous system
  general characteristics, 424
  involuntary actions, 356, 411
  nerve fibers, 424
  neurotransmitters, 428–31
  peripheral nervous system, 411
  sympathetic division, 424–26
Autoregulation, 790
Autosomal inheritance, 918, 920, 923
AV bundle, 567
Axial portion, 12, 207f
Axial skeleton, 192, 205t, 206
Axillary artery, **34,** 597, 601f
Axillary border, 226
Axillary nerves, 421
Axillary region, 23, 25f, 622
Axillary vein, **34,** 602f, 603, 607f
Axis, of cervical vertebra, 221
Axon(s)
  action potentials, 369
  definition, 353, 354
  functions, 164, 356
  myelination, 358, 359
  neurons, 164, 357f
  regeneration, 364
  use of term, 438
Axon terminal, 356
Azygos vein, **44,** 603, 607f

B

Babinski reflex, 394
Back, surface anatomy of, **342**
Back pressure, 579
Bacteria
  arthritis, 278
  cell structure, 76
  dental caries, 661
  endocarditis, 557
  infections, 453, 534t
  liver function, 675
  microflora, 652, 689
Balance
  acid-base, 819–24
  electrolytes, 811, 815–18, 820
  water, 811, 813–15
Baldness, 179
Balkan endemic nephropathy, 775
Ball-and-socket joint, 265, 267t
Banting, Frederick, 513
Bariatric surgery, 711
Baroreceptor(s), 439, 571
Baroreceptor reflexes, 569–70, 571f
Barrett's esophagus, 665
Basal body, 87
Basal conditions, 707
Basal metabolic rate (BMR), 500, 698, 707, 728
Basal nuclei, 398, 404, 405f, 412t
Base(s), 59–60. *See also* Acid-base balance
Basement membrane, 144, 145, 172, 174f
Base-pairing rules, for DNA, 127
Basilar artery, 593, 595f, 596f
Basilar membrane, 453
Basilic vein, 602f, 603, 607f
Basophils, 531–32, 535t
B cell(s), 532, 628–29, 630t, 631f, 632–33. *See also* B lymphocytes
B cell lymphoma, 637t
BeadChip tests, 546
Beaumont, William, 668
Bedsore, 173
Beeturia, 803
Bell-shaped curve, 924–26
Benign joint hypermobility syndrome, 263
Benign prostatic hypertrophy (BPH), 841
Benign tumor, 104
Beriberi, 713–14
Berylliosis, 749
Beryllium, 53t, 749
Best, Charles Herbert, 513
Beta-2 agonists, 489t
Beta endorphins, 374
Beta globulins, 535, 536t

Beta-hydroxybutyric acid, 825
Beta interferons, 360
Beta oxidation, 702
Beta radiation, 53, 56
Beta receptors, 430
Beta waves, 411
Bicarbonate buffer system, 821, 822t
Bicarbonate ions, 62t, 766, 788t
Biceps brachii muscle, 302–304, 305, 314f, 315f, 316, 317f, **342–44, 347, 350**
Biceps femoris muscle, 306f, 325f, 326, 328t, 330f, 331f, **343–44, 351–52**
Biceps-jerk reflex, 394
Bifidus factor, 904
Bilateral, as term of relative position, 21
Bile, 675–76, 943
Bile canaliculi, 673, 675f
Bile duct, 672f, 673, 674f, 675f, 677
Bile pigments, 675, 676
Bile salts, 675, 678
Bilirubin, 522, 530, 531f, 676, 940, 943
Biliverdin, 530, 531f, 676
Binding site, of hormone, 486
Biochemistry, 50, 51
Bioethics, and postmortem sperm retrieval, 876
Biogenic amines, 373t
Biomarker, 51
Biomineralization, 204
Biotin, 716, 717t, 719f
Bipolar neurons, 353, 360, 362t, 469
Birds
  basal nuclei, 404
  muscle fibers, 299
  neural stem cells, 365
Birth control, 862–66
Birth defects, 896–97
Birth process, 899–902
1,3-Bisphoglyceric acid, 944, 945f
Bitter receptors, 450
Blastocyst, 862, 879–82, 886t
Blastomere, 879
Blast-related brain injury, 405
Blebs, 109
Blood. *See also* Blood pressure; Blood vessels
  blood groups and transfusions, 544–48
  cells, 202–203, 524–34, 535t
  clotting, 183, 185f, 538–42, 543t, 712
  connective tissues, 162
  definition, 523
  fetal circulation, 895, 897–98, 899f
  hemostasis, 538–43
  laboratory tests, 940–42
  oxygen supply and cellular respiration, 294–95
  path through heart, 560–61
  plasma, 535–38
  supply to heart, 561–63
Blood cells, 202–203, 524–34, 535t
Blood coagulation, 538–41
Blood clotting, 183, 185f, 538–42, 543t, 712
Blood doping, 489
Blood gases, 536, 537
Bloodgen, 546
Blood platelets, 534
Blood pressure
  aging, 611
  arterial, 580, 582–85
  capillary blood flow, 578–79
  control of, 585–87, 588
  diet, 725
  heart action, 582–83
  homeostatic mechanism, 10
  kidney failure, 779, 789
  venous blood flow, 587, 589
Blood reservoirs, 580
Blood-testis barrier, 836
Blood typing, 546
Blood urea nitrogen (BUN), 538, 704, 940
Blood vessel(s), 173, 430t, 573–80, 581, 779, 780f, 781f. *See also* Arteries; Capillaries; Veins
Blood vessel spasm, 538
Blood volume, 523, 583
Bloom, David, 543
Blue diaper syndrome, 803
B lymphocytes, 628–29. *See also* B cells
Body, of uterus, 854
Body cavities, 12, 13f
Body covering, 14
Body fluids, 811–13
Body mass index (BMI), 708, 710
Body movement, and skeletal muscle actions, 301–303
Body odor, of males, 483
Body regions, and terminology, 22–24
Body sections, and terminology, 22
Body temperature. *See also* Cold environment; Heat
  elevated, 184
  energy expenditure from foods, 707
  homeostasis, 9–10
  life-span changes, 186
  ovulation, 863
  regulation of, 181–82, 183f
Body weight, and diet, 708–709. *See also* Obesity
Bolus, 664
Bombay phenotype, 547
Bomb calorimeter, 706–707
Bone. *See also* Skeletal system; *specific bones*
  age-related changes, 238, 240
  connective tissue, 160–62
  development and growth, 197–201
  function, 202–204
  heart rate, 572
  number, 205–206
  structure of skeletal system, 193–95
Bone marrow
  blood cell formation, 202–203
  lymphocytes, 629f
  stem cells, 526
  sternal puncture, 225
  transplants, 108
Bone mineral density, 204
Bone morphogenetic proteins, 201
Bone remodeling, 200
Bony processes, 194
Boron, 53t
"Botox," 186
Botulinum toxin, 186, 291
Boyle's law, 747
Brachial artery, **34,** 597, 601f
Brachialis muscle, 305f, 306f, 315f, 316, 317f, **344, 350**
Brachial plexus, **34,** 420–21, 422f
Brachial region, 23, 25f
Brachial veins, 602, 603, 607f
Brachiocephalic artery, **35, 37, 46–47,** 592, 593f, 595f, 596f, 601f
Brachiocephalic veins, **34, 36, 46,** 593f, 602, 603, 607f, 619f
Brachioradialis muscle, 305f, 316, 317f, 318f, **342, 350**
Bradycardia, 552, 572
Brain. *See also* Blast-related brain injury; Cerebellum; Cerebrum; Traumatic brain injury
  arteries, 593, 596
  development of, 398, 400t
  major parts of, 412t
  pain impulses, 443
  PET scans, 70–71
  veins, 602
Brain banks, 354
Brain death, 411
Brainstem, **39.** *See also* Cerebellum; Medulla oblongata; Pons
  basal nuclei, 405f
  communication between brain and spinal cord, 384, 389f
  sensory impulses, 395f, 401f
  structure of brain and, 398, 399f, 407–409, 412t
Brain tumors, 363
Brain waves, 411
Branches, of bronchial tree, 743–44
Breast, **31,** 862
Breast cancer, 637, 864–65
Breastfeeding, 904
Breath analysis, for alcohol, 762
Breathing. *See also* Respiration
  control of, 755–59
  mechanism, 747, 750–55
  reasons for, 726–27
Brewer's yeast, 725
"Brittle bone disease," 924
Broad ligament, 847, 849f, 853f, 854
Broca's area, 401f, 402
Bronchial arteries, 592
Bronchial asthma, 756
Bronchial tree, 743–46, 747t
Bronchioles, 430t, 745
Bronchitis, 739
Bronchomediastinal trunk, 617, 619f
Bronchopulmonary segment, 743
Bronchus, **36, 47,** 735
Brown-Séquard syndrome, 395
Buccal region, 23, 25f
Buccinator muscle, 306, 307f, **346**
Buffers, 60
Bulbospongiosus muscle, 320, 322, 323t
Bulbourethral gland, 801f, 834f, 841, 845t
Bulimia, 725, 728
Bulk elements, 52
Bundle branch block, 567
"Burning man syndrome," 82
Burns, healing of, 184, 186, 187f
Bursa, 260, 265, 271–72, 276–77
Bursitis, 278

C

Cacosmia, 451
Cadaver dissection, 3
Cadherins, 171
Caffeine, 796, 815, 904t
Calcaneal tendon, 306f, 330, 331f, **345, 352**
Calcaneal tuberosity, 236, 237f
Calcaneus, 236, 237f, 331f, **345**
Calcitonin, 487t, 491, 501, 503
Calcium
- blood clotting, 541t
- blood tests, 940
- bone loss and supplements, 240
- chemical constituents of cells, 62t
- electrolyte balance, 818, 819f
- elemental composition of human body, 52t
- heart action, 573
- ion channels, 82
- muscle contraction, 291
- nerve impulse conduction, 372
- nutrition, 717, 721t
- osteopenia and osteoporosis, 204
- parathyroid hormone, 503
- urine and urine tests, 788t, 943

Calcium channel blockers, 301, 588t
California Cryobank, 876
Calluses, 173
Calmodulin, 301, 491
Calorie, 698, 706, 707t, 723
Calorimeter, 698
Canaliculus (caniculi), 161, 192, 195
Cancer. *See also* Bone cancer; Breast cancer; Cervical cancer; Leukemia; Prostate cancer; Skin cancer; Testicular cancer
- angiogenesis, 574
- basement membranes, 145
- causes of death, 911t
- characteristics of cells, 104t
- chronic pain, 445
- definition, 104
- dietary supplements, 725
- digestive system, 691
- DNA microarrays, 933–34
- extracellular matrix, 154
- hypercalcemia, 818
- immune response, 644
- immunotherapy, 637
- microRNA-based diagnostic tests, 136
- steps in development, 105f

Canine teeth, 659
Cannabinoids, 444
Capacitation, of sperm cells, 841
Capillaries, 576–79, 580t, 784–85, 787f. *See also* Cardiovascular system; *specific capillaries*
Capitate, 230f
Capitulum, 226, 229f
Capsule, of connective tissue, 621
Carbaminohemoglobin, 766
Carbohydrates
- biomarkers, 51
- diet, 702, 706t
- metabolic pathways and storage, 124, 126f
- nutrition, 700–702
- organic chemicals in cells, 61–62, 69t
- size of molecules, 63f

Carbon, 52t, 53t. *See also* Carbon dioxide; Carbon monoxide
Carbonate, 62t
Carbon dioxide
- blood tests, 940
- chemical constituents of cells, 61, 62t
- citric acid cycle, 946f
- control of breathing, 757, 758, 763
- gas transport, 766–67
- molecular structure, 69f
- respiratory excretion, 822–23

Carbonic acid, 819, 822
Carbonic anhydrase, 766, 822
Carbon monoxide, 61, 764
Carboxypepidase, 671, 683t
Carcina, 735
Carcinomas, 150, 735
Cardia, 665, 666f
Cardiac center, of medulla oblongata, 408, 586
Cardiac conduction system, 565–67, 611
Cardiac control center, 569–70
Cardiac cycle, 564, 568–72
Cardiac muscle(s), 301, 302f
Cardiac muscle fibers, 565
Cardiac muscle tissue, 164, 165t
Cardiac output (CO), 585
Cardiac veins, 562f, 563
Cardioaccelerator reflex centers, 571, 586
Cardioinhibitor reflex, 571, 586
Cardiology, 24
Cardiovascular disease, molecular causes of, 608. *See also* Heart disease
Cardiovascular system
- actions, 564–73
- arterial system, 592–99, 610
- autonomic stimulation, 430t
- blood pressure, 580, 582–90
- blood vessels, 573–80
- definition, 553
- digestive system, 610, 692
- endocrine system, 516, 610
- exercise, 590, 611
- fetal blood and circulation, 895, 897–98, 899t
- integumentary system, 188, 610
- life-span changes, 606, 609, 611, 803, 805, 908t
- lymphatic system, 610, 645
- major functions of, 8t, 16, 17f
- muscular system, 333, 610
- nervous system, 378, 610
- paths of circulation, 590–92
- postnatal development, 905, 906t
- reproductive system, 610, 868
- respiratory system, 610, 768
- skeletal system, 239, 610
- structure of heart, 553–63
- urinary system, 610, 804
- venous system, 600, 602–605, 606f, 607f

Carina, 743, 745
Carotene, 176, 709, 710
Carotid artery, **43,** 595f, 601f
- external, 596–97
- internal, 597

Carotid body, 597, 758
Carotid canal, 210, 211f, 217t, **249–51**
Carotid sinus, 595f, 597
Carotene, 698
Carpal bones, 192, 206, 228f, 229, 231t, **344**
Carpal region, 23, 25f
Carpal tunnel syndrome, 424
Carpometacarpal joint, 270t
Carpus, 229
Cartilage, 158, 160
Cartilaginous cells, of epiphyseal plate, 199f
Cartilaginous joints, 262–63, 267t
CASA system, 843
Cat(s), and polydactyly, 231
Catabolism, 114, 115, 116, 126f
Catalysis, of enzymes, 117
Catalysts, 59
Cataract, 477
Catecholamines, 505
Cations, 55
Cauda equina, **40–41,** 389, 419
Caudate lobe, of liver, 673, 674f
Caudate nucleus, 404
Causalgia, 442
CCRS receptors, 76
CD4 helper T cells, 76, 630
Cecum, **34–35,** 651, 687
Celiac artery, **36,** 594f, 601f
Celiac disease, 686
Celiac region, 23, 25f, 686
Cell(s)
- aging process, 20
- blood, 202–203, 524–34, 535t
- cell cycle, 100–103
- chemical constituents of, 60–71
- composite, 76, 78f
- cytoplasm, 81–89
- death of, 106, 109
- differentiated, 76, 105
- division, 101, 103–105
- levels of organization, 4, 5f
- membrane, 76, 78–81, 82, 91t
- movements in and out of, 90–99
- nervous system, 356–59
- nucleus, 76, 89–90, 91f
- shape and function, 77f
- sizes of, 76, 77f
- stem and progenitor, 105–106
- structures and functions of parts, 91t

Cell body, 356, 357f
Cell cycle, 100–103
Cell division, 101, 103–105
Cell membrane, 76, 78–81, 82, 91t
Cell membrane potential, 365–71
Cell specialization, and proteomics, 144
Cell surface proteins, 81t
Cellular adhesion molecules (CAMs), 80, 81
Cellular immune response, 629–30, 631f
Cellular metabolism. *See also* Metabolism
- cellular respiration, 119, 120–24
- control of reactions, 117–19
- energy for, 119–20
- nucleic acids, 124–31
- processes of, 115–16

Cellular respiration, 119, 120–24, 294–95, 736, 944–47
Cellular turnover, 682

Cellulose, 701
Cementum, 660
Centers for Disease Control and Prevention (CDC), 523, 642
Central canal, 161, 195, 363, 389
Central nervous system (CNS), 354, 355f, 361–63, 364
Central sulcus, 399
Central vein, of liver, 673, 675f
Central venous pressure, 589–90
Centrioles, 86–87, 88f
Centromere, 101, 831
Centrosomes, 86–87, 91t
Cephalic phase, of gastic secretion, 668, 669t
Cephalic region, 23, 25f
Cephalic vein, 602f, 603, 607f
Cerebellar ataxia, 878
Cerebellar cortex, 410
Cerebellar peduncles, 410
Cerebellum, **38–39,** 398, 399f, 401f, 405f, 410, 412t
Cerebral aqueduct, 385, 386f, 407
Cerebral arterial circle, 593
Cerebral arteries, 596f, 597
Cerebral cortex, 400–401, 403f
Cerebral hemispheres, 398, 400f, 405f
Cerebral peduncles, 407
Cerebral vascular accident (CVA), 401, 588
Cerebrospinal fluid (CSF), 385–86, 387f, 388
Cerebrum, 1, **38–39,** 395f, 396f, 398–404, 412
Ceruminous glands, 182t, 450
Cervical cancer, 869
Cervical cap, 863, 866t
Cervical curvature, 219
Cervical enlargement, 387, 389f
Cervical nerves, 418
Cervical orifice, 853f, 854
Cervical plexuses, 420
Cervical region, 23, 25f, 622
Cervical vertebrae, **39,** 219f, 220–21, 222f, 223t
Cervix, 848f, 853f, 854, 877f
CFTR protein, 919
Chambers, of heart, 555–57
Chaperones, and protein synthesis, 132
Charles I (King of Spain), 797
Cheeks, 656, 660t
Chemical barriers
  birth control, 863
  infection, 626–27, 628t
Chemical bonds, 52, 55
Chemical digestion, 652
Chemical energy, 120
Chemical reactions, 58–59
Chemistry
  chemical constituents of cells, 60–71
  definition, 51
  hormones, 484–85
  structure of matter, 51–60
Chemoreceptors, 439, 757–58
Chemotaxis, 636t
Chemotherapy, for breast cancer, 865
Cheyne-Stokes breathing, 909
Chief cells, 666, 667f
Childbirth, 899–902
Children. *See also* Adolescence; Infants
  adipose tissue, 158
  development, 906, 908t
  ear infections, 453
  skull, **259**
  vaccines and immunity, 638
  vision screening programs, 465
Chin, 234t
China, and peanut allergies, 617
Chlamydia infection, 867t
Chloride, 62t, 82, 766, 788t, 818, 940
Chloride shift, 766
Chlorine, 52t, 719–20, 721t
Cholecystectomy, 679
Cholecystogram, 679
Cholecystokinin, 669, 672, 678t
Cholelithiasis, 679
Cholera, 816
Cholesterol, 52, 65f, 80, 97, 608, 702, 703, 940
Cholinergic neurotransmitters, 428
Chondrocyte, 143, 158, 160
Chondrodysplasia, 157t
Chondroitin, 261
Chordae tendineae, 557, 558f, 559f
Chorion, 875, 885, 890f
Chorionic villi, 885, 886
Chorionic villus sampling (CVS), 932, 933t
Choroid artery, 595f
  anterior, 597
Choroid coat, 437, 466, 470t
Choroid plexus, 363, 382, 385, 387f
Chromatids, 831
Chromatin, 90, 91t
"Chromatin remodeling," 132
Chromatophilic substance, 356
Chromium, 722, 723t
Chromosomes, 87, 90, 128f, 916, 918–20, 929–33. *See also* DNA; Gene(s); Genetic(s); RNA
Chronic glomerulonephritis, 779, 782
Chronic granulomatous disease, 548t
Chronic myeloid leukemia (CML), 536
Chronic obstructive pulmonary diseases (COPD), 758, 911t
Chronic pain, 442, 445
Chylomicrons, 684
Chyme, 651, 669, 670f, 671f, 673f, 683, 689
Chymopapain, 225
Chymotrypsin, 671, 683t
Cigarette smoke and smoking, 51, 736, 739, 896–97. *See also* Nicotine
Cilia, 87–88, 91t, 146, 854
Ciliary body, 466
Ciliary muscles, 466, 467
Ciliary processes, 466
Circadian rhythms, 512
Circle of Willis, 593
Circular fibers, 654
Circular muscles, 468
Circular set, 468
Circular sulcus, 400
Circulation, as characteristic of life, 8t
Circumcision, 843
Circumduction, of joint, 268f, 269
Circumflex arteries, 561, 562f, 597f
Circumflex ilian artery, deep, 598
Cisterna chyli, 620f
Cisternae, 83, 288
Citric acid, 944, 946f
Citric acid cycle, 120, 121f, 122, 123f, 294, 944
Classical pathway, of complement, 627
Clavicle, **31–32,** 192, 206, 225, 227f, 231t, 272f, 315f, **343**
Clavicular notch, 224f
Cleavage, 875, 879–82, 886t
Cleft palate, 213
Cleidocranial dysplasia, 225
Clitoris, 848f, 855–56, 857t
Clomid (clomiphene), 851
Clone, 629
Closed fracture, 202
Clostridium botulinum, 186, 291
Clotting, of blood, 183, 185f, 538–42, 712
Clubfoot, 238
Coagulation, of blood, 538–41
Cobalt, 52t, 54, 715, 721, 723t
Cocaine, 374t, 377, 837, 904t
Cocarboxylase, 713
Coccygeal nerves, 418f, 419
Coccygeus, 320, 323t
Coccyx, **41,** 206, 218, 219f, 222, 223t, 232f, 234t, **343,** 388f
Cochlea, 437, 453, 455, 456f, 457f
Cochlear branch, 416
Cochlear duct, 453
Cochlear implant, 455
Codominant gene, 922
Codons, 131, 135t
Coenzymes, 114, 119, 714
Cofactors, 118–19
Coitus interruptus, 862, 866t
Cold environment, and homeostasis, 9–10. *See also* Hypothermia
Cold receptors, 440
Colitis, 689
Collagen, 69f, 155, 157, 716
Collagenous fibers, 155, 156t
Collateral axon, 356
Collateral ganglia, 425–26
Collectin(s), 627
Collecting ducts, 618, 778f, 781–82, 783f, 786f, 797t
Colloid, 499
Colloid osmotic pressure, 535, 578, 812
Colon, **38, 48,** 687. *See also* Ascending colon; Descending colon
Colonoscopy, 691
Colony-stimulating factors (CSFs), 530, 630
Color
  eye, 926
  hair, 179
  skin, 174, 176, 179, 925f, 926
  vision, 438, 475, 928
Colorblindness, 438, 928
Colorectal cancer, 637, 691
Colostomy, 691
Colostrum, 902, 904, 905
Columnar epithelial cells, 145, 146
Coma, 409
Combined hormone contraceptives, 864–65, 866t
Comminuted fracture, 202f
Common carotid artery, **32–33, 35–37, 46–47,** 592, 593f, 595f, 596, 597f, 601f

Common fibular nerves, 423
Common hepatic duct, 673
Common iliac artery, **35–36,** 593, 594f, 898f
Common iliac vein, 607f
Communicating arteries
anterior, 596f
posterior, 596f, 597
Compact bone, 194, 195, 196f, 240
Compartments, 286, 811, 812–13
Compartment syndrome, 286
Compensated acidosis, 825
Complement, 627, 636t
Complementary base pairs, 127
Completely penetrant genotype, 924
Complete proteins, 705
Compliance, and lungs, 751
Composite cells, 76, 78f
Compound(s), 52, 55
Compound fracture, 202
Compound gland, 150, 152t
Computerized tomography (CT) scanning, 70–71
Concave surface, 471
Concentration(s), of acids and bases, 59–60
Concentration gradient, 92
Concentric muscle contraction, 298, 299f
Condom, 863, 866t
Conduction, 182
Conductive deafness, 459
Condylar canal, 211f
Condylar joint, 265–66, 267t
Condyle, 192, 208t, 226, 260. *See also* Lateral Condyle; Mandibular condyle; Medial condyle; Occipital condyle
Cones, 473, 474f
Confocal microscope, 79
Conformation, of amino acid, 65, 66
Congenital erythropoietic porphyria, 138f
Conjoined twins, 900
Conjunctiva, 463
Conjunctivitis, 463
Connecticut Agricultural College, 925f
Connective tissues, 146t, 152–62, 285–87
Constrictor muscles, 663
Consumer fraud, and genetic testing, 917
Continuous ambulatory peritoneal dialysis, 779
Continuous positive airway pressure, 757
Contraception, 830, 862–66
Contractility, and blood pressure, 586
Contraction, of muscles, 289–99, 300–301
Contralateral, as term of relative position, 21
Control center, 9
Conus medullaris, 388f, 389
Convection, 182
Convergence, and neuronal pools, 375
Convex surface, 471
Convulsions, 372
Coordination, and organ systems, 16
Copper, 52t, 721–22, 723t
Coproporphyria, 138t
Coracobrachialis muscle, **32–34,** 312, 314t, 315f
Coracohumeral ligament, 271, 272f
Coracoid process, 192, 226, 227f, 272f
Coriell Institute (New Jersey), 910
Cornea, 437, 465
Corniculate cartilages, 740–41
Coronal section, of skull, 215f
Coronal suture, 208, 209f, 210f, 215f, 218f, **245–46, 255**
Corona radiata, 849, 851f, 853f, 880f
Coronary angioplasty, 561, 574, 609
Coronary arteries, 561, 562f, 592, 601f
Coronary artery disease (CAD), 609
Coronary bypass surgery, 574, 609
Coronary embolism, 588
Coronary heart disease, 725
Coronary ligament, 673, 674f
Coronary sinus, 557, 562f, 563
Coronary stent, 609
Coronary thrombosis, 588
Coronoid fossa, 226, 229f
Coronoid process, 210f, 214, 229, 230f, **252, 273**
Corpora cavernosa, 856
Corpora quadrigemina, 407
Corpus albicans, 853f, 859
Corpus callosum, **39, 42,** 398, 399f, 404
Corpus cavernosum, 834f, 843
Corpus luteum, 853f, 858
Corpus spongiosum, 834f, 843
Cortex, of ovary, 847
Cortical nephrons, 784, 785f
Cortical radiate arteries, 779, 780f, 783f, 786f
Cortical spreading depression, 356
Corticospinal tract, 393, 395, 396t, 402
Corticosteroids, 181t
Corticotropes, 494
Corticotropin(s), 489t
Corticotropin-releasing hormone (CRH), 487t, 496, 514–15
Cortisol, 486f, 487t, 507–508, 509t
Costal cartilage, **44–45,** 224f, 227f, 598f
Costal region, 23, 25f
Costocervical arteries, 596
Coughing, 754, 755t
Countercurrent mechanism, 795–96, 797f
Covalent bond, 50, 56–57
Cow milk, 904
Coxal region, 23, 25f
Cox-2 inhibitors, 278, 691
Coyle, Ed, 699
Cramps, muscular, 296
Cranial branch, 417
Cranial cavity, 12, 14f, 213f, **257**
Cranial nerves, 411, 412, 414–17
Cranial region, 1
Cranial sutures, 262f
Cranium, 206, 208, 209f, 210, 212
Creatine, 294, 538, 726
Creatinine, 788t, 798, 940, 943
Creatinine clearance test, 798, 943
Cremasteric reflex, 394
Crest, 208t
Cribiform plate, 192, 212, 213f, 215f, **253, 257**
Cricoid cartilage, 735, 740, 741f
Crista ampullaris, 460, 462f
Cristae, 84
Crista galli, 192, 212, 213f, 215f, **253, 256–57**
Critical period, in development, 892
Croatian Institute for Brain Research, 354
Crohn disease, 637t, 691
Cross bridge cycling, 287, 288f, 291–92
Crossed extensor reflex, 391
Crossed fused ectopia, 802t
Cross over, of chromosomes, 832
Cross section, 22, 23f
Crown, of tooth, 659
Cruciate ligaments, 276, 277f
Crura, 830, 843, 856
Crural region, 23, 25f
Crying, 755t
Cryo-electron tomography, 171
Cryptorchidism, 835
Cryptosporidiosis, 626
Cubital region, 23, 25f
Cubital vein, 602f, 603
median, 607f
Cuboidal epithelium, 145, 146, 148
Cuboid bone, 237f
Cuneiform cartilages, 741
Cupula, 460
Curare, 374t
Curvature, of vertebral column, 218–19
Cushing syndrome, 509
Cutaneous carcinoma, 175
Cutaneous membranes, 163
Cuts, in skin, 183
Cyanide, 119, 124
Cyanocobalamin, 715, 717t
Cyanosis, 176, 526
Cyclic adenosine monophosphate (cAMP), 487–88, 490f
Cyclic guanosine monophosphate (cGMP), 474–75, 491
Cyclosporine, 904t
Cysteine, 66f, 135t, 705t, 719
Cystic duct, **35,** 672f, 674f, 677
Cystic fibrosis, 66, 82, 744, 919, 920, 923
Cystitis, 774, 800
Cytochrome(s), 946
Cytochrome oxidase, 946
Cytokines, 630, 637
Cytokinesis, 101–103
Cytoplasm, 76, 81–89
Cytoplasmic division, 101–103
Cytosine, 128f, 948f, 949f
Cytoskeleton, 82, 90f
Cytosol, 75, 76
Cytotoxic T cell, 630

D

Dalton, John, 438
Dark adapted eye, 475
Dartos muscle, 842
Deafness, 455, 459. *See also* Hearing
Deamination, 114, 704
Death. *See also* Apoptosis
cigarette smoking and secondhand smoke, 736
end of life and process of, 907–909
leading causes of in U.S., 911t
obesity as cause of, 710
Decellularization, and tissue engineering, 166
Decibels (dB), 458

Deciduous teeth, 651
Decomposition, and chemical reactions, 58
Decubitus ulcer, 173
Deep, as term of relative position, 21
Deep fascia, 287
Deep partial-thickness burn, 184
Deep vein thrombosis, 543
Defecation, 690
Defensins, 626–27
Defibrillators, 553
Dehydration, 810, 816
Dehydration synthesis, 115, 117f
Dehydrocholesterol, 200
Delayed-reaction allergy, 641
Delta waves, 411
Deltoid muscle, **31–32,** 304, 305, 306f, 311f, 312f, 313f, 314t, 315, **342–43, 347–48**
Deltoid tuberosity, 226, 229f
Denaturation, of protein, 66
Dendrites, 353, 354, 356, 357f
Dens (odontoid process), 221
Dense connective tissue, 155, 158, 159f, 162t
Dental caries, 651, 661
Dentate nucleus, 410
Dentin, 659
Deoxyhemoglobin, 526
Deoxyribonuclease (DNase), 744
Deoxyribonucleotide, 949f
Deoxyribose, 68f, 701
Department of Agriculture (USDA), 722, 724f
Depolarized membrane, 368
Depo-Provera (medroxyprogesterone acetate), 865, 866t
Deposition, and bone remodeling, 200
Depression
  of joint, 269f, 270
  psychiatric disorder, 373t
Dermal papilla, 174f, 177
Dermatology, 24
Dermatome, 419–20
Dermis, 170, 172, 174f, 177, 186
Descending aorta, **36–37,** 592, 595t
Descending colon, **34–36,** 604f, 687
Descending tract, 393–96
Desmosome, 144, 146t
Detached retina, 473
Detrusor muscle, 774, 800
Development. *See also* Children; Fetus; Growth; Infants
  abnormalities of urinary system, 802t
  adolescence, 906–907, 908t
  adulthood, 907, 908t
  brain, 398, 400t
  childhood, 906, 908t
  definition, 876
  embryonic stage, 885–92
  fetal stage, 892–94
  infancy, 905–906, 908t
  ossification timetable, 200t
  postnatal period, 904–906, 908t
  prenatal period, 879–903
  senescence, 907, 908t
Deviated septum, 737
Diabetes
  diabetes insipidus, 498, 814
  diabetes mellitus, 512, 513, 642t, 792, 825, 911t
  maturity-onset diabetes of the young (MODY), 510
Diabetic retinopathy, 477
Diacylglycerol (DAG), 491
Diagnostic Mabkits, 636–37
Diapedesis, 533
Diaphragm
  birth control, 863, 866t
  respiratory system, 12, **33–37, 40, 44–49,** 555f, 665f, 750–51
Diarrhea, 686, 825
Diarthrotic joints, 261, 267t
Diastole, 564
Diastolic pressure, 552, 580, 583
Diencephalon, 398, 399f, 400t, 404–405, 407, 408f, 412t
Diet. *See also* Nutrition
  acid-base balance, 823
  blood vessel disorders, 581
  calories, 707–708
  carbohydrates, 702, 706t
  endurance athletes, 699, 726
  healthy eating, 722–28
  lipids, 703–704, 706t
  obesity, 711
  pregnancy, 884, 885
  proteins, 705, 706, 823
  red blood cell production, 528–29
  vegetarian, 699t
  vitamins, 713t
Dietary supplements, 722, 725, 726
Diet drugs, 711
Differential white blood cell count (DIFF), 533–34, 942
Differentiated cells, 76, 105
Diffusion, 90, 92–93, 578, 760–62
Diffusional equilibrium, 92
Digestion, 8t, 652
Digestive system
  alimentary canal, 652–56
  autonomic stimulation, 430t
  cardiovascular system, 610, 692
  endocrine system, 516
  esophagus, 665
  hormones, 678t
  integumentary system, 188, 692
  large intestine, 686–90
  life-span changes, 690, 693, 908t
  liver, 673–78
  lymphatic system, 645, 692
  major functions of, 8t, 18f
  mouth, 656–60
  muscular system, 333, 692
  nervous system, 378, 692
  pancreas, 671–72
  pharynx, 661, 663–64
  reproductive system, 692, 868
  respiratory system, 692, 768
  salivary glands, 660–61
  skeletal system, 239, 692
  small intestine, 678–86
  stomach, 665–71
  urinary system, 692, 804
Digital artery, 597f
Digital region, 23, 25f
Dihydrotestosterone, 846
Dihydroxyacetone phosphate, 945f
Dipeptide, 116, 117f
Diplopia, 465
Disaccharides, 50, 62, 63f, 684f, 700
Disease. *See also* Cancer; Genetic disorders; Infection; Medical science; Viruses
  genetic mutations, 136–37
  stem cells, 108
Dislocation, of shoulder, 272
Dissecting aneurysm, 581
Disseminated intravascular clotting, 539
Distal, as term of relative position, 21
Distal convoluted tubule, 780f, 781, 783f, 786f, 797t
Distal phalanx, 230f, 237f, **344–45**
Diuresis, 792
Diuretics, 82t, 482, 489t, 498, 588, 796, 815
Divergence, of nerve impulses, 376
Diverticulosis, 691
DNA (deoxyribonucleic acid). *See also* Genetic(s)
  cellular metabolism, 124, 125–27
  definition, 68
  RNA compared to, 133t
  structure, 128f, 129f, 948–49
DNA chips, 546
DNA damage response, 137
DNA microarrays, 537, 933–34
DNA polymerase, 132
DNA profiling, 130
DNA tests, 917
Dogs, and sense of smell, 446
Dominant follicle, 849
Dominant inheritance, 920–22, 923
Dopa, 505
Dopamine, 373t, 404, 406
Dorsal arch vein, 602f
Dorsal branch, 420
Dorsal flexors, 328, 330
Dorsalis pedis artery, 599, 600f
Dorsalis pedis vein, 606f
Dorsal position, 1
Dorsal respiratory group, 756–57
Dorsal root, 419, 420f
Dorsal root ganglion, 419, 420f
Dorsal scapular nerve, 421
Dorsiflexion, 269
Dorsum region, 23, 25f
Double vision, 465
Down, John Langdon Haydon, 931
Down-regulation, of hormones, 485
Down syndrome, 884, 929, 931
Doxorubicin, 904t
Drug(s). *See also* ACE inhibitors; Antibiotics; Diuretics; Nonsteroidal anti-inflammatory drugs; Selective serotonin reuptake inhibitors; Tricylic antidepressants
  addiction, 377, 886
  delivery through skin, 173
  derived from plants, 115
  immunosuppressive, 641
  nanotechnology, 145
  obesity and diet drugs, 711
  smell and taste disorders, 451
Drug abuse, 886. *See also* Alcoholism; Cocaine; Heroin; Marijuana
Dual reuptake inhibitors, 374t

Duchenne muscular dystrophy, 137, 289, 923, 928
Ducts, of eye, 463
Ductus arteriosus, 897–98, 899t, 906t
Ductus deferens, **34,** 834f, 835, 836f, 839–40, 845t
Ductus venosus, 897, 898f, 899t, 906t
Duodenum, **35–36,** 666f, 667f, 670f, 671f, 672f, 673f, 680
Dural sinuses, 384
Dura mater, **42,** 384, 387f, 388f
Dwarfism, 497. *See also* Pituitary dwarfism
Dynamic equilibrium, 459, 460, 705–706
Dysgeusia, 450t
Dyslexia, 402
Dysosmia, 450t
Dystrophic epidermolysis bullosa, 157t
Dystrophin, 137, 289, 928

E

Ear, 458t. *See also* Hearing
Ear infection, 453
Eccentric muscle contraction, 298, 299f
Eccrine glands, 180, 182t, 430t
Ectoderm, 875, 885, 887f
Ectodermal cells, 885
Ectopic pregnancy, 882
Edema
- definition, 552, 810, 816
- filtration, 94
- heart failure, 579, 817
- hypoproteinemia, 816–17
- inflammation, 817
- lymph system, 620, 621, 817
- nephrotic syndrome, 793

Edmonton protocol, for diabetes, 513
Effectors, 9, 355, 393t
Efferent arteriole, 780f, 781, 783f, 786f
Efferent lymphatic vessels, 622
Efferent neurons, 389, 391f
Egg cells, 846, 849, 877f
Egg donor, 831, 879
Egypt (ancient), and diabetes, 513
Ehrlich, Paul, 636
Ejaculation, 830, 844, 845f
Ejaculatory duct, 834f, 840
Elastic cartilage, 160, 162t
Elastic connective tissue, 158, 159f, 162t
Elastic fibers, 155, 156t
Elastic recoil, of lung tissue, 751
Elastin, 155
Elbow joint, 270t, 272–73
Electrocardiogram, 552, 568, 569–70f
Electroencephalogram (EEG), 411
Electrolytes
- balance, 811, 815–18
- bile, 675
- chemistry, 59, 60
- definition, 50
- intestinal absorption, 685, 686t
- plasma, 538
- types of, 59t

Electron, 51t, 52
Electron shells, 55
Electron transport chain, 120, 121f, 122–24, 294, 945–47
Elements, 51–52, 53t, 939
Elevation, of joint, 269f, 270
Elimination, of urine, 798–803
Ellipsoidal joint, 265–66
Embolism, 522, 541
Embolus, 541
Embryo, 849, 888f. *See also* Fetus
Embryonic disc, 885
Embryonic stage, of development, 885–92
Embryo proper, 880
Emergency department, 2
Emission, of sperm cells, 844, 845f
Emotional upset, and breathing patterns, 758
Emphysema, 739, 756
Emulsification, 678
Enamel, of tooth, 659
Encephalitis, 382, 644
End-diastolic volume (EDV), 585
Endocarditis, 557, 564–65, 567
Endocardium, 555, 556f, 557t
Endochondral bones, 197–98, 199f
Endochondral ossification, 198
Endocrine glands, 150, 483, 484f, 485f
Endocrine system. *See also* Hormones
- adrenal glands, 504–509
- autonomic stimulation, 430t
- cardiovascular system, 516, 610
- definition, 483
- digestive system, 516, 692
- general characteristics of, 483–84
- hormonal secretions, 491–92
- hormone action, 484–91
- integumentary system, 188, 516
- life-span changes, 515, 908t
- lymphatic system, 516, 645
- major functions of, 8t, 16, 17f
- muscular system, 333, 516
- nervous system, 378, 516
- pancreas, 509–11
- parathyroid glands, 502–504
- pineal gland, 511–13
- pituitary gland, 492–99
- reproductive system, 516, 868
- respiratory system, 516, 768
- skeletal system, 239, 516
- stress, 513–15
- thyroid gland, 499–501
- urinary system, 516, 804

Endocrinology, 24
Endocytosis, 96–98, 100t, 792
Endoderm, 885, 887f
Endodermal cells, 885
Endolymph, 453
Endometriosis, 861
Endometrium, 853f, 854, 855, 859, 882f
Endomysium, 286
Endoneurium, 412, 413f
Endoplasmic reticulum (ER), 75, 82–83, 84f, 91t
Endorphins, 373t, 375, 444
Endoscopic papillotomy, 679
Endosteum, 194–95
Endothelium, of artery, 574, 575f, 576
End-stage renal disease (ESRD), 779
End-systolic volume (ESV), 585
Energy. *See also* Adenosine triphosphate
- metabolic reactions, 119–20
- muscle contraction, 294
- nutrition, 706–709, 728t

Energy balance, 708
Energy shells, 55
Enkephalins, 373t, 374, 443
Enriched nutrients, 712
Enterogastric reflex, 670, 671f
Enterokinase, 671, 683t
Environmental factors. *See also* Body temperature; Internal environment
- aging, 20
- cancer, 104–105
- height and skin color, 926
- human life span, 910

Environmental tobacco smoke (ETS), 736
Enzymes
- definition, 64, 114
- digestive system, 683t
- lysosomes, 86
- membrane proteins, 80, 81t
- metabolic reactions, 117, 118f
- peroxisomes, 86

Eosinophils, 531, 532f, 534t, 535t
Ependyma, 353, 362f, 363
Ephedrine, 431
Epicardium, 555, 556f, 557t
Epicondyle, 208t, 226. *See also* Lateral epicondyle; Medial epicondyle
Epicranial aponeurosis, 306, 307f
Epicranius, 306, 307f
Epidemiology, 24
Epidermal growth factor, 104
Epidermis, 170, 172, 173–74, 176, 186
Epidermolysis bulbosa, 89
Epididymis, **34, 41,** 834f, 835, 836f, 838, 839f, 845t
Epidural space, 384
Epigastric arteries, 598, 599f
Epigastric region, 22, 24f
Epiglottic cartilage, 740, 741f
Epiglottis, 663f, 664, 735, 738f, 740, 741f, 742
Epilepsy, 373t
Epimysium, 286
Epinephrine, 426, 429, 487t, 504–505, 506t
Epineurium, 412, 413f
Epiphyseal plate, 198–200, 201f
Epiphysis, 194
Episiotomy, 901
Epitestosterone, 489
Epithelial cells, 77f, 145, 456, 682, 891
Epithelial membranes, 163
Epithelial tissue, 143, 144–52, 153t
Equilibrium, 459
Erectile dysfunction, 844
Erection, of penis, 844
Erector spinae, 308, 310, **342**
Erythroblast(s), 527
Erythroblastosis fetalis, 547–48, 624
Erythrocystosis, 548t
Erythrocytes, 522, 526, 535t, 942
Erythropoiesis, 527
Erythropoietic protoporphyria, 138f
Erythropoietin (EPO), 489, 513, 522, 527, 777
Esophageal arteries, 592
Esophageal hiatus, 651, 665

Esophagus, **36–37, 39–40, 43–44, 47,** 653f, 663f, 665, 666f, 738f
Essential amino acids, 705
Essential fatty acids, 703
Essential hypertension, 588
Essential nutrients, 699
Estradiol, 857, 884
Estriol, 857
Estrogen(s), 201, 497, 508, 857, 858, 859t, 861, 865, 884
Estrogen replacement therapy (ERT), 861
Estrone, 857
Ethanol, 904t
Ethiopia, and human evolution, 193
Ethmoidal sinus, **42,** 212, 214f, **253, 255**
Ethmoid bone, 209f, 210t, 212, 213f, 215f, **245–47, 253, 255–57,** 738
Eukaryotic cells, 76
Eumelanin, 178, 179
Evaporation, 182, 814
Eversion, 269
Evertor, and leg muscles, 334
Evolution
  allergies, 639
  fossilized skulls, 193
  mitochondrial DNA, 84
  Neanderthals, 179
  pregnancy and "morning sickness," 885
Exchange reaction, 58
Excitation contraction coupling, 291
Excitatory postsynaptic potential (EPSP), 371
Excretion. *See also* Urinary system; Urine
  characteristics of life, 8t
  organ systems, 17–18
Excretory system, 18. *See also* Urinary system
Exercise. *See also* Athletes and athletic performance
  breathing, 760
  cardiovascular system, 590, 611
  energy expenditure, 707t
  joints, 280
  muscular system, 300, 334
  obesity, 711
  osteoporosis, 204
Exocrine glands, 150, 151f, 152t, 482, 483, 484f
Exocytosis, 98, 99f, 100t
Expiration, 747, 751–52
Expiratory reserve volume (ERV), 752, 754t
Extension, 267, 305
Extensor carpi radialis brevis muscle, **350**
Extensor carpi radialis longus muscle, 317f, 318f, 319t, 320, **342, 350**
Extensor carpi ulnaris muscle, **350**
Extensor digitorum muscles, 305, 318f, 319t, 320, 328, 329f, 330f, 332t, **342, 344–45, 350**
Extensor hallucis longus muscle, 329f, 330, 332t, 334
Extensor muscles, 315, 316, 320, 328
Extensor retinaculum muscle, 318f, 320, 329f, 330f, 334, **350**
External acoustic meatus, 210, 211f, **253,** 450
External anal sphincter, 320, 322f
External respiration, 736
External urethral orifice, 801, 802, 843
Exteroreceptive senses, 440
Extracellular fluid, 9, 10f, 810, 812, 820
Extracellular fluid compartment, 811
Extracellular matrix (ECM), 153, 154
Extracorporeal liver assist device (ELAD), 676
Extracorporeal membrane oxygenation, 746
Extracorporeal shock-wave lithotripsy (ESWL), 799
Extrafollicular cells (C cells), 499
Extrapyramidal tracts, 395
Extremophiles, 119
Extrinsic clotting mechanism, 539, 540f, 541t
Extrinsic muscles, 464–65
Eye, **42,** 209f, 926. *See also* Iris; Sight
Eyelid, 462

F

Facet, 208t
Facial artery, 595f, 596f, 597
Facial bones, 206, 209f
Facial expression, 305–307
Facial nerves, 415, 417t
Facial skeleton, 212–16
Facial vein, 602f, 607f
Facilitated diffusion, 92–93, 94f, 100t
Facilitation, and neuronal pools, 375
Factor V Leiden, 548t
FAD, 714, 944
$FADH_2$, 945, 946f
Fainting, 759
Falciform ligament, **33, 48,** 673
Falls, by elderly, 240, 280
False ribs, 223, 224f
False vocal cords, 741
Falx cerebelli, **42,** 384t, 410
Falx cerebri, **42,** 384t, 398
Familial amyloidosis, 567
Familial hypercholesterolemia, 608, 922
Familial hypertrophic cardiomyopathy, 608
Famine, 725
Fanconi anemia, 883
Farsightedness, 472
Fascia, 285, 286f, 287
Fascicles, 284, 286
Fasciculus cuneatus, 394, 395f, 396t, 401
Fasciculus gracilis, 394, 396t
Fast-twitch muscle fibers, 299
Fat(s), 65f, 126f, 684, 685f. *See also* Fatty acid(s)
Fatal familial insomnia, 409t, 723
Fat globules, 678
Fatigue, of muscles, 296
Fat-soluble vitamins, 709–12, 713t
Fatty acid(s)
  acid-base balance, 819
  anabolism, 115–16
  carbohydrate use, 701
  chemistry, 65f, 69f
  definition, 63, 64f
  intestinal absorption, 684, 685f, 686t
Fatty acid oxidases, 702–703
Feces, 690, 814
Federation of International Football Associations, 763
Female. *See also* Gender; Pregnancy
  age-related bone loss, 240
  differences in male skeleton, 234t
  pelvis, 233f, 234, 322f
  reproductive system, 18, 846–61
  urinary bladder and urethra, 801f
Femopatellar joint, 270t
Femoral arteries, **32–33, 35, 45, 350,** 598, 599f, 600–601f
Femoral nerves, **32–34,** 423
Femoral region, 23, 25f
Femoral vein, **31–33, 35, 45, 350,** 604, 606f, 607f
Femur, **37, 45,** 206, 234, 235f, 238t, 240, 275f, 277f
Ferritin, 530, 675, 940
Fertilization, 876, 877–78. *See also In vitro* fertilization; Infertility
Fetal alcohol syndrome, 896, 897
Fetal cell sorting, 932–33
Fetal stage, of development, 892–94
Fetus. *See also* Development; Embyro; Pregnancy
  birth process, 899–902
  blood and circulation, 895, 897–98, 899f
  cocaine exposure, 837
  lymphocytes, 628
  medical treatments, 895
  ovaries, 847
  pituitary gland, 493
  prenatal tests for chromosome abnormalities, 931–33
  skull, **258**
  spleen, 624
  testes, 834f
Fever, 184, 628
Fiberoptic bronchoscope, 745
Fiber-optic endoscope, 679
Fibrillation, 572
Fibrinogen, 535, 536t, 541t
Fibrinolysin, 522
Fibrin-stabilizing factor, 541t
Fibrin threads, 543t
Fibroblasts, 108, 153, 156t, 157f, 165f, 541
Fibrocartilage, 160, 161f, 162t, 203, 264f
Fibrocystic breast disease, 865
Fibrosis, 154
Fibrous coat, 798
Fibrous joints, 261–62, 267t, 279
Fibrous layer, of vagina, 855
Fibrous pericardium, 12, **48,** 554
Fibrous tunic, 465
Fibula, 206, 235f, 236, 238t, **344**
Fibular artery, 599, 600–601f
Fibular collateral ligament, 276, 277f
Fibularis brevis muscle, 329f, 330f, 331f, **352**
Fibularis longus muscle, 305f, 329f, 330f, 331f, 332t, 334, **344, 351–52**
Fibularis retinacula muscle, 330f, 331f, 334
Fibularis tertius muscle, 328, 329f, 330f, 332t
Fibular veins, 606f
"Fight or flight" response, 506, 514
Filtration, and diffusion or osmosis, 94, 100t
Filum terminale, 387f, 388f, 389
Fimbriae, 830, 848f, 849f, 852, 853f
Fingerprints, 177. *See also* DNA fingerprinting
First-degree burn, 184

First line of defense, 626, 823
First polar body, 849, 850f, 880f
Fissure, 208t, 398
Fissured fracture, 202f
Fixed cells, 153
Flaccid paralysis, 382, 397
Flagella, 87, 88, 91t
Flared ilium, 233f
Flat bones, 193, 194f
Flavine adenine dinucleotide. *See* FAD
Flavoproteins, 714
Flexion, 267, 305
Flexor carpi radialis, 317f, 318, 319t, **344**
Flexor carpi ulnaris, 317f, 318–19, **344**
Flexor digitorum, 319, 331f, 332
Flexor muscles, 316, 317f, 318–19, 326, 328
Flexor retinaculum, 317f, 331f, 334
Floaters, in eye, 477
Floating ribs, 223
Fluid compartments, 811
Fluoride, 661
Fluorine, 52t, 53t, 70, 722, 723t
Folic acid (folacin), 715–16, 717t
Folinic acid, 716
Follicle(s), 853f. *See also* Primordial follicle
  maturation, 849, 851
  ovarian, 830
  thyroid gland, 499
Follicle-stimulating hormone (FSH), 487t, 491, 492, 497, 499t, 845, 859t
Follicular cells, 637, 847, 853f
Fomix, 848f
Fontanel, 208t, 216, 218f, 262
Food. *See also* Diet; Food poisoning; Food pyramids; Nutrients; Nutrition
  energy values of, 706–707
  requirements for life, 7, 9t
Food and Drug Administration (FDA), 725, 726
Food poisoning, 790
Food pyramids, 722–23, 724f
Foot, 236–38, 328, 330, 332, 334, **345**
Foramen, definition, 208t. *See also* Infraorbital foramen; Jugular foramen; Mandibular foramen; Mastoid foramen; Vertebral foramen
Foramen lacerum, 211f, 213f, 217t, **249–51, 256–57**
Foramen magnum, 208, 211f, 213f, 215f, 217t, **249–51, 253, 256–57,** 389f
Foramen ovale, 211f, 213f, 217t, **249–51, 254, 256–57,** 897, 898f, 899t, 906t
Foramen rotundum, 211f, 213f, 217t, **254, 257**
Foramen spinosum, 211f, 213f, 217t, **249–51, 254, 256–57**
Forearm, 316, 317f, 319f, **344, 350**
Forebrain, 398, 400t
Forehead, 234t
Forensics, and DNA profiling, 130
Fornix, 399f, 854
Fortified nutrients, 712
Fossa, definition, 208t. *See also* Coronoid fossa; Iliac fossa; Incisive fossa; Infraspinous fossa; Mandibular fossa; Supraspinous fossa
Fossa ovalis, 905
Fourth ventricle, 385, 386f, 387f, 408f
Fovea, definition, 208t
Fovea capitis, 192, 236f, 260, 274
Fovea centralis, 469, 473
Fox, Michael J., 406
Fox P2 function, 404
Fractures, of bone, 202–203, 204, 240
"Fragility fracture," 204
Frank-Starling law of the heart, 586
Fraternal twins, 880
Free nerve endings, 440, 446t
Free radicals, 710, 909
Free (apical) surface, 144
Frenulum, 651
"Frequency combs," 762
Frontal bone, **39,** 208, 209f, 210t, 211f, 213f, 215f, 218f, **245–48, 253, 255–57, 341,** 738
Frontal eye field, 402
Frontalis muscle, 305f, 306, **346**
Frontal lobe, **42,** 399, 400f, 401f, 402, 403t
Frontal plane, 22, 23f
Frontal region, 23, 25f
Frontal sinus, **39, 42,** 208, 213f, 214f, 215f, **255,** 738f
Frontal suture, 216, 218f, **258**
Fructose, 701, 840, 944, 945f
Full-thickness burn, 186
Fulminant hepatitis, 677
Fumaric acid, 946f
Functional residual capacity (FRC), 753, 754t
Functional syncytium, 565
Fundus. 665–66, 667f, 854
Funiculus, 382

G

GABA, 373t
Gage, Phineas, 383
GAIT (Glucosamine/Chondroitin Arthritis Intervention Trial), 261
Galactose, 701
Gallbladder, **33–35, 45,** 604f, 653f, 672, 674f, 677–78, 679
Gallium-67, 54
Gallstones, 678
Gamete intrafallopian transfer (GIFT), 879
Gamma globulins, 535, 536t, 634
Gamma radiation, 53, 56
Ganglion (ganglia), 360, 363, 382, 414
Ganglion cells, 469
Gap junctions, 144, 146t
Gardasil, 869
Gas transport, and respiratory system, 762–67
Gastric arteries, 592, 594f
Gastric bypass surgery, 513, 711
Gastric folds, 666f, 667f
Gastric glands, 651, 666, 667f
Gastric juice, 666, 668t
Gastric lipase, 668, 683t
Gastric pits, 666, 667f
Gastric secretions, 666–69
Gastric veins, 603, 604f
Gastrin, 668, 678t
Gastritis, 668
Gastrocnemius muscle, 277f, 305f, 306f, 326f, 329f, 330f, 331f, 332t, **344–45, 352**
Gastroenterology, 24
Gastrula, 885, 886t
Gastrulation, 885
Gated channels, 366
Gender, and phenotype, 928. *See also* Female; Male
Gene(s), 125, 633. *See also* Chromosomes; DNA; Genetic(s); RNA
Gene expression, 125, 924, 933–34
Gene expression profiling, 865, 933, 934
General Accounting Office (U.S.), 917
General senses, 438, 440–46
General stress syndrome, 514, 515t
Gene therapy, 922. *See also* Genetic disorders; Genetic tests
Genetic(s). *See also* Chromosomes; DNA; Gene(s); Gene expression; Genome; Genomic(s); RNA
  aging, 20
  anatomy and physiology, 933–34
  cellular metabolism, 125–27
  definition, 916, 917
  expression of single genes, 924
  hair and skin pigmentation, 179
  human life span, 910
  modes of inheritance, 918–22
  multifactorial traits, 924–26
  obesity, 711
  sex chromosomes and sex determination, 927–28
Genetic code, 124, 130, 132
Genetic disorders, 929–33. *See also* Gene therapy
Genetic heterogeneity, 924
Genetic information, 135–38
Genetic Information Nondiscrimination Act (2008), 917
Genetic tests, 51, 656, 865, 917. *See also* Genetic disorders; Gene therapy; Laboratory tests
Genicular arteries, 598, 600f
Genital herpes, 867t
Genital region, 23, 25f
Genital warts, 867t, 869
Genome, 125, 139, 917
Genomic(s), 918
Genomic imprinting, 928
Genotype, 920
George III (King of England), 530
Geophagy, 722
Geriatrics, 24
Germinal epithelium, 830
Gerontology, 24
Ghrelin, 700
GI camera, 654
Gigantism, 497. *See also* Pituitary gigantism
Gingko biloba, 725
Glandular epithelium, 150, 152, 153t
Glans penis, 834f, 843
Glatiramer, 360
Glaucoma, 477
Gleevec, 536, 537
Glenohumeral ligaments, 270t, 271, 272f
Glenoid cavity, 192, 226, 227f, 260, 272f

Glenoid labrum, 271, 272f
Gliding joints, 266
Global Polio Eradication Initiative, 290
Globulins, 535, 536t, 940
Globus pallidus, 404
Glomerular capillary pressure, 788
Glomerular capsule, 781, 783f, 786f, 797t
Glomerular filtrate, 786, 788
Glomerular filtration, 785–88, 794
Glomerular filtration rate (GFR), 788–90, 803
Glomerulonephritis, 642t, 782, 817
Glomerulus, 774, 779, 781, 783f, 797t
Glossopharyngeal nerves, 416, 417t
Glottis, 741
Glucagon, 487t, 491, 509–10, 511t
Glucocorticoid, 507
Glucokinase, 510
Gluconeogenesis, 701, 704
Glucosamine, 261
Glucose
  aerobic respiration, 125f, 819
  anaerobic respiration, 819
  blood tests, 940
  diffusion into cell, 93
  homeostatic mechanism, 10–11
  metabolic pathways, 124
  molecular structure, 69f
  nutrition, 701
  oxygen debt, 295
  structural formulas, 63f
  urine, 788t, 792, 798
Glucose-6-phosphate, 945f
Glucosuria, 792
Glutamate, 373t
Glutamic acid, 135t, 705t
Glutamine, 135t, 705t
Gluteal arteries, 598, 599f
Gluteal nerves, 423
Gluteal region, 23, 25f
Gluteal tuberosity, 236f
Gluteal veins, 605
Gluten, 686
Gluteus maximus muscle, **45**, 306f, 322f, 323, 325f, 326f, 327t, **343, 348, 351**
Gluteus medius muscle, **37**, 323, 325f, 326f
Gluteus minimus muscle, 323, 327t
Glyceraldehyde-3-phosphate, 945, 946f
Glycerol, 63, 115–16, 685f, 686t
Glycine, 69f, 135t, 705t
Glycogen, 50, 62
Glycolysis, 120, 121f, 944
Glycoproteins, 485, 487t, 617
GM1 ganglioside, 397
Goblet cells, 146
Goiter, 496, 501t, 502f
Golgi apparatus, 83, 85f, 91t
Golgi tendon organs, 444–45, 446t
Gomphosis, 262, 267t
Gonadal arteries, 593, 594f, 601f
Gonadal veins, 604, 607f
Gonadotropes, 494
Gonadotropin(s), 489, 497, 845
Gonadotropin-releasing hormone (GnRH), 487t, 497, 845, 846, 858
Gonorrhea, 867t
Gout, 279t, 797
G protein, 488
Gracilis muscle, **34–37**, 305f, 306f, 324f, 325, 326f, 327t, 331f, **350**
Graft-versus-host disease (GVHD), 641, 643
Granulations, 183
Granulocyte(s), 530, 535t
Granulocyte colony stimulating factor (G-CSF), 104
Granulosa cells, 849, 851f
Grapheme-color type synesthesia, 447
Graves disease, 501t, 502f, 642t
Gravity, 585
Gray commissure, 389, 390f
Gray matter, in brain, **42,** 358, 390f, 391f
Gray rami, 426
Greater omentum, **33, 38, 48,** 681
Greater palatine foramen, 211f, 217t, **249, 251**
Greater pelvis, 233
Greater sciatic notch, 232f, 233
Greater trochanter, 236f, 275f
Greater tubercle, 226, 229f
Greater wing, of sphenoid bone, 211f, **254**
Greater trochanter, 236f
Greece (ancient), and diabetes, 513. *See also* Aristotle; Hippocrates
Greenstick fracture, 202f
Grollman, Arthur, 775
Ground substance, 153
Growth. *See also* Development
  characteristics of life, 8t
  definition, 876
  epiphyseal plate, 198–200
Growth factors, 104, 183
Growth hormone (GH), 201, 487t, 489, 492, 494, 497, 499t, 515. *See also* Human growth hormone
Growth hormone-releasing hormone (GHRH), 487t, 494
Guanine, 128f, 948f, 949f
Gubernaculum, 830, 833
Gynecology, 24
Gyrus, 398, 399f, 400f

H

Hageman factor, 540, 541t
Hair cells, 455, 457f
Hair follicle, 170, 178–79
Hair loss, 179
Hair shaft, 178
Hairy cell leukemia, 534t
Hamate, 230f
Hamstring muscles, 328t, **343**
Hand, 228f, 229–31, 316–20, **344, 350**
Haploid cell, 831
Hapten, 628
Hard palate, 212, 657, 664f, 738f
Harvard University, 354, 816
Hashimoto disease, 501t
Haversian canals, 160
Head, **346.** *See also* Skull
  arteries, 595f, 596–97
  of femur, 236f
  of fibula, 236
  muscles, 307f, 308, 310
  of radius, 228, 230f
  of rib, 223
  skeletal structures and use of term, 208t
  sperm cell, 837
  surface anatomy, **341**
  of ulna, 229, 230f
  veins, 602
Hearing, sense of, 450–58, 476–77, 459. *See also* Deafness
Heart, **34, 38, 43, 46.** *See also* Cardiovascular system; Heart disease
  aging, 609, 611
  blood pressure, 582–83
  blood supply to, 561–63, 593f
  cardiac conduction system, 565–67
  cardiac cycle, 564, 568–72
  cardiac muscle fibers, 565
  chambers and valves, 555–57
  coverings of, 554
  electrocardiogram, 568, 569–70f
  failure, 154
  path of blood through, 560–61
  size and location of, 553–54
  skeleton of, 559f, 560
  sounds of, 564–65
  tissue engineering, 166
  transplant, 566
  wall of, 555, 557t
Heartbeat, 564. *See also* Heart rate
Heartburn, 670
Heart disease, 574, 609, 911t. *See also* Cardiovascular disease
Heart failure, 154
Heart rate, 590. *See also* Heartbeat; Pulse
Heat
  body temperature regulation, 181–82
  muscles and production of, 296
  requirements for life, 7, 9t
Heatstroke, 811
Heavy chains (H-chains), 634
Heavy metals, 530
Height
  age-related loss of, 238, 280
  genetic and environmental factors, 925f, 926
Helicobacter pylori, 668
Helium, 53t, 55f
Helper T cells, 630, 631f
Hematocrit (HCT), 522, 523, 524f, 940
Hematology, 24
Hematoma, 203f
Hematopoietic growth factors, 526
Hematopoietic stem cells, 524, 525f, 526
Hematopoiesis, 192, 202
Heme, 531f
Hemiazygos veins, 603
Hemicellulose, 701
Hemispheric dominance, in brain, 403–404
Hemizygous gene, 928
Hemochromatosis, 717
Hemodialysis, 777, 779
Hemoglobin
  blood tests, 941
  definition, 522, 735
  fetal circulation, 897
  iron, 720
  oxygen transport, 762–63
  protein buffer system, 822

red blood cells, 526
structural formula, 531f
urine tests, 943
Hemolytic anemia, 529t, 642t
Hemolytic jaundice, 676
Hemolytic uremic syndrome (HUS), 790
Hemophilia, 548t, 928
Hemorrhagic disease of the newborn, 713
Hemorrhoids, 688
Hemostasis, 522, 538–43
Hensel, Abby and Brittany, 900
Heparin, 155, 522, 532, 542–43
Hepatic arteries, 592, 594f
Hepatic cells, 673
Hepatic coma, 675
Hepatic ducts, 672f, 673, 674f
Hepatic gland, 651
Hepatic lobules, 673
Hepatic portal system, 603
Hepatic portal vein, 603, 673, 674f, 898f
Hepatic sinusoids, 603, 673, 675f
Hepatic veins, 604, 607f
Hepatitis, 677
Hepatopancreatic ampulla, 671
Hepatopancreatic sphincter, 671
HERC2 gene, 926
Herd immunity, 638
Hereditary hemochromatosis, 548t
Hereditary osteoarthritis, 157t
Hernia
hiatal, 665
indirect inguinal, 835
strangulated, 835
Herniated disc, 225
Heroin, 375, 904t
Heterogeneity, genetic, 924
Heterozygous chromosomes, 916, 919
Hiatal hernia, 665
Hiccups, 745, 755t
High altitude, and breathing, 763
High-density lipoprotein (HDL), 684–85, 703, 940
Hilum, 621, 747, 776
Hindbrain, 398, 400t
Hinge joint, 266, 267t
Hip bone, **45,** 231–33, 238t
Hip fracture, 204, 240
Hip joint, 270t, 274, 275f
Hippocampus, 404
Hippocrates, 797, 803
Hirschsprung disease, 690
Histamine, 155, 373t, 532, 579, 639, 668
Histidine, 135t, 705t
Histology, 24, 143, 144
Histones, 127, 128f
History, of study of human body, 3. *See also* Egypt; Greece
HIV. *See* Human immunodeficiency virus
Holocrine glands, 150, 152f, 153t, 170
Homeostasis, 1, 8–12, 19f, 183f, 200
Homeostatic mechanisms, 9–11, 12
Homologous chromosomes, 832, 916, 919
Homo sapiens idaltu, 193
Hookworm, 534t
Horizontal cells, 469
Hormone(s). *See also* Androgens; Endocrine system; Estrogens; Growth hormone; Hormone replacement therapy; Steroid hormones
acne, 181
actions of, 485–91
bone growth and development, 201
chemistry, 484–85
control of secretions, 491–92
definition, 16, 482, 483
digestive tract, 678t
female reproductive functions, 857–61
male reproductive functions, 845–46
mammary glands, 903t
pregnancy, 883–84, 885t
smooth muscle contraction, 301
Hormone-receptor complex, 485
Hormone replacement therapy (HRT), 861
Horseshoe kidney, 802t
"Hot flashes," 860–61
Human(s)
evolution, 193
genome, 917
life span, 910–11
major elements in body, 52t
organization of body, 12–19
pheromones, 483
skin color, 174, 176
Human chorionic gonadotropin (hCG), 882, 883, 884, 885t
Human genome project, 136
Human growth hormone (HGH), 497. *See also* Growth hormone
Human immunodeficiency virus (HIV), 76, 98, 99f, 626, 638, 642–43, 644, 869. *See also* Acquired immune deficiency syndrome
Human leukocyte antigens (HLA), 630
Human menopausal gonadotropin (hMG), 861
Human Metabolome Database, 139
Human Oral Microbiome Database, 652
Human papilloma virus (HPV), 869
Humerus, **34,** 206, 226, 227f, 228f, 229f, 231t, 272f, **273,** 314f, 315
Humoral immune response, 632–33
Humoral immunity, 616
Hunter-gatherers, 3
Huntington disease (HD), 363, 373t, 922, 923
Hyaline cartilage, 143, 160, 162t
Hybrid fixator, 203
Hybridoma, 636
Hydrocephalus, 388
Hydrochloric acid, 668t, 669
Hydrogen
acid-base balance, 60t, 793, 819–24
atomic structure, 53t, 55f
chemical constituents of cells, 60t, 62t
elemental composition of human body, 52t
tubular secretion, 793
Hydrogenation, of unsaturated fat, 63
Hydrogen bond, 57–58, 948f
Hydrogen carriers, 944–45, 946f
Hydrolysis, 116, 125f, 702, 820
Hydrophilic molecules, 68
Hydrostatic pressure, 8, 578, 789, 812–13
Hydroxide ions, 820
Hydroxyapatite, 203
Hydroxycholecalciferol, 503, 711
17-Hydroxycorticosteroids, 943
Hydroxyurea, 527
Hygiene hypothesis, 756
Hymen, 854
Hyoid bone, 205t, 206, 207f, 664, 738f, 741f
Hypercalcemia, 503, 573, 818
Hyperextension, 267, 269
Hypergeusia, 450t
Hypericin, 725
Hyperkalemia, 573, 820
Hypermobility syndrome, 263
Hypernatremia, 820
Hyperosmia, 450t
Hyperoxia, 760
Hyperparathyroidism, 504t, 818
Hyperpolarized membrane, 368
Hypersensitivities, 639
Hypersomnia, 373t
Hypertension, 588. *See also* Blood pressure
Hyperthermia, 182, 184
Hyperthyroidism, 482, 501t
Hypertonic solution, 75, 94, 95f
Hypertrophic pyloric stenosis, 666
Hypertrophy, 284, 300
Hyperventilation, 758, 759, 825
Hypervitaminosis, 710, 712
Hypocalcemia, 503
Hypochondriac region, 22, 24f
Hypodermic injections, 173
Hypogastric region, 22, 24f
Hypogeusia, 450t
Hypoglossal branch, 417
Hypoglossal canal, 215f, 217t
Hypoglycemia, 509, 510
Hypokalemia, 820
Hyponatremia, 726, 816, 820
Hypoparathyroidism, 504t
Hypophyseal portal veins, 493
Hypopituitary dwarfism, 497
Hypoproteinemia, 810, 816–17
Hyposmia, 450t
Hypothalamus
diencephalon, 405
female reproductive functions, 857, 858
homeostasis, 9–10, 183f
hormones secreted by, 487t, 492, 496
male reproductive functions, 845, 846, 847f
pituitary gland, 493f
regulatory functions, 407
stress response, 514
Hypothermia, 182, 183f
Hypothyroidism, 501t
Hypotonic solution, 75, 94, 95f
Hypoxia, 526, 763
Hypoxia-inducible factor (HIF-1), 574
Hysterosalpingogram, 861
H zones, 287, 288

I

I bands, 287
Identical twins, 880
Idiotypes, 634
Ileocecal sphincter, 686

Iliac arteries, 598, 599f, 600–601f
internal, 898f
Iliac crest, **37,** 231, 232f, **343**
Iliac fossa, 231, 232f
Iliac region, 22, 24f
Iliac spine. *See* Inferior iliac spine; Superior iliac spine
Iliacus, **37,** 323, 324f, 327t, **350**
Iliac veins, 604, 605, 606f, 607f
Iliocostalis cervicis, 309f, 310t
Iliocostalis lumorum, 309f, 310t
Iliocostalis thoracis, 309f, 310t
Iliofemoral ligament, 274, 275f
Iliolumbar artery, 598, 599f
Iliotibial tract, 325f, 331f, **344**
Ilium, **35,** 231, 232f, 680
Illinois Masonic Medical Center, 883
Immediate-reaction allergy, 639, 640f
Immune complex reactions, 639
Immune system and immunity
allergic reactions, 639–41
antigens, 628
autoimmunity, 641–43
B cells and humoral immune response, 632–35
body defense against infection, 625–26
definition, 616
immune responses, 637
immunotherapy, 636
innate (nonspecific) defenses, 626–28
lymphocyte origins, 628–29
practical classification of, 638
T cells and cellular immune response, 629–30, 631f
transcytosis, 98
transplantation and tissue rejection, 641
Immunoglobulin(s), 633, 634–35
Immunoglobulin A (IgA), 635
Immunoglobulin D (IgD), 635
Immunoglobulin E (IgE), 635
Immunoglobulin G (IgG), 635
Immunoglobulin M (IgM), 635
Immunology, 24
Immunosuppressive drugs, 641
Immunotherapy, 636–37
Impact resection arthroplasty, 276
Implantable cardioverter defibrillator (ICD), 553
Implantation, of blastocyst, 882
Impulse processing, and nervous system, 374–76
"Inborn error of metabolism," 137–38
Incisive foramen, 211f, 217t
Incisive fossa, **249, 251**
Incisors, 659
Inclusions, 89
Incomplete dominance, 922
Incompletely penetrant genotype, 924
Incontinence, in elderly, 805
Incontinentia pigmenti, 928
Incus, 451–52
Indirect inguinal hernia, 835
Inert atoms, 55
Infants. *See also* Children; Development; Fetus
adipose tissue, 158
antibodies, 635
breastfeeding and formula, 904
dehydration, 798, 816
drugs of abuse and withdrawal symptoms, 886
kidneys, 798
postnatal period, 904–906, 908t
premature, 893
respiratory distress syndrome, 750
skull, 216, 218f
sleep apnea, 757
vitamin K deficiency, 713
Infection. *See also* Bacteria; Viruses
causes of death, 911t
of ear, 453
of urinary tract, 800
Inferior, as term of relative position, 21
Inferior articulating process, 220
Inferior iliac spine, 232f
Inferior peduncles, 410
Inferior rectus, 464, 465t
Infertility, 842–43, 851, 861. *See also In vitro* fertilization
Inflammation
antibodies, 636t
blood vessel disorders, 581
definition, 183, 616
edema, 817
endocrime system and homeostasis, 514
major actions of immune response, 627, 628t
urinary tract, 800
Inflammatory bowel disease, 691
Inflation reflex, 758
Influenza, 632, 638, 765, 911t
Infraglenoid tubercle, 227f
Infraorbital foramen, 209f, 217t, **247–48.** *See also* Supraorbital foramen
Infrapatellar bursa, 265, 277
Infraspinatus muscle, 314t, 316, **342, 348–49**
Infraspinous fossa, 226, 227f
Infraspinous muscle, 306f, 313f
Infundibulum, 405, 407, 492, 852, 853f, 877f
Inguinal canal, **33,** 593, 833
Inguinal hernia, 835
Inguinal region, 23, 25f, 622
Inhalation, 735
Inhaled particles, 749
Inheritance, modes of, 918–22. *See also* Genetic(s)
Inhibin, 845
Inhibitory postsynaptic potential (IPSP), 372
Initial segment, 368
Injectable contraception, 865
Injury, as cause of death, 911t. *See also* Inflammation; Spinal cord injuries; Traumatic brain injury
Innate (nonspecific) defenses, 626–28
Inner cell mass, 880
Inner ear, 453–57
Inner membrane, of mitochondria, 946
Inner tunic, 469
Inner vascular layer, 849
Innervation, of alimentary tube, 656
Innocence Project, 130
Inorganic chemicals, 60, 61
Inorganic mineral salts, 203
Insecticides, 293
Insertion, of skeletal muscle, 303–304
Insomnia, 373t, 409t
Inspiration, 747, 750–51, 759f
Inspiratory capacity (IC), 753, 754t
Inspiratory reserve volume (IRV), 752, 754t
Institute of Medicine, 712
Insula, 400
Insulin, 93, 487t, 489t, 510, 511t, 513, 515, 699–700
Insulin-like growth factor-1 (IGF-1), 495
Integral protein, 81t
Integration, of organ systems, 16
Integrin, 81
Integumentary system. *See also* Skin
autonomic stimulation, 430t
cardiovascular system, 188, 610
digestive system, 188, 692
endocrine system, 188, 516
life-span changes, 186–87, 908t
lymphatic system, 188, 645
major functions of, 8t, 14, 16f
muscular system, 188, 333
nervous system, 188, 378
reproductive system, 188, 868
respiratory system, 188, 768
skeletal system, 188, 239
urinary system, 188, 804
Intensity, of vocal sound, 741
Interatrial septum, 555
Intercalated disc, 143, 164, 284, 301, 302f
Intercarpal joint, 270t
Intercellular junctions, 144, 146t
Intercondylar eminence, 236f
Intercondylar fossa, 236f
Intercostal arteries, 592, 598, 601f
Intercostal muscles, 750–51, 752
external, **32–33, 37,** 312f
internal, **33, 37,** 312f
Intercostal nerves, 423
Intercostal trunks, 617, 619f
Intercostal veins, 603
Interferons, 626, 637
Interleukin(s), 530
Interleukin-1 (IL-1), 628, 630
Interleukin-2 (IL-2), 630
Interleukin-6 (IL-6), 334
Interlobar arteries, 779, 780f
Intermediary metabolism, 115
Intermediate cuneiform, 237f
Intermediate fibers, 299
Internal accessory sex organs, 837–42
Internal acoustic meatus, 213f, 215f, 217t, **255**
Internal anal sphincter, 320
Internal environment, 9, 10f
Internal respiration, 736
Interneurons, 361, 362t, 393t
Internodal atrial muscle, 566
Interosseous ligament, 261
Interosseous membrane, 261
Interphase, 75, 100
Interstitial cell(s), 835, 845t
Interstitial cell-stimulating hormone (ICSH), 845

Intertarsal joint, 270t
Intertubercular groove, 229f
Interventricular arteries
anterior, 561
posterior, 561, 562f
Interventricular foramina, 385, 386f
Interventricular septum, 555, 559f
Interventricular sulci, 556
Intervertebral discs, **37, 40–41,** 192, 206, 218, 219f
Intervertebral foramen, 219f, 220, 420
Intervertebral joint, 270t
Intestinal gas, 689
Intestinal gastrin, 678t
Intestinal glands, 681–82
Intestinal lipase, 683
Intestinal phase, of gastric secretion, 669
Intestinal somatostatin, 669
Intestinal trunk, 617, 619f
Intestinal villi, 681
Intracellular fluid, 810, 812
Intracellular fluid compartment, 811
Intracranial pressure (ICP), 388
Intracytoplasmic sperm injection (ICSI), 879
Intradermal injections, 173
Intralobular bronchioles, 743
Intramembranous bones, 192, 197
Intramembranous ossification, 197, 198t
Intramuscular injections, 173
Intrauterine devices (IUD), 866
Intrauterine growth retardation (IUGR), 897
Intravascular oxygenator, 746
Intrinsic clotting mechanism, 539, 540–41
Intrinsic factor, 528, 668, 715
Inulin, 798
Inulin clearance test, 798
Inversion, of joint, 269
Invertor, and leg muscles, 334
In vitro fertilization, 831, 841, 878–79, 883
Involution, and childbirth, 902
Iodide pump, 500
Iodine, 52t, 54, 722, 723t
Iodopsins, 475
Ion(s). *See also* Ammonium; Calcium; Carbonate; Chloride; Hydrogen; Phosphate; Sodium; Sulfate
bonding of atoms, 55
cell membrane potential, 365
channels, 82
definition, 51t
inorganic substances, 62t
Ionic bonds, 55
Ionizing radiation, 56
Ipsilateral, as term of relative position, 21
Iris, of eye, 430t, 437, 467, 926
Iron, 52t, 528, 720–21, 723t
Iron-binding capacity, 941
Iron deficiency anemia, 529t
Irregular bones, 193, 194f
Ischial spine, 232f, 233
Ischial tuberosity, 232f, 233, **343**
Ischiocavernosus muscle, 322, 323t
Ischiofemoral ligament, 274, 275f
Ischium, 232f, 233, 275f
Isocritic acid, 946f
Isograft, 641
Isoleucine, 135t, 705t
Isometric muscle contraction, 298–99
Isotonic glucose, 813
Isotonic muscle contraction, 284, 298, 299f
Isotonic solution, 75, 94, 95f
Isotope, 50, 52–53
Isotretinoin, 181, 897
Isthmus, 499

J

Jacobs syndrome, 929
Jaundice, 176, 177, 676
Jaw angle, 234t
Jejunum, **35,** 680
Johns Hopkins School of Medicine, 375
Joint(s). *See also* Skeletal system; Synovial joints
classification of, 261–63
disorders of, 278
life-span changes, 278–80
replacement of, 276
types of movements, 267–70
Joint capsule, 264, 276
Joint disorders, 278
Joint reflex, 760
Joint replacement, 276
Jugular foramen, 210, 211f, 213f, 215f, 217t, **249–51, 256, 257**
Jugular notch, 224f, **343**
Jugular trunk, 618, 619f
Jugular veins, **43, 46–47,** 593, 602, 603f, 607f, 619f
external, **33**
internal, **32–33, 36**
Juvenile rheumatoid arthritis, 279t
Juxtaglomerular apparatus, 782, 784
Juxtaglomerular cells, 782, 784
Juxtamedullary nephron, 774, 784, 785f, 786f

K

Kaposi sarcoma, 642
Karyokinesis, 101
Karyotype, 916, 918, 920f
Kawasaki disease, 279t
Kennedy, John F., 509
Keratin, 147–48, 170
Keratinization, 148, 173
α-Ketoglutaric acid, 946f
Ketone bodies, 703
Ketonuria, 810
Kidney(s), **36, 44–45,** 704, 776–85, 803, 825, 911t
Kidney stone, 799
Kilocalorie, 706
Klinefelter syndrome, 929
Knee joint, 270t, 274, 276–77, 278, 280f, **344**
Köhler, Georges, 636
Korotkoff's sound, 583
Kupffer cells, 604, 673
Kwashiorkor, 726, 727f
Kyphosis, 225

L

Labia majora, 848f, 855, 857t
Labia minora, 830, 848f, 855, 857t
Labor, and childbirth, 900
Laboratory tests, of clinical importance. *See also* Genetic tests; Prenatal tests
blood, 940–42
urine, 943
Labrum, 260
Labyrinth, 437, 453
Lacrimal apparatus, 463–64
Lacrimal bones, 209f, 210f, 214, 216t, **246–48**
Lacrimal gland, 306, 437, 463
Lacrimal sac, 463
Lactase, 682–83
Lacteal, 681
Lactic acid, 121, 122, 941
Lactic acid threshold, 295
Lactic dehydrogenase, 941
Lactobacillus strains, 652
Lactose, 684, 690, 701
Lactose intolerance, 684, 690
Lacunae, 158, 195, 875, 885
Lambdoid suture, 208, 210f, 211f, 215f, **246**
Lamellae, 160, 192
Lamellated corpuscles, 440, 446t
Laminectomy, 225
Landsteiner, Karl, 544
Lanugo, 875
Laparoscopy, 861
Large calorie, 706
Large intestine, **44,** 653f, 686–90, 691, 834f
Laryngopharynx, 663, 738f
Larynx, **32, 35, 39, 46,** 207f, 663f, 664, 738f, 740–41, 744f, 747t, 754
Latent period, 284, 296
Lateral, as term of relative position, 21
Lateral border, 226
Lateral canal, 460
Lateral condyle, 235, 236f, 277f
Lateral cuneiform, 237f
Lateral curvature, 225
Lateral epicondyle, 229f, 236f, **273, 342, 344**
Lateral funiculus, 390f
Lateral geniculate nucleus (LGN), 407
Lateral horn, 389
Lateral malleolus, 236, **345**
Lateral meniscus, 277f
Lateral pterygoid, 211f, 307, 308
Lateral rectus muscle, **42,** 415, 464, 465
Lateral spinothalamic tract, 395
Lateral sulcus, 399, 400f
Lateral ventricles, **38, 42,** 385, 386f
Latissimus dorsi muscle, **32–33,** 306f, 311f, 314t, 315, **342, 347–49**
Laughing, 754, 755t
Learning, and PET imaging, 71f
Leeches, 543
Leeuwenhoek, Anton van, 836
Left anterior descending artery, 561. *See also* Interventricular artery
Left hemisphere, 403, 404
Left lower quadrant (LLQ), of abdomen, 22, 24f

Left upper quadrant (LUQ), of abdomen, 22, 24f
Left ventricle, **44, 46,** 558f
Left ventricular assist device (LVAD), 566
Leg, muscles of, 324f, 325–28, 329f, 330f, 332t, 334, **345, 351**
Lens, 466–67
Lens fibers, 467
Leptin, 700
Lesser pelvis, 233–34
Lesser sciatic notch, 232f
Lesser trochanter, 236f, 275f
Lesser tubercle, 226, 229f
Lesser wing, of sphenoid bone, 211f, **254**
Leucine, 135t
Leukemia, 536–37, 934
Leukocytes, 522, 530, 533, 535t
Leukocytosis, 522, 533
Leukopenia, 533
Leukotrienes, 639
Levator ani, 320, 322f, 323t, **346**
Levator palpebrae superioris, 462, 465t
Levator scapulae, 311f, 312, 313t
Levers, 301–302, 304f
Lewy bodies, 406f
Lexical-gustatory synesthesia, 447
Life. *See also* Life-span changes
  characteristics of, 6, 8t
  maintenance of, 7–12
  span of human, 910–11
Life cycle, of red blood cells, 528f
Life expectancy, 911
Life span, of human, 910–11
Life-span changes. *See also* Age; Aging
  cardiovascular system, 606, 609, 611
  digestive system, 690, 693
  endocrine system, 515
  immune system, 644
  joints, 278–80
  muscular system, 334
  nervous system, 431
  nutrition, 728
  respiratory system, 767, 769
  senses, 476–77
  skeletal system, 238, 240
  skin, 186–87
  urinary system, 803, 805
  water imbalances, 816
Ligament(s)
  collagenous fibers, 155
  dense regular connective tissue, 158
  joint capsule, 264
  knee joint, 277f
  shoulder joint, 271, 272f
Ligamentum arteriosum, 593f, 905
Ligamentum capitis, 275f
Ligamentum teres, 905
Light adapted eye, 475
Light chains (L-chains), 634
Light microscope, 79
Light refraction, 471, 473
Lignin, 701
Limbic system, 407
Limited transport capacity, 791
Linea, 208t
Linea alba, **32,** 312f, 320, 321f, **347**
Linea aspera, 236f
Lingual artery, 595f, 596
Lingual frenulum, 657
Lingual tonsil, 651, 657, 738f
Linoleic acid, 703, 704
Linolenic acid, 703
Lip(s), 657, 660t
Lipase, 118
Lipids
  blood levels, 941
  cell membrane, 79, 80
  definition, 50, 698
  important groups of, 65t
  nutrition, 702–704, 706t
  structure and function, 62–64, 69t
Lipochromes, 926
Lipoid nephrosis, 793
Lissencephaly, 398
Lithium, 52f, 53t, 55f
Liver, **33, 34, 38, 40, 44–45, 49**
  carbohydrates, 701
  dietary iron, 721
  digestive system, 653f, 673–78
  fibrosis, 154
  kidneys, 777f
  lipids, 703
  transplant, 676
  veins from abdominal viscera, 604f
  vitamin B, 715
Liver fibrosis, 154
Lobar bronchi, 743
Lobes. *See also* Frontal lobe; Occipital lobe; Parietal lobe; Quadrate lobe; Temporal lobe
  of liver, **35, 48,** 673, 674f
  of lung, **34**
Lobules, of lung, 747
Local potential changes, 368
Long bones, 193, 194–95
Long head biceps brachii muscle, **32**
Longissimus capitis muscle, 309f, 310t
Longissumus cervicis muscle, 310t
Longissumus thoracis muscle, 309f, 310t
Longitudinal fibers, 654
Longitudinal fissure, 398, 400f, 405f, 410f
Longitudinal ligaments, anterior, 220
Longitudinal section, 22, 23f
Long Life Family Study, 910
Long-QT syndrome, 82
Long-term synaptic potentiation, 404
Loose connective tissue, 155
Lordosis, 225
Lorenzo's Oil (film 1992), 87
Low-density lipoproteins (LDL), 97, 608, 684, 703, 940
Lower esophageal sphincter, 665
Lower limbs, 205t, 206, 234–38, 598–99, 604–605, 606f
Lower motor neuron syndrome, 397
Lower respiratory tract, 737
LSD, 377
Lucid, Shannon, 585
Lumbar arteries, 593, 594f, 598, 601f
Lumbar curvature, 219
Lumbar enlargement, 387, 389f
Lumbar nerves, 418f, 419
Lumbar puncture, 388
Lumbar region, 22, 23, 24f, 25f
Lumbar trunk, 617, 619f
Lumbar veins, 604
Lumbar vertebrae, **37,** 219f, 222, 222f, 223t
Lumbar vertebral body, **40–41**
Lumbosacral plexus, 421, 423
Lumen, 653, 665f, 855
Lumpectomy, 865
Lunate, 230f
Lung(s), **33–35, 43–44, 46,** 747–52
Lung cancer, 637t, 739, 818
Luteinizing hormone (LH), 487t, 491, 492, 497, 499t, 845, 846, 858, 859t
Lyme disease, 274, 278, 279t
Lymph, 617, 619–20, 621
Lymphadenitis, 622
Lymphangitis, 622
Lymphatic capillaries, 617, 618f, 620
Lymphatic duct, 618, 619f
Lymphatic obstructions, 817
Lymphatic pathways, 617–19
Lymphatic system. *See also* Immune system and immunity
  cardiovascular system, 610, 645
  digestive system, 645, 692
  endocrine system, 516, 645
  integumentary system, 188, 645
  life-span changes, 644, 908t
  lymphatic pathways, 617–19
  lymph movement, 621
  lymph nodes, 617, 620f, 621–23, 626t
  major functions of, 8t, 16–17
  muscular system, 333, 645
  nervous system, 378, 645
  reproductive system, 645, 868
  respiratory system, 645, 768
  skeletal system, 239, 645
  thymus and spleen, 623–25, 626t
  tissue fluid and lymph, 619–20
  urinary system, 645, 804
Lymphatic trunks, 617–19, 620f
Lymphatic vessels, 617, 619f, 620f
Lymph nodes, 617, 620f, 621–23, 626t
Lymph nodules, 621
Lymphocytes, 532, 534, 535t, 625, 628–29
Lymph sinuses, 622
Lysine, 135t, 705
Lysis, 636t
Lysosomes, 75, 84, 86, 87, 91t, 97
Lysozyme, 464

M

Macromolecule, 4, 5f
Macronutrients, 699, 726
Macrophages, 143, 155, 156t, 522, 532, 625, 631f
Macula, 459, 461f
Macula densa, 782
Macula lutea, 437, 469, 473
Macular degeneration, 477
"Mad cow disease," 66
Magnesium
  atomic structure, 53t
  blood tests, 941
  chemical constituents of cells, 62t
  elemental composition of human body, 52t

nutrition, 720
urine, 788t
Magnetic resonance imagery (MRI), 6–7, 561
Mainstream smoke, 736
Major calyx, 774, 776, 778f
Major histocompatibility complex (MHC), 630
Major minerals, 717–20, 721t
Malabsorption, 686
Malaria, 534t, 922
Male. *See also* Gender
body odor, 483
differences in female skeleton, 234t
pelvis, 233f, 234, 322f
reproductive system, 18, 833–46
urinary bladder and urethra, 801f
Male climacteric, 846
Malic acid, 946f
Malignant hyperthermia, 811
Malignant tumor, 104
Malleus, 437, 451–52
Malnutrition, 698, 723, 725, 897
Maltase, 118, 682–83
Mammary glands, **31,** 182t, 620f, 830, 861–62, 902, 903t
Mammary region, 23, 25f
Mammatropes, 494
Mammillary bodies, 407
Mammogram, 864, 865
Mandible, **39,** 206, 208, 209f, 210f, 214, 215f, 216, 218f, **252, 255, 341,** 663f
Mandibular condyle, 210f, 214, **252**
Mandibular division, 415
Mandibular foramen, 214, 217t, **252**
Mandibular fossa, 210, 211f, **248–50, 253**
Mandibular ramus, **252**
Manganese, 52t, 721, 723t
Mania, 373t
Manometer, 388
Manubrium, 224
Maple syrup urine disease, 803
Marasmus, 726, 727f
Maraviroc, 76
Marburg virus, 523
Marfan syndrome, 157t, 581, 608, 924
Marginal artery, 561
Marijuana, 444
Marshall, Barry, 668
Massachusetts Superior Court, 876
Masseter, 305f, 307–308, **341, 346**
Mass movements, 689
Mast cells, 155, 156t, 635
Mastectomy, 865
Mastication, 307–308, 656
Mastitis, 830
Mastoid foramen, 211f. *See also* Stylomastoid foramen
Mastoid fontanel, 218f
Mastoiditis, 210
Mastoid process, 210, 215f, 234t, **246, 250, 253, 341**
Maternal serum markers, 933t
Matter, structure of, 51–60
Mature follicle, 849
Maturity-onset diabetes of the young (MODY), 510
Maxilla, **39,** 209f, 210f, 211f, 213f, 215f, 218f, **245, 248–49, 341**
Maxillary artery, 595f, 597
Maxillary bones, 212–13, 216t, 738
Maxillary division, 415
Maxillary sinuses, 212, 214f, 215f, **255**
Mayo Clinic Healthy Weight Pyramid, 724f
McCartney, Macie Hope, 895
McCune-Albright syndrome, 491
M cells, 98
Mean arterial pressure, 583
Mean corpuscular volume, 941
Meatus, definition, 208t. *See also* Auditory meatus; External acoustic meatus; Internal acoustic meatus
Mechanical barriers
birth control, 863
infection, 626, 628t
Mechanical digestion, 652
Mechanoreceptors, 439
Mechnikov, Ilya, 910
Medial, as term of relative position, 21
Medial border, 226
Medial condyle, 235, 236f, 277f
Medial cuneiform, 237f
Medial epicondyle, 229f, 236f, **273, 344**
Medial malleolus, 236f, **345**
Medial meniscus, 277f
Medial pterygoid, 211f, 307, 308
Medial rectus muscle, **42,** 464, 465
Median fissure, anterior, 389, 390f
Median nerves, 420
Median palatine suture, 211f, **249, 251**
Median plane, 22
Median sacral crest, 222, 223f
Mediastinal arteries, 592
Mediastinum, 12, 556f
Mediastinum testis, 835
Medical science. *See also* Disease; Drug(s); Space medicine; Surgery
human history, 3
laboratory tests of clinical importance, 940–43
treatment of dying, 907
Mediterranean Diet Pyramid, 724f
Medulla, of ovary, 847
Medulla oblongata, 398, 399f, 408–409, 410f, 412t, 755f
Medullary cavity, 194
Medullary respiratory center, 756, 757f
Megadosing, with vitamins, 710
Megakaryocytes, 526, 534
Megaloblastic anemia, 716
Meiosis, 100–101, 831–33
Melanin, 170, 174, 468, 926
Melanocytes, 174, 176f
Melanoma, 175
MELAS (mitochondrial encephalomyopathy, lactic acidosis, and strokelike episodes), 87
Melatonin, 512
Membrane
cell, 76, 78–81, 82, 91t
interosseous, 261
mitochondrial, 946, 947f
mucous, 163, 653
parietal, 1
pericardial, 1, 12
peritoneal, 14
placental, 886, 890f
pleural, 1, 12, 751
respiratory, 760–62
serous, 15f, 163
synovial, 163, 264, **273,** 275f, 277f
tectorial, 455, 456f
thoracic, 12–14
tissues, 163
Membrane potential, 366
Membranous labyrinth, 453, 454f, 455f
Membranous urethra, 801f, 802
Memory, and brain, 404
Memory cells, 630, 632, 637
Menarche, 857
Meningeal artery, 596f
Meningeal branch, 420
Meninges, 382, 384–85, 399f
Meningitis, 385
Meniscus, 265, 276, 277f, 278, 280f
Menopause, 240, 860–61
Menses, 830
Menstruation, 830, 857, 859
Mental foramen, 209f, 210f, 217t, **252**
Mental region, 23, 25f
Mental retardation, 929. *See also* Down syndrome
Merocine glands, 150, 152f, 153t
Mesangial cells, 781
Mesencephalon, 398, 400t, 407
Mesentery, **35, 49**
Mesenteric arteries, 599f, 601f
inferior, **36,** 593, 594f
superior, **36, 49,** 592, 594f
Mesenteric veins, 603, 604
superior, **36, 49**
Mesentery, 680, 681f
Mesoderm, 875, 885, 887f
Mesodermal cells, 885
Messenger RNA (mRNA), 130–31
Metabolic acidosis, 824, 825
Metabolic alkalosis, 824, 826
Metabolic pathways, 118
Metabolism. *See also* Cellular metabolism
characteristics of life, 6
definition, 1, 699
inborn errors of, 137–38
intermediary, 115
water of, 813
Metabolome, 139
Metacarpal bones, 206, 228f, 229, 230f, 231, **344**
Metacarpus, 229
Metaphase, 101, 102f, 103t, 832, 833
Metarterioles, 576, 577f
Metatarsal bones, 206, 235f, 236–38
Metatarsal joint, 270t
Metatarsophalangeal joint, 270t
Metatarsus, 236
Metencephalon, 400t
Methionine, 705, 719
Methotrexate, 904t
Metopic suture, 216
Microbiome, 652
Microchimerism, 643
Microfilaments, 88–89, 90f, 91f

Microflora, 652, 689
Microglia, 362f, 363
Micrographia, 406
Microgravity, 201, 585
Micrometers, and size of cells, 76
Micronutrients, 699
Microorganisms, 652. *See also* Bacteria; Viruses
MicroRNAs, 133, 136
Microscopes, 79, 836
Microtubules, 88, 89, 90f, 91t
Microvilli, 146, 148f, 681, 682f, 791
Micturition, 774, 802–803
Micturition reflex center, 802
Midbrain, 398, 399f, 400t, 407, 412t
Middle ear, 12, 205t, 451–53
Middle peduncles, 410
Middle phalanx, 230f, 237f, **344–45**
Middle tunic, 466–68
Midpiece, of sperm cell, 837
Migraine, 356
Milk, production and secretion of, 902–903, 904
Milstein, Cesar, 636
Mineral(s), and nutrition, 717–22, 726
Mineralocorticoid, 506
Minor calyx, 776, 778f
Minoxidil, 179
Minute ventilation, 754
Mitchell, S. Weir, 442
Mitochondria, 83–84, 86f, 87, 91t
Mitochondrial DNA, 84
Mitochondrial membrane, 946, 947f
Mitosis, 75, 100–101, 102f, 109, 910
Mitoxantrone, 360
Mitral valve, 557, 558f, 559f, 560t
Mitral valve prolapse (MVP), 557
Mixed-lineage leukemia, 537
Mixed nerves, 412, 413f
Mixing movements, in alimentary canal, 654, 656f
M lines, 287, 288f
Modiolus, 454f
Molding, infant skull, 216
Molars, 659
Molecular formula, 53
Molecule(s), 4, 5f, 51t, 53, 54, 68
Monoamine(s), 373
Monoamine oxidase, 374, 430–31
Monoamine oxidase inhibitors, 374t
Monoclonal antibodies, 636–37
Monocytes, 532, 534t, 535t, 627
Mononuclear phagocytic system, 627
Mononucleosis, 534t
Monosaccharides, 50, 62, 63f, 684f, 686t, 700, 701
Monosomy, 916, 929
Monounsaturated fatty acids, 63
Mons pubis, **31,** 830, 855
"Morning-after pill," 865
"Morning sickness," 885
Morphine, 375, 377
Morula, 875, 879, 886t
Motion sickness, 461
Motor areas, of cerebral cortex, 402
Motor end plate, 290
Motor nerves, 412
Motor neurons, 289, 353, 361, 362t, 391f, 393t
Motor speech area, 401f, 402
Motor unit, 290, 297–98
Mountain climbing, 763
Mouth, 653f, 656–60
Movement
  bone function, 202
  characteristics of life, 8t
  organ systems, 14
Mucin, 152
Mucosa, 653, 655t, 665f
Mucosal layer, of vagina, 855
Mucous cells, 660, 666, 667f, 668
Mucous coat, 798, 800
Mucous membranes, 163, 653
Mucous-secreting glands, 682
Muscular coat, 798
Muscular layer, of vagina, 855
Mucus, 152, 660, 668t
Multifactorial genetic traits, 924–26
Multiple endocrine neoplasia (MEN), 483
Multiple motor unit summation, 297–98
Multiple sclerosis (MS), 360, 637, 642t
Multipolar neurons, 353, 360, 362t
Multipotent adult progenitor cell, 526
Murmur, heart, 565
Muscarine, 429
Muscarinic receptors, 429
Muscle(s). *See* Muscle fiber; Muscle tissue; Muscular system; *specific muscles*
Muscle cells, 77f, 121
Muscle fatigue, 296
Muscle fiber, 177, 387, 412, 414
Muscle impulse, 290
Muscle relaxants, 82t
Muscle spindles, 444, 446t
Muscle strain, 289
Muscle tissue, 146t, 163–64
Muscle tone, 298
Muscular coat, 800
Muscular dystrophy, 289. *See also* Duchenne muscular dystrophy
Muscular layer, of alimentary canal, 654, 655t
Muscular responses, 296–99
Muscular system. *See also* Muscle fiber; Muscle tissue; Skeletal muscles
  cardiac muscle, 301
  cardiovascular system, 333, 610
  digestive system, 333, 692
  endocrine system, 333, 516
  eye, 465t
  integumentary system, 188, 333
  life-span changes, 334, 908t
  lymphatic system, 333, 645
  major functions of, 8t, 14, 16f
  muscular responses, 296–99
  nervous system, 333, 378
  reproductive system, 333, 868
  respiratory system, 333, 768
  skeletal system, 239, 333
  smooth muscles, 300–301
  urinary system, 333, 804
Musculocutaneous nerve, **34,** 420
Mutagen, 136
Mutant phenotype, 920
Mutations, genetic, 114, 135–37, 920
Myasthenia gravis (MG), 291, 293, 642t
Myelencephalon, 398, 400t
Myelin, 356, 357f, 358f, 359
Myelinated axons, 358, 359f, 364f, 371
Myelin sheath, 356
Myenteric plexus, 656
Myoblasts, 106
Myocardial infarction, 563
Myocardium, 552, 555, 556f, 557t
Myoepithelial cells, 902, 903f
Myofibrils, 89, 284, 287
Myoglobin, 294, 720
Myogram, 284
Myometrium, 853f, 854, 855
Myopia, 472
Myosin, 287, 288f, 289
Myosin chains, 608
Myositis ossificans, 328
Myostatin inhibitors, 489t

N

Nacre, 204
NAD$^+$, 944
Nails, of fingers, 177, 178f
Nanotechnology, 145
Narcolepsy, 409t
Nasal bones, 209f, 210f, 213f, 214, 215f, 216t, 218f, **245–47, 341**
Nasal cavity, 12, **38,** 213f, 737, 747t
Nasal conchae, 737, 738f
  inferior, **39,** 209f, 213f, 214, 216t, **247**
  middle, 212, 213f, 215f, **247, 253**
  superior, 212, 213f
Nasal region, 1, 23, 25f
Nasal septum, **42,** 737
Nash, Molly, 883
Nasolacrimal duct, 464
Nasopharynx, 663, 664f, 738f
Natalizumab, 360
National Human Neural Stem Cell Resource, 354
National Institute on Aging, 334, 910
National Institutes of Health, 261, 843
Natural killer (NK) cells, 627, 628t
Naturally acquired active immunity, 638, 639t
Naturally acquired passive immunity, 638, 639t
Nausea, 671
Navicular, 237f
Neanderthals, and genetic analysis, 179
Nearsightedness, 472
Neck
  anatomical, 226, 229f
  arteries, 595f, 596–97
  of bladder, 799–800
  of femur, 236f
  of rib, 223
  surface anatomy, **341**
  surgical, 226, 229f
  of tooth, 659
  veins, 602
Necrosis, 173
Negative feedback, 9, 492, 527
Negative nitrogen balance, 706
Neon, 53t
Neonatal period, of development, 875, 904–905, 908t

Neonatology, 24
Nephritis, 782
Nephrology, 24
Nephron, 774, 778f, 779, 781–85, 786f, 797t, 823
Nephron loop, 781, 783f, 786f, 795, 797t
Nephrotic syndrome, 793, 802t
Nerve(s), 354. *See also* Nerve fibers; Nerve impulses; Nervous system; Neuron(s)
Nerve cell, 77f. *See also* Neuron(s)
Nerve fibers, 387, 412, 424, 438
Nerve gas, 293
Nerve impulses, 16, 354, 369–70, 372
Nerve tracts, 389
Nervous system. *See also* Autonomic nervous system; Brain; Central nervous system; Peripheral nervous system
  cardiovascular system, 378, 610
  cell membrane potential, 365–71
  classification of cells, 359–64
  definition, 354
  description of cells, 356–59
  digestive system, 378, 692
  endocrine system, 378, 484t, 516
  general functions of, 355–56
  general senses, 440–46
  hormone secretion, 492
  impulse processing, 374–76
  integumentary system, 188, 378
  life-span changes, 431, 908t
  lymphatic system, 378, 645
  major functions of, 8t, 16, 17f
  meninges, 384–85
  muscular system, 333, 378
  receptors, sensation, and perception, 438–40
  reproductive system, 378, 868
  respiratory system, 378, 768
  skeletal system, 239, 378
  special senses, 446–76
  spinal cord, 387–96
  subdivisions of, 412t
  synapse, 365
  synaptic transmission, 371–74
  urinary system, 378, 804
  ventricles and cerebrospinal fluid, 385–86
Nervous tissues, 146t, 164–65
Nervous tunic, 465
Net filtration pressure, 788
Neural stem cells, 365
Neural tube, 398
Neural tube defect (NTD), 398, 716, 725
Neurilemma, 353, 356, 358f
Neurilemmal sheath, 356
Neurofibrils, 356, 357f, 358f
Neuroglia, 143, 164–65, 354, 355f, 361–64
Neurology, 24
Neuroma, 364
Neuromodulators, 374
Neuromuscular junction, 289–90
Neuron(s). *See also* Afferent neurons; Bipolar neurons; Efferent neurons; Interneurons; Motor neurons; Multipolar neurons; Receptors; Sensory neurons; Unipolar neurons
  blood glucose levels, 511
  classification of, 359–61, 362t
  definition, 143, 354
  pH of body fluids, 822
  structure and functions, 164, 165f, 355f, 357f
  synapse, 365
Neuronal pools, 374–75
Neuropathic pain, 442
Neuropeptides, 373t, 374, 700
Neurospheres, 354
Neutralization, 636t
Neutral solution, 810
Neurotransmitters, 289, 354, 372–74, 428–31
Neutron, 51t, 52
Neutrophils, 531, 532f, 534t, 535t, 627
New England Centenarian Study, 910
Niacin, 714, 717t
Niacinamide, 714
Nicotinamide adenine dinucleotide. *See* NAD^+
Nicotine, 374t, 377, 429, 904t. *See also* Cigarette smoke and smoking
Nicotinic receptors, 429, 714, 717t
Niemann-Pick type C disease, 608
Nigeria, and polio vaccine, 290
Nightblindness, 475, 710
Nipple, **31,** 862
Nitric oxide, 61, 373t, 587
Nitrogen, 52t, 53t
Nitrogen balance, 705–706
Nocturnal emissions, 845
Nodes of Ranvier, 356, 357f, 358f, 371
Nodule, 616
Nondisjunction, 929
Nonelectrolytes, 61
Nonprotein nitrogenous substances (NPNs), 537–38
Nonrespiratory air movements, 754–55
Nonsteroidal anti-inflammatory drugs (NSAIDs), 278, 445, 691
Nonsteroid hormones, 485, 486–88, 490–91
Norepinephrine, 301, 373t, 377, 426, 429, 430, 486f, 487t, 505, 506t
Normal range, of homeostasis, 12
Norovirus, 638
Nose, 737, 747t
Nuclear envelope, 89, 91t
Nuclear pores, 89
Nuclear waste, 56
Nucleases, 672, 683t
Nucleic acids, 68, 69t, 124–31, 820. *See also* DNA; RNA
Nucleolus, 90, 91t
Nucleotides, 68
Nucleus (nuclei)
  of atom, 50, 52
  basal, 398, 404, 405f, 412t
  brainstem, 407
  of cell, 76, 89–90, 91f
Nucleus cuneatus, 408
Nucleus gracilis, 408
Nucleus pulposus, 263
Nutrients. *See also* Diet; Food; Nutrition
  birth defects, 897
  definition, 698, 699
  intestinal absorption of, 686t
Nutrition. *See also* Diet; Food; Nutrients
  carbohydrates, 700–702
  definition, 699
  energy expenditures, 706–709
  healthy eating, 722–28
  height, 925f
  life-span changes, 728
  lipids, 702–704
  minerals, 717–22
  proteins, 704–706
  vitamins, 709–16
Nutritional edema, 706
Nystagmus, 473

O

Obesity
  definition, 698
  diet and weight, 708–709, 710–11
  hypertension, 588
Oblique fracture, 202f
Oblique muscles
  external, **31–33,** 305, 306f, 312f, 320, 321f, 322t, **347–48**
  inferior, 464, 465t
  internal, **32–33,** 312f, 320, 321f, 322t
  superior, 415, 464, 465t
Oblique popliteal ligament, 276, 277f
Oblique section, 22, 23f
O'Brien, Chloe, 883
Obsessive–compulsive disorder, 70–71
Obstetrics, 24. *See also* Pregnancy
Obstructive jaundice, 676
Ostructive sleep apnea syndrome, 409t
Obturator artery, 599f
Obturator foramen, **37,** 232f, 233, 234t
Obturator nerves, 423
OCA2 gene, 926
Occipital artery, 595f, 597
Occipital bone, 208, 210t, 211f, 213f, 215f, 218f, **246, 249–51, 253, 255–56, 341**
Occipital condyle, 208, 211f, **249–51, 253, 255**
Occipitalis, 306, **346**
Occipital lobe, **42,** 400, 401f, 402, 403t
Occipital region, 23, 25f
Occupational hazards, and birth defects, 897
Octet rule, 55
Oculomotor nerves, 414, 417t
Odontoid process, 192, 221
"Oldest old," 911
Olecranon fossa, 226
Olecranon process, 228f, 229, **273, 342–43**
Olfactory, definition, 437
Olfactory bulbs, 414, 447
Olfactory code, 448
Olfactory foramina, 212, 213f
Olfactory nerves, 414, 417t, 447–48
Olfactory organs, 446
Olfactory receptors, 414, 446
Olfactory stimulation, 448
Olfactory tracts, 447–48
Oligodendrocytes, 353, 358, 361–63
Oligomeganephronia, 802t

Olympic Committee, 489
Omega-3 fatty acids, 704
Oncogenes, 104
Oncology, 24
Oocytes, 831. *See also* Primary oocyte; Secondary oocyte
Oogenesis, 830, 849, 850f
Ophthalmic artery, 597
Ophthalmic division, 415
Ophthalmic vein, 602f
Ophthalmology, 24
Opiates, 375, 377
Opsin, 473
Opsonization, 636t
Optic canal, 209f, 211f, 213f, 217t, **257**
Optic chiasma, 382, 405, 475
Optic disc, 469
Optic foramina, 414
Optic nerve, **42,** 414, 417t, 466
Optic radiations, 476
Optic tracts, 405, 476
Oral cavity, 12, **39,** 656
Oral contraceptives, 859, 864–65, 904t
Oral glucose tolerance test, 512
Oral region, 23, 25f
Orbicularis oculi, 305f, 306, 307f, **346,** 437, 462, 465t
Orbicularis oris, 305f, 307f, **346**
Orbit, of skull, 234t, **253**
Orbital cavities, 12
Orbital fissure
  inferior, 209f, 217t, **247**
  superior, 209f, 211f, 213f, 217t, **247, 254, 257**
Orbital region, 1, 23, 25f
Orbital surface, **253**
Orchiectomy, 835
Oregon, and policies on dying and death, 907
Organ(s). *See also* Organ systems
  definition, 171
  levels of organization, 4, 5f
Organelle, 4, 5f
Organic chemicals, 60, 61–69
Organism, 4, 5f, 9t. *See also* Microorganisms
Organ systems. *See also* Cardiovascular system; Digestive system; Endocrine system; Integumentary system; Lymphatic system; Muscular system; Nervous system; Reproductive system; Respiratory system; Skeletal system; Urinary system
  description of, 14–19
  levels of organization, 4, 5f
  major functions of, 8t
Orgasm, 844, 856
Origin, of skeletal muscles, 303–304
Oropharynx, 663, 738f
Orthopedics, 25
Orthostatic intolerance, 585
Orthotic device, 308
Osmolality, 941, 943
Osmolarity, 816–17
Osmoles, 817
Osmoreceptor(s), 498, 813–14, 815
Osmoreceptor–ADH mechanism, 815
Osmosis, 93–94, 100t
Osmotic diuresis, 792
Osmotic pressure, 94, 535
Osseus labyrinth, 453, 454f
Osseus tissue, 143
Ossification, and bone development, 200t
Osteoarthritis, 278, 279t
Osteoblasts, 106, 160, 192, 197
Osteocalcin, 201
Osteoclasts, 192, 198, 203f
Osteocytes, 143, 161, 195, 197
Osteogenesis imperfecta, 157t, 197f, 924
Osteomalacia, 200, 712
Osteon, 161, 195, 196f, 240
Osteonectin, 201
Osteopenia, 204
Osteopontin, 201
Osteoporosis, 157, 203, 204, 725
Otic region, 23, 25f
Otitis media, 210, 453
Otolaryngology, 25
Otoliths, 459
Otosclerosis, 459
Outer ear, 450–51
Outer fibrous layer, 849
Outer tunic, 465–66
Oval window, 452
Ovarian follicles, 847, 851f
Ovarian ligament, 847, 849f, 853f
Ovary, **35–36,** 512, 847–49, 853f, 857t
Overnutrition, 723
Overweight, and obesity, 708
Ovulation, 851–52, 853f, 863
Oxaloacetic acid, 944, 946f
Oxidation, 120, 819
Oxygen
  atomic structure, 53t
  chemical constituents of cells, 61, 62t
  control of breathing, 757–58, 760
  diffusion into cell, 92
  electron transport chain, 946–47
  elemental composition of human body, 52t
  gas transport, 762–63, 764f, 767t
  muscular contraction, 294–95
  requirements for life, 7, 9t
Oxygen-carrying capacity, 527
Oxygen debt, 121, 295
Oxygen free radicals, 710
Oxygen saturation, 941
Oxyhemoglobin, 526, 763, 764f
Oxytocin, 301, 482, 486f, 487t, 493, 498, 499, 899–900, 901

P

Pacemaker, 566
Packed cell volume (PCV), 523
Paganini, Niccolò, 263
Pain
  ion channels, 82
  receptors, 439, 440
  sense of, 440–44
  treatment of, 445
Pain nerve pathways, 442–43
Palate, 234t, 657
Palatine bone, 209f, 211f, 213–14, 215f, **249, 251**
Palatine processes, 212
Palatine tonsils, 657, 738f
Paleontology, 193
Pallidotomy, 406
Palmaris longus, 317f, 319, **344**
Palmar region, 23, 25f
Palpebrae superioris, 437
Pancreas, **36, 44–45,** 487t, 492, 509–11, 604f, 653f, 671–72, 777f
Pancreatic acinar cells, 671
Pancreatic amylase, 671
Pancreatic duct, 671, 672f
Pancreatic glands, 430t
Pancreatic islets, 509–11
Pancreatic juice, 671–72
Pancreatic lipase, 671, 683t
Pancreatic secretion, 672
Pantothenic acid, 714–15, 717t
Papillae, 448, 657
Papillary muscle, 552, 557, 559f
Para-aminohippuric acid (PAH), 798
Paracrine secretions, 483
Paranasal sinuses, 214f
Parasomnias, 409t
Parasympathetic division, of autonomic nervous system, 424, 426, 428f
Parasympathetic impulses, 686, 856
Parathyroid gland, 482, 487t, 502–504, 515
Parathyroid hormone (PTH), 205f, 486f, 487t, 491, 502–503, 818, 819f, 884, 885t
Parathyroid hyperplasia, 483
Parietal bones, 208, 209f, 210t, 213f, 215f, 216t, 218f, **245–46, 255–56, 341**
Parietal cells, 666, 667f
Parietal lobe, 399, 400f, 401f, 402, 403t
Parietal membrane, 1
Parietal pericardium, 12, 554
Parietal peritoneum, 14, 777f, 799f
Parietal pleura, 12, 747
Parkinson disease, 319, 373t, 406
Parotid gland, **346,** 661, 662f, 663t
"Paroxysmal extreme pain disorder," 82
Partially complete protein, 705
Partial pressure, 757
Partial thromboplastin time (PTT), 541
Parturition, 899–902
Passive aging, 909
Passive processes, for movements in and out of cell, 100t
Patella, 206, 235, 238t, 277f, 324f, 329f, **344, 351–52**
Patellar dislocation, 235
Patellar ligament, 276, 277f, 324f, 328, 329f, **344, 352**
Patellar region, 23, 25f
Patellar reflex, 390, 392f
Patellar surface, 236f
Patellar tendon, 328
Patent ductus arteriosus (PDA), 905
Pathogen, 616
Pathologic fracture, 202
Pathology, 25
Pattern baldness, 179
PCP, 377
Peanut allergy, 617
Pectin, 701
Pectineus, 324f, 327t

Pectoral girdle, 205t, 206, 225–26, 227f, 310–12
Pectoralis major muscle, **31–32, 43,** 305, 312, 314t, **343, 347,** 862
Pectoralis minor muscle, **32,** 312, 313t
Pectoral nerves, 421
Pectoral region, 24, 25f
Pedal region, 24, 25f
Pediatrics, 25
Pedicels, 781
Pedicles, 220
Pedigrees (genetic), 920–21
Peeling, of sunburn, 184
Pellagra, 698, 714
Pelvic brim, 233
Pelvic cavity, 1, 12, 622
Pelvic diaphragm, 320
Pelvic girdle, 205t, 206, 231–34, 238t
Pelvic inflammatory disease, 869
Pelvic outlet, muscles of, 320, 322, 323t
Pelvic region, 24, 25f
Pelvic sacral foramen, **37**
Pelvis, 206, 231, 233–34, 598–99, 604–605, 606f
Penetrance, and gene expression, 924
Penetrating keratoplasty, 465
Penile urethra, 801f, 802
Penis, **33–34,** 801f, 834f, 843, 845t
Pepsin, 666, 668, 683t
Pepsinogen, 666, 668
Peptidases, 682–83
Peptide(s), 373, 487t
Peptide bond, 116, 117f
Peptide hormones, 485
Perception, 438–40
Percutaneous transluminal coronary angioplasty (PTCA), 609
Perforating canals, 195
Perforins, 627, 630
Pericardial arteries, 592
Pericardial cavity, 12, **43,** 554
Pericardial membranes, 1, 12
Pericardial sac, **33**
Pericardium, 1, **44,** 554, 556f
Perichondrium, 158
Peridontal ligament, 262, 660
Perilymph, 453, 454f
Perimetrium, 853f, 854, 855
Perimysium, 286
Perineal artery, 599f
Perineal region, 24, 25f
Perineurium, 412, 413f
Periodic Table, 939
Periosteum, 194
Peripheral, as term of relative position, 21
Peripheral chemoreceptors, 758
Peripheral nerve, 364, 412, 414
Peripheral nervous system
  definition, 353, 354, 411
  diagrammatic representation of, 355f
  nerve classification, 412, 414
  structure of nerves, 412
Peripheral neuropathy, 441
Peripheral protein, 80
Peripheral resistance (PR), 583–84, 585
Peristalsis, 300, 651, 654, 656f
Peristaltic rush, 686
Peristaltic waves, 664f, 670f, 686, 689, 798–99
Peritoneal cavity, 14
Peritoneal membranes, 14
Peritubular capillary, 780f, 783f, 784–85, 786f, 791
Permeability, of capillary, 576
Pernicious anemia, 529, 642t, 715
Peroxisomes, 86, 87, 91t
Perpendicular plate, 212, 215f, **253**
Persistent vegetative state, 383, 409
Peyer's patches, 621–22
P53 protein, 175
pH and pH scale, 59–60, 823, 824, 941, 943. *See also* Acid(s); Base(s)
Phagocytes, 96, 143
Phagocytosis, 75, 96–97, 100t, 627, 628t
Phalanges, of fingers and toes, 206, 228f, 230f, 231, 235f, 237f, 238. *See also* Distal phalanx; Middle phalanx; Proximal phalanx
Pharmacology, 25
Pharyngeal tonsils, 657, 738f
Pharynx, **43,** 653f, 661, 663–64, 740, 747t
Phencyclidine, 904t
Phenobarbitol, 904t
Phenotype, 916, 920, 928
Phenylalanine, 66f, 135t, 705t
Phenylpyruvic acid, 943
Pheochromocytoma, 483
Pheomelanin, 178, 179
Pheromones, 483
Phlebitis, 552, 581
Phosphatase, serum, 941
Phosphate, 62t, 788t, 818, 819f
Phosphate buffer system, 821, 822t
Phosphoenolpyruvic acid, 944, 945f
Phosphoglyceric acids, 944, 945f
Phospholipids, 64, 65f. *See also* Lipids
Phosphoproteins, 820
Phosphoric acid, 820
Phosphorus, 52t, 204, 717–18, 721t, 941
Phosphorylation, 119, 944
Photoreceptors, 437, 439, 475
Phrenic arteries, 592, 594f
Phrenic nerve, **47,** 420, 735, 750
Phrenic veins, 604
Physical stress, 201, 514
Physiological steady state, 92
Physiologic dead space, 753
Physiology, 1, 4, 54, 933–34
Pia mater, 385, 387f
Pica, 722
Pilocarpine, 431
Pineal cells, 511
Pineal gland, 407, 511–13
Pinocytosis, 75, 96, 100t
Piriformis, 323, 325f, 327t
Pisiform, 230f
Pitch, of vocal sound, 741
Pituitary dwarfism, 201. *See also* Dwarfism
Pituitary gigantism, 201. *See also* Gigantism
Pituitary gland, 487t, 492–99, 596f, 845, 857, 858, 859t
  posterior, 407, 487t, 489
Pivot joint, 266–67
Placenta, 512, 882, 884, 890f, 898f
Placental estrogens, 884, 885t
Placental lactogen, 884, 885t
Placental membrane, 886, 890f
Placental progesterone, 884, 885t
Plane joints, 266, 267t
Plant(s), secondary metabolites of as drugs, 115
Plantar arteries, 599
  lateral, 600f
  medial, 600f
Plantar flexion, 269
Plantar flexors, 330
Plantaris, 277f, 331f, 332
Plantar reflex, 394
Plantar region, 24, 25f
Plantar veins
  lateral, 606f
  medial, 606f
Plasma, 162, 523, 524f, 535–38, 789
Plasma cells, 633
Plasma colloid osmotic pressure, 619
Plasma electrolytes, 538
Plasma lipids, 537
Plasma nutrients, 536–37
Plasma proteins, 535–36
Plasma thromboplastin antecedent and component, 541t
Plasmin, 541
Plasminogen, 541
Platelet(s), 162, 535t
Platelet count, 534, 941
Platelet-derived growth factor (PDGF), 541, 781
Platelet plug, 538
Platysma, 306t, 307
Pleiotropy, 924
Pleural cavity, **36,** 747, 748f
Pleural membranes, 1, 12, 751
Plexus, 420
Plicae circulares, 682
Pluripotent cells, 106
Pneumocystis jirovecii, 642, 765
Pneumonia, 765, 911t
Pneumotaxic center, 757
Pneumothorax, 751
Podiatry, 25
Podocytes, 781
Polar molecule, 57, 68
Poliomyelitis, 290, 756
Polyclonal response, 633
Polycystic kidney disease, 802t
Polycythemia, 522
Polydactyly, 231, 924
Polygenic traits, 924–26
Polymerase chain reaction (PCR), 127, 132, 274
Polymorphonuclear leukocytes (PMNs), 531. *See also* Neutrophils
Polypeptide, 116
Polyploidy, 929
Polysaccharides, 62, 63f, 700
Polyunsaturated fatty acids, 50, 63
Pons, 398, 399f, 407, 410f, 412t, 755f
Pontine respiratory group, 757
Popliteal artery, 599, 600–601f
Popliteal region, 24, 25f
Popliteal veins, 604, 606f, 607f

Popliteus muscle, 277f
Pores, of skin, 170, 180
Porphyria cutanea tarda, 138f
Porphyria variegata, 138f, 530, 548t, 924
Porta hepatis, 673
Portal system, of pituitary gland, 493
Positive chemotaxis, 533
Positive energy balance, 708
Positive feedback mechanism, 11–12, 539, 900, 901f
Positive nitrogen balance, 706
Positron emission tomography (PET) imaging, 70–71
Posterior, as term of relative position, 21
Posterior canal, 460
Posterior cavity, 469
Posterior chamber, 467
Posterior fontanel, 218f
Posterior funiculus, 389, 390f
Posterior group, of thigh muscles, 323
Posterior horns, 389, 390f
Posterior longitudinal ligaments, 220
Posterior median sulcus, 389, 390f
Posterior sacral foramina, 222, 223f
Postganglionic fiber, 424
Postmortem sperm retrieval, 876
Postnatal period, of development, 875, 876, 904–909
Postpolio syndrome, 290
Postsynaptic neuron, 365
Posttraumatic stress disorder (PTSD), 405
Potassium. *See also* Sodium/potassium pump
  blood tests, 941
  chemical constituents of cells, 62t
  electrolyte balance, 818, 820
  elemental composition of human body, 52t
  heart action, 572, 573
  ion channels, 82, 365, 366, 370t
  nutrition, 718–19, 721t
  urine, 788t, 794t
PQ interval, 568
PR interval, 568
Prader-Willi syndrome, 497
Preactive dying, 908
Precapillary sphincters, 578
Precipitation, 636t
Preembryonic stage, of development, 895t
Preganglionic fiber, 424, 427f, 428f
Pregnancy. *See also* Fetus; Infertility
  definition, 876
  dietary iron, 721
  fertilization, 877–78
  fetal development, 892–94
  hormonal changes during, 883–84, 885t
  naturally acquired passive immunity, 638
  prenatal development, 879–903
  Rh blood group, 547
  transport of sex cells, 876–77
Pregnancy-associated plasma protein A (PAPP–A), 884
Pregnanediol, 861
Preimplantation genetic diagnosis (PGD), 883
Preload, and blood pressure, 586
Premature beat, 572
Premature infants, 893
Premenstrual dysphoric disorder (PMDD), 859
Premenstrual syndrome (PMS), 859
Prenatal period, of development, 875, 876, 879–903
Prenatal tests, for chromosome abnormalities, 931–33. *See also* Laboratory tests
Prepatellar bursa, 265, 276–77
Prepuce, 843
Presbycusis, 477
Presbyopia, 472
Prediabetes, 512
Premolars, 659
Prepuce, 834f
Pressure. *See also* Atmospheric pressure; Blood pressure; Hydrostatic pressure; Osmotic pressure
  characteristics of life, 8, 9t
  sense of, 440, 441f
Pressure ulcers, 173
Presynaptic neuron, 365
Prevention, of breast cancer, 865
Primary bronchi, 743
Primary follicle, 849, 853f
Primary germ layers, 885
Primary immune response, 637
Primary malnutrition, 723
Primary metabolites, 115
Primary oocyte, 847, 850f, 851f, 853f
Primary ossification center, 198
Primary sex organs, 833, 846
Primary spermatocyte, 836
Primary structure, of amino acid, 65–66, 67f
Primary teeth, 658, 659t
Prime mover, 304
Primordial follicle, 830, 847, 849, 851f, 852f, 853f
Prion, 66
Proaccelerin, 541t
Probenecid, 793
Probiotics, 652
Procaine, 372
Process, definition, 208t
Procollagen, 157
Products, of chemical reactions, 58
Progenitor cell, 105–106, 107f
Progesterone, 857, 859, 884
Projection, and sensation, 439
Prokaryotic cells, 76
Prolactin, 482, 487t, 495, 499t, 902
Prolactin release–inhibiting hormone (PRL), 487t, 492, 495
Prolactin-releasing factor (PRF), 487t, 495
Proline, 135t, 705t
Pronation, of joint, 268f, 269
Pronator quadratus, 316, 317f
Pronator teres, 316, 317f
Propelling movements, in alimentary canal, 654
Prophase, 75, 101, 102f, 103t, 832, 833
Proprioceptive senses, 440
Proprioceptors and proprioception, 414, 439, 444–46
Prosencephalon, 398, 400t
Prostacyclin (PGI), 542, 543t
Prostaglandins, 485, 486f, 491, 639, 840
Prostate cancer, 841
Prostate gland, **45,** 799f, 801f, 834f, 840, 841, 845t
Prostate specific antigen (PSA), 51, 841
Prostatic urethra, 801
Protease, 118
Protection, as bone function, 202
Protein(s). *See also* Amino acids
  anabolism, 116
  bone, 201
  catabolism, 126f
  cell membrane, 79, 80, 81t
  dietary deficiency, 706
  hormones, 487t
  nutrition, 704–706
  organic compounds, 69t
  structure and functions, 64–66, 67f
  synthesis, 131–34
  total serum, 941
  urine, 798
Protein buffer system, 821–22
Protein fibers, 153
Protein hormones, 485
Protein kinases, 488
Protein misfolding, 66
Proteome, 139
Proteomics, 144, 918, 933
Prothrombin, 539, 541t, 712
Prothrombin time (PT), 541
Proton, 51t, 52
Protraction, of joint, 269
Provitamins, 709
Proximal, as term of relative position, 21
Proximal convoluted tubule, 780f, 781, 783f, 786f, 797t
Proximal phalanx, 230f, 237f, **344–45**
Proximal tubule, 774
Pseudohermaphroditism, 846
Pseudostratified columnar epithelium, 143, 146, 148f, 153t
Psoas major, **37,** 323, 324f, 327t
Psoas minor, 324f
Psoriasis, 173, 637t
Psychiatry, 25. *See also* Depression; Mania; Obsessive–compulsive disorder; Posttraumatic stress disorder; Schizophrenia
Psychological stress, 514
Pteroylmonoglutamic acid, 716
Puberty, 830, 846, 862
Pubic arch, 232f, 233, 234t
Pubic crest, 232f
Pubic tubercle, 232f
Pubis, 232f, 233, 275f
Pubofemoral ligament, 275f
Pudendal artery, 598, 599f, 600f
Pudendal nerve, 423
Pudendal veins, 605
Pulled hamstrings, 326
Pulmonary arteries, **35, 40,** 557, 558f, 560f, 562f, 591f, 593f, 608f, 745f, 898f
Pulmonary circuit, 553, 590–92
Pulmonary edema, 592
Pulmonary embolism, 543, 581
Pulmonary trunk, **34–35,** 557, 558f, 562f, 898f

Pulmonary valve, 557, 558f, 559f, 560t
Pulmonary veins, **35,** 557, 558f, 560f, 562f, 591f, 592, 593f, 745f, 898f
Pulmonic sound, 565
Pulse, and blood pressure, 582, 584, 597. *See also* Heart rate
Punnett squares, 920, 921f
Pupil, of eye, 467
Purified protein derivative (PPD), 639
Purines, 126
Purkinje fibers, 567
Pus, 533
Putamen, 404
P wave, 568
Pyloric antrum, 666
Pyloric canal, 666
Pyloric sphincter, 651, 666, 667f, 670f, 672f
Pylorus, 666f, 667f
Pyramidal cells, 402
Pyramidal tracts, 395
Pyridostigmine bromide, 293
Pyridoxal, 715
Pyridoxamine, 715
Pyridoxine, 715
Pyrimidines, 126–27
Pyruvic acid, 944

Q

QRS complex, 568, 570f
Quadrate lobe, 673, 674f
Quadratus lumborum muscle, **37,** 308, 309f, 310
Quadriceps femoris muscle group, 277f, 324f, 328
Quaternary structure, of protein, 66, 67f
Quinlan, Karen Ann, 383
Q wave, 568

R

Radial artery, 597, 601f
Radial collateral ligament, 272, **273**
Radial muscles, 468
Radial nerves, 421
Radial set, 468
Radial tuberosity, 228, 230f
Radial veins, 602, 607f
Radiation, 182
Radioactive isotopes, 53, 54, 70, 904t
Radiocarpal joint, 270t
Radiology, 25
Radioulnar joints
  distal, 270t
  proximal, 270t
Radius, 206, 228, 230f, 231t, **273,** 303
Radon, 56
Ramus, 208t, 214
Ransome, Joseph, 438
Rapid eye movement (REM) sleep, 409
Rate-limiting enzyme, 118
Reactants, 58
Receptive relaxation, 654
Receptor(s). *See also* Adrenergic receptors; Alpha receptors; Beta receptors; CCRS receptors; Muscarinic receptors; Nicotinic receptors; Sensory receptors; Trace amine-associated receptors
  cell membrane, 80
  homeostasis, 9
  reflex arc, 393t
  sensory, 439
Receptor cells, 469
Receptor-mediated endocytosis, 97, 98, 100t
Receptor proteins, 81t
Recessive inheritance, 920–22, 923
Recommended Daily Allowance (RDA), 722
Recruitment, of motor units, 297–98
Rectal artery, 598, 599f
Rectal veins, 605
Rectouterine pouch, 848f
Rectum, **35, 37, 41, 45,** 651, 653f, 687, 688f, 799f, 848f
Rectus abdominis muscle, **31–32, 41, 44–45,** 305f, 312f, 320, 321f, 322t, **347**
Rectus femoris muscle, **32, 34–36, 45,** 305f, 324f, 325f, 327f, 328, **344, 350**
Red fibers, 299
Red blood cell(s), 79f, 95f, 162, 526–30, 531t. *See also* Anemia; Hemoglobin
Red blood cell count (RBCC), 527, 941
Red blood cell distribution width (RBCDW), 941
Red marrow, 203, 225
Red nucleus, 407
Red pulp, of spleen, 624, 625
Reeve, Christopher, 397
Referred pain, 441–42
Reflex arcs, 389, 391f, 393t
Reflex behavior, 390–91, 394, 754
Reflex sympathetic dystrophy, 442
Refraction disorders, 472
Refractory period, of nerve impulse, 370
Regeneration, and injuries to nervous system, 364, 397
Regulation
  of bile release, 678
  of cardiac cycle, 568–72
  of electrolyte intake and output, 815–18
  of gastric secretions, 668–69
  of glomerular filtration rate, 790
  of hydrogen ion concentration, 820–24
  hypothalamus, 407
  of male sex hormones, 846
  of pain impulses, 443–44
  of pancreatic secretion, 672
  of small intestine secretions, 683
  of urine concentration and volume, 794–96
  of water intake and output, 813–15
Relative position, terms of, 21
Relative refractory period, 370
Relaxation, of muscles, 293, 294t
Relaxin, 884, 885t
REM-sleep behavior disorder, 409t
Renal agenesis, 802t
Renal arteries, 593, 601f, 778f, 779, 780f, 898f
Renal baroreceptors, 790
Renal capsule, 776, 778f
Renal clearance, 798
Renal columns, 776, 778f
Renal corpuscle, 778f, 779, 783f, 797t
Renal cortex, 774, 776, 778f, 783f, 785f
Renal dysplasia, 802t
Renal excretion, of hydrogen ions, 823
Renal fascia, 777f
Renal hypoplasia, 802t
Renal medulla, 776, 778f, 783f, 785f
Renal papillae, 774, 776, 778f
Renal pelvis, 776, 776f, 780f
Renal plasma threshold, 792
Renal pyramids, 776, 778f
Renal sinus, 776, 778f
Renal tubules, 778f, 779, 797t, 803
Renal veins, 604, 607f, 778f, 779, 780f
Renin, 506, 588, 777, 790
Renin-angiotensin system, 506, 790
Repair enzymes, 137
Replication, of DNA, 127, 129f
Reproduction. *See also* Reproductive system
  characteristics of life, 8t
  organ systems, 18
Reproductive cycle, 857–59
Reproductive system
  autonomic stimulation, 430t
  cardiovascular system, 610, 868
  digestive system, 692, 868
  endocrine system, 516, 868
  hormonal control of female, 857–61
  hormonal control of male, 845–46
  integumentary system, 188, 868
  life-span changes, 908t
  lymphatic system, 645, 868
  major functions of, 8t
  meiosis, 831–33
  muscular system, 333, 868
  nervous system, 378, 868
  organs of female, 846–56
  organs of male, 833–45
  respiratory system, 768, 868
  skeletal system, 239, 868
  urinary system, 804, 868
"Reprogrammed" differentiated cells, 108
Reserpine, 374t, 431
Residual volume (RV), 752, 754t
Resorption, and bone remodeling, 200
Respiration. *See also* Aerobic respiration; Anaerobic respiration; Breathing; Cellular respiration; Respiratory system
  characteristics of life, 8t
  definition, 736
Respiratory acidosis, 824
Respiratory alkalosis, 824, 825–26
Respiratory areas, 755–57
Respiratory bronchioles, 743, 745f
Respiratory capacities, 752–53, 754t
Respiratory center, 409, 822
Respiratory cycle, 752
Respiratory disorder, 756
Respiratory distress syndrome, 750. *See also* Acute respiratory distress syndrome
Respiratory membrane, 760–62
Respiratory system
  alveolar gas exchanges, 759–62
  autonomic stimulation, 430t
  breathing mechanism, 747, 750–55

cardiovascular system, 610, 768
cigarette smoking, 739
control of breathing, 755–59
digestive system, 692, 768
endocrine system, 516, 768
excretion of carbon dioxide, 822–23
gas transport, 762–67
integumentary system, 188, 768
life-span changes, 767, 769, 908t
lymphatic system, 645, 768
major functions of, 8t, 17, 18f
muscular system, 333, 768
nervous system, 378, 768
organs of, 737–47
parts of, 747t
reasons for breathing, 726–27
reproductive system, 768, 868
skeletal system, 239, 768
urinary system, 768, 804
Respiratory syncytial virus, 637t
Respiratory tubes, 744–46
Respiratory volumes, 752–53, 754t
Responsiveness, as characteristic of life, 8t
Resting potential, 366, 367f
Resting tidal volume, 752
Restless legs syndrome, 409t
Rete testis, 835, 836f
Reticular connective tissue, 158, 159f, 162t
Reticular fibers, 156
Reticular formation, 408f, 409
Reticulocytes, 527
Reticulospinal tracts, 395, 396t, 402
Retina, 469, 470f, 473
Retinal, 709f, 710
Retinal pigment epithelium, 473
Retinol, 709f
Retraction, of joint, 269
Retrolental fibroplasia (RLF), 760
Retroperitoneal fat, **45**
Retroperitoneal positioning, of kidneys, 776, 777f
Reversible reaction, 59
R group, 65
Rh blood types, 546, 547–48
Rheumatic fever, 642t
Rheumatoid arthritis, 265, 278, 279t, 491, 637t, 642t, 934
Rhodopsin, 473, 474f, 475, 710
RhoGAM, 548
Rhombencephalon, 398, 400t
Rhomboid muscle, 306f, 311, 313t, **348**
Rhythmicity, of smooth muscles, 300
Rhythm method, of contraception, 863, 866t
Rib(s), **37, 43–46,** 222–23, 224f, 226f, 227f
Rib facet, 219f
Riboflavin, 714, 717t
Ribose, 68f, 701
Ribosomal RNA (rRNA), 132
Ribosomes, 75, 82, 91t
Ribozymes, 132
Rickets, 200, 712
Right atrium, **35, 46**
Right hemisphere, 403, 404
Right lower quadrant (RLQ), of abdomen, 22, 24f
Right lymphatic duct, 618, 619f, 620f
Right upper quadrant (RUQ), of abdomen, 22, 24f
Right ventricle, **35, 44–46,** 558f
Rigor mortis, 293
Rinne test, 459
Risk factors, for stroke, 588t
RNA (ribonucleic acid). *See also* Chromosomes; Genetic(s)
cellular metabolism, 130–31
DNA compared to, 133t
structure of, 68, 948–49
RNA interference (RNAi), 136
RNA polymerase, 131
Rods, of eye, 473, 474f
Root
of hair follicle, 178
of penis, 843
of tooth, 659
Root canals, 659
Rotation, of joint, 268f, 269
Rotator cuff, 271
Rotator muscles, 316
Rothman, Cappy, 876
Rough endoplasmic reticulum, 83, 84f
Round bones, 193
Round ligament of uterus, **35,** 849f, 853f, 854
Round window, 453
Rubella virus, 896
Rubrospinal tracts, 395, 396t, 402
Ruddy, Erin Zammett, 536f
"Rule of nines," for burns, 186, 187f
Ruptured disc, 225
R wave, 568

S

Saccharin, 701
Saccule, 459, 461f
Sacral artery, 599f
middle, 593, 594f
Sacral canal, 222, 223f, 232f
Sacral curvature, 219, 233f
Sacral foramina, anterior, 222, 223f
Sacral hiatus, 222, 223f, 232f
Sacral nerves, 418f, 419
Sacral promontory, 222, 223f, 232f, 233f
Sacral region, 24, 25f
Sacroiliac joints, 222, 231, 232f, 270t
Sacromeres, 287, 288f
Sacrum, **37, 41,** 206, 218, 219f, 222, 223f, 223t, 232f, 234t, **343,** 388f
Saddle joint, 266f, 267
Sagittal plane, 22, 23f
Sagittal section, of skull, 215f
Sagittal suture, 208, 218f, **246**
St. John's Wort, 725
St. Martin, Alexis, 668
Salicylic acid, 181t
Salivary amylase, 660, 683t
Salivary glands, 104, 430t, 653f, 660–61, 662f, 663t
Salivary secretions, 660
Salt(s), 59, 203–204. *See also* Bile salts; Sodium chloride
Saltatory conduction, 353, 371
Salt craving, 815
Salt receptors, 450
Saphenous veins, **31–32, 35,** 327f, 606f
greater, 581
lesser, 581
small, 605, 607f
Sarcolemma, 287
Sarcoplasm, 284, 287
Sarcoplasmic reticulum, 284, 288
Sartorius muscle, **31–33, 35–36, 45,** 305f, 306f, 324f, 325f, 326f, 327f, 328, 331f, **344, 350**
Satellite cells, 363
Saturated fats, 63
Saturated fatty acid, 63
Scab, 183, 185f
Scala tympani, 453
Scala vestibuli, 453
Scalp, **39, 42**
Scanning electron microscope (SEM), 79
Scanning probe microscope, 79
Scaphoid, 230f
Scapula, 206, 226, 227f, 231t, 272f, 313f, 315f
Scar tissue, 183, 185f
Schiavo, Terry, 383
Schizophrenia, 373t, 431
Schwann cells, 356, 357f, 358f, 363
Sciatica, 424
Sciatic nerves, 423
Scintillation counter, 54, 609
Sclera, 437, 465, 470t
Scleroderma, 279t, 643
Scoliosis, 225
Scrotum, **34, 41,** 834f, 842, 845t
Scurvy, 716
"Seat belt contusions," 2
Sebaceous glands, 153f, 170, 179–80, 181, 182t
Sebum, 180
Secondary bronchi, 743, 744f
Secondary follicle, 849
Secondary immune response, 637
Secondary malnutrition, 723, 725
Secondary metabolites, 115
Secondary oocyte, 849, 850f, 851f, 852f, 853f, 877, 880f, 881f
Secondary ossification centers, 198
Secondary pacemaker, 573
Secondary sex characteristics, 846, 847f, 857
Secondary spermatocytes, 836
Secondary structure, of amino acid, 66, 67f
Secondary teeth, 658, 659t
Second-degree burn, 184
Secondhand smoke, 736
Second line of defense, 626, 823
Second messengers, 487
Second polar body, 849, 850f
Secretin, 672, 673f, 678t
Secretion
autocrine, 483
control of hormonal, 491–92
gastric, 666–69
milk production, 902–903, 904
pancreatic, 672
paracrine, 483
salivary, 660

small intestine, 683
tubular, 786, 793–94
Sedimentation rate, erythrocyte, 942
Segmentation, of alimentary canal, 654, 656f
Selectin, 81
Selective estrogen receptor modulators (SERMs), 489t, 865
Selectively permeable cells, 79
Selective serotonin reuptake inhibitors (SSRIs), 374t, 859
Selenium, 722, 723t
Sellar joint, 267
Sella turcica, 211f, 212, 213f, 215f, **254–57,** 493f
Semen, 841
Semicircular canals, 453
Semilunar valve, 552
Semimembranosus muscle, 306f, 326, 327f, 328t, 331f, **351**
Seminal vesicle, 834f, 840, 845t
Seminiferous tubules, 835, 836f, 845t
Semispinalis capitis muscle, 308, 309f, 310
Semitendinosus muscle, 306f, 326, 327f, 328t, 331f, **343, 351**
Senescence, and development, 875, 907, 908t
Sensation, 438–40
Senses
equilibrium, 459–61
hearing, 450–58
life-span changes, 476–77
sight, 462–76
smell, 446–68
taste, 448–50
Sensoneural deafness, 459
Sensory adaptation, 440
Sensory area, of cerebrum, 401
Sensory impulses, from ear, 458t
Sensory nerves, 412
Sensory neurons, 353, 360–61, 362t, 393t
Sensory receptors, 177, 355, 389, 439, 446–47
Sensory speech area, 401
Septa, 835
9/11-associated air pollution, 749
Septicemia, 911t
Serine, 135t, 705t
Serology, 546
Serosa, 654, 655t
Serotonin, 373t, 443, 534
Serous cells, 660
Serous coat, 800
Serous fluid, 152
Serous layer, 654
Serous membranes, 15f, 163
Serratus anterior muscle, **31–32,** 305f, 312, 313t, 314f, **343, 347**
Serum, 541
Serum glutamic pyruvic transaminase, 941
Serum protrombin conversion accelerator, 541t
Sesamoid bones, 193, 194f, 230f, 237f
Set point, 9
Sex cells, transport of, 876–77
Sex chromosomes, 919, 927–28
Sex determination, 927
Sex hormones, 508
Sex-influenced inheritance, 928
Sex-limited trait, 928
Sexually transmitted diseases (STDs), 867, 869
Shark cartilage, 725
Shigatoxin, 790
Shock, and blood pressure, 789
Short bones, 193, 194f
Short head biceps brachii muscle, **32**
Shoulder, 270t, 271–72, 313f, 597, 602–603
"Siamese twins," 900
Sickle cell disease, 195, 527, 529t, 548t, 922
Side chain, 65
Sight, sense of, 462–76. *See also* Eye
Sigmoid colon, **35–36, 41, 49,** 687
Simethicone, 670
Simple columnar epithelium, 146, 148f, 153f
Simple cuboidal epithelium, 147f, 153f
Simple diffusion, 100t
Simple glands, 150, 152t
Simple squamous epithelium, 145, 146, 147f, 153t
Single nucleotide polymorphisms (SNPs), 135–36
Sinoatrial (SA) node, 565–66, 571f, 573
Sinuses, 208t, 738, 740f, 747t
Skeletal muscle. *See also* Muscular system
actions, 301–305
contraction, 289–96
fibers, 287–88
lymph flow, 621
major muscles, 305–34
structure, 285–89
tissue, 163, 165t, 302t
use and disuse of, 300
Skeletal system. *See also* Bone; Joints
cardiovascular system, 239, 610
definition, 193
digestive system, 239, 692
endocrine system, 239, 516
integumentary system, 188, 239
life-span changes, 238, 240, 908t
lower limbs, 234–38
lymphatic system, 239, 645
major functions of, 8t, 14, 16f
muscular system, 239, 333
nervous system, 239, 378
pectoral girdle, 225–26
pelvic girdle, 231–34
reproductive system, 239, 868
respiratory system, 239, 768
skull, 206, 208–17, 218f
upper limb, 226–31
urinary system, 239, 804
veretebral column, 218–22
Skin. *See also* Integumentary system
accessory structures, 177–81
body temperature regulation, 181–82, 183f
color, 174, 176, 179, 925f, 926
healing of wounds and burns, 183–86, 187f
lymphatic vessels, 619
organization, 205–206
tissues, 171–77
Skin cancer, 175, 712
Skin cells, 171
Skin glands, 179–81, 182t
Skin substitutes, 186
Skull, **42, 245–59**
bones of, 205t
divisions of skeleton, 206
fossilized and human evolution, 193
joints, 270t
male versus female, 234t
skeletal system, 206, 208–17, 218f
Sleep apnea, 757
Sleep paralysis, 409t
Sleep and sleep disorders, 373t, 409, 757
Sliding filament model, 291, 292
Slow-twitch muscle fibers, 299
Slow-wave sleep, 409
Small intestine, **33–34, 38, 40–41, 44, 48–49,** 653f, 655f, 678–86
Small saphenous vein, 605, 607f
Smell, sense of, 446–48, 451
Smoking. *See* Cigarette smoke and smoking; Nicotine
Smooth endoplasmic reticulum, 83, 84f
Smooth muscle(s), 300–301, 302t
Smooth muscle fibers, 300
Smooth muscle tissue, 163, 164f, 165t
Sneezing, 754, 755t
Social Security Administration, 876
Sodium. *See also* Salt(s); Sodium chloride; Sodium/potassium pump
atomic structure, 53t
blood tests, 942
chemical constituents of cells, 62t
diffusion and active transport, 95
electrolyte balance, 818, 820
elemental composition of human body, 52t
hypertension, 588
ion channels, 82, 365, 366, 367f, 369, 370t, 372
nutrition, 719, 721t
urine, 788t, 792, 794t
Sodium chloride, 795, 797f, 813
Sodium hydroxide, 59
Sodium/potassium pump, 96, 365, 366, 367f, 812–13
Soft palate, 657, 664
Soleus muscle, 305f, 306f, 329f, 330f, 331f, 332, **345, 351–52**
Somatic afferent fibers, 412, 414
Somatic efferent fibers, 412, 414
Somatic nervous system, 356, 411
Somatostatin, 487t, 494, 511, 668, 678t
Somatotropes, 494
Somatotropin (STH), 494
Sounds, of heart, 564–65
Sour receptors, 449–50
Space medicine, 201, 585
Spastic paralysis, 397
Specialized cells, 78f
Specialized connective tissues, 156
Special senses, 438, 446–76
Species resistance, 626, 628t
Speech, 755t. *See also* Motor speech area
Sperm, 831, 877. *See also* Spermatogenesis; Sperm cells
Spermatic arteries, 593
Spermatic cord, **33, 41,** 835

Spermatids, 836
Spermatogenesis, 830, 836, 837, 839f. *See also* Sperm; Sperm cells
Spermatogenic cells, 835
Spermatogonia, 836, 838f
Sperm cells, 833, 835–37, 841, 847f
Sperm donation, 878
Spermicide, 865, 866t
Sphenoidal fontanel, 218f
Sphenoidal sinus, **39, 42,** 212, 213f, 214f, 215f, **254,** 738f
Sphenoid bone, 209f–13f, 215f, 218f, **245–46, 248–51, 254–57,** 493f, 738
Sphenomandibularis, 308
Spheroidal joint, 265
Sphincter muscle, 306
Sphincter urethrae muscle, 322, 323t
Sphygmomanometer, 582–83
Spina bifida, 166, 398, 892
Spinal artery, 596f
Spinal branch, 417
Spinal cord, **38, 40, 43–45,** 385f, 387–96, 397, 399f, 405f, 596f
Spinal cord injuries, 397, 934
Spinalis capitis muscle, 309f, 310t
Spinalis cervicis muscle, 309f, 310t
Spinalis thoracis muscle, 309f, 310t
Spinal nerves, 387, 411, 412, 417–23
Spine. *See also* Spinal cord; Vertebrae
  definition, 208t
  of scapula, 226, 227f
Spinocerebellar tracts, 395, 396t
Spinothalamic tract, 393, 395, 396t
Spinous process, 220
Spiral fracture, 202f
Spiral lamina, 454f
Spiral organ, 455, 456f, 457f
Spirometer, 753
Spirometry, 752
Spleen, **34–36, 45,** 604f, 623–25, 626t, 777f
Splenic artery, 594f
Splenic vein, 604
Splenius capitis muscle, 308, 309f, 310, **346**
Spondyloarthropathy, 279t
Spondylolisthesis, 221
Spondylolysis, 220, 221
Spongy bone, 194, 195
Spontaneous fracture, 202
Spontaneous mutations, 136
Sprains, 278
Squamous epithelium, 143, 145
Squamous suture, 209f, 210, 215f, **246, 248**
SRY gene, 927
Stanford University, 375
Stapedius, 452
Stapes, 451–52
Starvation, 725–28
Static equilibrium, 459–60
Statins, 97
Steele, Allen, 278
Stem cells
  baldness therapy, 179
  blood cells and hematopoietic, 524, 525f, 526
  brain function and malfunction, 354
  cell pathways, 107f
  heart transplant, 566
  neural, 365
  structure and function, 105–106
  tooth extraction as source, 660
  treatment of disease, 108
Stereoscopic vision, 475
Sternal angle, 224
Sternal puncture, 225
Sternal region, 24, 25f
Sternoclavicular joint, 270t
Sternocleidomastoid muscle, **31–32,** 305, 306f, 307f, 308, 310, 312f, **341, 343, 346–47**
Sternocostal joint, 270t
Sternum, **33, 39–40, 43,** 206, 223–25, 227f, 555f, 598f
Steroid hormones, 485–86, 489
Steroid molecules, 64, 65f. *See also* Lipids
Stickler syndrome, 157t
Stomach, **33–35, 38, 40,** 604f, 653f, 665–71
Strabismus, 465
Strangulated hernia, 835
Stratified columnar epithelium, 150, 153t
Stratified cuboidal epithelium, 148, 149f, 153t
Stratified squamous epithelium, 143, 145, 147–48, 149f
Striated muscle, 143, 163
Stratum basale, 173, 174f
Stratum corneum, 173, 174f
Stratum granulosum, 173, 174f
Stratum lucidum, 173, 174f
Stratum spinosum, 173, 174f
Strausbaugh, Linda, 925f
Strength training, 334
Strep A infection, 279t
Streptococcus bacteria, 557
Stress and stress response
  endocrine system, 513–15
  hypertension, 588
Stretch receptors, 439, 444f
Stretch reflex, 444
Stroke, 401, 588, 911t
Strong acid, 820
Strong bases, 820
Structural formulas, of atoms, 57
Stuart-Prower factor, 541t
Styloid process, 210, 211f, 215f, 228, 229, 230f
Stylomastoid foramen, 211f, 217t, **249–51.** *See also* Mastoid foramen
Subacromial bursa, 272
Subarachnoid space, 385, 387f, 388f
Subatomic particles, 4, 5f
Subchondral plate, 264
Subclavian arteries, **36–37, 46–47,** 592, 593f, 595f, 596f, 597f, 601f
Subclavian trunk, 617, 619f, 620f
Subclavian veins, **33–35,** 602, 603, 607f
Subcoracoid bursa, 272
Subcutaneous fascia, 287
Subcutaneous injections, 173
Subcutaneous layer, 172–73
Subcutaneous tissue, **42**
Subdeltoid bursa, 271–72
Subdural hematoma, 384
Sublingual glands, 661, 662f, 663t
Submandibular glands, 661, 662f, 663t
Submucosa, of alimentary canal, 654, 655t, 665f
Submucosal plexus, 656
Submucous coat, 800
Subscapular bursa, 271, 272f
Subscapularis muscle, **33,** 314t, 315f, 316
Subscapular nerve, 421
Substance P, 373t, 374
Substantia nigra, 406
Substrate, 114, 117, 118f
Succinic acid, 946f
Succinyl-CoA, 946f
Sucralose, 701
Sucrase, 118, 682–83
Sudden cardiac arrest, 590
Sudden infant death, 757
Sugars, 62, 92, 701
Sulcus, 398, 399f, 400f, 401f
Sulfate, 62t, 788t, 818
Sulfur, 52t, 719, 721t, 819
Sulfuric acid, 819
Sunburn, 109f, 184
Sun poisoning, 184
Superficial, as term of relative position, 21
Superficial digital flexor, **344**
Superficial partial-thickness burn, 184
Superficial temporal arteries, 597
Superficial transversus perinei muscle, 320, 322f, 323t
Superior, as term of relative position, 21
Superior articulating processes, 220, 223f
Superior border, 226
Superior iliac spine
  anterior, **31, 33, 37,** 231, 232f
  posterior, 232f, 233, **343**
Superior peduncles, 410
Superior pubic ramus, 834f
Superior rectus muscle, 464, 465t
Superior region, 1
Superoxide dismutase (SOD), 909
Supination, of joint, 268f, 269
Supinator, 316, 317f
Support function
  of bone, 202
  of organ systems, 14
Suppression amblyopia, 465
Suprascapular notch, 227f
Supraglenoid tubercle, 227f
Supraorbital foramen, 208, 209f, 217t, **245, 247–48.** *See also* Infraorbital foramen
Supraorbital notch, 208, 209f, **245, 253, 341**
Supraorbital ridge, 234t
Suprapatellar bursa, 265, 276, 277f
Suprarenal arteries, 592, 594f, 601f, 778f
Suprarenal veins, 604, 778f
Supraspinatus muscle, 311f, 313f, 314t, 315
Supraspinous fossa, 226, 227f
Supratrochlear region, 622
Sural region, 24, 25f
Surface anatomy, 341–45
Surface tension, 750
Surfactant, 750, 905
Surgery. *See also* Medical science
  birth control, 866, 867f
  hypothermia, 183
  obesity, 711
Surgical neck, 226, 229f

Suspensory ligaments, 466, 847, 849f, 853f, 862
Sustained muscle contraction, 297, 298
Sustentacular cells, 835–36
Sutton, Josiah, 130
Sutural ligament, 262
Suture
bone, 205, 208t
joint, 260, 261–62, 267t
Swallowing mechanism, 664
Swanson, Alfred, 276
S wave, 568
Sweat, 814, 816
Sweat glands, 180
Sweet receptors, 449
Sympathetic chain ganglia, 425
Sympathetic division, of autonomic nervous system, 424–26, 427f
Sympathetic tone, 429
Sympathetic trunks, 425
Symphysis joints, 263, 267t, 279–80
Symphysis pubis, **36–37, 41,** 232f, 233, 264f, 270t, 799f, 848f
Synapse, 289, 353, 354. *See also* Synaptic transmission
Synapsis, 832
Synaptic cleft, 290, 356
Synaptic knobs, 356, 372f
Synaptic potentials, 371–72
Synaptic transmission, 365, 371–74
Synarthrotic joints, 261, 267t
Synchondrosis, 262–63, 267t, 279
Syncytium, 301, 552
Syndesmosis, 260, 261, 267t
Synergist, 284, 304
Synesthesia, 447
Synovial cavity, 264
Synovial fluid, 260, 264, 265
Synovial joints, 263–77, 280
Synovial membrane, 163, 264, **273,** 275f, 277f
Synthesis
chemical reactions, 50, 58
protein, 131–34
Synthetic folic acid, 716
Syphilis, 867t
Systemic capillaries, 560f
Systemic circuit, 553, 590, 592
Systemic lupus erythematosus, 279t, 642t
Systole, 564
Systolic pressure, 552, 580, 583

T

Tachycardia, 552, 572
Tactile corpuscles, 440, 446t
Tail, of sperm cell, 837
Tardive dyskinesia, 373t
Talocrural joint, 270t
Talus, 236, 237f
Tanning, and skin cancer, 175
Tapeworm, 534t
Tardive dyskinesia, 406
Target cells, 16, 483
Targeted muscle reinnervation, 300
Tarsal region, 24, 25f
Tarsals (tarsus), 206, 235f, 236, 237f, 238t, **345**
Taste, 448–50, 451
Taste buds, 448, 449f
Taste cells, 448, 449
Taste hairs, 448
Taste nerve pathways, 450
Taste pore, 448
Taste receptors, 448–49
Taste sensations, 449–50
Tattoo, 177
Taylor, Amillia, 893
Tay-Sachs disease, 87, 359
T cells, 532, 534t, 629–30, 631f. *See also* T lymphocytes
Tear gland, 430t, 464
Tectorial membrane, 455, 456f
Teeth, 658–60, 661, 690
Telencephalon, 398, 400t
Telomeres, 103
Telophase, 101, 102f, 103t, 833
Temperature. *See also* Body temperature; Cold environment; Heat
definition, 7
homeostatic mechanisms, 9–10
sense of, 440
Temporal artery, 601f
superior, 595f
Temporal bones, 209f, 210, 211f, 212, 213f, 215f, 218f, **245–46, 248–50, 253, 341**
Temporalis muscle, **42,** 306f, 307, 308, **341, 346**
Temporal lobe, **42,** 399, 400f, 401f, 402, 403t
Temporal process, 214
Temporal veins, 607f
Temporomandibular joint, 270t
Temporomandibular joint (TMJ) syndrome, 308
Tendinitis, 286
Tendons, 155, 158, 285, 286
Tenosynovitis, 286
Tensor fasciae latae muscle, **32, 35–36,** 305f, 323, 324f, 325f, 327t, **350**
Tensor tympani, 452
Tentorium cerebelli, 384t, 400
Teratogens, 892, 896–97
Teres major muscle, **33,** 306f, 311f, 313f, 314t, 315, **342**
Teres minor muscle, 306f, 311f, 313f, 314t, 316, **348–49**
Terminal bronchioles, 743, 744f, 745f
Terminal ganglia, 426
Terminology, anatomical, 20–25
Tertiary bronchi, 743, 744f
Tertiary structure, of amino acid, 66, 67f
Testes, **34, 38, 41,** 512, 833–37, 845t
Testicular cancer, 835
Testosterone, 201, 489, 833, 846, 857
Tetanic contraction, 284, 297, 298f, 372
Tetrahydrolipostatin, 711
Tetraploidy, 929
Texting, and muscular movements, 285
Thalamotomy, 406
Thalamus, **39, 42,** 395f, 405, 407, 410f
Thalassemia, 529t
Thalidomide, 896
Thallium-201, 54
Theca externa, 851f
Theca interna, 851f
Thermoreceptors, 183f, 437, 439, 440
Theta waves, 411
Thiamine, 713–14, 717t, 719t
Thigh, muscles of, 323–25, 327t, **343, 350–51**
Third-degree burn, 186
Third ventricle, **42,** 385, 386f, 387f
Thirst center, 813
Thoracic aorta, **47,** 592, 595t, 598f
Thoracic artery, internal, 598
Thoracic cage, 205t, 206, 222–25
Thoracic cavity, 12, 13f, 14f, **37,** 623
Thoracic curvature, 219, 225
Thoracic duct, 618, 619f, 620f
Thoracic membranes, 12–14
Thoracic nerves, 418
Thoracic outlet syndrome, 424
Thoracic veins, 603
Thoracic vertebrae, 219f, 220f, 221, 222f, 223t, 224f
Thoracic wall, 597–98, 603
Thoracodorsal nerve, 421
Thorax, 15f, **47, 342**
Threonine, 135t, 705t
Threshold potential, 368
Threshold stimulus, and muscular responses, 296
Thrombin, 539
Thrombocytes, 522, 534, 535t
Thrombocytopenia, 542
Thrombophlebitis, 581
Thrombopoietin (TPO), 526, 534, 542
Thrombus, 541
Thymine, 128f, 948f, 949f
Thymosins, 512
Thymus, 512, 623, 624f, 626t, 644
Thyrocervical arteries, 596
Thyrocervical axis, 595f
Thyroid arteries, 595f, 596
Thyroid cancer, 483
Thyroid cartilage, **33–34, 343,** 740, 741f
Thyroid gland, **32–34, 46**
aging, 515
bone calcium, 205f
endocrine system, 499–501
radioactive isotopes, 54
Thyroid hormone, 201, 485–86, 500–501
Thyroid-stimulating hormone (TSH), 487t, 490, 492, 496, 499t, 942
Thyrotropin-releasing hormone, 487t, 496
Thyroxine, 201, 487t, 500, 501, 942
Tibia, 206, 235, 238t, 277f, 329f, **345**
Tibial arteries, 599, 600–601f
Tibial collateral ligament, 276, 277f
Tibialis anterior muscle, 305f, 328, 329f, 332t, **344–45, 351**
Tibialis posterior muscle, 331f, 332t, 334, **345**
Tibial tuberosity, 235, 236f, **344**
Tibial veins, 604, 606f, 607f
Tibiofemoral joint, 270t
Tibiofibular joints, 270t
Tidal volume, 752, 754t
Tight junctions, 144, 146t
Time course, of hydrogen ion regulation, 823

Tinnitus, 477
Tissue(s)
connective, 152–62
definition, 144
engineering, 166, 186
epithelial, 144–52
fluid formation, 619
levels of organization, 4, 5f
life-span changes, 20
membranes, 163
muscle, 163–64, 302t
nervous, 164–65
skin, 171–77
transplantation and rejection of, 641
Tissue thrombloplastin activator (tPA), 539, 541t, 542
Titins, 287
T lymphocytes, 623. *See also* T cells
Tonsil(s), 621, 657
Tonsilitis, 657
Tongue, **38–39, 43,** 657, 664, 738f
Tonometer, 477
Tooth decay, 3
Total lung capacity (TLC), 753, 754t
Totipotent cell, 105
Touch, sense of, 440, 441f
Toxic chemicals, 451, 837
Toxicity, of minerals, 717, 719
Toxicology, 25
Trabecular bone, 194, 240
Trace amine-associated receptors, 483
Trace elements, 52, 720–22, 723t
Trachea, **33–36, 38–40, 46,** 664f, 738f, 741f, 742, 743f, 744f, 747t
Tracheostomy, 742, 743f
Transaminase, 942
Transcellular fluid, 811
Transcription, of RNA, 130, 135t
Transcriptome, 139
Transcutaneous electrical nerve stimulation (TENS), 308, 445
Transcytosis, 98, 99f, 100t
Transdermal patch, 173
Trans fats, 63
Transferrin, 530
Transfer RNA (tRNA), 131, 134f
Transforming growth factor B (TGF-B), 608f
Transfusions, of blood, 544–48
Transient ischemic attack (TIA), 401, 588
Transitional epithelium, 150, 151f, 153t
Translation, of RNA, 131, 135t
Translocation, of chromosomes, 929
Transmembrane protein, 80
Transmissible spongiform encephalopathies, 66
Transmission electron microscope (TEM), 79
Transplants
bone marrow, 108
heart, 566
kidney, 779
liver, 676
tissue rejection, 637t, 641
Transport, and organ systems, 16–17. *See also* Active transport; Electron transport chain; Gas transport; Limited transport capacity; Sex cells
Transverse colon, **34–35, 40, 45, 49,** 687
Transverse fissure, 398, 399f, 400f
Transverse foramina, 220
Transverse fracture, 202f
Transverse humeral ligament, 271, 272f
Transverse plane, 22, 23f
Transverse process, 220
Transverse tubules, 288
Tranversus abdominis muscle, **32–33, 37,** 312f, 320, 321f, 322t
Trapezius muscle, **31,** 305f, 306f, 311, 312f, 313t, 315f, **341–42, 348–49**
Trastuzumab, 865
Traumatic brain injury (TBI), 405
Tretinoin, 181t
Triad, 288
Triceps brachii muscle, 272f, 303, 306f, 313f, 314f, 316, 318f, **342, 347–50**
Triceps-jerk reflex, 394
Tricuspid valve, 557, 558f, 559f, 560t
Tricyclic antidepressants, 374t
Trigeminal nerves, 415, 417t
Trigeminal neuralgia, 415
Triglycerides, 63, 64f, 65f, 702, 942. *See also* Lipids
Trigone, 774, 799, 801f
Triidothyronine, 487t, 500, 501, 942
Triploid chromosome number, 929
Triplo-X, 929
Triptans, 356
Triquetrum, 230f
Trisomy, 916, 929, 930t
Tristearin, 62
Trochanter, 208t
Trochlea, 226, 229f, **273**
Trochlear nerves, 415, 417t
Trochlear notch, 229, 230f
Trochoid joint, 266–67
Trophoblast, 875, 880, 882f
Tropic hormones, 492
Tropomyosin, 287, 288f, 289
Troponin, 287, 288f, 289
True ribs, 223, 224f
Trypsin, 671, 683t
Trypsinogen, 671
Tryptophan, 135t, 374t, 705t, 714
Tubal ligation, 866, 867f
Tubal pregnancy, 882
Tubercle, 208t, 223, 765
Tuberculin skin test, 639
Tuberculosis, 534t, 639, 735, 765
Tuberosity, 208t
Tubular dysgenesis, 802t
Tubular glands, 150
Tubular reabsorption, 786, 790–92, 794
Tubular secretion, 786, 793–94
Tumor, 104
Tunic, of artery, 574, 575f, 576
Tunica albuginea, 835, 836f, 843, 847
Turner syndrome, 497, 929
T wave, 568
Twins, 880, 900
Twitch, of muscle, 296, 298f
Tumor suppressor genes, 104
Tympanic cavity, 451
Tympanic membrane, 437, 450
Tympanic reflex, 452–53
Tympanostomy tubes, 453
Typhoid fever, 534t
Tyrosine, 135t, 705t

U

Ulcer(s), 668. *See also* Pressure ulcers
Ulcerative colitis, 642t, 691
Ulna, 206, 227f, 228–29, 230f, 231t, **273**
Ulnar artery, 597, 601f
Ulnar collateral ligament, 272, **273**
Ulnar nerves, 420
Ulnar veins, 602, 607f
Ultrasonography, 6–7
Ultrasound, 931, 932f, 933t
Ultratrace elements, 52
Umami, 450
Umbilical arteries, 891, 898, 899t, 906t
Umbilical cord, 875, 891
Umbilical cord stem cells, 108
Umbilical region, 22, 24f, 25f
Umbilical vein, 891, 897, 898f, 899t, 906t
Umbilicus, **31, 343, 347**
Uncompensated acidosis, 825
Undernutrition, 723
UNICEF, 115
Unipolar neurons, 353, 360, 362t
Universal precautions, 523
Universal recipients and donors, 546
University of Connecticut, 925f
University of Maryland, 354
University of Pittsburgh, 910
University of Texas at Austin, 699
Unmyelinated axons, 358, 359f
Unsaturated fats, 63, 64
Unsaturated fatty acids, 63
Upper limbs, 205t, 206, 226–31, **342,** 597, 602–603
Upper motor neuron syndrome, 397
Upper respiratory tract, 737
Up-regulation, of hormones, 485
Uracil, 949f
Uranium, 56
Urbanization, and human health, 3
Urea, 704, 788t, 796–97
Ureter, **35–36,** 776f, 778f, 780f, 798–99, 801f, 834f, 853f
Urethra, **37, 45,** 776f, 799f, 800–802, 834f, 848f
Urethral glands, 801
Urethral sphincter, 800, 803
Uric acid, 788t, 797, 942, 943
Urinalysis, 803
Urinary bladder, **33–36, 38, 41,** 166, 776f, 799–800, 805, 834f, 848f. *See also* Urinary system; Urine
Urinary excretion, 786
Urinary excretion of odoriferous component of asparagus, 803
Urinary system. *See also* Urinary bladder; Urine
autonomic stimulation, 430t
cardiovascular system, 610, 804
description, 775
developmental abnormalities, 802t
digestive system, 692, 804
elimination of urine, 798–803

endocrine system, 516, 804
integumentary system, 188, 804
kidneys, 776–85
life-span changes, 803, 805, 908t
lymphatic system, 645, 804
major functions of, 8t, 17–18
muscular system, 333, 804
nervous system, 378, 804
reproductive system, 804, 868
respiratory system, 768, 804
skeletal system, 239, 804
urine formation, 785–98
Urinary tract infection (UTI), 800
Urine. *See also* Urinary bladder; Urinary system
composition of, 798
elimination of, 798–903
formation of, 785–98
laboratory tests, 943
regulation of water output, 814, 815
relative concentrations of components, 788t
Urogenital diaphragm, 320, 322, 834f
Urology, 25
USA Track and Field, 816
Uterine artery, 598
Uterine tube, **35,** 848f, 849f, 852, 853f, 854, 857t
Uterine veins, 605
Uterus, **35–36,** 848f, 849f, 854, 857t, 877f, 884
Utricle, 459, 461f
Uvula, 657, 738f

V

Vaccines, 290, 638, 644
Vagina, **37,** 853f, 854–55, 857t, 877f
Vaginal orifice, 848f, 854
Vaginal process, 835
Vaginal veins, 605
Vagus nerve, **47,** 416, 417t, 568
Valine, 135t, 705t
Valium, 374t, 377
Valves
of heart, 555–57, 558f, 559f, 560t
of veins, 579
Variably expressive phenotype, 924
Varicose veins, 581
Vasa recta, 785, 786f, 795, 797
Vascular endothelial growth factor (VEGF), 104, 574, 781
Vascular tunic, 465
Vasectomy, 866, 867f
Vasoconstriction, 576, 584f
Vasodilation, 576, 584f
Vasomotor center, of medulla oblongata, 409, 587
Vasomotor fibers, 576
Vasopressin, 482, 498
Vasospasm, 538
Vastus intermedius, **36,** 324f, 328
Vastus lateralis muscle, **34–36, 45,** 305f, 306f, 324f, 325f, 326f, 328, 330f, **344, 350**
Vastus medialis muscle, **34–35,** 305f, 324f, 327f, 328, **344, 350**
Vegetarian diets, 699t
Veins, 579–80. *See also* Cardiovascular system; *specific vein*
Vena cava, 608f
inferior, **37, 44–45, 47,** 557, 558f, 560f, 562f, 602, 604, 605, 606f, 607f, 777f, 778f, 898f
interior, **36**
superior, **35–36, 46,** 557, 558f, 560f, 562f, 593f, 602, 603f, 607f, 898f
Venipuncture, 603
Venoconstriction, 589
Venous blood flow, 587, 589
Venous pathways, 600, 602
Venous return, 585
Venous sinus, 602f
Ventilation, 736, 756
Ventral branch, 420
Ventral respiratory group, 757
Ventral root, 420
Ventricles. *See also* Fourth ventricle; Lateral ventricle; Left ventricle; Right ventricle; Third ventricle
heart, 555, 608f
nervous system, **40,** 363, 385–86, 387f, 388f, 398
Ventricular fibrillation, 572
Ventricular syncytium, 565
Venules, 579–80
Vermis, 410
Vernoff, Bruce and Gaby, 876
Vertebrae, 218, 219–21. *See also* Cervical vertebrae; Lumbar vertebrae; Thoracic vertebrae; Vertebral column
Vertebral arteries, **43,** 593, 595f, 596f, 601f
Vertebral body, **44–45,** 598f
Vertebral border, 226
Vertebral canal, 12
Vertebral column, **38,** 205t, 206, 218–22, 225, 308, 310. *See also* Spinal cord; Vertebrae
Vertebral compression, 240
Vertebral foramen, 220. *See also* Intervertebral foramen
Vertebral region, 24, 25f
Vertebral spine, **342**
Vertebral vein, 602
Vertebra prominens, 219f, 220
Vertebrocostal joint, 270t
Vertex position, 894
Very-low-density lipoprotein (VLDL), 684, 703
Vesical arteries, 598, 599f
superior, 905
Vesical veins, 605
Vesicles, 75, 83, 91t
Vesicoureteral reflux, 802t
Vestibular branch fibers, 416
Vestibular bulbs, 856
Vestibular glands, 856, 857t
Vestibular membrane, 453
Vestibule, 453, 656, 856, 857t
Vestibulocochlear nerves, 416, 417t
Veterans Administration, 261
Viagra (sildenafil), 844
Villi, 651. *See also* Microvilli
Viruses, 76, 360, 638. *See also* Human immunodeficiency virus (HIV); Human papilloma virus; Influenza; Marburg virus
Viscera, 12
Visceral afferent fibers, 412, 414
Visceral branch, 420
Visceral efferent fibers, 412
Visceral pain, 441–42
Visceral pericardium, 12, 554
Visceral peritoneum, 14, 654
Visceral pleura, 12, 747
Visceral senses, 446
Visceral smooth muscle, 300
Visceroreceptive senses, 440
Viscosity, and blood pressure, 584–85
Visible Human Project, 3
Vision. *See* Eye; Sight
Visual accessory organs, 462–65
Visual nerve pathways, 475–76
Visual pigments, 473–75
Visual receptors, 473
Vital capacity (VC), 753, 754t
Vital signs, 2
Vitamin(s)
definition, 119
nutrition, 709–16, 726
supplements, 717
Vitamin A, 200, 475, 709–11, 713t
Vitamin B, 529, 713–16, 717t
Vitamin C, 200, 529, 716, 717f, 719t
Vitamin D, 172, 186–87, 200, 204, 240, 502–503, 711–12, 713t, 777
Vitamin E, 712, 713t
Vitamin K, 539, 712, 713t
Vitreous body, 469
Vitreous humor, 437, 469
Vocal cords, 741, 742f
Volar arch arteries, 597
Volatile acid, 822
Voltage-gated sodium channels, 369
Voluntary muscle, 163, 284
Vomer bone, 209f, 211f, 214, 215f, 216t, **245, 247, 249–51**
Vomiting and vomiting center, 670–71, 825
Von Willebrand disease, 548t

W

Wall. *See also* Abdominal wall; Alveolar wall; Thoracic wall
of alimentary canal, 652–54, 655t
of heart, 555, 557t
of large intestine, 688
of small intestine, 681–82
Wandering cells, 153
Warfarin, 542
Warm receptors, 440
Warning signs, of breast cancer, 864
Warren, J. Robin, 668
Washington, George, 742
Water
balance, 811, 813–15, 816–17
chemical constituents of cells, 61, 62t
metabolism, 813
molecular structure, 69f
osmosis, 93–94

reabsorption in urinary system, 792, 794t
requirements for life, 7, 9
Water intoxication, 816
Water-soluble vitamins, 709, 713–16, 718t
Weak acid, 820
Weak base, 820
Weber test, 459
Weight lifting, 300
Wernicke's area, 401
Whiplash, 424
White, Tim, 193
White blood cell(s), 162, 530–34
White blood cell count (WBCC), 533, 942
White fibers, 299
White matter, in brain, **42,** 358, 390f, 391f
White pulp, of spleen, 624
White rami, 425
Whooping cough, 534t
Wild type genetic mutation, 920
Wilson disease, 803
"Winter vomiting disease," 638
Wisdom teeth, 658, 660
Withdrawal reflex, 390–91, 392f, 393f
Wnt4 gene, 927
Wolf's Paw (herb), 775
World Anti-doping Agency, 489
World Bank, 115
World Health Organization, 290, 523, 843
Wormian bones, 205–206
Wounds, in skin, 183–86
Wrist, 270t

X

Xenograft, 641
Xeroma pigmentosum, 137
Xeropthalmia, 710
Xerostomia, 690
Xiphoid process, 224, 225
X chromosome, 927–28
X-linked inheritance, 920, 923, 928
X-linked recessive disorders, 928
XO syndrome, 929
X-ray(s), 70
X-ray fluoroscopy, 609
XXY syndrome, 929
XYY syndrome, 929–30

Y

Yawning, 755
Y chromosome, 927–28
Yellow marrow, 203
Y-linked genes, 920
Yolk sac, 891
Yucca Mountain (Nevada), 56

Z

Ziconotide, 445
Zinc, 52t, 722, 723t
Z lines, 287, 288f
Zona pellucida, 830, 849, 851f, 853f, 877, 880f
Zone of calcified cartilage, 198
Zone of hypertrophic cartilage, 198
Zone of proliferating cartilage, 198
Zone of resting cartilage, 198
Zygomatic arch, 211f, 212, **246, 248, 341, 346**
Zygomatic bones, 209f, 210f, 211f, 214, 215f, 216t, 218f, **245–49**
Zygomatic process, 210f, 212, **253**
Zygomaticus major and minor, 305f, 306, 307f
Zygote, 849, 850f, 878, 886t
Zygote intrafallopian transfer (ZIFT), 879
Zymogen granules, 671

Aids to Understanding Words

acetabul-, vinegar cup: *acetabul*um
adip-, fat: *adip*ose tissue
aer-, air: *aer*obic respiration
af-, to: *af*ferent arteriole
agglutin-, to glue together: *agglutin*ation
alb-, white: *alb*inism
aliment-, food: *aliment*ary canal
allant-, sausage-shaped: *allant*ois
alveol-, small cavity: *alveol*us
an-, without: *an*aerobic respiration
ana-, up: *ana*bolic
andr-, man: *andr*ogens
angio-, vessel: *angio*tensin
anul-, ring: *anul*ar ligament
append-, to hang something: *append*icular
arth-, joint: *arth*rology
astr-, starlike: *astr*ocyte
ather-, porridge: *ather*osclerosis
aud-, to hear: *aud*itory
auto-, self: *auto*immune disease
ax-, axis: *ax*ial skeleton
bas-, base: *bas*al metabolic rate
bi -, two: *bi*polar neuron
bil-, bile: *bil*irubin
bio-, life: *bio*chemistry
-blast, budding: osteo*blast*
brady-, slow: *brady*cardia
bronch-, windpipe: *bronch*us
burs-, bag, purse: prepatellar *burs*a
calat-, something inserted: inter*calat*ed disc
calor-, heat: *calor*ie
calyc-, small cup: *calyc*es
canal-, channel: *canal*iculus
carcin-, spreading sore: *carcin*oma
cardi-, heart: peri*cardi*um
cari-, decay: dental *cari*es
carin-, keel-like: *carin*a
carot-, carrot: *carot*ene
carp-, wrist: *carp*als
cata-, down: *cata*bolic
cec-, blindness: *cec*um
cephal-, head: en*cephal*itis
cerebr-, brain: *cerebr*um
chiasm-, cross: optic *chiasm*a
chondr-, cartilage: *chondr*ocyte
chorio-, skin: *chorio*n
choroid, skinlike: *choroid* plexus
chromo-, color: *chromo*some
chym-, juice: *chym*e
-clast, broken: osteo*clast*
clav-, bar: *clav*icle
cleav-, to divide: *cleav*age
co-, with: *co*enzyme
cochlea, snail: *cochlea*
condyl-, knob: *condyl*e
contra-, against, counter: *contra*ception
corac-, beaklike: *corac*oid process
corn-, horn: *corn*ea
cort-, covering: *cort*ex
cran-, helmet: *cran*ial
cribr-, sievelike: *cribr*iform plate
cric-, ring: *cric*oid cartilage
-crin, to secrete: endo*crin*e
crist-, ridge: *crist*a galli
-crit, to separate: hemato*crit*
crur-, lower part: *crur*a
cut-, skin: sub*cut*aneous
cyst-, bladder: *cyst*itis
cyt-, cell: *cyt*oplasm
-cyt, cell: osteo*cyt*e
de-, undoing: *de*amination
decidu-, falling off: *decidu*ous
dendr-, tree: *dendr*ite
derm-, skin: *derm*is
detrus-, to force away: *detrus*or muscle
di-, two: *di*saccharide
diastol-, dilation: *diastol*e
diuret-, to pass urine: *diuret*ic
dors-, back: *dors*al
ect-, outside: *ect*oderm
edem-, swelling: *edem*a
ejacul-, to shoot forth: *ejacul*ation
embol-, stopper: *embol*us
-emia, a blood condition: hypoprotein*emia*
endo-, within: *endo*plasmic reticulum
ependym-, tunic: *ependym*a
epi-, upon: *epi*thelial tissue
erg-, work: syn*erg*ist
erythr-, red: *erythr*ocyte
exo-, outside: *exo*crine gland
extra-, outside: *extra*cellular
fasc-, bundle: *fasc*iculus
fimb-, fringe: *fimb*riae
flacc-, flabby: *flacc*id paralysis
follic-, small bag: hair *follic*le
fov-, pit: *fov*ea
frenul-, bridle, restraint: *frenul*um
funi-, small cord or fiber: *funi*culus
gangli-, swelling: *gangli*on
gastr-, stomach: *gastr*ic gland
-gen, be produced: aller*gen*
-genesis, origin: spermato*genesis*
germ-, to bud or sprout: *germ*inal
glen-, joint socket: *glen*oid cavity
-glia, glue: neuro*glia*
glom-, little ball: *glom*erulus
glyc-, sweet: *glyc*ogen
-gram, something written: electrocardio*gram*
gubern-, to steer, to guide: *gubern*aculum
hema-, blood: *hema*toma
hemo-, blood: *hemo*globin
hepa-, liver: *hepa*tic duct
hetero-, other, different: *hetero*zygous
hiat-, opening: esophageal *hiat*us
hist-, web, tissue: *hist*ology
hol-, entire, whole: *hol*ocrine gland
hom-, same, common: *hom*ologous chromosomes
homeo-, same: *homeo*stasis
horm-, impetus, impulse: *horm*one
humor-, fluid: *humor*al
hyal-, resemblance to glass: *hyal*ine cartilage
hyper-, above: *hyper*tonic
hypo-, below: *hypo*tonic
im-, (or in-), not: *im*balance
immun-, free: *immun*ity
inflamm-, to set on fire: *inflamm*ation
inhal-, to breathe in: *inhal*ation
inter-, between: *inter*phase
intra-, inside: *intra*membranous
iris, rainbow: *iris*
iso-, equal: *iso*tonic
juxta-, near to: *juxta*medullary nephron
karyo-, nucleus: *karyo*type
kerat-, horn: *kerat*in
labi-, lip: *labi*a